Biochemistry

Biochemistry

FOURTH EDITION

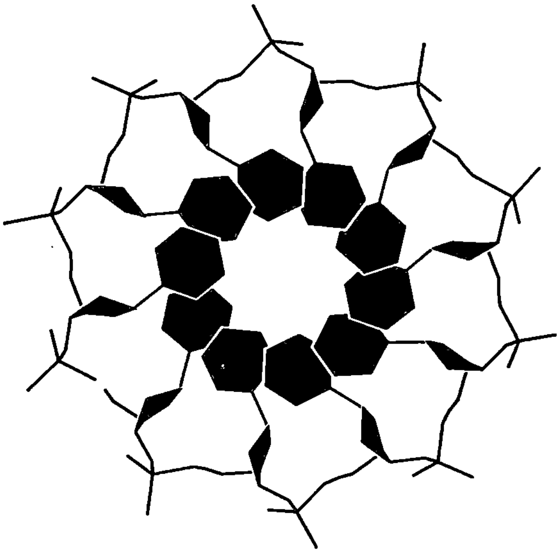

Lubert Stryer

STANFORD UNIVERSITY

W. H. Freeman and Company
New York

Library of Congress Cataloging-in-Publication Data

Stryer, Lubert.
Biochemistry/Lubert Stryer.—4th ed.

Includes index.
ISBN 0-7167-2009-4
1. Biochemistry. I. Title.
QP514.2.S66 1995
574.19'2—dc20 94-22832

Printed in the United States of America

Fourth printing 1997, KP

To my teachers

Paul F. Brandwein

Daniel L. Harris

Douglas E. Smith

Elkan R. Blout

Edward M. Purcell

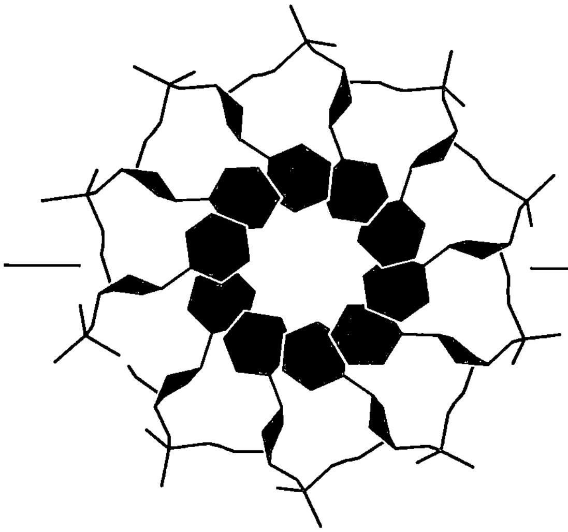

Biochemistry

Contents

Topics

PART I Molecular Design of Life 1

Chapter 1 Prelude 3

Chapter 2 Protein Structure and Function 17

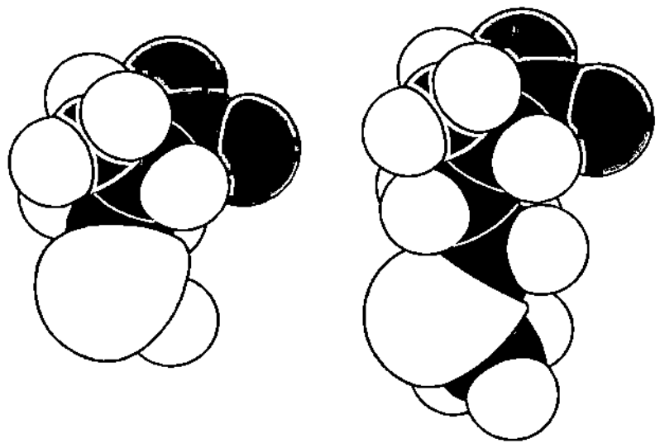

Chapter 3 Exploring Proteins 45

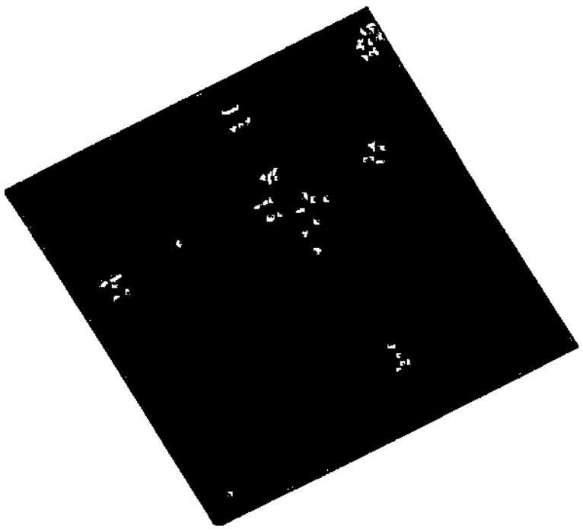

PART II Proteins: Conformation, Dynamics, and Function 145

Chapter 7 Portrait of an Allosteric Protein 147

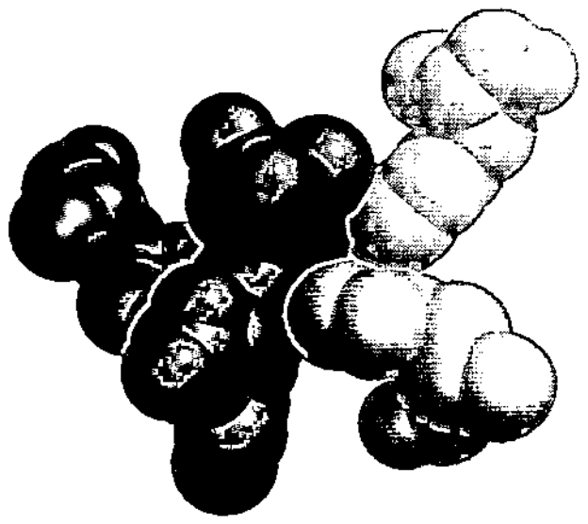

Chapter 8 Enzymes: Basic Concepts and Kinetics 181

Chapter 9 Catalytic Strategies 207

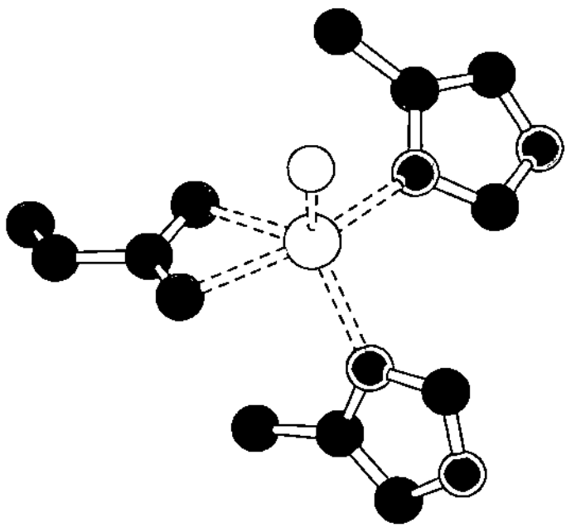

Chapter 10 Regulatory Strategies 237

Chapter 11 Membrane Structure and Dynamics 263

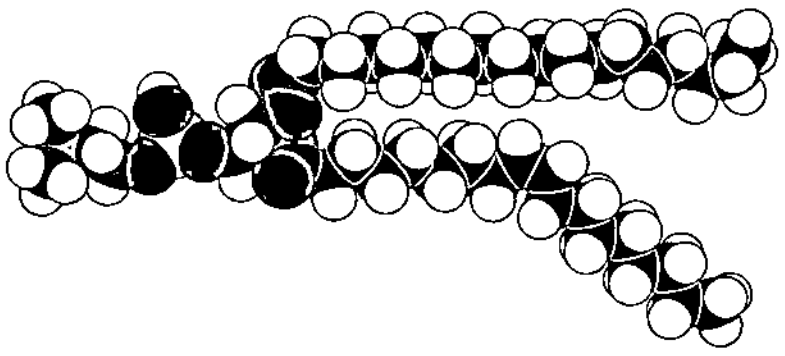

Chapter 12 Membrane Channels and Pumps 291

Chapter 13 Signal Transduction Cascades 325

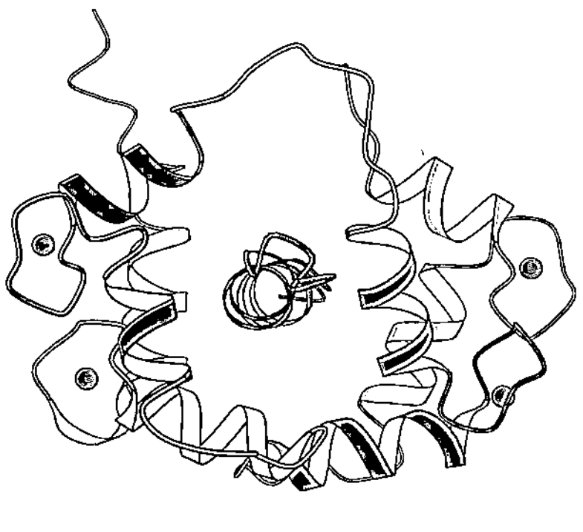

Chapter 14 Antibodies and T-Cell Receptors 361

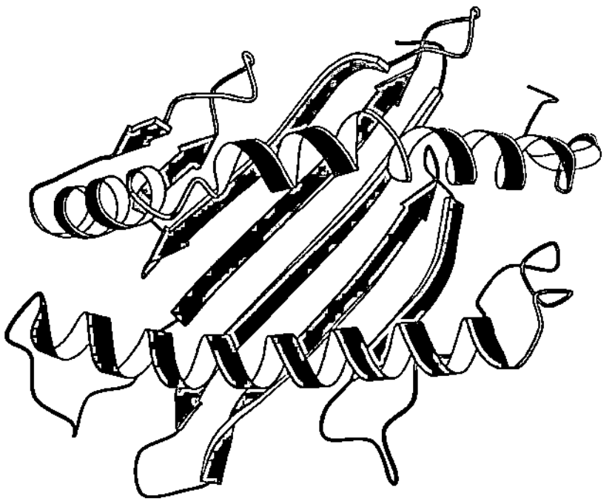

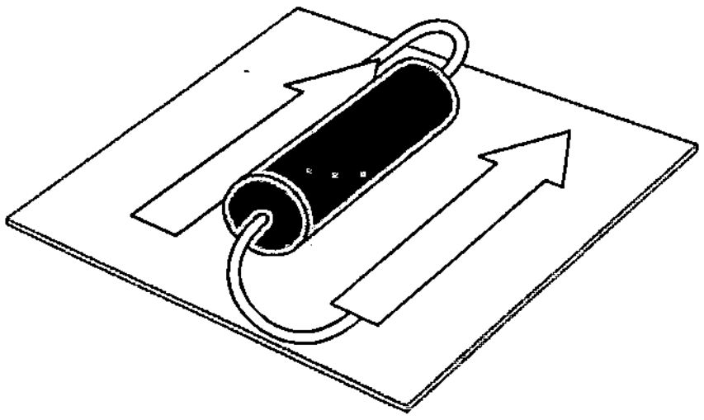

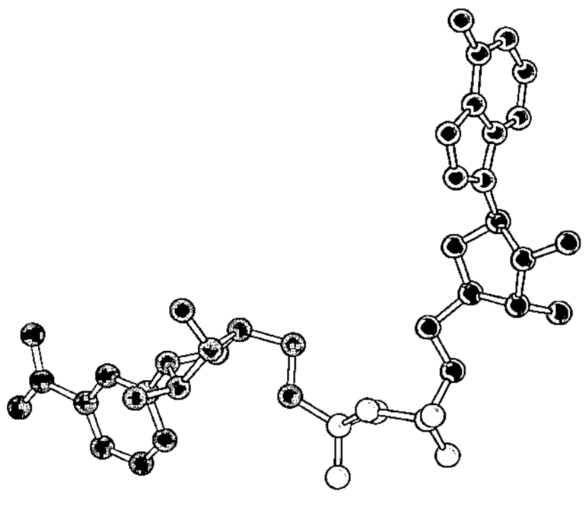

Chapter 20 Citric Acid Cycle 509

Chapter 21 Oxidative Phosphorylation 529

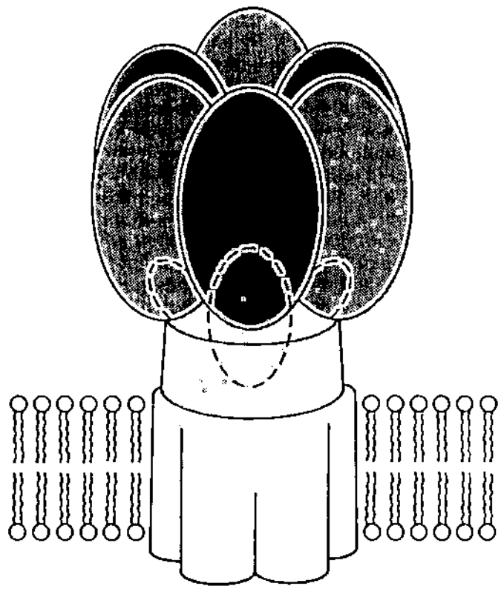

Chapter 22 Pentose Phosphate Pathway and Gluconeogenesis 559

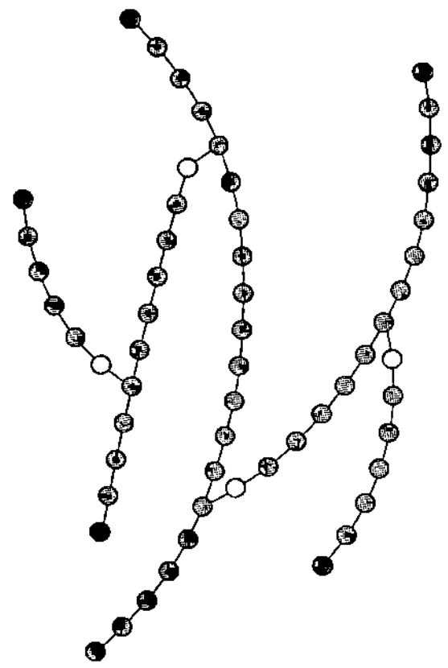

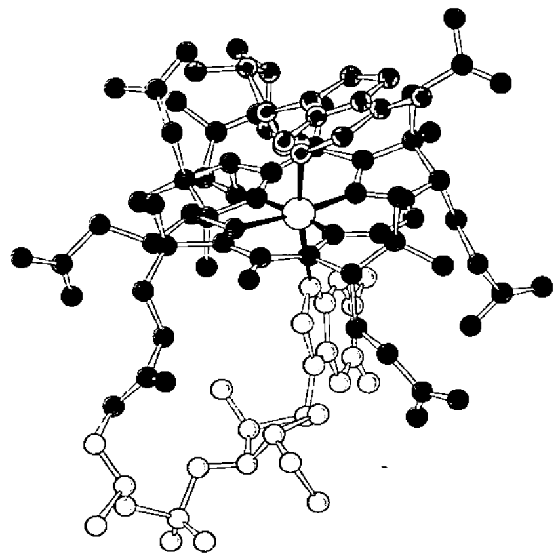

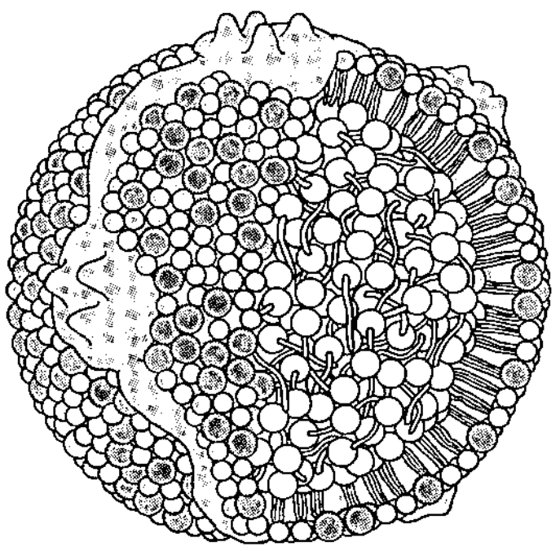

Chapter 28 Biosynthesis of Amino Acids and Heme 713

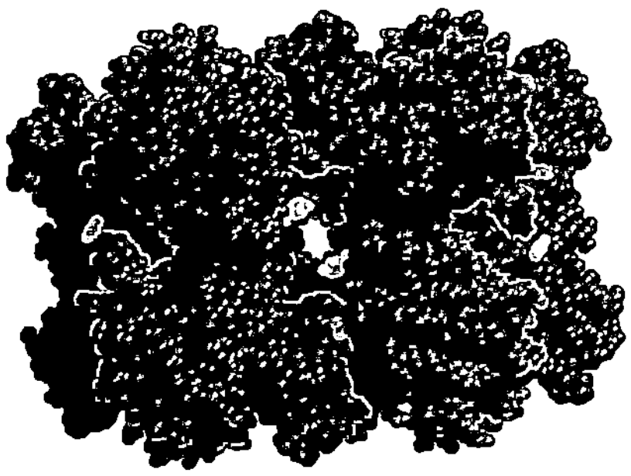

Chapter 29 Biosynthesis of Nucleotides 739

Chapter 30 Integration of Metabolism 763

PART V *Genes: Replication and Expression* *785*

Chapter 31 DNA Structure, Replication, and Repair 787

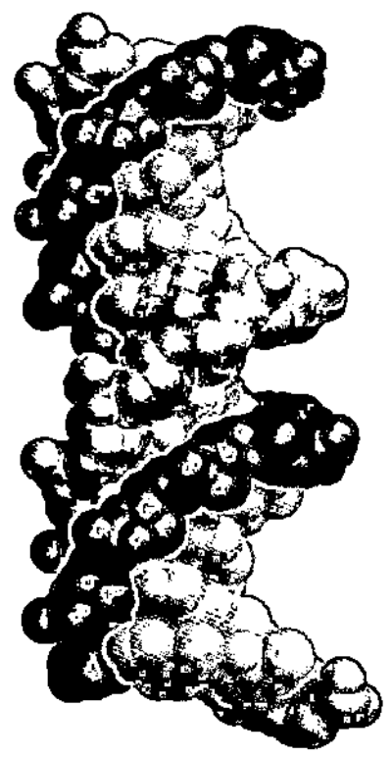

Chapter 32 Gene Rearrangements 819

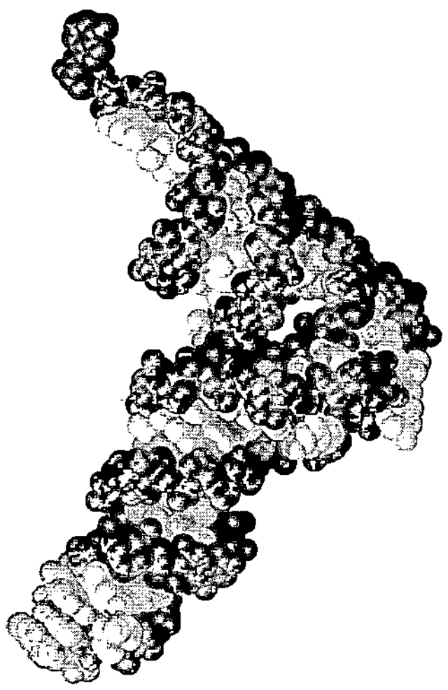

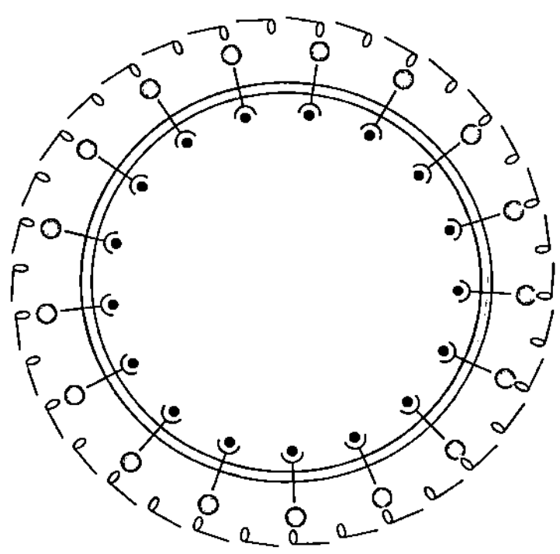

Chapter 36 Control of Gene Expression in Prokaryotes 949

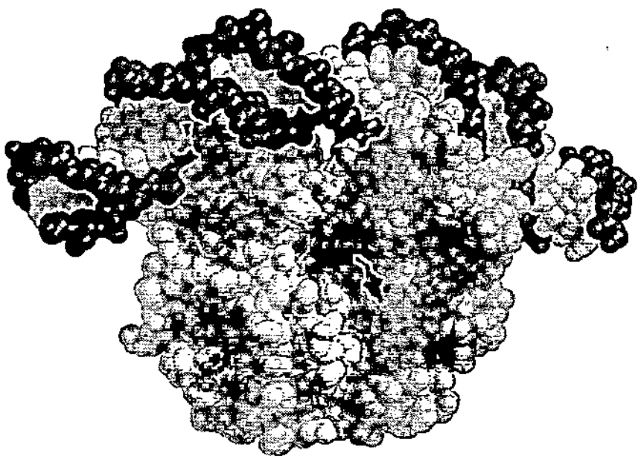

Chapter 37 Eukaryotic Chromosomes and Gene Expression 975

Preface

This is a wonderful time in biochemistry. Recombinant DNA technology, protein chemistry, and structural biology have come together to reveal the molecular mechanisms of fundamental biological processes. Many of our dreams of only a few years ago have been fulfilled. One of the fruits of this continuing harvest is a deeper understanding of protein form and function. We now see that proteins are highly sophisticated molecular machines that process energy, matter, and information. Their beautiful molecular ballet is coming into view. Indeed, we can even see the steps taken by a single molecular motor and the flow of ions through a single membrane channel.

These profound advances compel a change in how biochemistry is taught. I have expanded Part II, Proteins: Conformation, Dynamics, and Function, to bring our new understanding of proteins into the core of biochemistry. Membrane channels and pumps, signal transduction cascades, antibodies and T-cell receptors, and molecular motors are now presented in Part II rather than in a separate section at the end of the book. The proteins considered in these chapters reveal much about fundamental processes such as free-energy conversion, signal amplification, and molecular recognition. An understanding of proteins as energy-transforming and information-processing devices enriches the study of metabolism. Oxidative phosphorylation and photosynthesis, for example, are more readily understood and appreciated if the proton-pumping mechanism of bacteriorhodopsin is considered first. Likewise, the hormonal control of metabolism is easier to comprehend if G-protein and tyrosine-kinase cascades are presented first. Another noteworthy change is the new chapter on protein folding and design, an exciting area of inquiry that has broad import.

In 1962, the structures of only two proteins—myoglobin and hemoglobin—were known at atomic resolution. I was fortunate to see them at close range while I was a postdoctoral fellow at the Medical Research Council Laboratory of Molecular Biology in Cambridge, England. Max Perutz, John Kendrew, and the young crystallographers in the laboratory instilled in me a deep appreciation of molecular architecture and its importance in understanding the molecular basis of life. Now, three decades later, we see the fruition of structural biology. X-ray crystallography, joined by electron crystallography and nuclear magnetic resonance spectroscopy, have revealed the structures of hundreds of proteins. This wealth of structural information has illuminated and enriched our understanding of how proteins function as enzymes, transporters, motors, signal transducers, and gene regulators. Recurring structural modules have come into view, bringing insight into how proteins evolved. Structural biology has also contributed greatly to our understanding of how genetic information is replicated and expressed. DNA is now seen as a highly dynamic molecule that can twist, bend, and breathe. The structures of many complexes of DNA and RNA with polymerases, synthetases, transcriptional regulators, and other cognate proteins have been immensely informative. They are presented in depth in Part V, Genes: Replication and Expression. Throughout this edition, as in previous ones, I have sought to vividly depict the beautiful relations between molecular architecture and biological function, inspired by the experiences of my scientific childhood.

I am grateful to Elva Diaz, Richard Gumport, Roger Koeppe, Carl Rhodes, and John Tymoczko for their critical and perceptive review of the entire manuscript, and to Alexander Glazer for his sustained support and wise counsel. Clint Ballou, Ronald Bentley, Stuart Edelstein, Roger Koeppe II, Brian Law, Peter Rubenstein, Alan Sachs, Ihor Skrypka, Ronald Somerville, Mary Waltner, and Derrick Williamson provided valuable comments on major parts of this book. I have also gained much from the advice provided by Robert Baldwin, Jeremy Berg, Nicholas Cozzarelli, Marilyn Farquhar, Michel Goldberg, Peter Kim, Daniel Koshland, Jr., Richard Mathies, David McKay, George Palade, Peter Parham, Suzanne Pfeffer, Alexander Rich, Timothy Ryan, Robert Shulman, Paul Sigler, James Spudich, Thomas Steitz, Nigel Unwin, Max Vasquez, John Wagner, and William Weis. Joanne Tisch carefully scrutinized the proofs and helped in many other ways with the preparation of this edition, as with the last.

I wish to thank Richard Gumport, Ana Jonas, Richard Mintel, Carl Rhodes, and Roger Koeppe for having written a *Student Companion* to this work. This valuable study guide is an outgrowth of their devotion to teaching and the advancement of biochemical knowledge.

The flowering of structural biology has been accompanied and reinforced by rapid advances in molecular graphics. This edition contains hundreds of new drawings of biomolecules that were generated using a host of powerful tools for visualizing molecular architecture. I am indebted to Michael Levitt for introducing me to *MacImdad,* his interactive graphics program for personal computers, and for providing advice on molecular structure and graphics. Most of the new space-filling models shown in this edition were drawn using *MIDAS,* a program developed by Thomas Ferrin and Conrad Huang. Ball-and-stick models and schematic representations were drawn primarily with *MOLSCRIPT,* a program written by Per Kraulis. I have also benefited from *GRASP,* a program devised by Anthony Nicholls. David Hinds, Kevin Flaherty, and Peter David guided me through UNIX and helped me in many ways with both software and hardware on my Silicon Graphics IRIS computer. I am grateful to Paul Haeberli for making it possible to capture in print the striking molecular images seen on the IRIS screen. I have made extensive use of the Protein Data Bank, a rich and highly accessible repository of structural information maintained by Brookhaven National Laboratory. Many crystallographers shared with me the atomic coordinates of their recently solved structures.

Sonia DiVittorio played a key role in developing this edition and in guiding the preparation of new art. Her sensitive appreciation of biochemistry and dedication to excellence in publishing have made their mark. I am grateful to her for her outstanding contributions. Bill O'Neal, Jr., thoughtfully edited the manuscript, which gained much from his careful scrutiny. I am also indebted to Tomo Narashima and Ian Worpole for many fine drawings. Bill Page was the illustration coordinator. John Hatzakis provided an esthetic design, and Stan Hatzakis crafted each page with a fine eye for integrating word and picture. Georgia Lee Hadler expertly guided the flow and convergence of manuscript and thousands of pages of text and art proof. Julia DeRosa ably coordinated a challenging production process. Nancy Brooks provided valuable assistance at crucial times. Her command of the classics enlivened and enriched the preparation of this edition.

The twelve foreign-language editions of this book—Chinese, French, German, Greek, Italian, Japanese, Korean, Polish, Portuguese, Russian, Serbo-Croatian, and Spanish—have given me special pleasure. I am indebted to the translators for their devoted and scholarly contributions. They have made this book accessible to students in many lands and have fostered the universality of biochemistry. I have greatly enjoyed my contact with students and faculty around the world and look forward to continuing this stimulating and rewarding dialogue.

I was invited several years ago by a medical student newsletter to contribute to an issue entitled "Why Teach?" I responded by recounting my earliest recollection of school. During World War II, in the midst of uncertainty and strife, I attended a primary school in Shanghai that was taught by two Danish ladies in their small apartment. Their educational strategy was simple and highly effective: each student both taught and learned. I learned multiplication from a youngster who had mastered it only a week earlier, and in turn, I taught subtraction to a less advanced peer. Teaching remains my best way of learning. I also teach because the sharing of knowledge is joyful. Kindred souls are joined by the inseparable acts of teaching and learning.

Twenty-three years have passed since I embarked on the intellectual journey of writing a textbook of biochemistry. Each edition has been demanding in a different way. I am deeply grateful to my wife, Andrea, for her sustained support of this endeavor from its inception. Her wisdom has made all the difference. I have been nurtured too by our enlarging family. Two wonderful daughters have joined us—Daniel's wife, Stacy, and Michael's wife, Barri. The birth of our granddaughter Leah a few weeks ago has been an added blessing.

Lubert Stryer

NOVEMBER 1994

Prefaces

PREFACE TO THE THIRD EDITION

Biochemistry has been profoundly transformed by recombinant DNA technology. The genome is now an open book—any passage can be read. The cloning and sequencing of millions of bases of DNA have greatly enriched our understanding of genes and proteins. Indeed, recombinant DNA technology has led to the integration of molecular genetics and protein chemistry. The intricate interplay of genotype and phenotype is now being unraveled at the molecular level. One of the fruits of this harvest is insight into how the genome is organized and its expression is controlled. The molecular circuitry of growth and development is coming into view. The reading of the genome is also providing a wealth of amino acid sequence information that illuminates the entire protein landscape. Scarce proteins can be produced in abundance by transfected cells. Moreover, precisely designed novel proteins can be generated by site-specific mutagenesis to elucidate how proteins fold, catalyze reactions, transduce signals, transport ions, and interconvert different forms of free energy.

Our understanding of molecular evolution also has been greatly enriched by the recombinant DNA revolution. Families and superfamilies of proteins have come into view. Theme and variation at the level of proteins are vivid expressions of the underlying processes of gene duplication and divergence. The genes of complex proteins display the coming together in evolution of exons encoding functional modules. The many recurring structural and mechanistic motifs seen throughout nature testify to the fundamental unity of all forms of life. The discovery of catalytic RNA enables us to envision an RNA world early in the evolution of life, prior to the appearance of DNA and protein. The ubiquity of ribonucleotides in metabolism and the central roles played by them are reflections of their ancient origins—of the early RNA world, when RNA served both as gene and enzyme.

These remarkable advances compel a major change in the way biochemistry is taught. I have altered the architecture of this book to provide a new framework for the exposition of fundamental themes and principles of biochemistry. The book begins with a new part, entitled Molecular Design of Life, that provides an overview of the central molecules of life—DNA, RNA, and proteins—and their interplay. Recombinant DNA technology and other experimental methods for exploring proteins and genes are also presented in this part. This introduction prepares the reader for the detailed consideration of protein structure and function that follows. The teaching of metabolism is likewise enriched by this new organization. The other major structural changes are the addition of a chapter on carbohydrates and one on protein targeting. Many sections of the book have been extensively revised and hundreds of new illustrations have been added. I have tried to preserve the unity of biochemistry as an intellectual discipline. My goal has been to make this powerful language comprehensible and to share its beautiful imagery.

I am grateful to Alexander Glazer, Daniel Koshland, Jr., and Alexander Rich for having encouraged me to write this edition. I would not have embarked on this endeavor had it not been for their warm support and good counsel.

The planning of this edition unexpectedly took place in terrain quite different from Yale, Stanford, and Aspen, where the first two editions took form. In December 1985, my family and I went to Nepal to trek in the Everest region. After two rewarding days in Kathmandu, we were on the verge of boarding the plane to Lukla, only to be turned back with the disappointing news that the landing strip was closed because of snow. We then headed for the Annapurna region in four-wheel-drive vehicles but had to return a day later because the road vanished in the heavy rain. The inaccessibility of the high Himalayas led to an abrupt change of itinerary. We flew to Bangkok and arrived in 95-degree heat, carrying our parkas and arctic sleeping bags. Instead of hiking at 12,000 feet, we found ourselves at a hotel pool at sea level. This dislocation led to a totally unforeseen benefit. I was able to unhurriedly reflect on the remarkable development of biochemistry since I wrote my last edition. I had the leisure to think and dream and plan this book. Best of all, I was able to share my thoughts with my son Daniel, who was then a senior majoring in human biology, and gain from his insights.

Alexander Glazer, Richard Gumport, Roger Koeppe, James Rawn, Carl Rhodes, and Peter Rubenstein read the entire manuscript. I have benefited greatly from their scholarly and perceptive criticism. Steve Block, Daniel Branton, Simone Brutlag, Carolyn Cohen, Jeffrey Critchfield, Peter Cullis, Russell Doolittle, Marilyn Farquhar, Robert Fletterick, Robert Fox, Michel Goldberg, Jack Griffith, James Hageman, Stephen Harrison, Brian Holl, Leroy Hood, Horace Jackson, Gunther Kohlhaw, Arthur Kornberg, Roger Kornberg, Stephen Kron, Michael Levitt, Bo Malmström, Lynne Mercer, Albert Mildvan, Jeremy Nathans, Christopher Newgard, Marion O'Leary, George Palade, Peter Parham, Frederic Richards, Ed Rock, Gottfried Schatz, Gray Scrimgeour, Paul Sigler, Jeffrey Sklar, James Spudich, Thomas Steitz, Nigel Unwin, Ronald Vale, William Wickner, and Robley Williams also gave valuable advice and help.

The contributors of many striking and informative illustrations are acknowledged in the figure legends. I am also indebted to crystallographers who have deposited the atomic coordinates of their solved structures in the Protein Data Bank, a valuable resource maintained by Brookhaven National Laboratory. Many new figures depicting molecular structure were generated on our departmental molecular graphics computer facility. David Austen and William Hurja helped me use this excellent system.

I was able to concentrate on the writing of this book because my office was in the capable hands of Joanne Tisch. She played a critical role in preparing the manuscript and reading the proofs. Her sensitivity, intelligence, and good spirits lightened my load. The Medline bibliographic retrieval system of the National Library of Medicine greatly facilitated my search of the literature. The staff of the Lane Medical Library and Falconer Biology Library of Stanford University were most helpful in locating books and references.

Andrew Kudlacik edited this manuscript with a fine sense of style and meaning. Mike Suh skillfully integrated word and picture in the design of each page. Susan Moran kept a watchful and discerning eye over many thousands of pages of manuscript, figures, and proofs. I also wish to thank Tom Cardamone and Shirley Baty for many outstanding drawings.

I am grateful to my family for their sustained support of this endeavor, which was more arduous than anticipated. My sons, Michael and Daniel, now embarked on their own careers, cheered me from afar. My wife, Andrea, provided criticism, advice, and encouragement in just the right proportions. I have been nurtured, too, by many who have reached out to express their warmth and interest in continuing this dialogue of biochemistry. I feel very fortunate and privileged to partake in this process at such a wonderful time.

December 1987

PREFACE TO THE SECOND EDITION

The pace of discovery in biochemistry has been exceptionally rapid during the past several years. This progress has greatly enriched our understanding of the molecular basis of life and has opened many new areas of inquiry. The sequencing of DNA, the construction and cloning of new combinations of genes, the elucidation of metabolic control mechanisms, and the unraveling of membrane transport and transduction processes are some of the highlights of recent research. One of my aims in this edition has been to weave new knowledge into the fabric of the text. I have sought to enhance the book's teaching effectiveness by centering the exposition of new material on common themes wherever feasible and by citing recurring motifs. I have also tried to convey a sense of the intellectual power and beauty of the discipline of biochemistry.

I am indebted to Thomas Emery, Henry Epstein, Alexander Glazer, Roger Kornberg, Robert Martin, and Jeffrey Sklar for their counsel, criticism, and encouragement in the preparation of this edition. Robert Baldwin, Charles Cantor, Richard Caprioli, David Eisenberg, Alan Fersht, Robert Fletterick, Herbert Friedmann, Horace Jackson, Richard Keynes, Sung-Hou Kim, Aaron Klug, Arthur Kornberg, Daniel Koshland, Jr., Samuel Latt, Vincent Marchesi, David Nelson, Garth Nicolson, Vernon Oi, Robert Renthal, Carl Rhodes, Frederic Richards, James Rothman, Peter Sargent, Howard Schachman, Joachim Seelig, Eric Shooter, Elizabeth Simons, James Spudich, Theodore Steck, Thomas Steitz, Judit C.-P. Stenn, Robert Trelstad, Christopher Walsh, Simon Whitney, and Bernhard Witkop also gave valuable advice.

Patricia Mittelstadt edited both editions of this text. I deeply appreciate her critical and sustained contributions. I am indebted to Donna Salmon for her outstanding drawings. David Clayton, David Dressler, John Heuser, Lynne Mercer, Kenneth Miller, George Palade, Nigel Unwin, and Robley Williams generously provided many fine electron micrographs. Betty Hogan typed the manuscript and played an indispensable role in its preparation. Cary Leiden and Karen Marzotto carefully read the proofs. I also wish to thank Michael Graves for his excellent photographic work.

My wife, Andrea, and my sons, Michael and Daniel, have cheerfully allowed this text to become a member of the family. I am deeply grateful to them for their patience and buoyancy. Andrea provided much advice on style and design, as she did for the first edition.

I have been heartened by the many letters that I have received from readers of the first edition. Their comments and criticisms have enlightened, stimulated, and encouraged me. I look forward to a continuing dialogue with readers in the years ahead.

August 1980

PREFACE TO THE FIRST EDITION

This book is an outgrowth of my teaching of biochemistry to undergraduates, graduate students, and medical students at Yale and Stanford. My aim is to provide an introduction to the principles of biochemistry that gives the reader a command of its concepts and language. I also seek to give an appreciation of the process of discovery in biochemistry. My exposition of the principles of biochemistry is organized around several major themes:

1. Conformation—exemplified by the relationship between the three-dimensional structure of proteins and their biological activity

2. Generation and storage of metabolic energy

3. Biosynthesis of macromolecular precursors

4. Information—storage, transmission, and expression of genetic information

5. Molecular physiology—interaction of information, conformation, and metabolism in physiological processes

The elucidation of the three-dimensional structure of proteins, nucleic acids, and other biomolecules has contributed much in recent years to our understanding of the molecular basis of life. I have emphasized this aspect of biochemistry by making extensive use of molecular models to give a vivid picture of architecture and dynamics at the molecular level. Another stimulating and heartening aspect of contemporary biochemistry is its increasing interaction with medicine. I have presented many examples of this interplay. Discussions of molecular diseases such as sickle-cell anemia and of the mechanism of action of drugs such as penicillin enrich the teaching of biochemistry. Finally I have tried to define several challenging areas of inquiry in biochemistry today, such as the molecular basis of excitability.

In writing this book, I have benefited greatly from the advice, criticism, and encouragement of many colleagues and students. Leroy Hood, Arthur Kornberg, Jeffrey Sklar, and William Wood gave me invaluable counsel on its overall structure. Richard Caprioli, David Cole, Alexander Glazer, Robert Lehman, and Peter Lengyel read much of the manuscript and made many very helpful suggestions. I am indebted to Frederic Richards for sharing his thoughts on macromolecular conformation and for extensive advice on how to depict three-dimensional structures. Deric Bownds, Thomas Broker, Jack Griffith, Hugh Huxley, and George Palade made available to me many striking electron micrographs. I am also very thankful for the advice and criticism that were given at various times in the preparation of this book by Richard Dickerson, David Eisenberg,

Moises Eisenberg, Henry Epstein, Joseph Fruton, Michel Goldberg, James Grisolia, Richard Henderson, Harvey Himel, David Hogness, Dale Kaiser, Samuel Latt, Susan Lowey, Vincent Marchesi, Peter Moore, Allan Oseroff, Jordan Pober, Edward Reich, Russell Ross, Mark Smith, James Spudich, Joan Steitz, Thomas Steitz, and Alan Waggoner.

I am grateful to the Commonwealth Fund for a grant that enabled me to initiate the writing of this book. The interest and support of Robert Glaser, Terrance Keenan, and Quigg Newton came at a critical time. One of my aims in writing this book has been to achieve a close integration of word and picture and to illustrate chemical transformations and three-dimensional structures vividly. I am especially grateful to Donna Salmon, John Foster, and Jean Foster for their work on the drawings, diagrams, and graphs. Many individuals at Yale helped to bring this project to fruition. I particularly wish to thank Margaret Banton and Sharen Westin for typing the manuscript, William Pollard for photographing space-filling models, and Martha Scarf for generating the computer drawings of molecular structures on which many of the illustrations in this book are based. John Harrison and his staff at the Kline Science Library helped in many ways.

Much of this book was written in Aspen. I wish to thank the Aspen Center of Physics and the Given Institute of Pathobiology for their kind hospitality during several summers. I have warm memories of many stimulating discussions about biochemistry and molecular aspects of medicine that took place in the lovely garden of the Given Institute and while hiking in the surrounding wilderness areas. The concerts in Aspen were another source of delight, especially after an intensive day of writing.

I am deeply grateful to my wife, Andrea, and to my children, Michael and Daniel, for their encouragement, patience, and good spirit during the writing of this book. They have truly shared in its gestation, which was much longer than expected. Andrea offered advice on style and design and also called my attention to the remark of the thirteenth-century Chinese scholar Tai T'ung (*The Six Scripts: Principles of Chinese Writing*): "Were I to await perfection, my book would never be finished."

I welcome comments and criticisms from readers.

October 1974

Biochemistry

PART

I

Molecular Design of Life

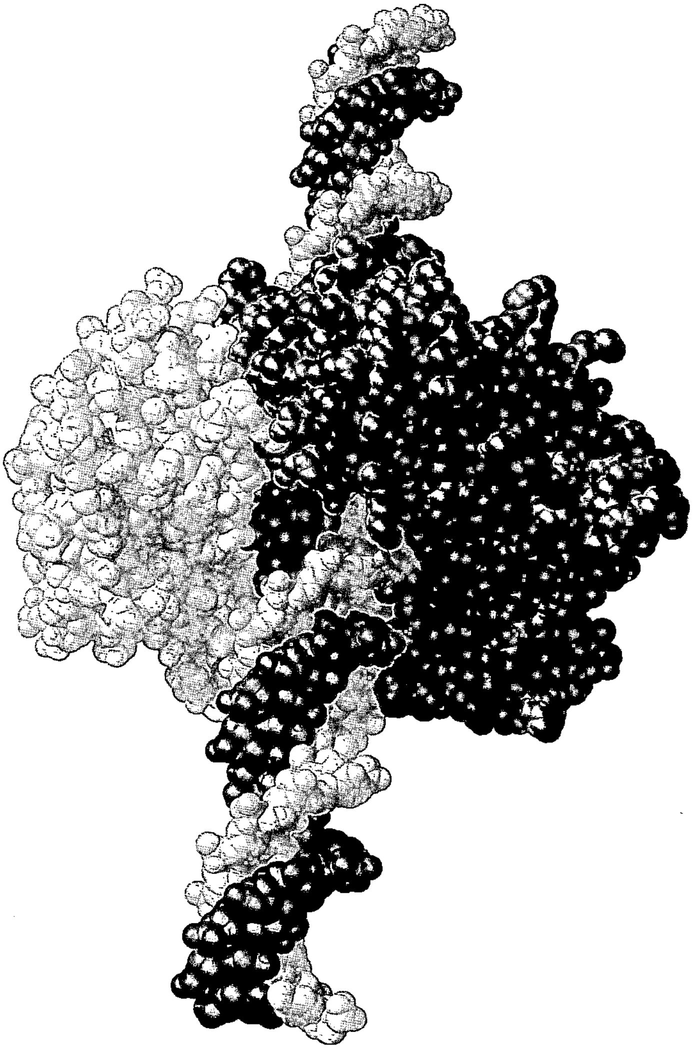

A double-helical DNA molecule (green and red strands) is encircled by a protein clamp (blue and yellow subunits) that serves as a docking site for DNA polymerase and the other components of the replication machinery. [Based on coordinates kindly provided by Dr. John Kuriyan. X.P. Kong, R. Onrust, M. O'Donnell, and J. Kuriyan. Cell 69(1992):425.]

1

Prelude

Biochemistry is the study of the molecular basis of life. There is much excitement and ferment in biochemistry throughout the world for several reasons.

First, the chemical mechanisms of many central processes of life are now understood. The discovery of the double-helical structure of deoxyribonucleic acid (DNA), the elucidation of the flow of information from gene to protein, the determination of the three-dimensional structure and mechanism of action of many protein molecules, and the unraveling of central metabolic pathways are some of the outstanding achievements of biochemistry. Much has also been learned about molecular machines that harvest energy, detect signals, and process information. The development of recombinant DNA technology has made the genome an open book. Its story, full of riches and surprises, can now be read.

Second, common molecular patterns and principles underlie the diverse expressions of life. Organisms as different as the bacterium *Escherichia coli* and human beings use the same building blocks to construct macromolecules. The flow of genetic information from DNA to ribonucleic acid (RNA) to protein is essentially the same in all organisms. Adenosine triphosphate (ATP), the universal currency of energy in biological systems, is generated in similar ways by all forms of life.

Third, biochemistry is profoundly influencing medicine. The molecular lesions causing sickle-cell anemia, cystic fibrosis, hemophilia, and many other genetic diseases have been elucidated. Knowledge of the underlying defects opens the door to the discovery and implementation of effective therapies. Biochemistry is also contributing richly to clinical diagnostics. For example, elevated levels of telltale enzymes in blood reveal whether a patient has recently had a myocardial infarction. DNA probes are coming into play in the precise diagnosis of inherited disorders, infectious diseases, and cancers. Genetically engineered strains of bacteria containing recombinant DNA are producing valuable proteins such as insulin, growth hormone, and stimulators of blood cell development. Furthermore, biochemistry makes possible the rational design of new drugs. Agriculture, too, is benefiting from recombinant DNA technology, which can produce designed changes in the genetic endowment of plants, such as the acquisition of greater resistance to insects.

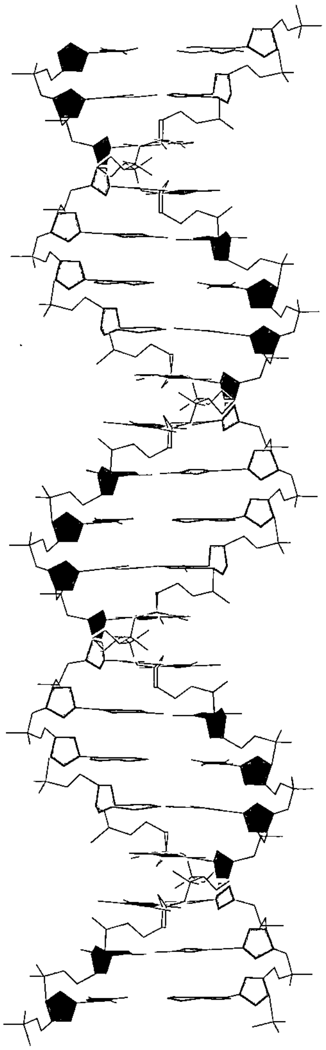

Figure 1-1
Model of the DNA double helix. The diameter of the helix is about 20 Å.

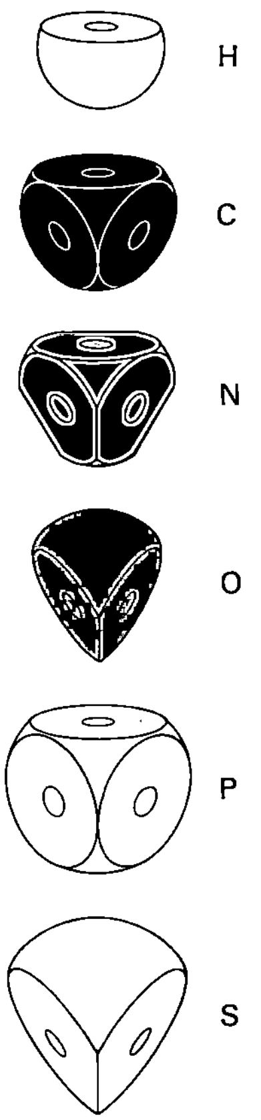

Figure 1-2
Space-filling models of hydrogen, carbon, nitrogen, oxygen, phosphorus, and sulfur atoms.

Fourth, the rapid development of powerful biochemical concepts and techniques in recent years has enabled investigators to tackle some of the most challenging and fundamental problems in biology and medicine. How does a fertilized egg give rise to cells as different as those in muscle, the brain, and the liver? How do cells find each other in forming a complex organ? How is the growth of cells controlled? What are the causes of cancer? What is the molecular mechanism of memory? What is the molecular basis of mental disorders such as Alzheimer's disease and schizophrenia?

MOLECULAR MODELS DEPICT THREE-DIMENSIONAL STRUCTURE

The interplay between the three-dimensional structure of biomolecules and their biological function is the unifying motif of this book. Three types of atomic models will be used to depict molecular architecture: space-filling, ball-and-stick, and skeletal. The *space-filling models* are the most realistic. The size and configuration of an atom in a space-filling model are determined by its bonding properties and van der Waals radius (Figure 1-2). The colors of the model atoms are set by convention:

Hydrogen, white	Nitrogen, blue	Phosphorus, yellow
Carbon, black	Oxygen, red	Sulfur, yellow

Space-filling models of several simple molecules are shown in Figure 1-3.

Ball-and-stick models are not as realistic as space-filling models, because the atoms are depicted as spheres of radius smaller than the van der Waals radius. However, the bonding arrangement is easier to see because the bonds are explicitly represented by sticks. In an illustration, the taper of a stick, representing parallax, tells which of a pair of bonded atoms is closer to the reader. More of a complex structure can be seen in a ball-and-stick model than in a space-filling model. An even simpler image is achieved with *skeletal models,* which show only the molecular framework. In these models, atoms are not shown explicitly. Rather, their positions are implied by the junctions and ends of bonds. Skeletal models are frequently used to depict large biological macromolecules, such as protein molecules having several thousand atoms. Space-filling, ball-and-stick, and

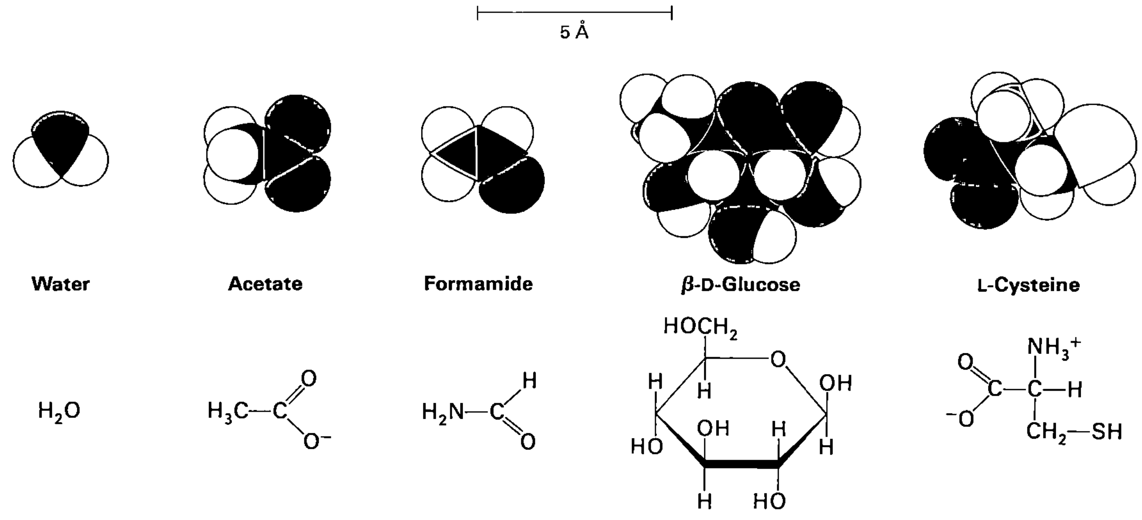

Figure 1-3
Space-filling models of water, acetate, formamide, glucose, and cysteine.

skeletal models of ATP are compared in Figure 1-4. Schematic representations of extended structures will also be used, as exemplified by a pair of intertwined ribbons to depict a DNA double helix.

SPACE, TIME, AND ENERGY

In considering molecular structure, it is important to have a sense of scale (Figure 1-5). The angstrom (Å) unit, which is equal to 10^{-10} meter (m) or 0.1 nanometer (nm), is customarily used as the measure of length at the atomic level. The length of a C–C bond, for example, is 1.54 Å. Small biomolecules, such as sugars and amino acids, are typically several angstroms long. Biological macromolecules, such as proteins, are at least tenfold larger. For example, hemoglobin, the oxygen-carrying protein in red blood cells, has a diameter of 65 Å. Another tenfold increase in size brings us to assemblies of macromolecules. Ribosomes, the sites of protein synthesis, have diameters of about 300 Å. The range from 100 Å (10 nm) to 1000 Å (100 nm) also encompasses most viruses. Cells are typically a hundred times as large, in the range of micrometers (μm). For example, a red blood cell is 7 μm (7×10^4 Å) long. It is important to note that the limit of resolution of the light microscope is about 2000 Å (0.2 μm), which corresponds to the size of many subcellular organelles. Mitochondria, the major generators of ATP in aerobic cells, can just be resolved by the light microscope. Most of our knowledge of biological structure in the range from 1 Å (0.1 nm) to 10^4 Å (1 μm) has come from x-ray crystallography and electron microscopy. Crystallographers usually use angstroms as the unit of length, whereas electron microscopists prefer nanometers or micrometers. We will use Å for lengths shorter than 100 Å (10 nm), and nm or μm for longer dimensions.

The molecules of life are constantly in flux. Chemical reactions in biological systems are catalyzed by enzymes, which typically convert substrate into product in milliseconds (ms, 10^{-3} s). Some enzymes act even more rapidly, in times as short as a few microseconds (μs, 10^{-6} s). Many conformational changes in biological macromolecules also are rapid. For example, the unwinding of the DNA double helix, which is essential for its

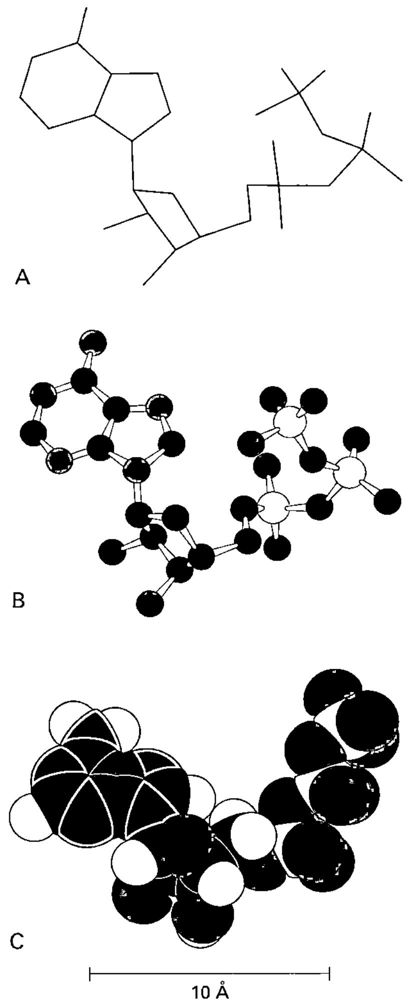

Figure 1-4
Comparison of (A) skeletal, (B) ball-and-stick, and (C) space-filling models of ATP. Hydrogen atoms are not shown in models A and B.

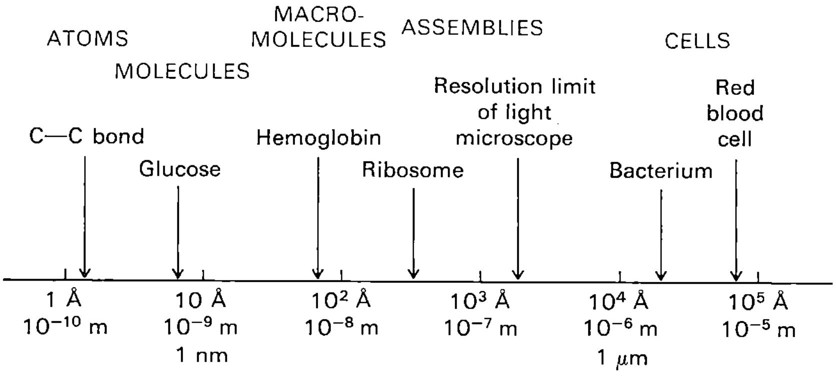

Figure 1-5
Dimensions of some biomolecules, assemblies, and cells.

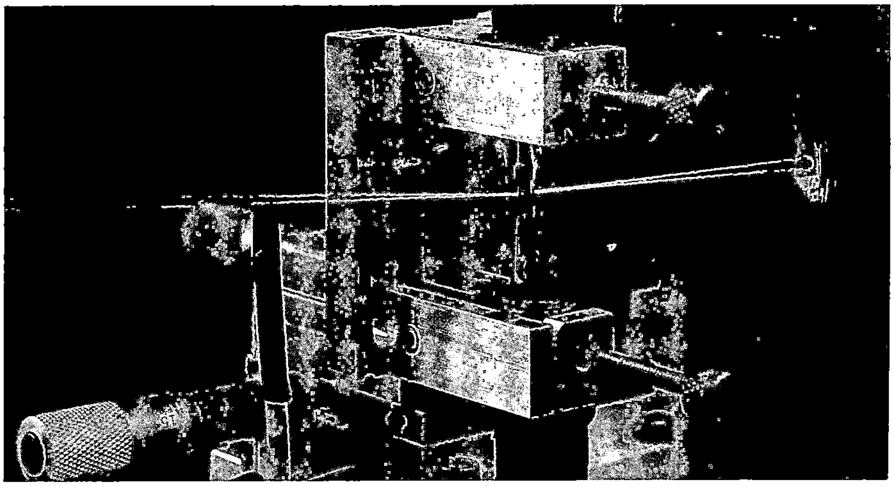

Figure 1-6
Very rapid reactions in biological systems can be probed using picosecond or even briefer light pulses that are generated by lasers. The green beam triggers a photosensitive molecule, and the red beam monitors the resulting conformational changes. [Courtesy of Dr. Robert Schoenlein, Dr. Richard Mathies, and Dr. Charles Shank.]

replication and expression, is a microsecond event. Many noncovalent interactions between groups in macromolecules are formed and broken in even shorter intervals. The rotation of one domain of a protein with respect to another can take place in nanoseconds (ns, 10^{-9} s). Even more rapid processes can be probed with very short light pulses from lasers (Figure 1-6). It is remarkable that the primary event in vision—a change in structure of the light-absorbing group—occurs within a few picoseconds (ps, 10^{-12} s) after the absorption of a photon (Figure 1-7). Likewise, the primary step in photosynthesis—the light-induced transfer of an electron—takes place in picoseconds.

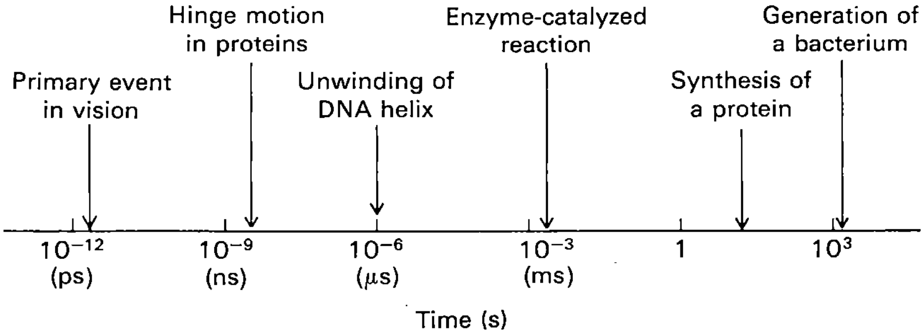

Figure 1-7
Typical times of some processes in biological systems.

We shall be concerned with energy changes in molecular events (Figure 1-8). The ultimate source of energy for life is the sun. The energy of a green photon, for example, is 57 kilocalories per mole (kcal/mol). An alternative unit of energy is the joule (J), which is equal to 0.239 calorie; 1 kcal/mol is equal to 4.184 kJ/mol. We will use the kilocalorie as the unit of energy because it is preferred by most biochemists. ATP, the universal currency of energy, has a usable energy content of about 12 kcal/mol. In contrast, the average energy of each vibrational degree of freedom in a molecule is much smaller, 0.6 kcal/mol at 25°C. This amount of energy is much less than that needed to dissociate covalent bonds (e.g., 83 kcal/mol for a C–C bond). Hence, the covalent framework of biomolecules is stable in the absence of enzymes and inputs of energy. On the other hand, noncovalent bonds in biological systems typically have an energy of only a few kilocalories per mole, so that thermal energy is enough to make and break them.

Energy content

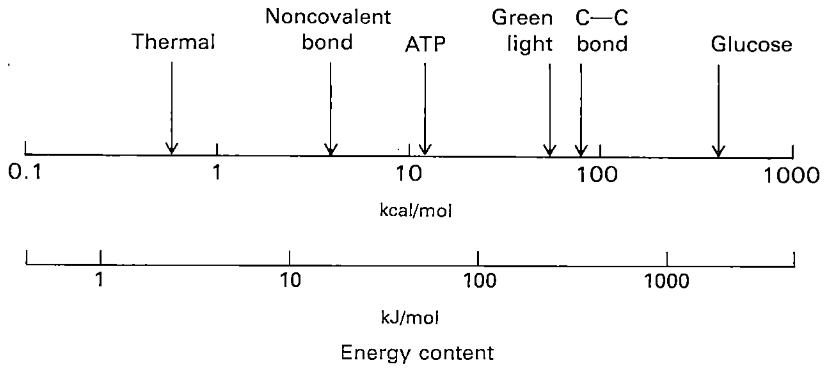

Figure 1-8
Some biologically important energies.

REVERSIBLE INTERACTIONS OF BIOMOLECULES ARE MEDIATED BY THREE KINDS OF NONCOVALENT BONDS

Reversible molecular interactions are at the heart of the dance of life. Weak, noncovalent forces play key roles in the faithful replication of DNA, the folding of proteins into intricate three-dimensional forms, the specific recognition of substrates by enzymes, and the detection of signal molecules. Indeed, all biological structures and processes depend on the interplay of noncovalent interactions as well as covalent ones. The three fundamental noncovalent bonds are *electrostatic bonds, hydrogen bonds,* and *van der Waals bonds.* They differ in geometry, strength, and specificity. Furthermore, these bonds are profoundly affected in different ways by the presence of water. Let us consider the characteristics of each:

1. *Electrostatic bonds.* A charged group on a substrate can attract an oppositely charged group on an enzyme. The force (F) of such an *electrostatic attraction* is given by Coulomb's law:

$$F = \frac{q_1 q_2}{r^2 D}$$

in which q_1 and q_2 are the charges of the two groups, r is the distance between them, and D is the dielectric constant of the medium. The attraction is strongest in a vacuum (where D is 1) and is weakest in a medium such as water (where D is 80). This kind of attraction is also called an *ionic bond, salt linkage, salt bridge,* or *ion pair.* The distance between oppositely charged atoms in an optimal electrostatic attraction is about 2.8 Å.

2. *Hydrogen bonds* can be formed between uncharged molecules as well as charged ones. In a hydrogen bond, *a hydrogen atom is shared by two other atoms.* The atom to which the hydrogen is more tightly linked is called the *hydrogen donor,* whereas the other atom is the *hydrogen acceptor.* The acceptor has a partial negative charge that attracts the hydrogen atom. In fact, a hydrogen bond can be considered an intermediate in the transfer of a proton from an acid to a base. It is reminiscent of a ménage à trois.

$$-CH_2-C\overset{\displaystyle O}{\underset{\displaystyle O^-}{\big<}}$$

Negatively charged
group of a substrate

$$\overset{+}{H_3N}-CH_2-$$

Positively charged
group of an enzyme

O—H ··· O

$$\diagdown N-H \cdots O=C\diagup$$

$$O-H \cdots O=C\diagup$$

Strong hydrogen bonds

O
⋮
O—H

Weak hydrogen bond

Hydrogen Hydrogen
donor acceptor

—O—H ··· N—
2.88 Å

Hydrogen Hydrogen
donor acceptor

—N—H ··· O—
3.04 Å

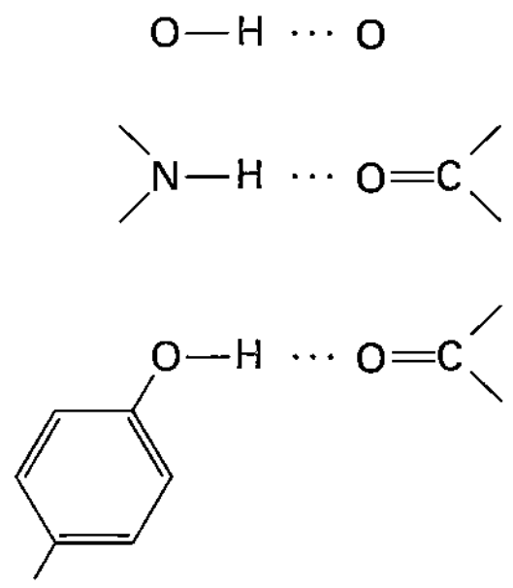

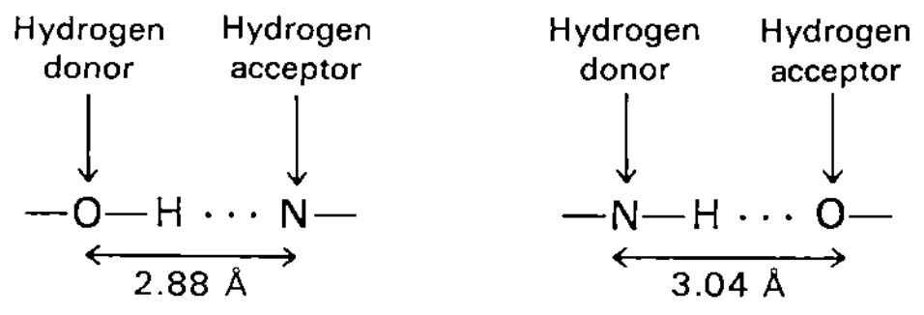

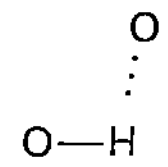

Table 1-1
Typical hydrogen-bond lengths

Bond	*Length (Å)*
O—H · · · O	2.70
O—H · · · O⁻	2.63
O—H · · · N	2.88
N—H · · · O	3.04
N⁺—H · · · O	2.93
N—H · · · N	3.10

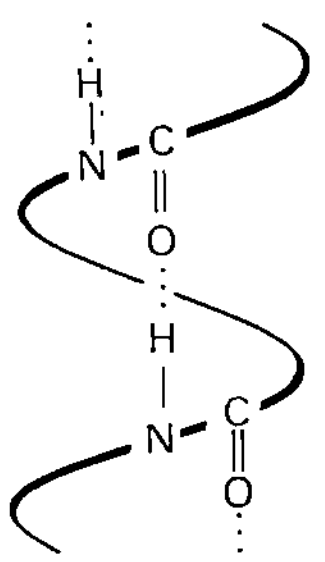

Figure 1-9
Schematic diagram of hydrogen bonding between an amide and a carbonyl group in an α helix of a protein.

The donor in a hydrogen bond in biological systems is an oxygen or nitrogen atom that has a covalently attached hydrogen atom. The acceptor is either oxygen or nitrogen. The kinds of hydrogen bonds formed and their bond lengths are given in Table 1-1. The bond energies range from about 3 to 7 kcal/mol. Hydrogen bonds are stronger than van der Waals bonds but much weaker than covalent bonds. The length of a hydrogen bond is intermediate between that of a covalent bond and a van der Waals bond. *An important feature of hydrogen bonds is that they are highly directional.* The strongest hydrogen bonds are those in which the donor, hydrogen, and acceptor atoms are colinear. The α helix, a recurring motif in proteins, is stabilized by hydrogen bonds between amide (–NH) and carbonyl (–CO) groups (Figure 1-9). Another example of the importance of hydrogen bonding is the DNA double helix, which is held together by hydrogen bonds between bases on opposite strands (see Figure 1-1).

3. *Van der Waals bonds,* a nonspecific attractive force, come into play when any two atoms are 3 to 4 Å apart. Though weaker and less specific than electrostatic and hydrogen bonds, van der Waals bonds are no less important in biological systems. The basis of a van der Waals bond is that the distribution of electronic charge around an atom changes with time. At any instant, the charge distribution is not perfectly symmetric. This transient asymmetry in the electronic charge around an atom encourages a similar asymmetry in the electron distribution around its neighboring atoms. The resulting attraction between a pair of atoms increases as they come closer, until they are separated by the van der Waals *contact distance* (Figure 1-10). At a shorter distance, very strong repulsive forces become dominant because the outer electron clouds overlap. The contact distance between an oxygen and carbon atom, for example, is 3.4 Å, which is obtained by adding 1.4 and 2.0 Å, the contact radii (Table 1-2) of the O and C atoms.

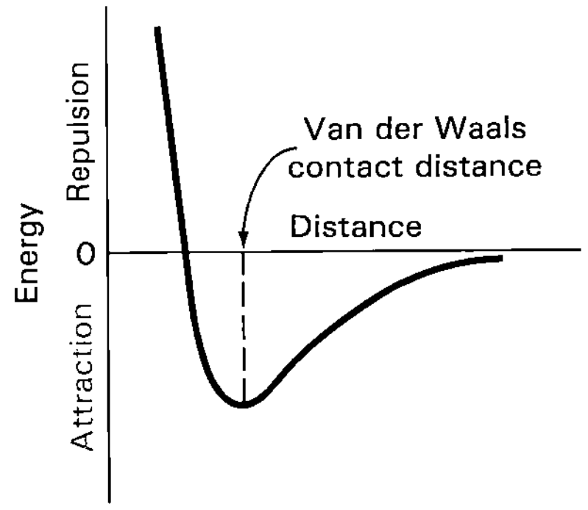

Figure 1-10
Energy of a van der Waals interaction as a function of the distance between two atoms.

Table 1-2
Van der Waals contact radii of atoms (Å)

Atom	*Radius*
H	1.2
C	2.0
N	1.5
O	1.4
S	1.85
P	1.9

The van der Waals bond energy of a pair of atoms is about 1 kcal/mol. It is considerably weaker than a hydrogen or electrostatic bond, which is in the range of 3 to 7 kcal/mol. A single van der Waals bond counts for very little because its strength is only a little more than the average thermal energy of molecules at room temperature (0.6 kcal/mol). Furthermore, the van der Waals force fades rapidly when the distance between a pair of atoms becomes even 1 Å greater than their contact distance. It becomes significant only when numerous atoms in one of a pair of molecules can simultaneously come close to many atoms of the other. This can happen only if the shapes of the molecules match. In other words, effective van der Waals interactions depend on *steric complementarity.* Though there is virtually no specificity in a single van der Waals interaction, *specificity arises when there is an opportunity to make a large number of van der Waals bonds simultaneously.* Repulsions between atoms closer than the van der Waals contact distance are as important as attractions for establishing specificity.

THE BIOLOGICALLY IMPORTANT PROPERTIES OF WATER ARE ITS POLARITY AND COHESIVENESS

Water profoundly influences all molecular interactions in biological systems. Two properties of water are especially important in this regard:

1. *Water is a polar molecule.* The shape of the molecule is triangular, not linear, and so there is an asymmetrical distribution of charge. The oxygen nucleus draws electrons away from the hydrogen nuclei, which leaves the region around those nuclei with a net positive charge. The water molecule is thus an electrically polar structure.

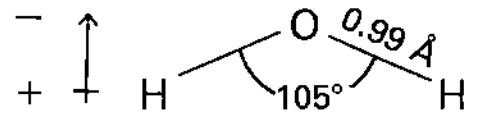

2. *Water is highly cohesive.* Neighboring water molecules have high affinity for each other. A positively charged region in one water molecule tends to orient itself toward a negatively charged region in one of its neighbors. Ice has a highly regular crystalline structure in which all potential hydrogen bonds are made (Figure 1-11). Liquid water has a partly ordered structure in which hydrogen bonded clusters of molecules are continually forming and breaking up. Each molecule is hydrogen bonded to an average of 3.4 neighbors in liquid water, compared with 4 in ice.

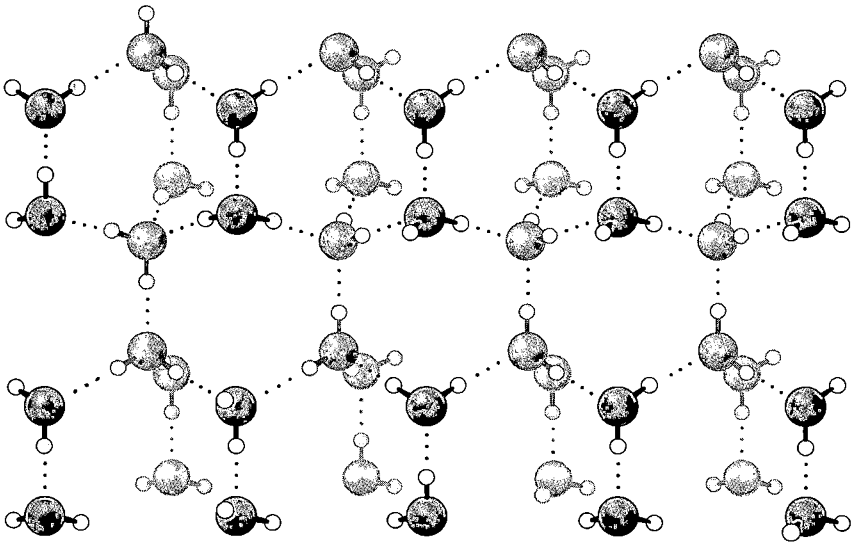

Figure 1-11
Structure of a form of ice. [After L. Pauling and P. Pauling. *Chemistry* (W.H. Freeman, 1975), p. 289.]

WATER SOLVATES POLAR MOLECULES AND HENCE WEAKENS IONIC AND HYDROGEN BONDS

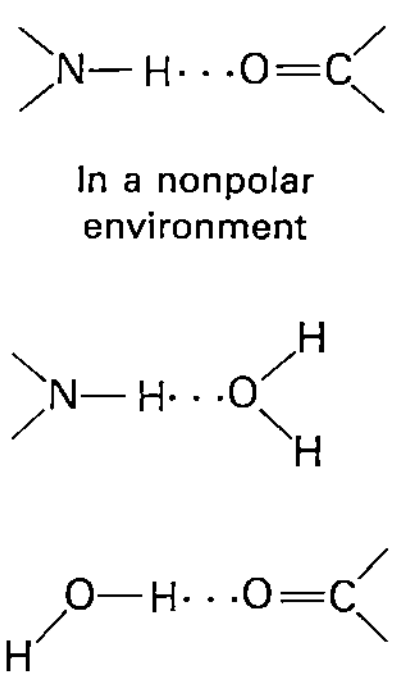

Figure 1-12
Water competes for hydrogen bonds.

The polarity and hydrogen-bonding capability of water make it a highly interacting molecule. Water is an excellent solvent for polar molecules. The reason is that water greatly weakens electrostatic forces and hydrogen bonding between polar molecules by competing for their attractions. For example, consider the effect of water on hydrogen bonding between a carbonyl and an amide group (Figure 1-12). The hydrogen atoms of water can replace the amide hydrogen group as hydrogen-bond donors, and the oxygen atom of water can replace the carbonyl oxygen as the acceptor. Hence, a strong hydrogen bond between a CO and an NH group forms only if water is excluded.

Water diminishes the strength of electrostatic attractions by a factor of 80, the dielectric constant of water, compared with the same interactions in a vacuum. Water has an unusually high dielectric constant (Table 1-3) because of its polarity and capacity to form oriented solvent shells around ions (Figure 1-13). These oriented solvent shells produce electric fields of their own, which oppose the fields produced by the ions. Consequently, electrostatic attractions between ions are markedly weakened by the presence of water.

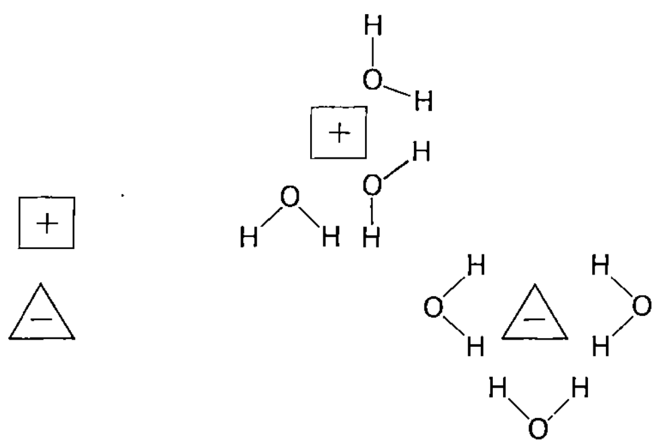

Electrostatic interaction in a nonpolar environment
Water surrounds the charged groups and attenuates their interaction

Figure 1-13
Water attenuates electrostatic attractions between charged groups.

The existence of life on earth depends critically on the capacity of water to dissolve a remarkable array of polar molecules that serve as fuels, building blocks, catalysts, and information carriers. High concentrations of these polar molecules can coexist in water, where they are free to diffuse and find each other. However, the excellence of water as a solvent poses a problem, for it also weakens interactions between polar molecules. *Biological systems have solved this problem by creating water-free microenvironments where polar interactions have maximal strength and specificity.* We shall see many examples of the critical importance of these specially constructed niches in protein molecules.

Table 1-3
Dielectric constants of some solvents at 20°C

Substance	*Dielectric constant*
Hexane	1.9
Benzene	2.3
Diethyl ether	4.3
Chloroform	5.1
Acetone	21.4
Ethanol	24
Methanol	33
Water	80
Hydrogen cyanide	116

HYDROPHOBIC ATTRACTIONS: NONPOLAR GROUPS TEND TO ASSOCIATE IN WATER

The sight of dispersed oil droplets coming together in water to form a single large oil drop is a familiar one. An analogous process occurs at the atomic level: *nonpolar molecules or groups tend to cluster together in water.*

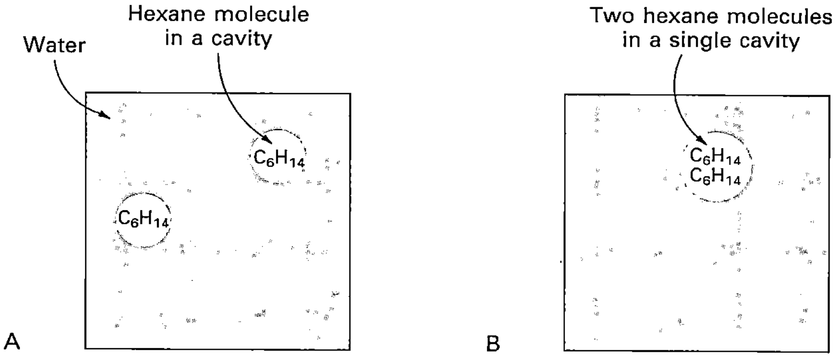

Figure 1-14
A schematic representation of two molecules of hexane in a small volume of water. (A) The hexane molecules occupy different cavities in the water structure, or (B) they occupy the same cavity, which is energetically more favored.

These associations are called *hydrophobic attractions.* In a figurative sense, water tends to squeeze nonpolar molecules together.

Let us examine the basis of hydrophobic attractions, which are a major driving force in the folding of macromolecules, the binding of substrates to enzymes, and the formation of membranes that define the boundaries of cells and their internal compartments. Consider the introduction of a single nonpolar molecule, such as hexane, into some water. A cavity in the water is created, which temporarily disrupts some hydrogen bonds between water molecules. The displaced water molecules then reorient themselves to form a maximum number of new hydrogen bonds. This is accomplished at a price: the number of ways of forming hydrogen bonds in the cage of water around the hexane molecule is much fewer than in pure water. The water molecules around the hexane molecule are much more ordered than elsewhere in the solution. Now consider the arrangement of two hexane molecules in water. Do they sit in two small cavities (Figure 1-14A) or in a single larger one (Figure 1-14B)? The experimental fact is that the two hexane molecules come together and occupy a single large cavity. This association releases some of the more ordered water molecules around the separated hexanes. In fact, the basis of a hydrophobic attraction is this enhanced freedom of released water molecules. *Nonpolar solute molecules are driven together in water not primarily because they have a high affinity for each other but because water bonds strongly to itself.*

DESIGN OF THIS BOOK

This book has five parts, each having a major theme:

I Molecular Design of Life
II Proteins: Conformation, Dynamics, and Function
III Metabolic Energy: Generation and Storage
IV Biosynthesis of Building Blocks
V Genes: Replication, Expression, and Repair

Part I is an overview of the central molecules of life—DNA, RNA, and proteins—and their interplay. We begin with proteins, which are unique in being able to recognize and bind a remarkably diverse array of molecules. Proteins

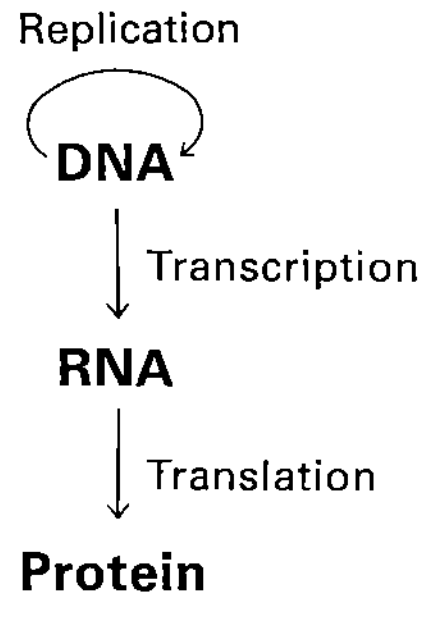

Figure 1-15
Flow of genetic information.

determine the pattern of chemical transformations in biological systems by catalyzing nearly all the necessary chemical reactions. We then turn to DNA, the repository of genetic information in all cells. The discovery of the DNA double helix led immediately to an understanding of how DNA replicates. The following chapter deals with the flow of genetic information from DNA to RNA to protein. The first step, called transcription, is the synthesis of RNA, and the second, called translation, is the synthesis of proteins according to instructions given by templates of messenger RNA. The genetic code, which specifies the relation between the sequence of 4 kinds of bases in DNA and RNA and the 20 kinds of amino acids in proteins, is beautiful in its simplicity. Three bases constitute a codon, the unit that specifies an amino acid. Translation is carried out by the coordinated interplay of more than a hundred kinds of protein and RNA molecules in an organized assembly called the ribosome. Experimental methods for exploring proteins and genes are also presented in Part I. Recombinant DNA technology is introduced here, and some examples of its power and generality in analyzing and altering both genes and proteins are given.

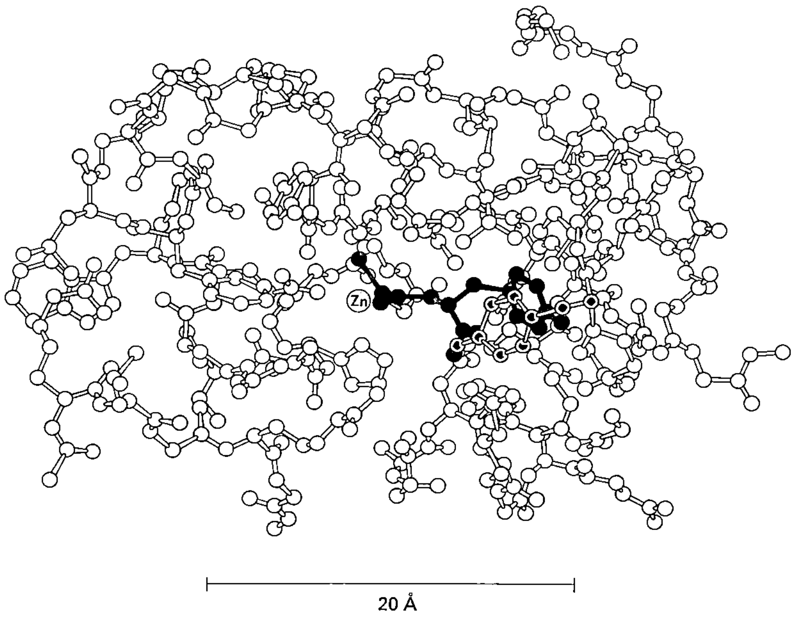

Figure 1-16
Structure of an enzyme-substrate complex. Glycyltyrosine (shown in red) is bound to carboxypeptidase A, a digestive enzyme. Several groups of the enzyme that are important for binding or catalysis are shown in yellow, blue, and green. Only a quarter of the enzyme is shown. [After W.N. Lipscomb. *Proc. Robert A. Welch Found. Conf. Chem. Res.* 15(1971):141.]

The interplay of three-dimensional structure and biological activity as exemplified by proteins is the major theme of Part II. This section begins with myoglobin and hemoglobin, the oxygen-carrying proteins in vertebrates, because these proteins illustrate many principles of protein architecture and function. Hemoglobin is especially interesting because its binding of oxygen is regulated by specific molecules in its environment. Sickle-cell anemia graphically illustrates how a single mutation can lead to a serious disease. We then turn to enzymes and consider how they recognize substrates and enhance reaction rates by factors of a million or more. The enzymes lysozyme, carboxypeptidase A, and chymotrypsin are examined closely because they reveal many general principles of catalysis. The regulation of enzymatic activity by specific control proteins and other signal molecules is considered next. Allosteric interactions between distant sites on proteins play key roles in the transmission and processing of information at the molecular level.

We then turn to membranes, which are organized sheetlike assemblies of lipids and proteins. Membranes are highly selective permeability barriers that give cells individuality by separating them from their environment. Three major classes of proteins in membranes—channels, pumps, and receptors—control the flow of matter and information between the external milieu and the inside of cells. The pumping of Na^+, K^+, and Ca^{2+} generates gradients that are at the heart of excitability of nerve and muscle. Receptors and enzymes too come into play in the sensing and processing of sensory and hormonal stimuli. How do bacteria detect nutrients in their environment and move toward them? How is a retinal rod cell triggered by a single photon? How are growth signals recognized, and how do molecular lesions in these pathways lead to cancer? The molecular circuitry of these signal transduction processes is rapidly being deciphered.

We continue our exploration of the rich and diverse protein landscape by taking a close look at the molecular basis of the immune response. How does an organism scan vast numbers of molecules to decide which ones are foreign? The key proteins in recognition—immunoglobulins and T-cell receptors—unleash protective responses when they encounter bacteria, viruses, and other pathogens. The next chapter deals with proteins as molecular motors. We are learning how chemical energy is harnessed to produce coordinated motion as in muscle contraction and the beating of cilia. The ATP-powered movement of myosin on actin tracks, and of kinesin on dynein tracks, is now being viewed and analyzed at the atomic level. Part II concludes with a discussion of one of the most challenging and important questions in biochemistry: how does the amino acid sequence of a protein specify its three-dimensional form? Recent advances in our understanding of protein folding have opened a new area of research, the design of novel proteins with tailor-made properties.

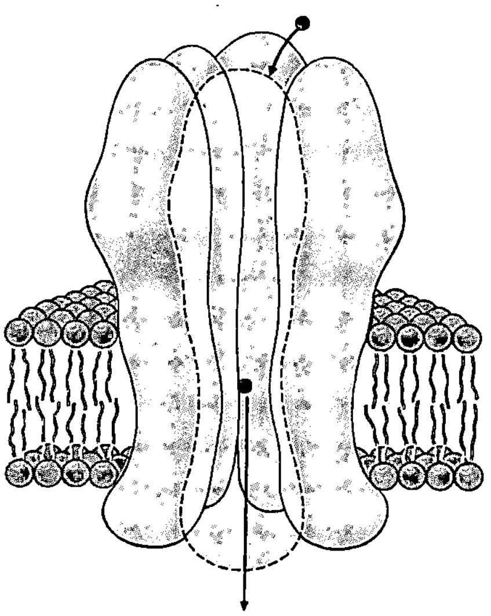

Figure 1-17
Acetylcholine receptor channel. The five subunits of this protein span the lipid bilayer to form a pore that conducts ions at synapses between nerve and muscle. [After N. Unwin. *Cell* 72(1993):37.]

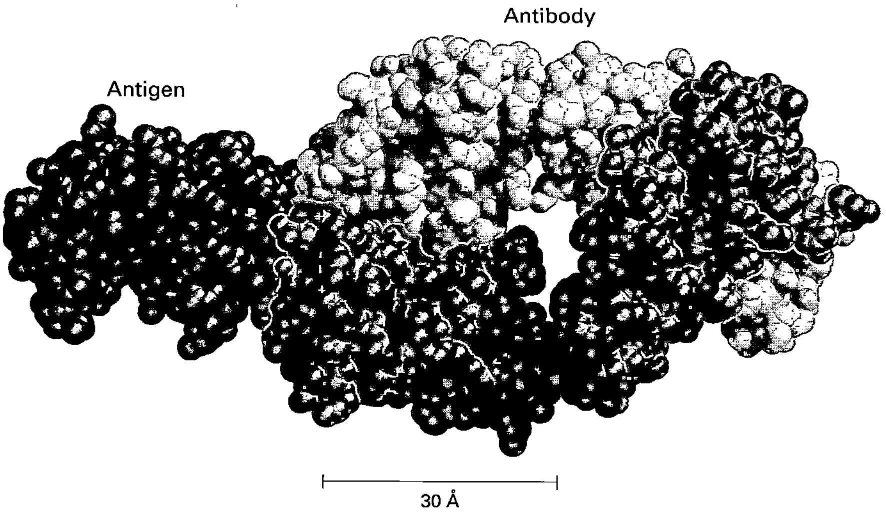

Figure 1-18
Three-dimensional structure of an antigen-antibody complex. The foreign protein is shown in red and an antibody fragment in yellow and blue. [Drawn from Protein Data Bank coordinate set 1fdl.pdb. T.O. Fischmann, G.A. Bentley, T.N. Bhat, G. Boulot, R.A. Mariuzza, S.E.V. Phillips, D. Tello, and R.J. Poljak. *J. Biol. Chem.* 266(1991):12915.]

Part III deals with the generation and storage of metabolic energy. First, the overall strategy of metabolism is presented. Cells convert energy from fuel molecules into ATP. In turn, ATP drives most energy-requiring processes in cells. In addition, reducing power in the form of nicotinamide adenine dinucleotide phosphate (NADPH) is generated for use in biosyn-

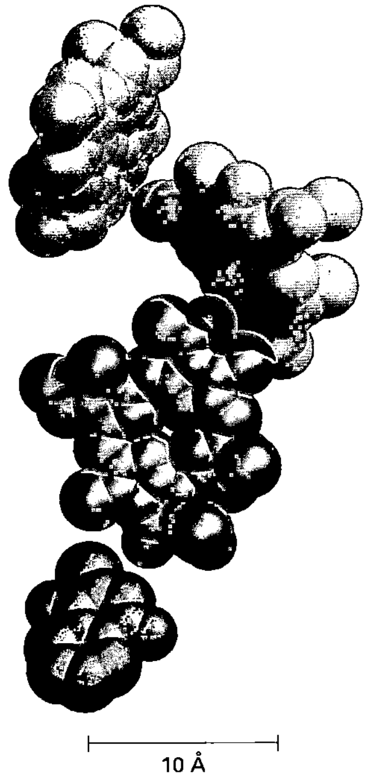

Figure 1-19
Structure of the electron transfer chain in the photosynthetic reaction center of a purple sulfur bacteria. The absorption of light by the bacteriochlorophyll molecule (green) on top leads to the ejection of an electron, which is transferred through two chlorophyll-like molecules (yellow and blue) to an acceptor (red). The light-absorbing chlorophyll shown here is one of a pair. Only a small portion of the reaction center is shown here. [Drawn from 1prc.pdb. J. Deisenhofer, O. Epp, K. Miki, R. Huber, and H. Michel. *Nature* 318(1965):618.]

theses. The metabolic pathways that carry out these reactions are then presented in detail. For example, the generation of ATP from glucose requires a sequence of three series of reactions—glycolysis, the citric acid cycle, and oxidative phosphorylation. The last two are also common to the generation of ATP from the oxidation of fats and some amino acids, the other major fuels. We see here an illustration of molecular economy. Two storage forms of fuel molecules, glycogen and triacylglycerols (neutral fats), are also discussed in Part III. The concluding topic of this part of the book is photosynthesis, in which the primary event is the light-activated transfer of an electron from one substance to another against a chemical potential gradient. As in oxidative phosphorylation, electron flow leads to the pumping of protons across a membrane, which in turn drives the synthesis of ATP. In essence, life is powered by proton batteries that are ultimately energized by the sun.

Part IV considers the biosynthesis of building blocks. These small recurring units serve as precursors of macromolecules (DNA, RNA, proteins, and polysaccharides) and as components of membranes. We begin with the synthesis of membrane lipids and steroids. The pathway for the synthesis of cholesterol, a 27-carbon steroid, is of particular interest because all its carbon atoms come from a 2-carbon precursor. The reactions leading to the synthesis of selected amino acids and the heme group are then discussed. Key steps in these biosynthetic pathways are feedback inhibited by their products and by other regulatory mechanisms to precisely adjust supply and demand. The biosynthesis of nucleotides, the activated precursors of DNA and RNA, is then considered. The final chapter in this part deals with the integration of metabolism. How are energy-yielding and energy-consuming reactions coordinated to meet the needs of an organism? The running of a marathon graphically illustrates how metabolic fuels are selected to achieve maximum power.

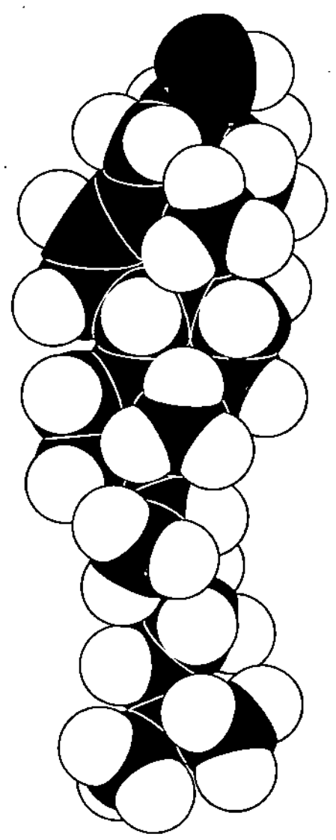

Figure 1-20
Model of cholesterol, a 27-carbon lipid present in many biological membranes.

The transmission and expression of genetic information constitute the central theme of Part V. The genetic role and structure of DNA were introduced in

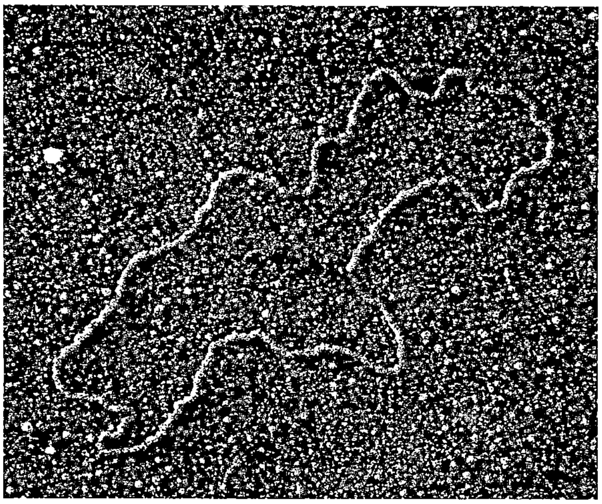

Figure 1-21
Electron micrograph of a DNA molecule. [Courtesy of Dr. Thomas Broker.]

Part I, as was the flow of genetic information. We now resume our consideration of this theme, enriched with a knowledge of proteins and metabolic transformations. The mechanisms of DNA replication and DNA repair are discussed first. A remarkable aspect of DNA replication is its very high accuracy—the error frequency in replicating a nucleotide is less than one in a million. The processes of genetic recombination and transposition, which produce new combinations of DNA, are then presented. We next turn to transcription and the processing of nascent transcripts to form functional RNA molecules. The mechanism of protein synthesis, in which transfer RNAs, messenger RNAs, and ribosomes interact, comes next. RNA molecules play key roles in the splicing of RNA and in protein synthesis. We then consider how proteins are specifically targeted to many different destinations. The signals that guide proteins from one intracellular compartment to another in eukaryotic cells are being decoded.

The next topic is the control of gene expression in bacteria. Many of their genes are clustered in units called operons that are coordinately switched in response to internal or environmental signals. Genetic and structural studies have revealed how the lactose and tryptophan operons of *E. coli* are controlled. We will also see how the interplay of repressors and activators of transcription determines the developmental fate of lambda (λ) phage, a virus that infects *E. coli*. The following chapter deals with the molecular architecture of eukaryotic chromosomes and regulation of their expression. The large amount of DNA in the genomes of eukaryotes—3×10^9 base pairs in humans—endows them with potentialities not present in prokaryotes and poses additional challenges. How is the DNA packaged inside the nucleus, and how is it selectively expressed? Multiple transcriptional regulators come together on DNA to activate the expression of dispersed genes. These control proteins contain structural motifs such as zinc fingers and leucine zippers that recur in all eukaryotes. Indeed, similar homeodomain proteins establish the segmentation pattern—the basic body plan—in the embryos of flies and humans. The molecular dance of development is being unveiled.

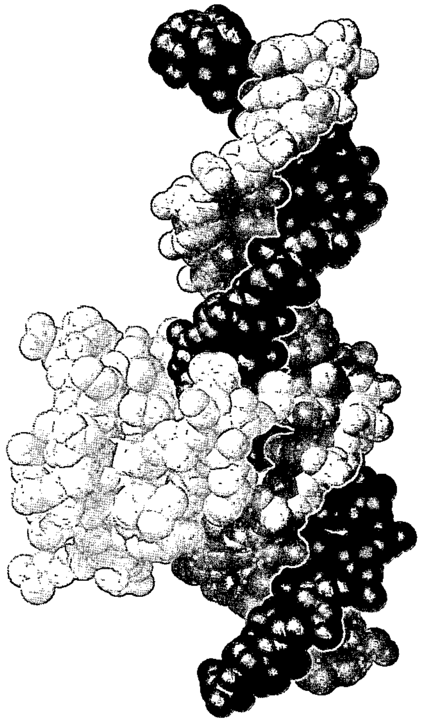

Figure 1-22
Three-dimensional structure of a homeodomain protein (yellow) bound to a target site on double-helical DNA (red and green). [Drawn from 1hdd.pdb. C.R. Kissingger, B. Liu, E. Martin-Blanco, T.B. Kornberg, and C.O. Pabo. *Cell* 63(1990):579.]

"Analogy would lead me one step further, namely, to the belief that all animals and plants have descended from some one prototype. . . . All living things have much in common, in their chemical composition, their germinal vesicles, their cellular structure, and their laws of growth and reproduction. . . . Probably all the organic beings which have ever lived on this earth have descended from some one primordial form, into which life was first breathed. . . . From so simple a beginning endless forms most beautiful and most wonderful have been, and are being, evolved."

CHARLES DARWIN
On the Origin of Species (1859)

Biochemistry reveals the deep connections between different forms of life. The sequences of protein, DNA, and RNA molecules are enormously informative. As words reveal the history of languages, these documents of evolutionary history shed light on processes that took place millions and even billions of years ago. Ancient molecules can be distinguished from those that came into being in recent times, and the pathway of evolution can be retraced. The identity of a 14-nucleotide ribosomal RNA sequence in all eubacteria, archaebacteria, and eukaryotes attests to their common origin. Likewise, the presence of the same ribonucleotide unit in several central molecules of metabolism—ATP, NADH, $FADH_2$, and coenzyme A—reflects the underlying kinship of all organisms. The discovery of catalytic RNA enables us to envision an RNA world early in the evolution of life, before the appearance of DNA and protein.

We see, too, that the mechanisms of many molecular processes have been conserved over a long span of evolution. The retinal receptor that captures light in the first step of vision in animals is strikingly similar to the receptor used by yeast to detect the presence of a potential mate. Furthermore, these sensory receptors are coupled to amplifier proteins that are variations on a common theme. Nor have we only retrospective clues to the evolutionary process. The rapid changes of influenza virus and human immunodeficiency virus seen in recent years are graphic indicators of evolution in action, of the power of mutation and selection in generating new forms.

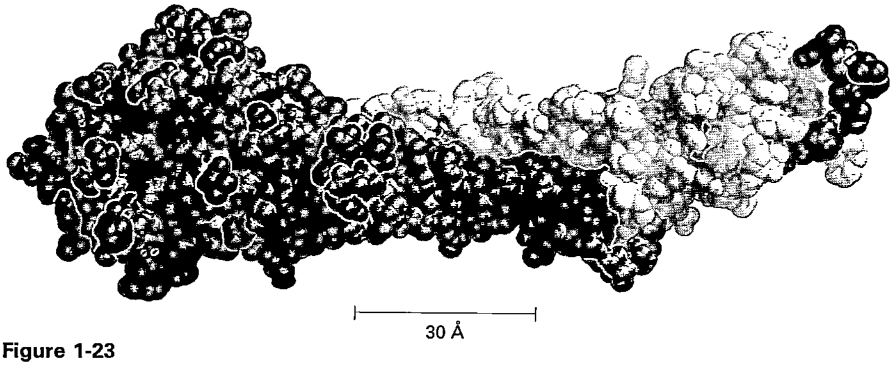

Figure 1-23
Influenza virus evades the immune response of hosts by changing its protein coat. The blue portion of the viral hemagglutinin protein faces the outside of the virus particle, and the yellow portion is located nearer the viral membrane. Surface residues that mutated in the interval from 1969 to 1983 are shown in red. [Drawn from 1hgg.pdb. N.K. Sauter, G.D. Glick, J.H. Brown, R.L. Crowther, S.-J. Park, M.B. Eisen, J.J. Skehel, J.R. Knowles, and D.C. Wiley. *Proc. Nat. Acad. Sci.* 89(1992):324.]

All forms of life arose from a common ancestor. All are subject to the same laws of physics and chemistry. Biochemistry is an intellectually coherent and beautiful discipline because of the underlying unity of life.

CHAPTER

2

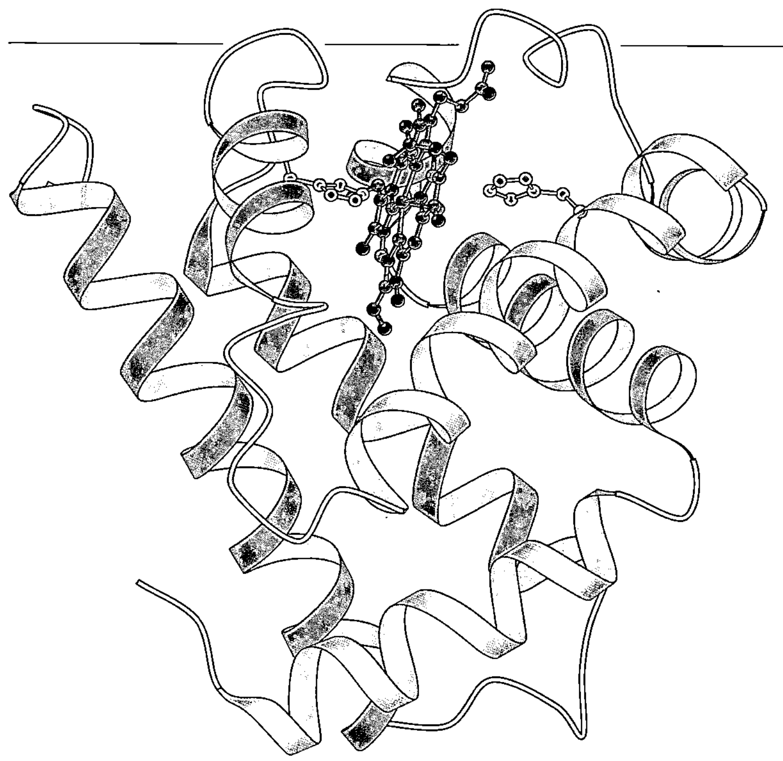

Protein Structure and Function

Proteins play crucial roles in virtually all biological processes. Their significance and the remarkable scope of their activity are exemplified in the following functions:

1. *Enzymatic catalysis.* Nearly all chemical reactions in biological systems are catalyzed by specific macromolecules called *enzymes.* Some of these reactions, such as the hydration of carbon dioxide, are quite simple. Others, such as the replication of an entire chromosome, are highly intricate. Enzymes exhibit enormous catalytic power. They usually increase reaction rates by at least a millionfold. Indeed, chemical transformations in vivo rarely proceed at perceptible rates in the absence of enzymes. Several thousand enzymes have been characterized, and many of them have been crystallized. The striking fact is that nearly all known enzymes are proteins. Thus, proteins are central in determining the pattern of chemical transformations in biological systems.

2. *Transport and storage.* Many small molecules and ions are transported by specific proteins. For example, hemoglobin transports oxygen in erythrocytes, whereas myoglobin, a related protein, transports oxygen in muscle. Iron is carried in the plasma of blood by transferrin and is stored in the liver as a complex with ferritin, a different protein.

Protein—
Derived from the Greek word *proteios,* which means "of the first rank." A word coined by Jöns J. Berzelius in 1838 to emphasize the importance of this class of molecules.

Figure 2-1
Photomicrograph of a crystal of hexokinase, a key enzyme in the utilization of glucose. [Courtesy of Dr. Thomas Steitz and Dr. Mark Yeager.]

Opening Image: *Structure of myoglobin, an oxygen-carrying protein in muscle. O_2 binds to the heme group (red), near two key histidine residues (blue). Two of the α helices are shown in green, and the others in orange. The same structural motif occurs in hemoglobin, the oxygen-carrying protein in blood. [Drawn from 1mbn.pdb. H.C. Watson. Prog. Stereochem. 4(1969):299.]*

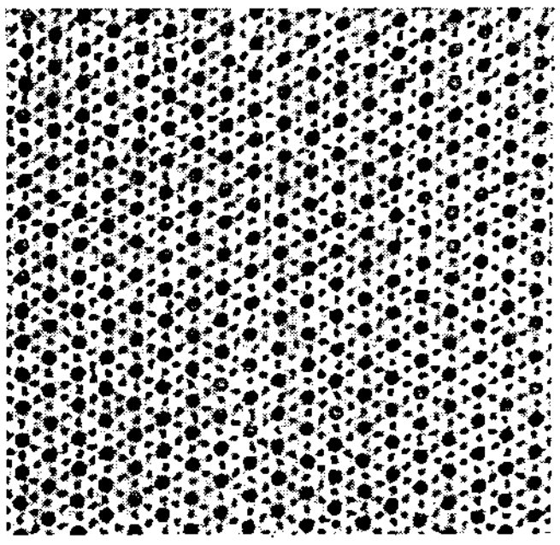

Figure 2-2
Electron micrograph of a cross section of insect flight muscle showing a hexagonal array of two kinds of protein filaments. [Courtesy of Dr. Michael Reedy.]

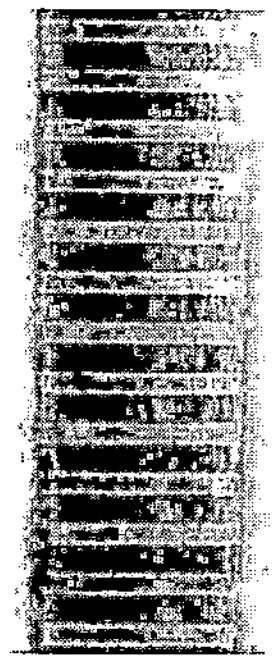

Figure 2-3
Electron micrograph of a fiber of collagen. [Courtesy of Dr. Jerome Gross and Dr. Romaine Bruns.]

3. *Coordinated motion.* Proteins are the major component of muscle. Muscle contraction is accomplished by the sliding motion of two kinds of protein filaments. On the microscopic scale, such coordinated motions as the movement of chromosomes in mitosis and the propulsion of sperm by their flagella also are produced by contractile assemblies consisting of proteins.

4. *Mechanical support.* The high tensile strength of skin and bone is due to the presence of collagen, a fibrous protein.

5. *Immune protection.* Antibodies are highly specific proteins that recognize and combine with such foreign substances as viruses, bacteria, and cells from other organisms. Proteins thus play a vital role in distinguishing between self and nonself.

6. *Generation and transmission of nerve impulses.* The response of nerve cells to specific stimuli is mediated by receptor proteins. For example, rhodopsin is the light-sensitive protein in retinal rod cells. Receptor proteins that can be triggered by specific small molecules, such as acetylcholine, are responsible for transmitting nerve impulses at synapses—that is, at junctions between nerve cells.

7. *Control of growth and differentiation.* Controlled sequential expression of genetic information is essential for the orderly growth and differentiation of cells. Only a small fraction of the genome of a cell is expressed at any one time. In bacteria, repressor proteins are important control elements that silence specific segments of the DNA of a cell. In higher organisms, growth and differentiation are controlled by growth factor proteins. For example, nerve growth factor guides the formation of neural networks. The activities of different cells in multicellular organisms are coordinated by hormones. Many of them, such as insulin and thyroid-stimulating hormone, are proteins. Indeed, proteins serve in all cells as sensors that control the flow of energy and matter.

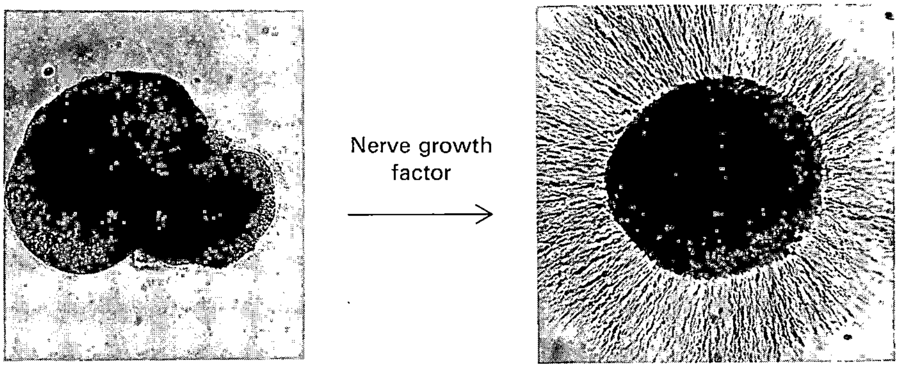

Figure 2-4
Photomicrograph of a ganglion showing the proliferation of nerves after addition of nerve growth factor, a complex of proteins. [Courtesy of Dr. Eric Shooter.]

PROTEINS ARE BUILT FROM A REPERTOIRE OF 20 AMINO ACIDS

Amino acids are the basic structural units of proteins. An α-amino acid consists of an amino group, a carboxyl group, a hydrogen atom, and a distinctive R group, all of which are bonded to an *α carbon* atom. This carbon atom is named α because it is adjacent to the carboxyl (acidic) group (Figure 2-5). An R group is referred to as a *side chain* for reasons that will be evident shortly.

$$
\begin{array}{ccc}
& \text{NH}_2 & & & \text{NH}_3{}^+ \\
& | & & & | \\
\text{H}\!-\!\text{C}\!-\!\text{COOH} & & & \text{H}\!-\!\text{C}\!-\!\text{COO}^- \\
& | & & & | \\
& \text{R} & & & \text{R}
\end{array}
$$

Un-ionized form of an Dipolar ion (or zwitterion)
amino acid form of an amino acid

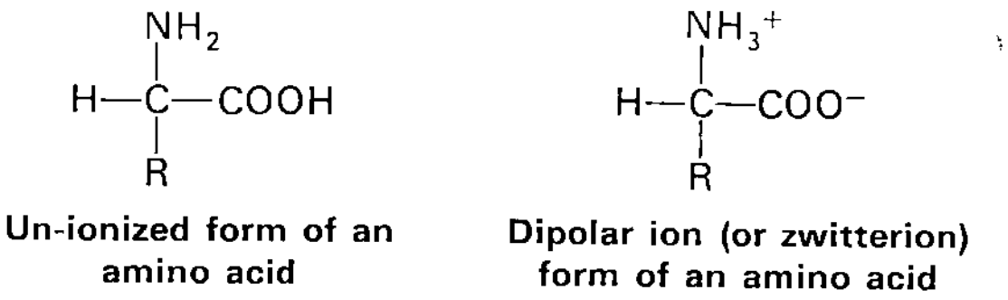

Figure 2-5
Structure of the un-ionized and zwitterion forms of an α-amino acid. The R group is the side chain.

Amino acids in solution at neutral pH are predominantly *dipolar ions* (or *zwitterions*) rather than un-ionized molecules. In the dipolar form of an amino acid, the amino group is protonated ($-\text{NH}_3{}^+$) and the carboxyl group is dissociated ($-\text{COO}^-$). The ionization state of an amino acid varies with pH (Figure 2-6). In acid solution (e.g., pH 1), the carboxyl group is un-ionized ($-\text{COOH}$) and the amino group is ionized ($-\text{NH}_3{}^+$). In alkaline solution (e.g., pH 11), the carboxyl group is ionized ($-\text{COO}^-$) and the amino group is un-ionized ($-\text{NH}_2$). For glycine, the pK of the carboxyl group is 2.3 and that of the amino group is 9.6. In other words, the midpoint of the first ionization is at pH 2.3, and that of the second is at pH 9.6. For a review of acid-base concepts and pH, see the appendix to this chapter.

$$
\begin{array}{ccccc}
\text{NH}_3{}^+ & & \text{NH}_3{}^+ & & \text{NH}_2 \\
| & & | & & | \\
\text{H}\!-\!\text{C}\!-\!\text{COOH} & \rightleftharpoons & \text{H}\!-\!\text{C}\!-\!\text{COO}^- & \rightleftharpoons & \text{H}\!-\!\text{C}\!-\!\text{COO}^- \\
| & \text{H+} & | & \text{H+} & | \\
\text{R} & & \text{R} & & \text{R}
\end{array}
$$

Predominant form Predominant form Predominant form
at pH 1 at pH 7 at pH 11

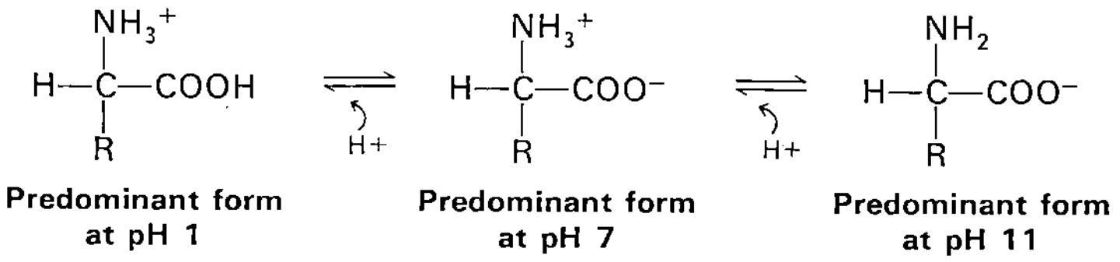

Figure 2-6
Ionization states of an amino acid depend on pH.

The tetrahedral array of four different groups about the α carbon atom confers optical activity on amino acids. The two mirror-image forms are called the L isomer and the D isomer (Figure 2-7). *Only L amino acids are constituents of proteins.* Hence, the designation of the optical isomer will be omitted and the L isomer implied in discussions of proteins herein, unless otherwise noted.

Twenty kinds of side chains varying in *size, shape, charge, hydrogen-bonding capacity,* and *chemical reactivity* are commonly found in proteins. Indeed, all proteins in all species, from bacteria to humans, are constructed from the same set of 20 amino acids. This fundamental alphabet of proteins is at least two billion years old. The remarkable range of functions mediated by proteins results from the diversity and versatility of these 20 kinds of building blocks. We shall explore ways in which this alphabet is used to create the intricate three-dimensional structures that enable proteins to carry out so many biological processes.

Let us look at this set of amino acids. The simplest one is *glycine,* which has just a hydrogen atom as its side chain (Figure 2-8). *Alanine* comes next, with a methyl group. Larger hydrocarbon side chains (three and

L Isomer

D Isomer

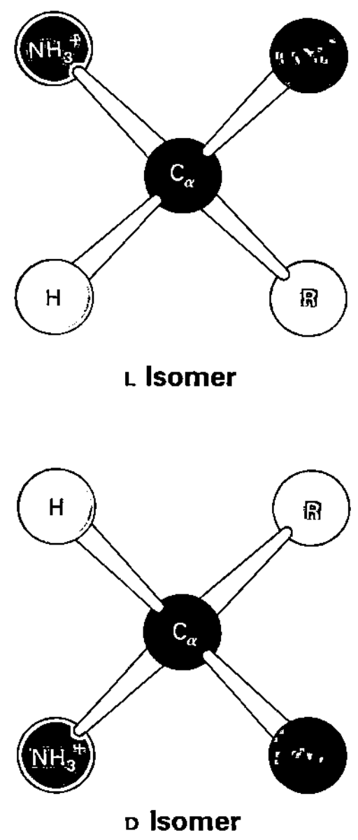

Figure 2-7
Absolute configurations of the L and D isomers of amino acids. R refers to the side chain. The L and D isomers are mirror images of each other.

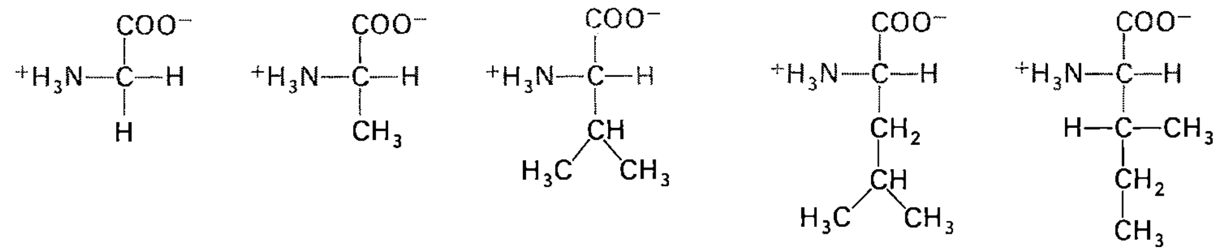

Glycine (Gly, G) | Alanine (Ala, A) | Valine (Val, V) | Leucine (Leu, L) | Isoleucine (Ile, I)

Figure 2-8
Amino acids having aliphatic side chains.

four carbons long) are found in *valine, leucine,* and *isoleucine.* These larger aliphatic side chains are *hydrophobic*—that is, they have an aversion to water and like to cluster. As will be discussed later, the three-dimensional structure of water-soluble proteins is stabilized by the coming together of hydrophobic side chains to avoid contact with water. The different sizes and shapes of these hydrocarbon side chains (Figure 2-9) enable them to pack together to form compact structures with few holes.

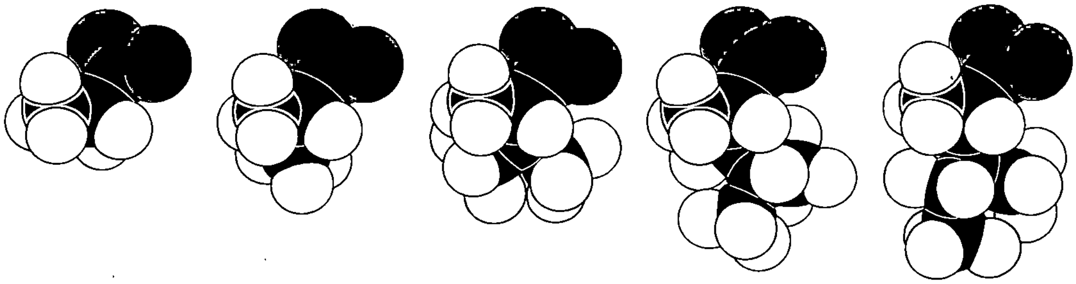

Gly Ala Val Leu Ile

Figure 2-9
Models of aliphatic amino acids.

Proline also has an aliphatic side chain, but it differs from other members of the set of 20 in that its side chain is bonded to both the nitrogen and the α carbon atoms. The resulting cyclic structure (Figure 2-10) markedly influences protein architecture. Proline, often found in the bends of folded protein chains, is not averse to being exposed to water.

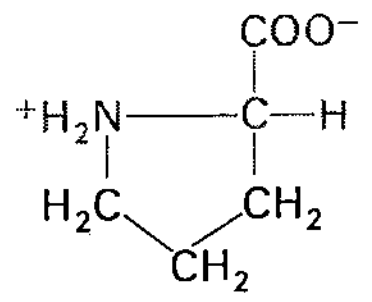

Proline (Pro, P)

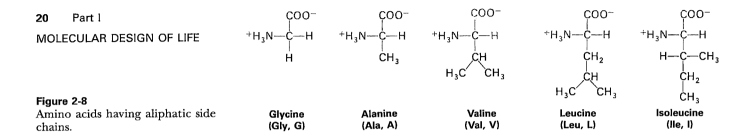

Figure 2-10
Proline differs from the other common amino acids in that its side chain is bonded to the backbone nitrogen atom as well as the α carbon atom.

Three amino acids with *aromatic side chains* are part of the fundamental repertoire (Figure 2-11). *Phenylalanine,* as its name indicates, contains a phenyl ring attached to a methylene ($-CH_2-$) group. The aromatic ring of *tyrosine* contains a hydroxyl group, which makes tyrosine less hydrophobic than phenylalanine. Moreover, this hydroxyl group is reactive, in contrast with the rather inert side chains of the other amino acids discussed thus far. *Tryptophan* has an indole ring joined to a methylene group; this side chain contains a nitrogen atom in addition to carbon and hydrogen atoms. Phenylalanine and tryptophan are highly hydrophobic. The aro-

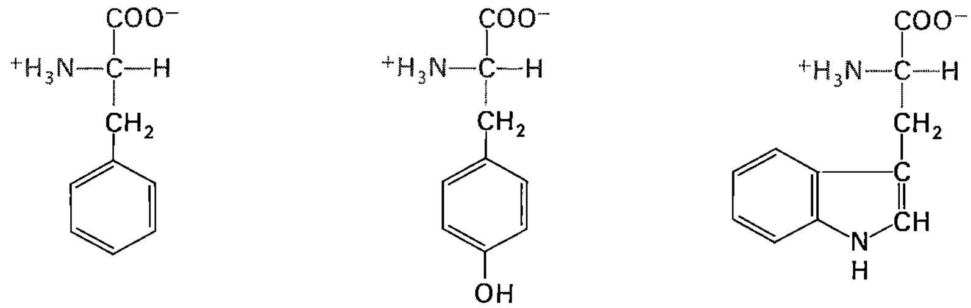

Figure 2-11
Phenylalanine, tyrosine, and tryptophan have aromatic side chains.

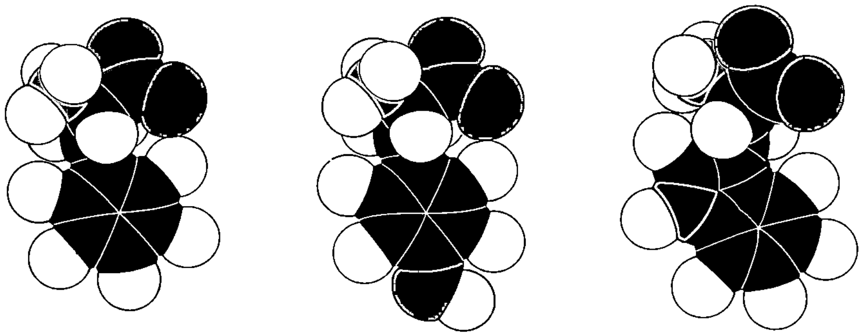

Figure 2-12
Models of the aromatic amino acids.

matic rings of phenylalanine, tryptophan, and tyrosine contain delocalized pi (π) electron clouds that enable them to interact with other π-systems and to transfer electrons.

A *sulfur atom* is present in the side chains of two amino acids (Figure 2-13). *Cysteine* contains a sulfhydryl group (–SH), and *methionine* contains a sulfur atom in a thioether linkage (–S–CH_3). Both these sulfur-containing side chains are hydrophobic. The sulfhydryl group of cysteine is highly reactive. As will be discussed shortly, cysteine plays a special role in stabilizing some proteins by forming disulfide links.

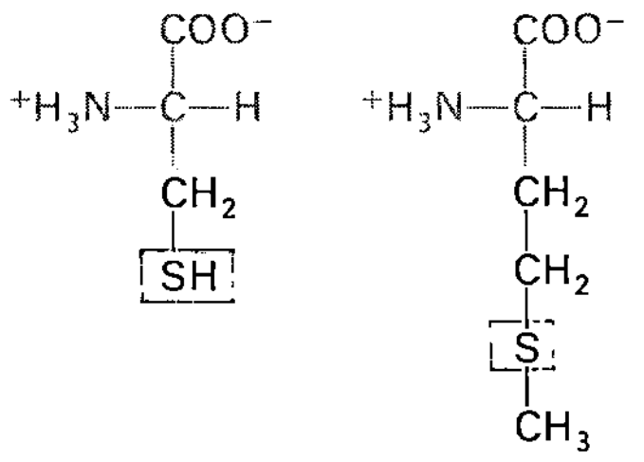

Figure 2-13
Cysteine and methionine have sulfur-containing side chains.

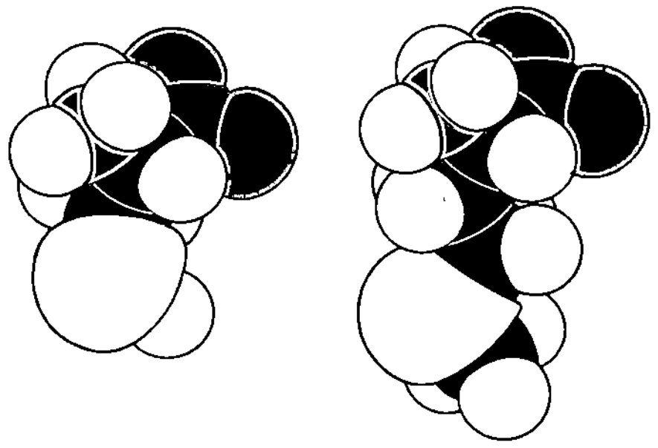

Figure 2-14
Models of cysteine and methionine.

Two amino acids, *serine* and *threonine*, contain aliphatic *hydroxyl groups* (Figure 2-15). Serine can be thought of as a hydroxylated version of alanine. The hydroxyl groups on serine and threonine make them much more *hydrophilic* (water-loving) and *reactive* than alanine and valine. Threonine, like isoleucine, contains two centers of asymmetry. All other amino acids in the basic set of 20, except for glycine, contain a single asymmetric center (the α carbon atom). Glycine is unique in being optically inactive.

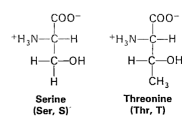

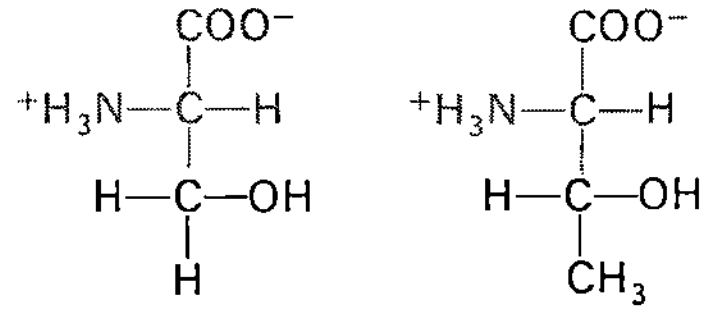

Figure 2-15
Serine and threonine have aliphatic hydroxyl side chains.

We turn now to amino acids with very polar side chains, which render them highly *hydrophilic*. *Lysine* and *arginine* are *positively charged* at neutral pH. *Histidine* can be uncharged or positively charged, depending on its local environment. Indeed, histidine is often found in the active sites of enzymes, where its imidazole ring can readily switch between these states to catalyze the making and breaking of bonds. These *basic amino acids* are depicted in Figure 2-16. The side chains of arginine and lysine are the longest ones in the set of 20.

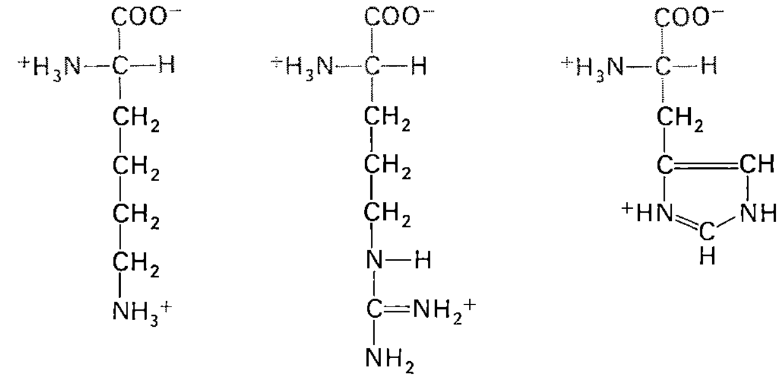

Lysine (Lys, K) **Arginine (Arg, R)** **Histidine (His, H)**

Figure 2-16
Lysine, arginine, and histidine have basic side chains.

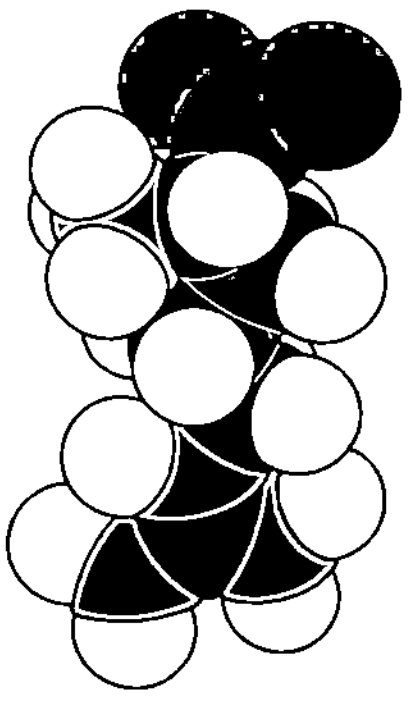

Arg

Figure 2-17
Model of arginine. The planar outer part of the side chain, consisting of three nitrogens bonded to a carbon atom, is called a guanidinium group.

The set of amino acids also contains two with *acidic side chains, aspartic acid* and *glutamic acid.* These amino acids are usually called *aspartate* and *glutamate* to emphasize that their side chains are nearly always negatively charged at physiological pH (Figure 2-18). Uncharged derivatives of glutamate and aspartate are *glutamine* and *asparagine,* which contain a terminal amide group in place of a carboxylate.

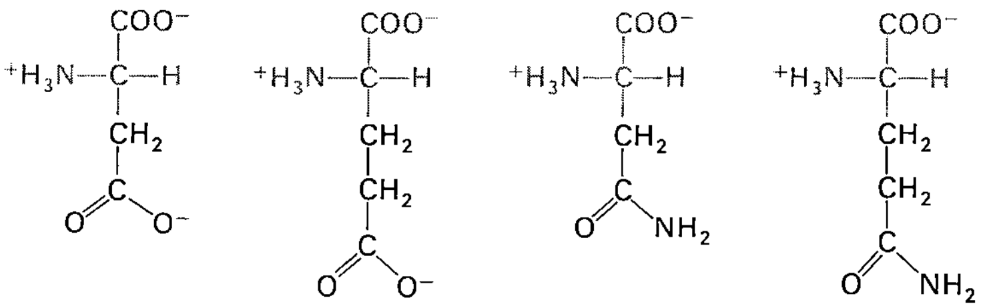

Aspartate (Asp, D) **Glutamate (Glu, E)** **Asparagine (Asn, N)** **Glutamine (Gln, Q)**

Figure 2-18
Acidic amino acids (aspartate and glutamate) and their amide derivatives (asparagine and glutamine).

Seven of the twenty amino acids have readily ionizable side chains. Equilibria and typical p*K* values for ionization of the side chains of arginine, lysine, histidine, aspartic and glutamic acids, cysteine, and tyrosine in proteins are given in Table 2-1. Two other groups in proteins, the terminal α-amino group and the terminal α-carboxyl group, can be ionized, and are also shown in the table.

Amino acids are often designated by either a three-letter abbreviation or a one-letter symbol to facilitate concise communication (Table 2-2). The abbreviations for amino acids are the first three letters of their names, except for tryptophan (Trp), asparagine (Asn), glutamine (Gln), and isoleucine (Ile). The symbols for the small amino acids are the first letters of their names (e.g., G for glycine and L for leucine); the other symbols have been agreed upon by convention. These abbreviations and symbols are an integral part of the vocabulary of biochemists.

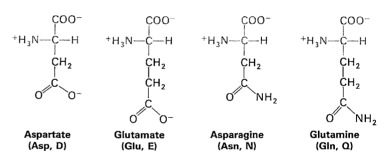

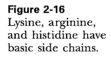

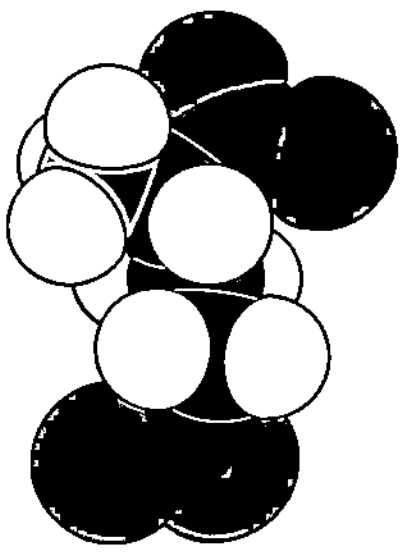

Glu

Figure 2-19
Model of glutamate.

Table 2-1
pK values of ionizable groups in proteins

Group	Acid ⇌ base + H⁺	Typical pK*
Terminal carboxyl	—COOH ⇌ —COO⁻ + H⁺	3.1
Aspartic and glutamic acid	—COOH ⇌ —COO⁻ + H⁺	4.4
Histidine	—CH₂—[imidazole, +HN NH] ⇌ —CH₂—[imidazole, N NH] + H⁺	6.5
Terminal amino	—NH₃⁺ ⇌ —NH₂ + H⁺	8.0
Cysteine	—SH ⇌ —S⁻ + H⁺	8.5
Tyrosine	—⟨C₆H₄⟩—OH ⇌ —⟨C₆H₄⟩—O⁻ + H⁺	10.0
Lysine	—NH₃⁺ ⇌ —NH₂ + H⁺	10.0
Arginine	—N(H)—C(=NH₂⁺)NH₂ ⇌ —N(H)—C(=NH)NH₂ + H⁺	12.0

*pK values depend on temperature, ionic strength, and the microenvironment of the ionizable group.

Table 2-2
Abbreviations for amino acids

Amino acid	Three-letter abbreviation	One-letter symbol
Alanine	Ala	A
Arginine	Arg	R
Asparagine	Asn	N
Aspartic acid	Asp	D
Asparagine or aspartic acid	Asx	B
Cysteine	Cys	C
Glutamine	Gln	Q
Glutamic acid	Glu	E
Glutamine or glutamic acid	Glx	Z
Glycine	Gly	G
Histidine	His	H
Isoleucine	Ile	I
Leucine	Leu	L
Lysine	Lys	K
Methionine	Met	M
Phenylalanine	Phe	F
Proline	Pro	P
Serine	Ser	S
Threonine	Thr	T
Tryptophan	Trp	W
Tyrosine	Tyr	Y
Valine	Val	V

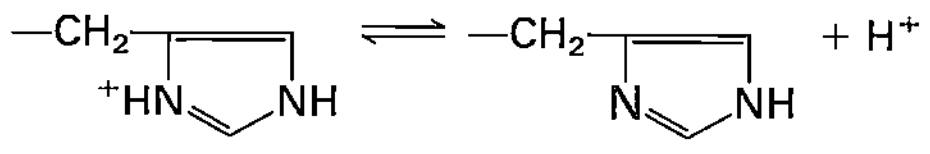

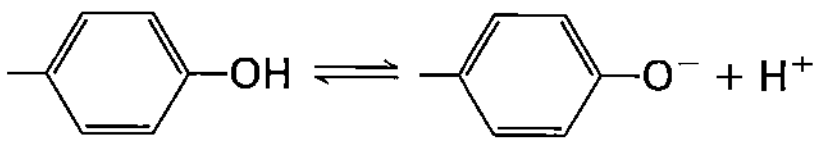

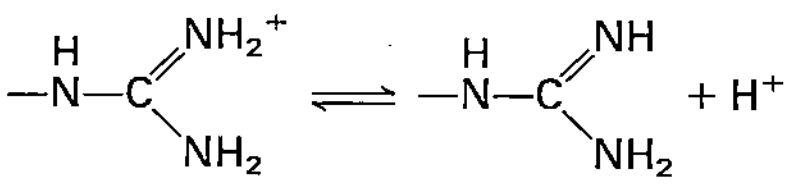

AMINO ACIDS ARE LINKED BY PEPTIDE BONDS TO FORM POLYPEPTIDE CHAINS

In proteins, the α-carboxyl group of one amino acid is joined to the α-amino group of another amino acid by a *peptide bond* (also called an *amide bond*). The formation of a dipeptide from two amino acids by loss of a water molecule is shown in Figure 2-20. The equilibrium of this reaction lies on the side of hydrolysis rather than synthesis. Hence, the biosynthesis of peptide bonds requires an input of free energy, whereas their hydrolysis is thermodynamically downhill.

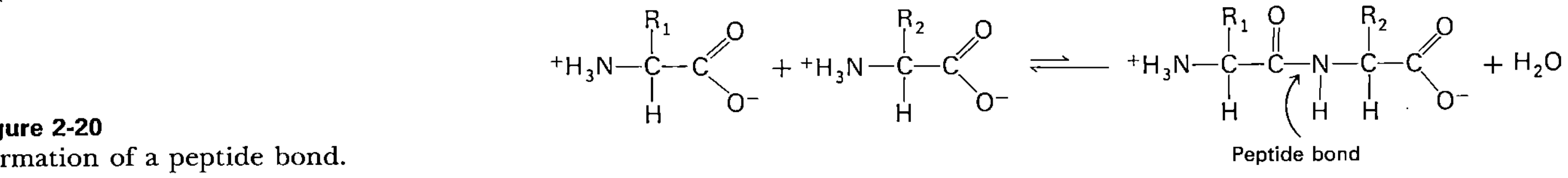

Figure 2-20
Formation of a peptide bond.

Many amino acids joined by peptide bonds form a *polypeptide chain,* which is unbranched (Figure 2-21). An amino acid unit in a polypeptide is called a *residue.* A polypeptide chain has direction because its building blocks have different ends—namely, the α-amino and the α-carboxyl groups. By convention, *the amino end is taken to be the beginning of a polypeptide chain,* and so the sequence of amino acids in a polypeptide chain is written starting with the amino-terminal residue. Thus, in the tripeptide Ala-Gly-Trp (AGW), alanine is the amino-terminal residue and tryptophan is the carboxy-terminal residue. Note that Trp-Gly-Ala (WGA) is a different tripeptide.

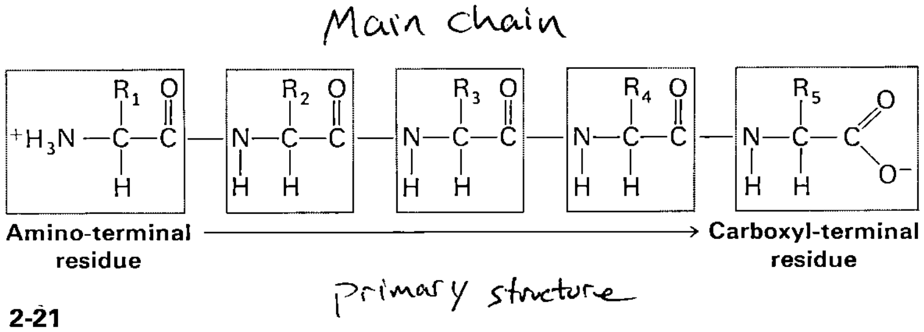

Figure 2-21
A pentapeptide. The constituent amino acid residues are outlined. The chain starts at the amino end.

A polypeptide chain consists of a regularly repeating part, called the *main chain,* and a variable part, comprising the distinctive *side chains* (Figure 2-22). The main chain is sometimes termed the *backbone.* Most natural polypeptide chains contain between 50 and 2000 amino acid residues. The mean molecular weight of an amino acid residue is about 110, and so the molecular weights of most polypeptide chains are between 5500 and 220,000. We can also refer to the mass of a protein, which is expressed in units of daltons; one *dalton* is equal to one atomic mass unit. A protein with a molecular weight of 50,000 has a mass of 50,000 daltons, or 50 kd (kilodaltons).

Dalton—
A unit of mass very nearly equal to that of a hydrogen atom (precisely equal to 1.0000 on the atomic mass scale). Named after John Dalton (1766–1844), who developed the atomic theory of matter.

Kilodalton (kd)—
A unit of mass equal to 1000 daltons.

Figure 2-22
A polypeptide chain is made up of a regularly repeating main chain (backbone) and distinctive side chains (R_1, R_2, R_3, shown in green).

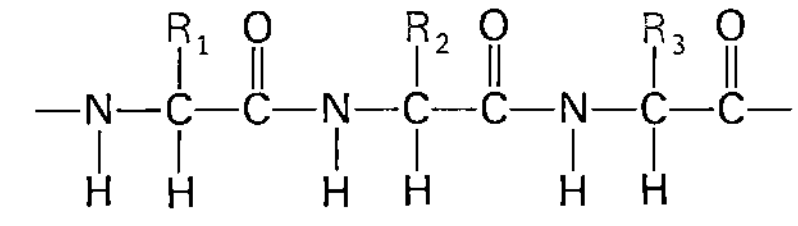

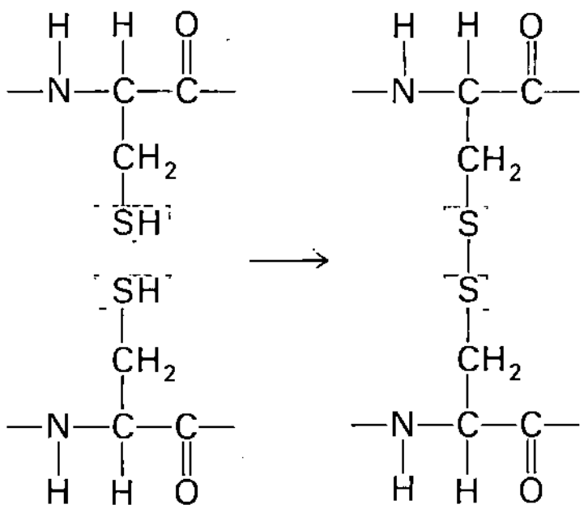

Cysteine Cystine

Figure 2-23
A disulfide bridge (–S–S–) is formed from the sulfhydryl groups (–SH) of two cysteine residues. The product is a cystine residue.

Some proteins contain *disulfide bonds.* These cross-links between chains or between parts of a chain are formed by the oxidation of cysteine residues. The resulting disulfide is called *cystine* (Figure 2-23). Intracellular proteins usually lack disulfide bonds, whereas extracellular proteins often contain several. Nonsulfur cross-links derived from lysine side chains are present in some proteins. For example, collagen fibers in connective tissue are strengthened in this way, as are fibrin blood clots.

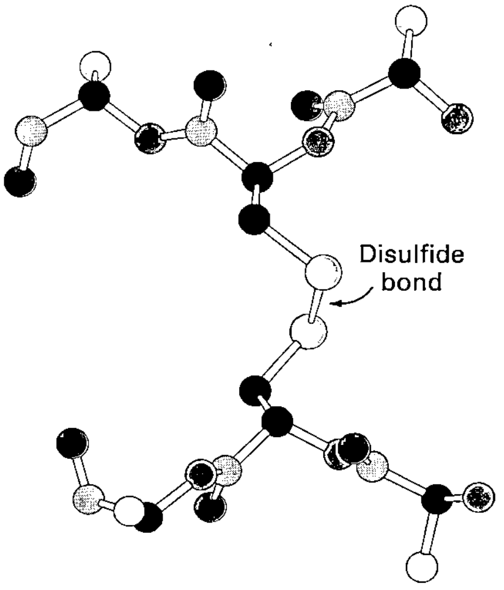

Figure 2-24
Model of a disulfide cross-link.

PROTEINS HAVE UNIQUE AMINO ACID SEQUENCES THAT ARE SPECIFIED BY GENES

In 1953, Frederick Sanger determined the amino acid sequence of insulin, a protein hormone (Figure 2-25). *This work is a landmark in biochemistry because it showed for the first time that a protein has a precisely defined amino acid sequence.* Moreover, it demonstrated that insulin consists only of L amino acids in peptide linkage between α-amino and α-carboxyl groups. This accomplishment stimulated other scientists to carry out sequence studies of a wide variety of proteins. Indeed, the complete amino acid sequences of more than 10,000 proteins are now known. The striking fact is that each protein has a unique, precisely defined amino acid sequence.

A series of incisive studies in the late 1950s and early 1960s revealed that the amino acid sequences of proteins are genetically determined. The sequence of nucleotides in DNA, the molecule of heredity, specifies a complementary sequence of nucleotides in RNA, which in turn specifies the amino acid sequence of a protein (p. 95). In particular, each of the 20 amino acids of the repertoire is encoded by one or more specific sequences of three nucleotides. Furthermore, proteins in all organisms are synthesized from their constituent amino acids by a common mechanism.

Amino acid sequences are important for several reasons. First, knowledge of the sequence of a protein is very helpful, indeed usually essential, in elucidating its mechanism of action (e.g., the catalytic mechanism of

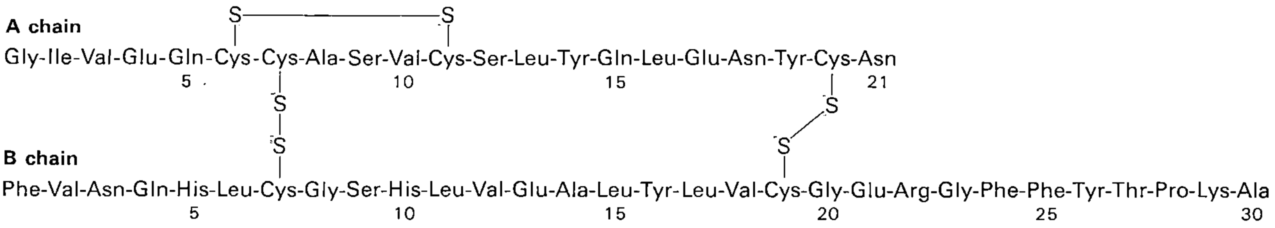

Figure 2-25
Amino acid sequence of bovine insulin.

an enzyme). Proteins with novel properties can be generated by varying the sequence of known structures. Second, analyses of relations between amino acid sequences and three-dimensional structures of proteins are uncovering the rules that govern the folding of polypeptide chains. Amino acid sequence is the link between the genetic message in DNA and the three-dimensional structure that performs a protein's biological function. Third, sequence determination is part of molecular pathology, a rapidly growing area of medicine. Alterations in amino acid sequence can produce abnormal function and disease. Fatal disease, such as sickle-cell anemia and cystic fibrosis, can result from a change in a single amino acid in a single protein. Fourth, the sequence of a protein reveals much about its evolutionary history. Proteins resemble one another in amino acid sequence only if they have a common ancestor. Consequently, molecular events in evolution can be traced from amino acid sequences; molecular paleontology is a flourishing area of research.

PROTEIN MODIFICATION AND CLEAVAGE CONFER NEW CAPABILITIES

Many of the set of 20 amino acids can be modified after synthesis of a polypeptide chain to enhance its capabilities. For example, the amino termini of many proteins are *acetylated*, which makes these proteins more resistant to degradation. In newly synthesized collagen, many proline residues become hydroxylated to form *hydroxyproline* (Figure 2-26). The added hydroxyl groups stabilize the collagen fiber. The biological significance of this modification is evident in scurvy, which results from insufficient hydroxylation of collagen because of a deficiency of vitamin C. Another specialized amino acid produced by a finishing touch is *γ-carboxyglutamate.* In vitamin K deficiency, insufficient carboxylation of glutamate in prothrombin, a clotting protein, can lead to hemorrhage. Most proteins, such as antibodies, that are secreted by cells acquire *carbohydrate units* on specific asparagine residues. The addition of sugars makes a protein more hydrophilic. In contrast, a protein can be made more hydrophobic by the addition of a *fatty acid* to an α-amino group or a cysteine sulfhydryl group.

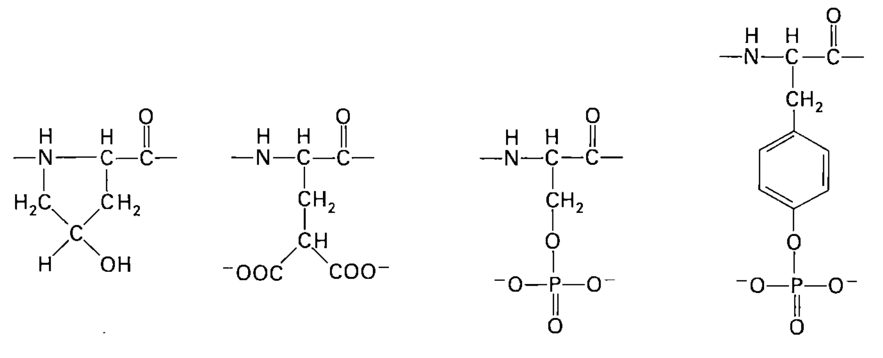

Hydroxyproline γ-Carboxyglutamate *O*-Phosphoserine *O*-Phosphotyrosine

Figure 2-26
Modified amino acid residues found in some proteins: hydroxyproline, γ-carboxyglutamate, *O*-phosphoserine, and *O*-phosphotyrosine. Groups added after the polypeptide chain is synthesized are shown in red.

Many hormones, such as epinephrine (adrenaline), alter the activities of enzymes by stimulating the phosphorylation of the hydroxyl amino acids serine and threonine; *phosphoserine* and *phosphothreonine* are the most ubiquitous modified amino acids in proteins. Growth factors such as insulin act by triggering the phosphorylation of the hydroxyl group of tyrosine residues to form *phosphotyrosine.* The phosphate groups on these three modified amino acids can readily be removed, enabling them to act as reversible switches in regulating cellular processes. Indeed, some cancer-producing genes (oncogenes) act by stimulating excessive phosphorylation of tyrosine residues on proteins that control cell proliferation.

Many proteins are cleaved and trimmed after synthesis. For example, digestive enzymes are synthesized as inactive precursors that can be stored safely in the pancreas. After being released into the intestine, these precursors become activated by peptide bond cleavage. In blood clotting, soluble fibrinogen is converted into insoluble fibrin by peptide bond cleavage. A number of polypeptide hormones, such as adrenocorticotropic hormone, arise from the splitting of a single large precursor protein, a molecular cornucopia. Likewise, the proteins of poliovirus are produced by cleavage of a giant polyprotein precursor. Several key proteins of the human immunodeficiency virus-1 (HIV-1), which causes acquired immune deficiency syndrome (AIDS), arise by specific cleavage of a long precursor. We shall encounter many more examples of modification and cleavage as essential features of protein formation and function. Indeed, these finishing touches account for much of the versatility, precision, and elegance of protein action and regulation.

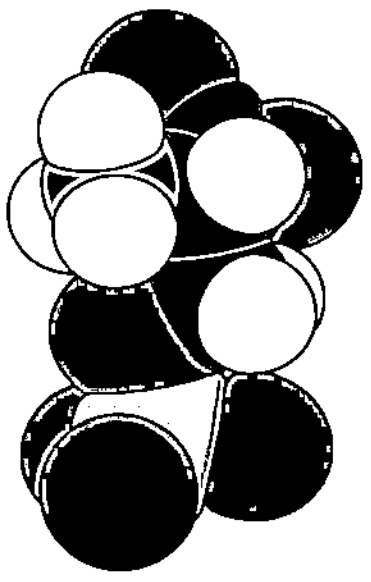

Figure 2-27
Model of phosphoserine.

THE PEPTIDE UNIT IS RIGID AND PLANAR

A striking characteristic of proteins is that they have well-defined three-dimensional structures. A stretched-out or randomly arranged polypeptide chain is devoid of biological activity, as will be discussed shortly. *Function arises from conformation,* which is the three-dimensional arrangement of atoms in a structure. Amino acid sequences are important because they specify the conformation of proteins.

In the late 1930s, Linus Pauling and Robert Corey began x-ray crystallographic studies of the precise structure of amino acids and peptides. Their aim was to obtain a set of standard bond distances and bond angles for these building blocks and then use this information to predict the conformation of proteins. One of their important findings was that *the peptide unit is rigid and planar* (Figure 2-28). The hydrogen of the substituted amino group is nearly always trans (opposite) to the oxygen of the carbonyl group. The only exceptions are X-Pro peptide bonds (*X* denotes any residue), which can be cis (on the same side) or trans. The bond between the carbonyl carbon atom and the nitrogen atom of the peptide unit is not free to rotate because this link has partial double-bond character (Figure 2-29). The length of this bond is 1.32 Å, which is between that

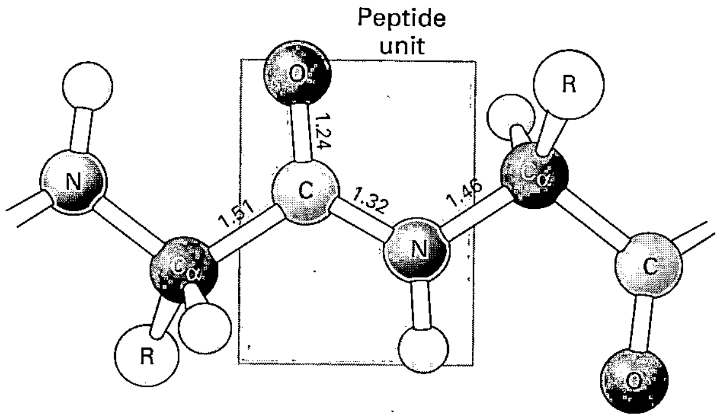

Figure 2-28
The peptide unit is a rigid planar array of four atoms (N, H, C, and O). Standard bond distances (in Å) are shown.

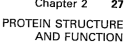

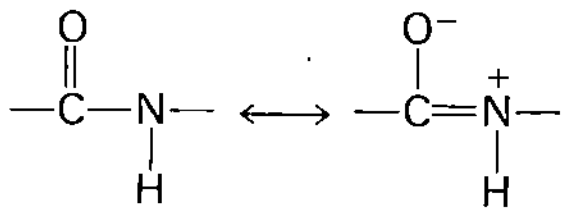

Figure 2-29
The peptide unit is planar because the carbon–nitrogen bond has partial double-bond character.

of a C–N single bond (1.49 Å) and a C=N double bond (1.27 Å). In contrast, the link between the α carbon atom and the carbonyl carbon atom is a pure single bond. The bond between the α carbon atom and the peptide nitrogen atom also is a pure single bond. Consequently, *there is a large degree of rotational freedom about these bonds on either side of the rigid peptide unit* (Figure 2-30). The rigidity of the peptide bond enables proteins to have well-defined three-dimensional forms. The freedom of rotation on either side of the peptide unit is equally important because it allows proteins to fold in many different ways.

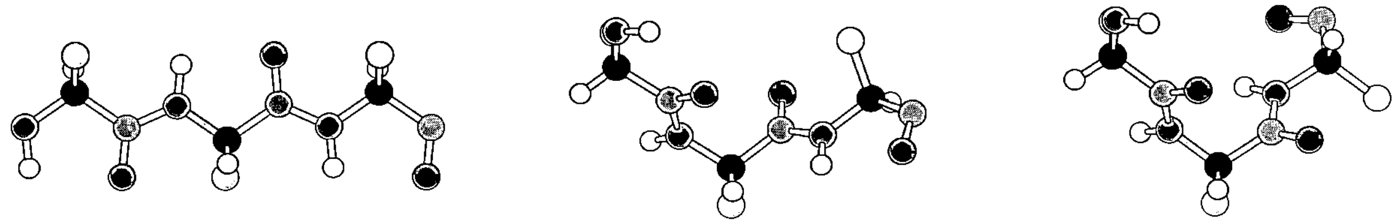

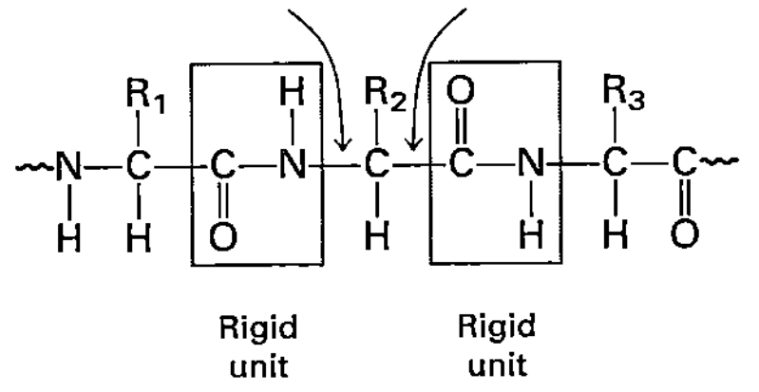

Figure 2-30
There is considerable freedom of rotation about the bonds joining the α carbon atom (C_α) to the main-chain nitrogen (N) and carbonyl carbon (C) atoms. Three allowed conformations of a tripeptide unit are shown in the upper part of the figure.

POLYPEPTIDE CHAINS CAN FOLD INTO REGULAR STRUCTURES SUCH AS THE α HELIX

Can a polypeptide chain fold into a regularly repeating structure? To answer this question, Pauling and Corey evaluated a variety of potential polypeptide conformations by building precise molecular models. They adhered closely to the experimentally observed bond angles and distances for amino acids and small peptides. In 1951, they proposed two periodic polypeptide structures, called the *α helix* (alpha helix) and the *β pleated sheet* (beta pleated sheet).

The α helix is a rodlike structure. The tightly coiled polypeptide main chain forms the inner part of the rod, and the side chains extend outward in a helical array (Figures 2-31 and 2-32). The α helix is stabilized by hydrogen bonds between the NH and CO groups of the main chain. The CO group of each amino acid is hydrogen bonded to the NH group of the amino acid that is situated four residues ahead in the linear sequence (Figure 2-33). Thus, *all the main-chain CO and NH groups are hydrogen bonded.* Each residue is related to the next one by a rise of 1.5 Å along the helix axis and a rotation of 100°, which gives 3.6 amino acid residues per turn of helix. Thus, amino acids spaced three and four apart in the linear sequence are spatially quite close to one another in an α helix. In contrast, amino acids two apart in the linear sequence are situated on opposite sides of the helix and so are unlikely to make contact. The *pitch* of the α helix, which is equal to the product of the translation (1.5 Å) and the number of residues per turn (3.6), is 5.4 Å. The *screw sense* of a helix can be right-handed (clockwise) or left-handed (counterclockwise); the α helices found in proteins are right-handed.

The α-helical content of proteins ranges widely, from nearly none to almost 100%. For example, the digestive enzyme chymotrypsin is virtually devoid of α helix. In contrast, about 75% of myoglobin and hemoglobin is α-helical. Single-stranded α helices are usually less than 45 Å long. However, two or more α helices can entwine to form very stable structures, which can have a length of 1000 Å (100 nm or 0.1 μm) or more. Such *α-helical coiled coils* (Figure 2-34) are found in myosin and tropomyosin in muscle (p. 395), fibrin in blood clots, and keratin in hair. The helical

Angstrom (Å)—
A unit of length equal to 10^{-10} meter.

$$1\ \text{Å} = 10^{-10}\ \text{m} = 10^{-8}\ \text{cm} = 10^{-4}\ \mu\text{m} = 10^{-1}\ \text{nm}$$

Named for Anders J. Ångström (1814–1874), a spectroscopist.

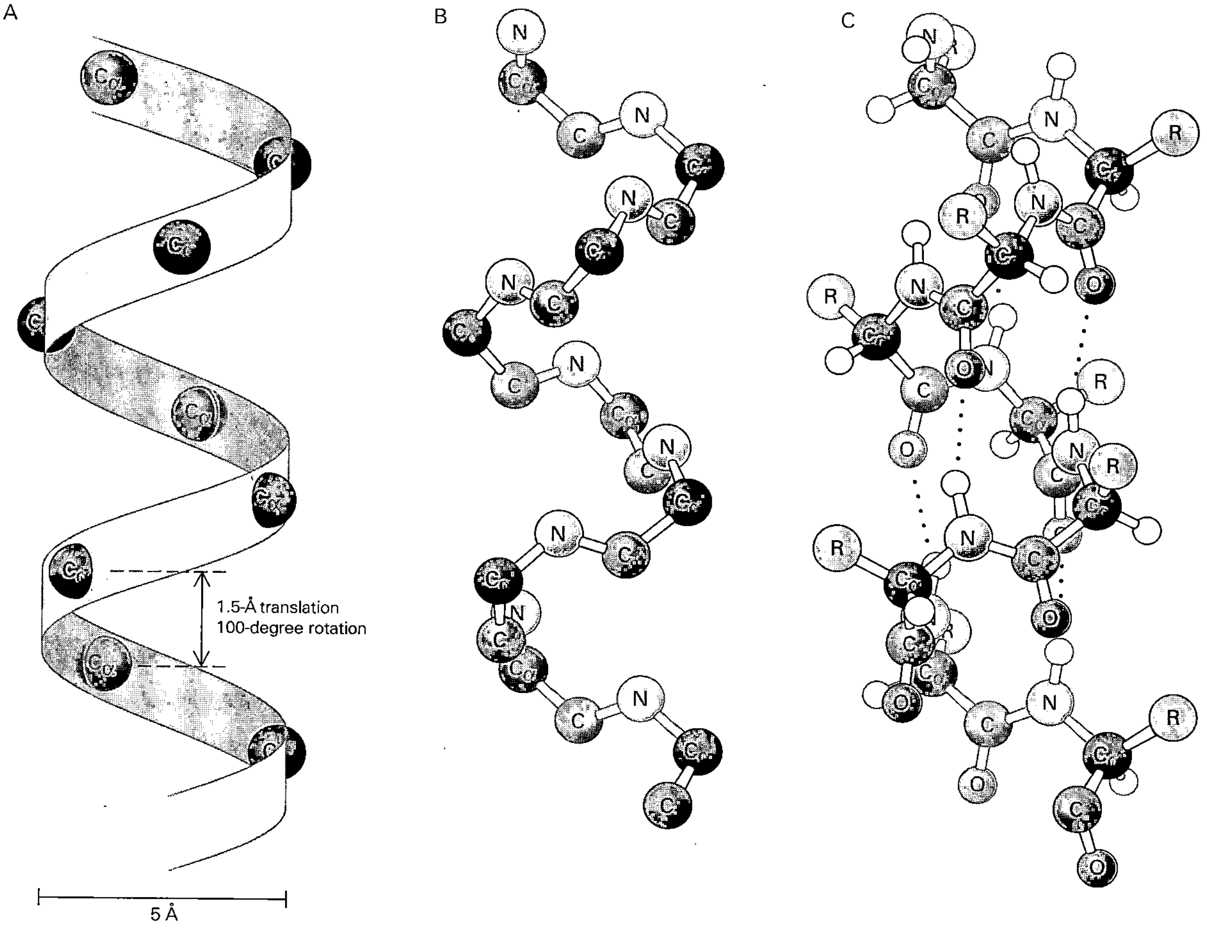

Figure 2-31
Models of a right-handed α helix. (A) Only the α carbon atoms are shown on a helical ribbon; (B) only the backbone nitrogen (N), α carbon (C_α), and carbonyl carbon (C) atoms are shown; (C) entire helix. Hydrogen bonds (denoted in part C by red dots) between NH and CO groups stabilize the helix.

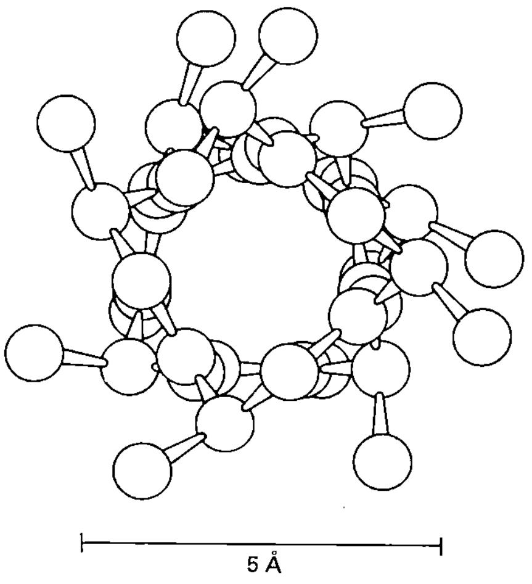

Figure 2-32
Cross-sectional view of an α helix. Note that the side chains (shown in green) are on the outside of the helix. The van der Waals radii of the atoms are larger than shown here; hence, there is actually almost no free space inside the helix.

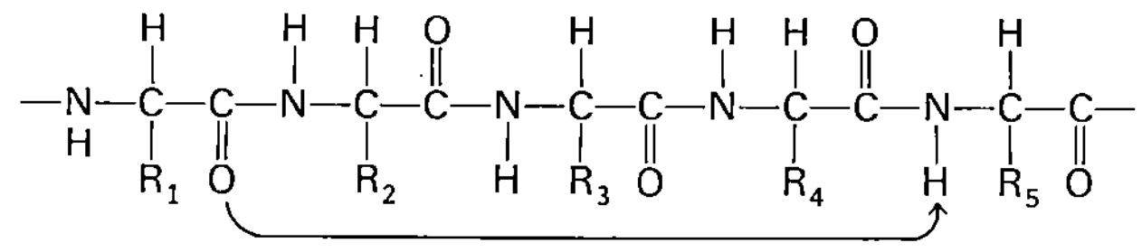

Figure 2-33
In the α helix, the CO group of residue n is hydrogen bonded to the NH group of residue $(n + 4)$.

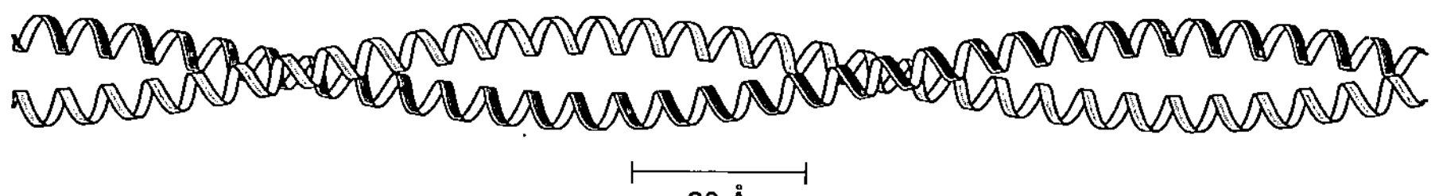

Figure 2-34
α-Helical coiled coil. The two helices wind around each other to form a superhelix. The chains in most naturally occurring coiled coils have the same amino-to-carboxyl direction.

The sharp quills of porcupines are made of keratin, a protein rich in α-helical coiled coils.

"When we consider that the fibrous proteins of the epidermis, the keratinous tissues, the chief muscle protein, myosin, and now the fibrinogen of the blood all spring from the same peculiar shape of molecule, and are therefore probably all adaptations of a single root idea, we seem to glimpse one of the great coordinating facts in the lineage of biological molecules."

K. Bailey, W.T. Astbury, and K.M. Rudall
Nature, 1943

cables in these proteins serve a mechanical role in forming stiff bundles of fibers, as in porcupine quills. The cytoskeleton (internal scaffolding) of cells is rich in intermediate filaments, which also are two-stranded α-helical coiled coils. The forces stabilizing this recurring motif will be discussed in later chapters (pp. 396 and 423).

The structure of the α helix was deduced by Pauling and Corey six years before it was actually seen in the x-ray reconstruction of the structure of myoglobin. *The elucidation of the structure of the α helix is a landmark in molecular biology because it demonstrated that the conformation of a polypeptide chain can be predicted if the properties of its components are rigorously and precisely known.*

β PLEATED SHEETS ARE STABILIZED BY HYDROGEN BONDING BETWEEN β STRANDS

Pauling and Corey discovered another periodic structural motif, which they named the *β pleated sheet* (β because it was the second structure they elucidated, the α helix having been the first). The β pleated sheet differs markedly from the rodlike α helix. A polypeptide chain in a β pleated sheet, called a *β strand,* is almost fully extended (Figure 2-35) rather than being tightly coiled as in the α helix. The axial distance between adjacent amino acids is 3.5 Å, in contrast with 1.5 Å for the α helix. Another difference is that the β pleated sheet is stabilized by hydrogen bonds between NH and CO groups in *different* polypeptide strands, whereas in the α helix the hydrogen bonds are between NH and CO groups in the *same* strand. Adjacent chains in a β pleated sheet can run in the same direction *(parallel β sheet)* or in opposite directions *(antiparallel β sheet)* (Figure 2-36). For example, silk fibroin consists almost entirely of stacks of antiparallel β sheets. Such β sheet regions are a recurring structural motif in many proteins. Structural units comprising from two to five parallel or antiparallel β strands are especially common.

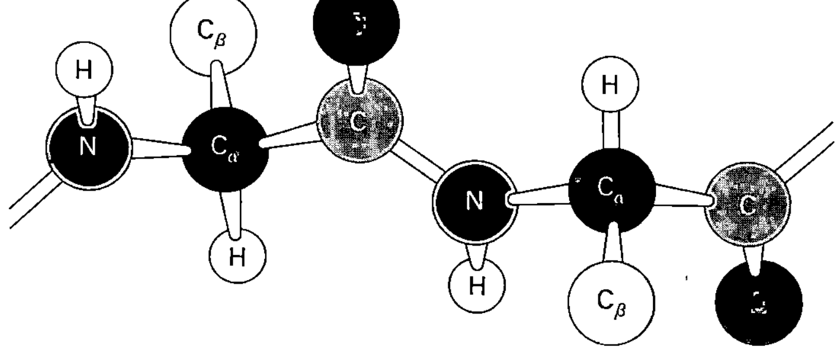

Figure 2-35
Conformation of a dipeptide unit in a β strand. The polypeptide chain is almost fully stretched out.

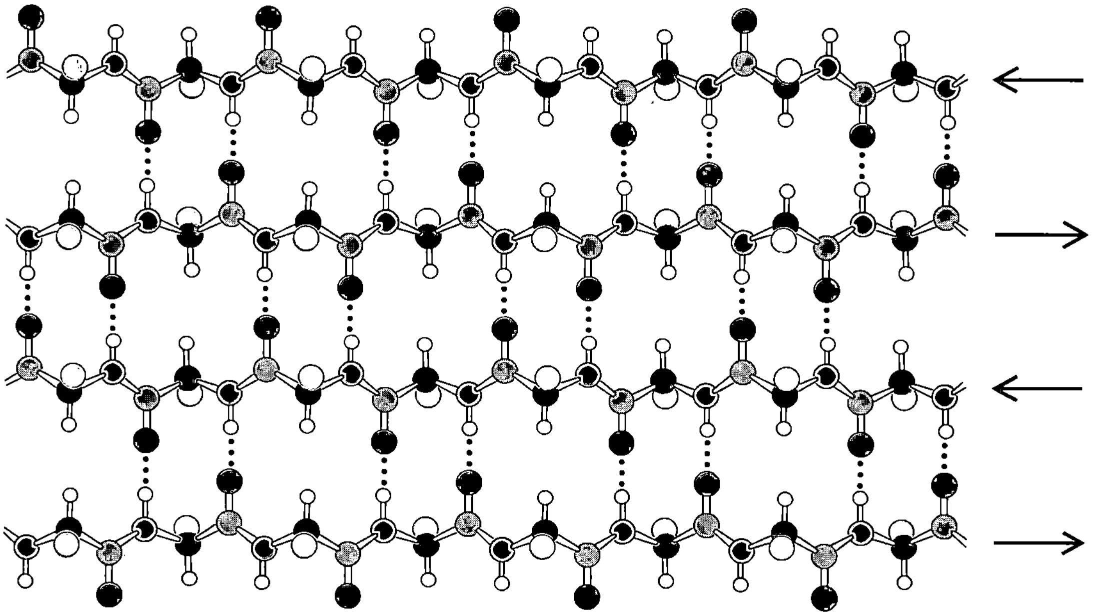

Figure 2-36
Antiparallel β pleated sheet. Adjacent β strands run in opposite directions. Hydrogen bonds between NH and CO groups of adjacent strands stabilize the structure. The side chains (shown in green) are above and below the plane of the sheet.

POLYPEPTIDE CHAINS CAN REVERSE DIRECTION BY MAKING HAIRPIN TURNS

Most proteins have compact, globular shapes owing to reversals in the direction of their polypeptide chains. Many of these reversals are accomplished by a common structural element called the *β turn.* The essence of this hairpin turn is that the CO group of residue n of a polypeptide is hydrogen-bonded to the NH group of residue $n + 3$ (Figure 2-37). Thus, a polypeptide chain can abruptly reverse its direction. β Turns often connect antiparallel β strands, hence their name. They are also known as *reverse turns* or *hairpin bends.*

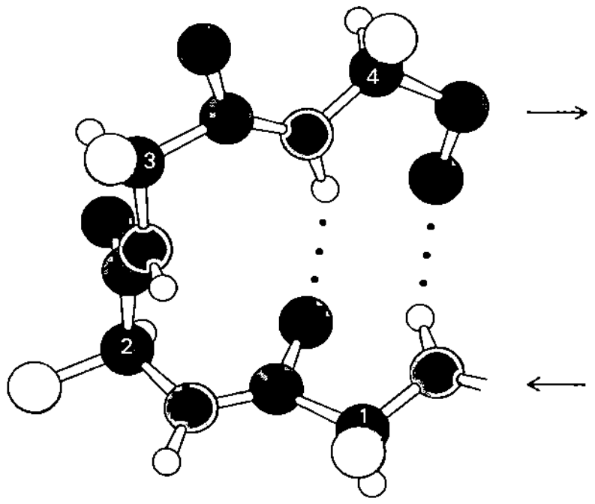

Figure 2-37
Structure of a β turn. The NH and CO groups of residue 1 of the tetrapeptide shown here are hydrogen bonded, respectively, to the CO and NH groups of residue 4, which results in a hairpin turn.

THE TRIPLE-STRANDED COLLAGEN HELIX IS STABILIZED BY PROLINE AND HYDROXYPROLINE

A special type of helix is present in collagen, the most abundant protein of mammals. Collagen is the main fibrous component of skin, bone, tendon, cartilage, and teeth. This extracellular protein contains three helical polypeptide chains, each nearly 1000 residues long. The amino acid sequence of collagen is remarkably regular: *nearly every third residue is glycine.* Proline is also present to a much greater extent than in most other proteins. Furthermore, collagen contains 4-hydroxyproline, which is rarely found elsewhere. The sequence glycine-proline-hydroxyproline (Gly-Pro-Hyp) recurs frequently (Figure 2-38).

13
-Gly-Pro-Met-Gly-Pro-Ser-Gly-Pro-Arg-
22
-Gly-Leu-Hyp-Gly-Pro-Hyp-Gly-Ala-Hyp-
31
-Gly-Pro-Gln-Gly-Phe-Gln-Gly-Pro-Hyp-
40
-Gly-Glu-Hyp-Gly-Glu-Hyp-Gly-Ala-Ser-
49
-Gly-Pro-Met-Gly-Pro-Arg-Gly-Pro-Hyp-
58
-Gly-Pro-Hyp-Gly-Lys-Asn-Gly-Asp-Asp-

Figure 2-38
Amino acid sequence of part of a collagen chain. Every third residue is glycine in a region spanning more than a thousand residues. Collagen also has an abundance of proline and hydroxyproline residues.

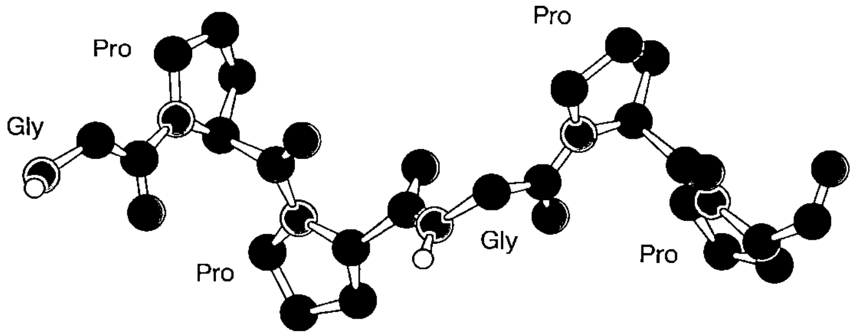

Figure 2-39
Conformation of a single strand of the collagen triple helix. The sequence shown here is Gly-Pro-Pro-Gly-Pro-Pro.

Collagen is a rod-shaped molecule, about 3000 Å long and only 15 Å in diameter. The helical motif of its three chains (Figure 2-39) is entirely different from that of the α helix. Hydrogen bonds within a strand are absent. Instead, *each of the three collagen helices is stabilized by steric repulsion of the pyrrolidone rings of the proline and hydroxyproline residues.* The pyrrolidone rings keep out of each other's way when the polypeptide chain assumes this helical form, which is much more open than the tightly coiled α helix. The three strands wind around each other to form a *superhelical cable* (Figure 2-40). The *rise* (axial distance) per residue in this superhelix is 2.9 Å, and there are nearly three residues per turn. *The three strands are hydrogen bonded to each other.* The hydrogen donors are the peptide NH groups of glycine residues, and the hydrogen acceptors are the peptide CO groups of residues on the other chains. The hydroxyl groups of hydroxyproline residues also participate in hydrogen bonding, as will be discussed further when we consider the action of vitamin C (p. 454).

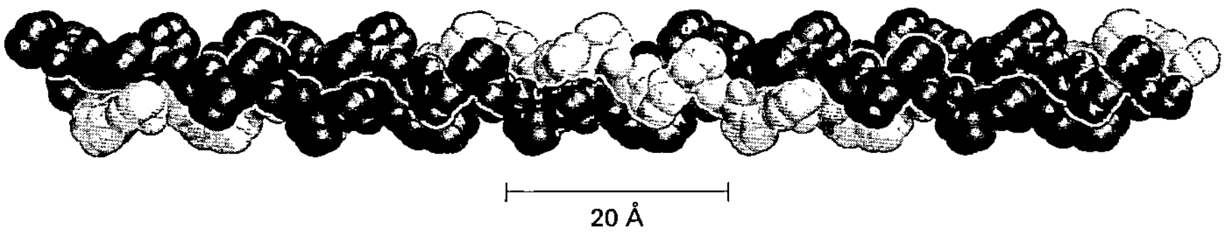

Figure 2-40
Space-filling model of the collagen triple helix. The three strands are shown in different colors. [Drawn from coordinates reported in K. Okuyama, K. Okuyama, S. Arnott, M. Takayanagi, and M. Kakudo. *J. Mol. Biol.* 152(1981):427.]

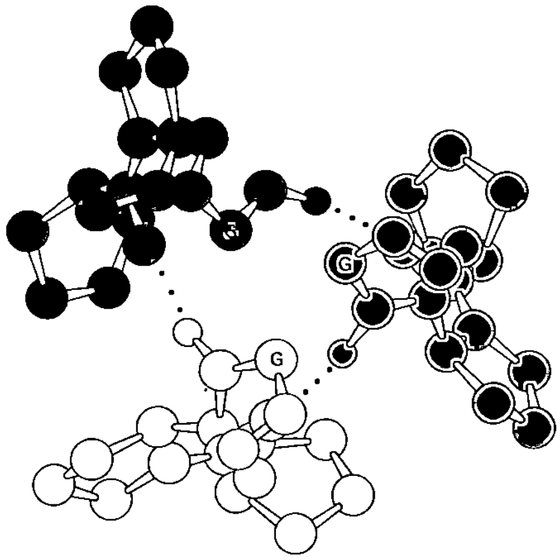

Figure 2-41
Cross section of a model of collagen. Each strand of the triple helix is shown in a different color. Each is hydrogen bonded to the other two (··· denotes a hydrogen bond). The α carbon atom of a glycine residue in each strand is labeled G. Every third residue must be glycine because there is no space near the helix axis (center) for a larger amino acid residue. Note that the pyrrolidone rings are on the outside.

We can now understand why glycine occupies every third position in a stretch of a thousand residues that form the helical region of collagen. The inside of the triple-stranded helical cable is very crowded (Figure 2-41). Indeed, *the only residue that can fit in an interior position is glycine.* Because there are three residues per turn of helix, every third residue on each strand must be glycine. The amino acid residue on either side of glycine is located on the outside of the cable, where there is room for the bulky rings of proline and hydroxyproline residues.

PROTEINS ARE RICH IN HYDROGEN-BONDING POTENTIALITY

What are the forces that determine the three-dimensional architecture of proteins? As was explained in Chapter 1, all reversible molecular interactions in biological systems are mediated by three kinds of forces: *electrostatic bonds, hydrogen bonds,* and *van der Waals bonds.* We have already seen hydrogen bonds between main-chain NH and CO groups at work in forming α helices, β sheets, and collagen cables. In fact, side chains of 11 of the

20 fundamental amino acids also can participate in hydrogen bonding. It is convenient to group these amino acids according to their hydrogen-bonding potentialities:

1. The side chains of tryptophan and arginine can serve as *hydrogen bond donors only.*

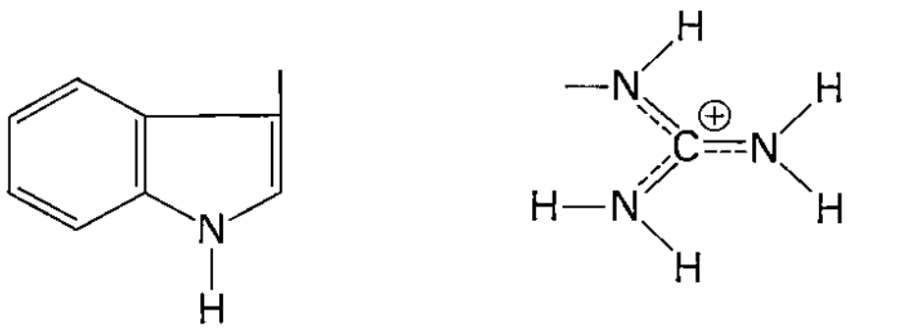

Hydrogen donor group of tryptophan **Hydrogen donor groups of arginine**

2. Like the peptide unit itself, the side chains of asparagine, glutamine, serine, and threonine can serve as *hydrogen bond donors and acceptors.*

3. The hydrogen-bonding capabilities of lysine (and the terminal amino group), aspartic and glutamic acid (and the terminal carboxyl group), tyrosine, and histidine vary with pH. These groups can serve as both acceptors *and* donors within a certain pH range, and as acceptors *or* donors (but not both) at other pH values, as shown for aspartate and glutamate in Figure 2-42. *The hydrogen-bonding modes of these ionizable residues are pH-dependent.*

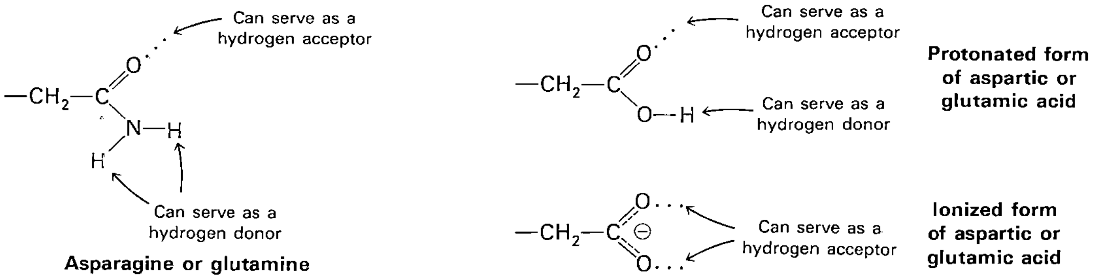

Figure 2-42
Hydrogen-bonding groups of several side chains in proteins.

WATER-SOLUBLE PROTEINS FOLD INTO COMPACT STRUCTURES WITH NONPOLAR CORES

Let us now see how these forces shape the structure of proteins. X-ray crystallographic studies have revealed the detailed three-dimensional structures of more than 300 proteins, as will be discussed in subsequent chapters. We begin here with a preview of *myoglobin,* the first protein to be seen in atomic detail.

Myoglobin, the oxygen carrier in muscle, is a single polypeptide chain of 153 amino acids and has a mass of 18 kd. The capacity of myoglobin to bind oxygen depends on the presence of *heme,* a nonpolypeptide *prosthetic (helper) group* consisting of protoporphyrin and a central iron atom. *Myoglobin is an extremely compact molecule* (Figure 2-43). Its overall dimensions are 45 × 35 × 25 Å, an order of magnitude less than if it were fully stretched out. *Myoglobin is built primarily of α helices,* of which there are eight. About 70% of the main chain is folded into α helices, and much of

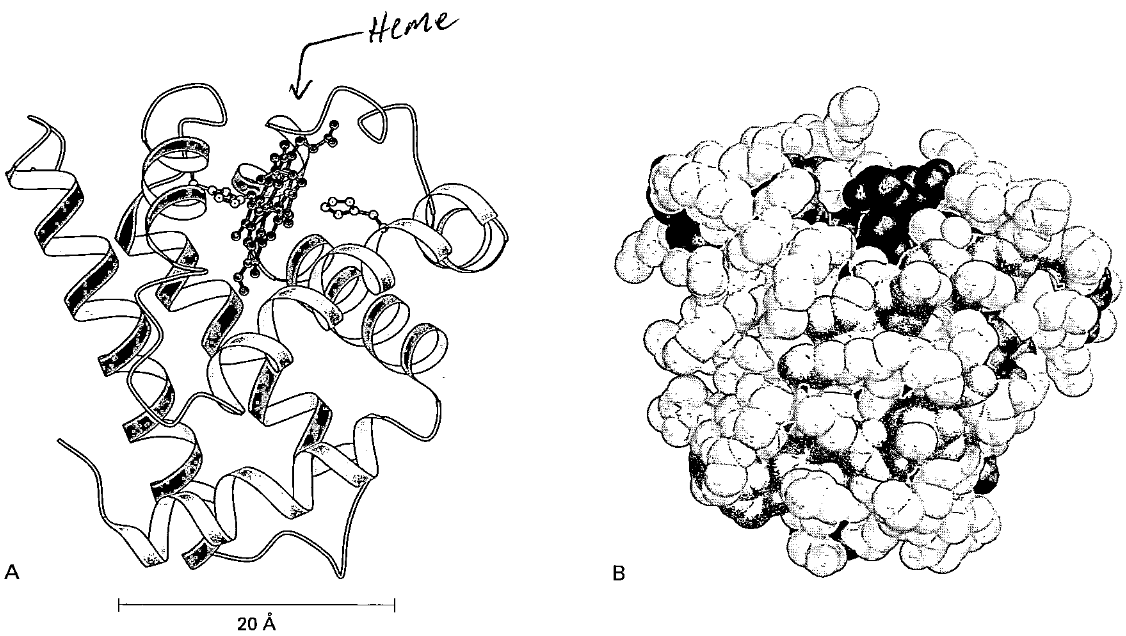

Figure 2-43
Three-dimensional structure of myoglobin, an oxygen-carrying protein in muscle. (A) A schematic drawing of the backbone conformation. The two α helices in front are shown in green, and the others in orange. The heme group is depicted in red, and two critical histidine residues in blue. (B) A space-filling model of myoglobin in the same orientation. [Drawn from 1mbn.pdb. H.C. Watson. *Prog. Stereochem.* 4(1969):299. Part A was drawn using the graphics program *MOLSCRIPT,* described in P. Kraulis, *J. Appl. Cryst.* 24(1991):946. Part B was drawn using *MIDAS,* described in T. Ferrin, C. Huang, L. Jarvis, and R. Langridge. *J. Mol. Graphics* 6(1988):13.]

the rest of the chain forms turns between helices. Four of the turns contain proline, which tends to disrupt α helices because of steric hindrance by its rigid five-membered ring.

The folding of the main chain of myoglobin, like that of other proteins, is complex and devoid of symmetry. However, a unifying principle emerges from the distribution of side chains. The striking fact is that *the interior consists almost entirely of nonpolar residues* such as leucine, valine, methionine, and phenylalanine (Figure 2-44). Polar residues such as aspartate, glutamate, lysine, and arginine are absent from the inside of myoglobin. The only polar residues inside are two histidines, which play critical roles in the binding of heme oxygen. The outside of myoglobin, on the other hand, consists of both polar and nonpolar residues. The space-filling model also shows that there is very little empty space inside.

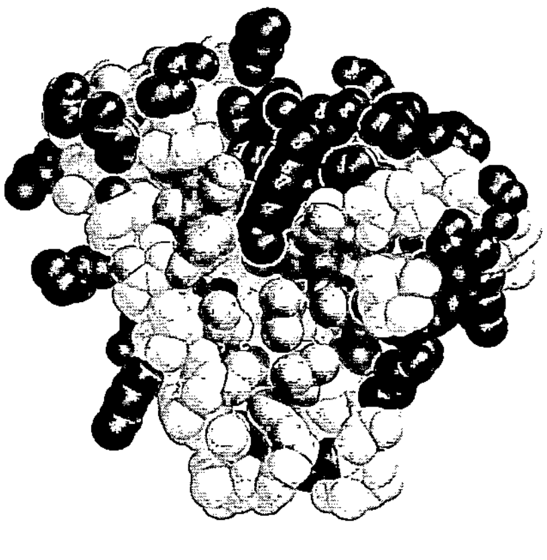

Figure 2-44
View of the inside of myoglobin. The front of a model shown in the same orientation as part B of Figure 2-43 has been sliced off. An exposed face is shown here. Nonpolar side chains are shown in yellow, polar ones in blue, other side chains in white, and the heme group in red. The inside of the protein is distinctly nonpolar.

This contrasting distribution of polar and nonpolar residues reveals a key facet of protein architecture. In an aqueous environment, protein folding is driven by the strong tendency of hydrophobic residues to be excluded from water. Recall that water is highly cohesive and that hydrophobic groups are thermodynamically more stable when clustered in the interior of the molecule than when extended into the aqueous surroundings (p. 9). *The polypeptide chain therefore folds spontaneously so that its hydrophobic side chains are buried and its polar, charged chains are on the surface.* The fate of the main chain accompanying the hydrophobic side chains is important, too. An unpaired peptide NH or CO markedly prefers water to a nonpolar milieu. The secret of burying a segment of main chain in a hydrophobic environment is to pair all the NH and CO groups by hydrogen bonding. This pairing is neatly accomplished in an α helix or β sheet. Van der Waals bonds between tightly packed hydrocarbon side chains also contribute to the stability of proteins. We can now understand why the set of 20 amino acids contains several that differ subtly in size and shape (see Figure 2-9). Nature can choose among them to fill the interior of a protein neatly and thereby maximize van der Waals interactions, which require intimate contact.

Ribonuclease A, a pancreatic enzyme that hydrolyzes RNA, exemplifies a rather different mode of protein folding. This single polypeptide chain of 124 residues is folded mainly into β strands, in contrast with myoglobin, which contains α helices and lacks β strands. Ribonuclease, like myoglobin, contains a tightly packed, highly nonpolar interior. This enzyme is further stabilized by four disulfide bonds. The structure of a protein can be symbolized in highly schematic form by depicting β strands as broad arrows, α helices as helical ribbons, and connecting regions as strings. This representation, devised by Jane Richardson, is very useful for concisely representing relations of these elements in proteins, especially large ones, and for detecting structural motifs that recur in different proteins. A ribbon drawing of ribonuclease is shown in Figure 2-45.

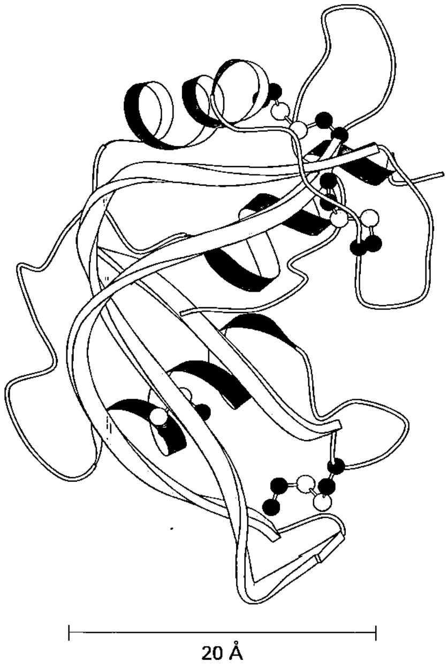

Figure 2-45
Ribonuclease A, a digestive enzyme, has a complex three-dimensional structure. In this schematic diagram, α helices are shown in red, and β strands in green. The four disulfide bonds in the protein are also drawn, with sulfur atoms shown in yellow. [Drawn from 3rn3.pdb. A. Wlodawer, N. Borrakoti, D.S. Moss, and B. Howlin. *Acta Cryst.* 42B(1986):379.]

Integral membrane proteins, those that span biological membranes, are designed differently from proteins that are soluble in aqueous solution. The permeability barrier of membranes is formed by lipids, which are highly hydrophobic. Thus, the part of a membrane protein that spans this region must have a hydrophobic exterior. As will be discussed in Chapter 11, the transmembrane portion of a membrane protein usually consists of bundles of α helices with nonpolar side chains (such as those of leucine and phenylalanine) facing out from the surface of the protein.

THERE ARE FOUR BASIC LEVELS OF STRUCTURE IN PROTEIN ARCHITECTURE

Four levels of structure are frequently cited in discussions of protein architecture. *Primary structure* is the amino acid sequence. *Secondary structure* refers to the spatial arrangement of amino acid residues that are near one another in the linear sequence. Some of these steric relationships are of a regular kind, giving rise to a periodic structure. The α helix and β strand are elements of secondary structure. *Tertiary structure* refers to the spatial arrangement of amino acid residues that are far apart in the linear sequence, and to the pattern of disulfide bonds. The dividing line between secondary and tertiary structure is a matter of taste.

Proteins containing more than one polypeptide chain exhibit an additional level of structural organization. Each polypeptide chain in such a protein is called a *subunit*. *Quaternary structure* refers to the spatial arrangement of subunits and the nature of their contacts. Hemoglobin, for example, consists of two α and two β chains (Figure 2-46). The subunit interfaces in this tetramer participate in transmitting information between distant binding sites for O_2, CO_2, and H^+. Viruses make the most of a limited amount of genetic information by forming coats that use the same kind of subunit repetitively in a symmetric array (Figure 2-47).

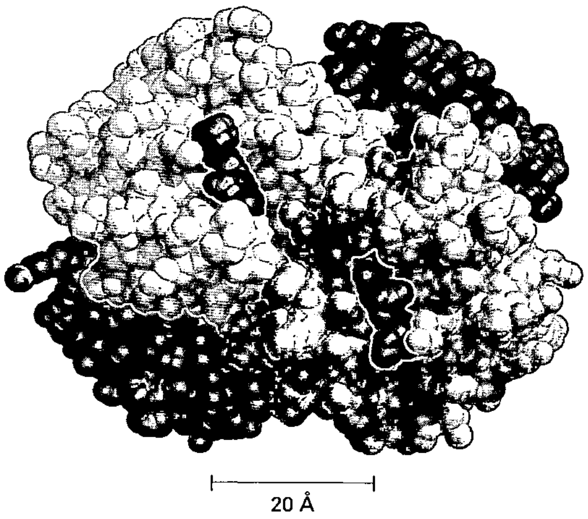

Figure 2-46
Quaternary structure of hemoglobin. The two identical α subunits are shown in yellow and orange; and the two identical β subunits in blue and green. Heme groups are shown in red; two of them can be seen edge-on. [Drawn from 4hhb.pdb. G. Fermi, M.F. Perutz, B. Shaanan, and R. Fourme. *J. Mol. Biol.* 175(1984):159.]

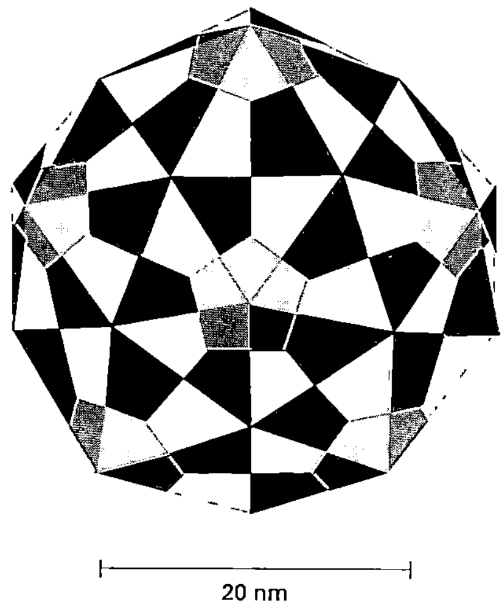

Figure 2-47
The highly symmetric coat of poliovirus contains 60 copies of each of four subunits. The infectious RNA molecule is inside this shell. Three kinds of subunits can be seen in this view. Human rhinovirus 14, the cause of the common cold, has a similar architecture. Both viruses have the symmetry of an icosahedron, one of the five regular Platonic forms. [After J.M. Hogle, M. Chow, and D.J. Filman. The structure of poliovirus. Copyright © 1987 by Scientific American, Inc. All rights reserved.]

Studies of protein conformation, function, and evolution have revealed the importance of two other levels of organization. *Supersecondary structure* refers to clusters of secondary structure. For example, a β strand separated from another β strand by an α helix is found in many proteins; this motif is called a *$\beta\alpha\beta$ unit* (Figure 2-48). Some polypeptide chains fold

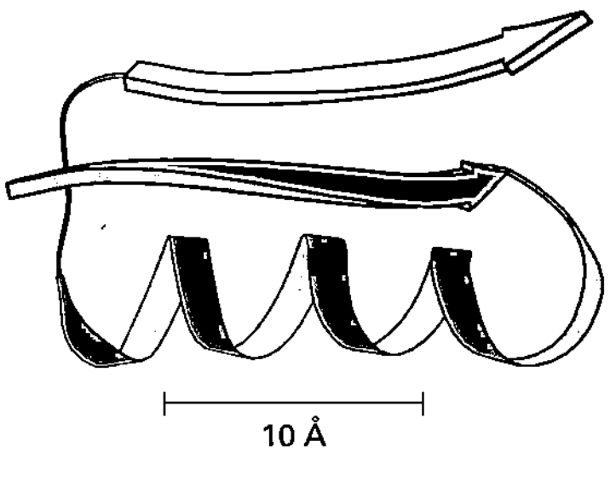

Figure 2-48
Schematic drawing of the backbone of a $\beta\alpha\beta$ unit, a recurring motif in proteins. The β strands are shown in blue and green, and the α helix between them is shown in red. [Drawn from 6adh.pdb. H. Eklund, J-P. Samama, L. Wallen, C-I. Branden, A. Akeson, and T.A. Jones. *J. Mol. Biol.* 146(1981):561.]

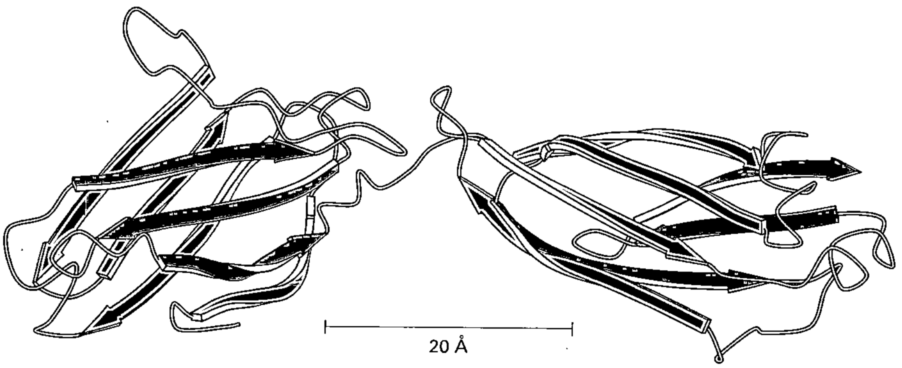

Figure 2-49
The light chain of an antibody molecule consists of two similar domains. A sandwich of β sheets (shown in blue and red) forms the core of each domain. [Drawn from 7fab.pdb. F.A. Saul and R.J. Poljak. *Proteins* 14(1992):363.]

into two or more compact regions that may be joined by a flexible segment of polypeptide chain, rather like pearls on a string. These compact globular units, called *domains,* range in size from about 100 to 400 amino acid residues. For example, a 25-kd L chain of an antibody is folded into two domains (Figure 2-49). Indeed, these domains resemble each other, which suggests that they arose by duplication of a primordial gene. An important principle has emerged from analyses of genes and proteins in higher eukaryotes: *protein domains are often encoded by distinct parts of genes called exons* (p. 114).

THE AMINO ACID SEQUENCE OF A PROTEIN SPECIFIES ITS THREE-DIMENSIONAL STRUCTURE

Insight into the relation between the amino acid sequence of a protein and its conformation came from the classic work of Christian Anfinsen on ribonuclease. As was mentioned earlier, ribonuclease is a single polypeptide chain consisting of 124 amino acid residues (Figure 2-50). Its four disulfide bonds can be cleaved reversibly by reducing them with a reagent such as β-*mercaptoethanol,* which forms mixed disulfides with cysteine side chains (Figure 2-51). In the presence of a large excess of β-

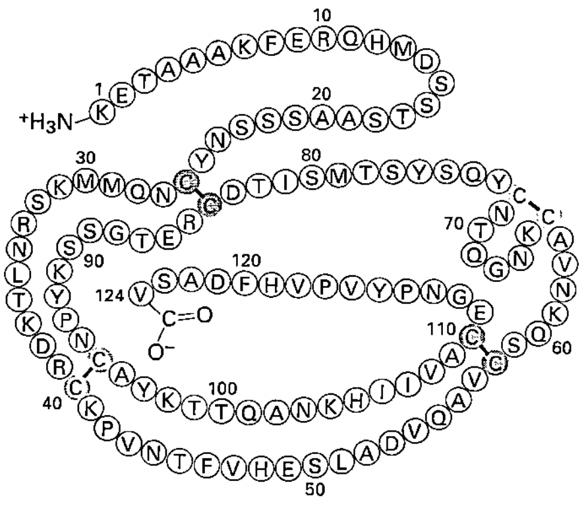

Figure 2-50
Amino acid sequence of bovine ribonuclease. The four disulfide bonds are shown in color. [After C.H.W. Hirs, S. Moore, and W.H. Stein. *J. Biol. Chem.* 235(1960):633.]

Oxidized protein Mixed disulfide Reduced protein

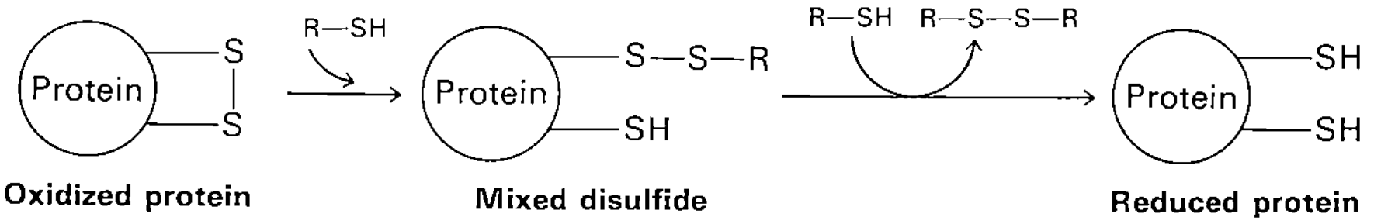

Figure 2-51
Reduction of the disulfide bonds in a protein by an excess of a sulfhydryl reagent (R–SH) such as β-mercaptoethanol.

mercaptoethanol, the mixed disulfides also are reduced, so that the final product is a protein in which the disulfides (cystines) are fully converted into sulfhydryls (cysteines). However, it was found that ribonuclease at 37°C and pH 7 cannot be readily reduced by β-mercaptoethanol unless the protein is partly unfolded by agents such as *urea* or *guanidine hydrochloride.* Although the mechanism of action of these agents is not fully understood, it is evident that they disrupt noncovalent interactions. Most polypeptide chains devoid of cross-links assume a *random-coil conformation* in

$$HO-CH_2-CH_2-SH$$
β-Mercaptoethanol

$$H_2N-\overset{\overset{\displaystyle O}{\|}}{C}-NH_2$$
Urea

$$H_2N-\overset{\overset{\displaystyle NH_2^+\ Cl^-}{\|}}{C}-NH_2$$
Guanidine hydrochloride

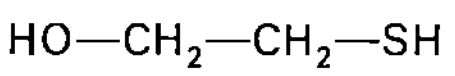

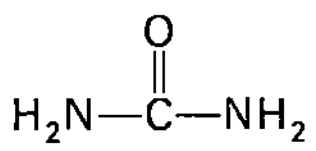

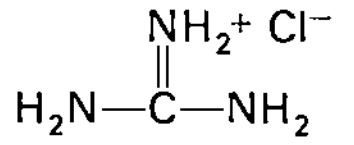

8 M urea or 6 M guanidine HCl, as evidenced by physical properties such as viscosity and optical activity. When ribonuclease was treated with β-mercaptoethanol in 8 M urea, the product was a fully reduced, randomly coiled polypeptide chain *devoid of enzymatic activity*. In other words, ribonuclease was *denatured* by this treatment (Figure 2-52).

Anfinsen then made the critical observation that the denatured ribonuclease, freed of urea and β-mercaptoethanol by dialysis, slowly regained enzymatic activity. He immediately perceived the significance of this chance finding: the sulfhydryls of the denatured enzyme became oxidized by air, and the enzyme spontaneously refolded into a catalytically active form. Detailed studies then showed that nearly all the original enzymatic activity was regained if the sulfhydryls were oxidized under suitable conditions (Figure 2-53). All the measured physical and chemical properties of the refolded enzyme were virtually identical with those of the native enzyme. These experiments showed that *the information needed to specify the complex three-dimensional structure of ribonuclease is contained in its amino acid sequence.* Subsequent studies have established the generality of this central principle of molecular biology: *sequence specifies conformation.*

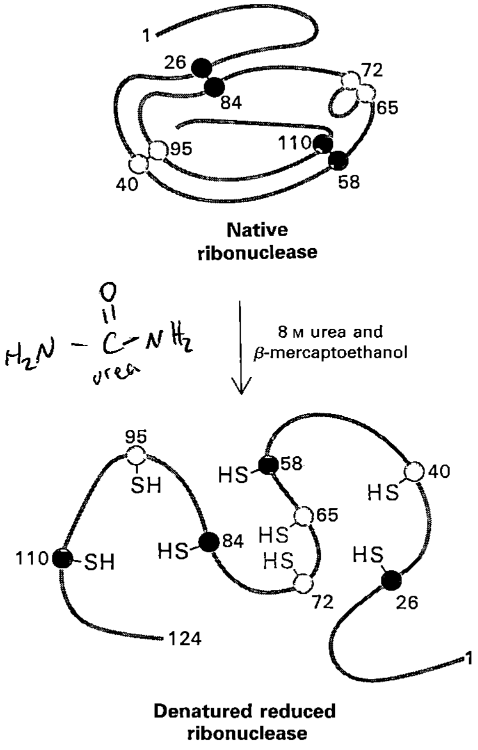

Figure 2-52
Reduction and denaturation of ribonuclease.

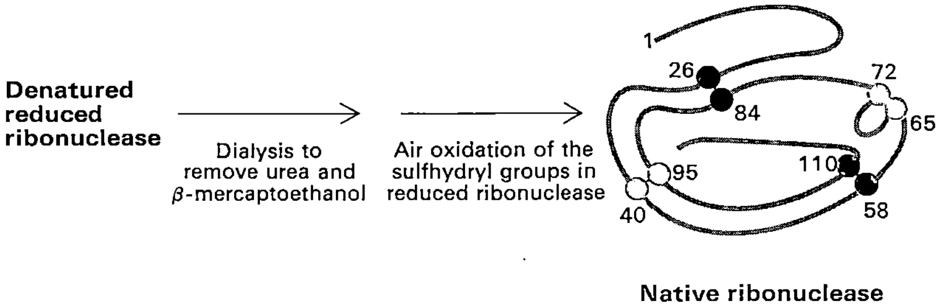

Figure 2-53
Renaturation of ribonuclease.

A quite different result was obtained when reduced ribonuclease was reoxidized while it was still in 8 M urea. This preparation was then dialyzed to remove the urea. Ribonuclease reoxidized in this way had only 1% of the enzymatic activity of the native protein. Why was the outcome of this experiment different from the one in which reduced ribonuclease was reoxidized in a solution free of urea? The reason is that wrong disulfide pairings were formed when the random-coil form of the reduced molecule was reoxidized. There are 105 different ways of pairing eight cysteines to form four disulfides; only one of these combinations is enzymatically active. The 104 wrong pairings have been picturesquely termed "scrambled" ribonuclease. Anfinsen then found that scrambled ribonuclease spontaneously converted into fully active, native ribonuclease when trace amounts of β-mercaptoethanol were added to an aqueous solution of the protein (Figure 2-54). The added β-mercaptoethanol catalyzed the rearrangement of disulfide pairings until the native structure was regained, which took about 10 hours. This process was driven entirely by the decrease in free energy as the scrambled conformations were converted into the stable, native conformation of the enzyme. Thus, *the native form of ribonuclease is the thermodynamically most stable structure.* However, it is important to note that the folding of many proteins is assisted by enzymes that catalyze the attainment of the lowest energy state (p. 429).

Anfinsen (1964) wrote:

> It struck me recently that one should really consider the sequence of a protein molecule, about to fold into a precise geometric form, as a line of melody written in canon form and so designed by Nature to fold back upon itself, creating harmonic chords of interaction consistent with biological function. One might carry the analogy further by suggesting that the kinds of chords formed in a protein with scrambled disulfide bridges, such as I mentioned earlier, are dissonant, but that, by giving an opportunity for rearrangement by the addition of mercaptoethanol, they modulate to give the pleasing harmonics of the native molecule. Whether or not some conclusion can be drawn about the greater thermodynamic stability of Mozart's over Schoenberg's music is something I will leave to the philosophers of the audience.

How are the harmonic chords of interaction created in the conversion of an unfolded polypeptide chain into a folded protein? Can the conformation of a protein be deduced from its amino acid sequence? These challenging problems, which are central to biochemistry, will be discussed further in Chapter 16.

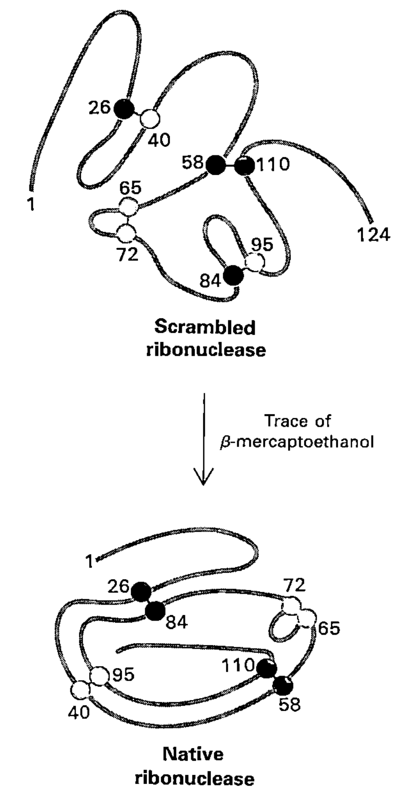

Figure 2-54
Formation of native ribonuclease from scrambled ribonuclease in the presence of a trace of β-mercaptoethanol.

SPECIFIC BINDING AND TRANSMISSION OF STRUCTURAL CHANGES ARE AT THE HEART OF PROTEIN ACTION

The first step in the action of a protein is its binding of another molecule. *Proteins as a class of macromolecules are unique in being able to recognize and interact with highly diverse molecules.* For example, myoglobin tightly binds a heme group when its polypeptide chain is partly folded. The acquisition of heme enables myoglobin to carry out its biological function, which is to reversibly bind O_2. Proteins also combine with other proteins to produce highly ordered arrays, such as the contractile filaments in muscle. The binding of foreign molecules to antibody proteins enables the immune system to distinguish between self and nonself. Furthermore, the expression of many genes is controlled by the binding of proteins that recognize specific DNA sequences. Proteins are able to interact specifically with such a wide range of molecules because they are highly proficient at forming *complementary surfaces and clefts.* The rich repertoire of side chains on these surfaces and in these clefts enables proteins to form hydrogen bonds, electrostatic bonds, and van der Waals bonds with other molecules.

As was mentioned in the introduction to this chapter, nearly all reactions in biological systems are catalyzed by proteins called *enzymes.* We can now appreciate why proteins play the unique role of determining the pattern of chemical transformations. *The catalytic power of proteins comes from their capacity to bind substrate molecules in precise orientations and to stabilize transition states in the making and breaking of chemical bonds.* This basic principle can be made concrete by a simple example of enzymatic catalysis, the hydration of carbon dioxide by *carbonic anhydrase.*

$$H_2O + CO_2 \longrightarrow HCO_3^- + H^+$$

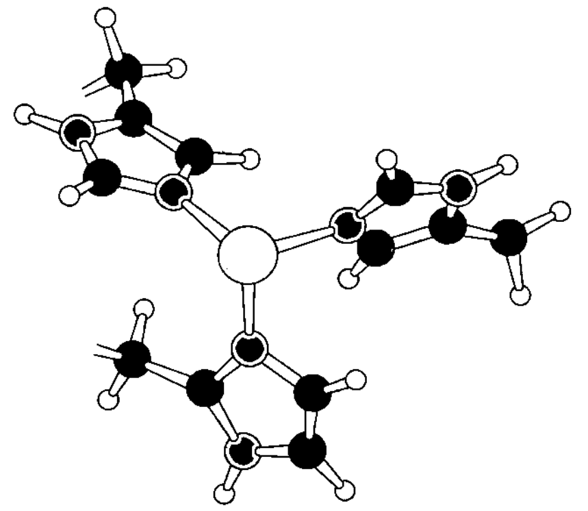

Figure 2-55
Model showing the coordination of a zinc ion (yellow) to three nitrogen atoms of histidine side chains in the catalytic site of carbonic anhydrase.

How does carbonic anhydrase accelerate this reaction by a factor of more than a million? Part of this catalytic enhancement is due to the action of a zinc ion that is coordinated to the imidazole groups of three histidine residues in the enzyme (Figure 2-55). The zinc ion is located at the bottom of a deep cleft some 15 Å from the surface of the protein.

Nearby is a group of residues that recognizes and binds carbon dioxide. Water bound to the zinc ion is rapidly converted into hydroxide ion, which is precisely positioned to attack the carbon dioxide molecule bound next to it (Figure 2-56). The zinc ion helps to orient the CO_2 as

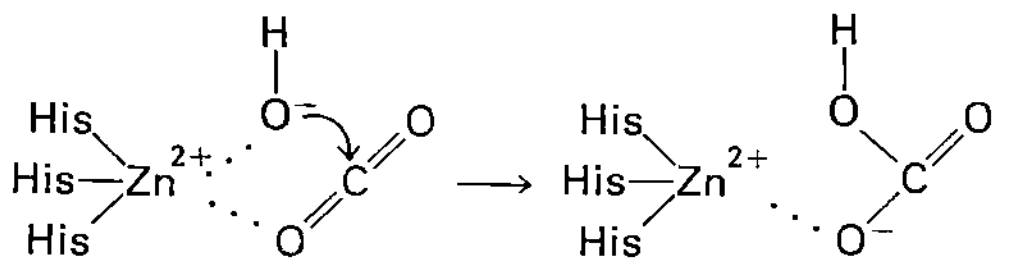

Figure 2-56
Essence of the catalytic mechanism of carbonic anhydrase. Hydroxide ion and carbon dioxide are precisely positioned for the facile formation of bicarbonate by their binding to Zn^{2+}.

well as to provide a very high local concentration of OH^-. Carbonic anhydrase, like other enzymes, is a potent catalyst because *it brings substrates into close proximity and optimizes their orientation for reaction.* Another recurring catalytic device is the *use of charged groups to polarize substrates and stabilize transition states.* Some enzymes even form covalent bonds with substrates. We shall consider enzymatic mechanisms in detail in later chapters.

Some of the most interesting and important proteins contain multiple binding sites that communicate with each other. A conformational change induced by the binding of a molecule to one site in a protein can alter other sites more than 20 Å away. Thus, proteins can be built to serve as *molecular switches* to receive, integrate, and transmit signals. Many proteins contain regulatory sites called *allosteric sites* that control their binding of other molecules and alter their catalytic rates (Figure 2-57). For example, the binding of oxygen to the heme groups of hemoglobin is altered by the binding of H^+ and CO_2 to distant sites in the protein. This dependence of oxygen binding on pH and carbon dioxide concentration makes hemoglobin a very efficient oxygen transporter (p. 159). Allosteric control mediated by conformational changes in protein molecules is central to the regulation of metabolism. In the nervous system, channel proteins sense multiple stimuli and act as logic gates, akin to an *AND* gate on a computer chip.

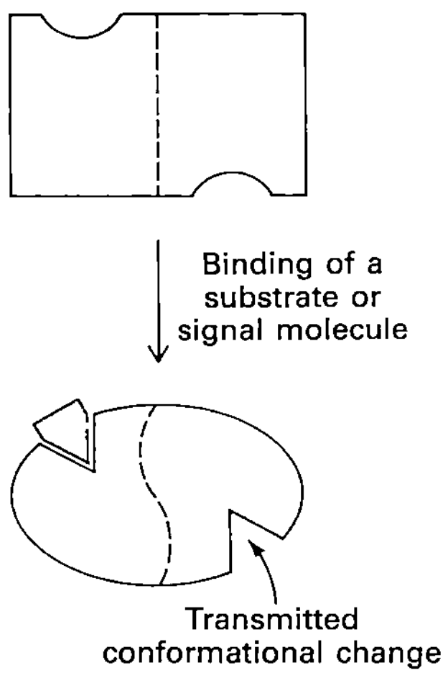

Figure 2-57
Schematic diagram of an allosteric interaction in a protein. The binding of a small molecule or macromolecule to a site in the protein leads to conformational changes that are propagated to distant sites.

Proteins containing pairs of sites that are coupled to each other by conformational changes have the capacity to convert energy from one form to another. Suppose that a protein has a catalytic site that hydrolyzes adenosine triphosphate (ATP) to adenosine diphosphate (ADP), an energetically favored reaction (Figure 2-58). The change from a bound triphosphate to a diphosphate group induces a change at the catalytic site that is transmitted to a different binding site some distance away on the same protein. The role of this second site is to bind another protein when ADP is bound to the first site and to release it when ATP is again bound to the first site. Indeed, enzymes with these properties function as molecular motors that convert chemical bond energy into directed movement, as in muscle contraction (p. 399).

Figure 2-58
ATP binding and hydrolysis can control the affinity of one protein for another. The pair of proteins shown here interact strongly when ADP is bound to one of them but not when ATP is bound. Cycles of binding and release coupled to ATP hydrolysis can produce directional movement, as in muscle contraction.

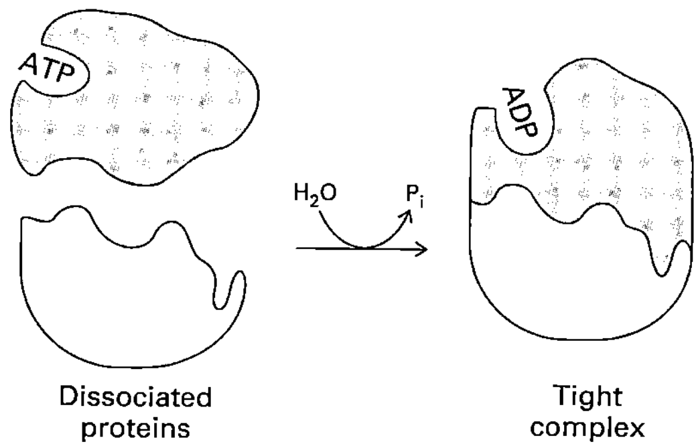

SUMMARY

Proteins play key roles in virtually all biological processes. Nearly all catalysts in biological systems are proteins called enzymes. Hence, proteins determine the pattern of chemical transformations in cells. Proteins mediate a wide range of other functions, such as transport and storage, coordinated motions, mechanical support, immune protection, excitability, integration of metabolism, and control of growth and differentiation.

The basic structural units of proteins are amino acids. All proteins in all species from bacteria to humans are constructed from the same set of 20 amino acids. The side chains of these building blocks differ in size, shape, charge, hydrogen-bonding capacity, and chemical reactivity. They can be grouped as follows: (1) aliphatic side chains—glycine, alanine, valine, leucine, isoleucine, and proline; (2) hydroxyl aliphatic side chains—serine and threonine; (3) aromatic side chains—phenylalanine, tyrosine, and tryptophan; (4) basic side chains—lysine, arginine, and histidine; (5) acidic side chains—aspartic acid and glutamic acid; (6) amide side chains—asparagine and glutamine; and (7) sulfur side chains—cysteine and methionine.

Many amino acids, usually more than a hundred, are joined by peptide bonds to form a polypeptide chain. A peptide bond links the α-carboxyl group of one amino acid and the α-amino group of the next one. Disulfide cross-links can be formed by cysteine residues. Some proteins are covalently modified and cleaved after their synthesis. Proteins have unique amino acid sequences that are genetically determined. The critical determinant of the biological function of a protein is its conformation, which is the three-dimensional arrangement of the atoms of a molecule. Three regularly repeating conformations of polypeptide chains are known: the α helix, the β pleated sheet, and the collagen helix. Segments of the α helix and the β pleated sheet are found in many proteins, as are hairpin β turns.

The amino acid sequence of a protein specifies its three-dimensional structure, as was first shown for ribonuclease. Reduced, unfolded ribonuclease spontaneously forms the correct disulfide pairings and regains full enzymatic activity when oxidized by air after removal of mercaptoethanol and urea. The strong tendency of hydrophobic residues to be herded together by water drives the folding of soluble proteins. Proteins are stabilized by many reinforcing hydrogen bonds and van der Waals interactions as well as by hydrophobic interactions.

Proteins are a unique class of macromolecules in being able to specifically recognize and interact with highly diverse molecules. The repertoire of 20 kinds of side chains enables proteins to fold into distinctive structures and form complementary surfaces and clefts. The catalytic power of enzymes comes from their capacity to bind substrates in precise orientations and to stabilize transition states in the making and breaking of chemical bonds. Conformational changes transmitted between distant sites in protein molecules are at the heart of the capacity of proteins to transduce energy and information.

APPENDIX: Acid-Base Concepts

Ionization of water

Water dissociates into hydronium (H_3O^+) and hydroxyl (OH^-) ions. For simplicity, we refer to the hydronium ion as a hydrogen ion (H^+) and write the equilibrium as

$$H_2O \rightleftharpoons H^+ + OH^-$$

The equilibrium constant K_{eq} of this dissociation is given by

$$K_{eq} = \frac{[H^+][OH^-]}{[H_2O]} \qquad (1)$$

in which the terms in brackets denote molar concentrations. Because the concentration of water (55.5 M) is changed little by ionization, expression 1 can be simplified to give

$$K_w = [H^+][OH^-] \qquad (2)$$

in which K_w is the ion product of water. At 25°C, K_w is 1.0×10^{-14}.

Note that the concentrations of H^+ and OH^- are reciprocally related. If the concentration of H^+ is high, then the concentration of OH^- must be low, and vice versa. For example, if $[H^+] = 10^{-2}$ M, then $[OH^-] = 10^{-12}$ M.

Definition of acid and base

An acid is a proton donor. A base is a proton acceptor.

$$\text{Acid} \rightleftharpoons H^+ + \text{base}$$

$$\underset{\text{Acetic acid}}{CH_3\text{–}COOH} \rightleftharpoons H^+ + \underset{\text{Acetate}}{CH_3\text{–}COO^-}$$

$$\underset{\text{Ammonium ion}}{NH_4^+} \rightleftharpoons H^+ + \underset{\text{Ammonia}}{NH_3}$$

The species formed by the ionization of an acid is its conjugate base. Conversely, protonation of a base yields its conjugate acid. Acetic acid and acetate ion are a conjugate acid-base pair.

Definition of pH and p*K*

The pH of a solution is a measure of its concentration of H^+. The pH is defined as

$$\text{pH} = \log_{10}\frac{1}{[H^+]} = -\log_{10}[H^+] \qquad (3)$$

The ionization equilibrium of a weak acid is given by

$$HA \rightleftharpoons H^+ + A^-$$

The apparent equilibrium constant K for this ionization is

$$K = \frac{[H^+][A^-]}{[HA]} \qquad (4)$$

The pK of an acid is defined as

$$\text{p}K = -\log K = \log\frac{1}{K} \qquad (5)$$

Inspection of equation 4 shows that the pK of an acid is the pH at which it is half dissociated, when $[A^-] = [HA]$.

Henderson-Hasselbalch equation

What is the relationship between pH and the ratio of acid to base? A useful expression can be derived from equation 4. Rearrangement of that equation gives

$$\frac{1}{[H^+]} = \frac{1}{K}\frac{[A^-]}{[HA]} \qquad (6)$$

Taking the logarithm of both sides of equation 6 gives

$$\log\frac{1}{[H^+]} = \log\frac{1}{K} + \log\frac{[A^-]}{[HA]} \qquad (7)$$

Substituting pH for $\log 1/[H^+]$ and pK for $\log 1/K$ in equation 7 yields

$$\text{pH} = \text{p}K + \log\frac{[A^-]}{[HA]} \qquad (8)$$

which is commonly known as the Henderson-Hasselbalch equation.

The pH of a solution can be calculated from equation 8 if the molar proportion of A^- to HA and the pK of HA are known. Consider a solution of 0.1 M acetic acid and 0.2 M acetate ion. The pK of acetic acid is 4.8. Hence, the pH of the solution is given by

$$\text{pH} = 4.8 + \log\frac{0.2}{0.1} = 4.8 + \log 2$$
$$= 4.8 + 0.3 = 5.1$$

Conversely, the pK of an acid can be calculated if the molar proportion of A^- to HA and the pH of the solution are known.

Buffering power

An acid-base conjugate pair (such as acetic acid and acetate ion) has an important property: it resists changes in the pH of a solution. In other words, it acts as a *buffer.* Consider the addition of OH^- to a solution of acetic acid (HA):

$$HA + OH^- \rightleftharpoons A^- + H_2O$$

A plot of the dependence of the pH of this solution on the amount of OH^- added is called a *titration curve* (Figure 2-59). Note that there is an inflection point in the curve at pH 4.8, which is the pK of acetic acid. In the vicinity of this pH, a relatively large amount of OH^- produces little change in pH. In general, a weak acid is most effective in buffering against pH changes in the vicinity of its p*K* value.

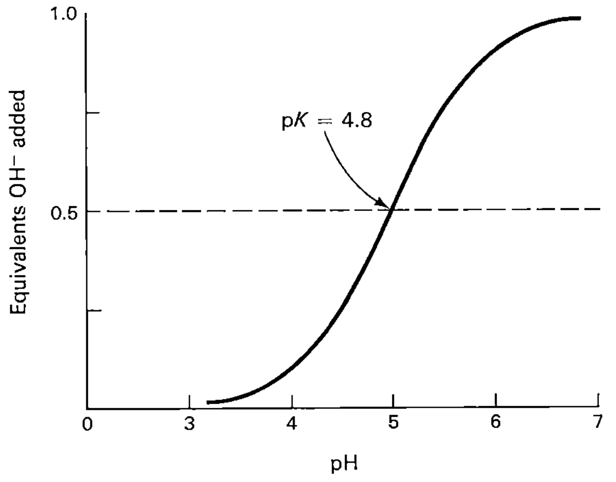

Figure 2-59
Titration curve of acetic acid.

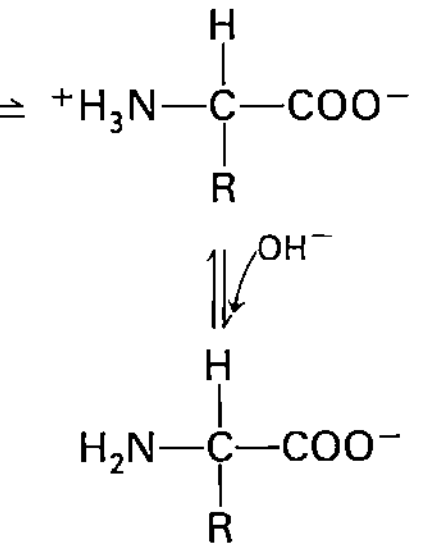

Figure 2-60
Titration of the α-carboxyl and α-amino groups of an amino acid.

Table 2-3
p*K* values of some amino acids

Amino acid	p*K* values (25°C)		
	α-COOH group	α-NH_3^+ group	Side chain
Alanine	2.3	9.9	
Glycine	2.4	9.8	
Phenylalanine	1.8	9.1	
Serine	2.1	9.2	
Valine	2.3	9.6	
Aspartic acid	2.0	10.0	3.9
Glutamic acid	2.2	9.7	4.3
Histidine	1.8	9.2	6.0
Cysteine	1.8	10.8	8.3
Tyrosine	2.2	9.1	10.9
Lysine	2.2	9.2	10.8
Arginine	1.8	9.0	12.5

After J.T. Edsall and J. Wyman. *Biophysical Chemistry* (Academic Press, 1958), ch. 8.

p*K* values of amino acids

An amino acid such as glycine contains two ionizable groups: an α-carboxyl group and a protonated α-amino group. As base is added, these two groups are titrated (Figure 2-60). The p*K* of the α-COOH group is 2.4, whereas that of the α-NH_3^+ group is 9.8. The p*K* values of these groups in other amino acids are similar (Table 2-3). Some amino acids, such as aspartic acid, also contain an ionizable side chain. The p*K* values of ionizable side chains in amino acids range from 3.9 (aspartic acid) to 12.5 (arginine).

SELECTED READINGS

Where to start

Doolittle, R.F., 1985. Proteins. *Sci. Amer.* 253(4):88–99. [A fine overview that emphasizes molecular evolution.]

Richardson, J.S., Richardson, D.C., Tweedy, N.B., Gernert, K.M., Quinn, T.P., Hecht, M.H., Erickson, B.W., Yan, Y., McClain, R.D., Donlan, M.E., and Suries, M.C., 1992. Looking at proteins: Representations, folding, packing, and design. *Biophys. J.* 63:1186–1220. [A beautifully illustrated account of protein folding and design.]

Richards, F.M., 1991. The protein folding problem. *Sci. Amer.* 264:54–57.

Karplus, M., and McCammon, J. A., 1986. The dynamics of proteins. *Sci. Amer.* 254(4):42–51.

Books

Branden, C., and Tooze, J., 1991. *Introduction to Protein Structure.* Garland. [An exceptionally well illustrated account of the three-dimensional structure of proteins, with emphasis on recurring structural motifs.]

Perutz, M.F., 1992. *Protein Structure: New Approaches to Disease and Therapy.* W.H. Freeman. [A highly readable and engaging account of how x-ray crystallography has enriched our understanding of disease processes.]

Hendrickson, W.A., and Wüthrich, K., 1993. *Macromolecular Structures 1994.* Current Biology. [This annual volume contains descriptions and illustrations of biological macromolecules whose structures were elucidated at atomic resolution during the preceding year.]

Creighton, T.E., 1992. *Proteins: Structures and Molecular Principles* (2nd ed.). W.H. Freeman. [A valuable volume that emphasizes conformation and folding.]

Kyte, J., 1994. *Structure in Protein Chemistry.* Garland.

Cantor, C.R., and Schimmel, P.R., 1980. *Biophysical Chemistry.* W.H. Freeman. [Chapters 2 and 5 in Part I and Chapters 20 and 21 in Part III give an excellent account of the principles of protein conformation.]

Lesk, A.M., 1991. *Protein Architecture: A Practical Approach.* IRL Press.

Schultz, G.E., and Schirmer, R.H., 1979. *Principles of Protein Structure.* Springer-Verlag.

Neurath, H., and Hill, R.L. (eds.), 1976. *The Proteins* (3rd ed.). Academic Press. [A multivolume treatise that contains many fine articles.]

Covalent modification of proteins

Wold, F., 1981. In vivo chemical modification of proteins (post-translational modification). *Ann. Rev. Biochem.* 50:783–814.

Glazer, A.N., DeLange, R.J., and Sigman, D.S., 1975. *Chemical Modification of Proteins.* North-Holland.

Conformation of proteins

Chothia, C., and Finkelstein, A.V., 1990. The classification and origin of protein folding patterns. *Ann. Rev. Biochem.* 59:1007–1039.

Richardson, J.S., 1981. The anatomy and taxonomy of protein structure. *Adv. Protein Chem.* 34:167–339. [An incisive account of three-dimensional architecture, with emphasis on supersecondary structural motifs.]

Thornton, J.M., and Gardner, S.P., 1989. Protein motifs and data-base searching. *Trends Biochem. Sci.* 14:300–304.

Protein folding

Anfinsen, C.B., 1973. Principles that govern the folding of protein chains. *Science* 181:223–230.

Baldwin, R.L., 1989. How does protein folding get started? *Trends Biochem. Sci.* 14:291–294.

Gething, M-J., and Sambrook, J., 1992. Protein folding in the cell. *Nature* 355:33–45.

Molecular graphics

Ferrin, T., Huang, C., Jarvis, L., and Langridge, R., 1988. The MIDAS display system. *J. Mol. Graphics* 6:13–27. [Nearly all the new space-filling drawings of proteins in this edition were prepared using this highly versatile program.]

Kraulis, P., 1991. MOLSCRIPT: A program to produce both detailed and schematic plots of protein structures. *J. Appl. Cryst.* 24:946–950. [Nearly all the new ribbon and ball-and-stick drawings of proteins in this edition were prepared using this excellent program.]

Richardson, D.C., and Richardson, J.S., 1994. Kinemages—Simple macromolecular graphics for interactive teaching and publication. *Trends Biochem. Sci.* 19:135–138.

PROBLEMS

1. *Shape and dimension.*
   (a) Tropomyosin, a 70-kd muscle protein, is a two-stranded α-helical coiled coil. What is the length of the molecule?
   (b) Suppose that a 40-residue segment of a protein folds into a two-stranded antiparallel β structure with a 4-residue hairpin turn. What is the longest dimension of this motif?

2. *Contrasting isomers.* Poly-L-leucine in an organic solvent such as dioxane is α-helical, whereas poly-L-isoleucine is not. Why do these amino acids with the same number and kinds of atoms have different helix-forming tendencies?

3. *Active again.* A mutation that changes an alanine residue in the interior of a protein to a valine is found to lead to a loss of activity. However, activity is regained when a second mutation at a different position changes an isoleucine residue to a glycine. How might this second mutation lead to a restoration of activity?

4. *Shuffle test.* An enzyme that catalyzes disulfide-sulfhydryl exchange reactions, called protein disulfide isomerase (PDI), has been isolated. Inactive scrambled ribonuclease is rapidly converted into enzymatically active ribonuclease by PDI. In contrast, insulin is rapidly inactivated by PDI. What does this important observation imply about the relation between the amino acid sequence of insulin and its three-dimensional structure?

5. *Stretching a target.* A protease is an enzyme that catalyzes the hydrolysis of peptide bonds of target proteins. How might a protease bind a target protein so that its main chain becomes fully extended in the vicinity of the vulnerable peptide bond?

6. *Often irreplaceable.* Glycine is a highly conserved amino acid residue in the evolution of proteins. Why?

7. *Potential partners.* Identify the groups in a protein that can form hydrogen bonds or electrostatic bonds with an arginine side chain at pH 7.

8. *Permanent waves.* The shape of hair is determined in part by the pattern of disulfide bonds in keratin, its major protein. How can curls be induced?

CHAPTER

3

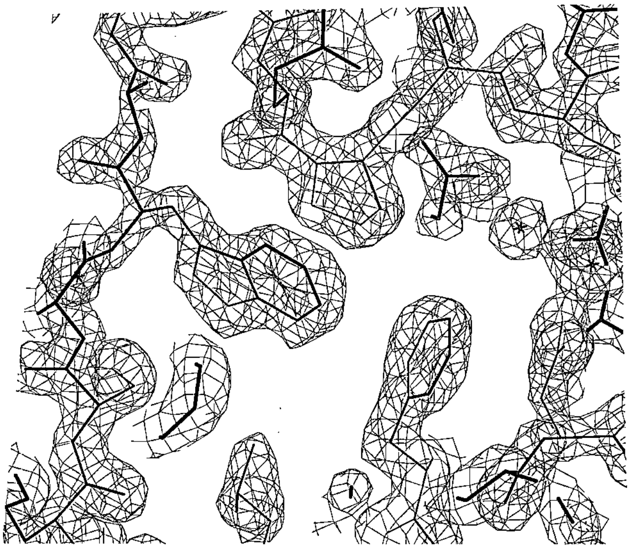

Exploring Proteins

In the preceding chapter, we saw that proteins play crucial roles in nearly all biological processes—in catalysis, transport, coordinated motion, excitability, and control of growth and differentiation. This remarkable range of functions arises from the folding of proteins into many distinctive three-dimensional structures that bind highly diverse molecules. One of the major goals of biochemistry is to determine how amino acid sequences specify the conformations of proteins. We also want to learn how proteins bind specific substrates and other molecules, mediate catalysis, and transduce energy and information. An indispensable step in these studies is the purification of the protein of interest.

The three key approaches to analyzing and purifying proteins are electrophoresis, chromatography, and ultracentrifugation. The amino acid sequence can be determined once a protein is purified. The strategy is to divide and conquer, to obtain specific fragments that can readily be sequenced. Automated peptide sequencing and the application of recombinant DNA methods are providing a wealth of amino acid sequence data that are opening new vistas. The physiological context of a protein also needs to be known to fully understand how it acts. Antibodies are choice probes for locating proteins in vivo and measuring their quantities. X-ray crystallography and nuclear magnetic resonance spectroscopy are the principal techniques for elucidating three-dimensional structure, the key determinant of function. We close with the synthesis of peptides, which makes feasible the synthesis of new drugs, antigens for inducing the formation of specific antibodies, and functional fragments of proteins.

The exploration of proteins by this array of physical and chemical techniques has greatly enriched our understanding of the molecular basis of life and makes it possible to tackle some of the most challenging questions of biology in molecular terms.

Photomicrograph of crystals of adolase. [Courtesy of Dr. David Eisenberg.]

Opening Image: *Electron-density map of a protein determined by x-ray crystallographic analysis. The blue polyhedra show regions of high electron density. The superposed red skeletal framework is the molecular interpretation of the map. Only a small part of the map of the whole protein is shown here. [Courtesy of Dr. William Weis.]*

PROTEINS CAN BE SEPARATED BY GEL ELECTROPHORESIS AND DISPLAYED

A molecule with a net charge will move in an electric field. This phenomenon, termed *electrophoresis,* offers a powerful means of separating proteins and other macromolecules, such as DNA and RNA. The velocity of migration (v) of a protein (or any molecule) in an electric field depends on the electric field strength (E), the net charge on the protein (z), and the frictional coefficient (f).

$$v = \frac{Ez}{f} \qquad (1)$$

The electric force Ez driving the charged molecule toward the oppositely charged electrode is opposed by the viscous drag fv arising from friction between the moving molecule and the medium. The frictional coefficient f depends on both the mass and shape of the migrating molecule and the viscosity (η) of the medium. For a sphere of radius r,

$$f = 6\pi\eta r \qquad (2)$$

Electrophoretic separations are nearly always carried out in gels (or on solid supports such as paper) rather than in free solution for two reasons. First, gels suppress convective currents produced by small temperature gradients, a requirement for effective separation. Second, gels serve as molecular sieves that enhance separation (Figure 3-1). Molecules that are small compared with the pores in the gel readily move through the gel, whereas molecules much larger than the pores are almost immobile. Intermediate-size molecules move through the gel with various degrees of facility. Polyacrylamide gels are choice supporting media for electrophoresis because they are chemically inert and are readily formed by the polymerization of acrylamide. Moreover, their pore sizes can be controlled by choosing various concentrations of acrylamide and methylenebisacrylamide (a cross-linking reagent) at the time of polymerization (Figure 3-2).

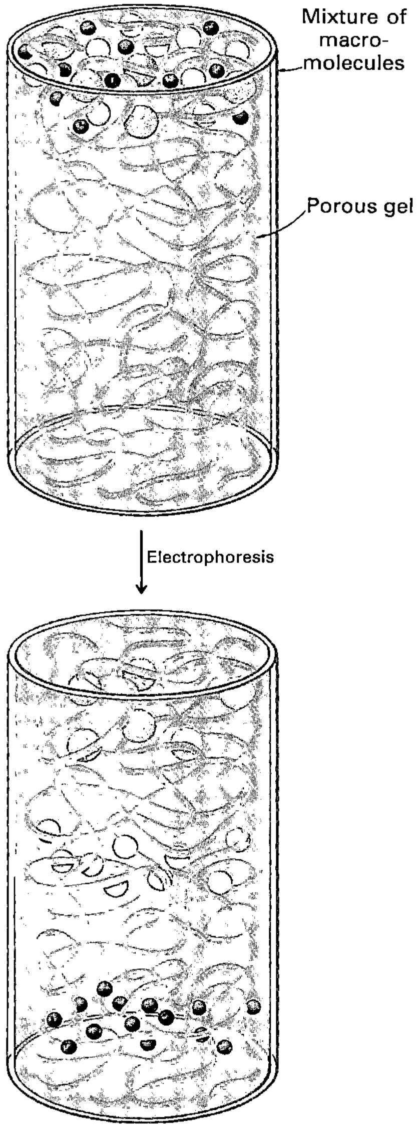

Figure 3-1
Sieving action of a porous polyacrylamide gel.

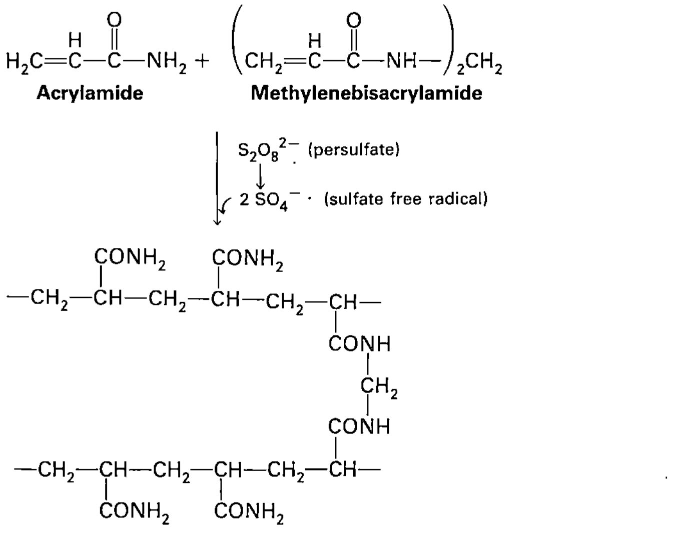

Figure 3-2
Formation of a polyacrylamide gel. The pore size can be controlled by adjusting the concentration of activated monomer (red) and cross-linker (green).

Proteins can be separated largely on the basis of mass by electrophoresis in a polyacrylamide gel under denaturing conditions. The mixture of proteins is first dissolved in a solution of sodium dodecyl sulfate (SDS), an anionic detergent that disrupts nearly all noncovalent interactions in native proteins. Mercaptoethanol or dithiothreitol is also added to reduce disulfide bonds. Anions of SDS bind to main chains at a ratio of about one SDS for every two amino acid residues, which gives a complex of SDS with a denatured protein a large net negative charge that is roughly proportional to the mass of the protein. The negative charge acquired on binding SDS is usually much greater than the charge on the native protein; this native charge is thus rendered insignificant. The SDS complexes with the denatured proteins are then electrophoresed on a polyacrylamide gel, typically in the form of a thin vertical slab (Figure 3-3). The direction of electrophoresis is from top to bottom. Finally, the proteins in the gel can be visualized by staining them with silver or a dye such as Coomassie blue, which reveals a series of bands (Figure 3-4). Radioactive labels can be detected by placing a sheet of x-ray film over the gel, a procedure called *autoradiography.*

$H_3C—(CH_2)_{10}—CH_2OSO_3^-\ Na^+$

Sodium dodecyl sulfate (SDS)

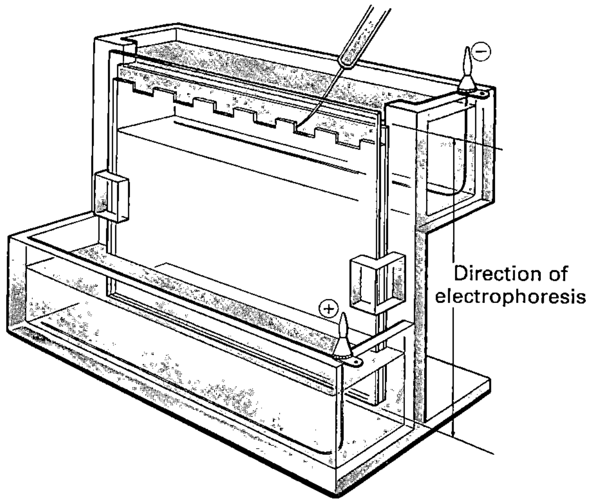

Figure 3-3
Loading of samples for electrophoresis. Typically, several samples are electrophoresed on one flat polyacrylamide gel. A microliter syringe is used to place solutions of proteins in the wells of the slab. A cover is then placed over the gel chamber and 200 volts is applied. The negatively charged SDS–protein complexes migrate in the direction of the anode, at the bottom of the gel.

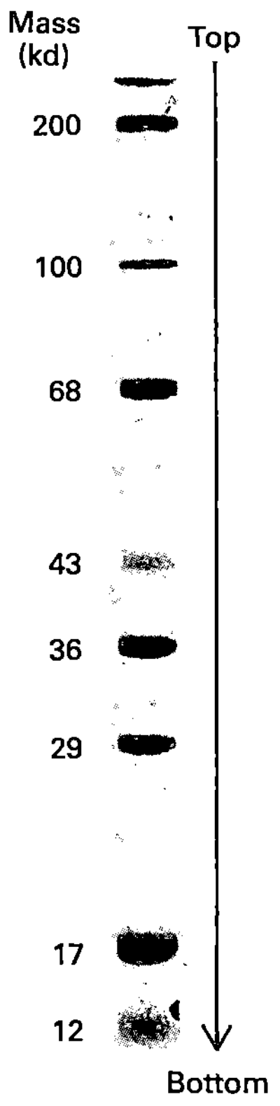

Figure 3-4
Proteins electrophoresed on an SDS–polyacrylamide gel can be visualized by staining with Coomassie blue.

Small proteins move rapidly through the gel, whereas large ones stay at the top, near the point of application of the mixture. The mobility of most polypeptide chains under these conditions is linearly proportional to the logarithm of their mass (Figure 3-5). This empirical relationship is not obeyed by some proteins; for example, some carbohydrate-rich proteins and membrane proteins migrate anomalously. SDS–polyacrylamide gel electrophoresis is rapid, sensitive, and capable of a high degree of resolution. Electrophoresis and staining take several hours. As little as 0.1 μg (~2 pmol) of a protein gives a distinct band when stained with Coomassie blue, and even less (~0.02 μg) can be detected with a silver stain. Proteins that differ in mass by about 2% (e.g., 40 and 41 kd, arising from a difference of about 10 residues) can usually be distinguished.

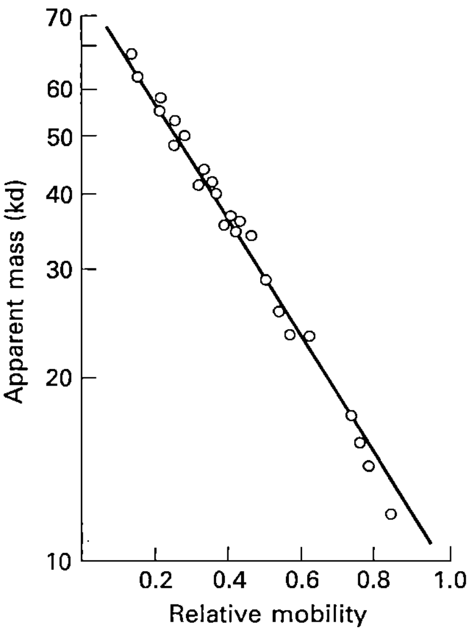

Figure 3-5
The electrophoretic mobility of many proteins in SDS–polyacrylamide gels is inversely proportional to the logarithm of their mass. [After K. Weber and M. Osborn, *The Proteins,* vol. 1, 3rd ed. (Academic Press, 1975), p. 179.]

Proteins can also be separated electrophoretically on the basis of their relative contents of acidic and basic residues. The *isoelectric point (pI)* of a protein is the pH at which its net charge is zero. At this pH, its electrophoretic mobility is zero because z in equation 1 is equal to zero. For example, the pI of cytochrome c, a highly basic electron transport protein, is 10.6, whereas that of serum albumin, an acidic protein in blood, is 4.8. Suppose that a mixture of proteins is electrophoresed in a pH gradient in a gel in the absence of SDS. Each protein will move until it reaches a position in the gel at which the pH is equal to the pI of the protein. This method of separating proteins according to their isoelectric point is called *isoelectric focusing.* The pH gradient in the gel is formed first by electrophoresing a mixture of *polyampholytes* (small multicharged polymers) having many pI values. Isoelectric focusing can readily resolve proteins that differ in pI by as little as 0.01, which means that proteins differing by one net charge can be separated.

Isoelectric focusing can be combined with SDS–polyacrylamide gel electrophoresis to obtain very high-resolution separations. A single sample is first subjected to isoelectric focusing. This single-lane gel is then placed horizontally on top of an SDS–polyacrylamide slab. The proteins are thus spread across the top of the polyacrylamide gel according to how far they migrated during isoelectric focusing. They are then electrophoresed again in a perpendicular direction (vertically) to yield a two-dimensional pattern of spots. In such a gel, proteins have been separated in the horizontal direction on the basis of isoelectric point, and in the vertical direction on the basis of mass. It is remarkable that more than a thousand different proteins in the bacterium *Escherichia coli* can be resolved in a single experiment by two-dimensional electrophoresis (Figure 3-6).

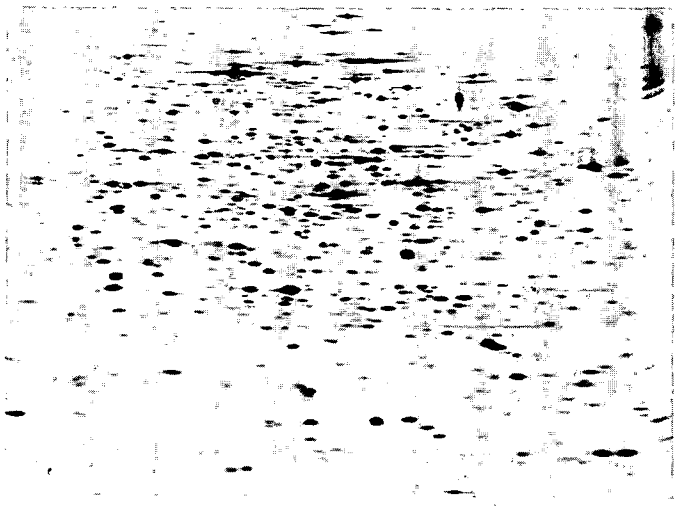

Figure 3-6
Two-dimensional electrophoresis of the proteins from *E. coli.* More than a thousand different proteins from this bacterium have been resolved. These proteins were separated according to their isoelectric pH in the horizontal direction and their apparent mass in the vertical direction. [Courtesy of Dr. Patrick H. O'Farrell.]

PROTEINS CAN BE PURIFIED ACCORDING TO SIZE, SOLUBILITY, CHARGE, AND BINDING AFFINITY

Proteins can readily be visualized and differentiated by the electrophoretic methods described above. These gel techniques can also be used to obtain small quantities (micrograms) of purified polypeptides. However, they do not provide large amounts of purified proteins in their native state. Substantial quantities of purified proteins, of the order of many milligrams, are needed to elucidate fully their three-dimensional structure and their mechanism of action. Several thousand proteins have been purified in active form on the basis of such characteristics as *size, solubility,*

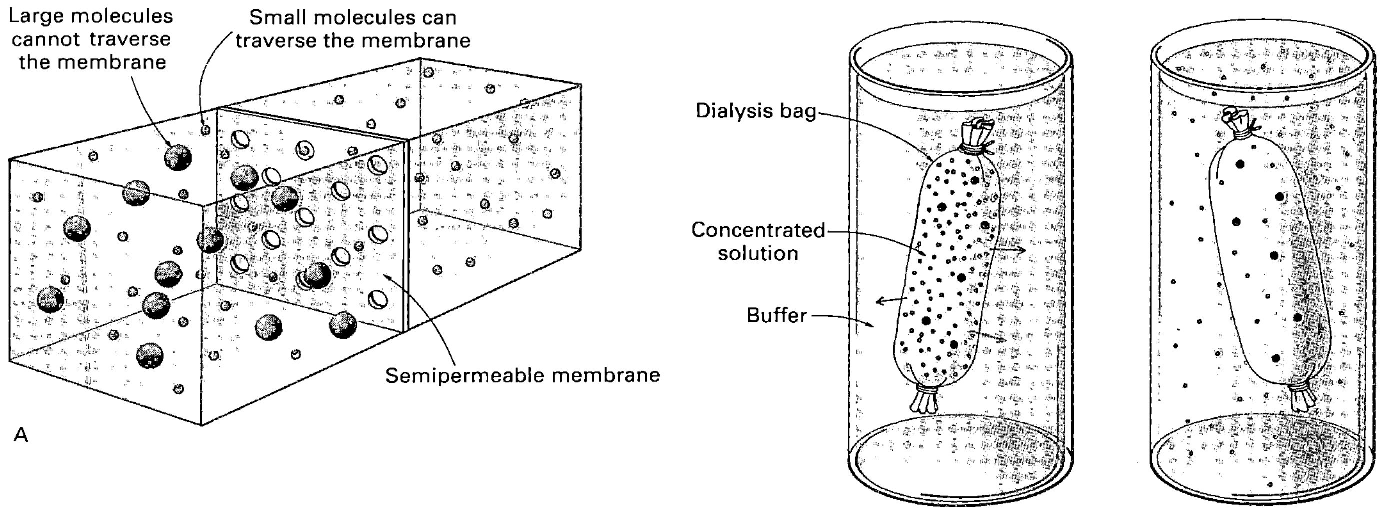

Figure 3-7
Separation of molecules on the basis of size by dialysis. (A) Principle. (B) Application. Protein molecules (red) are retained within the dialysis bag, whereas small molecules (blue) diffuse into the surrounding medium.

charge, and *specific binding affinity.* At each step in purification, the preparation is assayed for a distinctive property of the protein of interest (e.g., enzymatic activity) to assess the efficacy of the procedure.

Proteins can be separated from small molecules by dialysis through a semipermeable membrane, such as a cellulose membrane with pores (Figure 3-7). Molecules having dimensions significantly greater than the pore diameter are retained inside the dialysis bag, whereas smaller molecules and ions traverse the pores of such a membrane and emerge in the dialysate outside the bag.

More discriminating separations on the basis of size can be achieved by the technique of *gel-filtration chromatography* (Figure 3-8). The sample is applied to the top of a column consisting of porous beads made of an insoluble but highly hydrated polymer such as dextran or agarose (which are carbohydrates) or polyacrylamide. Sephadex, Sepharose, and Bio-gel are commonly used commercial preparations of these beads, which are typically 100 μm (0.1 mm) in diameter. Small molecules can enter these beads, but large ones cannot. The result is that small molecules are distributed both in the aqueous solution inside the beads and between them, whereas large molecules are located only in the solution between the beads. *Large molecules flow more rapidly through this column and emerge first because a smaller volume is accessible to them* (see Figure 3-8). It should be noted that the order of emergence of molecules from a column of porous beads is the reverse of the order in gel electrophoresis, in which a *continuous* polymer framework impedes the movement of large molecules (see Figure 3-4). Much larger quantities of protein can be separated by gel-filtration chromatography than by gel electrophoresis, but at the price of lower resolution.

The solubility of most proteins is lowered at high salt concentrations. This effect, called *salting out,* is very useful, though not well understood. The dependence of solubility on salt concentration differs from one protein to another. Hence, salting out can be used to fractionate proteins. For example, 0.8 M ammonium sulfate precipitates fibrinogen, a blood-clotting protein, whereas 2.4 M is needed to precipitate serum albumin. Salting out is also useful for concentrating dilute solutions of proteins.

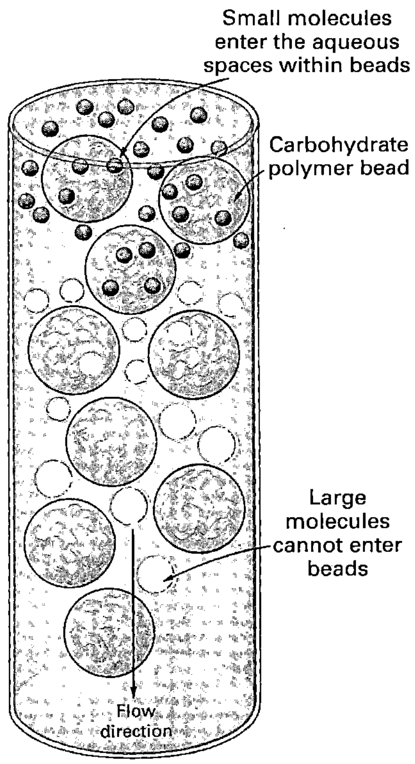

Figure 3-8
Separation of three proteins differing in size by gel-filtration chromatography. Large proteins emerge sooner than small ones because they cannot enter the internal volume of the beads.

Positively charged
protein binds to
negatively charged
bead

Negatively charged
protein flows
through

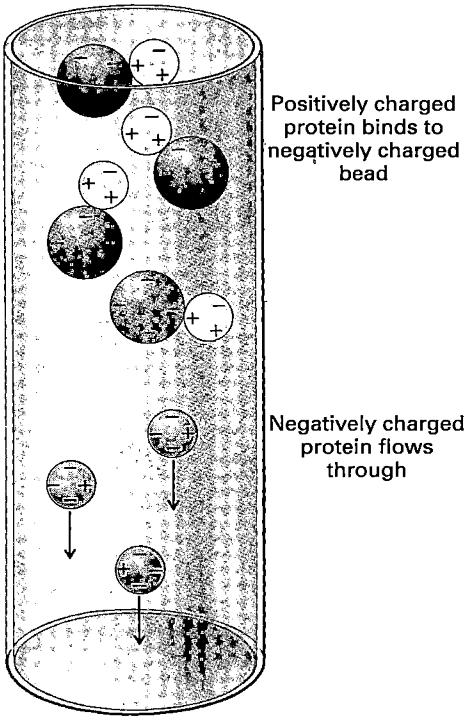

Figure 3-9
Ion-exchange chromatography separates proteins mainly according to their net charge.

Proteins can be separated on the basis of their net charge by *ion-exchange chromatography*. If a protein has a net positive charge at pH 7, it will usually bind to a column of beads containing carboxylate groups, whereas a negatively charged protein will not (Figure 3-9). A positively charged protein bound to such a column can then be eluted (released) by increasing the concentration of sodium chloride or another salt in the eluting buffer. Sodium ions compete with positively charged groups on the protein for binding to the column. Proteins that have a low density of net positive charge will tend to emerge first, followed by those having a higher charge density. Factors other than net charge, such as affinity for the supporting matrix, can also influence the behavior of proteins on ion-exchange columns. Negatively charged proteins (anionic proteins) can be separated by chromatography on positively charged diethyl-aminoethyl–cellulose (DEAE–cellulose) columns. Conversely, positively charged proteins (cationic proteins) can be separated on negatively charged carboxymethyl–cellulose (CM–cellulose) columns.

$$\boxed{\begin{array}{c}\text{Cellulose}\\\text{or}\\\text{agarose}\end{array}}\!-\!CH_2\!-\!CH_2\!-\!\overset{+}{N}\!\!\begin{array}{c}H\\\diagup\\[-4pt]\diagdown\end{array}\!\!\begin{array}{c}C_2H_5\\[6pt]C_2H_5\end{array}$$

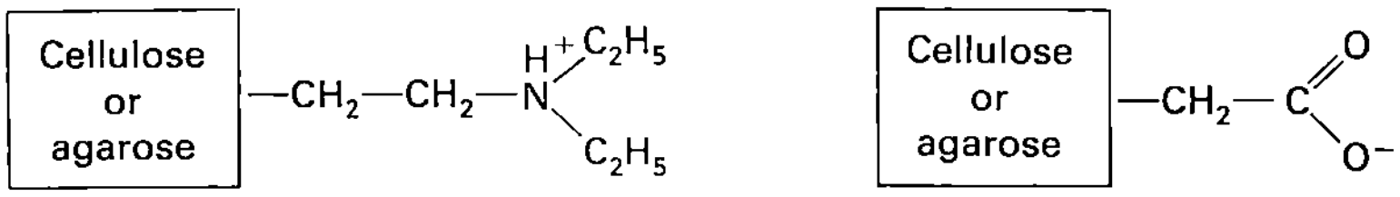

Diethylaminoethyl (DEAE) group
(Protonated form)

$$\boxed{\begin{array}{c}\text{Cellulose}\\\text{or}\\\text{agarose}\end{array}}\!-\!CH_2\!-\!C\!\!\begin{array}{c}\diagup O\\[-2pt]\diagdown O^-\end{array}$$

Carboxymethyl (CM) group
(Ionized form)

Affinity chromatography is another powerful and generally applicable means of purifying proteins. This technique takes advantage of the high affinity of many proteins for specific chemical groups. For example, the plant protein concanavalin A can be purified by passing a crude extract through a column of beads containing covalently attached glucose residues. Concanavalin A binds to such a column because it has affinity for glucose, whereas most other proteins do not. The bound concanavalin A can then be released from the column by adding a concentrated solution of glucose. The glucose in solution displaces the column-attached glucose residues from binding sites on concanavalin A (Figure 3-10). In general, affinity chromatography can be effectively used to isolate a protein that recognizes group X by (1) covalently attaching X or a derivative of it to a column, (2) adding a mixture of proteins to this column, which is then washed with buffer to remove unbound proteins, and (3) eluting the desired protein by adding a high concentration of a soluble form of X.

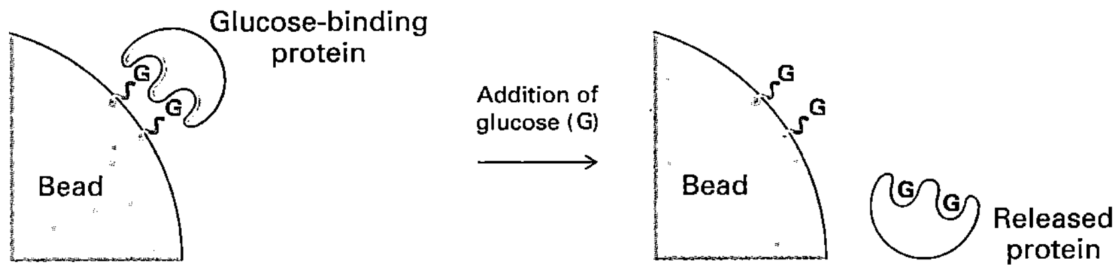

Figure 3-10
Affinity chromatography of concanavalin A (shown in yellow) on a solid support containing covalently attached glucose residues (G).

ULTRACENTRIFUGATION IS VALUABLE FOR SEPARATING BIOMOLECULES AND DETERMINING THEIR MASS

Centrifugation is a powerful and generally applicable method for separating and analyzing cells, organelles, and biological macromolecules. A particle moving in a circle of radius r at an angular velocity ω is subject to a centrifugal (outward) field equal to $\omega^2 r$. The *centrifugal force* F_c on this particle is equal to the product of its effective mass m' and the centrifugal field.

$$F_c = m'\omega^2 r = m(1 - \bar{v}\rho)\omega^2 r \quad (3)$$

The effective mass m' is less than the mass m because the displaced fluid exerts an opposing force. This buoyancy factor is equal to $(1 - \bar{v}\rho)$, where $\bar{v}$ is the partial specific volume of the particle (the reciprocal of its density) and ρ is the density of the solution. A particle moves in this field at a constant velocity v when F_c is equal to the viscous drag vf, where f is the frictional coefficient of the particle. Hence, the migration velocity (sedimentation velocity) of the particle is

$$v = \frac{F_c}{f} = \frac{m(1 - \bar{v}\rho)\omega^2 r}{f} \quad (4)$$

Note that this expression for movement in a centrifugal field is analogous to equation 1 for movement in an electric field.

Equation 4 shows that the sedimentation velocity is directly proportional to the strength of the centrifugal field. Hence, it is possible to define a measure of sedimentation that depends on the properties of the particle and solution but is independent of how fast the sample is spun. The *sedimentation coefficient s*, defined as the velocity divided by the centrifugal field, is equal to

$$s = \frac{v}{\omega^2 r} = \frac{m(1 - \bar{v}\rho)}{f} \quad (5)$$

Sedimentation coefficients are usually expressed in *Svedberg units.* A svedberg (S) is equal to 10^{-13} seconds. For example, suppose that a 150-kd antibody protein is spun in an ultracentrifuge at a radius of 8 cm at 75,000 revolutions per minute (rpm). The centrifugal field under these conditions is 4.9×10^8 cm/s^2, which is about 500,000 times the strength of the earth's gravitational field (g). If the velocity of the protein in this field is 3.4×10^{-4} cm/s, then its sedimentation coefficient is 7S.

Several important conclusions can be drawn from equation 5:

1. The sedimentation velocity of a particle depends in part on its mass. A more massive particle always sediments more rapidly than does a less massive one of the same shape and density.

2. Shape is important too because it affects the viscous drag. The frictional coefficient f of a compact particle is smaller than that of an extended particle of the same mass. A parachutist with a defective unopened parachute falls much more quickly than one with a functioning opened parachute. Hence, elongated particles sediment more slowly than do spherical ones of the same mass.

3. A dense particle moves more rapidly than a less dense one because the opposing buoyant force is smaller for the denser particle ($\bar{v}\rho$ is smaller for the denser particle).

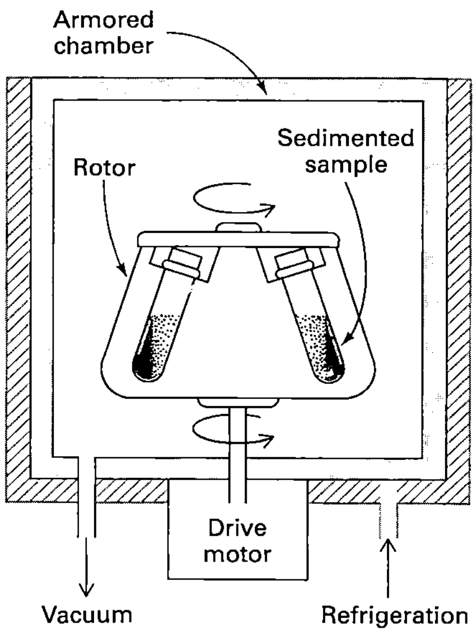

Figure 3-11
Ultracentrifuge rotor containing two sample tubes.

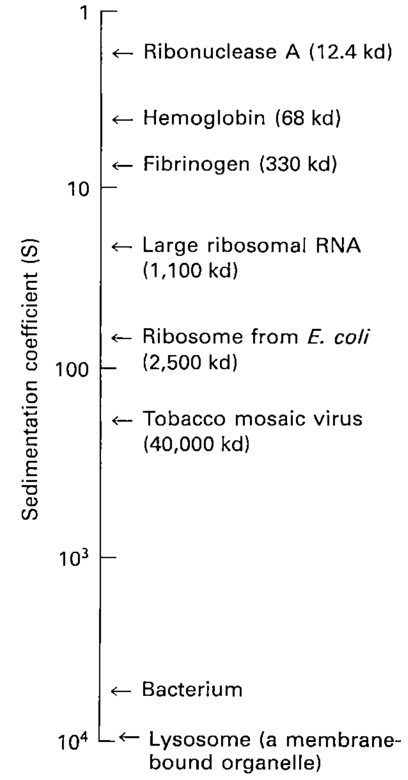

Figure 3-12
Span of S values of biomolecules and cells.

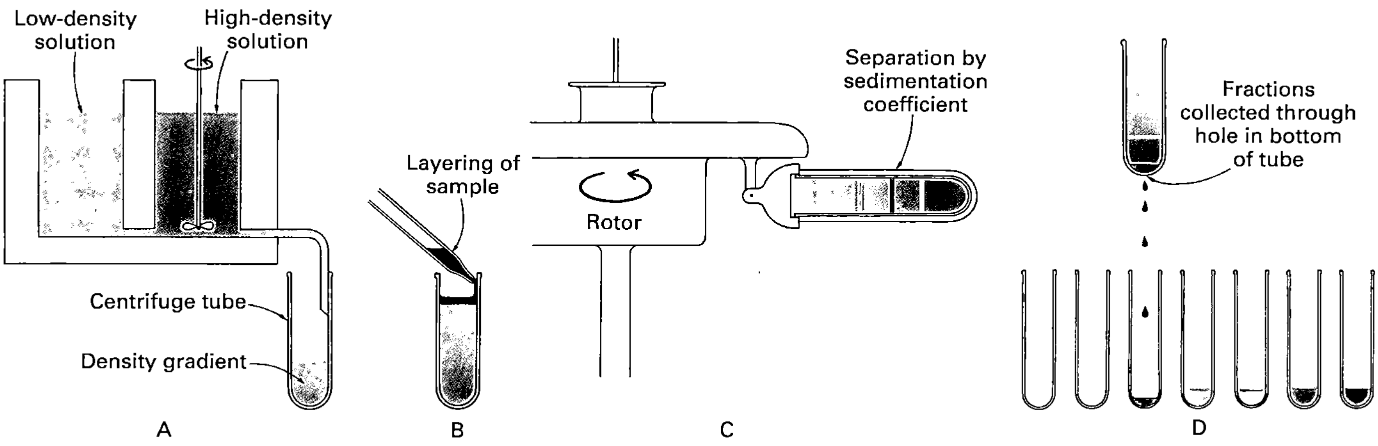

Figure 3-13
Zonal centrifugation. The steps are (A) forming of a density gradient, (B) layering the sample on top of the gradient, (C) placing the tube in a swinging-bucket rotor and centrifuging it, and (D) collecting the samples. [After D. Freifelder, *Physical Biochemistry,* 2nd ed. (W.H. Freeman, 1982), p. 397.]

4. The sedimentation velocity depends also on the density of the solution (ρ). Particles sink when $\bar{v}\rho < 1$, float when $\bar{v}\rho > 1$, and do not move when $\bar{v}\rho = 1$.

Let us see how centrifugation can be used to separate proteins with different sedimentation coefficients (Figure 3-13). The first step in *zonal centrifugation* (also called *band centrifugation*) is to form a density gradient in a centrifuge tube by mixing different proportions of a low-density solution (such as 5% sucrose) and a high-density one (such as 20% sucrose). The role of the density gradient here is to prevent convective flow. A small volume of a solution containing the mixture of proteins to be separated is then layered on top of the density gradient. When the rotor is spun, proteins move through the gradient and separate according to their sedimentation coefficients. Centrifugation is stopped before the fastest protein reaches the bottom of the tube. The separated bands of protein can be harvested by making a hole in the bottom of the tube and collecting drops. The drops can be assayed for protein content and catalytic activity or another functional property. This sedimentation-velocity technique readily separates proteins differing in sedimentation coefficient by a factor of two or more.

The mass (molecular weight) of a protein can be directly determined by *sedimentation equilibrium,* in which a sample is centrifuged at relatively low speed so that sedimentation is counterbalanced by diffusion. Under these conditions, a smooth gradient of protein concentration develops. The dependence of concentration on distance from the rotation axis reveals the mass of the particle. *The sedimentation-equilibrium technique for determining mass is rigorous and can be applied under nondenaturing conditions in which the native structure of multimeric proteins is preserved.* In contrast, SDS–polyacrylamide gel electrophoresis (p. 48) provides an *estimate* of the mass of dissociated polypeptide chains under *denaturing* conditions.

THE MASS OF PROTEINS CAN BE PRECISELY DETERMINED BY ELECTROSPRAY MASS SPECTROMETRY

The very low volatility of proteins was a barrier for many years to using mass spectrometry, an established analytical technique in organic chemistry. This difficulty has been circumvented by the recent introduction of *electrospray ionization mass spectrometry.* A protein sample in an acidic volatile solvent is sprayed into a mass spectrometer. The solvent surrounding individual droplets evaporates rapidly in the vacuum chamber of the instrument, leaving unfragmented bare protein molecules carrying multiple positive charges. These charged protein molecules are accelerated by an electric field and then deflected by a magnetic field. They are separated according to *m/z*, the ratio of their mass to their charge. The mass spectrum of a pure protein shows a set of peaks corresponding to different numbers of bound protons (Figure 3-14A). Because adjacent peaks in the mass spectrum arise from proteins containing $n-1$, n, and $n+1$ bound protons, the mass of the parent protein molecule can be deduced (Figure 3-14B). Mass spectrometric determinations are accurate to about 1 part in 10^4, which means that the mass of a 10-kd protein can be measured to within 1 dalton.

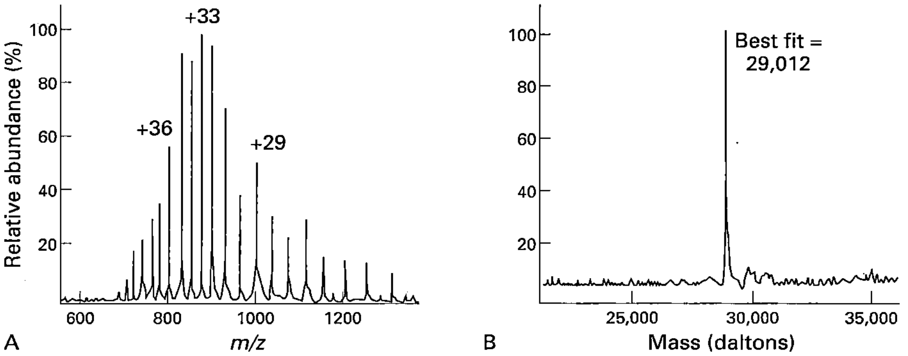

Figure 3-14
Determination of the mass of a protein by electrospray mass spectrometry. (A) The *m/z* spectrum of carbonic anhydrase. Adjacent peaks arise from protein molecules that differ by 1 in the number of bound protons. (B) The deduced mass distribution of the sample shows a major peak with a mass of 29,012 daltons. [Courtesy of Dr. Ian Jardine.]

AMINO ACID SEQUENCES CAN BE DETERMINED BY AUTOMATED EDMAN DEGRADATION

We turn to the elucidation of amino acid sequence. Let us consider how the sequence of a short peptide, such as

Ala-Gly-Asp-Phe-Arg-Gly

could be established. First, the *amino acid composition* of the peptide is determined. The peptide is hydrolyzed into its constituent amino acids by heating it in 6 N HCl at 110°C for 24 hours. Stanford Moore and William Stein showed that amino acids in hydrolysates can be separated by ion-exchange chromatography on columns of sulfonated polystyrene and quantitated by reacting them with *ninhydrin.* Amino acids treated this way give an intense blue color, except for proline, which gives a yellow color because it contains a secondary amino group. The concentration of amino acid in a solution is proportional to the optical absorbance of the solution after heating it with ninhydrin. This technique can detect a microgram (10 nmol) of an amino acid, which is about the amount present in a thumbprint. As little as a nanogram (10 pmol) of an amino acid can be detected by means of *fluorescamine,* which reacts with the α-amino group to form a highly fluorescent product (Figure 3-15). The identity of

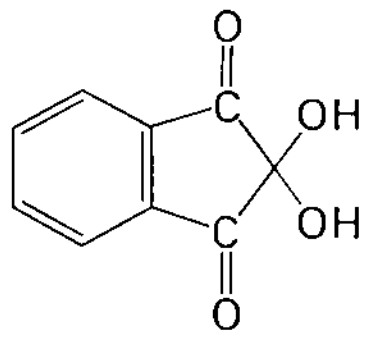

Ninhydrin

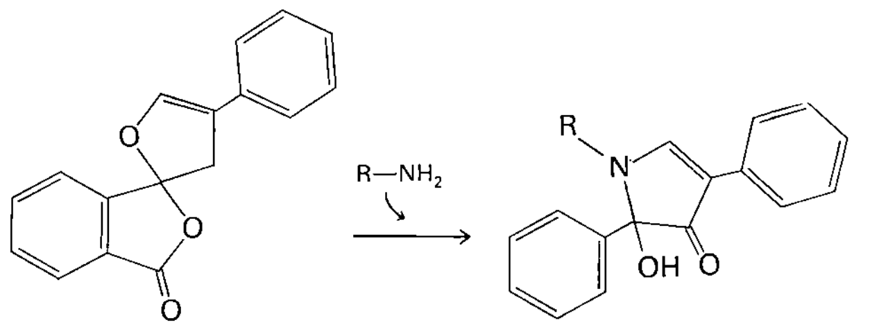

Figure 3-15
Reaction of fluorescamine with the α-amino group of an amino acid to form a fluorescent derivative.

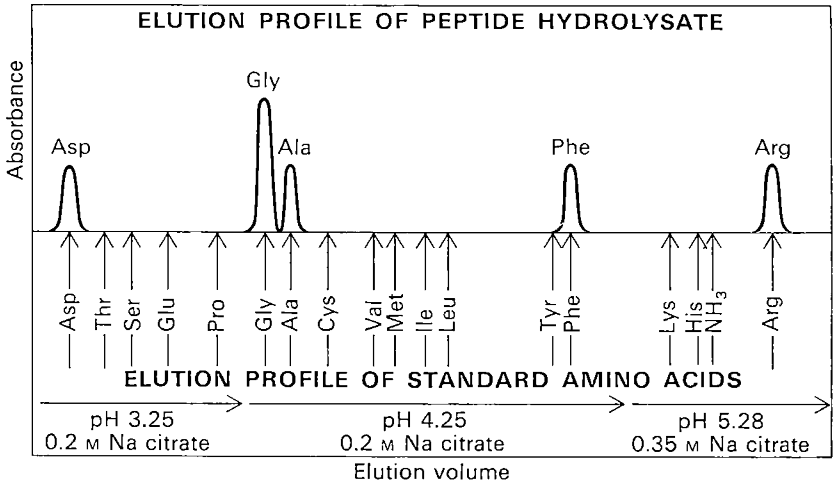

Figure 3-16
Different amino acids in a peptide hydrolysate can be separated by ion-exchange chromatography on a sulfonated polystyrene resin (such as Dowex-50). Buffers of increasing pH are used to elute the amino acids from the column. Aspartate, which has an acidic side chain, is first to emerge, whereas arginine, which has a basic side chain, is the last.

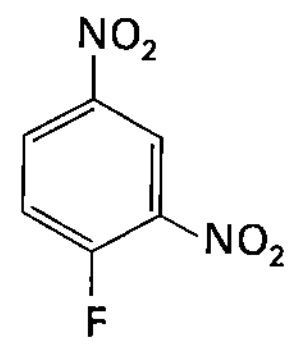

Fluorodinitrobenzene

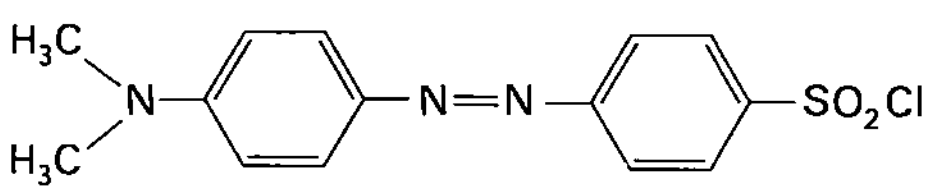

Dabsyl chloride

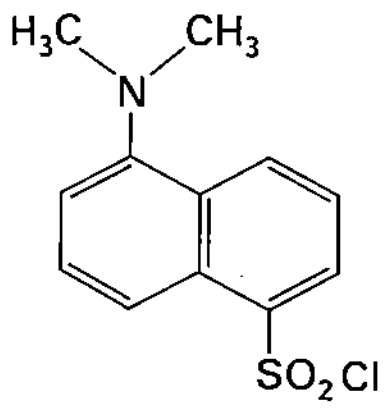

Dansyl chloride

the amino acid is revealed by its elution volume, which is the volume of buffer used to remove the amino acid from the column (Figure 3-16). A comparison of the chromatographic patterns of our sample hydrolysate with that of a standard mixture of amino acids would show that the amino acid composition of the peptide is

(Ala, Arg, Asp, Gly_2, Phe)

The parentheses denote that this is the amino acid composition of the peptide, not its sequence.

The amino-terminal residue of a protein or peptide can be identified by labeling it with a compound that forms a stable covalent link. *Fluorodinitrobenzene (FDNB)* was first used for this purpose by Frederick Sanger. *Dabsyl chloride* is now commonly used because it forms intensely colored derivatives that can be detected with high sensitivity. It reacts with an uncharged α-NH_2 group to form a sulfonamide derivative that is stable under conditions that hydrolyze peptide bonds (Figure 3-17). Hydrolysis of our sample dabsyl–peptide in 6 N HCl would yield a dabsyl–amino acid, which could be identified as dabsyl–alanine by its chromatographic properties. *Dansyl chloride* too is a valuable labeling reagent because it forms fluorescent sulfonamides.

Although the dabsyl method for determining the amino-terminal residue is sensitive and powerful, it cannot be used repeatedly on the same

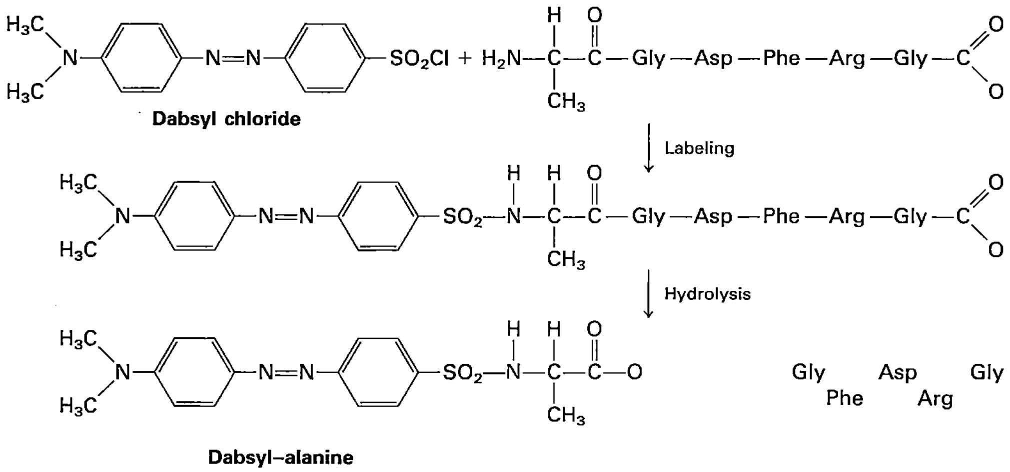

Figure 3-17
Determination of the amino-terminal residue of a peptide. Dabsyl chloride labels the peptide, which is then hydrolyzed using HCl. The dabsyl–amino acid (dabsyl–alanine in this example) is identified by its chromatographic characteristics.

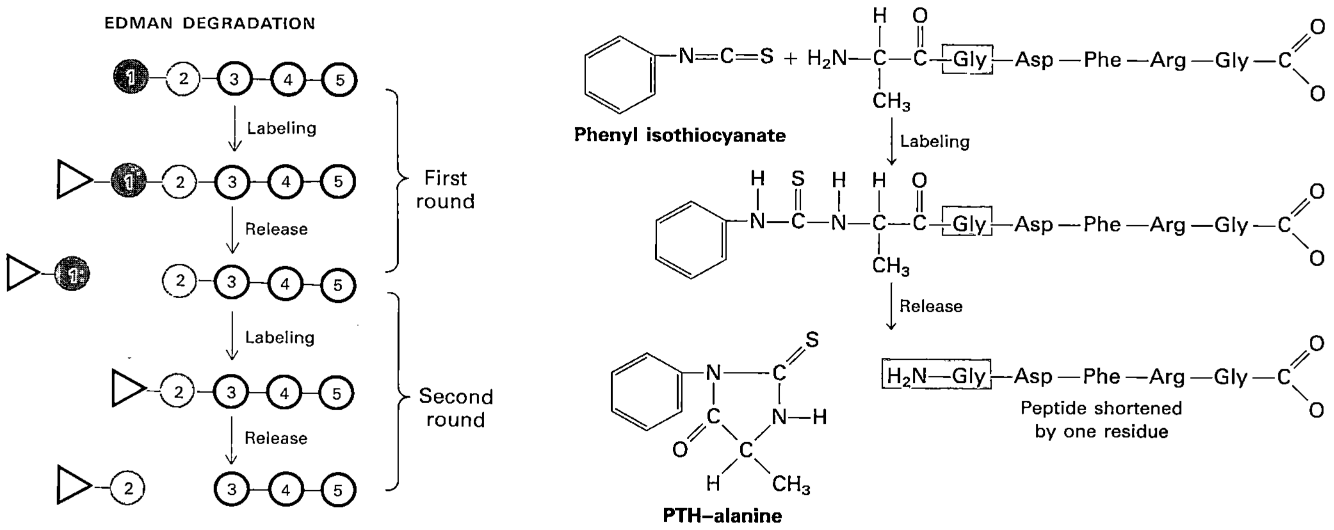

peptide because the peptide is totally degraded in the acid-hydrolysis step. Pehr Edman devised a method for labeling the amino-terminal residue and cleaving it from the peptide without disrupting the peptide bonds between the other amino acid residues. The *Edman degradation* sequentially removes one residue at a time from the amino end of a peptide (Figure 3-18). *Phenyl isothiocyanate* reacts with the uncharged terminal amino group of the peptide to form a phenylthiocarbamoyl derivative. Then, under mildly acidic conditions, a cyclic derivative of the terminal amino acid is liberated, which leaves an intact peptide shortened by one amino acid. The cyclic compound is a phenylthiohydantoin (PTH)–amino acid, which can be identified by chromatographic procedures. Furthermore, the amino acid composition of the shortened peptide

(Arg, Asp, Gly_2, Phe)

can be compared with that of the original peptide

(Ala, Arg, Asp, Gly_2, Phe)

The difference between these analyses is one alanine residue, which shows that alanine is the amino-terminal residue of the original peptide. The Edman procedure can then be repeated on the shortened peptide. The amino acid analysis after the second round of degradation is

(Arg, Asp, Gly, Phe)

showing that the second residue from the amino end is glycine. This conclusion can be confirmed by chromatographic identification of PTH–glycine obtained in the second round of the Edman degradation. Three more rounds of the Edman degradation will reveal the complete sequence of the original peptide.

Analyses of protein structures have been markedly accelerated by the development of *sequenators,* which are automated instruments for the determination of amino acid sequence. In a liquid-phase sequenator, a thin film of protein in a spinning cylindrical cup is subjected to the Edman degradation. The reagents and extracting solvents are passed over the immobilized film of protein, and the released PTH–amino acid is identified by *high-pressure liquid chromatography* (also called *high-performance liquid chromatography,* HPLC; Figure 3-19). One cycle of the Edman degradation is carried out in less than two hours. By repeated degradations, the amino

Figure 3-18
The Edman degradation. The labeled amino-terminal residue (PTH–alanine in the first round) can be released without hydrolyzing the rest of the peptide. Hence, the amino-terminal residue of the shortened peptide (Gly-Asp-Phe-Arg-Gly) can be determined in the second round. Three more rounds of the Edman degradation reveal the complete sequence of the original peptide.

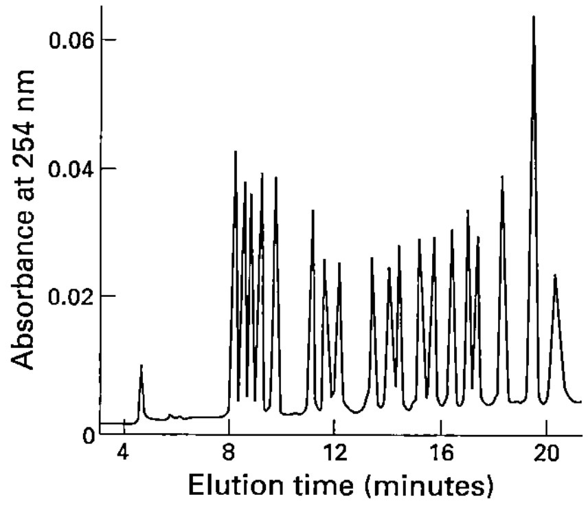

Figure 3-19
PTH–amino acids can be rapidly separated by high-pressure liquid chromatography (HPLC). In this HPLC profile, a mixture of PTH–amino acids is clearly resolved into its components. An unknown amino acid can be identified by its elution position relative to the known ones.

acid sequence of some fifty residues in a protein can be determined. Gas-phase sequenators can analyze picomole quantities of peptides and proteins. This high sensitivity makes it feasible to analyze the sequence of a protein sample eluted from a single band of an SDS–polyacrylamide gel.

PROTEINS CAN BE SPECIFICALLY CLEAVED INTO SMALL PEPTIDES TO FACILITATE ANALYSIS

Peptides much longer than about 50 residues cannot be reliably sequenced by the Edman method because not quite all peptides in the reaction mixture release the amino acid derivative at each step. If the efficiency of release of each round were 98%, the proportion of "correct" amino acid released after 60 rounds would be only 0.3 (0.98^{60})—a hopelessly impure mix. This obstacle can be circumvented by specifically cleaving a protein into peptides not much longer than 50 residues. In essence, the strategy is to *divide and conquer.*

Specific cleavage can be achieved by chemical or enzymatic methods. For example, Bernhard Witkop and Erhard Gross discovered that *cyanogen bromide (CNBr)* splits polypeptide chains only on the carboxyl side of methionine residues (Figure 3-20). A protein that has 10 methionines will usually yield 11 peptides on cleavage with CNBr. Highly specific cleavage is also obtained with trypsin, a proteolytic enzyme from pancreatic juice. Trypsin cleaves polypeptide chains on the carboxyl side of arginine and lysine residues (Figure 3-21). A protein that contains 9 lysines and 7 arginines will usually yield 17 peptides on digestion with trypsin. Each of these tryptic peptides, except for the carboxyl-terminal peptide of the protein, will end with either arginine or lysine. Several other ways of specifically cleaving polypeptide chains are given in Table 3-1.

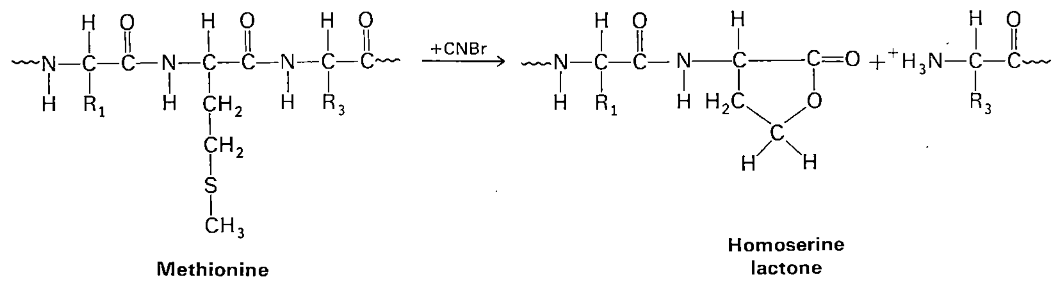

Figure 3-20
Cyanogen bromide cleaves polypeptides on the carboxyl side of methionine residues.

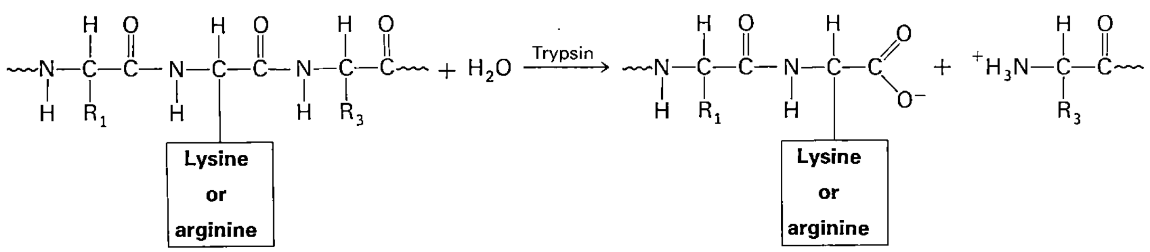

Figure 3-21
Trypsin hydrolyzes polypeptide on the carboxyl side of arginine and lysine residues.

Table 3-1
Specific cleavage of polypeptides

Reagent	*Cleavage site*
Chemical cleavage	
Cyanogen bromide	Carboxyl side of methionine residues
O-Iodosobenzoate	Carboxyl side of tryptophan residues
Hydroxylamine	Asparagine–glycine bonds
2-Nitro-5-thiocyanobenzoate	Amino side of cysteine residues
Enzymatic cleavage	
Trypsin	Carboxyl side of lysine and arginine residues
Clostripain	Carboxyl side of arginine residues
Staphylococcal protease	Carboxyl side of aspartate and glutamate residues (glutamate only under certain conditions)

The peptides obtained by specific chemical or enzymatic cleavage are separated by chromatography. The sequence of each purified peptide is then determined by the Edman method. At this point, the amino acid sequences of segments of the protein are known, but the order of these segments is not yet defined. The necessary additional information is obtained from *overlap peptides* (Figure 3-22). A second enzyme is used to split the polypeptide chain at other linkages. For example, chymotrypsin cleaves preferentially on the carboxyl side of aromatic and some other bulky nonpolar residues. Because these chymotryptic peptides overlap two or more tryptic peptides, they can be used to establish the order of the peptides. The entire amino acid sequence of the polypeptide chain is then known.

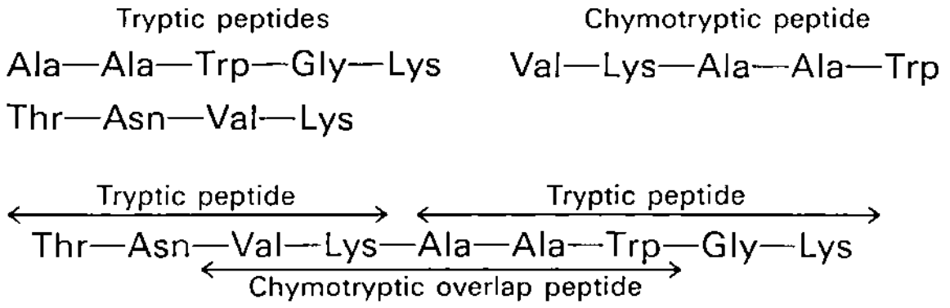

Figure 3-22
The peptide obtained by chymotrypic digestion overlaps two tryptic peptides, which thus establishes their order.

These methods apply to a protein consisting of a single polypeptide chain devoid of disulfide bonds. Additional steps are necessary if a protein has disulfide bonds or more than one chain. For a protein made up of two or more polypeptide chains held together by noncovalent bonds, denaturing agents, such as urea or guanidine hydrochloride, are used to dissociate the chains. The dissociated chains must be separated before sequence determination can begin. Polypeptide chains linked by disulfide bonds are first separated by reduction with thiols such as β-mercaptoethanol or dithiothreitol. To prevent the cysteine residues from recombining, they are then alkylated with iodoacetate to form stable *S*-carboxymethyl derivatives (Figure 3-23).

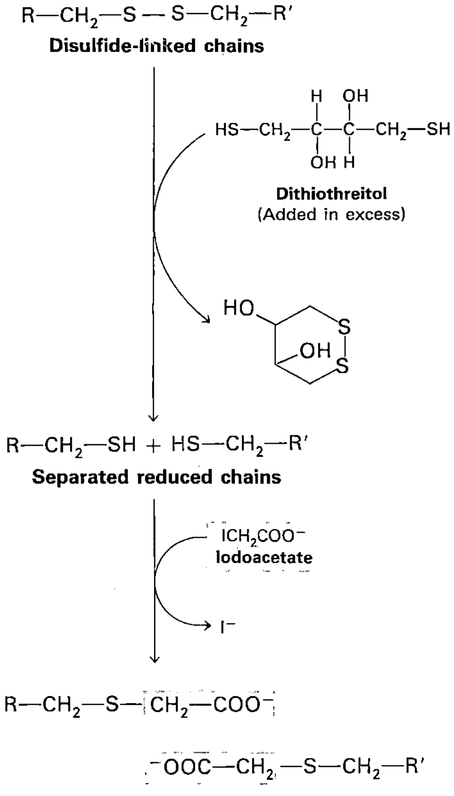

Figure 3-23
Polypeptides linked by disulfide bonds can be separated by reduction with dithiothreitol followed by alkylation.

The positions of disulfide bonds can be determined by a *diagonal electrophoresis* technique (Figure 3-24). First, the protein is specifically cleaved into peptides under conditions in which the disulfides stay intact. The mixture of peptides is applied to a corner of a sheet of paper and electrophoresed in a single lane along one side. The resulting sheet is exposed to vapors of performic acid, which cleaves disulfides and converts them into cysteic acid residues. Peptides originally linked by disulfides are now independent and also more acidic because of the formation of an SO_3^- group. This mixture is electrophoresed in the perpendicular direction under the same conditions as in the first electrophoresis. Peptides that were devoid of disulfides will have the same mobility as before, and consequently all will be located on a single diagonal line. In contrast, the newly formed peptides containing cysteic acid will usually migrate differently from their parent disulfide-linked peptide and hence will lie off the diagonal.

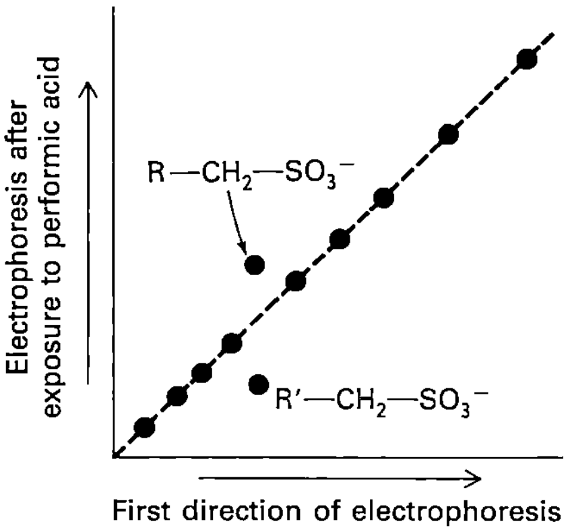

Figure 3-24
Detection of peptides joined by disulfides by diagonal electrophoresis. The mixture of peptides is electrophoresed in a single lane in one direction (horizontal) before treatment with performic acid, and then in the perpendicular direction (vertical).

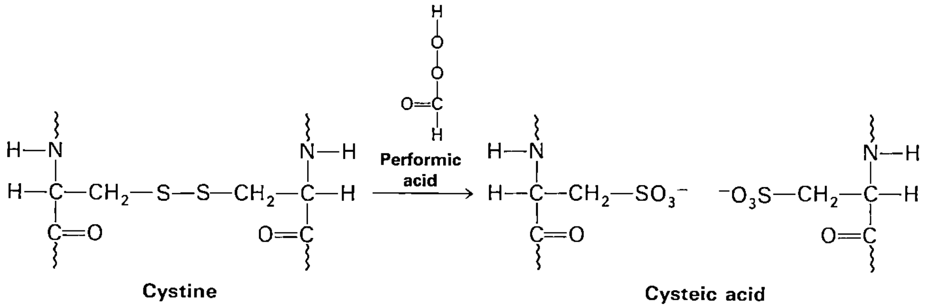

RECOMBINANT DNA TECHNOLOGY HAS REVOLUTIONIZED PROTEIN SEQUENCING

Hundreds of proteins have been sequenced by Edman degradation of peptides derived from specific cleavages. Protein sequence determination is a demanding and time-consuming process. The elucidation of the sequence of large proteins, those with more than 1000 residues, usually requires heroic effort. Fortunately, a complementary experimental approach based on recombinant DNA technology has become available. As will be discussed in Chapter 6, long stretches of DNA can be cloned and sequenced. The sequence of the four kinds of bases in DNA—adenine (A), thymine (T), guanine (G), and cytosine (C)—directly reveals the amino acid sequence of the protein encoded by the gene or the corresponding messenger RNA molecule (Figure 3-25). The amino acid sequence deduced by reading the DNA sequence is that of the *nascent* protein, the direct product of the translational machinery in ribosomes that links amino acids in a sequence specified by a messenger RNA template.

DNA sequence	GGG	TTC	TTG	GGA	GCA	GCA	GGA	AGC	ACT	ATG	GGC	GCA
Amino acid sequence	Gly	Phe	Leu	Gly	Ala	Ala	Gly	Ser	Thr	Met	Gly	Ala

Figure 3-25
The complete nucleotide sequence of HIV-1 (human immunodeficiency virus), the cause of AIDS, was determined within a year after the isolation of the virus. A portion of the DNA sequence specified by the RNA genome of the virus is shown here with the corresponding amino acid sequence (deduced from a knowledge of the genetic code).

As discussed previously, many proteins are modified after synthesis (p. 26). Some have their ends trimmed, and others arise by cleavage of a larger initial polypeptide chain. Cysteine residues in some proteins are oxidized to form disulfide links, which can be either within a chain or between polypeptide chains. Specific side chains of some proteins are altered. For example, proteins targeted to membranes usually contain carbohydrate units attached to specific asparagine side chains. Amino acid sequences derived from DNA sequences are rich in information, but they do not disclose such posttranslational modifications. Chemical analyses of proteins themselves are needed to delineate the nature of these changes, which are critical for the biological activities of most proteins. *Thus, DNA sequencing and protein chemical analyses are complementary approaches toward elucidating the structural basis of protein function.* Recombinant DNA technology is producing a wealth of amino acid sequence information at a remarkable rate. The sequences of more than 10,000,000 residues in more than 30,000 proteins are now known and readily retrievable from computer databases. At the present pace, the number of known sequences is expected to double in less than two years.

AMINO ACID SEQUENCES PROVIDE MANY KINDS OF INSIGHTS

Amino acid sequences can be highly informative in a variety of ways:

1. *The sequence of a protein of interest can be compared with all other known sequences to ascertain whether significant similarities exist. Does this protein belong to one of the established families?* For example, myoglobin and hemoglobin belong to the globin family. Chymotrypsin and trypsin are members of the serine protease family, a clan of proteolytic enzymes that have a common catalytic mechanism based on a reactive serine residue. A search for kinship between a newly sequenced protein and several thousand previously sequenced ones takes only a few minutes on a personal computer. Quite unexpected results sometimes emerge from such comparisons. For example, a viral protein that produces cancer in susceptible hosts was found to be nearly identical with a normal cellular growth factor (p. 354). This startling finding advanced the understanding of both oncogenic viruses (cancer-producing viruses) and the normal cell cycle. Comparison of amino acid sequences has also revealed that many larger proteins of higher organisms are built of domains that have come together by the fusion of gene segments. Proteins with new properties have arisen from novel combinations of these modules.

2. *Comparison of sequences of the same protein in different species yields a wealth of information about evolutionary pathways.* Genealogical relations between species can be inferred from sequence differences between their proteins, and the time of divergence of two evolutionary lines can be estimated because of the clocklike nature of random mutations. For example, a comparison of serum albumins of primates indicates that human beings and African apes diverged only 5 million years ago, not 30 million years ago as was previously thought. These sequence analyses have opened a new perspective on the fossil record and the pathway of human evolution.

3. *Amino acid sequences can be searched for the presence of internal repeats.* Many proteins apparently have arisen by duplication of a primordial gene followed by its diversification. For example, calmodulin, a ubiquitous calcium sensor in eukaryotes, contains four similar calcium-binding mod-

ules that arose by gene duplication (Figure 3-26). Antibody molecules too are built of a series of similar domains. The amino acid sequences of proteins express their evolutionary history.

Repeating units in calmodulin

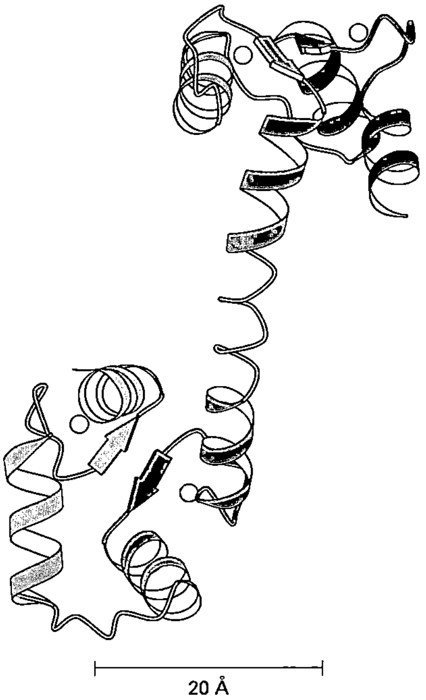

Figure 3-26
Calmodulin, a calcium sensor, contains four similar units in a single polypeptide chain. These calcium-binding motifs are shown in shades of red, and the bound calcium ions in green. [Drawn from 3cln.pdb. Y.S. Babu, C.E. Bugg, and W.J. Cook. *J. Mol. Biol.* 204(1988):191.]

4. *Amino acid sequences contain signals that determine the destination of proteins and control their processing.* Many proteins destined for export from a cell or for a membrane location contain a signal sequence, a stretch of about 20 hydrophobic residues near the amino terminus. Potential sites for the addition of carbohydrate units to asparagine residues can be identified by finding Asn-X-Ser and Asn-X-Thr in the sequence. Pairs of basic residues, such as Arg-Arg, mark potential sites of proteolytic cleavage, as in proinsulin, the precursor of insulin.

5. *Sequence data provide a basis for preparing antibodies specific for a protein of interest.* Specific antibodies can be very useful in determining the amount of a protein, ascertaining its distribution within a cell, and cloning its gene (p. 61).

6. *Amino acid sequences are also valuable for making DNA probes that are specific for the genes encoding the corresponding proteins* (p. 131). Protein sequencing is an integral part of molecular genetics, just as DNA cloning is central to the analysis of protein structure and function.

PROTEINS CAN BE QUANTITATED AND LOCALIZED BY HIGHLY SPECIFIC ANTIBODIES

The purification of a protein or the determination of its sequence opens the door to the application of immunologic methods. An *antibody* is a protein synthesized by an animal in response to the presence of a foreign substance, called an *antigen* (Chapter 14). Antibodies (also called *immunoglobulins*) have specific affinity for the antigens that elicited their synthesis. Proteins, polysaccharides, and nucleic acids are effective antigens. Antibodies can also be elicited by small molecules, such as synthetic peptides, provided that the small molecules are attached to a macromolecular carrier. The group recognized by an antibody is called an *antigenic determinant* (or *epitope*). Animals have a very large repertoire of antibody-producing cells, each producing antibody of a single specificity. An antigen acts by stimulating the proliferation of the small number of cells that were already forming complementary antibody. The major class of antibody in blood plasma is *immunoglobulin G,* a 150-kd protein containing two identical sites for the binding of antigen (Figure 3-27).

Antibodies that recognize a particular protein can be obtained by injecting the protein into a rabbit twice, three weeks apart. Blood is drawn from the immunized rabbit several weeks later and centrifuged. The resulting serum, called an *antiserum,* usually contains the desired antibody. The antiserum or its immunoglobulin G fraction can be used directly. Alternatively, antibody molecules specific for the antigen can be purified by affinity chromatography. Antibodies produced in this way are *polyclonal*—that is, they are products of many different populations of antibody-producing cells and hence differ somewhat in their precise specificity and affinity for the antigen. The discovery of a means of producing *monoclonal antibodies* of virtually any desired specificity was a major break-

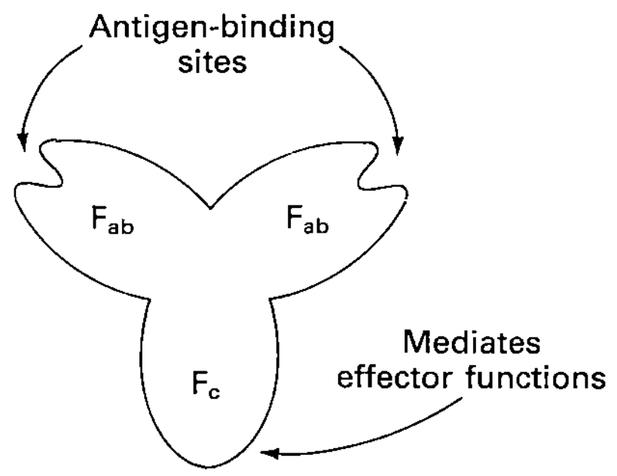

Figure 3-27
Diagram of immunoglobulin G (IgG), the major class of antibody molecules in blood plasma. IgG contains two antigen-binding F_{ab} units and an F_c unit that mediates effector functions such as the lysis of cell membranes.

through (p. 366). Monoclonal antibodies, in contrast with polyclonal ones, are homogeneous because they are synthesized by a population of identical cells (a clone). Each such population is descended from a single *hybridoma cell* formed by fusing an antibody-producing cell with a tumor cell that has the capacity for unlimited proliferation.

Closely related proteins can be distinguished by antibodies; indeed, a difference of just one residue on the surface can be detected. Antibodies can be used as exquisitely specific analytic reagents to quantitate the amount of a protein or other antigen. In a *solid-phase immunoassay,* antibody specific for a protein of interest is attached to a polymeric support such as a sheet of polyvinylchloride (Figure 3-28). A drop of cell extract or a sample of serum or urine is laid on the sheet, which is washed after formation of the antibody-antigen complex. Antibody specific for a different site on the antigen is then added, and the sheet is again washed. This second antibody carries a radioactive or fluorescent label so that it can be detected with high sensitivity. The amount of second antibody bound to the sheet is proportional to the quantity of antigen in the sample. The sensitivity of the assay can be enhanced even further if the second antibody is attached to an enzyme such as alkaline phosphatase. This enzyme can rapidly convert many molecules of an added colorless substrate into colored products, or nonfluorescent substrates into intensely fluorescent products (Figure 3-29). Less than a nanogram (10^{-9} g) of a protein can readily be measured by such an *enzyme-linked immunosorbent assay* (ELISA), which is rapid and convenient. For example, pregnancy can be detected within a few days after conception by immunoassaying urine for the presence of human chorionic gonadotropin (hCG), a 37-kd protein hormone produced by the placenta.

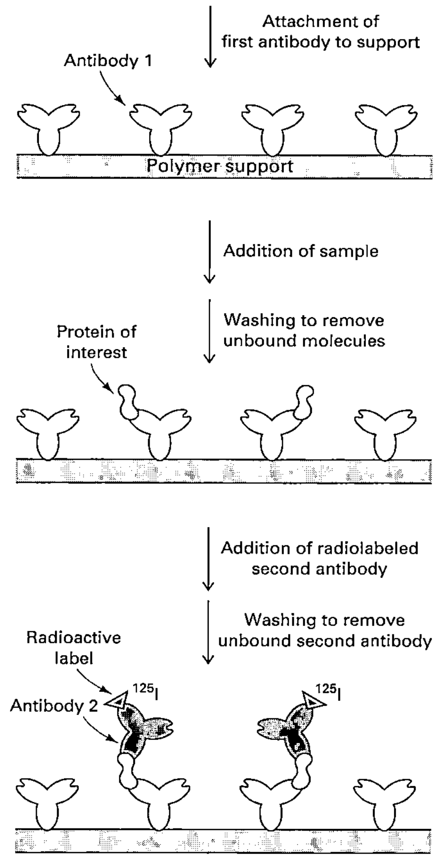

Figure 3-28
Solid-phase immunoassay. The steps are coupling of specific antibody to a solid support, addition of the sample, washing to remove soluble compounds, and addition of a radiolabeled second antibody specific for a different site on the protein being detected.

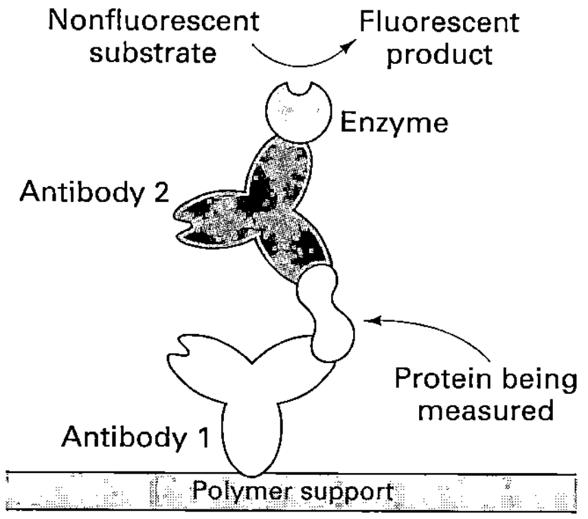

Figure 3-29
Enzyme-linked immunosorbent assay (ELISA). The steps are the same as in the immunoassay described in Figure 3-28 except that an enzyme instead of a radiolabel is attached to the second antibody. An intensely colored or fluorescent compound is formed by the catalytic action of this enzyme.

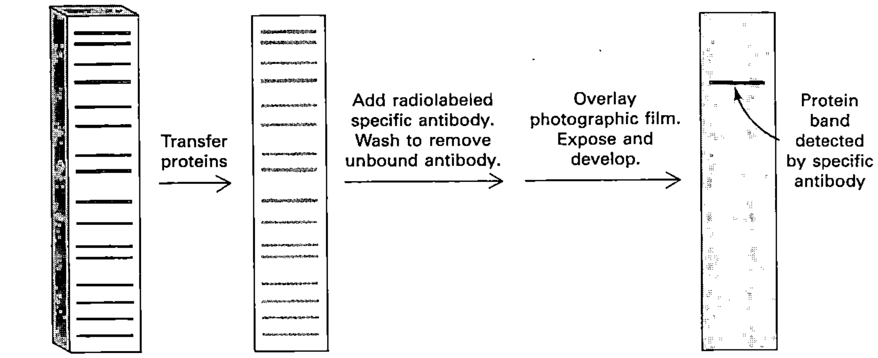

Figure 3-30
Detection of a protein on a gel by Western blotting. Proteins on an SDS–polyacrylamide gel are transferred to a polymer sheet and stained with radioactive antibody. A red band corresponding to the protein of interest appears in the autoradiogram.

Very small quantities of a protein of interest in a cell or in body fluid can be detected by an immunoassay technique called *Western blotting* (Figure 3-30). A sample is electrophoresed on an SDS–polyacrylamide gel. The resolved proteins on the gel are transferred (by blotting) to a polymer sheet to make them more accessible for reaction with a subsequently added antibody that is specific for the protein of interest. The antibody-antigen complex on the sheet then can be detected by rinsing the sheet with a second antibody specific for the first (e.g., goat antibody that recognizes mouse antibody). A radioactive label on the second antibody produces a dark band on x-ray film (an autoradiogram). Alternatively, an enzyme on the second antibody generates a colored product, as in the ELISA method. Western blotting makes it possible to find a protein in a complex mixture, the proverbial needle in a haystack. This technique is used advantageously in the cloning of genes (p. 141).

Antibodies are also valuable in determining the spatial distribution of antigens. Cells can be stained with fluorescent-labeled antibodies and examined by *fluorescence microscopy* to reveal the localization of a protein of interest. For example, arrays of parallel bundles are evident in cells stained with antibody specific for actin, a protein that polymerizes into filaments (Figure 3-31). Actin filaments are constituents of the cytoskeleton, the internal scaffolding of cells that controls their shape and movement. The finest resolution of fluorescence microscopy is about 0.2 μm (200 nm or 2000 Å) because of the wavelength of visible light. Finer spatial resolution can be achieved by electron microscopy using antibodies

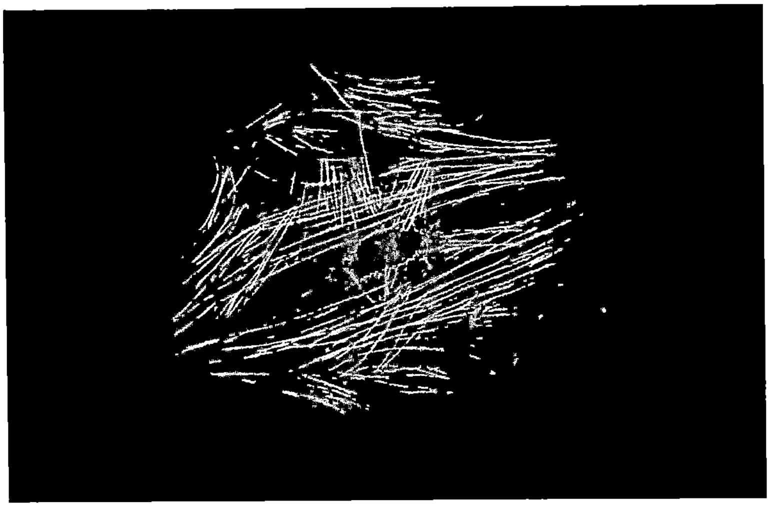

Figure 3-31
Fluorescence micrograph of actin filaments in a cell stained with an antibody specific to actin. [Courtesy of Dr. Elias Lazarides.]

tagged with electron-dense markers. For example, ferritin conjugated to an antibody can readily be visualized by electron microscopy because it contains an electron-dense core of iron hydroxide. Clusters of gold can also be conjugated to antibodies to make them highly visible under the electron microscope. *Immunoelectron microscopy* can define the position of antigens to a resolution of 10 nm (100 Å) or finer (Figure 3-32).

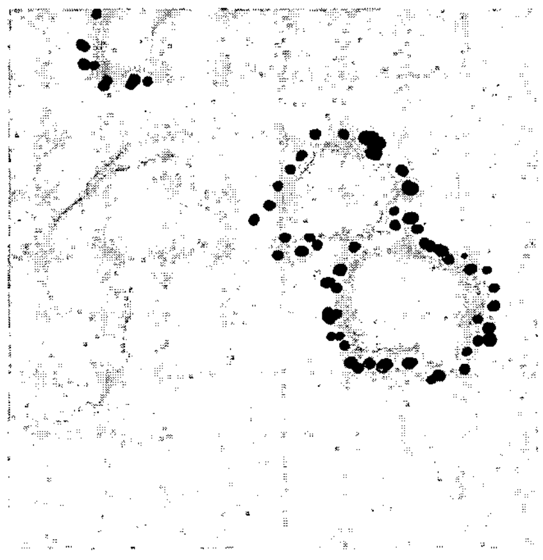

Figure 3-32
The opaque particles (15 nm (150 Å) diameter) in this electron micrograph are clusters of gold atoms bound to antibody molecules. These membrane vesicles from the synapses of neurons contain a channel protein that is recognized by the specific antibody. [Courtesy of Dr. Peter Sargent.]

CIRCULAR DICHROISM IS A SENSITIVE INDICATOR OF THE MAIN-CHAIN CONFORMATION OF PROTEINS

We turn now to experimental approaches for elucidating the conformation of proteins and begin with circular dichroism, an expression of optical activity. Proteins are optically active because they are *dissymmetric*—that is, they cannot be superimposed on their mirror images. Two kinds of dissymmetry are present: *configurational* and *conformational*. Amino acid residues other than glycine have intrinsic optical activity because of the L configuration about their α carbon atom. Threonine and isoleucine possess an additional asymmetric center. Electronic interactions between different residues in a protein also contribute to optical activity. The right-handed screw sense of an α helix, for example, gives rise to a large conformational contribution to optical activity.

Optical activity originates in the absorption bands of dissymmetric compounds. The absorption coefficients for left and right circularly polarized light are different if the electrons participating in the transition to the excited state sense a dissymmetric environment. This expression of optical activity is called *circular dichroism*. The related effect, called *optical rotation*, is the change in the direction of polarization of linearly polarized light on passing through an optically active solution. Circular dichroism (CD) spectra rather than optical rotatory spectra are measured because they are simpler to interpret. The magnitude of the CD, $\Delta\varepsilon$, is usually less than $10\ cm^{-1}\ M^{-1}$ even for a strongly optically active absorption band, compared with an ordinary absorbance coefficient ε of the order $10^4\ cm^{-1}\ M^{-1}$. Hence, sensitive instrumentation is required to measure CD.

The far-ultraviolet CD spectrum of a protein is sensitive to its main-chain conformation. The informative region is between 170 to 240 nm, where the peptide amide group has multiple overlapping absorption bands. Measurements of synthetic polypeptides and proteins of known structure have defined the CD spectra of α helices, β structure motifs, and random coils (Figure 3-33). *The α helix makes a dominant contribution with its negative CD bands at 208 and 222 nm, and its positive band at 192 nm.* A randomly arranged polypeptide chain, by contrast, has a negative CD band centered at 199 nm. Estimates of the α helix content of proteins derived from CD spectra agree well with values obtained from x-ray crystallographic studies. The content of β structures can also be estimated from CD spectra, but the uncertainty is greater because β structures are less regular than the α helices and contribute less to the CD spectrum.

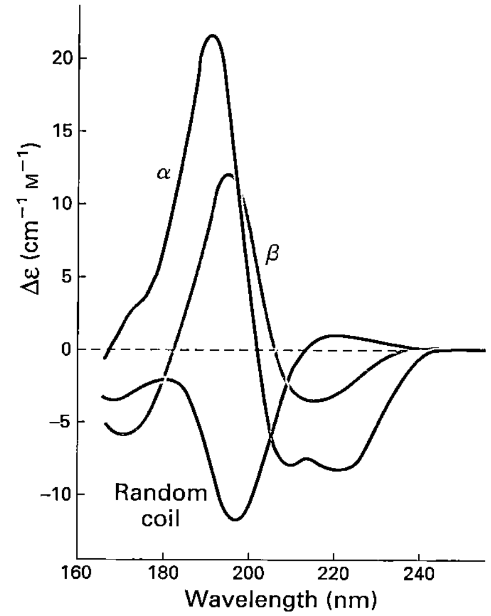

Figure 3-33
Circular dichroism spectra of an α helix (red), a β pleated sheet (green), and (C) a random-coil region (black) of a polypeptide chain. [After W.C. Johnson, Jr. *Proteins: Structure, Function, and Genetics* 7(1990):205.]

X-RAY CRYSTALLOGRAPHY REVEALS THREE-DIMENSIONAL STRUCTURE IN ATOMIC DETAIL

The understanding of protein structure and function has been greatly enriched by x-ray crystallography, a technique that can reveal the precise three-dimensional positions of most of the atoms in a protein molecule.

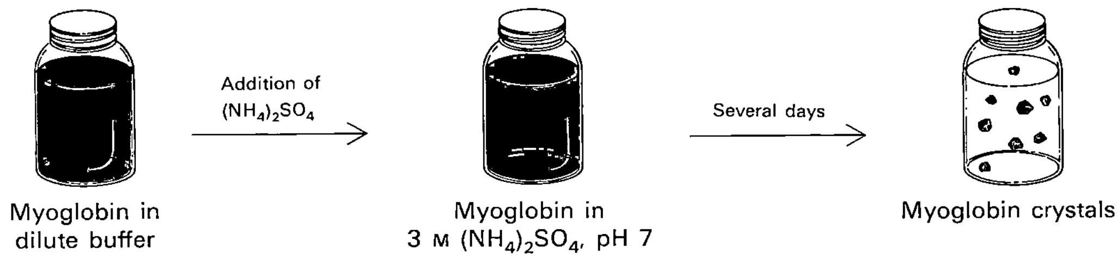

Figure 3-34
Crystallization of myoglobin.

Let us consider some basic aspects of this powerful method. First, crystals of the protein of interest are needed because the technique requires that all molecules be precisely oriented. Crystals can often be obtained by adding ammonium sulfate or another salt to a concentrated solution of protein to reduce its solubility. For example, myoglobin crystallizes in 3 M ammonium sulfate (Figure 3-34). Slow salting out favors the formation of highly ordered crystals instead of amorphous precipitates. Some proteins crystallize readily, whereas others do so only after much effort has been expended in finding the right conditions. Crystallization is an art; the best practitioners have great perseverance and patience, as well as a golden touch. Increasingly large and complex proteins are being crystallized. For example, polio virus, an 8500-kd assembly of 240 protein subunits surrounding an RNA core, has been crystallized and its structure solved by x-ray methods.

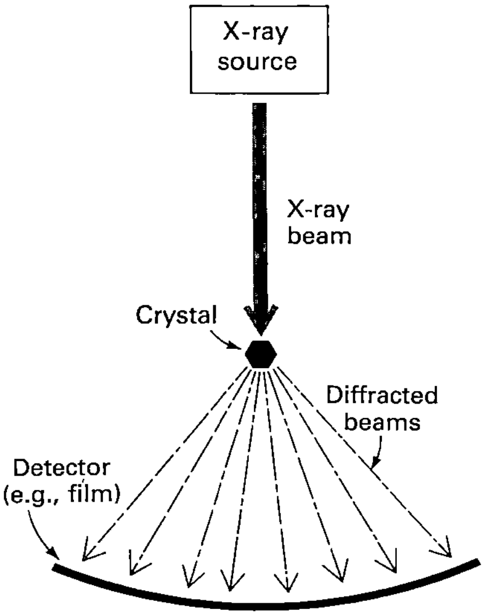

Figure 3-35
Essence of an x-ray crystallographic experiment: an x-ray beam, a crystal, and a detector.

The three components in an x-ray crystallographic analysis are a *source of x-rays,* a *protein crystal,* and a *detector* (Figure 3-35). A beam of x-rays of wavelength 1.54 Å is produced by accelerating electrons against a copper target. A narrow beam of x-rays strikes the protein crystal. Part of it goes straight through the crystal; the rest is *scattered* in various directions. The scattered (or *diffracted*) beams can be detected by x-ray film, the blackening of the emulsion being proportional to the intensity of the scattered x-ray beam, or by a solid-state electronic detector. The basic physical principles underlying the technique are

1. *Electrons scatter x-rays.* The amplitude of the wave scattered by an atom is proportional to its number of electrons. Thus, a carbon atom scatters six times as strongly as a hydrogen atom.

2. *The scattered waves recombine.* Each atom contributes to each scattered beam. The scattered waves reinforce one another at the film or detector if they are in phase (in step) there, and they cancel one another if they are out of phase.

3. *The way in which the scattered waves recombine depends only on the atomic arrangement.*

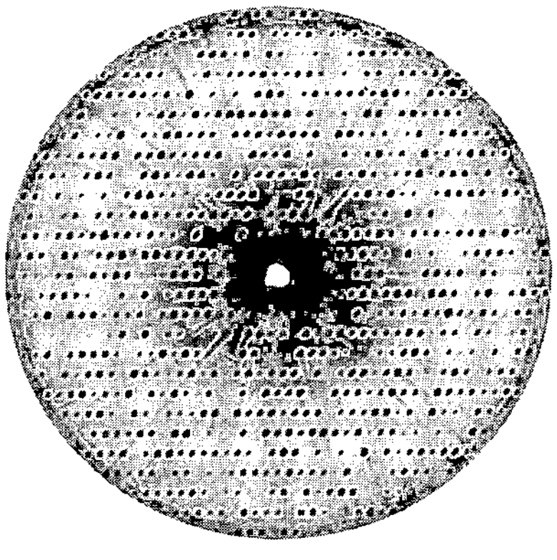

Figure 3-36
X-ray precession photograph of a myoglobin crystal.

The protein crystal is mounted in a capillary and positioned in a precise orientation with respect to the x-ray beam and the film. Precessional motion of the crystal results in an x-ray photograph consisting of a regular array of spots called *reflections.* The x-ray photograph shown in Figure 3-36 is a two-dimensional section through a three-dimensional array of 25,000 spots. The intensity of each spot is measured. These *intensities* are the basic experimental data of an x-ray crystallographic analysis. The next step is to reconstruct an image of the protein from the observed intensities. In light microscopy or electron microscopy, the diffracted beams are focused by lenses to directly form an image. However, lenses for focusing x-rays do not exist. Instead, the image is formed by applying a mathematical relation called a *Fourier transform.* For each spot, this operation yields a wave of electron density, whose amplitude is proportional to the square root of the observed intensity of the spot. Each wave also has a *phase*—that is, the timing of its crests and troughs relative to those of other waves. The phase of each wave determines whether it reinforces or cancels the waves contributed by the other spots. These phases can be deduced from the well-understood diffraction patterns produced by heavy-atom reference markers such as uranium or mercury at specific sites in the protein.

The stage is then set for the calculation of an electron-density map, which gives the density of electrons at a large number of regularly spaced points in the crystal. This three-dimensional electron-density distribution is represented by a series of parallel sections stacked on top of each other. Each section is a transparent plastic sheet (or a layer in a computer image) on which the electron-density distribution is represented by contour lines (Figure 3-37), like the contour lines used in geological survey maps to depict altitude (Figure 3-38). The next step is to interpret the electron-density map. A critical factor is the *resolution* of the x-ray analysis, which is determined by the number of scattered intensities used in the Fourier synthesis. The fidelity of the image depends on the resolution of the Fourier synthesis, as shown by the optical analogy in Figure 3-39. A resolution of 6 Å reveals the course of the polypeptide chain but few other structural details. The reason is that polypeptide chains pack together so that their centers are between 5 and 10 Å apart. Maps at higher resolution are needed to delineate groups of atoms, which lie from 2.8 to 4.0 Å apart, and individual atoms, which are between 1.0 and 1.5 Å apart. The ultimate resolution of an x-ray analysis is determined by the degree of perfection of the crystal. For proteins, this limiting resolution is usually about 2 Å.

The structures of more than 300 proteins have been elucidated at atomic resolution. Knowledge of their detailed molecular architecture has provided insight into how proteins recognize and bind other mole-

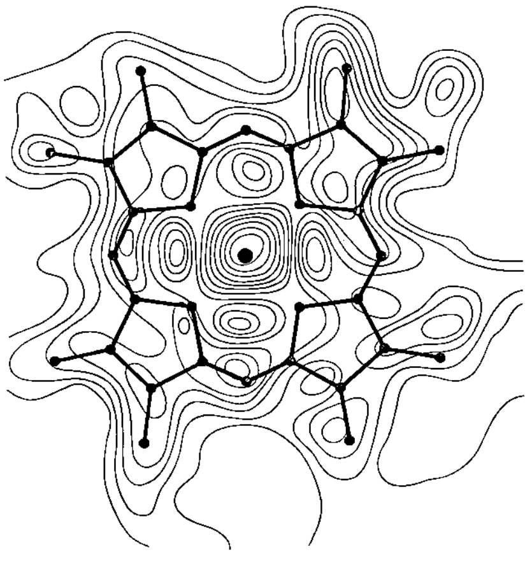

Figure 3-37
Section from the electron-density map of myoglobin showing the heme group. The peak of the center of this section corresponds to the position of the iron atom. [From J.C. Kendrew. The three-dimensional structure of a protein molecule. Copyright © 1961 by Scientific American, Inc. All rights reserved.]

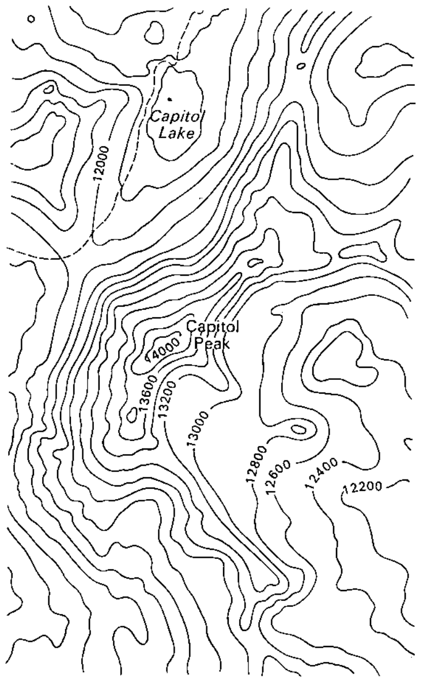

Figure 3-38
Section from a U.S. Geological Survey map of the Capitol Peak Quadrangle, Colorado.

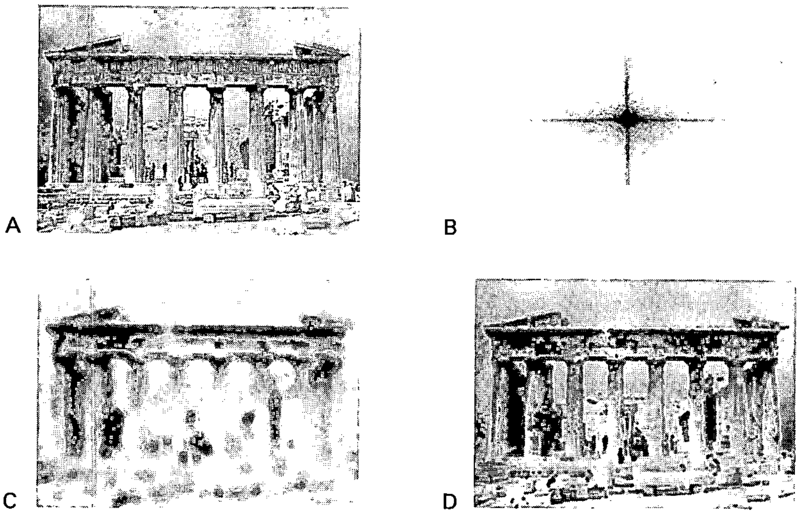

Figure 3-39
Effect of resolution on the quality of a reconstructed image, as shown by an optical analog of x-ray diffraction: (A) a photograph of the Parthenon; (B) an optical diffraction pattern of the Parthenon; (C and D) images reconstructed from the pattern in B. More data were used to obtain D than C, which accounts for the higher quality of the image in D. [Courtesy of Dr. Thomas Steitz (part A) and Dr. David DeRosier (part B).]

cules, how they function as enzymes, how they fold, and how they evolved. This extraordinarily rich harvest is continuing at a rapid pace and profoundly influencing the entire field of biochemistry.

Table 3-2
Biologically important nuclei giving NMR signals

Nucleus	Natural abundance (% by weight of the element)
^{1}H	99.984
^{2}H	0.016
^{13}C	1.108
^{14}N	99.635
^{15}N	0.365
^{17}O	0.037
^{23}Na	100.0
^{25}Mg	10.05
^{31}P	100.0
^{35}Cl	75.4
^{39}K	93.1

NUCLEAR MAGNETIC RESONANCE (NMR) SPECTROSCOPY CAN REVEAL THE STRUCTURE OF PROTEINS IN SOLUTION

X-ray crystallography is being complemented by nuclear magnetic resonance (NMR) spectroscopy, which is unique in being able to reveal the atomic structure of macromolecules *in solution*. Certain atomic nuclei, such as ordinary hydrogen (^{1}H), are intrinsically magnetic (Table 3-2). The spinning of the positively charged proton, like that of any rotating charged particle, generates a magnetic moment. This moment can take either of two orientations (called α and β) when an external magnetic field is applied (Figure 3-40). The energy difference between these states is proportional to the strength of the imposed field. The α state has a slightly lower energy and hence is slightly more populated (by a factor of the order of 1.00001 in a typical experiment) because it is aligned with

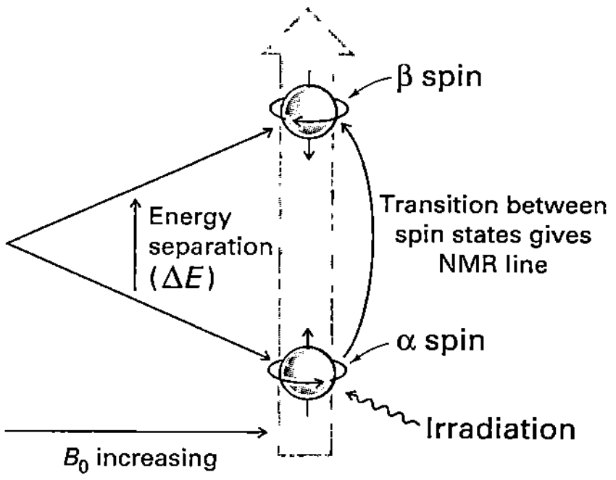

Figure 3-40
Essence of NMR spectroscopy. The energies of the two orientations of a nucleus of spin $\frac{1}{2}$ (such as ^{31}P and ^{1}H) depend on the strength of the applied magnetic field. Absorption of electromagnetic radiation of appropriate frequency (v_0) induces a transition from the lower to the upper level.

the field. A transition from the lower (α) to the upper (β) state occurs when a nucleus absorbs electromagnetic radiation of appropriate frequency. The resonance frequency v_0 of an isolated nucleus is

$$v_0 = \frac{\gamma H_0}{2\pi}$$

where H_0 is the strength of the steady magnetic field and γ is a constant (called the magnetogyric ratio) for a given nucleus. For example, the resonance frequency for ^{1}H in a 100 kilogauss (10 tesla) field is 426 megaherz (MHz), which lies in the radiofrequency region of the spectrum. A plot of the energy absorbed versus frequency would show a peak at 426 MHz.

Nuclear magnetic resonance (NMR) spectroscopy is a very informative technique because the local magnetic field is not identical with the applied field B_0 for all nuclei in the sample. The flow of electrons around a magnetic nucleus generates a small local magnetic field that opposes the external field. The degree of shielding from B_0 depends on the surrounding electron density. Consequently, *nuclei in different environments absorb energy at slightly different resonance frequencies*, an effect termed the *chemical shift*. These separations are expressed in fractional units δ (parts per million, *ppm*) relative to a standard compound, such as a water-soluble deriv-

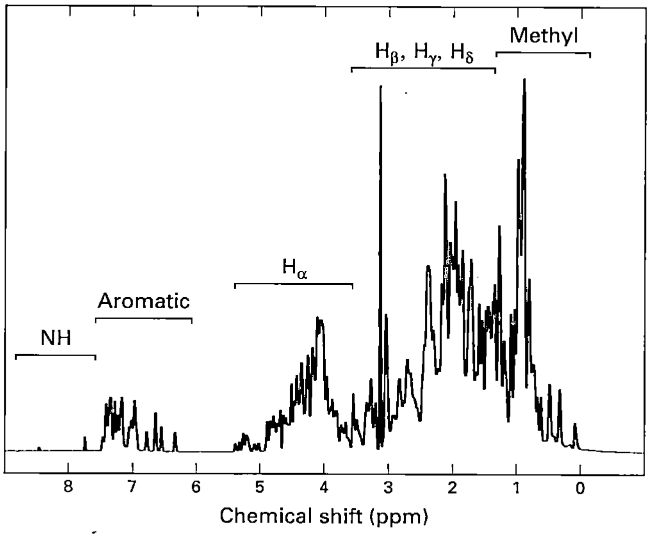

Figure 3-41
One-dimensional proton NMR spectrum of calmodulin, a calcium sensor. The usual range of chemical shifts of particular protons in proteins are marked. [Courtesy of Dr. Mitsu Ikura.]

ative of tetramethysilane. For example, a $-CH_3$ proton typically exhibits a δ of 1 ppm, compared with a δ of 7 ppm for an aromatic proton. The chemical shifts of most protons in protein molecules fall between 1 and 9 ppm (Figure 3-41). Absorption peaks in an NMR spectrum are usually called *lines.* A particular proton usually gives rise to more than one NMR line because of the influence of nonequivalent neighboring protons, an effect called *spin-spin coupling.* Hydrogen atoms separated by three or fewer covalent bonds are coupled to one another in this way.

The transient magnetization induced in a sample by a radio-frequency pulse decays with time; the sample relaxes to its equilibrium state. Relaxation processes are highly informative about macromolecular structure and dynamics because they are very sensitive to both geometry and motion. Especially revealing is the *nuclear Overhauser effect* (*NOE*), an interaction between nuclei that is proportional to the inverse sixth power of the distance between them. Magnetization is transferred from an excited nucleus to an unexcited one if they are less than about 5 Å apart (Figure 3-42A). *A two-dimensional nuclear Overhauser enhancement spectroscopy (NOESY) spectrum graphically displays pairs of protons that are in close proximity.* The diagonal of a NOESY spectrum corresponds to a one-dimensional chemical shift spectrum. The off-diagonal peaks provide crucial new information: *they identify pairs of protons that are less than 5 Å apart* (Figure

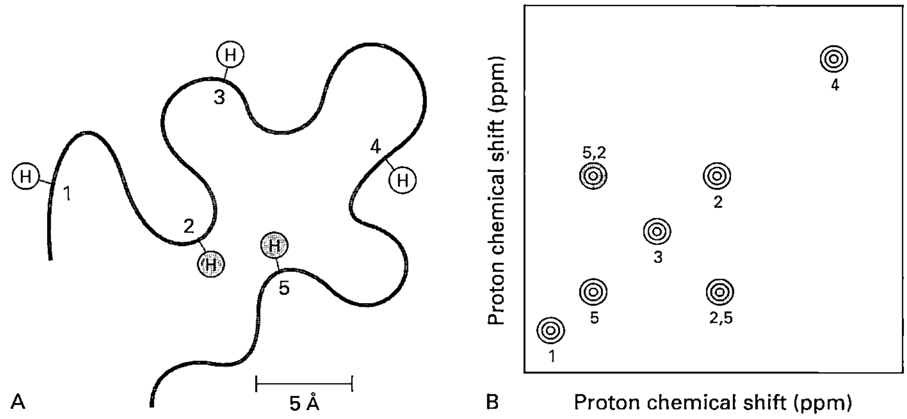

Figure 3-42
The nuclear Overhauser effect (NOE) identifies pairs of protons that are in close proximity. (A) Schematic diagram of a polypeptide chain highlighting five particular protons. Protons 2 and 5 are in close proximity (~4 Å apart), whereas other pairs are farther apart. (B) A highly simplified NOESY spectrum. The diagonal shows five peaks corresponding to the five protons in part (A). The peak above the diagonal and the symmetrically related one below reveal that proton 2 is close to proton 5.

3-42B). Overlapping peaks in NOESY spectra can usually be resolved by obtaining NMR spectra of proteins that are labeled with ^{15}N and ^{13}C. Irradiation of these nuclei separates NOE peaks along additional axes, an approach termed *multidimensional NMR spectroscopy*. The three-dimensional structure of a protein can be reconstructed from a large number of such proximity relations.

NMR spectroscopy and x-ray crystallography are the only two techniques that can reveal the three-dimensional structure of proteins and other biomolecules in atomic detail. X-ray methods give the highest-resolution images, but crystals are required. NMR methods, in contrast, are effective with proteins in solution, provided that highly concentrated solutions (~1 mM, or 15 mg/ml for a 15-kd protein) can be obtained. The upper bound on size is about 30 kd because NMR peaks of larger proteins cannot be adequately resolved and assigned at this time. However, much can be accomplished within this limit because protein domains are usually smaller than 30 kd. Furthermore, NMR spectroscopy provides a wealth of information about dynamics. Thus, NMR and x-ray techniques nicely complement each other in structural studies (Figure 3-43).

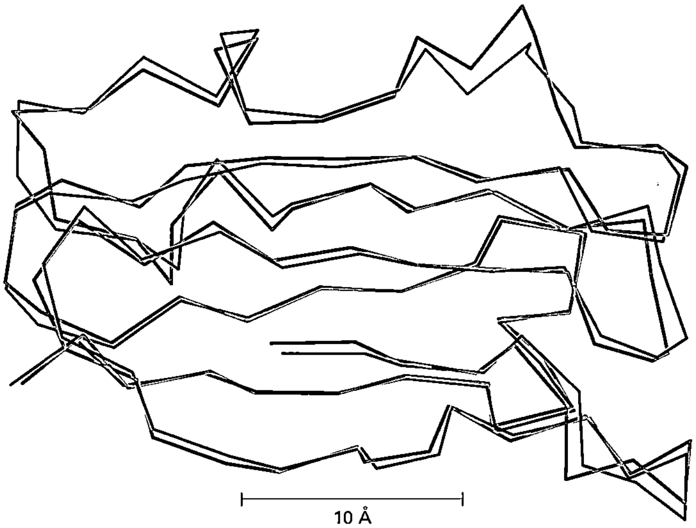

Figure 3-43
Comparison of the structures of plastocyanin, an electron-carrying protein in photosynthesis, determined by x-ray crystallography (blue) and NMR spectroscopy (red). [Drawn from 4pcy.pdb and 9pcy.pdb. W.B. Church, J.M. Guss, J.J. Potter, and H.C. Freeman. *J. Biol. Chem.* 261(1986):234, and J.M. Moore, C.A. Lepre, G.P. Gippert, W.J. Chazin, D.A. Case, and P.E. Wright. *J. Mol. Biol.* 221(1993):533.]

PEPTIDES CAN BE SYNTHESIZED BY AUTOMATED SOLID-PHASE METHODS

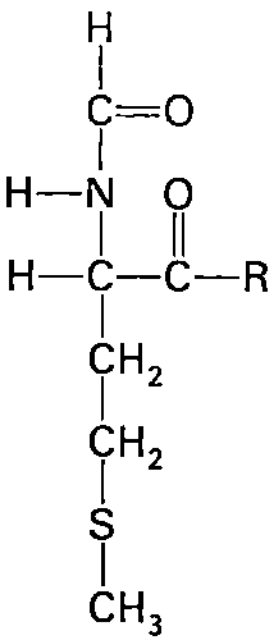

fMet peptide

Synthesizing peptides of defined sequence is important for several reasons. First, *studying synthetic peptides can help define the rules governing the three-dimensional structure of proteins.* We can ask whether a particular sequence by itself folds into an α helix, β strand, or hairpin turn or behaves as a random coil. Second, *synthetic peptides can be used to isolate receptors for many hormones and other signal molecules.* For example, white blood cells are attracted to bacteria by formylmethionyl peptides that come from the breakdown of bacterial proteins. Synthetic formylmethionyl peptides have been useful in identifying the cell-surface receptor for this class of peptides. Synthetic peptides can be attached to agarose beads to prepare

$$S\text{———————————}S$$
$$^+H_3N\text{—Cys—Tyr—Phe—Glu—Asp—Cys—Pro—Arg—Gly—}C\diagdown\substack{O\\NH_2}$$

1 2 3 4 5 6 7 8 9

8-Arginine vasopressin
A **(Antidiuretic hormone, ADH)**

$$S\text{————————————}S$$
$$CH_2\text{—}CH_2\text{—}\underset{\substack{\|\\O}}{C}\text{—Tyr—Phe—Glu—Asp—Cys—Pro—}_D\text{-Arg—Gly—}C\diagdown\substack{O\\NH_2}$$

B **1-Desamino-8-D-arginine vasopressin**

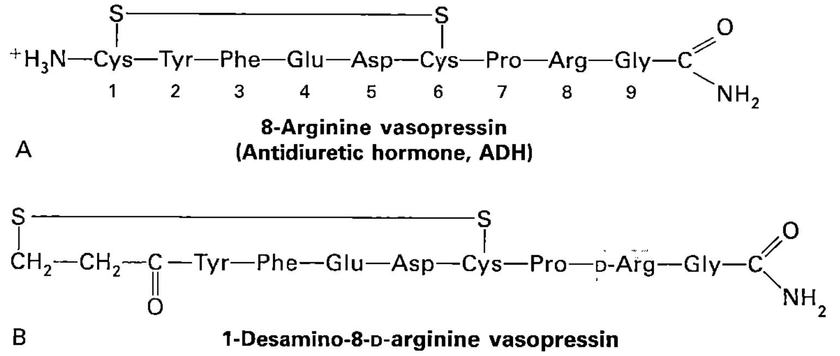

Figure 3-44
Structural formulas of (A) vasopressin, a peptide hormone that stimulates water resorption, and (B) 1-desamino-8-D-arginine vasopressin, a more stable synthetic analog of this antidiuretic hormone.

affinity chromatography columns for the purification of receptor proteins that specifically recognize the peptides. Third, *synthetic peptides can serve as drugs*. Vasopressin is a peptide hormone that stimulates the reabsorption of water in the distal tubules of the kidney, leading to the formation of a more concentrated urine. Patients with diabetes insipidus are deficient in *vasopressin* (also called *antidiuretic hormone*), and so they excrete large volumes of urine (more than 5 liters per day) and are continually thirsty because of this massive loss of fluid. This defect can be treated by administering 1-desamino-8-D-arginine vasopressin, a synthetic analog of the missing hormone (Figure 3-44). This synthetic peptide is degraded in vivo much more slowly than vasopressin and, additionally, does not increase the blood pressure. Fourth, *synthetic peptides can serve as antigens to stimulate the formation of specific antibodies*.

Peptides are synthesized by linking an amino group to a carboxyl group that has been activated by reacting it with a reagent such as *dicyclohexylcarbodiimide* (DCC) (Figure 3-45). The attack of a free amino group on the activated carboxyl leads to the formation of a peptide bond and the release of dicylohexylurea. A unique product is formed only if a single amino group and a single carboxyl group are available for reaction. Hence, it is necessary to *block* (protect) all other potentially reactive groups. For example, the α-amino group of the component containing the activated carboxyl group can be blocked with a *tert*-butyloxycarbonyl (*t*-Boc) group. This *t*-Boc protecting group can be subsequently removed by exposing the peptide to dilute acid, which leaves peptide bonds intact.

Peptides can be readily synthesized by a *solid-phase method* devised by R. Bruce Merrifield. Amino acids are added stepwise to a growing peptide chain that is linked to an insoluble matrix, such as polystyrene beads. A major advantage of this solid-phase method is that the desired product at

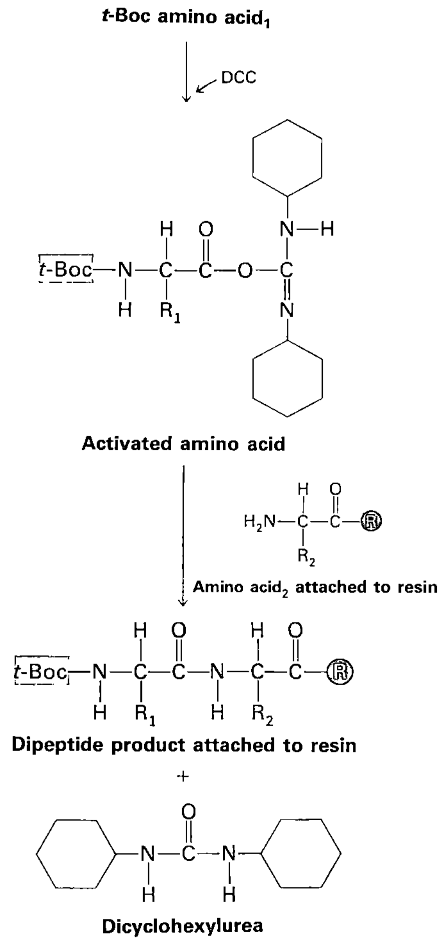

Figure 3-45
Dicyclohexylcarbodiimide is used to activate carboxyl groups for the formation of peptide bonds.

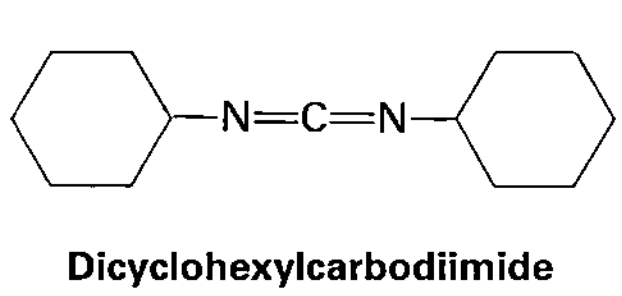

Dicyclohexylcarbodiimide
(DCC)

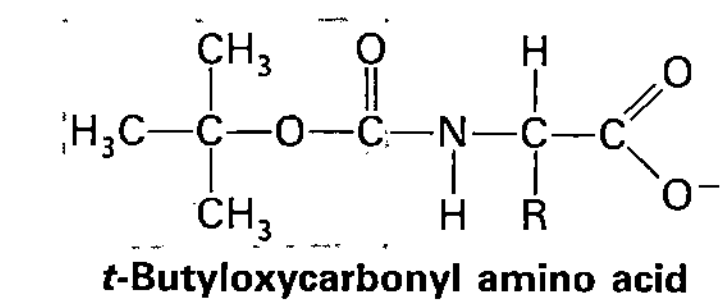

t-Butyloxycarbonyl amino acid
(t-Boc amino acid)

each stage is bound to beads that can be rapidly filtered and washed, and so the need to purify intermediates is obviated. All reactions are carried out in a single vessel, which eliminates losses due to repeated transfers of products. The carboxyl-terminal amino acid of the desired peptide sequence is first anchored to the polystyrene beads (Figure 3-46). The *t*-Boc protecting group of this amino acid is then removed. The next amino acid (in the protected *t*-Boc form) and dicyclohexylcarbodiimide, the coupling agent, are added together. After formation of the peptide bond, excess reagents and dicyclohexylurea are washed away, which leaves the beads with the desired dipeptide product. Additional amino acids are linked by the same sequence of reactions. At the end of the synthesis, the peptide is released from the beads by adding hydrofluoric acid (HF), which cleaves the carboxyl ester anchor without disrupting peptide bonds. Protecting groups on potentially reactive side chains, such as that of lysine, are also removed at this time. This cycle of reactions can readily be automated, which makes it feasible to routinely synthesize peptides containing about 50 residues in good yield and purity. In fact, Merrifield has synthesized interferons (155 residues) that have antiviral activity and ribonuclease (124 residues) that is catalytically active.

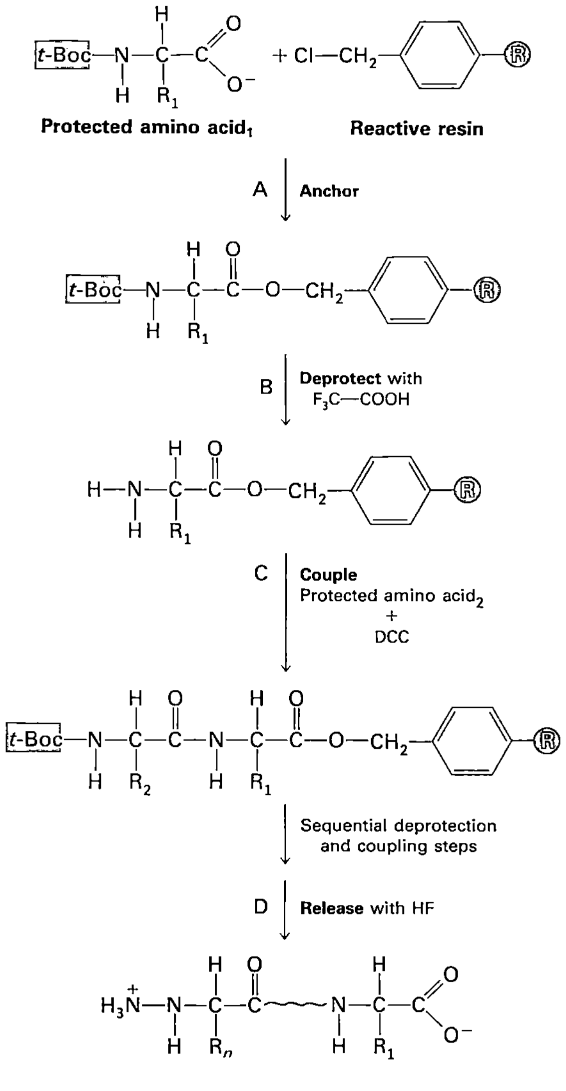

Figure 3-46
Sequence of steps in solid-phase peptide synthesis. (A) Anchoring of the C-terminal amino acid, (B) deprotection of the amino terminus, and (C) coupling of the next residue. Steps (B) and (C) are repeated for each added amino acid. Finally, (D) the completed peptide is released from the resin.

SUMMARY

The purification of a protein is an essential step in elucidating its structure and function. Proteins can be separated from each other and from other molecules on the basis of such characteristics as size, solubility, charge, and binding affinity. SDS–polyacrylamide gel electrophoresis separates the polypeptide chains of proteins under denaturing conditions largely according to mass. Proteins can also be separated electrophoretically on the basis of net charge by isoelectric focusing in a pH gradient. Ultracentrifugation and gel-filtration chromatography resolve proteins according to size, whereas ion-exchange chromatography separates them mainly on the basis of net charge. The high affinity of many proteins for specific chemical groups is exploited in affinity chromatography, in which proteins bind to columns containing beads bearing covalently linked substrates, inhibitors, or other specifically recognized groups. The mass of a protein can be precisely determined by sedimentation equilibrium measurements or by electrospray mass spectrometry. Circular dichroism, an expression of optical activity, provides information about the secondary structure of proteins.

The amino acid composition of a protein can be ascertained by hydrolyzing it into its constituent amino acids in 6 N HCl at 110°C. They can be separated by ion-exchange chromatography and quantitated by reacting them with ninhydrin or fluorescamine. Amino acid sequences can be determined by Edman degradation, which removes one amino acid at a time from the amino end of a peptide. Phenylisothiocyanate reacts with the terminal amino group to form a phenylthiocarbamoyl derivative, which cyclizes under mildly acidic conditions to give a phenylthiohydantoin (PTH)–amino acid and a peptide shortened by one residue. Automated repeated Edman degradations by a sequenator can analyze sequences of about 50 residues. Longer polypeptide chains are broken into shorter ones for analysis by specifically cleaving them with a reagent such as cyanogen bromide, which splits peptide bonds on the carboxyl side of methionine residues. Enzymes such as trypsin, which cleaves on the carboxyl side of lysine and arginine residues, are also very useful in splitting proteins.

Recombinant DNA techniques have revolutionized amino acid sequencing. The nucleotide sequence of DNA molecules reveals the amino acid sequence of nascent proteins encoded by them but does not disclose posttranslational modifications. Amino acid sequences are rich in information concerning the kinship of proteins, their evolutionary relations, and diseases produced by mutations. Knowledge of a sequence provides valuable clues to conformation and function. Proteins can be detected and quantitated by highly specific antibodies; monoclonal antibodies are especially useful because they are homogeneous. Enzyme-linked immunosorbent assays (ELISA) and Western blots of SDS–polyacrylamide gels are used extensively. Proteins can also be localized within cells by immunofluorescence microscopy and immunoelectron microscopy.

Our understanding of how proteins fold, recognize other molecules, and catalyze chemical reactions has been greatly enriched by x-ray crystallography. Electrons scatter x-rays; the way in which the scattered waves recombine depends only on the atomic arrangement. The three-dimensional structures of more than 300 proteins are now known in atomic detail. Nuclear magnetic resonance (NMR) spectroscopy complements x-ray crystallography by revealing the structure and dynamics of proteins in solution. The chemical shift of nuclei depends on their local environment. Furthermore, the spins of neighboring nuclei interact with each other in ways that provide definitive structural information.

Polypeptide chains can be synthesized by automated solid-phase methods in which the carboxyl end of the growing chain is linked to an insoluble support. The α-carboxyl group of the incoming amino acid is activated by dicyclohexylcarbodiimide and joined to the α-amino group of the growing chain. Synthetic peptides can serve as drugs and as antigens to stimulate the formation of specific antibodies. They also provide insight into relations between amino acid sequence and conformation.

SELECTED READINGS

Where to start

Hunkapiller, M.W., and Hood, L.E., 1983. Protein sequence analysis: automated microsequencing. *Science* 219:650–659.

Merrifield, B., 1986. Solid phase synthesis. *Science* 232:341–347.

Doolittle, R.F., 1989. Similar amino acid sequences revisited. *Trends Biochem. Sci.* 14:244–245.

Sanger, F., 1988. Sequences, sequences, sequences. *Ann. Rev. Biochem.* 57:1–28. [Reminiscences of a great biochemist who expresses gratitude for having been given the chance to fulfill some of his wildest dreams.]

Books

Creighton, T.E., 1992. *Proteins: Structure and Molecular Properties* (2nd ed.). W.H. Freeman. [Rich in information and illuminating examples.]

Kyte, J., 1994. *Structure in Protein Chemistry.* Garland.

Deutscher, M. (ed.), 1990. *Guide to Protein Purification.* Academic Press.

Scopes, R., 1982. *Protein Purification: Principles and Practice.* Springer-Verlag.

Perutz, M., 1992. *Protein Structure: New Approaches to Disease and Therapy.* W.H. Freeman. [Chapter 1, entitled "Diffraction without Tears," is an engaging introduction, written by a master. A book to be savored.]

Methods in Enzymology. Academic Press. [The more than 200 volumes of this series are a treasure house of experimental procedures. Especially pertinent to this chapter are volumes 91, 114, and 115 on protein structure and x-ray crystallography; volumes 92 and 121 on immunochemical techniques; volumes 182 and 245 on protein purification; volume 193 on mass spectrometry; volumes 176 and 177 on NMR spectroscopy; and volume 183 on sequence comparisons and molecular evolution.]

Physical chemistry of proteins

Cantor, C.R., and Schimmel, P.R., 1980. *Biophysical Chemistry.* W.H. Freeman. [An outstanding exposition of fundamental principles and experimental methods.]

Freifelder, D., 1982. *Physical Biochemistry: Applications to Biochemistry and Molecular Biology.* W.H. Freeman. [Contains a lucid discussion of ultracentrifugation.]

Tanford, C., 1961. *Physical Chemistry of Macromolecules.* Wiley. [Chapter 6 gives an excellent account of transport processes.]

Ultracentrifugation and mass spectrometry

Schachman, H.K., 1959. *Ultracentrifugation in Biochemistry.* Academic Press.

Bowen, T.J., 1970. *An Introduction to Ultracentrifugation.* Wiley-Interscience.

Arnott, D., Shabanowitz, J., and Hunt, D.F., 1993. Mass spectrometry of proteins and peptides: sensitive and accurate mass measurement and sequence analysis. *Clin. Chem.* 39:2005–2010.

Chait, B.T., and Kent, S.B.H., 1992. Weighing naked proteins: Practical, high-accuracy mass measurement of peptides and proteins. *Science* 257:1885–1894.

Jardine, I., 1990. Molecular weight analysis of proteins. *Meth. Enzymol.* 193:441–455. [This volume contains a wealth of information on mass spectrometry of proteins and peptides.]

Edmonds, C.G., Loo, J.A., Loo, R.R., Udseth, H.R., Barinaga, C.J., and Smith, R.D., 1991. Application of electrospray ionization mass spectrometry and tandem mass spectrometry in combination with capillary electrophoresis for biochemical investigations. *Biochem. Soc. Trans.* 19:943–947.

Amino acid sequence determination

Hunkapiller, M.W., Strickler, J.E., and Wilson, K.J., 1984. Contemporary methodology for protein structure determination. *Science* 226:304–311.

Aebersold, R., Pipes, G.D., Wettenhall, R.E., Nika, H., and Hood, L.E., 1990. Covalent attachment of peptides for high sensitivity solid-phase sequence analysis. *Analyt. Biochem.* 187:56–65.

Hewick, R.M., Hunkapiller, M.W., Hood, L.E., and Dreyer, W.J., 1981. A gas-liquid solid phase peptide and protein sequenator. *J. Biol. Chem.* 256:7990–7997.

Konigsberg, W.H., and Steinman, H.M., 1977. Strategy and methods of sequence analysis. *In* Neurath, H., and Hill, R.L. (eds.), *The Proteins* (3rd ed.), vol. 3, pp. 1–178. Academic Press.

Stein, S., and Udenfriend, S., 1984. A picomole protein and peptide chemistry: some applications to the opioid peptides. *Analyt. Chem.* 136:7–23.

Moore, S., and Stein, W.H., 1973. Chemical structures of pancreatic ribonuclease and deoxyribonuclease. *Science* 180:458–464. [A classic account of pioneering research in protein chemistry.]

Spectroscopy

Wüthrich, K., 1989. Protein structure determination in solution by nuclear magnetic resonance spectroscopy. *Science* 243:45–50.

Clore, G.M., and Marius, A.M., 1991. Structures of larger proteins in solution: three- and four-dimensional heteronuclear NMR spectroscopy. *Science* 252:1390–1399.

Wüthrich, K., 1986. *NMR of Proteins and Nucleic Acids.* Wiley-Interscience.

Johnson, W.C., Jr., 1990. Protein secondary structure and circular dichroism: a practical guide. *Proteins* 7:205–214.

X-ray crystallography

Matthews, B.W., 1977. X-ray structure of proteins. *In* Neurath, H., and Hill, R.L. (eds.), *The Proteins* (3rd ed.), vol. 3, pp. 404–590. Academic Press.

Holmes, K.C., and Blow, D.M., 1965. *The Use of X-ray Diffraction in the Study of Protein and Nucleic Acid Structure.* Wiley-Interscience.

Glusker, J.P., and Trueblood, K.N., 1972. *Crystal Structure Analysis: A Primer.* Oxford University Press. [A lucid and concise introduction to x-ray crystallography in general.]

Sequence comparisons and molecular evolution

Doolittle, R.F., 1992. Reconstructing history with amino acid sequences. *Protein Sci.* 1:191–200.

Lipman, D.J., and Pearson, W.R., 1985. Rapid and sensitive protein similarity searches. *Science* 227:1435–1441. [Description of an algorithm that searches for similarities between an amino acid sequence and a large database of previously determined sequences. This program can be run on a personal computer.]

Pearson, W.R., 1990. Rapid and sensitive sequence comparison with FASTP and FASTA. *Meth. Enzymol.* 183:63–98.

Feng, D.F., and Doolittle, R.F., 1990. Progressive alignment and phylogenetic tree construction of protein sequences. *Meth. Enzymol.* 183:375–387.

Doolittle, R.F., 1990. Searching through sequence databases. *Meth. Enzymol.* 183:99–110.

Wilson, A.C., 1985. The molecular basis of evolution. *Sci. Amer.* 253(4):164.

Karlin, S., Zuker, M., and Brocchieri, L., 1994. Measuring residue associations in protein structures. Possible implications for protein folding. *J. Mol. Biol.* 239:227–248.

PROBLEMS

1. *Valuable reagents.* The following reagents are often used in protein chemistry:

CNBr	Dabsyl chloride
Urea	6 N HCl
β-Mercaptoethanol	Ninhydrin
Trypsin	Phenyl isothiocyanate
Performic acid	Chymotrypsin

Which one is the best suited for accomplishing each of the following tasks?
   (a) Determination of the amino acid sequence of a small peptide.
   (b) Identification of the amino-terminal residue of a peptide (of which you have less than 0.1 μg).
   (c) Reversible denaturation of a protein devoid of disulfide bonds. Which additional reagent would you need if disulfide bonds were present?
   (d) Hydrolysis of peptide bonds on the carboxyl side of aromatic residues.
   (e) Cleavage of peptide bonds on the carboxyl side of methionines.
   (f) Hydrolysis of peptide bonds on the carboxyl side of lysine and arginine residues.

2. *Acid-base relations.* What is the ratio of base to acid at pH 4, 5, 6, 7, and 8 for an acid with a pK of 6?

3. *Finding an end.* Anhydrous hydrazine has been used to cleave peptide bonds in proteins. What are the reaction products? How might this technique be used to identify the carboxyl-terminal amino acid?

4. *Crafting a new breakpoint.* Ethyleneimine reacts with cysteine side chains in proteins to form *S*-aminoethyl derivatives. The peptide bonds on the carboxyl side of these modified cysteine residues are susceptible to hydrolysis by trypsin. Why?

5. *Spectrometry.* The absorbance A of a solution is defined as

$$A = \log_{10} (I_0/I)$$

in which I_0 is the incident light intensity and I is the transmitted light intensity. The absorbance is related to the molar absorption coefficient (extinction coefficient) ε (in $\text{cm}^{-1}\ \text{M}^{-1}$), concentration c (in M), and path length l (in cm) by

$$A = \varepsilon l c$$

The absorption coefficient of myoglobin at 580 nm is 15,000 $\text{cm}^{-1}\ \text{M}^{-1}$. What is the absorbance of a 1 mg/ml solution across a 1-cm path? What percentage of the incident light is transmitted by this solution?

6. *A slow mover.* Tropomyosin, a 93-kd muscle protein, sediments more slowly than does hemoglobin (65 kd). Their sedimentation coefficients are 2.6S and 4.31S, respectively. Which structural feature of tropomyosin accounts for its slow sedimentation?

7. *Sedimenting spheres.* What is the dependence of the sedimentation coefficient S of a spherical protein on its mass? How much more rapidly does an 80-kd protein sediment than does a 40-kd protein?

8. *Size estimate.* The relative electrophoretic mobilities of a 30-kd protein and a 92-kd protein used as standards on an SDS–polyacrylamide gel are 0.80 and 0.41, respectively. What is the apparent mass of a protein having a mobility of 0.62 on this gel?

9. *A new partnership?* The gene encoding a protein with a single disulfide bond undergoes a mutation that changes a serine residue into a cysteine residue. You want to find out whether the disulfide pairing in this mutant is the same as in the original protein. Propose an experiment to directly answer this question.

10. *Helix-coil transitions.* (a) Circular dichroism measurements have shown that poly-L-lysine is a random coil at pH 7 but becomes α-helical as the pH is raised above 10. Account for this pH-dependent conformational transition. (b) Predict the pH dependence of the helix-coil transition of poly-L-glutamate.

11. *Sorting cells.* Fluorescence-activated cell sorting (FACS) is a powerful technique for separating cells according to their content of particular molecules. For example, a fluorescent-labeled antibody specific for a cell-surface protein can be used to detect cells containing such a molecule. Suppose that you want to isolate cells that possess a receptor enabling them to detect bacterial degradation products. However, you do not yet have an antibody directed against this receptor. Which fluorescent-labeled molecule would you prepare to identify such cells?

12. *Mirror images.* Suppose that a protease is synthesized by the solid-phase method from D rather than L amino acids. How would the sedimentation, electrophoretic, and circular dichroism properties of this enzyme compare with those of the native form? What prediction can you make about the relation of peptide substrates of the D and L enzymes?

13. *Peptides on a chip.* Large numbers of different peptides can be synthesized in a small area on a solid support. This high-density array can then be probed with a fluorescent-labeled protein to find out which peptides are recognized. The binding of an antibody to an array of 1024 different peptides occupying a total area the size of a thumbnail is shown in Figure 3-47. How would you synthesize such a peptide array? [Hint: Use light instead of acid to deprotect the terminal amino group in each round of synthesis.]

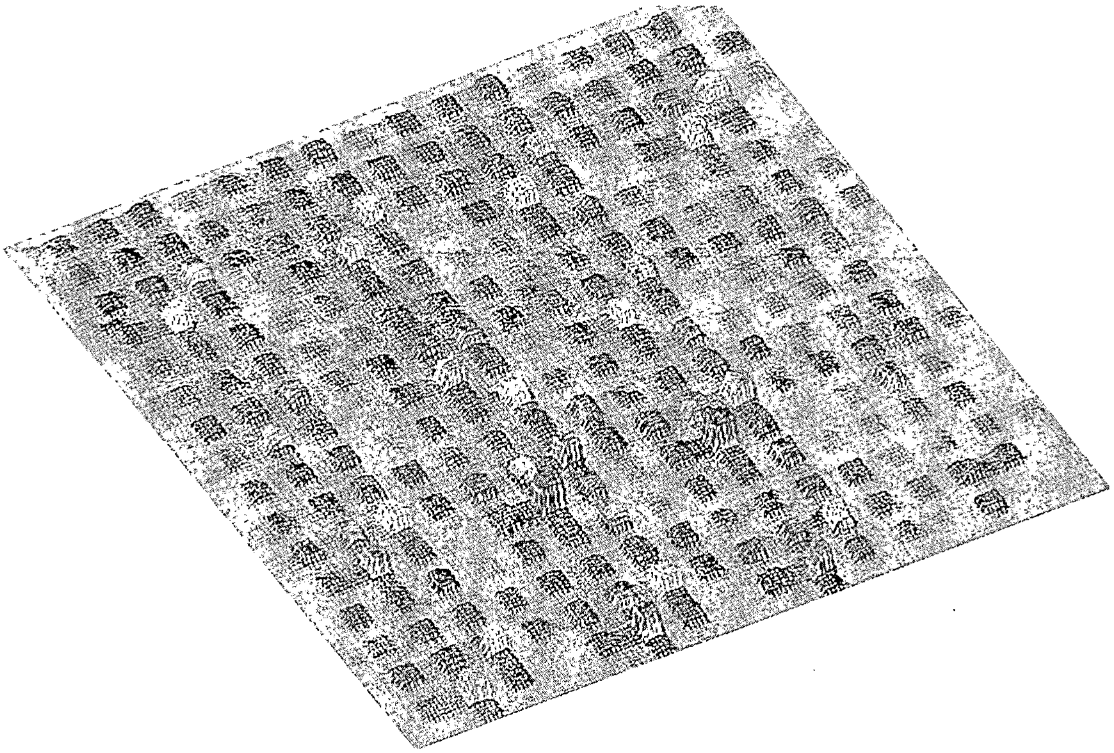

Figure 3-47
Fluorescence scan of an array of 1024 peptides in a 1.6-cm^2 area. Each synthesis site is a 400-μm by 400-μm square. A fluorescent-labeled monoclonal antibody was added to the chip to identify peptides that are recognized. The height and color of each square denote the fluorescence intensity. [After S.P.A. Fodor, J.L. Read, M.C. Pirrung, L. Stryer, A.T. Lu, and D. Solas. *Science* 251(1991):767.]

CHAPTER

4

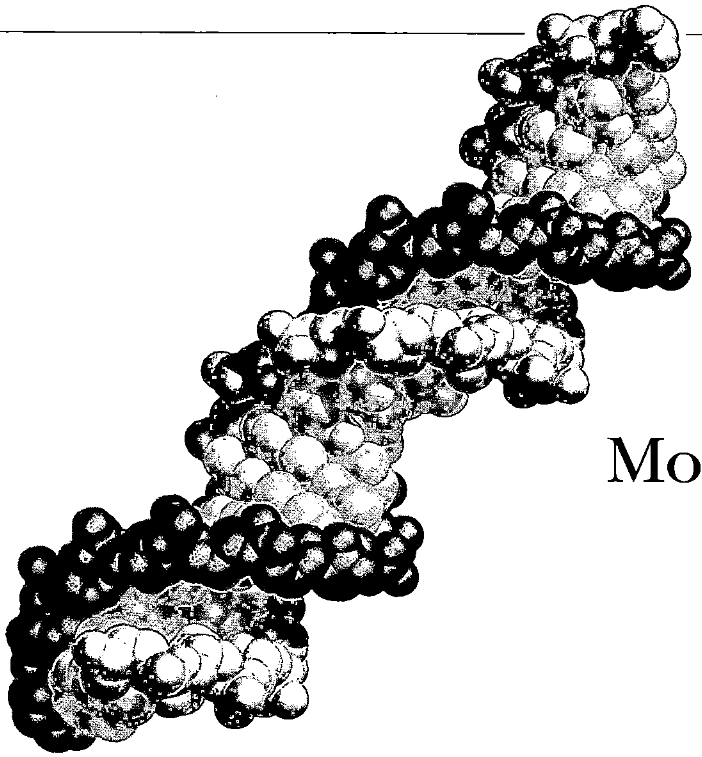

DNA and RNA: Molecules of Heredity

DNA is a very long, threadlike macromolecule made up of a large number of deoxyribonucleotides, each composed of a base, a sugar, and a phosphate group. *The bases of DNA molecules carry genetic information, whereas their sugar and phosphate groups perform a structural role.* This chapter presents the key experiments that revealed that DNA is the genetic material, then describes the DNA double helix. When this structure was discovered, the complementary nature of its two chains immediately suggested that each is a template for the other in DNA replication. DNA polymerases are the enzymes that replicate DNA by taking instructions from DNA templates. These exquisitely specific enzymes replicate DNA with an error frequency of less than 1 in 100 million nucleotides. The genes of all cells and many viruses are made of DNA. Some viruses, however, use RNA (ribonucleic acid) as their genetic material. This chapter concludes with examples of the genetic role of RNA in plant viruses and animal tumor viruses.

DNA CONSISTS OF FOUR KINDS OF BASES JOINED TO A SUGAR-PHOSPHATE BACKBONE

DNA is a polymer of deoxyribonucleotide units. A *nucleotide* consists of a nitrogenous base, a sugar, and one or more phosphate groups. The sugar in a deoxyribonucleotide is *deoxyribose.* The *deoxy* prefix indicates that this

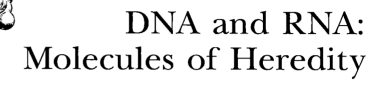

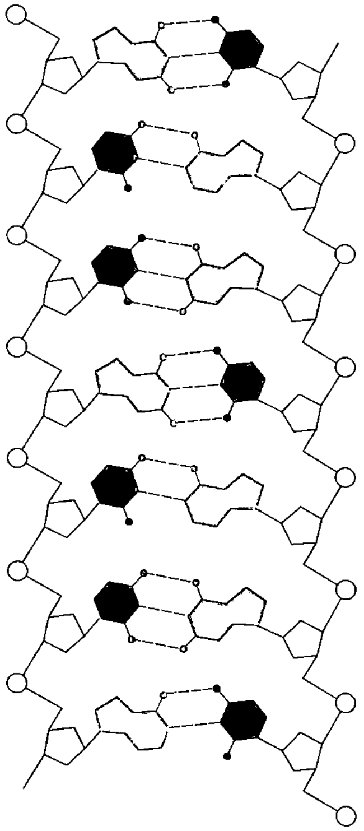

Figure 4-1
Schematic diagram of the structure of DNA. The sugar-phosphate backbone is shown in black, and the bases are shown in color. [After A. Kornberg. The synthesis of DNA. Copyright © 1968 by Scientific American, Inc. All rights reserved.]

Opening Image: *Model of double-helical DNA. The sugar-phosphate backbone is shown in dark colors, and the bases in light colors.*

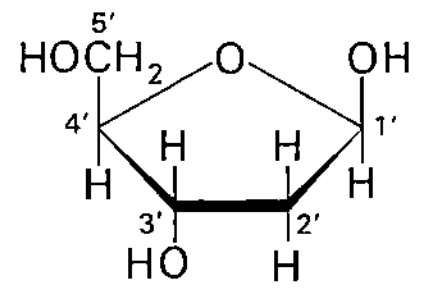

β-D-2-Deoxyribose

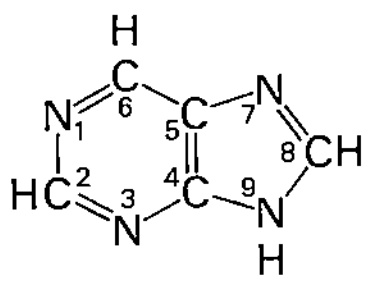

Purine

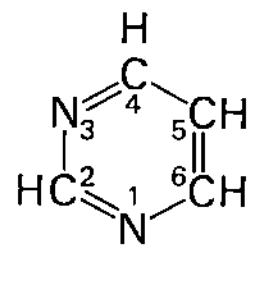

Pyrimidine

sugar lacks an oxygen atom that is present in ribose, the parent compound. The nitrogenous base is a derivative of *purine* or *pyrimidine.*

The purines in DNA are *adenine (A)* and *guanine (G)*, and the pyrimidines are *thymine (T)* and *cytosine (C)*.

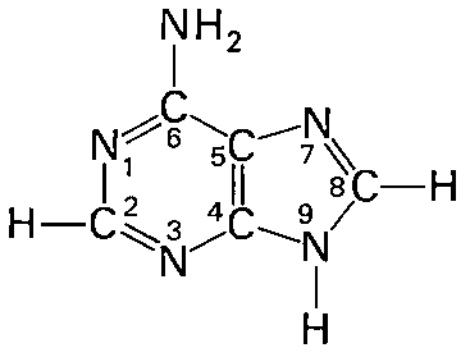

Adenine
(A)

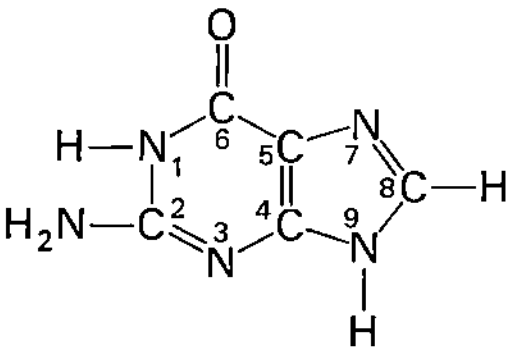

Guanine
(G)

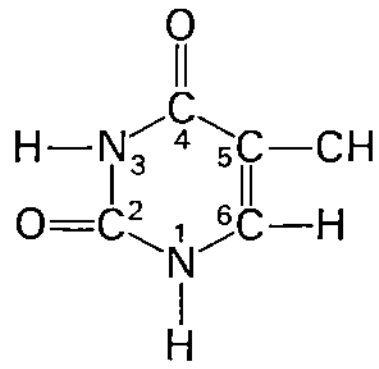

Thymine
(T)

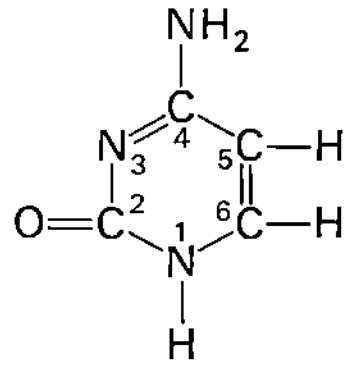

Cytosine
(C)

A *nucleoside* consists of a purine or pyrimidine base bonded to a sugar. The four nucleoside units in DNA are called *deoxyadenosine, deoxyguanosine, deoxythymidine,* and *deoxycytidine.* In a deoxyribonucleoside, N-9 of a purine or N-1 of a pyrimidine is attached to C-1 of deoxyribose. The configuration of this *N*-glycosidic linkage is β (the base lies above the plane of the sugar). A *nucleotide* is a phosphate ester of a nucleoside. The most common site of esterification in naturally occurring nucleotides is the hydroxyl group attached to C-5 of the sugar. Such a compound is called a *nucleoside 5′-phosphate* or a *5′-nucleotide.* For example, *deoxyadenosine 5′-triphosphate (dATP)* is an activated precursor in the synthesis of DNA; the nucleotide is activated by the presence of two phosphoanhydride bonds in its triphosphate unit. A primed number denotes an atom of the sugar, whereas an unprimed number denotes an atom of the purine or pyrimidine ring. The prefix *d* in dATP indicates that the sugar is deoxyribose to distinguish this compound from ATP, in which the sugar is ribose.

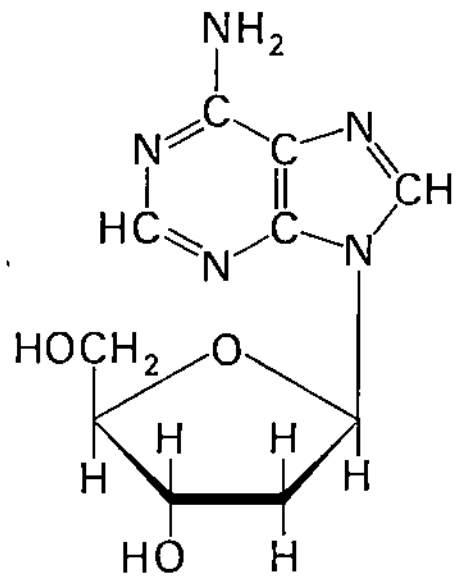

Deoxyadenosine
(A nucleoside)

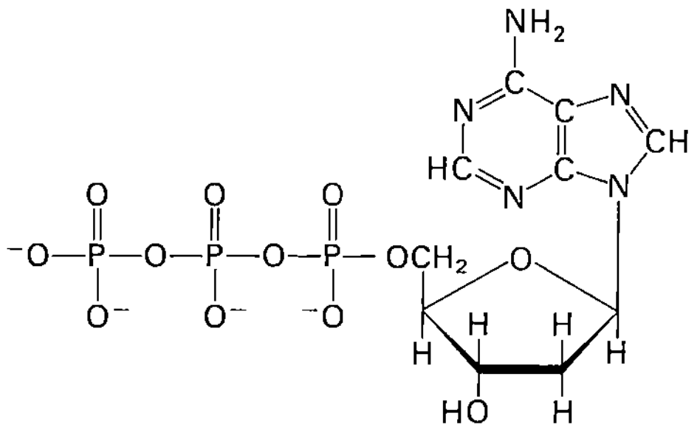

Deoxyadenosine 5′-triphosphate
(dATP)
(A nucleotide)

The *backbone* of DNA, which is invariant throughout the molecule, consists of deoxyriboses linked by phosphate groups. Specifically, the 3′-hydroxyl of the sugar moiety of one deoxyribonucleotide is joined to the 5′-hydroxyl of the adjacent sugar by a phosphodiester bridge. The *variable part* of DNA is its *sequence of four kinds of bases (A, G, C, and T).* The corresponding nucleotide units are called *deoxyadenylate, deoxyguanylate, deoxycytidylate,* and *deoxythymidylate.* The structure of a DNA chain is shown in Figure 4-2.

The structure of a DNA chain can be concisely represented in the following way. The symbols for the four principal deoxyribonucleosides are

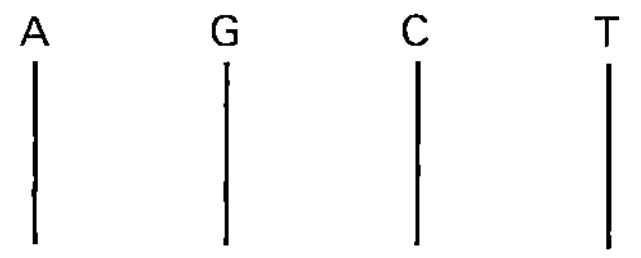

The bold line refers to the sugar, whereas A, G, C, and T represent the bases. In the diagram below, the Ⓟ within the diagonal line denotes a phosphodiester bond. This diagonal line joins the middle of one bold line with the end of another. These junctions refer to the 3′-OH and 5′-OH, respectively. In this example, the symbol Ⓟ indicates that deoxyadenylate is linked to deoxycytidylate by a phosphodiester bridge. Specifically, the 3′-OH of deoxyadenylate is joined through a phosphoryl group to the 5′-OH of deoxycytidylate.

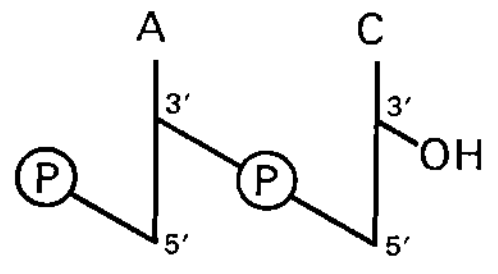

Now suppose that deoxyguanylate becomes linked to the deoxycytidylate unit of this dinucleotide. The resulting trinucleotide can be represented by

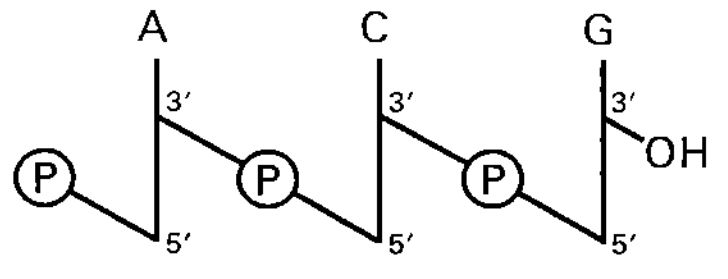

An even more abbreviated notation for this trinucleotide is pApCpG or pACG.

A DNA chain has polarity. One end of the chain has a 5′-OH group and the other a 3′-OH group, neither of which is linked to another nucleotide. By convention, the symbol ACG means that the unlinked 5′-OH group is on deoxyadenylate, whereas the unlinked 3′-OH group is on deoxyguanylate. Thus, *the base sequence is written in the 5′ → 3′ direction.* Recall that the amino acid sequence of a protein is written in the amino → carboxyl direction. Note that ACG and GCA refer to different compounds, just as Glu-Phe-Ala differs from Ala-Phe-Glu.

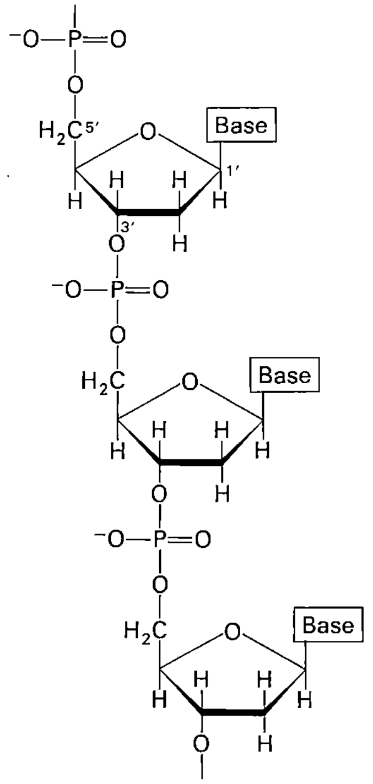

Figure 4-2
Structure of part of a DNA chain.

TRANSFORMATION OF PNEUMOCOCCI BY DNA REVEALED THAT GENES ARE MADE OF DNA

The pneumococcus bacterium played an important part in the discovery of the genetic role of DNA. A pneumococcus is normally surrounded by a slimy, glistening polysaccharide capsule. This outer layer is essential for the pathogenicity of the bacterium, which causes pneumonia in humans and other susceptible mammals. Mutants devoid of a polysac-

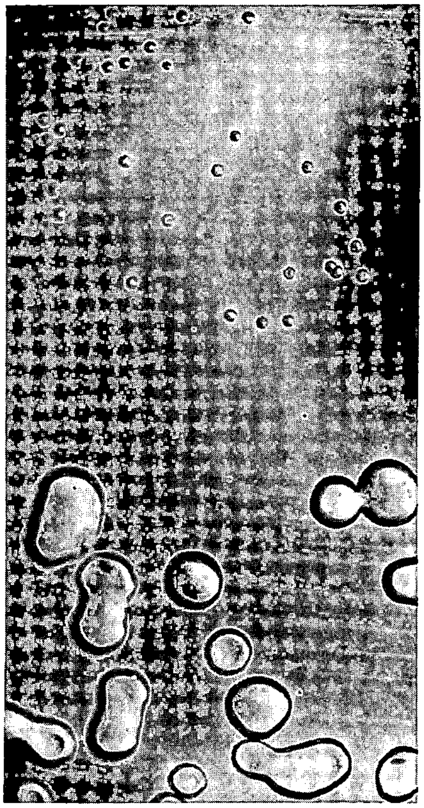

Figure 4-3
Transformation of nonpathogenic R pneumococci (small colonies) to pathogenic S pneumococci (large, glistening colonies) by DNA from heat-killed S pneumococci. [From O.T. Avery, C.M. MacLeod, and M. McCarty. *J. Exp. Med.* 79 (1944):158.]

charide coat are not pathogenic. The normal bacterium is referred to as the *S form* (because it forms smooth colonies in a culture dish), whereas mutants without capsules are called *R forms* (because they form rough colonies). R mutants lack an enzyme needed for the synthesis of capsular polysaccharide.

In 1928, Fred Griffith discovered that a nonpathogenic R mutant could be *transformed* into the pathogenic S form in the following way. He injected mice with a mixture of live R and heat-killed S pneumococci. The striking finding was that this mixture was lethal to the mice, whereas either live R or heat-killed S pneumococci alone were not. The blood of the dead mice contained live S pneumococci. Thus, the heat-killed S pneumococci had somehow transformed live R pneumococci into live S pneumococci. This change was permanent: the transformed pneumococci yielded pathogenic progeny of the S form. It was subsequently found that this R → S transformation can occur in vitro. Some of the cells in a growing culture of the R form were transformed into the S form by the addition of a *cell-free extract* of heat-killed S pneumococci. This finding set the stage for the elucidation of the chemical nature of the "transforming principle."

The cell-free extract of heat-killed S pneumococci was fractionated and the transforming activity of its components assayed. In 1944, Oswald Avery, Colin MacLeod, and Maclyn McCarty published their discovery that *"a nucleic acid of the deoxyribose type is the fundamental unit of the transforming principle of Pneumococcus Type III."* The experimental basis for their conclusion was (1) the purified, highly active transforming principle gave an elemental chemical analysis that agreed closely with that calculated for DNA; (2) the optical, ultracentrifugal, diffusive, and electrophoretic properties of the purified material were like those of DNA; (3) there was no loss of transforming activity upon extraction of protein or lipid; (4) the polypeptide-cleaving enzymes trypsin and chymotrypsin did not affect transforming activity; (5) ribonuclease (known to digest ribonucleic acid) had no effect on the transforming principle; and (6) in contrast, transforming activity was lost following the addition of deoxyribonuclease.

This work is a landmark in the development of biochemistry. Until 1944, it was generally assumed that chromosomal proteins carry genetic information and that DNA plays a secondary role. This prevailing view was decisively shattered by the rigorously documented finding that *purified DNA has genetic specificity.* Avery gave a vivid description of this research and of its implications in a letter that he wrote in 1943 to his brother, a medical microbiologist at another university (Figure 4-4).

Further support for the genetic role of DNA came from the studies of a virus that infects the bacterium *Escherichia coli.* The T2 bacteriophage consists of a core of DNA surrounded by a protein coat (Figure 4-5). In 1951, Roger Herriott suggested that "the virus may act like a little hypodermic needle full of transforming principles; the virus as such never enters the cell; only the tail contacts the host and perhaps enzymatically cuts a small hole through the outer membrane and then the nucleic acid of the virus head flows into the cell." In 1952, this idea was tested by Alfred Hershey and Martha Chase in the following way. Phage DNA was labeled with the radioisotope ^{32}P, whereas the protein coat was labeled with ^{35}S. These labels are highly specific because DNA does not contain sulfur and the protein coat is devoid of phosphorus. A sample of an *E. coli* culture was infected with labeled phage, which became attached to the bacteria during a short incubation period. The suspension was spun for a few minutes in a Waring Blendor at 10,000 rpm. This treatment subjected the phage-infected cells to very strong shearing forces, which severed the

For the past two years, first with MacLeod and now with Dr. McCarty, I have been trying to find out what is the chemical nature of the substance in the bacterial extract which induces this specific change. The crude extract of Type III is full of capsular polysaccharide, G (somatic) carbohydrate, nucleoproteins, free nucleic acids of both the yeast and thymus type, lipids, and other cell constituents. Try to find in the complex mixtures the active principle! Try to isolate and chemically identify the particular substance that will by itself, when brought into contact with the R cell derived from Type II, cause it to elaborate Type III capsular polysaccharide and to acquire all the aristocratic distinctions of the same specific type of cells as that from which the extract was prepared! Some job, full of headaches and heartbreaks. But at last perhaps we have it.

. . . if we prove to be right—and of course that is a big if—then it means that both the chemical nature of the inducing stimulus is known and the chemical structure of the substance produced is also known, the former being thymus nucleic acid, the latter Type III polysaccharide, and both are thereafter reduplicated in the daughter cells and after innumerable transfers without further addition of the inducing agent and the same active and specific transforming substance can be recovered far in excess of the amount originally used to induce the reaction. Sounds like a virus—may be a gene. But with mechanisms I am not now concerned. One step at a time and the first step is what is the chemical nature of the transforming principle? Some one else can work out the rest. Of course the problem bristles with implications. It touches the biochemistry of the thymus type of nucleic acids which are known to constitute the major part of chromosomes but have been thought to be alike regardless of origin and species. It touches genetics, enzyme chemistry, cell metabolism and carbohydrate synthesis. But today it takes a lot of well documented evidence to convince anyone that the sodium salt of deoxyribose nucleic acid, protein free, could possibly be endowed with such biologically active and specific properties and that is the evidence we are now trying to get. It is lots of fun to blow bubbles but it is wiser to prick them yourself before someone else tries to.

Figure 4-4
Part of a letter from Oswald Avery to his brother Roy, written in May 1943. [From R.D. Hotchkiss. In *Phage and the Origins of Molecular Biology*, J. Cairns, G.S. Stent, and J.D. Watson, eds. (Cold Spring Harbor Laboratory, 1966), pp. 185–186.]

connections between the viruses and bacteria. The resulting suspension was centrifuged at a speed sufficient to throw the bacteria to the bottom of the tube. Thus, the pellet contained the infected bacteria, whereas the supernatant contained smaller particles. These fractions were analyzed for ^{32}P and ^{35}S to determine the location of the phage DNA and the protein coat. The results of these experiments were

1. Most of the phage DNA was found in the bacteria.

2. Most of the phage protein was found in the supernatant.

3. The blender treatment had almost no effect on the competence of the infected bacteria to produce progeny virus.

Additional experiments showed that less than 1% of the ^{35}S was transferred from the parental phage to the progeny phage. In contrast, 30% of the parental ^{32}P appeared in the progeny. These simple, incisive experiments led to the conclusion that "*a physical separation of the phage T2 into genetic and non-genetic parts is possible. . . .* The sulfur-containing protein of resting phage particles is confined to a protective coat that is responsible for the adsorption of bacteria, and functions as an instrument for the injection of the phage DNA into the cell. This protein probably has no

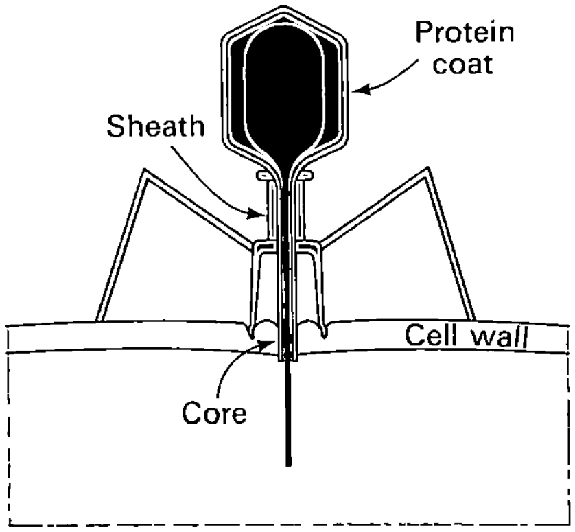

Figure 4-5
Diagram of a T2 bacteriophage injecting its DNA into a bacterial cell. [After W.B. Wood and R.S. Edgar. Building a bacterial virus. Copyright © 1967 by Scientific American, Inc. All rights reserved.]

function in the growth of intracellular phage. The DNA has some function. Further chemical inferences should not be drawn from the experiments presented."

The cautious tone of this conclusion did not diminish its impact. The genetic role of DNA soon became a generally accepted fact. The experiments of Hershey and Chase strongly reinforced what Avery, MacLeod, and McCarty had found eight years earlier in a different system. Additional support came from studies of the DNA content of single cells, which showed that in a given species *the DNA content is the same for all cells that have a diploid set of chromosomes. Haploid cells were found to have half as much DNA.*

THE DISCOVERY OF THE DNA DOUBLE HELIX BY WATSON AND CRICK REVOLUTIONIZED BIOLOGY

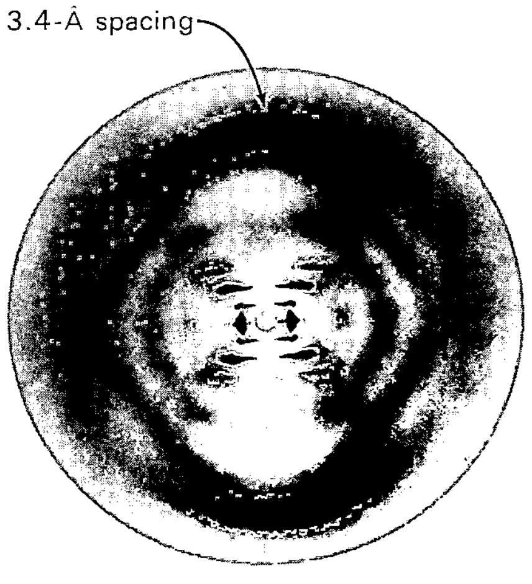

Figure 4-6
X-ray diffraction photograph of a hydrated DNA fiber. The central cross is diagnostic of a helical structure. The strong arcs on the meridian arise from the stack of base pairs, which are 3.4 Å apart. [Courtesy of Dr. Maurice Wilkins.]

In 1953, James Watson and Francis Crick deduced the three-dimensional structure of DNA and immediately inferred its mechanism of replication. This brilliant accomplishment ranks as one of the most significant in the history of biology because it led the way to an understanding of gene function in molecular terms. Watson and Crick analyzed x-ray diffraction photographs of DNA fibers taken by Rosalind Franklin and Maurice Wilkins (Figure 4-6) and derived a structural model that has proved to be essentially correct. The important features of their model of DNA are

1. Two helical polynucleotide chains are coiled around a common axis. The chains run in opposite directions (Figure 4-7).

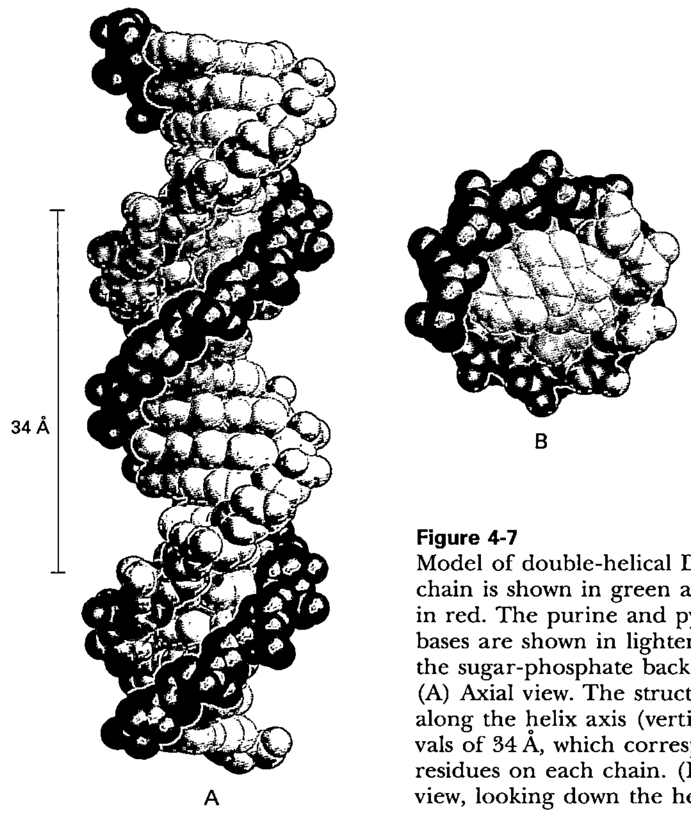

Figure 4-7
Model of double-helical DNA. One chain is shown in green and the other in red. The purine and pyrimidine bases are shown in lighter colors than the sugar-phosphate backbone.
(A) Axial view. The structure repeats along the helix axis (vertical) at intervals of 34 Å, which corresponds to 10 residues on each chain. (B) Radial view, looking down the helix axis.

2. The purine and pyrimidine bases are on the inside of the helix, whereas the phosphate and deoxyribose units are on the outside (Figure 4-8). The planes of the bases are perpendicular to the helix axis. The planes of the sugars are nearly at right angles to those of the bases.

3. The diameter of the helix is 20 Å. Adjacent bases are separated by 3.4 Å along the helix axis and related by a rotation of 36 degrees. Hence, the helical structure repeats after 10 residues on each chain, that is, at intervals of 34 Å.

4. The two chains are held together by hydrogen bonds between pairs of bases. Adenine is always paired with thymine; guanine is always paired with cytosine (Figures 4-9 and 4-10).

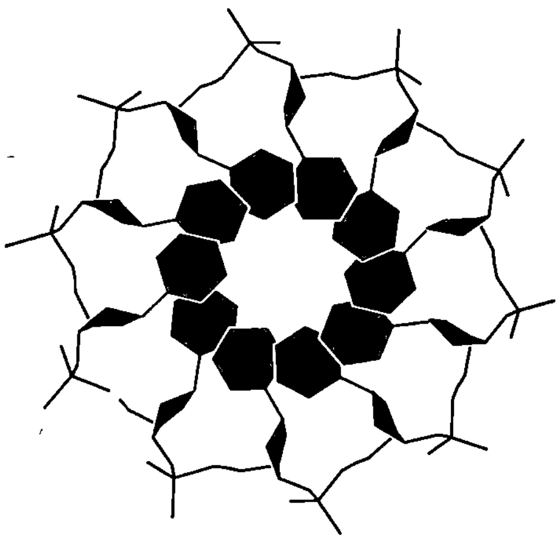

Figure 4-8
Diagram of one of the strands of a DNA double helix, viewed down the helix axis. The bases (all pyrimidines here) are inside, whereas the sugar-phosphate backbone is outside. The tenfold symmetry is evident. The bases are shown in blue and the sugars in red.

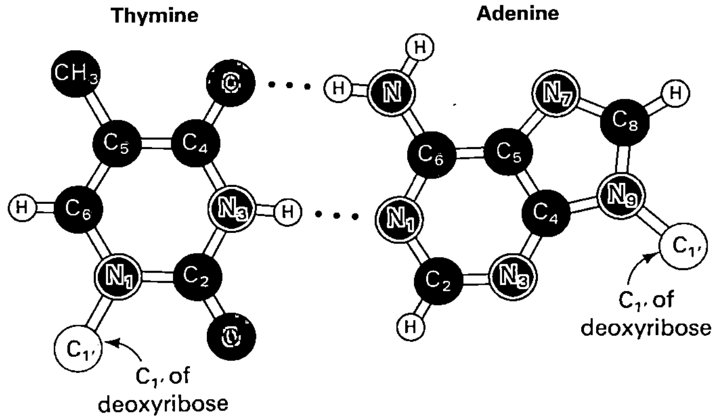

Figure 4-9
Model of an adenine-thymine base pair.

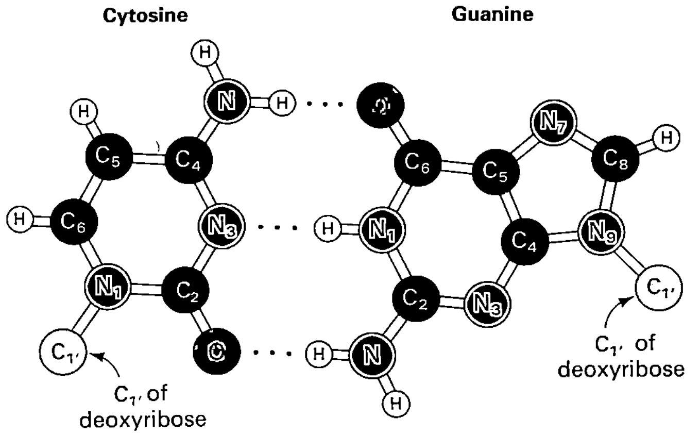

Figure 4-10
Model of a guanine-cytosine base pair.

5. The sequence of bases along a polynucleotide chain is not restricted in any way. *The precise sequence of bases carries the genetic information.*

The most important aspect of the DNA double helix is the specificity of the pairing of bases. Watson and Crick deduced that adenine must pair with thymine, and guanine with cytosine, because of steric and hydrogen-bonding factors. The steric restriction is imposed by the regular helical nature of the sugar-phosphate backbone of each polynucleotide chain. The glycosidic bonds

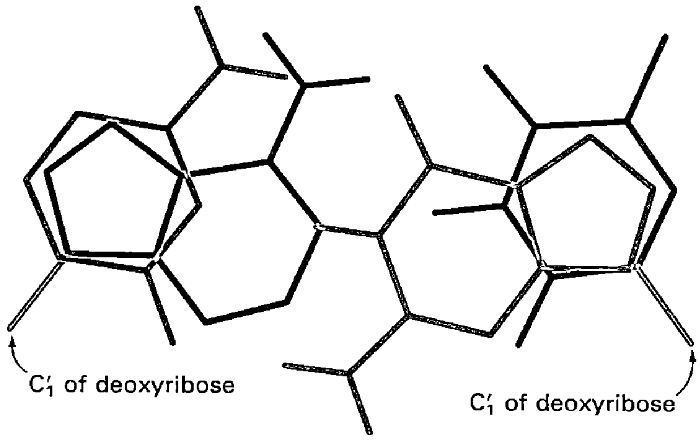

Figure 4-11
Superposition of a G·C base pair (blue) on an A·T base pair (red). Note that the positions of the glycosidic bonds (green) and of the C-1′ atom of deoxyribose are virtually identical for the two base pairs.

that are attached to a bonded pair of bases are very nearly 10.8 Å apart (Figure 4-11). A purine-pyrimidine base pair fits perfectly in this space. In contrast, there is insufficient room for two purines. There is more than enough space for two pyrimidines, but they would be too far apart to form hydrogen bonds. Hence, one member of a base pair in a DNA helix must always be a purine and the other a pyrimidine because of steric factors. The base pairing is further restricted by hydrogen-bonding requirements. The hydrogen atoms in the purine and pyrimidine bases have well-defined positions. Adenine cannot pair with cytosine because there would be two hydrogens near one of the bonding positions and none at the other. Likewise, guanine cannot pair with thymine. In contrast, adenine forms two hydrogen bonds with thymine, whereas guanine forms three with cytosine (see Figures 4-9 and 4-10). The orientations and distances of these hydrogen bonds are optimal for achieving strong attraction between the bases.

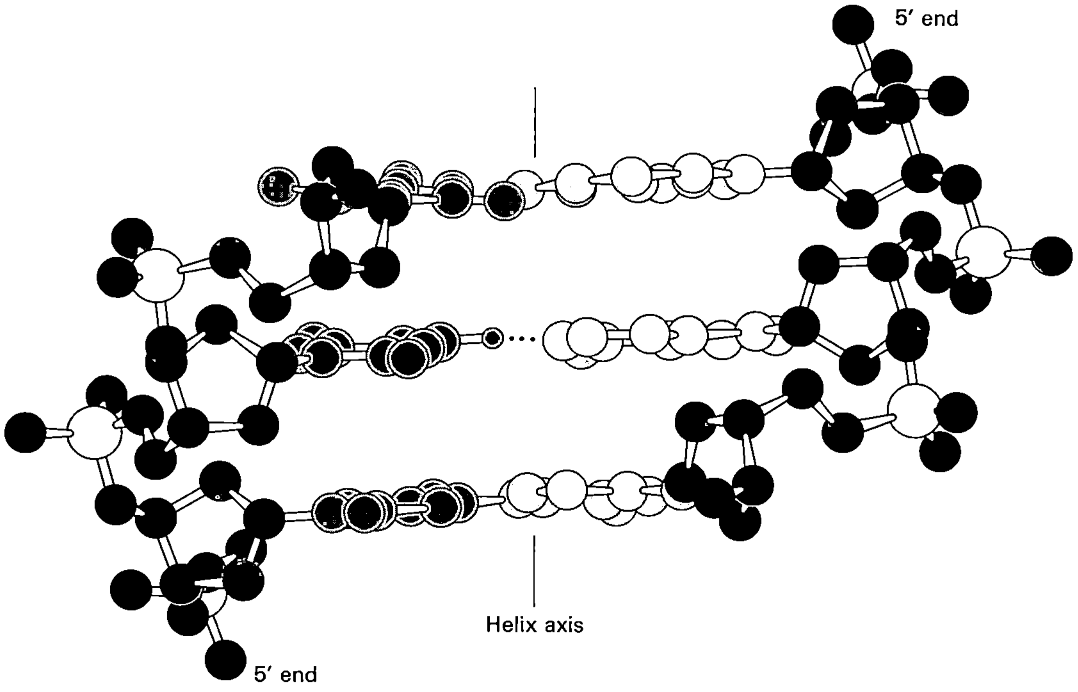

Figure 4-12
Model of a double-helical DNA molecule showing three base pairs (gray). Note that the two strands run in opposite directions. B form DNA, the classical Watson-Crick double helix, is depicted here. In this form, the planes of the bases are perpendicular to the helix axis.

This base-pairing scheme was strongly supported by the results of earlier studies of the base compositions of DNAs from different species. In 1950, Erwin Chargaff found that the *ratios of adenine to thymine and of guanine to cytosine were nearly 1.0 in all species studied.* The meaning of these equivalences was not evident until the Watson-Crick model was proposed. Only then could it be seen that they reflect an essential facet of DNA structure and function—the specificity of base pairing.

The double-helical structure of DNA shown in Figure 4-12 is *the B form of DNA (B-DNA)*. As will be discussed in a later chapter (p. 788), DNA can assume different helical forms, such as A-DNA and Z-DNA. Under physiologic conditions, most of the DNA is in the Watson-Crick B form.

THE COMPLEMENTARY CHAINS ACT AS TEMPLATES FOR EACH OTHER IN DNA REPLICATION

The double-helical model immediately suggested a mechanism for the replication of DNA. Watson and Crick published their hypothesis a month after they had presented their structural model in a beautifully simple and lucid paper:

> If the actual order of the bases on one of the pairs of chains were given, one could write down the exact order of the bases on the other one, because of the specific pairing. Thus one chain is, as it were, the complement of the other, and it is this feature which suggests how the deoxyribonucleic acid molecule might duplicate itself.
>
> Previous discussions of self-duplication have usually involved the concept of a template, or mould. Either the template was supposed to copy itself directly or it was to produce a "negative," which in its turn was to act as a template and produce the original "positive" once again. In no case has it been explained in detail how it would do this in terms of atoms and molecules.
>
> Now our model for deoxyribonucleic acid is, in effect, a *pair* of templates, each of which is complementary to the other. We imagine that prior to duplication the hydrogen bonds are broken, and the two chains unwind and separate. Each chain then acts as a template for the formation onto itself of a new companion chain, so that eventually we shall have *two* pairs of chains, where we only had one before. Moreover, the sequence of the pairs of bases will have been duplicated exactly.

DNA REPLICATION IS SEMICONSERVATIVE

Watson and Crick proposed that one of the strands of each daughter DNA molecule is newly synthesized, whereas the other is passed on unchanged from the parent DNA molecule. This distribution of parental atoms is called *semiconservative.* A critical test of this hypothesis was carried out by Matthew Meselson and Franklin Stahl. The parent DNA was labeled with ^{15}N, the heavy isotope of nitrogen, to make it denser than ordinary DNA. This was accomplished by growing *E. coli* for many generations in a medium that contained $^{15}NH_4Cl$ as the sole nitrogen source. The bacteria were abruptly transferred to a medium that contained ^{14}N, the ordinary isotope of nitrogen. The question asked was: What is the distribution of ^{14}N and ^{15}N in the DNA molecules after successive rounds of replication?

The distribution of ^{14}N and ^{15}N was revealed by the newly developed technique of *density-gradient equilibrium sedimentation.* A small amount of DNA was dissolved in a concentrated solution of cesium chloride having a density close to that of the DNA ($\sim$1.7 g/cm^3). This solution was centrifuged until it was nearly at equilibrium. The opposing processes of sedimentation and diffusion created a gradient in the concentration of cesium chloride across the centrifuge cell. The result was a stable density gradient, ranging from 1.66 to 1.76 g/cm^3. The DNA molecules in this density gradient were driven by centrifugal force into the region where the solution's density was equal to their own. High-molecular-weight DNA yielded a narrow band that was detected by its absorption of ultraviolet light. A mixture of ^{14}N DNA and ^{15}N DNA molecules gave clearly separate bands because they differ in density by about 1% (Figure 4-13).

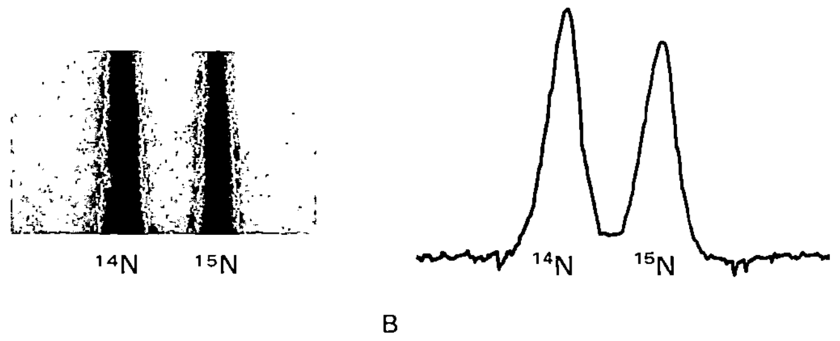

Figure 4-13
Resolution of ^{14}N DNA and ^{15}N DNA by density-gradient centrifugation: (A) ultraviolet absorption photograph of a centrifuge cell; (B) densitometric tracing of the absorption photograph. [From M. Meselson and F.W. Stahl. *Proc. Nat. Acad. Sci.* 44(1958):671.]

DNA was extracted from the bacteria at various times after they were transferred from a ^{15}N to a ^{14}N medium. Analysis of these samples by the density-gradient technique showed that there was a single band of DNA after one generation (Figure 4-14). The density of this band was precisely halfway between those of ^{14}N DNA and ^{15}N DNA. *The absence of ^{15}N DNA indicated that parental DNA was not preserved as an intact unit on replication.* The absence of ^{14}N DNA indicated that all the daughter DNA molecules derived some of their atoms from the parent DNA. This proportion had to be half because the density of the hybrid DNA band was halfway between those of ^{14}N DNA and ^{15}N DNA.

After two generations, there were equal amounts of two bands of DNA. One was hybrid DNA, and the other was ^{14}N DNA. Meselson and Stahl concluded from these incisive experiments *"that the nitrogen of a DNA molecule is divided equally between two physically continuous subunits; that, following duplication, each daughter molecule receives one of these; and that the subunits are conserved through many duplications."* Their results agreed perfectly with the Watson-Crick model for DNA replication (Figure 4-15).

THE DOUBLE HELIX CAN BE REVERSIBLY MELTED

The two strands of a DNA helix readily come apart when the hydrogen bonds between its paired bases are disrupted. This can be accomplished by heating a solution of DNA or by adding acid or alkali to ionize its bases. The unwinding of the double helix is called *melting* because it occurs abruptly at a certain temperature. The *melting temperature* (T_m) is defined

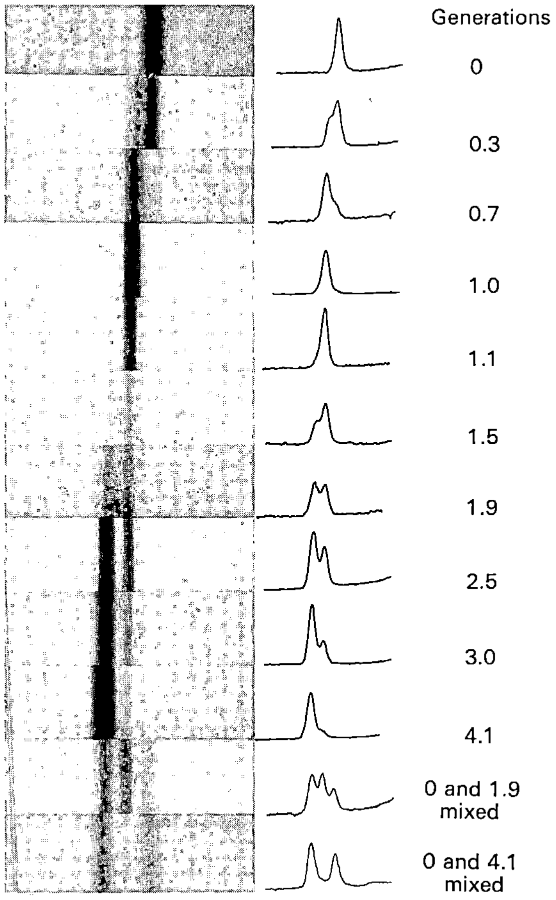

Figure 4-14
Detection of semiconservative replication in *E. coli* by density-gradient centrifugation. The position of a band of DNA depends on its content of ^{14}N and ^{15}N. After 1.0 generation, all the DNA molecules are hybrids containing equal amounts of ^{14}N and ^{15}N. No parental DNA (^{15}N) is left after 1.0 generation. [From M. Meselson and F.W. Stahl. *Proc. Nat. Acad. Sci.* 44(1958):671.]

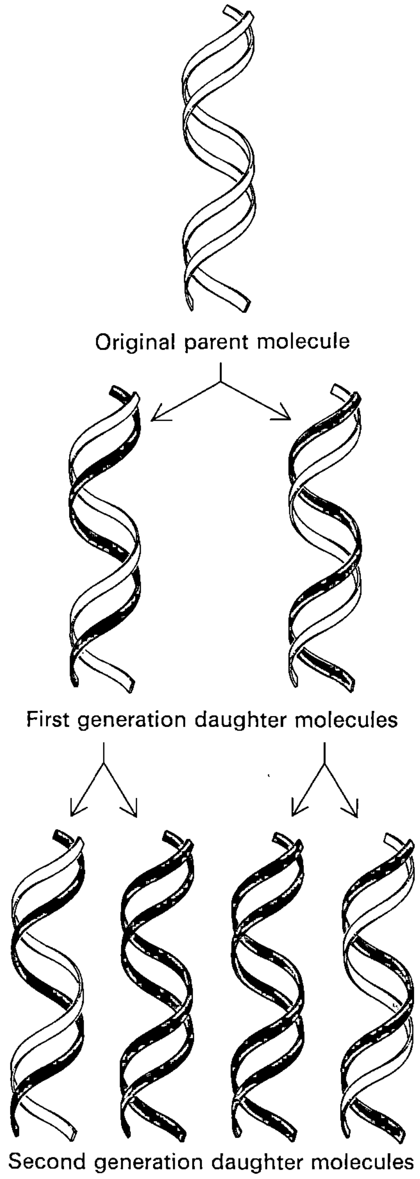

Figure 4-15
Schematic diagram of semiconservative replication. Parental DNA is shown in green and newly synthesized DNA in red. [After M. Meselson and F.W. Stahl. *Proc. Nat. Acad. Sci.* 44(1958):671.]

as the temperature at which half the helical structure is lost. The abruptness of the transition indicates that the DNA double helix is a *highly cooperative structure,* held together by many reinforcing bonds; it is stabilized by the stacking of bases as well as by base pairing. The melting of DNA is readily monitored by measuring its absorbance of light at wavelength 260 nm. The unstacking of the base pairs results in increased absorbance, an effect called *hyperchromism* (Figure 4-16).

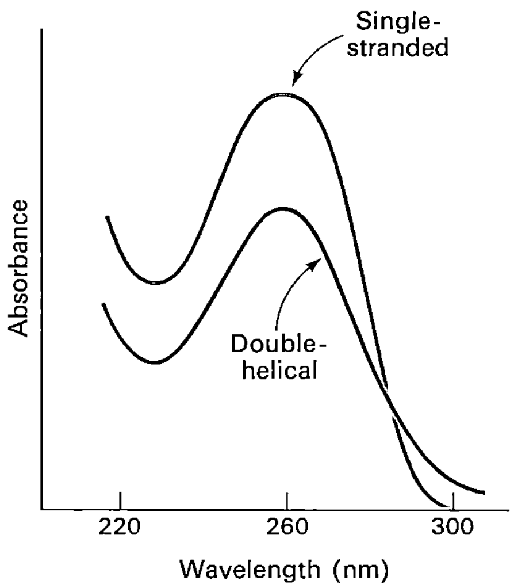

Figure 4-16
The absorbance of a DNA solution at wavelength 260 nm increases when the double helix is melted into single strands.

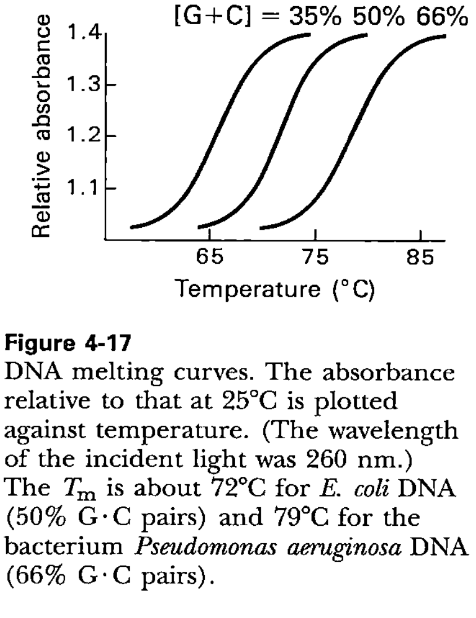

Figure 4-17
DNA melting curves. The absorbance relative to that at 25°C is plotted against temperature. (The wavelength of the incident light was 260 nm.) The T_m is about 72°C for *E. coli* DNA (50% G·C pairs) and 79°C for the bacterium *Pseudomonas aeruginosa* DNA (66% G·C pairs).

The melting temperature of a DNA molecule depends markedly on its base composition. DNA molecules rich in G·C base pairs have a higher T_m than those having an abundance of A·T base pairs (Figure 4-17). In fact, the T_m of DNA from many species varies linearly with G·C content, rising from 77°C to 100°C as the fraction of G·C pairs increases from 20% to 78%. G·C base pairs are more stable than A·T pairs because their bases are held together by three hydrogen bonds rather than by two. In addition, adjacent G·C base pairs interact more strongly with one another than do adjacent A·T base pairs. Hence, *the AT-rich regions of DNA are the first to melt* (Figure 4-18). The double helix is melted in vivo by the action of specific proteins (p. 805).

Separated complementary strands of DNA spontaneously reassociate to form a double helix when the temperature is lowered below T_m. This

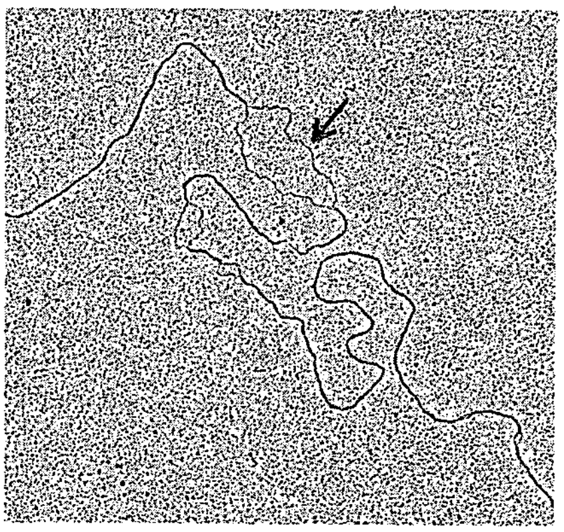

Figure 4-18
Electron micrograph of a DNA molecule partly unwound by alkali. Single-stranded regions appear as loops that stain less intensely than double-stranded segments. These unwound regions are rich in A·T base pairs. One of them is marked by an arrow. [From R.B. Inman and M. Schnos. *J. Mol. Biol.* 49(1970):93.]

renaturation process is sometimes called *annealing*. The facility with which double helices can be melted and then reassociated is crucial for the biological functions of DNA. It is also important to appreciate that DNA is a flexible molecule even at temperatures much below T_m.

DNA MOLECULES ARE VERY LONG

A striking characteristic of naturally occurring DNA molecules is their length. DNA molecules must be very long to encode the large number of proteins present in even the simplest cells. The *E. coli* chromosome, for example, is a single molecule of double-helical DNA consisting of four million base pairs. The mass of this DNA molecule is 2.6 gigadaltons (2.6×10^9 daltons). It has a *highly asymmetric shape* when taken out of the cell. The length of *E. coli* DNA is 14×10^6 Å, but its diameter is only 20 Å. The 1.4-mm length of this DNA molecule corresponds to a macroscopic dimension, whereas its width of 20 Å is on the atomic scale. Bruno Zimm found that the largest chromosome of *Drosophila melanogaster* contains a single DNA molecule of 6.2×10^7 base pairs, which has a length of 2.1 cm. Such highly asymmetric molecules are very susceptible to cleavage by shearing forces. Unless special precautions are taken in their handling, they easily break into segments whose masses are a thousandth of that of the original molecule.

DNA molecules from many bacteria and viruses have been directly visualized by electron microscopy (Figure 4-19). The dimensions of some of these DNA molecules are given in Table 4-1. It should be noted that even the smallest DNA molecules are highly elongate. The DNA from polyoma virus, for example, consists of 5100 base pairs and has a length of 1.7 μm (17,000 Å). Hemoglobin, which is roughly spherical, has a diameter of 65 Å; collagen, one of the longest proteins, has a length of 3000 Å. These comparisons emphasize the remarkable length and asymmetry of DNA molecules.

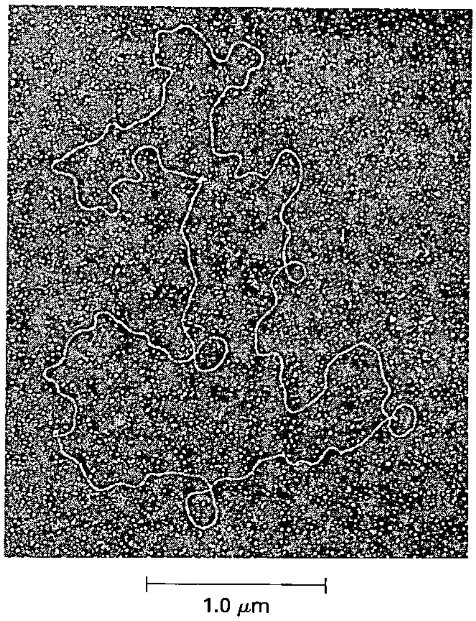

Figure 4-19
Electron micrograph of a circular DNA molecule from λ bacteriophage. [Courtesy of Dr. Thomas Broker.]

Table 4-1
Sizes of DNA genomes

Organism	*Base pairs (in thousands, or kb)*	*Length (μm)*
Viruses		
Polyoma or SV40	5.1	1.7
λ phage	48.6	17
T2 phage	166	56
Vaccinia	190	65
Bacteria		
Mycoplasma	760	260
E. coli	4,000	1,360
Eukaryotes		
Yeast	13,500	4,600
Drosophila	165,000	56,000
Human	2,900,000	990,000

After A. Kornberg and T.A. Baker. *DNA Replication* (2nd ed.). (W.H. Freeman, 1992), p. 20.

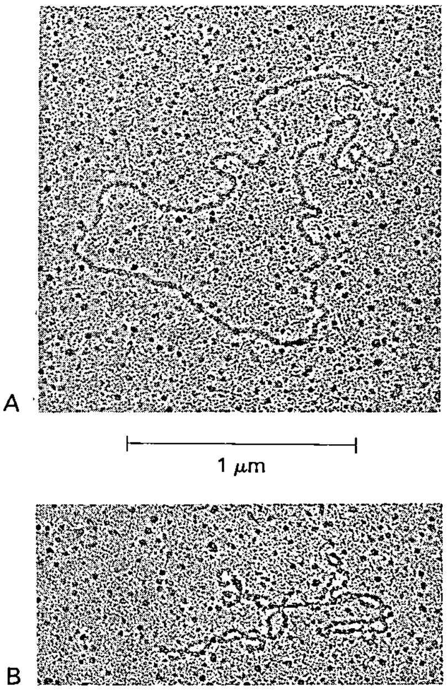

Figure 4-20
Electron micrographs of DNA from mitochondria: (A) relaxed circular form; (B) supercoiled circular form. [Courtesy of Dr. David Clayton.]

MANY DNA MOLECULES ARE CIRCULAR AND SUPERCOILED

Electron microscopy has shown that intact DNA molecules from many sources are circular (see Figure 4-19). The finding that *E. coli* has a circular chromosome was anticipated by genetic studies that revealed that *the gene-linkage map of this bacterium is circular.* The term *circular* refers to the continuity of the DNA chain, not to its geometrical form. DNA molecules in vivo necessarily have a very compact shape. Note that the length of the *E. coli* chromosome is about a thousand times as long as the greatest diameter of the bacterium.

Not all DNA molecules are circular. DNA from the T7 bacteriophage, for example, is *linear.* The DNA molecules of some viruses, such as the λ bacteriophage, *interconvert between linear and circular forms.* The linear form is present inside the virus particle, whereas the circular form is present in the host cell.

A new property appears in the conversion of a linear DNA duplex into a closed circular molecule. The axis of the double helix can itself be twisted to form a *superhelix.* A circular DNA without any superhelical turns is known as a *relaxed molecule.* Supercoiling is biologically important for two reasons. First, *a supercoiled DNA has a more compact shape than its relaxed counterpart* (Figure 4-20). Supercoiling is critical for the packaging of DNA in the cell. Second, *supercoiling affects the capacity of the double helix to unwind, and thereby affects its interactions with other molecules.* These topological features of DNA will be discussed further in a later chapter (p. 794).

DNA IS REPLICATED BY POLYMERASES THAT TAKE INSTRUCTIONS FROM TEMPLATES

We turn now to the molecular mechanism of DNA replication. In 1958, Arthur Kornberg and his colleagues isolated an enzyme from *E. coli* that catalyzes the synthesis of DNA. They named the enzyme *DNA polymerase;* it is now called *DNA polymerase I* because other DNA polymerases have since been found. DNA replication is mediated by the intricate and coordinated interplay of more than 20 proteins. We focus here on DNA polymerase I to illustrate some new principles.

DNA polymerase I is a 103-kd single polypeptide chain. *It catalyzes the step-by-step addition of deoxyribonucleotide units to a DNA chain:*

$$(\text{DNA})_{n \text{ residues}} + \text{dNTP} \rightleftharpoons (\text{DNA})_{n+1} + \text{PP}_i$$

(The abbreviation dNTP denotes any deoxyribonucleoside triphosphate, and PP_i denotes the pyrophosphate group.) DNA polymerase I requires the following components to synthesize a chain of DNA (Figure 4-21):

1. All four of the activated precursors—the *deoxyribonucleoside 5′-triphosphates dATP, dGTP, dTTP, and dCTP*—must be present. Mg^{2+} ion is also required.

2. DNA polymerase I adds deoxyribonucleotides to the 3′-hydroxyl terminus of a preexisting DNA chain. In other words, a *primer chain* with a free 3′-OH group is required.

3. A *DNA template* is essential. The template can be single- or double-stranded DNA. Double-stranded DNA is an effective template only if its sugar-phosphate backbone is broken at one or more sites.

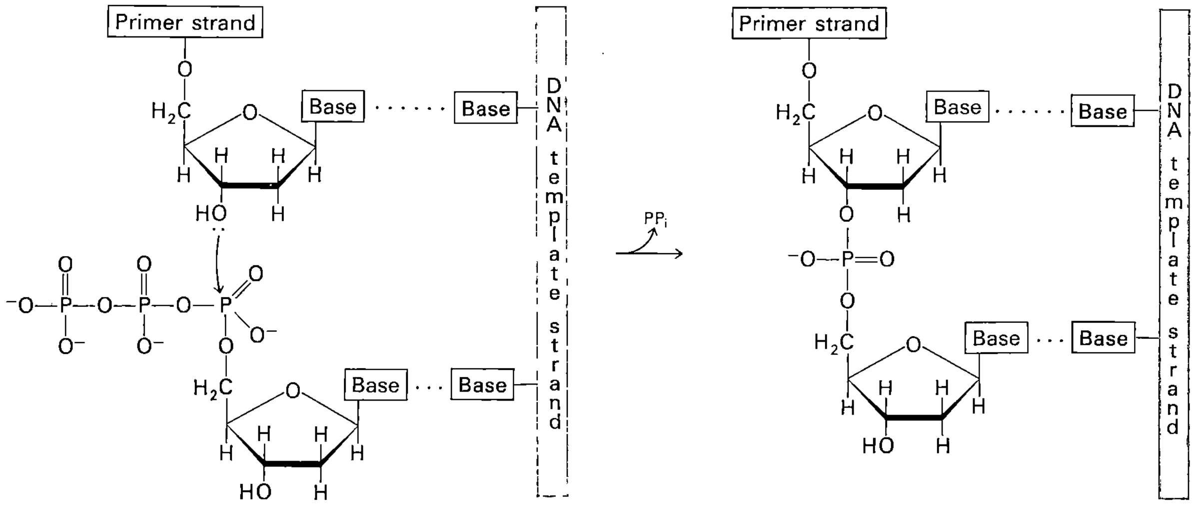

Figure 4-21
Chain-elongation reaction catalyzed by DNA polymerases.

The chain-elongation reaction catalyzed by DNA polymerase is *a nucleophilic attack of the 3′-OH terminus of the primer on the innermost phosphorus atom of a deoxyribonucleoside triphosphate.* A phosphodiester bridge is formed and pyrophosphate is concomitantly released. The subsequent hydrolysis of pyrophosphate by inorganic pyrophosphatase, a ubiquitous enzyme, drives the polymerization forward. *Elongation of the DNA chain proceeds in the 5′ → 3′ direction* (Figure 4-22).

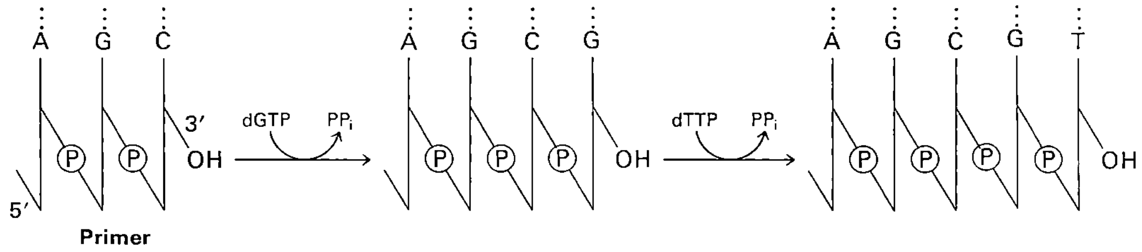

Figure 4-22
DNA polymerases catalyze elongation of DNA chains in the 5′ → 3′ direction.

DNA polymerase catalyzes the formation of a phosphodiester bond only if the base on the incoming nucleotide is complementary to the base on the template strand. The probability of making a covalent link is very low unless the incoming base forms a Watson-Crick type of base pair with the base on the template strand. Thus, DNA polymerase is a *template-directed enzyme.* The enzyme takes instructions from the template and synthesizes a product with a base sequence complementary to that of the template. Indeed, DNA polymerase I was the first template-directed enzyme to be discovered. Another striking feature of DNA polymerase I is that it corrects mistakes in DNA by removing mismatched nucleotides. These properties of DNA polymerase I contribute to the remarkably high fidelity of DNA replication, which has an error rate of less than 10^{-8} per base pair (p. 801).

SOME VIRUSES HAVE SINGLE-STRANDED DNA DURING PART OF THEIR LIFE CYCLE

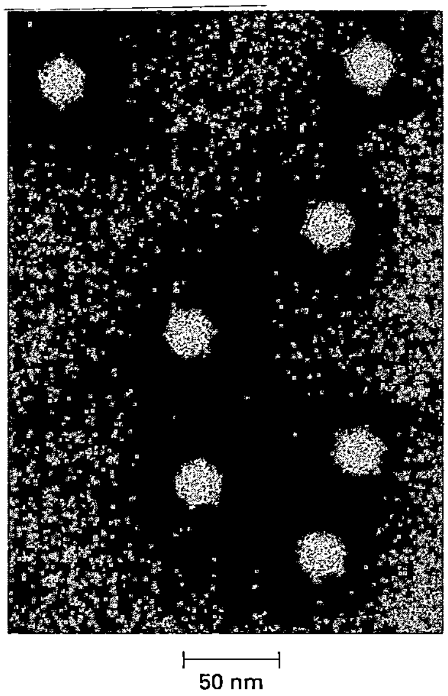

Figure 4-23
Electron micrograph of ϕX174 virus particles. [Courtesy of Dr. Robley Williams.]

Not all DNA is double-stranded. Robert Sinsheimer discovered that the DNA in *ϕX174, a small virus that infects* E. coli, *is single-stranded.* Several experimental results led to this unexpected conclusion. First, the base ratios of ϕX174 DNA do not conform to the rule that [A] = [T] and [G] = [C]. Second, a solution of ϕX174 DNA is much less viscous than a solution of the same concentration of *E. coli* DNA. The hydrodynamic properties of ϕX174 DNA are like those of a randomly coiled polymer. In contrast, the DNA double helix behaves hydrodynamically as a quite rigid rod. Third, the amino groups of the bases of ϕX174 DNA react readily with formaldehyde, whereas the bases in double-helical DNA are virtually inaccessible to this reagent.

The finding of this single-stranded DNA raised doubts concerning the universality of the semiconservative replicative scheme proposed by Watson and Crick. However, it was soon shown that ϕX174 DNA is single-stranded for only a part of the life cycle of the virus. Sinsheimer found that infected *E. coli* cells contain a *double-stranded form of ϕX174 DNA.* This double-helical DNA is called the *replicative form* because it serves as the template for the synthesis of the DNA of the progeny virus. The generality of the Watson-Crick scheme for replication was reinforced by the finding of this double-stranded viral DNA intermediate.

THE GENES OF SOME VIRUSES ARE MADE OF RNA

Genes in all prokaryotic and eukaryotic organisms are made of DNA. In viruses, genes are made of either DNA or RNA (ribonucleic acid). RNA, like DNA, is a long, unbranched polymer consisting of nucleotides joined by 3′ → 5′ phosphodiester bonds (Figure 4-24). The covalent structure of RNA differs from that of DNA in two respects. As indicated by its name, the sugar units in RNA are *riboses* rather than deoxyriboses. Ribose con-

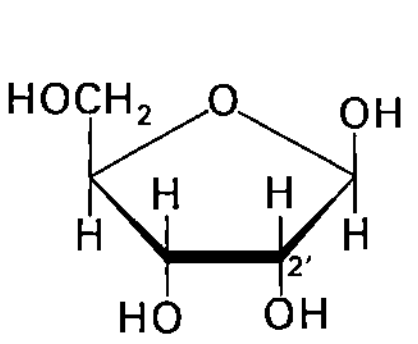

Ribose

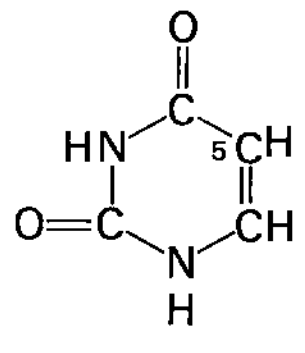

Uracil

tains a 2′-hydroxyl group not present in deoxyribose. The other difference is that one of the four major bases in RNA is *uracil (U)* instead of thymine (T). Uracil, like thymine, can form a base pair with adenine, but it lacks the methyl group present in thymine. RNA molecules can be single-stranded or double-stranded. RNA cannot form a double helix of the B-DNA type because of steric interference by the 2′-hydroxyl groups of its ribose units. However, RNA can adopt a modified double-helical form in which the base pairs are tilted about 20 degrees away from the perpendicular to the helix axis, a structure like A-DNA (p. 790).

Tobacco mosaic virus, which infects the leaves of tobacco plants, is one of the best-characterized RNA viruses. It consists of a single strand of RNA (6390 nucleotides) surrounded by a protein coat of 2130 identical subunits. The protein can be separated from the RNA by treatment of the

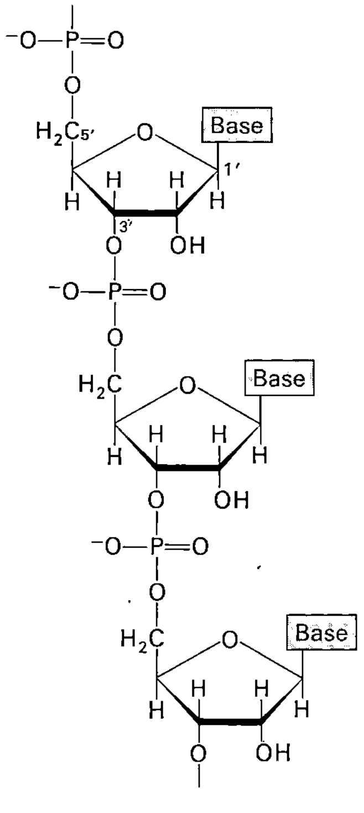

Figure 4-24
Structure of part of an RNA chain.

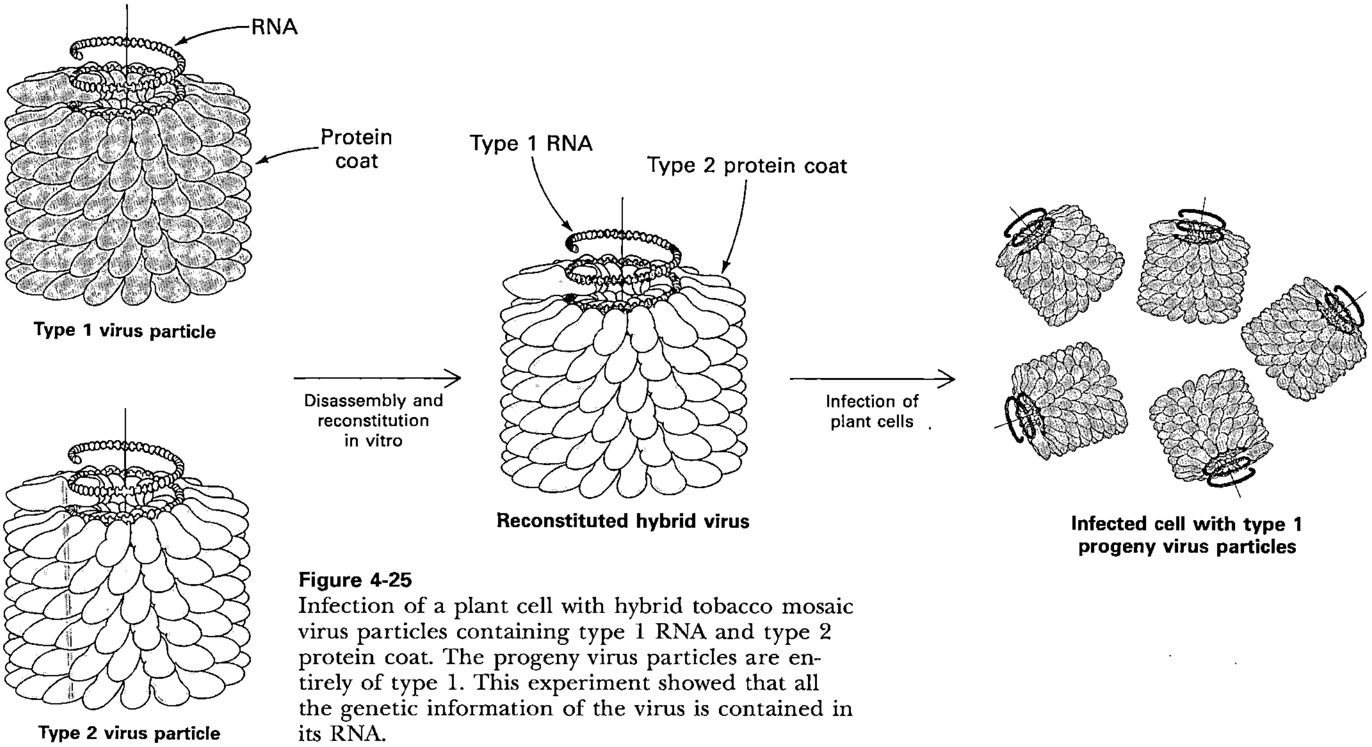

Figure 4-25
Infection of a plant cell with hybrid tobacco mosaic virus particles containing type 1 RNA and type 2 protein coat. The progeny virus particles are entirely of type 1. This experiment showed that all the genetic information of the virus is contained in its RNA.

virus with phenol. *The isolated viral RNA is infective, whereas the viral protein is not.* Synthetic hybrid virus particles (virions) provide additional evidence that the genetic specificity of the virus resides exclusively in its RNA. A variety of strains of tobacco mosaic virus are known. A synthetic hybrid virus was prepared from the RNA of strain 1 and the protein of strain 2. Another was prepared from the RNA of strain 2 and the protein of strain 1. After infection, *the progeny virus always consisted of RNA and protein corresponding to the RNA in the infecting hybrid virus* (Figure 4-25).

Tobacco mosaic virus replicates in an infected plant cell by first synthesizing a (−) RNA strand that is complementary to the (+) RNA strand in the virus particle. The (−) RNA strand then serves as the template for the synthesis of a large number of (+) RNA strands that become packaged in new virus particles that are released by the cell. These syntheses are catalyzed by RNA polymerases that take instructions from RNA templates (*RNA-directed RNA polymerases*). The (+) strand of viral RNA, by convention, is the one that serves as the template (messenger RNA) for the synthesis of proteins.

Virion—
The complete extracellular form of a virus consisting of DNA or RNA surrounded by a coat.

Capsid—
The protein coat surrounding viral DNA or RNA.

RNA TUMOR VIRUSES AND OTHER RETROVIRUSES REPLICATE THROUGH DOUBLE-HELICAL DNA INTERMEDIATES

A number of RNA viruses produce malignant tumors in susceptible animal hosts. *Rous sarcoma virus* is one of the best-studied members of this group of *RNA tumor viruses,* which contain a single strand of RNA (p. 834). A striking feature of RNA tumor viruses is that they replicate through *DNA* intermediates (Figure 4-26). Human immunodeficiency virus-1 (HIV-1), the cause of AIDS, employs the same genetic strategy. RNA tumor virus and HIV-1 are known as *retroviruses* because genetic information flows from RNA to DNA rather than from DNA to RNA.

Infection begins with the delivery of viral (+) RNA into the host cell. This (+) RNA is the template for the synthesis of a complementary (−) DNA strand by *reverse transcriptase,* a viral enzyme that also gains entry into the cell. Reverse transcriptase is an *RNA-directed DNA polymerase.* Reverse transcriptase is an apt name because RNA serves as a template for the synthesis of DNA, the reverse of the usual direction of information flow. The (−) DNA then serves as a template for the synthesis of (+) DNA (Figure 4-26). The resulting double-helical DNA version of the viral genome becomes incorporated into the chromosomal DNA of the host and is replicated along with the normal cellular DNA in the course of cell division. At some later time, the integrated viral genome is expressed to form viral (+) RNA and viral proteins, which assemble into new virus particles. We shall return in a later chapter (p. 834) to this unusual genetic strategy.

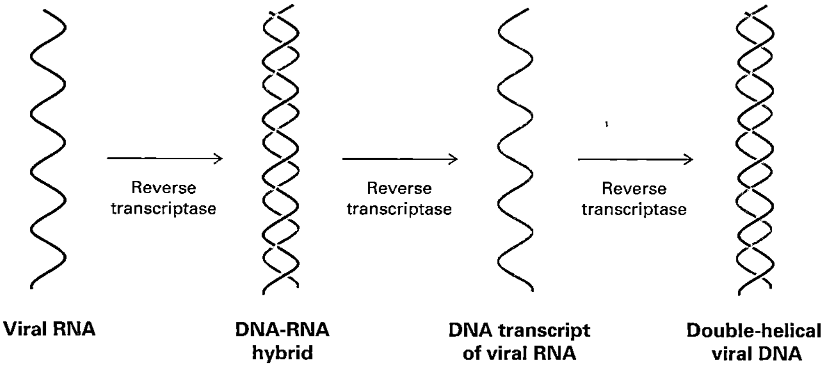

Figure 4-26
Information flows from RNA to DNA in the life cycle of retroviruses. DNA complementary to viral RNA is synthesized by reverse transcriptase, an enzyme brought into the cell by the infecting virus particle. Reverse transcriptase catalyzes the synthesis of (−) DNA, the digestion of (+) RNA, and the subsequent synthesis of (+) DNA.

SUMMARY

DNA is the molecule of heredity in all prokaryotic and eukaryotic organisms. In viruses, the genetic material is either DNA or RNA. All cellular DNA consists of two very long, helical polynucleotide chains coiled around a common axis. The two strands of the double helix run in opposite directions. The sugar-phosphate backbone of each strand is on the outside of the double helix, whereas the purine and pyrimidine bases are inside. The two chains are held together by hydrogen bonds between pairs of bases. Adenine (A) is always paired with thymine (T), and guanine (G) is always paired with cytosine (C). Hence, one strand of a double helix is the complement of the other. Genetic information is encoded in the precise sequence of bases along a strand. Most DNA molecules are circular. The axis of the double helix of a circular DNA can itself be coiled to form a superhelix. Supercoiled DNA is more compact than is relaxed DNA.

In the replication of DNA, the two strands of a double helix unwind and separate as new chains are synthesized. Each parent strand acts as a template for the formation of a new complementary strand. Thus, the replication of DNA is semiconservative—each daughter molecule receives one strand from the parent DNA molecule. The replication of DNA is a complex process carried out by many proteins, including several DNA polymerases. The activated precursors in the synthesis of DNA are the four deoxyribonucleoside 5′-triphosphates. The new strand is synthesized in the 5′ → 3′ direction by a nucleophilic attack by the 3′-hydroxyl terminus of the primer strand on the innermost phosphorus atom of the incoming deoxyribonucleoside triphosphate. Most important, DNA polymerases catalyze the formation of a phosphodiester bond only if the base on the incoming nucleotide is complementary to the base on the template strand. In other words, DNA polymerases are template-directed enzymes.

Some viruses have single-stranded DNA during a part of their life cycle. The DNA in ϕX174, a small virus that infects *E. coli,* is single-stranded. In the infected host, however, a complementary strand is made, to form a double-helical replicative intermediate. The genes of some viruses, such as tobacco mosaic virus, are made of single-stranded RNA. An RNA-directed RNA polymerase mediates the replication of this viral RNA. Retroviruses, exemplified by RNA tumor viruses and HIV-1, have an unusual genetic strategy: information flows from RNA to DNA, the reverse of normal. Their single-stranded RNA is transcribed into double-stranded DNA by reverse transcriptase, an RNA-directed DNA polymerase.

SELECTED READINGS

Where to start

Felsenfeld, G., 1985. DNA. *Sci. Amer.* 253(4):58–67. [Also printed in *The Molecules of Life,* an excellent series of articles. W.H. Freeman, 1985.]

Darnell, J.E., Jr., 1985. RNA. *Sci. Amer.* 253(4):26–36. [Also printed in *The Molecules of Life.*]

Dickerson, R.E., 1983. The DNA helix and how it is read. *Sci. Amer.* 249(6):94–111.

Crick, F.H.C., 1954. The structure of the hereditary material. *Sci. Amer.* 191(4):54–61. [A wonderful account of the discovery of the double helix. Reprinted in *Scientific American: Science in the 20th Century,* 1991.]

Discovery of the major concepts

Avery, O.T., MacLeod, C.M., and McCarty, M., 1944. Studies on the chemical nature of the substance inducing transformation of pneumococcal types. Induction of transformation by a deoxyribonucleic acid fraction isolated from Pneumococcus Type III. *J. Exp. Med.* 79:137–158.

Hershey, A.D., and Chase, M., 1952. Independent functions of viral protein and nucleic acid in growth of bacteriophage. *J. Gen. Physiol.* 36:39–56.

Chargaff, E., 1950. Chemical specificity of nucleic acids and mechanism of their enzymatic degradation. *Experentia* 6:201–209.

Watson, J.D., and Crick, F.H.C., 1953*a*. Molecular structure of nucleic acid. A structure for deoxyribose nucleic acid. *Nature* 171:737–738.

Watson, J.D., and Crick, F.H.C., 1953*b*. Genetic implications of the structure of deoxyribonucleic acid. *Nature* 171:964–967.

Kornberg, A., 1960. Biologic synthesis of deoxyribonucleic acid. *Science* 131:1503–1508.

Meselson, M., and Stahl, F.W., 1958. The replication of DNA in *Escherichia coli. Proc. Nat. Acad. Sci.* 44:671–682.

Taylor, J.H. (ed.), 1965. *Selected Papers on Molecular Genetics.* Academic Press. [Contains the classic papers listed above.]

Zubay, G.L. (ed.), 1968. *Papers in Biochemical Genetics.* Holt, Rinehart, and Winston.

DNA structure

Saenger, W., 1984. *Principles of Nucleic Acid Structure.* Springer-Verlag. [An outstanding advanced account of the three-dimensional structure of nucleotides, DNA, and RNA. Contains many excellent illustrations.]

Dickerson, R.E., Drew, H.R., Conner, B.N., Wing, R.M., Fratini, A.V., and Kopka, M.L., 1982. The anatomy of A-, B-, and Z-DNA. *Science* 216:475–485.

Cantor, C.R., and Schimmel, P.R., 1980. *Biophysical Chemistry* (3 vols.). W.H. Freeman. [Chapters 3 and 6 (in Part I) and Chapters 22, 23, and 24 (in Part III) give an excellent account of the conformation of nucleic acids.]

DNA replication

Kornberg, A., and Baker, T.A., 1992. *DNA Replication* (2nd ed.). W.H. Freeman. [An outstanding and highly readable book.]

Reminiscences and historical accounts

Watson, J.D., 1968. *The Double Helix.* Atheneum. [A lively, personal account of the discovery of the structure of DNA and its biological implications.]

McCarty, M., 1985. *The Transforming Principle: Discovering that Genes Are Made of DNA.* Norton. [A warm and lucid account of one of the major discoveries of this century by a sensitive participant.]

Cairns, J., Stent, G.S., and Watson, J.D. (eds.), 1966. *Phage and the Origins of Molecular Biology.* Cold Spring Harbor Laboratory. [A fascinating collection of reminiscences by some of the architects of molecular biology.]

Olby, R., 1974. *The Path to the Double Helix.* University of Washington Press.

Portugal, F.H., and Cohen, J.S., 1977. *A Century of DNA: A History of the Discovery of the Structure and Function of the Genetic Substance.* MIT Press.

Judson, H., 1979. *The Eighth Day of Creation.* Simon and Schuster.

PROBLEMS

1. *Complements.* Write the complementary sequence (in the standard 5′ → 3′ notation) for

   (a) GATCAA. (c) ACGCGT.
   (b) TCGAAC. (d) TACCAT.

2. *Compositional constraint.* The composition (in mole-fraction units) of one of the strands of a double-helical DNA is [A] = 0.30 and [G] = 0.24.
   (a) What can you say about [T] and [C] for the same strand?
   (b) What can you say about [A], [G], [T], and [C] of the complementary strand?

3. *Lost DNA.* The DNA of a deletion mutant of λ bacteriophage has a length of 15 μm instead of 17 μm. How many base pairs are missing from this mutant?

4. *An unseen pattern.* What result would Meselson and Stahl have obtained if the replication of DNA were conservative (i.e., the parental double helix stayed together)? Give the expected distribution of DNA molecules after 1.0 and 2.0 generations for conservative replication.

5. *A fortunate circumstance.* Griffith used heat-killed S pneumococci to transform R mutants. Studies years later showed that double-stranded DNA is needed for efficient transformation and that high temperatures melt the DNA double helix. Why were Griffith's experiments nevertheless successful?

6. *A matter of competence.* Strains of *Bacillus subtilis* that can be transformed by foreign DNA are termed *competent.* Others, termed *noncompetent,* are insusceptible to transformation. How might these strains differ from each other?

7. *A propitious choice.* Bacteriophage M13 infects *E. coli* differently from the way bacteriophage T2 does. The M13 protein coat is removed in the inner membrane of the bacterial cell, where it is sequestered and subsequently used for the envelopment of progeny DNA. Why would M13 have been much less suitable than T2 was for the experiments carried out by Hershey and Chase?

8. *Tagging DNA.*
   (a) Suppose that you want to radioactively label DNA but not RNA in dividing and growing bacterial cells. Which radioactive molecule would you add to the culture medium?
   (b) Suppose that you want to prepare DNA in which the backbone phosphorus atoms are uniformly labeled with ^{32}P. Which precursors should be added to a solution containing DNA polymerase I and primed template DNA? Specify the position of radioactive atoms in these precursors.

9. *Finding a template.* A solution contains DNA polymerase I and the Mg^{2+} salts of dATP, dGTP, dCTP, and dTTP. The DNA molecules listed below are added to aliquots of this solution. Which of them would lead to DNA synthesis?
   (a) A single-stranded closed circle containing 1000 nucleotide units.
   (b) A double-stranded closed circle containing 1000 nucleotide pairs.
   (c) A single-stranded closed circle of 1000 nucleotides base paired to a linear strand of 500 nucleotides with a free 3′-OH terminus.
   (d) A double-stranded linear molecule of 1000 nucleotide pairs with a free 3′-OH at each end.

10. *The right start.* Suppose that you want to assay reverse transcriptase activity. If polyriboadenylate is the template in the assay, what should you use as the primer? Which radioactive nucleotide should you use to follow chain elongation?

11. *Essential degradation.* Reverse transcriptase has ribonuclease activity as well as polymerase activity. What is the role of its ribonuclease activity?

12. *Virus hunting.* You have purified a virus that infects turnip leaves. Treatment of a sample with phenol removes viral proteins. Application of the residual material to scraped leaves results in the formation of progeny virus particles. You infer that the infectious substance is a nucleic acid. Propose a simple and highly sensitive means of determining whether the infectious nucleic acid is DNA or RNA.

13. *Mutagenic consequences.* Spontaneous deamination of cytosine bases in DNA occurs at low but measurable frequency. Cytosine is converted into uracil by loss of its amino group. After this conversion, which base pair occupies this position in each of the daughter strands resulting from one round of replication? Two rounds of replication?

14. *Eons ago.* The atmosphere of the primitive earth before the emergence of life contained N_2, NH_3, H_2, HCN, CO, and H_2O. Which of these compounds is the most likely precursor of most of the atoms in adenine? Why?

15. *Information content.*
    (a) How many different 8-mer sequences of DNA are there? [Hint: There are 16 possible dinucleotides and 64 possible trinucleotides.]
    (b) How many bits of information are stored in an 8-mer DNA sequence? In the *E. coli* genome? In the human genome?
    (c) Compare each of these values with the amount of information that can be stored on a personal computer diskette. A byte is equal to 8 bits.

CHAPTER

5

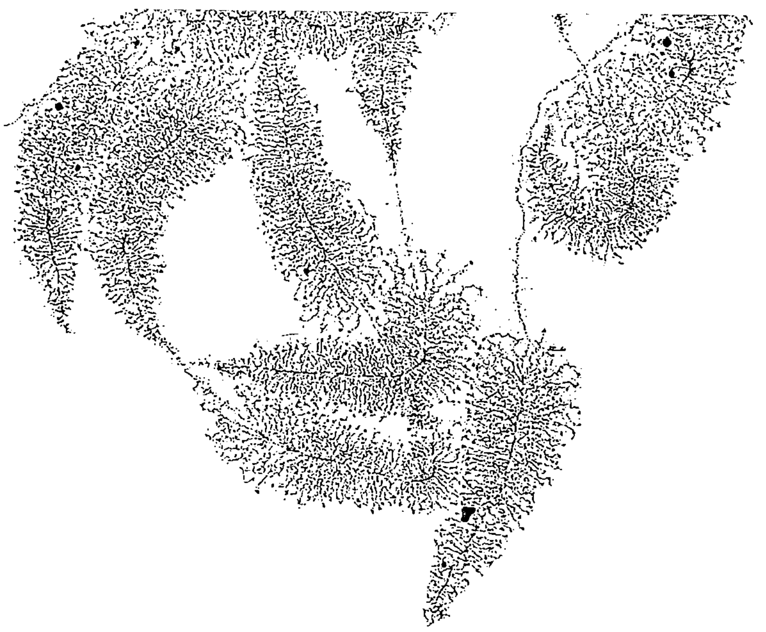

Flow of Genetic Information

We turn now from the storage and transmission of genetic information to its expression. Genes specify the kinds of proteins that are made by cells. However, DNA is not the direct template for protein synthesis. Rather, the templates for protein synthesis are RNA (ribonucleic acid) molecules. This chapter begins with an account of the discovery that a class of RNA molecules called *messenger RNAs (mRNAs)* are the information-carrying intermediates in protein synthesis. Other RNA molecules, such as transfer RNA (tRNA) and ribosomal RNA (rRNA) are part of the protein-synthesizing machinery. All forms of cellular RNA are synthesized by RNA polymerases that take instructions from DNA templates. This process of *transcription* is followed by *translation,* the synthesis of proteins according to instructions given by mRNA templates. Thus, the flow of genetic information in normal cells is

$$\text{DNA} \xrightarrow{\text{Transcription}} \text{RNA} \xrightarrow{\text{Translation}} \text{protein}$$

This brings us to the genetic code, the relation between the sequence of bases in DNA (or its mRNA transcript) and the sequence of amino acids in a protein. The code, which is nearly the same in all organisms, is beautiful in its simplicity. A sequence of three bases, called a *codon,* specifies an amino acid. Codons in mRNA are read sequentially by tRNA molecules, which serve as adaptors in protein synthesis. Protein synthesis takes place on ribosomes, which are complex assemblies of rRNAs and more than 50 kinds of proteins. Newly synthesized proteins contain signals that enable them to be targeted to specific destinations. The last theme to be considered is the interrupted character of most eukaryotic genes, which are mosaics of nucleic acid sequences called *introns* and *exons.* Both are transcribed, but introns are cut out of newly synthesized RNA molecules, leaving mature RNA molecules with continuous exons. The existence of introns and exons has profound implications for evolution.

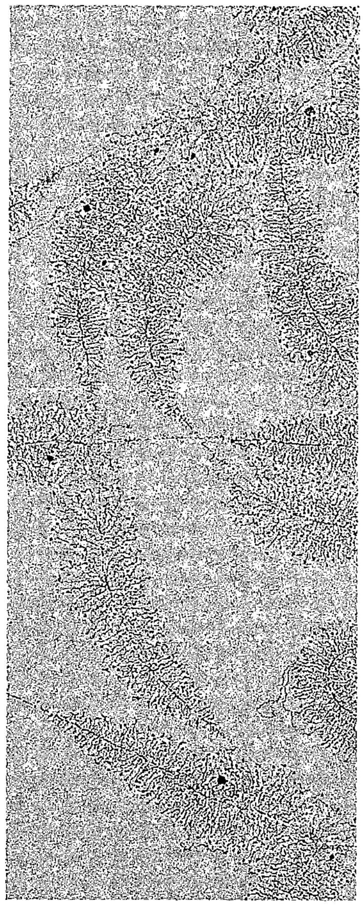

Opening Image: *Transcription of DNA to form precursors of ribosomal RNA. This electron micrograph shows a tandem array of nascent RNA molecules emerging from a thread of DNA. [From O.L. Miller, Jr., and B.R. Beatty. Portrait of a gene. J. Cell Physiol. 74 (supp. 1, 1969):225.]*

SEVERAL KINDS OF RNA PLAY KEY ROLES IN GENE EXPRESSION

RNA is a long, unbranched macromolecule consisting of nucleotides joined by 3′ → 5′ phosphodiester bonds (see Figure 4-24). As the name indicates, the sugar unit in RNA is ribose. The four major bases in RNA are adenine (A), uracil (U), guanine (G), and cytosine (C). Adenine can

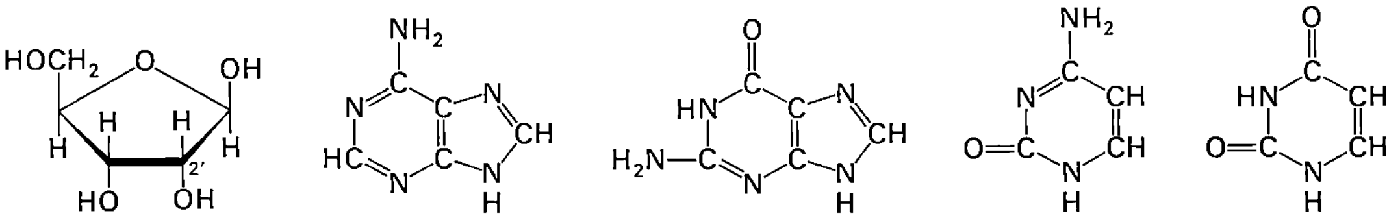

Ribose **Adenine** **Guanine** **Cytosine** **Uracil**

pair with uracil, and guanine with cytosine. The number of nucleotides in RNA ranges from as few as seventy-five to many thousands. *RNA molecules are usually single-stranded,* except in some viruses. Consequently, an RNA molecule need not have complementary base ratios. In fact, the proportion of adenine differs from that of uracil, and the proportion of guanine differs from that of cytosine, in most RNA molecules. However, *RNA molecules do contain regions of double-helical structure that are produced by the formation of hairpin loops* (Figure 5-1). In these regions, A pairs with U, and G pairs with C. The base pairing in RNA hairpins is frequently imperfect. G can also form a base pair with U, but it is less strong than the G·C base pair (p. 887). Some of the apposing bases may not be complementary at all, and one or more bases along a single strand may be looped out to facilitate the pairing of the others. The proportion of helical regions in different kinds of RNA varies over a wide range; a value of 50% is typical.

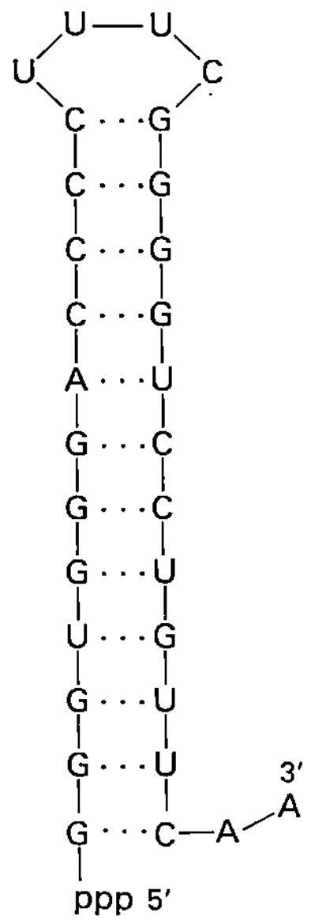

Figure 5-1
RNA can fold back on itself to form double-helical regions.

Cells contain several kinds of RNA (Table 5-1). *Messenger RNA (mRNA)* is the template for protein synthesis. An mRNA molecule is produced for each gene or group of genes that is to be expressed. Consequently, mRNA is a very heterogeneous class of molecules. In *Escherichia coli,* the average length of an mRNA molecule is about 1.2 kb. *Transfer RNA (tRNA)* carries amino acids in an activated form to the ribosome for peptide-bond formation, in a sequence determined by the mRNA template. There is at least 1 kind of tRNA for each of the 20 amino acids. Transfer RNA consists

Table 5-1
RNA molecules in *E. coli*

Type	*Relative amount (%)*	*Sedimentation coefficient (S)*	*Mass (kd)*	*Number of nucleotides*
Ribosomal RNA (rRNA)	80	23	1.2×10^3	3700
		16	0.55×10^3	1700
		5	3.6×10^1	120
Transfer RNA (tRNA)	15	4	2.5×10^1	75
Messenger RNA (mRNA)	5	Heterogeneous		

of about 75 nucleotides (having a mass of about 25 kd), which makes it the smallest of the RNA molecules. *Ribosomal RNA* (*rRNA*), the major component of ribosomes, plays both a catalytic and structural role in protein synthesis (p. 889). In *E. coli,* there are three kinds of rRNA, called *23S, 16S,* and *5S RNA* because of their sedimentation behavior. One molecule of each of these species of rRNA is present in each ribosome. Ribosomal RNA is the most abundant of the three types of RNA. Transfer RNA comes next, followed by messenger RNA, which constitutes only 5% of the total RNA. Eukaryotic cells contain additional small RNA molecules. *Small nuclear RNA* (*snRNA*) molecules, for example, participate in the splicing of RNA exons. A small RNA molecule in the cytosol plays a role in the targeting of newly synthesized proteins to intracellular compartments and extracellular destinations.

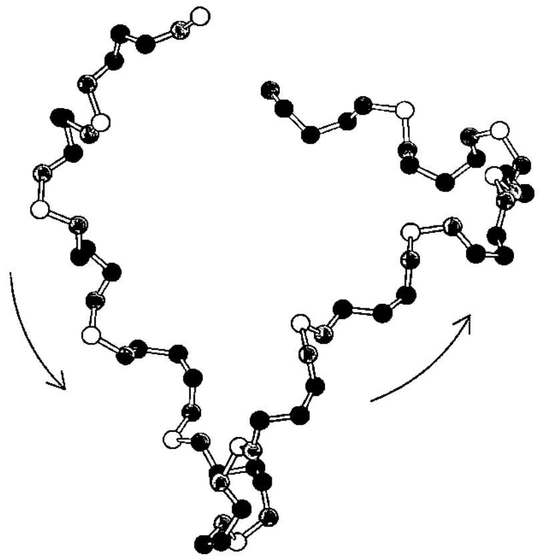

Figure 5-2
Course of the sugar-phosphate backbone of an RNA hairpin. The atoms shown are the phosphorus (yellow) and bridging oxygens (red) of phosphodiester bonds, and the 3′, 4′, and 5′ carbons (black) of ribose. The green arrows mark the 5′ → 3′ direction of the chain. [Drawn from 6tna.pdb.]

DISCOVERY OF MESSENGER RNA, THE INFORMATION CARRIER IN PROTEIN SYNTHESIS

The concept of mRNA was formulated by Francois Jacob and Jacques Monod in a classic paper published in 1961. Because proteins are synthesized in the cytoplasm rather than in the nucleus of eukaryotic cells, it was evident to them that genes must specify a chemical intermediate, which they called the "structural messenger." What is the nature of this carrier of genetic information? An important clue came from their studies of the control of protein synthesis in *E. coli* (Chapter 36). Certain enzymes in *E. coli,* such as those that participate in the uptake and utilization of lactose, are inducible—that is, the amount of these enzymes increases more than a thousandfold if lactose or an analog of this sugar is present. The kinetics of induction were very revealing. The addition of an inducer elicited maximal synthesis of lactose-utilizing proteins within a few minutes. Furthermore, removal of inducer resulted in the cessation of synthesis in an equally short time. Hence, Jacob and Monod surmised that *the messenger must be a very short-lived intermediate* with the following properties:

1. The messenger should be a polynucleotide.

2. The base composition of the messenger should reflect the base composition of the DNA that specifies it.

3. The messenger should be very heterogeneous in size because genes (or groups of genes) vary in length. They correctly assumed that three nucleotides code for one amino acid and calculated that the molecular weight of a messenger should be at least a half million.

4. The messenger should be transiently associated with ribosomes, the sites of protein synthesis.

5. The messenger should be synthesized and degraded very rapidly.

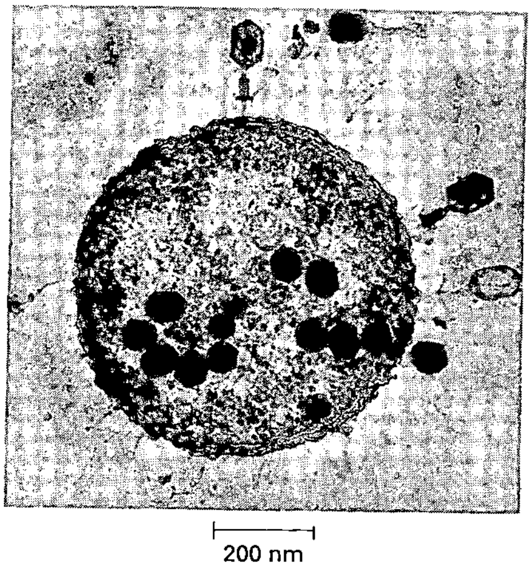

Figure 5-3
Electron micrograph of an *E. coli* cell infected by T2 viruses. [Courtesy of Dr. Lee Simon.]

Their hypothesis was tested in a choice system—*E. coli* infected with T2 bacteriophage (Figure 5-3). Nearly all proteins made by a bacterium after infection are genetically determined by the phage it harbors. The synthesis of these proteins is not accompanied by an overall synthesis of RNA. In particular, neither ribosomal RNA nor transfer RNA is synthesized after infection. However, a minor RNA fraction with a short half-life does appear soon after infection. In fact, the nucleotide composition of this RNA is like that of the phage DNA.

How does infection lead to the synthesis of a new set of proteins? One possibility a priori is that phage DNA specifies a new set of ribosomes. In

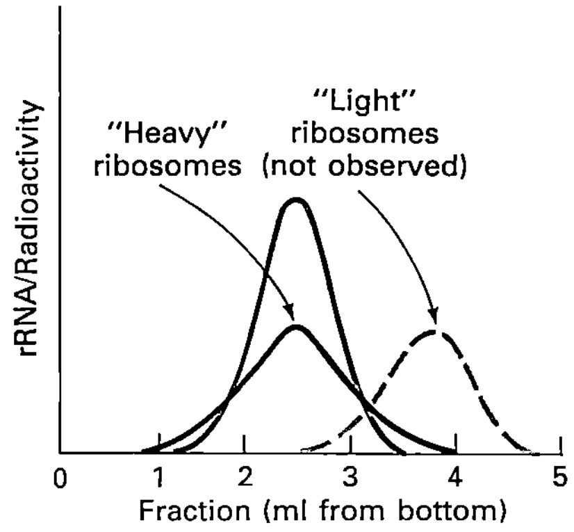

Figure 5-4
Density-gradient centrifugation of ribosomes of normal and T2 phage-infected bacteria. ^{32}P-labeled RNA (red) synthesized *after* infection banded with ribosomes formed before infection ("heavy" ribosomes). Ribosomes were not synthesized after infection, as evidenced by the absence of "light" ribosomes.

this model, genes control the synthesis of specialized ribosomes, and each ribosome can make only one kind of protein. An alternative model, the one proposed by Jacob and Monod, is that ribosomes are nonspecialized structures that receive coding information from the gene in the form of an unstable messenger RNA.

Sydney Brenner, Francois Jacob, and Matthew Meselson crafted an incisive experiment to determine *whether new ribosomes are synthesized after infection or whether new RNA joins preexisting ribosomes.* Bacteria were grown in a medium containing heavy isotopes (^{15}N and ^{13}C), infected with phage, and then immediately transferred to a medium containing light isotopes (^{14}N and ^{12}C). Ribosomes synthesized before and after infection could be separated by density-gradient centrifugation because their densities would differ ("heavy" and "light," respectively). Furthermore, any new RNA was to be detected through the introduction of the radioisotope ^{32}P- or ^{14}C-uracil; new protein, by ^{35}S.

This elegant isotope-labeling experiment (Figure 5-4) showed that

1. *Ribosomes are not synthesized after infection,* as evidenced by the absence of "light" ribosomes.

2. RNA *is* synthesized after infection. Most of the radioactively labeled RNA emerges in the "heavy" ribosome peak. Thus, *most of the new RNA is associated with preexisting ribosomes.* Additional experiments showed that this new RNA turns over rapidly during the growth of phage.

3. The radioisotope ^{35}S appears transiently in the "heavy" ribosome peak, which indicates that *new proteins are synthesized in preexisting ribosomes.*

Hence, *ribosomes are nonspecialized structures which synthesize, at a given time, the protein dictated by the messenger they happen to contain.* Studies of uninfected bacterial cells gave the same result. The bold hypothesis that *messenger RNA (mRNA) is the information carrier between gene and protein* became an established fact in a very short time.

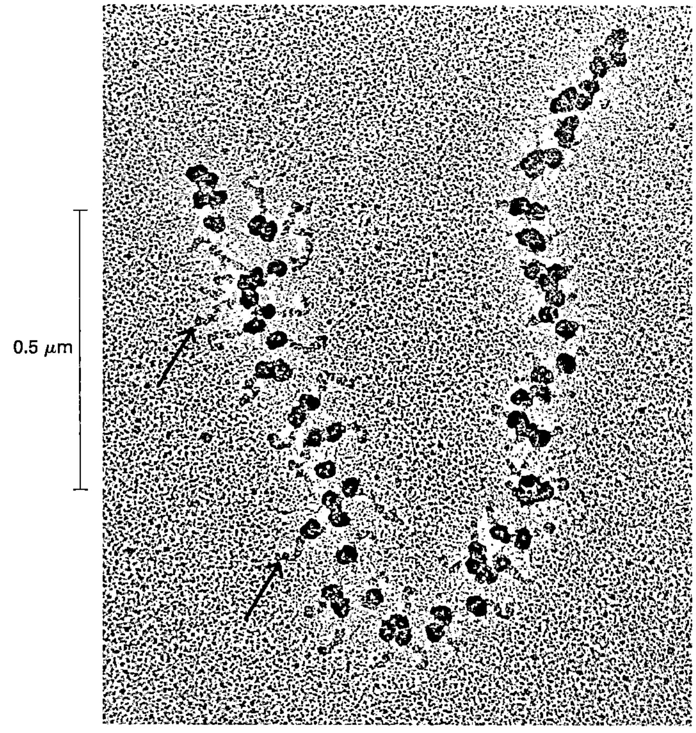

Figure 5-5
Translation of an mRNA. This electron micrograph shows a series of ribosomes on an mRNA molecule. Nascent polypeptide chains (marked by arrows) emerge from the ribosomes. [After C. Francke, J.-E. Edström, A.W. McDowall, and O.L. Miller, Jr. *EMBO J.* 1(1982):59.]

HYBRIDIZATION STUDIES SHOWED THAT MESSENGER RNA IS COMPLEMENTARY TO ITS DNA TEMPLATE

In 1961, Sol Spiegelman developed a new technique called *hybridization* to answer the following question: Is the RNA synthesized after infection with T2 phage complementary to T2 DNA? It was known from the work of Julius Marmur and Paul Doty that heating double-helical DNA above its melting temperature results in the formation of single strands. If the mixture is cooled slowly, these strands reassociate to form a double-helical structure with biological activity. Furthermore, double helices are formed only from strands derived from the same or from closely related organisms. This observation suggested to Spiegelman that a double-stranded DNA–RNA hybrid might be formed from a mixture of single-stranded DNA and RNA if their base sequences were complementary (Figure 5-6). This idea was tested in the following way:

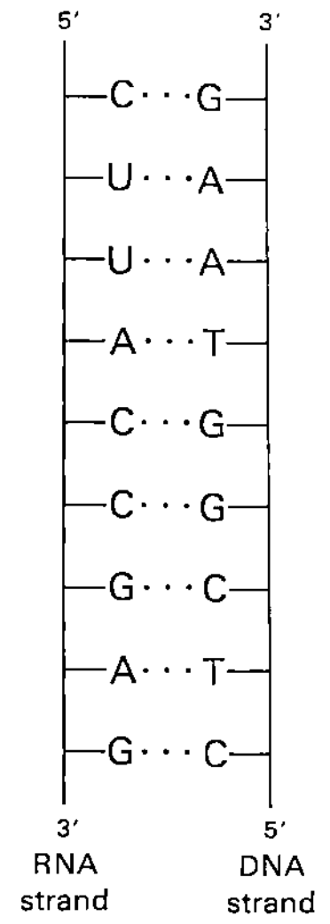

Figure 5-6
An RNA–DNA hybrid can be formed if the RNA and DNA have complementary sequences.

1. A sample of T2 mRNA (the RNA synthesized after infection of *E. coli* with T2 phage) was prepared with a ^{32}P label. T2 DNA labeled with ^{3}H was prepared separately.

2. A mixture of the T2 mRNA and T2 DNA was heated to 100°C, which melted the double-helical DNA into single strands. This solution of single-stranded RNA and DNA was slowly cooled to room temperature.

3. The cooled mixture was analyzed by density-gradient centrifugation. The samples were centrifuged for several days in swinging-bucket rotors. The plastic sample tubes were then punctured at the bottom, and drops were collected for analysis.

Analysis of the density gradient after centrifugation revealed that *T2 mRNA formed a hybrid with T2 DNA* (Figure 5-7). In contrast, T2 mRNA did

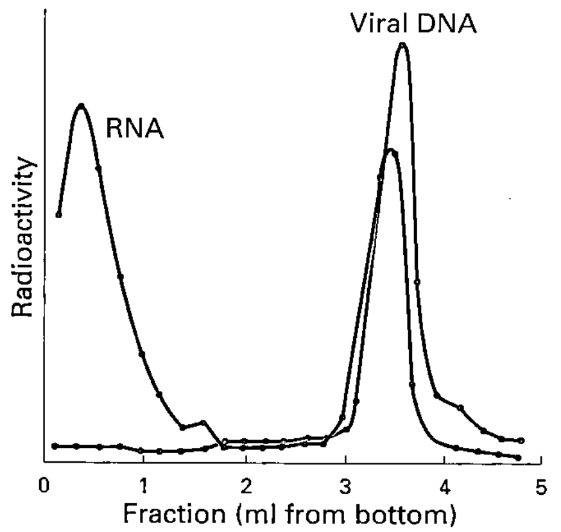

Figure 5-7
The RNA produced after *E. coli* has been infected with T2 phage is complementary to the viral DNA. In this hybridization experiment, RNA was labeled with ^{32}P, whereas T2 DNA was labeled with ^{3}H. The distribution of radioactivity in a cesium chloride density gradient shows that much of the RNA synthesized after infection bands with the T2 DNA. The mRNA–DNA double-stranded hybrid that is formed is slightly denser than double-helical DNA. [After S. Spiegelman. Hybrid nucleic acids. Copyright © 1964 by Scientific American, Inc. All rights reserved.]

not hybridize with DNA derived from a variety of bacteria and unrelated viruses, even if their base ratios were like those of T2 DNA. Subsequent experiments showed that the mRNA fraction of uninfected cells hybridized with the DNA derived from that particular organism but not from unrelated ones. These incisive experiments revealed that *the base sequence of mRNA is complementary to that of its DNA template. Furthermore, a powerful tool was developed for tracing the flow of genetic information in cells and for determining whether two nucleic acid molecules are similar.*

ALL CELLULAR RNA IS SYNTHESIZED BY RNA POLYMERASES

The concept of mRNA stimulated the search for an enzyme that synthesizes RNA according to instructions given by a DNA template. In 1960, Jerard Hurwitz and Samuel Weiss independently discovered such an enzyme, which they named *RNA polymerase.* The enzyme from *E. coli* requires the following components for the synthesis of RNA:

1. *Template.* The preferred template is *double-stranded DNA.* Single-stranded DNA can also serve as a template. RNA, whether single- or double-stranded, is not an effective template, nor are RNA–DNA hybrids.

2. *Activated precursors.* All four *ribonucleoside triphosphates*—ATP, GTP, UTP, and CTP—are required.

3. *Divalent metal ion.* Mg^{2+} or Mn^{2+} are effective. Mg^{2+} meets this requirement in vivo.

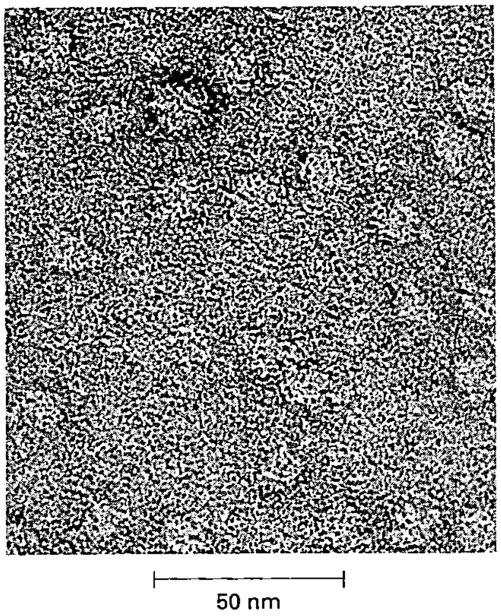

Figure 5-8
Electron micrograph of RNA polymerase from *E. coli.* [Courtesy of Dr. Robley Williams and Dr. Michael Chamberlin.]

RNA polymerase catalyzes the initiation and elongation of RNA chains. The reaction catalyzed by this enzyme is

$$(\text{RNA})_{n \text{ residues}} + \text{ribonucleoside triphosphate} \rightleftharpoons (\text{RNA})_{n+1 \text{ residues}} + \text{PP}_i$$

The synthesis of RNA is like that of DNA in several respects (Figure 5-9). First, the direction of synthesis is 5′ → 3′. Second, the mechanism of elongation is similar: the 3′-OH group at the terminus of the growing chain makes a nucleophilic attack on the innermost phosphate of the incoming nucleoside triphosphate. Third, the synthesis is driven forward by the hydrolysis of pyrophosphate. In contrast with DNA polymerase, RNA polymerase does not require a primer. Another difference is that the DNA template is fully conserved in RNA synthesis, whereas it is semiconserved in DNA synthesis. Also, RNA polymerase lacks the nuclease capability used by DNA polymerase to excise mismatched nucleotides.

All three types of cellular RNA—mRNA, tRNA, and rRNA—are synthesized in *E. coli* by the same RNA polymerase according to instructions given by a DNA template. In mammalian cells, there is a division of labor among several different kinds of RNA polymerases. We shall return to these RNA polymerases in Chapter 33.

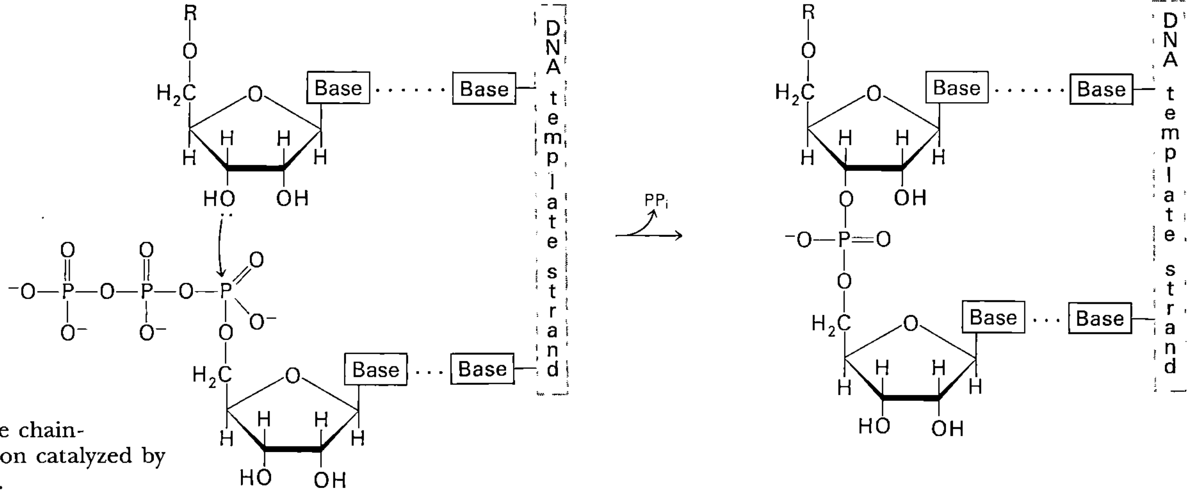

Figure 5-9
Mechanism of the chain-elongation reaction catalyzed by RNA polymerase.

RNA POLYMERASE TAKES INSTRUCTIONS FROM A DNA TEMPLATE

RNA polymerase, like the DNA polymerases described in the preceding chapter, takes instructions from a DNA template. The earliest evidence was the finding that the *base composition* of newly synthesized RNA is the complement of that of the DNA template strand, as exemplified by the RNA synthesized from a template of single-stranded ϕX174 DNA (Table 5-2). Hybridization experiments also revealed that RNA synthesized by RNA polymerase is complementary to its DNA template. The strongest evidence for the fidelity of transcription came from base-sequence studies showing that the RNA sequence is the precise complement of the DNA template sequence (Figure 5-10).

Table 5-2
Base composition (percentage) of RNA synthesized from a viral DNA template

DNA template (plus strand of ϕX174)		*RNA product*	
A	25	25	U
T	33	32	A
G	24	23	C
C	18	20	G

5′-GCGGCGACGCGCAGUUAAUCCCACAGCCGCCAGUUCCGCUGGCGGCAUUUU-3′ **mRNA**

3′-CGCCGCTGCGCGTCAATTAGGGTGTCGGCGGTCAAGGCGACCGCCGTAAAA-5′ **Template strand of DNA**

5′-GCGGCGACGCGCAGTTAATCCCACAGCCGCCAGTTCCGCTGGCGGCATTTT-3′ **Coding strand of DNA**

Figure 5-10
The base sequence of mRNA (red) is the complement of that of the DNA template strand (green). The sequence shown here is from the tryptophan operon, a segment of DNA containing the genes for five enzymes that catalyze the synthesis of tryptophan. The other strand of DNA (black) is called the coding strand because it has the same sequence as the RNA transcript except for T in place of U.

TRANSCRIPTION BEGINS NEAR PROMOTER SITES AND ENDS AT TERMINATOR SITES

DNA templates contain regions called *promoter sites* that specifically bind RNA polymerase and determine where transcription begins. In *bacteria,* two sequences on the 5′ (upstream) side of the first nucleotide to be transcribed are important (Figure 5-11A). One of them, called the *Pribnow box,* has the consensus sequence TATAAT and is centered at −10 (10 nucleotides on the 5′ side of the first one transcribed, which is denoted by +1). The other, called the *−35 region,* has the consensus sequence TTGACA. The first nucleotide transcribed is usually a purine.

Consensus sequence—
The base sequences of promoter sites are not all identical. However, they do possess common features, which can be represented by an idealized consensus sequence. Each base in the consensus sequence TATAAT is found in a majority of prokaryotic promoters. Nearly all promoter sequences differ from this consensus sequence at only two or fewer bases.

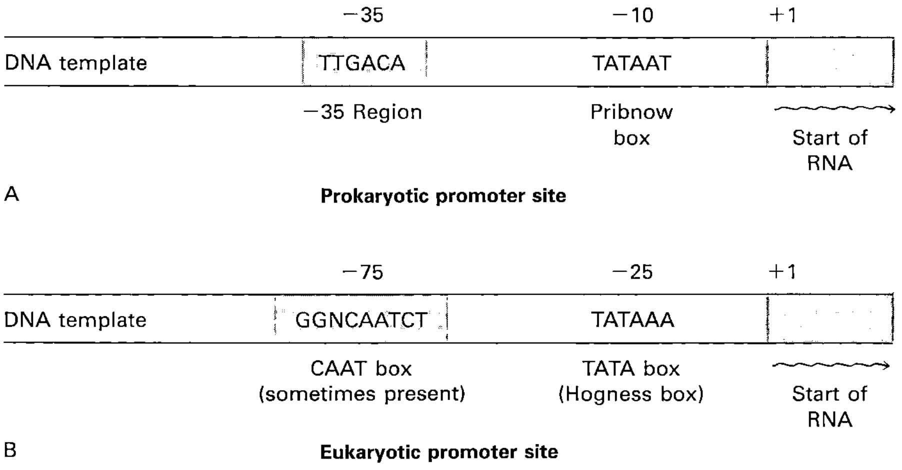

Figure 5-11
Promoter sites for transcription in (A) prokaryotes and (B) eukaryotes. Consensus sequences are shown. The first nucleotide to be transcribed is numbered +1. The adjacent nucleotide on the 5′ side is numbered −1. The sequences shown are those of the coding strand of DNA.

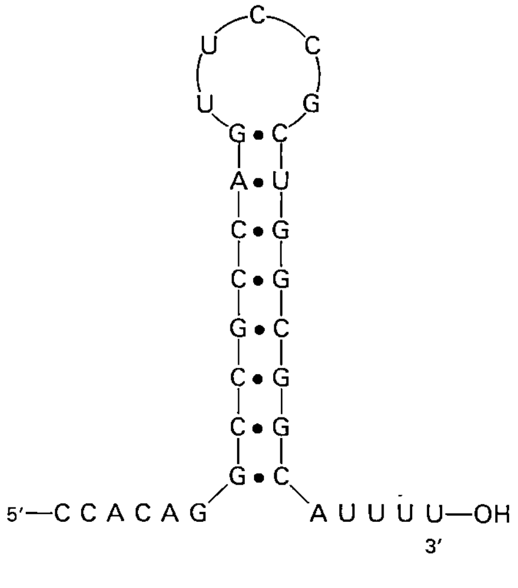

Figure 5-12
Base sequence of the 3′ end of an mRNA transcript in *E. coli*. A stable hairpin structure is followed by a sequence of U residues.

Eukaryotic genes encoding proteins have promoter sites with a TATAAA consensus sequence centered at about −25 (Figure 5-11B). This *TATA box* (also called the *Hogness box*) is like the prokaryotic Pribnow box except that it is farther upstream. Many eukaryotic promoters also have a *CAAT box* with a GGN*CAAT*CT consensus sequence centered at about −75. Transcription of eukaryotic genes is further stimulated by *enhancer sequences,* which can be quite distant (up to several kilobases) from the start site, on either its 5′ or its 3′ side.

RNA polymerase proceeds along the DNA template and transcribes one of its strands until a terminator sequence is reached. This sequence codes for a termination signal, which in *E. coli* is a *base-paired hairpin* on the newly synthesized RNA molecule (Figure 5-12). This hairpin is formed by base pairing of self-complementary sequences that are rich in G and C. Nascent RNA spontaneously dissociates from RNA polymerase when this hairpin is followed by a string of U residues. Alternatively, RNA synthesis can be terminated by the action of *rho,* a protein. Less is known about the termination of transcription in eukaryotes. A more detailed discussion of the initiation and termination of transcription will be given in Chapter 33. The important point now is that *discrete start and stop signals for transcription are encoded in the DNA template.*

TRANSFER RNA IS THE ADAPTOR MOLECULE IN PROTEIN SYNTHESIS

We have seen that mRNA is the template for protein synthesis. How then does it direct amino acids to become joined in the correct sequence? In 1958, Francis Crick wrote

> One's first naive idea is that the RNA will take up a configuration capable of forming twenty different "cavities," one for the side chain of each of the twenty amino acids. If this were so, one might expect to be able to play the problem backwards—that is, to find the configuration of RNA by trying to form such cavities. All attempts to do this have failed, and on physical-chemical grounds the idea does not seem in the least plausible.

He observed that RNA does not have the knobby hydrophobic surfaces that could distinguish valine from leucine and isoleucine, nor does it have properly positioned charged groups to discriminate between positively and negatively charged amino acid side chains. Crick then proposed an entirely different mechanism for recognition of the mRNA template.

> RNA presents mainly a sequence of sites where hydrogen bonding could occur. One would expect, therefore, that whatever went onto the template in a *specific* way did so by forming hydrogen bonds. It is therefore a natural hypothesis that the amino acid is carried to the template by an adaptor molecule, and that the adaptor is the part which actually fits onto the RNA. In its simplest form one would require twenty adaptors, one for each amino acid.

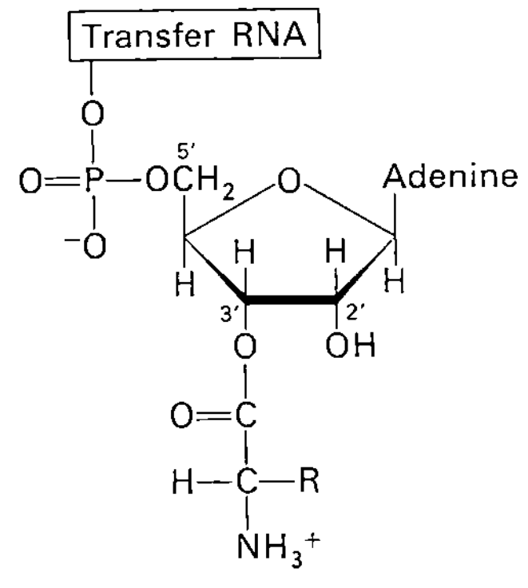

Figure 5-13
Mode of attachment of an amino acid (shown in red) to a tRNA molecule. The amino acid is esterified to the 3′-hydroxyl group of the terminal adenosine of tRNA.

This highly innovative hypothesis soon became an established fact. *The adaptor in protein synthesis is transfer RNA (tRNA).* The structure and reactions of these remarkable molecules will be considered in detail in Chapter 34. For the moment it suffices to note that tRNA contains an *amino acid attachment site* and a *template-recognition site.* A tRNA molecule carries a specific amino acid in an activated form to the site of protein synthesis. The carboxyl group of this amino acid is esterified to the 3′- or 2′-hydroxyl group of the ribose unit at the 3′ end of the tRNA chain (Figure

5-13). The esterified amino acid may migrate between the 2′- and 3′-hydroxyl groups during protein synthesis. The joining of an amino acid to a tRNA to form an *aminoacyl-tRNA* is catalyzed by a specific enzyme called an *aminoacyl-tRNA synthetase* (or *activating enzyme*). This esterification reaction is driven by ATP. There is at least one specific synthetase for each of the 20 amino acids. The template-recognition site on tRNA is a sequence of three bases called the *anticodon* (Figure 5-14). The anticodon on tRNA recognizes a complementary sequence of three bases on mRNA, called the *codon.*

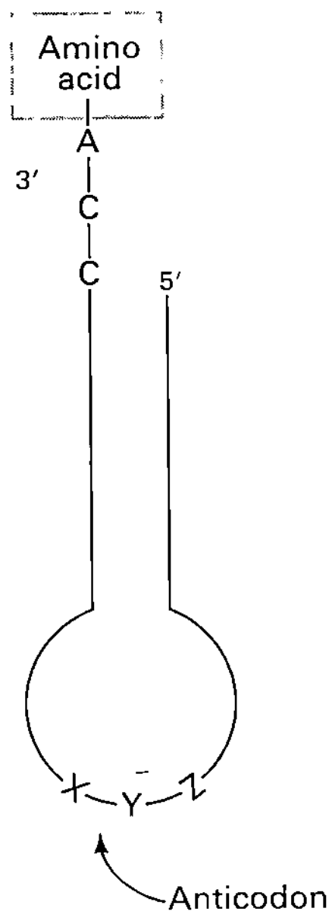

Figure 5-14
Symbolic diagram of an aminoacyl-tRNA showing the amino acid attachment site and the anticodon, which is the template-recognition site.

AMINO ACIDS ARE CODED BY GROUPS OF THREE BASES STARTING FROM A FIXED POINT

The *genetic code* is the relation between the sequence of bases in DNA (or its RNA transcripts) and the sequence of amino acids in proteins. Experiments by Francis Crick, Sydney Brenner, and others established the following features of the genetic code by 1961:

1. *What is the coding ratio?* A single-base code can specify only four kinds of amino acids because there are four kinds of bases in DNA. Sixteen kinds of amino acids can be specified by a two-base code ($4 \times 4 = 16$), whereas 64 kinds of amino acids can be determined by a three-base code ($4 \times 4 \times 4 = 64$). Proteins are built from a basic set of 20 amino acids, and so it was evident from this simple calculation that three or more bases are probably needed to specify one amino acid. Genetic experiments then showed that *an amino acid is in fact coded by a group of three bases.* This group of bases is called a *codon.*

2. *Is the code nonoverlapping or overlapping?* In a nonoverlapping triplet code, each group of three bases in a sequence ABCDEF . . . specifies only one amino acid—ABC specifies the first, DEF the second, and so forth—whereas in a completely overlapping triplet code, ABC specifies the first amino acid, BCD the second, CDE the third, and so forth.

These alternatives were distinguished by studies of the sequence of amino acids in mutants. Suppose that the base C is mutated to C′. In a nonoverlapping code, only amino acid 1 will be changed. In a completely overlapping code, amino acids 1, 2, and 3 will all be altered by a mutation of C to C′. Amino acid sequence studies of tobacco mosaic virus mutants and abnormal hemoglobins showed that alterations usually affected only a single amino acid. Hence, it was concluded that the *genetic code is nonoverlapping.*

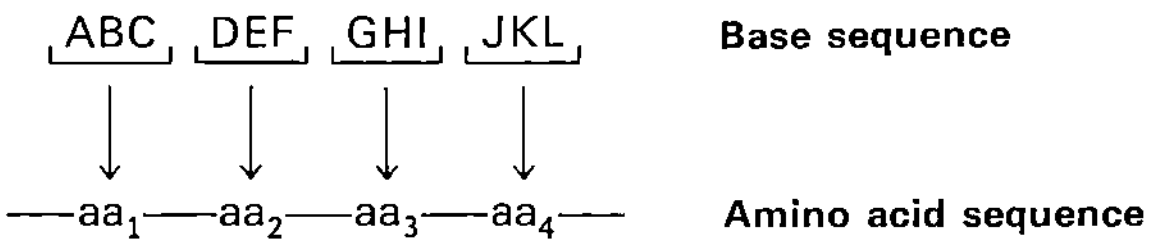

3. *How is the correct group of three bases read?* One possibility a priori is that one base (denoted as Q) serves as a "comma" between groups of three bases.

. . . QABCQDEFQGHIQ JKLQ . . .

This turned out not to be the case. Rather, *the sequence of bases is read sequentially from a fixed starting point.* There are no commas.

Start
↓
ABC ¦ DEF ¦ GHI ¦ JKL ¦ MNO

$aa_1—aa_2—aa_3—aa_4—aa_5$

Suppose that a mutation deletes base G.

Start ↓ G (deleted)

ABC ¦ DEF ¦ HIJ ¦ KLM ¦ NOP

$aa_1—aa_2—aa_3—aa_4—aa_5$

Normal (aa1–aa2) Altered (aa3–aa5)

The first two amino acids in the resulting polypeptide chain will be normal, but the rest of the base sequence will be read incorrectly because *the reading frame has been shifted* by the deletion of G. Suppose instead that a base Z has been added between F and G.

Start ↓ Insertion

ABC ¦ DEF ¦ ZGH ¦ IJK ¦ LMN

$aa_1—aa_2—aa_3—aa_4—aa_5$

Normal (aa1–aa2) Altered (aa3–aa5)

This addition also disrupts the reading frame starting at the codon for amino acid 3. In fact, genetic studies of addition and deletion mutants revealed many of the features of the genetic code.

4. As mentioned earlier, there are 64 possible base triplets and 20 amino acids. *Is there just 1 triplet for each of the 20 amino acids, or are some amino acids coded by more than 1 triplet?* Genetic studies showed that most of the 64 triplets do code for amino acids. Subsequent biochemical studies demonstrated that 61 of the 64 triplets specify particular amino acids. Thus, *for most amino acids, there is more than one code word.* In other words, the genetic code is *degenerate.*

DECIPHERING THE GENETIC CODE: SYNTHETIC RNA CAN SERVE AS A MESSENGER

What is the relation between 64 kinds of code words and the 20 kinds of amino acids? In principle, this question can be directly answered by comparing the amino acid sequence of a protein with the corresponding base sequence of its gene or mRNA. However, this approach was not experimentally feasible in 1961 because the base sequences of genes and mRNA molecules were entirely unknown. The breaking of the genetic code seemed remote until Marshall Nirenberg discovered that *the addition of polyuridylate [poly(U)] to a cell-free protein-synthesizing system led to the synthesis of polyphenylalanine.* Poly(U) evidently served as a messenger RNA. The first code word was deciphered: UUU codes for phenylalanine. This remarkable experiment pointed the way to the complete elucidation of the genetic code.

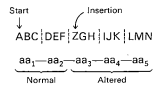

Let us consider Nirenberg's landmark experiment in more detail. The two essential components were a cell-free system that actively synthesized proteins, and a synthetic polyribonucleotide that served as messenger RNA. Proteins were synthesized when the system was supplemented with ATP, GTP, and amino acids. At least one of the added amino acids was radioactive so that its incorporation into protein could be detected. This mixture was incubated at 37°C for about an hour. Trichloroacetic acid was then added to terminate the reaction and precipitate the proteins, leaving the free amino acids in the supernatant. The precipitate was washed and its radioactivity was then measured to determine how much labeled amino acid was incorporated into newly synthesized protein.

A crucial feature of this system is that protein synthesis can be halted by the addition of deoxyribonuclease, which destroys the template for the synthesis of new mRNA (Figure 5-15). The mRNA present at the time of addition of deoxyribonuclease has a short life, and so protein synthesis stops within a few minutes. Nirenberg then found that protein synthesis resumed on addition of a crude fraction of mRNA. Thus, here was a *cell-free protein-synthesizing system that was responsive to the addition of mRNA.*

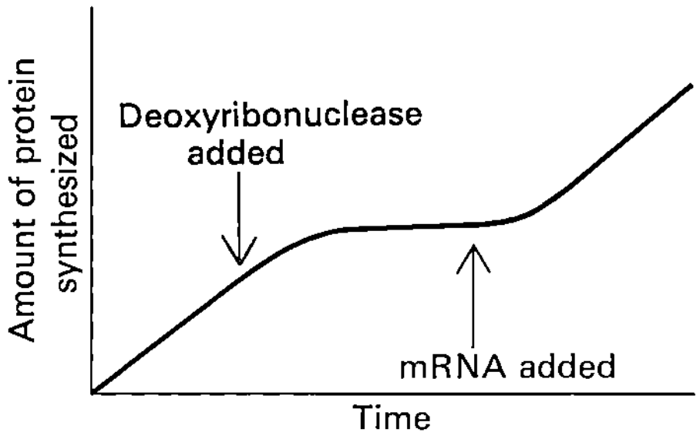

Figure 5-15
Protein synthesis in a cell-free system stops a few minutes after the addition of deoxyribonuclease and resumes following the addition of mRNA.

The other critical component in this experiment was a synthetic polyribonucleotide—namely, poly(U). This polymer was synthesized by using *polynucleotide phosphorylase.* This enzyme, discovered in 1955 by Marianne Grunberg-Manago and Severo Ochoa, catalyzes the synthesis of polyribonucleotides from ribonucleoside diphosphates:

$$(\text{RNA})_n + \text{ribonucleoside diphosphate} \rightleftharpoons (\text{RNA})_{n+1} + \text{P}_i$$

Polynucleotide phosphorylase, like RNA polymerase, catalyzes RNA synthesis. However, the activated substrates here are ribonucleoside diphosphates rather than triphosphates. Orthophosphate is a product instead of pyrophosphate. Hence, the reaction catalyzed by polynucleotide phosphorylase cannot be driven to the right by the hydrolysis of pyrophosphate. Indeed, the equilibrium in vivo is in the direction of RNA degradation, not synthesis. A critical difference is that *polynucleotide phosphorylase does not utilize a template,* which made it a most valuable experimental tool in deciphering the genetic code. The RNA synthesized had a composition dictated only by the ratios of the ribonucleotides present in the incubation mixture and a sequence that was essentially random. Thus, poly(U) was synthesized by incubating high concentrations of UDP with the enzyme. Copolymers of U and A with a random sequence were prepared by including ADP in this mixture.

A variety of synthetic ribonucleotides were added to the cell-free protein-synthesizing system, and the incorporation of ^{14}C-labeled L-phenylalanine was measured. The results were striking.

Polyribonucleotide added	*^{14}C counts per minute*
None	44
Poly(A)	50
Poly(C)	38
Poly(U)	39,800

The same experiment was carried out with a different ^{14}C-labeled amino acid in each incubation mixture. It was found that *poly(A) led to the synthesis of polylysine* and that *poly(C) led to the synthesis of polyproline.* Thus, three code words were deciphered.

Code word	*Amino acid*
UUU	Phenylalanine
AAA	Lysine
CCC	Proline

TRINUCLEOTIDES PROMOTE THE BINDING OF SPECIFIC TRANSFER RNA MOLECULES TO RIBOSOMES

The use of mixed copolymers as templates revealed the *composition but not the sequence of codons* corresponding to particular amino acids. Valine, for example, is coded by 2 U and 1 G, but is the triplet UUG, UGU, or GUU? This question was answered by (1) measuring the codon-dependent binding of specific tRNA molecules to ribosomes and (2) using synthetic polyribonucleotides with an ordered sequence as templates.

In 1964, Nirenberg discovered that *trinucleotides promote the binding of specific tRNA molecules to ribosomes.* For example, the addition of pUUU (the *p* indicates that the 5′ end of the trinucleotide is phosphorylated) led to the binding of phenylalanine tRNA, whereas pAAA markedly enhanced the binding of lysine tRNA, as did pCCC for proline tRNA. Dinucleotides were ineffective in stimulating the binding of tRNA to ribosomes. These studies showed that a *trinucleotide (like a triplet in mRNA) specifically binds the particular tRNA molecule for which it is the code word.* A simple and rapid binding assay was devised (Figure 5-16): tRNA molecules bound to ribosomes are retained by a cellulose nitrate filter, whereas unbound tRNA passes through the filter. The kind of tRNA retained by the filter was identified by using a particular ^{14}C-labeled amino acid.

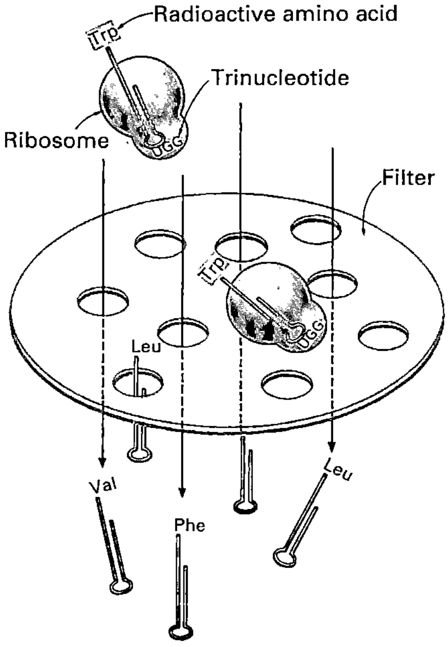

Figure 5-16
Filter-binding assay for detecting the binding of a trinucleotide to a specific tRNA molecule. This complex binds to a ribosome, which adheres to the filter. In contrast, tRNA alone passes through the filter. The tRNA molecules corresponding to a particular amino acid are specifically labeled by enzymatic esterification with a radioactive amino acid. In the example shown here, pUGG binds to tRNA specific for tryptophan.

All 64 trinucleotides were synthesized by organic-chemical or enzymatic techniques. For each trinucleotide, the binding of tRNA molecules corresponding to all 20 amino acids was assayed. For example, it was found that pUUG stimulated the binding of leucine tRNA only, pUGU stimulated the binding of cysteine tRNA only, and pGUU stimulated the binding of valine tRNA only. Hence, it was concluded that the codons UUG, UGU, and GUU correspond to leucine, cysteine, and valine, respectively. For a few codons, no tRNA was strongly bound, whereas, for a few others, more than one kind of tRNA was bound. For most codons, clear-cut binding results were obtained. In fact, about 50 codons were deciphered by this simple and elegant experimental approach.

COPOLYMERS WITH A DEFINED SEQUENCE WERE ALSO INSTRUMENTAL IN BREAKING THE CODE

At about the same time, H. Gobind Khorana succeeded in synthesizing polyribonucleotides with a defined repeating sequence. These regular copolymers were used as templates in the cell-free protein-synthesizing system. Let us examine some of the results. A copolymer consisting of an alternating sequence of two bases P and Q,

PQP ¦ QPQ ¦ PQP ¦ QPQ ¦ PQP ¦. . .

contains two codons, PQP and QPQ. The polypeptide product should therefore contain an alternating sequence of two kinds of amino acids (abbreviated as aa_1 and aa_2).

$$aa_1—aa_2—aa_1—aa_2—aa_1—$$

Whether aa_1 or aa_2 is amino-terminal in the polypeptide product depends on whether the reading frame starts at P or Q. When poly(UG) was the template, a polypeptide with an alternating sequence of cysteine and valine was synthesized.

UGU¦GUG¦UGU¦GUG¦UGU¦GUG

Cys—Val—Cys—Val—Cys—Val

This result unequivocally confirmed the triplet nature of the code and showed that either UGU or GUG codes for cysteine and that the other triplet codes for valine. When this result was considered together with tRNA binding data, it was evident that UGU codes for cysteine and that GUG codes for valine. The polypeptides synthesized in response to several other alternating copolymers of two bases were

Template	*Product*	*Codon assignments*	
Poly(UC)	Poly(Ser-Leu)	UCU ⟶ Ser	CUC ⟶ Leu
Poly(AG)	Poly(Arg-Glu)	AGA ⟶ Arg	GAG ⟶ Glu
Poly(AC)	Poly(Thr-His)	ACA ⟶ Thr	CAC ⟶ His

Now consider a template consisting of three bases in a repeating sequence, poly(PQR). If the reading frame starts at P, the resulting polypeptide should contain only one kind of amino acid, the one coded by the triplet PQR.

Start
↓
PQR¦PQR¦PQR¦PQR¦PQR¦PQR. . .

$aa_1—aa_1—aa_1—aa_1—aa_1—aa_1$

However, if the reading frame starts at Q, the polypeptide synthesized should contain a different amino acid, the one coded by the triplet QRP.

Start
↓
P¦QRP¦QRP¦QRP¦QRP¦QRP¦QRP. . .

$aa_2—aa_2—aa_2—aa_2—aa_2—aa_2$

If the reading frame starts at R, a third kind of polypeptide should be formed, containing only the amino acid coded by the triplet RPQ.

Start
↓
PQ¦RPQ¦RPQ¦RPQ¦RPQ¦RPQ¦RPQ. . .

$aa_3—aa_3—aa_3—aa_3—aa_3—aa_3$

Thus, the expected products are three different homopolypeptides. In fact, this was observed for most templates consisting of repeating sequences of three nucleotides. For example, poly(UUC) led to the synthesis of polyphenylalanine, polyserine, and polyleucine. This result, taken together with the outcome of other experiments, showed that UUC codes for phenylalanine, UCU codes for serine, and CUU codes for leucine. The polypeptides synthesized in response to other templates of this type are given in Table 5-3. Note that poly(GUA) and poly(GAU) each elicited the synthesis of two rather than three homopolypeptides. The reason will be evident shortly.

Table 5-3
Homopolypeptides synthesized using messengers containing repeating trinucleotide sequences

Messenger	*Homopolypeptides synthesized*
Poly(UUC)	Phe, Ser, Leu
Poly(AAG)	Lys, Arg, Glu
Poly(UUG)	Leu, Cys, Val
Poly(CCA)	Gln, Asn, Thr
Poly(GUA)	Val, Ser
Poly(UAC)	Tyr, Thr, Leu
Poly(AUC)	Ile, Ser, His
Poly(GAU)	Asp, Met

Khorana also synthesized several copolymers with repeating tetranucleotides, such as poly(UAUC). This template led to the synthesis of a polypeptide with the repeating sequence Tyr-Leu-Ser-Ile, irrespective of the reading frame.

UAU | CUA | UCU | AUC | UAU | CUA | UCU | AUC

Tyr—Leu—Ser—Ile—Tyr—Leu—Ser—Ile

Four codon assignments were deduced from this result.

A very different result was obtained when poly(GUAA) was the template. The only products were dipeptides and tripeptides. Why not longer chains? The reason is that one of the triplets present in this polymer—namely, UAA—codes not for an amino acid but rather for the termination of protein synthesis:

GUA | AGU | AAG | UAA | GUA | A. . .

Val—Ser—Lys—Stop

Only di- and tripeptides were formed also when poly(AUAG) was the template, because UAG is a second signal for chain termination.

AUA | GAU | AGA | UAG | AUA | G. . .

Ile—Asp—Arg—Stop

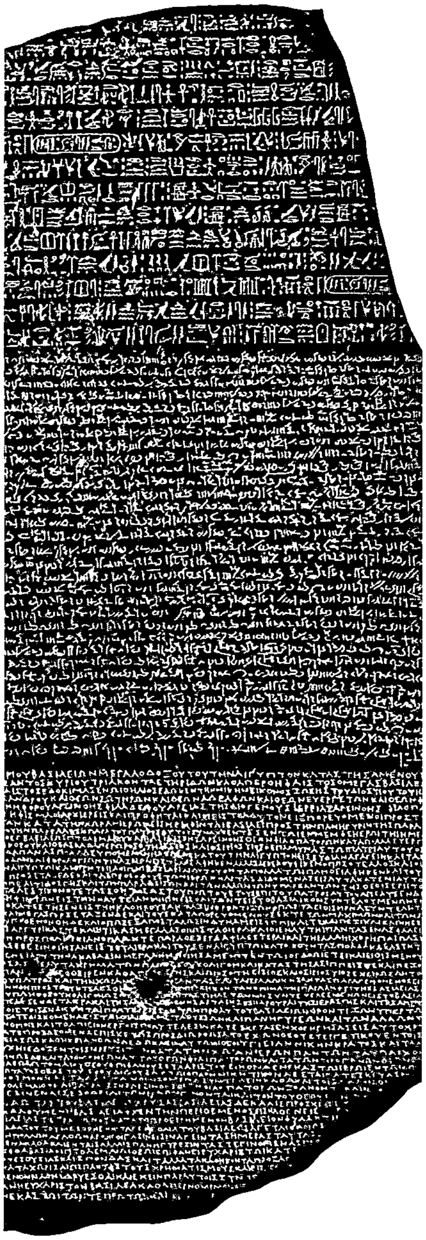

Rosetta stone, inscribed in hieroglyphs, demotic, and Greek. [© Archiv/PhotoResearchers, Inc.]

Now let us look again at Table 5-3. Two rather than three homopolypeptides were synthesized with poly(GUA) as a template because the third reading frame corresponds to the sequence

G | UAG | UAG | UAG . . .

Stop—Stop—Stop

which is a repeating sequence of termination signals. What about poly(GAU)? Again, only two homopolypeptides were synthesized with this template, because the third reading frame corresponds to

GA | UGA | UGA | UGA | U. . .

Stop—Stop—Stop

UGA is yet another signal for chain termination. In fact, *UAG, UAA, and UGA are the only three codons that do not specify an amino acid.*

Khorana's synthesis of polynucleotides with a defined sequence was an outstanding accomplishment. The use of these polymers as templates for protein synthesis, together with Nirenberg's studies of the trinucleotide-stimulated binding of tRNA to ribosomes, resulted in the complete elucidation of the genetic code by 1966, an outcome that would have been deemed a quixotic dream only six years earlier.

MAJOR FEATURES OF THE GENETIC CODE

All 64 codons have been deciphered (Table 5-4). Sixty-one triplets correspond to particular amino acids, whereas three code for chain termination. Because there are 20 amino acids and 61 triplets that code for them, it is evident that the code is highly *degenerate.* In other words, *many amino acids are designated by more than one triplet.* Only tryptophan and methionine are coded by just one triplet. The other 18 amino acids are coded by two or more. Indeed, leucine, arginine, and serine are specified by six codons each. The number of codons for a particular amino acid is correlated with its frequency of occurrence in proteins.

Codons that specify the same amino acid are called *synonyms.* For example, CAU and CAC are synonyms for histidine. Note that synonyms are not distributed haphazardly throughout the table of the genetic code (see Table 5-4). An amino acid specified by two or more synonyms occupies a single box (unless there are more than four synonyms). The amino acids in a box are specified by codons that have the same first two bases but differ in the third base, as exemplified by GUU, GUC, GUA, and GUG. Thus, *most synonyms differ only in the last base of the triplet.* Inspection of the code shows that XYC and XYU always code for the same amino acid, whereas XYG and XYA usually code for the same amino acid. The structural basis for these equivalences of codons will become evident when we consider the nature of the anticodons of tRNA molecules (p. 887).

What is the biological significance of the extensive degeneracy of the genetic code? If the code were not degenerate, 20 codons would

Table 5-4
The genetic code

First position (5′ end)	*Second position*				*Third position (3′ end)*
	U	C	A	G	
U	Phe	Ser	Tyr	Cys	U
	Phe	Ser	Tyr	Cys	C
	Leu	Ser	Stop	Stop	A
	Leu	Ser	Stop	Trp	G
C	Leu	Pro	His	Arg	U
	Leu	Pro	His	Arg	C
	Leu	Pro	Gln	Arg	A
	Leu	Pro	Gln	Arg	G
A	Ile	Thr	Asn	Ser	U
	Ile	Thr	Asn	Ser	C
	Ile	Thr	Lys	Arg	A
	Met	Thr	Lys	Arg	G
G	Val	Ala	Asp	Gly	U
	Val	Ala	Asp	Gly	C
	Val	Ala	Glu	Gly	A
	Val	Ala	Glu	Gly	G

Note: This table identifies the amino acid encoded by each triplet. For example, the codon 5′ AUG 3′ on mRNA specifies methionine, whereas CAU specifies histidine. UAA, UAG, and UGA are termination signals. AUG is part of the initiation signal, in addition to coding for internal methionines.

designate amino acids and 44 would lead to chain termination. The probability of mutating to chain termination would therefore be much higher with a nondegenerate code. Chain-termination mutations usually lead to inactive proteins, whereas substitutions of one amino acid for another are usually rather harmless. Thus, *degeneracy minimizes the deleterious effects of mutations.* Degeneracy of the code may also be significant in permitting DNA base composition to vary over a wide range without altering the amino acid sequence of the proteins encoded by the DNA. The G + C content of bacterial DNA ranges from less than 30% to more than 70%. DNA rich in G·C has a higher melting temperature than DNA rich in A·T. As might be expected, the DNA of algae residing in hot springs has a high content of G·C. DNA molecules with quite different G + C contents could code for the same proteins if different synonyms were consistently used.

MESSENGER RNA CONTAINS START AND STOP SIGNALS FOR PROTEIN SYNTHESIS

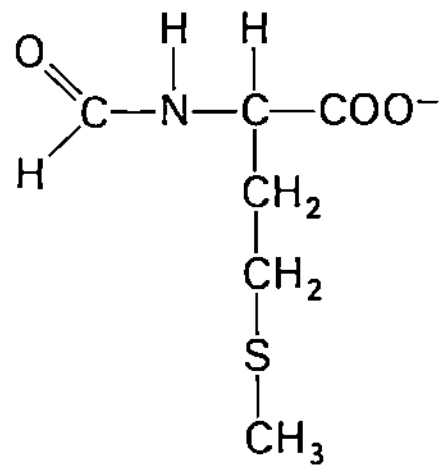

Formylmethionine (fMet)

It has already been mentioned that *UAA, UAG, and UGA designate chain termination.* These codons are read not by tRNA molecules but rather by specific proteins called *release factors* (p. 901). The start signal for protein synthesis is more complex. Polypeptide chains in bacteria start with a modified amino acid—namely, formylmethionine (fMet). A specific tRNA, the initiator tRNA, carries fMet. This fMet-tRNA recognizes the codon AUG or, less frequently, GUG. However, AUG is also the codon for an internal methionine, and GUG is the codon for an internal valine. Hence, the signal for the first amino acid in a prokaryotic polypeptide chain must be more complex than for all subsequent ones. *AUG (or GUG) is only part of the initiation signal* (Figure 5-17). In bacteria, the initiating AUG (or GUG) is preceded several nucleotides away by a purine-rich sequence that base-pairs with a complementary sequence in a ribosomal RNA molecule (p. 895). In eukaryotes, the AUG closest to the 5′ end of an mRNA is usually the start signal for protein synthesis. This particular AUG is read by an initiator tRNA conjugated to methionine.

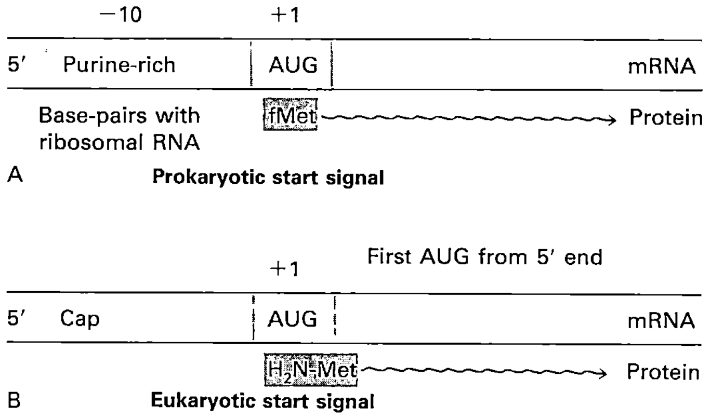

Figure 5-17
Start signals for the initiation of protein synthesis in (A) prokaryotes and (B) eukaryotes. In eukaryotic mRNAs the 5′ end, called a *cap,* contains modified bases (Chapter 33).

THE GENETIC CODE IS NEARLY UNIVERSAL

The genetic code was deciphered by studies of trinucleotides and synthetic mRNA templates in cell-free systems derived from bacteria. Is the genetic code the same in all organisms? Analyses of spontaneous and specifically designed mutations in viruses, bacteria, and higher organisms have been highly informative. The base sequences of many wild-type and mutant genes are known, as are the amino acid sequences of their encoded proteins. In each case, the nucleotide change in the gene and the amino acid change in the protein are as predicted by the genetic code. Furthermore, mRNAs can be correctly translated by the protein-synthesizing machinery of very different species. For example, human hemoglobin mRNA is correctly translated by a wheat germ extract. As will be discussed in the next chapter, bacteria efficiently express recombinant DNA molecules encoding human proteins such as insulin. These experimental findings strongly suggested that the genetic code is universal.

A surprise was encountered when the sequence of human mitochondrial DNA became known. Human mitochondria read UGA as a codon for tryptophan rather than as a stop signal (Table 5-5). Furthermore, AGA and AGG are read as stop signals rather than as codons for arginine, and AUA is read as a codon for methionine instead of isoleucine. Mitochondria of other species, such as those of yeast, also have a genetic code that differs slightly from the standard one. Mitochondria can have a different genetic code from the rest of the cell because mitochondrial DNA encodes a distinct set of tRNAs. Do any cellular protein-synthesizing systems deviate from the standard genetic code? Ciliated protozoa differ from most organisms in reading UAA and UAG as codons for amino acids rather than as stop signals; UGA is their sole termination signal. Thus, *the genetic code is nearly but not absolutely universal.* Variations clearly exist in mitochondria and in species, such as ciliates, that branched off very early in eukaryotic evolution. It is interesting to note that two of the codon reassignments in human mitochondria diminish the information content of the third base of the triplet (e.g., both AUA and AUG specify methionine). Most variations from the standard genetic code are in the direction of a simpler code.

Table 5-5
Distinctive codons of human mitochondria

Codon	*Standard code*	*Mitochondrial code*
UGA	Stop	Trp
UGG	Trp	Trp
AUA	Ile	Met
AUG	Met	Met
AGA	Arg	Stop
AGG	Arg	Stop

Why has the code remained nearly invariant through billions of years of evolution, from bacteria to humans? A mutation that altered the reading of mRNA would change the amino acid sequence of most, if not all, proteins synthesized by that particular organism. Many of these changes would undoubtedly be deleterious, and so there would be strong selection against a mutation with such pervasive consequences.

THE SEQUENCES OF GENES AND THEIR ENCODED PROTEINS ARE COLINEAR

Before the genetic code was deciphered, high-resolution gene-mapping studies carried out by Seymour Benzer in the 1950s had revealed that genes are unbranched structures. This important result was in harmony with the established fact that DNA consists of a linear sequence of base pairs. Polypeptide chains also have unbranched structures. The following question was therefore asked: *Is there a linear correspondence between a gene and its polypeptide product?*

Charles Yanofsky approached this problem in 1967 by using mutants of *E. coli* that produced an altered enzyme molecule. Numerous bacteria harboring mutations in the gene for the α chain of tryptophan synthetase (an enzyme catalyzing the final step in the synthesis of tryptophan) were isolated. The positions of the mutations on the genetic map for the α chain were determined by measuring frequencies of recombination. Some of the mutations fell very close to each other on the genetic map, whereas others were far apart. The next task was to determine the location of the amino acid substitution produced by each of these mutations. First, the amino acid sequence of the wild-type α chain was determined. Then, the position and nature of the change in each mutant protein was identified. *The order of the mutations on the genetic map proved to be the same as the order of the corresponding changes in the amino acid sequence of the polypeptide product* (Figure 5-18). The same result was subsequently obtained in many other systems. Thus, *genes and their polypeptide products are colinear.*

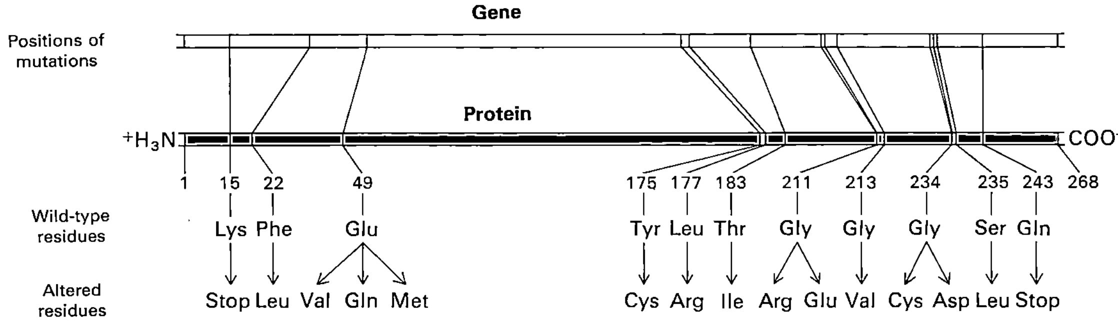

Figure 5-18
Colinearity of the gene and the amino acid sequence of the α chain of tryptophan synthetase. The locations of mutations in DNA were determined by genetic-mapping techniques. The positions of the altered amino acids in the mutant proteins are in the same order as the corresponding mutations. [After C. Yanofsky. Gene structure and protein structure. Copyright © 1967 by Scientific American, Inc. All rights reserved.]

MOST EUKARYOTIC GENES ARE MOSAICS OF INTRONS AND EXONS

In bacteria, polypeptide chains are encoded by a continuous array of triplet codons in DNA. It was assumed for many years that genes in higher organisms also are continuous. This view was unexpectedly shattered in 1977, when investigators in several laboratories discovered that several genes are *discontinuous.* The mosaic nature of eukaryotic genes was revealed by electron microscopic studies of hybrids formed between mRNA and a segment of DNA containing the corresponding gene (Figure 5-19).

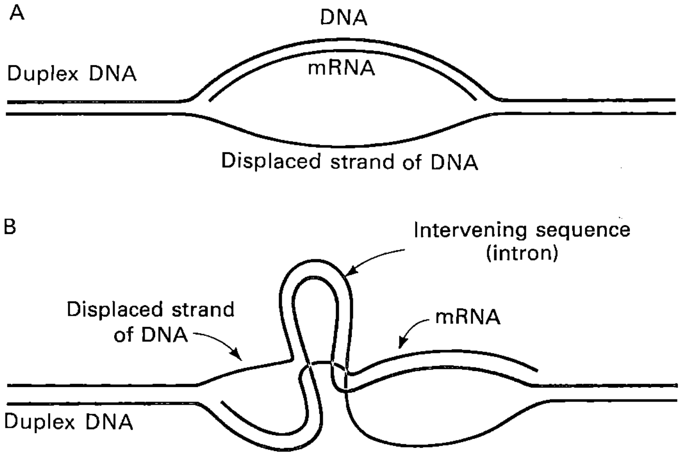

Figure 5-19
Detection of intervening sequences by electron microscopy. An mRNA molecule (shown in red) is hybridized to genomic DNA containing the corresponding gene. (A) A single loop of single-stranded DNA (shown in blue) is seen if the gene is continuous. (B) Two loops of single-stranded DNA (blue) and a loop of double-stranded DNA (blue and green) are seen if the gene contains an intervening sequence. Additional loops are evident if more than one intervening sequence is present.

For example, the gene for the β chain of hemoglobin is interrupted within its amino acid coding sequence by a long *intervening sequence* of 550 base pairs and a short one of 120 base pairs. Thus, the *β-globin gene is split into three coding sequences.*

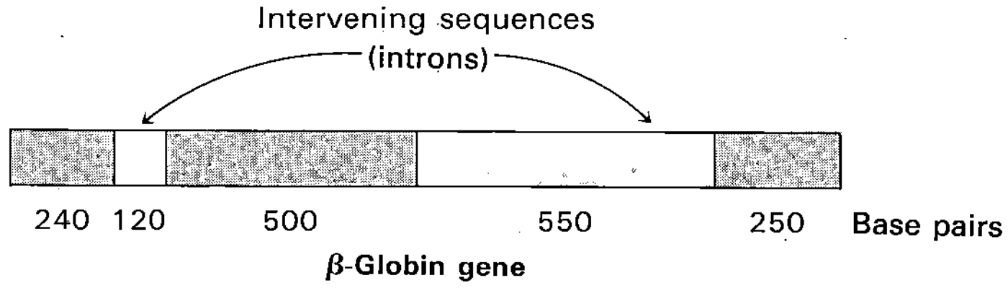

At what stage in gene expression are intervening sequences removed? Newly synthesized RNA chains isolated from nuclei are much larger than the mRNA molecules derived from them. In fact, the primary transcript of the β-globin gene contains two regions that are not present in the mRNA. *These intervening sequences in the 15S primary transcript are excised, and the coding sequences are simultaneously linked by a precise splicing enzyme to form the mature 9S mRNA* (Figure 5-20). Regions that are removed from the primary transcript are called *introns* (for *int*ervening sequences), whereas those that are retained in the mature RNA are called *exons* (for *ex*pressed regions).

Another split eukaryotic gene is the one for ovalbumin in chickens, which is made up of eight exons separated by seven long introns (Figure 5-21). Even more striking is the collagen gene, which contains more than 40 exons. A common feature in the expression of these genes is that their exons are ordered in the same sequence in mRNA as in DNA. Thus, *split genes, like continuous genes, are colinear with their polypeptide products.*

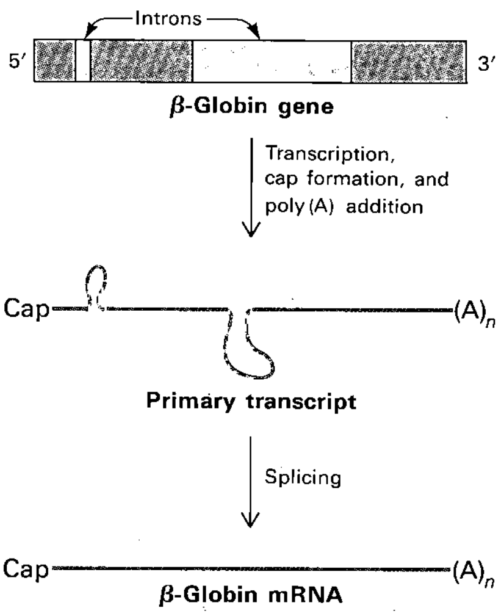

Figure 5-20
Transcription of the β-globin gene and removal of the intervening sequences in the primary RNA transcript. Cap formation and the addition of poly(A) are discussed in Chapter 33.

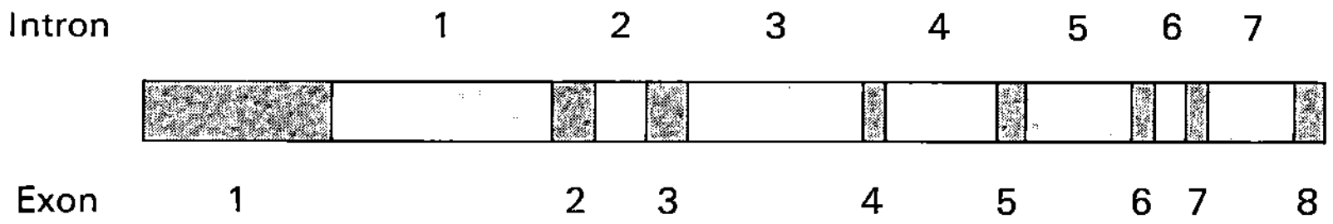

Figure 5-21
Structure of the chick ovalbumin gene. The introns (noncoding regions) are shown in yellow and the exons (translated regions) in blue.

Splicing is a facile complex operation that is carried out by *spliceosomes,* which are assemblies of proteins and small RNA molecules (p. 862). This enzymatic machinery recognizes signals in the nascent RNA that specify the splice sites. *Introns nearly always begin with GU and end with an AG that is preceded by a pyrimidine-rich tract (Figure 5-22). This consensus sequence is part of the signal for splicing.*

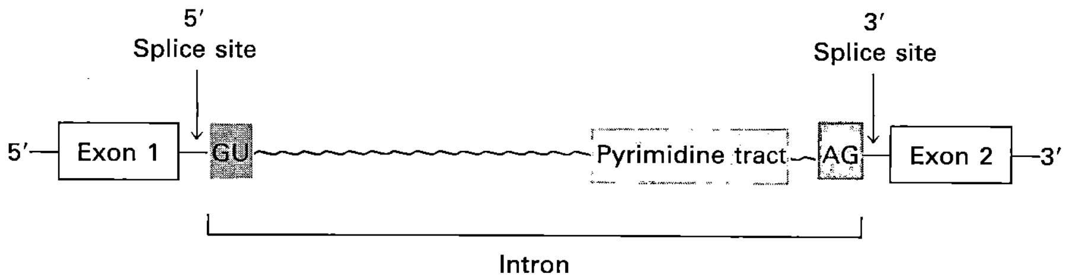

Figure 5-22
Consensus sequence for the splicing of mRNA precursors.

MANY EXONS ENCODE PROTEIN DOMAINS

Most genes of higher eukaryotes, such as birds and mammals, are split. Lower eukaryotes, such as yeast, have a much higher proportion of continuous genes. In eubacteria, such as *E. coli,* no split genes have been found. Have introns been inserted into genes in the evolution of higher organisms? Or have introns been removed from genes to form the streamlined genomes of prokaryotes and simple eukaryotes? Comparisons of the DNA sequences of genes encoding proteins that are highly conserved in evolution strongly suggest that *introns were present in ancestral genes and were lost in the evolution of organisms that have become optimized for very rapid growth, such as eubacteria and yeast.* The positions of introns in some genes are at least one billion years old. Furthermore, a common mechanism of splicing developed before the divergence of fungi, plants, and vertebrates, as shown by the finding that mammalian cell extracts can splice yeast RNA.

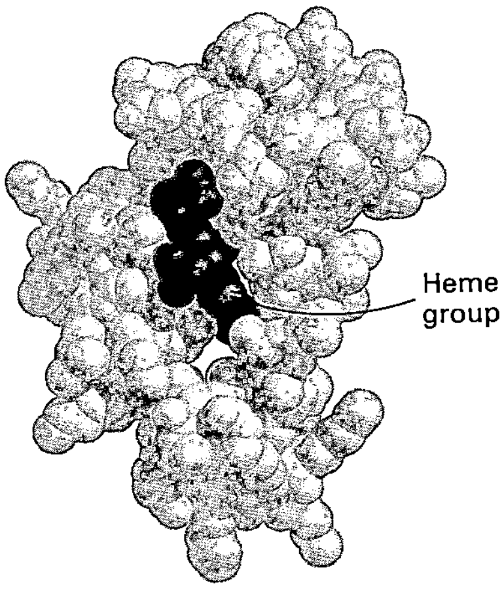

Figure 5-23
The core of myoglobin is encoded by the central exon of its gene. The polypeptide chain encoded by this exon is shown in blue and the oxygen-binding heme group in red. [Drawn from 1mbn.pdb. H.C. Watson. *Prog. Stereochem.* 4(1969):299.]

Many exons encode discrete structural and functional units of proteins. For example, the central exon of myoglobin and hemoglobin genes encodes a heme-binding region (Figure 5-23) that can reversibly bind O_2 (p. 153). Other exons specify α-helical segments that anchor proteins in cell membranes. An entire domain of a protein may be encoded by a single exon. An attractive hypothesis is that *new proteins arose in evolution by the rearrangement of exons encoding discrete structural elements, binding sites, and catalytic sites.* Exon shuffling is a rapid and efficient means of generating novel genes because it preserves functional units, but allows them to interact in new ways (Figure 5-24). Introns are extensive regions in which DNA can break and recombine with no deleterious effect on encoded proteins. In contrast, the exchange of sequences between different exons usually leads to loss of function.

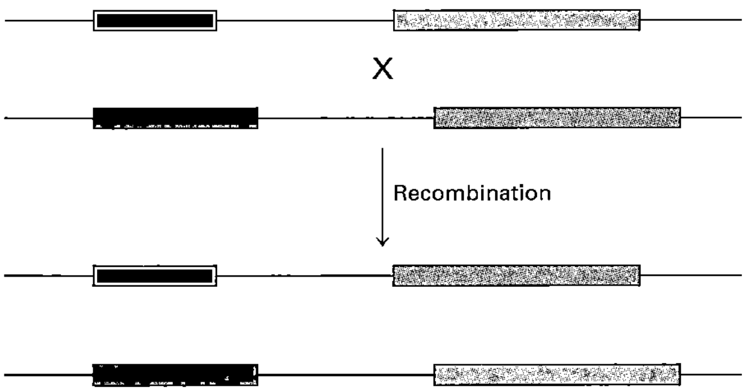

Figure 5-24
Exons can readily be shuffled by recombination of DNA to expand the genetic repertoire.

Another advantage conferred by split genes is the potentiality for generating a series of related proteins by splicing a nascent RNA transcript in different ways. For example, a precursor of an antibody-producing cell forms an antibody that is anchored in the cell's plasma membrane (Figure 5-25). Stimulation of such a cell by a specific foreign antigen that is recognized by the attached antibody leads to cell differentiation and proliferation. The activated antibody-producing cells then splice their nascent RNA transcript in an alternative manner to form soluble antibody molecules that are secreted rather than retained on the cell surface. We see here a clear-cut example of a benefit conferred by the complex arrangement of in-

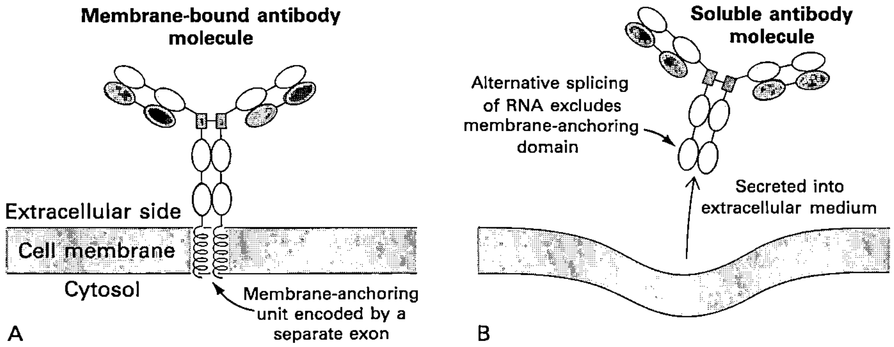

Figure 5-25
Alternative splicing generates mRNAs that are templates for different forms of a protein: (A) a membrane-bound antibody on the surface of a lymphocyte, and (B) its soluble counterpart, exported from the cell. The membrane-bound antibody is anchored to the plasma membrane by a helical segment (highlighted in yellow) that is encoded by its own exon.

trons and exons in higher organisms. *Alternative splicing is a facile means of forming a set of proteins that are variations of a basic motif according to a developmental program.*

RNA PROBABLY CAME BEFORE DNA AND PROTEINS IN EVOLUTION

The flow of genetic information from DNA to RNA to protein depends on the intricate interplay of enzymes and other proteins with nucleic acids. Likewise, the replication of DNA and RNA is mediated by the interaction of polymerases and other proteins with nucleic acid templates. How did nucleic acid molecules early in the evolution of life replicate in the absence of enzymes? A likely solution to this enigma comes from the discovery in 1983 that *RNA molecules as well as proteins can be enzymes.* Thomas Cech found that the precursor of a ribosomal RNA in *Tetrahymena* (a ciliated protozoan) undergoes self-splicing (p. 864). The intron in this precursor RNA molecule is precisely removed by a catalytic activity of the RNA itself. This liberated intron then loses a short 5′-terminal sequence to form a 395-nucleotide RNA molecule that catalyzes the transformation of *other* RNA molecules. This intron-derived RNA catalyzes the cleavage and joining of RNA chains at specific sites without itself being consumed (p. 231). Hence, it is a true enzyme. Protein catalysts, known for a century, are now joined by RNA catalysts (*ribozymes*).

Cech's revolutionary discovery enables us to envision an RNA world early in the evolution of life before the appearance of DNA and protein. Walter Gilbert has proposed that RNA molecules first catalyzed their own replication and developed a repertoire of enzymatic activities. In the next stage, RNA molecules began to synthesize proteins, which emerged as superior enzymes because their 20 side chains are more versatile than the four bases of RNA. Finally, DNA was formed by reverse transcription of RNA. DNA replaced RNA as the genetic material because its double helix is a more stable and reliable store of genetic information than is single-stranded RNA. At this point, RNA was left with roles it has retained to this day as the information carrier (mRNA) and adapter (tRNA) in protein synthesis and as critical components (rRNAs) of ribosomes and other assemblies that mediate gene expression. The present intricate (indeed, baroque) mechanism of information transfer from gene to protein is an ancient epic that probably began when RNA alone wrote the script, directed the action, and played all the key parts.

SUMMARY

The flow of genetic information in normal cells is from DNA to RNA to protein. The synthesis of RNA from a DNA template is called transcription, whereas the synthesis of a protein from an RNA template is termed translation. Cells contain several kinds of RNA: messenger RNA (mRNA), transfer RNA (tRNA), ribosomal RNA (rRNA), and small nuclear RNA (snRNA). Most RNA molecules are single-stranded, but many contain extensive double-helical regions that arise from the folding of the chain into hairpins. The smallest RNA molecules are the tRNAs, which contain as few as 75 nucleotides, whereas the largest ones are some mRNAs, which may have more than 5000 nucleotides. All cellular RNA is synthesized by RNA polymerase according to instructions given by DNA templates. The activated intermediates are ribonucleoside triphosphates. The direction of RNA synthesis is $5' \rightarrow 3'$, like that of DNA synthesis. RNA polymerase differs from DNA polymerase in not requiring a primer. Another difference is that the DNA template is fully conserved in RNA synthesis, whereas it is semiconserved in DNA synthesis. Many RNA molecules are cleaved and chemically modified after transcription.

The base sequence of a gene is colinear with the amino acid sequence of its polypeptide product. The genetic code is the relation between the sequence of bases in DNA (or its RNA transcript) and the sequence of amino acids in proteins. Amino acids are coded by groups of three bases (called codons) starting from a fixed point. Sixty-one of the 64 codons specify particular amino acids, whereas the other three codons (UAA, UAG, and UGA) are signals for chain termination. Thus, for most amino acids there is more than one code word. In other words, the code is degenerate. Codons specifying the same amino acid are called synonyms. Most synonyms differ only in the last base of the triplet. The genetic code, which is nearly the same in all organisms, was deciphered after the discovery that the polyribonucleotide poly(U) codes for polyphenylalanine. Various synthetic polyribonucleotides then were used as mRNAs in cell-free protein-synthesizing systems. Natural mRNAs contain start and stop signals for translation, just as genes do for directing where transcription begins and ends.

Most genes in higher eukaryotes are discontinuous. Coding sequences (exons) in these split genes are separated by intervening sequences (introns), which are removed in the conversion of the primary transcript into mRNA and other functional mature RNA molecules. For example, the β-globin gene in mammals contains two introns. Nascent RNA molecules contain signals that specify splice sites. Split genes, like continuous genes, are colinear with their polypeptide products. A striking feature of many exons is that they encode functional domains in proteins. New proteins probably arose in evolution by the shuffling of exons. Introns may have been present in primordial genes, but were lost in the evolution of such fast-growing organisms as bacteria and yeast. The discovery that certain RNA molecules undergo self-splicing and can serve as enzymes suggests that RNA came before DNA and proteins in evolution.

SELECTED READINGS

Where to start

Crick, F.H.C., 1966. The genetic code III. *Sci. Amer.* 215(4):55–62. [A view of the code when it was almost completely elucidated.]

Chambon, P., 1981. Split genes. *Sci. Amer.* 244(5):60–71.

Yanofsky, C., 1967. Gene structure and protein structure. *Sci. Amer.* 216(5):80–94. [Presents the evidence for colinearity.]

Cech, T.R., 1990. Self-splicing and enzymatic activity of an intervening sequence RNA from Tetrahymena. *Biosci. Rep.* 10:239–61. [Nobel Lecture account of the discovery of catalytic RNA.].

Books

Berg, P., and Singer, M., 1991. *Genes and Genomes.* University Science Books.

Lodish, H., Baltimore, D., Berk, A., Zipursky, L., and Matsudaira, P., 1995. *Molecular Cell Biology* (3rd ed.). Scientific American Books.

Lewin, B., 1994. *Genes* (5th ed.). Oxford.

Watson, J.D., Hopkins, N.H., Roberts, J.W., Steitz, J.A., and Weiner, A.M., 1987. *Molecular Biology of the Gene* (4th ed.). Benjamin/Cummings.

Discovery of messenger RNA

Jacob, F., and Monod, J., 1961. Genetic regulatory mechanisms in the synthesis of proteins. *J. Mol. Biol.* 3:318–356.

Brenner, S., Jacob, F., and Meselson, M., 1961. An unstable intermediate carrying information from genes to ribosomes for protein synthesis. *Nature* 190:576–581.

Hall, B.D., and Spiegelman, S., 1961. Sequence complementarity of T2-DNA and T2-specific RNA. *Proc. Nat. Acad. Sci.* 47:137–146.

Genetic code

Crick, F.H.C., Barnett, L., Brenner, S., and Watts-Tobin, R.J., 1961. General nature of the genetic code for proteins. *Nature* 192:1227–1232.

Khorana, H.G., 1968. Nucleic acid synthesis in the study of the genetic code. In *Nobel Lectures: Physiology or Medicine* (1963–1970), pp. 341–369. American Elsevier (1973).

Nirenberg, M., 1968. The genetic code. In *Nobel Lectures: Physiology or Medicine* (1963–1970), pp. 372–395. American Elsevier (1973).

Crick, F.H.C., 1958. On protein synthesis. *Symp. Soc. Exp. Biol.* 12:138–163. [A brilliant anticipatory view of the problem of protein synthesis. The adaptor hypothesis is presented in this article.]

Woese, C.R., 1967. *The Genetic Code.* Harper & Row.

Colinearity of gene and protein

Yanofsky, C., Carlton, B.C., Guest, J.R., Helinski, D.R., and Henning, U., 1964. On the colinearity of gene structure and protein structure. *Proc. Nat. Acad. Sci.* 51:266–272.

Sarabhai, A.S., Stretton, O.W., Brenner, S., and Bolle, A., 1964. Colinearity of gene with polypeptide chain. *Nature* 201:13–17.

Introns, exons, and split genes

Sharp, P.A., 1988. RNA splicing and genes. *J. Amer. Med. Assoc.* 260:3035–3041.

Dorit, R.L., Schoenbach, L., and Gilbert, W., 1990. How big is the universe of exons? *Science* 250:1377–1382.

Cochet, M., Gannon, F., Hen, R., Maroteaux, L., Perrin, F., and Chambon, P., 1979. Organization and sequence studies of the 17-piece chicken conalbumin gene. *Nature* 282:567–574.

Tilghman, S.M., Tiemeier, D.C., Seidman, J.G., Peterlin, B.M., Sullivan, M., Maijel, J.V., and Leder, P., 1978. Intervening sequence of DNA identified in the structural portion of a mouse β-globin gene. *Proc. Nat. Acad. Sci.* 75:725–729.

Catalytic activity of RNA

Altman, S., 1990. Enzymatic cleavage of RNA by RNA. *Biosci. Rep.* 10:317–337. [Nobel Lecture account of the discovery of catalytic RNA.]

Zaug, A.J., and Cech, T.R., 1986. The intervening sequence RNA of *Tetrahymena* is an enzyme. *Science* 231:470–475.

Cech, T.R., 1990. Self-splicing of group I introns. *Ann. Rev. Biochem.* 59:543–568.

Cech, T.R., 1993. Catalytic RNA: structure and mechanism. *Biochem. Soc. Trans.* 21:229–234.

Molecular evolution

Gesteland, R.F., and Atkins, J.F. (eds.), 1993. *The RNA World.* Cold Spring Harbor Laboratory Press.

Wilson, A.C., 1985. The molecular basis of evolution. *Sci. Amer.* 253(4):164–173.

Gilbert, W., 1986. The RNA world. *Nature* 319:618.

Sharp, P.A., 1985. On the origin of RNA splicing and introns. *Cell* 42:397–400.

Marchionni, M., and Gilbert, W., 1986. The triosephosphate isomerase gene from maize: Introns antedate the plant–animal divergence. *Cell* 46:133–141.

Osawa, S., Juke , T.H., Watanabe, K., and Muto, A., 1992. Recent evidence for evolution of the genetic code. *Microbiol. Rev.* 56:229–264.

PROBLEMS

1. *Key polymerases.* Compare DNA polymerase I and RNA polymerase from *E. coli* in regard to each of the following features:
   (a) Activated precursors.
   (b) Direction of chain elongation.
   (c) Conservation of the template.
   (d) Need for a primer.

2. *Encoded sequences.*
   (a) Write the sequence of the mRNA molecule synthesized from a DNA template strand having the sequence

   5′-ATCGTACCGTTA-3′

   (b) What amino acid sequence is encoded by the following base sequence of an mRNA molecule? Assume that the reading frame starts at the 5′ end.

   5′-UUGCCUAGUGAUUGGAUG-3′

   (c) What is the sequence of the polypeptide formed on addition of poly(UUAC) to a cell-free protein-synthesizing system?

3. *A tougher chain.* RNA is readily hydrolyzed by alkali, whereas DNA is not. Why?

4. *A potent blocker.* How does cordycepin (3′-deoxyadenosine) block the synthesis of RNA?

5. *Silent RNA.* The code word GGG could not be deciphered in the same way as was UUU, CCC, and AAA because poly(G) does not act as a template. Poly(G) forms a triple-stranded helical structure. Why is it an ineffective template?

6. *Two from one.* Khorana synthesized by organic-chemical methods two complementary deoxyribonucleotides, each with nine residues: $d(TAC)_3$ and $d(GTA)_3$. Partially overlapping duplexes that formed on mixing these oligonucleotides then served as templates for the synthesis by DNA polymerase of long, repeating double-helical DNA chains. The next step was to obtain long polyribonucleotide chains with a sequence complementary to only one of the two DNA strands. How did he obtain only poly(UAC)? Only poly(GUA)?

7. *Back to the bench.* A protein chemist told a molecular geneticist that he had found a new mutant hemoglobin in which aspartate replaced lysine. The molecular geneticist expressed surprise and sent his friend scurrying back to the laboratory.
   (a) Why was the molecular geneticist dubious about the reported amino acid substitution?
   (b) Which amino acid substitutions would have been more palatable to the molecular geneticist?

8. *Triple entendre.* The RNA transcript of a region of G4 phage DNA contains the sequence 5′-AAAUGAGGA-3′. This sequence encodes three different polypeptides. What are they?

9. *Valuable synonyms.* Proteins generally have low contents of Met and Trp, intermediate ones of His and Cys, and high ones of Leu and Ser. What is the relation between the number of codons of an amino acid and its frequency of occurrence in proteins? What might be the selective advantage of this relation?

10. *A new translation.* A transfer RNA with a UGU anticodon is enzymatically conjugated to ^{14}C-labeled cysteine. The cysteine unit is then chemically modified to alanine (using Raney nickel, which removes the sulfur atom of cysteine). The altered aminoacyl-tRNA is added to a protein-synthesizing system containing normal components except for this tRNA. The mRNA added to this mixture contains the following sequence:

    5′-UUUUGCCAUGUUUGUGCU-3′

    What is the sequence of the corresponding radiolabeled peptide?

11. *Fire and ice.* Valine is specified by four codons. How might the relative frequencies of their usage in an alga isolated from a volcanic hot spring differ from those of an alga isolated from an Antarctic bay?

12. *Eons apart.* The amino acid sequences of a yeast protein and a human protein carrying out the same function are found to be 60% identical. However, the corresponding DNA sequences are only 45% identical. Account for this differing degree of identity.

CHAPTER

6

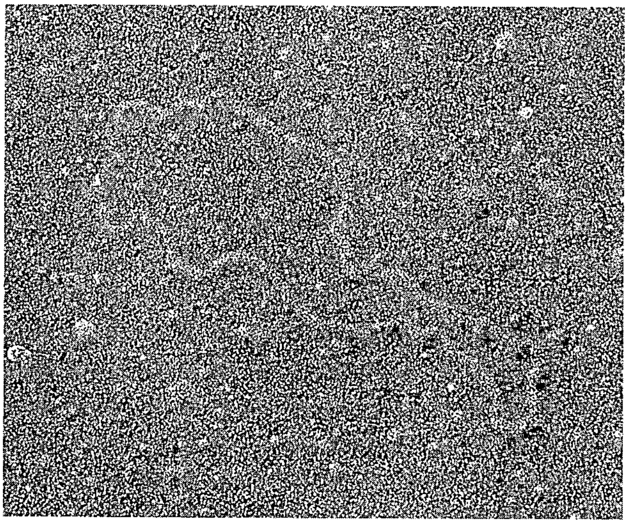

Exploring Genes

Recombinant DNA technology, a new approach to exploring the central molecules of life, came into being in the 1970s. It has revolutionized biochemistry by providing powerful means of analyzing and altering genes and proteins. The genetic endowment of organisms can now be precisely changed in designed ways. Recombinant DNA technology is a fruit of several decades of basic research on DNA, RNA, and viruses. It depends, first, on having enzymes that can cut, join, and replicate DNA and reverse-transcribe RNA. Earlier chapters have already discussed DNA polymerases and reverse transcriptase. This chapter begins with restriction enzymes, which cut very long DNA molecules into specific fragments that can be manipulated. The availability of many kinds of restriction enzymes and of DNA ligases, enzymes that join DNA strands, makes it feasible to treat DNA sequences as modules and to move them at will from one DNA molecule to another. Thus, recombinant DNA technology is based on nucleic acid enzymology.

A second foundation is the base-pairing language that mediates nucleic acid recognition. We have already seen that hybridization with complementary DNA or RNA probes is a sensitive and powerful means of detecting specific nucleotide sequences. In recombinant DNA technology, base pairing is used to construct new combinations of DNA as well as to detect and amplify particular sequences. This revolutionary technology is also critically dependent on the existence of viruses and detailed knowledge concerning their interplay with susceptible hosts. Viruses are the ultimate parasites. They efficiently deliver their own DNA (or RNA) into hosts, subverting them either to replicate the viral genome and produce viral protein or to incorporate viral DNA into the host genome. Likewise, plasmids, which are accessory chromosomes, have been indispensable in recombinant DNA technology.

This chapter also introduces some of the benefits of these new methods. For example, the discovery of restriction enzymes led to the development of techniques for the rapid sequencing of DNA. A wealth of information

Figure 6-1
Crystals of human insulin produced by bacteria harboring recombinant DNA encoding the hormone. [From R.E. Chance, E.P. Kroeff, and J.A. Hoffmann. In *Insulins, Growth Hormone, and Recombinant DNA Technology,* J.L. Gueriguian, ed. (Raven Press, 1981), p. 77.]

***Opening Image:** Electron micrograph of pSC101, the first plasmid vector used in the cloning of DNA. [Courtesy of Dr. Stanley N. Cohen.]*

concerning gene architecture, the control of gene expression, and protein structure has come from the sequencing of more than 100 million bases in DNA molecules from viruses, bacteria, and higher organisms. DNA molecules can also be synthesized de novo. The automated solid-phase synthesis of DNA provides highly specific probes and synthetic tailor-made genes. A single copy of a specific DNA sequence in a complex mixture can be amplified more than a millionfold by the polymerase chain reaction (PCR). This powerful technique is being used to detect pathogens and genetic diseases, determine the source of a hair left at the scene of a crime, and resurrect genes from fossils millions of years old. The final part of this chapter deals with the construction and cloning of novel combinations of genes. New genes can be efficiently expressed by host cells, as exemplified by the production of human insulin by bacteria. Moreover, specific mutations can be made in vitro to engineer proteins in designed ways.

RESTRICTION ENZYMES SPLIT DNA INTO SPECIFIC FRAGMENTS

Restriction enzymes, also called *restriction endonucleases*, recognize specific base sequences in double-helical DNA and cleave both strands of the duplex at specific places. To biochemists, these exquisitely precise scalpels are marvelous gifts of nature. They are indispensable for analyzing chromosome structure, sequencing very long DNA molecules, isolating genes, and creating new DNA molecules that can be cloned. Restriction enzymes were discovered by Werner Arber, Hamilton Smith, and Daniel Nathans in the late 1960s.

Restriction enzymes are found in a wide variety of prokaryotes. Their biological role is to cleave foreign DNA molecules. The cell's own DNA is not degraded because the sites recognized by its own restriction enzymes are methylated. Many restriction enzymes recognize specific sequences of four to eight base pairs and hydrolyze a phosphodiester bond in each strand in this region. A striking characteristic of these cleavage sites is that they possess *twofold rotational symmetry*. In other words, the recognized sequence is *palindromic* and the cleavage sites are symmetrically positioned. For example, the sequence recognized by a restriction enzyme from *Streptomyces achromogenes* is

Cleavage site
↓
5′ C—C—G—C—G—G 3′

3′ G—G—C—G—C—C 5′
↑
Cleavage Symmetry axis
site

In each strand, the enzyme cleaves the C-G phosphodiester bond on the 3′ side of the symmetry axis.

More than 90 restriction enzymes have been purified and characterized. Their names consist of a three-letter abbreviation for the host organism (e.g., Eco for *Escherichia coli*, Hin for *Haemophilus influenzae*, Hae for *Haemophilus aegyptius*) followed by a strain designation (if needed) and a roman numeral (if more than one restriction enzyme is produced). The specificities of several of these enzymes are shown in Figure 6-2. Note that the cuts may be staggered or even.

Palindrome—
A word, sentence, or verse that reads the same from right to left as it does from left to right.

Radar
Madam, I'm Adam
Able was I ere I saw Elba
Roma tibi subito motibus
ibit amor

Derived from the Greek *palindromos,* "running back again."

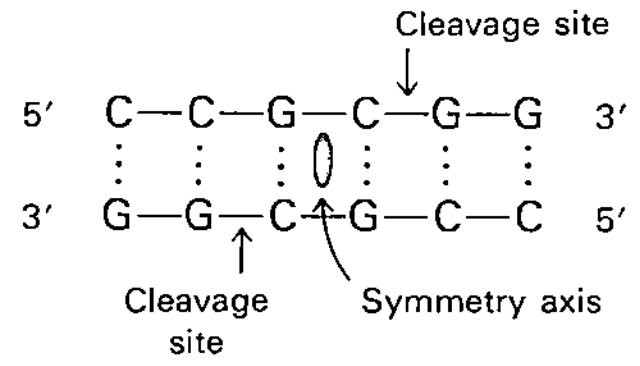

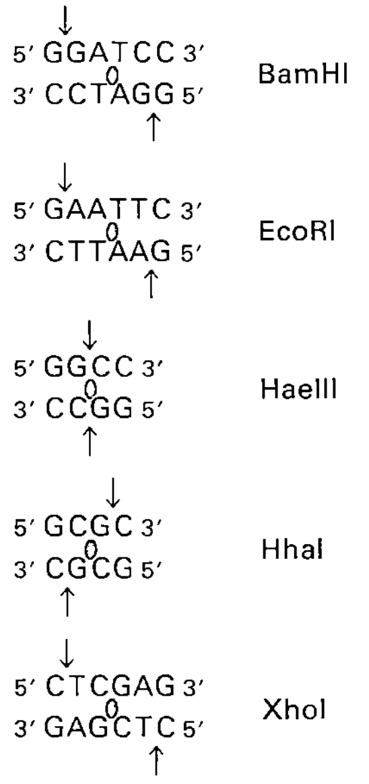

Figure 6-2
Specificities of some restriction endonucleases. The base-pair sequences that are recognized by these enzymes contain a twofold axis of symmetry. The two strands in these regions are related by a 180-degree rotation around the axis marked by the green symbol. The cleavage sites are denoted by red arrows. The abbreviated name of each restriction enzyme is given at the right of the sequence that it recognizes.

Restriction enzymes are used to cleave DNA molecules into specific fragments that are more readily analyzed and manipulated than the parent molecule. For example, the 5.1-kilobase (kb) circular duplex DNA of the tumor-producing SV40 virus is cleaved at 1 site by EcoRI, 4 sites by HpaI, and 11 sites by HindIII. A piece of DNA produced by the action of one restriction enzyme can be specifically cleaved into smaller fragments by another restriction enzyme. The pattern of such fragments can serve as a *fingerprint* of a DNA molecule, as will be discussed shortly. Indeed, complex chromosomes containing hundreds of millions of base pairs can be mapped by using a series of restriction enzymes.

Kilobase (kb)—
A unit of length equal to 1000 base pairs of a double-stranded nucleic acid molecule (or 1000 bases of a single-stranded molecule).
One kilobase of double-stranded DNA has a contour length of 0.34 μm and a mass of about 660 kd.

RESTRICTION FRAGMENTS CAN BE SEPARATED BY GEL ELECTROPHORESIS AND VISUALIZED

Small differences between related DNA molecules can be readily detected because their restriction fragments can be separated and displayed by gel electrophoresis. In many types of gels, the electrophoretic mobility of a DNA fragment is inversely proportional to the logarithm of the number of base pairs, up to a certain limit. Polyacrylamide gels are used to separate fragments containing up to about 1000 base pairs, whereas more porous agarose gels are used to resolve mixtures of larger fragments (up to about 20 kb). An important feature of these gels is their high resolving power. In certain kinds of gels, fragments differing in length by just one nucleotide out of several hundred can be distinguished. Moreover, entire chromosomes containing millions of nucleotides can now be separated on agarose gels by applying pulsed electric fields in different directions (p. 977). Bands or spots of radioactive DNA in gels can be visualized by autoradiography. Alternatively, a gel can be stained with ethidium bromide, which fluoresces an intense orange when bound to double-helical DNA (Figure 6-3). A band containing only 50 ng of DNA can readily be seen.

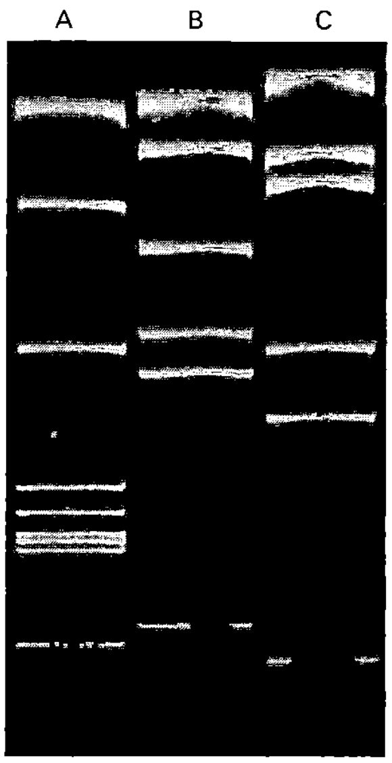

Figure 6-3
Gel electrophoresis pattern showing the fragments produced by cleaving SV40 DNA with each of three restriction enzymes. These fragments were made fluorescent by staining the gel with ethidium bromide. [Courtesy of Dr. Jeffrey Sklar.]

A restriction fragment containing a specific base sequence can be identified by hybridizing it with a labeled complementary DNA strand (Figure 6-4). A mixture of restriction fragments is separated by electrophoresis through an agarose gel, denatured to form single-stranded DNA, and transferred to a nitrocellulose sheet. The positions of the DNA fragments in the gel are preserved in the nitrocellulose sheet, where they can be hybridized with a ^{32}P-labeled single-stranded *DNA probe*. Autoradiogra-

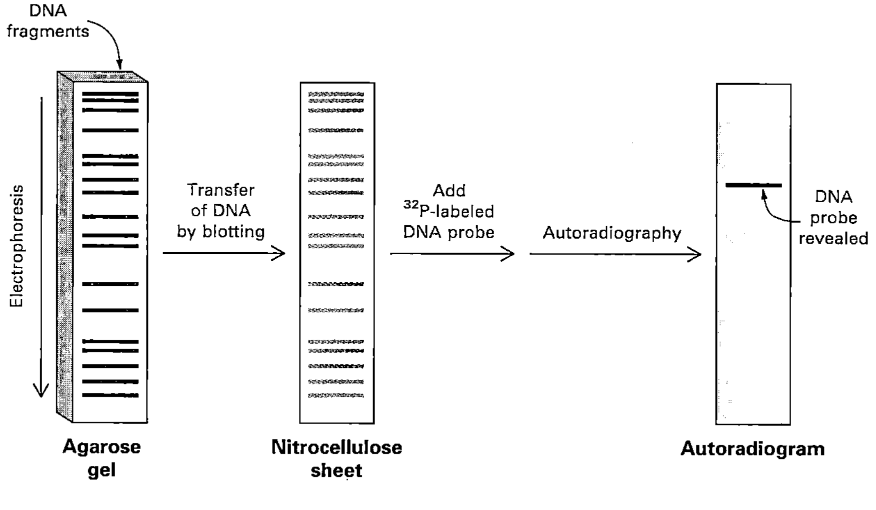

Figure 6-4
Southern blotting. A DNA fragment containing a specific sequence can be identified by separating a mixture of fragments by electrophoresis, transferring them to nitrocellulose, and hybridizing with a ^{32}P-labeled probe complementary to the sequence. The fragment containing the sequence is then visualized by autoradiography.

> *Restriction-fragment-length polymorphism (RFLP)*—
> Southern blotting can be used to follow the inheritance of selected genes. Mutations within restriction sites change the sizes of restriction fragments and hence the positions of bands in Southern blot analyses. The existence of genetic diversity in a population is termed polymorphism. The detected mutation may itself cause disease or it may be closely linked to one that does. Genetic diseases such as sickle-cell anemia (p. 169), cystic fibrosis, and Huntington's chorea can be detected by RFLP analyses.

phy then reveals the position of the restriction fragment with a sequence complementary to that of the probe. A particular fragment in the midst of a million others can readily be identified in this way, like finding a needle in a haystack. This powerful technique is known as *Southern blotting* because it was devised by E.M. Southern. Likewise, RNA molecules can be separated by gel electrophoresis, and specific sequences can be identified by hybridization following transfer to nitrocellulose. This analogous technique for the analysis of RNA has been whimsically termed *Northern blotting.* A further play on words accounts for the term *Western blotting,* which refers to a technique for detecting a particular protein by staining with specific antibody (p. 62). Southern, Northern, and Western blots are also known as *DNA, RNA,* and *protein blots.*

DNA CAN BE SEQUENCED BY SPECIFIC CHEMICAL CLEAVAGE (MAXAM-GILBERT METHOD)

The analysis of DNA structure and its relation to gene expression has also been markedly facilitated by the development of powerful techniques for the sequencing of DNA molecules. The *chemical cleavage method* devised by Allan Maxam and Walter Gilbert starts with a single-stranded DNA that is labeled at one end with ^{32}P. Polynucleotide kinase is usually used to add ^{32}P at the 5′-hydroxyl terminus. The labeled DNA is then broken preferentially at one of the four nucleotides. The conditions are chosen so that an average of one break is made per chain. In the reaction mixture for a given base, then, each broken chain yields a radioactive fragment extending from the ^{32}P label to one of the positions of that base, and such fragments are produced for every position of the base. For example, if the sequence is

5′-^{32}P-GCTACGTA-3′

the *radioactive* fragments produced by specific cleavage on the 5′ side of each of the four bases would be

Cleavage at A:	^{32}P-GCT
	^{32}P-GCTACGT
Cleavage at G:	^{32}P-GCTAC
Cleavage at C:	^{32}P-G
	^{32}P-GCTA
Cleavage at T:	^{32}P-GC
	^{32}P-GCTACG

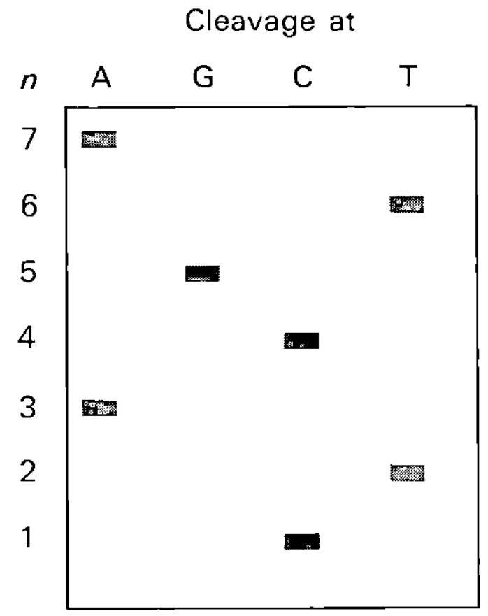

Figure 6-5
Schematic diagram of a gel showing the radioactive fragments formed by specific cleavage of 5′-^{32}P-GCTACGTA-3′ at each of the four bases (A, G, C, and T lanes). The number of nucleotides (n) in a fragment is shown on the left. The base sequence is read in ascending order.

The fragments in each mixture are then separated by polyacrylamide gel electrophoresis, which can resolve DNA molecules differing in length by just one nucleotide. The next step is to look at an autoradiogram of this gel. In this idealized example (Figure 6-5), the lowest band would be in the C lane, and the next one up in the T lane, followed by one in the A lane. Hence, the sequence of the first three nucleotides detected is 5′-CTA-3′ (the identity of the G at the 5′ end is not revealed). Reading all seven bands in ascending order gives the sequence 5′-CTACGTA-3′. Thus, *the autoradiogram of a gel produced from four different chemical cleavages displays a pattern of bands from which the sequence can be read directly.*

In practice, DNA is specifically cleaved by reagents that modify and then remove certain bases from their sugars. Purines are damaged by *dimethylsulfate,* which methylates guanine at N-7 and adenine at N-3. The glycosidic bond of a methylated purine is readily broken by heating at neutral pH, which leaves the sugar without a base. Subsequent heating in alkali leads to cleavage of the backbone at G, whereas treatment with dilute acid causes cleavage at both A and G. Cytosine and thymine are split off by *hydrazine*. The backbone is then cleaved at both C and T by *piperidine*. Hydrazinolysis in the presence of 2 M NaCl spares thymine. Treatment with piperidine then leads to cleavage only at C. Thus, DNA can be cleaved at only G, A and G, C and T, and only C, and the sequence can be deduced by comparing the four corresponding lanes in an autoradiogram (Figure 6-6).

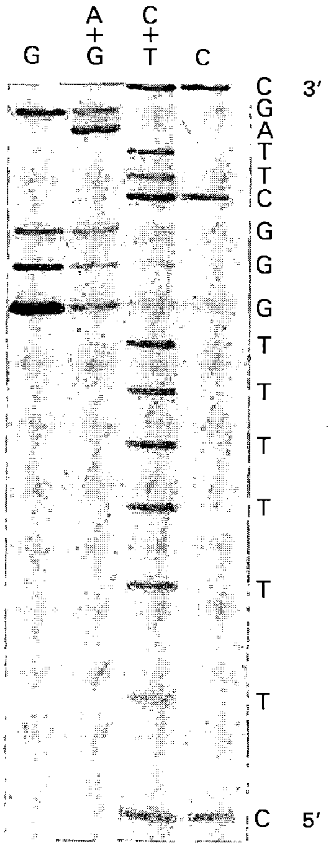

Figure 6-6
Autoradiogram of a gel showing labeled fragments produced by chemical cleavage. The 5′ end of the DNA strand was labeled with ^{32}P. The shortest fragment is at the bottom of the gel. Hence, the base sequence is 5′-CTTTTTTGGGCTTAGC-3′. [Courtesy of Dr. David Dressler.]

DNA IS USUALLY SEQUENCED BY CONTROLLED TERMINATION OF REPLICATION (SANGER DIDEOXY METHOD)

DNA can also be sequenced by generating fragments through the *controlled interruption of enzymatic replication,* a method developed by Frederick Sanger and co-workers. In fact, this is now the method of choice because of its simplicity. DNA polymerase I is used to copy a particular sequence of a single-stranded DNA. The synthesis is primed by a complementary fragment, which may be obtained from a restriction enzyme digest or synthesized chemically. In addition to the four deoxyribonucleoside triphosphates (radioactively labeled), the incubation mixture contains a *2′,3′-dideoxy analog* of one of them. The incorporation of this analog blocks further growth of the new chain because it lacks the 3′-hydroxyl terminus needed to form the next phosphodiester bond. Hence, fragments of various lengths are produced in which the dideoxy analog is at the 3′ end (Figure 6-7). Four such sets of *chain-terminated fragments* (one

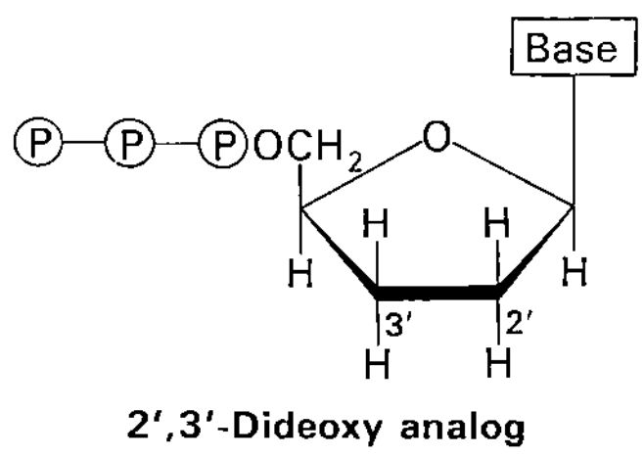

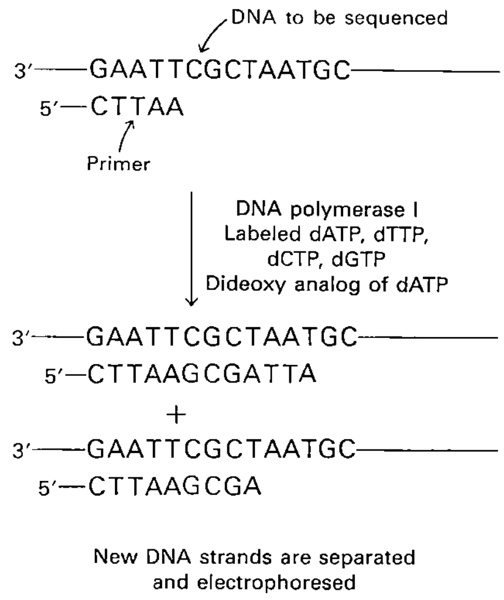

Figure 6-7
Strategy of the chain-termination method for the sequencing of DNA. Fragments are produced by adding the 2′,3′-dideoxy analog of a dNTP to each of four polymerization mixtures. For example, the addition of the dideoxy analog of dATP (shown in red) results in fragments ending in A.

for each dideoxy analog) are then electrophoresed, and the base sequence of the new DNA is read from the autoradiogram of the four lanes.

Fluorescence detection is a highly effective alternative to autoradiography. A fluorescent tag is attached to an oligonucleotide primer—a differently colored one in each of the four chain-terminating reaction mixtures (e.g., a blue emitter for termination at A and a red one for termination at C). The reaction mixtures are combined and electrophoresed together. The separated bands of DNA are then detected by their fluorescence as they emerge from the gel; the sequence of their colors directly gives the

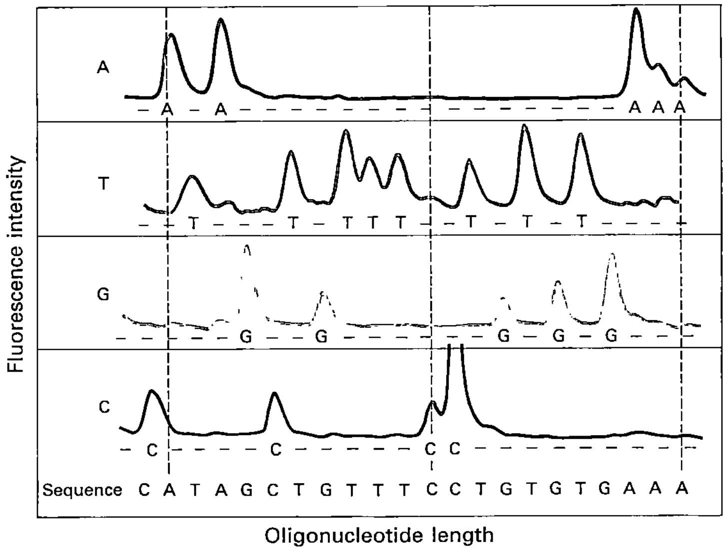

Figure 6-8
Fluorescence detection of oligonucleotide fragments produced by the dideoxy method. Each of the four chain-terminating mixtures is primed with a tag that fluoresces at a different wavelength (e.g., blue for A). The sequence determined by fluorescence measurements at four wavelengths is shown at the bottom of the figure. [From L.M. Smith, J.Z. Sanders, R.J. Kaiser, P. Hughes, C. Dodd, C.R. Connell, C. Heiner, S.B.H. Kent, and L.E. Hood. *Nature* 321(1986):674.]

base sequence (Figure 6-8). Sequences of up to 500 bases can be determined in this way. Fluorescence detection is attractive because it eliminates the use of radioactive reagents and can readily be automated.

The complete sequence of the 5386 bases in ϕX174 DNA was determined by Sanger and co-workers in 1977, just a quarter century after Sanger's pioneering elucidation of the amino acid sequence of a protein. This accomplishment is a landmark in molecular biology because it revealed the total information content of a DNA genome. This tour de force was followed several years later by the determination of the sequence of human mitochondrial DNA, a double-stranded circular DNA molecule containing 16,569 base pairs. It encodes 2 ribosomal RNAs, 22 transfer RNAs, and 13 proteins. This achievement was followed by the sequencing of the 48,513 base pairs of the DNA of λ phage, a virus that infects *E. coli.* Recently, an entire yeast chromosome, 310,000 base pairs long, was sequenced by a consortium of 35 laboratories. The wealth of information derived from the sequencing of more than 1.5×10^6 bases of DNA molecules from diverse sources will be considered in detail in later chapters. A major goal now is the sequencing of the entire human genome, which contains 3.5×10^9 base pairs.

DNA PROBES AND GENES CAN BE SYNTHESIZED BY AUTOMATED SOLID-PHASE METHODS

DNA strands, like polypeptides (p. 70), can be synthesized by the sequential addition of activated monomers to a growing chain that is linked to an insoluble support. The activated monomers are protonated *deoxyribonucleoside 3′-phosphoramidites* (Figure 6-9). In step 1, the 3′-phosphorus atom of this incoming unit becomes joined to the 5′-oxygen of the growing chain to form a *phosphite triester.* The 5′-OH of the activated monomer is unreactive because it is blocked by a dimethoxytrityl (DMT) protecting group. Likewise, amino groups on the purine and pyrimidine bases are blocked. Coupling is carried out under anhydrous conditions because water reacts with phosphoramidites. In step 2, the phosphite triester (in which P is trivalent) is oxidized by iodine to form a *phosphotriester* (in which P is pentavalent). In step 3, the DMT protecting group on the 5′-OH of the growing chain is removed by addition of dichloroacetic acid, which leaves

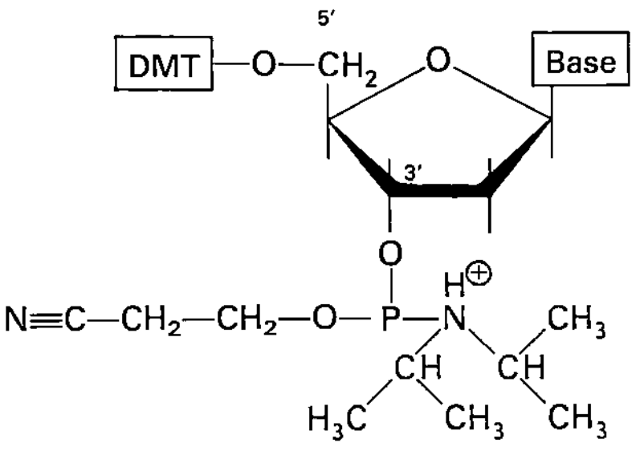

Protonated phosphoramidite
(The 5′-hydroxyl is blocked by a dimethoxytrityl protecting group)

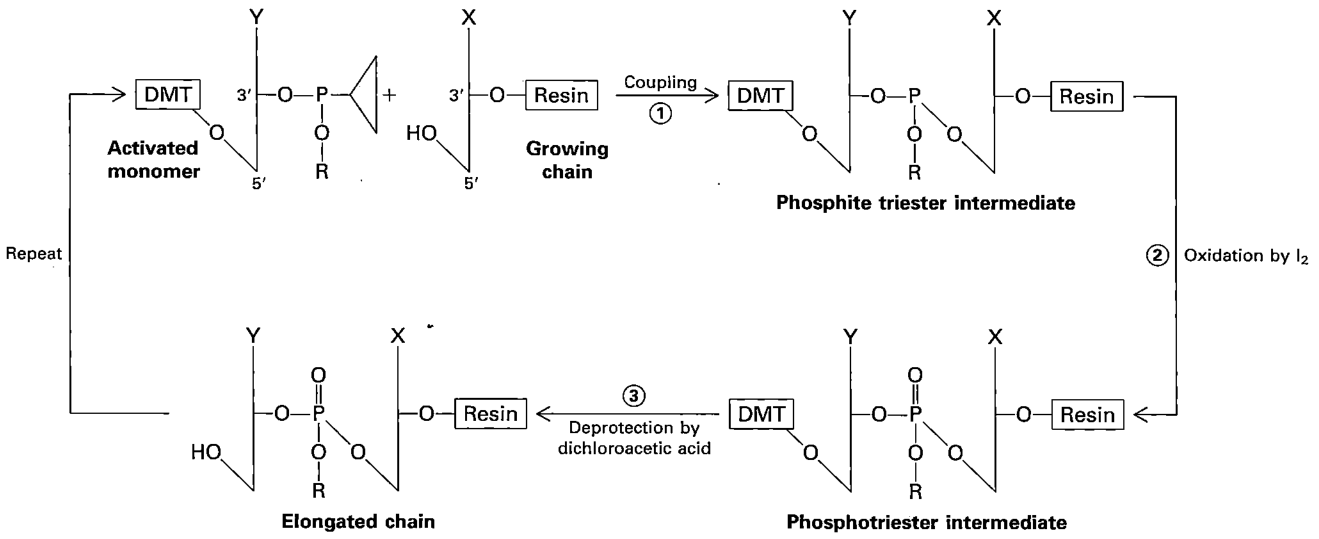

Figure 6-9
Solid-phase synthesis of a DNA chain by the phosphite triester method. The activated monomer added to the growing chain is a deoxyribonucleoside 3′-phosphoramidite containing a DMT (dimethoxytrityl) protecting group on its 5′-oxygen atom, a β-cyanoethyl (R) protecting group on its 3′-phosphoryl oxygen, and a protecting group on the purine of pyrimidine. The phosphoramidite group is denoted by a green triangle. The addition of NH_3 at the end of the synthesis removes all protecting groups and releases the oligonucleotide from the solid support.

other protecting groups intact. The DNA chain is now elongated by one unit and ready for another cycle of addition. Each cycle takes only about 10 minutes and elongates more than 98% of the chains.

This solid-phase approach is ideal for the synthesis of DNA, as it is for polypeptides, because the desired product stays on the insoluble support until the final release step. All the reactions occur in a single vessel, and excess soluble reagents can be added to drive reactions to completion. At the end of each step, soluble reagents and by-products are washed away from the glass beads that bear the growing chains. At the end of the synthesis, NH_3 is added to remove all protecting groups and release the oligonucleotide from the solid support. Because elongation is never 100% complete, the new DNA chains are of diverse lengths—the desired chain is the longest one. The sample can be purified by high-performance liquid chromatography or by electrophoresis on polyacrylamide gels. DNA chains up to 100 nucleotides long can readily be synthesized by this automated method.

The ability to rapidly synthesize DNA chains of any selected sequence opens many experimental avenues. For example, an oligonucleotide labeled at one end with ^{32}P or a fluorescent tag can be used to search for a complementary sequence in a very long DNA molecule or even in a genome consisting of many chromosomes (Figure 6-10). The use of labeled oligonucleotides as *DNA probes* is powerful and general. For example, a DNA probe that is base-paired to a known complementary sequence in a chromosome can serve as the starting point of an exploration of adjacent uncharted DNA. Such a probe can be used as a *primer* to initiate the replication of neighboring DNA by DNA polymerase. One of the most exciting applications of the solid-phase approach is the *synthesis of new tailor-made genes.* New proteins with novel properties can now be produced in abundance by expressing synthetic genes. *Protein engineering* has become a reality.

Figure 6-10
The locations of genes on chromosomes can be mapped by hybridizing complementary DNA probes that are tagged with fluorescent labels. The sites of six genes on human chromosome 5 are shown in this image, which was reconstructed from digital fluorescence scans. Each probe appears twice because the DNA in this metaphase chromosome is already duplicated. [After T. Ried, A. Baldini, T.C. Rand, and D.C. Ward. *Proc. Nat. Acad. Sci.* 89(1992):1388.]

DNA CAN BE SEQUENCED BY HYBRIDIZATION TO OLIGONUCLEOTIDE ARRAYS (DNA CHIPS)

The synthesis of *arrays of oligonucleotides* makes it feasible to sequence DNA in a new way. *Sequencing by hybridization (SBH)* uses a set of oligonucleotide probes of defined sequence to search for complementary sequences on a longer target DNA molecule. Suppose that a target 12-mer DNA with the sequence

5′-AGCCTAGCTGAA-3′

is mixed with a complete set of octanucleotide probes. Only 5 of the 65,536 (4^8) octamers would form hybrids if perfect complementarity were required. The sequence of the target 12-mer can then be reconstructed by aligning the overlapping sequences of the hybridizing oligomers:

```
3′-TCGGATCG-5′
     CGGATCGA
       GGATCGAC
         GATCGACT
           ATCGACTT
3′-TCGGATCGACTT-5′
5′-AGCCTAGCTGAA-3′
```

In practice, the hybridization pattern is more complex because oligonucleotides that are not perfectly complementary to the target will still form duplexes, though with lower affinity.

High-density arrays of oligonucleotides on a solid support can be formed using light to direct the pattern of synthesis. A photolabile protecting group is used in place of the acid-labile dimethoxytrityl group. Illumination through a photolithographic mask releases the photolabile protecting group in selected regions. An array of all 65,536 octamers has been synthesized in a 1.6 cm^2 area, about the size of a thumbnail. The hybridization of fluorescent-labeled target DNA to complementary probes on the surface of a *DNA chip* can readily be visualized by fluorescence microscopy (Figure 6-11).

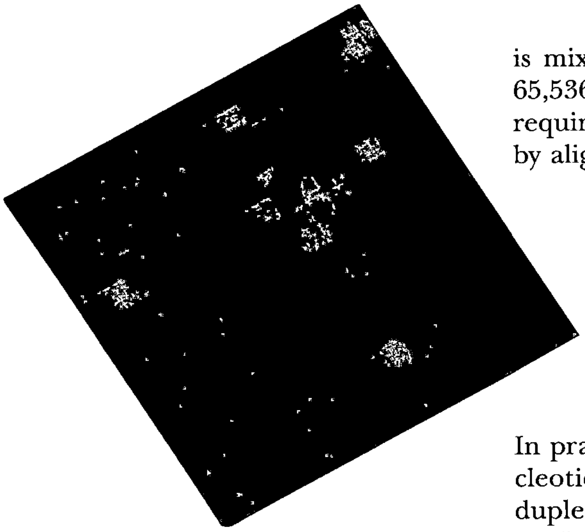

Figure 6-11
DNA sequencing by hybridization. The principle is illustrated here by the hybridization of a fluorescent-labeled oligonucleotide of known sequence (5′-GCGGCGGC-3′) to a light-generated array of 256 octanucleotides on a solid support. The central four bases in the sequence 3′-CGXXXXCG-5′ are varied on this chip. The fluorescence intensity of the chip after hybridization is denoted by the height and color of each pixel. The single red pixel has the sequence 3′-CGCCGCCG-5′, the perfect complement of the fluorescent-labeled target DNA. The weakly fluorescent pixels have sequences that differ from the perfect complement at one of the central four bases. [After A.C. Pease, D. Solas, E.J. Sullivan, M.T. Cronin, C.P. Holmes, and S.P.A. Fodor. *Proc. Nat. Acad. Sci.* 91(1994):5022–5026.

RESTRICTION ENZYMES AND DNA LIGASE ARE KEY TOOLS IN FORMING RECOMBINANT DNA MOLECULES

The pioneering work of Paul Berg, Herbert Boyer, and Stanley Cohen in the early 1970s led to the development of recombinant DNA technology, which has revolutionized biochemistry. New combinations of unrelated genes can be constructed in the laboratory by applying recombinant DNA techniques. These novel combinations can be cloned—amplified manyfold—by introducing them into suitable cells, where they are replicated by the DNA synthesizing machinery of the host. The inserted genes are often transcribed and translated in their new setting. What is most striking is that the genetic endowment of the host can be permanently altered in a designed way.

Let us begin by seeing how novel DNA molecules can be constructed in the laboratory. A DNA fragment of interest is covalently joined to a DNA *vector*. The essential feature of a vector is that it can replicate autonomously in an appropriate host. Plasmids (naturally occurring circles of DNA that act as accessory chromosomes) and λ phage (a virus) are choice vectors for cloning in *E. coli*. The vector can be prepared for splicing by cleaving it at a single specific site with a restriction enzyme. For example, the plasmid pSC101, a 9.9-kb double-helical circular DNA molecule, is split at a unique site by the EcoRI restriction enzyme. The staggered cuts made by this enzyme produce *complementary single-stranded ends,* which have specific affinity for each other and hence are known as *cohesive ends.* Any DNA fragment can be inserted into this plasmid if it has the same cohesive ends. Such a fragment can be prepared from a larger piece of DNA by using the same restriction enzyme as was used to open the plasmid DNA (Figure 6-12). The single-stranded ends of the fragment are then complementary to those of the cut plasmid. The DNA fragment and the cut plasmid can be annealed and then joined by *DNA ligase,* which catalyzes the formation of a phosphodiester bond at a break in a DNA chain. DNA ligase requires a free 3′-OH group and a 5′-phosphate group. Furthermore, the chains joined by ligase must be in a double helix. An energy source, such as ATP or NAD^+, is required for the joining reaction, as will be discussed in Chapter 31.

Chimeric DNA—
A recombinant DNA molecule containing unrelated genes. From *chimera,* a mythological creature with the head of a lion, the body of a goat, and the tail of a serpent:

"a thing of immortal make, not human, lion-fronted and snake behind, a goat in the middle, and snorting out the breath of the terrible flame of bright fire . . . ."

Iliad (6.179)

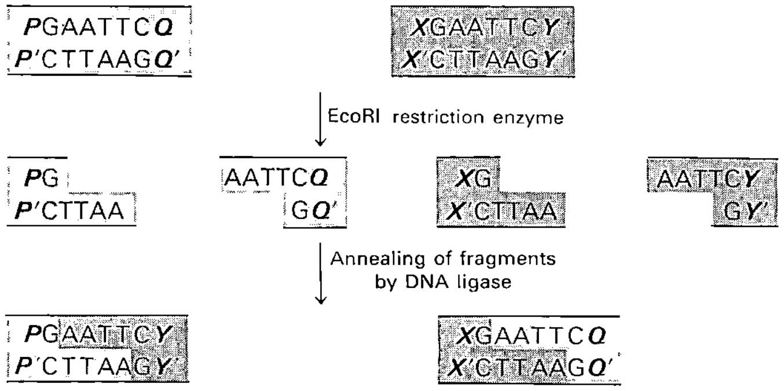

Figure 6-12
Joining of DNA molecules by the cohesive-end method. One of the parental DNA molecules (shown in green) carries genes *P* and *Q* separated by a restriction site, and the other (shown in red) carries *X* and *Y*. One of the recombinant molecules contains *P* and *Y*, and the other contains *Q* and *X*.

This cohesive-end method for joining DNA molecules can be made general by using a *short, chemically synthesized DNA linker* that can be cleaved by restriction enzymes. First, the linker is covalently joined to the ends of a DNA fragment or vector. For example, the 5′ ends of a decameric linker and a DNA molecule are phosphorylated by polynucleotide kinase and then joined by the ligase from T4 phage (Figure 6-13). This ligase can form a covalent bond between blunt-ended (flush-ended) double-helical DNA molecules. Cohesive ends are produced when these terminal extensions are cut by an appropriate restriction enzyme. Thus, *cohesive ends*

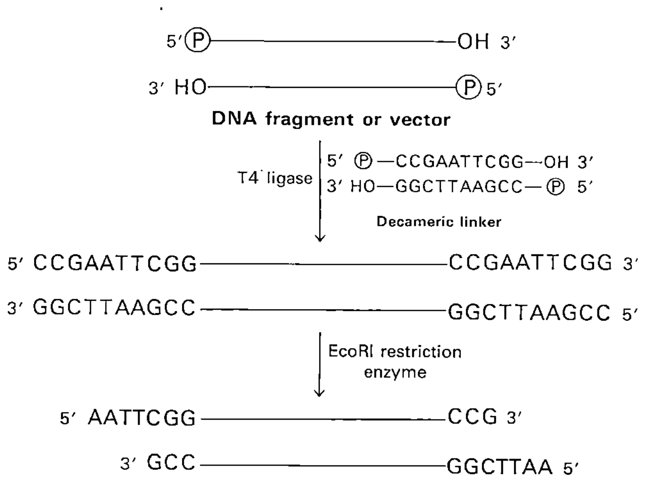

Figure 6-13
Formation of cohesive ends by the addition and cleavage of a chemically synthesized linker.

corresponding to a particular restriction enzyme can be added to virtually any DNA molecule. We see here the fruits of combining enzymatic and synthetic chemical approaches in crafting new DNA molecules.

PLASMIDS AND LAMBDA PHAGE ARE CHOICE VECTORS FOR DNA CLONING IN BACTERIA

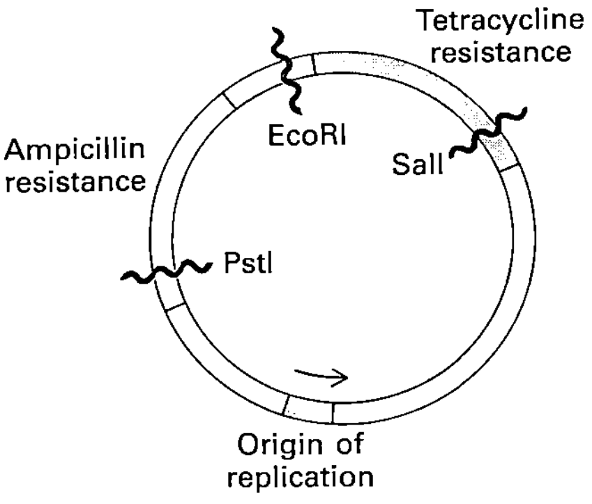

Figure 6-14
Genetic map of pBR322, a plasmid with two genes for antibiotic resistance. pBR322, like all other plasmids, is a circular duplex DNA.

Many plasmids and bacteriophages have been ingeniously modified to enhance the delivery of recombinant DNA molecules into bacteria and to facilitate the selection of bacteria harboring them. *Plasmids* are circular duplex DNA molecules occurring naturally in some bacteria and ranging in size from two to several hundred kilobases. They carry genes for the inactivation of antibiotics, the production of toxins, and the breakdown of natural products. These *accessory chromosomes* can replicate independently of the host chromosome. In contrast with the host genome, they are dispensable under certain conditions. A bacterial cell may have no plasmids at all, or it may house as many as 20 copies of a plasmid.

One of the most useful plasmids for cloning is pBR322, which contains genes for resistance to tetracycline and ampicillin (an antibiotic like penicillin). This plasmid can be cleaved at a variety of unique sites by different endonucleases, and DNA fragments inserted. Insertion of DNA at the EcoRI restriction site does not alter either of the genes for antibiotic resistance (Figure 6-14). However, insertion at the HindIII, SalI, or BamHI restriction site inactivates the gene for tetracycline resistance, an effect called *insertional inactivation.* Cells containing pBR322 with a DNA insert at one of these restriction sites are resistant to ampicillin but sensitive to tetracycline, and so they can be readily *selected.* Cells that failed to take up the vector are sensitive to both antibiotics, whereas cells containing pBR322 without a DNA insert are resistant to both.

Lambda (λ) phage is another widely used vector (Figure 6-15). This bacteriophage enjoys a choice of life styles: it can destroy its host or it can become part of its host (Chapter 36). In the *lytic pathway,* viral functions are fully expressed: viral DNA and proteins are quickly produced and packaged into virus particles, leading to the lysis (destruction) of the host cell and the sudden appearance of about 100 progeny virus particles, or *virions.* In the *lysogenic pathway,* the phage DNA becomes inserted into the host-cell genome and can be replicated together with host-cell DNA for

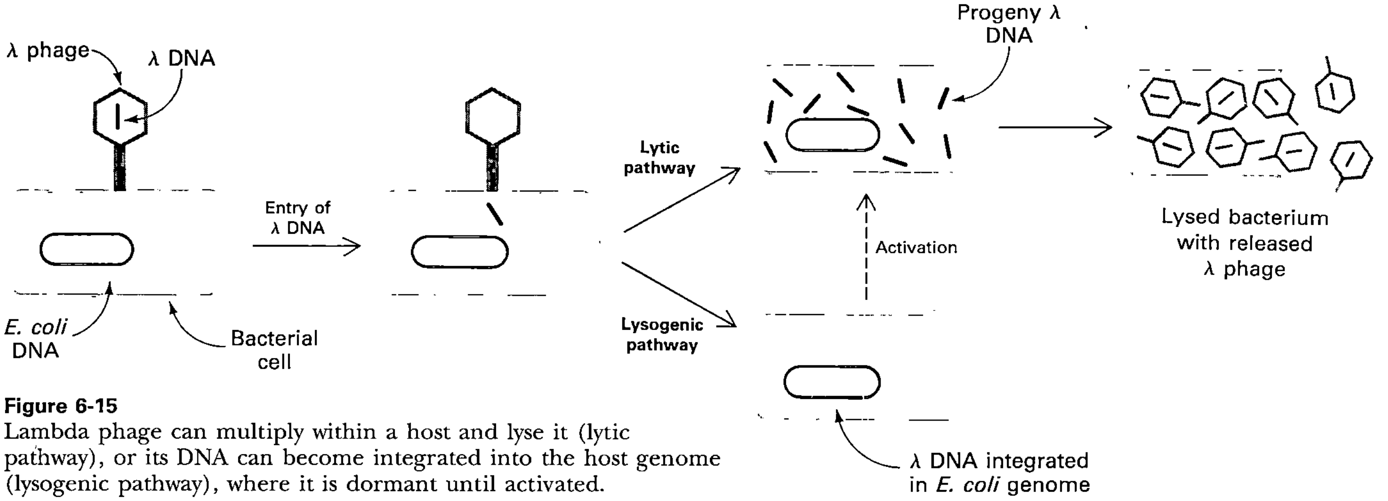

Figure 6-15
Lambda phage can multiply within a host and lyse it (lytic pathway), or its DNA can become integrated into the host genome (lysogenic pathway), where it is dormant until activated.

many generations, remaining inactive. Certain environmental changes can trigger the expression of this dormant viral DNA, which leads to the formation of progeny virus and lysis of the host. Large segments of the 48-kb DNA of λ phage are not essential for productive infection and can be replaced by foreign DNA.

Mutant λ phages designed for cloning have been constructed. An especially useful one called λgt-λβ contains only two EcoRI cleavage sites instead of the five normally present (Figure 6-16). After cleavage, the middle segment of this λ DNA molecule can be removed. The two remaining pieces of DNA have a combined length equal to 72% of a normal genome length. This amount of DNA is too little to be packaged into a λ particle. The range of lengths that can be readily packaged is from 75% to 105% of a normal genome length. However, *a suitably long DNA insert (such as 10 kb) between the two ends of λ DNA enables such a recombinant DNA molecule (93% of normal length) to be packaged.* Nearly all infective λ particles formed in this way will contain an inserted piece of foreign DNA. Another advantage of using these modified viruses as vectors is that they enter bacteria much more easily than do plasmids. A variety of λ mutants have been constructed for use as cloning vectors. One of them, called a *cosmid,* can serve as a vector for large DNA inserts (up to ~45 kb).

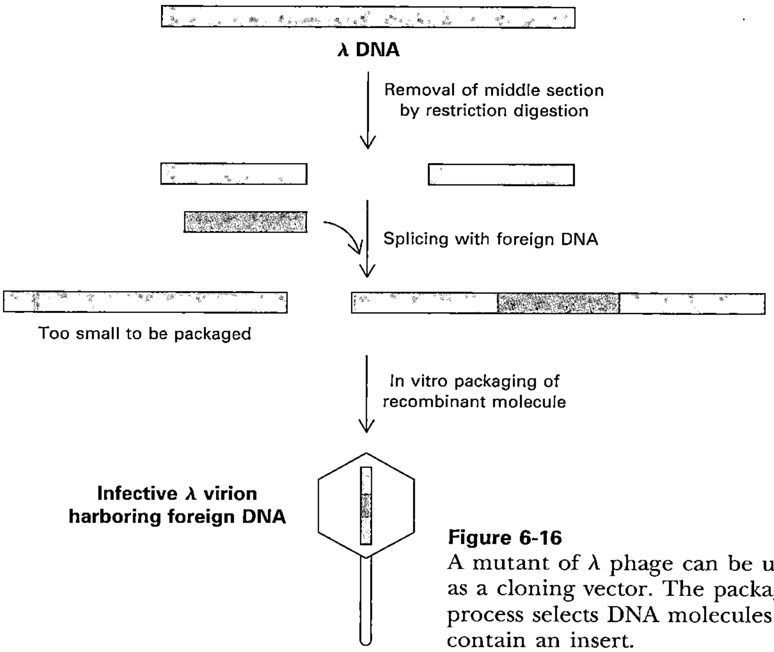

Figure 6-16
A mutant of λ phage can be used as a cloning vector. The packaging process selects DNA molecules that contain an insert.

M13 phage is another very useful vector for cloning DNA. This filamentous virus is 900 nm long and only 9 nm wide (Figure 6-17). Its 6.4-kb single-stranded circle of DNA is protected by a coat of 2710 identical protein subunits. M13 enters *E. coli* through the bacterial sex pilus, a protein appendage. The single-stranded DNA in the virus particle (called the (+) strand) is replicated through an intermediate circular double-stranded replicative form (RF) containing (+) and (−) strands, much as in φX174 (p. 90). Only the (+) strand is packaged into new virus particles. About a thousand progeny M13 are produced per generation. A striking feature of M13 is that it does not kill its bacterial host. Consequently, large quantities of M13 can be grown and easily harvested (1 gram from 10 liters of culture fluid).

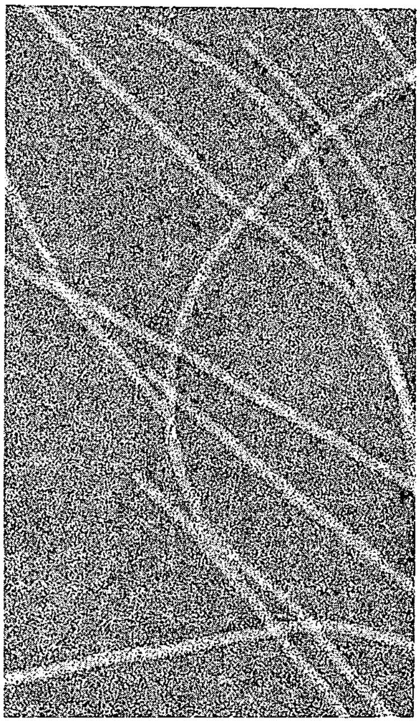

Figure 6-17
Electron micrograph of M13 filamentous phage. [Courtesy of Dr. Robley Williams.]

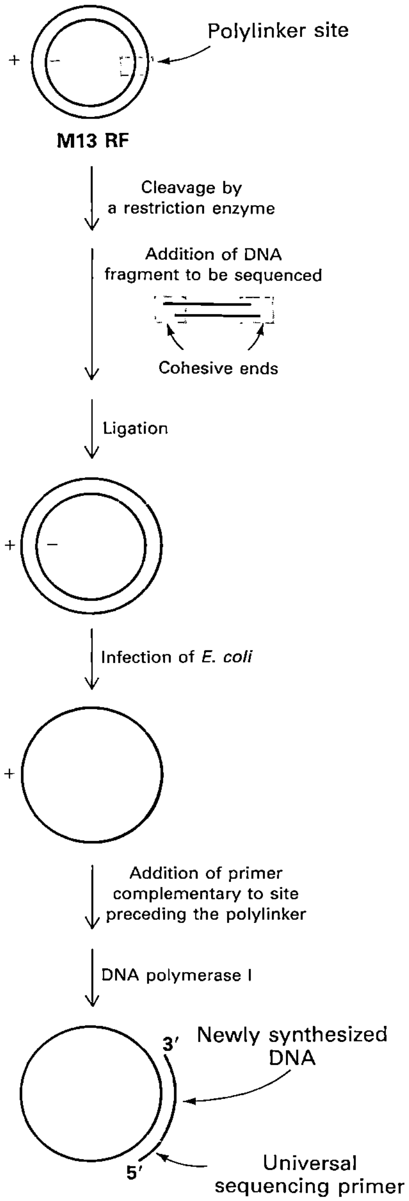

Figure 6-18
M13 phage DNA is very useful in sequencing DNA fragments by the dideoxy method. A double-stranded DNA fragment is inserted into M13 RF DNA. Synthesis of a new strand is primed by an oligonucleotide that is complementary to a sequence near the inserted DNA.

An M13 vector is prepared for cloning by cutting its circular RF DNA at a single site with a restriction enzyme. The cut is made in a *polylinker* region that contains a series of closely spaced recognition sites for restriction enzymes; each such site occurs only once in the vector. A double-stranded foreign DNA fragment produced by cleavage with the same restriction enzyme is then ligated to the cut vector (Figure 6-18). The foreign DNA can be inserted in two different orientations because the ends of both DNA molecules are the same. Hence, half the new (+) strands packaged into virus particles will contain one of the strands of the foreign DNA, and half will contain the other strand. Infection of *E. coli* by a single virus particle will yield a large amount of single-stranded M13 DNA containing the same strand of the foreign DNA. An oligonucleotide that hybridizes adjacent to the polylinker region is used as a primer for sequencing the insert. This oligomer is called a *universal sequencing primer* because it can be used to sequence *any* insert. M13 is ideal for sequencing but not for long-term propagation of recombinant DNA, because inserts longer than about 1 kb are not stably maintained.

SPECIFIC GENES CAN BE CLONED FROM A DIGEST OF GENOMIC DNA

Ingenious cloning and selection methods have made feasible the isolation of a specific DNA segment several kilobases long out of a genome containing more than 3×10^6 kb. Let us see how a gene that occurs just once in a human genome can be cloned. A sample containing many molecules of total genomic DNA is first mechanically sheared or partly digested by restriction enzymes into large fragments (Figure 6-19). This nearly random population of overlapping DNA fragments is then separated by gel electrophoresis to isolate a set about 15 kb long. Synthetic linkers are attached to the ends of these fragments, cohesive ends are formed, and

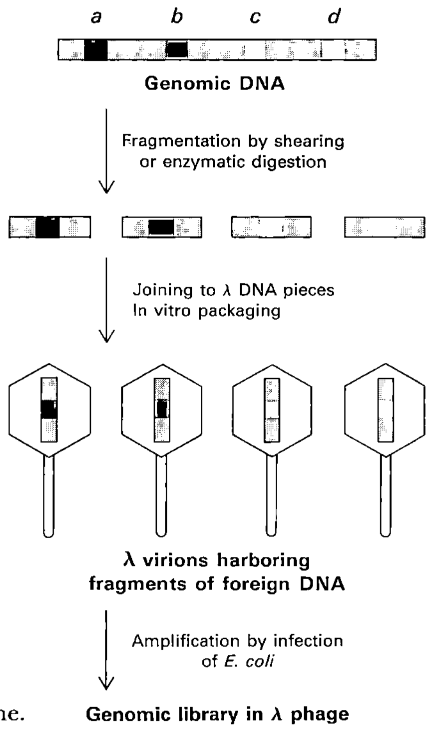

Figure 6-19
Creating a genomic library from a digest of a whole eukaryotic genome.

the fragments are then inserted into a vector, such as λ phage DNA, prepared with the same cohesive ends. *E. coli* are then infected with these recombinant phages. The resulting lysate contains fragments of human DNA housed in a sufficiently large number of virus particles to ensure that nearly the entire genome is represented. These phage constitute a *genomic library*. Phage can be propagated indefinitely, and so the library can be used repeatedly over long periods.

This genomic library is then screened to find the very small portion of phage harboring the gene of interest. A calculation shows that a 99% probability of success requires screening about 500,000 clones; hence, a very rapid and efficient screening process is essential. This can be accomplished by DNA hybridization.

A dilute suspension of the recombinant phage is first plated on a lawn of bacteria (Figure 6-20). Where each phage particle has landed and infected a bacterium, a *plaque* containing identical phage develops on the plate. A replica of this master plate is then made by applying a sheet of nitrocellulose. Infected bacteria and phage DNA released from lysed cells adhere to the sheet in a pattern of spots corresponding to the plaques. Intact bacteria on this sheet are lysed with NaOH, which also serves to denature the DNA so that it becomes accessible for hybridization with a ^{32}P-labeled probe. *The presence of a specific DNA sequence in a single spot on the replica can be detected by using a radioactive complementary DNA or RNA molecule as a probe.* Autoradiography then reveals the positions of spots harboring recombinant DNA. The corresponding plaques are picked out of the intact master plate and grown. A million clones can readily be screened in a day by a single investigator.

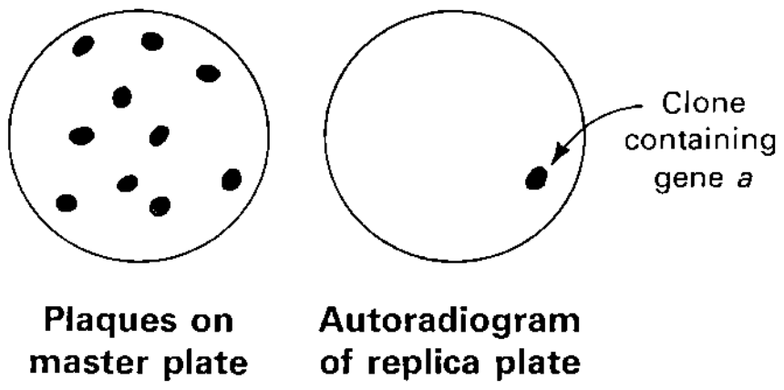

Figure 6-20
Screening a genomic library for a specific gene. Here, a plate is tested for plaques containing gene *a* of Figure 6-19.

This method makes it possible to isolate virtually any gene, *provided that a probe is available.* How does one obtain a specific probe? One approach is to *start with the corresponding mRNA from cells in which it is abundant.* For example, precursors of red blood cells contain large amounts of mRNAs for hemoglobin, and plasma cells are rich in mRNAs for antibody molecules. The mRNAs from these cells can be fractionated by size to enrich for the one of interest. As will be described shortly, a DNA complementary to this mRNA can be synthesized in vitro and cloned to produce a highly specific probe.

Alternatively, *a probe for a gene can be prepared if part of the amino acid sequence of the protein encoded by the gene is known.* A problem arises because a given peptide sequence can be encoded by a number of oligonucleotides. Thus, for this purpose, peptide sequences containing tryptophan and methionine are preferred, because these amino acids are specified by a single codon, whereas other amino acid residues have between two and six codons (p. 109). The strategy is to choose a peptide that has a high proportion of tryptophan and methionine. For example, even for the pentapeptide

Trp-Tyr-Met-Cys-Met

there are four possible DNA coding sequences because tyrosine and cysteine can each be specified by either of two codons.

3′ ACC-ATA-TAC-ACA-TAC 5′
ACC-ATG-TAC-ACA-TAC
ACC-ATA-TAC-ACG-TAC
ACC-ATG-TAC-ACG-TAC
Trp-Tyr-Met-Cys-Met

All the coding DNA sequences (or their complements) are synthesized by the solid-phase method and made radioactive by phosphorylating their 5′ ends with ^{32}P-orthophosphate. The replica plate is exposed to a mixture of these probes and autoradiographed to identify clones with a complementary DNA sequence. Positive clones are then sequenced to determine which ones have a sequence matching that of the protein of interest. Some of them may contain the desired gene or a significant segment of it.

LONG STRETCHES OF DNA CAN BE EFFICIENTLY ANALYZED BY CHROMOSOME WALKING

A typical genomic DNA library housed in λ phage vectors consists of DNA fragments about 15 kb long. However, many eukaryotic genes are much longer—for example, the dystrophin gene, which is mutated in Duchenne muscular dystrophy, is 2000 kb long. How can such long stretches of DNA be analyzed? The development of cosmids helped because these chimeras of plasmids and λ phages can house 45-kb inserts. Much larger pieces of DNA can now be propagated in *yeast artificial chromosomes (YACs)*. YACs contain a centromere, an *autonomous replication sequence (ARS,* where replication begins), a pair of telomeres (normal ends of eukaryotic chromosomes), selectable marker genes, and a cloning site (Figure 6-21). Genomic DNA is partially digested using a restriction endonuclease that cuts, on the average, at distant sites. The fragments are then separated by pulse field gel electrophoresis, and the large ones (~450 kb) are eluted and ligated into YACs. Artificial chromosomes bearing inserts of 100 to 1000 kb are efficiently replicated in yeast cells.

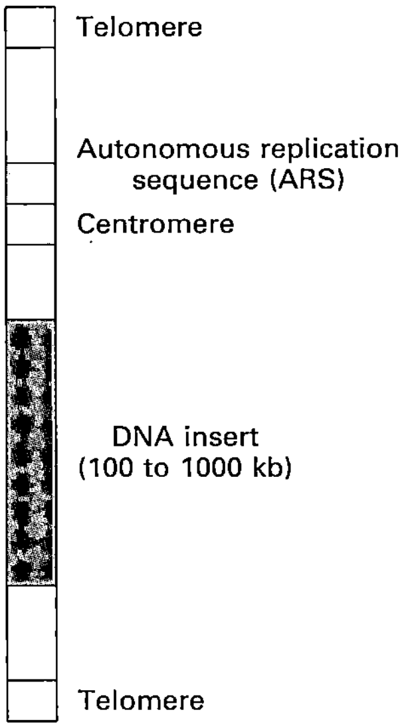

Figure 6-21
Schematic diagram of a yeast artificial chromosome (YAC). DNA inserts as large as 1000 kb can be propagated in this vector.

Equally important in analyzing large genes is the capacity to scan long regions of DNA. The fragments in a cosmid or YAC library are produced by random cleavage of many DNA molecules, and so some of the fragments overlap one another. Suppose that a fragment containing region *A* selected by hybridization with a complementary probe *A′* also contains region *B* (Figure 6-22). A new probe *B′* can be prepared by cleaving this fragment and subcloning region *B*. If the library is screened again with probe *B′*, new fragments containing region *B* will be found. Some will contain a previously unknown region *C*. Hence, we now have information about a segment of DNA encompassing regions *A*, *B*, and *C*. This process of subcloning and rescreening is called *chromosome walking*. Long stretches of DNA can be analyzed in this way, provided that each of the new probes is complementary to a unique region.

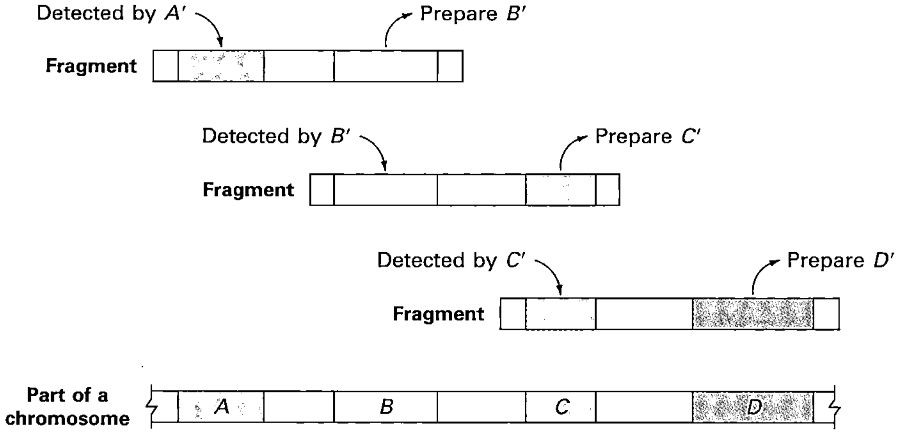

Figure 6-22
Chromosome walking. Long regions of unknown DNA can be explored starting with a known base sequence by subcloning and rescreening. New probes are designed on the basis of the DNA sequences that have been determined.

SELECTED DNA SEQUENCES CAN BE GREATLY AMPLIFIED BY THE POLYMERASE CHAIN REACTION (PCR)

In 1984, Kary Mullis devised an ingenious method called the *polymerase chain reaction (PCR)* for amplifying specific DNA sequences. Consider a DNA duplex consisting of regions *ABCDE*. Millions of copies of *C* (the *target*) can readily be obtained by PCR if the sequences of *B* and *D* (the *flanking sequences*) are known. Let us denote one strand of this duplex by *a-b-c-d-e* and the complementary one by *a′-b′-c′-d′-e′*. PCR is carried out by adding the following components to a solution containing the target sequence: (1) a pair of primers, *b* and *d′*, (2) all four deoxyribonucleoside triphosphates (dNTPs), and (3) a heat-stable DNA polymerase. A PCR cycle consists of three steps (Figure 6-23).

1. *Strand separation.* The two strands of the parent DNA molecule are separated by heating the solution to 95°C for 15 s.

2. *Hybridization of primers.* The solution is then abruptly cooled to 54°C to allow each primer to hybridize to a DNA strand. Primer *b* hybridizes to *b′* on one strand, and primer *d′* hybridizes to *d* on the complementary strand. Parent DNA duplexes are not formed because the primers are present in large excess. Primers are typically 20 to 30 nucleotides long.

3. *DNA synthesis.* The solution is then heated to 72°C, the optimal temperature for *Taq* DNA polymerase. This heat-stable polymerase comes from *Thermus aquaticus,* a thermophilic bacterium. Elongation of both primers occurs in the direction of the target sequence because the 3′ end of primer *d′* faces *c*, and the 3′ end of primer *b* faces *c′*. Polymerization is allowed to proceed for 30 s. One of the new DNA strands is *b-c-d-e* and the other is *a′-b′-c′-d′*. Hence, both strands of the target are replicated.

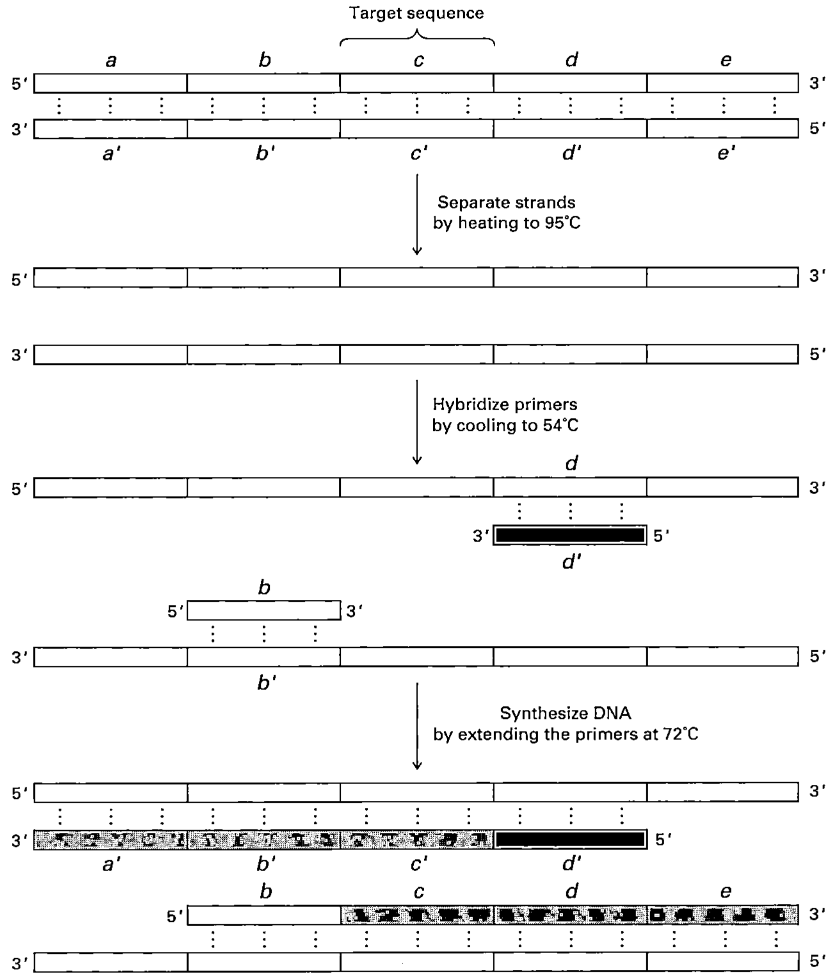

Figure 6-23
Polymerase chain reaction (PCR). A cycle consists of three steps: strand separation, hybridization of primers, and extension of primers by DNA synthesis. The reactions are carried out in a closed vessel. The cycle is driven by changes in temperature. Sequences on one strand of the original DNA are denoted by *abcde* and those on the complementary strand by *a′b′c′d′e′*. Primer *b* is shown in yellow and primer *d′* in blue; new DNA is shown in red.

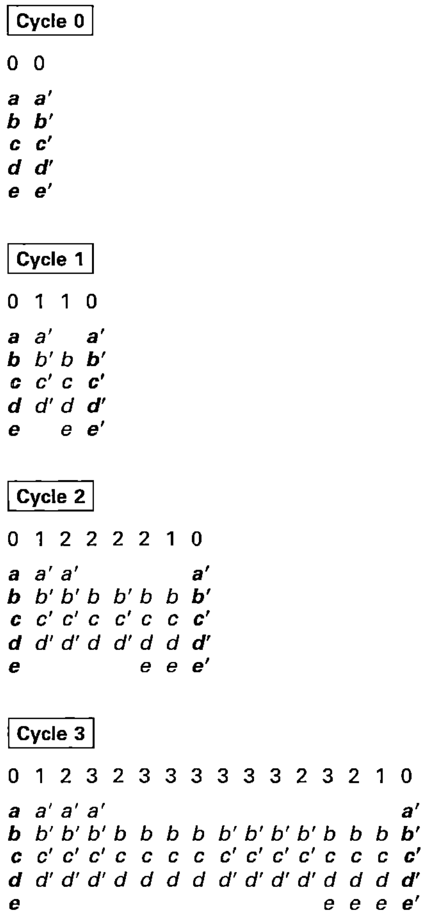

Figure 6-24
Products of three cycles of the polymerase chain reaction. The addition of primers *b* and *d'* results in an exponential amplification of the target sequence *c* and its complement *c'* (both shown in green). The numerals denote the cycle in which the sequence was produced.

These three steps—*strand separation, hybridization of primers, and DNA synthesis*—can be carried out repetitively just by changing the temperature of the reaction mixture. The thermostability of the polymerase makes it feasible to carry out PCR in a closed container; no reagents are added after the first cycle. A key feature of PCR is that *all the new DNA strands serve as templates in successive cycles.* Specifically, *b-c-d-e* formed in the first cycle serves as the template for the synthesis of *b'-c'-d'* in the second cycle and subsequent ones. Likewise, *a'-b'-c'-d'* serves as the template for the synthesis of *b-c-d*. At the end of the third cycle, *b-c-d* and *b'-c'-d'* comprise half the DNA strands (Figure 6-24). *This DNA, consisting of the target sequence flanked by the primers, increases exponentially in subsequent cycles, whereas the other DNA sequences in the mixture increase only linearly.* Hence, nearly all the DNA after a few cycles is *BCD*. Ideally, after *n* cycles, this sequence is amplified 2^n-fold. The amplification is a millionfold after 20 cycles, and a billionfold after 30 cycles, which can be carried out in less than an hour.

Several features of this remarkable method for amplifying DNA are noteworthy. First, the sequence of the target *C* need not be known. All that is required is knowledge of the flanking sequences *B* and *D*. Second, the target can be much larger than the primers. Targets as large as 10 kb have been amplified by PCR. Third, primers do not have to be perfectly matched to flanking sequences to amplify targets. Knowing the sequence of a gene makes it possible to search for variations on the theme. Families of genes are being discovered by PCR. Fourth, PCR is highly specific because of the stringency of hybridization at high temperature. The only DNA that is amplified is that situated between primers that have hybridized. A gene comprising less than a millionth of the total DNA of a higher organism is accessible by PCR. Fifth, PCR is exquisitely sensitive. *A single DNA molecule can be amplified and detected.*

PCR IS A POWERFUL TECHNIQUE IN MEDICAL DIAGNOSTICS, FORENSICS, AND MOLECULAR EVOLUTION

DNA sequencing and cloning have been greatly simplified by the application of PCR. Furthermore, the scope and power of recombinant DNA technology have been markedly enhanced by this innovative technique. PCR can provide valuable diagnostic information in medicine. Bacteria and viruses can be readily detected using specific primers. For example, PCR can reveal the presence of human immunodeficiency virus-1 (HIV-1) in individuals who have not mounted an immune response to this pathogen and therefore would be missed with an antibody assay. Finding *Mycobacterium tuberculosis* bacilli in tissue specimens is slow and laborious. With PCR, as few as 10 tubercle bacilli per million human cells can readily be detected. PCR is also a promising method for the early detection of cer-

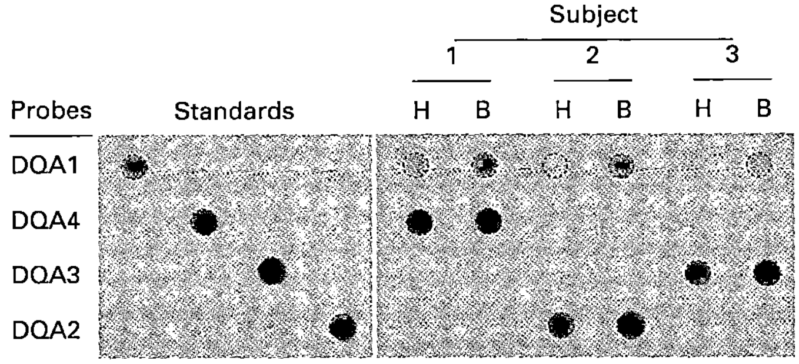

Figure 6-25
The HLA type of a person can be determined by PCR analysis of a single hair (H) or from a sample of blood (B). A 245-bp region of genomic DNA was amplified by PCR and immobolized on a nylon membrane. These "dots" from different people were then hybridized with *allele-specific oligonucleotide probes (ASOs)*. Hybrids were then detected using an enzyme (horseradish peroxidase) to generate a visible product. The three subjects tested have different HLA types. [After R. Higuchi, C.H. von Beroldingen, G.F. Sensabaugh, and H.A. Erlich. *Nature* 332(1988):543.]

tain cancers. Mutations of certain growth-control genes, such as the *RAS* genes, can be identified by PCR. The capacity to greatly amplify selected regions of DNA can also be highly informative in monitoring cancer chemotherapy. Ideally, treatment should be stopped when cancerous cells have been eliminated and resumed immediately if a relapse occurs. PCR is ideal for detecting leukemias caused by chromosomal rearrangements.

PCR is also making an impact in forensics and legal medicine. The DNA profile of an individual is highly distinctive because many genetic loci are highly variable within a population. For example, organ transplants are rejected when the HLA types (human leukocyte antigen types) of the donor and recipient are not sufficiently matched (p. 380). PCR amplification of multiple genes is being used to establish biological parentage in disputed paternity and immigration cases. Analyses of blood stains and semen samples by PCR have been informative in numerous assault and rape cases. Hairs are often found at the scene of crimes. The root of a single shed hair contains enough DNA for typing by PCR (Figure 6-25).

Furthermore, PCR enables us to reconstruct DNA from ancient samples by amplifying the rare surviving fragments. Portions of some genes of 2400-year-old Egyptian mummies and of 7500-year-old archaeological remains have recently been read using PCR. The HLA types of ancient humans provide glimpses into the population dynamics of their communities. The oldest DNA analyzed thus far comes from a termite embalmed in amber (Figure 6-26). The DNA in this sample is 25–30 million years old! The sequences of ribosomal RNA genes of this fossil termite from the Miocene epoch provide insights into the evolution of termites and cockroaches. DNA from ancient plants have also been revealing. A key chloroplast gene from a magnolia fossil of the same epoch has been resurrected. Molecular archaeology and molecular paleontology have sprung to life. PCR has given ancient remains a new voice to tell their tale.

Figure 6-26
This fossil termite encased in amber was the source of 25–30-million-year-old DNA. PCR analysis of this ancient DNA revealed the sequence of a gene for ribosomal RNA. Comparisons of DNA sequences of extinct and living species can reveal missing links in evolution. [After R. DeSalle, J. Gatesy, W. Wheeler, and D. Grimaldi. *Science* 257(1992):1933.]

COMPLEMENTARY DNA (cDNA) PREPARED FROM mRNA CAN BE EXPRESSED IN HOST CELLS

Can mammalian DNA be cloned and expressed by *E. coli*? Recall that most mammalian genes are mosaics of introns and exons. These interrupted genes cannot be expressed by bacteria, which lack the machinery to splice introns out of the primary transcript. However, this difficulty can be circumvented by causing bacteria to take up recombinant DNA that is complementary to mRNA. For example, proinsulin, a precursor of insulin, is synthesized by bacteria harboring plasmids that contain DNA complementary to mRNA for proinsulin (Figure 6-27). Indeed, much of the insulin used today by millions of diabetics is produced by bacteria.

Figure 6-27
Synthesis of proinsulin, a precursor of insulin, by transformed (genetically altered) clones of *E. coli*. The clones contain the mammalian proinsulin gene.

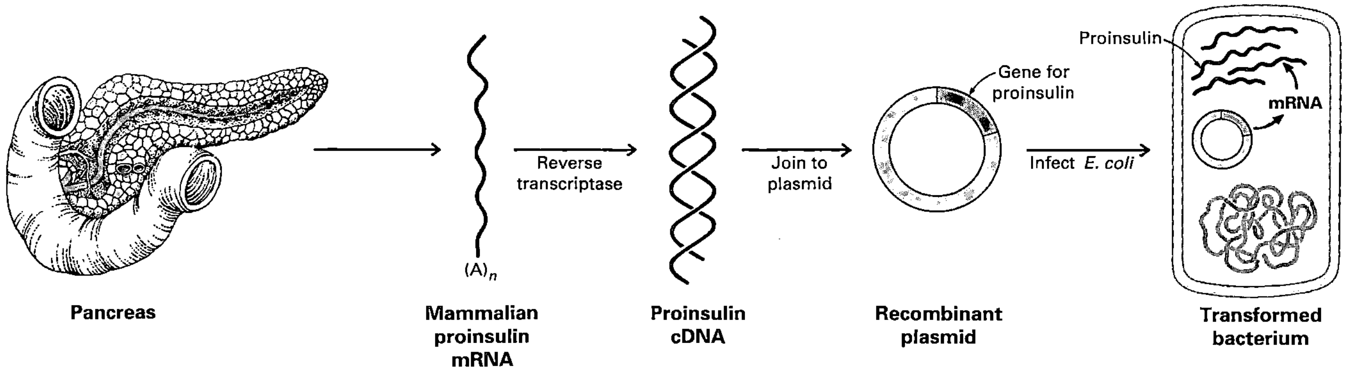

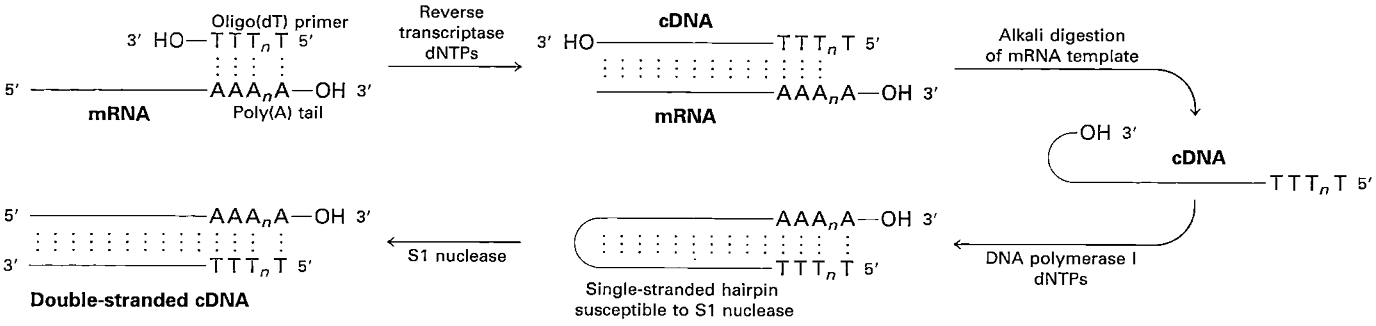

Figure 6-28
Forming a cDNA duplex from mRNA by using reverse transcriptase.

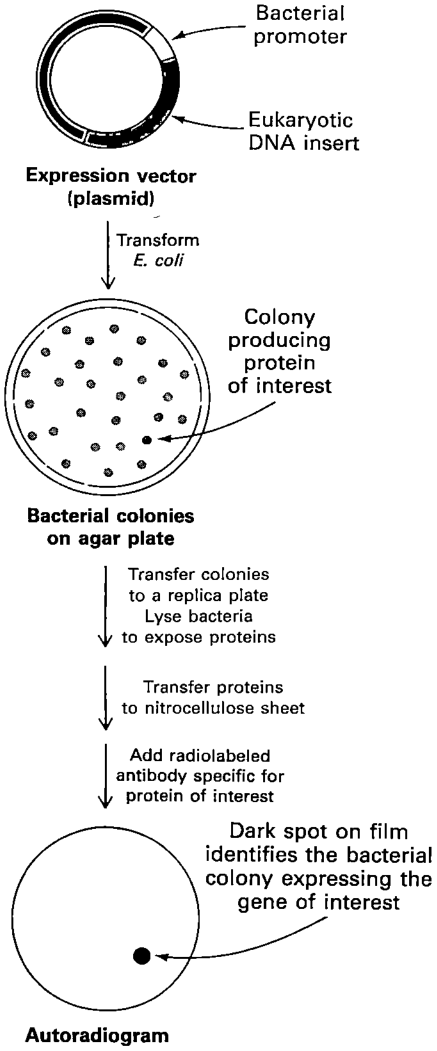

Figure 6-29
Screening of expression vector products by staining with specific antibody.

The key to forming *complementary DNA (cDNA)* is the enzyme *reverse transcriptase.* As was discussed earlier (p. 92), retroviruses use this enzyme to form a DNA–RNA hybrid as the first step in replicating their genomic RNA. Reverse transcriptase synthesizes a DNA strand complementary to an RNA template if it is provided with a primer that is base-paired to the RNA and contains a free 3′-OH group. We can use this enzyme to synthesize DNA from mRNA by providing an oligo(dT) primer that pairs with the poly(A) sequence at the 3′ end of most eukaryotic mRNA molecules (Figure 6-28). The rest of the cDNA strand is then synthesized in the presence of the four deoxyribonucleoside triphosphates. The RNA strand of this RNA–DNA hybrid is subsequently hydrolyzed by raising the pH. Unlike RNA, DNA is resistant to alkaline hydrolysis. The 3′ end of the newly synthesized DNA strand then forms a hairpin loop and primes the synthesis of the opposite DNA strand. The hairpin loop is removed by digestion with S1 nuclease, which recognizes unpaired nucleotides. Synthetic linkers can be added to this double-helical DNA for ligation to a suitable vector.

cDNA molecules can be inserted into vectors that favor their efficient expression in hosts such as *E. coli.* Such plasmids or phages are called *expression vectors.* To maximize transcription, the cDNA is inserted into the vector in the correct reading frame near a strong bacterial promoter. In addition, these vectors assure efficient translation by encoding a ribosome-binding site on the mRNA near the initiation codon. *cDNA clones can be screened on the basis of their capacity to direct the synthesis of a foreign protein in bacteria.* A radioactive antibody specific for the protein of interest can be used to identify the colonies of bacteria that harbor the corresponding cDNA vector (Figure 6-29). As described earlier, spots of bacteria on a replica plate are lysed to release proteins, which bind to an applied nitrocellulose filter. A ^{125}I-labeled antibody specific for the protein of interest is added, and autoradiography reveals the location of the desired colonies on the master plate. This *immunochemical screening* approach can be used whenever a protein is expressed and corresponding antibody is available.

NEW GENES INSERTED INTO EUKARYOTIC CELLS CAN BE EFFICIENTLY EXPRESSED

Bacteria are ideal hosts for the amplification of DNA molecules. They can also serve as factories for the production of a wide range of prokaryotic and eukaryotic proteins. However, posttranslational modifications such as specific cleavages of polypeptides and attachment of carbohydrate units are not carried out by bacteria because they lack the necessary enzymes. Thus, many eukaryotic genes can be correctly expressed only in eukaryotic host cells. Another motivation for introducing recombinant

DNA molecules into cells of higher organisms is to gain insight into how their genes are organized and expressed: How are genes turned on and off in embryological development? How does a fertilized egg give rise to an organism with highly differentiated cells that are organized in space and time? These central questions of biology can now be fruitfully approached by expressing foreign genes in mammalian cells.

Recombinant DNA molecules can be introduced into animal cells in several ways. In one, foreign DNA molecules *precipitated by calcium phosphate* are taken up by animal cells. A small fraction of the imported DNA becomes stably integrated into the chromosomal DNA. The efficiency of incorporation is low, but the method is useful because it is easy to apply. In another method, DNA is *microinjected* into cells. A fine-tipped (0.1-μm-diameter) glass micropipet containing a solution of foreign DNA is inserted into a nucleus (Figure 6-30). A skilled investigator can inject hundreds of cells per hour. About 2% of injected mouse cells are viable and contain the new gene. In a third method, *viruses* can be used to bring new genes into animal cells. The most effective vectors are *retroviruses* (RNA tumor viruses) (Figure 6-31). As discussed earlier (p. 91), these viruses replicate through DNA intermediates, the reverse of the normal flow of information (hence the prefix *retro*). A striking feature of the life cycle of a retrovirus is that the double-helical DNA form of its genome, produced by the action of reverse transcriptase, becomes randomly incorporated into host chromosomal DNA. This DNA version of the viral genome, called *proviral DNA,* can be efficiently expressed by the host cell and replicated along with normal cellular DNA. Retroviruses do not usually kill their hosts. Foreign genes have been efficiently introduced into mammalian cells by infecting them with vectors derived from *Maloney murine leukemia virus,* which can accept inserts as long as 6 kb. Some genes introduced by this retroviral vector into the genome of a transformed host cell are efficiently expressed.

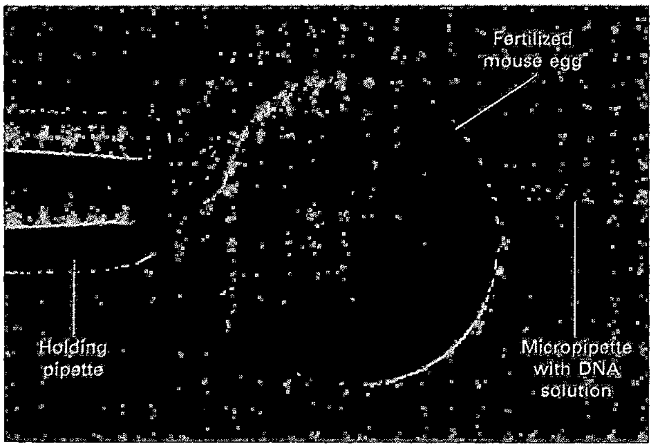

Figure 6-30
Microinjection of cloned plasmid DNA into the male pronucleus of a fertilized mouse egg.

Two other viral vectors are extensively used. *Vaccinia virus,* a large DNA-containing virus, replicates in the cytoplasm of mammalian cells, where it shuts down host-cell protein synthesis. *Baculovirus* infects insect cells, which can be conveniently cultured. Insect larvae infected with this virus can serve as efficient protein factories. Vectors based on these large-genome viruses have been engineered to efficiently express DNA inserts.

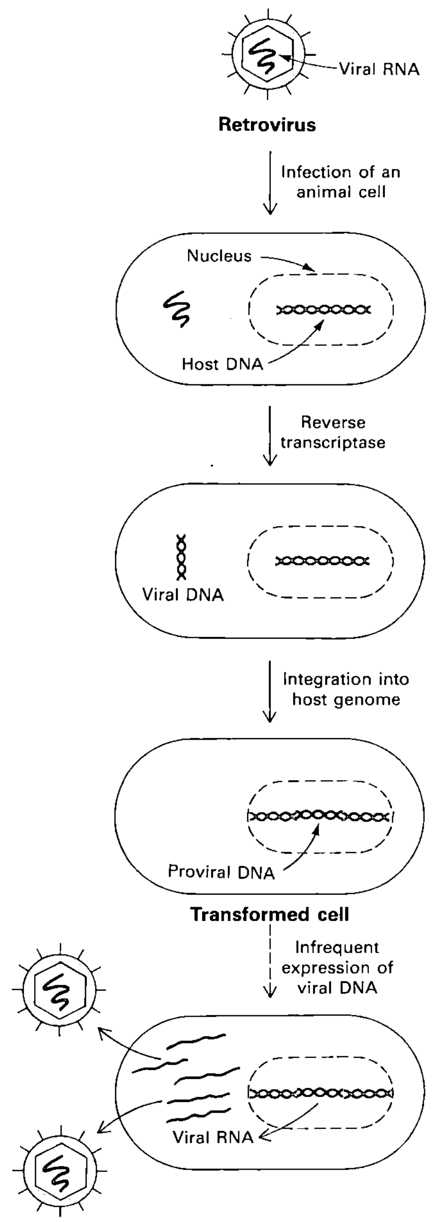

Figure 6-31
Life cycle of a retrovirus.

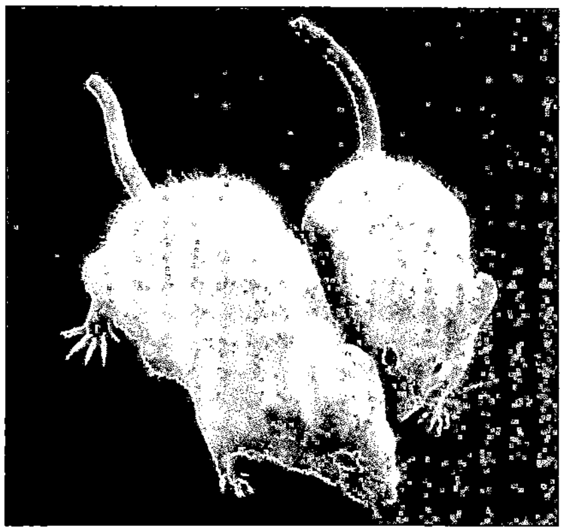

Figure 6-32
Injection of the gene for growth hormone into a fertilized mouse egg gave rise to giant mouse (left), about twice the weight of his sibling (right). [Courtesy of Dr. Ralph Brinster.]

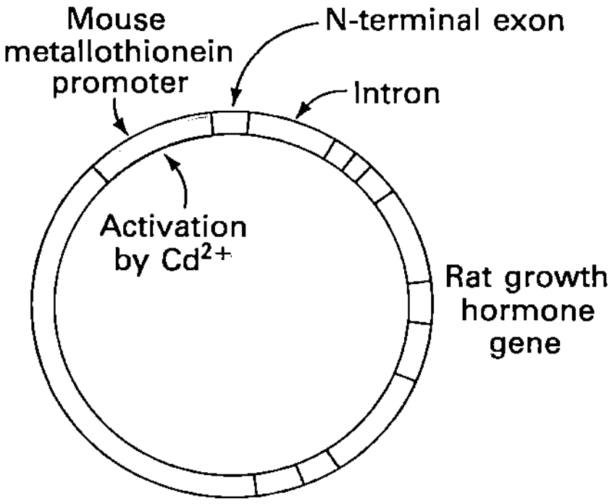

Figure 6-33
The gene for rat growth hormone (shown in yellow) was inserted into a plasmid next to the metallothionein promoter, which is activated by the addition of heavy metals, such as cadmium ion.

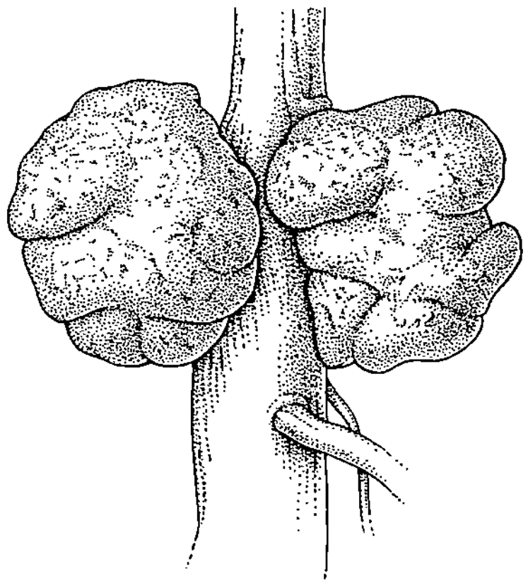

Figure 6-34
Crown gall, a plant tumor, is caused by a bacterium (*Agrobacterium tumefaciens*) that carries a tumor-inducing plasmid (Ti plasmid).

TRANSGENIC ANIMALS HARBOR AND EXPRESS GENES THAT WERE INTRODUCED INTO THEIR GERM LINE

Genetically engineered giant mice (Figure 6-32) illustrate the expression of foreign genes in mammalian cells. *Growth hormone (somatotropin),* a 21-kd protein, is normally synthesized by the pituitary gland. A deficiency of this hormone produces dwarfism, and an excess leads to gigantism. The gene for rat growth hormone was placed next to the mouse metallothionein promoter on a plasmid (Figure 6-33). This promoter is normally located on a chromosome, where it controls the transcription of metallothionein, a cysteine-rich protein that has high affinity for heavy metals. The synthesis of this protective protein by the liver is induced by heavy-metal ions such as cadmium. Several hundred copies of this plasmid were microinjected into the male pronucleus of a fertilized mouse egg, which was then inserted into the uterus of a foster mother mouse. A number of mice that developed from these microinjected eggs contained the gene for rat growth hormone, as shown by Southern blots of their DNA. These *transgenic mice,* containing multiple copies (~30 per cell) of the rat growth hormone gene, grew much more rapidly than control mice. The level of growth hormone in these mice was 500 times as high as in normal mice, and their body weight at maturity was twice normal. The foreign DNA had been transcribed and its five introns correctly spliced out to form functional mRNA. *These experiments strikingly demonstrate that a foreign gene under the control of a new promoter can be integrated and efficiently expressed in mammalian cells.*

TUMOR-INDUCING (Ti) PLASMIDS CAN BE USED TO INTRODUCE NEW GENES INTO PLANT CELLS

The common soil bacterium *Agrobacterium tumefaciens* infects and introduces foreign genes into plants (Figure 6-34). A lump of tumor tissue called a *crown gall* grows at the site of infection. Crown galls synthesize opines, a group of amino acid derivatives that are metabolized by the infecting bacteria. In essence, the metabolism of the plant cell is diverted to satisfy the highly distinctive appetite of the intruder. The instructions for the synthesis of opines and the switch to the tumor state come from *Ti plasmids (tumor-inducing plasmids)* that are carried by *Agrobacterium.* A small portion of the Ti plasmid becomes integrated into the genome of infected plant cells; this 20-kb segment is called *T-DNA* (transferred DNA) (Figure 6-35).

Ti plasmid derivatives can be used as vectors to deliver foreign genes into plant cells. First, a segment of foreign DNA is inserted into the

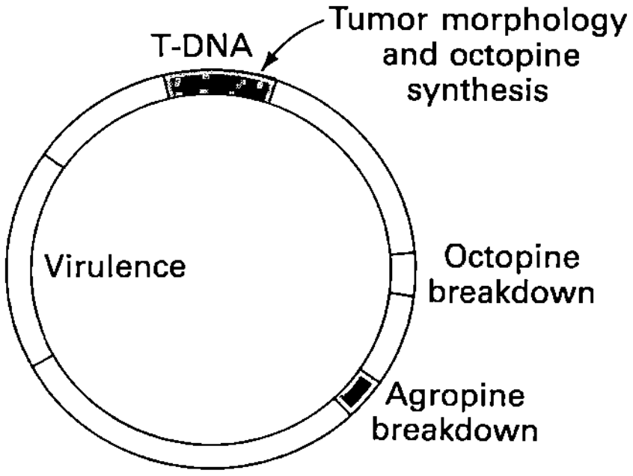

Figure 6-35
Agrobacteria containing Ti plasmids can deliver foreign genes into some plant cells. [After M. Chilton. A vector for introducing new genes into plants. Copyright © 1983 by Scientific American, Inc. All rights reserved.]

T-DNA region of a small plasmid by restriction enzymes and ligases. This synthetic plasmid is added to *Agrobacterium* colonies harboring naturally occurring Ti plasmids. By recombination, Ti plasmids containing the foreign gene are formed. These Ti vectors hold great promise for exploring the genomes of plant cells and modifying plants to improve their agricultural value and crop yield. However, they are not suitable for transforming all types of plants. Ti-plasmid transfer works with dicots (broad-leaved plants such as grapes) and a few kinds of monocots but not with economically important cereal monocots.

Foreign DNA can be introduced into cereal monocots as well as dicots by applying intense electric fields, a technique called *electroporation* (Figure 6-36). First, the cellulose wall surrounding plant cells is removed by adding cellulase; this produces *protoplasts,* plant cells with exposed plasma membranes. Electric pulses then are applied to a suspension of protoplasts and plasmid DNA. Because high electric fields make membranes transiently permeable to large molecules, plasmid DNA molecules enter the cells. The cell wall is then allowed to reform, which results in viable plant cells. Maize cells and carrot cells have been stably transformed in this way with plasmid DNA that includes genes for resistance to antibiotics. Moreover, the plasmid DNA is efficiently expressed by the transformed cells. Electroporation is also an effective means of delivering foreign DNA into animal cells. DNA can also be introduced into plant cells by bombarding them with 1-μm-diameter tungsten pellets coated with DNA. *DNA particle guns* have been used to transform corn and wheat.

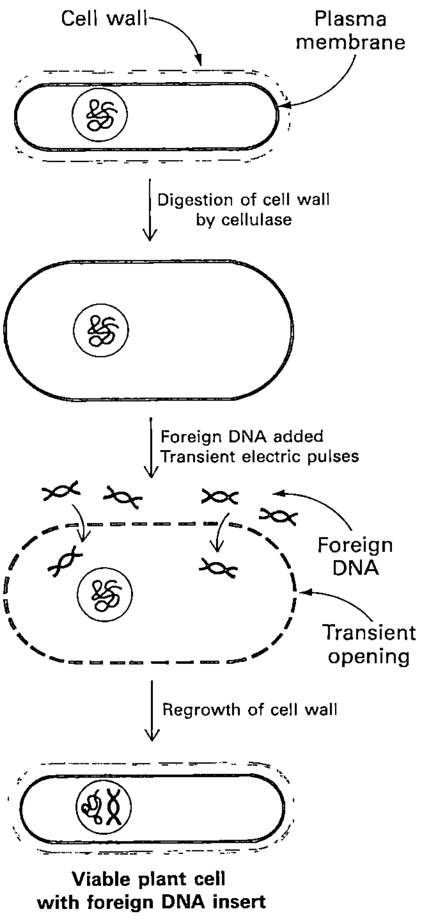

Figure 6-36
Foreign DNA can be introduced into plant cells by electroporation, applying intense electric fields to make their plasma membranes transiently permeable.

NOVEL PROTEINS CAN BE ENGINEERED BY SITE-SPECIFIC MUTAGENESIS

Much has been learned about genes and proteins by selecting genes from the repertoire offered by nature. In the classic genetic approach, mutations are generated randomly throughout the genome, and those exhibiting a particular phenotype are selected. Analysis of these mutants then reveals which genes are altered, and DNA sequencing identifies the precise nature of the changes. *Recombinant DNA technology now makes it feasible to create specific mutations in vitro.* We can construct new genes with designed properties by making three kinds of directed changes: *deletions, insertions,* and *substitutions.*

A specific deletion can be produced by cleaving a plasmid at two sites with a restriction enzyme and religating to form a smaller circle. This simple approach usually removes a large block of DNA. A smaller deletion can be made by cutting a plasmid at a single site. The ends of the linear DNA are then digested with an exonuclease that removes nucleotides from both strands. The shortened piece of DNA is then religated to form a circle that is missing a short length of DNA about the restriction site.

Mutant proteins with single amino acid substitutions can readily be produced by *oligonucleotide-directed mutagenesis* (Figure 6-37). Suppose that

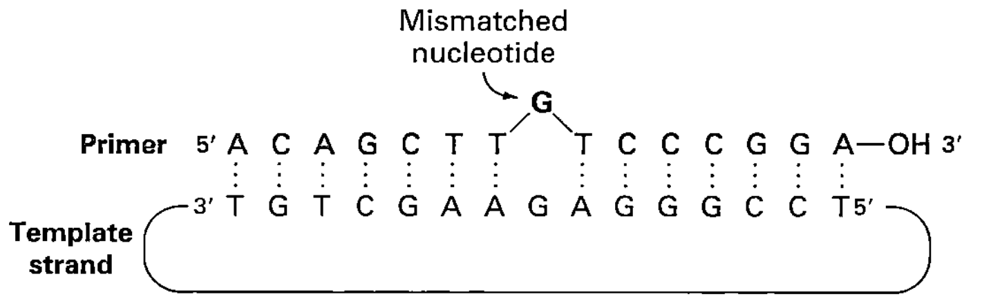

Figure 6-37
Oligonucleotide-directed mutagenesis. A primer containing a mismatched nucleotide is used to produce a desired point mutation.

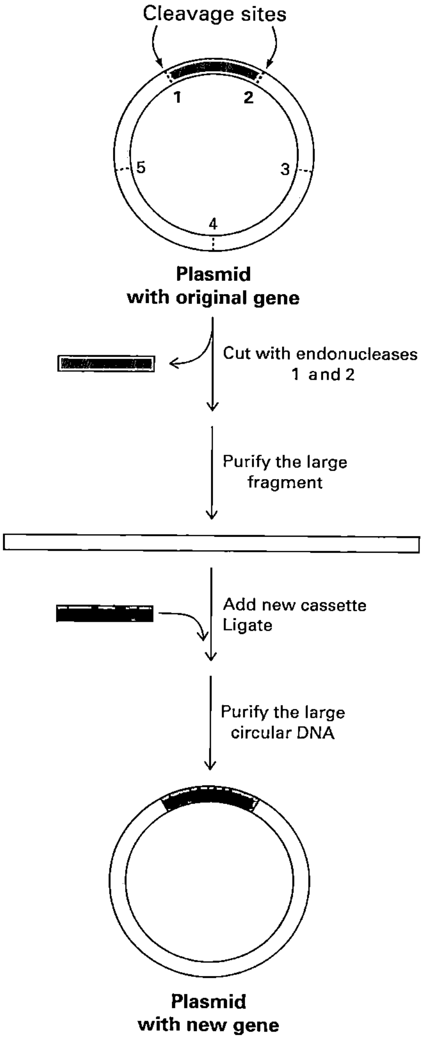

Figure 6-38
Cassette mutagenesis. DNA is cleaved at a pair of unique restriction sites by two different restriction endonucleases. A synthetic oligonucleotide with ends that are complementary to these sites (the *cassette*) is then ligated to the cleaved DNA. The method is highly versatile because the inserted DNA can have any desired sequence.

we want to replace a particular serine residue with cysteine. This mutation can be made if (1) we have a plasmid containing the gene or cDNA for this protein and (2) we know the base sequence around the site to be altered. If the serine of interest is encoded by TCT, we need to change the C to a G to get cysteine, which is encoded by TGT. The key to this mutation is to prepare an oligonucleotide primer that is complementary to this region of the gene except that it contains TGT instead of TCT. The two strands of the plasmid are separated, and the primer is then annealed to the complementary strand. (The mismatch of 1 base pair out of 15 is tolerable if the annealing is carried out at an appropriate temperature. An attractive feature of nucleic acid hybridization is that its *stringency*—the required closeness of the match—can be experimentally controlled by choice of temperature and ionic strength.) The primer is then elongated by DNA polymerase, and the double-stranded circle is closed by adding DNA ligase. Subsequent replication of this duplex yields two kinds of progeny plasmid, half with the original TCT sequence and half with the mutant TGT sequence. Expression of the plasmid containing the new TGT sequence will produce a protein with the desired substitution of serine for cysteine at a unique site. We will encounter many examples of the use of oligonucleotide-directed mutagenesis to precisely alter regulatory regions of genes and to produce proteins with tailor-made features.

Cassette mutagenesis is another valuable approach. Plasmid DNA is cut with a pair of restriction enzymes to remove a short segment (Figure 6-38). A synthetic double-stranded oligonucleotide (the *cassette*) with cohesive ends that are complementary to the ends of the cut plasmid is then added and ligated. Each plasmid now contains the desired mutation. It is convenient to introduce into the plasmid unique restriction sites spaced about 40 nucleotides apart so that mutations can readily be made anywhere in the sequence.

Novel proteins can also be created by splicing gene segments that encode domains that are not associated in nature. For example, a gene for an antibody can be joined to a gene for a toxin to produce a chimeric protein that kills cells that are recognized by the antibody. These *immunotoxins* are being evaluated as anticancer agents. Furthermore, entirely new genes can be synthesized de novo by the solid-phase method. Noninfectious coat proteins of viruses can be produced in large amounts by recombinant DNA methods. They can serve as *synthetic vaccines* that are safer than conventional vaccines prepared by inactivating pathogenic viruses. A subunit of the hepatitis B virus produced in yeast is proving to be an effective vaccine against this debilitating viral disease.

RECOMBINANT DNA TECHNOLOGY HAS OPENED NEW VISTAS

The analysis of the molecular basis of life has been revolutionized by recombinant DNA technology. Complex chromosomes are rapidly being mapped and dissected into units that can be manipulated and deciphered. The amplification of genes by cloning has provided abundant quantities of DNA for sequencing. Genes are now open books that can be read. New insights are emerging, as exemplified by the discovery of introns in eukaryotic genes. Central questions of biology, such as the molecular basis of development, are now being fruitfully explored. DNA and RNA sequences provide a wealth of information about evolution. Biochemists now move back and forth between gene and protein and feel at home in both areas of inquiry.

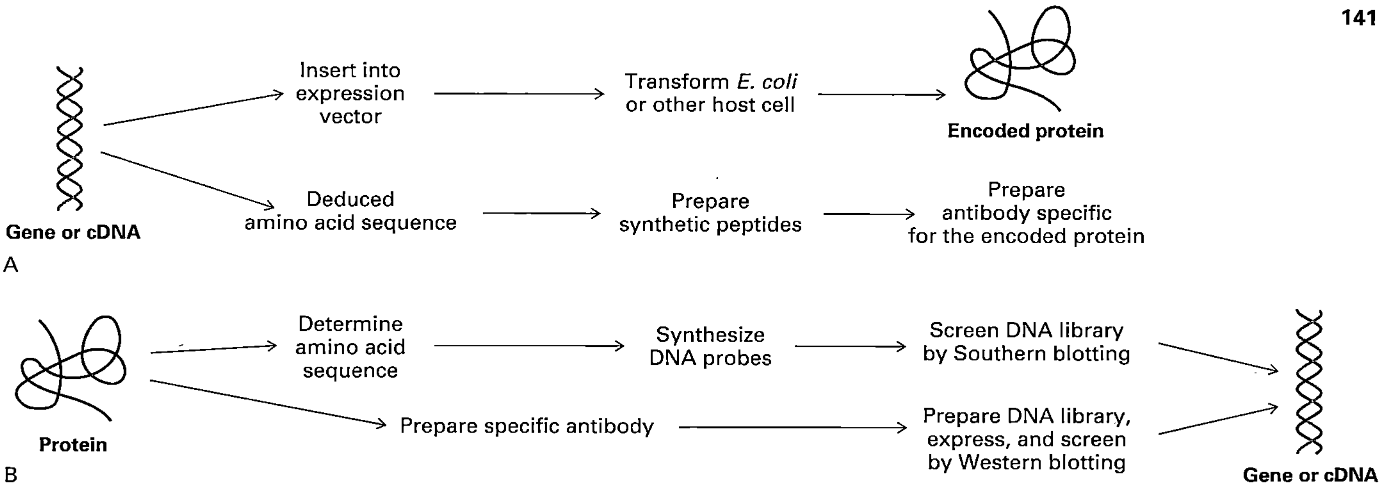

Figure 6-39
The techniques of protein chemistry and nucleic acid chemistry are mutually reinforcing. (A) From DNA (or RNA) to protein, and (B) from protein to DNA.

Analyses of genes and cDNA can reveal the existence of previously unknown proteins, which can be isolated and purified (Figure 6-39A). Conversely, purification of a protein can be the starting point for the isolation and cloning of its gene or cDNA (Figure 6-39B). Very small amounts of protein or nucleic acid suffice because of the sensitivity of recently developed microchemical techniques and the amplification afforded by gene cloning and the polymerase chain reaction. The powerful techniques of protein chemistry, nucleic acid chemistry, immunology, and molecular genetics are highly synergistic.

New kinds of proteins can be created by altering genes in specific ways. Site-specific mutagenesis opens the door to understanding how proteins fold, recognize other molecules, catalyze reactions, and process information. Large amounts of protein can be obtained by expressing cloned genes or cDNAs in bacteria or eukaryotic cells. Hormones such as insulin and antiviral agents such as interferon are being produced by bacteria. Tissue plasminogen activator, which is administered to patients after a heart attack, is made in large quantities in mammalian cells. A new pharmacology that is beginning to profoundly alter medicine has come into being. Recombinant DNA technology is also providing highly specific diagnostic reagents, such as DNA probes for the detection of genetic diseases, infections, and cancers. Human gene therapy has been successfully initiated. White blood cells deficient in adenosine deaminase, an essential enzyme, are taken from patients and returned after being transformed in vitro to correct the genetic error. Agriculture, too, is benefiting from genetic engineering. Transgenic crops with increased resistance to insects, herbicides, and drought have been produced.

SUMMARY

The recombinant DNA revolution in biology is rooted in the repertoire of enzymes that act on nucleic acids. Restriction enzymes, a group of key reagents, are endonucleases that recognize specific base sequences in double-helical DNA and cleave both strands of the duplex. Specific fragments of DNA formed by the action of restriction enzymes can be separated and displayed by gel electrophoresis. The pattern of these restriction fragments is a fingerprint of a DNA molecule. A DNA fragment containing a particular sequence can be identified by hybridizing it with a labeled single-stranded DNA probe (Southern blotting). Analysis of DNA molecules has also advanced by the development of rapid sequencing techniques. DNA can be sequenced by chemical cleavage at particular bases (Maxam-Gilbert method) or by controlled interruption of replication (Sanger dideoxy method). The fragments produced are separated by gel electrophoresis and visualized by autoradiography of a ^{32}P label at the 5′ end or by fluorescent tags. The recent sequencing of an entire yeast chromosome (310 kb) demonstrates the power of these techniques in illuminating large segments of the genome.

DNA probes for hybridization reactions, as well as new genes, can be synthesized by the sequential addition of deoxyribonucleoside 3′-phosphoramidites to a growing chain that is linked to an insoluble support. DNA chains a hundred nucleotides long can readily be synthesized by this automated solid-phase method. Oligonucleotide arrays synthesized on solid supports (DNA chips) can be used to sequence DNA by hybridization. The polymerase chain reaction (PCR) makes it possible to greatly amplify specific segments of DNA in vitro. The region amplified is determined by the pair of primers that are added to the target DNA along with a thermostable DNA polymerase and deoxyribonucleoside triphosphates. The exquisite sensitivity of PCR makes it a choice technique in detecting pathogens and cancer markers, genotyping individuals, and reading DNA from fossils that are millions of years old.

New genomes can be constructed in the laboratory, introduced into host cells, and expressed. Novel DNA molecules are made by joining fragments that have complementary cohesive ends produced by the action of a restriction enzyme. DNA ligase seals gaps in DNA chains if they are within a double helix. Plasmids (circular accessory chromosomes) and λ phage are choice vectors for cloning DNA in bacteria. Yeast artificial chromosomes (YACs) can propagate large pieces of DNA, up to a million base pairs long. Specific genes can be cloned from a digest of DNA. This genomic library can be screened with a complementary DNA or RNA probe. Alternatively, one can form cDNA from mRNA by the action of reverse transcriptase. cDNAs inserted into expression vectors that contain strong promoters are efficiently expressed by bacteria. Foreign DNA can be carried into mammalian cells by a retrovirus or be directly injected. The production of giant mice by injecting the gene for rat growth hormone into fertilized mouse eggs vividly shows that mammalian cells can be genetically altered in a designed way. New DNA can be brought into plant cells by the soil bacterium *Agrobacterium tumefaciens,* which harbors Ti (tumor-inducing) plasmids. DNA can also be introduced into plant cells by applying intense electric fields, which render them transiently permeable to very large molecules, or by bombarding them with DNA-coated microparticles. Agriculture is benefiting from the engineering of transgenic plants with valuable properties such as increased resistance to insects.

Novel proteins can be engineered by generating specific mutations in vitro. A mutant protein with a single amino acid substitution can be produced by priming DNA replication with an oligonucleotide encoding the new amino acid. Plasmids can be engineered to permit facile insertion of a DNA cassette containing any desired mutation. The techniques of protein and nucleic acid chemistry are highly synergistic. Investigators now move back and forth between gene and protein with great facility. Recombinant DNA technology is beginning to profoundly alter medicine by providing new diagnostic and therapeutic agents and revealing molecular mechanisms of disease.

SELECTED READINGS

Where to start

Berg, P., 1981. Dissections and reconstructions of genes and chromosomes. *Science* 213:296–303.

Gilbert, W., 1981. DNA sequencing and gene structure. *Science* 214:1305–1312.

Sanger, F., 1981. Determination of nucleotide sequences in DNA. *Science* 214:1205–1210.

Mullis, K.B., 1990. The unusual origin of the polymerase chain reaction. *Sci. Amer.* 262(4):56–65.

Books on recombinant DNA technology

Watson, J.D., Gilman, M., Witkowski, J., and Zoller, M., 1992. *Recombinant DNA* (2nd ed.). Scientific American Books. [A highly lucid and well-illustrated account with many interesting examples.]

Sambrook, J., Fritsch, E.F., and Maniatis, T., 1989. *Molecular Cloning: A Laboratory Manual* (2nd ed.). Cold Spring Harbor Laboratory.

Zyskind, J.W., and Bernstein, S.I., 1992. *Recombinant DNA Laboratory Manual.* Academic Press.

Mullis, K.B., Ferrê, F., and Gibbs, R.A. (eds.), 1994. *The Polymerase Chain Reaction.* Birkhaüser.

Grierson, D. (ed.), 1991. *Plant Genetic Engineering.* Chapman and Hall.

Methods in Enzymology. Academic Press. [Many volumes in this series deal with recombinant DNA technology. In particular, see volumes 152–155, 185, 208, 211, 212, and 216–218.]

DNA sequencing and synthesis

Hunkapiller, T., Kaiser, R.J., Koop, B.F., and Hood, L., 1991. Large-scale and automated DNA sequence determination. *Science* 254:59–67.

Sanger, F., Nicklen, S., and Coulson, A.R., 1977. DNA sequencing with chain-terminating inhibitors. *Proc. Nat. Acad. Sci.* 74:5463–5467.

Maxam, A.M., and Gilbert, W., 1977. A new method for sequencing DNA. *Proc. Nat. Acad. Sci.* 74:560–564.

Smith, L.M., Sanders, J.Z., Kaiser, R.J., Hughes, P., Dodd, C., Connell, C.R., Heiner, C., Kent, S.B.H., and Hood, L.E., 1986. Fluorescence detection in automated DNA sequence analysis. *Nature* 321:674–679.

Caruthers, M.H., Beaton, G., Wu, J.V., and Wiesler, W., 1992. Chemical synthesis of deoxyoligonucleotides and deoxyoligonucleotide analogs. *Meth. Enzymol.* 211:3–20.

Pease, A.C., Solas, D., Sullivan, E.J., Cronin, M.T., Holmes, C.P., and Fodor, S.P.A., 1994. Light-generated oligonucleotide arrays for rapid DNA sequence analysis. *Proc. Nat. Acad. Sci.* 91:5022–5026.

Origins of DNA cloning

Nathans, D., 1979. Restriction endonucleases, simian virus 40, and the new genetics. *Science* 206:903–909.

Jackson, D.A., Symons, R.H., and Berg, P., 1972. Biochemical method for inserting new genetic information into DNA of simian virus 40: Circular SV40 DNA molecules containing lambda phage genes and the galactose operon of *Escherichia coli. Proc. Nat. Acad. Sci.* 69:2904–2909.

Lobban, P.E., and Kaiser, A.D., 1973. Enzymatic end-to-end joining of DNA molecules. *J. Mol. Biol.* 78:453–471.

Cohen, S.N., Chang, A., Boyer, H., and Helling, R., 1973. Construction of biologically functional bacterial plasmids in vitro. *Proc. Nat. Acad. Sci.* 70:3240–3244.

Cohen, S.N., 1985. DNA cloning: Historical perspectives. *Biogenetics of Neurohormonal Peptides,* pp. 3–14. Academic Press.

Watson, J.D., and Tooze, J., 1981. *The DNA Story.* W.H. Freeman.

Therapeutic proteins

Gilbert, W., and Villa-Komaroff, L., 1980. Useful proteins from recombinant bacteria. *Sci. Amer.* 242(4):74–94.

Johnson, I.S., 1983. Human insulin from recombinant DNA technology. *Science* 219:632–637.

Medin, J.A., Hunt, L., Gathy, K., Evans, R.K., and Coleman, M.S., 1990. Efficient, low-cost protein factories: Expression of human adenosine deaminase in baculovirus-infected insect larvae. *Proc. Nat. Acad. Sci.* 87:2760–2764.

Winter, G., and Milstein, C., 1991. Man-made antibodies. *Nature* 349:293–299.

Brown, F., 1990. From Jenner to genes—the new vaccines. *Lancet* 335:587–590.

Introduction of genes into animal cells

Anderson, W.F., 1992. Human gene therapy. *Science* 256:808–813.

Verma, I., 1990. Gene therapy. *Sci. Amer.* 263(5):68–84.

Brinster, R.L., and Palmiter, R.D., 1986. Introduction of genes into the germ lines of animals. *Harvey Lectures* 80:1–38. [The production of large mice by injection of the rat growth hormone gene is discussed in this review.]

Grosveld, F., and Kollias, G. (ed.), 1992. *Transgenic Animals.* Academic Press.

Genetic engineering of plants

Gasser, C.S., and Fraley, R.T., 1992. Transgenic crops. *Sci. Amer.* 266(6):62–69.

Gasser, C.S., and Fraley, R.T., 1989. Genetically engineering plants for crop improvement. *Science* 244:1293–1299.

Shimamoto, K., Terada, R., Izawa, T., and Fujimoto, H., 1989. Fertile transgenic rice plants regenerated from transformed protoplasts. *Nature* 338:274–276.

Chilton, M-D., 1983. A vector for introducing new genes into plants. *Sci. Amer.* 248(6):50.

Polymerase chain reaction (PCR)

Arnheim, N., and Erlich, H., 1992. Polymerase chain reaction strategy. *Ann. Rev. Biochem.* 61:131–156.

Erlich, H.A., Gelfand, D., and Sninsky, J.J., 1991. Recent advances in the polymerase chain reaction. *Science* 252:1643–1651.

Erlich, H.A. (ed.), 1989. *PCR Technology: Principles and Applications for DNA Amplification.* Stockton Press.

Kirby, L.T. (ed.), 1990. *DNA Fingerprinting. An Introduction.* Stockton Press.

Eisenstein, B.I., 1990. The polymerase chain reaction. A new method for using molecular genetics for medical diagnosis. *New Engl. J. Med.* 322:178–183.

Pääbo, S., 1993. Ancient DNA. *Sci. Amer.* 269(5):86–92.

Higuchi, R., von Beroldingen, C.H., Sensabaugh, G.F., and Erlich, H.A., 1988. DNA typing from single hairs. *Nature* 332:543–546.

Hagelberg, E., Gray, I.C., Jeffreys, A.J., 1991. Identification of the skeletal remains of a murder victim by DNA analysis. *Nature* 352:427–429.

Lawlor, D.A., Dickel, C.D., Hauswirth, W.W., and Parham, P., 1991. Ancient HLA genes from 7500-year-old archaeological remains. *Nature* 349:785–788.

Soltis, P.S., Soltis, D.E., and Smiley, C.J., 1992. An rbcL sequence from a Miocene Taxodium (bald cypress). *Proc. Nat. Acad. Sci.* 89:449–451.

DeSalle, R., Gatesby, J., Wheeler, W., and Grimaldi, D., 1992. DNA sequences from a fossil termite in oligo-miocene amber and their phylogenetic implications. *Science* 257:1933–1936.

PROBLEMS

1. *Reading sequences.*
   (a) An autoradiogram of a gel containing four lanes of DNA fragments produced by ideal chemical cleavage at each of the four bases is shown in Figure 6-40. The DNA contained a ^{32}P label at its 5′ end. What is its sequence?
   (b) Suppose that the Sanger dideoxy method shows that the template strand sequence is 5′-TGCAATGGC-3′. Sketch the gel pattern that would lead to this conclusion.

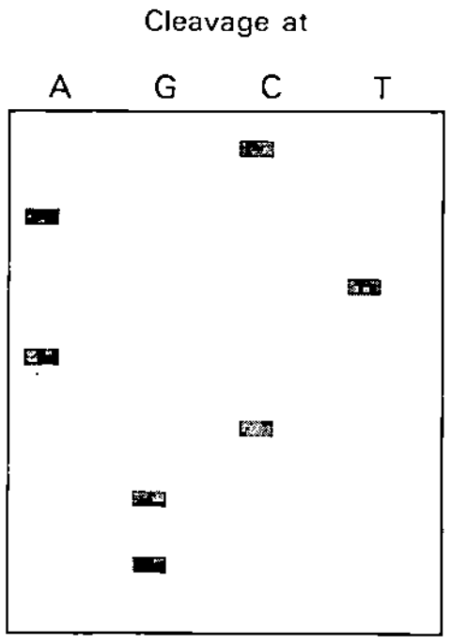

Figure 6-40

2. *The right template.* Ovalbumin is the major protein of egg white. The chicken ovalbumin gene contains eight exons separated by seven introns. Should one use ovalbumin cDNA or ovalbumin genomic DNA to form the protein in *E. coli*? Why?

3. *The right cuts.* Suppose that a human genomic library is prepared by exhaustive digestion of human DNA with the EcoRI restriction enzyme. Fragments averaging about 4 kb in length would be generated.
   (a) Is this procedure suitable for cloning large genes? Why?
   (b) Is this procedure suitable for mapping extensive stretches of the genome by chromosome walking? Why?

4. *A revealing cleavage.* Sickle-cell anemia arises from a mutation in the gene for the β chain of human hemoglobin. The change from GAG to GTG in the mutant eliminates a cleavage site for the restriction enzyme MstII, which recognizes the target sequence CCTGAGG. These findings form the basis of a diagnostic test for the sickle-cell gene. Propose a rapid procedure for distinguishing between the normal and the mutant gene. Would a positive result prove that the mutant contains GTG in place of GAG?

5. *Autocatalysis.* Thomas Cech showed that the ribosomal RNA (rRNA) precursor in the ciliate protozoan *Tetrahymena* can self-splice without binding any *Tetrahymena* protein. He cloned in a plasmid a region of *Tetrahymena* DNA consisting of the intron and flanking sequences present in the precursor RNA. Suggest how this plasmid was used to establish that *Tetrahymena* proteins are not required for the splicing of the ribosomal RNA precursor.

6. *Many melodies from one cassette.* Suppose that you have isolated an enzyme that digests paper pulp and have obtained its cDNA. The goal is to produce a mutant that is effective at high temperature. You have engineered a pair of unique restriction sites in the cDNA that flank a 30-bp coding region. Propose a rapid technique for generating many different mutants in this region.

7. *Terra incognita.* PCR is typically used to amplify DNA that lies between two known sequences. Suppose that you want to explore DNA on both sides of a single known sequence. Devise a variation of the usual PCR protocol that would enable you to amplify entirely new genomic terrain.

8. *A puzzling ladder.* A gel pattern displaying PCR products shows four strong bands. The four pieces of DNA have lengths that are approximately in the ratio of 1:2:3:4. The largest band is cut out of the gel, and PCR is repeated with the same primers. Again, a ladder of four bands is evident in the gel. What does this reveal about the structure of the encoded protein?

9. *Landmarks in the genome.* Many laboratories around the world are mapping the human genome. It is essential that the results be merged at an early stage to provide a working physical map of each chromosome. In particular, we need to know whether a YAC studied in one laboratory overlaps a YAC studied in another when only a small proportion of each (less than 5%) has been sequenced. Propose a simple test for overlap based on the *transfer of information but not of materials* between the two laboratories.

PART

II

Proteins: Conformation, Dynamics, and Function

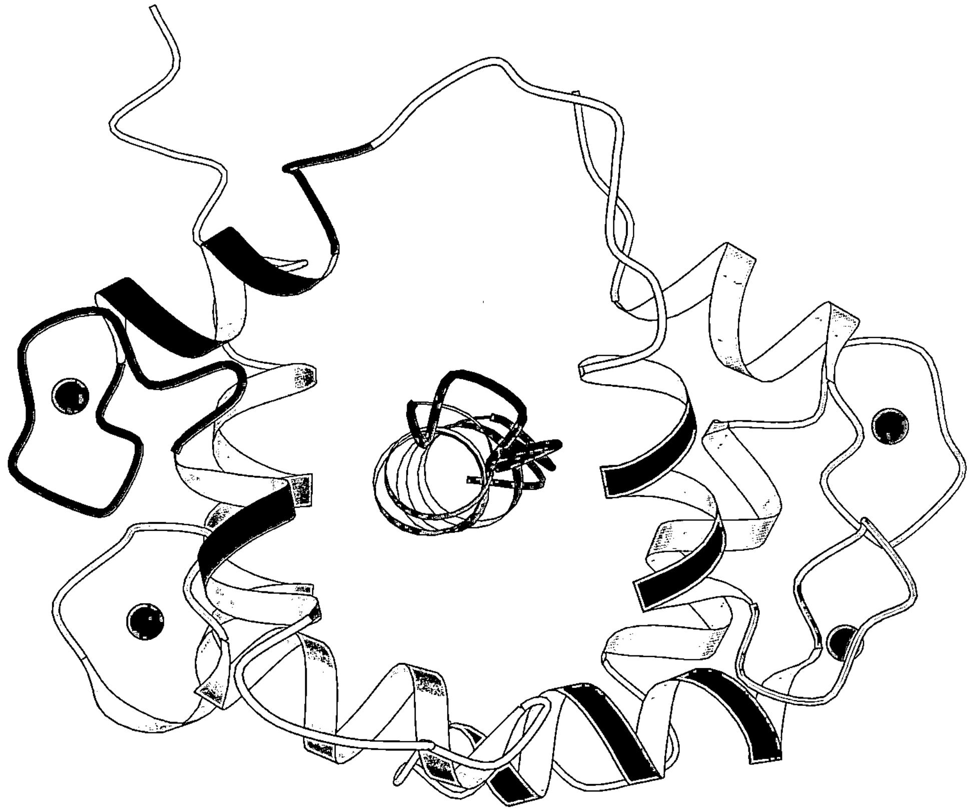

Calcium-calmodulin wraps around myosin light-chain kinase (purple), a target protein, to switch on its enzymatic activity. The four calcium-binding modules (EF hands) are shown in blue, green, yellow, and red, and the bound Ca^{2+} ions in orange. This structure was solved by NMR spectroscopy. [Drawn from coordinates kindly provided by Dr. Mitsu Ikura. M. Ikura, G.M. Clore, A.M. Groneborn, G. Zhu, C.B. Klee, and A. Bax. Science *256(1992):632.]*

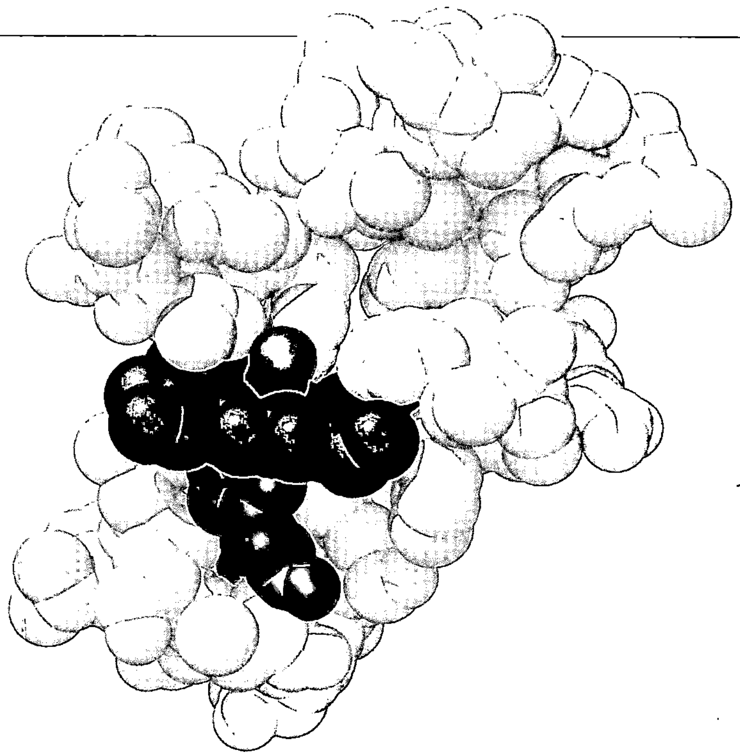

CHAPTER

7

Portrait of an Allosteric Protein

The transition from anaerobic to aerobic life was a major step in evolution because it uncovered a rich reservoir of energy. Eighteen times as much energy is extracted from glucose in the presence of oxygen as in its absence. Vertebrates have evolved two principal mechanisms for supplying their cells with a continuous and adequate flow of oxygen. The first is a circulatory system that actively delivers oxygen to cells. The second is the use of *oxygen-carrying molecules* to overcome the limitation imposed by the low solubility of oxygen in water. The oxygen carriers in vertebrates are the proteins *hemoglobin* and *myoglobin.* Hemoglobin, which is contained in red blood cells, serves as the oxygen carrier in blood and also plays a vital role in the transport of carbon dioxide and hydrogen ion. Myoglobin, which is located in muscle, provides a reserve supply of oxygen and facilitates the movement of oxygen within muscle.

Part II of this book begins with myoglobin and hemoglobin because they illustrate many important principles of protein conformation, dynamics, and function. Their three-dimensional structures, known in atomic detail, reveal much about how proteins fold, bind other molecules, and integrate information. The binding of O_2 by hemoglobin is regulated by H^+, CO_2, and organic phosphates. These regulators greatly affect the oxygen-binding properties of hemoglobin by binding to sites on the protein far from where O_2 is bound. Indeed, interactions between spatially distinct sites, termed *allosteric interactions,* occur in many proteins. Allo-

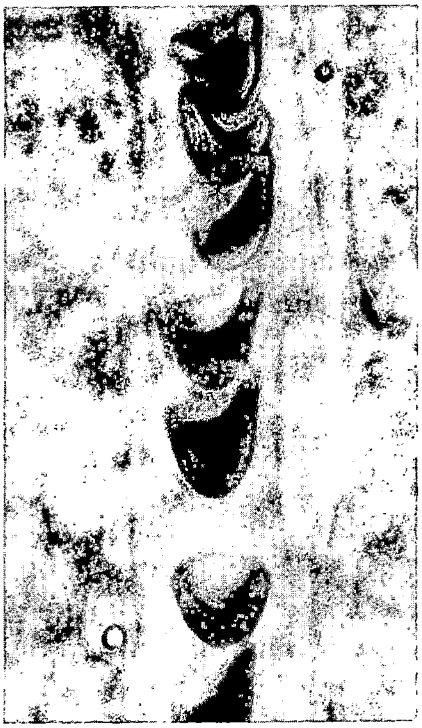

Figure 7-1
Erythrocytes flowing through a small blood vessel. [From P.I. Brånemark. *Intravascular Anatomy of Blood Cells in Man.* (Basel: S. Karger AG, 1971).]

Opening Image: *Model of an oxygen-binding site in hemoglobin. O_2 (green) is bound on one side of a heme group (red). A histidine residue (blue) is bound on the other side. [Drawn from 1hho.pdb. B. Shaanan. J. Mol. Biol. 171(1983):31.]*

steric effects play a critical role in controlling and integrating molecular events in biological systems. Hemoglobin is the best-understood allosteric protein, and so it is rewarding to examine its structure and function in some detail. Furthermore, the discovery of mutant hemoglobins first revealed that disease can arise from a change of a single amino acid in a protein. The concept of molecular disease, now an integral part of medicine, came from studies of a mutant hemoglobin. Hemoglobin has also been a rich source of insight into the molecular basis of evolution.

OXYGEN BINDS TO A HEME PROSTHETIC GROUP

The capacity of myoglobin or hemoglobin to bind oxygen depends on the presence of a nonpolypeptide unit, namely, a *heme group*. The heme also gives myoglobin and hemoglobin their distinctive color. Indeed, many proteins require tightly bound, specific nonpolypeptide units for their biological activities. Such a unit is called a *prosthetic group*. A protein without its characteristic prosthetic group is termed an *apoprotein*.

The heme consists of an organic part and an iron atom. The organic part, *protoporphyrin*, is made up of four *pyrrole* rings. The four pyrroles are linked by methene bridges to form a tetrapyrrole ring. Four methyl, two vinyl, and two propionate side chains are attached.

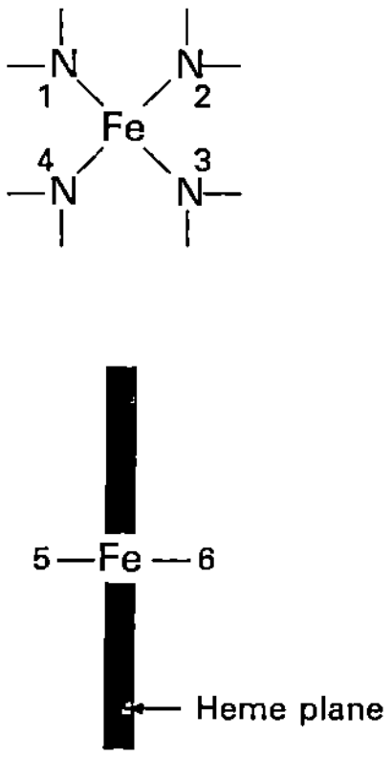

Figure 7-2
The iron atom in heme can form six bonds.

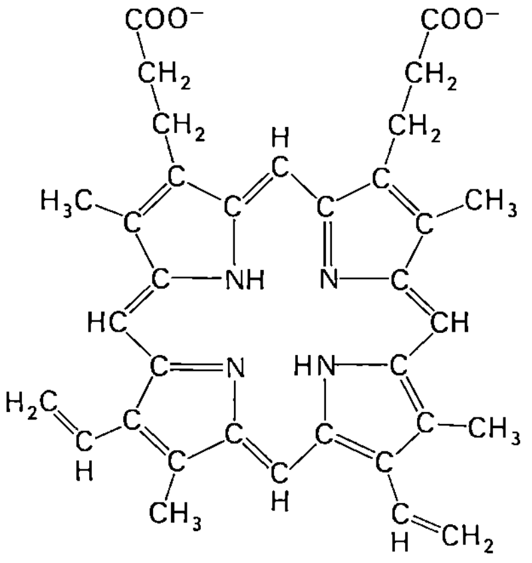

Protoporphyrin IX

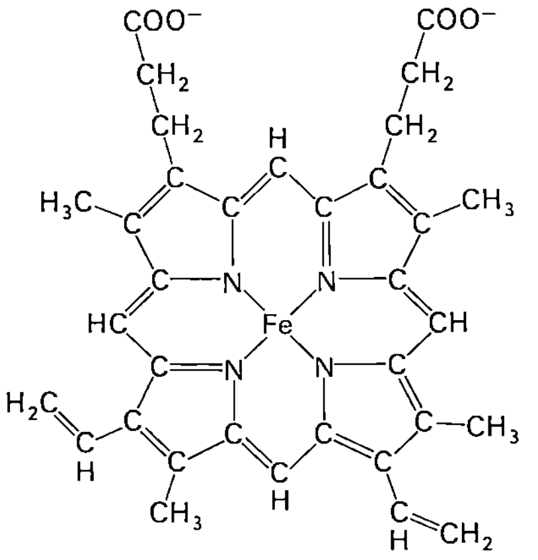

Heme
(Fe–protoporphyrin IX)

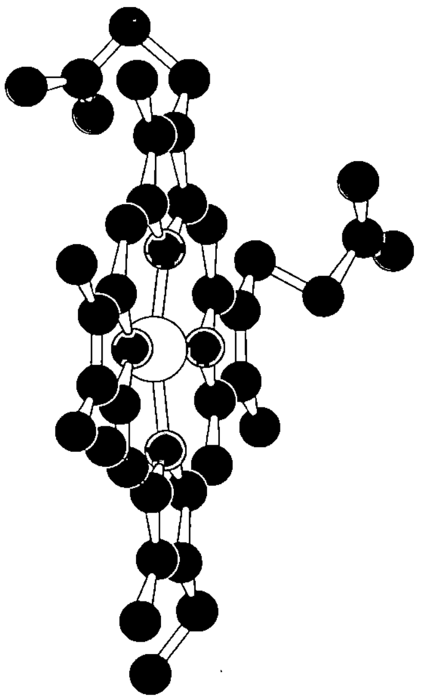

Figure 7-3
The heme group in myoglobin (yellow for Fe, blue for N, red for O, and black for C).

The iron atom in heme binds to the four nitrogens in the center of the protoporphyrin ring (Figures 7-2 and 7-3). The iron can form two additional bonds, one on either side of the heme plane. These bonding sites are termed the *fifth* and *sixth coordination positions*. The iron atom can be in the ferrous (+2) or the ferric (+3) oxidation state, and the corresponding forms of hemoglobin are called *ferrohemoglobin* and *ferrihemoglobin* (also called *methemoglobin*). Only ferrohemoglobin, the +2 oxidation state, can bind oxygen. The same nomenclature applies to myoglobin.

MYOGLOBIN HAS A COMPACT STRUCTURE AND A HIGH CONTENT OF α HELICES

The elucidation of the three-dimensional structure of myoglobin by John Kendrew and of hemoglobin by Max Perutz are landmarks in biochemistry. These studies came to fruition in the late 1950s and showed that x-ray crystallography (p. 64) can reveal the structure of molecules as large as proteins. The determination of the three-dimensional structures of these proteins was a great stimulus to the field of protein crystallography. Kendrew chose myoglobin for x-ray analysis because it is relatively small, easily prepared in quantity, and readily crystallized. It had the additional advantage of being closely related to hemoglobin, which was already being studied by his colleague Perutz. Myoglobin from the skeletal muscle of the sperm whale was selected because it is stable and forms excellent crystals. The skeletal muscle of diving mammals, such as whales, seals, and porpoises, is particularly rich in myoglobin, which serves as a store of oxygen during a dive.

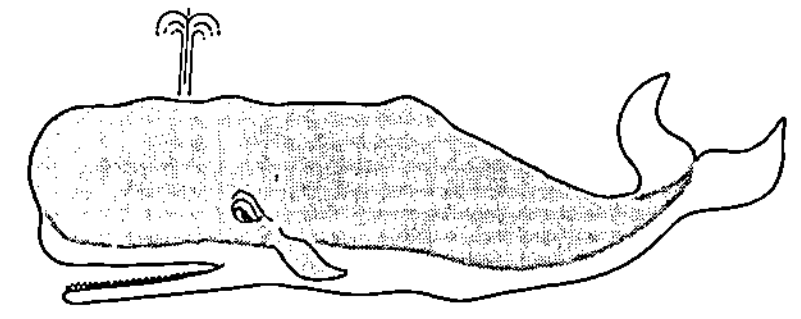

A sperm whale.

In 1957, Kendrew and his colleagues saw what no one had ever seen before: a three-dimensional picture of a protein molecule in all its complexity. The structure was more intricate than had been imagined. Its meaning was revealed two years later, when a high-resolution image was obtained. A wealth of structural detail emerged. The positions of 1200 of the 1260 nonhydrogen atoms were clearly defined to a precision of better than 0.3 Å. The course of the main chain and the position of the heme group are shown in Figure 7-4 (also see Figure 2-43 on p. 34).

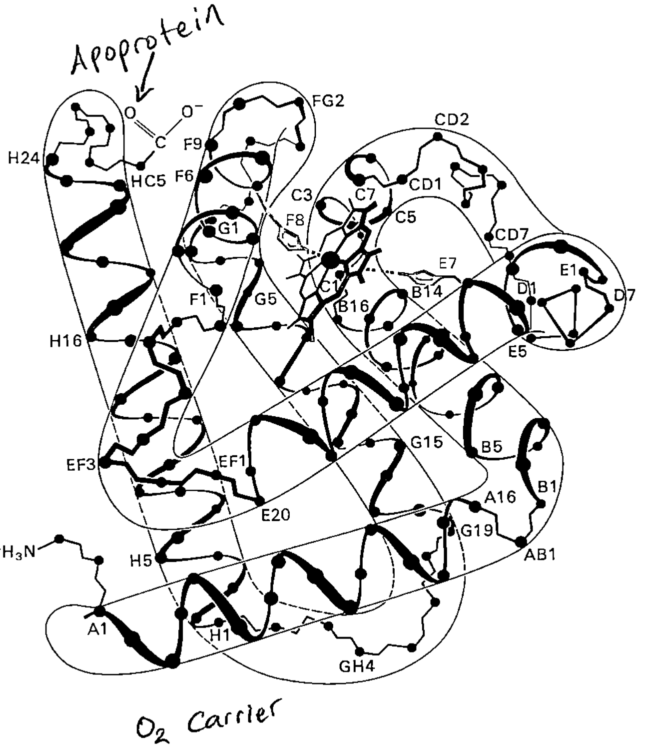

Figure 7-4
Model of myoglobin at high resolution. Only the α carbon atoms are shown. The heme group is shown in red, and two key histidine residues in green. [After R.E. Dickerson. In *The Proteins,* vol. 2, 2nd ed., H. Neurath, ed. (Academic Press, 1964), p. 634.]

Val-	Leu-	Ser-	Glu-	Gly-	Glu-	Trp-	Gln-	Leu-	Val-	10
NA1	NA2	A1	A2	A3	A4	A5	A6	A7	A8	
Leu-	His-	Val-	Trp-	Ala-	Lys-	Val-	Glu-	Ala-	Asp-	20
A9	A10	A11	A12	A13	A14	A15	A16	AB1	B1	
Val-	Ala-	Gly-	His-	Gly-	Gln-	Asp-	Ile-	Leu-	Ile-	30
B2	B3	B4	B5	B6	B7	B8	B9	B10	B11	
Arg-	Leu-	Phe-	Lys-	Ser-	His-	Pro-	Glu-	Thr-	Leu-	40
B12	B13	B14	B15	B16	C1	C2	C3	C4	C5	
Glu-	Lys-	Phe-	Asp-	Arg-	Phe-	Lys-	His-	Leu-	Lys-	50
C6	C7	CD1	CD2	CD3	CD4	CD5	CD6	CD7	CD8	
Thr-	Glu-	Ala-	Glu-	Met-	Lys-	Ala-	Ser-	Glu-	Asp-	60
D1	D2	D3	D4	D5	D6	D7	E1	E2	E3	
Leu-	Lys-	Lys-	His-	Gly-	Val-	Thr-	Val-	Leu-	Thr-	70
E4	E5	E6	E7	E8	E9	E10	E11	E12	E13	
Ala-	Leu-	Gly-	Ala-	Ile-	Leu-	Lys-	Lys-	Lys-	Gly-	80
E14	E15	E16	E17	E18	E19	E20	EF1	EF2	EF3	
His-	His-	Glu-	Ala-	Glu-	Leu-	Lys-	Pro-	Leu-	Ala-	90
EF4	EF5	EF6	EF7	EF8	F1	F2	F3	F4	F5	
Gln-	Ser-	His-	Ala-	Thr-	Lys-	His-	Lys-	Ile-	Pro-	100
F6	F7	F8	F9	FG1	FG2	FG3	FG4	FG5	G1	
Ile-	Lys-	Tyr-	Leu-	Glu-	Phe-	Ile-	Ser-	Glu-	Ala-	110
G2	G3	G4	G5	G6	G7	G8	G9	G10	G11	
Ile-	Ile-	His-	Val-	Leu-	His-	Ser-	Arg-	His-	Pro-	120
G12	G13	G14	G15	G16	G17	G18	G19	GH1	GH2	
Gly-	Asp-	Phe-	Gly-	Ala-	Asp-	Ala-	Gln-	Gly-	Ala-	130
GH3	GH4	GH5	GH6	H1	H2	H3	H4	H5	H6	
Met-	Asn-	Lys-	Ala-	Leu-	Glu-	Leu-	Phe-	Arg-	Lys-	140
H7	H8	H9	H10	H11	H12	H13	H14	H15	H16	
Asp-	Ile-	Ala-	Ala-	Lys-	Tyr-	Lys-	Glu-	Leu-	Gly-	150
H17	H18	H19	H20	H21	H22	H23	H24	HC1	HC2	
Tyr-	Gln-	Gly								153
HC3	HC4	HC5								

Figure 7-5
Amino acid sequence of sperm whale myoglobin. The label below each residue in the sequence refers to its position in an α-helical region or a nonhelical region. For example, B4 is the fourth residue in the B helix; EF7 is the seventh residue in the nonhelical region between the E and F helices. [After A.E. Edmundson. *Nature* 205(1965):883; H.C. Watson. *Prog. Stereochem.* 4(1969):299–333.]

Some important features of myoglobin are

1. Myoglobin is *extremely compact.* The overall dimensions are about 45 × 35 × 25 Å.

2. About 75% of the main chain is in an *α-helical conformation.* The eight major helical segments, all right-handed, are referred to as A, B, C, . . . , H. The first residue in helix A is designated A1, the second A2, and so forth (Figure 7-5). Five nonhelical segments lie between helices (named CD, e.g., if located between the C and D helices). Myoglobin has two other nonhelical regions: two residues at the amino-terminal end (named NA1 and NA2) and five residues at the carboxyl-terminal end (named HC1 through HC5).

3. Four of the helices are terminated by a *proline* residue, whose five-membered ring does not fit within a straight stretch of α helix.

4. The main-chain peptide groups are *planar,* and the carbonyl group of each is *trans* to the NH. Also, the bond angles and distances are like those in dipeptides and other organic compounds (p. 27).

5. The inside and outside are well defined. There is little empty space inside (Figure 7-6). *The interior consists almost entirely of nonpolar residues* such as leucine, valine, methionine, and phenylalanine. In contrast, glutamic and aspartic acids, glutamine, asparagine, lysine, and arginine are absent from the interior of the protein. Residues that have both a polar and a nonpolar part, such as threonine, tyrosine, and tryptophan, are oriented so that their nonpolar portions point inward. The only polar residues inside myoglobin are two histidines, which have a critical function at the binding site. The outside of the protein has both polar and nonpolar residues.

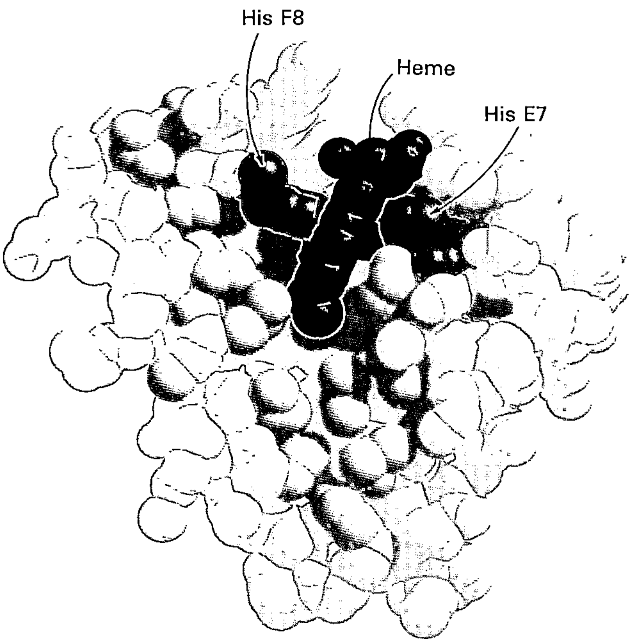

Figure 7-6
Space-filling model of myoglobin showing the close packing of residues in the interior of myoglobin. The heme group is shown in red, the proximal (F8) histidine in blue, the distal (E7) histidine in green, and nonpolar residues in yellow. [Drawn from 1mbn.pdb. H.C. Watson. *Prog. Stereochem.* 4(1969):299.]

A HINDERED HEME ENVIRONMENT IS ESSENTIAL FOR REVERSIBLE OXYGENATION

The heme group is located in a crevice in the myoglobin molecule. The highly polar propionate side chains of the heme are on the surface of the molecule. At physiological pH, these carboxylic acid groups are ionized. The rest of the heme is inside the molecule, where it is surrounded by nonpolar residues except for two histidines. The iron atom of the heme is directly bonded to one of these histidines, namely, residue F8 (Figures 7-7 and 7-8). This histidine, which occupies the fifth coordination position, is called the *proximal histidine.* The iron atom is about 0.3 Å out of the plane of the porphyrin, on the same side as histidine F8. *The oxygen-binding site is on the other side of the heme plane, at the sixth coordination position.* A second histidine residue (E7), termed the *distal histidine,* is near the heme but not bonded to it. A section of an electron-density map displaying the heme is shown in Figure 7-9.

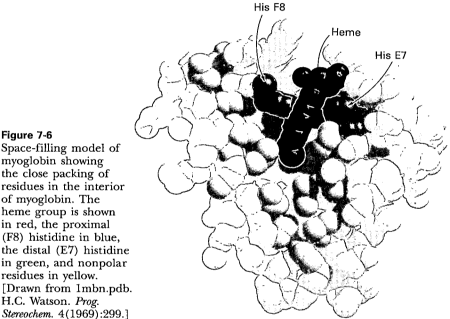

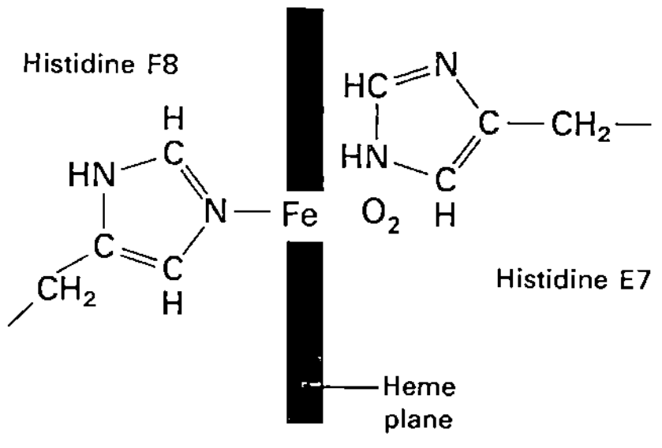

Figure 7-7
Schematic diagram of the oxygen-binding site in myoglobin.

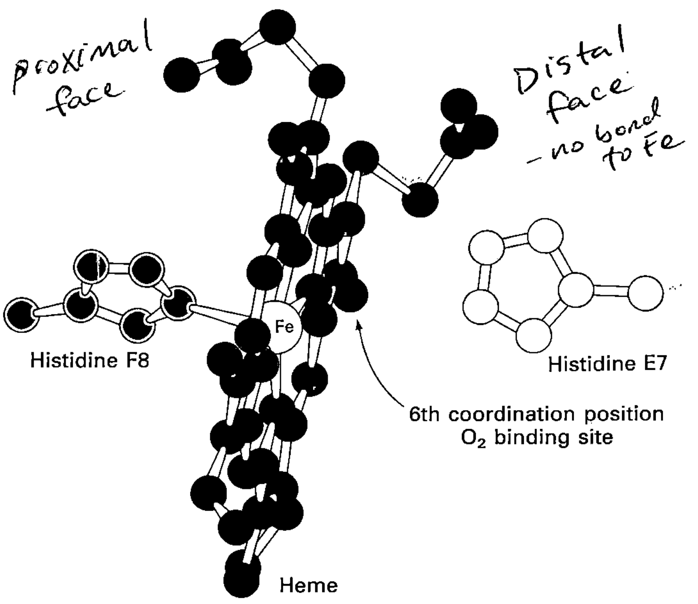

Figure 7-8
Model of the oxygen-binding site in myoglobin showing the heme group, the proximal histidine (F8), and the distal histidine (E7).

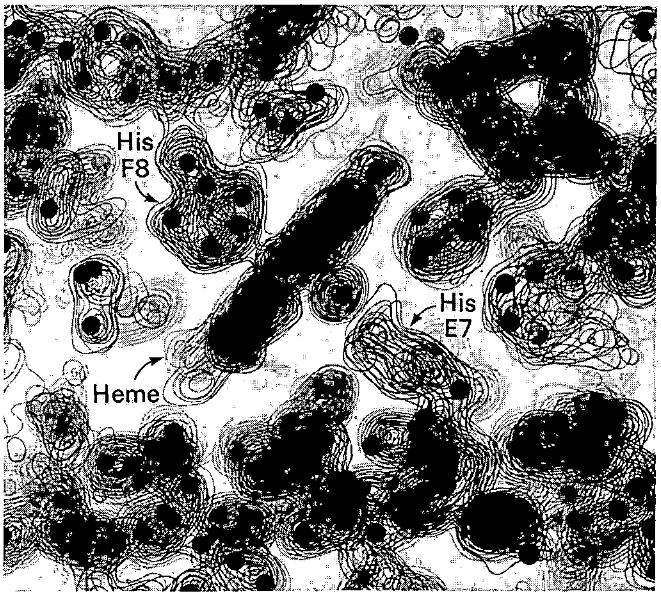

Figure 7-9
Section of an electron-density map of myoglobin near the oxygen-binding site. Electron density extending across the lower part of the map is the E helix. [Courtesy of Dr. John Kendrew.]

The conformations of the three physiologically pertinent forms of myoglobin—deoxymyoglobin, oxymyoglobin, and ferrimyoglobin—are very similar except at the sixth coordination position (Table 7-1). In deoxymyoglobin, it is empty; in oxymyoglobin, it is occupied by O_2; in ferrimyoglobin, it is occupied by water. The axis of the bound O_2 is at an angle to the iron–oxygen bond (Figure 7-10).

The oxygen-binding site comprises only a small fraction of the volume of the myoglobin molecule. Indeed, oxygen is directly bonded only to the iron atom of the heme. Why is the polypeptide portion of myoglobin needed for oxygen transport and storage? The answer lies in the oxygen-binding properties of an isolated heme group. In water, a free ferrous heme group can bind oxygen, but it does so for only a fleeting moment. The reason is that O_2 very rapidly oxidizes the ferrous heme to ferric heme, which cannot bind oxygen. A complex of O_2 sandwiched between two hemes is an intermediate in this reaction. *In myoglobin, the heme group is much less susceptible to oxidation because two myoglobin molecules cannot readily associate to form a heme–O_2–heme complex.* The formation of this sandwich is sterically hindered by the distal histidine and other residues surrounding the sixth coordination site.

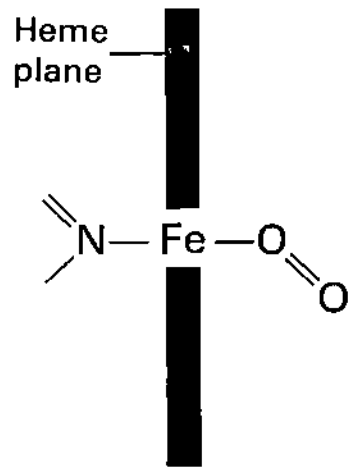

Figure 7-10
Bent, end-on orientation of O_2 in oxymyoglobin. The angle between the O_2 axis and the Fe–O bond is 121 degrees.

Table 7-1
Heme environment

Form	Oxidation state of Fe	Occupant	
		5th coordination position	6th coordination position
Deoxymyoglobin	+2	His F8	Empty
Oxymyoglobin	+2	His F8	O_2
Ferrimyoglobin	+3	His F8	H_2O

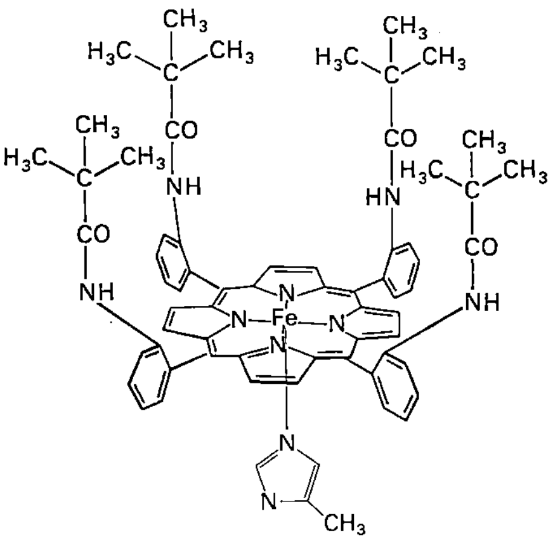

Figure 7-11
Structural formula of a picket-fence iron porphyrin. [After J. Collman, J.I. Brauman, E. Rose, and K.S. Suslick. *Proc. Nat. Acad. Sci.* 75(1978):1053.]

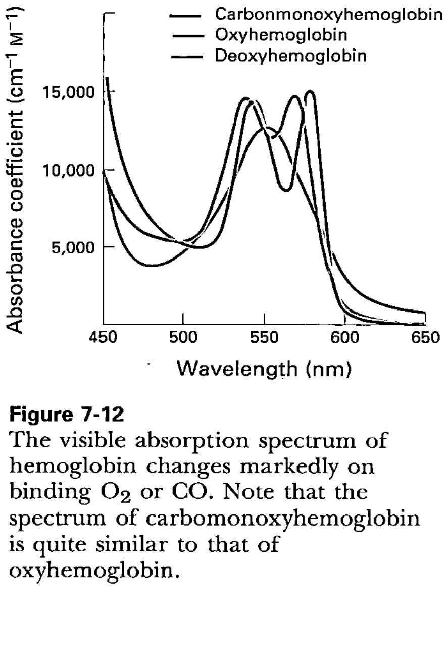

Figure 7-12
The visible absorption spectrum of hemoglobin changes markedly on binding O_2 or CO. Note that the spectrum of carbomonoxyhemoglobin is quite similar to that of oxyhemoglobin.

The strongest evidence for the importance of steric factors in determining the rate of oxidation of heme comes from studies of synthetic model compounds. James Collman synthesized *picket-fence iron porphyrin complexes* (Figure 7-11) that mimic the oxygen-binding sites of myoglobin and hemoglobin. These compounds have a protective enclosure for binding O_2 on one side of the porphyrin ring, whereas the other side is left unhindered so that it can bind a base. In fact, when the base is a substituted imidazole (like histidine), the oxygen affinity of the picket-fence compound (see Figure 7-11) is like that of myoglobin. Furthermore, the picket fence stabilizes the ferrous form of this iron porphyrin and thus enables it to reversibly bind oxygen for long periods. The critical difference between this model compound and free heme is the presence of the picket fence, which blocks the formation of the sandwich dimer.

CARBON MONOXIDE BINDING IS DIMINISHED BY THE PRESENCE OF THE DISTAL HISTIDINE

Carbon monoxide is a poison because it combines with ferromyoglobin and ferrohemoglobin and thereby blocks oxygen transport. An isolated heme in solution typically binds CO some 25,000 times as strongly as O_2. However, the binding affinity of myoglobin and hemoglobin for CO is only about 200 times as great as for O_2. How do these proteins suppress the innate preference of heme for carbon monoxide? The answer comes from x-ray crystallographic and infrared spectroscopic studies of complexes of CO and O_2 with myoglobin and synthetic iron porphyrins. In the very tightly bound complexes of CO with isolated iron porphyrins, the Fe, C, and O atoms are in a linear array (Figure 7-13). In carbonmonoxymyoglobin, by contrast, the CO axis is at an angle to the Fe–C bond. The linear binding of CO is prevented mainly by the steric hindrance of the distal histidine. On the other hand, the O_2 axis is at an angle to the Fe–O bond in oxymyoglobin. Thus, *the protein forces CO to bind at an angle rather than in line. This bent geometry in the globins weakens the interaction of CO with the heme.*

The decreased affinity of myoglobin and hemoglobin for CO is biologically important. Carbon monoxide was a potential hazard long before the emergence of industrialized societies because it is produced endogenously (within cells) in the breakdown of heme (p. 735). The level of endogenously formed carbon monoxide is such that about 1% of the sites in myoglobin and hemoglobin are blocked by CO, a tolerable degree of

Figure 7-13
Structural basis of the diminished affinity of myoglobin and hemoglobin for carbon monoxide: (A) linear mode of binding of CO to isolated iron porphyrins; (B) bent mode of binding of CO to myoglobin and hemoglobin, in which the distal histidine (E7) prevents CO from binding linearly and so the affinity for CO is markedly reduced; (C) bent mode of binding of O_2 in myoglobin and hemoglobin. Isolated iron porphyrins also bind O_2 in a bent mode.

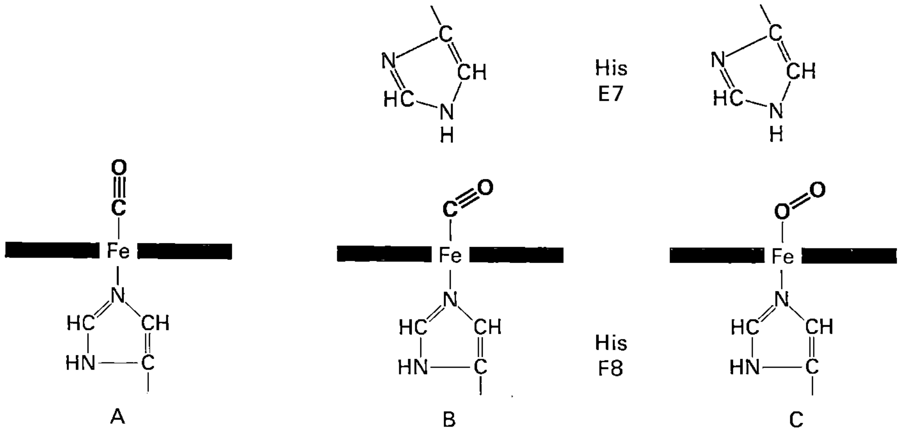

inhibition. However, endogenously produced CO would cause massive poisoning if the affinity of these proteins for CO was like that of isolated iron porphyrins. This challenge was solved by the evolution of heme proteins that discriminate between O_2 and CO by sterically imposing a bent and hence weaker mode of binding for CO.

Thus, myoglobin has created a special microenvironment that confers distinctive properties on its prosthetic group. In general, *the function of a prosthetic group is modulated by its polypeptide environment.* For example, the same heme group has quite a different function in cytochrome *c*, a protein in the terminal oxidation chain in the mitochondria of all aerobic organisms. In cytochrome *c*, the heme is a reversible carrier of electrons rather than of oxygen. Heme has yet another function in the enzyme catalase, where it catalyzes the conversion of hydrogen peroxide into water and oxygen.

THE CENTRAL EXON OF MYOGLOBIN ENCODES A FUNCTIONAL HEME-BINDING UNIT

As was discussed in Chapter 5, most eukaryotic genes are mosaics of exons (coding sequences) and introns (noncoding intervening sequences). It was mentioned that exons encode discrete structural and functional units of proteins (p. 114). Myoglobin illustrates this general principle nicely. The gene for myoglobin consists of three exons: an amino-terminal one encoding residues 1 to 30 (NA1 to B11), a central one encoding residues 31 to 105 (B12 to G6), and a carboxyl-terminal one encoding residues 106 to 153 (G7 to HC5). A look at the structure of native myoglobin shows that nearly the entire heme-binding site is specified by the central exon (Figure 7-14). The regions encoded by the other two exons make very few contacts with the heme. What are the functional properties of an isolated polypeptide consisting of residues 31 to 105? The answer to this question is not yet known, but the properties of another fragment are revealing. Digestion of apomyoglobin with clostripain, an arginine-specific protease, yields a polypeptide containing residues 32 to 139, which corresponds to the central exon and part of the C-terminal

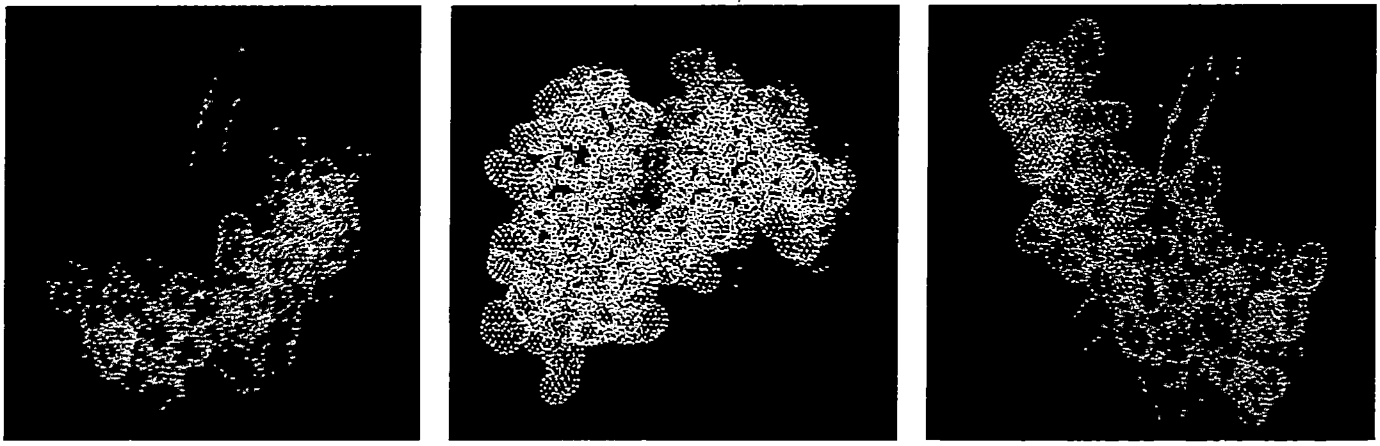

A B C

Figure 7-14
Region of myoglobin encoded by (A) exon 1, (B) exon 2, and (C) exon 3. The central exon encodes a functional oxygen-binding module. The heme group is shown in red.

one. This fragment binds heme. Furthermore, this complex, called *mini-myoglobin,* binds O_2 and CO reversibly. Indeed, the rates of association and dissociation are nearly the same as those of the intact protein.

These findings indicate that the conformation of mini-myoglobin is very similar to that of native myoglobin. Because the amino-terminal exon and the last 14 residues of the C-terminal exon are not essential for reversible oxygenation, it is evident that the central exon contains much of the information for binding heme and maintaining the native fold of an oxygen-binding protein. The exon-intron organization of the constituent chains of hemoglobin is very similar to that of myoglobin. Amino acid sequence comparisons suggest that the genes for myoglobin and hemoglobin diverged some 700 million years ago. Thus, the central exon of these oxygen carriers is an ancient piece of DNA that encoded a functional heme-binding module eons ago.

HEMOGLOBIN CONSISTS OF FOUR POLYPEPTIDE CHAINS

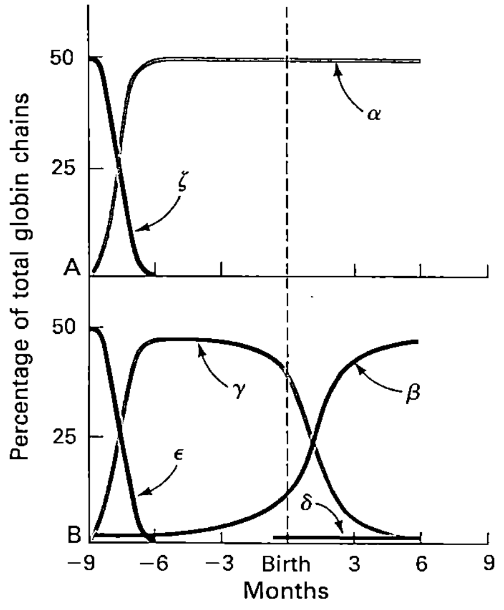

Figure 7-15
Expression of hemoglobin genes in human development: (A) α and ζ genes; (B) β, γ, δ, and ε genes.

We turn now to hemoglobin, the oxygen transporter in erythrocytes. Vertebrate hemoglobins consist of four polypeptide chains, two of one kind and two of another. The four chains are held together by noncovalent attractions. Each contains a heme group and a single oxygen-binding site. *Hemoglobin A,* the principal hemoglobin in adults, consists of two alpha (α) chains and two beta (β) chains. Adults also have a minor hemoglobin (~2% of the total hemoglobin) called *hemoglobin* A_2, which contains delta (δ) chains in place of the β chains of hemoglobin A. Thus, the subunit composition of hemoglobin A is $\alpha_2\beta_2$, and that of hemoglobin A_2 is $\alpha_2\delta_2$.

Embryos and fetuses have distinctive hemoglobins. Shortly after conception, embryos synthesize zeta (ζ) chains (which are α-like chains) and epsilon (ε) chains (which are β-like). In the course of development, ζ is replaced by α, and ε is replaced by gamma (γ), and then by β (Figure 7-15). The major hemoglobin during the latter two-thirds of fetal life, *hemoglobin F,* has the subunit composition $\alpha_2\gamma_2$. The α and ζ chains contain 141 residues; the β, γ, and δ chains contain 146 residues. Why does hemoglobin consist of multiple polypeptides, and why do they differ? We will see shortly that subunit interactions are at the heart of hemoglobin's capacity to transport O_2, CO_2, and H^+in a physiologically responsive way.

X-RAY ANALYSIS OF HEMOGLOBIN: A LABOR OF LOVE OVER A QUARTER CENTURY

Perutz's elucidation of the three-dimensional structure of hemoglobin, a monumental accomplishment, began in 1936. He left Austria that year to pursue graduate work in England and decided on hemoglobin as his thesis subject. The largest structure that had been solved then was the dye phthalocyanin, which contains 58 atoms. In choosing to tackle a molecule one hundred times as large, as Perutz wrote years later, it was little wonder that "my fellow students regarded me with a pitying smile. . . . Fortunately, the examiners of my doctoral thesis did not insist on a determination of the structure, otherwise I should have had to remain a graduate student for twenty-three years." Fortunately, too, Lawrence Bragg became director of the Cavendish Laboratory in Cambridge at this time. In 1913, he and his father were the first to use x-ray crystallography to solve structures. Three decades later, he was eager to see the technique applied to proteins. Bragg wrote, "I was frank about the outlook. It was like multiplying a zero probability that success would be achieved by an infinity of importance if the structure came out; the result of this mathematic operation was anyone's guess." Success came in 1959, when Perutz obtained a low-resolution electron-density image of horse oxyhemoglobin, followed several years later by high-resolution maps of both human and horse oxyhemoglobin and deoxyhemoglobin. The three-dimensional structures of human and horse hemoglobins are very similar.

The hemoglobin molecule is nearly spherical, with a diameter of 55 Å. The four chains are packed together in a tetrahedral array (Figure 7-16). The heme groups are located in crevices near the exterior of the molecule, one in each subunit. The four oxygen-binding sites are far apart; the distance between the two closest iron atoms is 25 Å. Each α chain is in contact with both β chains. In contrast, there are few interactions between the two α chains or between the two β chains.

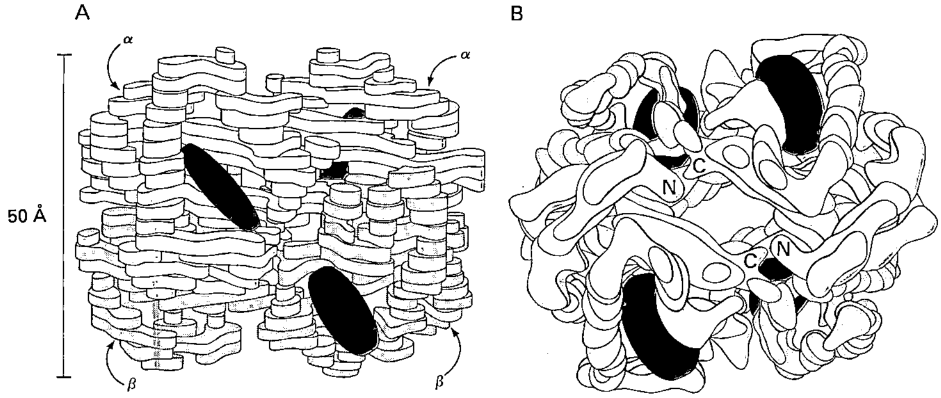

Figure 7-16
Model of hemoglobin at low resolution. The α chains in this model are yellow, the β chains blue, and the heme groups red. View (A) is at right angles to view (B); the top of (A) is visible in (B). [After M.F. Perutz. The hemoglobin molecule. Copyright © 1964 by Scientific American, Inc. All rights reserved.]

HEMOGLOBIN SUBUNITS CLOSELY RESEMBLE MYOGLOBIN IN THREE-DIMENSIONAL STRUCTURE

The three-dimensional structures of myoglobin and the α and β chains of human hemoglobin are strikingly similar (Figure 7-17). The eight helices in each chain of hemoglobin are virtually superposable on those of myoglobin. This close resemblance in the folding of their main chains was

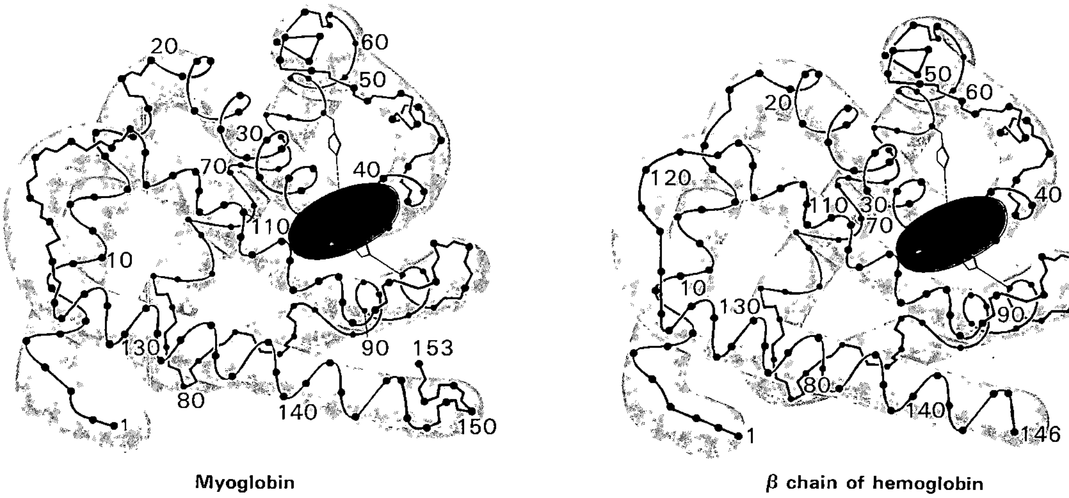

Figure 7-17
Comparison of the conformations of the main chain of myoglobin and the β chain of hemoglobin. The similarity of their conformations is evident. [From M.F. Perutz. The hemoglobin molecule. Copyright © 1964 by Scientific American, Inc. All rights reserved.]

Figure 7-18
Comparison of the amino acid sequences of sperm whale myoglobin and the α and β chains of human hemoglobin, for residues F1 to F9. The amino acid sequences of these three polypeptide chains are much less alike than are their three-dimensional structures.

	Helix position								
Human hemoglobin	F1	F2	F3	F4	F5	F6	F7	F8	F9
α chain	-Leu-	Ser-	Ala-	Leu	-Ser-	Asp-	Leu-	His	-Ala-
β chain	-Phe-	Ala-	Thr-	Leu	-Ser-	Glu-	Leu-	His	-Cys-
Sperm whale myoglobin	-Leu-	Lys-	Pro-	Leu	-Ala-	Gln-	Ser-	His	-Ala-

Only triple identities in this sequence of nine residues (F4 Leu and F8 His, boxed)

unexpected because their amino acid sequences are rather different. In fact, these three chains are identical at only 24 of 141 positions. Hence, *quite different amino acid sequences can specify very similar three-dimensional structures* (Figure 7-18).

It is evident that the three-dimensional form of sperm whale myoglobin and of the α and β chains of human hemoglobin has broad biological significance. In fact, this motif, called the *globin fold,* is common to all known vertebrate myoglobins and hemoglobins. *The intricate folding of the polypeptide chain, first discovered in myoglobin, is nature's fundamental design for an oxygen carrier: it places the heme in an environment that enables it to carry oxygen reversibly.* The genes for myoglobin and for the α, β, and other chains of hemoglobin are variations on a fundamental theme. This family of genes almost certainly arose by gene duplication and diversification.

> *Hemo*—
> A prefix from the Greek word meaning "blood."
>
> *Myo*—
> A prefix from the Greek word meaning "muscle."
>
> *Globin*—
> A protein belonging to the myoglobin-hemoglobin family.

The amino acid sequences of hemoglobins from more than 60 species (ranging from lamprey eels to humans) are known. A comparison of these sequences shows considerable variability at most positions. However, nine positions have the same residue in most species studied thus far (Table 7-2). These highly conserved residues are especially important for

Table 7-2
Highly conserved amino acid residues in hemoglobins

Position	*Amino acid*	*Role*
F8	Histidine	Proximal heme-linked histidine
E7	Histidine	Distal histidine near the heme
CD1	Phenylalanine	Heme contact
F4	Leucine	Heme contact
B6	Glycine	Allows the close approach of the B and E helices
C2	Proline	Helix termination
HC2	Tyrosine	Cross-links the H and F helices
C4	Threonine	Uncertain
H10	Lysine	Uncertain

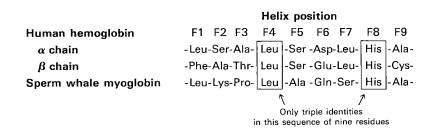

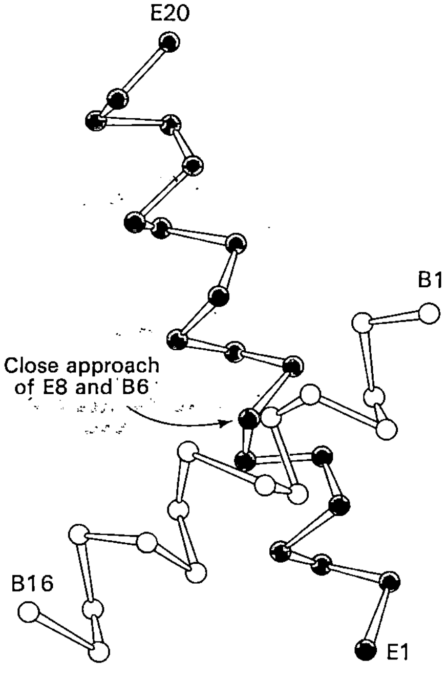

Figure 7-19
Crossing of the B and E helices in myoglobin. Residue B6 is almost invariably glycine because there is no space for a larger side chain.

the function of the hemoglobin molecule. For example, the invariant F8 histidine is directly bonded to the heme iron. Several of them directly affect the oxygen-binding site. Another invariant residue, tyrosine HC2, stabilizes the structure by forming a hydrogen bond between the H and F helices. Glycine B6 is invariant because of its small size: a side chain larger than a hydrogen atom would not allow the B and E helices to approach each other as closely as they do (Figure 7-19). Proline C2 may be essential because it defines one end of the C helix.

The amino acid residues in the interior of hemoglobins vary considerably. However, the change is always of one nonpolar residue for another (as from alanine to isoleucine). Thus, *the striking nonpolar character of the interior of the molecule is conserved.* The nonpolar core is important in binding the heme group and in stabilizing the three-dimensional structure of each subunit (see Figure 7-6).

In contrast, surface residues are highly variable. Indeed, few are consistently positively or negatively charged. It might be thought that proline residues would be preserved because of their role as helix breakers. However, this is not so. Only one proline is invariant, yet the lengths and directions of the helices in all globins are very similar. Obviously, there are other ways of terminating or bending α helices. For example, an α helix can be destabilized by hydrogen bonding between the OH group of a serine or threonine residue and a main-chain carbonyl group.

ALLOSTERIC INTERACTIONS ENABLE HEMOGLOBIN TO COORDINATELY TRANSPORT O_2, CO_2, AND H^+

The α and β subunits of hemoglobin have the same structural design as myoglobin. However, new properties of profound biological importance emerge when different subunits come together to form a tetramer. *Hemoglobin is a much more intricate and sentient molecule than is myoglobin.* Hemoglobin transports H^+ and CO_2 in addition to O_2. Furthermore, the oxygen-binding properties of hemoglobin are regulated by interactions between separate, nonadjacent sites. *Hemoglobin is an allosteric protein, whereas myoglobin is not.* This difference is expressed in three ways:

1. The binding of O_2 to hemoglobin enhances the binding of additional O_2 to the same hemoglobin molecule. In other words, O_2 binds cooperatively to hemoglobin. In contrast, the binding of O_2 to myoglobin is not cooperative.

2. The affinity of hemoglobin for oxygen depends on pH, whereas that of myoglobin is independent of pH. The CO_2 molecule also affects the oxygen-binding characteristics of hemoglobin. Both H^+ and CO_2 promote the release of bound O_2. Reciprocally, O_2 promotes the release of bound H^+ and CO_2.

3. The oxygen affinity of hemoglobin is further regulated by organic phosphates such as 2,3-bisphosphoglycerate (BPG). The result is that hemoglobin has a lower affinity for oxygen than does myoglobin.

Torr—
A unit of pressure equal to that exerted by a column of mercury 1 mm high at 0°C and standard gravity (1 mm Hg). Named after Evangelista Torricelli (1608–1647), the inventor of the mercury barometer.

OXYGEN BINDS COOPERATIVELY TO HEMOGLOBIN

The saturation Y is defined as the fractional occupancy of all the oxygen-binding sites in a solution. The value of Y can range from 0 (all sites empty) to 1 (all sites filled). A plot of Y versus pO_2, the partial pressure of oxygen, is called an *oxygen dissociation curve.* The oxygen dissociation curves of myoglobin and hemoglobin differ in two ways (Figure 7-20). First, for any given pO_2 Y is higher for myoglobin than for hemoglobin. This means that *myoglobin has a higher affinity for oxygen than does hemoglobin.* Oxygen affinity can be characterized by a quantity P_{50}, which is the partial pressure of oxygen at which 50% of sites are filled (i.e., at which $Y = 0.5$). For myoglobin, P_{50} is typically 1 torr, whereas for hemoglobin, P_{50} is 26 torrs.

The second difference is that *the oxygen dissociation curve of myoglobin is hyperbolic, whereas that of hemoglobin is sigmoidal.* Let us consider these curves in quantitative terms, starting with the one for myoglobin because it is simpler. The binding of oxygen to myoglobin (Mb) can be described by a simple equilibrium.

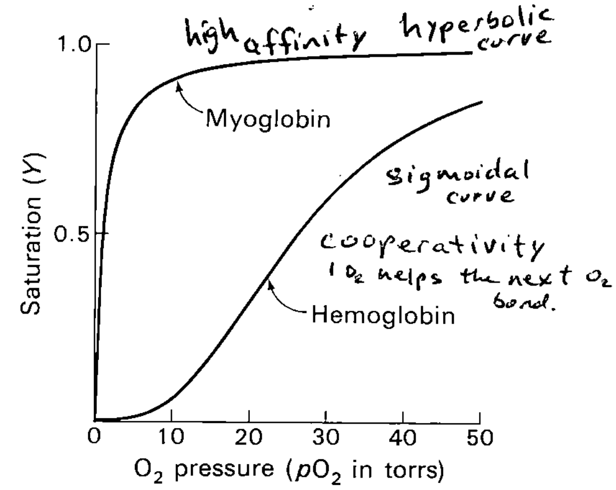

Figure 7-20
Oxygen dissociation curves of myoglobin and hemoglobin. Saturation of the oxygen-binding sites is plotted as a function of the partial pressure of oxygen surrounding the solution.

$$MbO_2 \rightleftharpoons Mb + O_2 \tag{1}$$

The equilibrium constant K for the dissociation of oxymyoglobin is

$$K = \frac{[Mb][O_2]}{[MbO_2]} \tag{2}$$

in which $[MbO_2]$ is the concentration of oxymyoglobin, [Mb] is the concentration of deoxymyoglobin, and $[O_2]$ is the concentration of uncombined oxygen, all in moles per liter. Then the saturation Y is

$$Y = \frac{[MbO_2]}{[MbO_2] + [Mb]} \tag{3}$$

Substitution of equation 2 into equation 3 yields

$$Y = \frac{[O_2]}{[O_2] + K} \tag{4}$$

Because oxygen is a gas, it is convenient to express its concentration in terms of pO_2, the partial pressure of oxygen (in torrs) in the atmosphere surrounding the solution, and to use P_{50} in place of K. Equation 4 then becomes

$$Y = \frac{pO_2}{pO_2 + P_{50}} \tag{5}$$

Equation 5 plots as a hyperbola. In fact, the oxygen dissociation curve calculated from equation 5, taking P_{50} to be 1 torr, closely matches the experimentally observed curve for myoglobin (see Figure 7-20). In contrast, the sigmoidal curve for hemoglobin cannot be matched by any curve described by equation 5.

THE HILL COEFFICIENT IS A MEASURE OF COOPERATIVITY

In 1913, Archibald Hill showed that the curve obtained from the oxygen-binding data for hemoglobin agrees with the equation derived for the *hypothetical* equilibrium

$$Hb(O_2)_n \rightleftharpoons Hb + nO_2 \tag{6}$$

This expression yields

$$Y = \frac{(pO_2)^n}{(pO_2)^n + (P_{50})^n} \tag{7}$$

which can be rearranged to give

$$\frac{Y}{1 - Y} = \left(\frac{pO_2}{P_{50}}\right)^n \tag{8}$$

This equation states that the ratio of oxyheme (Y) to deoxyheme ($1 - Y$) is equal to the nth power of the ratio of pO_2 to P_{50}. Taking the logarithms of both sides of equation 8 gives

$$\log \frac{Y}{1 - Y} = n \log pO_2 - n \log P_{50} \tag{9}$$

A plot of $\log [Y/(1 - Y)]$ versus $\log pO_2$, called a *Hill plot*, approximates a straight line. Its slope n at the midpoint of the binding ($Y = 0.5$) is called the *Hill coefficient*. The value of n increases with the degree of cooperativity; the maximum possible value of n is equal to the number of binding sites.

Myoglobin gives a Hill plot with $n = 1.0$ (Figure 7-21), which means that O_2 molecules bind independently of each other, as indicated in equation 1. In contrast, the Hill coefficient of 2.8 for hemoglobin indicates that *the binding of oxygen in hemoglobin is cooperative.* Binding at one heme facilitates the binding of oxygen at the other hemes on the same tetramer. Conversely, the unloading of oxygen at one heme facilitates the unloading of oxygen at the others. In other words, the heme groups of a hemoglobin molecule communicate with each other. The mechanism of cooperative binding of oxygen by hemoglobin, sometimes called *heme-heme interaction,* will be discussed shortly.

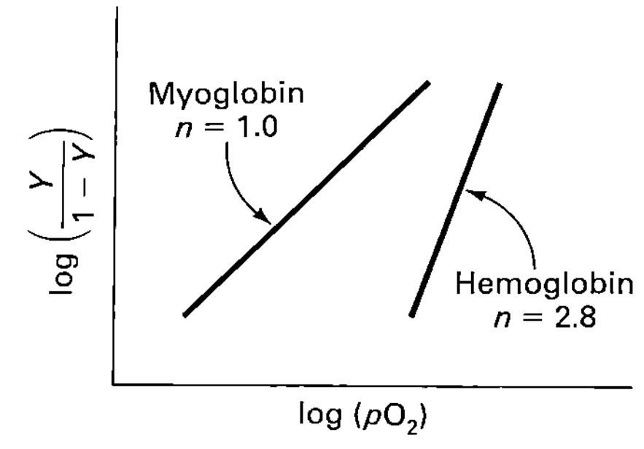

Figure 7-21
Hill plot for the binding of O_2 to myoglobin and hemoglobin. The slope of 2.8 for hemoglobin indicates that it binds oxygen cooperatively, in contrast with myoglobin, which has a slope of 1.0.

THE COOPERATIVE BINDING OF OXYGEN MAKES HEMOGLOBIN A MORE EFFICIENT OXYGEN TRANSPORTER

What is the biological significance of the cooperative binding of oxygen by hemoglobin? The oxygen saturation of hemoglobin changes more rapidly with changes in the partial pressure of O_2 than it would if the oxygen-binding sites were independent of each other. Let us consider a specific example (Figure 7-22). Assume that the alveolar pO_2 is 100 torrs, and that the pO_2 in the capillary of an active muscle is 20 torrs. Let $P_{50} = 26$ torrs, and take $n = 2.8$. Then Y in the alveolar capillaries will be 0.98, and Y in the muscle capillaries will be 0.32. The oxygen delivered will be proportional to the difference in Y, which is 0.66. Let us now make the same calculation for a hypothetical oxygen carrier for which P_{50} is also 26 torrs, but in which the binding of oxygen is not cooperative ($n = 1$). Then $Y_{\text{alveoli}} = 0.79$, and $Y_{\text{muscle}} = 0.43$, and so the difference in Y is equal to 0.36. Thus, *the cooperative binding of oxygen by hemoglobin enables it to deliver 1.83 times as much oxygen under typical physiological conditions as it would if the sites were independent.*

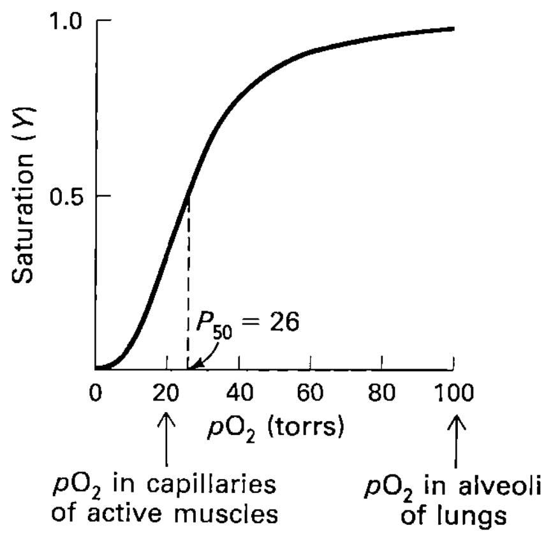

Figure 7-22
Oxygen dissociation curve of hemoglobin. Typical values for pO_2 in the capillaries of active muscle and in the alveoli of the lung are marked on the horizontal axis. Note that P_{50} for hemoglobin under physiological conditions lies between these values.

O_2 binding ↓ w/[H^+] ↑
" " " w/[CO_2] ↑

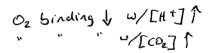

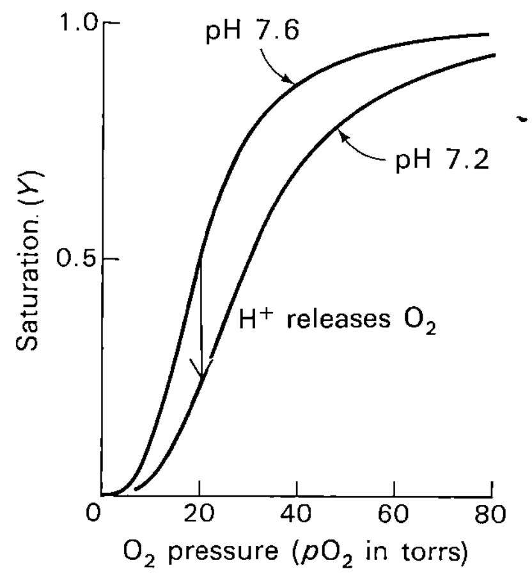

Figure 7-23
Effect of pH on the oxygen affinity of hemoglobin. Lowering the pH from 7.6 to 7.2 results in the release of O_2 from oxyhemoglobin.

H^+ AND CO_2 PROMOTE THE RELEASE OF O_2: THE BOHR EFFECT

Myoglobin shows no change in oxygen binding over a broad range of pH, nor does CO_2 have an appreciable effect. In hemoglobin, however, acidity enhances the release of oxygen. In the physiological range, a lowering of pH shifts the oxygen dissociation curve to the right, so that the oxygen affinity is decreased (Figure 7-23). Increasing the concentration of CO_2 (at constant pH) also lowers the oxygen affinity. In rapidly metabolizing

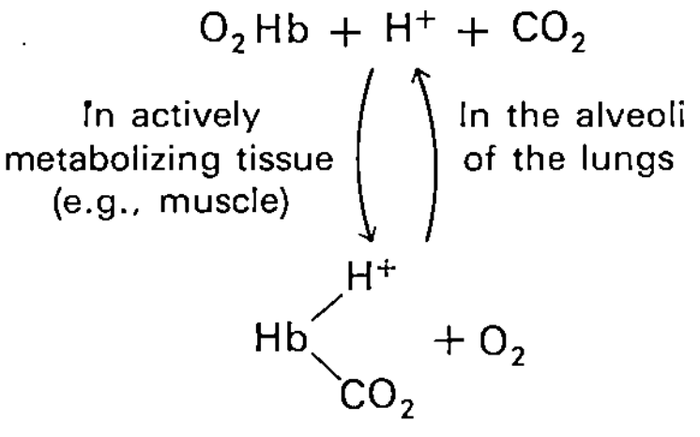

Figure 7-24
Summary of the Bohr effect. The actual mechanism and stoichiometry are more complex than indicated in this diagram.

tissue, such as contracting muscle, much CO_2 and acid are produced. *The presence of higher levels of CO_2 and H^+ in the capillaries of such metabolically active tissue promotes the release of O_2 from oxyhemoglobin.* This important mechanism for meeting the higher oxygen needs of metabolically active tissues was discovered by Christian Bohr in 1904.

The reciprocal effect, discovered 10 years later by J.S. Haldane, occurs in the alveolar capillaries of the lungs. The high concentration of O_2 there unloads H^+ and CO_2 from hemoglobin, just as the high concentration of H^+ and CO_2 in active tissues drives off O_2. These linkages between the binding of O_2, H^+, and CO_2 are known as the *Bohr effect* (Figure 7-24).

BPG LOWERS THE OXYGEN AFFINITY OF HEMOGLOBIN

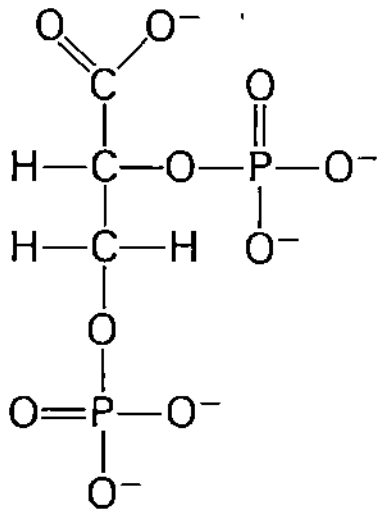

2,3-Bisphosphoglycerate (BPG)
(2,3-Diphosphoglycerate, DPG)

The oxygen affinity of hemoglobin within red cells is lower than that of hemoglobin in free solution. As early as 1921, Joseph Barcroft wondered, "Is there some third substance present . . . which forms an integral part of the oxygen-hemoglobin complex?" Indeed there is. Reinhold Benesch and Ruth Benesch showed in 1967 that *2,3-bisphosphoglycerate* (*BPG*, also known as *2,3-diphosphoglycerate, DPG*) binds to hemoglobin and has a large effect on its affinity for oxygen. This highly anionic organic phosphate is present in human red cells at about the same molar concentration as hemoglobin. In the absence of BPG, the P_{50} of hemoglobin is 1 torr, like that of myoglobin. In its presence, P_{50} becomes 26 torrs (Figure 7-25). Thus, *BPG lowers the oxygen affinity of hemoglobin by a factor of 26, which is essential in enabling hemoglobin to unload oxygen in tissue capillaries.* BPG diminishes the oxygen affinity of hemoglobin by binding to deoxyhemoglobin but not to the oxygenated form.

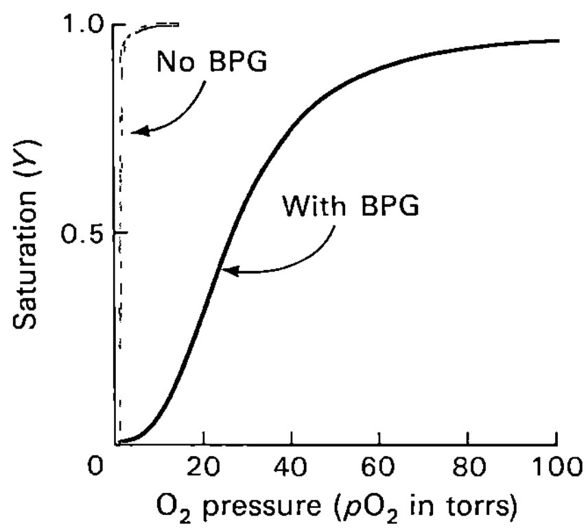

Figure 7-25
2,3-Bisphosphoglycerate (BPG) decreases the oxygen affinity of hemoglobin.

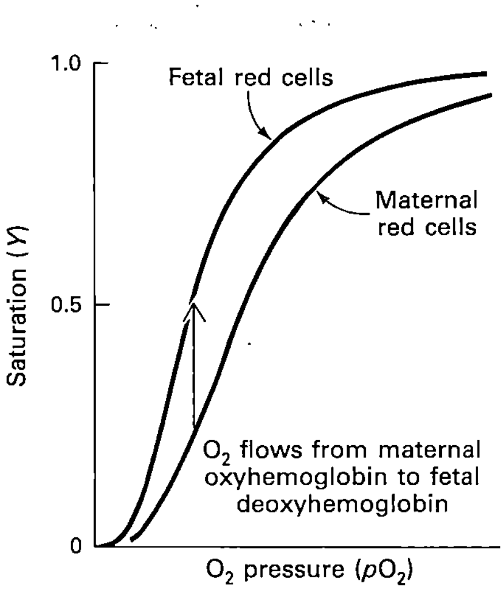

Figure 7-26
Fetal red blood cells have a higher oxygen affinity than do maternal red blood cells. The reason is that, in the presence of BPG, the oxygen affinity of fetal hemoglobin is higher than that of maternal hemoglobin.

FETAL HEMOGLOBIN HAS A HIGHER OXYGEN AFFINITY THAN MATERNAL HEMOGLOBIN

Fetuses have their own kind of hemoglobin, called *hemoglobin F* ($\alpha_2\gamma_2$), which differs from adult hemoglobin A ($\alpha_2\beta_2$), as mentioned previously. An important property of hemoglobin F is that it has a higher oxygen affinity under physiological conditions than does hemoglobin A (Figure 7-26). The higher oxygen affinity of hemoglobin F optimizes the transfer of oxygen from the maternal to the fetal circulation. Hemoglobin F is oxygenated at the expense of hemoglobin A on the other side of the placental circulation. The higher oxygen affinity of fetal blood was known for many years, but an understanding of its basis could come only after the discovery of BPG. *Hemoglobin F binds BPG less strongly than does hemoglobin A and consequently has a higher oxygen affinity.* In the absence of BPG, the oxygen affinity of hemoglobin F is actually lower than that of hemoglobin A. We see here a clear-cut biological advantage of having distinct forms of a protein, called *isoforms* or *isotypes,* in different tissues. Indeed, many proteins have multiple isoforms, but the reasons are usually less evident than for hemoglobin.

THE QUATERNARY STRUCTURE OF HEMOGLOBIN CHANGES MARKEDLY ON OXYGENATION

Let us now consider the structural basis of these allosteric effects. Hemoglobin can be dissociated into its constituent chains. The properties of the isolated α chain are very much like those of myoglobin. The α chain by itself has a high oxygen affinity, a hyperbolic oxygen dissociation curve, and oxygen-binding characteristics that are insensitive to pH, CO_2 concentration, and BPG level. The isolated β chains readily associate to form a tetramer (β_4). Like the α chain and myoglobin, β_4 lacks the allosteric properties of hemoglobin and has a high oxygen affinity. In short, *the allosteric properties of hemoglobin arise from interactions between its subunits. The functional unit of hemoglobin is a tetramer consisting of two kinds of polypeptide chains.*

In 1938, Felix Haurowitz found that crystals of deoxyhemoglobin shattered when they were exposed to oxygen. Deoxymyoglobin crystals, in contrast, can bind and release oxygen without losing their form. The shattering of hemoglobin crystals suggested that the protein undergoes a major conformational change on binding O_2. Indeed, x-ray crystallographic studies carried out years later showed that *oxy- and deoxyhemoglobin differ markedly in quaternary structure* (Figure 7-27).

The oxygenated molecule is more compact. For instance, the distance between the iron atoms of the β chains decreases from 40 to 33 Å on oxygenation. The changes in the contacts between the α and β chains are of special interest (Figure 7-28). In the transition from oxy- to deoxyhemoglobin, large structural changes take place at two of the four contact

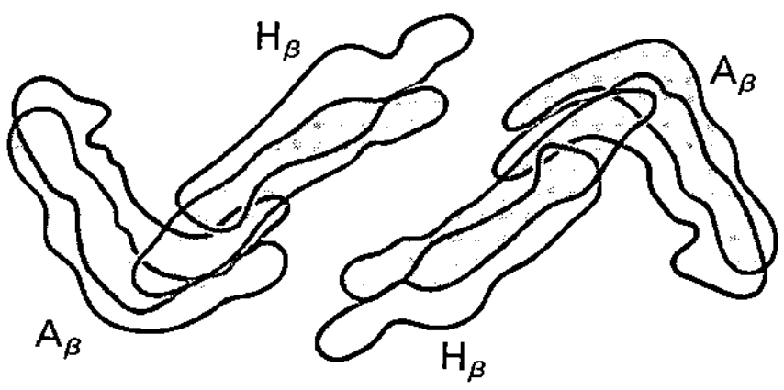

Figure 7-27
Projection of part of the electron-density maps of oxyhemoglobin (shown in red) and deoxyhemoglobin (shown in blue) at a resolution of 5.5 Å. The A and H helices of the two β chains of hemoglobin are shown here. The center of the diagram corresponds to the central cavity of the molecule. It shows one of the conformational changes accompanying oxygenation—a movement of the H helices toward each other. [After M.F. Perutz. *Nature* 228(1970):738.]

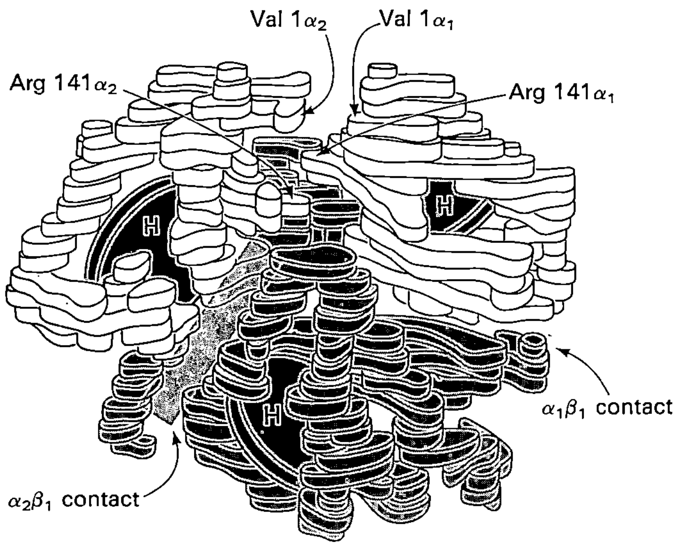

Figure 7-28
Model of oxyhemoglobin at low resolution showing the two kinds of interfaces between α and β chains. The α chains are white, the β chains gray. Three hemes (red) can be seen in this view of the molecule. The $\alpha_2\beta_1$ contact region is shown in blue, the $\alpha_1 b_1$ contact region in yellow. [After M.F. Perutz and L.F. TenEyck. *Cold Spring Harbor Symp. Quant. Biol.* 36(1971):296.]

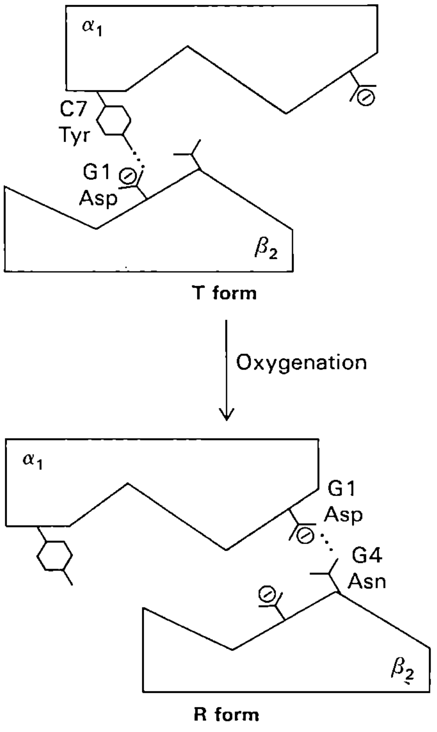

Figure 7-30
The $\alpha_1\beta_2$ interface switches from the T to the R form on oxygenation. The dovetailed construction of this interface allows the subunits to readily adopt either of the two forms.

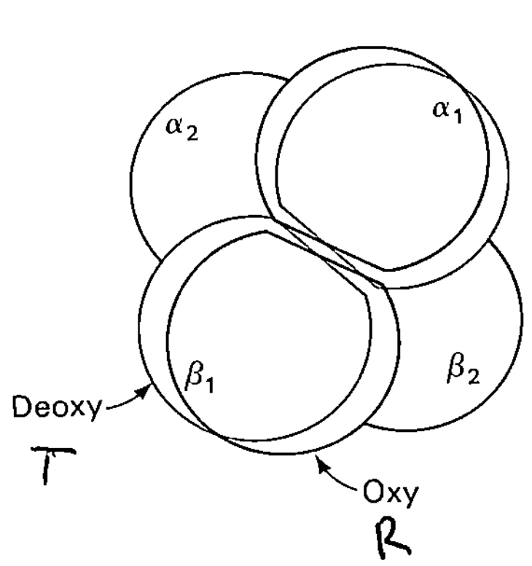

Figure 7-29
Schematic diagram showing the change in quaternary structure on oxygenation. One pair of $\alpha\beta$ subunits shifts with respect to the other by a rotation of 15 degrees and a translation of 0.8 Å. The oxy form of the rotated $\alpha\beta$ subunit is shown in red and the deoxy form in blue. [After J. Baldwin and C. Chothia. *J. Mol. Biol.* 129(1979):192. Copyright by Academic Press, Inc. (London) Ltd.]

regions (the $\alpha_1\beta_2$ contact and the identical $\alpha_2\beta_1$ contact) but not at the others (the $\alpha_1\beta_1$ contact and the identical $\alpha_2\beta_2$ contact). Furthermore, the $\alpha_1\beta_1$ pair rotates relative to the other pair by 15 degrees. Some atoms at the interface between these pairs shift by as much as 6 Å (Figure 7-29).

In fact, *the $\alpha_1\beta_2$ contact region is designed to act as a switch between two alternative structures.* The two forms of this dovetailed interface are stabilized by different sets of hydrogen bonds (Figure 7-30). This interface is closely connected to the heme groups, and so structural changes in it affect the hemes. Reciprocally, structural changes at the hemes affect this interface. Most residues in it are the same in all species because of its key role in mediating allosteric interactions. Nearly all mutations in this interface diminish cooperative oxygen binding, whereas changes in the $\alpha_1\beta_1$ interface do not.

In oxyhemoglobin, the carboxyl-terminal residues of all four chains have almost complete freedom of rotation. In deoxyhemoglobin, by contrast, these terminal groups are anchored. The terminal carboxylates and the side chains of the C-terminal residues participate in salt links (electrostatic interactions) that tie the tetramer. Deoxyhemoglobin is a tauter, more constrained molecule than oxyhemoglobin because of the presence of eight additional salt links. *The quaternary structure of deoxyhemoglobin is termed the T (tense or taut) form; that of oxyhemoglobin, the R (relaxed) form.* The designations R and T are generally used to describe alternative quaternary structures of an allosteric protein, the T form having a lower affinity for the substrate (p. 166).

OXYGEN BINDING SWITCHES QUATERNARY STRUCTURE BY MOVING IRON INTO THE PLANE OF THE PORPHYRIN

The conformational changes discussed thus far take place at some distance from the heme. Now let us see what happens at the heme group itself on oxygenation. In deoxyhemoglobin, the iron atom is about 0.4 A out of the porphyrin plane toward the proximal histidine, so that the heme group is domed (convex) in the same direction (Figure 7-31). On oxygenation, the iron atom moves into the plane of the porphyrin to form a strong bond with O_2, and the heme becomes more planar. Structural studies of many synthetic iron porphyrins have shown that the iron atom is out of plane in five-coordinated compounds, whereas it is in plane or nearly so in six-coordinated complexes.

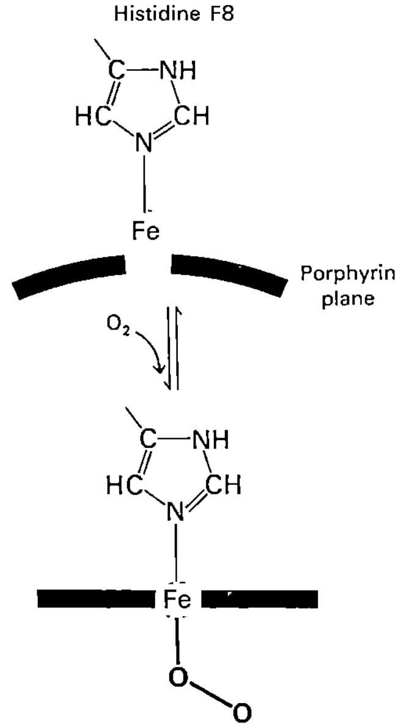

Figure 7-31
The iron atom moves into the plane of the heme on oxygenation. The proximal histidine (F8) is pulled along with the iron atom and becomes less tilted.

How does the movement of the iron atom into the plane of the heme favor the switch in quaternary structure from T to R? The iron atom pulls the proximal histidine with it when it moves into the plane of the porphyrin. This movement of histidine F8 shifts the F helix, the EF corner, and the FG corner (Figure 7-32). These conformational changes are in turn transmitted to the subunit interfaces, where they rupture interchain salt links to switch the protein to the R form. Thus, *a structural change (oxygenation) within a subunit is translated into structural changes at the interfaces between subunits. The binding of oxygen at one heme site is thereby communicated to parts of the molecule that are far away.*

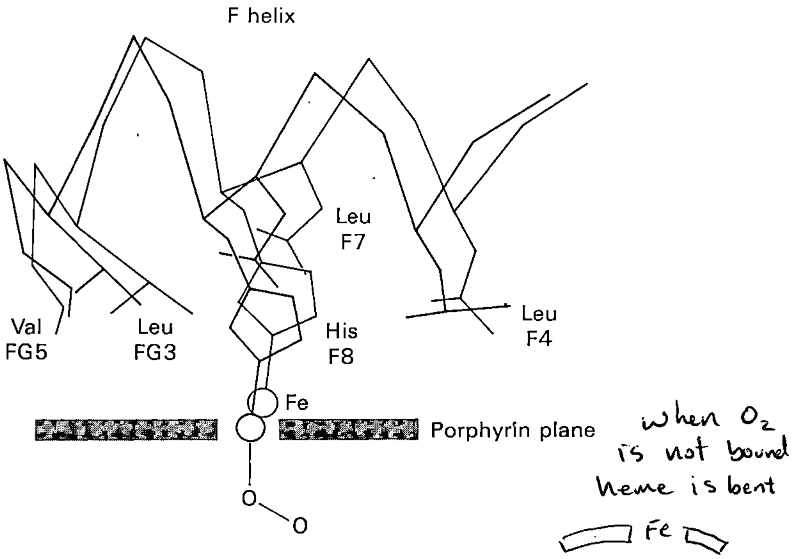

Figure 7-32
Conformational changes induced by the movement of the iron atom on oxygenation. The oxygenated structure is shown in red and the deoxygenated structure in blue. [After J. Baldwin and C. Chothia. *J. Mol. Biol.* 129(1979):192.]

BPG DECREASES OXYGEN AFFINITY BY CROSS-LINKING DEOXYHEMOGLOBIN

How does BPG lower the oxygen affinity of hemoglobin? Only one molecule of BPG is bound to deoxyhemoglobin. This is an unusual stoichiometry—an $\alpha_2\beta_2$ tetramer would be expected to possess at least two binding sites for a small molecule. The presence of just one binding site immediately suggested that *BPG binds on the symmetry axis of the hemoglobin molecule in the central cavity,* where the four subunits are near one another (Figure 7-33). Indeed, x-ray analysis confirmed this proposal and showed that the

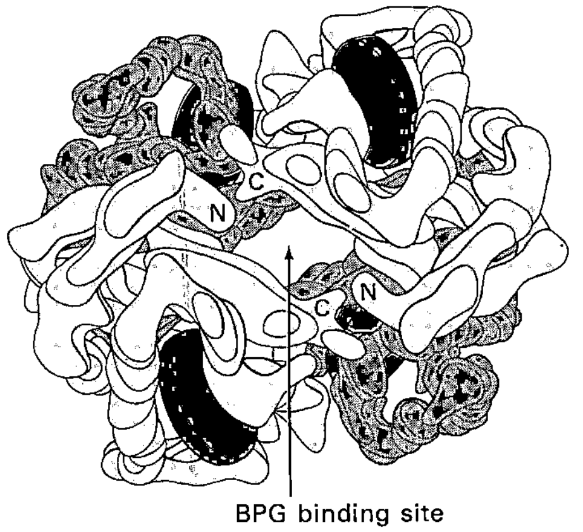

Figure 7-33
The binding site for BPG is in the central cavity of deoxyhemoglobin. [After M.F. Perutz. The hemoglobin molecule. Copyright © 1964 by Scientific American, Inc. All rights reserved.]

binding site for BPG is constituted by multiple positively charged residues on *each* β chain: the α-amino group, His 2, Lys 82, and His 143. These groups interact with the strongly negatively charged BPG, which carries nearly four negative charges at physiologic pH. *BPG is stereochemically complementary to this constellation of positively charged groups facing the central cavity of the hemoglobin molecule* (Figure 7-34). Why does BPG bind more weakly to fetal than to adult hemoglobin? Residue 143 in hemoglobin F is an uncharged serine instead of the positively charged histidine in hemoglobin A.

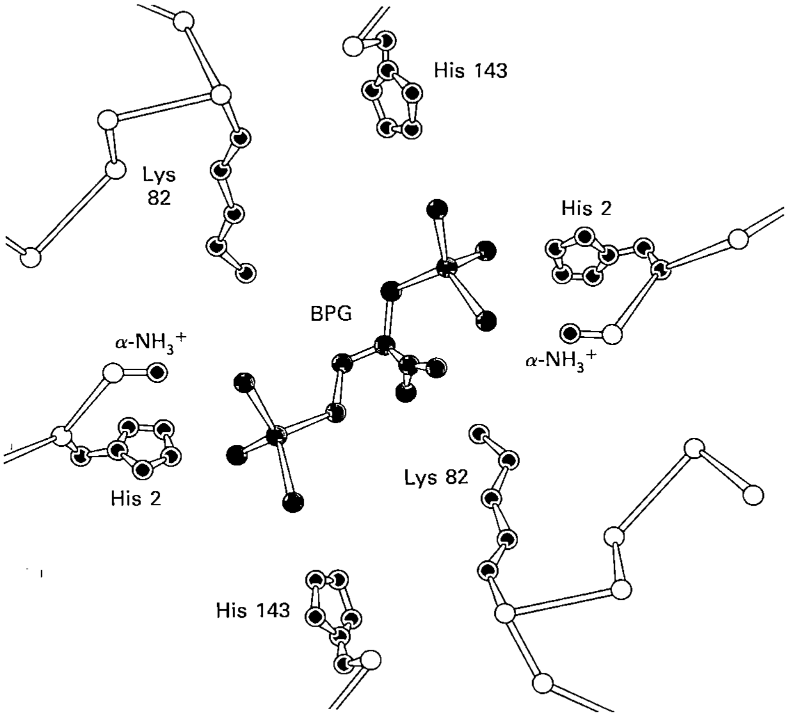

Figure 7-34
Mode of binding of BPG to human deoxyhemoglobin. BPG interacts with three positively charged groups on each β chain. [After A. Arnone. *Nature* 237(1972):148.]

On oxygenation, BPG is extruded because the central cavity becomes too small. Specifically, the gap between the H helices of the β chains becomes narrowed (see Figure 7-27). Also, the distance between the α-amino groups increases from 16 to 20 Å, which prevents them from simultaneously binding the phosphates of a BPG molecule. The reason why BPG decreases oxygen affinity is now evident. *BPG stabilizes the deoxyhemoglobin quaternary structure by cross-linking the* β *chains.* In other words, BPG shifts the equilibrium toward the T form. As mentioned previously, the carboxyl-terminal residues of deoxyhemoglobin form eight salt links that must be broken for oxygenation to occur. The binding of BPG contributes additional cross-links that must be broken, and so the oxygen affinity of hemoglobin is diminished.

CO_2 BINDS TO THE TERMINAL AMINO GROUPS OF HEMOGLOBIN AND LOWERS ITS OXYGEN AFFINITY

In aerobic metabolism, about 0.8 equivalents of CO_2 are produced per O_2 consumed. Most of the CO_2 is transported as *bicarbonate,* which is formed within red cells by the action of *carbonic anhydrase.*

$$CO_2 + H_2O \rightleftharpoons HCO_3^- + H^+$$

Much of the H^+ generated by this reaction is taken up by deoxyhemoglobin as part of the Bohr effect. The remainder of the CO_2 is carried by hemoglobin in the form of *carbamate,* because the un-ionized form of the α-amino groups of hemoglobin can react reversibly with CO_2.

$$\mathrm{R{-}NH_2 + CO_2 \rightleftharpoons R{-}\overset{H}{\underset{}{N}}{-}\underset{\underset{O}{\|}}{C}{-}O^- + H^+}$$

The bound carbamates form salt bridges that stabilize the T form. *Hence, the binding of CO_2 lowers the oxygen affinity of hemoglobin.*

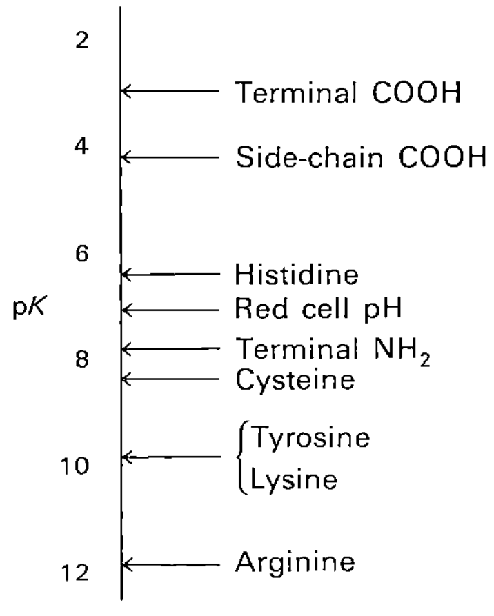

Figure 7-35
Typical p*K*s of some functional groups in proteins. The environment of a particular group can make its actual p*K* higher or lower than the value given here.

DEOXYGENATION INCREASES THE AFFINITY OF SEVERAL PROTON-BINDING SITES

About 0.5 H^+ is taken up by hemoglobin for each molecule of O_2 that is released. This uptake of H^+ helps to buffer the pH in metabolically active tissues. It indicates that deoxygenation increases the affinities of some sites for H^+. Specifically, the p*K*s of some groups must be *raised* in the transition from oxy- to deoxyhemoglobin; an increase in p*K* means stronger binding of H^+. Which groups have their p*K*s raised? The potential sites and typical p*K* values are given in Figure 7-35. The side-chain carboxylates of glutamate and aspartate normally have p*K*s of about 4. It is unlikely that their p*K*s would be raised to at least 7, which would be necessary for a group to participate in the Bohr effect. On the other hand, the normal p*K*s of the side chains of tyrosine, lysine, and arginine are usually more than 10, and so it seems unlikely that they could be lowered enough to allow uptake of H^+. Hence, the most plausible candidates are the terminal amino group and the side chains of histidine and cysteine, which normally have p*K* values near 7.

X-ray and chemical studies suggest that three groups account for much of the Bohr effect: the side chains of histidines β146 and α122 and the α-amino group of the α chain. In oxyhemoglobin, histidine β146 rotates freely, whereas, in deoxyhemoglobin, this terminal residue participates in a number of interactions. Of particular significance is the interaction of its imidazole ring with the negatively charged aspartate 94 in the FG corner of the same β chain. The close proximity of this negatively charged group enhances the likelihood that the imidazole group binds a proton (Figure 7-36). In other words, the proximity of aspartate 94 raises the p*K* of histidine 146. Thus, *in the transition from oxy- to deoxyhemoglobin, histidine 146 acquires a greater affinity for H^+* because its local environment becomes more negatively charged. The immediate environments of the other two groups implicated in the Bohr effect also are more negatively charged in deoxyhemoglobin because of changes of quaternary structure induced by the release of O_2.

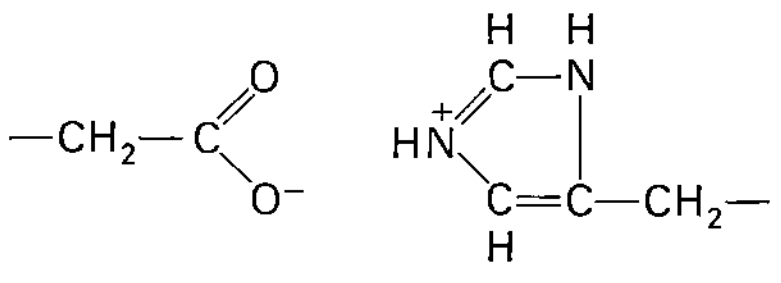

Asp 94 **His 146**

Figure 7-36
Aspartate 94 raises the p*K* of histidine 146 in deoxyhemoglobin but not in oxyhemoglobin. The proximity of the negative charge on aspartate 94 favors protonation of histidine 146 in deoxyhemoglobin.

IN THE SEQUENTIAL MODEL FOR ALLOSTERIC INTERACTIONS SUBUNITS CHANGE CONFORMATION ONE AT A TIME

We have seen the structures of fully deoxygenated and fully oxygenated hemoglobin. These splendid images provide insight into how the binding of O_2, CO_2, H^+, and BPG influence each other. They also whet our appetite for a deeper understanding of allosteric transitions. In particular, how does hemoglobin switch from the deoxy to the oxy structure when it binds successive O_2 molecules? What is the actual allosteric mechanism?

Two models—the *sequential model* and the *concerted model*—offer contrasting views of how cooperative interactions occur. The simplest form of the sequential model, which was developed by Daniel Koshland, Jr., makes three assumptions:

1. Only two conformational states, T and R, are accessible to any subunit.

2. The binding of ligand switches the conformation of the subunit to which it is bound but not that of its neighbors. In other words, the T to R transition in a subunit is *induced* by the binding of ligand to that particular subunit.

3. The conformational change elicited by the binding of substrate in one subunit can increase or decrease the binding affinity of the other subunits in the same molecule. A T subunit with an R neighbor has higher affinity for ligand than does a T subunit with a T neighbor because the subunit interfaces are different in TR and TT.

The sequential model is schematically shown in Figure 7-37. The T state is symbolized by a square, and the R state by a circle. Deoxyhemoglo-

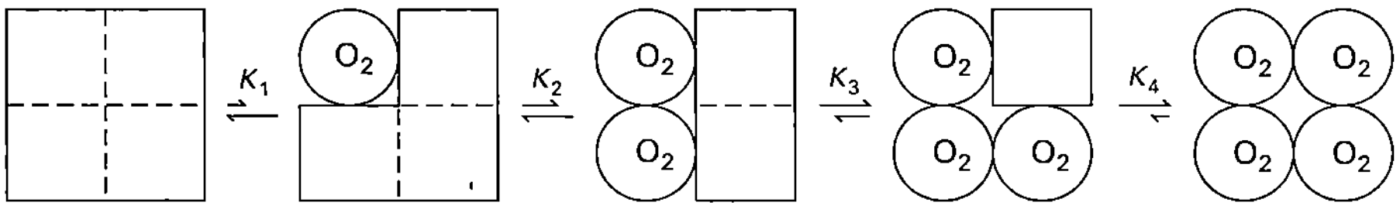

Figure 7-37
Simple sequential model for a tetrameric allosteric protein. The binding of a ligand to a subunit changes the conformation of that particular subunit from the T (square) to the R (circle) form. This transition increases the affinity of the other subunits for the ligand.

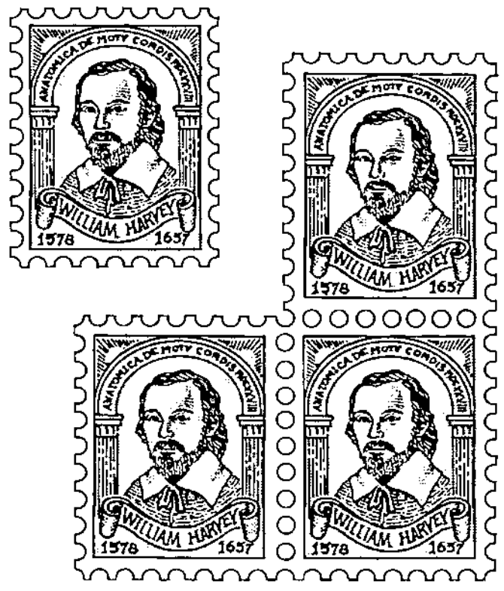

Figure 7-38
Postage stamp analogy of the simple sequential model. Two perforated edges (i.e., sets of salt bridges) must be torn to remove the first stamp. Only one perforated edge must be torn to remove the second stamp, and one edge again to remove the third stamp. The fourth stamp is then free. According to this model, the affinity of hemoglobin for successive O_2 molecules increases because fewer salt bridges need be broken.

bin is in the T_4 state. The binding of O_2 to one of the subunits changes its conformation from T to R, but leaves the other subunits in the T form. The oxygen-binding affinity of unoccupied sites in RT_3 is higher than in T_4 because some salts links have been broken on binding the first O_2. R_2T_2 and R_3T, which have higher oxygen affinities than does RT_3, are formed when the second and third O_2 bind. Finally, R_4 is produced on binding the fourth O_2. One can think of the sequential model in terms of a postage stamp analogy (Figure 7-38).

IN THE CONCERTED MODEL FOR ALLOSTERIC INTERACTIONS ALL SUBUNITS CHANGE CONFORMATION TOGETHER

A different view of allosteric interactions was proposed in 1965 by Jacques Monod, Jeffries Wyman, and Jean-Pierre Changeux. The essence of their elegant and incisive model is that *symmetry is conserved in allosteric transitions.* The assumptions of their model, commonly known as the *concerted* or *MWC model* (Figure 7-39), are

1. The protein interconverts between two conformations, R and T. All the subunits of a particular molecule must be in the T form, or all must be in the R form. Hybrids such as TR are forbidden.

2. Ligands bind with low affinity to the T form, and with high affinity to the R form.

3. The binding of each ligand increases the probability that *all* subunits in that molecule are in the R form. The allosteric transition is said to be *concerted* because all subunits change in unison from T to R, or vice versa. Stated in terms of the postage-stamp analogy, either *all four* perforated edges or *none* are broken.

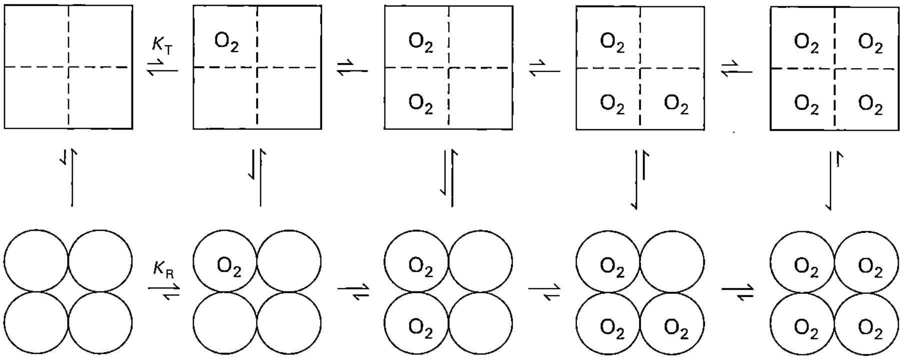

Figure 7-39
Concerted (Monod-Wyman-Changeux, or MWC) model for a tetrameric allosteric protein. The squares denote the T form, and the circles denote the R form. The ratio of T to R forms in the absence of ligand is L. The dissociation constants for the binding of ligand to the T and R states are K_T and K_R.

It is instructive to consider the concerted model in quantitative terms. In the absence of ligand, the two allowed states are symbolized as T_0 and R_0, and L_0 is the ratio of their concentrations.

$$L_0 = [T_0]/[R_0] \tag{10}$$

The dissociation constant of the ligand-protein complex is K_T for a subunit in the T state, and K_R for a subunit in the R state. The affinity of a subunit for the ligand depends solely on whether it is in the T or R state and not on whether the sites on neighboring subunits are occupied. Now let us define c as the ratio of dissociation constants

$$c = K_R/K_T \tag{11}$$

The binding of a ligand to an oligomer changes the [T]/[R] ratio by the factor c. When two ligands are bound, the [T]/[R] ratio is changed by c^2. For j substrates bound, the [T]/[R] ratio is given by

$$[T_j]/[R_j] = c^j L_0 \tag{12}$$

In other words, part of the binding energy of the ligand is used to drive the transition from the T to the R state.

We can now calculate the dependence of Y, the fraction of active sites containing a bound ligand, on the ligand concentration S. Let n be the number of subunits (each containing a binding site for ligand), and $\alpha = S/K_R$. Y is then given by

$$Y = \frac{\alpha(1+\alpha)^{n-1} + Lc\alpha(1+c\alpha)^{n-1}}{(1+\alpha)^n + L(1+c\alpha)^n} \tag{13}$$

and f_R, the fraction of molecules in the R form, by

$$f_R = \frac{(1+\alpha)^n}{(1+\alpha)^n + L(1+c\alpha)^n} \tag{14}$$

Homotropic effects—
Effects due to allosteric interactions between *identical* ligands, as exemplified by the cooperative binding of O_2 to hemoglobin.

Heterotropic effects—
Effects due to allosteric interactions between *different* ligands, as exemplified by the decrease in O_2 binding elicited by H^+, CO_2, and BPG.

We are now ready to apply the concerted model to the cooperative binding of oxygen by hemoglobin. The observed sigmoidal oxygen dissociation curve can be fit with $L_0 = 9000$ and $c = 0.014$ (Figure 7-40A and B). In other words, for deoxyhemoglobin, the equilibrium overwhelmingly favors the T state by a factor of 9000. The c value of 0.014 indicates that oxygen binds 71-fold more strongly to a site in the R state than to one in the T state (the α and β chains are assumed to be equivalent). *Part of the binding energy of O_2 is used to shift the conformational equilibrium from T to R.* In fact, the conformational equilibrium is changed c-fold by the binding of each O_2. The [T]/[R] ratio therefore changes from 9000 to 126, 1.76, .025, and .00035 on binding the first, second, third, and fourth O_2 molecules, respectively. Thus, *the population of hemoglobin molecules switches from being predominantly T (low affinity) to predominantly R (high affinity) when, on average, slightly more than two O_2 are bound.* When half the sites are occupied, the major species are fully deoxygenated and fully oxygenated hemoglobin (Figure 7-40C).

The actual allosteric mechanism of hemoglobin is more complex than envisioned by either the simple sequential or the concerted model. These models should be regarded as limiting cases. Actual allosteric processes combine, in varying degrees, elements of both. It is remarkable that so much of the allosteric behavior of hemoglobin can be described by the concerted model with just three parameters (L, K_R, and K_T).

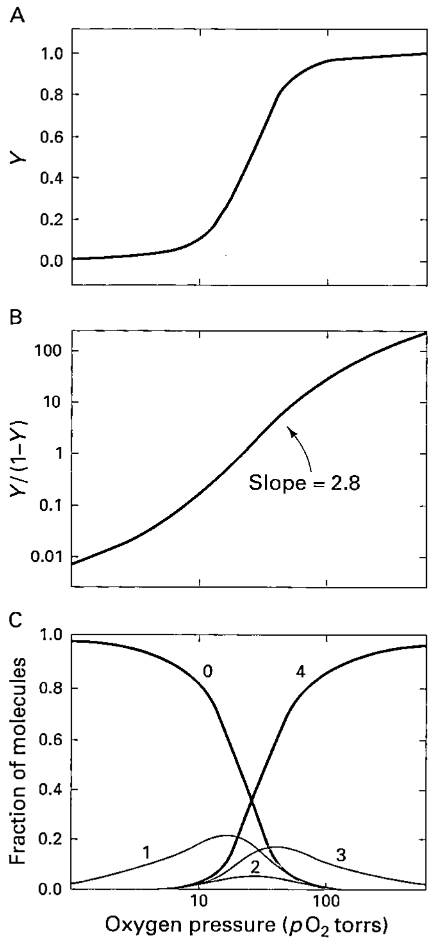

Figure 7-40
Oxygen-binding properties of hemoglobin calculated according to the concerted model with $L_0 = 9000$, $c = 0.014$, and $P_{50} = 26$ torrs. (A) Fractional saturation (Y). (B) Hill plot. The slope is 2.8 at $Y = 0.5$. (C) Fraction of molecules containing 0, 1, 2, 3, and 4 bound O_2 molecules.

COMMUNICATION WITHIN A PROTEIN MOLECULE

We have seen that the binding of O_2, H^+, CO_2, and BPG by hemoglobin are linked. These molecules are bound to spatially distinct sites that communicate with each other by means of complex conformational changes within the protein. The binding sites are separate because these molecules are structurally very different. *The interplay between these different sites is mediated primarily by changes in quaternary structure.* In fact, most known allosteric proteins consist of multiple polypeptide chains. The contact region between two chains can serve as a switch that transmits conformational changes from one subunit to another. An allosteric protein does not have fixed properties. Rather, its functional characteristics are regulated by specific molecules in its environment. Consequently, allosteric interactions have immense importance in cellular function. *In the evolutionary transition from myoglobin to hemoglobin, a macromolecule capable of perceiving information from its environment has emerged.*

SICKLE-SHAPED RED BLOOD CELLS WERE FOUND IN A CASE OF SEVERE ANEMIA

In 1904, James Herrick, a Chicago physician, examined a 20-year-old black college student who had been admitted to the hospital because of a cough and fever. The patient felt weak and dizzy and had a headache. For about a year he had been experiencing palpitations and shortness of breath. On physical examination, the patient appeared rather well developed physically. There was a tinge of yellow in the whites of his eyes, and the visible mucous membranes were pale. His heart was distinctly enlarged. Examination of the blood showed that the patient was markedly anemic. The number of red cells was half of what is normal.

The patient's blood smear contained unusual red cells, which were described by Herrick in these terms: "The shape of the red cells was very irregular, but *what especially attracted attention was the large number of thin, elongated, sickle-shaped and crescent-shaped forms.*" The treatment was supportive, consisting of rest and nourishing food. The patient left the hospital four weeks later, less anemic and feeling much better. However, his blood still exhibited a "tendency to the peculiar crescent-shape in the red corpuscles though this was by no means as noticeable as before."

Herrick was puzzled by the clinical picture and laboratory findings. Indeed, he waited six years before publishing the case history and then candidly asserted that "not even a definite diagnosis can be made." He noted the chronic nature of the disease, and the diversity of abnormal physical and laboratory findings: cardiac enlargement, a generalized swelling of lymph nodes, jaundice, anemia, and evidence of kidney damage. He concluded that the disease could not be explained on the basis of an organic lesion in any one organ. He singled out the abnormal blood picture as the key finding and titled his case report *Peculiar Elongated and Sickle-Shaped Red Blood Corpuscles in a Case of Severe Anemia.* Herrick suggested that "some unrecognized change in the composition of the corpuscle itself may be the determining factor."

Figure 7-41
Red cells from the blood of a patient with sickle-cell anemia, as viewed under a light microscope. [Courtesy of Dr. Frank Bunn.]

SICKLE-CELL ANEMIA IS A GENETICALLY TRANSMITTED, CHRONIC, HEMOLYTIC DISEASE

Other cases of this disease, called *sickle-cell anemia,* were found soon after the publication of Herrick's description. Indeed, sickle-cell anemia is not a rare disease. It is a significant public health problem wherever there is a substantial black population. The incidence of sickle-cell anemia among blacks is about 4 per 1000. In the past, it has usually been a fatal disease, often before age 30, as a result of infection, renal failure, cardiac failure, or thrombosis. Sickled red cells become trapped in the small blood vessels, which impairs the circulation and leads to damage of multiple organs. Sickled cells are more fragile than normal ones. They hemolyze readily and consequently have a shorter life than normal cells, which leads to a severe anemia. The chronic course of the disease is punctuated by crises in which the proportion of sickled cells is especially high. During such a crisis, the patient may go into shock.

Sickle-cell anemia is genetically transmitted. *Patients with sickle-cell anemia are homozygous for an abnormal gene* located on an autosomal chromosome. Offspring who receive the abnormal gene from one parent but its normal allele from the other have *sickle-cell trait. Such heterozygous people are usually not symptomatic.* Only 1% of the red cells in a heterozygote's venous circulation are sickled, in contrast with about 50% in a homozygote.

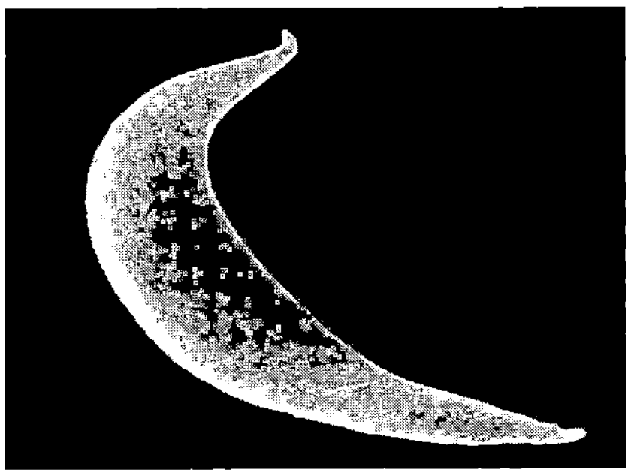

Figure 7-42
Scanning electron micrograph of an erythrocyte from a patient with sickle-cell anemia. [Courtesy of Dr. Jerry Thornwaite and Dr. Robert Leif.]

THE SOLUBILITY OF DEOXYGENATED SICKLE HEMOGLOBIN IS ABNORMALLY LOW

Herrick correctly surmised the location of the defect in sickle-cell anemia. Red cells from a patient with this disease will sickle on a microscope slide in vitro if the concentration of oxygen is reduced. In fact, the hemoglobin in these cells is itself defective. Deoxygenated sickle-cell hemoglobin has an abnormally low solubility. A fibrous precipitate is formed when a concentrated solution of sickle-cell hemoglobin is deoxygenated. This precipitate deforms red cells and gives them their sickle shape. Sickle-cell hemoglobin is commonly referred to as *hemoglobin S (Hb S)* to distinguish it from hemoglobin A (Hb A), the normal adult hemoglobin.

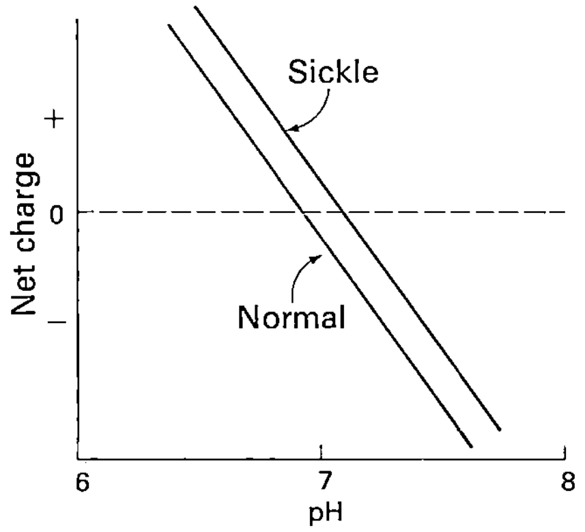

Figure 7-43
Net charge of sickle-cell hemoglobin and of normal hemoglobin as a function of pH, as measured by their electrophoretic mobilities. The isoelectric point of a molecule is the pH at which its mobility is 0.

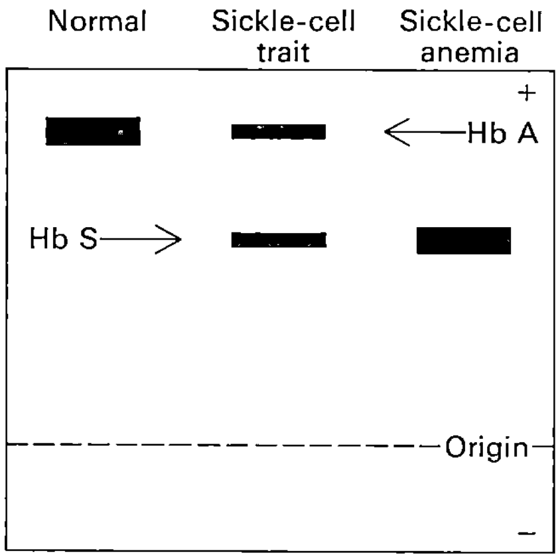

Figure 7-44
Gel electrophoresis pattern of hemoglobin isolated from a normal person, from a person with sickle-cell trait, and from a person with sickle-cell anemia.

DISCOVERY OF A MOLECULAR DISEASE: A SINGLE AMINO ACID IN THE β CHAIN IS MUTATED

In 1949, Linus Pauling and his associates examined the physical-chemical properties of hemoglobin from normal people and from those with sickle-cell trait or sickle-cell anemia. Their experimental approach was to search for differences between these hemoglobins by electrophoresis (Figure 7-43). They found that the isoelectric point (p. 48) of sickle-cell hemoglobin is higher than that of normal hemoglobin in both the oxygenated and the deoxygenated state.

	pI		
	Normal	*Sickle-cell*	*Difference*
Oxyhemoglobin	6.87	7.09	0.22
Deoxyhemoglobin	6.68	6.91	0.23

These observations suggested that *there is a difference in the number or kind of ionizable groups in the two hemoglobins.* An estimate was made from the acid-base titration curve of hemoglobin in the neighborhood of pH 7. A change of 1 pH unit in the hemoglobin solution produces a change of about 13 charges. The difference in isoelectric pH of 0.23 therefore corresponds to about three charges per hemoglobin molecule. It was concluded that *sickle-cell hemoglobin has between two and four more net positive charges per molecule than does normal hemoglobin.*

Patients with sickle-cell anemia (who are homozygous for the sickle gene) have hemoglobin S but no hemoglobin A. In contrast, people with sickle-cell trait (who are heterozygous for the sickle gene) have both kinds of hemoglobin in approximately equal amounts (Figure 7-44). Thus, Pauling's study revealed *"a clear case of a change produced in a protein molecule by an allelic change in a single gene."* This was the first demonstration of a *molecular disease.*

The precise difference between normal and sickle-cell hemoglobin was identified in 1954, when Vernon Ingram devised a new technique for detecting amino acid substitutions in proteins. The hemoglobin molecule was split into smaller units for analysis, because it was anticipated that it would be easier to detect an altered amino acid in a peptide containing about 20 residues than in a protein 10 times as large. Trypsin was used to specifically cleave hemoglobin on the carboxyl side of its lysine and arginine residues. Because an $\alpha\beta$ half of hemoglobin contains a total of 27 lysine and arginine residues, *tryptic digestion* produced 29 different peptides. The next step was to separate the peptides in two dimensions (Figure 7-45). The mixture of peptides was placed in a spot at one corner of a large sheet of filter paper. *Electrophoresis* was carried out in one direc-

Figure 7-45
Fingerprinting. A mixture of peptides produced by proteolytic cleavage is resolved by electrophoresis in the horizontal direction followed by chromatography in the vertical direction.

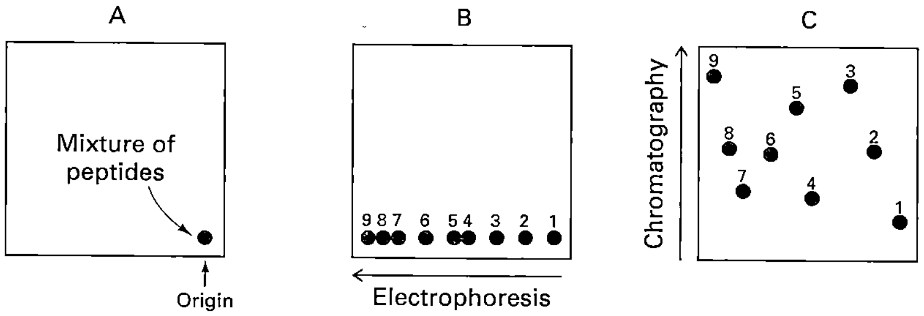

tion, followed by *chromatography* in the perpendicular direction. Finally, the peptide spots were made visible by staining the filter paper with ninhydrin. This sequence of steps—selective cleavage of a protein into small peptides, followed by their separation in two dimensions—is called *fingerprinting.*

The fingerprints of hemoglobins A and S were highly revealing (Figure 7-46). When they were compared, *all but one of the peptide spots matched.* The spot that differed was eluted from each fingerprint and shown to be a single peptide consisting of eight amino acids. Amino acid analysis indicated that this peptide in hemoglobin S differed from the one in hemoglobin A by a single amino acid. Ingram determined the sequence of this peptide and showed that *hemoglobin S contains valine instead of glutamate at position 6 of the β chain:*

Hemoglobin A	Val-His-Leu-Thr-Pro-Glu-Glu-Lys-
Hemoglobin S	Val-His-Leu-Thr-Pro-Val-Glu-Lys-
	β1 2 3 4 5 6 7 8

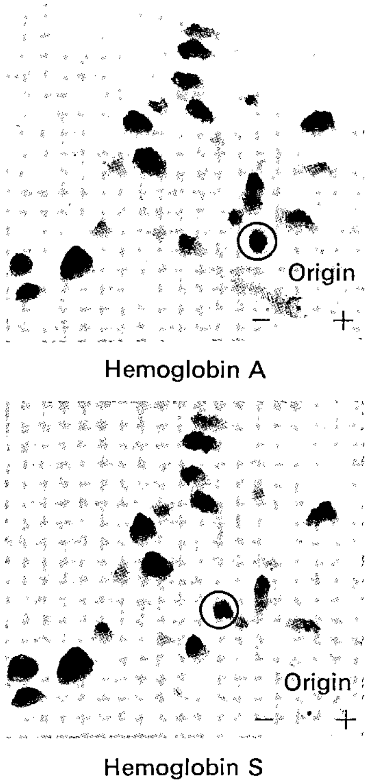

Figure 7-46
Comparison of the ninhydrin-stained fingerprints of hemoglobin A and hemoglobin S. The position of the peptide that is different in these hemoglobins is encircled in red. [Courtesy of Dr. Corrado Baglioni.]

STICKY PATCHES ON DEOXYGENATED HEMOGLOBIN S LEAD TO THE FORMATION OF FIBROUS PRECIPITATES

The substitution of valine for glutamate at position 6 of the β chains places a nonpolar residue on the outside of hemoglobin S (Figure 7-47). The oxygen affinity and allosteric properties of hemoglobin are virtually unaffected by this change. *However, this alteration markedly reduces the solubility of the deoxygenated but not the oxygenated form of hemoglobin.* The reason is that the valine side chain of hemoglobin S interacts with a complementary sticky patch on another hemoglobin molecule (Figure 7-48). The complementary site, formed by phenylalanine β85 and leucine β88, is exposed in deoxygenated but not in oxygenated hemoglobin. Thus, *sickling occurs when there is a high concentration of the deoxygenated form of hemoglobin S.*

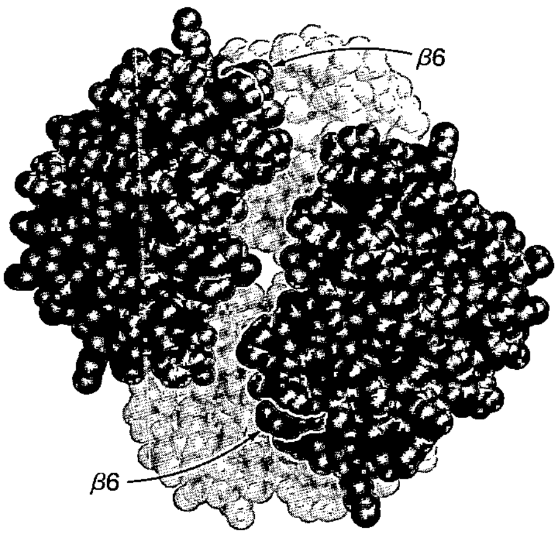

Figure 7-47
The positions of the amino acid changes in hemoglobin S (glutamate to valine at residue 6 of each β chain) are shown in red in this model of deoxyhemoglobin. Note that β6 is located at the surface of the protein. The α chains are shown in yellow, and the β chains in blue. [Drawn from 4hhb.pdb. G. Fermi, M.F. Perutz, B. Shaanan, and R. Fourme. *J. Mol. Biol.* 175(1984):159.]

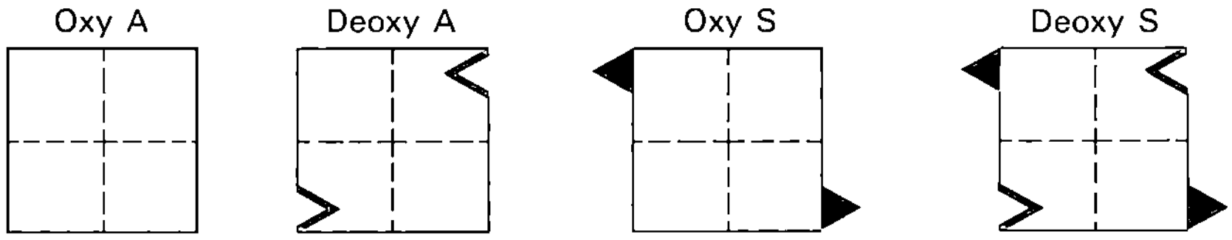

Figure 7-48
The red triangle represents the sticky patch that is present on both oxy- and deoxyhemoglobin S but not on either form of hemoglobin A. The complementary site is represented by an indentation that can accommodate the triangle. This complementary site is present in deoxyhemoglobin S and is probably also present in deoxyhemoglobin A.

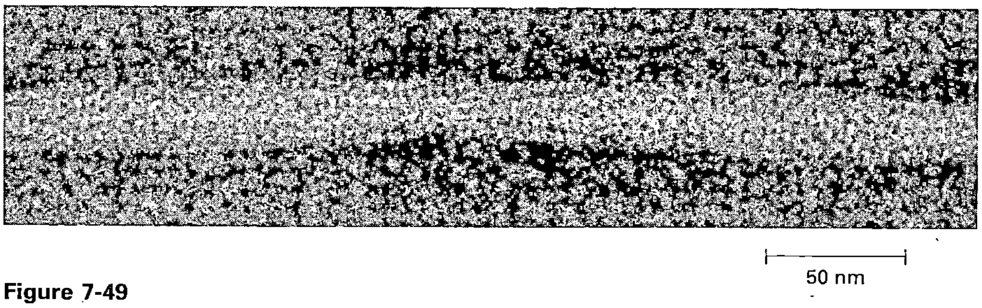

Figure 7-49
Electron micrograph of a negatively stained fiber of deoxyhemoglobin S. [From G. Dykes, R.H. Crepeau, and S.J. Edelstein. *Nature* 272(1978):509.]

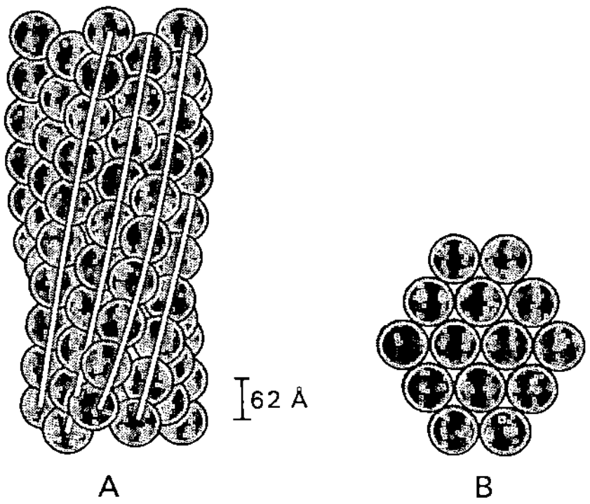

Figure 7-50
Fourteen-stranded helical model of the deoxyhemoglobin S fiber: (A) axial view; (B) cross-sectional view. Each circle represents a hemoglobin S molecule. [After G. Dykes, R.H. Crepeau, and S.J. Edelstein. *Nature* 272(1978):509.]

Further insight into sickle-cell disease comes from examining the nature of the precipitate formed by deoxygenated hemoglobin S, which deforms the red cell. Electron microscopy shows that the precipitate consists of fibers having a diameter of 21.5 nm (Figure 7-49). Each fiber is a fourteen-stranded helix made of seven strongly interacting pairs of hemoglobin molecules (Figure 7-50). Multiple polar interactions, in addition to the critical one between sticky patches, stabilize the fiber.

What determines whether a red cell becomes sickled during its passage through the capillary circulation, which takes about a second? *The striking experimental finding is that the rate of fiber formation is proportional to about the tenth power of the effective concentration of deoxyhemoglobin S. Thus, fiber formation is a highly concerted reaction.* These kinetic data indicate that nucleation is the rate-limiting phase in fiber formation. The fiber grows rapidly once a critical cluster of about 10 deoxyhemoglobin S molecules has been formed. The important clinical implication is that *kinetic,* as well as thermodynamic, factors are important in sickling. A red cell that is supersaturated with deoxyhemoglobin S will not sickle if the lag time for fiber formation is longer than the transit time from the peripheral capillaries to the alveoli of the lungs, where reoxygenation occurs.

These facts account for several clinical characteristics of sickle-cell anemia. A vicious cycle is set up when sickling occurs in a small blood vessel. The blockage of the vessel creates a local region of low oxygen concentration. Hence, more hemoglobin goes into the deoxy form and so more sickling occurs. The very strong dependence of the polymerization rate on the concentration of deoxyhemoglobin S also accounts for the fact that people with sickle-cell trait are usually asymptomatic. The concentration of deoxyhemoglobin S in the red cells of these heterozygotes is about half that in homozygotes, and so their rate of fiber formation is about a thousandfold slower (30 s, compared with 30 msec).

THE HIGH INCIDENCE OF THE SICKLE GENE IS DUE TO THE PROTECTION CONFERRED AGAINST MALARIA

The frequency of the sickle gene is as high as 40% in certain parts of Africa. Until recently, most homozygotes have died before reaching adulthood, and so there must have been strong selective pressure to maintain the high incidence of the gene. James Neel proposed that the heterozygote enjoys advantages not shared by either the normal homozygote or the sickle-cell homozygote. Indeed, sickle-cell trait confers a small but highly significant degree of protection against the most lethal form of malaria, perhaps by accelerating the destruction of infected erythrocytes. In a malaria-infested region, the reproductive fitness of a person with sickle-cell trait is about 15% higher than that of someone with normal hemoglobin. The incidence of malaria and the frequency of the sickle gene in Africa are strongly correlated (Figure 7-51). We see here a striking example of *balanced polymorphism*—an allele that is highly deleterious in a homozygote persists because the heterozygous state is advantageous.

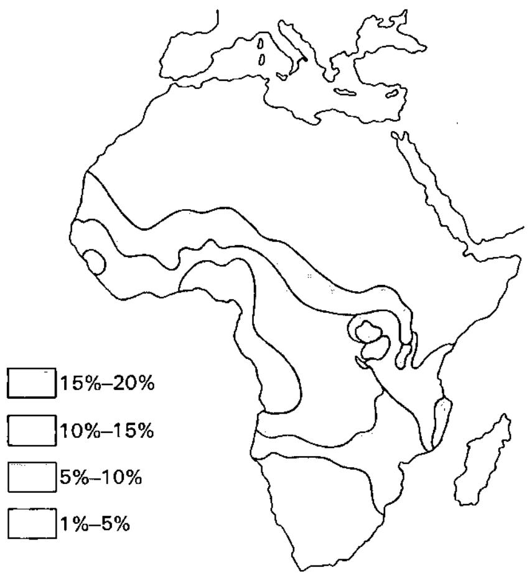

Figure 7-51
Frequency of the sickle-cell gene in Africa. High frequencies are restricted to regions where malaria is a major cause of death. [After A.C. Allison. Sickle cells and evolution. Copyright © 1956 by Scientific American, Inc. All rights reserved.]

FETAL DNA CAN BE ANALYZED FOR THE PRESENCE OF THE SICKLE-CELL GENE

The substitution of valine for glutamate at $\beta6$ in hemoglobin S results from a change in a single base, T for A. This mutation can readily be detected by cleaving DNA with a restriction enzyme that recognizes the sequence in this immediate region. The target for the restriction endonuclease MstII is the palindromic sequence CCTNAGG (where N denotes any base), which is present in the gene for the β chain of hemoglobin A (β^A gene) but not in the one for hemoglobin S (β^S). Because of the absence of this target site in the β^S gene, complete digestion of the gene by MstII produces a 1.3-kb fragment instead of the normal 1.1-kb fragment (Figure 7-52A). The fragments in the digested sample of DNA are separated by gel electrophoresis and visualized by Southern blotting (p. 122) with a ^{32}P-labeled DNA probe complementary to the 1.1-kb fragment. The 1.3-kb fragment is also stained by this probe because it contains the 1.1-kb sequence. An autoradiogram reveals whether the β^A gene, the β^S gene, or both are present in the sample (Figure 7-52B).

An attractive feature of this restriction enzyme method is that the DNA sample can come from *any* fetal cell. In contrast, a hemoglobin sample could be obtained only from red blood cells or their precursors. A sample of amniotic fluid is obtained from the fetus more readily and with less potential hazard than are blood-forming cells. DNA analyses can also be performed on biopsy samples of chorionic villi obtained early in pregnancy, at about eight weeks gestation. These techniques for obtaining and analyzing genomic DNA are generally applicable. The number of genetic diseases that can be detected early in pregnancy by restriction enzyme cleavage of fetal DNA followed by Southern blotting with highly

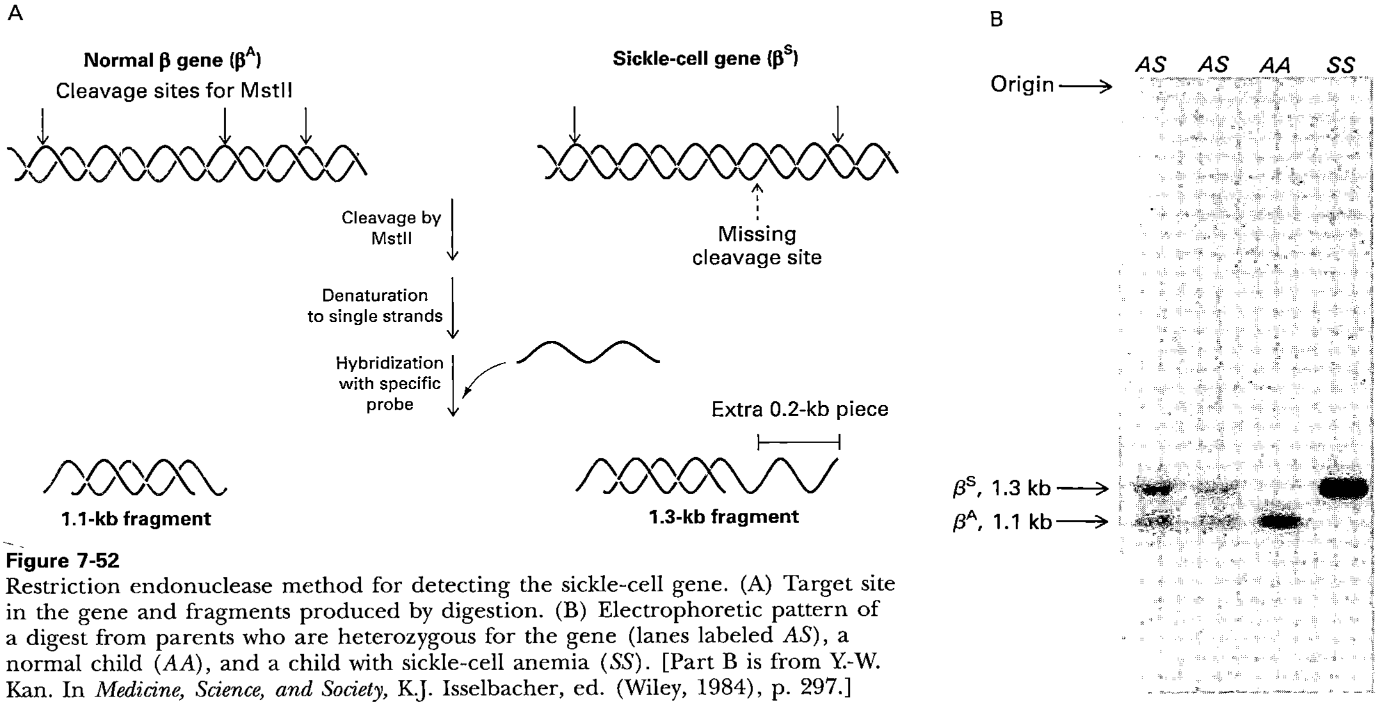

Figure 7-52
Restriction endonuclease method for detecting the sickle-cell gene. (A) Target site in the gene and fragments produced by digestion. (B) Electrophoretic pattern of a digest from parents who are heterozygous for the gene (lanes labeled *AS*), a normal child (*AA*), and a child with sickle-cell anemia (*SS*). [Part B is from Y.-W. Kan. In *Medicine, Science, and Society,* K.J. Isselbacher, ed. (Wiley, 1984), p. 297.]

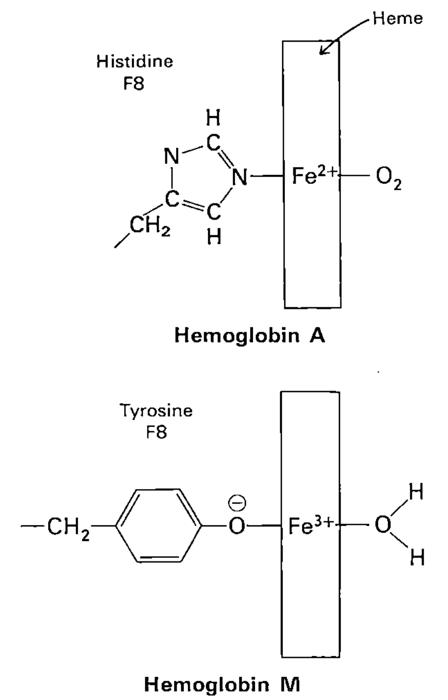

Figure 7-53
Substitution of tyrosine for the proximal histidine (F8) results in the formation of a hemoglobin M. The negatively charged oxygen atom of tyrosine is coordinated to the iron atom, which is in the ferric state. Water rather than O_2 is bound at the sixth coordination position.

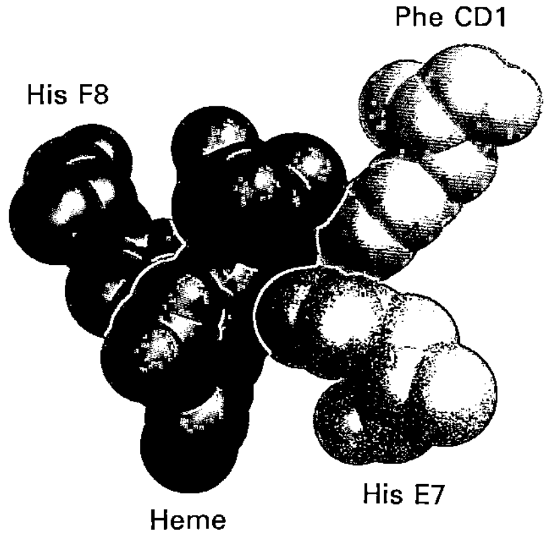

Figure 7-54
Location of phenylalanine CD1, in normal hemoglobin. The aromatic ring of this residue is in contact with the heme. In hemoglobin Hammersmith, serine replaces this phenylalanine residue; this markedly weakens the binding of heme.

specific probes is increasing rapidly. Amplification of fetal DNA by the polymerase chain reaction (p. 133) has facilitated the detection of genetic abnormalities. Parents of fetuses who are at risk can now make informed decisions as to whether to terminate pregnancy.

MOLECULAR PATHOLOGY OF HEMOGLOBIN

More than 300 abnormal hemoglobins have been discovered by the examination of patients with clinical symptoms and by electrophoretic surveys of normal populations. In northern Europe, 1 of 300 people is heterozygous for a variant of hemoglobin A. The frequency of any one mutant allele is less than 10^{-4}, which is lower by several orders of magnitude than the frequency of the sickle allele in regions where malaria is endemic. In other words, most abnormal hemoglobins do not confer a selective advantage on the person. They are almost always neutral or harmful.

Abnormal hemoglobins are of several types:

1. *Altered exterior.* Nearly all substitutions on the surface of the hemoglobin molecule are harmless. Hemoglobin S is a striking exception.

2. *Altered active site.* The defective subunit cannot bind oxygen because of a structural change near the heme that directly affects oxygen binding. For example, substitution of tyrosine for the histidine proximal or distal to heme stabilizes the heme in the ferric form, which can no longer bind oxygen (Figure 7-53). The tyrosine side chain is ionized in this complex with the ferric ion of the heme. Mutant hemoglobins characterized by a permanent ferric state of two of the hemes are called *hemoglobin M.* The letter M signifies that the altered chains are in the *methemoglobin* (ferrihemoglobin) form. These patients are usually cyanotic. The disease has been seen only in heterozygous form, because homozygosity would almost certainly be lethal.

3. *Altered tertiary structure.* The polypeptide chain is prevented from folding into its normal conformation. These hemoglobins are usually unstable. For example, in hemoglobin Hammersmith, phenylalanine at CD1, adjacent to the heme, is replaced by serine (Figure 7-54). The affinity of this hemoglobin for its heme groups is much lower than normal. Amino acid substitutions at sites far from the heme also can prevent hemoglobin from folding into its normal conformation. An instructive mutant is hemoglobin Riverdale-Bronx, which has arginine in place of glycine at B6. Recall that glycine occupies this position in nearly all known normal myoglobins and hemoglobins (p. 156). This mutant hemoglobin does not fold normally because arginine at B6 is too large to fit into the narrow space between the B and E helices (see Figure 7-19).

4. *Altered quaternary structure.* Some mutations at subunit interfaces lead to the loss of allosteric properties. These hemoglobins usually have an abnormal oxygen affinity. The $\alpha_1\beta_2$ contact region, which changes markedly on oxygenation, is especially vulnerable to mutation.

THALASSEMIAS ARE GENETIC DISORDERS OF HEMOGLOBIN SYNTHESIS

The genetic diseases considered thus far are ones in which hemoglobin molecules are produced in essentially normal amounts but have impaired function because of the change of a single amino acid residue. The *thalassemias,* a different class of genetic disorders, are characterized by *defective synthesis of one or more hemoglobin chains.* The chain that is affected is denoted by a prefix, as in α-thalassemia. The name of this group of diseases comes from the Greek word *thalassa,* meaning "sea," because of the high incidence of some forms of thalassemia among people living near the Mediterranean Sea. Indeed, some 20% of the population in parts of Italy are carriers of a gene for this disease. The geographic distribution of some thalassemia genes parallels that of malaria, which suggests that the heterozygote benefits from the presence of the gene, as in sickle-cell trait.

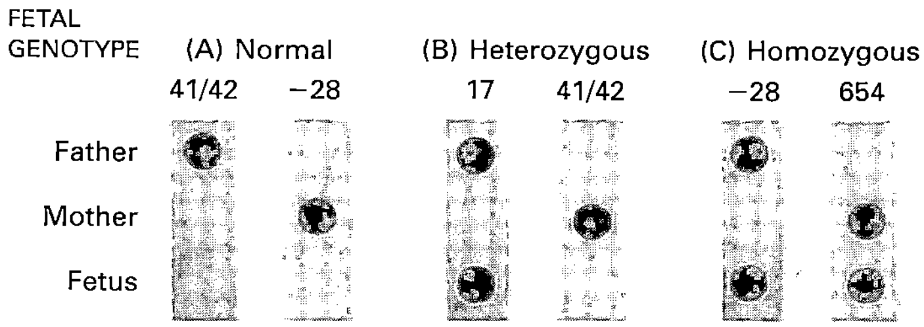

Figure 7-55
Prenatal diagnosis of thalassemia by oligonucleotide hybridization. Parts A, B, and C show analyses of the DNA of three sets of parents and their fetuses. Each parent is a heterozygote for β-thalassemia at the codon noted above the corresponding lane. Fetus A is normal, fetus B is a heterozygote, and C is a homozygote for β-thalassemia. [Courtesy of Dr. Yuet Wai Kan. S.P. Cai, C.A. Chang, J.Z. Zhang, R.K. Saiki, H.A. Erlich, and Y.-W. Kan. *Blood* 73(1989)372.]

Thalassemias are produced by many different mutations that lead to the absence or deficiency of a globin chain in a variety of ways:

1. *The gene is missing.*

2. The gene is present but *RNA synthesis or processing is impaired.* For example, a mutation in the TATA box of the promoter site (p. 102) decreases the amount of RNA synthesized. Mutations within introns or near the exon-intron boundary can lead to aberrant splicing of the primary RNA transcript.

3. Globin mRNA is produced in normal amounts, but it encodes a grossly abnormal protein. For example, mutation of an amino acid codon to a stop codon (say TGG for tryptophan to TGA) will result in *premature termination* of protein synthesis. The deletion or addition of a nucleotide, a *frameshift mutation,* leads to an entirely different amino acid sequence on the distal side of the mutation. Most of these hemoglobin chains are very unstable and are rapidly degraded.

IMPACT OF THE DISCOVERY OF MOLECULAR DISEASES

Analyses of mutations affecting oxygen transport have had a major impact on molecular biology, medicine, and genetics. Their significance is threefold:

1. *They are sources of insight into relations between the structure and function of normal hemoglobin.* Mutations of a single amino acid residue are highly specific chemical modifications provided by nature. They shed light on facets of the protein that are critical for function.

2. *The discovery of mutant hemoglobins has revealed that disease can arise from a change of a single amino acid in one kind of polypeptide chain.* The concept of molecular disease, which is now an integral part of medicine, had its origins in the incisive studies of sickle-cell hemoglobin. The thalassemias have provided striking illustrations of the consequences of aberrant transcription, RNA splicing, and protein synthesis.

3. *The finding of mutant hemoglobins has enhanced our understanding of evolutionary processes.* Mutations are the raw material of evolution; the studies of sickle-cell anemia have shown that a mutation may be both beneficial and harmful.

On disease and evolution—
"Subjectively, to evolve must most often have amounted to suffering from a disease. And these diseases were of course molecular. The appearance of the concept of good and evil, interpreted by man as his painful expulsion from Paradise, was probably a molecular disease that turned out to be evolution."

E. Zuckerkandl and L. Pauling
In *Horizons in Biochemistry,*
M. Kasha and B. Pullman, eds.
(Academic Press, 1964),
pp. 189–225

SUMMARY

Myoglobin and hemoglobin are the oxygen-carrying proteins in vertebrates. Myoglobin facilitates the transport of oxygen in muscle and serves as a reserve store of oxygen, whereas hemoglobin is the oxygen carrier in blood. These proteins contain tightly bound heme, a substituted porphyrin with a central iron atom. The ferrous (+2) state of heme binds O_2, whereas the ferric (+3) state does not.

Myoglobin, a single polypeptide chain of 153 residues (18 kd), has a compact shape. The inside of myoglobin consists almost exclusively of nonpolar residues. About 75% of the polypeptide chain is α-helical. The single ferrous heme group is located in a nonpolar niche, which protects it from oxidation to the ferric form. The iron atom of the heme is directly bonded to a nitrogen atom of a histidine side chain. This proximal histidine occupies the fifth coordination position. The sixth coordination position on the other side of the heme plane is the binding site for O_2. The nearby distal histidine diminishes the binding of CO at the oxygen-binding site and inhibits the oxidation of heme to the ferric state. The central exon of the myoglobin gene encodes nearly the entire heme-binding site.

Hemoglobin consists of four polypeptide chains, each with a heme group. Hemoglobin A, the predominant hemoglobin in adults, has the subunit structure $\alpha_2\beta_2$. The three-dimensional structure of the α and β chains of hemoglobin is strikingly similar to that of myoglobin. Yet new properties appear in tetrameric hemoglobin that are not present in monomeric myoglobin. Hemoglobin transports H^+ and CO_2 in addition to O_2. Furthermore, their binding is regulated by allosteric interactions, which are interactions between separate sites on the same protein. Indeed, hemoglobin is the best-understood allosteric protein.

Hemoglobin exhibits three kinds of allosteric effects. First, the oxygen-binding curve of hemoglobin is sigmoidal, which means that the binding of oxygen is cooperative. The binding of oxygen to one heme facilitates the binding of oxygen to the other hemes in the same molecule. Second, H^+ and CO_2 promote the release of O_2 from hemoglobin, an effect that is physiologically important in enhancing the release of O_2 in metabolically active tissues such as muscle. Conversely, O_2 promotes the release of H^+ and CO_2 in the alveolar capillaries of the lungs. These allosteric linkages between the binding of H^+, CO_2, and O_2 are known as the Bohr effect. Third, the affinity of hemoglobin for O_2 is further regulated by 2,3-bisphosphoglycerate (BPG), a small molecule with a very high density of negative charge. BPG binds tightly to deoxyhemoglobin but not to oxyhemoglobin. Hence, BPG lowers the oxygen affinity of hemoglobin. Fetal hemoglobin ($\alpha_2\gamma_2$) has a higher oxygen affinity than adult hemoglobin because it binds BPG less tightly.

The allosteric properties of hemoglobin arise from interactions between its α and β subunits. The T (tense) quaternary structure is constrained by salt links between different subunits, giving it a low affinity for O_2. These intersubunit salt links are absent from the R (relaxed) form, which has a high affinity for O_2. On oxygenation, the iron atom moves into the plane of the heme, pulling the proximal histidine with it. This motion cleaves some of the salt links, and the equilibrium is shifted from T to R. BPG stabilizes the deoxy state by binding to positively charged groups around the central cavity of hemoglobin. Carbon dioxide, another allosteric agent, binds to the terminal amino groups of all four chains by forming readily reversible carbamate linkages. The hydrogen ions participating in the Bohr effect are bound to several pairs of sites that have a more negatively charged environment in the deoxy than in the oxy state.

Two limiting models for allosteric interactions have been proposed. In the simple sequential model, the binding of a ligand induces a conformational change in the subunit to which it is bound but not in neighboring subunits. The affinities of neighboring subunits are changed by the altered subunit interface. In the concerted model, the symmetry of the oligomer is conserved; all subunits in a particular protein must be in the R or T form. Binding of ligand increases the proportion of molecules that are in the R form. The actual allosteric mechanism of hemoglobin is more complex than envisioned by either model.

Gene mutation resulting in the change of a single amino acid in a single protein can produce disease. The best-understood molecular disease is sickle-cell anemia. Hemoglobin S, the abnormal hemoglobin in this disease, consists of two normal α chains and two mutant β chains. In hemoglobin S, glutamate at residue 6 of the β chain is replaced by valine. This substitution of a nonpolar side chain for a polar one drastically reduces the solubility of deoxyhemoglobin S, which leads to the formation of fibrous precipitates that deform the red cell and give it a sickle shape. The resulting destruction of red cells produces a chronic hemolytic anemia. Sickle-cell anemia arises when a person is homozygous for the mutant sickle gene. The heterozygous condition, called sickle-cell trait, is relatively asymptomatic. About 1 of 10 blacks in the United States is heterozygous for the sickle gene, and as many as 4 of 10 in some parts of Africa. This high incidence of a gene that is harmful in homozygotes is due to its beneficial effect in heterozygotes—sickle-cell trait confers a measure of protection against the most lethal form of malaria. This is an example of balanced polymorphism. Sickle-cell anemia can be diagnosed in utero by restriction endonuclease digestion of a sample of fetal DNA, followed by Southern blotting.

Several hundred mutant hemoglobins have been found by studies of the hemoglobin of patients with hematologic symptoms and by surveys of normal populations. Several classes of mutant hemoglobins are known: (1) Most substitutions on the surface of hemoglobin are harmless. Hemoglobin S is a striking exception. (2) Most substitutions near the heme impair the oxygen-binding site. For example, replacement of the proximal or distal histidine with tyrosine locks the heme in the ferric state, which cannot bind oxygen. (3) Many alterations in the interior of the molecule distort the tertiary structure and produce unstable hemoglobins. (4) Many changes at subunit interfaces lead to changes in oxygen affinity and the loss of allosteric properties. Thalassemias, a different class of genetic disorders, are characterized by the absence or defect of a globin chain. Some underlying causes are (1) absence of a globin gene, (2) impaired transcription or defective RNA processing, and (3) premature termination of protein synthesis or a shift in the reading frame.

SELECTED READINGS

Where to start

Kendrew, J.C., 1961. The three-dimensional structure of a protein molecule. *Sci. Amer.* 205(6):96–11.

Perutz, M.F., Fermi, G., Luisi, B., Shaanan, B., and Liddington, R.C., 1987. Stereochemistry of cooperative mechanisms in hemoglobin. *Acc. Chem. Res.* 20:309–321.

Ranney, H.M., 1992. The spectrum of sickle cell disease. *Hospital Practice (Office Edition)* 27:133–137.

Dickerson, R.E., and Geis, I., 1983. *Hemoglobin: Structure, Function, Evolution and Pathology.* Benjamin/Cummings. [A beautifully illustrated account with a broad perspective.]

Structure and reactivity

Perutz, M.F., 1978. Hemoglobin structure and respiratory transport. *Sci. Amer.* 239(6):92–125.

Perutz, M.F., 1987. Molecular anatomy, physiology, and pathology of hemoglobin. *In* Stamatayonnopoulos, C. (ed.), *Molecular Basis of Blood Diseases.* Saunders.

Fermi, G., and Perutz, M.F., 1981. *Atlas of Molecular Structures in Biology. 2. Haemoglobin and Myoglobin.* Clarendon Press.

Fermi, G., Perutz, M.F., Shaanan, B., and Fourme, R., 1984. The crystal structure of human deoxyhaemoglobin at 1.74 Å resolution. *J. Mol. Biol.* 175:159–174.

Go, M., 1981. Correlation of DNA exonic regions with protein structural units in haemoglobin. *Nature* 291:90–92.

De Sanctis, G., Falcioni, G., Giardina, B., Ascoli, F., and Brunori, M., 1986. Mini-myoglobin: Preparation and reaction with oxygen and carbon monoxide. *J. Mol. Biol.* 188:73–76.

Collman, J.P., Brauman, J.I., Halbert, T.R., and Suslick, K.S., 1976. Nature of O_2 and CO binding to metalloporphyrins and heme proteins. *Proc. Nat. Acad. Sci.* 73:3333–3337. [Presents the structural basis for the decreased binding of carbon monoxide by myoglobin and hemoglobin.]

Interaction of hemoglobin with H^+, CO_2, and BPG

Kilmartin, J.V., 1976. Interaction of haemoglobin with protons, CO_2, and 2,3-diphosphoglycerate. *Brit. Med. Bull.* 32:209–222.

Benesch, R., and Benesch, R.E., 1969. Intracellular organic phosphates as regulators of oxygen release by haemoglobin. *Nature* 221:618–622.

Tyuma, I., and Shimizu, K., 1970. Effect of organic phosphates on the difference in oxygen affinity between fetal and adult human hemoglobin. *Fed. Proc.* 29:1112–1114.

Allosteric mechanism of hemoglobin

Perutz, M.F., 1989. Mechanisms of cooperativity and allosteric regulation in proteins. *Quart. Rev. Biophys.* 22:139–237.

Ackers, G.K., Doyle, M.L., Myers, D., and Daugherty, M.A., 1992. Molecular code for cooperativity in hemoglobin. *Science* 255:54–63.

Ho, C., 1992. Proton nuclear magnetic resonance studies of hemoglobin: Cooperative interactions and partially ligated intermediates. *Adv. Prot. Chem.* 43:153–312.

Baldwin, J., and Chothia, C., 1979. Hemoglobin: The structural changes related to ligand binding and its allosteric mechanism. *J. Mol. Biol.* 129:175–220.

Gelin, B.R., Lee, A.W.-M., and Karplus, M., 1983. Hemoglobin tertiary structure change on ligand binding: Its role in the cooperative mechanism. *J. Mol. Biol.* 171:489–559.

Allosteric models

Monod, J., Wyman, J., and Changeux, J.-P., 1965. On the nature of allosteric transitions: A plausible model. *J. Mol. Biol.* 12:88–118. [Presentation of the concerted model.]

Koshland, D.L., Jr., Nemethy, G., and Filmer, D., 1966. Comparison of experimental binding data and theoretical models in proteins containing subunits. *Biochemistry* 5:365–385. [Presentation of the simple sequential model.]

Wyman, J., and Gill, S.J., 1990. *Binding and Linkage. Functional Chemistry of Biological Macromolecules.* University Books. [A rigorous and fundamental treatment of the thermodynamics of allosteric interactions.]

Sickle-cell anemia

Edelstein, S.J., 1986. *The Sickled Cell. From Myths to Molecules.* Harvard University Press. [A fascinating account that brings together anthropology, biochemistry, structural biology, and molecular genetics.]

Herrick, J.B., 1910. Peculiar elongated and sickle-shaped red blood corpuscles in a case of severe anemia. *Arch. Intern. Med.* 6:517–521.

Pauling, L., Itano, H.A., Singer, S.J., and Wells, I.C., 1949. Sickle cell anemia: A molecular disease. *Science* 110:543–548.

Ingram, V.M., 1957. Gene mutation in human haemoglobin: The chemical difference between normal and sickle cell haemoglobin. *Nature* 180:326–328.

Rodgers, D.W., Crepeau, R.H., and Edelstein, S.J., 1987. Pairings and polarities of the 14 strands in sickle cell hemoglobin fibers. *Proc. Nat. Acad. Sci.* 84:6157–6161.

Eaton, W.A., and Hofrichter, J., 1990. Sickle cell hemoglobin polymerization. *Adv. Prot. Chem.* 40:63–279.

Molecular genetics of hemoglobin

Kan, Y.W., 1992. Development of DNA analysis for human diseases. Sickle cell anemia and thalassemia as a paradigm. *J. Amer. Med. Assoc.* 267:1532–1536.

Embury, S.H., Scharf, S.J., Saiki, R.K., Gholson, M.A., Golbus, M., Arnheim, N., and Erlich, H.A., 1987. Rapid prenatal diagnosis of sickle cell anemia by a new method of DNA analysis. *New Engl. J. Med.* 316:656–661.

Scriver, C.R., Beaudet, A.L., Sly, W.S., and Valle, D. (eds.), 1989. *Metabolic Basis of Inherited Disease* (6th ed.). McGraw-Hill. [Chapter 93 contains an excellent discussion of hemoglobinopathies.]

Honig, G.R., and Adams, J.G., 1986. *Human Hemoglobin Genetics.* Springer-Verlag.

Saiki, R.K., Scharf, S., Faloona, F., Mullis, K.B., Horn, G.T., Erlich, H.A., and Arnheim, N., 1985. Enzymatic amplification of beta-globin genomic sequences and restriction site analysis for diagnosis of sickle cell anemia. *Science* 230:1350–1354.

Bunn, H.F., and Forget, B.G., 1986. *Hemoglobin: Molecular, Genetic and Clinical Aspects.* Saunders.

Site-specific mutagenesis and genetic engineering

Mouneimne, Y., Barhoumi, R., Myers, T., Slogoff, S., and Nicolau, C., 1990. Stable rightward shifts of the oxyhemoglobin dissociation curve induced by encapsulation of inositol hexaphosphate in red blood cells using electroporation. *FEBS Lett.* 275:117–120.

Egeberg, K.D., Springer, B.A., Sligar, S.G., Carver, T.E., Rohlfs, R.J., and Olson, J.S., 1990. The role of Val68(E11) in ligand binding to sperm whale myoglobin. Site-directed mutagenesis of a synthetic gene. *J. Biol. Chem.* 265:11788–11795.

Baudin, C.V., Pagnier, J., Marden, M., Bohn, B., Lacaze, N., Kister, J., Schaad, O., Edelstein, S.J., and Poyart, C., 1990. Enhanced polymerization of recombinant human deoxyhemoglobin $\beta 6$ Glu $\rightarrow$ Ile. *Proc. Nat. Acad. Sci.* 87:1845–1849.

PROBLEMS

1. *Hemoglobin content.* The average volume of a red blood cell is 87 μm^3. The mean concentration of hemoglobin in red cells is 34 g/100 ml.
 (a) What is the weight of the hemoglobin contained in a red cell?
 (b) How many hemoglobin molecules are there in a red cell?
 (c) Could the hemoglobin concentration in red cells be much higher than the observed value? (Hint: Suppose that a red cell contained a crystalline array of hemoglobin molecules in a cubic lattice with 65 Å sides.)

2. *Iron content.* How much iron is there in the hemoglobin of a 70-kg adult? Assume that the blood volume is 70 ml/kg of body weight and that the hemoglobin content of blood is 16 g/100 ml.

3. *Oxygenating myoglobin.* The myoglobin content of some human muscles is about 8 g/kg. In sperm whale, the myoglobin content of muscle is about 80 g/kg.
 (a) How much O_2 is bound to myoglobin in human muscle and in that of sperm whale? Assume that the myoglobin is saturated with O_2.
 (b) The amount of oxygen dissolved in tissue water (in equilibrium with venous blood) at 37°C is about 3.5×10^{-5} M. What is the ratio of oxygen bound to myoglobin to that directly dissolved in the water of sperm whale muscle?

4. *Release kinetics.* The equilibrium constant K for the binding of oxygen to myoglobin is 10^{-6} M, where K is defined as

$$K = \frac{[Mb][O_2]}{[MbO_2]}$$

 The rate constant for the combination of O_2 with myoglobin is 2×10^7 $M^{-1} s^{-1}$.
 (a) What is the rate constant for the dissociation of O_2 from oxymyoglobin?
 (b) What is the mean duration of the oxymyoglobin complex?

5. *Tuning oxygen affinity.* What is the effect of each of the following treatments on the oxygen affinity of hemoglobin A in vitro?
 (a) Increase in pH from 7.2 to 7.4.
 (b) Increase in pCO_2 from 10 to 40 torrs.
 (c) Increase in [BPG] from 2×10^{-4} to 8×10^{-4} M.
 (d) Dissociation of $\alpha_2\beta_2$ into monomer subunits.

6. *Avian and reptilian counterparts.* The erythrocytes of birds and turtles contain a regulatory molecule different from BPG. This substance is also effective in reducing the oxygen affinity of human hemoglobin stripped of BPG. Which of the following substances would you predict to be most effective in this regard?

(a) Glucose 6-phosphate. (d) Malonate.
(b) Inositol hexaphosphate. (e) Arginine.
(c) HPO_4^{2-}. (f) Lactate.

7. *Tuning proton affinity.* The pK of an acid depends partly on its environment. Predict the effect of the following environmental changes on the pK of a glutamic acid side chain.
 (a) A lysine side chain is brought into close proximity.
 (b) The terminal carboxyl group of the protein is brought into close proximity.
 (c) The glutamic acid side chain is shifted from the outside of the protein to a nonpolar site inside.

8. *Linkage relations.* The concept of linkage is crucial for the understanding of many biochemical processes. Consider a protein molecule P that can bind A or B or both:

$$\begin{array}{ccc}
P & \overset{K_A}{\rightleftharpoons} & PA \\
K_B \updownarrow & & \updownarrow K_{AB} \\
PB & \underset{K_{BA}}{\rightleftharpoons} & PAB
\end{array}$$

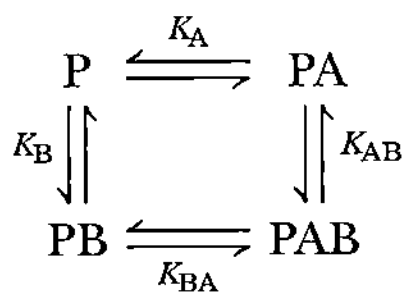

 The dissociation constants for these equilibria are defined as

$$K_A = \frac{[P][A]}{[PA]} \qquad K_B = \frac{[P][B]}{[PB]}$$

$$K_{BA} = \frac{[PB][A]}{[PAB]} \qquad K_{AB} = \frac{[PA][B]}{[PAB]}$$

 (a) Suppose $K_A = 5 \times 10^{-4}$ M, $K_B = 10^{-3}$ M, and $K_{BA} = 10^{-5}$ M. Is the value of the fourth dissociation constant K_{AB} defined? If so, what is it?
 (b) What is the effect of [A] on the binding of B? What is the effect of [B] on the binding of A?

9. *Poison puzzle.* Carbon monoxide, an odorless gas, combines with hemoglobin to form CO–hemoglobin. Crystals of CO–hemoglobin are isomorphous with those of oxyhemoglobin. Each heme in hemoglobin can bind one carbon monoxide molecule, but O_2 and CO cannot simultaneously bind to the same heme. The binding affinity for CO is about 200 times as great as that for oxygen. Exposure for 1 hour to a CO concentration of 0.1% in inspired air leads to the occupancy by CO of about half the heme sites in hemoglobin, a proportion that is frequently fatal.

 An interesting problem was posed (and partly solved) by J.S. Haldane and J.G. Priestley in 1935:

 If the action of CO were simply to diminish the oxygen-carrying power of the hemoglobin, without other modification of its properties, the symptoms of CO poisoning would be very difficult to understand in the light of other knowledge. Thus, a person whose blood is half-saturated with CO is practically helpless, as we have just seen; but a person whose hemoglobin percentage is simply diminished to half by anemia may be going about his work as usual.

What is the key to this seeming paradox?

10. *Optimal transport.* A protein molecule P reversibly binds a small molecule L. The dissociation constant K for the equilibrium

$$P + L \rightleftharpoons PL$$

is defined as

$$K = \frac{[P][L]}{[PL]}$$

The protein transports the small molecule from a region of high concentration $[L_A]$ to one of low concentration $[L_B]$. Assume that the concentrations of unbound small molecules remain constant. The protein goes back and forth between regions A and B.

(a) Suppose $[L_A] = 10^{-4}$ M and $[L_B] = 10^{-6}$ M. What value of K yields maximal transport? One way of solving this problem is to write an expression for ΔY, the change in saturation of the ligand-binding site in going from region A to B. Then, take the derivative of ΔY with respect to K.

(b) Treat oxygen transport by hemoglobin in a similar way. What value of P_{50} would give a maximal ΔY? Assume that the P in the lungs is 100 torrs, whereas P in the tissue capillaries is 20 torrs. Compare your calculated value of P_{50} with the physiological value of 26 torrs.

11. *On the mutant trail.* A hemoglobin with an abnormal electrophoretic mobility is detected in a screening program. Fingerprinting after tryptic digestion reveals that the amino acid substitution is in the β chain. The normal amino-terminal tryptic peptide (Val-His-Leu-Thr-Pro-Glu-Glu-Lys) is missing. A new tryptic peptide consisting of six amino acid residues is found. Valine is the amino-terminal residue of this peptide.

(a) Which amino acid substitutions are consistent with these data?

(b) Which single-base changes in DNA sequence could give these amino acid substitutions? The DNA sequence encoding the normal amino-terminal region is GTGCACCTGACTCCTGAG-GAGAAG.

(c) How should the electrophoretic mobility of this hemoglobin compare with those of Hb A and Hb S at pH 8?

12. *Allosteric transition.* Consider an allosteric protein that obeys the concerted model. Suppose that the ratio of T to R formed in the absence of ligand is 10^5, $K_T = 2$ mM, and $K_R = 5$ μM. The protein contains four binding sites for ligand. What is the fraction of molecules in the R form when 0, 1, 2, 3, and 4 ligands are bound?

13. *Triple hit.* Some mutations in a hemoglobin gene affect all three of the hemoglobins A, A_2, and F, whereas others affect only one of them. Why?

14. *Saving grace.* Hemoglobin A inhibits the formation of long fibers of hemoglobin S and the subsequent sickling of the red cell upon deoxygenation. Why does hemoglobin A have this effect?

15. *Carbamoylation.* Cyanate was a promising antisickling drug until clinical trials uncovered its toxic side effects, such as damage to peripheral nerves. Cyanate carbamoylates the terminal amino groups of hemoglobin. It behaves as a reactive analog of CO_2.

$$R-NH_2 \ + \ ^-NCO \ + \ H^+ \longrightarrow R-\overset{\overset{\displaystyle H}{|}}{N}-\overset{\overset{\displaystyle}{\underset{\underset{\displaystyle O}{\|}}{C}}}-NH_2$$

| Terminal amino group | Cyanate | Carbamoylated derivative |

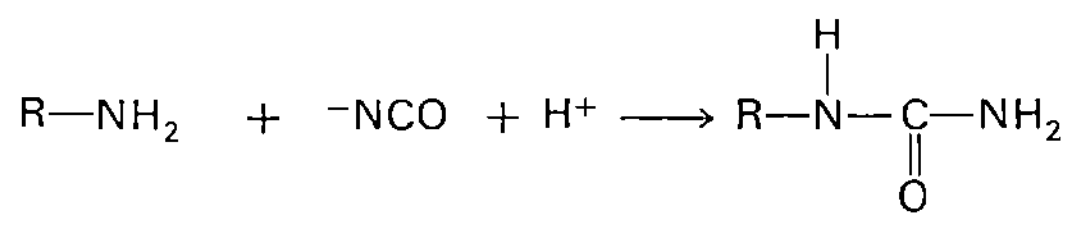

Predict the resulting change in oxygen affinity. Why does carbamoylation have an antisickling effect?

16. *A changing slope.* The Hill plot for hemoglobin has a slope of about 2.8 at the midpoint ($Y = 0.5$). However, it is important to note that the slope of the Hill plot is not constant (see Figure 7-40B). At both very low and very high oxygen pressure, the slope approaches 1. Why?

CHAPTER

8

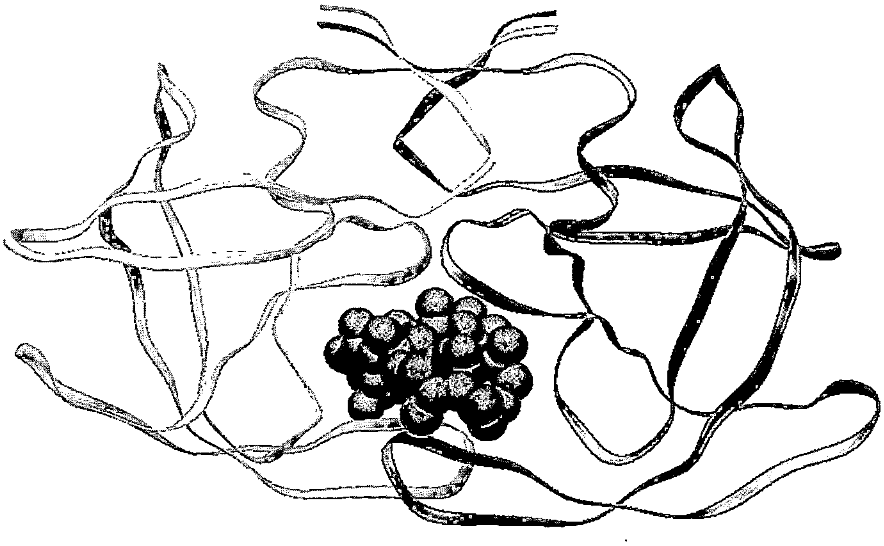

Enzymes: Basic Concepts and Kinetics

Enzymes, the catalysts of biological systems, are remarkable molecular devices that determine the pattern of chemical transformations. They also mediate the transformation of different forms of energy. The most striking characteristics of enzymes are their *catalytic power* and *specificity*. Furthermore, the actions of many enzymes are regulated. Nearly all known enzymes are proteins. The discovery of catalytically active RNA molecules, however, indicates that proteins do not have an absolute monopoly on catalysis.

Proteins as a class of macromolecules are highly effective in catalyzing diverse chemical reactions because of their capacity to specifically bind a very wide range of molecules. By utilizing the full repertoire of intermolecular forces, enzymes bring substrates together in an optimal orientation, the prelude to making and breaking chemical bonds. In essence, they catalyze reactions by stabilizing transition states, the highest-energy species in reaction pathways. By doing this selectively, an enzyme determines which one of several potential chemical reactions actually occurs. Enzymes can also act as molecular switches in regulating catalytic activity and transforming energy because of their capacity to couple the actions of separate binding sites.

ENZYMES HAVE IMMENSE CATALYTIC POWER

Enzymes accelerate reactions by factors of at least a million. Indeed, most reactions in biological systems do not occur at perceptible rates in the absence of enzymes. Even a reaction as simple as the hydration of carbon

Opening Image: *Structure of HIV-1 protease, a dimer of identical subunits (yellow and blue). Protease activity is blocked by a peptide inhibitor (red) that binds to the active site at the interface between subunits. [Drawn by Dr. Anthony Nicholls from 4hvp.pdb. M. Miller, J. Schneider, B.K. Sethyanarayana, M.V. Toth, G.R. Marshall, L. Clawson, L. Selk, S.B.H. Kent, and A. Wlodawer. Science 246(1989):1149.]*

dioxide is catalyzed by an enzyme, namely, carbonic anhydrase (p. 39). The transfer of CO_2 from the tissues into the blood and then to the alveolar air would be less complete in the absence of this enzyme. In fact, carbonic anhydrase is one of the fastest enzymes known. Each enzyme molecule can hydrate 10^5 molecules of CO_2 per second. This catalyzed reaction is 10^7 times faster than the uncatalyzed one.

ENZYMES ARE HIGHLY SPECIFIC

Enzymes are highly specific both in the reaction catalyzed and in their choice of reactants, which are called *substrates*. An enzyme usually catalyzes a single chemical reaction or a set of closely related reactions. Side reactions leading to the wasteful formation of by-products rarely occur in enzyme-catalyzed reactions, in contrast with uncatalyzed ones. The degree of specificity for substrate is usually high and sometimes virtually absolute.

Let us consider *proteolytic enzymes* as an example. The reaction catalyzed by these enzymes is the hydrolysis of a peptide bond.

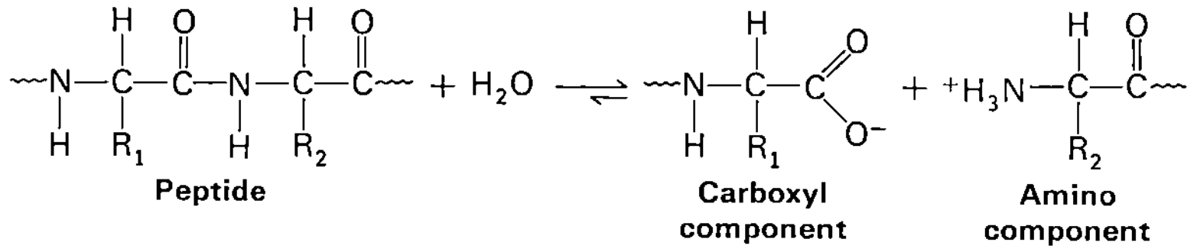

Most proteolytic enzymes also catalyze a different but related reaction, namely, the hydrolysis of an ester bond.

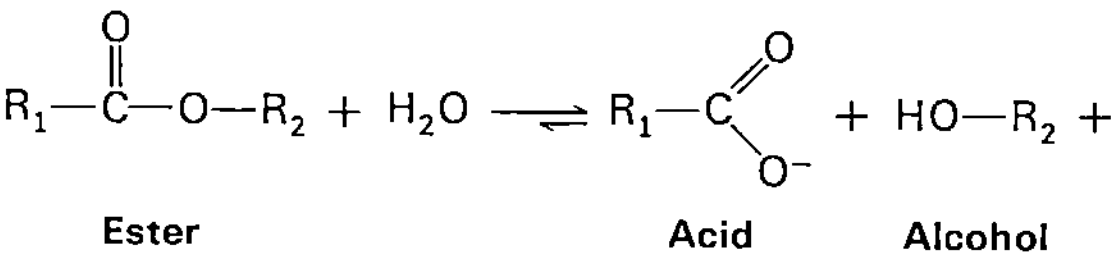

Proteolytic enzymes differ markedly in their degree of substrate specificity. Subtilisin, which comes from certain bacteria, is quite undiscriminating about the nature of the side chains adjacent to the peptide bond to be cleaved. Trypsin (p. 56) is quite specific in that it catalyzes the splitting of peptide bonds on the carboxyl side of lysine and arginine residues only (Figure 8-1A). Thrombin, an enzyme that participates in blood clotting, is even more specific than trypsin. It catalyzes the hydrolysis of Arg-Gly bonds in particular peptide sequences only (Figure 8-1B).

DNA polymerase I, a template-directed enzyme (p. 88), is a highly specific catalyst. The sequence of nucleotides in the DNA strand that is being synthesized is determined by the sequence of nucleotides in another DNA strand that serves as a template. DNA polymerase I is remarkably precise in carrying out the instructions given by the template. The wrong nucleotide is inserted into a new DNA strand less than once in a million times because DNA polymerase proofreads the nascent product and corrects its mistakes.

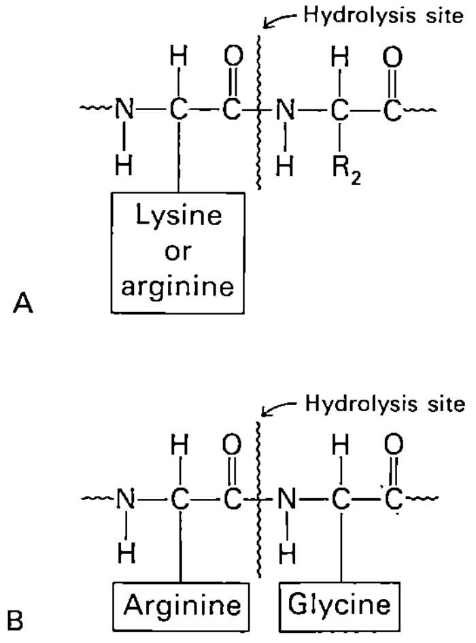

Figure 8-1
Comparison of the specificities of (A) trypsin and (B) thrombin. Trypsin cleaves on the carboxyl side of arginine and lysine residues. Thrombin cleaves Arg-Gly bonds in particular sequences only.

THE CATALYTIC ACTIVITIES OF MANY ENZYMES ARE REGULATED

The enzyme that catalyzes the first step in a biosynthetic pathway is usually inhibited by the ultimate product (Figure 8-2). The biosynthesis of isoleucine in bacteria illustrates this type of control, which is called *feedback inhibition.* Threonine is converted into isoleucine in five steps, the first of which is catalyzed by threonine deaminase. This enzyme is inhibited when the concentration of isoleucine reaches a sufficiently high level. Isoleucine inhibits by binding to the enzyme at a regulatory site, which is distinct from the catalytic site. This inhibition is mediated by an *allosteric interaction,* which is rapidly reversible. When the level of isoleucine drops sufficiently, threonine deaminase becomes active again, and consequently isoleucine is synthesized once more.

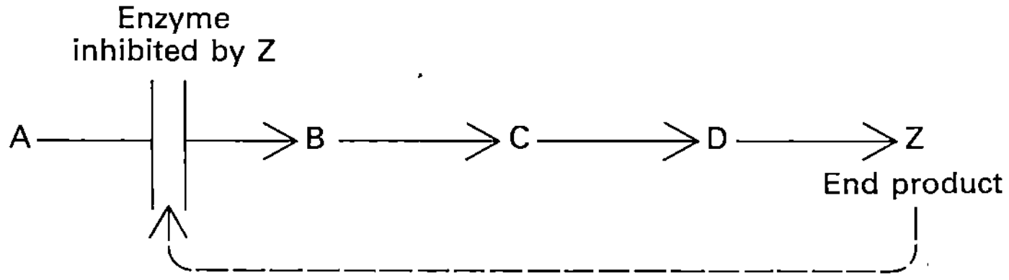

Figure 8-2
Feedback inhibition of the first enzyme in a pathway by reversible binding of the final product.

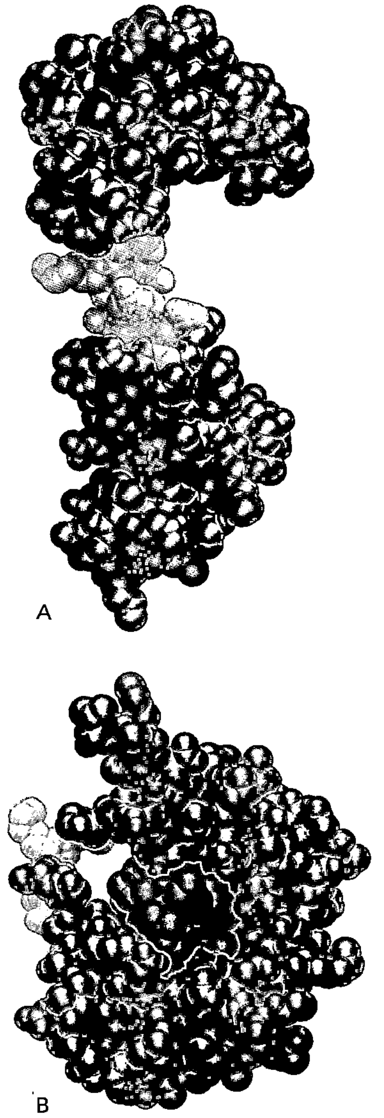

Figure 8-3
Calmodulin, a ubiquitous calcium sensor in eukaryotes, regulates the activities of many intracellular proteins. (A) Space-filling model of calmodulin (blue) containing four bound calcium ions. (B) Calcium–calmodulin wraps around myosin light-chain kinase (red), a target protein, to switch on its enzymatic activity. [(A) Drawn from 3cln.pdb. Y.S. Babu, C.E. Bugg, and W.J. Cook. *J. Mol. Biol.* 204(1988):191. (B) Coordinates kindly provided by Dr. Mitsu Ikura. M. Ikura, G.M. Clore, A.M. Groneborn, G. Zhu, C.B. Klee, and A. Bax. *Science* 256(1992):632.]

Enzymes are also controlled by *regulatory proteins,* which can either stimulate or inhibit. The activities of many enzymes are regulated by *calmodulin,* a 17-kd protein that serves as a calcium sensor in nearly all eukaryotic cells (Figure 8-3). The binding of Ca^{2+} to multiple sites in calmodulin induces a major conformational change that converts it from an inactive to an active form. Activated calmodulin then binds to many enzymes and other target proteins in the cell and modifies their activities (p. 349)

Covalent modification is a third mechanism of enzyme regulation. Many enzymes are controlled by the *reversible attachment of phosphoryl groups* to specific serine and threonine residues. The activities of enzymes that synthesize and degrade glycogen are coordinated in this way (p. 595). The phosphorylation of tyrosine residues on growth factor receptors is a critical step in inducing cell differentiation and proliferation (p. 350). Specific enzymes called *kinases* catalyze the attachment of phosphoryl groups; *phosphatases* catalyze their removal by hydrolysis.

Some enzymes are synthesized in an inactive precursor form, which is activated at a physiologically appropriate time and place. The digestive

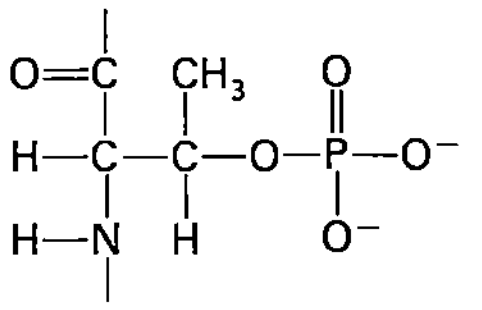

Phosphothreonine residue

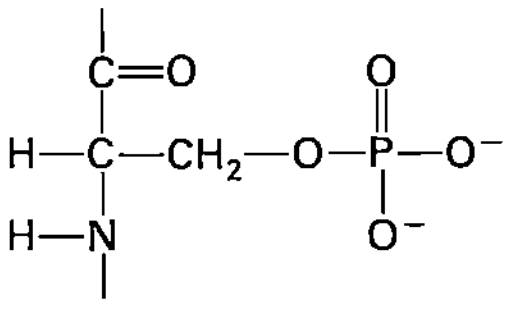

Phosphoserine residue

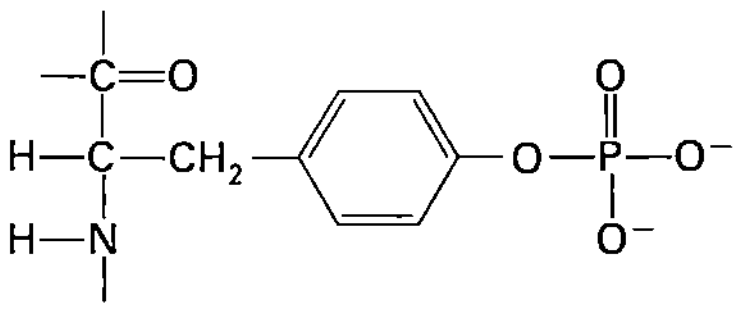

Phosphotyrosine residue

enzymes exemplify this kind of control, which is called *proteolytic activation.* For example, trypsinogen is synthesized in the pancreas and is activated by peptide-bond cleavage in the small intestine to form the active enzyme trypsin (Figure 8-4). This type of control is also repeatedly used in the sequence of enzymatic reactions that lead to the clotting of blood. The enzymatically inactive precursors of proteolytic enzymes are called *zymogens.*

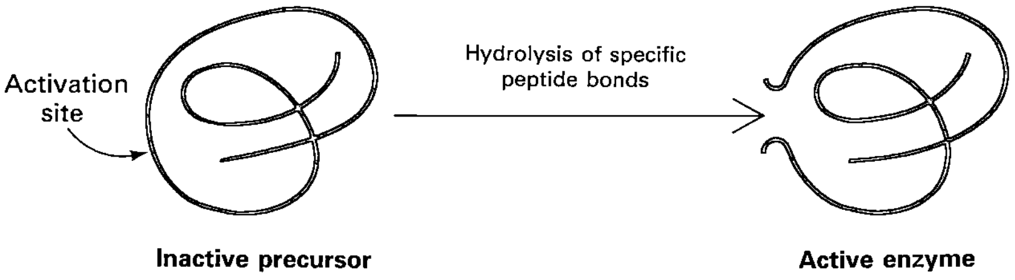

Figure 8-4
Activation of a zymogen by hydrolysis of specific peptide bonds.

ENZYMES TRANSFORM DIFFERENT FORMS OF ENERGY

In many biochemical reactions, *the energy of the reactants is converted with high efficiency into a different form.* For example, in photosynthesis, light energy is converted into chemical-bond energy. In mitochondria, the free energy contained in small molecules derived from food is converted into a different currency, the free energy of adenosine triphosphate (ATP). The chemical-bond energy of ATP is then utilized in many different ways. In muscle contraction, the energy of ATP is converted by myosin into mechanical energy. Membranes of cells and organelles contain pumps that utilize ATP to transport molecules and ions against chemical and electrical gradients. The molecular mechanisms of these energy-transducing enzymes are being unraveled. We will see in subsequent chapters how unidirectional cycles of discrete steps—binding, chemical transformation, and release—lead to the conversion of one form of energy into another.

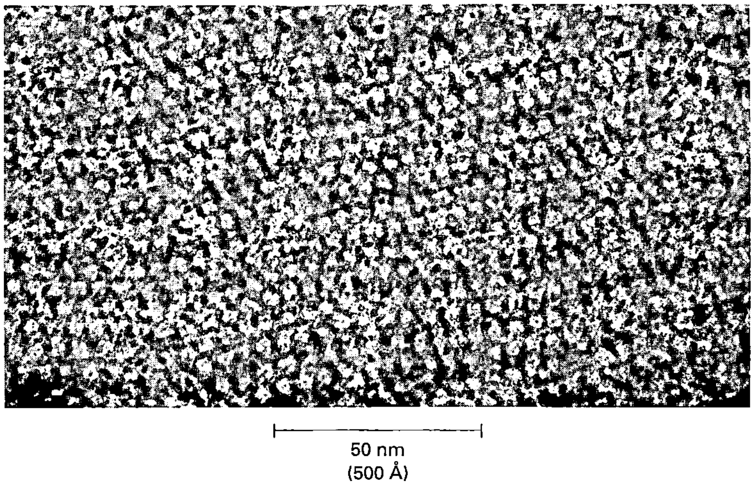

Figure 8-5
Electron micrograph of sodium-potassium pump molecules in a plasma membrane. These densely packed enzyme molecules catalyze the ATP-driven flux of Na^+ and K^+ out of and into cells. [Courtesy of Dr. Guido Zampighi.]

FREE ENERGY IS THE MOST USEFUL THERMODYNAMIC FUNCTION IN BIOCHEMISTRY

Let us review some key thermodynamic relations. In thermodynamics, a *system* is the matter within a defined region. The matter in the rest of the universe is called the *surroundings. The first law of thermodynamics states that the total energy of a system and its surroundings is a constant.* In other words, energy is conserved. The mathematical expression of the first law is

$$\Delta E = E_B - E_A = Q - W \quad (1)$$

in which E_A is the energy of a system at the start of a process and E_B at the end of the process, Q is the heat absorbed by the system, and W is the work done by the system. An important feature of equation 1 is that *the change in energy of a system depends only on the initial and final states and not on the path of the transformation.*

The first law of thermodynamics cannot be used to predict whether a reaction can occur spontaneously. Some reactions do occur spontaneously even when ΔE is positive (the energy of the system increases). In such cases, the system absorbs heat from its surroundings. It is evident that a function different from ΔE is required. One such function is the *entropy* (S), which is a measure of the *degree of randomness or disorder of a system.* The entropy of a system increases (ΔS is positive) when it becomes more disordered (Figure 8-6). *The second law of thermodynamics states that a process can occur spontaneously only if the sum of the entropies of the system and its surroundings increases.*

$$(\Delta S_{\text{system}} + \Delta S_{\text{surroundings}}) > 0 \text{ for a spontaneous process} \quad (2)$$

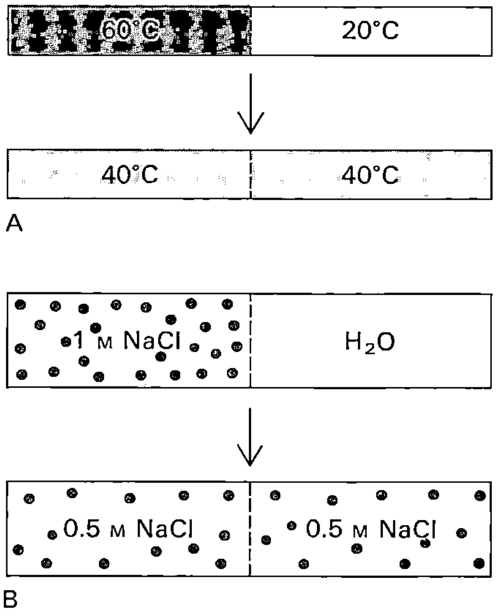

Figure 8-6
Processes that are driven by an increase in the entropy of a system: (A) diffusion of heat; and (B) diffusion of a solute.

Note that the entropy of a system can decrease during a spontaneous process, provided that the entropy of the surroundings increases so that their sum is positive. For example, the formation of a highly ordered biological structure is thermodynamically feasible because the decrease in the entropy of such a system is more than offset by an increase in the entropy of its surroundings.

One difficulty in using entropy as a criterion of whether a biochemical process can occur spontaneously is that the entropy changes of chemical reactions are not readily measured. Furthermore, the criterion of spontaneity given in equation 2 requires that both the entropy change of the surroundings and that of the system of interest be known. These difficulties are obviated by using a different thermodynamic function called the *free energy,* which is denoted by the symbol G (or F, in the older literature). In 1878, Josiah Willard Gibbs created the free-energy function by combining the first and second laws of thermodynamics. The basic equation is

$$\Delta G = \Delta H - T\Delta S \quad (3)$$

in which ΔG is the change in free energy of a system undergoing a transformation at constant pressure (P) and temperature (T), ΔH is the change in enthalpy (heat content) of this system, and ΔS is the change in entropy of this system. Note that the properties of the surroundings do not enter into this equation. The enthalpy change is given by

$$\Delta H = \Delta E + P\Delta V \quad (4)$$

The volume change, ΔV, is small for nearly all biochemical reactions, and so ΔH is nearly equal to ΔE. Hence,

$$\Delta G \cong \Delta E - T\Delta S \quad (5)$$

Thus, the ΔG of a reaction depends both on the change in internal energy and on the change in entropy of the system.

The change in free energy (ΔG) of a reaction, in contrast with the change in internal energy (ΔE) of a reaction, is a valuable criterion of whether it can occur spontaneously:

1. *A reaction can occur spontaneously only if* ΔG *is negative.*

2. A system is at equilibrium and no net change can take place if ΔG is zero.

3. A reaction cannot occur spontaneously if ΔG is positive. An input of free energy is required to drive such a reaction.

Two additional points need to be emphasized here. First, the ΔG of a reaction depends only on the free energy of the products (the final state) minus that of the reactants (the initial state). *The* ΔG *of a reaction is independent of the path (or molecular mechanism) of the transformation.* The mechanism of a reaction has no effect on ΔG. For example, the ΔG for the oxidation of glucose to CO_2 and H_2O is the same whether it occurs by combustion in vitro or by a series of many enzyme-catalyzed steps in a cell. Second, *the* ΔG *provides no information about the rate of a reaction.* A negative ΔG indicates that a reaction can occur spontaneously, but it does not signify whether it will proceed at a perceptible rate. As will be discussed shortly (p. 188), the rate of a reaction depends on the *free energy of activation* ($\Delta G^{\ddagger}$), which is unrelated to ΔG.

STANDARD FREE-ENERGY CHANGE OF A REACTION AND ITS RELATION TO THE EQUILIBRIUM CONSTANT

Consider the reaction

$$A + B \rightleftharpoons C + D$$

The ΔG of this reaction is given by

$$\Delta G = \Delta G^{\circ} + RT \log_e \frac{[C][D]}{[A][B]} \qquad (6)$$

in which ΔG° is the *standard free-energy change,* R is the gas constant, T is the absolute temperature, and [A], [B], [C], and [D] are the molar concentrations (more precisely, the activities) of the reactants. ΔG° is the free-energy change for this reaction under standard conditions—that is, when each of the reactants A, B, C, and D is present at a concentration of 1.0 M (for a gas, the standard state is usually chosen to be 1 atmosphere). Thus, the ΔG of a reaction depends on the *nature* of the reactants (expressed in the ΔG° term of equation 6) and on their *concentrations* (expressed in the logarithmic term of equation 6).

A convention has been adopted to simplify free-energy calculations for biochemical reactions. The standard state is defined as having a pH of 7. Consequently, when H^+ is a reactant, its activity has the value 1 (corresponding to a pH of 7) in equations 6 and 9. The activity of water also is taken to be 1 in these equations. The *standard free-energy change at pH 7,* denoted by the symbol $\Delta G^{\circ\prime}$, will be used throughout this book. The *kilocalorie* (abbreviated *kcal*) will be used as the unit of energy.

The relation between the standard free energy and the equilibrium constant of a reaction can be readily derived. At equilibrium, $\Delta G = 0$. Equation 6 then becomes

Units of energy—
A *calorie* (cal) is equivalent to the amount of heat required to raise the temperature of 1 gram of water from 14.5°C to 15.5°C.
A *kilocalorie* (kcal) is equal to 1000 cal.
A *joule* (J) is the amount of energy needed to apply a 1-newton force over a distance of 1 meter. A *kilojoule* (kJ) is equal to 1000 J.
1 kcal = 4.184 kJ

$$0 = \Delta G^{\circ\prime} + RT\log_e \frac{[\mathrm{C}][\mathrm{D}]}{[\mathrm{A}][\mathrm{B}]} \quad (7)$$

and so

$$\Delta G^{\circ\prime} = -RT\log_e \frac{[\mathrm{C}][\mathrm{D}]}{[\mathrm{A}][\mathrm{B}]} \quad (8)$$

The equilibrium constant under standard conditions, K'_{eq}, is defined as

$$K'_{eq} = \frac{[\mathrm{C}][\mathrm{D}]}{[\mathrm{A}][\mathrm{B}]} \quad (9)$$

Substituting equation 9 into equation 8 gives

$$\Delta G^{\circ\prime} = -RT\log_e K'_{eq} \quad (10)$$

$$\Delta G^{\circ\prime} = -2.303\, RT\log_{10} K'_{eq} \quad (11)$$

which can be rearranged to give

$$K'_{eq} = 10^{-\Delta G^{\circ\prime}/(2.303RT)} \quad (12)$$

Substituting $R = 1.987 \times 10^{-3}$ kcal mol^{-1} deg^{-1} and $T = 298$ K (corresponding to 25°C) gives

$$K'_{eq} = 10^{-\Delta G^{\circ\prime}/1.36} \quad (13)$$

when $\Delta G^{\circ\prime}$ is expressed in kcal/mol. Thus, the standard free energy and the equilibrium constant of a reaction are related by a simple expression. For example, an equilibrium constant of 10 gives a standard free-energy change of −1.36 kcal/mol (−5.69 kJ/mol) at 25°C (Table 8-1).

Table 8-1
Relation between $\Delta G^{\circ\prime}$ and K'_{eq} (at 25°C)

K'_{eq}	$\Delta G^{\circ\prime}$	
	kcal/mol	*kJ/mol*
10^{-5}	6.82	28.53
10^{-4}	5.46	22.84
10^{-3}	4.09	17.11
10^{-2}	2.73	11.42
10^{-1}	1.36	5.69
1	0	0
10	−1.36	−5.69
10^{2}	−2.73	−11.42
10^{3}	−4.09	−17.11
10^{4}	−5.46	−22.84
10^{5}	−6.82	−28.53

Let us calculate $\Delta G^{\circ\prime}$ and ΔG for the isomerization of dihydroxyacetone phosphate to glyceraldehyde 3-phosphate as an example. This reaction occurs in glycolysis (p. 488). At equilibrium, the ratio of glyceraldehyde 3-phosphate to dihydroxyacetone phosphate is 0.0475 at 25°C (298 K) and pH 7. Hence, $K'_{eq} = 0.0475$. The standard free-energy change for this reaction is then calculated from equation 11

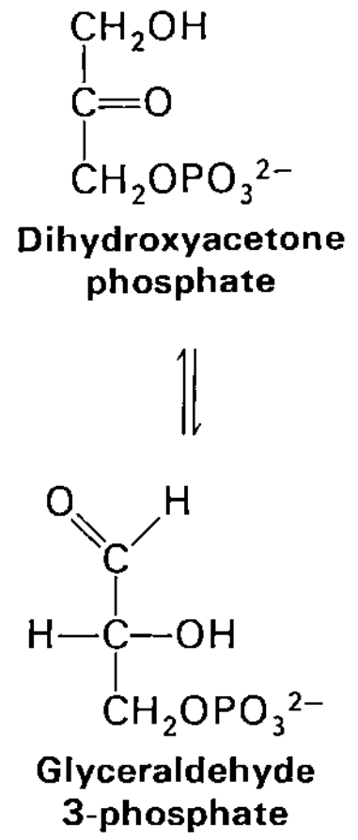

Dihydroxyacetone phosphate ⇌ **Glyceraldehyde 3-phosphate**

$$\begin{aligned}\Delta G^{\circ\prime} &= -2.303\, RT\log_{10} K'_{eq}\\ &= -2.303 \times 1.987 \times 10^{-3} \times 298 \times \log_{10}(0.0475)\\ &= +1.8 \text{ kcal/mol}\end{aligned}$$

Now let us calculate ΔG for this reaction when the initial concentration of dihydroxyacetone phosphate is 2×10^{-4} M and the initial concentration of glyceraldehyde 3-phosphate is 3×10^{-6} M. Substituting these values into equation 6 gives

$$\begin{aligned}\Delta G &= 1.8 \text{ kcal/mol} + 2.303\, RT\log_{10} \frac{3 \times 10^{-6}\ \mathrm{M}}{2 \times 10^{-4}\ \mathrm{M}}\\ &= 1.8 \text{ kcal/mol} - 2.5 \text{ kcal/mol}\\ &= -0.7 \text{ kcal/mol}\end{aligned}$$

This negative value for the ΔG indicates that the isomerization of dihydroxyacetone phosphate to glyceraldehyde 3-phosphate can occur spontaneously when these species are present at the concentrations stated above. Note that ΔG for this reaction is negative although $\Delta G^{\circ\prime}$ is positive. *It is important to stress that whether the* ΔG *for a reaction is larger, smaller, or the same as* $\Delta G^{\circ\prime}$ *depends on the concentrations of the reactants.* The criterion of spontaneity for a reaction is ΔG, not $\Delta G^{\circ\prime}$.

ENZYMES CANNOT ALTER REACTION EQUILIBRIA

An enzyme is a catalyst, and consequently it cannot alter the equilibrium of a chemical reaction. This means that an enzyme accelerates the forward and reverse reaction by precisely the same factor. Consider the interconversion of A and B. Suppose that in the absence of enzyme the forward rate constant (k_F) is $10^{-4}\ s^{-1}$ and the reverse rate constant (k_R) is $10^{-6}\ s^{-1}$. The equilibrium constant K is given by the ratio of these rate constants:

$$A \underset{10^{-6}\,s^{-1}}{\overset{10^{-4}\,s^{-1}}{\rightleftharpoons}} B$$

$$K = \frac{[B]}{[A]} = \frac{k_F}{k_R} = \frac{10^{-4}}{10^{-6}} = 100$$

The equilibrium concentration of B is 100 times that of A, whether or not enzyme is present. However, it would take more than an hour to approach this equilibrium without enzyme, whereas equilibrium would be attained within a second in the presence of a suitable enzyme. *Enzymes accelerate the attainment of equilibria but do not shift their position.*

> "I think that enzymes are molecules that are complementary in structure to the activated complexes of the reactions that they catalyze, that is, to the molecular configuration that is intermediate between the reacting substances and the products of reaction for these catalyzed processes. The attraction of the enzyme molecule for the activated complex would thus lead to a decrease in its energy and hence to a decrease in the energy of activation of the reaction and to an increase in the rate of reaction."
>
> LINUS PAULING
> *Nature* 161:707 (1948)

ENZYMES ACCELERATE REACTIONS BY STABILIZING TRANSITION STATES

A chemical reaction of substrate S to form product P goes through a *transition state* $S^{\ddagger}$ that has a higher free energy than either S or P.

$$\underset{\text{Substrate}}{S} \overset{K^{\ddagger}}{\rightleftharpoons} \underset{\text{Transition state}}{S^{\ddagger}} \overset{V}{\longrightarrow} \underset{\text{Product}}{P}$$

The transition state is the most seldom occupied species along the reaction pathway because it has the highest free energy. The *Gibbs free energy of activation,* symbolized by $\Delta G^{\ddagger}$, is equal to the difference in free energy between the transition state and the substrate. The double dagger (‡) denotes a thermodynamic quantity of a transition state.

$$\Delta G^{\ddagger} = G_{S^{\ddagger}} - G_S$$

The reaction rate V is proportional to the concentration of $S^{\ddagger}$, which depends on $\Delta G^{\ddagger}$ because it is in equilibrium with S.

$$[S^{\ddagger}] = [S]e^{-\Delta G^{\ddagger}/RT}$$

$$V = \nu[S^{\ddagger}] = \frac{kT}{h}[S]e^{-\Delta G^{\ddagger}/RT}$$

In these equations, k is Boltzmann's constant and h is Planck's constant. The value of kT/h at 25°C is $6.2 \times 10^{12}\ s^{-1}$. Suppose that the free energy of activation is 6.82 kcal/mol. The ratio $[S^{\ddagger}]/[S]$ is then 10^{-5} (see Table 8-1); we have assumed that $[S] = 1$, and so the reaction rate V is $6.2 \times 10^{7}\ s^{-1}$. As Table 8-1 shows, a decrease of 1.36 kcal/mol in $\Delta G^{\ddagger}$ results in a tenfold faster V.

Enzymes accelerate reactions by decreasing $\Delta G^{\ddagger}$, *the activation barrier.* The combination of substrate and enzyme creates a new reaction pathway whose transition state energy is lower than that of the reaction in the absence of enzyme (Figure 8-7). *The essence of catalysis is specific binding of the transition state,* to be discussed later in this chapter and the next.

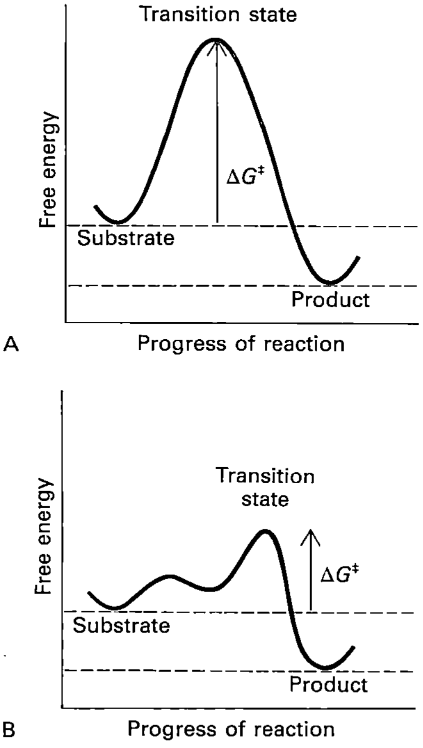

Figure 8-7
Enzymes accelerate reactions by decreasing $\Delta G^{\ddagger}$, the free energy of activation. The free-energy profiles of uncatalyzed (A) and catalyzed (B) reactions are compared.

FORMATION OF AN ENZYME-SUBSTRATE COMPLEX IS THE FIRST STEP IN ENZYMATIC CATALYSIS

Much of the catalytic power of enzymes comes from their bringing substrates together in favorable orientations in *enzyme-substrate (ES)* complexes. The substrates are bound to a specific region of the enzyme called the *active site.* Most enzymes are highly selective in their binding of substrates. Indeed, the catalytic specificity of enzymes depends in part on the specificity of binding. Furthermore, the activities of some enzymes are controlled at this stage.

The existence of ES complexes has been shown in a variety of ways:

1. At a constant concentration of enzyme, the reaction rate increases with increasing substrate concentration until a maximal velocity is reached (Figure 8-8). In contrast, uncatalyzed reactions do not show this saturation effect. In 1913, Leonor Michaelis interpreted the *maximal velocity of an enzyme-catalyzed reaction* in terms of the formation of a discrete ES complex. At a sufficiently high substrate concentration, the catalytic sites are filled and so the reaction rate reaches a maximum. Though indirect, this is the most general evidence for the existence of ES complexes.

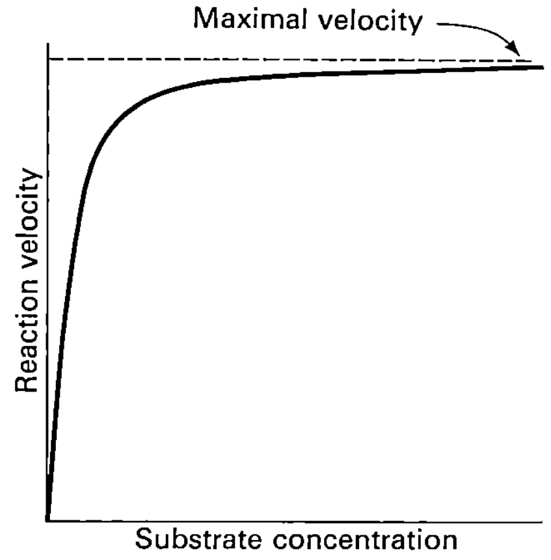

Figure 8-8
Velocity of an enzyme-catalyzed reaction as a function of the substrate concentration.

2. ES complexes have been directly visualized by *electron microscopy,* as in the micrograph of DNA polymerase I bound to its DNA template (Figure 8-9). *X-ray crystallography* has provided high-resolution images of substrates and substrate analogs bound to the active sites of many enzymes. In the next chapter, we shall take a close look at several of these complexes. Moreover, x-ray studies carried out at low temperatures (to slow reactions down) are providing revealing views of intermediates in enzymatic reactions. Also, light can be used to generate a substrate from a photolabile precursor that is bound to the active site in the crystal. Dynamic images of these nascent ES complexes can be obtained with fast pulses of x-rays.

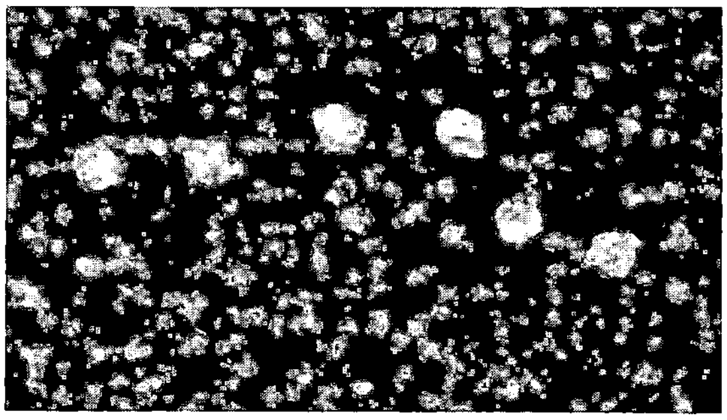

Figure 8-9
Electron micrograph of DNA polymerase I molecules (white spheres) bound to a threadlike synthetic DNA template. [Courtesy of Dr. Jack Griffith.]

3. The *spectroscopic characteristics* of many enzymes and substrates change upon formation of an ES complex just as the absorption spectrum of deoxyhemoglobin changes markedly when it binds oxygen or when it is oxidized to the ferric state, as described previously (see Figure 7-12, on p. 152). These changes are particularly striking if the enzyme contains a colored prosthetic group. Tryptophan synthetase, a bacterial enzyme that contains a pyridoxal phosphate prosthetic group, affords a nice illustration. This enzyme catalyzes the synthesis of L-tryptophan from L-serine and indole. The addition of L-serine to the enzyme produces a marked increase in the fluorescence of the pyridoxal phosphate group (Figure 8-10). The subsequent addition of indole, the second substrate, quenches

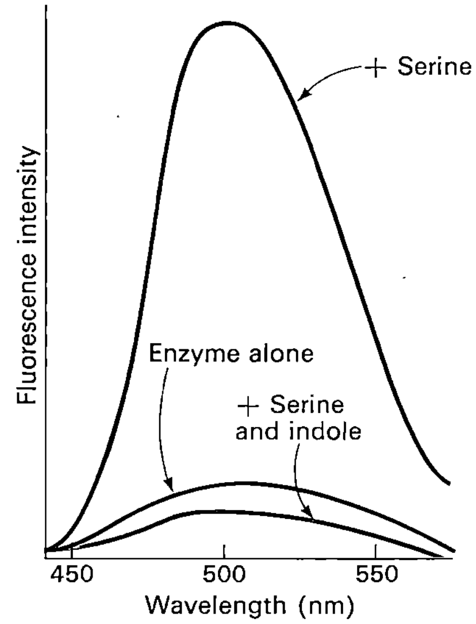

Figure 8-10
Fluorescence intensity of the pyridoxal phosphate group at the active site of tryptophan synthetase changes upon addition of serine and indole, the substrates.

this fluorescence to a level even lower than that of the enzyme alone. Thus, fluorescence spectroscopy reveals the existence of an enzyme-serine complex and of an enzyme-serine-indole complex. Other spectroscopic techniques, such as nuclear magnetic resonance and electron spin resonance, also are highly informative about ES interactions.

ACTIVE SITES OF ENZYMES HAVE SOME COMMON FEATURES

The active site of an enzyme is the region that binds the substrates (and the prosthetic group, if any) and contains the residues that directly participate in the making and breaking of bonds. These residues are called the *catalytic groups.* Although enzymes differ widely in structure, specificity, and mode of catalysis, a number of generalizations concerning their active sites can be stated:

1. *The active site takes up a relatively small part of the total volume of an enzyme.* Most of the amino acid residues in an enzyme are not in contact with the substrate. This raises the intriguing question of why enzymes are so big. Nearly all enzymes are made up of more than 100 amino acid residues, which gives them a mass greater than 10 kd and a diameter of more than 25 Å.

2. *The active site is a three-dimensional entity* formed by groups that come from different parts of the linear amino acid sequence—indeed, residues far apart in the linear sequence may interact more strongly than adjacent residues in the amino acid sequence, as has already been seen for myoglobin and hemoglobin. In lysozyme, an enzyme that will be discussed in more detail in the next chapter, the important groups in the active site are contributed by residues numbered 35, 52, 62, 63, and 101 in the linear sequence of 129 amino acids (Figure 8-11).

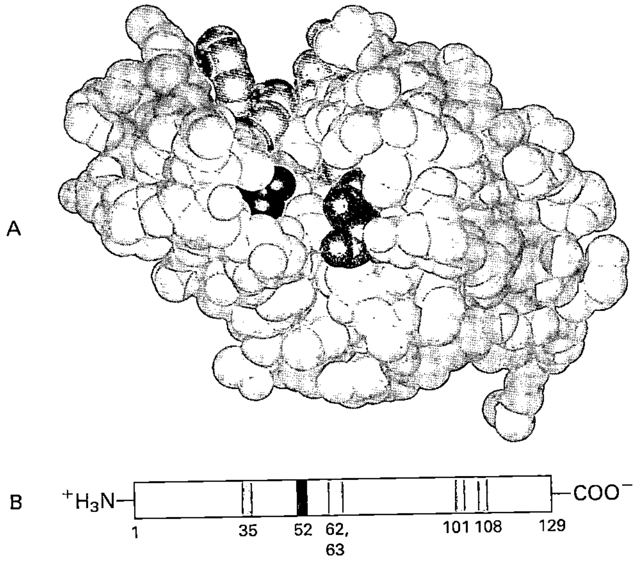

Figure 8-11
(A) Space-filling model of lysozyme. Two catalytically critical residues are shown in red and green. Several other residues in the active site are shown in yellow. (B) Schematic diagram of lysozyme showing that the active site is formed by residues that come from different segments of the polypeptide chain. [(A) Drawn from 6lyz. R. Diamond. *J. Mol. Biol.* 82(1974):371.]

3. *Substrates are bound to enzymes by multiple weak attractions.* ES complexes usually have equilibrium constants that range from 10^{-2} to 10^{-8} M, corresponding to free energies of interaction ranging from about −3 to −12 kcal/mol. The noncovalent interactions in ES complexes are much weaker than covalent bonds, which have energies between −50 and −110 kcal/mol. As was discussed in Chapter 1 (p. 7), reversible interac-

tions of biomolecules are mediated by electrostatic bonds, hydrogen bonds, van der Waals forces, and hydrophobic interactions. Van der Waals forces become significant in binding only when numerous substrate atoms can simultaneously come close to many enzyme atoms. Hence, the enzyme and substrate should have complementary shapes. The directional character of hydrogen bonds between enzyme and substrate often enforces a high degree of specificity.

4. *Active sites are clefts or crevices.* In all enzymes of known structure, substrate molecules are bound to a cleft or crevice. Water is usually excluded unless it is a reactant. The nonpolar character of much of the cleft enhances the binding of substrate. However, the cleft may also contain polar residues. It creates a microenvironment in which certain of these residues acquire special properties essential for catalysis. The internal positions of these polar residues are biologically crucial exceptions to the general rule that polar residues are exposed to water.

5. *The specificity of binding depends on the precisely defined arrangement of atoms in an active site.* To fit into the site, a substrate must have a matching shape. Emil Fischer's metaphor of the lock and key (Figure 8-13), expressed in 1890, has proved to be highly stimulating and fruitful. However, it is now evident that the shapes of the active sites of many enzymes are markedly modified by the binding of substrate, as was postulated by Daniel E. Koshland, Jr., in 1958. The active sites of these enzymes assume shapes that are complementary to that of the substrate only *after* the substrate is bound. This process of dynamic recognition is called *induced fit* (Figure 8-14).

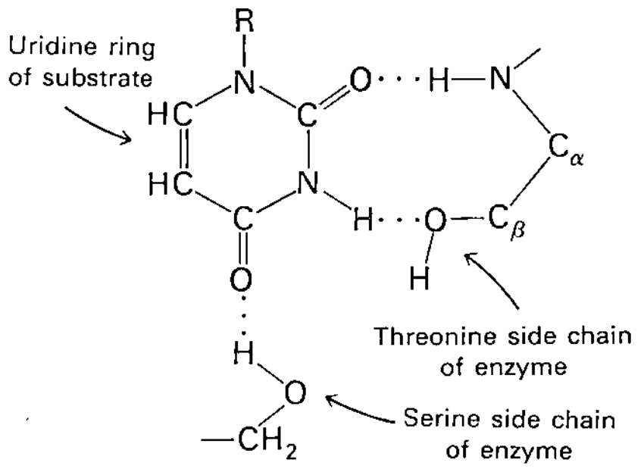

Figure 8-12
Hydrogen bond interactions in the binding of a uridine substrate to ribonuclease. [After F.M. Richards, H.W. Wyckoff, and N. Allewell. In *The Neurosciences: Second Study Program,* F.O. Schmidt, ed. (Rockefeller University Press, 1970), p. 970.]

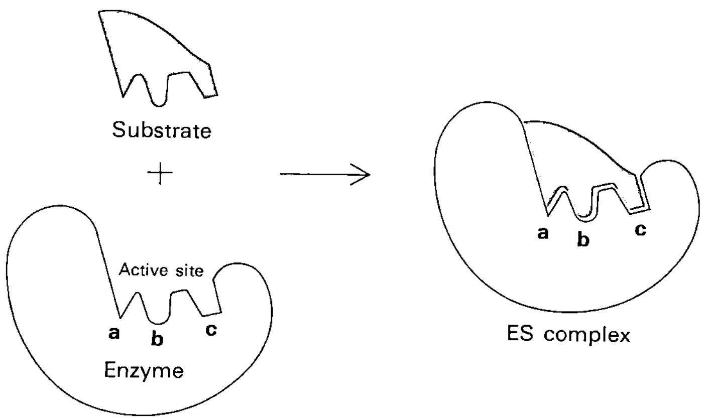

Figure 8-13
Lock-and-key model of the interaction of substrates and enzymes. The active site of the unbound enzyme is complementary in shape to that of the substrate.

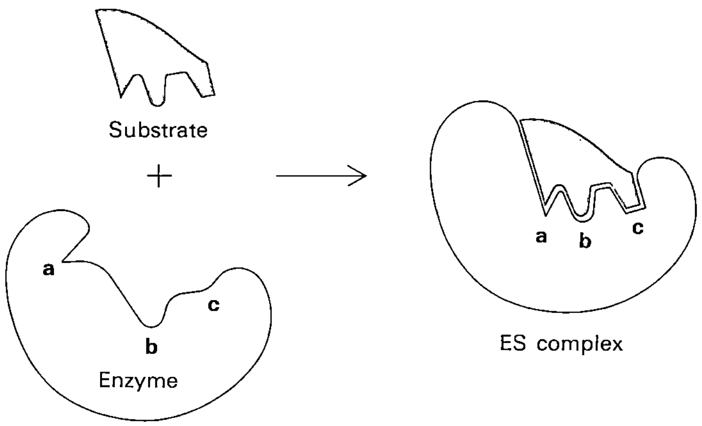

Figure 8-14
Induced-fit model of the interaction of substrates and enzymes. The enzyme changes shape upon binding substrate. The active site has a shape complementary to that of the substrate only after the substrate is bound.

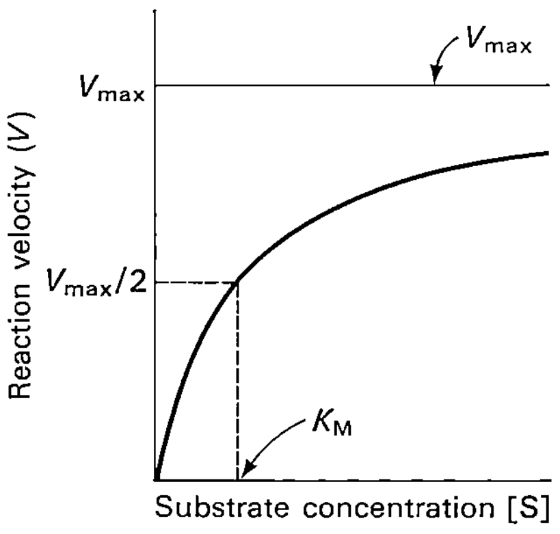

Figure 8-15
A plot of the reaction velocity V as a function of the substrate concentration [S] for an enzyme that obeys Michaelis-Menten kinetics (V_{max} is the maximal velocity and K_M is the Michaelis constant).

THE MICHAELIS-MENTEN MODEL ACCOUNTS FOR THE KINETIC PROPERTIES OF MANY ENZYMES

For many enzymes, the rate of catalysis V varies with the substrate concentration [S] in a manner shown in Figure 8-15. V is defined as the number of moles of product formed per second. At a fixed concentration of enzyme, V is almost linearly proportional to [S] when [S] is small. At high [S], V is nearly independent of [S]. In 1913, Leonor Michaelis and Maud Menten proposed a simple model to account for these kinetic characteristics. The critical feature in their treatment is that a specific ES complex is a necessary intermediate in catalysis. The model proposed, which is the simplest one that accounts for the kinetic properties of many enzymes, is

$$E + S \underset{k_2}{\overset{k_1}{\rightleftharpoons}} ES \xrightarrow{k_3} E + P \tag{14}$$

An enzyme E combines with S to form an ES complex, with a rate constant k_1. The ES complex has two possible fates. It can dissociate to E and S, with a rate constant k_2, or it can proceed to form product P, with a rate constant k_3. It is assumed that almost none of the product reverts to the initial substrate, a condition that holds in the initial stage of a reaction before the concentration of product is appreciable.

We want an expression that relates the rate of catalysis to the concentrations of substrate and enzyme and the rates of the individual steps. Our starting point is that the catalytic rate is equal to the product of the concentration of the ES complex and k_3.

$$V = k_3[ES] \tag{15}$$

Now we need to express [ES] in terms of known quantities. The rates of formation and breakdown of ES are given by

$$\text{Rate of formation of ES} = k_1[E][S] \tag{16}$$

$$\text{Rate of breakdown of ES} = (k_2 + k_3)[ES] \tag{17}$$

We are interested in the catalytic rate under steady-state conditions. In a *steady state,* the concentrations of intermediates stay the same while the concentrations of starting materials and products are changing. This occurs when the rates of formation and breakdown of the ES complex are equal. On setting the right-hand sides of equations 16 and 17 equal,

$$k_1[E][S] = (k_2 + k_3)[ES] \tag{18}$$

By rearranging equation 18,

$$[ES] = \frac{[E][S]}{(k_2 + k_3)/k_1} \tag{19}$$

Equation 19 can be simplified by defining a new constant, K_M, called the *Michaelis constant,*

$$K_M = \frac{k_2 + k_3}{k_1} \tag{20}$$

and substituting it into equation 19, which then becomes

$$[ES] = \frac{[E][S]}{K_M} \tag{21}$$

Now let us examine the numerator of equation 21. The concentration of uncombined substrate [S] is very nearly equal to the total substrate concentration, provided that the concentration of enzyme is much lower than that of the substrate. The concentration of uncombined enzyme [E] is equal to the total enzyme concentration [E_T] minus the concentration of the ES complex.

$$[E] = [E_T] - [ES] \tag{22}$$

On substituting this expression for [E] in equation 21,

$$[ES] = ([E_T] - [ES])[S]/K_M \tag{23}$$

Solving equation 23 for [ES] gives

$$[ES] = [E_T]\frac{[S]/K_M}{1 + [S]/K_M} \tag{24}$$

or

$$[ES] = [E_T]\frac{[S]}{[S] + K_M} \tag{25}$$

By substituting this expression for [ES] into equation 15, we get

$$V = k_3[E_T]\frac{[S]}{[S] + K_M} \tag{26}$$

The maximal rate V_{max} is attained when the catalytic sites on the enzyme are saturated with substrate—that is, when [S] is much greater than K_M—so that $[S]/([S] + K_M)$ approaches 1. Thus,

$$V_{max} = k_3[E_T] \tag{27}$$

Substituting equation 27 into equation 26 yields the *Michaelis-Menten equation.*

$$V = V_{max}\frac{[S]}{[S] + K_M} \tag{28}$$

This equation accounts for the kinetic data given in Figure 8-15. At very low substrate concentration, when [S] is much less than K_M, $V = [S]V_{max}/K_M$; that is, the rate is directly proportional to the substrate concentration. At high substrate concentration, when [S] is much greater than K_M, $V = V_{max}$; that is, the rate is maximal, independent of substrate concentration.

The meaning of K_M is evident from equation 28. When $[S] = K_M$, then $V = V_{max}/2$. Thus, *K_M is equal to the substrate concentration at which the reaction rate is half its maximal value.*

V_{max} AND K_M CAN BE DETERMINED BY VARYING THE SUBSTRATE CONCENTRATION

The Michaelis constant, K_M, and the maximal rate, V_{max}, can be readily derived from rates of catalysis measured at different substrate concentrations if an enzyme operates according to the simple scheme given in equation 14. It is convenient to transform the Michaelis-Menten equation into one that gives a straight-line plot. This can be done by taking the reciprocal of both sides of equation 28 to give

$$\frac{1}{V} = \frac{1}{V_{max}} + \frac{K_M}{V_{max}} \cdot \frac{1}{[S]} \tag{29}$$

A plot of $1/V$ versus $1/[S]$, called a *Lineweaver-Burk plot*, yields a straight line with an intercept of $1/V_{max}$ and a slope of K_M/V_{max} (Figure 8-16). Alternatively, K_M and V_{max} can be obtained by fitting the data to equation 28 using a computer program.

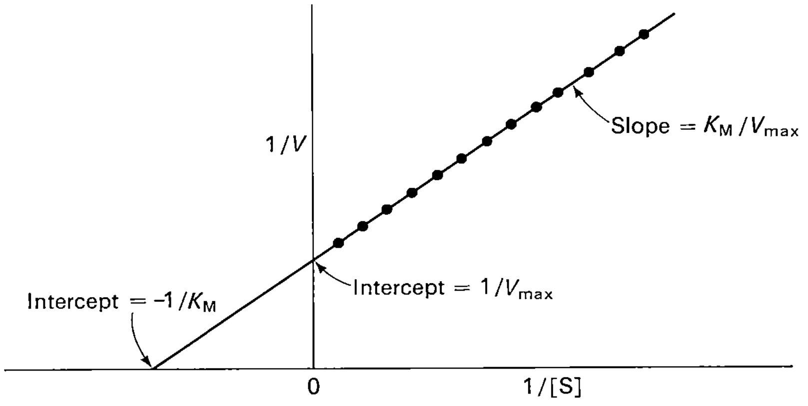

Figure 8-16
A double-reciprocal plot of enzyme kinetics: $1/V$ is plotted as a function of $1/[S]$. The slope is K_M/V_{max}, the intercept on the vertical axis is $1/V_{max}$, and the intercept on the horizontal axis is $-1/K_M$.

SIGNIFICANCE OF K_M AND V_{max} VALUES

The K_M values of enzymes range widely (Table 8-2). For most enzymes, K_M lies between 10^{-1} and 10^{-7} M. The K_M value for an enzyme depends on the particular substrate and also on environmental conditions such as pH, temperature, and ionic strength. The Michaelis constant K_M has two meanings. First, K_M is the concentration of substrate at which half the active sites are filled. Once the K_M is known, the fraction of sites filled f_{ES} at any substrate concentration can be calculated from

$$f_{ES} = \frac{V}{V_{max}} = \frac{[S]}{[S] + K_M} \tag{30}$$

Second, K_M is related to the rate constants of the individual steps in the catalytic scheme given in equation 14. In equation 20, K_M is defined as $(k_2 + k_3)/k_1$. Consider a limiting case in which k_2 is much greater than k_3. This means that dissociation of the ES complex to E and S is much more rapid than formation of E and product. Under these conditions $(k_2 \gg k_3)$,

Table 8-2
K_M values of some enzymes

Enzyme	*Substrate*	K_M (μM)
Chymotrypsin	Acetyl-L-tryptophanamide	5000
Lysozyme	Hexa-*N*-acetylglucosamine	6
β-Galactosidase	Lactose	4000
Threonine deaminase	Threonine	5000
Carbonic anhydrase	CO_2	8000
Penicillinase	Benzylpenicillin	50
Pyruvate carboxylase	Pyruvate	400
	HCO_3^-	1000
	ATP	60
Arginine-tRNA synthetase	Arginine	3
	tRNA	0.4
	ATP	300

$$K_M = \frac{k_2}{k_1} \tag{31}$$

The dissociation constant of the ES complex is given by

$$K_{ES} = \frac{[E][S]}{[ES]} = \frac{k_2}{k_1} \tag{32}$$

In other words, K_M *is equal to the dissociation constant of the ES complex if* k_3 *is much smaller than* k_2. When this condition is met, K_M is a measure of the strength of the ES complex: a high K_M indicates weak binding; a low K_M indicates strong binding. It must be stressed that K_M indicates the affinity of the ES complex only when k_2 is much greater than k_3.

The *turnover number* of an enzyme is *the number of substrate molecules converted into product by an enzyme molecule in a unit time when the enzyme is fully saturated with substrate.* It is equal to the kinetic constant k_3. The maximal rate V_{max} reveals the turnover number of an enzyme if the concentration of active sites $[E_T]$ is known, because

$$V_{max} = k_3[E_T] \tag{33}$$

For example, a 10^{-6} M solution of carbonic anhydrase catalyzes the formation of 0.6 M H_2CO_3 per second when it is fully saturated with substrate. Hence, k_3 is $6 \times 10^5\ s^{-1}$. This turnover number is one of the largest known. Each round of catalysis occurs in a time equal to $1/k_3$, which is 1.7 μs for carbonic anhydrase. The turnover numbers of most enzymes with their physiological substrates fall in the range from 1 to 10^4 per second (Table 8-3).

Table 8-3
Maximum turnover numbers of some enzymes

Enzyme	*Turnover number (per second)*
Carbonic anhydrase	600,000
3-Ketosteroid isomerase	280,000
Acetylcholinesterase	25,000
Penicillinase	2,000
Lactate dehydrogenase	1,000
Chymotrypsin	100
DNA polymerase I	15
Tryptophan synthetase	2
Lysozyme	0.5

KINETIC PERFECTION IN ENZYMATIC CATALYSIS: THE k_{cat}/K_M CRITERION

When the substrate concentration is much greater than K_M, the rate of catalysis is equal to k_3, the turnover number, as described in the preceding section. However, most enzymes are not normally saturated with substrate. Under physiological conditions, the $[S]/K_M$ ratio is typically between 0.01 and 1.0. When $[S] \ll K_M$, the enzymatic rate is much less than k_3 because most of the active sites are unoccupied. Is there a number that characterizes the kinetics of an enzyme under these conditions? Indeed there is, as can be shown by combining equations 15 and 21 to give

$$V = \frac{k_3}{K_M}[E][S] \tag{34}$$

When $[S] \ll K_M$, the concentration of free enzyme, [E], is nearly equal to the total concentration of enzyme $[E_T]$, and so

$$V = \frac{k_3}{K_M}[S][E_T] \tag{35}$$

Thus, when $[S] \ll K_M$, the enzymatic velocity depends on the value of k_3/K_M and on [S].

Are there any physical limits on the value of k_3/K_M? Note that this ratio depends on k_1, k_2, and k_3, as can be shown by substituting for K_M.

$$k_3/K_M = \frac{k_3 k_1}{k_2 + k_3} < k_1 \tag{36}$$

Suppose that the rate of formation of product (k_3) is much faster than the rate of dissociation of the ES complex (k_2). The value of k_3/K_M then approaches k_1. Thus the ultimate limit on the value of k_3/K_M is set by k_1, the rate of formation of the ES complex. *This rate cannot be faster than the diffusion-controlled encounter of an enzyme and its substrate.* Diffusion limits the value of k_1 so that it cannot be higher than between 10^8 and 10^9 M^{-1} s^{-1}. Hence, the upper limit on k_3/K_M is between 10^8 and 10^9 M^{-1} s^{-1}.

This restriction also pertains to enzymes having more complex reaction pathways than that of equation 14. Their maximal catalytic rate when substrate is saturating, denoted by k_{cat}, depends on several rate constants rather than on k_3 alone. The pertinent parameter for these enzymes is k_{cat}/K_M. In fact, the k_{cat}/K_M ratios of the enzymes carbonic anhydrase, acetylcholinesterase, and triosephosphate isomerase are between 10^8 and 10^9 M^{-1} s^{-1}, which shows that they have attained *kinetic perfection. Their catalytic velocity is restricted only by the rate at which they encounter substrate in the solution.* Any further gain in catalytic rate can come only by decreasing the time for diffusion. Indeed, some series of enzymes are associated into organized assemblies (p. 517) so that the product of one enzyme is very rapidly found by the next enzyme. In effect, products are channeled from one enzyme to the next, much as in an assembly line. Thus, the limit imposed by the rate of diffusion in solution can be partly overcome by confining substrates and products in the limited volume of a multienzyme complex.

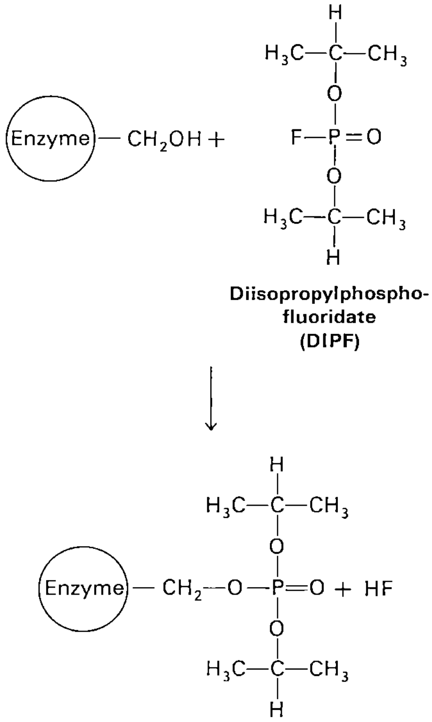

Figure 8-17
Inactivation of acetylcholinesterase by diisopropylphosphofluoridate (DIPF).

ENZYMES CAN BE INHIBITED BY SPECIFIC MOLECULES

The inhibition of enzymatic activity by specific small molecules and ions is important because it serves as a major control mechanism in biological systems. Also, many drugs and toxic agents act by inhibiting enzymes. Furthermore, inhibition can be a source of insight into the mechanism of enzyme action: residues critical for catalysis can often be identified by using specific inhibitors. The value of transition state analogs has already been discussed.

Enzyme inhibition can be either reversible or irreversible. An *irreversible inhibitor* dissociates very slowly from its target enzyme because it becomes very tightly bound to the enzyme, either covalently or noncovalently. The action of nerve gases on acetylcholinesterase, an enzyme that plays an important role in the transmission of nerve impulses, exemplifies irreversible inhibition. Diisopropylphosphofluoridate (DIPF), one of these agents, reacts with a critical serine residue at the active site to form an inactive diisopropylphosphoryl enzyme (Figure 8-17). Alkylating reagents, such as iodoacetamide, irreversibly inhibit the catalytic activity of some enzymes by modifying cysteine and other side chains (Figure 8-18).

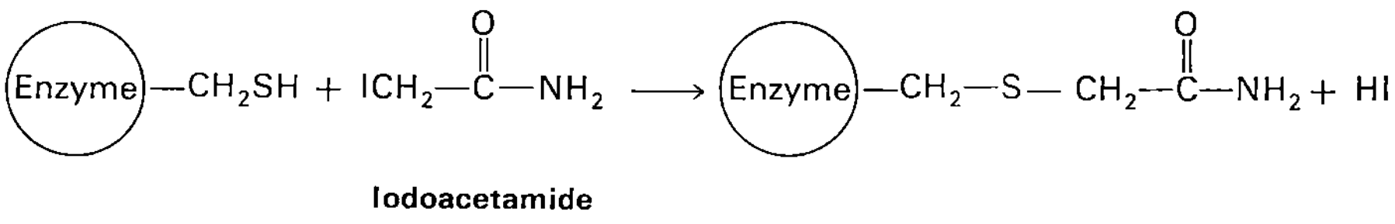

Figure 8-18
Inactivation of an enzyme with a critical cysteine residue by iodoacetamide.

Reversible inhibition, in contrast with irreversible inhibition, is characterized by a rapid dissociation of the enzyme-inhibitor complex. In *competitive inhibition,* the enzyme can bind substrate (forming an ES complex) or inhibitor (EI) but not both (ESI). Many competitive inhibitors resemble

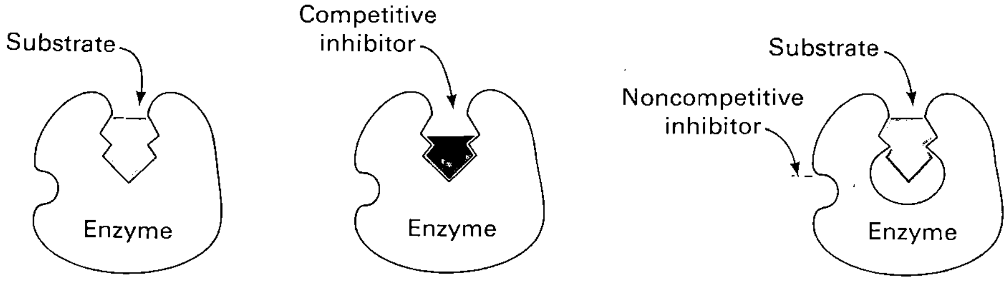

Figure 8-19
Distinction between a competitive and a noncompetitive inhibitor: (top) enzyme-substrate complex; (middle) a competitive inhibitor prevents the substrate from binding; (bottom) a noncompetitive inhibitor does not prevent the substrate from binding.

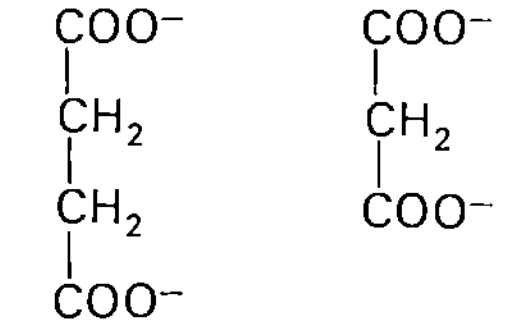

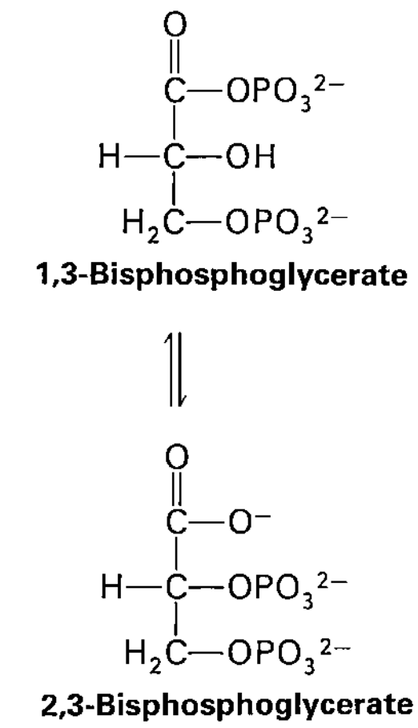

the substrate and bind to the active site of the enzyme (Figure 8-19). The substrate is thereby prevented from binding to the same active site. *A competitive inhibitor diminishes the rate of catalysis by reducing the proportion of enzyme molecules bound to a substrate.* A classic example of competitive inhibition is the action of malonate on succinate dehydrogenase, an enzyme that removes two hydrogen atoms from succinate (p. 512). Malonate differs from succinate in having one rather than two methylene groups. A physiologically important example of competitive inhibition is found in the formation of 2,3-bisphosphoglycerate (BPG, p. 160) from 1,3-bisphosphoglycerate. Bisphosphoglycerate mutase, the enzyme catalyzing this isomerization, is competitively inhibited by even low levels of 2,3-bisphosphoglycerate. In fact, it is not uncommon for an enzyme to be competitively inhibited by its own product because of the product's structural resemblance to the substrate. Competitive inhibition can be overcome by increasing the concentration of substrate.

In *noncompetitive inhibition,* which is also reversible, the inhibitor and substrate can bind simultaneously to an enzyme molecule (Figure 8-19). Hence, their binding sites do not overlap. A noncompetitive inhibitor acts by decreasing the turnover number rather than by diminishing the proportion of enzyme molecules that are bound to substrate. Noncompetitive inhibition, in contrast with competitive inhibition, cannot be overcome by increasing the substrate concentration. A more complex pattern, called *mixed inhibition,* is produced when an inhibitor both affects the binding of substrate and alters the turnover number of the enzyme.

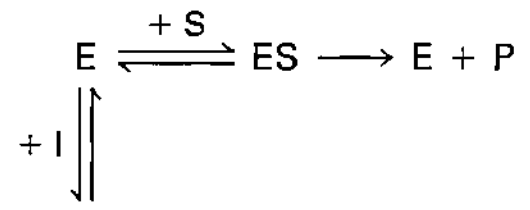

COMPETITIVE AND NONCOMPETITIVE INHIBITION ARE KINETICALLY DISTINGUISHABLE

Let us return to enzymes that exhibit Michaelis-Menten kinetics. Measurements of the rates of catalysis at different concentrations of substrate and inhibitor serve to distinguish between competitive and noncompetitive inhibition. In *competitive inhibition,* the intercept of the plot of $1/V$ versus $1/[S]$ is the same in the presence and absence of inhibitor, although the slope is different (Figure 8-20). This reflects the fact that V_{max} is not altered by a competitive inhibitor. *The hallmark of competitive inhibition is that it can be overcome by a sufficiently high concentration of substrate.* At a sufficiently high concentration, virtually all the active sites are filled by substrate, and the enzyme is fully operative. The increase in the slope of the $1/V$ versus $1/[S]$ plot indicates the strength of binding of competitive inhibitor. In the presence of a competitive inhibitor, equation 29 is replaced by

$$\begin{array}{cc} COO^- & COO^- \\ | & | \\ CH_2 & CH_2 \\ | & | \\ CH_2 & COO^- \\ | & \\ COO^- & \end{array}$$

Succinate **Malonate**

$$\begin{array}{c} O \\ \| \\ C\!-\!OPO_3^{2-} \\ | \\ H\!-\!C\!-\!OH \\ | \\ H_2C\!-\!OPO_3^{2-} \end{array}$$

1,3-Bisphosphoglycerate

$$\begin{array}{c} O \\ \| \\ C\!-\!O^- \\ | \\ H\!-\!C\!-\!OPO_3^{2-} \\ | \\ H_2C\!-\!OPO_3^{2-} \end{array}$$

2,3-Bisphosphoglycerate

$$E \underset{}{\overset{+\,S}{\rightleftarrows}} ES \longrightarrow E + P$$
$$+ I \Big\updownarrow$$
$$EI$$

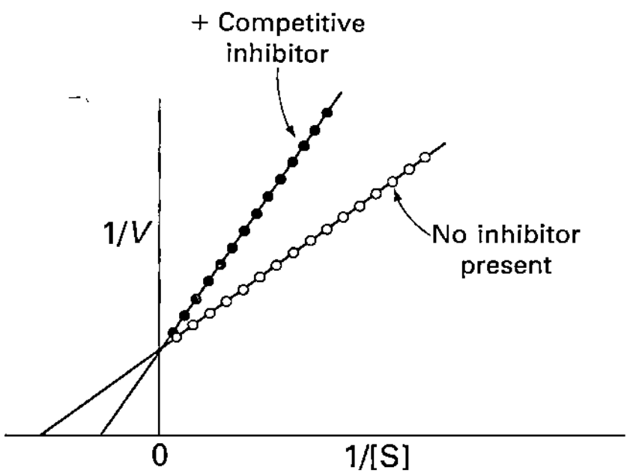

Figure 8-20
A double-reciprocal plot of enzyme kinetics in the presence (•••••) and absence (-ο-ο-ο-ο-) of a competitive inhibitor; V_{max} is unaltered, whereas K_M is increased.

$$\frac{1}{V} = \frac{1}{V_{max}} + \frac{K_M}{V_{max}}\left(1 + \frac{[I]}{K_i}\right)\left(\frac{1}{[S]}\right) \tag{37}$$

in which [I] is the concentration of inhibitor and K_i is the dissociation constant of the enzyme-inhibitor complex.

$$E + I \rightleftharpoons EI$$

$$K_i = \frac{[E][I]}{[EI]} \tag{38}$$

In other words, the slope of the plot is increased by the factor $(1 + [I]/K_i)$ in the presence of a competitive inhibitor. Consider an enzyme with a K_M of 10^{-4} M. In the absence of inhibitor, $V = V_{max}/2$ when $[S] = 10^{-4}$ M. In the presence of 2×10^{-3} M competitive inhibitor that is bound to the enzyme with a K_i of 10^{-3} M, the apparent K_M will be 3×10^{-4} M. Substitution of these values into equation 29 gives $V = V_{max}/4$.

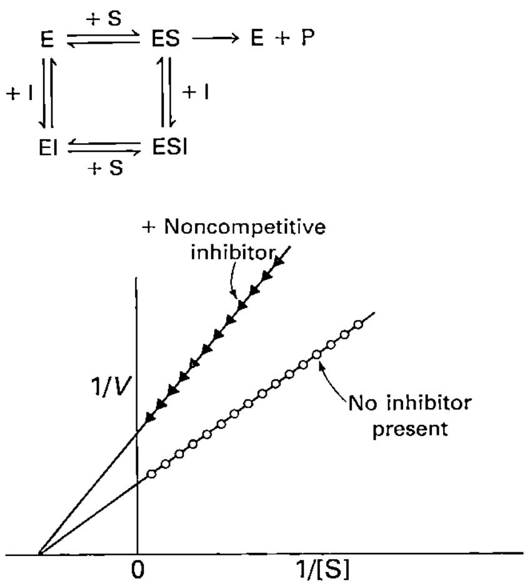

Figure 8-21
A double-reciprocal plot of enzyme kinetics in the presence (-◂◂◂◂◂-) and absence (-o-o-o-o-o-) of a noncompetitive inhibitor; K_M is unaltered by the noncompetitive inhibitor, whereas V_{max} is decreased.

In *noncompetitive inhibition* (Figure 8-21), V_{max} is decreased to V^I_{max}, and so the intercept on the vertical axis is increased. The new slope, which is equal to K_M/V^I_{max}, is larger by the same factor. In contrast with V_{max}, K_M is not affected by this kind of inhibition. *Noncompetitive inhibition cannot be overcome by increasing the substrate concentration.* The maximal velocity in the presence of a noncompetitive inhibitor, V^I_{max}, is given by

$$V^I_{max} = \frac{V_{max}}{1 + [I]/K_i} \tag{39}$$

ALLOSTERIC ENZYMES DO NOT OBEY MICHAELIS-MENTEN KINETICS

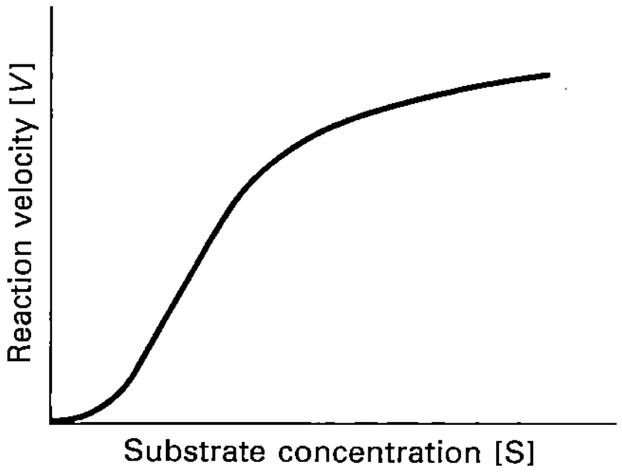

Figure 8-22
Sigmoidal dependence of reaction velocity on substrate concentration for an allosteric enzyme.

The Michaelis-Menten model has greatly affected the development of enzyme chemistry. Its virtues are simplicity and broad applicability. However, the kinetic properties of many enzymes cannot be accounted for by the Michaelis-Menten model. An important group consists of the *allosteric enzymes,* which often display sigmoidal plots (Figure 8-22) of the reaction velocity V versus substrate concentration [S], rather than the hyperbolic plots predicted by the Michaelis-Menten equation (equation 28). Recall that the oxygen-binding curve of myoglobin is hyperbolic, whereas that of hemoglobin is sigmoidal. The binding of enzymes to substrates is analogous. In allosteric enzymes, the binding of substrate to one active site can affect the properties of other active sites in the same enzyme molecule. A possible outcome of this interaction between subunits is that the binding of substrate becomes cooperative, which would give a sigmoidal plot of V versus [S]. A Hill plot (p. 159) for such an enzyme would have a slope greater than 1.0. In addition, the activity of an allosteric enzyme may be altered by regulatory molecules that are bound to sites other than the catalytic sites, just as oxygen binding to hemoglobin is affected by BPG, H^+, and CO_2.

Consider an allosteric enzyme consisting of two identical subunits, each containing an active site (Figure 8-23A). The T (tense) form has low affinity and the R (relaxed) form has high affinity for substrate. Two limiting models for the allosteric process were presented in the previous chapter (p. 166). In the simple sequential model, the binding of substrate to one of the subunits induces a T → R transition in that subunit but not in the other (Figure 8-23A). The affinity of the other subunit for substrate is increased because the subunit interface has been altered by the binding

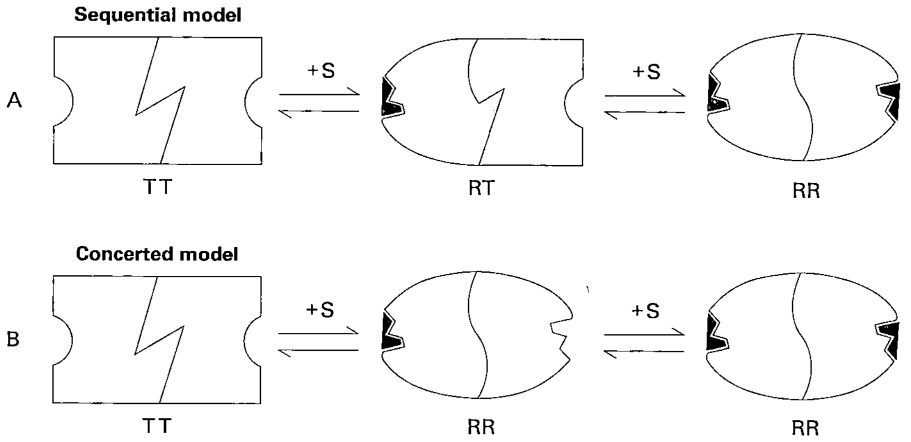

Figure 8-23
Comparison of allosteric models. (A) Simple sequential model. (B) Concerted model.

of the first substrate molecule. In the concerted model, the binding of substrate to one of the subunits increases the probability that both switch from the T to the R form (Figure 8-23B). Symmetry is conserved in the concerted model but not in the sequential model.

The effects of allosteric activators and inhibitors can be accounted for quite simply by the concerted model. An allosteric inhibitor binds preferentially to the T form, whereas an allosteric activator binds preferentially to the R form (Figure 8-24). Consequently, *an allosteric inhibitor shifts the R → T conformational equilibrium toward T, whereas an allosteric activator shifts it toward R.* The result is that an allosteric activator increases the binding of substrate to the enzyme, whereas an allosteric inhibitor decreases substrate binding.

It is noteworthy that most allosteric interactions alter f_{ES} (the fraction of enzyme molecules containing bound substrate, also termed Y) rather than V_{max}. For such enzymes, the dependence of f_{ES} on [S] is sigmoidal rather than hyperbolic (Figure 8-25). Allosteric activators shift this curve to the left (to higher saturation), whereas allosteric inhibitors shift it to the right (to lower saturation).

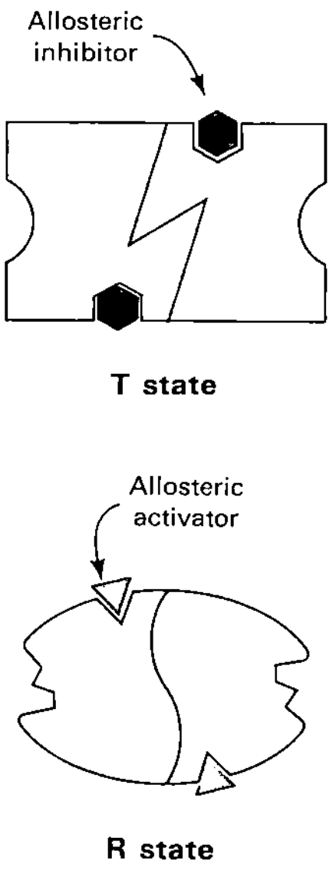

Figure 8-24
In the concerted model, an allosteric inhibitor (represented by a hexagon) stabilizes the T state, whereas an allosteric activator (represented by a triangle) stabilizes the R state.

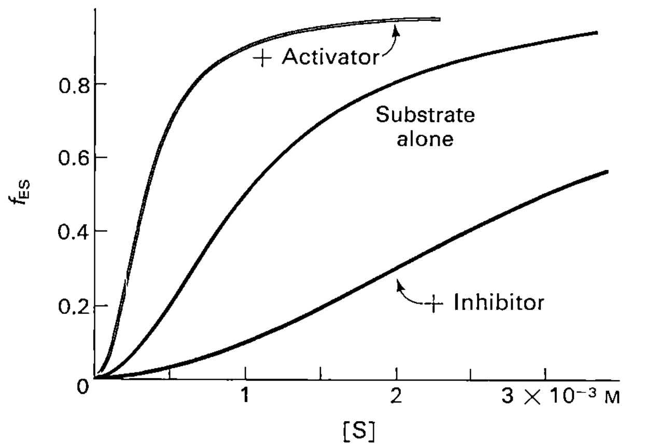

Figure 8-25
Dependence of f_{ES}, the fraction of catalytic sites containing bound substrate, on the substrate concentration for an allosteric enzyme. An allosteric activator shifts the curve to the left, whereas an allosteric inhibitor shifts it to the right by changing L_0 (the equilibrium constant that gives the ratio of T to R forms in the absence of substrate). The activator decreases L_0 from 10^4 to 10^3, whereas the inhibitor increases L_0 to 10^5.

TRANSITION STATE ANALOGS ARE POTENT INHIBITORS OF ENZYMES

We turn now to compounds that provide intimate views of the catalytic process itself. Linus Pauling proposed in 1946 that compounds resembling the transition state of a catalyzed reaction should be very effective inhibitors of enzymes. These mimics are called *transition state analogs.* The

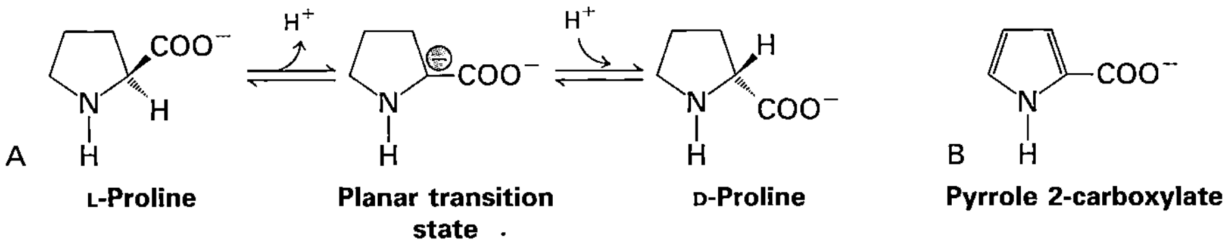

Figure 8-26
(A) The isomerization of L-proline to D-proline by proline racemase, a bacterial enzyme, proceeds through a planar transition state in which the α carbon is trigonal rather than tetrahedral. (B) Pyrrole 2-carboxylate is a transition state analog because it has a planar geometry.

inhibition of proline racemase is an instructive example. *The racemization of proline proceeds through a transition state in which the tetrahedral α carbon atom has become trigonal by loss of a proton* (Figure 8-26). In the trigonal form, all three bonds are in the same plane; C_α also carries a net negative charge. This symmetric carbanion can be reprotonated on one side to give the L isomer or on the other side to give the D isomer. This picture is supported by the finding that pyrrole 2-carboxylate binds to the racemase 160 times as tightly as does proline. *The α carbon atom of this inhibitor, like that of the transition state, is trigonal.* An analog that also carries a negative charge on C_α would be expected to bind even more tightly, but it has not been feasible to synthesize a stable compound of this kind. In general, highly potent and specific inhibitors of enzymes can be produced by synthesizing compounds that more closely resemble the transition state than the substrate itself. The inhibitory power of transition state analogs underscores the essence of catalysis: *selective binding of the transition state.*

CATALYTIC ANTIBODIES CAN BE FORMED BY USING TRANSITION STATE ANALOGS AS IMMUNOGENS

In 1969, William Jencks proposed that antibodies specific for the transition state of a chemical reaction should have catalytic power. This incisive prediction was realized in 1986 when the laboratories of Richard Lerner and Peter Schultz found that *catalytic antibodies could be produced by using transition state analogs as immunogens.* The preparation of an antibody that catalyzes the insertion of a metal ion into a porphyrin nicely illustrates this experimental approach. Ferrochelatase, the final enzyme in the biosynthetic pathway for the production of heme, catalyzes the insertion of Fe^{2+} into protoporphyrin IX. The nearly planar porphyrin must be bent for iron to enter (Figure 8-27A).

What kind of immunogen might elicit the production of an antibody that would catalyze this metallation? The clue came from studies showing that *N*-methylprotoporphyrin is a potent inhibitor of ferrochelatase. This compound resembles the transition state because N-*alkylation forces the porphyrin to be bent.* Moreover, it was known that *N*-alkylporphyrins chelate metal ions 10^4 times faster than their unalkylated counterparts. Bending increases the exposure of the pyrrole nitrogen lone pairs of electrons to solvent, which facilitates chelation.

Indeed, an effective catalyst was produced using an *N*-alkylporphyrin as the immunogen (Figure 8-27B). *The resulting antibody distorted a planar porphyrin to facilitate the entry of a metal ion.* Eighty porphyrin molecules were metallated per hour per antibody molecule, a rate only tenfold less than that of ferrochelatase. The uncatalyzed reaction was 2500 times as

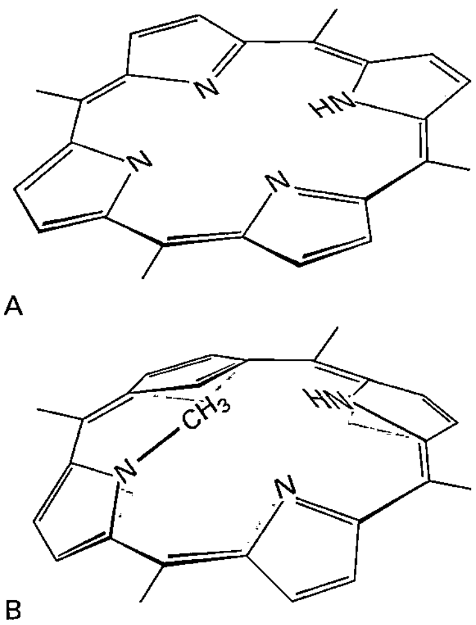

Figure 8-27
(A) The insertion of a metal ion into a porphyrin proceeds through a transition state in which the porphyrin is bent. (B) *N*-methylmesoporphyrin, a bent porphyrin, is an effective immunogen because it resembles the transition state of the reaction.

slow as the antibody-catalyzed reaction. Antibodies catalyzing many other kinds of chemical reactions—exemplified by ester and amide hydrolysis, amide bond formation, transesterification, photoinduced cleavage, photoinduced dimerization, decarboxylation, and oxidization—have been produced using similar strategies. *The power of transition state analogs is now evident: (1) They provide insight into catalytic mechanisms, (2) they can serve as potent and specific inhibitors of enzymes, and (3) they can be used as immunogens to generate a wide range of novel catalysts.*

> "If complementarity between the active site and the transition state contributes significantly to enzymatic catalysis, it should be possible to synthesize an enzyme by constructing such an active site. One way to do this is to prepare an antibody to a haptenic group which resembles the transition state of a given reaction. The combining sites of such antibodies should be complementary to the transition state and should cause an acceleration by forcing bound substrates to resemble the transition state."
>
> W.P. Jencks
> *Catalysis in Chemistry and Enzymology* (McGraw-Hill, 1969), p. 288.

PENICILLIN IRREVERSIBLY INACTIVATES A KEY ENZYME IN BACTERIAL CELL WALL SYNTHESIS

Penicillin was discovered by Alexander Fleming in 1928, when he observed by chance that bacterial growth was inhibited by a contaminating mold (*Penicillium*). Fleming was encouraged to find that an extract from the mold was not toxic when injected into animals. However, on trying to concentrate and purify the antibiotic, he found that "penicillin is easily destroyed, and to all intents and purposes we failed. We were bacteriologists—not chemists—and our relatively simple procedures were unavailing." Ten years later, Howard Florey, a pathologist, and Ernst Chain, a biochemist, carried out an incisive series of studies that led to the isolation, chemical characterization, and clinical use of this antibiotic.

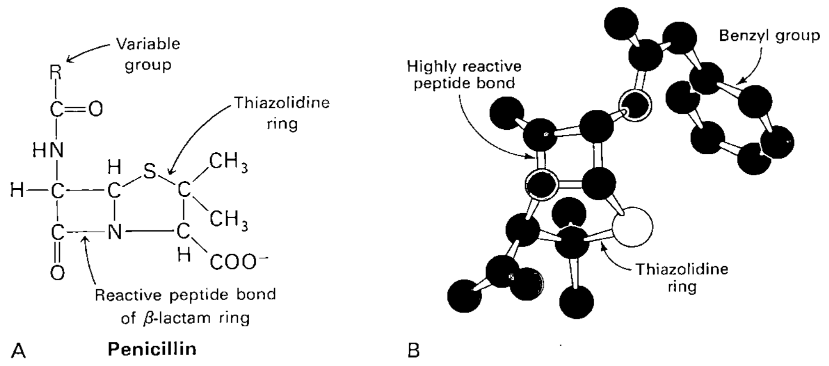

Figure 8-28
(A) Structural formula of penicillin and (B) model of benzyl penicillin. The reactive site of penicillin is the peptide bond of its β-lactam ring.

Penicillin consists of a thiazolidine ring fused to a *β-lactam* ring, to which a variable R group is attached by a peptide bond. In benzyl penicillin, for example, R is a benzyl group (Figure 8-28). This structure can undergo a variety of rearrangements, which accounts for the instability first encountered by Fleming. In particular, the β-lactam ring is very labile. Indeed, this property is closely tied to the antibiotic action of penicillin, as will be evident shortly.

How does penicillin inhibit bacterial growth? In 1957, Joshua Lederberg showed that bacteria ordinarily susceptible to penicillin could be grown in its presence if a hypertonic medium were used. The organisms obtained in this way, called *protoplasts,* are devoid of a cell wall and consequently lyse when transferred to a normal medium. Hence, it was inferred that penicillin interferes with the synthesis of the bacterial cell wall. The cell wall macromolecule, called a *peptidoglycan,* consists of linear polysaccharide chains that are cross-linked by short peptides (Figure 8-29). The

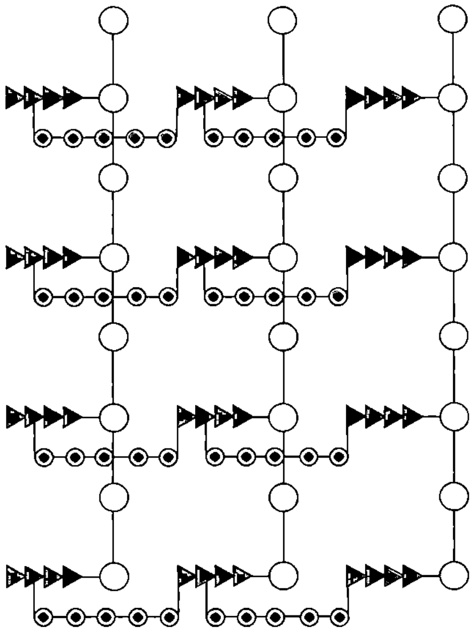

Figure 8-29
Schematic diagram of the peptidoglycan in *Staphylococcus aureus*. The sugars are shown in yellow, the tetrapeptides in red, and the pentaglycine bridges in blue. The cell wall is a single, enormous bag-shaped macromolecule because of extensive cross-linking.

enormous bag-shaped peptidoglycan confers mechanical support and prevents bacteria from bursting from their high internal osmotic pressure.

In 1965, James Park and Jack Strominger independently deduced that penicillin blocks the last step in cell wall synthesis, namely the cross-linking of different peptidoglycan strands. In the formation of the cell wall of *Staphylococcus aureus*, the amino group at one end of a pentaglycine chain attacks the peptide bond between two D-alanine residues in another peptide unit (Figure 8-30). A peptide bond is formed between glycine and

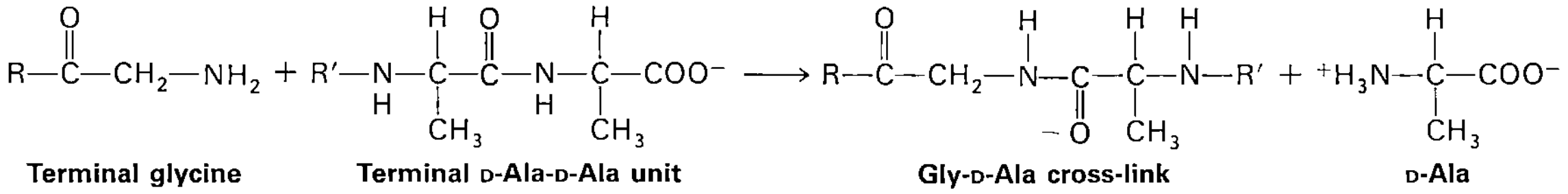

Terminal glycine residue of pentaglycine bridge · **Terminal D-Ala-D-Ala unit** · **Gly-D-Ala cross-link** · **D-Ala**

Figure 8-30
The amino group of the pentaglycine bridge in the *S. aureus* cell wall attacks the peptide bond between two D-Ala residues to form a cross-link.

one of the D-alanine residues, and the other D-alanine residue is released. This cross-linking reaction is catalyzed by *glycopeptide transpeptidase*. Bacterial cell walls are unique in containing D amino acids, which form cross-links by a mechanism different from that used to synthesize proteins.

Penicillin inhibits the cross-linking transpeptidase by the Trojan horse stratagem. The transpeptidase normally forms an *acyl intermediate* with the penultimate D-alanine residue of the D-Ala-D-Ala-peptide (Figure 8-31).

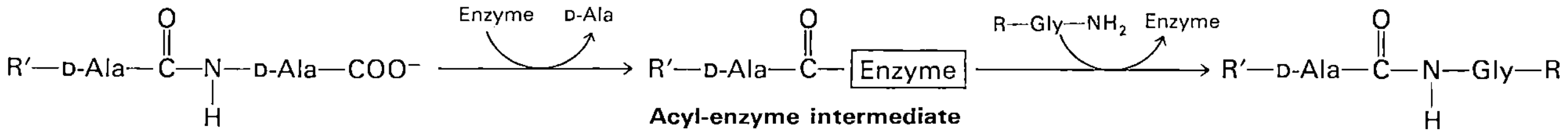

Figure 8-31
An acyl-enzyme intermediate is formed in the transpeptidation reaction.

This covalent acyl-enzyme intermediate then reacts with the amino group of the terminal glycine in another peptide to form the cross-link. Penicillin is welcomed into the active site of the transpeptidase because it mimics the D-Ala-D-Ala moiety of the normal substrate. Bound penicillin then forms a covalent bond with a serine residue at the active site of the enzyme (Figure 8-32). *This penicilloyl-enzyme does not react further. Hence, the transpeptidase is irreversibly inhibited.*

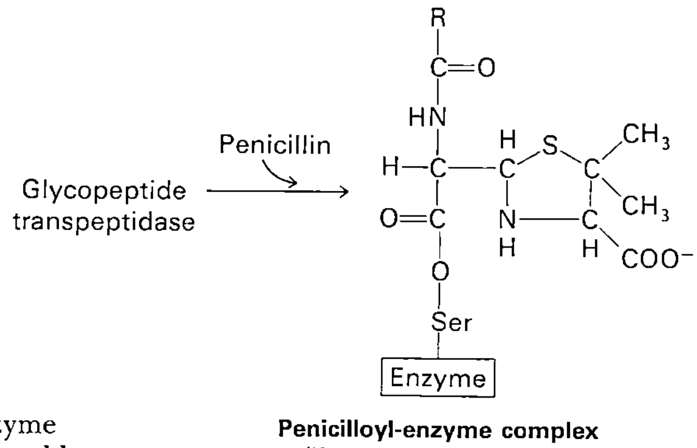

Penicilloyl-enzyme complex
(Enzymatically inactive)

Figure 8-32
Formation of a penicilloyl-enzyme complex, which is indefinitely stable.

Why is penicillin such an effective inhibitor of the transpeptidase? First, the four-membered β-lactam ring of penicillin is strained, which makes it highly reactive. Second, the conformation of this part of penicillin is probably very similar to that of the transition state of the normal substrate, a species that interacts strongly with the enzyme (Figure 8-33). In other words, *penicillin is a transition state analog, a striking example of molecular mimicry executed with perfection.*

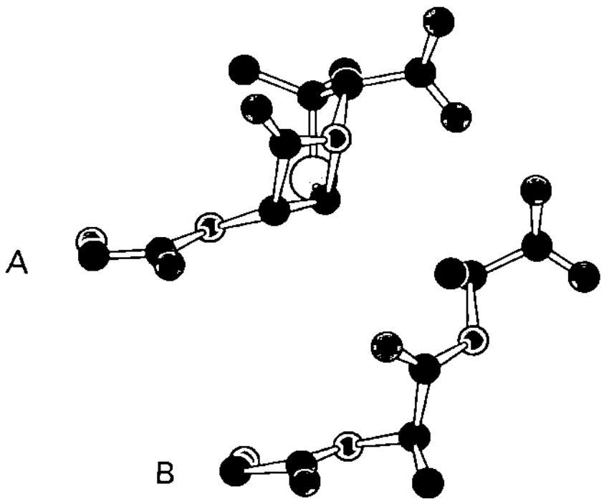

Figure 8-33
The conformation of penicillin in the vicinity of its reactive peptide bond (A) resembles the postulated conformation of the transition state of R–D-Ala–D-Ala (B) in the transpeptidation reaction. [After B. Lee. *J. Mol. Biol.* 61(1971):464.]

SUMMARY

Free energy is the most valuable thermodynamic function for determining whether a reaction can occur and for understanding the energetics of catalysis. Free energy is a measure of the capacity of a system to do useful work at constant temperature and pressure. A reaction can occur spontaneously only if the change in free energy (ΔG) is negative. The ΔG of a reaction is independent of path and depends only on the nature of the reactants and their activities (which can sometimes be approximated by their concentrations). The free-energy change of a reaction that occurs when reactants and products are at unit activity is called the standard free-energy change ($\Delta G°$). Biochemists usually use $\Delta G°'$, the standard free-energy change at pH 7.

The catalysts in biological systems are enzymes, and nearly all of them are proteins. Enzymes are highly specific and have great catalytic power. They enhance reaction rates by factors of at least 10^6. Enzymes do not alter reaction equilibria. Rather, they serve as catalysts by reducing the free energy of activation of chemical reactions. Enzymes accelerate reactions by providing a new reaction pathway in which the transition state (the highest-energy species) has a lower free energy and hence is more accessible than in the uncatalyzed reaction. The first step in catalysis is the formation of an enzyme-substrate complex. Substrates are bound to enzymes at active-site clefts from which water is largely excluded when the substrate is bound. The specificity of enzyme-substrate interactions arises mainly from hydrogen bonding, which is directional, and the shape of the active site, which rejects molecules that do not have a sufficiently complementary shape. The recognition of substrates by enzymes is a dynamic process accompanied by conformational changes at active sites.

The Michaelis-Menten model accounts for the kinetic properties of some enzymes. In this model, an enzyme (E) combines with a substrate (S) to form an enzyme-substrate (ES) complex, which can proceed to form a product (P) or to dissociate into E and S.

$$\mathrm{E} + \mathrm{S} \underset{k_2}{\overset{k_1}{\rightleftharpoons}} \mathrm{ES} \xrightarrow{k_3} \mathrm{E} + \mathrm{P}$$

The rate V of formation of product is given by the Michaelis-Menten equation

$$V = V_{\mathrm{max}} \frac{[\mathrm{S}]}{[\mathrm{S}] + K_{\mathrm{M}}}$$

in which V_{max} is the rate when the enzyme is fully saturated with substrate, and K_{M}, the Michaelis constant, is the substrate concentration at which the reaction rate is half maximal. The maximal rate, V_{max}, is equal to the product of k_3 and the total concentration of enzyme. The kinetic constant k_3, called the turnover number, is the number of substrate molecules converted into product per unit time at a single catalytic site when the enzyme is fully saturated with substrate. Turnover numbers for most enzymes are between 1 and 10^4 per second.

Enzymes can be inhibited by specific small molecules or ions. In irreversible inhibition, the inhibitor is covalently linked to the enzyme or bound so tightly that its dissociation from the enzyme is very slow. For example, penicillin irreversibly inactivates an essential enzyme in the formation of bacterial cell walls by mimicking the normal substrate in the cross-linking reaction. In contrast, reversible inhibition is characterized by a rapid equilibrium between enzyme and inhibitor. A competitive inhibitor prevents the substrate from binding to the active site. It reduces the reaction velocity by diminishing the proportion of enzyme molecules that are bound to substrate. In noncompetitive inhibition, the inhibitor decreases the turnover number. Competitive inhibition can be distinguished from noncompetitive inhibition by determining whether the inhibition can be overcome by raising the substrate concentration.

The catalytic activity of many enzymes is regulated in vivo. Allosteric interactions, which are defined as interactions between spatially distinct sites, are particularly important in this regard. The enzyme catalyzing the first step in a biosynthetic pathway is usually inhibited by the final product. Enzymes are also controlled by regulatory proteins such as calmodulin, which senses the intracellular Ca^{2+} level. Covalent modifications such as phosphorylation of serine, threonine, and tyrosine side chains are a third means of modulating enzymatic activity. The conversion of an inactive precursor protein into an active enzyme by peptide-bond cleavage, a process termed proteolytic activation, is another recurring device.

The essence of catalysis is selective stabilization of the transition state. Hence, enzymes bind the transition state more tightly than the substrate. Transition state analogs are stable compounds that mimic key features of this highest energy species. They are potent and specific inhibitors of enzymes. Pyrrole 2-carboxylate, for example, blocks proline racemase because it has a planar geometry about its α carbon atom, like that of the transition state in the racemization reaction. Transition state analogs are also valuable as immunogens in generating catalytic antibodies. A bent porphyrin has been used as the immunogen to produce an antibody that catalyzes the insertion of metal ions into porphyrins.

SELECTED READINGS

Where to start

Koshland, D.E., Jr., 1987. Evolution of catalytic function. *Cold Spring Harbor Symp. Quant. Biol.* 52:1–7. [An engaging account of the importance of protein flexibility and subunit interactions.]

Jencks, W.P., 1987. Economics of enzyme catalysis. *Cold Spring Harbor Symp. Quant. Biol.* 52:65–73. [An incisive statement of the thermodynamic principles governing enzyme action.]

Cech, T.R., 1986. RNA as an enzyme. *Sci. Amer.* 255(5):64–75.

Lerner, R.A., and Tramontano, A., 1988. Catalytic antibodies. *Sci. Amer.* 258(3):58–70.

Books on enzymes

Fersht, A., 1983. *Enzyme Structure and Mechanism* (2nd ed.). W.H. Freeman. [A concise and lucid introduction to enzyme action, with emphasis on physical principles.]

Walsh, C., 1979. *Enzymatic Reaction Mechanisms.* W.H. Freeman. [An excellent account of the chemical basis of enzyme action. This book demonstrates that the large number of enzyme-catalyzed reactions in biological systems can be grouped into a small number of types of chemical reactions.]

Page, M.I., and Williams, A. (eds.), 1987. *Enzyme Mechanisms.* Royal Society of Chemistry. [Contains many valuable articles on enzyme models, transition state analogs, and classes of enzyme-catalyzed reactions.]

Bender, M.L., Bergeron, R.J., and Komiyama, M., 1984. *The Bioorganic Chemistry of Enzymatic Catalysis.* Wiley-Interscience.

Dugas, H., and Penney, C., 1981. *Bioorganic Chemistry: A Chemical Approach to Enzyme Action.* Springer-Verlag.

Jencks, W.P., 1969. *Catalysis in Chemistry and Enzymology.* McGraw-Hill.

Abelson, J.N., and Simon, M.I. (eds.), 1992. *Methods in Enzymology*. Academic Press. [More than 200 volumes have been published in this continuing series. A treasure house of useful experimental methods.]

Boyer, P.D. (ed.), 1970. *The Enzymes* (3rd ed.). Academic Press. [This multivolume treatise on enzymes contains a wealth of information. Volumes 1 and 2 (available in a paperback edition) deal with general aspects of enzyme structure, mechanism, and regulation. Volume 3 and subsequent ones contain detailed and authoritative articles on individual enzymes.]

Books on thermodynamics

Edsall, J.T., and Gutfreund, H., 1983. *Biothermodynamics: The Study of Biochemical Processes at Equilibrium.* Wiley.

Klotz, I.M., 1967. *Energy Changes in Biochemical Reactions.* Academic Press. [A concise introduction, full of insight.]

Transition state stabilization

Pauling, L., 1948. Nature of forces between large molecules of biological interest. *Nature* 161:707–709. [A visionary statement of the importance of strain in enzymatic catalysis.]

Leinhard, G.E., 1973. Enzymatic catalysis and transition-state theory. *Science* 180:149–154.

Kraut, J., 1988. How do enzymes work? *Science* 242:533–540. [A critical analysis of the principle of transition state stabilization.]

Transition state analogs and other enzyme inhibitors

Waxman, D.J., and Strominger, J.L., 1983. Penicillin-binding proteins and the mechanism of action of β-lactam antibiotics. *Ann. Rev. Biochem.* 52:825–869.

Abraham, E.P., 1981. The beta-lactam antibiotics. *Sci. Amer.* 244:76–86.

Walsh, C.T., 1984. Suicide substrates, mechanism-based enzyme inactivators: Recent developments. *Ann. Rev. Biochem.* 53:493–535.

Catalytic antibodies

Lerner, R.A., Benkovic, S.J., and Schultz, P.G., 1991. At the crossroads of chemistry and immunology: Catalytic antibodies. *Science* 252:659–667. [A stimulating presentation of how catalytic antibodies can serve as sources of insight into biological catalysis.]

Cochran, A.G., and Schultz, P.G., 1990. Antibody-catalyzed porphyrin metallation. *Science* 249:781–783.

Schultz, P.G., 1988. The interplay of chemistry and biology in the design of enzymatic catalysts. *Science* 240:426–433. [Both catalytic antibodies and hybrid enzymes formed by site-specific insertion of synthetic catalytic groups are discussed in this informative review.]

Enzyme kinetics and mechanisms

Fersht, A.R., Leatherbarrow, R.J., and Wells, T.N.C., 1986. Binding energy and catalysis: A lesson from protein engineering of the tyrosyl-tRNA synthetase. *Trends Biochem. Sci.* 11:321–325.

Jencks, W.P., 1975. Binding energy, specificity, and enzymic catalysis: The Circe effect. *Advan. Enzymol.* 43:219–410.

Knowles, J.R., and Albery, W.J., 1976. Evolution of enzyme function and the development of catalytic efficiency. *Biochemistry* 15:5631–5640.

Molecular interactions and binding

Richards, F.M., Wyckoff, H.W., and Allewell, N., 1970. The origin of specificity in binding: A detailed example in a protein-nucleic acid interaction. *In* Schmitt, F.O. (ed.), *The Neurosciences: Second Study Program,* pp. 901–912. Rockefeller University Press.

Davidson, N., 1967. Weak interactions and the structure of biological macromolecules. *In* Quarton, G.C., Melnechuk, T., and Schmitt, F.O. (eds.), *The Neurosciences: A Study Program,* pp. 46–56. Rockefeller University Press.

Allosteric models

Monod, J., Changeux, J.-P., and Jacob, F., 1963. Allosteric proteins and cellular control systems. *J. Mol. Biol.* 6:306–329. [A classic paper that introduced the concept of allosteric interactions.]

Koshland, D.L., Jr., Nemethy, G., and Filmer, D., 1966. Comparison of experimental binding data and theoretical models in proteins containing subunits. *Biochemistry* 5:365–385. [The simple sequential model is presented in this influential article.]

Wyman, J., and Gill, S.J., 1990. *Binding and Linkage. Functional Chemistry of Biological Macromolecules.* University Books. [A rigorous and fundamental treatment of the thermodynamics of allosteric interactions.]

PROBLEMS

1. *Hydrolytic driving force.* The hydrolysis of pyrophosphate to orthophosphate is important in driving forward biosynthetic reactions such as the synthesis of DNA. This hydrolytic reaction is catalyzed in *Escherichia coli* by a pyrophosphatase that has a mass of 120 kd and consists of six identical subunits. For this enzyme, a unit of activity is defined as the amount of enzyme that hydrolyzes 10 μ moles of pyrophosphate in 15 minutes at 37°C under standard assay conditions. The purified enzyme has a V_{max} of 2800 units per milligram of enzyme.
   - (a) How many moles of substrate are hydrolyzed per second per milligram of enzyme when the substrate concentration is much greater than K_M?
   - (b) How many moles of active site are there in 1 mg of enzyme? Assume that each subunit has one active site.
   - (c) What is the turnover number of the enzyme? Compare this value with others mentioned in this chapter.
2. *Destroying the Trojan horse.* Penicillin is hydrolyzed and thereby rendered inactive by penicillinase (also known

as β-lactamase), an enzyme present in some resistant bacteria. The mass of this enzyme in *Staphylococcus aureus* is 29.6 kd. The amount of penicillin hydrolyzed in 1 minute in a 10-ml solution containing 10^{-9} g of purified penicillinase was measured as a function of the concentration of penicillin. Assume that the concentration of penicillin does not change appreciably during the assay.

[Penicillin]	*Amount hydrolyzed (nanomoles)*
1	0.11
3	0.25
5	0.34
10	0.45
30	0.58
50	0.61

(a) Plot $1/V$ versus $1/[S]$ for these data. Does penicillinase appear to obey Michaelis-Menten kinetics? If so, what is the value of K_M?
(b) What is the value of V_{max}?
(c) What is the turnover number of penicillinase under these experimental conditions? Assume one active site per enzyme molecule.

3. *Counterpoint.* Penicillinase (β-lactamase) hydrolyzes penicillin. Compare penicillinase with glycopeptide transpeptidase.

4. *Mode of inhibition.* The kinetics of an enzyme are measured as a function of substrate concentration in the presence and absence of 2 mM inhibitor (I).

[S] (μM)	*Velocity (μmoles/min)*	
	No inhibitor	*Inhibitor*
3	10.4	4.1
5	14.5	6.4
10	22.5	11.3
30	33.8	22.6
90	40.5	33.8

(a) What are the values of V_{max} and K_M in the absence of inhibitor? In its presence?
(b) What type of inhibition is this?
(c) What is the binding constant of this inhibitor?
(d) If $[S] = 10$ μM and $[I] = 2$ mM, what fraction of the enzyme molecules have a bound substrate? A bound inhibitor?
(e) If $[S] = 30$ μM, what fraction of the enzyme molecules have a bound substrate in the presence and absence of 2 mM inhibitor? Compare this ratio with the ratio of the reaction velocities under the same conditions.

5. *A different mode.* The kinetics of the enzyme discussed in problem 4 are measured in the presence of a different inhibitor. The concentration of this inhibitor is 100 μM.
(a) What are the values of V_{max} and K_M in the presence of this inhibitor? Compare them with those obtained in problem 4.
(b) What type of inhibition is this?
(c) What is the dissociation constant of this inhibitor?

[S] (μM)	*Velocity (μmoles/min)*	
	No inhibitor	*Inhibitor*
3	10.4	2.1
5	14.5	2.9
10	22.5	4.5
30	33.8	6.8
90	40.5	8.1

(d) If $[S] = 30$ μM, what fraction of the enzyme molecules have a bound substrate in the presence and absence of 100 μM inhibitor?

6. *A fresh view.* The plot of $1/V$ versus $1/[S]$ is sometimes called a Lineweaver-Burk plot. Another way of expressing the kinetic data is to plot V versus $V/[S]$, which is known as an Eadie-Hofstee plot.
(a) Rearrange the Michaelis-Menten equation to give V as a function of $V/[S]$.
(b) What is the significance of the slope, the vertical intercept, and the horizontal intercept in a plot of V versus $V/[S]$?
(c) Make a sketch of a plot of V versus $V/[S]$ in the absence of an inhibitor, in the presence of a competitive inhibitor, and in the presence of a noncompetitive inhibitor.

7. *Potential donors and acceptors.* The hormone progesterone contains two ketone groups. At pH 7, which side chains of the receptor might form hydrogen bonds with progesterone? (Assume that the side chains in the receptor protein have the same p*K*s as in the amino acids in aqueous solution.)

8. *Competing substrates.* Suppose that two substrates A and B compete for an enzyme. Derive an expression relating the ratio of the rates of utilization of A and B, V_A/V_B, to the concentrations of these substrates and their values of k_3 and K_M. (Hint: Express V_A as a function of k_3/K_M for substrate A, and do the same for V_B.) Is specificity determined by K_M alone?

9. *A tenacious mutant.* Suppose that a mutant enzyme binds a substrate 100-fold as tightly as does the native enzyme. What is the effect of this mutation on catalytic rate if the binding of the transition state is unaffected?

CHAPTER

9

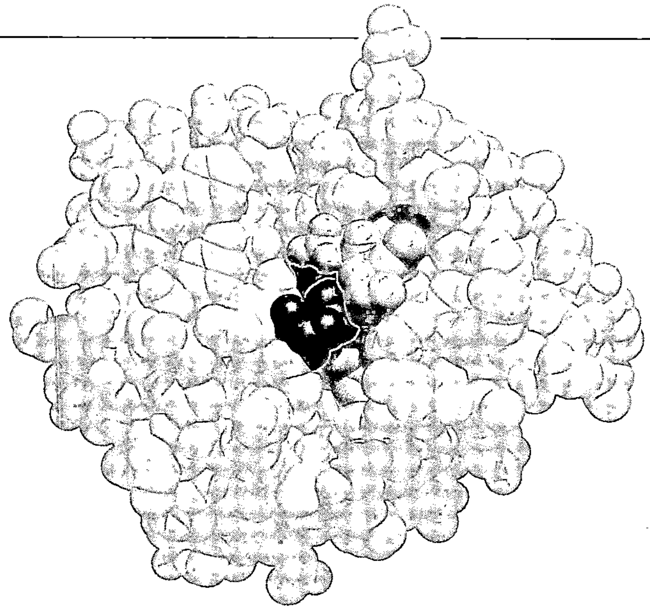

Catalytic Strategies

What are the sources of the catalytic power and selectivity of enzymes? This chapter presents the catalytic strategies used by several well-understood enzymes: lysozyme, ribonuclease A, carboxypeptidase, and chymotrypsin. The three-dimensional structures of these enzymes are known in atomic detail, as is their mode of binding of substrate analogs and inhibitors. The reactions catalyzed are relatively simple—namely, the hydrolysis of glycosidic, phosphodiester, and peptide bonds. Each of these enzymes digests a macromolecular substrate and exemplifies an important category of hydrolytic reactions. The actions of these enzymes illustrate many important principles of catalysis. In particular, we shall see how enzymes facilitate the formation of transition states. Potent and specific enzyme inhibitors are being designed on the basis of our emerging knowledge of enzyme structure and mechanism. A strategy for inhibiting the HIV-1 protease, which is essential for replication of the AIDS virus, is presented as an example. The chapter ends with a brief account of the discovery of catalytic RNA and its biological significance.

FLEMING'S DISCOVERY OF LYSOZYME

In 1922, Alexander Fleming, a bacteriologist in London, had a cold. He was not one to waste a moment and consequently used his cold as an opportunity to do an experiment. He allowed a few drops of his nasal mucus to fall on a culture plate containing bacteria. He was excited to find some time later that the bacteria near the mucus had been dissolved away. Fleming showed that the antibacterial substance was an enzyme, which he named *lysozyme*—*lyso* because of its capacity to lyse bacteria and *zyme* because it was an enzyme. He also discovered a small round bacterium that was particularly susceptible to lysozyme; he named it *Micrococcus lysodeikticus* because it was a displayer of lysis (*deiktikos* means "able to

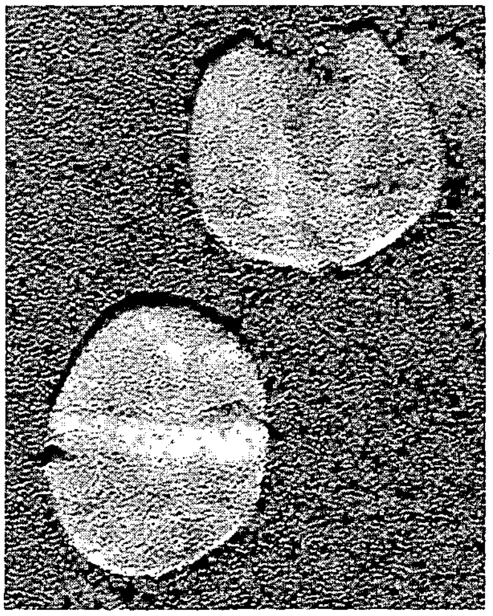

Figure 9-1
Electron micrograph of the isolated cell wall of *Micrococcus lysodeikticus*. [Courtesy of Dr. Nathan Sharon.]

Opening Image: *Model of chymotrypsin, a digestive enzyme. A substrate analog (yellow) is bound next to three key catalytic residues (red, blue, and green). [Drawn from 7gch.pdb. K. Brady, A. Wei, D. Ringe, and R.H. Abeles. Biochemistry 29(1990):7600.]*

show" in Greek). Fleming found that tears are a rich source of lysozyme. Volunteers provided tears after they suffered a few squirts of lemon—an "ordeal by lemon." The *St. Mary's Hospital Gazette* published a cartoon showing children coming for a few pennies to Fleming's laboratory, where one attendant administered beatings while another collected their tears! Fleming was disappointed to find that lysozyme was not effective against the most harmful bacteria. But seven years later, he did discover a highly effective antibiotic, penicillin—a striking illustration of Pasteur's comment that chance favors the prepared mind.

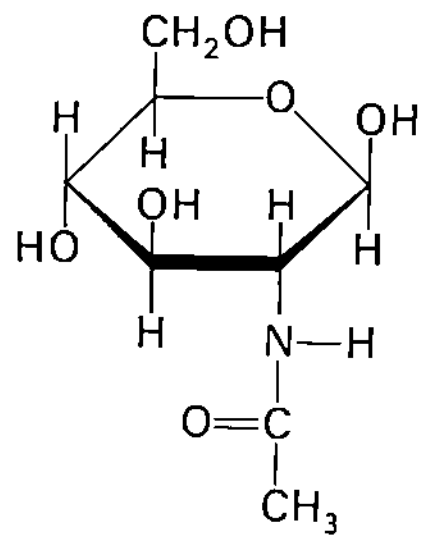

N-Acetylglucosamine (NAG)

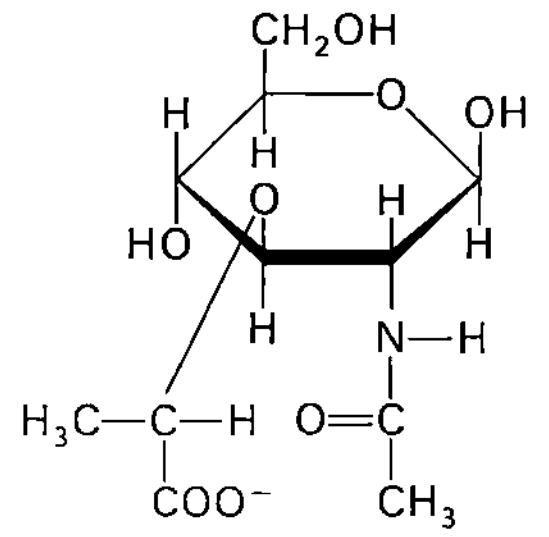

N-Acetylmuramate (NAM)

Figure 9-2
Sugar residues in the polysaccharide of bacterial cell walls.

LYSOZYME CLEAVES BACTERIAL CELL WALLS

Lysozyme dissolves certain bacteria by cleaving the polysaccharide component of their cell walls. The function of the cell wall in bacteria is to confer mechanical support. A bacterial cell devoid of a cell wall usually bursts because of the high internal osmotic pressure. We saw in the preceding chapter (p. 202) that penicillin blocks cell wall synthesis by inactivating a transpeptidase that catalyzes the formation of essential cross-links between peptide units. Let us now focus on lysozyme's target, the polysaccharide portion of cell walls.

The cell wall polysaccharide is made up of two kinds of sugars: N-*acetylglucosamine (NAG)* and N-*acetylmuramate (NAM)*. NAM and NAG are derivatives of glucosamine in which the amino group is acetylated (Figure 9-2). In NAM, a lactyl side chain is attached to C-3 of the sugar ring by an ether bond. In bacterial cell walls, NAM and NAG are joined by *glycosidic linkages* between C-1 of one sugar and C-4 of the other. The oxygen atom in a glycosidic bond can be located either above or below the plane of the sugar ring. In the α configuration, the oxygen is below the plane of the sugar; in the β configuration, it is above (see Chapter 18 for a more detailed discussion of the properties and nomenclature of sugars). All glycosidic bonds of the cell wall polysaccharide have a *β configuration* (Figure 9-3). NAM and NAG alternate in sequence. Thus, the cell wall

Figure 9-3
NAM is linked to NAG by a $\beta(1 \rightarrow 4)$ glycosidic bond.

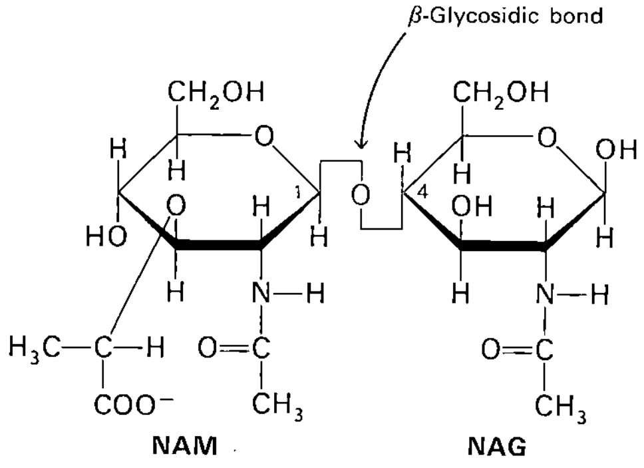

polysaccharide is an alternating polymer of NAM and NAG residues joined by $\beta(1 \rightarrow 4)$ glycosidic linkages. Note that polysaccharide chains, like polypeptide and polynucleotide chains, have directionality. Different polysaccharide chains are cross-linked by short peptides that are attached to NAM residues (see Figure 8-29, on p. 201).

Lysozyme, a glycosidase, hydrolyzes the glycosidic bond between C-1 of NAM and C-4 of NAG (Figure 9-4). The other glycosidic bond, between C-1 of NAG and C-4 of NAM, is not cleaved. *Chitin*, a polysaccharide found in the shell of crustaceans, is also a substrate for lysozyme. Chitin consists only of NAG residues joined by $\beta(1 \rightarrow 4)$ glycosidic links.

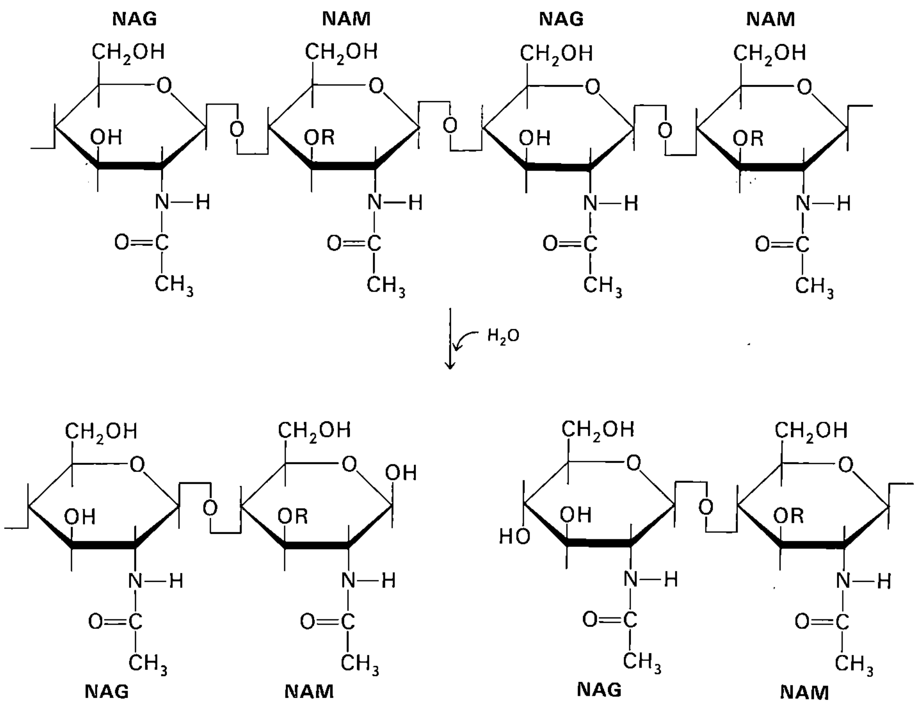

Figure 9-4
Lysozyme hydrolyzes the glycosidic bond (shown in green) between NAM and NAG (R refers to the lactyl group of NAM).

LYSOZYME IS A COMPACT PROTEIN WITH A COMPLEX FOLD

Lysozyme is a relatively small enzyme (14.6 kd). The one obtained from chicken egg white, a rich source, is a single polypeptide chain of 129 residues. This highly stable protein is cross-linked by four disulfide bridges. In 1965, David Phillips and his colleagues determined the three-dimensional structure of lysozyme. Their high-resolution electron-density map was the first for an enzyme molecule. Lysozyme is a compact molecule, roughly ellipsoidal in shape, with dimensions 45 × 30 × 30 Å. The folding of this protein is complex (Figure 9-5). There is much less α helix than in myoglobin and hemoglobin. In a number of regions, the polypeptide chain is in an extended β sheet conformation. The interior of lysozyme, like that of myoglobin and hemoglobin, is almost entirely nonpolar. Hydrophobic interactions play an important role in the folding of lysozyme, as they do for most proteins.

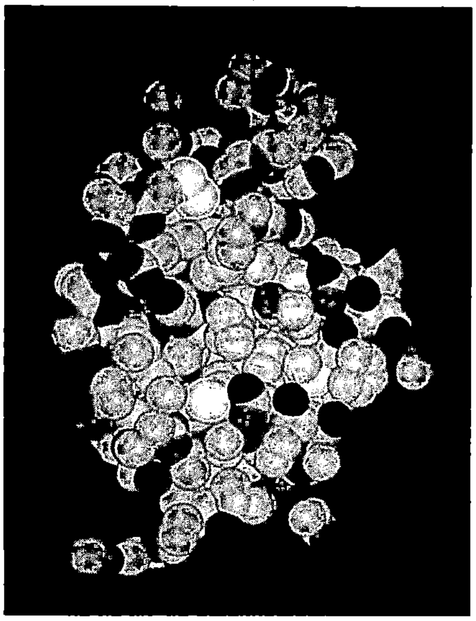

Figure 9-5
Space-filling model of lysozyme.

FINDING THE ACTIVE SITE IN LYSOZYME

Knowledge of the detailed three-dimensional structure of lysozyme did not immediately reveal its catalytic mechanism. Even the location of the active site was not obvious on looking at the electron-density map. Lysozyme does not contain a prosthetic group and thus lacks a built-in marker at its active site, in contrast with such proteins as myoglobin and hemoglobin. The essential information needed to identify the active site, specify the mode of substrate binding, and elucidate the enzymatic mechanism came from an x-ray crystallographic study of the interaction of lysozyme with inhibitors. When the three-dimensional structure of a protein is known, the mode of binding of small molecules can often be determined quite readily by x-ray crystallographic methods. These experiments are feasible because protein crystals are quite porous. Rather large inhibitor molecules can diffuse in the channels between different protein molecules and find their way to specific binding sites. The electron density corresponding to the additional molecule can be calculated directly from the intensities of the x-ray reflections (using the phases already determined for the native protein crystal) if the crystal structure is not markedly altered. This technique is called the *difference Fourier method.*

Ideally, one would like to use the difference Fourier method to elucidate the structure of an enzyme-substrate (ES) complex undergoing catalysis. However, under ordinary conditions, the conversion of bound substrate into product is much more rapid than the diffusion of new substrate into the crystal, so that ES complexes in the crystal are fleeting and rare. This difficulty can sometimes be circumvented by cooling the crystal (e.g., to −50°C) to slow the catalytic process. This experimental approach is called *cryoenzymology.* Another approach is to use light to convert a bound precursor into a substrate at the active site. The entire diffraction pattern of a crystal containing such an uncaged substrate can be collected in a few seconds using an intense x-ray beam. Alternatively, much information can be derived from a study of a complex of an enzyme with an unreactive (or very slowly reactive) analog of its substrate. For lysozyme, this was achieved with the trimer of *N*-acetylglucosamine (tri-NAG, or NAG_3). Oligomers of *N*-acetylglucosamine consisting of fewer than five residues are hydrolyzed very slowly or not at all. However, they do bind to the active site of the enzyme. Indeed, tri-NAG is a potent competitive inhibitor of lysozyme.

MODE OF BINDING OF TRI-*N*-ACETYLGLUCOSAMINE, A COMPETITIVE INHIBITOR

The x-ray study of the tri-NAG complex with lysozyme showed the location of the active site, revealed the interactions responsible for the specific binding of substrate, and led to the proposal of a detailed enzymatic mechanism. Tri-NAG binds to lysozyme in a cleft at the surface of the enzyme and occupies about half the cleft. Tri-NAG is bound to lysozyme by hydrogen bonds and van der Waals interactions. Electrostatic interactions cannot occur because tri-NAG lacks ionic groups.

The *hydrogen bonds* between tri-NAG and lysozyme are shown in Figure 9-6. The carboxylate group of aspartate 101 is hydrogen-bonded to sugar residues A and B. Four hydrogen bonds are made with residue C: the NH of the indole ring of tryptophan 62 is hydrogen-bonded to the oxygen attached to C-6 (the ring moves 0.75 Å when tri-NAG binds to the enzyme); the adjacent amino acid residue, tryptophan 63, is similarly hydro-

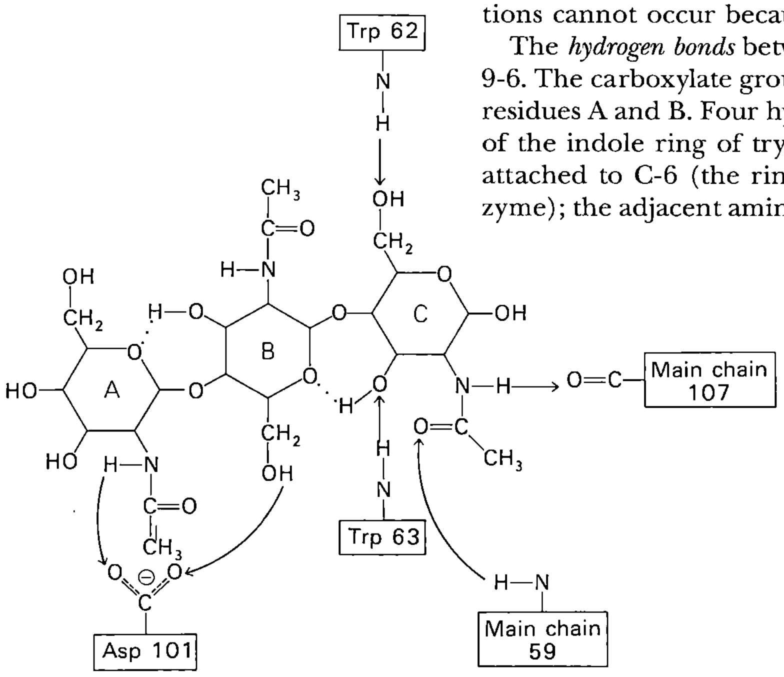

Figure 9-6
Hydrogen bonds between tri-NAG and lysozyme are marked by arrows in this highly schematic diagram. The hydrogen-binding groups of tri-NAG are shown in blue; those of lysozyme, in red.

gen-bonded to the oxygen attached to C-3; and two other hydrogen bonds are formed with peptide groups of the main chain. Tri-NAG also makes many *van der Waals contacts* with the enzyme. For example, sugar residue B fits closely to the indole ring of tryptophan 57.

FROM STRUCTURE TO THE ENZYMATIC MECHANISM OF LYSOZYME

1. *How would a substrate bind?* The mode of binding of a rapidly cleaved substrate cannot be established directly by x-ray crystallography. However, the structure of an enzyme complex with a competitive inhibitor provides clues to how a substrate binds. The finding that tri-NAG fills only half the cleft in lysozyme was a very suggestive starting point. It seemed likely that additional sugar residues, which would fill the rest of the cleft, were required for formation of a reactive ES complex. There was space for three additional sugar residues. This was encouraging, because it was known that the hexamer of *N*-acetylglucosamine (hexa-NAG) is rapidly hydrolyzed by the enzyme.

Three additional sugar residues, named D, E, and F, were fitted into the cleft by careful model building (Figure 9-7). Residues E and F went in nicely, making a number of good hydrogen bonds and van der Waals contacts. However, *residue D would not fit unless it was distorted.* Its C-6 and O-6 atoms came too close to several groups on the enzyme unless the ring was distorted from its normal conformation, which has the appearance of a chair (p. 214).

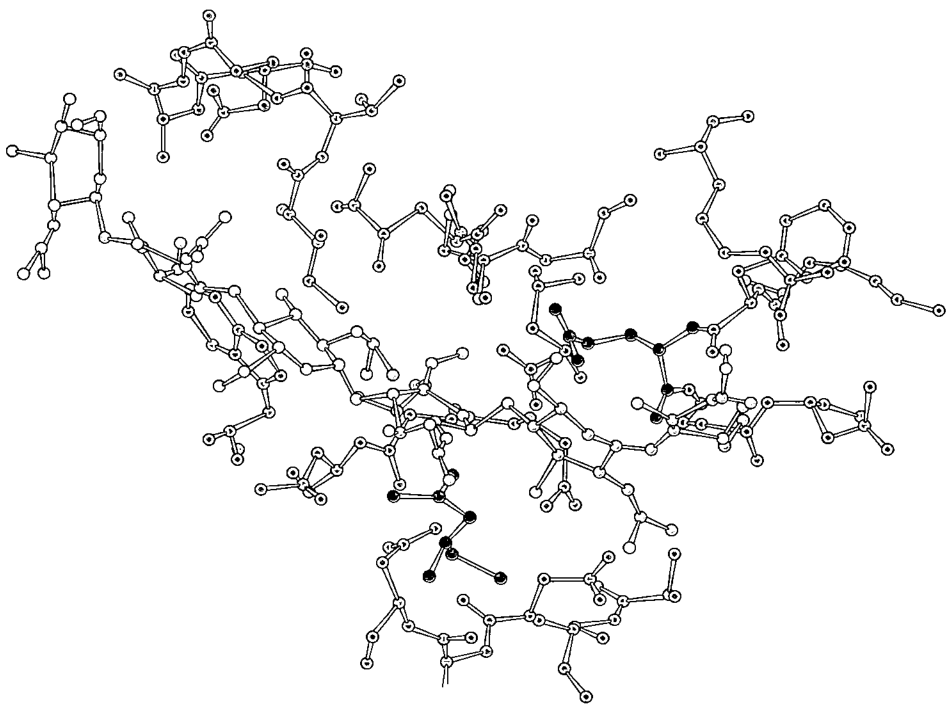

Figure 9-7
Mode of binding of hexa-NAG (shown in yellow) to lysozyme. The locations of sugar residues A, B, and C (left) are those observed in the tri-NAG–lysozyme complex, whereas those of residues D, E, and F (right) are inferred by model building. The two residues shown in red directly participate in catalysis.

2. *Which bond is cleaved?* The rate of hydrolysis of oligomers of *N*-acetylglucosamine increases strikingly as the number of residues is increased from four to five—that is, from NAG_4 to NAG_5 (Table 9-1). The cleavage rate increases further when a sixth residue is added (NAG_6), but it stays the same as the number of residues is increased to eight. This finding is

Table 9-1
Effectiveness of oligomers of *N*-acetylglucosamine as substrates

Substrate	*Relative rate of hydrolysis*
NAG_2	0
NAG_3	1
NAG_4	8
NAG_5	4,000
NAG_6	30,000
NAG_8	30,000

consistent with the crystallographic result showing that the active-site cleft would be fully occupied by six sugar residues.

Which bond of hexa-NAG is cleaved? Because tri-NAG is stable, the A–B bond (i.e., the glycosidic bond between the A and B sugar residues) cannot be the site of cleavage. Similarly, the B–C bond cannot be the site of cleavage. A second crucial piece of evidence is that site C cannot be occupied by NAM. NAG fits nicely into site C, but NAM is too large because of its lactyl side chain. The bond cleaved in bacterial cell walls is the NAM-NAG link. Hence, the C–D bond cannot be the site of cleavage if the bacterial cell wall polysaccharide binds to the enzyme in the same manner as hexa-NAG. The inability of NAM to fit into site C excludes yet another cleavage site: the E–F bond. The cell wall polysaccharide is an alternating polymer of NAM and NAG, and so NAM cannot occupy site E if it cannot occupy site C.

These observations eliminated the A–B, B–C, C–D, and E–F bonds as possible sites of enzymatic cleavage of the hexamer substrate. Hence, *the D–E bond was the only remaining candidate for the cleavage site* (Figure 9-8).

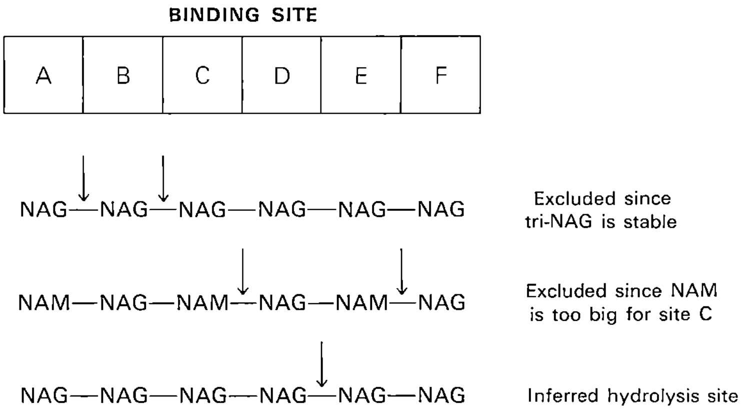

Figure 9-8
Steps in deducing that the glycosidic bond between sugar residues D and E is the one cleaved by lysozyme.

3. *Which groups on the enzyme participate in catalysis?* The inference that it is the D–E bond that is split was an important step toward defining the groups on the enzyme that carry out the hydrolysis reaction. However, it was necessary to localize the site of cleavage even more precisely: on which side of the glycosidic oxygen atom is the bond cleaved? This question was answered by carrying out enzymatic hydrolysis in the presence of water enriched with ^{18}O, the stable heavy isotope of oxygen (Figure 9-9). The sugars isolated contained ^{18}O attached to C-1 of the D sugar. In contrast, the hydroxyl group attached to C-4 of the E sugar contained the ordinary isotope of oxygen. Hence, *the bond split is the one between C-1 of residue D and the oxygen of the glycosidic linkage to residue E.* This experiment illustrates the value of isotopes in elucidating enzymatic mechanisms. In the absence of an isotopic marker, it would have been very difficult, if not impossible, to establish the precise site of cleavage.

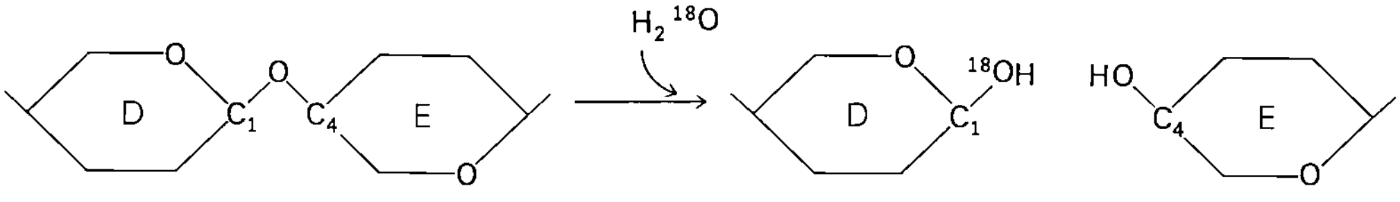

Figure 9-9
Hydrolysis in ^{18}O water showed that lysozyme cleaves the C_1–O bond rather than the O–C_4 bond. (Only the skeletons of the D and E residues are shown here.)

A search was then made for possible catalytic groups close to the glycosidic bond that is cleaved. A *catalytic group* is one that directly participates in making or breaking covalent bonds. The donation or abstraction of a hydrogen ion is a critical step in most enzymatic reactions. Hence, the most likely candidates are groups that can serve as *proton donors or acceptors*. The only plausible catalytic residues near the glycosidic bond that is cleaved by lysozyme are aspartic acid 52 and glutamic acid 35. The aspartic acid residue is on one side of the glycosidic linkage, and the glutamic acid residue is on the other. These two acidic side chains have markedly different environments. Aspartic acid 52 is in a distinctly polar environment, where it serves as a hydrogen bond acceptor in a complex network of hydrogen bonds. In contrast, glutamic acid 35 lies in a nonpolar region. Dissociation of a proton from a carboxyl group is less favored in a nonpolar than in a polar environment. Hence, at pH 5, the pH optimum for the hydrolysis of chitin by lysozyme, *aspartic acid 52 is probably in the ionized COO^- form (aspartate), whereas glutamic acid 35 is in the un-ionized COOH form.* The nearest oxygen atom of each of these acid groups is located about 3 Å away from the glycosidic linkage (Figure 9-10).

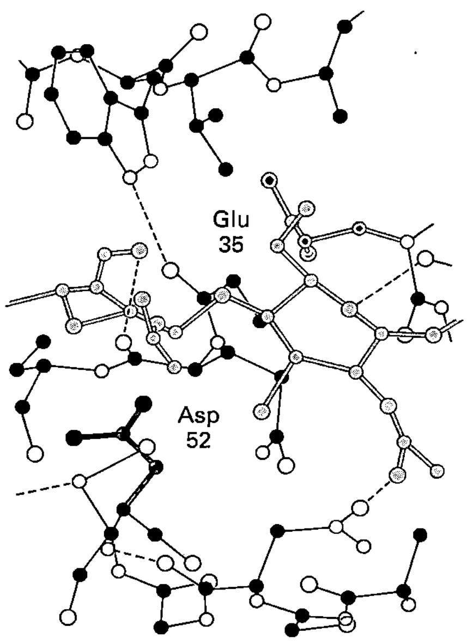

Figure 9-10
Structure of part of the active site of lysozyme. The D (left) and E (right) rings of the hexa-NAG substrate are shown in yellow. The side chains of aspartate 52 (red) and glutamic acid 35 (green) are in close proximity. [After W.N. Lipscomb. *Proc. Robert A. Welch Found. Conf. Chem. Res.* 15(1971):150.]

A CARBONIUM ION INTERMEDIATE IS CRITICAL FOR CATALYSIS

Phillips and his colleagues proposed a detailed catalytic mechanism for lysozyme based on the preceding structural data. The essential steps are

1. The –COOH group of glutamic acid 35 donates an H^+ to the glycosidic oxygen atom between rings D and E. The transfer of a proton cleaves the bond between C-1 of the D ring and the glycosidic oxygen atom (Figure 9-11).

2. This creates a positive charge on C-1 of the D ring. This transient species is called a *carbonium ion* because it contains a positively charged carbon atom. It is also known as an *oxocarbonium ion* because some of the positive charge is shared by the ring oxygen atom.

3. The dimer of NAG consisting of residues E and F diffuses away from the enzyme.

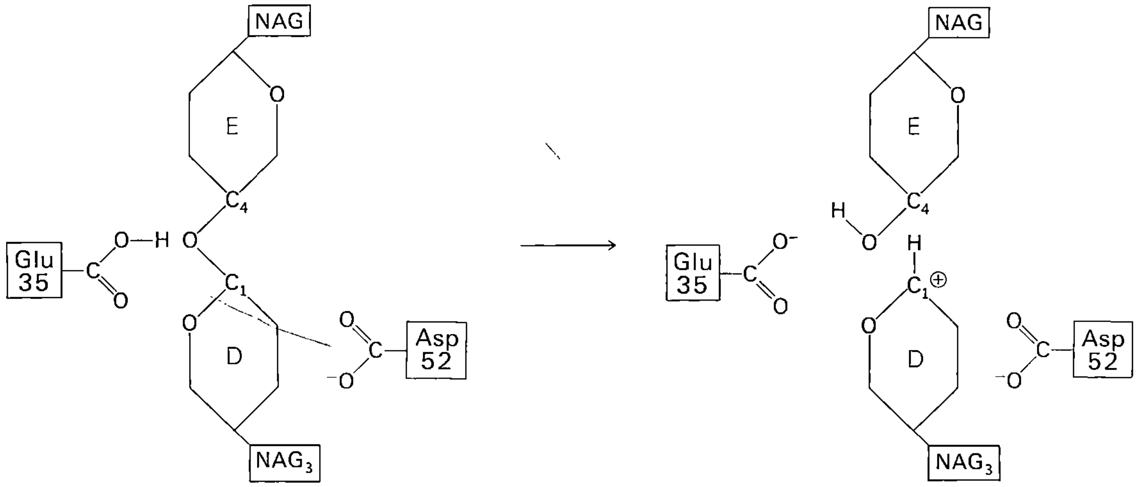

Figure 9-11
The first step in catalysis by lysozyme is the transfer of an H^+ from Glu 35 to the oxygen atom of the glycosidic bond. The glycosidic bond is thereby cleaved, and a carbonium ion intermediate is formed.

4. The carbonium ion intermediate then reacts with OH^- (or H_2O) from the solvent (Figure 9-12). Glutamic acid 35 becomes reprotonated, and tetra-NAG, consisting of residues A, B, C, and D, diffuses away from the enzyme. Lysozyme is then ready for another round of catalysis.

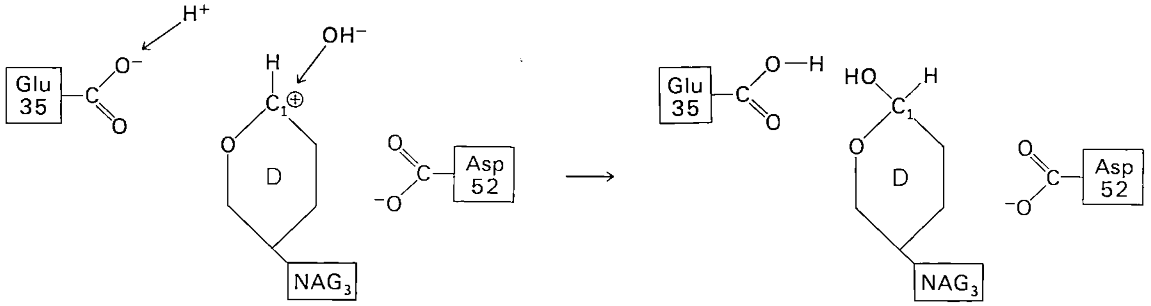

Figure 9-12
The cleavage reaction is completed by the addition of OH^- to the carbonium ion intermediate and H^+ to the side chain of Glu 35.

The critical elements of this proposed catalytic scheme are

1. *General acid catalysis.* A proton is transferred from glutamic acid 35, which is un-ionized and optimally located 3 Å away from the glycosidic oxygen atom. The term *general acid* indicates that the source of the donor is a donor group rather than free H^+.

2. *Promotion of the formation of the carbonium ion intermediate.* The enzymatic reaction is markedly facilitated by two different factors that stabilize the carbonium ion intermediate:

a. The *electrostatic* factor is the presence of a negatively charged group 3 Å away from the carbonium ion intermediate. Aspartate 52, which is in the negatively charged carboxylate form, electrostatically stabilizes the positive charge on C-1 of ring D.

b. The *geometrical* factor is the distortion of ring D (Figure 9-13). Hexa-NAG fits best into the active-site cleft if sugar residue D is distorted out of its customary chair conformation into a half-chair form. This distortion enhances catalysis because the half-chair geometry markedly promotes the formation of the carbonium ion. In the half-chair form, the planarity of carbon atoms 1, 2, and 5 and the ring oxygen atom enables the positive charge to be shared by resonance between C-1 and the ring oxygen atom.

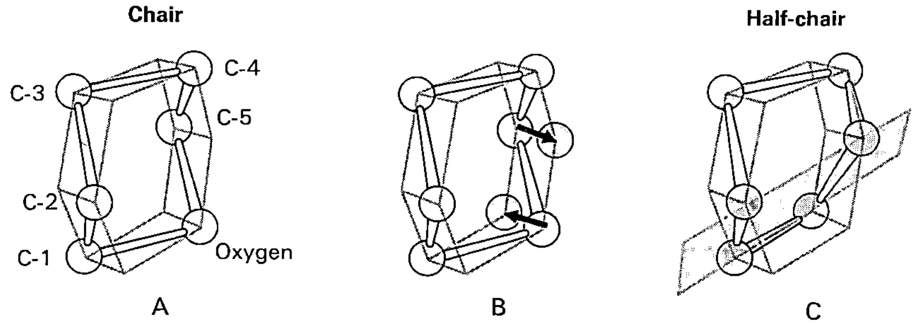

Figure 9-13
Distortion of the D ring of the substrate of lysozyme into a half-chair form: (A) a sugar residue in the normal chair form; (B) on binding to lysozyme, the ring oxygen atom and C-5 of sugar residue D move so that C-1, C-2, C-5, and O are in the same plane, as shown in part C. [After D.C. Phillips. The three-dimensional structure of an enzyme molecule. Copyright © 1966 by Scientific American, Inc. All rights reserved.]

THE PROPOSED MECHANISM IS SUPPORTED BY MANY LINES OF EXPERIMENTAL EVIDENCE

The mode of binding of substrate and the mechanism of catalysis proposed on the basis of the crystallographic studies have been tested in a variety of chemical experiments. All the experimental results thus far support the crystallographic hypothesis. A number of experimental findings are especially pertinent in this regard:

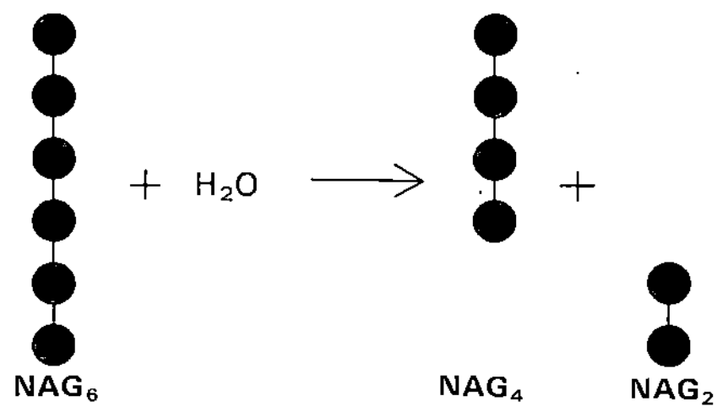

Figure 9-14
Hexa-NAG is hydrolyzed to tetra-NAG and di-NAG.

1. *Cleavage pattern.* Hexa-NAG is split into tetra-NAG and di-NAG, which confirms the crystallographic hypothesis that cleavage occurs between the fourth and fifth residues of a hexamer (Figure 9-14).

2. *Transition state analogs.* Evidence for the role of distortion of residue D comes from studies of a transition state analog of the substrate—that is, a compound that has the geometry of the catalyzed transition state *before* it is bound to the enzyme, as well as when it is bound. The D ring of the pure lactone analog of tetra-NAG (Figure 9-15) has a half-chair conformation. When bound to lysozyme, the C-1, C-2, C-4, C-5, and O atoms of the D ring of this analog are coplanar. This sofa conformation (Figure 9-16) is similar to the postulated half-chair form for the transition state. In fact, the binding of this lactone analog of tetra-NAG to subsites A through D of lysozyme is 3600 times as strong as the binding of tetra-NAG itself. This finding suggests that *the distortion of the D ring of a normal substrate could accelerate catalysis several thousandfold.*

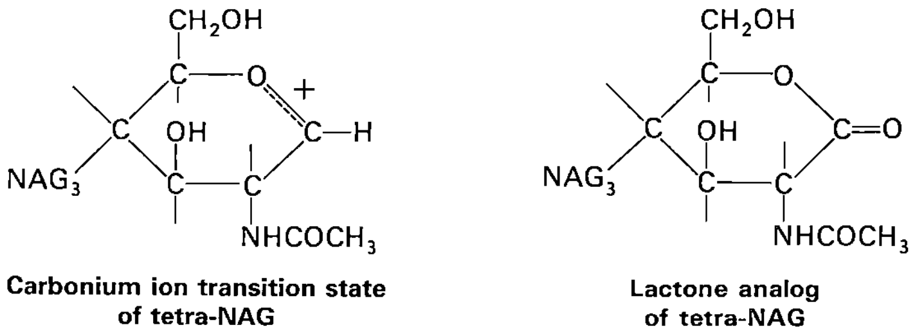

Figure 9-15
The lactone analog of tetra-NAG resembles the transition state in the reaction catalyzed by lysozyme because its D ring has a conformation like that of a half-chair form.

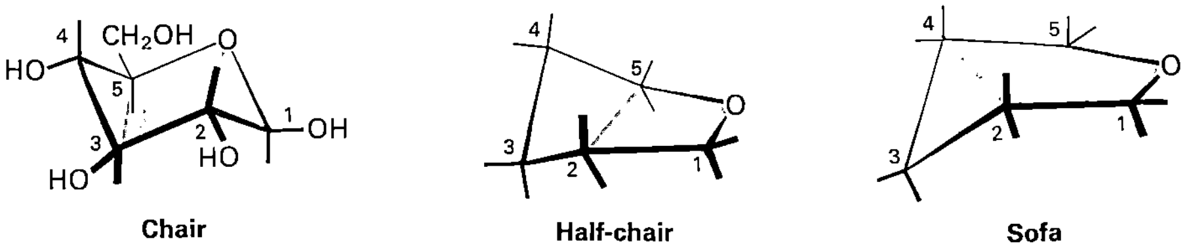

Figure 9-16
Comparison of the chair, half-chair, and sofa conformations of a sugar ring.

3. *Dependence of the catalytic rate on pH.* The rate of hydrolysis of chitin (poly-NAG) is most rapid at pH 5 (Figure 9-17). The enzymatic activity drops sharply on either side of this optimal pH. The decrease on the alkaline side is due to the ionization of glutamic acid 35, whereas the decrease in rate on the acid side reflects the protonation of aspartate 52. Lysozyme is active only when glutamic acid 35 is un-ionized and aspartate 52 is ionized.

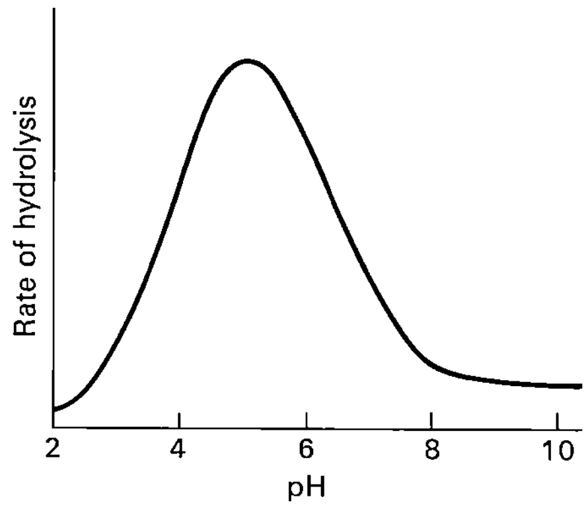

Figure 9-17
The rate of hydrolysis of chitin (poly-NAG) by lysozyme is maximal near pH 5.

4. *Selective chemical modification.* Esterification of aspartate 52 leads to a total loss of catalytic activity. This critical residue is protected from chemical modification by bound substrate.

5. *X-ray evidence.* The structure of NAM-NAG-NAM bound to lysozyme was recently solved at high resolution. This trisaccharide occupies the B-C-D sites rather than the A-B-C sites because NAM cannot fit into site C. The atomic coordinates clearly show that NAM in site D is distorted from the chair to the sofa conformation. Thus, *the bound sugar is forced into a conformation resembling that of the transition state.*

A CYCLIC PHOSPHATE INTERMEDIATE IS FORMED IN THE HYDROLYSIS OF RNA BY RIBONUCLEASE A

Let us now turn to the catalytic action of *ribonuclease A,* a digestive enzyme secreted by the pancreas. The folding of this enzyme (p. 35) and its binding of a uracil base (p. 191) have been mentioned. Ribonuclease A has been intensively studied for many years because it is small (124 residues, 13.7 kd), plentiful, and stable. The enzyme can be cleaved by subtilisin at a single peptide bond to yield *ribonuclease S,* a catalytically active complex consisting of an S-peptide moiety (residues 1–20) and an S-protein moiety (residues 21–124) bound together by multiple noncovalent interactions. Neither the S-peptide nor the S-protein alone is active. The catalytic properties of ribonuclease A and S are nearly the same, and so we shall refer to them as *ribonuclease.*

Ribonuclease catalyzes the hydrolysis of phosphodiester bonds in RNA chains. The bonds cleaved are ones between phosphorus and 5′-oxygen atoms (Figure 9-18). One of the new termini formed has a free 5′-OH group, and the other has a free 3′-phosphate group. Ribonuclease-catalyzed hydrolysis, like base-catalyzed hydrolysis in the absence of enzyme, proceeds through a *2′,3′-cyclic phosphate intermediate.* In the first stage, cleavage of the scissile bond forms a cyclic phosphate terminus and a free 5′-OH. In the second stage, water reacts with the cyclic phosphate to yield a free 3′-phosphate end. The formation of the cyclic phosphate intermediate is reversible, whereas its hydrolysis is virtually irreversible. This cyclic intermediate can readily be isolated because it is formed much more rapidly than it is hydrolyzed.

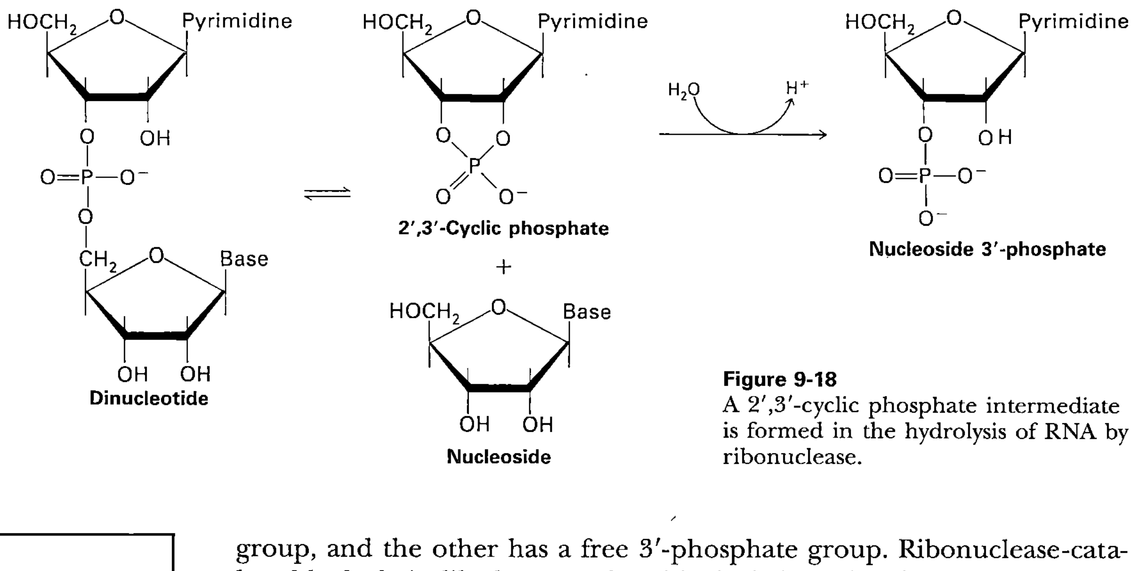

Figure 9-18
A 2′,3′-cyclic phosphate intermediate is formed in the hydrolysis of RNA by ribonuclease.

Scissile bond—
The bond in a substrate molecule that is cleaved by an enzyme.
From the Latin word *scissus,* meaning "torn asunder." The same root is seen in the word *scissors.*

The nucleotide on the 3′ side of the scissile bond must be a pyrimidine, because a purine ring is too large to be accommodated in the active site without distorting it. A uracil or cytosine ring binds to the active site by multiple, precisely directed hydrogen bonds (p. 191). In contrast, ribonuclease is indifferent to the nature of all other bases in its RNA substrate because binding is primarily electrostatic. A series of nine positively charged lysine and arginine side chains form salt bridges with the negatively charged phosphate backbone of RNA. Single-stranded DNA also binds to ribonuclease, but it is not hydrolyzed because DNA lacks a 2′-hydroxyl group and hence cannot form a 2′,3′-cyclic intermediate.

Much was deduced about the catalytic mechanism of ribonuclease from chemical studies before its three-dimensional structure was visualized. A plot of the catalytic rate versus pH is bell-shaped, like that of lysozyme (see Figure 9-17). Because the pH optimum is about 7, it was proposed that two histidine residues participate in catalysis, one in the basic form and the other in the acidic form. Chemical modification studies supported this notion. Reaction of ribonuclease with iodoacetate, an alkylating reagent, led to the carboxymethylation of the imidazole ring of histidine 119 or histidine 12 but not both in the same enzyme molecule (Figure 9-19). Modification of either histidine side chain inactivated the enzyme. Substrates or competitive inhibitors protected ribonuclease from modification and inactivation by iodoacetate. These findings suggested that *histidines 12 and 119 are near each other in the active site and act in concert as proton donors and acceptors in catalyzing the formation of the cyclic intermediate and its subsequent hydrolysis.*

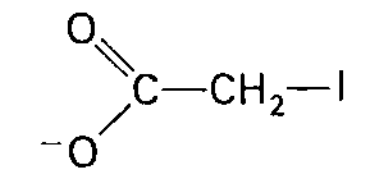

Iodoacetate

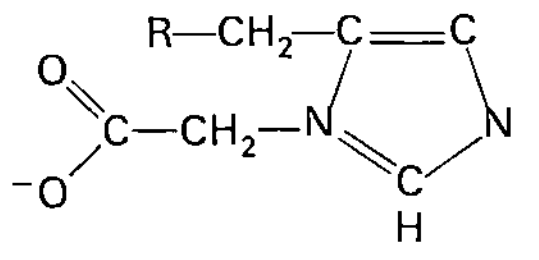

1-Carboxymethyl-His 119

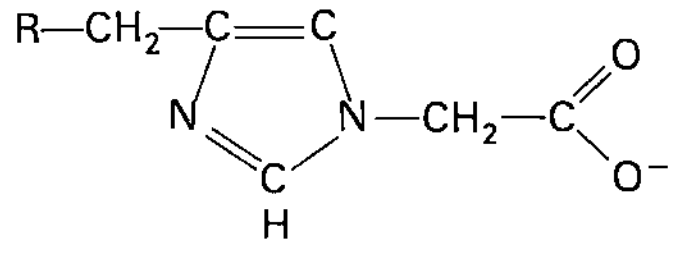

3-Carboxymethyl-His 12

Figure 9-19
Iodoacetate alkylates two histidine residues in ribonuclease. The products are 3-carboxymethyl-histidine 12 and 1-carboxymethyl-histidine 119.

PHOSPHORUS IS PENTACOVALENT IN THE TRANSITION STATE FOR RNA HYDROLYSIS

X-ray crystallographic studies carried out by Frederic Richards and Harold Wyckoff showed that both histidines are at the active site, just where expected. They solved the three-dimensional structure of a complex of ribonuclease with a substrate analog, namely, the phosphonate analog of UpC, in which a methylene group ($-CH_2-$) replaces the oxygen atom in the scissile bond. This analog, though sterically very similar to UpC, cannot be hydrolyzed by ribonuclease. The high-resolution structure of this enzyme-analog complex (Figure 9-20) has been very valuable in formulating a detailed catalytic mechanism.

Phosphonate—
A compound containing a C–P bond. Phosphonates can serve as nonhydrolyzable analogs of esters.

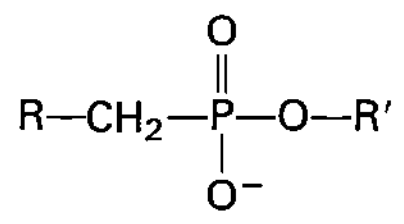

Phosphonate

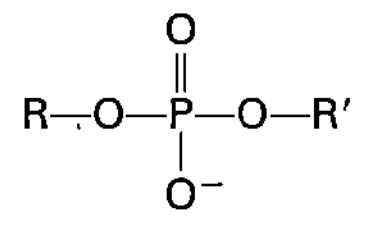

Phosphate ester

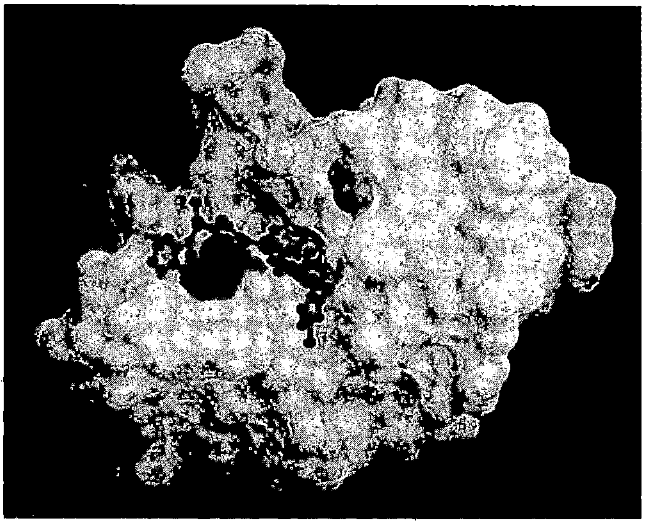

Figure 9-20
Three-dimensional structure of a complex of ribonuclease with a substrate analog. The phosphonate analog of UpC is shown in orange. The side chains shown in color are histidine 12, green; histidine 119, blue; and lysine 41, pink. [Courtesy of Dr. Frederic Richards, Dr. Harold Wyckoff, and Dr. Art Perlo.]

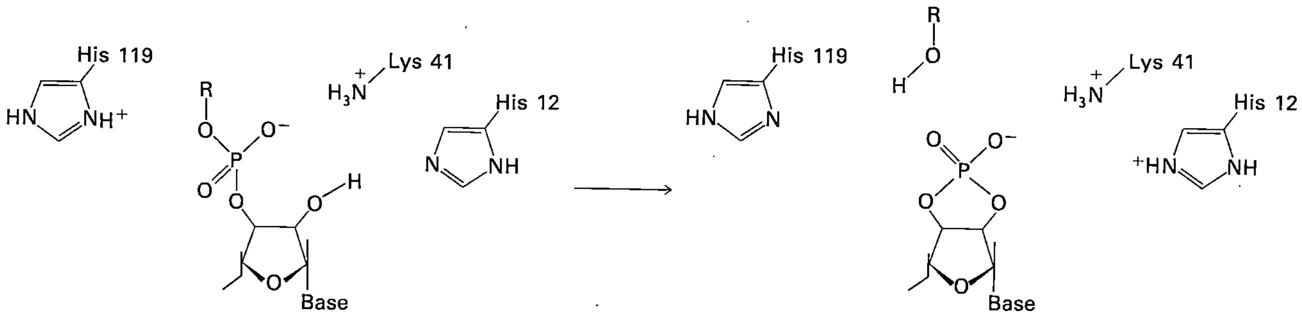

Figure 9-21
Proposed mechanism for the formation of the cyclic phosphate intermediate in catalysis by ribonuclease. A pentacovalent transition state is formed. Hydrolysis of the cyclic phosphate intermediate is essentially the reverse of this reaction, with water substituting for ROH.

Three residues play a critical role in catalysis: *histidine 12, histidine 119,* and *lysine 41* (Figure 9-21). The reaction begins with an attack by the 2′-O on the phosphorus atom of the scissile bond, in the following way. First, the un-ionized form of histidine 12 accepts a proton from the 2′-OH, which enhances the nucleophilicity of this oxygen atom. At the same time, the protonated form of histidine 119 begins to donate its proton to the 5′-O of the adjacent ribose, and the 2′-O begins to form a bond with P, which becomes transiently bonded to five oxygen atoms. This pentacovalent transition state is stabilized electrostatically by the nearby positively charged side chain of lysine 41. The bond between P and the 5′-O breaks when the proton from histidine 119 is completely transferred to this oxygen atom. At the same time, a bond between P and the 2′-O becomes fully formed, producing the 2′,3′-cyclic intermediate.

Hydrolysis of this cyclic intermediate, the second stage of the reaction, is nearly a reversal of the first stage. A difference is that water replaces the 5′-O component that was removed. Histidine 12 is now the proton donor and histidine 119 is the proton acceptor. The capacity of histidine to act as either an acid or base at physiologic pH accounts for its appearance in the active sites of many enzymes.

The geometry of the pentacovalent transition state is interesting. The tetrahedral geometry of phosphorus in RNA changes to that of a *trigonal bipyramid* when phosphorus becomes pentacovalent. Phosphorus occupies the center, three oxygen atoms lie in the equatorial plane, and two at the apices of the pyramid (Figure 9-22). In the formation of the cyclic intermediate, the outgoing 5′-O is at one apex, and the incoming 2′-O is at the other. In the hydrolysis of this intermediate, a water O is at one apex and the 2′-O is at the other. *In each stage, one apex is occupied by the attacking nucleophile and the other by the leaving group.* This geometry of attacking and leaving groups is termed *in-line.* We shall see later that many phosphoryl transfer reactions proceed by an in-line mechanism. For example, in the synthesis of DNA and RNA, the 3′-O is the attacking nucleophile and the pyrophosphate unit is the leaving group.

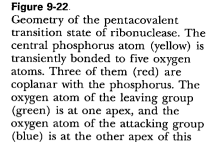

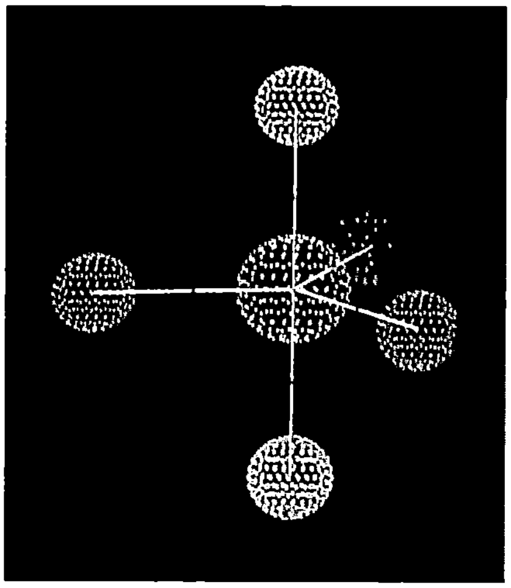

Figure 9-22
Geometry of the pentacovalent transition state of ribonuclease. The central phosphorus atom (yellow) is transiently bonded to five oxygen atoms. Three of them (red) are coplanar with the phosphorus. The oxygen atom of the leaving group (green) is at one apex, and the oxygen atom of the attacking group (blue) is at the other apex of this trigonal bipyramid.

CARBOXYPEPTIDASE A IS A ZINC-CONTAINING PROTEOLYTIC ENZYME

Let us now turn to carboxypeptidase A, a digestive enzyme that hydrolyzes the carboxyl-terminal peptide bond in polypeptide chains. Hydrolysis occurs most readily if the carboxyl-terminal residue has an aromatic or a bulky aliphatic side chain (Figure 9-23).

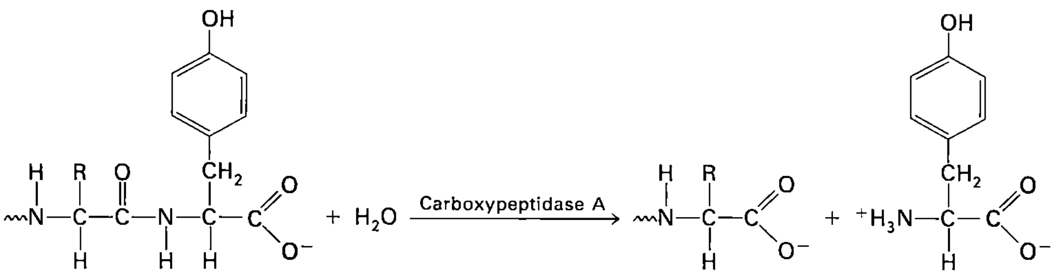

Figure 9-23
Reaction catalyzed by carboxypeptidase A.

The three-dimensional structure of carboxypeptidase A at a resolution of 2 Å was solved by William Lipscomb in 1967. This enzyme is a single polypeptide chain of 307 amino acid residues. Carboxypeptidase A has a compact shape that approximates an ellipsoid of dimensions 50 × 42 × 38 Å. The enzyme contains regions of α helix (38%) and of β pleated sheet (17%). A tightly bound zinc ion is essential for enzymatic activity. This zinc ion is located in a groove near the surface of the molecule, where it is coordinated to two histidine side chains, a glutamate side chain, and a water molecule (Figure 9-24). A large pocket near the zinc ion accommodates the side chain of the terminal residue of the peptide substrate.

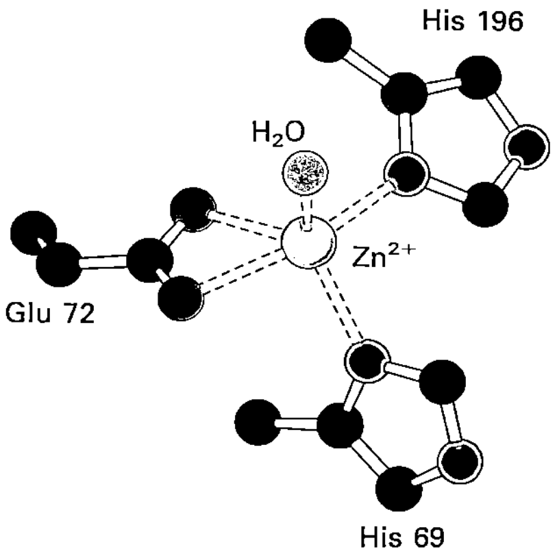

Figure 9-24
A zinc ion is coordinated to two histidine side chains, a glutamate side chain, and a water molecule at the active site of carboxypeptidase A. [Drawn from 5cpa.pdb. D.C. Rees, M. Lewis, and W.N. Lipscomb. *J. Mol. Biol.* 168(1983):367.]

Two facets of the catalytic mechanism of carboxypeptidase A are particularly noteworthy:

1. *Induced fit.* The binding of substrate is accompanied by many changes in the structure of the enzyme.

2. *Activation of water.* The zinc ion at the active site markedly increases the reactivity of the bound water molecule.

BINDING OF SUBSTRATE INDUCES LARGE STRUCTURAL CHANGES AT THE ACTIVE SITE OF CARBOXYPEPTIDASE A

The mode of binding of substrate and the catalytic mechanism of carboxypeptidase A have been deduced from x-ray and chemical studies carried out by numerous investigators. High-resolution x-ray analyses of complexes with substrate analogs, slowly hydrolyzed substrates, transition state analogs, and stable intermediates have contributed much to our understanding of this enzyme. The binding of glycyltyrosine, a slowly hydrolyzed substrate, is accompanied by a large structural rearrangement of the active site (Figure 9-25), as envisioned by Daniel Koshland, Jr., in his induced-fit model of enzyme action. Especially striking is the 12-Å movement of the phenolic hydroxyl group of tyrosine 248, a distance equal to about a quarter of the diameter of this protein. This motion is accomplished primarily by a facile rotation about a single carbon–carbon bond. The hydroxyl group of tyrosine 248 moves from the surface of the molecule to the vicinity of the terminal carboxylate of the substrate. An important consequence of this motion is that it closes the active-site cavity and extrudes water from it.

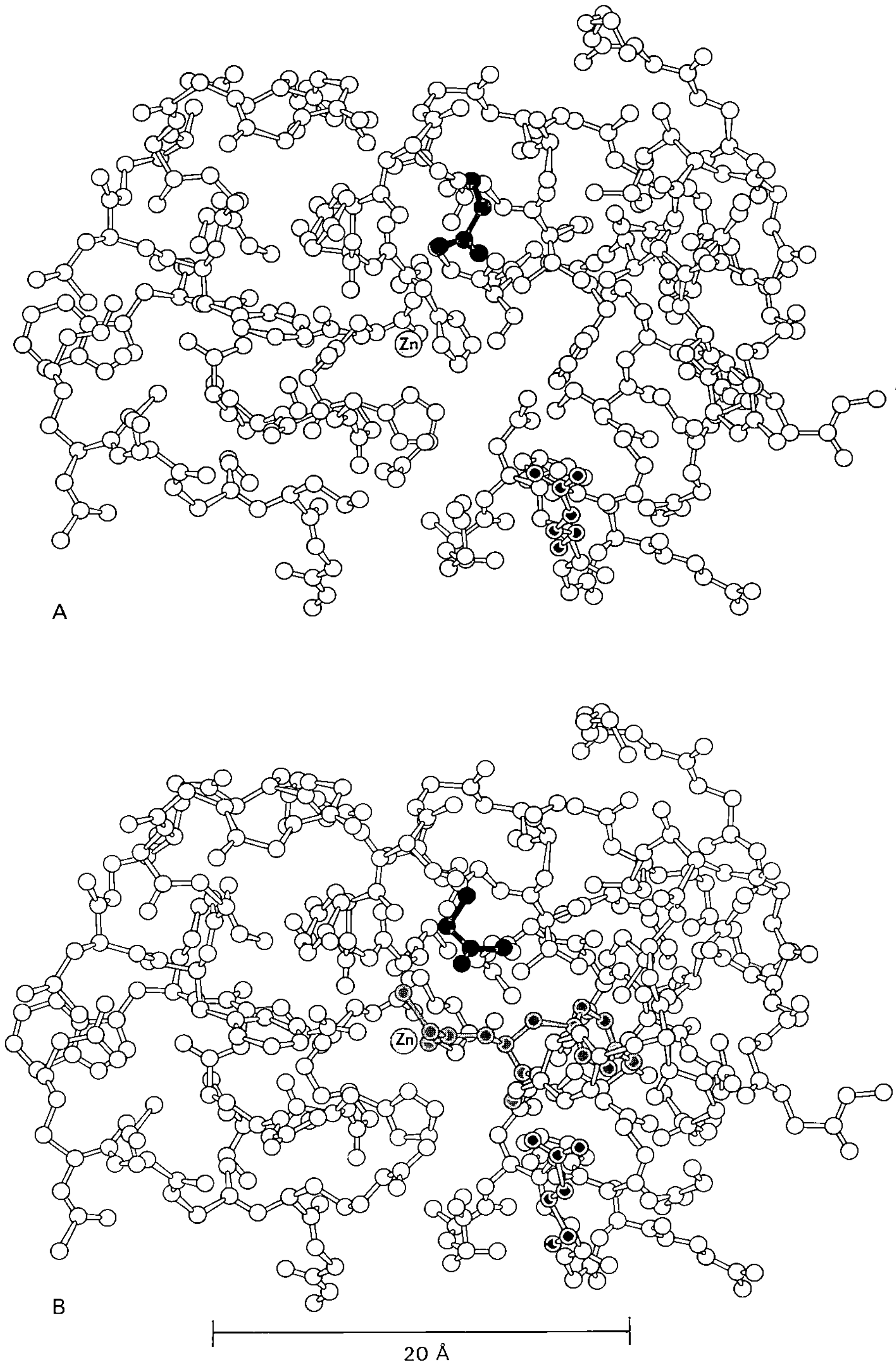

Figure 9-25
The structure of carboxypeptidase A changes upon binding substrate:
(A) enzyme alone (Arg 145 is shown in blue, Glu 270 in red, Tyr 248 in green, and Zn^{2+} in yellow);
(B) enzyme-substrate complex (glycyltyrosine, the substrate, is shown in gray). [After W.N. Lipscomb. *Proc. Robert A. Welch Found. Conf. Chem. Res.* 15(1971):140–141.]

How does carboxypeptidase A recognize the terminal carboxylate group of a peptide substrate? Three interactions are important (Figure 9-26). The induced-fit movement of tyrosine 248 enables its phenolic hydroxyl group to become hydrogen-bonded to this group. Furthermore, the guanidinium group of arginine 145 moves 2 Å in this transition so that it can form a strong salt bridge with the terminal COO^-, which is also hydrogen-bonded to a side-chain amide (Asn 144). In addition, the terminal side chain of the substrate sits in a hydrophobic pocket of the enzyme, which explains why carboxypeptidase A requires an aromatic or bulky nonpolar residue at this position. Two interactions involving the penultimate residue are also important. The carbonyl oxygen of the scissile peptide bond interacts with the guanidinium group of arginine 127. Finally, tyrosine 248 is also hydrogen-bonded to the peptide NH of the penultimate residue.

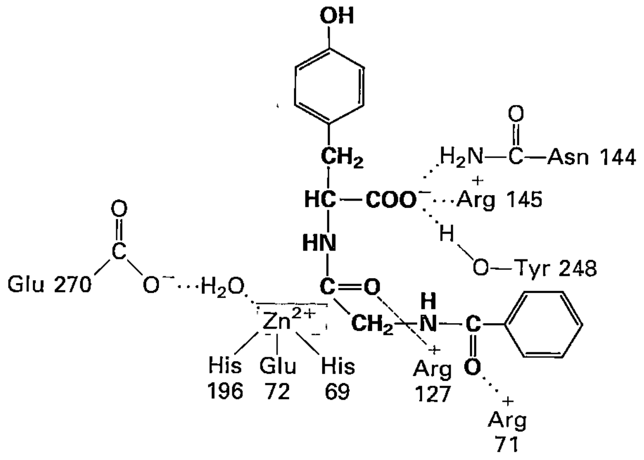

Figure 9-26
Mode of binding of *N*-benzoylglycyltyrosine to carboxypeptidase A. [After D.W. Christianson and W.N. Lipscomb. *Acc. Chem. Res.* 22(1989):62–69.]

THE ZINC ION AT THE ACTIVE SITE ACTIVATES A WATER MOLECULE

Proteins and peptides are stable at neutral pH in the absence of a protease because water does not readily attack peptide bonds. A key catalytic task of carboxypeptidase A is to activate water. This is accomplished by the bound zinc ion with assistance from the adjacent carboxylate of glutamate 270. In fact, *the zinc-bound H_2O behaves much like an OH^- ion.* Recall that the zinc ion at the active site of carbonic anhydrase plays a similar role (p. 39).

The first step in catalysis is the attack of the activated water molecule on the carbonyl group of the scissile peptide bond (Figure 9-27). Specifically, the nucleophilic oxygen atom of activated H_2O attacks the carbonyl carbon atom. Glutamate 270 simultaneously accepts a proton from the H_2O. *A negatively charged tetrahedral intermediate is formed.* This intermediate is stabilized by electrostatic interactions with Zn^{2+} and the positively charged side chain of arginine 127.

The next step is the transfer of a proton from the COOH group of glutamate 270 to the peptide NH group. The peptide bond is concomitantly cleaved, and the reaction products diffuse away from the active site. The need for substrate-induced structural changes in the active site of carboxypeptidase A can now be appreciated. The bound substrate is surrounded on all sides by catalytic groups of the enzyme. This arrangement promotes catalysis in three ways: (1) activation of H_2O by Zn^{2+}, (2) proton abstraction and donation by glutamate 270, and (3) electrostatic sta-

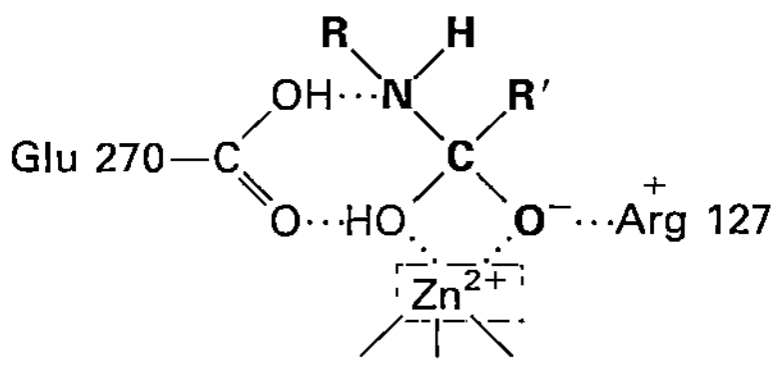

Figure 9-27
Proposed tetrahedral transition state in the peptide-bond hydrolysis by carboxypeptidase A.

bilization by arginine 127. It is evident that a substrate could not enter such an array of catalytic groups (nor could a product leave) unless the enzyme were flexible. *A flexible protein provides a much larger repertoire of potentially catalytic conformations than does a rigid one.*

CHYMOTRYPSIN IS A SERINE PROTEASE

Chymotrypsin, like carboxypeptidase A, is a mammalian digestive enzyme. However, its specificity and catalytic mechanism are quite different, which makes it rewarding to examine chymotrypsin, too, in some detail. The biological role of chymotrypsin is to catalyze the hydrolysis of proteins in the small intestine. It is selective for peptide bonds on the *carboxyl side* of the *aromatic side chains* of tyrosine, tryptophan, and phenylalanine, and of large *hydrophobic residues such as methionine.* Chymotrypsin has further significance as a member of an important family of proteins, the *serine proteases.* Trypsin, another digestive enzyme, and thrombin, a clotting enzyme, are other members of this pervasive clan. Part of the substrate becomes covalently attached to the enzyme during catalysis by this family of proteases.

Chymotrypsin, a 25-kd enzyme, consists of three polypeptide chains connected by two interchain disulfide bonds. As will be discussed in the next chapter, chymotrypsin is synthesized as a single-chain inactive precursor called *chymotrypsinogen.* The three-dimensional structure of the enzyme was solved at 2-Å resolution (Figure 9-28) by David Blow. The molecule is a compact ellipsoid of dimensions 51 × 40 × 40 Å. Chymotrypsin contains several antiparallel β pleated sheet regions and little α helix. All charged groups are on the surface of the molecule except for three that play a critical role in catalysis.

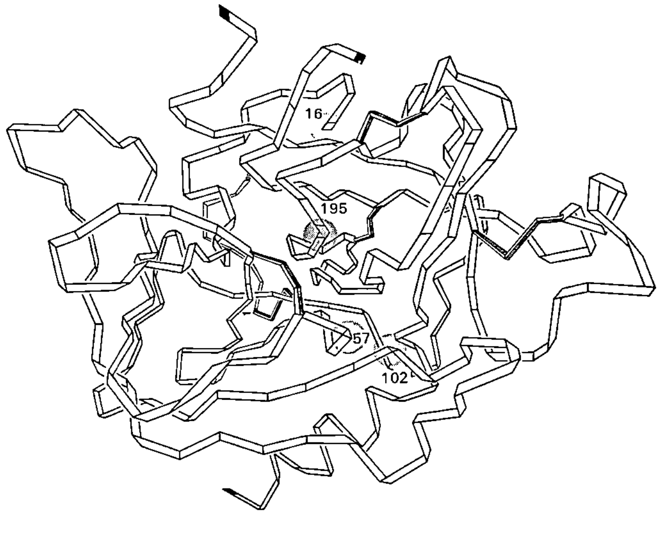

Figure 9-28
Three-dimensional structure of α-chymotrypsin. Only the α carbon atoms are shown. Catalytically important residues are marked in color. [Courtesy of Dr. David Blow.]

PART OF THE SUBSTRATE IS COVALENTLY BOUND TO CHYMOTRYPSIN DURING CATALYSIS

Chymotrypsin, like many proteases, hydrolyzes *ester bonds* in addition to peptide bonds. Although unimportant physiologically, ester-bond hydrolysis is of interest because of its close relationship to peptide-bond hydrolysis. Indeed, much of our knowledge of the catalytic mechanism of chymotrypsin comes from studies of the hydrolysis of simple esters. Chymotrypsin catalyzes the hydrolysis of peptide or ester bonds in two distinct stages. This was first revealed by studies of the kinetics of hydrolysis of *p*-nitrophenyl acetate. When large amounts of enzyme are used, there is an initial *rapid burst* of *p*-nitrophenol product, followed by its formation at a much *slower steady-state rate* (Figure 9-29).

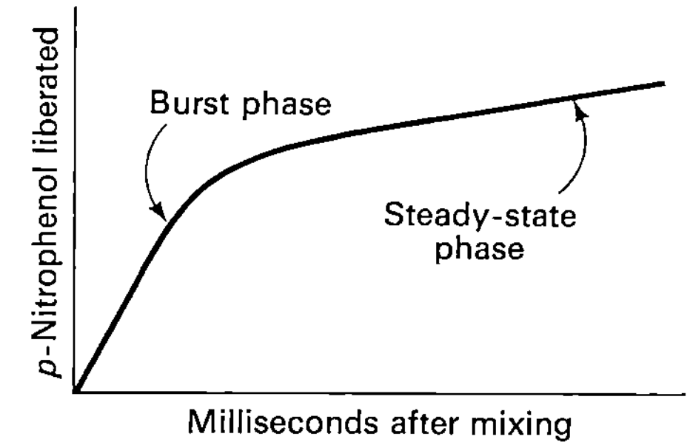

Figure 9-29
Two phases in the formation of *p*-nitrophenol are evident following the mixing of chymotrypsin and *p*-nitrophenyl acetate.

The first step is the combination of *p*-nitrophenyl acetate with chymotrypsin to form an enzyme-substrate (ES) complex. The ester bond of the substrate is then cleaved. One of the products, *p*-nitrophenol, is released from the enzyme, whereas the acetyl group of the substrate becomes covalently attached to the enzyme (Figure 9-30).

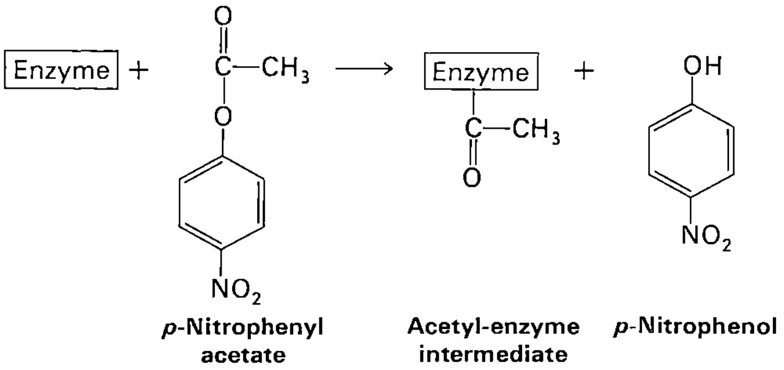

Figure 9-30
Acylation: formation of the acetyl-enzyme intermediate in catalysis by chymotrypsin.

Water then attacks the acetyl-enzyme complex to yield acetate ion and regenerate the enzyme (Figure 9-31).

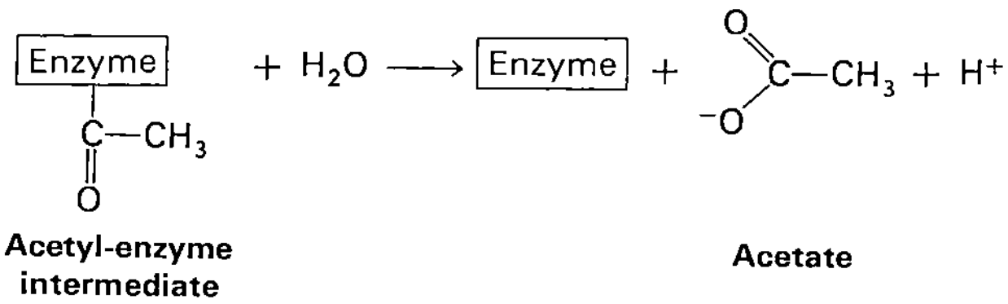

Figure 9-31
Deacylation: hydrolysis of the acetyl-enzyme intermediate.

The initial rapid burst of *p*-nitrophenol production corresponds to the formation of the acetyl-enzyme complex. This step is called *acylation*. The slower steady-state production of *p*-nitrophenol corresponds to the hydrolysis of the acetyl-enzyme complex to regenerate the free enzyme. This second step, called *deacylation*, is much slower than the first, so that it determines the overall rate of hydrolysis of esters by chymotrypsin. In fact, the acetyl-enzyme complex is sufficiently stable to be isolated under appropriate conditions. The catalytic mechanism of chymotrypsin can thus be represented by the adjacent scheme, in which P_1 is the amine (or alcohol) component of the substrate, $E{-}P_2$ is the covalent intermediate, and P_2 is the acid component of the substrate.

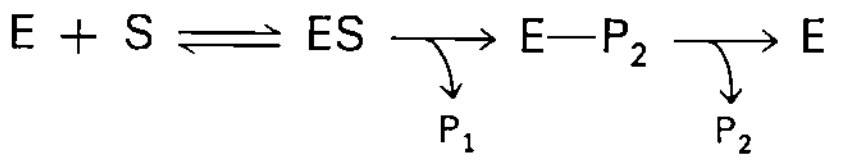

A distinctive feature of this mechanism is the appearance of a covalent intermediate. In the particular reaction discussed above, an acetyl group is covalently bonded to the enzyme. In general, the group attached to chymotrypsin at the $E{-}P_2$ stage is an acyl group. Thus, $E{-}P_2$ is an *acyl-enzyme intermediate.*

THE ACYL GROUP IS ATTACHED TO AN UNUSUALLY REACTIVE SERINE RESIDUE ON THE ENZYME

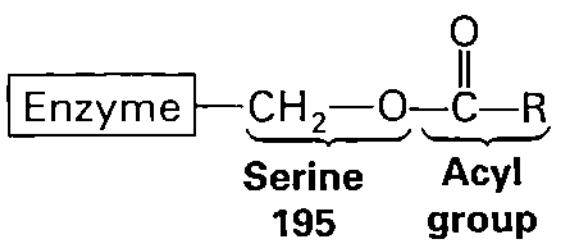

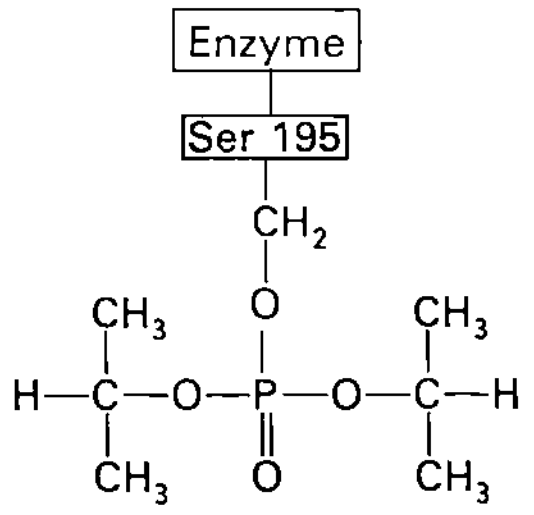

Diisopropylphosphoryl derivative

The site of attachment of the acyl group was identified following the isolation of E–P_2, which is quite stable at pH 3. The acyl group is linked to the oxygen atom of a specific serine residue, namely, serine 195. This serine residue is unusually reactive. It can be specifically labeled with *organic fluorophosphates,* such as diisopropylphosphofluoridate (DIPF). DIPF reacts only with serine 195 to form an inactive *diisopropylphosphoryl-enzyme complex,* which is indefinitely stable. The remarkable reactivity of serine 195 is highlighted by the fact that the other 27 serine residues in chymotrypsin are untouched by DIPF. Proteolytic enzymes containing a highly reactive serine, as evidenced by their susceptibility to DIPF, are known as *serine proteases.*

THE CATALYTIC ROLE OF HISTIDINE 57 WAS DEMONSTRATED BY AFFINITY LABELING

The importance of a second residue in catalysis was shown by *affinity labeling.* The strategy was to react chymotrypsin with a molecule that (1) specifically binds to the active site because it resembles a substrate and then (2) forms a stable covalent bond with a group on the enzyme that is in close proximity. These criteria are met by tosyl-L-phenylalanine chloromethyl ketone (TPCK), whose structure is shown in Figure 9-32. The phenylalanine side chain of TPCK enables it to bind specifically to chymotrypsin. The reactive group in TPCK is the chloromethyl ketone, which alkylates one of the ring nitrogens of histidine 57. TPCK is positioned to react with this residue because of its specific binding to the active site of the enzyme. The TPCK derivative of chymotrypsin is enzymatically inactive.

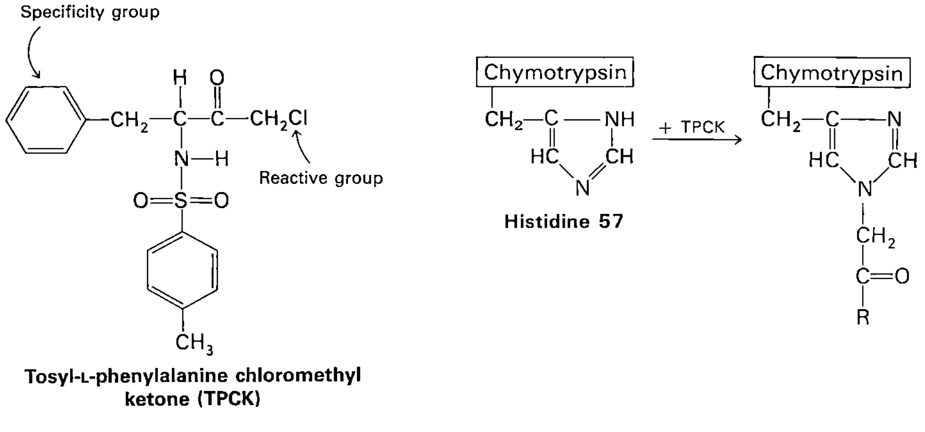

Figure 9-32
Affinity labeling of a histidine residue in chymotrypsin by tosyl-L-phenylalanine chloromethyl ketone (TPCK), a reactive substrate analog.

Three lines of evidence indicated that histidine 57 is part of the active site. First, the affinity-labeling reaction was highly stereospecific; the D isomer of TPCK was totally ineffective. Second, the reaction was inhibited when a competitive inhibitor of chymotrypsin, β-phenylpropionate, was present. Third, the rate of inactivation by TPCK varied with pH in nearly the same way as did the rate of catalysis.

SERINE, HISTIDINE, AND ASPARTATE FORM A CATALYTIC TRIAD IN CHYMOTRYPSIN

The catalytic activity of chymotrypsin depends on the unusual properties of serine 195. A $-CH_2OH$ group is ordinarily quite unreactive under physiologic conditions. What makes it so reactive in the active site of chymotrypsin? A convincing explanation has emerged from x-ray studies of the three-dimensional structure of the enzyme. As was foreseen by affinity-labeling studies, histidine 57 is adjacent to serine 195 (Figure 9-33). The carboxylate group of aspartate 102, buried in the protein, also is next to histidine 57. These three residues form a *catalytic triad.*

In the absence of substrate, histidine 57 is unprotonated (Figure 9-34). However, it is poised to accept the proton from the serine 195 –OH group when this oxygen atom carries out a nucleophilic attack on the substrate. The role of the $-COO^-$ group of aspartate 102 is to stabilize the positively charged form of histidine 57 in the transition state. In addition, aspartate 102 orients histidine 57 and ensures that it is in the appropriate tautomeric form to accept a proton from serine 195.

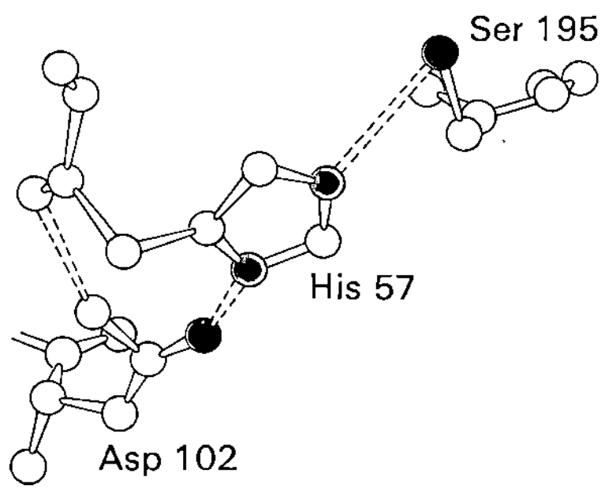

Figure 9-33
Conformation of the serine-histidine-aspartate catalytic triad in chymotrypsin. [After D.M. Blow and T.A. Steitz. X-ray diffraction studies of enzymes. *Ann. Rev. Biochem.* 39(1970): 86. Copyright © 1970 by Annual Reviews Inc. All rights reserved.]

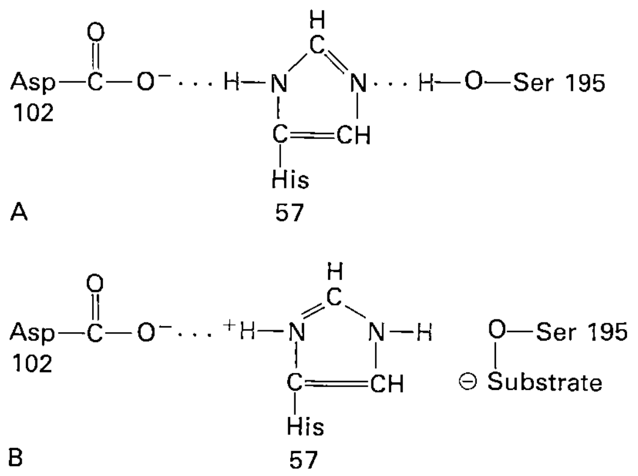

Figure 9-34
Role of the catalytic triad in chymotrypsin: (A) enzyme alone; (B) on addition of a substrate, a proton is transferred from serine 195 to histidine 57. The positively charged imidazole ring is stabilized by electrostatic interaction with negatively charged aspartate 102.

Crystallographic studies of complexes of chymotrypsin with substrate analogs have also shown the location of the site of specific recognition and the likely orientation of the susceptible peptide bond. Formyl-L-tryptophan binds to chymotrypsin with its indole side chain fitted neatly into a pocket near serine 195 (Figure 9-35). This deep cleft accounts for the specificity of chymotrypsin for aromatic and other bulky hydrophobic side chains.

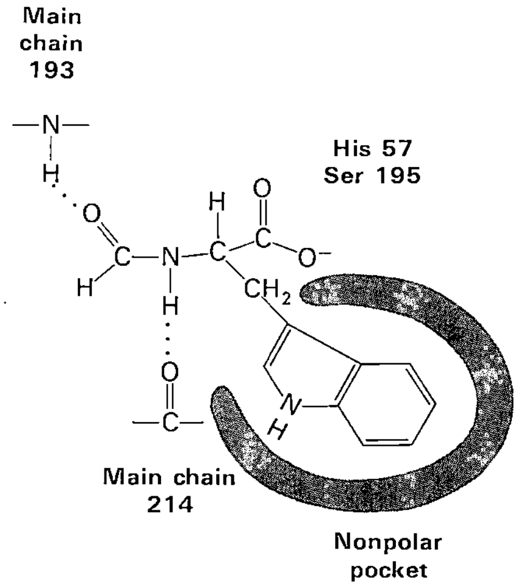

Figure 9-35
Schematic representation of the binding of formyl-L-tryptophan, a substrate analog, to chymotrypsin.

A TRANSIENT TETRAHEDRAL INTERMEDIATE IS FORMED DURING CATALYSIS BY CHYMOTRYPSIN

A plausible catalytic mechanism for chymotrypsin has been deduced from extensive x-ray crystallographic and chemical data. In this mechanism, *histidine 57 and serine 195 participate directly in the cleavage of the susceptible peptide bond of the substrate.* Hydrolysis of the peptide bond starts with an attack by the oxygen atom of the hydroxyl group of serine 195 on the carbonyl carbon atom of the susceptible peptide bond. The carbon–oxygen bond of this carbonyl group becomes a single bond, and the oxygen atom acquires a net negative charge. The four atoms now bonded to the carbonyl carbon are arranged as in a tetrahedron. The formation of this *transient tetrahedral intermediate* from a planar amide group is made

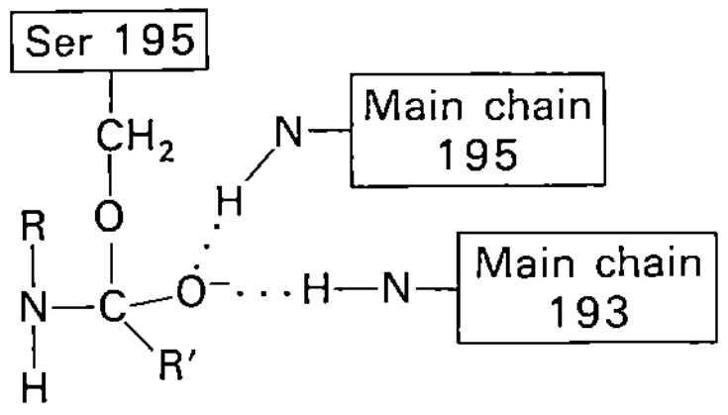

Figure 9-36
The tetrahedral transition state in the acylation reaction of chymotrypsin. The hydrogen bonds formed by two NH groups from the main chain of the enzyme are critical in stabilizing this species.

possible by hydrogen bonds between the negatively charged carbonyl oxygen atom (called an *oxyanion*) and two main-chain NH groups (Figure 9-36). This site is called the *oxyanion hole.*

The other essential event in the formation of this tetrahedral transition state is the transfer of a proton from serine 195 to histidine 57 (Figure 9-37). This proton transfer is markedly facilitated by the presence of the catalytic triad. Aspartate 102 precisely orients the imidazole ring of histidine 57 and partly neutralizes the charge that develops on it during the transition state. The proton held by the protonated form of histidine 57 is

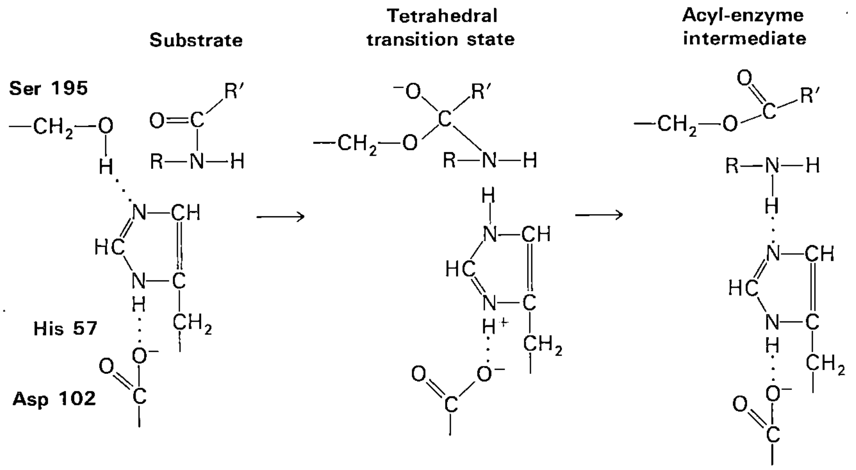

Figure 9-37
First stage in the hydrolysis of a peptide by chymotrypsin: *acylation.* A tetrahedral transition state is formed, in which the peptide bond is cleaved. The amine component then rapidly diffuses away, leaving an acyl-enzyme intermediate.

then donated to the nitrogen atom of the susceptible peptide bond, which thus is cleaved. At this stage, the amine component is hydrogen-bonded to histidine 57, whereas the acid component of the substrate is esterified to serine 195. The amine component diffuses away, completing the *acylation stage* of the hydrolytic reaction.

The next stage, *deacylation* (Figure 9-38), begins when a water molecule takes the place occupied earlier by the amine component of the substrate. In essence, *deacylation is the reverse of acylation, with H_2O substituting for the amine component.* First, histidine 57 draws a proton away from water. The resulting OH^- ion immediately attacks the carbonyl carbon atom of the acyl group that is attached to serine 195. As in acylation, a transient

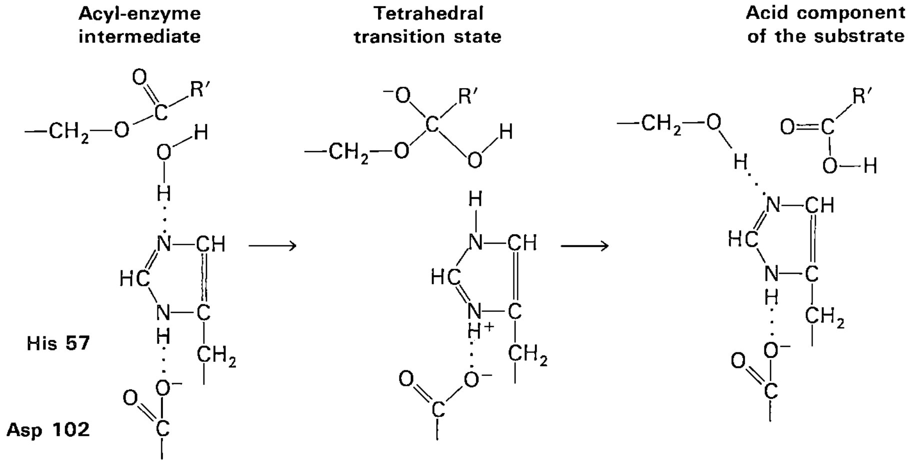

Figure 9-38
Second stage in the hydrolysis of a peptide by chymotrypsin: *deacylation.* The acyl-enzyme intermediate is hydrolyzed by water. Note that deacylation is essentially the reverse of acylation, with water in the role of the amine component of the original substrate.

tetrahedral intermediate is formed. Histidine 57 then donates the proton to the oxygen atom of serine 195, which then releases the acid component of the substrate. This acid component diffuses away and the enzyme is ready for another round of catalysis.

TRYPSIN AND ELASTASE ARE VARIATIONS ON THE CHYMOTRYPSIN THEME

Trypsin and elastase, two other digestive enzymes, are like chymotrypsin in many respects: (1) About 40% of the amino acid sequences of these three enzymes are identical. The degree of identity is even higher (~60%) for residues located in the interior of the enzymes. (2) X-ray studies have shown that their three-dimensional structures are very similar (Figure 9-39). (3) A serine-histidine-aspartate catalytic triad is present in all three. (4) The serine residue in this triad is modified by fluorophosphates (such as DIPF), causing a loss of catalytic activity. The amino acid sequence around this serine is the same in all three enzymes: Gly-Asp-Ser-Gly-Gly-Pro. (5) All three enzymes have nearly identical catalytic mechanisms. The catalytic triad and oxyanion hole in each promotes the formation of a tetrahedral transition state. A covalent acyl-enzyme intermediate is formed by all three during catalysis. (6) As will be discussed in the next chapter, these enzymes are secreted by the pancreas as inactive precursors that become activated by cleavage of a single peptide bond.

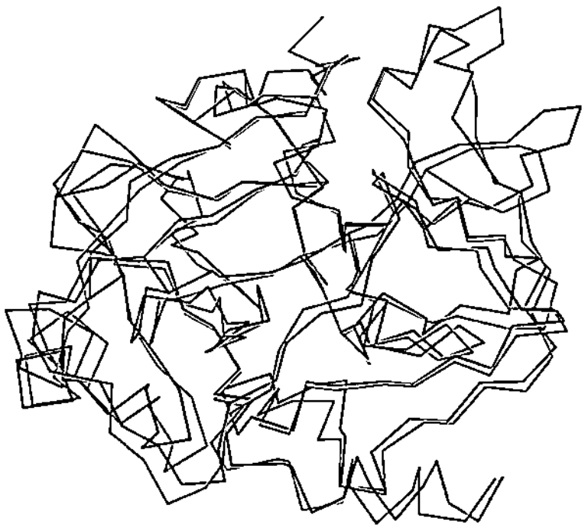

Figure 9-39
The main-chain conformation of elastase (red) is very similar to that of chymotrypsin (blue). The mean deviation of the superposed α carbon atoms of the two enzymes is 1.7 Å. [Drawn from 2cha.pdb and 1est.pdb. J.J. Birktoft and D.M. Blow. *J. Mol. Biol.* 68(1972):187, and L. Sawyer, D.M. Shotton, J.W. Campbell, P.L. Wendell, H. Muirhead, H.C. Watson, R. Diamond, and R.C. Ladner. *J. Mol. Biol.* 118(1978):137.]

Although similar in structure and mechanism, these enzymes differ markedly in substrate specificity. Chymotrypsin requires an aromatic or bulky nonpolar side chain on the amino side of the scissile bond. Trypsin requires a lysine or arginine residue. Elastase cannot cleave either of these kinds of substrates. Its specificity is directed toward the smaller uncharged side chains. X-ray studies have shown that *these different specificities are due to quite small structural differences in the binding site* (Figure 9-40). In chymotrypsin, a nonpolar pocket serves as a niche for the aromatic or bulky nonpolar side chain. In trypsin, one residue in this pocket is different from chymotrypsin: a serine is replaced by an aspartate. This aspartate in the nonpolar pocket of trypsin can form a strong electrostatic bond with a positively charged lysine or arginine side chain of a substrate. In elastase, the pocket does not exist, because the two glycine residues lining it in chymotrypsin are replaced by the much bulkier valine and threonine.

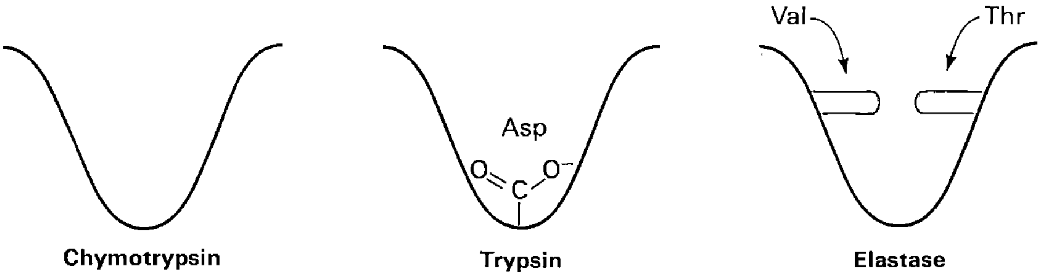

Figure 9-40
A highly simplified representation of part of the substrate-binding site in chymotrypsin, trypsin, and elastase.

SERINE, ZINC, THIOL, AND ASPARTYL PROTEASES ARE THE MAJOR CLASSES OF PROTEOLYTIC ENZYMES

Thus far, consideration has been given to two classes of proteolytic enzymes: the *zinc proteases,* exemplified by carboxypeptidase A, and the *serine proteases.* The *thiol proteases* (also called *sulfhydryl* or *cysteine proteases*) are

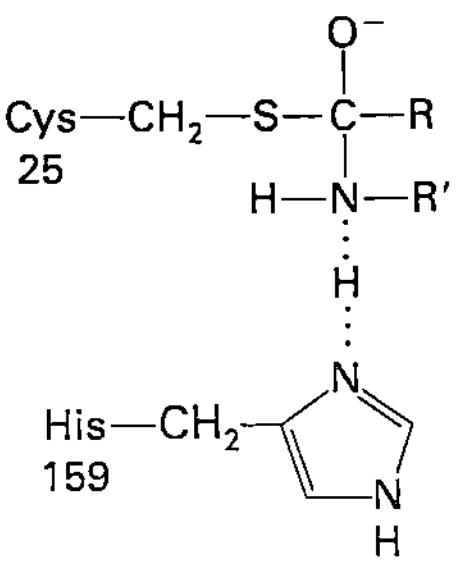

Figure 9-41
Transition state in catalysis by papain, a thiol protease. The –SH group of a cysteine residue (shown in red) plays a role like that of the –OH group of serine 195 in chymotrypsin. Another similarity is that the nucleophilicity of the –SH group is enhanced by the adjacent histidine (shown in green), which serves as a proton acceptor.

another widely distributed group. Papain, derived from papaya, contains an active-site cysteine that acts like serine 195 in chymotrypsin. Catalysis proceeds through a thiol ester intermediate and is facilitated by a nearby histidine side chain (Figure 9-41). Papain is widely used as a meat tenderizer. The thiol protease cathepsin B degrades proteins in lysosomes, the membrane-enveloped organelles that carry out intracellular digestion in animal cells.

The last major class of proteolytic enzymes are the *aspartyl proteases,* which are also known as *carboxyl* or *acid proteases.* They are discussed below. *Indeed, all known proteolytic enzymes belong to one of these four major classes. There appear to be only a limited number of efficient ways of catalyzing the cleavage of peptide bonds.*

THE CATALYTIC ARRAY OF PEPSIN CONSISTS OF AN ACTIVATED WATER BETWEEN TWO ASPARTATES

The best-understood member of the aspartyl protease class is *pepsin,* the principal proteolytic enzyme in gastric juice. The optimum pH for catalysis by this 35-kd single-chain protein is between 2 and 3. X-ray crystallographic studies by Michael James have shown that the active site contains a water molecule flanked by two aspartates (Figure 9-42). One must be ionized and the other un-ionized for the enzyme to be active. A network of interactions changes the p*K* of one of the carboxylates from a customary value of about 4.5 to the unusually low value of 1.2. In other words, the protein environment makes this aspartate a very strong acid.

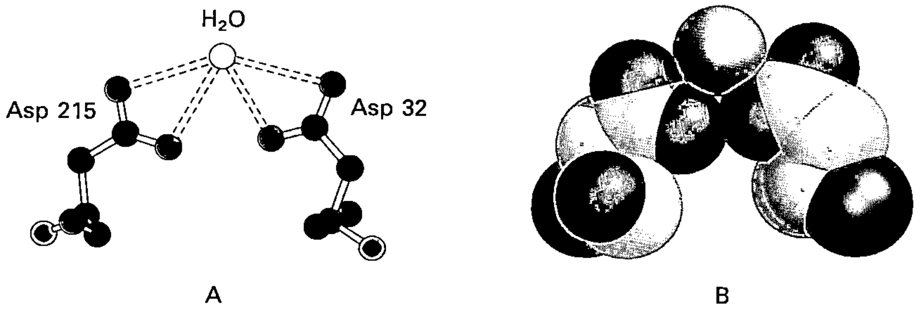

Figure 9-42
A water molecule at the active site of pepsin is activated by a pair of aspartates. (A) Ball-and-stick model. (B) Space-filling model. The oxygen atom of the activated water molecule is shown in green. [Drawn from 4pep.pdb. A.R. Sielecki, A.A. Fedorov, A. Boodhoo, N.S. Andreeva, and M.N. James. *J. Mol. Biol.* 214(1990):143.]

Pepsin consists of two structurally similar lobes (Figure 9-43), which suggests that it evolved by gene duplication and fusion. The substrate sits in a deep groove between them. One of the key aspartates comes from the amino-terminal lobe and the other from the carboxy-terminal lobe. *These aspartates play a dual role in catalysis: (1) They activate the water molecule positioned between them and (2) serve as proton acceptors and donors.*

The aspartyl protease class is diverse and broadly distributed in nature, as exemplified by (1) *renin,* an enzyme that raises blood pressure in mammals by converting angiotensinogen to angiotensin I, (2) *chymosin* (also called *rennin*), an enzyme from the fourth stomach of cows, which has

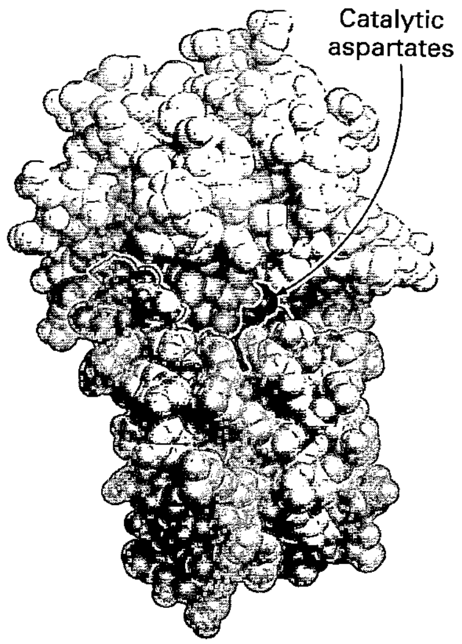

Figure 9-43
Pepsin consists of two similar domains (shown in yellow and blue). Each contributes one of the two catalytic aspartates (red). [Drawn from 4pep.pdb. A.R. Sielecki, A.A. Fedorov, A. Boodhoo, N.S. Andreeva, and M.N. James. *J. Mol. Biol.* 214(1990):143.]

been used by humans for many centuries to make cheese, (3) *rhizopus pepsin* and *penicillopepsin,* which enable fungi to digest decaying plant matter, and (4) *HIV-1 protease,* which is essential for the replication of the AIDS virus, as will be discussed shortly. Most aspartyl proteases are inhibited by very low (~nM) concentrations of *pepstatin,* a microbial peptide that acts as a transition state analog because it contains a tetrahedral hydroxymethylene unit.

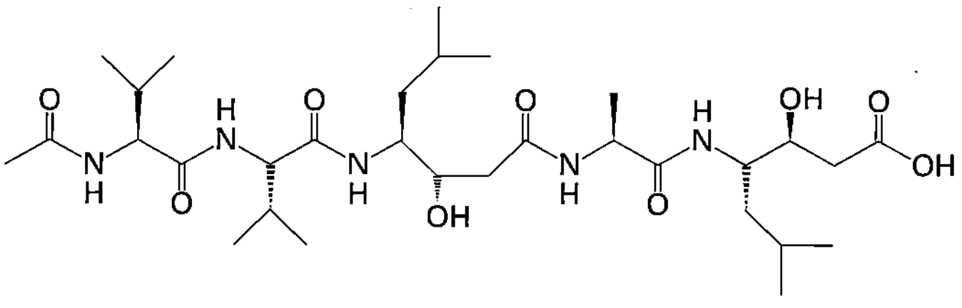

N-Acetylpepstatin

Wood engraving depicting the making of cheese in medieval Germany. [Johannes Stumpf, 1550.]

AN ASPARTYL PROTEASE IS ESSENTIAL FOR REPLICATION OF HUMAN IMMUNODEFICIENCY VIRUS (HIV)

The replication of human immunodeficiency virus-1 (HIV-1), the cause of acquired human immunodeficiency syndrome (AIDS), requires proteolytic cleavage of a *polyprotein* that contains several key viral proteins. Proteolysis is carried out by a virus-encoded protein that is itself contained within the polyprotein. The first task of the *HIV-1 protease* is to liberate itself from the confines of the polyprotein. It then hydrolyzes additional peptide bonds in the polyprotein to free the other essential viral proteins. The presence of an Asp-Thr-Gly sequence in the HIV-1 protease suggested that it is an aspartyl protease because this sequence recurs in the active sites of this class of enzymes. However, HIV-1 protease (11 kd, 99 residues) is not quite half the size of eukaryotic aspartyl proteases and contains only one such sequence.

How then is an active site formed? Answering this question depended on having an abundant supply of the enzyme. Large quantities were produced by expression of the cDNA in *Escherichia coli* and by solid-phase chemical synthesis. X-ray crystallographic studies then revealed that *HIV-1 protease is a symmetric dimer with the critical aspartate pair at the subunit interface* (Figure 9-44). Indeed, the viral enzyme looks very much like

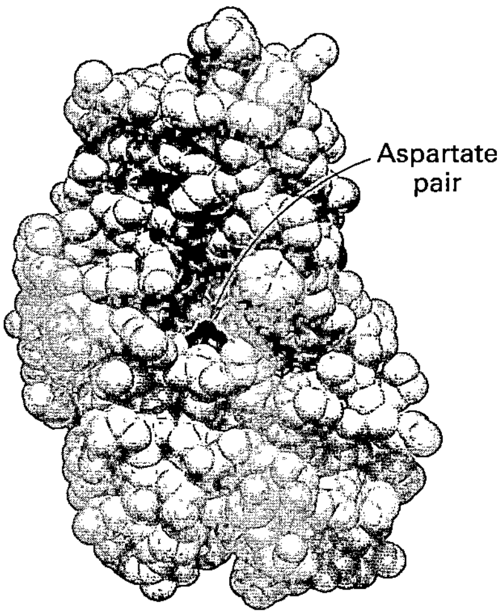

Figure 9-44
HIV-1 protease is a dimer of identical chains. One of them is shown in green, and the other in blue. The active site is located at their interface. Each chain contributes an aspartate (red) that is critical for catalysis. [Drawn from 4hvp.pdb. M. Miller, J. Schneider, B.K. Sethyanarayana, M.V. Toth, G.R. Marshall, L. Clawson, L. Selk, S.B.H. Kent, and A. Wlodawer. *Science* 246(1989):1149.]

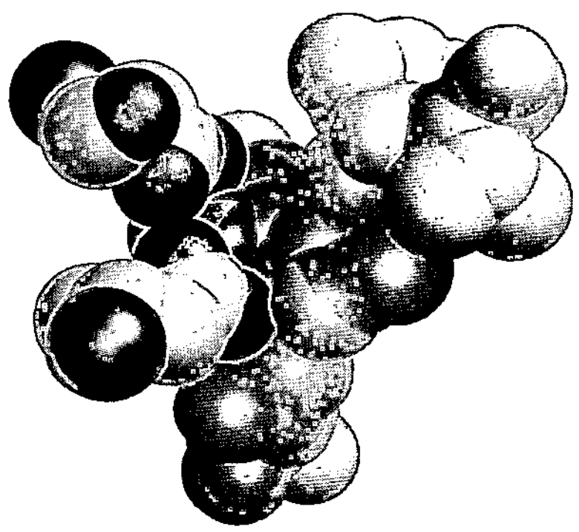

Figure 9-45
Interaction of an HIV-1 protease inhibitor (yellow) with the critical aspartate pair at the active site. The oxygen atom (green) of the hydroxyl group of the inhibitor lies between the carboxylate groups of the enzyme. [Drawn from 8hvp.pdb. M. Jaskolski, A.G. Tomasselli, T.K. Sawyer, D.G. Staples, R.L. Heinrikson, J. Schneider, S.B.H. Kent, and A. Wlodawer. *Biochemistry* 30(1991):1600.]

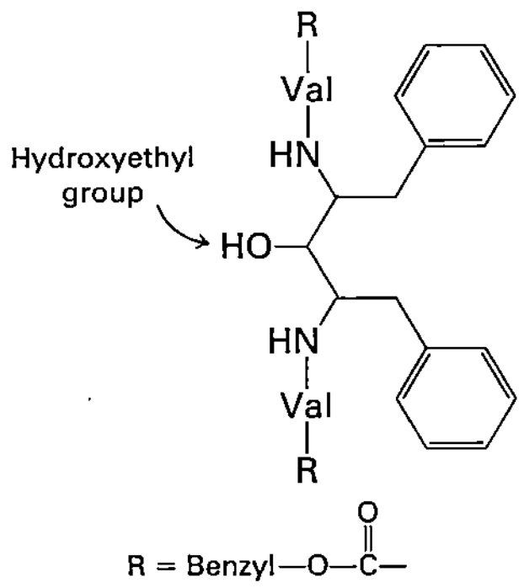

Figure 9-46
Structure of a potent symmetric inhibitor of HIV-1 protease. The portion shown in red has a geometry like that of the transition state.

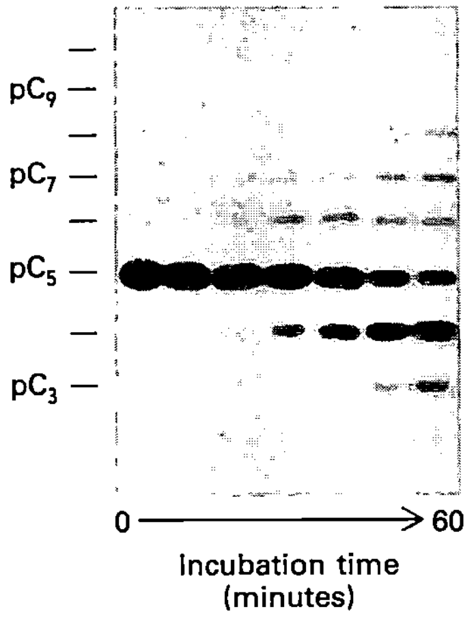

Figure 9-47
Catalysis of the cleavage and elongation of pentacytidylate (C_5) by L19 RNA. The molar ratio of substrate to enzyme was 6. The 5′ end of C_5 was labeled with ^{32}P phosphate so that the reaction products could be visualized by autoradiography. [After A.J. Zaug and T.R. Cech. *Science* 231(1985):471.]

pepsin and other eukaryotic aspartyl proteases. The degree of similarity is especially high in the vicinity of the active site.

Mutation of either aspartate at the active site gives a totally inactive protease. Furthermore, this mutant virus is not infective. The essentiality of the protease for viral multiplication makes it an attractive target for the design of drugs to treat AIDS. The first step in designing an anti-AIDS drug is to elucidate the mode of binding of substrate analogs and infer the geometry of the transition state. A clue came from the finding that the protease cleaves peptide bonds between a hydrophobic residue and proline when the next residue is also hydrophobic. High-affinity inhibitors were then produced by replacing proline with a *hydroxyethyl group* to mimic the tetrahedral transition state. In fact, *N*-acetylpepstatin and related inhibitors bearing such a group are potent inhibitors of HIV-1 protease ($K_i \sim 1$ nM). X-ray analysis has shown that they bind in an extended conformation. As expected, their hydroxyethyl group interacts intimately with the critical aspartate pair of the enzyme.

An anti-AIDS drug that blocks the HIV-1 protease must not appreciably inhibit important human aspartyl proteases such as renin. How might such selectivity be achieved? The viral protease differs from host aspartyl proteases in being precisely twofold symmetric. Hence, an inhibitor with the same symmetry as the viral protease should be highly specific. Crystal structures of asymmetric inhibitors bound to HIV-1 protease pointed the way in the design of a highly potent symmetric inhibitor that contains a hydroxyethyl group (Figure 9-46). This inhibitor is 10^4-fold more effective in blocking the HIV-1 protease than host proteases because *its symmetry and geometry match that of its target.* It is evident that drug design has entered a new stage. Many promising drug candidates now come from our knowledge of three-dimensional structure and reaction mechanism.

RNA MOLECULES CAN BE POTENT ENZYMES

Thousands of proteins have been purified and shown to possess enzymatic activity. It was believed for more than 50 years that all enzymes are proteins. However, experiments carried out in the 1980s revealed that *RNA molecules, too, can be highly active enzymes.* A striking example is provided by L19 RNA, a shortened form of an intron released by the self-splicing of a ribosomal RNA precursor in *Tetrahymena thermophila,* a protozoan (p. 865). Thomas Cech discovered that L19 RNA catalyzes the cleavage and joining of oligoribonucleotide substrates in a highly specific manner. Pentacytidylate (C_5) is converted by L19 RNA into longer and shorter oligomers (Figure 9-47). Specifically, C_5 is degraded to C_4 and C_3.

At the same time, C_6 and longer oligomers are formed. Thus, L19 RNA is both a *ribonuclease* and an *RNA polymerase.*

This RNA enzyme acts much more rapidly on C_5 than on U_5 and not at all on A_5 and G_5. It obeys Michaelis-Menten kinetics with a K_M of 42 μM and a k_{cat} of 0.033 s^{-1} for C_5. The essential need for a 2′-OH in the substrate is demonstrated by the finding that deoxy-C_5 is a competitive inhibitor of C_5 with a K_i of 260 μM. Thus, L19 RNA displays several hallmarks of classic enzymatic catalysis: *a high degree of substrate specificity, Michaelis-Menten saturation kinetics, and susceptibility to competitive inhibition.*

How does this 395-nucleotide RNA molecule act both as a ribonuclease and a polymerase? Its three-dimensional structure is not yet known, but some inferences about its catalytic mechanism can be made (Figure 9-48). Pentacytidylate binds to a specific site in L19 RNA. Several of the

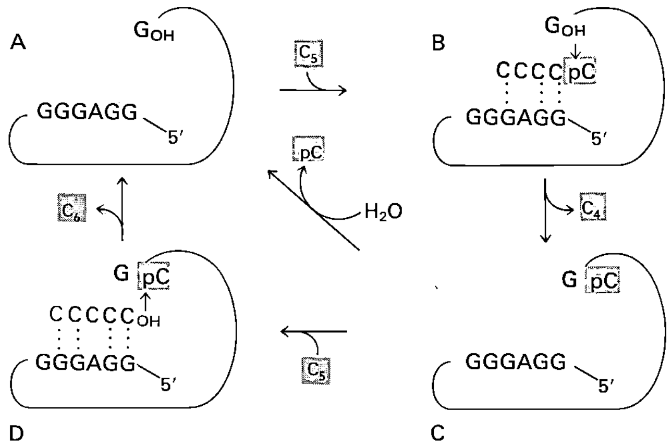

Figure 9-48
Proposed catalytic mechanism of L19 RNA. (A) Enzyme alone. (B) C_5 is hydrogen-bonded to the enzyme, and the terminal pC is linked covalently to the terminal G of the enzyme. The remaining C_4 moiety is free to dissociate from the enzyme. (C) Hydrolysis of GpC restores the enzyme to its original state. (D) Alternatively, the covalently attached pC can be attacked by a second C_5 molecule to form C_6. [After A.J. Zaug and T.R. Cech. *Science* 231(1985):473.]

five C bases of this substrate are probably base-paired to a G-rich sequence of this RNA enzyme. The phosphodiester bond between the fourth and fifth C of bound C_5 is then attacked by the 3′-OH of the terminal G residue of this enzyme. A phosphodiester bond is formed between the terminal G of the enzyme and the terminal C of the substrate to produce a –GpC *covalent intermediate.* C_4 is concomitantly released during this *transesterification reaction.* This highly labile new phosphodiester bond can be attacked by water to release pC and regenerate the enzyme. This series of three steps accounts for the *ribonuclease* activity of L19 RNA.

Alternatively, the –GpC enzyme intermediate can bind another C_5 molecule, which then attacks –GpC to produce a C_6 product and regenerate the enzyme. This step accounts for the *polymerase* action of L19 RNA. Thus, the covalent intermediate –GpC can be attacked either by OH^- or by the 3′-OH group of a second substrate. Hydrolysis is favored by a high pH, whereas transesterification to form C_6 is promoted by a high concentration of the C_5 substrate. C_6 formed in this way can be elongated to C_7 and longer oligomers by successive transesterification reactions. It seems likely that the phosphodiester bond of –GpC is unusually reactive because it is activated electrostatically to form a pentacovalent transition state.

How does the catalytic efficacy of this RNA enzyme compare with that of protein enzymes? The k_{cat}/K_M value for L19 RNA is ~10^3 s^{-1} M^{-1}, five orders of magnitude less than that for the most catalytically proficient proteins (p. 195). However, it is similar to that for ribonuclease A. *The rate of hydrolysis of C_5 by this RNA enzyme is about 10^{10} times the uncatalyzed rate. It is evident that RNA molecules, like proteins, can be very effective catalysts.* RNA molecules can form precise three-dimensional structures that bind specific substrates and stabilize transition states. However, RNAs differ from proteins in not being able to form large nonpolar niches. RNAs are also much less versatile than are proteins because they contain only 4 different building blocks instead of 20. Hence, *while most enzymes are proteins, it is evident that proteins do not have a monopoly on catalysis.* RNA enzymes (sometimes called *ribozymes*) are very well suited for recognizing and transforming nucleic acids because they share a common base-pairing language with their substrates.

The finding of enzymatic activity in L19 RNA and in the RNA component of ribonuclease P (p. 879), an enzyme that participates in the processing of tRNA precursors, has opened new areas of inquiry and changed the way we think about molecular evolution. The discovery that RNA can be a catalyst as well as an information carrier suggested that an RNA world occurred early in the evolution of life, before the appearance of DNA and protein (p. 115).

SUMMARY

Enzymes can assume conformations that are complementary to the transition states of reactions catalyzed by them. This chapter showed how several well-understood hydrolytic enzymes bind substrates and facilitate the formation of transition states. Lysozyme, a small glycosidase, cleaves bacterial cell walls. Its substrate is an alternating polymer of *N*-acetylglucosamine (NAG) and *N*-acetylmuramic acid (NAM) residues joined by $\beta(1 \rightarrow 4)$ glycosidic linkages. Lysozyme hydrolyzes the glycosidic bond between C-1 of NAM and C-4 of NAG. Oligomers of NAG consisting of six or more residues are also good substrates.

The lysozyme molecule has a cleft to which substrates are bound by many hydrogen bonds and van der Waals interactions. The critical groups in catalysis are the un-ionized carboxyl of glutamic acid 35 and the ionized carboxylate of aspartate 52. Both groups are about 3 Å from the scissile bond between residues D and E of a hexameric substrate. Glutamate 35 cleaves this bond by donating an H^+ to the bridge oxygen between residues D and E. C-1 of the D ring then becomes positively charged. This transient species, called a carbonium ion, then reacts with OH^- from the solvent, and glutamic acid 35 becomes protonated again. Lysozyme is ready for another round of catalysis after the products diffuse away. Catalysis is markedly enhanced by two factors that promote the formation of the carbonium ion intermediate. The electrostatic factor is the proximity of the negatively charged side chain of aspartate 52. The geometrical factor is the distortion, in the transition state, of ring D into a half-chair form, which enables the positive charge to be shared between C-1 and the ring oxygen atom.

Ribonuclease A, a digestive enzyme secreted by the pancreas, hydrolyzes RNA on the 3′ side of pyrimidine residues. The reaction begins with a nucleophilic attack by the 2′-O of ribose on the phosphorus atom of the scissile bond. The un-ionized form of histidine 12 accepts a proton to enhance the nucleophilicity of the 2′-O. The protonated form of histidine 119 simultaneously donates a proton to the 5′-O to cleave the scissile bond and form a 2′,3′-cyclic phosphate intermediate. In the transition state, phosphorus becomes pentacovalent and acquires the geometry of a trigonal bipyramid, in which three oxygens lie in the equatorial plane and two at the apices. One apex is occupied by the attacking nucleophile (the

2′-O) and the other by the leaving group (the 5′-O). The negatively charged phosphate group in the transition state is stabilized by a salt bridge with the positively charged side chain of lysine 41. Hydrolysis of the cyclic phosphate intermediate, the second stage of the reaction, is essentially a reversal of the first, in which H_2O replaces the 5′-OH component that was removed.

Carboxypeptidase A, a proteolytic enzyme, recognizes C-terminal residues that contain an aromatic or bulky aliphatic side chain; it hydrolyzes the terminal peptide bond. Substrate binding induces large structural changes at the active site, which convert it from a water-filled region into a hydrophobic one. Carboxypeptidase A illustrates induced fit in catalysis. A tightly bound zinc ion plays a key role in catalysis by activating a water molecule, which then attacks the carbonyl carbon atom of the scissile peptide bond. A negatively charged tetrahedral intermediate is stabilized by electrostatic interactions with Zn^{2+} and an arginine side chain. A glutamate residue adjacent to the carbonyl carbon atom accepts a proton from the activated water and then donates it to the peptide NH to cleave the bond. Carboxypeptidase A is a member of the zinc protease class, which includes the bacterial enzyme thermolysin.

In chymotrypsin and other serine proteases, a highly reactive serine 195 plays a critical role in catalysis. The first stage in the hydrolysis of a peptide substrate is acylation, the formation of a covalent acyl-enzyme intermediate, in which the carboxyl component of the substrate is esterified to the hydroxyl group of serine 195. The nucleophilicity of the serine –OH is markedly enhanced by histidine 57, which accepts a proton from serine as serine attacks the carbonyl carbon atom of the substrate. The resulting positively charged histidine is stabilized by electrostatic interaction with negatively charged aspartate 102. Serine, histidine, and aspartate form a catalytic triad that is at the heart of the catalytic action of all serine proteases. The negative charge on the tetrahedral transition state is also stabilized by hydrogen bonding to two main-chain NH groups in the oxyanion hole. The second stage, deacylation, is in essence a reverse of the first, with H_2O substituting for the amine component. Chymotrypsin, trypsin, elastase, several clotting factors, and other vertebrate serine proteases probably arose from a common ancestral gene.

The two other major classes of proteolytic enzymes are the thiol proteases and the aspartyl proteases. Papain, a thiol protease from papaya, contains an active-site cysteine that acts like serine 195 in chymotrypsin. Pepsin, the best understood aspartyl protease, contains an activated water molecule between an un-ionized and an ionized aspartate. This catalytic cluster is located in a deep groove between the two similar domains of the enzyme. The HIV-1 protease, another member of this class, is required for replication of the virus that causes AIDS. This viral-encoded proteolytic enzyme is a dimer of identical subunits; each contributes one of the two essential aspartates. Knowledge of the three-dimensional structure and catalytic mechanism of aspartyl proteases provides a basis for the design of transition state analogs that specifically inhibit HIV-1 protease and may be of value in treating AIDS.

The discovery of RNA enzymes in the 1980s came as a surprise because it was previously thought that proteins have a monopoly on catalysis. An RNA molecule derived from the intron of a self-splicing RNA from a ciliated protozoan acts as a ribonuclease and an RNA polymerase. This RNA enzyme displays a high degree of specificity, Michaelis-Menten kinetics, and susceptibility to competitive inhibition. RNA enzymes probably preceded protein enzymes in evolution.

SELECTED READINGS

Where to start

Phillips, D.C., 1966. The three-dimensional structure of an enzyme molecule. *Sci. Amer.* 215(5):78–90. [A superb article on the three-dimensional structure and catalytic mechanism of lysozyme.]

Cech, T.R., 1990. Self-splicing and enzymatic activity of an intervening sequence RNA from Tetrahymena. *Biosci. Rep.* 10:239–261. [A fascinating account of the discovery of catalytic RNA is given in this Nobel Lecture.]

Lolis, E., and Petsko, G.A., 1990. Transition-state analogues in protein crystallography: Probes of the structural source of enzyme catalysis. *Ann. Rev. Biochem.* 59:597–630.

Books

Fersht, A., 1985. *Enzyme Structure and Mechanism.* W.H. Freeman. [Lucid discussions of catalytic mechanisms are given in Chapters 2 and 15.]

Page, M.I., and Williams, A. (eds.), 1987. *Enzyme Mechanisms.* Royal Society of Chemists. [Contains many informative and concise articles on the major kinds of catalytic mechanisms in biochemical systems.]

Boyer, P.D. (ed.). *The Enzymes* (3rd ed.). Academic Press. [This multivolume treatise continues to be a valuable reference. For example, see vol. 7, pp. 666–868, for a fine article on lysozyme.]

Lysozyme

Strynadka, N.C., and James, M.N., 1991. Lysozyme revisited: Crystallographic evidence for distortion of an *N*-acetylmuramic acid residue bound in site D. *J. Mol. Biol.* 220:401–424.

Cheetham, J.C., Artymiuk, P.J., and Phillips, D.C., 1992. Refinement of an enzyme complex with inhibitor bound at partial occupancy. Hen egg-white lysozyme and tri-*N*-acetylchitotriose at 1.75 Å resolution. *J. Mol. Biol.* 224:613–628.

Johnson, L.N., Cheetham, J., McLaughlin, P.J., Acharya, K.R., Barford, D., and Phillips, D.C., 1988. Protein-oligosaccharide interactions: Lysozyme, phosphorylase, amylases. *Curr. Top. Microbiol. Immunol.* 139:81–134.

Kumagai, I., Sunada, F., Takeda, S., and Miura, K., 1992. Redesign of the substrate-binding site of hen egg white lysozyme based on the molecular evolution of C-type lysozymes. *J. Biol. Chem.* 267:4608–4612.

Ford, L.O., Johnson, L.N., Mackin, P.A., Phillips, D.C., and Tjian, R., 1974. Crystal structure of a lysozyme-tetrasaccharide lactone complex. *J. Mol. Biol.* 88:349–371.

Ribonuclease

Richards, F.M., and Wyckoff, H., 1971. Bovine pancreatic ribonuclease. *In* Boyer, P.D. (ed.), *The Enzymes* (3rd ed.), vol. 4, pp. 647–806. Academic Press.

McPherson, A., Brayer, G., Cascio, D., and Williams, R., 1986. The mechanism of binding of a polynucleotide chain to pancreatic ribonuclease. *Science* 232:765–768.

Borah, B., Chen, C.W., Egan, W., Miller, M., Wlodawer, A., and Cohen, J.S., 1985. Nuclear magnetic resonance and neutron diffraction studies of the complex of ribonuclease A with uridine vanadate, a transition-state analogue. *Biochemistry* 24:2058–2067.

Zinc proteases

Christianson, D.W., and Lipscomb, W.N., 1989. Carboxypeptidase A. *Acc. Chem. Res.* 22:62–69.

Kim, H., and Lipscomb, W.N., 1991. Comparison of the structures of three carboxypeptidase A–phosphonate complexes determined by x-ray crystallography. *Biochemistry* 30:8171–8180.

Shoham, G., Christianson, D.W., and Oren, D.A., 1988. Complex between carboxypeptidase A and a hydrated ketomethylene substrate analogue. *Proc. Nat. Acad. Sci.* 85:684–688.

Makinen, M.W., Wells, G.B., and Kang, S.O., 1984. Structure and mechanism of carboxypeptidase A. *Advan. Inorg. Biochem.* 6:1–69. [An alternative catalytic mechanism involving the formation of a mixed anhydride intermediate is presented in this review.]

Matthews, B.W., 1988. Structural basis of the action of thermolysin and related zinc peptidases. *Acc. Chem. Res.* 21:333–340.

Monzingo, A.F., and Matthews, B.W., 1984. Binding of *N*-carboxymethyl dipeptide inhibitors to thermolysin determined by x-ray crystallography: A novel class of transition-state analogues for zinc peptidases. *Biochemistry* 23:5724–5729.

Christianson, D.W., 1991. Structural biology of zinc. *Adv. Prot. Chem.* 42:281–355.

Serine proteases

Steitz, T.A., and Shulman, R.G., 1982. Crystallographic and NMR studies of the serine proteases. *Ann. Rev. Biochem. Biophys.* 11:419–444.

Stroud, R.M., 1974. A family of protein-cutting proteins. *Sci. Amer.* 231(1):24–88.

Blow, D.M., 1976. Structure and mechanism of chymotrypsin. *Acc. Chem. Res.* 9:145–152.

Kossiakoff, A.A., and Spencer, S.A., 1981. Direct determination of the protonation states of aspartic acid-102 and histidine-57 in the tetrahedral intermediate of the serine proteases: Neutron structure of trypsin. *Biochemistry* 20:6462–6474.

Thiol proteases

Kamphuis, I.G., Drenth, J., and Baker, E.N., 1985. Thiol proteases. Comparative studies based on the high-resolution structures of papain and actinidin, and on amino acid sequence information for cathepsins B and H, and stem bromelain. *J. Mol. Biol.* 182:317–329.

Musil, D., Zucic, D., Turk, D., Engh, R.A., Mayr, I., Huber, R., Popovic, T., Turk, V., Towatari, T., and Katunuma, N., 1991. The refined 2.15 Å x-ray crystal structure of human liver cathepsin B: The structural basis for its specificity. *EMBO J.* 10:2321–2330.

Matsumoto, K., Yamamoto, D., Ohishi, H., Tomoo, K., Ishida,

T., Inoue, M., Sadatome, T., Kitamura, K., and Mizuno, H., 1989. Mode of binding of E-64-c, a potent thiol protease inhibitor, to papain as determined by x-ray crystal analysis of the complex. *FEBS Lett.* 245:177–180.

Aspartyl proteases

Davies, D.R., 1990. The structure and function of the aspartic proteinases. *Ann. Rev. Biophys. Biophys. Chem.* 19:189–215.

Sielecki, A.R., Fedorov, A.A., Boodhoo, A., Andreeva, N.S., and James, M.N., 1990. Molecular and crystal structures of monoclinic porcine pepsin refined at 1.8 Å resolution. *J. Mol. Biol.* 214:143–170.

Jaskolski, M., Tomasselli, A.G., Sawyer, T.K., Staples, D.G., Heinrikson, R.L., Schneider, J., Kent, S.B., and Wlodawer, A., 1991. Structure at 2.5-Å resolution of chemically synthesized human immunodeficiency virus type 1 protease complexed with a hydroxyethylene-based inhibitor. *Biochemistry* 30:1600–1609.

Swain, A.L., Miller, M.M., Green, J., Rich, D.H., Schneider, J., Kent, S.B., and Wlodawer, A., 1990. X-ray crystallographic structure of a complex between a synthetic protease of human immunodeficiency virus 1 and a substrate-based hydroxyethylamine inhibitor. *Proc. Nat. Acad. Sci.* 87:8805–8809.

Fitzgerald, P.M., McKeever, B.M., VanMiddlesworth, J.F., Springer, J.P., Heimbach, J.C., Leu, C.T., Herber, W.K., Dixon, R.A., and Darke, P.L., 1990. Crystallographic analysis of a complex between human immunodeficiency virus type 1 protease and acetyl-pepstatin at 2.0-Å resolution. *J. Biol. Chem.* 265:14209–14219.

Wlodawer, A., Miller, M., Jaskolski, M., Sathyanarayana, B.K., Baldwin, E., Weber, I.T., Selk, L.M., Clawson, L., Schneider, J., and Kent, S.B., 1989. Conserved folding in retroviral proteases: Crystal structure of a synthetic HIV-1 protease. *Science* 245:616–621.

Design of enzyme inhibitors

Cushman, D.W., and Ondetti, M.A., 1991. History of the design of captopril and related inhibitors of angiotensin converting enzyme. *Hypertension* 17:589–592.

Erickson, J., Neidhart, D.J., VanDrie, J., Kempf, D.J., Wang, X.C., Norbeck, D.W., Plattner, J.J., Rittenhouse, J.W., Turon, M., and Wideburg, N., 1990. Design, activity, and 2.8 Å crystal structure of a C2 symmetric inhibitor complexed to HIV-1 protease. *Science* 249:527–533.

Sielecki, A.R., Hayakawa, K., Fujinaga, M., Murphy, M.E., Fraser, M., Muir, A.K., Carilli, C.T., Lewicki, J.A., Baxter, J.D., and James, M.N., 1989. Structure of recombinant human renin, a target for cardiovascular-active drugs, at 2.5 Å resolution. *Science* 243:1346–1351.

Catalytic RNA

Zaug, A.J., and Cech, T.R., 1986. The intervening sequence RNA of *Tetrahymena* is an enzyme. *Science* 231:431–475.

Cech, T.R., Herschlag, D., Piccirilli, J.A., and Pyle, A.M., 1992. RNA catalysis by a group I ribozyme. Developing a model for transition state stabilization. *J. Biol. Chem.* 267:17479–17482.

Cech, T.R., 1993. Catalytic RNA: Structure and mechanism. *Biochem. Soc. Trans.* 21:229–234.

Pyle, A.M., and Cech, T.R., 1991. Ribozyme recognition of RNA by tertiary interactions with specific ribose 2′-OH groups. *Nature* 350:628–631.

Altman, S., 1990. Enzymatic cleavage of RNA by RNA. *Biosci. Rep.* 10:317–337. [This Nobel Lecture presents the discovery and characterization of the catalytic RNA subunit of the enzyme ribonuclease P of *E. coli.*]

PROBLEMS

1. *Kinetic forecast.* Predict the relative rates of hydrolysis by lysozyme of these oligosaccharides (G stands for an *N*-acetylglucosamine residue, and M for *N*-acetylmuramic acid):
   (a) M-M-M-M-M-M
   (b) G-M-G-M-G-M
   (c) M-G-M-G-M-G

2. *Occupancy forecast.* Predict on the basis of the data given in Figure 9-49 which of the sugar binding sites, A to F, on lysozyme will be occupied in the most prevalent complex with each of these oligosaccharides (same abbreviations as in problem 1).
   (a) G-G
   (b) G-M
   (c) G-G-G-G

3. *A revealing tag.* Suppose that hexa-NAG is synthesized so that the glycosidic oxygen between its D and E sugar residues is labeled with ^{18}O. Where will this isotope appear in the products formed by hydrolysis with lysozyme?

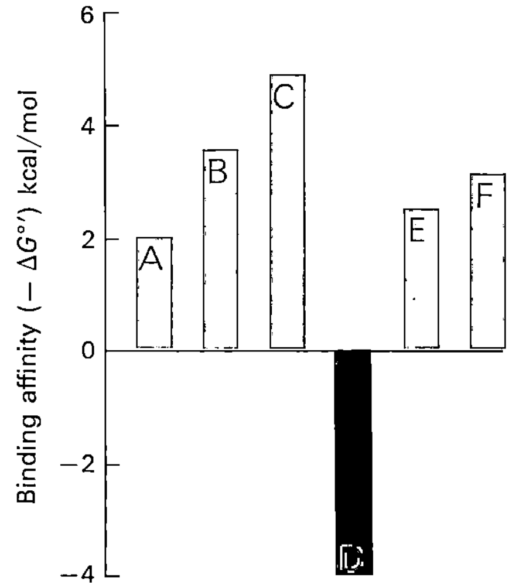

Figure 9-49
Contributions made by the six sugars of hexa-NAG to the standard free energy of binding to lysozyme.

4. *Affinity accounting.* An analog of tetra-NAG containing –H in place of $-CH_2OH$ at C-5 of residue D binds to lysozyme much more strongly than does tetra-NAG. Propose a structural basis for this difference in binding affinity.

5. *Metal coordination.* Compare the coordination of the zinc atom in carboxypeptidase A with that of the iron atom in oxymyoglobin and oxyhemoglobin.
   (a) Which atoms are directly bonded to these metal ions?
   (b) Which side chains contribute these metal-binding groups?
   (c) Which other side chains in proteins can bind Zn^{2+} or Fe^{2+}?

6. *Detective work.* In Agatha Christie's *Murder on the Orient Express,* Hercule Poirot's task as a detective was compounded by the presence of *so many* plausible suspects. Likewise, the catalytic mechanism of carboxypeptidase A was difficult to unravel because of the presence of several potential catalytic groups at the active site. For example, the hydroxyl group of tyrosine 248 was postulated to be the proton donor to the NH group of the substrate in the cleavage step. Design an experiment that critically tests this hypothesis.

7. *On the thiol trail.* What is a simple means of determining whether a recently discovered proteolytic enzyme is a thiol protease?

8. *A critical donation.* The transfer of a proton from an enzyme to its substrate is often a key step in catalysis.
   (a) Does this occur in the proposed catalytic mechanisms for ribonuclease A, chymotrypsin, lysozyme, and carboxypeptidase A?
   (b) If so, in each case, identify the proton donor.

9. *Variation on an inhibitory theme.* Recall that TPCK, an affinity-labeling reagent, inactivates chymotrypsin by alkylating histidine 57.
   (a) Design an affinity-labeling reagent for trypsin that resembles TPCK.
   (b) How would you test its specificity?

10. *An aldehyde inhibitor.* Elastase is specifically inhibited by an aldehyde derivative of one of its substrates:

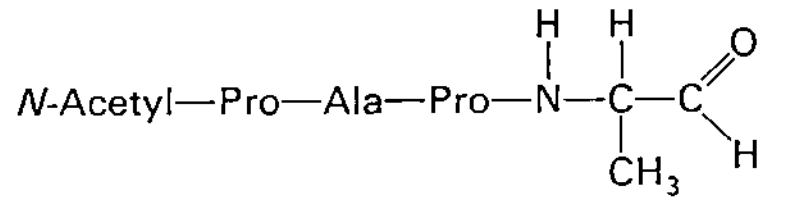

In fact, this aldehyde is an analog of the transition state for catalysis by elastase.
   (a) Which residue at the active site of elastase is most likely to form a covalent bond with this aldehyde?
   (b) What type of covalent link would be formed?

11. *A perfect dimer.* The HIV-1 protease, like other retroviral proteases, is a dimer of identical subunits rather than a single chain twice as large. What is the selective advantage to the virus of a dimeric arrangement?

12. *Designing anti-AIDS drugs.* Symmetric inhibitors of the HIV-1 protease are attractive candidates for anti-AIDS drugs because their symmetry matches that of the target. Propose another mechanism of selectively inhibiting HIV-1 protease.

13. *An antihypertensive drug.* Angiotensin II, an octapeptide formed by the kidneys, raises blood pressure. This pressor hormone is generated by the action of two proteolytic enzymes. First, angiotensinogen, a 14-residue peptide, is cleaved by *renin* to yield angiotensin I, a decapeptide. Then, the last two residues of angiotensin I are removed in a single step by the hydrolytic action of angiotensin converting enzyme (ACE) to form angiotensin II. This hormone rapidly elevates blood pressure by constricting arterioles. ACE is a zinc-containing protease that is mechanistically similar to carboxypeptidase A except that it is cleaves the penultimate rather than the last peptide bond. Captopril, a potent inhibitor (K_i = 0.2 nM) of ACE, is widely used to treat hypertension. How might captopril inhibit ACE?

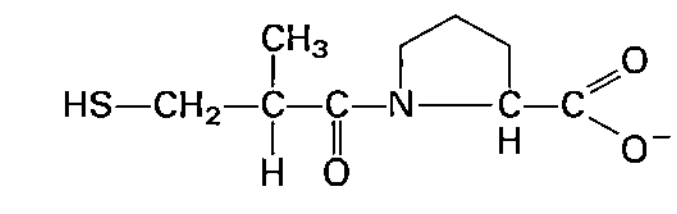

Captopril

14. *Recapitulation.* List the factors contributing to the catalytic power of enzymes.

CHAPTER

10

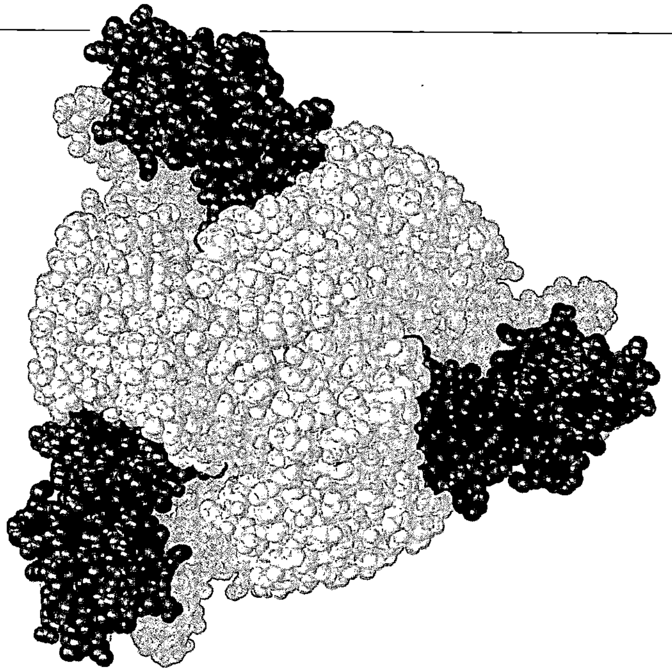

Regulatory Strategies

The capacity of proteins to specifically bind virtually any biomolecule and stabilize its transition state endows enzymes with great catalytic power. This chapter considers another remarkable property of enzymes, the precise regulation of their catalytic activity. Their dynamic character, particularly their inherent ability to transmit conformational changes between spatially distinct sites within a molecule, is exploited in many enzymes to make them *catalytically active at the right time and in the right place.* Enzymatic activity is regulated in four principal ways:

1. *Allosteric control.* We saw in Chapter 7 that the binding of O_2, H^+, and CO_2 to hemoglobin is beautifully coordinated by conformational changes that are transmitted across subunits. Allosteric regulation by small molecules is also profoundly significant in controlling enzymatic activity. This chapter examines one of the best-understood allosteric enzymes, *aspartate transcarbamoylase* from *Escherichia coli.* Catalysis of the first step in pyrimidine biosynthesis by this enzyme is feedback-inhibited by cytidine triphosphate, the final product. The presence of separable regulatory and catalytic subunits has made aspartate transcarbamoylase a choice object of inquiry. Moreover, its three-dimensional structure is known in atomic detail and its allosteric mechanism has been delineated.

2. *Stimulation and inhibition by control proteins.* Stimulatory or inhibitory proteins, as well as small molecules, regulate the activity of enzymes. *Calmodulin,* for example, senses the intracellular concentration of Ca^{2+} and activates multiple proteins inside a cell when calcium levels rise. Blood clotting is markedly accelerated by *antihemophilic factor,* a protein that enhances the activity of a serine protease, as will be discussed later in this

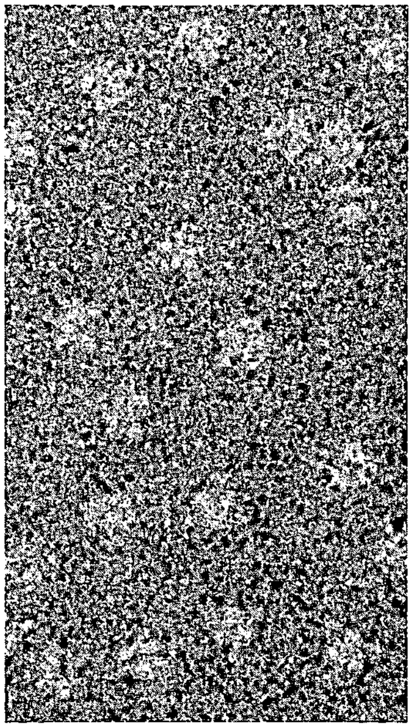

Figure 10-1
Electron micrograph of aspartate transcarbamoylase. [Courtesy of Dr. Robley C. Williams.]

Opening Image: *Aspartate transcarbamoylase, an allosteric enzyme that catalyzes the first step in pyrimidine synthesis, consists of six catalytic (blue) and six regulatory (red) subunits. Conformational changes induced by the binding of substrate and allosteric effectors are transmitted over distances as large as 80 Å. [Drawn from coordinates kindly provided by Dr. Raymond Stevens.]*

chapter. Conversely, the catalytic activity of some enzymes, such as protein kinase A, is held in check by inhibitory subunits. Cyclic AMP, an intracellular messenger in the action of many hormones, activates this kinase by unleashing its catalytic subunits.

3. *Reversible covalent modification.* The catalytic properties of many enzymes are markedly altered by the covalent attachment of phosphoryl groups. For example, glycogen phosphorylase (the enzyme that releases sugar units from glycogen, an energy store) is activated by phosphorylation of a specific serine residue. ATP serves as the phosphoryl donor in these reactions, which are catalyzed by *protein kinases.* The removal of phosphoryl groups by hydrolysis is catalyzed by *protein phosphatases.* The structure, specificity, and control of *protein kinase A,* a ubiquitous eukaryotic enzyme that regulates diverse target proteins, will be considered in this chapter. ATP also donates its AMP unit to reversibly control the activity of bacterial enzymes, such as glutamine synthetase.

4. *Proteolytic activation.* The enzymes controlled by the mechanisms cited above cycle between active and inactive states. A different regulatory motif is used to *irreversibly* convert an inactive enzyme into an active one. Many enzymes are activated by the hydrolysis of a few or even one peptide bond in inactive precursors called *zymogens* or *proenzymes.* This regulatory mechanism generates digestive enzymes such as chymotrypsin, trypsin, and pepsin. This chapter will also consider the mechanism of blood clotting, a remarkable cascade of zymogen activations. Active digestive and clotting enzymes are switched off by the irreversible binding of specific inhibitory proteins that are irresistible lures to their molecular prey.

ASPARTATE TRANSCARBAMOYLASE IS FEEDBACK-INHIBITED BY THE FINAL PRODUCT OF THE PYRIMIDINE PATHWAY

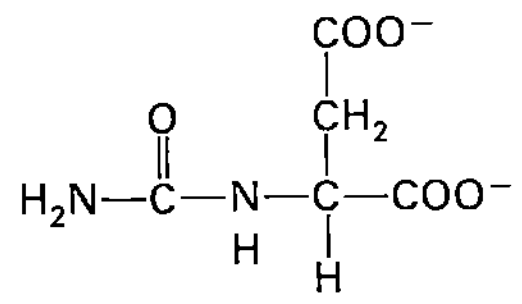

N-Carbamoylaspartate

Pyrimidine biosynthesis begins with the formation of N-*carbamoylaspartate* from aspartate and carbamoyl phosphate. This reaction is catalyzed by *aspartate transcarbamoylase (ATCase),* an especially interesting regulatory enzyme. The enzyme from *E. coli* has been studied in detail by many investigators.

John Gerhart and Arthur Pardee discovered that *ATCase is feedback-inhibited by cytidine triphosphate (CTP),* the final product of this biosynthetic pathway. Furthermore, they found that the binding of carbamoyl phosphate and aspartate is cooperative, as reflected in the sigmoidal dependence of reaction velocity on substrate concentration (Figure 10-2). Cooperative binding serves to switch on the synthesis of *N*-carbamoylaspartate over a rather narrow range of concentration of substrates. CTP inhibits ATCase by decreasing its affinity for substrates without affecting its V_{max}. The extent of inhibition, which may reach 90%, depends on the concentrations of the substrates. In contrast, *ATP is an activator of ATCase.* It enhances the affinity of the enzyme for its substrates, but also leaves V_{max} unaltered. ATP competes with CTP for binding to regulatory sites. Consequently, high levels of ATP prevent CTP from inhibiting the enzyme.

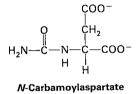

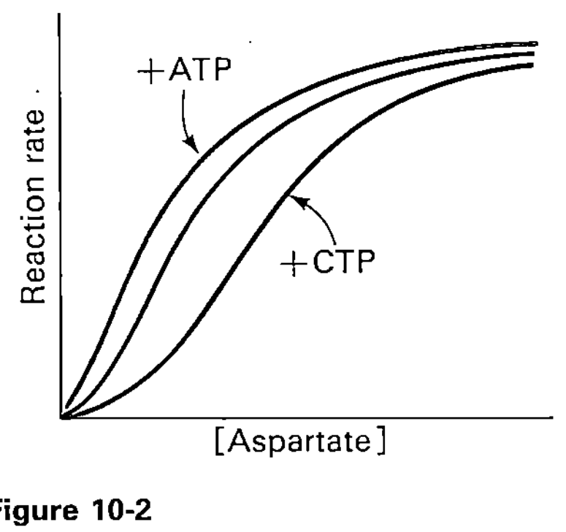

Figure 10-2
Allosteric effects in aspartate transcarbamoylase. ATP is an activator, whereas CTP is an inhibitor. [After J.C. Gerhart. *Curr. Top. Cell Regul.* 2(1970):275.]

The biological significance of the regulation of ATCase by CTP and ATP is twofold. First, activation by ATP signals that energy is available for DNA replication and leads to the synthesis of needed pyrimidine nucleotides. Second, feedback inhibition by CTP assures that *N*-carbamoylaspartate and subsequent intermediates in the pathway are not needlessly formed when pyrimidines are abundant.

ASPARTATE TRANSCARBAMOYLASE CONSISTS OF SEPARABLE CATALYTIC AND REGULATORY SUBUNITS

The regulatory properties of ATCase vanish when the enzyme is treated with a mercurial such as *p*-hydroxymercuribenzoate, which reacts with sulfhydryl groups. ATP and CTP have no effect on the catalytic activity of mercurial-treated ATCase. Furthermore, the binding of substrates becomes noncooperative. By contrast, the modified enzyme has the same maximal catalytic activity as does the native enzyme.

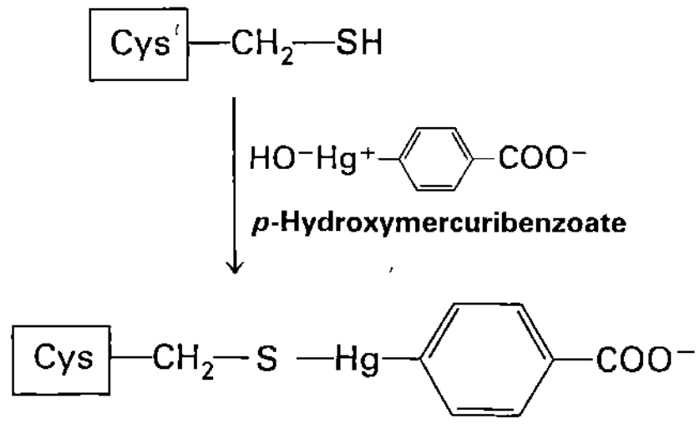

Ultracentrifugation studies carried out by Gerhart and Howard Schachman showed that *mercurials dissociate ATCase into two kinds of subunits* (Figure 10-3). The sedimentation coefficient of the native enzyme is 11.6S, whereas that of the dissociated subunits is 2.8S and 5.8S. The subunits can be readily separated by ion-exchange chromatography because they differ markedly in charge, or by centrifugation in a sucrose density gradient because they differ in size. Furthermore, the attached *p*-mercuribenzoate groups can be removed from the separated subunits by adding an excess of mercaptoethanol.

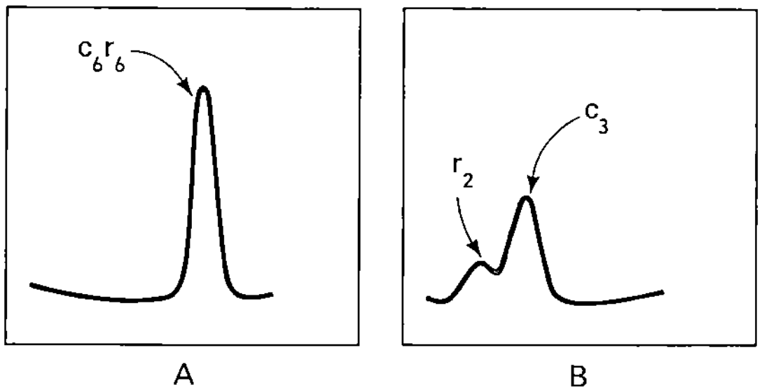

Figure 10-3
Sedimentation velocity patterns of (A) native ATCase and (B) the enzyme dissociated by a mercurial into regulatory and catalytic subunits. [After J.C. Gerhart and H.K. Schachman. *Biochemistry* 4(1965):1054.]

The large subunit, called the *catalytic subunit,* is catalytically active but unresponsive to ATP and CTP. The small subunit is devoid of catalytic activity but can bind CTP or ATP. Hence, it is called the *regulatory subunit.* The catalytic subunit (c_3) consists of three c chains (34 kd each), and the regulatory subunit consists of two r chains (17 kd each). The catalytic and regulatory subunits combine rapidly when they are mixed. The resulting complex has the same structure, c_6r_6, as that of the native enzyme.

$$3\ r_2 + 2\ c_3 \longrightarrow r_6c_6$$

Furthermore, the reconstituted enzyme has the same allosteric properties as those of the native enzyme. Thus, *ATCase is composed of discrete catalytic and regulatory subunits, which interact in the native enzyme to produce its allosteric behavior.* ATCase is a choice object of inquiry because its catalytic and regulatory functions can readily be separated and brought back together.

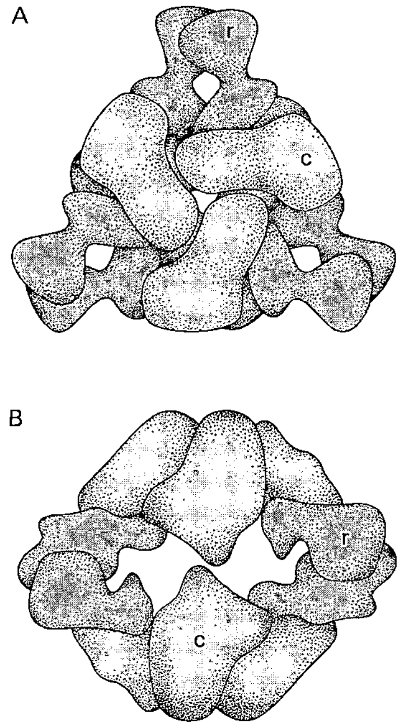

Figure 10-4
Subunit arrangement in aspartate transcarbamoylase. The six catalytic chains (c) are shown in blue and the regulatory chains (r) in red. (A) View looking down the threefold symmetry axis. (B) A perpendicular view. [After K.L. Krause, K.W. Volz, and W.N. Lipscomb. *Proc. Nat. Acad. Sci.* 82(1985):1643.]

X-RAY ANALYSES HAVE REVEALED THE STRUCTURE OF ATCase AND ITS COMPLEX WITH A BISUBSTRATE ANALOG

William Lipscomb and co-workers have elucidated the three-dimensional structure of ATCase at a resolution of 2.6 Å. They also have solved the structure of complexes of ATCase with CTP, the allosteric inhibitor, and with substrate analogs. The design of this enzyme is distinctive (Figure 10-4). One of the catalytic trimers (c_3) lies above and the other catalytic trimer lies below an equatorial belt of three regulatory dimers (r_2). The enzyme contains a large central cavity, which is accessible through several channels. The diameter of a sphere encompassing ATCase is about 13 nm (130 Å), twice the size of hemoglobin.

A view of the path of the main chain of half the enzyme molecule (c_3r_3) is shown in Figure 10-5. We are looking down the threefold symmetry axis of the molecule. The three identical cr units in this view are related by a rotation of 120 degrees. Each r chain on the periphery consists of two domains. The outer one contains the binding site for CTP and the inner one interacts with a catalytic chain in the same trimer. The inner domain of the r chain also interacts with a c chain in the opposite trimer. Thus, as shown in Figure 10-4A, each r chain is in contact with two c chains, and each c chain is in contact with two r chains. The inner lobe of the r chain also contains a zinc ion that plays a structural role by coordinating the sulfur atoms of four cysteine residues.

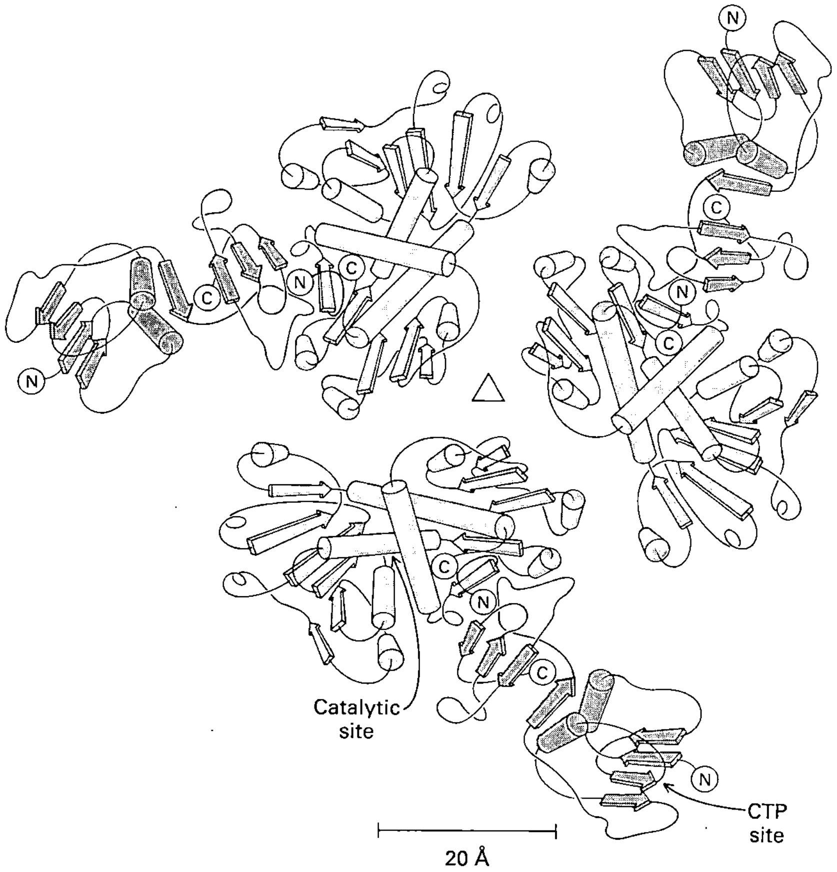

Figure 10-5
Three-dimensional structure of half (c_3r_3) of the ATCase molecule. The c chains are shown in blue and the r chains in red. The catalytic sites are far from the regulatory sites. The green triangle marks the threefold symmetry axis of ATCase. The relation of this half molecule to the other one is depicted in Figure 10-4. [After E.R. Kantrowitz, S.C. Patra-Landis, and W.N. Lipscomb. *Trends Biochem. Sci.* 5(1980):152.]

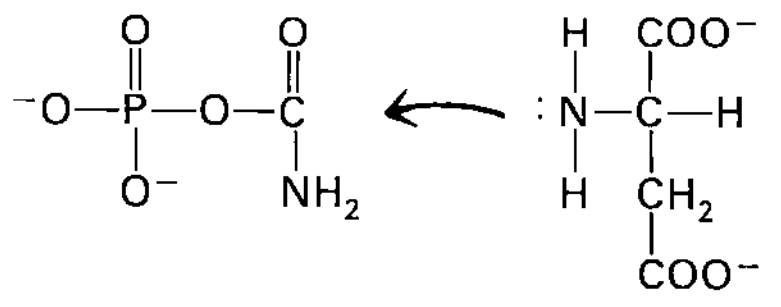

Bound substrates

A

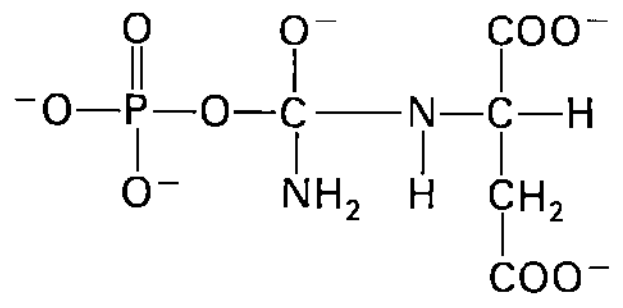

Transition state

B

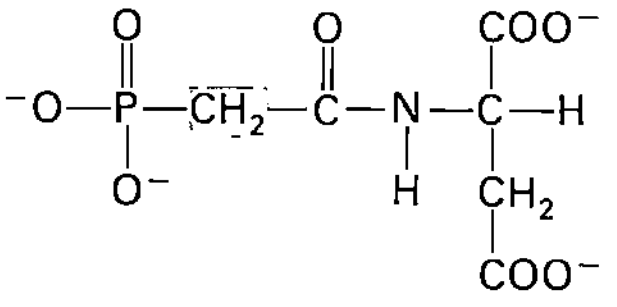

***N*-(Phosphonacetyl)-L-aspartate (PALA)**

C

Figure 10-6
(A) Nucleophilic attack of the amino group of aspartate on the carbonyl carbon atom of carbamoyl phosphate. (B) Transition state. (C) PALA resembles the bound pair of substrates and the transition state.

Where is the active site? In the formation of *N*-carbamoylaspartate, carbamoyl phosphate binds first by multiple electrostatic and hydrogen-bond interactions. Aspartate then binds, and its amino group attacks the carbonyl carbon atom of carbamoyl phosphate (Figure 10-6A). A tetrahedral transition state seems likely (Figure 10-6B). *N*-(phosphonacetyl)-L-aspartate (PALA) (Figure 10-6C), a potent inhibitor, resembles the two-substrate complex and the transition state. PALA binds tightly to ATCase with a dissociation constant of about 10 nM. This unreactive *bisubstrate analog* has proved to be invaluable in studies of ATCase. X-ray analysis of crystals of ATCase containing PALA shows that each of the six active sites is located near the interface between catalytic chains. Because it has a high negative charge, PALA binds electrostatically to four arginines and one lysine in an active site (Figure 10-7). It is also bound to the enzyme by many hydrogen bonds. An active site is formed by residues belonging to two catalytic chains. Specifically, many residues from one chain join with a serine and a lysine residue from an adjacent chain to form a binding pocket for PALA.

The hydrogen bonding of the imidazole side chain of histidine 134 to the carbonyl oxygen atom of PALA is also noteworthy. The carbonyl oxygen atom of carbamoyl phosphate becomes negatively charged when the amino group of aspartate attacks the carbonyl carbon atom. The protonated form of histidine 134 probably stabilizes this negatively charged atom in the tetrahedral transition state.

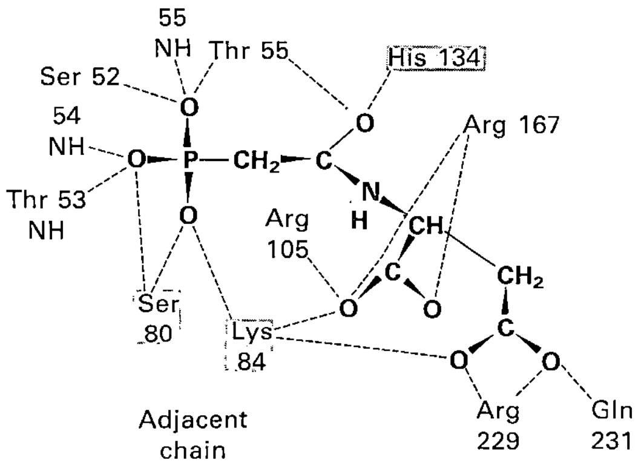

Figure 10-7
Mode of binding of PALA (shown in red) to the catalytic site of ATCase. Not all the electrostatic and hydrogen bonds formed are shown in this schematic diagram. Histidine 134 (shown in blue) may stabilize the negative charge on the carbonyl oxygen atom in the transition state. This binding site is at the interface of catalytic chains. Residues contributed by the adjacent catalytic chain are marked in yellow. [After K.W. Volz, K.L. Krause, and W.N. Lipscomb. *Biochem. Biophys. Res. Comm.* 136(1986):823.]

ALLOSTERIC INTERACTIONS IN ATCase ARE MEDIATED BY LARGE CHANGES IN QUATERNARY STRUCTURE

Ultracentrifugation studies showed that the binding of substrate leads to a 3% *decrease* in the sedimentation coefficient of ATCase. This finding indicated that ATCase *expands* on binding substrates. X-ray studies of unliganded ATCase and its complex with PALA have revealed that large changes in quaternary structure accompany the binding of this bisubstrate analog (Figure 10-8). The catalytic trimers move apart 12 Å and turn by 10 degrees. In addition, each regulatory dimer turns 15 degrees about its twofold axis. The binding of PALA also leads to substantial changes in the tertiary structure of each catalytic chain (Figure 10-9).

The active site at the interface of catalytic chains is formed by side chains from both. Each catalytic chain contains a carbamoyl phosphate–binding domain (N-terminal) and an aspartate-binding domain (C-terminal). Binding of the two substrates brings these domains together by 2 Å and in-

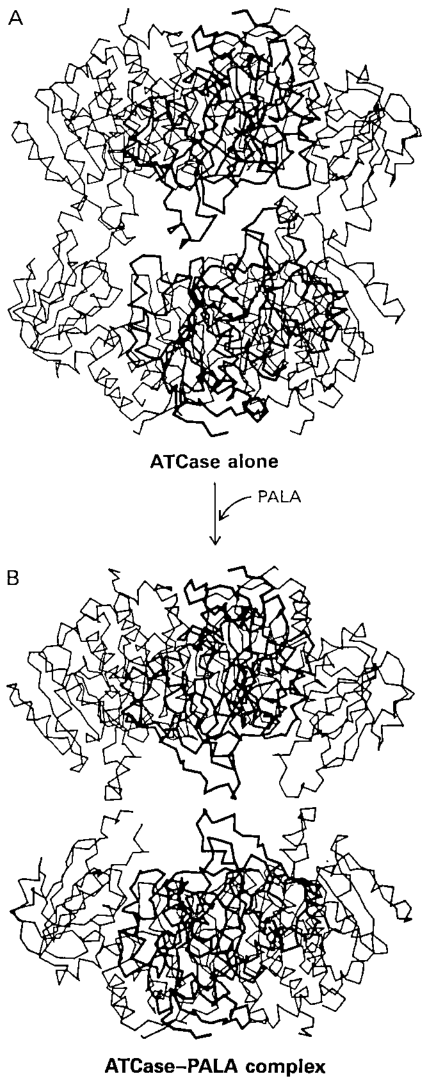

Figure 10-8
Allosteric transition of the catalytic chains of ATCase. (A) Unliganded T state and (B) liganded R state. The regulatory chains have been omitted for clarity. The catalytic trimers move apart 12 Å and turn 10 degrees in the transition to the high-affinity R state. [After K.L. Krause, K.W. Volz, and W.N. Lipscomb. *Proc. Nat. Acad. Sci.* 82(1985):1643.]

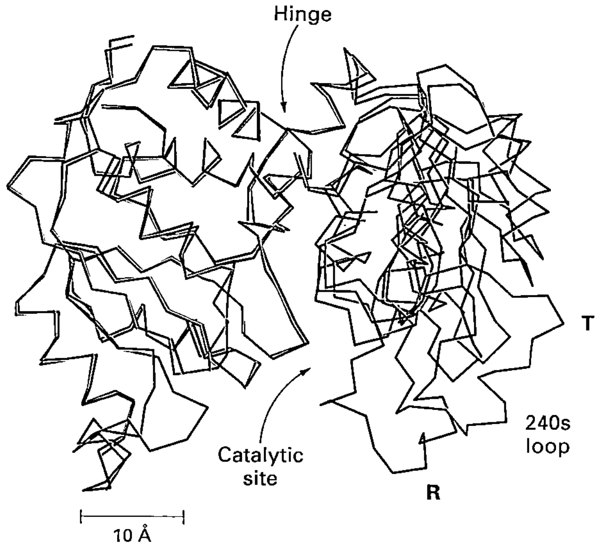

Figure 10-9
Change in tertiary structure of a catalytic chain in the transition from the T state (shown in blue) to the R state (shown in red). Movement about a hinge brings the two domains closer together in the T → R transition. Note that the carbamoyl phosphate-binding domain (left) changes little, in contrast with the aspartate-binding domain (right). The 240s loop undergoes the largest change in position. The active site is located in a cleft between the domains. [Drawn from 8atc.pdb and 5atl.pdb. R.C. Stevens, J.E. Gouaux, and W.N. Lipscomb. *Biochemistry* 29(1990): 7691. The hinge movement is discussed in E.S. Huang, E.P. Rock, and S. Subbiah. *Curr. Biol.* 3(1993):740.]

duces them to switch to the R conformation. A loop consisting of residues 230 to 245, called the *240s loop*, undergoes a major reorientation in this transition (see Figure 10-9). In the T state, side chains from the 240s loop bridge the upper and lower catalytic trimers. This constraining interaction is disrupted by substrate binding. The movement of the 240s loop liberates several groups, enabling them to interact with substrate and promote catalysis.

THE BINDING OF SUBSTRATES TO ATCase LEADS TO A HIGHLY CONCERTED ALLOSTERIC TRANSITION

Two limiting models for allosteric interactions were presented in the chapter on hemoglobin (p. 166). Recall that the concerted model assumes that all subunits in a molecule switch from the T to the R state in unison. The central tenet of the concerted model is that symmetry is preserved in allosteric transitions. In contrast, the simple sequential model assumes that the T → R transition occurs only in subunits containing bound ligand.

Which model best accounts for the allosteric properties of ATCase? The simple sequential model predicts that the fraction of catalytic chains in the R state (f_R) is equal to the fraction containing bound substrate (Y). The concerted model, in contrast, predicts that f_R increases more rapidly than Y as the substrate concentration is increased (see equations 13 and 14 on p. 167). These predictions were tested by measuring f_R and Y as a function of the concentration of succinate, an unreactive analog of aspartate, in the presence of a saturating concentration of carbamoyl phosphate. The value of Y was determined from the change in absorbance at 280 nm, and the value of f_R from the sedimentation coefficient. As was mentioned earlier, the T → R transition expands the enzyme, which increases its frictional coefficient and thereby decreases its sedimentation coefficient. A decisive result was obtained (Figure 10-10). *The change in* f_R *leads the change in* Y *on addition of succinate, as predicted by the concerted model.*

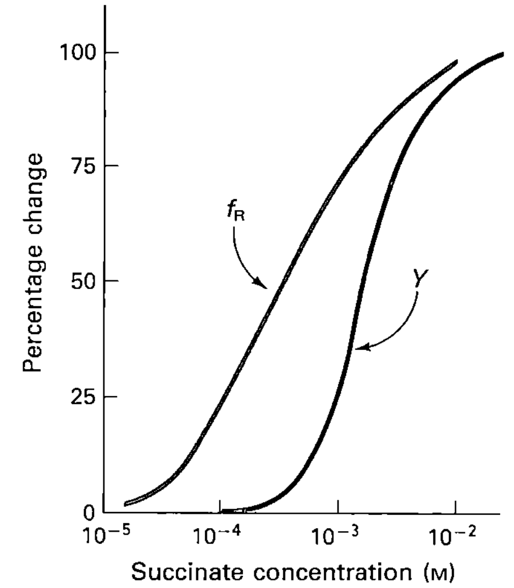

Figure 10-10
Dependence of the fraction of ATCase in the R state (f_R) and the fractional occupancy of substrate-binding sites (Y) on the concentration of succinate, a substrate analog. The change in f_R precedes the change in Y, as predicted by the concerted model. [From M.W. Kirschner and H.K. Schachman. *Biochemistry* 12(1966):2997.]

Spectroscopic studies of ATCase containing colored reporter groups provided further insight into the nature of the allosteric transition. ATCase was reacted with nitromethane to form a colored nitrotyrosine group (λ_{max} = 430 nm) in each of its catalytic chains (Figure 10-11). An essential lysine at each catalytic site was also modified to block the binding of substrate. Catalytic trimers from this doubly modified enzyme were then combined with native trimers to form a hybrid enzyme. The binding of succinate to the functional catalytic sites of the native c_3 moiety changed the visible absorption spectrum of nitrotyrosine residues in the *other* c_3 moiety of the hybrid enzyme (Figure 10-11C). Thus, the binding of substrate analog to the active sites of one trimer altered the structure of the other trimer.

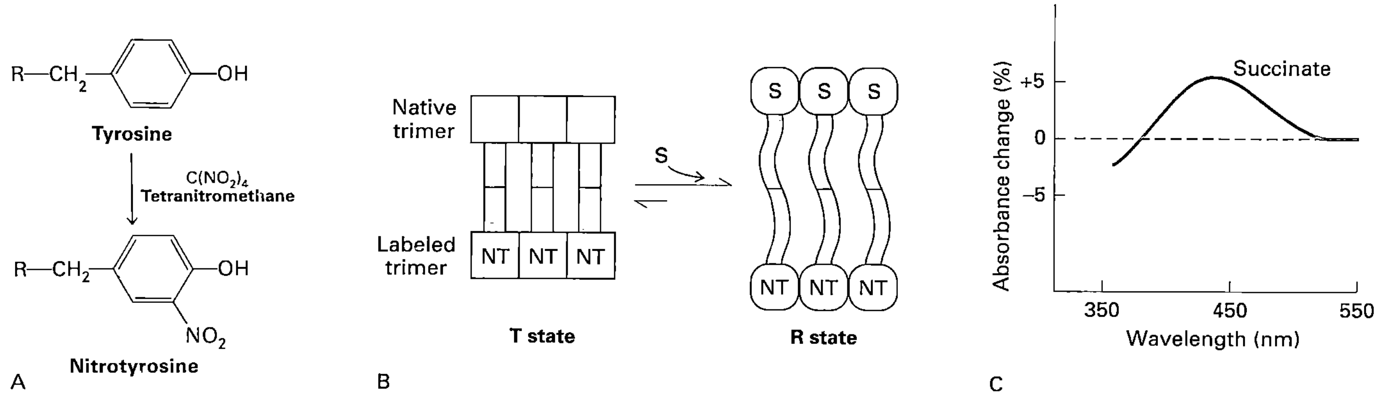

Figure 10-11
Long-range allosteric effects in ATCase. (A) Nitration of tyrosine to form a colored nitrotyrosine group that served as a reporter. (B) A hybrid ATCase molecule containing a nitrotyrosylated c_3 trimer (red), a native c_3 trimer, and native r subunits was prepared. The binding of succinate (S) to native catalytic sites in one trimer changed the environment of nitrotyrosyl groups in the other trimer, as evidenced by the change in the nitrotyrosyl absorption spectrum (C). Succinate binding led to a concerted T → R transition. [After H.K. Schachman. *J. Biol. Chem.* 263(1988):18583.]

ATP AND CTP REGULATE ATCase ACTIVITY BY SHIFTING THE EQUILIBRIUM BETWEEN T AND R FORMS

According to the concerted model, an allosteric activator shifts the conformational equilibrium of all subunits toward the R state, whereas an allosteric inhibitor shifts it toward the T state. A different ATCase hybrid was constructed to test this prediction. Normal regulatory subunits were combined with nitrotyrosine-containing catalytic subunits. The addition of ATP in the absence of substrate increased the absorbance at 430 nm (Figure 10-12), the same change as was elicited by the addition of succinate (see Figure 10-11). Thus, ATP (an allosteric activator) shifted the equilibrium to the R form. Conversely, CTP in the absence of substrate

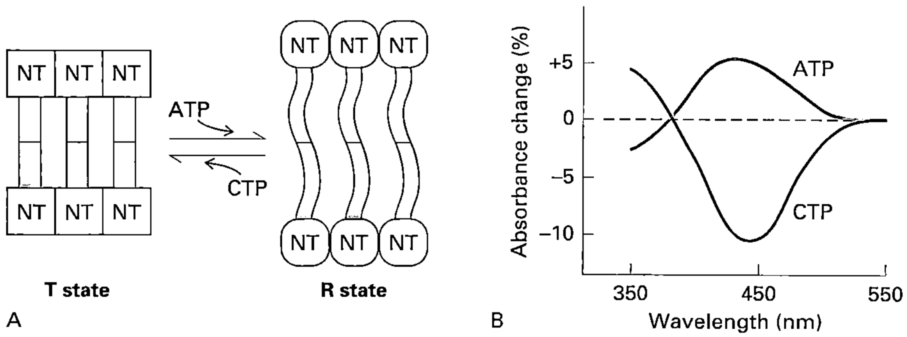

Figure 10-12
The binding of ATP and CTP to the regulatory subunits of ATCase leads to global conformational changes. (A) A hybrid ATCase containing nitrotyrosylated catalytic subunits and native regulatory subunits was formed. (B) The addition of ATP, an activator, shifted the allosteric equilibrium of the enzyme to the R state, as evidenced by the change in the nitrotyrosine absorption spectrum. Conversely, CTP, an inhibitor, shifted the allosteric equilibrium to the T state. [After H.K. Schachman. *J. Biol. Chem.* 263(1988):18583.]

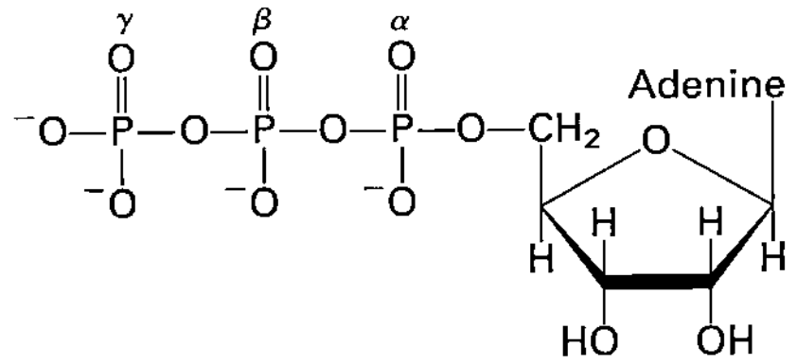

**Adenosine triphosphate
(ATP)**

decreased the absorbance at 430 nm. Hence, this allosteric inhibitor shifted the equilibrium to the T form. Sedimentation velocity experiments provided further evidence that ATP and CTP, like succinate, induce global conformational changes. Thus, *the concerted model accounts for the ATP-induced and CTP-induced (heterotropic), as well as for the substrate-induced (homotropic), allosteric interactions of ATCase.*

PHOSPHORYLATION IS A HIGHLY EFFECTIVE MEANS OF SWITCHING THE ACTIVITY OF TARGET PROTEINS

The activity of many enzymes, membrane channels, and other target proteins is regulated by phosphorylation, the most common and ubiquitous type of reversible covalent modification. The enzymes catalyzing these reactions are called *protein kinases*. The terminal (γ) phosphoryl group of ATP is transferred to specific *serine* and *threonine* residues by one class of protein kinases, and to specific *tyrosine* residues by another. The acceptors in protein phosphorylation reactions are located inside cells, where ATP is abundant. Proteins that are entirely extracellular are not regulated by reversible phosphorylation.

$$R{-}OH + ATP \rightleftharpoons R{-}O{-}\overset{\overset{\displaystyle O}{\|}}{\underset{\underset{\displaystyle O^-}{|}}{P}}{-}O^- + ADP + H^+$$

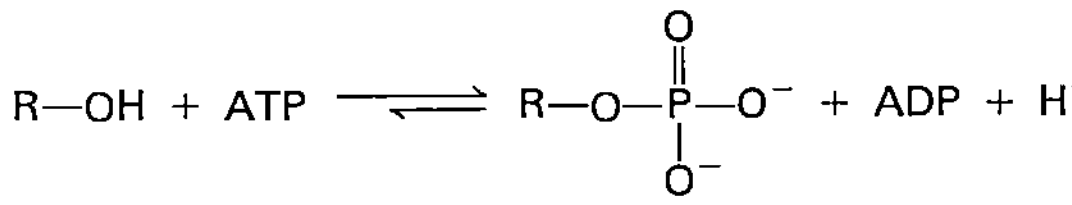

Protein phosphatases reverse the effects of kinases by catalyzing the hydrolysis of phosphoryl groups attached to proteins. The unmodified hydroxyl-containing side chain is regenerated and orthophosphate (P_i) is produced.

$$R{-}O{-}\overset{\overset{\displaystyle O}{\|}}{\underset{\underset{\displaystyle O^-}{|}}{P}}{-}O^- + H_2O \rightleftharpoons R{-}OH + P_i$$

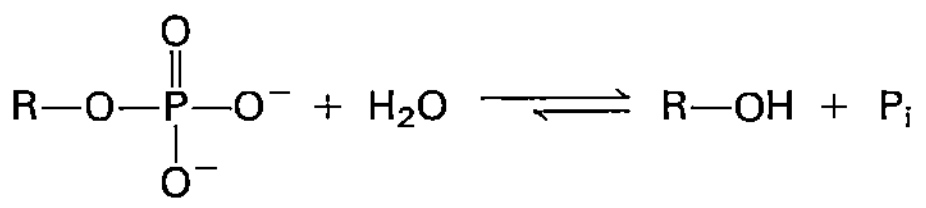

Both phosphorylation and dephosphorylation are essentially irreversible under physiologic conditions. Note that the net outcome of the two reactions is the hydrolysis of ATP to ADP and P_i, which has a ΔG of -12 kcal/mol under cellular conditions (p. 448). This highly favorable free-energy change assures that target proteins cycle unidirectionally between unphosphorylated and phosphorylated forms (Figure 10-13). The rate of cycling depends on the activities of kinases and phosphatases. We will see shortly how a major protein kinase is switched on.

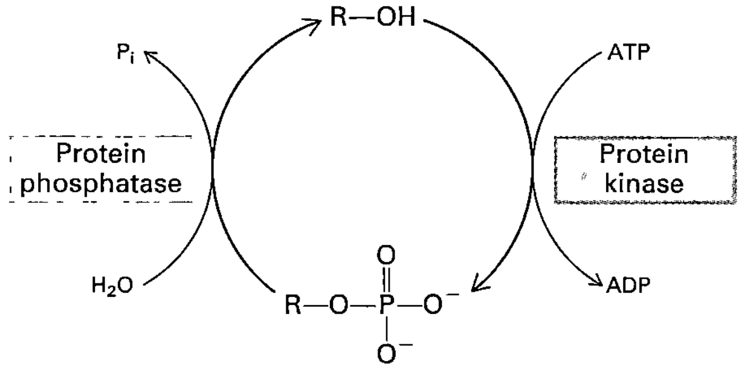

Figure 10-13
Targets of protein kinases cycle unidirectionally between dephosphorylated and phosphorylated forms. The reactions catalyzed by kinases and phosphatases are essentially irreversible under physiologic conditions.

Phosphorylation is a highly effective means of controlling the activity of proteins for structural, thermodynamic, and kinetic reasons:

1. A phosphoryl group adds two negative charges to a modified protein. Electrostatic interactions in the unmodified protein can be disrupted and new ones can be formed. Such structural changes can markedly alter substrate binding and catalytic activity.

2. A phosphoryl group can form three hydrogen bonds. The tetrahedral geometry makes these interactions highly directional.

3. The free-energy of phosphorylation is large. Of the −12 kcal/mol provided by ATP, about half is consumed in making phosphorylation irreversible; the other half is conserved in the phosphorylated protein. Recall that a free-energy change of 1.36 kcal/mol corresponds to a factor of 10 in an equilibrium constant (p. 187). Hence, phosphorylation can change the conformational equilibrium between different functional states by a large factor, of the order of 10^4.

4. Phosphorylation and dephosphorylation can occur in less than a second or over a span of hours. The kinetics can be adjusted to meet the timing needs of physiologic process.

5. Phosphorylation often evokes *highly amplified* effects. A single activated kinase can phosphorylate hundreds of target proteins in a short interval. Furthermore, a single target enzyme can transform a large number of substrate molecules.

The degree of specificity of protein kinases varies. *Dedicated protein kinases* phosphorylate a single protein or several closely related ones. *Multifunctional protein kinases,* on the other hand, modify many different targets: they have a wide reach and can coordinate diverse processes. Comparisons of amino acid sequences of many phosphorylation sites show that multifunctional kinases recognize related sequences. For example, the *consensus motif* recognized by protein kinase A is Arg-Arg-X-*Ser*-Z or Arg-Arg-X-*Thr*-Z, where X is a small residue, Z is a large hydrophobic one, and *Ser* or *Thr* is the site of phosphorylation. It should be noted that this sequence is not absolutely required. Lysine, for example, can substitute for one of the arginines, though not as effectively. Short synthetic peptides containing a consensus motif are nearly always phosphorylated by serine-threonine protein kinases. Thus, *the primary determinant of specificity is the linear sequence surrounding the serine or threonine residue.* However, distant residues can contribute to specificity. Also, the accessibility of the phosphorylation site may be affected by the conformation of the protein.

CYCLIC AMP ACTIVATES PROTEIN KINASE A (PKA) BY UNLEASHING ITS CATALYTIC SUBUNITS

Many hormones trigger the formation of cyclic AMP, which is formed by cyclization of ATP. Cyclic AMP serves as an intracellular messenger in mediating their physiologic actions, as will be discussed in Chapter 13. The striking finding is that *most effects of cyclic AMP in eukaryotic cells are mediated by the activation of a single protein kinase. This key enzyme is called protein kinase A (PKA) or cAMP-dependent protein kinase (cAPK).* We will refer to it as PKA. The kinase alters the activities of target proteins by phosphorylating specific serine or threonine residues.

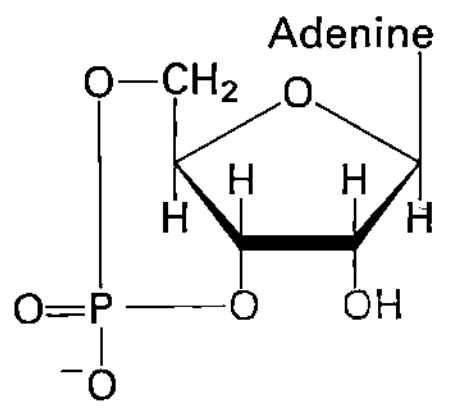

Cyclic AMP (cAMP)

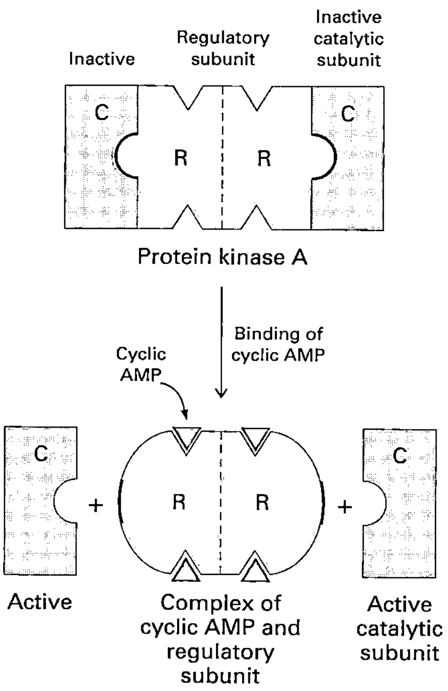

Figure 10-14
The binding of four molecules of cAMP activates protein kinase A by dissociating the inhibited holoenzyme (R_2C_2) into a regulatory subunit (R_2) and two catalytically active subunits (2C). The pseudosubstrate sequence of each R chain is depicted in red.

PKA is activated by cAMP concentrations of the order of 10 nM. What is the activation mechanism? The enzyme in muscle consists of two kinds of subunits: a 49-kd regulatory (R) subunit, which has high affinity for cyclic AMP, and a 38-kd catalytic (C) subunit. In the absence of cyclic AMP, the regulatory and catalytic subunits form an R_2C_2 complex that is enzymatically inactive (Figure 10-14). The binding of two molecules of cyclic AMP to each of the regulatory subunits leads to the dissociation of R_2C_2 into an R_2 subunit and two C subunits. These free catalytic subunits are then enzymatically active. Thus, *the binding of cyclic AMP to the regulatory subunit relieves its inhibition of the catalytic subunit.* The presence of distinct regulatory and catalytic subunits in this protein kinase is reminiscent of aspartate transcarbamoylase.

How does cAMP activate the kinase? Each R chain contains the sequence Arg-Arg-Gly-*Ala*-Ile, which matches the consensus sequence for phosphorylation except for the presence of alanine in place of serine. In the R_2C_2 complex, this *pseudosubstrate sequence* of R occupies the catalytic site of C, thereby preventing the entry of protein substrates. The binding of cAMP to the R chains allosterically moves the pseudosubstrate sequences out of the catalytic sites. The released C chains are then free to bind and phosphorylate substrate proteins.

ATP AND THE TARGET PROTEIN BIND TO A DEEP CLEFT IN THE CATALYTIC SUBUNIT OF PROTEIN KINASE A

The three-dimensional structure of the catalytic subunit of PKA containing a bound 20-residue peptide inhibitor has recently been determined by x-ray crystallography. The 350-residue catalytic subunit is bilobed (Figure 10-15). A deep cleft between the lobes is filled by ATP and part of the inhibitor. The small lobe makes many contacts with Mg^{2+}–ATP, whereas the large lobe binds peptide and contributes the key catalytic residues. *This structure has broad significance because residues 40 to 280 constitute a conserved catalytic core that is shared by more than a hundred different protein kinases.* For example, in nearly all known protein kinases, a glycine-rich loop (*Gly*-Thr-*Gly*-Ser-Phe-*Gly*) participates in ATP binding. The catalytic loop contains an aspartate and an asparagine that are nearly invariant in this family of enzymes.

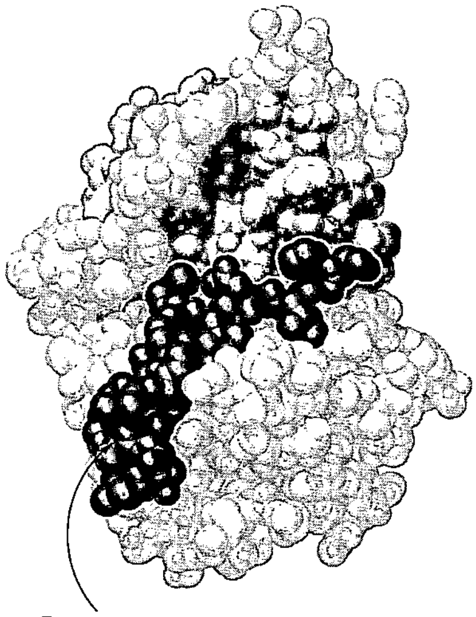

Figure 10-15
Three-dimensional structure of a complex of the catalytic subunit of protein kinase A and an inhibitor bearing a pseudosubstrate sequence. The inhibitor (red) binds in a cleft between the domains (green and blue) of the enzyme. The bound ATP is sequestered in the active site and cannot be seen. [Drawn from 1atp.pdb. J. Zheng, D.R. Knighton, L.F. TenEyck, R. Karlsson, N.-H. Xuong, S.S. Taylor, and J.M. Sowadski. *Biochemistry* 32(1993):2154.]

The bound peptide in this crystal occupies the active site because it contains the pseudosubstrate sequence Arg-Arg-Asn-*Ala*-Ile. The structural basis for recognition of this consensus motif is now evident (Figure 10-16). The guanidinium nitrogen atoms of the first arginine are 3 Å away from the carboxylate side chain of a glutamate residue of the enzyme. The second arginine likewise forms ion pairs with two carboxylates. The nonpolar side chain of isoleucine, which matches Z in the consensus sequence, fits snugly in a hydrophobic groove formed by two leucine residues of the enzyme.

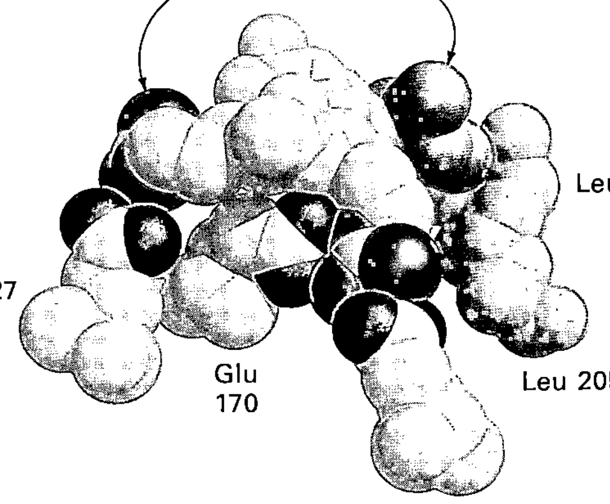

Figure 10-16
Mode of binding of a pseudosubstrate sequence to the active site of protein kinase A. The two arginine side chains (nitrogen atoms in blue) of the pseudosubstrate form salt bridges with three glutamate carboxylates (oxygen atoms in red). Hydrophobic interactions are also important in the recognition of substrate. The isoleucine residue (green) of the pseudosubstrate is in contact with a pair of leucines (yellow) of the enzyme. [Drawn from 1atp.pdb. J. Zheng, D.R. Knighton, L.F. TenEyck, R. Karlsson, N.-H. Xuong, S.S. Taylor, and J.M. Sowadski. *Biochemistry* 32(1993):2154.]

MANY ENZYMES ARE ACTIVATED BY SPECIFIC PROTEOLYTIC CLEAVAGE

We turn now to a different mechanism of enzyme regulation. Enzymes such as lysozyme acquire full enzymatic activity as they spontaneously fold into their characteristic three-dimensional forms. In contrast, many other enzymes are synthesized as inactive precursors that are subsequently activated by cleavage of one or a few specific peptide bonds. The inactive precursor is called a *zymogen* (or a *proenzyme*). Proteolytic activation, in contrast with reversible regulation by phosphorylation, can occur outside cells because a sustained energy source (ATP) is not needed. Another noteworthy difference is that proteolytic activation, in contrast with allosteric control and reversible covalent modification, can occur just once in the life of an enzyme molecule.

Activation of enzymes and other proteins by specific proteolysis recurs frequently in biological systems. For example:

1. The *digestive enzymes* that hydrolyze proteins are synthesized as zymogens in the stomach and pancreas (Table 10-1).

Table 10-1
Gastric and pancreatic zymogens

Site of synthesis	*Zymogen*	*Active enzyme*
Stomach	Pepsinogen	Pepsin
Pancreas	Chymotrypsinogen	Chymotrypsin
Pancreas	Trypsinogen	Trypsin
Pancreas	Procarboxypeptidase	Carboxypeptidase
Pancreas	Proelastase	Elastase

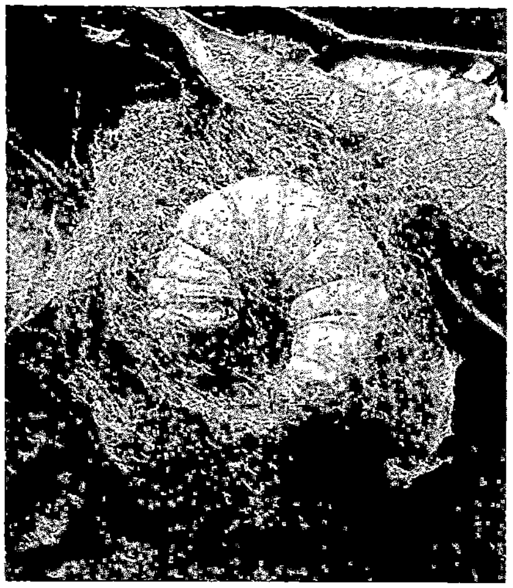

Figure 10-17
The cocoon spun by the larva of the silk moth *Bombyx mori* is digested at pupation by cocoonase, a proteolytic enzyme derived from a catalytically inactive precursor. [Stephen Dalton, NHPA.]

2. *Blood clotting* is mediated by a cascade of proteolytic activations that assures a rapid and amplified response to trauma.

3. Some protein hormones are synthesized as inactive precursors. For example, *insulin* is derived from *proinsulin* by proteolytic removal of a peptide.

4. The fibrous protein *collagen,* the major constituent of skin and bone, is derived from *procollagen,* a soluble precursor.

5. Many *developmental processes* are controlled by the activation of zymogens. In the metamorphosis of a tadpole into a frog, large amounts of collagen are resorbed from the tail in the course of a few days. Likewise, much collagen is broken down in a mammalian uterus after delivery. The conversion of *procollagenase* to *collagenase,* the active protease, is precisely timed in these remodeling processes. *Cocoonase,* a key enzyme in insect development, is also derived from an inactive precursor (Figure 10-17).

The mechanism of activation of three major digestive enzymes—chymotrypsin, trypsin, and pepsin—will be considered first. We shall then turn to blood clotting, which is mediated by a series of proteolytic activations culminating in the conversion of soluble fibrinogen into an insoluble fibrin clot. Proteolytic activation, in contrast with allosteric control, is irreversible. Hence, another means of switching off the activities of these enzymes is needed. Two specific inhibitory proteins that bind very tightly to proteolytic enzymes and block their activity will be discussed.

CHYMOTRYPSINOGEN IS ACTIVATED BY SPECIFIC CLEAVAGE OF A SINGLE PEPTIDE BOND

Chymotrypsin is a digestive enzyme that hydrolyzes proteins in the small intestine. Its inactive precursor, *chymotrypsinogen,* is synthesized in the pancreas, as are several other zymogens and digestive enzymes. Indeed, the pancreas is one of the most active organs in synthesizing proteins. The enzymes and zymogens are synthesized in the acinar cells of the pancreas and stored inside membrane-bound granules (Figure 10-18). The zymogen granules accumulate at the apex of the acinar cell; when the cell is stimulated by a hormonal signal or a nerve impulse, the contents of the granules are released into a duct leading into the duodenum.

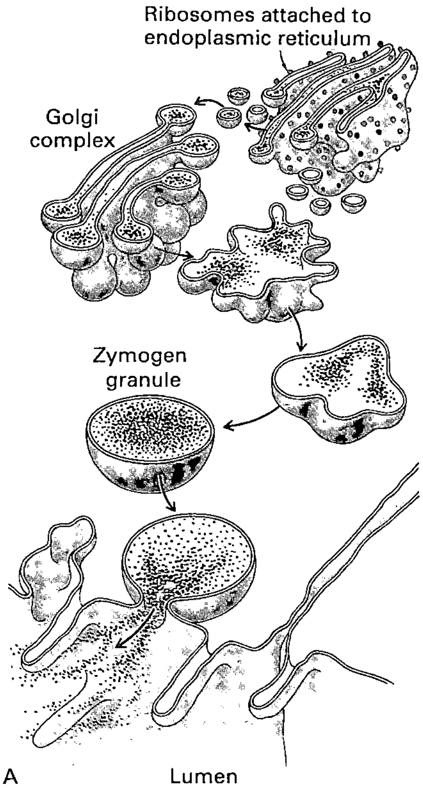

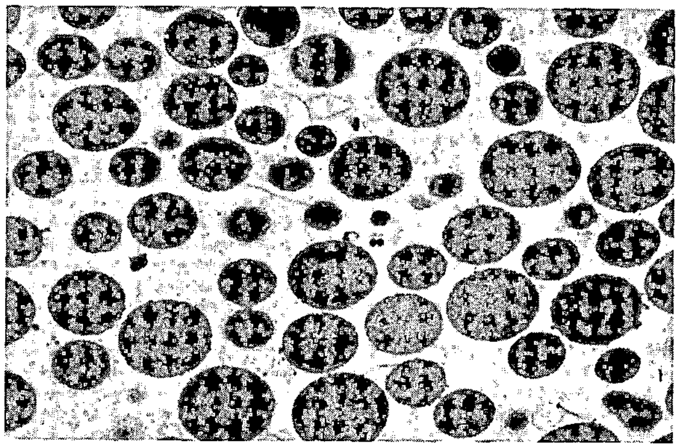

Figure 10-18
Secretion of zymogens by an acinar cell of the pancreas. (A) Diagrammatic representation of the secretory process. (B) Zymogen granules appear in electron micrographs as very dense bodies because they have a high concentration of protein. [Courtesy of Dr. George Palade.]

Chymotrypsinogen, a single polypeptide chain consisting of 245 amino acid residues, is virtually devoid of enzymatic activity. It is converted into a fully active enzyme when the peptide bond joining arginine 15 and isoleucine 16 is cleaved by trypsin (Figure 10-19). The resulting active enzyme, called π-chymotrypsin, then acts on other π-chymotrypsin molecules. Two dipeptides are removed to yield α-chymotrypsin, the stable form of the enzyme. The three resulting chains in α-chymotrypsin remain linked to each other by two interchain disulfide bonds. The additional cleavages made in converting π- into α-chymotrypsin are superfluous, because π-chymotrypsin is already fully active. The striking feature of this activation process is that *cleavage of a single specific peptide bond transforms the protein from a catalytically inactive form into one that is fully active.*

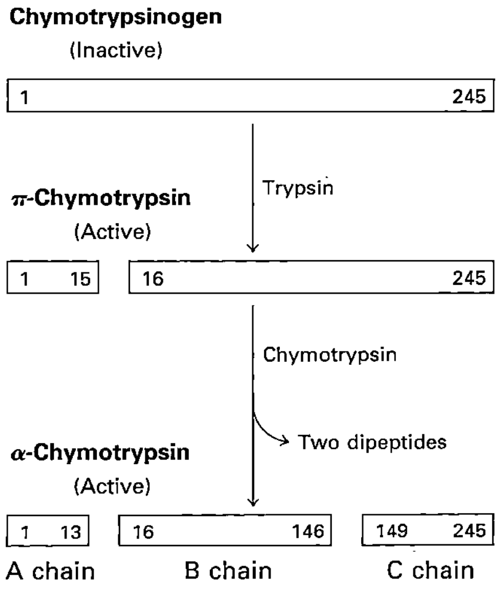

Figure 10-19
Proteolytic activation of chymotrypsinogen. The three chains of α-chymotrypsin are linked by two interchain disulfide bonds (A to B, and B to C).

PROTEOLYTIC ACTIVATION OF CHYMOTRYPSINOGEN LEADS TO THE FORMATION OF A SUBSTRATE BINDING SITE

How does cleavage of a single peptide bond activate the zymogen? Elucidation of the three-dimensional structure of chymotrypsinogen by Joseph Kraut revealed the key conformational changes:

1. The newly formed *amino-terminal group of isoleucine 16 turns inward and interacts with aspartate 194* in the interior of the chymotrypsin molecule (Figure 10-20). Protonation of this amino group stabilizes the active form of chymotrypsin, as shown by the dependence of enzyme activity on pH.

2. This electrostatic interaction triggers a number of conformational changes. Methionine 192 moves from a deeply buried position in the zymogen to the surface of the molecule in the enzyme, and residues 187 and 193 become more extended. These changes result in the formation of the *substrate specificity site* for aromatic and bulky nonpolar groups. One side of this site is made up of residues 189 through 192. *This cavity for part of the substrate is not fully formed in the zymogen.*

3. The tetrahedral transition state in catalysis by chymotrypsin is stabilized by hydrogen bonds between the negatively charged carbonyl oxygen atom of the substrate and two NH groups of the main chain of the enzyme (see Figure 9-36, on p. 226). One of these NH groups is not appropriately located in chymotrypsinogen, and so *the oxyanion hole is incomplete in the zymogen.*

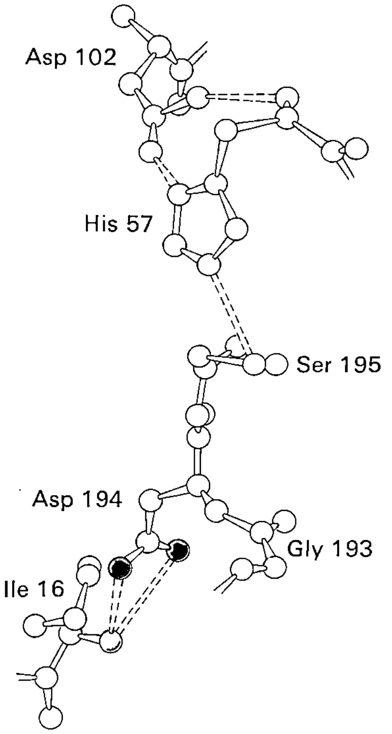

Figure 10-20
Environment of aspartate 194 and isoleucine 16 in chymotrypsin. The electrostatic interaction between the carboxylate of Asp 194 (red) and the α-NH_2 group of Ile 16 (blue) is essential for the activity of chymotrypsin. These groups are adjacent to the Ser-His-Asp catalytic triad (shown on top). [After D.M. Blow and T.A. Steitz. X-ray diffraction studies of enzymes. *Ann. Rev. Biochem.* 39(1970):86. Copyright © 1970 by Annual Reviews Inc. All rights reserved.]

4. The conformational changes elsewhere in the molecule are very small. Thus, *the switching on of enzymatic activity in a protein can be accomplished by discrete, highly localized conformational changes that are triggered by the hydrolysis of a single peptide bond.*

A HIGHLY MOBILE REGION BECOMES ORDERED IN THE ACTIVATION OF TRYPSINOGEN

The structural changes accompanying the activation of trypsinogen are somewhat different from those in the activation of chymotrypsinogen. X-ray analyses have shown that the conformation of four stretches of polypeptide, constituting about 15% of the molecule, changes markedly on activation. *These regions, called the activation domain, are very floppy in the zymogen, whereas they have a well-defined conformation in trypsin.* Furthermore, the oxyanion hole (p. 226) in trypsinogen is too far from histidine 57 to promote the formation of the tetrahedral transition state.

The digestion of proteins in the duodenum requires the concurrent action of several proteolytic enzymes, because each is specific for a limited number of side chains. Thus, the zymogens must be switched on at the same time. Coordinated control is achieved by the action of *trypsin as the common activator of all the pancreatic zymogens*—trypsinogen, chymotrypsinogen, proelastase, and procarboxypeptidase. How is the trypsin produced? The cells that line the duodenum secrete an enzyme, *enteropeptidase,* that hydrolyzes a unique lysine–isoleucine peptide bond in trypsinogen as it enters the duodenum from the pancreas (Figure 10-21). The small amount of trypsin produced in this way activates more trypsinogen and the other zymogens. Thus, *the formation of trypsin by enteropeptidase is the master activation step.*

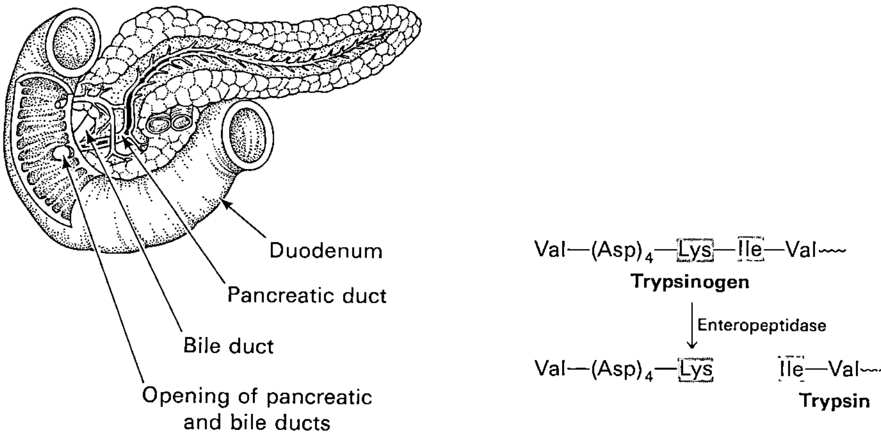

Figure 10-21
Trypsinogen is activated by enteropeptidase as it enters the duodenum.

PEPSINOGEN CLEAVES ITSELF IN AN ACIDIC ENVIRONMENT TO FORM HIGHLY ACTIVE PEPSIN

Pepsin digests proteins in the highly acidic environment of the stomach. Its pH optimum of 2 is very unusual for an enzyme. Recall that pepsin and other members of the aspartyl protease family contain two aspartate residues in their catalytic center, one in the –COOH form and the other in the $-COO^-$ form. *Pepsinogen,* the zymogen, contains an amino-terminal precursor segment of 44 residues that is proteolytically removed in the formation of pepsin. This activation occurs spontaneously below pH 5. The essential event is the cleavage of the peptide bond between leucine 16 and isoleucine 17 in the precursor segment. The rate of formation of pepsin is independent of the concentration of pepsinogen, which shows that activation is *intramolecular.*

The precursor segment is highly basic, whereas pepsin is highly acidic. X-ray crystallographic analyses have shown that the active site is fully formed in pepsinogen but is blocked at neutral pH by residues of the precursor segment. Six lysine and arginine side chains of the precursor segment form salt bridges with the carboxylate side chains of glutamate and aspartate residues of the pepsin moiety. Most important, a lysine side chain of the precursor interacts electrostatically with the pair of aspartates at the active site (Figure 10-22). When the pH is lowered, salt bridges between the precursor segment and the pepsin moiety are broken because several carboxylates become protonated. During the ensuing conformational rearrangement, the now exposed catalytic site hydrolyzes the peptide bond between the precursor and pepsin moieties.

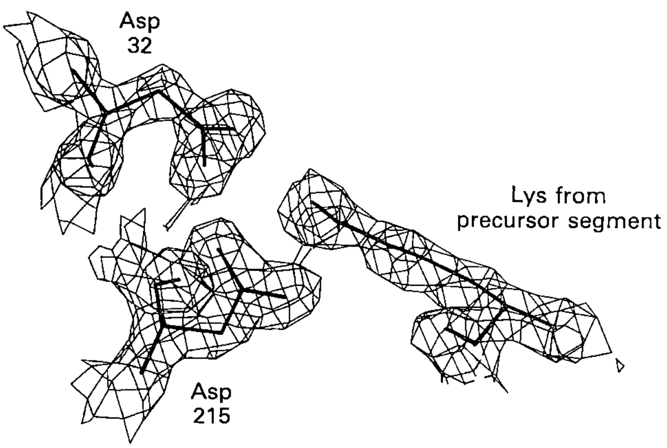

Figure 10-22
The active site of pepsinogen is blocked at neutral pH by an electrostatic interaction between a lysine side chain from the precursor segment (shown in blue) and a critical pair of aspartate carboxylates (shown in red). The polygons depict electron-density contour surfaces. [After M.N.G. James and A.R. Sielecki. *Nature* 319(1986):35.]

PANCREATIC TRYPSIN INHIBITOR BINDS VERY TIGHTLY TO THE ACTIVE SITE OF TRYPSIN

The conversion of a zymogen into a protease by cleavage of a single peptide bond is a precise means of switching on enzymatic activity. However, this activation step is irreversible, and so a different mechanism is needed to stop proteolysis. This is accomplished by specific protease inhibitors. For example, *pancreatic trypsin inhibitor,* a 6-kd protein, inhibits trypsin by binding very tightly to its active site. The dissociation constant of the complex is 0.1 pM, which corresponds to a standard free energy of binding of about −18 kcal/mol. In contrast with nearly all known protein assemblies, this complex is not dissociated into its constituent chains by treatment with 8 M urea or 6 M guanidine hydrochloride.

The reason for the exceptional stability of the complex is that pancreatic trypsin inhibitor is a very effective substrate analog. X-ray analyses have shown that the side chain of lysine 15 of this inhibitor binds to the aspartate side chain in the specificity pocket of the enzyme (Figure 10-23). In addition, there are many hydrogen bonds between the main chain

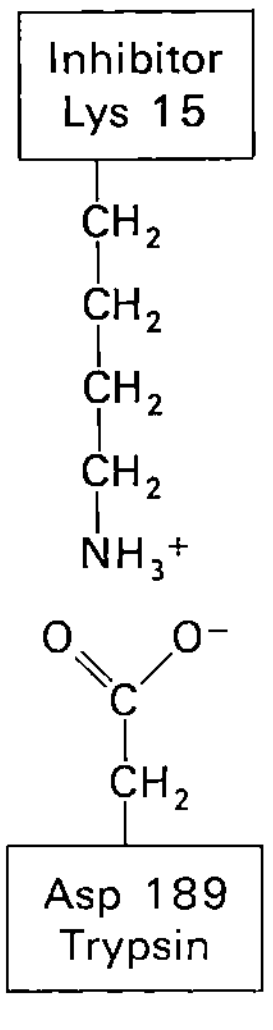

Figure 10-23
A prominent feature of the interaction of pancreatic trypsin inhibitor with trypsin is an electrostatic bond between lysine 15 and aspartate 189. The $-NH_3^+$ group of lysine 15 is also hydrogen-bonded to several oxygen atoms in the specificity pocket of trypsin.

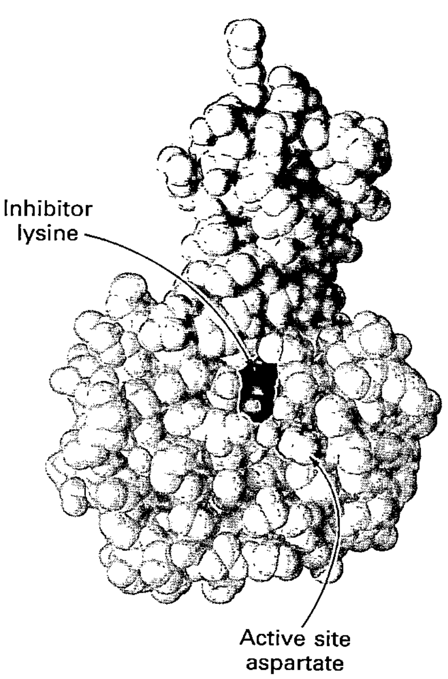

Figure 10-24
Structure of a complex of trypsin (blue) and pancreatic trypsin inhibitor (yellow). Lys 15 (red) of the inhibitor penetrates into the active site of the enzyme and forms a salt bridge with Asp 189 (green) in the active site. [Drawn from 2ptc.pdb. M. Marquart, J. Walter, J. Deisenhofer, W. Bode, and R. Huber. *Acta Cryst.* 39B(1983):480.]

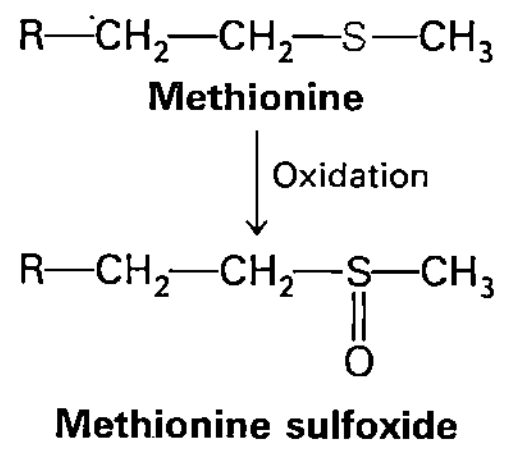

Figure 10-25
Oxidation of methionine to the sulfoxide.

of trypsin and that of its inhibitor, as in an antiparallel β pleated sheet. This array of hydrogen bonds is like that made by a true substrate. Furthermore, the carbonyl group of lysine 15 and the surrounding atoms of the inhibitor fit snugly in the oxyanion hole of the enzyme. The scissile peptide bond in this complex is not planar; the carbonyl carbon atom is on its way to becoming tetrahedral, a key facet of the transition state (p. 225). Indeed, the peptide bond between lysine 15 and alanine 16 in pancreatic trypsin inhibitor is cleaved, but at a very slow rate. The half-life of this trypsin-inhibitor complex is several months. *The inhibitor has very high affinity for trypsin because its structure is almost perfectly complementary to that of the active site* (Figure 10-24). The restricted flexibility of the inhibitor in the binding site contributes to the blocking of catalysis and accounts for the very slow dissociation of the complex.

INSUFFICIENT α_1-ANTITRYPSIN ACTIVITY LEADS TO DESTRUCTION OF THE LUNGS AND EMPHYSEMA

α_1-Antitrypsin (also called *α_1-antiproteinase*), a 53-kd plasma protein, protects tissues from digestion by elastase, a secretory product of neutrophils (white blood cells that engulf bacteria). *Antielastase* would be a more accurate name for this inhibitor, because it blocks elastase much more effectively than it blocks trypsin. Like pancreatic trypsin inhibitor, α_1-antitrypsin blocks the action of target enzymes by binding nearly irreversibly to their active site. Genetic disorders leading to deficient α_1-antitrypsin show that this inhibitor is physiologically important. For example, the substitution of lysine for glutamate at residue 53 in the type Z mutant slows the secretion of this inhibitor from liver cells. Serum levels of the inhibitor are about 15% of normal in people homozygous for this defect. The consequence is that unrestrained excess elastase destroys alveolar walls in the lungs by digesting elastic fibers and other connective tissue proteins.

The resulting clinical condition is called *emphysema* (also known as *destructive lung disease*). People with emphysema must breathe much harder to exchange the same volume of air because their alveoli are much less resilient than normal. Cigarette smoking markedly increases the likelihood that even a type Z heterozygote will develop emphysema. The reason is that smoke oxidizes methionine 358 of the inhibitor (Figure 10-25), a residue essential for binding elastase. Indeed, this methionine side chain is the bait that selectively traps elastase. The *methionine sulfoxide* oxidation product, in contrast, does not lure elastase, a striking consequence of the insertion of just one oxygen atom into a protein.

CLOTTING OCCURS BY A CASCADE OF ZYMOGEN ACTIVATIONS

Blood clots are formed by a *series of zymogen activations.* In this enzymatic *cascade,* the activated form of one clotting factor catalyzes the activation of the next (Figure 10-26). Very small amounts of the initial factors suffice to trigger the cascade because of the catalytic nature of the activation process. The numerous steps yield a *large amplification,* assuring a rapid response to trauma. The clotting cascade illustrates some general principles of how reaction sequences are triggered and controlled. Many familiar themes are evident. For example, several of the activated clotting factors are serine proteases.

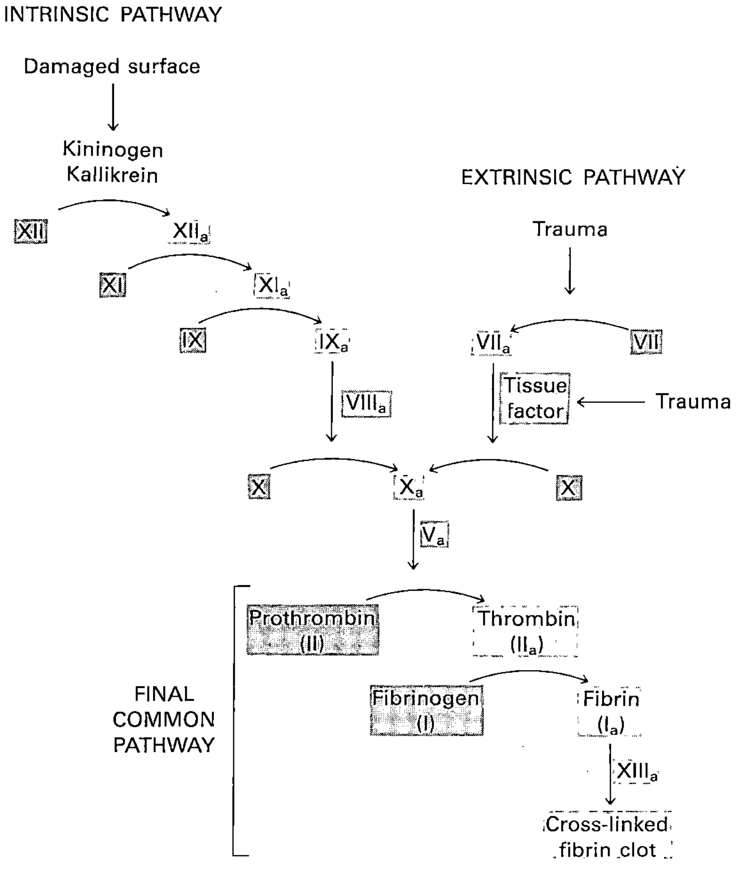

Figure 10-26
Blood clotting cascade. A fibrin clot is formed by the interplay of the intrinsic, extrinsic, and final common pathways. The intrinsic pathway begins with the activation of Factor XII (Hageman factor) by contact with abnormal surfaces produced by injury. The extrinsic pathway is triggered by trauma, which activates Factor VII and releases a lipoprotein called tissue factor from blood vessels. Inactive forms of clotting factors are shown in red and their activated counterparts in green. Stimulatory proteins that are not themselves enzymes are shown in blue. A striking feature of this process is that the activated form of one clotting factor catalyzes the activation of the next factor.

In 1863, Joseph Lister showed that blood stayed fluid in the excised jugular vein of an ox, but that it rapidly clotted when it was transferred to a glass vessel. This unphysiologic surface activated a reaction sequence that has become known as the *intrinsic clotting pathway*. Clotting can also be triggered by substances that are released from tissues as a consequence of trauma to them. This *extrinsic clotting pathway* and the intrinsic one converge on a common sequence of final steps to form a *fibrin clot*. The intrinsic and extrinsic pathways interact with each other in vivo. Indeed, both are needed for proper clotting, as evidenced by clotting disorders caused by a deficiency of a single protein in one of the pathways.

FIBRINOGEN IS CONVERTED BY THROMBIN INTO A FIBRIN CLOT

The best-characterized part of the clotting process is the conversion of fibrinogen into fibrin by thrombin, a proteolytic enzyme. Fibrinogen is made up of three globular units connected by two rods (Figure 10-27). This 340-kd protein consists of pairs of three kinds of chains: Aα, Bβ, γ. The rod regions are triple-stranded α-helical coiled coils, a recurring motif in proteins (p. 28). Thrombin cleaves four *arginine–glycine peptide bonds* in the central globular region of fibrinogen to release an A peptide of 18 residues from each of the two α chains and a B peptide of 20 residues from each of the two β chains. These A and B peptides are called *fibrinopeptides*. A fibrinogen molecule devoid of these fibrinopeptides is called *fibrin monomer* and has the subunit structure $(\alpha\beta\gamma)_2$.

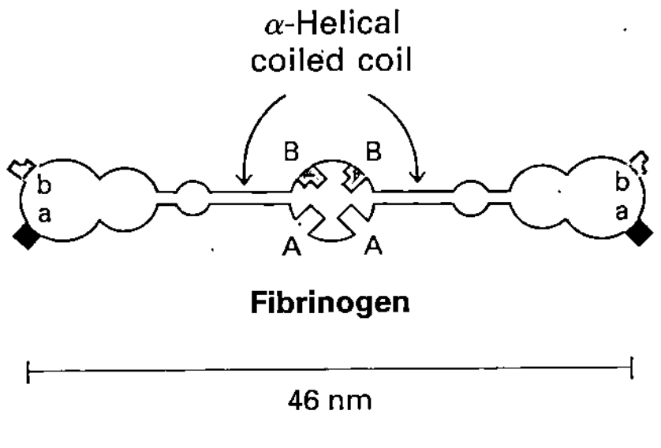

Figure 10-27
Schematic diagram of a fibrinogen molecule. The rod regions are triple-stranded α-helical coiled coils. Thrombin cleaves fibrinopeptides A and B from the central globule of fibrinogen. Each of the two end domains of fibrinogen contains sites a and b that are complementary to the newly exposed sites in the central globule. [After J.W. Weisel. *Biophys. J.* 50(1986):1080.]

Fibrin monomers spontaneously assemble into ordered fibrous arrays called *fibrin*. Electron micrographs and low-angle x-ray patterns show that fibrin has a periodic structure that repeats every 23 nm (Figure 10-28). Because fibrinogen is about 46 nm long, it seems likely that the fibrin

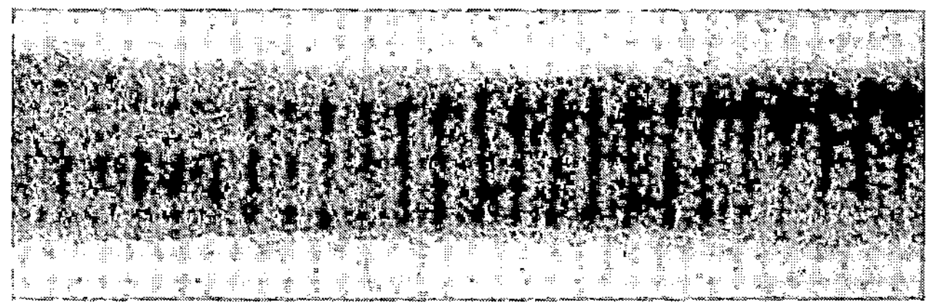

Figure 10-28
Electron micrograph of fibrin. The 23-nm period along the fiber axis is half the length of a fibrinogen molecule. [Courtesy of Dr. Henry Slayter.]

monomers associate to form a half-staggered array. Why do fibrin monomers aggregate, whereas their parent fibrinogen molecules stay in solution? The fibrinopeptides of all vertebrate species studied thus far have a *large net negative charge*. Aspartate and glutamate residues are found in abundance. An unusual negatively charged derivative of tyrosine, namely, *tyrosine-O-sulfate*, is found in fibrinopeptide B. The presence of these and other negatively charged groups in the fibrinopeptides keeps fibrinogen molecules apart. *Their release by thrombin gives fibrin monomers a different surface-charge pattern, leading to their specific aggregation.* The newly formed clot is stabilized by the formation of amide bonds between the side chains of lysine and glutamine residues in different monomer units. This cross-linking reaction is catalyzed by *transglutaminase (Factor XIII$_a$)*.

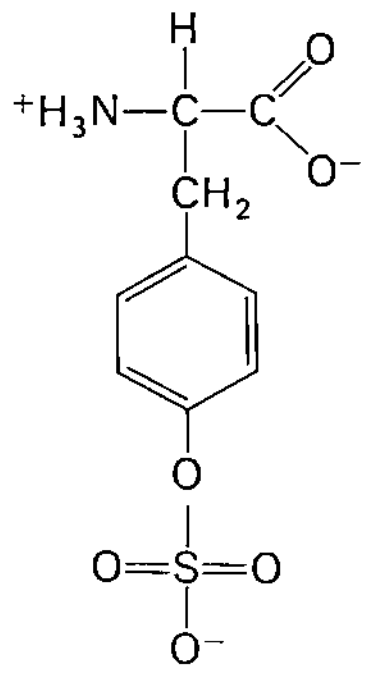

Tyrosine-O-sulfate

THROMBIN IS HOMOLOGOUS TO TRYPSIN

The specificity of thrombin for arginine–glycine bonds suggested that it might resemble trypsin. Indeed it does, as shown by amino acid sequence and x-ray crystallographic studies. Thrombin has a mass of 34 kd and consists of two chains. The A chain of 49 residues exhibits no detectable homology to the pancreatic enzymes. The B chain, however, is quite similar in sequence to trypsin, chymotrypsin, and elastase (Figure 10-29). The sequence around its active-site serine is Gly-Asp-Ser-Gly-Gly-Pro, the same as that in the pancreatic serine proteases. X-ray crystallographic studies have shown that thrombin, like trypsin, contains a catalytic triad, an oxyanion hole, and an aspartate residue at the bottom of its substrate-binding cleft. This negatively charged group forms an electrostatic bond with a positively charged arginine side chain of its substrate. However, thrombin is much more specific than is trypsin. It cleaves only certain arginine–glycine bonds, whereas trypsin cleaves most peptide bonds following arginine or lysine residues.

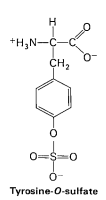

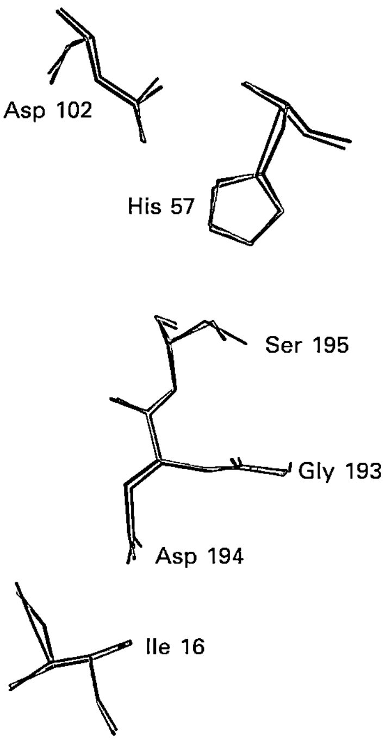

Figure 10-29
The conformation of the B chain of thrombin (red) is similar to that of trypsin (blue), as shown in this superposition of their active sites. The salt bridge between Ile 16 and Asp 194 is also shown. The orientation here is nearly the same as in Figure 10-20. [Drawn from 1abi.pdb and 2ptc.pdb. M. Marquart, J. Walter, J. Deisenhofer, W. Bode, and R. Huber. *Acta Cryst.* 39B(1983):480; X. Qiu, K.P. Padmanabhan, V.E. Carperos, A. Tulinsky, T. Kline, J.M. Maraganore, and J.W. Fenton, III. *Biochemistry* 31(1992):11689.]

Thrombin, like the pancreatic serine proteases, is synthesized as a zymogen called *prothrombin.* Proteolytic cleavage of the Arg 274–Thr 275 bond releases a 32-kd fragment from the 66-kd zymogen (Figure 10-30). Cleavage of the Arg 323–Ile 324 bond then yields active thrombin. An ion pair like the one between the positively charged amino group of Ile 16 and the negatively charged aspartate 194 in chymotrypsin (see Figure 10-20) stabilizes the active form of thrombin.

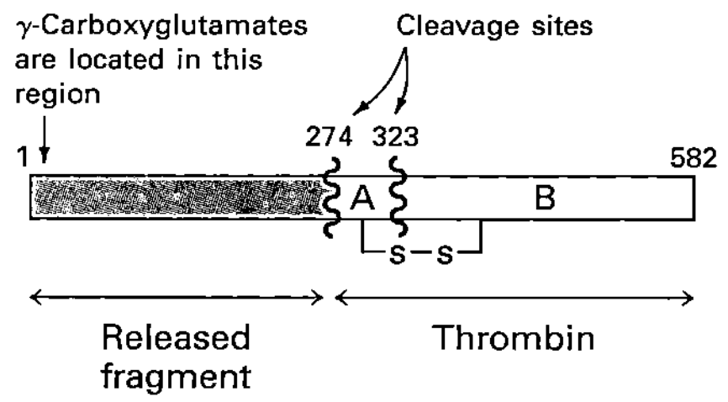

Figure 10-30
Structure of prothrombin. Cleavage of two peptide bonds yields thrombin. The released amino-terminal fragment of prothrombin is shown in blue. All the γ-carboxyglutamate residues are in this fragment. The A and B chains of thrombin, which are joined by a disulfide, are shown in yellow.

VITAMIN K IS REQUIRED FOR THE SYNTHESIS OF PROTHROMBIN AND OTHER CALCIUM-BINDING PROTEINS

Vitamin K (Figure 10-31) had been known for many years to be essential for the synthesis of prothrombin and several other clotting factors. Subsequent studies of the abnormal prothrombin synthesized in the absence of vitamin K or in the presence of vitamin K antagonists, such as dicoumarol, revealed the mode of action of this vitamin. *Dicoumarol* is found in spoiled sweet clover and causes a fatal hemorrhagic disease in cattle fed on this hay. This coumarin derivative is used clinically as an *anticoagulant* to prevent thromboses in patients prone to clot formation. Dicoumarol and such related vitamin K antagonists as *warfarin* also serve as effective rat poisons. Cows fed dicoumarol synthesize an abnormal prothrombin that does not bind Ca^{2+}, in contrast with normal prothrombin. This difference was puzzling for some time because abnormal prothrombin has the same number of amino acid residues and gives the same amino acid analysis after acid hydrolysis as does normal prothrombin.

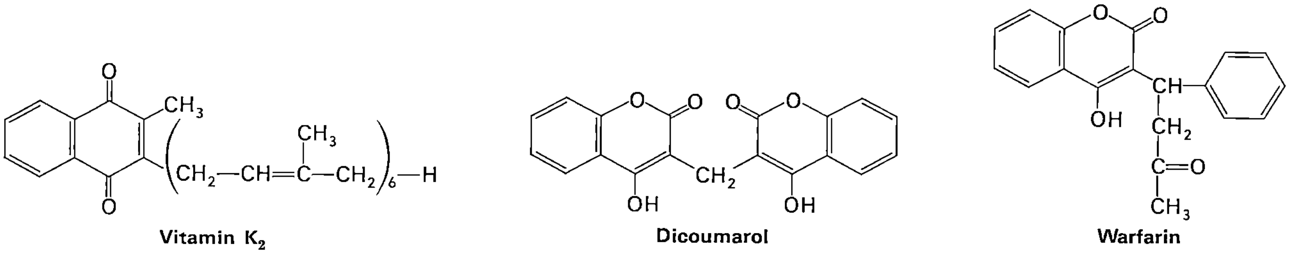

Figure 10-31
Formulas of vitamin K_2 and of two antagonists, dicoumarol and warfarin.

Fragmentation of normal prothrombin showed that its capacity to bind Ca^{2+} resides in its amino-terminal region. The electrophoretic mobility of an amino-terminal peptide from abnormal prothrombin was then found to be markedly different. Nuclear magnetic resonance studies revealed that normal prothrombin contains *γ-carboxyglutamate,* a previously unknown residue that evaded detection because its second carboxyl group is lost on acid hydrolysis. The abnormal prothrombin formed following administration of anticoagulants lacks this modified amino acid. In fact, the first 10 glutamate residues in the amino-terminal region of prothrombin are carboxylated to γ-carboxyglutamate by a vitamin K–dependent enzyme system.

The vitamin K–dependent carboxylation reaction converts glutamate, a weak chelator of Ca^{2+}, into γ-carboxyglutamate, a much stronger chelator. The binding of Ca^{2+} by prothrombin anchors it to phospholipid membranes derived from blood platelets following injury. The functional significance of the binding of prothrombin to phospholipid surfaces is that it brings prothrombin into close proximity with two proteins that mediate its con-

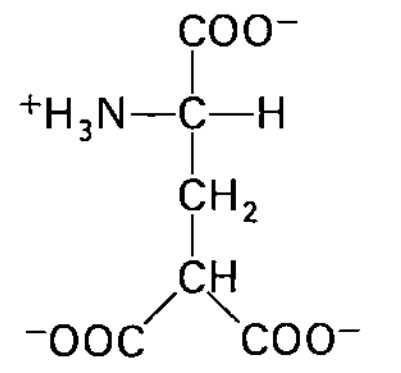

γ-Carboxyglutamate

> *An account of an hemorrhagic disposition existing in certain families—*
> "About seventy or eighty years ago, a woman by the name of Smith settled in the vicinity of Plymouth, New Hampshire, and transmitted the following idiosyncrasy to her descendants. It is one, she observed, to which her family is unfortunately subject, and has been the source not only of great solicitude, but frequently the cause of death. If the least scratch is made on the skin of some of them, as mortal a hemorrhage will eventually ensue as if the largest wound is inflicted. . . . It is a surprising circumstance that the males only are subject to this strange affection, and that all of them are not liable to it. . . . Although the females are exempt, they are still capable of transmitting it to their male children."
>
> JOHN OTTO (1803)

version into thrombin—Factor X_a (a serine protease) and Factor V (a stimulatory protein). The amino-terminal fragment of prothrombin, which contains the Ca^{2+}-binding sites, is released in this activation step. Thrombin freed in this way from the phospholipid surface can then activate fibrinogen in the plasma.

HEMOPHILIA REVEALED AN EARLY STEP IN CLOTTING

How is Factor X activated? Biochemical studies of early steps in clotting are more difficult than those of late ones because the early clotting factors are present in only small amounts. The concentration of Factor X in blood is only 0.01 mg/ml, compared with 3 mg/ml for fibrinogen. The concentrations of some of the earlier factors are even lower. Furthermore, these proteins are highly labile. Some of the important breakthroughs in the elucidation of the pathways of clotting have therefore come from studies of patients with bleeding disorders.

Classic hemophilia, the best-known clotting defect, is genetically transmitted as a sex-linked recessive characteristic. Heterozygous females are asymptomatic carriers. A famous carrier of this disease was Queen Victoria, who transmitted it to the royal families of Prussia, Spain, and Russia. *In classic hemophilia, Factor VIII (antihemophilic factor) of the intrinsic pathway is missing or has markedly reduced activity. VIII is not itself a protease or a zymogen.* Rather, VIII markedly stimulates the activation of X, the final protease of the intrinsic pathway, by IX_a, a serine protease. Thus, activation of the intrinsic pathway is severely impaired in hemophilia.

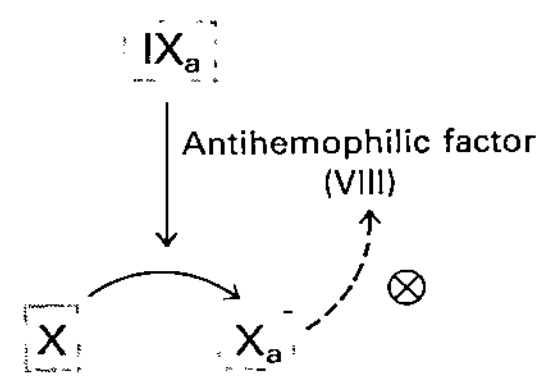

Figure 10-32
Antihemophilic factor (VIII) stimulates the activation of Factor X by IX_a. It is interesting to note that the activity of VIII is markedly increased by limited proteolysis by thrombin and X_a. This positive feedback amplifies the clotting signal and accelerates clot formation once a threshold is reached.

ANTIHEMOPHILIC FACTOR PRODUCED BY RECOMBINANT DNA TECHNOLOGY IS THERAPEUTICALLY EFFECTIVE

In the past, hemophiliacs had been treated with transfusions of a concentrated plasma fraction containing VIII. This therapy carried the risk of infection. Indeed, many hemophiliacs contracted hepatitis and AIDS. A safer preparation of VIII was urgently needed. The cloning of the gene for VIII began with the purification of several milligrams of protein from 25,000 liters of cows' blood, an arduous task necessitated by the low concentration of VIII in plasma (less than 0.05 μg/ml). Determination of the amino acid sequences of several peptides of this very large protein (330 kd) led to the synthesis of a set of oligonucleotide probes and the isolation of the gene from a human genomic library. A hamster cell that has incorporated this 186-kb gene (Figure 10-33) into its genome secretes large amounts of human Factor VIII, which is then purified. Indeed, VIII produced by recombinant DNA technology is replacing plasma concentrates in treating hemophilia. Another benefit of these gene-cloning studies is that DNA probes can now be used to diagnose hemophilia in fetuses

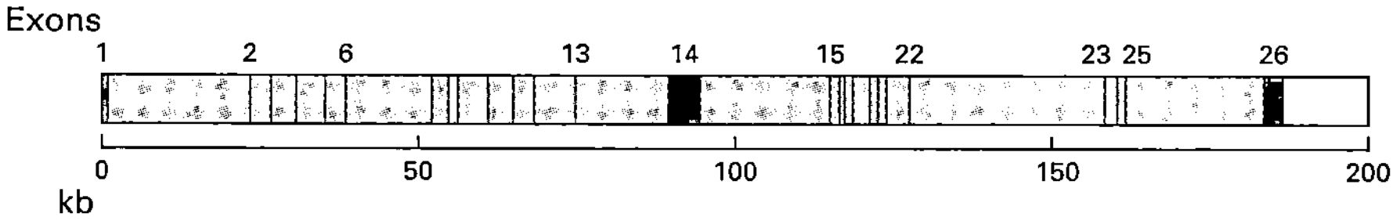

Figure 10-33
The human gene for Factor VIII consists of 26 exons spanning 186 kb. [After R.M. Lawn and G.A. Vehar. The molecular genetics of hemophilia. Copyright © 1986 by Scientific American, Inc. All rights reserved.]

(as described for sickle-cell anemia, p. 173). These probes have also revealed that classic hemophilia is produced by many kinds of missense and chain-termination mutations in the Factor VIII gene.

THROMBIN AND OTHER SERINE PROTEASES IN CLOTTING ARE IRREVERSIBLY INHIBITED BY ANTITHROMBIN III

It is evident that the clotting process must be precisely regulated. There is a fine line between hemorrhage and thrombosis. Clotting must occur rapidly yet remain confined to the area of injury. What are the mechanisms that normally limit clot formation to the site of injury? The lability of clotting factors contributes significantly to the control of clotting. Activated factors are short-lived because they are diluted by blood flow, removed by the liver, and degraded by proteases. For example, the stimulatory proteins V_a and VIII_a are digested by protein C, a protease that is switched on by the action of thrombin. Thus, thrombin triggers the deactivation of the clotting cascade in addition to catalyzing the formation of fibrin.

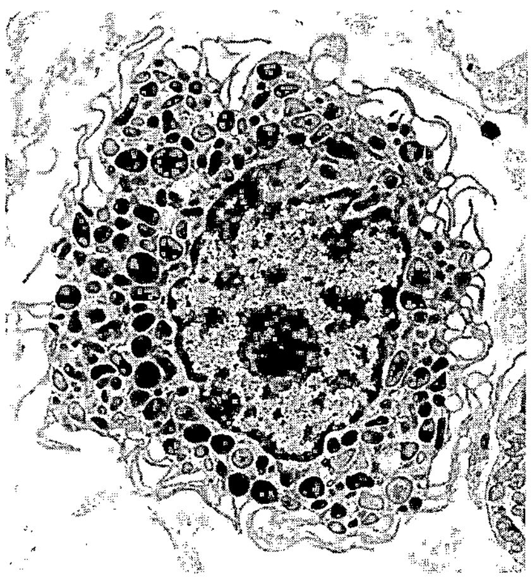

Figure 10-34
Electron micrograph of a mast cell. Heparin and other molecules in the dense granules are released into the extracellular space when the cell is triggered to secrete. [Courtesy of Lynne Mercer.]

Specific inhibitors of clotting factors are also critical in the termination of clotting. The most important one is *antithrombin III,* a plasma protein that inactivates thrombin by forming an irreversible complex with it. Antithrombin III resembles α_1-antitrypsin except that it inhibits thrombin much more strongly than it inhibits elastase. Antithrombin III also blocks other serine proteases in the clotting cascade, namely, XII_a, XI_a, IX_a, and X_a. The inhibitory action of antithrombin III is enhanced by *heparin,* a negatively charged polysaccharide (p. 474) found in mast cells near the walls of blood vessels and on the surfaces of endothelial cells. Heparin acts as an *anticoagulant* by increasing the rate of formation of irreversible complexes between antithrombin III and the serine protease clotting factors. Antitrypsin and antithrombin are *serpins,* a family of *ser*ine *p*rotease *in*hibitors.

The amount of thrombin and antithrombin in blood is precisely balanced to achieve rapid clotting that is limited to the site of injury. This fine balance was disturbed in a 14-year-old boy who died of a bleeding disorder because of a mutation in his α_1-antitrypsin, which normally inhibits elastase (p. 227). Methionine 358 in the binding pocket for elastase was replaced by arginine. *This change of a single amino acid residue dramatically altered the specificity of this inhibitory protein. It became an antithrombin because of the lure of arginine 358.* Consequently, his thrombin activity was reduced. α_1-Antitrypsin activity normally increases markedly following injury to counteract excess elastase arising from stimulated neutrophils. During one of these episodes, this patient's thrombin activity dropped to such a low level that a fatal hemorrhage ensued. We see here a striking example of how a change of a single residue in a protein can dramatically alter specificity, an intimation of what probably occurred many times in evolution. This bleeding disorder further illustrates the critical importance of having the right amount of a protease inhibitor.

FIBRIN CLOTS ARE LYSED BY PLASMIN

Clots are not permanent structures. On the contrary, they are designed to be dissolved when the structural integrity of damaged areas is restored. Fibrin is split by *plasmin,* a serine protease that hydrolyzes peptide bonds in the triple-stranded connector rod regions. Plasmin molecules can diffuse through aqueous channels in the porous fibrin clot to cut the accessible connector rods. Plasmin is formed by proteolytic activation of *plasminogen,* an inactive precursor. This conversion is carried out by *tissue-type plasminogen activator* (TPA), a 72-kd protein made of several types of recurring domains: a finger module (found in fibronectin, an extracellular matrix protein), a growth factor module (found in proteins that stimulate cell proliferation), two kringle modules (triply disulfide-bonded units having the shape of a classic Scandinavian pastry), and a serine protease module (Figure 10-35). The kringle modules bind TPA to fibrin clots, where it swiftly activates adhering plasminogen. In contrast, free plasminogen is activated very slowly by TPA.

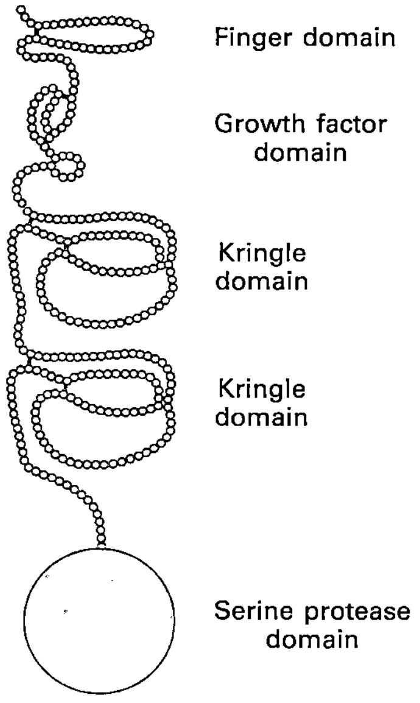

Figure 10-35
Modular structure of tissue-type plasminogen activator (TPA). [After B.A. McCullen and K. Fujikawa. *J. Biol. Chem.* 260(1985):5333.]

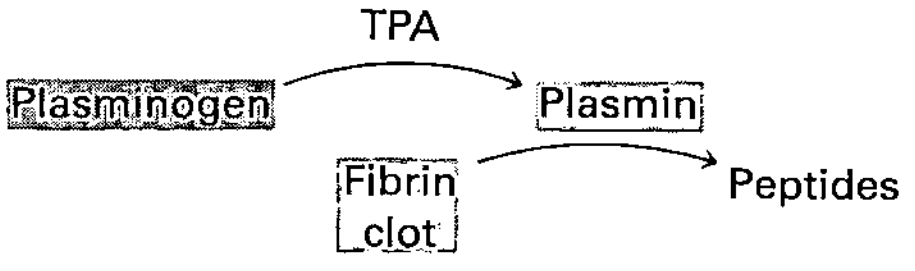

Figure 10-36
Plasmin, formed from plasminogen by the proteolytic action of tissue-type plasminogen activator (TPA), lyses fibrin clots.

The gene for TPA has been cloned and expressed in cultured mammalian cells. Indeed, several hundred thousand patients have been treated with TPA produced by recombinant DNA methods. Clinical studies have shown that TPA administered intravenously within an hour of the formation of a blood clot in a coronary artery markedly increases the likelihood of surviving a heart attack.

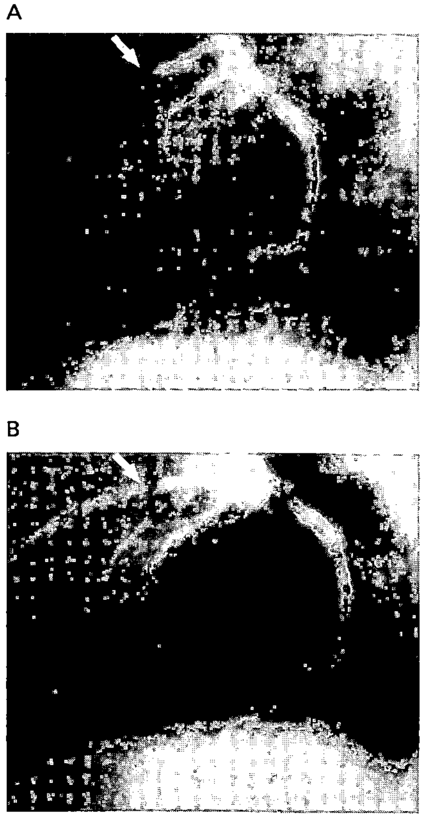

Figure 10-37
Tissue-type plasminogen activator leads to the dissolution of blood clots. X-ray of blood vessels in the heart (A) before and (B) three hours after the administration of TPA. The position of the clot is marked by the arrow in (A). [After F. Van de Werf, P.A. Ludbrook, S.R. Bergmann, A.J. Tiefenbrunn, K.A.A. Fox, H. de Geest, M. Verstraete, D. Collen, and B.E. Sobel. *New Engl. J. Med.* 310(1984):609.]

SUMMARY

The activities of many enzymes are regulated by allosteric interactions, the binding of stimulatory and inhibitory proteins, reversible covalent modification, and proteolytic activation. Aspartate transcarbamoylase (ATCase), one of the best-understood allosteric enzymes, catalyzes the synthesis of *N*-carbamoylaspartate, the first intermediate in the synthesis of pyrimidines. ATCase is feedback-inhibited by CTP, the final product of the pathway. This inhibition is reversed by ATP. ATCase consists of separable catalytic (c_3) subunits (which bind the substrates) and regulatory (r_2) subunits (which bind CTP and ATP). The inhibitory effect of CTP, the stimulatory action of ATP, and the cooperative binding of substrates are mediated by large changes in quaternary structure. On binding substrates, the c_3 subunits of the c_6r_6 enzyme move apart and reorient themselves. This allosteric transition is highly concerted, as postulated by the Monod-Wyman-Changeux (MWC) model. All subunits of an ATCase molecule simultaneously switch from the T (low-affinity) to the R (high-affinity) state, or in the opposite direction.

Phosphorylation, the most common type of reversible covalent modification, is a potent means of controlling the activity of enzymes and other proteins. One class of protein kinases phosphorylates specific serine and threonine residues on proteins, and another class phosphorylates specific tyrosine residues. Dedicated protein kinases phosphorylate a single target, whereas multifunctional protein kinases modify many different ones. Signals can be highly amplified by phosphorylation because a single kinase can act on many target molecules. The regulatory actions of protein kinases are reversed by protein phosphatases, which catalyze the hydrolysis of attached phosphoryl groups.

Cyclic AMP (cAMP) serves as an intracellular messenger in the transduction of many hormonal and sensory stimuli. cAMP switches on protein kinase A (PKA), a major multifunctional kinase, by binding to the regulatory subunit of the enzyme, thereby releasing and activating PKA's catalytic subunits. In the absence of cAMP, the catalytic sites of PKA are occupied by pseudosubstrate sequences from the regulatory subunit.

The activation of an enzyme by proteolytic cleavage of one or a few peptide bonds is a recurring control mechanism. The inactive precursor is called a zymogen (or a proenzyme). Trypsin converts chymotrypsinogen, a zymogen, into active chymotrypsin by hydrolyzing a single peptide bond. The newly formed amino terminus turns inward and forms a salt bridge with an aspartate in the interior. This electrostatic interaction triggers a series of local conformational changes that result in the creation of a pocket for the binding of the aromatic (or bulky nonpolar) side chain of the substrate. In addition, an oxyanion hole that stabilizes the transition state is formed. The activation of trypsinogen by enteropeptidase or trypsin leads to the formation of a binding pocket for the substrate. A molecule of pepsinogen activates itself in acidic media by hydrolyzing a particular peptide bond.

Zymogen activations play a major role in mediating blood clotting. A striking feature of the clotting process is that it occurs by a cascade of zymogen conversions, in which the activated form of one clotting factor catalyzes the activation of the next precursor. Many of the activated clotting factors are serine proteases. Clotting involves the interplay of two reaction sequences, called the intrinsic and extrinsic pathways. They converge on a final common pathway that results in the formation of a fibrin clot. Fibrinogen, a highly soluble molecule in the plasma, is converted by thrombin into fibrin by the hydrolysis of four arginine–glycine bonds.

The resulting fibrin monomer spontaneously forms long, insoluble fibers called fibrin. Several clotting proteins have a calcium-binding domain containing γ-carboxyglutamate residues formed after translation in a reaction that requires vitamin K. Zymogen activation is also essential in the lysis of clots. Plasminogen is converted to plasmin, a serine protease that cleaves fibrin, by tissue-type plasminogen activator (TPA).

The action of irreversible protein inhibitors is exemplified by α_1-antitrypsin, a plasma protein that inhibits elastase released by neutrophils. A deficiency of this inhibitor leads to the destruction of alveoli in the lungs and emphysema. Blood clotting is held in check by antithrombin III, an inhibitor of thrombin and several other serine proteases in the clotting cascade. Antihemophilic factor (Factor VIII), the missing component in hemophilia, illustrates the action of stimulatory proteins. The production of Factor VIII and TPA by recombinant DNA methods demonstrates the power of bringing protein chemistry and molecular genetics together to solve challenging medical problems.

SELECTED READINGS

Where to start

Kantrowitz, E.R., and Lipscomb, W.N., 1990. *Escherichia coli* aspartate transcarbamoylase: The molecular basis for a concerted allosteric transition. *Trends Biochem. Sci.* 15:53–59.

Schachman, H.K., 1988. Can a simple model account for the allosteric transition of aspartate transcarbamoylase? *J. Biol. Chem.* 263:18583–18586.

Neurath, H., 1989. Proteolytic processing and physiological regulation. *Trends Biochem. Sci.* 14:268–271.

Bode, W., and Huber, R., 1992. Natural protein proteinase inhibitors and their interaction with proteinases. *Eur. J. Biochem.* 204:433–451.

Aspartate transcarbamoylase and allosteric interactions

Schachman, H.K., 1974. Anatomy and physiology of a regulatory enzyme: Aspartate transcarbamylase. *Harvey Lectures* 68:67–113. [A classical account, soon after the initial discoveries.]

Eisenstein, E., Markby, D.W., and Schachman, H.K., 1990. Heterotropic effectors promote a global conformational change in aspartate transcarbamoylase. *Biochemistry* 29:3724–3731.

Werner, W.E., and Schachman, H.K., 1989. Analysis of the ligand-promoted global conformational change in aspartate transcarbamoylase. Evidence for a two-state transition from boundary spreading in sedimentation velocity experiments. *J. Mol. Biol.* 206:221–230.

Lahue, R.S., and Schachman, H.K., 1986. Communication between polypeptide chains in aspartate transcarbamoylase. Conformational changes at the active sites of unliganded chains resulting from ligand binding to other chains. *J. Biol. Chem.* 261:3079–3084.

Foote, J., and Schachman, H.K., 1985. Homotropic effects in aspartate transcarbamylase. What happens when the enzyme binds a single molecule of the bisubstrate analog *N*-phosphonacetyl-L-aspartate? *J. Mol. Biol.* 186:175–184.

Newell, J.O., Markby, D.W., and Schachman, H.K., 1989. Cooperative binding of the bisubstrate analog *N*-(phosphonacetyl)-L-aspartate to aspartate transcarbamoylase and the heterotropic effects of ATP and CTP. *J. Biol. Chem.* 264:2476–2481.

Stevens, R.C., Reinisch, K.M., and Lipscomb, W.N., 1991. Molecular structure of *Bacillus subtilis* aspartate transcarbamoylase at 3.0 Å resolution. *Proc. Nat. Acad. Sci.* 88:6087–6091.

Stevens, R.C., Gouaux, J.E., and Lipscomb, W.N., 1990. Structural consequences of effector binding to the T state of aspartate carbamoyltransferase: Crystal structures of the unligated and ATP- and CTP-complexed enzymes at 2.6-Å resolution. *Biochemistry* 29:7691–7701.

Gouaux, J.E., and Lipscomb, W.N., 1990. Crystal structures of phosphonoacetamide ligated T and phosphonoacetamide and malonate ligated R states of aspartate carbamoyltransferase at 2.8-Å resolution and neutral pH. *Biochemistry* 29:389–402.

Protein kinase A

Taylor, S.S., Knighton, D.R., Zheng, J., Sowadski, J.M., Gibbs, C.S., and Zoller, M.J., 1993. A template for the protein kinase family. *Trends Biochem. Sci.* 18:84–89.

Gibbs, C.S., Knighton, D.R., Sowadski, J.M., Taylor, S.S., and Zoller, M.J., 1992. Systematic mutational analysis of cAMP-dependent protein kinase identifies unregulated catalytic subunits and defines regions important for the recognition of the regulatory subunit. *J. Biol. Chem.* 267:4806–4814.

Knighton, D.R., Zheng, J.H., TenEyck, L., Ashford, V.A., Xuong, N.H., Taylor, S.S., and Sowadski, J.M., 1991. Crystal structure of the catalytic subunit of cyclic adenosine monophosphate-dependent protein kinase. *Science* 253:407–414.

Knighton, D.R., Zheng, J.H., TenEyck, L., Xuong, N.H., Taylor, S.S., and Sowadski, J.M., 1991. Structure of a peptide inhibitor bound to the catalytic subunit of cyclic adenosine monophosphate-dependent protein kinase. *Science* 253:414–420.

Adams, S.R., Harootunian, A.T., Buechler, Y.J., Taylor, S.S., and Tsien, R.Y., 1991. Fluorescence ratio imaging of cyclic AMP in single cells. *Nature* 349:694–697.

Zymogen activation

Neurath, H., 1986. The versatility of proteolytic enzymes. *J. Cell. Biochem.* 32:35–49.

Bode, W., and Huber, R., 1986. Crystal structure of pancreatic serine endopeptidases. *In* Desnuelle, P., Sjostrom, H., and Noren, O. (eds.), *Molecular and Cellular Basis of Digestion,* pp. 213–234. Elsevier. [Excellent presentation of the three-dimensional structure and activation mechanism.]

Huber, R., and Bode, W., 1978. Structural basis of the activation and action of trypsin. *Acc. Chem. Res.* 11:114–122.

Stroud, R.M., Kossiakoff, A.A., and Chambers, J.L., 1977. Mechanism of zymogen activation. *Ann. Rev. Biophys. Bioeng.* 6:177–193.

Sielecki, A.R., Fujinaga, M., Read, R.J., and James, M.N., 1991. Refined structure of porcine pepsinogen at 1.8 Å resolution. *J. Mol. Biol.* 219:671–692.

Protease inhibitors

Carrell, R., and Travis, J., 1985. α_1-Antitrypsin and the serpins: Variation and countervariation. *Trends Biochem. Sci.* 10:20–24.

Carp, H., Miller, F., Hoidal, J.R., and Janoff, A., 1982. Potential mechanism of emphysema: α_1-Proteinase inhibitor recovered from lungs of cigarette smokers contains oxidized methionine and has decreased elastase inhibitory capacity. *Proc. Nat. Acad. Sci.* 79:2041–2045.

Owen, M.C., Brennan, S.O., Lewis, J.H., and Carrell, R.W., 1983. Mutation of antitrypsin to antithrombin. *New Engl. J. Med.* 309:694–698. [Case report and molecular analysis of a methionine-to-arginine mutation in α_1-antiproteinase that led to a fatal bleeding disorder.]

Travis, J., and Salvesen, G.S., 1983. Human plasma proteinase inhibitors. *Ann. Rev. Biochem.* 52:655–709.

Clotting cascade

Hedner, U., and Davie, E.W., 1989. Introduction to hemostasis and the vitamin-K dependent coagulation factors. *In* Scriver, C.R., Beaudet, A.L., Sly, W.S., and Valle, D. (eds.), *The Metabolic Basis of Inherited Disease* (6th ed.), pp. 2107–2134. McGraw-Hill. [An excellent introduction to the clotting cascade. Also see the following chapters 85 to 89 for accounts of the clinical consequences of genetic defects in fibrinogen, antihemophilic factor, and antithrombin.]

Doolittle, R.F., 1981. Fibrinogen and fibrin. *Sci. Amer.* 245(12):126–135.

Lawn, R.M., and Vehar, G.A., 1986. The molecular genetics of hemophilia. *Sci. Amer.* 254(3):48–65.

Wintrobe, M.W. (ed.), 1980. *Blood, Pure and Eloquent.* McGraw-Hill. [The title of this wonderful collection of essays comes from a poem by John Donne. They vividly illustrate how studies of blood and its disorders have opened new areas of inquiry. In particular, see Oscar Ratnoff's essay (Chapter 18) "Why Do People Bleed?"]

Masie, R.K., 1967. *Nicholas and Alexandra.* Dell. [A fascinating account of the interplay of hemophilia and Russian history.]

Rao, S.P., Poojary, M.D., Elliott, B.J., Melanson, L.A., Oriel, B., and Cohen, C., 1991. Fibrinogen structure in projection at 18 Å resolution. Electron density by coordinated cryo-electron microscopy and x-ray crystallography. *J. Mol. Biol.* 222:89–98.

Stubbs, M.T., Oschkinat, H., Mayr, I., Huber, R., Angliker, H., Stone, S.R., and Bode, W., 1992. The interaction of thrombin with fibrinogen. A structural basis for its specificity. *Eur. J. Biochem.* 206:187–195.

Martin, P.D., Robertson, W., Turk, D., Huber, R., Bode, W., and Edwards, B.F., 1992. The structure of residues 7–16 of the A alpha-chain of human fibrinogen bound to bovine thrombin at 2.3-Å resolution. *J. Biol. Chem.* 267:7911–7920.

Rydel, T.J., Tulinsky, A., Bode, W., and Huber, R., 1991. Refined structure of the hirudin-thrombin complex. *J. Mol. Biol.* 221:583–601.

Suttie, J.W., 1985. Vitamin K-dependent carboxylase. *Ann. Rev. Biochem.* 54:459–477.

PROBLEMS

1. *Activity profile.* A histidine residue in the active site of aspartate transcarbamoylase is thought to be important in stabilizing the transition state of the bound substrates. Predict the pH dependence of the catalytic rate, assuming that this interaction is essential and dominates the pH-activity profile of the enzyme.

2. *Allosteric switching.* A substrate binds 100 times as tightly to the R state of an allosteric enzyme as to its T state. Assume that the concerted (MWC) model applies to this enzyme.
   (a) By what factor does the binding of one substrate molecule per enzyme molecule alter the ratio of the concentrations of enzyme molecules in the R and T states?
   (b) Suppose that L, the ratio of [T] to [R] in the absence of substrate, is 10^7 and that the enzyme contains four binding sites for substrate. What is the ratio of enzyme molecules in the R state to that in the T state in the presence of saturating amounts of substrate, assuming that the concerted model is obeyed?

3. *Negative cooperativity.* You have isolated a dimeric enzyme that contains two identical active sites. The binding of substrate to one active site decreases the substrate affinity of the other active site. Which allosteric model best accounts for this negative cooperativity?

4. *Paradoxical at first glance.* Recall that phosphonacetyl-L-aspartate (PALA) is a potent inhibitor of ATCase be-

cause it mimics the two physiologic substrates. However, low concentrations of this unreactive bisubstrate analog *increase* the reaction velocity. This stimulatory action is particularly striking in the ATCase-catalyzed breakdown of *N*-carbamoyl phosphate by arsenate, a nonphysiological reaction. On addition of PALA, the reaction rate increases until an average of three PALA are bound per enzyme molecule (Figure 10-38). This maximal velocity is 17-fold greater than in the absence of PALA. The reaction rate then decreases to nearly zero on adding three more molecules of PALA per enzyme. Why do low concentrations of PALA activate ATCase?

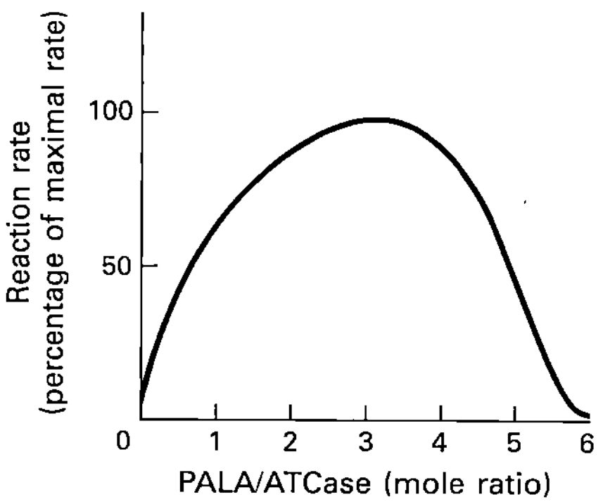

Figure 10-38
Effect of PALA on ATCase rate.

5. *Zymogen activation.* When very low concentrations of pepsinogen are added to acidic media, how does the half-time for activation depend on zymogen concentration?

6. *A revealing assay.* Suppose that you have just examined a young boy with a bleeding disorder highly suggestive of classic hemophilia (Factor VIII deficiency). Because of the late hour, the laboratory that carries out specialized coagulation assays is closed. However, you happen to have a sample of blood from a classic hemophiliac you admitted to the hospital an hour earlier. What is the simplest and most rapid test you can perform to determine whether your present patient is also deficient in Factor VIII activity?

7. *Counterpoint.* The synthesis of Factor X, like that of prothrombin, requires vitamin K. Factor X also contains γ-carboxyglutamate residues in its amino-terminal region. However, activated Factor X, in contrast with thrombin, retains this region of the molecule. What is a likely functional consequence of this difference between the two activated species?

8. *A discerning inhibitor.* Antithrombin III forms an irreversible complex with thrombin but not with prothrombin. What is the most likely reason for this difference in reactivity?

9. *Repeating heptads.* Each of the three types of chains of fibrin contains repeating heptapeptide units (*abcdefg*) in which residues *a* and *d* are hydrophobic. Propose a reason for this regularity.

10. *Drug design.* A drug company has decided to prepare by recombinant DNA methods a modified α_1-antitrypsin that will be more resistant to oxidation than is the naturally occurring inhibitor. Which single amino acid substitution would you recommend to them?

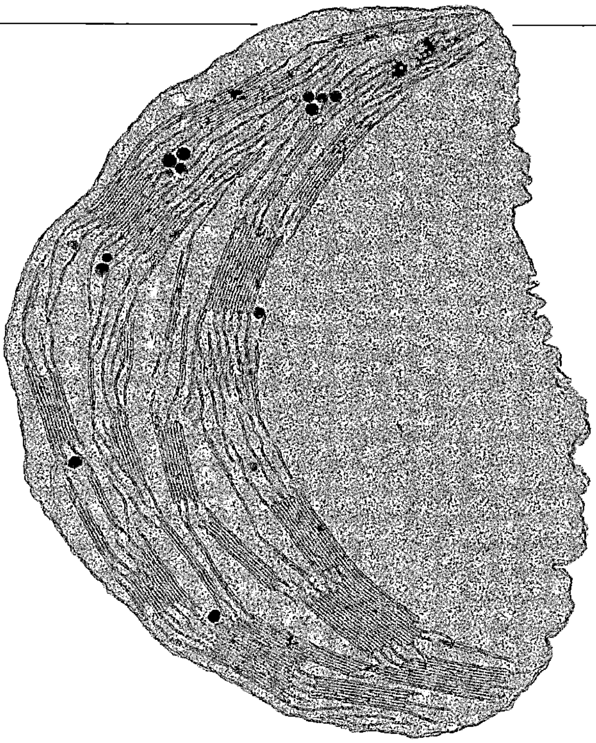

CHAPTER

11

Membrane Structure and Dynamics

We now turn to biological membranes, which are organized, sheetlike assemblies consisting mainly of proteins and lipids. The functions carried out by membranes are indispensable for life. Plasma membranes give cells their individuality by separating them from their environment. *Membranes are highly selective permeability barriers* rather than impervious walls because they contain specific *channels* and *pumps.* These transport systems regulate the molecular and ionic composition of the intracellular medium. Eukaryotic cells also contain internal membranes that form the boundaries of organelles such as mitochondria, chloroplasts, and lysosomes. Functional specialization in the course of evolution has been closely linked to the formation of such compartments.

Membranes also control the flow of information between cells and their environment. They contain *specific receptors for external stimuli.* The movement of bacteria toward food, the response of target cells to hormones such as insulin, and the perception of light are processes in which the primary event is the detection of a signal by a specific receptor in a membrane. In turn, *some membranes generate signals,* which can be chemical or electrical, as in the transmission of nerve impulses. Thus, membranes play a central role in biological communication.

The two most important *energy conversion processes* in biological systems are carried out by membrane systems that contain ordered arrays of enzymes and other proteins. *Photosynthesis,* in which light is converted into chemical-bond energy, occurs in the inner membranes of chloroplasts

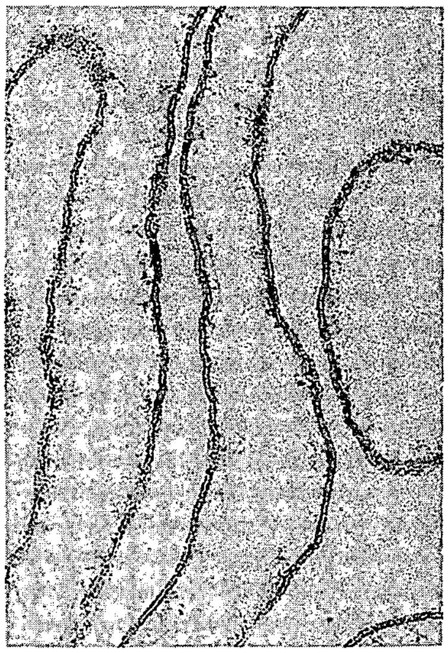

Figure 11-1
Electron micrograph of a preparation of plasma membranes from red blood cells. These membranes are seen "on edge," in cross section. [Courtesy of Dr. Vincent Marchesi.]

Opening Image: *Electron micrograph of a whole chloroplast from a spinach leaf. The stacked thylakoid membranes are the sites of energy conversion in photosynthesis. [Courtesy of Dr. Kenneth Miller.]*

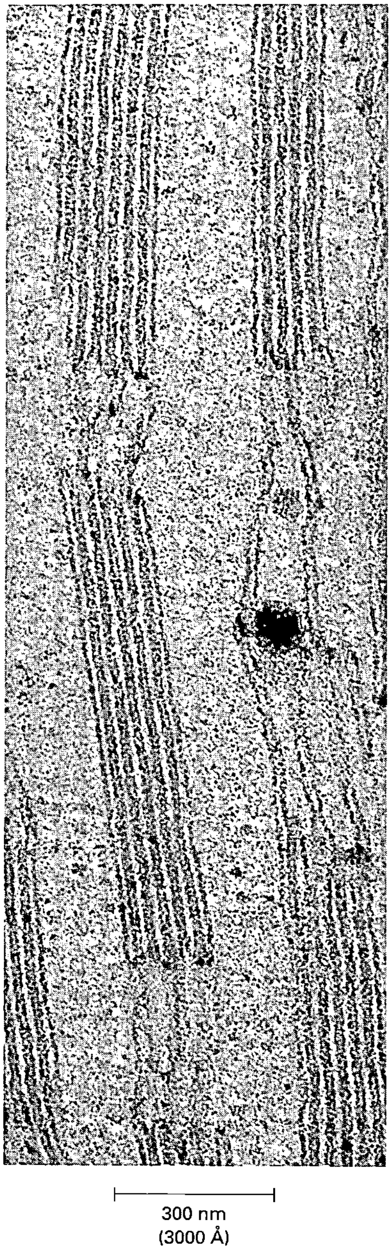

Figure 11-2
Light is converted into chemical-bond energy by photosynthetic assemblies in the thylakoid membranes of chloroplasts. [Courtesy of Dr. M.C. Ledbetter.]

(Figure 11-2), whereas *oxidative phosphorylation,* in which adenosine triphosphate (ATP) is formed by the oxidation of fuel molecules, takes place in the inner membranes of mitochondria. These and other membrane processes will be discussed in detail in later chapters. This chapter deals with some essential features that are common to most biological membranes.

MANY COMMON FEATURES UNDERLIE THE DIVERSITY OF BIOLOGICAL MEMBRANES

Membranes are as diverse in structure as they are in function. However, they do have in common a number of important attributes:

1. Membranes are *sheetlike structures,* only a few molecules thick, that form *closed boundaries* between different compartments. The thickness of most membranes is between 60 Å (6 nm) and 100 Å (10 nm).

2. Membranes consist mainly of *lipids* and *proteins.* Their mass ratio ranges from 1:4 to 4:1. Membranes also contain *carbohydrates* that are linked to lipids and proteins.

3. Membrane lipids are relatively small molecules that have both a *hydrophilic* and a *hydrophobic* moiety. These lipids spontaneously form *closed bimolecular sheets* in aqueous media. These *lipid bilayers* are barriers to the flow of polar molecules.

4. *Specific proteins mediate distinctive functions of membranes.* Proteins serve as pumps, channels, receptors, energy transducers, and enzymes. Membrane proteins are embedded in lipid bilayers, which create suitable environments for their action.

5. Membranes are *noncovalent assemblies.* The constituent protein and lipid molecules are held together by many noncovalent interactions, which are cooperative.

6. Membranes are *asymmetric.* The two faces of biological membranes always differ from each other.

7. Membranes are *fluid structures.* Lipid molecules diffuse rapidly in the plane of the membrane, as do proteins, unless they are anchored by specific interactions. In contrast, they do not rotate across the membrane. Membranes can be regarded as *two-dimensional solutions of oriented proteins and lipids.*

8. Most membranes are *electrically polarized,* with the inside negative (typically −60 millivolts). Membrane potential plays a key role in transport, energy conversion, and excitability, as will be discussed in the next chapter.

PHOSPHOLIPIDS ARE THE MAJOR CLASS OF MEMBRANE LIPIDS

Lipids differ markedly from the other groups of biomolecules considered thus far. By definition, lipids are water-insoluble biomolecules that are highly soluble in organic solvents such as chloroform. Lipids have a variety of biological roles: they serve as fuel molecules, highly concentrated energy stores, signal molecules, and components of membranes. The first three roles of lipids will be discussed in later chapters. Here, our focus is on lipids as membrane constituents. The three major kinds of membrane lipids are *phospholipids, glycolipids,* and *cholesterol.*

Let us start with phospholipids, because they are abundant in all biological membranes. Phospholipids are derived from either *glycerol,* a three-carbon alcohol, or *sphingosine,* a more complex alcohol. Phospholipids derived from glycerol are called *phosphoglycerides.* A phosphoglyceride consists of a glycerol backbone, two fatty acid chains, and a phosphorylated alcohol.

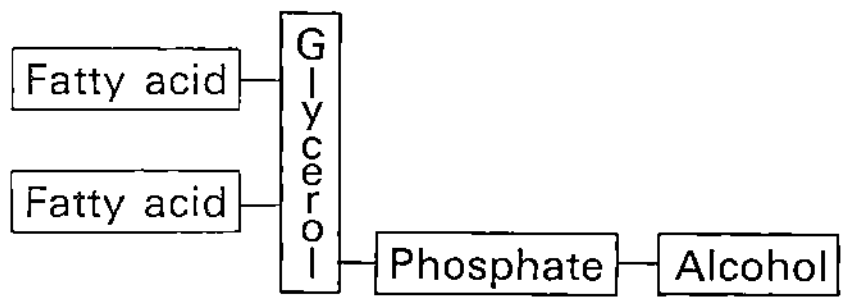

Components of a phosphoglyceride

The *fatty acid chains* in phospholipids and glycolipids usually contain an even number of carbon atoms, typically between 14 and 24. The 16- and 18-carbon fatty acids are the most common ones. In animals, the hydrocarbon chain in fatty acids is unbranched. Fatty acids may be saturated or unsaturated. The configuration of double bonds in unsaturated fatty acids is nearly always cis. As will be discussed shortly, the length and the degree of unsaturation of fatty acid chains in membrane lipids have a profound effect on membrane fluidity. The structures of the ionized form of two common fatty acids—palmitic acid (C_{16}, saturated) and oleic acid (C_{18}, one double bond)—are shown in Figure 11-3. They will be referred to as palmitate and oleate to emphasize the fact that they are ionized under physiologic conditions. The nomenclature of fatty acids is presented in Chapter 24.

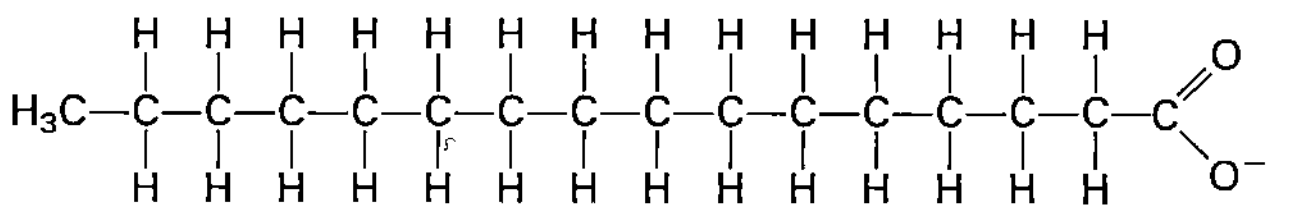

Palmitate
(Ionized form of palmitic acid)

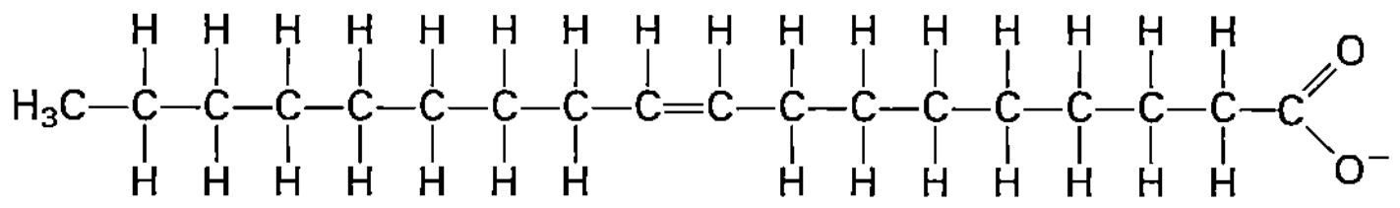

Oleate
(Ionized form of oleic acid)

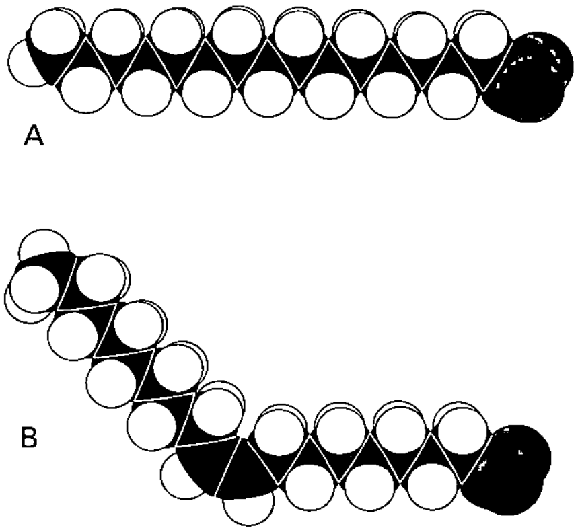

Figure 11-3
Space-filling models of (A) palmitate (C_{16}, saturated) and (B) oleate (C_{18}, unsaturated). The cis double bond in oleate produces a bend in the hydrocarbon chain.

In phosphoglycerides, the hydroxyl groups at C-1 and C-2 of glycerol are esterified to the carboxyl groups of two fatty acid chains. The C-3 hydroxyl group of the glycerol backbone is esterified to phosphoric acid. The resulting compound, called *phosphatidate* (or *diacylglycerol 3-phosphate*), is the simplest phosphoglyceride. Only small amounts of phosphatidate are present in membranes. However, it is a key intermediate in the biosynthesis of the other phosphoglycerides. The absolute configuration of the glycerol 3-phosphate moiety of membrane lipids is shown in Figure 11-4.

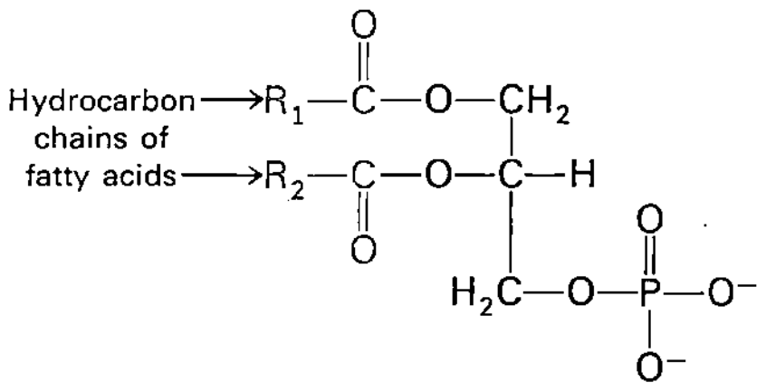

Phosphatidate (Diacylglycerol 3-phosphate)

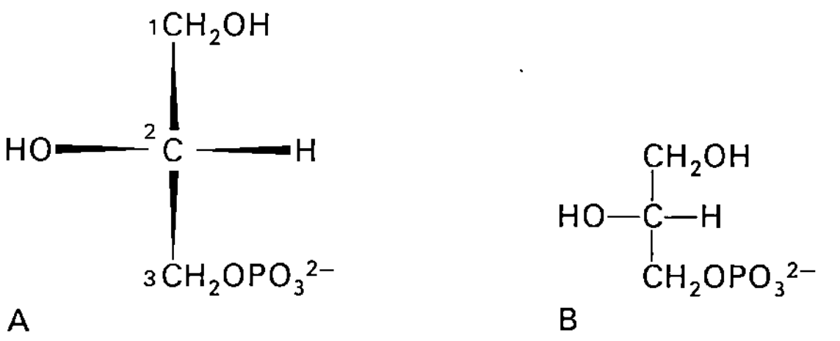

Figure 11-4
Absolute configuration of the glycerol 3-phosphate moiety of membrane lipids: (A) H and OH, attached to C-2, are in front of the plane of the page, whereas C-1 and C-3 are behind it. (B) Fischer representation of this structure. In a Fischer projection, horizontal bonds denote bonds in front, whereas vertical bonds denote bonds behind the plane of the page.

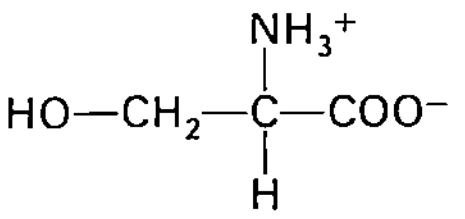

The major phosphoglycerides are derivatives of phosphatidate. The phosphate group of phosphatidate becomes esterified to the hydroxyl group of one of several alcohols. The common alcohol moieties of phosphoglycerides are serine, ethanolamine, choline, glycerol, and inositol.

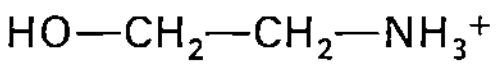

$$HO-CH_2-\underset{\underset{H}{|}}{\overset{\overset{NH_3^+}{|}}{C}}-COO^-$$

Serine

$$HO-CH_2-CH_2-NH_3^+$$

Ethanolamine

$$HO-CH_2-CH_2-\overset{+}{N}(CH_3)_3$$

Choline

$$HO-CH_2-\underset{\underset{OH}{|}}{\overset{\overset{H}{|}}{C}}-CH_2-OH$$

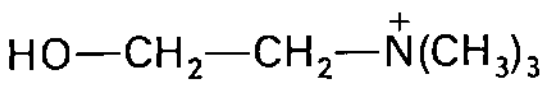

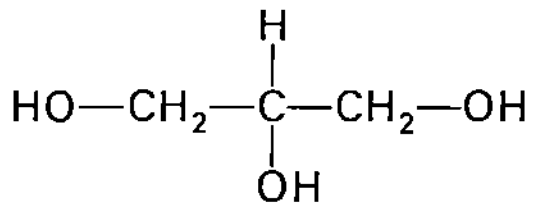

Glycerol

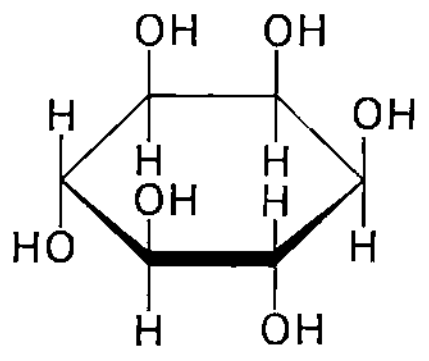

Inositol

Now let us link some of these components to form phosphatidyl choline, a phosphoglyceride found in most membranes of higher organisms.

$$H_3C-(CH_2)_{14}-\overset{\overset{O}{\|}}{C}-O-CH_2$$

$$H_3C-(CH_2)_7-\underset{H}{C}=\underset{H}{C}-(CH_2)_7-\overset{\overset{}{}}{\underset{\underset{O}{}}{C}}-O-\overset{}{C}-H$$

$$H_2C-O-\overset{\overset{O}{\|}}{\underset{\underset{O^-}{|}}{P}}-O-CH_2-CH_2-\overset{+}{N}\overset{-CH_3}{\underset{-CH_3}{-CH_3}}$$

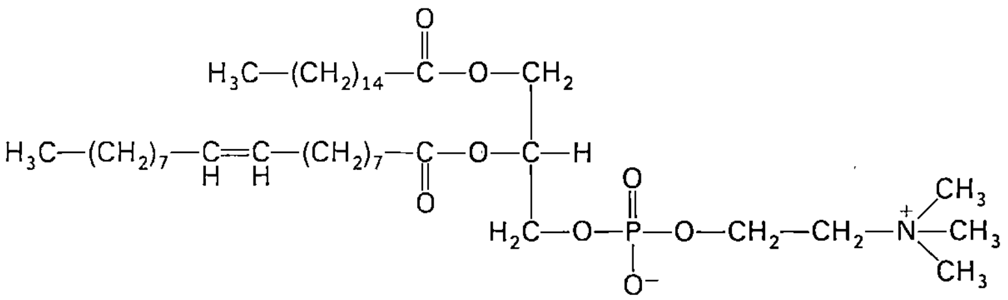

A phosphatidyl choline
(1-Palmitoyl-2-oleoyl-phosphatidyl choline)

The structural formulas of phosphatidyl choline and the other principal phosphoglycerides—namely, phosphatidyl ethanolamine, phosphatidyl serine, phosphatidyl inositol, and diphosphatidyl glycerol—are given in Figure 11-5.

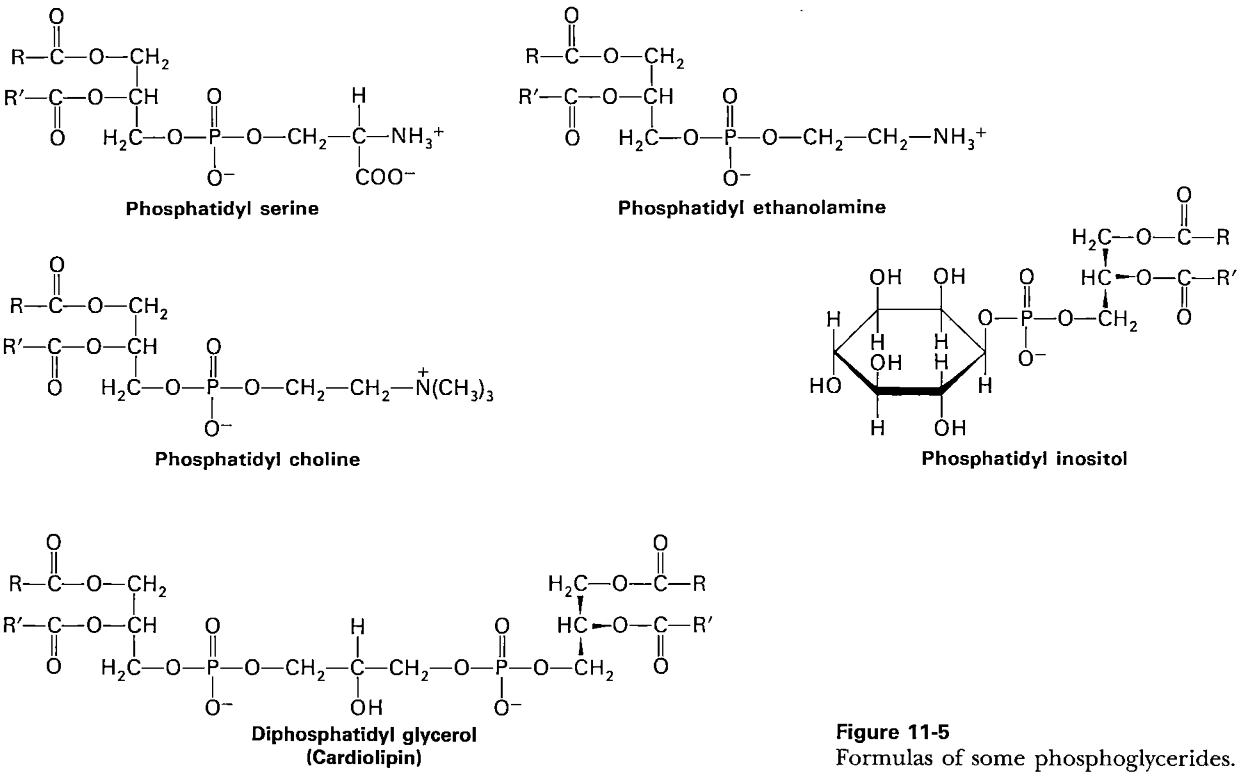

Figure 11-5
Formulas of some phosphoglycerides.

Sphingomyelin is the only phospholipid in membranes that is not derived from glycerol. Instead, the backbone in sphingomyelin is *sphingosine,* an amino alcohol that contains a long, unsaturated hydrocarbon chain. In sphingomyelin, the amino group of the sphingosine backbone is linked to a fatty acid by an amide bond. In addition, the primary hydroxyl group of sphingosine is esterified to phosphoryl choline. As will be shown shortly, the conformation of sphingomyelin resembles that of phosphatidyl choline.

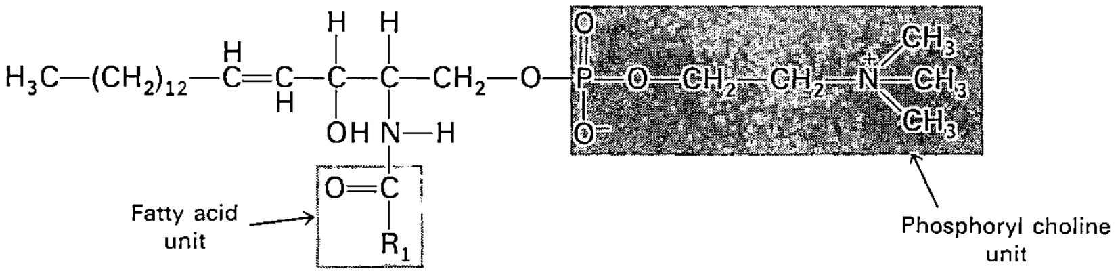

Sphingomyelin

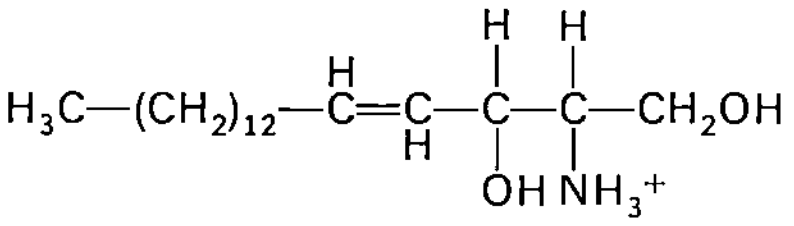
Sphingosine

MANY MEMBRANES ALSO CONTAIN GLYCOLIPIDS AND CHOLESTEROL

Glycolipids, as their name implies, are *sugar-containing lipids.* In animal cells, glycolipids, as with sphingomyelin, are derived from sphingosine. The amino group of the sphingosine backbone is acylated by a fatty acid, as in sphingomyelin. Glycolipids differ from sphingomyelin in the nature of the unit that is linked to the primary hydroxyl group of the sphingosine backbone. In glycolipids, one or more sugars (rather than phosphoryl choline) are attached to this group. The simplest glycolipid is *cerebroside,* in which there is only one sugar residue, either glucose or galactose. More complex glycolipids, such as *gangliosides,* contain a branched chain of as many as seven sugar residues.

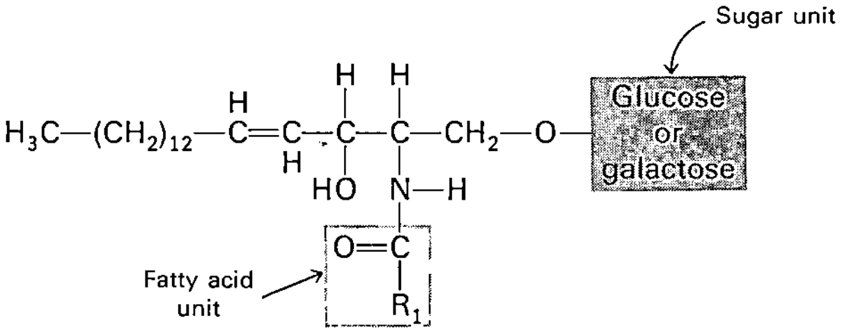

Cerebroside
(A glycolipid)

Another important lipid in some membranes is *cholesterol.* This sterol is present in eukaryotes but not in most prokaryotes. The oxygen atom in its 3-OH group comes from O_2. Cholesterol evolved after the earth's atmosphere became aerobic. The plasma membranes of eukaryotic cells are usually rich in cholesterol, whereas the membranes of their organelles typically have smaller amounts of this neutral lipid.

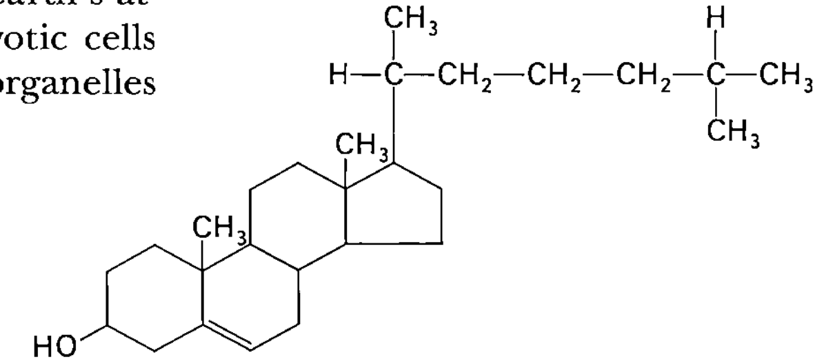
Cholesterol

Table 11-1
Hydrophobic and hydrophilic units of membrane lipids

Membrane lipid	*Hydrophobic unit*	*Hydrophilic unit*
Phosphoglycerides	Fatty acid chains	Phosphorylated alcohol
Sphingomyelin	Fatty acid chain and hydrocarbon chain of sphingosine	Phosphoryl choline
Glycolipid	Fatty acid chain and hydrocarbon chain of sphingosine	One or more sugar residues
Cholesterol	Entire molecule except for OH group	OH group at C-3

MEMBRANE LIPIDS ARE AMPHIPATHIC MOLECULES CONTAINING A HYDROPHILIC AND A HYDROPHOBIC MOIETY

The repertoire of membrane lipids is extensive, perhaps even bewildering at first sight. However, they possess a critical common structural theme: *membrane lipids are amphipathic molecules* (amphiphilic molecules). They contain both a *hydrophilic* and a *hydrophobic* moiety (Table 11-1).

Let us look at a space-filling model of a phosphoglyceride, such as phosphatidyl choline (Figure 11-6). Its overall shape is roughly rectangular. The two fatty acid chains are approximately parallel to each other, whereas the phosphoryl choline moiety points in the opposite direction. Sphingomyelin has a similar conformation (Figure 11-7). The sugar group of a glycolipid occupies nearly the same position as the phosphoryl choline unit of sphingomyelin. Therefore, the following shorthand has been adopted to represent these membrane lipids. The hydrophilic unit, also called the *polar head group,* is represented by a circle, whereas the hydrocarbon tails are depicted by straight or wavy lines (Figure 11-8).

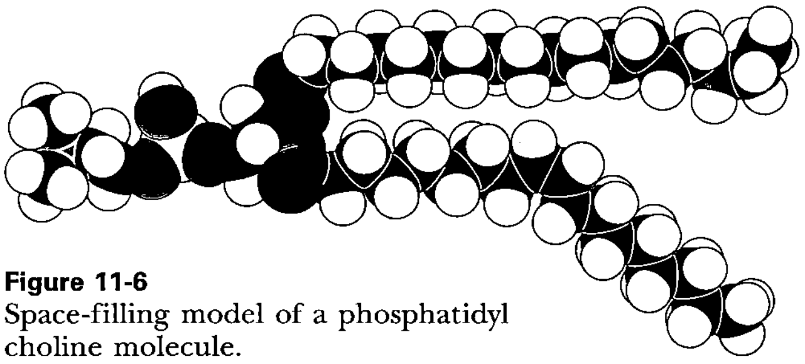

Figure 11-6
Space-filling model of a phosphatidyl choline molecule.

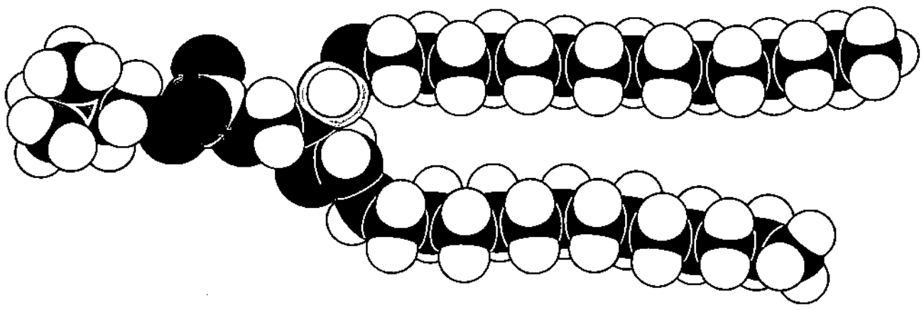

Figure 11-7
Space-filling model of a sphingomyelin molecule.

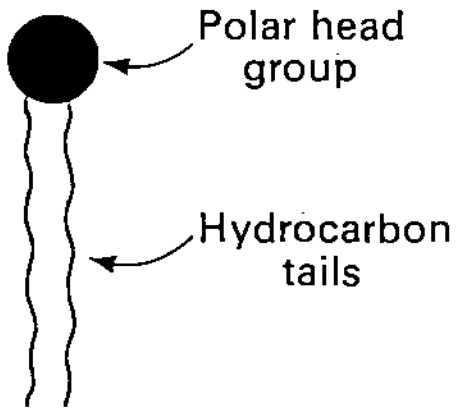

Figure 11-8
Symbol for a phospholipid or glycolipid molecule.

AMPHIPATHIC LIPIDS FORM ORIENTED MONOLAYERS AT AIR-WATER INTERFACES

In 1773, Benjamin Franklin wrote a letter to a scientist friend about the effect of oil on water. Franklin was fascinated by the ancient observation that waves in a storm can be smoothed by pouring oil into the sea. In the letter, Franklin described an experiment he carried out to learn more about this curious phenomenon. He saw one day that the water in a large pond in an English village was rough because of a strong wind, and lost no time in dropping a small quantity of olive oil (a trioleate ester of glycerol) on the surface. The effect was striking:

> the oil, though no more than a tea spoonful, produced an instant calm over a space several yards square, which spread amazingly, and extended itself gradually till it reached the lee side, making all that quarter of the pond, perhaps half an acre, as smooth as a looking-glass.

Franklin, a fine experimenter and keen observer, further remarked that

> one circumstance struck me with particular surprise. This was the *sudden, wide, and forcible spreading of a drop of oil on the face of the water*. . . . If a drop of oil is put on a highly polished marble table, or on a looking-glass that lies horizontally, the drop remains in place, spreading very little. But when put on water, it spreads instantly, many feet round, becoming so thin as to produce the prismatic colors, for a considerable space, and beyond them so much thinner as to be invisible, except in its effect of smoothing the waves at a much greater distance.

How thin? Charles Tanford, in his delightful book *Ben Franklin Stilled the Waves,* notes that Franklin could have made the simple calculation but did not. A teaspoon of oil ($\sim$2 cm^3) spread over a half acre ($\sim 2 \times 10^7$ cm^2) forms a layer only 10^{-7} cm (10 Å or 1 nm) thick. Franklin had almost determined the dimension of a molecule! And why did the oil spread so forcibly? A full understanding came more than a century later from experiments carried out by Lord Rayleigh, a great English physicist and baron; Agnes Pockels, an intuitive, self-taught German scientist who worked at home in her kitchen; and Irving Langmuir, an innovative physical chemist who worked in an American industrial research laboratory. Their studies revealed that *oil spread on water forms a layer just one molecule thick.* This monolayer extends only a nanometer or two above the water. In 1917, Langmuir deduced that *the oil molecules in a monolayer are oriented so that their polar head groups are in contact with water, whereas their nonpolar lipid moieties project into air, where they make contact with the lipid units of their neighbors* (Figure 11-10). Both the water-loving and water-hating passions of the amphipathic oil molecules are thereby fulfilled.

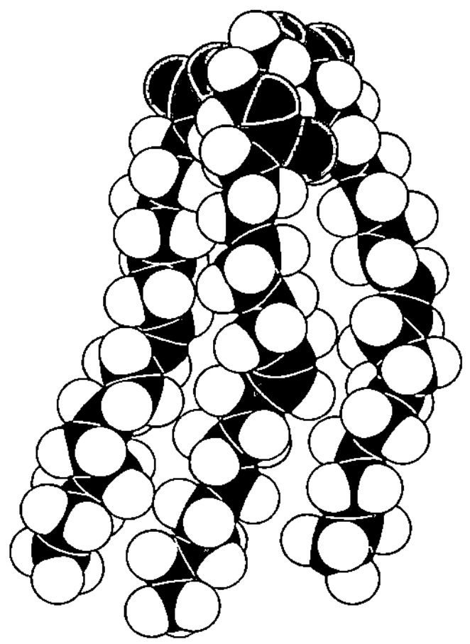

Figure 11-9
Space-filling model of the trioleate ester of glycerol.

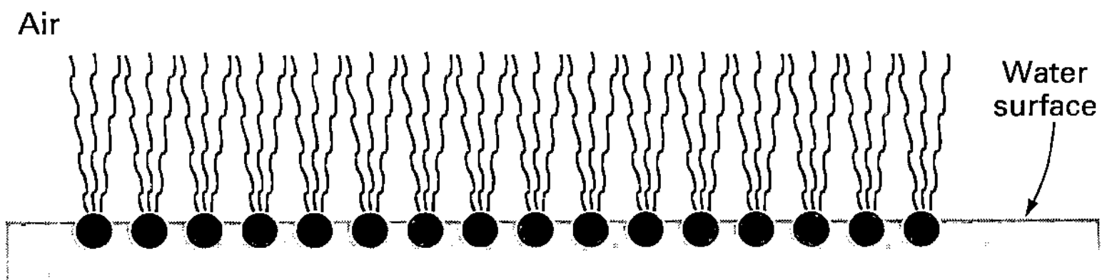

Figure 11-10
Monolayer of oil molecules at an air-water interface.

PHOSPHOLIPIDS AND GLYCOLIPIDS READILY FORM BIMOLECULAR SHEETS IN AQUEOUS MEDIA

Now let us consider the arrangement of phospholipids and glycolipids *within* an aqueous medium. Their polar head groups, like those of Franklin's olive oil, like to be in contact with water, whereas their hydrocarbon tails will try to avoid water. How can this be done *inside* water? One way is to form a *micelle,* a globular structure in which polar head groups are surrounded by water and hydrocarbon tails are sequestered inside, facing one another (Figure 11-11). Alternatively, the strongly opposed tastes of

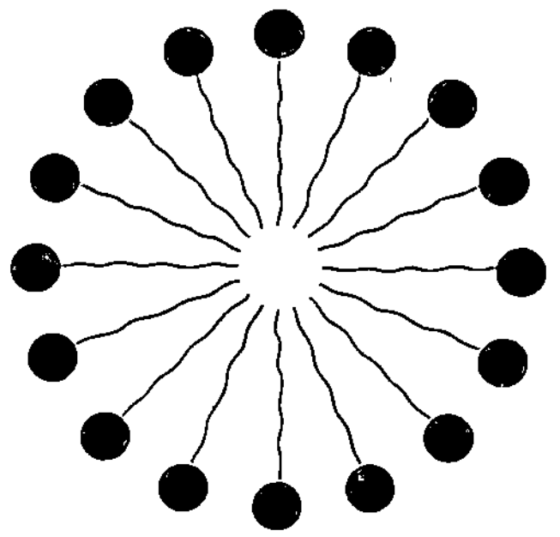

Figure 11-11
Diagram of a section of a micelle formed from ionized fatty acid molecules. Most phospholipids do not form micelles.

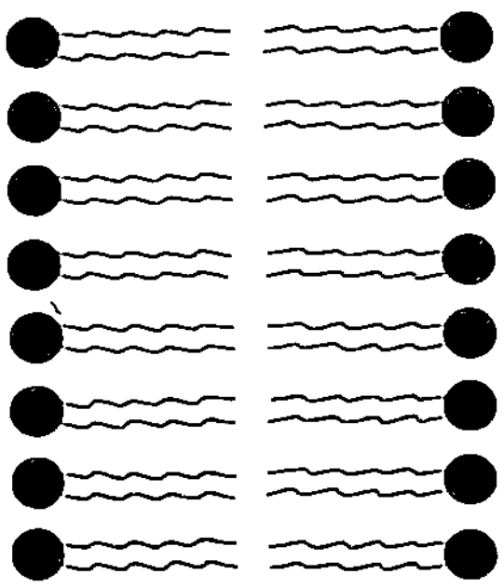

Figure 11-12
Diagram of a section of a bilayer membrane formed from phospholipid molecules.

the hydrophilic and hydrophobic moieties of membrane lipids can be satisfied by forming a *bimolecular sheet,* which is also called a *lipid bilayer* (Figure 11-12).

The favored structure for most phospholipids and glycolipids in aqueous media is a bimolecular sheet rather than a micelle. The reason is that their two fatty acyl chains are too bulky to fit into the interior of a micelle. In contrast, salts of fatty acids (such as sodium palmitate, a constituent of soap), which contain only one chain, readily form micelles. *The formation of bilayers instead of micelles by phospholipids is of critical biological importance.* A micelle is a limited structure, usually less than 20 nm (200 Å) in diameter. In contrast, a bimolecular sheet can have macroscopic dimensions, such as a millimeter (10^6 nm or 10^7 Å). Phospholipids and glycolipids are key membrane constituents because they readily form extensive bimolecular sheets. Furthermore, these sheets serve as permeability barriers, yet they are quite fluid.

The formation of lipid bilayers is a *self-assembly process.* In other words, the structure of a bimolecular sheet is inherent in the structure of the constituent lipid molecules, specifically in their amphipathic character.

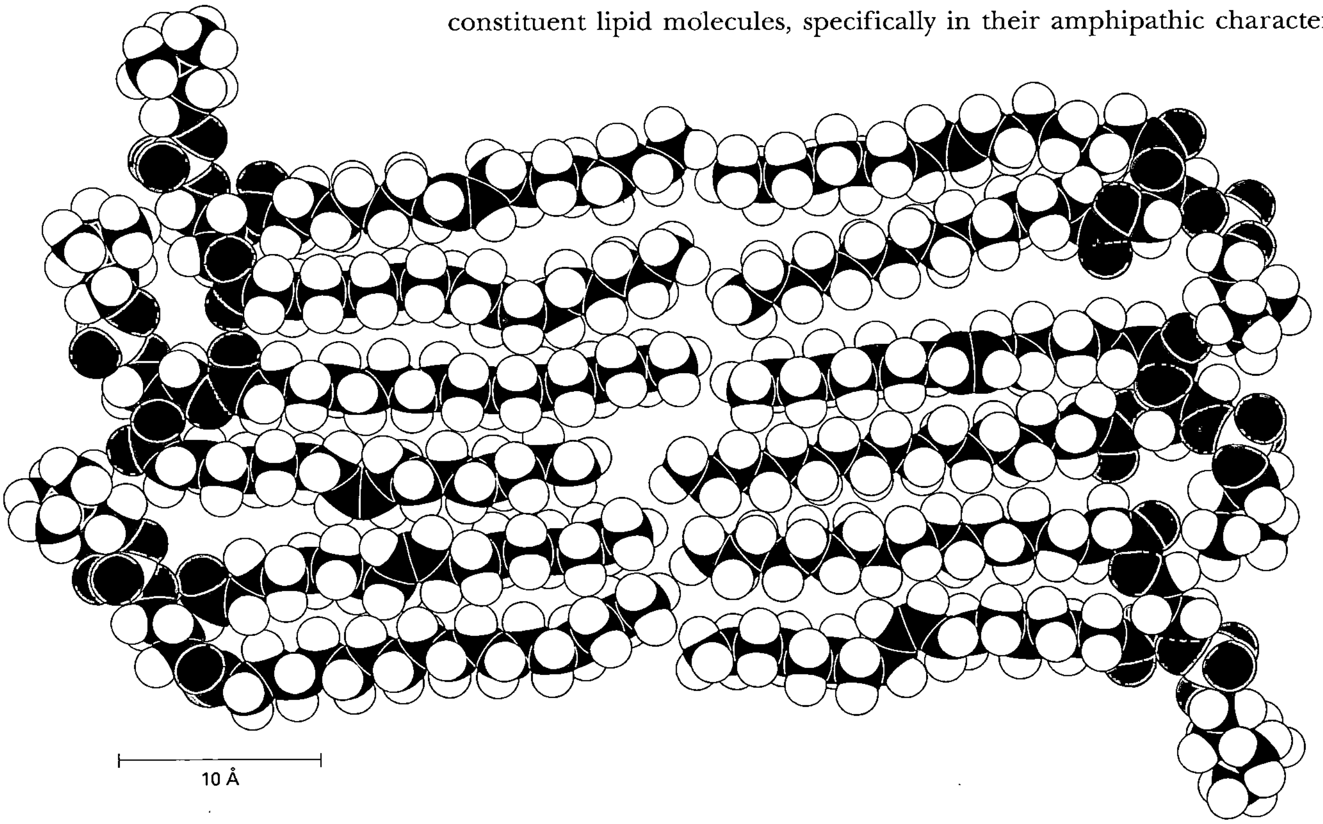

Figure 11-13
Space-filling model of a section of a highly fluid phospholipid bilayer membrane.

The formation of lipid bilayers from glycolipids and phospholipids is a rapid and spontaneous process in water. *Hydrophobic interactions are the major driving force for the formation of lipid bilayers.* Recall that hydrophobic interactions also play a dominant role in the folding of proteins in aqueous solution. Water molecules are released from the hydrocarbon tails of membrane lipids as these tails become sequestered in the nonpolar interior of the bilayer. Furthermore, there are *van der Waals attractive forces* between the hydrocarbon tails. These forces favor close packing of the tails. Finally, there are *electrostatic and hydrogen-bonding attractions between the polar head groups and water molecules.* Thus, lipid bilayers are stabilized by the full array of forces that mediate molecular interactions in biological systems.

LIPID BILAYERS ARE NONCOVALENT, COOPERATIVE STRUCTURES

Another important feature of lipid bilayers is that they are *cooperative structures.* They are held together by many *reinforcing, noncovalent interactions.* Phospholipids and glycolipids cluster together in water to minimize the number of exposed hydrocarbon chains. A pertinent analogy is the huddling together of sheep in the cold to minimize the area of exposed body surface. Clustering is also favored by the van der Waals attractive forces between adjacent hydrocarbon chains. These energetic factors have three significant biological consequences: (1) lipid bilayers have an inherent tendency to be *extensive,* (2) lipid bilayers will tend to *close on themselves* so that there are no edges with exposed hydrocarbon chains, which results in the formation of a compartment, and (3) lipid bilayers are *self-sealing* because a hole in a bilayer is energetically unfavorable.

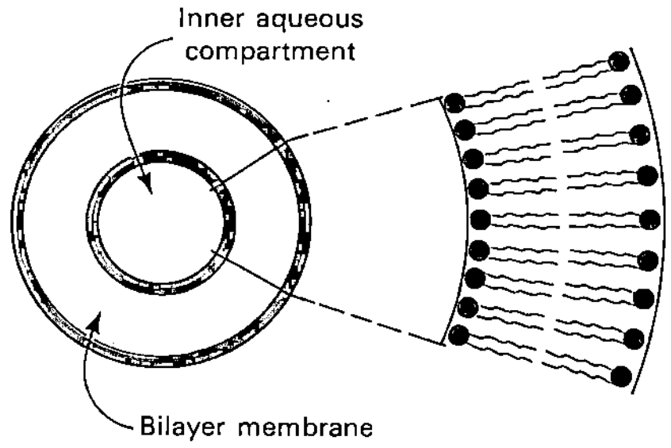

Figure 11-14
Diagram of a lipid vesicle.

LIPID VESICLES (LIPOSOMES) AND PLANAR BILAYER MEMBRANES ARE VALUABLE MODEL SYSTEMS

The permeability of lipid bilayers has been measured in two well-defined synthetic systems: lipid vesicles and planar bilayer membranes. These model systems have been sources of insight into a major function of biological membranes—namely, their role as permeability barriers. *Lipid vesicles* (also known as *liposomes*) are aqueous compartments enclosed by a lipid bilayer (Figure 11-14). They can be formed by suspending a suitable lipid, such as phosphatidyl choline, in an aqueous medium. This mixture is then *sonicated* (i.e., agitated by high-frequency sound waves) to give a dispersion of closed vesicles that are quite uniform in size. Alternatively, vesicles can be prepared by rapidly mixing a solution of lipid in ethanol with water. This can be accomplished by injecting the lipid through a fine needle into an aqueous solution. Vesicles formed by these methods are nearly spherical in shape and have a diameter of about 50 nm (500 Å). Larger vesicles (of the order of 1 μm, or 10^4 Å, in diameter) can be prepared by slowly evaporating the organic solvent from a suspension of phospholipid in a mixed solvent system.

Ions or molecules can be trapped in the aqueous compartment of lipid vesicles by forming the vesicles in the presence of these substances (Figure 11-15). For example, if vesicles 50 nm in diameter are formed in a 0.1 M glycine solution, about 2000 molecules of glycine will be trapped in each inner aqueous compartment. These glycine-containing vesicles can be separated from the surrounding solution of glycine by dialysis or by gel-filtration chromatography. The permeability of the bilayer membrane to glycine can then be determined by measuring the rate of efflux of glycine from the inner compartment of the vesicle to the ambient solution. These lipid vesicles are valuable not only for permeability studies. They fuse with the plasma membrane of many kinds of cells and can thus be used to introduce a wide variety of impermeant substances into cells. The selective fusion of lipid vesicles with particular kinds of cells is a promising means of controlling the delivery of drugs to target cells.

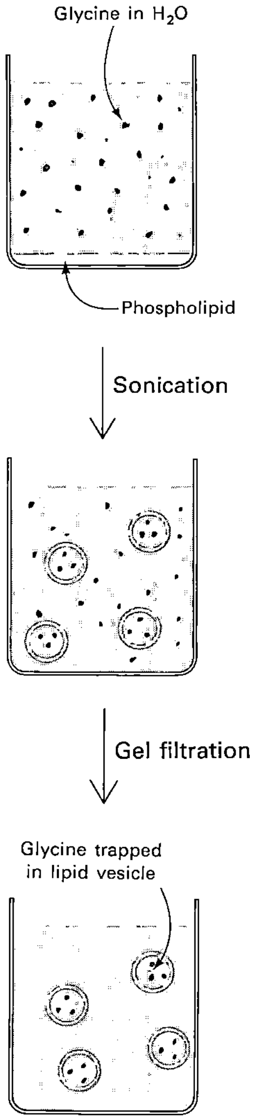

Figure 11-15
Preparation of a suspension of lipid vesicles containing glycine molecules.

Another well-defined synthetic membrane is a *planar bilayer membrane.* This structure can be formed across a 1-mm hole in a partition between two aqueous compartments. Such a membrane is very well suited for electrical studies because of its large size and simple geometry. Paul Mueller and Donald Rudin showed that a large bilayer membrane can be readily formed in the following way. A fine paintbrush is dipped into a mem-

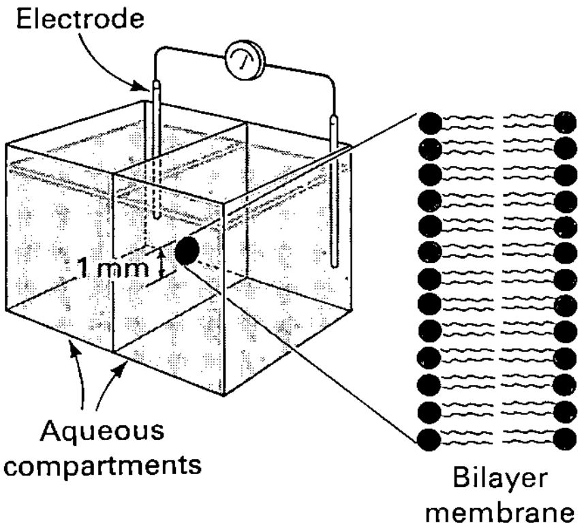

Figure 11-16
Experimental arrangement for the study of planar bilayer membranes. A bilayer membrane is formed across a 1-mm hole in a septum that separates two aqueous compartments.

brane-forming solution, such as phosphatidyl choline in decane. The tip of the brush is then stroked across a hole (1 mm in diameter) in a partition between two aqueous media. The lipid film across the hole thins spontaneously; the excess lipid forms a torus at the edge of the hole. A planar bilayer membrane consisting primarily of phosphatidyl choline is formed within a few minutes. The electrical conduction properties of this macroscopic bilayer membrane are readily studied by inserting electrodes into each aqueous compartment (Figure 11-16). For example, its permeability to ions is determined by measuring the current across the membrane as a function of the applied voltage.

LIPID BILAYERS ARE HIGHLY IMPERMEABLE TO IONS AND MOST POLAR MOLECULES

Permeability studies of lipid vesicles and electrical conductance measurements of planar bilayers have shown that *lipid bilayer membranes have a very low permeability for ions and most polar molecules.* Water is a conspicuous exception to this generalization; it readily traverses such membranes. The range of measured permeability coefficients is very wide (Figure 11-17). For example, Na^+ and K^+ traverse these membranes 10^9 times more slowly than does H_2O. Tryptophan, a zwitterion at pH 7, crosses the membrane 10^3 times more slowly than indole, a structurally related molecule that lacks ionic groups. As was discovered by Charles Overton in 1901, *the permeability coefficients of small molecules are correlated with their solubility in a nonpolar solvent relative to their solubility in water.* This relationship suggests that a small molecule might traverse a lipid bilayer membrane in the following way: first, it sheds its solvation shell of water; then, it becomes dissolved in the hydrocarbon core of the membrane; finally, it diffuses through this core to the other side of the membrane, where it becomes resolvated by water. An ion such as Na^+ traverses membranes very slowly because the removal of its coordination shell of water molecules is highly unfavored energetically.

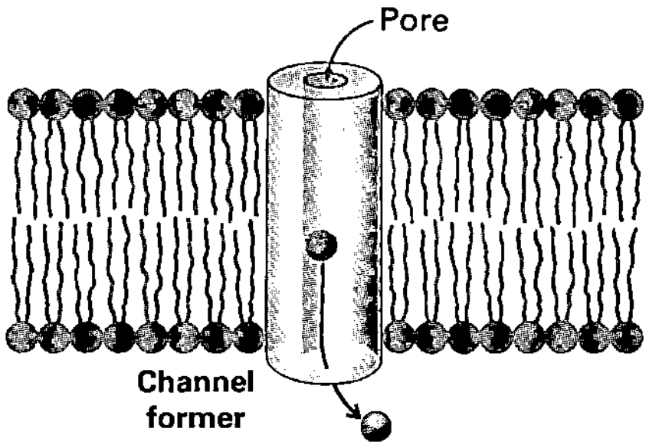

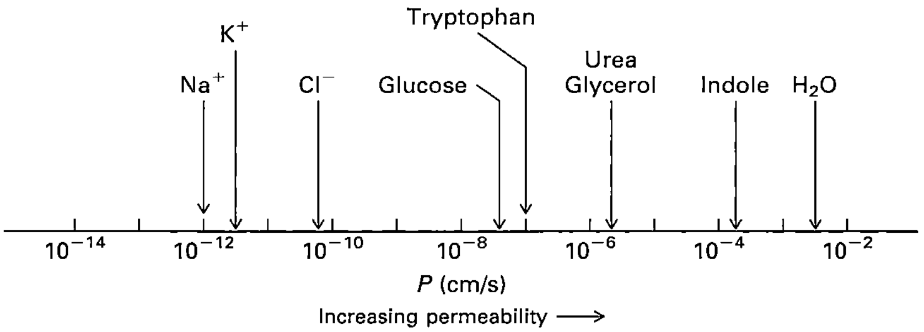

Figure 11-17
The permeability coefficients (P) of ions and molecules in lipid bilayer membranes span a wide range of values.

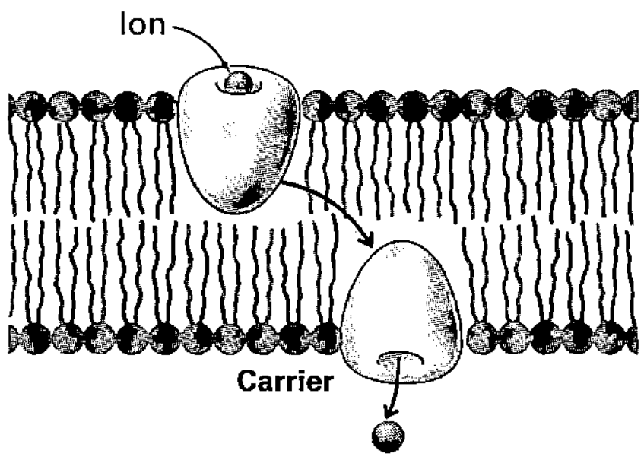

Figure 11-18
Schematic diagram comparing a carrier transport antibiotic with a channel-forming one.

TRANSPORT ANTIBIOTICS ARE CARRIERS OR CHANNELS

Some microorganisms secrete compounds that disable other species in their environs by making their membranes abnormally permeable. These small molecules, called *transport antibiotics,* are valuable experimental tools because of their specificity, in addition to being sources of insight into how ions are transported. They act in two different ways, as *carriers* and as *channel formers* (Figure 11-18). Carriers bind an ion on one side of a membrane and shuttle it across to the other side. Channel formers, in contrast, form a continuous aqueous pore that traverses the hydrocarbon core of the membrane.

Carrier antibiotics are donut-shaped molecules. *A single metal ion is coordinated to several oxygen atoms that surround a central cavity* (Figure 11-19). The number of oxygen atoms that bind the metal ion is typically six or eight. *The periphery of the carrier consists of hydrocarbon groups.* The roles of the central oxygen atoms and the hydrocarbon exterior are evident. In an aqueous medium, a metal ion such as K^+ binds several water molecules through their oxygen atoms. A carrier competes with water for the ion by chelating it to several appropriately arranged oxygen atoms in its central cavity. The hydrocarbon periphery makes the ion-carrier complex soluble in the interior of lipid bilayer. In essence, *these antibiotics catalyze the transport of ions across membranes by making them soluble in lipid.*

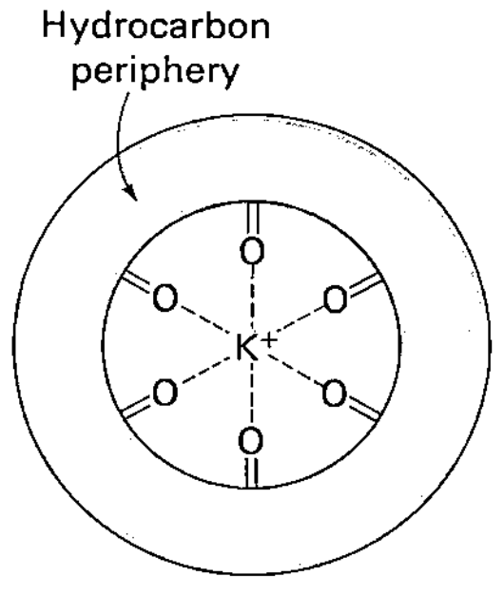

Figure 11-19
Schematic diagram of the chelation of K^+ by oxygen atoms of a carrier transport antibiotic.

Valinomycin, a cyclic molecule containing four repeating residues that are alternately joined by ester and peptide bonds (Figure 11-20), is a well-understood carrier antibiotic.

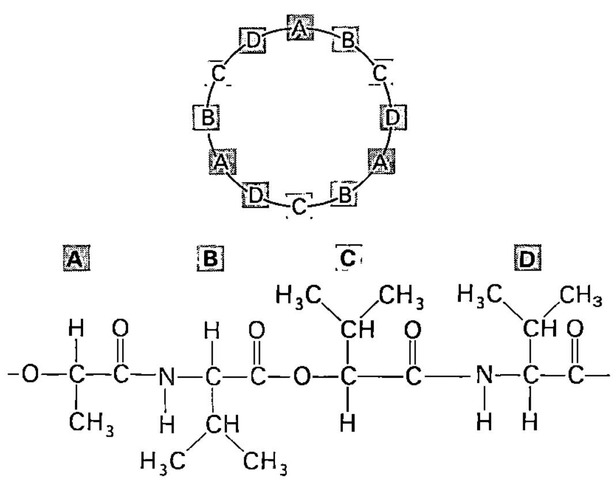

Figure 11-20
Valinomycin is a cyclic molecule made up of L-lactate (A), L-valine (B), D-hydroxyisovalerate (C), and D-valine (D) residues.

The three-dimensional structures of valinomycin and its complex with K^+ are shown in Figure 11-21. The K^+ ion is coordinated to six oxygen atoms, which are arranged octahedrally around the center of the molecule. These oxygen atoms come from the carbonyl groups of the six valine residues in the antibiotic. The methyl and isopropyl side chains constitute the hydrocarbon periphery of valinomycin.

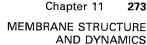

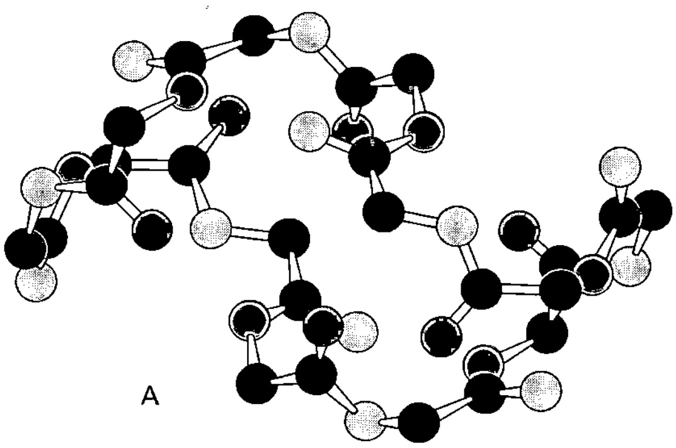

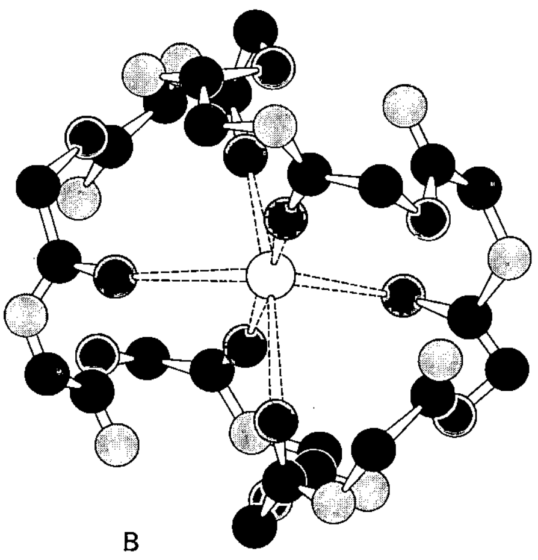

Figure 11-21
Models of (A) valinomycin and (B) its complex with K^+. The conformation of the antibiotic changes upon binding K^+. The carbonyl oxygen atoms that chelate K^+ are shown in dark red, and the other oxygen atoms in light red. [Drawn from atomic coordinates kindly provided by Dr. William Duax (for valinomycin) and by Dr. Larry Steinrauf (for the valinomycin–K^+ complex).]

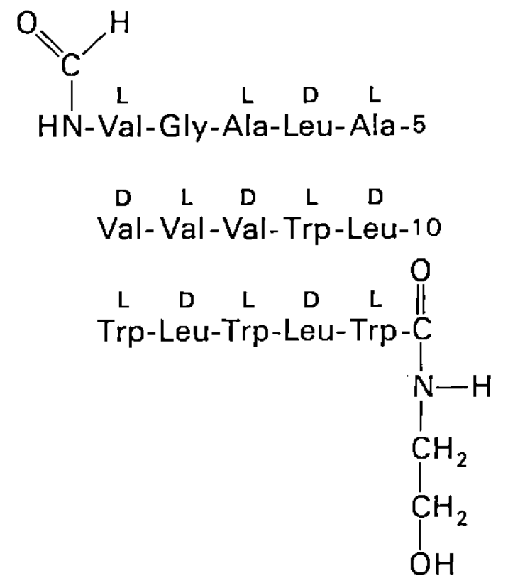

Figure 11-22
Gramicidin A contains alternating L and D amino acid residues.

The flexibility of the valinomycin molecule is also noteworthy. Chelation of K^+ is a step-by-step process in which water molecules in the hydration shell are successively displaced by oxygen atoms of the antibiotic. The activation barrier for binding is small because new bonds are being made as old ones are being broken. Likewise, the energy of activation for release, the reverse process, is small. Hence, valinomycin picks up and unloads K^+ many times a second. The importance of flexibility is evident here, as in enzyme action.

THE FLOW OF IONS THROUGH A SINGLE CHANNEL IN A MEMBRANE CAN BE DETECTED

Gramicidin A, in contrast with valinomycin, acts as a channel former. This intensively studied transport antibiotic is an unusual 15-residue peptide containing alternating L and D residues (Figure 11-22). The conductance of a planar bilayer membrane containing a very small amount of gramicidin A is not constant. Denis Haydon discovered that the conductance of this membrane to Na^+ fluctuates with time in a quantized manner (Figure 11-23). These steps in conductance arise from the spontaneous open-

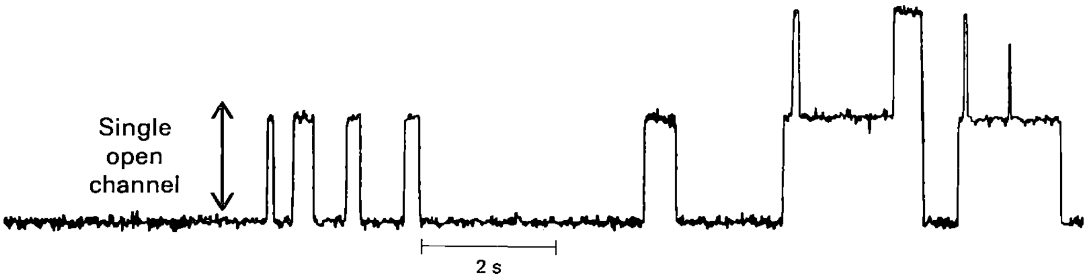

Figure 11-23
The conductance of a lipid bilayer membrane containing a few molecules of gramicidin A fluctuates in a step-by-step manner. The smallest step in conductance arises from a single open channel. [Courtesy of Dr. Olaf Andersen.]

ing and closing of gramicidin channels. A single channel stays open for about a second. This channel is highly permeable to monovalent cations but not to divalent cations or to anions. In fact, *more than 10^7 cations can traverse a single channel in a second.* This transport rate is only a factor of 10 less than that of diffusion through pure water. In contrast, the maximal transport rate of diffusional carriers is less than 10^3 ions per second because rotation and translation of these transporters in the membrane take at least a millisecond.

Spectroscopic and chemical-modification studies revealed that a transmembrane channel is formed from two molecules of gramicidin A (Figure 11-24). In the membrane, this conducting helical dimer is in equilibrium with nonconducting monomers. In fact, the steps in conductance in Figure 11-23 correspond to the formation and dissociation of dimers. The dimer contains a 4-Å-diameter aqueous channel that is surrounded by polar peptide groups. In contrast, the hydrophobic side chains are at the periphery of the channel, in contact with hydrocarbon chains of membrane phospholipids. The carbonyl groups surrounding the aqueous pore transiently coordinate the cation as it passes through the channel. The alternation of L and D residues places all the side chains on the outside of the cylinder, leaving a central hole.

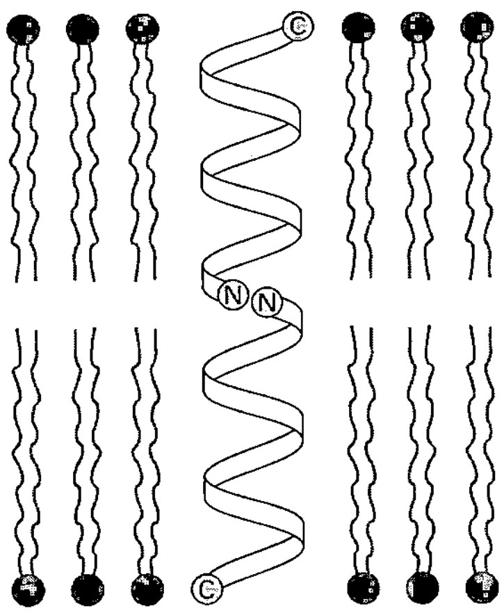

Figure 11-24
Schematic diagram of a gramicidin A channel formed by the association of two polypeptides at their *N*-formyl ends. Each chain is folded into a β helix, which resembles a rolled-up β pleated sheet. This model was proposed by Dan Urry. [After S. Weinstein, B.A. Wallace, E.R. Blout, J.S. Morrow, and W. Veatch. *Proc. Nat. Acad. Sci.* 76(1979):4230.]

PROTEINS CARRY OUT MOST MEMBRANE PROCESSES

We now turn to membrane proteins, which are responsible for most of the dynamic processes carried out by membranes. Membrane lipids form a permeability barrier and thereby establish compartments, whereas *specific proteins mediate nearly all other membrane functions.* Membrane lipids create the appropriate environment for the action of such proteins.

Membranes differ in their protein content. Myelin, a membrane that serves as an insulator around certain nerve fibers, has a low content of protein (18%). Lipid, the major molecular species in myelin, is well suited for insulation. In contrast, the plasma membranes of most other cells are much more active. They contain many pumps, gates, receptors, and enzymes. The protein content of these plasma membranes is typically 50%. Membranes involved in energy transduction, such as the internal membranes of mitochondria and chloroplasts, have the highest content of protein, typically 75%.

The protein components of a membrane can be readily visualized by *SDS–polyacrylamide gel electrophoresis.* The membrane to be analyzed is first solubilized in a 1% solution of sodium dodecyl sulfate (SDS), an amphipathic molecule. This detergent disrupts most protein-protein and protein-lipid interactions. The sample is then electrophoresed in an acrylamide gel containing SDS. As discussed earlier (p. 47), the electrophoretic mobility of many proteins in this SDS-containing gel depends on their mass rather than on their net charge. The gel electrophoresis patterns of three membranes—the plasma membrane of erythrocytes, the photoreceptor membrane of retinal rod cells, and the sarcoplasmic reticulum membrane of muscle—are shown in Figure 11-25. It is evident that these three membranes have very different protein compositions. In general, *membranes performing different functions contain different repertoires of proteins.*

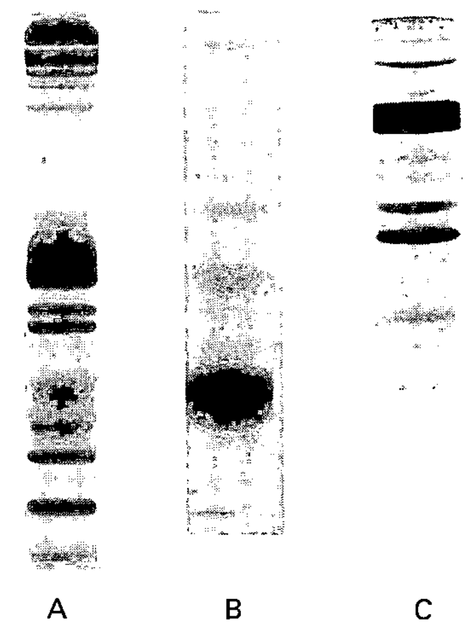

Figure 11-25
SDS–acrylamide gel patterns of (A) the plasma membrane of erythrocytes, (B) the photoreceptor membranes of retinal rod cells, and (C) the sarcoplasmic reticulum membrane of muscle cells. [Courtesy of Dr. Theodore Steck (A) and Dr. David MacLennan (C).]

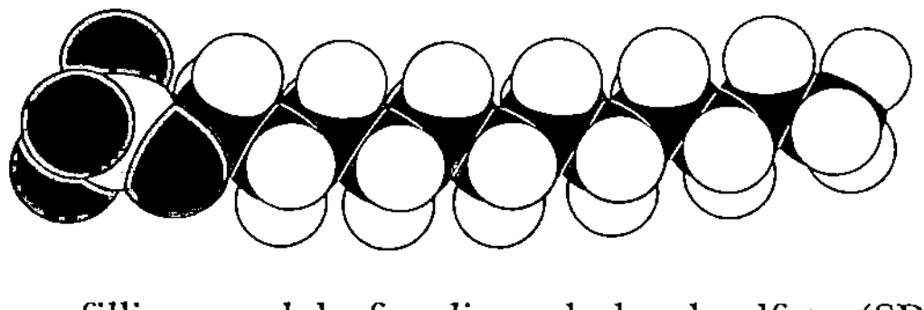

Space-filling model of sodium dodecyl sulfate (SDS).

MANY MEMBRANE PROTEINS SPAN THE LIPID BILAYER

Some membrane proteins can be solubilized by relatively mild means, such as extraction by a solution of high ionic strength (e.g., 1 M NaCl). Other membrane proteins are bound much more tenaciously and can be solubilized only by using a detergent or an organic solvent. Membrane proteins can be classified as being either *peripheral* or *integral* on the basis of this difference in dissociability (Figure 11-26). Integral proteins interact extensively with the hydrocarbon chains of membrane lipids, and so they can be released only by agents that compete for these nonpolar interactions. In fact, nearly all known integral membrane proteins span the lipid bilayer. In contrast, peripheral proteins are bound to membranes primarily by electrostatic and hydrogen-bond interactions. These polar interactions can be disrupted by adding salts or by changing the pH. Many peripheral membrane proteins are bound to the surfaces of integral proteins, either on the cytosolic or extracellular side of the membrane. Others are anchored to the lipid bilayer by a covalently attached hydrophobic chain, such as a fatty acid (p. 934).

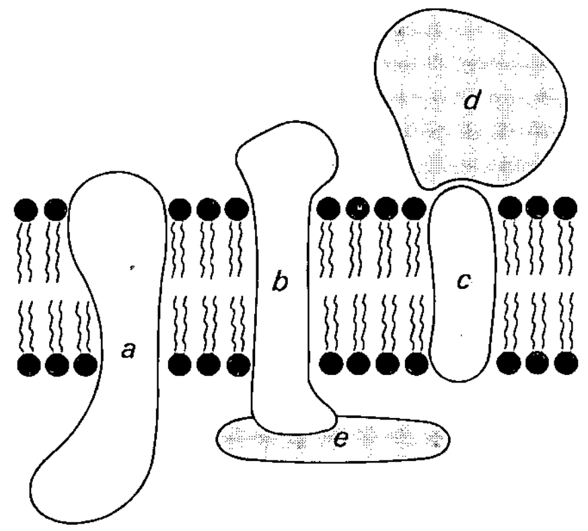

Figure 11-26
Integral membrane proteins (*a*, *b*, *c*) interact extensively with the hydrocarbon region of the bilayer. Nearly all known integral membrane proteins traverse the lipid bilayer. Peripheral membrane proteins (*d*) and (*e*) bind to the surface of integral membrane proteins.

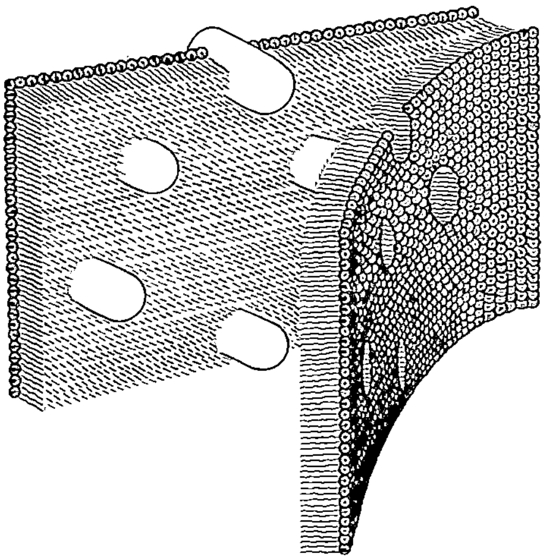

Figure 11-27
Technique of freeze-fracture electron microscopy. The cleavage plane passes through the middle of the bilayer membrane. [After S.J. Singer. *Hospital Practice* 8(1973):81.]

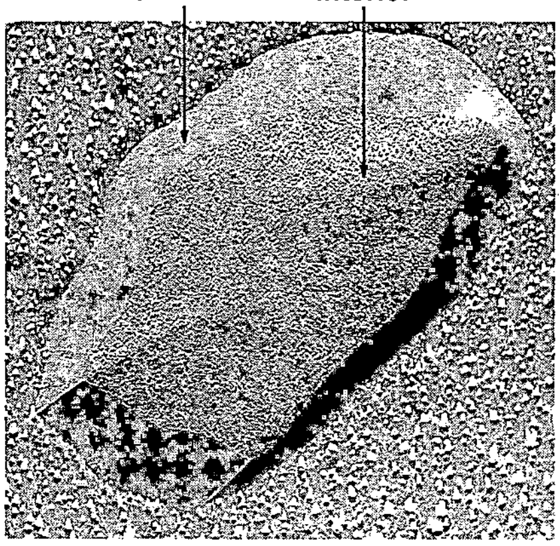

Figure 11-28
Freeze-etch electron micrograph of the plasma membrane of a red blood cell. The interior of the membrane, which has been exposed by fracture of the membrane, is rich in globular particles that have a diameter of about 75 Å. These particles are integral membrane proteins. [Courtesy of Dr. Vincent Marchesi.]

Freeze-fracture electron microscopy is a valuable technique for ascertaining whether proteins are located in the interior of biological membranes. Cells or membrane fragments are rapidly frozen to the temperature of liquid nitrogen. The frozen membrane is then fractured by the impact of a microtome knife. Cleavage usually occurs along a plane in the middle of the bilayer, between its leaflets (Figure 11-27). Hence, extensive regions *within* the lipid bilayer are exposed. These exposed regions can then be shadowed with carbon and platinum, which produces a replica of the interior of the bilayer. The external surfaces of membranes can also be viewed by combining freeze-fracture and deep-etching techniques. First, the interior of a frozen membrane is exposed by fracturing. The ice that covers one of the adjacent membrane surfaces is sublimed away; this process is termed *deep-etching.* The combined technique, which is called *freeze-etching electron microscopy,* provides a view of the interior of a native membrane and of both its surfaces.

Freeze-etching studies provided the first direct evidence for the presence of integral proteins in many biological membranes. The interior of erythrocyte membrane, for example, is dense with globular particles approximately 75 Å in diameter (Figure 11-28). The inside of the sarcoplasmic reticulum membrane is also rich in globular particles. In contrast, synthetic bilayers formed from phosphatidyl choline yield smooth fracture faces. Also, the fracture faces of large areas of myelin membranes are smooth, as might be expected for a relatively inert membrane that serves primarily as an insulator.

LIPIDS AND MANY MEMBRANE PROTEINS DIFFUSE RAPIDLY IN THE PLANE OF THE MEMBRANE

Biological membranes are not rigid structures. On the contrary, lipids and many membrane proteins are constantly in lateral motion. The rapid movement of membrane proteins has been visualized by means of fluorescence microscopy. Human cells and mouse cells in culture can be induced to fuse with each other. The resulting hybrid cell is called a *heterokaryon.* Part of the plasma membrane of this heterokaryon comes from a mouse cell, the rest from a human cell. Do the membrane proteins derived from the mouse and human cells stay segregated in the heterokaryon or do they intermingle? This question was answered by using fluorescent-labeled antibodies as markers that could be followed by light microscopy. An antibody specific for mouse membrane proteins was labeled to show a green fluorescence, and an antibody specific for human membrane proteins was labeled to show a red fluorescence (Figure 11-29). In a newly formed heterokaryon, half the surface displayed green fluorescing patches, the other half red. However, in less than an hour (at 37°C), the red and green fluorescing patches became completely intermixed. This

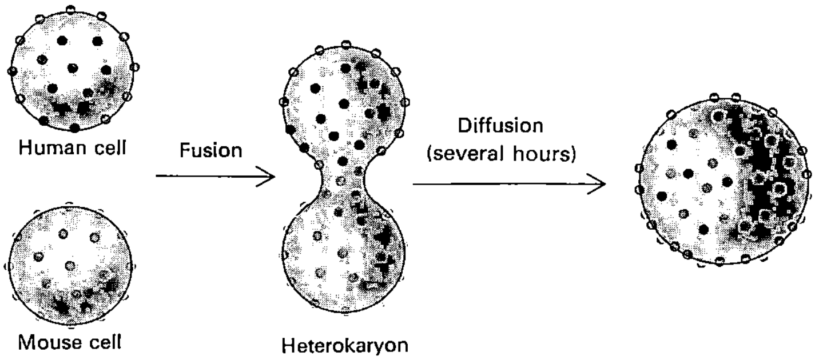

Figure 11-29
Diagram showing the fusion of a mouse cell and a human cell, followed by diffusion of membrane components in the plane of the plasma membrane. The green and red fluorescing markers are completely intermingled after several hours.

experiment revealed that a *membrane protein can diffuse through a distance of several microns in approximately one minute.*

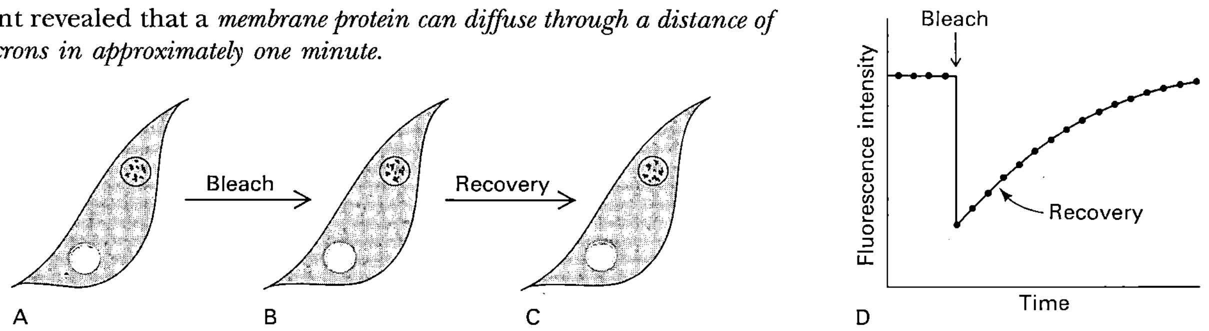

Figure 11-30
Fluorescence photobleaching recovery technique: (A) fluorescence from a labeled cell-surface component in a small illuminated region of a cell; (B) the fluorescent molecules are bleached by an intense light pulse; (C) the fluorescence intensity recovers as bleached molecules diffuse out of the illuminated region and unbleached molecules diffuse into it; the rate of recovery (D) depends on the diffusion coefficient.

A more general and quantitative method for measuring the lateral mobility of membrane molecules in intact cells is the *fluorescence photobleaching recovery technique* (Figure 11-30). First, a cell-surface component is specifically labeled with a fluorescent chromophore. A small region of the cell surface (~3 μm^2) is viewed through a fluorescence microscope. The fluorescent molecules in this region are then destroyed by a very intense light pulse from a laser. The fluorescence of this region is subsequently monitored as a function of time using a light level sufficiently low to prevent further bleaching. If the labeled component is mobile, bleached molecules leave and unbleached molecules enter the illuminated region, which results in an increase in the fluorescence intensity. The rate of recovery of fluorescence depends on the lateral mobility of the fluorescent-labeled component, which can be expressed in terms of a diffusion coefficient D. The average distance s traversed in two dimensions in time t depends on D according to

$$s = (4Dt)^{1/2}$$

The diffusion coefficient of lipids in a variety of membranes is about 1 $\mu m^2\ s^{-1}$. Thus, a phospholipid molecule diffuses an average distance of 2 μm in 1 s. This means that a *lipid molecule can travel from one end of a bacterium to the other in a second.* The magnitude of the observed diffusion coefficient indicates that the viscosity of the membrane is about one hundred times that of water, rather like that of olive oil.

In contrast, proteins vary markedly in their lateral mobility. *Some proteins are nearly as mobile as lipids, whereas others are virtually immobile.* For example, the photoreceptor protein rhodopsin, a very mobile protein, has a diffusion coefficient of 0.4 $\mu m^2\ s^{-1}$. The rapid movement of rhodopsin is essential for fast signaling. At the other extreme is fibronectin, a peripheral glycoprotein that interacts with the extracellular matrix. For fibronectin, D is less than $10^{-4}\ \mu m^2\ s^{-1}$. Fibronectin has a very low mobility because it is anchored to actin filaments on the other side of the plasma membrane through *integrin.* This transmembrane protein links the extracellular matrix to the cytoskeleton.

Fibronectin—
An adhesive cell-surface protein.
Derived from the Latin roots *fibra,* "fiber," and *nectere* "to bind or connect."

MEMBRANE PROTEINS DO NOT ROTATE ACROSS BILAYERS AND MEMBRANE LIPIDS DO SO VERY SLOWLY

The spontaneous rotation of lipids from one face of a membrane to the other is a very slow process, in contrast with their movement parallel to the plane of the bilayer. The transition of a molecule from one membrane surface to the other is called *transverse diffusion,* or *flip-flop,* whereas diffusion in the plane of a membrane is termed *lateral diffusion* (Figure 11-31). The flip-flop of phospholipid molecules in phosphatidyl choline vesicles has been directly measured by electron spin resonance tech-

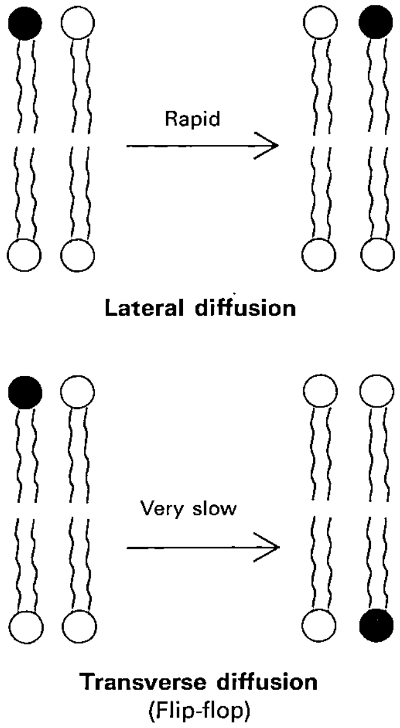

Figure 11-31
Lateral diffusion of lipids is much more rapid than transverse diffusion (flip-flop).

niques, which showed that *a phospholipid molecule flip-flops once in several hours* (see problem 5, p. 290, for the experimental design). Thus, a phospholipid molecule takes about 10^9 times as long to flip-flop across a membrane as it takes to diffuse a distance of 50 Å in the lateral direction. The free-energy barriers to the flip-flopping of protein molecules are even larger than for lipids because proteins have more extensive polar regions. In fact, *the flip-flop of a protein molecule has not been observed.* Hence, *membrane asymmetry can be preserved for long periods.*

BIOLOGICAL MEMBRANES ARE FLUID MOSAICS OF LIPIDS AND PROTEINS

In 1972, S. Jonathan Singer and Garth Nicolson proposed a fluid mosaic model for the overall organization of biological membranes. The essence of their model is that *membranes are two-dimensional solutions of oriented lipids and globular proteins* (Figure 11-32).

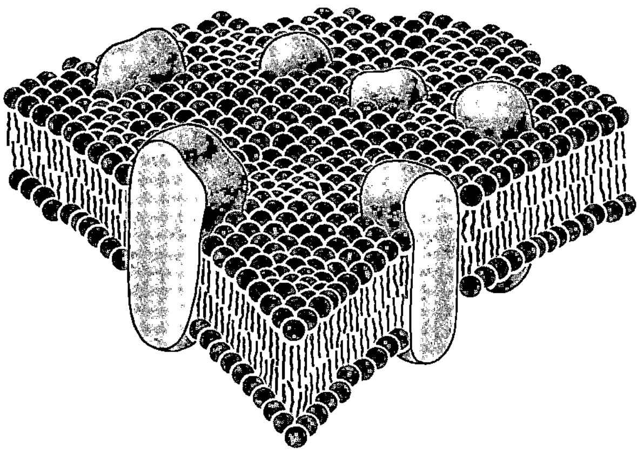

Figure 11-32
Fluid mosaic model. [After S.J. Singer and G.L. Nicolson. *Science* 175(1972):723. Copyright © 1972 by the American Association for the Advancement of Science.]

This proposal is supported by a wide variety of experimental observations. The major features of the model are

1. Most of the membrane phospholipid and glycolipid molecules are arranged in a bilayer. This lipid bilayer has a dual role: it is both a *solvent* for integral membrane proteins and a *permeability barrier.*

2. A small proportion of membrane lipids interact specifically with particular membrane proteins and may be essential for their function.

3. Membrane proteins are free to diffuse laterally in the lipid matrix unless restricted by special interactions, whereas they are not free to rotate from one side of a membrane to the other.

ALL BIOLOGICAL MEMBRANES ARE ASYMMETRIC

Membranes are structurally and functionally asymmetric. The outer and inner surfaces of *all known biological membranes have different components and different enzymatic activities.* A clear-cut example is the pump that regulates the concentration of Na^+ and K^+ ions in cells (p. 310). This transport protein is located in the plasma membrane of nearly all cells in higher organisms. The Na^+-K^+ pump is oriented so that it pumps Na^+ out of the cell and K^+ into it (Figure 11-33). Furthermore, ATP must be on the inside of the cell to drive the pump. Ouabain, a specific inhibitor of the pump, is effective only if it is located outside.

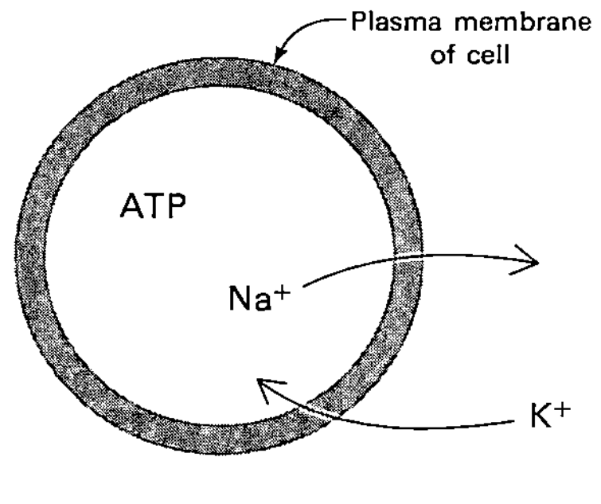

Figure 11-33
Asymmetry of the Na^+-K^+ transport system in plasma membranes.

As will be discussed in detail in Chapter 35, membrane proteins have a unique orientation because they are synthesized and inserted into the membrane in an asymmetric manner. This absolute asymmetry is preserved because membrane proteins do not rotate from one side of the membrane to the other, and because *membranes are always synthesized by growth of preexisting membranes.* Lipids, too, are asymmetrically distributed as a consequence of their mode of biosynthesis, but this asymmetry is usually not absolute, except in the case of glycolipids. In the membrane of red blood cells, sphingomyelin and phosphatidyl choline are preferentially located in the outer leaflet of the bilayer, whereas phosphatidyl ethanolamine and phosphatidyl serine are mainly in the inner leaflet. Large amounts of cholesterol are present in both leaflets.

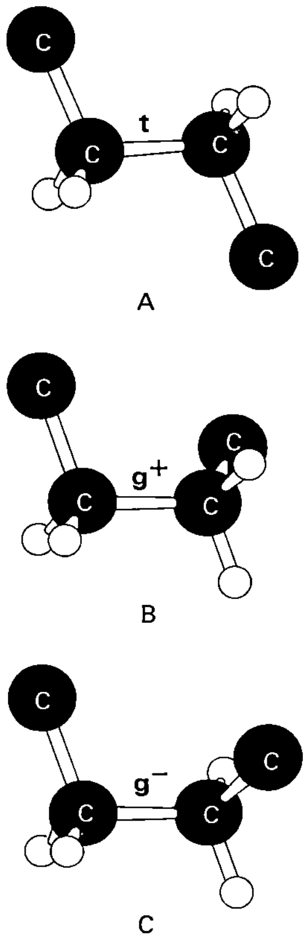

Figure 11-34
Conformation of C–C bonds in fatty acyl chains: (A) trans (t) conformation; (B and C) a 120-degree rotation yields a gauche (g) conformation, which can be either g^+ (clockwise rotation) or g^- (counterclockwise rotation).

MEMBRANE FLUIDITY IS CONTROLLED BY FATTY ACID COMPOSITION AND CHOLESTEROL CONTENT

The fatty acyl chains in bilayer membranes can exist in an ordered, rigid state or in a relatively disordered, fluid state. In the ordered state, all the C–C bonds have a trans conformation, whereas in the disordered state, some are in the gauche conformation (Figure 11-34).

The transition from the rigid (all trans) to the fluid (partly gauche) state occurs rather abruptly as the temperature is raised above T_m, the melting temperature. *This transition temperature depends on the length of the fatty acyl chains and on their degree of unsaturation.* The rigid state is favored by the presence of saturated fatty acyl residues because their straight hydrocarbon chains interact very favorably with each other (Figure 11-35A). On the other hand, *a cis double bond produces a bend in the hydrocarbon chain. This bend interferes with a highly ordered packing of fatty acyl chains, and so* T_m *is lowered* (Figure 11-35B). The length of the fatty acyl chain also affects the transition temperature. Long hydrocarbon chains interact more strongly than do short ones. Specifically, each additional $-CH_2-$ group makes a favorable contribution of about −0.5 kcal/mol to the free energy of interaction of two adjacent hydrocarbon chains.

Prokaryotes regulate the fluidity of their membranes by varying the number of double bonds and the length of their fatty acyl chains. For example, the ratio of saturated to unsaturated fatty acyl chains in the *Escherichia coli* membrane decreases from 1.6 to 1.0 as the growth temperature is lowered from 42°C

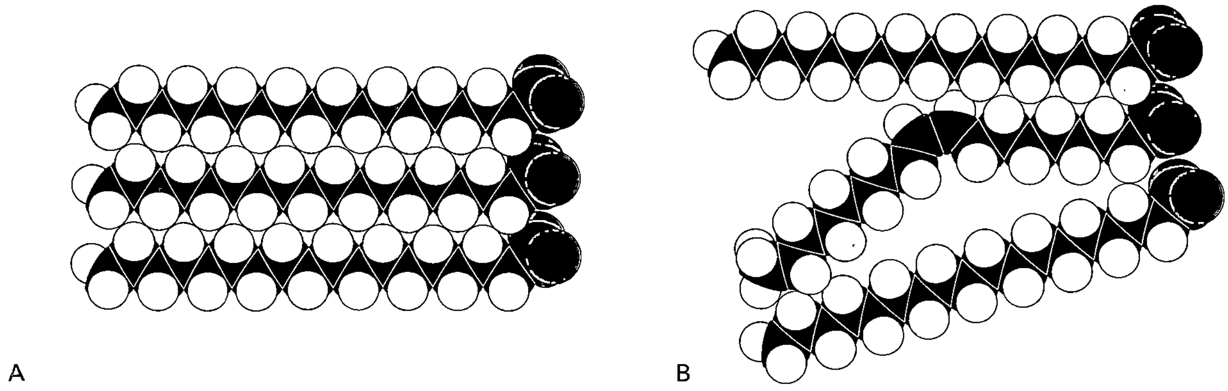

Figure 11-35
The highly ordered packing of fatty acid chains is disrupted by the presence of cis double bonds. These space-filling models show the packing of (A) three molecules of stearate (C_{18}, saturated) and (B) a molecule of oleate (C_{18}, unsaturated) between two molecules of stearate.

to 27°C. This decrease in the proportion of saturated residues prevents the membrane from becoming too rigid at the lower temperature. *In eukaryotes, cholesterol also is a key regulator of membrane fluidity.* Cholesterol contains a bulky steroid nucleus with a hydroxyl group at one end and a flexible hydrocarbon tail at the other end (Figure 11-36). Cholesterol inserts into bilayers with its long axis perpendicular to the plane of the membrane. The hydroxyl group of cholesterol hydrogen-bonds to a carbonyl oxygen atom of a phospholipid head group, whereas the hydrocarbon tail of cholesterol is located in the nonpolar core of the bilayer. Cholesterol prevents the crystallization of fatty acyl chains by fitting between them. In fact, high concentrations of cholesterol abolish phase transitions of bilayers. An opposite effect of cholesterol is to sterically block large motions of fatty acyl chains, which makes membranes less fluid. Thus, *cholesterol moderates the fluidity of membranes.*

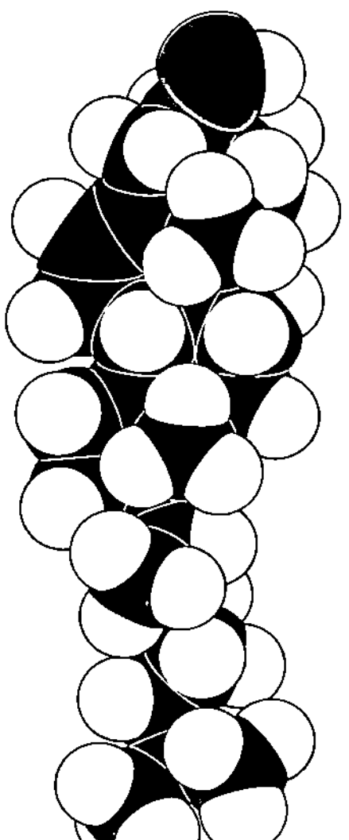

Figure 11-36
Space-filling model of cholesterol.

CARBOHYDRATE UNITS ARE LOCATED ON THE EXTRACELLULAR SIDE OF PLASMA MEMBRANES

Membranes of eukaryotic cells usually have a carbohydrate content of between 2% and 10% contributed by the sugar residues of their *glycolipids* and *glycoproteins.* As mentioned earlier, the glycolipids of higher organisms are derivatives of sphingosine with one or more attached sugar residues. In membrane glycoproteins, sugars are attached either to the amide nitrogen atom in the side chain of asparagine (termed an N-*linkage*) or to the oxygen atom in the side chain of serine or threonine (termed an O-*linkage*). The sugar directly attached to one of these side chains is usually *N*-acetylglucosamine or *N*-acetylgalactosamine (Figure 11-37). The amino acid sequence adjacent to an *N*-linked sugar is either Asn-X-Ser or Asn-X-Thr (X is any residue).

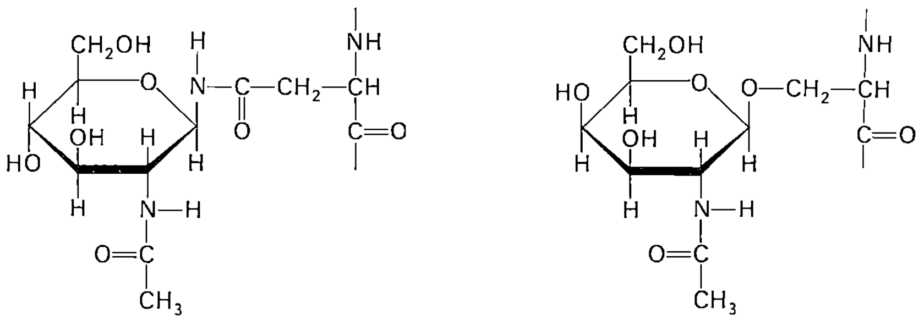

A **N-Acetylglucosamine linked to an asparagine residue**

B **N-Acetylgalactosamine linked to a serine residue**

Figure 11-37
Linkage of a sugar to (A) an asparagine side chain (an *N*-linkage) and (B) a serine side chain (an *O*-linkage).

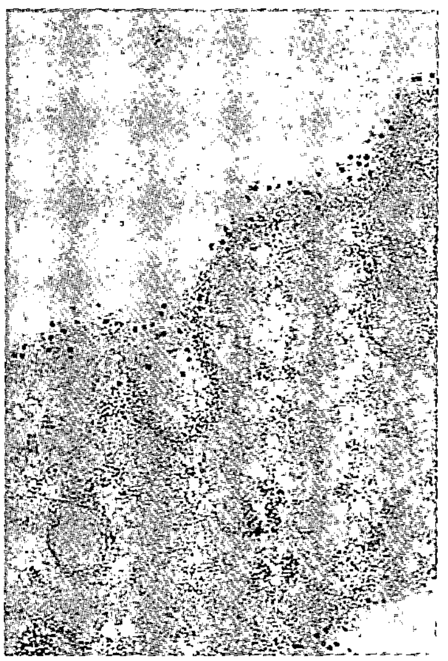

Figure 11-38
Electron micrograph showing the labeling of carbohydrate groups on the surface of a capillary endothelial cell by concanavalin A-ferritin. This lectin recognizes terminal mannosyl residues. [Courtesy of Dr. Richard Pino.]

The location of these carbohydrate groups in membranes can be revealed by specific labeling techniques. *Lectins,* which are plant proteins with high affinity for specific sugar residues, are valuable probes for this purpose. For example, *concanavalin A* binds to internal and nonreducing terminal α-mannosyl residues, whereas *wheat-germ agglutinin* binds to terminal *N*-acetylglucosamine residues. These lectins can be readily seen in electron micrographs if they are conjugated to an electron-dense marker such as ferritin, a 460-kd protein with a central core of about 4000 Fe^{3+} ions (Figure 11-38). Concanavalin A binds specifically to the outer surface of the erythrocyte membrane and not to the cytosolic surface. Studies of the binding of a wide variety of lectins to many kinds of cell membranes have shown that *sugar residues of membrane glycolipids and glycoproteins are always located on the extracellular side of mammalian plasma membranes* (Figure 11-39). The reason for this asymmetry will become evident when we consider how proteins are targeted to membranes (Chapter 35).

Carbohydrate groups orient glycoproteins in membranes. Because sugars are very hydrophilic, the sugar residues of a glycoprotein or of a glycolipid will strongly prefer to be located near the aqueous surface rather than in the hydrocarbon core. The cost in free energy of inserting an oligosaccharide chain into the hydrocarbon core of a membrane is very high. Consequently, there is a high barrier to the rotation of a glycoprotein from one side of a membrane to the other. The carbohydrate moieties of membrane glycoproteins help to maintain the asymmetric character of biological membranes. Carbohydrates on cell surfaces can also be important in *cell-cell recognition* (p. 478).

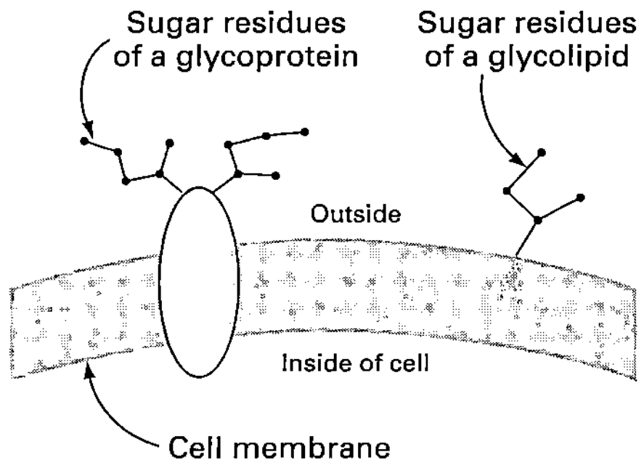

Figure 11-39
Sugar residues of membrane glycoproteins and glycolipids are located on the extracellular surface of mammalian plasma membranes.

FUNCTIONAL MEMBRANE SYSTEMS CAN BE RECONSTITUTED FROM PURIFIED COMPONENTS

Peripheral membrane proteins can be removed from a membrane by changing the pH and salt concentration. The integral membrane proteins in this stripped preparation are then released from their interactions with each other and with membrane lipids by adding a detergent. An ideal detergent has three properties: (1) It should dissociate the protein of interest from other membrane components, (2) it should not unfold the protein, and (3) it should be readily removable following purification. The balance is a delicate one. For example, sodium dodecyl sulfate is highly effective in dissociating proteins from other membrane components, but it denatures them and is hard to remove. Gentler detergents are used, such as sodium cholate and octyl glucoside. An early step in purifying and reconstituting a membrane protein is to test a series of detergents to determine which one is best suited for the particular task.

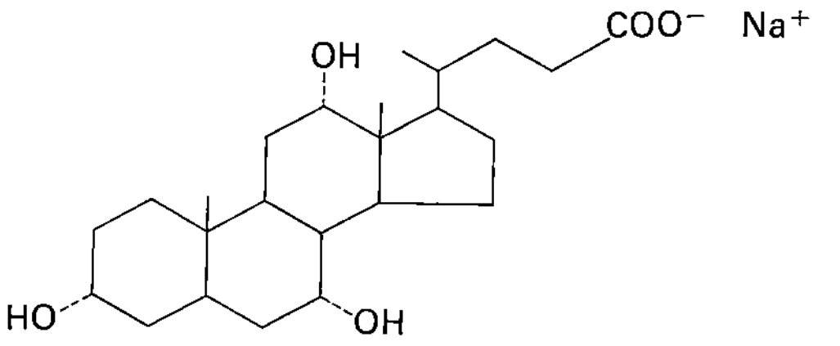

Sodium cholate

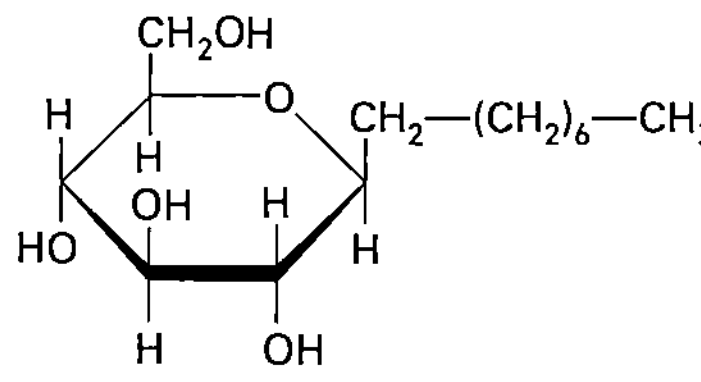

Octyl-β-glucoside

An excess of detergent is added so that not more than one protein molecule is present in a detergent micelle (Figure 11-40). The components of this mixture of proteins, phospholipids, and detergent are then separated by chromatographic and sedimentation methods similar to those used to purify water-soluble proteins. For example, ion-exchange chromatography is effective when nonionic (e.g., Triton X-100) or zwitterionic (e.g., DDAO) detergents are used. Some purified membrane

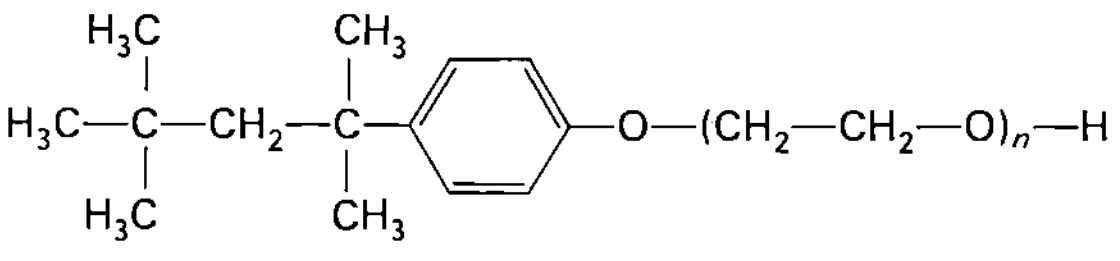

Polyoxyethylene *p*-*t*-octyl phenol
(Triton X series)

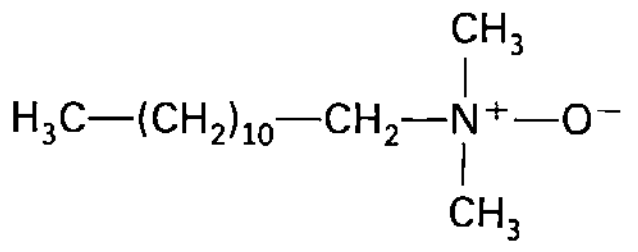

Dodecyldimethylamine oxide
(DDAO, Ammonyx LO)

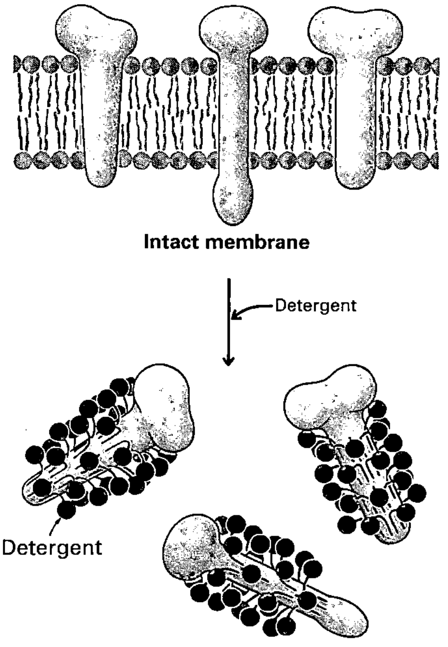

Figure 11-40
Solubilization of integral membrane proteins by the addition of detergent.

proteins while in detergent solution are functionally active. For example, rhodopsin, a photoreceptor protein from the rod cells of the retina, retains its 500-nm absorption band in detergents. Furthermore, excitation of the light-absorbing group of this receptor triggers the same structural change while in detergent as in the intact membrane.

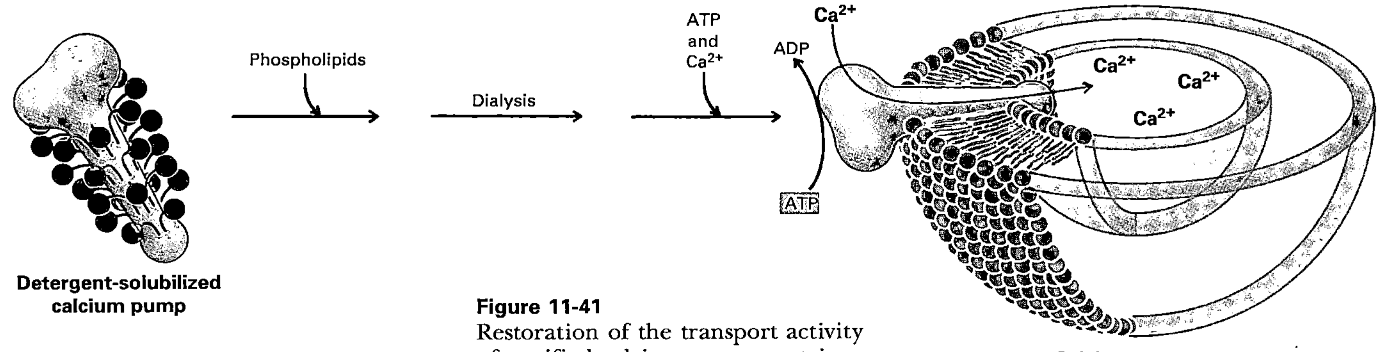

Figure 11-41
Restoration of the transport activity of purified calcium pump protein (Ca^{+2}-ATPase).

However, some of the most important properties of membrane proteins are expressed only in a membrane. For example, activity of membrane channels can be detected only in an intact membrane. Likewise, the effect of a transmembrane voltage or pH gradient can be measured only when the protein of interest is located in a membrane separating two compartments. These *vectorial properties* of membrane proteins, which are at the heart of their biological importance, become manifest only when they are situated in a membrane. Hence, the second stage in the study of a membrane protein is to return it to the bilayer (Figure 11-41). This can be accomplished by adding synthetic phospholipids to the purified protein in detergent solution and dialyzing away the detergent. Phospholipid vesicles containing the protein form spontaneously as the concentration of detergent decreases during dialysis. An important difference between such a reconstituted membrane and the native one is that the proteins in the vesicle are usually oriented randomly. However, vectorial functions can be assayed by the addition of substrates and ions to one side of the membrane only. For example, vesicles containing a calcium pump protein (called the Ca^{2+}-ATPase) accumulate Ca^{2+} in their inner aqueous space if ATP, the energy source, and Ca^{2+} are added to the outside (see Figure 11-41). As will be discussed in subsequent chapters, *the reconstitution of functionally active membrane systems from purified components is a powerful experimental approach to the elucidation of membrane processes.*

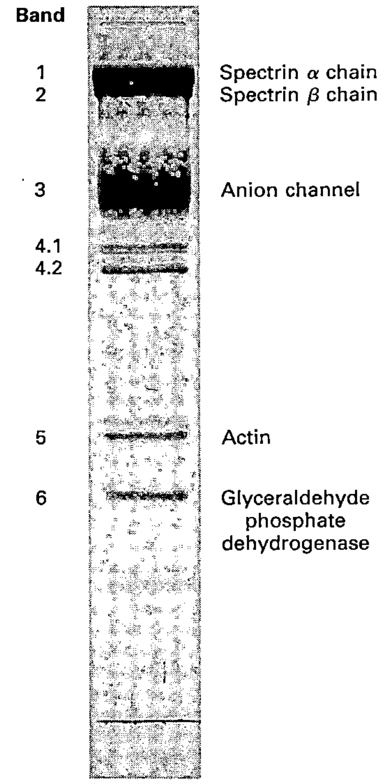

Figure 11-42
SDS-polyacrylamide gel pattern of the erythrocyte membrane. This gel was stained with Coomassie blue. [Courtesy of Dr. Vincent Marchesi.]

GLYCOPHORIN, A TRANSMEMBRANE PROTEIN, FORMS A CARBOHYDRATE COAT AROUND RED CELLS

Erythrocytes have been choice objects of inquiry in studies of membranes because of their ready availability and relative simplicity. They lack organelles and thus have only a single membrane, the plasma membrane. Nearly all the cytoplasmic contents of these cells can be released by osmotic hemolysis to give *ghosts,* which are quite pure plasma membranes. Coomassie staining of an SDS–polyacrylamide gel shows more than six prominent bands (Figure 11-42). When such a gel is stained with the periodic acid–Schiff reagent (PAS reagent), several related carbohydrate-rich proteins become evident. One of them, *glycophorin A*, is noteworthy because it accounts for much of the cell-surface carbohydrate. Moreover, it has been intensively studied. Glycophorin A is a single polypeptide chain with 16 attached oligosaccharide units. Indeed, 60% of the mass of this glycoprotein is carbohydrate, and so its name (derived from the Greek for "to carry sugar") is apt.

Vincent Marchesi determined the amino acid sequence of glycophorin A, the first integral membrane protein to be sequenced (Figure 11-43).

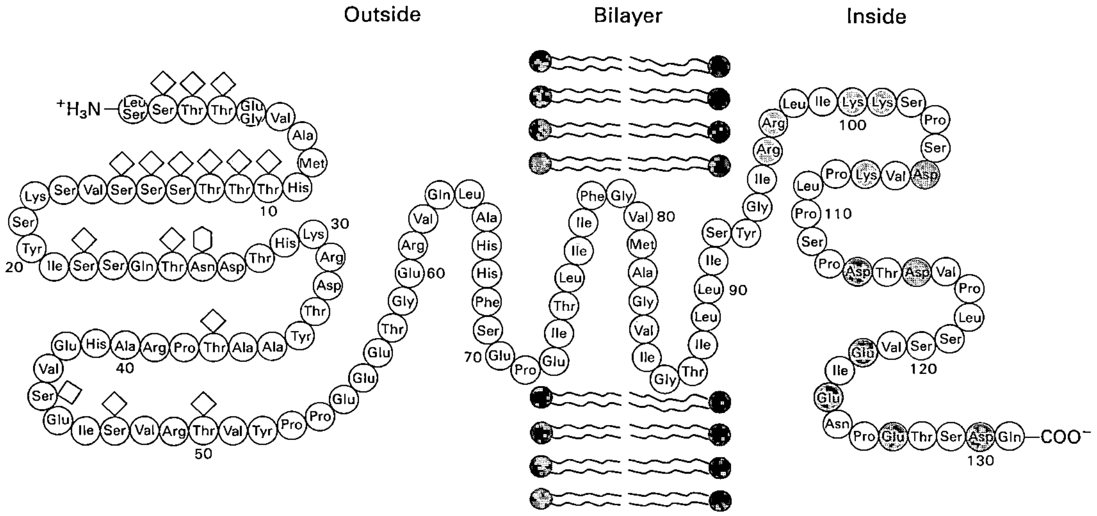

Figure 11-43
Amino acid sequence and transmembrane disposition of glycophorin A from the red-cell membrane. The fifteen *O*-linked carbohydrate units are shown in light green and the *N*-linked unit in dark green. The hydrophobic residues (yellow) buried in the bilayer form a transmembrane α helix. The carboxyl-terminal part of the molecule, located on the cytosolic side of the membrane, is rich in negatively charged (red) and positively charged (blue) residues. [Courtesy of Dr. Vincent Marchesi.]

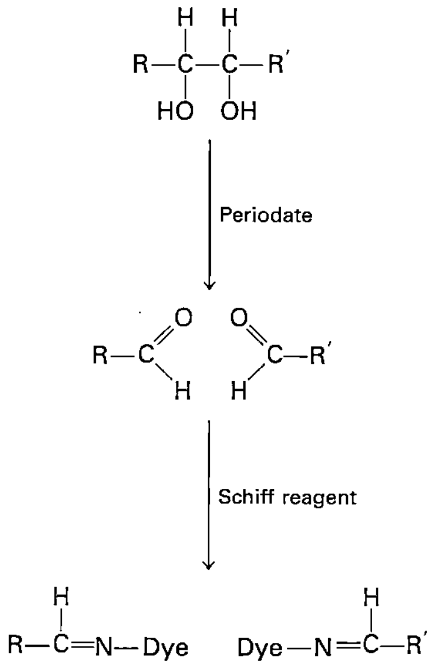

Ferritin-labeled antibodies specific for a determinant near the amino terminus labeled the extracellular side of the membrane, whereas those specific for the carboxyl-terminal region labeled the cytosolic side. Hence, glycophorin A spans the plasma membrane, with all molecules pointed in the same direction. Proteolytic digestion and chemical-modification studies showed that glycophorin consists of three parts: (1) an amino-terminal region containing all the carbohydrate units, which is located on the extracellular face of the membrane; (2) a hydrophobic middle region, which is buried within the hydrocarbon core of the membrane; and (3) a carboxyl-terminal region rich in polar and ionized side chains, which is exposed on the cytosolic face of the red-cell membrane. The hydrophobic membrane-spanning region is α-helical.

Glycophorin A contains about 100 sugar residues arranged in 16 units. Fifteen of them are attached to the hydroxyl group of serine or threonine residues (*O*-linked), and one of them is linked to the amide NH of an asparagine side chain (*N*-linked). These carbohydrate units are rich in sialic acid (p. 476), a negatively charged sugar. The carbohydrate units of glycophorins give red cells a very hydrophilic, anionic coat, which enables them to circulate without adhering to other cells and vessel walls.

Periodic acid–Schiff (PAS) reagent—
Detection of carbohydrate units by periodate oxidation, which generates aldehydes, followed by addition of a dye containing an amino or hydrazide group.

$$R-\underset{HO}{\overset{H}{C}}-\underset{OH}{\overset{H}{C}}-R'$$

↓ Periodate

$$R-\underset{H}{\overset{O}{C}} \quad \underset{H}{\overset{O}{C}}-R'$$

↓ Schiff reagent

$$R-\underset{H}{\overset{}{C}}=N-Dye \qquad Dye-N=\underset{H}{\overset{}{C}}-R'$$

TRANSMEMBRANE HELICES CAN BE ACCURATELY PREDICTED FROM AMINO ACID SEQUENCES

Table 11-2
Polarity scale for identifying transmembrane helices

Amino acid residue	*Transfer free energy (kcal/mol)*
Phe	3.7
Met	3.4
Ile	3.1
Leu	2.8
Val	2.6
Cys	2.0
Trp	1.9
Ala	1.6
Thr	1.2
Gly	1.0
Ser	0.6
Pro	−0.2
Tyr	−0.7
His	−3.0
Gln	−4.1
Asn	−4.8
Glu	−8.2
Lys	−8.8
Asp	−9.2
Arg	−12.3

Note: The free energies are for the transfer of an amino acid residue in an α helix from the membrane interior (assumed to have a dielectric constant of 2) to water. After D.M. Engelman, T.A. Steitz, and A. Goldman. *Ann. Rev. Biophys. Biophys. Chem.* 15(1986):330.

The presence of a highly nonpolar sequence of 19 residues in glycophorin A (shown in yellow in Figure 11-43) first suggested that this segment of the protein traverses the lipid bilayer and is folded into an α helix. Studies of synthetic polypeptides had shown that α helices are more stable in nonpolar media than in water, which competes for hydrogen bonding with main-chain NH and CO groups. *One approach to identifying transmembrane helices is to ask whether a postulated helical segment prefers to be in a hydrocarbon milieu or in water.* Specifically, we want to calculate the free-energy change when a helical segment is transferred from the interior of a membrane to water. Free-energy estimates for amino acid residues are given in Table 11-2. For example, the transfer of a poly-L-arginine helix from the interior of a membrane to water would be highly favorable (−12.3 kcal/mol per arginine in the helix), whereas the transfer of a poly-L-phenylalanine helix would be unfavorable (+3.7 kcal/mol per phenylalanine in the helix).

The hydrocarbon core of membranes is typically 30 Å wide, which can be traversed by an α helix consisting of 20 residues. We can take the amino acid sequence of a protein and calculate the free-energy change in transferring a hypothetical α helix of residues 1 to 20 from the membrane interior to water. The same calculation can be made for residues 2 to 21, 3 to 22, and so forth until we reach the end of the sequence. The span of 20 residues chosen for this calculation is called the *window*. For glycophorin A, a plot of the free-energy change versus the position of the first residue in the window shows a clear-cut maximum of +40 kcal/mol when residue 73 is the first in the window (Figure 11-44). This *hydropathy plot* correctly identifies the transmembrane α helix of glycophorin. *In general, a peak of* + *20 kcal/mol or more in a hydropathy plot based on a window of 20 residues indicates that a polypeptide segment could be a membrane-spanning α helix.* By this criterion, glycophorin has only one membrane-spanning helix, in agreement with experimental findings. Indeed, hydropathy plots correctly predicted the positions of all 11 transmembrane helices in a photosynthetic reaction center (p. 287). It should be noted, however, that a peak in a hydropathy plot does not prove that a segment is a transmem-

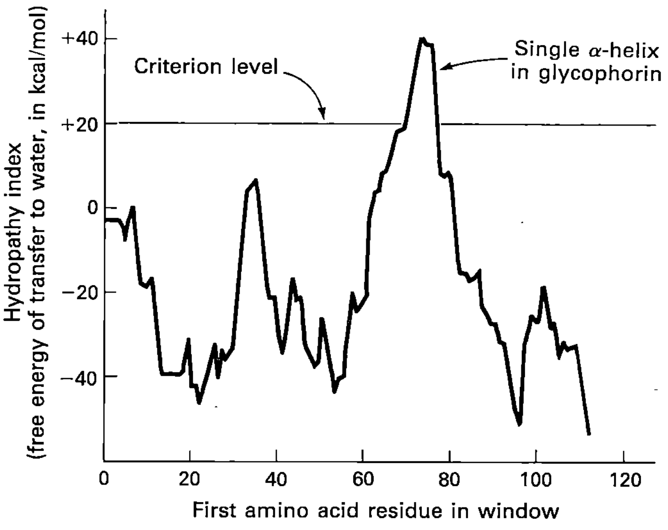

Figure 11-44
Hydropathy plot for glycophorin. The free energy for transferring a helix of 20 residues from the membrane to water is plotted as a function of the position of the first residue of the helix in the sequence of the protein. Peaks of greater than +20 kcal/mol in hydropathy plots are indicative of potential transmembrane helices. [After D.M. Engelman, T.A. Steitz, and A. Goldman. Identifying nonpolar transbilayer helices in amino acid sequences of membrane proteins. *Ann. Rev. Biophys. Biophys. Chem.* 15(1986):343. Copyright © 1986 by Annual Reviews, Inc. All rights reserved.]

brane helix. Even soluble proteins may have such highly nonpolar regions. Conversely, some membrane proteins have amphipathic transmembrane helices that escape detection by these plots. However, it is clear that hydropathy plots are very valuable in suggesting how membrane proteins fold and in stimulating experiments to test such models.

SPECTRIN FORMS A MEMBRANE CYTOSKELETON THAT MAKES ERYTHROCYTES RESISTANT TO SHEAR

An erythrocyte is exposed to powerful shearing forces and undergoes large changes in shape during its lifetime of 120 days. It travels hundreds of kilometers in this period, much of it through narrow passageways. The mechanical stability and resilience of the erythrocyte membrane come from a partnership between the plasma membrane and an underlying meshwork called the *membrane skeleton* (Figure 11-45). The major constituent of this infrastructure is *spectrin,* which consists of a 260-kd α chain and a 225-kd β chain (Figure 11-46). Amino acid and cDNA sequence analyses have revealed that each spectrin chain consists of repeats of a 106-residue unit that folds into a triple-stranded α-helical coiled coil. These two chains are aligned in parallel and twisted about each other to form a flexible rod that is about 100 nm (1000 Å) long. Two of these $\alpha\beta$ units join end to end to form a tetramer with a length of 200 nm.

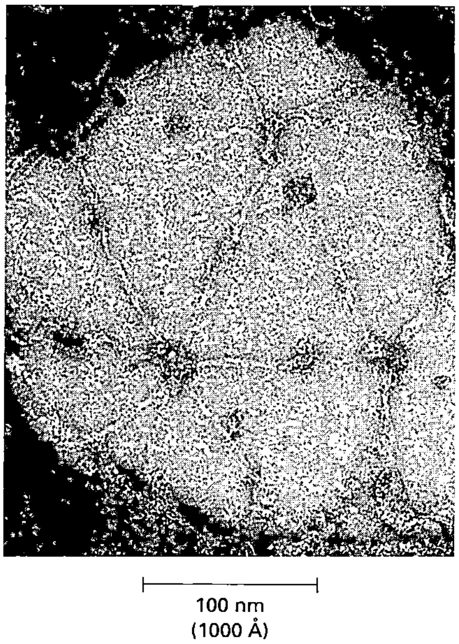

Figure 11-45
Electron micrograph of a portion of the erythrocyte membrane skeleton that has been spread tenfold to display the hexagonal and pentagonal network of spectrin and other protein molecules. [From T.J. Byers and D. Branton. *Proc. Nat. Acad. Sci.* 82(1985):6153.]

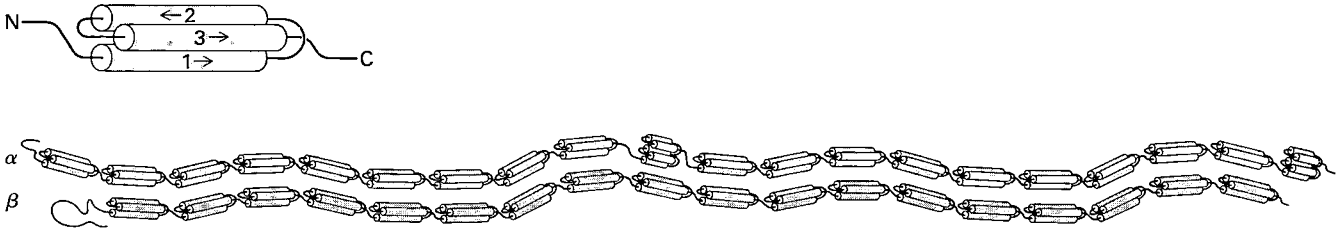

Figure 11-46
Schematic diagram of spectrin. The α chain is shown in blue and the β chain in red. The repeating segment of 106 residues folds into a triple-stranded α-helical bundle. These segments are flexibly joined in both chains. [Courtesy of Dr. Vincent Marchesi.]

Spectrin does not bind directly to the plasma membrane. Instead, different parts of spectrin interact with *ankyrin* and *protein 4.1* (Figure 11-47). Ankyrin forms a cross-link between the β chain of spectrin and the cytosolic domain of the *anion channel,* a transporter that mediates the exchange of HCO_3^- for Cl^- across the red-cell membrane. This exchanger plays a key role in the transport of CO_2 and the buffering of red-cell pH, which accounts for its abundance. The anion channel constitutes about a third of the total membrane protein. Protein 4.1 promotes the binding of actin filaments to the carboxyl-terminal portions of both spectrin chains and links this spectrin-actin complex to the cytosolic face of glycophorin. Several spectrin tetramers insert at each protein 4.1 junction complex to form a continuous meshwork underlying the plasma membrane.

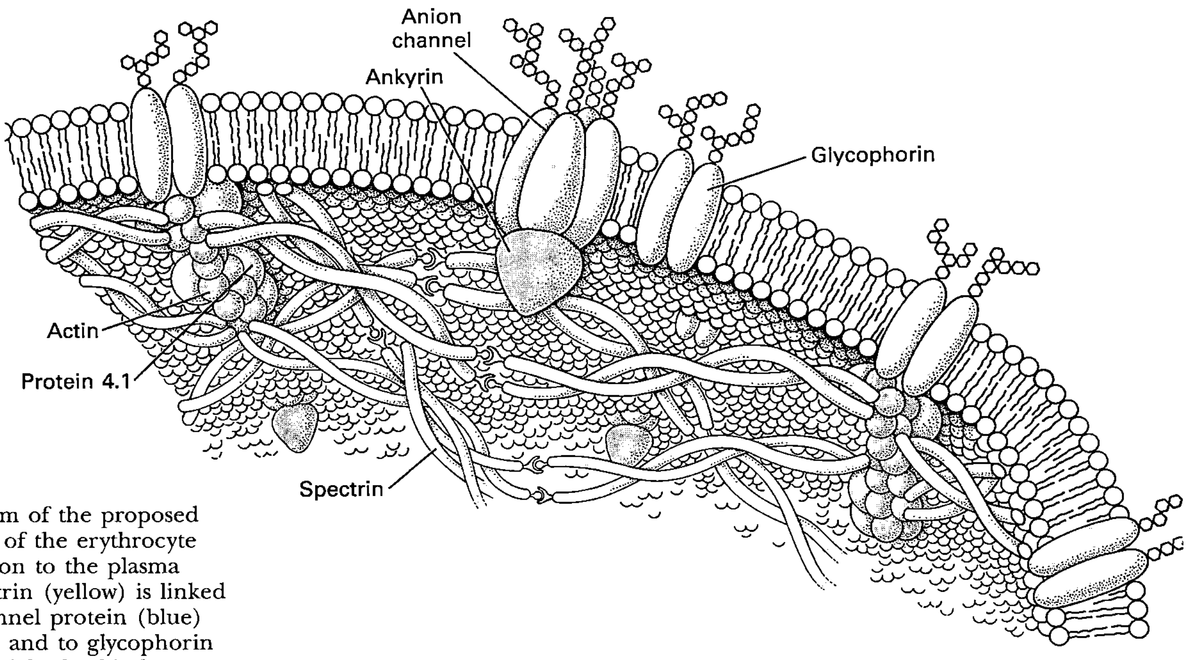

Figure 11-47
Schematic diagram of the proposed mode of binding of the erythrocyte membrane skeleton to the plasma membrane. Spectrin (yellow) is linked to the anion channel protein (blue) by ankyrin (red), and to glycophorin by protein 4.1, which also binds an actin filament. [After S.B. Shohet and S.E. Lux. *Hospital Practice* 19(Nov. 1984):90. Redrawn on the basis of a drawing by Robert Margulies.]

Variations on this theme are found in other tissues. *Fodrin* in epithelial cells closely resembles spectrin. *Synapsin-1,* a protein once thought to be specific to neurons, is homologous to protein 4.1. *Actin,* a component of contractile assemblies, is present in nearly all eukaryotic cells. *It seems likely that membrane skeletons like the one formed by spectrin in erythrocytes play key roles in altering and stabilizing the shapes of many kinds of cells.*

ELECTRON-MICROSCOPIC AND X-RAY ANALYSES OF CRYSTALLINE MEMBRANE PROTEINS ARE HIGHLY REVEALING

One of the goals of membrane research is to obtain high-resolution images of membrane systems to understand how they carry out their biological roles. X-ray crystallographic studies of proteins such as hemoglobin, lysozyme, and chymotrypsin have provided insight into how water-soluble proteins fold, bind other molecules, and catalyze reactions. Likewise, high-resolution views of membrane channels, pumps, and receptors are needed. An important start has been made with membrane proteins that form two-dimensional crystals, ordered lattices in the plane of the membrane. Extensive crystalline sheets are suitable for structural analysis by electron microscopy, as was first shown for the *purple membrane* of *Halobacterium halobium* (a salt-loving bacterium). This specialized region of the cell membrane contains *bacteriorhodopsin,* a 25-kd protein, which converts the energy of light into a transmembrane proton gradient that is used to synthesize ATP (p. 318). The experimental strategy is to take electron micrographs with a very low dose of electrons to minimize radiation damage. This is feasible with a crystalline array of many thousands of bacteriorhodopsin molecules. Also, unstained samples are used to provide a high-fidelity image. Electron micrographs are taken at various tilt angles to obtain a series of different views of the structure projected onto a plane. The final stage is to recombine the information from these projections by Fourier techniques. This technique is called *electron crystallography.*

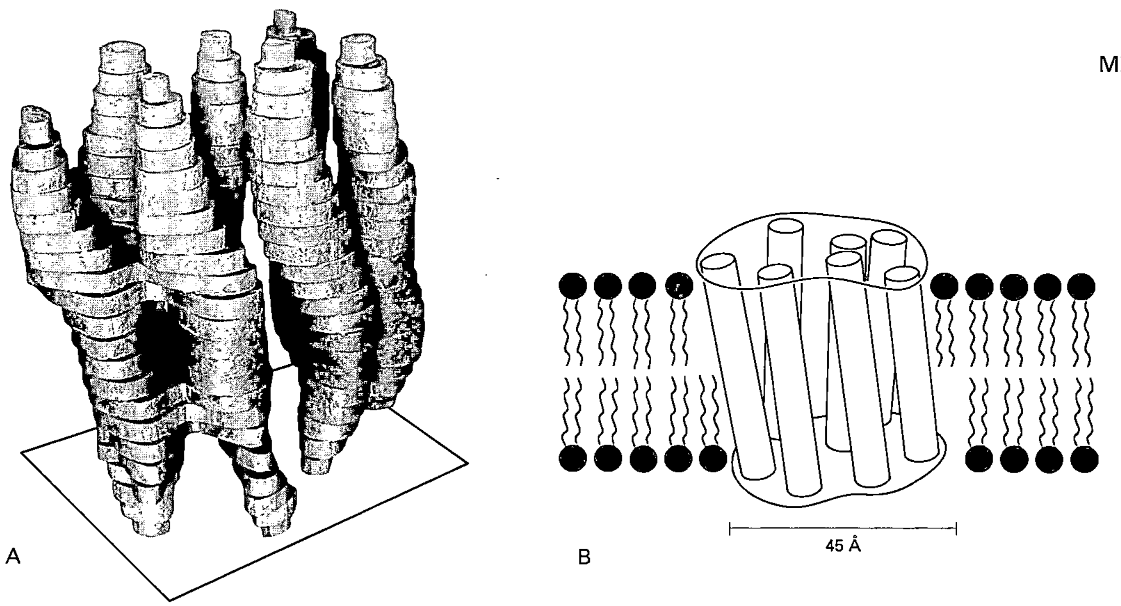

Figure 11-48
(A) Model of bacteriorhodopsin constructed from a 7-Å three-dimensional map. (B) Interpretive diagram showing the arrangement of α-helical segments in the lipid bilayer. The connections between helices were not evident at this resolution. [Courtesy of Dr. Richard Henderson and Dr. Nigel Unwin.]

In 1975, Richard Henderson and Nigel Unwin used this method to obtain the first picture of a membrane protein (Figure 11-48). Their 7-Å-resolution map showed that *bacteriorhodopsin contains seven closely packed α helices extending nearly perpendicular to the plane of the membrane for most of its 45-Å width.* Lipid bilayer regions fill the spaces between the protein molecules. A higher-resolution map subsequently revealed the connections between α helices, the location of the light-absorbing group, and the positions of most of the side chains (p. 319). These structural studies have led to a deeper understanding of how bacteriorhodopsin, the simplest known energy-converting protein, uses light to pump protons across a membrane. Electron crystallography has also provided revealing views of the acetylcholine receptor channel, which participates in the transmission of nerve impulses (p. 293), and gap junctions, which serve as channels between cells (p. 307).

The photosynthetic reaction center from a purple bacterium (*Rhodopseudomonas viridis*) has been crystallized in *three dimensions* from a detergent solution, and its structure has been solved by *x-ray analysis* at 3-Å resolution. This landmark study by Johann Deisenhofer, Hartmut Michel, and Robert Huber provided a highly detailed and informative image of an energy-converting membrane assembly (Figure 11-49). Moreover, it demonstrated that membrane proteins, like water-soluble proteins, can form highly ordered three-dimensional crystals that are amenable to x-ray analysis at high resolution. The 150-kd reaction center consists of four polypeptide chains, of which three traverse the membrane. The membrane-spanning segments are α-helical, as in bacteriorhodopsin. The function of the reaction center is to convert light into a separation of charge, a process mediated by its 12 prosthetic groups and an iron atom. We shall return to this remarkable assembly when we consider the primary energy-converting step in photosynthesis (p. 669).

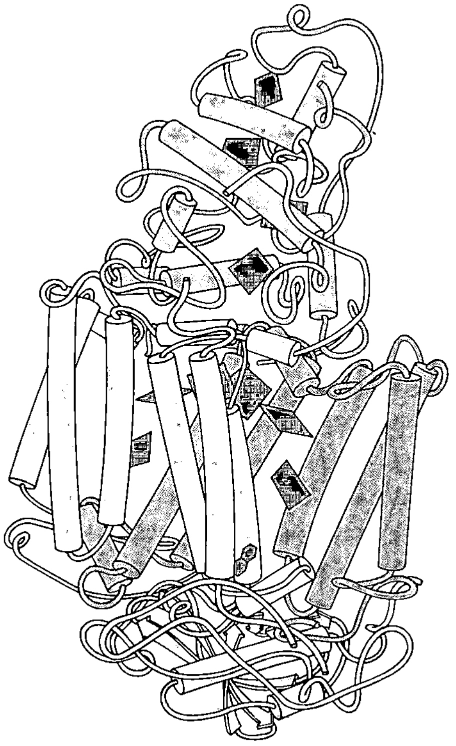

Figure 11-49
Three-dimensional structure of the photosynthetic reaction center from a purple bacterium. Three of the four proteins (shown in yellow, green, and red) traverse the lipid bilayer and consist largely of transmembrane α helices. The prosthetic groups of this energy-converting protein complex are shown in gray. [After a drawing kindly provided by Dr. Jane Richardson. Based on atomic coordinates provided by Dr. Johann Deisenhofer, Dr. Hartmut Michel, and Dr. Robert Huber.]

SUMMARY

Biological membranes are sheetlike structures, typically 75 Å thick, that are composed of protein and lipid molecules held together by noncovalent interactions. Membranes are highly selective permeability barriers. They create closed compartments, which may be entire cells or organelles within a cell. Pumps and gates in membranes regulate the molecular and ionic compositions of these compartments. Membranes also control the flow of information between cells. For example, many membranes contain receptors for hormones such as insulin. Furthermore, membranes are intimately involved in such energy conversion processes as photosynthesis and oxidative phosphorylation.

The major classes of membrane lipids are phospholipids, glycolipids, and cholesterol. Phosphoglycerides, a type of phospholipid, consist of a glycerol backbone, two fatty acid chains, and a phosphorylated alcohol. The fatty acid chains usually contain between 14 and 24 carbon atoms; they may be saturated or unsaturated. Phosphatidyl choline, phosphatidyl serine, and phosphatidyl ethanolamine are major phosphoglycerides. Sphingomyelin, a different type of phospholipid, contains a sphingosine backbone instead of glycerol. Glycolipids are sugar-containing lipids derived from sphingosine. A common feature of these membrane lipids is that they are amphipathic molecules. They spontaneously form extensive bimolecular sheets in aqueous solutions because they contain both a hydrophilic and a hydrophobic moiety. These lipid bilayers are highly impermeable to ions and most polar molecules, yet they are quite fluid, which enables them to act as a solvent for membrane proteins.

Distinctive membrane functions such as transport, communication, and energy transduction are mediated by specific proteins. Integral membrane proteins span the lipid bilayer, whereas peripheral membrane proteins are bound to membrane surfaces by electrostatic and hydrogen bond interactions. Membranes are structurally and functionally asymmetric, as exemplified by the directionality of ion transport systems and the restriction of sugar residues to the external surface of mammalian plasma membranes. Membranes are dynamic structures in which proteins and lipids diffuse rapidly in the plane of the membrane (lateral diffusion), unless restricted by special interactions. In contrast, the rotation of lipids from one face of a membrane to the other (transverse diffusion, or flip-flop) is usually very slow. Proteins do not rotate across bilayers; hence, membrane asymmetry can be preserved. The degree of fluidity of a membrane partly depends on the chain length of its lipids and the extent to which their constituent fatty acids are unsaturated.

The permeability of lipid bilayers has been measured in two well-defined model systems, lipid vesicles (liposomes) and planar bilayer membranes. Transport antibiotics increase the permeability of membranes by acting as ion carriers or channel formers. Valinomycin, a cyclic molecule, shuttles K^+ across membranes by providing a hydrophobic shell around the sequestered ion. Gramicidin A, in contrast, dimerizes to form a transmembrane pore that is permeable to monovalent cations.

The erythrocyte membrane, one of the most intensively studied and best-understood membrane systems, contains two abundant transmembrane proteins. Glycophorin A, which bears many covalently attached sugar units, gives red cells a negatively charged carbohydrate coat. The anion channel mediates the exchange of bicarbonate and chloride ions. These integral membrane proteins are linked by protein 4.1 and ankyrin to a flexible meshwork consisting mainly of spectrin. This membrane skeleton enables erythrocytes to resist strong shearing forces.

300 nm
(3000 Å)

Freeze-fracture electron micrograph of unilamellar phospholipid vesicles. [Courtesy of Dr. P.R. Cullis.]

Three-dimensional images of membrane proteins can be reconstructed from electron micrographs of two-dimensional crystalline arrays. Bacteriorhodopsin, a light-driven proton pump in the plasma membrane of halobacteria, contains seven transmembrane α helices. X-ray analyses of three-dimensional crystals of the photosynthetic reaction center of a purple bacterium have revealed the atomic structure of this energy-converting assembly of four proteins.

SELECTED READINGS

Where to start

Bretscher, M.S., 1985. The molecules of the cell membrane. *Sci. Amer.* 253(4):100–108.

Unwin, N., and Henderson, R., 1984. The structure of proteins in biological membranes. *Sci. Amer.* 250(2):78–94.

Deisenhofer, J., and Michel, H., 1989. The photosynthetic reaction centre from the purple bacterium *Rhodopseudomonas viridis. EMBO J.* 8:2149–2170. [This Nobel Lecture provides a lucid account of how the three-dimensional structure of a membrane protein was elucidated in atomic detail.]

Singer, S.J., and Nicolson, G.L., 1972. The fluid mosaic model of the structure of cell membranes. *Science* 175:720–731.

Books

Gennis, R.B., 1989. *Biomembranes: Molecular Structure and Function.* Springer-Verlag. [An informative and lucid account of the fundamentals of membrane structure, dynamics, and function. The cited references are a fine entrée into a large literature.]

Vance, D.E., and Vance, J.E. (eds.), 1991. *Biochemistry of Lipids, Lipoproteins, and Membranes.* Elsevier. [Contains many excellent articles on lipid structure, dynamics, metabolism, and assembly into membranes.]

Overton, C.E., 1901. *Studies of Narcosis and a Contribution to General Pharmacology.* [A translation of this classic work with illuminating commentary has been published by R.L. Lipnick, 1991, Chapman and Hall. Overton discovered that membrane permeability is correlated with a high solubility in a nonpolar solvent relative to that in water.]

Racker, E., 1985. *Reconstitutions of Transporters, Receptors, and Pathological States.* Academic Press. [An engaging personal account of the reconstitution of membrane assemblies. Many strategies and examples are presented.]

Tanford, C., 1980. *The Hydrophobic Effect: Formation of Micelles and Biological Membranes,* 2nd ed. Wiley-Interscience. [A lucid statement of thermodynamic aspects of membrane structure.]

Tanford, C., 1989. *Ben Franklin Stilled the Waves.* Duke University Press. [A delightful and perceptive account of Franklin's experiment on the spreading of oil on water and its significance for biology. The subtitle of this book is *An Informal History of Pouring Oil on Water with Reflections on the Ups and Downs of Scientific Life in General.*]

Membrane lipids and dynamics

Bloom, M., Evans, E., and Mouritsen, O.G., 1991. Physical properties of the fluid lipid-bilayer component of cell membranes: A perspective. *Quart. Rev. Biophys.* 24:293–397.

Elson, E.L., 1986. Membrane dynamics studied by fluorescence correlation spectroscopy and photobleaching recovery. *Soc. Gen. Physiol. Ser.* 40:367–383.

Zachowski, A., and Devaux, P.F., 1990. Transmembrane movements of lipids. *Experientia* 46:644–656.

Devaux, P.F., 1992. Protein involvement in transmembrane lipid asymmetry. *Ann. Rev. Biophys. Biomol. Struct.* 21:417–439.

Silvius, J.R., 1992. Solubilization and functional reconstitution of biomembrane components. *Ann. Rev. Biophys. Biomol. Struct.* 21:323–348.

Yeagle, P.L., Albert, A.D., Boesze, B.K., Young, J., and Frye, J., 1990. Cholesterol dynamics in membranes. *Biophys. J.* 57:413–424.

Frye, C.D., and Edidin, M., 1970. The rapid intermixing of cell surface antigens after formation of mouse-human heterokaryons. *J. Cell Sci.* 7:319–335.

Kornberg, R.D., and McConnell, H.M., 1971. Inside-outside transitions of phospholipids in vesicle membranes. *Biochemistry* 10:1111–1120.

Ostro, M.J., 1987. Liposomes. *Sci. Amer.* 256(1):102–110. [Potential applications of liposomes for drug delivery are discussed.]

Conformation of membrane proteins

Lipowsky, R., 1991. The conformation of membranes. *Nature* 349:475–481.

Altenbach, C., Marti, T., Khorana, H.G., and Hubbell, W.L., 1990. Transmembrane protein structure: Spin labeling of bacteriorhodopsin mutants. *Science* 248:1088–1092.

Fasman, G.D., and Gilbert, W.A., 1990. The prediction of transmembrane protein sequences and their conformation: An evaluation. *Trends Biochem. Sci.* 15:89–92.

Jennings, M.L., 1989. Topography of membrane proteins. *Ann. Rev. Biochem.* 58:999–1027.

Engelman, D.M., Steitz, T.A., and Goldman, A., 1986. Identifying non-polar transbilayer helices in amino acid sequences of membrane proteins. *Ann. Rev. Biophys. Biophys. Chem.* 15:321–353.

Erythrocyte membrane

Bennett, V., 1992. Ankyrins. Adaptors between diverse plasma membrane proteins and the cytoplasm. *J. Biol. Chem.* 267:8703–8706.

Marchesi, V.T., 1985. The cytoskeletal system of red blood cells. *Hospital Practice* 20(11):113–131.

Marchesi, S.L., Conboy, J., Agre, P., Letsinger, J.T., Marchesi, V.T., Speicher, D.W., and Mohandas, N., 1990. Molecular analysis of insertion/deletion mutations in protein 4.1 in elliptocytosis. I. Biochemical identification of rearrangements in the spectrin/actin binding domain and functional characterizations. *J. Clin. Invest.* 86:516–523.

Davies, K.A., and Lux, S.E., 1989. Hereditary disorders of the red cell membrane skeleton. *Trends Genet.* 5:222–227.

Lux, S.E., John, K.M., Kopito, R.R., and Lodish, H.F., 1989. Cloning and characterization of band 3, the human erythrocyte anion-exchange protein (AE1). *Proc. Nat. Acad. Sci.* 86:9089–9093.

Lux, S.E., John, K.M., and Bennett, V., 1990. Analysis of cDNA for human erythrocyte ankyrin indicates a repeated structure with homology to tissue-differentiation and cell-cycle control proteins. *Nature* 344:36–42.

Electron and x-ray crystallography of membrane proteins

Henderson, R., Baldwin, J.M., Ceska, T.A., Zemlin, F., Beckmann, E., and Downing, K.H., 1990. Model for the structure of bacteriorhodopsin based on high-resolution electron cryo-microscopy. *J. Mol. Biol.* 213:899–929.

Henderson, R., and Unwin, P.N.T., 1975. Three-dimensional model of purple membrane obtained by electron microscopy. *Nature* 257:28–32. [The first view of a membrane protein.]

Amos, L.A., Henderson, R., and Unwin, P.N.T., 1982. Three-dimensional structure determination by electron microscopy of two-dimensional crystals. *Prog. Biophys. Mol. Biol.* 39:183–231.

Deisenhofer, J., and Michel, H., 1991. High-resolution structures of photosynthetic reaction centers. *Ann. Rev. Biophys. Biophys. Chem.* 20:247–266.

Reconstitution of membrane assemblies

Madden, T.D., 1986. Current concepts in membrane protein reconstitution. *Chem. Phys. Lipids* 40:207–222.

Thomas, T.C., and McNamee, M.G., 1990. Purification of membrane proteins. *Meth. Enzymol.* 182:499–520.

Helenius, A., Sarvas, M., and Simons, K., 1981. Asymmetric and symmetric membrane reconstitution by detergent elimination. *Eur. J. Biochem.* 116:27–31.

Fleischer, S., and Fleischer, B. (eds.), 1990. *Methods in Enzymology,* vol. 192. Academic Press. [This volume and preceding ones in the series subtitled *Membranes* contain excellent articles on experimental methods in membrane research, including procedures for the preparation of model and reconstituted membranes. Also see volumes 31, 32, 52, 56, 81, 88, 96–98, 125–127, 156–157, 172–174, and 191.]

PROBLEMS

1. *Population density.* How many phospholipid molecules are there in a 1-μm^2 region of a phospholipid bilayer membrane? Assume that a phospholipid molecule occupies 70 $Å^2$ of the surface area.

2. *Lipid diffusion.* What is the average distance traversed by a membrane lipid in 1 μs, 1 ms, and 1 s? Assume a diffusion coefficient of 10^{-8} cm^2/s.

3. *Protein diffusion.* The diffusion coefficient D of a rigid spherical molecule is given by

$$D = \frac{kT}{6\pi\eta r}$$

in which η is the viscosity of the solvent, r is the radius of the sphere, k is the Boltzman constant (1.38×10^{-16} erg/deg), and T is the absolute temperature. What is the diffusion coefficient at 37°C of a 100-kd protein in a membrane that has an effective viscosity of 1 poise (1 poise = 1 erg s/cm^3)? What is the average distance traversed by this protein in 1 μs, 1 ms, and 1 s? Assume that this protein is an unhydrated, rigid sphere of density 1.35 g/cm^3.

4. *Cold sensitivity.* The conductance of a lipid bilayer membrane containing a carrier antibiotic decreased abruptly when the temperature was lowered from 40°C to 36°C. In contrast, there was little change in conductance of the same bilayer membrane when it contained a channel-forming antibiotic. Why?

5. *Flip-flop.* The transverse diffusion of phospholipids in a bilayer membrane was investigated using a paramagnetic analog of phosphatidyl choline, called *spin-labeled phosphatidyl choline.*

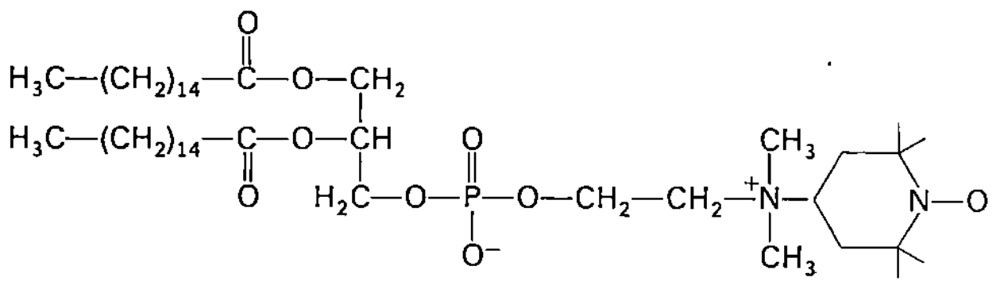

The nitroxide (NO) group in spin-labeled phosphatidyl choline gives a distinctive paramagnetic resonance spectrum. This spectrum disappears when nitroxides are converted into amines by reducing agents such as ascorbate.

Lipid vesicles containing phosphatidyl choline (95%) and the spin-labeled analog (5%) were prepared by sonication and purified by gel-filtration chromatography. The outside diameter of these liposomes was about 25 nm. The amplitude of the paramagnetic resonance spectrum decreased to 35% of its initial value within a few minutes of the addition of ascorbate. There was no detectable change in the spectrum within a few minutes after the addition of a second aliquot of ascorbate. However, the amplitude of the residual spectrum decayed exponentially with a half-time of 6.5 hr. How would you interpret these changes in the amplitude of the paramagnetic spectrum?

CHAPTER

12

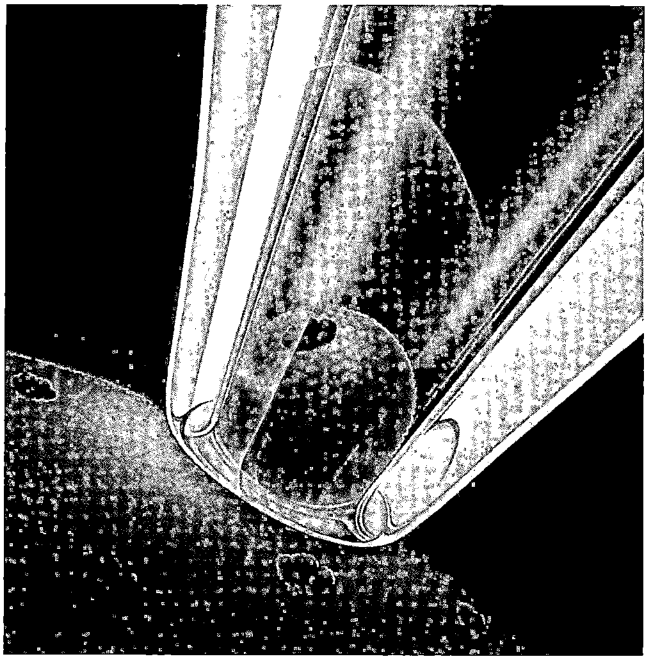

Membrane Channels and Pumps

The lipid bilayer of biological membranes, as was discussed in the preceding chapter, is intrinsically impermeable to ions and polar molecules. Permeability is conferred by two classes of allosteric proteins, *channels* and *pumps*. Channels enable ions to flow rapidly through membranes in a thermodynamically downhill direction. Pumps, by contrast, use a source of free energy such as ATP or light to drive the uphill transport of ions or molecules. We begin by examining the *acetylcholine receptor*, a channel that mediates the transmission of nerve signals across synapses. We will focus on three questions: (1) What is the architecture of the channel? Specifically, how is a transmembrane pore formed? (2) How is the channel opened and then shut? In other words, how is the channel gated? (3) Some ions readily flow through the channel, whereas others are rejected. How is ionic selectivity achieved?

We then turn to a pair of channels—the sodium channel and the potassium channel—that mediate action potentials in nerve axon membranes. These channels are opened by membrane depolarization rather than by the binding of an allosteric effector. They are *voltage-gated*, in contrast with the acetylcholine receptor channel, which is *ligand-gated*. Our focus here will be on how the sodium channel and the potassium channel respond to the voltage across the plasma membrane. A new principle of allosteric switching will become evident. These channels are also of inter-

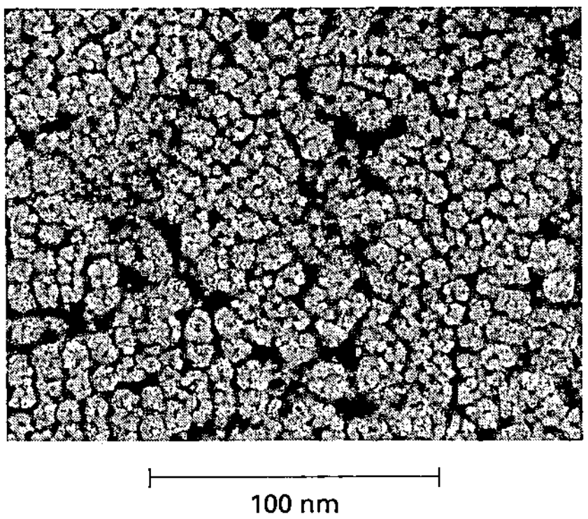

Figure 12-1
Electron micrograph of acetylcholine receptors in a postsynaptic membrane. [Courtesy of Dr. John Heuser and Dr. Shelly Salpeter.]

Opening Image: *The flow of ions through a single membrane channel (shown in red) can be detected by patch clamping, a technique that has revolutionized the study of these molecular assemblies. [After E. Neher and B. Sakmann. The patch clamp technique. Copyright © 1992 by Scientific American, Inc. All rights reserved.]*

est because they swiftly and deftly distinguish between quite similar ions (e.g., Na^+ and K^+). This is an exciting time in the study of channels because powerful techniques from different fields have come together to shed light on their structure and dynamics. It is remarkable that the flow of ions through a single channel in a membrane can readily be detected. The cloning and expression of genes for many kinds of channels have opened new vistas in our understanding of these elegant molecular devices. Furthermore, electron crystallography is providing tantalizing views of their molecular architecture.

Rapid progress is also being made on understanding how membrane pumps carry out uphill transport. In essence, pumps convert one currency of free energy into another. First, we will consider the ATP-driven sodium-potassium pump of animal cells, which plays a key role in maintaining cell volume and the intracellular ionic composition. The Na^+-K^+ ATPase also generates the Na^+ and K^+ gradients that are at the heart of the excitability of nerve and muscle. Phosphorylation and dephosphorylation of the pump are coupled to changes in orientation and affinity of its ion-binding sites. A different mechanism of active transport, one that utilizes the gradient of one ion to drive the active transport of another, will be illustrated by the sodium-calcium exchanger. This pump plays an important role in extruding Ca^{2+} from cells. Bacteria often use a proton gradient to concentrate fuel molecules inside the cell, as exemplified by the action of lactose permease. The chapter concludes with a view of the action of bacteriorhodopsin, a light-driven pump that generates a proton gradient in a salt-loving bacterium. We catch here our first glimpse of how a biological energy transduction process operates at the atomic level.

THE ACETYLCHOLINE RECEPTOR CHANNEL MEDIATES THE TRANSMISSION OF NERVE SIGNALS ACROSS SYNAPSES

Nerve impulses are communicated across most synapses by small, diffusible molecules called *neurotransmitters*, such as acetylcholine. The presynaptic membrane of a synapse is separated from the postsynaptic membrane by a gap of about 50 nm, called the *synaptic cleft*. The end of the presynaptic axon is filled with *synaptic vesicles*, each containing about 10^4 acetylcholine molecules (Figure 12-2). The arrival of a nerve impulse leads to the synchronous export of the contents of some 300 vesicles, which raises the acetylcholine concentration in the cleft from 10 nM to 500 μM in less than a millisecond.

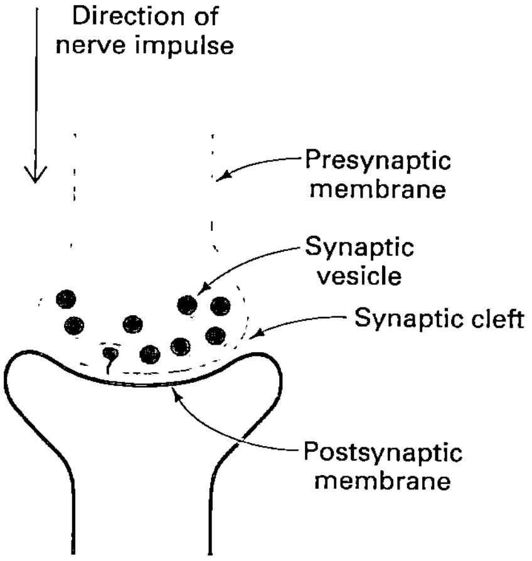

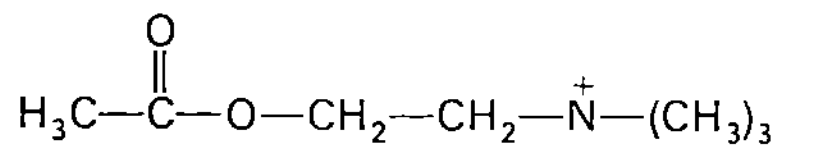

The binding of acetylcholine to the postsynaptic membrane markedly changes its ionic permeabilities (Figure 12-3). *The conductance of both Na^+ and K^+ increases greatly within 0.1 msec, leading to a large inward current of Na^+ and a smaller outward current of K^+.* The inward Na^+ current depolarizes the postsynaptic membrane and triggers an action potential. *Acetyl-*

Figure 12-2
Schematic diagram of a synapse.

$$H_3C-\overset{\overset{\textstyle O}{\|}}{C}-O-CH_2-CH_2-\overset{+}{N}-(CH_3)_3$$

Acetylcholine

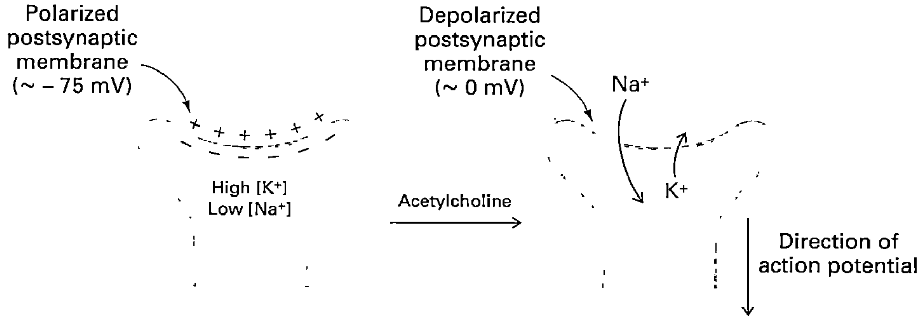

Figure 12-3
Acetylcholine depolarizes the postsynaptic membrane by increasing the conductance of Na^+ and K^+.

choline opens a single kind of cation channel, which is almost equally permeable to Na^+ and K^+. The influx of Na^+ is much larger than the efflux of K^+ because the electrochemical gradient across the membrane is steeper for Na^+. This change in ion permeability is mediated by the *nicotinic acetylcholine receptor*; for brevity, we shall refer to it as the *acetylcholine receptor.* Acetylcholine also activates the *muscarinic acetylcholine receptor,* which has a different design and function (p. 346).

The acetylcholine receptor channel is the best-understood ligand-gated channel, and so it is rewarding to examine it in some detail. The *electric organ* of *Torpedo,* an electric fish, is a choice source because its electroplaxes (voltage-generating cells) are very rich in cholinergic postsynaptic membranes. The receptor is very densely packed in these membranes (~20,000 per μm^2). Another exotic biological material has been invaluable in the isolation of acetylcholine receptors. Snake neurotoxins such as *α-bungarotoxin* (from the venom of a Formosan snake) and *cobratoxin* block neuromuscular transmission. These small (7-kd) basic proteins bind specifically and very tightly to acetylcholine receptors and hence can be used as tags.

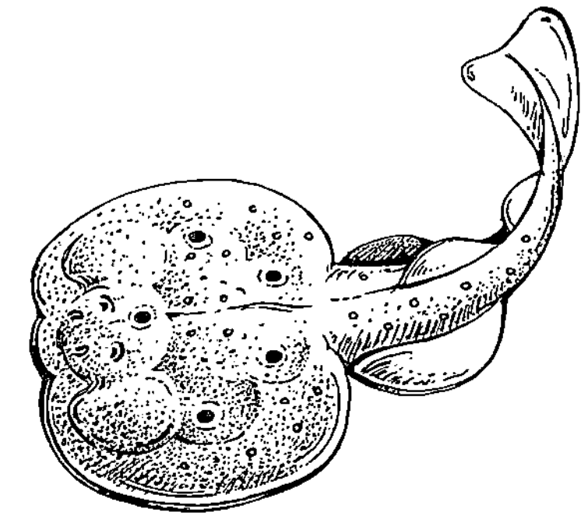

Torpedo marmarota, an electric fish, contains an electric organ that is rich in acetylcholine receptor channels.

The acetylcholine receptor of the electric organ has been solubilized by adding a nonionic detergent to a postsynaptic membrane preparation and purified by affinity chromatography on a column bearing covalently attached cobratoxin. This 268-kd receptor is a pentamer of four kinds of subunits, $\alpha_2\beta\gamma\delta$. Each α chain contains one binding site for acetylcholine. The cloning and sequencing of the cDNAs for these subunits (50–58 kd) showed that they have similar sequences. It seems likely that the genes for the α, β, γ, and δ subunits arose by duplication and divergence of a common ancestral gene. The amino-terminal half of each chain forms a hydrophilic extracellular (synaptic) domain. The carboxy-terminal half, by contrast, contains four predominantly hydrophobic segments that span the bilayer membrane.

THE FIVE SUBUNITS OF THE ACETYLCHOLINE RECEPTOR ARE SYMMETRICALLY POSITIONED AROUND THE PORE

The structure of the acetylcholine receptor from the electric organ of *Torpedo* has been elucidated by Nigel Unwin. His three-dimensional electron crystallographic analysis at 9-Å resolution revealed that the five subunits of the closed form of the receptor are arranged regularly around its central axis (Figure 12-4). The structure has approximate *fivefold (pentago-*

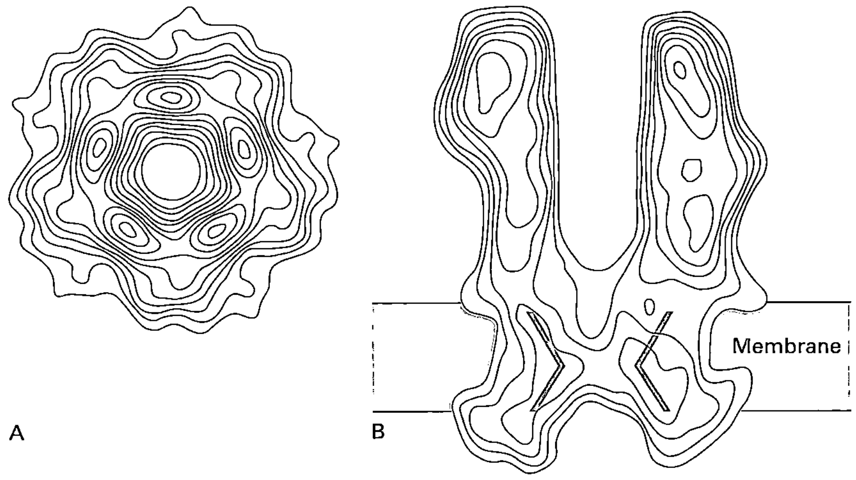

Figure 12-4
Electron-density map of the closed form of the acetylcholine receptor channel. The structure was determined by electron crystallography at 9-Å resolution. (A) View down the fivefold symmetry axis of the channel. The pore diameter at the synaptic mouth of the channel is 22 Å. (B) View along the pore axis. The position of the bilayer membrane is shown in yellow. The axes of two transmembrane helices lining the pore are marked in green. [Courtesy of Dr. Nigel Unwin.]

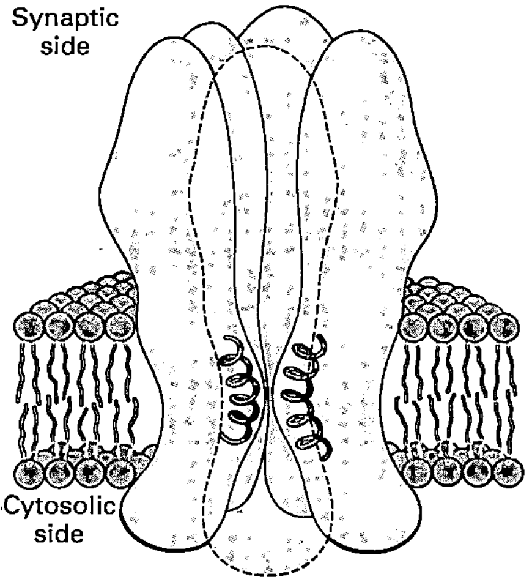

Figure 12-5
Schematic diagram of the closed form of the acetylcholine receptor channel. The narrowest part of the pore is occluded by side chains coming from five helices. [Courtesy of Dr. Nigel Unwin.]

nal) symmetry, in harmony with the similarity of its five constituent subunits. The receptor is cylindrical with a mean diameter of about 65 Å. All five rod-shaped subunits span the membrane. The receptor protrudes about 60 Å on the synaptic side of the membrane and about 20 Å on the cytosolic side (Figure 12-5). The pore of the channel is along its symmetry axis. The two acetylcholine binding sites are at the synaptic end of the α subunits, far from each other and from the pore.

The mouth of the channel on the synaptic surface of the receptor is very wide (22 Å). At the level of the outer phospholipid head groups, the lumen abruptly narrows to less than 10 Å. It widens again to about 20 Å at the level of the inner phospholipid head groups. Thus, *the channel is tripartite* (see Figure 12-5). *It contains an extracellular entrance domain, a transmembrane domain arranged around a narrow pore, and a cytosolic entrance domain.* The pore is lined by five α helices, one from each subunit.

PATCH-CLAMP CONDUCTANCE MEASUREMENTS REVEAL THE ACTIVITIES OF SINGLE CHANNELS

The study of ion channels has been revolutionized by the *patch-clamp technique,* which was introduced by Erwin Neher and Bert Sakmann in 1976. A clean glass pipette with a tip diameter of about 1 μm is pressed against an intact cell to form a seal (Figure 12-6). Slight suction leads to the formation of a very tight seal so that the resistance between the inside of the pipette and the bathing solution is many gigaohms (one gigaohm is equal to 10^9 ohms). Thus, a gigaohm seal (called a *gigaseal*) ensures that an electric current flowing through the pipette is identical with the current flowing through the membrane covered by the pipette. The gigaseal makes possible high-resolution current measurements while a known volt-

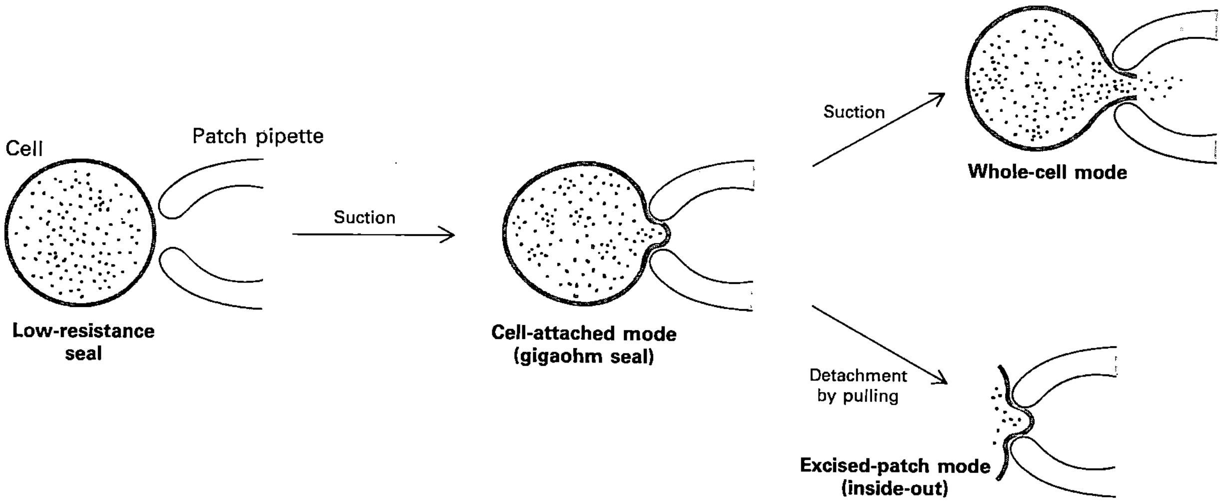

Figure 12-6
The patch-clamp technique for monitoring channel activity is highly versatile. A high-resistance seal (gigaseal) is formed between the pipette and a small patch of plasma membrane. This configuration is called *cell-attached.* The breaking of the membrane patch by increased suction produces a low-resistance pathway between the pipette and interior of the cell. The activity of the channels in the entire plasma membrane can be monitored in this *whole-cell mode.* To prepare a membrane in the *excised-patch mode,* the pipette is pulled away from the cell. A piece of plasma membrane with its cytosolic side now facing the medium is monitored by the patch pipette.

age is applied across the membrane. In fact, patch clamping increased the precision of such measurements a hundredfold. *The flow of ions through a single channel and transitions between different states of a channel can be monitored with a time resolution of microseconds.* Furthermore, the activity of channels in their native membrane environment, even in intact cells, can be directly observed. New vistas are opened whenever a single biomolecule can be seen in action.

The activity of a single acetylcholine receptor channel is graphically displayed in patch-clamp recordings of postsynaptic membranes of skeletal muscle (Figure 12-7). The addition of acetylcholine is followed by transient openings of the channel. The current i flowing through an open channel is 4 pA (picoamperes) when the membrane potential V is -100 mV. An ampere is the flow of 6.24×10^{18} charges per second. Hence, *2.5×10^7 Na^+ ions per second flow through an open channel.* The conductance g of a channel is equal to $i/(V - E_r)$, where E_r is the reversal potential at which there is no net flux; g is expressed in units of siemens (the reciprocal of an ohm), i in amperes, and V in volts. If $E_r = 0$, a current of 4 pA at a potential of 100 mV corresponds to a conductance of 40 pS (picosiemens).

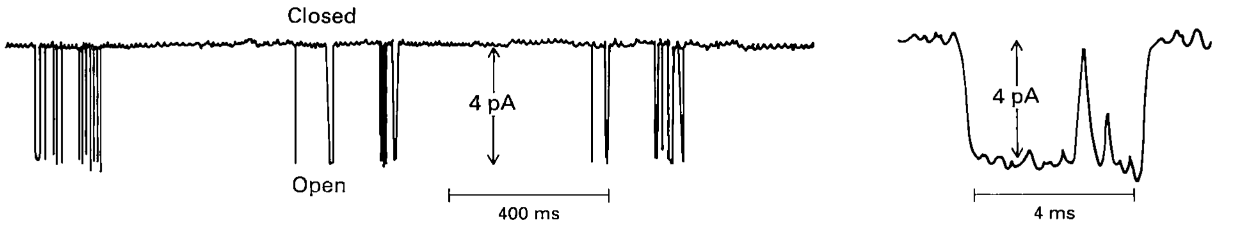

Figure 12-7
Patch-clamp recordings showing the conductance of an acetylcholine receptor channel in the presence of acetylcholine. The channel undergoes frequent transitions between open and closed states. [Courtesy of Dr. D. Colquhoun and Dr. B. Sakmann.]

XENOPUS OOCYTES EXPRESS MICROINJECTED mRNAs ENCODING SUBUNITS OF THE ACETYLCHOLINE RECEPTOR

The cloning of cDNAs for the subunits of the acetylcholine receptor has led to new ways of exploring how this channel functions. mRNAs transcribed in vitro from these cDNAs are microinjected into *Xenopus* oocytes, which lack their own acetylcholine receptor (Figure 12-8). Polypeptide

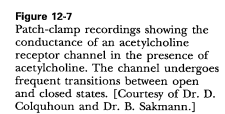

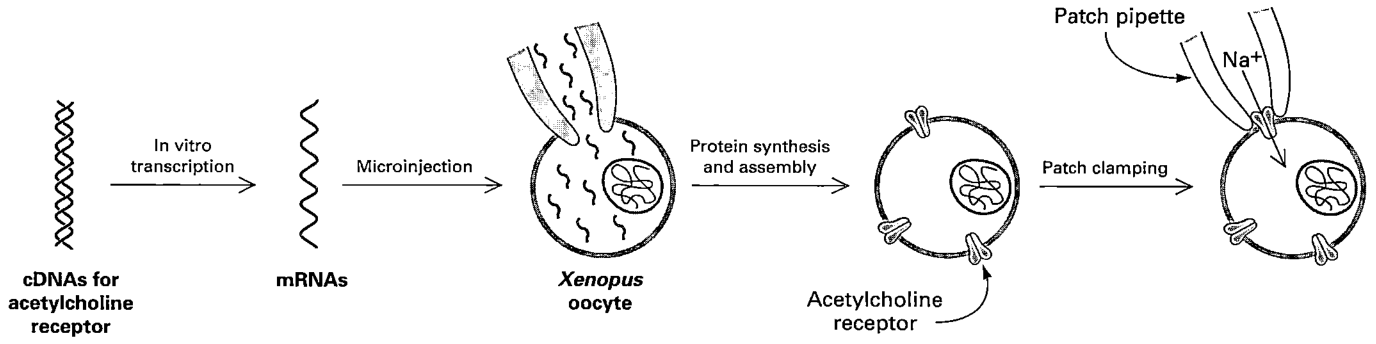

Figure 12-8
Xenopus oocytes can be used to express mRNAs derived from cDNAs. The steps in expression are in vitro transcription of the cDNA, microinjection of the resulting mRNA, translation of the mRNA by the oocyte, and targeting of the channel to the plasma membrane. Microinjection of the mRNAs for the four kinds of subunits of the acetylcholine receptor results in the formation of functional channels, as evidenced by patch-clamping studies of membrane conductance.

chains translated from these injected mRNAs are targeted to the plasma membrane of the oocyte. Patch-clamp measurements then reveal whether functional acetylcholine receptors are formed. In fact, fully active channels are formed only when all four kinds of subunits are expressed. Partly active channels are assembled if either the γ or δ chain is missing; one of these chains probably substitutes for the other in forming a pentameric channel. The oocyte expression system is now widely used to delineate relations between the structure and function of ion channels. Biochemistry, molecular genetics, and electrophysiology have come together.

THE BINDING OF TWO ACETYLCHOLINE MOLECULES TRANSIENTLY OPENS A CATION-SELECTIVE PORE

Measurements of the dependence of channel opening on the concentration of acetylcholine give a Hill coefficient of 1.97. Hence, *two molecules of acetylcholine must bind to the receptor to open the channel* (Figure 12-9). Patch-clamp studies have revealed that acetylcholine (A) binds to the closed state (C) of the receptor to form AC and A_2C, which then undergoes a transition to the open state A_2O. The rate constant for association of acetylcholine with either site of the receptor is $10^8\ \text{M}^{-1}\ \text{s}^{-1}$, close to the diffusion-controlled limit. Hence, the binding of acetylcholine under physiologic conditions occurs in less than ~100 μs. Likewise, the intramolecular transition from the closed A_2C to the open A_2O state is very rapid (30 μs). Thus, *the channel opens swiftly in response to a sudden increase in the acetylcholine concentration in the synaptic cleft.*

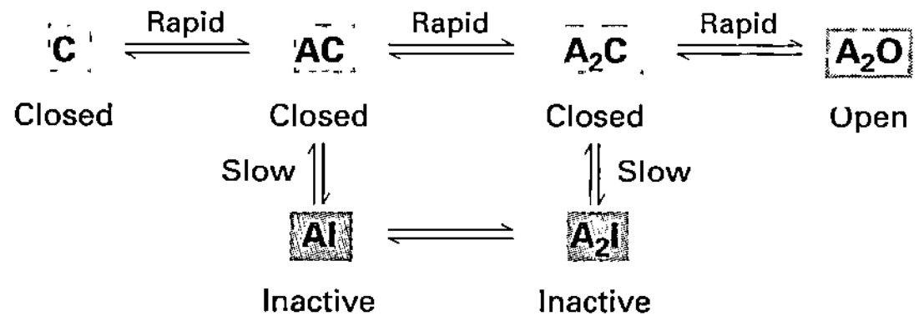

Figure 12-9
Multiple conformational states of the acetylcholine receptor. The abbreviations used are A, acetylcholine; C, closed state of the receptor; O, open state; I, inactive state. The binding of two molecules of acetylcholine rapidly opens the channel. If the concentration of acetylcholine stays high, the channel spontaneously closes (desensitizes) by making a transition to an inactive state.

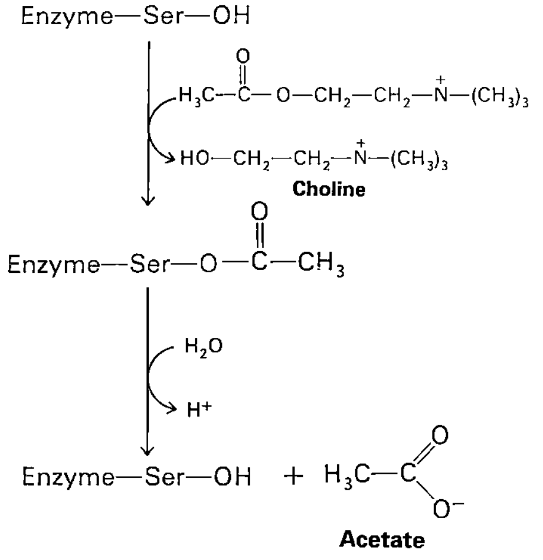

Figure 12-10
Catalytic mechanism of acetylcholinesterase.

The channel stays open under physiologic conditions for only about a millisecond because acetylcholine in the synaptic cleft is rapidly hydrolyzed to acetate and choline by *acetylcholinesterase* (Figure 12-10). This enzyme is anchored to the postsynaptic membrane by a covalently attached glycolipid group. Acetylcholinesterase has a very high turnover number (25,000 s^{-1}). Indeed, acetylcholinesterase has attained kinetic perfection (p. 195), as indicated by its k_{cat}/K_M value of $2 \times 10^8\ \text{M}^{-1}\ \text{s}^{-1}$. The catalytic prowess of acetylcholinesterase enables synapses to transmit action potentials at high frequency. *Organic fluorophosphates,* such as diisopropylphosphofluoridate (DIPF), inhibit acetylcholinesterase by forming *very stable covalent phosphoryl-enzyme complexes.* The phosphoryl group becomes bonded to the active-site serine, as in serine proteases that have reacted with DIPF (p. 196). Many organic phosphate compounds have been synthesized for use as *agricultural insecticides* or as *nerve gases* for chemical warfare.

What is the structural basis of channel opening? A comparison of the structures of the closed and open forms would be highly revealing. However, the open form has not been imaged thus far because it is so short-lived. If the concentration of acetylcholine remains high, the channel spontaneously closes in less than a second, a process called *desensitization* (see Figure 12-9). A plausible mechanism for channel opening can nevertheless be inferred from available data. As was noted earlier, the pore is lined by five α helices, one from each subunit. The amino acid sequences of these helices point to the presence of alternating ridges of small polar or neutral residues (serine, threonine, glycine) and large nonpolar ones (isoleucine, leucine, phenylalanine). In the closed state, the large residues occlude the channel by forming a tight hydrophobic ring (Figure 12-11). Indeed, each subunit has a bulky leucine at a critical position—

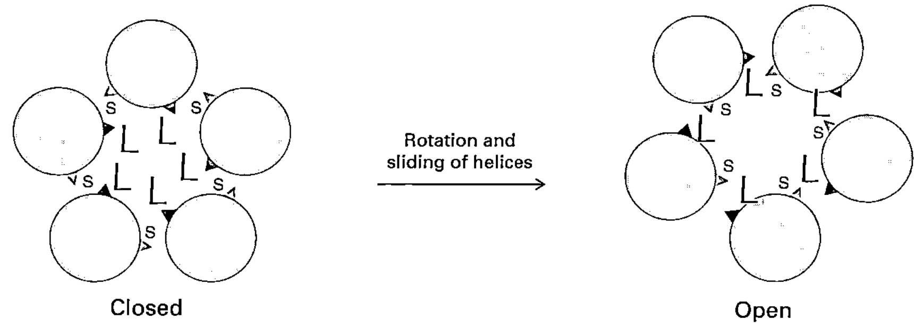

Figure 12-11
Large hydrophobic side chains (L) occlude the pore of the closed form of the acetylcholine receptor channel. Channel opening is probably mediated by tilting the helices that line the pore. Large residues move away from the pore and small ones (S) take their place. [After N. Unwin. *Neuron* 3(1989):665.]

the bend in the α helix. The binding of acetylcholine could allosterically tilt these helices to shift the positions of the ridges. The pore would then be open because it would be lined by small polar residues rather than by large hydrophobic ones.

Monovalent or divalent cations, but not anions, readily flow through the open form of the acetylcholine receptor channel. What makes the channel cation-selective? The amino acid sequences of the pore-forming helices are again highly informative. *Three rings of negatively charged residues are present* (Figure 12-12). One of them is located within the transmembrane region of the pore, and the other two flank the entrances to this narrow segment. Anions such as Cl^- cannot enter the pore because they are repelled by these negatively charged rings. Studies of the permeability of a series of organic cations differing in size (e.g., alkylammonium ions) suggest that the narrowest part of the pore has a diameter of 6.5 Å.

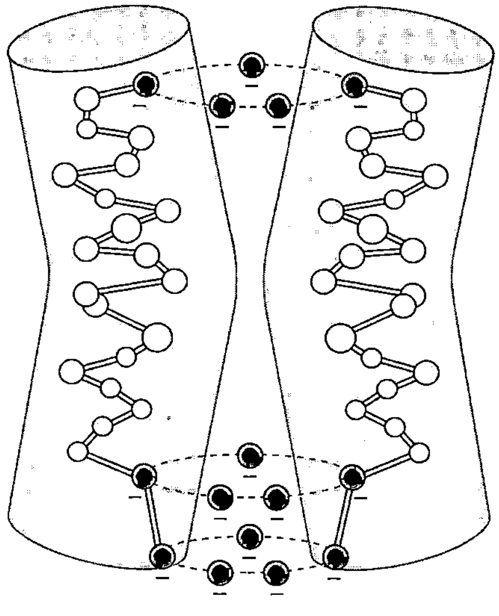

Figure 12-12
The selectivity of the acetylcholine receptor channel for cations is imposed by three rings of negatively charged side chains (red) in the pore. [After J.-P. Changeux. Chemical Signaling in the Brain. Copyright © 1993 by Scientific American, Inc. All rights reserved.]

ACTION POTENTIALS ARE MEDIATED BY TRANSIENT CHANGES IN Na^+ AND K^+ PERMEABILITY

We turn now from ligand-gated channels to voltage-gated channels. A nerve impulse is an electrical signal produced by the flow of ions across the plasma membrane of a neuron. The interior of a neuron, like that of most other cells, has a high concentration of K^+ and a low one of Na^+. These ionic gradients are generated by an ATP-driven pump (p. 310). In the resting state, the membrane of an axon is much more permeable to

K^+ than to Na^+, and so the membrane potential is largely determined by the ratio of the internal to the external concentration of K^+ (Figure 12-13A). The membrane potential of −60 mV in unstimulated axons is near the −75-mV level (the K^+ equilibrium potential) that would be given by a membrane permeable only to K^+. A nerve impulse, or *action potential,* is generated when the membrane potential is depolarized beyond a critical threshold value (i.e., from −60 to −40 mV). The membrane potential becomes positive within about a millisecond and attains a value of about +30 mV before turning negative again. This amplified depolarization is propagated along the nerve terminal. The giant axons of squids played an important role in the discovery of the nature of the action potential. Electrodes can be readily inserted into these unusually large axons, which have a diameter of about a millimeter, and so squid axons have been favorite objects of inquiry.

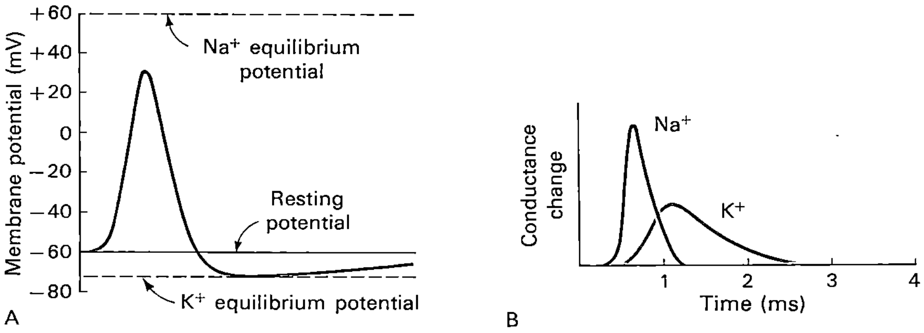

Figure 12-13
Depolarization of an axon membrane results in an action potential. Time course of (A) the change in membrane potential and (B) the change in Na^+ and K^+ conductances.

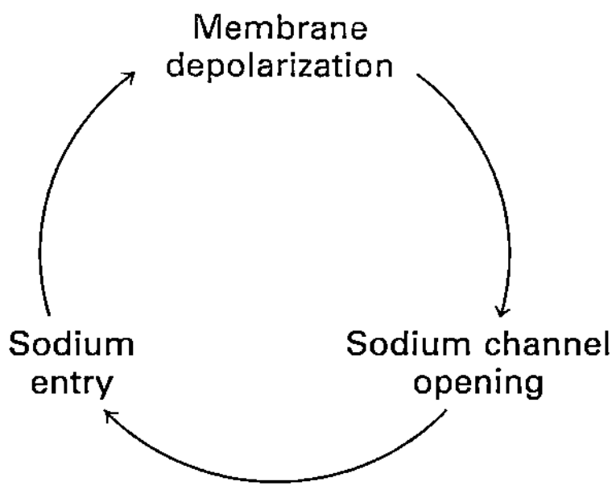

Positive-feedback loop in the generation of an action potential

Ingenious experiments carried out by Alan Hodgkin and Andrew Huxley revealed that *action potentials arise from large, transient changes in the permeability of the axon membrane to* Na^+ *and* K^+ *ions* (Figure 12-13B). Two kinds of voltage-sensitive channels, one selectively permeable to Na^+ and the other to K^+, were defined. The conductance of the membrane to Na^+ changes first. Depolarization of the membrane beyond the threshold level leads to an opening of Na^+ channels. Sodium ions begin to flow into the cell because of the large electrochemical gradient across the plasma membrane. The entry of Na^+ further depolarizes the membrane, and so more gates for Na^+ are opened. This positive feedback between depolarization and Na^+ entry leads to a very rapid and large change in membrane potential, from about −60 mV to +30 mV in a millisecond.

Sodium channels spontaneously close and potassium channels begin to open at about this time (Figure 12-13B). Consequently, potassium ions flow outward, and so the membrane potential returns to a negative value. At about 2 ms, the membrane potential is −75 mV, the K^+ equilibrium potential. The resting level of −60 mV is restored in a few milliseconds as the K^+ conductance decreases to the value characteristic of the unstimulated state. Only a very small proportion of the sodium and potassium ions in a nerve cell, of the order of one in a million, flows across the plasma membrane during the action potential. Clearly, the action potential is a very efficient means of signaling over large distances.

Hodgkin and Huxley's pioneering studies of nerve action potentials posed three key questions about membrane channels:

1. What is the molecular basis of the *ionic selectivity* of channels? The sodium channel is 11 times as permeable to Na^+ as to K^+. The potassium channel is even more discriminating; it is more than 100 times as permeable to K^+ as to Na^+.

2. What is the mechanism of *voltage gating*? How does a channel sense voltage and alter the flow of ions through its pore?

3. What is the mechanism of *inactivation*? How does a channel become inactive during an action potential and subsequently return to a closed but activatable state?

TETRODOTOXIN AND SAXITOXIN ARE POTENT NEUROTOXINS BECAUSE THEY BLOCK SODIUM CHANNELS

The purification of the sodium channel was made possible by the use of tetrodotoxin, a highly potent poison from the puffer (fugu) fish. Tetrodotoxin (TTX) blocks the conduction of nerve impulses along axons and in excitable membranes of nerve fibers, which leads to respiratory paralysis. The lethal dose for a mouse is about 10 ng. TTX is a very useful label because it binds very tightly ($K_i \sim 1$ nM) to the sodium channel and blocks the flow of Na^+ ions. Moreover, it is highly specific. TTX has no effect on the potassium channel or other channels. *Saxitoxin* (STX), which is produced by a marine dinoflagellate, acts in the same way. Shellfish, particularly clams and mussels, feeding on such dinoflagellates during a "red tide" are poisonous. Indeed, a contaminated mussel may contain enough saxitoxin to kill fifty humans! A common feature of TTX and STX is the presence of a guanido group (Figure 12-14). This positively charged group of the toxin interacts with a negatively charged carboxylate at the mouth of the channel on the extracellular side of the membrane. TTX and STX block the flow of Na^+ by plugging the entrance to the pore.

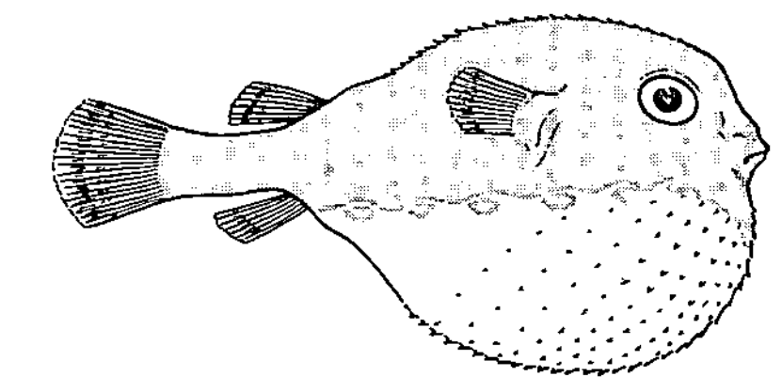

A puffer fish, regarded as a culinary delicacy in Japan.

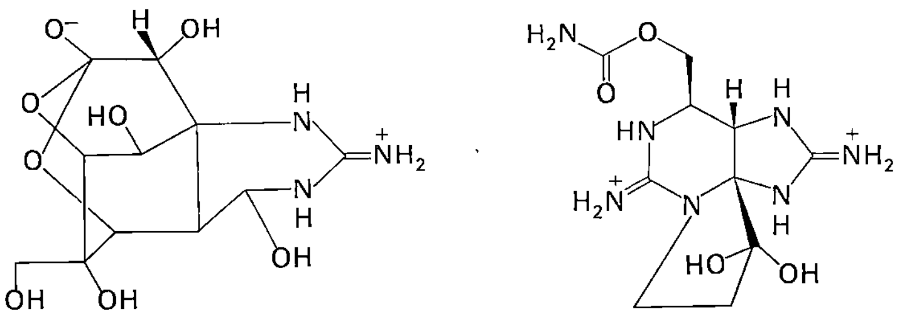

Tetrodotoxin (TTX) **Saxitoxin (STX)**

Figure 12-14
Poisons of the Na^+ channel. Tetrodotoxin has one guanido group (blue), whereas saxitoxin has two.

PURIFIED SODIUM CHANNELS RECONSTITUTED IN LIPID BILAYERS ARE FUNCTIONALLY ACTIVE

The sodium channel was first purified from the electric organ of *Electrophorus electricus*, a rich source. The 750-V discharge of the electric organ of this electric eel arises from the synchronous firing of thousands of electroplax cells connected in series (Figure 12-15). Electroplaxes are evolved from muscle cells; they have retained their electrically excitable outer membrane but not their contractile apparatus. The first step in purification was the addition of a nonionic detergent (Triton X-100) to solubilize the integral membrane proteins of electroplaxes. Radioactive tetrodotoxin was used as a specific tag to locate the sodium channel in

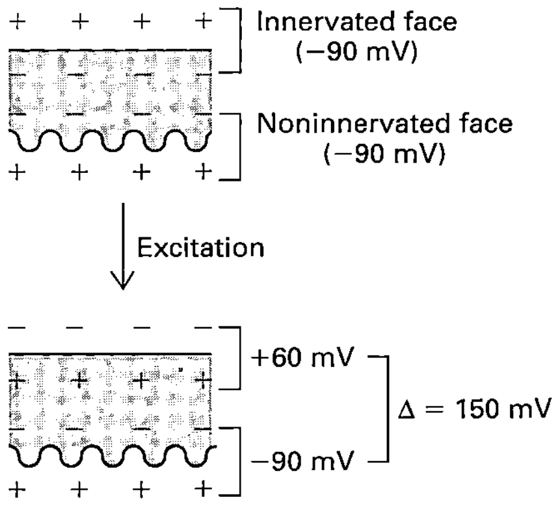

Figure 12-15
Generation of a voltage by an electroplax cell of the electric eel. The innervated face is rich in voltage-gated sodium channels, which open on depolarization.

subsequent steps. The channel was separated from other proteins in the extract on the basis of charge (by ion-exchange chromatography), carbohydrate content (by affinity chromatography on an immobilized lectin column), and size (by sucrose density sedimentation). About 5 mg of channel protein was obtained from a 300-g electric organ.

The channel from eel electric organ is a single 260-kd chain. The one from rat brain contains a 36-kd β_1 chain and a 33-kd β_2 chain in addition to a 260-kd α chain. Does the reconstituted α subunit of the brain sodium channel have channel activity? This question was answered by incorporating the α subunit into phospholipid vesicles, which were then fused with a planar bilayer membrane to facilitate patch clamping. Single-channel recordings revealed that the reconstituted α subunit is indeed a functional channel (Figure 12-16A). The conductance in 0.5 M NaCl is 25 pS, similar to that of native channels. As expected, reconstituted channels are blocked by TTX applied to the extracellular side. The reconstituted channel is *Na^+-selective,* favoring Na^+ over K^+ by a factor of 8. Furthermore, channel opening is highly *voltage-dependent* (Figure 12-16B). Depolarization opens channels. Finally, reconstituted channels, like native ones, undergo *inactivation.*

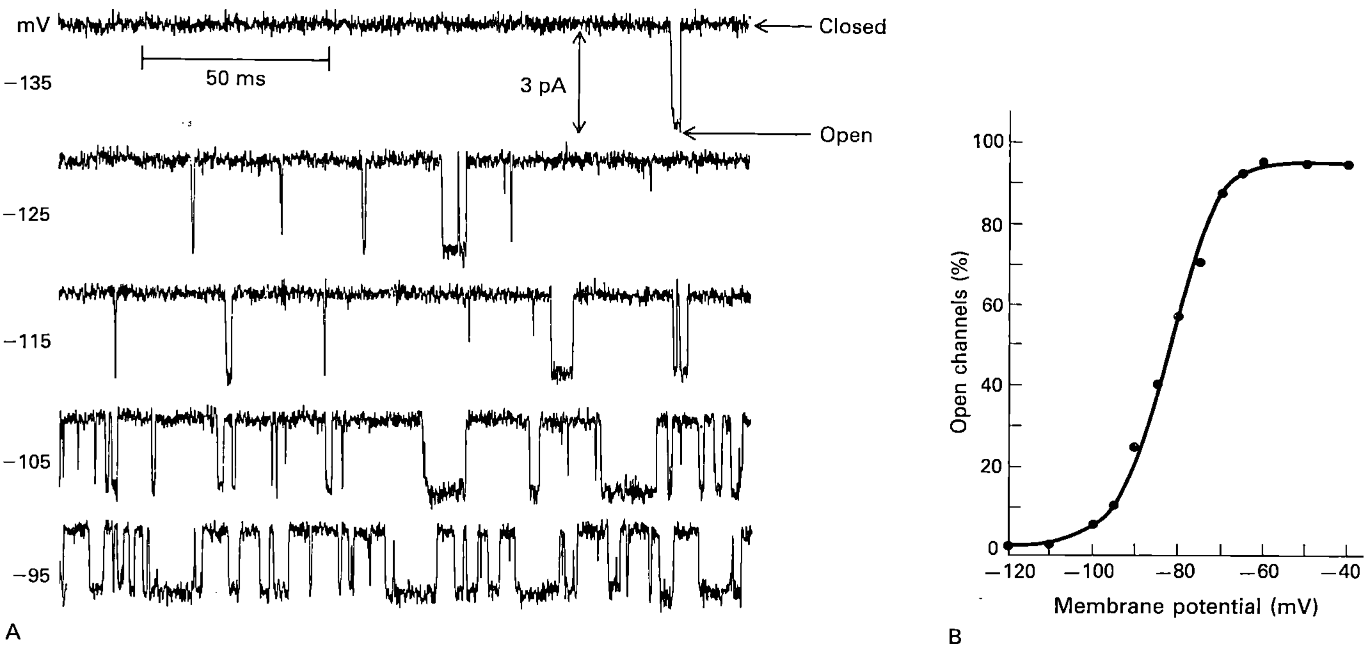

Figure 12-16
Voltage dependence of reconstituted sodium channels. (A) Conductance recordings showing transitions between the closed and open states of the channel at several membrane voltages. (B) Dependence of the fraction of open channels on the applied membrane voltage. [Courtesy of Dr. Mauricio Montal.]

THE FOUR REPEATING UNITS OF THE SODIUM CHANNEL FORM A HIGHLY DISCRIMINATING PORE

The availability of purified protein enabled Shosaku Numa and co-workers to clone and sequence the cDNA for the sodium channel from the eel electric organ. They began by screening an electroplax cDNA expression library with antibody specific for the electroplax sodium channel. Positive clones were further probed with oligonucleotides that encoded short peptide sequences known to be present in the channel protein. The cDNA for the eel sodium channel was then used as a probe to identify by low-stringency hybridization (p. 131) the corresponding cDNA in a rat brain library. Indeed, cDNAs for three related α subunits of rat brain sodium channel were found in this way.

The eel and rat cDNA sequences are 61% similar, which indicates that the amino acid sequence of the sodium channel has been conserved over a long evolutionary period. Most interesting, *the channel contains four internal repeats having similar amino acid sequences* (Figure 12-17). These repeats probably arose by gene duplication and divergence. Hydrophobicity profiles indicate that each homology unit contains five hydrophobic segments (S1, S2, S3, S5, and S6). Each repeat also contains a highly positively charged S4 segment; arginine or lysine residues are present at nearly every third residue. Numa proposed that segments S1–S6 are membrane-spanning α helices. The S4 segments, as will be discussed shortly, are in fact the voltage sensors of the channel.

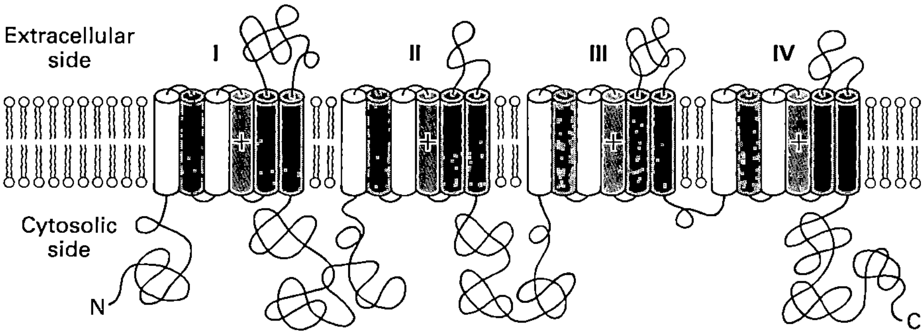

Figure 12-17
The Na^+ channel is a single polypeptide chain with four repeating units (I–IV). Each repeat probably folds into six transmembrane helices. The helix labeled + contains many lysine and arginine residues and is involved in voltage-gating. [After M. Noda, T. Ikeda, T. Kayono, H. Suzuki, H. Takeshima, M. Kurasaki, H. Takahashi, and S. Numa. *Nature* 320(1986):188.]

The sodium channel favors the passage of Na^+ over K^+ by a factor of 11. How is this selectivity achieved? A definitive answer to this question will have to await a high-resolution view of the three-dimensional structure of the channel. However, electrophysiological studies of the relative permeabilities of alkali cations and organic cations provide some valuable clues. The dependence of permeability on ionic size (Table 12-1) indicates that the channel is narrow. *Ions having diameters greater than 5 Å are excluded* (Figure 12-18).

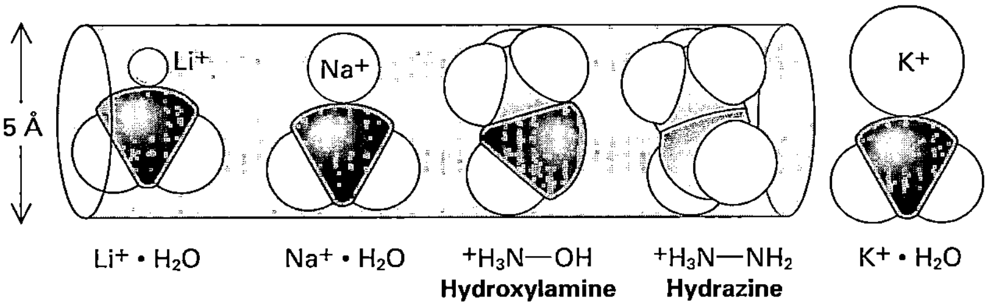

Figure 12-18
The ionic selectivity of the sodium channel partly depends on steric factors. Na^+ and Li^+, together with a water molecule, fit in the channel, as do hydroxylamine and hydrazine. In contrast, K^+ with a water molecule is too large. [After R.D. Keynes. Ion-channels in the nerve-cell membrane. © 1979 by Scientific American, Inc. All rights reserved.]

Table 12-1
Relative permeabilities of the sodium and potassium channels in axon membranes

Ion	*Na^+ channel*	*K^+ channel*
Li^+	0.93	<0.01
Na^+	1.00	<0.01
K^+	0.09	1.00
Rb^+	<0.01	0.91
Cs^+	<0.01	<0.08
NH_4^+	0.16	0.13
$OHNH_3^+$	0.94	<0.03
$H_2NNH_3^+$	0.59	<0.03
$H_3CNH_3^+$	<0.01	<0.02

Conductance is not governed by size alone. For example, methylamine ($H_3CNH_3^+$) has nearly the same dimensions as hydrazine ($H_2NNH_3^+$) and hydroxylamine ($HONH_3^+$), yet it is much less permeant. A likely reason for this difference is that the methyl group of methylamine, in contrast with the amino group of hydrazine or the hydroxyl group of hydroxylamine, cannot form a hydrogen bond with an oxygen atom at the constriction of the pore. Hence, methylamine does not fit. Another significant finding is that the conductance of the sodium channel to all permeant cations decreases markedly when the pH is lowered. In fact, the relative permeability follows a titration curve for an acid with a pK of 5.2, which suggests that the active form of the channel contains a negatively charged carboxylate group. Thus, *the sodium channel selects for Na^+ by providing a negatively charged site with a small radius* (Figure 12-19). A K^+ ion cannot readily pass through this constriction, called the *selectivity filter,* because it is larger than Na^+. The ionic radius of K^+ is 1.33 Å, compared with 0.95 Å for Na^+.

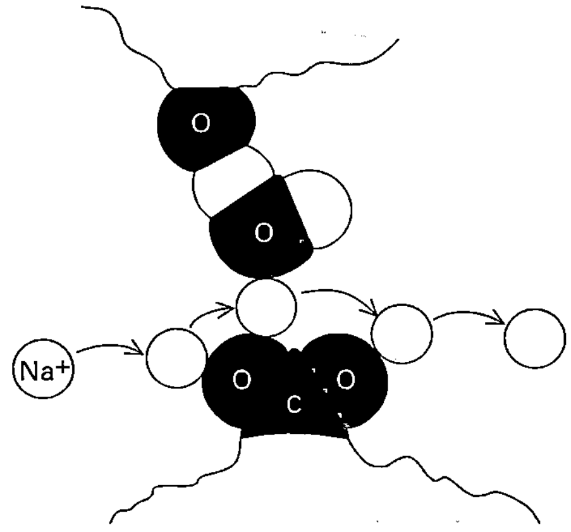

Figure 12-19
Proposed selectivity filter of the sodium channel. A Na^+ ion (green) crosses the narrowest part of the pore bound to a single water molecule. A side-chain carboxylate interacts electrostatically with Na^+ at this site. [After B. Hille. *J. Gen. Physiol.* 66(1975):535.]

A CHANNEL IS VOLTAGE-SENSITIVE IF CHARGED GROUPS MOVE ACROSS THE BILAYER IN OPENING THE PORE

How does a voltage-gated channel sense the membrane potential? Consider a channel that can exist in a closed state A or an open state B. *The equilibrium between these states depends on membrane voltage if there is a net movement of charge across the bilayer in the transition from A to B.* The reason is that electrical work is performed whenever charge moves across a voltage gradient. The charged groups of the protein that move partly or completely across the bilayer in opening the channel are called *gating charges.*

A quantitative relationship between channel opening and membrane voltage can readily be derived. Let V_0 denote the voltage at which the open and closed states have the same free energy and hence [A] = [B]. For a mole of channels, the free energy obtained in taking z gating charges from voltage V_0 to V is $zF(V - V_0)$, where F is the faraday constant. The voltage dependence of the equilibrium constant K for the transition between closed and open states is then given by

$$K = \frac{[\text{closed}]}{[\text{open}]} = e^{-zF(V - V_0)/RT}$$

where R is the gas constant, and T is the absolute temperature. The fraction f of open channels is equal to

$$f_{\text{open}} = \frac{1}{1 + e^{-zF(V - V_0)/RT}}$$

In these equations, F is 2.306×10^4 cal volt^{-1} mol^{-1}, and R is 1.987 cal mol^{-1} deg^{-1}; V and V_0 are expressed in volts. The steepness of the voltage-dependence curve depends only on the gating charge z. The value of z is readily obtained by plotting log (f/(1 − f)) versus V; the slope is proportional to z (Figure 12-20). V_0 is given by the x-intercept.

Hodgkin and Huxley's classic studies revealed that z is about +6 for opening of the sodium channel. A movement of six positive charges from the cytosolic to the extracellular side of the plasma membrane would give the observed voltage dependence. Alternatively, six negative charges could move from the extracellular to the cytosolic side to give the same effect. An identical voltage dependence would be obtained if 12 charges moved halfway across the membrane or if a larger number of partial charges (dipoles) moved the same distance. *What counts is the sum of charge times the fractional distance traversed across the lipid bilayer.* Movement of charge outside the bilayer does not contribute significantly to voltage dependence, because the electric field of the membrane drops sharply outside the low-dielectric lipid region.

We can now calculate the free-energy change produced by the movement of six gating charges in an action potential. The membrane potential changes from about −60 mV to nearly +40 mV, and so ΔV is equal to 0.1 V. The electrical work performed is given by $zF\Delta V$, which is equal to 6×23.1 kcal volt^{-1} mol^{-1} $\times$ 0.1 V, or 13.8 kcal/mol. In general, a free-energy input of 1.36 kcal/mol can change the equilibrium constant of a transition by a factor of 10 (p. 187). Hence, the movement of six gating charges through a potential of 0.1 V can alter the ratio of closed to open states by a very large factor, of the order of 10^{10}. *It is evident that the movement of gating charges in response to changes in membrane potential is a very effective means of switching the conformation of a protein.*

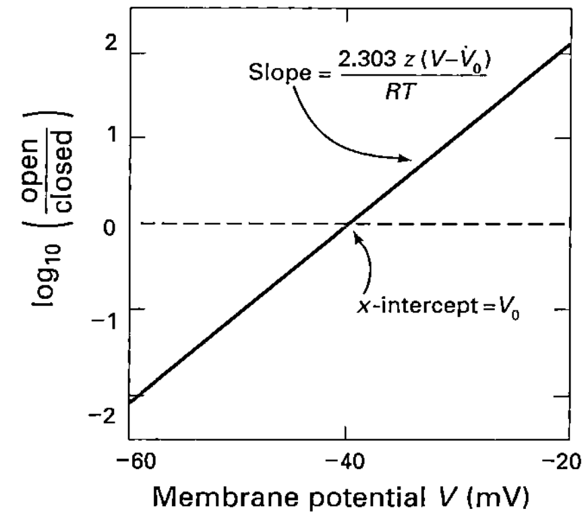

Figure 12-20
Voltage dependence of the sodium channel. The logarithm of the ratio of open to closed channels is plotted versus membrane potential. The x-intercept gives the value of V_0 (−40 mV), and the slope gives the value of z (+6.0).

FOUR POSITIVELY CHARGED TRANSMEMBRANE HELICES SERVE AS VOLTAGE SENSORS

What is the molecular identity of the gating charges that determine whether the sodium channel is closed or open? As was mentioned earlier, charged groups participate in voltage-dependent gating only if they move partly or totally across the lipid bilayer. The S4 segments attracted attention soon after the sequence of the sodium channel was determined because they contain 4, 5, 6, and 8 arginine or lysine residues in repeats I, II, III, and IV, respectively. Furthermore, these positively charged residues are located at every third residue, which suggests that they emerge on one side of a transmembrane α helix (Figure 12-21). According to this model,

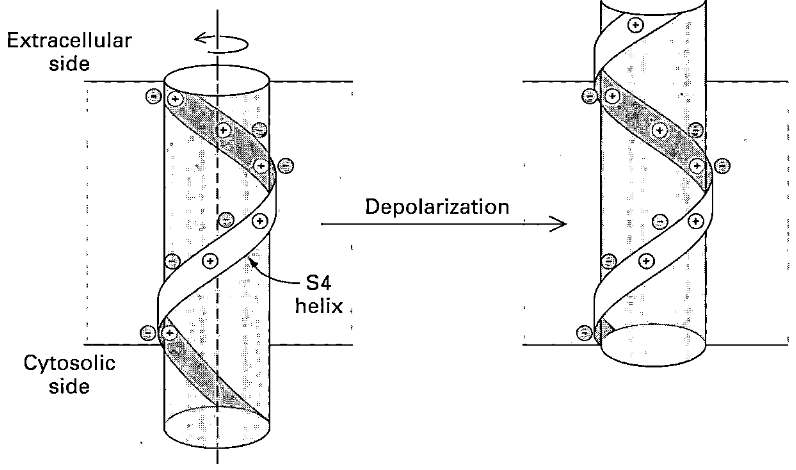

Figure 12-21
Postulated movement of the S4 voltage sensor in the voltage-induced opening of the sodium channel. Depolarization induces an outward movement and rotation of this positively charged helix. The positively charged arginine and lysine side chains (blue) of the S4 helix are thought to be paired with negatively charged side chains from other parts of the channel. Depolarization breaks a salt bridge and leads to the net outward movement of one (shown here) or two positive charges. [After W.A. Catterall. *Trends Neurosci.* 9(1986):7.]

these positively charged residues on each S4 are paired at the resting membrane potential with negative charges on other transmembrane helices. *Depolarization is thought to lead to a spiral motion of S4 accompanied by the net movement of one or two positive charges to the extracellular side of the membrane.* Such a translation of the four S4 segments could open a gate by moving a steric barrier to ion flow. Site-specific mutagenesis studies support the proposal that the S4 segments serve as voltage sensors. Mutations of S4 arginine and lysine residues to glutamine make the channel less sensitive to voltage (z becomes smaller). Furthermore, as will be discussed shortly, the S4 motif occurs in other voltage-gated channels.

Studies of charge movements leading to the opening of sodium channels—called *gating currents*—have revealed the existence of at least *three closed states* preceding the open one (Figure 12-22). The equivalent of two charges moves across the bilayer in each of these depolarization-induced transitions, which occur in less than 25 μs. The *open state* lasts only about 1 ms, even if the membrane is kept depolarized. The open state has a short life because it spontaneously converts into an *inactive state* that cannot be opened irrespective of the membrane potential. Inactivation is mediated by the cytosolic loop between repeats III and IV (see Figure 12-17). The return to a closed but activatable state requires repolarization. Thus, *the sodium channel cycles unidirectionally during an action potential.* It opens only once during each action potential. The existence of an absolutely refractory period assures that an action potential propagates in a single direction. Another noteworthy feature of the sodium channel is that it contains a *built-in timer* that determines the duration of ion flow. A nerve membrane can conduct hundreds of action potentials per second only if the channel stays open for not much longer than a millisecond.

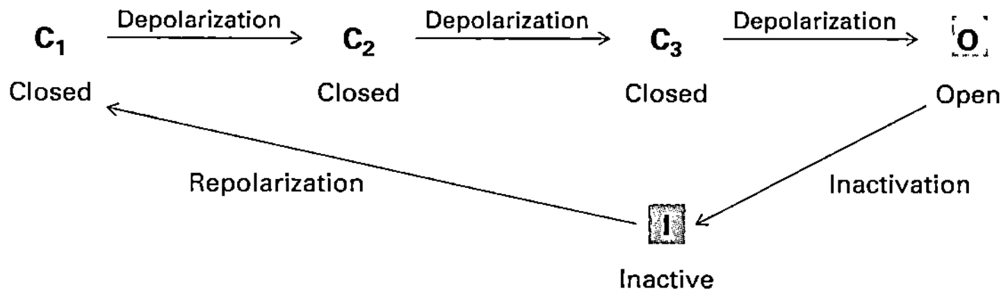

Figure 12-22
The sodium channel cycles unidirectionally during each action potential. Depolarization opens the channel, which then rapidly inactivates. The closed but activatable form is restored when the membrane repolarizes.

POTASSIUM CHANNELS HAVE S4 VOLTAGE SENSORS LIKE THOSE OF SODIUM CHANNELS

Open potassium channels drive the membrane potential toward the K^+ equilibrium potential, which is -98 mV for mammalian muscle. They help restore the resting state after depolarization (see Figure 12-13). In addition, they diminish excitability by their hyperpolarizing action. These stabilizing and restoring functions are mediated by an extensive repertoire of potassium channels differing in kinetics, sensitivity to voltage, and modulation by calcium ion and a variety of neurotransmitters.

Potassium channels resisted biochemical study for many years because of their very low abundance and the absence of known high-affinity ligands comparable to TTX. The breakthrough came from genetic studies of mutant fruit flies that shake violently when anesthetized with ether.

These *Shaker* mutants also have poor motor control when not anesthetized. A fast transient potassium channel that normally spaces successive action potentials in rhythmic firing is defective in *Shaker* flies. The subsequent mapping and cloning of the *Shaker* gene, which contains 21 exons over a span of 65 kb, was a tour de force. Alternative splicing of the transcripts of this complex gene gives rise to at least five mRNAs that produce a family of channels with distinctive properties.

Shaker cDNA encodes a 70-kd protein that associates to form a tetrameric channel. The potassium channel, like the sodium channel, exhibits fourfold symmetry. Furthermore, the hydrophobicity profile suggests that a *Shaker* subunit, like a domain of the sodium channel, contains six transmembrane helices. Most striking, each *Shaker* subunit contains an S4 voltage sensor. Thus, *potassium channels and sodium channels have a similar overall architecture and essentially the same voltage-sensing mechanism* (Figure 12-23). Calcium channels, too, have approximate fourfold symmetry and S4 voltage sensors.

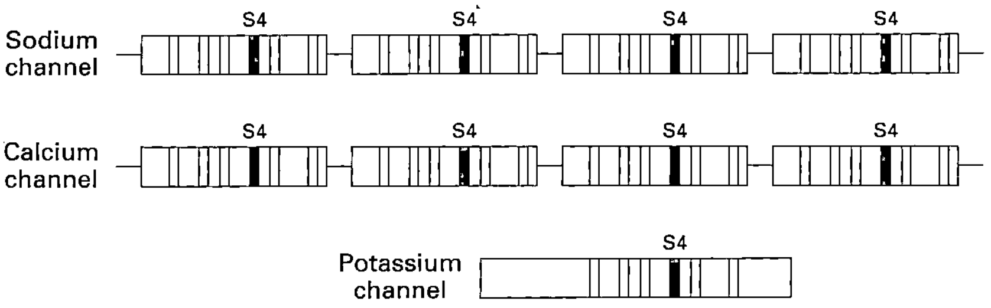

Figure 12-23
Structural similarity of sodium, calcium, and potassium channels. They exhibit approximate fourfold symmetry. Each subunit or domain of these homologous channels contains an S4 voltage sensor (red) and five highly hydrophobic segments (yellow). [After L.-Y. Jan and Y.-N. Jan. *Cell* 56(1989):13.]

POTASSIUM CHANNELS ARE HIGHLY SELECTIVE

Potassium channels are 100-fold more permeable to K^+ than to Na^+. How is this high degree of selectivity achieved? The narrow diameter (3 Å) of the selectivity filter of the potassium channel enables it to reject ions having a radius larger than 1.5 Å. However, a bare Na^+ is small enough (Table 12-2) to pass through the pore. Indeed, the ionic radius of Na^+ is substantially smaller than that of K^+. How then is Na^+ rejected? We need to consider the free-energy cost of dehydrating these ions; they cannot pass through the channel bearing a retinue of water molecules. The key point is that the free-energy cost of dehydrating Na^+ (72 kcal/mol) is much larger than for K^+ (55 kcal/mol). *The channel pays most of the cost of dehydrating K^+ by providing compensating interactions with oxygen atoms lining the selectivity filter.* However, these oxygens are positioned so that they do not interact too favorably with Na^+. The higher cost of dehydrating Na^+ would be unrecovered, and so Na^+ would be rejected. The ionic radii of oxygen, potassium, and sodium are 1.4, 1.33, and 0.95 Å, respectively. Hence a ring of oxygens positioned so that the K^+–O distance is 2.73 Å (1.4 + 1.33 Å) would be optimal for interacting with K^+ compared with Na^+, which would like to have shorter Na^+–O bonds (0.95 + 1.4 = 2.35 Å). Thus, *the potassium channel avoids closely embracing Na^+, giving Na^+ no choice but to stay hydrated and hence be impermeant.*

Table 12-2
Properties of alkali cations

Ion	*Ionic radius (Å)*	*Hydration free energy (kcal/mol)*
Li^+	0.60	−98
Na^+	0.95	−72
K^+	1.33	−55
Rb^+	1.48	−51
Cs^+	1.69	−47

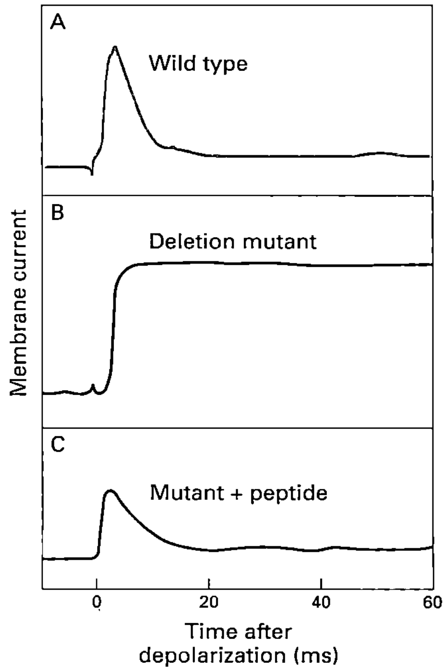

Figure 12-24
The amino-terminal region of the potassium chain is critical for inactivation. (A) The wild-type *Shaker* potassium channel displays rapid inactivation following opening. (B) A mutant channel lacking residues 6 to 46 does not inactivate. (C) Inactivation can be restored by adding a peptide consisting of residues 1 to 20 at a concentration of 100 μM. [After W.N. Zagotta, T. Hoshi, and R.W. Aldrich. *Science* 250(1990):568.]

INACTIVATION OCCURS BY OCCLUSION OF THE PORE: THE BALL-AND-CHAIN MODEL

The potassium channel, like the sodium channel, undergoes inactivation within milliseconds of channel opening (Figure 12-24A). Exposure of the cytoplasmic side of either channel to trypsin produces a trimmed channel that stays persistently open following depolarization. A second clue concerning the mechanism of inactivation came from the finding that alternatively spliced variants of the potassium channel have markedly different inactivation kinetics; these variants differed from each other only near their amino terminus. A mutant *Shaker* channel lacking 42 amino acids near the N terminus opened in response to depolarization but did not inactivate (Figure 12-24B). Most revealing, inactivation was restored by adding a synthetic peptide corresponding to the first 20 residues of the native channel.

These experiments strongly support the *ball-and-chain model* for channel inactivation that was proposed years earlier (Figure 12-25). The first 20 residues of the potassium channel form a cytoplasmic unit (the *ball*) that is attached to a flexible segment of the polypeptide (the *chain*). When the channel is closed, the ball rotates freely in the aqueous space. When the channel opens, the ball finds a complementary site in the pore and occludes it. This site is not available in the closed state. Hence, the channel opens for an interval before it undergoes inactivation by occlusion. Shortening the chain speeds inactivation because the ball finds its target more quickly. Conversely, lengthening the chain slows inactivation. Thus, the duration of the open state is controlled by the length and flexibility of the tether.

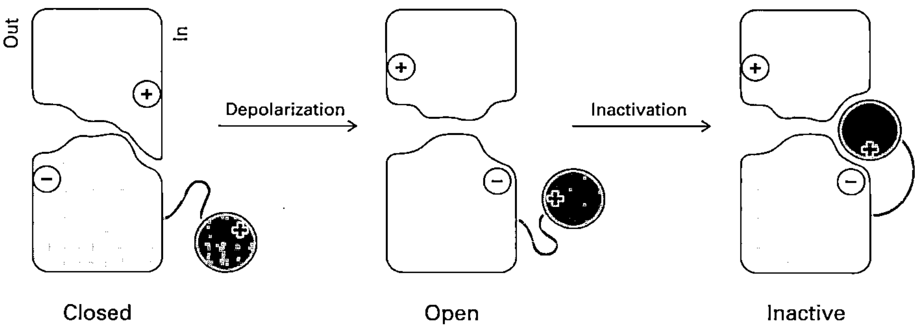

Figure 12-25
Ball-and-chain model for channel inactivation. The inactivation domain (the "ball," red) is tethered to the channel by a flexible chain (green). In the closed state, the ball is located in the cytosol. Depolarization opens the channel and creates a negatively charged binding site for the positively charged ball near the mouth of the pore. Movement of the ball into this site inactivates the channel by occluding it. [After C.M. Armstrong and F. Bezanilla. *J. Gen. Physiol.* 70(1977):567.]

GAP JUNCTIONS ALLOW IONS AND SMALL MOLECULES TO FLOW BETWEEN COMMUNICATING CELLS

The channels we have considered thus far—the acetylcholine receptor, the sodium channel, and the potassium channel—have narrow pores and are moderately to highly selective as to which ions are permeant. They are closed in the resting state and have short open-state lifetimes, typically a millisecond, that enable them to transmit high-frequency neural signals.

We turn now to a channel with a very different purpose. *Gap junctions,* also known as *cell-to-cell channels,* serve as passageways between the interiors of contiguous cells. Gap junctions are clustered in discrete regions of the plasma membranes of apposed cells. Electron micrographs of sheets of gap junctions show them tightly packed in a regular hexagonal array (Figure 12-26). A 20-Å central hole, the lumen of the channel, is prominent. A tangential view (Figure 12-27) shows that these channels span the intervening space, or gap, between apposed cells (hence, the name *gap junction*). The width of the gap between the cytosols of the two cells is about 35 Å.

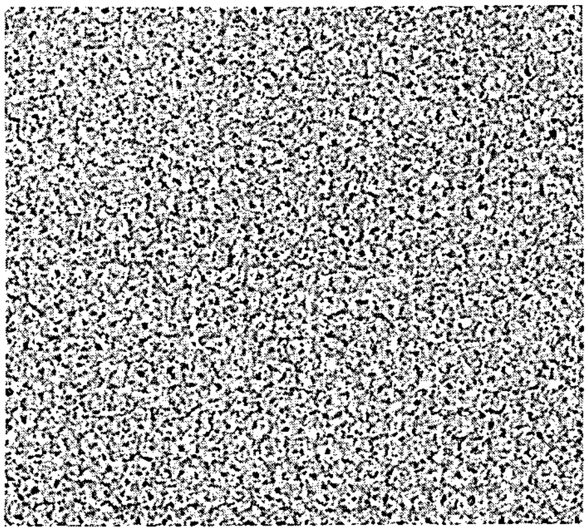

Figure 12-26
Electron micrograph of a sheet of isolated gap junctions. The cylindrical connexons form a hexagonal lattice with a unit-cell length of 85 Å. The densely stained central hole has a diameter of about 20 Å. [Courtesy of Dr. Nigel Unwin and Dr. Guido Zampighi.]

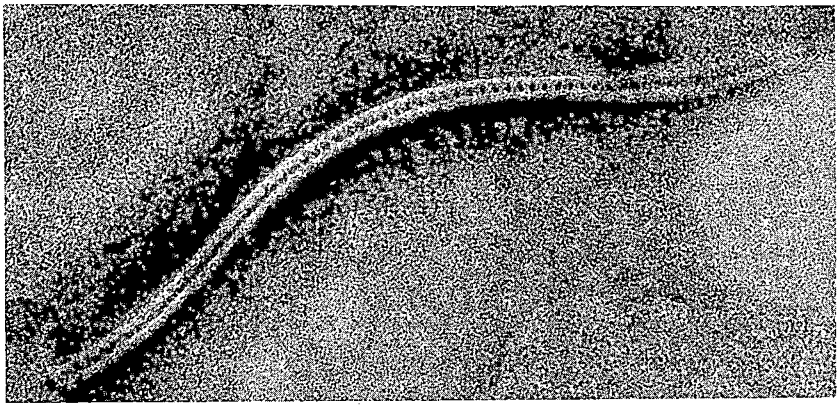

Figure 12-27
Electron micrograph of a tangential view of apposed cell membranes that are joined by gap junctions. [From E.L. Hertzberg and N.B. Gilula. *J. Biol. Chem.* 254(1979):2143.]

Small hydrophilic molecules as well as ions can pass through gap junctions. The bore size of the junctions was determined by microinjecting a series of fluorescent molecules into cells and observing their passage into adjoining cells. All polar molecules with a mass of less than about 1 kd can readily pass through these cell-to-cell channels. Thus, *inorganic ions and most metabolites (e.g., sugars, amino acids, and nucleotides) can flow between the interiors of cells joined by gap junctions.* In contrast, proteins, nucleic acids, and polysaccharides are too large to traverse these channels. *Gap junctions are important for intercellular communication.* Cells in some excitable tissues, such as heart muscle, are coupled by the rapid flow of ions through these junctions, which assure a rapid and synchronous response to stimuli. Gap junctions are also essential for the nourishment of cells that are distant from blood vessels, as in lens and bone. Moreover, communicating channels are important in development and differentiation. For example, a pregnant uterus is transformed from a quiescent protector of the fetus to a forceful ejector at the onset of labor; the formation of functional gap junctions at that time creates a syncytium of muscle cells that contract in synchrony.

A cell-to-cell channel is made of 12 molecules of *connexin,* a 32-kd transmembrane protein. The six hexagonally arrayed connexins form a half-channel, called a *connexon* or *hemichannel.* Two connexons join end to end

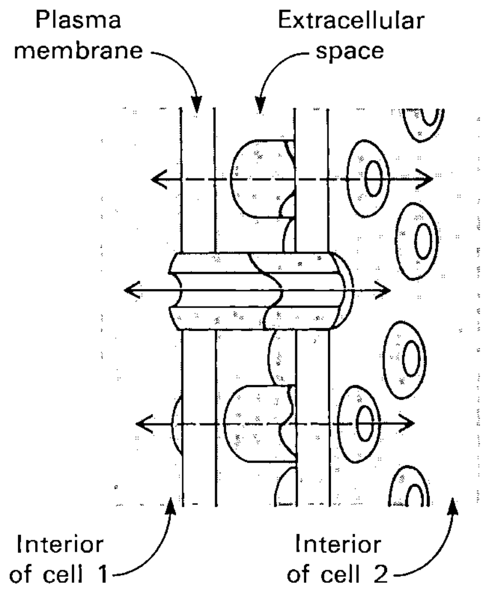

Figure 12-28
Schematic diagram of a gap junction. [Courtesy of Dr. Werner Loewenstein.]

in the intercellular space to form a functional channel between the communicating cells. Cell-to-cell channels differ from other membrane channels in three respects: (1) they traverse *two* membranes rather than one, (2) they connect cytosol to cytosol, rather than to the extracellular space or the lumen of an organelle, and (3) the connexons forming a channel are synthesized by different cells. Gap junctions form readily when cells are brought together. A cell-to-cell channel, once formed, tends to stay open for seconds to minutes. They are closed by high concentrations of calcium ion (mM) and by low pH. *The closing of gap junctions by Ca^{2+} and H^+ serves to seal normal cells from traumatized or dying neighbors.* The Hill coefficient for Ca^{2+}-induced closure is 3 and that of H^+-induced closure is 4.5, indicating that channel closing is a cooperative event. Gap junctions are also controlled by membrane potential and by hormone-induced phosphorylations.

The connexin subunits of a connexon are cylindrical, about 25 Å in diameter and 75 Å long. Each protrudes from the bilayer only slightly on the cytoplasmic side (7 Å) compared with the extracellular side (17 Å). The central hole is about 20 Å wide in the extracellular region but is narrower within the membrane. On exposure to Ca^{2+}, the subunits of each connexon rotate and slide so that the long axis of each becomes more aligned with the axis of the channel (Figure 12-29). A rather small change in tilt—only 7.5 degrees—leads to a quite large displacement—about 9 Å—at the cytoplasmic end of each subunit. Hence, the channel-lining faces of a pair of subunits on opposite sides of a channel move 18 Å toward each other, which closes the channel. A noteworthy feature of this mechanism is that switching is achieved without gross distortion of the individual connexin chains. In both the open and closed states, the polar faces of the subunits remain exposed to water, and the nonpolar ones to the hydrophobic core of the bilayer. Switching is facile because the free energies of the open and closed states do not differ markedly. A relatively small conformational change due to the binding of Ca^{2+} at one end of a subunit is amplified into a much larger one at the other end by the cooperative sliding and tilting of subunits.

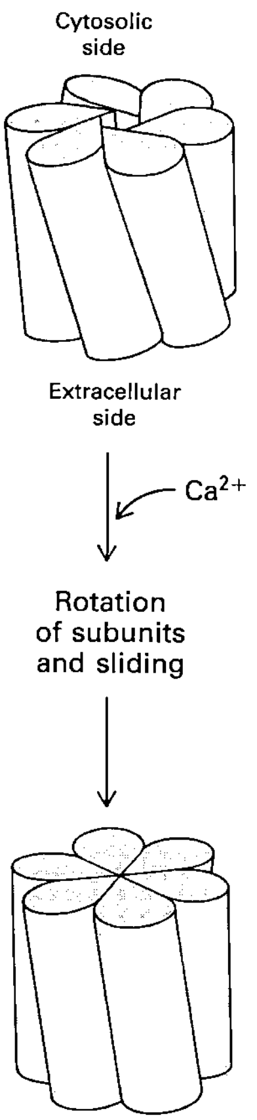

Figure 12-29
Model for the closure of gap junctions by Ca^{2+}. [After a drawing kindly provided by Dr. Nigel Unwin and Dr. Guido Zampighi.]

MEMBRANE CHANNELS EXHIBIT MANY RECURRING MOTIFS

Common structural and mechanistic themes underlie the diversity of membrane channels:

1. Channels are constructed by the association of multiple homologous subunits or domains. Most channels are formed from four, five, or six subunits (Figure 12-30).

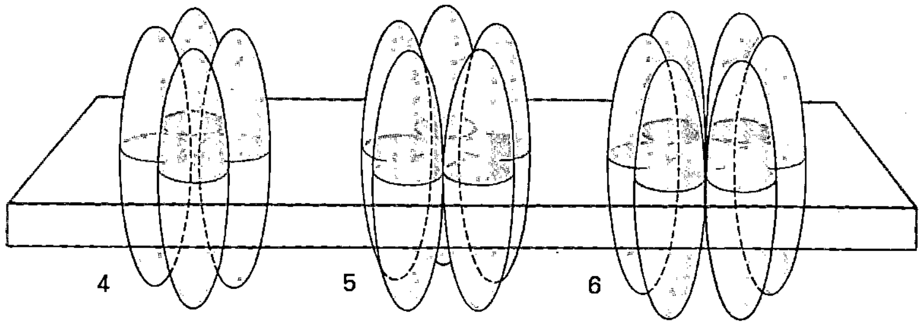

Figure 12-30
Many membrane channels are made of four, five, or six subunits (or domains). The central symmetry axis delineates a gated, water-filled pathway for the ions. The diameter of the pore increases with the number of subunits. The lipid bilayer is shown in yellow. [After N. Unwin. *Neuron* 3(1989):665.]

2. The pore lies along the symmetry axis of the channel.

3. The pore is lined by α helices or β strands. Side chains emerging from these structural elements establish the selectivity of the channel. The acetylcholine receptor channel, for example, is selective for cations because its pore contains negatively charged rings. The homologous γ-aminobutyric acid receptor ($GABA_A$ receptor) conducts anions such as Cl^- but not cations because it contains positively charged rings.

4. The degree of selectivity also depends on the diameter of the narrowest part of the pore. Channels formed from four subunits (the sodium channel and the potassium channel) have the smallest pores (diameters of 5 Å and 3 Å, respectively) and are the most selective. Gap junctions, formed from six subunits, have a minimum diameter of about 16 Å and are least selective. The acetylcholine receptor channel, made of five subunits, has a diameter of 6.5 Å and an intermediate selectivity.

5. Channels are allosteric proteins that can be gated by membrane potential, allosteric effectors, or covalent modification. The binding of a neurotransmitter such as acetylcholine provides a free-energy input that opens the gate. Alternatively, the gate can be opened by the depolarization-induced movement of gating charges within the bilayer. The tilting of helices lining the pore is a facile mechanism for gating.

ACTIVE TRANSPORT REQUIRES A COUPLED INPUT OF FREE ENERGY

Channels allow ions to flow either way. The direction of net flux is determined by the thermodynamic gradient, which allows Na^+ and Ca^{2+} to enter, and K^+ to leave cells. How are large gradients of these ions generated, and how are fuel molecules concentrated in cells? These processes are carried out by *active transport systems* that differ in an essential way from channels, which are passive transport devices. *Whether a transport process is passive or active depends on the change in free energy of the transported species.* Consider an uncharged solute molecule. The free-energy change in transporting this species from side 1 where it is present at a concentration of c_1 to side 2 where it is present at concentration c_2 is

$$\Delta G = RT\log_e \frac{c_2}{c_1} = 2.303\, RT\log_{10} \frac{c_2}{c_1}$$

For a charged species, the electrical potential across the membrane must also be considered. The sum of the concentration and electrical terms is called the *electrochemical potential.* The free-energy change is then given by

$$\Delta G = RT\log_e \frac{c_2}{c_1} + ZF\Delta V$$

in which Z is the electrical charge of the transported species, ΔV is the potential in volts across the membrane, and F is the faraday (23.062 kcal V^{-1} mol^{-1}).

A transport process must be active when ΔG is positive, whereas it can be passive when ΔG is negative. *Active transport requires a coupled input of free energy, whereas passive transport can occur spontaneously.* For example, consider the transport of an uncharged molecule from $c_1 = 10^{-3}$ mM to $c_2 = 10^{-1}$ mM.

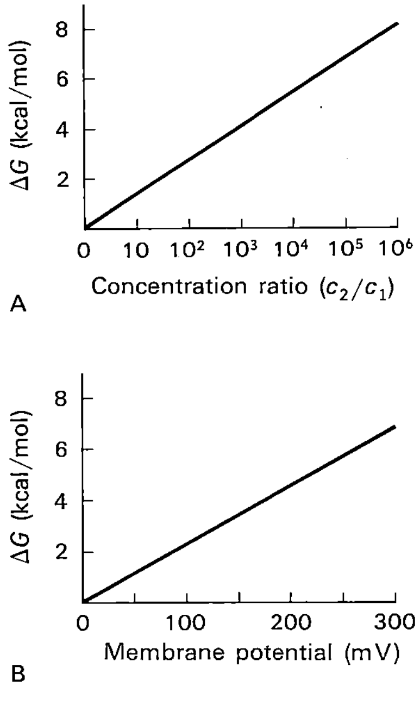

Figure 12-31
Free-energy change in transporting (A) an uncharged solute from a compartment at concentration c_1 to one at c_2 and (B) a singly charged species across a membrane to the side having the same charge as that of the transported ion. Note that the free-energy change imposed by a membrane potential of 59 mV is equivalent to that imposed by a concentration ratio of 10 for a singly charged ion at 25°C.

$$\Delta G = 2.303\ RT \log_{10} \frac{10^{-1}}{10^{-3}}$$
$$= 2.303 \times 1.99 \times 298 \times 2$$
$$= +2.7\ \text{kcal/mol}$$

At 25°C (298 K), ΔG is +2.7 kcal/mol, indicating that this transport process is active and hence requires an input of free energy. It could be driven, for example, by the hydrolysis of ATP, which yields −12 kcal/mol under typical cellular conditions.

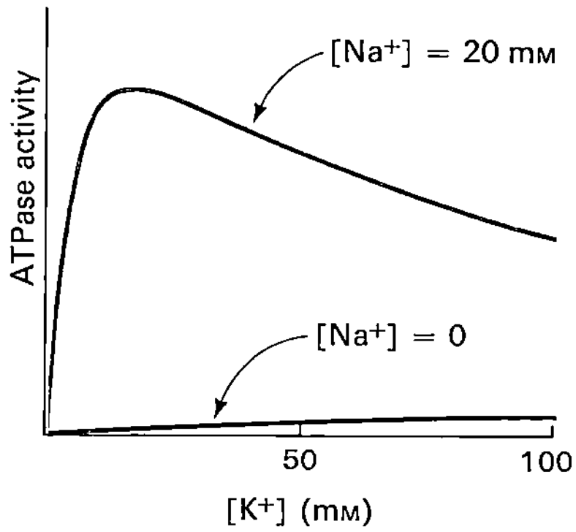

Figure 12-32
The Na^+-K^+ ATPase, an integral part of the Na^+-K^+ pump, hydrolyzes ATP only if both Na^+ and K^+ are present, in addition to Mg^{2+}.

ATP HYDROLYSIS DRIVES THE PUMPING OF SODIUM AND POTASSIUM IONS ACROSS THE PLASMA MEMBRANE

Most animal cells have a high concentration of K^+ and a low one of Na^+ relative to the external medium. These ionic gradients are generated by a specific transport system that is called the *Na^+-K^+ pump* because the movement of these ions is linked. The active transport of Na^+ and K^+ is of great physiologic significance. Indeed, more than a third of the ATP consumed by a resting animal is used to pump these ions. The Na^+-K^+ gradient in animal cells controls cell volume, renders nerve and muscle cells electrically excitable, and drives the active transport of sugars and amino acids.

In 1957, Jens Skou discovered an enzyme that hydrolyzes ATP only if Na^+ and K^+ are present in addition to Mg^{2+}, which is required by all ATPases (Figure 12-32). Hence, this enzyme was named the *Na^+-K^+ ATPase.*

$$\text{ATP} + H_2O \xrightarrow{Na^+,\ K^+,\ Mg^{2+}} \text{ADP} + P_i + H^+$$

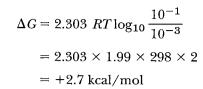

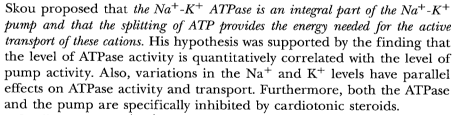

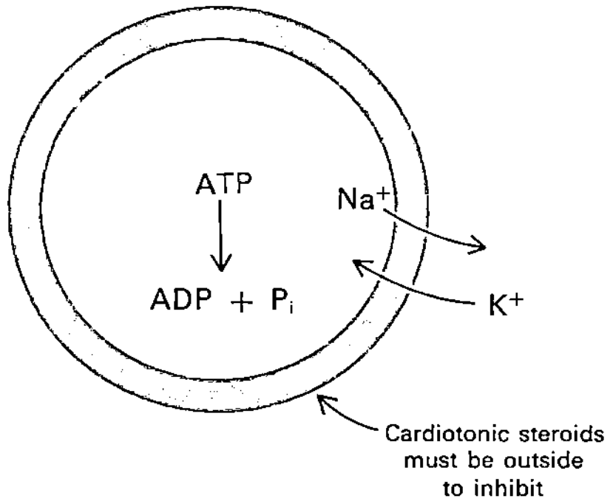

Figure 12-33
The Na^+-K^+ pump is oriented in the plasma membrane.

Skou proposed that *the Na^+-K^+ ATPase is an integral part of the Na^+-K^+ pump and that the splitting of ATP provides the energy needed for the active transport of these cations.* His hypothesis was supported by the finding that the level of ATPase activity is quantitatively correlated with the level of pump activity. Also, variations in the Na^+ and K^+ levels have parallel effects on ATPase activity and transport. Furthermore, both the ATPase and the pump are specifically inhibited by cardiotonic steroids.

Studies of the Na^+-K^+ pump in erythrocyte ghosts have revealed the orientation of this ATPase and pump. In a hypotonic salt solution, an erythrocyte swells and holes appear in its membrane. The inside of the swollen erythrocyte equilibrates with the external medium, so that hemoglobin diffuses out to leave a pale cell (hence, a ghost). If the external medium is then made isotonic, the membrane again becomes a permeability barrier. Thus, the molecular and ionic composition inside a ghost can be controlled by resealing it in an appropriate medium. Transport and enzymatic studies of such erythrocyte ghosts have shown that Na^+ must be inside, whereas K^+ must be outside to activate the ATPase and to be transported across the membrane. ATP is an effective substrate only when it is inside the cell. Likewise, vanadate must be in the cytosol to inhibit the pump and ATPase. In contrast, cardiotonic steroids inhibit the pump and the ATPase only when they are located outside the cell.

The pump consists of two kinds of subunits, α (112 kd) and β (35 kd), that are associated in the membrane as an $\alpha_2\beta_2$ tetramer (Figure 12-34). Hydrophobicity plots and membrane localization studies suggest that the

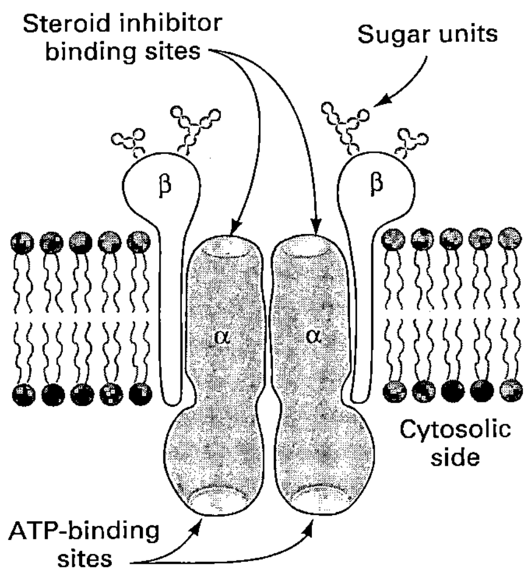

Figure 12-34
Schematic diagram of the subunit structure and orientation of the Na^+-K^+ pump.

α chain contains at least eight transmembrane helices. Much of the α chain, including its ATPase site, is located on the cytosolic side of the membrane. The small portion of α on the extracellular side contains the binding site for cardiotonic steroid inhibitors. The β chain, which contains a single transmembrane helix, does not appear to be essential for ATPase or transport activity.

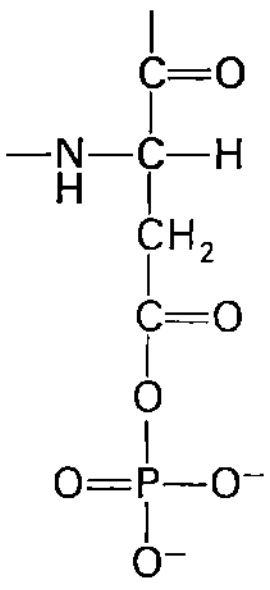

β-Aspartyl phosphate
(E-P intermediate)

How does ATP drive the active transport of Na^+ and K^+? An important clue was the finding that the ATPase is phosphorylated by ATP in the presence of Na^+ and Mg^{2+}. The site of phosphorylation is the side chain of a specific aspartate residue. This *β-aspartyl phosphoryl intermediate* (E-P) is then hydrolyzed if K^+ is present. Phosphorylation does not require K^+, whereas dephosphorylation does not require Na^+ or Mg^{2+}.

$$E + ATP \xrightleftharpoons{Na^+,\, Mg^{2+}} E\text{-}P + ADP$$

$$E\text{—}P + H_2O \xrightarrow{K^+} E + P_i$$

Na^+-dependent phosphorylation and K^+-dependent dephosphorylation are not the only critical reactions. The pump also interconverts between at least two different conformations, denoted by E_1 and E_2. Thus, at least four conformational states—E_1, E_1-P, E_2-P, and E_2—participate in the transport of Na^+ and K^+ and concomitant hydrolysis of ATP (Figure 12-35). *Three Na^+ and two K^+ are transported per ATP hydrolyzed in this unidirectional cycle,* which occurs in 10 ms. Thus, 300 Na^+ and 200 K^+ are transported by each pump molecule when it operates at maximal velocity.

An important feature of the pump is that *ATP is not hydrolyzed unless Na^+ and K^+ are transported.* In other words, the system is coupled so that ATP is not wasted. Tight coupling is a general characteristic of biological assemblies that mediate energy conversion. We shall see that electrons do not normally flow through the mitochondrial electron transport chain unless ATP is concomitantly generated (p. 552). The coupling of ATP hydrolysis and muscle contraction (p. 400) also illustrates this principle. The Na^+-K^+ pump can be reversed to *synthesize* ATP by exposing it to steep ionic gradients. This is achieved by incubating red cells in a medium containing a much higher than normal concentration of Na^+ and a much lower than normal concentration of K^+.

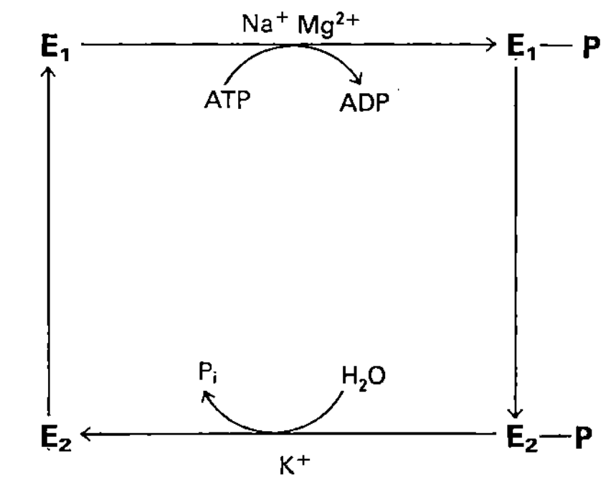

Figure 12-35
Enzymatic cycle of the Na^+-K^+ ATPase.

THE BINDING CAVITY ALTERNATELY FACES THE INSIDE AND OUTSIDE OF THE CELL IN EACH TRANSPORT CYCLE

How do phosphorylation and dephosphorylation of the ATPase lead to transport of Na^+ and K^+ across the membrane? Not enough is known about the three-dimensional structure of the pump to formulate a detailed mechanism of how it acts. Nevertheless, it is instructive to consider a simple model for a pump that was proposed years ago by Oleg Jardetzky. In this model, a protein must fulfill three structural conditions to function as a pump:

1. It must contain a cavity large enough to admit a small molecule or ion.

2. It must be able to assume two conformations, such that the cavity is open to the inside in one form and to the outside in the other.

3. The affinity for the transported species must be different in the two conformations.

Let us apply this model to Na^+ and K^+ transport (Figure 12-36). The two conformations are the E_1 and E_2 forms described earlier. It is assumed that (1) the ion-binding cavity in E_1 faces the inside of the cell, whereas in E_2 the cavity faces the outside, and that (2) E_1 has a high affinity for Na^+, whereas E_2 has a high affinity for K^+. The model also makes use of two established facts: (1) Na^+ triggers phosphorylation, whereas K^+ triggers dephosphorylation, and (2) E_2 is stabilized by phosphorylation, whereas E_1 is stabilized by dephosphorylation.

In the postulated reaction cycle, the binding of Na^+ inside (step 1) triggers the phosphorylation of E_1 by ATP (step 2). Phosphorylation switches the conformation to E_2, everting the ion-binding sites so that they now face the outside (step 3). E_2-P has low affinity for Na^+, which is released outside (steps 4 and 5). In contrast, E_2-P has high affinity for K^+,

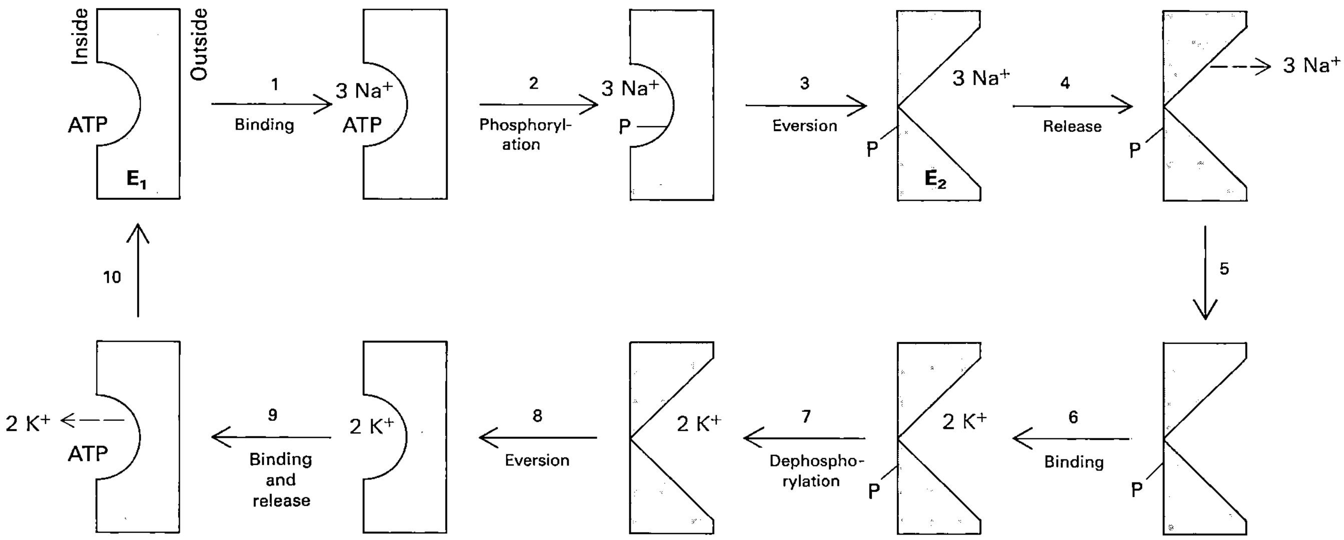

Figure 12-36
Schematic diagram of a proposed mechanism for the Na^+-K^+ pump. The upper sequence of reactions depicts the extrusion of three Na^+ ions, whereas the lower reactions show the entry of two K^+ ions. The E_1 (yellow) and E_2 (blue) forms are shown here as having very different conformations. The actual conformational differences may be quite small.

which binds on the outside (step 6). The binding of K^+ triggers the dephosphorylation of E_2-P (step 7). The enzyme devoid of a covalently attached phosphoryl group is not stable in the E_2 form. It everts to E_1 (step 8), which has low affinity for K^+. The binding of ATP accelerates the release of K^+ (steps 9 and 10) to complete the cycle. The actual reaction is undoubtedly more complex. For example, eversion proceeds through an occluded intermediate in which the binding cavity is inaccessible from either side of the membrane.

In Figure 12-36, E_1 and E_2 are depicted as having very different shapes. However, it must be stressed that the conformational differences between them need not be large. A shift of a few atoms by a distance of 2 Å might suffice to alter the relative affinities of the cavity for Na^+ and K^+ and change its orientation. There is ample precedent for assuming that phosphorylation can readily induce changes of this magnitude. Recall that the noncovalent binding of bisphosphoglycerate to hemoglobin markedly changes its oxygen affinity (p. 160).

The change in free energy accompanying the transport of Na^+ and K^+ can be calculated. Suppose that the concentration of Na^+ outside and inside the cell is 143 and 14 mM, respectively, and that of K^+ is 4 and 157 mM. At a membrane potential of −50 mV, the free-energy change in transporting 3 moles of Na^+ out and 2 moles of K^+ into the cell is +9.9 kcal. The hydrolysis of a single ATP per transport cycle provides sufficient free energy, about −12 kcal/mol under cellular conditions, to drive the uphill transport of these ions.

DIGITALIS SPECIFICALLY INHIBITS THE Na^+-K^+ PUMP BY BLOCKING ITS DEPHOSPHORYLATION

Certain steroids derived from plants are potent inhibitors ($K_i \sim$ 10 nM) of the Na^+-K^+ pump. Digitoxigenin and ouabain are members of this class of inhibitors, which are known as *cardiotonic steroids* because of their profound effects on the heart (Figure 12-37). These compounds inhibit the dephosphorylation of E_2-P when applied on the *extracellular face* of the membrane. Likewise, K^+ must be present on the extracellular side of the membrane to activate dephosphorylation.

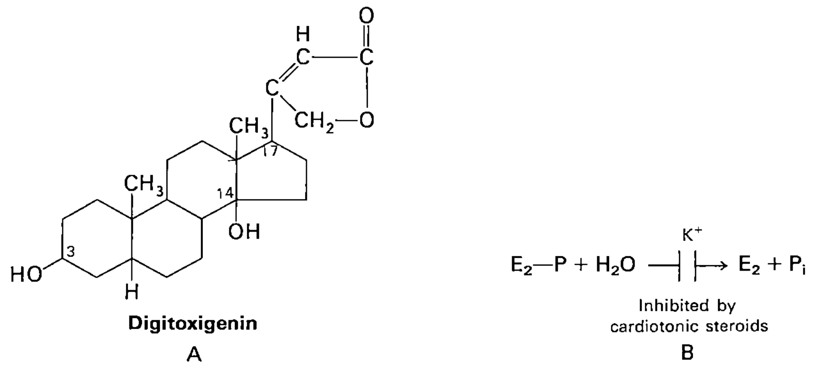

Figure 12-37
Cardiotonic steroids such as digitoxigenin inhibit the Na^+-K^+ pump by blocking the dephosphorylation of E_2-P.

Digitalis, a mixture of cardiotonic steroids derived from the dried leaf of the foxglove plant (*Digitalis purpurea*), is of great clinical significance. Digitalis increases the force of contraction of heart muscle, which makes it a choice drug in the treatment of congestive heart failure. Inhibition of the Na^+-K^+ pump by digitalis leads to a higher level of Na^+ inside the cell. The diminished Na^+ gradient results in slower extrusion of Ca^{2+} by the sodium-calcium exchanger (p. 316). The subsequent increase in the intracellular level of Ca^{2+} enhances the contractility of cardiac muscle. It is interesting to note that digitalis was effectively used long before the discovery of the Na^+-K^+ ATPase. In 1785, William Withering, a physician and botanist, published *An Account of the Foxglove and Some of its Medical Uses,* in which he described how he first learned about the use of digitalis to cure congestive heart failure (Figure 12-38).

"In the year 1775, my opinion was asked concerning a family receipt for the cure of the dropsy. I was told that it had long been kept a secret by an old woman in Shropshire, who had sometimes made cures after the more regular practitioners had failed. . . . This medicine was composed of twenty or more different herbs; but it was not very difficult for one conversant in these subjects to perceive that the active herb could be no other than the foxglove. . . . It has a power over the motion of the heart to a degree yet unobserved in any other medicine, and this power may be converted to salutary ends."

WILLIAM WITHERING

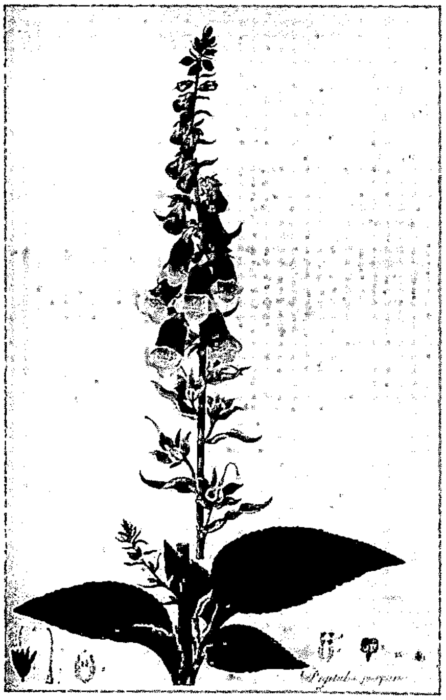

Figure 12-38
Foxglove (*Digitalis purpurea*) is the source of digitalis, one of the most widely used drugs. The quotation is from *An Account of the Foxglove and Some of its Medical Uses: with Practical Remarks on Dropsy, and Other Diseases,* by William Withering, M.D., Physician to the General Hospital at Birmingham. This figure of a foxglove plant was drawn by William Curtis, botanist to the Society of Apothecaries and a fellow of the Linnean Society, and appeared as the frontispiece to Withering's book, published in London, 1785. [Reproduced by permission of the Linnean Society of London.]

CALCIUM IONS ARE PUMPED OUT OF THE CYTOSOL BY RELATED ATPases

Calcium ion is a key intracellular messenger in many eukaryotic signal transduction processes (p. 347). The cytosolic Ca^{2+} level of an unexcited cell is typically 100 nM, compared with an extracellular concentration of 1.5 mM. The influx of Ca^{2+} is thermodynamically favorable under all physiological conditions. The large electrochemical gradient of Ca^{2+} across the plasma membrane is maintained by two transport systems: an ATP-driven calcium pump called the *Ca^{2+}-ATPase* and a *sodium-calcium exchanger* that uses the Na^+ gradient across the plasma membrane as the energy source.

The calcium level in many intracellular compartments is much higher than in the cytosol. The *endoplasmic reticulum (ER), an extensive membrane-enclosed network,* serves as a large and readily mobilizable internal store of Ca^{2+} (p. 344). A Ca^{2+}-ATPase in the ER membrane pumps Ca^{2+} into this compartment. The best-understood ATP-driven calcium pump is the one in the *sarcoplasmic reticulum (SR)* of skeletal muscle, a specialized form of the ER. Muscle contraction is triggered by an abrupt rise in the cytosolic calcium level (p. 402). Relaxation depends on the rapid removal of Ca^{2+} from the cytosol by a Ca^{2+}-ATPase that comprises more than 80% of the integral membrane protein of the SR and occupies nearly half its surface area (Figure 12-39). The density of the pump in the SR membrane is about 25,000 per μm^2, amongst the highest of any membrane protein. The plasma membrane contains a related ATPase that is present at much lower density.

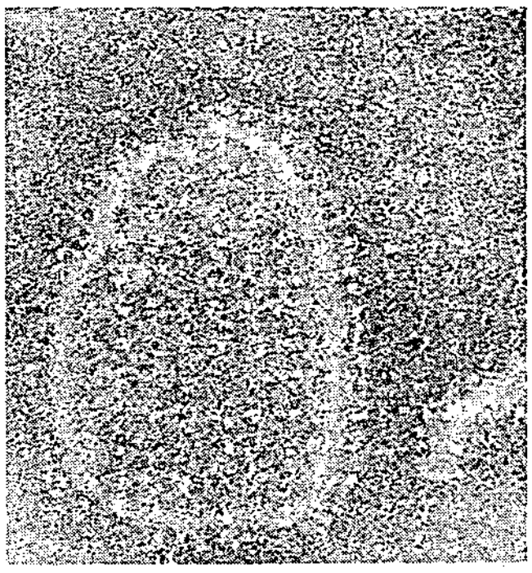

Figure 12-39
Membrane vesicle formed from purified Ca^{2+}-ATPase. The globular particles on the surface of the membrane are part of the ATPase molecule, which extends across the membrane. [From P.S. Stewart and D.H. MacLennan. *J. Biol. Chem.* 249(1974):987.]

In both the plasma membrane and SR Ca^{2+}-ATPase, a specific aspartate residue is transiently phosphorylated by ATP in a Ca^{2+}-dependent reaction.

$$E + ATP \xrightleftharpoons{Ca^{2+},\, Mg^{2+}} E\text{–}P + ADP$$

$$E\text{—}P + H_2O \xrightarrow{Mg^{2+}} E + P_i$$

A cycle of conformational changes driven by phosphorylation and dephosphorylation transports two Ca^{2+} for each ATP hydrolyzed by the SR pump. This Ca^{2+}-ATPase cycle resembles the Na^+-K^+ ATPase cycle in several respects: (1) The enzyme exists in two major conformations, E_1 and E_2. Two Ca^{2+} ions from the cytosol bind to E_1 and trigger its phosphorylation. (2) A β-aspartyl phosphate E-P intermediate is formed. A 10-residue sequence containing this aspartate is identical in the two pumps. These enzymes are members of a class of transporters called *P-type ion-motive ATPases* (*P* refers to the phosphorylated intermediate). (3) Phosphorylation favors E_2, and dephosphorylation favors E_1. (4) The ion-binding sites evert and their affinity for Ca^{2+} decreases more than a thousandfold in the transition from E_1 to E_2. Eversion proceeds through an occluded state. (5) The transition from E_2 to E_1 is markedly accelerated by ATP. (6) Vanadate inhibits Ca^{2+} transport and ATPase activity by stabilizing the E_2 form.

The SR Ca^{2+}-ATPase is a single 110-kd polypeptide. cDNA sequencing, electron microscopy, and chemical studies of this pump indicate that it has a tripartite structure (Figure 12-40). A large cytosolic headpiece is joined by a stalk to a domain made of 10 transmembrane helices. Only a small part of the protein protrudes on the lumenal side of the SR. The headpiece contains a *nucleotide-binding domain* (which binds ATP), a *phosphorylation domain* (which contains the transiently phosphorylated aspartate), and a *β sheet domain.* Site-specific mutagenesis studies suggest that the two transported Ca^{2+} are bound near the center of the membrane. One of the challenges now is to determine how phosphorylation of a single aspartate triggers eversion of the calcium-binding site.

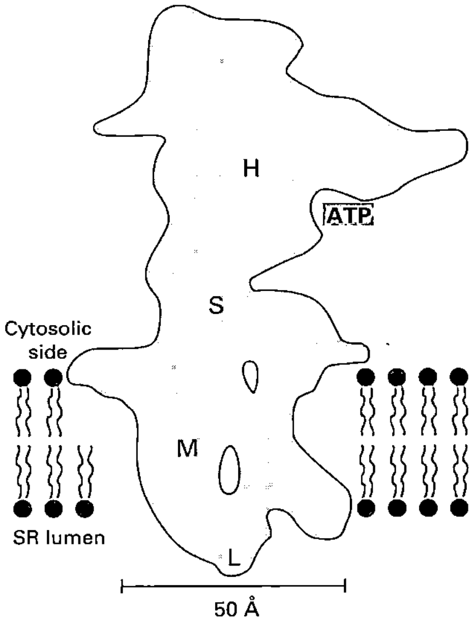

Figure 12-40
Schematic diagram of the sarcoplasmic reticulum Ca^{2+}-ATPase based on an electron crystallographic analysis at 14-Å resolution. Most of the 110-kd protein is located on the cytosolic side of the membrane. The head region (H) contains domains for ATP binding and phosphorylation. A 25-Å stalk (S) joins the head to the transmembrane domain (M), which contains 10 helical segments that traverse the bilayer. Only a small part of the pump (L) is located in the lumen of the SR. [After C. Toyoshima, H. Sasabe, and D.L. Stokes. *Nature* 362(1993):469.]

A functional Ca^{2+} pump has been reconstituted from purified Ca^{2+}-ATPase and phospholipids. The Ca^{2+}-ATPase was isolated from sarcoplasmic reticulum membranes after it had been solubilized by cholate, a detergent (p. 281). This solubilized enzyme was added to phospholipids obtained from soybeans. Membrane vesicles formed spontaneously when the detergent was removed by dialysis. *These reconstituted vesicles pumped in Ca^{2+} at a rapid rate (30 s^{-1}) when provided with ATP and Mg^{2+}.* The pump is tightly coupled so that ATP is hydrolyzed only when Ca^{2+} is transported. Moreover, the reconstituted pump can be operated in reverse.

Calcium gradients can drive the synthesis of ATP. Vesicles loaded with Ca^{2+} are placed in a medium containing P_i, Mg^{2+}, and EGTA (a chelator of calcium). Under these conditions, E_2-P is formed, but there is no flow of calcium out of the vesicle. The addition of ADP to the external medium then leads to calcium efflux and the synthesis of ATP. In fact, one ATP is formed per two Ca^{2+} transported (Figure 12-41). Thus, the Ca^{2+}-ATPase is a reversible molecular machine; it can transduce phosphoryl potential into an ion gradient, or vice versa.

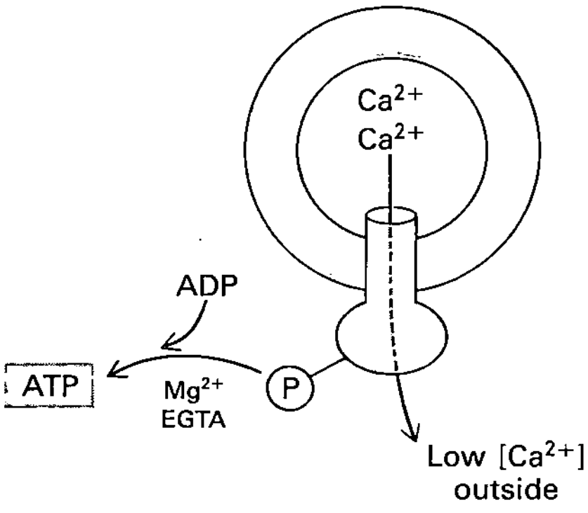

Figure 12-41
The calcium pump can be reversed in vitro to synthesize ATP. EGTA sequesters external Ca^{2+} to maintain a large gradient across the membrane.

THE SODIUM GRADIENT ACROSS THE PLASMA MEMBRANE CAN BE TAPPED TO PUMP IONS AND MOLECULES

Many active transport processes are not directly driven by the hydrolysis of ATP. Instead, the uphill flow of an ion or molecule is coupled to the downhill flow of another ion. The *sodium-calcium exchanger* in the plasma membrane uses the electrochemical gradient of Na^+ to pump Ca^{2+} out of the cell (Figure 12-42). Three Na^+ enter the cell for each Ca^{2+} that is extruded. The cost of transport by this exchanger is paid by the Na^+-K^+ pump, which generates the requisite sodium gradient. *The exchanger has lower affinity for Ca^{2+} than does the Ca^{2+}-ATPase, but its capacity to extrude Ca^{2+} is greater.* The cytosolic Ca^{2+} level can be lowered to several micromolar by the exchanger; submicromolar Ca^{2+} levels are attained by the subsequent action of the Ca^{2+}-ATPase.

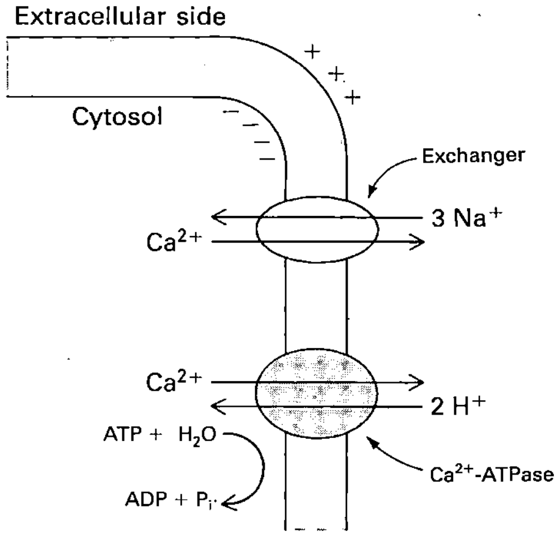

Figure 12-42
The Na^+-Ca^{2+} exchanger (green) works in parallel with the Ca^{2+}-ATPase (red) to maintain a low cytosolic level of Ca^{2+}. Extrusion of Ca^{2+} by the exchanger is driven by the sodium-motive force across the plasma membrane.

The exchanger can extrude about 2000 Ca^{2+} per second, compared with only 30 per second for the Ca^{2+}-ATPase pump. These transport rates are orders of magnitude lower than the $\sim 10^7\ s^{-1}$ flux through many channels. In general, *active transport is much slower than transport through channels.* In an exchanger or pump, the binding site for the transported species is not simultaneously accessible from both sides of the membrane. Rather, eversion of the site is coupled to a phosphorylation-dephosphorylation cycle or to the binding and transport of another ion. In contrast, an open channel has a pore through which ions can flow at any instant from either side; the direction of ion movement is determined solely by the electrochemical gradient.

The sodium gradient across the plasma membrane can be tapped to drive the entry as well as the extrusion of molecules or ions. A *symporter* carries the transported species in the same direction across a membrane, whereas an *antiporter* carries them in opposite directions. For example, glucose is pumped into some animal cells by the simultaneous entry of Na^+ (Figure 12-43). The entry of Na^+ provides a free-energy input of 2.2 kcal/mol under typical cellular conditions (external $[Na^+]$ = 117 mM, internal $[Na^+]$ = 30 mM, and membrane potential = −60 mV). This is sufficient to generate a 36.7-fold concentration gradient of an uncharged molecule such as glucose.

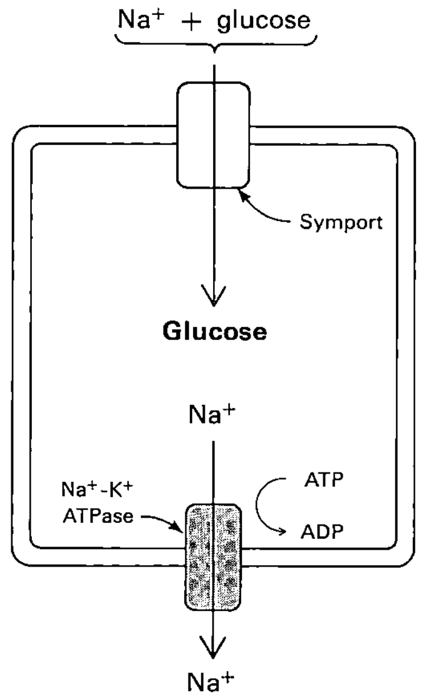

Figure 12-43
A Na^+ gradient drives the active transport of glucose. This symport system is present in the plasma membrane of intestinal cells and kidney cells.

PROTON FLOW ACROSS THE CELL MEMBRANE DRIVES MANY BACTERIAL TRANSPORT PROCESSES

Symporters and antiporters are ancient molecular machines. Many bacterial transport systems are driven by the flow of protons across the plasma membrane. The best-understood symport system in bacteria is the one for lactose in *Escherichia coli* (Figure 12-44). This denizen of the mammalian lower intestinal tract has evolved a highly efficient mechanism for concentrating lactose, a sugar that is abundant in milk (p. 471). This disaccharide is actively transported by *lactose permease,* a 47-kd integral membrane protein. Vesicles formed from bacterial membranes are able to actively transport lactose and many other substances. Their simplicity makes them attractive for mechanistic studies. Analysis of such vesicles showed that an imposed pH gradient, with the outside acidic, leads to the accumulation of lactose inside. Alternatively, lactose can be pumped in the absence of a pH gradient if the membrane potential is negative inside. In both cases, protons move down their thermodynamic gradient. Hence, the driving force for the active transport of lactose is the electrochemical gradient of H^+ across the plasma membrane. This free-energy source, called the *proton-motive force,* plays a central role in ATP generation in nearly all organisms (p. 544) and in many bacterial energy conversion processes (p. 412).

Reconstituted vesicles containing only purified lactose permease in a phosphatidyl choline bilayer are fully active in pumping lactose. The permease contains a specific proton-binding site in addition to a lactose-binding site. *The energetically uphill transport of one lactose molecule is coupled to the downhill transport of one proton.* Lactose and a proton bind first to sites that face the outside (Figure 12-45). The binding of the pair, but not of lactose or H^+ alone, leads to the eversion of their sites. The bound proton

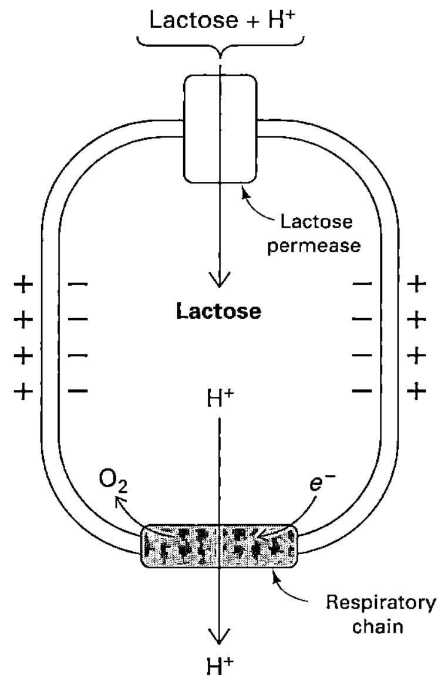

Figure 12-44
A proton gradient drives the active transport of some sugars and amino acids into bacteria. This proton gradient is generated by the flow of electrons through the respiratory chain.

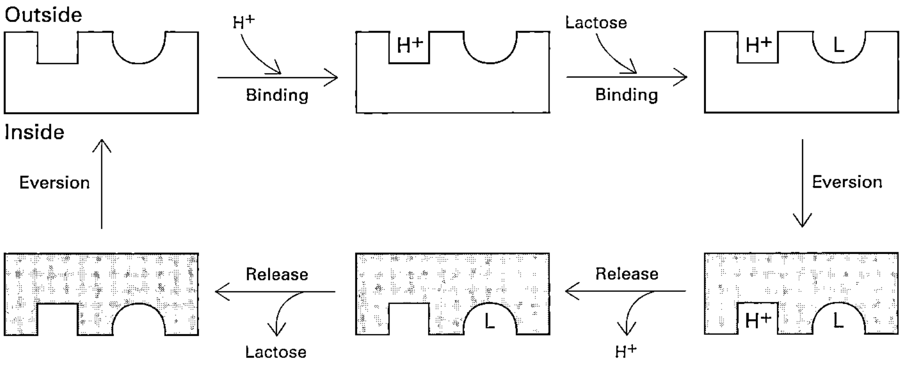

Figure 12-45
Lactose permease pumps lactose into bacterial cells by drawing on the proton-motive force. The binding sites evert when a lactose molecule and a proton are bound to external sites. After these species are released inside the cell, the binding sites again evert to complete the transport cycle.

is then released on the inside, followed by the dissociation of lactose. The binding sites on the empty permease now evert to again face the outside. This eversion is very slow if either partner lingers on the permease. *The transport cycle has directionality because the proton-motive force is greater on the outside than on the inside.* The binding of H^+ to the outside and the release of H^+ on the inside induces the uptake and dissociation of lactose against a concentration gradient.

BACTERIORHODOPSIN IS A LIGHT-DRIVEN PROTON PUMP IN HALOBACTERIA

The active transport processes discussed thus far are driven by an existing cellular energy source, such as ATP, the sodium-motive force, or the proton-motive force. We turn now to an energy-harvesting process in which *light is converted into a proton gradient. Halobacterium halobium* is a salt-loving archaebacterium that inhabits natural salt lakes and areas where seawater is evaporated to produce salt, such as the enclosed shallows in San Francisco Bay. In the presence of O_2, halobacteria oxidize fuel molecules to generate ATP. When oxygen is scarce, they switch to a photosynthetic mode and synthesize large amounts of a 26-kd integral membrane protein called *bacteriorhodopsin* because it contains *retinal* as its light-absorbing group. Retinal, derived from β-carotene, was already known to be the chromophore of *rhodopsin,* the photoreceptor protein in the eyes of higher organisms (p. 333). Retinal is joined to the ε-amino group of a specific lysine residue of bacteriorhodopsin by a Schiff base linkage. In 1973, Walther Stoeckenius and Dieter Oesterhelt discovered that *bacteriorhodopsin is a light-driven proton pump.* Light isomerizes the all-trans retinal chromophore of the protein to the 13-cis form (Figure 12-46). As retinal cycles back to the all-trans form, a proton is pumped from the cytosol to the outside. In essence, *light is converted into proton-motive force, which is then used to synthesize ATP.*

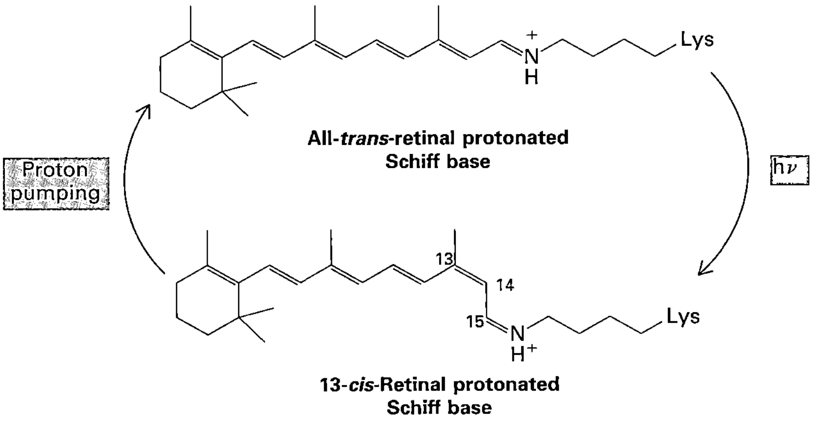

Figure 12-46
Bacteriorhodopsin is a light-driven proton pump. Light (hν) triggers the photoisomerization of the retinal chromophore from the all-trans to the 13-cis form. A proton is pumped from the cytosol to the extracellular side in the isomerization back to the all-trans form.

Bacteriorhodopsin is a very expressive molecule: its color graphically reveals many of its key transitions. A cycle of conformational transitions (Figure 12-47) is initiated by the absorption of a photon by bacteriorhodopsin, denoted as BR_{568} because its absorption spectrum peaks at 568 nm. Light isomerizes BR_{568} to form K_{590}, a transient red-absorbing intermediate. This energy-trapping conformational transition occurs very rapidly, in less than a picosecond (1 ps = 10^{-12} s). A series of intermediates with distinctive absorption spectra (L_{550}, M_{412}, N_{550}, and O_{640}) then appear in the microsecond to millisecond time range. The yellow color of the M_{412} species is especially revealing. *M_{412} absorbs maximally at a much shorter wavelength than do the others because its Schiff base linkage is deprotonated.* The Schiff base is reprotonated in forming N_{550}. Indeed, *deprotonation and reprotonation of the Schiff base are at the heart of the proton pumping mechanism.* The retinal group then isomerizes back to the all-trans form to regenerate BR_{568}, which is ready to absorb another photon. In bright light, some 50 H^+ are pumped per second by each bacteriorhodopsin molecule.

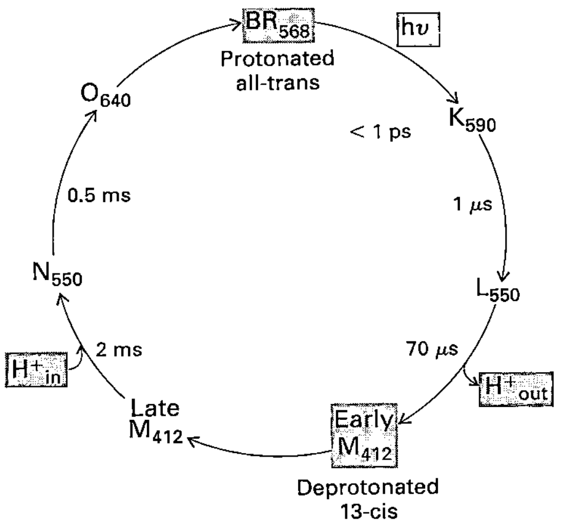

Figure 12-47
Photocycle of bacteriorhodopsin.

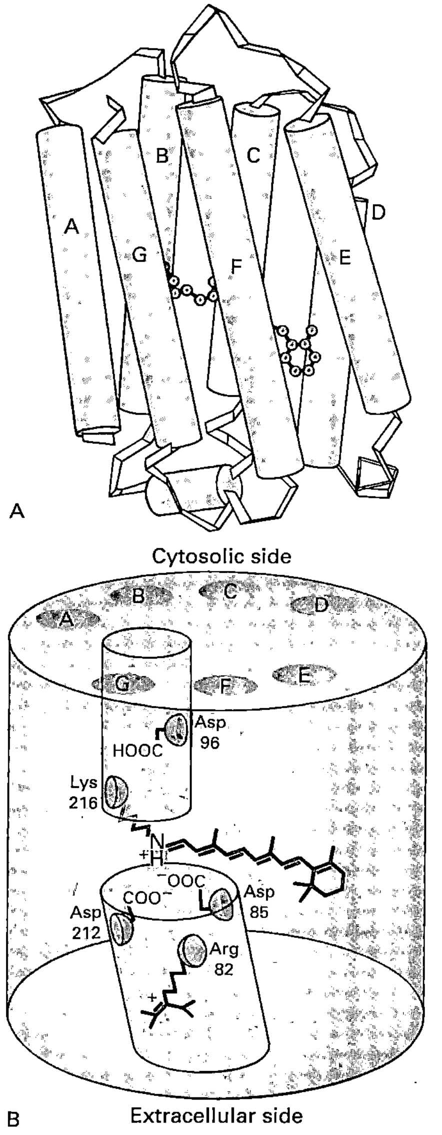

Figure 12-48
Three-dimensional structure of bacteriorhodopsin determined by electron crystallography at 3.5-Å resolution. (A) Schematic diagram of the course of the main chain and the location of bound retinal (red). (B) Arrangement of the key groups in light-driven proton pumping. [After R. Henderson, J.M. Baldwin, T.A. Ceska, F. Zemlin, F. Beckman, and K.H. Downing. *J. Mol. Biol.* 213(1990): 899.]

How does the light-induced isomerization of retinal drive the uptake of a proton from the inside to the outside of the cell? The dynamics of this pump are coming into view as a result of incisive structural, spectroscopic, and site-specific mutagenesis studies. In 1990, Richard Henderson and co-workers solved the three-dimensional structure of bacteriorhodopsin (BR_{568}) at 3.5-Å resolution. The course of the entire polypeptide chain and the location of retinal were revealed (Figure 12-48A). Retinal is located inside the protein, where it is surrounded by side chains from six of the seven helices (labeled A to G). The alternating single- and double-bond chain of this photosensitive chromophore is held in place by three tryptophans. Retinal is joined by a protonated Schiff base linkage to Lys 216. The negatively charged oxygens of two aspartates (Asp 85 and 212) serve as counterions for the positively charged nitrogen atom of the Schiff base. Retinal blocks the free flow of protons across the membrane. It sits at the junction of two chambers that would form a channel if they were in contact. Arg 82, Asp 85, and Asp 212 are situated in the half-channel facing the outside, and Asp 96 is located in the half-channel facing the cytosol. The two half-channels are never directly in contact. Rather, retinal shuttles a proton from the inner half-channel to the outer one.

All the players are now on stage, ready to be set in motion by light. Photoisomerization of the all-trans form of retinal to 13-*cis*-retinal moves the Schiff base by 2 Å (Figure 12-49). The nitrogen atom is now optimally positioned to transfer its proton to Asp 85. *Protonation of Asp 85 induces the neighboring Arg 82 to transfer its proton to the external medium.* Half the photocycle is completed in reaching M_{412}. A conformational transition involving many protein atoms then takes place at M_{412} without altering its color. Eversion of the Schiff base nitrogen returns it to a more polar environment. *It now faces the cytosol and picks up a proton from Asp 96, which in turn acquires a proton from the cytosol.* At this stage (N_{550}), a proton has been pumped from the inside to the outside of the cell; the energy of a photon has been converted into that of a proton gradient. The remaining tasks are to reisomerize the retinal to the all-trans form and restore the protein conformation to that of BR_{568}. *We have here our first glimpse of how a biological energy transduction process operates at the atomic level.*

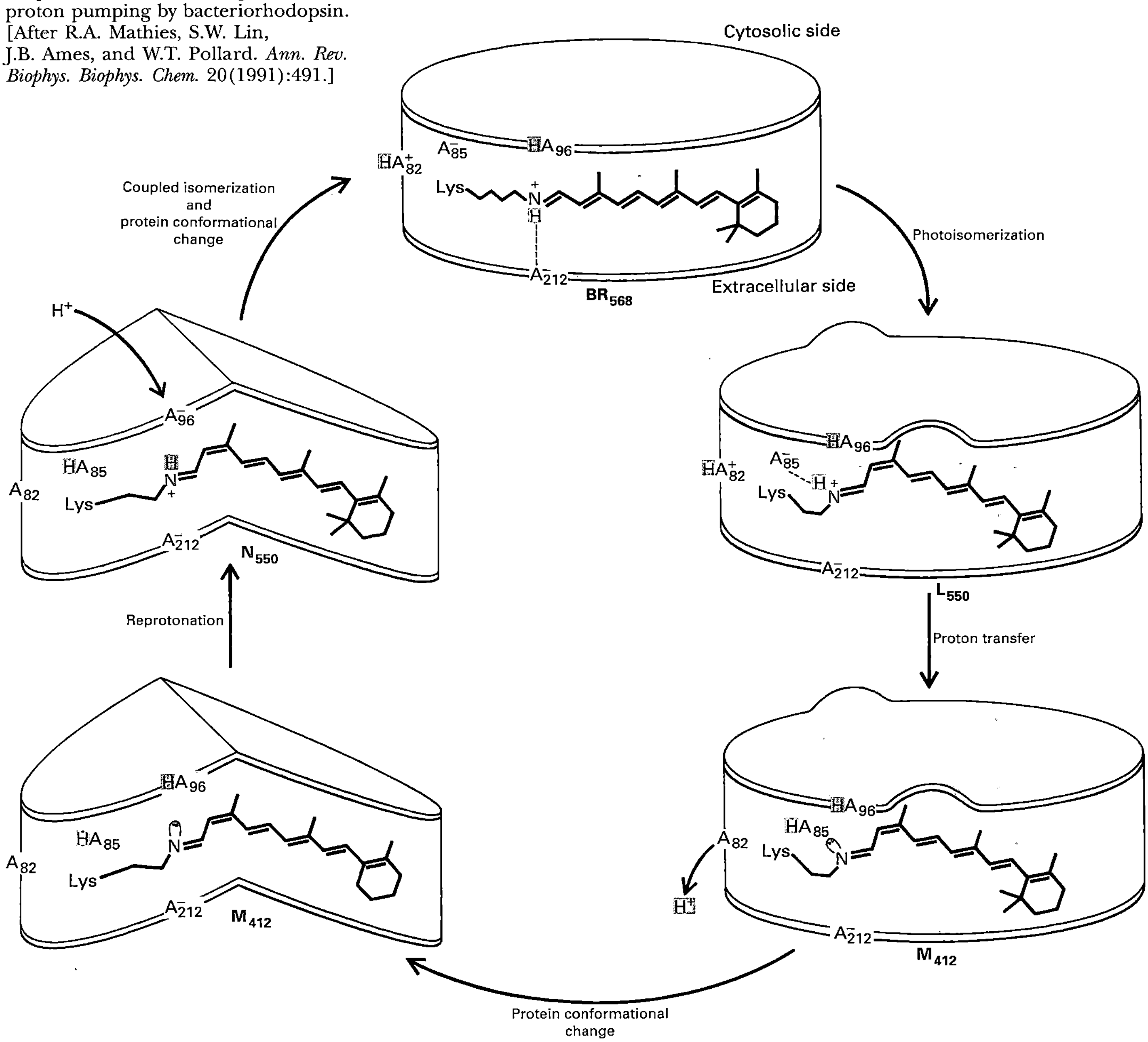

Figure 12-49
Proposed mechanism of light-driven proton pumping by bacteriorhodopsin. [After R.A. Mathies, S.W. Lin, J.B. Ames, and W.T. Pollard. *Ann. Rev. Biophys. Biophys. Chem.* 20(1991):491.]

SUMMARY

The flow of ions and molecules across biological membranes is controlled by two classes of transmembrane proteins, channels and pumps. Channels form continuous polar pathways across membranes that allow ions to move down their electrochemical gradients. The flux can be as high as 10^7 ions per second. Channels are built from identical or homologous subunits, or from homologous domains of a single polypeptide chain. The number of subunits is typically four, five, or six. The pore of the channel is along the symmetry axis of the oligomer. All channels have a closed and an open state, and some also have an inactive state which is controlled by a built-in timer. The transition between the closed and open form is allosterically regulated by voltage, the binding of another molecule, or covalent modification. The high degree of specificity of many channels is mediated by intimate interactions between the transported ion and residues lining the pore at a region called the selectivity filter. Ion flow through a single channel can be directly observed by the patch-clamp technique.

The acetylcholine receptor channel, which mediates synaptic transmission, is the best-understood ligand-gated channel. The pore of this cation-specific pentameric channel ($\alpha_2\beta\gamma\delta$) is along its fivefold symmetry axis. The binding of two acetylcholines opens a channel. The rapid lowering of the acetylcholine concentration in the synaptic cleft by the hydrolytic action of acetylcholinesterase leads to channel closure. The channel becomes desensitized if the concentration of acetylcholine stays high.

Action potentials are mediated by transient increases in the permeability of Na^+ and then of K^+. The sodium channel contains four internal repeats of similar sequence, which associate symmetrically to form a pore. The channel selects for Na^+ by providing a negatively charged site with a small radius. Depolarization opens the sodium channel by moving its four S4 voltage sensors. In general, a channel is voltage-dependent if charge moves across the bilayer on opening. Soon after opening, the sodium channel spontaneously inactivates. The closed but activatable state is restored only after the membrane is repolarized. Potassium channels, which diminish excitability when open, are tetramers of subunits that contain S4 voltage sensors. They select K^+ over Na^+ by taking advantage of the higher cost of dehydrating Na^+. Inactivation of the potassium channel occurs by occlusion of its pore by an amino-terminal ball and chain. The S4 motif and overall architectural plan also recur in voltage-gated calcium channels.

Ionic gradients are generated across cell membranes and fuel molecules are concentrated by active transport systems that differ in essential ways from channels, which are passive transport devices. A transport process is passive if ΔG for the transported species is negative, whereas it is active if ΔG is positive. The free-energy change depends on the concentration ratio of the transported species and on the membrane potential if it is charged. Active transport requires a coupled input of free energy, which can be provided by a source such as ATP, the electrochemical gradient of Na^+ or H^+ (sodium-motive force or proton-motive force), or light. Active transporters, called pumps, differ from channels in an essential way: a continuous pore from one side of the membrane to the other is never present. Rather, they undergo a cycle of conformational transitions that simultaneously change the orientation and affinity of the binding site for the transported species. A small portion of the transport protein is transiently everted during each transport cycle, which is driven unidirectionally by the expenditure of free energy. The maximal transport rate of pumps is $\sim 10^4\ s^{-1}$, much less than for channels.

The most prevalent transport system in animal cells is the Na^+-K^+ pump, which hydrolyzes a molecule of ATP to pump out three Na^+ ions and pump in two K^+ ions. When Na^+ is bound to the cytosolic side of the pump, ATP phosphorylates an aspartate side chain. Phosphorylation everts the binding site and releases Na^+ on the extracellular side. The subsequent binding of K^+ induces the hydrolysis of the attached phosphoryl group, which everts the binding site to its initial orientation and releases K^+ inside the cell. Cardiotonic steroids such as digitalis bind to an external site and inhibit dephosphorylation. Calcium ion is pumped out of the cytosol by related ATPases. The one in the sarcoplasmic reticulum membrane of muscle transports two Ca^{2+} per ATP hydrolyzed.

Some transport systems are driven by ionic gradients rather than by the hydrolysis of ATP. An antiporter carries a pair of transported species in opposite directions, whereas a symporter carries them in the same direction. The sodium-calcium exchanger is an antiporter that uses the sodium gradient to pump calcium out of cells. The active transport of glucose and amino acids into some animal cells is mediated by symporters that also are driven by the sodium-motive force. Bacteria generally use H^+ instead of Na^+ to drive symporters and antiporters. For example, the active transport of lactose into *E. coli* by lactose permease is coupled to the movement of a proton into the bacterium.

Halobacteria contain bacteriorhodopsin, the simplest known light-driven proton pump. Structural, spectroscopic, and site-specific mutagenesis studies have provided our first glimpse of an energy transduction process at the atomic level. This 26-kd protein contains a covalently attached retinal chromophore that cycles between an all-trans and a 13-cis form. Photoisomerization of all-trans-retinal is followed by the deprotonation of the Schiff base linkage and the transport of one H^+ from the inside to the outside of the cell. Retinal everts and changes its affinity for protons during the photocycle.

SELECTED READINGS

Where to start

Unwin, N., 1993. Neurotransmitter action: Opening of ligand-gated ion channels. *Cell* 72:31–41.

Neher, E., and Sakmann, B., 1992. The patch clamp technique. *Sci. Amer.* 266(3):44–51. [A lucid exposition of an experimental method that revolutionized the study of ion channels.]

Miller, C., 1989. Genetic manipulation of ion channels: A new approach to structure and mechanism. *Neuron* 2:1195–1205.

Sakmann, B., 1992. Elementary steps in synaptic transmission revealed by currents through single ion channels. *Science* 256:503–512.

Books

Hille, B., 1992. *Ionic Channels of Excitable Membranes* (2nd ed.). Sinauer. [An outstanding presentation of concepts, experimental methods, and channel properties. A book to savor.]

Läuger, P., 1991. *Electrogenic Ion Pumps.* Sinauer. [A rigorous exposition of fundamental physical principles underlying the operation of molecular pumps. The focus is on proton pumps and ATP-driven pumps.]

Stein, W.D., 1990. *Channels, Carriers, and Pumps. An Introduction to Membrane Transport.* Academic Press.

Miller, C. (ed.), 1986. *Ion Channel Reconstitution.* Plenum. [Contains many informative articles on channel purification and reconstitution of activity. The sodium channel and the acetylcholine receptor are discussed in detail.]

Cramer, W.A., and Knaff, D.B., 1991. *Energy Transduction in Biological Membranes. A Textbook of Bioenergetics.* Springer-Verlag.

Hodgkin, A., 1992. *Chance & Design: Reminiscences of Science in Peace and War.* Cambridge University Press.

Voltage-gated ion channels

Stuhmer, W., 1991. Structure-function studies of voltage-gated ion channels. *Ann. Rev. Biophys. Biophys. Chem.* 20:65–78.

Numa, S., 1989. A molecular view of neurotransmitter receptors and ionic channels. *Harvey Lectures* 83:121–165.

Guy, H.R., and Conti, F., 1990. Pursuing the structure and function of voltage-gated channels. *Trends Neurosci.* 13:201–206.

Stuhmer, W., Conti, F., Suzuki, H., Wang, X.D., Noda, M., Yahagi, N., Kubo, H., and Numa, S., 1989. Structural parts involved in activation and inactivation of the sodium channel. *Nature* 339:597–603.

Miller, C., 1991. 1990: Annus mirabilis of potassium channels. *Science* 252:1092–1096.

Isacoff, E., Papazian, D., Timpe, L., Jan, Y.-N., and Jan, L.-Y., 1990. Molecular studies of voltage-gated potassium channels. *Cold Spring Harbor Symp. Quant. Biol.* 55:9–17.

Hoshi, T., Zagotta, W.N., and Aldrich, R.W., 1990. Biophysical and molecular mechanisms of *Shaker* potassium channel inactivation. *Science* 250:533–538.

Papazian, D.M., Timpe, L.C., Jan, Y.-N., and Jan, L.-Y., 1991. Alteration of voltage-dependence of *Shaker* potassium channel by mutations in the S4 sequence. *Nature* 349:305–310.

Miller, R.J., 1992. Voltage-sensitive Ca^{2+} channels. *J. Biol. Chem.* 267:1403–1406.

Catterall, W.A., 1991. Excitation-contraction coupling in vertebrate skeletal muscle: A tale of two calcium channels. *Cell* 64:871–874.

Ligand-gated ion channels

Unwin, N., 1995. Acetylcholine receptor channel imaged in the open state. *Nature* 373: 37–43. [An elegant recent study revealing the mechanism of channel opening induced by the binding of acetylcholine.]

Galzi, J-L., Revah, F., Bessis, A., and Changeux, J-P., 1991. Functional architecture of the nicotinic acetylcholine receptor: From electric organ to brain. *Ann. Rev. Pharmacol.* 31:37–72.

Colquhoun, D., and Sakmann, B., 1981. Fluctuations in the microsecond time range of the current, through single acetylcholine receptor ion channels. *Nature* 294:464–466.

Jackson, M.B., 1989. Perfection of a synaptic receptor: Kinetics and energetics of the acetylcholine receptor. *Proc. Nat. Acad. Sci.* 86:2199–2203.

Toyoshima, C., and Unwin, N., 1990. Three-dimensional structure of the acetylcholine receptor by cryoelectron microscopy and helical image reconstruction. *J. Cell. Biol.* 6:2623–2635.

Unwin, N., Toyoshima, C., and Kubalek, E., 1988. Arrangement of the acetylcholine receptor subunits in the resting and desensitized states, determined by cryoelectron microscopy of crystallized *Torpedo* postsynaptic membranes. *J. Cell. Biol.* 107:1123–1138.

Imoto, K., Busch, C., Sakmann, B., Mishina, M., Konno, T., Nakai, J., Bujo, H., Mori, Y., Fukuda, K., and Numa, S., 1988. Rings of negatively charged amino acids determine the acetylcholine receptor channel conductance. *Nature* 335:645–648.

Langosch, D., Becker, C.M., and Betz, H., 1990. The inhibitory glycine receptor: A ligand-gated chloride channel of the central nervous system. *Eur. J. Biochem.* 194:1–8.

Seeburg, P.H., Wisden, W., Verdoorn, T.A., Pritchett, D.B., Werner, P., Herb, A., Luddens, H., Sprengel, R., and Sakmann, B., 1990. The $GABA_A$ receptor family: Molecular and functional diversity. *Cold Spring Harbor Symp. Quant. Biol.* 55:29–40.

Verdoorn, T.A., Burnashev, N., Monyer, H., Seeburg, P.H., and Sakmann, B., 1991. Structural determinants of ion flow through recombinant glutamate receptor channels. *Science* 252:1715–1718.

ATP-driven ion pumps

Pedersen, P.L., and Carafoli, E., 1987. Ion motive ATPases. *Trends Biochem. Sci.* 12:146–150 and 12:186–189. [A lucid overview of ATP-driven ion pumps, emphasizing common motifs.]

MacLennan, D.H., 1990. Molecular tools to elucidate problems in excitation-contraction coupling. *Biophys. J.* 58:1355–1365.

Carafoli, E., 1992. The Ca^{2+} pump of the plasma membrane. *J. Biol. Chem.* 267:2115–2118.

Racker, E., 1972. Reconstitution of a calcium pump with phospholipids and a purified Ca^{2+}-adenosine triphosphatase from sarcoplasmic reticulum. *J. Biol. Chem.* 247:8198–8200.

Estes, J.W., and White, P.D., 1965. William Withering and the purple foxglove. *Sci. Amer.* 212(6):110–117. [An interesting historical account of digitalis, a specific inhibitor of Na^+-K^+ ATPases.]

Kijima, Y., Ogunbunmi, E., and Fleischer, S., 1991. Drug action of thapsigargin on the Ca^{2+} pump protein of sarcoplasmic reticulum. *J. Biol. Chem.* 266:22912–22918.

Symporters and antiporters

Kaback, H.R., Bibi, E., and Roepe, P.D., 1990. β-Galactoside transport in *E. coli*: A functional dissection of *lac* permease. *Trends Biochem. Sci.* 8:309–314.

Kaback, H.R., 1990. Membrane vesicles, bioenergetics, molecules, and mechanisms, in *The Bacteria,* Vol. XII, pp. 151–202. Academic Press.

Hilgemann, D.W., Nicoll, D.A., and Philipson, K.D., 1991. Charge movement during Na^+ translocation by native and cloned cardiac Na^+/Ca^{2+} exchangers. *Nature* 352:715–718.

Hediger, M.A., Turk, E., and Wright, E.M., 1989. Homology of the human intestinal Na^+/glucose and *Escherichia coli* Na^+/proline cotransporters. *Proc. Nat. Acad. Sci.* 86:5748–5752.

Silverman, M., 1991. Structure and function of hexose transporters. *Ann. Rev. Biochem.* 60:757–794.

Kopito, R.R., 1990. Molecular biology of the anion exchanger family. *Int. Rev. Cytol.* 123:177–199.

Bacteriorhodopsin

Henderson, R., Baldwin, J.M., Ceska, T.A., Zemlin, F., Beckmann, E., and Downing, K.H., 1990. Model for the structure of bacteriorhodopsin based on high-resolution electron cryo-microscopy. *J. Mol. Biol.* 213:899–929.

Mathies, R.A., Lin, S.W., Ames, J.B., and Pollard, W.T., 1991. From femtoseconds to biology: Mechanism of bacteriorhodopsin's light-driven proton pump. *Ann. Rev. Biophys. Biophys. Chem.* 20:491–518.

Khorana, H.G., 1988. Bacteriorhodopsin, a membrane protein that uses light to translocate protons. *J. Biol. Chem.* 263:7439–7442.

Oesterhelt, D., and Tittor, J., 1989. Two pumps, one principle: Light-driven ion transport in halobacteria. *Trends Biochem. Sci.* 14:57–61.

Conformational transitions and energetics

Jencks, W.P., 1989. Utilization of binding energy and coupling rules for active transport and other coupled vectorial processes. *Meth. Enzymol.* 171:145–164.

Tanford, C., 1983. Mechanism of free energy coupling in active transport. *Ann. Rev. Biochem.* 52:379–409.

Jardetzky, O., 1966. Simple allosteric model for membrane pumps. *Nature* 211:969.

PROBLEMS

1. *Concerted opening.* Suppose that a channel obeys the concerted allosteric model (MWC model, p. 167). The binding of ligand to the R state (the open form) is 20 times as tight as to the T state (the closed form). In the absence of ligand, the ratio of closed to open channels is 10^5. If the channel is a tetramer, what is the fraction of open channels when 1, 2, 3, and 4 ligands are bound?

2. *Respiratory paralysis.* Tabun and sarin have been used as chemical warfare agents, and parathion has been employed as an insecticide. What is the molecular basis of their lethal actions?

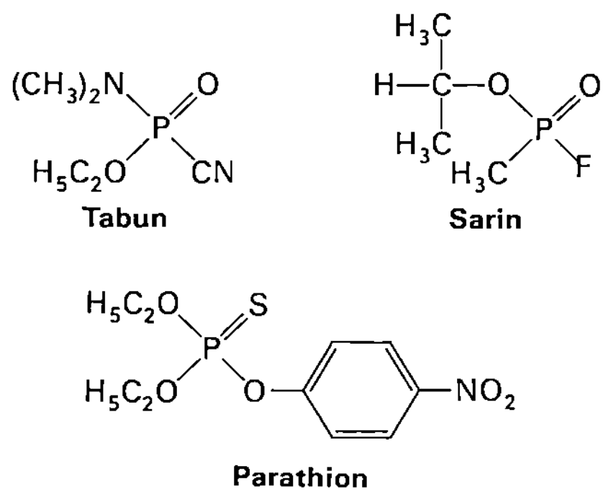

3. *Ligand-induced channel opening.* The ratio of open to closed forms of the acetylcholine receptor channel containing 0, 1, and 2 bound acetylcholines is 5×10^{-6}, 1.2×10^{-3}, and 14, respectively.
   (a) By what factor is the open/closed ratio increased by the binding of the first acetylcholine? The second acetylcholine?
   (b) What are the corresponding free-energy contributions to channel opening at 25°C?
   (c) Can the allosteric transition be accounted for by the MWC concerted model?

4. *Voltage-induced channel opening.* The fraction of open channels at 5 mV steps beginning at −45 mV and ending at +5 mV at 20°C is 0.02, 0.04, 0.09, 0.19, 0.37, 0.59, 0.78, 0.89, 0.95, 0.98, and 0.99. (a) At what voltage are half the channels open? (b) What is the value of the gating charge? (c) How much free energy is contributed by the movement of the gating charge in the transition from −45 mV to +5 mV?

5. *Frog poison.* Batrachotoxin (BTX) is a steroidal alkaloid from the skin of a poisonous Colombian frog. In the presence of BTX, sodium channels in an excised patch stay persistently open when the membrane is depolarized. They close when the membrane is repolarized. Which transition is blocked by BTX?

This poisonous Colombian frog *(Phyllobates terribilis)* contains batrachotoxin in its skin, the source of the poison used on blowgun darts. [Tom McHugh, Photo Researchers.]

6. *Valium target.* γ-Aminobutyric acid (GABA) opens channels that are specific for chloride ions. The $GABA_A$ receptor channel is pharmacologically important because it is the target of Valium, which is used to diminish anxiety.
   (a) The extracellular concentration of Cl^- is 123 mM and the intracellular concentration is 4 mM. In which direction does Cl^- flow through an open channel when the membrane potential is in the −60 mV to +30 mV range?
   (b) What is the effect of chloride channel opening on the excitability of a neuron?
   (c) The hydropathy profile of the $GABA_A$ receptor resembles that of the acetylcholine receptor. Predict the number of subunits in this chloride channel.

7. *The price of extrusion.* What is the free-energy cost of pumping Ca^{2+} out of a cell when the cytosolic concentration is 0.4 μM, the extracellular concentration is 1.5 mM, and the membrane potential is −60 mV?

8. *Rapid transit.* A channel exhibits current steps of 5 pA at a membrane potential of −50 mV. The mean open time is 1 ms, and the reversal potential is 0 mV.
   (a) What is the conductance of this channel?
   (b) How many univalent ions flow through the channel during its mean open time?
   (c) How long does it take an ion to pass through the channel?

9. *Pumping protons.* Design an experiment to show that lactose permease can be reversed in vitro to pump protons.

CHAPTER

13

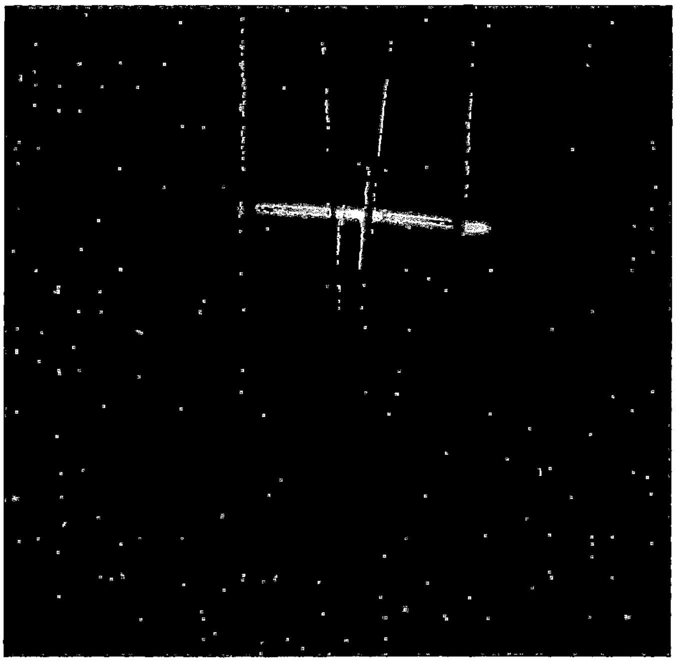

Signal Transduction Cascades

Cells are highly responsive to signals from their environment. The movement of bacteria toward nutrients, the detection of light by retinal cells, the release of fuel molecules by hormones signaling hunger, and the induction of differentiation by growth factors exemplify this fundamental property. The sensing and processing of stimuli are mediated by *signal transduction cascades.* These molecular circuits in cells are constructed from *receptors, enzymes, channels,* and *regulatory proteins.* They detect, amplify, and integrate diverse external signals.

We begin with bacterial chemotaxis, the movement of bacterial cells toward attractants and away from repellents. Bacteria, like people, perpetually seek greener pastures. A sensory system sends signals to flagellar motors that determine whether a bacterium swims smoothly in a straight line or takes a tumble to change its course. Adaptation automatically resets the sensitivity to enable a bacterium to respond to small changes in concentration rather than to absolute values. We then turn to visual excitation in vertebrates, an exquisitely sensitive process—a rod cell in the retina can be excited by a single photon. Light triggers a highly amplified enzymatic cascade that culminates in the closure of membrane channels. Several themes introduced here recur in the action of many hormones. The effect of epinephrine, like that of light, is mediated by a seven-helix receptor and a G protein. The consequent activation of adenylate cyclase leads to the synthesis of cyclic AMP, which in turn stimulates a protein kinase that phosphorylates multiple targets. Cyclic AMP is a key cytosolic messenger in many physiologic processes.

Opening Image: *A retinal rod cell can be triggered by a single photon. This micrograph shows a piece of retina from which the outer segment of a rod cell has been drawn into a suction pipette. The rod cell is excited by light (green) shining through a slit. [Courtesy of Dr. Denis Baylor.]*

Our focus will then shift to calcium ion, another ubiquitous messenger in eukaryotes. Stimulation of many cells leads to a rise in the cytosolic calcium level, which is detected by calmodulin and other calcium sensors that control enzymes and channels. The final topic of this chapter is the receptor tyrosine kinases. Insulin, epidermal growth factor, and other growth factors bind to the extracellular domain of these transmembrane enzymes and trigger their autophosphorylation and activation. Receptor tyrosine kinases are central to the control of growth and differentiation. Indeed, mutations of these receptors or their partners often lead to cancer.

CHEMORECEPTORS ON BACTERIA DETECT ATTRACTANTS AND REPELLENTS AND SEND SIGNALS TO FLAGELLA

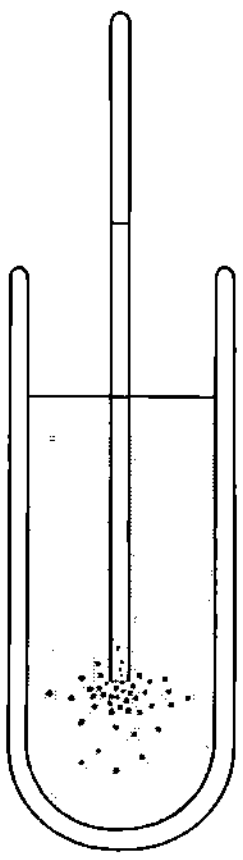

Figure 13-1
Chemotaxis of bacteria to a capillary containing an attractant such as glucose.

In the late nineteenth century, Wilhelm Pfeffer, a German botanist, showed that motile bacteria cluster near the mouth of a thin capillary tube containing an attractant, such as a sugar (Figure 13-1). In contrast, they move away from the tube if it contains a repellent, such as phenol, a harmful substance. This directed movement of bacteria toward some specific substances and away from others is called *chemotaxis.* In the 1960s, Julius Adler began to investigate the molecular basis of bacterial chemotaxis. Biochemical, genetic, and structural analyses carried out by him and many other investigators have revealed much about this process. Chemotaxis begins with the binding of attractants and repellents to receptor proteins called *chemoreceptors* that span the plasma membrane. Some of these molecules, such as aspartate, bind directly to chemoreceptors, whereas others are carried by binding proteins. Information from chemoreceptors is transmitted to a *central processing system* that analyzes and integrates many stimuli. This sensory processing system then sends a signal to the flagellar motors that determines whether a bacterium swims smoothly in a straight line or abruptly alters its course by tumbling.

THE DIRECTION OF FLAGELLAR ROTATION DETERMINES WHETHER BACTERIA SWIM SMOOTHLY OR TUMBLE

Bacteria swim by rotating flagella that emerge from their surfaces (Figure 13-2). An *Escherichia coli* or *Salmonella typhimurium* bacterium has about six flagella. These thin helical filaments (15 nm in diameter and 10 μm long) are built from *flagellin* subunits. A bacterial flagellum is an extracellular appendage that is rotated by a motor located at the junction of the flagellum and cell envelope (p. 411). It is powered by the proton-motive force across the plasma membrane. An intriguing feature of this motor is that it can rotate clockwise or counterclockwise, and switch between them almost instantaneously.

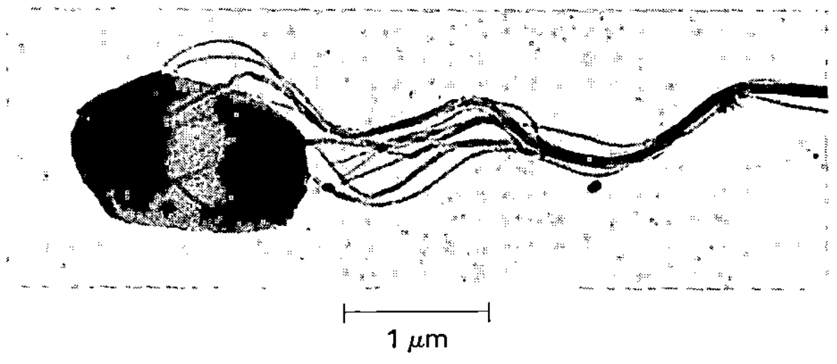

Figure 13-2
Electron micrograph of *S. typhimurium* showing flagella in a bundle.
[Courtesy of Dr. Daniel Koshland, Jr.]

An *E. coli* bacterium typically swims smoothly in nearly a straight line for about a second (Figure 13-3). In this interval, it travels a distance of about 30 μm, which is equal to about 15 body lengths. The bacterium then abruptly alters its course by tumbling (Figure 13-4). The average change in direction is about 60 degrees. What determines whether a bacterium swims smoothly or tumbles? When the flagella rotate counterclockwise, the helical filaments form a coherent bundle and the cell swims smoothly. When the flagella rotate clockwise, in contrast, the bundle flies apart because the screw sense of the helical flagella does not match the direction of rotation. Each flagellum then pulls in a different direction, and so the cell tumbles.

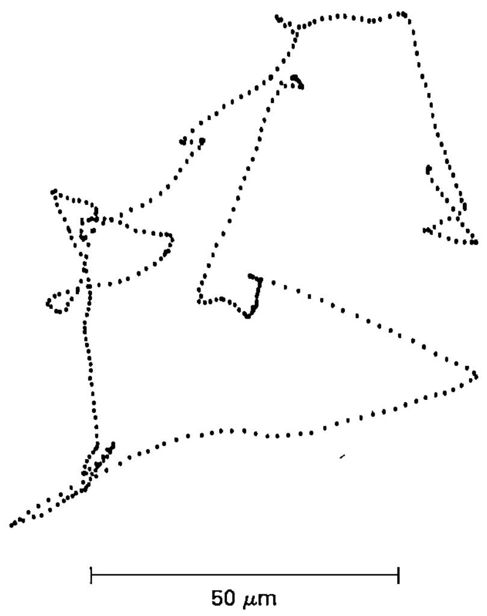

Figure 13-3
A projection of the track of an *E. coli* bacterium obtained with a microscope that automatically follows its motion in three dimensions. The points show the locations of the bacterium at 80-ms intervals. [From H.C. Berg. *Nature* 254(1975):390.]

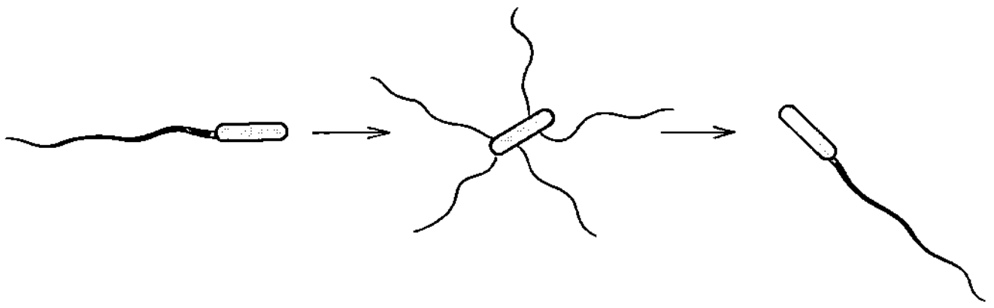

Figure 13-4
Tumbling is caused by an abrupt reversal of the flagellar motor, which disperses the flagellar bundle. A second reversal of the motor restores smooth swimming, but (most likely) in a different direction. [After a drawing kindly provided by Dr. Daniel Koshland, Jr.]

BACTERIA DETECT TEMPORAL GRADIENTS RATHER THAN INSTANTANEOUS SPATIAL GRADIENTS

When a bacterium moves toward an increasing concentration of an attractant, tumbling becomes less frequent. In contrast, when it moves away from an attractant, tumbling becomes more frequent. Repellents have opposite effects on the tumbling frequency. Thus, if a bacterium is moving toward an attractant or away from a repellent it will swim smoothly for a longer time than if it is moving in the opposite direction. Tumbling serves to reorient a misdirected bacterium. *The regulation of tumbling frequency is central to chemotaxis.*

Does a bacterium compare the concentration of an attractant at one end of the cell with that at the other end or does it compare the concentration of attractant now with that of a few moments ago? In other words, is the sensing mechanism spatial or temporal? Daniel Koshland, Jr., and Robert Macnab answered this important question by carrying out an ingeniously simple experiment. They rapidly mixed a suspension of bacteria in a medium devoid of an attractant with a solution containing an attractant, and they then observed the tumbling frequency of the bacteria. The striking finding was that tumbling was suppressed within a second after mixing. The bacteria swam for long distances in a straight line in this rapidly mixed solution, which was devoid of a spatial gradient. Hence, this experiment demonstrated that *bacteria possess a temporal sensing mechanism.* In other words, a bacterium detects a spatial gradient of attractant not by comparing the concentration at its head and tail, but by *traveling through space and comparing its observations through time.*

A bacterium decides whether or not to tumble by comparing the concentrations of attractants and repellents sensed in the past second with those encountered in the previous three seconds. If the concentration of a repellent increases or if that of an attractant decreases, tumbling is enhanced. In essence, *bacterial chemotaxis is a biased random walk:* highly purposeful behavior arises from the *selection* of random movements.

FOUR KINDS OF CHEMORECEPTORS TRANSMIT SIGNALS ACROSS THE PLASMA MEMBRANE

Chemotaxis begins with the binding of molecules to four kinds of receptors in the plasma membrane (Figure 13-5). In *E. coli* and other gram-negative bacteria, the plasma membrane is surrounded by an outer membrane that is permeable to small molecules; a periplasmic space separates the plasma membrane from the outer membrane. The chemoreceptors are encoded by the *tsr, tar, trg,* and *tap* genes and were originally called *methyl-accepting chemotaxis proteins (MCPs)* because they are reversibly methylated. Repellents and some attractants bind directly to these receptors, whereas other attractants bind to periplasmic proteins, which then interact with the receptors. For example, the attractant aspartate binds directly to the tar protein, which is also known as the *aspartate receptor.*

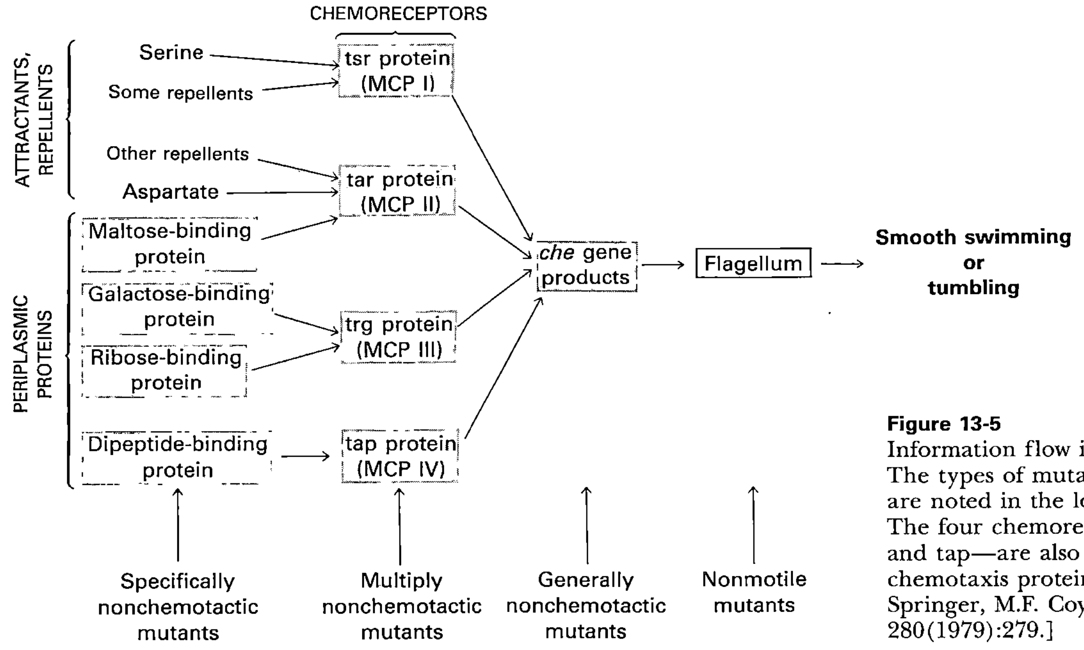

Figure 13-5
Information flow in bacterial chemotaxis. The types of mutants that have been isolated are noted in the lower part of the diagram. The four chemoreceptors (blue)—tsr, tar, trg, and tap—are also known as methyl-accepting chemotaxis proteins (MCPs). [After M.S. Springer, M.F. Coy, and J. Adler. *Nature* 280(1979):279.]

The four chemoreceptors have the same design. Each ~60-kd protein contains four regions: (1) a periplasmic domain that binds an attractant or a repellent, (2) a transmembrane segment consisting of two α helices, (3) a cytosolic region that interacts with components of the central processing system, and (4) a cytosolic region that can be reversibly methylated at several sites (Figure 13-6). Their cytosolic domains are highly conserved because they interact with the same components of the processing system. In fact, the cytosolic domains of tsr and tar contain an identical 48-residue sequence. In contrast, their periplasmic domains are distinctive because they bind different attractants, repellents, and attractant-protein complexes. A chimeric receptor formed from the amino-

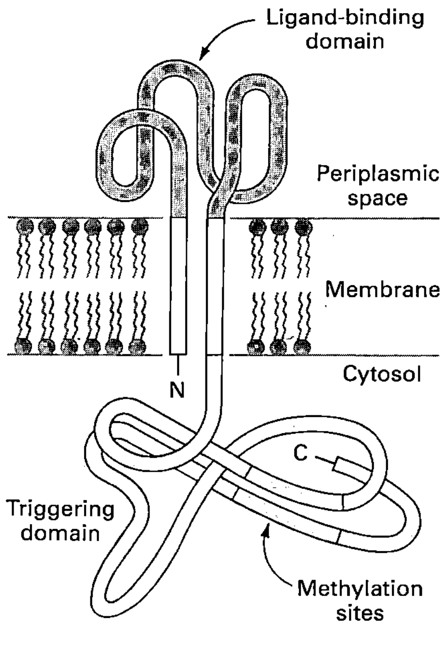

Figure 13-6
Schematic diagram of a chemoreceptor. Chemotactic signals are transmitted across the plasma membrane by these transmembrane proteins. Repellents, attractants, and transport proteins bind to their periplasmic domain. Their cytosolic domain communicates with a central processing system that controls the direction of flagellar rotation. Reversible methylation of their cytosolic domain is central to adaptation. [After D.E. Koshland, Jr., A.F. Russo, and N.I. Gutterson. *Cold Spring Harbor Symp. Quant. Biol.* 48(1983):805.]

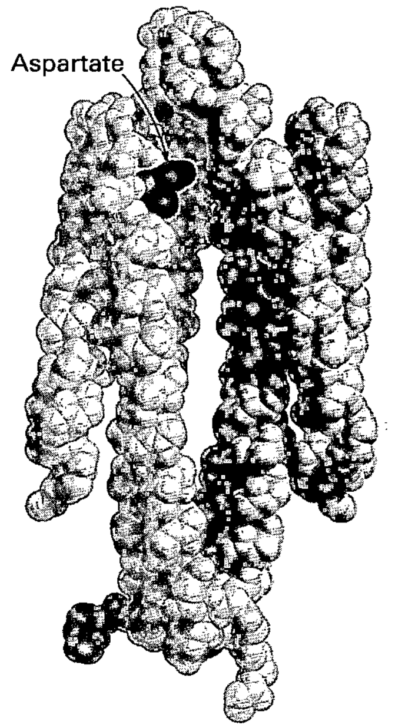

Figure 13-7
Three-dimensional structure of the ligand-binding domain of tar, the aspartate receptor in chemotaxis. One of the chains of the dimer is shown in blue, and the other in yellow. Each forms a four-helix bundle. A single aspartate (red) is bound near the twofold axis of the dimer. The lowest part of the dimer is located near the periplasmic surface of the membrane. [Drawn from coordinates kindly provided by Dr. Sung-Hou Kim and Dr. Joanne Yeh. M.V. Milburn, G.G. Privé, D.L. Milligan, W.G. Scott, J. Yeh, J. Jancarik, D.E. Koshland, Jr., and S.-H. Kim. *Science* 254(1991):1342.]

terminal half of tar and the carboxyl-terminal half of tsr is triggered by aspartate and other attractants that are normally sensed by tar. Thus, these chemoreceptors transmit signals across the membrane in essentially the same way.

The three-dimensional structure of the periplasmic domain of the aspartate receptor has recently been determined. This ligand-binding domain dimerizes to form a highly elongated structure (Figure 13-7). Each subunit is a *four-helix bundle* that is 20 Å in diameter and more than 70 Å long. The binding sites for aspartate are at the interface between subunits, more than 60 Å from the membrane surface. It will be interesting to see the other portions of this receptor and learn how the binding of aspartate changes signaling by the cytosolic region.

A RECEPTOR-TRIGGERED PHOSPHORYLATION CASCADE CONTROLS THE DIRECTION OF FLAGELLAR ROTATION

How do receptors bias the direction of flagellar rotation? Genetic analyses revealed the existence of a common processing system that is encoded by *che* genes and provided valuable clues concerning its molecular circuitry. Eight *che* loci have been identified: *A, B, C, D, W, R, Y,* and *Z.* Mutations in these genes impair chemotaxis to all attractants and repellents, in contrast with mutations in a particular kind of receptor, which block responses to a subset of these agents (see Figure 13-5). Bacteria devoid of all cytosolic Che proteins (known as "gutted" strains) swim smoothly all the time because their flagella rotate counterclockwise (CCW) continuously. The addition of CheY protein to a gutted strain by controlled expression of a CheY gene leads to increased clockwise (CW) rotation. Furthermore, flagella rotate exclusively CW in mutants that express abnormally high levels of CheY protein in the presence of normal amounts of the other Che proteins. These findings indicated that *the direction of flagellar rotation is controlled by the CheY protein.*

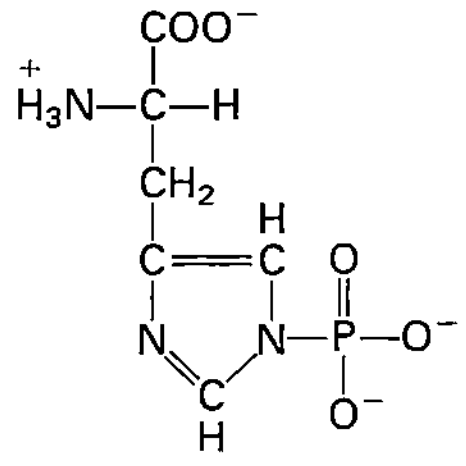

N-3-Phosphohistidine

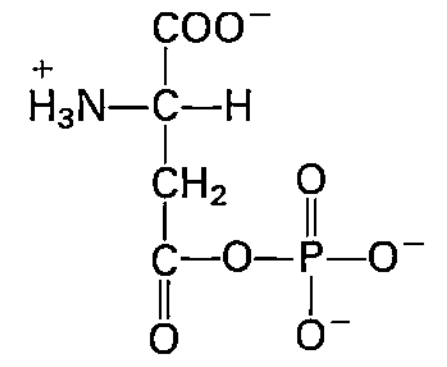

β-Aspartyl phosphate

The essentiality of ATP for both the excitation and the adaptation phases of chemotaxis (p. 331) provided a tantalizing hint as to how CheY might be controlled. A specific aspartate of CheY can become phosphory-

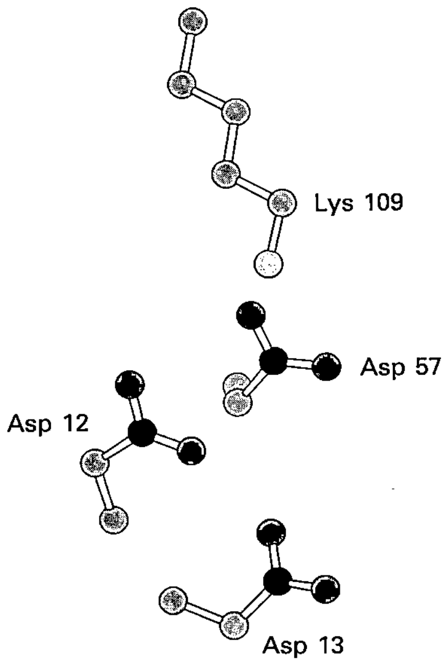

Figure 13-8
The acid pocket of CheY contains a cluster of three aspartates. In the dephosphorylated form shown here, Asp 57 interacts with Lys 109. Phosphorylation of Asp 57 alters this interaction to produce propagated conformational changes that make P–CheY an inducer of tumbling. [Drawn from 3chy.pdb. K. Volz and P. Matsumura. *J. Biol. Chem.* 266(1991):15511.]

lated (Figure 13-8). Most significant, *phosphorylated CheY induces tumbling, whereas dephosphorylated CheY has virtually no effect on the flagellar switch.* Phosphorylated CheY is a transient species. The attached phosphoryl group is spontaneously hydrolyzed in a reaction that is increased 100-fold by CheZ.

How does CheY acquire a phosphate group and how is its phosphorylation controlled by chemoreceptors? Mel Simon made a breakthrough when he found that *purified CheA protein undergoes a slow autophosphorylation.* The site of phosphorylation is N-3 of a histidine residue. This autophosphorylation is greatly accelerated by the binding of CheA, together with CheW, to an unoccupied receptor or a repellent-receptor complex but not to an attractant-receptor complex (Figure 13-9). The phosphoryl group attached to CheA can then be transferred to CheY. Hence, *repellents increase the level of phosphorylated CheA and, in turn, of phosphorylated CheY. Attractants have the opposite effect.* In this way, repellents increase CW rotation and tumbling, whereas attractants increase CCW rotation and smooth swimming.

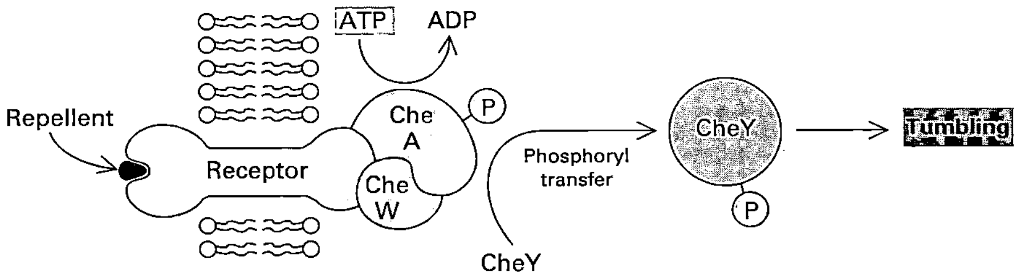

Figure 13-9
Flow of phosphoryl groups in chemotaxis. A repellent bound to a chemoreceptor stimulates the autophosphorylation of CheA. The phosphoryl group of CheA is then transferred to CheY. Phosphorylated CheY induces tumbling.

It is interesting to note that CheA belongs to a *family of regulatory proteins that enable bacteria to sense their environment and internal metabolic state.* For example, NtrB, which is a homolog of CheA, controls the expression of genes that respond to the nitrogen balance of the cell. All homologs of CheA are *autophosphorylating kinases* that regulate the activity of target proteins by phosphorylating them. These sensor-effector pairs are called *two-component regulatory systems.*

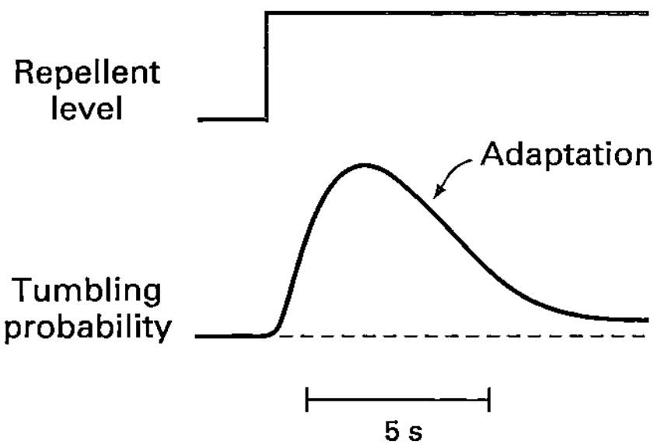

Figure 13-10
Adaptation in chemotaxis. A step increase in the concentration of a repellent leads to a rapid increase in the probability of CW rotation and, hence, tumbling. Within a few seconds, the probability of CW rotation returns to the basal state.

ADAPTATION TO CHEMOTACTIC STIMULI IS MEDIATED BY REVERSIBLE METHYLATION OF CHEMORECEPTORS

Consider the response of a bacterium to a sudden increase in the concentration of a repellent. The level of phosphorylated CheY (P–CheY) rapidly increases, which induces CW rotation and tumbling. After a few seconds, however, the probability of CW rotation returns to the original level (Figure 13-10). The resetting of sensitivity is called *adaptation.* How is it achieved in chemotaxis? A clue came from the finding that adaptation is severely impaired in mutants lacking both the CheR and the CheB proteins. *CheR* encodes a *methyltransferase* and *CheB* encodes a *methylesterase.* Four glutamate side chains on each chemoreceptor can be methylated by

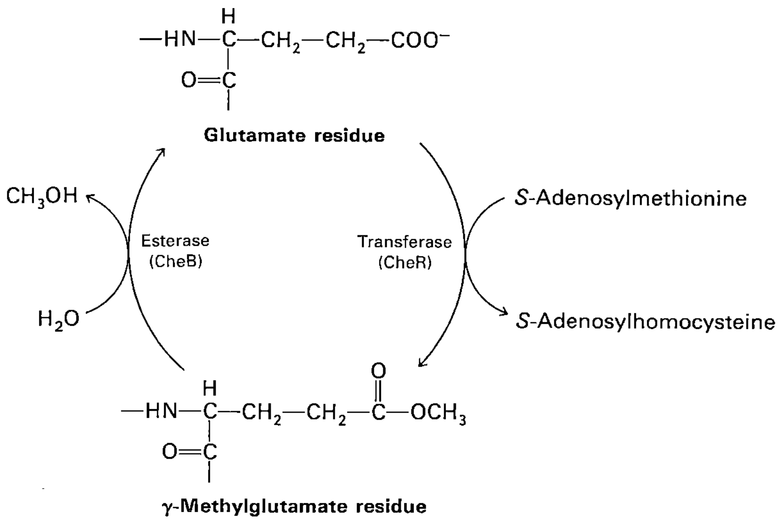

Figure 13-11
Reversible methylation of glutamate residues in the cytosolic domain of chemoreceptors.

CheB using *S*-adenosylmethionine (p. 721) as the methyl donor (Figure 13-11). These γ-methyl groups are subsequently removed by the hydrolytic action of CheB. As will be evident shortly, *reversible methylation is central to adaptation.*

Methylation mediates adaptation through a regulatory circuit based on four interactions:

1. The binding of an attractant to a receptor promotes methylation, whereas the binding of a repellent inhibits methylation. In the absence of attractant or repellent, a receptor contains, on average, two methyl groups.

2. Methylation increases the potency of a receptor in triggering the autophosphorylation of CheA and the consequent phosphorylation of CheY.

3. CheB as well as CheY is phosphorylated by P–CheA. The site of phosphorylation is again an aspartate side chain.

4. The methylesterase activity of CheB is markedly increased by phosphorylation.

Now let us see how these links between methylation and phosphorylation lead to adaptation. The addition of a repellent leads to a transiently higher level of P–CheA and P–CheY, which induces CW rotation and tumbling (Figure 13-12). On a slower time scale, CheB becomes phosphorylated. The consequent increase in methylesterase activity and occupancy of the receptor by repellent lead to a decrease in the degree of

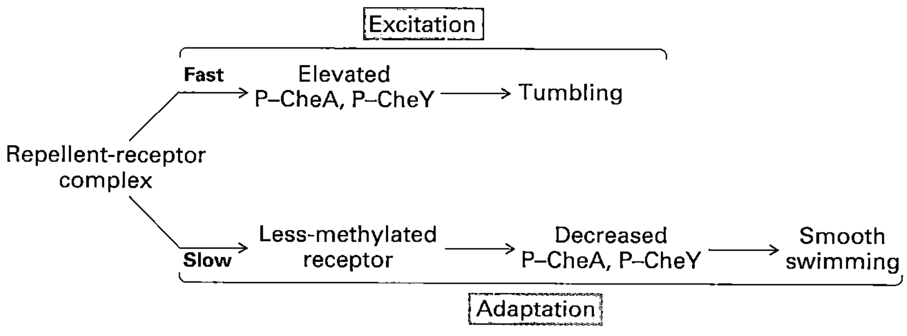

Figure 13-12
Excitation and adaptation in chemotaxis are coupled. An increase in the concentration of repellent leads to increased autophosphorylation of CheA. CheY is rapidly phosphorylated to generate an excitatory signal (CW rotation, tumbling), and CheB is slowly phosphorylated to generate a negative-feedback signal that underlies adaptation.

methylation of the receptor. Because less P–CheA and less P–CheY are then formed, the probability of CW rotation decreases and the bacterium tumbles less. The response to a step increase of an attractant is the reverse of that for a repellent. In both cases, *methylation serves as a negative feedback signal that brings the sensory system back to the unstimulated state.*

The chemoreceptors themselves are impressive miniature information-processing devices. Their output depends on (1) whether repellent is bound, (2) whether attractant is bound, and (3) whether 0, 1, 2, 3, or 4 methyl groups are attached. A receptor can be in one of 20 states (2 × 2 × 5) that differ in their triggering of phosphorylation and their susceptibility to methylation. Another noteworthy feature of this sensory system is that excitation is automatically followed by adaptation because phosphorylation and methylation are linked. *Bacterial chemotaxis graphically shows that highly purposeful behavior can arise from the biasing of random events.*

A RETINAL ROD CELL CAN BE EXCITED BY A SINGLE PHOTON

We now turn to a transduction cascade in higher organisms that *converts light into atomic motion and then into a nerve signal.* Vertebrates have two kinds of photoreceptor cells, called *rods* and *cones* because of their distinctive shapes. *Cones* function in bright light and are responsible for color vision, whereas *rods* function in dim light but do not perceive color. A

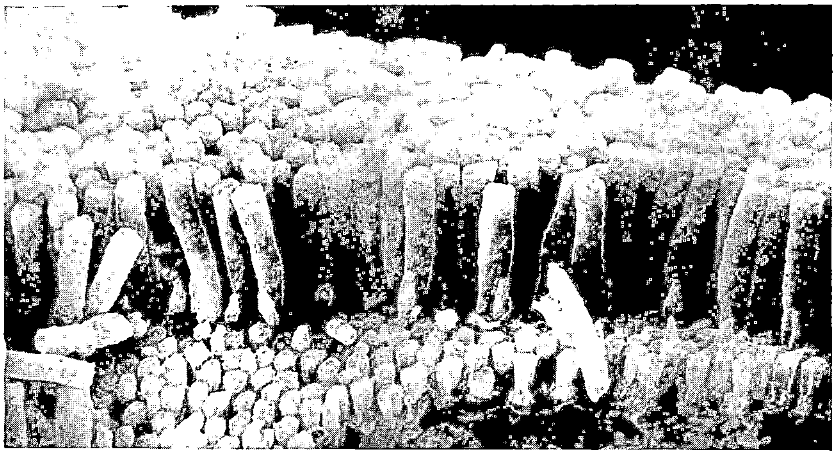

Figure 13-13
Scanning electron micrograph of retinal rod cells. [Courtesy of Dr. Deric Bownds.]

human retina contains about three million cones and a hundred million rods. Rods and cones form synapses with bipolar cells, which in turn interact with other nerve cells in the retina. The electrical signals generated by the photoreceptors are processed by an intricate array of nerve cells within the retina and then transmitted to the brain by the fibers of the optic nerve. Thus, the retina has a dual function: to transform light into nerve impulses and to integrate visual information.

In 1938, Selig Hecht discovered through psychophysical studies that *a human rod cell can be excited by a single photon.* Let us explore the molecular basis of this exquisite sensitivity. Rods are slender, elongated structures; in humans, they have a diameter of 1 μm and a length of 40 μm. The major functions of a rod cell are highly compartmentalized (Figure 13-14). *The outer segment of a rod is specialized for photoreception.* It contains a stack of about 1000 *discs,* which are closed, flattened sacs about 16 nm

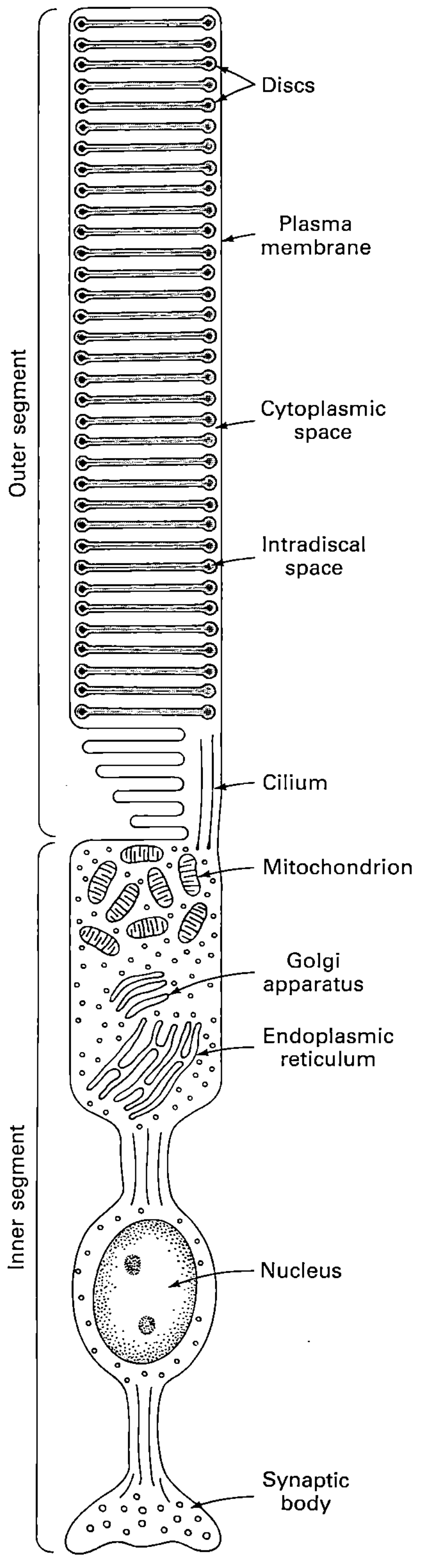

Figure 13-14
Schematic diagram of a retinal rod cell.

thick. These membranous structures are densely packed with photoreceptor proteins. The disc membranes are separate from the plasma membrane of the outer segment. A slender, immotile cilium joins the outer segment and the *inner segment,* which is rich in mitochondria and ribosomes. The inner segment generates ATP at a very rapid rate to power the transduction process and actively synthesizes proteins. The discs in the outer segment have a life of a month and are continually renewed. The inner segment also contains the nucleus, which is next to the *synaptic body.* Many transmitter-rich vesicles are present in this synaptic region.

The plasma membrane of a rod cell contains cation-specific channels that are open in the dark. Sodium ions rapidly flow into the outer segment in the dark because the channels are highly permeable to Na^+ and the electrochemical gradient is large. This gradient is maintained by Na^+-K^+ ATPase pumps located in the inner segment. *Light blocks these cation-specific channels in the outer segment.* Consequently, the influx of Na^+ decreases, and the plasma membrane becomes hyperpolarized—more negative on the inside. This *light-induced hyperpolarization* (Figure 13-15) is then transmitted by the plasma membrane from the outer segment to the synaptic body. *A single photon absorbed by a dark-adapted rod closes hundreds of cation-specific channels and leads to a hyperpolarization of about 1 mV,* which is sensed by the synapse and conveyed to other neurons of the retina.

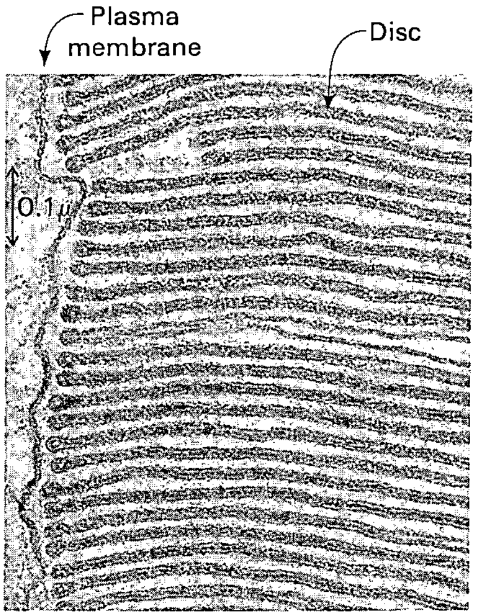

Electron micrograph of a rod outer segment, showing the stacks of discs. [From J.E. Dowling. The organization of vertebrate visual receptors. In *Molecular Organization and Biological Function.* J.M. Allen, ed. (Harper & Row, 1967).]

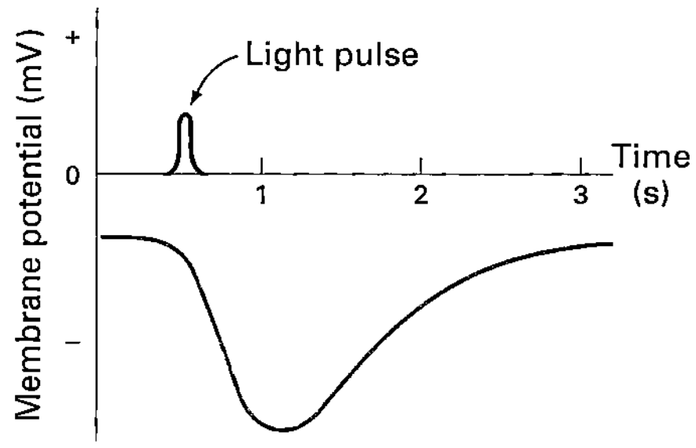

Figure 13-15
Light hyperpolarizes the plasma membrane of a retinal rod cell.

RHODOPSIN, THE PHOTORECEPTOR PROTEIN OF RODS, IS A MEMBER OF THE SEVEN-HELIX RECEPTOR FAMILY

How does light lead to the closure of membrane channels and the consequent hyperpolarization? Light must be absorbed to stimulate a photoreceptor cell. In addition, the light-absorbing group (called a *chromophore*) must undergo a conformational change after it has absorbed a photon. The photosensitive molecule in the discs of rod cells is *rhodopsin,* which consists of *opsin,* a protein, and *11-*cis*-retinal,* a prosthetic group (Figure 13-16). The precursor of 11-*cis*-retinal is *all-*trans*-retinol* (vitamin A), which cannot be synthesized de novo by mammals. A deficiency of vitamin A leads to *night blindness* and eventually to the deterioration of the

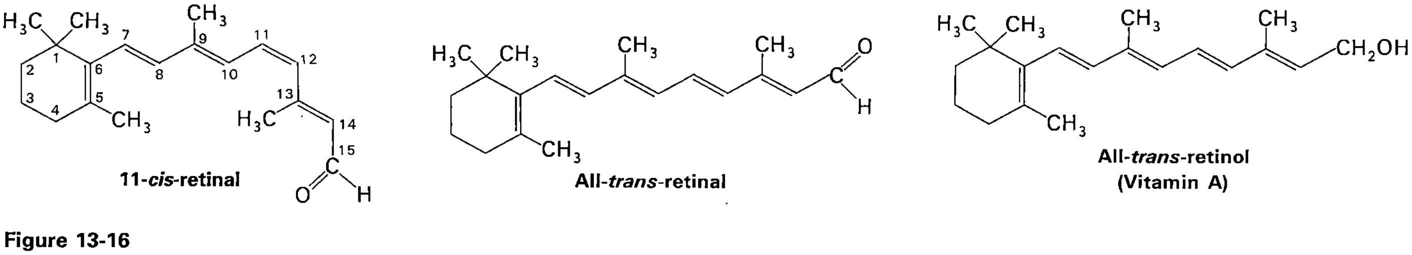

Figure 13-16
Structure of 11-*cis*-retinal, all-*trans*-retinal, and all-*trans*-retinol (vitamin A).

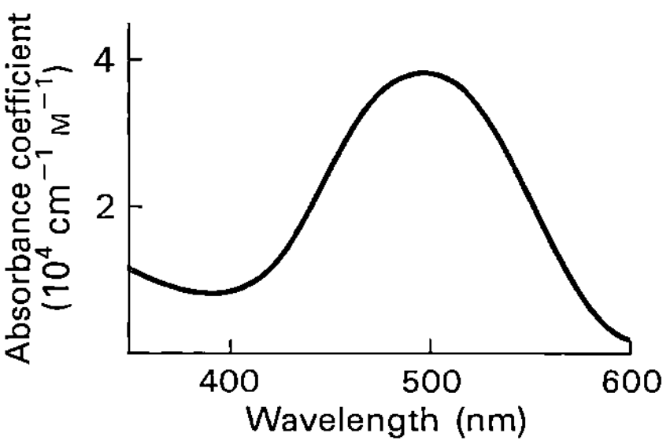

Figure 13-17
Absorption spectrum of rhodopsin.

outer segments of rods. All-*trans*-retinol (see Figure 13-16) is converted into 11-*cis*-retinal in several steps.

The color of rhodopsin and its responsiveness to light depend on the presence of 11-*cis*-retinal. This very effective chromophore gives rhodopsin a broad absorption band in the visible region of the spectrum with a peak at 500 nm, which nicely matches the solar output. The intensity of the visible absorption band of rhodopsin is also noteworthy. The integrated absorption strength of the visible absorption band of rhodopsin (Figure 13-17) approaches the maximum value attainable by organic compounds. 11-*cis*-Retinal has these favorable chromophoric properties because it is a *polyene*. Its six alternating single and double bonds constitute a long, unsaturated electron network. Recall that retinal also serves as the chromophore in bacteriorhodopsin, a light-driven proton pump (p. 318).

11-*cis*-Retinal is covalently attached to the protein by a *Schiff base linkage.* The aldehyde group of 11-*cis*-retinal is linked to the ε-amino group of a lysine residue (Lys 296). An unprotonated Schiff base would absorb maximally at 380 nm, compared with 440 nm or a longer wavelength for a protonated one. The 500-nm absorption maximum (λ_{max}) of rhodopsin clearly shows that its Schiff base is protonated and that additional interactions shift λ_{max} to the red.

$$R-C\overset{O}{\underset{H}{\big\langle}} + H_2N-(CH_2)_4-Opsin \overset{H^+}{\rightleftharpoons} R-C\overset{H}{=}\overset{+}{N}\underset{H}{-}(CH_2)_4-Opsin + H_2O$$

11-*cis*-Retinal **Lysine side chain** **Protonated Schiff base**

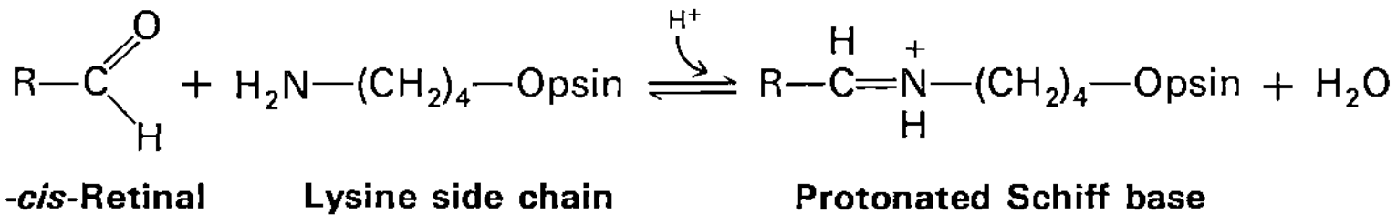

Rhodopsin, a 40-kd membrane protein, contains seven transmembrane α helices (Figure 13-18). Its amino terminus is on the intradiscal side of the membrane, and its carboxyl terminus is on the cytosolic side. 11-*cis*-Retinal lies in a pocket of the protein, near the center of the membrane, with its long axis nearly parallel to the plane of the membrane. As will be discussed shortly, the cytosolic side contains a region that transmits the

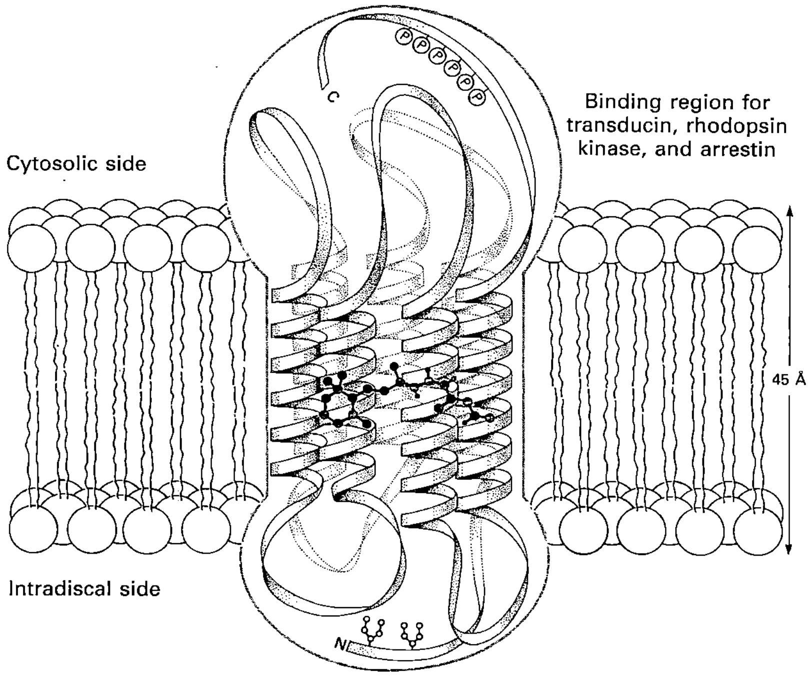

Figure 13-18
Model of rhodopsin. The seven-helix motif of this transmembrane receptor occurs in other sensory and hormonal receptors. The 11-*cis*-retinal chromophore is located near the center of the bilayer. The amino-terminal segment, located on the intradiscal side of the membrane, contains two *N*-linked oligosaccharide units. The cytosolic domain transmits the excitation signal to transducin. Phosphorylation of multiple serines and threonines (marked P) in the carboxyl-terminal tail deactivates photoexcited rhodopsin. [After E.A. Dratz and P.A. Hargrave. *Trends Biochem. Sci.* 8(1983):128 and L. Stryer. *Sci. Amer.* 255(7):42.]

Cytosolic side

Binding region for transducin, rhodopsin kinase, and arrestin

45 Å

Intradiscal side

excitation signal to an enzymatic cascade. The excited receptor is deactivated by the phosphorylation of serine and threonine residues in the adjacent carboxyl-terminal tail. *This seven-helix motif recurs in many eukaryotic membrane receptors from yeast to humans.*

VISUAL EXCITATION IS TRIGGERED BY PHOTOISOMERIZATION OF 11-*cis*-RETINAL

In 1958, George Wald and his co-workers discovered that *light isomerizes the 11-*cis*-retinal group of rhodopsin to all-*trans*-retinal.* This isomerization, the initial event in visual excitation, markedly alters the geometry of retinal (Figure 13-19). The Schiff base linkage of retinal moves approximately 5 Å in relation to the ring portion of the chromophore. In essence, *a photon has been converted into atomic motion.*

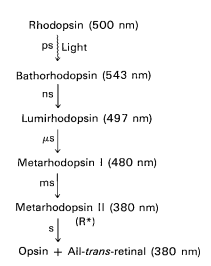

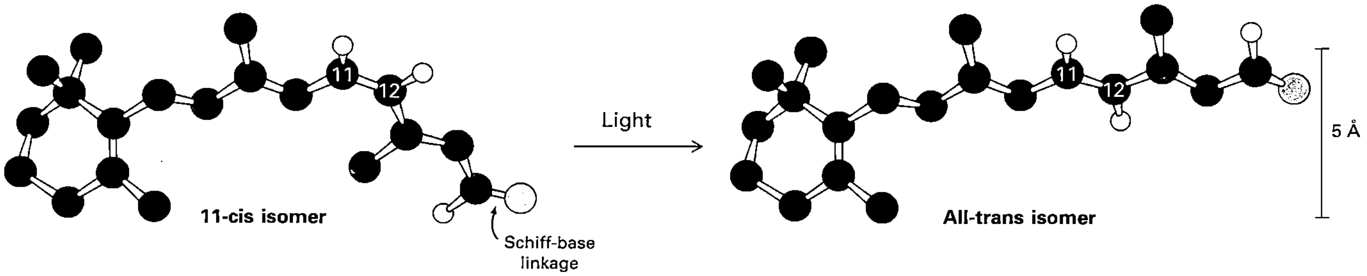

Figure 13-19
The primary event in visual excitation is the isomerization of the 11-cis isomer of the Schiff base of retinal to the all-trans form. The double bond between C11 and C12 is shown in green; hydrogen atoms attached to these carbons are shown in yellow.

Much of the isomerization of retinal takes place within a few picoseconds of the absorption of a photon, as shown by the appearance of a new absorption band following an intense laser pulse. This photolytic intermediate, called *bathorhodopsin,* contains a strained all-trans form of the chromophore. Both retinal and the protein continue to change their conformations, as reflected in the formation of a series of transient intermediates with distinctive spectral properties (Figure 13-20). The Schiff base linkage becomes deprotonated in the transition from metarhodopsin I to II, which takes about a millisecond. Metarhodopsin II, called *photoexcited rhodopsin,* triggers an enzymatic cascade, as will be discussed shortly. The unprotonated Schiff base in metarhodopsin II is hydrolyzed in about a minute to yield opsin and all-*trans*-retinal, which diffuses away from the protein because it does not fit into the binding site for the 11-cis isomer. All-*trans*-retinal is reduced to all-*trans*-retinol, which is then oxidized and isomerized in the dark to 11-*cis*-retinal. Finally, rhodopsin is regenerated by the binding of 11-*cis*-retinal and the formation of a protonated Schiff base linkage.

Rhodopsin (500 nm)
↓ ps, Light
Bathorhodopsin (543 nm)
↓ ns
Lumirhodopsin (497 nm)
↓ μs
Metarhodopsin I (480 nm)
↓ ms
Metarhodopsin II (380 nm)
↓ s (R*)
Opsin + All-*trans*-retinal (380 nm)

Figure 13-20
Intermediates in the photolysis of rhodopsin. The wavelength of the absorption maximum of each species and the time constant of each transition are given. R* denotes photoexcited rhodopsin.

PHOTOEXCITED RHODOPSIN ACTIVATES A G PROTEIN CASCADE LEADING TO CYCLIC GMP HYDROLYSIS

The closure of cation-specific channels and the consequent hyperpolarization are highly amplified responses of the outer segment. The flow of more than a *million* sodium ions is blocked by the absorption of a *single*

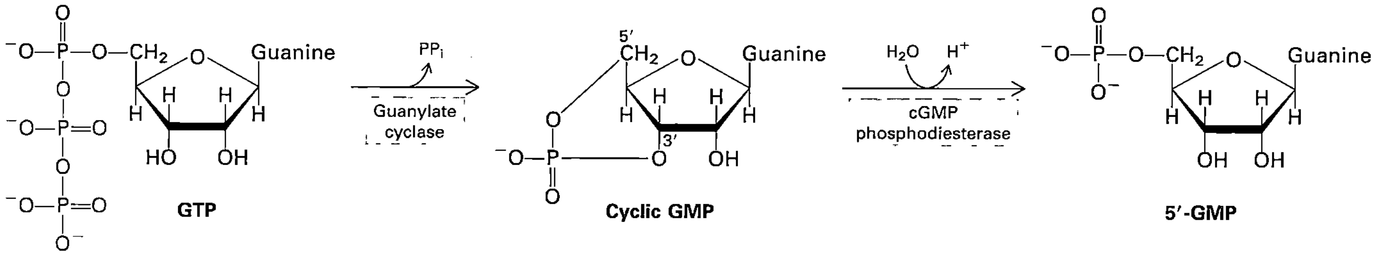

Figure 13-21
Cyclic GMP, the excitatory messenger in vision, is formed by guanylate cyclase and degraded by a specific phosphodiesterase.

photon by a dark-adapted rod. What is the mechanism of this remarkable amplification? In the dark, cation-specific channels in the plasma membrane are kept open by *cyclic GMP,* a cyclic nucleotide derived from GTP (Figure 13-21). *Photoexcited rhodopsin triggers an enzymatic cascade resulting in the hydrolysis of cyclic GMP.* In outline, the flow of information in visual excitation is from *photoexcited rhodopsin (R*)* to *transducin (T_α–GTP)* to a *phosphodiesterase* (PDE*) that hydrolyzes cyclic GMP (cGMP). The light-induced drop in the concentration of cGMP then closes channels.

$$R \xrightarrow{\text{Light}} R^* \longrightarrow T_\alpha\text{-GTP} \longrightarrow PDE^* \left\langle \begin{array}{l} cGMP \longrightarrow \boxed{\text{Open channels}} \\ \\ 5'\text{-GMP} \qquad \boxed{\text{Closed channels}} \end{array} \right.$$

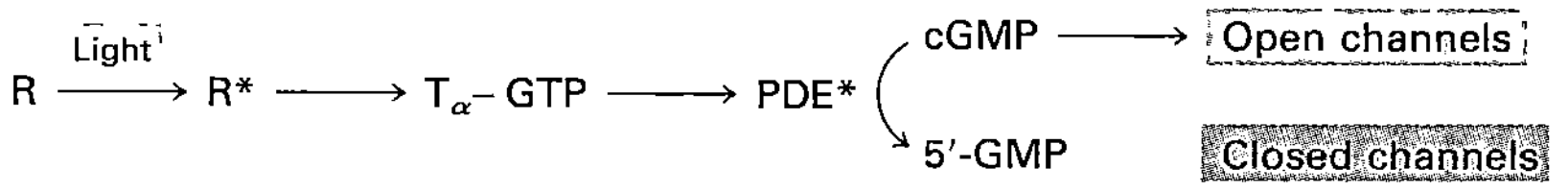

Transducin, the signal-coupling protein in visual excitation, interconverts between an inactive GDP state and an active GTP state. It consists of α (39 kd), β (36 kd), and γ (8 kd) subunits; the guanyl nucleotide binding site is on the α subunit. In the dark, transducin is in the GDP form, which is inactive. *Photoexcited rhodopsin activates transducin by forming a complex with it and catalyzing the exchange of GTP for bound GDP* (Figure 13-22). The binding of GTP to transducin leads to the release of R*, which enables it to catalyze the activation of another molecule of transducin. *A single R* catalyzes the activation of 500 molecules of transducin, the first stage of amplification in vision.*

The binding of GTP also leads to the dissociation of T_α–GTP from $T_{\beta\gamma}$. T_α–GTP, the active form of transducin, then switches on the phosphodies-

Figure 13-22
Light-activated transducin cycle in visual excitation. [After L. Stryer, J.B. Hurley, and B.K.-K. Fung. *Trends Biochem. Sci.* 6(1981):245.]

T: transducin
R: rhodopsin
R*: photoexcited rhodopsin
R*-P: multiply phosphorylated photoexcited rhodopsin
PDE$_i$: inhibited form of the phosphodiesterase specific for cGMP
PDE*: activated form of the phosphodiesterase
K: rhodopsin kinase
A: arrestin

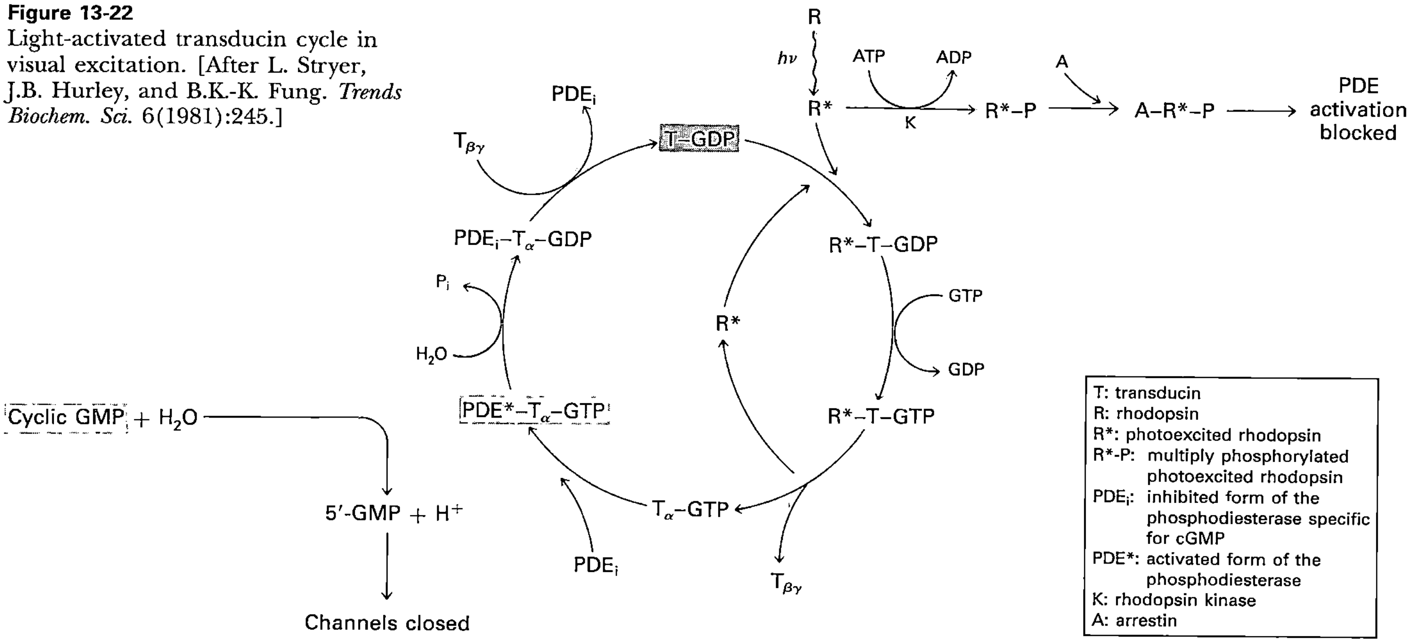

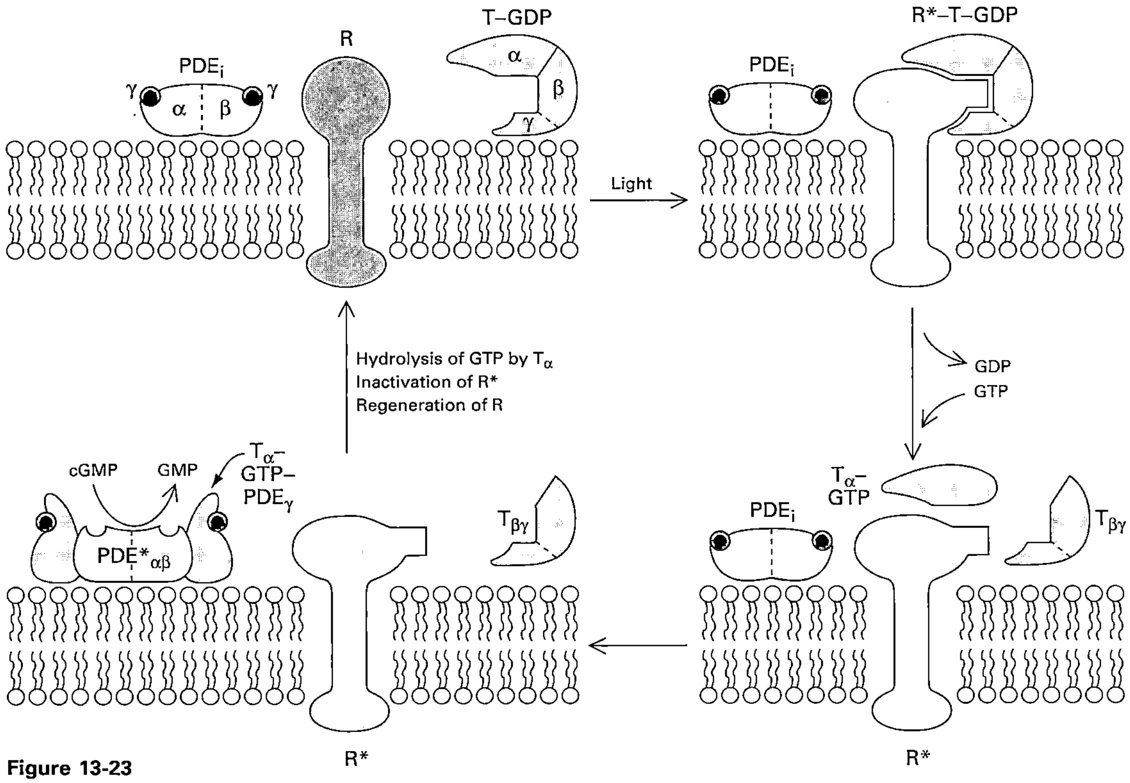

Figure 13-23
Schematic diagram of the cyclic GMP cascade of vision. [After L. Stryer. *Cold Spring Harbor Symp. Quant. Biol.* 48(1983):841.]

terase (PDE) by relieving an inhibitory constraint. In the dark, the two catalytic subunits (α and β) of PDE are held in check by a pair of inhibitory subunits (γ) (Figure 13-23). T_α–GTP activates PDE by prying away its γ subunits. Activated PDE has great catalytic prowess: it has a turnover number of 4200 s^{-1} and a k_{cat}/K_M value of $6 \times 10^7\ M^{-1}\ s^{-1}$, near the diffusion-controlled limit. *The hydrolysis of cGMP by the phosphodiesterase is the second stage of amplification.*

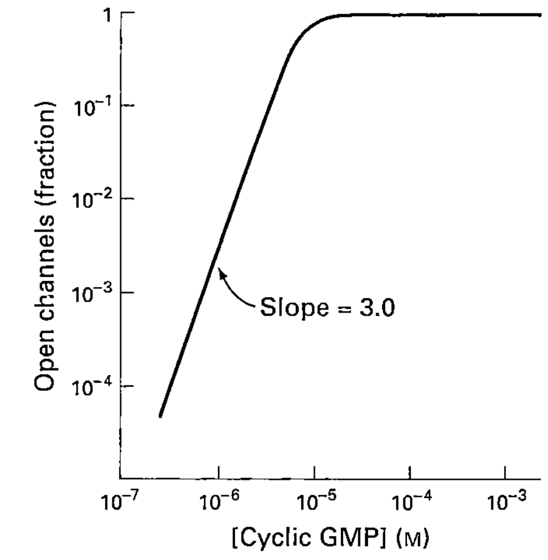

Figure 13-24
Cooperative opening of the cation-specific channel of retinal rod cells by cGMP. The initial slope of this double logarithmic plot of the dependence of channel opening on the cGMP concentration is 3.0. Hence, channel opening requires the binding of at least three molecules of cGMP. [After D.A. Baylor. *Invest. Ophthalmol. Vis. Sci.* 28(1987):34.]

CYCLIC GMP HYDROLYSIS CLOSES CATION-SPECIFIC CHANNELS TO GENERATE A NERVE SIGNAL

How does the light-triggered hydrolysis of cGMP lead to the closure of cation-specific channels in the plasma membrane? The answer came from patch-clamp studies (see Figure 12-6 on p. 294) of the plasma membrane of rod outer segments. Channels in an excised patch of this membrane were opened by the addition of cGMP (or a hydrolysis-resistant cGMP analog) to its cytosolic face. Other nucleotides were ineffective, and ATP was not required. Hence, *opening of the channel by cGMP is direct rather than mediated by covalent modification or by the binding of a cytosolic protein.*

The channel is a multimer of 80-kd subunits. Channel opening is highly cooperative with respect to cGMP binding. The Hill coefficient of 3.0 indicates that channel opening requires the binding of at least three molecules of cGMP (Figure 13-24). Part of the free energy of binding is used to change the conformation of the channel from the closed to the

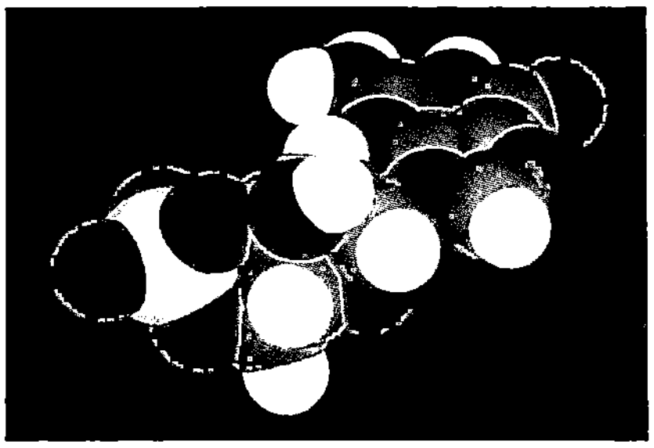

Model of cyclic GMP.

open form, as in the acetylcholine receptor channel (p. 296) and other ligand-gated channels. The high degree of cooperativity increases the sensitivity of the channel to small changes in the concentration of cGMP, which enables it to act as a switch. The channel opens and closes in times of milliseconds in response to physiologic changes in the level of cGMP. In essence, *the plasma membrane is a cyclic GMP electrode.* A similar channel, gated by cyclic AMP, plays a key role in olfaction.

THE LIGHT-INDUCED LOWERING OF THE CALCIUM LEVEL COORDINATES RECOVERY AND ADAPTATION

How does the system return to the dark state? The α subunit of transducin has a built-in *GTPase activity* that hydrolyzes bound GTP to GDP (see Figure 13-22). Hydrolysis occurs in less than a second when transducin is bound to the phosphodiesterase. The phosphodiesterase becomes inactive when transducin is converted into the T_α–GDP form. Hydrolysis of T_α–GTP to T_α–GDP is necessary but not sufficient for deactivation of the phosphodiesterase. R* must also be deactivated so that it does not continue to trigger the activation of transducin. *Rhodopsin kinase* catalyzes the phosphorylation of R* at multiple serine and threonine residues in its carboxyl-terminal region (see Figure 13-18). *Arrestin,* an inhibitory protein, then binds to phosphorylated R* to block the binding of transducin and prevent further activation of the phosphodiesterase.

The restoration of the dark state also requires the synthesis of cGMP from GTP, a reaction catalyzed by *guanylate cyclase.*

$$\text{GTP} \xrightarrow[\text{cyclase}]{\text{Guanylate}} \text{cyclic GMP} + \text{PP}_i$$

Calcium ion plays a key role in controlling this enzyme. In the dark, Ca^{2+} as well as Na^+ enters the rod outer segment through cGMP-gated channels in the plasma membrane (Figure 13-25). Calcium influx is balanced by its efflux through an exchanger. The thermodynamically favorable influx of four Na^+ and efflux of one K^+ pumps one Ca^{2+} out of the cell against a steep electrochemical gradient. Following illumination, the entry of Ca^{2+} through the cGMP-gated channel stops but its export by the exchanger continues. Consequently, the cytosolic level of Ca^{2+} drops from about 500 nM to 50 nM following illumination. *This light-induced lowering of the calcium level markedly stimulates guanylate cyclase* (Figure 13-26). Newly synthesized cGMP reopens the channels to restore the dark state.

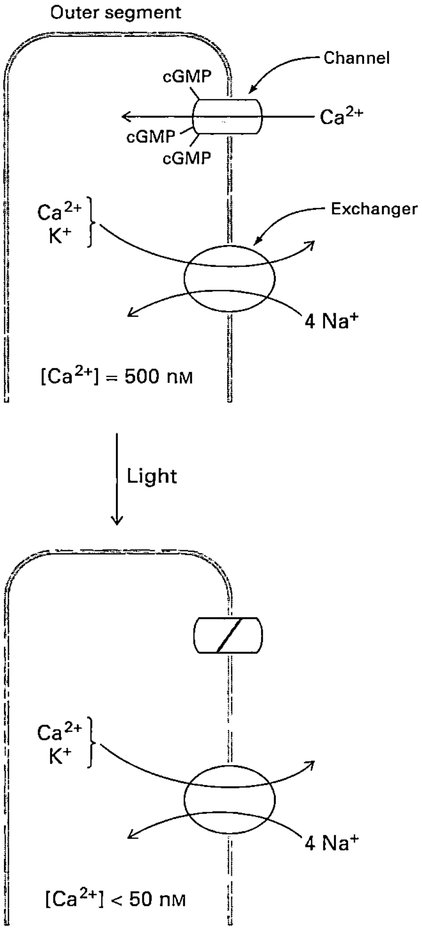

Figure 13-25
Light lowers the cytosolic calcium level in retinal rod cells by blocking the entry of Ca^{2+} through the cGMP-gated channel.

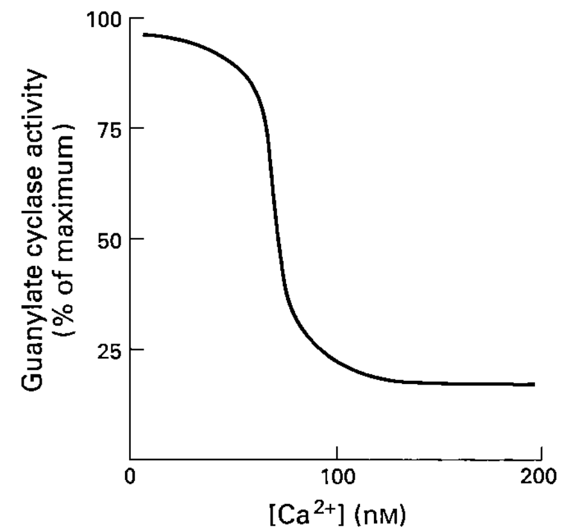

Figure 13-26
Guanylate cyclase activity in retinal rods is markedly stimulated by the light-induced lowering of the cytosolic Ca^{2+} level. [After K.-W. Koch and L. Stryer. *Nature* 334(1988):64.]

Thus, recovery following a light pulse is mediated by the GTPase activity of transducin, the inactivation of R* by rhodopsin kinase and arrestin, and the activation of a calcium-regulated guanylate cyclase. Excitation automatically sets in motion the recovery process, by closing channels to lower the calcium level (Figure 13-27). In essence, *cyclic GMP is the excitatory messenger, and calcium ion is the remembrance of photons past.* Electrophysiological studies suggest that calcium ion plays a key role in adaptation as well as recovery. The visual system, like the bacterial chemotaxis system discussed earlier, continuously resets its sensitivity so that it can respond to incremental stimuli. *Adaptation* enables a rod cell to perceive contrast over more than a 10,000-fold range of background light intensity. The challenge now is to unravel the molecular circuitry of this key facet of vision.

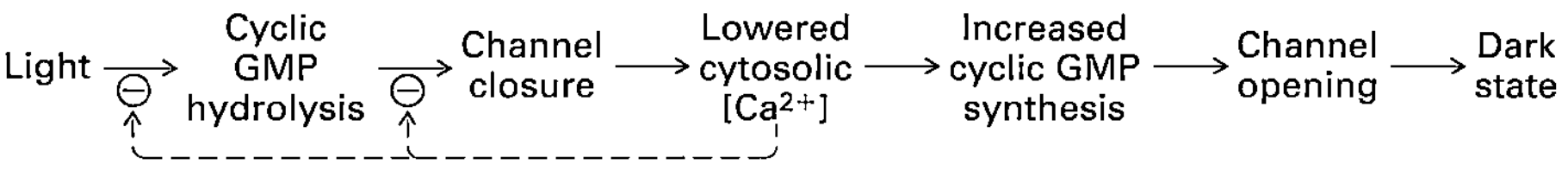

Figure 13-27
Visual excitation is followed by recovery and adaptation. The light-induced lowering of the cytosolic Ca^{2+} level serves as a coordinating signal. The minus signs denote negative feedback.

COLOR VISION IS MEDIATED BY THREE CONE RECEPTORS THAT ARE HOMOLOGS OF RHODOPSIN

The remarkable Thomas Young was a practicing physician, a professor of physics, and a distinguished Egyptologist. He proved the wave nature of light and deciphered many of the hieroglyphs of the Rosetta stone. In 1802, Young proposed that color vision is mediated by *three fundamental receptors.* Spectrophotometric studies of intact retinas carried out more than a century and a half later revealed that there are indeed *three types of cone cells: blue-, green-, and red-absorbing.* The absorption spectra of their three photoreceptor proteins have been measured by illuminating the outer segment of cones with a beam of light having a diameter of only 1 μm. The responses of different cone cells to monochromatic light of different wavelengths have shown that they belong to three groups: some are excited maximally by blue light, others by green light, and the rest by red light. In goldfish, the absorption maxima (λ_{max}) of the three color receptors are 455, 530, and 625 nm, and in humans, 426, 530, and ~560 nm. Recall that λ_{max} of rhodopsin is 500 nm.

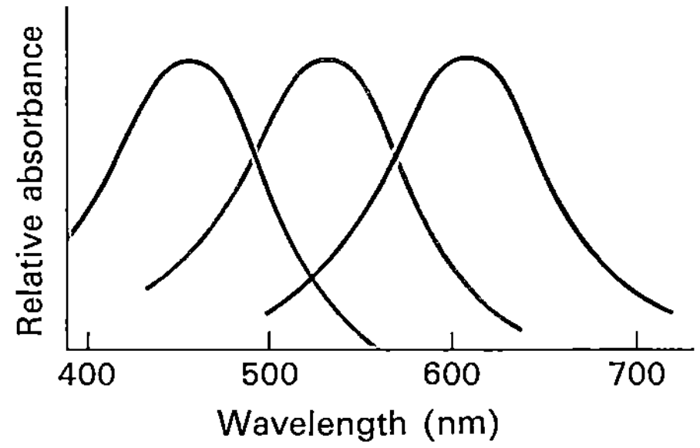

Figure 13-28
Absorption spectra of the three receptors mediating color vision in goldfish.

The photoreceptor proteins of cone cells, too, are seven-helix receptors, and they also contain 11-*cis*-retinal as their chromophore. How is the absorption spectrum of 11-*cis*-retinal tuned by the protein environment? Studies of site-specific mutants and of different primate cone receptors have revealed that three hydroxyl-containing residues located near retinal determine the spectral difference between the green and red receptors (Figure 13-29). *Replacement of a nonpolar residue with a polar one (e.g., alanine to serine) at each of these positions shifts λ_{max} to the red by about 10 nm.*

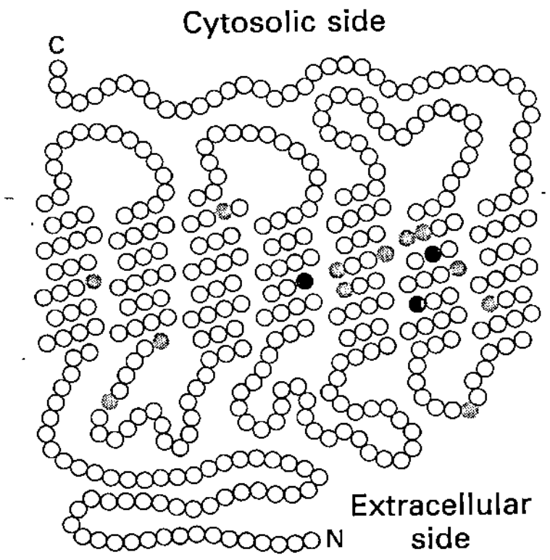

Figure 13-29
The amino acid sequences of the green- and red-absorbing photoreceptor proteins of human cone cells are very similar. Open circles denote identical residues and colored circles mark residues that are different. Three residues (dark red) account for most of the 30-nm difference between their absorption maxima. [Courtesy of Dr. Jeremy Nathans and Dr. David Hogness.]

CYCLIC AMP, A SECOND MESSENGER IN THE ACTION OF MANY HORMONES, IS FORMED BY ADENYLATE CYCLASE

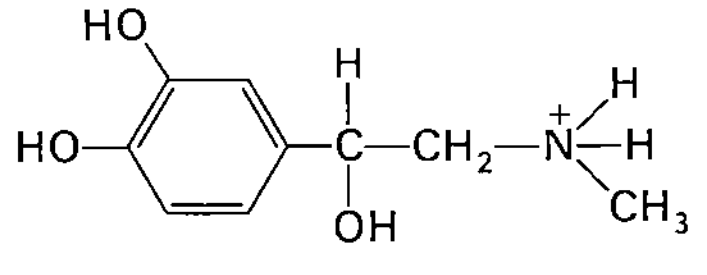

Epinephrine

$^{+}H_3N$-His-Ser-Glu-Gly-Thr-
-Phe-Thr-Ser-Asp-Tyr-
-Ser-Lys-Tyr-Leu-Asp-
-Ser-Arg-Arg-Ala-Gln-
-Asp-Phe-Val-Gln-Trp-
-Leu-Met-Asn-Thr-COO^-

Glucagon

The pioneering studies of Earl Sutherland in the 1950s led to a major breakthrough in our understanding of how hormones act at the molecular level. The initial aim was to determine how epinephrine (a catecholamine derived from tyrosine) and glucagon (a 29-residue peptide) elicit the breakdown of glycogen, a storage form of glucose, by the liver. Sutherland discovered that these hormones bind to receptors on the surface of liver cells, where they trigger the formation of *cyclic AMP (cAMP)*, a small nucleotide derived from ATP. The hormones themselves do not enter the liver cell. Rather, all their intracellular effects are mediated by cAMP. In essence, the hormones serve as *first messengers* (from one cell to another), and cAMP as the *second messenger* (within a cell).

Cyclic AMP is synthesized by the cyclization of ATP (Figure 13-30). The 3′-OH group of the ribose unit attacks the α-phosphoryl group of ATP to form a phosphodiester bond, with the concomitant release of pyrophosphate. *Adenylate cyclase,* a 120-kd integral membrane protein with multiple transmembrane segments, catalyzes this intramolecular reaction. The synthesis of cAMP, like that of cGMP, is slightly endergonic ($\Delta G^{\circ\prime}$ = +1.6 kcal/mol). Both are driven by the subsequent hydrolysis of pyrophosphate, a reaction catalyzed by *pyrophosphatase.* cAMP is hydrolyzed to 5′-AMP by a specific *phosphodiesterase.* The hydrolysis of both cAMP and cGMP is highly exergonic ($\Delta G^{\circ\prime}$ = −12 kcal/mol).

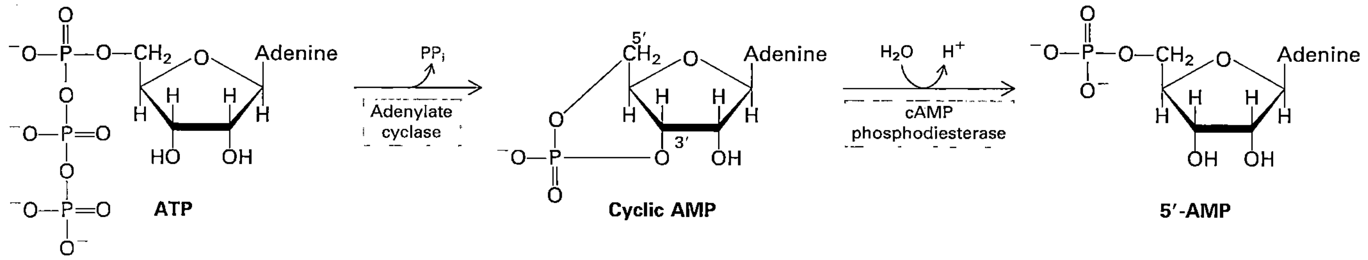

Figure 13-30
Enzyme-catalyzed synthesis and degradation of cyclic AMP.

Several experimental criteria have been used to determine whether cyclic AMP serves as a second messenger in mediating the action of a particular hormone.

1. Adenylate cyclase in a target cell should be stimulated by hormones affecting that cell. The cyclase activity of unresponsive cells should not be elevated by addition of the hormone.

2. The change in cAMP level in a target cell should precede or occur at the same time as the final outcome of hormonal stimulation. Variations in hormone levels should be matched by variations in the concentration of cAMP.

3. Inhibitors of the phosphodiesterase should act synergistically with hormones that employ cAMP as a second messenger.

4. The biological effects of a hormone should be mimicked by the addition of cAMP or a related compound to the target cells. In practice, cAMP cannot be readily used in this way because it penetrates cells poorly. However, less polar derivatives of cAMP, such as dibutyryl cAMP, are permeant; hydrolysis of the ester bonds inside the cell then liberates cAMP.

Experiments based on these criteria have revealed that *cyclic AMP is a second messenger for many hormones in addition to epinephrine and glucagon*

(Table 13-1). Cyclic AMP affects a very wide range of cellular processes. For example, it enhances the degradation of storage fuels, increases the secretion of acid by the gastric mucosa, leads to the dispersion of melanin pigment granules, diminishes the aggregation of blood platelets, and induces the opening of chloride channels.

Table 13-1
Hormones using cyclic AMP as a second messenger

Calcitonin
Chorionic gonadotropin
Corticotropin
Epinephrine
Follicle-stimulating hormone
Glucagon
Lipotropin
Luteinizing hormone
Melanocyte-stimulating hormone
Norepinephrine
Parathyroid hormone
Thyroid-stimulating hormone
Vasopressin

SEVEN-HELIX RECEPTORS ACTIVATE ADENYLATE CYCLASE THROUGH A STIMULATORY G PROTEIN (G_S)

Epinephrine (also called *adrenaline*) triggers the adenylate cyclase cascade by binding to the *β-adrenergic receptor,* a 64-kd protein that spans the plasma membrane of target cells. The purification of this receptor by Robert Lefkowitz led to the cloning and sequencing of its gene. Its hydrophobicity profile pointed to the presence of seven transmembrane helices (Figure 13-31). We have already seen this *seven-helix motif* in rhodopsin and the color vision receptors. The amino-terminal region of the β-adrenergic receptor, like that of rhodopsin, contains *N*-linked oligosaccharides and lies on the extracellular side of the membrane. The carboxyl-terminal region, which contains serines and threonines that are reversibly phosphorylated, resides on the cytosolic side. Bound epinephrine, like retinal, sits in a pocket formed by transmembrane helices.

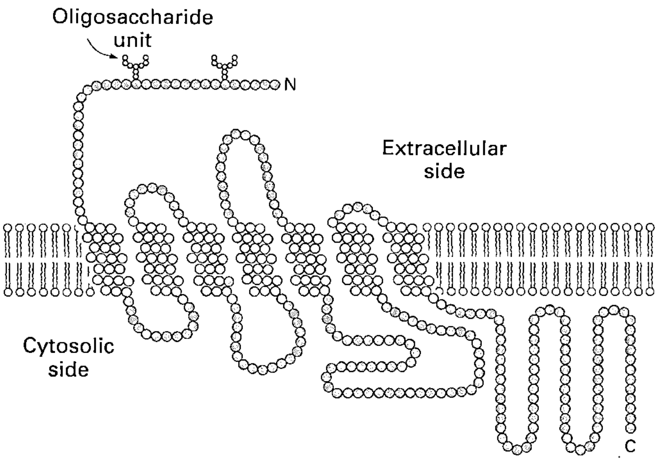

Figure 13-31
Seven-helix motif of the β-adrenergic receptor. Transmembrane helices are shown in yellow. Two *N*-linked oligosaccharide units (green) are located on the extracellular side of the plasma membrane. A loop on the cytosolic side participates in activating G_S, the stimulatory G protein. Phosphorylation of multiple serine and threonine residues in the carboxyl-terminal tail prevents the receptor from interacting with the G protein. [After H.G. Dohlman, M.G. Caron, and R.J. Lefkowitz. *Biochemistry* 26(1987):2660.]

How does the binding of epinephrine lead to the activation of adenylate cyclase? A key clue was Martin Rodbell's finding that GTP in addition to hormone is essential for activation. Equally revealing was the observation that hormone stimulates GTP hydrolysis. These findings led to the discovery that *a guanyl nucleotide–binding protein is an intermediary in the activation process.* This signal-coupling protein was named *G protein* (*G* for guanyl nucleotide). The hormone-receptor complex does not directly stimulate adenylate cyclase. Rather, the activated receptor stimulates the G protein, which carries the excitation signal to adenylate cyclase.

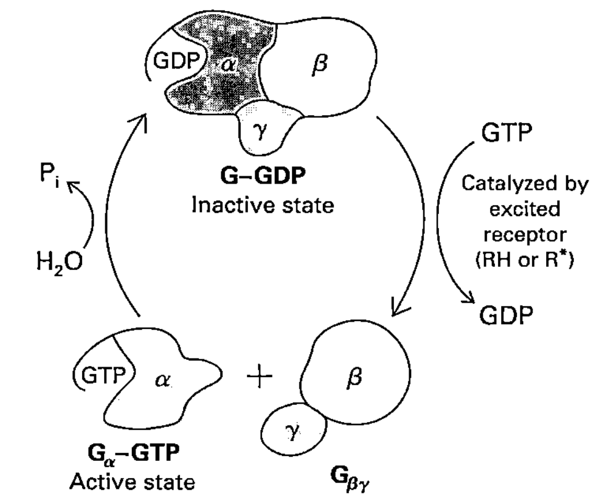

Figure 13-32
G proteins interconvert between an inactive GDP form and an active GTP form. The exchange of GTP for bound GDP is catalyzed by an activated receptor (e.g., RH, hormone-receptor complex; R*, photoexcited rhodopsin). G_α–GTP activates the effector protein. Hydrolysis of bound GTP brings the G protein back to the inactive state. The cycle is driven by the phosphoryl potential of GTP.

The G protein controlling adenylate cyclase is called the *stimulatory G protein* (G_S). How does G_S stimulate adenylate cyclase? Alfred Gilman purified G_S and found that it consists of α (45 kd), β (35 kd), and γ (7 kd) subunits. G_S, like transducin (p. 336), interconverts between a GDP form and a GTP form (Figure 13-32). The GTP form of G_S activates adenylate cyclase, whereas the GDP form does not. In the absence of hormone, nearly all of G_S is in the inactive GDP form. *The binding of hormone to the receptor triggers the exchange of GTP for bound GDP:* the hormone-receptor complex (but not the unoccupied receptor) binds to the G protein, induces the release of bound GDP, and allows GTP to enter. The α subunit bearing GTP ($G_{S\alpha}$–GTP) dissociates from the $\beta\gamma$ subunit ($G_{\beta\gamma}$). Adenylate cyclase is then activated by $G_{S\alpha}$–GTP. Thus, *the flow of information is from the hormone-receptor complex to G_S and then to adenylate cyclase* (Figure 13-33). Many $G_{S\alpha}$ are formed for each bound hormone, giving an *amplified response.* The activation of adenylate cyclase by epinephrine and that

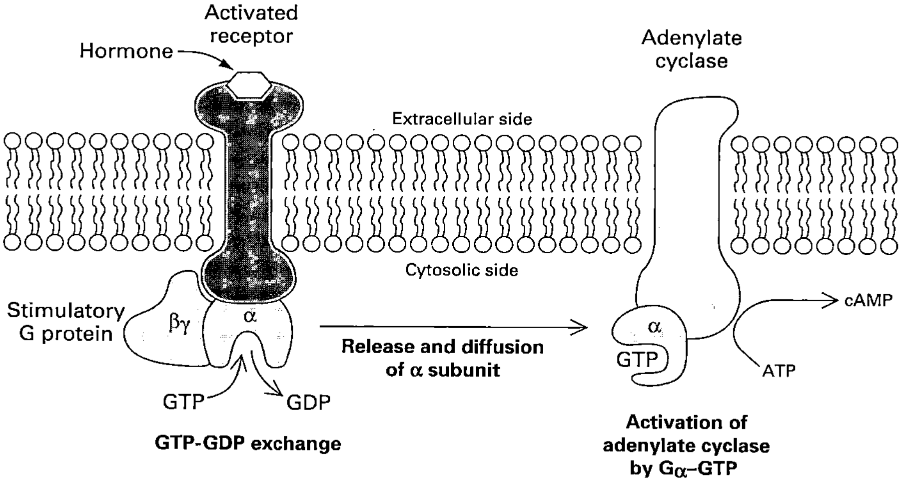

Figure 13-33
The activation of adenylate cyclase by the binding of a hormone to its specific receptor is mediated by G_S, the stimulatory G protein. A single hormone-receptor complex catalyzes the formation of many molecules of G_S. Hydrolysis of GTP bound to the α subunit of G_S terminates the activation of adenylate cyclase.

of cGMP phosphodiesterase by photoexcited rhodopsin (p. 337) are very similar in mechanism. G_S and transducin are members of the same family of signal-coupling proteins.

How is the activation of adenylate cyclase switched off? Recall that transducin possesses a built-in deactivation mechanism. Likewise, *G_S has intrinsic GTPase activity.* GTP bound to the α subunit of G_S is hydrolyzed on the time scale of minutes to GDP. Hydrolysis of GTP bound by $G_{S\alpha}$ is necessary but not sufficient for deactivation of adenylate cyclase.

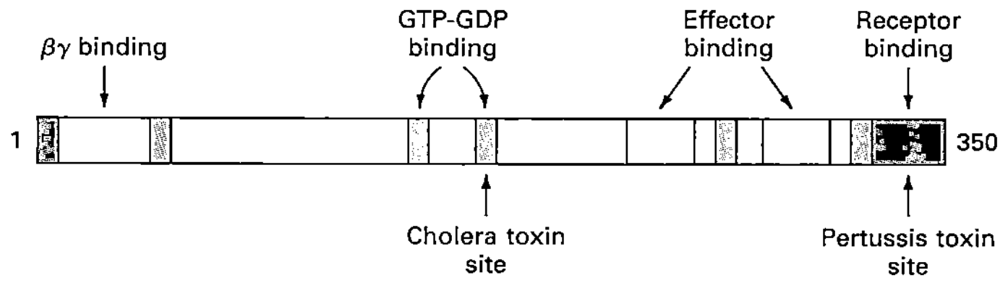

Functional regions of G_α

> *Hormones*—
> Chemical messengers that coordinate the activities of different cells in multicellular organisms. The term *hormone* (from the Greek, "to spur on") was first used in 1904 by William Bayliss and Ernest Starling to describe the action of secretin, a molecule secreted by the duodenum that stimulates the flow of pancreatic juice. Several very fruitful concepts emerged from their work: (1) Hormones are molecules synthesized by specific tissues (*glands*), (2) they are secreted directly into the blood, which carries them to their sites of action, and (3) they specifically alter the activities of responsive tissues (target organs or *target cells*).

The hormone-receptor complex too must be rendered inactive to return to the unstimulated state. The hormone-receptor complex, like photoexcited rhodopsin, is deactivated by phosphorylation. *β-Adrenergic receptor kinase,* which is similar to rhodopsin kinase, phosphorylates the carboxyl-terminal tail of the hormone-receptor complex but not of the unoccupied receptor. Finally, the binding of *β-arrestin,* a homolog of retinal arrestin, caps the phosphorylated receptor to further diminish its capacity to trigger the activation of G_S. Phosphorylation and the binding of β-arrestin account for the *desensitization (adaptation)* of the receptor following prolonged exposure to epinephrine. The adenylate cyclase cascade, like many other signal transduction processes, is designed to respond to *changes* in the strength of stimuli rather than to their absolute level. *Adaptation is advantageous because it enables receptors to respond to incremental stimuli over a wide range of background levels.*

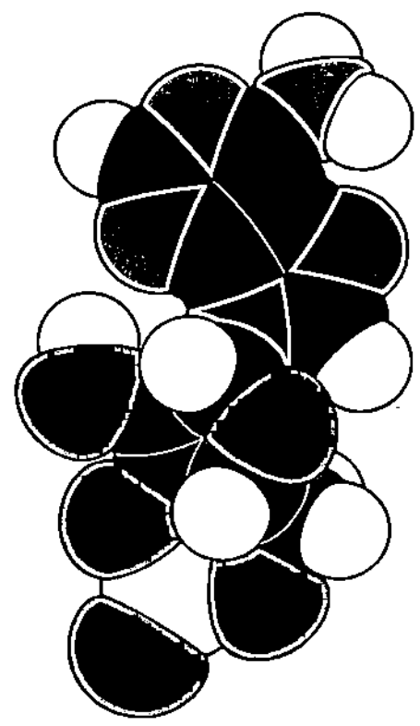

Model of cyclic AMP.

CYCLIC AMP STIMULATES THE PHOSPHORYLATION OF MANY TARGET PROTEINS BY PROTEIN KINASE A (PKA)

How does cyclic AMP influence so many cellular processes? Is there a common denominator for its diverse effects (see Table 13-1)? Indeed there is. *Most effects of cyclic AMP in eukaryotic cells are mediated by activation of a single protein kinase. This key enzyme is called protein kinase A (PKA).* As was discussed earlier (p. 246), PKA consists of two regulatory (R) chains and two catalytic (C) chains. In the absence of cAMP, the R_2C_2 complex is catalytically inactive. The binding of cAMP to the regulatory chains releases the catalytic chains, which are enzymatically active on their own. Activated PKA then phosphorylates specific serine and threonine residues in many targets to alter their activity. The significance and far reach of PKA are seen in the following examples of its actions:

1. In *glycogen metabolism* (p. 595), phosphorylation of two enzymes by PKA leads to the breakdown of this polymeric store of glucose and to the inhibition of further glycogen synthesis.

2. Epithelial cells contain a chloride channel called the *cystic fibrosis transmembrane regulator (CFTR).* This channel is opened by PKA-catalyzed phosphorylation of the regulatory domain of CFTR. Regulation of this channel is defective in *cystic fibrosis,* the most common lethal genetic disorder of Caucasians.

3. PKA stimulates the expression of specific genes by phosphorylating a transcriptional activator called the *cAMP-response element binding protein (CREB)* (p. 1004).

4. *Synaptic transmission* between pairs of neurons in *Aplysia* (a marine snail) is enhanced by serotonin, a neurotransmitter that is released by adjacent interneurons. Serotonin binds to a seven-helix receptor to trigger an adenylate cyclase cascade (Figure 13-34). The rise in cAMP level activates PKA, which closes potassium channels by phosphorylating them. Closure of potassium channels increases the excitability of the target cell.

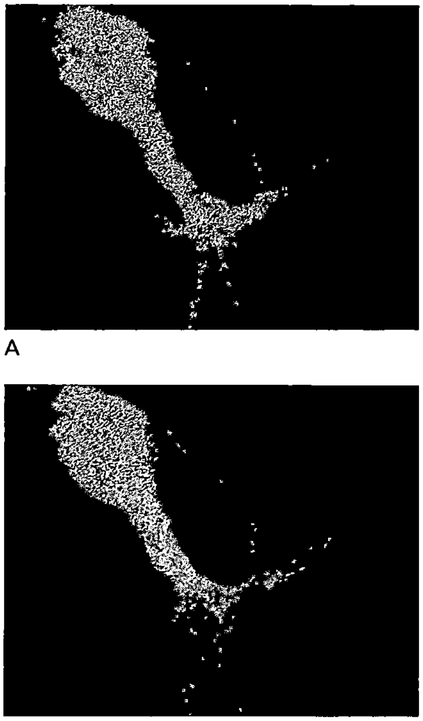

Figure 13-34
Serotonin, a neurotransmitter, markedly increased the cAMP level in an *Aplysia* sensory neuron. The cAMP level was monitored using fluorescent-labeled PKA that was microinjected into the neuron. Blue denotes a low cAMP level; yellow and red denote a high level. The free cAMP level increased from (A) less than 50 nM in the unstimulated neuron to (B) more than 1 μM 19 s after the addition of serotonin to the external medium. [After B.J. Backsai, B. Hochner, M. Mahau-Smith, S.R. Adams, B.-K. Kaang, E.R. Kandel, and R.Y. Tsien. *Science* 260(1993):222.]

RECEPTOR-TRIGGERED HYDROLYSIS OF PHOSPHATIDYL INOSITOL BISPHOSPHATE GENERATES TWO MESSENGERS

We turn now to another ubiquitous cascade that is mediated by G proteins and evokes many responses (Table 13-2). The *phosphoinositide cascade,* like the adenylate cyclase cascade, converts extracellular signals into intracellular ones. The intracellular messengers formed by activation of this pathway arise from *phosphatidyl inositol 4,5-bisphosphate* (PIP_2), a phospholipid

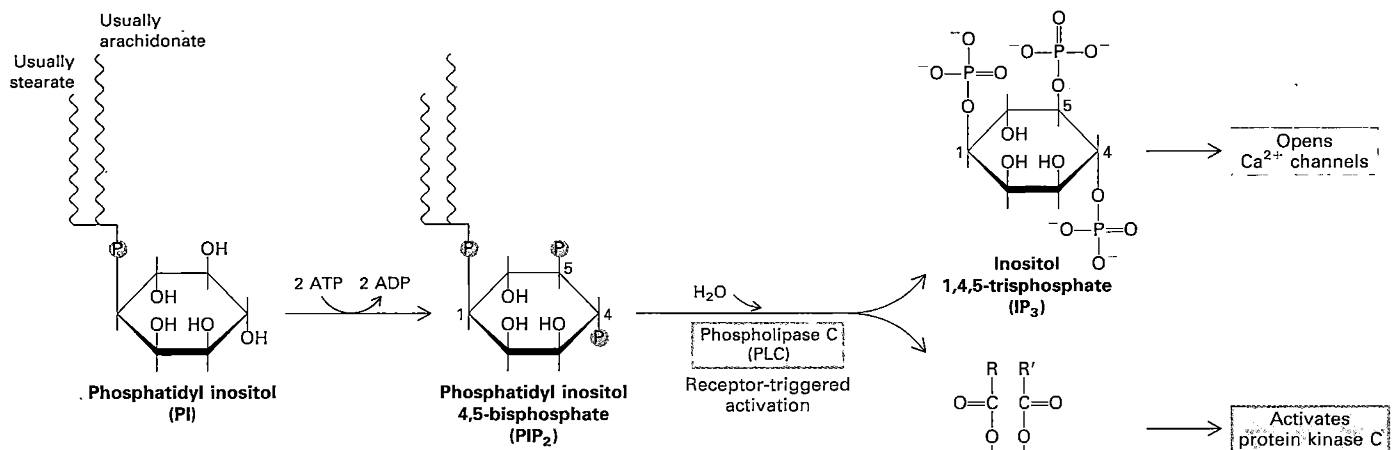

Figure 13-35
Synthesis and receptor-triggered hydrolysis of phosphatidyl inositol 4,5-bisphosphate (PIP_2). Two messengers are formed.

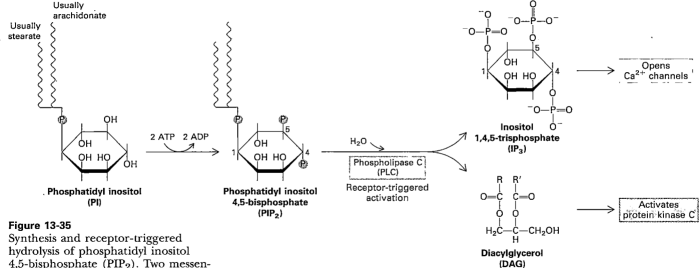

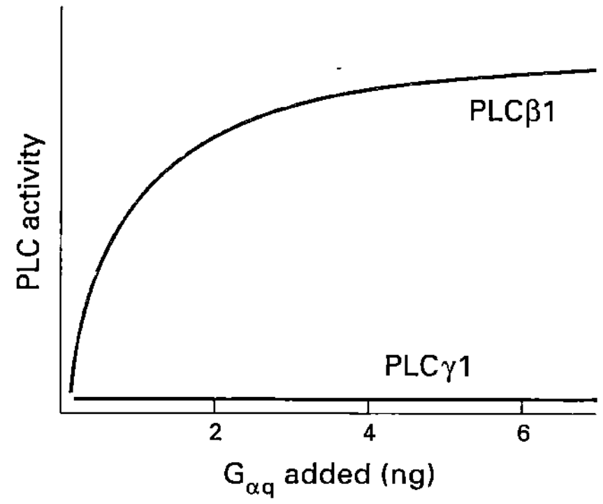

Figure 13-36
The GTP form of the α subunit of G_q activates the β1 isoform but not the γ1 isoform of phospholipase C. [After S.J. Taylor, H.Z. Chae, S.G. Rhee, and J.H. Exton. *Nature* 350(1991):516.]

in the plasma membrane (Figure 13-35). The binding of a hormone such as vasopressin to a cell-surface receptor leads to the activation of *phospholipase C.* This membrane-bound enzyme hydrolyzes the phosphodiester bond linking the phosphorylated inositol unit to the acylated glycerol moiety. Two messengers—*inositol 1,4,5-trisphosphate* (IP_3) and *diacylglycerol*—are formed by the cleavage of PIP_2.

Four kinds of mammalian phospholipase C (PLC)—α, β, γ, and δ—have been studied in detail. They range in mass from 61 to 154 kd, and are quite dissimilar apart from a few common sequences. PLCs are cytosolic enzymes that act on membrane-inserted phosphoinositide substrates. Their enzymatic activity increases markedly when the calcium level is raised from 100 nM to 1 μM. How do receptors control the activity of PLCs? A G protein that specifically activates a PLC has recently been purified. *G_q increases the catalytic activity of the β1 isoform of PLC and increases its affinity for Ca^{2+}* (Figure 13-36). Several other PLC isoforms are also activated by G proteins. By contrast, PLCγ1 is activated by a receptor tyrosine kinase (p. 353) rather than by a G protein.

INOSITOL 1,4,5-TRISPHOSPHATE (IP_3) OPENS CHANNELS TO RELEASE CALCIUM ION FROM INTRACELLULAR STORES

Table 13-2
Effects mediated by the phosphoinositide cascade

Glycogenolysis in liver cells
Histamine secretion by mast cells
Serotonin release by blood platelets
Aggregation of blood platelets
Insulin secretion by pancreatic islet cells
Epinephrine secretion by adrenal chromaffin cells
Smooth muscle contraction
Visual transduction in invertebrate photoreceptors

The effects of IP_3 have been delineated by microinjecting it into cells and by adding it to cells whose plasma membrane has been made permeable. Michael Berridge and co-workers found that *IP_3 causes the rapid release of Ca^{2+} from intracellular stores—the endoplasmic reticulum and, in smooth muscle cells, the sarcoplasmic reticulum.* The elevated level of Ca^{2+} in the cytosol then triggers processes such as smooth muscle contraction, glycogen breakdown, and vesicle release. The injection of IP_3 into *Xenopus* oocytes suffices to activate many of the early events of fertilization.

The cDNA for the IP_3-gated channel encodes a very large protein (313 kd) containing multiple membrane-spanning segments. Electron-microscopic and chemical cross-linking studies indicate that four subunits associate to form a channel. Channel opening requires the binding of at least three molecules of IP_3 to sites on the cytosolic side of the membrane. *The highly cooperative opening of calcium channels by nanomolar concentrations of IP_3 enables cells to detect and amplify very small changes in the concentration of* *this messenger.* Furthermore, *IP_3 is a short-lived messenger because it is rapidly converted into derivatives that do not open the channel.* The lifetime of IP_3 in most cells is less than a few seconds and in olfactory neurons is as short as 100 ms (Figure 13-37). IP_3 can be degraded to inositol by the sequential action of phosphatases, or it can be phosphorylated to inositol 1,3,4,5-tetrakisphosphate, which is then converted to inositol by an alternative route. Lithium ion, widely used to treat manic-depressive disorders, may act by inhibiting the recycling of inositol 1,3,4-trisphosphate.

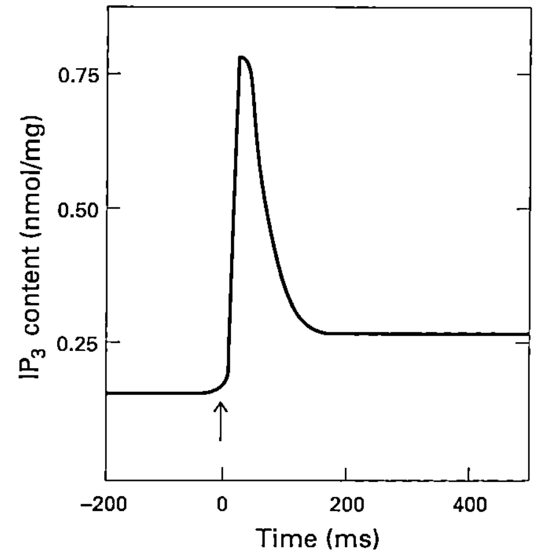

Figure 13-37
IP_3 is rapidly formed and degraded in olfactory cilia following the addition of pyrazine, an odorant present in bell peppers. The arrow marks the time of addition. Olfactory cilia contain a phosphoinositide cascade that is triggered by the binding of odorants to seven-helix receptors. [After I. Boekhoff, E. Tareilus, J. Strotmann, and H. Breer. *EMBO J.* 9(1990):2453.]

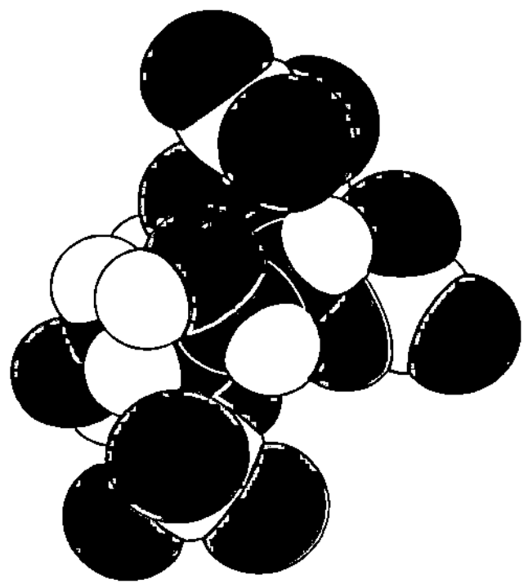

Figure 13-38
Model of inositol 1,4,5-trisphosphate (IP_3). This messenger molecule has a very high density of negatively charged and polar groups.

DIACYLGLYCEROL ACTIVATES PROTEIN KINASE C (PKC), WHICH PHOSPHORYLATES MANY TARGET PROTEINS

Diacylglycerol, the other messenger formed by the receptor-triggered hydrolysis of PIP_2, activates *protein kinase C (PKC).* This *multifunctional protein kinase* phosphorylates serine and threonine residues in many target proteins. Yasutomi Nishizuka found that PKC is enzymatically active only in the presence of Ca^{2+} and phosphatidyl serine. *Diacylglycerol increases the affinity of PKC for Ca^{2+} and thereby renders it active at physiologic levels of this ion.*

Many mammalian and *Drosophila* cDNAs for PKC have been cloned. They encode proteins having a mass of about 80 kd. All contain an N-terminal *regulatory domain* and a C-terminal *catalytic domain.* Proteolysis at the junction of these domains yields a persistently active catalytic fragment and a regulatory fragment that binds Ca^{2+} and diacylglycerol. How does the regulatory domain inhibit the catalytic domain of intact PKC? The regulatory domain of this enzyme, like the R subunit of protein kinase A, contains a *pseudosubstrate sequence* that is rich in positively charged residues.

-RFARKG**A**LRQKNVHEVKN-

An effective substrate has serine or threonine in place of the marked alanine. In the absence of diacylglycerol, the pseudosubstrate sequence occupies the substrate binding site. Binding of diacylglycerol disrupts this interaction and enables a protein substrate to enter.

Diacylglycerol, like IP_3, acts transiently because it is rapidly metabolized. It can be phosphorylated to phosphatidate (p. 265), or it can be hydrolyzed to glycerol and its constituent fatty acids. Arachidonate, the C_{20}-polyunsaturated fatty acid that usually occupies the 2-position on the

Figure 13-39
Model of diacylglycerol. This messenger, in contrast with IP_3, is quite nonpolar.

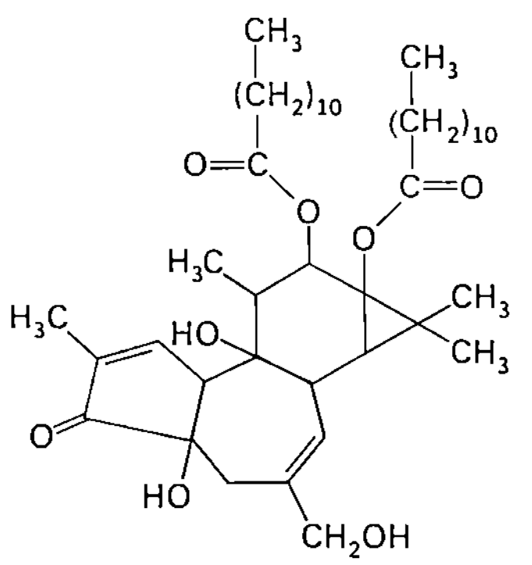

A phorbol ester

glycerol moiety of PIP_2, is the precursor of a series of 20-carbon hormones such as the prostaglandins (p. 624). Thus, *the phosphoinositide pathway gives rise to many molecules that have signaling roles.*

The importance of protein kinase C in controlling cell division and proliferation is revealed by the action of *phorbol esters.* These polycyclic alcohol derivatives from croton oil are carcinogenic; they are known as *tumor promoters.* Phorbol esters activate protein kinase C because they resemble diacylglycerol. The activation is persistent because phorbol esters, unlike diacylglycerol, are not readily degraded.

SEVEN-HELIX RECEPTORS AND G PROTEINS PLAY KEY ROLES IN DIVERSE SIGNAL TRANSDUCTION PROCESSES

We have seen that seven-helix receptors (also called *serpentine receptors*) and G proteins play key roles in sensory and hormonal transduction processes, as exemplified by vision, the adenylate cyclase cascade, and the phosphoinositide cascade. Indeed, they are crucial for signaling in all eukaryotes and have a long evolutionary history. In yeast, the receptors for the two kinds of mating factor peptides are members of the seven-helix family. These pheromone receptors are coupled to G proteins that determine the developmental fate of the cell. Some examples of the wide range of signals detected by seven-helix receptors are given in Table 13-3.

The common themes of these ancient and elegant signal-transducing couples are

1. Seven-helix receptors always act through G proteins. Reciprocally, G proteins are activated only by seven-helix receptors.

2. Seven-helix receptors contain a binding pocket that is located near the center of the bilayer. Many different kinds of ligands can be bound to these sites. The size of the repertoire of seven-helix receptors in olfactory neurons is striking: several hundred receptors differing in odorant-binding specificity enable us to smell a very large number of compounds.

3. Binding of a specific ligand to a seven-helix receptor induces conformational changes that are transmitted to loops on the cytosolic side of the membrane. The third cytosolic loop is especially important in conveying the signal to cognate G proteins.

4. Activated seven-helix receptors switch on G proteins by catalyzing the exchange of GTP for bound GDP. They do so by opening the guanyl nucleotide binding site of the G protein. The sole role of the activated receptor is to greatly accelerate the exchange of GTP for GDP.

Table 13-3
Physiologic processes mediated by G proteins

Stimulus	*Receptor*	*G protein*	*Effector*	*Physiologic response*
Epinephrine	β-Adrenergic receptor	G_s	Adenylate cyclase	Glycogen breakdown
Serotonin	Serotonin receptor	G_s	Adenylate cyclase	Behavioral sensitization and learning in *Aplysia*
Light	Rhodopsin	Transducin	cGMP phosphodiesterase	Visual excitation
Odorants	Olfactory receptors	G_{olf}	Adenylate cyclase	Olfaction
fMet peptide	Chemotactic receptor	G_q	Phospholipase C	Chemotaxis
Acetylcholine	Muscarinic receptor	G_i	Potassium channel	Slowing of pacemaker activity

After L. Stryer and H.R. Bourne. *Ann. Rev. Cell Biol.* 2(1986):393.

5. Seven-helix receptors bearing a specific ligand are deactivated by phosphorylation of serine and threonine residues in their C-terminal tail.

6. The binding of GTP to a G protein leads to its activation. The G protein dissociates into G_α–GTP and $G_{\beta\gamma}$ subunits, and the activated receptor is released. The excitation signal is usually carried forward by G_α–GTP (Figure 13-40). In some cases, $G_{\beta\gamma}$ switches on the target.

7. The α subunit of G proteins hydrolyzes bound GTP to return the G protein to the inactive form. In some systems, this built-in timer is accelerated by the target protein. The proportion of G protein in the active state is determined both by the rate of receptor-catalyzed GTP-GDP exchange and by the rate of hydrolysis of bound GTP.

8. Many G proteins are covalently modified by cholera toxin or pertussis toxin. These toxins catalyze the transfer of an adenine diphosphate ribose group from NAD^+ (p. 449) to an arginine or cysteine side chain of G_α. Modification by cholera toxin stabilizes the GTP form of the G protein, whereas modification by pertussis toxin stabilizes the GDP form.

Figure 13-40
Three-dimensional structure of the α subunit of transducin. GTPγS (red), a hydrolysis-resistant analog of GTP, is bound in a deep cleft between the GTPase domain (yellow) and the highly helical domain (blue). The GTPase domain of transducin is like that of ras, a GTP-binding protein that participates in growth control cascades. [Drawn from 1tnd.pdb. J.P. Noel, H.E. Hamm, and P.B. Sigler. *Nature* 366(1993):654.]

CALCIUM ION IS A UBIQUITOUS CYTOSOLIC MESSENGER

Calcium ion is an intracellular messenger in many eukaryotic signal-transducing pathways, such as vision (p. 338), the phosphoinositide cascade, and the regulation of muscle contraction (p. 402). Why has nature chosen this ion to mediate so many signaling processes? *The intracellular level of Ca^{2+} must be kept low because phosphate esters are highly abundant and calcium phosphates are quite insoluble.* As was discussed in the preceding chapter, all cells have transport systems—the Ca^{2+}-ATPase and the sodium-calcium exchanger—for the extrusion of Ca^{2+}. The cytosolic level of Ca^{2+} in unexcited cells is typically 100 nM, several orders of magnitude less than the concentration in the extracellular milieu. This steep gradient presents cells with a matchless opportunity: *the cytosolic Ca^{2+} concentration can be abruptly raised for signaling purposes by transiently opening calcium channels in the plasma membrane or in an intracellular membrane.*

A second property of Ca^{2+} that makes it a highly suitable intracellular messenger is that it can bind tightly to proteins. Negatively charged oxygens (from the side chains of glutamate and aspartate) and uncharged oxygens (main-chain carbonyls and side-chain oxygens from glutamine and asparagine) bind well to Ca^{2+} (Figure 13-41). *The capacity of Ca^{2+} to be coordinated to multiple ligands—six to eight oxygen atoms—enables it to cross-link different segments of a protein and induce large conformational changes.* Furthermore, *the binding of Ca^{2+} can be highly selective.* Mg^{2+}, a potential competitor, can be spurned because it does not have appreciable affinity for uncharged oxygen atoms. Another important difference between these ions is that Mg^{2+} prefers to form small and symmetric coordination shells, whereas Ca^{2+} can form asymmetric complexes having a larger radius. Thus, Ca^{2+} is well suited for binding to irregularly shaped crevices in proteins and can be selected over Mg^{2+} even when the latter is a thousandfold more abundant.

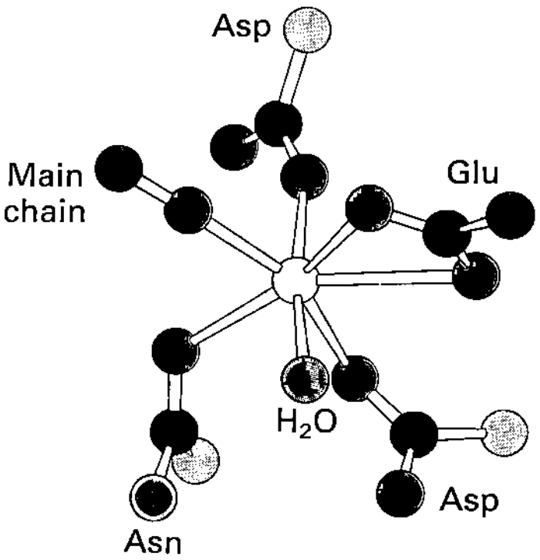

Figure 13-41
Mode of binding of Ca^{2+} (green) to calmodulin, a ubiquitous calcium sensor in eukaryotes. Calcium is coordinated to six oxygen atoms (red) of the protein and one (purple) from water.

Our understanding of the role of calcium in cellular processes has been greatly enhanced by the use of calcium-specific reagents. *Ionophores* such as *A23187* and *ionomycin* can traverse a lipid bilayer because they have a hydrophobic periphery. They can be used to introduce Ca^{2+} into cells and organelles. Many physiologic responses that are normally triggered by the binding of hormones to cell-surface receptors can also be elicited by

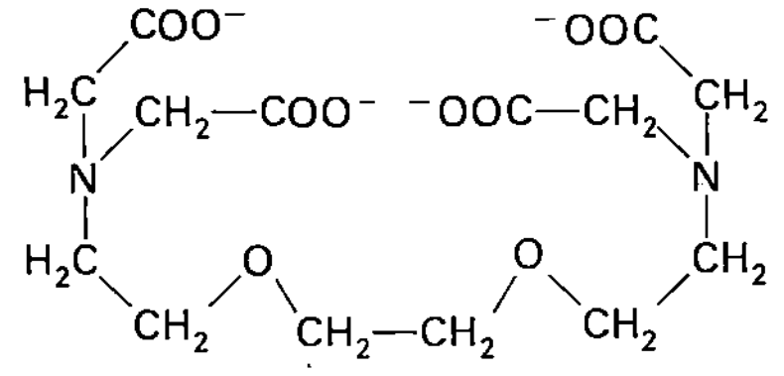

Figure 13-42
EGTA specifically binds Ca^{2+} with high affinity.

using calcium ionophores to raise the cytosolic calcium level. Conversely, the concentration of unbound calcium in a cell can be made very low (nanomolar or less) by introducing a calcium-specific chelator such as *EGTA* (Figure 13-42). A process postulated to be mediated by a rise in the cytosolic level of Ca^{2+} should be blocked by lowering the free calcium level with a chelator.

The concentration of free Ca^{2+} in intact cells can be monitored using polycyclic chelators such as *Fura-2* and *Fluo-3*. The fluorescence properties of these indicators change markedly when Ca^{2+} is bound. Calcium concentrations in the nanomolar to micromolar range can also be measured using *aequorin,* a protein produced by a luminous jellyfish (*Aequorea forskalea*). This bioluminescent protein contains an activated chromophore that emits light on binding Ca^{2+}; the rate of emission depends on the free Ca^{2+} level.

EF HANDS ARE RECURRING CALCIUM-BINDING MODULES

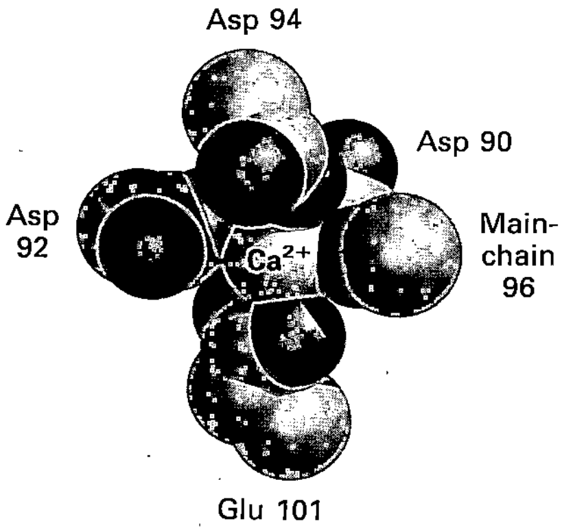

Figure 13-43
Mode of binding of Ca^{2+} (green) to a cluster of six oxygen atoms (red) provided by parvalbumin, a calcium buffer in muscle. The seventh ligand, a water oxygen atom, is not shown here. [Drawn from 5cpv.pdb. A.L. Swain, R.H. Kretsinger, and E.L. Amma. *J. Biol. Chem.* 264(1989):16620.]

X-ray crystallographic studies have provided informative views of calcium-binding proteins. The first to be elucidated was *parvalbumin,* a 12-kd protein that serves as a high-affinity calcium buffer in carp muscle. It contains two similar Ca^{2+}-binding sites, which are formed by a helix, a loop, and another helix. Seven oxygen atoms are coordinated to each Ca^{2+}: a carboxylate oxygen from each of three aspartates, two carboxylate oxygens from a glutamate, a main-chain carbonyl oxygen, and an oxygen of a bound water molecule (Figure 13-43). One of the binding sites is formed by helices E and F, which are positioned like the forefinger and thumb of the right hand (Figure 13-44). The Ca^{2+}-binding site is formed by a loop between these helices. Robert Kretsinger named this structural motif the *EF hand* and proposed that the two Ca^{2+}-binding sites of parvalbumin arose by duplication of a primordial gene encoding a calcium-binding loop. John Collins then noted that the amino acid sequences of parvalbumin and troponin C (a regulator of muscle contraction, p. 403) are similar and suggested that troponin C also contains EF hands. *Subsequent x-ray analyses have shown that EF hands recur in troponin C and many other intracellular calcium-binding proteins.* Indeed, sequences diagnostic of the EF hand motif are present in more than 100 proteins.

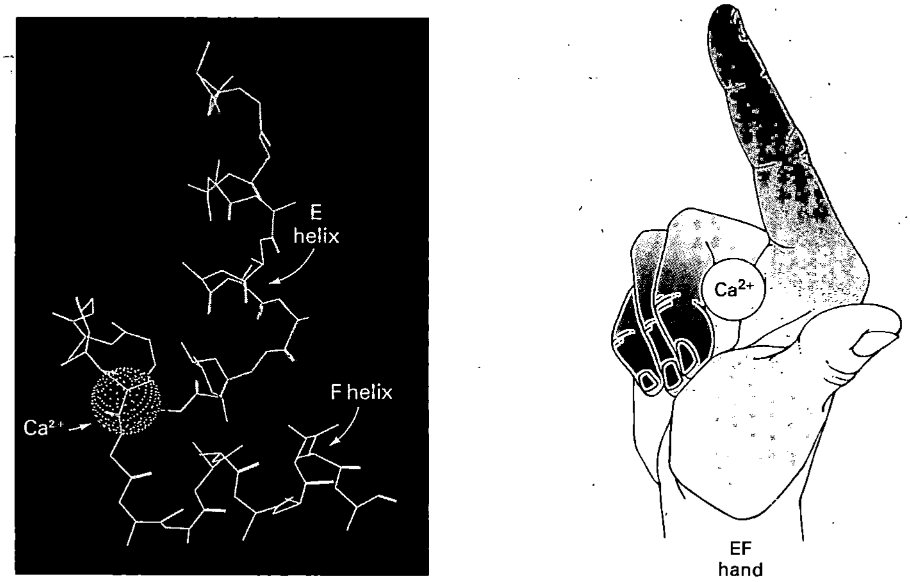

Figure 13-44
The binding sites for Ca^{2+} in many calcium-sensing proteins are formed by a helix-loop-helix unit. This recurring motif is called an EF hand.

THE CALCIUM-BOUND FORM OF CALMODULIN STIMULATES MANY ENZYMES AND TRANSPORTERS

Calmodulin, a 17-kd member of the EF hand family, serves as a calcium sensor in nearly all eukaryotic cells. It consists of two similar globular lobes joined by a long α helix, which gives it a length of 65 Å (Figure 13-45). Each lobe contains two EF hands that are 11 Å apart. Indeed, EF hands always come in pairs because the calcium-binding loop of one hand stabilizes the loop of the other by antiparallel β sheet hydrogen bonding and hydrophobic interactions. Calmodulin contains four calcium-binding sites. At each EF hand, Ca^{2+} is bound to six oxygen atoms of the protein and one of a water molecule (see Figure 13-41).

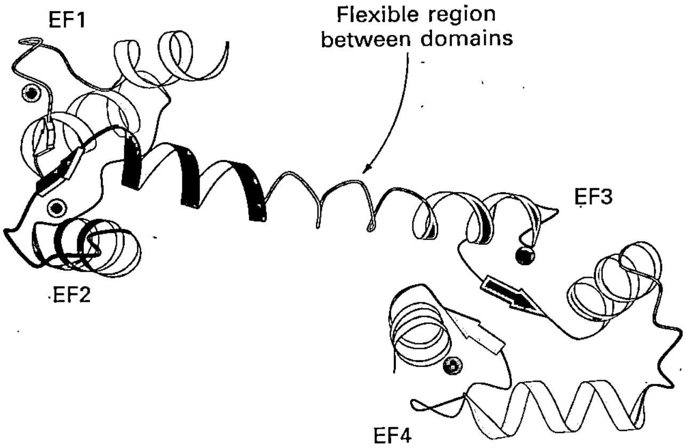

Figure 13-45
Calmodulin contains four EF hands (shown in different colors). A calcium ion (orange) is bound to each. An EF hand interacts with another EF hand through a short antiparallel β sheet (denoted by pairs of arrows). The helix joining the two domains of the proteins is flexible near the center. [Drawn from 3cln.pdb. Y.S. Babu, C.E. Bugg, and W.J. Cook. *J. Mol. Biol.* 204(1988):191.]

Calmodulin is activated by the binding of three or four Ca^{2+}, which occurs when the cytosolic calcium level is raised above about 500 nM by the opening of calcium channels in the plasma membrane or in the membrane surrounding an internal store. Ca^{2+}–calmodulin stimulates a wide array of enzymes, pumps, and other target proteins. Two targets are especially noteworthy. The *multifunctional calmodulin-dependent protein kinase II (CaM kinase II)* phosphorylates many different proteins. It regulates fuel metabolism, ionic permeability, neurotransmitter synthesis, and neurotransmitter release. The binding of Ca^{2+}–calmodulin to CaM kinase II activates this multimeric enzyme and enables it to phosphorylate target proteins (Figure 13-46).

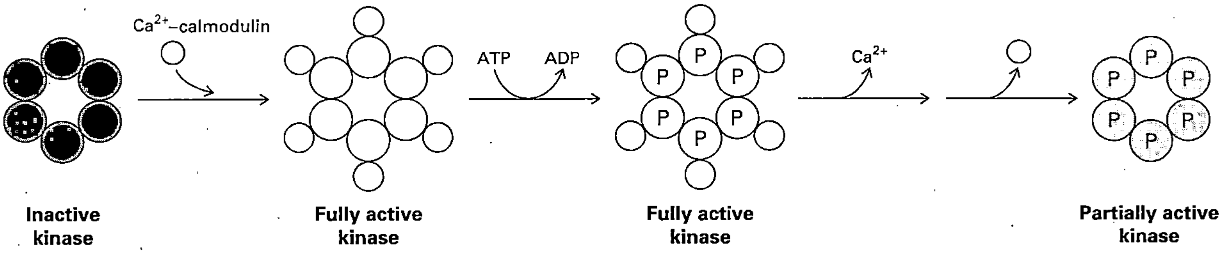

Figure 13-46
CaM kinase II, a multimeric enzyme, is activated by the binding of multiple molecules of Ca^{2+}–calmodulin. Half the dodecamer is shown in this schematic diagram. Autophosphorylation gives an enzyme that is partially active in the absence of Ca^{2+}. This autonomous enzyme is taken back to the basal state by a phosphatase. [After H. Schulman, P.I. Hanson, and T. Meyer. *Cell Calcium* 13(1992):401.]

In addition, the activated enzyme phosphorylates itself, which makes it partially active in phosphorylating exogenous substrates even when Ca^{2+}–calmodulin is not bound. In essence, autophosphorylation of CaM kinase II is the memory of a previous calcium pulse. The plasma membrane *Ca^{2+}-ATPase pump* is another important target of Ca^{2+}–calmodulin. Stimulation of the pump by Ca^{2+}–calmodulin drives the calcium level down to restore the low-calcium basal state.

How does calmodulin recognize many different targets? Comparisons of the amino acid sequences of calmodulin-binding domains of target proteins suggested that *calmodulin recognizes positively charged, amphipathic α helices.* Indeed, amphipathic peptides composed of only leucine, lysine, and tryptophan residues bind very tightly to Ca^{2+}–calmodulin but not to the calcium-free protein. Each of the lobes of Ca^{2+}–calmodulin contains large hydrophobic patches that are flanked by negatively charged regions, which are nicely complementary to the positively charged amphiphilic α helices of its targets. The binding of Ca^{2+} makes these patches accessible. *The central helix of calmodulin serves as a flexible tether that allows the two lobes to come together to form a contiguous binding site that can accommodate a wide variety of target proteins.* The dumbbell shape of the calcium-free protein changes into a globular form in which the target peptide sits in a hydrophobic channel surrounded by many side chains from both domains (Figure 13-47; also see Figure 8-3 on p. 183).

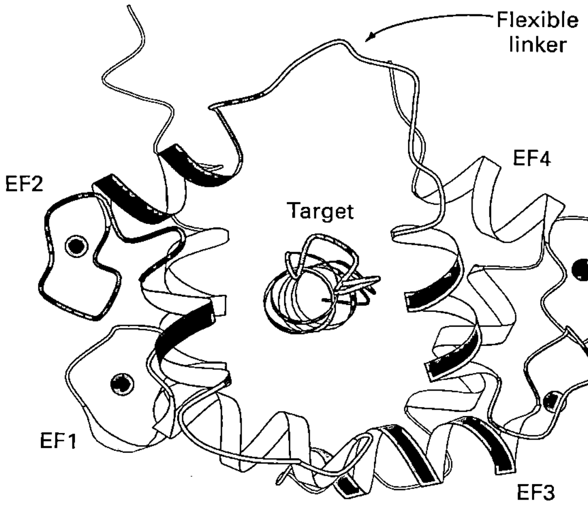

Figure 13-47
Calcium–calmodulin wraps around myosin light-chain kinase (purple), a target protein, to switch on its enzymatic activity. The four EF hands of calmodulin are shown in the same colors as in Figure 13-45. The bound Ca^{2+} ions are shown in orange. [Drawn from coordinates kindly provided by Dr. Mitsu Ikura. M. Ikura, G.M. Clore, A.M. Groneborn, G. Zhu, C.B. Klee, and A. Bax. *Science* 256(1992):632.]

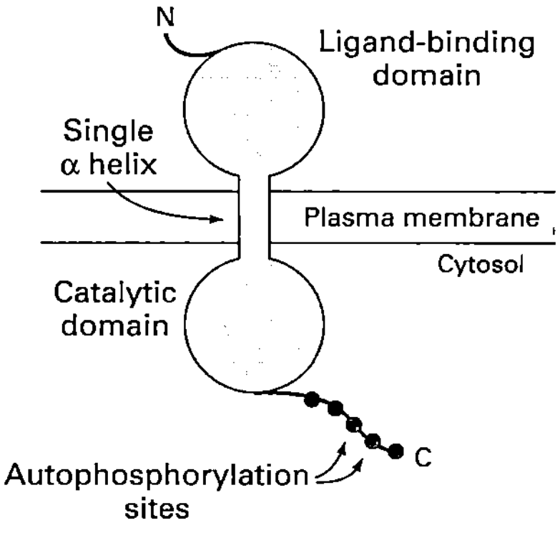

Figure 13-48
Schematic diagram of a receptor tyrosine kinase. The extracellular domain contains a binding site for a growth factor, and the cytosolic domain contains a tyrosine kinase catalytic site. The two domains are connected by a single transmembrane helix.

MEMBRANE-SPANNING RECEPTOR TYROSINE KINASES CONTROL CELL GROWTH AND DIFFERENTIATION

We continue with the theme of protein phosphorylation in signal transduction and consider a new class of proteins, *receptor tyrosine kinases* (Figure 13-48). The binding of growth factors such as *insulin, epidermal growth factor,* and *platelet-derived growth factor* to the extracellular domain of these transmembrane receptors switches on the kinase activity of their catalytic domain, which is located on the cytosolic side of the membrane. Receptor tyrosine kinases, in contrast with seven-helix receptors, are themselves enzymes. By phosphorylating tyrosine residues in target proteins, they directly activate proteins that induce cell growth and differentiation. Mutations causing persistent tyrosine kinase activity in these receptors lead to cancer. Indeed, such mutants account for a high proportion of known *oncogenes* (cancer-producing genes).

The cloning of many receptor tyrosine kinases by Axel Ullrich and co-workers has revealed several structural patterns and a common mechanism of action. Four classes of receptor tyrosine kinases are evident (Figure 13-49). Type 1 is exemplified by the *epidermal growth factor receptor*, type 2 by the *insulin receptor*, type 3 by the *platelet-derived growth factor receptor*, and type 4 by the *fibroblast growth factor receptor*. Types 1, 3, and 4 are monomeric, whereas type 2 is an $\alpha_2\beta_2$ tetramer. The transmembrane chains of these receptors cross the lipid bilayer only once.

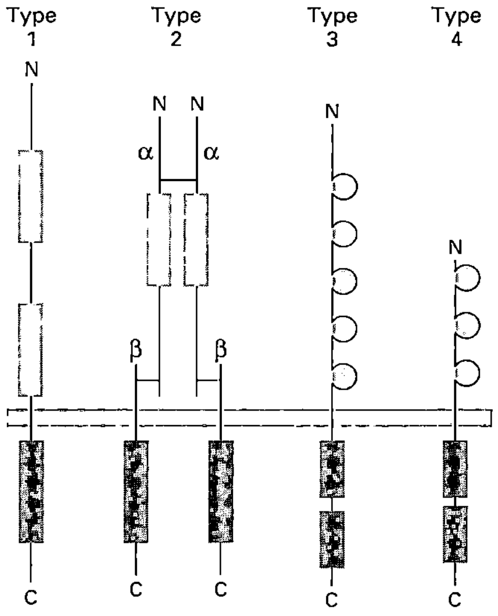

Figure 13-49
Four classes of receptor tyrosine kinases. Cysteine-rich extracellular domains (green) are present in type 1 and 2 receptors. Antibody-like domains (blue) are present in type 3 and 4 receptors, and their kinase domains are interrupted by an insert. Kinase domains are shown in red. In type 2 receptors, each α chain is disulfide-bonded to a β chain and the other α chain. The N terminus of all chains is on the extracellular side.

RECEPTOR TYROSINE KINASES ARE ACTIVATED BY LIGAND-INDUCED DIMERIZATION

How are receptor tyrosine kinases activated by growth factors? Let us examine the best-understood mechanism. *Epidermal growth factor (EGF)* was discovered serendipitously by Stanley Cohen while he was studying nerve growth factor (NGF), a protein that stimulates the development of sympathetic neurons and some sensory neurons. He found that the injection of crude preparations of NGF into newborn mice led to precocious eyelid opening and tooth eruption. These unexpected effects pointed to the existence of a new growth factor, which turned out to be EGF, a 6-kd polypeptide that stimulates the growth of epidermal and epithelial cells (Figure 13-50). This 53-residue growth factor is produced by the cleavage of an *EGF precursor*, a very large (1168-residue) transmembrane protein.

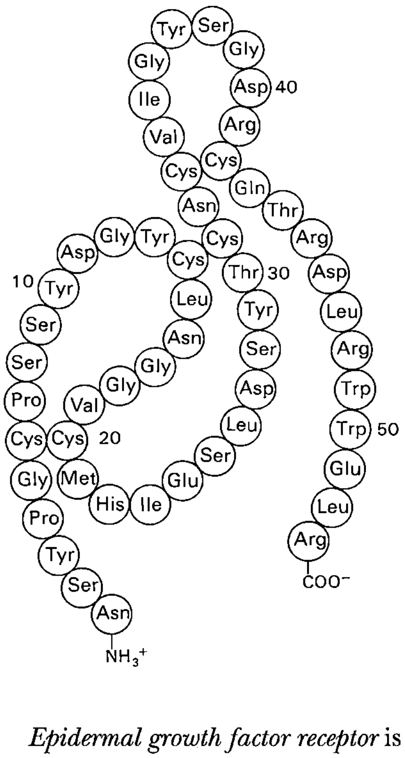

Figure 13-50
Amino acid sequence of epidermal growth factor (EGF). [After C.R. Savage, Jr., J.H. Hash, and S. Cohen. *J. Biol. Chem.* 248(1973):7669.]

Epidermal growth factor receptor is a single polypeptide chain consisting of 1186 residues. The receptor is monomeric and enzymatically inactive in the absence of the growth factor. *Binding of EGF to the extracellular domain causes the receptor to dimerize and undergo autophosphorylation* (Figure 13-51). The catalytic site of one chain phosphorylates five tyrosines located near the C terminus of the other chain in the dimer. *Autophosphorylation enhances the capacity of the receptor to phosphorylate other targets.* As will be evident shortly, *phosphorylated tyrosines serve as docking sites for target proteins.*

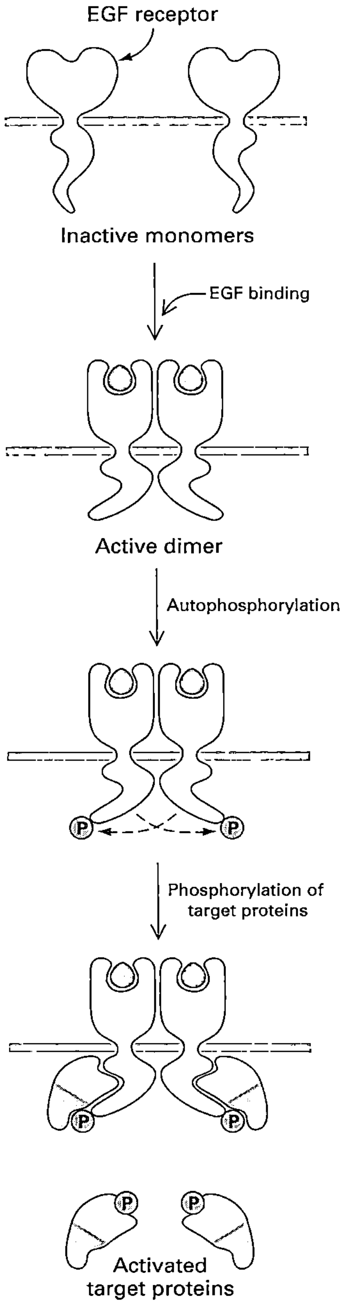

Figure 13-51
The binding of EGF (green) leads to dimerization of the EGF receptor and switching on of its tyrosine kinase activity. Autophosphorylation enables the receptor to bind and phosphorylate target proteins. Phosphorylated tyrosines (red) on the activated EGF receptor are recognized by conserved sequences on target proteins called SH2 domains (blue). Other receptor tyrosine kinases act similarly. [After J. Schlessinger and A. Ullrich. *Neuron* 9(1992):383.]

The binding of platelet-derived growth factor and fibroblast growth factor likewise dimerizes their receptors and leads to their autophosphorylation. What about the insulin receptor, which is a disulfide-bonded dimer of $\alpha\beta$ pairs, whether or not insulin is bound (see Figure 13-49)? The α chains are on the extracellular side, whereas the β chains cross the membrane once. Each $\alpha\beta$ subunit is derived from a single-chain precursor. The insulin receptor resembles the EGF receptor in sequence and overall architecture. Is the binding of growth factor signaled across the membrane portion of both receptors in the same way? This question was answered by synthesizing a gene that encoded a *chimeric receptor*—the extracellular portion came from the insulin receptor, and the membrane-spanning and cytosolic parts from the EGF receptor (Figure 13-52). The striking result was that the binding of insulin induced tyrosine kinase activity, as evidenced by rapid autophosphorylation. Hence, *the insulin receptor and the EGF receptor employ a common mechanism of signal transmission across the plasma membrane.* Dimerization is necessary but not sufficient for activation. Binding of the growth factor must also bring the subunits of the dimer into apposition so that autophosphorylation occurs. All receptor tyrosine kinases probably use essentially the same activation mechanism.

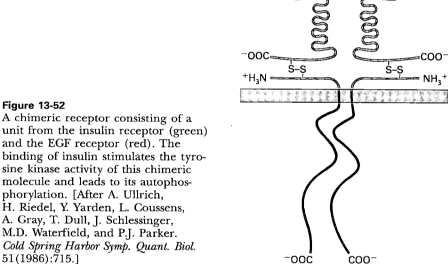

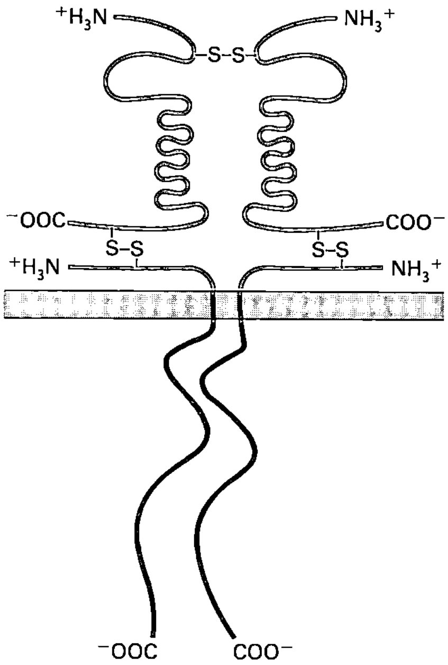

Figure 13-52
A chimeric receptor consisting of a unit from the insulin receptor (green) and the EGF receptor (red). The binding of insulin stimulates the tyrosine kinase activity of this chimeric molecule and leads to its autophosphorylation. [After A. Ullrich, H. Riedel, Y. Yarden, L. Coussens, A. Gray, T. Dull, J. Schlessinger, M.D. Waterfield, and P.J. Parker. *Cold Spring Harbor Symp. Quant. Biol.* 51(1986):715.]

Is the kinase activity of this family of receptors essential for the induction of cell division and differentiation? The sequence GXGXXG occurs near the N terminus of the catalytic domain. An invariant lysine that is essential for kinase activity is located about 20 residues on the carboxyl side of this sequence. *Substitution of the essential lysine by any other residue blocks the capacity of the receptor to activate all known targets. Kinase activity is essential for the transmission of growth signals.*

ACTIVATED TYROSINE KINASE RECEPTORS ARE RECOGNIZED BY SH2 DOMAINS OF TARGET PROTEINS

Phosphorylated tyrosine residues of activated receptors have another important role, which was revealed through studies of the *src protein,* a 60-kd membrane-associated tyrosine kinase. Src contains a tyrosine kinase catalytic domain but lacks an extracellular ligand-binding domain. Activated receptor tyrosine kinases bind src and phosphorylate tyrosine residues on the protein, which switches on its catalytic activity. How do src and other growth-control proteins recognize the activated form of a receptor tyrosine kinase? They contain a 100-residue module called *SH2 (src homology region 2)*. This recurring motif plays a key role in the recognition process (Figure 13-53).

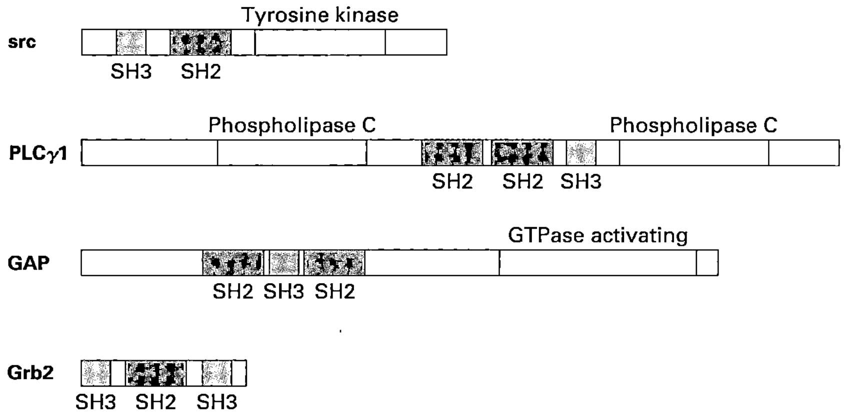

Figure 13-53
SH2 (red) and SH3 (blue) domains recur in signaling proteins. Some of them contain other functional modules (yellow). [After J. Schlessinger and A. Ullrich. *Neuron* 9(1992):383.]

SH2 domains bind phosphorylated peptides (Figure 13-54). In fact, *they direct src and other proteins containing them to activated receptor tyrosine kinases.* The activation of the γ1 isoform of phospholipase C (PLCγ1) nicely illustrates the role of SH2 domains in the transmission of growth stimuli. Most isoforms of PLC are controlled by seven-helix receptors through G proteins. However, PLCγ1 is activated by the EGF receptor, which leads to the mobilization of Ca^{2+} and the activation of protein kinase C. PLCγ1, in contrast with other isoforms of PLC, contains SH2 domains. They bind to phosphorylated tyrosines near the C terminus of the activated receptor. *This docking places a key tyrosine of PLCγ1 in just the right orientation to enable it to be phosphorylated by the catalytic domain of the EGF receptor.* Phosphorylation of PLCγ1 switches on its phospholipase activity. Moreover, it enables

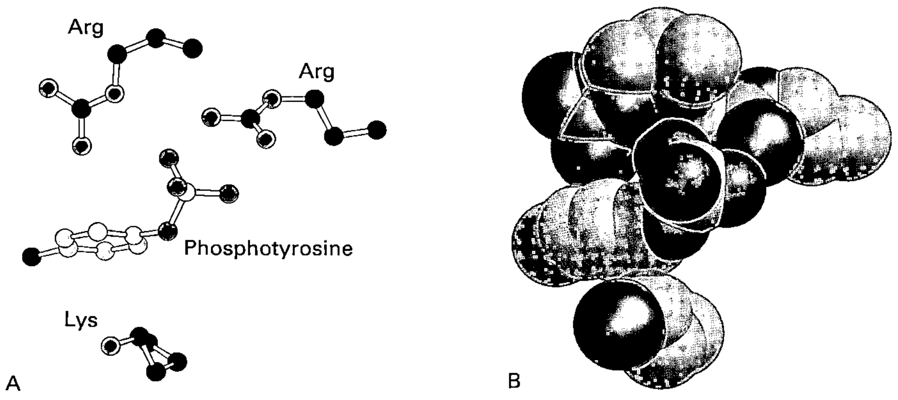

Figure 13-54
Recognition of a phosphorylated tyrosine peptide by an SH2 domain. Two arginines and a lysine from an SH2 domain make intimate contact with both the phosphate group and the aromatic ring (green) of phosphotyrosine. (A) Ball-and-stick model. (B) Space-filling model. [Drawn from coordinates kindly provided by Dr. John Kuriyan. G. Waksman, S. Shoelson, N. Pant, D. Cowburn, and J. Kuriyan. *Cell* 72(1993):1.]

the SH3 domain of PLCγ1 to bind to the membrane cytoskeleton, which positions it to hydrolyze membrane-bound PIP_2 to IP_3 and diacylglycerol. The activated EGF receptor phosphorylates a number of other proteins that bear SH2 domains to induce the proliferation of epithelial cells.

MANY CANCER-PRODUCING GENES (ONCOGENES) ENCODE ALTERED SIGNAL TRANSDUCTION PROTEINS

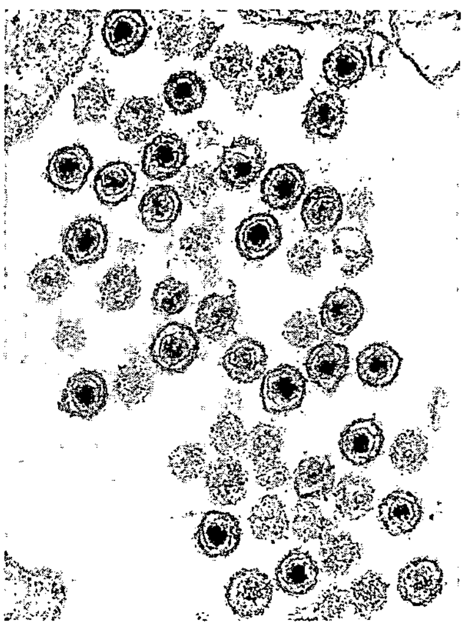

Figure 13-55
Electron micrograph of avian sarcoma virus (Rous sarcoma virus), a retrovirus. The densely stained virions are near the surface of an infected chicken cell. [Courtesy of Dr. Samuel Dales.]

In 1909, a poultry farmer in rural upstate New York observed a tumor in a hen and took the sick fowl to Peyton Rous, an investigator at the Rockefeller Institute in New York City. Rous prepared a cell-free filtrate from this connective-tissue tumor and injected it into normal chickens. The striking result was that the recipients developed highly malignant tumors of the same kind, called *sarcomas.* Rous also found that the tumor-causing agent in the filtrate, now known as *Rous sarcoma virus (RSV)* or *avian sarcoma virus (ASV),* could be propagated by serial passage through chickens. Fifty-five years later, Rous received the Nobel prize in recognition of this seminal discovery.

Studies of mutant viruses decades later also revealed that only one viral gene—the *src* gene—is required for transformation (the induction of cancer). The *src* gene acquired its name from its capacity to direct the synthesis of *sar*coma-producing protein. Does the viral *src* gene have a cellular counterpart? Hybridization studies carried out by Michael Bishop and Harold Varmus revealed that normal chicken cells contain a *src* gene that is very similar to the viral one. The cellular gene is called c-*src* to distinguish it from the viral v-*src.* As was mentioned earlier, the src protein is a cytosolic tyrosine kinase. The 19 C-terminal residues of the cellular protein are replaced by 12 unrelated residues in the viral protein. *The viral version of src is constitutively active because it lacks the inhibitory C-terminal region that is present in the cellular version.* The viral version of src does not need to be phosphorylated by receptor tyrosine kinases to transmit growth signals. Hence, it is intrinsically cancer-producing, and so v-*src* is an *oncogene.*

Many other oncogenes have been discovered (Table 13-4). Most of them encode receptor tyrosine kinases or participants in cascades triggered by them. For example, the v-*erbB* oncogene of avian erythroblastosis virus is derived from the gene for the EGF receptor. The last 34 residues

Table 13-4
Retroviral oncogenes

Class	*Oncogene*	*Retrovirus*
Tyrosine kinases	*abl*	Abelson murine leukemia virus
	erbB	Avian erythroblastosis virus
	src	Avian sarcoma virus (Rous sarcoma virus)
Growth factors	*sis*	Simian sarcoma virus
Growth factor receptors	*erbB*	Avian erythroblastosis virus
Guanyl nucleotide-binding	Ha-*ras*	Harvey murine sarcoma virus
proteins	Ki-*ras*	Kirsten murine sarcoma virus
Nuclear proteins	*fos*	FBJ osteosarcoma virus
	myb	Avian myeloblastosis virus
	myc	Avian myelocytomatosis virus

and nearly all the extracellular domain are missing (Figure 13-56). The result is a constitutively active membrane-spanning tyrosine kinase. A different means of generating a permanently active tyrosine kinase is illustrated by the *neu* oncogene, originally identified in a rat *neu*roblastoma tumor. The human analog *HER2* is often found in breast and ovarian carcinomas, and is correlated with a poor prognosis. A point mutation in the transmembrane region activates the neu protein by inducing its dimerization in the absence of ligand. *Oncogenic neu is persistently autophosphorylated.* Furthermore, *it forms a complex with PLCγ1 and activates the phospholipase in the absence of an external signal.*

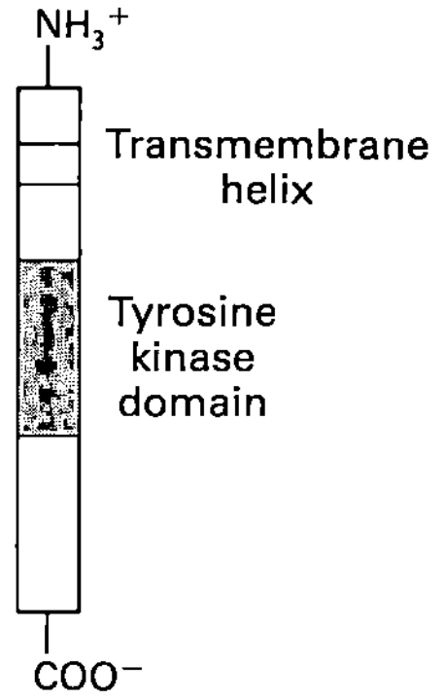

Figure 13-56
The viral oncogene v-*erbB* encodes a truncated form of the EGF receptor. This oncogene product has a permanently stimulated tyrosine kinase activity.

MUTATIONS OF RAS THAT BLOCK ITS GTPase ACTIVITY LEAD TO CANCER

Impaired GTPase activity in a regulatory protein can also lead to cancer. Mammalian cells contain three 21-kd *ras proteins* (H-, K-, and N-ras) that cycle between GTP and GDP forms. The GTP forms of these proteins stimulate cell growth and differentiation in mammals. Ras proteins are members of the *small GTPase protein family* (20 to 35 kd). They differ from G proteins in being smaller and monomeric but share several mechanistic and structural motifs. The GTPase domain of ras (Figure 13-57) is like its counterpart in transducin.

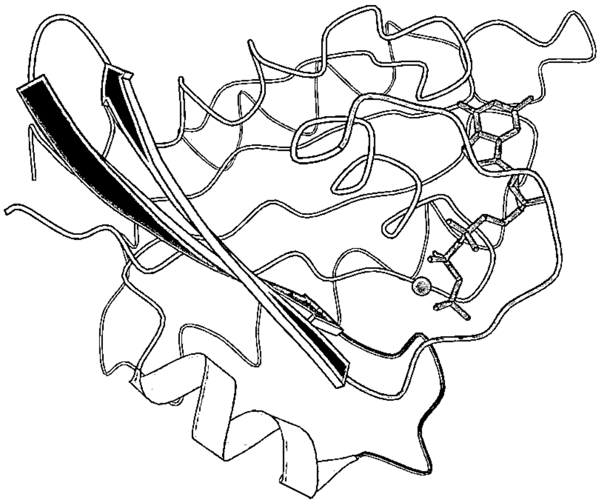

Figure 13-57
Three-dimensional structure of ras protein containing a hydrolysis-resistant analog of GTP (purple). Mg^{2+} (orange sphere) is coordinated to the β and γ phosphoryl groups of the GTP analog. The loop shown in blue is critical in binding and hydrolyzing GTP. It also participates in the conformational change induced by GTP hydrolysis. The helix shown in green, which interacts with effector proteins, shifts position when GDP is bound. [Drawn from 5p21.pdb. E.F. Pai, U. Krengel, G.A. Petsko, R.S. Goody, W. Kabsch, and A. Wittinghofer. *EMBO J.* 9:2351(1990).]

The ras proteins were discovered through the transforming action of a viral counterpart, *v-ras,* that is encoded by murine sarcoma viruses. This viral oncogene product differs from the normal cellular one in having diminished GTPase activity, which keeps it persistently activated. Indeed, a single amino acid change (glycine to valine) converts the normal gene into an oncogenic one. The hydrolysis of bound GTP to GDP in normal ras proteins is greatly accelerated by a 116-kd *GTPase activating protein (GAP).* GAP binds to the oncogenic ras protein but fails to speed its hydrolytic rate, which is only 0.01 s^{-1}. Consequently, oncogenic ras is nearly always in the GTP form.

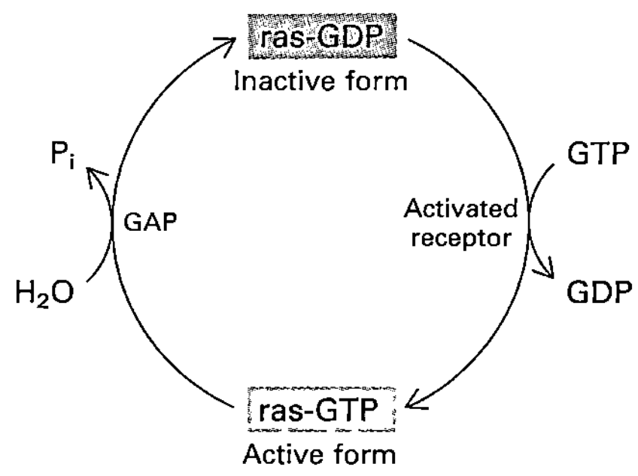

RAS PLAYS A KEY ROLE IN THE CONTROL OF CELL FATE

Studies of an interesting *Drosophila* mutant have provided a revealing view of ras in action. Each structural unit of the compound eye of the fruit fly contains 8 photoreceptor cells, named R1 to R8 (Figure 13-58), and 12 nonneuronal cells. A mutation called *sevenless (sev)* prevents R7 from developing into a photoreceptor cell; instead, it becomes a lens cell. The sev protein on R7 is normally activated by a membrane-bound protein on the adjacent R8 cell. The gene encoding this activator was evocatively named *bride-of-sevenless* and less romantically abbreviated as *boss*. How does *boss* direct R7 to become a photoreceptor cell? The cloning of the *sev* gene showed that it encodes a very large protein (2554 residues) with two putative transmembrane segments and a tyrosine kinase domain. The target of the sev kinase in the R7 cell is a protein encoded by the *son-of-sevenless (sos)* gene; they are bridged by the drk protein. The sos protein in turn activates ras by stimulating GTP-GDP exchange. Thus, *the ras protein is a participant in a Drosophila differentiation pathway that is triggered by activation of a receptor tyrosine kinase through a cell-cell interaction* (Figure 13-59).

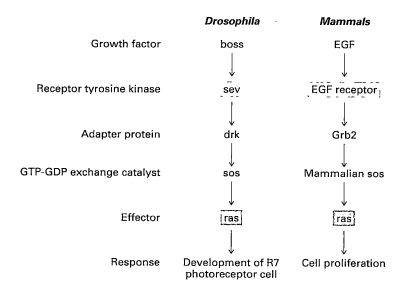

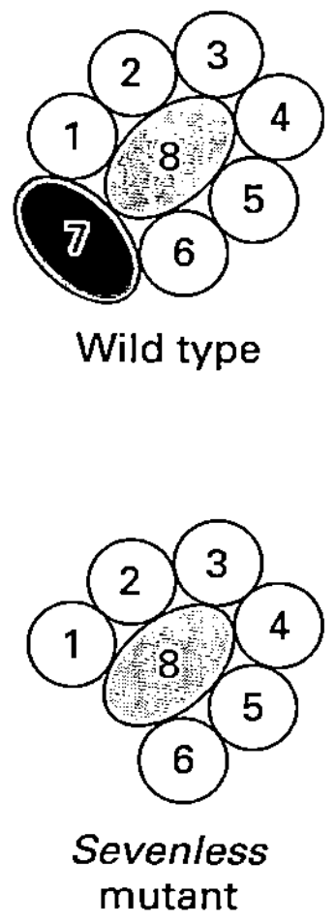

Figure 13-58
Array of photoreceptors in an ommatidium of the compound eye of *Drosophila*. In the wild type, eight cells (R1 to R8) develop. In a *sevenless* mutant, R7 is missing. The sev protein, a receptor tyrosine kinase on R7, is normally activated by a cell-surface protein on R8.

	Drosophila	***Mammals***
Growth factor	boss ↓	EGF ↓
Receptor tyrosine kinase	sev ↓	EGF receptor ↓
Adapter protein	drk ↓	Grb2 ↓
GTP-GDP exchange catalyst	sos ↓	Mammalian sos ↓
Effector	ras ↓	ras ↓
Response	Development of R7 photoreceptor cell	Cell proliferation

Figure 13-59
In *Drosophila* and mammals, receptor tyrosine kinases activate ras through similar proteins. The core of growth-control cascades is highly conserved in evolution.

How does ras control growth and differentiation in mammals? The exciting recent finding is that mammals have a receptor tyrosine kinase pathway strikingly similar to that of flies (Figure 13-59). A mammalian homolog of son-of-sevenless carries the growth signal to ras, which then acts on serine-threonine kinases and other targets. The core of the growth-control pathway is at least 800 million years old. The GTP-GDP switch in ras and G proteins is even more ancient. This elegant molecular device, perfected early in evolution, now serves a variety of cellular functions, including the translocation of macromolecules in protein synthesis, the transduction of hormonal and sensory stimuli, and the regulation of cell proliferation.

SUMMARY

The two best-understood signal transduction processes are bacterial chemotaxis and vertebrate vision. In bacterial chemotaxis, motile cells move toward attractants (such as aspartate) and away from repellents (such as fatty acids). Information from chemoreceptors in the cell membrane is processed by a transduction cascade and transmitted to the rotatory motors of the bacterial flagella. Counterclockwise rotation leads to smooth swimming in a nearly straight line, whereas clockwise rotation induces tumbling, which serves to change the direction of movement. Tumbling is suppressed when a bacterium moves toward an attractant or away from a repellent. Four kinds of receptors transiently bias the direction of flagellar rotation by activating a phosphorylation cascade. Repellent-receptor complexes stimulate the autophosphorylation of CheA. The phosphoryl group is then transferred to CheY, which thereby converts it into a form that induces tumbling. Attractants have the opposite effect. The response to an increase in concentration of attractant or repellent is transient. Adaptation enables bacteria to sense gradients rather than absolute concentrations of attractants and repellents. Reversible methylation of chemotaxis receptors is critical for adaptation. Bacterial chemotaxis reveals that highly purposeful behavior can result from the biasing of random events.

A retinal rod cell can be excited by a single photon. The chromophore of rhodopsin, the photoreceptor protein, is 11-*cis*-retinal, which is derived from all-*trans*-retinol (vitamin A). Rhodopsin is a member of the seven-helix family of membrane receptors that includes the β-adrenergic receptor, olfactory receptors, and yeast mating factor receptors. Visual excitation is triggered by the photoisomerization of 11-*cis*-retinal to the all-trans form. Photoexcited rhodopsin (R*) activates an enzymatic cascade leading to the highly amplified hydrolysis of cyclic GMP (cGMP). A single R* activates hundreds of molecules of transducin, a G protein, by catalyzing the exchange of GTP for bound GDP. The α subunit of transducin bearing GTP then activates a phosphodiesterase specific for cGMP. The lowering of the cGMP level directly closes cation-specific channels in the plasma membrane. Transducin hydrolyzes bound GTP to return to the inactive GDP state. In addition, rhodopsin kinase and arrestin deactivate R*, and guanylate cyclase synthesizes cGMP to restore the dark state. The cytosolic level of Ca^{2+} coordinates recovery and adaptation. Color vision is mediated by three cone receptors that are homologs of rhodopsin.

Cyclic AMP (cAMP) serves as an intracellular messenger in the action of many hormones (e.g., epinephrine). The binding of such a hormone to a seven-helix receptor in the plasma membrane triggers the exchange of GTP for GDP bound to the stimulatory G protein (G_S). $G_{S\alpha}$-GTP activates adenylate cyclase, an integral membrane protein. cAMP then activates protein kinase A (PKA) by binding to its regulatory subunit, thus unleashing its catalytic chains. PKA, a multifunctional kinase, alters the activity of many target proteins by phosphorylating serine and threonine residues. The phosphoinositide cascade is also mediated by seven-helix receptors and G proteins. The receptor-triggered activation of phospholipase C generates two intracellular messengers by hydrolysis of phosphatidyl inositol 4,5-bisphosphate. Inositol trisphosphate (IP_3) opens calcium channels in the endoplasmic and sarcoplasmic reticulum membranes, leading to an elevated level of Ca^{2+} in the cytosol. Diacylglycerol activates protein kinase C, which phosphorylates serine and threonine residues in target proteins. Calcium ion, a ubiquitous intracellular messenger in eukaryotes, acts by binding to calmodulin and other calcium sensors. Calmodulin contains four calcium-binding modules called EF hands that recur in other proteins. Ca^{2+}–calmodulin activates target proteins by binding to positively charged amphipathic helices.

Membrane-spanning receptor tyrosine kinases control cell growth and differentiation. They contain an extracellular domain that binds a growth factor, a transmembrane region, and a cytosolic domain that contains a catalytic site. The best-understood receptor is the one for epidermal growth factor (EGF). The binding of EGF to the receptor induces its dimerization and autophosphorylation. Five tyrosine residues in the carboxyl-terminal tail become autophosphorylated, enabling the receptor to phosphorylate target proteins such as the $\gamma1$ isoform of phospholipase C. Phosphorylated tyrosines in activated receptor tyrosine kinases generally serve as docking sites for SH2 domains that are present in numerous signaling proteins.

Many oncogenes (cancer-producing genes) arise from mutations that give rise to altered proteins in cascades that transduce growth and differention signals. Rous sarcoma virus causes cancer in fowl because it encodes a constitutively active form of the src protein, a cytosolic tyrosine kinase. Mutations of genes for receptor tyrosine kinases often generate oncogenes, as exemplified by *HER2*, which is found in a high proportion of human breast carcinomas. The ras protein, a key intermediate in tyrosine kinase cascades, cycles between an inactive GDP form and an active GTP form. Oncogenic ras has abnormally low GTPase activity, which keeps it persistently active.

SELECTED READINGS

Where to start

Stock, J.B., Lukat, G.S., and Stock, A.M., 1991. Bacterial chemotaxis and the molecular logic of intracellular signal transduction networks. *Ann. Rev. Biophys. Biophys. Chem.* 20:109–136.

Stryer, L., 1987. The molecules of visual excitation. *Sci. Amer.* 255(7):42–50.

Linder, M.E., and Gilman, A.G., 1992. G proteins. *Sci. Amer.* 267(1):56–65.

Bourne, H.R., Sanders, D.A., and McCormick, F., 1990. The GTPase superfamily: A conserved switch for diverse cell functions. *Nature* 348:125–132. [Also see their 1991 article in *Nature* 349:117–127.]

Fantl, W.J., Johnson, D.E., and Williams, L.T., 1993. Signalling by receptor tyrosine kinases. *Ann. Rev. Biochem.* 62:453–481.

Bacterial chemotaxis

Koshland, D.E., Jr., 1988. Chemotaxis as a model second-messenger system. *Biochemistry* 27:5829–5834.

Adler, J., 1976. The sensing of chemicals by bacteria. *Sci. Amer.* 234(4):40–47.

Berg, H.C., and Purcell, E.M., 1977. Physics of chemoreception. *Biophys. J.* 20:193–219. [A beautiful paper, rich in ideas and insight. Also see H.C. Berg's *Random Walks in Biology* (Princeton, 1983).]

Macnab, R., and Koshland, D.E., Jr., 1972. The gradient-sensing mechanism in bacterial chemotaxis. *Proc. Nat. Acad. Sci.* 69:2509–2512. [These experiments revealed that bacteria detect a temporal gradient rather than an instantaneous spatial one.]

Bourret, R.B., Borkovich, K.A., and Simon, M.I., 1991. Signal transduction pathways involving protein phosphorylation in prokaryotes. *Ann. Rev. Biochem.* 60:401–441.

Ninfa, E.G., Stock, A., Mowbray, S., and Stock, J., 1991. Reconstitution of the bacterial chemotaxis signal transduction system from purified components. *J. Biol. Chem.* 266:9764–9770.

Vision

Schnapf, J.L., and Baylor, D.A., 1987. How photoreceptor cells respond to light. *Sci. Amer.* 256(4):40–47. [A lucid account of the electrophysiology of rods and cones.]

Hargrave, P.A., 1991. Seven-helix receptors. *Curr. Opin. Struct. Biol.* 1:575–581.

Fung, B.-K., Hurley, J.B., and Stryer, L., 1981. Flow of information in the light-triggered cyclic nucleotide cascade of vision. *Proc. Nat. Acad. Sci.* 78:152–156.

Noel, J.P., Hamm, H., and Sigler, P.B., 1993. The 2.2 Å crystal structure of transducin-α complexed with GTPγS. *Nature* 366:654–663.

Fesenko, E.E., Kolesnikov, S.S., and Lyubarsky, A.L., 1985. Induction by cyclic GMP of cationic conductance in plasma membrane of retinal rod outer segment. *Nature* 313:310–313.

Nathans, J., 1989. The genes for color vision. *Sci. Amer.* 260(2):42–49.

Wald, G., 1968. The molecular basis of visual excitation. *Nature* 219:800–807. [This Nobel Lecture contains an interesting account of the discovery of the primary event in vision.]

Adenylate cyclase cascades

Dohlman, H.G., Thorner, J., Caron, M.G., and Lefkowitz, R.J., 1991. Model systems for the study of seven-transmembrane segment receptors. *Ann. Rev. Biochem.* 60:653–688.

Krupinski, J., Coussen, F., Bakalyar, H.A., Tang, W.J., Feinstein, P.G., Orth, K., Slaughter, C., Reed, R.R., and Gilman, A.G., 1989. Adenylyl cyclase amino acid sequence: Possible channel- or transporter-like structure. *Science* 244:1558–1564.

Adams, S.R., Harootunian, A.T., Buechler, Y.J., Taylor, S.S., and Tsien, R.Y., 1991. Fluorescence ratio imaging of cyclic AMP in single cells. *Nature* 349:694–697.

Buck, L., and Axel, R., 1991. A novel multigene family may encode odorant receptors: A molecular basis for odor recognition. *Cell* 65:175–187.

Phosphoinositide cascade

Berridge, M.J., 1993. Inositol trisphosphate and calcium signalling. *Nature* 361:315–325.

Sternweis, P.C., and Smrcka, A.V., 1992. Regulation of phospholipase C by G proteins. *Trends Biochem. Sci.* 17:502–506.

Mikoshiba, K., 1993. Inositol 1,4,5-trisphosphate receptor. *Trends Pharm. Sci.* 14:86–89.

Nishizuka, Y., 1992. Intracellular signaling by hydrolysis of phospholipids and activation of protein kinase C. *Science* 258:607–614.

Rhee, S.G., and Choi, K.D., 1992. Regulation of inositol phospholipid-specific phospholipase C isozymes. *J. Biol. Chem.* 267:12393–12396.

Calcium sensors and signaling

Kretsinger, R.H., 1987. Calcium coordination and the calmodulin fold: Divergent versus convergent evolution. *Cold Spring Harbor Symp. Quant. Biol.* 52:499–510.

Ikura, M., Clore, G.M., Groneborn, A.M., Zhu, G., Klee, C.B., and Bax, A., 1992. Solution structure of a calmodulin-target peptide complex by multidimensional NMR. *Science* 256:632–638.

Cohen, P., and Klee, C.B. (eds.), 1988. *Calmodulin.* Elsevier. [A valuable collection of authoritative articles on many facets of calmodulin and its interactions.]

Hanson, P.I., and Schulman, H., 1992. Neuronal Ca^{2+}/calmodulin-dependent protein kinases. *Ann. Rev. Biochem.* 61:559–601.

Tsien, R.Y., 1988. Fluorescence measurement and photochemical manipulation of cytosolic free calcium. *Trends Neurosci.* 11:419–424. [This issue of *TINS* contains many excellent articles on calcium in signaling.]

Woods, N.M., Cuthbertson, S.R., and Cobbold, P.H., 1986. Repetitive transient rises in cytoplasmic free calcium in hormone-stimulated hepatocytes. *Nature* 319:600–602.

Meyer, T. , and Stryer, L., 1991. Calcium spiking. *Ann. Rev. Biophys. Biophys. Chem.* 20:153–174.

Receptor tyrosine kinases

Schlessinger, J., and Ullrich, A., 1992. Growth factor signaling by receptor tyrosine kinases. *Neuron* 9:383–391.

Cantley, L.C., Auger, K.R., Carpenter, C., Duckworth, B., Graziani, A., Kapeller, R., and Soltoff, S., 1991. Oncogenes and signal transduction. *Cell* 64:281–302.

Koch, C.A., Anderson, D., Moran, M.F., Ellis, C., and Pawson, T., 1991. SH2 and SH3 domains: Elements that control interactions of cytoplasmic signaling proteins. *Science* 252:668–674.

Waksman, G., Shoelson, S.E., Pant, N., Cowburn, D., and Kuriyan, J., 1993. Binding of a high affinity phosphotyrosyl peptide to the Src SH2 domain: Crystal structures of the complexed and peptide-free forms. *Cell* 72:779–790.

Rhee, S.G., 1991. Inositol phospholipid-specific phospholipase C: Interaction of the γl isoform with tyrosine kinase. *Trends Biochem. Sci.* 16:297–301.

Weiner, D.B., Liu, J., Cohen, J.A., Williams, W.V., and Greene, M., 1989. A point mutation in the *neu* oncogene mimics ligand induction of receptor aggregation. *Nature* 339:230–231.

Fischer, E.H., Charbonneau, H., and Tonks, N.K., 1991. Protein tyrosine phosphatases: A diverse family of intracellular and transmembrane enzymes. *Science* 253:401–406.

Ras protein

Schlessinger, J., 1993. How receptor tyrosine kinases activate ras. *Trends Biochem. Sci.* 18:273–275.

Simon, M.A., Dodson, G.S., and Rubin, G.M., 1993. An SH3-SH2-SH3 protein is required for p21Ras1 activation and binds to sevenless and Sos proteins in vitro. *Cell* 73:169–177.

Tong, L.A., de Vos, A.M., Milburn, M.V., and Kim, S.H., 1991. Crystal structures at 2.2 Å resolution of the catalytic domains of normal ras protein and an oncogenic mutant complexed with GDP. *J. Mol. Biol.* 217:503–516.

Pai, E.F., Krengel, U., Petsko, G.A., Goody, R.S., Kabsch, W., and Wittinghofer, A., 1990. Refined crystal structure of the triphosphate conformation of H-ras p21 at 1.35 Å resolution: Implications for the mechanism of GTP hydrolysis. *EMBO J.* 9:2351–2359.

Schlichting, I., Almo, S.C., Rapp, G., Wilson, K., Petratos, K., Lentfer, A., Wittinghofer, A., Kabsch, W., Pai, E.F., and Petsko, G.A., 1990. Time-resolved x-ray crystallographic study of the conformational change in Ha-Ras p21 protein on GTP hydrolysis. *Nature* 345:309–315.

PROBLEMS

1. *The richness of phosphorylation.* Compare and contrast the role of phosphorylation in bacterial chemotaxis, vision, and growth-control cascades.

2. *Attractant response.* What is the sequence of molecular changes in the bacterial chemotaxis system following a step increase in the concentration of an attractant such as aspartate?

3. *Eeyore and Candide.* Bacterial chemotaxis mutants have distinctive personalities.
   (a) A phenotype reminiscent of Eeyore, the persistent pessimist in Milne's *Winnie the Pooh,* arises from the loss of which *che* gene?
   (b) The loss of a different *che* gene gives a phenotype resembling Voltaire's Candide, eternally optimistic, convinced that "this is the best of all possible worlds." Which gene?

4. *Truncated rods.* A retinal rod outer segment can be severed near its base to produce a fragment that carries out phototransduction if provided with GTP and ATP. Small molecules and proteins can be diffused into a truncated rod because its base is not sealed. Predict the effect of each of the following manipulations on the response to a light flash that excites a few rhodopsin molecules:
   (a) Addition of an inhibitor of rhodopsin kinase.
   (b) Addition of an inhibitor of the cGMP phosphodiesterase.
   (c) Addition of EGTA.
   (d) Addition of GTPγS (a hydrolysis-resistant analog of GTP).

5. *Color blindness.* In 1794, John Dalton presented a lecture entitled "Extraordinary Facts Relating to the Vision of Colours" to the Manchester Literary and Philosophical Society. Dalton, who later developed the atomic theory, emphasized how his own perception of color differed from that of most people: ". . . that part of the image which others call red appears to me little more than a shade or defect of light; after that the orange, yellow, and green seem one colour which descends pretty uniformly from intense to a rare yellow, making what I should call different shades of yellow. . . . " Dalton, like 8% of males and 0.6% of females generally, was red-green color blind. This form of color blindness is rarer in females because it is caused by a recessive sex-linked mutation. Recent molecular genetic studies have shown that genes for the red and green receptors are next to each other on the X chromosome. In some color-blind men, the green gene is lost. In others, a hybrid of the red and green genes is found in place of the red one. Predict the absorption spectrum of such a hybrid gene. Why would it lead to altered color vision?

6. *Inhibitory constraints.* Many proteins in signal transduction pathways are activated by the removal of an inhibitory constraint. Give three examples of this recurring mechanism.

7. *Fulminating diarrhea.* Cholera toxin catalyzes the transfer of an adenosine diphosphate ribose unit to an arginine residue of the stimulatory G protein (Figure 13-60). This modification, termed *ADP-ribosylation,* blocks the hydrolysis of GTP bound to the α subunit of G_S. How does cholera toxin induce a diarrhea that can be lethal if body fluids are not quickly replaced? [Hint: Phosphorylation of chloride channels in intestinal epithelial cells increases the efflux of Cl^- and water.]

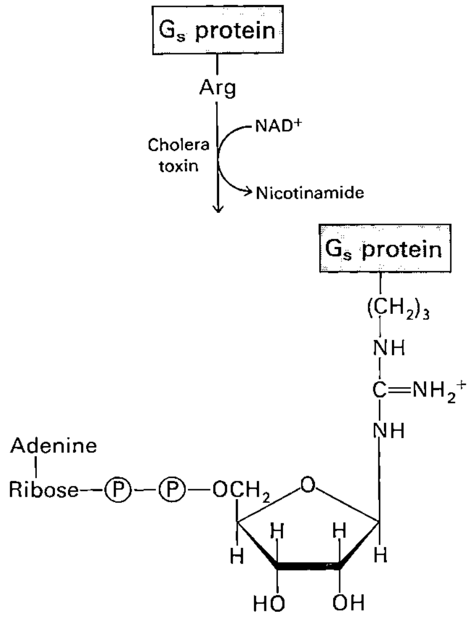

Figure 13-60
Cholera toxin catalyzes the ADP-ribosylation of G_S and some other G proteins.

8. *Calcium pump regulation.* The Ca^{2+}-ATPase pump in the plasma membrane of eukaryotic cells is activated by Ca^{2+}–calmodulin. The C-terminal region of this 138-kd pump contains a sequence similar to one in calmodulin.

Ca^{2+}-ATPase	Glu-Glu-Glu-Ile-Phe-Glu
Calmodulin	Glu-Glu-Glu-Ile-Arg-Glu

How might this sequence make the pump responsive to activation by Ca^{2+}–calmodulin?

9. *Reversing oncogenesis.* Receptor tyrosine kinases are potential drug targets in cancer therapy. How might the effect of an oncogenic mutation be reversed by a small molecule? Propose a search strategy for such a drug.

CHAPTER

14

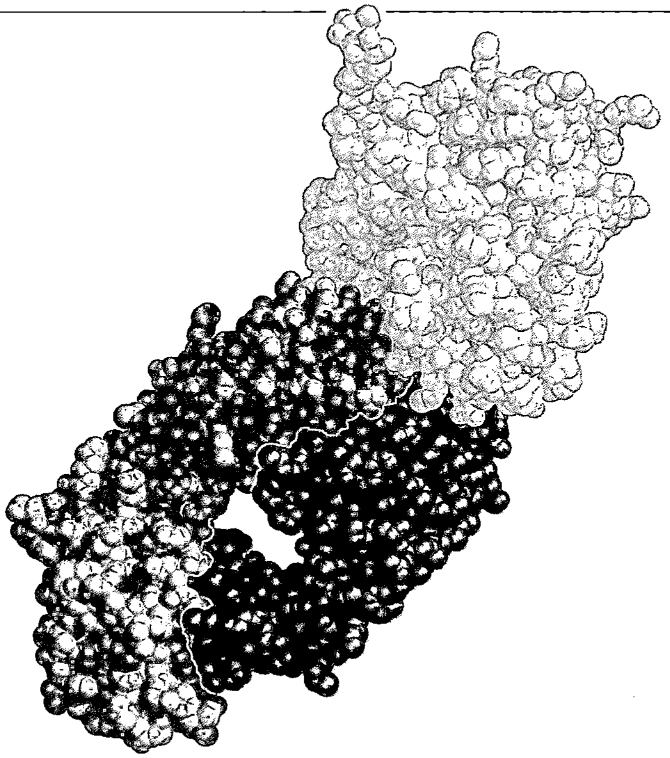

Antibodies and T-Cell Receptors

The vertebrate immune system is a vast network of molecules and cells having a single goal: *to distinguish between self and nonself.* Its primary function is to protect vertebrates against microorganisms—viruses, bacteria, and parasites. It continuously scans countless numbers of molecular units to decide which ones are foreign and initiate their destruction. The immune system learns from experience and remembers its encounters. Its hallmarks are *specificity* and *memory.* The immune system uses two different but related strategies. The recognition elements of the *humoral immune response* are soluble proteins called *antibodies (immunoglobulins),* which are produced by plasma cells. In the *cellular immune response,* T lymphocytes kill cells that display foreign motifs on their surface. They also stimulate the humoral response by helping B cells, the precursors of plasma cells. Both killing and helping are initiated by the binding of *T-cell receptors* to peptides presented by *MHC proteins* on the surface of target cells.

The molecular basis of these remarkable processes is rapidly being unraveled. Immunology is deepening our understanding of disease processes and providing new diagnostic and therapeutic approaches. Monoclonal antibodies, for example, serve as highly specific detectors of pathogenic microorganisms, tissue damage, and cancer. The immune system is subverted by a retrovirus in acquired immune deficiency syndrome (AIDS), a devastating disease. Immunology is also providing insight into how cell development is triggered and how cells stimulate and suppress

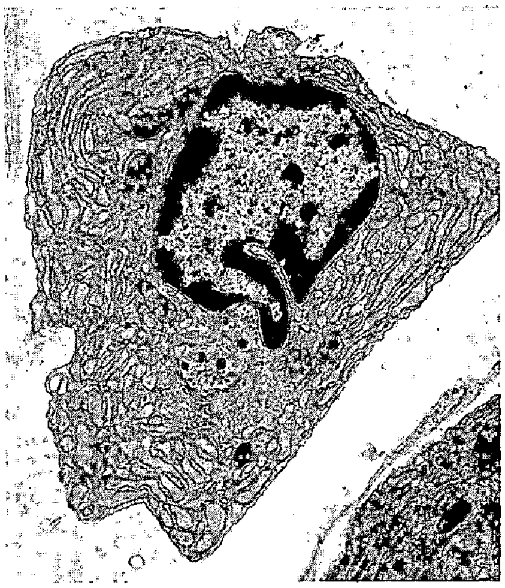

Figure 14-1
Electron micrograph of a plasma cell. The rough endoplasmic reticulum of this secretory cell is highly developed. [Courtesy of Lynne Mercer.]

Opening Image: *Three-dimensional structure of an antibody-antigen complex. An antibody fragment (F_{ab}) (blue and red) is bound to neuraminidase (yellow), a key enzyme of influenza virus. [Drawn from 1nca.pdb. W.R. Tulip, J.N. Varghese, W.G. Laver, R.G. Webster, and P.M. Colman. J. Mol. Biol. 227(1992):122.]*

one another. The immunoglobulin fold is a fundamental structural motif that is used by all multicellular organisms to mediate cell-cell adhesion and recognition. Links between molecular immunology and developmental biology have been forged. It is a time of ferment and excitement.

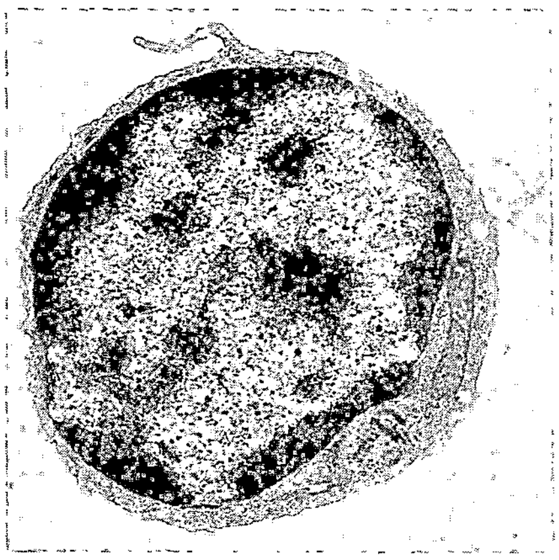

Electron micrograph of a B lymphocyte. [Courtesy of Lynne Mercer.]

SPECIFIC ANTIBODIES ARE SYNTHESIZED FOLLOWING EXPOSURE TO AN ANTIGEN

Antibodies (or immunoglobulins) are proteins synthesized by an animal in response to the presence of a foreign substance. They are secreted by *plasma cells*, which are derived from *B lymphocytes* (*B cells*). These soluble proteins are the recognition elements of the *humoral immune response* (*humor* is the Latin word for "liquid"). Each antibody has specific affinity for the foreign material that stimulated its synthesis. A foreign macromolecule capable of eliciting antibody formation is called an *antigen* (or *immunogen*). Proteins, polysaccharides, and nucleic acids are usually effective antigens. The specific affinity of an antibody is not for the entire macromolecular antigen, but for a particular site on it called the *antigenic determinant* (or *epitope*).

Most small foreign molecules do not stimulate antibody formation. However, they can elicit the formation of specific antibody if they are attached to macromolecules. The macromolecule is then the *carrier* of the attached chemical group, which is called a *haptenic determinant*. The small foreign molecule by itself is called a *hapten*. Antibodies elicited by attached haptens will bind unattached haptens as well.

Animals can make specific antibodies against virtually any foreign chemical group. The dinitrophenyl group (DNP) is particularly effective in eliciting antibody formation and therefore has been used extensively as a haptenic determinant. Antibodies specific for DNP (termed *anti-DNP antibody*) can be obtained in the following way:

1. DNP groups are covalently attached to a carrier protein such as bovine serum albumin (BSA) by reaction of fluorodinitrobenzene with lysine and other nucleophilic side chains (Figure 14-2).

2. DNP–BSA, the immunogen (antigen), is injected into a rabbit. The level of anti-DNP antibody in the serum of the rabbit starts to rise a few days later (Figure 14-3). These *early antibody molecules* belong to the *immunoglobulin M (IgM)* class and have a mass of nearly 1000 kd.

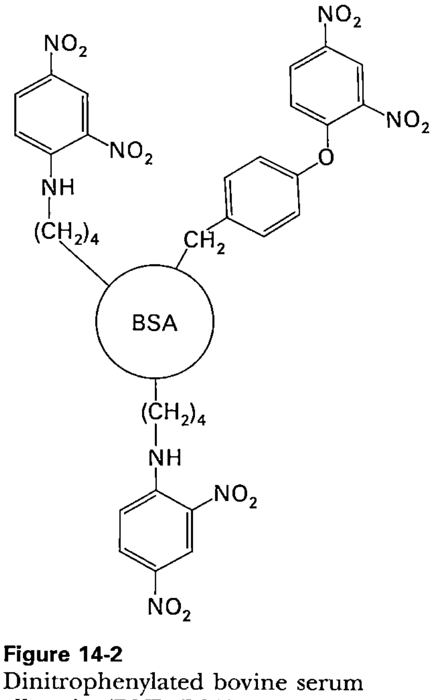

Figure 14-2
Dinitrophenylated bovine serum albumin (DNP–BSA) is an effective immunogen.

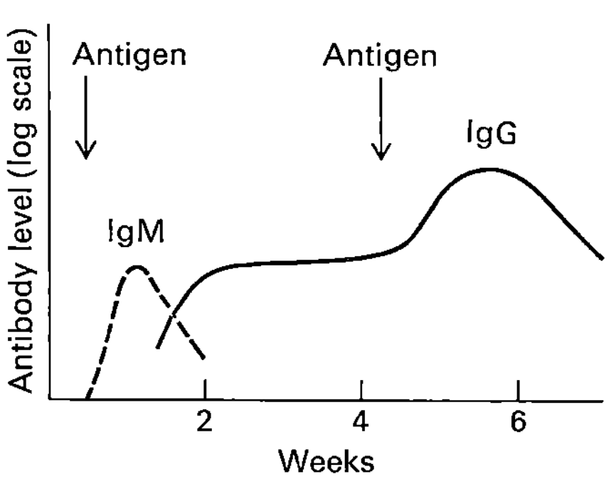

Figure 14-3
Kinetics of the appearance of immunoglobulins M and G (IgM and IgG) in the serum following immunization.

3. Approximately 10 days after the injection of immunogen, the amount of immunoglobulin M decreases, and there is a concurrent increase in the amount of anti-DNP antibody of a different class, called *immunoglobulin G* (IgG), which has a mass of 150 kd.

4. The level of anti-DNP antibody of the immunoglobulin G class reaches a plateau approximately three weeks after the injection of immunogen. A booster dose of DNP–BSA given at that time produces a further increase in the level of anti-DNP antibody in the rabbit's serum.

5. Blood is drawn from the immunized rabbit. The separated serum (called an *antiserum* because it is obtained after immunization) may contain as much as 1 mg/ml of anti-DNP antibody. Nearly all of it is of the IgG class, the principal immunoglobulin in serum.

6. The next step is to separate anti-DNP antibody from antibodies of other specificities and from other serum proteins. The distinguishing feature of anti-DNP antibody is its very high affinity for DNP. Consequently, it can be separated by affinity chromatography using a column consisting of dinitrophenyl groups covalently attached to an insoluble carbohydrate matrix.

ANTIBODIES ARE FORMED BY SELECTION RATHER THAN BY INSTRUCTION

Enzymes acquired their specificities over millions of years of evolution. Specific antibodies appear in the serum of an animal only a few weeks after exposure to virtually any foreign determinant. How are specific antibodies formed so rapidly? The *instructive theory,* which was proposed by Linus Pauling in 1940, postulated that the antigen acts as a template that directs the folding of the nascent antibody chain. In this model, antibody molecules of a given amino acid sequence have the potential for forming combining sites of many different specificities. The particular one formed depends on the antigen present at the time of folding. A contrasting hypothesis, called the *selective theory* or *clonal selection theory,* was proposed and developed by Macfarlane Burnet, Niels Jerne, David Talmage, and Joshua Lederberg in the 1950s. It postulated that the *combining site of an antibody molecule is completely determined before it ever encounters antigen.* In the selective theory, the antigen affects only the *amount* of specific antibody produced.

The instructive theory predicted that an antibody would lose its specificity if it were unfolded and then refolded in the absence of antigen. This prediction contrasted with experimental results obtained in Christian Anfinsen's laboratory on the refolding of ribonuclease. Recall that unfolded ribonuclease spontaneously assumes the three-dimensional structure, specificity, and catalytic activity of the native enzyme upon removal of the denaturing agent (p. 38). A substrate need not be present during the refolding of ribonuclease. Anfinsen's important experiment on ribonuclease provided the impetus for devising a similar test for antibodies (Figure 14-4). A fragment containing the antigen-binding site of anti-DNP antibody was unfolded in 7 M guanidine hydrochloride. The denatured fragment had no detectable affinity for DNP haptens. Hydrodynamic and optical measurements showed that its conformation approached that of a random coil. On removal of guanidine hydrochloride by dialysis, *the chain refolded in the absence of hapten into a compact unit that regained specific affinity for DNP.* This decisive finding contradicted a critical prediction of the instructive theory. By contrast, the selective theory was strongly supported by the observation that *antibody-producing cells can synthesize large amounts of specific antibody in the complete absence of the corresponding antigen.*

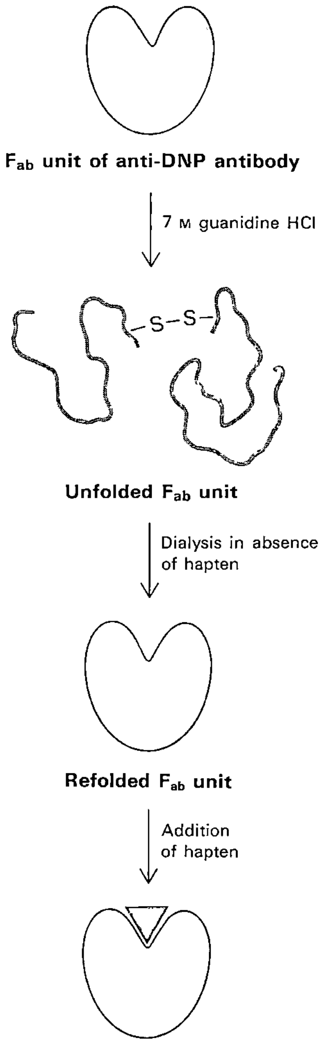

Figure 14-4
The F_{ab} fragment of anti-DNP antibody binds DNP after being unfolded and refolded in the absence of this hapten. This experiment showed that specificity is inherent in the amino acid sequence.

The clonal selection theory provides a unifying view of the immune response. The essential features of this visionary hypothesis, now firmly established, are

1. Each antibody-producing cell makes antibody of a single kind. *The commitment to synthesize a particular kind of antibody is made before the cell ever encounters antigen.*

2. The specificity of an antibody is determined by its amino acid sequence. Each cell has a distinctive base sequence in its DNA that determines the amino acid sequence of its immunoglobulin chains.

3. As it begins to mature, each antibody-producing cell makes small amounts of a specific antibody, which appears on its surface. An immature cell is killed if it encounters a molecule that binds to the antibody it expresses. Such cells are eliminated during fetal life. Hence, an animal does not usually make antibody against its own macromolecules—it is *self-tolerant.* In contrast, *a mature cell is stimulated if it encounters antigen. Large amounts of antibody are then synthesized, and the cell is triggered to divide.* The descendants of this cell are a *clone.* They have the same genetic makeup as the cell that was initially triggered by its encounter with antigen, and therefore all of them make antibody of the same specificity.

4. A clone tends to persist after the disappearance of the antigen. These cells retain the capacity to be stimulated by antigen if it reappears, which provides for long-term *immunological memory* and a quick response to reinfection.

On immunity and the plague of Athens—
"All speculation as to its origins and its causes, if causes can be found adequate to produce so great a disturbance, I leave to other writers. . . . For myself, I shall simply set down its nature, and explain the symptoms by which it may be recognized by the student, if it should ever break out again. This I can the better do, as I had the disease myself, and watched its operation in the case of others. . . .
Yet it was with those who had recovered from the disease that the sick and the dying found most compassion. These knew what it was from experience, and had now no fear for themselves; *for the same man was never attacked twice*—never at least fatally. And such persons not only received the congratulations of others, but themselves also, in the elation of the moment, *half entertained the vain hope that they were for the future safe from any disease whatsoever.* . . ."

THUCYDIDES (460–400 B.C.)
The History of the Peloponnesian War

The essence of the clonal selection theory, as expressed by Burnet, is that

> no combining site is in any evolutionary sense adapted to a particular antigenic determinant. The pattern of the combining site is there and if it happens to fit, in the sense that the affinity of adsorption to a given antigenic determinant is above a certain value, immunologically significant reaction will be initiated.

The selection of a preexisting pattern that is randomly generated is not a new theme in biology. Indeed, it is at the heart of Darwin's theory of evolution. It is interesting to note that instructive theories have preceded selective theories: Lamarck's before Darwin's, antigen-directed binding before clonal selection.

ANTIBODIES CONSIST OF DISTINCT ANTIGEN-BINDING AND EFFECTOR UNITS

Immunoglobulin G, the major antibody in serum, has a mass of 150 kd. A fruitful approach in studying such a large protein is to split it into fragments that retain activity. In 1959, Rodney Porter showed that immunoglobulin G can be cleaved into three active 50-kd fragments by the limited proteolytic action of papain.

$$\underset{(150\text{ kd})}{\text{IgG}} \xrightarrow{\text{Papain}} \underset{(50\text{ kd each})}{2\,\text{F}_{\text{ab}}} + \underset{(50\text{ kd})}{\text{F}_{\text{c}}}$$

Two of these fragments bind antigen. They are called F_{ab} (*ab* stands for "antigen-binding," *F* for "fragment"). Each F_{ab} contains one combining site for hapten or antigen, and it has the same binding affinity for hapten as does the whole molecule. However, F_{ab} cannot cross-link an antigen that contains multiple determinants because it is univalent (i.e., contains only one combining site), in contrast with intact immunoglobulin G, which contains two *identical* antigen-binding sites (Figure 14-5).

The other fragment (also 50 kd), called F_c because it crystallizes readily, does not bind antigen, but it has other important biological activities. F_c mediates protective responses termed *effector functions,* such as *complement fixation.* The binding of an antibody to an antigenic determinant on the surface of a foreign cell leads to a cascade of reactions that lyses the intruder (p. 376). Antibody-antigen complexes are also taken up by *macrophages,* which contain receptors for the F_c unit on their surface. Multivalent antigens induce the cross-linking of antibodies bound to the *F_c receptors,* which in turn triggers *phagocytosis.* These complexes are internalized and delivered to lysosomes for digestion. F_c units are also recognized by membrane transport systems, such as the one that ferries antibodies across the placenta to the fetus.

Electron-microscopic studies showed that immunoglobulin G has the shape of the letter *Y* and that the hapten binds near the ends of the F_{ab} units. Furthermore, the F_c and the two F_{ab} units of the intact antibody are joined by a hinge that allows rapid variation in the angle between the F_{ab} units through a wide range (Figure 14-6). This kind of mobility, called *segmental flexibility,* enhances the formation of antibody-antigen complexes by enabling both combining sites on an antibody to bind a multivalent antigen (i.e., an antigen having multiple determinants). The distance between combining sites at the tips of the F_{ab} units can be adjusted to match the distance between specific determinants on the antigen (e.g., a virus with many repeating subunits). The binding affinity is much higher when both combining sites of an antibody, rather than just one, are complexed to a multivalent antigen.

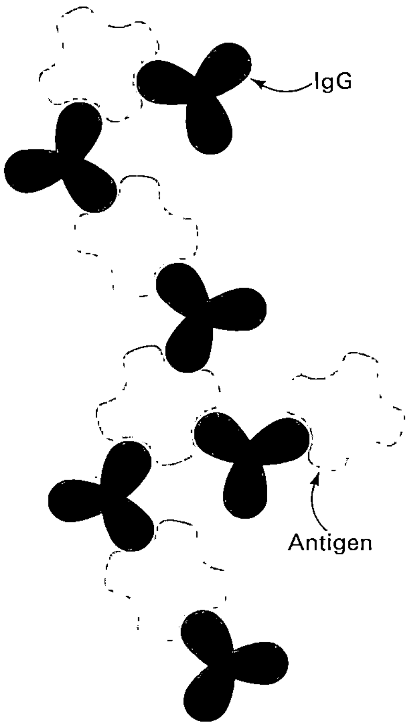

Figure 14-5
Schematic diagram of a lattice formed by the cross-linking of IgG (blue) and a multivalent antigen (yellow).

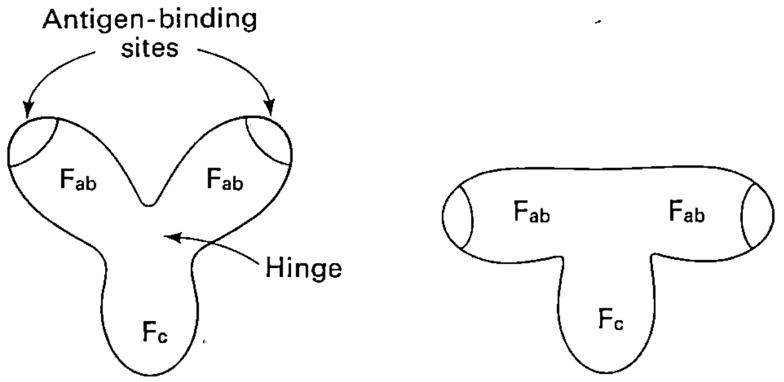

Figure 14-6
Immunoglobulin G has the shape of the letter *Y.* It contains a hinge that confers segmental flexibility.

MONOCLONAL ANTIBODIES WITH VIRTUALLY ANY DESIRED SPECIFICITY CAN READILY BE PREPARED

Antibodies differ from enzymes in an important respect. Most naturally occurring antibodies of a given specificity, such as anti-DNP antibody, are not a single molecular species. Analyses of the binding of DNP haptens to a preparation of anti-DNP antibody revealed a wide range of binding affinities—the dissociation constants ranged from about 0.1 nM to 1 μM. Furthermore, a large number of bands were evident when anti-DNP antibody was subjected to isoelectric focusing. *Antibody molecules having a common specificity are normally heterogeneous because they are produced by many different kinds of antibody-producing cells.*

Heterogeneity is a barrier to elucidating the molecular mechanism of a biological process. The hurdle was first overcome by taking advantage of *multiple myeloma,* a malignant disorder of antibody-producing cells. In this cancer, a single transformed lymphocyte or plasma cell divides uncontrollably. Consequently, a very large number of *cells of a single kind* are produced. They are a *clone* because they are descended from the same cell and have identical properties. Large amounts of *immunoglobulin of a single kind* are secreted by these tumors. Myeloma immunoglobulins are normal in structure but they are homogeneous. A myeloma can be transplanted from one mouse to another, where it continues to proliferate. Furthermore, these antibody-producing tumors synthesize the same kind of homogeneous antibody generation after generation.

A limitation in studying myeloma immunoglobulins is that their corresponding antigens are usually not known. In 1975, Cesar Milstein and Georges Köhler discovered that *large amounts of homogeneous antibody of nearly any desired specificity can be obtained by fusing an antibody-producing cell with a myeloma cell.* A mouse is immunized with an antigen, and its spleen is removed several weeks later (Figure 14-7). A mixture of lymphocytes and plasma cells from this spleen is fused in vitro with myeloma cells. Each of the resulting hybrid cells, called *hybridoma cells,* indefinitely produces homogeneous antibody specified by the parent cell from the spleen. Hybridoma cells can then be screened to determine which ones produce antibody having the desired specificity. These positive cells can be grown

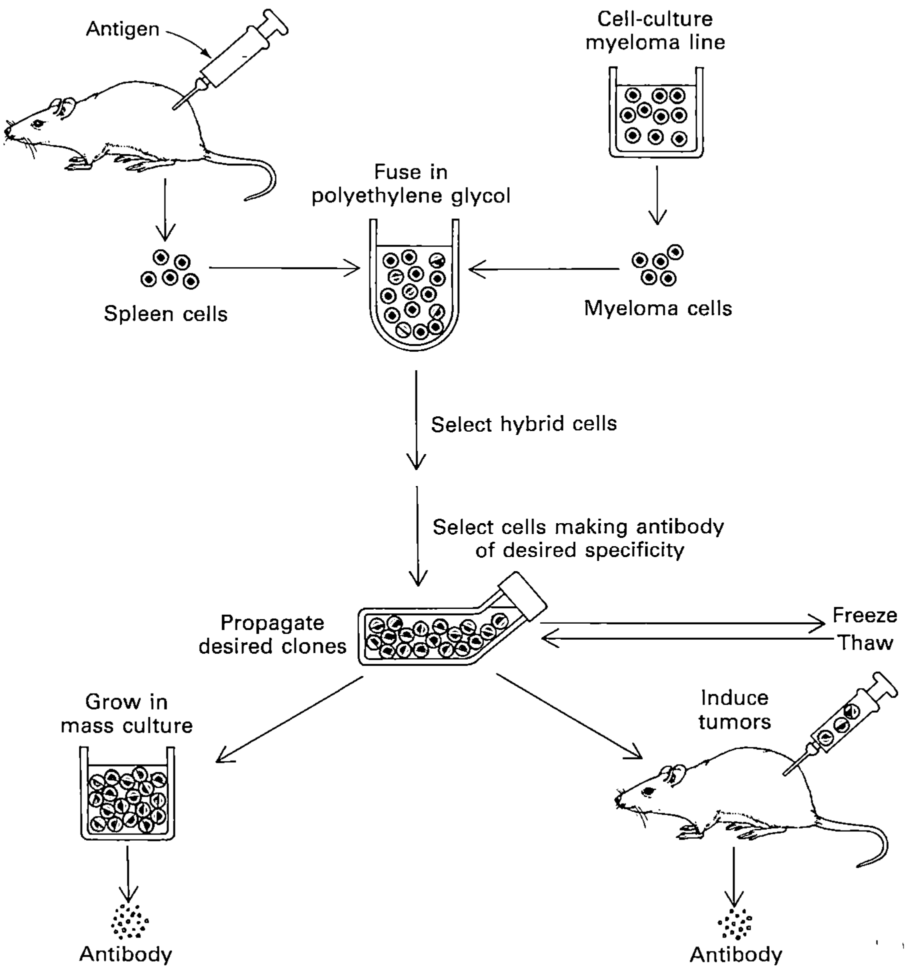

Figure 14-7
Preparation of monoclonal antibodies. Hybridoma cells are formed by fusion of antibody-producing cells and myeloma cells. These cells but not the parent ones are allowed to proliferate by growing them in a selective medium. Hybrid cells are then screened to determine which ones produce antibody of the desired specificity. [After C. Milstein. Monoclonal antibodies. Copyright © 1980 by Scientific American, Inc. All rights reserved.]

in culture medium or injected into mice to induce myelomas. Alternatively, the cells can be frozen and stored for long periods.

The hybridoma method of producing *monoclonal antibodies* has opened new vistas in biology and medicine. *Large amounts of homogeneous antibodies with tailor-made specificities can readily be prepared. They provide insight into relations between antibody structure and specificity. Moreover, monoclonal antibodies can serve as precise analytical and preparative reagents.* For example, a pure antibody can be obtained against an antigen that has not yet been isolated. This feature is especially advantageous in studies of cell-surface proteins. Subsets of lymphocytes can be isolated by fluorescence-activated cell sorting using monoclonal antibodies that are specific for distinctive protein markers on these cells. Proteins that guide development have been identified using monoclonal antibodies as tags. Monoclonal antibodies attached to solid supports can be used as affinity columns to purify scarce proteins. This method has been used to purify interferon (an antiviral protein) 5000-fold from a crude mixture. *Clinical laboratories are using monoclonal antibodies in many assays.* For example, the detection in blood of isozymes that are normally localized in the heart points to a myocardial infarction. Blood transfusions have been made safer by antibody screening of donor blood for viruses that cause AIDS, hepatitis, and other infectious diseases. Monoclonal antibodies are also being evaluated for use as therapeutic agents, as in the treatment of cancer. Furthermore, the vast repertoire of antibody specificity can be tapped to generate catalytic antibodies having novel features not found in naturally occurring enzymes.

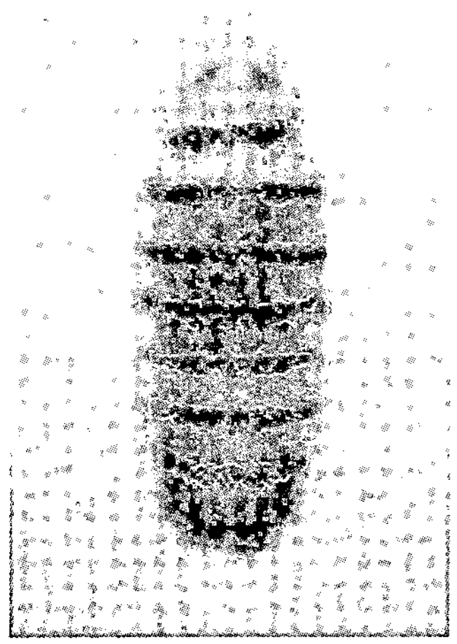

Figure 14-8
Fluorescence micrograph of a developing *Drosophila* embryo showing the expression pattern of *engrailed,* a key gene in specifying the body plan. The embryo was stained with a fluorescent-labeled monoclonal antibody specific for the DNA-binding protein encoded by *engrailed.* [Courtesy of Dr. Nipam Patel and Dr. Corey Goodman.]

LIGHT (L) AND HEAVY (H) CHAINS OF ANTIBODIES CONSIST OF A VARIABLE (V) AND A CONSTANT (C) REGION

Let us now turn to the molecular basis of antibody specificity. Immunoglobulin G consists of two kinds of polypeptide chains, a 25-kd *light (L)* chain and a 50-kd *heavy (H)* chain. The subunit composition is L_2H_2. Each L chain is bonded to an H chain by a disulfide bond, and the H chains are bonded to each other by at least one disulfide bond (Figure 14-9). The fragments produced by papain digestion were then related to this model. Papain cleaves the H chains on the carboxyl-terminal side of the disulfide bond that links the L and H chains. Thus, F_{ab} consists of the entire L chain and the amino-terminal half of the H chain, whereas F_c consists of the carboxyl-terminal halves of both H chains.

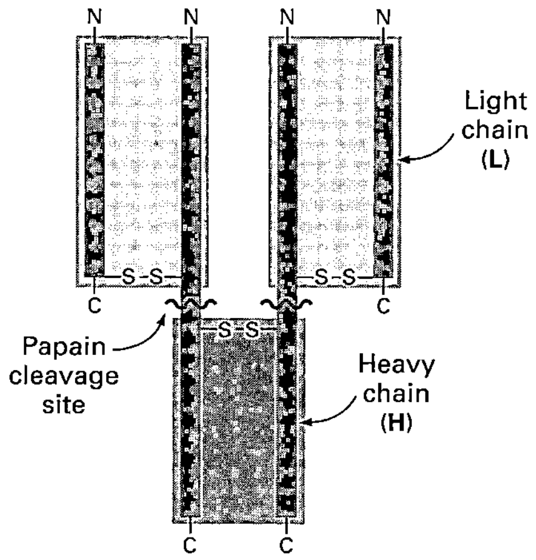

Figure 14-9
The subunit structure of IgG is L_2H_2. The F_{ab} units obtained by digestion with papain are shaded blue, whereas the F_c unit is shaded red.

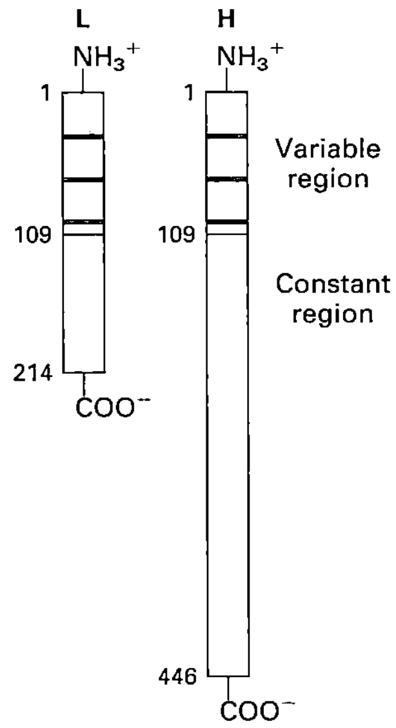

Figure 14-10
The light (L) chains and the heavy (H) chains of immunoglobulins consist of a variable region (green) and a constant region (yellow). Hypervariable regions (complementarity-determining regions, CDRs) are shown in red.

The first complete amino acid sequences of myeloma L chains, which contain about 214 amino acids, were determined in 1965. These studies showed that Bence-Jones proteins, which are dimers of L chains (first isolated by Henry Bence-Jones in 1847), from different patients have different amino acid sequences. Most striking, these differences in sequence are confined to the amino-terminal half of the polypeptide chain. Each Bence-Jones protein studied has a unique amino acid sequence from positions 1 to 108. In contrast, the sequences of many of these proteins are the same starting at position 109. Thus, *the L chain consists of a variable region (residues 1 to 108) and a constant region (residues 109 to 214)* (Figure 14-10). This remarkable finding had no precedent.

The H chain, consisting of 446 amino acid residues, is about twice as long as the L chain. Again, analyses of different myeloma immunoglobulins showed that all the differences in sequence are located in the 108 residues at the amino-terminal end. Thus, *the heavy chain, like the light chain, consists of a variable region and a constant region.* The variable region of the heavy chain has the same length as that of the light chain, whereas the constant region is about three times as long.

Three segments in the L chain and three in the H chain display far more variability than do other residues in the variable regions of these chains. In 1970, Elvin Kabat proposed that these *hypervariable segments* form the antigen-binding site and that antibody specificity is determined by the nature of their amino acid residues. X-ray crystallographic studies subsequently confirmed this incisive hypothesis, as will be discussed shortly. Hypervariable regions are also called *complementarity-determining regions (CDRs),* because they determine antibody specificity.

IMMUNOGLOBULINS CONSIST OF HOMOLOGOUS DOMAINS

The complete amino acid sequence of an immunoglobulin molecule was determined by Gerald Edelman in 1968. Several features of the sequence (Figure 14-11) were unanticipated: (1) The variable region of the light chain (V_L) is similar in sequence to the variable region of the heavy chain (V_H). (2) The constant region of the heavy chain (C_H) consists of equal thirds (C_H1, C_H2, and C_H3) that are similar in sequence. (3) Furthermore, the constant region of the light chain (C_L) closely resembles the three domains of the constant region of the heavy chain. (4) An intrachain disulfide bond is located in the same position in each domain of both the L and H chains.

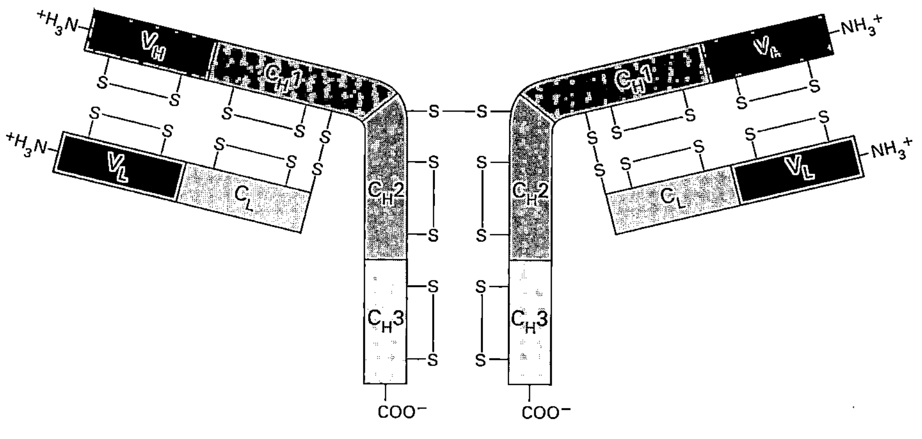

Figure 14-11
IgG consists of homologous domains. Each contains a conserved disulfide bond. [After G.M. Edelman. The structure and function of antibodies. Copyright © 1970 by Scientific American, Inc. All rights reserved.]

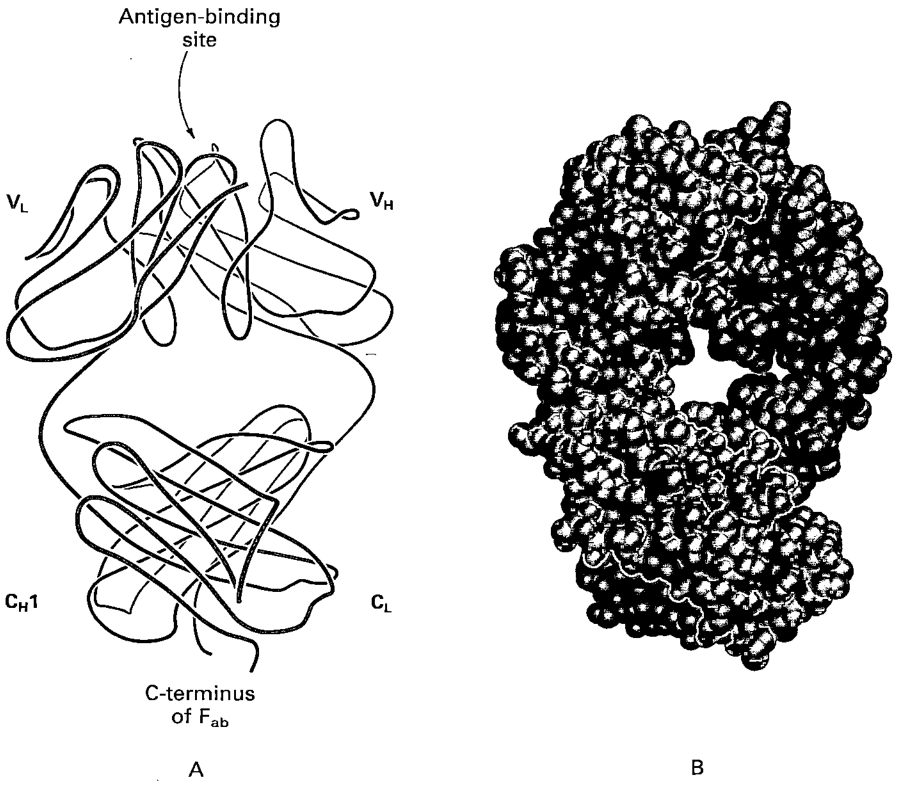

Figure 14-12
Structure of the $F_{ab'}$ unit of an antibody molecule. The L chain is shown in blue and the H chain in red. (A) Schematic diagram showing the course of the main chain. (B) Space-filling model. [Drawn from 1fve.pdb. C. Eigenbrot, M. Randal, L. Presta, P. Carter, and A.A. Kossiakoff. *J. Mol. Biol.* 229(1993):969.]

These similarities suggested that immunoglobulins are folded into compact domains, each subserving a distinctive molecular function. It also seemed likely that these domains resemble each other in tertiary structure. Indeed they do, as shown by x-ray crystallographic studies. The three-dimensional structure of the $F_{ab'}$ fragment (which is nearly the same as the F_{ab} fragment) of a human immunoglobulin has been solved at 2.0-Å resolution. This fragment consists of a *tetrahedral array of four globular subunits—V_L, V_H, C_L, and C_H1—which are strikingly similar in three-dimensional structure* (Figure 14-12). The structure of the F_c unit, which mediates effector functions, has also been solved at high resolution (Figure 14-13). The two C_H2 and two C_H3 domains of F_c are like the constant domains of the F_{ab} units.

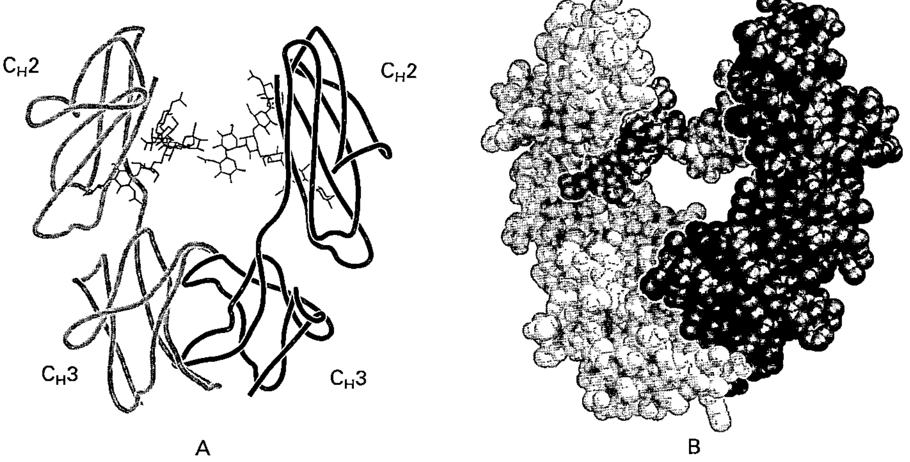

Figure 14-13
Structure of the F_c unit of an antibody molecule. The H chains are shown in red and yellow, and the attached carbohydrate units in green and blue. (A) Schematic diagram of the main chain and carbohydrate units. (B) Space-filling model. [Drawn from 1fc1.pdb. J. Deisenhofer. *Biochemistry* 20(1981):2361.]

THE IMMUNOGLOBULIN FOLD, A SANDWICH OF β SHEETS, IS A VERSATILE FRAMEWORK FOR GENERATING DIVERSITY

Let us examine more closely the structural core of immunoglobulin domains. A common structural feature is the presence of *two broad sheets of antiparallel β strands* (Figure 14-14). Many hydrophobic side chains are tightly packed between these sheets, which are joined by a disulfide bond. This recurring structural motif is called the *immunoglobulin fold.* The C domains contain three β strands in one sheet and four in the apposed one. The V domains are similar except for the addition of two β strands in one of the loops.

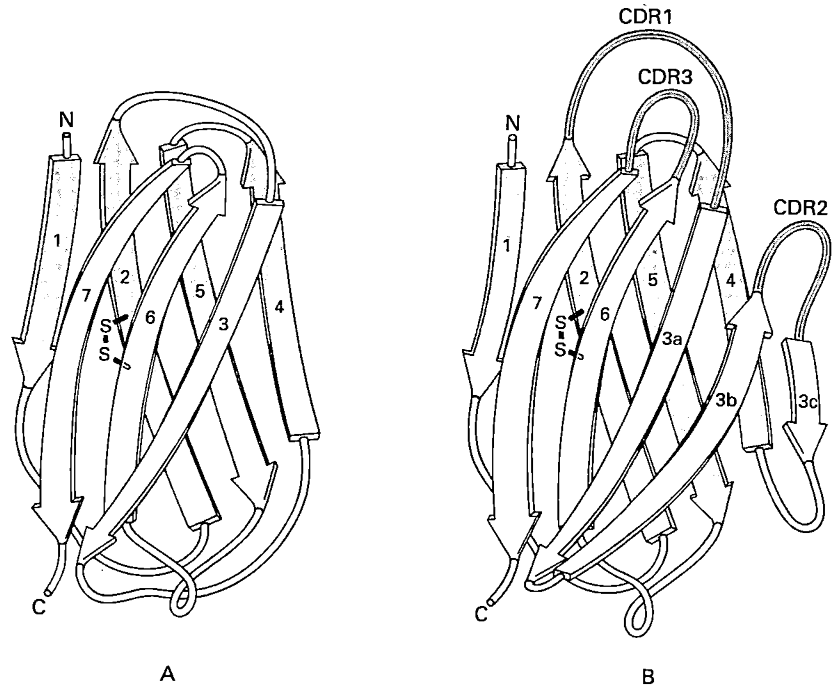

Figure 14-14
Immunoglobulin domains consist of two sheets of antiparallel β strands (yellow and blue). The sheets are bridged by a conserved disulfide bond. (A) A constant region domain contains three β strands in one sheet and four in the other. (B) A variable region domain contains two additional β strands (green). Three key loops (red) compose the complementarity-determining regions (CDRs) that form part of the antigen-binding site. [After C. Branden and J. Tooze. *Introduction to Protein Structure* (Garland, 1991), p. 185.]

The positions of the complementarity-determining regions (CDRs) of V_L and V_H are striking. X-ray analyses have shown that these hypervariable sequences, the sources of antibody specificity, are located in loops at one end of the sandwich (see Figure 14-14B). CDR1 emerges from the loop between strands 2 and 3, and CDR3 from the loop between strands 6 and 7. The reason for the extra two strands in V domains is now evident: the loop between them displays CDR2. *The immunoglobulin core serves as a framework that allows almost infinite variation of the loops.* Superposition of the structures of many different V domains shows that diverse CDRs emerge from a highly conserved framework. Moreover, V_L and V_H associate so that their CDRs come together to form a binding site (Figure 14-15). Virtually any V_L can pair with any V_H. Hence, *a very large number of different binding sites can be constructed by their combinatorial association.*

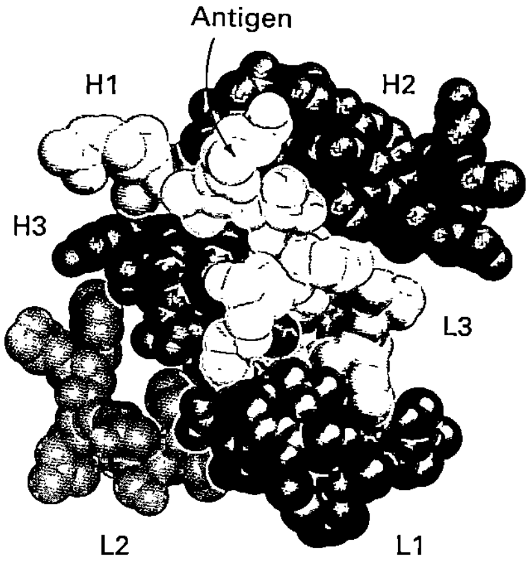

Figure 14-15
A peptide antigen (white) makes contact with five of six CDRs (shown in different colors) in the combining site of an antipeptide antibody. L1, L2, and L3 are L chain CDRs; H1, H2, and H3 are H chain CDRs. [Drawn from 2igf.pdb. R.L. Stanfield, T.M. Fieser, R.A. Lerner, and I.A. Wilson. *Science* 248(1990):712.]

X-RAY ANALYSES HAVE REVEALED HOW ANTIBODIES BIND HAPTENS AND ANTIGENS

X-ray crystallographic studies of more than 14 haptens and antigens bound to F_{ab} have provided further insight into the structural basis of antibody specificity. Phosphorylcholine, for example, binds to a cavity that is lined by residues from five CDRs, two from the L chain and three from the H chain (Figure 14-16). The positively charged trimethylammonium group of phosphorylcholine is buried inside the wedge-shaped cavity, where it interacts electrostatically with two negatively charged glutamates. The negatively charged phosphate group of the hapten binds to the positively charged guanidinium group of an arginine residue at the mouth of the crevice and to a nearby lysine. The phosphate group is also hydrogen bonded to the hydroxyl group of a tyrosine residue and to the guanido group of the arginine side chain. Numerous van der Waals interactions, such as those made by a tryptophan side chain, also stabilize this complex.

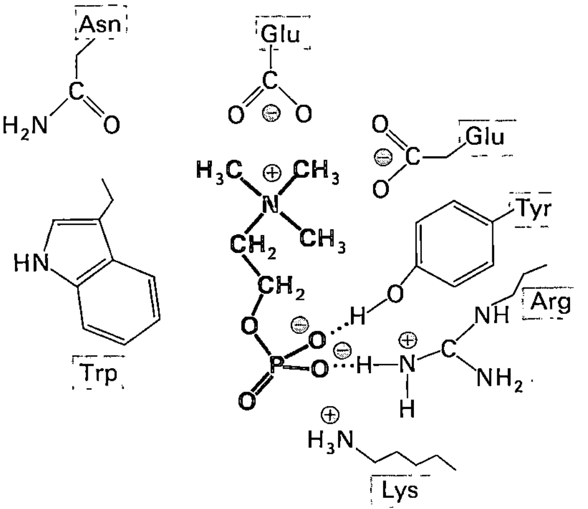

Figure 14-16
Mode of binding of phosphorylcholine (shown in green) to an antibody combining site. [After E.A. Padlan, D.R. Davies, S. Rudikoff, and M. Potter. *Immunochemistry* 13(1976):945.]

The binding of phosphorylcholine does not significantly change the structure of the antibody. In contrast, other haptens produce significant conformational changes. The binding of a trinucleotide of deoxythymidylate to an antibody that recognizes single-stranded DNA expanded the combining site (Figure 14-17). Many residues were displaced more than 2 Å in this transition. Indeed, the indole ring of a tryptophan residue moved 4 Å to stack with the central thymine of the trinucleotide. *Induced fit plays a role in the formation of many antibody-antigen complexes.* A malleable combining site can accommodate many more kinds of ligands than can a rigid one. Thus, induced fit increases the repertoire of antibody specificities.

Figure 14-17
The combining site of an antibody expands on binding a trinucleotide antigen. The L chain is shown in blue, the H chain in red, and the trinucleotide $(dT)_3$ in yellow. The dots depict the solvent-exposed surfaces of these units. (A) Antibody alone. (B) Expanded combining site of the complex. (C) Antibody-antigen complex. [Courtesy of Dr. Allen Edmundson.]

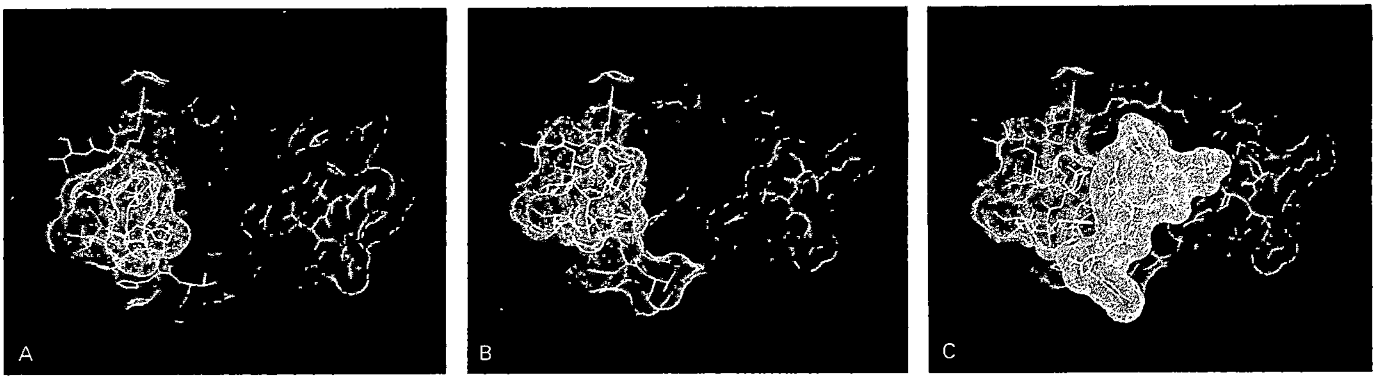

How do large antigens interact with antibodies? An antibody directed against lysozyme binds two polypeptide segments that are widely separated in the primary structure, residues 18 to 27 and 116 to 129. All six CDRs of the antibody make contact with this epitope (Figure 14-18). The V_L and V_H domains change little on binding. However, they slide 1 Å apart to favor more intimate contact with lysozyme. Only one water molecule remains in the region of contact, which is extensive (about 30 × 20 Å). The formation of 12 hydrogen bonds and numerous van der Waals interactions contributes to the high affinity (K_d = 20 nM) of the antibody-antigen interaction. The apposed surfaces are rather flat. The only exception is the side chain of glutamine 121 of lysozyme, which penetrates deeply into the antibody combining site, where it forms a hydrogen bond with a main-chain carbonyl oxygen and is surrounded by three aromatic side chains. Indeed, many antibody combining sites are rich in aromatic residues. *The interactions of haptens and antigens with antibodies are much like those of substrates with enzymes.* Numerous weak electrostatic, hydrogen bond, and van der Waals interactions, reinforced by hydrophobic interactions, combine to give strong and specific binding.

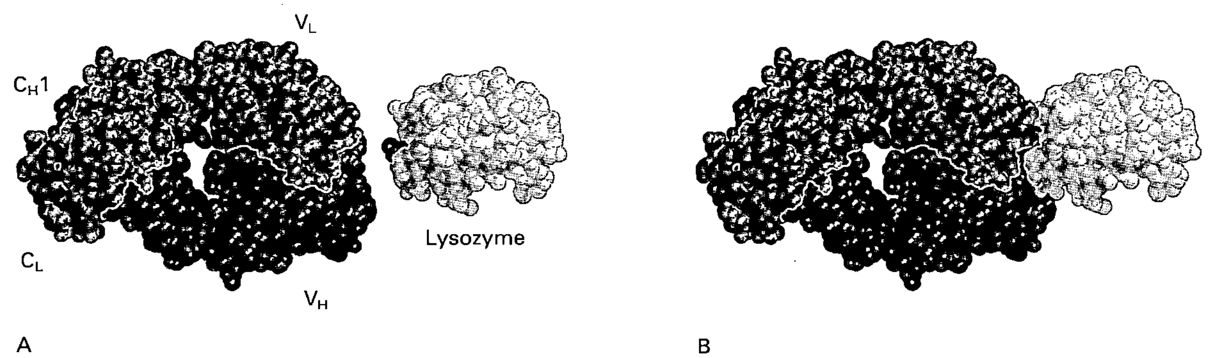

Figure 14-18
Structure of lysozyme bound to an F_{ab} molecule. The H chain is shown in red, the L chain in blue, and lysozyme in yellow. The antibody-antigen complex is shown separated (A) and together (B). A glutamine side chain (purple) of lysozyme penetrates deeply into the antibody. [Drawn from 1fdl.pdb. T.O. Fischmann, G.A. Bentley, T.N. Bhat, G. Boulot, R.A. Mariuzza, S.E.V. Phillips, D. Tello, and R.J. Poljak. *J. Biol. Chem.* 266(1991):12915.]

VARIABLE AND CONSTANT REGIONS ARE ENCODED BY SEPARATE GENES THAT BECOME JOINED

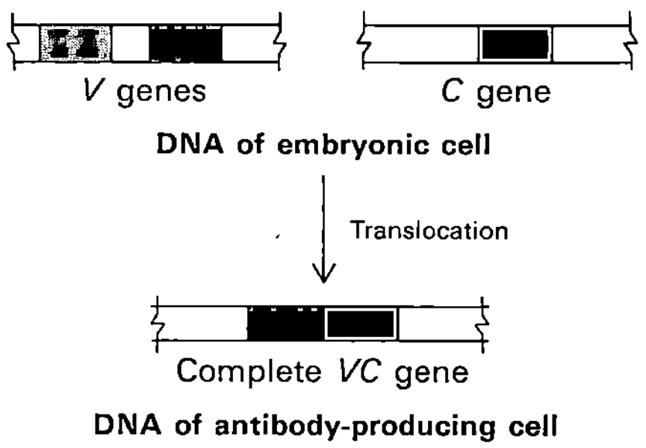

Figure 14-19
A *V* gene is translocated near a *C* gene in the differentiation of an antibody-producing cell.

The discovery of distinct variable and constant regions in the L and H chains raised the possibility that immunoglobulin genes, as well as their polypeptide products, have an unusual architecture. In 1965, William Dreyer and Claude Bennett proposed that multiple *V* genes are separate from a single *C* gene in embryonic (germ-line) DNA. According to their model, one of these *V* genes becomes joined to the *C* gene during differentiation of the antibody-producing cell. A critical test of this novel hypothesis had to await the isolation of pure immunoglobulin mRNA and the development of techniques for analyzing complex mammalian genomes. Twenty years later, Susumu Tonegawa found that V *and* C *genes are indeed far apart in embryonic (germ-line) DNA but are closely associated in the DNA of antibody-producing cells. Thus, immunoglobulin genes are translocated during the differentiation of lymphocytes* (Figure 14-19).

SEVERAL HUNDRED GENES ENCODE THE VARIABLE REGIONS OF L AND H CHAINS

An animal can synthesize large amounts of specific antibody against virtually any foreign determinant within a few weeks after being exposed to it. The number of different kinds of antibodies that can be made by an animal is very large, at least a million. We have seen that antibody specificity is determined by the amino acid sequences of the variable regions of both light and heavy chains. This brings us to the key question: *How are different variable-region sequences generated?* More specifically, we can ask:

1. *When* is diversity generated? During the life of an individual animal (somatic) or in the course of evolution (genetic)?

2. *How* is diversity generated? By a random evolutionary process, somatic recombination, or somatic mutation?

Recombinant DNA technology provided the means to obtain definitive answers to these questions. A key finding is that mice have *several hundred genes for the variable regions of* κ *light chains* (V_κ; κ is the major family of light chains) *and of heavy chains* (V_H). For 300 V_κ genes and 300 V_H genes, a maximum of 9×10^4 different specificities could be generated by the combination (300×300) of L and H chains. This calculated upper limit of 9×10^4 falls far short of the number of different kinds of antibodies that an animal can produce ($>10^6$). *The number of variable-region genes in the germ line is too small to be the sole source of antibody diversity. Clearly, some of this diversity must be generated during the lifetime of an animal in the differentiation of its lymphocytes.*

J (JOINING) GENES AND *D* (DIVERSITY) GENES INCREASE ANTIBODY DIVERSITY

Sequencing studies carried out by Tonegawa, Philip Leder, and Leroy Hood revealed that V *genes in embryonic cells do not encode the entire variable region of L and H chains.* The germ-line *V* gene stops at the codon for amino acid residue 95 rather than 108, which is the end of the variable region of the polypeptide chain. Where is the DNA that encodes the last 13 residues of the V region? For L chains in embryonic cells, this stretch of DNA is located in an unexpected place: near the *C* gene. It is called the *J* gene because it joins the *V* and *C* genes in a differentiated cell. In fact, *a tandem array of four* J *genes is located near the* C *gene in embryonic cells* (Figure 14-20). In the differentiation of an antibody-producing cell, *a* V *gene becomes spliced to a* J *gene to form a complete gene for the variable region.* *J* genes are important contributors to antibody diversity because they encode part of the last hypervariable segment (CDR3). In forming a continuous V_κ gene, any of several hundred *V* genes can become linked to any of four *J* genes. Thus, *somatic recombination of these gene segments amplifies the diversity already present in the germ line.* Recombination between these genes can occur at one of several bases near the codon for residue 95. In fact, *additional diversity is generated by allowing* V *and* J *genes to become spliced in different joining frames.* The immune system clearly takes delight in sweet disorder!

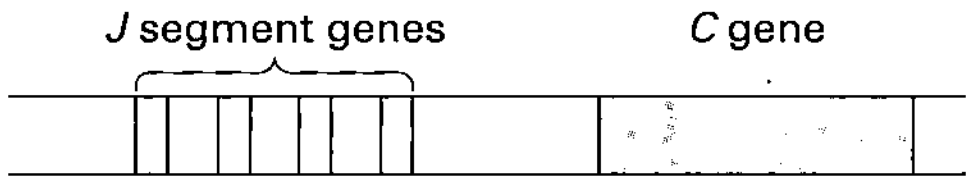

Figure 14-20
A tandem array of *J* genes, which encode part of the last hypervariable segment of the V region of the light chain, is located near the *C* gene.

The variable region of heavy chains is encoded by yet another gene segment, called *D* for "diversity." Some fifteen *D* segments lie between hundreds of V_H and five J_H segments (Figure 14-21). A *D* segment joins a J_H segment; a V_H segment then becomes linked to DJ_H. More kinds of antigen-binding patches and clefts can be formed by the H than by the L chain because it is encoded by three rather than two gene segments. Moreover, CDR3 of the H chain is diversified by the action of ***terminal deoxyribonucleotidyl transferase,*** a special polymerase that does not use a template. This enzyme inserts extra nucleotides between V_H and *D*.

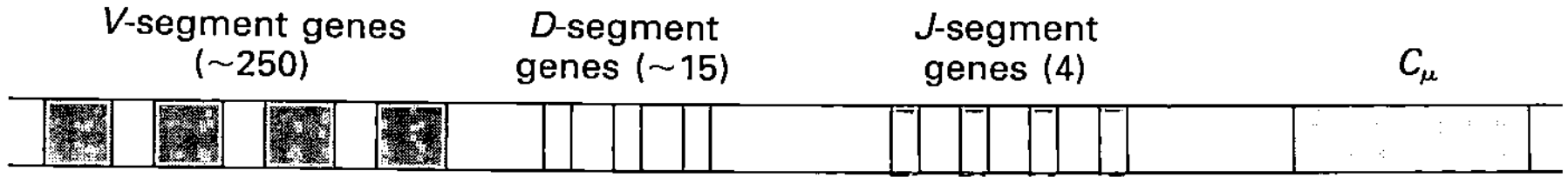

Figure 14-21
The variable region of the heavy chain is encoded by *V*-, *D*-, and *J*-segment genes.

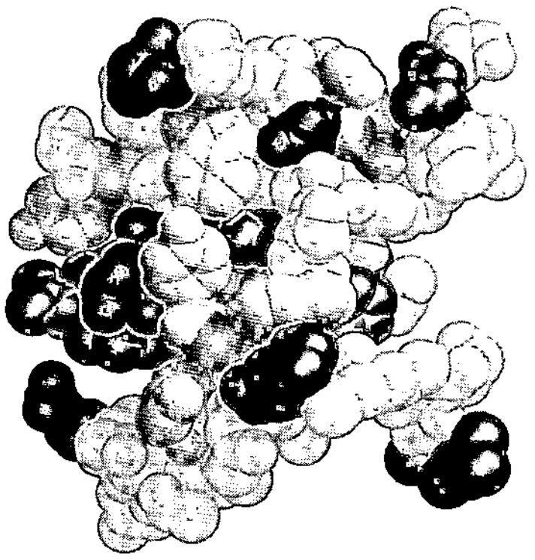

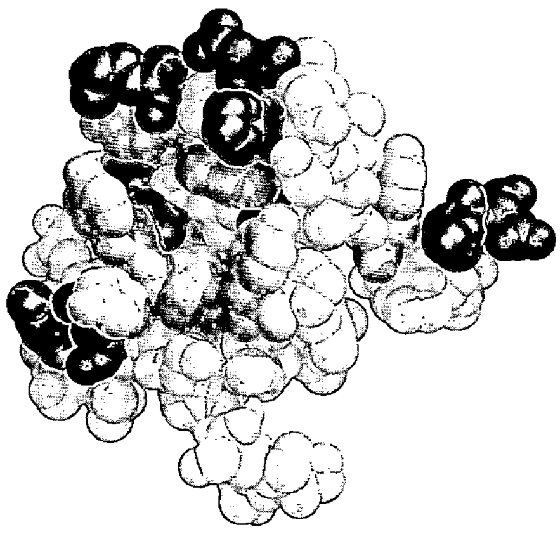

MORE THAN 10^8 ANTIBODIES CAN BE FORMED BY COMBINATORIAL ASSOCIATION AND SOMATIC MUTATION

Let us recapitulate the sources of antibody diversity. The germ line contains a rather large repertoire of variable-region genes. For κ light chains, there are about 250 *V*-segment genes and 4 *J*-segment genes. There are at least three frames for the joining of each *V* and *J*. Hence, a total of 250 $\times$ 4 $\times$ 3, or 3000, kinds of complete V_κ genes can be formed by the combinations of *V* and *J*. A larger number of heavy-chain genes can be formed because of the existence of *D* segments. For 250 *V*, 15 *D*, and 5 *J* gene segments that can be joined in three frames, the number of complete V_H genes that can be formed is 56,250. The association of 3000 kinds of L chains with 56,250 kinds of H chains would yield 1.7 $\times$ 10^8 different antibody specificities. This calculated number is sufficiently large to account for the remarkable range of antibodies that can be synthesized by an animal.

Figure 14-22
A gallery of complementarity-determining regions of antibodies. The five CDRs shown here have very different patterns of positively charged (blue), negatively charged (red), and aromatic (yellow) residues. [Drawn from 1fdl.pdb, 2fbj.pdb, 6fab.pdb, 3hfm.pdb, and 1nca.pdb.]

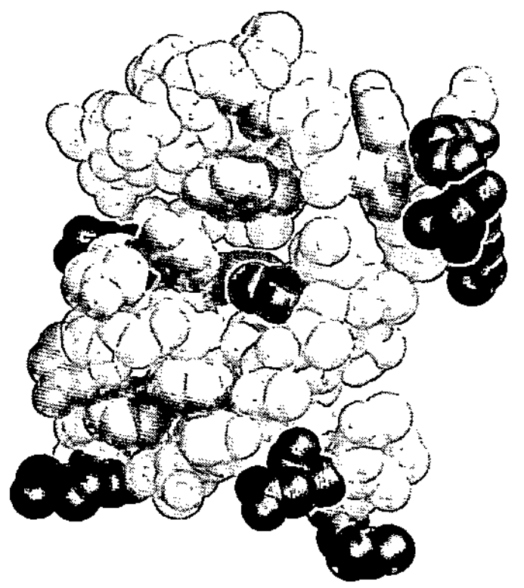

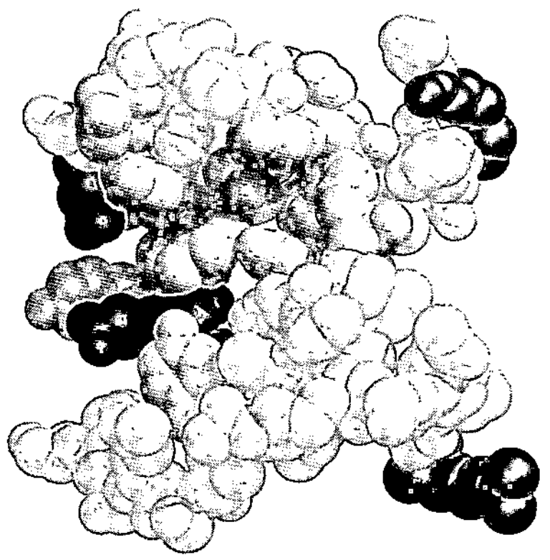

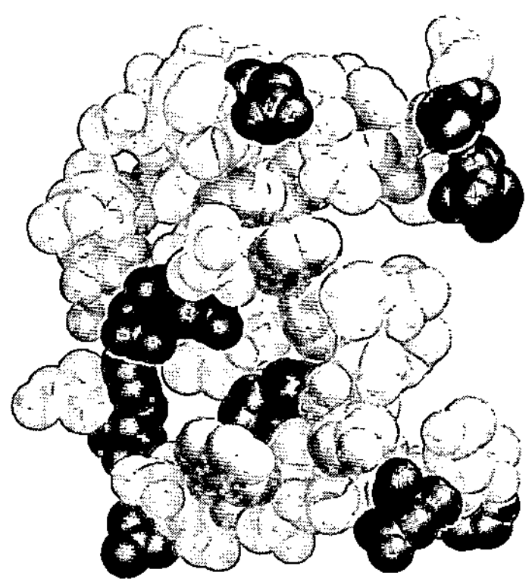

Somatic mutation further increases the diversity of heavy chains. Indeed, *affinity maturation*—the 1000-fold increase in binding affinity seen in the course of a typical humoral immune response—arises from somatic mutation. The generation of an expanded repertoire leads to the selection of antibodies that more precisely fit the antigen. Thus, *nature draws on each of the three proposed sources of diversity—a germ-line repertoire, somatic recombination, and somatic mutation—to form the rich variety of antibodies that protect an organism from foreign incursions.*

FIVE CLASSES OF IMMUNOGLOBULINS MEDIATE DISTINCT EFFECTOR FUNCTIONS

Thus far, we have focused on immunoglobulin G. There are in fact five classes of immunoglobulins (Table 14-1), each consisting of heavy and light chains. The constant regions of the heavy chains differ from one class to another, whereas those of the light chains are the same (either λ or κ). The heavy chains in immunoglobulin G are called γ chains, whereas those in immunoglobulins A, M, D, and E are called α, μ, δ, and ϵ, respectively. These different heavy chains, variations on a fundamental theme, give the five classes of immunoglobulins distinct biological characteristics. *Isotypes* are immunoglobulins belonging to the same class (i.e., sharing the same constant regions). *Different isotypes mediate different effector functions.*

Table 14-1
Properties of immunoglobulin classes

Class	*Serum concentration (mg/ml)*	*Mass (kd)*	*Sedimentation coefficient(s)*	*Light chains*	*Heavy chains*	*Chain structure*
IgG	12	150	7	κ or λ	γ	$\kappa_2\gamma_2$ or $\lambda_2\gamma_2$
IgA	3	180–500	7, 10, 13	κ or λ	α	$(\kappa_2\alpha_2)_n$ or $(\lambda_2\alpha_2)_n$
IgM	1	950	18–20	κ or λ	μ	$(\kappa_2\mu_2)_5$ or $(\lambda_2\mu_2)_5$
IgD	0.1	175	7	κ or λ	δ	$\kappa_2\delta_2$ or $\lambda_2\delta_2$
IgE	0.001	200	8	κ or λ	ϵ	$\kappa_2\epsilon_2$ or $\lambda_2\epsilon_2$

Note: n = 1, 2, or 3. IgM and oligomers of IgA also contain J chains that join immunoglobulin molecules. IgA in secretions has an additional secretory piece.

As mentioned previously, *immunoglobulin M (IgM)* is the first class of antibodies to appear in the serum after injection of an antigen. The presence of 10 combining sites enables IgM to bind especially tightly to antigens containing multiple identical epitopes, such as bacterial cell surfaces. The association constant for multivalent binding is termed *avidity* to distinguish it from *affinity,* which denotes the binding strength of a single combining site. The first antibodies produced in an immune response have a lower affinity for antigen than those formed later. The presence of 10 combining sites in IgM compared with 2 in IgG enables IgM to bind many multivalent antigens that would slip away from IgG.

Immunoglobulin G (IgG) is the principal antibody in the serum. IgG was originally called *γ-globulin* because of its solubility and electrophoretic characteristics. Subclasses of IgG (e.g., IgGl and IgG2) differ in their degree of segmental flexibility and their capacity to trigger complement fixation and other effector functions. *Immunoglobulin A (IgA)* is the major class of antibodies in external secretions, such as saliva, tears, bronchial mucus, and intestinal mucus. Thus, IgA serves as a first line of defense against bacterial and viral antigens. IgA is transported across epithelial

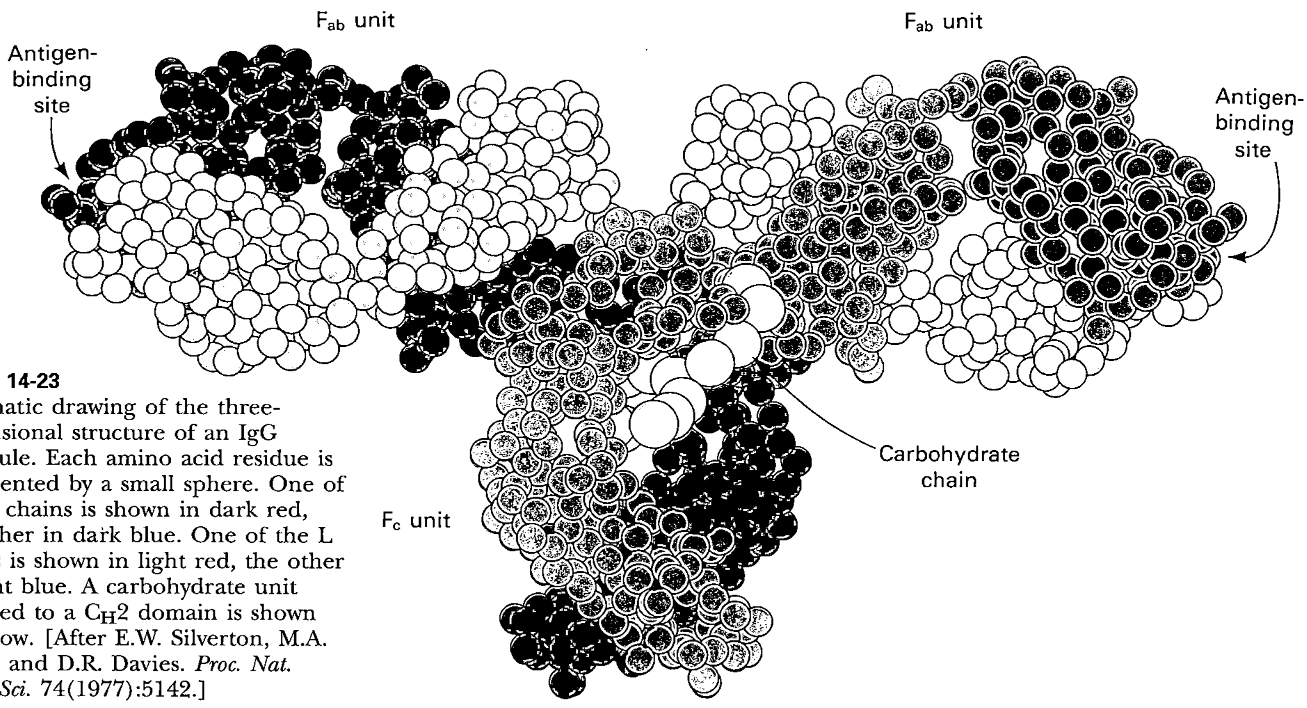

Figure 14-23
Schematic drawing of the three-dimensional structure of an IgG molecule. Each amino acid residue is represented by a small sphere. One of the H chains is shown in dark red, the other in dark blue. One of the L chains is shown in light red, the other in light blue. A carbohydrate unit attached to a C_H2 domain is shown in yellow. [After E.W. Silverton, M.A. Navia, and D.R. Davies. *Proc. Nat. Acad. Sci.* 74(1977):5142.]

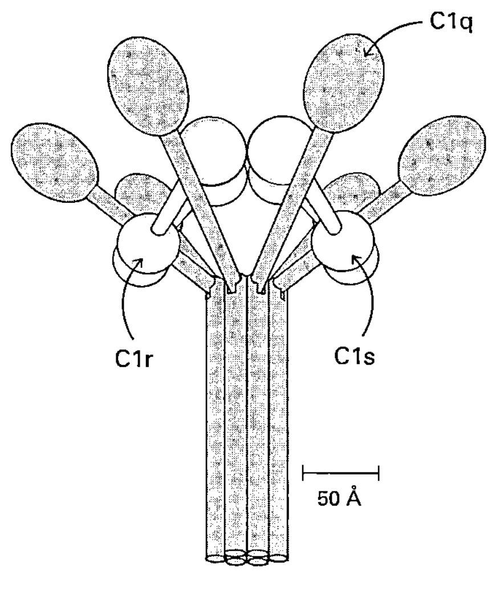

Figure 14-24
Model of the C1 component of complement. C1q is shown in red, C1r in blue, and C1s in green. The globular heads of C1q emerge from long stems consisting of triple-stranded helices of the type found in collagen. Binding of the F_c units of antibody-antigen complexes to the globular heads of C1q triggers the complement cascade by activating C1r and then C1s. [After G.J. Arlaud, M.G. Colomb, and J. Gagnon. *Immunol. Today* 8(1987):1076.]

cells from the blood side to the extracellular side by a receptor called *secretory component.* The benefits provided by *immunoglobulin D (IgD)* are not yet known. *Immunoglobulin E (IgE)* is important in conferring protection against parasites, but IgE also causes allergic reactions. IgE-antigen complexes cross-link receptors on the surfaces of mast cells to trigger a cascade that leads to the release of granules. Histamine (p. 724), one of the agents released, induces smooth muscle contraction and stimulates the secretion of mucus.

ANTIGEN-ANTIBODY COMPLEXES TRIGGER THE COMPLEMENT CASCADE TO LYSE TARGET CELLS

Complement fixation is one of the most important effector functions of antibodies. The complement system lyses microorganisms and infected cells by forming holes in their plasma membrane. More than 15 soluble proteins in blood participate in this precisely regulated cascade. The triggering of lysis by antibody-antigen complexes is mediated by the *classical pathway,* beginning with the activation of C1, the first component of the pathway. Each of the six globular heads of C1q (Figure 14-24) contains a binding site for the F_c moiety of an antibody molecule. C1 is not activated by IgG alone nor by a single antigen-IgG complex. Rather, a cluster of several F_c units serves as the trigger when they are brought together by the formation of a multivalent antibody-antigen complex. Subclasses of IgG differ in their effectiveness in triggering complement. C1 can also be activated by the binding of a multivalent antigen to a single IgM molecule, which exposes several of its F_c units.

C1r and C1s are located in a cavity formed by the heads and stems of C1q. *The simultaneous binding of several F_c units to the globular heads of C1q converts C1r from a zymogen into a protease.* This activation step, in which C1r undergoes an internal cleavage, is reminiscent of the activation of pepsinogen by acid (p. 250). It seems likely that the binding of F_c units to C1q pulls an inhibitory segment out from Clr. Activated Clr in turn splits a peptide bond in Cls to convert it into an active protease. A series of zymogen conversions analogous to those of the clotting cascade is triggered by activated Cls. In the penultimate step, a *membrane attack complex* (consisting of C5b, C6, C7, and C8) penetrates the membrane bearing the antigen. *This cluster catalyzes the entry of some 16 molecules of C9, which associate to form large pores (11 nm) in the membrane* (Figure 14-25). The result is lysis and destruction of the target cell.

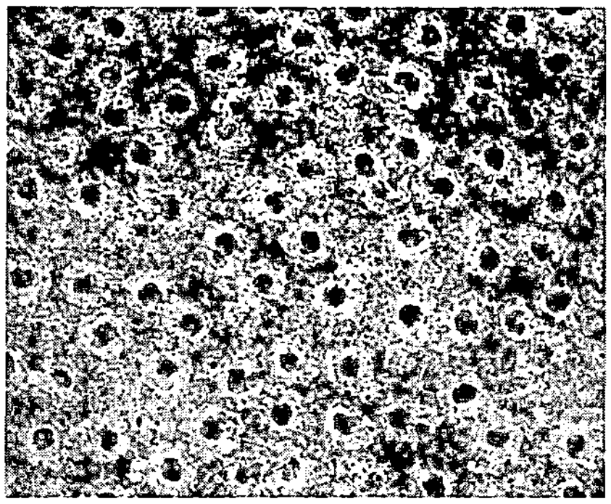

Figure 14-25
Electron micrograph of an erythrocyte membrane that was lysed by complement. The 11-nm-diameter pores are formed by oligomers of C9, the final component of the cascade. [Courtesy of Dr. Eckhard Podack.]

DIFFERENT CLASSES OF ANTIBODIES ARE FORMED BY THE HOPPING OF V_H GENES

An antibody-producing cell first synthesizes IgM and then makes either IgG, IgA, IgD, or IgE of the *same specificity*. In this switch, the light chain is unchanged. Furthermore, the variable region of the heavy chain stays the same. *Only the constant region of the heavy chain changes.* This step in the differentiation of an antibody-producing cell is called *class switching* (Figure 14-26).

In embryonic mouse cells, the genes for the constant region of each class of heavy chain—called C_μ, C_δ, C_γ, C_ε, *and* C_α*—are next to each other* (Figure 14-27).

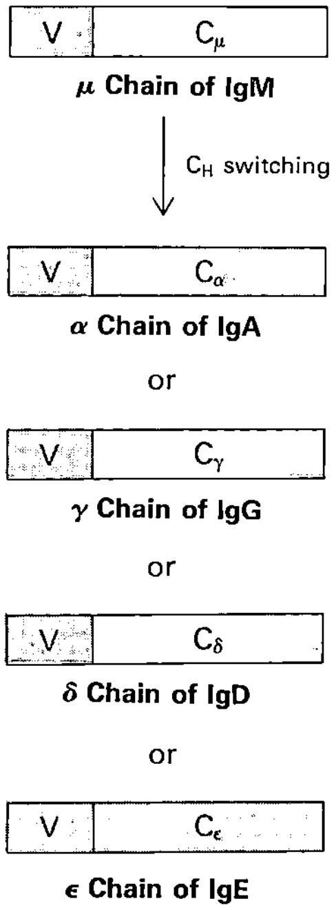

Figure 14-26
Synthesis of different classes of immunoglobulins. The V_H region is first associated with C_μ and then with another C region to form an H chain of a different class.

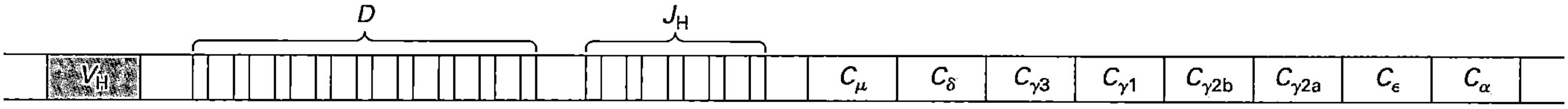

Figure 14-27
The genes for the constant regions of μ, δ, γ, ε, and α chains are next to each other in germ-line DNA.

There are eight in all, including four genes for the constant regions of γ chains. *A complete gene for the heavy chains of IgM antibody is formed by the translocation of a* V_H *gene segment to a* DJ_H *gene segment* (Figure 14-28). This joining brings V_H, D, J_H, and C_μ together to form a functional gene, which is then transcribed. Several introns are excised in the processing of the primary transcript to form mRNA for the μ chain.

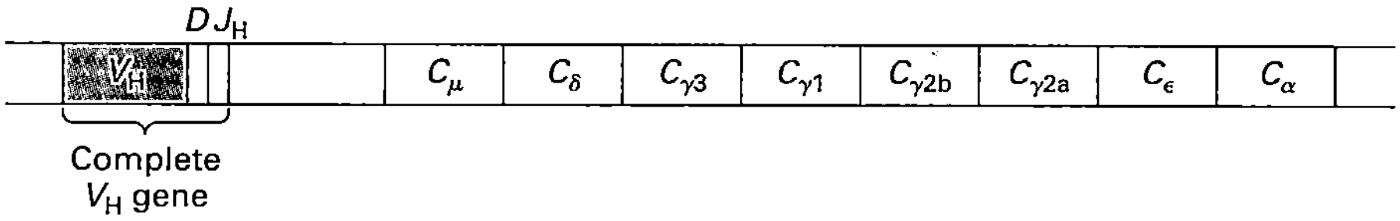

Figure 14-28
A complete V_H gene is formed by the coming together of *V*, *D*, and *J* segments near the *C* gene for the μ chain.

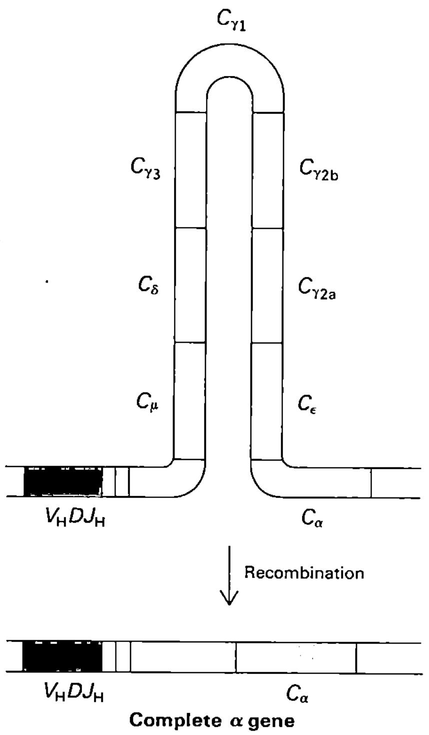

Figure 14-29
Structural basis of C_H switching. The V_HDJ_H gene moves from its position near C_μ to one near C_α by intrachromosomal recombination.

How are other heavy chains formed? *Class switching is mediated by the movement of a* VDJ *gene from a site near one* C *gene to a site near another* C *gene* (Figure 14-29). The mechanism of recombination will be discussed in a later chapter (p. 833). The important point now is that antigen-binding specificity is conserved in class switching because the entire V_HDJ_H gene is translocated. For example, the antigen-combining specificity of IgA produced by a particular cell is the same as that of IgM synthesized at an earlier stage of its development. *The biological significance of* C_H *switching is that a whole recognition domain (the variable domain) is shifted from the early constant region* (C_μ) *to one of several other constant regions that mediate different effector functions.*

KILLER T CELLS DESTROY INFECTED CELLS THAT DISPLAY FOREIGN PEPTIDES BOUND TO CLASS I MHC PROTEINS

The immune system employs two systems of recognition elements, soluble antibodies and cell-attached T-cell receptors. We have seen that antibodies tag foreign cells for destruction by the complement system and phagocytosis by macrophages. Antibodies can also block the action of proteins required for the invasiveness of pathogens. Why then is a second set of receptors and effectors needed? The reason is that antibodies, though highly effective against extracellular pathogens, confer little protection against microorganisms that are predominantly intracellular, such as viruses and mycobacteria (which cause tuberculosis and leprosy). These pathogens are shielded from antibodies by the host cell membrane (Figure 14-30).

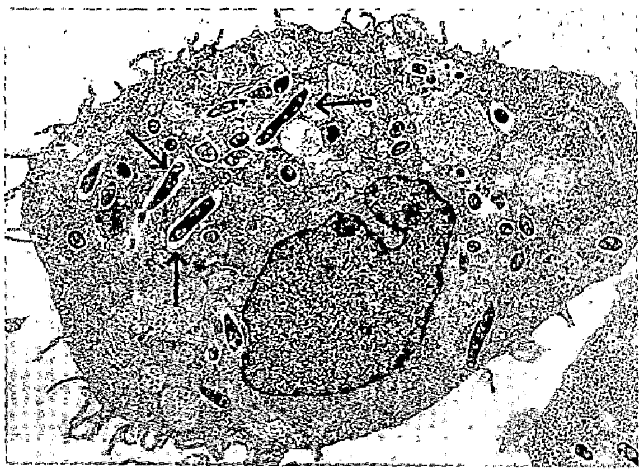

Figure 14-30
Electron micrograph showing mycobacteria (arrows) inside an infected macrophage. [Courtesy of Dr. Stanley Falkow.]

A different and more subtle strategy, T-cell–mediated immunity, has evolved to cope with intracellular pathogens. *T cells continually scan the surfaces of all cells and kill those that exhibit foreign markings.* The task is not simple; intracellular microorganisms are not so obliging as to intentionally leave telltale traces on the surface of their host. Quite the contrary, successful pathogens are masters of the art of camouflage. Vertebrates have devised an ingenious mechanism—*cut and display*—to reveal the presence of stealthy intruders. *Nearly all vertebrate cells exhibit on their surface a sample of peptides derived from the digestion of proteins in their cytosol.* These peptides are displayed by integral membrane proteins that are encoded by the *major histocompatibility complex (MHC).* Specifically, peptides derived from cytosolic proteins are bound to *class I MHC proteins* (p. 381).

How are these peptides generated and delivered to the plasma membrane? The process starts in the cytosol with the degradation of proteins, normal ones as well as those of pathogens (Figure 14-31). Digestion is carried out by very large multisubunit assemblies called *proteasomes.* The resulting peptide fragments are transported from the cytosol into the lumen of the endoplasmic reticulum (ER), a membrane-enclosed network, by an ATP-driven pump. In the ER, peptides combine with nascent *class I MHC proteins;* these complexes are then targeted to the plasma membrane.

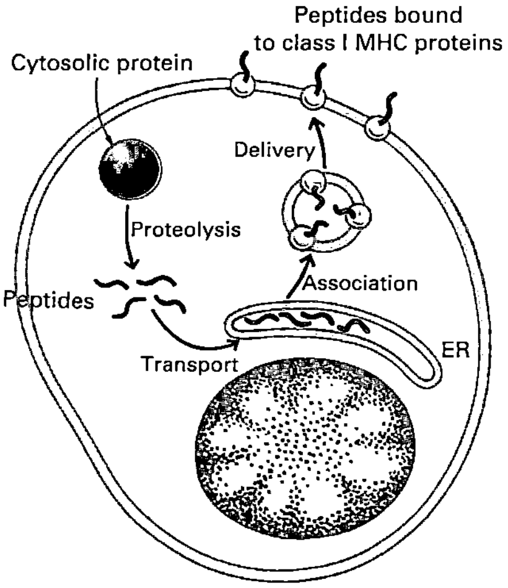

Figure 14-31
Class I MHC proteins on the surface of most cells display peptides that are formed by proteolysis of cytosolic proteins.

MHC proteins embedded in the plasma membrane tenaciously grip their bound peptides so that they can be touched and scrutinized by T-cell receptors on the surface of a killer cell. *Foreign peptides bound to class I MHC proteins signal that a cell is infected and mark it for destruction by killer T cells* (Figure 14-32). An assembly consisting of the foreign peptide–MHC complex, the T-cell receptor, and numerous accessory proteins triggers a cascade that lyses the infected cell by forming large pores in the plasma membrane. Infected cells are sacrificed to save the organism.

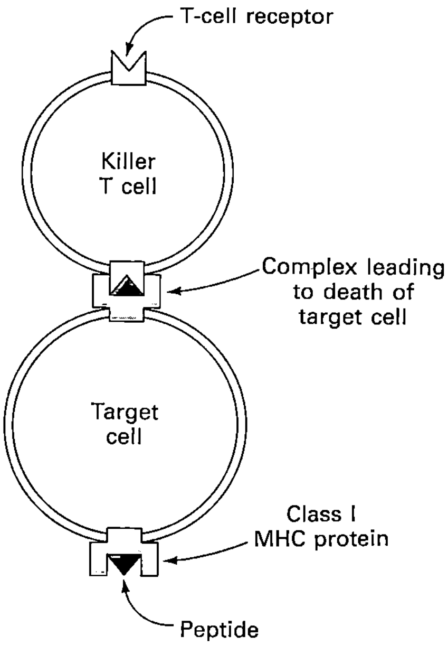

Figure 14-32
T-cell receptors on killer T cells recognize foreign peptides bound to class I MHC proteins (green) on infected target cells.

HELPER T CELLS STIMULATE B CELLS THAT DISPLAY FOREIGN PEPTIDES BOUND TO CLASS II MHC PROTEINS

Not all T cells are killers. *Helper T cells, a different class, stimulate the proliferation of specific B lymphocytes and thereby serve as partners in determining which kinds of immunoglobulins are produced.* The essentiality of helper T cells is graphically revealed by the devastation wrought by the AIDS virus, which destroys these cells. Helper T cells, like killer T cells, detect foreign peptides that are presented on cell surfaces by MHC proteins. However, the source of the peptides, the MHC proteins that bind them, and the transport pathway are different.

Helper T cells recognize peptides that are bound to class II, but not class I, MHC proteins. Their helping action is focused on B cells, macrophages, and dendritic cells because class II proteins, unlike class I proteins, are not expressed by most cells. Another important difference is that peptides presented by class II MHC proteins do not come from the cytosol. Rather, they arise from the degradation of proteins that have

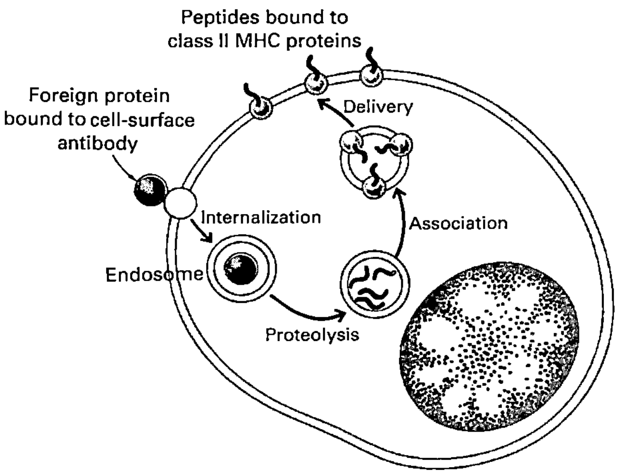

Figure 14-33
Class II MHC proteins on antigen-presenting cells (e.g., macrophages) display peptides that are formed by digestion of external proteins following internalization.

been internalized by endocytosis (p. 935). Consider, for example, a virus particle that is captured by a cluster of membrane-bound immunoglobulins on the surface of a B cell (Figure 14-33). This complex is delivered to an endosome, a membrane-enclosed acidic compartment, where it is digested. The resulting peptides become associated with class II MHC proteins, which are sent to the cell surface. Peptides from the cytosol cannot reach class II proteins, whereas peptides from endosomal compartments cannot reach class I proteins. This segregation of displayed peptides is biologically critical. Association of a foreign peptide with a class II MHC protein signals that a cell has *encountered* a pathogen and serves as a call for *help*. In contrast, association with a class I MHC protein signals that a cell has *succumbed* to a pathogen and is a call for *destruction*.

MAJOR HISTOCOMPATIBILITY COMPLEX (MHC) PROTEINS ARE HIGHLY DIVERSE PRESENTERS OF ANTIGENS

MHC class I and II proteins, the presenters of peptides to T cells, were discovered because of their role in *transplantation rejection*. A tissue transplanted from one person to another, or from one wild mouse to another, is usually rejected by the immune system. In contrast, tissues transplanted from one identical twin to another, or between mice of an inbred strain, are accepted. Genetic analyses revealed that rejection occurs when tissues are transplanted between individuals having different genes in the major histocompatibility complex, a cluster of more than 75 genes playing key roles in immunity. The 3500-kb span of the MHC is about the length of the entire *Escherichia coli* chromosome. The MHC encodes class I proteins (presenters to killer T cells) and class II proteins (presenters to helper T cells), as well as class III proteins (components of the complement cascade) and many other proteins that play key roles in immunity.

Humans express six different class I genes (three from each parent) and six different class II genes. The three loci for class I genes are called HLA-A, HLA-B, and HLA-C; those for class II genes are called HLA-DP, HLA-DQ, and HLA-DR. The abbreviation HLA stands for *human leukocyte antigen*. These loci are *highly polymorphic*: many alleles of each are present in the population. For example, more than 40 HLA-A, 70 HLA-B, and 30 HLA-C alleles are known; the numbers increase each year. Hence, the likelihood that two unrelated individuals have identical class I and II proteins is very small ($<10^{-4}$), accounting for transplantation rejection unless there is close matching of genotypes.

Class I MHC proteins consist of a 44-kd α chain noncovalently bound to a 12-kd polypeptide called *β_2-microglobulin.* The α chain has three extracellular domains—α_1, α_2, and α_3—a transmembrane segment, and a cytosolic tail (Figure 14-34). Differences between class I proteins are located mainly in the α_1 and α_2 domains, which form the peptide-binding site. The α_3 domain, which interacts with a constant β_2-microglobulin (β_2m), is largely conserved. *Class II MHC proteins* have a similar overall plan. They consist of a 33-kd α chain and a noncovalently bound 30-kd β chain. Each contains two extracellular domains, a transmembrane segment, and a short cytosolic tail (Figure 14-35). The peptide-binding site is formed by the α_1 and β_1 domains, the most variable regions of the protein.

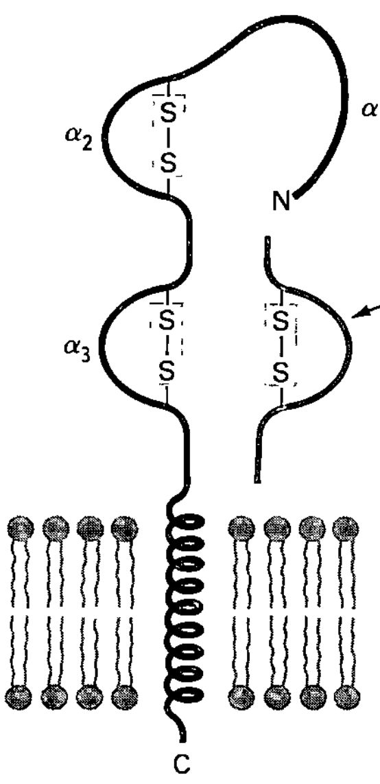

Figure 14-34
Schematic diagram of the domain structure of a class I MHC protein.

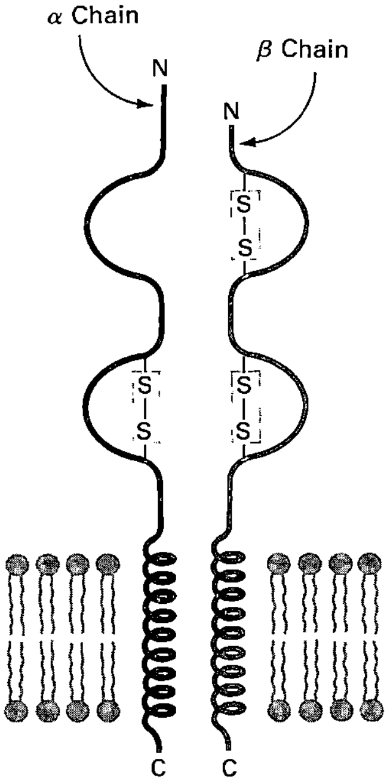

Figure 14-35
Schematic diagram of the domain structure of a class II MHC protein.

Why are MHC proteins so highly variable? *Their diversity makes possible the presentation of a very wide range of peptides to T cells.* A *particular* class I or class II molecule may not be able to bind any of the peptide fragments of a viral protein. The likelihood of a fit is markedly increased by having several kinds (usually six) of each class of presenters in each individual. If all members of a species had identical class I or class II molecules, they would be much more vulnerable to infection. Mutant pathogens that evade presentation would undoubtedly arise and thrive.

PEPTIDES PRESENTED BY MHC PROTEINS OCCUPY A DEEP GROOVE THAT IS FLANKED BY α HELICES

The three-dimensional structure of a human MHC class I protein (HLA-A2) was solved in 1987 by Don Wiley, Jack Strominger, and their coworkers. Cleavage by papain of the HLA heavy chain several residues before the transmembrane segment yielded a soluble heterodimeric fragment that could be crystallized in aqueous solution. The 70-Å-long protein consists of two sets of structurally similar domains (Figure 14-36).

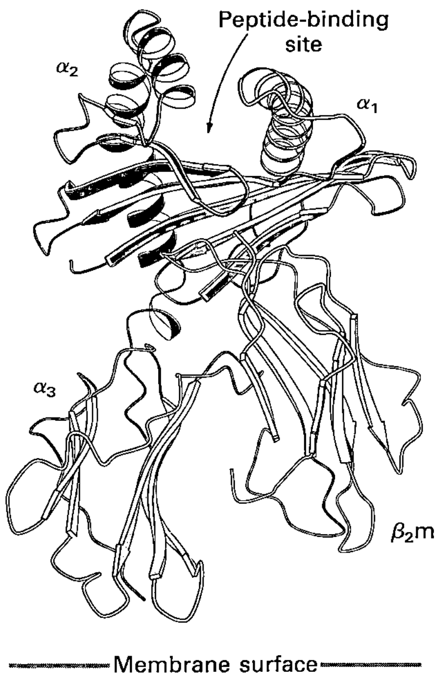

Figure 14-36
Three-dimensional structure of a class I MHC protein. The α_1 domain is shown in blue, the α_2 domain in red, the α_3 domain in green, and β_2m in yellow. The peptide-binding site is formed by α_1 and α_2. In the intact protein, a transmembrane helix follows α_3. [Drawn from 1hsa.pdb. D.R. Madden, J.C. Gorga, J.L. Strominger, and D.C. Wiley. *Cell* 70(1992):1035.]

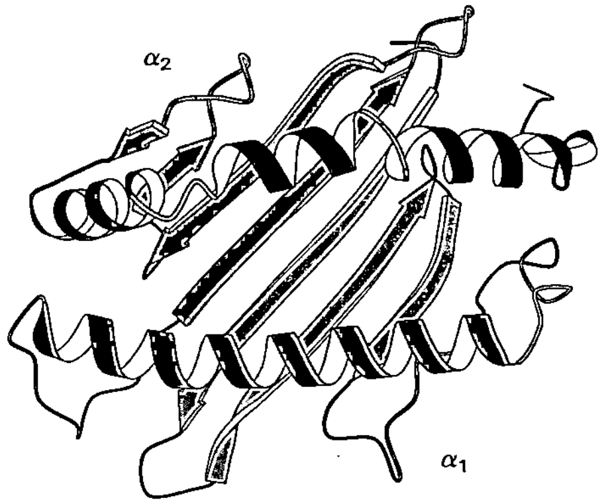

Figure 14-37
Binding groove of a class I MHC protein. The floor of the binding site is formed by a β sheet. The groove is flanked by a pair of α helices. The groove is formed by residues from the α_1 (blue) and α_2 (red) domains. [Drawn from 1hsa.pdb. D.R. Madden, J.C. Gorga, J.L. Strominger, and D.C. Wiley. *Cell* 70(1992):1035.]

The α_3-β_2m pair is closest to the membrane, whereas the α_1-α_2 pair is farthest away. The α_3 and β_2m fold into β sheet sandwiches like immunoglobulin constant domains, but the two domains are paired differently than in antibodies. The α_1 and α_2 domains, by contrast, exhibit a novel architecture. They associate intimately to form a deep groove that serves as the peptide-binding site (Figure 14-37). The floor of the groove, which is about 25 Å long and 10 Å wide, is formed by eight β strands, four from each domain. A helix contributed by the α_1 domain forms one side, and a helix contributed by the α_2 domain forms the other side.

The groove in HLA-B7 is filled by a nine-residue peptide in an extended conformation (Figure 14-38). Pockets in the cleft bind four side chains and both the amino and carboxy termini of the displayed peptide. Position 2 (P2) of these peptides is always arginine, P3 is a hydrophobic residue, and P6 is a nonpolar or a small polar residue. The other residues are highly variable. The sequences of peptides bound to HLA-B7 suggest that *many millions of different peptides could be presented by this particular class I MHC protein.* What makes it so eclectic in its tastes? The protein stretches its peptide quarry (rather like a medieval rack) and accommodates its side chains in pockets that are quite tolerant of diversity (except at P2). Only four of the nine residues seem to matter. A remarkable feature of this interaction is its stability. Once bound, a peptide is not released, even in days. MHC class I and class II proteins are designed to hold tenaciously valuable samples from inside the cell.

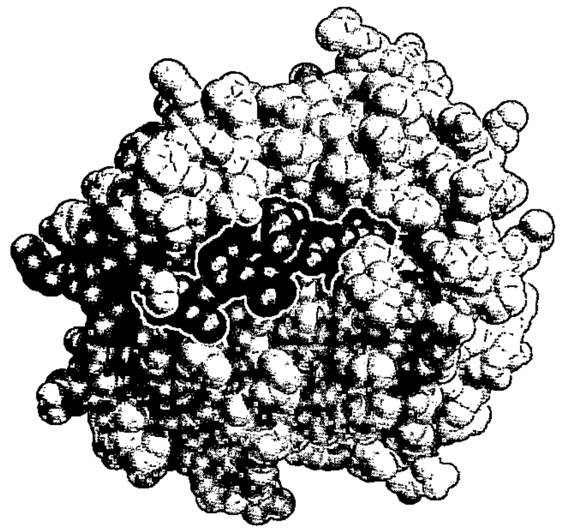

Figure 14-38
Binding of a peptide to a class I MHC protein. The 9-mer peptide is shown in red, the α chain of the MHC protein in blue, and β_2m in yellow. [Drawn from 1hhi.pdb. D.R. Madden, D.N. Garboczi, and D.C. Wiley. *Cell* 75(1993):693-708.]

T-CELL RECEPTORS ARE ANTIBODY-LIKE PROTEINS CONTAINING VARIABLE AND CONSTANT REGIONS

We are now ready to consider the receptor that recognizes peptides displayed by MHC proteins on target cells. The T-cell receptor consists of a 43-kd α chain (T_α) joined by a disulfide bond to a 43-kd β chain (T_β) (Figure 14-39). Both chains span the plasma membrane and have a short carboxyl-terminal region on the cytosolic side. A small proportion of T cells express a receptor consisting of γ and δ chains in place of α and β. T_α and T_β, like immunoglobulin L and H chains, consist of *variable* and *constant* regions. Indeed, *these domains of the T-cell receptor are homologous in sequence and three-dimensional structure to the V and C domains of immunoglobu-*

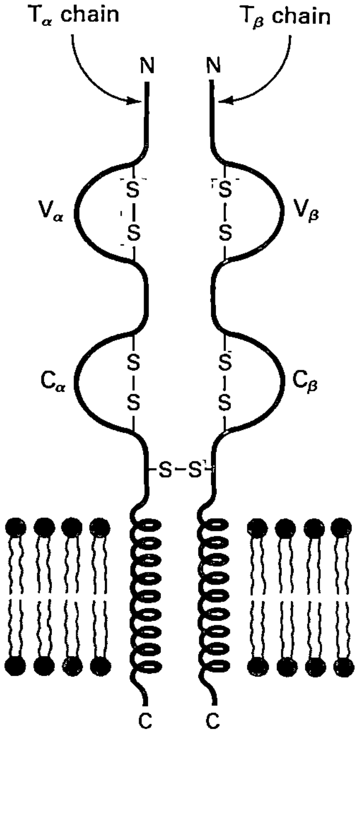

Figure 14-39
Schematic diagram of a T-cell receptor. Each polypeptide chain of this disulfide-linked heterodimer consists of a variable and a constant region.

lins. Furthermore, hypervariable sequences present in the V regions of T_α and T_β form the binding site for the epitope.

Moreover, the genetic architecture of these homologous proteins is similar to that of immunoglobulins (Figure 14-40). The variable region of T_α is encoded by about 100 *V* segment genes and 50 *J* (joining) segment genes. T_β is encoded by two *D* (diversity) gene segments in addition to about 30 *V* and 12 *J* segments. Again, diversity of component genes and different reading frames increase the number of kinds of proteins formed. *At least 10^7 different specificities could arise from combinations of this repertoire of genes.* Thus, T-cell receptors, like immunoglobulins, can recognize a very large number of different epitopes. All the receptors on a particular T cell have the same specificity.

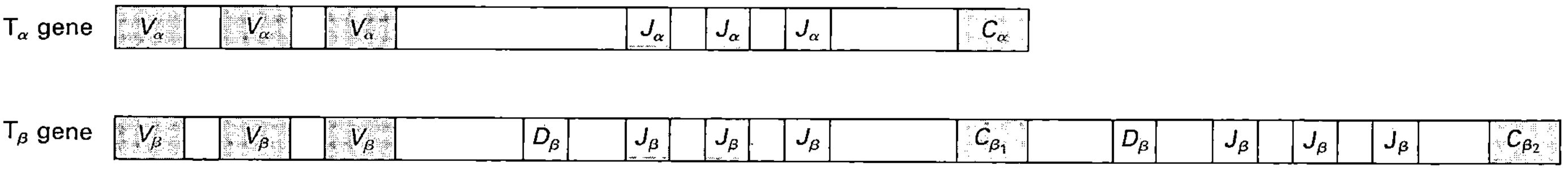

Figure 14-40
The T-cell receptor is encoded by multiple *V*-, *D*-, and *J*-gene segments. *C* genes follow *J*-segment genes.

How do T cells recognize their targets? The three-dimensional structure of the T-cell receptor is not yet known. However, a plausible working hypothesis is depicted in Figure 14-41. The variable regions of the α and β chains of the T-cell receptor form a combining site that recognizes a *combined epitope*—foreign peptide bound to an MHC class I or II protein on the target cell. Neither the foreign peptide alone nor the MHC protein alone forms a complex with the T-cell receptor. This dual recognition process assures that T cells devote all their attention to other cells and are not diverted by soluble molecules.

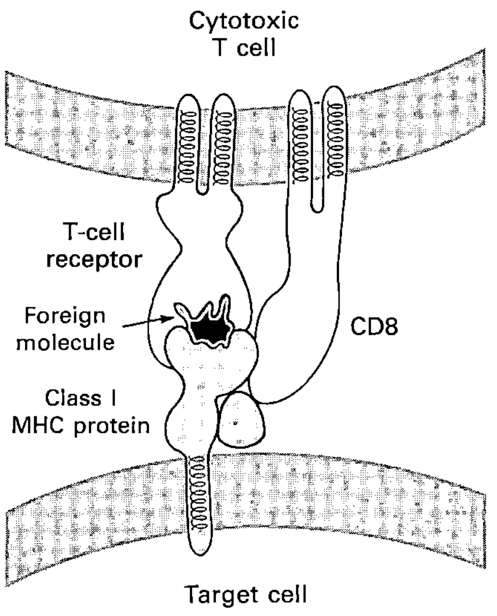

Figure 14-41
Model of antigen recognition by a T cell. A class I MHC protein (blue) presents a foreign molecule (red) to a T-cell receptor (green). The T cell contains other cell-surface proteins that bind markers on target cells. CD8 (yellow), which is present on cytotoxic T cells, binds to class I MHC proteins. [After J. Goverman, T. Hunkapiller, and L. Hood. *Cell* 45(1986):475.]

CD8 ON KILLER T CELLS AND CD4 ON HELPER T CELLS ACT IN CONCERT WITH T-CELL RECEPTORS

The T-cell receptor does not act alone in recognizing and profoundly altering the fate of target cells. Five integral membrane proteins form a *CD3 complex* that is associated with the T-cell receptor. CD45, a membrane-bound tyrosine phosphatase, is also associated with the T-cell re-

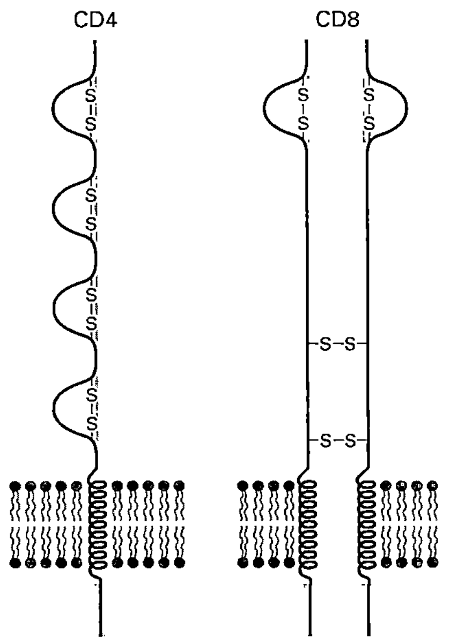

Figure 14-42
CD4 and CD8 are T-cell proteins that participate in recognizing MHC-peptide complexes of target cells. They contain extracellular domains (blue) that resemble immunoglobulin domains. The cytosolic tails (yellow) of CD4 and CD8 bind p56^{lck}, a tyrosine kinase that plays a key role in T-cell signaling. [After D.L. Leahy, R. Axel, and W.A. Hendrickson. *Cell* 68(1992):1145.]

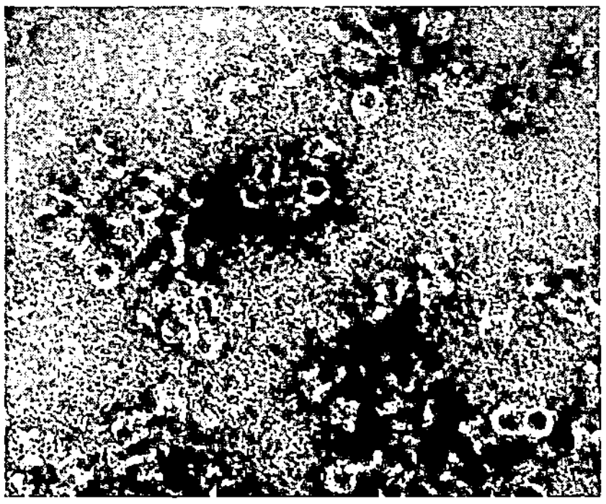

Figure 14-43
Electron micrograph showing pores in the membrane of a cell that was lysed by a cytotoxic (killer) T cell. The pores are formed by the polymerization of perforin, a protein akin to the C9 component of complement. [Courtesy of Dr. Eckhard Podack.]

ceptor. The abbreviation *CD* stands for "cluster of differentiation," a cell-surface marker that identifies a lineage or stage of differentiation. Antibodies specific for particular CD proteins have been invaluable in following the development of leukocytes and in discovering new interactions. Cells expressing different sets of CD proteins can readily be sorted once they are tagged with fluorescent-labeled antibodies.

Killer and helper T cells have the same T-cell receptors. Why then does the T-cell receptor on a killer cell interact only with peptide bound to class I MHC proteins, whereas the receptor on a helper cell interacts only with class II? *The molecules that direct T cells to different targets are the integral membrane proteins CD8 and CD4, which serve as coreceptors.* Each chain in the CD8 dimer contains a domain that resembles an immunoglobulin variable domain (Figure 14-42). CD4, which is monomeric, contains four such immunoglobulin-like domains. Thus, *immunoglobulin folds are present in many molecules of immunity: T-cell receptors, class I and II MHC proteins, CD3, CD4, and CD8, in addition to immunoglobulins themselves.* Indeed, the significance of the immunoglobulin fold extends far beyond the immune system. Recall that several receptor tyrosine kinases contain this motif (p. 351). Furthermore, many cell-adhesion molecules (CAMs) are built of these domains. *The immunoglobulin fold plays a fundamental role in mediating cell-cell interactions in invertebrates as well as vertebrates.*

A killer cell expresses CD8 but not CD4, whereas a helper cell expresses CD4 but not CD8. What do these coreceptors recognize on target cells? CD8 interacts primarily with the relatively constant α_3 domain of class I MHC proteins, whereas CD4 binds the constant part of class II proteins. CD8 and CD4 are important not only in directing T cells to their appropriate targets. These coreceptors also play a critical role in signal transduction. The cytosolic tails of CD8 and CD4 contain docking sites for p56^{lck}, a cytosolic tyrosine kinase akin to the src protein (p. 353). Binding of several T-cell receptors to peptide-MHC complexes on a target cell leads to the stimulation of this cytosolic tyrosine kinase. The consequent activation of the γ1 isoform of phospholipase C switches on the phosphoinositide cascade (p. 344). One consequence of the activation of killer T cells is the secretion of *perforin.* This 70-kd protein lyses the target cell by polymerizing to form transmembrane pores 10 nm wide (Figure 14-43). Perforin is similar to C9, the final component in the complement cascade. By contrast, activated helper cells secrete *interleukins* (protein hormones that act locally) to stimulate the proliferation of B cells. The mechanism of T-cell activation and the interplay of T cells with target cells are exciting and important areas of inquiry.

HUMAN IMMUNODEFICIENCY VIRUSES SUBVERT THE IMMUNE SYSTEM BY DESTROYING HELPER T CELLS

In 1981, the first cases of a new disease called *acquired immune deficiency syndrome (AIDS)* were recognized. The victims died of rare infections because their immune systems were crippled. The cause was identified two years later by Luc Montagnier. AIDS is produced by *human immunodeficiency virus (HIV).* Two kinds are known: HIV-1 and the much less common HIV-2. Both are related to human T-cell lymphotrophic virus I (HTLV-I), the cause of a rare leukemia. Like other *retroviruses,* HIV contains a single-stranded RNA genome that is replicated through a double-stranded DNA intermediate (p. 92). This viral DNA becomes integrated into the genome of the host cell. In fact, viral genes are transcribed only when they are integrated into the host DNA.

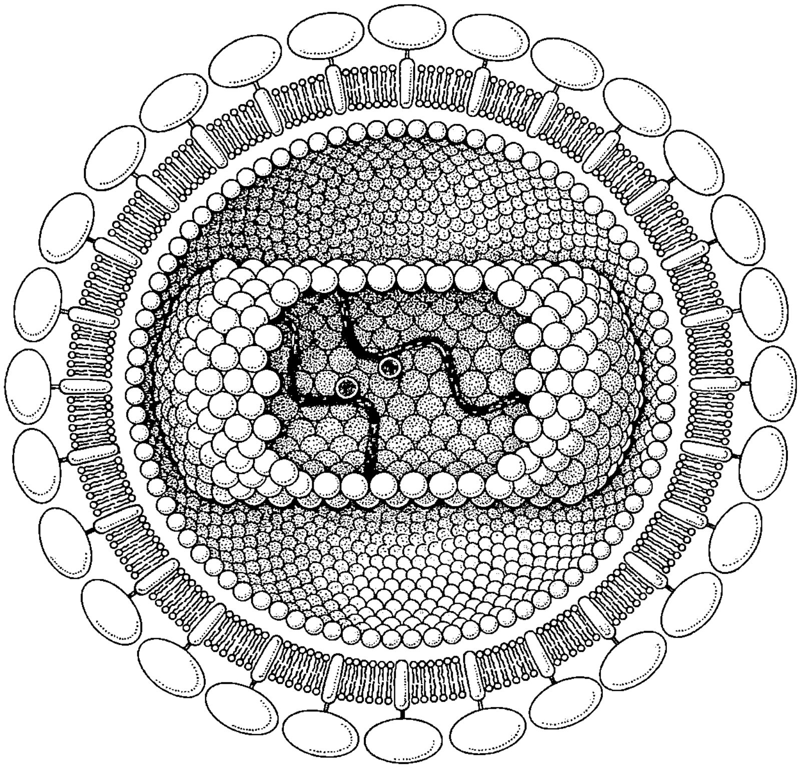

Figure 14-44
Schematic diagram of human immunodeficiency virus (HIV), the cause of AIDS. The membrane envelope glycoproteins gp41 and gp120 are shown in dark and light green. The core of the virus contains two kinds of protein subunits, p18 (orange) and p24 (white), an RNA genome (red), and several molecules of reverse transcriptase (blue). [After R.C. Gallo. The AIDS virus. Copyright © 1987 by Scientific American, Inc. All rights reserved.]

The HIV virion is enveloped by a lipid bilayer membrane containing two glycoproteins: gp41 spans the membrane and is associated with gp120, which is located on the external face (Figure 14-44). The core of the virus contains two copies of the RNA genome and associated transfer RNAs, and several molecules of reverse transcriptase. They are surrounded by many copies of two proteins called p18 and p24. *The host cell for HIV is the helper T cell.* The gp120 molecules on the membrane of HIV bind to CD4 molecules on the surface of helper T cells. This interaction allows the associated viral gp41 to insert its amino-terminal head into the host cell membrane. The viral membrane and the helper cell membrane fuse, and the viral core is released directly into the cytosol (Figure 14-45).

Infection by HIV leads to the lysis of helper T cells. The permeability of the host plasma membrane is markedly increased by the insertion of viral glycoproteins and the budding of virus particles. The influx of ions and water disrupts the ionic balance, causing osmotic lysis. Cytosolic calcium levels also become abnormally high. Viral mRNAs also interfere with the

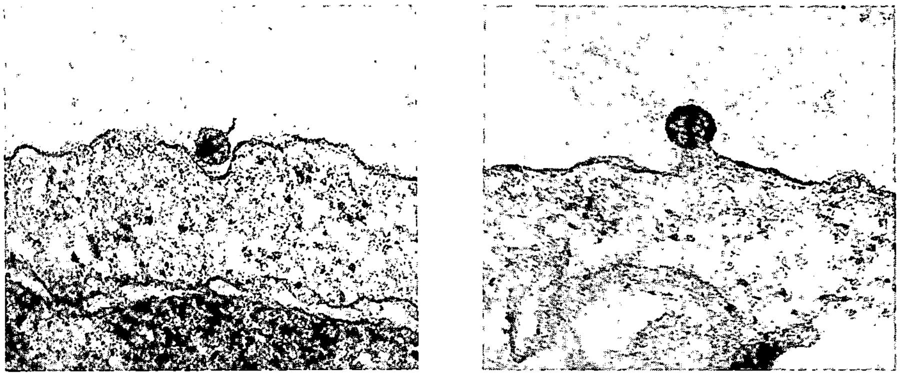

A B

Figure 14-45
Entry of HIV into a T-lymphocyte precursor. The viral membrane fuses with the plasma membrane of the target cell. [After B.S. Stein, S.D. Gowda, J.D. Lifson, R.C. Penhallow, K.G. Bensch, and E.G. Engleman. *Cell* 49(1987):664.]

expression of host-cell transcripts. Infected cells can fuse many normal CD4-bearing cells to form giant multinucleated cells. In addition to depleting helper T cells, HIV markedly impairs their function. For example, soluble gp120 binds to CD4 on helper T cells, blocking their interaction with peptide–class II MHC complexes on targets. Also, the binding of soluble gp120 to CD4 molecules on uninfected cells may trigger their destruction by the complement pathway.

The development of an effective AIDS vaccine is made difficult by the antigenic diversity of HIV strains. HIV constantly changes its coat proteins, both gp120 and gp41, to evade detection by the host. Indeed, the mutation rate of HIV is 65 times higher than that of influenza virus, a skilled practitioner of camouflage. The aim now is to define relatively conserved sequences in these HIV proteins and use them as immunogens. An alternative approach is to use soluble fragments of CD4 produced by recombinant DNA technology to block gp120. Ideally, one would like to have a small molecule that binds very tightly and specifically to the region of gp120 that makes contact with CD4. The recent elucidation of the three-dimensional structure of CD4 provides a starting point for the design of such inhibitors.

SUMMARY

The immune system employs two systems of recognition elements to distinguish between self and nonself: soluble antibodies and cell-attached T-cell receptors. Antibodies are populations of protein molecules (immunoglobulins) that are synthesized by an animal in response to a foreign macromolecule, called an antigen or immunogen. Antibodies have high affinity for antigens that induced their formation. Small foreign molecules (haptens) elicit the formation of specific antibody if they are attached to a macromolecule. Antibody synthesis occurs by selection rather than by instruction. An antigen binds to the surface of lymphocytes already committed to making antibodies specific for that antigen. The combination of antigen and surface receptors triggers cell division and the synthesis of large amounts of specific antibody. Antibodies directed against a specific determinant are usually heterogeneous, because they are the products of many antibody-producing cells. The antibodies produced by a single cell or by a clone are homogeneous. Monoclonal antibodies are formed by hybridoma cells obtained by fusing an immortal myeloma tumor cell with a lymphocyte that produces an antibody having the desired specificity.

Five classes of antibodies are formed. Immunoglobulin G (IgG) is the principal one in serum, but IgM is the first class to appear following exposure to an antigen. IgA is the major class in external secretions, and IgE protects against parasites; the role of IgD is not known. Antibodies consist of light (L) and heavy (H) chains. Immunoglobulin G, which has the subunit structure L_2H_2, contains two binding sites for antigen. Immunoglobulin G can be enzymatically cleaved into two F_{ab} fragments, which bind antigen but do not form a precipitate, and one F_c fragment, which mediates effector functions such as the binding of complement. L and H chains consist of a variable (V) region (an amino-terminal sequence of about 108 residues) and a constant (C) region. Antigen-binding sites are formed by three hypervariable segments (complementarity-determining regions, or CDRs) from the L chain and three from the H chain.

Antibody molecules are folded into compact domains of about 108 residues that have homologous sequences. The immunoglobulin fold, a sandwich of antiparallel β pleated sheets stapled by a disulfide bond, is a highly versatile framework. Loops between β strands can be varied without changing the structural core. Haptens and antigens bind to crevices, pockets, and extensive flat surfaces formed by the CDRs. The array of interactions used to achieve complementarity and high affinity is like that employed in enzyme-substrate complexes. The constant-region domains mediate effector functions. Antigen-antibody complexes trigger the complement cascade, a series of zymogen conversions culminating in the formation of a membrane attack complex, which lyses target cells. They also activate phagocytosis by macrophages and induce granule release by mast cells.

Variable and constant regions are encoded by separate genes that become joined during cell differentiation. There are several hundred genes for the V regions of L and H chains. A complete gene for the V region of a light chain is formed by the recombination of an incomplete *V* segment gene with one of several *J* (joining) segment genes encoding part of the last hypervariable segment. A tandem array of *J* genes is located near the *C* gene. Heavy-chain V-region genes are formed by the recombination of *V*-, *D*- (diversity), and *J*-segment genes. More than 10^8 different specificities can be generated by combining different *V*, *D*, and *J* gene segments, somatic mutation of segments, and combining different L and H chains. Different classes of antibodies are formed by the translocation of a V_H gene from the *C* gene of one class to that of another.

T-cell–mediated immunity evolved to cope with intracellular pathogens, which are shielded from antibodies by the host cell membrane. T cells continually scan the surfaces of all cells and kill those that exhibit foreign markings. Vertebrate cells exhibit on their surface a sample of peptides derived from the digestion of proteins in their cytosol. Peptides generated by the action of proteasomes are transported into the lumen of the endoplasmic reticulum. They combine with nascent class I major histocompatibility complex (MHC) proteins, which tenaciously bind them in a long groove and present them on the cell surface for scrutiny by T cells. Class I MHC proteins are highly polymorphic, which protects the population from viral mutants that might otherwise resist presentation. Killer T cells detect foreign peptides bound to class I MHC proteins.

T-cell receptors are antibody-like proteins containing variable and constant regions. The interplay of germ-line diversity and somatic recombination generates a very large repertoire of T-cell receptors. The foreign peptide–MHC complex triggers a cascade initiated by the T-cell receptor in concert with CD3 and CD8. An associated cytosolic tyrosine kinase ($p56^{lck}$) plays a key role in signaling.

Helper T cells, a different set, stimulate the proliferation of specific B lymphocytes and thereby serve as partners in determining which kinds of immunoglobulins are produced by B cells. Foreign peptides bound to class II MHC proteins trigger helper T cells. These peptides come from the digestion of proteins that are endocytosed by a B cell or macrophage. They signal that a cell has encountered a pathogen and serve as a call for help. In contrast, complexes of foreign peptides bound to class I proteins indicate that a cell has succumbed to a pathogen and must be destroyed to save the organism. Helper T cells recognize class II rather than class I MHC proteins because they contain CD4 in place of CD8. When activated, they stimulate the proliferation of B cells. HIV-1, a retrovirus that causes AIDS, subverts the immune system by destroying helper T cells. The gp120 protein in the membrane envelope of HIV-1 binds to CD4 on the surface of the helper cell to induce fusion and gain viral entry into the host cell.

SELECTED READINGS

Where to start

Nossal, G.J.V., 1993. Life, death, and the immune system. *Sci. Amer.* 269(3):53–62. [A highly informative primer on immunology. The other articles in this issue also are valuable.]

Tonegawa, S., 1985. The molecules of the immune system. *Sci. Amer.* 253(4):122–131.

Leder, P., 1982. The genetics of antibody diversity. *Sci. Amer.* 246(5):102–115.

Grey, H.M., Sette, A., and Buus, S., 1989. How T cells see antigen. *Sci. Amer.* 261(5):56–64.

Hunkapiller, T., Goverman, J., Koop, B.F., and Hood, L., 1989. Implications of the diversity of the immunoglobulin gene superfamily. *Cold Spring Harbor Symp. Quant. Biol.* 54:15–29.

Books

Kuby, J., 1994. *Immunology* (2nd ed.). W.H. Freeman. [A stimulating account of the experimental foundations of immunology.]

Abbas, A.K., Lichtman, A.H., and Pober, J.S., 1992. *Cellular and Molecular Immunology* (2nd ed). Saunders. [A concise and highly informative introduction that emphasizes the interplay of immunology and medicine.]

Cold Spring Harbor Symposia on Quantitative Biology, 1989. Volume 54. Immunological Recognition. [An outstanding collection of articles on antibody, T-cell, and MHC recognition processes. Lymphocyte signaling is also presented in detail.]

Nisinoff, A., 1985. *Introduction to Molecular Immunology* (2nd ed.). Sinauer. [A concise and perceptive introduction.]

Weir, D.M. (ed.), 1986. *Handbook of Experimental Immunology.* Oxford University Press.

Annual Review of Immunology. [Contains excellent review articles on many facets of molecular and cellular immunology. Volume 1 was published in 1983.]

Structure of antibodies and antibody-antigen complexes

Davies, D.R., Padlan, E.A., and Sheriff, S., 1990. Antibody-antigen complexes. *Ann. Rev. Biochem.* 59:439–473.

Poljak, R.J., 1991. Structure of antibodies and their complexes with antigens. *Mol. Immunol.* 28:1341–1345.

Marquart, M., Deisenhofer, J., Huber, R., and Palm, W., 1980. Crystallographic refinement and atomic models of the intact immunoglobulin molecule Kol and its antigen-binding fragment at 3.0 Å and 1.9 Å resolution. *J. Mol. Biol.* 141:369–391.

Herron, J.N., He, X.M., Ballard, D.W., Blier, P.R., Pace, P.E., Bothwell, A.L., Voss, E.J., and Edmundson, A.B., 1991. An autoantibody to single-stranded DNA: Comparison of the three-dimensional structures of the unliganded Fab and a deoxynucleotide-Fab complex. *Proteins* 11:159–175.

Silverton, E.W., Navia, M.A., and Davies, D.R., 1977. Three-dimensional structure of an intact human immunoglobulin. *Proc. Nat. Acad. Sci.* 74:5140–5144.

Valentine, R.C., and Green, N.M., 1967. Electron microscopy of an antibody-hapten complex. *J. Mol. Biol.* 27:615–617.

Padlan, E.A., Silverton, E.W., Sheriff, S., Cohen, G.H., Smith, G.S., and Davies, D.R., 1989. Structure of an antibody-antigen complex: Crystal structure of the HyHEL-10 Fab-lysozyme complex. *Proc. Nat. Acad. Sci.* 86:5938–5942.

Rini, J., Schultze-Gahmen, U., Wilson, I.A., 1992. Structural evidence for induced fit as a mechanism for antibody-antigen recognition. *Nature* 255:959–965.

Fischmann, T.O., Bentley, G.A., Bhat, T.N., Boulot, G., Mariuzza, R.A., Phillips, S.E., Tello, D., and Poljak, R.J., 1991. Crystallographic refinement of the three-dimensional structure of the Fab D1.3-lysozyme complex at 2.5-Å resolution. *J. Biol. Chem.* 266:12915–12920.

Effector functions of immunoglobulins

Reid, K.B.M., and Law, S.K.A., 1988. *Complement.* IRL Press.

Burton, D.R., 1990. Antibody: The flexible adaptor molecule. *Trends Biochem. Sci.* 15:64–69.

Sim, R.B., and Reid, K.B., 1991. C1: Molecular interactions with activating systems. *Immunol. Today* 12:307–311.

Generation of antibody diversity

Tonegawa, S., 1988. Somatic generation of immune diversity. *Biosci. Rep.* 8:3–26. [This Nobel Lecture provides an excellent account of how much of the diversity of immunoglobulins is generated.]

Honjo, T., and Habu, S., 1985. Origin of immune diversity: Genetic variation and selection. *Ann. Rev. Biochem.* 54:803–830.

Berek, C., Griffiths, G.M., and Milstein, C., 1985. Molecular events during maturation of the immune response to oxazolone. *Nature* 316:412–418.

Applications of monoclonal antibodies

Milstein, C., 1980. Monoclonal antibodies. *Sci. Amer.* 243(4):66–74.

Waldmann, T.A., 1991. Monoclonal antibodies in diagnosis and therapy. *Science* 252:1657–1661.

Lerner, R.A., and Tramontano, A., 1988. Catalytic antibodies. *Sci. Amer.* 258(3):58–60.

MHC proteins and antigen processing

Bjorkman, P.J., and Parham, P., 1990. Structure, function, and diversity of class I major histocompatibility complex molecules. *Ann. Rev. Biochem.* 59:253–288.

Goldberg, A.L., and Rock, K.L., 1992. Proteolysis, proteasomes, and antigen presentation. *Nature* 357:375–379.

Madden, D.R., Gorga, J.C., Strominger, J.L., and Wiley, D.C., 1992. The three-dimensional structure of HLA-B27 at 2.1 Å resolution suggests a general mechanism for tight binding to MHC. *Cell* 70:1035–1048.

Brown, J.H., Jardetzky, T.S., Gorga, J.C., Stern, L.J., Urban, R.G., Strominger, J.L., and Wiley, D.C., 1993. Three-dimensional structure of the human class II histocompatibility antigen HLA-DR1. *Nature* 364:33–39.

Saper, M.A., Bjorkman, P.J., and Wiley, D.C., 1991. Refined structure of the human histocompatibility antigen HLA-A2 at 2.6 Å resolution. *J. Mol. Biol.* 219:277–319.

Madden, D.R., Gorga, J.C., Strominger, J.L., and Wiley, D.C., 1991. The structure of HLA-B27 reveals nonamer self-peptides bound in an extended conformation. *Nature* 353:321–325.

Bjorkman, P.J., Saper, M.A., Samraoui, B., Bennett, W.S., Strominger, J.L., and Wiley, D.C., 1987. Structure of the human class I histocompatibility antigen, HLA-A2. *Nature* 329:506–512.

T-cell receptors and signaling complexes

Janeway, C.J., 1992. The T cell receptor as a multicomponent signalling machine: CD4/CD8 coreceptors and CD45 in T cell activation. *Ann. Rev. Immunol.* 10:645–674.

Podack, E.R., and Kupfer, A., 1991. T-cell effector functions: mechanisms for delivery of cytotoxicity and help. *Ann. Rev. Cell Biol.* 7:479–504.

Davis, M.M., 1990. T cell receptor gene diversity and selection. *Ann. Rev. Biochem.* 59:475-496.

Wang, J.H., Yan, Y.W., Garrett, T.P., Liu, J.H., Rodgers, D.W., Garlick, R.L., Tarr, G.E., Husain, Y., Reinherz, E.L., and Harrison, S.C., 1990. Atomic structure of a fragment of human CD4 containing two immunoglobulin-like domains. *Nature* 348:411–418.

Leahy, D.J., Axel, R., and Hendrickson, W.A., 1992. Crystal structure of a soluble form of the human T cell coreceptor CD8 at 2.6 Å resolution. *Cell* 68:1145–1162.

Sefton, B.M., 1991. The lck tyrosine protein kinase. *Oncogene* 6:683–686.

HIV and AIDS

Fauci, A.S., 1988. The human immunodeficiency virus: Infectivity and mechanisms of pathogenesis. *Science* 239:617–622.

Gallo, R.C., and Montagnier, L., 1988. AIDS in 1988. *Sci. Amer.* 259:41–48.

Haseltine, W.A., 1992. The molecular biology of the human immunodeficiency virus type 1. *FASEB J.* 5:2349–2360. [Also see the other articles on AIDS in this issue of the journal.]

Discovery of major concepts

Ada, G.L., and Nossal, G., 1987. The clonal selection theory. *Sci. Amer.* 257(2):62–69.

Porter, R.R., 1973. Structural studies of immunoglobulins. *Science* 180:713–716.

Edelman, G.M., 1973. Antibody structure and molecular immunology. *Science* 180:830–840.

Kohler, G., 1986. Derivation and diversification of monoclonal antibodies. *Science* 233:1281–1286.

Milstein, C., 1986. From antibody structure to immunological diversification of immune response. *Science* 231:1261–1268.

Jerne, N.K., 1973. The immune system. *Sci. Amer.* 229(1):52–60.

Janeway, C.A., Jr., 1989. Approaching the asymptote? Evolution and revolution in immunology. *Cold Spring Harbor Symp. Quant. Biol.* 54:1–13.

PROBLEMS

1. *Energetics and kinetics.* Suppose that the dissociation constant of an F_{ab}-hapten complex is 3×10^{-7} M at 25°C.
   (a) What is the standard free energy of binding?
   (b) Immunologists often speak of affinity (K_a), the reciprocal of the dissociation constant, in comparing antibodies. What is the affinity of this F_{ab}?
   (c) The rate constant of release of hapten from the complex is 120 s^{-1}. What is the rate constant for association? What does the magnitude of this value imply about the extent of structural change in the antibody on binding hapten?

2. *Sugar niche.* An antibody specific for dextran, a polysaccharide of glucose residues, was tested for its binding of glucose oligomers. The full binding affinity was obtained when the oligomer contained six glucose residues. How does the size of this site compare with that of the active site of lysozyme?

3. *A brilliant emitter.* Certain naphthalene derivatives exhibit a weak yellow fluorescence when they are in a highly polar environment (such as water) and an intense blue fluorescence when they are in a markedly nonpolar environment (such as hexane). The binding of ε-dansyl-lysine to specific antibody is accompanied by a marked increase in its fluorescence intensity and a shift in color from yellow to blue. What does this finding reveal about the hapten-antibody complex?

4. *Avidity versus affinity.* The standard free energy of binding of F_{ab} derived from an antiviral IgG is −7 kcal/mol at 25°C.
   (a) Calculate the dissociation constant of this interaction.

(b) Predict the dissociation constant of the intact IgG assuming that both combining sites of the antibody can interact with viral epitopes and that the free energy cost of assuming a favorable hinge angle is +3 kcal/mol.

5. *Miniantibody.* The F_{ab} fragment of an antibody molecule has essentially the same affinity for a monovalent hapten as does intact IgG.
   (a) What is the smallest unit of an antibody that could retain the specificity and binding affinity of the whole protein?
   (b) Design a compact single-chain protein that is likely to specifically bind antigen with high affinity.

6. *Turning on B cells.* B lymphocytes, the precursors of plasma cells, are triggered to proliferate by the binding of multivalent antigens to receptors on their surface. The cell-surface receptors are transmembrane immunoglobulins. Univalent antigens, by contrast, do not activate B cells.
   (a) What do these findings reveal about the mechanism of B-cell activation?
   (b) How might antibodies be used to activate B cells?

7. *Switching to a soluble antibody.* An unactivated B cell contains membrane-bound immunoglobulin (the μ_m form of IgM), whereas an activated cell secretes soluble antibody of the same specificity (μ_s). The H chains of μ_m differ from those of μ_s in possessing a transmembrane segment and a short cytosolic tail. The L chains of μ_m and μ_s are identical. Furthermore, μ_m and μ_s are encoded by the same genes. Propose a mechanism for generating membrane-bound and soluble antibodies of the same specificity.

8. *An ingenious cloning strategy.* In the cloning of the gene for the α chain of the T-cell receptor, T-cell cDNAs were hybridized with B-cell mRNAs. What was the purpose of this hybridization step? Can the principle be applied generally?

CHAPTER

15

Molecular Motors

Coordinated movement is central to life. Three eukaryotic motility systems that are driven by ATP will be considered first. In higher eukaryotes, *muscle contraction* is mediated by the sliding of interdigitating *myosin* and *actin* filaments. Indeed, most cells of eukaryotes, from yeast to humans, are capable of active movement because of the interaction of myosin and actin. The beating of *cilia* and *flagella* depends on the interplay of a different pair of proteins—*dynein* and *tubulin.* Microtubules made of tubulins also form the mitotic spindle, which organizes the movement of chromosomes in cell division. Furthermore, microtubules serve as tracks for the movement of vesicles and organelles within cells, as in the transport of secretory vesicles along the axons of neurons. *Kinesin* is one of several proteins mediating the *movement of vesicles on microtubules.* The binding of ATP to myosin, dynein, and kinesin induces conformational transitions in these motor proteins. These structural changes are reversed by the hydrolysis of bound ATP and release of ADP and P_i, which advance the motor protein on the actin or tubulin track.

Bacteria use a different kind of motor and energy source to move. They are propelled by the *rotation of flagellar motors* located in their cytoplasmic membrane. These motors are powered by *proton-motive force* rather than by ATP. How do ATP-driven and proton-driven conformational cycles in these molecular engines lead to coordinated motion? The concerted use of a wide range of experimental approaches drawn from protein chemistry, structural biology, and molecular genetics is lifting the veil from the mystery of directed movement.

Opening Image: *Leonardo da Vinci's "Study of a rearing horse" for The Battle of Anghiari (c. 1504). [The Royal Collection © 1994 Her Majesty Queen Elizabeth II.]*

MUSCLE CONTAINS INTERACTING THICK AND THIN PROTEIN FILAMENTS

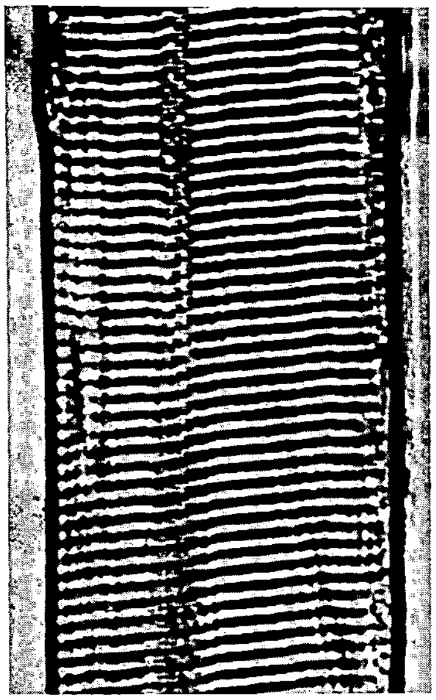

Figure 15-1
Phase-contrast light micrograph of a skeletal muscle fiber about 50 μm in diameter. The A bands are dark and the I bands are light. [Courtesy of Dr. Hugh Huxley.]

We begin with the structural basis of contraction in vertebrate striated muscle, the best-understood force-generating process. Vertebrate muscle that is under voluntary control has a *striated* appearance when examined under the light microscope (Figure 15-1). It consists of multinucleate cells that are bounded by an electrically excitable plasma membrane. A muscle cell contains many parallel *myofibrils,* each about 1 μm in diameter, that are immersed in the cytosol. An electron micrograph of a longitudinal section of a myofibril displays a wealth of structural detail (Figures 15-2A and 15-3). The functional unit, called a *sarcomere,* repeats every 2.3 μm (23,000 Å) along the fibril axis. A dark *A band* and a light *I band* alternate regularly. The central region of the A band, termed the *H zone,* is less dense than the rest of the band. The I band is bisected by a very dense, narrow *Z line.*

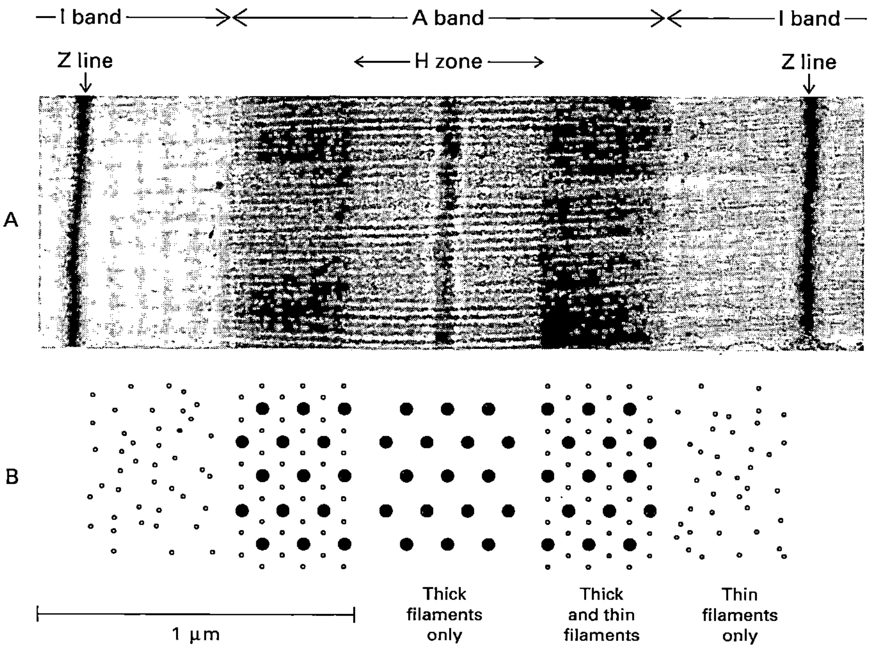

Figure 15-2
(A) Electron micrograph of a longitudinal section of a skeletal muscle myofibril, showing a single sarcomere. (B) Schematic diagrams of cross sections are shown below the corresponding regions in the micrograph. [Courtesy of Dr. Hugh Huxley.]

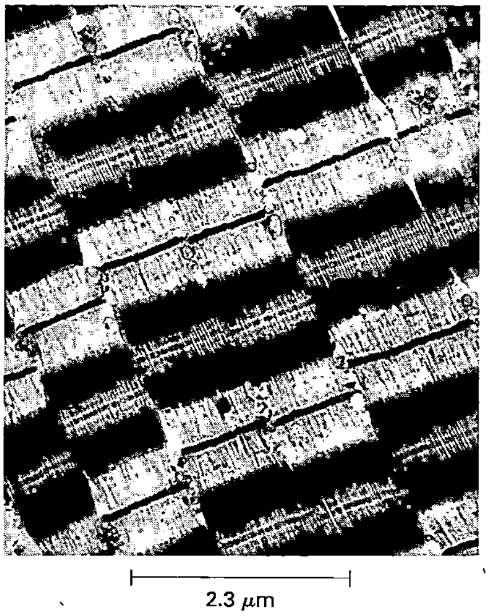

Figure 15-3
Electron micrograph of a longitudinal section of a skeletal muscle fiber, showing several myofibrils extending from upper left to lower right. [Courtesy of Dr. Hugh Huxley.]

The underlying molecular plan of a sarcomere is revealed by cross sections of a myofibril, which show that there are *two kinds of interacting protein filaments* (Figure 15-2B). The *thick filaments* have diameters of about 15 nm (150 Å), whereas the *thin filaments* have diameters of about 9 nm (90 Å). The thick filaments are primarily *myosin.* The thin filaments contain *actin, tropomyosin,* and the *troponin complex.* Each thin filament has three neighboring thick filaments and each thick filament is encircled by six thin filaments. The thick and thin filaments interact by *cross-bridges,* which are domains of myosin molecules. In muscle depleted of ATP, cross-bridges emerge at regular intervals from the thick filaments and bridge a gap of 13 nm between the surfaces of thick and thin filaments (Figures 15-4 and 15-5).

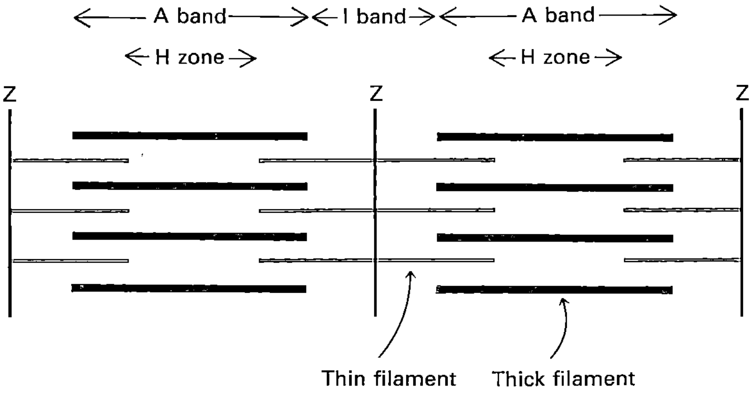

Figure 15-4
Schematic diagram of the structure of striated muscle, showing overlapping arrays of thick and thin filaments. [Courtesy of Dr. Hugh Huxley.]

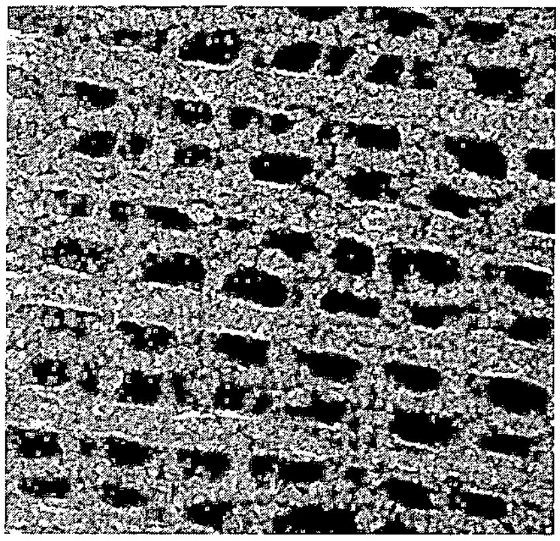

Figure 15-5
Electron micrograph showing cross-bridges between thick and thin filaments. [Courtesy of Dr. John Heuser.]

THICK AND THIN FILAMENTS SLIDE PAST EACH OTHER IN MUSCLE CONTRACTION

Muscle shortens by as much as a third of its original length as it contracts. How is shortening achieved? In the 1950s, Andrew Huxley and Ralph Niedergerke, and Hugh Huxley and Jean Hanson, independently proposed a *sliding-filament model* on the basis of x-ray, light-microscopic, and electron-microscopic studies. The essential features of their imaginative and incisive model are

1. The lengths of the thick and thin filaments do not change during muscle contraction.

2. Instead, the length of the sarcomere decreases because the overlap between the two types of filaments increases. *Thick and thin filaments slide past each other during contraction* (Figure 15-6).

3. The force of contraction is generated by a process that *actively moves* one type of filament past neighboring filaments of the other type.

The sliding-filament model is strongly supported by measurements of the A and I bands and the H zone in stretched, resting, and contracted muscle. The length of the A band is constant, which means that the thick filaments do not change size. The distance between the Z line and the adjacent edge of the H zone is also constant, which indicates that the thin filaments do not change size. In contrast, the size of the H zone and also of the I band decreases on contraction, because the thick and thin filaments overlap more.

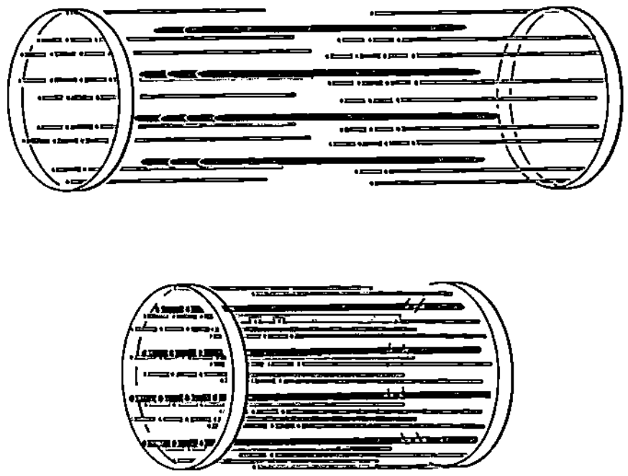

Figure 15-6
Sliding-filament model. [From H.E. Huxley. The mechanism of muscular contraction. Copyright © 1965 by Scientific American, Inc. All rights reserved.]

MYOSIN FORMS THICK FILAMENTS, HYDROLYZES ATP, AND REVERSIBLY BINDS ACTIN

Myosin has three biologically critical activities. First, myosin molecules spontaneously assemble into filaments in solutions of physiologic ionic strength and pH. In fact, the *thick filament* consists mainly of myosin molecules. Second, myosin is an enzyme. In 1939, Vladimir Engelhardt and Militsa Lyubimova discovered that myosin is an *ATPase.*

$$ATP + H_2O \longrightarrow ADP + P_i + H^+$$

This overall reaction, which occurs in a series of discrete steps, provides the free energy for muscle contraction. Third, myosin binds to the polymerized form of *actin (F-actin),* the major constituent of the thin filament. Indeed, this interaction is critical for the generation of the force that moves the thick and thin filaments past each other. Myosin can be regarded as a *mechanoenzyme* because it catalyzes the conversion of chemical-bond energy into mechanical energy.

Actin and myosin form a complex called *actomyosin* when these proteins are mixed in solution. Complex formation is accompanied by a large increase in viscosity. In the 1940s, Albert Szent-Györgyi showed that this increase is reversed by the addition of ATP. His observation revealed that ATP dissociates actomyosin into actin and myosin. He also prepared threads of actomyosin in which the molecules were oriented by flow. A striking result was obtained when the threads were immersed in a solution containing ATP, K^+, and Mg^{2+}. *The actomyosin threads contracted, whereas threads formed from myosin alone did not.* These incisive experiments suggested that *the force of muscle contraction arises from the interplay of myosin, actin, and ATP.*

The ATPase activity of myosin is markedly enhanced by F-actin. Indeed, Albert Szent-Györgyi named actin for its capacity to activate ATP hydrolysis by myosin. Actin increases the turnover number of myosin 200-fold, from 0.05 s^{-1} to 10 s^{-1}. ATP bound to myosin is rapidly hydrolyzed, but ADP and P_i are slow to leave. Actin increases the turnover number of myosin by binding to the myosin-ADP-P_i complex and accelerating the release of products (Figure 15-7). Actomyosin then binds ATP, which leads to the dissociation of actin and myosin. The resulting ATP-myosin complex is ready for another round of catalysis. These reactions, like those of all known ATPases, require Mg^{2+}.

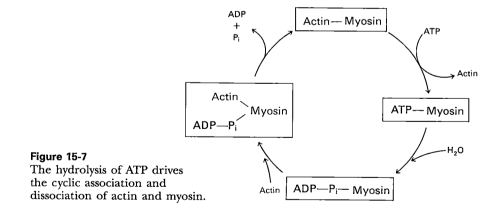

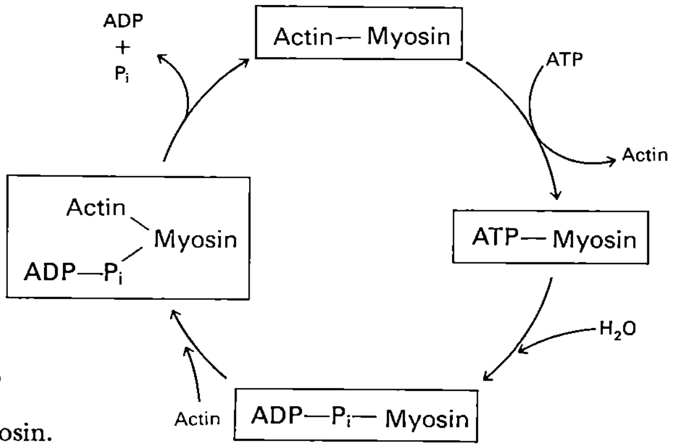

Figure 15-7
The hydrolysis of ATP drives the cyclic association and dissociation of actin and myosin.

MYOSIN CONSISTS OF TWO GLOBULAR HEADS JOINED TO A LONG α-HELICAL COILED-COIL TAIL

Myosin is a large protein (520 kd) made of six polypeptide chains: two identical *heavy chains* (each 220 kd) and two pairs of *light chains* (each ~20 kd). Electron micrographs show that the molecule consists of a double-headed globular region joined to a very long rod (Figure 15-8). The rod is a two-stranded α-helical coiled coil formed by the heavy chains

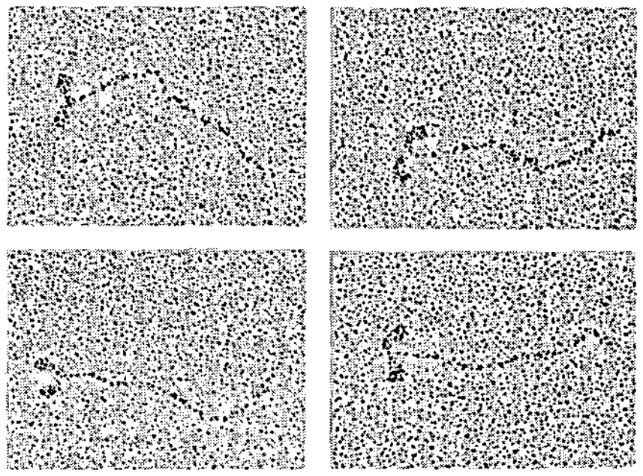

Figure 15-8
Electron micrographs of myosin molecules. [Courtesy of Dr. Paula Flicker, Dr. Theo Walliman, and Dr. Peter Vibert.]

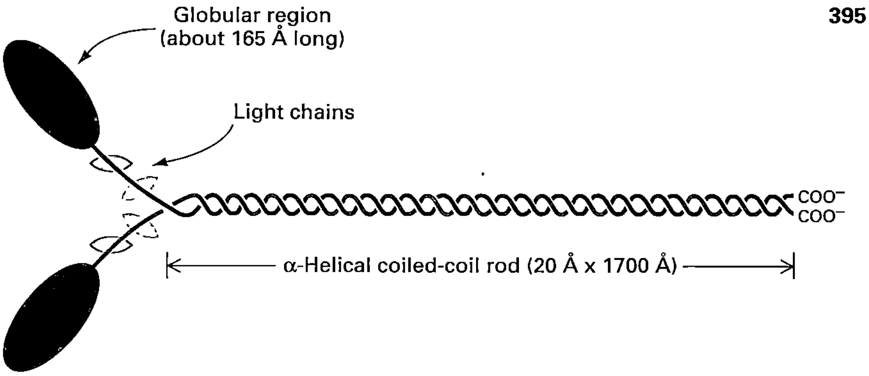

Figure 15-9
Schematic diagram of a myosin molecule.

(Figure 15-9). The heavy chain of each head binds two different light chains. They have modulatory roles and are called, for historical reasons, the *essential light chain (ELC)* and the *regulatory light chain (RLC)*.

Studies of fragments of myosin produced by limited proteolysis have provided insight into the function of this very large protein molecule. Proteolytic dissection is often rewarding, as exemplified by the cleavage of antibodies into F_{ab} and F_c fragments (p. 364). In fact, the cleavage of myosin into active fragments came first. Myosin can be split by trypsin into two partially functional fragments, called *light meromyosin (LMM)* and *heavy meromyosin (HMM)* (Figure 15-10).

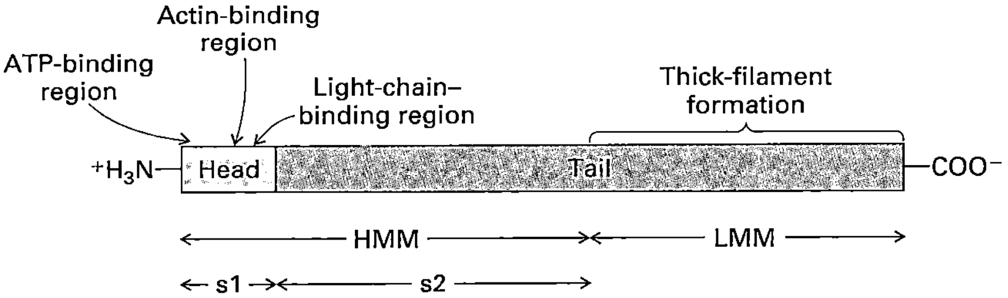

Figure 15-10
Roles of different regions of the myosin molecule. The protein can be cleaved into fragments that retain some functions of the intact protein. The actual lengths of S2 and HMM relative to S1 are about double those shown here.

LMM, like myosin, forms filaments but lacks ATPase activity and does not combine with actin. Heavy meromyosin, by contrast, catalyzes the hydrolysis of ATP and binds to actin but does not form filaments. HMM can be split further into two identical globular subfragments (each called *S1*) and one rod-shaped subfragment (called *S2*). S1 contains an ATPase site, an actin-binding site, and two light-chain binding sites. Indeed, the S1 moieties are the force-generating units of myosin.

The three-dimensional structure of S1 has recently been determined at 2.8-Å resolution by Ivan Rayment and co-workers. S1 is 165 Å long, 65 Å wide, and 40 Å thick (Figure 15-11). The heavy chain of S1 contains an N-terminal 25-kd segment, a central 50-kd segment, and a C-terminal 20-kd segment. The ATPase site is formed by the 25-kd and 50-kd seg-

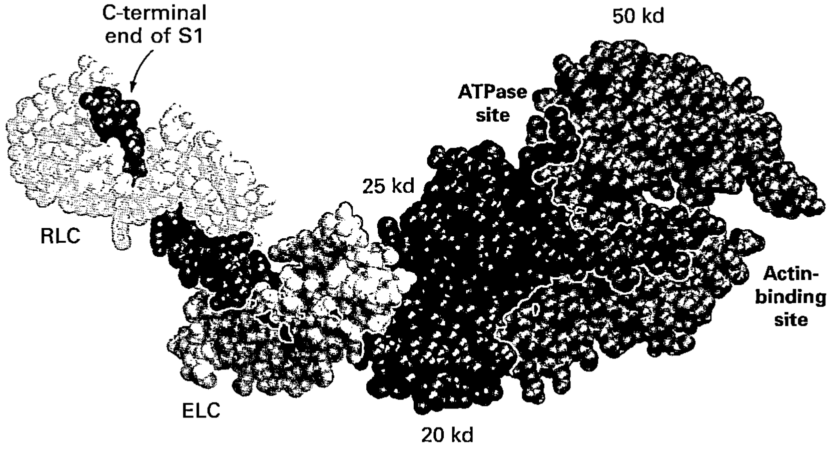

Figure 15-11
Three-dimensional structure of the S1 head of myosin determined by x-ray crystallography at 2.8-Å resolution. The nucleotide-binding site is empty in this crystal form. The heavy chain of S1 consists of a 25-kd N-terminal domain (purple), a 50-kd central domain (blue), and a 20-kd C-terminal domain (red). The actin-binding site and the ATPase site are on opposite sides of the 50-kd domain. Two light chains (RLC, yellow, and ELC, green) are bound to the long C-terminal α helix of the heavy chain. [Courtesy of Dr. Ivan Rayment and Dr. Hazel Holden.]

ments, whereas the actin-binding site is formed by the 50-kd and 20-kd segments. Nearly all known myosins share the catalytic-site sequence

Gly-Glu-Ser-Gly-Ala-Gly-Lys-Thr

which is similar to the sequences found in the active sites of other ATPases. The two light chains wrap around the 20-kd segment, which forms an 85-Å-long α helix that spans much of S1.

The two S1 heads are joined at a hinge to the tail of myosin, which is a two-stranded coiled coil (Figure 15-12). The formation of this very long rod (1700 Å, or 170 nm) is favored by the absence of proline over a span of more than a thousand residues and by the abundance of leucine, alanine, and glutamate. The two strands are in register, pointing the same way. Their axes are about 10 Å apart, enabling the side chains of the two strands to interact intimately to reinforce the helical structure. Regularities in the amino acid sequence of the tail promote the formation of a coiled-coil rod. The tail consists of *repeating seven-residue units (abcdefg)* in which *a* and *d* are usually hydrophobic (Figure 15-13). Residues *a* and *d* form a zigzag pattern of knobs and holes that interlock with those of the other strand to form a tight-fitting hydrophobic core. In contrast, residues *b*, *c*, and *f*, which are located on the periphery of the coiled coil, tend to be charged.

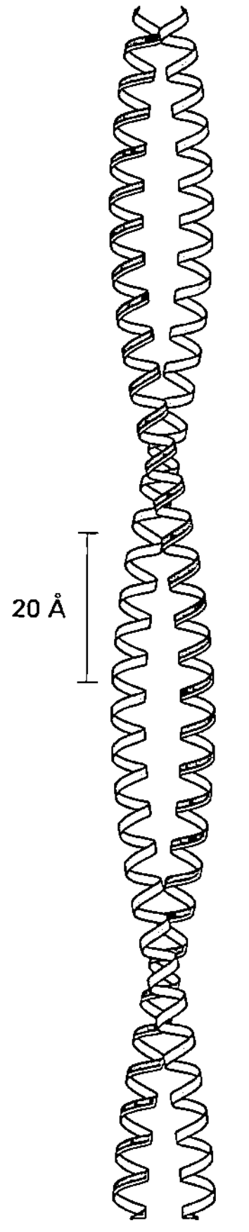

Figure 15-12
Model of a two-stranded α-helical coiled coil.

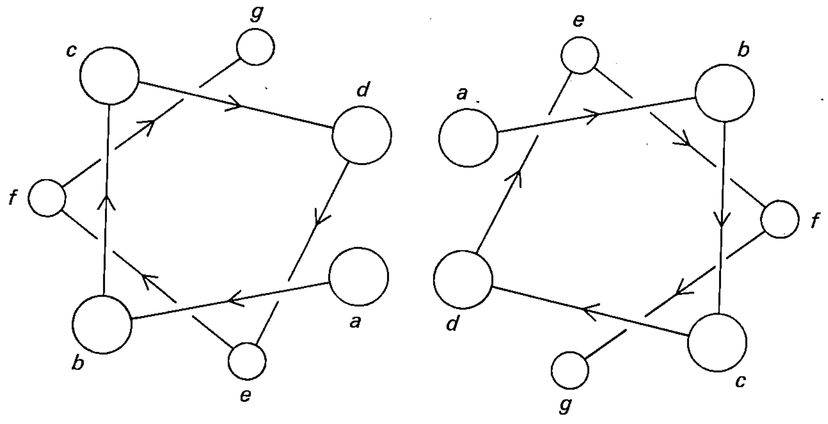

Figure 15-13
Schematic diagram of a cross section of an α-helical coiled coil showing the positions of side chains. Residues *a* and *d* (yellow) of each strand pack tightly to form a hydrophobic core. Residues *b*, *c*, and *f* (green) on the periphery tend to be charged.

Actin, a ubiquitous protein in eukaryotes, is the major constituent of thin filaments. In solutions of low ionic strength, actin is a 42-kd monomer called *G-actin* because of its globular shape. As the ionic strength is increased to the physiologic level, G-actin polymerizes into a fibrous form, *F-actin,* which closely resembles the thin filaments of intact muscle. Actin, like myosin, is an ATPase. However, ATP hydrolysis by actin does not power muscle contraction. Rather, the ATP-ADP cycle of actin participates in filament assembly and disassembly.

The structures of G-actin and F-actin have been solved by Wolfgang Kabsch and Kenneth Holmes. G-actin is a bipartite protein (Figure 15-14). ATP (or ADP) is bound in a cleft between the two domains of the monomer. In F-actin, each monomer is related to the next by a rotation

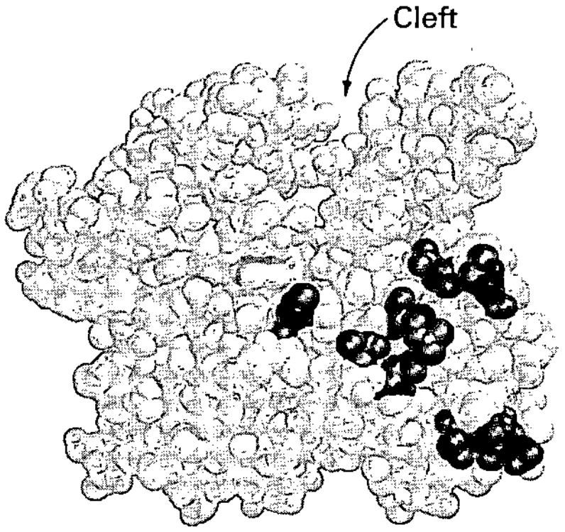

Figure 15-14
Space-filling model of G-actin. ATP (yellow) is bound in a cleft between the two domains. Residues known to participate in the binding of the S1 heads of myosin are shown in dark red. [Drawn from 1atn.pdb. W. Kabsch, H.G. Mannherz, D. Suck, E.F. Pai, and K.C. Holmes. *Nature* 347(1990):37.]

of 166 degrees and a translation of 27.5 Å, which gives rise to a double-stranded appearance (Figure 15-15). Indeed, thin filaments appear in electron micrographs rather like two strings of beads wound around each other. Each monomer in the filament makes contact with four others, which accounts for the high degree of cooperativity of polymerization. Actin monomers assemble into filaments when a *critical concentration* is reached; the value for ATP–actin is 20-fold lower than for ADP–actin.

If S1 (or myosin) is added to F-actin filaments (or thin filaments) in the absence of ATP, a distinctive arrowhead pattern appears (Figure 15-16).

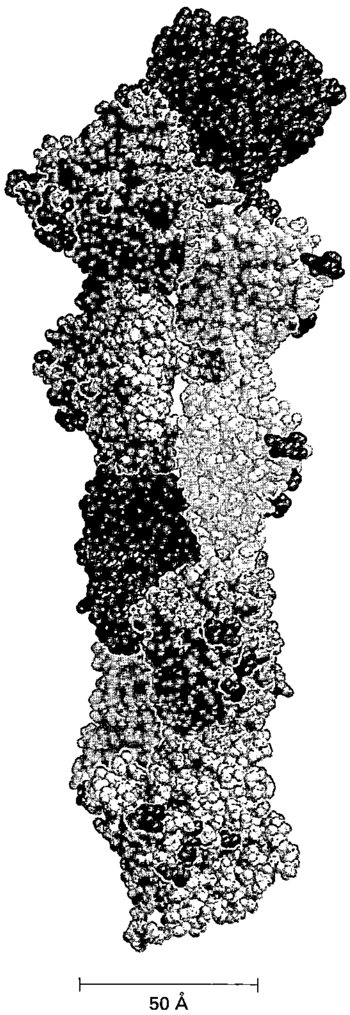

Figure 15-15
Structure of F-actin. The identical subunits of this helical assembly are depicted in several colors to show the interactions of an actin monomer with its four neighbors. All monomers point in the same direction. The helix repeats after 13 subunits. The myosin-binding sites (dark red) are at the periphery of the filament. [Drawn from coordinates kindly provided by Dr. Ronald Milligan. K.C. Holmes, D. Popp, W. Gerhard, and W. Kabsch. *Nature* 347(1990):44.]

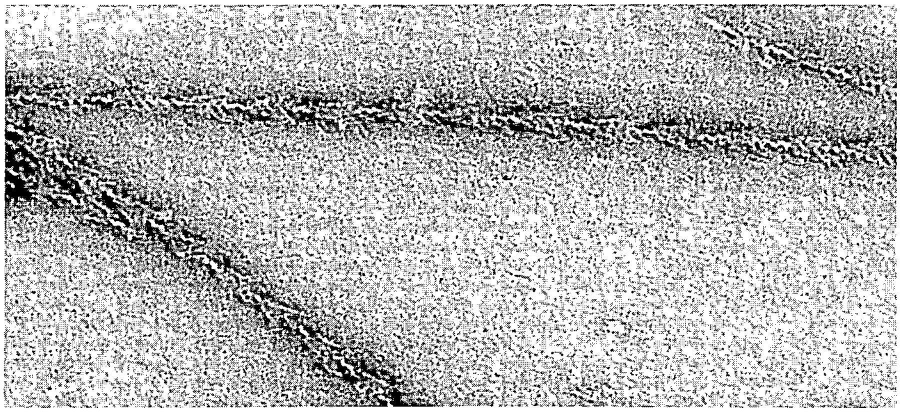

Figure 15-16
Electron micrograph of filaments of F-actin decorated with the S1 moiety of HMM. The "arrowheads" all point the same way. [Courtesy of Dr. James Spudich.]

These structures are picturesquely called *decorated filaments.* The long axis of S1 is oriented at about 45 degrees with respect to the helix axis of F-actin (Figure 15-17). *The arrows on a decorated filament always point the same way for its entire length.* Thus, *all actin units of a thin filament have the same directionality.*

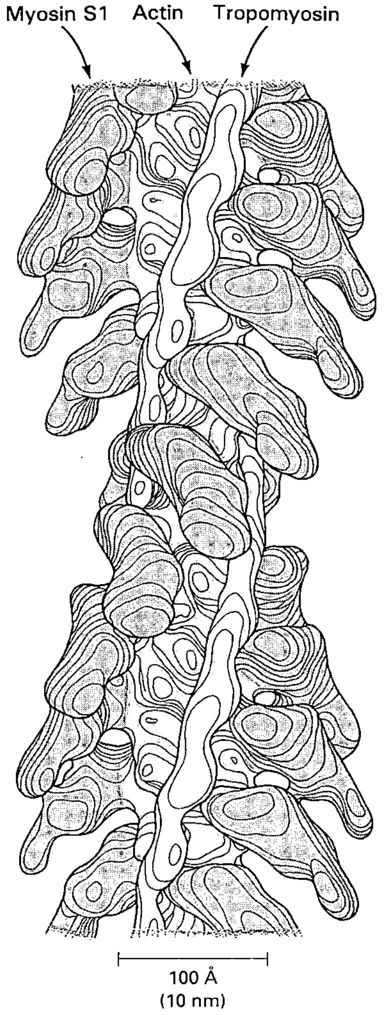

Figure 15-17
Structure of a reconstituted thin filament decorated with the S1 fragment of myosin. This image, obtained by three-dimensional analyses of electron micrographs of unstained frozen filaments, corresponds to the structure of the intact system at the end of the power stroke. The location of the troponin complex is not defined by this analysis. The pitch of the helical array is 37 nm. [After a drawing kindly provided by Dr. Ronald Milligan and Dr. Paula Flicker.]

THE POLARITY OF THICK AND THIN FILAMENTS REVERSES IN THE MIDDLE OF A SARCOMERE

A thick filament dissociated from muscle has a diameter of 16 nm (160 Å) and a length of 1.5 μm (15,000 Å). Cross-bridges emerge from the filament axis in a regular helical array at intervals of 14.3 nm along the filament axis because of regularities in the amino acid sequence of myosin's tail. Midway along the length of a thick filament is a 150-nm bare region devoid of projecting cross-bridges (Figure 15-18). *The myosin molecules on one side of the bare zone point one way, whereas those on the other side point in the opposite direction.* Thus, *a thick filament is bipolar, whereas a thin filament is unipolar.*

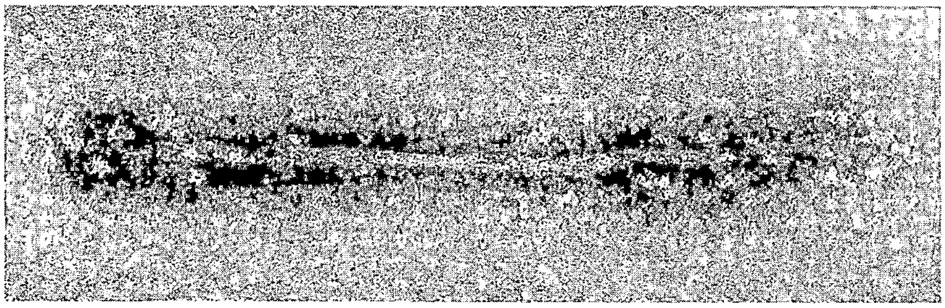

Figure 15-18
Electron micrograph of a reconstituted thick filament. The projections on either side of the bare zone are the cross-bridges. [Courtesy of Dr. Hugh Huxley.]

In muscle depleted of ATP, myosin cross-bridges decorate actin filament so that the arrows on all filaments point away from the Z line. Thus, *all the thin filaments on one side of a Z line have the same orientation, whereas those on the opposite side have the reverse polarity.* The structural polarity of both the thick and the thin filaments is crucial for coherent motion. The sliding force developed by the interaction of individual actin and myosin units adds up because the interacting units have the same relative orientation. Furthermore, *the absolute direction of actin and myosin molecules reverses halfway between Z lines* (Figure 15-19). Consequently, the two thin filaments that bind the cross-bridges of a thick filament are drawn toward each other, which shortens the distance between Z lines (Figure 15-20).

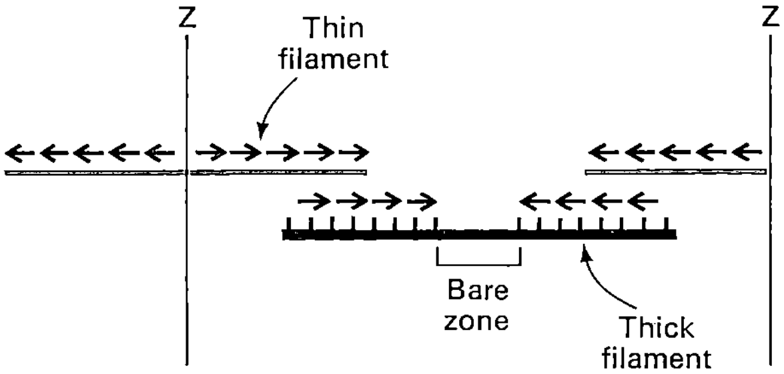

Figure 15-19
The polarity of the thick and thin filaments reverses halfway between the Z lines.

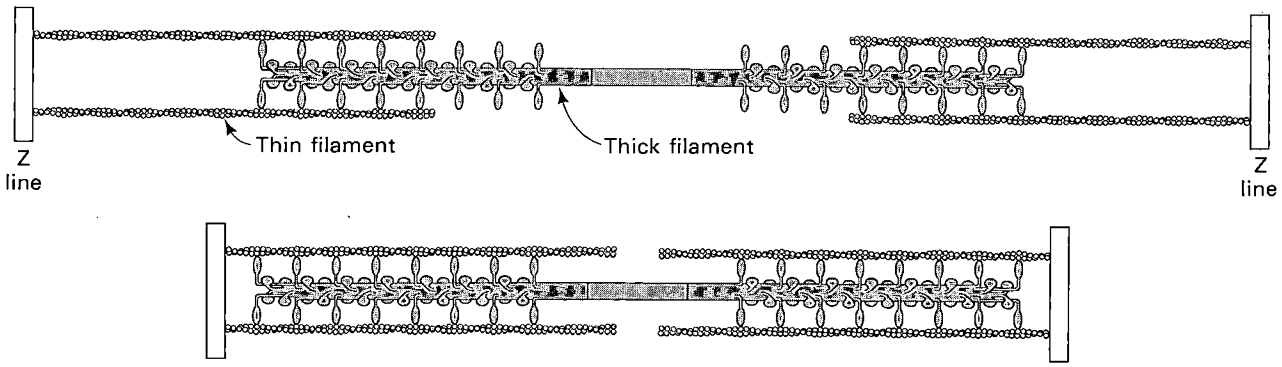

Figure 15-20
Schematic diagram showing the interaction of thick and thin filaments in skeletal muscle contraction. [After a diagram kindly provided by Dr. James Spudich.]

THE POWER STROKE IN CONTRACTION IS DRIVEN BY CONFORMATIONAL CHANGES IN THE MYOSIN S1 HEAD

How does the cyclic formation and dissociation of complexes between actin and S1 lead to a sliding of the thin and thick filaments? A plausible mechanism for the generation of force, pieced together from many biochemical, biophysical, and structural experiments, is shown in Figure 15-21. The ATP-ADP cycle in myosin is thought to produce directional movement in the following way:

1. In resting muscle (see part A of Figure 15-21), S1 heads are unable to interact with actin units in thin filaments because of steric interference by tropomyosin, a regulatory protein (p. 402). In this state, the hydrolysis products ADP and P_i are still bound to myosin.

2. When muscle is stimulated, tropomyosin shifts position. S1 heads can then reach out from the thick filament and attach to actin units on thin filaments (B).

3. The binding of myosin–ADP–P_i to actin leads to the release of P_i. The subsequent dissociation of ADP induces a major conformational change in S1 (C). The change in orientation of S1 relative to actin constitutes the *power stroke* of muscle contraction—the thin filament is pulled a distance of about 100 Å. ADP is released from myosin at the end of the power stroke.

4. The subsequent binding of ATP to myosin leads to the rapid release of actin. The S1 head is again detached from the thin filament (D).

5. Finally, the bound ATP is hydrolyzed by the free myosin head, which resets it for the next interaction with the thin filament (E).

The essence of the process is a cyclic change both in the shape of the myosin S1 head and in its affinity for actin. The binding of ATP to the catalytic site, near the center of S1, leads to the detachment of S1 from actin at a

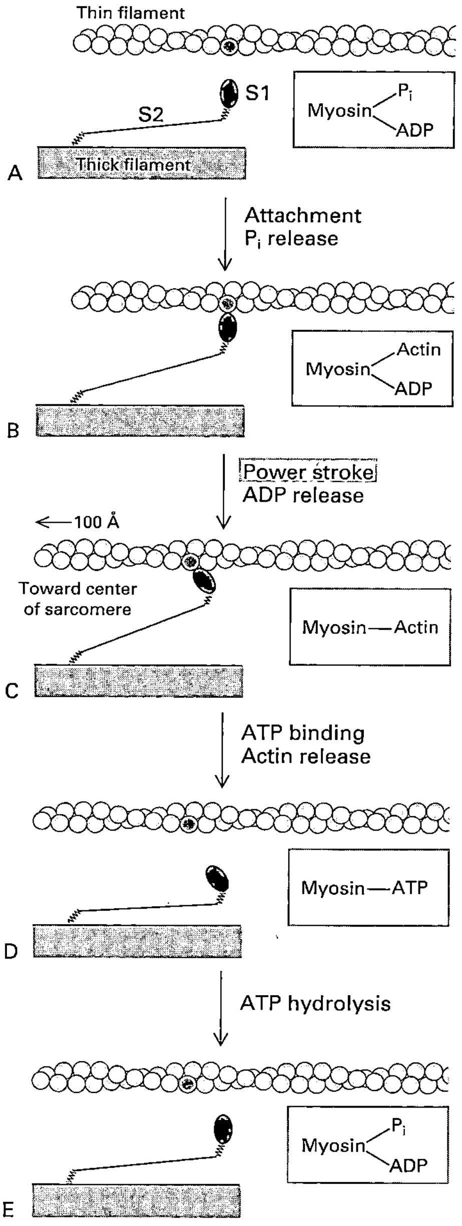

Figure 15-21
Proposed mechanism for the generation of force by the interaction of an S1 unit of a myosin filament with an actin filament. In the power stroke, the thin filament moves relative to the thick filament when S1 undergoes conformational changes accompanying the release of ADP.

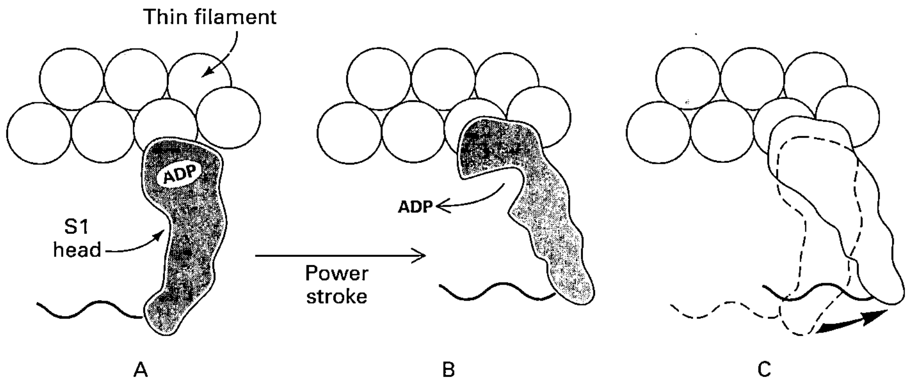

Figure 15-22
Schematic diagram showing the inferred change in the shape of the S1 head and its relation to the thin filament (A) before and (B) after the power stroke. The outline of S1 in these states is compared in (C). [After I. Rayment, H.M. Holden, M. Whittaker, C.B. Yohn, M. Lorenz, K.C. Holmes, and R.A. Milligan. *Science* 261(1993):58.]

distant site. S1 is designed so that it strongly binds either ATP or actin but not both. Given a choice, S1 markedly prefers ATP to actin. The active site cleft then closes around the bound ATP (Figure 15-22). Cleft closure and the consequent hydrolysis of ATP lead to domain movements that change the curvature of S1. The release of the γ phosphate accompanying the binding of S1 to actin initiates the power stroke. When ADP is released, the curvature of S1 again changes. The result is a ~100-Å (10-nm) movement of the thin filament relative to the thick filament with each ATP-ADP cycle.

It is noteworthy that myosin does not hydrolyze ATP at an appreciable rate unless it interacts with actin. Unproductive hydrolysis of ATP is thereby prevented. Recall that ATP-driven pumps, such as the Na^+-K^+ ATPase, likewise do not hydrolyze ATP unless useful work is performed (p. 311). Equally significant, the affinity of actin for myosin depends on the nature of the bound nucleotide. Actin interacts strongly with myosin when ADP is bound to myosin or when myosin's nucleotide site is empty. In contrast, actin binds weakly to the ATP complex of myosin. The control of protein-protein interactions by bound nucleotides is a recurring theme in biochemistry. In an earlier chapter, we saw that the binding of GTP to transducin, an amplifier protein in vision, leads to its release from photoexcited rhodopsin (p. 337).

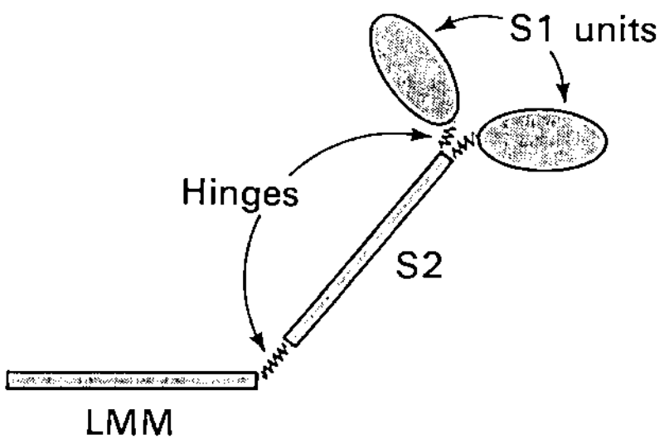

Figure 15-23
Two kinds of hinges in myosin—one between S1 and S2 and the other between S2 and the LMM region—are important in allowing the S1 unit to alter its contacts with actin during the power stroke.

Myosin contains two kinds of hinges that enable its S1 heads to attach and detach from actin and to change their orientation when bound. One type of hinge is located between each S1 head and the S2 rod, and the other is positioned between S2 and the LMM unit of myosin (Figure 15-23). These hinges are flexible regions of polypeptide chain, vulnerable to cleavage by proteolytic enzymes. Indeed, the enzymatic splitting of myosin into LMM, S2, and S1 pieces expresses the fact that *myosin is constructed of domains joined by hinges.* The hinge between S1 and S2 enables S1 to interact with actin in one orientation when ADP is bound and in a different one when ADP is released. The other hinge, between S2 and LMM, allows considerable variation in the position of S1 relative to the thick filament, which permits S1 to interact precisely with actin. Hence, the system works over a wide range of lateral spacings of the thick and thin filaments. Thus, *segmental flexibility* plays a critical role in muscle contraction, as in the binding of antibodies to multivalent antigens (p. 365).

MYOSIN-COATED BEADS MOVE UNIDIRECTIONALLY ON ORIENTED ACTIN CABLES

One of the goals of biochemical studies is to reconstitute biological function in vitro using purified components. Reconstitution tests our understanding of biological processes. Furthermore, reconstituted systems are well suited for mechanistic studies because of their relative simplicity. We have previously discussed reconstituted systems that carry out processes such as ATP-driven ion pumping (p. 310) and light-driven proton transport (p. 318). An assay system for direct observation of movements produced by interactions between myosin and actin was devised by Michael Sheetz and James Spudich. They made use of *Nitella,* a giant alga that contains cables of actin fixed to the cytoplasmic face of rows of chloroplasts. A key feature of these cables, which mediate cytoplasmic streaming, is that actin filaments in them point in the same direction. The striking finding is that *myosin-coated beads move unidirectionally along these actin cables in the presence of ATP* (Figure 15-24), in the same direction as the cytosol in cytoplasmic streaming. Beads coated with skeletal muscle myosin move

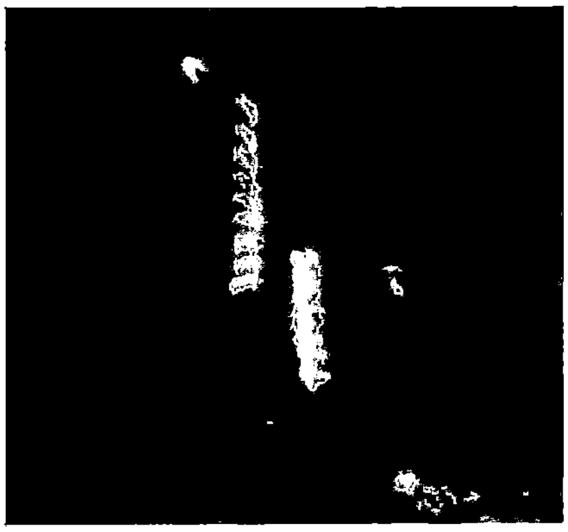

Figure 15-24
Movement of myosin-coated beads on actin cables. Yellow-fluorescent beads coated with myosin were added to an opened *Nitella* cell. On addition of ATP, the beads moved on actin cables lying above the chloroplasts, which fluoresce red. A series of exposures taken at intervals of 1 s indicate that the velocity of movement was 3 μm/s. [Courtesy of Dr. James Spudich and Dr. Michael Sheetz.]

with a speed of 5 μm/s, which is approximately the speed of contraction of intact sarcomeres in vivo. It has been estimated that 25 myosin heads on each bead give rise to this movement. In essence, myosin molecules "walk" on actin cables (Figure 15-25). Alternatively, actin filaments can move on a glass surface containing bound myosin molecules if ATP is present.

The velocity of movement depends on the type of myosin used in the assay, not on the type of actin. Smooth muscle myosin moves at 0.4 μm/s and *Dictyostelium* myosin at 1 μm/s, considerably slower than skeletal muscle myosin, which is also the fastest in vivo. Thus, *the velocity depends on the nature of the motor (myosin) rather than of the track (actin).* Furthermore, heavy meromyosin moves at nearly the same velocity as does intact myosin, indicating that the distal 85 nm of the myosin tail does not participate in the generation of contractile force. In fact, actin filaments move on films coated with S1. Thus, no portion of the α-helical tail of myosin is essential for movement. *The force of muscle contraction is generated entirely within the S1 head.*

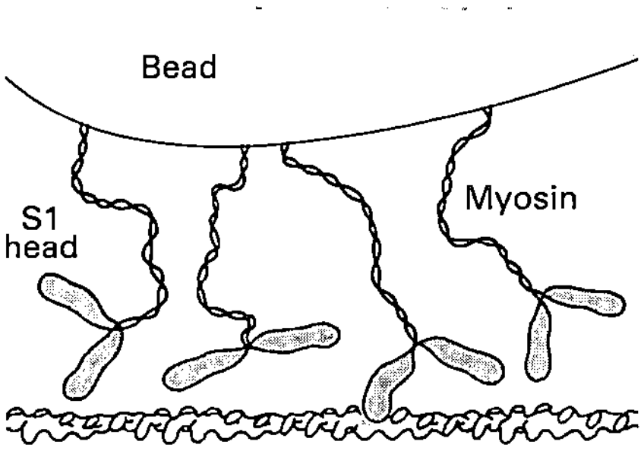

Figure 15-25
Schematic diagram of an in vitro motility system. Myosin molecules attached to beads "walk" unidirectionally along actin filaments as ATP is hydrolyzed. [After a drawing kindly provided by Dr. James Spudich.]

This in vitro motility system has opened the door to the direct observation of steps taken by a *single* myosin molecule (Figure 15-26A). A second advance was needed to overcome the difficulty of seeing discrete steps amidst random Brownian motion. Beads can be held in place in solution using intensely focused laser beams as *optical traps.* An actin filament with

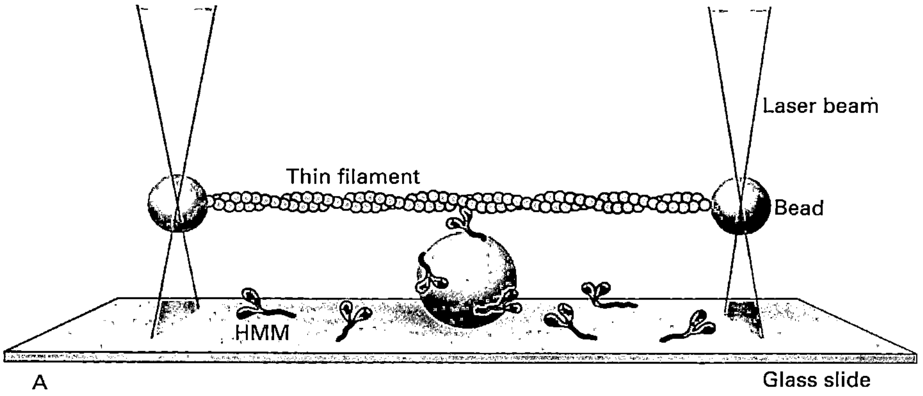

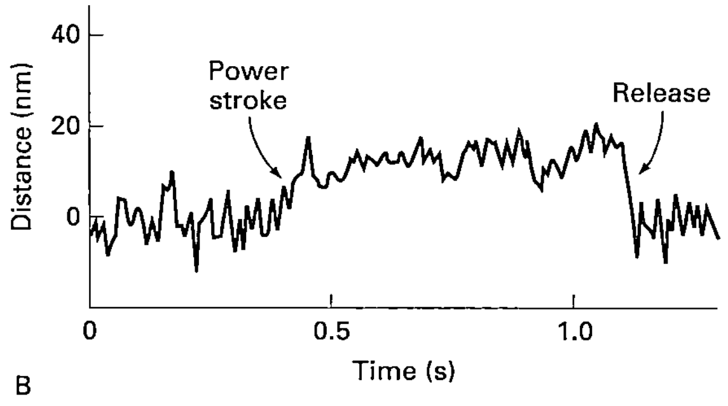

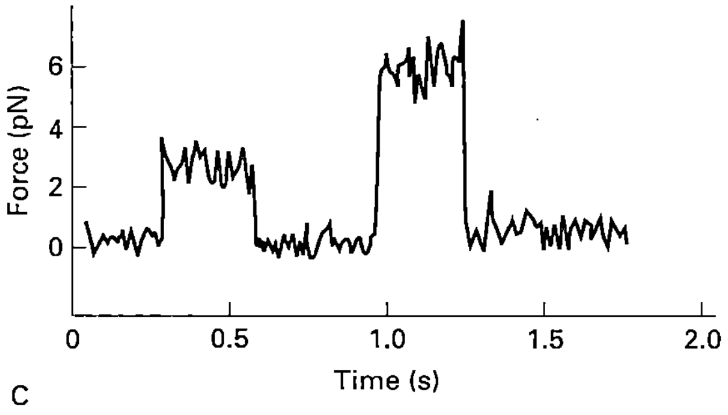

Figure 15-26
Measurement of the action of a single myosin molecule. (A) An actin filament (blue) is placed above a heavy meromyosin (HMM) fragment (red) that projects from a bead on a glass slide. A bead attached to each end of the actin filament is held in an optical trap produced by a focused, intense infrared laser beam (green). The position of the beads can be measured with nanometer precision. (B) Recording of the displacement of the actin filament induced by HMM in the presence of 1 μM ATP. (C) Recording of the force exerted by a myosin molecule when the actin filament is prevented from moving. The displacement and force transients are briefer at higher ATP levels. [After J.T. Finer, R.M. Simmons, and J.A. Spudich. *Nature* 368(1994):113.]

a bead at each end can be positioned next to a myosin molecule that projects from a surface. The movement of the filament can then be detected (Figure 15-26B). *The average step size is about 11 nm,* which strongly supports the swinging-cross-bridge model of muscle contraction. Furthermore, the average force exerted by a single myosin molecule is about 4 piconewtons (pN) (Figure 15-26C). The dependence of the duration of steps and forces on ATP concentration suggests that one ATP is hydrolyzed during each cross-bridge cycle.

TROPONIN AND TROPOMYOSIN MEDIATE THE REGULATION OF MUSCLE CONTRACTION BY CALCIUM ION

How is muscle contraction controlled? In resting muscle, the myosin–ADP–P_i complex is poised to interact with actin but is restrained from making productive contact. What is the nature of this restraint and how is it relieved? Setsuro Ebashi discovered that actin and myosin are held in check by *tropomyosin* and the *troponin complex* when the cytosolic calcium level is low. Their inhibitory action is suspended when the calcium level is transiently raised. In resting (relaxed) muscle, Ca^{2+} is sequestered in the sarcoplasmic reticulum (SR), an extensive intracellular compartment, by an active transporter (p. 314). ATP-driven calcium pumps in the SR membrane lower the concentration of Ca^{2+} in the cytosol to less than 1 μM. A nerve impulse leads to the release of Ca^{2+} from the sacs of the sarcoplasmic reticulum, which raises the cytosolic concentration to about 10 μM and leads to muscle contraction.

Tropomyosin and the troponin complex are located in the thin filament and constitute about a third of its mass. Tropomyosin is a two-

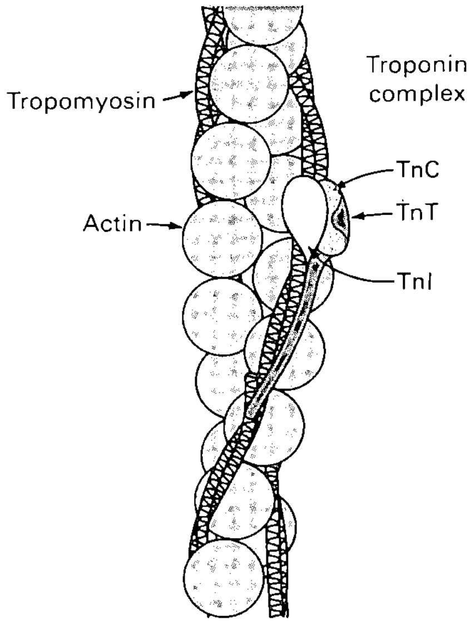

Figure 15-27
Model of a thin filament. The troponin complex consists of three components: TnI (white), TnC (green), and TnT (red). In relaxed muscle (low Ca^{2+}), tropomyosin (yellow) prevents actin from interacting with myosin S1 units. [After C. Cohen. Protein switch of muscle contraction. Copyright © 1975 by Scientific American, Inc. All rights reserved. The arrangement of the troponin complex is based on a drawing kindly provided by Dr. Larry Smillie.]

stranded α-helical rod. This highly elongated 70-kd protein is aligned nearly parallel to the long axis of the thin filament (Figure 15-27). Troponin is a complex of three polypeptide chains: TnC (18 kd), TnI (24 kd), and TnT (37 kd). TnC binds calcium ions, TnI binds to actin, and TnT binds to tropomyosin. The troponin complex is located in the thin filaments at intervals of 385 Å, a period set by the length of a tropomyosin unit. Each troponin complex, through tropomyosin, regulates the interactions of some seven actin units. They inhibit muscle contraction when the cytosolic calcium level is low but not when it is high.

Troponin C, the calcium-sensing component of the complex, consists of two homologous domains. An amino-terminal and a carboxyl-terminal globular domain are connected by a long eight-turn α-helix (Figure 15-28). Each domain contains two binding sites for Ca^{2+}. The ones in the carboxyl-terminal domain have high affinity for Ca^{2+} ($K = 0.1\ \mu\text{M}$), whereas those in the amino-terminal domain have low affinity ($K = 10\ \mu\text{M}$). TnC is a member of the *EF hand superfamily of calcium-binding proteins* that was discussed in an earlier chapter (p. 348). The structure of TnC closely resembles that of calmodulin, a ubiquitous calcium sensor.

In resting muscle, the high-affinity sites of TnC are occupied by Ca^{2+} but the low-affinity ones are empty. When Ca^{2+} is released from the sarcoplasmic reticulum, it occupies the low-affinity sites, which changes the conformation of the amino-terminal domain, making it more similar to the carboxyl-terminal domain. This conformational change in TnC is transmitted to the other components of the troponin complex and then to tropomyosin. It seems likely that TnT, which has a highly elongated shape, controls the position of tropomyosin on the thin filament, near the interface between actin and the S1 head of myosin. When the Ca^{2+} level is low, tropomyosin sterically blocks the binding of S1 to actin.

The essentiality of the troponin-tropomyosin couple in the regulation of contraction is graphically illustrated by in vitro motility assays. The movement of myosin-coated beads on actin cables is insensitive to Ca^{2+}, when only these two proteins are present. The addition of troponin and tropomyosin makes the beads responsive to Ca^{2+}. Movement is blocked if Ca^{2+} is absent and restored by the addition of micromolar levels of Ca^{2+}.

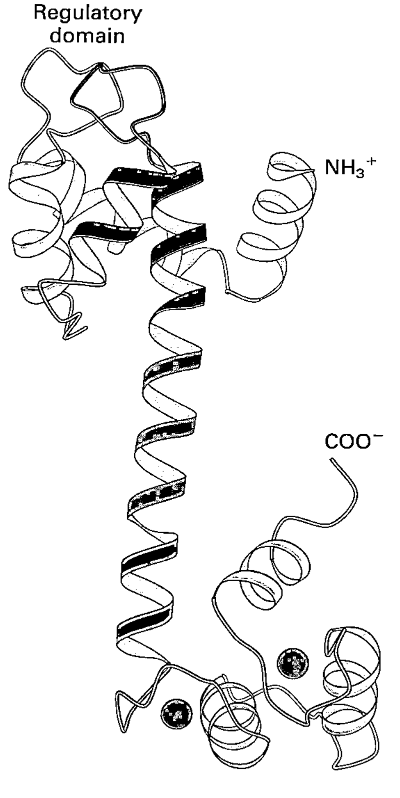

Figure 15-28
Schematic diagram of the structure of troponin C. The amino-terminal regulatory domain (top) is separated from a homologous carboxyl-terminal domain (bottom) by a long α helix (gray). The four EF hands (calcium-binding sites) are shown in green, red, blue, and yellow. The sites in the regulatory domain are unoccupied in this crystal form. The two Ca^{2+} bound to high-affinity sites in the carboxyl-terminal domain are shown in purple. [Drawn from 5tnc.pdb. O. Herzberg and M.N.G. James. *Nature* 313(1985):655.]

Thus, a reconstituted system containing only four proteins—myosin, actin, troponin, and tropomyosin—exhibits calcium-regulated movement. It is evident that *Ca^{2+} controls muscle contraction by an allosteric mechanism in which the flow of information is*

$$Ca^{2+} \longrightarrow \text{troponin} \longrightarrow \text{tropomyosin} \longrightarrow \text{actin} \longrightarrow \text{myosin}$$

ACTIN AND MYOSIN HAVE CONTRACTILE ROLES IN NEARLY ALL EUKARYOTIC CELLS

It has long been known that many nonmuscle cells can move and change shape. The migration of cells in the development of embryos, the movement of macrophages to injured tissues, and the retraction of clots by blood platelets vividly exemplify the universality of cell motility. What is the molecular basis of cell movement? An early clue came from work on *Physarum polycephalum,* a plasmodial slime mold, which contains a streaming mass of cytoplasm. Extracts of these primitive cells had properties like that of actomyosin from striated muscle. The addition of ATP produced a rapid drop in viscosity, which was followed by a slow rise accompanying the hydrolysis of ATP. Furthermore, this slime mold contains a large amount of actin that is very similar to muscle actin. It forms thin filaments and interacts with myosin. Most interesting, *slime mold actin* reacts with S1 heads from *vertebrate skeletal muscle* to give decorated filaments like those formed from vertebrate actin and myosin. In fact, slime mold actin differs from rabbit muscle actin at only 17 of 375 amino acid residues. Thus, *actin is a highly conserved, ancient protein of eukaryotes.*

Likewise, myosin has been isolated from this slime mold and from many other eukaryotic cells. Myosins are much more diverse than actin. Two classes have been found in nonmuscle cells. *Myosin I molecules are single-headed, whereas myosin II molecules are double-headed, like those in muscle.* The head of myosin I, like that of myosin II, contains a binding site for ATP and another one for actin. Their amino acid sequences are about 40% identical. Myosin I contains at least one light chain, whereas each head in myosin II has two associated light chains. Myosin I lacks the long α-helical coiled coil of myosin II and hence cannot form bipolar filaments. In contrast, myosin I possesses a carboxyl-terminal lipid-binding domain that enables it to associate with membranes. Myosin I can transport membrane-bound organelles on actin tracks (Figure 15-29). The coiled-coil tails of myosin II molecules in nonmuscle cells determine the kind of cargo that is carried by the motor. The assembly and functional properties of nonmuscle myosins are controlled by phosphorylation of serine and threonine residues in their tails.

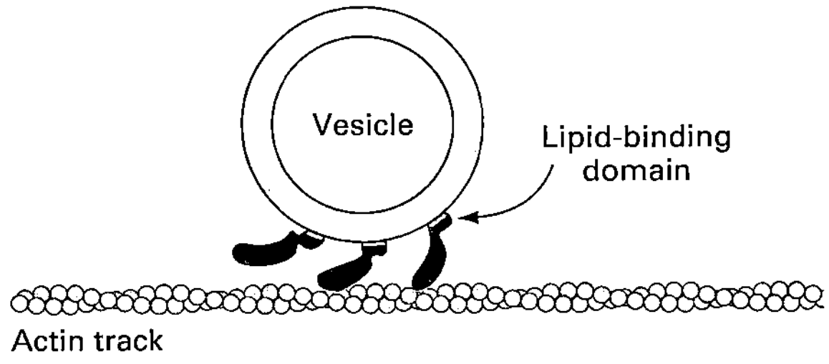

Figure 15-29
The short tail of myosin I contains a lipid-binding domain that enables it to carry vesicles on actin tracks. [After T.D. Pollard, S.K. Doberstein, and H.G. Zot. *Ann. Rev. Physiol.* 53(1991):653.]

THE BEATING OF CILIA AND FLAGELLA IS PRODUCED BY THE DYNEIN-INDUCED SLIDING OF MICROTUBULES

Eukaryotic cells have an internal scaffolding called the *cytoskeleton* that gives them their distinctive shapes. The cytoskeleton also enables cells to transport vesicles, undergo changes in shape, and migrate. This dynamic structure is formed by three classes of filamentous assemblies: *microfilaments, intermediate filaments,* and *microtubules.* As was mentioned earlier, microfilaments (~7-nm diameter) are made of actin. Intermediate filaments (7- to 11-nm diameter), present in epithelial cells of animals, contain *two- or three-stranded α-helical coiled-coil cores* flanked by diverse sequences on each side. The *keratins* of hair are prominent intermediate filament proteins.

Microtubules, the third class of cytoskeletal elements, are hollow cylindrical structures built from two kinds of similar 50-kd subunits, *α- and β-tubulin.* Their outer diameter of about 30 nm clearly distinguishes them from microfilaments and intermediate filaments. The rigid wall of a microtubule is made of a helical array of alternating α- and β-tubulin subunits (Figure 15-30). A microtubule can be regarded as being made of 13 protofilaments that run parallel to its long axis (Figure 15-31).

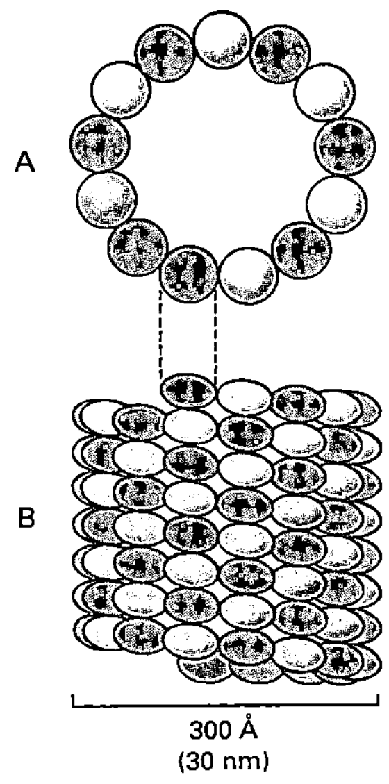

Figure 15-30
Schematic diagram showing the helical pattern of tubulin subunits in a microtubule: (A) cross-sectional view showing the arrangement of 13 protofilaments; (B) surface lattice of α and β subunits. [After J.A. Snyder and J.R. McIntosh. *Ann. Rev. Biochem.* 45(1976):706.]

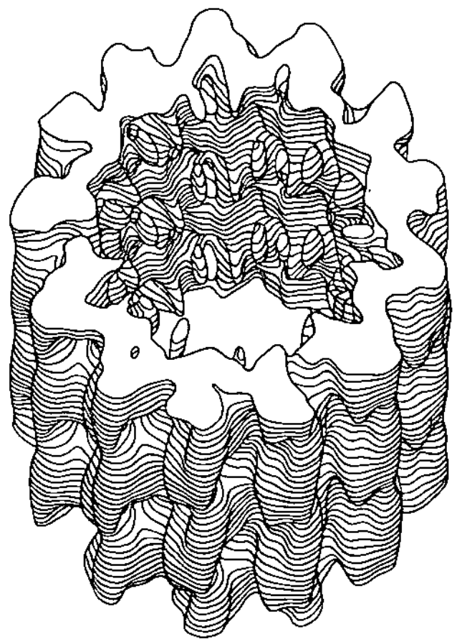

Figure 15-31
Image of a microtubule obtained from x-ray diffraction analysis of fibers. The resolution of the analysis was 18 Å. [After a drawing kindly provided by Dr. Lorena Beese, Dr. Gerald Stubbs, and Dr. Carolyn Cohen.]

Microtubules are major components of eukaryotic *cilia* and *flagella.* These hairlike organelles protrude from the surface of many cells. Cilia act as oars to move a stream of liquid parallel to a stationary cell's surface. For example, the coordinated beating of cilia on cells lining the respiratory passages serves to sweep out foreign particles. Free cells such as sperm and protozoa are propelled by either cilia or flagella. Electron-microscopic studies have revealed that cilia and flagella from nearly all eukaryotes have the same fundamental design: a bundle of fibers called an *axoneme* is surrounded by a membrane that is continuous with the plasma membrane. In fact, the fibers in an axoneme are microtubules: a peripheral group of nine pairs of microtubules surrounds two singlet microtubules (Figure 15-32). This recurring motif is known as a *9 + 2 array.* Flagella differ from cilia in being much longer.

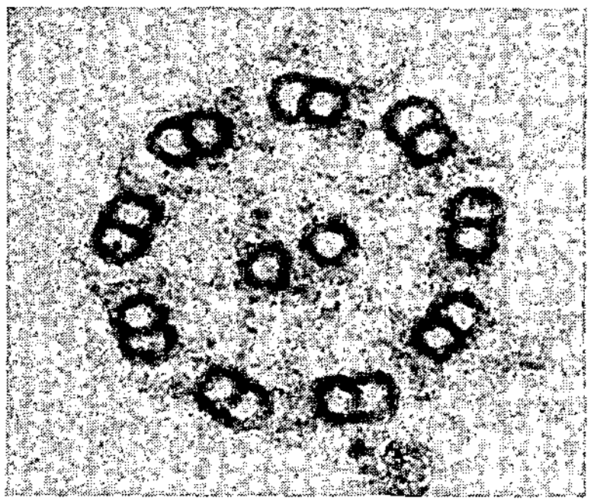

Figure 15-32
Electron micrograph of a cross section of a flagellar axoneme. Nine outer microtubule doublets surround two singlets. [Courtesy of Dr. Joel Rosenbaum.]

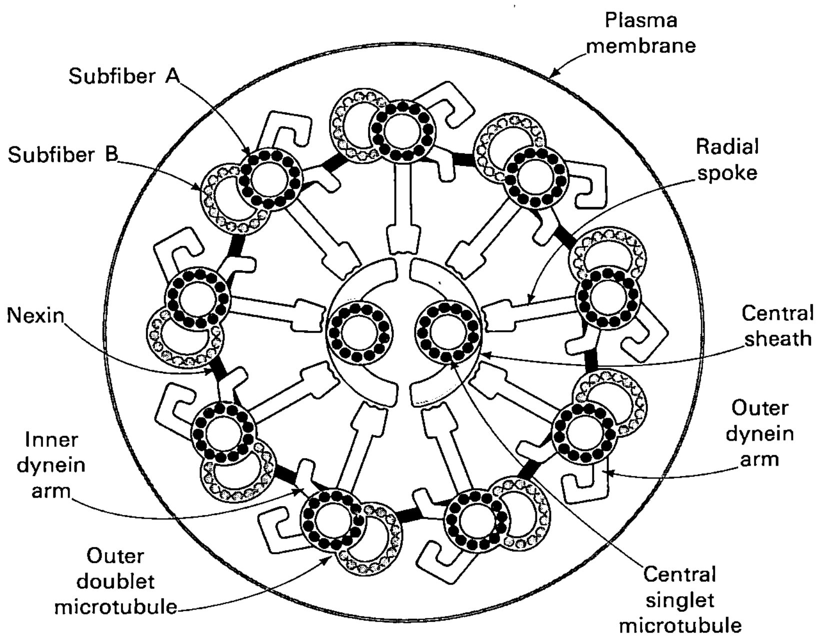

Figure 15-33
Schematic diagram of the structure of an axoneme.

A schematic diagram of the structure of an axoneme is shown in Figure 15-33. Each of the nine outer doublets looks like a figure eight, 37 nm × 25 nm. The smaller of the pair, *subfiber A*, is joined to a central sheath of the cilium by radial spokes. Microtubule doublets are also held together by *nexin* links. Two *arms* emerge from each subfiber A. All the arms in a given cilium point in the same direction. Treatment of cilia with detergent and then with salt at high concentration removes the plasma membrane surrounding them and solubilizes an ATPase called *dynein*. The outer fibers retain their ninefold cylindrical arrangement but are devoid of arms. They can be restored by the addition of dynein under suitable ionic conditions. Hence, *the arms of subfibers A are dyneins with ATPase activity*. In the absence of ATP, dynein arms bind tightly to subfiber B.

Dynein is a remarkably large protein containing one, two, or three heads, depending on the source, and multiple associated polypeptides (Figure 15-35). A 1000-to-2000-kd dynein is comparable in mass to an entire ribosome. The head region, like that of myosin, acts as a cross-

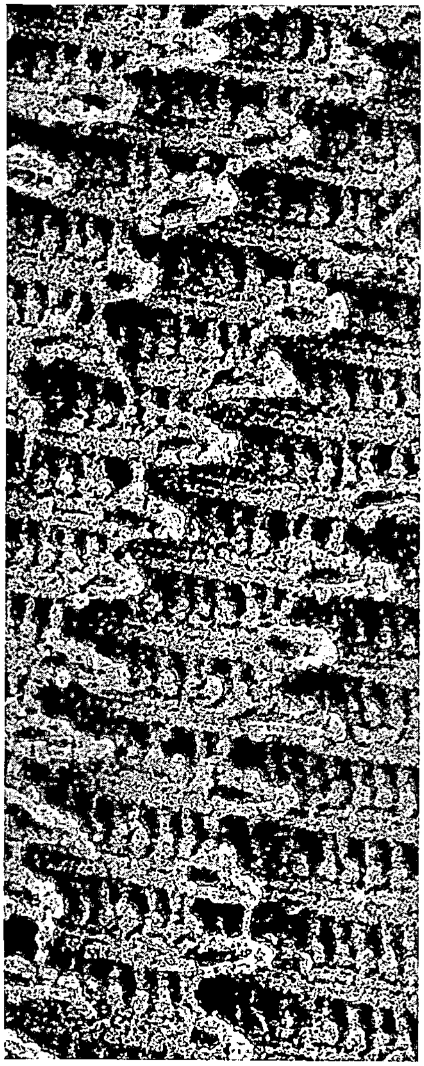

Figure 15-34
Freeze-fracture electron micrograph of microtubules in the motile axostyle (central supporting rod) of a protozoan. The microtubules are linked by oblique cross-bridges that resemble the dynein arms in the cilia of higher organisms. [Courtesy of Dr. John Heuser.]

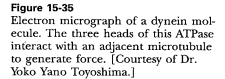

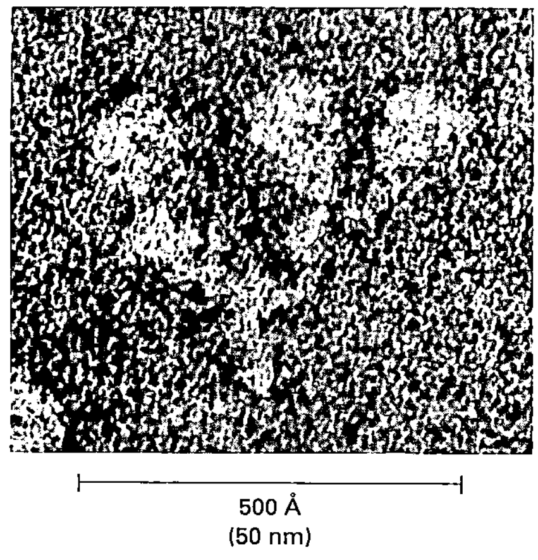

Figure 15-35
Electron micrograph of a dynein molecule. The three heads of this ATPase interact with an adjacent microtubule to generate force. [Courtesy of Dr. Yoko Yano Toyoshima.]

bridge. The amino acid sequence of dynein is unrelated to that of myosin, but its ATPase cycle is similar. The binding of ATP to dynein releases it from the B subfiber. Hydrolysis of bound ATP yields dynein–ADP–P_i, which binds again to the B subfiber. The association of dynein with the B subfiber induces the release of P_i, as in the attachment of S1 to actin. The power stroke probably occurs in this step.

How does the ATPase cycle of dynein cause cilia and flagella to beat? Peter Satir and Ian Gibbons showed that the *outer doublets of the axoneme slide past each other to produce bending.* The force between adjacent doublets is generated by *dynein cross-bridges.* The dynein arms on subfiber A of one doublet walk along subfiber B of an adjacent doublet as ATP is being hydrolyzed, as in the movement of myosin cross-bridges on an actin filament in skeletal muscle. In an intact cilium, the radial spokes resist this sliding motion, which instead is converted into a local bending. Nexin, a highly extensible protein, keeps adjacent doublets together during the sliding process. *An important difference between muscle and cilia is that myosin forms bipolar filaments that cross-link antiparallel actin filaments, whereas dynein drives the sliding of parallel adjacent microtubules. Hence, the sarcomere shortens but the axoneme bends.*

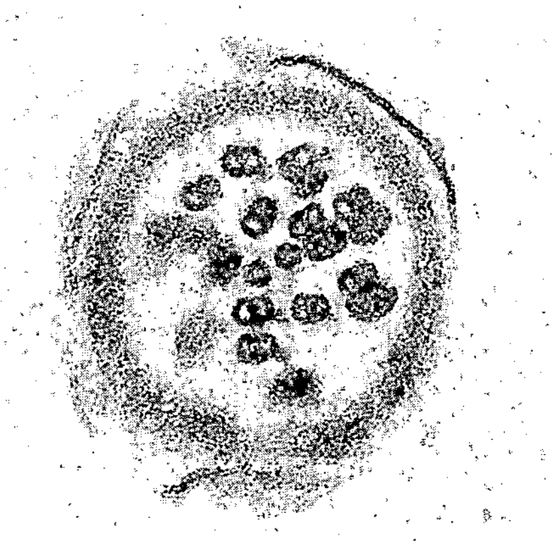

Figure 15-36
Electron micrograph of a cilium from a patient with the immotile-cilia syndrome. The arrangement of microtubules in this cilium is disordered. [Courtesy of Dr. Bjorn Afzelius.]

Defective cilia have been found in a group of patients with chronic pulmonary disorders. Bjorn Afzelius showed that the cilia in their respiratory tracts are immotile. Males with this genetic defect are also infertile because their sperm are unable to move. This disease, called the *immotile-cilia syndrome,* arises from a variety of molecular lesions. The most common one is the absence of outer and inner dynein arms. Other defects producing this disease are the absence of spokes, nexin links, and central microtubules. An intriguing finding is that many patients with immotile cilia have a complete left-right reversal of internal organs (situs inversus), which suggests that microtubular arrays play a critical role in establishing left-right asymmetry early in embryogenesis.

THE RAPID GTP-DRIVEN ASSEMBLY AND DISASSEMBLY OF MICROTUBULES IS CENTRAL TO MORPHOGENESIS

Microtubules are also important in determining the shapes of cells and in separating daughter chromosomes in mitosis (Figure 15-37). Microtubules in cells are formed by the addition of α- and β-tubulin molecules to preexisting filaments or nucleation centers. De novo polymerization would be too slow because multiple subunits must come together simultaneously to form the 13-protofilament helix. Instead, microtubules in cells radiate from *centrosomes* and from the *poles* of mitotic spindles (Figure 15-38). These sites of the initiation of microtubule growth are called

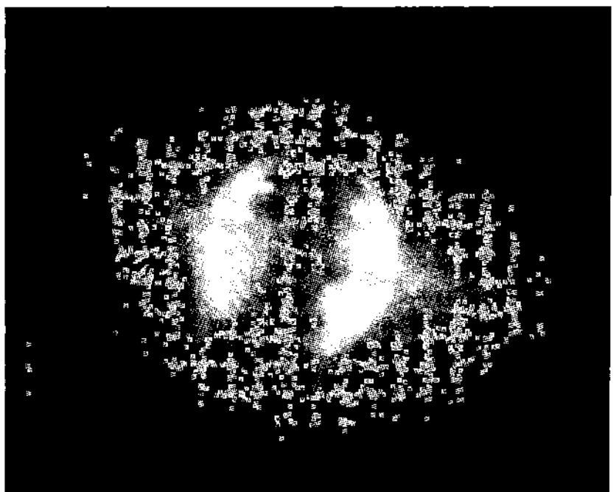

Figure 15-37
Fluorescence micrograph of microtubules in the mitotic spindle of a mouse fibroblast. A fluorescent-labeled antitubulin antibody was used to display these filaments. [Courtesy of Dr. Marc Kirschner.]

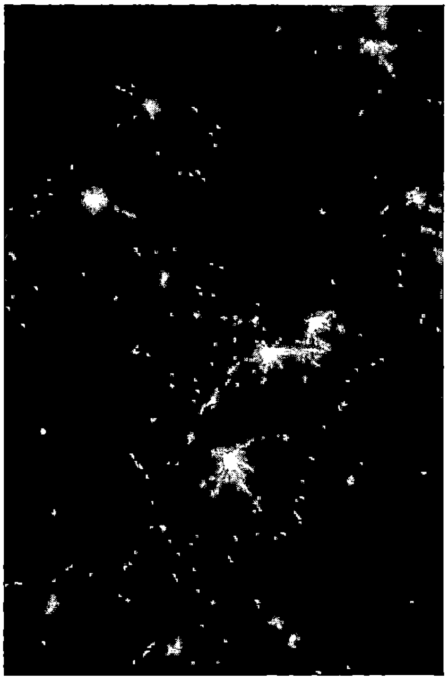

Figure 15-38
Fluorescence micrograph showing the in vivo nucleation of microtubules from centrosomes. [Courtesy of Dr. Tim Mitchison and Dr. Marc Kirschner.]

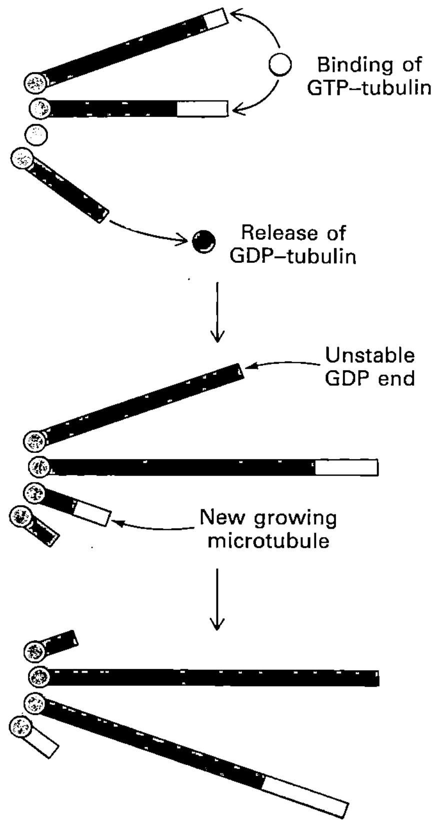

Figure 15-39
Some microtubules grow and others simultaneously regress because their ends are different. Microtubules with GDP–tubulin (red) at their ends are unstable, in contrast with those having GTP–tubulin (green) at their ends. Tubulin has an inherent GTPase activity that converts bound GTP to GDP. Nucleation sites, which bind the minus ends of microtubules, are shown in blue.

microtubule-organizing centers (MTOC). The minus end of a microtubule is the one bound to an MTOC, and the plus end is the free one.

In fact, *most microtubules undergo rapid assembly and disassembly.* Fluorescent tubulins injected into cells become incorporated into mitotic spindles in about 15 seconds, and into other microtubules in several minutes. The hydrolysis of GTP bound to tubulin is at the heart of the rapid turnover of microtubules. The critical concentration for the formation of microtubules is much lower for GTP–tubulin than for GDP–tubulin. GTP–tubulin adds to the plus end of microtubules. The bound nucleotide is hydrolyzed seconds later. A GDP–tubulin unit located within a microtubule stays there, but one situated at an exposed end dissociates from the filament. Hence, microtubules become longer when the rate of addition of GTP–tubulin to the plus end is faster than the rate of hydrolysis of bound GTP. Marc Kirschner and Tim Mitchison found that some microtubules in a population of filaments lengthen while others simultaneously shorten. This property, called *dynamic instability,* arises from random fluctuations in whether the plus end contains GTP–tubulin or GDP–tubulin (Figure 15-39).

The dynamic instability of microtubules is exploited in cell development. The formation of the mitotic spindle is an instructive example. Microtubules join each pole of the mitotic spindle to kinetochores, the binding sites at the centromeres of daughter chromosomes (Figure 15-40). The nucleation sites at the two poles do not send out microtubules precisely aimed at kinetochore targets. Instead, hundreds of randomly oriented microtubules emanate from the poles. Microtubules reaching kinetochores are stabilized, whereas the others fall apart because they have an exposed plus end. Thus, *dynamic instability produces a large repertoire of structures; those engaged in constructive interactions become stabilized.* We see here, as in the operations of the immune system (p. 364), *the importance of selection in generating pattern out of random variation.*

Colchicine, an alkaloid from the autumn crocus, inhibits cell processes that depend on functioning microtubules by blocking their polymerization. For example, dividing cells are arrested at metaphase by the action of colchicine because microtubules are essential for moving chromosomes. Colchicine also inhibits the movement of vesicles along microtubule tracks; hence, many secretory processes are blocked by this com-

Figure 15-40
The dynamic instability of microtubules is exploited in the formation of the mitotic spindle. Microtubule growth in many directions is nucleated from a pole of the spindle (blue). Microtubules reaching the kinetochores at centromeres (yellow) of chromosomes are stabilized. The other microtubules regress when spontaneous fluctuations give rise to GDP ends (red) as a result of hydrolysis of bound GTP (green). [After M. Kirschner and T. Mitchison. *Cell* 45(1986):339.]

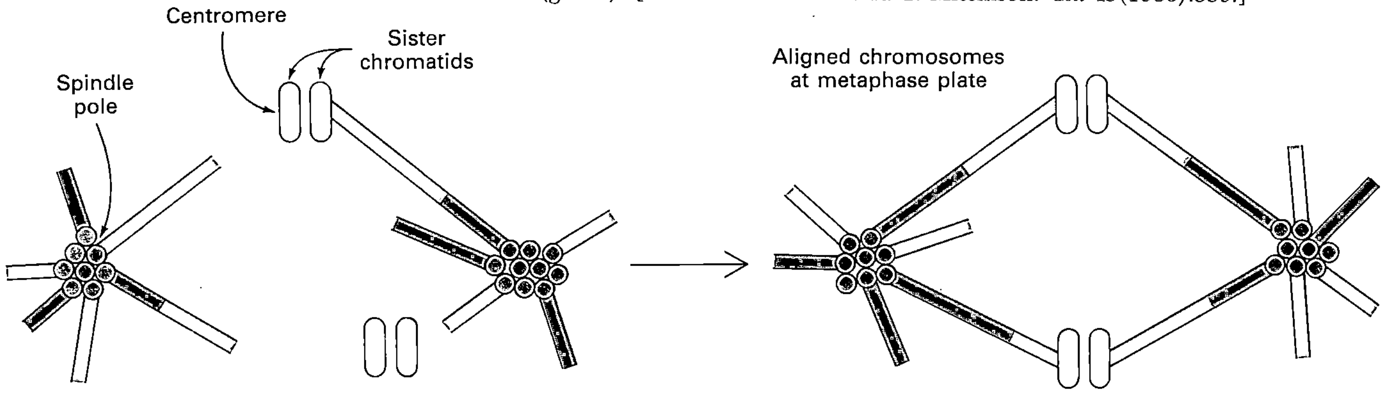

pound. Colchicine, a potent anti-inflammatory agent, has been used for several centuries to treat acute attacks of gout (p. 756). *Taxol,* which comes from the bark of the Pacific yew tree, has the opposite action. It stabilizes tubulin in microtubules and promotes polymerization. Taxol, an anticancer drug, blocks the proliferation of rapidly dividing cells by interfering with the mitotic spindle.

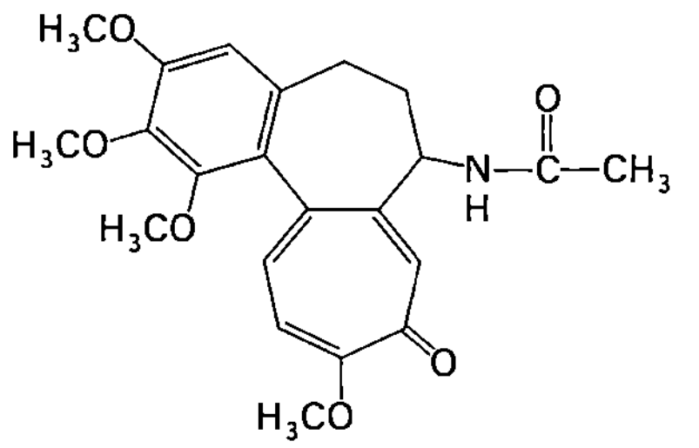

Colchicine

KINESIN MOVES VESICLES AND ORGANELLES UNIDIRECTIONALLY ALONG MICROTUBULE TRACKS

Microtubules also participate in the movement of vesicles and organelles in nearly all eukaryotic cells (Figure 15-41). The development of video-enhanced contrast microscopy made it feasible to view the motions of 30-nm vesicles in vivo. Vesicle transport is most striking in neurons, where it occurs rapidly over long distances. Vesicles move from the cell body to the nerve terminal at rates of up to 5 μm/s, enabling them to traverse a distance of a meter in about a day. Indeed, vesicles also move rapidly in extruded axoplasm (the cytoplasm of nerve axons). *The tracks for these movements are single microtubules.*

The development of a reconstituted in vitro system that transports vesicles led to the discovery of a new ATP-driven molecular engine. Vesicles and organelles are transported by *kinesin,* a large water-soluble protein consisting of two 110-kd subunits and two 70-kd subunits. Kinesin is a highly elongated protein, about 110 nm long (Figure 15-42). The amino-terminal globular domain of each heavy chain contains an ATP-binding site and a microtubule-binding site. This motor domain is followed by an α-helical coiled-coil stalk. The carboxyl-terminal region, together with the light chains, binds to a specific receptor in the membrane of a vesicle or organelle. Thus, one end of kinesin interacts with a microtubule track and the other end grasps the cargo to be transported. Observations of the movements of asymmetric organelles, such as mitochondria, suggest that they are hoisted and carried rather than rolled along microtubule tracks by kinesin molecules.

Organelles and vesicles containing kinesin move from the minus end of a microtubule (at a microtubule organizing center, such as a centrosome) to the plus end. Hence, ***kinesin produces movement from the center of a cell to its periphery (anterograde transport).*** Beads coated with kinesin move in the same direction as do vesicles (Figure 15-43). The ATPase activity of kinesin is stimulated 50-fold by the addition of microtubules, just as the ATPase activity of myosin is enhanced by microfilaments. *ATP hydrolysis by kinesin is tightly coupled to its binding to a microtubule and to the generation of force.* Kinesin differs from myosin and dynein in that ATP promotes its binding to, rather than its release from, its partner protein. ATP is hydrolyzed by kinesin while it is attached to a microtubule; hydrolysis enables kinesin to release its hold so that it can take another step in the walk along the microtubule. *A motor can thus have either higher or lower affinity for its*

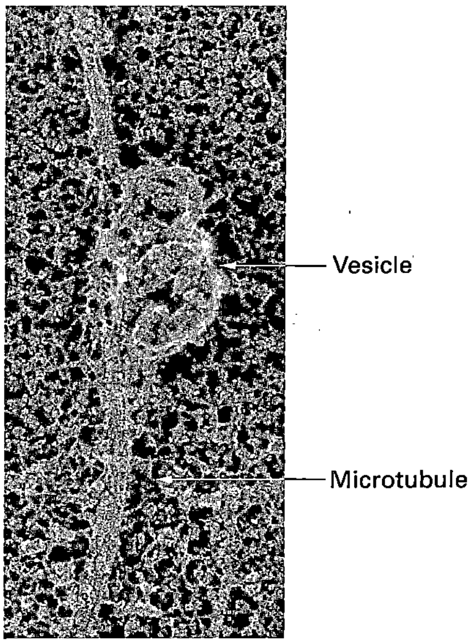

Figure 15-41
Freeze-fracture electron micrograph of a vesicle on a microtubule. [Courtesy of Dr. Ronald Vale.]

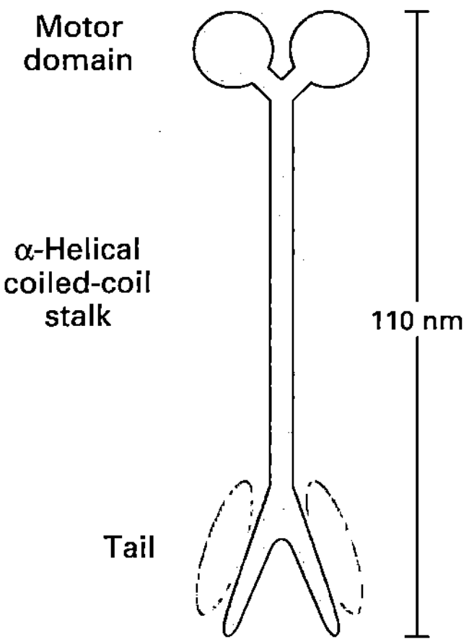

Figure 15-42
Kinesin consists of two heads, an α-helical coiled-coil stalk, and a bifurcated tail. Two 70-kd light chains (yellow) are associated with the two 110-kd heavy chains (green). [After R.D. Vale. *Trends Biochem. Sci.* 17(1992):300.]

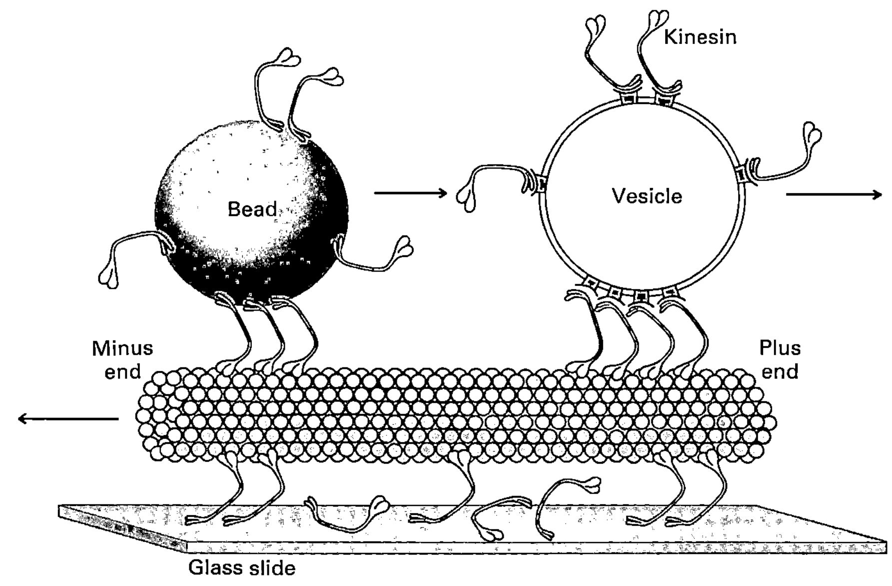

Figure 15-43
Schematic diagram showing movements mediated by kinesin, an ATP-driven motor, on microtubule tracks. Vesicles and beads containing kinesin molecules on their surface move toward the plus end of microtubules. Kinesin is bound to vesicles by receptors (shown in red) in their membrane. Conversely, microtubules can move on glass slides containing bound kinesin. [After a drawing kindly provided by Dr. Ronald Vale.]

partner protein when bound ATP has been hydrolyzed. In either case, the key to energy transduction is the large change in affinity of the pair of proteins during the ATP-ADP cycle.

A different engine carries cargo from the periphery of a cell to its center (retrograde transport). *Cytoplasmic dynein, a motor protein related to axonemal dynein in flagella and cilia, is the retrograde transporter.* The polarity of microtubules serves as a compass for the navigation of particles in the cytoplasm. The direction of movement is determined by the motor rather than by the track.

A SINGLE KINESIN MOTOR CAN MOVE A VESICLE ON A MICROTUBULE TRACK

How many kinesins are needed to move a vesicle or organelle? Analyses of the movement of kinesin-coated beads on microtubule tracks, and conversely, of microtubules on kinesin-coated glass, have answered this important question. As was mentioned earlier, an intensely focused laser beam can be used as a micromanipulator *(optical tweezer)* to position particles. Beads containing a small number of bound kinesin molecules were placed on microtubules and their subsequent motions were monitored. Several outcomes were observed: (1) some beads failed to bind; (2) others bound, moved a distance, and diffused away from the microtubule before reaching the end; and (3) the rest walked the full length of the track. The dependence of these outcomes on the average number of bound kinesins in a population of beads revealed that *a single kinesin molecule can propel a bead along a microtubule.* Likewise, a single kinesin molecule adsorbed to glass can move a microtubule. However, a particle with a single kinesin drifts away from the track after moving a distance of 1 to 2 μm because kinesin transiently detaches from tubulin during each ATPase cycle. A bead containing two kinesin molecules stays on the track for a much longer distance because only rarely are both simultaneously detached. Thus, *a small number of kinesin molecules suffice to move a vesicle over a long distance on a microtubule track.* Recent optical trap experiments have revealed that a kinesin molecule takes 8-nm (80-Å) steps (Figure 15-44), a size consistent with the dimensions of this motor protein.

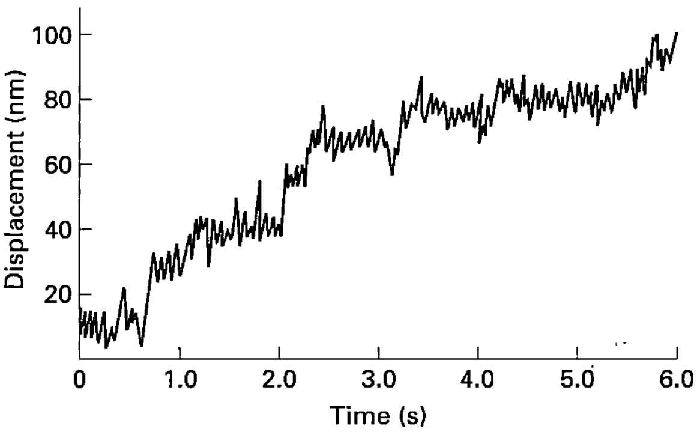

Figure 15-44
Direct observation of the movement of a bead carried by a single kinesin motor on a microtubule track. Multiple steps were taken in this six-second interval. The average step size is about 80 Å. [After K. Svoboda, C.F. Schmidt, B.J. Schnapp, and S.M. Block. *Nature* 365(1993):721.]

The movement of actin-coated filaments on myosin-coated glass surfaces and that of microtubules on kinesin coatings differ in an important respect. Actin filaments move faster on surfaces containing a high rather than a low density of myosin. By contrast, microtubules move at a velocity independent of kinesin density. The reason for this difference is that *a myosin molecule spends most of its time free from an actin track, whereas kinesin spends most of its time gripping a microtubule.* The kinetic properties of myosin have been optimized for fast movement, which is favored by having very brief periods of attachment to actin. In contrast, kinesin has been designed to stay on track most of the time so that vesicles containing only a few motor molecules can be transported steadily; the price paid is a lower speed.

BACTERIA SWIM BY ROTATING THEIR FLAGELLA

In an earlier chapter, we saw that bacteria move toward attractants and away from repellents (p. 326). They swim by rotating their flagella. An *Escherichia coli* or *Salmonella typhimurium* bacterium has about six flagella that emerge from random positions on its surface. These thin helical filaments, which are 15 nm in diameter and 10 μm long, are built from 53-kd *flagellin* subunits. When flagella rotate counterclockwise, they form a coherent bundle that transmits a propulsive force (Figure 15-45). The result is smooth swimming in a nearly straight line. In contrast, when flagella rotate clockwise, the bundle flies apart because the helical filaments can no longer dovetail into each other. Such a bacterium tumbles transiently, and then swims in a different direction when counterclockwise rotation is restored. Tumbling redirects a bacterium that is headed in an unpromising direction.

Bacteria swim at a velocity of about 25 μm/s, which corresponds to about 10 body lengths per second. A human sprinting at a proportional rate would set an enviable new record by running the 100-meter dash in 5.5 seconds. Furthermore, bacteria are very efficient swimmers. Less than 1% of their total energy production is consumed by their flagellar motors. Their performance is all the more impressive given their small size in relation to the viscosity of water. Bacteria cannot coast, not even for a microsecond. The viscous resistance encountered by a bacterium is akin to what we would experience if we tried to swim in molasses.

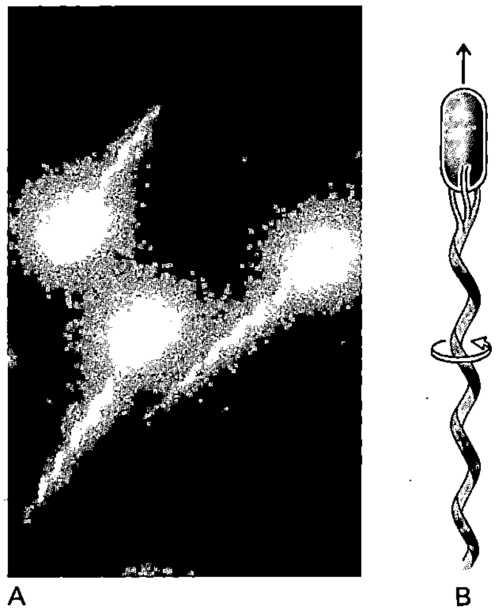

Figure 15-45
Bacteria such as *E. coli* and *S. typhimurium* swim by rotating their flagella. When the motors rotate counterclockwise, the flagella form a propulsive bundle. (A) Flash photograph of several swimming bacteria. (B) Schematic diagram. [Courtesy of Dr. Robert Macnab.]

PROTON FLOW DRIVES BACTERIAL FLAGELLAR ROTATION

A bacterial flagellum, in contrast with a eukaryotic flagellum or cilium, is an extracellular appendage that transmits but does not generate motile force. The propulsive force is generated instead by a motor that is located within the cytoplasmic membrane of the cell. The flagella, though passive, are intricate assemblies. Genetic studies revealed that they are encoded by some 40 genes. The motor itself is encoded by just two genes, *motA* and *motB*. How does this motor work? An early clue came from the finding that *ATP is not required for flagellar rotation.* In contrast, all known eukaryotic motors are powered by ATP or GTP. A bacterial cell depleted of energy sources can be induced to swim by placing it in a medium that is more acidic than its cytosol. In fact, *the energy source for flagellar rotation is the proton-motive force across the cytoplasmic membrane.* The magnitude of this force depends on both the membrane potential and the pH gradient across the membrane.

Bacterial flagella typically rotate at a velocity of 100 revolutions per second. Each revolution is driven by the flow of about 1000 protons across the membrane. *Rotatory force is generated by the interplay of the motA and motB proteins,* which are located in the cytoplasmic membrane to harness the proton-motive force (Figure 15-46). The motA protein forms *proton channel complexes* across this membrane; these channels are elastically linked to the cell wall. MotB proteins are located at the periphery of the M ring, where they interact with motA.

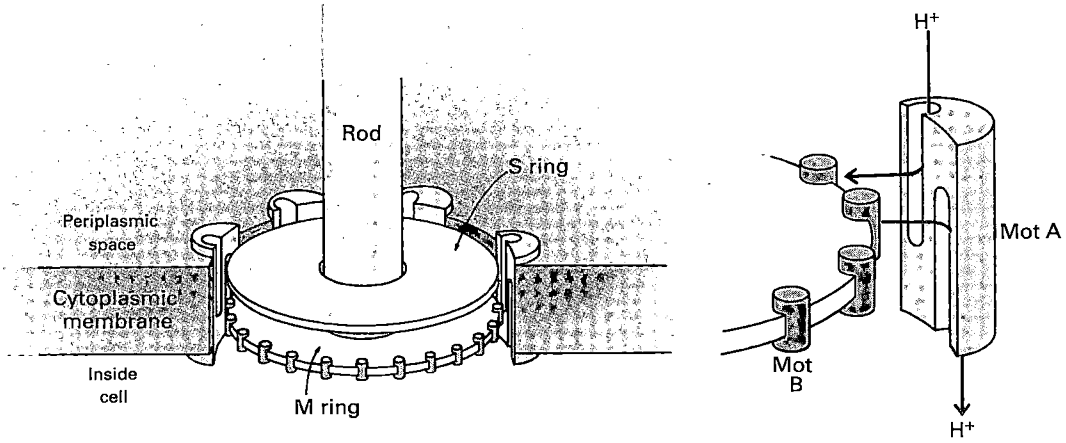

Figure 15-46
Schematic diagram of the rotatory motor of a bacterium. The motA protein (blue) spans the cytoplasmic membrane of the bacterial cell. Protons flow through the motA half-channels and bind to the motB proteins (red) at the periphery of the M ring (yellow). Rotation of the M ring enables the bound protons to be released into a second set of half-channels in motA. [After F.M. Harold. *The Vital Force: A Study of Bioenergetics* (W.H. Freeman, 1986), p. 150.]

How does proton flux through a channel complex lead to rotation of the M ring bearing the flagellar rod? A plausible mechanism has been proposed by Howard Berg and co-workers (Figure 15-47). In their model, the M ring has a set of proton-accepting sites (provided by motB) that are equally spaced around its periphery. This ring interacts with eight channel complexes formed by motA. Each channel complex contains two *half-channels;* one of them is accessible from the cytosol, and the other from the extracellular side. The half-channels are never in contact with each other. Rather, *a proton is transferred from a half-channel to a site on the M ring. This proton can be donated to the other half-channel only if the M ring rotates.*

How could proton shuttling between motA and motB drive the rotation of the M ring? Suppose that (1) neither side of a channel complex can move past a proton-accepting site if the site is *occupied,* and (2) the center of the complex cannot move past a proton-accepting site if the site is *empty.* Now consider the channel complex shown in part A of Figure 15-47. Because site 2 is empty and site 3 is filled, the complex can move to the right but not to the left. If random motion carries it one step to the

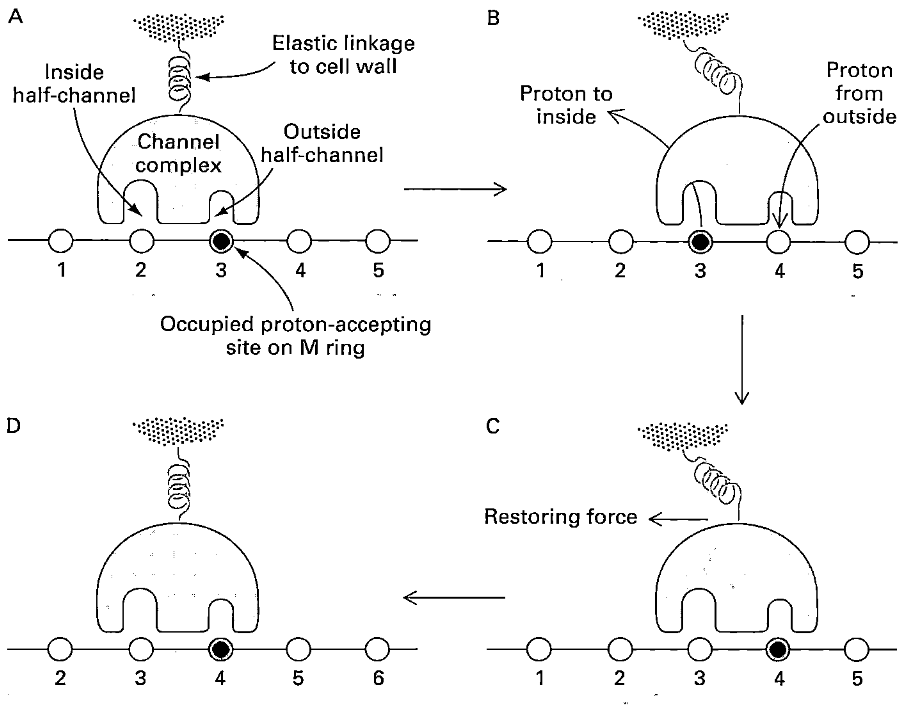

Figure 15-47
Proton shuttling between the motA channel complex (blue) and the motB proteins (circular sites) drives the rotation of the M ring (yellow). Protons flow unidirectionally because of the pH gradient and membrane potential across the cytoplasmic membrane. [After M. Meister, S.R. Caplan, and H.C. Berg. *Biophys. J.* 55(1989):905.]

right, as shown in (B), the proton on site 3 can move into the cytosol, and a proton from the outside can move into the adjacent empty site 4, to give (C). Finally, the M ring rotates counterclockwise to relieve the force exerted by the elastic linkage of the channel complex. The configuration shown in (D) is identical with that of (A) except that the M ring has rotated counterclockwise a distance equal to the spacing of the proton-accepting sites.

Channel complexes in swimming bacteria move more often to the right than the left because a ring site in contact with the outer half-channel is more likely to be protonated than a site in contact with the cytosolic half-channel. The reason for this asymmetry is that the concentration of protons at the two sites differs because the proton potential outside is higher than inside under physiologic conditions. In general, *proton gradients are readily interconvertible sources of free energy.* Two examples were presented in an earlier chapter. Lactose permease couples the energetically uphill transport of lactose into bacterial cells to the downhill flux of protons (p. 317). Bacteriorhodopsin harvests the energy of light by generating a proton gradient (p. 318). In the operation of the flagellar motor, we see *the transduction of a proton gradient into rotatory motion.*

SUMMARY

Vertebrate striated muscle consists of two kinds of interacting protein filaments. The thick filaments contain myosin, whereas the thin filaments contain actin, tropomyosin, and troponin. The hydrolysis of ATP bound to myosin drives the sliding of these filaments past each other. Myosin is a large protein (520 kd) composed of two heavy chains and four light chains. The heavy chains are folded to form two elongated S1 heads that are joined to a double-stranded α-helical coiled-coil rod. Two light chains that have modulatory roles are bound to each S1 head. The S1 heads and part of the rod form cross-bridges that interact with actin to generate the contractile force. The remainder of the myosin molecule forms the backbone of the thick filament. Actin is a globular protein (42 kd) that polymerizes to form thin filaments. Both thin and thick filaments have direction, which reverses halfway between the Z lines.

ATP binding and hydrolysis drives the cyclic formation and dissociation of complexes between the myosin cross-bridges of thick filaments and the actin units of thin filaments, which shortens the distance between Z lines. The power stroke of muscle contraction comes from conformational changes within the S1 head accompanying the release of ADP from the nucleotide-binding site. Hinges between domains of myosin are important in enabling S1 heads to reversibly attach and detach from actin and to change their orientation when bound. ATP-driven movement can be reconstituted in simple, well-defined systems. Myosin-coated beads move unidirectionally on actin cables. Optical trap experiments have shown that the step taken by a single myosin molecule is about 11 nm.

The contraction of skeletal muscle is regulated by Ca^{2+}. The interaction of actin and myosin is inhibited by the troponin complex and tropomyosin when the level of Ca^{2+} is low. Nerve excitation triggers the release of Ca^{2+} from the sarcoplasmic reticulum. The binding of Ca^{2+} to troponin C, a member of the EF hand family of calcium sensors, alters the interaction of tropomyosin with actin to allow myosin to bind actin and generate the contractile force.

Actin and myosin are ancient proteins, as shown by their presence in yeast and slime molds. In fact, these proteins have contractile roles in nearly all eukaryotic cells. Actin, which is particularly abundant, forms 7-nm-diameter microfilaments. They participate in a wide range of cellular movements, as exemplified by cell migration in development and the movement of macrophages to injured tissues. Single-headed myosin I molecules transport membrane-bound organelles on actin tracks. The diverse tails of double-headed myosin II molecules specify the cargo that is carried by these motors in nonmuscle cells. The dynamic cytoskeletons of eukaryotic cells also contain intermediate filaments (~10-nm diameter) and microtubules (~30-nm diameter). Intermediate filaments are formed from proteins that have a common α-helical coiled-coil core but are otherwise quite different.

Microtubules have multiple architectural and contractile roles in nearly all eukaryotic cells. These hollow fibers are built from globular α- and β-tubulin subunits. Colchicine inhibits movements mediated by microtubules by blocking their polymerization. A eukaryotic cilium or flagellum contains nine microtubule doublets around two singlets. The outer doublets are cross-linked by dynein, an ATPase. The dynein-induced sliding of adjacent microtubule doublets produces a local bending of the cilium and causes it to beat. Microtubules also serve as tracks for the movements of vesicles and organelles. The rapid GTP-driven assembly and disassembly of microtubules are central to their actions in cell development, as exemplified by the formation of the mitotic spindle. Microtubules spontaneously disassemble if they do not reach a target that stabilizes their ends. Two kinds of ATP-driven motors, kinesin and cytoplasmic dynein, carry vesicles and organelles on microtubule tracks. Kinesin moves particles from the center of a cell to its periphery (anterograde transport), whereas cytoplasmic dynein moves them in the opposite direction (retrograde transport). A single kinesin molecule can move a bead on a microtubule track.

Bacteria swim by rotating their flagella. These passive extracellular appendages are rotated by reversible motors located in the cytoplasmic membrane. Rotation is powered by proton-motive force rather than by hydrolysis of ATP or GTP. Each revolution is driven by the flow of about 1000 protons across the membrane. The motor is formed from two gene products. The motA protein forms proton channel complexes, whereas the motB protein provides proton-binding sites at the periphery of the M ring. Protons cannot flow directly through motA, because its two half-channels are not contiguous. Rather, protons are transferred from motA to motB. After the M ring rotates, protons are transferred from motB to the other side of motA. The operation of the flagellar motor shows how a proton gradient can be used as a free-energy source.

SELECTED READINGS

Where to start

Cook, R., 1990. Force generation in muscle. *Curr. Opin. Cell Biol.* 2:62–66.

Rayment, I., Holden, H.M., Whittaker, M., Yohn, C.B., Lorenz, M., Holmes, K.C., and Milligan, R.A., 1993. Structure of the actin-myosin complex and its implications for muscle contraction. *Science* 261:58–65.

Finer, J.T., Simmons, R.M., and Spudich, J.A., 1994. Single myosin molecule mechanics: Piconewton forces and nanometre steps. *Nature* 368:113–119.

Vale, R.D., 1992. Microtubule motors: Many new models off the assembly line. *Trends Biochem. Sci.* 17:300–304.

Books

Sugi, H. (ed.), 1992. *Muscle Contraction and Cell Motility.* Springer-Verlag.

Philosophical Transactions of the Royal Society of London. Series B: Biological Sciences, 1992. Nucleoside triphosphatases in energy transduction, cellular regulation and information transfer in biological systems. Volume 336, pp. 1–112. [This symposium volume contains many excellent articles on motor proteins and the cytoskeleton.]

Squire, J.M., 1986. *Muscle: Design, Diversity, and Disease.* Benjamin/Cummings.

Pollack, G.H., and Sugi, H. (eds.), 1984. *Contractile Mechanisms in Muscle.* Plenum.

Development of major concepts

Huxley, H.E., 1965. The mechanism of muscular contraction. *Sci. Amer.* 213(6):18–27.

Huxley, H.E., 1971. The structural basis of muscle contraction. *Proc. Roy. Soc. London Ser. B* 178:131–149.

Huxley, A.F., 1975. The origin of force in skeletal muscle. *Ciba Found. Symp.* 31:271–290.

Summers, K.E., and Gibbons, I.R., 1971. ATP-induced sliding of tubules in trypsin-treated flagella of sea-urchin sperm. *Proc. Nat. Acad. Sci.* 68:3092–3096.

Myosin and actin

Rayment, I., Rypniewski, W.R., Schmidt-Bäse, K., Smith, R., Tomchick, D.R., Benning, M.M., Winkelmann, D.A., Wesenberg, G., and Holden, H.M., 1993. Three-dimensional structure of myosin subfragment-1: A molecular motor. *Science* 261:50–58.

Kabsch, W., Mannherz, H.G., Suck, D., Pai, E.F., and Holmes, K.C., 1990. Atomic structure of the actin:DNase I complex. *Nature* 347:37–44.

Kabsch, W., and Vandekerckhove, J., 1992. Structure and function of actin. *Ann. Rev. Biophys. Biomol. Struct.* 21:49–76.

Holmes, K.C., Popp, D., Gebhard, W., and Kabsch, W., 1990. Atomic model of the actin filament. *Nature* 347:44–49.

Cooper, J.A., 1991. The role of actin polymerization in cell motility. *Ann. Rev. Physiol.* 53:585–605.

Spudich, J.A., and Warrick, H.M., 1991. A tale of two motors. *Curr. Opin. Struct. Biol.* 1:264–269.

Pollard, T.D., Doberstein, S.K., and Zot, H.G., 1991. Myosin-I. *Ann. Rev. Physiol.* 53:653–681.

Microtubules, dynein, and kinesin

Kirschner, M., and Mitchison, T., 1986. Beyond self-assembly: From microtubules to morphogenesis. *Cell* 45:329–342.

Vallee, R., 1991. Cytoplasmic dynein: Advances in microtubule-based motility. *Trends Cell Biol.* 1:25–29.

Schroer, T.A., and Sheetz, M.P., 1991. Functions of microtubule-based motors. *Ann. Rev. Physiol.* 53:629–652.

Vallee, R.B., and Shpetner, H.S., 1990. Motor proteins of cytoplasmic microtubules. *Ann. Rev. Biochem.* 59:909–932.

Vale, R.D., and Goldstein, L.S., 1990. One motor, many tails: An expanding repertoire of force-generating enzymes. *Cell* 60:883–885.

Witman, G.B., 1992. Axonemal dyneins. *Curr. Opin. Cell Biol.* 4:74–79.

Afzelius, B., 1985. The immotile-cilia syndrome: A microtubule-associated defect. *CRC Crit. Rev. Biochem.* 19:63–87.

In vitro motility systems

Kron, S.J., Toyoshima, Y.Y., Uyeda, T.Q., and Spudich, J.A., 1991. Assays for actin sliding movement over myosin-coated surfaces. *Meth. Enzymol.* 196:399–416.

Kron, S.J., Drubin, D.G., Botstein, D., and Spudich, J.A., 1992. Yeast actin filaments display ATP-dependent sliding movement over surfaces coated with rabbit muscle myosin. *Proc. Nat. Acad. Sci.* 89:4466–4470.

Zot, H.G., Doberstein, S.K., and Pollard, T.D., 1992. Myosin-I moves actin filaments on a phospholipid substrate: Implications for membrane targeting. *J. Cell Biol.* 116:367–376.

Vale, R.D., Schnapp, B.J., Reese, T.S., and Sheetz, M.P., 1985. Organelle, bead, and microtubule translocations promoted by soluble factors from the squid giant axon. *Cell* 40:559–569.

Yang, J.T., Saxton, W.M., Stewart, R.J., Raff, E.C., and Goldstein, L.S., 1990. Evidence that the head of kinesin is sufficient for force generation and motility in vitro. *Science* 249:42–47.

Force generation

Irving, M., Lombardi, V., Piazzesi, G., and Ferenczi, M.A., 1992. Myosin head movements are synchronous with the elementary force-generating process in muscle. *Nature* 357:156–158.

Huxley, H.E., Kress, M., Faruqi, A.F., and Simmons, R.M., 1988. X-ray diffraction studies on muscle during rapid shortening and their implications concerning cross-bridge behaviour. *Advan. Exp. Med. Biol.* 226:347–352.

Howard, J., Hudspeth, A.J., and Vale, R.D., 1989. Movement of microtubules by single kinesin molecules. *Nature* 342:154–158.

Block, S.M., Goldstein, L.S., and Schnapp, B.J., 1990. Bead movement by single kinesin molecules studied with optical tweezers. *Nature* 348:348–352.

Svoboda, K., Schmidt, C.F., Schnapp, B.J., and Block, S.M., 1993. Direct observation of kinesin stepping by optical trapping interferometry. *Nature* 365:721–727.

Eisenberg, E., and Hill, T.L., 1985. Muscle contraction and free energy transduction in biological systems. *Science* 227:999–1006.

Bacterial flagellar motor

Purcell, E.M., 1977. Life at low Reynolds number. *Am. J. Physiol.* 45:3–11. [An incisive account of how bacteria swim in a world dominated by viscous forces. This delightful essay, full of insight, is a classic.]

Meister, M., Caplan, S.R., and Berg, H.C., 1989. Dynamics of a tightly coupled mechanism for flagellar rotation. Bacterial motility, chemiosmotic coupling, protonmotive force. *Biophys. J.* 55:905–914.

Blair, D.F., and Berg, H.C., 1990. The MotA protein of *E. coli* is a proton-conducting component of the flagellar motor. *Cell* 60:439–449.

Macnab, R.M., and Parkinson, J.S., 1991. Genetic analysis of the bacterial flagellum. *Trends Genet.* 7:196–200.

PROBLEMS

1. *Diverse motors.* Skeletal muscle, eukaryotic cilia, and bacterial flagella use different molecular strategies for the conversion of free energy into coherent motion. Compare and contrast these motility systems with respect to (a) the free-energy source, (b) the number of essential components and their identity, and (c) cellular localization.

2. *Load-dependent ATPase activity.* Experiments carried out more than a half century ago showed that the ATPase rate of skeletal muscle is slowed when a large load is applied. Propose a molecular explanation for this observation.

3. *Rigor mortis.* Why does the body stiffen after death?

4. *Theme and variation.* Calmodulin is typically activated by Ca^{2+} with an apparent K of about 1 μM. Troponin C, however, has a K of about 10 μM. What is a likely advantage of troponin's lower calcium affinity? What price must be paid for this gain?

5. *Smooth muscle.* Smooth muscle, in contrast with skeletal muscle, is not regulated by a troponin-tropomyosin mechanism. Instead, vertebrate smooth muscle contraction is controlled by the degree of phosphorylation of its light chains. Phosphorylation induces contraction, and dephosphorylation leads to relaxation. Smooth muscle contraction is triggered by an increase in the cytosolic Ca^{2+} level. Propose a mechanism for this action of Ca^{2+} based on your knowledge of other signal transduction processes (Chapter 13).

6. *Mushroom poison.* The importance of rapid turnover of microfilaments in cell motility is highlighted by the action of phalloidin, a toxic cyclic peptide from *Amanita phalloides*, a highly poisonous mushroom. Propose four ways in which phalloidin could, in principle, interfere with microfilament function.

7. *Vesicle hauling.* Consider the action of a single kinesin molecule in moving a vesicle along a microtubule track. The force F required to drag a spherical particle of radius a at a velocity v in a medium having a viscosity η is

$$F = 6\pi\eta a v$$

Suppose that a 2-μm bead is carried at a velocity of 0.5 μm/s in an aqueous medium ($\eta = .01$ poise $= 0.01\ \mathrm{g\ cm^{-1}\ s^{-1}}$)

(a) What is the magnitude of the force exerted by the kinesin molecule? Express the value in dynes (1 dyne $= 1\ \mathrm{g\ cm\ s^{-2}}$).

(b) How much work is performed in 1 s? Express the value in ergs (an erg is a dyne cm).

(c) A kinesin motor hydrolyzes 40 ATP per second. What is the energy content of the ATP, expressed in ergs? Compare this value with the actual work done.

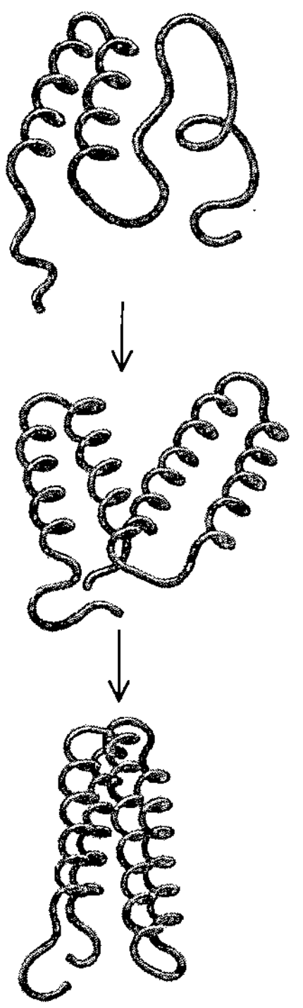

Protein Folding and Design

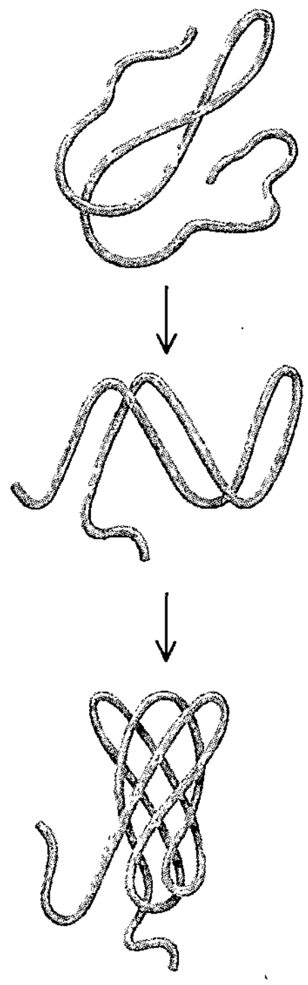

How does the amino acid sequence of a protein specify its three-dimensional structure? And how does an unfolded polypeptide chain acquire the form of the native protein? These closely related questions are the core of the *protein folding problem,* one of the most challenging and important areas of inquiry in biochemistry. We begin with a simple and revealing calculation showing that proteins cannot fold by a random search of all possible conformations. Rather, *protein folding occurs by a progressive stabilization of intermediates.* An unfolded polypeptide chain rapidly condenses into a *molten globule,* which possesses much of the secondary but not the well-packed tertiary structure of the native state. The formation of this compact species is driven by the need to sequester hydrophobic side chains.

We then consider the formation of α helices, β strands, and β turns because they are formed early in protein folding and play a key role in generating molten globules. Combinations of these elements, called *folding units,* are successively formed. Experimental methods for charting the folding process in atomic detail are then presented. Partially folded intermediates of lysozyme and bovine pancreatic trypsin inhibitor have been trapped and characterized. They reveal that *protein folding is guided by interactions that also stabilize the final folded state.* Folding units stabilize other flickering structures to form subdomains and then domains.

The folding of most proteins in cells is assisted by enzymes. Protein disulfide isomerase catalyzes the shuffling of disulfide bonds to find optimal pair-

Opening Image: *Simulation of the folding of protein molecules. α Helix (top) and β sheet (side) formation are key events in the compaction of polypeptide chains. [After a drawing kindly provided by Dr. Jane Richardson.]*

ings. Peptidyl prolyl isomerase also plays a key role in folding by accelerating the cis-trans isomerization of prolyl peptide bonds. In the crowded milieu of the cell, nascent proteins would get tangled together were it not for the action of molecular chaperones. These protein guides, many of which are powered by ATP, minimize protein aggregation and dissociate any aggregates that have formed.

We then turn to the prediction of three-dimensional structure from amino acid sequence, a formidable challenge. An encouraging start has been made in deciphering how one-dimensional information is converted into three-dimensional architecture, the determinant of biological activity. The pattern of hydrophilic and hydrophobic residues in amino acid sequences is highly influential in specifying the final conformation. This chapter concludes with a consideration of protein design, an emerging and promising field. Protein design tests our understanding of basic principles and creates useful new molecules.

PROTEINS FOLD BY PROGRESSIVE STABILIZATION OF INTERMEDIATES RATHER THAN BY RANDOM SEARCH

How are the harmonic chords of interaction, to use Anfinsen's metaphor (p. 39), created in the conversion of an unfolded polypeptide chain into a folded protein? One possibility a priori would be that all possible conformations are searched to find the energetically most favorable one. How long would such a random search take? Consider a small protein with 100 residues. Cyrus Levinthal calculated that if each residue can assume three different positions, the total number of structures is 3^{100}, which is equal to 5×10^{47}. If it takes 10^{-13} s to convert one structure into another, the total search time would be $5 \times 10^{47} \times 10^{-13}$ s, which is equal to 5×10^{34} s, or 1.6×10^{27} years! Clearly, it would take much too long for even a small protein to fold properly by randomly trying out all possible conformations. The enormous difference between calculated and actual folding times is called *Levinthal's paradox.*

The way out of this dilemma is to recognize the power of *cumulative selection.* Richard Dawkins, in *The Blind Watchmaker,* asked how long it would take a monkey poking randomly at a typewriter to reproduce Hamlet's remark to Polonius, "Methinks it is like a weasel." An astronomically large number of keystrokes, of the order of 10^{40}, would be required. However, suppose that we preserved each correct character and allowed the monkey to retype only the wrong ones. In this case, only a few thousand keystrokes, on average, would be needed. The crucial difference between these cases is that the first employs a completely random search, whereas, in the second, *partially correct intermediates are retained* (Figure 16-1).

The essence of protein folding is the retention of partially correct intermediates. However, the protein folding problem is much more difficult than the one presented to our simian Shakespeare. First, proteins are only marginally stable. The free-energy difference between the folded and unfolded states of a typical 100-residue protein is 10 kcal/mol. The average stabilization per residue is only 0.1 kcal/mol, which is less than random thermal energy ($RT = 0.6$ kcal/mol at room temperature). This means that cor-

```
 200  ?T(\G{+s]x[A.N5~,#ATxSGpn`e0@
 400  oDr'Jh7s]DFR:W4l'u+^v6zpJseOi
 600  e2ih'8zs]n527x8l8d_ih=Hldseb]
 800  S#dh>}/s]]tZqC%lP%DK<|!^aseZ]
1000  VOth>nLs]ut/isjl_kwojjwMasef]
1200  juth+nvs_it_is[lukh?SCw=ase5]
1400  Iifhdn4s_it_isOl/ks/IxwLase~]
1600  M?thinrs_it_is]Xk?T"_woasel]
1800  MSthinWs_it_is_lwkN7OKw(asel]
2000  Mhthin_s_it_is_likv,aww_asel]
2200  MMthinns_it_is_lik+5avwlasel]
2400  MethinXs_it_is_likydaqw)asel]
2600  Methin4s_it_is_lik2das_weasel]
2800  Methinhs_it_is_likeOaTweasel]
2883  Methinks_it_is_like_a_weasel]
```

```
 200  )z~hg)W4{(cu!kO{d6jS!NlEyUx}p
 400  "W hi\kR.<&CfA%4-Y1G!iT$6({|6
 600  .L=hinkm4(uMGP^lAWoE6klwW=yiS
 800  AthinkaPa_vYH_liR\Hb,Uo4\-"(
1000  OFthinksP)@fZO_li8v] /+Eln26B
1200  6ithinksMVt]-V_likm+g1#K~}BFk
1400  vxthinksaEt]Ow_like.S1Geutks]
1600  :Othinks<it]MC_likesN2[eaVe4]
1800  uxthinksqit]Or_likeQh)weaoew]
2000  Y/thinks_it_ld_like7alwea)e&]
2200  Methinks_it_lW_like_a[weawel]
2400  Methinks_it_is_like_a;weasel]
2431  Methinks_it_is_like_a_weasel]
```

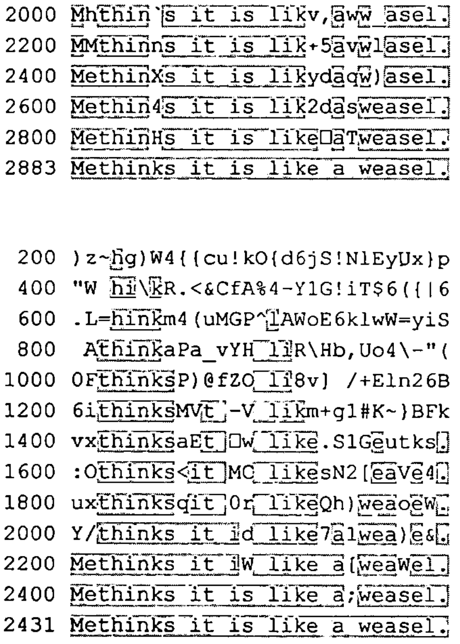

Figure 16-1
A monkey randomly poking a typewriter could write a line from Shakespeare's *Hamlet,* provided that correct keystrokes were retained. Two computer simulations are shown. The cumulative number of keystrokes is given at the left of each line.

rect intermediates, especially those formed early in folding, can be lost. The analogy is that the monkey would be quite free to undo its correct keystrokes. Second, the criterion of correctness is not a residue-by-residue scrutiny of conformation by an omniscient observer but rather the total free energy of the transient species. Intermediates can be scored only by their free energies. Third, some intermediates, called *kinetic traps,* have a favorable free energy but are not on the path to the final folded form. No wonder then that protein folding is such an intriguing problem for both theoreticians and experimentalists.

MOLTEN GLOBULES CONTAINING NATIVE SECONDARY STRUCTURE ARE FORMED EARLY IN FOLDING

Most small proteins can be reversibly unfolded by raising the temperature or by adding a denaturing agent such as H^+, urea, or guanidine hydrochloride. When several different indices of protein conformation were monitored under equilibrium conditions, they often changed together as a function of temperature or concentration of denaturant (Figure 16-2). Furthermore, the transition was the same whether approached from one extreme or the other (e.g., starting at low or high temperature). These findings implied that only two conformational states are present in appreciable amounts—a *native state N* and an *unfolded state U*—and that they are in rapid equilibrium

$$U \rightleftharpoons N$$

Thus, at the midpoint of the thermal or denaturant-induced transition, half the molecules are fully unfolded and the other half fully native.

Other proteins exhibit an additional state under certain denaturing conditions. α-Lactalbumin, a 14-kd protein found in milk, undergoes an acid-induced conformational transition. At pH 4, α-lactalbumin has nearly the same secondary structure as the native protein, as evidenced by its far-ultraviolet circular dichroism spectra (Figure 16-3). Recall that circular dichroism distinguishes between α helix, β sheet, and random-coil structures (p. 63). However, spectral studies in the near ultraviolet showed that many of the interactions of aromatic side chains were disrupted by acid treatment, as they are by 6 M guanidine hydrochloride (GuHCl) and other strong denaturants. Hydrodynamic studies of the pH 4 species showed that it was nearly as condensed as the native form. By contrast, a strong denaturant produces a random coil, which is greatly expanded. These equilibrium studies of α-lactalbumin pointed to the existence of a *molten globule* state, which contains *native secondary but not tertiary structure.* The term *globule* emphasizes its *condensed state,* and the adjective *molten* emphasizes the fluctuating nature of the interactions between secondary-structure units. Indeed, the molten globule state M is a readily observed intermediate between the unfolded and native forms of many proteins.

$$U \rightleftharpoons M \rightleftharpoons N$$

What is the driving force for the formation of molten globules? As was discussed in an earlier chapter (p. 34), proteins are stabilized by hydrophobic interactions. A completely unfolded polypeptide chain in water (in the absence of a denaturing agent) is unstable because many nonpolar side chains are in contact with the aqueous medium. These hydrophobic groups spontaneously come together to avoid water—this condensation is called a *hydrophobic collapse* (Figure 16-4).

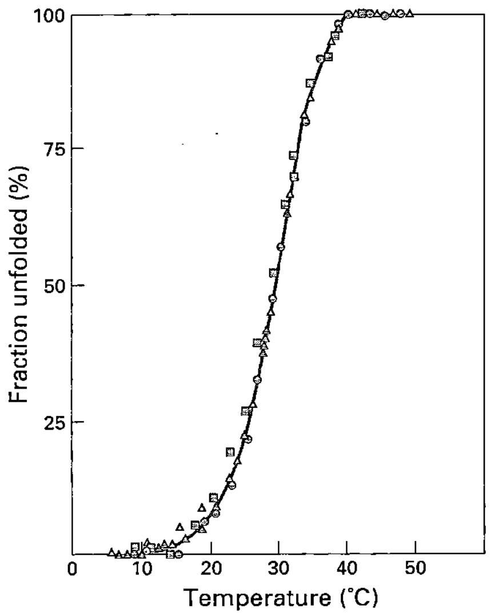

Figure 16-2
The thermal transition of ribonuclease is highly cooperative and fully reversible. The changes in viscosity (□), optical activity (○), and ultraviolet absorbance (△) parallel each other. The sample was heated (blue symbols), cooled for 16 hours, and heated again (red symbols). [After T. Creighton. *Proteins: Structures and Molecular Properties,* 2nd ed. (W.H. Freeman, 1992), p. 288. Based on A. Ginsburg and W.R. Carroll. *Biochemistry* 4(1965):2159.]

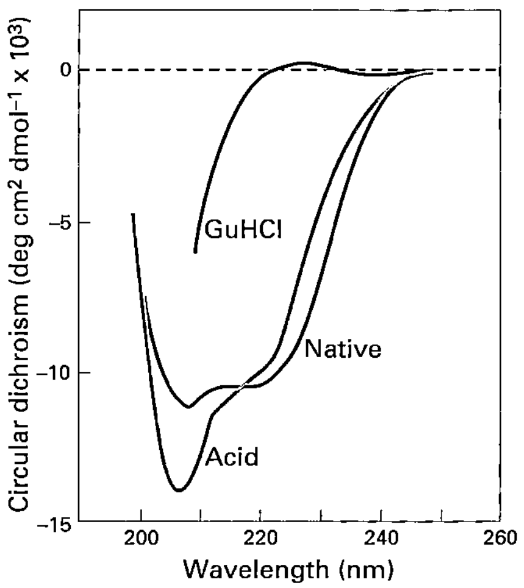

Figure 16-3
Circular dichroism spectra of native α-lactalbumin (red), the acid-unfolded form (green), and the guanidine HCl–unfolded form (black). The molten globule state in acid has much of the secondary structure of the native protein. [After K. Kuwajima, Y. Hiraoka, M. Ikeguchi, and S. Sugai. *Biochemistry* 24(1985):874.]

Figure 16-4
A molten globule formed by helices A (blue), G (red), and H (yellow) is an early intermediate in the folding of myoglobin. The structure of this unit in the final folded state is shown here. The interactions between helices in the molten globule are loose. (A) Schematic diagram. (B) Space-filling model. [Drawn from 1mbn.pdb. H.C. Watson. *Prog. Stereochem.* 4(1969):299. B.D. Barrick and R.L. Baldwin. *Protein Sci.* 2(1993):869.]

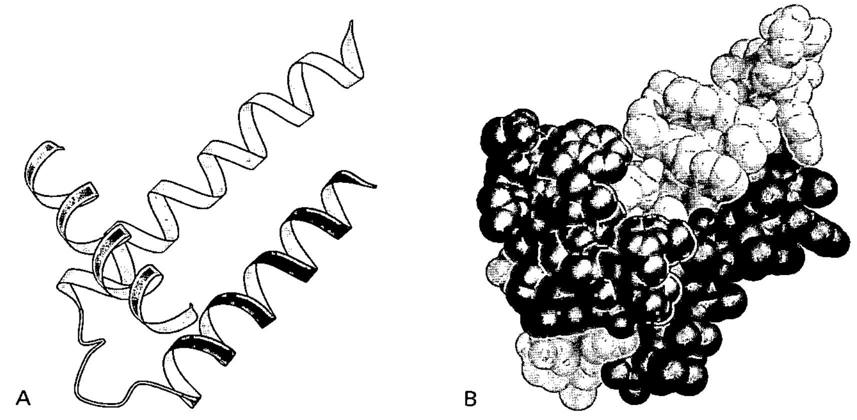

Molten globule formation is also driven by the formation of stretches of secondary structure. Many short peptides in water have considerable helix-forming potential. The nonpolar interior of a molten globule stabilizes nascent helices by shielding their main-chain amides and carbonyl groups from the competing attractions of water. Also, many helices contain a hydrophobic face. The sequestration of these residues within the core of the molten globule also promotes helix formation. Thus, *hydrophobic collapse and acquisition of stable secondary structure are mutually reinforcing events in the formation of molten globules.*

RAMACHANDRAN PLOTS DISPLAY ALLOWED CONFORMATIONS OF THE MAIN CHAIN

The formation of α helix and β strands is at the heart of the folding process. We therefore need to understand the factors that influence the secondary structure of the polypeptide. Let us begin by considering the range of backbone conformations that are accessible to proteins. Recall that the main chain can rotate on either side of each rigid peptide unit (p. 28). The degree of rotation at the bond between the nitrogen and α carbon atoms of the main chain is called *phi* (ϕ), and that between the α carbon and carbonyl carbon atoms is called *psi* (ψ) (Figure 16-5). The

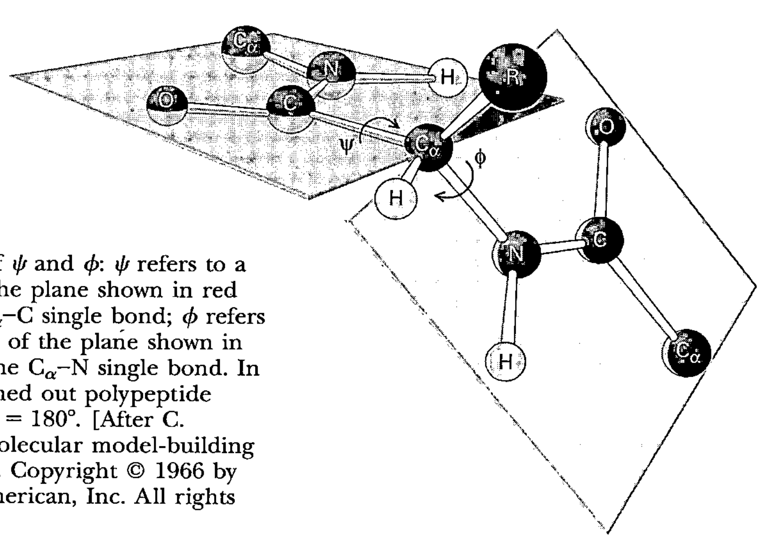

Figure 16-5
Definition of ψ and ϕ: ψ refers to a rotation of the plane shown in red about the C_α–C single bond; ϕ refers to a rotation of the plane shown in blue about the C_α–N single bond. In a fully stretched out polypeptide chain, $\psi = \phi = 180°$. [After C. Levinthal. Molecular model-building by computer. Copyright © 1966 by Scientific American, Inc. All rights reserved.]

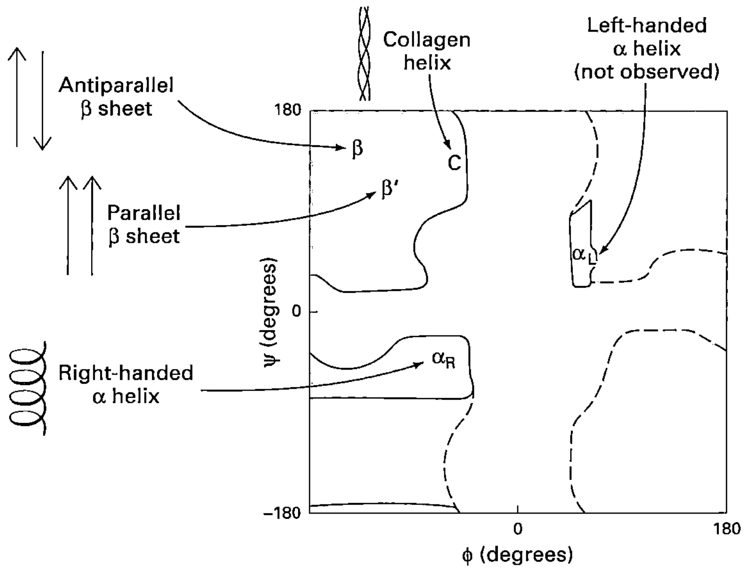

Figure 16-6
Ramachandran plot showing allowed values of ϕ and ψ for L-alanine residues (green regions). Additional conformations are accessible to glycine (yellow regions) because it has a very small side chain.

conformation of the main chain is completely defined when ϕ and ψ are specified for each residue in the chain. G.N. Ramachandran recognized that a residue in a polypeptide chain cannot have just *any* pair of values of ϕ and ψ. Certain combinations are virtually forbidden because of steric hindrance. By assuming that atoms behave as hard spheres, allowed ranges of ϕ and ψ can be predicted and visualized in steric contour diagrams called *Ramachandran plots.* Such a plot for poly-L-alanine (or any amino acid except glycine and proline) shows three separate allowed ranges (the green regions in Figure 16-6). One of them contains ϕ-ψ values that generate the antiparallel β sheet, the parallel β sheet, and the collagen helix. A second region has ϕ-ψ values that produce the right-handed α helix; a third, the left-handed α helix. Though sterically allowed, left-handed α helices are not found in proteins because they are energetically much less favored than right-handed ones.

For glycine, these three allowed regions are larger, and a fourth appears (shown in yellow in Figure 16-6) because a hydrogen atom causes less steric hindrance than a methyl group. *Glycine enables the polypeptide backbone to make turns that would not be possible with another residue. Proline too is special* (Figure 16-7). The five-membered ring of proline prevents rotation about the C_α–N bond, which fixes ϕ at about −65°. Hence, a proline residue has a markedly restricted range of allowed conformations. The residue on the amino-terminal side of a proline is also constrained because of steric hindrance imposed by the five-membered ring. In particular, X in an X-Pro sequence is unlikely to be in an α helix. Proline also disfavors α helix formation because it lacks an amide H atom for hydrogen bonding.

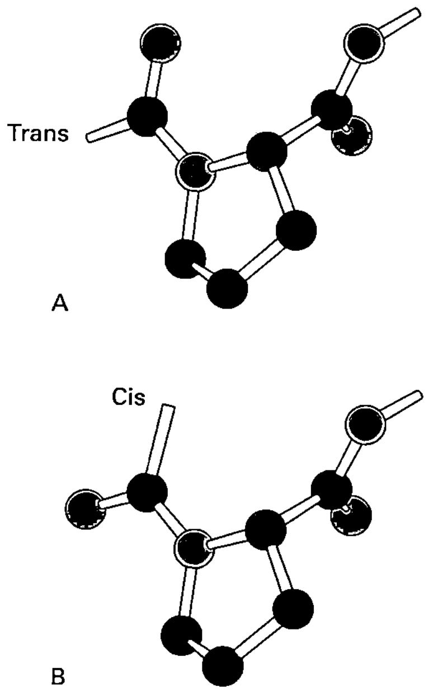

Figure 16-7
Proline can form an (A) trans or (B) cis peptide bond with the preceding residue in the chain. Another distinctive feature of proline is the presence of a five-membered ring, which locks ϕ (the angle between N and C_α) at about −65°.

AMINO ACID RESIDUES HAVE DIFFERENT PROPENSITIES FOR FORMING α HELICES, β SHEETS, AND β TURNS

What determines whether a particular sequence in a protein forms an α helix, a β strand, or a β turn (reverse turn)? The frequency of occurrence of amino acid residues in these secondary structures (Table 16-1) has provided insight into relations between sequence and conformation. *Residues such as glutamate, alanine, and leucine promote α helix formation, whereas valine and isoleucine favor β strand formation.* Glycine, asparagine, and proline, on the other hand, have a propensity for forming reverse turns, in which the polypeptide chain abruptly changes direction by 180 degrees (see Figure 2-37 on p. 31).

Table 16-1
Relative frequencies of occurrence of amino acid residues in the secondary structures of proteins

Amino acid	*α helix*	*β sheet*	*β turn*
Ala	1.29	0.90	0.78
Cys	1.11	0.74	0.80
Leu	1.30	1.02	0.59
Met	1.47	0.97	0.39
Glu	1.44	0.75	1.00
Gln	1.27	0.80	0.97
His	1.22	1.08	0.69
Lys	1.23	0.77	0.96
Val	0.91	1.49	0.47
Ile	0.97	1.45	0.51
Phe	1.07	1.32	0.58
Tyr	0.72	1.25	1.05
Trp	0.99	1.14	0.75
Thr	0.82	1.21	1.03
Gly	0.56	0.92	1.64
Ser	0.82	0.95	1.33
Asp	1.04	0.72	1.41
Asn	0.90	0.76	1.28
Pro	0.52	0.64	1.91
Arg	0.96	0.99	0.88

After T.E. Creighton. *Proteins: Structures and Molecular Properties,* 2nd ed. (W.H. Freeman, 1992), p. 256.

Studies of the conformation of synthetic polypeptides by Elkan Blout, Gerald Fasman, and Ephraim Katchalski, and analyses of the three-dimensional structures of proteins, have revealed some reasons for these preferences. Branching at the β carbon atom, as occurs in valine and isoleucine, tends to destabilize the α helix because of steric hindrance. These hydrophobic residues can readily be accommodated in β sheets, where their side chains project out of the plane containing the main chain. Serine, aspartate, and asparagine tend to disrupt α helices because their side chains contain a hydrogen bond donor and acceptor in close proximity to the main chain, where they compete for main-chain NH and CO groups. Proline is not usually found inside α helices for steric reasons mentioned above, but it can be accommodated as the N-terminal residue of a helical segment. Glycine can readily fit into an α helix, but it does not favor helix formation because it has such a variety of conformational options.

Can one predict the secondary structure of proteins using this knowledge of the conformational preferences of amino acid residues? Predictions of secondary structure based on local sequence (e.g., a stretch of six or fewer residues) have proved to be about 60% accurate. What stands in the way of more accurate prediction? Note that the conformational preferences of amino acid residues (see Table 16-1) are not tipped all the way to one structure. For example, glutamate, one of the strongest helix formers, prefers α helix over β strand by only a factor of 2. The preference ratios of most other residues are smaller.

Indeed, a particular pentapeptide sequence can adopt an α-helical conformation in one protein and a β strand conformation in another (Figure 16-8). Hence, some amino acid sequences do not uniquely determine secondary structure. *The context is often crucial in determining the conformational outcome.* Tertiary interactions—interactions between residues that are far apart in the linear sequence—may be decisive in specifying the secondary structure of some segments.

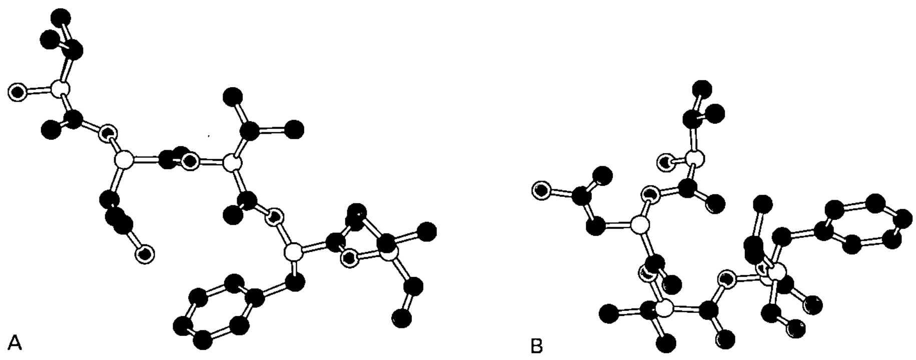

Figure 16-8
The same pentapeptide sequence (VNTFV) has very different conformations in (A) ribonuclease and (B) erythrocruorin, an insect hemoglobin. α Carbons are shown in green, and other atoms in standard colors. [Drawn from 2rns.pdb and 1ecd.pdb. W. Kabsch and C. Sander. *Proc. Nat. Acad. Sci.* 81(1984):1075.]

By contrast, *other sequences contain the information needed to specify a particular secondary structure.* A peptide corresponding to residues 1 to 13 of ribonuclease A, a helical region of the enzyme, is 25% α-helical in water at low temperature. This peptide exemplifies an *autonomous folding unit*—interactions directing its folding are inherent in its amino acid sequence. Robert Baldwin has found that stable α helices can be formed by relatively short peptides. For example, a 16-mer containing 13 alanines and, to confer water-solubility, 3 lysines, is 80% α-helical.

Acetyl-AAAAKAAAAKAAAAKA-NH_2

FOLDING MOTIFS (SUPERSECONDARY STRUCTURES) ARE FORMED FROM α HELICES AND β STRANDS

More than 60% of the polypeptide backbone consists of α helices and β strands. Proteins contain an abundance of these regularly repeating structures to fulfill two essential requirements. First, proteins must be compact to minimize exposure of hydrophobic residues to water. Second, most of their buried polar groups (such as main-chain NH and CO groups) must be hydrogen bonded to compensate for the cost of removing them from favorable interactions with water. The shape of α helices and β strands determines how they interact to form compact domains. The cylindrical contour of α helices enables them to stack around a central core or to form layered structures. β Strands associate to form β sheets, which can stack with one another or with α helices to form layered structures. *Motifs that arise by the association of α helices or β strands play a fundamental role in protein folding.* They are called *folding motifs* or *supersecondary structures.*

Let us first examine the packing of α helices. The side chains of an α helix form ridges that are separated by shallow grooves. *Two helices can neatly pack together if the ridges of one fit into the grooves of the other.* The two preferred orientations are those in which the angle between the axes of adjacent helices is +20° or −50° (Figure 16-9). The *four-helix bundle,* one of the simplest folding motifs, utilizes the +20° packing mode (Figure 16-10). In this bundle, helical segments that are adjacent in the amino acid sequence are *antiparallel* to one another. Antiparallel packing of neighboring structural units is generally favored because it costs less energy to bend a connecting loop by 180° than by 360°. We have already encountered the −50° packing mode in myoglobin and hemoglobin. In general, one side of a helix faces the hydrophobic interior, and the other is exposed to solvent. *This arrangement creates a hydrophobic core whose diameter is about 11 Å, the length of two side chains.*

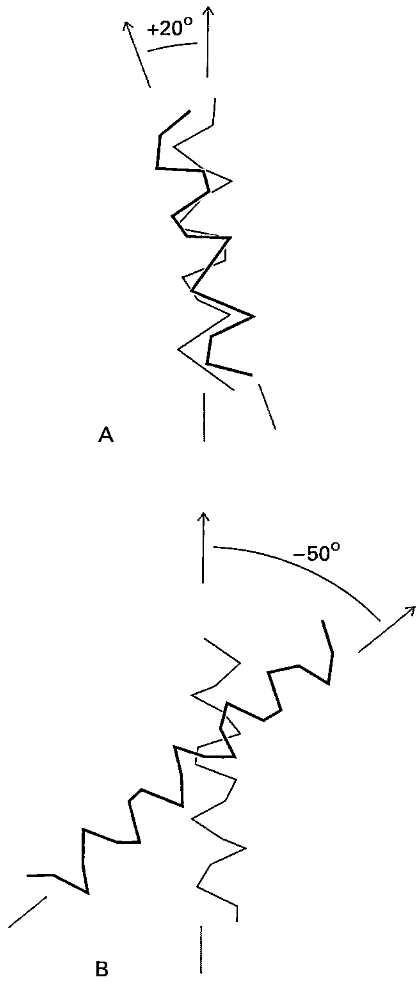

Figure 16-9
Two modes of packing of α helices. (A) +20°. (B) −50°. [After C. Chothia and A.V. Finkelstein. *Ann. Rev. Biochem.* 59(1990):1007.]

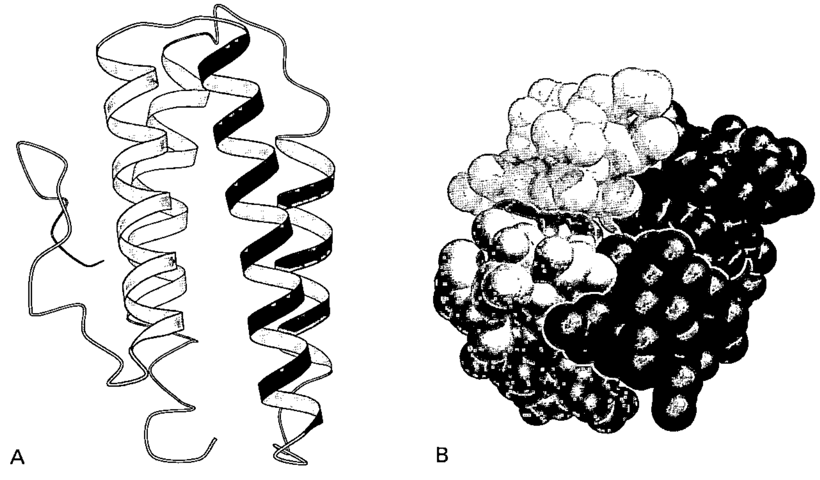

Figure 16-10
The four-helix bundle is the major structural motif of myohemerythrin, an oxygen-binding protein in worms. (A) Schematic diagram of the main-chain conformation. (B) Space-filling model of the four-helix bundle, shown in cross section. [Drawn from 2mhr.pdb. S. Sheriff, W.A. Hendrickson, and L.J. Smith. *J. Mol. Biol.* 197(1987):273.]

β Strands also readily form folding motifs. The interacting strands can be parallel or antiparallel. The simplest *β* motif is a *β hairpin,* which consists of *two antiparallel strands that are joined by a short loop* (Figure 16-11A). An extensive sheet can be formed from several *β* hairpins (Figure 16-11B). More complex patterns can be formed with longer loops. For example, four *β* strands can come together in an antiparallel fashion (Figure 16-11C) to form what is called a *Greek key motif* because the pattern occurs in ancient Greek ornaments. One side of an antiparallel *β* sheet typically faces the solvent, and the other faces a hydrophobic core, as in immunoglobulins (see Figure 14-14 on p. 370). The loops in these structures are almost always in contact with solvent.

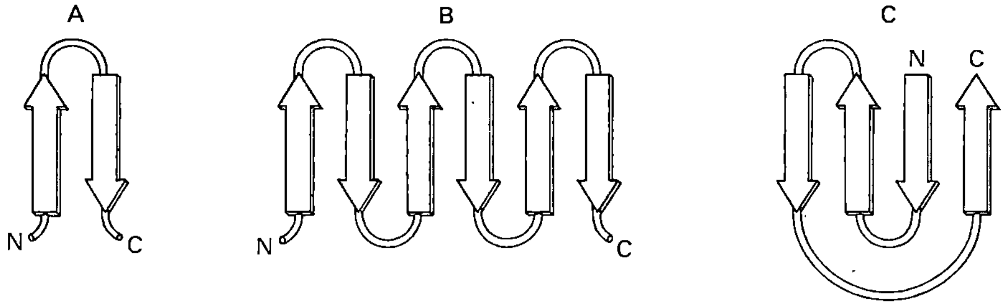

Figure 16-11
Schematic diagram of *β* structures. (A) *β* Hairpin. (B) *β* Sheet. (C) Greek key. [After C. Branden and J. Tooze. *Introduction to Protein Structure* (Garland, 1991), p. 21.]

β Strands can be paired in a parallel fashion if a crossover connection is made. The connector has to link opposite ends of the sheet and hence must be much longer than is required for antiparallel association. The crossover connector is usually an *α* helix (Figure 16-12). A supersecondary structure of this kind is called a *βαβ motif.* The hydrophobic side of the *β* strands packs tightly against the hydrophobic face of the *α* helix.

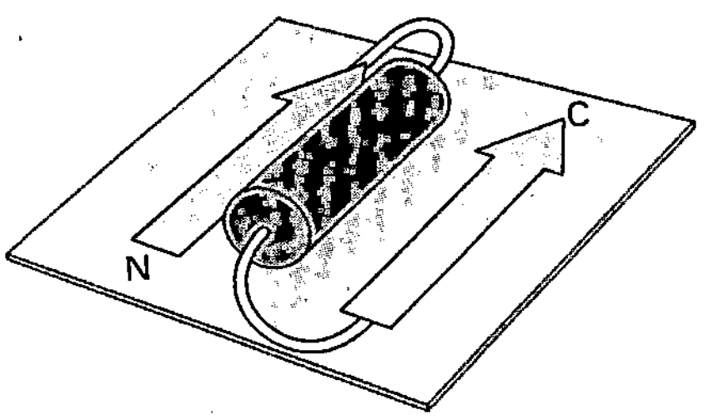

Figure 16-12
Schematic diagram of a *βαβ* motif. The *β* strands are shown in green, and the *α* helix joining them is shown in red. [After C. Branden and J. Tooze. *Introduction to Protein Structure* (Garland, 1991), p. 24.]

PARTIALLY FOLDED INTERMEDIATES CAN BE DETECTED, TRAPPED, AND CHARACTERIZED

To understand the folding process, we need to visualize partially folded states, preferably at atomic resolution. During the past decade, important advances have been made in identifying folding intermediates and in opening experimental windows on the folding process itself:

1. *Rapid-kinetic studies.* Folding is fast: many proteins fold in less than a second. Equilibrium studies are convenient but miss most of the action. Folding intermediates are generated by abruptly changing the external conditions to favor the refolding of an unfolded protein. Refolding can be observed starting at about 10 ms after mixing. For example, a solution containing an unfolded protein in 5 M guanidine hydrochloride is rapidly mixed with a buffer to lower the concentration of the denaturing agent. A progression of distinctive intermediates between the U and N states can then be detected by monitoring several spectroscopic properties, such as fluorescence and circular dichroism. In the folding of cytochrome *c,* an electron transport protein, 44% of the *α* helix content of the native protein is formed within 4 ms of mixing, the resolution time of the apparatus (Figure 16-13). The other 56% of *α* helix is formed progressively in the 10-ms to 1-s time range. By contrast, well-packed tertiary structure, as reflected in the circular dichroism of aromatic side chains, is formed much later. Several seconds are required for the finishing touches in tertiary structure, which are essential for enzymatic activity.

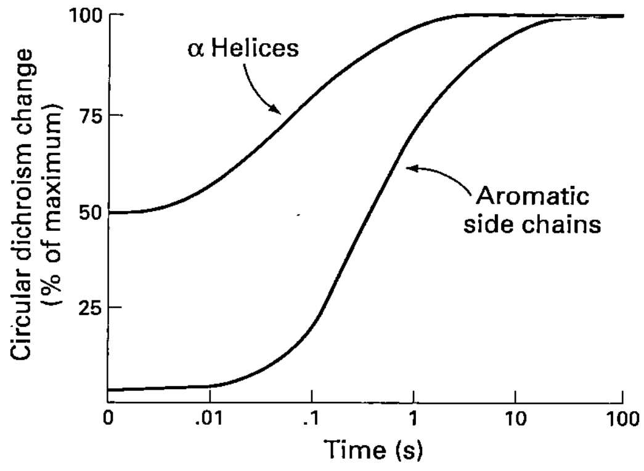

Figure 16-13
α Helix formation is more rapid than tertiary-structure rearrangements of aromatic side chains in the folding of cytochrome *c*. The kinetics of these structural changes were determined from circular dichroism measurements at 222 nm and 289 nm, respectively. [After G. Elöve, A.F. Chaffotte, H. Roder, and M.E. Goldberg. *Biochemistry* 31(1992):6879.]

2. *Trapping of disulfide-bonded intermediates.* Proteins containing disulfide bonds can be unfolded by reducing these cross-links. Folding is then initiated by removal of the reducing agent, which allows disulfides to reform by oxidation in air (Figure 16-14). Intermediates can be trapped covalently by blocking uncombined cysteines with iodoacetate, an alkylating agent (p. 57). After chromatographic separation, the disulfide bond pattern in each intermediate can be determined.

Figure 16-14
The sequence of formation of disulfide bonds in proteins can be determined by trapping free cysteine residues with iodoacetate. The *S*-carboxymethyl derivative of cysteine is stable.

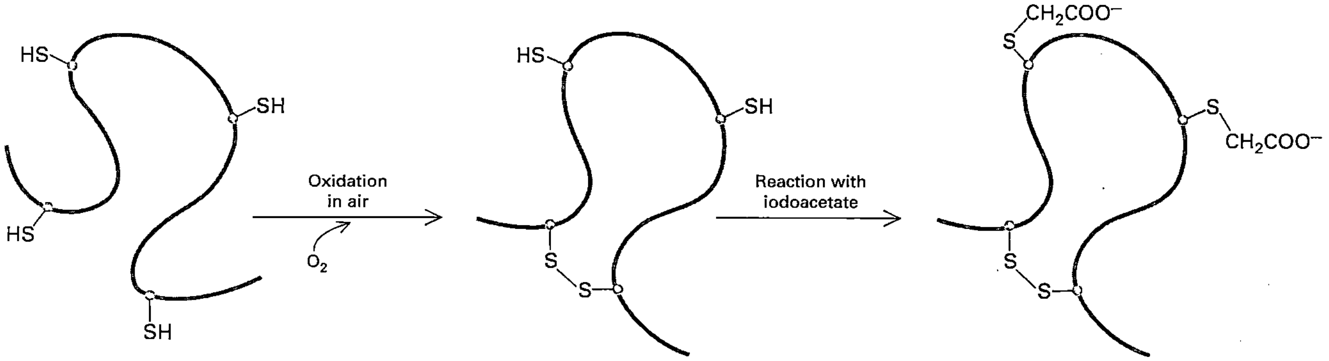

3. *Pulsed-label NMR studies.* Pulsed hydrogen-deuterium exchange and NMR spectroscopy have recently been brought together to unravel, at the atomic level, the intricacies of protein folding. A protein is unfolded in a D_2O-denaturant solution to change amide NH groups to ND groups (Figure 16-15). Refolding is then initiated by diluting the sample in D_2O to lower the concentration of denaturant. At first, most of the ND groups are accessible for exchange, but with time, they become protected as a result of the acquisition of secondary and tertiary structure. The degree

Figure 16-15
Pulsed hydrogen-deuterium exchange can be used to monitor the acquisition of secondary structure in protein folding.

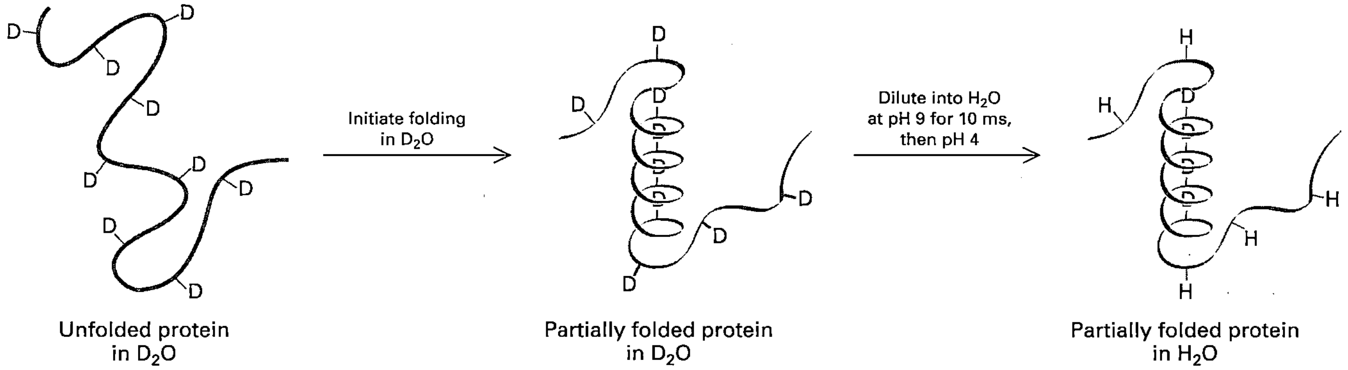

of protection is probed by diluting the sample in H_2O to exchange H for D at open sites. A high pH (e.g., 9) is chosen to promote rapid exchange (~1 ms for accessible protons). Labeling is stopped some 10 ms later by lowering the pH to 4. The exchange of amide protons is intrinsically very slow at this pH. The pattern of labeling is determined at leisure by two-dimensional NMR spectroscopy. Because individual NH resonances can be assigned, the precise locations of exchangeable sites are identified. In general, sites protected from rapid H-D exchange are located in α helices, β sheets, or β turns. In contrast, exchangeable amides do not have fixed hydrogen bond partners and hence are located in regions which have not yet folded, or which fluctuate very rapidly between folded and unfolded forms. By varying the time for folding before the pulse, *we can see the progression of protection and hence the acquisition of secondary structure, residue by residue.*

4. *Synthesis of folding kernels.* The entire polypeptide chain is not required for the folding of some segments. For example, two fragments from bovine pancreatic trypsin inhibitor can associate to form a 30-residue disulfide-bonded peptide having secondary and tertiary structure like that of the intact protein. By removing parts of a protein, we can discern the roles of different regions in the folding process.

5. *Site-specific mutagenesis.* Specific residues in a protein can be genetically altered to delineate their contribution to both stability and folding.

6. *Protein design.* One of the most stringent tests of our understanding of protein folding is the engineering of novel proteins with distinctive functions. Encouraging starts have been made in synthesizing new scaffolds, metal-binding proteins, channels, and catalysts.

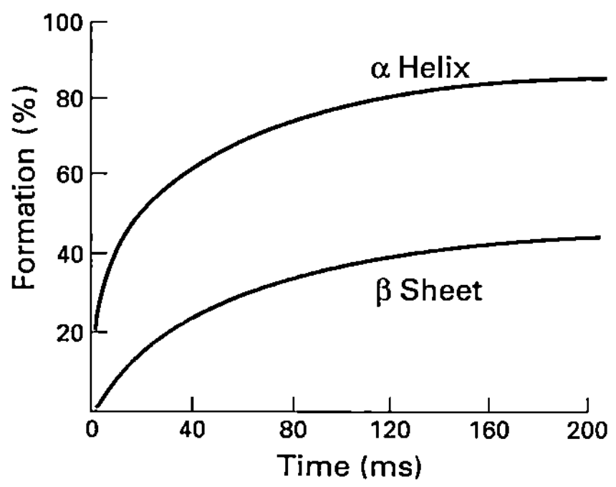

Figure 16-16
In the refolding of lysozyme, helix A is formed before the β sheet. Proton exchangeability was measured at different times after the initiation of folding. [After S.E. Radford, C.M. Dobson, and P.A. Evans. *Nature* 358(1992):303.]

LYSOZYME FOLDS ALONG MULTIPLE PARALLEL PATHWAYS

Pulsed exchange NMR spectroscopy has provided an informative picture of the folding of hen egg white lysozyme, a 129-residue enzyme that was discussed in detail earlier (p. 207). Lysozyme contains four α helices and a β sheet. Amides in helices become protected within about 60 ms of the onset of folding, whereas amides in the β sheet become protected 10 times more slowly (Figure 16-16). *It is evident that lysozyme does not fold in a single cooperative event. Rather, different regions become stabilized at different times.* The α helices form one domain of lysozyme, and the β sheet forms the other (Figure 16-17). Many of the tertiary interactions of the native

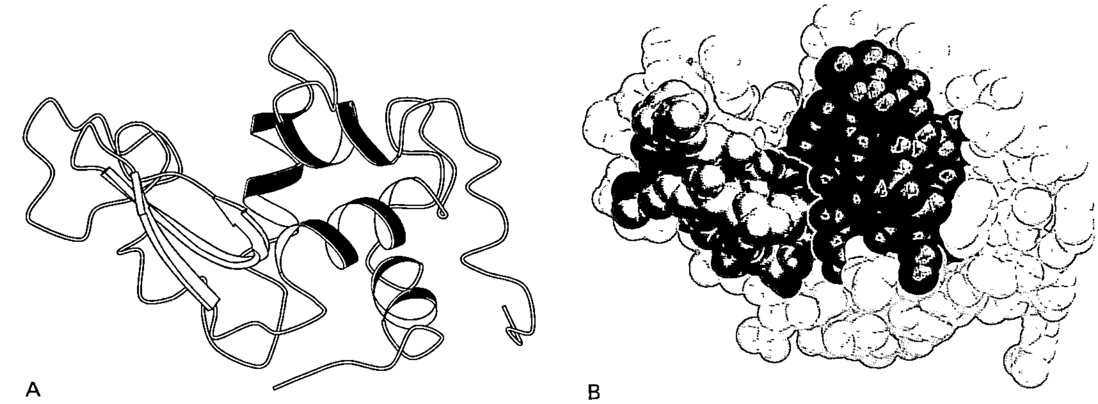

Figure 16-17
Structure of lysozyme showing the α helix domain (red) and the β sheet domain (green). The active site is located at the interface of these domains. (A) Schematic diagram. (B) Space-filling model. [Drawn from 6lyz.pdb. R. Diamond. *J. Mol. Biol.* 82(1974):371.]

protein are made after the helices and sheet are constructed and require the cooperative interplay of the two domains. Furthermore, detailed analysis of the folding kinetics shows that different lysozyme molecules in the same solution fold by kinetically distinct pathways. Folding does not occur along a single pathway with a unique sequence of intermediates. Rather, *unfolded lysozyme molecules refold along multiple parallel paths.*

NATIVE DISULFIDE-BONDED INTERMEDIATES PREDOMINATE IN THE FOLDING OF A TRYPSIN INHIBITOR

Another view of protein folding comes from studies of bovine pancreatic trypsin inhibitor (BPTI), a 6-kd protein that inhibits trypsin by inserting a lysine residue into the specificity pocket of the enzyme (see Figure 10-23 on p. 251). BPTI has been a choice protein for folding studies because of its simplicity. This 58-residue single-polypeptide chain is stabilized by three disulfide bonds, denoted by [5-55], [30-51], and [14-38] (Figure 16-18). BPTI can be reversibly unfolded by reducing these disulfides; the reduced protein does not inhibit trypsin. Addition of an oxidizing agent leads to disulfide bond formation and the restoration of inhibitory activity. An attractive feature of BPTI is that the state of its disulfide bonds, and hence its biological activity, can be controlled by using small thiols and disulfide derivatives, such as the reduced and oxidized forms of glutathione (p. 731). Furthermore, intermediates containing one, two, or three disulfide bonds can be isolated at different times after refolding is allowed to begin.

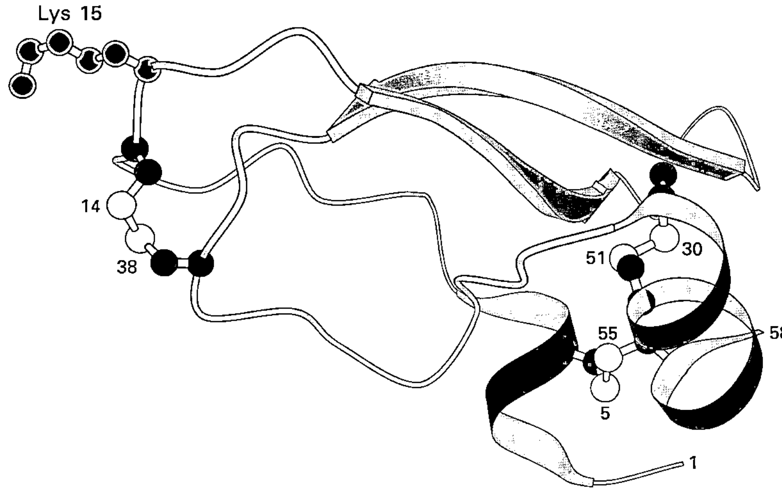

Figure 16-18
Schematic diagram of the structure of bovine pancreatic trypsin inhibitor (BPTI). The α helices are shown in red and the β strands in green. The three disulfide bonds (sulfur atoms in yellow) are marked. BPTI inhibits trypsin by inserting Lys 15 (blue) into the specificity pocket of the enzyme. [Drawn from 6pti.pdb. A. Wlodawer, J. Nachman, G.L. Gilliland, W. Gallagher, and C. Woodward. *J. Mol. Biol.* 198(1987):469.]

What is the sequence of disulfide bond formation in the refolding of the fully reduced protein? A total of 75 different species containing 1, 2, or 3 disulfide bonds are feasible. Only a small fraction of all potential pairings correspond to those of the native folded protein. Do the major intermediates in refolding contain native or nonnative disulfide bonds? These questions were answered in the following way. Refolding of the fully reduced, unfolded protein was initiated by adding oxidized glutathione and lowering the concentration of denaturing agent by dilution. At various times, aliquots were removed, and acid was added to shift the equilibrium from RS^- to RSH to block disulfide bond formation and rearrangement. The different species were then separated by high-pressure liquid chromatography (Figure 16-19), and their disulfide bond pattern was as-

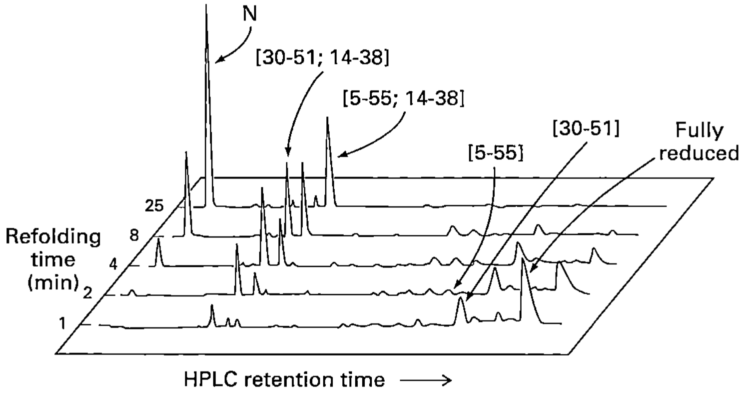

Figure 16-19
Intermediates in the folding of BPTI, starting with the fully reduced protein, can be trapped and characterized. Disulfide bond formation was quenched at the indicated times by addition of acid. The identities of the HPLC peaks were determined after free sulfhydryls were reacted with iodoacetate to prevent rearrangements. Only native disulfide bonds are present in the major species. [After J.S. Weissman and P.S. Kim. *Science* 253(1991):1390.]

certained after their free sulfhydryl groups were reacted with iodoacetate to form stable *S*-carboxymethyl adducts. These analyses revealed the identity of intermediates quenched at different times after the onset of refolding.

Reduced BPTI is oxidized initially to many one-disulfide intermediates, which rapidly rearrange to form two major species, [30-51] and [5-55] (Figure 16-20). The [5-55] intermediate inhibits trypsin, but it is much less stable than the native protein. A second disulfide bond forms very rapidly in both species. The productive intermediate, containing [30-51] and [14-38], does not readily form the third disulfide to complete folding. Though Cys 5 and Cys 55 are next to each other, their oxidation is very slow because they are buried. The [14-38] disulfide breaks to allow the [5-55] disulfide to form; this rearrangement takes several minutes. This new two-disulfide species rapidly acquires a third disulfide bond ([14-38]) to complete the folding process. Most significant, *all major intermediates in the folding of BPTI contain only native disulfide bonds. Thus, the same interactions that stabilize the final folded structure also guide the protein in reaching this conformation.*

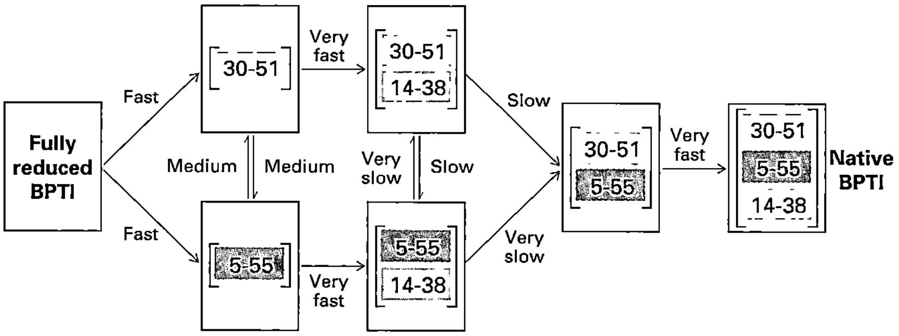

Figure 16-20
Folding pathway of BPTI in vitro. The very fast reactions occur in milliseconds, whereas the very slow ones occur in months. The species containing 5-55 and 14-38 disulfide bonds is kinetically trapped in the absence of enzymes. [After J.S. Weissman and P.S. Kim. *Science* 253(1991):1390.]

SUBDOMAINS CAN FOLD INTO NATIVE STRUCTURES

Is the entire polypeptide chain required for folding or can smaller portions adopt the native conformation? It has been appreciated for some time that domains can fold independently of one another. For example, an immunoglobulin F_{ab} domain can fold on its own. Recent studies indicate that *subdomains too can serve as structural kernels.* A synthetic peptide

P_α corresponding to residues 43 to 58 of BPTI, and another peptide P_β corresponding to residues 20 to 33, were synthesized. Cys 55 was replaced with alanine to ensure that only the [30-51] disulfide formed when an oxidant was added to a solution containing the two peptides. This 30-residue disulfide-bonded peptide pair was designed to mimic a key intermediate in the folding of BPTI. In fact, the peptide pair at 4°C adopts nearly the same secondary and tertiary structure as occurs in the intact native protein. Circular dichroism and NMR studies showed that the C-terminal helix, the central antiparallel β sheet, and the hydrophobic core of the native protein are present in the peptide pair (Figure 16-21).

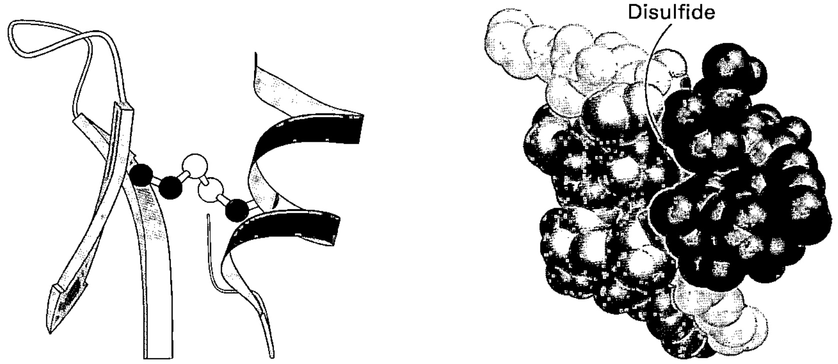

Figure 16-21
Structure of the $P_\alpha P_\beta$ peptide pair in intact BPTI. P_α (residues 43 to 58) forms an α helix (red), whereas P_β (residues 20 to 33) forms a β hairpin (green). (A) Schematic diagram. (B) Space-filling model. The 30-51 disulfide bond is buried. Only a small part of a sulfur atom is evident at the interface of the two peptides. [Drawn from 6pti.pdb. A. Wlodawer, J. Nachman, G.L. Gilliland, W. Gallagher, and C. Woodward. *J. Mol. Biol.* 198(1987):469.]

Likewise, a peptide pair joined by the [5-55] disulfide folds into an essentially native conformation at low temperature. Subdomains mimic portions of the native protein except that they are less stable because some reinforcing interactions are missing. Thus, the folding pathway for BPTI is largely determined by the propensity to form native-like subdomains in two key intermediates, the [30-51] and the [5-55] single-disulfide species. These studies of peptide models reinforce the conclusion that *the process of folding is guided by interactions that also stabilize the final native structure.*

PROTEIN FOLDING IN VIVO IS CATALYZED BY ISOMERASES AND CHAPERONE PROTEINS

The refolding of many proteins in vitro is much slower and less efficient than in vivo. The reason for this difference is that *protein folding in vivo is assisted by catalysts.*

1. The formation of correct disulfide pairings in nascent proteins is catalyzed by *protein disulfide isomerase (PDI).* PDI binds the polypeptide backbone of protein substrates. The enzyme preferentially interacts with peptides that contain cysteine residues but is otherwise undiscriminating. The broad substrate specificity of PDI enables it to speed the folding of diverse disulfide-containing proteins. By shuffling disulfide bonds, PDI enables proteins to quickly find the thermodynamically most stable pairings amongst those that are accessible. What is the catalytic mechanism of this enzyme? PDI contains two -Cys-Gly-His-Cys- sequences. The thiols of these cysteines are highly reactive at physiologic pH because they have a lower pK (7.3) than do most thiols in proteins (8.5). A disulfide bond of

the protein substrate is attacked by an RS^- group of the enzyme to form a covalent enzyme-substrate intermediate. The liberated thiol of the substrate is now free to attack another disulfide bond to form a different pairing (Figure 16-22). *PDI is especially important in accelerating disulfide interchange in kinetically trapped folding intermediates,* such as the BPTI intermediate containing [5-55] and [14-38] disulfides (see Figure 16-20). The isomerase accelerates disulfide shuffling 6000-fold in this intermediate.

2. Peptide bonds in proteins are nearly always in the trans configuration. The exceptions are X-Pro peptide bonds (X denotes any residue), of which 6% are cis (Figure 16-23; also see Figure 16-7). Prolyl isomerization is the rate-limiting step in the folding of many proteins in vitro.

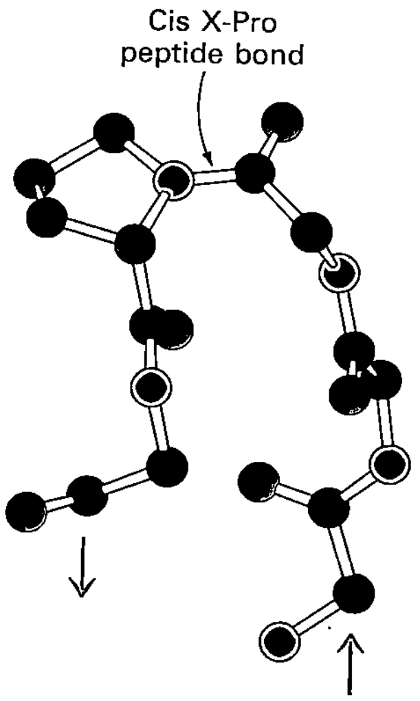

Figure 16-23
Cis-proline is well suited to making hairpin turns in polypeptide chains. [Drawn from 2rns.pdb. E.E. Kim, R. Varadarajan, H.W. Wyckoff, and F.M. Richards. *Biochemistry* 31(1992):12304.]

Spontaneous isomerization is slow because the bond between the carbonyl carbon and the amide nitrogen has partial double bond character (p. 27). *Peptidyl prolyl isomerases (PPIases) accelerate cis-trans isomerization more than 300-fold by twisting the peptide bond so that the C, O, and N atoms are no longer planar.* In this transition state, the C–N bond has more single bond character because resonance is minimized. Hence, the activation barrier for isomerization is lowered.

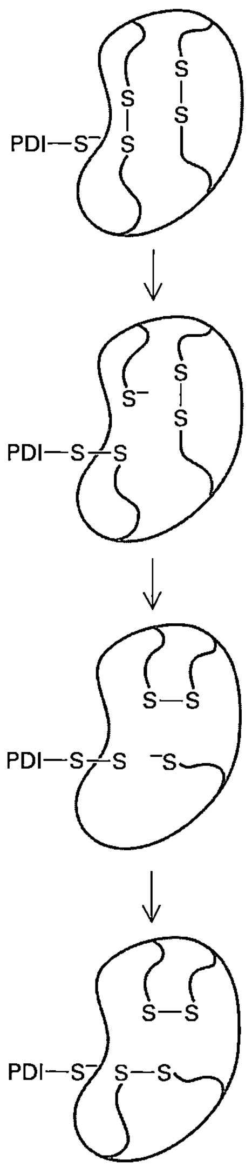

Figure 16-22
Protein disulfide isomerase (PDI) catalyzes disulfide interchange. A thiolate anion of the enzyme plays a key role in catalysis.

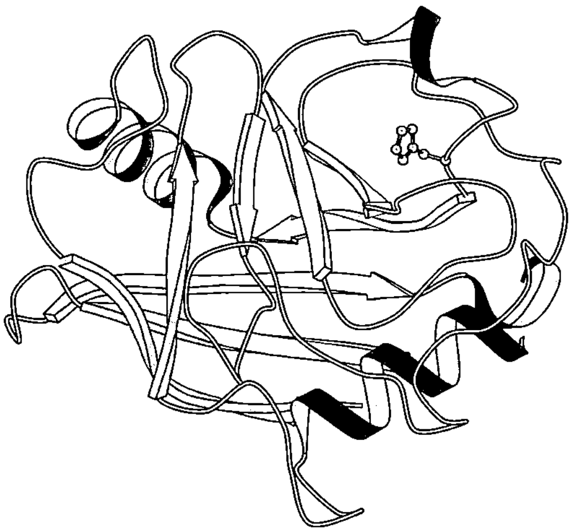

Figure 16-24
Schematic diagram of the structure of cyclophilin A, a peptidyl prolyl isomerase (PPIase). Polypeptide substrates bind to the surface of this β barrel. The staves of the barrel are formed by eight antiparallel β strands. α Helices are shown in red and β strands in green. A critical histidine residue at the catalytic site is shown in blue. [Drawn from 2cpl.pdb. H. Ke, L.D. Zydowsky, J. Liu, and C.T. Walsh. *Proc. Nat. Acad. Sci.* 88:9483-9487.]

3. Unfolded and partially folded proteins tend to aggregate when present at high concentration. The refolding of denatured proteins in vitro is usually carried out in dilute solution to minimize intermolecular contact and aggregation. Protein folding in vivo, however, effectively occurs in a very crowded milieu. A nascent protein in the cytosol or endoplasmic reticulum is surrounded by many potential molecular distractions. How are improper intermolecular interactions prevented in vivo? The answer is that several classes of proteins, picturesquely termed *molecular chaperones,* assist protein folding by inhibiting improper attachments and prying apart illegitimate liaisons. *Chaperones reversibly bind unfolded segments of polypeptide that might otherwise serve as centers of aggregation or diversion.* The hydrolysis of ATP powers the untangling of trapped intermediates. We will return to these fascinating proteins in a later chapter (p. 919).

MANY PROTEINS HAVE BEEN SELECTED IN EVOLUTION TO BE MARGINALLY STABLE

Thermodynamic studies have shown that the native state of a protein under physiologic conditions is typically only 5 to 10 kcal/mol more stable than the unfolded form. Is this relatively small difference inherent in the nature of proteins? Or have proteins been selected to be marginally stable? Site-specific mutagenesis studies have shown that proteins can be strengthened to resist thermal denaturation by the introduction of disulfide bonds. A suitably positioned disulfide can raise the melting temperature (T_m) of a protein such as T4 phage lysozyme by 4°C or more. Indeed, a mutant containing three additional disulfides has a T_m that is 23°C higher than the wild-type enzyme. Proteins that have a cavity in their hydrophobic core can be stabilized by substituting a large nonpolar residue for a small one (e.g., leucine for valine). Conversely, proteins can be destabilized by removing disulfide bonds, introducing bulky substituents that cannot readily be accommodated in the hydrophobic core, or replacing polar residues that make favorable hydrogen bonds or salt links.

Studies of the thermal stability of collagens have also been revealing. Recall that collagen is the major protein of connective tissue and that it forms a triple-stranded helix that is very different from the α helix (see Figure 2-40 on p. 32). The collagen triple helix is stabilized by hydrogen bonding between strands and by the steric locking effect of its proline and hydroxyproline residues, which keep ϕ near $-60°$. Collagens from different species have different melting temperatures (Table 16-2). The pattern is informative: collagens from icefish have the lowest T_m, whereas those from warm-blooded mammals have the highest. The T_m is raised by increasing the content of proline and hydroxyproline. In general, the

Table 16-2
Dependence of thermal stability on proline and hydroxyproline content

Source	Proline plus hydroxyproline (per 1000 residues)	Thermal stability (°C) T_s	Thermal stability (°C) T_m	Body temperature (°C)
Calf skin	232	65	39	37
Shark skin	191	53	29	24–28
Cod skin	155	40	16	10–14

Note: T_s is the shrinkage temperature of a collagen fiber, and T_m is the melting temperature of an isolated collagen molecule.

melting temperature of a collagen is only slightly above the body temperature of its source.

It is evident that protein stability can be precisely controlled. Indeed, many proteins have been selected to be marginally stable because it is biologically advantageous:

1. *Protein flexibility is at the heart of protein action.* Proteins must undergo conformational change to bind substrates, interact with other proteins, and carry out catalysis. Allosteric interactions are mediated by structural transitions that enable noncontiguous sites to communicate with each other. Likewise, proteins rapidly switch between different conformational states in serving as signal transducers and energy converters.

2. *Nascent proteins must not be allowed to fall into deep traps en route to the native state.* Intermediates in protein folding should be short-lived—the activation barrier to the next productive state must not be large.

3. *Partial unfolding enables proteins to be transported across membrane barriers.* The transport of a protein from one cellular compartment to another (e.g., from the cytosol into a mitochondrion) can occur only if the protein can unfold sufficiently to enable it to thread through a membrane channel (p. 916).

4. *Proteins are constructed so that they can readily be taken apart.* For example, collagen molecules are degraded in a controlled manner to remodel tissues. Large amounts of collagen in the tail fin of a tadpole are rapidly broken down during its metamorphosis to a frog (Figure 16-25). A specific collagenase cuts collagen across its three strands to form a 750-residue fragment and a 250-residue fragment. These fragments, but not the 1000-residue native molecule, spontaneously unfold at their ends and thereby become susceptible to degradation by other proteases. This controlled breakdown would not occur if the T_m of collagen and its fragments were much higher than the body temperature.

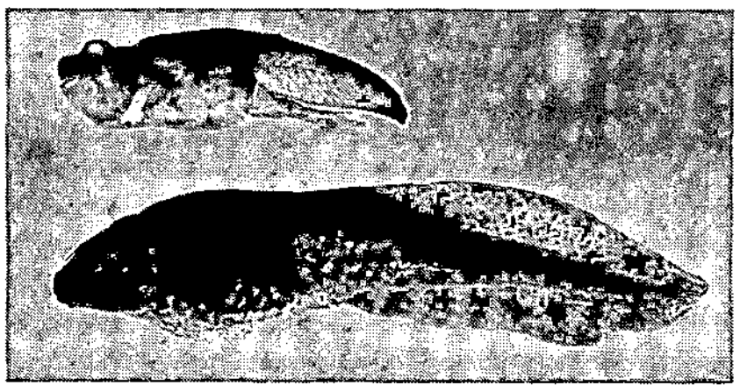

Figure 16-25
Collagen is rapidly removed from the tail fin of a tadpole during metamorphosis. [From C.M. Lapiere and J. Gross. In *Mechanisms of Hard Tissue Destruction,* R.F. Sognnaes, ed., p. 665. Copyright 1963 by the American Association for the Advancement of Science.]

VERY DIFFERENT AMINO ACID SEQUENCES CAN GENERATE STRIKINGLY SIMILAR PROTEIN FOLDS

The oxygen-carrying proteins of mammals, insects, and plants have very different amino acid sequences. Human hemoglobin, insect erythrocruorin, and root nodule leghemoglobin exhibit less than 20% sequence identity. However, *the three-dimensional structure of these evolutionarily distant globins is remarkably similar* (Figure 16-26). Indeed, a comparison of the sequences of 226 globins shows that only two residues—the proximal heme-linked histidine and a phenylalanine that is sandwiched against the heme group—are identical in all. *It is evident that many different sequences can give rise to essentially the same main-chain conformation.*

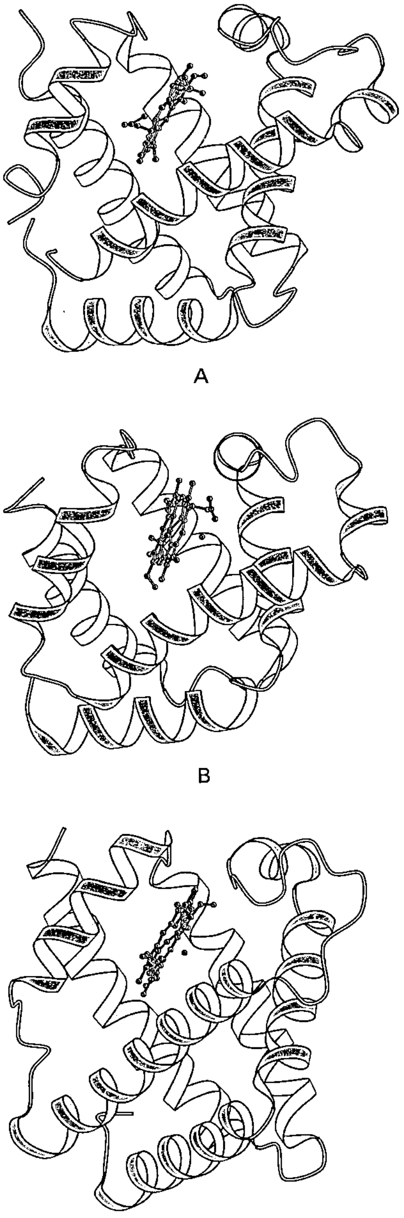

A

B

C

Figure 16-26
Proteins with different sequences (< 20% identity) can have very similar folds, as exemplified by the oxygen-carrying globins of mammals, insects, and plants. (A) Human hemoglobin. (B) Insect erythrocruorin. (C) Root nodule leghemoglobin.

What then determines whether a protein will adopt a globin fold? *The main-chain conformation appears to be specified by the pattern of nonpolar and polar residues.* Thirty-two positions in the sequence are nearly always hydrophobic; all these side chains are inside the protein. Surprisingly large changes in the volume of these residues can be tolerated. Helices can shift their position by several angstroms and their orientation by as much as 30 degrees to compensate for changes in the volume of the hydrophobic core. Another 32 positions are nearly always occupied by a charged or polar residue, or by Gly or Ala. These hydrophilic residues are located on the surface of the protein. *The distinctive hydrophilic versus hydrophobic pattern of these conserved residues, which comprise nearly half the total, distinguishes globins from all other proteins.*

Mutagenesis studies of λ repressor, a phage protein that controls gene expression (p. 959), were carried out to learn the range of allowable sequence changes in a functionally important region. A helix of one monomer packs against the corresponding helix of another monomer to form a dimer that binds to specific sites on DNA and silences gene expression (Figure 16-27A). Two or three residues at a time were simultaneously mutated by using mixtures of nucleotides to synthesize codons for all twenty amino acids at these positions. Functional proteins were then selected and sequenced. The important finding was that *most positions can be changed with retention of function* (Figure 16-27B). In particular, solvent-accessible residues can readily be substituted, whereas buried ones are much more conserved. For example, the activity of the repressor was unaffected by the replacement of an external glutamate with any of 12 other amino acids.

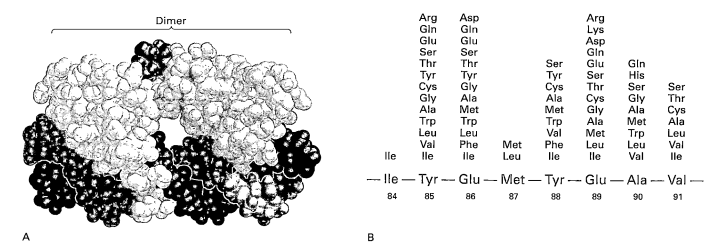

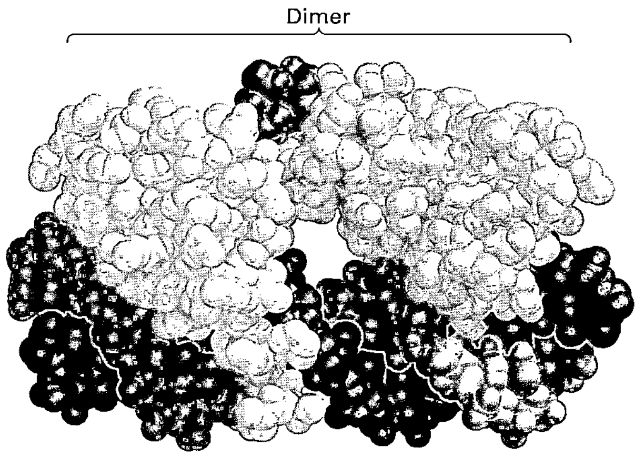

A

B

	Arg	Asp			Arg		
	Gln	Gln			Lys		
	Glu	Glu			Asp		
	Ser	Ser			Gln		
	Thr	Thr		Ser	Glu	Gln	
	Tyr	Tyr		Tyr	Ser	His	
	Cys	Gly		Cys	Thr	Ser	Ser
	Gly	Ala		Ala	Cys	Gly	Thr
	Ala	Met		Met	Gly	Ala	Cys
	Trp	Trp		Trp	Ala	Met	Ala
	Leu	Leu		Val	Met	Trp	Leu
	Val	Phe	Met	Phe	Leu	Leu	Val
Ile	Ile	Ile	Leu	Ile	Ile	Val	Ile
— Ile —	Tyr —	Glu —	Met —	Tyr —	Glu —	Ala —	Val —
84	85	86	87	88	89	90	91

Figure 16-27
Structure and mutagenesis of the repressor protein of λ phage.
(A) λ Repressor binds as a dimer (yellow and blue) to DNA (green and red). Helix 5 (shown in darker color) of one subunit interacts with helix 5 of the other subunit. (B) Functionally acceptable residues of a key part of helix 5, as determined by mutagenesis. [(A) Drawn from 1lmb.pdb. L.J. Beamer and C.O. Pabo. *J. Mol. Biol.* 227(1992):177. (B) After J.F. Reidhaar-Olson and R.T. Sauer. *Science* 241(1988)53.]

AN ENCOURAGING START HAS BEEN MADE IN PREDICTING THE THREE-DIMENSIONAL STRUCTURE OF PROTEINS

One of the goals of protein chemistry is to be able to predict the three-dimensional structure of a protein, given its amino acid sequence. Prediction is difficult because (1) a polypeptide chain has a vast number of potential conformations and (2) proteins are only marginally stable, which means that the free-energy difference between the unfolded and folded states is small. *No simple code relates the one-dimensional information of a sequence to the rich three-dimensional form of the folded protein.* The folding problem is inherently complex.

Significant progress is now being made in this challenging area of inquiry for several reasons. First, the structures of an increasingly large number of proteins are being solved by x-ray crystallography and NMR spectroscopy. More than 100 different protein folds are now known. Second, DNA sequencing is providing a wealth of amino acid sequence information. By comparing the sequences of homologous proteins, we can learn far more about the rules governing structure than from a single sequence. Third, sophisticated computer programs are taking advantage of the rapidly enlarging sequence and structure databases. Subtle patterns and relationships can be deduced and displayed. Fourth, predictions of structure can be rapidly tested. Theory and experiment have come together and are mutually reinforcing.

The first step in structure prediction is to ask whether the sequence of a new protein is similar to one whose three-dimensional structure is already known. *If two proteins are more than 40% identical in sequence, their backbone conformations are very likely to be nearly the same in the region of partial identity.* A pair of proteins differing markedly in sequence can have essentially the same backbone structure *if their hydrophobicity patterns are alike.* The structural similarity of root nodule leghemoglobin and human hemoglobin is a striking illustration (see Figure 16-26). *Functional motifs in proteins can also be identified by scanning amino acid sequences.* Calcium-binding EF hands (p. 348), for example, can be found by searching for a distinctive pattern of hydrophobic and oxygen-containing side chains in successive 29-residue segments of an amino acid sequence (Figure 16-28). X-ray crystallographic studies have shown that recoverin, a calcium sensor in vision, indeed contains EF hands (Figure 16-29), as was predicted from its amino acid sequence.

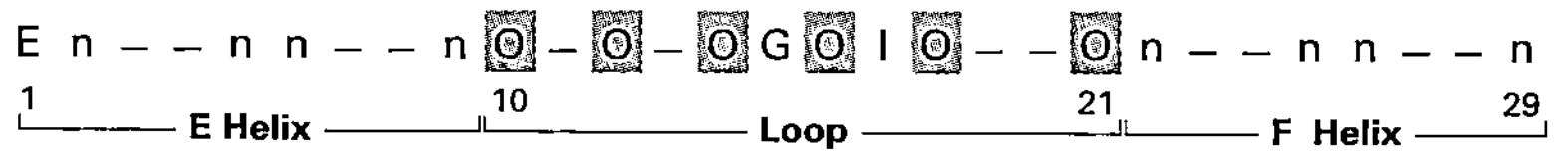

Figure 16-28
Consensus sequence for calcium-binding EF hands in proteins. EF hands can be detected by scanning amino acid sequences for this 29-residue motif. The symbol n (yellow) denotes a nonpolar side chain, and O (red) denotes an oxygen-containing side chain (Asp, Asn, Glu, Gln, Ser, Thr). The residues marked in red form the calcium-binding loop.

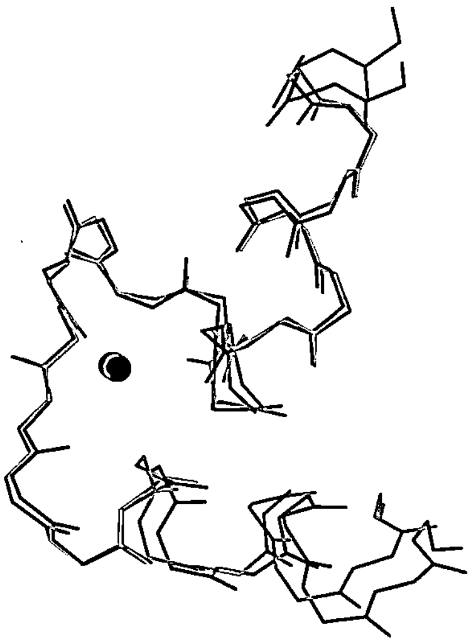

Figure 16-29
The EF hands of calmodulin (blue) and recoverin (red) are very similar in structure. Bound Ca^{2+} is shown as a sphere. The mean difference in the position of main-chain atoms is 0.8 Å. [Drawn from 3cln.pdb. Y.S. Babu, C.E. Bugg, and W.J. Cook. *J. Mol. Biol.* 204(1988):191; and 1rec.pdb. K.M. Flaherty, S. Zozulya, L. Stryer, and D.B. McKay. *Cell* 75(1993):709.]

Investigators are now tackling the most demanding problem, the prediction of structure in the absence of prior three-dimensional information about a related protein. By aligning many homologous sequences from different species, one can discern the pattern of essential hydrophobic and hydrophilic residues. This pattern is the starting point in identifying folding units. The accurate prediction of much of the backbone structure of the catalytic domain of protein kinase A before its crystal structure was solved is indicative of the rapid progress that is being made.

PROTEIN DESIGN TESTS OUR GRASP OF BASIC PRINCIPLES AND CREATES USEFUL NEW MOLECULES

The design and synthesis of novel proteins and of variations on nature's themes are important for several reasons:

1. De novo synthesis tests our understanding of fundamental principles. The ultimate criterion of the validity of a theory is its predictive

power. Can we construct proteins from scratch? And do these new constructs have the structural and functional properties that theory and modeling lead us to expect?

2. Naturally occurring proteins are remarkable catalysts, recognition devices, amplifiers, structural materials, and energy-transducing machines. However, their properties are not always optimal for medical, agricultural, or industrial applications. For example, greater thermal stability or resistance to extremes of pH is often desired. Can we use our knowledge of protein folding and function to improve and enhance what nature has already generated?

3. The most challenging task is to construct proteins with novel functions, to expand nature's repertoire in entirely new directions. For example, it would be very useful to construct catalysts for chemical reactions that have no counterpart in the biological world.

The de novo design of proteins is under way. The formation of a *parallel two-stranded rod* was a significant start. The inspiration came from the rod regions of myosin and tropomyosin. These highly elongated muscle proteins are *α-helical coiled coils,* in which each strand slowly winds around the other (see Figure 15-12 on p. 396). Recall that *coiled coils* contain seven-residue repeats called *heptads,* whose sequence is denoted by *abcdefg.* Sites *a* and *d* are nearly always occupied by large nonpolar residues, whereas the other sites are occupied by polar residues. The reason for this repeating arrangement is that sites *a* and *d* face the inside of the rod, whereas the other sites are exposed to solvent. These principles served as the basis for the design of a *heterodimeric coiled coil* consisting of a pair of 30-residue peptides, *α* and *β*.

b c d e f g a b c d e f g a b c d e f g a b c d e f g a b c
α Ac–A Q L E K E L Q A L E K E N A Q L E W E L Q A L E K E L A Q–NH_2
β Ac–A Q L K K K L Q A L K K K N A Q L K W K L Q A L K K K L A Q–NH_2

Both contain leucine at nearly all positions *a* and *d* to provide hydrophobic stabilization. They differ at positions *e* and *g*, where *α* contains glutamate, and *β* contains lysine (Figure 16-30). The reason for making *α* anionic and *β* cationic at these sites was to selectively stabilize the *αβ* dimer by electrostatic attraction. By contrast, *αα* and *ββ* were expected to be destabilized by electrostatic repulsion. Indeed, *α and β associate to form a stable parallel α-helical coiled coil.* Circular dichroism spectra showed that *αβ* is highly helical, whereas the individual peptides are random coils. The dissociation constant of *αβ* is 30 nM, indicating that the strands have high affinity for each other. Formation of the heterodimer is favored over

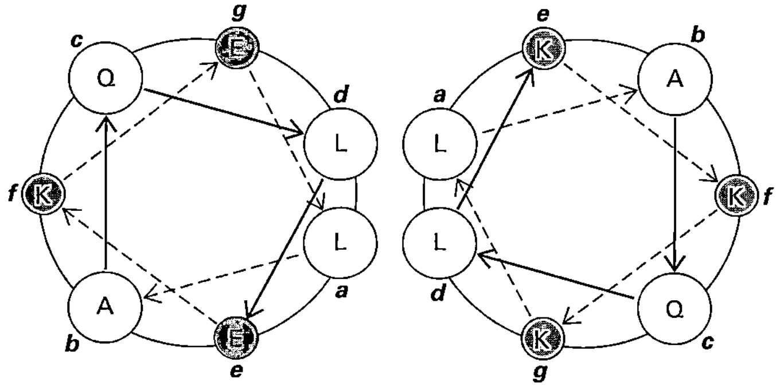

Figure 16-30
Helical wheel representation of an *α*-helical coiled coil. The location of each of the heptad sites is shown. The left strand (*α* peptide) has the sequence AQLEKEL . . . , and the right strand (*β* peptide) has the sequence AQLKKKL . . . , both starting at site *b*. The coiled coil is stabilized by hydrophobic interactions between leucines (yellow), and electrostatic interactions between glutamates (red) and lysines (blue). [After E.K. O'Shea, K.J. Lumb, and P.S. Kim. *Current Biol.* 3(1993):658.]

that of the homodimers by a factor of 10^5. Two proteins that do not normally associate could be brought together by attaching α to one of them and β to the other.

How can one construct a *water-soluble globular protein?* One approach is to build the protein from α helices because they are stabilized by short-range interactions. β Sheets are more difficult to make because they are held together by hydrogen bonds between distant segments of the sequence. Furthermore, nature has already shown that simple proteins can be built from *four antiparallel α helices,* as exemplified by myohemerythrin (see Figure 16-10). The strands in a four-helix bundle do not wind around each other as in a coiled coil. Rather, pairs of helices pack tightly in a limited region and then diverge because the angle between their axes is about 20°. *The key to making a four-helix bundle protein is to join four amphipathic helices through short linkers that can make hairpin turns* (Figure 16-31A).

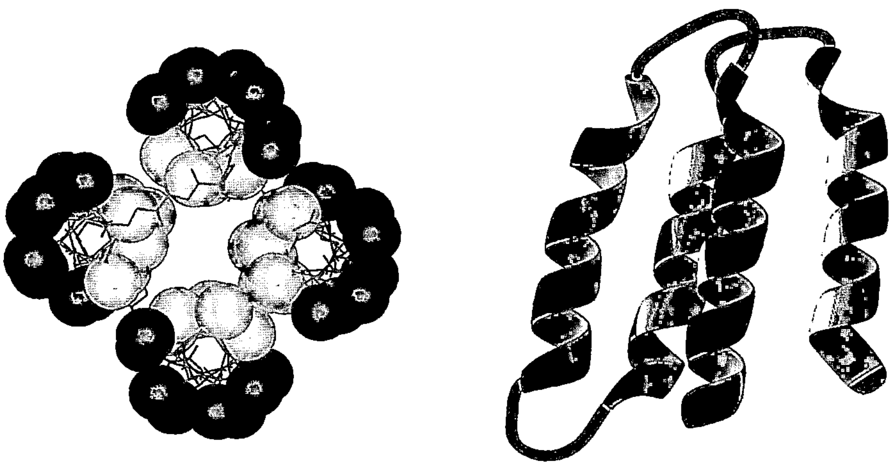

A B

Figure 16-31
Design of a four-helix bundle. (A) Helical wheel representation of the four helices. Yellow circles denote nonpolar residues, and red circles denote polar residues. The connections between helices and their ends are shown in blue. (B) Ribbon diagram of a four-helix bundle. [After S. Kamtekar, J.M. Schiffer, H. Xiong, J.M. Babik, and M.H. Hecht. *Science* 262(1993):1680.]

Amphipathic helices are required for two reasons: (1) The bundle is stabilized by hydrophobic interactions between the nonpolar faces of the four helices. (2) Each helix has a polar face to confer water-solubility and to prevent the protein from forming multimers. A 79-residue protein containing four potentially α-helical 14-residue segments joined by glycine-rich linkers was designed by students and faculty in a graduate seminar. The corresponding gene was synthesized and expressed in *Escherichia coli.* The resulting protein, named Felix, indeed behaves as a four-helix bundle as evidenced by its hydrodynamic properties, circular dichroism spectra, and susceptibility to reversible denaturation by guanidine hydrochloride. Subsequent molecular genetic studies have shown that many different amino acid sequences can give rise to four-helix bundles. The common theme is the pattern of hydrophilic and hydrophobic residues (Figure 16-31B).

A metal-binding protein with a β sheet framework has been designed and synthesized (Figure 16-32). The scaffold is like the one in the variable region of the immunoglobulin heavy chain (V_H). Recall that the immunoglobulin fold, a sandwich of β sheets, is a versatile framework for generating diversity (p. 370). The complementary-determining regions form loops that come together to form the antigen-binding site. A 61-residue protein was crafted from a spartan version of V_H consisting of just six antiparallel β strands connected by short loops. A zinc-binding site was fashioned by placing a histidine in one loop, and two histidines in the facing loop. These residues were chosen because an array of three histidines forms the zinc-binding site at the catalytic center of carbonic anhydrase (see Figure 2-55 on p. 39). The designed protein indeed folds into a β sandwich and binds Zn^{2+} with high affinity.

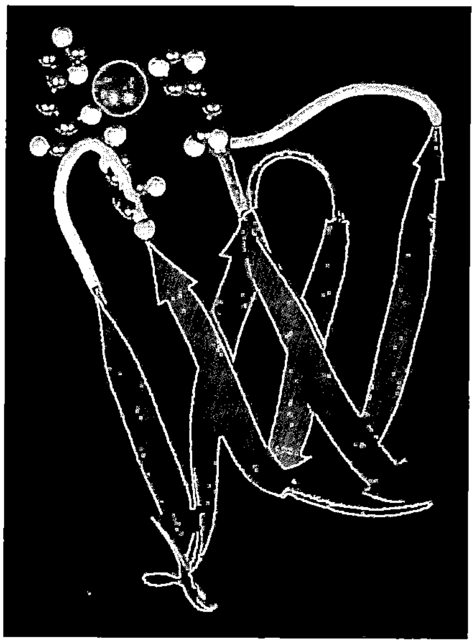

Figure 16-32
Schematic diagram of a metal-binding protein with a β sheet framework. The zinc ion shown in blue is coordinated to three histidine residues that emerge from loops between β strands. [Courtesy of Dr. Maurizio Sollazo and co-workers at IRBM, Italy. A. Pessi, E. Bianchi, A. Crameri, S. Venturini, A. Tramontano, and M. Sollazzo. *Nature* 362(1993):367.]

SUMMARY

The richness and complexity of protein architecture pose two fundamental questions: What is the relationship between amino acid sequence and three-dimensional structure? And what is the process by which an unfolded chain folds into the native biologically active form of the protein? A simple calculation shows that a random search of all possible conformations of a polypeptide chain to find the energetically most favorable state would take much too long ($\sim 10^{27}$ years). Rather, protein folding takes place in times of seconds to minutes by the progressive stabilization of intermediates that resemble parts of the final folded state. The native form of a protein is typically only 10 kcal/mol more stable than the unfolded form, which means that the average contribution of a single residue to stability is less than the magnitude of thermal energy. The native form of a protein is the consequence of many mutually reinforcing interactions. Many proteins have been selected to be marginally stable because flexibility is essential for protein folding, function, and removal.

Kinetic studies of the folding of many proteins have shown that the molten globule is an early intermediate. Molten globules contain much of the secondary structure but little of the well-packed tertiary structure of the native form. Interactions between secondary-structure units fluctuate rapidly in these intermediates. The driving forces for molten globule formation are hydrophobic interactions leading to compaction and the acquisition of secondary structure.

The formation of α helices and β strands is at the heart of the folding process. The conformational possibilities of the main chain are depicted in Ramachandran plots, which show allowed combinations of ϕ and ψ angles. Amino acid residues have different propensities for forming α helices, β strands, and turns. Alanine and leucine, for example, favor α helix formation, whereas valine and isoleucine favor β sheet formation. Glycine and proline promote the formation of turns. More than 60% of the secondary structure of most proteins consists of α helices and β strands. These structural elements come together to form folding motifs (supersecondary structure) such as four-helix bundles, β hairpins, and $\beta\alpha\beta$ units.

The detection and characterization of partially folded states has been vital in advancing our understanding of protein architecture and folding. Rapid-kinetic circular dichroism measurements have shown that α helices can form within a few milliseconds. The formation of defined tertiary structure is usually much slower. Disulfide-bonded intermediates can be trapped to follow the sequence of formation of these cross-links. Pulsed hydrogen-deuterium (H-D) exchange followed by NMR spectral measurements has enabled us to see the progressive acquisition of secondary structure, residue by residue, starting in times of milliseconds. Fragments of proteins have been synthesized to define autonomous folding units. For example, 15-residue peptides can fold into α helices. Site-specific mutagenesis has also provided a wealth of information about the contributions of specific residues to folding and stability.

Pulsed H-D exchange experiments have shown that different regions of proteins become stabilized at different times during folding. Lysozyme, for example, folds along multiple parallel paths; the α-helical domain is formed before the β sheet domain. The trapping of disulfide-bonded species in the folding of bovine pancreatic trypsin inhibitor has shown that all major intermediates contain only native disulfide bonds. Most important, protein folding is generally guided by interactions that stabilize the final native state.

Protein folding in cells is assisted by catalysts. Protein disulfide isomerase (PDI) accelerates disulfide interchange reactions several thousand-fold in kinetically trapped intermediates. Peptidyl prolyl isomerases (PPIases) markedly accelerate the isomerization of X-Pro peptide bonds, which can be cis or trans; nearly all other peptide bonds are trans. Unfolded and partially folded proteins tend to aggregate when present at high concentration. Chaperone proteins inhibit improper attachments and pry apart illegitimate liaisons. The unraveling action of many chaperones is driven by the hydrolysis of ATP.

The design and construction of proteins is under way. Parallel α-helical coiled coils can be formed by dimerization of polypeptides containing heptad repeats; hydrophobic stabilization is provided by interactions between leucines at interior positions of the two strands. Antiparallel packing of four amphipathic helical segments residing in a single polypeptide chain gives a four-helix bundle that is globular and water-soluble. A β sheet sandwich can also serve as a framework in making a globular protein. A metal-binding site was crafted by placing three histidines in loops at one end of the six-stranded β sheet. Protein design is important because it tests our understanding of fundamental principles and expands nature's repertoire for medical, industrial, and agricultural uses.

SELECTED READINGS

Where to start

Richards, F.M., 1991. The protein folding problem. *Sci. Amer.* 264:54–57.

Baldwin, R.L., 1989. How does protein folding get started? *Trends Biochem. Sci.* 14:291–294.

Gething, M.-J., and Sambrook, J., 1992. Protein folding in the cell. *Nature* 355:33–45.

DeGrado, W.F., Wasserman, Z.R., and Lear, J.D., 1989. Protein design, a minimalist approach. *Science* 243:622–628.

Richardson, J.S., Richardson, D.C., Tweedy, N.B., Gernert, K.M., Quinn, T.P., Hecht, M.H., Erickson, B.W., Yan, Y., McClain, R.D., Donlan, M.E., and Surfes, M.C., 1992. Looking at proteins: Representations, folding, packing, and design. *Biophys. J.* 63:1186–1209. [A beautifully illustrated and stimulating account of basic principles.]

Reviews

Kim, P.S., and Baldwin, R.L., 1990. Intermediates in the folding reactions of small proteins. *Ann. Rev. Biochem.* 59:631–660.

Kuwajima, K., 1989. The molten globule state as a clue for understanding the folding and cooperativity of globular-protein structure. *Proteins* 6:87–103.

Scholtz, J.M., and Baldwin, R.L., 1992. The mechanism of alpha helix formation by peptides. *Ann. Rev Biophys. Biomol. Struct.* 21:95–118.

Goldberg, M.E., 1991. Investigating protein conformation, dynamics and folding with monoclonal antibodies. *Trends Biochem. Sci.* 16:358–362.

Creighton, T.E., 1990. Protein folding. *Biochem. J.* 270:1–16.

Bowie, J.U., Reidhaar, O.J., Lim, W.A., and Sauer, R.T., 1990. Deciphering the message in protein sequences: Tolerance to amino acid substitutions. *Science* 247:1306–1310.

Seckler, R., and Jaenicke, R., 1992. Protein folding and protein refolding. *FASEB J.* 6:2545–2552.

Goldenberg, D.P., 1992. Native and non-native intermediates in the BPTI folding pathway. *Trends Biochem. Sci.* 17:257–261.

Books

Creighton, T.E., 1992. *Proteins: Structure and Molecular Properties* (2nd ed.). W.H. Freeman.

Branden, C., and Tooze, J., 1991. *Introduction to Protein Structure.* Garland.

Creighton, T.E. (ed.), 1992. *Protein Folding.* W.H. Freeman.

Gierasch, L.M., and King, J. (eds.), 1990. *Protein Folding: Deciphering the Second Half of the Genetic Code.* American Association for the Advancement of Science.

Lesk, A.M., 1991. *Protein Architecture: A Practical Approach.* IRL Press.

Langone, J.J. (ed.), 1991. *Meth. Enzymol.,* vol. 202. [Deals with protein design and modeling.]

Dawkins, R., 1987. *The Blind Watchmaker.* Norton. [Chapter 3 presents an engaging and perceptive account of the power of cumulative selection. The monkey at the typewriter analogy is given there.]

Folding intermediates

Englander, S.W., and Kallenbach, N.R., 1984. Hydrogen exchange and structural dynamics of proteins and nucleic acids. *Quart. Rev. Biophys.* 16:521–655.

Radford, S.E., Dobson, C.M., and Evans, P.A., 1992. The folding of hen lysozyme involves partially structured intermediates and multiple pathways. *Nature* 358:302–307.

Weissman, J.S., and Kim, P.S., 1991. Reexamination of the folding of BPTI: Predominance of native intermediates. *Science* 253:1386–1393.

Leopold, P.E., Montal, M., and Onuchic, J.N., 1992. Protein folding funnels: A kinetic approach to the sequence-structure relationship. *Proc. Nat. Acad. Sci.* 89:8721–8725.

Zwanzig, R., Szabo, A., and Bagchi, B., 1992. Levinthal's paradox. *Proc. Nat. Acad. Sci.* 89:20–22. [Presents a simple theory showing how stabilization of correct intermediates can greatly reduce the search time.]

Chaffotte, A.F., Guillou, Y., and Goldberg, M.E., 1992. Kinetic resolution of peptide bond and side chain far-uv circular dichroism during the folding of hen egg white lysozyme. *Biochemistry* 31:9694–9702.

Serrano, L., Matouschek, A., and Fersht, A.R., 1992. The folding of an enzyme. VI. The folding pathway of barnase: Comparison with theoretical models. *J. Mol. Biol.* 224:847–859.

Daggett, V., and Levitt, M., 1992. A model of the molten globule state from molecular dynamics simulations. *Proc. Nat. Acad. Sci.* 89:5142–5146.

Staley, J.P., and Kim, P.S., 1990. Role of a subdomain in the folding of bovine pancreatic trypsin inhibitor. *Nature* 344:685–688.

Structural motifs and folding determinants

Chothia, C., and Finkelstein, A.V., 1990. The classification and origin of protein folding patterns. *Ann. Rev. Biochem.* 59:1007–1039.

Richardson, J.S., 1981. The anatomy and taxonomy of protein structure. *Advan. Prot. Chem.* 34:167–339.

Richardson, J.S., 1977. β-Sheet topology and the relatedness of proteins. *Nature* 268:495–500.

Bashford, D., Chothia, C., and Lesk, A.M., 1987. Determinants of a protein fold. Unique features of the globin amino acid sequence. *J. Mol. Biol.* 196:199–216.

Protein stability

Wilson, K.P., Malcolm, B.A., and Matthews, B.W., 1992. Structural and thermodynamic analysis of compensating mutations within the core of chicken egg white lysozyme. *J. Biol. Chem.* 267:10842–10849.

Sauer, R.T., and Lim, W.A., 1991. The role of internal packing interactions in determining the structure and stability of a protein. *J. Mol. Biol.* 219:359–376.

Lee, C., and Levitt, M., 1991. Accurate prediction of the stability and activity effects of site-directed mutagenesis on a protein core. *Nature* 352:448–491.

Isomerases and chaperones in protein folding

Schmid, F.X., 1991. Catalysis and assistance of protein folding. *Curr. Opin. Struct. Biol.* 1:36–41.

Noiva, R., and Lennarz, W.J., 1992. Protein disulfide isomerase: A multifunctional protein resident in the lumen of the endoplasmic reticulum. *J. Biol. Chem.* 267:3553–3556.

Schonbrunner, E.R., and Schmid, F.X., 1992. Peptidyl-prolyl cis-trans isomerase improves the efficiency of protein disulfide isomerase as a catalyst of protein folding. *Proc. Nat. Acad. Sci.* 89:4510–4513.

Ellis, R.J., and van der Vies, S.M., 1991. Molecular chaperones. *Ann. Rev. Biochem.* 60:321–347.

Protein design

Cohen, C., and Parry, D.A., 1990. Alpha-helical coiled coils and bundles: How to design an alpha-helical protein. *Proteins* 7:1–15.

Hecht, M.H., Richardson, J.S., Richardson, D.C., and Ogden, R.C., 1990. De novo design, expression, and characterization of Felix: A four-helix bundle protein of native-like sequence. *Science* 249:884–891.

Regan, L., and DeGrado, W.F., 1988. Characterization of a helical protein designed from first principles. *Science* 241:976–978.

Lear, J.D., Wasserman, Z.R., and DeGrado, W.F., 1988. Synthetic amphiphilic peptide models for protein ion channels. *Science* 240:1177–1181.

Grove, A., Iwamoto, T., Montal, M.S., Tomich, J.M., and Montal, M., 1992. Synthetic peptides and proteins as models for pore-forming structure of channel proteins. *Meth. Enzymol.* 207:510–525.

Pessi, A., Bianchi, E., Crameri, A., Venturini, S., Tramontano, A., and Sollazzo, M., 1993. A designed metal-binding protein with a novel fold. *Nature* 362:367–369.

Kametekar, S., Schiffer, J.M., Xiong, H., Babik, J.M., and Hecht, M.H., 1993. Protein design by binary patterning of polar and nonpolar amino acids. *Science* 262:1680–1685.

Regan, L., 1993. The design of metal-binding sites in proteins. *Ann. Rev. Biophys. Biomol. Struct.* 22:257–281.

PROBLEMS

1. *Motif hunt.* Calmodulin contains four EF hands. Find them in the amino acid sequence shown below.

1	MADQLTEEQIAEFKEAFSLFDKDGDGTITT	30
31	KELGTVMRSLGQNPTEAELQDMINEVDADG	60
61	NGTIDFPEFLTMMARKMKDTDSEEEIREAF	90
91	RVFDKDGNGYISAAELRHVMTNLGEKLTDE	120
121	EVDEMIREADIDGDGQVNYEEFVQMMTAK	149

2. *Multiple alanine mutagenesis.* A solvent-exposed α helix of bacteriophage T4 lysozyme contains three alanine residues. A variant with alanine at 10 consecutive positions of this helix folds normally and has nearly the same enzymatic activity as the wild type. Also, the backbone conformation of this multiple mutant is essentially identical with that of the wild type.
   (a) What does this mutant reveal about the information content of the amino acid sequence of the wild type enzyme?
   (b) Substitution of a buried leucine of this helix with alanine lowers the melting temperature of the protein by 8°C. Why?
   (c) In contrast, replacement of an exposed serine with alanine raises the melting temperature by 1°C. Why?
   (d) How might multiple alanine mutagenesis be used to simplify the protein folding problem?

3. *Circularly permuted sequences.* Phosphoribosyl anthranilate isomerase, an enzyme that participates in the synthesis of tryptophan (p. 725), folds as an $\alpha\beta$ barrel in which the amino and carboxy termini are in close proximity. A mutant was constructed in which these termini were joined and new ones were generated at a different site. This mutant has essentially the same sequence as the wild type except that the order of secondary structure elements with respect to the ends of the chain is different. This circularly permuted variant was enzymatically active. What does this finding reveal about the folding process?

4. *Bispecific antibodies.* The rearrangement of *V, D,* and *J* segments to form a functional immunoglobulin gene occurs on only one chromosome because the protein product suppresses the rearrangement of the other chromosome. Hence, all antigen-binding sites made by a single cell are the same. However, antibodies containing two binding sites with different specificities can be produced in the laboratory.
   (a) How might two F_{ab} units be joined to produce a bispecific antibody?
   (b) How might a bispecific antibody be used therapeutically?

5. *Immunosuppression.* Cyclosporin A (Figure 16-33), a cyclic 11-residue peptide from fungi, is widely used to inhibit graft rejection following organ transplantation. This immunosuppressant acts by blocking the activation of cytotoxic T lymphocytes. The target is calcineurin, a calcium-sensitive protein phosphatase. However, cyclosporin must be activated by cyclophilin (see Figure 16-24) before it can inhibit the phosphatase. How might cyclophilin activate cyclosporin?

6. *Molecular design.* Propose a design for a membrane channel.

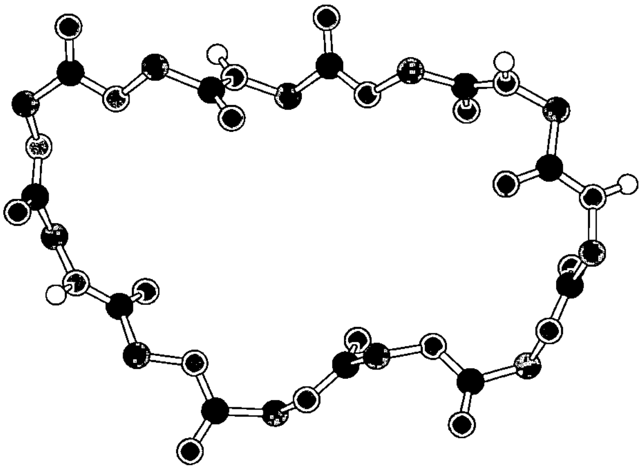

Figure 16-33
Model of cyclosporin A (CsA). The main chain of this 11-residue cyclic peptide is shown here. [Drawn from 2cys.pdb. C. Weber, G. Wider, B. Von Freyberg, R. Traber, W. Braun, H. Widmer, and K. Wuthrich. *Biochemistry* 30(1991):6563.]

PART

III

Metabolic Energy: Generation and Storage

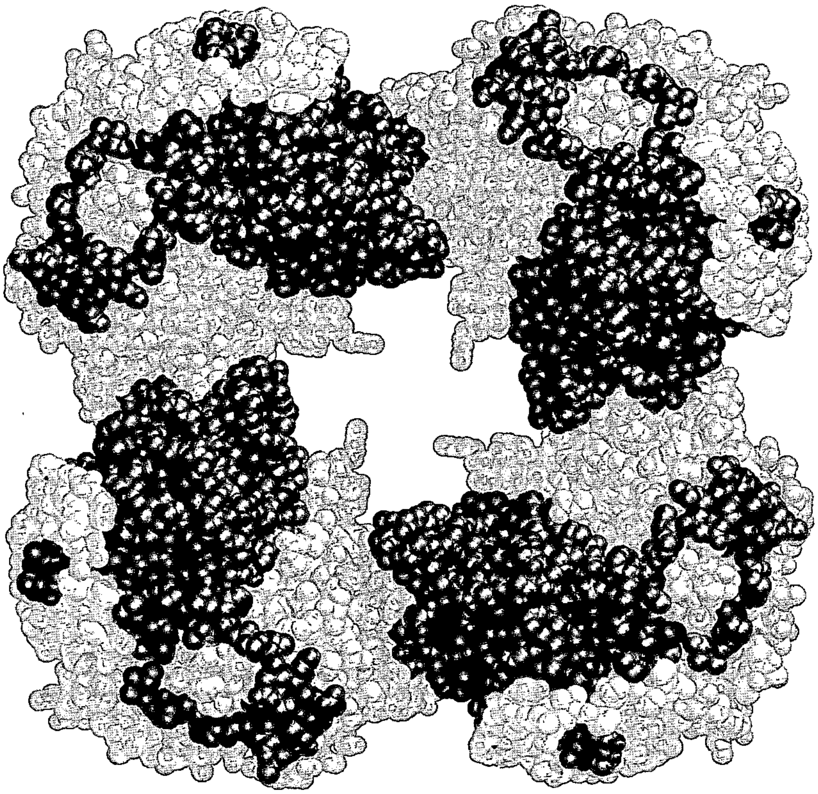

Structure of the transacetylase core of the pyruvate dehydrogenase complex. This multienzyme assembly catalyzes the irreversible funneling of the product of glycolysis into the citric acid cycle, the final common pathway for the oxidation of fuel molecules. The core consists of 24 identical chains. Four of the eight trimers can be seen in this view. [Drawn from 1eaa.pdb. A. Mattevi, G. Oblomolova, E. Schulze, K.H. Kalk, A.H. Westphal, A. de Kok, and W.G.J. Hol. Science 255(1992):1544.]

CHAPTER

17

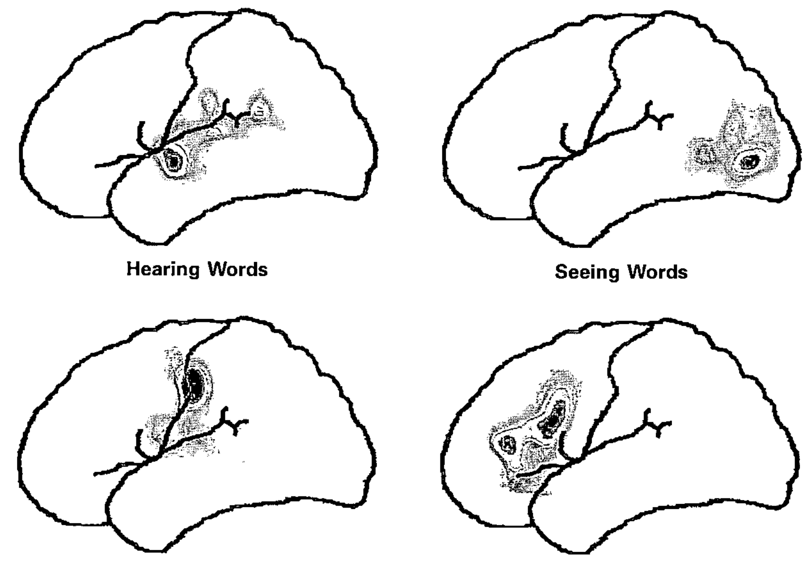

Metabolism: Basic Concepts and Design

The concepts of conformation and dynamics developed in Part II—especially those dealing with the specificity and catalytic power of enzymes, the regulation of their catalytic activity, and the transport of molecules and ions across membranes—enable us to turn now to two major questions of biochemistry:

1. *How do cells extract energy and reducing power from their environment?*
2. *How do cells synthesize the building blocks of their macromolecules?*

These processes are carried out by a highly integrated network of chemical reactions, which are collectively known as *metabolism.*

More than a thousand chemical reactions take place in even as simple an organism as *Escherichia coli.* The array of reactions may seem overwhelming at first glance. However, closer scrutiny reveals that metabolism has a *coherent design containing many common motifs.* The number of reactions in metabolism is large, but the number of *kinds* of reactions is relatively small. The mechanisms of these reactions are usually quite simple. For example, a double bond is often formed by dehydration of an alcohol. Furthermore, a group of about a hundred molecules plays a central role in all forms of life. Metabolic pathways are also regulated in common ways. The purpose of this chapter is to introduce some general principles and motifs of metabolism.

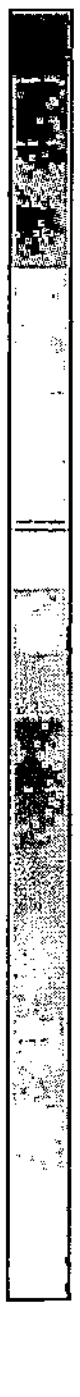

Opening Image: *Metabolic images of the human brain. Specific cognitive acts increase the metabolic activity and blood flow in specific regions of the brain (yellow and red denote highest activity). These images were obtained by positron emission tomography (PET scanning), a technique that detects the spatial distribution of an isotope such as* ^{15}O *following intravenous administration of* $H_2{}^{15}O$*. The scale at the right shows the color coding of the relative increase in blood flow. [Courtesy of Dr. Marcus Raichle.]*

A THERMODYNAMICALLY UNFAVORABLE REACTION CAN BE DRIVEN BY A FAVORABLE ONE

As was discussed earlier (p. 185), free energy is the most useful thermodynamic concept in biochemistry. A reaction can occur spontaneously only if ΔG, the change in free energy, is negative. Recall that ΔG for the formation of products C and D from substrates A and B is given by

$$\Delta G = \Delta G^\circ + RT \log_e \frac{[\mathrm{C}][\mathrm{D}]}{[\mathrm{A}][\mathrm{B}]}$$

Thus, the ΔG of a reaction depends on the *nature* of the reactants (expressed by the ΔG° term, the standard free-energy change) and on their *concentrations* (expressed by the logarithmic term). The standard free-energy change at pH 7 is denoted by $\Delta G^{\circ\prime}$.

An important thermodynamic fact is that *the overall free-energy change for a chemically coupled series of reactions is equal to the sum of the free-energy changes of the individual steps.* Consider the reactions

$$\begin{array}{ll} \mathrm{A} \rightleftharpoons \mathrm{B} + \mathrm{C} & \Delta G^{\circ\prime} = +5 \text{ kcal/mol} \\ \underline{\mathrm{B} \rightleftharpoons \mathrm{D} \qquad\quad} & \underline{\Delta G^{\circ\prime} = -8 \text{ kcal/mol}} \\ \mathrm{A} \rightleftharpoons \mathrm{C} + \mathrm{D} & \Delta G^{\circ\prime} = -3 \text{ kcal/mol} \end{array}$$

Under standard conditions, A cannot be spontaneously converted into B and C because ΔG is positive. However, the conversion of B into D under standard conditions is thermodynamically feasible. Because free-energy changes are additive, the conversion of A into C and D has a $\Delta G^{\circ\prime}$ of −3 kcal/mol, which means that it can occur spontaneously under standard conditions. Thus, *a thermodynamically unfavorable reaction can be driven by a thermodynamically favorable reaction that is coupled to it.* In this example, the reactions are coupled by the *shared chemical intermediate B.*

An uphill reaction can be coupled to a downhill reaction in two other principal ways. An *activated protein conformation* can store free energy, which can then be used to drive a thermodynamically unfavorable reaction. We have already encountered many examples of proteins as energy conversion devices. Molecular motors such as myosin, kinesin, and dynein convert the phosphoryl potential of ATP into mechanical energy (Chapter 15). The active transport of Na^+ and K^+ across membranes is driven by the phosphorylation of the sodium-potassium pump by ATP and its subsequent dephosphorylation (p. 311). This reaction cycle changes the affinity of the pump for the transported ions and the orientation of its ion-binding sites with respect to the inside and outside of the cell. The harvesting of photons by bacteriorhodopsin to pump protons across the cell's membrane further illustrates the diversity and power of proteins to transduce free energy (p. 318).

Ionic gradients across membranes also serve as a versatile means of coupling uphill reactions to downhill ones. The electrochemical potential of Na^+, for example, can be tapped to pump Ca^{2+} out of cells or to transport nutrients such as sugars and amino acids into cells (Chapter 12). As will be discussed in subsequent chapters, proton gradients produced by the oxidation of fuel molecules or by photosynthesis ultimately power the synthesis of most of the ATP in cells. Membrane-bound proteins mediate all these energy conversion processes by undergoing cycles of conformational transitions.

ATP IS THE UNIVERSAL CURRENCY OF FREE ENERGY IN BIOLOGICAL SYSTEMS

Living things require a continual input of free energy for three major purposes: the performance of mechanical work in muscle contraction and other cellular movements, the active transport of molecules and ions, and the synthesis of macromolecules and other biomolecules from simple precursors. The free energy used in these processes, which maintain an organism in a state that is far from equilibrium, is derived from the environment. *Chemotrophs* obtain this energy by the oxidation of foodstuffs, whereas *phototrophs* obtain it by trapping light energy. Part of the free energy derived from the oxidation of foodstuffs and from light is transformed into a highly accessible form before it is used for motion, active transport, and biosynthesis. The free-energy donor in most energy-requiring processes is *adenosine triphosphate (ATP)*. The central role of ATP in energy exchanges in biological systems was perceived in 1941 by Fritz Lipmann and by Herman Kalckar.

ATP is a nucleotide consisting of an adenine, a ribose, and a triphosphate unit (Figure 17-1). The active form of ATP is usually a complex of ATP with Mg^{2+} or Mn^{2+}. In considering the role of ATP as an energy carrier, we can focus on its triphosphate moiety. *ATP is an energy-rich molecule because its triphosphate unit contains two phosphoanhydride bonds.* A large amount of free energy is liberated when ATP is hydrolyzed to adenosine diphosphate (ADP) and orthophosphate (P_i) or when ATP is hydrolyzed to adenosine monophosphate (AMP) and pyrophosphate (PP_i).

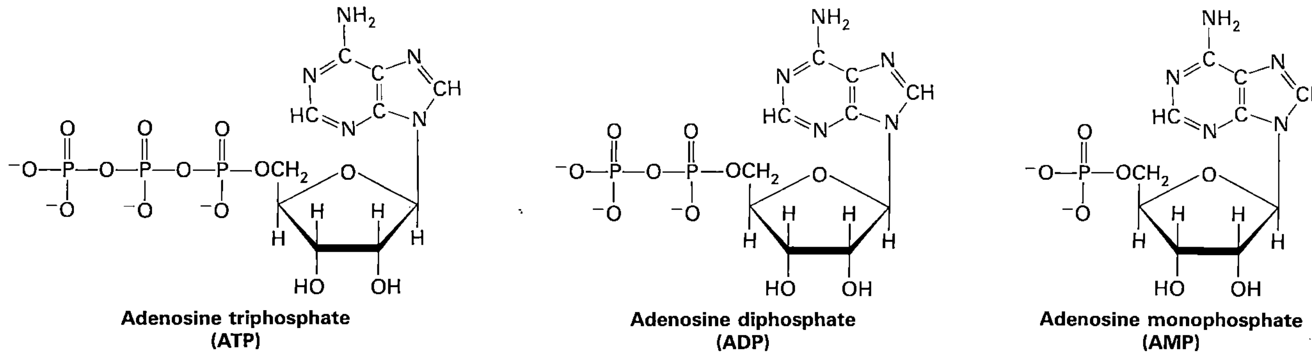

Figure 17-1
Structure of ATP, ADP, and AMP. These adenylates consist of adenine (blue), a ribose (black), and a tri-, di-, or monophosphate unit (red). The innermost phosphorus atom of ATP is designated P_α, the middle one P_β, and the outermost one P_γ.

$$ATP + H_2O \rightleftharpoons ADP + P_i + H^+ \qquad \Delta G^{\circ\prime} = -7.3 \text{ kcal/mol}$$

$$ATP + H_2O \rightleftharpoons AMP + PP_i + H^+ \qquad \Delta G^{\circ\prime} = -7.3 \text{ kcal/mol}$$

The $\Delta G^{\circ\prime}$ for these reactions depends on the ionic strength of the medium and on the concentrations of Mg^{2+} and Ca^{2+}. We shall use a value of −7.3 kcal/mol. Under typical cellular conditions, the actual ΔG for these hydrolyses is approximately −12 kcal/mol.

ATP, AMP, and ADP are interconvertible. The enzyme *adenylate kinase* (also called *myokinase*) catalyzes the reaction:

$$ATP + AMP \rightleftharpoons ADP + ADP$$

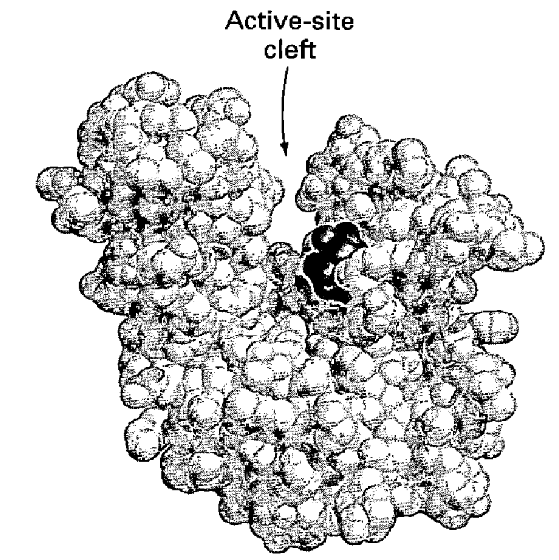

Figure 17-2
Structure of adenylate kinase (blue) containing a bound AMP (red). The active-site cleft closes on binding ATP, the second substrate. [Drawn from 1ak3.pdb. K. Diederichs and G.E. Schultz. *Biochemistry* 29(1990):8138.]

The free energy liberated in the hydrolysis of ATP is harnessed to drive reactions that require an input of free energy, such as muscle contraction. In turn, ATP is formed from ADP and P_i when fuel molecules are oxidized in chemotrophs or when light is trapped by phototrophs. *This ATP-ADP cycle is the fundamental mode of energy exchange in biological systems.*

Some biosynthetic reactions are driven by nucleoside triphosphates that are analogous to ATP, namely, guanosine triphosphate (GTP), uridine triphosphate (UTP), and cytidine triphosphate (CTP). The diphosphate forms of these nucleotides are denoted by GDP, UDP, and CDP, and the monophosphate forms by GMP, UMP, and CMP. Enzymes can catalyze the transfer of the terminal phosphoryl group from one nucleotide to another, as in the reactions

$$\text{ATP} + \text{GDP} \rightleftharpoons \text{ADP} + \text{GTP}$$

$$\text{ATP} + \text{GMP} \rightleftharpoons \text{ADP} + \text{GDP}$$

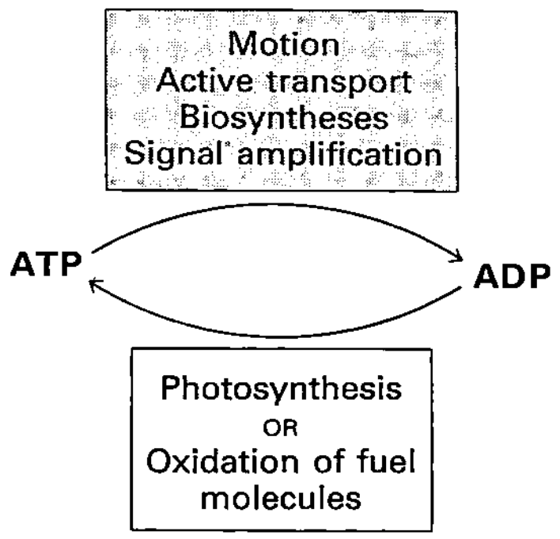

Figure 17-3
The ATP-ADP cycle is the fundamental mode of energy exchange in biological systems.

ATP IS CONTINUOUSLY FORMED AND CONSUMED

ATP serves as the principal *immediate donor of free energy* in biological systems rather than as a long-term storage form of free energy. In a typical cell, an ATP molecule is consumed within a minute following its formation. *The turnover of ATP is very high.* For example, a resting human consumes about 40 kg of ATP in 24 hours. During strenuous exertion, the rate of utilization of ATP may be as high as 0.5 kg per minute. Motion, active transport, signal amplification, and biosynthesis can occur only if ATP is continuously regenerated from ADP (Figure 17-3). Phototrophs harvest the free energy in light to generate ATP, whereas chemotrophs form ATP by the oxidation of fuel molecules. The energy-conserving event in both processes is the pumping of protons across a membrane to generate a proton-motive force. The proton gradient then powers the synthesis of ATP.

STRUCTURAL BASIS OF THE HIGH PHOSPHORYL TRANSFER POTENTIAL OF ATP

Let us compare the standard free energy of hydrolysis of ATP with that of a phosphate ester, such as glycerol 3-phosphate:

$$\text{ATP} + H_2O \rightleftharpoons \text{ADP} + P_i + H^+ \qquad \Delta G^{\circ\prime} = -7.3 \text{ kcal/mol}$$

$$\text{Glycerol 3-phosphate} + H_2O \rightleftharpoons \text{glycerol} + P_i \qquad \Delta G^{\circ\prime} = -2.2 \text{ kcal/mol}$$

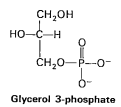

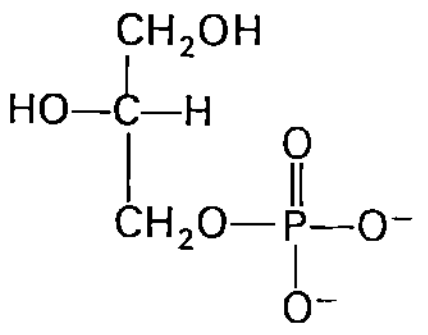

Glycerol 3-phosphate

The magnitude of $\Delta G^{\circ\prime}$ for the hydrolysis of glycerol 3-phosphate is much smaller than that of ATP. This means that ATP has a stronger tendency to transfer its terminal phosphoryl group to water than does glycerol 3-phosphate. In other words, ATP has a higher *phosphoryl potential (phosphoryl group-transfer potential)* than does glycerol 3-phosphate.

What is the structural basis of the high phosphoryl potential of ATP? The structures of both ATP and its hydrolysis products, ADP and P_i, must be examined to answer this question because $\Delta G^{\circ\prime}$ depends on the *difference* in free energies of the products and reactants. Two factors are important: *electrostatic repulsion* and *resonance stabilization.* At pH 7, the triphosphate unit of ATP carries about four negative charges. These charges repel one another strongly because they are in close proximity. The repulsion between them is reduced when ATP is hydrolyzed. Furthermore, ADP and P_i enjoy greater resonance stabilization than does ATP. Orthophosphate has a number of resonance forms of similar energy (Figure 17-4), whereas the γ-phosphoryl group of ATP has only a few. Forms like that shown in Figure 17-5 are unlikely because the two phosphorus atoms compete for electron pairs on oxygen. Also, there is a positive charge on an oxygen atom adjacent to a positively charged phosphorus atom, which is electrostatically unfavorable.

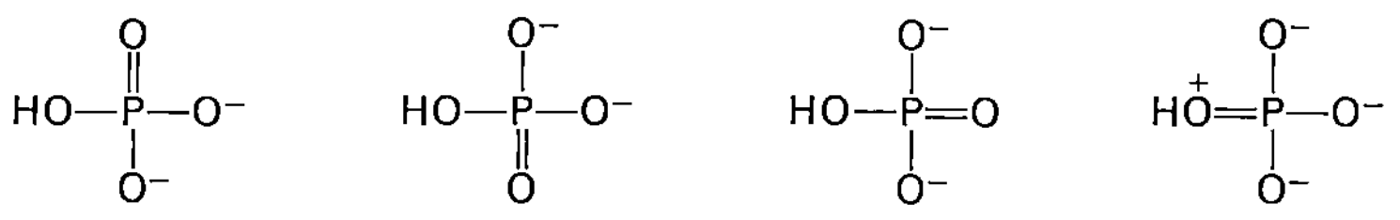

Figure 17-4
Significant resonance forms of orthophosphate.

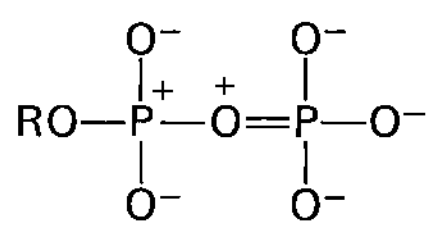

Figure 17-5
An improbable resonance form of the terminal portion of ATP.

Various other compounds in biological systems have a high phosphoryl potential. In fact, some of them, such as phosphoenolpyruvate, acetyl phosphate, and creatine phosphate (Figure 17-6), have a higher transfer potential than does ATP. This means that phosphoenolpyruvate can transfer its phosphoryl group to ADP to form ATP. Indeed, this is one of the ways in which ATP is generated in the breakdown of sugars. It is significant that ATP has a phosphoryl transfer potential that is intermediate among the biologically important phosphorylated molecules (Table 17-1). *This intermediate position enables ATP to function efficiently as a carrier of phosphoryl groups.*

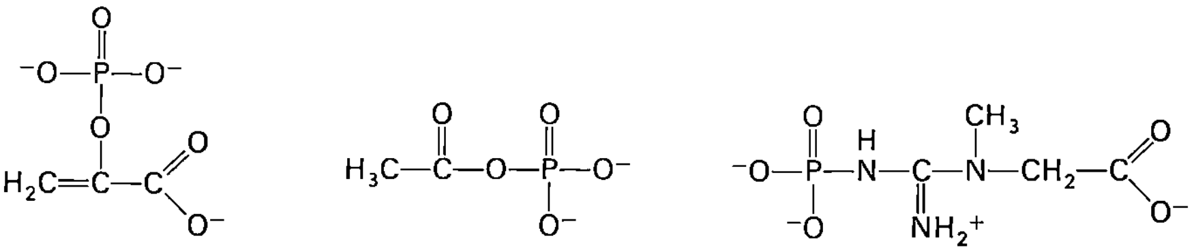

Figure 17-6
Some compounds with a higher phosphoryl transfer potential than that of ATP.

ATP is often called a high-energy phosphate compound, and its phosphoanhydride bonds are referred to as high-energy bonds. There is nothing special about the bonds themselves. *They are high-energy bonds in the sense that much free energy is released when they are hydrolyzed,* for the reasons given above. Lipmann's term "high-energy bond" and his symbol ~P (squiggle P) for a compound having a high phosphoryl transfer potential are vivid, concise, and useful notations. In fact, Lipmann's squiggle did much to stimulate interest in bioenergetics.

Table 17-1
Free energies of hydrolysis of some phosphorylated compounds

Compound	$\Delta G^{\circ\prime}$ *(kcal/mol)*
Phosphoenolpyruvate	−14.8
Carbamoyl phosphate	−12.3
Acetyl phosphate	−10.3
Creatine phosphate	−10.3
Pyrophosphate	−8.0
ATP (to ADP)	−7.3
Glucose 1-phosphate	−5.0
Glucose 6-phosphate	−3.3
Glycerol 3-phosphate	−2.2

Note: Phosphoenolpyruvate has the highest phosphoryl group-transfer potential of the compounds listed. 1 kcal/mol corresponds to 4.184 kJ/mol.

CREATINE PHOSPHATE IS A RESERVOIR OF ~P IN MUSCLE

The amount of ATP in muscle suffices to sustain contractile activity for less than a second. Vertebrate muscle contains a reservoir of high-potential phosphoryl groups in the form of *creatine phosphate* (also called *phos-*

phocreatine; see Figure 17-6), which can readily transfer its phosphoryl group to ATP. This reaction is catalyzed by *creatine kinase.*

$$\text{Creatine phosphate} + \text{ADP} + \text{H}^+ \rightleftharpoons \text{ATP} + \text{creatine}$$

At pH 7, the standard free energy of hydrolysis of creatine phosphate is −10.3 kcal/mol, compared with −7.3 kcal/mol for ATP. Hence, the standard free-energy change in forming ATP from creatine phosphate is −3 kcal/mol, which corresponds to an equilibrium constant of 162.

$$K = \frac{[\text{ATP}][\text{creatine}]}{[\text{ADP}][\text{creatine phosphate}]} = 10^{-\Delta G^{\circ\prime}/(2.303\,RT)} = 10^{3/1.36} = 162$$

In resting muscle, typical concentrations of these metabolites are [ATP] = 4 mM, [ADP] = 0.013 mM, [creatine phosphate] = 25 mM, and [creatine] = 13 mM. *The abundance of creatine phosphate and its high phosphoryl transfer potential relative to that of ATP make it a highly effective ~P buffer.* Creatine phosphate maintains a high concentration of ATP during periods of muscular exertion. Indeed, creatine phosphate is the major source of ~P for a runner during the first four seconds of a 100-meter sprint.

ATP HYDROLYSIS SHIFTS THE EQUILIBRIA OF COUPLED REACTIONS BY A FACTOR OF 10^8

An understanding of the role of ATP in energy coupling can be enhanced by considering a chemical reaction that is thermodynamically unfavorable without an input of free energy. Suppose that the standard free energy of the conversion of A into B is +4 kcal/mol.

$$\text{A} \rightleftharpoons \text{B} \qquad \Delta G^{\circ\prime} = +4\ \text{kcal/mol}$$

The equilibrium constant K'_{eq} of this reaction at 25°C is related to $\Delta G^{\circ\prime}$ by

$$K'_{eq} = \frac{[\text{B}]_{eq}}{[\text{A}]_{eq}} = 10^{-\Delta G^{\circ\prime}/1.36} = 1.15 \times 10^{-3}$$

Thus, net conversion of A into B cannot occur when the molar ratio of B to A is equal to or greater than 1.15×10^{-3}. However, A can be converted into B when the [B]/[A] ratio is higher than 1.15×10^{-3} if the reaction is coupled to the hydrolysis of ATP. The new overall reaction is

$$\text{A} + \text{ATP} + \text{H}_2\text{O} \rightleftharpoons \text{B} + \text{ADP} + \text{P}_i + \text{H}^+ \qquad \Delta G^{\circ\prime} = -3.3\ \text{kcal/mol}$$

Its standard free-energy change of −3.3 kcal/mol is the sum of $\Delta G^{\circ\prime}$ for the conversion of A into B (+4 kcal/mol) and for the hydrolysis of ATP (−7.3 kcal/mol). The equilibrium constant of this coupled reaction is

$$K'_{eq} = \frac{[\text{B}]_{eq}}{[\text{A}]_{eq}} \cdot \frac{[\text{ADP}]_{eq}[\text{P}_i]_{eq}}{[\text{ATP}]_{eq}} = 10^{3.3/1.36}$$
$$= 2.67 \times 10^2$$

At equilibrium, the ratio of [B] to [A] is given by

$$\frac{[\text{B}]_{eq}}{[\text{A}]_{eq}} = K'_{eq}\frac{[\text{ATP}]_{eq}}{[\text{ADP}]_{eq}[\text{P}_i]_{eq}}$$

The ATP-generating system of cells maintains the [ATP]/[ADP] [P_i] ratio at a high level, typically of the order of 500. For this ratio,

$$\frac{[B]_{eq}}{[A]_{eq}} = 2.67 \times 10^2 \times 500 = 1.34 \times 10^5$$

which means that the hydrolysis of ATP enables A to be converted into B until the [B]/[A] ratio reaches a value of 1.34×10^5. This equilibrium ratio is strikingly different from the value of 1.15×10^{-3} for the reaction A → B that does not include ATP hydrolysis. In other words, the coupled hydrolysis of ATP has changed the equilibrium ratio of B to A by a factor of about 10^8.

We see here the thermodynamic essence of ATP's action as an energy-coupling agent. Cells maintain a high level of ATP by using oxidizable substrates or light as sources of free energy. The hydrolysis of an ATP molecule in a coupled reaction then changes the equilibrium ratio of products to reactants by a very large factor, of the order of 10^8. More generally, the hydrolysis of n ATP molecules changes the equilibrium ratio of a coupled reaction (or sequence of reactions) by a factor of 10^{8n}. For example, the hydrolysis of three ATP molecules in a coupled reaction changes the equilibrium ratio by a factor of 10^{24}. Thus, a thermodynamically unfavorable reaction sequence can be converted into a favorable one by coupling it to the hydrolysis of a sufficient number of ATP molecules in a new reaction. It should also be emphasized that A and B in the preceding coupled equation may be interpreted very generally, not only as different chemical species. For example, A and B may represent different conformations of a protein, as in muscle contraction. Alternatively, A and B may refer to the concentrations of an ion or molecule on the outside and inside of a cell, as in the active transport of a nutrient.

NADH AND $FADH_2$ ARE THE MAJOR ELECTRON CARRIERS IN THE OXIDATION OF FUEL MOLECULES

Chemotrophs derive free energy from the oxidation of fuel molecules, such as glucose and fatty acids. In aerobic organisms, the ultimate electron acceptor is O_2. However, electrons are not transferred directly from fuel molecules and their breakdown products to O_2. Instead, these substrates transfer electrons to special carriers, which are either *pyridine nucleotides* or *flavins*. The reduced forms of these carriers then transfer their high-potential electrons to O_2 by means of an electron transport chain located in the inner membrane of mitochondria. The proton gradient formed as a result of this flow of electrons then drives the synthesis of ATP from ADP and P_i. This process, called *oxidative phosphorylation* (Chapter 21), is the major source of ATP in aerobic organisms. Alternatively, the high-potential electrons derived from the oxidation of fuel molecules can be used in biosyntheses that require *reducing power* in addition to ATP.

Nicotinamide adenine dinucleotide (NAD^+) is a major electron acceptor in the oxidation of fuel molecules (Figure 17-7). The reactive part of NAD^+ is its nicotinamide ring, a pyridine derivative. *In the oxidation of a substrate, the nicotinamide ring of NAD^+ accepts a hydrogen ion and two electrons, which are equivalent to a hydride ion.* The reduced form of this carrier is called *NADH.* In the oxidized form, the nitrogen atom is tetravalent and carries a positive charge, as indicated by NAD^+. In the reduced form, NADH, the nitrogen atom is trivalent.

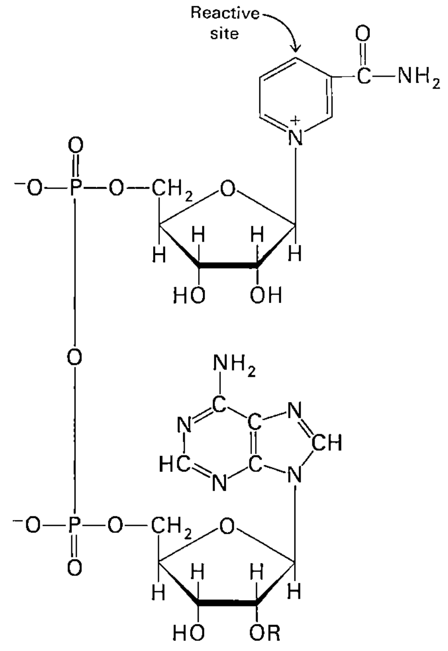

Figure 17-7
Structure of the oxidized form of nicotinamide adenine dinucleotide (NAD^+) and of nicotinamide adenine dinucleotide phosphate ($NADP^+$). In NAD^+, R = H; in $NADP^+$, R = PO_3^{2-}.

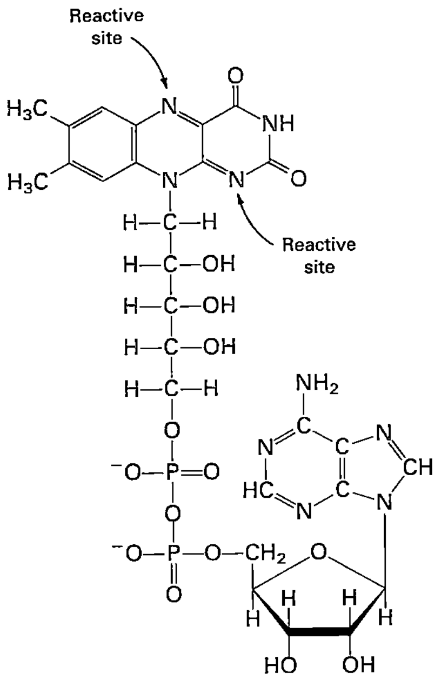

Figure 17-8
Structure of the oxidized form of flavin adenine dinucleotide (FAD). This electron carrier consists of a flavin mononucleotide (FMN) unit (shown in green) and an AMP unit (shown in red).

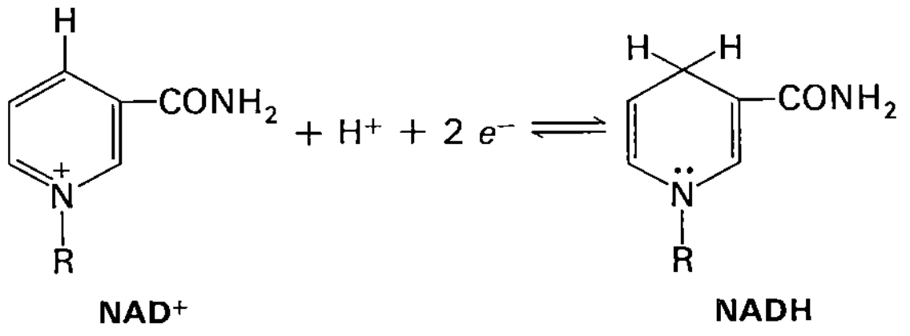

NAD^+ is the electron acceptor in many reactions of the type

$$NAD^+ + R{-}CH(OH){-}R' \rightleftharpoons NADH + R{-}C(=O){-}R' + H^+$$

In this dehydrogenation, one hydrogen atom of the substrate is directly transferred to NAD^+, whereas the other appears in the solvent as a proton. Both electrons lost by the substrate are transferred to the nicotinamide ring.

The other major electron carrier in the oxidation of fuel molecules is *flavin adenine dinucleotide* (Figure 17-8). The abbreviations for the oxidized and reduced forms of this carrier are FAD and $FADH_2$, respectively. FAD is the electron acceptor in reactions of the type

$$FAD + R{-}CH_2{-}CH_2{-}R' \rightleftharpoons FADH_2 + R{-}CH{=}CH{-}R'$$

The reactive part of FAD is its isoalloxazine ring (Figure 17-9). FAD, like NAD^+, can accept two electrons. In doing so, FAD, unlike NAD^+, takes up a proton as well as a hydride ion. These electron carriers and flavin mononucleotide (FMN), an electron carrier related to FAD, will be discussed further in Chapter 21.

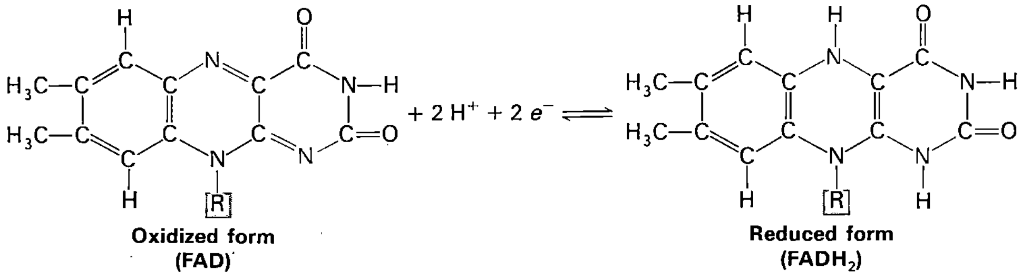

Figure 17-9
Structures of the reactive parts of FAD and $FADH_2$.

NADPH IS THE MAJOR ELECTRON DONOR IN REDUCTIVE BIOSYNTHESES

In most biosyntheses, the precursors are more oxidized than the products. Hence, reductive power is needed in addition to ATP. For example, in the biosynthesis of fatty acids, the keto group of an added C_2 unit is reduced to a methylene group in several steps. This sequence of reactions requires an input of four electrons.

Chapter 17 **451**

METABOLISM

$$R—CH_2—\overset{\overset{\displaystyle O}{\|}}{C}—R' + 4\ H^+ + 4\ e^- \longrightarrow R—CH_2—CH_2—R' + H_2O$$

The electron donor in most reductive biosyntheses is NADPH, the reduced form of nicotinamide adenine dinucleotide phosphate ($NADP^+$; see Figure 17-7). NADPH differs from NADH in that the 2'-hydroxyl group of its adenosine moiety is esterified with phosphate. NADPH carries electrons in the same way as NADH. However, *NADPH is used almost exclusively for reductive biosyntheses, whereas NADH is used primarily for the generation of ATP*. The extra phosphate group on NADPH is a tag that directs this reducing agent to discerning biosynthetic enzymes. The biological significance of the distinction between NADPH and NADH will be discussed later (p. 559).

NADH, NADPH, and $FADH_2$ react slowly with O_2 in the absence of catalysts. Likewise, ATP is hydrolyzed slowly (in times of many hours or even days) in the absence of a catalyst. These molecules are kinetically quite stable in the face of a large thermodynamic driving force for reaction with O_2 (in the case of the electron carriers) and H_2O (in the case of ATP). *The stability of these molecules in the absence of specific catalysts is essential for their biological function because it enables enzymes to control the flow of free energy and reductive power.*

COENZYME A IS A UNIVERSAL CARRIER OF ACYL GROUPS

Coenzyme A is another central molecule in metabolism. In 1945, Lipmann found that a heat-stable cofactor was required in many enzyme-catalyzed acetylations. This cofactor was named *coenzyme A (CoA)*, the A standing for *acetylation*. It was isolated, and its structure was determined several years later (Figure 17-10). The terminal sulfhydryl group in CoA is

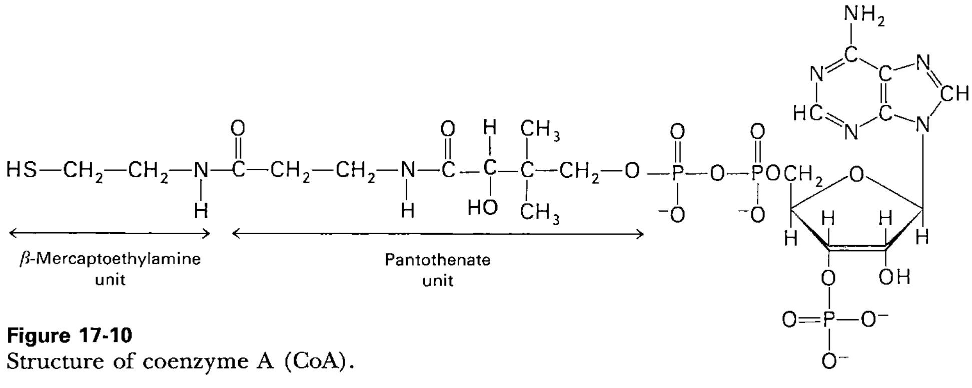

Figure 17-10
Structure of coenzyme A (CoA).

the reactive site. Acyl groups are linked to CoA by a thioester bond. The resulting derivative is called an *acyl CoA*. An acyl group often linked to CoA is the acetyl unit; this derivative is called *acetyl CoA*. The $\Delta G^{\circ\prime}$ for the hydrolysis of acetyl CoA has a large negative value:

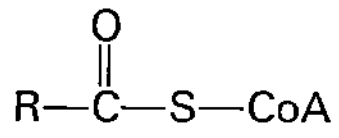

Acetyl CoA + H_2O $\rightleftharpoons$ acetate + CoA + H^+

$$\Delta G^{\circ\prime} = -7.5\ \text{kcal/mol}$$

$$R—\overset{\overset{\displaystyle O}{\|}}{C}—S—CoA$$
Acyl CoA

$$H_3C—\overset{\overset{\displaystyle O}{\|}}{C}—S—CoA$$
Acetyl CoA

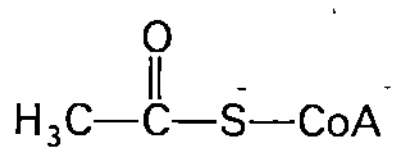

The hydrolysis of a thioester is thermodynamically more favorable than that of an oxygen ester because the double-bond character of the C–O bond does not extend significantly to the C–S bond. Consequently, *acetyl CoA has a high acetyl potential (acetyl group–transfer potential).* Acetyl CoA carries an activated acetyl group, just as ATP carries an activated phosphoryl group.

ACTIVATED CARRIERS EXEMPLIFY THE MODULAR DESIGN AND ECONOMY OF METABOLISM

Coenzyme A is highly versatile. It carries activated acyl groups ranging in size from 2-carbon to 24-carbon or even longer units. Coenzyme A can deliver activated acyl groups for degradation and energy generation, or for biosynthetic purposes. For example, some proteins acquire long-chain fatty acyl groups to enable them to anchor into bilayer membranes. Likewise, NADH, NADPH, and $FADH_2$ mediate many different electron transfers, and *S*-adenosylmethionine participates in methylations, as we have already seen in bacterial chemotaxis (p. 331). Indeed, *most interchanges of activated groups in metabolism are accomplished by a rather small set of carriers* (Table 17-2). The existence of a recurring set of activated carriers in all organisms is one of the unifying motifs of biochemistry. Furthermore, it illustrates the modular design of metabolism. A small set of molecules has been selected in the course of evolution to carry out a very wide range of tasks. Metabolism is readily comprehended because of the economy and elegance of its underlying design.

Table 17-2
Some activated carriers in metabolism

Carrier molecule	*Group carried in activated form*
ATP	Phosphoryl
NADH and NADPH	Electrons
$FADH_2$	Electrons
$FMNH_2$	Electrons
Coenzyme A	Acyl
Lipoamide	Acyl
Thiamine pyrophosphate	Aldehyde
Biotin	CO_2
Tetrahydrofolate	One-carbon units
S-Adenosylmethionine	Methyl
Uridine diphosphate glucose	Glucose
Cytidine diphosphate diacylglycerol	Phosphatidate
Nucleoside triphosphates	Nucleotides

MOST WATER-SOLUBLE VITAMINS ARE COMPONENTS OF COENZYMES

Lipmann has commented that "doctors like to prescribe vitamins and millions of people take them, but it requires a good deal of biochemical sophistication to understand why they are needed and how the organism uses them." Vitamins are organic molecules that are needed in small amounts in the diets of some higher animals. These molecules serve nearly the same roles in all forms of life, but higher animals have lost the capacity to synthesize them. Vitamins can be grouped according to whether they are soluble in water or in nonpolar solvents. The *water-soluble vitamins* are ascorbic acid (vitamin C) and a series known as the vitamin B complex (Figure 17-11). Ascorbate, the ionized form of ascorbic acid, serves as a reducing agent (an antioxidant), as will be discussed shortly. The vitamin B series are components of coenzymes (Table 17-3). For example, riboflavin (vitamin B_2) is a precursor of FAD, and pantothenate is a component of coenzyme A.

Figure 17-11
Structures of some water-soluble vitamins.

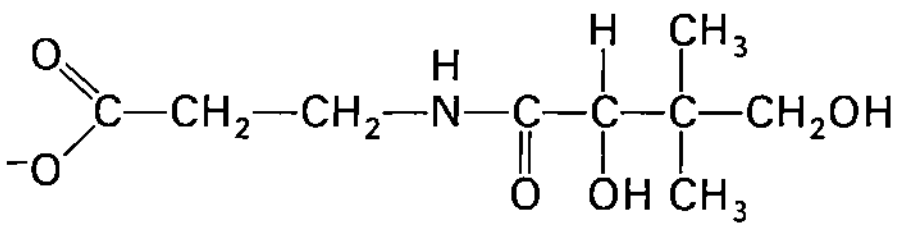

Pantothenate

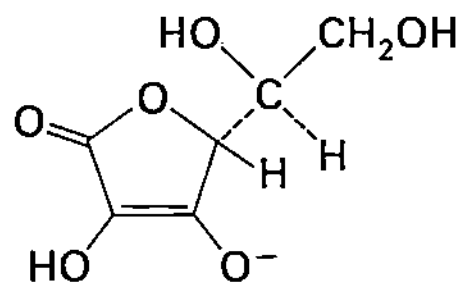

L-Ascorbate (Vitamin C)

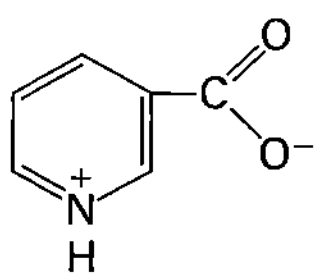

Nicotinate (Niacin)

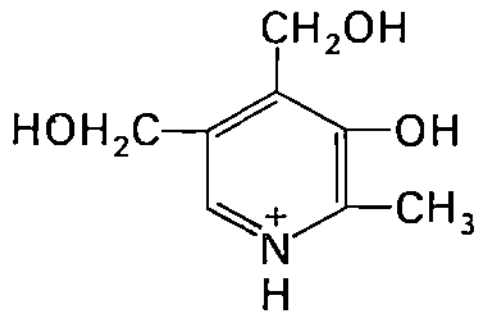

Pyridoxine (A form of vitamin B_6)

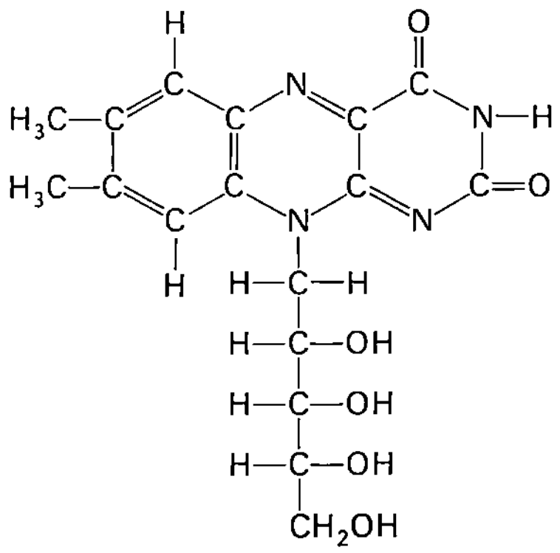

Riboflavin (Vitamin B_2)

Table 17-3
Coenzyme derivatives of some water-soluble vitamins

Vitamin	*Coenzyme derivative*
Thiamine (vitamin B_1)	Thiamine pyrophosphate
Riboflavin (vitamin B_2)	Flavin adenine dinucleotide and flavin mononucleotide
Nicotinate (niacin)	Nicotinamide adenine dinucleotide
Pyridoxine, pyridoxal, and pyridoxamine (vitamin B_6)	Pyridoxal phosphate
Pantothenate	Coenzyme A
Biotin	Covalently attached to carboxylases
Folate	Tetrahydrofolate
Cobalamin (vitamin B_{12})	Cobamide coenzymes

FAT-SOLUBLE VITAMINS PARTICIPATE IN DIVERSE PROCESSES SUCH AS BLOOD CLOTTING AND VISION

Much is also known about the molecular actions of *fat-soluble vitamins*, which are designated by the letters A, D, E, and K (Figure 17-12). Vitamin K, which is required for normal blood clotting (*K* stands for *k*oagulation), participates in the carboxylation of glutamate residues to γ-carboxyglutamate, which makes it a much stronger chelator of Ca^{2+} (p. 255). Vitamin A (retinol) is the precursor of retinal, the light-sensitive group in rhodopsin and other visual pigments (p. 333). A deficiency of this vitamin leads to night blindness. Furthermore, young animals require

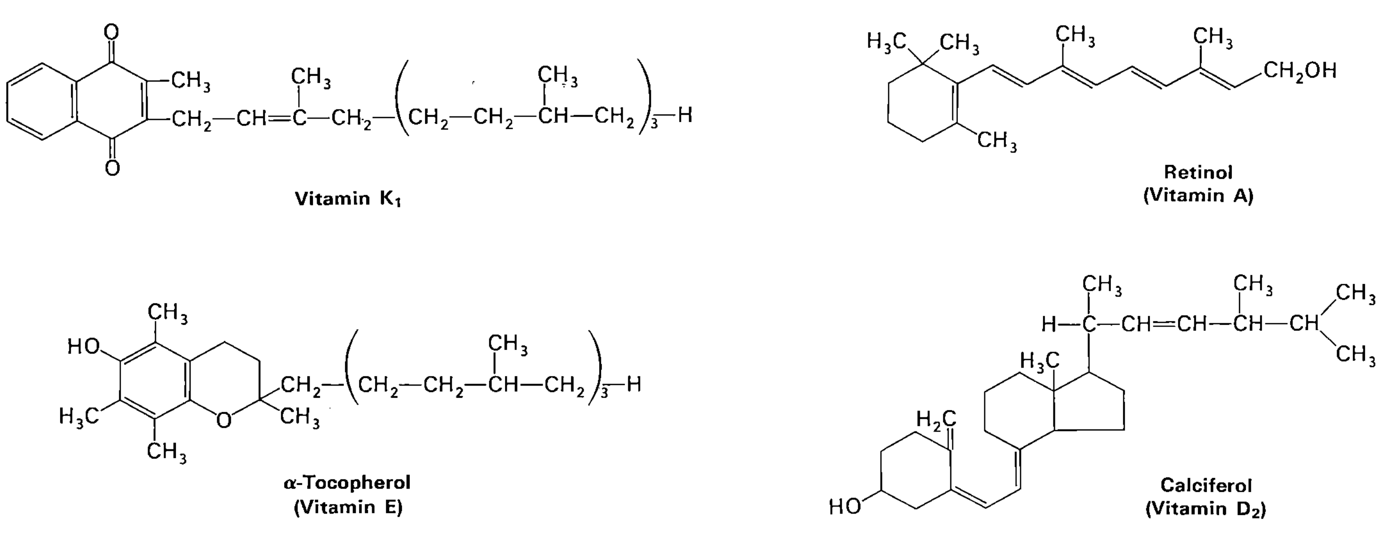

Figure 17-12
Structures of some fat-soluble vitamins.

vitamin A for growth. Retinoic acid, which contains a terminal carboxylate in place of the alcohol terminus of retinol, activates the transcription of specific genes that mediate growth and development (p. 1002). The metabolism of calcium and phosphorus is regulated by a hormone that is derived from vitamin D (p. 707). A deficiency in vitamin D impairs bone formation in growing animals. Infertility in rats is a consequence of vitamin E (α-tocopherol) deficiency. This vitamin protects unsaturated membrane lipids from oxidation.

ASCORBATE (VITAMIN C) IS REQUIRED FOR THE HYDROXYLATION OF PROLINE RESIDUES IN COLLAGEN

In an early chapter, we considered the triple-helical structure of collagen, the major protein of connective tissue, and noted that it contains *4-hydroxyproline*, an amino acid rarely found elsewhere (p. 31). How is this unusual amino acid formed and what is its role? Radioactive labeling studies showed that prolines on the amino side of glycine residues in nascent collagen chains become hydroxylated. The oxygen atom that becomes attached to C-4 of proline comes from O_2. The other oxygen atom of O_2 is taken up by α-ketoglutarate, which is converted into succinate (Figure 17-13). This complex reaction is catalyzed by *prolyl hydroxylase,* a *dioxygenase.* It is assisted by an Fe^{2+} ion, which is tightly bound to it and needed to activate O_2. The enzyme also converts α-ketoglutarate into succinate without hydroxylating proline. In this partial reaction, an Fe^{3+}–O^- complex is formed, which inactivates the enzyme. How is the active enzyme regenerated? *Ascorbate (vitamin C)* comes to the rescue by reducing the ferric ion of the inactivated enzyme. In the recovery process, ascorbate is oxidized to dehydroascorbic acid (Figure 17-14). Thus, ascorbate serves here as a specific *antioxidant.*

Primates and guinea pigs are unable to synthesize ascorbic acid and hence must acquire it from their diets. The importance of ascorbate becomes strikingly evident in *scurvy.* A vivid description of this dietary deficiency disease was given by Jacques Cartier in 1536, when it afflicted his men as they were exploring the Saint Lawrence River:

> Some did lose all their strength, and could not stand on their feet. . . . Others also had all their skins spotted with spots of blood of a purple colour:

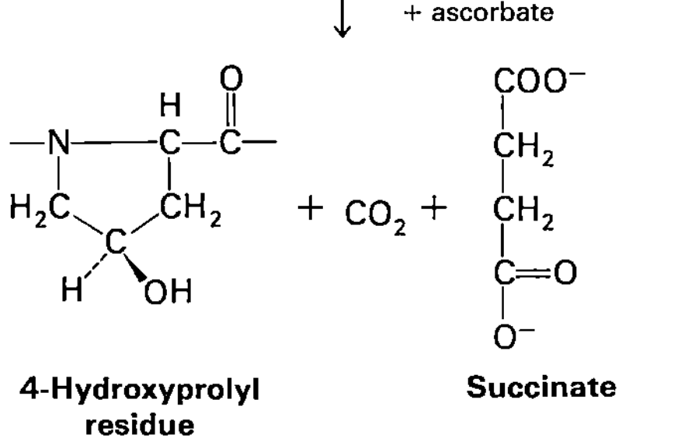

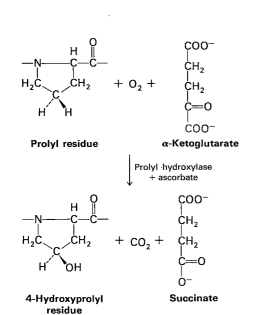

Figure 17-13
Hydroxylation of a proline residue at C-4 by the action of prolyl hydroxylase, an enzyme that activates molecular oxygen.

then did it ascend up to their ankles, knees, thighs, shoulders, arms, and necks. Their mouths became stinking, their gums so rotten, that all the flesh did fall off, even to the roots of the teeth, which did also almost all fall out.

The means of preventing scurvy was succinctly stated by James Lind, a Scottish physician, in 1753:

> Experience indeed sufficiently shows that as greens or fresh vegetables, with ripe fruits, are the best remedies for it, so they prove the most effectual preservatives against it.

Lind urged the inclusion of lemon juice in the diet of sailors, and some 40 years later the admiralty took his advice. After limes were substituted for lemons in 1865, British sailors began to be known as "limeys."

Why does impaired hydroxylation have such devastating consequences? *Collagen synthesized in vitro in the absence of ascorbate has a lower melting temperature than does the normal protein.* Studies of the thermal stability of synthetic polypeptides have been especially informative. The T_m of the poly(Pro-Pro-Gly) triple helix is 24°C, compared with 58°C for the poly(Pro-Hyp-Gly) triple helix (*Hyp* stands for "hydroxyproline"). Hydroxyproline stabilizes the collagen triple helix by forming interstrand hydrogen bonds. The abnormal fibers formed by insufficiently hydroxylated collagen contribute to the skin lesions and blood vessel fragility seen in scurvy.

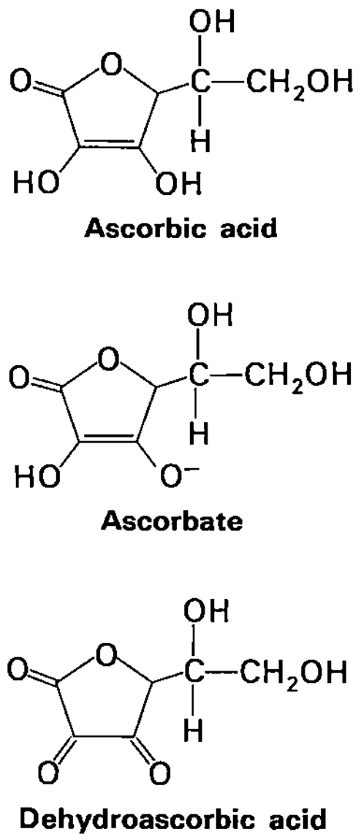

Figure 17-14
Formulas of ascorbic acid (vitamin C) and ascorbate (its ionized form). The pK_a of the acidic hydroxyl group of ascorbic acid is 4.2. Dehydroascorbic acid is the oxidized form of ascorbate.

STAGES IN THE EXTRACTION OF ENERGY FROM FOODSTUFFS

Let us take an overview of the process of energy generation in higher organisms before considering it in detail in subsequent chapters. Hans Krebs described three stages in the generation of energy from the oxidation of foodstuffs (Figure 17-15).

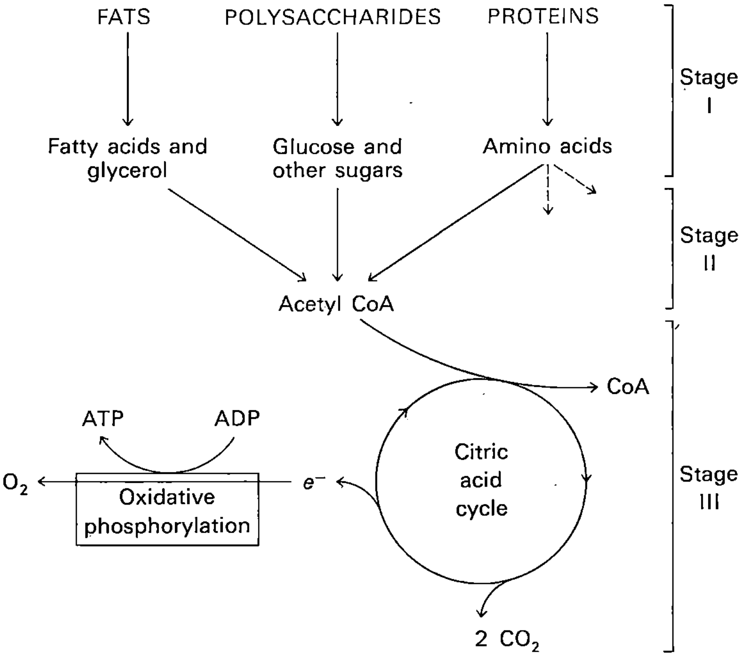

Figure 17-15
Stages in the extraction of energy from foodstuffs.

In the first stage, large molecules in food are broken down into smaller units. Proteins are hydrolyzed to their 20 kinds of constituent amino acids, polysaccharides are hydrolyzed to simple sugars such as glucose, and fats are hydrolyzed to glycerol and fatty acids. No useful energy is generated in this phase.

In the second stage, these numerous small molecules are degraded to a few simple units that play a central role in metabolism. In fact, most of them—sugars, fatty acids, glycerol, and several amino acids—are converted into the acetyl unit of acetyl CoA. Some ATP is generated in this stage, but the amount is small compared with that obtained from the complete oxidation of the acetyl unit of acetyl CoA.

The third stage consists of the citric acid cycle and oxidative phosphorylation, which are the final common pathways in the oxidation of fuel molecules. Acetyl CoA brings acetyl units into this cycle, where they are completely oxidized to CO_2. Four pairs of electrons are transferred (three to NAD^+ and one to FAD) for each acetyl group that is oxidized. Then, ATP is generated as electrons flow from the reduced forms of these carriers to O_2 in a process called *oxidative phosphorylation.* More than 90% of the ATP generated by the degradation of foodstuffs is formed in this third stage.

METABOLIC PROCESSES ARE REGULATED IN THREE PRINCIPAL WAYS

Even the simplest bacterial cell has the capacity for carrying out more than a thousand interdependent reactions. It is evident that this complex network must be rigorously regulated. At the same time, metabolic control must be flexible, because the external environment of cells is not constant. Metabolism is regulated by controlling (1) *the amounts of enzymes,* (2) *their catalytic activities,* and (3) *the accessibility of substrates.* The amount of a particular enzyme depends on both its rate of synthesis and its rate of degradation. The level of most enzymes is adjusted primarily by changing the *rate of transcription* of genes encoding them. Lactose, for example, induces a more than 50-fold increase in the rate of synthesis of β-galactosidase, an enzyme required for the breakdown of this disaccharide (Chapter 36).

The catalytic activity of enzymes is controlled in several ways. *Reversible allosteric control* is especially important. For example, the first reaction in many biosynthetic pathways is allosterically inhibited by the ultimate product of the pathway. The inhibition of aspartate transcarbamoylase by cytidine triphosphate (p. 238) is a well-understood example of *feedback inhibition.* Another recurring mechanism is *reversible covalent modification.* For example, glycogen phosphorylase, the enzyme catalyzing the breakdown of glycogen, a storage form of sugar, is activated by phosphorylation of a particular serine residue when glucose is scarce (p. 590). Hormones such as epinephrine trigger signal transduction cascades that lead to highly amplified changes in metabolic patterns. As was discussed earlier (p. 343), cyclic AMP and calcium ion serve as intracellular messengers that coordinate the activities of many target proteins.

Metabolism is also regulated by controlling the *flux of substrates.* Insulin, for example, promotes the entry of glucose into many kinds of cells. The transfer of substrates from one compartment of a cell to another (e.g., from the cytosol to the mitochondria) can also serve as a control point.

An important general principle of metabolism is that *biosynthetic and degradative pathways are almost always distinct.* This separation is necessary for energetic reasons, as will be evident in subsequent chapters. It also facilitates the control of metabolism. In eukaryotes, metabolic regulation and flexibility are also enhanced by compartmentation. For example, fatty acid oxidation occurs in mitochondria, whereas fatty acid synthesis occurs in the cytosol (the soluble part of the cytoplasm). *Compartmentation segregates opposed reactions.*

Many reactions in metabolism are controlled by the *energy status* of the cell. One index of the energy status is the *energy charge,* which is proportional to the mole fraction of ATP plus half the mole fraction of ADP, given that ATP contains two anhydride bonds, whereas ADP contains one. Hence, the energy charge is defined as

$$\text{Energy charge} = \frac{[\text{ATP}] + \frac{1}{2}[\text{ADP}]}{[\text{ATP}] + [\text{ADP}] + [\text{AMP}]}$$

The energy charge can have a value ranging from 0 (all AMP) to 1 (all ATP). Daniel Atkinson has shown that *ATP-generating (catabolic) pathways are inhibited by a high energy charge, whereas ATP-utilizing (anabolic) pathways are stimulated by a high energy charge.* In plots of the reaction rates of such pathways versus the energy charge, the curves are steep near an energy charge of 0.9, where they usually intersect (Figure 17-16). It is evident that

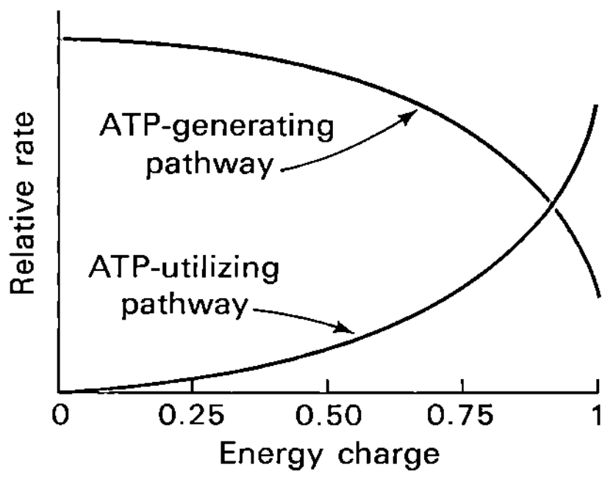

Figure 17-16
Effect of the energy charge on the relative rates of a typical ATP-generating (catabolic) pathway and a typical ATP-utilizing (anabolic) pathway.

the control of these pathways is designed to maintain the energy charge within rather narrow limits. In other words, *the energy charge, like the pH of a cell, is buffered.* The energy charge of most cells is in the range of 0.80 to 0.95. An alternative index of the energy status is the *phosphorylation potential,* which is defined as

$$\text{Phosphorylation potential} = \frac{[\text{ATP}]}{[\text{ADP}][\text{P}_i]}$$

The phosphorylation potential, in contrast with the energy charge, depends on the concentration of P_i and is directly related to the free energy available from ATP.

NUCLEAR MAGNETIC RESONANCE SPECTROSCOPY REVEALS METABOLIC EVENTS IN INTACT ORGANISMS

A century ago, Thomas Huxley said, "What an enormous revolution would be made in biology, if physics or chemistry could supply the physiologist with a means of making out the molecular structure of living tissues comparable to that which spectroscopy affords to the inquirer into the nature of the heavenly bodies." Recent advances in nuclear magnetic resonance (NMR) spectroscopy are bringing this hope closer to realization. Metabolic processes in skeletal muscle, heart, and brain of intact organisms can now be explored noninvasively by NMR methods. As was discussed earlier, nuclei in different environments absorb energy at slightly different resonance frequencies, an effect termed the *chemical shift* (p. 66). For example, the chemical shifts of the phosphorus atom in $H_2PO_4^-$ and HPO_4^{2-} differ by 2.4 ppm (parts per million). However, a single peak is observed from solutions of orthophosphate because $H_2PO_4^-$ and HPO_4^{2-} interconvert rapidly. The position of this peak depends on the ratio of these species, and hence can serve as an indicator of intracellular pH (Figure 17-17).

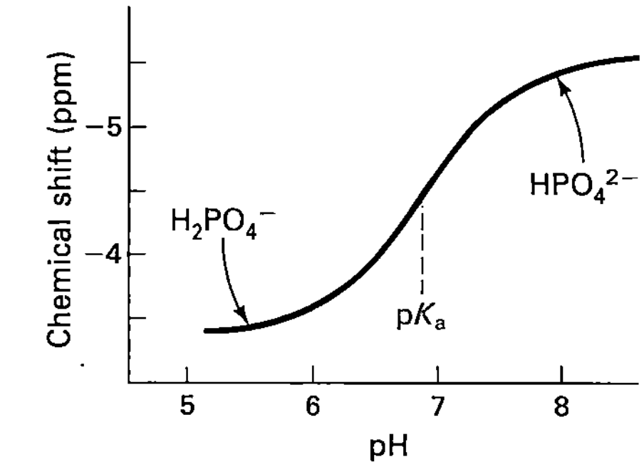

Figure 17-17
The chemical shift of the ^{31}P NMR line of orthophosphate depends on pH. The shifts are expressed relative to the position of the creatine phosphate resonance. [After D.G. Gadian, G.K. Radda, R.E. Richards, and P.J. Seeley. In *Biological Applications of Magnetic Resonance,* R.G. Shulman, ed. (Academic Press, 1979), p. 475.]

The ^{31}P NMR spectra of the forearm muscle of a human subject before and during exercise are shown in Figure 17-18. Five peaks are evident in this strikingly simple spectrum from a complex organ. Three of them arise primarily from the α, β, and γ phosphorus atoms of ATP. The other two come from the phosphorus atoms of creatine phosphate and orthophosphate. ADP and other phosphate compounds do not contribute appreciably to these spectra because they are present at much lower concentration or are not free to rotate. Slowly rotating nuclei such as the phosphorus atoms of nucleotides tightly bound to proteins and those of phospholipid bilayers and nucleic acids have very broad lines that raise the background of such NMR spectra but do not appear as discrete peaks. The spectrum observed after 19 minutes of exercise shows that *the amount of creatine phosphate has decreased markedly, whereas that of orthophosphate has increased markedly. In contrast, the ATP level stayed nearly constant because it was buffered by the creatine phosphate.*

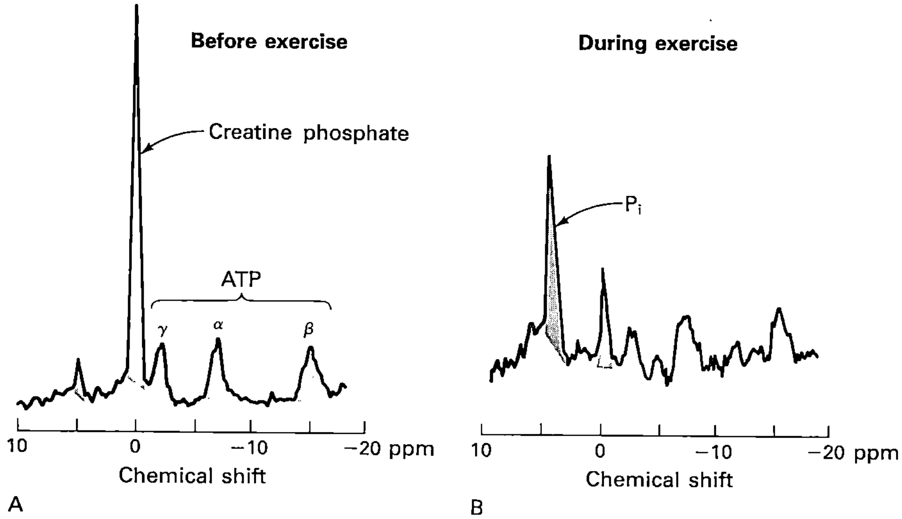

Figure 17-18
Effect of exercise on the level of ATP, creatine phosphate, and orthophosphate in the forearm muscle of a human subject. (A) ^{31}P NMR spectrum before exercise, and (B) after 19 minutes of exercise. Spectrum A was collected in 256 s and spectrum B in 64 s. Note that the three phosphorus atoms in ATP have different chemical shifts. [After G.K. Radda. *Science* 233(1986):641.]

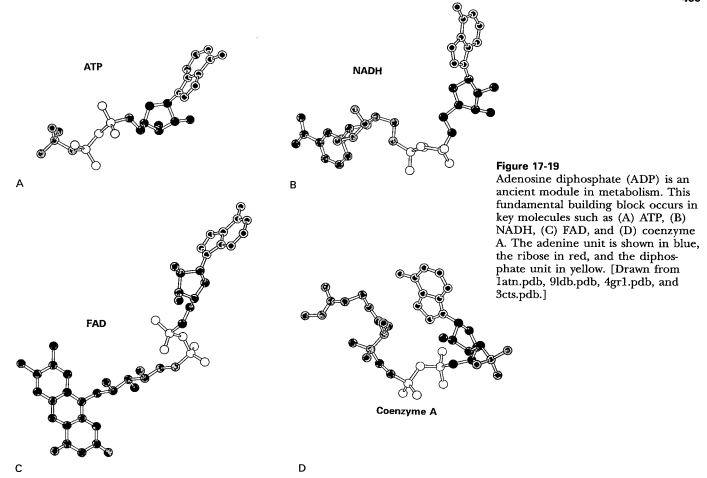

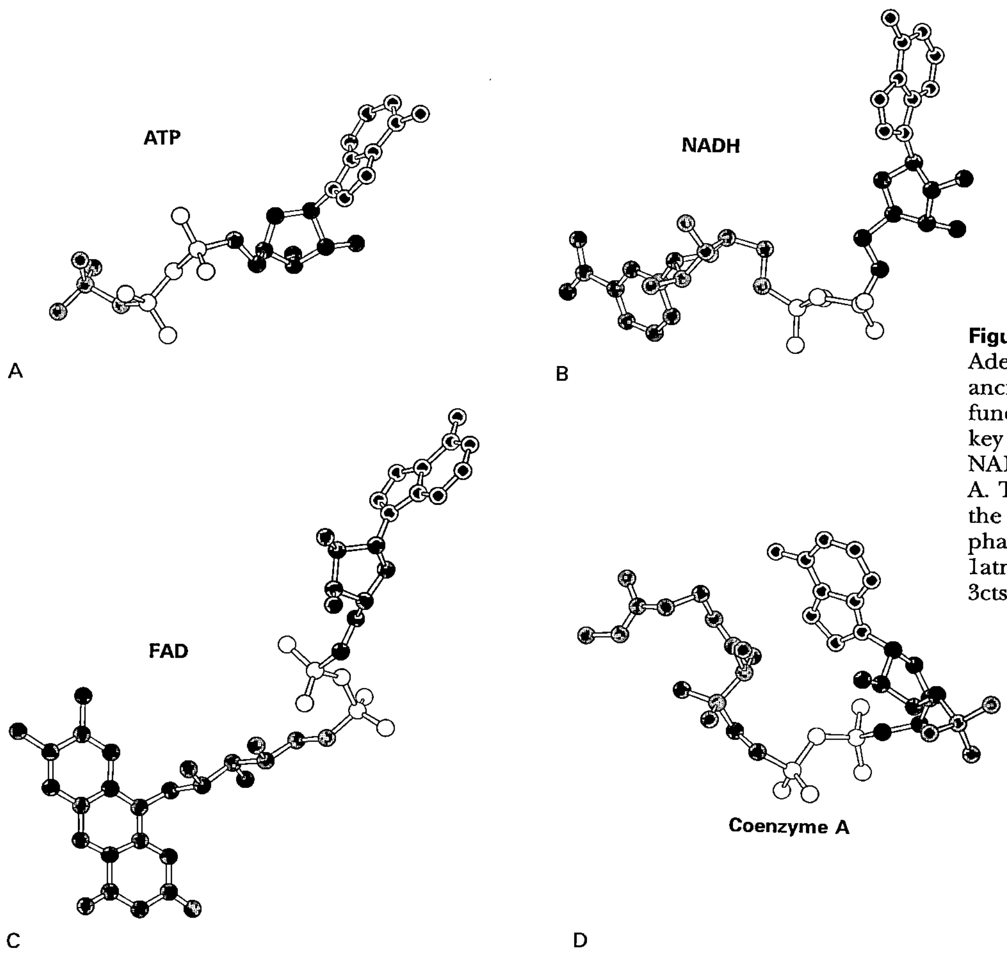

Figure 17-19
Adenosine diphosphate (ADP) is an ancient module in metabolism. This fundamental building block occurs in key molecules such as (A) ATP, (B) NADH, (C) FAD, and (D) coenzyme A. The adenine unit is shown in blue, the ribose in red, and the diphosphate unit in yellow. [Drawn from 1atn.pdb, 9ldb.pdb, 4gr1.pdb, and 3cts.pdb.]

THE CENTRAL ROLE OF RIBONUCLEOTIDES IN METABOLISM REFLECTS THEIR ANCIENT ORIGINS

Many of the central molecules of metabolism in all forms of life are ribonucleotides. Why do activated carriers such as ATP, NADH, $FADH_2$, and coenzyme A contain adenosine phosphate units? A likely explanation is that RNA came before proteins and DNA in evolution (p. 115). The earliest catalysts most probably were RNA molecules, termed *ribozymes*. Non-RNA units such as the isoalloxazine ring were recruited to serve as efficient carriers of activated electrons and chemical units, a function not readily performed by RNA itself. We can picture the adenine ring of $FADH_2$ binding to a uracil unit in a niche of a ribozyme by base pairing. When proteins replaced RNA as the major catalysts to achieve greater versatility, the ribonucleotide coenzymes stayed essentially unchanged because they were already well suited to their metabolic roles. The nicotinamide unit of NADH, for example, can readily transfer electrons irrespective of whether the adenine unit interacts with a base in a ribozyme or with amino acid residues in a protein enzyme. That molecules and motifs of metabolism are common to all forms of life testifies to their common origin and to the retention of functioning modules over billions of years of evolution. Our understanding of metabolism, like that of other biological processes, is enriched by inquiry into how these beautifully integrated patterns of reactions came into being.

SUMMARY

All cells extract energy from their environment and convert foodstuffs into cellular components by a highly integrated network of chemical reactions called metabolism. Most of the central molecules of metabolism are the same in all forms of life. Ribonucleotides such as ATP and NADH are especially prominent, reflecting their ancient origins. Moreover, many metabolic patterns are essentially the same in bacteria, plants, and animals.

The most valuable thermodynamic concept for understanding bioenergetics is free energy. A reaction can occur spontaneously only if the change in free energy (ΔG) is negative. A thermodynamically unfavorable reaction can be driven by a thermodynamically favorable one. ATP, the universal currency of energy in biological systems, is an energy-rich molecule because it contains two phosphoanhydride bonds. The repulsion between the negatively charged phosphate groups is reduced when ATP is hydrolyzed. Also, ADP and P_i are stabilized by resonance more than is ATP. The hydrolysis of ATP shifts the equilibrium of a coupled reaction by a factor of about 10^8.

The basic strategy of metabolism is to form ATP, NADPH, and building blocks for biosyntheses. ATP is consumed in muscle contraction and other motions of cells, active transport, signal transduction processes, and biosyntheses. NADPH, which carries two electrons at a high potential, provides reducing power in the biosynthesis of cell components from more-oxidized precursors. ATP and NADPH are continuously generated and consumed.

Vitamins are small biomolecules that are needed in small amounts in the diet of higher animals. The water-soluble ones are vitamin C (ascorbate, an antioxidant) and the vitamin B complex (components of coenzymes). Ascorbate is required for the hydroxylation of proline residues in collagen, which stabilizes the triple helix. The fat-soluble ones are vitamin A (a precursor of retinal), D (a regulator of calcium and phosphorus metabolism), E (an antioxidant in membranes), and K (a participant in the carboxylation of glutamate).

There are three stages in the extraction of energy from foodstuffs by aerobic organisms. In the first stage, large molecules are broken down into smaller ones, such as amino acids, sugars, and fatty acids. In the second stage, these small molecules are degraded to a few simple units that have a pervasive role in metabolism. One of them is the acetyl unit of acetyl CoA, a carrier of activated acyl groups. The third stage of metabolism is the citric acid cycle and oxidative phosphorylation, in which ATP is generated as electrons flow to O_2, the ultimate electron acceptor, and fuels are completely oxidized to CO_2. Most transfers of activated groups in metabolism are mediated by a recurring set of carriers.

Metabolism is regulated in a variety of ways. The amounts of some critical enzymes are controlled by regulation of the rate of protein synthesis and degradation. In addition, the catalytic activities of many enzymes are regulated by allosteric interactions (as in feedback inhibition) and by covalent modification. The movement of many substrates into cells and subcellular compartments is also controlled. Distinct pathways for biosynthesis and degradation also contribute to metabolic regulation. The energy charge, which depends on the relative amounts of ATP, ADP, and AMP, plays a role in metabolic regulation. A high energy charge inhibits ATP-generating (catabolic) pathways, whereas it stimulates ATP-utilizing (anabolic) pathways. The level of ATP and other key metabolites in living organisms can be monitored noninvasively by nuclear magnetic resonance.

SELECTED READINGS

Where to start

Radda, G.K., 1992. Control, bioenergetics, and adaptation in health and disease: Noninvasive biochemistry from nuclear magnetic resonance. *FASEB J.* 6:3032–3038.

McGrane, M.M., Yun, J.S., Patel, Y.M., and Hanson, R.W., 1992. Metabolic control of gene expression: In vivo studies with transgenic mice. *Trends Biochem. Sci.* 17:40–44.

DeLuca, H.F., 1988. The vitamin D story: A collaborative effort of basic science and clinical medicine. *FASEB J.* 2:224–236.

Shulman, R.G., 1983. NMR spectroscopy of living cells. *Sci. Amer.* 248(1):86–93.

Books

Krebs, H.A., and Kornberg, H.L., 1957. *Energy Transformations in Living Matter.* Springer-Verlag. [Includes a valuable appendix by K. Burton containing thermodynamic data.]

Linder, M.C. (ed.), 1991. *Nutritional Biochemistry and Metabolism* (2nd ed.). Elsevier.

Gottschalk, G., 1986. *Bacterial Metabolism* (2nd ed.). Springer-Verlag.

Nicholls, D.G., and Ferguson, S.J., 1992. *Bioenergetics 2.* Academic Press.

Martin, B.R., 1987. *Metabolic Regulation, A Molecular Approach.* Blackwell Scientific Publications.

Thermodynamics

Edsall, J.T., and Gutfreund, H., 1983. *Biothermodynamics: The Study of Biochemical Processes at Equilibrium.* Wiley. [A concise account with many informative examples.]

Klotz, I.M., 1967. *Energy Changes in Biochemical Reactions.* Academic Press. [A concise introduction, full of insight.]

Hill, T.L., 1977. *Free Energy Transduction in Biology.* Academic Press.

Alberty, R.A., 1993. Levels of thermodynamic treatment of biochemical reaction systems. *Biophys. J.* 65:1243–1254.

Alberty, R.A., and Goldberg, R.N., 1992. Standard thermodynamic formation properties for the adenosine 5′-triphosphate series. *Biochemistry* 31:10610–10615.

Alberty, R.A., 1968. Effect of pH and metal ion concentration on the equilibrium hydrolysis of adenosine triphosphate to adenosine diphosphate. *J. Biol. Chem.* 243:1337–1343.

Jencks, W.P., 1970. Free energies of hydrolysis and decarboxylation. *In* Sober, H.A., *Handbook of Biochemistry* (2nd ed.), pp. J181–J186. Chemical Rubber Co.

Goldberg, R.N., 1984. *Compiled Thermodynamic Data Sources for Aqueous and Biochemical Systems: An Annotated Bibliography (1930–1983).* National Bureau of Standards Special Publication 685, U.S. Government Printing Office.

Regulation of metabolism

Newsholme, E.A., and Start, C., 1973. *Regulation in Metabolism.* Wiley. [An excellent treatment of metabolic regulation in mammals.]

Atkinson, D.E., 1977. *Cellular Energy Metabolism and Its Regulation.* Academic Press. [Contains a detailed account of the concept of energy charge and of other aspects of metabolic control.]

Erecińska, M., and Wilson, D.F., 1978. Homeostatic regulation of cellular energy metabolism. *Trends Biochem. Sci.* 3:219–223. [Review of the regulatory role of the phosphorylation potential in mitochondria.]

Metabolic imaging

Avison, M.J., Hetherington, H.P., and Shulman, R.G., 1986. Applications of NMR to studies of tissue metabolism. *Ann. Rev. Biophys. Biophys. Chem.* 15:377.

Mark, F., Jeffrey, H., Rajagopal, A., Malloy, C.R., and Sherry, A.D., 1991. ^{13}C-NMR: A simple yet comprehensive method for analysis of intermediary metabolism. *Trends Biochem. Sci.* 16:5–10.

Belliveau, J.W., Kennedy, D.J., McKinstry, R.C., Buchbinder, B.R., Weisskoff, R.M., Cohen, M.S., Vevea, J.M., Brady, T.J., and Rosen, B.R., 1991. Functional mapping of the human visual cortex by magnetic resonance imaging. *Science* 254:716–719.

Gruetter, R., Novotny, E.J., Boulware, S.D., Rothman, D.L., Mason, G.F., Shulman, G.I., Shulman, R.G., and Tamborlane, W.V., 1992. Direct measurement of brain glucose concentrations in humans by ^{13}C NMR spectroscopy. *Proc. Nat. Acad. Sci.* 89:1109–1112.

Hausser, K.H., and Kalbitzer, H.R., 1991. *NMR in Medicine and Biology.* Springer-Verlag.

Boivin, M.J., Giordani, B., Berent, S., Amato, D.A., Lehtinen, S., Koeppe, R.A., Buchtel, H.A., Foster, N.L., and Kuhl, D.E., 1992. Verbal fluency and positron emission tomographic mapping of regional cerebral glucose metabolism. *Cortex* 28:231–239.

Raichle, M.E., 1990. Anatomical explorations of mind: Studies with modern imaging techniques. *Cold Spring Harbor Symp. Quant. Biol.* 55:983–986.

Ascorbate (vitamin C) and scurvy

Kivirikko, K.I., Myllylä, R., and Pihlajaneiemi, T., 1989. Protein hydroxylation: Prolyl 4-hydroxylase, an enzyme with four cosubstrates and a multifunctional subunit. *FASEB J.* 3:1609–1617.

Berg, R.A., and Prockop, D.J., 1973. The thermal transition of a non-hydroxylated form of collagen: Evidence for a role for hydroxyproline in stabilizing the triple-helix of collagen. *Biochem. Biophys. Res. Comm.* 52:115–120.

Sauberlich, H.E., 1994. Pharmacology of vitamin C. *Ann. Rev. Nutrit.* 14:371–392.

Jukes, T.H., 1992. Antioxidants, nutrition, and evolution. *Prevent. Med.* 21:270–276.

Nishikimi, M., and Udenfriend, S., 1977. Scurvy as an inborn error of ascorbic acid biosynthesis. *Trends Biochem. Sci.* 2:111–112.

Major, R.H. (ed.), 1945. *Classic Descriptions of Disease* (3rd ed.). Thomas. [Cartier's description of scurvy is given on p. 587.]

Historical aspects

Kalckar, H.M., 1991. 50 years of biological research—from oxidative phosphorylation to energy requiring transport regulation. *Ann. Rev. Biochem.* 60:1–37.

Kalckar, H.M. (ed.), 1969. *Biological Phosphorylations.* Prentice-Hall. [A valuable collection of many classic papers on bioenergetics.]

Fruton, J.S., 1972. *Molecules and Life.* Wiley-Interscience. [Perceptive and scholarly essays on the interplay of chemistry and biology since 1800. Metabolism and bioenergetics are among the topics treated in detail.]

Lipmann, F., 1971. *Wanderings of a Biochemist.* Wiley-Interscience. [Contains reprints of some of the author's classic papers and several delightful essays.]

PROBLEMS

1. *Flow of ~P.* What is the direction of each of the following reactions when the reactants are initially present in equimolar amounts? Use the data given in Table 17-1.
   (a) ATP + creatine $\rightleftharpoons$ creatine phosphate + ADP
   (b) ATP + glycerol $\rightleftharpoons$ glycerol 3-phosphate + ADP
   (c) ATP + pyruvate $\rightleftharpoons$ phosphoenolpyruvate + ADP
   (d) ATP + glucose $\rightleftharpoons$ glucose 6-phosphate + ADP

2. *A proper inference.* What information do the $\Delta G^{\circ\prime}$ data given in Table 17-1 provide about the relative rates of hydrolysis of pyrophosphate and acetyl phosphate?

3. *A potent donor.* Consider the reaction

   ATP + pyruvate $\rightleftharpoons$ phosphoenolpyruvate + ADP

   (a) Calculate $\Delta G^{\circ\prime}$ and K'_{eq} at 25°C for this reaction, using the data given in Table 17-1.
   (b) What is the equilibrium ratio of pyruvate to phosphoenolpyruvate if the ratio of ATP to ADP is 10?

4. *Isomeric equilibrium.* Calculate $\Delta G^{\circ\prime}$ for the isomerization of glucose 6-phosphate to glucose 1-phosphate. What is the equilibrium ratio of glucose 6-phosphate to glucose 1-phosphate at 25°C?

5. *~CH_3.* The formation of acetyl CoA from acetate is an ATP-driven reaction:

   Acetate + ATP + CoA $\rightleftharpoons$ acetyl CoA + AMP + PP_i

   (a) Calculate $\Delta G^{\circ\prime}$ for this reaction, using data given in this chapter.
   (b) The PP_i formed in the above reaction is rapidly hydrolyzed in vivo because of the ubiquity of inorganic pyrophosphatase. The $\Delta G^{\circ\prime}$ for the hydrolysis of PP_i is −8 kcal/mol. Calculate the $\Delta G^{\circ\prime}$ for the overall reaction. What effect does the hydrolysis of PP_i have on the formation of acetyl CoA?

6. *Acid strength.* The p*K* of an acid is a measure of its proton group–transfer potential.
   (a) Derive a relation between ΔG° and p*K*.
   (b) What is the ΔG° for the ionization of acetic acid, which has a p*K* of 4.8?

7. *Activated sulfate.* Fibrinogen contains tyrosine-*O*-sulfate. Propose an activated form of sulfate that could react in vivo with the aromatic hydroxyl group of a tyrosine residue in a protein to form tyrosine-*O*-sulfate.

8. *Dynamics revealed.* Two chemical species give rise to a single peak in an NMR spectrum if their rate of interconversion is fast compared with the difference between their resonance frequencies. For example, orthophosphate gives a single peak at pH values at which both $H_2PO_4^-$ and HPO_4^{2-} are present (see Figure 17-17). The chemical shifts of these forms of orthophosphate differ by 2.4 ppm relative to a standard compound that resonates at 129 MHz.
   (a) Calculate the difference between the resonance frequencies of these species.
   (b) What is the minimum interconversion rate of these species?
   (c) What is the minimum rate constant for the association of H^+ with HPO_4^{2-}?

9. *Raison d'être.* The muscles of some invertebrates are rich in *arginine phosphate* (phosphoarginine). Propose a function for this amino acid derivative. How would you test your hypothesis?

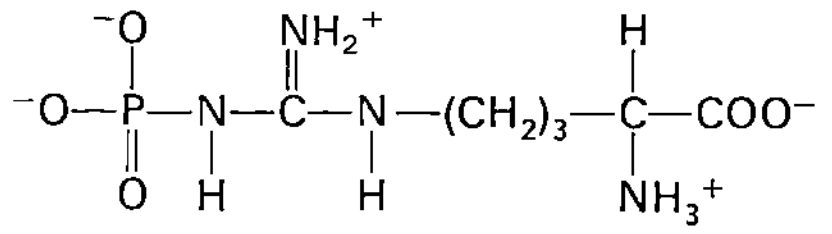

Arginine phosphate

10. *Recurring motif.* What is the structural feature common to ATP, FAD, NAD^+, and CoA?

CHAPTER

18

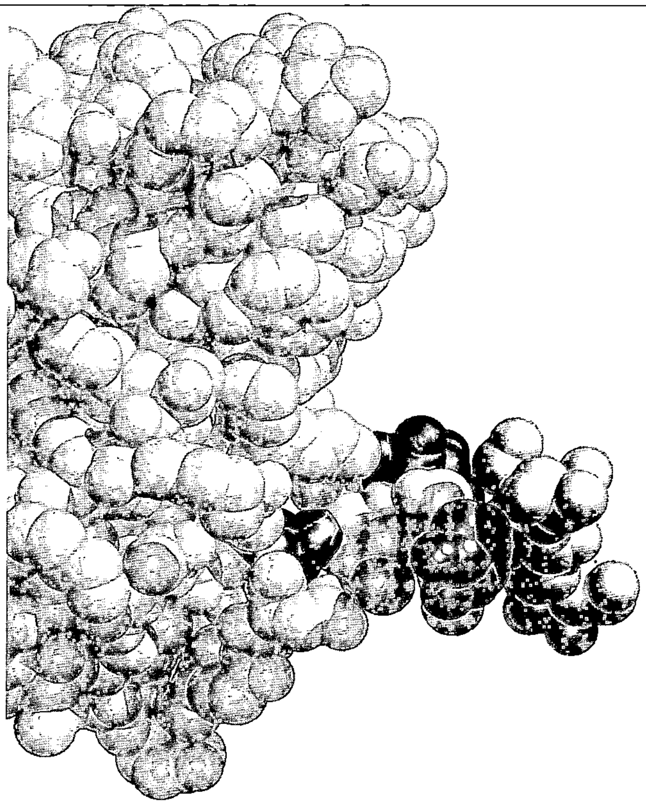

Carbohydrates

Before embarking on our metabolic journey, let us take an overview of carbohydrates, one of the four major classes of biomolecules. We have already considered the other three—proteins, nucleic acids, and lipids. Carbohydrates are aldehyde or ketone compounds with multiple hydroxyl groups. They make up most of the organic matter on earth because of their multiple roles in all forms of life. First, carbohydrates serve as *energy stores, fuels, and metabolic intermediates.* Starch in plants and glycogen in animals are polysaccharides that can be rapidly mobilized to yield glucose, a prime fuel for the generation of energy. ATP, the universal currency of free energy, is a phosphorylated sugar derivative, as are many coenzymes. Second, ribose and deoxyribose sugars form part of the *structural framework of RNA and DNA.* The conformational flexibility of these sugar rings is important in the storage and expression of genetic information. Third, polysaccharides are *structural elements in the cell walls of bacteria and plants, and in the exoskeletons of arthropods.* In fact, cellulose, the main constituent of plant cell walls, is one of the most abundant organic compounds in the biosphere. Fourth, carbohydrates are *linked to many proteins and lipids.* For example, the sugar units of glycophorin, an integral membrane protein, give red blood cells a highly polar anionic coat.

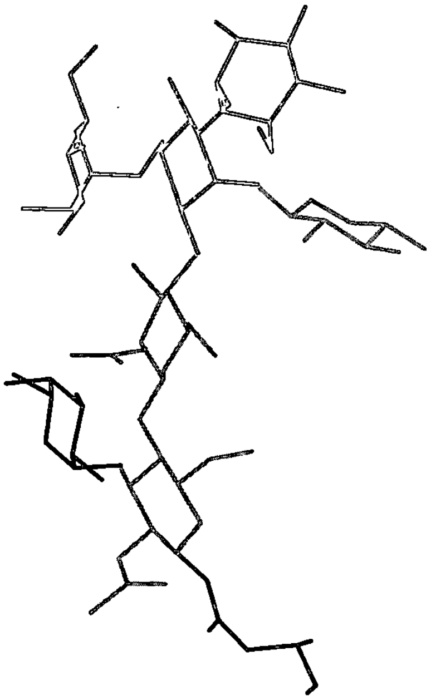

Opening Image: *Structure of a carbohydrate-bearing protein. The seven-residue oligosaccharide projects from the surface of the protein. Four kinds of carbohydrate units—N-acetylglucosamine (yellow), mannose (green), fucose (red), and xylose (orange)—are present in this branched chain. The oligosaccharide chain is attached to an asparagine residue (blue) of the protein. [Drawn from 1lte.pdb. B. Shaanan, H. Lis, and N. Sharon. Science 254(1991):862.]*

Recent studies have revealed that *carbohydrate units on cell surfaces play key roles in cell-cell recognition processes.* Fertilization begins with the binding of a sperm to a specific oligosaccharide on the surface of an egg. The adhesion of leukocytes to the lining of injured blood vessels and the return of lymphocytes to their sites of origin in lymph nodes further illustrate the importance of carbohydrates in recognition processes. Carbohydrates have entered the limelight as information-rich molecules, full of significance in development and repair.

MONOSACCHARIDES ARE ALDEHYDES OR KETONES WITH MULTIPLE HYDROXYL GROUPS

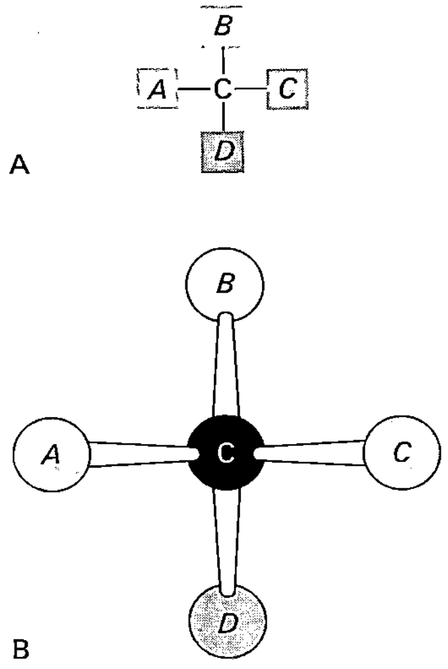

Figure 18-1
(A) Fischer representation of a tetrahedral carbon atom with substituents *A, B, C,* and *D;* and (B) a model, showing the stereochemistry denoted by this projection.

Monosaccharides, the simplest carbohydrates, are aldehydes or ketones that have two or more hydroxyl groups; the empirical formula of many is $(CH_2O)_n$. The smallest ones, for which $n = 3$, are glyceraldehyde and dihydroxyacetone. They are *trioses.* Glyceraldehyde is also an *aldose* because it contains an aldehyde group, whereas dihydroxyacetone is a *ketose* because it contains a keto group.

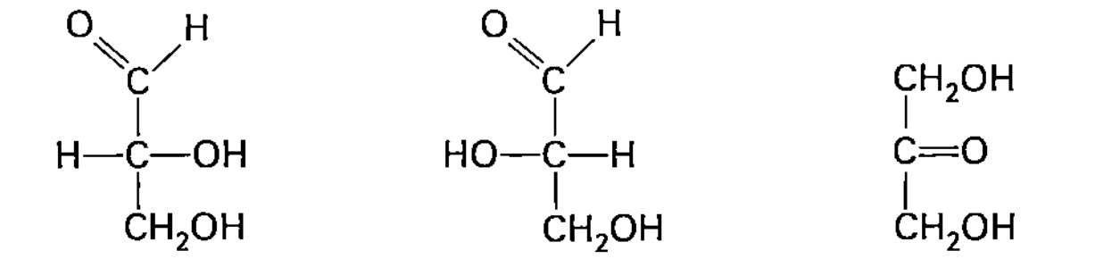

D-Glyceraldehyde (An aldose) **L-Glyceraldehyde** (An aldose) **Dihydroxyacetone** (A ketose)

Glyceraldehyde has a single asymmetric carbon. Thus, there are two stereoisomers of this three-carbon aldose, D-glyceraldehyde and L-glyceraldehyde. The prefixes D and L designate the absolute configuration. Recall that in a Fischer projection of a molecule, atoms joined to an asymmetric carbon atom by horizontal bonds are in front of the plane of the page, and those joined by vertical bonds are behind (Figure 18-1).

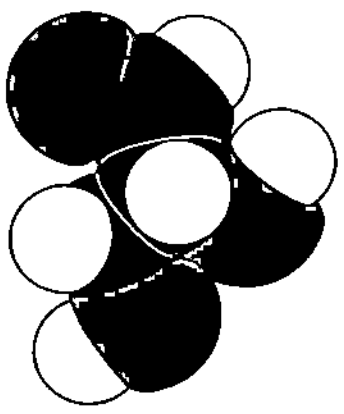

Figure 18-2
Model showing the absolute configuration of D-glyceraldehyde. Fischer's arbitrary assignment of the D configuration to this stereoisomer was shown years later, by x-ray crystallography, to be correct.

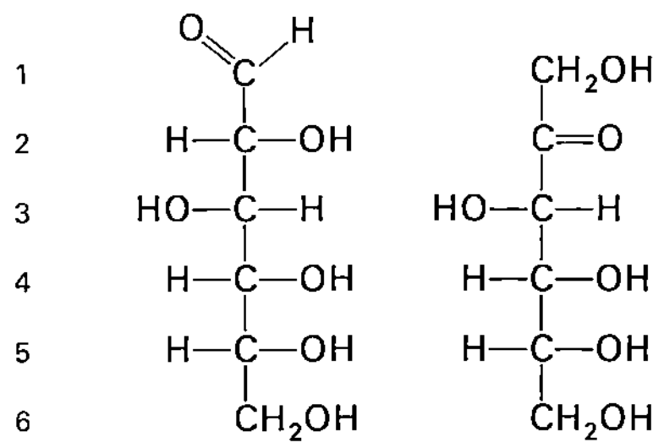

D-Glucose (An aldose) **D-Fructose** (A ketose)

Aldoses with 4, 5, 6, and 7 carbon atoms are called *tetroses, pentoses, hexoses,* and *heptoses.* Two common hexoses are D-*glucose* (an aldose) and D-*fructose* (a ketose). For sugars with more than one asymmetric carbon atom, the symbols D and L refer to the absolute configuration of the asymmetric carbon farthest from the aldehyde or keto group. These hexoses belong to the D series because their configuration at C-5 is the same as that in D-glyceraldehyde.

In general, a molecule with n asymmetric centers and no plane of symmetry has 2^n stereoisomeric forms. For aldotrioses, $n = 1$, and so there are two stereoisomers, D- and L-glyceraldehyde. They are *enantiomers* (mir-

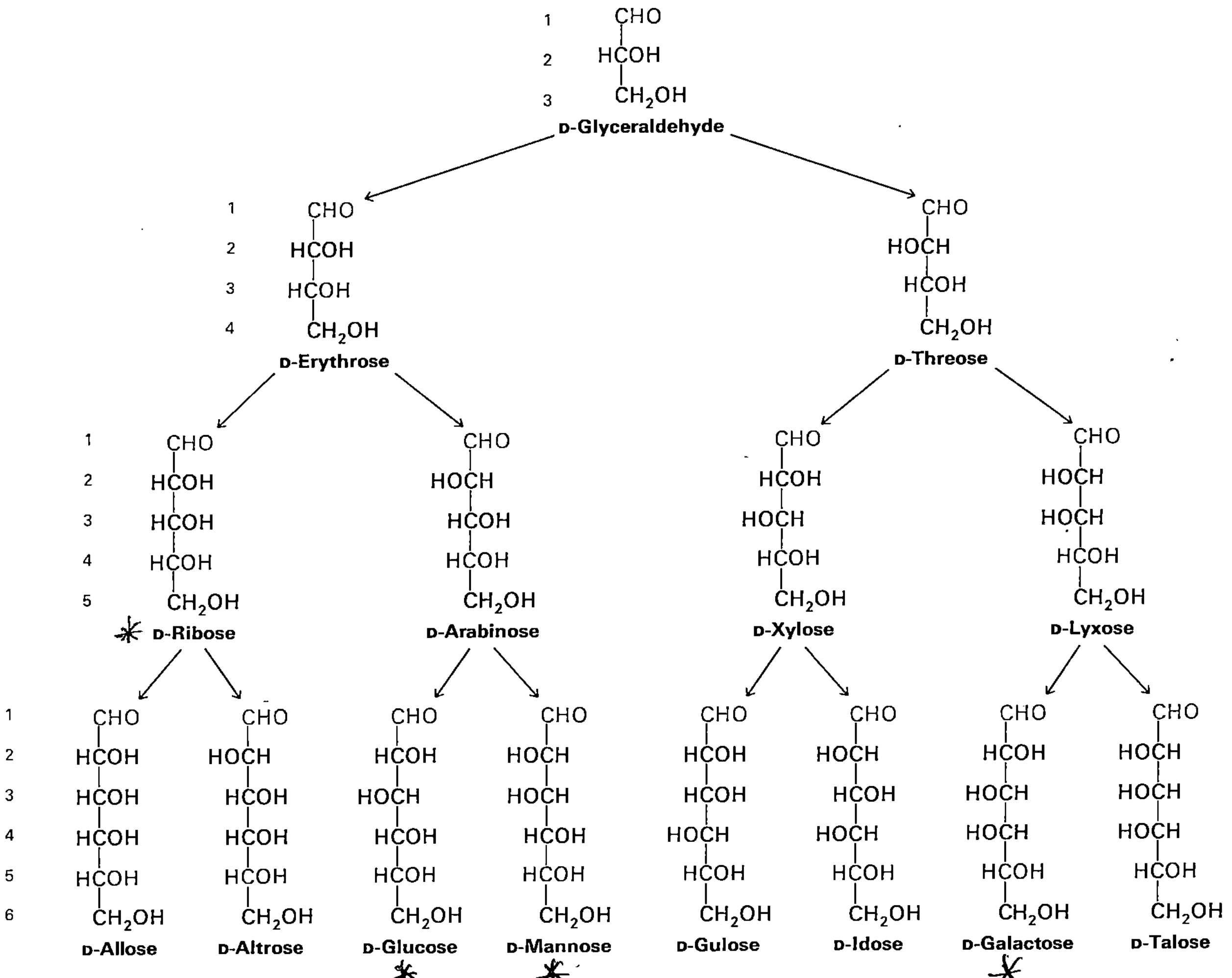

Figure 18-3
Stereochemical relations of D aldoses containing three, four, five, and six carbon atoms. These sugars are D aldoses because they contain an aldehyde group (shown in green) and have the configuration of D-glyceraldehyde at their farthest asymmetric center (shown in red).

* Know these carbos

ror images) of each other. Addition of an HCOH group gives four aldotetroses because $n = 2$. Two of them are D sugars and the other two are the enantiomeric L sugars. Let us follow the D sugar series (Figure 18-3). One of these four-carbon aldoses is D-erythrose and the other is D-threose. They have the same configuration at C-3 (because they are D sugars) but opposite configurations at C-2. They are *diastereoisomers*, not enantiomers, because they are not mirror images of each other.

The five-carbon aldoses have three asymmetric centers, which give 8 (2^3) stereoisomers, 4 in the D series. D-*Ribose* belongs to this group. The six-carbon aldoses have four asymmetric centers, and so there are 16 (2^4) stereoisomers, 8 in the D series. D-Glucose, D-mannose, and D-galactose are abundant six-carbon aldoses. Note that D-glucose and D-mannose differ only in configuration at C-2. D Sugars differing in configuration at a single asymmetric center are *epimers*. Thus, D-glucose and D-mannose are epimers at C-2; D-glucose and D-galactose are epimers at C-4. Emil Fischer's elucidation in 1891 of the configuration of D-glucose was a remarkable achievement that greatly stimulated the field of organic chemistry.

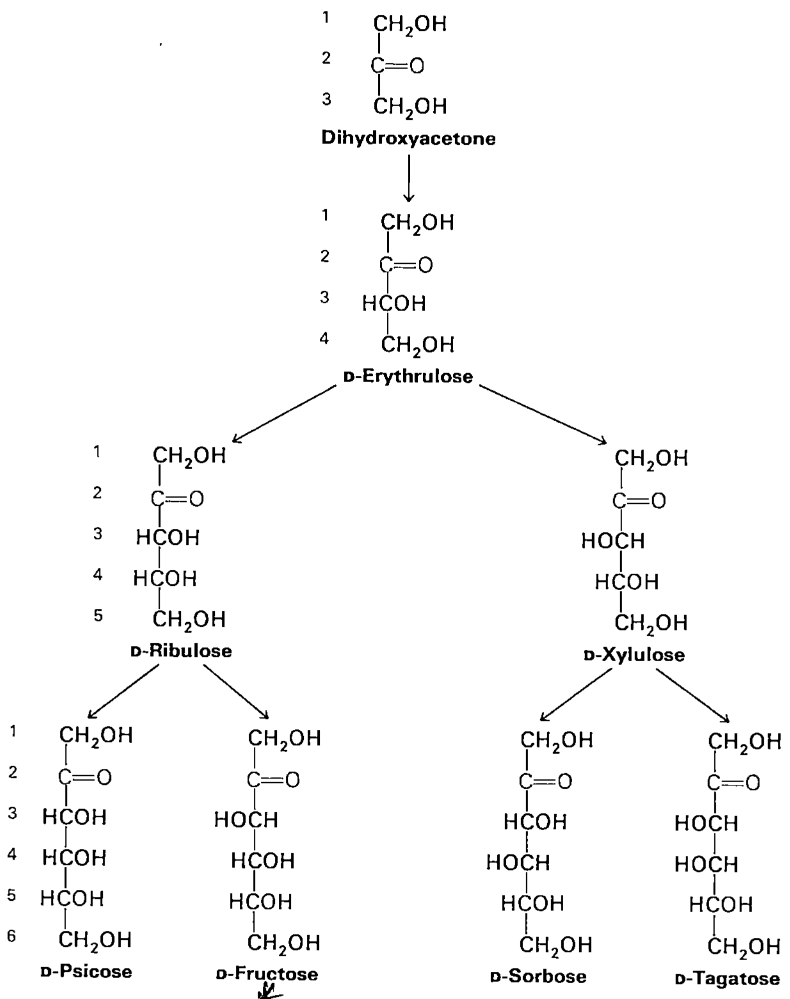

Figure 18-4
Stereochemical relations of D ketoses containing three, four, five, and six carbon atoms. These sugars are D ketoses because they contain a keto group (shown in green) and have the configuration of D-glyceraldehyde at their farthest asymmetric center (shown in red).

The stereochemical relations of D ketoses are shown in Figure 18-4. Dihydroxyacetone, the simplest of these sugars, is optically inactive. D-Erythrulose is the sole four-carbon D ketose because ketoses have one fewer asymmetric center than do aldoses with the same number of carbon atoms. Hence, there are two five-carbon and four six-carbon D ketoses. D-*Fructose* is the most abundant ketohexose.

PENTOSES AND HEXOSES CYCLIZE TO FORM FURANOSE AND PYRANOSE RINGS

The predominant forms of glucose and fructose in solution are not open chains. Rather, the open-chain forms of these sugars cyclize into rings. In general, an aldehyde can react with an alcohol to form a *hemiacetal*.

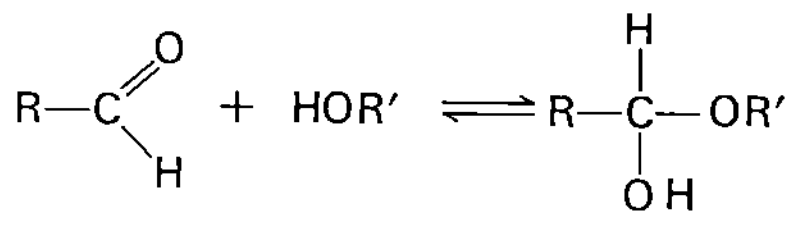

Aldehyde **Alcohol** **Hemiacetal**

The C-1 aldehyde in the open-chain form of glucose reacts with the C-5 hydroxyl group to form an *intramolecular hemiacetal*. The resulting six-membered ring is called *pyranose* because of its similarity to *pyran*.

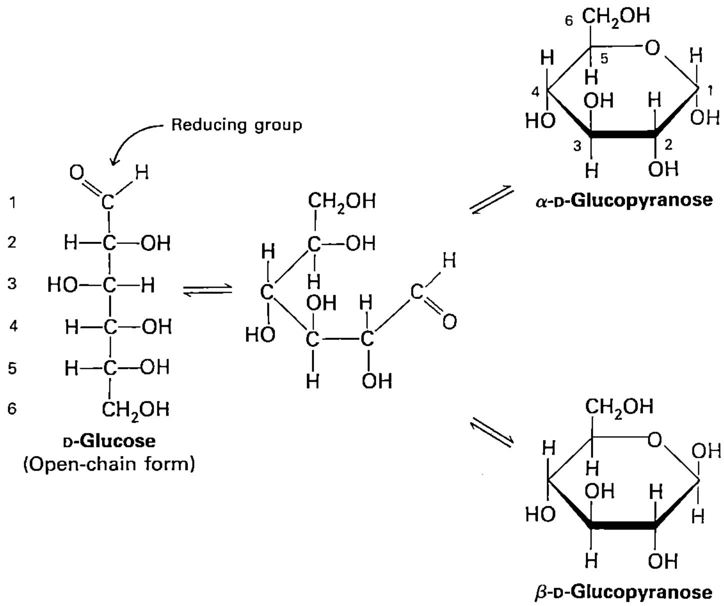

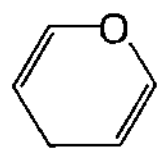
Pyran

Similarly, a ketone can react with an alcohol to form a *hemiketal*.

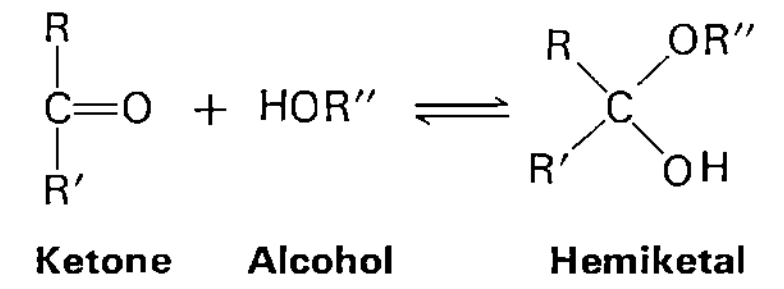

The C-2 keto group in the open-chain form of fructose can react with the C-5 hydroxyl group to form an *intramolecular hemiketal*. This five-membered ring is called *furanose* because of its similarity to *furan*.

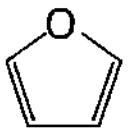
Furan

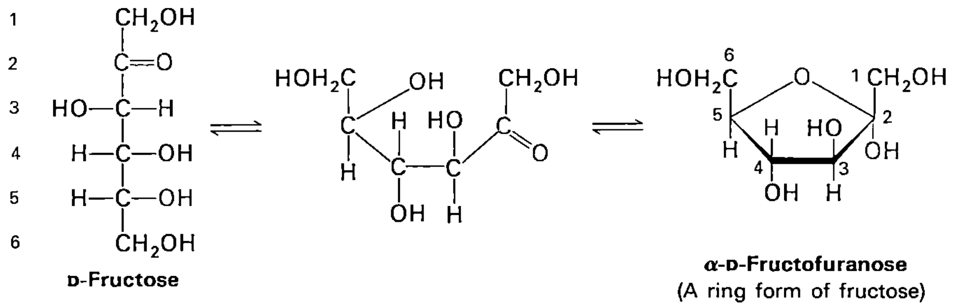

The depictions of glucopyranose and fructofuranose on this page are *Haworth projections*. In such a projection, the carbon atoms in the ring are not explicitly shown. The approximate plane of the ring is perpendicular to the plane of the paper, with the heavy line on the ring projecting toward the reader.

An additional asymmetric center is created when glucose cyclizes. C-1, the carbonyl carbon atom in the open-chain form, becomes an asymmetric center in the ring form of the sugar. Two ring structures can be formed: α-D-glucopyranose and β-D-glucopyranose. For D sugars drawn as Haworth projections, *the designation α means that the hydroxyl group attached to C-1 is below the plane of the ring; β means that it is above the plane of the ring.* The C-1 carbon is called the *anomeric carbon atom,* and so the α and β forms are *anomers.*

The same nomenclature applies to the furanose ring form of fructose, except that α and β refer to the hydroxyl groups attached to C-2, the anomeric carbon atom. Fructose also forms pyranose rings. In fact, the pyranose form predominates in fructose free in solution, whereas the furanose form is the major one in most of its derivatives.

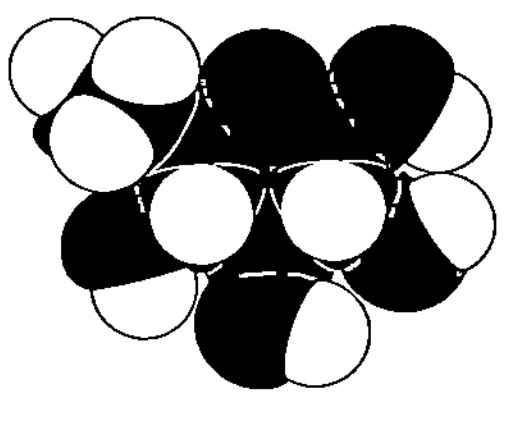

Figure 18-5
Model of β-D-glucopyranose.

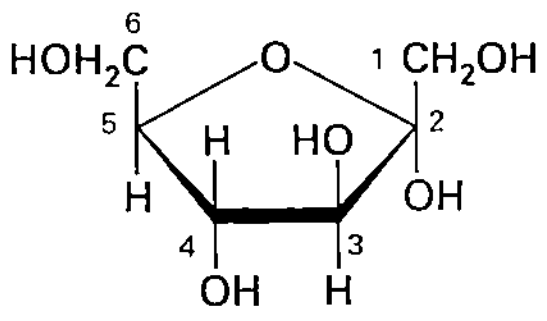

α-D-Fructofuranose

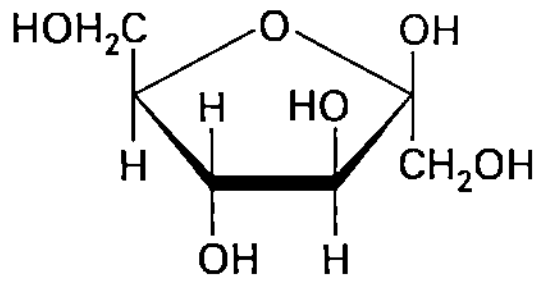

β-D-Fructofuranose

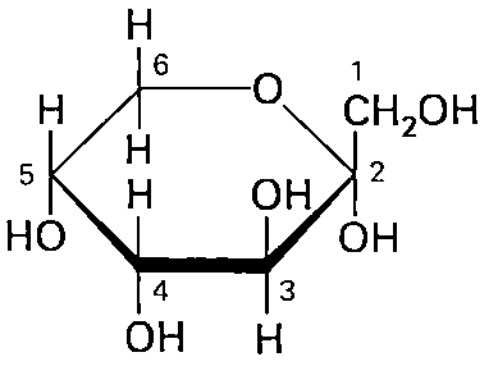

α-D-Fructopyranose

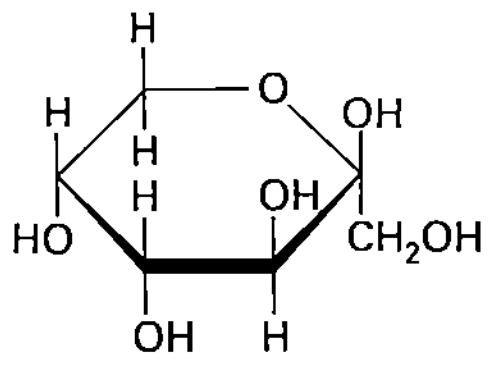

β-D-Fructopyranose

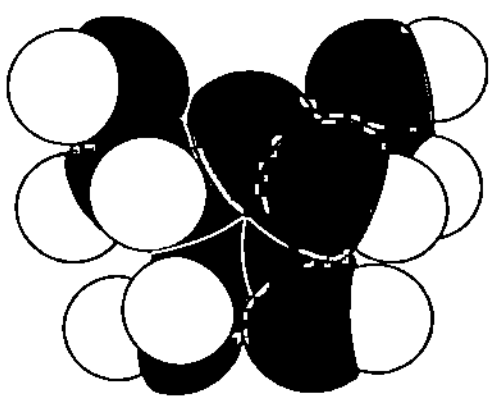

Figure 18-6
Model of β-D-fructofuranose.

Five-carbon sugars such as D-ribose and 2-deoxy-D-ribose form furanose rings, as was exemplified by the structure of these units in RNA and DNA.

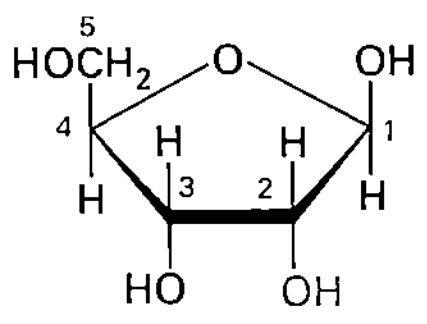

β-D-Ribofuranose

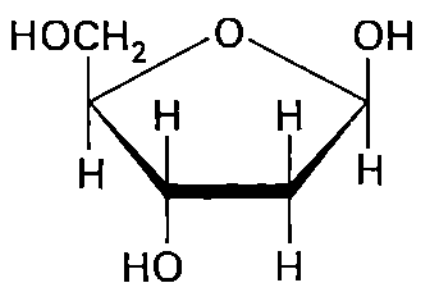

2-Deoxy-β-D-ribofuranose

Reducing sugars—
Sugars containing a free aldehyde or keto group reduce indicators such as cupric ion (Cu^{2+}) complexes to the cuprous form (Cu^{+}). The reducing agent in these reactions is the open-chain form of the aldose or ketose. The reducing end of a sugar is the one containing a free aldehyde or keto group.

In water, α-D-glucopyranose and β-D-glucopyranose interconvert through the open-chain form to give an equilibrium mixture. This interconversion was detected many years ago by following changes in *optical rotation* and was called *mutarotation.* An equilibrium mixture contains about one-third α anomer, two-thirds β anomer, and very little ($<1\%$) of the open-chain form. Likewise, the α and β anomers of both the pyranose and the furanose forms of fructose interconvert through the open-chain form. Some cells contain *mutarotases* that accelerate the interconversion of anomeric sugars. We shall use the terms *glucose* and *fructose* to refer to the equilibrium mixture of the open-chain and ring forms.

CONFORMATION OF PYRANOSE AND FURANOSE RINGS

The six-membered pyranose ring, like cyclohexane, cannot be planar because of the tetrahedral geometry of its saturated carbon atoms. Instead, pyranose rings adopt *chair* and *boat* conformations (Figure 18-7). The substituents on the ring carbon atoms have two orientations: *axial* and *equatorial.* Axial bonds are nearly perpendicular to the average plane of the ring, whereas equatorial bonds are nearly parallel to this plane. As can be seen from Figure 18-7 (or even better, from actual molecular models in your hands), axial substituents emerge above and below the average plane of the ring, whereas equatorial substituents emerge at the periphery. Axial substituents sterically hinder each other if they emerge on the same side of the ring (e.g., 1,3-diaxial groups). In contrast, equatorial substituents are less crowded. *The chair form of β-D-glucopyranose predominates because all axial positions are occupied by hydrogen atoms.* The bulkier $-OH$ and $-CH_2OH$ groups emerge at the less hindered periphery. In contrast, the boat form of glucose is highly disfavored because it is so sterically hindered.

Furanose rings, like pyranose rings, are not planar. They can be puckered so that four atoms are nearly coplanar and the fifth is about 0.5 Å away from this plane. This conformation is called an *envelope form* because the structure resembles an opened envelope with the back flap raised (Figure 18-8). In the ribose moiety of most biomolecules, either C-2 or C-3 is out of plane on the same side as C-5. These conformations are called $C_{2'}$-*endo* and $C_{3'}$-*endo*. As will be discussed further in Chapter 31, the sugars in RNA are in the $C_{3'}$-*endo* form, whereas the sugars in the Watson-Crick DNA double helix are in the $C_{2'}$-*endo* form. Furanose rings can interconvert rapidly between different conformational states. They are more flexible than pyranose rings, which may account for their selection as components of RNA and DNA.

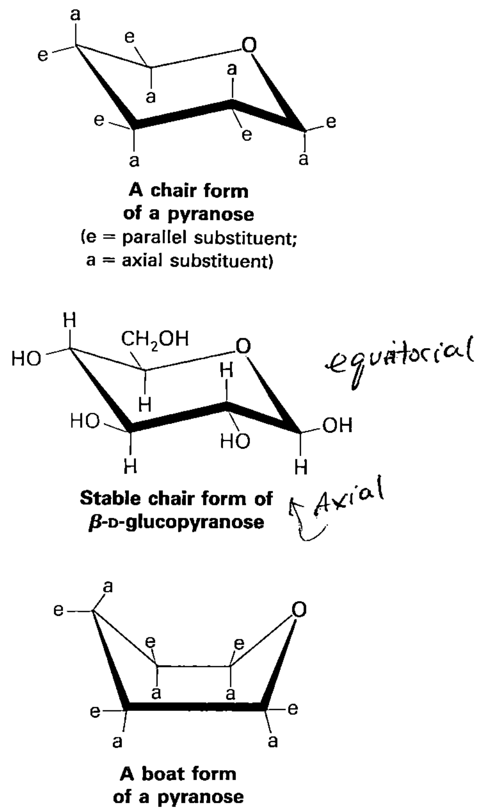

Figure 18-7
Chair and boat conformations of pyranose rings. For sugars with large equatorial substituents, the chair form is energetically more favorable because it is less hindered.

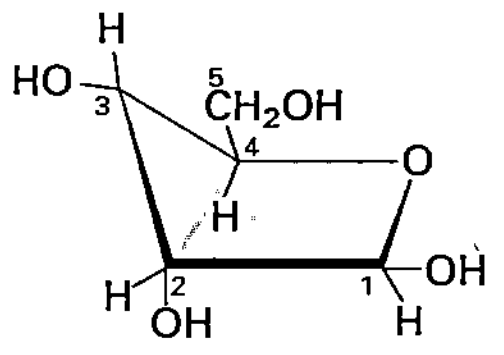

Figure 18-8
An envelope form of β-D-ribose. The $C_{3'}$-*endo* conformation is shown. C-3 is out of plane on the same side as C-5.

CARBOHYDRATES ARE JOINED TO ALCOHOLS AND AMINES BY GLYCOSIDIC BONDS

When glucose is warmed in anhydrous methanol containing HCl, its anomeric carbon atom reacts with the hydroxyl group of the alcohol to form two acetals, *methyl α-D-glucopyranoside* and *methyl β-D-glucopyranoside.* Acid facilitates removal of the $-OH$ group by protonating the anomeric carbon atom.

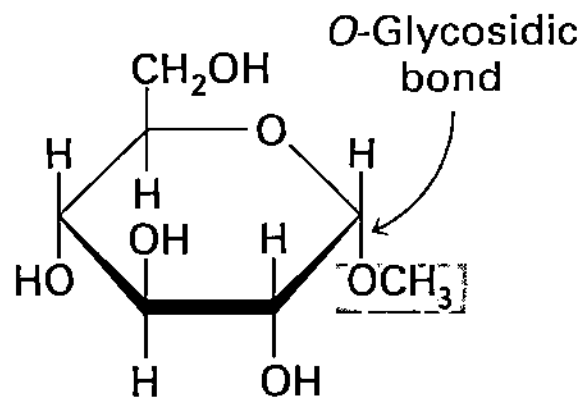

Methyl α-D-glucopyranoside

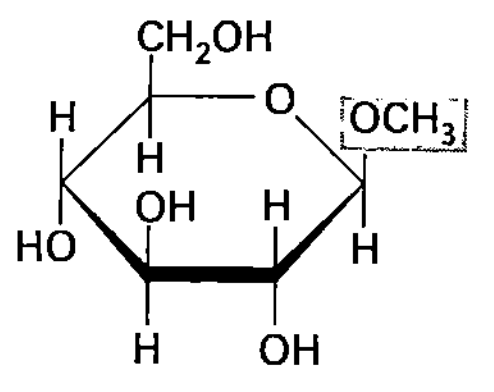

Methyl β-D-glucopyranoside

The new bond between C-1 of glucose and the oxygen atom of methanol is called a *glycosidic bond*—specifically, an *O*-glycosidic bond. Sugars can be linked to each other by *O*-glycosidic bonds to form disaccharides and polysaccharides. In cellulose, for example, D-glucose residues are joined

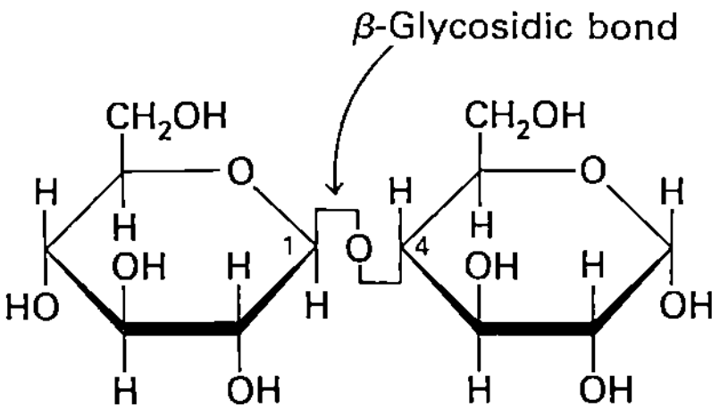

Cellobiose
(β-D-Glucopyranosyl-(1→4)-α-D-glucopyranose)

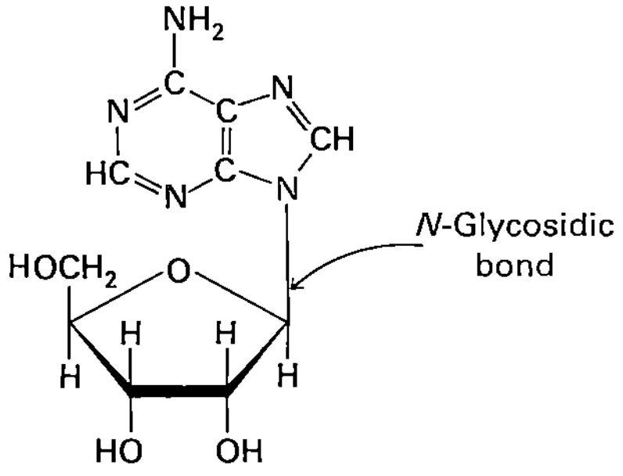

Adenosine

by glycosidic linkages between C-1 of one sugar and the hydroxyl oxygen atom on C-4 of an adjacent sugar. The glycosidic bonds in cellulose have a β configuration. In other words, the bond emerging from C-1 lies above the plane of the ring when viewed in the standard orientation. Hence, glucose units in cellulose are joined by β(1→4) glycosidic linkages, which can be concisely denoted by the abbreviation β-1,4. Recall that *N*-acetylmuramate and *N*-acetylglucosamine sugars in bacterial cell wall polysaccharides also are joined by β-1,4 linkages (p. 208). By contrast, amylose is an α-1,4 polymer of glucose.

The anomeric carbon atom of a sugar can be linked to the nitrogen atom of an amine by an *N*-glycosidic bond. The crucial biological importance of this type of glycosidic linkage is evident in such central biomolecules as nucleotides, RNA, and DNA. *N*-glycosidic linkages in virtually all naturally occurring biomolecules have the β configuration.

PHOSPHORYLATED SUGARS ARE KEY INTERMEDIATES IN ENERGY GENERATION AND BIOSYNTHESES

Phosphorylated sugars are another important class of derivatives. In the next chapter, we shall see that the first step in glycolysis, the breakdown of glucose to obtain energy, is its conversion to *glucose 6-phosphate.* The transfer of a phosphoryl group from ATP to the C-6 hydroxyl group of glucose is catalyzed by hexokinase. Several subsequent intermediates in this metabolic pathway, such as dihydroxyacetone phosphate and glyceraldehyde 3-phosphate, are phosphorylated sugars. In fact, one of the strategies of glycolysis is to form three-carbon intermediates that can transfer their phosphate groups to ADP to achieve a net synthesis of ATP.

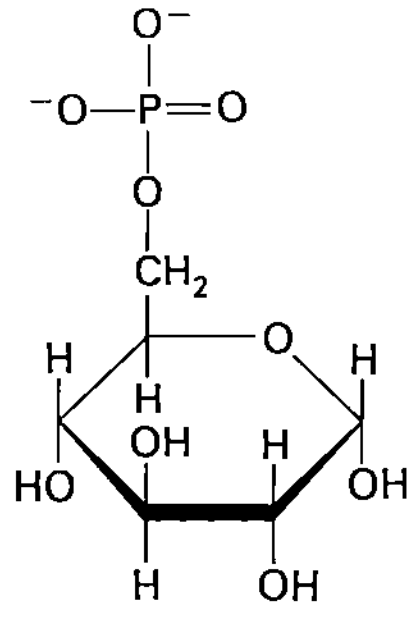

Glucose 6-phosphate

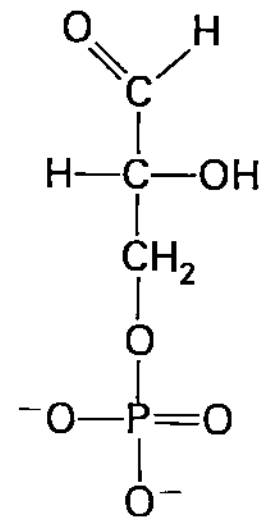

Glyceraldehyde 3-phosphate

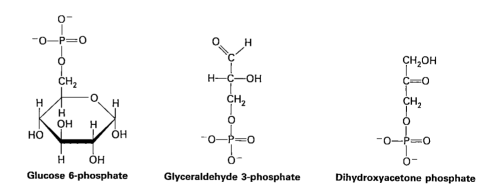

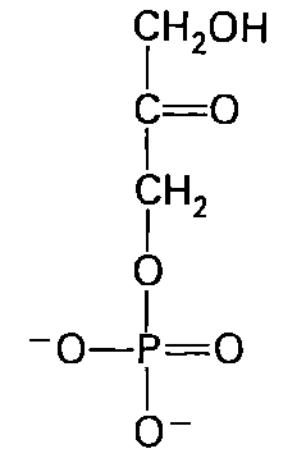

Dihydroxyacetone phosphate

Phosphorylation also serves to make sugars anionic. The p*K* values of a sugar phosphate group are about 2.1 and 6.8. Hence, at an intracellular pH of 7.4, the net charge of a sugar phosphate such as glucose 6-phosphate is about −1.8. Such a group can have strong electrostatic interactions with the active site of an enzyme. The negative charge contributed by phosphorylation also prevents these sugars from spontaneously crossing lipid bilayer membranes. Phosphorylation helps to retain biomolecules inside cells, an effect whimsically characterized as "the importance of being ionized."

Another function of phosphorylation is the *creation of reactive intermediates* for the formation of *O*- and *N*-glycosidic linkages. For example, a multiply phosphorylated derivative of ribose plays key roles in the biosyn-

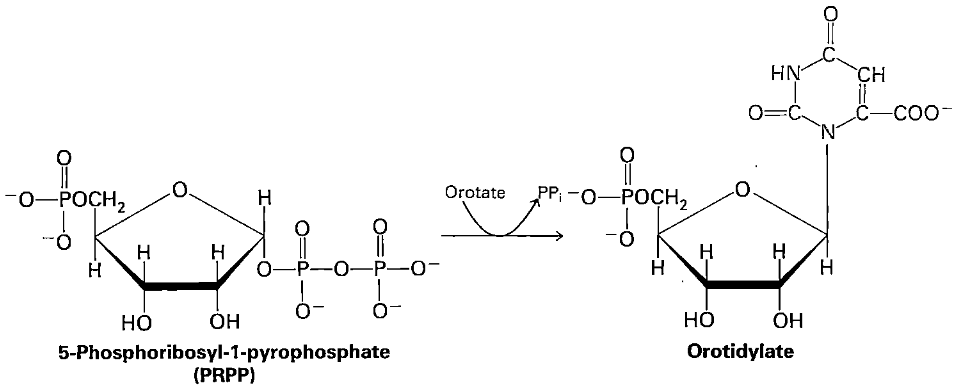

Figure 18-9
The *N*-glycosidic linkage of pyrimidine nucleotides is formed by displacement of the pyrophosphate group of PRPP, the activated intermediate. The configuration of the glycosidic bond is inverted in this reaction.

theses of purine and pyrimidine nucleotides (p. 741). The pyrophosphate group of 5-phosphoribosyl-1-pyrophosphate (PRPP) is displaced by a nitrogen atom of a free pyrimidine (orotate) to form a pyrimidine nucleotide (orotidylate) (Figure 18-9).

SUCROSE, LACTOSE, AND MALTOSE ARE THE COMMON DISACCHARIDES

Disaccharides consist of two sugars joined by an *O*-glycosidic bond. Three abundant disaccharides are sucrose, lactose, and maltose (Figure 18-10). *Sucrose* (common table sugar) is obtained commercially from cane or beet. The anomeric carbon atoms of a glucose unit and a fructose unit are joined in this disaccharide; the configuration of this glycosidic linkage is α for glucose and β for fructose. *Consequently, sucrose lacks a free reducing group (an aldehyde or ketone end), in contrast with most other sugars.* The hydrolysis of sucrose to glucose and fructose is catalyzed by *sucrase* (also called *invertase* because hydrolysis changes the optical activity from dextro- to levorotatory).

Lactose, the disaccharide of milk, consists of galactose joined to glucose by a β-1,4 glycosidic linkage (see Figure 18-10 and Figure 18-11). Lactose

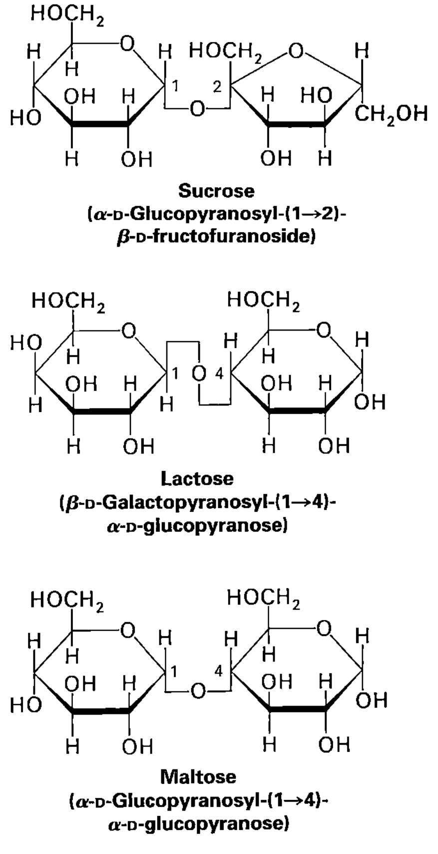

Figure 18-10
Formulas of three common disaccharides: sucrose, lactose, and maltose. The α configuration of the anomeric carbon atom at the reducing end of maltose and lactose is shown here.

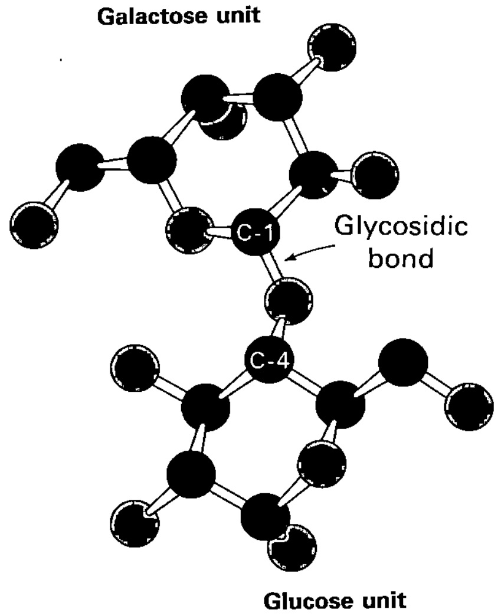

Figure 18-11
Model of lactose. Galactose is linked to glucose by a β-1,4 glycosidic bond.

is hydrolyzed to these monosaccharides by *lactase* in humans (by *β-galactosidase* in bacteria). In *maltose,* two glucose units are joined by an α-1,4 glycosidic linkage. Maltose comes from the hydrolysis of starch and is in turn hydrolyzed to glucose by *maltase.* Sucrase, lactase, and maltase are located on the outer surface of epithelial cells lining the small intestine. These cells have many fingerlike folds called *microvilli* that markedly increase their surface area for digestion and absorption of nutrients.

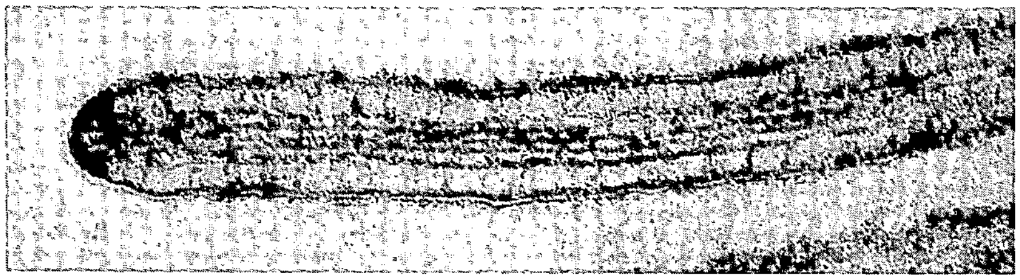

Figure 18-12
Electron micrograph of a microvillus projecting from an intestinal epithelial cell. Lactase and other enzymes that hydrolyze carbohydrates are present on the outer face of the plasma membrane. The filaments inside the microvillus contain actin, a contractile protein. [From M.S. Mooseker and L.G. Tilney. *J. Cell Biol.* 67(1975):725.]

MOST ADULTS ARE INTOLERANT OF MILK BECAUSE THEY ARE DEFICIENT IN LACTASE

Nearly all infants and children are able to digest lactose. In contrast, a majority of the adults in the world are deficient in *lactase,* which makes them intolerant of milk. In a lactase-deficient adult, lactose accumulates in the lumen of the small intestine after ingestion of milk because there is no mechanism for the uptake of this disaccharide. The large osmotic effect of the unabsorbed lactose leads to an influx of fluid into the small intestine. Hence, the clinical symptoms of lactose intolerance are abdominal distention, nausea, cramping, pain, and a watery diarrhea. Lactase deficiency appears to be inherited as an autosomal recessive trait and is usually first expressed in adolescence or young adulthood. The prevalence of lactase deficiency in human populations varies greatly. For example, 3% of Danes are deficient in lactase, compared with 97% of Thais. Human populations that do not consume milk in adulthood generally have a high incidence of lactase deficiency, which is also characteristic of other mammals. Milk treated with lactase is available for consumption by lactose-intolerant people. The capacity of humans to digest lactose in adulthood seems to have evolved since the domestication of cattle some ten thousand years ago.

GLYCOGEN, STARCH, AND DEXTRAN ARE MOBILIZABLE STORES OF GLUCOSE

Animal cells store glucose in the form of glycogen. As will be discussed in detail in Chapter 23, glycogen is a very large, branched polymer of glucose residues. Most of the glucose units in glycogen are linked by α-1,4-glycosidic bonds. The branches are formed by α-1,6-glycosidic bonds, which occur about once in ten units (Figure 18-13). These branches serve to increase the solubility of glycogen and make its sugar units accessible. They are released from the many nonreducing ends of this highly branched carbohydrate store.

The nutritional reservoir in plants is *starch,* of which there are two forms. *Amylose,* the unbranched type of starch, consists of glucose residues in α-1,4 linkage. *Amylopectin,* the branched form, has about one α-1,6 linkage per thirty α-1,4 linkages, and so it is like glycogen except for its lower degree of branching.

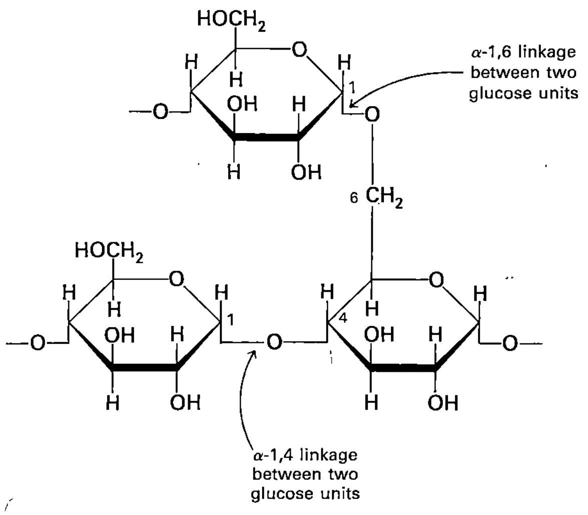

Figure 18-13
A branch in glycogen is formed by an α-1,6 glycosidic linkage.

More than half the carbohydrate ingested by humans is starch. Both amylopectin and amylose are rapidly hydrolyzed by *α-amylase,* which is secreted by the salivary glands and the pancreas. α-Amylase, an endoglycosidase, hydrolyzes internal α-1,4 linkages to yield *maltose, maltotriose,* and *α-dextrin.* Maltose consists of two glucose residues in α-1,4 linkage (p. 471), and maltotriose of three such residues. α-Dextrin is made up of several glucose units joined by an α-1,6 linkage in addition to α-1,4 linkages. Maltose and maltotriose are hydrolyzed to glucose by *maltase,* whereas α-dextrin is hydrolyzed to glucose by *α-dextrinase.* Malt contains *β-amylase,* an enzyme that hydrolyzes starch into maltose by sequential removal of disaccharide units from nonreducing ends. Malt derived from barley or other grains is used to make beer.

Dextran, a storage polysaccharide in yeasts and bacteria, also consists only of glucose residues, but differs from glycogen and starch in that nearly all linkages are α-1,6. Occasional branches are formed by α-1,2, α-1,3, or α-1,4 linkages, depending on the species.

Contemporary drawing depicting a 16th century German brewery. Barley was steeped in water for several days and then allowed to germinate in a warm damp room. β-Amylase and maltase produced by the sprouts then digest the starch into glucose, which is fermented. [*Der Bierbreuwer,* by Jost Ammon.]

CELLULOSE, THE MAJOR STRUCTURAL POLYMER OF PLANTS, CONSISTS OF LINEAR CHAINS OF GLUCOSE UNITS

Cellulose, the other major polysaccharide of plants, serves a structural rather than a nutritional role. *Cellulose is one of the most abundant organic compounds in the biosphere.* Some 10^{15} kg of cellulose is synthesized and degraded on earth each year! It is an unbranched polymer of glucose residues joined by β-1,4 linkages. The β configuration allows cellulose to form very long straight chains (Figure 18-14). Each glucose residue is related to the next by a rotation of 180°, and the ring oxygen atom of one is hydrogen-bonded to the 3-OH group of the next. Fibrils are formed by

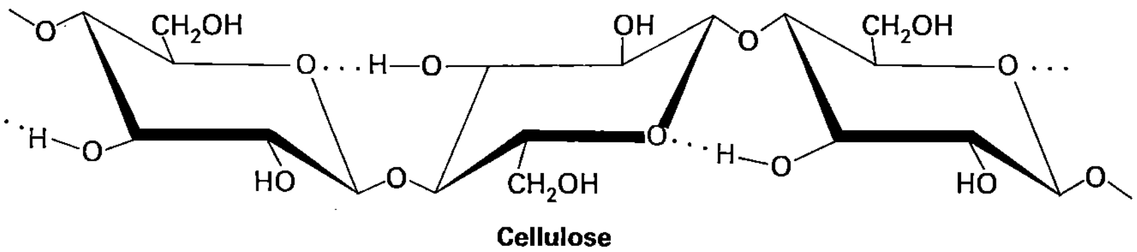

Figure 18-14
Schematic diagram showing the conformation of cellulose. The structure is stabilized by hydrogen bonds between adjacent glucose units in the same strand. In fibrils of cellulose, hydrogen bonds are formed between different strands as well.

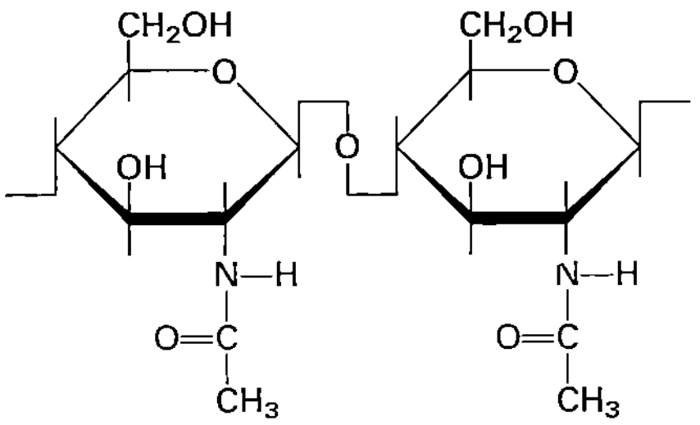

Chitin
(NAG-β(1→4)-NAG repeat)

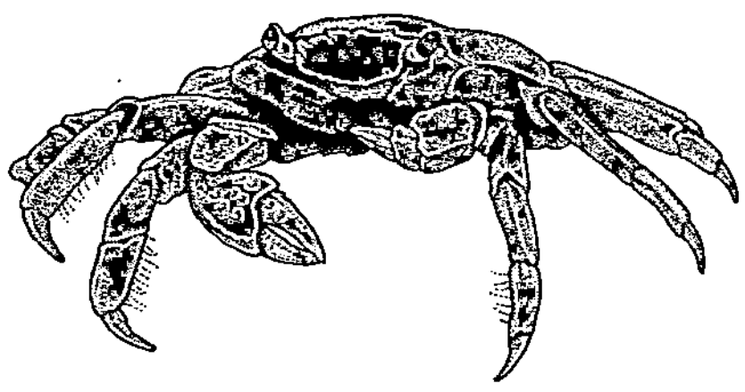

A Sally lightfoot crab. The exoskeleton of this arthropod is rich in chitin, one of the most abundant biopolymers on earth.

parallel chains. The α-1,4 linkages in glycogen and starch produce a very different molecular architecture. A hollow helix is formed instead of a straight chain. These differing consequences of the α and β linkages are biologically important. *The straight chain formed by β linkages is optimal for the construction of fibers having a high tensile strength. In contrast, the open helix formed by α linkages is well suited to forming an accessible store of sugar.*

Mammals lack *cellulases* and therefore cannot digest wood and vegetable fibers. However, some ruminants harbor cellulase-producing bacteria in their digestive tracts and thus can digest cellulose. Fungi and protozoa also secrete cellulases. In fact, the digestion of wood by termites depends on protozoa in their gut, a mutually beneficial association.

The exoskeletons of insects and crustacea contain *chitin,* which consists of *N*-acetylglucosamine residues in β-1,4 linkage. Chitin forms long straight chains that serve a structural role. Thus, chitin is like cellulose except that the substituent at C-2 is an acetylated amino group instead of a hydroxyl group.

GLYCOSAMINOGLYCANS ARE ANIONIC POLYSACCHARIDE CHAINS MADE OF REPEATING DISACCHARIDE UNITS

A different kind of repeating polysaccharide is present on the cell surface and in the extracellular matrix of animals. Many *glycosaminoglycans* are made of *disaccharide repeating units* containing a derivative of an *amino sugar,* either glucosamine or galactosamine. At least one of the sugars in the repeating unit has a *negatively charged carboxylate or sulfate group.* Chondroitin sulfate, keratan sulfate, heparin, heparan sulfate, dermatan sulfate, and hyaluronate are the major glycosaminoglycans (Figure 18-15). Heparin is synthesized as a nonsulfated proteoglycan, which is then deacetylated and sulfated. Incomplete modification leads to a mixture of variously sulfated sequences. One of them acts as an anticoagulant by binding specifically to antithrombin, which accelerates its sequestration of thrombin (p. 257). Heparan sulfate is like heparin except that it has fewer *N*- and *O*-sulfate groups and more *N*-acetyl groups.

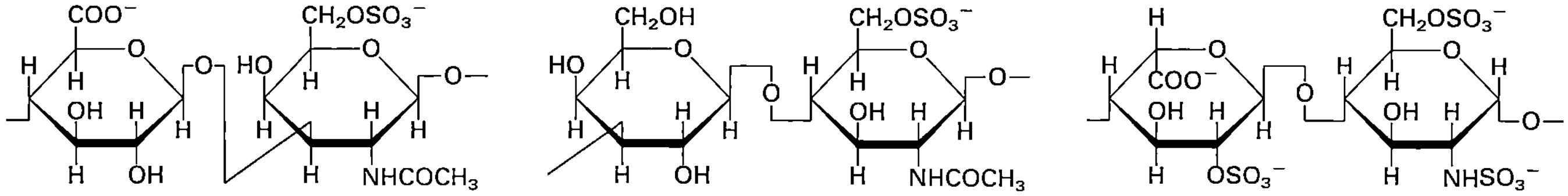

Chondroitin 6-sulfate **Keratan sulfate** **Heparin**

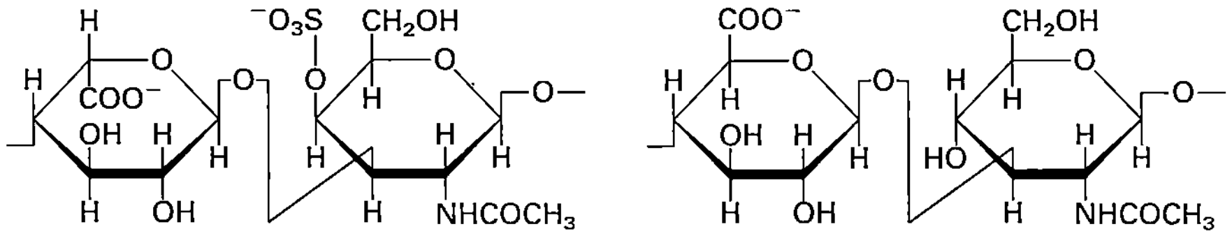

Dermatan sulfate **Hyaluronate**

Figure 18-15
Structural formulas of the repeating disaccharide units of some major glycosaminoglycans. Negatively charged groups are shown in red, and amino groups in blue.

Proteoglycans are proteins containing one or more covalently linked glycosaminoglycan chains. The best-characterized member of this diverse class is the proteoglycan in the extracellular matrix of cartilage (Figure 18-16). Keratan sulfate and chondroitin sulfate chains are covalently

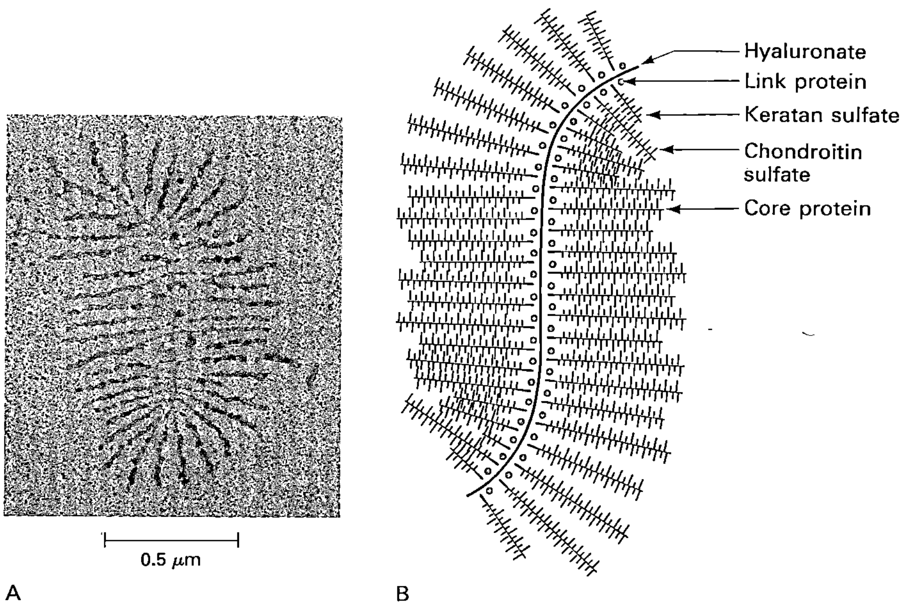

Figure 18-16
(A) Electron micrograph of a proteoglycan from cartilage. Proteoglycan monomers emerge laterally at regular intervals from the opposite sides of an elongated central filament of hyaluronate. (B) Schematic diagram. [(A) Courtesy of Dr. Lawrence Rosenberg. From J.A. Buckwalter and L. Rosenberg. *Coll. Rel. Res.* 3(1983): 489–504. (B) After L. Rosenberg. In *Dynamics of Connective Tissue Macromolecules,* M. Burleigh and R. Poole, eds. (North Holland, 1975), p. 105.]

attached to a polypeptide backbone called the *core protein.* About 140 of these proteins are noncovalently bound at intervals of 30 nm to a very long filament of hyaluronate. This interaction is promoted by a small *link protein.* The entire complex has a mass of 2×10^6 daltons and a length of about 2 μm. Cartilage can cushion compressive forces because these highly hydrated polyanions spring back after being deformed. Proteoglycans in other tissues have quite different roles. For example, the heparan sulfate and chondroitin sulfate chains of *syndecan,* a membrane-inserted proteoglycan, bind collagen and thereby mediate the adhesion of connective tissue cells to the extracellular matrix. Glycosaminoglycan chains also serve as docking sites for fibroblast growth factor and other proteins that stimulate cell proliferation.

OLIGOSACCHARIDES ARE ATTACHED TO INTEGRAL MEMBRANE PROTEINS AND MANY SECRETED PROTEINS

More complex and diverse carbohydrate units are displayed by many integral membrane proteins and secreted proteins. Glycophorin A, for example, forms a coat around red blood cells (p. 282). As was mentioned in Chapter 11, carbohydrates are attached to either the side-chain oxygen atom of serine or threonine residues by *O*-glycosidic linkages or to the side-chain nitrogen of asparagine residues by *N*-glycosidic linkages. *N*-linked oligosaccharides contain a *common pentasaccharide core* consisting of three mannose and two *N*-acetylglucosamine residues. Additional sugars are attached to this common core in many different ways to form the great variety of oligosaccharide patterns found in glycoproteins. Examples are shown in Figure 18-17 and in the space-filling model on page 463. In the *high-mannose type,* additional mannose residues are linked to

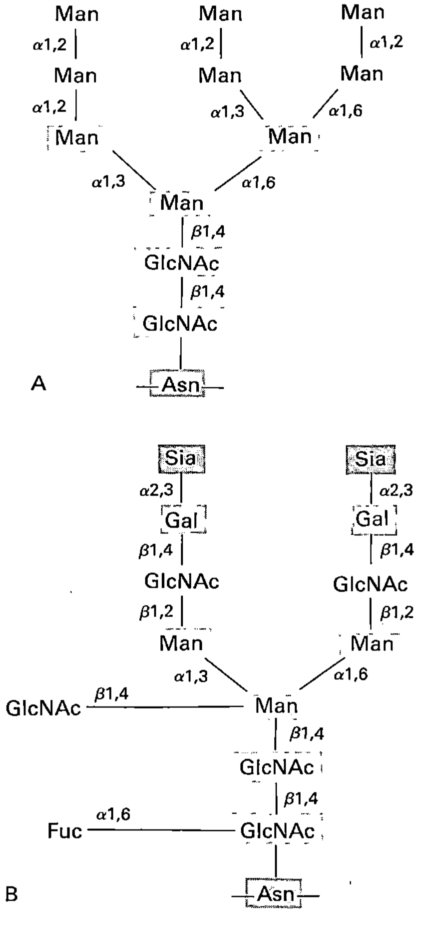

Figure 18-17
N-linked oligosaccharide units in glycoproteins contain a common core (shown in yellow) of three mannose and two *N*-acetylglucosamine residues. Additional sugars are added to form many different patterns, of which only two are shown here. (A) High-mannose type. (B) Complex type. [After R. Kornfeld and S. Kornfeld. *Ann. Rev. Biochem.* 54(1985):633.]

the core. In the *complex type,* different combinations of *N*-acetylglucosamine, galactose, sialic acid, and L-fucose residues are built on the core. Fucose (6-deoxy-L-galactose) is derived from mannose. Sialic acid (*N*-acetylneuraminate), also derived from mannose, is a nine-carbon sugar with a carboxylate group. The formulas of sugars commonly found in oligosaccharide units of glycoproteins are given in Figure 18-18.

Abbreviations for sugars—

Fuc	Fucose
Gal	Galactose
GalNAc	*N*-Acetylgalactosamine
Glc	Glucose
GlcNAc	*N*-Acetylglucosamine
Man	Mannose
Sia	Sialic acid
NeuNAc	*N*-Acetylneuraminate (sialic acid)

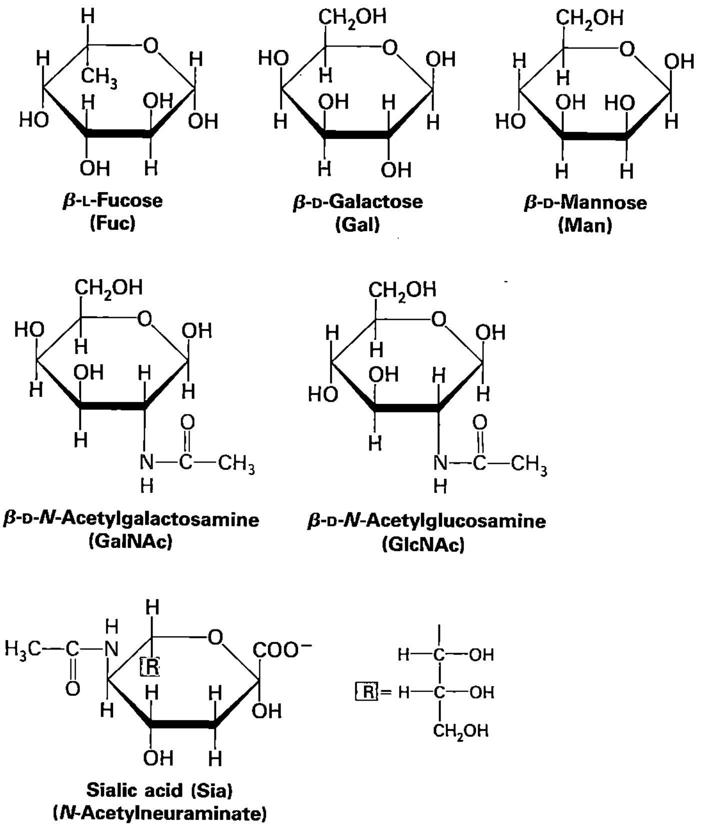

Figure 18-18
Sugar residues commonly found in glycoproteins.

A carbohydrate sequence can be concisely represented by the abbreviated names of the residues and the linkages between them. For example,

$$\text{Gal}\beta 1 \longrightarrow 4\text{GlcNAc}\beta 1 \longrightarrow 6\text{Gal}$$

denotes an oligosaccharide in which galactose is joined by a β-1,4 glycosidic linkage to *N*-acetylglucosamine, which in turn is joined by a β-1,6 glycosidic linkage to galactose. Oligosaccharide chains, like polypeptide and polynucleotide chains, have directionality. The residue at the right end of the sequence is the reducing terminus.

CARBOHYDRATE-BINDING PROTEINS CALLED LECTINS MEDIATE MANY BIOLOGICAL RECOGNITION PROCESSES

The diversity and complexity of the carbohydrate units of glycoproteins suggest that they are rich in information and functionally important. Nature does not construct complex patterns when simple ones suffice. Cellulose and starch, for example, are built solely from glucose units. Glycosaminoglycan chains, a step up in complexity, are linear arrays of disaccharide repeats. Other glycoproteins, by contrast, contain multiple types of residues joined by many kinds of glycosidic linkages. *An enormous number of patterns of surface sugars are possible* because (1) different monosaccharides can be joined to each other through any of several hydroxyl groups, (2) the C-1 linkage can have either an α or a β configuration, and (3) extensive branching is possible. Indeed, *many more different oligosaccharides can be formed from four sugars than oligopeptides from four amino acids.*

Why all this intricacy and diversity? It is becoming evident that carbohydrates are information-rich molecules that guide many biological processes. For example, the terminal carbohydrate residues on a glycoprotein can serve as a signal that directs liver cells to remove the protein from the blood. The best-understood cell-surface receptor for this purpose is the *asialoglycoprotein receptor.* Many newly synthesized glycoproteins, such as immunoglobulins and peptide hormones, contain carbohydrate units with terminal sialic acid residues (see Figure 18-17B). In the course of hours or days, depending on the particular protein, these groups are removed by sialylases protruding from the surface of blood vessels. The exposed galactose residues of the trimmed glycoproteins are detected by asialoglycoprotein receptors in the plasma membranes of liver cells (Figure 18-19). The complex of the asialoglycoprotein and its receptor is then

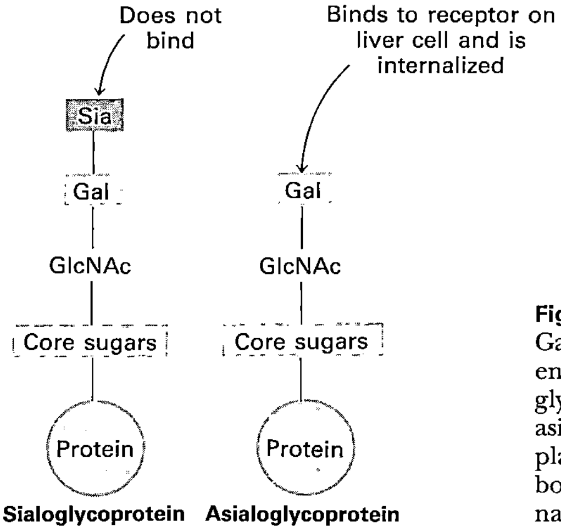

Figure 18-19
Galactose residues at the nonreducing ends of oligosaccharide units of glycoproteins are recognized by the asialoglycoprotein receptors in the plasma membranes of liver cells. The bound glycoproteins are then internalized and removed from the blood.

internalized by the liver cell by a process called *endocytosis* (p. 935). *In essence, these oligosaccharide units mark the passage of time, to indicate when the proteins carrying them should be removed from circulation.* The rate of removal of sialic acid from glycoproteins is controlled by the structure of the protein itself. Hence, proteins can be designed to have lifetimes ranging from a few hours to many weeks, depending on physiologic needs.

Plants contain many specific carbohydrate-binding proteins called *lectins* (from the Latin word *legere,* meaning "to select"). For example, *concanavalin A* (from the jack bean) binds to internal and nonreducing terminal α-mannose residues. *Wheat germ agglutinin, peanut lectin,* and *phytohemagglutinin* (from red kidney bean) recognize disaccharide or oligosaccharide units (Figure 18-20). All known lectins contain two or more binding sites for carbohydrate units, which accounts for their ability to

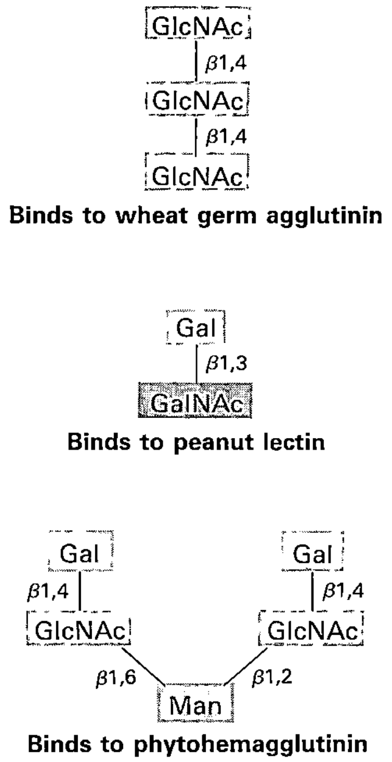

Figure 18-20
Sugar units recognized by wheat germ agglutinin, peanut lectin, and phytohemagglutinin.

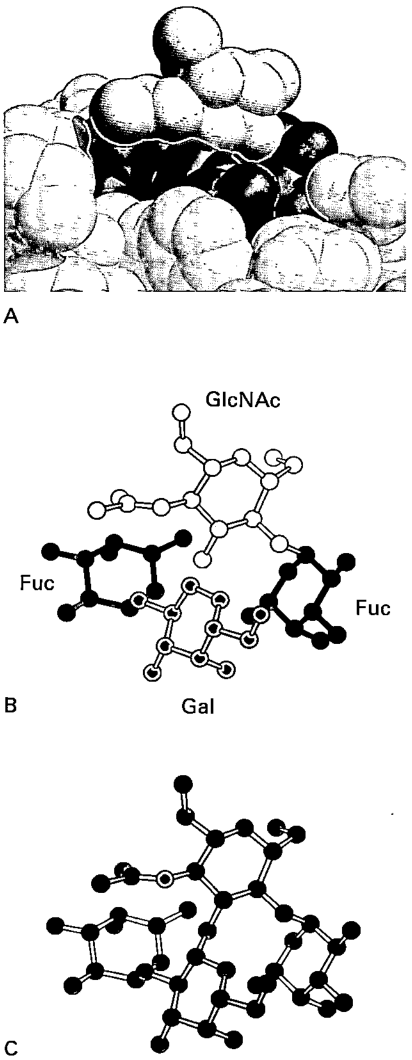

Figure 18-21
(A) Space-filling model of a red blood cell oligosaccharide (the Lewis B blood group determinant) bound to a lectin. The galactose residue is shown in blue, the two fucose residues in red, and *N*-acetylglucosamine in yellow. (B) Ball-and-stick model of the tetrasaccharide in the same orientation. The colors are the same as in part A. (C) Ball-and-stick model using standard colors for the atoms. [Drawn from 1led.pdb. L.T.J. Delbaerre, M. Vandonselaar, L. Prasad, J.W. Quail, K.S. Wilson, and Z. Dauter. *J. Mol. Biol.* 230(1993):950.]

agglutinate (cross-link) erythrocytes and other cells. Lectins are very useful probes of cell surfaces because of their capacity to recognize specific oligosaccharide patterns. What is their physiologic role? A hint comes from the finding that a lectin participates in binding a nitrogen-fixing bacterium (*Rhizobium trifolii*) to the surface of the root hairs of clover. This lectin cross-links receptors on the cell wall of root hairs to bacterial capsular polysaccharides and lipopolysaccharides.

Bacteria, too, contain lectins. The adherence of *Escherichia coli* to epithelial cells of the gastrointestinal tract is mediated by bacterial lectins that recognize oligosaccharide units on the surface of target cells. These lectins are located on slender hairlike appendages called *fimbriae (pili)*. *Neisseria gonorrhoeae* infects human genital or oral epithelial cells but not those of other tissues or species because their surfaces lack carbohydrates that are recognized by this pathogen. Some viruses also gain entry into host cells by adhering to cell-surface carbohydrates. Influenza virus, for example, contains a *hemagglutinin* protein (see Figure 1-23 on p. 16) that recognizes sialic acid residues on cells lining the respiratory tract.

Carbohydrates are also critical in the interaction of sperm with ovulated eggs. In mammals, ovulated eggs are surrounded by an extracellular coat called the *zona pellucida (ZP)*. *O*-linked oligosaccharides that are attached to ZP3, a glycoprotein in this coat, are recognized by a receptor on the surface of sperm. These essential oligosaccharides contain an α-linked galactose at their nonreducing end. Binding triggers the release of sperm enzymes—proteases and hyaluronidase—that dissolve the zona pellucida to allow sperm entry.

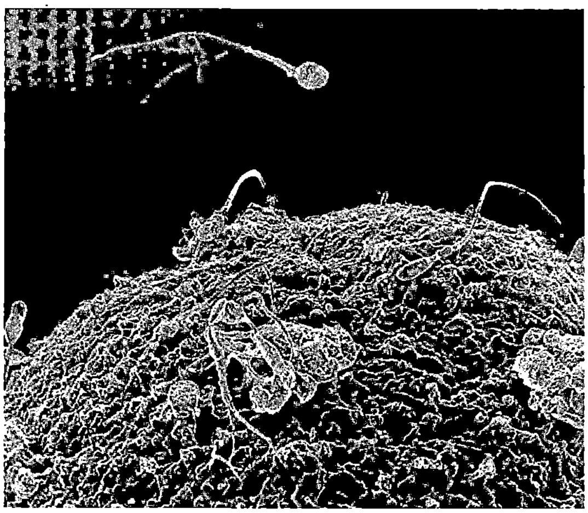

Figure 18-22
Scanning electron micrograph showing the interaction of sperm with an egg. A carbohydrate on the surface of the egg is recognized by the sperm. [From Lennart Nilsson, *How Was I Born,* Delacorte Press.]

CELL ADHESION IS DIRECTED BY THE INTERPLAY OF SELECTINS AND THEIR CARBOHYDRATE PARTNERS

Lymphocytes circulating in the blood tend to migrate to lymphoid sites from which they were originally derived. This return, called *homing,* is mediated by specific interactions between receptors on the surface of lymphocytes and carbohydrates on the endothelial lining of lymph nodes. Several experiments pointed to the importance of carbohydrates in this process: (1) Homing can be blocked by phosphomannans (polymers of mannose 6-phosphate) or by fucoidin (a polymer of fucose

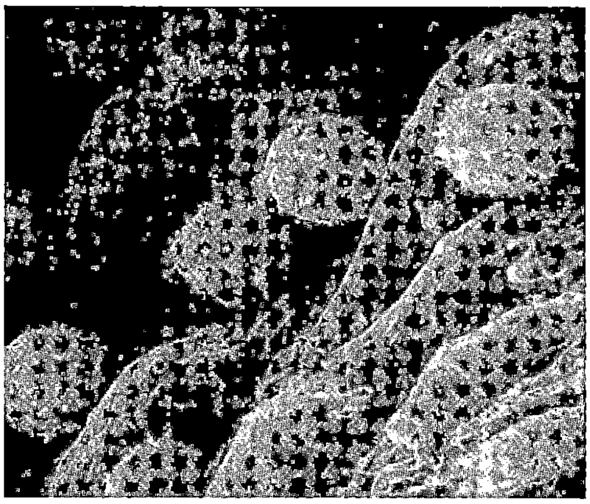

Figure 18-23
Scanning electron micrograph showing lymphocytes adhering to the endothelial lining of a lymph node. Adhesive proteins called L-selectins on the lymphocyte surface bind specifically to carbohydrates on the lining of lymph node vessels. [Courtesy of Dr. Eugene Butcher.]

4-phosphate). (2) Enzymatic removal of sialic acid residues from the endothelial surface of blood vessels prevents the adhesion of lymphocytes. (3) Lymphocytes bind selectively to polyacrylamide gels containing covalently attached phosphomannans or fucoidin.

The *homing receptor* of lymphocytes has recently been isolated. This integral membrane protein contains an extracellular N-terminal *lectin domain* that recurs in other vertebrate cell-surface receptors and plasma proteins. Indeed, the binding of neutrophils and other leukocytes to sites of injury in the inflammatory response is mediated by proteins akin to the homing receptor. These carbohydrate-binding adhesive proteins are called *selectins*. The *L, E,* and *P* forms of members of this family bind specifically to carbohydrates on *l*ymph-node vessels, *e*ndothelium, and activated blood *p*latelets, respectively. Each contains a conserved 120-residue carbohydrate-recognition domain (CRD) that complexes Ca^{2+} together with the specific carbohydrate (Figure 18-24). New therapeutic agents that control inflammation and block abnormal clotting are likely to emerge from a deeper understanding of how selectins bind and distinguish different carbohydrates.

The development of the nervous system requires highly specific adhesion of neurons. Cell-surface carbohydrates may serve as guides in forming the intricate circuitry of the brain. The diversity of carbohydrates in the nervous system and their selective spatial expression suggest that they are rich in information that can be rapidly scanned by receptor proteins. The deciphering of the carbohydrate code will undoubtedly open new vistas in biology.

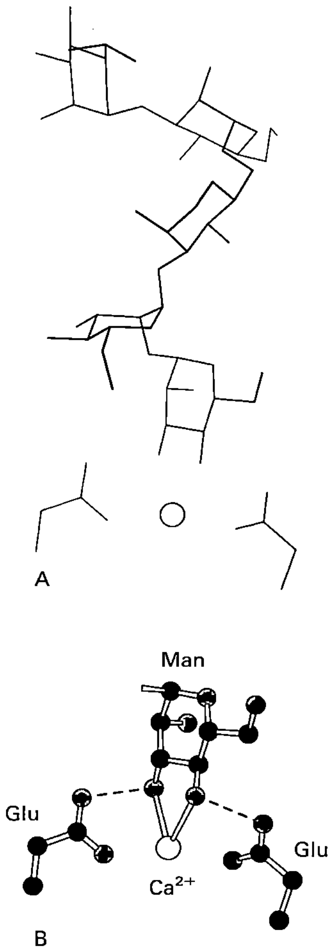

Figure 18-24
Binding of a mannose-rich oligosaccharide to the carbohydrate-recognition domain of a mammalian lectin. (A) Skeletal model showing five mannosyl residues (blue) at the binding site of a C-type (calcium-dependent) mannose-binding protein. A calcium ion (green) and two glutamate residues (red) play a key role in binding the oligosaccharide. (B) Ball-and-stick model showing the coordination of two hydroxyl oxygen atoms of a mannose residue to Ca^{2+} (green); oxygens are shown in red, and carbons in black. The mannose residue is also hydrogen-bonded (blue dashed lines) to the side-chain oxygens of two glutamates. Protein-bound water molecules also participate in binding the oligosaccharide. [Drawn from 2msb.pdb. W.I. Weis, K. Drickamer, and W.A. Hendrickson. *Nature* 360(1992):127.]

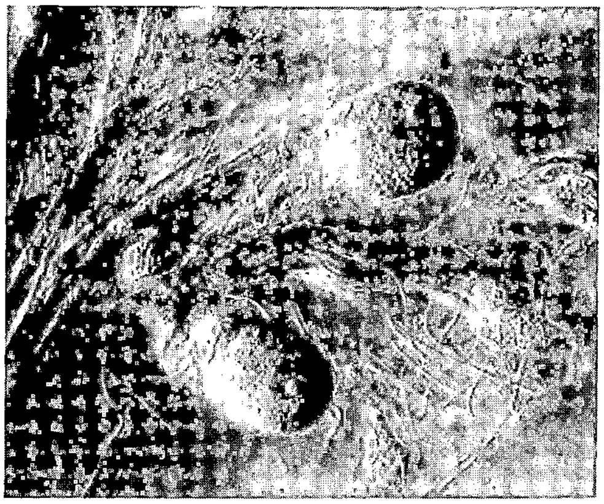

Figure 18-25
Light micrograph showing neurons interacting in the development of the nervous system. Cell-surface carbohydrates have been implicated in the formation of adhesions between interacting cells. [Courtesy of Dr. Peter Sargent.]

SUMMARY

Carbohydrates are aldehydes or ketones with two or more hydroxyl groups. Aldoses are carbohydrates with an aldehyde group (as in glyceraldehyde and glucose), whereas ketoses contain a keto group (as in dihydroxyacetone and fructose). A sugar belongs to the D series if the absolute configuration of its asymmetric carbon farthest from the aldehyde or keto group is the same as that of D-glyceraldehyde. Most naturally occurring sugars belong to the D series. The C-1 aldehyde in the open-chain form of glucose reacts with the C-5 hydroxyl group to form a six-membered pyranose ring. The C-2 keto group in the open-chain form of fructose reacts with the C-5 hydroxyl group to form a five-membered furanose ring. Pentoses such as ribose and deoxyribose also form furanose rings. An additional asymmetric center is formed at the anomeric carbon atom (C-1 in aldoses and C-2 in ketoses) in these cyclizations. The hydroxyl group attached to the anomeric carbon atom is below the plane of the ring (viewed in the standard orientation) in the α anomer, whereas it is above the ring in the β anomer. Not all the atoms in the rings lie in the same plane. Rather, pyranose rings usually adopt the chair conformation, and furanose rings the envelope conformation.

Sugars are joined to alcohols and amines by glycosidic bonds from the anomeric carbon atom. For example, *O*-glycosidic bonds link sugars to one another in disaccharides and polysaccharides. *N*-glycosidic bonds link sugars to purines and pyrimidines in nucleotides, RNA, and DNA. Phosphate-sugar esters such as glucose 6-phosphate are important metabolic intermediates. Phosphorylation also generates reactive intermediates for synthesizing *O*- and *N*-glycosidic linkages.

Sucrose, lactose, and maltose are the common disaccharides. Sucrose (common table sugar), obtained from cane or beet, consists of α-glucose and β-fructose joined by a glycosidic linkage between their anomeric carbons. Lactose (in milk) consists of galactose joined to glucose by a β-1,4 linkage. Maltose (from starch) consists of two glucoses joined by an α-1,4 linkage. Starch is a polymeric form of glucose in plants, and glycogen serves a similar role in animals. Most of the glucose units in starch and glycogen are in α-1,4 linkage. Glycogen has more branch points formed by α-1,6 linkages than does starch, which makes glycogen more soluble. Cellulose, the major structural polymer of plant cell walls, consists of glucose units joined by β-1,4 linkages. These β linkages give rise to long straight chains that form fibrils with high tensile strength. In contrast, the α linkages in starch and glycogen lead to open helices, in keeping with their roles as mobilizable energy stores. Cell surfaces and extracellular matrices of animals contain polymers of repeating disaccharides called glycosaminoglycans. One of the units in each repeat is a derivative of glucosamine or galactosamine. These highly anionic carbohydrates have a high density of carboxylate or sulfate groups. Proteins bearing covalently linked glycosaminoglycans are termed proteoglycans.

The oligosaccharide units on integral membrane proteins are linked either to the side-chain oxygen atom of serine or threonine residues or to the side-chain amide nitrogen of asparagine residues. *N*-linked oligosaccharides contain a common core consisting of three mannose and two *N*-acetylglucosamine residues. Additional sugars are attached to this core to form diverse patterns. Carbohydrates are important in molecular targeting and cell-cell recognition. The asialoglycoprotein receptor of liver cells removes from circulation glycoproteins that display terminal galactose residues because of the loss of sialic acid. Plants contain many soluble carbohydrate-binding proteins called lectins. The binding of a nitrogen-

Model of heparan sulfate, a cell-surface glycosaminoglycan that participates in recognition.

fixing microorganism to the root hairs of a plant is mediated by a lectin that recognizes cell-surface sugars on the partners in this symbiotic relationship. Cell adhesion in vertebrates is guided by the interplay of carbohydrate-binding proteins called selectins and their sugar targets. The homing of lymphocytes to their sites of origin and the adhesion of leukocytes to the lining of blood vessels in inflammation are mediated by selectins that recognize different oligosaccharide units. A small number of carbohydrate residues can be joined in many different ways to form highly diverse patterns that can be distinguished by the lectin domains of protein receptors.

SELECTED READINGS

Where to start

Sharon, N., and Lis, H., 1993. Carbohydrates in cell recognition. *Sci. Amer.* 268(1):82–89.

Lasky, L.A., 1992. Selectins: Interpreters of cell-specific carbohydrate information during inflammation. *Science* 258:964–969.

Weiss, P., and Ashwell, G., 1989. The asialoglycoprotein receptor: Properties and modulation by ligand. *Prog. Clin. Biol. Res.* 300:169–184.

Sharon, N., 1980. Carbohydrates. *Sci. Amer.* 245(5):90–116.

Books

Rees, D.A., 1977. *Polysaccharide Shapes.* Wiley. [A concise and lucid account of the conformation of sugars and polysaccharides.]

El Khadem, H.S., 1988. *Carbohydrate Chemistry.* Academic Press.

Ginsburg, V., and Robbins, P.W. (eds.), 1981. *Biology of Carbohydrates,* Vols. 1–3. Wiley.

McGee, H., 1984. *On Food and Cooking: The Science and the Lore of the Kitchen.* Collier. [A delightful and informative account of the biochemistry of foods and cooking.]

Fukuda, M. (ed.), 1992. *Cell Surface Carbohydrates and Cell Development.* CRC Press. [Contains many excellent articles on the biosynthesis and biological role of carbohydrates in cell-cell recognition.]

Preiss, J. (ed.), 1980. *The Biochemistry of Plants,* vol. 3: *Carbohydrates: Structure and Function.* Academic Press. [Contains many fine articles on the structure, biosynthesis, and function of carbohydrates in plants.]

Lactase deficiency

Semenza, G., and Auricchio, S., 1989. Small-intestinal disaccharidases. *In* Scriver, C.R., Beaudet, A.L., Sly, W.S., and Valle, D. (eds.), *The Metabolic Basis of Inherited Disease* (6th ed.), pp. 2975–2997. McGraw-Hill. [Also see the article by G. Flatz on the genetics of lactase activity, pp. 2999–3006.]

Structure of carbohydrate-binding proteins

Vyas, N.K., 1991. Atomic features of protein-carbohydrate interactions. *Curr. Opin. Struct. Biol.* 1:732–740.

Weis, W.I., Drickamer, K., and Hendrickson, W.A., 1992. Structure of a C-type mannose-binding protein complexed with an oligosaccharide. *Nature* 360:127–134.

Wright, C.S., 1992. Crystal structure of a wheat germ agglutinin/glycophorin-sialoglycopeptide receptor complex. Structural basis for cooperative lectin-cell binding. *J. Biol. Chem.* 267:14345–14352.

Shaanan, B., Lis, H., and Sharon, N., 1991. Structure of a legume lectin with an ordered *N*-linked carbohydrate in complex with lactose. *Science* 254:862–866.

Glycoproteins

Kjellén, L., and Lindahl, U., 1991. Proteoglycans: Structures and interactions. *Ann. Rev. Biochem.* 60:443–475.

Yanagishita, M., and Hascall, V., 1992. Cell surface heparan sulfate proteoglycans. *J. Biol. Chem.* 267:9451–9454.

Ruoslahti, E., 1989. Proteoglycans in cell regulation. *J. Biol. Chem.* 264:13369–13372.

Carbohydrates in recognition processes

Sharon, N., and Lis, H., 1989. Lectins as cell recognition molecules. *Science* 246:227–234.

Wassarman, P.M., 1990. Profile of a mammalian sperm receptor. *Development* 108:1–17.

Turner, M.L., 1992. Cell adhesion molecules: A unifying approach to topographic biology. *Biol. Rev. Cambridge Phil. Soc.* 67:359–377.

Feizi, T., 1992. Blood group–related oligosaccharides are ligands in cell-adhesion events. *Biochem. Soc. Trans.* 20:274–278.

Jessell, T.M., Hynes, M.A., and Dodd, J., 1990. Carbohydrates and carbohydrate-binding proteins in the nervous system. *Ann. Rev. Neurosci.* 13:227–255.

Brandley, B.K., 1991. Cell surface carbohydrates in cell adhesion. *Sem. Cell. Biol.* 2:281–287.

Plant cell walls

McNeil, M., Darvill, A.G., Fry, S.C., and Albersheim, P., 1984. Structure and function of the primary cell walls of plants. *Ann. Rev. Biochem.* 53:625–663.

Albersheim, P., and Darvill, A.G., 1985. Oligosaccharins. *Sci. Amer.* 253(3):58–64. [Oligosaccharins, fragments of plant cell walls, control functions such as development and defense against disease.]

PROBLEMS

1. *Word origin.* Account for the origin of the term *carbohydrate.*

2. *Couples.* Indicate whether each of the following pairs of sugars consists of anomers, epimers, or an aldose-ketose pair:
   (a) D-glyceraldehyde and dihydroxyacetone.
   (b) D-glucose and D-mannose.
   (c) D-glucose and D-fructose.
   (d) α-D-glucose and β-D-glucose.
   (e) D-ribose and D-ribulose.
   (f) D-galactose and D-glucose.

3. *Tollen's test.* Glucose and other aldoses are oxidized by an aqueous solution of a silver-ammonia complex. What are the reaction products?

4. *Mutarotation.* The specific rotations ($[\alpha]_D$) of the α and β anomers of D-glucose are +112 degrees and +18.7 degrees. $[\alpha]_D$ is defined as the observed rotation of light of wavelength 589 nm (the D line of a sodium lamp) passing through 10 cm of a 1 g/ml solution of a sample. When a crystalline sample of α-D-glucopyranose is dissolved in water, the specific rotation decreases from 112 degrees to an equilibrium value of 52.7 degrees. At equilibrium, what are the proportions of the α and β anomers? Assume that the concentration of the open-chain form is negligible.

5. *Telltale adduct.* Glucose reacts slowly with hemoglobin and other proteins to form covalent compounds. Why is glucose reactive? What is the nature of the adduct formed?

6. *Periodate cleavage.* Compounds containing hydroxyl groups on adjacent carbon atoms undergo carbon–carbon bond cleavage when treated with periodate ion (IO_4^-). How can this reaction be used to distinguish between pyranosides and furanosides?

7. *Oxygen source.* Does the oxygen atom attached to C-1 in methyl α-D-glucopyranoside come from glucose or methanol?

8. *Sugar lineup.* Identify the four sugars shown below.

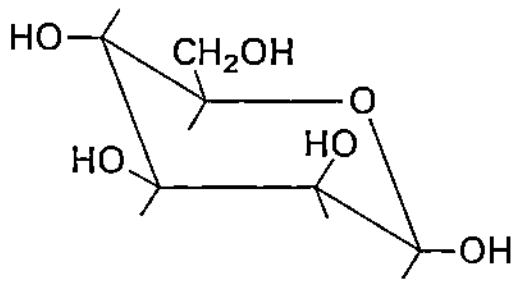

A

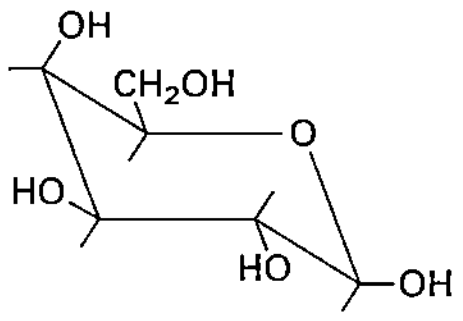

B

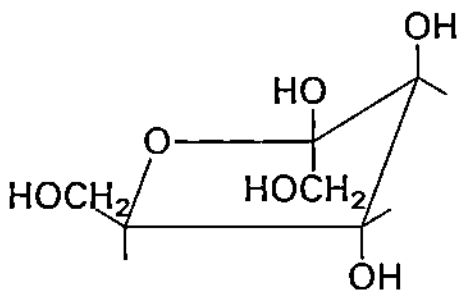

C

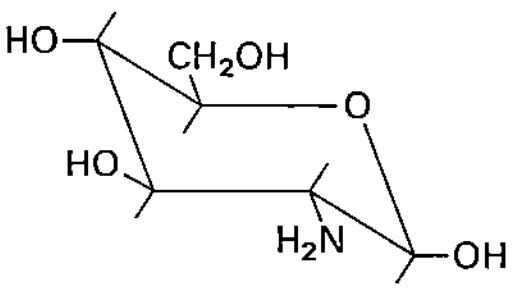

D

9. *Cellular glue.* A trisaccharide unit of a cell-surface glycoprotein is postulated to play a critical role in mediating cell-cell adhesion in a particular tissue. Design a simple experiment to test this hypothesis.

CHAPTER

19

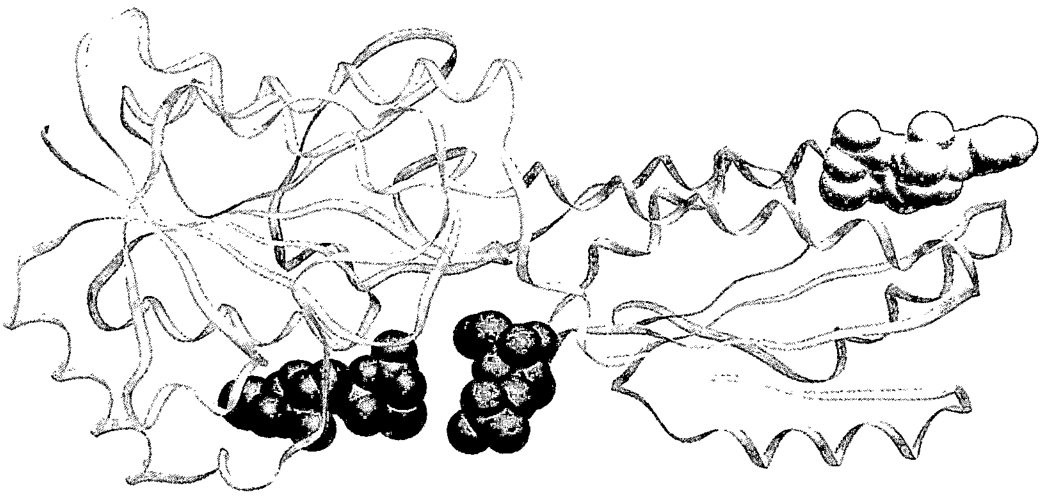

Glycolysis

We begin our consideration of the generation of metabolic energy with glycolysis, a nearly universal pathway in biological systems. *Glycolysis is the sequence of reactions that converts glucose into pyruvate with the concomitant production of a relatively small amount of ATP.* In aerobic organisms, glycolysis is the prelude to the citric acid cycle and the electron transport chain, which together harvest most of the energy contained in glucose. Under aerobic conditions, pyruvate enters mitochondria, where it is completely oxidized to CO_2 and H_2O. If the supply of oxygen is insufficient, as in actively contracting muscle, pyruvate is converted into lactate. Under anaerobic conditions, yeast transforms pyruvate into ethanol. The formation of ethanol and lactate from glucose are examples of fermentations.

Glycolysis—
Derived from the Greek stem *glyk-*, "sweet," and the word *lysis*, "dissolution."

The elucidation of glycolysis has a rich history. Indeed, the development of biochemistry and the delineation of this central pathway went hand in hand. A key discovery was made by Hans Buchner and Eduard Buchner in 1897, quite by accident. They were interested in manufacturing cell-free extracts of yeast for possible therapeutic use. These extracts had to be preserved without using antiseptics such as phenol, and so they decided to try sucrose, a commonly used preservative in kitchen chemistry. They obtained a startling result: sucrose was rapidly fermented into alcohol by the yeast juice. The significance of this finding was immense. *The Buchners demonstrated for the first time that fermentation could occur outside living cells.* The accepted view of their day, asserted by Louis Pasteur in 1860, was that fermentation is inextricably tied to living cells. The chance

Fermentation—
An ATP-generating process in which organic compounds act as both donors and acceptors of electrons. Fermentation can occur in the absence of O_2. Discovered by Pasteur, who described fermentation as *"la vie sans l'air"* ("life without air").

Opening Image: *Three-dimensional structure of phosphofructokinase, a key enzyme in glycolysis. Fructose 6-phosphate (red), a substrate, is next to ADP (blue), which occupies the ATP-binding site. The bacterial enzyme is allosterically activated by ADP (green) at a distant site. Only one of the four subunits of the tetrameric enzyme is shown here. [Drawn from 4pfk.pdb. T. Schirmer and P.R. Evans. Nature 343(1990):140. The image was generated by Dr. Anthony Nicholls using GRASP, described in A. Nicholls, K. Sharp, and B. Honig. Proteins 11(1991):281.]*

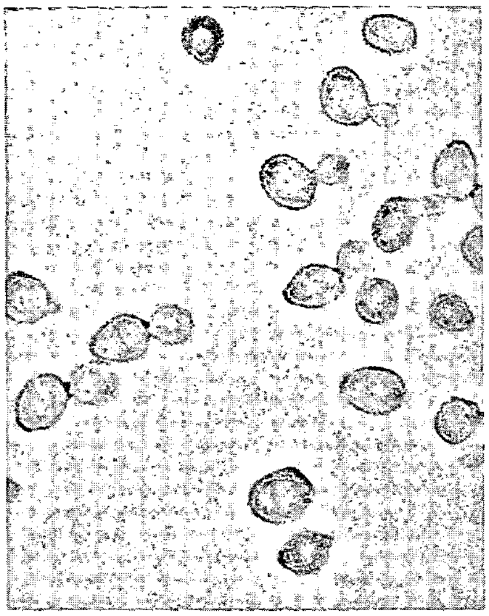

Light micrograph of yeast cells. [Courtesy of Dr. Randy Schekman.]

discovery of the Buchners refuted this vitalistic dogma and opened the door to modern biochemistry. *Metabolism became chemistry.*

The next critical contribution was made by Arthur Harden and William Young in 1905. They added yeast juice to a solution of glucose and found that fermentation started almost immediately but soon ceased unless inorganic phosphate was added. They deduced that *inorganic phosphate became linked to a sugar,* and they proceeded to isolate a hexose diphosphate, which was later shown to be fructose 1,6-bisphosphate, a key intermediate in glycolysis. Furthermore, Harden and Young discovered that yeast juice lost its activity if it was dialyzed or heated to 50°C. However, activity was restored when inactive *dialyzed* juice was mixed with inactive *heated* juice. Thus, activity depended on the presence of two kinds of substances: a heat-labile, nondialyzable component (called *zymase*) and a heat-stable, dialyzable fraction (called *cozymase*). We now know that "zymase" consists of a number of enzymes, whereas "cozymase" consists of metal ions, adenosine triphosphate (ATP), adenosine diphosphate (ADP), and coenzymes such as nicotinamide adenine dinucleotide (NAD^+).

Studies of muscle extracts then showed that many of the reactions of lactic fermentation were the same as those of alcoholic fermentation. *This was an exciting discovery because it revealed an underlying unity in biochemistry.* The complete glycolytic pathway was elucidated by 1940, largely because of the pioneering contributions of Gustav Embden, Otto Meyerhof, Carl Neuberg, Jacob Parnas, Otto Warburg, Gerty Cori, and Carl Cori. Glycolysis is also known as the *Embden-Meyerhof pathway.*

Enzyme—
A term coined by Friedrich Wilhelm Kühne in 1878 to designate catalytically active substances that had previously been called ferments. Derived from the Greek words *en,* "in," and *zyme,* "leaven."

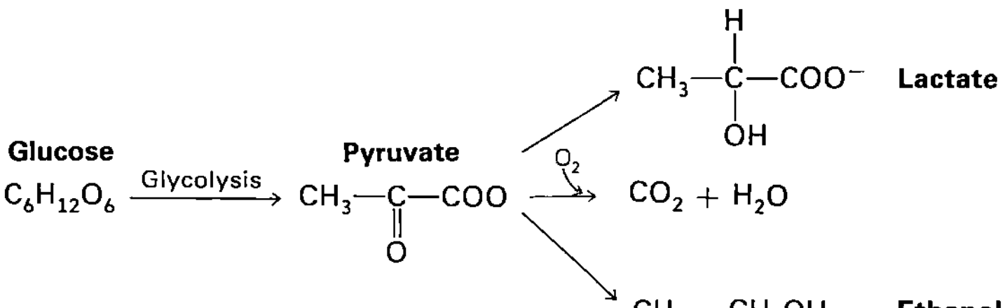

Figure 19-1
Some fates of glucose.

AN OVERVIEW OF KEY STRUCTURES AND REACTIONS

Learning the sequence of events in a metabolic pathway is easier with a firm grasp of the structures of the reactants and an understanding of the types of reactions taking place. Glycolytic intermediates have either six or three carbons. The *six-carbon units* are derivatives of *glucose* and *fructose.* The *three-carbon units* are derivatives of *dihydroxyacetone, glyceraldehyde, glycerate,* and *pyruvate.*

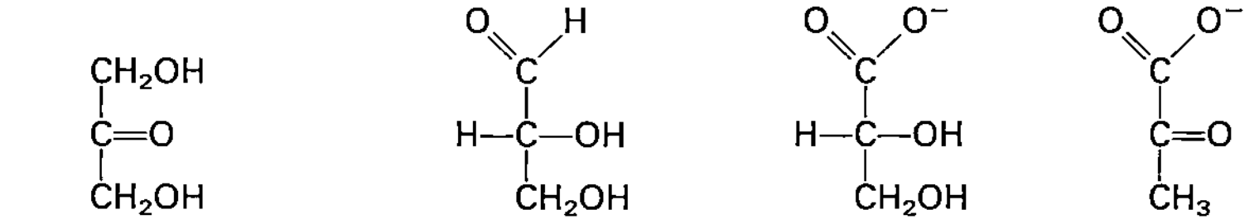

Dihydroxyacetone **Glyceraldehyde** **Glycerate** **Pyruvate**

All intermediates between glucose and pyruvate are *phosphorylated.* The phosphoryl groups in these compounds are linked as either *esters* or *anhydrides.*

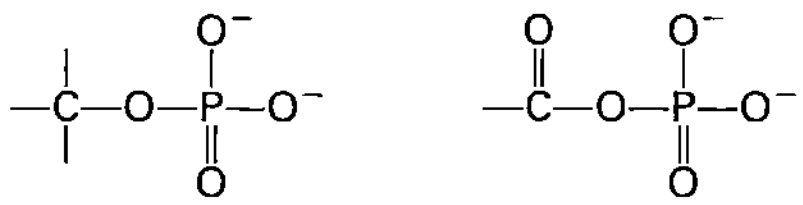

Ester **Anhydride**

Now let us look at some of the kinds of reactions that occur in glycolysis and in many other metabolic pathways:

1. *Phosphoryl transfer.* A phosphoryl group is transferred from ATP to a glycolytic intermediate, or from the intermediate to ADP, by a *kinase.*

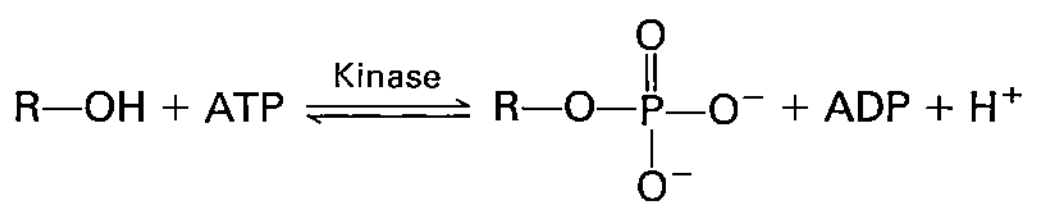

2. *Phosphoryl shift.* A phosphoryl group is shifted from one oxygen atom to another within a molecule by a *mutase.*

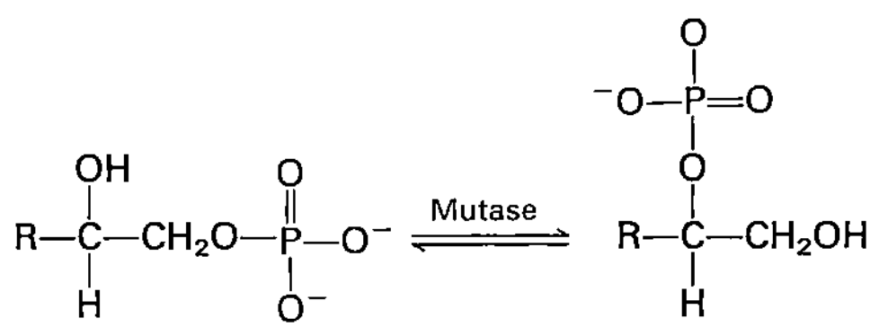

3. *Isomerization.* A ketose is converted into an aldose, or vice versa, by an *isomerase.*

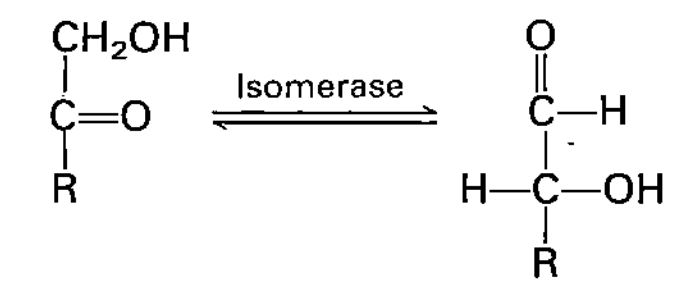

Ketose **Aldose**

4. *Dehydration.* A molecule of water is eliminated by a *dehydratase.*

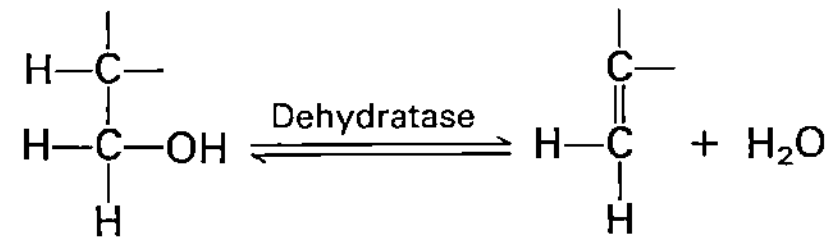

5. *Aldol cleavage.* A carbon–carbon bond is split in a reversal of an aldol condensation by an *aldolase.*

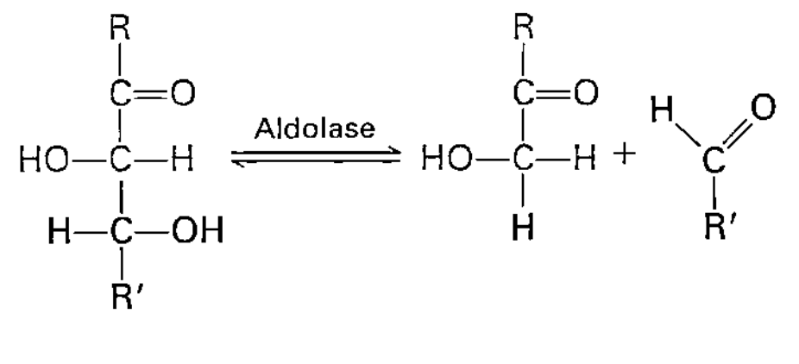

FORMATION OF FRUCTOSE 1,6-BISPHOSPHATE FROM GLUCOSE

We now start our journey down the glycolytic pathway. The reactions in this pathway take place in the cell cytosol. The first stage, which is the conversion of glucose into fructose 1,6-bisphosphate, consists of three steps: a phosphorylation, an isomerization, and a second phosphorylation reaction. *The strategy of these initial steps in glycolysis is to trap the substrate*

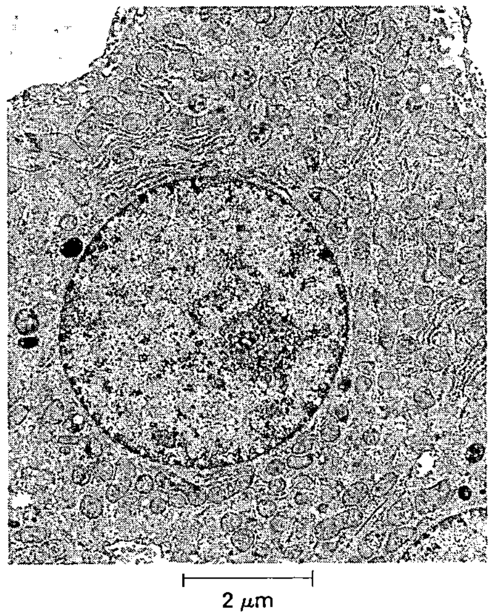

Electron micrograph of a liver cell. Glycolysis takes place in the cytosol. [Courtesy of Dr. Anne Hubbard.]

in the cell and form a compound that can be readily cleaved into phosphorylated three-carbon units. ATP is subsequently extracted from the three-carbon units.

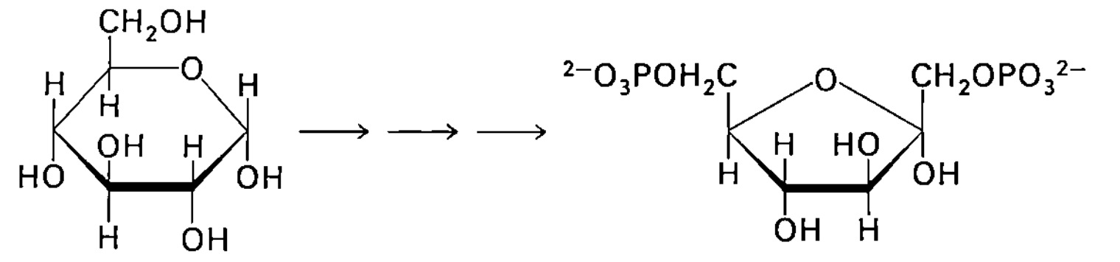

Glucose **Fructose 1,6-bisphosphate**

Glucose enters most cells through specific transport proteins and has one principal fate: *it is phosphorylated by ATP to form glucose 6-phosphate.* The transfer of the phosphoryl group from ATP to the hydroxyl group on C-6 of glucose is catalyzed by *hexokinase.*

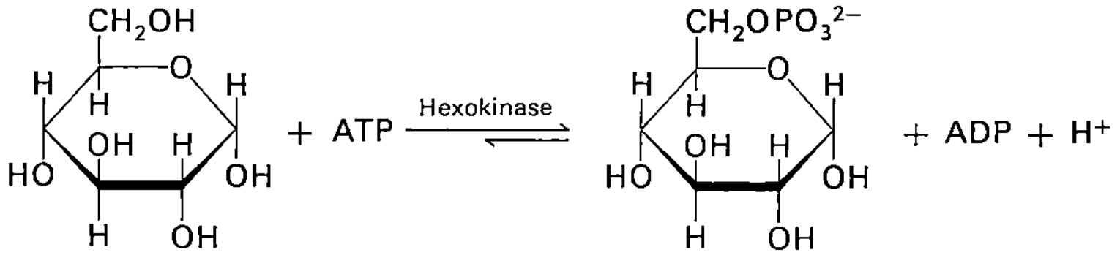

Glucose **Glucose 6-phosphate**

Phosphoryl transfer is a basic reaction in biochemistry. As was discussed earlier (p. 485), *kinases* are enzymes that catalyze the transfer of a phosphoryl group from ATP to an acceptor. Hexokinase, then, catalyzes the transfer of a phosphoryl group from ATP to a variety of six-carbon sugars (*hexoses*), such as glucose and mannose. *Hexokinase, like all other kinases, requires Mg^{2+} (or another divalent metal ion such as Mn^{2+}) for activity.* The divalent metal ion forms a complex with ATP. The structures of two possible Mg^{2+}–ATP complexes are shown in Figure 19-2.

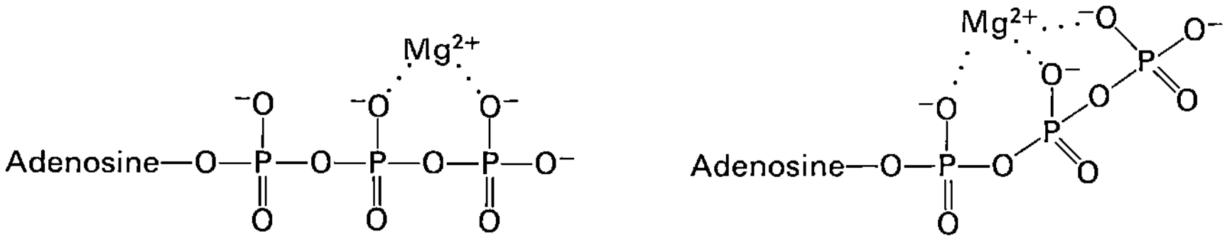

Figure 19-2
Modes of binding Mg^{2+} to ATP.

The next step in glycolysis is the *isomerization of glucose 6-phosphate to fructose 6-phosphate.* The *six-membered pyranose ring* of glucose 6-phosphate is converted into the *five-membered furanose ring* of fructose 6-phosphate. Recall that the open-chain form of glucose has an aldehyde group on C-1, whereas the open-chain form of fructose has a keto group on C-2. The aldehyde on C-1 reacts with the hydroxyl group on C-5 to form the pyranose ring, whereas the keto group on C-2 reacts with the C-5 hydroxyl to form the furanose ring. Thus, the isomerization of glucose 6-phosphate to fructose 6-phosphate is a *conversion of an aldose into a ketose.*

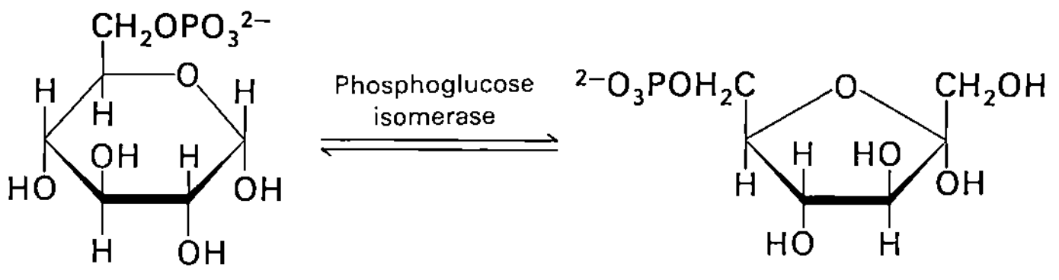

Glucose 6-phosphate **Fructose 6-phosphate**

The open-chain representations of these sugars show the essence of this reaction.

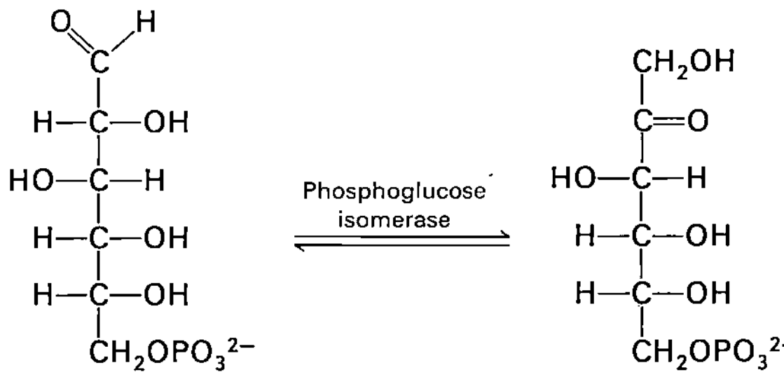

Glucose 6-phosphate (An aldose) — **Fructose 6-phosphate** (A ketose)

A second phosphorylation reaction follows the isomerization step. *Fructose 6-phosphate is phosphorylated by ATP to fructose 1,6-bisphosphate.* This compound was formerly known as *fructose 1,6-diphosphate. Bis*phosphate means two separate phosphate groups, whereas *di*phosphate (as in adenosine diphosphate) means two phosphate groups joined by an anhydride bond. Hence, the name *fructose 1,6-bisphosphate* should be used.

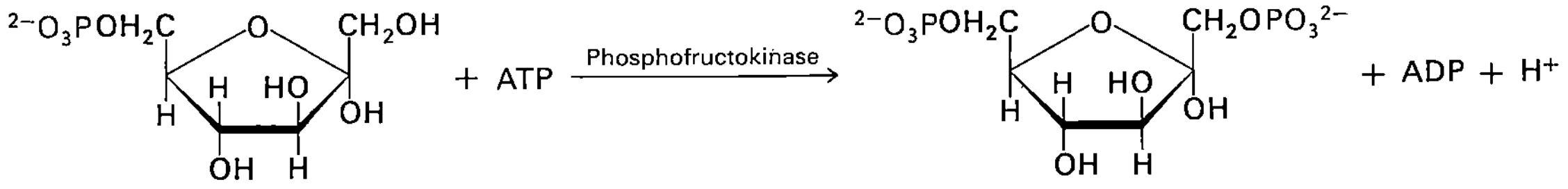

Fructose 6-phosphate — **Fructose 1,6-bisphosphate**

This reaction is catalyzed by *phosphofructokinase,* an allosteric enzyme. The pace of glycolysis is critically dependent on the level of activity of this enzyme, which is allosterically controlled by ATP and several other metabolites (p. 493).

FORMATION OF GLYCERALDEHYDE 3-PHOSPHATE BY CLEAVAGE AND ISOMERIZATION

The second stage of glycolysis consists of four steps, starting with the splitting of fructose 1,6-bisphosphate into *glyceraldehyde 3-phosphate* and *dihydroxyacetone phosphate.* The remaining steps in glycolysis involve three-carbon units rather than six-carbon units.

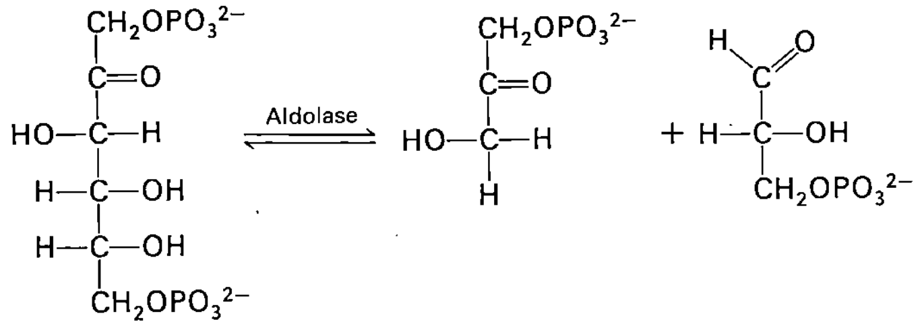

Fructose 1,6-bisphosphate — **Dihydroxyacetone phosphate** — **Glyceraldehyde 3-phosphate**

This reaction is catalyzed by *aldolase.* This enzyme derives its name from the nature of the reverse reaction, an aldol condensation.

Glyceraldehyde 3-phosphate is on the direct pathway of glycolysis. Dihydroxyacetone phosphate is not, but it can be readily converted into glycer-

Aldol condensation—
The combination of two carbonyl compounds (e.g., an aldehyde and a ketone) to form an aldol (a β-hydroxy-carbonyl compound).

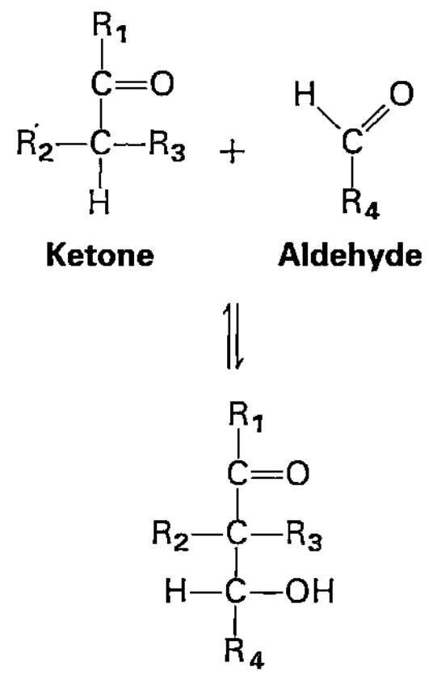

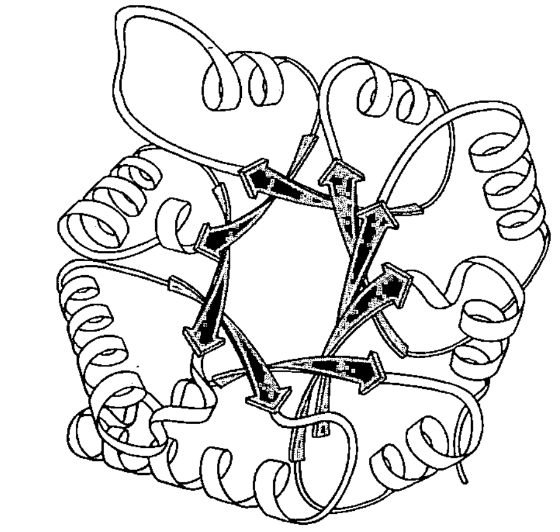

Figure 19-3
Triose phosphate isomerase consists of a central core of eight parallel β strands (red) surrounded by eight α helices (green). Connecting regions are shown in yellow. This structural motif, called an $\alpha\beta$ barrel, is also found in one of the domains of pyruvate kinase. [After a drawing kindly provided by Dr. Jane Richardson.]

aldehyde 3-phosphate. These compounds are isomers: dihydroxyacetone phosphate is a ketose, whereas glyceraldehyde 3-phosphate is an aldose. The isomerization of these three-carbon phosphorylated sugars is catalyzed by *triose phosphate isomerase* (Figure 19-3). This reaction is rapid and reversible. At equilibrium, 96% of the triose phosphate is dihydroxyacetone phosphate. However, the reaction proceeds readily from dihydroxyacetone phosphate to glyceraldehyde 3-phosphate because of efficient removal of this product by subsequent reactions.

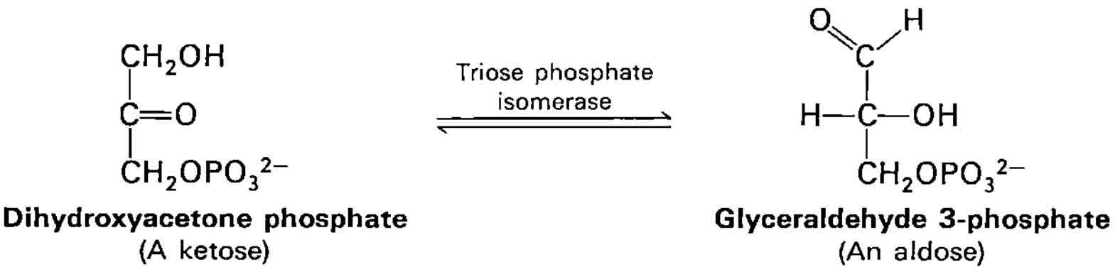

Thus, two molecules of glyceraldehyde 3-phosphate are formed from one molecule of fructose 1,6-bisphosphate by the sequential action of aldolase and triose phosphate isomerase. The economy of metabolism is evident in this reaction sequence. The isomerase funnels dihydroxyacetone into the main glycolytic pathway—a separate set of reactions are not needed.

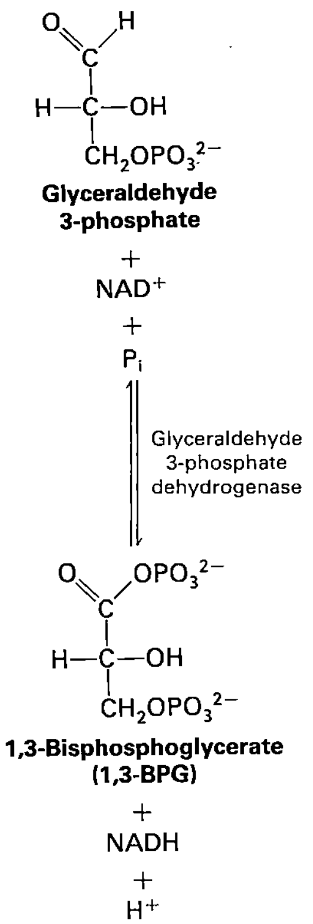

ENERGY CONSERVATION: PHOSPHORYLATION IS COUPLED TO THE OXIDATION OF GLYCERALDEHYDE 3-PHOSPHATE

The preceding steps in glycolysis have transformed one molecule of glucose into two molecules of glyceraldehyde 3-phosphate. No energy has yet been extracted. On the contrary, two molecules of ATP have been invested thus far. We come now to a series of steps that harvest some of the energy contained in glyceraldehyde 3-phosphate. The initial reaction in this sequence is the *conversion of glyceraldehyde 3-phosphate into 1,3-bisphosphoglycerate (1,3-BPG),* a reaction catalyzed by *glyceraldehyde 3-phosphate dehydrogenase.* In the earlier literature, 1,3-BPG was known as 1,3-diphosphoglycerate (1,3-DPG).

A *high-potential phosphorylated compound* is generated in this oxidation-reduction reaction. The aldehyde group at C-1 is converted into an *acyl phosphate,* which is a *mixed anhydride* of phosphoric acid and a carboxylic

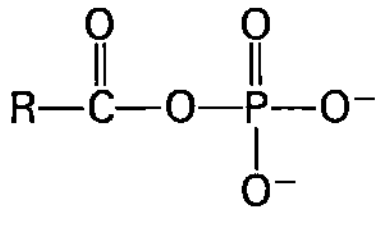

Acyl phosphate

acid. The energy for the formation of this anhydride, which has a high phosphoryl group–transfer potential, comes from the oxidation of the aldehyde group. Note that C-1 in 1,3-BPG is at the oxidation level of a carboxylic acid. NAD^+ (p. 450) is the electron acceptor in this oxidation. The mechanism of this complex reaction, which couples oxidation and phosphorylation, will be discussed later in this chapter (p. 501).

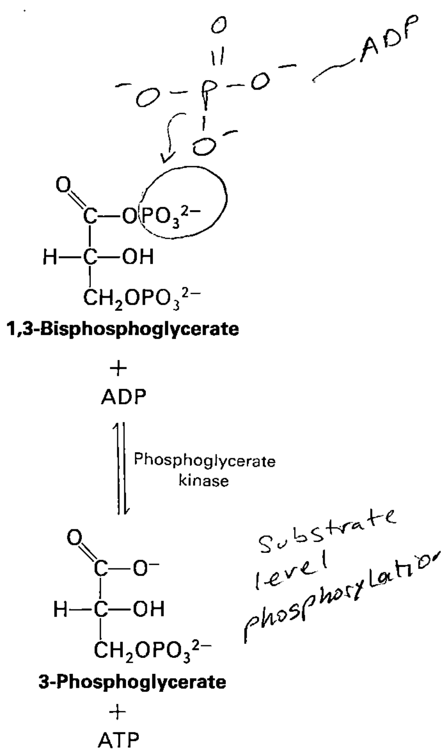

FORMATION OF ATP FROM 1,3-BISPHOSPHOGLYCERATE

In the next step, the high phosphoryl transfer potential of 1,3-BPG is used to generate ATP. Indeed, this is the first ATP-generating reaction in glycolysis. *Phosphoglycerate kinase* catalyzes the transfer of the phosphoryl group from the acyl phosphate of 1,3-BPG to ADP. ATP and 3-phosphoglycerate are the products.

Thus, the outcomes of the reactions catalyzed by glyceraldehyde 3-phosphate dehydrogenase and phosphoglycerate kinase are

1. Glyceraldehyde 3-phosphate, an aldehyde, is oxidized to 3-phosphoglycerate, a carboxylic acid.
2. NAD^+ is concomitantly reduced to NADH.
3. ATP is formed from P_i and ADP.

FORMATION OF PYRUVATE AND THE GENERATION OF A SECOND ATP

In the last stage of glycolysis, 3-phosphoglycerate is converted into pyruvate, and a second molecule of ATP is formed.

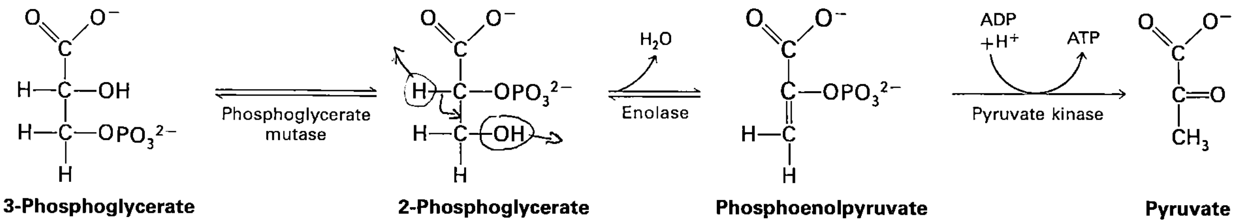

The first reaction is a rearrangement. The position of the phosphoryl group shifts in the conversion of *3-phosphoglycerate into 2-phosphoglycerate,* a reaction catalyzed by *phosphoglycerate mutase.* In general, a *mutase* is an enzyme that catalyzes the intramolecular shift of a chemical group, such as a phosphoryl group.

In the second reaction, an *enol* is formed by the dehydration of 2-phosphoglycerate. *Enolase* catalyzes the formation of *phosphoenolpyruvate.* This dehydration reaction markedly elevates the group transfer potential of the phosphoryl group. An *enol phosphate* has a high phosphoryl transfer potential, whereas the phosphate ester of an ordinary alcohol has a low one. The reasons for this difference will be discussed later (p. 504).

In the last reaction, *pyruvate* is formed, and ATP is generated concomitantly. The virtually irreversible transfer of a phosphoryl group from phosphoenolpyruvate to ADP is catalyzed by *pyruvate kinase.*

ENERGY YIELD IN THE CONVERSION OF GLUCOSE INTO PYRUVATE

The net reaction in the transformation of glucose into pyruvate is

$$\text{Glucose} + 2\,P_i + 2\,\text{ADP} + 2\,\text{NAD}^+ \longrightarrow 2\,\text{pyruvate} + 2\,\text{ATP} + 2\,\text{NADH} + 2\,H^+ + 2\,H_2O$$

Thus, *two molecules of ATP are generated in the conversion of glucose into two molecules of pyruvate.* A summary of the steps in which ATP is consumed or formed is given in Table 19-1. Recall that a pair of three-carbon units are formed from fructose 1,6-bisphosphate. The reactions of glycolysis are summarized in Figure 19-4 and Table 19-2.

Figure 19-4
The glycolytic pathway.

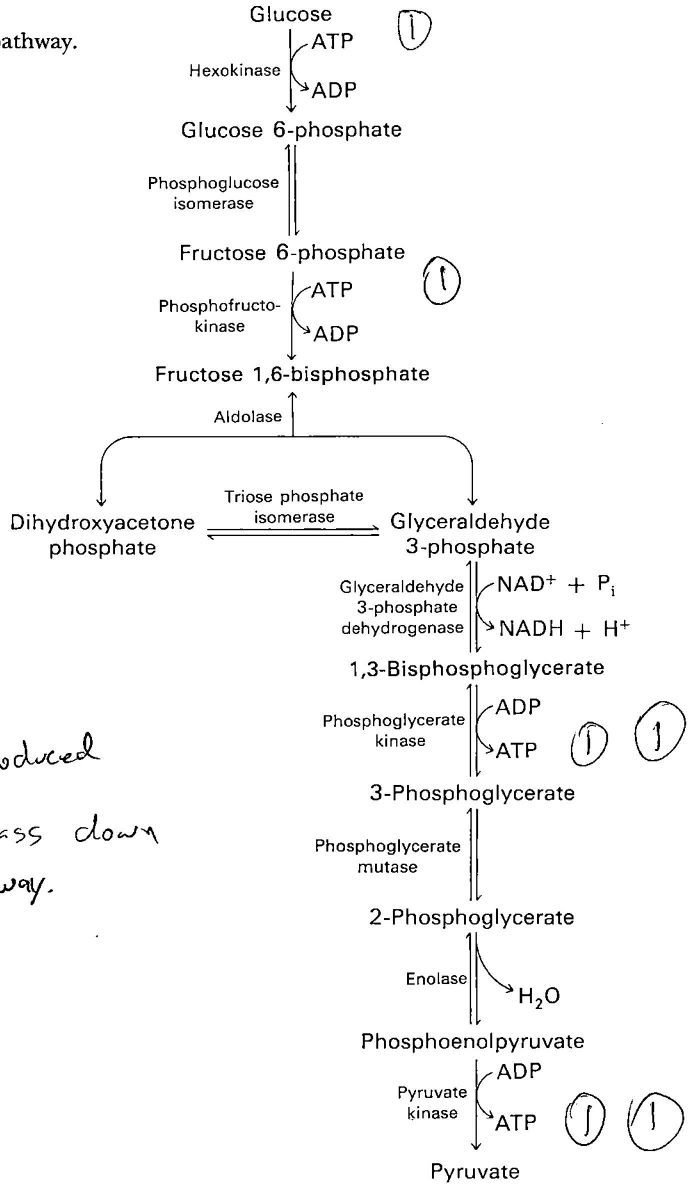

+2 ATPs produced with each pass down glycolytic pathway.

Table 19-1
Consumption and generation of ATP in glycolysis

Reaction	*ATP change per glucose*
Glucose ⟶ glucose 6-phosphate	−1
Fructose 6-phosphate ⟶ fructose 1,6-bisphosphate	−1
2 1,3-Bisphosphoglycerate ⟶ 2 3-phosphoglycerate	+2
2 Phosphoenolpyruvate ⟶ 2 pyruvate	+2
Net	+2

Table 19-2
Reactions of glycolysis

Step	*Reaction*	*Enzyme*	*Type**	ΔG°′	ΔG
1	Glucose + ATP ⟶ glucose 6-phosphate + ADP + H^+	Hexokinase	a	−4.0	−8.0
2	Glucose 6-phosphate ⇌ fructose 6-phosphate	Phosphoglucose isomerase	c	+0.4	−0.6
3	Fructose 6-phosphate + ATP ⟶ fructose 1,6-bisphosphate + ADP + H^+	Phosphofructokinase	a	−3.4	−5.3
4	Fructose 1,6-bisphosphate ⇌ dihydroxyacetone phosphate + glyceraldehyde 3-phosphate	Aldolase	e	+5.7	−0.3
5	Dihydroxyacetone phosphate ⇌ glyceraldehyde 3-phosphate	Triose phosphate isomerase	c	+1.8	+0.6
6	Glyceraldehyde 3-phosphate + P_i + NAD^+ ⇌ 1,3-bisphosphoglycerate + NADH +H^+	Glyceraldehyde 3-phosphate dehydrogenase	f	+1.5	−0.4
7	1,3-Bisphosphoglycerate + ADP ⇌ 3-phosphoglycerate + ATP	Phosphoglycerate kinase	a	−4.5	+0.3
8	3-Phosphoglycerate ⇌ 2-phosphoglycerate	Phosphoglycerate mutase	b	+1.1	+0.2
9	2-Phosphoglycerate ⇌ phosphoenolpyruvate +H_2O	Enolase	d	+0.4	−0.8
10	Phosphoenolpyruvate + ADP + H^+ ⟶ pyruvate +ATP	Pyruvate kinase	a	−7.5	−4.0

*Reaction type: (a) phosphoryl transfer; (b) phosphoryl shift; (c) isomerization; (d) dehydration; (e) aldol cleavage; (f) phosphorylation coupled to oxidation.

Note: $\Delta G^{\circ\prime}$ and ΔG are expressed in kcal/mol. ΔG, the actual free-energy change, has been calculated from $\Delta G^{\circ\prime}$ and known concentrations of reactants under typical physiologic conditions. Glycolysis can proceed only if the ΔG values of all reactions are negative. The small positive ΔG values of three of the above reactions indicate that the concentrations of metabolites in vivo in cells undergoing glycolysis are not precisely known.

ENTRY OF FRUCTOSE AND GALACTOSE INTO GLYCOLYSIS

Let us consider how two other abundant sugars—fructose and galactose—can be funneled into the glycolytic pathway. Recall that the hydrolysis of sucrose (common table sugar) yields fructose and glucose, and that hydrolysis of lactose (milk sugar) gives galactose and glucose. Fructose itself is present in many foods (e.g., honey). A typical daily intake of fructose is 100 grams. Much of it is metabolized by the liver, using the *fructose 1-phosphate pathway* (Figure 19-5). The first step is the phosphorylation of fructose to fructose 1-phosphate by fructokinase. Fructose 1-phosphate is then split into glyceraldehyde and dihydroxyacetone phosphate. This aldol cleavage is catalyzed by a specific fructose 1-phosphate aldolase. Glyceraldehyde is then phosphorylated to glyceraldehyde 3-phosphate by triose kinase so that it too can enter glycolysis.

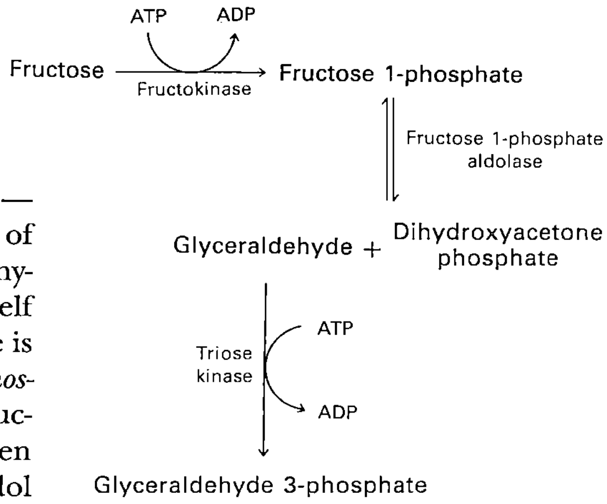

Figure 19-5
Fructose enters the glycolytic pathway via the fructose 1-phosphate pathway.

Alternatively, *fructose can be phosphorylated to fructose 6-phosphate by hexokinase.* However, the affinity of hexokinase for glucose is 20 times as high as it is for fructose. Little fructose 6-phosphate is formed in the liver because of the abundance of glucose relative to fructose in this organ. In contrast, adipose tissue has much more fructose than glucose. Hence, the formation of fructose 6-phosphate is not competitively inhibited to an appreciable extent, and most of the fructose in adipose tissue is metabolized through fructose 6-phosphate.

Galactose is converted into glucose 6-phosphate in four steps. The first reaction in the *galactose-glucose interconversion pathway* is the phosphorylation of galactose to galactose 1-phosphate by *galactokinase.*

$$\text{Galactose} + \text{ATP} \longrightarrow \text{galactose 1-phosphate} + \text{ADP} + \text{H}^+$$

Galactose 1-phosphate then acquires a uridyl group from uridine diphosphate glucose (UDP-glucose), an intermediate in the synthesis of glycosidic linkages (p. 586). The products of this reaction, which is catalyzed by *galactose 1-phosphate uridyl transferase,* are UDP-galactose and glucose 1-phosphate.

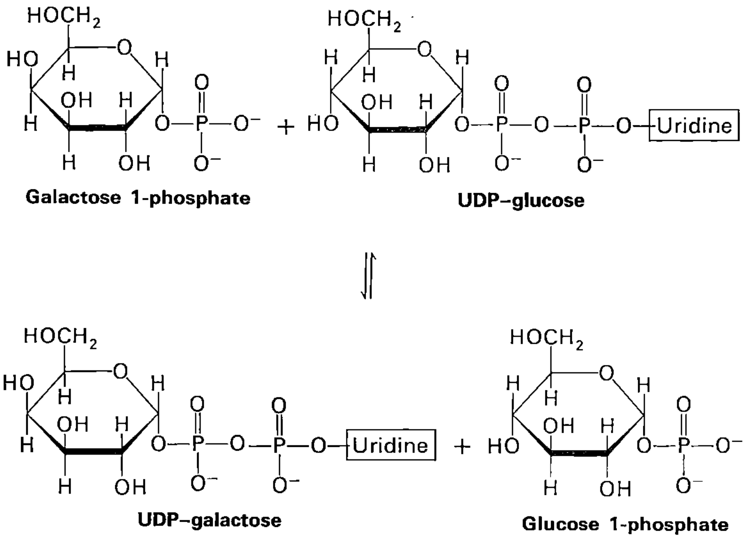

The galactose moiety of UDP is then epimerized to glucose. The configuration of the hydroxyl group at C-4 is inverted by *UDP-galactose 4-epimerase.* The sum of the reactions catalyzed by galactokinase, the transferase, and the epimerase is

$$\text{Galactose} + \text{ATP} \longrightarrow \text{glucose 1-phosphate} + \text{ADP} + \text{H}^+$$

Note that UDP-glucose is not consumed in the conversion of galactose to glucose because it is regenerated from UDP-galactose by the epimerase. This reaction is reversible, and the product of the reverse direction is also important. *The conversion of UDP-glucose into UDP-galactose is essential for the synthesis of galactosyl residues in complex polysaccharides and glycoproteins if the amount of galactose in the diet is inadequate to meet these needs.*

Finally, glucose 1-phosphate, formed from galactose, is isomerized to glucose 6-phosphate by *phosphoglucomutase.* We shall return to this reaction when we consider the synthesis and degradation of glycogen, which proceeds through glucose 1-phosphate (p. 584).

GALACTOSE IS HIGHLY TOXIC IF THE TRANSFERASE IS MISSING

The absence of galactose 1-phosphate uridyl transferase causes *galactosemia,* a severe disease that is inherited as an autosomal recessive trait. The metabolism of galactose in people who have this disease is blocked at galactose 1-phosphate. Afflicted infants fail to thrive. Vomiting or diarrhea occurs when milk is consumed, and enlargement of the liver and jaundice are common. Furthermore, many galactosemics become mentally retarded. The blood galactose level is markedly elevated, and galactose is found in the urine. The absence of the transferase in red blood cells is a definitive diagnostic criterion.

Galactosemia is treated by the exclusion of galactose from the diet. A galactose-free diet leads to a striking regression of virtually all the clinical symptoms, except for mental retardation, which may not be reversible. Continued galactose intake may lead to death in some patients. *The damage in galactosemia is caused by an accumulation of toxic substances, rather than by the absence of an essential compound.* Patients are able to synthesize UDP–galactose from UDP–glucose because their epimerase activity is normal. One of the toxic substances is *galactitol,* which is formed by reduction of galactose. The presence of aldose reductase in the lens of the eye causes galactitol to accumulate there, which leads to the entry of water and the development of cataracts.

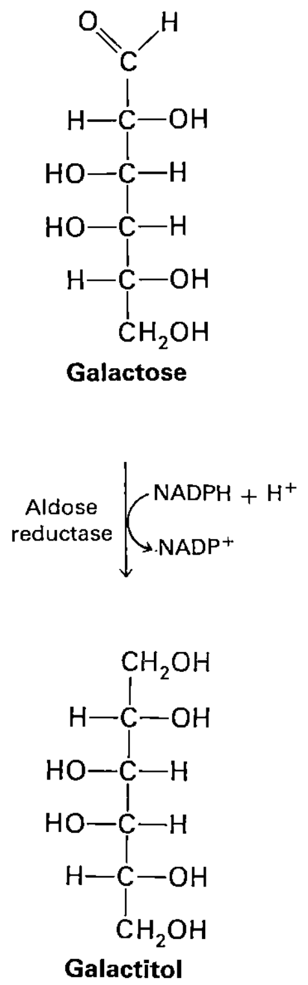

PHOSPHOFRUCTOKINASE IS THE KEY ENZYME IN THE CONTROL OF GLYCOLYSIS

The glycolytic pathway has a dual role: it degrades glucose to generate ATP, and it provides building blocks for synthetic reactions, such as the formation of long-chain fatty acids. The rate of conversion of glucose into pyruvate is regulated to meet these two major cellular needs. *In metabolic pathways, enzymes catalyzing essentially irreversible reactions are potential sites of control.* In glycolysis, the reactions catalyzed by hexokinase, phosphofructokinase, and pyruvate kinase are virtually irreversible; hence, they would be expected to have regulatory as well as catalytic roles. In fact, each of them serves as a control site. Their activities are regulated by the reversible binding of allosteric effectors or by covalent modification. Also, the amounts of these key enzymes are varied by transcriptional control to meet changing metabolic needs. Reversible allosteric control, regulation by phosphorylation, and transcriptional control typically occur in times of milliseconds, seconds, and hours, respectively.

Phosphofructokinase is the most important control element in the glycolytic pathway of mammals. The enzyme from liver (a 340-kd tetramer) is *inhibited by high levels of ATP,* which lower its affinity for fructose 6-phosphate. A high concentration of ATP converts the hyperbolic binding curve of fructose 6-phosphate into a sigmoidal one (Figure 19-6). This allosteric effect is elicited by the binding of ATP to a specific regulatory site that is distinct from the catalytic site. The inhibitory action of ATP is reversed by AMP, and so *the activity of the enzyme increases when the ATP/AMP ratio is lowered.* In other words, *glycolysis is stimulated as the energy charge falls.* A second control feature comes into play when the pH drops appreciably. The inhibition of phosphofructokinase by H^+ prevents excessive formation of lactate (p. 497) and a precipitous drop in blood pH (acidosis).

Glycolysis also furnishes carbon skeletons for biosyntheses, and so phosphofructokinase should also be regulated by a signal indicating whether

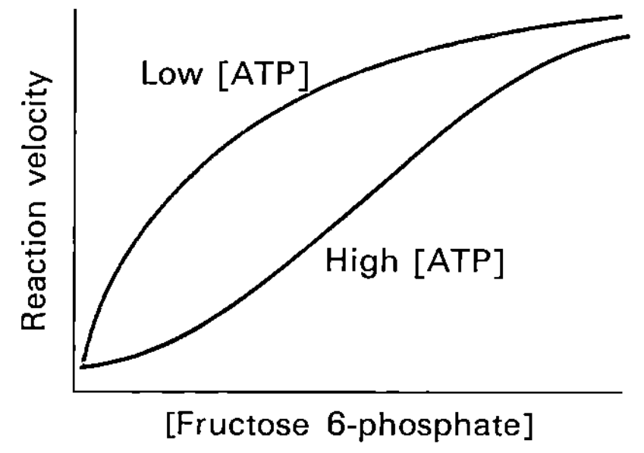

Figure 19-6
Allosteric regulation of phosphofructokinase. A high level of ATP inhibits the enzyme by decreasing its affinity for fructose 6-phosphate. AMP diminishes and citrate enhances the inhibitory effect of ATP.

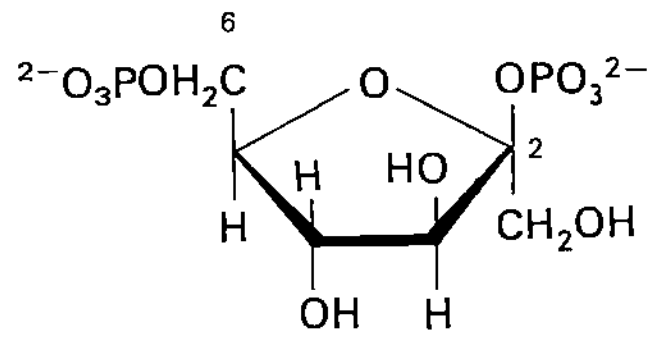

Fructose 2,6-bisphosphate (F-2,6-BP)

building blocks are abundant or scarce. Indeed, *phosphofructokinase is inhibited by citrate,* an early intermediate in the citric acid cycle (p. 510). A high level of citrate means that biosynthetic precursors are abundant and so additional glucose should not be degraded for this purpose. Citrate inhibits phosphofructokinase by enhancing the inhibitory effect of ATP.

A new regulator of glycolysis was discovered in 1980 by Henri-Géry Hers and Emile Van Schaftingen. They found that *fructose 2,6-bisphosphate,* a previously unknown metabolite, is a potent activator of phosphofructokinase. Fructose 2,6-bisphosphate (F-2,6-BP) activates phosphofructokinase in liver by increasing its affinity for fructose 6-phosphate and diminishing the inhibitory effect of ATP (Figure 19-7). In essence, *F-2,6-BP is an allosteric activator that shifts the conformational equilibrium of this tetrameric enzyme from the T state to the R state.*

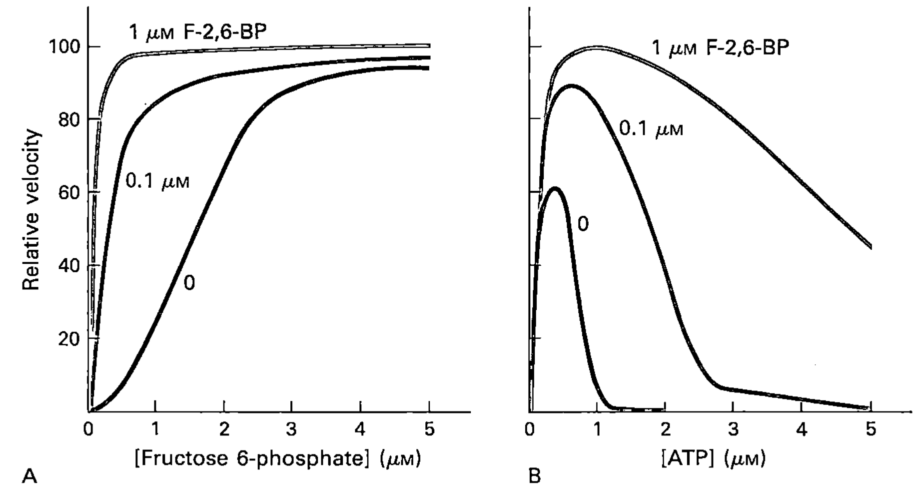

Figure 19-7
Phosphofructokinase is activated by fructose 2,6-bisphosphate. (A) The sigmoidal dependence of velocity on substrate concentration becomes hyperbolic in the presence of 1 μM fructose 2,6-bisphosphate, and (B) the inhibitory effect of ATP is reversed. [After H.-G. Hers and E. Van Schaftingen. *Proc. Nat. Acad. Sci.* 78(1981):2862.]

A REGULATED BIFUNCTIONAL ENZYME SYNTHESIZES AND DEGRADES FRUCTOSE 2,6-BISPHOSPHATE

Fructose 2,6-bisphosphate is formed by the phosphorylation of fructose 6-phosphate, in a reaction catalyzed by phosphofructokinase 2 (PFK2), a different enzyme from phosphofructokinase (PFK) (Figure 19-8). F-2,6-BP is hydrolyzed to fructose 6-phosphate by a specific phosphatase, fructose bisphosphatase 2 (FBPase2). The striking finding is that *both PFK2 and FBPase2 are present in a single 55-kd polypeptide chain.* This bifunctional enzyme contains an N-terminal *regulatory domain,* followed by a *kinase domain* and a *phosphatase domain.* It is interesting to note that PFK2 resem-

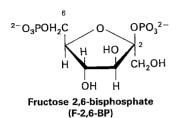

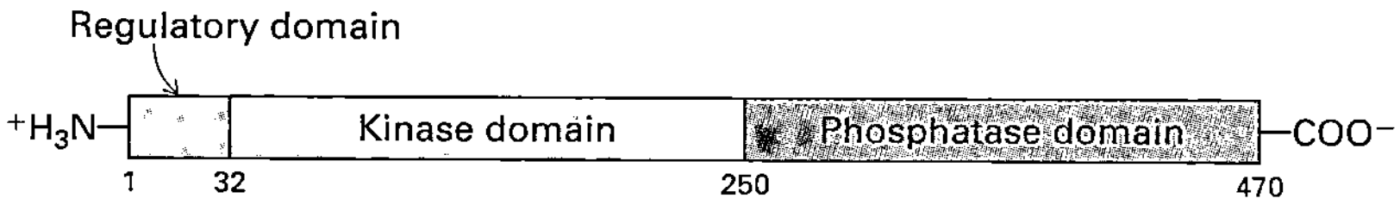

Domain structure of bifunctional enzyme

bles phosphofructokinase, whereas FBPase2 resembles phosphoglycerate mutase. The bifunctional enzyme probably arose by the fusion of genes encoding the kinase and phosphatase domains.

Fructose 6-phosphate accelerates the synthesis of F-2,6-BP and inhibits its hydrolysis. Hence, *an abundance of fructose 6-phosphate leads to a higher concentration of F-2,6-BP, which in turn stimulates phosphofructokinase.* Such a process is called *feedforward stimulation.* Furthermore, the activities of

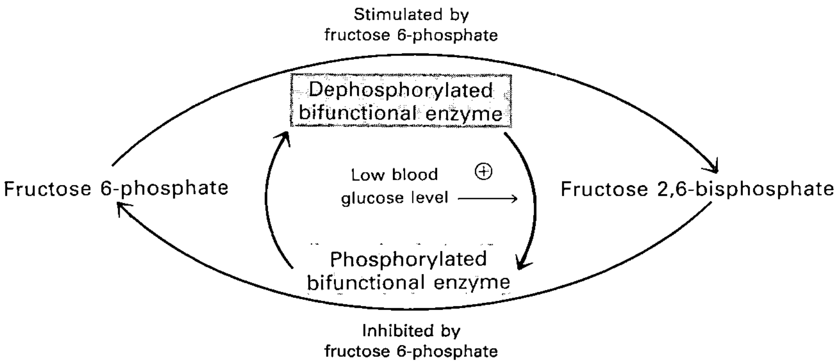

Figure 19-8
Control of the synthesis and degradation of fructose 2,6-bisphosphate. Fructose 6-phosphate accelerates the formation of F-2,6-BP and inhibits its hydrolysis. A low blood glucose level leads to a higher level of the phosphorylated bifunctional enzyme and hence to a lower level of F-2,6-BP.

PFK2 and FBPase2 are reciprocally controlled by *phosphorylation of a single serine residue.* When glucose is scarce, a rise in the blood level of the hormone glucagon triggers a cyclic AMP cascade leading to the phosphorylation of this bifunctional enzyme. This covalent modification activates FBPase2 and inhibits PFK2, lowering the level of F-2,6-BP. Conversely, when glucose is abundant, the enzyme loses its attached phosphate group, which leads to a rise in the level of F-2,6-BP and the consequent acceleration of glycolysis. This coordinated control is facilitated by having the kinase and phosphatase domains on the same polypeptide chain as the regulatory domain. We shall return to this elegant switch when we consider the integration of carbohydrate metabolism (p. 766).

HEXOKINASE AND PYRUVATE KINASE ALSO SET THE PACE OF GLYCOLYSIS

Hexokinase, the enzyme catalyzing the first step of glycolysis, is inhibited by glucose 6-phosphate. When phosphofructokinase is inactive, the concentration of fructose 6-phosphate rises. In turn, the level of glucose 6-phosphate rises because it is in equilibrium with fructose 6-phosphate. Hence, *the inhibition of phosphofructokinase leads to the inhibition of hexokinase.* However, liver possesses *glucokinase,* a specialized isoform of hexokinase that is not inhibited by glucose 6-phosphate. Glucokinase phosphorylates glucose only when it is abundant because it has a much higher K_M for glucose than does hexokinase (5 mM, compared with 0.1 mM). The role of glucokinase is to provide glucose 6-phosphate for the synthesis of glycogen, a storage form of glucose (p. 586). The high K_M of glucokinase in the liver gives brain and muscle first call on glucose when its supply is limited.

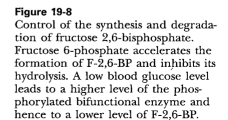

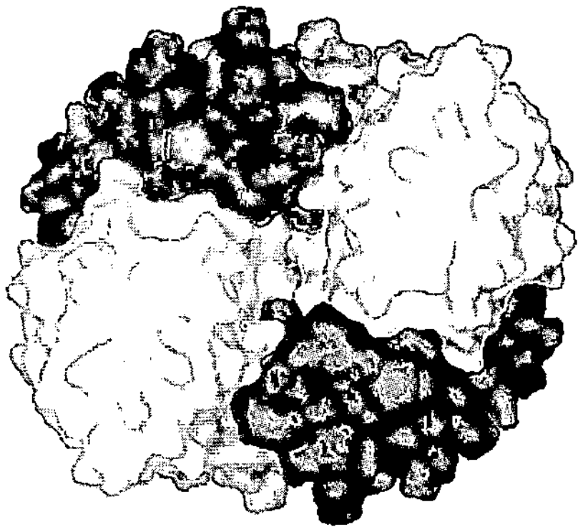

Structure of a bacterial phosphofructokinase. The four identical subunits are shown in different colors in a model depicting the surface of the enzyme. [Courtesy of Dr. Anthony Nicholls. Drawn from 4pfk.pdb. T. Schirmer and P.R. Evans. *Nature* 343(1990):140.]

Why is phosphofructokinase rather than hexokinase the pacemaker of glycolysis? The reason becomes evident on noting that glucose 6-phosphate is not solely a glycolytic intermediate. Glucose 6-phosphate can also be converted into glycogen or it can be oxidized by the pentose phosphate pathway (p. 559) to form NADPH. The first irreversible reaction unique to the glycolytic pathway, called the *committed step,* is the phosphorylation of fructose 6-phosphate to fructose 1,6-bisphosphate. Thus, it is highly appropriate for phosphofructokinase to be the primary control site in glycolysis. In general, *the enzyme catalyzing the committed step in a metabolic sequence is the most important control element in the pathway.*

Pyruvate kinase, the enzyme catalyzing the third irreversible step in glycolysis, controls the outflow from this pathway. This final step yields ATP and pyruvate, a central metabolic intermediate that can be oxidized further or used as a building block. Several forms of pyruvate kinase (a tetramer of 57-kd subunits) encoded by different genes are present in mammals: the L type predominates in liver, and the M type in muscle and brain. These variations on a common theme, called *isoenzymes* or *isozymes,* have essentially the same architectural plan and catalytic mechanism but differ in how they are regulated. Both the L and M forms bind phosphoenolpyruvate cooperatively. Fructose 1,6-bisphosphate, the product of the preceding irreversible step in glycolysis, activates pyruvate kinase to enable it to keep pace with the oncoming high flux of intermediates. ATP allosterically inhibits pyruvate kinase to slow glycolysis when the energy charge is high (Figure 19-9).

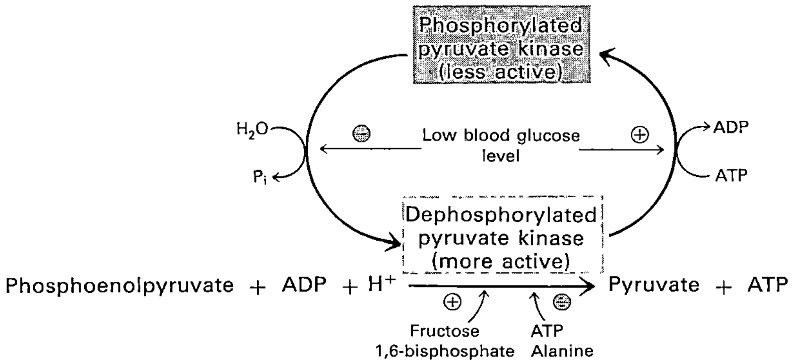

Figure 19-9
Control of the catalytic activity of pyruvate kinase.

Alanine (synthesized in one step from pyruvate, p. 630) also allosterically inhibits pyruvate kinase, in this case to signal that building blocks are abundant. The catalytic properties of the L form—but not of the M form—are also controlled by reversible phosphorylation. When the blood glucose level is low, glucagon triggers a cyclic AMP cascade that leads to the phosphorylation of pyruvate kinase, which diminishes its activity. A rise in the cytosolic calcium level induced by hormones such as vasopressin also leads to phosphorylation and inhibition of pyruvate kinase. *These hormone-triggered phosphorylations, like that of the bifunctional enzyme controlling the levels of fructose 2,6-bisphosphate, prevent the liver from consuming glucose when it is more urgently needed by brain and muscle* (p. 575). We see here a clear-cut example of how isoenzymes contribute to the metabolic diversity of different organs.

DIVERSE FATES OF PYRUVATE: ETHANOL, LACTATE, OR ACETYL COENZYME A

The sequence of reactions from glucose to pyruvate is similar in all organisms and in all kinds of cells. In contrast, the fate of pyruvate is variable. Three reactions of pyruvate are of prime importance:

1. *Ethanol* is formed from pyruvate in yeast and several other microorganisms. The first step is the decarboxylation of pyruvate.

$$\text{Pyruvate} + H^+ \longrightarrow \text{acetaldehyde} + CO_2$$

This reaction is catalyzed by *pyruvate decarboxylase.* Thiamine pyrophosphate, the coenzyme here, also participates in reactions catalyzed by decarboxylases (p. 516) and other enzymes (p. 566).

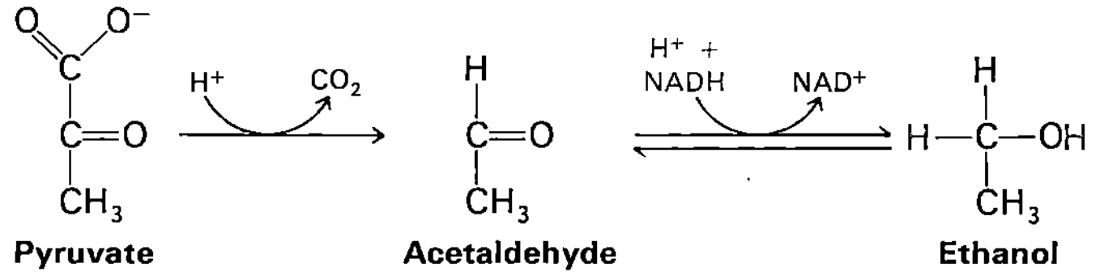

The second step is the reduction of acetaldehyde to ethanol by NADH, in a reaction catalyzed by *alcohol dehydrogenase.*

$$\text{Acetaldehyde} + \text{NADH} + H^+ \rightleftharpoons \text{ethanol} + \text{NAD}^+$$

The active site of alcohol dehydrogenase contains a zinc ion that is coordinated to the sulfur atoms of two cysteine residues and a histidine nitrogen atom (Figure 19-10). As in carboxypeptidase A (p. 221), Zn^{2+} polarizes the carbonyl group of the substrate to stabilize the transition state (Figure 19-11).

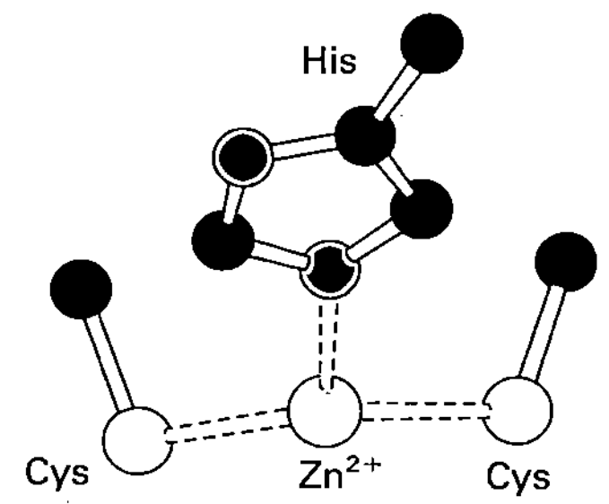

Figure 19-10
Model of the active site of liver alcohol dehydrogenase showing the coordination of the zinc ion. [Drawn from 2ohx.pdb; coordinates deposited by S. Al-Karadaghi and E.S. Cedergren-Zeppezauer.]

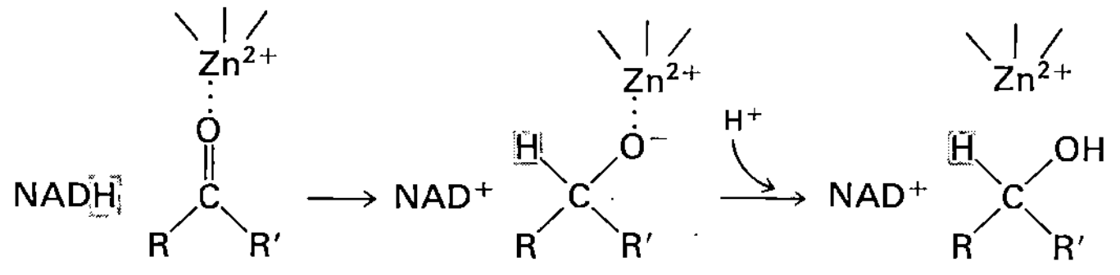

Figure 19-11
Catalytic mechanism of alcohol dehydrogenase.

The conversion of glucose into ethanol is called *alcoholic fermentation.* The net result of this anaerobic process is

$$\text{Glucose} + 2\,P_i + 2\,\text{ADP} + 2\,H^+ \longrightarrow 2\,\text{ethanol} + 2\,CO_2 + 2\,\text{ATP} + 2\,H_2O$$

It is important to note that NAD^+ and NADH do not appear in this equation, even though they are crucial for the overall reaction. NAD^+ generated in the reduction of acetaldehyde to ethanol is consumed in the oxidation of glyceraldehyde 3-phosphate. Thus, *there is no net oxidation-reduction in the conversion of glucose into ethanol.*

2. *Lactate* is normally formed from pyruvate in a variety of microorganisms. The reaction also occurs in the cells of higher organisms when the amount of oxygen is limiting, as in muscle during intense activity. The reduction of pyruvate by NADH to form lactate is catalyzed by *lactate dehydrogenase.*

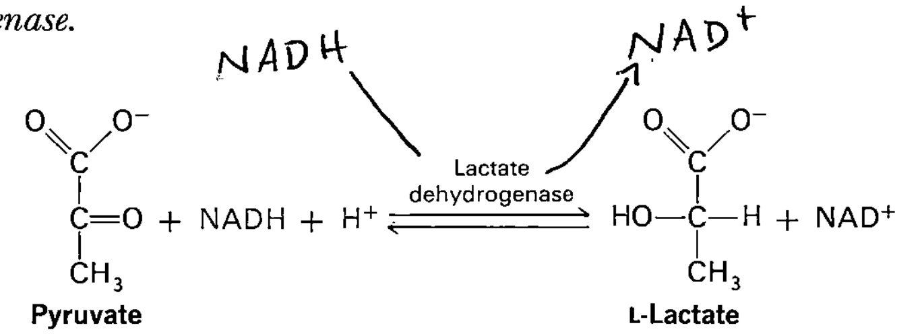

this rxn is heavily dependent on conc. of reactants. (direction of reaction)

high[NADH] / low[NAD] →

low[NADH] / high[NAD] ←

The overall reaction in the conversion of glucose into lactate is

$$\text{Glucose} + 2\,P_i + 2\,\text{ADP} \longrightarrow 2\,\text{lactate} + 2\,\text{ATP} + 2\,H_2O$$

As in alcoholic fermentation, there is no net oxidation-reduction. The NADH formed in the oxidation of glyceraldehyde 3-phosphate is consumed in the reduction of pyruvate. *The regeneration of NAD$^+$ in the reduction of pyruvate to lactate or ethanol sustains the continued operation of glycolysis under anaerobic conditions.* If NAD$^+$ were not regenerated, glycolysis could not proceed beyond glyceraldehyde 3-phosphate, which means that no ATP would be generated. In effect, the formation of lactate by aerobic organisms buys time, as will be discussed in Chapter 22.

3. Only a small fraction of the energy of glucose is released in its anaerobic conversion into lactate (or ethanol). Much more energy can be extracted aerobically by means of the citric acid cycle and the electron transport chain. The entry point to this oxidative pathway is *acetyl coenzyme A (acetyl CoA),* which is formed inside mitochondria by the oxidative decarboxylation of pyruvate.

$$\text{Pyruvate} + \text{NAD}^+ + \text{CoA} \longrightarrow \text{acetyl CoA} + \text{CO}_2 + \text{NADH}$$

This reaction, which is catalyzed by the pyruvate dehydrogenase complex, will be discussed in detail in the next chapter. The NAD$^+$ required for this reaction and for the oxidation of glyceraldehyde 3-phosphate is regenerated when NADH ultimately transfers its electrons to O_2 through the electron transport chain in mitochondria.

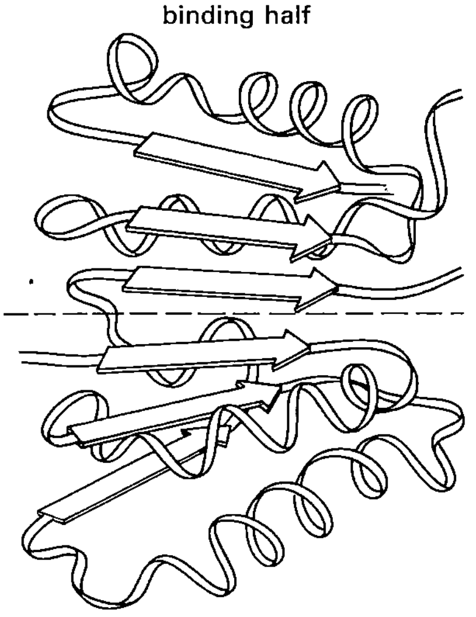

Figure 19-12
Schematic diagram of the NAD$^+$-binding region in dehydrogenases. The nicotinamide-binding half (shaded green) is structurally similar to the adenine-binding half (shaded yellow) of the site. [After M.G. Rossmann, A. Liljas, C.-I. Brändén, and L.J. Banaszak. In *The Enzymes,* vol. 10, 3rd ed. (Academic Press, 1975), p. 68.]

THE BINDING SITE FOR NAD$^+$ IS VERY SIMILAR IN MANY DEHYDROGENASES

Lactate dehydrogenase from skeletal muscle, a 140-kd tetramer, and alcohol dehydrogenase, an 84-kd dimer, have quite different three-dimensional structures. However, their NAD$^+$-binding domains are strikingly similar (Figure 19-12). This nucleotide-binding region is made up of four α helices and a sheet of six parallel β strands. The conformations of NAD$^+$ bound to lactate dehydrogenase and to alcohol dehydrogenase also are nearly the same. The adenine moiety of NAD$^+$ is bound in a hydrophobic crevice. In contrast, the nicotinamide unit is bound so that the reactive side of the ring is in a polar environment, whereas the other side makes contact with hydrophobic residues of the enzyme. The bound NAD$^+$ has an extended conformation (Figure 19-13). The three-dimen-

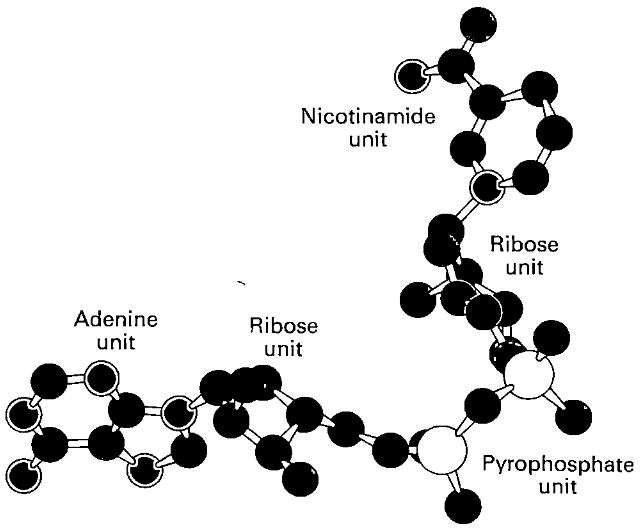

Figure 19-13
Model of NAD$^+$. The conformation shown here is the one found in the complex of NAD$^+$ and lactate dehydrogenase.

sional structures of two other NAD⁺-requiring enzymes—glyceraldehyde 3-phosphate dehydrogenase and malate dehydrogenase (an enzyme of the citric acid cycle, p. 512)—are also known at high resolution. Their NAD⁺-binding domains closely resemble those of lactate dehydrogenase and alcohol dehydrogenase. *The NAD⁺-binding region common to these four enzymes is a fundamental structural motif of NAD⁺-linked dehydrogenases.*

INDUCED FIT IN HEXOKINASE: GLUCOSE CLOSES THE ACTIVE-SITE CLEFT

X-ray crystallographic studies of yeast hexokinase have revealed that the binding of glucose leads to a large conformational change in the enzyme. Hexokinase consists of two lobes, which come together when glucose is bound (Figure 19-14). Glucose induces a 12-degree rotation of one lobe with respect to the other, resulting in movements of the polypeptide backbone of as much as 8 Å. The cleft between the lobes closes, and the bound glucose becomes surrounded by protein, except for its 5-hydroxymethyl group.

The closing of the cleft in hexokinase is a striking example of the role of *induced fit* in enzyme action, as originally proposed by Koshland (p. 191). The glucose-induced structural changes are likely to be significant in two ways. First, the environment around the glucose becomes much more nonpolar, which encourages the donation of the terminal phosphoryl group of ATP. Second, the embracing of glucose by hexokinase enables the enzyme to discriminate against H_2O as a substrate. If hexokinase were rigid, a water molecule occupying the binding site for the $-CH_2OH$ of glucose would attack the γ-phosphoryl group of ATP. In other words, a rigid kinase would necessarily be an ATPase as well as a kinase. The undesirable ATPase activity is prevented by making hexokinase enzymatically active only when glucose closes the cleft. It is interesting to note that pyruvate kinase, phosphoglycerate kinase, and phosphofructokinase also contain clefts between lobes that close when substrate is bound. *Substrate-induced cleft closing is likely to be a general feature of kinases.*

Glucose

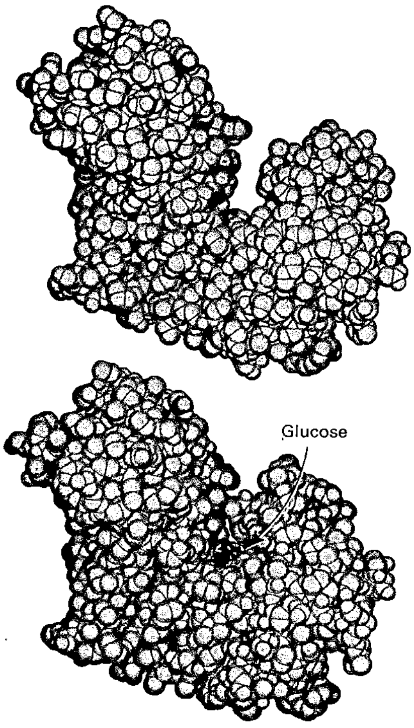

Figure 19-14
The conformation of hexokinase changes markedly on binding glucose (shown in red). The two lobes of the enzyme come together and surround the substrate. [Courtesy of Dr. Thomas Steitz.]

ALDOLASE FORMS A SCHIFF BASE WITH DIHYDROXYACETONE PHOSPHATE

Let us turn now to the catalytic mechanism of aldolase. For convenience, we will examine the condensation of dihydroxyacetone phosphate and glyceraldehyde 3-phosphate to form fructose 1,6-bisphosphate, the reverse of the glycolytic reaction. First, dihydroxyacetone phosphate forms a protonated Schiff base with a specific lysine residue in the active site of animal aldolases.

$$E-NH_2 + \begin{array}{c} CH_2OPO_3^{2-} \\ | \\ O=C \\ | \\ CH_2OH \end{array} + H^+ \rightleftharpoons \begin{array}{c} CH_2OPO_3^{2-} \\ | \\ E-\overset{+}{N}=C \\ | \quad | \\ H \quad CH_2OH \end{array} + H_2O$$

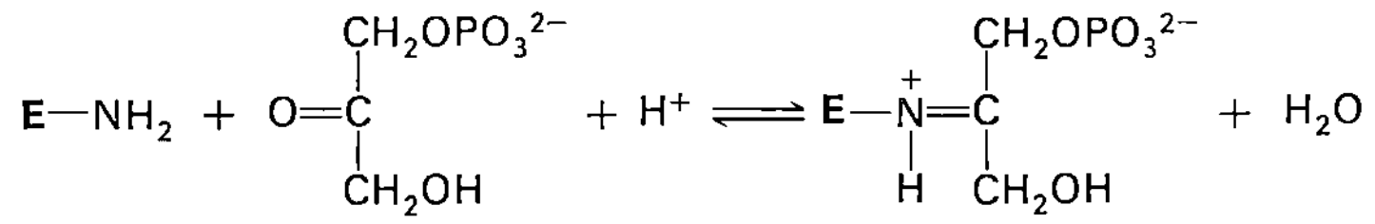

Protonated Schiff base

In this reaction, a nucleophile (the amino group) attacks the carbonyl group to form a tetrahedral intermediate, which then dehydrates. *The resulting protonated Schiff base plays a critical role in catalysis. It promotes the formation of the enolate anion of dihydroxyacetone phosphate by serving as an electron sink (a potent electron acceptor).*

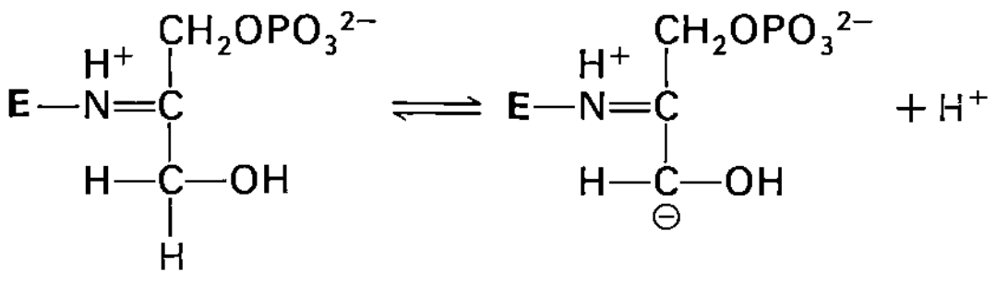

Enolate anion

The enolate anion then adds to the aldehyde group of glyceraldehyde 3-phosphate to form a protonated Schiff base of fructose 1,6-bisphosphate.

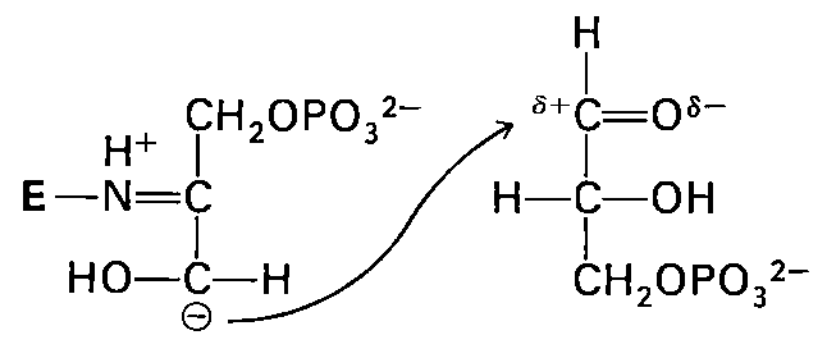

Enolate anion   Glyceraldehyde 3-phosphate

The Schiff base is deprotonated and hydrolyzed to yield fructose 1,6-bisphosphate and the regenerated enzyme.

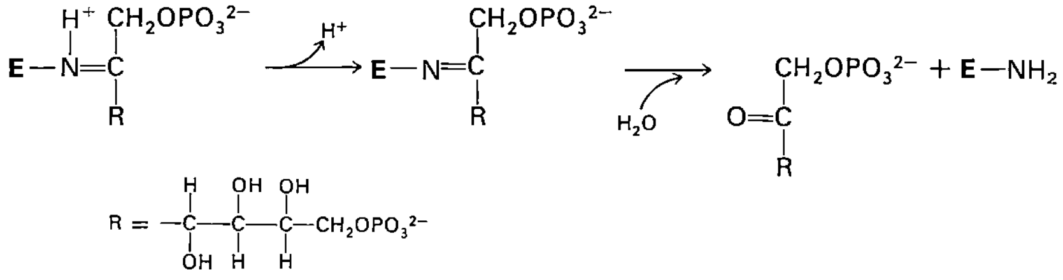

The pathway for the cleavage of fructose 1,6-bisphosphate is simply the reverse of the one for its formation.

KINETIC PERFECTION IN CATALYSIS: TRIOSE PHOSPHATE ISOMERASE IN ACTION

Much is known about the catalytic mechanism of triosephosphate isomerase (TIM), an especially interesting enzyme. TIM catalyzes the transfer of a hydrogen atom from C-1 to C-2 in converting dihydroxyacetone phosphate into glyceraldehyde 3-phosphate, an intramolecular oxidation-reduction. This isomerization of a ketose into an aldose proceeds through an *enediol intermediate* (Figure 19-15). When the reaction is carried out in D_2O, deuterium becomes incorporated into C-2. This finding rules out a

Figure 19-15
Catalytic mechanism of triosephosphate isomerase.

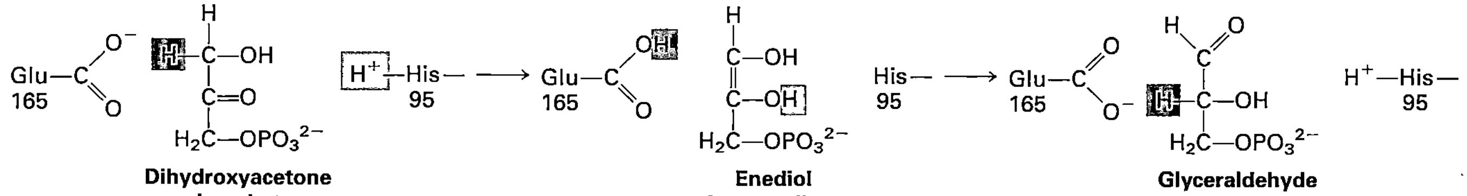

Dihydroxyacetone phosphate   Enediol intermediate   Glyceraldehyde 3-phosphate

direct transfer of a hydride ion (:H$^-$) from C-1 to C-2. Rather, the first step is the removal of a proton from C-1 by a basic group of the protein to form an enediol. Isotope labeling studies of the reverse reaction provided another key clue. Some of the tritium attached to C-2 of specifically labeled glyceraldehyde 3-phosphate emerged in C-1 of dihydroxyacetone phosphate, whereas the rest exchanged with protons of water. This finding revealed that *abstraction and donation of a proton are mediated by the same catalytic base.*

X-ray crystallographic studies then showed that glutamate 165 plays this role (Figure 19-16). However, a carboxylate group by itself is not basic enough to pull a proton away from a carbon atom adjacent to a carbonyl group. Histidine 95 assists catalysis by donating a proton to the C-2 carbonyl group. The phosphate group of the substrate is held in place by a salt bridge with lysine 12 and hydrogen bonds with two main-chain NH groups. The positively charged end of an α helix dipole is also trained on the dianionic phosphate moiety.

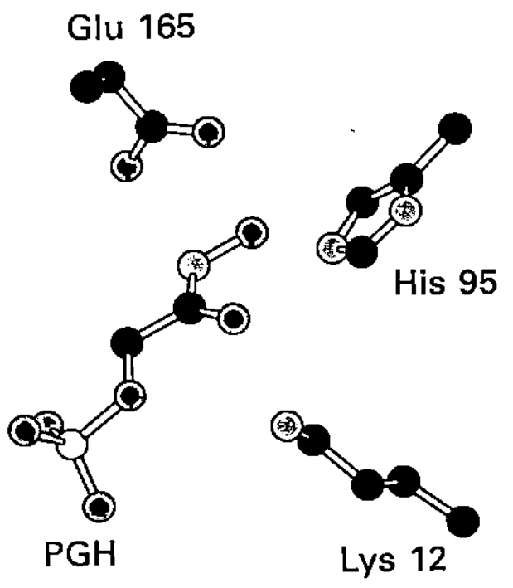

Figure 19-16
Active site of triosephosphate isomerase. Phosphoglycolhydroxamate (PGH), a substrate analog, is tightly bound. Glu 165 and His 95 play key catalytic roles. Lys 12 forms a salt bridge with the phosphate group of the substrate analog and polarizes the C-2 oxygen atom. [Drawn from 7tim.pdb. R.C. Davenport, P.A. Bash, B.A. Seaton, M. Karplus, G.A. Petsko, and D. Ringe. *Biochemistry* 30(1991):5821.]

Two other features of this isomerase are noteworthy. First, TIM displays great catalytic prowess. It accelerates isomerization by a factor of 10^{10} compared with that obtained with a simple base catalyst such as acetate ion. Indeed, the k_{cat}/K_M ratio for isomerization of glyceraldehyde 3-phosphate is 2×10^8 M^{-1} s^{-1}, which is close to the diffusion-controlled limit. In other words, the rate-limiting step in catalysis is the diffusion-controlled encounter of substrate and enzyme. TIM is an example of a *kinetically perfect enzyme* (p. 195). Second, TIM suppresses an undesired side reaction, the decomposition of the enediol intermediate into methylglyoxal and P_i (Figure 19-17). In solution, this unfavorable reaction is 100 times faster than isomerization. Hence, TIM must prevent the enediol from leaving the enzyme. This labile intermediate is sequestered in the active site by the movement of a loop of 10 residues. This lid shuts the active site when the enediol is present, and reopens when isomerization is completed. *We see here a striking example of the channeling of catalysis by induced fit.*

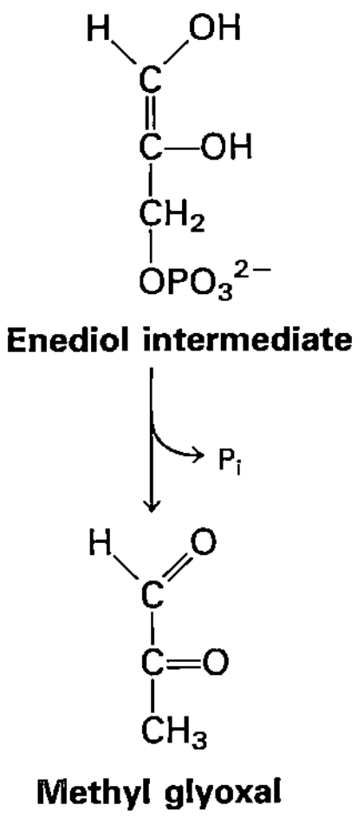

Figure 19-17
In solution, the enediol intermediate in the isomerization of dihydroxyacetone phosphate decomposes into methylglyoxal and P_i. The isomerase blocks this undesired side reaction.

A THIOESTER IS FORMED IN THE OXIDATION OF GLYCERALDEHYDE 3-PHOSPHATE

Glyceraldehyde 3-phosphate dehydrogenase catalyzes the oxidative phosphorylation of its aldehyde substrate.

$$\text{Glyceraldehyde 3-phosphate} + P_i + NAD^+ \longrightarrow 1{,}3\text{-BPG} + NADH + H^+$$

The conversion of an aldehyde into an acyl phosphate entails both *oxidation* and *phosphorylation*. Oxidation requires the *removal of a hydride ion* (:H⁻), which is a hydrogen nucleus and two electrons. Removal of a hydride ion from an aldehyde is energetically costly because of the dipolar character of the carbonyl group: the carbon atom of the carbonyl group already has a partial positive charge. Removal of the hydride ion is greatly facilitated by making the carbon atom less positively charged. This is accomplished by the *addition of a nucleophile*, represented here by X⁻:

$$H^+ + \overset{\delta-}{O}{=}\overset{R}{\underset{\delta+}{C}}{-}H + X^- \longrightarrow HO{-}\overset{R}{\underset{X}{C}}{-}H \xrightarrow[NAD^+\quad NADH\ (reduced)]{} O{=}\overset{R}{\underset{X}{C}} + {:}H^- + H^+$$

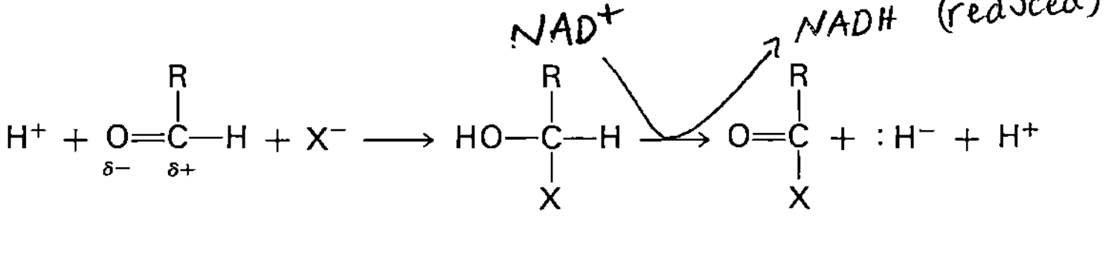

The hydride ion readily leaves the addition compound because the carbon atom no longer carries a large positive charge. Furthermore, some of the free energy of the oxidation is preserved in the acyl intermediate. Addition of orthophosphate to this acyl intermediate yields an acyl phosphate, which has a high group transfer potential.

$$R{-}\overset{O}{\overset{\|}{C}}{-}X + {}^-O{-}\overset{O}{\underset{OH}{\overset{\|}{P}}}{-}O^- \longrightarrow R{-}\overset{O}{\overset{\|}{C}}{-}O{-}\overset{O}{\underset{O^-}{\overset{\|}{P}}}{-}O^- + XH$$

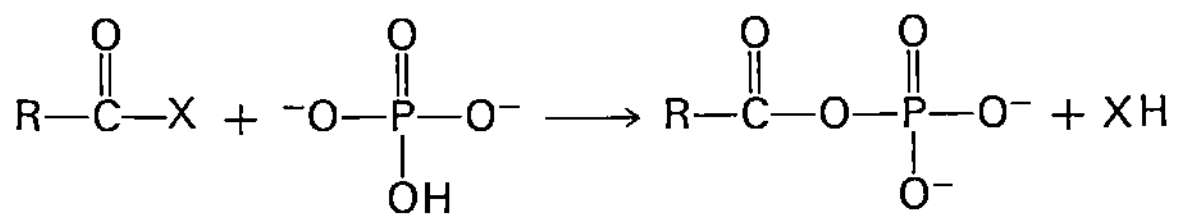

Now let us see how glyceraldehyde 3-phosphate dehydrogenase carries out these steps (Figure 19-18). *The nucleophile X⁻ is the sulfhydryl group of a cysteine residue at the active site.* The aldehyde substrate reacts with the ionized form of this sulfhydryl group to form a hemithioacetal. The next step is the *transfer of a hydride ion to a molecule of NAD⁺ that is tightly bound to the enzyme.* The products of this reaction are the reduced coenzyme NADH and a thioester. *This thioester is an energy-rich intermediate,* corresponding to the acyl intermediate mentioned earlier. NADH dissociates from the enzyme, and another NAD⁺ binds to the active site. Orthophosphate then attacks the thioester to form 1,3-BPG, a high-potential phosphoryl donor.

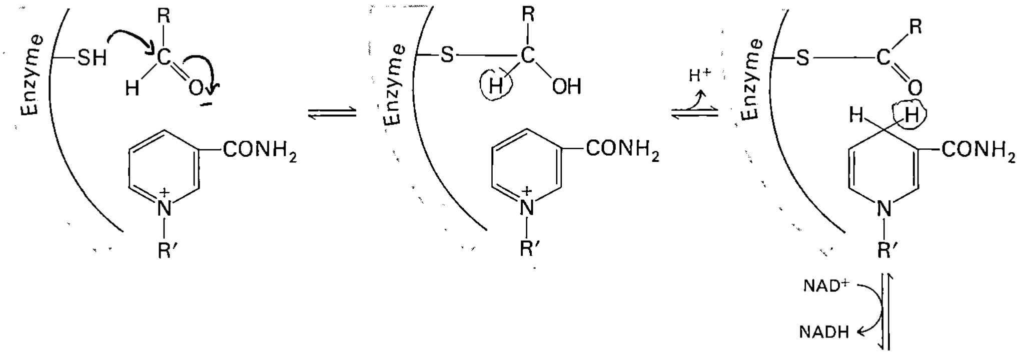

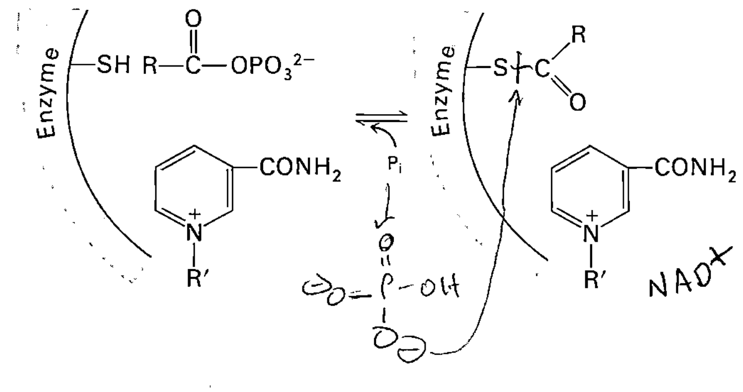

Figure 19-18
Catalytic mechanism of glyceraldehyde 3-phosphate dehydrogenase.

A crucial aspect of the formation of 1,3-BPG from glyceraldehyde 3-phosphate is that a thermodynamically unfavorable reaction, the formation of an acyl phosphate from a carboxylate, is driven by a thermodynamically favorable reaction, the oxidation of an aldehyde.

$$R\text{—}\overset{\overset{\displaystyle O}{\|}}{C}\text{—}H + NAD^+ + H_2O \rightleftharpoons R\text{—}\overset{\overset{\displaystyle O}{\|}}{C}\text{—}O^- + NADH + 2\ H^+$$

These two reactions are *coupled by the thioester intermediate,* which preserves much of the free energy released in the oxidation reaction. We see here the *use of a covalent enzyme-bound intermediate as a mechanism of energy coupling.*

ARSENATE, AN ANALOG OF PHOSPHATE, POISONS BY UNCOUPLING OXIDATION AND PHOSPHORYLATION

Arsenate (AsO_4^{3-}) closely resembles P_i in structure and reactivity. In the reaction catalyzed by glyceraldehyde 3-phosphate dehydrogenase, arsenate can replace phosphate in attacking the energy-rich thioester intermediate. The product of this reaction, 1-arseno-3-phosphoglycerate, is unstable, in contrast with 1,3-bisphosphoglycerate. 1-Arseno-3-phosphoglycerate and other acyl arsenates are rapidly and spontaneously hydrolyzed. Hence, the net reaction in the presence of arsenate is

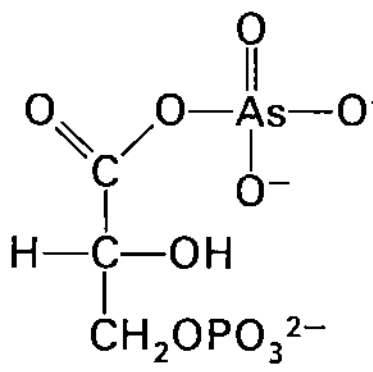

1-Arseno-3-phosphoglycerate

$$\text{Glyceraldehyde 3-phosphate} + NAD^+ + H_2O \longrightarrow \text{3-phosphoglycerate} + NADH + 2\ H^+$$

Note that glycolysis proceeds in the presence of arsenate but that the ATP normally formed in the conversion of 1,3-bisphosphoglycerate into 3-phosphoglycerate is lost. Thus, *arsenate uncouples oxidation and phosphorylation by forming a highly labile acyl arsenate.* Arsenic is a potent poison because arsenate generally substitutes for phosphate in phosphoryl transfer reactions. Also, arsenite (AsO_2^-) forms adducts with thiols. One likely reason for the choice of phosphorus over arsenic in the evolution of biomolecules is the greater kinetic stability of its energy-rich compounds.

2,3-BISPHOSPHOGLYCERATE, AN ALLOSTERIC EFFECTOR OF HEMOGLOBIN, ARISES FROM 1,3-BISPHOSPHOGLYCERATE

We have seen that fructose 2,6-bisphosphate, a regulatory molecule, arises from a glycolytic intermediate. Another regulatory molecule coming from this pathway is *2,3-bisphosphoglycerate* (2,3-BPG), a controller of oxygen transport in erythrocytes (p. 160). Red blood cells have a high concentration of 2,3-BPG, typically 4 mM, in contrast with most other cells, which have only trace amounts. The synthesis and degradation of 2,3-BPG are a short detour from the glycolytic pathway (Figure 19-19).

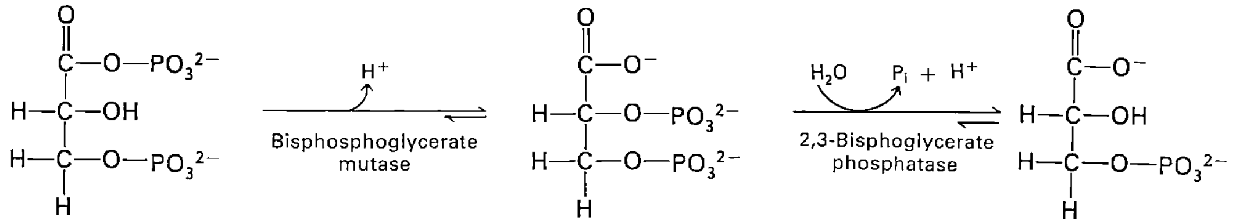

1,3-Bisphosphoglycerate (1,3-BPG) **2,3-Bisphosphoglycerate (2,3-BPG)** **3-Phosphoglycerate**

Figure 19-19
Synthesis and degradation of 2,3-bisphosphoglycerate.

Bisphosphoglycerate mutase converts 1,3-BPG into 2,3-BPG. 2,3-BPG is a potent competitive inhibitor of its own formation. The concentration of this allosteric effector also depends on its hydrolysis to 3-phosphoglycerate, a reaction catalyzed by *2,3-bisphosphoglycerate phosphatase.* Both the synthesis and degradation of 2,3-BPG are nearly irreversible.

The mutase reaction has an interesting mechanism. *3-Phosphoglycerate* is an obligatory participant although it does not appear in the overall stoichiometry. The mutase binds 1,3-BPG and 3-phosphoglycerate simultaneously. In this ternary complex, a phosphoryl group is transferred from position 1 of 1,3-BPG to position 2 of 3-phosphoglycerate (Figure 19-20).

Table 19-3
Typical concentrations of glycolytic intermediates in erythrocytes

Intermediate	μM
Glucose	5000
Glucose 6-phosphate	83
Fructose 6-phosphate	14
Fructose 1,6-bisphosphate	31
Dihydroxyacetone phosphate	138
Glyceraldehyde 3-phosphate	19
1,3-Bisphosphoglycerate	1
2,3-Bisphosphoglycerate	4000
3-Phosphoglycerate	118
2-Phosphoglycerate	30
Phosphoenolpyruvate	23
Pyruvate	51
Lactate	2900
ATP	1850
ADP	138
P_i	1000

After S. Minakami and H. Yoshikawa. *Biochem. Biophys. Res. Comm.* 18(1965):345.

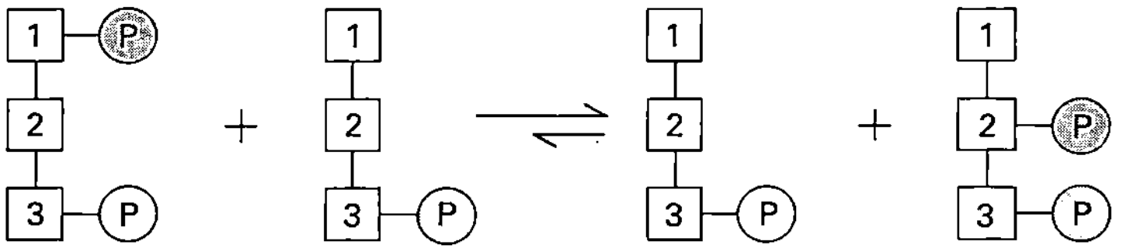

Figure 19-20
3-Phosphoglycerate participates in the conversion of 1,3-bisphosphoglycerate into 2,3-bisphosphoglycerate.

2,3-BPG is important not only in red blood cells. Phosphoglycerate mutase, the enzyme catalyzing the interconversion of 3-phosphoglycerate and 2-phosphoglycerate, requires a catalytic amount of 2,3-BPG to be active. 2,3-BPG phosphorylates a histidine residue at the catalytic site to maintain the active form of the enzyme. Thus, 2,3-bisphosphoglycerate plays an essential role in basic metabolism. It seems likely that 2,3-BPG was present eons before it was recruited by red cells to control the oxygen affinity of hemoglobin.

ENOL PHOSPHATES ARE POTENT PHOSPHORYL DONORS

Because it is an *acyl phosphate,* 1,3-BPG has a high group transfer potential. A different kind of high-energy phosphate compound is generated several steps later in glycolysis. Phosphoenolpyruvate, an *enol phosphate,* is formed by the dehydration of 2-phosphoglycerate. The $\Delta G^{\circ\prime}$ of hydrolysis of a phosphate ester of an ordinary alcohol is -3 kcal/mol, whereas that of phosphoenolpyruvate is -14.8 kcal/mol. Why does phosphoenolpyruvate have such a high phosphoryl group–transfer potential? The answer is that the enol formed upon transfer of the phosphoryl group undergoes a conversion into a ketone—namely, pyruvate.

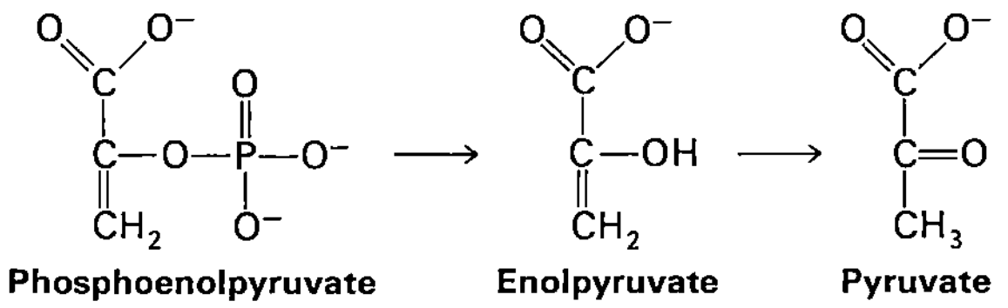

The $\Delta G^{\circ\prime}$ of the enol-ketone conversion is very large, of the order of -10 kcal/mol. Thus, *the high phosphoryl group–transfer potential of phosphoenolpyruvate arises primarily from the large driving force of the subsequent enol-ketone conversion.*

A FAMILY OF TRANSPORTERS ENABLES GLUCOSE TO ENTER AND LEAVE ANIMAL CELLS

The thermodynamically downhill movement of glucose across the plasma membrane of animal cells is mediated by several *glucose transporters*. The members of this protein family, named GLUT1 to 5, consist of a single polypeptide chain about 500 residues long. The common structural theme is the presence of 12 transmembrane segments (Figure 19-21). The binding site for glucose alternately faces the inside and outside of the cell when occupied by a sugar; this eversion is accomplished by conformational changes within the transporter and not by a rotation of the whole protein.

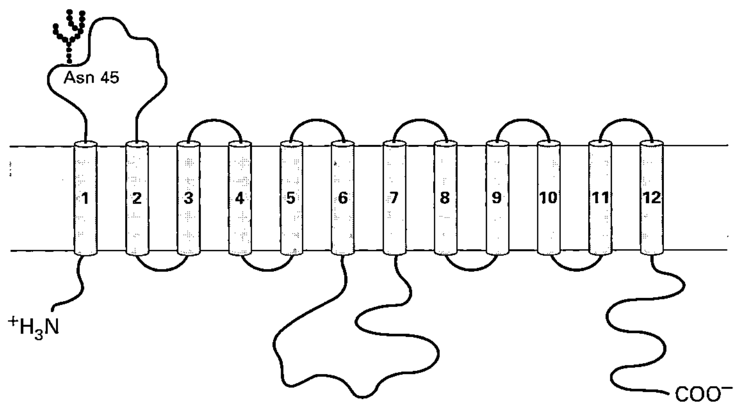

Figure 19-21
Model of a mammalian glucose transporter. The hydrophobicity profile of the protein points to the presence of 12 transmembrane α helices. Chemical labeling and tryptic cleavage studies support this postulated topography. [From M. Muekler, C. Caruso, S.A. Baldwin, M. Panico, M. Blench, H.R. Morris, W.J. Allard, G.E. Lienhard, and H.F. Lodish. *Science* 229(1985):941.]

The members of this family have distinctive roles:

1. GLUT1 and 3, present in nearly all mammalian cells, are responsible for basal glucose uptake. Their K_M for glucose is about 1 mM, significantly less than the normal serum glucose level, which typically ranges from 4 mM to 8 mM. Hence, GLUT1 and 3 continually transport glucose at an essentially constant rate.

2. GLUT5, present in the small intestine, works in tandem with the Na^+-glucose symporter in the absorption of glucose from the gut. The symporter pumps glucose into the intestinal epithelial cell. GLUT5 in the plasma membrane on the opposite side of the cell then releases glucose into the bloodstream.

3. GLUT2, present in liver and pancreatic β cells, is distinctive in having a very high K_M for glucose (15–20 mM). Hence, the rate of entry of glucose into these tissues is proportional to the blood glucose level. The pancreas can thereby sense the glucose level and accordingly adjust the rate of insulin secretion. The high K_M of GLUT2 also assures that glucose rapidly enters liver cells only in times of plenty. When the blood glucose level is low, glucose preferentially enters brain and other tissues because their glucose transporters have a lower K_M than that of liver.

4. GLUT4, which has a K_M of 5 mM, mediates the entry of glucose into muscle and fat cells. Insulin, which signals the fed state, leads to a rapid increase in the number of GLUT4 transporters in the plasma membrane. Hence, insulin promotes the uptake of glucose by muscle and fat.

This family of transporters vividly illustrates how isoforms of a single protein can profoundly shape the metabolic character of cells and contribute to their diversity and functional specialization.

SUMMARY

Glycolysis is the set of reactions that converts glucose into pyruvate. In aerobic organisms, glycolysis is the prelude to the citric acid cycle and the electron transport chain, where most of the free energy in glucose is harvested. The 10 reactions of glycolysis occur in the cytosol. In the first stage, glucose is converted into fructose 1,6-bisphosphate by a phosphorylation, an isomerization, and a second phosphorylation reaction. Two molecules of ATP are consumed per molecule of glucose in these reactions, which are the prelude to the net synthesis of ATP. In the second stage, fructose 1,6-bisphosphate is cleaved by aldolase into dihydroxyacetone phosphate and glyceraldehyde 3-phosphate, which are readily interconvertible. Glyceraldehyde 3-phosphate is then oxidized and phosphorylated to form 1,3-bisphosphoglycerate, an acyl phosphate with a high phosphoryl transfer potential. 3-Phosphoglycerate is then formed as an ATP is generated. In the last stage of glycolysis, phosphoenolpyruvate, a second intermediate with a high phosphoryl transfer potential, is formed by a phosphoryl shift and a dehydration. Another ATP is generated as phosphoenolpyruvate is converted into pyruvate. There is a net gain of two molecules of ATP in the formation of two molecules of pyruvate from one molecule of glucose.

The electron acceptor in the oxidation of glyceraldehyde 3-phosphate is NAD^+, which must be regenerated for glycolysis to continue. In aerobic organisms, the NADH formed in glycolysis transfers its electrons to O_2 through the electron transport chain, which thereby regenerates NAD^+. Under anaerobic conditions, NAD^+ is regenerated by the reduction of pyruvate to lactate. In some microorganisms, NAD^+ is normally regenerated by the synthesis of lactate or ethanol from pyruvate. These two processes are examples of fermentations.

The glycolytic pathway has a dual role: it degrades glucose to generate ATP, and it provides building blocks for the synthesis of cellular components. The rate of conversion of glucose into pyruvate is regulated to meet these two major cellular needs. Under physiologic conditions, the reactions of glycolysis are readily reversible except for the ones catalyzed by hexokinase, phosphofructokinase, and pyruvate kinase. Phosphofructokinase, the most important control element in glycolysis, is inhibited by high levels of ATP and citrate, and it is activated by AMP and fructose 2,6-bisphosphate. In liver, this bisphosphate signals that glucose is abundant. Hence, phosphofructokinase is active when either energy or building blocks are needed. Hexokinase is inhibited by glucose 6-phosphate, which accumulates when phosphofructokinase is inactive. Pyruvate kinase, the other control site, is allosterically inhibited by ATP and alanine, and it is activated by fructose 1,6-bisphosphate. Consequently, pyruvate kinase is maximally active when the energy charge is low and glycolytic intermediates accumulate. Pyruvate kinase, like the bifunctional enzyme controlling the level of fructose 2,6-bisphosphate, is regulated by phosphorylation. A low level of glucose in the blood promotes the phosphorylation of liver pyruvate kinase, which diminishes its activity and thereby decreases glucose consumption in the liver.

Induced fit enhances the catalytic efficiency of hexokinase and triosephosphate isomerase by preventing undesired side reactions. The isomerase is an example of a catalytically perfect enzyme limited only by the diffusion-controlled encounter of enzyme and substrate. A thioester intermediate conserves some of the free energy of oxidation of glyceraldehyde 3-phosphate; attack by P_i yields an energy-rich acyl phosphate. Arsenate, an analog of phosphate, uncouples oxidation and phosphorylation.

The thermodynamically downhill entry of glucose into animal cells is mediated by a family of glucose transporters named GLUT1 to 5. The binding site for glucose in this plasma membrane protein, when occupied, alternately faces the inside and outside of the cell. The different K_M values and differential regulation of this family of transporters shape the metabolic character of cells in different organs.

SELECTED READINGS

Where to start

Boiteux, A., and Hess, B., 1981. Design of glycolysis. *Phil. Trans. Roy. Soc. Lond. B* 293:5–22. [A stimulating presentation of the regulation of glycolysis in a symposium volume entitled *The Enzymes of Glycolysis: Structure, Activity, and Evolution.*]

Pilkis, S.J., and Granner, D.K., 1992. Molecular physiology of the regulation of hepatic gluconeogenesis and glycolysis. *Ann. Rev. Physiol.* 54:885–909.

Knowles, J.R., 1991. Enzyme catalysis: Not different, just better. *Nature* 350:121–124. [An engaging account of the catalytic strategy of triosephosphate isomerase.]

Granner, D., and Pilkis, S., 1990. The genes of hepatic glucose metabolism. *J. Biol. Chem.* 265:10173–10176.

Books

Ochs, R.S., Hanson, R.W., and Hall, J. (eds.), 1985. *Metabolic Regulation.* Elsevier. [A collection of essays originally published in *Trends in Biochemical Sciences (TIBS),* a highly readable and informative journal.]

Fersht, A., 1985. *Enzyme Structure and Mechanism* (2nd ed.). W.H. Freeman. [Contains succinct accounts of the catalytic and regulatory mechanisms of numerous enzymes participating in metabolism.]

Newsholme, E.A., and Start, C., 1973. *Regulation in Metabolism.* Wiley. [Chapters 3 and 6 provide excellent accounts of the control of carbohydrate metabolism.]

Structure of glycolytic enzymes

Anderson, C.M., Zucker, F.H., and Steitz, T.A., 1979. Space-filling models of kinase clefts and conformation changes. *Science* 204:375–380. [A well-illustrated and lucid article showing that kinases generally have a deep cleft that closes on binding substrate.]

Schirmer, T., and Evans, P.R., 1990. Structural basis of the allosteric behaviour of phosphofructokinase. *Nature* 343:140–145.

Gamblin, S.J., Cooper, B., Millar, J.R., Davies, G.J., Littlechild, J.A., and Watson, H.C., 1990. The crystal structure of human muscle aldolase at 3.0 Å resolution. *FEBS Lett.* 262:282–286.

Harlos, K., Vas, M., and Blake, C.F., 1992. Crystal structure of the binary complex of pig muscle phosphoglycerate kinase and its substrate 3-phospho-D-glycerate. *Proteins* 12:133–144.

Catalytic mechanisms

Rose, I.A., 1981. Chemistry of proton abstraction by glycolytic enzymes (aldolase, isomerases, and pyruvate kinase). *Phil. Trans. Roy. Soc. Lond. B* 293:131–144.

Knowles, J.R., and Albery, W.J., 1977. Perfection in enzyme catalysis: The energetics of triosephosphate isomerase. *Acc. Chem. Res.* 10:105–111.

Bash, P.A., Field, M.J., Davenport, R.C., Petsko, G.A., Ringe, D., and Karplus, M., 1991. Computer simulation and analysis of the reaction pathway of triosephosphate isomerase. *Biochemistry* 30:5826–5832.

Boyer, P. D. (ed.), 1972. *The Enzymes* (3rd ed.). Academic Press. [Volumes 5 through 9 contain authoritative reviews of each of the glycolytic enzymes. Alcohol dehydrogenase and lactate dehydrogenase are discussed in vol. 10, and pyruvate kinase in vol. 18.]

Regulation of glycolysis

Pilkis, S.J., and Claus, T.H., 1991. Hepatic gluconeogenesis/glycolysis: Regulation and structure/function relationships of substrate cycle enzymes. *Ann. Rev. Nutrit.* 11:465–515.

Hers, H.-G., and Van Schaftingen, E., 1982. Fructose 2,6-bisphosphate two years after its discovery. *Biochem. J.* 206:1–12.

Middleton, R.J., 1990. Hexokinases and glucokinases. *Biochem. Soc. Trans.* 18:180–183.

Sugar transporters

Silverman, M., 1991. Structure and function of hexose transporters. *Ann. Rev. Biochem.* 60:757–794.

Thorens, B., Charron, M.J., and Lodish, H.F., 1990. Molecular physiology of glucose transporters. *Diabetes Care* 13:209–218.

Meadow, N.D., Fox, D.K., and Roseman, S., 1990. The bacterial phosphoenolpyruvate:glycose phosphotransferase system. *Ann. Rev. Biochem.* 59:497–542.

Genetic diseases

Scriver, C.R., Beaudet, A.L., Sly, W.S., and Valle, D. (eds.), 1989. *The Metabolic Basis of Inherited Disease* (6th ed.). McGraw-Hill. [Chapters 11 and 13 deal with disorders of fructose and galactose metabolism. Deficiencies of glycolytic enzymes in erythrocytes are discussed in Chapter 94.]

Historical aspects

Kalckar, H.M. (ed.), 1969. *Biological Phosphorylations: Development of Concepts.* Prentice-Hall. [Contains many of the classical papers on glycolysis.]

Fruton, J.S., 1972. *Molecules and Life: Historical Essays on the Interplay of Chemistry and Biology.* Wiley-Interscience. [Includes a meticulously documented account of the elucidation of the nature of fermentation and how it led to enzyme chemistry.]

PROBLEMS

1. *Kitchen chemistry*. Sucrose is commonly used to preserve fruits. Why is glucose not well suited for preserving foods?

2. *Tracing carbon atoms*. Glucose labeled with ^{14}C at C-1 is incubated with the glycolytic enzymes and necessary cofactors.
   (a) What is the distribution of ^{14}C in the pyruvate that is formed? (Assume that the interconversion of glyceraldehyde 3-phosphate and dihydroxyacetone phosphate is very rapid compared with the subsequent step.)
   (b) If the specific activity of the glucose substrate is 10 mCi/mM, what is the specific activity of the pyruvate that is formed?

3. *Lactic fermentation*. Write a balanced equation for the conversion of glucose into lactate.
   (a) Calculate the standard free-energy change of this reaction using the data given in Table 19-2 (p. 491) and the fact that $\Delta G^{\circ\prime}$ is -6 kcal for the reaction

   $$\text{Pyruvate} + \text{NADH} + \text{H}^+ \rightleftharpoons \text{lactate} + \text{NAD}^+$$

   (b) What is the free-energy change ($\Delta G'$, not $\Delta G^{\circ\prime}$) of this reaction when the concentrations of reactants are: glucose, 5 mM; lactate, 0.05 mM; ATP, 2 mM; ADP, 0.2 mM; and P_i, 1 mM?

4. *High potential*. What is the equilibrium ratio of phosphoenolpyruvate to pyruvate under standard conditions when [ATP]/[ADP] = 10?

5. *Hexose-triose equilibrium*. What are the equilibrium concentrations of fructose 1,6-bisphosphate, dihydroxyacetone phosphate, and glyceraldehyde 3-phosphate when 1 mM fructose 1,6-bisphosphate is incubated with aldolase under standard conditions?

6. *Double labeling*. 3-Phosphoglycerate labeled uniformly with ^{14}C is incubated with 1,3-BPG labeled with ^{32}P at C-1. What is the radioisotope distribution of the 2,3-BPG that is formed on addition of BPG mutase?

7. *An informative analog*. Xylose has the same structure as glucose except that it has a hydrogen atom at C-6 in place of a hydroxymethyl group. The rate of ATP hydrolysis by hexokinase is markedly enhanced by the addition of xylose. Why?

8. *The far reach of glycolysis*. Oxygen transport can be affected in genetic disorders of glycolysis in red cells.
   (a) How are glycolysis and oxygen transport linked?
   (b) How is oxygen affinity altered by a deficiency of hexokinase?
   (c) How is oxygen affinity altered by a deficiency of pyruvate kinase?

9. *Distinctive sugars*. The intravenous infusion of fructose into healthy volunteers leads to a two- to fivefold increase in the level of lactate in the blood, a far greater increase than that observed following the infusion of the same amount of glucose.
   (a) Why is glycolysis more rapid following the infusion of fructose?
   (b) Fructose has been used in place of glucose for intravenous feeding. Why is this use of fructose unwise?

10. *Catalytic metal*. Aldolases in prokaryotes contain a tightly bound divalent metal ion that is essential for catalysis. Propose a catalytic function for this metal ion.

11. *Contrasting inactivators*. Prokaryotic aldolases are inactivated by ethylenediaminetetraacetate (EDTA), a chelator of divalent metal ions, whereas animal aldolases discussed in this chapter are inactivated by sodium borohydride. Account for this difference.

12. *Metabolic mutants*. Predict the effect of each of the following on the pace of glycolysis in liver cells:
    (a) Loss of the allosteric site for ATP in phosphofructokinase.
    (b) Loss of the binding site for citrate in phosphofructokinase.
    (c) Loss of the phosphatase domain of the bifunctional enzyme that controls the level of fructose 2,6-bisphosphate.
    (d) Loss of the binding site for fructose 1,6-bisphosphate in pyruvate kinase.

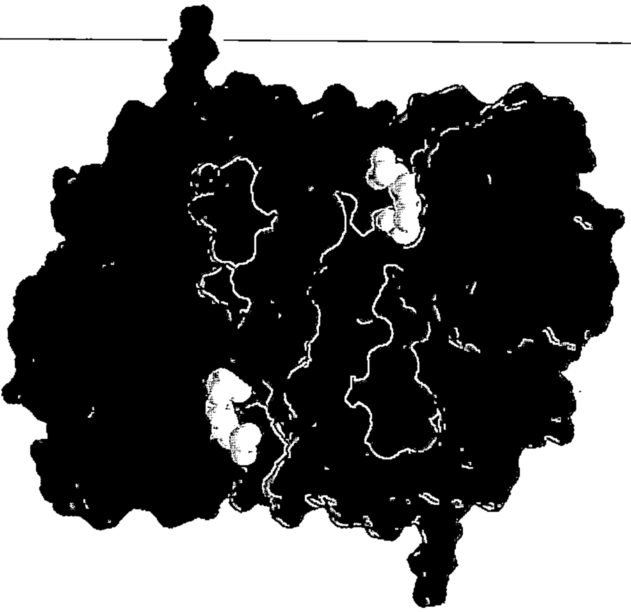

Citric Acid Cycle

The preceding chapter considered the glycolytic pathway, in which glucose is converted into pyruvate. Under aerobic conditions, the next step in the generation of energy from glucose is the oxidative decarboxylation of pyruvate to form acetyl coenzyme A (acetyl CoA). This activated acetyl unit is then completely oxidized to CO_2 by the *citric acid cycle,* a series of reactions that is also known as the *tricarboxylic acid cycle* or the *Krebs cycle.* The citric acid cycle is the *final common pathway for the oxidation of fuel molecules*—amino acids, fatty acids, and carbohydrates. Most fuel molecules enter the cycle as acetyl CoA. The cycle also provides intermediates for biosyntheses. In eukaryotes, the reactions of the citric acid cycle occur inside mitochondria, in contrast with those of glycolysis, which occur in the cytosol.

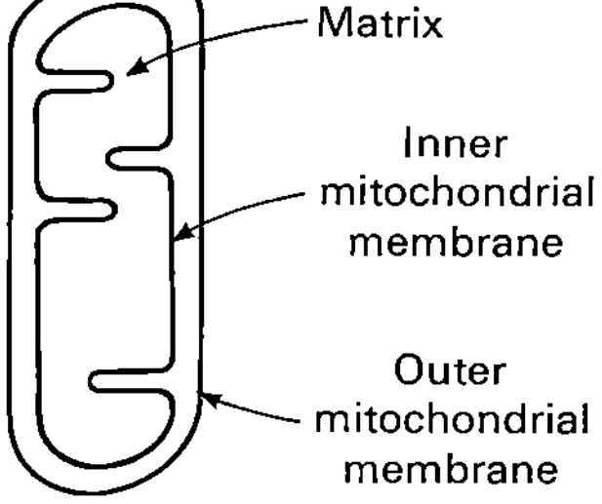

Figure 20-1
Schematic diagram of a mitochondrion. The oxidative decarboxylation of pyruvate and the sequence of reactions in the citric acid cycle occur within the mitochondrial matrix.

FORMATION OF ACETYL COENZYME A FROM PYRUVATE

The oxidative decarboxylation of pyruvate to form acetyl CoA, which occurs in the mitochondrial matrix, is the link between glycolysis and the citric acid cycle.

$$\text{Pyruvate} + \text{CoA} + \text{NAD}^+ \longrightarrow \text{acetyl CoA} + \text{CO}_2 + \text{NADH}$$

This irreversible funneling of the product of glycolysis into the citric acid cycle is catalyzed by the *pyruvate dehydrogenase complex.* This very large multienzyme complex is a highly integrated array of three kinds of enzymes, which will be discussed in detail later in this chapter (p. 514).

***Opening Image:** Model of citrate synthase showing the solvent-accessible surface. This enzyme catalyzes the first step in the citric acid cycle, the joining of a four-carbon unit and a two-carbon unit to form citrate. The two identical subunits (red and blue) interdigitate. The active sites are near the subunit interfaces. Activated two-carbon units are brought to the enzyme by coenzyme A (yellow), which binds near the surface. [Drawn by Dr. Anthony Nicholls from 5cts.pdb. M. Karpusas, B. Branchaud, and S.J. Remington. Biochemistry 29(1990):2213.]*

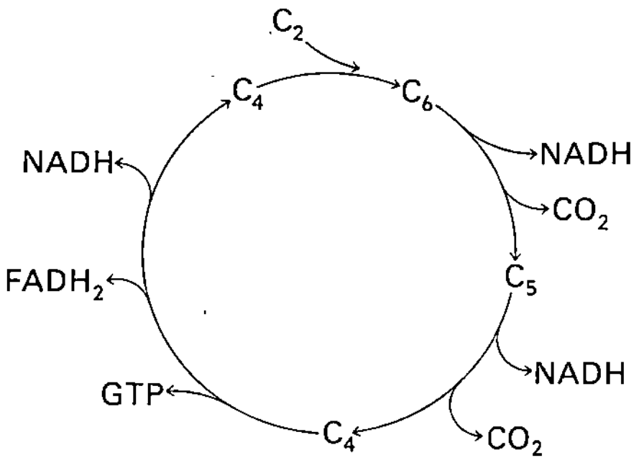

Figure 20-2
An overview of the citric acid cycle.

AN OVERVIEW OF THE CITRIC ACID CYCLE

The overall pattern of the citric acid cycle is shown in Figure 20-2. A four-carbon compound (oxaloacetate) condenses with a two-carbon acetyl unit to yield a six-carbon tricarboxylic acid (citrate). An isomer of citrate is then oxidatively decarboxylated. The resulting five-carbon compound (α-ketoglutarate) is oxidatively decarboxylated to yield a four-carbon compound (succinate). Oxaloacetate is then regenerated from succinate. Two carbon atoms enter the cycle as an acetyl unit and two carbon atoms leave the cycle in the form of two molecules of CO_2. An acetyl group is more reduced than CO_2, and so oxidation-reduction reactions must take place in the citric acid cycle. In fact, there are four such reactions. Three hydride ions (hence, six electrons) are transferred to three NAD^+ molecules, whereas one pair of hydrogen atoms (hence, two electrons) is transferred to a flavin adenine dinucleotide (FAD) molecule. These electron carriers yield nine molecules of adenosine triphosphate (ATP) when they are oxidized by O_2 in the electron transport chain. In addition, one high-energy phosphate bond is formed in each round of the citric acid cycle itself.

OXALOACETATE CONDENSES WITH ACETYL COENZYME A TO FORM CITRATE

The cycle starts with the joining of a four-carbon unit, oxaloacetate, and a two-carbon unit, the acetyl group of acetyl CoA. Oxaloacetate reacts with acetyl CoA and H_2O to yield citrate and CoA.

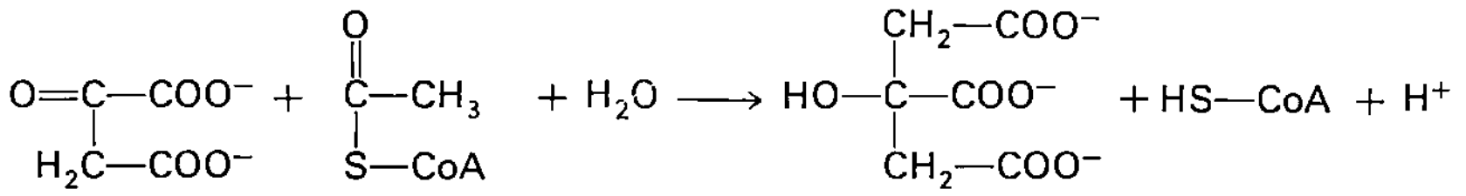

$$\underset{\text{Oxaloacetate}}{\begin{array}{c}O=C-COO^- \\ | \\ H_2C-COO^-\end{array}} + \underset{\text{Acetyl CoA}}{\begin{array}{c}O \\ || \\ C-CH_3 \\ | \\ S-CoA\end{array}} + H_2O \longrightarrow \underset{\text{Citrate}}{\begin{array}{c}CH_2-COO^- \\ | \\ HO-C-COO^- \\ | \\ CH_2-COO^-\end{array}} + HS-CoA + H^+$$

$$\underset{\text{Citryl CoA}}{\begin{array}{c}O \\ || \\ CH_2-C-S-CoA \\ | \\ HO-C-COO^- \\ | \\ CH_2-COO^-\end{array}}$$

This reaction, which is an aldol condensation followed by a hydrolysis, is catalyzed by *citrate synthase*. Oxaloacetate first condenses with acetyl CoA to form *citryl CoA*, which is then hydrolyzed to citrate and CoA. Hydrolysis of citryl CoA pulls the overall reaction far in the direction of the synthesis of citrate.

CITRATE IS ISOMERIZED INTO ISOCITRATE

Citrate must be isomerized into isocitrate to enable the six-carbon unit to undergo oxidative decarboxylation. The isomerization of citrate is accomplished by a *dehydration* step followed by a *hydration* step. The result is an interchange of an H and an OH. The enzyme catalyzing both steps is called *aconitase* because cis-*aconitate* is an intermediate.

$$\underset{\text{Citrate}}{\begin{array}{c}COO^- \\ | \\ H-C-H \\ | \\ ^-OOC-C-OH \\ | \\ CH_2 \\ | \\ COO^-\end{array}} \underset{\xleftarrow{\quad}}{\overset{H_2O}{\rightleftharpoons}} \underset{\textit{cis}\text{-Aconitate}}{\begin{array}{c}H \\ | \\ ^-OOC-C \\ || \\ ^-OOC-C \\ | \\ CH_2 \\ | \\ COO^-\end{array}} \underset{\xleftarrow{\quad}}{\overset{H_2O}{\rightleftharpoons}} \underset{\text{Isocitrate}}{\begin{array}{c}COO^- \\ | \\ H-C-OH \\ | \\ ^-OOC-C-H \\ | \\ CH_2 \\ | \\ COO^-\end{array}}$$

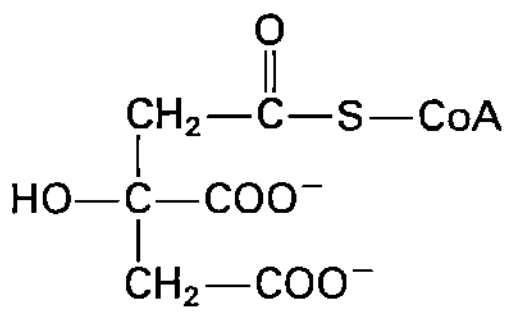

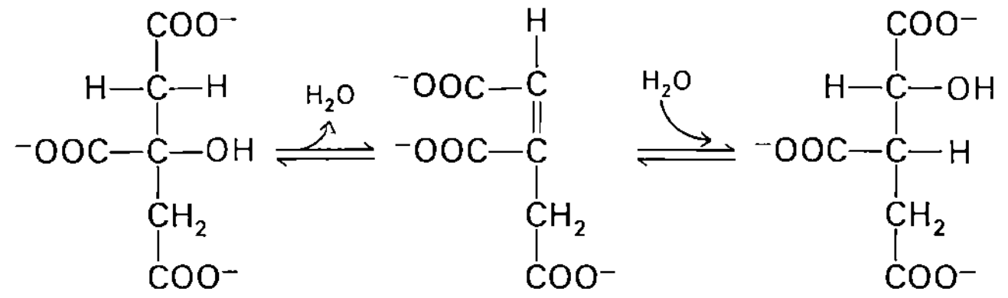

Aconitase contains iron that is not bonded to heme. Rather, its four iron atoms are complexed to four inorganic sulfides and four cysteine sulfur atoms (Figure 20-3). This Fe-S cluster binds citrate and participates in dehydrating and rehydrating the bound substrate. Proteins with Fe-S clusters are known as *iron-sulfur proteins* or *nonheme iron proteins.*

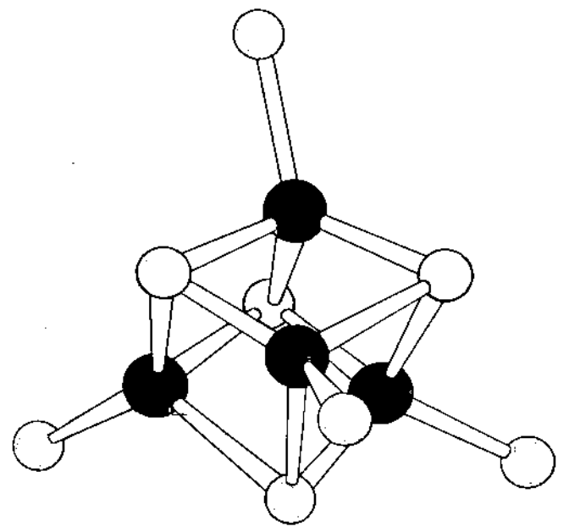

Figure 20-3
An iron-sulfur cluster in aconitase. Each of the four iron atoms (red) in this 4Fe-4S cubic array is bonded to three inorganic sulfides (green) and a cysteine sulfur atom (yellow). [After H. Beinert and A.J. Thomson. *Arch. Biochem. Biophys.* 222(1983):358.]

ISOCITRATE IS OXIDIZED AND DECARBOXYLATED TO α-KETOGLUTARATE

We come now to the first of four oxidation-reduction reactions in the citric acid cycle. The oxidative decarboxylation of isocitrate is catalyzed by *isocitrate dehydrogenase.*

$$\text{Isocitrate} + \text{NAD}^+ \rightleftharpoons \alpha\text{-ketoglutarate} + \text{CO}_2 + \text{NADH}$$

The intermediate in this reaction is oxalosuccinate, an unstable β-keto acid. While bound to the enzyme, it loses CO_2 to form α-ketoglutarate. The rate of formation of α-ketoglutarate is important in determining the overall rate of the cycle, as will be discussed later (p. 525).

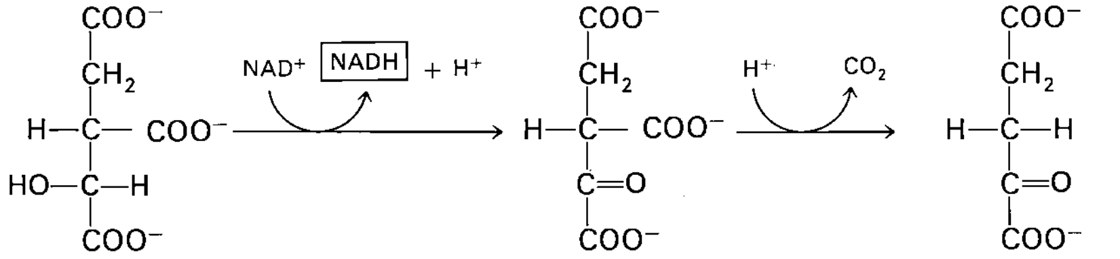

Isocitrate **Oxalosuccinate** **α-Ketoglutarate**

SUCCINYL COENZYME A IS FORMED BY THE OXIDATIVE DECARBOXYLATION OF α-KETOGLUTARATE

The conversion of isocitrate into α-ketoglutarate is followed by a second oxidative decarboxylation reaction, the formation of succinyl CoA from α-ketoglutarate.

$$\alpha\text{-Ketoglutarate} + \text{NAD}^+ + \text{CoA} \longrightarrow \text{succinyl CoA} + \text{CO}_2 + \text{NADH}$$

This reaction is catalyzed by the *α-ketoglutarate dehydrogenase complex,* an organized assembly consisting of three kinds of enzymes. *The mechanism of this reaction is very similar to that of the conversion of pyruvate into acetyl CoA* (p. 514).

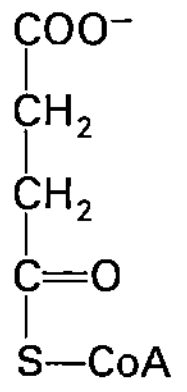

Succinyl CoA

A HIGH-ENERGY PHOSPHATE BOND IS GENERATED FROM SUCCINYL COENZYME A

The succinyl thioester of CoA has an energy-rich bond. The $\Delta G^{\circ\prime}$ for hydrolysis of succinyl CoA is about −8 kcal/mol, which is comparable with that of ATP (−7.3 kcal/mol). *The cleavage of the thioester bond of succinyl CoA is coupled to the phosphorylation of guanosine diphosphate (GDP).*

$$\text{Succinyl CoA} + \text{P}_\text{i} + \text{GDP} \rightleftharpoons \text{succinate} + \text{GTP} + \text{CoA}$$

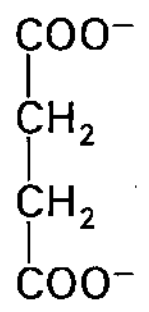

Succinate

This readily reversible reaction ($\Delta G^{\circ\prime} = -0.8$ kcal/mol) is catalyzed by *succinyl CoA synthetase.* It is the only step in the citric acid cycle that directly yields a high-energy phosphate bond. GTP itself is used as a phosphoryl donor in protein synthesis (p. 896) and signal transduction processes

(p. 336). Alternatively, its γ phosphoryl group can readily be transferred to adenosine diphosphate (ADP) to form ATP, in a reaction catalyzed by *nucleoside diphosphokinase.*

$$\text{GTP} + \text{ADP} \rightleftharpoons \text{GDP} + \text{ATP}$$

OXALOACETATE IS REGENERATED BY OXIDATION OF SUCCINATE

Reactions of four-carbon compounds constitute the final stage of the citric acid cycle (Figure 20-4). Succinate is converted into oxaloacetate in three steps: an oxidation, a hydration, and a second oxidation reaction. Oxaloacetate is thereby regenerated for another round of the cycle, as energy is trapped in the form of $FADH_2$ and NADH.

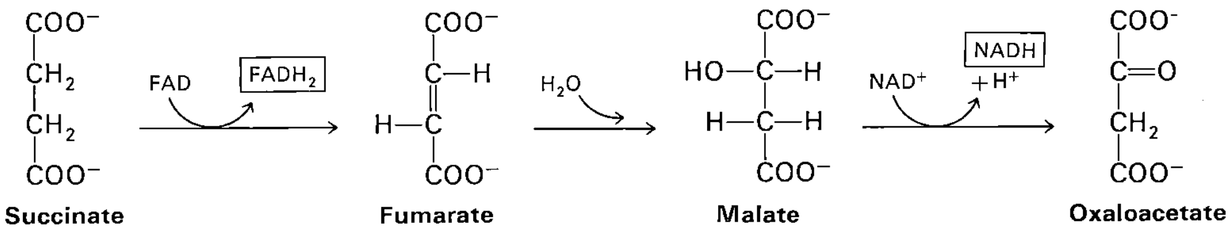

Figure 20-4
Final stage of the citric acid cycle: from succinate to oxaloacetate.

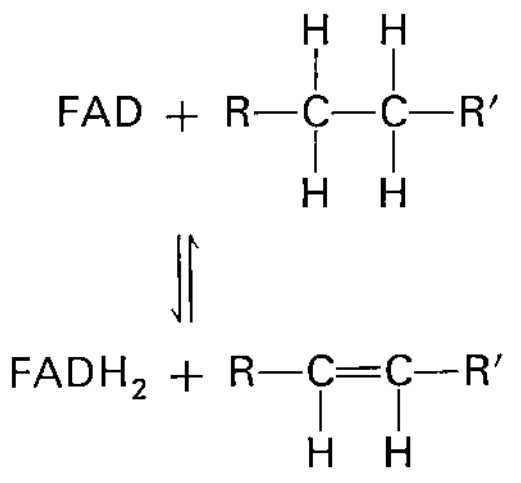

Succinate is oxidized to fumarate by *succinate dehydrogenase.* The hydrogen acceptor is FAD rather than NAD^+, which is used in the other three oxidation reactions in the cycle. FAD is the hydrogen acceptor in this reaction because the free-energy change is insufficient to reduce NAD^+. FAD is nearly always the electron acceptor in oxidations that remove two hydrogen atoms from a substrate. In succinate dehydrogenase, the isoalloxazine ring of FAD is covalently attached to a histidine side chain of the enzyme (denoted E-FAD).

$$\text{E-FAD} + \text{succinate} \rightleftharpoons \text{E-FADH}_2 + \text{fumarate}$$

Succinate dehydrogenase, like aconitase, is an *iron-sulfur protein* (also called a *nonheme iron protein*). Indeed, succinate dehydrogenase contains three different kinds of iron-sulfur clusters, 2Fe-2S (two irons bonded to two inorganic sulfides), 3Fe-4S, and 4Fe-4S. We shall consider the role of these iron-sulfur clusters in the electron transfer reactions of oxidative phosphorylation (p. 535) and photosynthesis (p. 663). Succinate dehydrogenase, consisting of a 70-kd and a 27-kd subunit, differs from other enzymes in the citric acid cycle in being imbedded in the inner mitochondrial membrane. In fact, *succinate dehydrogenase is directly linked to the electron transport chain.* The $FADH_2$ produced by the oxidation of succinate does not dissociate from the enzyme, in contrast with NADH produced in other oxidation-reduction reactions. Rather, two electrons from $FADH_2$ are transferred directly to Fe-S clusters of the enzyme. The ultimate acceptor of these electrons is molecular oxygen, as will be discussed in the next chapter.

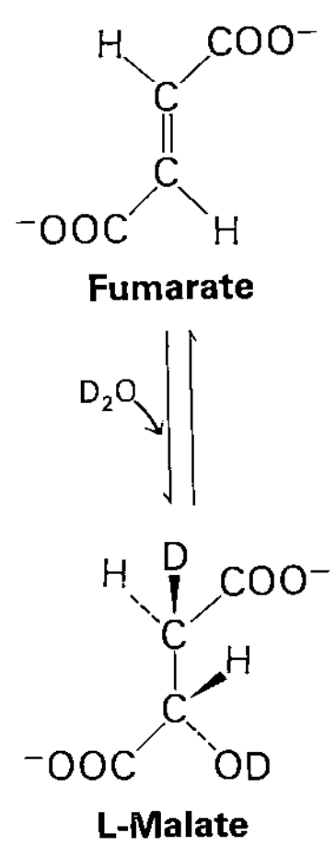

The next step is the hydration of fumarate to form L-malate. *Fumarase* catalyzes a stereospecific trans addition of H and OH, as shown by deuterium-labeling studies. The OH group adds to only one side of the double bond of fumarate; hence, only the L isomer of malate is formed.

Finally, malate is oxidized to form oxaloacetate. This reaction is catalyzed by *malate dehydrogenase,* and NAD^+ is again the hydrogen acceptor.

$$\text{Malate} + \text{NAD}^+ \rightleftharpoons \text{oxaloacetate} + \text{NADH} + \text{H}^+$$

STOICHIOMETRY OF THE CITRIC ACID CYCLE

The net reaction of the citric acid cycle is

$$\text{Acetyl CoA} + 3\ NAD^+ + FAD + GDP + P_i + 2\ H_2O \longrightarrow 2\ CO_2 + 3\ NADH + FADH_2 + GTP + 2\ H^+ + CoA$$

Let us recapitulate the reactions that give this stoichiometry (Figure 20-5 and Table 20-1):

1. Two carbon atoms enter the cycle in the condensation of an acetyl unit (from acetyl CoA) with oxaloacetate. Two carbon atoms leave the cycle in the form of CO_2 in the successive decarboxylations catalyzed by isocitrate dehydrogenase and α-ketoglutarate dehydrogenase. As will be discussed shortly, the two carbon atoms that leave the cycle are different from the ones that entered in that round.

2. Four pairs of hydrogen atoms leave the cycle in the four oxidation reactions. Two NAD^+ molecules are reduced in the oxidative decarboxylations of isocitrate and α-ketoglutarate, one FAD molecule is reduced in the oxidation of succinate, and one NAD^+ molecule is reduced in the oxidation of malate.

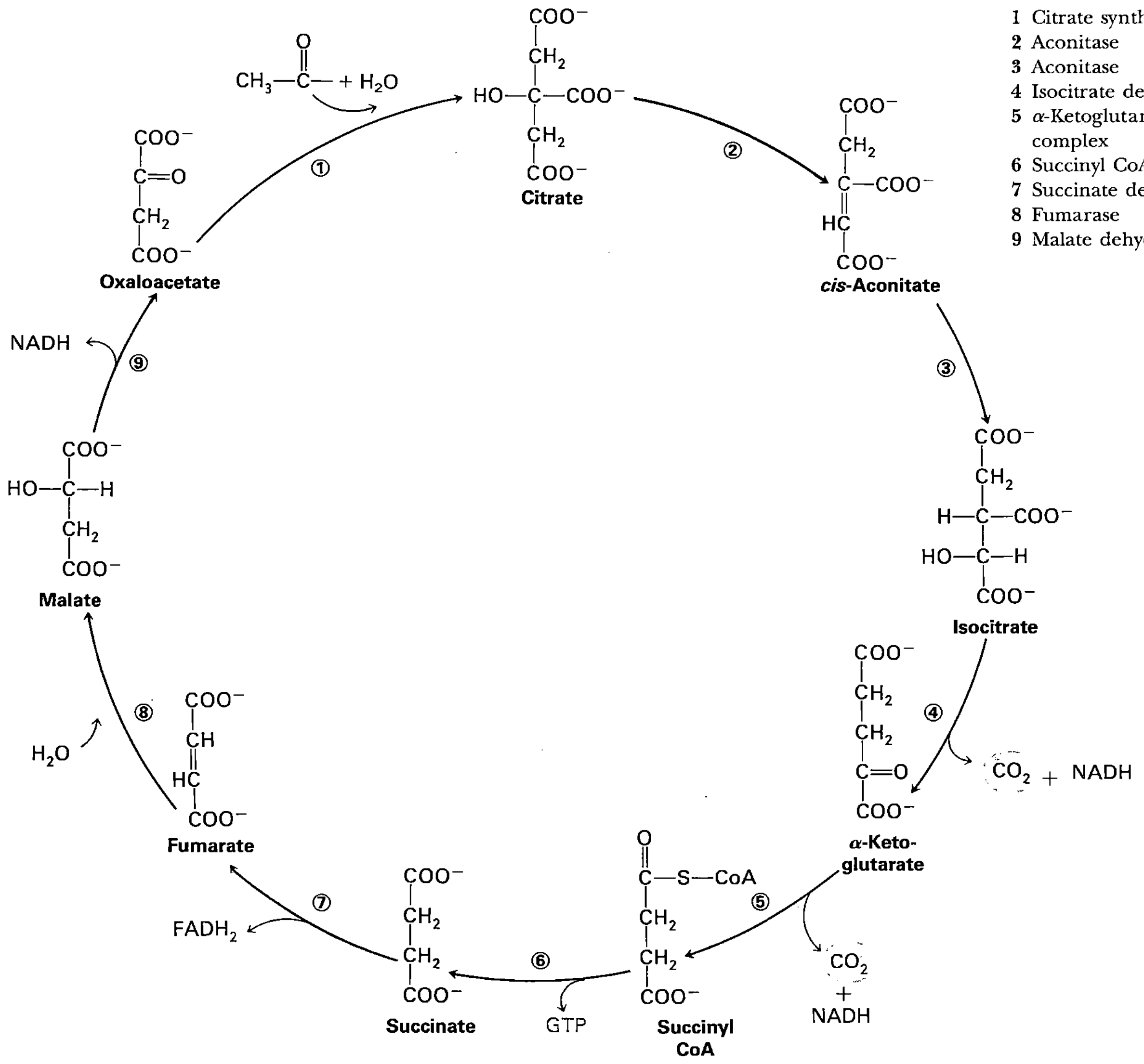

Figure 20-5
Citric acid cycle. This series of reactions is catalyzed by the following enzymes, as numbered in the diagram:

1 Citrate synthase
2 Aconitase
3 Aconitase
4 Isocitrate dehydrogenase
5 α-Ketoglutarate dehydrogenase complex
6 Succinyl CoA synthetase
7 Succinate dehydrogenase
8 Fumarase
9 Malate dehydrogenase

Table 20-1
Citric acid cycle

Step	Reaction	Enzyme	Prosthetic group	Type*	$\Delta G^{\circ\prime}$
1	Acetyl CoA + oxaloacetate + H_2O $\longrightarrow$ citrate + CoA + H^+	Citrate synthase		a	−7.5
2	Citrate $\rightleftharpoons$ *cis*-aconitate + H_2O	Aconitase	Fe-S	b	+2.0
3	*cis*-Aconitate + H_2O $\rightleftharpoons$ isocitrate	Aconitase	Fe-S	c	−0.5
4	Isocitrate + NAD^+ $\rightleftharpoons$ α-ketoglutarate + CO_2 + NADH	Isocitrate dehydrogenase		d + e	−2.0
5	α-Ketoglutarate + NAD^+ + CoA $\rightleftharpoons$ succinyl CoA + CO_2 + NADH	α-Ketoglutarate dehydrogenase complex	Lipoic acid, FAD, TPP	d + e	−7.2
6	Succinyl CoA + P_i + GDP $\rightleftharpoons$ succinate + GTP + CoA	Succinyl CoA synthetase		f	−0.8
7	Succinate + FAD (enzyme-bound) $\rightleftharpoons$ fumarate + $FADH_2$ (enzyme-bound)	Succinate dehydrogenase	FAD, Fe-S	e	~0
8	Fumarate + H_2O $\rightleftharpoons$ L-malate	Fumarase		c	−0.9
9	L-Malate + NAD^+ $\rightleftharpoons$ oxaloacetate + NADH + H^+	Malate dehydrogenase		e	+7.1

*Reaction type: (a) condensation; (b) dehydration; (c) hydration; (d) decarboxylation; (e) oxidation; (f) substrate-level phosphorylation.

3. One high-energy phosphate bond (in the form of GTP) is generated from the energy-rich thioester linkage in succinyl CoA.

4. Two water molecules are consumed: one in the synthesis of citrate by the hydrolysis of citryl CoA, the other in the hydration of fumarate.

As will be discussed in the next chapter, the NADH and $FADH_2$ formed in the citric acid cycle are oxidized by the electron transport chain. The transfer of electrons from these carriers to O_2, the ultimate electron acceptor, leads to the pumping of protons across the inner mitochondrial membrane. The proton-motive force then powers the generation of ATP; the stoichiometry is about 2.5 ATP per NADH, and 1.5 ATP per $FADH_2$. Only one high-energy phosphate bond per acetyl unit is directly formed in the citric acid cycle. Nine more high-energy phosphate bonds are generated when three NADH and one $FADH_2$ are oxidized by the electron transport chain.

Molecular oxygen does not participate directly in the citric acid cycle. However, the cycle operates only under aerobic conditions because NAD^+ and FAD can be regenerated in the mitochondrion only by the transfer of electrons to molecular oxygen. *Glycolysis has both an aerobic and an anaerobic mode, whereas the citric acid cycle is strictly aerobic.* Recall that glycolysis can proceed under anaerobic conditions because NAD^+ is regenerated in the conversion of pyruvate into lactate.

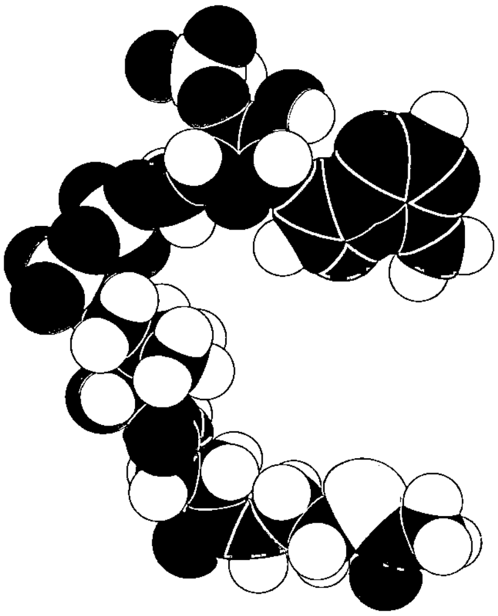

Figure 20-6
Model of acetyl CoA.

THE PYRUVATE DEHYDROGENASE COMPLEX IS A MULTIMERIC ASSEMBLY OF THREE KINDS OF ENZYMES

We now turn to several important and interesting reaction mechanisms. The oxidative decarboxylation of pyruvate in the formation of acetyl CoA is catalyzed by the *pyruvate dehydrogenase complex,* an organized assembly of three kinds of enzymes (Table 20-2). The net reaction catalyzed is

$$\text{Pyruvate} + \text{CoA} + \text{NAD}^+ \longrightarrow \text{acetyl CoA} + \text{CO}_2 + \text{NADH}$$

The mechanism of this reaction is more complex than might be suggested by its stoichiometry. *Thiamine pyrophosphate (TPP), lipoamide,* and *FAD* serve as catalytic cofactors, in addition to CoA and NAD^+, the stoichiometric cofactors.

Table 20-2
Pyruvate dehydrogenase complex of *E. coli*

Enzyme	*Abbreviation*	*Number of chains*	*Prosthetic group*	*Reaction catalyzed*
Pyruvate dehydrogenase component	E_1	24	TPP	Oxidative decarboxylation of pyruvate
Dihydrolipoyl transacetylase	E_2	24	Lipoamide	Transfer of the acetyl group to CoA
Dihydrolipoyl dehydrogenase	E_3	12	FAD	Regeneration of the oxidized form of lipoamide

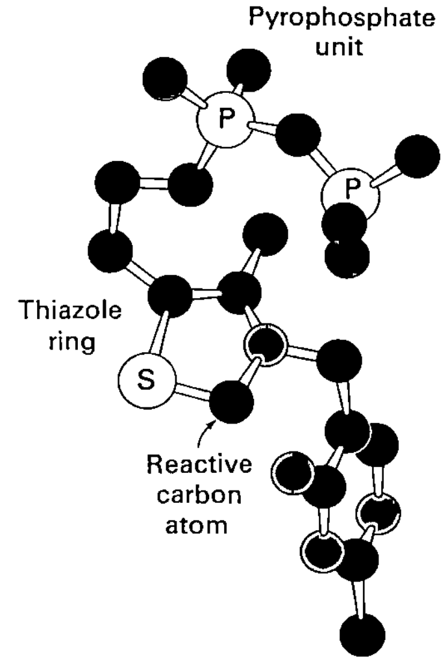

Figure 20-7
Model of thiamine pyrophosphate.

There are four steps in the conversion of pyruvate into acetyl CoA. First, pyruvate is *decarboxylated* after it combines with TPP. This reaction is catalyzed by the *pyruvate dehydrogenase component* (E_1) of the multienzyme complex.

$$\text{Pyruvate} + \text{TPP} \longrightarrow \text{hydroxyethyl-TPP} + CO_2$$

A key feature of TPP, the prosthetic group of the pyruvate dehydrogenase component, is that the carbon atom between the nitrogen and sulfur atoms in the thiazole ring is much more acidic than most =CH– groups.

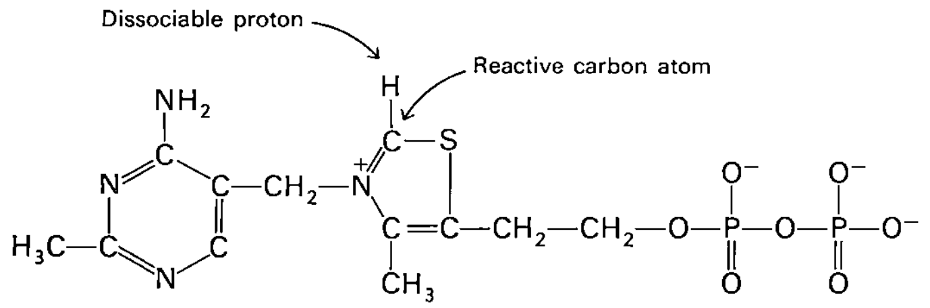

Thiamine pyrophosphate (TPP)

It ionizes to form a *carbanion,* which readily adds to the carbonyl group of pyruvate.

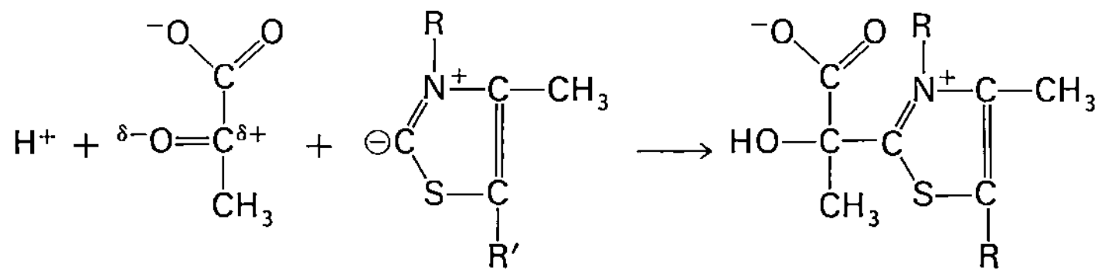

The positively charged ring nitrogen of TPP then acts as an electron sink to stabilize the formation of a negative charge, which is necessary for decarboxylation.

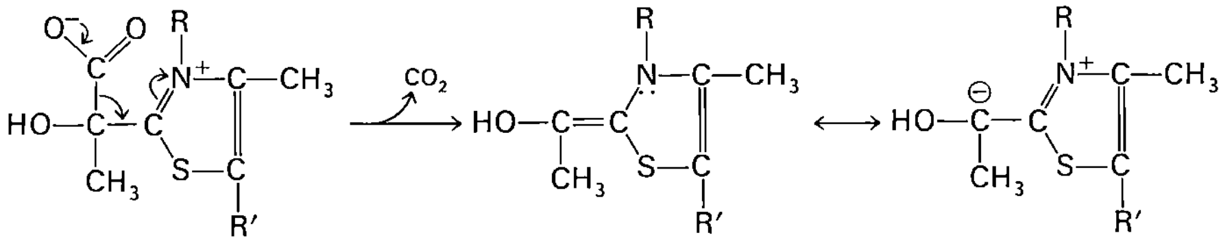

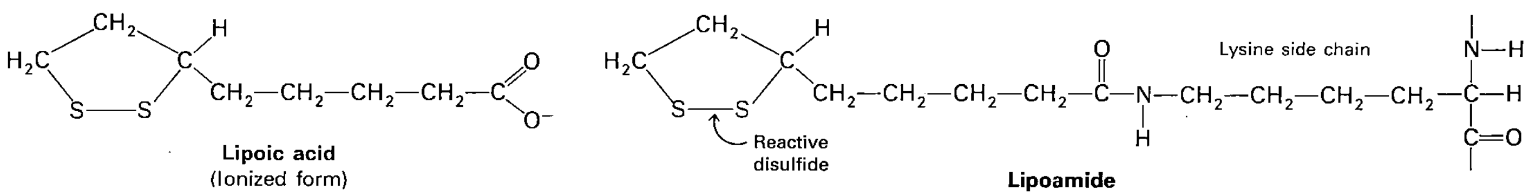

Figure 20-8
Structures of lipoic acid and lipoamide. Lipoic acid is covalently attached to a specific lysine side chain of dihydrolipoyl transacetylase (E_2). Note that this prosthetic group is at the end of a long, flexible chain, which enables it to rotate from one active site to another in the enzyme complex.

Protonation then gives *hydroxyethyl thiamine pyrophosphate.*

Second, the hydroxyethyl group attached to TPP is *oxidized* to form an acetyl group and concomitantly *transferred to lipoamide.* The oxidant in this reaction is the disulfide group of lipoamide, which is converted into the sulfhydryl form. This reaction, also catalyzed by the pyruvate dehydrogenase component, yields *acetyllipoamide.*

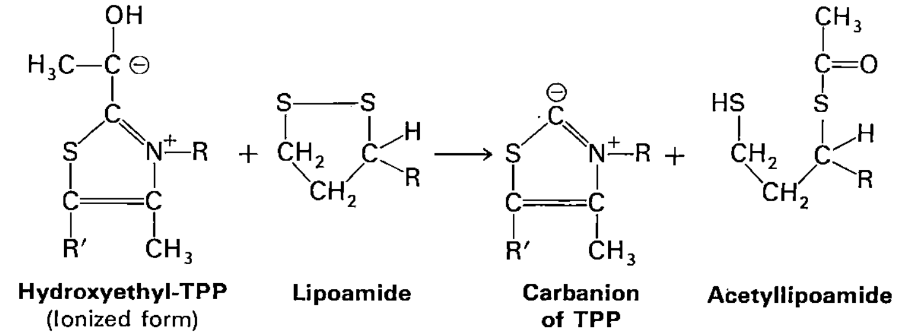

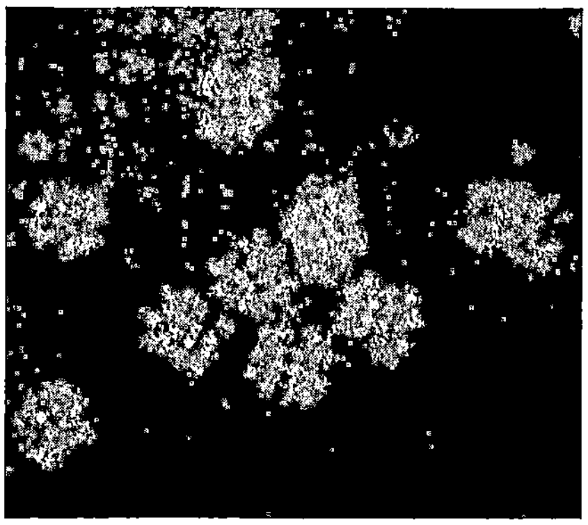

Figure 20-9
Electron micrograph of the pyruvate dehydrogenase complex from *E. coli.* [Courtesy of Dr. Lester Reed.]

Third, *the acetyl group is transferred from acetyllipoamide to CoA to form acetyl CoA.* Dihydrolipoyl transacetylase (E_2) catalyzes this reaction. The energy-rich thioester bond is preserved as the acetyl group is transferred to CoA.

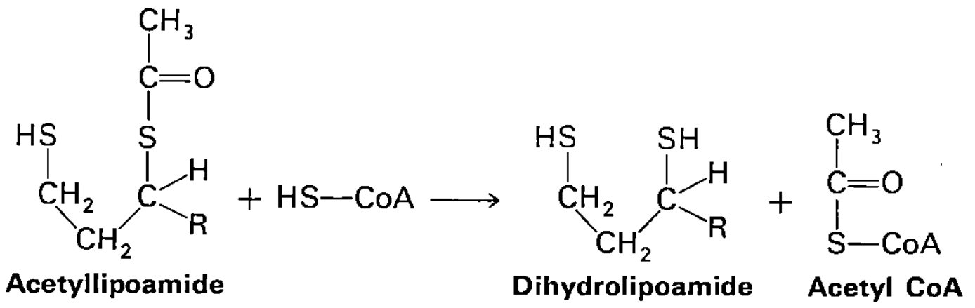

Fourth, *the oxidized form of lipoamide is regenerated by dihydrolipoyl dehydrogenase (E_3).* Two electrons are transferred to an FAD prosthetic group of the enzyme and then to NAD^+. The unusual electron transfer potential of FAD bound to this protein enables it to transfer electrons to NAD^+.

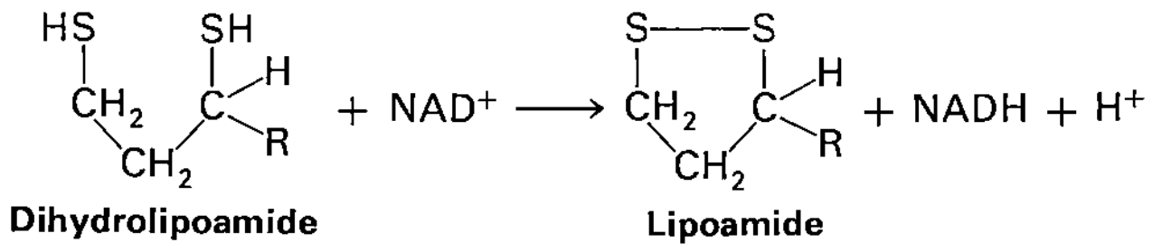

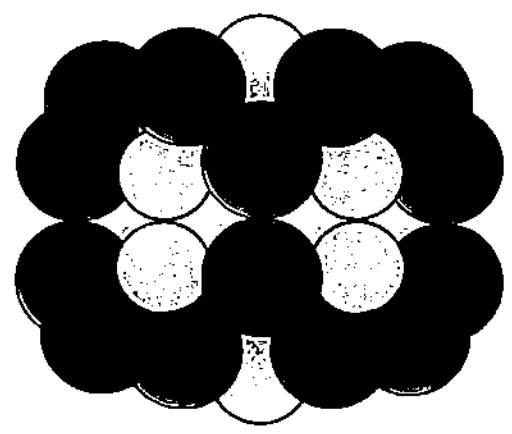

Figure 20-10
Model of the pyruvate dehydrogenase complex from *E. coli.* The transacetylase core (E_2) is shown in yellow, the pyruvate dehydrogenase component (E_1) in red, and dihydrolipoyl dehydrogenase (E_3) in green. [After a drawing kindly provided by Dr. Lester Reed.]

The studies of Lester Reed have been sources of insight into the structure and assembly of the pyruvate dehydrogenase complex. The enzyme complex from *Escherichia coli* has been studied intensively. It has a mass of about 5 million daltons and consists of 60 polypeptide chains (see Table 20-2), which makes it even larger than a ribosome. A polyhedral structure with a diameter of about 30 nm is evident in electron micrographs (Figure 20-9). *The transacetylase polypeptide chains (E_2) form the core of this complex.* The pyruvate dehydrogenase units and lipoyl dehydrogenase units are bound to the outside of the transacetylase core (Figure 20-10).

X-ray studies have shown that eight trimers of E_2 come together to form a hollow cube (Figure 20-11). The internal cavity of this cage is connected to the outside by large channels that cross the faces of the cube. Electron microscopic studies of the interactions of E_1 and E_3 with the E_2 core indicate that their active sites are far apart (~40 Å).

The structural integration of three kinds of enzymes makes possible the coordinated catalysis of a complex reaction (Figure 20-12). All the intermediates in the oxidative decarboxylation of pyruvate are tightly bound to the complex. The proximity of one enzyme to another *increases the overall reaction rate* and *minimizes side reactions.* The activated intermediates are transferred from one active site to another by the lipoamide prosthetic group of the transacetylase (see Figure 20-8). The attachment of the lipoyl group to the ε-amino group of a lysine residue on the transacetylase provides a flexible arm for the reactive ring. The high degree of mobility of the lipoyl domain and its 14-Å molecular string enables the lipoyl moiety of a transacetylase subunit (E_2) to interact with the thiamine pyrophosphate unit of an adjacent pyruvate dehydrogenase subunit (E_1) and with the flavin unit of an adjacent dihydrolipoyl dehydrogenase (E_3). Furthermore, the acetyl group can be shuttled from one lipoyl unit to another.

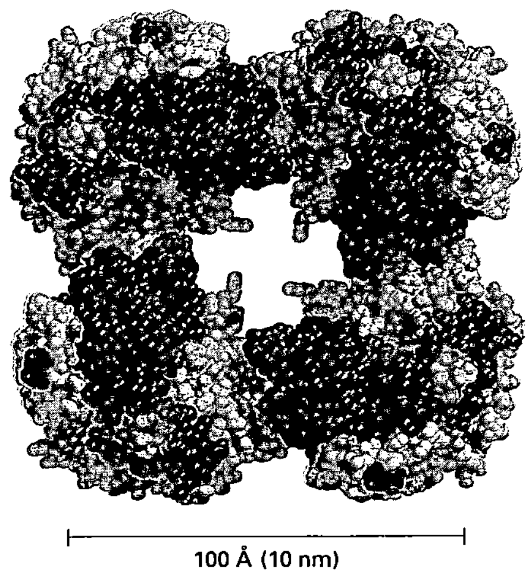

Figure 20-11
Structure of the E_2 transacetylase core of the pyruvate dehydrogenase complex of *Azobacter vinelandii,* a gram-negative bacterium. The view here is down the fourfold axis of the core. The three tightly associated subunits are depicted in different colors. [Drawn from 1eaa.pdb A. Mattevi, G. Oblomolova, E. Schulze, K.H. Kalk, A.H. Westphal, A. de Kok, and W.G.J. Hol. *Science* 255(1992):1544.]

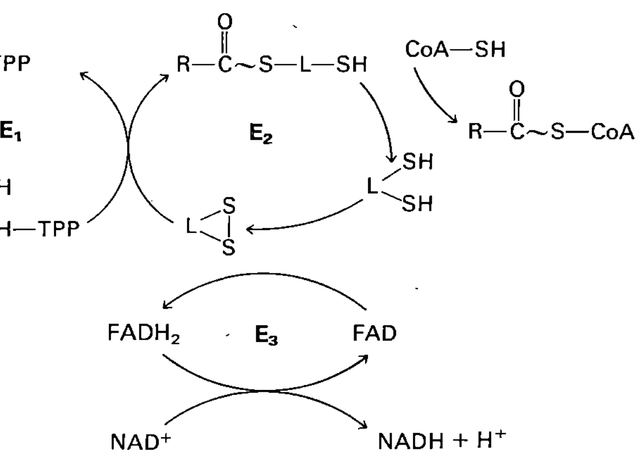

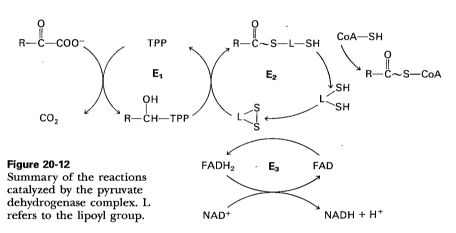

Figure 20-12
Summary of the reactions catalyzed by the pyruvate dehydrogenase complex. L refers to the lipoyl group.

VARIATION ON A MULTIENZYME THEME: THE α-KETOGLUTARATE DEHYDROGENASE COMPLEX

The oxidative decarboxylation of α-ketoglutarate closely resembles that of pyruvate, also an α-keto acid.

$$\alpha\text{-Ketoglutarate} + \text{CoA} + \text{NAD}^+ \longrightarrow \text{succinyl CoA} + \text{CO}_2 + \text{NADH}$$

$$\text{Pyruvate} + \text{CoA} + \text{NAD}^+ \longrightarrow \text{acetyl CoA} + \text{CO}_2 + \text{NADH}$$

The same cofactors are participants: TPP, lipoamide, CoA, FAD, and NAD^+. In fact, *the oxidative decarboxylation of α-ketoglutarate is catalyzed by an enzyme complex that is structurally similar to the pyruvate dehydrogenase complex.* The α-ketoglutarate dehydrogenase complex contains three kinds of enzymes: an α-ketoglutarate dehydrogenase component (E_1'), a transsuccinylase one (E_2'), and a dihydrolipoyl dehydrogenase one (E_3'). Again, E_1' binds to E_2', and E_2' binds to E_3', but E_1' does not bind directly to E_3'. Thus, *transsuccinylase* (like transacetylase) *is the core of the complex.* The α-ketoglutarate dehydrogenase component and transsuccinylase are different from the corresponding enzymes in the pyruvate dehydrogenase complex. However, *the dihydrolipoyl dehydrogenase parts* of the two complexes are identical.

It was noted earlier that chymotrypsin, trypsin, thrombin, and elastase are homologous enzymes. Here we see that the pyruvate and α-ketoglutarate dehydrogenase complexes are *homologous enzyme assemblies.* The structural and mechanistic motifs that coordinate catalysis in the entrée to the citric acid cycle are used again later in the cycle. Furthermore, a related multienzyme complex participates in the degradation of leucine, valine, and isoleucine (p. 646). The common thread is the transfer of an α-keto group to coenzyme A.

BERIBERI IS CAUSED BY A DEFICIENCY OF THIAMINE

Beriberi, a neurologic and cardiovascular disorder, is caused by a dietary deficiency of thiamine (also called *vitamin* B_1). The disease has been and continues to be a serious health problem in the Far East because rice, the major food, has a rather low content of thiamine. This deficiency is partially ameliorated if the whole rice grain is soaked in water before milling—some of the thiamine in the husk then leaches into the rice kernel. The problem is exacerbated if the rice is polished, because only the outer layer contains appreciable amounts of thiamine. Beriberi is also occasionally seen in alcoholics who are severely malnourished. The disease is characterized by neurologic and cardiac symptoms. Damage to the peripheral nervous system is expressed in terms of pain in the limbs, weakness of the musculature, and distorted skin sensation. The heart may be enlarged and the cardiac output inadequate.

Which biochemical processes might be affected by a deficiency of thiamine? *TPP is the prosthetic group of three important enzymes: pyruvate dehydrogenase, α-ketoglutarate dehydrogenase, and transketolase.* Transketolase transfers two-carbon units from one sugar to another; its role in the pentose phosphate pathway will be discussed in Chapter 22. *The common feature of enzymatic reactions utilizing TPP is the transfer of an activated aldehyde unit.* In beriberi, the levels of pyruvate and α-ketoglutarate in the blood are higher than normal. The increase in the level of pyruvate in the blood is especially pronounced after ingestion of glucose. A related finding is that the activities of the pyruvate and α-ketoglutarate dehydrogenase complexes in vivo are abnormally low. The low transketolase activity of red cells in beriberi is a reliable diagnostic indicator of the disease.

Beriberi—
A vitamin-deficiency disease first described in 1630 by Jacob Bonitus, a Dutch physician working in Java:

"A certain very troublesome affliction, which attacks men, is called by the inhabitants Beriberi (which means sheep). I believe those, whom this same disease attacks, with their knees shaking and the legs raised up, walk like sheep. It is a kind of paralysis, or rather Tremor: for it penetrates the motion and sensation of the hands and feet indeed sometimes of the whole body."

CITRATE SYNTHASE UNDERGOES A LARGE CONFORMATIONAL CHANGE ON BINDING OXALOACETATE

As was mentioned earlier, citrate synthase catalyzes the condensation of oxaloacetate and acetyl CoA to form citryl CoA, which is then hydrolyzed to citrate and CoA.

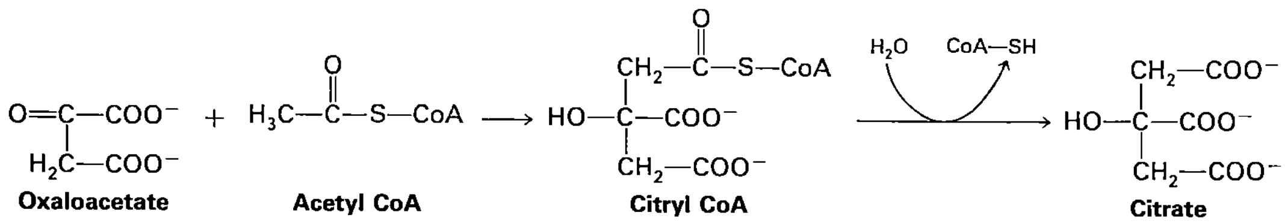

Mammalian citrate synthase is a dimer of identical 49-kd subunits. Each active site is located in a cleft between the large and small domain of a subunit, adjacent to the subunit interface (see the image on p. 509). X-ray

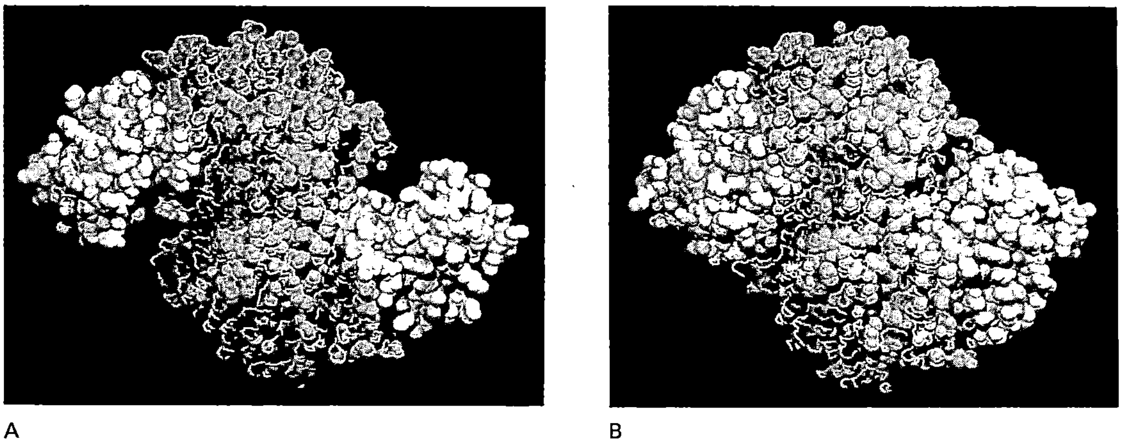

crystallographic studies of citrate synthase and its complexes with several substrates and inhibitors have revealed that the enzyme undergoes large conformational changes during catalysis. Oxaloacetate binds first, followed by acetyl CoA. The reason for ordered binding is that *oxaloacetate induces a major structural rearrangement leading to the creation of a binding site for acetyl CoA.* The open form of the enzyme observed in the absence of ligands is converted into a closed form (Figure 20-13) by the binding of oxaloacetate. In each subunit, the small domain rotates 19 degrees relative to the large domain. *Movements as large as 15 Å are produced by the rotation of α helices elicited by quite small shifts of side chains around bound oxaloacetate.* This conformational transition is reminiscent of the cleft closure in hexokinase induced by the binding of glucose (p. 499) and the subunit rearrangement in hemoglobin caused by the binding of O_2 (p. 161).

Figure 20-13
Citrate synthase undergoes a large conformational change on binding oxaloacetate. The small domain of each subunit of the homodimer is shown in yellow, and the large domain in blue. (A) Open form of enzyme alone. (B) Closed form of the liganded enzyme. [After S.J. Remington, G. Wiegand, and R. Huber. *J. Mol. Biol.* 158(1982):111.]

The synthase catalyzes the condensation reaction by bringing the substrates into close proximity, orienting them, and polarizing certain bonds. Two histidines and an aspartate are key players (Figure 20-14).

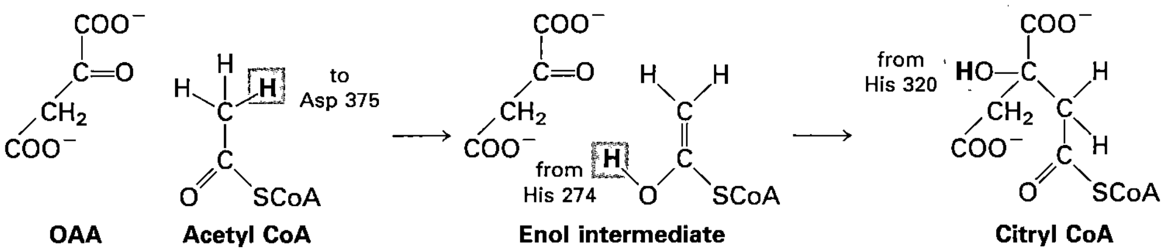

Figure 20-14
Mechanism of synthesis of citryl CoA by citrate synthase. The condensation of oxaloacetate (OAA) and acetyl CoA proceeds through an enol intermediate. The subsequent hydrolysis of citryl CoA yields citrate and CoA.

One of the histidines (His 274) donates a proton to the carbonyl oxygen of acetyl CoA to promote removal of a methyl proton by Asp 375. Oxaloacetate is activated by the transfer of a proton from His 320 to its carbonyl carbon atom. The subsequent attack of the enol of acetyl CoA on the carbonyl carbon of oxaloacetate results in the formation of a C–C bond. The newly formed citryl CoA induces additional structural changes in the enzyme. The active site becomes completely enclosed. His 274 participates again as a proton donor to hydrolyze the thioester. CoA leaves the enzyme, followed by citrate, and the enzyme returns to the initial open conformation.

Citrate synthase is designed to hydrolyze *citryl* CoA but not *acetyl* CoA, which would be wasteful. How is this discrimination accomplished? First, acetyl CoA does not bind to the enzyme until oxaloacetate is bound and ready for condensation. Second, the catalytic residues crucial for hydrolysis of the thioester linkage are not appropriately positioned until citryl CoA is formed. As with hexokinase (p. 499) and triosephosphate isomerase (p. 501), *induced fit prevents an undesirable side reaction.*

SYMMETRIC MOLECULES MAY REACT ASYMMETRICALLY

Let us follow the fate of a particular carbon atom in the citric acid cycle. Suppose that oxaloacetate were labeled with ^{14}C in the carboxyl carbon farthest from the keto group. Analysis of the α-ketoglutarate formed would show that none of the radioactive label had been lost. Decarboxylation of α-ketoglutarate would then yield succinate devoid of radioactivity. All the label would be in the released CO_2.

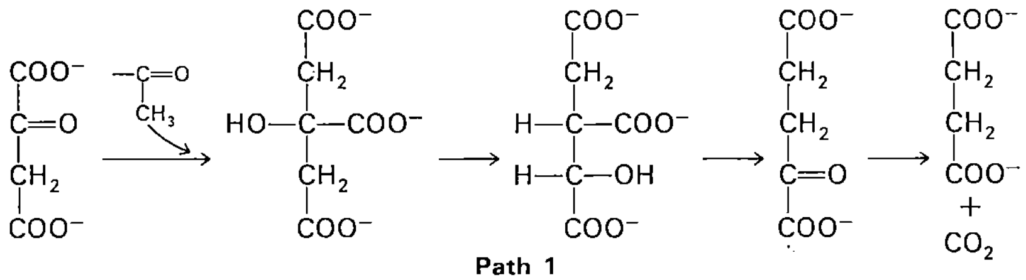

Path 1

The finding that *all* the label emerges in the CO_2 came as a surprise. Citrate is a symmetric molecule. Consequently, it was assumed that the two $-CH_2COO^-$ groups in it would react identically. Thus, for every citrate undergoing the reactions shown in path 1, it was thought that another citrate molecule would react as shown in path 2. If so, then only *half* the label should have emerged in the CO_2.

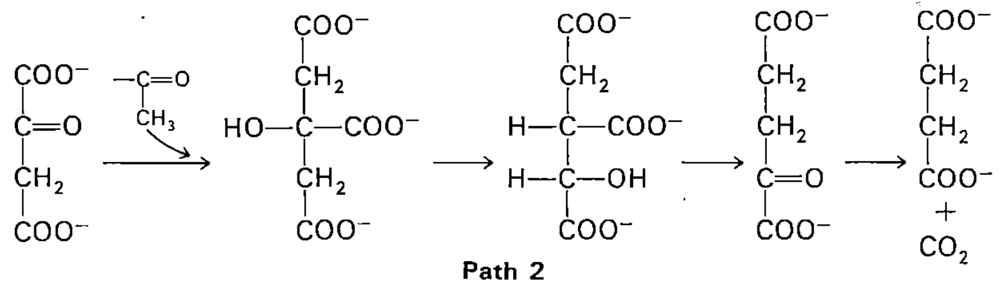

Path 2
(Does not occur)

Table 20-3
Commonly used radioisotopes

Isotope	*Half-life*
^3H	12.26 years
^{14}C	5730 years
^{22}Na	2.62 years
^{32}P	14.28 days
^{35}S	87.9 days
^{42}K	12.36 hours
^{45}Ca	163 days
^{59}Fe	45.6 days
^{125}I	60.2 days
^{203}Hg	46.9 days

The interpretation of these experiments, which were carried out in 1941, was that citrate (or any other symmetric compound) could not be an intermediate in the formation of α-ketoglutarate because of the asymmetric fate of the label. This view seemed compelling until Alexander Ogston incisively pointed out in 1948 that it is a fallacy to assume that the two identical groups of a symmetric molecule cannot be distinguished: "On the contrary, it is possible that *an asymmetric enzyme which attacks a symmetrical compound can distinguish between its identical groups*. . . the asymmetrical occurrence of isotope in a product cannot be taken as conclusive evidence against its arising from a symmetrical precursor."

Let us examine Ogston's assertion. For simplicity, consider a molecule in which two hydrogen atoms, a group X, and a different group Y are bonded to a tetrahedral carbon atom. Let us label one hydrogen A, the other B. Now suppose that an enzyme binds three groups of this substrate—X, Y, and H—at three complementary sites. Can H_A be distinguished from H_B? Figure 20-15 shows X, Y, and H_A bound to three points on the enzyme. In contrast, X, Y, and H_B cannot be bound to this active site; two of these three groups can be bound, but not all three. Thus, H_A and H_B will have different fates.

It should be noted that H_A and H_B are sterically not equivalent even though the molecule $CXYH_2$ is optically inactive. Similarly, the $-CH_2COO^-$ groups in citrate are sterically not equivalent even though citrate is optically inactive. *The symmetry rules that determine whether a compound has indistinguishable substituents are different from those that determine whether it is optically inactive:* (1) a molecule is optically inactive if it can be superimposed on its mirror image; (2) a molecule has indistinguishable substituents only if these groups can be brought into coincidence by a rotation that leaves the rest of the structure invariant.

Sterically nonequivalent groups such as H_A and H_B will almost always be distinguished in enzymatic reactions. The essence of the differentiation of these groups is that the enzyme holds the substrate in a specific orientation. Attachment at three points, as depicted in Figure 20-16, is a readily visualized way of achieving a particular orientation of the substrate, but it is not the only means of doing so.

The terms *chiral* and *prochiral* are now extensively used to describe the stereochemistry of molecules. A *chiral* molecule has handedness and hence is optically active. A *prochiral* molecule (or center within a molecule), such as citrate or $CXYH_2$, lacks handedness and is optically inactive, but its identical substituents are distinguishable, and it can become chiral in one step. A prochiral molecule (such as $CXYH_AH_B$) is transformed into a chiral one ($CXYZH_B$) when one of its identical atoms or groups (H_A in this example) is replaced. The prefixes R and S are used to unambiguously designate the configuration of chiral and prochiral centers, as described in the appendix to this chapter (p. 526).

HYDROGEN IS STEREOSPECIFICALLY TRANSFERRED BY NAD⁺ DEHYDROGENASES

In the 1950s, Birgit Vennesland and Frank Westheimer carried out elegant experiments on the stereospecificity of hydrogen transfer by NAD⁺ dehydrogenases. Ethanol labeled with two deuterium atoms at C-1 was the substrate in a reaction catalyzed by alcohol dehydrogenase. They found that the reduced coenzyme contained one atom of deuterium per molecule, whereas the other deuterium atom was in acetaldehyde. Thus, none of the deuterium was lost to the solvent, which shows that deuterium was directly transferred from the substrate to NAD⁺.

$$H_3C-\overset{\overset{\displaystyle D}{|}}{\underset{\underset{\displaystyle D}{|}}{C}}-OH + NAD^+ \longrightarrow CH_3-C\overset{\displaystyle O}{\underset{\displaystyle D}{\diagup}} + \begin{array}{l}\text{Reduced NAD} \\ \text{containing} \\ \text{1 deuterium}\end{array} + H^+$$

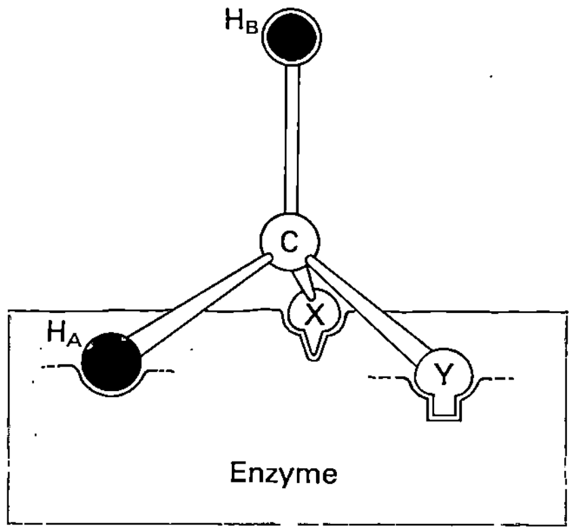

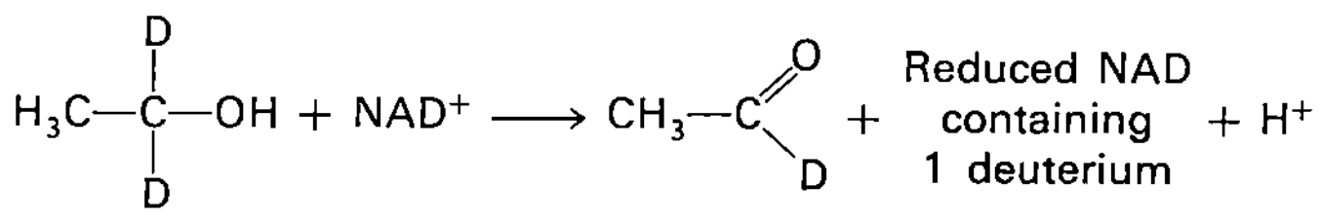

The deuterated reduced coenzyme formed in this reaction was then used to reduce acetaldehyde. The striking result was that *all the deuterium was transferred from the coenzyme to the substrate.* None remained in NAD⁺.

$$CH_3-C\overset{\displaystyle O}{\underset{\displaystyle H}{\diagup}} + H^+ + \begin{array}{l}\text{Reduced NAD} \\ \text{containing} \\ \text{1 deuterium}\end{array} \longrightarrow \underset{\substack{\text{(Contains} \\ \text{1 deuterium)}}}{CH_3CHDOH} + \underset{\substack{\text{(Contains} \\ \text{no deuterium)}}}{NAD^+}$$

Figure 20-15
H_A and H_B are sterically not equivalent if the substrate $CXYH_2$ is bound to the enzyme at three points.

Chirality—
"I call any geometrical figure, or any group of points, *chiral,* and say it has *chirality,* if its image in a plane mirror, ideally realized, cannot be brought to coincide with itself."

LORD KELVIN (1893)

Derived from the Greek word *cheir,* "hand."

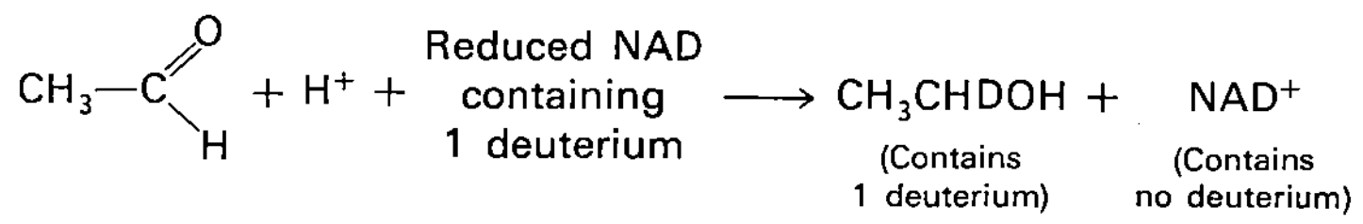

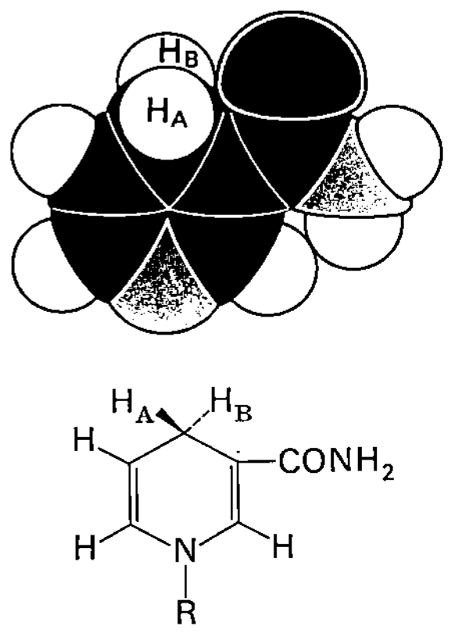

NADH

Figure 20-16
Model and formula of the nicotinamide moiety of NADH showing that H_A and H_B are on opposite sides of the ring. In the RS nomenclature, H_A is *pro-R* and H_B is *pro-S*.

These reactions revealed that the transfer catalyzed by alcohol dehydrogenase is stereospecific. *The positions occupied by the two hydrogen atoms at C-4 in NADH are not equivalent* (Figure 20-16). One of them (H_A) is in front of the nicotinamide plane; the other (H_B) is behind. In other words, C-4 is a *prochiral center*. Alcohol dehydrogenase, a chiral reagent, distinguishes between positions A and B at C-4. Deuterium is transferred from deuterated ethanol to position A only.

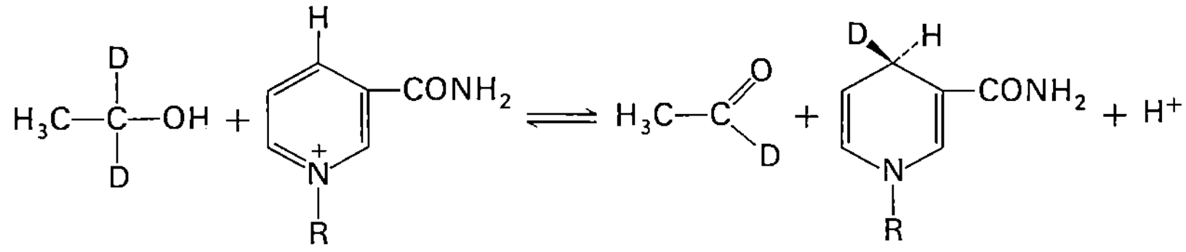

In the reverse reaction, the deuterium atom at position A is removed and directly transferred to acetaldehyde. Some dehydrogenases, such as glyceraldehyde phosphate dehydrogenase, transfer deuterium to position B because the nicotinamide ring of bound NAD^+ is rotated by 180 degrees relative to its orientation in alcohol dehydrogenase and other type A dehydrogenases.

THE CITRIC ACID CYCLE IS A SOURCE OF BIOSYNTHETIC PRECURSORS

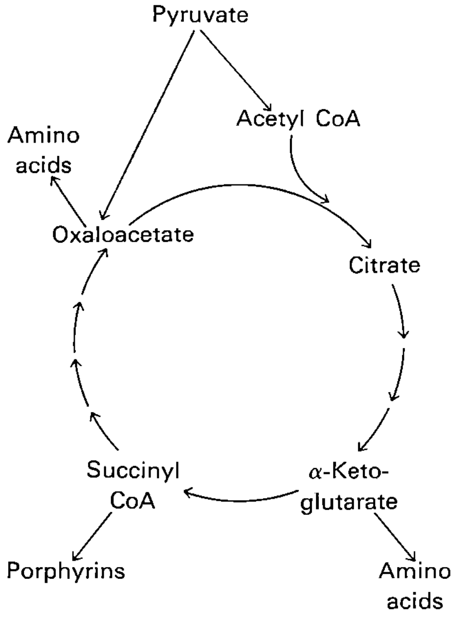

Figure 20-17
Biosynthetic roles of the citric acid cycle. Intermediates drawn off for biosyntheses (shown by red arrows) are replenished by the formation of oxaloacetate from pyruvate.

Thus far, discussion has focused on the citric acid cycle as the *major degradative pathway for the generation of ATP.* The citric acid cycle also *provides intermediates for biosyntheses* (Figure 20-17). For example, a majority of the carbon atoms in porphyrins come from *succinyl CoA.* Many of the amino acids are derived from *α-ketoglutarate and oxaloacetate.* The biosyntheses of these compounds will be discussed in subsequent chapters. The important point now is that *citric acid cycle intermediates must be replenished if any are drawn off for biosyntheses.* Suppose that oxaloacetate is converted into amino acids for protein synthesis. The citric acid cycle will cease to operate unless new oxaloacetate is formed, because acetyl CoA cannot enter the cycle unless it condenses with oxaloacetate. Even though oxaloacetate is used catalytically, a minimal level must be maintained to allow the cycle to function.

How is oxaloacetate replenished? Mammals lack the enzymes for the net conversion of acetyl CoA into oxaloacetate or another citric acid cycle intermediate. Rather, oxaloacetate is formed by the carboxylation of pyruvate, in a reaction catalyzed by pyruvate carboxylase (p. 571).

$$\text{Pyruvate} + CO_2 + \text{ATP} + H_2O \longrightarrow \text{oxaloacetate} + \text{ADP} + P_i + 2\ H^+$$

The synthesis of oxaloacetate by carboxylation of pyruvate is an example of an *anaplerotic reaction* (from the Greek, meaning to "fill up").

THE GLYOXYLATE CYCLE ENABLES PLANTS AND BACTERIA TO GROW ON ACETATE

Many bacteria and plants are able to grow on acetate or other compounds that yield acetyl CoA. They make use of a metabolic pathway that converts two-carbon acetyl units into four-carbon units (succinate) for energy production and biosyntheses. This reaction sequence, called the *glyoxylate*

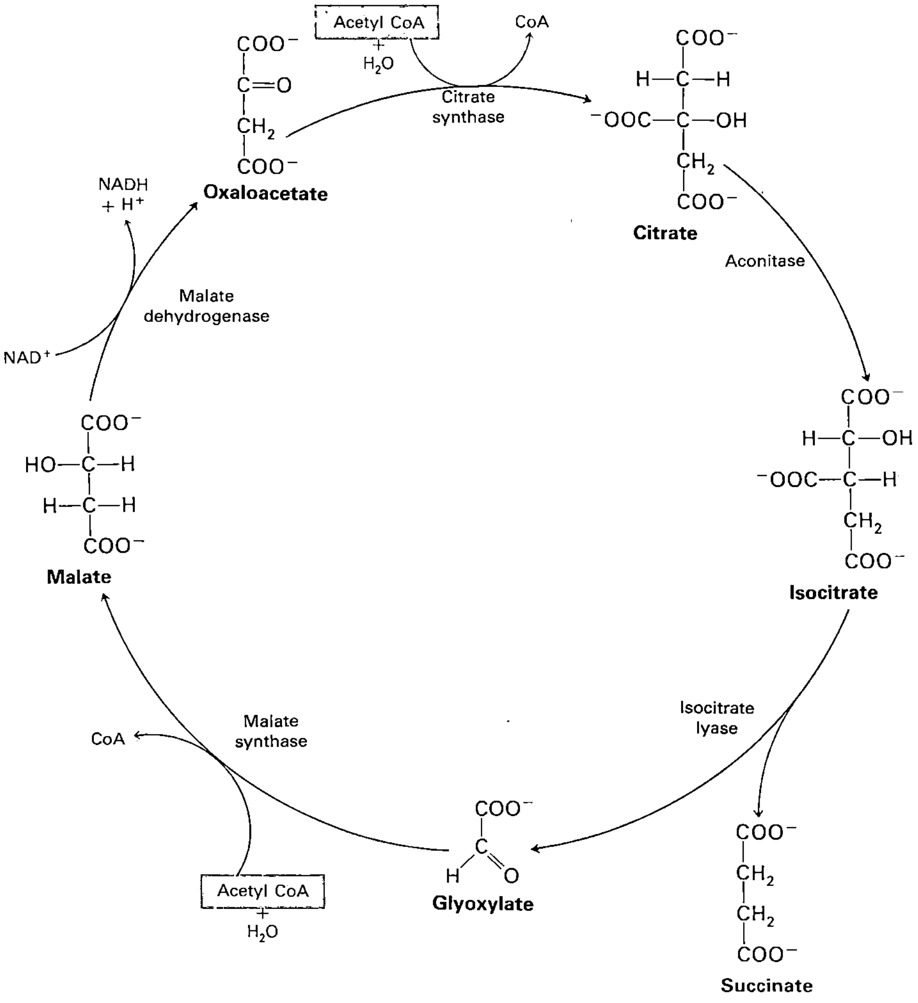

Figure 20-18
Glyoxylate cycle of plants and bacteria. The reactions of this cycle are the same as those of the citric acid cycle except for the ones catalyzed by isocitrate lyase and malate synthase.

cycle, bypasses the two decarboxylation steps of the citric acid cycle. Another key difference is that two molecules of acetyl CoA enter per turn of the glyoxylate cycle, compared with one in the citric acid cycle. The glyoxylate cycle (Figure 20-18), like the citric acid cycle, begins with the condensation of acetyl CoA and oxaloacetate to form citrate, which is then isomerized to isocitrate. Instead of being decarboxylated, isocitrate is cleaved by *isocitrate lyase* into succinate and glyoxylate. The subsequent steps regenerate oxaloacetate from glyoxylate. Acetyl CoA condenses with glyoxylate to form malate, in a reaction catalyzed by *malate synthase,* which resembles citrate synthase. Finally, malate is oxidized to oxaloacetate, as in the citric acid cycle. The sum of these reactions is

$$2\ \text{Acetyl CoA} + \text{NAD}^+ + 2\ \text{H}_2\text{O} \longrightarrow \text{succinate} + 2\ \text{CoA} + \text{NADH} + 2\ \text{H}^+$$

In plants, these reactions occur in organelles called *glyoxysomes.*

Bacteria and plants can synthesize acetyl CoA from acetate and CoA by an ATP-driven reaction that is catalyzed by *acetyl CoA synthetase.*

$$\text{Acetate} + \text{CoA} + \text{ATP} \longrightarrow \text{acetyl CoA} + \text{AMP} + \text{PP}_i$$

Pyrophosphate is then hydrolyzed to orthophosphate, and so 2 ~P are consumed in the activation of acetate. We shall return to this type of activation reaction in protein synthesis, where it is used to link amino acids to transfer RNAs (p. 880).

ISOCITRATE DEHYDROGENASE IN BACTERIA IS DEACTIVATED BY PHOSPHORYLATION AT THE ACTIVE SITE

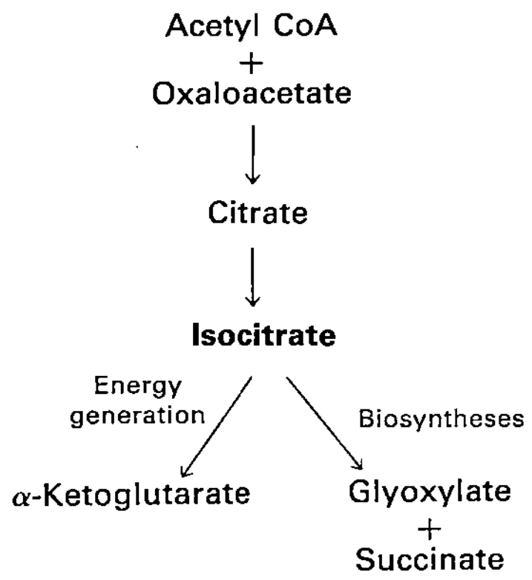

Figure 20-19
Isocitrate has two major fates in some bacteria and plants.

Isocitrate has two major fates in some bacteria and plants. When energy is needed, it is oxidatively decarboxylated to α-ketoglutarate. In contrast, when energy is abundant, isocitrate is split into succinate and glyoxylate. The citric acid cycle and the glyoxylate cycle compete for isocitrate at this key branch point (Figure 20-19).

In times of plenty, *isocitrate dehydrogenase is switched off by phosphorylation, which serves to funnel isocitrate into the glyoxylate pathway for the formation of biosynthetic intermediates.* The mechanism of deactivation is simple: phosphorylation of Ser 113 at the active site directly blocks the binding of isocitrate (Figure 20-20). The kinase that phosphorylates isocitrate dehydrogenase and the phosphatase that removes the phosphate group from the modified enzyme are located on the same polypeptide chain. Indeed, these opposing reactions are catalyzed by the same active site. Their velocities are reciprocally controlled by several metabolites. AMP, for example, activates the phosphatase and inhibits the kinase.

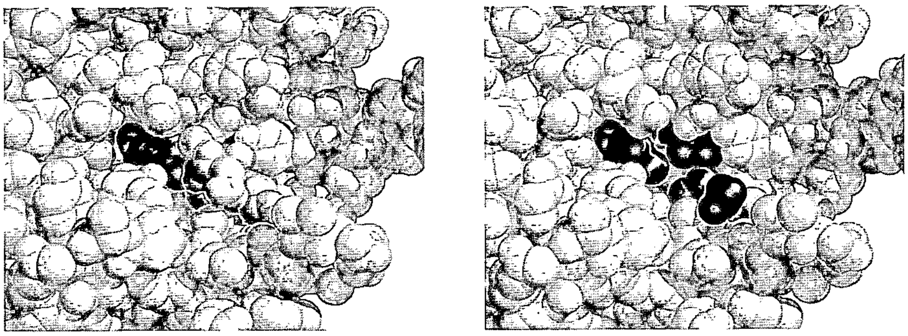

A B

Figure 20-20
Models of the active and inactive forms of isocitrate dehydrogenase from *E. coli.* (A) In the dephosphorylated form, isocitrate (yellow) is hydrogen-bonded to Ser 113 (green) and interacts electrostatically with three arginine residues (blue). A bound Mg^{2+} is shown in purple. (B) In the inactive phosphorylated form, isocitrate cannot bind because it is sterically blocked and repelled by the negatively charged phosphoryl group (red) attached to Ser 113. Inhibition by phosphorylation is direct—the two forms of the enzyme have nearly the same conformation. [Drawn from 5icd.pdb and 4icd.pdb. J.B. Hurley, A.M. Dean, P.E. Thorness, D.E. Koshland, Jr., and R.M. Stroud. *J. Biol. Chem.* 265 (1990):3599.]

THE PYRUVATE DEHYDROGENASE COMPLEX IS REGULATED BY REVERSIBLE PHOSPHORYLATION

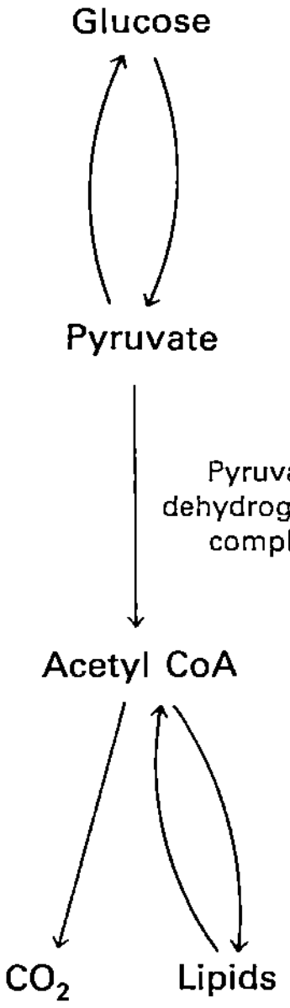

Figure 20-21
The formation of acetyl CoA from pyruvate is a key irreversible step.

The formation of acetyl CoA from pyruvate is a key irreversible step in animals because they are unable to convert acetyl CoA into glucose. The oxidative decarboxylation of pyruvate to acetyl CoA commits the carbon atoms of glucose to two principal fates: oxidation to CO_2 by the citric acid cycle, with the concomitant generation of energy, or incorporation into lipid (Figure 20-21). As expected of an enzyme at a critical decision point in metabolism, the activity of the pyruvate dehydrogenase complex is stringently controlled. *Phosphorylation of the pyruvate dehydrogenase component (E_1) by a specific kinase switches off the activity of the complex. Deactivation is reversed by the action of a specific phosphatase.* Both the kinase and phosphatase are bound to the transacetylase component (E_2), again highlighting the structural and mechanistic importance of this core. Increasing the

NADH/NAD$^+$, acetyl CoA/CoA, or ATP/ADP ratio promotes phosphorylation and hence deactivation of the complex. Pyruvate activates the dehydrogenase by inhibiting the kinase, whereas Ca^{2+} does so by stimulating the phosphatase. Thus, *pyruvate dehydrogenase is switched off when the energy charge is high and biosynthetic intermediates are abundant.* By contrast, hormones such as vasopressin and α_1-adrenergic agonists stimulate pyruvate dehydrogenase by triggering a rise in the cytosolic Ca^{2+} level, which in turn elevates the mitochondrial Ca^{2+} level. Insulin also accelerates the conversion of pyruvate into acetyl CoA by stimulating the dephosphorylation of the complex. In turn, glucose is funneled into pyruvate.

CONTROL OF THE CITRIC ACID CYCLE

The rate of the citric acid cycle is also precisely adjusted to meet an animal cell's needs for ATP (Figure 20-22). *The synthesis of citrate from oxaloacetate and acetyl CoA, the entry of two-carbon units into the cycle, is an important control point.* ATP is an allosteric inhibitor of citrate synthase. The effect of ATP is to increase the K_M for acetyl CoA. Thus, as the level of ATP increases, less of this enzyme is saturated with acetyl CoA and so less citrate is formed.

A second control point is isocitrate dehydrogenase. This enzyme is allosterically stimulated by ADP, which enhances its affinity for substrates. The binding of isocitrate, NAD$^+$, Mg^{2+}, and ADP is mutually cooperative. In contrast, NADH inhibits isocitrate dehydrogenase by directly displacing NAD$^+$. ATP, too, is inhibitory. It is important to note that several steps in the cycle require NAD$^+$ or FAD, which are abundant only when the energy charge is low.

A third control site in the citric acid cycle is α-ketoglutarate dehydrogenase. Some aspects of its control are like those of the pyruvate dehydrogenase complex, as might be expected from their homology. α-Ketoglutarate dehydrogenase is inhibited by succinyl CoA and NADH, the products of the reaction it catalyzes. Also, α-ketoglutarate dehydrogenase is inhibited by a high energy charge. Thus, *the funneling of two-carbon fragments into the citric acid cycle and the rate of the cycle are reduced when the cell has a high level of ATP.* Control is achieved by a variety of complementary mechanisms that operate at multiple sites (see Figure 20-22).

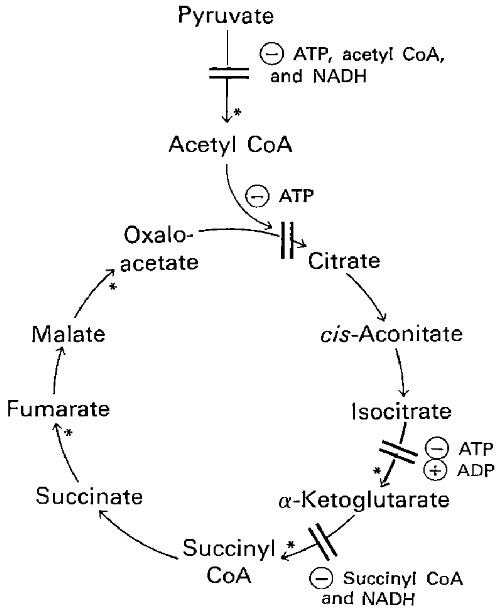

Figure 20-22
Control of the citric acid cycle and the oxidative decarboxylation of pyruvate: * indicates steps that require an electron acceptor (NAD$^+$ or FAD) that is regenerated by the respiratory chain.

SUMMARY

The citric acid cycle is the final common pathway for the oxidation of fuel molecules. It also serves as a source of building blocks for biosyntheses. Most fuel molecules enter the cycle as acetyl CoA. The link between glycolysis and the citric acid cycle is the oxidative decarboxylation of pyruvate to form acetyl CoA. In eukaryotes, this reaction and those of the cycle occur inside mitochondria, in contrast with glycolysis, which occurs in the cytosol.

The cycle starts with the condensation of oxaloacetate (C_4) and acetyl CoA (C_2) to give citrate (C_6), which is isomerized to isocitrate (C_6). Oxidative decarboxylation of this intermediate gives α-ketoglutarate (C_5). The second molecule of CO_2 comes off in the next reaction, in which α-ketoglutarate is oxidatively decarboxylated to succinyl CoA (C_4). The thioester bond of succinyl CoA is cleaved by P_i to yield succinate, and a high-energy phosphate bond in the form of GTP is concomitantly generated. Succinate is oxidized to fumarate (C_4), which is then hydrated to form malate (C_4). Finally, malate is oxidized to regenerate oxaloacetate (C_4). Thus, two carbon atoms from acetyl CoA enter the cycle, and two carbon atoms leave the cycle as CO_2 in the successive decarboxylations catalyzed by isocitrate dehydrogenase and α-ketoglutarate dehydrogenase. In the four oxidation-reductions in the cycle, three pairs of electrons are transferred to NAD^+ and one pair to FAD. These reduced electron carriers are subsequently oxidized by the electron transport chain to generate nine molecules of ATP. In addition, one high-energy phosphate bond is directly formed in the citric acid cycle. Hence, a total of 10 high-energy phosphate bonds are generated for each two-carbon fragment that is completely oxidized to H_2O and CO_2.

The citric acid cycle operates only under aerobic conditions because it requires a supply of NAD^+ and FAD. These electron acceptors are regenerated when NADH and $FADH_2$ transfer their electrons to O_2 through the electron transport chain, with the concomitant production of ATP. Consequently, the rate of the citric acid cycle depends on the need for ATP. The regulation of three enzymes in the cycle is also important for control. A high energy charge diminishes the activities of citrate synthase, isocitrate dehydrogenase, and α-ketoglutarate dehydrogenase. The irreversible formation of acetyl CoA from pyruvate is another important regulatory point. The activity of the pyruvate dehydrogenase complex is stringently controlled by reversible phosphorylation. These mechanisms complement each other in reducing the rate of formation of acetyl CoA when the energy charge of the cell is high and when biosynthetic intermediates are abundant.

APPENDIX: The RS Designation of Chirality

The absolute configuration of any chiral center can be unambiguously specified using the RS notation introduced in 1956 by Robert Cahn, Christopher Ingold, and Vladimir Prelog. Consider the chiral compound CHFClBr (Figure 20-23A). The first step in using the RS designation is to assign a *priority sequence* to the four substituents by applying the rule that *an atom with a higher atomic number has a higher priority than an atom with a lower atomic number.* Hence, the priority sequence of these four substituents is (a) Br, (b) Cl, (c) F, and (d) H, with Br having the highest priority (a). The next step is to *orient the molecule so that the group of lowest priority (d) points away from the viewer.* For CHFClBr, this means that the molecule should be oriented so that H is away from us (behind the plane of the page in Figure 20-23B). *The final step is to ask whether the path from a to b to c under these conditions is clockwise or counterclockwise.* If it is clockwise (right-handed), the configuration is R (from the Latin *rectus,* "right"). If it is counterclockwise (left-

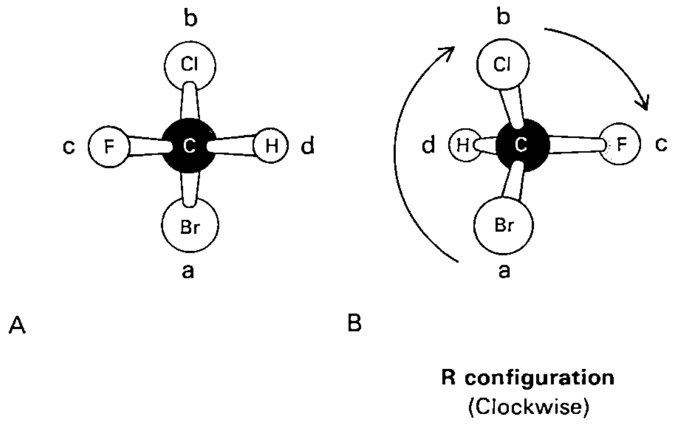

Figure 20-23
The CHFClBr stereoisomer shown in part A is rotated to place H behind the plane of the page so that the rule can be applied. In B, the direction of rotation from a to b to c is clockwise. Hence, this stereoisomer has the R configuration.

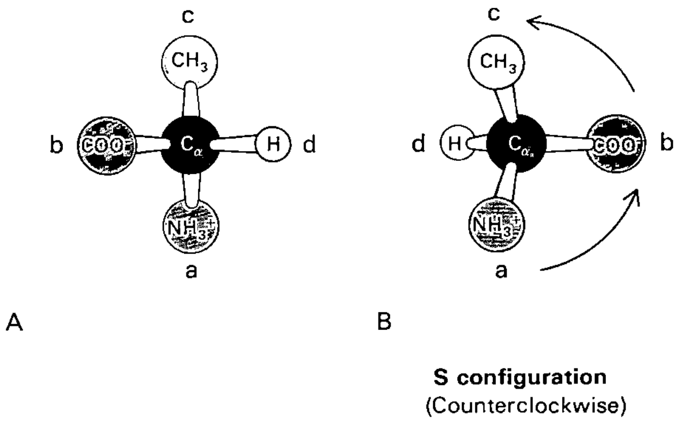

Figure 20-24
L-Alanine has an S configuration because, as can be seen in B, the direction of rotation from a to b to c is counterclockwise.

handed), the configuration is S (from the Latin *sinister*, "left"). The configuration of the stereoisomer of CHFClBr shown in Figure 20-23 is R.

Now let us consider the RS designation of alanine (Figure 20-24A). The four atoms bonded to the α carbon are N, C, C, and H. The priority sequence of the methyl carbon and the carboxyl carbon is determined by going outward to the next set of atoms. *It is useful to note the following priority sequence for biochemically important groups:* –SH (highest), –OR, –OH, –NHR, $-NH_2$, –COOR, –COOH, –CHO, $-CH_2OH$, $-C_6H_5$, $-CH_3$, –T, –D, –H (lowest). Thus, the priority sequence of the four groups attached to the α carbon of alanine is (a) $-NH_3^+$, (b) $-COO^-$, (c) $-CH_3$, and (d) –H. The next step is to orient L-alanine so that its lowest-priority group (–H) is behind the plane of the page (Figure 20-24B). The path from (a) $-NH_3^+$ to (b) $-COO^-$ to (c) $-CH_3$ is then counterclockwise (left-handed), and so L-alanine has an S configuration.

SELECTED READINGS

Where to start

Baldwin, J.E., and Krebs, H., 1981. The evolution of metabolic cycles. *Nature* 291:381–382. [A concise essay, rich in insight, with emphasis on the citric acid cycle.]

Reed, L.J., and Hackert, M.L., 1990. Structure-function relationships in dihydrolipoamide acyltransferases. *J. Biol. Chem.* 265:8971–8974.

Mattevi, A., Obmolova, G., Schulze, E., Kalk, K.H., Westphal, A.H., De Kok, A., and Hol, W.G., 1992. Atomic structure of the cubic core of the pyruvate dehydrogenase multienzyme complex. *Science* 255:1544–1550.

LaPorte, D.C., 1993. The isocitrate dehydrogenase phosphorylation cycle: Regulation and enzymology. *J. Cell. Biochem.* 51:14–18.

Books

Roche, T.E., and Patel, M.S. (eds.), 1989. *Alpha-keto Acid Dehydrogenase Complexes: Organization, Regulation, and Biomedical Ramifications.* Annals of the New York Academy of Sciences, vol. 573.

Lowenstein, J.M. (ed.), 1969. *Citric Acid Cycle: Control and Compartmentation.* Marcel Dekker.

Kay, J., and Weitzman, P.D.J. (eds.), 1987. *Krebs' Citric Acid Cycle—Half a Century and Still Turning.* Biochemical Society Symposium 54.

Gottschalk, G., 1986. *Bacterial Metabolism* (2nd ed.). Springer-Verlag. [An excellent account of the metabolic diversity of bacteria and the richness of their energy-generating pathways.]

Stereospecificity

Popják, G., 1970. Stereospecificity of enzymic reactions. *In* Boyer, P.D. (ed.), *The Enzymes* (3rd ed.), vol. 2, pp. 115–215. Academic Press. [An excellent review containing a discussion of the stereochemistry of the citric acid cycle.]

Ogston, A.G., 1948. Interpretation of experiments on metabolic processes using isotopic tracer elements. *Nature* 162:963.

Bentley, R., 1969. *Molecular Asymmetry in Biology,* vols. 1 and 2. Academic Press. [These volumes contain a wealth of information about stereospecificity in biochemical reactions. Chapter 2 of vol. 1 discusses nomenclature, and Chapter 4 discusses prochirality.]

Pyruvate dehydrogenase complex

Perham, R.N., 1991. Domains, motifs, and linkers in 2-oxo acid dehydrogenase multienzyme complexes: A paradigm in the design of a multifunctional protein. *Biochemistry* 30:8501–8512.

Patel, M.S., and Roche, T.E., 1990. Molecular biology and biochemistry of pyruvate dehydrogenase complexes. *FASEB J.* 4:3224–3233.

Green, J.D., Perham, R.N., Ullrich, S.J., and Appella, E., 1992. Conformational studies of the interdomain linker peptides in the dihydrolipoyl acetyltransferase component of the pyruvate dehydrogenase multienzyme complex of *Escherichia coli. J. Biol. Chem.* 267:23484–23488.

Randall, D.D., Miernyk, J.A., Fang, T.K., Budde, R.J., and Schuller, K.A., 1989. Regulation of the pyruvate dehydrogenase complexes in plants. *Ann. N.Y. Acad. Sci.* 573:192–205.

Structure of citric acid cycle enzymes

Remington, S.J., 1992. Structure and mechanism of citrate synthase. *Curr. Top. Cell. Regul.* 33:209–229.

Karpusas, M., Branchaud, B., and Remington, S.J., 1990. Proposed mechanism for the condensation reaction of citrate synthase: 1.9-Å structure of the ternary complex with oxaloacetate and carboxymethyl coenzyme A. *Biochemistry* 29:2213–2219.

Lauble, H., Kennedy, M.C., Beinert, H., and Stout, C.D., 1992. Crystal structures of aconitase with isocitrate and nitroisocitrate bound. *Biochemistry* 31:2735–2748.

Barnes, S.J., and Weitzman, P.D., 1986. Organization of citric acid cycle enzymes into a multienzyme cluster. *FEBS Lett.* 201:267–270. [Fumarase, malate dehydrogenase, citrate synthase, aconitase, and isocitrate dehydrogenase are loosely associated in a multienzyme cluster.]

Singer, T.P., and Johnson, M.K., 1985. The prosthetic groups of succinate dehydrogenase: 30 years from discovery to identification. *FEBS Lett.* 190:189–198.

Srere, P.A., 1992. The molecular physiology of citrate. *Curr. Top. Cell. Regul.* 33:261–275.

Regulation

Reed, L.J., Damuni, Z., and Merryfield, M.L., 1985. Regulation of mammalian pyruvate and branched-chain α-keto acid dehydrogenase complexes by phosphorylation-dephosphorylation. *Curr. Top. Cell Regul.* 27:41–49.

Williamson, J.R., and Cooper, R.H., 1980. Regulation of the citric acid cycle in mammalian systems. *FEBS Lett.* 117(Suppl.):K73–K85.

Randle, P.J., 1978. Pyruvate dehydrogenase complex: Meticulous regulator of glucose disposal in animals. *Trends Biochem. Sci.* 3:217–219.

Hurley, J.H., Dean, A.M., Sohl, J.L., Koshland, D.J., and Stroud, R.M., 1990. Regulation of an enzyme by phosphorylation at the active site. *Science* 249:1012–1016.

Evolutionary aspects

Gest, H., 1987. Evolutionary roots of the citric acid cycle in prokaryotes. *Biochem. Soc. Symp.* 54:3–16.

Weitzman, P.D.J., 1981. Unity and diversity in some bacterial citric acid cycle enzymes. *Advan. Microbiol. Physiol.* 22:185–244.

Discovery of the citric acid cycle

Krebs, H.A., and Johnson, W.A., 1937. The role of citric acid in intermediate metabolism in animal tissues. *Enzymologia* 4:148–156.

Krebs, H.A., 1970. The history of the tricarboxylic acid cycle. *Perspect. Biol. Med.* 14:154–170.

Krebs, H.A., and Martin, A., 1981. *Reminiscences and Reflections.* Clarendon Press.

PROBLEMS

1. *Flow of carbon atoms.* What is the fate of the radioactive label when each of the following compounds is added to a cell extract containing the enzymes and cofactors of the glycolytic pathway, the citric acid cycle, and the pyruvate dehydrogenase complex? (The ^{14}C label is printed in red.)

 (a) $H_3C-\overset{\overset{O}{\|}}{C}-COO^-$ (c) $H_3C-\overset{\overset{O}{\|}}{C}-COO^-$

 (b) $H_3C-\underset{\underset{O}{\|}}{C}-COO^-$ (d) $H_3C-\underset{\underset{O}{\|}}{C}-S-CoA$

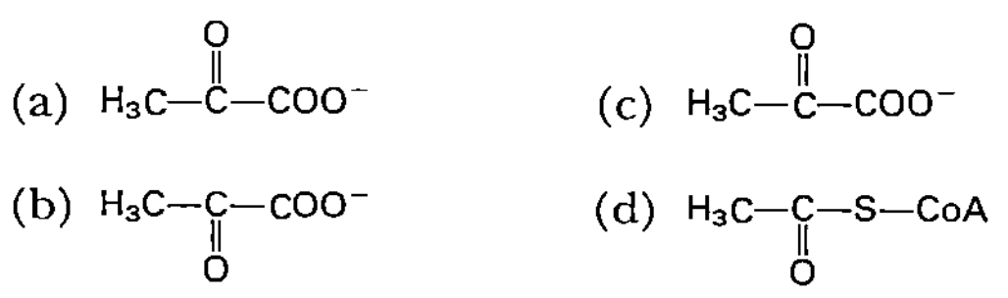

 (e) Glucose 6-phosphate labeled at C-1.

2. $C_2 + C_2 \rightarrow C_4$.
 (a) Which enzymes are required to get *net synthesis* of oxaloacetate from acetyl CoA?
 (b) Write a balanced equation for the net synthesis.
 (c) Do mammalian cells contain the requisite enzymes?

3. *Driving force.* What is the $\Delta G^{\circ\prime}$ for the complete oxidation of the acetyl unit of acetyl CoA by the citric acid cycle?

4. *Probing stereospecificity.* A sample of deuterated reduced NAD was prepared by incubating H_3C-CD_2-OH and NAD^+ with alcohol dehydrogenase. This reduced coenzyme was added to a solution of 1,3-BPG and glyceraldehyde 3-phosphate dehydrogenase. The NAD^+ formed by this second reaction contained one atom of deuterium, whereas glyceraldehyde 3-phosphate, the other product, contained none. What does this experiment reveal about the stereospecificity of glyceraldehyde 3-phosphate dehydrogenase?

5. *A potent inhibitor.* Thiamine thiazolone pyrophosphate binds to pyruvate dehydrogenase about 20,000 times as strongly as does thiamine pyrophosphate, and it competitively inhibits the enzyme. Why?

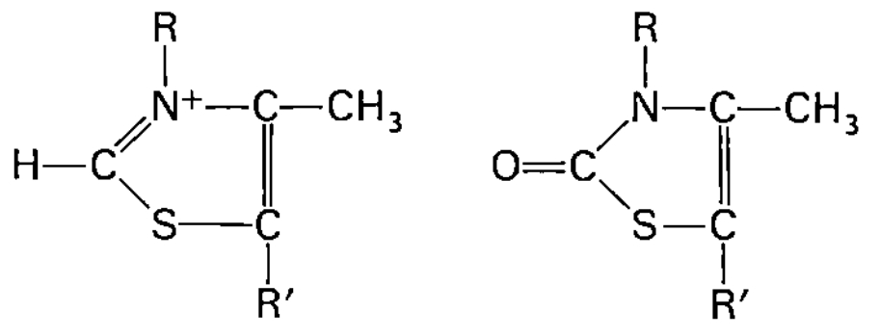

TPP Thiazolone analog
 of TPP

6. *Making oxaloacetate.* The oxidation of malate by NAD^+ to form oxaloacetate is a highly endergonic reaction under standard conditions $(\Delta G^{\circ\prime} = +7 \text{ kcal/mol})$. The reaction proceeds readily under physiologic conditions.
 (a) Why?
 (b) Assuming an $[NAD^+]/[NADH]$ ratio of 8 and a pH of 7, what is the lowest [malate]/[oxaloacetate] ratio at which oxaloacetate can be formed from malate?

7. *Theme and variation.* Propose a reaction mechanism for the condensation of acetyl CoA and glyoxylate in the glyoxylate cycle of plants and bacteria.

8. *Phosphatase action.* Hydrolysis of the phosphoryl group attached to Ser 113 in the inactivated form of isocitrate dehydrogenase proceeds very slowly if the phosphatase is added without ADP. Propose a mechanism for phosphatase action that accounts for the accelerating effect of ADP. [Hint: recall that the kinase and phosphatase activities of the modifying enzyme are mediated by the same active site.]

9. *Déjà vu.* Which other regulatory enzyme in metabolism contains kinase and phosphatase activities on the same polypeptide chain? How does it differ from the one that acts on bacterial isocitrate dehydrogenase?

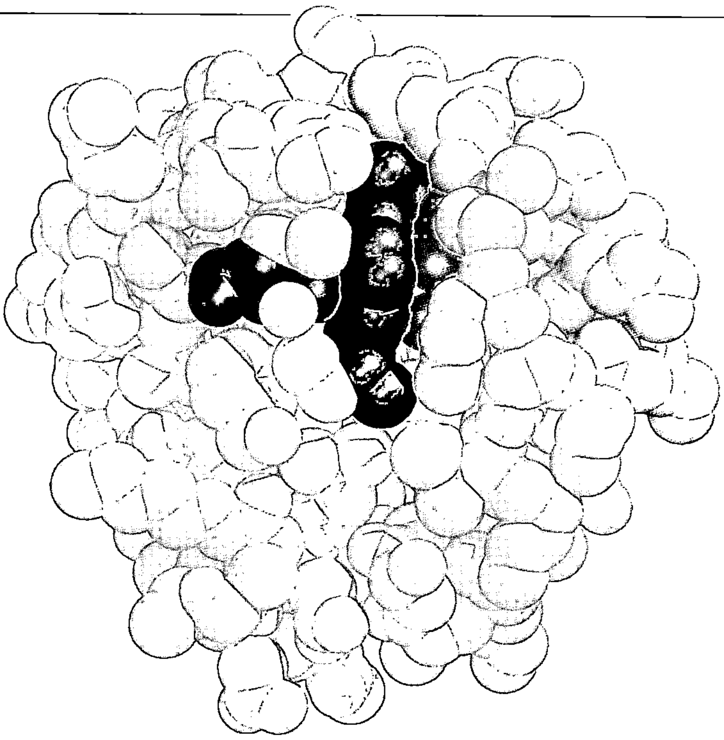

CHAPTER

21

Oxidative Phosphorylation

The NADH and $FADH_2$ formed in glycolysis, fatty acid oxidation, and the citric acid cycle are energy-rich molecules because each contains a pair of electrons having a high transfer potential. When these electrons are donated to molecular oxygen, a large amount of free energy is liberated, which can be used to generate ATP. *Oxidative phosphorylation is the process in which ATP is formed as a result of the transfer of electrons from NADH or* $FADH_2$ *to* O_2 *by a series of electron carriers.* This is the major source of ATP in aerobic organisms. For example, oxidative phosphorylation generates 26 of the 30 molecules of ATP that are formed when glucose is completely oxidized to CO_2 and H_2O.

Oxidative phosphorylation is conceptually simple and mechanistically complex. The flow of electrons from NADH or $FADH_2$ to O_2 through protein complexes located in the inner membrane of mitochondria leads to the pumping of protons out of the mitochondrial matrix. A proton-motive force is generated consisting of a pH gradient and a transmembrane electric potential. ATP is synthesized when protons flow back to the mitochondrial matrix through an enzyme complex. Thus, *oxidation and phosphorylation are coupled by a proton gradient across the inner mitochondrial membrane* (Figure 21-2).

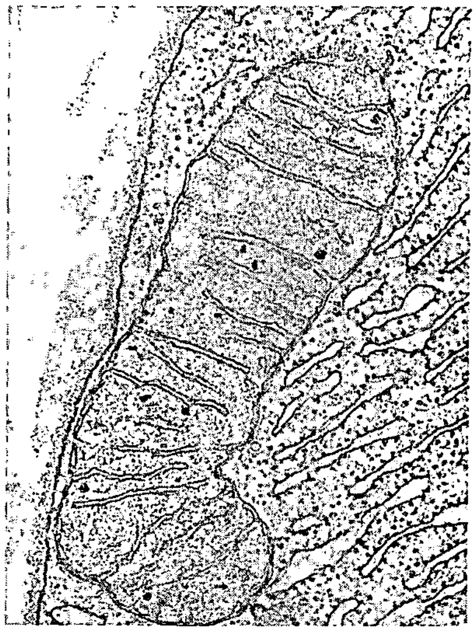

Figure 21-1
Electron micrograph of a mitochondrion. [Courtesy of Dr. George Palade.]

Opening Image: *Structure of cytochrome c, an electron carrier in the respiratory chain of mitochondria. The heme group is shown in red, a coordinated histidine in blue and methionine in green, and a highly conserved sequence in yellow. The conformation of cytochrome c has remained nearly constant over more than a billion years of evolution. [Drawn from 5cyt.pdb. T. Takano and R.E. Dickerson. J. Mol. Biol. 153(1981):79.]*

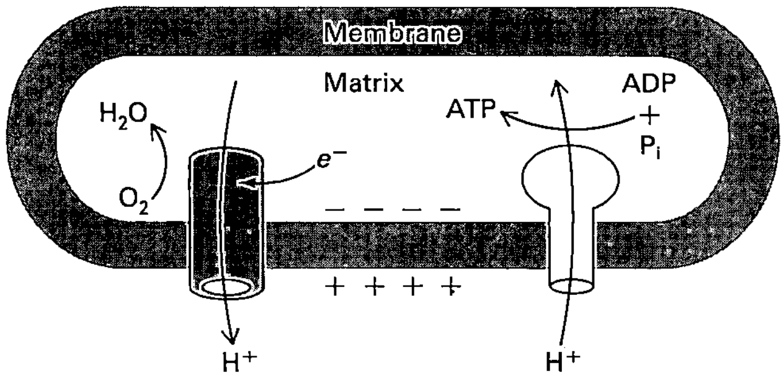

Figure 21-2
Essence of oxidative phosphorylation. Oxidation and ATP synthesis are coupled by transmembrane proton fluxes.

In essence, in oxidative phosphorylation, electron-motive force is converted into proton-motive force and then into phosphoryl potential. The first phase is carried out by three electron-driven proton pumps—*NADH-Q reductase, cytochrome reductase,* and *cytochrome oxidase.* These large transmembrane complexes contain multiple oxidation-reduction centers, such as flavins, quinones, iron-sulfur clusters, hemes, and copper ions. These three complexes are far more complex than bacteriorhodopsin, a light-driven proton pump discussed in an earlier chapter (p. 318). The second phase of oxidative phosphorylation is carried out by *ATP synthase,* an ATP-synthesizing assembly that is driven by the flow of protons back into the mitochondrial matrix. Oxidative phosphorylation vividly shows that *proton gradients are an interconvertible currency of free energy in biological systems.*

OXIDATIVE PHOSPHORYLATION IN EUKARYOTES OCCURS IN MITOCHONDRIA

> *Respiration*—
> An ATP-generating process in which an inorganic compound (such as O_2) serves as the ultimate electron acceptor. The electron donor can be either an organic compound or an inorganic one.

Mitochondria are oval-shaped organelles, typically about 2 μm in length and 0.5 μm in diameter. Eugene Kennedy and Albert Lehninger discovered a half century ago that *mitochondria contain the respiratory assembly, the enzymes of the citric acid cycle, and the enzymes of fatty acid oxidation.* Electron microscopic studies by George Palade and Fritjof Sjöstrand revealed that mitochondria have two membrane systems: an *outer membrane* and an extensive, highly folded *inner membrane.* The inner membrane is folded into a series of internal ridges called *cristae.* Hence, there are two compartments in mitochondria: the *intermembrane space* between the outer and inner membranes, and the *matrix,* which is bounded by the inner membrane (Figure 21-3). Oxidative phosphorylation takes place in the inner mitochondrial membrane, in contrast with most of the reactions of the citric acid cycle and fatty acid oxidation, which occur in the matrix.

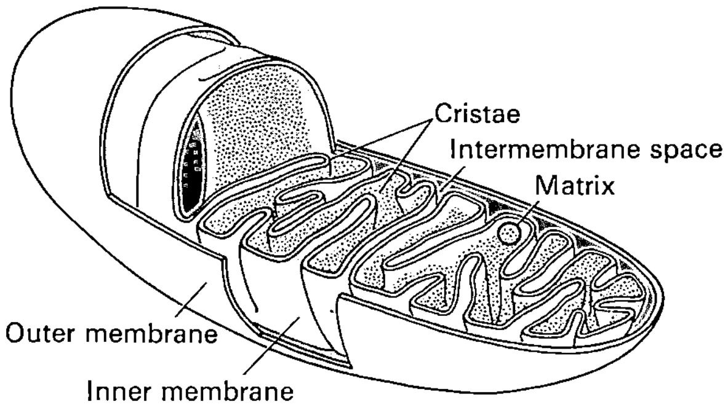

Figure 21-3
Diagram of a mitochondrion. [After *Biology of the Cell* by Stephen L. Wolfe. © 1972 by Wadsworth Publishing Company, Inc., Belmont, California 94002. Adapted by permission of the publisher.]

The outer membrane is quite permeable to most small molecules and ions because it contains many copies of *porin,* a transmembrane protein with a large pore. In contrast, the inner membrane is intrinsically impermeable to nearly all ions and polar molecules. A large family of transporters shuttles metabolites such as ATP and citrate across the inner mitochondrial membrane. The two faces of this membrane will be referred to as the *matrix side* and the *cytosolic side* (because it is freely accessible to most small molecules in the cytosol). They are also called the *N* and *P* sides because the membrane potential is negative on the matrix side and positive on the cytosolic side.

In prokaryotes, the electron-driven proton pumps and ATP-synthesizing complex are located in the cytoplasmic membrane, the inner of two membranes. The outer membrane of bacteria, like that of mitochondria, is permeable to most small metabolites because of the presence of porin.

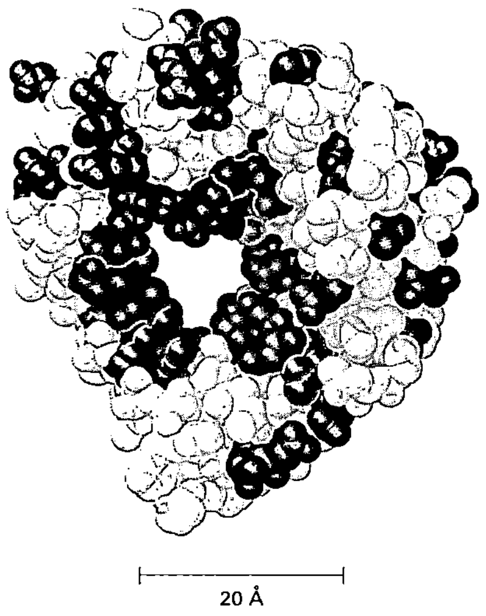

Porin forms large channels in the outer membrane of gram-negative bacteria, as it does in mitochondria. Small molecules (less than ~600 daltons) readily traverse these pores. Acidic residues are shown in red, basic ones in blue, and Ca^{2+} in green. [Drawn from 2por.pdb. M.S. Weiss, U. Abele, J. Weckesser, W. Welte, E. Schiltz, and G.E. Schultz. *Science* 254(1991):1627.]

REDOX POTENTIALS AND FREE-ENERGY CHANGES

In oxidative phosphorylation, the *electron transfer potential* of NADH or $FADH_2$ is converted into the *phosphoryl transfer potential (phosphoryl potential)* of ATP. We need quantitative expressions for these forms of free energy. The measure of phosphoryl transfer potential is already familiar to us: it is given by $\Delta G^{\circ\prime}$ for hydrolysis of the activated phosphate compound. The corresponding expression for the electron transfer potential is E_0', the *reduction potential* (also called the *redox potential* or *oxidation-reduction potential*).

The reduction potential is an electrochemical concept. Consider a substance that can exist in an oxidized form X and a reduced form X^-. Such a pair is called a *redox couple* (Figure 21-4). The reduction potential of this

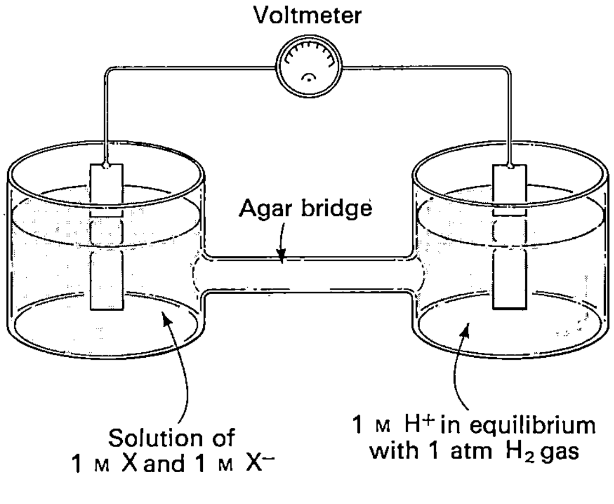

Figure 21-4
Apparatus for the measurement of the standard oxidation-reduction potential of a redox couple. Electrons, but not X or X^-, can flow through the agar bridge.

couple can be determined by measuring the electromotive force generated by a *sample half-cell* connected to a *standard reference half-cell.* The sample half-cell consists of an electrode immersed in a solution of 1 M oxidant (X) and 1 M reductant (X^-). The standard reference half-cell consists of an electrode immersed in a 1 M H^+ solution that is in equilibrium with H_2 gas at 1-atmosphere pressure. The electrodes are connected to a voltmeter, and electrical continuity between the half-cells is established by an agar bridge. Electrons then flow from one half-cell to the other. If the reaction proceeds in the direction

$$X^- + H^+ \longrightarrow X + \tfrac{1}{2} H_2$$

the reactions in the half-cells are

$$X^- \longrightarrow X + e^-$$

$$H^+ + e^- \longrightarrow \tfrac{1}{2} H_2$$

Thus, electrons flow from the sample half-cell to the standard reference half-cell, and consequently the sample-cell electrode is negative with respect to the standard-cell electrode. *The reduction potential of the $X{:}X^-$ couple is the observed voltage at the start of the experiment* (when X, X^-, and H^+ are 1 M). *The reduction potential of the $H^+{:}H_2$ couple is defined to be 0 V (volts).*

The meaning of the reduction potential is now evident. A negative reduction potential means that a substance has lower affinity for electrons than does H_2, as in the above example. A positive reduction potential means that a substance has higher affinity for electrons than does H_2. These comparisons refer to standard conditions, namely 1 M oxidant, 1 M reductant, 1 M H^+, and 1 atmosphere H_2. Thus, *a strong reducing agent (such as NADH) is poised to donate electrons and has a negative reduction potential, whereas a strong oxidizing agent (such as O_2) is ready to accept electrons and has a positive reduction potential.*

The reduction potentials of many biologically important redox couples are known (Table 21-1). This table is like those presented in chemistry texts except that a hydrogen ion concentration of 10^{-7} M (pH 7) instead of 1 M (pH 0) is the standard state adopted by biochemists. This difference is denoted by the prime in E_0'. Recall that the prime in $\Delta G^{\circ\prime}$ denotes a standard free-energy change at pH 7.

Table 21-1
Standard reduction potentials of some reactions

Oxidant	*Reductant*	n	E_0' *(V)*
Succinate + CO_2	α-Ketoglutarate	2	−0.67
Acetate	Acetaldehyde	2	−0.60
Ferredoxin (oxidized)	Ferredoxin (reduced)	1	−0.43
2 H^+	H_2	2	−0.42
NAD^+	NADH + H^+	2	−0.32
$NADP^+$	NADPH + H^+	2	−0.32
Lipoate (oxidized)	Lipoate (reduced)	2	−0.29
Glutathione (oxidized)	Glutathione (reduced)	2	−0.23
Acetaldehyde	Ethanol	2	−0.20
Pyruvate	Lactate	2	−0.19
Fumarate	Succinate	2	0.03
Cytochrome *b* (+3)	Cytochrome *b* (+2)	1	0.07
Dehydroascorbate	Ascorbate	2	0.08
Ubiquinone (oxidized)	Ubiquinone (reduced)	2	0.10
Cytochrome *c* (+3)	Cytochrome *c* (+2)	1	0.22
Fe (+3)	Fe (+2)	1	0.77
$\tfrac{1}{2} O_2$ + 2 H^+	H_2O	2	0.82

Note: E_0' is the standard oxidation-reduction potential (pH 7, 25°C) and *n* is the number of electrons transferred. E_0' refers to the partial reaction written as

$$\text{Oxidant} + e^- \longrightarrow \text{reductant}$$

The free-energy change of an oxidation-reduction reaction can be readily calculated from the difference in reduction potentials of the reactants. For example, consider the reduction of pyruvate by NADH.

(a) Pyruvate + NADH + H^+ $\rightleftharpoons$ lactate + NAD^+

The reduction potential of the NAD^+ : NADH couple is −0.32 V, whereas that of the pyruvate : lactate couple is −0.19 V. By convention, reduction potentials (as in Table 21-1) refer to partial reactions written as reductions: oxidant + e^- → reductant. Hence,

(b) Pyruvate + 2 H^+ + 2 e^- ⟶ lactate $E_0' = -0.19$ V
(c) NAD^+ + H^+ + 2 e^- ⟶ NADH $E_0' = -0.32$ V

To obtain (a) from (b) and (c), we need to reverse the direction of reaction (c) so that NADH appears on the left side of the arrow. In doing so, the sign of E_0' must be changed.

(b) Pyruvate + 2 H^+ + 2 e^- ⟶ lactate   −0.19 V
(d) NADH ⟶ NAD^+ + H^+ + 2 e^- +0.32 V

Reactions (b) and (d) can be added to get the desired reaction (a) and a $\Delta E_0'$ of +0.13 volt.

Now we can calculate the $\Delta G°'$ for the reduction of pyruvate by NADH. The standard free-energy change $\Delta G°'$ is related to the change in reduction potential $\Delta E_0'$ by

$$\Delta G°' = -nF\Delta E_0'$$

in which n is the number of electrons transferred, F is a proportionality constant called the *faraday* (23.06 kcal V^{-1} mol^{-1}), $\Delta E_0'$ is in volts, and $\Delta G°'$ is in kilocalories per mole. Note that $\Delta G°'$ is the amount of *free energy* that can be obtained per mole from a transformation, whereas $\Delta E_0'$ is the difference in *potential* between two states. Hence, $\Delta E_0'$ must be multiplied by nF to determine the free-energy yield. For the reduction of pyruvate, $n = 2$, and so

$$\begin{aligned}\Delta G°' &= -2 \times 23.06 \times 0.13 \\ &= -6 \text{ kcal/mol}\end{aligned}$$

A *positive* $\Delta E_0'$ (but a *negative* $\Delta G°'$) signifies an *exergonic reaction* under standard conditions.

A 1.14-VOLT POTENTIAL DIFFERENCE BETWEEN NADH AND O_2 DRIVES ELECTRON TRANSPORT THROUGH THE CHAIN

The driving force of oxidative phosphorylation is the electron transfer potential of NADH or $FADH_2$ relative to that of O_2. Let us calculate the $\Delta E_0'$ and $\Delta G°'$ of the oxidation of NADH by O_2. The pertinent partial reactions are

(a) $\frac{1}{2}$ O_2 + 2 H^+ + 2 e^- ⟶ H_2O $E_0' = +0.82$ V

(b) NAD^+ + H^+ + 2 e^- ⟶ NADH $E_0' = -0.32$ V

Subtracting reaction (b) from reaction (a) yields

(c) $\frac{1}{2}$ O_2 + NADH + H^+ $\rightleftharpoons$ H_2O + NAD^+ $\Delta E_0' = +1.14$ V

The free energy of oxidation of this reaction is then given by

$$\begin{aligned}\Delta G°' &= -nF\Delta E_0' = -2 \times 23.06 \times 1.14 \\ &= -52.6 \text{ kcal/mol}\end{aligned}$$

Table 21-2
Components of the mitochondrial electron transport chain

Enzyme complex	*Mass (kd)*	*Subunits*	*Prosthetic group*	*Oxidant or reductant*		
				Matrix side	*Hydrocarbon core*	*Cytosolic side*
NADH-Q reductase	880	≥34	FMN Fe-S	NADH	Q	
Succinate-Q reductase	140	4	FAD Fe-S	Succinate	Q	
Cytochrome reductase	250	10	Heme *b*-562 Heme *b*-566 Heme c_1 Fe-S		Q	Cyt *c*
Cytochrome oxidase	160	10	Heme *a* Heme a_3 Cu_A and Cu_B			Cyt *c*

Sources: J.W. DePierre and L. Ernster, *Ann. Rev. Biochem.* 46(1977):215, Y. Hatefi, *Ann. Rev. Biochem.* 54(1985):1015, and J.E. Walker, *Quart. Rev. Biophys.* 25(1992):253.

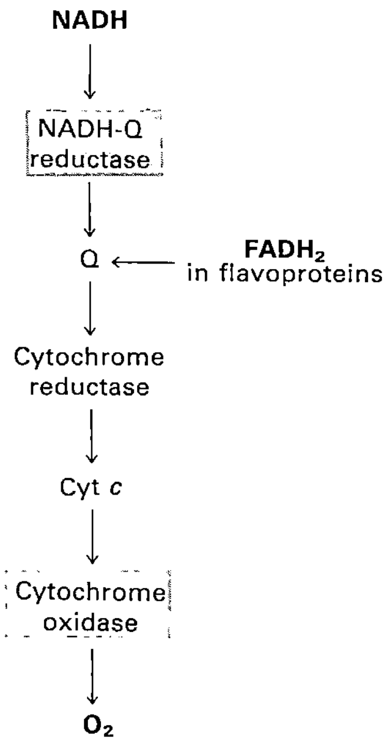

Figure 21-5
Sequence of electron carriers in the respiratory chain. Protons are pumped by the three complexes shown in color.

THE RESPIRATORY CHAIN CONSISTS OF THREE PROTON PUMPS LINKED BY TWO MOBILE ELECTRON CARRIERS

Electrons are transferred from NADH to O_2 through a chain of three large protein complexes called *NADH-Q reductase, cytochrome reductase,* and *cytochrome oxidase* (Figure 21-5 and Table 21-2). Electron flow within these transmembrane complexes leads to the pumping of protons across the inner mitochondrial membrane. As will be discussed shortly, the electron-carrying groups in these enzymes are *flavins, iron-sulfur clusters, hemes,* and *copper ions.* Electrons are carried from NADH-Q reductase to cytochrome reductase, the second complex of the chain, by the reduced form of *ubiquinone (Q),* a hydrophobic quinone that diffuses rapidly within the inner mitochondrial membrane. Ubiquinone also carries electrons from $FADH_2$ (produced, for example, by the oxidation of succinate in the citric acid cycle) to cytochrome reductase. Cytochrome *c,* a small protein, shuttles electrons from cytochrome reductase to cytochrome oxidase, the final component in the chain. NADH-Q reductase, succinate-Q reductase (p. 537), cytochrome reductase, and cytochrome oxidase are also called *Complex I, II, III,* and *IV,* respectively. Succinate-Q reductase (II), in contrast with the other complexes, does not pump protons.

THE HIGH-POTENTIAL ELECTRONS OF NADH ENTER THE RESPIRATORY CHAIN AT NADH-Q REDUCTASE

The electrons of NADH enter the chain at *NADH-Q reductase* (also called *NADH dehydrogenase* or *Complex I*), a large enzyme (880 kd) consisting of at least 34 polypeptide chains. This proton pump, like the other two in the respiratory chain, is encoded by two genomes, nuclear and mitochondrial. The initial step is the binding of NADH and the transfer of its two high-potential electrons to the *flavin mononucleotide (FMN)* prosthetic group of this complex to give the reduced form, $FMNH_2$.

$$NADH + H^+ + FMN \longrightarrow FMNH_2 + NAD^+$$

FMN can also accept one electron (or $FMNH_2$ can donate one electron) by forming a semiquinone radical intermediate (Figure 21-6).

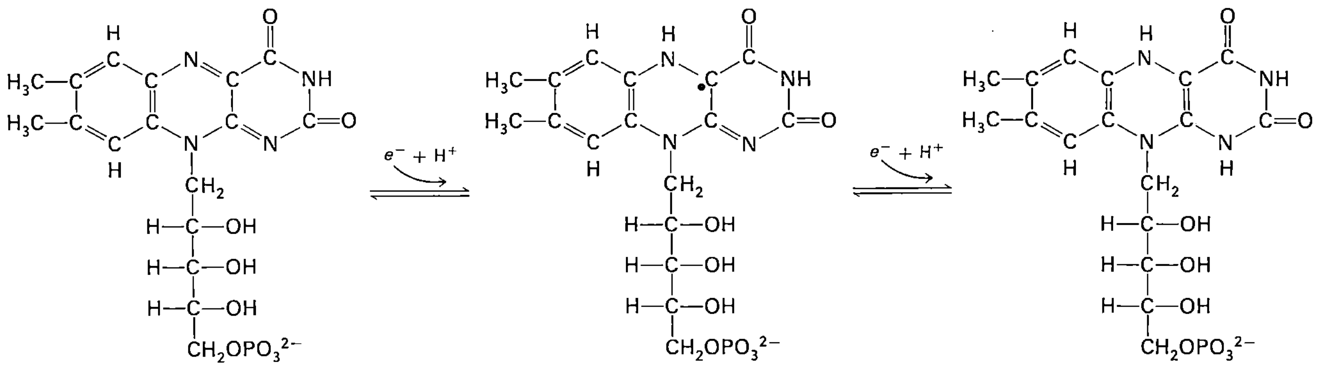

Flavin mononucleotide (FMN) · **Semiquinone intermediate** · **Reduced flavin mononucleotide ($FMNH_2$)**

Figure 21-6
The reduction of flavin mononucleotide (FMN) to $FMNH_2$ proceeds through a semiquinone intermediate.

Electrons are then transferred from $FMNH_2$ to a series of *iron-sulfur clusters* (abbreviated as Fe-S), the second type of prosthetic group in NADH-Q reductase. Fe-S clusters in *iron-sulfur proteins* (also called *nonheme iron proteins*) play a critical role in a wide range of reduction reactions in biological systems. Several types of Fe-S clusters are known (Figure 21-7). In the simplest kind, a single iron atom is tetrahedrally coordinated to the sulfhydryl groups of four cysteine residues of the protein. A second kind, denoted by [2Fe-2S], contains two iron atoms and two inorganic sulfides, in addition to four cysteine residues. A third type, designated as [4Fe-4S], contains four iron atoms, four inorganic sulfides, and four cysteine residues. NADH-Q reductase contains both [2Fe-2S] and [4Fe-4S] clusters. Iron atoms in these Fe-S complexes cycle between Fe^{2+} (reduced) or Fe^{3+} (oxidized) states.

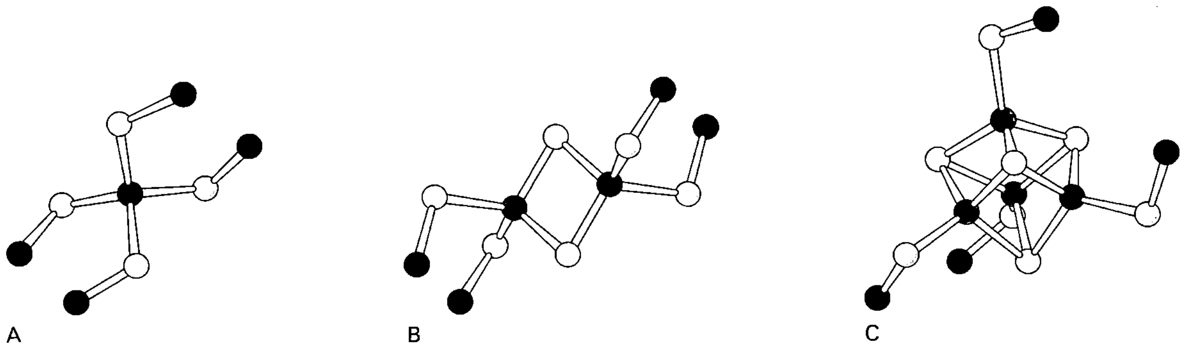

Figure 21-7
Molecular models of iron-sulfur complexes: (A) cluster containing one Fe; (B) [2Fe-2S] cluster; (C) [4Fe-4S] cluster. Iron atoms are shown in red, cysteine sulfur atoms in yellow, and inorganic sulfur atoms in green. [Drawn from atomic coordinates of model compounds kindly provided by Dr. Jeremy Berg.]

Electrons in the iron-sulfur clusters of NADH-Q reductase are then shuttled to *coenzyme Q*, also known as *ubiquinone (Q)* because it is ubiquitous in biological systems. Q is a quinone derivative with a long isoprenoid tail. The number of five-carbon isoprene units in Q depends on the species. The most common form in mammals contains 10 isoprene units

Oxidized form of coenzyme Q
(Q, ubiquinone)

Semiquinone
intermediate
(Q·⁻)

Reduced form of coenzyme Q
(QH₂, ubiquinol)

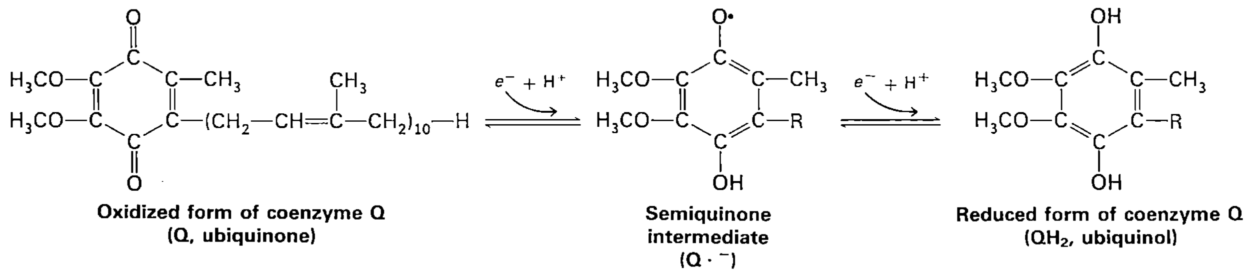

Figure 21-8
The reduction of ubiquinone (Q) to ubiquinol (QH₂) proceeds through a semi-quinone anion intermediate (Q·⁻).

(Q₁₀) (Figure 21-8). For simplicity, we will omit the subscript from this abbreviation. Ubiquinone is reduced to a *free-radical semiquinone anion* by the uptake of a single electron. Reduction of this enzyme-bound intermediate by a second electron yields *ubiquinol (QH₂)*.

$$
\begin{array}{c}
\text{NADH} \\
\text{NAD}^+
\end{array}
\begin{array}{c}
\text{FMN} \\
\text{FMNH}_2
\end{array}
\begin{array}{c}
\text{Reduced Fe-S} \\
\text{Oxidized Fe-S}
\end{array}
\begin{array}{c}
\text{Q} \\
\text{QH}_2
\end{array}
$$

NADH-Q reductase

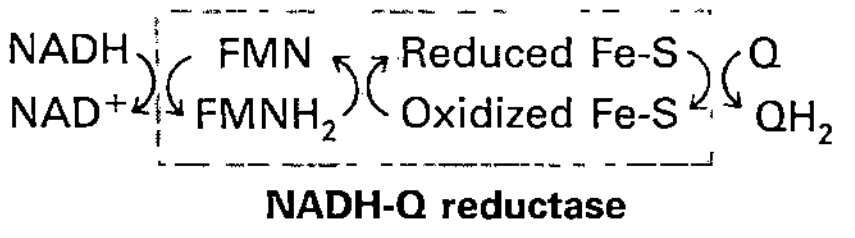

The flow of two electrons from NADH to QH₂ through NADH-Q reductase leads to the pumping of four H⁺ from the matrix to the cytosolic side of the inner mitochondrial membrane. The mechanism of conversion of electron potential into proton-motive force is not known. The complexity of NADH-Q reductase, which is larger than a ribosome, is daunting. A significant start has been made in elucidating the molecular architecture of this energy-transducing assembly. Electron microscopic studies of crystalline arrays have revealed that NADH-Q reductase is L-shaped, and biochemical studies have shown that the enzyme is bipartite (Figure 21-9). The hydrophobic horizontal arm, which is buried in the membrane, contains the subunits that are encoded by mitochondrial DNA. The vertical arm, which contains the peripheral membrane proteins of the complex, projects into the matrix. NADH binds to a site on the vertical arm, and transfers its electrons to FMN. They flow within the vertical unit to three [4Fe-4S] centers and then to a tightly bound Q. The pair of electrons on bound QH₂ are transferred to a [2Fe-2S] center, and finally, to a mobile Q in the hydrophobic core of the membrane. The challenge is to delineate the conformational changes induced by these electron transfers and learn how they are coupled to proton pumping.

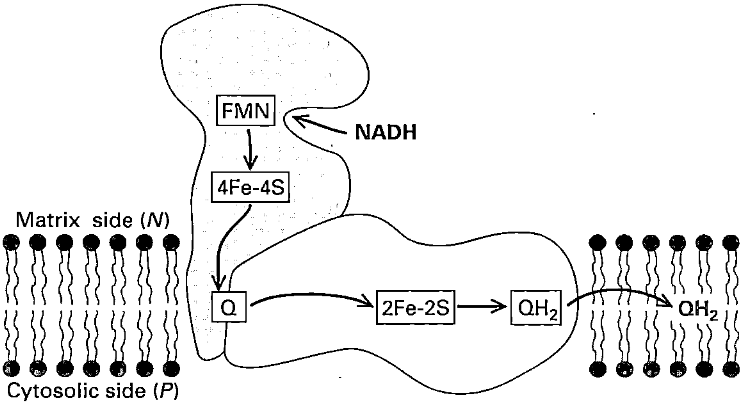

Figure 21-9
Model of NADH-Q reductase. The highly hydrophobic subunits encoded by the mitochondrial genome are located in the unit shown in yellow. The peripheral membrane proteins of this large electron-driven proton pump are in the unit shown in blue. This schematic diagram shows the postulated sequence of electron carriers between NADH and QH₂. [After G. Hofhaus, H. Weiss, and K. Leonard. *J. Mol. Biol.* 221(1991):1027.]

MITOCHONDRIAL DISEASES ARE BEING DISCOVERED

Several mutations in NADH-Q reductase lead to *Leber's hereditary optic neuropathy,* a maternally inherited form of blindness that usually strikes in midlife. Some of these mutations impair NADH utilization, whereas others block electron transfer to Q. The reason for maternal non-Mendelian inheritance is that the affected subunits of the reductase are encoded by mitochondrial DNA. A human egg harbors several hundred thousand molecules of mitochondrial DNA, whereas a sperm contributes only a few hundred and thus has little effect on the genotype. This neuropathy is one of many *mitochondrial diseases* that are being uncovered. Organs that are highly dependent on oxidative phosphorylation, such as the nervous system and the heart, are most vulnerable to mutations of mitochondrial DNA. The accumulation of mitochondrial mutations over several decades may contribute to aging and degenerative disorders.

UBIQUINOL (QH_2) IS ALSO THE ENTRY POINT FOR ELECTRONS FROM $FADH_2$ OF FLAVOPROTEINS

Recall that $FADH_2$ is formed in the citric acid cycle in the oxidation of succinate to fumarate by succinate dehydrogenase (p. 512). This enzyme is part of the *succinate-Q reductase complex (Complex II),* an integral membrane protein of the inner mitochondrial membrane. $FADH_2$ does not leave the complex. Rather, its electrons are transferred to Fe-S centers and then to Q for entry into the electron transport chain. Likewise, the $FADH_2$ moieties of *glycerol phosphate dehydrogenase* (p. 549) and *fatty acyl CoA dehydrogenase* (p. 609) transfer their high-potential electrons to Q to form QH_2, the reduced state. The isoprenoid tail makes Q highly nonpolar, which enables it to diffuse rapidly in the hydrocarbon core of the inner mitochondrial membrane. Succinate-Q reductase complex and other enzymes that transfer electrons from $FADH_2$ to Q, in contrast with NADH-Q reductase, are not proton pumps because the free-energy change of the catalyzed reaction is too small. Consequently, less ATP is formed from the oxidation of $FADH_2$ than from NADH.

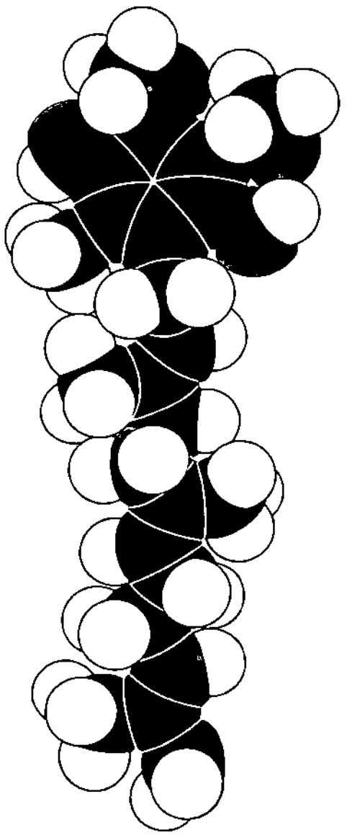

Figure 21-10
Model of ubiquinol. Only 3 of the 10 isoprenoid units are shown. These five-carbon units make the molecule very nonpolar.

ELECTRONS FLOW FROM UBIQUINOL TO CYTOCHROME *C* THROUGH CYTOCHROME REDUCTASE

The second of the three proton pumps in the respiratory chain is *cytochrome reductase* (also called *ubiquinol–cytochrome* c *reductase, cytochrome, cytochrome* bc$_1$ *complex,* or Complex III). *A cytochrome is an electron-transferring protein that contains a heme prosthetic group.* The key role of the brightly colored cytochromes in respiration was discovered in 1925 by David Keilin. Their iron atoms alternate between a reduced ferrous (+2) state and an oxidized ferric (+3) state during electron transport. The function of cytochrome reductase is to catalyze the transfer of electrons from QH_2 to cytochrome *c*, a water-soluble protein, and concomitantly pump protons across the inner mitochondrial membrane. The flow of a pair of electrons through this complex leads to the effective net transport of

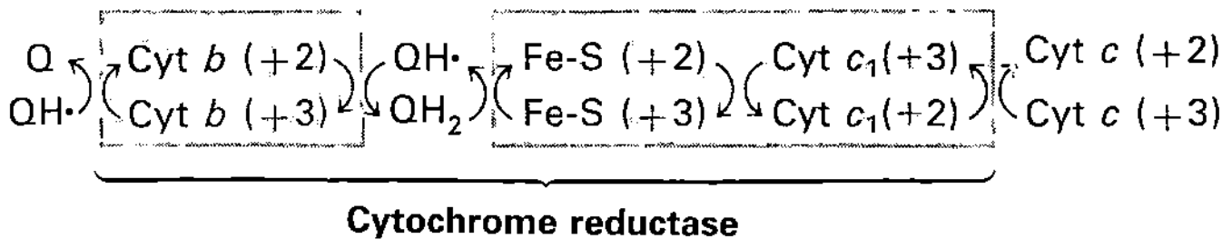

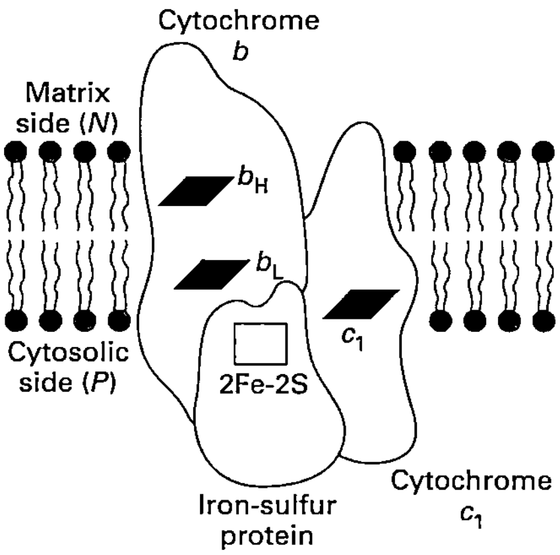

Figure 21-11
Model of a portion of cytochrome reductase. Four key redox centers are shown. Cytochrome b contains two hemes b_H and b_L. The iron-sulfur protein contains a 2Fe-2S center, and cytochrome c_1 contains a heme. Electrons from QH_2 enter the complex at a site between b_L and 2Fe-2S. They are finally transferred from the c_1 heme to cytochrome c, which approaches the complex from the cytosolic side. [After B.L. Trumpower. *J. Biol. Chem.* 265(1990):11409.]

2 H^+ to the cytosolic side, half the yield obtained with NADH-Q reductase because of a smaller thermodynamic driving force.

Cytochrome reductase itself contains two types of cytochromes, named b and c_1 (Figure 21-11). The reductase also contains an Fe-S protein and several other polypeptide chains. The prosthetic group of cytochromes b, c_1, and c is iron-protoporphyrin IX, the same *heme* as in myoglobin and hemoglobin (p. 148). The hemes of cytochromes c and c_1, in contrast with those in b, are covalently attached to the protein (Figure 21-12). The linkages are thioethers formed by the addition of the sulfhydryl groups of two cysteine residues to the vinyl groups of the heme.

A

$$R—CH{=}CH_2 \quad + \quad HS—CH_2—R' \quad \longrightarrow \quad R—\overset{\overset{\displaystyle CH_3}{|}}{C}H—S—CH_2—R'$$

Vinyl group of the heme Cysteine residue of the protein Thioether linkage

B

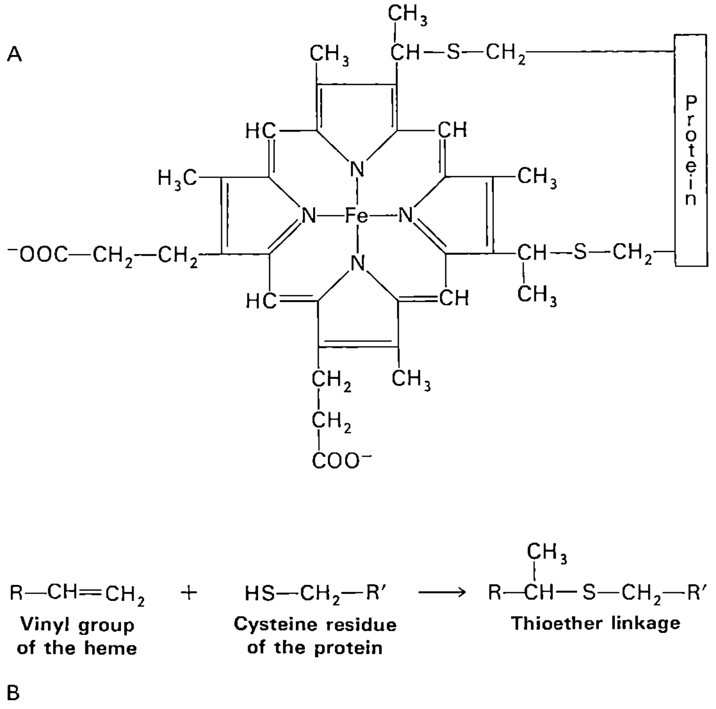

Figure 21-12
(A) The heme in cytochromes c and c_1 is covalently attached to two cysteine side chains by (B) thioether linkages.

Ubiquinol transfers one of its two high-potential electrons to the Fe-S cluster in the reductase. This electron is then shuttled sequentially to cytochrome c_1 and cytochrome c, which carries it away from this complex (Figure 21-13A). This one-electron transfer converts ubiquinol (QH_2) into the semiquinone anion ($Q \cdot \bar{}$). What about the electron residing in the semiquinone? This is where cytochrome b with its two heme groups enters the act. Its identical hemes have different electron affinities because they are in different polypeptide environments. Heme b_L, which is located near the cytosolic face of the membrane, has lower affinity for an electron than does heme b_H, which is near the matrix side. Hemes b_L and b_H are also known as b-*566* and b-*560* because they absorb light maximally at these wavelengths. $Q \cdot \bar{}$ rapidly transfers its electron to b_L to form Q, which is free to diffuse in the membrane. Heme b_L then reduces b_H, which in turn reduces a bound Q near the cytosolic side to form $Q \cdot \bar{}$.

At this stage, the task of the complex is only half complete because only one of the two electrons from QH_2 has been transferred to cytochrome c. The other electron is waiting in the wings, in the form of bound $Q \cdot \bar{}$. A

second molecule of QH_2 then reacts with the complex in the same way as the first. One of its electrons is transferred to the Fe-S cluster and then to cytochromes c_1 and c. The other electron goes from the newly formed $Q\cdot^-$ to b_L and then b_H. However, this time, b_H reduces bound $Q\cdot^-$ rather than Q, to complete the Q cycle (Figure 21-13B). Thus, two QH_2 are oxidized to form two Q, and one Q is reduced to QH_2, during a complete cycle. Why this complexity? The formidable problem solved here is to efficiently funnel electrons from a two-electron carrier (QH_2) to a one-electron carrier (cytochrome c). *The cytochrome* b *component of the reductase is in essence a recyling device that enables both electrons of QH_2 to be effectively used.*

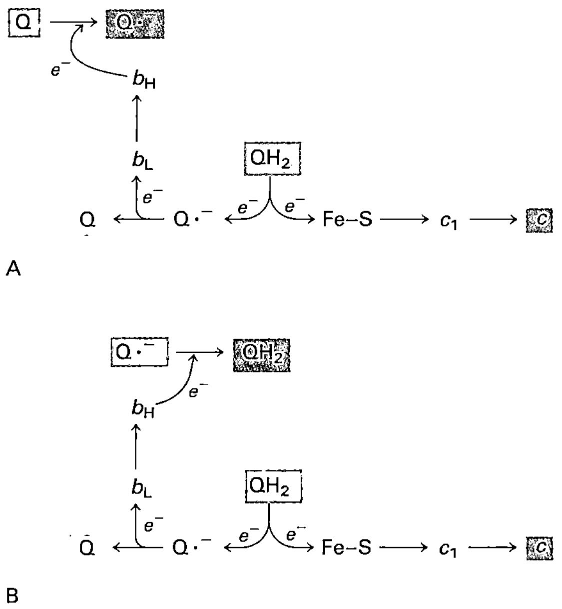

Figure 21-13
Pathway of electron transfer in cytochrome reductase. The input quinones are shaded green, and the reduced species are shaded yellow and red. (A) In the first stage, one electron from QH_2 reduces cytochrome c, and the other reduces a tightly bound Q to $Q\cdot^-$. (B) In the second stage, one electron from another QH_2 reduces a second molecule of cytochrome c. The other electron reduces tightly bound $Q\cdot^-$ to QH_2. The net result of the Q cycle is the transfer of a pair of electrons from QH_2 to two molecules of cytochrome c.

CYTOCHROME OXIDASE CATALYZES THE TRANSFER OF ELECTRONS FROM CYTOCHROME *c* TO O_2

Cytochrome oxidase, the last of the three proton-pumping assemblies of the respiratory chain, catalyzes the transfer of electrons from ferrocytochrome c (the reduced form) to molecular oxygen, the final acceptor.

$$4\,\text{Cyt}\ c\,(+2) + 4\,H^+ + O_2 \longrightarrow 4\,\text{cyt}\ c\,(+3) + 2\,H_2O$$

Four electrons are funneled into O_2 to completely reduce it to H_2O and concomitantly pump protons from the matrix to the cytosolic side of the inner mitochondrial membrane.

This reaction is carried out by a complex of 10 subunits, of which 3 (called *subunits I, II,* and *III*) are encoded by the mitochondrion's own genome. Cytochrome oxidase contains two *heme A* groups and two *copper ions.* Heme A differs from the heme in cytochrome c and c_1 in several ways: (1) a formyl group replaces a methyl group, (2) a C_{15} hydrocarbon chain replaces one of the vinyl groups, and (3) the heme is not covalently attached to the protein (Figure 21-14). The two heme A molecules, though chemically identical, have different properties because they are located in different parts of cytochrome oxidase. One of them is called *heme* a and the other *heme* a_3. Likewise, the two copper ions, called Cu_A and Cu_B, are distinct because they are bound differently by the protein.

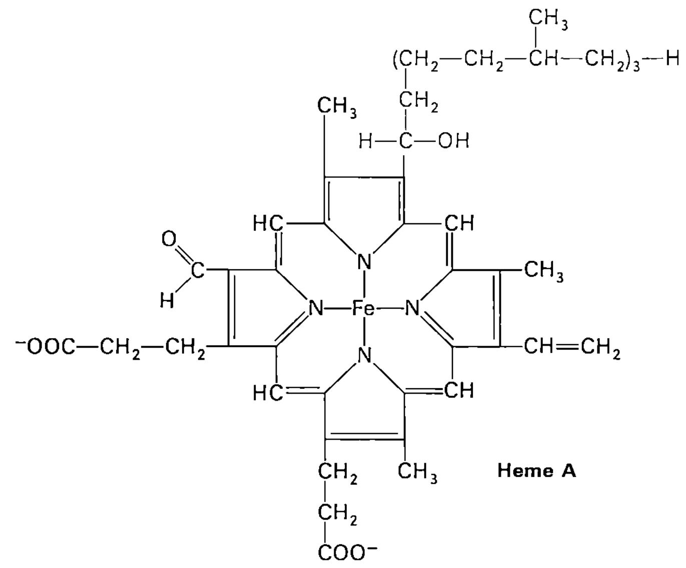

Figure 21-14
Heme A, which is present in cytochrome oxidase, differs from other hemes in having a formyl group (red) and a long hydrophobic side chain (green).

Heme *a* is close to Cu_A in subunit II, and heme a_3 is next to Cu_B in subunit I. The oxidation-reduction units of cytochrome oxidase known as *cytochromes* a and a_3 are located in subunits II and I, respectively.

Ferrocytochrome *c* donates its electron to the heme a–Cu_A cluster. An electron is then transferred to the heme a_3–Cu_B cluster, where O_2 is reduced in a series of steps to two molecules of H_2O. Molecular oxygen is an ideal terminal electron acceptor. Its high affinity for electrons provides a large thermodynamic driving force for oxidative phosphorylation. Moreover, O_2, in contrast with other strong electron acceptors (such as F_2), reacts very slowly unless activated by a catalyst. However, *danger lurks in the reduction of* O_2. The transfer of four electrons leads to safe products (two molecules of H_2O), but partial reduction generates hazardous compounds. In particular, *superoxide anion,* a potentially destructive compound (p. 533), is formed by the transfer of a single electron to O_2.

$$O_2 + e^- \rightleftharpoons O_2^- \quad \text{(Superoxide anion)}$$

The strategy for the safe reduction of O_2 is clear: *the catalyst must not release partly reduced intermediates.* Cytochrome oxidase meets this crucial criterion by binding O_2 between the Fe^{2+} and Cu^+ ions of its a_3–Cu_B center. The cycle (Figure 21-15) starts with the center fully oxidized: iron is in the +3 state and copper in the +2 state. The first electron relayed

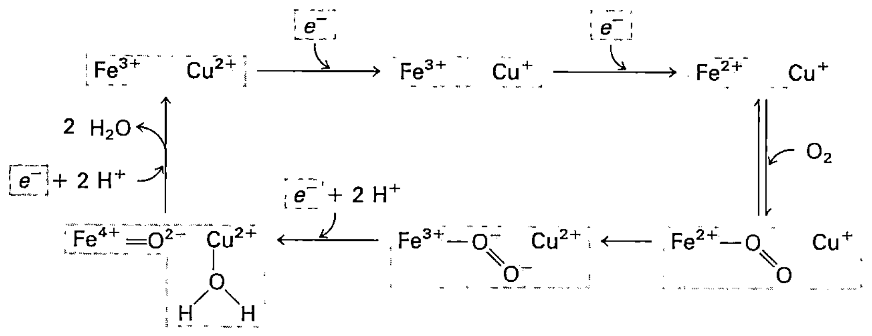

Figure 21-15
Proposed reaction cycle for the four-electron reduction of O_2 by cytochrome oxidase. [After M. Wikström and J.E. Morgan. *J. Biol. Chem.* 267(1992):10266.]

from ferrocytochrome *c* reduces the Cu^{2+}, and the second electron reduces the Fe^{3+}. The iron ion of this fully reduced center then binds molecular oxygen, which abstracts an electron from both ions to form a *peroxy* intermediate. The input of an electron and the uptake of two H^+ lead to cleavage of the bound peroxide. One oxygen atom is bound in the -2 state to iron in the *ferryl* $+4$ state, and the other oxygen is bound as water to Cu^{2+}. The input of the fourth electron and the uptake of two more H^+ result in the safe release of two molecules of H_2O and the regeneration of the oxidized bimetallic center. *Proton pumping occurs in the final two transitions, from the peroxy intermediate to water.* Four protons are translocated to the cytosolic side of the membrane when a pair of electrons flows through the oxidase.

ELECTROSTATIC INTERACTIONS ARE CRITICAL IN THE DOCKING OF CYTOCHROME *c* WITH ITS REACTION PARTNERS

Much more is known about the structure and interactions of cytochrome *c* than of any other electron-carrying protein. The solubility of this peripheral membrane protein in water has facilitated its purification and crystallization. Cytochrome *c* consists of a single polypeptide chain of 104 amino acid residues and a covalently attached heme group. The three-dimensional structures of the ferrous and ferric forms of cytochrome *c* from tuna have been elucidated at high resolution by Richard Dickerson (Figure 21-16). The protein is roughly spherical, with a diameter of 34 Å.

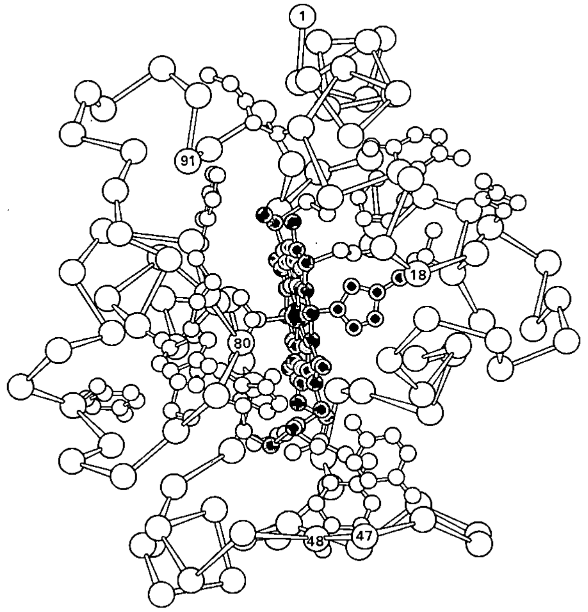

Figure 21-16
Three-dimensional structure of reduced cytochrome *c* from tuna. The heme group (red), methionine 80 (green), histidine 18 (blue), and the α carbon atoms are shown. [After T. Takano, O.B. Kallai, R. Swanson, and R.E. Dickerson. *J. Biol. Chem.* 248(1973):5244.]

The heme group is surrounded by many tightly packed hydrophobic side chains. The iron atom is bonded to the sulfur atom of a methionine residue and to the nitrogen atom of a histidine residue (Figure 21-17). The hydrophobic character of the heme environment makes the reduction potential of cytochrome *c* more positive (corresponding to a higher electron affinity) than that of the same heme complex in an aqueous milieu. It is energetically more costly to remove an electron from the heme in cytochrome *c* than from a heme in water because the dielectric constant near the iron atom is lower in cytochrome *c*.

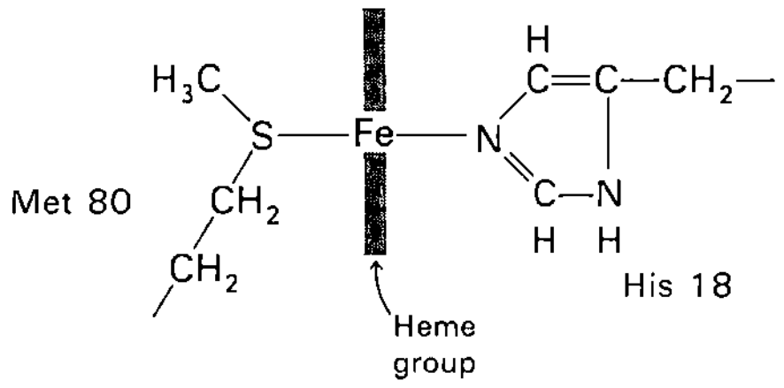

Figure 21-17
The iron atom of the heme group in cytochrome *c* is bonded to a methionine sulfur atom and a histidine nitrogen atom.

As was mentioned earlier, cytochrome *c* carries electrons from cytochrome reductase (the second proton-pumping complex) to cytochrome oxidase (the third). How does cytochrome *c* interact with its reductase and then with its oxidase? An important clue came from the distribution of charged residues on the surface of this highly basic protein. Cytochrome *c* molecules from all species studied thus far have clusters of lysine side chains around the heme crevice on one face of the protein. This array of positive charges on the surface of cytochrome *c* is central to the recognition and binding of the reductase and oxidase, which are negatively charged. For example, the interaction of cytochrome *c* with cytochrome oxidase is impaired if lysine 13 is modified. Furthermore, polylysine, a highly cationic synthetic polypeptide, competes with cytochrome *c* in binding to the oxidase and reductase. *Nearly every collision between cytochrome* c *and its reaction partners is productive because electrostatic forces bring complementary positively and negatively charged groups into apposition.*

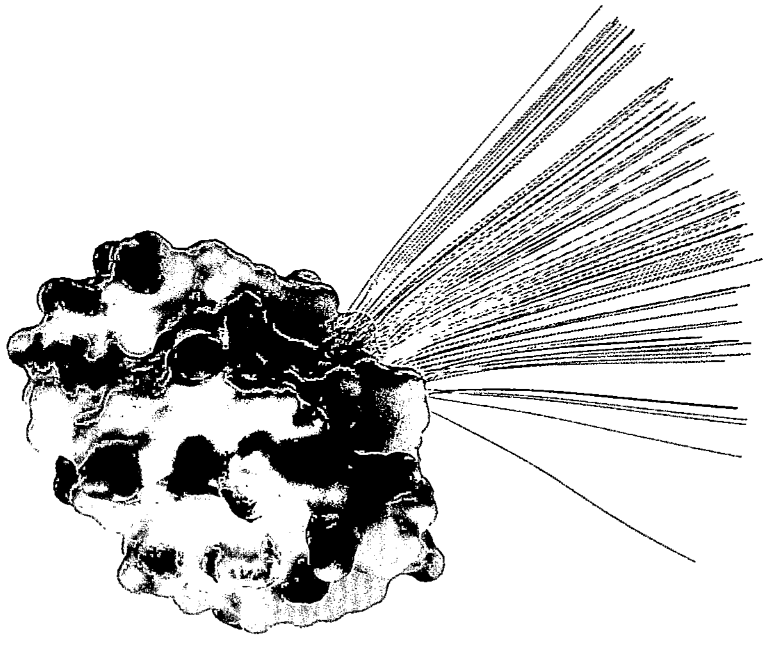

Figure 21-18
The surface of cytochrome *c* contains a positively charged cluster of residues (shown in blue) that is essential for the docking of its reaction partners. Only a small part of the surface of this highly basic protein is anionic (red). The electrostatic lines of force that attract its negatively charged partner are shown in green. [Drawn by Dr. Anthony Nicholls from 5cyt.pdb. T. Takano and R.E. Dickerson. *J. Mol. Biol.* 153(1981):79.]

ELECTRONS CAN BE TRANSFERRED BETWEEN GROUPS THAT ARE NOT IN CONTACT

How does cytochrome *c* accept an electron from its reductase, which it then donates to its oxidase? *The donor and acceptor heme groups need not be in contact. Electrons can be transferred between groups that are quite far apart.* For example, the rate of electron transfer in a particular model system decreased exponentially from about 10^{12} s^{-1} at van der Waals contact distance (~4 Å) to 1 s^{-1} at a separation of 25 Å. In general, the transfer rate depends on the change in redox potential, the distance between the donor and acceptor, and their relative orientation. Also, the intervening medium may play a role in guiding electron transfer.

The recently solved crystal structure of cytochrome *c* bound to cytochrome *c* peroxidase, a reaction partner, provides a direct view of a functional electron transfer complex. The hemes of these proteins are not touching—they are 19 Å apart, separated by a four-residue stretch of the main chain culminating in a tryptophan side chain (Figure 21-19). As anticipated, a positively charged cluster formed by several lysines of cytochrome *c* is important in binding the peroxidase.

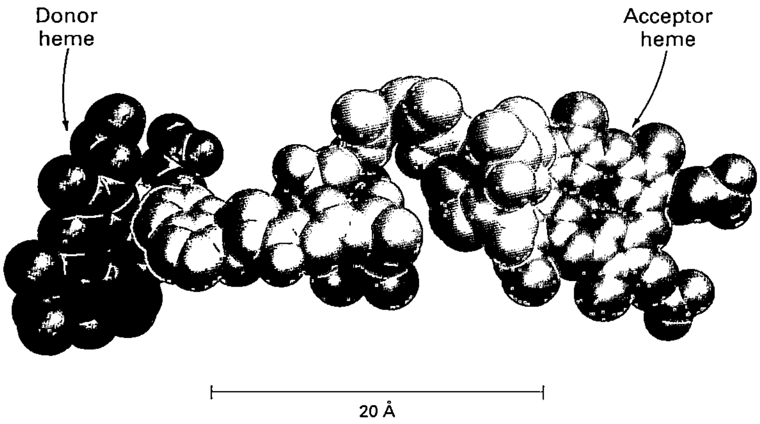

Figure 21-19
Space-filling model of the heme groups of a complex between cytochrome *c* (the electron donor) and cytochrome *c* peroxidase (the electron acceptor). The heme group of the donor is shown in red and that of the acceptor in purple. The intervening atoms of the donor are shown in blue and those of the acceptor in yellow. Note that the hemes are not in contact; the heme irons are 29.9 Å apart. [Drawn from 2pcc.pdb. H. Pelletier and J. Kraut. *Science* 258(1992):1748.]

THE CONFORMATION OF CYTOCHROME *c* HAS REMAINED ESSENTIALLY CONSTANT FOR MORE THAN A BILLION YEARS

Cytochrome *c* is present in all organisms having mitochondrial respiratory chains: plants, animals, and eukaryotic microorganisms. This electron carrier evolved more than 1.5 billion years ago, before the divergence of plants and animals. Its function has been conserved throughout this period, as evidenced by the fact that *the cytochrome* c *of any eukaryotic species reacts in vitro with the cytochrome oxidase of any other species tested thus far.* For example, wheat germ cytochrome *c* reacts with human cytochrome oxidase. A second criterion of the conservation of function is that the reduction potentials of all cytochrome *c* molecules studied are close to +0.25 V. Third, the absorption spectra of cytochrome *c* molecules of many species are virtually indistinguishable. Fourth, some prokaryotic cytochromes, such as cytochrome c_2 from a photosynthetic bacterium and cytochrome c_{550} from a denitrifying bacterium, closely resemble cytochrome *c* from tuna heart mitochondria (Figure 21-20).

The amino acid sequences of cytochrome *c* from more than 80 widely ranging eukaryotic species have been determined by Emil Smith, Emanuel Margoliash, and others. The striking finding is that *26 of 104 residues have been invariant for more than one and a half billion years of evolution.* The reasons for the constancy of many of these residues are evident now that the three-dimensional structure of the molecule is known. As might be expected, the heme ligands methionine 80 and histidine 18 are invariant, as are the two cysteines that are covalently bonded to the heme. A sequence of 11 residues from number 70 to number 80 is nearly the same in all cytochrome *c* molecules. Many hydrophobic residues in contact with the heme are invariant. Furthermore, most of the glycine residues in cytochrome *c* have been preserved. As discussed earlier (p. 32), glycine is important because it is small. The compact folding of the polypeptide chain requires glycine at certain sites. Several invariant lysine and arginine residues are located in the positively charged clusters on the surface of the molecule. One of these clusters interacts with cytochrome *c* reductase, and another with cytochrome oxidase.

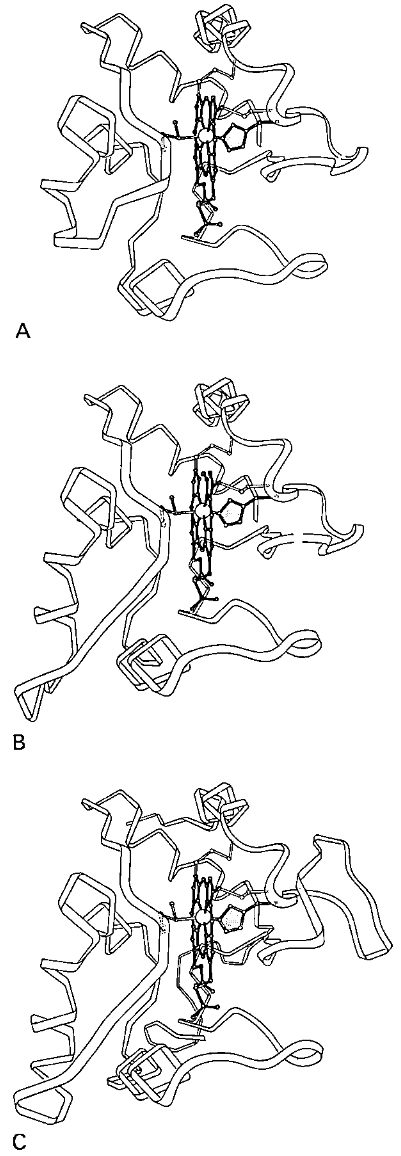

Figure 21-20
Conservation of the three-dimensional structure of cytochrome *c* in evolution is exemplified by the similarity in conformation of (A) cytochrome *c* from tuna heart mitochondria, (B) cytochrome c_2 from *Rhodospirillum rubrum*, a photosynthetic bacterium, and (C) cytochrome c_{550} from *Paracoccus denitrificans*, a denitrifying bacterium. [After F.R. Salamme. Reproduced, with permission, from the *Annual Review of Biochemistry*, Vol. 46. © 1977 by Annual Reviews, Inc.]

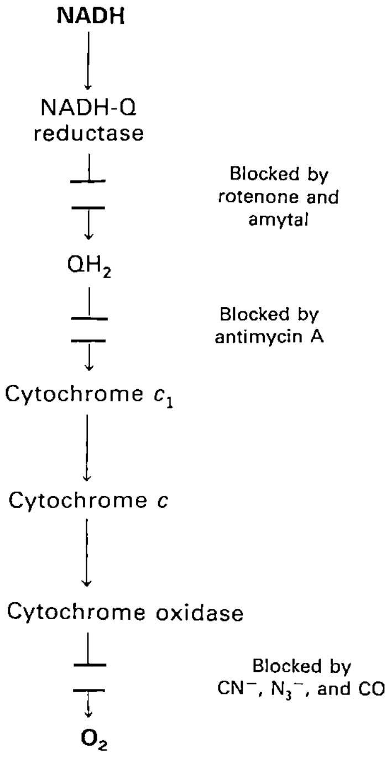

Figure 21-21
Sites of action of some inhibitors of electron transport.

ELECTRON TRANSFERS IN THE RESPIRATORY CHAIN CAN BE BLOCKED BY SPECIFIC INHIBITORS

Specific inhibitors of electron transport were invaluable in revealing the sequence of electron carriers in the respiratory chain. For example, *rotenone* and *amytal* block electron transfer in NADH-Q reductase and thereby prevent the utilization of NADH as a substrate (Figure 21-21). In contrast, electron flow resulting from the oxidation of succinate is unimpaired because these electrons enter through QH_2, beyond the block. *Antimycin A* interferes with electron flow from cytochrome b_H in cytochrome reductase. Furthermore, electron flow in cytochrome oxidase can be blocked by CN^-, N_3^-, and *CO*. Cyanide and azide react with the ferric form of heme a_3, whereas carbon monoxide inhibits the ferrous form.

OXIDATION AND PHOSPHORYLATION ARE COUPLED BY A PROTON-MOTIVE FORCE

Thus far, emphasis has been given to the flow of electrons from NADH to O_2, an exergonic process.

$$\text{NADH} + \tfrac{1}{2}\text{O}_2 + \text{H}^+ \rightleftharpoons \text{H}_2\text{O} + \text{NAD}^+ \qquad \Delta G^{\circ\prime} = -52.6 \text{ kcal/mol}$$

This free energy of oxidation is used to synthesize ATP, an endergonic process.

$$\text{ADP} + \text{P}_i + \text{H}^+ \rightleftharpoons \text{ATP} + \text{H}_2\text{O} \qquad \Delta G^{\circ\prime} = +7.3 \text{ kcal/mol}$$

The synthesis of ATP is carried out by a molecular assembly in the inner mitochondrial membrane. This enzyme complex (p. 546) has been called the *mitochondrial ATPase* or H^+*-ATPase* because it was discovered through its catalysis of the hydrolytic reaction. *ATP synthase,* its preferred name, emphasizes its actual role in the mitochondrion.

How is the oxidation of NADH coupled to the phosphorylation of ADP? It was first suggested that electron transfer leads to the formation of a covalent high-energy intermediate that serves as the precursor of ATP. This chemical-coupling hypothesis was based on the mechanism of ATP formation in glycolysis (e.g., the formation of 1,3-bisphosphoglycerate as a high-energy intermediate, p. 489). An alternative proposal was that the free energy of oxidation is trapped in an activated protein conformation, which then drives the synthesis of ATP. Investigators in many laboratories tried for several decades to isolate these putative energy-rich intermediates, but none were found.

A radically different mechanism, *the chemiosmotic hypothesis,* was postulated by Peter Mitchell in 1961. He proposed that electron transport and ATP synthesis are coupled by a proton gradient across the inner mitochondrial membrane rather than by a covalent high-energy intermediate or an activated protein conformation. In his model, the transfer of electrons through the respiratory chain leads to the pumping of protons from the matrix to the other side of the inner mitochondrial membrane. The H^+ concentration becomes higher on the cytosolic side, and an electrical potential with that side positive is generated (Figure 21-22). *Mitchell postulated that this proton-motive force drives the synthesis of ATP by the ATPase complex.* In essence, *the primary energy-conserving event induced by electron transport is the generation of a proton-motive force across the inner mitochondrial membrane.*

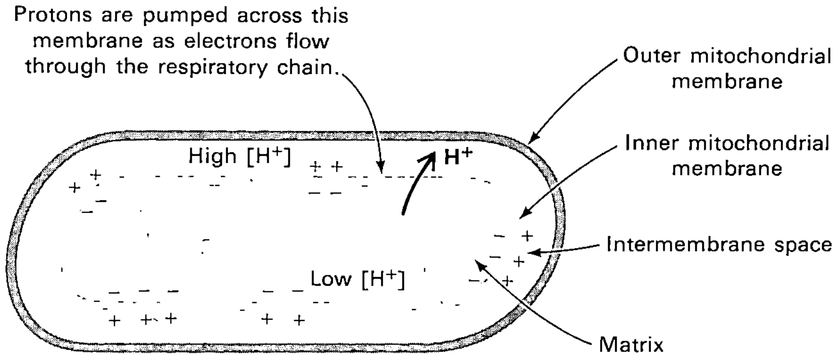

Figure 21-22
Electron transfer through the respiratory chain leads to the pumping of protons from the matrix to the cytosolic side of the inner mitochondrial membrane. The pH gradient and membrane potential constitute a proton-motive force that is used to drive ATP synthesis.

Mitchell's highly innovative hypothesis that oxidation and phosphorylation are coupled by a proton gradient is now supported by a wealth of evidence:

1. Electron transport generates a proton gradient across the inner mitochondrial membrane. The pH outside is 1.4 units lower than inside, and the membrane potential is 0.14 V, the outside being positive. The proton-motive force Δp (in volts) consists of a membrane-potential contribution (E_m) and a chemical gradient contribution (ΔpH). In the following equation, R is the gas constant, T is the absolute temperature, and F is the energy change as a mole of electrons passes through an electric potential of 1 volt.

$$\Delta p = E_m - \frac{2.3\ RT}{F}\Delta \mathrm{pH} = E_m - 0.06\ \Delta \mathrm{pH}$$

$$= 0.14 - 0.06(-1.4) = 0.224\ \mathrm{V}$$

This total proton-motive force of 0.224 V corresponds to a free energy of 5.2 kcal per mole of protons.

2. ATP is synthesized when a pH gradient is imposed on mitochondria or chloroplasts (p. 665) in the absence of electron transport.

3. Bacteriorhodopsin, a purple-membrane protein from halobacteria, pumps protons when illuminated (p. 318). Synthetic vesicles containing this bacterial protein and a purified ATPase from beef heart mitochondria synthesize ATP when illuminated (Figure 21-23). In this key experiment, carried out by Walther Stoeckenius and Efraim Racker, bacteriorhodopsin replaces the respiratory chain, which shows that the respiratory chain and ATP synthase are biochemically separate systems, linked only by a proton-motive force.

4. NADH-Q reductase, cytochrome reductase, and cytochrome oxidase pump protons out of the matrix. Their return drives ATP formation by ATP synthase.

5. A closed compartment is essential for oxidative phosphorylation. ATP synthesis coupled to electron transfer does not occur in soluble preparations or in membrane fragments lacking well-defined inside and outside compartments.

6. Substances that carry protons across the inner mitochondrial membrane dissipate the proton gradient and thereby uncouple oxidation and phosphorylation (p. 553).

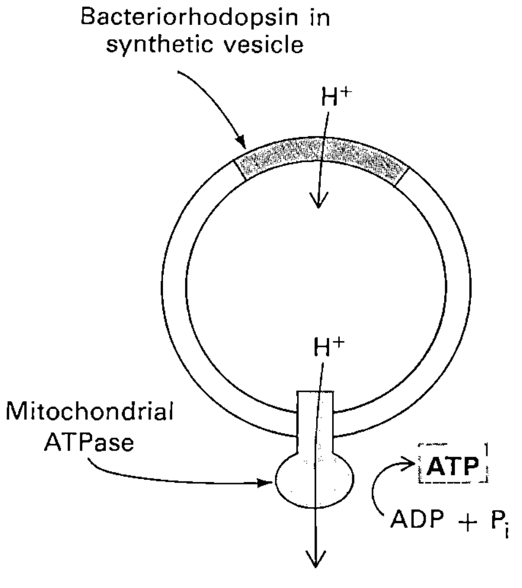

Figure 21-23
ATP is synthesized when reconstituted membrane vesicles containing bacteriorhodopsin (a light-driven proton pump) and ATP synthase are illuminated. The orientation of ATP synthase in this reconstituted membrane is the reverse of that in the mitochondrion.

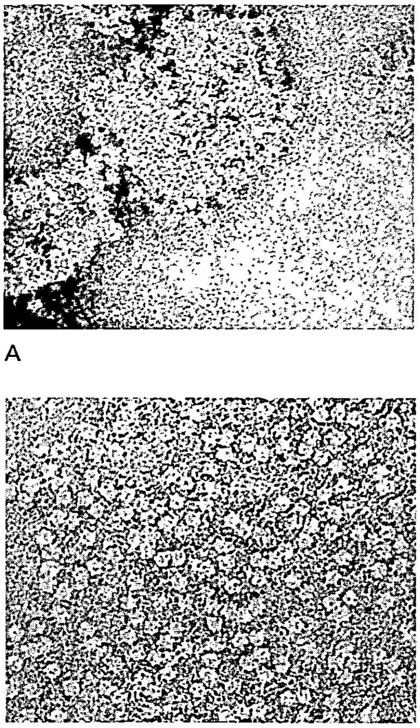

A

B

Figure 21-24
Electron micrographs of (A) a submitochondrial particle showing F_1 projections on its surface, and (B) isolated F_1 subunits. [Courtesy of Dr. Efraim Racker.]

ATP IS SYNTHESIZED BY AN ENZYME COMPLEX MADE OF A PROTON-CONDUCTING F_0 UNIT AND A CATALYTIC F_1 UNIT

We turn now to the utilization of the proton-motive force to synthesize ATP. *Submitochondrial particles* prepared by sonic disruption of the inner mitochondrial membrane have been valuable in studies of the mechanism of oxidative phosphorylation. *ATP synthase* is seen in electron micrographs of these inside-out vesicles as spherical projections from their outer surface (Figure 21-24). In intact mitochondria, these 85-Å-diameter projections are on the matrix side of the inner mitochondrial membrane. In 1960, Efraim Racker found that these knobs can be removed by mechanical agitation. The stripped submitochondrial particles can transfer electrons through their electron transport chain, but they can no longer synthesize ATP. In contrast, the separated 85-Å spheres do catalyze the hydrolysis of ATP. Most interesting, Racker found that the addition of these ATPase spheres to the stripped submitochondrial particles restored their capacity to synthesize ATP. These spheres are referred to as F_1. The normal role of the F_1 unit is to catalyze the synthesis of ATP. The ATPase activity exhibited by solubilized F_1 (in the absence of a proton gradient) is the reverse of its normal action. F_1 consists of five kinds of polypeptide chains with the stoichiometry $\alpha_3\beta_3\gamma\delta\varepsilon$. This assembly has a mass of 378 kd (Table 21-3).

The other major unit of ATP synthase is F_0, a hydrophobic segment that spans the inner mitochondrial membrane. *F_0 is the proton channel of the complex.* It consists of four kinds of polypeptide chains. The 8-kd chain, of which there are six per F_1, probably forms the transmembrane pore for protons. The stalk between F_0 and F_1 contains several other proteins (see Table 21-3). One of them renders the complex sensitive to *oligomycin,* an antibiotic that blocks ATP synthesis by interfering with the utilization of the proton gradient. *ATP synthase* is also known as *F_0F_1-ATPase.*

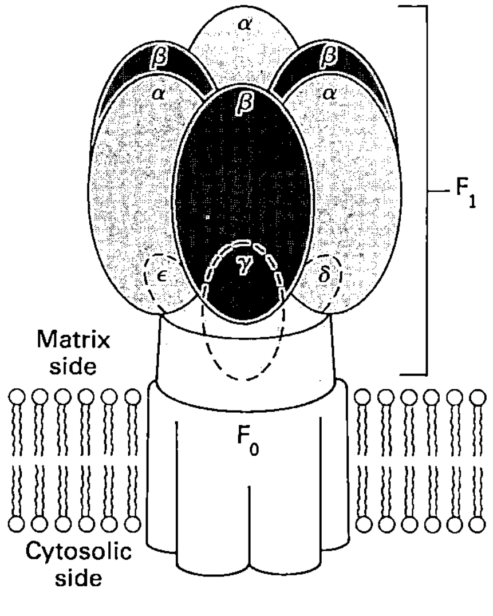

Figure 21-25
Schematic diagram of ATP synthase showing the proton-conducting F_0 unit (green) and the ATP-synthesizing F_1 unit (red). [After F.M. Harold. *The Vital Force: A Study of Bioenergetics* (W.H. Freeman, 1986), p. 238.]

Table 21-3
Components of the mitochondrial ATP-synthesizing complex

Subunits	*Mass (kd)*	*Role*	*Location*
F_1	378	Contains catalytic site for ATP synthesis	Spherical headpiece on matrix side
α	56		
β	52		
γ	34		
δ	14		
ε	6		
F_0	25	Contains proton channel	Transmembrane
	21		
	12		
	8		
F_1 inhibitor	10	Regulates proton flow and ATP synthesis	Stalk between F_0 and F_1
Oligomycin-sensitivity-conferring protein (OSCP)	23		
Fc_2 (F_6)	8		

Sources: J.W. DePierre and L. Ernster, *Ann. Rev. Biochem.* 46(1977):216, and Y. Kagawa, in L. Ernster (ed.), *Bioenergetics* (Elsevier, 1984), p. 151.

PROTON FLOW THROUGH ATP SYNTHASE LEADS TO THE RELEASE OF TIGHTLY BOUND ATP

ATP synthase catalyzes the formation of ATP from ADP and orthophosphate.

$$ADP^{3-} + P_i^{2-} + H^+ \rightleftharpoons ATP^{4-} + H_2O$$

The actual substrates are Mg^{2+} complexes of ADP and ATP, as in all known phosphoryl transfer reactions with these nucleotides. A terminal oxygen atom of ADP attacks the phosphorus atom of P_i to form a pentacovalent intermediate, which then dissociates into ATP and H_2O (Figure 21-26). The attacking oxygen atom of ADP and the departing oxygen atom of P_i occupy the apices of a trigonal bipyramid. This *in-line* geometry of attacking and leaving groups is also seen in the catalytic mechanism of ribonuclease A (p. 218).

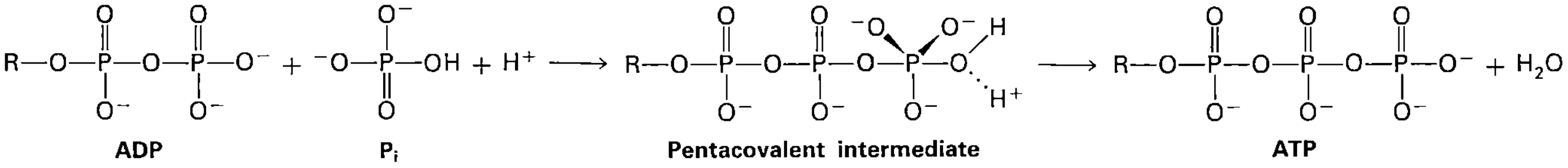

Figure 21-26
A pentacovalent phosphoryl intermediate is formed in the synthesis of ATP. The attacking oxygen atom of ADP and the leaving oxygen atom of P_i are at opposite apices of a trigonal bipyramid.

How does the flow of protons drive the synthesis of ATP? One possibility a priori is that the energized protons flowing through F_0 are funneled to the catalytic site on F_1, where they remove an oxygen from P_i to shift the equilibrium toward ATP synthesis. However, isotopic exchange experiments unexpectedly revealed that *enzyme-bound ATP forms readily in the absence of a proton-motive force.* When ADP and P_i were added to ATP synthase in $H_2{}^{18}O$, ^{18}O became incorporated into P_i through the synthesis of ATP and its subsequent hydrolysis (Figure 21-27).

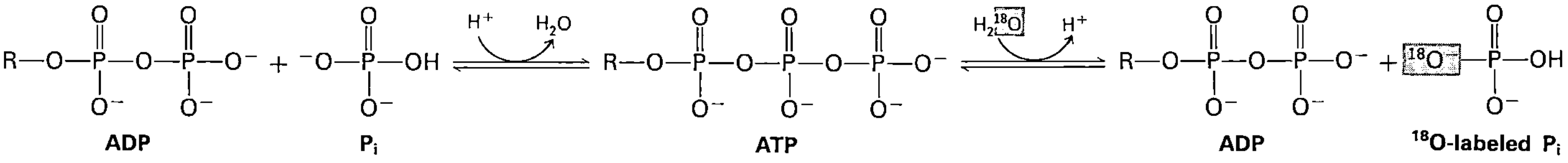

Figure 21-27
Isotope exchange experiment showing that enzyme-bound ATP is formed from ADP and P_i in the absence of a proton-motive force.

The rate of incorporation of ^{18}O into P_i showed that about equal amounts of bound ATP and ADP are in equilibrium at the catalytic site, even in the absence of a proton gradient. However, ATP does not leave the catalytic site unless protons flow through the enzyme. Paul Boyer showed that *the role of the proton gradient is not to form ATP but to release it from the synthase.* He also found that the nucleotide-binding sites of this enzyme interact with each other. The binding of ADP and P_i to one site promotes the release of ATP from another. In other words, ATP synthase exhibits *catalytic cooperativity.*

Boyer has proposed a *binding-change mechanism* for proton-driven ATP synthesis. In this working hypothesis (Figure 21-28), the three catalytic β subunits are intrinsically identical but are not functionally equivalent at any particular moment. One catalytic site is in the O form, which is open and has very low affinity for substrates. The second is in the L form, which binds them loosely and is catalytically inactive. The third is in the T form, which binds them tightly and is active. Consider an enzyme molecule with ATP bound to the T site. ADP and P_i then bind to the L site. Energy input by proton flux converts the T site into an O site, the L site into a T site, and the O site into an L site. These conversions permit ATP to be released from the new O site and enable another molecule of ATP to be formed from ADP and P_i at the new T site (former L site). Protons flow from the F_0 to the F_1 side of the membrane only when O, L, and T interconvert. These conformational transitions are linked, most likely by changes in subunit interactions.

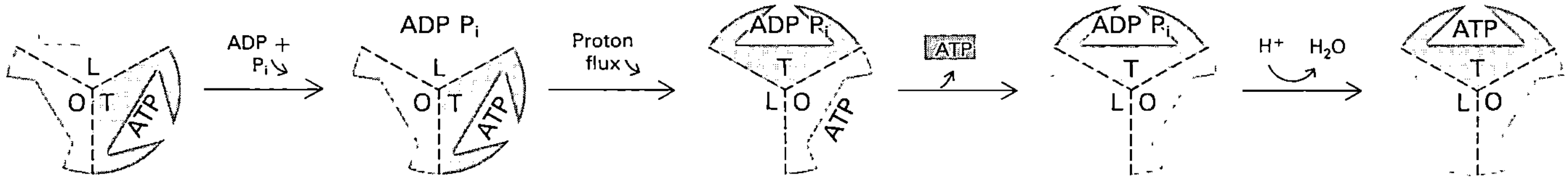

Figure 21-28
Binding-change mechanism for ATP synthase. The three catalytic sites cycle through three conformational states: O (open), L (loose binding), and T (tight binding). Proton flux through the synthase drives this interconversion of states. The essence of this proposed mechanism is that proton flux leads to the release of tightly bound ATP. [After R.L. Cross, D. Cunningham, and J.K. Tamura. *Curr. Top. Cell. Regul.* 24(1984):336.]

ATP synthase is driven by proton-motive force, which is the sum of contributions by the pH gradient and the membrane potential, as expressed in the equation on page 553. A pH gradient, for example, could cause aspartate residues to be in the –COOH form at pH 4.5 (when facing the cytosolic side) and in the $-COO^-$ form at pH 7.5 (when facing the matrix side). Then protonation on one side and deprotonation on the other could drive a reaction cycle unidirectionally to achieve net ATP synthesis. How would membrane potential achieve the same end? Suppose that the membrane potential is 0.18 V (the cytosolic side positive) and the pH is 7.5 on both sides. The concentration of H^+ within the F_0 channel will not be uniform because H^+ will be attracted to the matrix side of the membrane, which is negative relative to the other side (Figure 21-29). The potential difference of 0.18 V leads to a 1000-fold higher concentration of H^+ at the F_0-F_1 junction than at the entrance of the F_0 channel. Thus, *a positive membrane potential leads to ATP synthesis by producing a high local concentration of* H^+ at the gate between F_0 and F_1; specifically, 0.18 V produces the same concentration gradient as does a 3 pH unit difference between the two sides. *The translocation of three* H^+ *through the synthase leads to the formation of one ATP.*

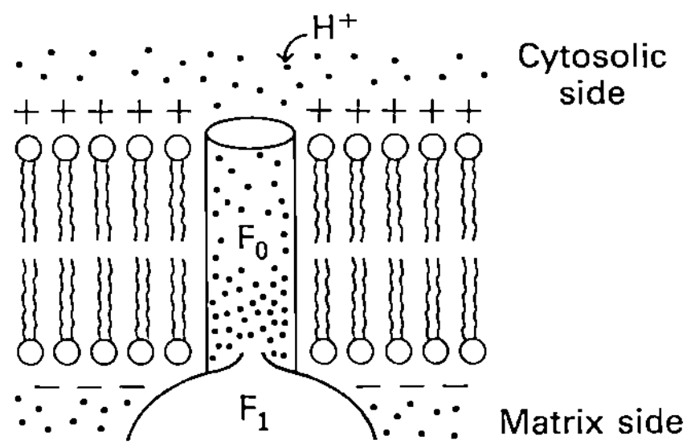

Figure 21-29
A positive membrane potential leads to a high local concentration of H^+ at the junction of the F_0 and F_1 subunits of ATP synthase.

ELECTRONS FROM CYTOSOLIC NADH ENTER MITOCHONDRIA BY SHUTTLES

The inner mitochondrial membrane is quite impermeable to NADH and NAD^+. But NADH is formed by glycolysis in the cytosol, in the oxidation of glyceraldehyde 3-phosphate, and NAD^+ must be regenerated for glycolysis to continue. How then does cytosolic NADH get oxidized by the respiratory chain? The solution is that *electrons from NADH,* rather than NADH itself, are carried across the mitochondrial membrane. One carrier is *glycerol 3-phosphate,* which readily traverses the outer mitochondrial membrane. The first step in this shuttle (Figure 21-30) is the transfer of electrons from NADH to dihydroxyacetone phosphate to form glycerol 3-phosphate.

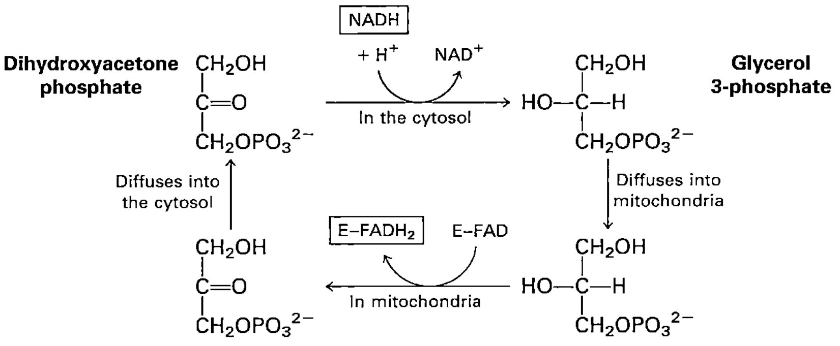

Figure 21-30
Glycerol phosphate shuttle.

This reaction, catalyzed by glycerol 3-phosphate dehydrogenase, occurs in the cytosol. Glycerol 3-phosphate is reoxidized to dihydroxyacetone phosphate on the outer surface of the inner mitochondrial membrane—an electron pair from glycerol 3-phosphate is transferred to the FAD prosthetic group of the mitochondrial glycerol dehydrogenase. This enzyme differs from its cytosolic counterpart: it uses FAD rather than NAD^+ as the electron acceptor, and it is a transmembrane protein. The dihydroxyacetone phosphate formed in the oxidation of glycerol 3-phosphate then diffuses back into the cytosol to complete the shuttle.

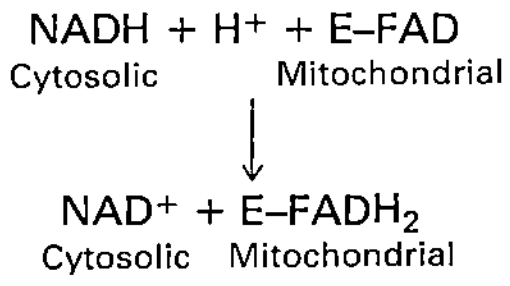

Glycerol phosphate shuttle

The reduced flavin inside the mitochondria transfers its electrons to the electron carrier Q, which then enters the respiratory chain as QH_2. As will be discussed shortly, *1.5 rather than 2.5 ATP are formed when cytosolic NADH transported by the glycerol phosphate shuttle is oxidized by the respiratory chain.* At first glance, this shuttle may seem to waste one ATP per cycle. The yield is lower because FAD rather than NAD^+ is the electron acceptor in mitochondrial glycerol 3-phosphate dehydrogenase. The use of FAD enables electrons from cytosolic NADH to be transported into mitochondria against an NADH concentration gradient. The price of this transport is one ATP per two electrons. This glycerol phosphate shuttle is especially prominent in insect flight muscle, which can sustain a very high rate of oxidative phosphorylation.

In heart and liver, electrons from cytosolic NADH are brought into mitochondria by the *malate-aspartate shuttle,* which is mediated by two membrane carriers and four enzymes. Electrons are transferred from NADH in the cytosol to oxaloacetate, forming malate, which traverses the inner mitochondrial membrane and is then reoxidized by NAD^+ in the matrix to form NADH. The resulting oxaloacetate does not readily cross the inner mitochondrial membrane, and so a transamination reaction (p. 630) is needed to form aspartate, which can be transported to the cytosolic side. This shuttle, in contrast with the glycerol phosphate shuttle, is readily reversible. Consequently, NADH can be brought into mitochondria by the malate-aspartate shuttle only if the $NADH/NAD^+$ ratio is higher in the cytosol than in the mitochondrial matrix.

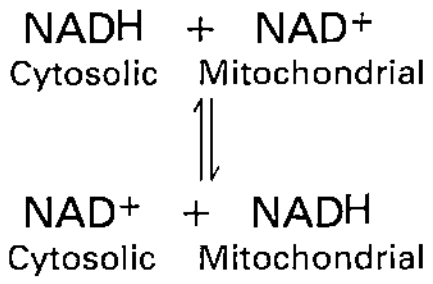

Malate-aspartate shuttle

THE ENTRY OF ADP INTO MITOCHONDRIA IS COUPLED TO THE EXIT OF ATP BY THE ATP-ADP TRANSLOCASE

ATP and ADP do not diffuse freely across the inner mitochondrial membrane. Rather, a specific transport protein, the *ATP-ADP translocase* (also called the *adenine nucleotide carrier*), enables these highly charged molecules to traverse this permeability barrier. Most important, the flows of ATP and ADP are coupled. *ADP enters the mitochondrial matrix only if ATP exits, and vice versa.* The reaction catalyzed by the translocase, which acts as an antiporter, is

$$ADP_c{}^{3-} + ATP_m{}^{4-} \longrightarrow ADP_m{}^{3-} + ATP_c{}^{4-}$$

in which the subscript c denotes the cytosolic side and m denotes the matrix side of the inner mitochondrial membrane.

The ATP-ADP translocase, a dimer of identical 30-kd subunits, contains a single nucleotide binding site that alternately faces the matrix and cytosolic sides of the membrane. ATP and ADP (both devoid of Mg^{2+}) are bound with nearly the same affinity. *In the presence of a positive membrane potential, the rate of binding-site eversion from the matrix to the cytosolic side is more rapid for ATP than for ADP because ATP has one more negative charge.* Hence, ATP is transported out of the matrix about 30 times more rapidly than is ADP, which leads to a higher phosphoryl potential on the outer side than on the matrix side. The translocase does not evert at an appreciable rate unless a nucleotide is bound. This design feature assures that the entry of ADP into the matrix is precisely coupled to the exit of ATP. The other side of the coin is that *the membrane potential is decreased by the exchange of ATP for ADP, which results in a net transfer of one negative charge out of the matrix.* ATP-ADP exchange is energetically expensive; about a quarter of the energy yield from electron transfer by the respiratory chain is consumed to regenerate the membrane potential that is tapped by this exchange process.

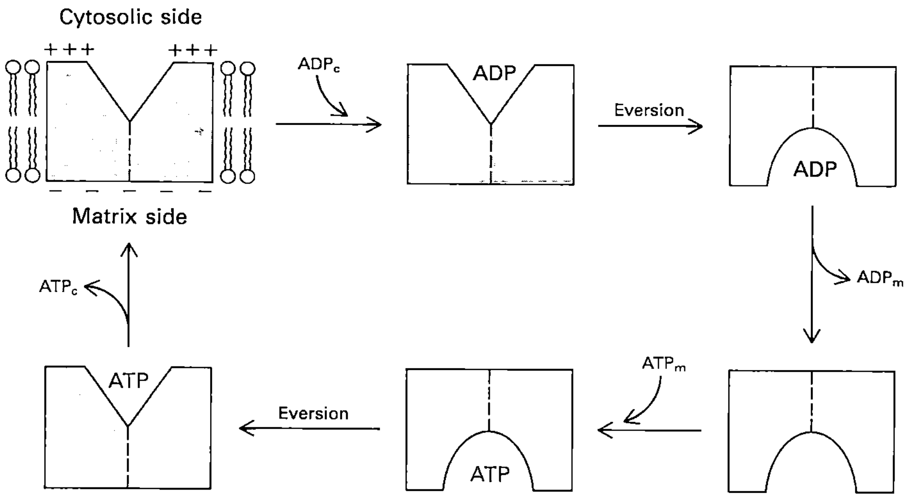

Figure 21-31
The mitochondrial ATP-ADP translocase catalyzes the coupled entry of ADP and exit of ATP into and from the matrix. The reaction cycle is driven by membrane potential. The actual conformational change corresponding to eversion of the binding site could be quite small.

The translocase is highly abundant, constituting about 14% of the inner mitochondrial membrane protein. It is specifically inhibited by very low concentrations of *atractyloside,* a plant glycoside, or of *bongkrekic acid,* an antibiotic from a mold. Atractyloside binds to the translocase when its nucleotide site faces outward, whereas bongkrekic acid binds when this site faces the matrix. Oxidative phosphorylation stops soon after either inhibitor is added, showing that the ATP-ADP translocase is essential.

MITOCHONDRIAL TRANSPORTERS FOR METABOLITES HAVE A COMMON TRIPARTITE MOTIF

The ATP-ADP translocase is but one of many mitochondrial transporters for ions and charged metabolites. For historical reasons, these transmembrane proteins are called carriers. Recall that some of them function as symporters and others as antiporters (p. 316). The *phosphate carrier*, which works in concert with the ATP-ADP translocase, mediates the electroneutral exchange of P_i for OH^- (or, indistinguishably, the electroneutral symport of P_i and H^+). The combined action of these two transporters leads to the exchange of cytosolic ADP and P_i for matrix ATP at the cost of an influx of one H^+. The *dicarboxylate carrier* enables malate, succinate, and fumarate to be exported from mitochondria in exchange for P_i. The *tricarboxylate carrier* transports citrate and a proton in exchange for malate. Pyruvate in the cytosol enters the mitochondrial matrix in exchange for OH^- (or together with H^+) by means of the *pyruvate carrier*. These mitochondrial transporters and more than five others have a common structural motif. They are constructed from three tandem repeats of a 100-residue module, each containing two putative transmembrane segments (Figure 21-32).

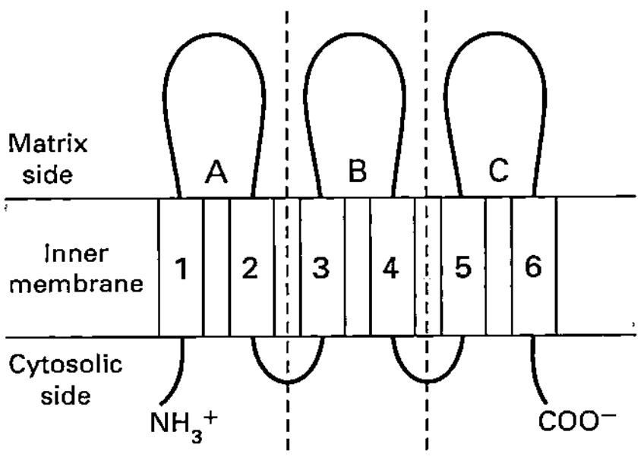

Figure 21-32
Many mitochondrial transporters consist of three similar ~100-residue units. These proteins contain six putative membrane-spanning segments. [After J.E. Walker. *Curr. Opin. Struct. Biol.* 2(1992):519.]

THE COMPLETE OXIDATION OF GLUCOSE YIELDS ABOUT 30 ATP

We can now estimate how many ATP are formed when glucose is completely oxidized to CO_2. The number of ATP (or GTP) formed in glycolysis and the citric acid cycle is unequivocally known because it is determined by the stoichiometries of chemical reactions. In contrast, the ATP yield of oxidative phosphorylation is less certain because the stoichiometries of proton pumping, ATP synthesis, and metabolite transport processes need not be integer numbers or even have fixed values. The best current estimate of the number of H^+ pumped from the matrix to the cytosolic side of the membrane by NADH-Q reductase, cytochrome reductase, and cytochrome oxidase per electron pair is 4, 2, and 4, respectively. The synthesis of an ATP is driven by the flow of about 3 H^+ through ATP synthase. An additional H^+ is consumed in transporting ATP from the matrix to the cytosol. Hence, about 2.5 cytosolic ATP are generated as a result of the flow of a pair of electrons from NADH to O_2. For electrons that enter at the level of cytochrome reductase, such as those from the oxidation of succinate or cytosolic NADH, the yield is about 1.5 ATP per electron pair. Hence, as tallied in Table 21-4, *about 30 ATP are formed when glucose is completely oxidized to CO_2;* this value supersedes the traditional estimate of 36 ATP. Most of the ATP, 26 out of 30 formed, is generated by oxidative phosphorylation.

Table 21-4
ATP yield from the complete oxidation of glucose

Reaction sequence	*ATP yield per glucose*
Glycolysis: Glucose into pyruvate (in the cytosol)	
Phosphorylation of glucose	−1
Phosphorylation of fructose 6-phosphate	−1
Dephosphorylation of 2 molecules of 1,3-BPG	+2
Dephosphorylation of 2 molecules of phosphoenolpyruvate	+2
2 NADH are formed in the oxidation of 2 molecules of glyceraldehyde 3-phosphate	
Conversion of pyruvate into acetyl CoA (inside mitochondria)	
2 NADH are formed	
Citric acid cycle (inside mitochondria)	
2 molecules of guanosine triphosphate are formed from 2 molecules of succinyl CoA	+2
6 NADH are formed in the oxidation of 2 molecules each of isocitrate, α-ketoglutarate, and malate	
2 $FADH_2$ are formed in the oxidation of 2 molecules of succinate	
Oxidative phosphorylation (inside mitochondria)	
2 NADH formed in glycolysis; each yields 1.5 ATP (assuming transport of NADH by the glycerol phosphate shuttle)	+3
2 NADH formed in the oxidative decarboxylation of pyruvate; each yields 2.5 ATP	+5
2 $FADH_2$ formed in the citric acid cycle; each yields 1.5 ATP	+3
6 NADH formed in the citric acid cycle; each yields 2.5 ATP	+15
NET YIELD PER GLUCOSE	+30

Source: The ATP yield of oxidative phosphorylation is based on values given in P.C. Hinkle, M.A. Kumar, A. Resetar, and D.L. Harris. *Biochemistry* 30(1991):3576. The current value of 30 ATP per glucose supersedes the earlier one of 36 ATP. The stoichiometries of proton pumping, ATP synthesis, and metabolite transport should be regarded as estimates. About two more ATP are formed per glucose oxidized when the malate-aspartate shuttle rather than the glycerol phosphate shuttle is used.

THE RATE OF OXIDATIVE PHOSPHORYLATION IS DETERMINED BY THE NEED FOR ATP

Under most physiologic conditions, electron transport is tightly coupled to phosphorylation. *Electrons do not usually flow through the electron transport chain to O_2 unless ADP is simultaneously phosphorylated to ATP.* Oxidative phosphorylation requires a supply of NADH (or other source of electrons at high potential), O_2, ADP, and P_i. The most important factor in determining the rate of oxidative phosphorylation is the *level of ADP*. The rate of oxygen consumption by mitochondria increases markedly when ADP is added and then returns to its initial value when the added ADP has been converted into ATP (Figure 21-33).

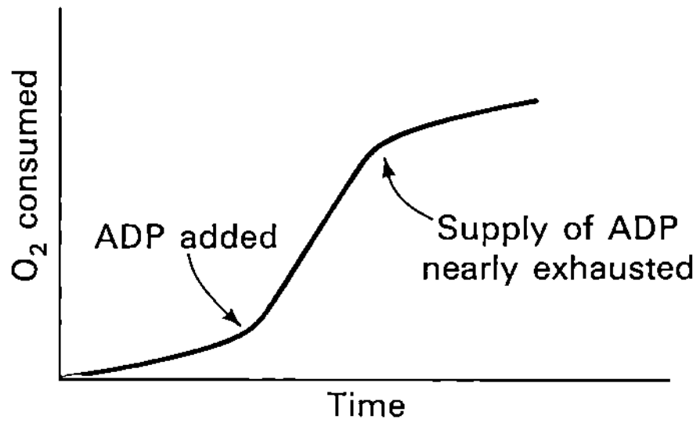

Figure 21-33
Respiratory control. Electrons are transferred to O_2 only if ADP is concomitantly phosphorylated to ATP.

The regulation of the rate of oxidative phosphorylation by the ADP level is called *respiratory control.* The level of ADP likewise affects the rate of the citric acid cycle because of its need for NAD^+ and FAD. The physiologic significance of this regulatory mechanism is evident. The ADP level increases when ATP is consumed, and so oxidative phosphorylation is coupled to the utilization of ATP. *Electrons do not flow from fuel molecules to O_2 unless ATP needs to be synthesized.* We see here another example of the regulatory significance of the energy charge.

THE PROTON GRADIENT CAN BE SHORT-CIRCUITED TO GENERATE HEAT

The tight coupling of electron transport and phosphorylation in mitochondria is disrupted by 2,4-dinitrophenol (DNP) and some other acidic aromatic compounds (Figure 21-34). These substances carry protons across the inner mitochondrial membrane. In the presence of these uncouplers, electron transport from NADH to O_2 proceeds normally, but ATP is not formed by the mitochondrial ATPase because the proton-motive force across the inner mitochondrial membrane is dissipated. The loss of respiratory control leads to increased oxygen consumption and oxidation of NADH. In contrast, DNP has no effect on phosphorylations involving high-energy chemical intermediates (e.g., 1,3-bisphosphoglycerate). DNP and other uncouplers are very useful in metabolic studies because of their specific effect on oxidative phosphorylation.

The uncoupling of oxidative phosphorylation can be biologically useful. *It is a means of generating heat to maintain body temperature in hibernating animals, some newborn animals (including humans), and mammals adapted to cold.* Brown adipose tissue, which is very rich in mitochondria, is specialized for this process of *thermogenesis*. The inner mitochondrial membrane of these mitochondria contains a large amount of *thermogenin* (also called the *uncoupling protein*), a dimer of 33-kd subunits that resembles the ATP-ADP translocase. Thermogenin forms a pathway for the flow of protons from the cytosol to the matrix. In essence, *thermogenin generates heat by short-circuiting the mitochondrial proton battery.* This dissipative proton pathway is activated by free fatty acids liberated from triacylglycerols in response to hormonal signals. The skunk cabbage uses an analogous mechanism to heat its floral spikes, increasing the evaporation of odoriferous molecules that attract insects to fertilize its flowers.

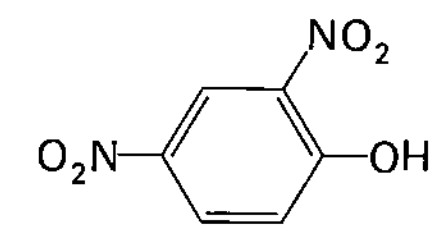

2,4-Dinitrophenol (DNP)

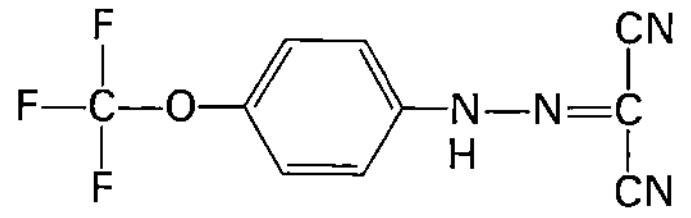

Carbonylcyanide-*p*-trifluoromethoxyphenylhydrazone

Figure 21-34
Formulas of two uncouplers of oxidative phosphorylation. These lipid-soluble substances can carry protons across the inner mitochondrial membrane. The dissociable proton is shown in red.

TOXIC DERIVATIVES OF O_2 SUCH AS SUPEROXIDE RADICAL ARE SCAVENGED BY PROTECTIVE ENZYMES

As was mentioned earlier, the one-electron reduction of O_2 yields $O_2\cdot^-$, *superoxide anion,* a potentially destructive radical. Cytochrome oxidase and other proteins that reduce O_2 have been designed not to release $O_2\cdot^-$. However, a small amount of superoxide anion is unavoidably formed, as in the oxidation of a ferroheme (Fe^{2+}) group of hemoglobin to ferriheme (Fe^{3+}) by O_2. Protonation of superoxide anion yields *hydroperoxyl radical ($HO_2\cdot$),* which can react spontaneously with another superoxide anion to form *hydrogen peroxide* (H_2O_2).

$$O_2 \xrightarrow{e^-} \underset{\text{Superoxide anion}}{O_2\cdot^-} \xrightarrow{H^+} \underset{\text{Hydroperoxyl radical}}{HO_2\cdot} \xrightarrow{HO_2\cdot \;\; O_2} \underset{\text{Hydrogen peroxide}}{H_2O_2}$$

Superoxide anion can be scavenged by *superoxide dismutase,* an enzyme that catalyzes the conversion of two of these radicals into hydrogen peroxide and molecular oxygen.

$$O_2\cdot^- + O_2\cdot^- \xrightarrow[\text{Superoxide dismutase}]{2\,H^+} H_2O_2 + O_2$$

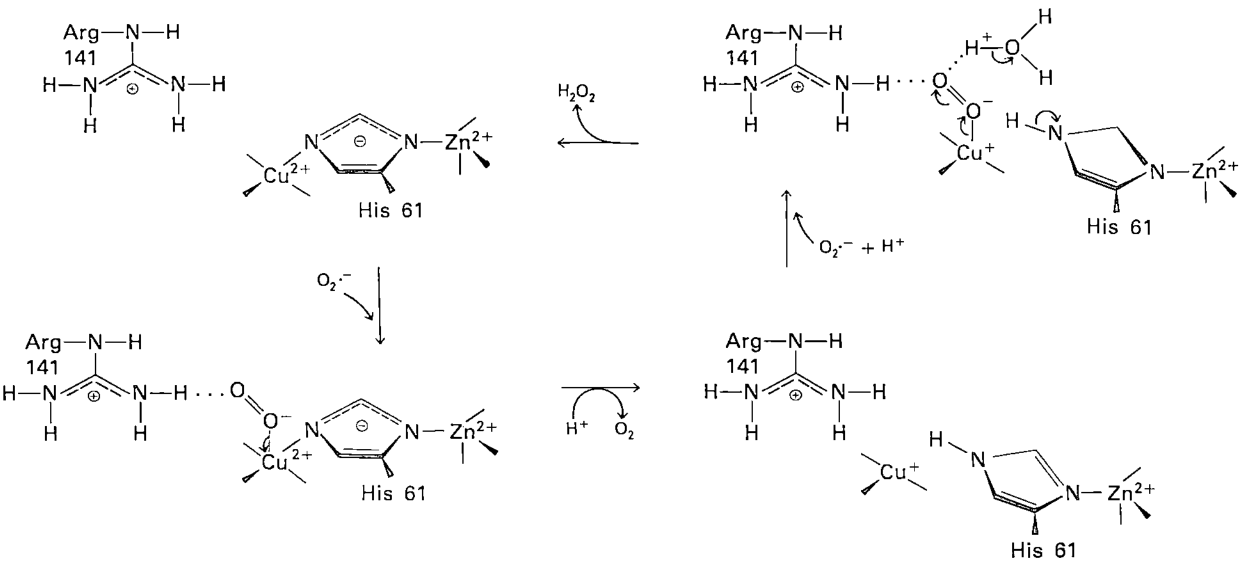

Figure 21-35
Proposed catalytic mechanism of superoxide dismutase. A copper ion at the active site cycles between +1 and +2 oxidation states. The histidine side chain coordinated to Cu^{2+} during part of the cycle is also bonded to Zn^{2+}. The other key participant is an arginine side chain. [After J.A. Tainer, E.D. Getzoff, J.S. Richardson, and D.C. Richardson. *Nature* 306(1983):286.]

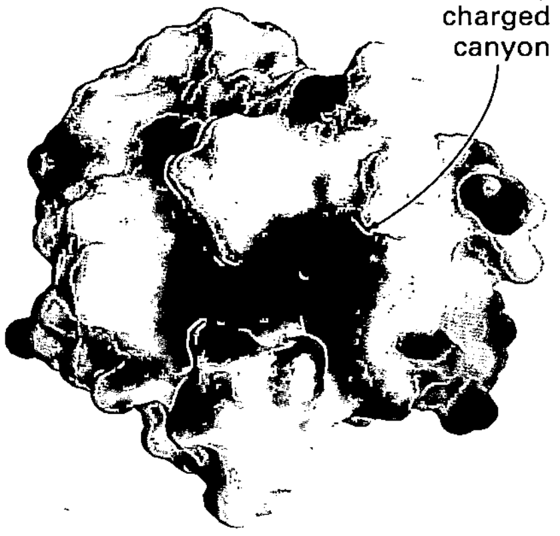

Figure 21-36
The active site of superoxide dismutase is located in a positively charged canyon (blue) that is surrounded by a ring of negatively charged residues (red). The superoxide anion is electrostatically attracted to the positively charged active site. [Drawn from 3sod.pdb. J.A. Tainer, E.D. Getzoff, J.S. Richardson, and D.C. Richardson. *Nature* 306(1983):284.]

The active site of the cytosolic enzyme in eukaryotes contains a copper ion and a zinc ion coordinated to the side chain of a histidine residue (Figure 21-35). The negatively charged superoxide is guided electrostatically to a very positively charged catalytic site at the bottom of a channel (Figure 21-36). $O_2\cdot^-$ binds to Cu^{2+} and the guanido group of an arginine residue. An electron is transferred from superoxide to cupric ion to form Cu^+ and O_2, which is released. A second superoxide enters the active site and binds to Cu^+, arginine, and H_3O^+. The bound $O_2\cdot^-$ acquires an electron from Cu^+ and two protons from its binding partners to form H_2O_2 and regenerate the Cu^{2+} state of the enzyme.

Superoxide dismutase is present in all aerobic organisms. *Escherichia coli* mutants lacking this enzyme are highly vulnerable to oxidative damage. Superoxide dismutase is an essential molecular adaptation to the challenge posed by an oxygen atmosphere. The importance of this enzyme is highlighted by the recent finding that mutations of superoxide dismutase cause *amyotrophic lateral sclerosis*. In this debilitating neurological disorder, motor neurons in the brain and spinal cord degenerate.

The hydrogen peroxide formed by superoxide dismutase and by the uncatalyzed reaction of hydroperoxy radicals is scavenged by *catalase*, a ubiquitous heme protein that catalyzes the dismutation of hydrogen peroxide into water and molecular oxygen.

$$H_2O_2 + H_2O_2 \xrightarrow{\text{Catalase}} 2\,H_2O + O_2$$

Catalase has great catalytic prowess. Its k_{cat}/K_M ratio of $4 \times 10^7\ \text{M}^{-1}\ \text{s}^{-1}$ places it in the elite group of enzymes that have attained kinetic perfection (p. 195). *Peroxidases*, which also are heme enzymes, catalyze an analogous reaction in which an alkyl peroxide is reduced to water and an alcohol by a reductant (AH_2). For example, the reductant could be cytochrome *c* or ascorbate.

$$ROOH + AH_2 \xrightarrow{\text{Peroxidase}} ROH + H_2O + A$$

POWER TRANSMISSION BY PROTON GRADIENTS: A CENTRAL MOTIF OF BIOENERGETICS

The main concept presented in this chapter is that mitochondrial electron transfer and ATP synthesis are linked by a transmembrane proton gradient. ATP synthesis in bacteria and chloroplasts (p. 665) also is driven by proton gradients. In fact, proton gradients power a variety of energy-requiring processes such as the active transport of Ca^{2+} by mitochondria, the entry of some amino acids and sugars into bacteria (p. 317), the rotation of bacterial flagella (p. 412), and the transfer of electrons from NADH to NADPH. Proton gradients can also be used to generate heat, as in hibernation. It is evident that *proton gradients are a central interconvertible currency of free energy in biological systems* (Figure 21-37). Mitchell has noted that the proton-motive force is a marvelously simple and effective store of free energy because it requires only a thin closed lipid membrane between two aqueous phases.

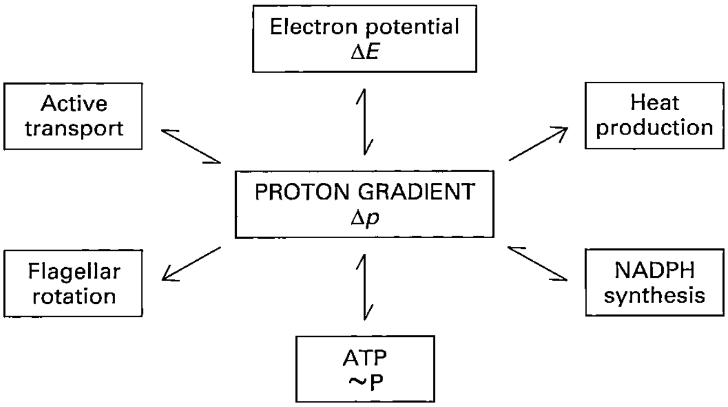

Figure 21-37
The proton gradient is an interconvertible form of free energy.

SUMMARY

In oxidative phosphorylation, the synthesis of ATP is coupled to the flow of electrons from NADH or $FADH_2$ to O_2 by a proton gradient across the inner mitochondrial membrane. Electron flow through three asymmetrically oriented transmembrane complexes results in the pumping of protons out of the mitochondrial matrix and the generation of a membrane potential. ATP is synthesized when protons flow back to the matrix through a channel in an ATP-synthesizing complex, called ATP synthase (also known as F_0F_1-ATPase or H^+-ATPase). Oxidative phosphorylation exemplifies a fundamental theme of bioenergetics: the transmission of free energy by proton gradients.

The electron carriers in the respiratory assembly of the inner mitochondrial membrane are flavins, iron-sulfur complexes, quinones, heme groups of cytochromes, and copper ions. Electrons from NADH are transferred to the FMN prosthetic group of NADH-Q reductase, the first of three complexes. This reductase also contains Fe-S centers. The electrons emerge in QH_2, the reduced form of ubiquinone (Q). This highly mobile hydrophobic carrier transfers its electrons to cytochrome reductase, a complex that contains cytochromes b and c_1 and an Fe-S center. This second complex reduces cytochrome c, a water-soluble peripheral membrane protein. Cytochrome c, like Q, is a mobile carrier of electrons, which it then transfers to cytochrome oxidase. This third complex contains cytochromes a and a_3 and two copper ions. A heme iron and a copper ion in this oxidase transfer electrons to O_2, the ultimate acceptor, to form H_2O.

The flow of electrons through each of these complexes leads to the pumping of protons from the matrix side to the cytosolic side of the inner mitochondrial membrane. A proton-motive force consisting of a pH gradient (cytosolic side acidic) and a membrane potential (cytosolic side positive) is generated. The flow of protons back to the matrix side through ATP synthase drives ATP synthesis. The enzyme complex consists of a hydrophobic F_0 unit that conducts protons through the membrane and a hydrophilic F_1 unit that catalyzes ATP synthesis sequentially at three sites. Protons flowing through ATP synthase release tightly bound ATP.

The flow of two electrons through NADH-Q reductase, cytochrome reductase, and cytochrome oxidase generates a gradient sufficient to synthesize 1, 0.5, and 1 molecule of ATP, respectively. Hence, 2.5 ATP are formed per NADH oxidized in the mitochondrial matrix, whereas only 1.5 ATP are made per $FADH_2$ oxidized because its electrons enter the chain at QH_2, after the first proton pumping site. Also, only 1.5 ATP are generated by the oxidation of cytosolic NADH because its electrons enter the respiratory chain at the second site owing to the action of the glycerol phosphate shuttle. The entry of ADP into the mitochondrial matrix is coupled to the exit of ATP by the ATP-ADP translocase, a transporter driven by membrane potential. About 30 ATP are generated when a molecule of glucose is completely oxidized to CO_2 and H_2O.

Electron transport is normally tightly coupled to phosphorylation. NADH and $FADH_2$ are oxidized only if ADP is simultaneously phosphorylated to ATP. This coupling, called respiratory control, can be disrupted by uncouplers such as DNP, which dissipate the proton gradient by carrying protons across the inner mitochondrial membrane.

SELECTED READINGS

Where to start

Dickerson, R.E., 1980. Cytochrome *c* and the evolution of energy metabolism. *Sci. Amer.* 242(3):137-153.

Trumpower, B.L., 1990. The protonmotive Q cycle. Energy transduction by coupling of proton translocation to electron transfer by the cytochrome bc_1 complex. *J. Biol. Chem.* 265:11409–11412.

Babcock, G.T., and Wikstrom, M., 1992. Oxygen activation and the conservation of energy in cell respiration. *Nature* 356:301–309.

Pedersen, P.L., and Amzel, L.M., 1993. ATP synthases. Structure, reaction center, mechanism, and regulation of one of nature's most unique machines. *J. Biol. Chem.* 268:9937–9940.

Fridovich, I., 1989. Superoxide dismutases. An adaptation to a paramagnetic gas. *J. Biol. Chem.* 264:7761–7764.

Books

Nicholls, D.G., and Ferguson, S.J., 1992. *Bioenergetics 2.* Academic Press. [A concise and lucid account of the experimental basis of our current understanding of oxidative phosphorylation.]

Harold, F.M., 1986. *The Vital Force: A Study of Bioenergetics.* W.H. Freeman. [A sensitive and scholarly account of bioenergetics emphasizing the development of ideas. Chapter 7 deals with mitochondria and oxidative phosphorylation.]

Ernster, L. (ed.), 1992. *Molecular Mechanisms in Bioenergetics.* Elsevier.

Tzagoloff, A., 1982. *Mitochondria.* Plenum.

Spiro, T.G. (ed.), 1982. *Iron-Sulfur Proteins.* Wiley-Interscience.

NADH-Q reductase

Walker, J.E., 1992. The NADH:ubiquinone oxidoreductase (complex I) of respiratory chains. *Quart. Rev. Biophys.* 25:253–324.

Weiss, H., Friedrich, T., Hofhaus, G., and Preis, D., 1991. The respiratory-chain NADH dehydrogenase (complex I) of mitochondria. *Eur. J. Biochem.* 197:563–576.

Hofhaus, G., Weiss, H., and Leonard, K., 1991. Electron microscopic analysis of the peripheral and membrane parts of mitochondrial NADH dehydrogenase (complex I). *J. Mol. Biol.* 221:1027–1043.

Cytochromes

Dickerson, R.E., 1972. The structure and history of an ancient protein. *Sci. Amer.* 226(4):58–72. [A beautifully illustrated account of the conformation and evolution of cytochrome *c.* Offprint 1245.]

Trumpower, B.L., 1990. Cytochrome bc_1 complexes of microorganisms. *Microbiol. Rev.* 54:101–129.

Pelletier, H., and Kraut, J., 1992. Crystal structure of a complex between electron transfer partners, cytochrome *c* peroxidase and cytochrome *c. Science* 258:1748–1755.

Wikstrom, M., and Morgan, J.E., 1992. The dioxygen cycle. Spectral, kinetic, and thermodynamic characteristics of ferryl and peroxy intermediates observed by reversal of the cytochrome oxidase reaction. *J. Biol. Chem.* 267:10266–10273.

Zhang, Y.Z., Ewart, G., and Capaldi, R.A., 1991. Topology of subunits of the mammalian cytochrome *c* oxidase: Relationship to the assembly of the enzyme complex. *Biochemistry* 30:3674–3681.

Capaldi, R.A., 1990. Structure and function of cytochrome *c* oxidase. *Ann. Rev. Biochem.* 59:569–596.

ATP synthase

Abrahams, J.P., Leslie, A.G.W., Lutter, R., and Walker, J.E., 1994. Structure at 2.8 Å resolution of F_1-ATPase, from bovine heart mitochondria. *Nature* 370:621–628.

Hinkle, P.C., Kumar, M.A., Resetar, A., and Harris, D.L., 1991. Mechanistic stoichiometry of mitochondrial oxidative phosphorylation. *Biochemistry* 30:3576–3582.

Thomas, P.J., Bianchet, M., Garboczi, D.N., Hullihen, J., Amzel, L.M., and Pedersen, P.L., 1992. ATP synthase: Structure-function relationships. *Biochim. Biophys. Acta* 1101:228–231.

Senior, A.E., 1990. The proton-translocating ATPase of *Escherichia coli*. *Ann. Rev. Biophys. Biophys. Chem.* 19:7–41.

Penefsky, H.S., and Cross, R.L., 1991. Structure and mechanism of F_0F_1-type ATP synthases and ATPases. *Advan. Enzymol. Mol. Biol.* 64:173–214.

Boyer, P.D., 1993. The binding change mechanism for ATP synthase—Some probabilities and possibilities. *Biochim. Biophys. Acta* 1140:215–250.

Fillingame, R.H., 1992. H^+ transport and coupling by the F_0 sector of the ATP synthase: Insights into the molecular mechanism of function. *J. Bioenerg. Biomemb.* 24:485–491.

Chance, B., Leigh, J.S., Jr., Kent, J., McCully, K., Nioka, S., Clark, B.J., Maris, J.M., and Graham, T., 1986. Multiple controls of oxidative metabolism in living tissues as studied by phosphorus magnetic resonance. *Proc. Nat. Acad. Sci.* 83:9458–9462.

Translocators

Walker, J.E., 1992. The mitochondrial transporter family. *Curr. Opin. Struct. Biol.* 2:519–526.

Klingenberg, M., 1992. Structure-function of the ADP/ATP carrier. *Biochem. Soc. Trans.* 20:547–550.

Klingenberg, M., 1990. Mechanism and evolution of the uncoupling protein of brown adipose tissue. *Trends Biochem. Sci.* 15:108–112.

Superoxide dismutase

Roberts, V.A., Fisher, C.L., Redford, S.M., McRee, D.E., Parge, H.E., Getzoff, E.D., and Tainer, J.A., 1991. Mechanism and atomic structure of superoxide dismutase. *Free Radical Res. Comm.* 1:269–278.

Tainer, J.A., Getzoff, E.D., Richardson, J.S., and Richardson, D.C., 1983. Structure and mechanism of Cu,Zn superoxide dismutase. *Nature* 306:284–287.

Rosen, D.R., *et al.*, 1993. Mutations in Cu/Zn superoxidase dismutase gene are associated with familial amyotrophic lateral sclerosis. *Nature* 362:59–62.

Mitochondrial diseases

Wallace, D.C., 1992. Diseases of the mitochondrial DNA. *Ann. Rev. Biochem.* 61:1175–1212.

Majander, A., Huoponen, K., Savontaus, M.L., Nikoskelainen, E., and Wikstrom, M., 1991. Electron transfer properties of NADH:ubiquinone reductase in the ND1/3460 and the ND4/11778 mutations of the Leber hereditary optic neuroretinopathy (LHON). *FEBS Lett.* 292:289–292.

Benecke, R., Strumper, P., and Weiss, H., 1992. Electron transfer complex I defect in idiopathic dystonia. *Ann. Neurol.* 32:683–686.

Historical aspects

Mitchell, P., 1979. Keilin's respiratory chain concept and its chemiosmotic consequences. *Science* 206:1148–1159. [Mitchell reviews the evolution of the chemiosmotic concept in this Nobel Lecture.]

Mitchell, P., 1976. Vectorial chemistry and the molecular mechanics of chemiosmotic coupling: Power transmission by proticity. *Trans. Biochem. Soc.* 4:399–430.

Racker, E., 1980. From Pasteur to Mitchell: A hundred years of bioenergetics. *Fed. Proc.* 39:210–215.

Keilin, D., 1966. *The History of Cell Respiration and Cytochromes.* Cambridge University Press.

Kalckar, H.M. (ed.), 1969. *Biological Phosphorylations: Development of Concepts.* Prentice-Hall. [A collection of classical papers on oxidative phosphorylation and other aspects of bioenergetics.]

Kalckar, H.M., 1991. Fifty years of biological research—From oxidative phosphorylation to energy requiring transport and regulation. *Ann. Rev. Biochem.* 60:1–37.

Fruton, J.S., 1972. *Molecules and Life: Historical Essays on the Interplay of Chemistry and Biology.* Wiley-Interscience. [Includes an excellent account of cellular respiration, starting on p. 262.]

PROBLEMS

1. *Energy harvest.* What is the yield of ATP when each of the following substrates is completely oxidized to CO_2 by a mammalian cell homogenate? Assume that glycolysis, the citric acid cycle, and oxidative phosphorylation are fully active.
   (a) Pyruvate.
   (b) Lactate.
   (c) Fructose 1,6-bisphosphate.
   (d) Phosphoenolpyruvate.
   (e) Galactose.
   (f) Dihydroxyacetone phosphate.

2. *Reference states.* The standard oxidation-reduction potential for the reduction of O_2 to H_2O is given as 0.82 V

in Table 21-1. However, the value given in textbooks of chemistry is 1.23 V. Account for this difference.

3. *Potent poisons.* What is the effect of each of the following inhibitors on electron transport and ATP formation by the respiratory chain?
   (a) Azide.
   (b) Atractyloside.
   (c) Rotenone.
   (d) DNP.
   (e) Carbon monoxide.
   (f) Antimycin A.

4. *P:O ratios.* The number of molecules of inorganic phosphate incorporated into organic form per atom of oxygen consumed, termed the *P:O ratio,* was frequently used as an index of oxidative phosphorylation.
   (a) What is the relation of the P:O ratio to the ratio of the number of protons translocated per electron pair ($H^+/2e^-$) and the ratio of the number of protons needed to synthesize an ATP and transport it to the cytosol ($\sim P/H^+$)?
   (b) What are the P:O ratios for electrons donated by matrix NADH and by succinate?

5. *Thermodynamic constraint.* Compare the $\Delta G^{\circ\prime}$ for the oxidation of succinate by NAD^+ and by FAD. Use the data given in Table 21-1, and assume that E_0' for the FAD/$FADH_2$ redox couple is nearly 0 V. Why is FAD rather than NAD^+ the electron acceptor in the reaction catalyzed by succinate dehydrogenase?

6. *Cyanide antidote.* The immediate administration of nitrite is a highly effective treatment for cyanide poisoning. What is the basis for the action of this antidote? (Hint: Nitrite oxidizes ferro- hemoglobin to ferrihemoglobin.)

7. *Chiral clue.* ATPγS, a slowly hydrolyzed analog of ATP, can be used to probe the mechanism of phosphoryl transfer reactions. Chiral ATPγS has been synthesized containing ^{18}O in a specific γ position and ordinary ^{16}O elsewhere in the molecule. Hydrolysis of this chiral molecule by ATP synthase in ^{17}O-enriched water yields inorganic [^{16}O,^{17}O,^{18}O]thiophosphate having the absolute configuration shown in Figure 21-38. In contrast, hydrolysis of this chiral ATPγS by a calcium-pumping ATPase from muscle gives thiophosphate of the opposite configuration. What is the simplest interpretation of these data?

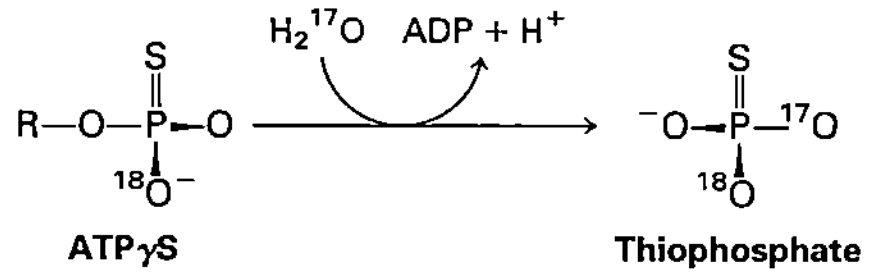

Figure 21-38
Stereochemistry of the hydrolysis of chiral ATPγS by ATP synthase.

8. *Currency exchange.* For a proton-motive force of 0.2 V (matrix negative), what is the maximum [ATP]/[ADP][P_i] ratio compatible with ATP synthesis? Calculate this ratio three times, assuming that the number of protons translocated per ATP formed is 2, 3, and 4 and that the temperature is 25°C.

9. *Runaway mitochondria.* Suppose that the mitochondria of a patient oxidized NADH irrespective of whether ADP was present. The P:O ratio for oxidative phosphorylation by these mitochondria was less than normal. Predict the likely symptoms of this disorder.

10. *An essential residue.* Conduction of protons by the F_0 unit of ATP synthase is blocked by modification of a single side chain by dicyclohexylcarbodiimide. What are the most likely targets of action of this reagent? How might you use site-specific mutagenesis to determine whether this residue is essential for proton conduction?

11. *Recycling device.* The cytochrome *b* component of cytochrome reductase enables both electrons of QH_2 to be effectively utilized in generating a proton-motive force. Cite another recycling device in metabolism that brings a potentially dead-end reaction product back into the mainstream.

12. *Crossover point.* The precise site of action of a respiratory-chain inhibitor can be revealed by the *crossover technique.* Britton Chance devised elegant spectroscopic methods for determining the proportions of the oxidized and reduced forms of each carrier. This is feasible because they have distinctive absorption spectra, as illustrated for cytochrome *c* (Figure 21-39). You are given a new inhibitor and find that its addition to respiring mitochondria causes the carriers between NADH and QH_2 to become more reduced, and those between cytochrome *c* and O_2 to become more oxidized. Where does your inhibitor act?

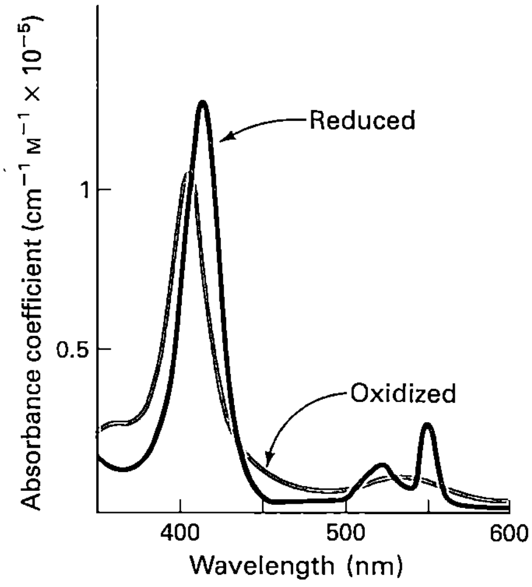

Figure 21-39
Absorption spectra of the oxidized and reduced forms of cytochrome *c.*

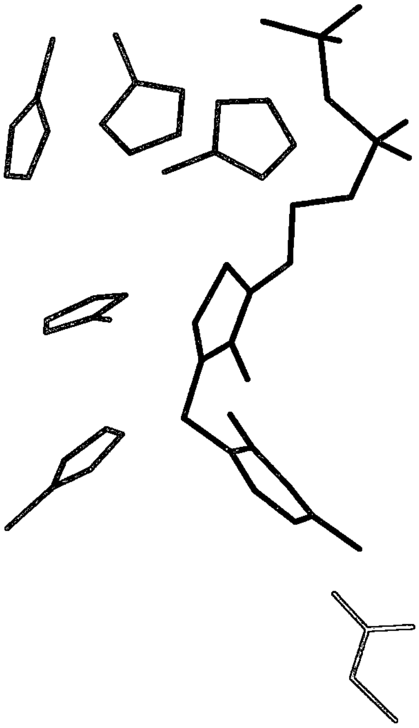

Pentose Phosphate Pathway and Gluconeogenesis

The preceding chapters on glycolysis, the citric acid cycle, and oxidative phosphorylation were primarily concerned with the generation of ATP, starting with glucose as a fuel. We now turn to the generation of a different type of metabolic energy—reducing power. Some of the high-potential electrons of fuel molecules must be conserved for biosynthetic purposes rather than transferred to O_2 to generate ATP. *The currency of readily available reducing power in cells is NADPH.* The phosphoryl group on C-2 of one of the ribose units of NADPH distinguishes it from NADH. As mentioned previously (p. 451), there is a *fundamental distinction between NADPH and NADH in most biochemical reactions. NADH is oxidized by the respiratory chain to generate ATP, whereas NADPH serves as an electron donor (hydride ion donor) in reductive biosyntheses.* This chapter also deals with the synthesis of glucose from noncarbohydrate precursors, by a process called *gluconeogenesis.*

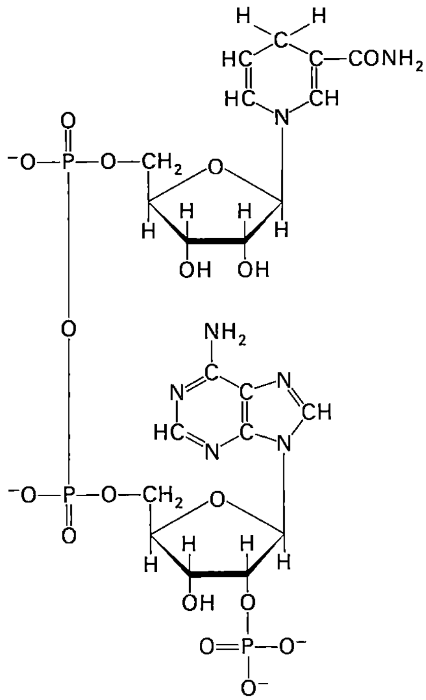

Reduced nicotinamide adenine dinucleotide phosphate (NADPH)

THE PENTOSE PHOSPHATE PATHWAY GENERATES NADPH AND SYNTHESIZES FIVE-CARBON SUGARS

In the pentose phosphate pathway, NADPH is generated when glucose 6-phosphate is oxidized to ribose 5-phosphate. This five-carbon sugar and its derivatives are components of such important biomolecules as ATP, CoA, NAD^+, FAD, RNA, and DNA.

$$\text{Glucose 6-phosphate} + 2\ NADP^+ + H_2O \longrightarrow \text{ribose 5-phosphate} + 2\ NADPH + 2\ H^+ + CO_2$$

Opening Image: *Active site cluster in transketolase, an enzyme that rearranges the carbon skeletons of sugar molecules. The thiamine pyrophosphate prosthetic group (red) enables the enzyme to transfer two-carbon fragments. Five histidine residues (blue) and a glutamate (green) participate in binding and catalysis. [Drawn from coordinates kindly provided by Dr. Y. Linqvist. Y. Linqvist, G. Schneider, U. Ermler, and M. Sundstrom. EMBO J. 11(1992):2373.]*

The pentose phosphate pathway also catalyzes the interconversion of three-, four-, five-, six-, and seven-carbon sugars in a series of nonoxidative reactions. All these reactions occur in the cytosol. In plants, part of the pentose phosphate pathway also participates in the formation of hexoses from CO_2 in photosynthesis (p. 675). The pentose phosphate pathway is sometimes called the *pentose shunt*, the *hexose monophosphate pathway*, or the *phosphogluconate oxidative pathway*.

TWO NADPH ARE GENERATED IN THE CONVERSION OF GLUCOSE 6-PHOSPHATE INTO RIBULOSE 5-PHOSPHATE

The oxidative branch of the pentose phosphate pathway starts with the dehydrogenation of glucose 6-phosphate at C-1, a reaction catalyzed by *glucose 6-phosphate dehydrogenase* (Figure 22-1). This enzyme is highly specific for $NADP^+$; the K_M for NAD^+ is about a thousand times as great as that for $NADP^+$. The product is *6-phosphoglucono-δ-lactone*, which is an intramolecular ester between the C-1 carboxyl group and the C-5 hydroxyl group. The next step is the hydrolysis of 6-phosphoglucono-δ-lactone by a specific *lactonase* to give *6-phosphogluconate*. This six-carbon sugar is then oxidatively decarboxylated by *6-phosphogluconate dehydrogenase* to yield *ribulose 5-phosphate*. $NADP^+$ is again the electron acceptor.

Know this enzyme
glucose 6 (P) Dehydrogenase

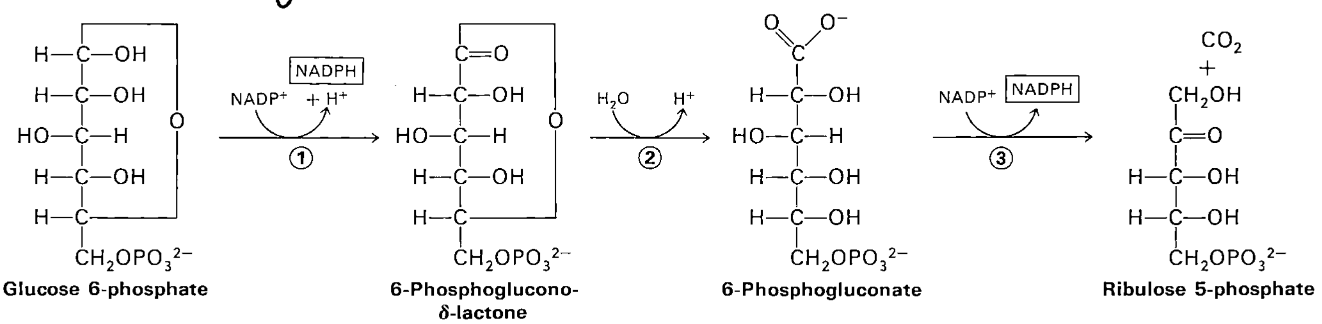

Figure 22-1
Oxidative branch of the pentose phosphate pathway. These three reactions are catalyzed by glucose 6-phosphate dehydrogenase, lactonase, and 6-phosphogluconate dehydrogenase.

RIBULOSE 5-PHOSPHATE IS ISOMERIZED TO RIBOSE 5-PHOSPHATE THROUGH AN ENEDIOL INTERMEDIATE

The final step in the synthesis of ribose 5-phosphate is the isomerization of ribulose 5-phosphate by *phosphopentose isomerase*.

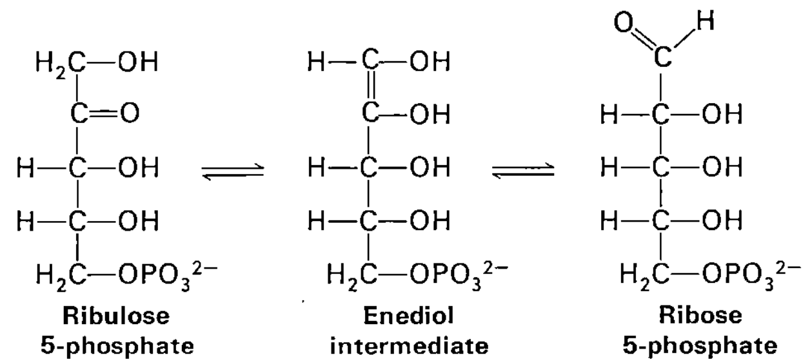

This reaction is similar to the glucose 6-phosphate → fructose 6-phosphate and the dihydroxyacetone phosphate → glyceraldehyde 3-phosphate reactions in glycolysis. *All three ketose-aldose isomerizations proceed through an enediol intermediate.*

THE PENTOSE PHOSPHATE PATHWAY AND GLYCOLYSIS ARE LINKED BY TRANSKETOLASE AND TRANSALDOLASE

The preceding reactions yield two NADPH and one ribose 5-phosphate for each glucose 6-phosphate oxidized. However, many cells need NADPH for reductive biosyntheses much more than ribose 5-phosphate for incorporation into nucleotides and nucleic acids. In these cases, ribose 5-phosphate is converted into glyceraldehyde 3-phosphate and fructose 6-phosphate by *transketolase* and *transaldolase. These enzymes create a reversible link between the pentose phosphate pathway and glycolysis by catalyzing these three reactions.*

$$C_5 + C_5 \underset{}{\overset{\text{Transketolase}}{\rightleftharpoons}} C_3 + C_7$$

$$C_7 + C_3 \underset{}{\overset{\text{Transaldolase}}{\rightleftharpoons}} C_4 + C_6$$

$$C_5 + C_4 \underset{}{\overset{\text{Transketolase}}{\rightleftharpoons}} C_3 + C_6$$

The net result of these reactions is the *formation of two hexoses and one triose from three pentoses.*

Transketolase transfers a two-carbon unit, whereas transaldolase transfers a three-carbon unit. The sugar that donates the two- or three-carbon unit is always a ketose, whereas the acceptor is always an aldose.

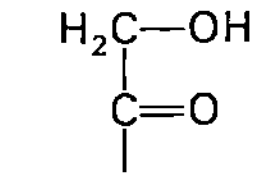

Transferred by transketolase

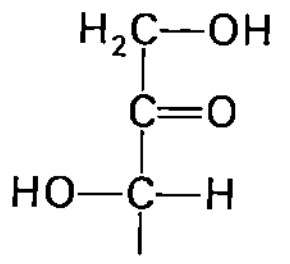

Transferred by transaldolase

The first of the three reactions linking the pentose phosphate pathway and glycolysis is the formation of *glyceraldehyde 3-phosphate* and *sedoheptulose 7-phosphate* from two pentoses.

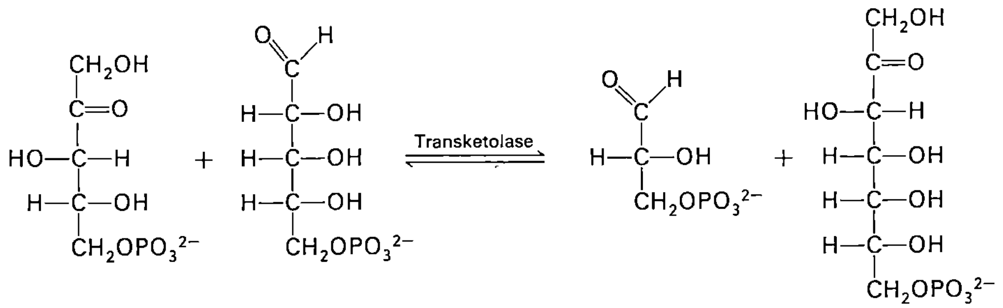

Xylulose 5-phosphate **Ribose 5-phosphate** **Glyceraldehyde 3-phosphate** **Sedoheptulose 7-phosphate**

The donor of the two-carbon unit in this reaction is xylulose 5-phosphate, an epimer of ribulose 5-phosphate. A ketose is a substrate of transketolase only if its hydroxyl group at C-3 has the configuration of xylulose rather than ribulose. Ribulose 5-phosphate is converted into the appropriate epimer for the transketolase reaction by *phosphopentose epimerase.*

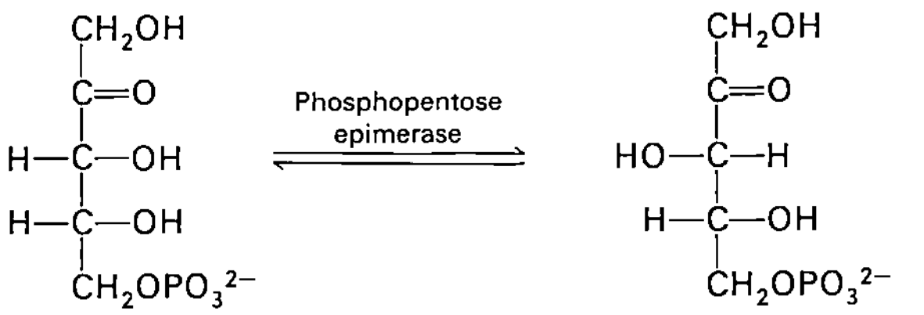

Ribulose 5-phosphate **Xylulose 5-phosphate**

Glyceraldehyde 3-phosphate and sedoheptulose 7-phosphate then react to form *fructose 6-phosphate* and *erythrose 4-phosphate.* This synthesis of a four-carbon sugar and a six-carbon sugar is catalyzed by *transaldolase.*

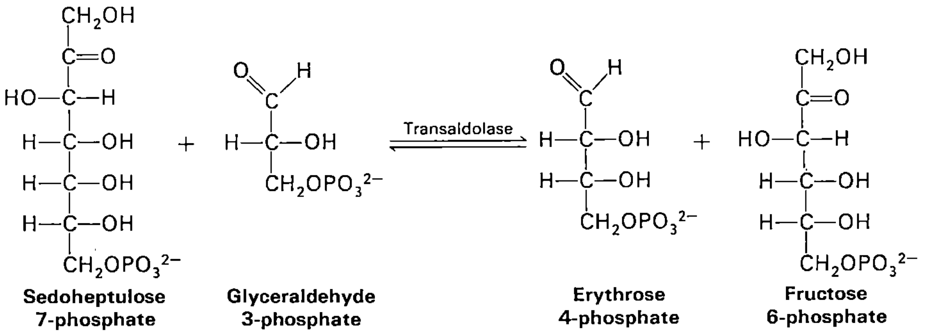

In the third reaction, transketolase catalyzes the synthesis of *fructose 6-phosphate* and *glyceraldehyde 3-phosphate* from erythrose 4-phosphate and xylulose 5-phosphate.

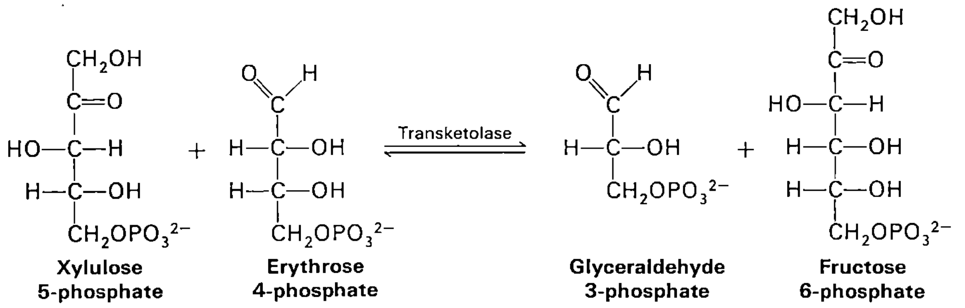

The sum of these reactions is

2 Xylulose 5-phosphate + ribose 5-phosphate $\rightleftharpoons$
2 fructose 6-phosphate + glyceraldehyde 3-phosphate

Xylulose 5-phosphate can be formed from ribose 5-phosphate by the sequential action of phosphopentose isomerase and phosphopentose epimerase, and so the net reaction starting from ribose 5-phosphate is

3 Ribose 5-phosphate $\rightleftharpoons$
2 fructose 6-phosphate + glyceraldehyde 3-phosphate

Thus, *excess ribose 5-phosphate formed by the pentose phosphate pathway can be completely converted into glycolytic intermediates.* It is evident that the carbon skeletons of sugars can be extensively rearranged to meet physiologic needs.

THE RATE OF THE PENTOSE PHOSPHATE PATHWAY IS CONTROLLED BY THE LEVEL OF NADP$^+$

The first reaction in the oxidative branch of the pentose phosphate pathway, the dehydrogenation of glucose 6-phosphate, is essentially irreversible. In fact, this reaction is rate-limiting under physiologic conditions and serves as the control site. The most important regulatory factor is the level of NADP$^+$, the electron acceptor in the oxidation of glucose 6-phosphate to 6-phosphoglucono-δ-lactone. Also, NADPH competes

Table 22-1
Pentose phosphate pathway

Reaction	Enzyme
Oxidative branch	
Glucose 6-phosphate + $NADP^+$ ⟶ 6-phosphoglucono-δ-lactone + NADPH + H^+	Glucose 6-phosphate dehydrogenase
6-Phosphoglucono-δ-lactone + H_2O ⟶ 6-phosphogluconate + H^+	Lactonase
6-Phosphogluconate + $NADP^+$ ⟶ ribulose 5-phosphate + CO_2 + NADPH	6-Phosphogluconate dehydrogenase
Nonoxidative branch	
Ribulose 5-phosphate ⇌ ribose 5-phosphate	Phosphopentose isomerase
Ribulose 5-phosphate ⇌ xylulose 5-phosphate	Phosphopentose epimerase
Xylulose 5-phosphate + ribose 5-phosphate ⇌ sedoheptulose 7-phosphate + glyceraldehyde 3-phosphate	Transketolase
Sedoheptulose 7-phosphate + glyceraldehyde 3-phosphate ⇌ fructose 6-phosphate + erythrose 4-phosphate	Transaldolase
Xylulose 5-phosphate + erythrose 4-phosphate ⇌ fructose 6-phosphate + glyceraldehyde 3-phosphate	Transketolase

with $NADP^+$ in binding to the enzyme. The ratio of $NADP^+$ to NADPH in the cytosol of a liver cell from a well-fed rat is about 0.014, several orders of magnitude lower than the ratio of NAD^+ to NADH, which is 700 under the same conditions. The marked effect of the $NADP^+$ level on the rate of the oxidative branch ensures that NADPH generation is tightly coupled to its utilization in reductive biosyntheses. The nonoxidative branch of the pentose phosphate pathway is controlled primarily by the availability of substrates.

THE FLOW OF GLUCOSE 6-PHOSPHATE DEPENDS ON THE NEED FOR NADPH, RIBOSE 5-PHOSPHATE, AND ATP

Let us follow the fate of glucose 6-phosphate in four different situations (see Figure 22-2):

Mode 1. *Much more ribose 5-phosphate than NADPH is required.* For example, rapidly dividing cells need ribose 5-phosphate for the synthesis of nucleotide precursors of DNA. Most of the glucose 6-phosphate is converted into fructose 6-phosphate and glyceraldehyde 3-phosphate by the glycolytic pathway. Transaldolase and transketolase then convert two molecules of fructose 6-phosphate and one molecule of glyceraldehyde 3-phosphate into three molecules of ribose 5-phosphate by a reversal of the reactions described earlier. The stoichiometry of mode 1 is

$$\text{5 Glucose 6-phosphate} + \text{ATP} \longrightarrow \text{6 ribose 5-phosphate} + \text{ADP} + H^+$$

Mode 2. *The needs for NADPH and ribose 5-phosphate are balanced.* The predominant reaction under these conditions is the formation of two NADPH and one ribose 5-phosphate from glucose 6-phosphate by the oxidative branch of the pentose phosphate pathway. The stoichiometry of mode 2 is

$$\text{Glucose 6-phosphate} + 2\ NADP^+ + H_2O \longrightarrow \text{ribose 5-phosphate} + 2\ NADPH + 2\ H^+ + CO_2$$

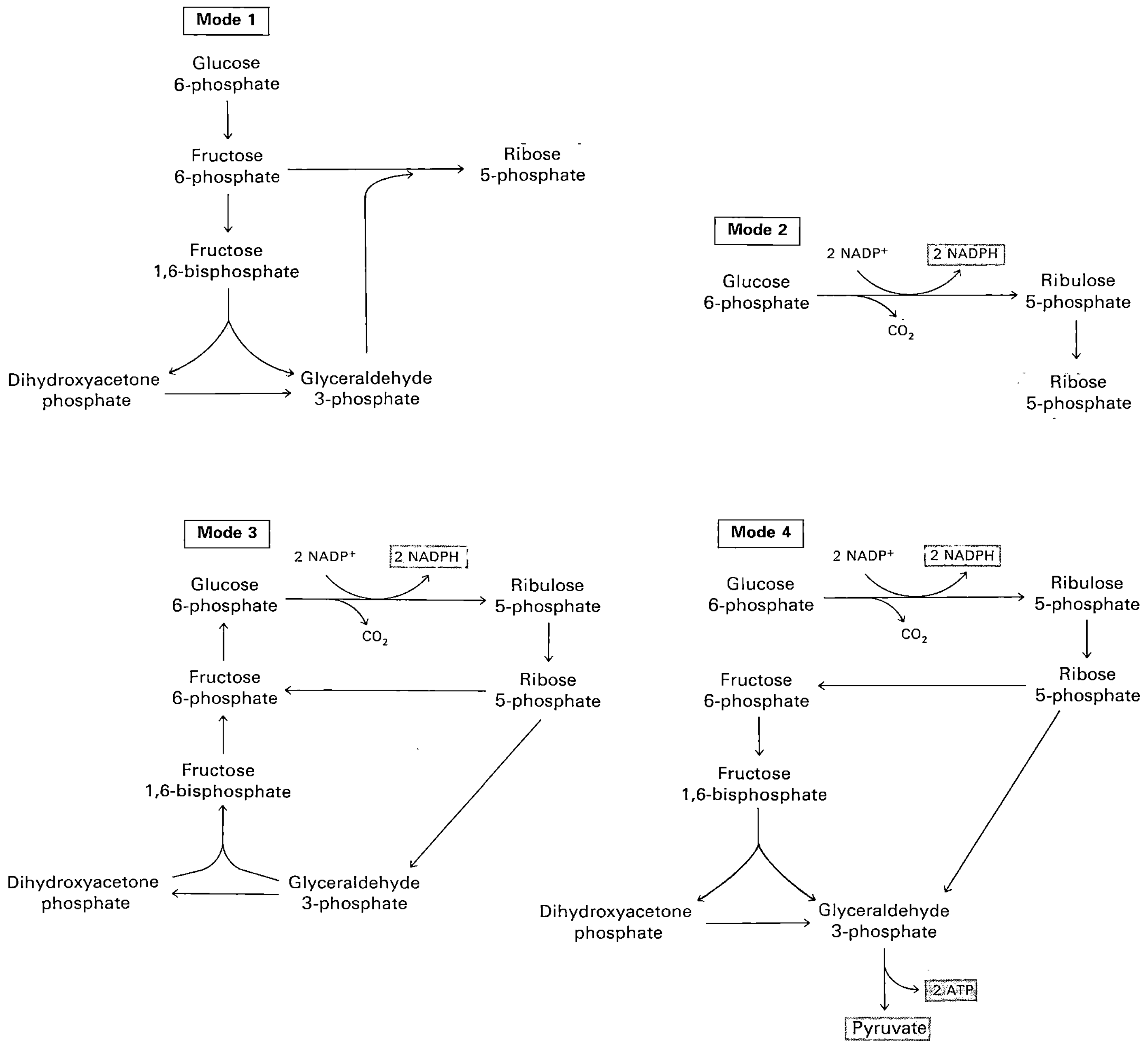

Figure 22-2
Four modes of the pentose phosphate pathway. Major products are shown in color.

Mode 3. *Much more NADPH than ribose 5-phosphate is required; glucose 6-phosphate is completely oxidized to CO_2.* For example, adipose tissue requires a high level of NADPH for the synthesis of fatty acids. Three groups of reactions are active in this situation. First, two NADPH and one ribose 5-phosphate are formed by the oxidative branch of the pentose phosphate pathway. Then, ribose 5-phosphate is converted into fructose 6-phosphate and glyceraldehyde 3-phosphate by transketolase and transaldolase. Finally, glucose 6-phosphate is resynthesized from fructose 6-phosphate and glyceraldehyde 3-phosphate by the gluconeogenic pathway (discussed later in this chapter). The stoichiometries of these three sets of reactions are

$$6 \text{ Glucose 6-phosphate} + 12 \text{ NADP}^+ + 6 \text{ H}_2\text{O} \longrightarrow 6 \text{ ribose 5-phosphate} + 12 \text{ NADPH} + 12 \text{ H}^+ + 6 \text{ CO}_2$$

$$\text{6 Ribose 5-phosphate} \longrightarrow \text{4 fructose 6-phosphate} + \text{2 glyceraldehyde 3-phosphate}$$

$$\text{4 Fructose 6-phosphate} + \text{2 glyceraldehyde 3-phosphate} + H_2O \longrightarrow \text{5 glucose 6-phosphate} + P_i$$

The sum of these reactions, which mode 3 comprises, is

$$\text{Glucose 6-phosphate} + 12\ NADP^+ + 7\ H_2O \longrightarrow 6\ CO_2 + 12\ NADPH + 12\ H^+ + P_i$$

Thus, *the equivalent of a glucose 6-phosphate can be completely oxidized to* CO_2 *with the concomitant generation of NADPH.* In essence, ribose 5-phosphate produced by the pentose phosphate pathway is recycled into glucose 6-phosphate by transketolase, transaldolase, and some of the enzymes of the gluconeogenic pathway.

Mode 4. *Much more NADPH than ribose 5-phosphate is required; glucose 6-phosphate is converted into pyruvate.* Alternatively, ribose 5-phosphate formed by the oxidative branch of the pentose phosphate pathway can be converted into pyruvate. Fructose 6-phosphate and glyceraldehyde 3-phosphate derived from ribose 5-phosphate enter the glycolytic pathway rather than reverting to glucose 6-phosphate. In this mode, *ATP and NADPH are concomitantly generated, and five of the six carbons of glucose 6-phosphate emerge in pyruvate.*

$$\text{3 Glucose 6-phosphate} + 6\ NADP^+ + 5\ NAD^+ + 5\ P_i + 8\ ADP \longrightarrow \text{5 pyruvate} + 3\ CO_2 + 6\ NADPH + 5\ NADH + 8\ ATP + 2\ H_2O + 8\ H^+$$

Pyruvate formed by these reactions can be oxidized to generate more ATP or it can be used as a building block in a variety of biosyntheses.

THE PENTOSE PHOSPHATE PATHWAY IS MUCH MORE ACTIVE IN ADIPOSE TISSUE THAN IN MUSCLE

Radioactive-labeling experiments can provide estimates of how much glucose 6-phosphate is metabolized by the pentose phosphate pathway and how much is metabolized by the combined action of glycolysis and the citric acid cycle. For example, one aliquot of a tissue homogenate is incubated with glucose labeled with ^{14}C at C-1, and another with glucose labeled with ^{14}C at C-6. The radioactivity of the CO_2 produced by the two samples is then compared. The rationale of this experiment is that only C-1 is decarboxylated by the pentose phosphate pathway, whereas C-1 and C-6 are decarboxylated equally when glucose is metabolized by the glycolytic pathway, the pyruvate dehydrogenase complex, and the citric acid cycle. The reason for the equivalence of C-1 and C-6 in the latter set of reactions is that glyceraldehyde 3-phosphate and dihydroxyacetone phosphate are rapidly interconverted by triose phosphate isomerase.

This experimental approach has shown that *the activity of the pentose phosphate pathway is very low in skeletal muscle, whereas it is very high in adipose tissue.* These findings support the idea that a major role of the pentose phosphate pathway is to generate NADPH for reductive biosyntheses. Large amounts of it are consumed by adipose tissue in the reductive synthesis of fatty acids from acetyl CoA (p. 621), as in the mammary gland during lactation.

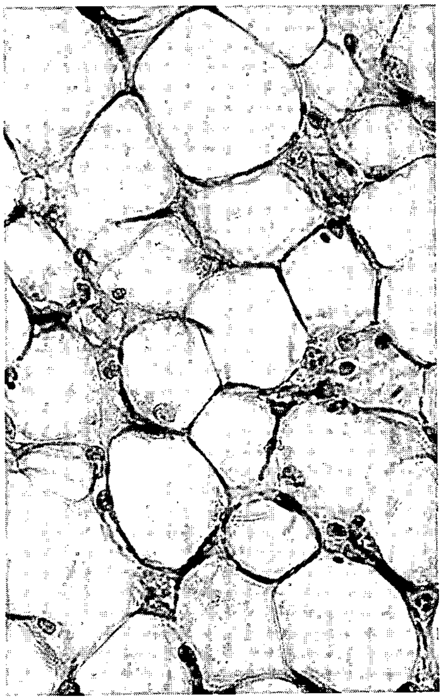

Figure 22-3
Light micrograph of adipose tissue.

TPP, THE PROSTHETIC GROUP OF TRANSKETOLASE, TRANSFERS A TWO-CARBON ACTIVATED ALDEHYDE

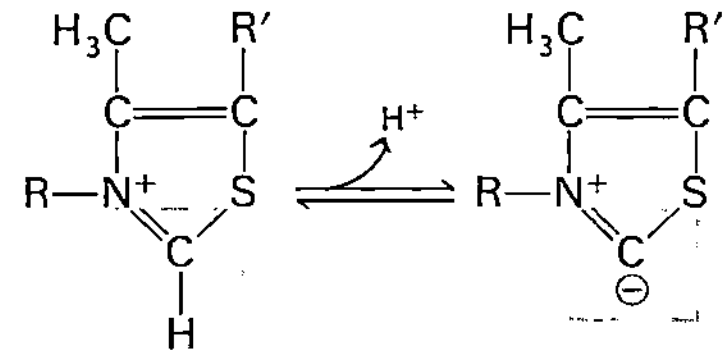

(TPP) Thiamine pyrophosphate   Carbanion

Transketolase contains a tightly bound thiamine pyrophosphate (TPP) as its prosthetic group. This prosthetic group has been encountered before, in the decarboxylation of pyruvate by the pyruvate dehydrogenase complex (p. 515). The mechanism of catalysis of transketolase is similar in that an *activated aldehyde unit is transferred to an acceptor.* The acceptor in the transketolase reaction is an aldose, whereas in the pyruvate dehydrogenase reaction it is lipoamide. In both reactions, the site of addition of the keto substrate is the *thiazole ring* of the prosthetic group. The C-2 carbon atom of bound TPP readily ionizes to give a *carbanion.* The negatively charged carbon atom of this reactive intermediate adds to the carbonyl group of the ketose substrate (e.g., xylulose 5-phosphate, fructose 6-phosphate, and sedoheptulose 7-phosphate).

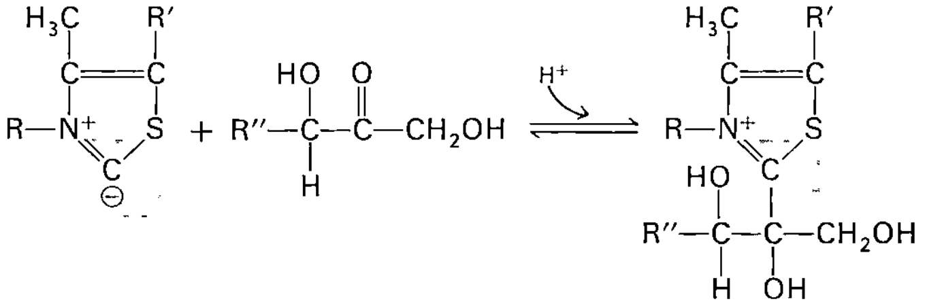

Carbanion   Ketose substrate   Addition compound

This addition compound loses its R″–CHOH moiety to yield a negatively charged, *activated glycoaldehyde* unit. The positively charged nitrogen in the thiazole ring acts as an *electron sink* to promote the development of a negative charge on the activated intermediate.

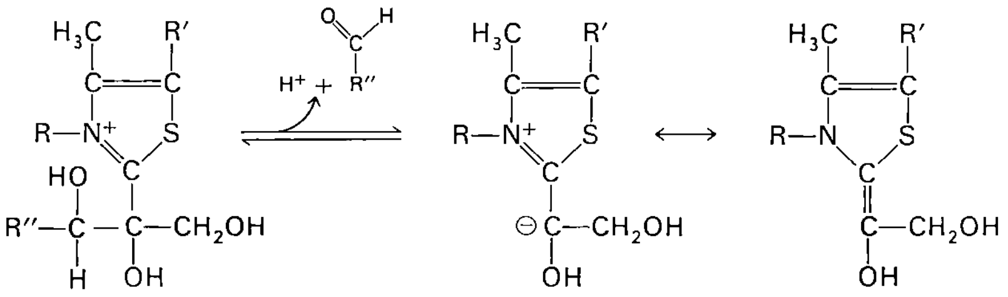

Addition compound   Resonance forms of the activated glycoaldehyde unit

The carbonyl group of a suitable aldehyde acceptor then condenses with the activated glycoaldehyde unit to form a new ketose, which is released from the enzyme.

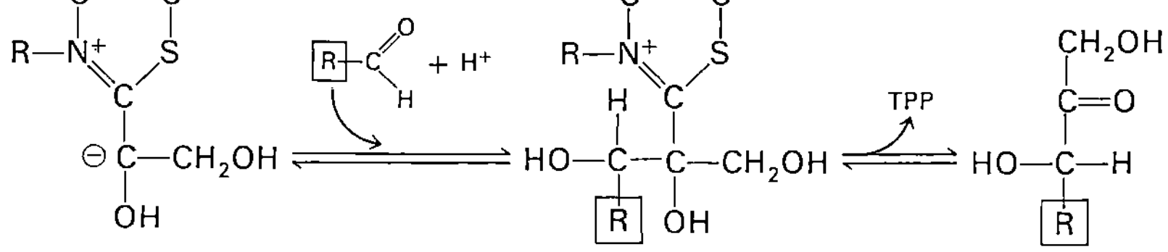

Activated glycoaldehyde   Addition compound   Ketose product

ACTIVATED DIHYDROXYACETONE IS CARRIED BY TRANSALDOLASE AS A SCHIFF BASE

Transaldolase transfers a three-carbon *dihydroxyacetone* unit from a ketose donor to an aldose acceptor. Transaldolase, in contrast with transketolase, does not contain a prosthetic group. Rather, *a Schiff base is formed between the carbonyl group of the ketose substrate and the ε-amino group of a lysine residue at the active site of the enzyme.* This kind of covalent enzyme-substrate (ES) intermediate is like that formed in fructose bisphosphate aldolase in the glycolytic pathway (p. 499).

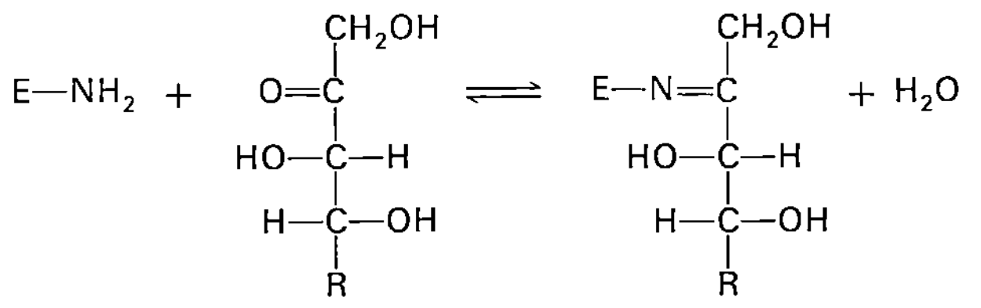

Ketose substrate Schiff base

The Schiff base becomes protonated, the bond between C-3 and C-4 is split, and an aldose is released.

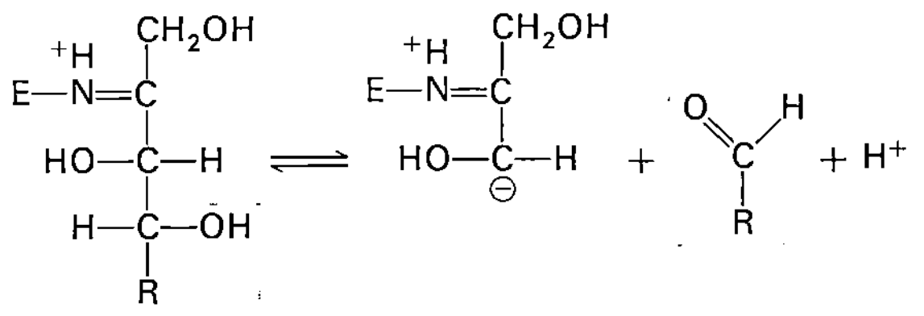

Protonated Schiff base Carbanion Aldose

The negative charge on the Schiff base carbanion moiety is stabilized by resonance. *The positively charged nitrogen atom of the Schiff base acts as an electron sink.* This nitrogen atom plays the same role in transaldolase as does the thiazole ring nitrogen in transketolase.

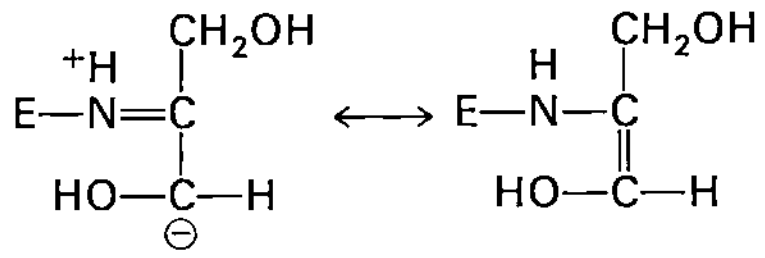

Resonance forms of the Schiff base carbanion

The Schiff base carbanion is stable until a suitable aldose becomes bound. The dihydroxyacetone moiety then reacts with the carbonyl group of the aldose. The ketose product is released by hydrolysis of the Schiff base.

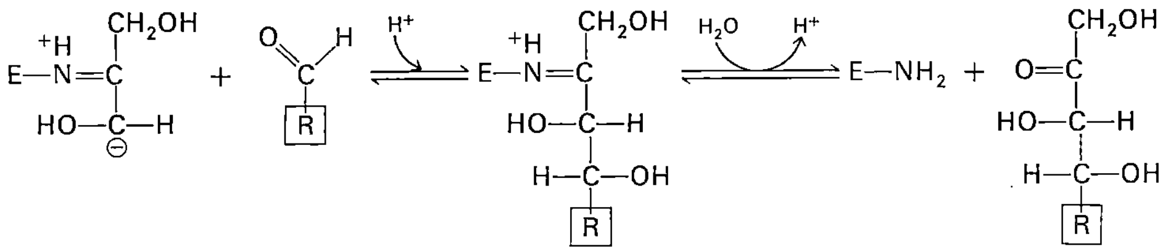

Carbanion Aldose substrate Schiff base Ketose product

GLUCOSE 6-PHOSPHATE DEHYDROGENASE DEFICIENCY CAUSES A DRUG-INDUCED HEMOLYTIC ANEMIA

An antimalarial drug, pamaquine, was introduced in 1926. Most patients tolerated the drug well, but a few developed severe symptoms within a few days after therapy was started. The urine turned black, jaundice developed, and the hemoglobin content of the blood dropped sharply. In some cases, massive destruction of red blood cells caused death.

This drug-induced hemolytic anemia was shown 30 years later to be caused by *a deficiency of glucose 6-phosphate dehydrogenase,* the enzyme catalyzing the first step in the oxidative branch of the pentose phosphate pathway. This defect, which affects hundreds of millions of people, is inherited as a sex-linked trait. Red blood cells are affected because they lack mitochondria and hence have no alternative means of generating NADPH (p. 560). The major role of NADPH in red cells is to reduce the disulfide form of *glutathione* to the sulfhydryl form, a reaction catalyzed by *glutathione reductase.*

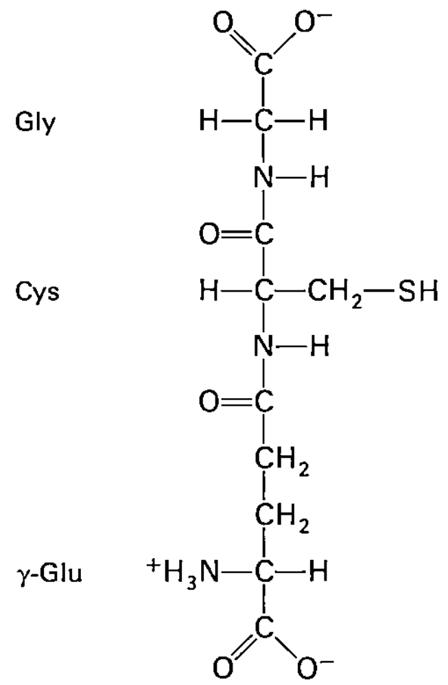

Reduced glutathione
(γ-Glutamylcysteinylglycine)

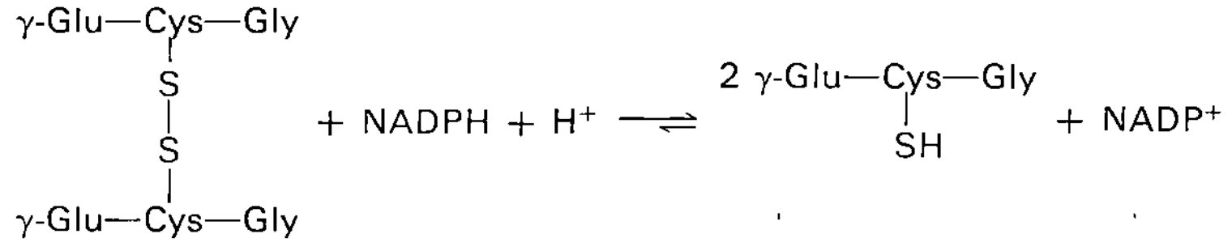

Oxidized glutathione (GSSG)

Reduced glutathione (GSH)

The reduced form of glutathione, a tripeptide with a free sulfhydryl group, serves as a *sulfhydryl buffer* that maintains the cysteine residues of hemoglobin and other red-cell proteins in the reduced state. The reduced form also plays a role in detoxification by reacting with hydrogen peroxide and organic peroxides, as will be discussed in a later chapter (p. 731). Normally, the ratio of the reduced to oxidized form of glutathione in red cells is 500.

$$2\ \text{GSH} + \text{R—O—OH} \longrightarrow \text{GSSG} + \text{H}_2\text{O} + \text{ROH}$$

Reduced glutathione is essential for maintaining the normal structure of red cells and for keeping hemoglobin in the ferrous state. Cells with a lowered level of reduced glutathione are more susceptible to hemolysis. Drugs such as pamaquine (used years ago to treat malaria) distort the surface of red cells in the absence of reduced glutathione, which makes them more liable to destruction and removal by the spleen. These drugs also increase the rate of formation of toxic peroxides, which are normally eliminated by reaction with reduced glutathione. The ingestion of broad beans (also known as fava beans, *Vicia fava*) can likewise induce a hemolytic anemia in dehydrogenase-deficient people. The deficiency is otherwise quite benign.

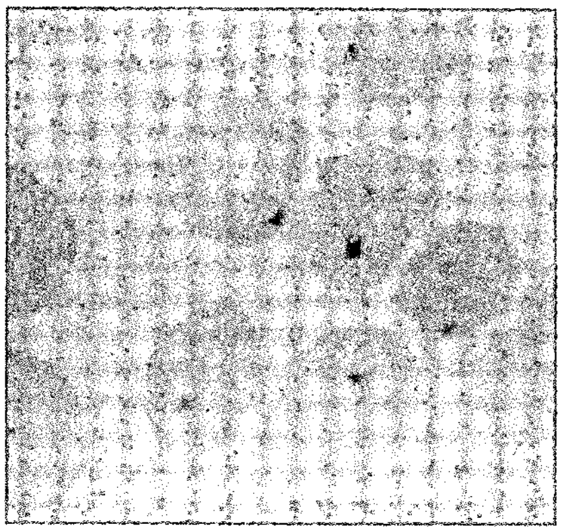

Figure 22-4
Light micrograph of red blood cells from a person deficient in glucose 6-phosphate dehydrogenase. The dark particles, called Heinz bodies, inside the cells are clumps of denatured protein that adhere to the plasma membrane and stain with basic dyes. Red blood cells in these people are highly susceptible to oxidative damage. [Courtesy of Dr. Stanley Schrier.]

The incidence of the most common form of glucose 6-phosphate dehydrogenase deficiency, characterized by a tenfold reduction in enzymatic activity in red cells, is 11% among Americans of African heritage. This high frequency suggests that the deficiency may be advantageous under certain environmental conditions. Indeed, *glucose 6-phosphate dehydrogenase deficiency protects against falciparum malaria.* The parasites causing this disease require reduced glutathione and the products of the pentose phosphate pathway for optimal growth. Thus, glucose 6-phosphate dehydrogenase deficiency and sickle-cell trait (p. 172) are parallel mechanisms of protection against malaria, which accounts for their high frequency in malaria-infested regions of the world. The occurrence of this dehydrogenase deficiency also clearly demonstrates that *atypical reactions to drugs may have a genetic basis. We see here once again the interplay of heredity and environment in the production of disease.* Galactosemia (p. 493) and phenylketonuria (p. 649) are other striking examples.

GLUTATHIONE REDUCTASE TRANSFERS ELECTRONS FROM NADPH TO OXIDIZED GLUTATHIONE THROUGH FAD

The regeneration of reduced glutathione is catalyzed by *glutathione reductase,* a dimer of 50-kd subunits. The electrons from NADPH are not directly transferred to the disulfide bond in oxidized glutathione. Rather, they are transferred from NADPH to a tightly bound flavin adenine dinucleotide (FAD), then to a disulfide bridge between two cysteine residues in the subunit, and finally to oxidized glutathione. Each subunit consists of three structural domains: an FAD-binding domain, an $NADP^+$-binding domain, and an interface domain (Figure 22-5). The FAD domain and $NADP^+$ domain resemble each other and are similar to nucleotide-binding domains in other dehydrogenases. FAD and $NADP^+$ are bound in an extended form, with their isoalloxazine and nicotinamide rings next to each other. It is interesting to note that the binding site for oxidized glutathione is formed by the FAD domain of one subunit and the interface domain of the other subunit.

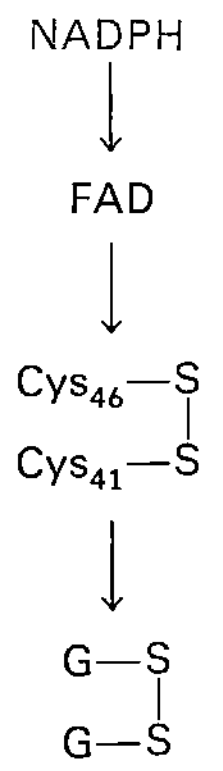

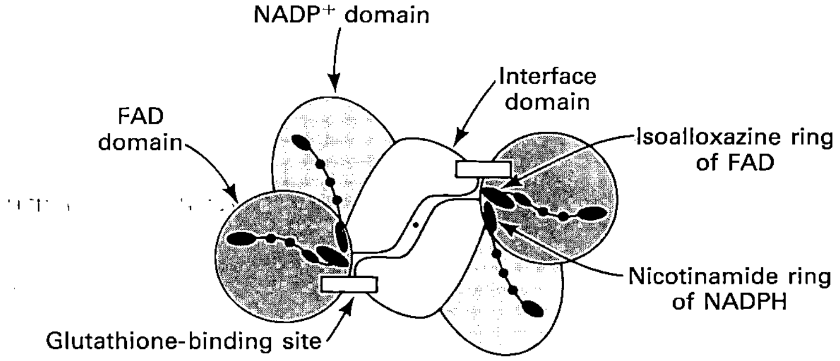

Figure 22-5
Schematic diagram of the domain structure of glutathione reductase. Each subunit in this dimeric enzyme consists of an $NADP^+$ domain, an FAD domain, and an interface domain. Glutathione is bound to the FAD domain of one subunit and the interface domain of another. [After G.E. Schultz, R.H. Schirmer, W. Sachsenheimer, and E.F. Pai. *Nature* 273(1978):123.]

GLUCOSE CAN BE SYNTHESIZED FROM NONCARBOHYDRATE PRECURSORS

We now turn to the *synthesis of glucose from noncarbohydrate precursors,* a process called *gluconeogenesis.* This metabolic pathway is very important because the brain is highly dependent on glucose as the primary fuel. Erythrocytes, too, require glucose. The daily glucose requirement of the brain in a typical adult is about 120 g, which accounts for most of the 160 g of glucose needed by the whole body. The amount of glucose present in body fluids is about 20 g, and that readily available from glycogen, a storage form of glucose (p. 581), is approximately 190 g. Thus, the direct glucose reserves are sufficient to meet the needs for glucose for about a day. In a longer period of starvation, glucose must be formed from noncarbohydrate sources for survival. Gluconeogenesis is also important during periods of intense exercise.

The gluconeogenic pathway converts pyruvate into glucose. Noncarbohydrate precursors of glucose enter the pathway chiefly at pyruvate, oxaloacetate, and dihydroxyacetone phosphate (Figure 22-6). The major noncarbohydrate precursors are *lactate, amino acids,* and *glycerol.* Lactate is formed by

Gluconeogenesis—
Formation of glucose from noncarbohydrate sources. Derived from the Greek words *glykys,* "sweet," *neo,* "new," and *genesis,* "origin" or "generation." The prefixes "gluco" and "glyco" are derived from the same Greek root, *glykys;* they differ only in transliteration. Glyconeogenesis is an older (and etymologically more accurate) term for gluconeogenesis.

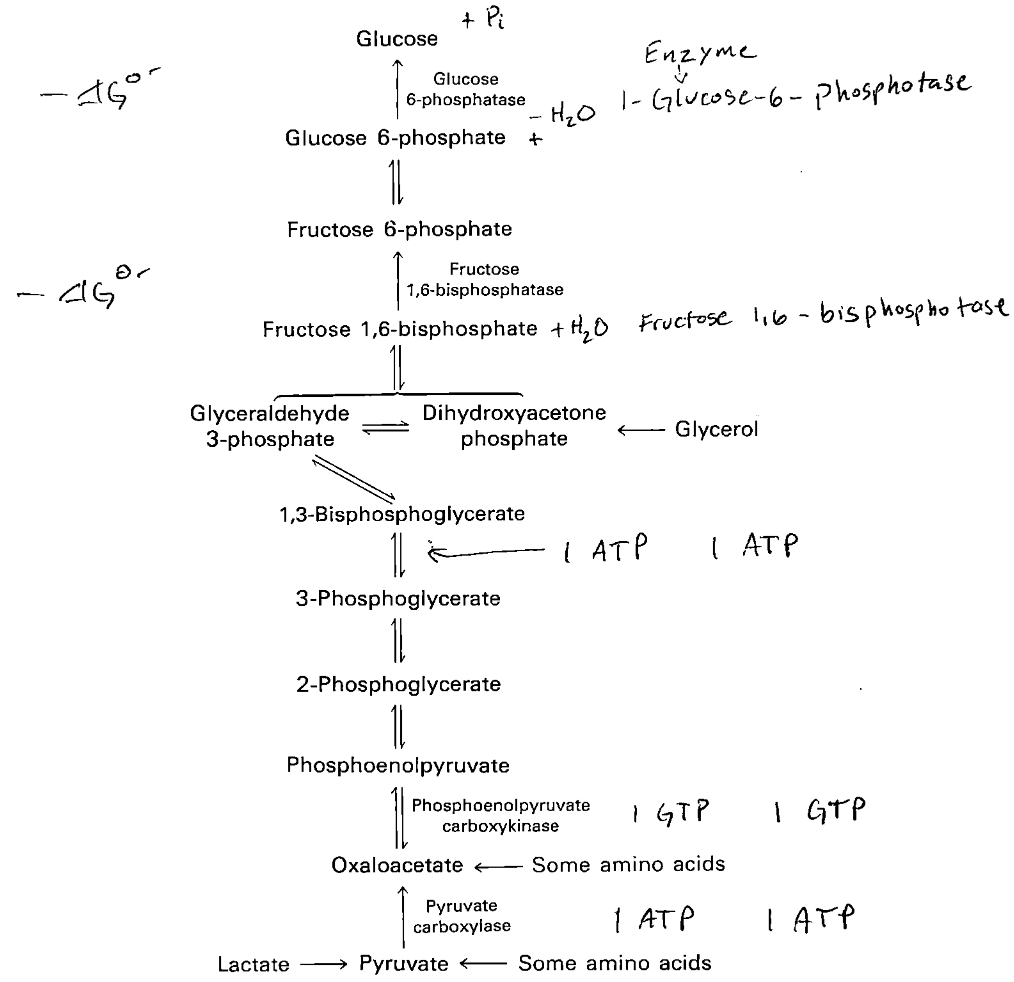

Figure 22-6
Pathway of gluconeogenesis. The distinctive reactions of this pathway are shown in red. The other reactions are common to glycolysis. The enzymes of gluconeogenesis are located in the cytosol, except for pyruvate carboxylase (in mitochondria) and glucose 6-phosphatase (membrane-bound inside the endoplasmic reticulum). The entry points for lactate, glycerol, and amino acids are shown.

active skeletal muscle when the rate of glycolysis exceeds the metabolic rate of the citric acid cycle and the respiratory chain (p. 497). Amino acids are derived from proteins in the diet and, during starvation, from the breakdown of proteins in skeletal muscle (p. 776). The hydrolysis of triacylglycerols (p. 605) in fat cells yields glycerol and fatty acids. Glycerol is a precursor of glucose, but animals cannot convert fatty acids into glucose, for reasons that will be discussed later (p. 613).

The major site of gluconeogenesis is the *liver*. Gluconeogenesis also occurs in the cortex of the *kidney*, but the total amount of glucose formed there is about one-tenth of that formed in the liver because of the kidney's smaller mass. Very little gluconeogenesis takes place in the brain, skeletal muscle, or heart muscle. Rather, *gluconeogenesis in the liver and kidney helps to maintain the glucose level in the blood so that brain and muscle can extract sufficient glucose from it to meet their metabolic demands.*

GLUCONEOGENESIS IS NOT A REVERSAL OF GLYCOLYSIS

In glycolysis, glucose is converted into pyruvate; in gluconeogenesis, pyruvate is converted into glucose. However, gluconeogenesis is not a reversal of glycolysis. Several different reactions are required because the thermodynamic equilibrium of glycolysis lies far on the side of pyruvate formation. The actual ΔG for the formation of pyruvate from glucose is about -20 kcal/mol under typical cellular conditions (p. 491). Most of the decrease in free energy in glycolysis takes place in the three essentially irreversible steps catalyzed by hexokinase, phosphofructokinase, and pyruvate kinase.

$$\text{Glucose} + \text{ATP} \xrightarrow{\text{Hexokinase}} \text{glucose 6-phosphate} + \text{ADP}$$

$$\text{Fructose 6-phosphate} + \text{ATP} \xrightarrow{\text{Phosphofructokinase}} \text{fructose 1,6-bisphosphate} + \text{ADP}$$

$$\text{Phosphoenolpyruvate} + \text{ADP} \xrightarrow{\text{Pyruvate kinase}} \text{pyruvate} + \text{ATP}$$

In gluconeogenesis, these virtually irreversible reactions of glycolysis are bypassed by the following new steps:

1. *Phosphoenolpyruvate is formed from pyruvate by way of oxaloacetate.* First, pyruvate is carboxylated to oxaloacetate at the expense of an ATP. Then, oxaloacetate is decarboxylated and phosphorylated to yield phosphoenolpyruvate, at the expense of a second high-energy phosphate bond.

$$\text{Pyruvate} + CO_2 + \text{ATP} + H_2O \longrightarrow \text{oxaloacetate} + \text{ADP} + P_i + 2\ H^+$$

$$\text{Oxaloacetate} + \text{GTP} \rightleftharpoons \text{phosphoenolpyruvate} + \text{GDP} + CO_2$$

The first reaction is catalyzed by *pyruvate carboxylase,* and the second by *phosphoenolpyruvate carboxykinase.* The sum of these reactions, which occur in mitochondria, is

$$\text{Pyruvate} + \text{ATP} + \text{GTP} + H_2O \longrightarrow \text{phosphoenolpyruvate} + \text{ADP} + \text{GDP} + P_i + 2\ H^+$$

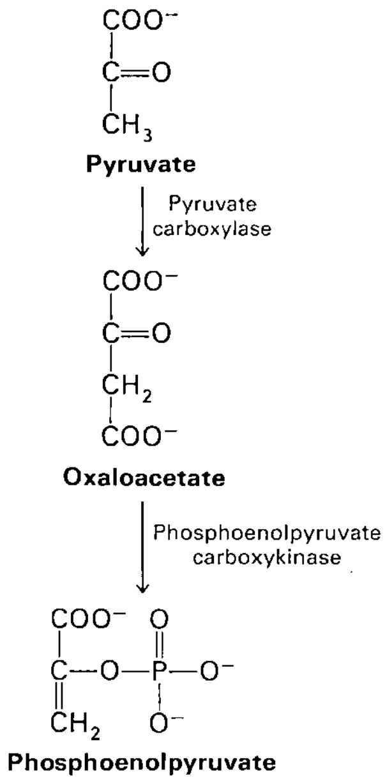

This two-step pathway for the formation of phosphoenolpyruvate from pyruvate is thermodynamically feasible, because $\Delta G^{\circ\prime}$ is +0.2 kcal/mol in contrast with +7.5 kcal/mol for the reaction catalyzed by pyruvate kinase. This much more favorable $\Delta G^{\circ\prime}$ results from the input of an additional high-energy phosphate bond in the carboxylation step.

2. *Fructose 6-phosphate is formed from fructose 1,6-bisphosphate by hydrolysis of the phosphate ester at C-1.* Fructose 1,6-bisphosphatase catalyzes this exergonic hydrolysis.

$$\text{Fructose 1,6-bisphosphate} + H_2O \longrightarrow \text{fructose 6-phosphate} + P_i$$

3. *Glucose is formed by hydrolysis of glucose 6-phosphate,* in a reaction catalyzed by glucose 6-phosphatase.

$$\text{Glucose 6-phosphate} + H_2O \longrightarrow \text{glucose} + P_i$$

The final step in the generation of glucose is not carried out in the cytosol. Rather, glucose 6-phosphate is transported into the lumen of the endoplasmic reticulum, where it is hydrolyzed by *glucose 6-phosphatase,* a membrane-bound enzyme (Figure 22-7). An associated Ca^{2+}-binding sta-

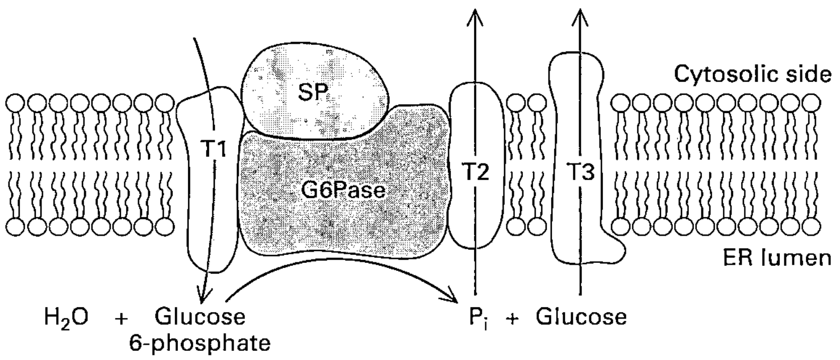

Figure 22-7
Schematic diagram of endoplasmic reticulum (ER) proteins that participate in the generation of glucose from cytosolic glucose 6-phosphate. T1 transports glucose 6-phosphate into the lumen of the ER, whereas T2 and T3 transport P_i and glucose, respectively, back into the cytosol. Glucose 6-phosphatase is stabilized by a Ca^{2+}-binding protein (SP). [After A. Burchell and I.D. Waddell. *Biochim. Biophys. Acta* 1092(1991):129.]

bilizing protein is essential for phosphatase activity. Glucose and P_i are then shuttled back to the cytosol by a pair of transporters. The glucose transporter in the endoplasmic reticulum membrane is like those found in the plasma membrane (p. 505). It is striking that five proteins are needed to transform cytosolic glucose 6-phosphate into glucose. Glucose 6-phosphatase is not present in brain and muscle; hence, glucose cannot be formed by these organs.

Table 22-2
Enzymatic differences between glycolysis and gluconeogenesis

Glycolysis	*Gluconeogenesis*
Hexokinase	Glucose 6-phosphatase
Phosphofructokinase	Fructose 1,6-bisphosphatase
Pyruvate kinase	Pyruvate carboxylase Phosphoenolpyruvate carboxykinase

BIOTIN IS A MOBILE CARRIER OF ACTIVATED CO_2

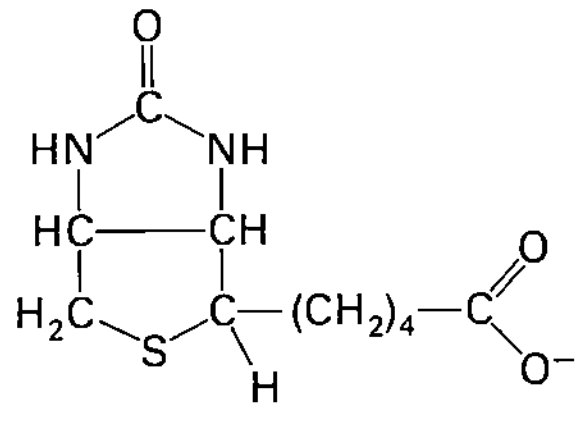

Biotin

The finding that mitochondria can form oxaloacetate from pyruvate led to the discovery of pyruvate carboxylase by Merton Utter in 1960. This enzyme is of especial interest because of its catalytic and allosteric properties. Pyruvate carboxylase contains a covalently attached prosthetic group, *biotin*, which serves as a *carrier of activated CO_2*. The carboxyl terminus of biotin is linked to the ε-amino group of a specific lysine residue by an amide bond.

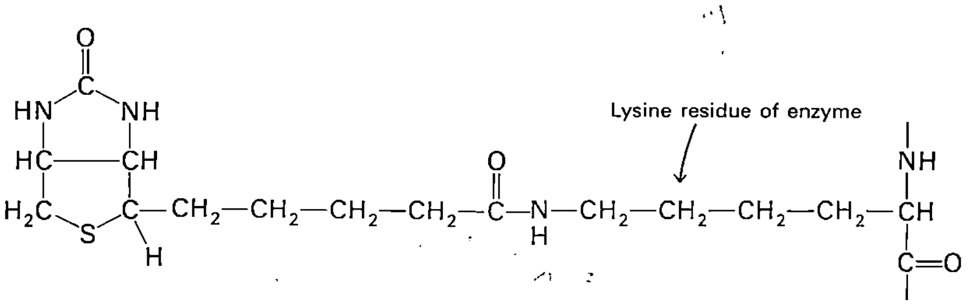

Note that biotin is attached to pyruvate carboxylase by a *long, flexible chain* like that of lipoamide in the pyruvate dehydrogenase complex. The carboxylation of pyruvate occurs in two stages:

$$\text{Biotin-enzyme} + \text{ATP} + \text{HCO}_3^- \xrightleftharpoons[]{\text{Mg}^{2+},\ \text{acetyl CoA}} \text{CO}_2 \sim \text{biotin-enzyme} + \text{ADP} + \text{P}_i$$

$$\text{CO}_2 \sim \text{biotin-enzyme} + \text{pyruvate} \xrightleftharpoons[]{\text{Mn}^{2+}} \text{biotin-enzyme} + \text{oxaloacetate}$$

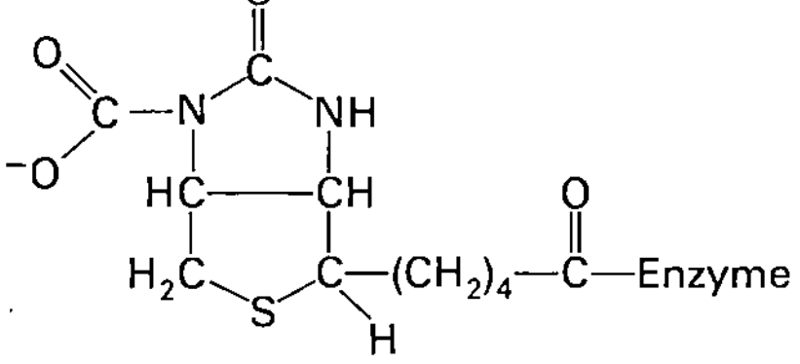

Carboxybiotin-enzyme intermediate

The carboxyl group in the carboxybiotin-enzyme intermediate is bonded to the N-1 nitrogen atom of the biotin ring. This carboxyl group is *activated*. The $\Delta G^{\circ\prime}$ for its cleavage

$$\text{CO}_2 \sim \text{biotin-enzyme} + \text{H}^+ \longrightarrow \text{CO}_2 + \text{biotin-enzyme}$$

is -4.7 kcal/mol, which enables carboxybiotin to transfer CO_2 to acceptors without the input of additional free energy.

The activated carboxyl group is then transferred from carboxybiotin to pyruvate to form oxaloacetate. The long, flexible link between biotin and the enzyme enables this prosthetic group to rotate from one active site of the enzyme (the ATP-bicarbonate site) to the other (the pyruvate site).

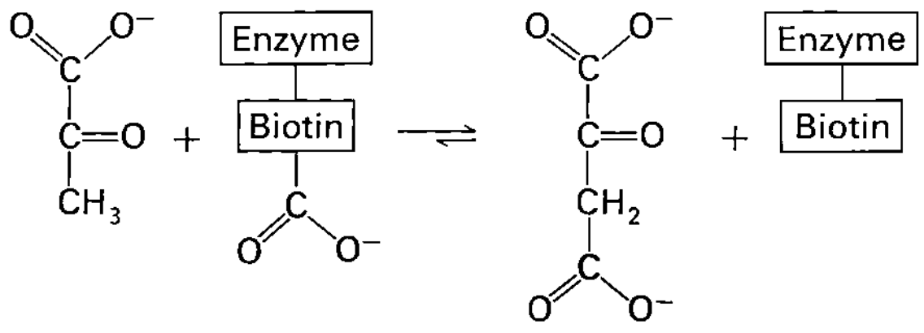

Pyruvate Oxaloacetate

PYRUVATE CARBOXYLASE IS ACTIVATED BY ACETYL CoA

The activity of pyruvate carboxylase depends on the presence of acetyl CoA. *Biotin is not carboxylated unless acetyl CoA (or a closely related acyl CoA) is bound to the enzyme.* The second partial reaction is not affected by acetyl CoA. The allosteric activation of pyruvate carboxylase by acetyl CoA is an important physiologic control mechanism. Oxaloacetate, the product of the pyruvate carboxylase reaction, is both a stoichiometric intermediate in gluconeogenesis and a catalytic intermediate in the citric acid cycle. *A high level of acetyl CoA signals the need for more oxaloacetate.* If there is a surplus of ATP, oxaloacetate will be consumed in gluconeogenesis. If there is a deficiency of ATP, oxaloacetate will enter the citric acid cycle upon condensing with acetyl CoA.

Thus, not only is pyruvate carboxylase important in gluconeogenesis, but it also plays a *critical role in maintaining the level of citric acid cycle intermediates.* These intermediates need to be replenished because they are consumed in some biosynthetic reactions, such as heme synthesis. This role of pyruvate carboxylase is termed *anaplerotic,* meaning "to fill up."

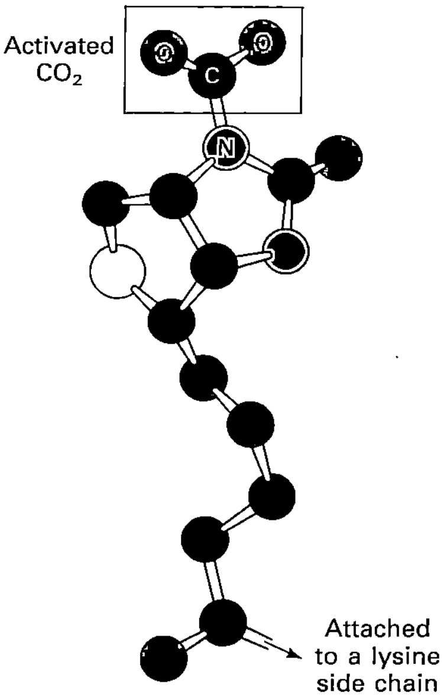

Figure 22-8
Molecular model of carboxybiotin.

OXALOACETATE IS SHUTTLED INTO THE CYTOSOL AND CONVERTED INTO PHOSPHOENOLPYRUVATE

Pyruvate carboxylase is a mitochondrial enzyme, whereas the other enzymes of gluconeogenesis leading to the formation of glucose 6-phosphate are cytoplasmic. Oxaloacetate, the product of the pyruvate carboxylase reaction, is transported across the mitochondrial membrane in the form of *malate* (Figure 22-9). Oxaloacetate is reduced to malate inside the mitochondrion by an NADH-linked malate dehydrogenase. Malate is transported across the mitochondrial membrane and is reoxidized to oxaloacetate by an NAD^+-linked malate dehydrogenase in the cytosol.

Finally, oxaloacetate is simultaneously *decarboxylated* and *phosphorylated* by phosphoenolpyruvate carboxykinase in the cytosol.

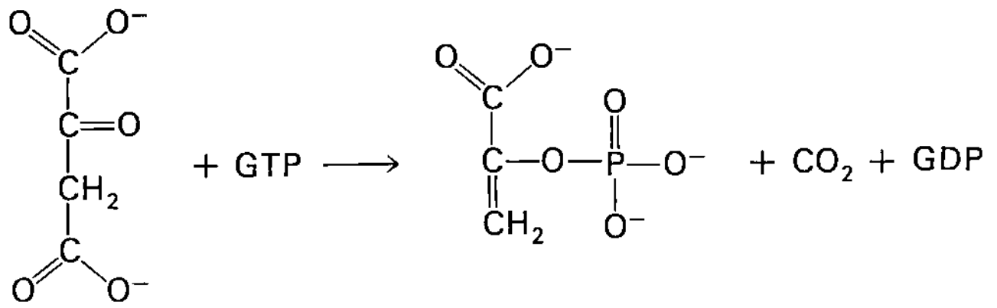

Oxaloacetate Phosphoenolpyruvate

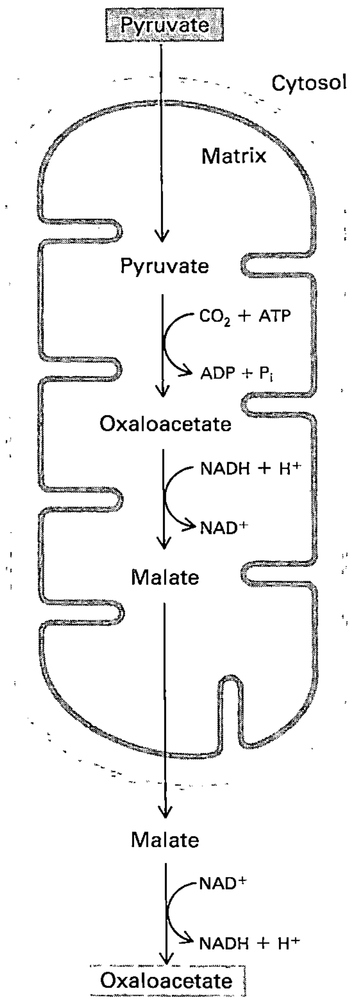

Figure 22-9
Oxaloacetate utilized in the cytosol for gluconeogenesis is formed in the mitochondrial matrix by carboxylation of pyruvate. Oxaloacetate leaves the mitochondrion in the form of malate, which is reoxidized to oxaloacetate in the cytosol.

The CO_2 that was added to pyruvate by pyruvate carboxylase comes off in this step. In fact, the phosphorylation reaction is made energetically feasible by the concomitant decarboxylation. *Decarboxylations often drive reactions that would otherwise be highly endergonic.* This device is used in the citric acid cycle and the pentose phosphate pathway, and we shall see it used in fatty acid synthesis (p. 617).

SIX HIGH-ENERGY PHOSPHATE BONDS ARE SPENT IN SYNTHESIZING GLUCOSE FROM PYRUVATE

The stoichiometry of gluconeogenesis is

$$2\ \text{Pyruvate} + 4\ \text{ATP} + 2\ \text{GTP} + 2\ \text{NADH} + 6\ H_2O \longrightarrow \text{glucose} + 4\ \text{ADP} + 2\ \text{GDP} + 6\ P_i + 2\ \text{NAD}^+ + 2\ H^+ \qquad \Delta G^{\circ\prime} = -9\ \text{kcal/mol}$$

In contrast, the stoichiometry for the reversal of glycolysis is

$$2\ \text{Pyruvate} + 2\ \text{ATP} + 2\ \text{NADH} + 2\ H_2O \longrightarrow \text{glucose} + 2\ \text{ADP} + 2\ P_i + 2\ \text{NAD}^+ \qquad \Delta G^{\circ\prime} = +20\ \text{kcal/mol}$$

Note that *six* high-energy phosphate bonds are used to synthesize glucose from pyruvate in gluconeogenesis, whereas only *two* molecules of ATP are generated in glycolysis in the conversion of glucose into pyruvate. Thus, the extra price of gluconeogenesis is four high-energy phosphate bonds per glucose synthesized from pyruvate. The four extra high-energy phosphate bonds are needed to turn an energetically unfavorable process (the reversal of glycolysis, $\Delta G^{\circ\prime} = +20$ kcal/mol) into a favorable one (gluconeogenesis, $\Delta G^{\circ\prime} = -9$ kcal/mol). Another way of looking at this energetic difference between glycolysis and gluconeogenesis is to recall that the input of an ATP equivalent changes the equilibrium constant of a reaction by a factor of about 10^8 (p. 448). Hence, the input of four additional high-energy bonds in gluconeogenesis changes the equilibrium constant by a factor of 10^{32}, which makes the conversion of pyruvate into glucose thermodynamically feasible.

GLUCONEOGENESIS AND GLYCOLYSIS ARE RECIPROCALLY REGULATED

Gluconeogenesis and glycolysis are coordinated so that one pathway is relatively inactive while the other is highly active. If both sets of reactions were highly active at the same time, the net result would be the hydrolysis of four ~P (two ATP plus two GTP) per reaction cycle. Both glycolysis and gluconeogenesis are highly exergonic under cellular conditions, and so there is no thermodynamic barrier to such cycling. However, the *amounts* and *activities* of the distinctive enzymes of each pathway are controlled so that both pathways are not highly active at the same time. The rate of glycolysis is also determined by the concentration of glucose, and the rate of gluconeogenesis by the concentrations of lactate and other precursors of glucose.

The interconversion of fructose 6-phosphate and fructose 1,6-bisphosphate (F-1,6-BP) is stringently controlled (Figure 22-10). As discussed in a previous chapter (p. 493), phosphofructokinase is stimulated by AMP, whereas it is inhibited by ATP and citrate. Fructose 1,6-bisphosphatase, on the other hand, is inhibited by AMP and activated by citrate. A high level of AMP indicates that the energy charge is low and signals the need for ATP generation. Conversely, a high level of ATP and citrate indicates that the energy charge is high and that biosynthetic intermediates are abundant. Under these conditions, glycolysis is nearly switched off and gluconeogenesis is promoted.

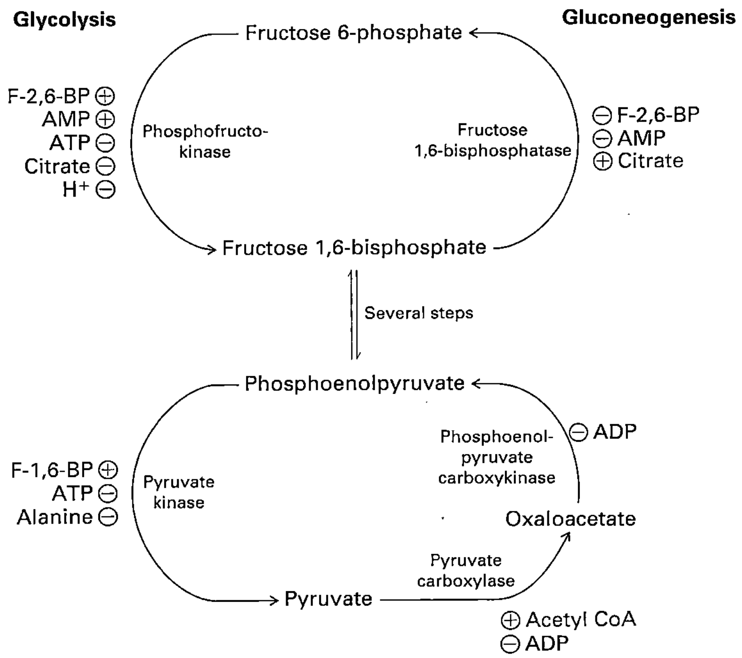

Figure 22-10
Reciprocal regulation of gluconeogenesis and glycolysis in the liver. The level of F-2,6-BP is high in the fed state and low in starvation. Another key control is the inhibition of pyruvate kinase by phosphorylation during starvation.

Phosphofructokinase and fructose 1,6-bisphosphatase are also reciprocally controlled by *fructose 2,6-bisphosphate (F-2,6-BP)*, a signal molecule derived from fructose 6-phosphate (p. 494). The level of F-2,6-BP is low during starvation, and high in the fed state, because of the antagonistic effects of glucagon and insulin on the production and degradation of this signal molecule. Recall that F-2,6-BP is formed and hydrolyzed by a bifunctional enzyme that is regulated by phosphorylation. *F-2,6-BP strongly stimulates phosphofructokinase and inhibits fructose 1,6-bisphosphatase.* Hence, glycolysis is accelerated and gluconeogenesis is diminished in the fed state. During starvation, gluconeogenesis predominates because the level of F-2,6-BP is very low. Glucose formed by the liver under these conditions is essential for the viability of brain and muscle.

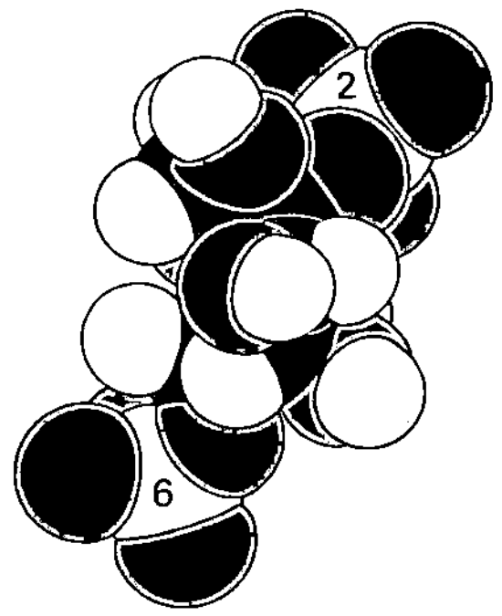

Fructose 2,6-bisphosphate (F-2,6-BP)

Model of fructose 2,6-bisphosphate, a key regulator of glycolysis and gluconeogenesis.

The interconversion of phosphoenolpyruvate and pyruvate is also precisely regulated. Recall that pyruvate kinase is controlled by allosteric effectors and by phosphorylation (p. 496). The enzyme in liver is inhibited by high levels of ATP and alanine, which signal that the energy charge is high and that building blocks are abundant. Conversely, pyruvate carboxylase, which catalyzes the first step in gluconeogenesis from pyruvate, is activated by acetyl CoA and inhibited by ADP. Likewise, phosphoenolpyruvate carboxykinase is inhibited by ADP, when the energy charge of a cell is low. Hence, gluconeogenesis is favored when the cell is rich in biosynthetic precursors and ATP. Pyruvate kinase, a key glycolytic enzyme, is also switched off when glucose is more urgently needed by brain and muscle. Glucagon triggers a cAMP cascade that leads to the phosphorylation and consequent inhibition of pyruvate kinase during starvation.

The amounts as well as the activities of these key enzymes are regulated. Hormones affect gene expression primarily by changing the rate of transcription. The degradation of some mRNAs also appears to be regulated. Transcriptional control occurs in times of hours to days, whereas allosteric control is much swifter, in times of seconds to minutes. Insulin, which rises following feeding, stimulates the expression of phosphofructokinase, pyruvate kinase, and the bifunctional enzyme that makes and degrades F-2,6-BP. Glucagon, which rises during starvation, inhibits the expression of these enzymes, and instead stimulates the production of two key gluconeogenic enzymes, phosphoenolpyruvate carboxykinase and fructose 1,6-bisphosphatase. The richness and complexity of hormonal control are graphically displayed by the promoter of the carboxykinase gene, which contains regulatory sequences that respond to insulin, glucagon, glucocorticoids, and thyroid hormone (Figure 22-11). The precise role of these regulatory elements and their interplay in metabolic regulation can now be studied in vivo in transgenic animals that harbor specifically engineered genes.

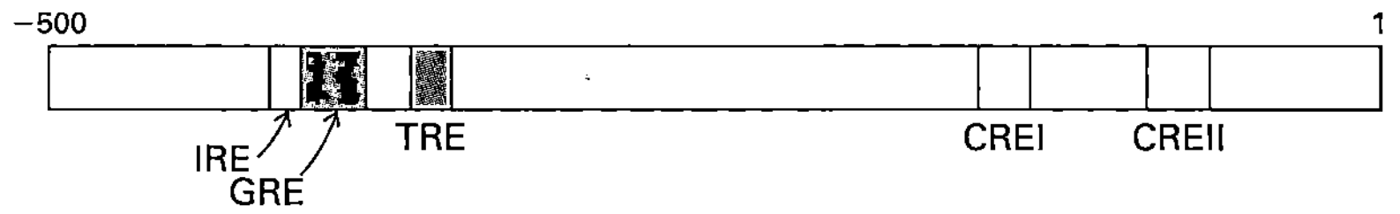

Figure 22-11
The promoter of the phosphoenolpyruvate carboxykinase gene, a 500-bp region, contains regulatory sequences that mediate the actions of several hormones. IRE, insulin response element; GRE, glucocorticoid response element; TRE, thyroid hormone response element; CREI and CREII, cAMP response elements. [After M.M. McGrane, J.S. Jun, Y.M. Patel, and R.W. Hanson. *Trends Biochem. Sci.* 17(1992):40.]

SUBSTRATE CYCLES AMPLIFY METABOLIC SIGNALS AND PRODUCE HEAT

A pair of reactions such as the phosphorylation of fructose 6-phosphate to fructose 1,6-bisphosphate and its hydrolysis back to fructose 6-phosphate is called a *substrate cycle.* As already mentioned, both reactions are not simultaneously fully active in most cells because of reciprocal allosteric controls. However, isotope labeling studies have shown that fructose 6-phosphate is phosphorylated to fructose 1,6-bisphosphate during gluconeogenesis. A limited degree of cycling also occurs in other pairs of opposed irreversible reactions. This cycling was regarded as an imperfection in metabolic control, and so substrate cycles have sometimes been called *futile cycles.* However, it now seems likely that substrate cycles are biologically important. One possibility is that *substrate cycles amplify metabolic signals.* Suppose that the rate of conversion of A into B is 100 and of B into A is 90, giving an initial net flux of 10. Assume that an allosteric effector increases the A → B rate by 20% to 120 and reciprocally decreases the B → A rate by 20% to 72. The new net flux is 48, and so a 20% change in the rates of the opposing reactions has led to a 380% increase in the net flux. In the example shown in Figure 22-12, this amplification is made possible by the rapid hydrolysis of ATP.

The other potential biological role of substrate cycles is the *generation of heat produced by the hydrolysis of ATP.* A striking example is provided by bumblebees, which have to maintain a thoracic temperature of about 30°C in order to fly. A bumblebee is able to maintain this high thoracic temperature and forage for food even when the ambient temperature is only 10°C because phosphofructokinase and fructose 1,6-bisphosphatase in its flight muscle are simultaneously highly active; the continuous hydrolysis of ATP generates heat. This bisphosphatase is not inhibited by AMP, which suggests that the enzyme is specially designed for the generation of heat. In contrast, the honeybee has almost no fructose 1,6-bisphosphatase activity in its flight muscle and consequently cannot fly when the ambient temperature is low.

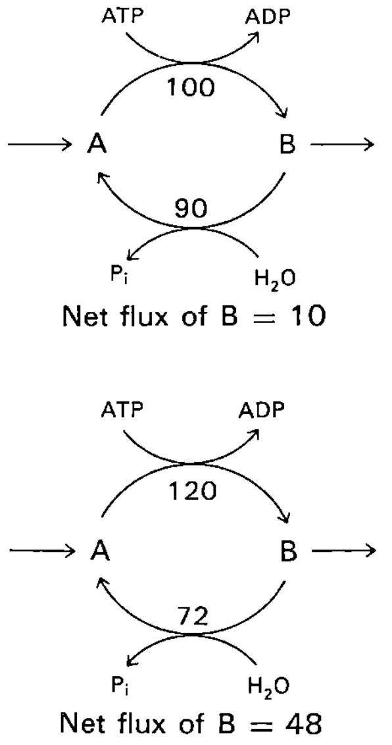

Figure 22-12
An ATP-driven substrate cycle operating at two different rates. A small change in the rates of the two opposing reactions results in a large change in the *net* flux of product B.

LACTATE AND ALANINE FORMED BY CONTRACTING MUSCLE ARE CONVERTED INTO GLUCOSE BY THE LIVER

The major raw materials of gluconeogenesis are lactate and alanine produced by active skeletal muscle and erythrocytes. The rate of production of pyruvate by glycolysis exceeds the rate of oxidation of pyruvate by the citric acid cycle in contracting skeletal muscle under anaerobic conditions, as during vigorous exercise. Under these conditions, moreover, the rate of formation of NADH by glycolysis is greater than the rate of its oxidation by the respiratory chain. Continued glycolysis depends on the availability of NAD^+ for the oxidation of glyceraldehyde 3-phosphate. The accumulation of both NADH and pyruvate is reversed by lactate dehydrogenase, which oxidizes NADH to NAD^+ as it reduces pyruvate to lactate.

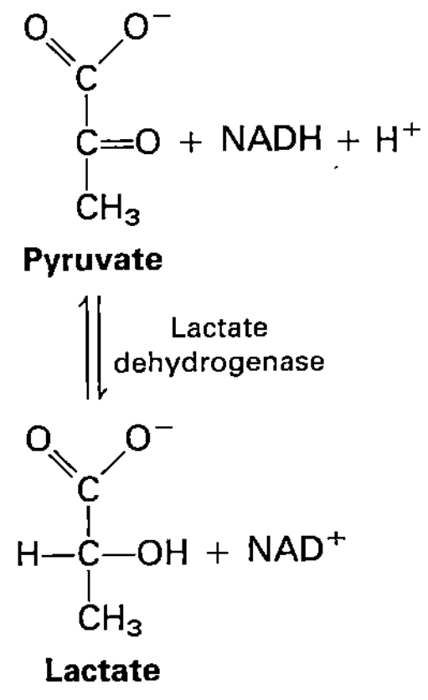

Lactate is a dead end in metabolism. It must be converted back into pyruvate before it can be metabolized. The only purpose of the reduction of pyruvate to lactate is to regenerate NAD^+ so that glycolysis can proceed in active skeletal muscle and erythrocytes. Note that the conversion of glucose to lactate does not involve a net oxidation-reduction. Rather, one of the carbon atoms in lactate is more oxidized than in glucose, and another is more reduced. *The formation of lactate buys time and shifts part of the metabolic burden from muscle to liver.*

The plasma membrane of most cells is highly permeable to lactate and pyruvate. Both substances diffuse out of active skeletal muscle into the blood and are carried to the liver. Much more lactate than pyruvate is carried because of the high $NADH/NAD^+$ ratio in contracting skeletal muscle. The lactate that enters the liver is oxidized to pyruvate, a reaction favored by the low $NADH/NAD^+$ ratio in the cytosol of liver cells. Pyruvate is then converted into glucose by the gluconeogenic pathway in liver. Glucose then enters the blood and is taken up by skeletal muscle. Thus, *liver furnishes glucose to contracting skeletal muscle, which derives ATP from the glycolytic conversion of glucose into lactate. Glucose is then synthesized from lactate by the liver. These reactions constitute the Cori cycle* (Figure 22-13). Recent studies have shown that alanine, like lactate, is a major precursor of glucose. In muscle, alanine is formed from pyruvate by transamination (p. 630); the reverse reaction occurs in the liver.

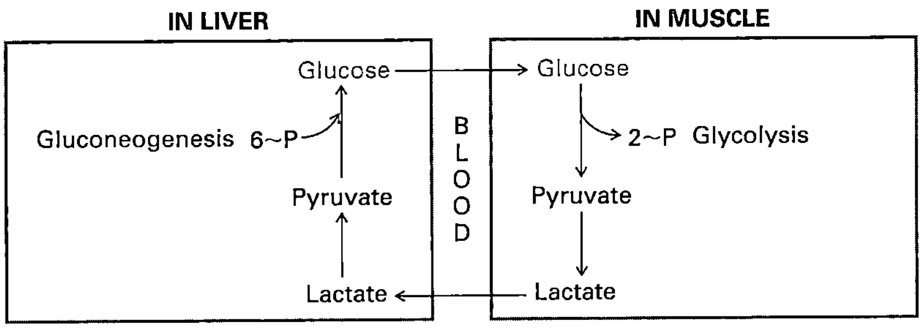

Figure 22-13
The Cori cycle. Lactate formed by active muscle is converted into glucose by the liver. This cycle shifts part of the metabolic burden of active muscle to the liver.

The interconversions of pyruvate and lactate are facilitated by differences in the catalytic properties of lactate dehydrogenase enzymes in different tissues. Lactate dehydrogenase is a tetramer of two kinds of 35-kd subunits encoded by similar genes: the H type predominates in the heart, and the homologous M type in skeletal muscle and the liver. These subunits associate to form five types of tetramers: H_4, H_3M_1, H_2M_2, H_1M_3, and M_4. These species are called *isozymes* (or *isoenzymes*). The H_4

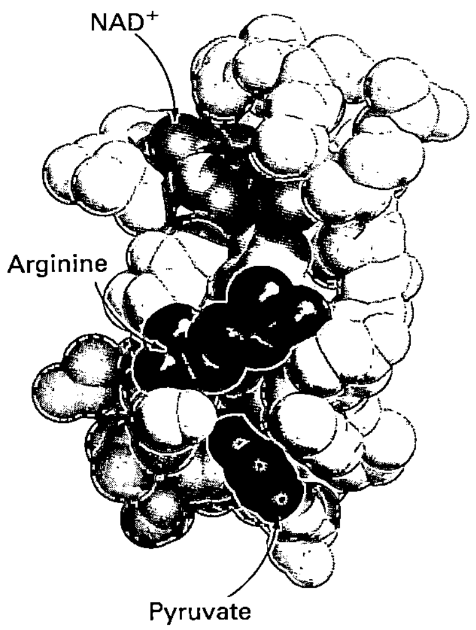

Figure 22-14
Model of the active site of lactate dehydrogenase. Bound NAD^+ is shown in yellow and pyruvate in red. An arginine residue that forms a salt bridge with the pyrophosphate moiety of NAD^+ is shown in blue. [Drawn from 3ldh.pdb. J.L. White, M.L. Hackert, M. Buehner, M.J. Adams, G.C. Ford, P.J. Lenz, Jr., I.E. Smiley, S.J. Steindel, and M.G. Rossman. *J. Mol. Biol.* 102(1976):759.]

isozyme (type 1) has higher affinity for substrates than does the M_4 isozyme (type 5). They also differ in that H_4 is allosterically inhibited by high levels of pyruvate, whereas M_4 is not. The other isozymes have intermediate properties depending on the ratio of the two kinds of chains. It has been proposed that H_4 is designed to oxidize lactate to pyruvate, which is then utilized as a fuel by the heart. The aerobic metabolism of the heart enables it to funnel pyruvate into the citric acid cycle. In contrast, M_4 is optimized to operate in the reverse direction, to convert pyruvate to lactate to allow glycolysis to proceed under anaerobic conditions. We see here an example of how gene duplication and divergence generate a series of homologous enzymes that foster metabolic cooperation between organs.

SUMMARY

The pentose phosphate pathway generates NADPH and ribose 5-phosphate in the cytosol. NADPH is used in reductive biosyntheses, whereas ribose 5-phosphate is used in the synthesis of RNA, DNA, and nucleotide coenzymes. The pentose phosphate pathway starts with the dehydrogenation of glucose 6-phosphate to form a lactone, which is hydrolyzed to give 6-phosphogluconate and then oxidatively decarboxylated to yield ribulose 5-phosphate. $NADP^+$ is the electron acceptor in both of these oxidations. The last step is the isomerization of ribulose 5-phosphate (a ketose) to ribose 5-phosphate (an aldose). A different mode of the pathway is active when cells need much more NADPH than ribose 5-phosphate. Under these conditions, ribose 5-phosphate is converted into glyceraldehyde 3-phosphate and fructose 6-phosphate by transketolase and transaldolase. These two enzymes create a reversible link between the pentose phosphate pathway and glycolysis. Xylulose 5-phosphate, sedoheptulose 7-phosphate, and erythrose 4-phosphate are intermediates in these interconversions. In this way, 12 NADPH can be generated for each glucose 6-phosphate that is completely oxidized to CO_2.

Only the nonoxidative branch of the pathway is appreciably active when much more ribose 5-phosphate than NADPH needs to be synthesized. Under these conditions, fructose 6-phosphate and glyceraldehyde 3-phosphate (formed by the glycolytic pathway) are converted into ribose 5-phosphate without the formation of NADPH. Alternatively, ribose 5-phosphate formed by the oxidative branch can be converted into pyruvate through fructose 6-phosphate and glyceraldehyde 3-phosphate. In this mode, ATP and NADPH are generated, and five of the six carbons of glucose 6-phosphate emerge in pyruvate. The interplay of the glycolytic and pentose phosphate pathways enables the levels of NADPH, ATP, and building blocks such as ribose 5-phosphate and pyruvate to be continuously adjusted to meet cellular needs.

Gluconeogenesis is the synthesis of glucose from noncarbohydrate sources, such as lactate, amino acids, and glycerol. Several of the reactions that convert pyruvate into glucose are common to glycolysis. Gluconeogenesis, however, requires four new reactions to bypass the essential irreversibility of the corresponding reactions in glycolysis. Pyruvate is carboxylated in mitochondria to oxaloacetate, which in turn is decarboxylated and phosphorylated in the cytosol to phosphoenolpyruvate. Two high-energy phosphate bonds are consumed in these reactions, which are catalyzed by pyruvate carboxylase and phosphoenolpyruvate carboxykinase. Pyruvate carboxylase contains a biotin prosthetic group. The other distinctive reactions of gluconeogenesis are the hydrolyses of fructose 1,6-

bisphosphate and glucose 6-phosphate, which are catalyzed by specific phosphatases. The major raw materials for gluconeogenesis by the liver are lactate and alanine produced from pyruvate by active skeletal muscle. The formation of lactate during intense muscular activity buys time and shifts part of the metabolic burden from muscle to liver.

Gluconeogenesis and glycolysis are reciprocally regulated so that one pathway is relatively inactive while the other is highly active. Phosphofructokinase and fructose 1,6-bisphosphatase are key control points. Fructose 2,6-bisphosphate, an intracellular signal molecule present at higher levels when glucose is abundant, activates glycolysis and inhibits gluconeogenesis by regulating these enzymes. Pyruvate kinase and pyruvate carboxylase are regulated by other effectors so that both are not maximally active at the same time. Allosteric regulation and reversible phosphorylation, which are rapid, are complemented by transcriptional control, which occurs in hours or days. A moderate degree of simultaneous activity of opposing enzymes, called substrate cycling, can be beneficial by amplifying metabolic signals.

SELECTED READINGS

Where to start

Horecker, B.L., 1976. Unravelling the pentose phosphate pathway. *In* Kornberg, A., Cornudella, L., Horecker, B.L., and Oro, J. (eds.), *Reflections on Biochemistry,* pp. 65–72. Pergamon.

Pilkis, S.J., and Granner, D.K., 1992. Molecular physiology of the regulation of hepatic gluconeogenesis and glycolysis. *Ann. Rev. Physiol.* 54:885–909.

McGrane, M.M., Yun, J.S., Patel, Y.M., and Hanson, R.W., 1992. Metabolic control of gene expression: In vivo studies with transgenic mice. *Trends Biochem. Sci.* 17:40–44. [This article focuses on the promoter of phosphoenolpyruvate carboxykinase, a key enzyme in gluconeogenesis.]

Melendez-Hevia, E., and Isidoro, A., 1985. The game of the pentose phosphate cycle. *J. Theor. Biol.* 117:251–263. [What is the simplest way of interconverting five- and six-carbon sugars? The pentose phosphate pathway is analyzed here in terms of a mathematical game of optimization.]

Books and general reviews

Wood, T., 1985. *The Pentose Phosphate Pathway.* Academic Press.

Pontremoli, S., and Grazi, E., 1969. Hexose monophosphate oxidation. *Compr. Biochem.* 17:163–189.

Pontremoli, S., and Grazi, E., 1968. Gluconeogenesis. *In* Dickens, F., Randle, P.J., and Whelan, W.J. (eds.), *Carbohydrate Metabolism and Its Disorders,* vol. 1, pp. 259–295. Academic Press.

Meister, A., and Anderson, M.E., 1983. Glutathione. *Ann. Rev. Biochem.* 52:711–760.

Enzymes and reaction mechanisms

Lindqvist, Y., Schneider, G., Ermler, U., and Sundstrom, M., 1992. Three-dimensional structure of transketolase, a thiamine diphosphate dependent enzyme, at 2.5 Å resolution. *EMBO J.* 11:2373–2379.

Karplus, P.A., and Schulz, G.E., 1989. Substrate binding and catalysis by glutathione reductase as derived from refined enzyme: Substrate crystal structures at 2 Å resolution. *J. Mol. Biol.* 210:163–180.

Horecker, B.L., 1964. Transketolase and transaldolase. *Compr. Biochem.* 15:48–70.

Scrutton, M.C., and Young, M.R., 1972. Pyruvate carboxylase. *In* Boyer, P.D. (ed.), *The Enzymes* (3rd ed.), vol. 6, pp. 1–35. Academic Press.

Regulation

Newsholme, E.A., and Start, C., 1973. *Regulation in Metabolism.* Wiley. [Chapter 6 deals with the control of glycolysis and gluconeogenesis in liver and kidney cortex.]

Newsholme, E.A., Chaliss, R.A.J., and Crabtree, B., 1984. Substrate cycles: Their role in improving sensitivity in metabolic control. *Trends Biochem. Sci.* 9:277–280.

Burchell, A., and Waddell, I.D., 1991. The molecular basis of the hepatic microsomal glucose-6-phosphatase system. *Biochim. Biophys. Acta* 1092:129–137.

Berthon, H.A., Kuchel, P.W., and Nixon, P.F., 1992. High control coefficient of transketolase in the nonoxidative pentose phosphate pathway of human erythrocytes: NMR, antibody, and computer simulation studies. *Biochemistry* 31:12792–12798.

Genetic diseases

Luzzatto, L., and Mehta, A., 1989. Glucose 6-phosphate dehydrogenase deficiency. *In* Scriver, C.R., Beaudet, A.L., Sly, W.S., and Valle, D. (eds.), *The Metabolic Basis of Inherited Disease* (6th ed.), pp. 2237–2265. McGraw-Hill.

Luzzatto, L., Usanga, E.A., and Reddy, S., 1969. Glucose 6-phosphate dehydrogenase deficient red cells: Resistance to infection by malarial parasites. *Science* 164:839–842.

Burchell, A., 1990. Molecular pathology of glucose 6-phosphatase. *FASEB J.* 4:2978–2988.

PROBLEMS

1. *Tracing glucose.* Glucose labeled with ^{14}C at C-6 is added to a solution containing the enzymes and cofactors of the oxidative branch of the pentose phosphate pathway. What is the fate of the radioactive label?

2. *Recurring decarboxylations.* Which reaction in the citric acid cycle is most analogous to the oxidative decarboxylation of 6-phosphogluconate to ribulose 5-phosphate? What kind of enzyme-bound intermediate is formed in both reactions?

3. *Carbon shuffling.* Ribose 5-phosphate labeled with ^{14}C at C-1 is added to a solution containing transketolase, transaldolase, phosphopentose epimerase, phosphopentose isomerase, and glyceraldehyde 3-phosphate. What is the distribution of the radioactive label in the erythrose 4-phosphate and fructose 6-phosphate that are formed in this reaction mixture?

4. *Synthetic stoichiometries.* What is the stoichiometry of synthesis of
   (a) Ribose 5-phosphate from glucose 6-phosphate without the concomitant generation of NADPH?
   (b) NADPH from glucose 6-phosphate without the concomitant formation of pentose sugars?

5. *Developmental transition.* During fetal development, the proportion of H chains in lactate dehydrogenase increases and that of M chains decreases. Propose a selective advantage of this shift in isozyme pattern during development.

6. *Trapping a reactive lysine.* Design a chemical experiment to identify the lysine residue that forms a Schiff base at the active site of transaldolase.

7. *Reductive power.* What ratio of NADPH to $NADP^+$ is required to sustain [GSH] = 10 mM and [GSSG] = 1 mM? Use the redox potentials given on page 532.

8. *Metabolic mutant.* What are the likely consequences of a genetic disorder rendering fructose 1,6-bisphosphatase in liver less sensitive to regulation by fructose 2,6-bisphosphate?

9. *Carriers on strings.* A lysine side chain serves as a flexible linker that enables the activated carboxyl group of carboxybiotin to rotate from one active site to another. Name another activated carrier on a molecular string and cite its function.

10. *Biotin snatcher.* Avidin, a 70-kd protein in egg white, has very high affinity for biotin (Figure 22-15). In fact, it is a highly specific inhibitor of biotin enzymes. Which of the following conversions would be blocked by the addition of avidin to a cell homogenate?
    (a) Glucose → pyruvate.
    (b) Pyruvate → glucose.
    (c) Oxaloacetate → glucose.
    (d) Glucose → ribose 5-phosphate.
    (e) Pyruvate → oxaloacetate.
    (f) Ribose 5-phosphate → glucose.

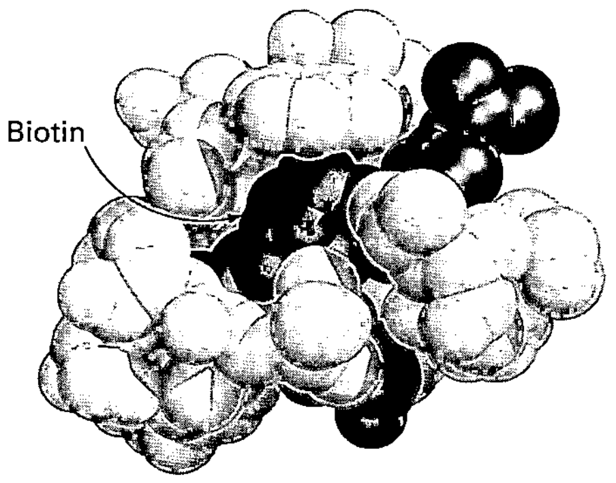

Figure 22-15
Biotin (red) is bound to a hydrophobic box of avidin. Aromatic residues lining the box are shown in yellow. Biotin also forms numerous hydrogen bonds. Two avidin residues that serve as acceptors are shown in blue. [Drawn from 2avi.pdb. O. Livnah, E.A. Bayer, M. Wilchek, and J.L. Sussman. *Proc. Nat. Acad. Sci.* 90(1993):5076.]

CHAPTER

23

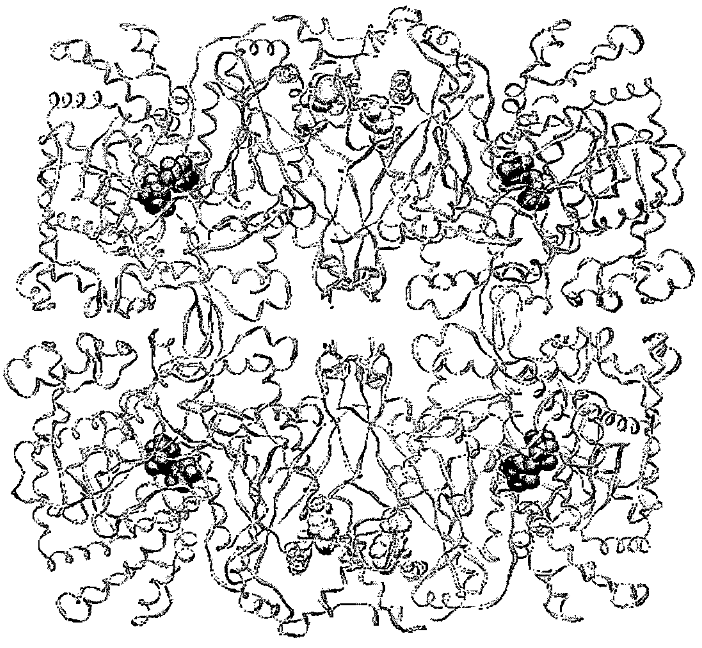

Glycogen Metabolism

Glycogen is a *readily mobilized storage form of glucose.* It is a very large, branched polymer of glucose residues (Figure 23-1). Most of the

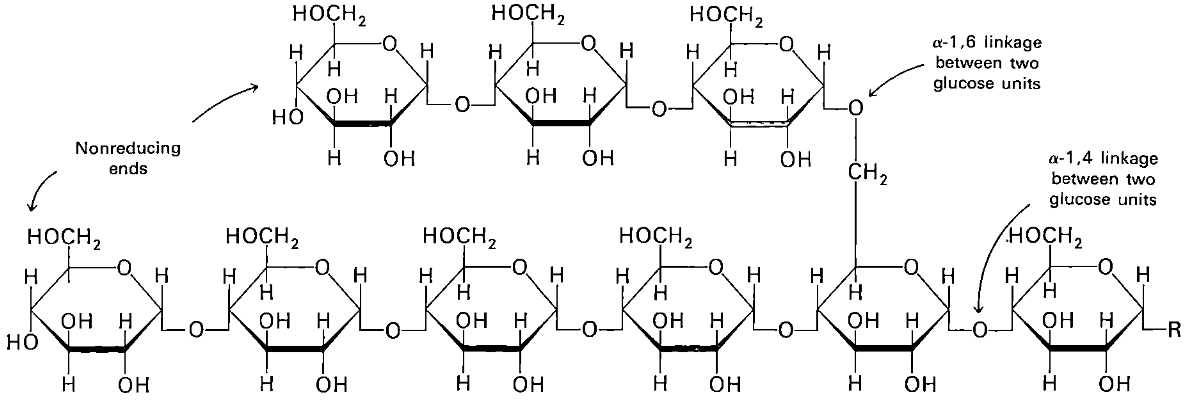

glucose residues in glycogen are linked by α-1,4 glycosidic bonds. Branches at about every tenth residue are created by α-1,6 glycosidic bonds. Recall that α glycosidic linkages form open helical polymers, whereas β linkages produce nearly straight strands that form structural fibrils, as in cellulose (p. 473).

Figure 23-1
Structure of two outer branches of a glycogen particle. The residues at the nonreducing ends are shown in red. The residue that starts a branch is shown in green. The rest of the glycogen molecule is represented by R.

Opening Image: *Structure of glycogen phosphorylase, the enzyme that releases glucose 1-phosphate from glycogen. The b form shown here is a tetramer of identical subunits. The main chain of two subunits is shown in gray, and of the other two, in blue. Each subunit contains a bound pyridoxal phosphate (red) at the catalytic site, and the allosteric activator AMP (green) at a distant site. [Drawn by Dr. Anthony Nicholls from 1pyg.pdb. S.R. Sprang, S.G. Withers, E.J. Goldsmith, R.J. Fletterick, and N.B. Madsen. Science 254(1991):1367.]*

The presence of glycogen greatly increases the amount of glucose that is immediately available between meals and during muscular activity. Glucose is virtually the only fuel used by the brain, except during prolonged starvation. The glucose in the body fluids of an average 70-kg man has an energy content of only 40 kcal, whereas the total body glycogen has an energy content of more than 600 kcal, even after an overnight fast. The two major sites of glycogen storage are the liver and skeletal muscle. The concentration of glycogen is higher in liver than in muscle, but more glycogen is stored in skeletal muscle because of its much greater mass. Glycogen is present in the cytosol in the form of granules ranging in diameter from 10 to 40 nm (Figure 23-2). These granules contain regulatory proteins as well as the enzymes that catalyze the synthesis and degradation of glycogen.

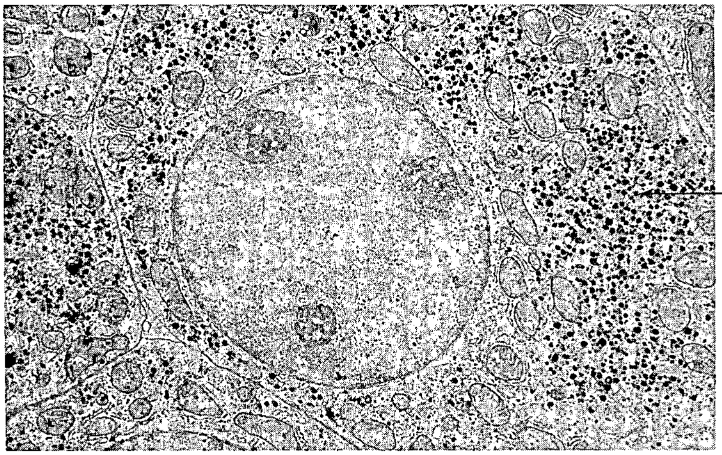

Figure 23-2
Electron micrograph of a liver cell. The dense particles in the cytoplasm are glycogen granules. [Courtesy of Dr. George Palade.]

The synthesis and degradation of glycogen in mammals are considered here in some detail for several reasons. First, these processes are important because they *regulate the blood glucose level* and provide a *reservoir of glucose* for strenuous muscular activity. Second, the synthesis and degradation of glycogen occur by *different reaction pathways,* which illustrates an important principle of biochemistry. Third, the hormonal regulation of glycogen metabolism is mediated by mechanisms that are of general significance. The role of *cyclic AMP* in the coordinated control of glycogen synthesis and breakdown is well understood and is a source of insight into the action of many hormones. The enzymes of glycogen metabolism are regulated by *reversible phosphorylation,* a recurring control device in all biological systems. The effects of hormones such as *epinephrine, glucagon,* and *insulin* on glycogen metabolism are now understood at the molecular level. Fourth, a number of inherited enzyme defects resulting in impaired glycogen metabolism have been characterized. Some of these *glycogen-storage diseases* are lethal in infancy, whereas others have a relatively mild clinical course.

PHOSPHORYLASE CATALYZES THE PHOSPHOROLYTIC CLEAVAGE OF GLYCOGEN INTO GLUCOSE 1-PHOSPHATE

The pathway of glycogen breakdown was elucidated by the incisive studies of Carl Cori and Gerty Cori. They showed that glycogen is *cleaved by orthophosphate (P_i)* to yield a new kind of phosphorylated sugar, which they identified as *glucose 1-phosphate.* The Coris also isolated and crystallized *glycogen phosphorylase,* the enzyme that catalyzes this reaction.

$$\underset{(n \text{ residues})}{\text{Glycogen}} + P_i \rightleftharpoons \text{glucose 1-phosphate} + \underset{(n-1 \text{ residues})}{\text{glycogen}}$$

Phosphorylase catalyzes the sequential removal of glycosyl residues from the nonreducing end of the glycogen molecule (the end with a free 4′-OH group). The glycosidic linkage between C-1 of the terminal residue and C-4 of the adjacent one is split by orthophosphate. Specifically, the bond between the C-1 carbon atom and the glycosidic oxygen atom is cleaved by orthophosphate, and the α configuration at C-1 is retained. Glucose 1-phosphate released from glycogen can readily be converted into glucose 6-phosphate (p. 585), a major metabolic intermediate.

Phosphorolysis—
The cleavage of a bond by orthophosphate (in contrast to *hydrolysis,* which refers to cleavage by water).

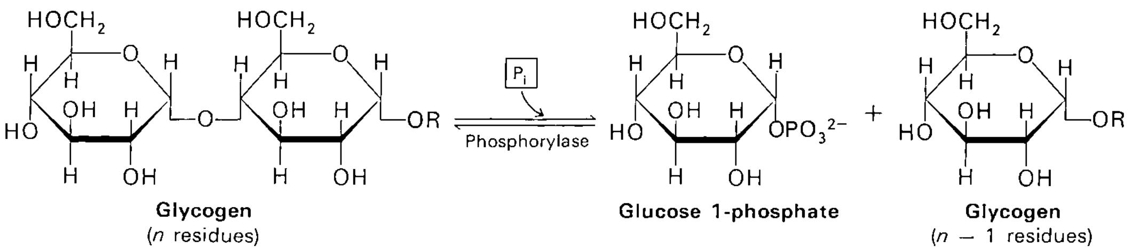

The reaction catalyzed by phosphorylase is readily reversible in vitro. At pH 6.8, the equilibrium ratio of orthophosphate to glucose 1-phosphate is 3.6. The $\Delta G^{\circ\prime}$ for this reaction is small because a glycosidic bond is replaced by a phosphate ester bond that has a nearly equal transfer potential. However, phosphorolysis proceeds far in the direction of glycogen breakdown in vivo because the [P_i]/[glucose 1-phosphate] ratio is usually greater than 100.

The phosphorolytic cleavage of glycogen is energetically advantageous because the released sugar is phosphorylated. In contrast, a hydrolytic cleavage would yield glucose, which would have to be phosphorylated at the expense of an ATP to enter the glycolytic pathway. An additional advantage of phosphorolytic cleavage for muscle cells is that glucose 1-phosphate, ionized under physiologic conditions, cannot diffuse out of the cell, whereas glucose can. We see here an example of *the importance of a metabolite's being ionized.*

A DEBRANCHING ENZYME ALSO IS NEEDED FOR THE BREAKDOWN OF GLYCOGEN

Glycogen is degraded to a limited extent by phosphorylase alone. However, the α-1,6 glycosidic bonds at the branch points are not susceptible to cleavage by phosphorylase. Indeed, phosphorylase stops cleaving α-1,4 linkages when it reaches a terminal residue four away from a branch point. About 1 in 10 residues is branched. The action of phosphorylase on two outer branches of a glycogen particle is shown in Figure 23-3. Five α-1,4 glycosidic bonds on one branch and three on the other are cleaved by phosphorylase. Cleavage by phosphorylase stops at this stage because terminal residues *a* and *d* are four away from the branch point *h.* A new enzymatic activity is required. *A transferase shifts a block of three glycosyl residues from one outer branch to the other.* The α-1,4 glycosidic link between *c*

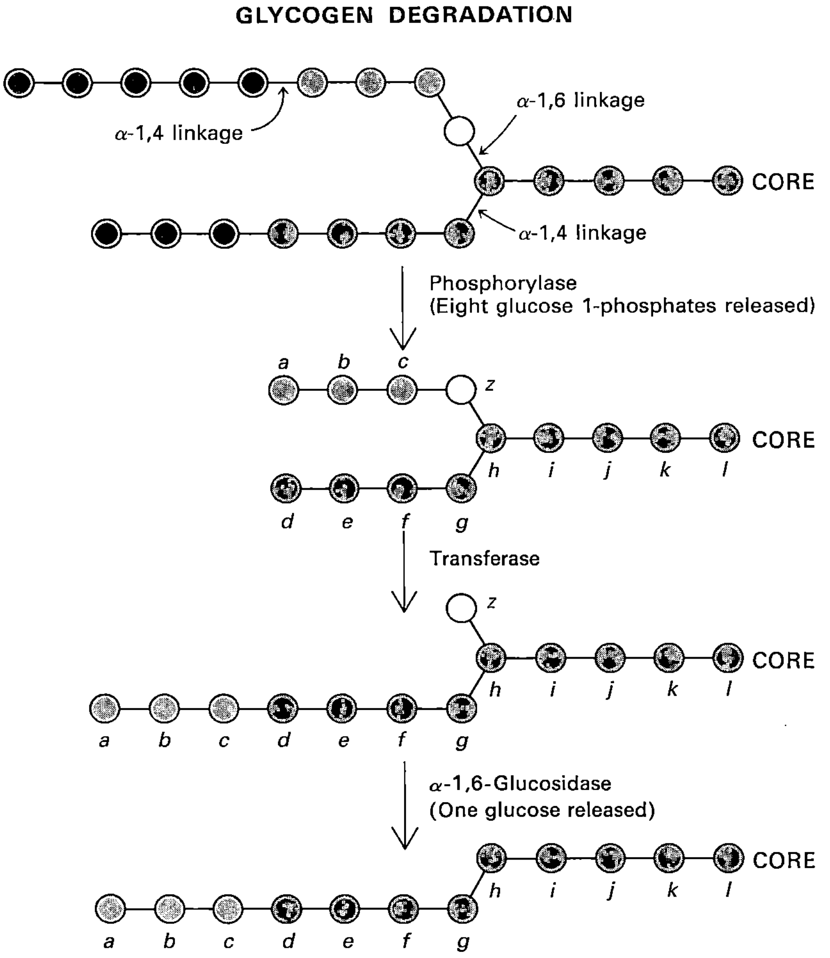

Figure 23-3
Steps in the degradation of glycogen.

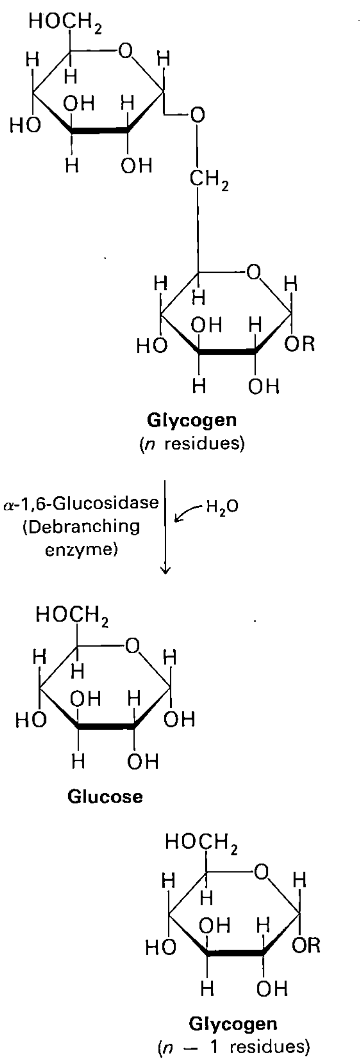

Figure 23-4
Hydrolysis of glycogen at an α-1,6 branch point by the α-1,6-glucosidase (debranching enzyme).

and *z* is broken, and a new α-1,4 link between *c* and *d* is formed. This transfer exposes residue *z* to the action of a third degradative enzyme, an *α-1,6-glucosidase,* which is also known as the *debranching enzyme.* This enzyme hydrolyzes the α-1,6 glycosidic bond between residues *z* and *h* (Figure 23-4).

Thus, the transferase and the α-1,6-glucosidase convert the branched structure into a linear one, which paves the way for further cleavage by phosphorylase. The hydrolysis of *z* renders all the residues *a* through *l* susceptible to phosphorylase. It is noteworthy that the transferase and the α-1,6-glucosidase are present in a single 160-kd polypeptide chain.

PHOSPHOGLUCOMUTASE CONVERTS GLUCOSE 1-PHOSPHATE INTO GLUCOSE 6-PHOSPHATE

Glucose 1-phosphate formed in the phosphorolytic cleavage of glycogen must be converted into glucose 6-phosphate to enter the metabolic mainstream. This phosphoryl shift is catalyzed by *phosphoglucomutase.* The equilibrium mixture contains 95% glucose 6-phosphate. The catalytic site of an active mutase molecule contains a phosphorylated serine residue. The phosphoryl group is transferred to the C-6 hydroxyl group of glucose

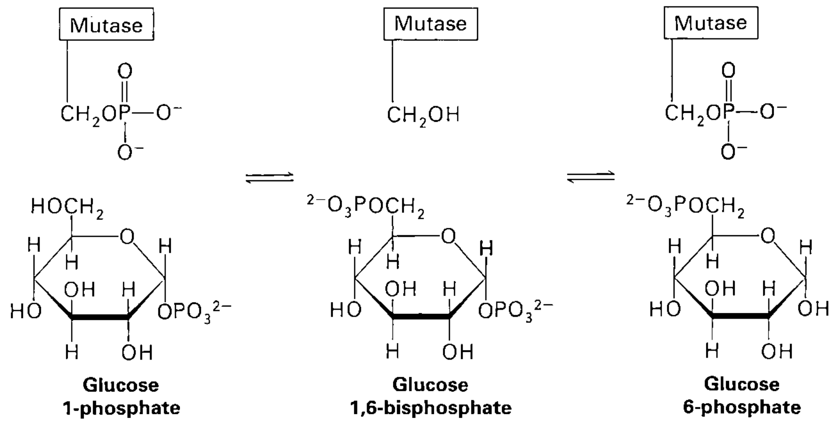

1-phosphate to form glucose 1,6-bisphosphate. The C-1 phosphoryl group of this intermediate is then shuttled to the same serine residue, resulting in the formation of glucose 6-phosphate and the regeneration of the phosphoenzyme.

These reactions are like those of *phosphoglyceromutase,* a glycolytic enzyme (p. 504). The role of 2,3-bisphosphoglycerate (2,3-BPG) in the interconversion of 2-phosphoglycerate and 3-phosphoglycerate is like that of glucose 1,6-bisphosphate in the interconversion of the phosphoglucoses. A phosphoenzyme intermediate participates in both reactions.

LIVER CONTAINS GLUCOSE 6-PHOSPHATASE, A HYDROLYTIC ENZYME ABSENT FROM MUSCLE

A major function of the liver is to maintain a relatively constant level of glucose in the blood. The liver releases glucose into the blood during muscular activity and between meals. The released glucose is taken up primarily by the brain and by skeletal muscle. Phosphorylated glucose, in contrast with glucose, cannot readily diffuse out of cells. The liver contains a hydrolytic enzyme, *glucose 6-phosphatase,* that enables glucose to leave that organ. This enzyme, located on the lumenal side of the smooth endoplasmic reticulum membrane, is essential for gluconeogenesis. Recall that glucose 6-phosphate is transported into the ER; glucose and P_i formed by hydrolysis are then shuttled back into the cytosol (p. 571).

$$\text{Glucose 6-phosphate} + H_2O \longrightarrow \text{glucose} + P_i$$

Glucose 6-phosphatase is also present in the kidneys and intestine, but *it is absent from muscle and the brain.* Consequently, glucose 6-phosphate is retained by muscle and the brain, which need large amounts of this fuel for the generation of ATP. In contrast, glucose is not a major fuel for the liver. Rather, the liver stores and releases glucose primarily for the benefit of other tissues (p. 773).

GLYCOGEN IS SYNTHESIZED AND DEGRADED BY DIFFERENT PATHWAYS

Three lines of evidence indicated that glycogen is synthesized in vivo by a pathway different from that of its degradation. First, the thermodynamics of the phosphorylase-catalyzed reaction do not permit synthesis at physi-

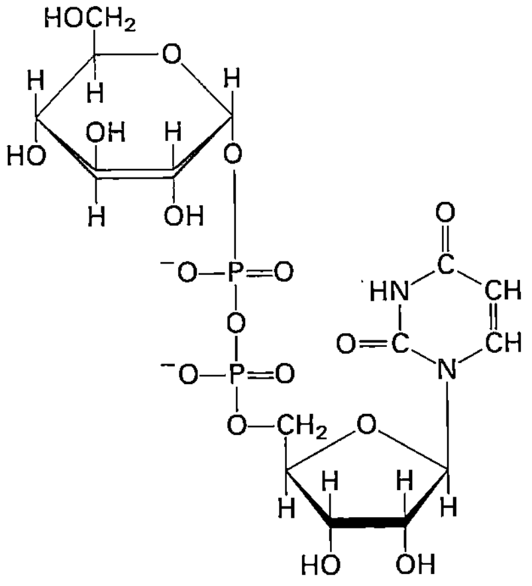

Uridine diphosphate glucose (UDP-glucose)

ologic [P_i]/[glucose 1-phosphate] ratios (p. 583). Second, hormones that lead to an increase in phosphorylase activity always elicit glycogen breakdown. Third, patients who lack muscle phosphorylase entirely (in McArdle's disease, see p. 599) are able to synthesize muscle glycogen. In 1957, Luis Leloir and his co-workers showed that glycogen is synthesized by a pathway that utilizes *uridine diphosphate glucose (UDP–glucose)* rather than glucose 1-phosphate as the activated glucose donor.

$$\text{Synthesis: Glycogen}_n + \text{UDP–glucose} \longrightarrow \text{glycogen}_{n+1} + \text{UDP}$$

$$\text{Degradation: Glycogen}_{n+1} + P_i \longrightarrow \text{glycogen}_n + \text{glucose 1-phosphate}$$

We now know that *biosynthetic and degradative pathways in biological systems are almost always distinct.* Glycogen metabolism provided the first example of this important principle. *Separate pathways afford much greater flexibility, both in energetics and in control.* The cell is not at the mercy of mass action; glycogen can be synthesized despite a high ratio of P_i to glucose 1-phosphate.

UDP-GLUCOSE IS AN ACTIVATED FORM OF GLUCOSE

UDP–glucose, the glucose donor in the biosynthesis of glycogen, is an *activated form of glucose,* just as ATP and acetyl CoA are activated forms of orthophosphate and acetate, respectively. The C-1 carbon atom of the glucosyl unit of UDP–glucose is activated because its hydroxyl group is esterified to the diphosphate moiety of UDP.

UDP–glucose is synthesized from glucose 1-phosphate and uridine triphosphate (UTP) in a reaction catalyzed by *UDP–glucose pyrophosphorylase.* The pyrophosphate liberated in this reaction comes from the outer two phosphoryl residues of UTP.

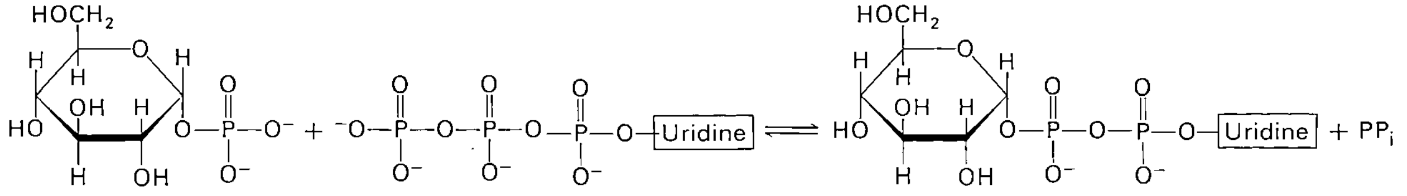

Glucose 1-phosphate

UDP–glucose

This reaction is readily reversible. However, pyrophosphate is rapidly hydrolyzed in vivo to orthophosphate by an inorganic pyrophosphatase. The essentially irreversible hydrolysis of pyrophosphate drives the synthesis of UDP–glucose.

$$\text{Glucose 1-phosphate} + \text{UTP} \rightleftharpoons \text{UDP–glucose} + PP_i$$
$$PP_i + H_2O \longrightarrow 2\,P_i$$

$$\text{Glucose 1-phosphate} + \text{UTP} + H_2O \longrightarrow \text{UDP–glucose} + 2\,P_i$$

The synthesis of UDP–glucose exemplifies a recurring theme in biochemistry: *many biosynthetic reactions are driven by the hydrolysis of pyrophosphate.* Another aspect of this reaction has broad significance. Nucleoside diphosphate sugars serve as glycosyl donors in the biosynthesis of many disaccharides and polysaccharides (p. 676).

GLYCOGEN SYNTHASE CATALYZES THE TRANSFER OF GLUCOSE FROM UDP-GLUCOSE TO A GROWING CHAIN

New glucosyl units are added to the nonreducing terminal residues of glycogen. The activated glucosyl unit of UDP–glucose is transferred to the hydroxyl group at a C-4 terminus of glycogen to form an α-1,4 glycosidic linkage. In elongation, UDP is displaced by the terminal hydroxyl group of the growing glycogen molecule. This reaction is catalyzed by *glycogen synthase.*

Synthase—
An enzyme catalyzing a synthetic reaction in which two units are joined without the direct participation of ATP (or another nucleoside triphosphate).

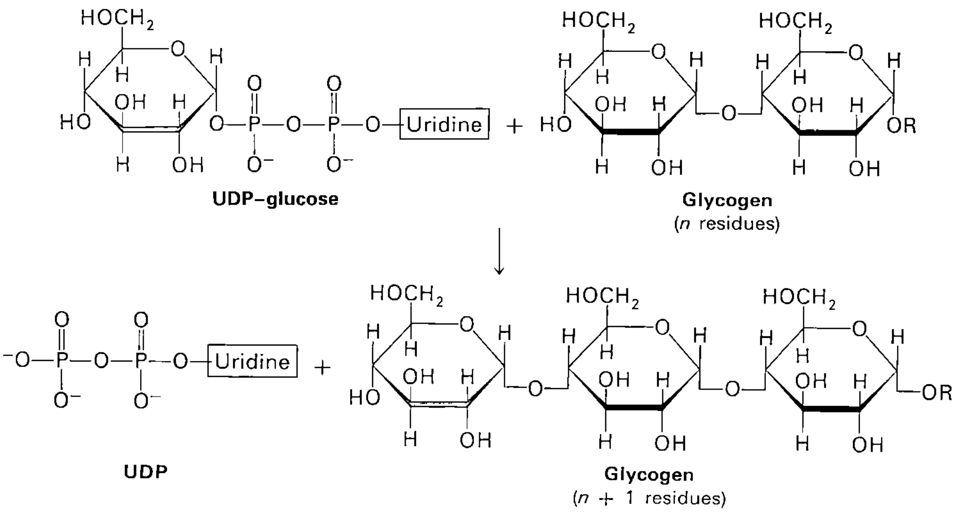

Glycogen synthase can add glucosyl residues only if the polysaccharide chain already contains more than four residues. Thus, glycogen synthesis requires a *primer.* This priming function is carried out by *glycogenin,* a 37-kd protein bearing an oligosaccharide of α-1,4 glucose units. C-1 of the first unit of this chain is covalently attached to the phenolic hydroxyl group of a specific tyrosine in glycogenin. How is this chain formed? Glycogenin autocatalyzes the addition of some eight glucose units; UDP–glucose is the donor in this autoglycosylation. At this point, glycogen synthase, which is tightly bound to glycogenin, takes over. Most significant, *glycogen synthase is catalytically efficient only when it is bound to glycogenin.* This dependence has two important consequences: (1) The number of glycogen granules is determined by the number of molecules of glycogenin. (2) Elongation stops when the synthase is no longer in contact with glycogenin, which forms the core of the particle. *The synthase-glycogenin interaction limits the size of glycogen granules.* We see here a simple and elegant molecular device for setting the size of a biological structure.

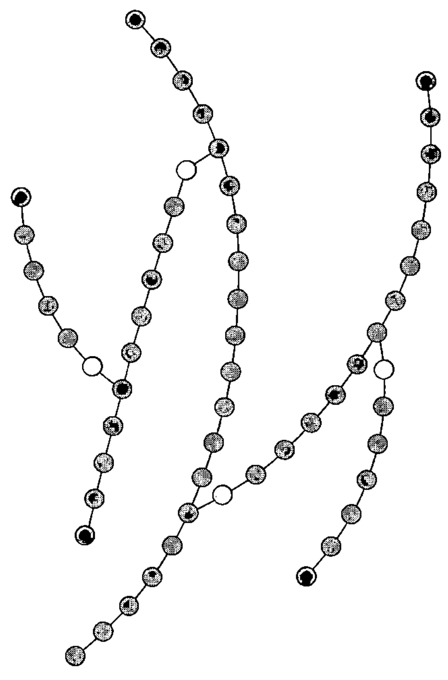

Figure 23-5
Diagram of a cross section of a glycogen molecule. Nonreducing ends are shown in red, and the start of new branches in green (residues are differentiated by the same colors as in Figure 23-1 on p. 581.)

A BRANCHING ENZYME FORMS α-1,6 LINKAGES

Glycogen synthase catalyzes only the synthesis of α-1,4 linkages. Another enzyme forms the α-1,6 linkages that make glycogen a branched polymer. *Branching is important because it increases the solubility of glycogen.* Furthermore, branching creates a large number of terminal residues, the sites of action of glycogen phosphorylase and synthase (Figure 23-5). Thus, *branching increases the rate of glycogen synthesis and degradation.*

Branching occurs after a number of glucosyl residues are joined in α-1,4 linkage by glycogen synthase. A branch is created by the breaking of an α-1,4 link and the formation of an α-1,6 link: this reaction is different from debranching. A block of residues, typically seven in number, is transferred to a more interior site. The *branching enzyme* that catalyzes this reaction is quite exacting. The block of seven or so residues must include the nonreducing terminus and come from a chain at least eleven residues long. In addition, the new branch point must be at least four residues away from a preexisting one.

GLYCOGEN IS A VERY EFFICIENT STORAGE FORM OF GLUCOSE

What is the cost of converting glucose 6-phosphate into glycogen and back into glucose 6-phosphate? The pertinent reactions have already been described, except for reaction 5 below, which is the regeneration of UTP. UDP is phosphorylated by ATP in a reaction catalyzed by *nucleoside diphosphokinase.*

(1) Glucose 6-phosphate $\longrightarrow$ glucose 1-phosphate
(2) Glucose 1-phosphate + UTP $\longrightarrow$ UDP–glucose + PP_i
(3) $PP_i + H_2O \longrightarrow 2\ P_i$
(4) UDP–glucose + $\text{glycogen}_n \longrightarrow \text{glycogen}_{n+1}$ + UDP
(5) UDP + ATP $\longrightarrow$ UTP + ADP

Sum: Glucose 6-phosphate + ATP + $\text{glycogen}_n + H_2O \longrightarrow \text{glycogen}_{n+1} + \text{ADP} + 2\ P_i$

Thus, one high-energy phosphate bond is spent in incorporating glucose 6-phosphate into glycogen. The energy yield from the breakdown of glycogen is highly efficient. About 90% of the residues are phosphorolytically cleaved to glucose 1-phosphate, which is converted at no cost into glucose 6-phosphate. The other 10% are branch residues, which are hydrolytically cleaved. One ATP is then used to phosphorylate each of these glucose molecules to glucose 6-phosphate. The complete oxidation of glucose 6-phosphate yields about 31 molecules of ATP and storage consumes slightly more than one ATP per glucose 6-phosphate, and so *the overall efficiency of storage is nearly 97%.*

PYRIDOXAL PHOSPHATE PARTICIPATES IN THE PHOSPHOROLYTIC CLEAVAGE OF GLYCOGEN

We turn now to the catalytic mechanism and allosteric regulation of glycogen phosphorylase. Both the glycogen substrate and the glucose 1-phosphate product have an α configuration at C-1 (the designation α means that the oxygen atom attached to C-1 is below the plane of the ring, see p. 468). This retention of configuration is a valuable clue to the catalytic mechanism of phosphorylase. A direct attack of phosphate on C-1 of a sugar would invert the configuration at this carbon because the reaction would proceed through a pentacovalent transition state (p. 218). The finding that the glucose 1-phosphate formed has an α rather than β configuration strongly suggests that an even number of steps (most simply, two) are required. A *carbonium ion intermediate* seems most likely, as in the lysozyme-catalyzed hydrolysis of bacterial cell wall polysaccharides (p. 213).

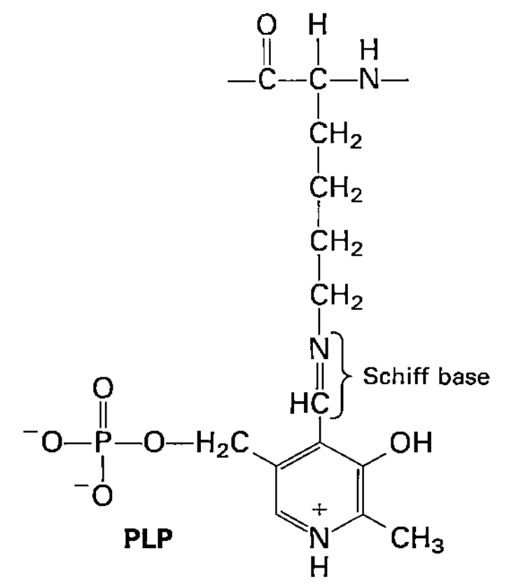

Figure 23-6
Pyridoxal 5′-phosphate (red) forms a Schiff base with a lysine residue (blue) at the active site of phosphorylase.

A second clue to the catalytic mechanism of phosphorylase is its requirement for *pyridoxal 5′-phosphate (PLP)*, a derivative of pyridoxine (vitamin B_6, p. 453). The aldehyde group of this coenzyme forms a Schiff base with a specific lysine side chain of the enzyme (Figure 23-6). X-ray diffraction and nuclear magnetic resonance studies indicate that the reacting orthophosphate group lies between the 5′-phosphate group of PLP and the glycogen substrate (Figure 23-7). *The 5′-phosphate of PLP acts in tandem with orthophosphate, by serving as a proton donor and as a proton acceptor (as a general acid-base catalyst).* Orthophosphate (in the HPO_4^{2-} form) donates a proton to O-4 of the departing glycogen chain and simultaneously acquires a proton from PLP. The carbonium ion intermediate formed in this step is then attacked by orthophosphate to form α-glucose 1-phosphate. We shall see in Chapter 25 (p. 631) that PLP acts quite differently in other enzymes. The special challenge faced by phosphorylase is to cleave glycogen *phosphorolytically* rather than hydrolytically to gain one ~P. This requires that water be excluded from the active site, hence the special role of PLP as a general acid-base catalyst.

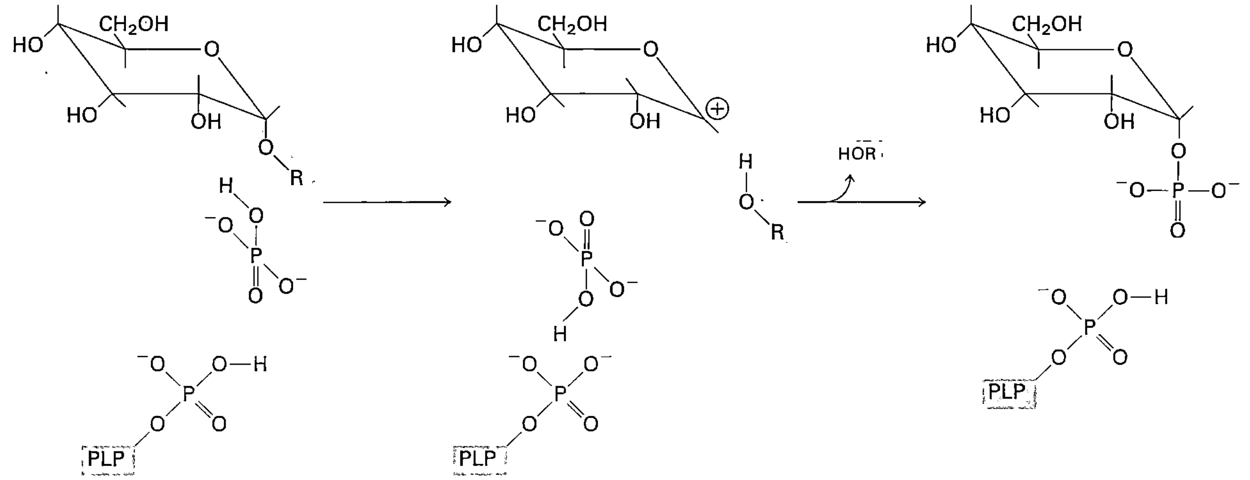

Figure 23-7
Catalytic mechanism of glycogen phosphorylase. PLP denotes pyridoxal phosphate linked as a Schiff base to the active site. The products are glucose 1-phosphate and HOR, the glycogen chain containing one less glucosyl unit. [After P.J. McLaughlin, D.I. Stuart, H.W. Klein, N.G. Oikonomakos, and L.N. Johnson. *Biochemistry* 23(1984):5871.]

PHOSPHORYLASE IS REGULATED BY ALLOSTERIC INTERACTIONS AND REVERSIBLE PHOSPHORYLATION

The existence of separate pathways for the synthesis and degradation of glycogen requires that they be rigorously controlled. ATP would be wastefully hydrolyzed if both sets of reactions were fully active at the same time. In fact, glycogen metabolism is precisely controlled by multiple interlocking mechanisms. *Phosphorylase is regulated by several allosteric effectors that signal the energy state of the cell and by reversible phosphorylation, which is responsive to hormones such as insulin, epinephrine, and glucagon.*

Skeletal muscle phosphorylase, a dimer or tetramer of 97-kd subunits, exists in two interconvertible forms: an *active* phosphorylase *a* and a *usually inactive* phosphorylase *b* (Figure 23-8). Phosphorylase *b* is converted into phosphorylase *a* by the phosphorylation of a single serine residue (serine 14) in each subunit. This covalent modification is catalyzed by *phosphorylase kinase,* an enzyme discovered by Edmond Fischer and Edwin Krebs.

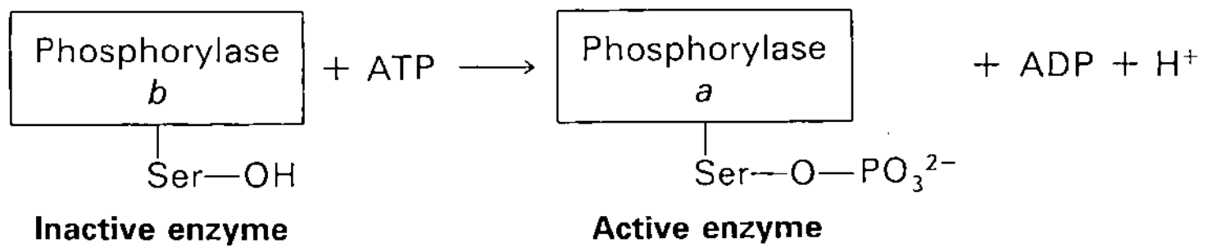

Phosphorylase *a* is deactivated by hydrolysis of the phosphoserine residue in each subunit, a reaction catalyzed by *protein phosphatase 1,* a major cellular phosphatase.

Muscle phosphorylase *b* is active only in the presence of high concentrations of AMP, which acts allosterically—it binds to the nucleotide binding site and alters the conformation of phosphorylase *b.* ATP acts as a negative allosteric effector by competing with AMP. Glucose 6-phosphate also inhibits phosphorylase *b,* primarily by binding to the AMP site. Under most physiologic conditions, *phosphorylase* b *is inactive because of the inhibitory effects of ATP and glucose 6-phosphate.* Phosphorylase *b* is active only when the energy charge of the muscle cell is low. In contrast, *phosphorylase* a *is fully active,* irrespective of the levels of AMP, ATP, and glucose 6-phosphate. In resting muscle, nearly all the enzyme is in the inactive *b* form. During exercise, the elevated level of AMP leads to the activation of phosphorylase *b.* As will be described, increased levels of epinephrine and electrical stimulation of muscle result in the formation of the active *a* form.

The regulation of phosphorylase can be viewed in terms of the concerted allosteric model (see Figure 23-8). The *b* form by itself is almost entirely in the inactive T state; the binding of AMP shifts the allosteric equilibrium to the active R state. Phosphorylation of *b* to *a* puts the enzyme almost entirely in the R state.

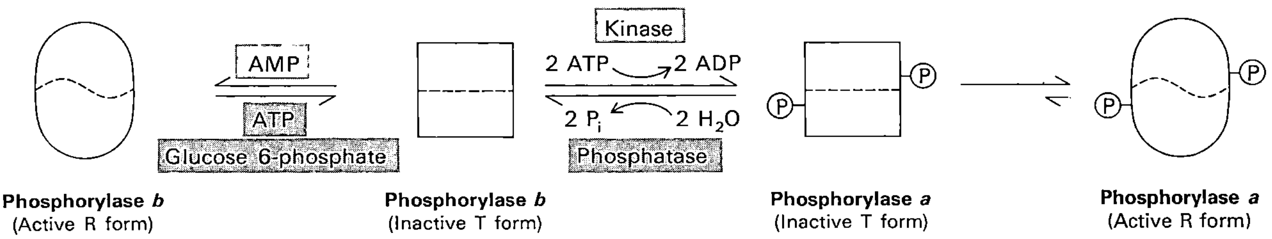

Figure 23-8
Control of glycogen phosphorylase in skeletal muscle. The enzyme can adopt a catalytically inactive T (tense) conformation or an active R (relaxed) conformation. The R → T equilibrium of phosphorylase *a* is far on the side of the active R state. In contrast, phosphorylase *b* is mostly in the inactive T state, unless the level of AMP is high and the levels of ATP and glucose 6-phosphate are low. Under most physiologic conditions, the proportion of active enzyme is determined by the rates of phosphorylation and dephosphorylation.

The allosteric regulation of liver phosphorylase differs from that of the muscle enzyme in two significant respects: (1) AMP does not activate liver phosphorylase *b*. (2) Liver phosphorylase *a* is deactivated by the binding of glucose. In other words, glucose shifts the allosteric equilibrium of the *a* form from R to T. The purpose of glycogen degradation in liver is to form glucose for *export to other tissues* when the blood glucose level is low. Hence, liver phosphorylase is responsive to glucose but not to AMP, an indicator of its own metabolic status. As will be discussed in a later chapter (p. 773), glucose is not a major fuel for the liver. By contrast, skeletal muscle needs glucose for bursts of activity. The activation of muscle phosphorylase *b* by AMP leads to the rapid mobilization of glycogen. Furthermore, the absence of glucose 6-phosphatase in muscle (p. 572) assures that glucose 6-phosphate derived from glycogen remains within the cell for energy generation.

STRUCTURAL CHANGES AT THE SUBUNIT INTERFACE ARE TRANSMITTED TO THE CATALYTIC SITES

X-ray crystallographic studies of both the *a* and *b* forms of glycogen phosphorylase from muscle are sources of insight into the catalytic and control mechanisms of this key enzyme in metabolism. The structures of both the T and the R states of the *a* and *b* forms have been solved. The 841 amino acid residues of the monomeric subunit are compactly folded into an *amino-terminal domain* (480 residues) with a *glycogen binding unit* (60 residues) and a *carboxyl-terminal domain* (361 residues) (Figure 23-9). The

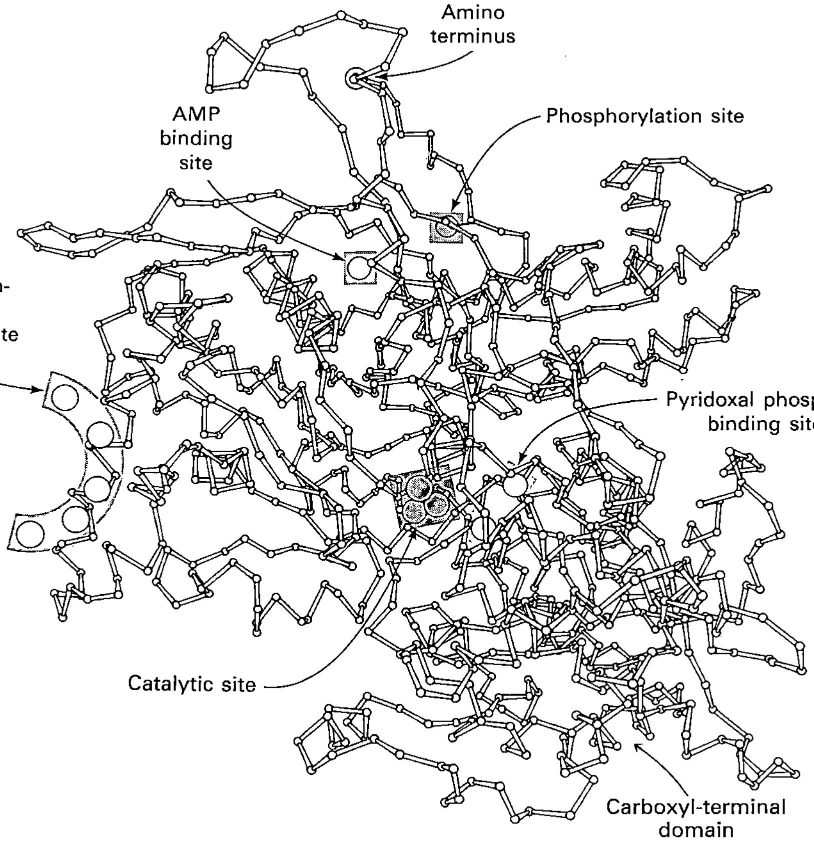

Figure 23-9
α Carbon diagram of a subunit of phosphorylase *a* showing the locations of the catalytic site (red), pyridoxal phosphate binding site (yellow), glycogen-particle binding site (green), AMP allosteric site (blue), and the phosphorylation site (gray). [Courtesy of Dr. Robert Fletterick, Dr. Neil Madsen, and Dr. Peter Kasvinsky.]

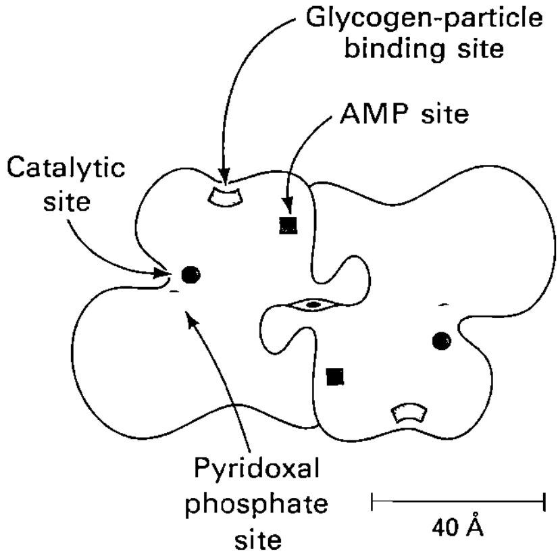

Figure 23-10
Schematic diagram of the phosphorylase *b* dimer viewed down the molecular twofold axis. [After I.T. Weber, L.N. Johnson, K.S. Wilson, D.G.R. Yeates, D.L. Wild, and J.A. Jenkins. *Nature* 274(1978):433.]

catalytic site is located in a deep crevice formed by residues from N- and C-terminal domains. This shielding of the active site from the aqueous milieu favors phosphorolysis over hydrolysis. *Pyridoxal phosphate* is positioned at the active site so that its phosphate group is next to the attacking orthophosphate. The *glycogen binding site* is 30 Å away from the catalytic site. Phosphorylase adheres to glycogen particles because it contains this docking site. The large separation between the docking site and the catalytic site enables the enzyme to phosphorolyze many terminal residues without having to dissociate and reassociate after each catalytic cycle. In other words, phosphorylase is *highly processive*.

AMP, an allosteric activator of muscle phosphorylase *b*, binds to a site next to the interface between subunits, far from the catalytic site and the glycogen binding site (Figure 23-10). Each of the adenine, ribose, and phosphate moieties of AMP binds a distinct segment of polypeptide chain (Figure 23-11). AMP also makes contact with two residues from the other subunit. *The binding of AMP alters the contacts between subunits at their interface and leads to long-range structural changes throughout the dimer.* Most important, both catalytic sites, which are more than 30 Å away from the AMP sites, are switched to the catalytically active state. We see here, as in hemoglobin (p. 161) and aspartate transcarbamoylase (p. 241), the critical role of subunit interfaces in transmitting allosteric interactions. In the T to R transition, a tunnel is opened making the active site accessible to the glycogen substrate.

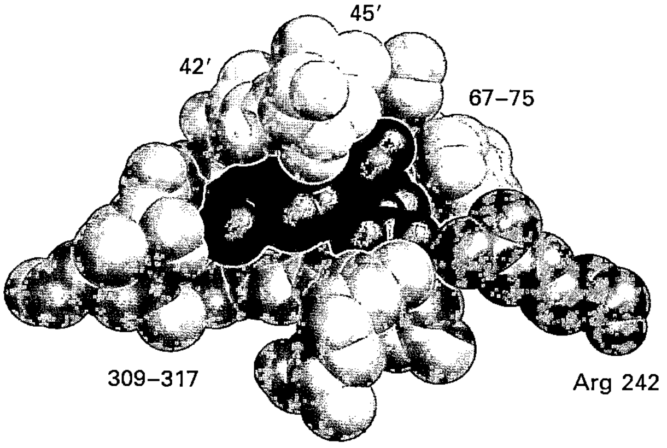

Figure 23-11
AMP (red) is located at the subunit interface of phosphorylase *b*. The nucleotide interacts with three regions (blue, green, and purple) of one subunit, and a different region (yellow) of the other subunit. The phosphate group of AMP forms a salt bridge with arginine 242. [Drawn from 1pyg.pdb. S.R. Sprang, S.G. Withers, E.J. Goldsmith, R.J. Fletterick, and N.B. Madsen. *Science* 254(1991):1367.]

Another critical change induced by AMP is the creation of a binding site for orthophosphate, which forms salt bridges with lysine and arginine side chains, and a hydrogen bond with a main-chain NH. In the T state, a negatively charged aspartate occupies the position of the essential arginine. ATP, which blocks activation by AMP, binds to the same site but in a different manner. However, ATP does not induce activation because its two additional phosphoryl groups prevent it from moving the three polypeptide segments the way AMP does.

The *phosphorylation site* in the conversion of phosphorylase *b* into *a* is serine 14, which is strategically located at the subunit interface. Phosphorylation leads to a dramatic change in the conformation of the 19 amino-terminal residues: in *b*, they are highly mobile, whereas in *a* they have a precise conformation and interact with other residues of both subunits. Specifically, the negatively charged phosphoryl group on serine 14 interacts electrostatically with two positively charged arginine side chains, one on each chain (Figure 23-12). These structural changes, like those induced by AMP, are transmitted throughout the protein, to both catalytic sites. One subunit of the dimer rotates 10 degrees with respect to the other in the T to R transition. The flexibility of the amino-terminal region in the *b* form is reminiscent of the very floppy activation domain in trypsinogen, which adopts a well-defined conformation on conversion into trypsin (p. 250).

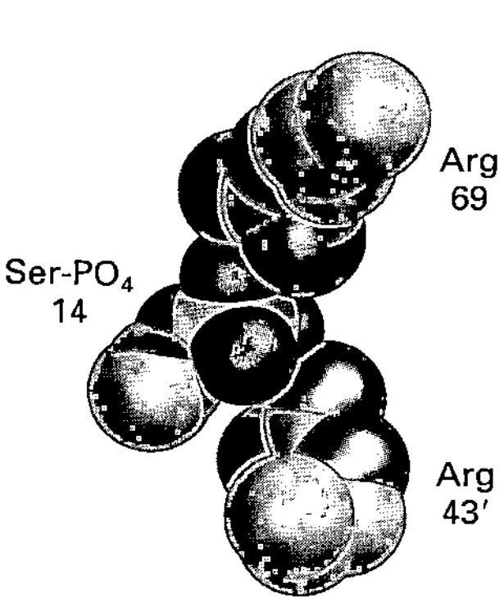

Figure 23-12
The phosphoryl group attached to serine 14 of phosphorylase *a* cross-links arginine residues of different subunits. [Drawn from 1gpa.pdb. D. Barford, S.-H. Hu, and L.N. Johnson. *J. Mol. Biol.* 218(1991):233.]

PHOSPHORYLASE KINASE IS ACTIVATED BY PHOSPHORYLATION AND CALCIUM ION

Phosphorylase kinase, the enzyme that activates phosphorylase by catalyzing the phosphorylation of serine 14, is a very large protein. The subunit composition of the kinase in skeletal muscle is $(\alpha\beta\gamma\delta)_4$, and its mass is 1200 kd. This kinase is under dual control. It is converted from *a low-activity form into a high-activity one by phosphorylation.* The enzyme catalyzing the activation of phosphorylase kinase is *protein kinase A (PKA),* which is switched on by cyclic AMP (p. 343). As will be discussed, hormones such as epinephrine induce the breakdown of glycogen by activating a cyclic AMP cascade.

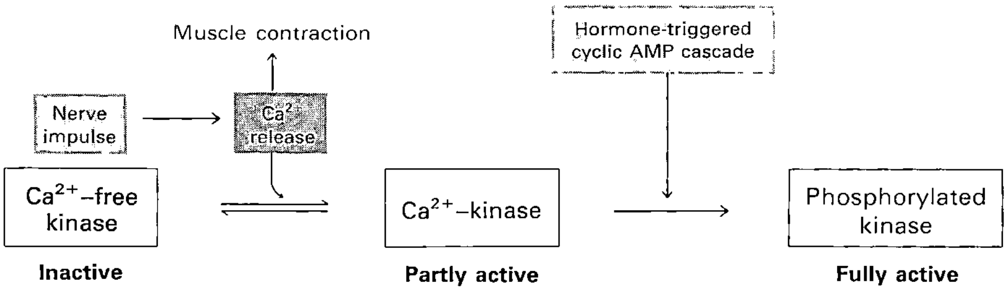

Phosphorylase kinase can be partly activated in a different way, by Ca^{2+} levels of the order of 1 μM. Its δ subunit is *calmodulin,* a calcium sensor that stimulates many enzymes in eukaryotes (p. 349). This mode of activation of the kinase is important in muscle, where contraction is triggered by the release of Ca^{2+} from the sarcoplasmic reticulum (p. 402). Hormones such as vasopressin raise the cytosolic Ca^{2+} level by a different mechanism. They activate the *phosphoinositide cascade* (p. 343) by binding to a seven-helix receptor in the plasma membrane, which in turn activates a G protein that stimulates phospholipase C. The consequent rise in the level of inositol 1,4,5-trisphosphate induces the release of Ca^{2+} from endoplasmic reticulum stores. This route for the activation of phosphorylase kinase, and hence of phosphorylase itself, is important in the liver.

GLYCOGEN SYNTHASE IS INACTIVATED BY THE PHOSPHORYLATION OF A SPECIFIC SERINE RESIDUE

The synthesis of glycogen is closely coordinated with its degradation. The activity of glycogen synthase, like that of phosphorylase, is regulated by covalent modification. The active *a* form of the synthase is converted into a usually inactive *b* form by phosphorylation. The phosphorylated *b* form requires a high level of glucose 6-phosphate for activity, whereas the dephosphorylated *a* form is active in its presence or absence. Thus, *phosphorylation has opposite effects on the enzymatic activities of glycogen synthase and phosphorylase.* Glycogen synthase is phosphorylated at multiple sites by protein kinase A and several other kinases. The unphosphorylated *a* form has a net charge of −13, compared with −31 for the fully phosphorylated *b* form. The nine phosphorylation sites are located in two clusters, near the N and C termini (Figure 23-13).

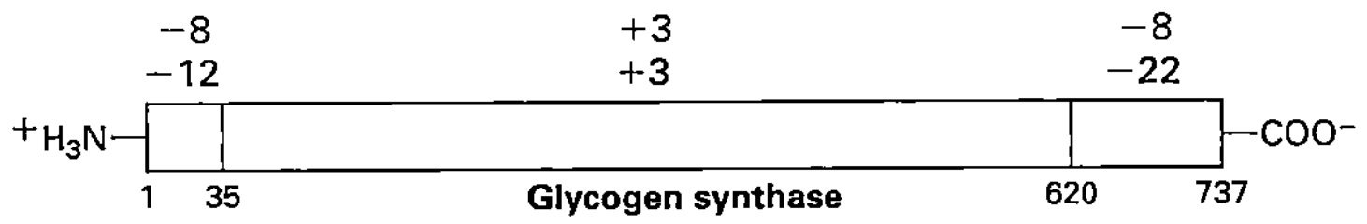

Figure 23-13
Glycogen synthase has a highly asymmetric charge distribution. Phosphorylation markedly changes the net charge of the amino- and carboxyl-terminal regions (yellow) of the enzyme. The net charge of these regions and the interior of the enzyme before and after complete phosphorylation are shown in green and red, respectively. [After M.F. Browner, K. Nakano, A.G. Bang, and R.J. Fletterick. *Proc. Nat. Acad. Sci.* 86(1989):1443.]

A CYCLIC AMP CASCADE COORDINATELY CONTROLS GLYCOGEN SYNTHESIS AND BREAKDOWN

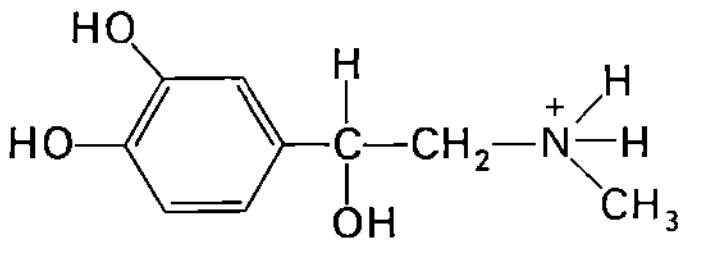

Epinephrine

Glycogen metabolism is profoundly affected by several hormones. *Insulin,* a polypeptide hormone (p. 25), induces the synthesis of glycogen. High levels of insulin in the blood signal the fed state, whereas low levels signal a fasted state (p. 773). *Glucagon* and *epinephine,* by contrast, trigger the breakdown of glycogen. Muscular activity or its anticipation leads to the release of *epinephrine,* a catecholamine derived from tyrosine, from the adrenal medulla. Epinephrine markedly stimulates glycogen breakdown in muscle and, to a lesser extent, in liver. The liver is more responsive to *glucagon,* a polypeptide hormone that is secreted by the α cells of the pancreas when the blood sugar level is low.

5 10
^+H_3N-His-Ser-Glu-Gly-Thr-Phe-Thr-Ser-Asp-Tyr-

15 20
-Ser-Lys-Tyr-Leu-Asp-Ser-Arg-Arg-Ala-Gln-

25 29
-Asp-Phe-Val-Gln-Trp-Leu-Met-Asn-Thr-COO^-

Glucagon

How do hormones trigger the breakdown of glycogen? The breakthrough was made 40 years ago by Earl Sutherland, who began his studies in the laboratory of Carl Cori and Gerty Cori. He discovered that phosphorylase is activated by phosphorylation and inactivated by dephosphorylation, the first example of enzyme regulation by covalent modification. The next exciting finding was that epinephrine and glucagon activated phosphorylase in a cell-free homogenate, just as they did in liver slices. Specific hormone effects had not previously been observed in cell-free systems. Sutherland then discovered that a heat-stable activator was produced when the particulate fraction (which contained the plasma membrane but not phosphorylase) was incubated with hormones and nucleoside triphosphates. The activator was shown to be a new adenine ribonucleotide, namely, cyclic AMP. The pieces of the puzzle neatly came together when Edwin Krebs and Donal Walsh discovered that cyclic AMP (cAMP) activates a protein kinase.

The entire signal transduction pathway from hormone to glycogen degradation is now understood in molecular terms (Figure 23-14):

1. Epinephrine and glucagon bind to seven-helix receptors in the plasma membrane of target cells and trigger the activation of the stimulatory G protein (G_S, p. 341). Recall that these receptors are members of the rhodopsin family.

2. The GTP form of the α subunit of G_S activates adenylate cyclase, a transmembrane enzyme that catalyzes the formation of cAMP from ATP (p. 342).

3. The elevated cytosolic level of cAMP activates *protein kinase A (PKA)*. This kinase is inactive in the absence of cAMP because of the inhibitory action of its regulatory subunits. The binding of cAMP to the regulatory subunits unleashes the catalytic subunits (p. 245).

4. PKA phosphorylates both phosphorylase kinase and glycogen synthase. Phosphorylation by this cyclic AMP–dependent kinase *switches on phosphorylase* (by activating phosphorylase kinase) and simultaneously *deactivates glycogen synthase* (directly). Glycogen synthase is also phosphorylated by phosphorylase kinase, which further assures that glycogen is not being synthesized while it is being degraded. Thus, *PKA coordinately controls glycogen degradation and synthesis.*

5. *The effects of hormones are highly amplified by the cyclic AMP cascade.* The receptor-triggered activation of the stimulatory G protein, the formation of cAMP by adenylate cyclase, and the covalent modification of phosphorylase by phosphorylase kinase are high-gain catalytic processes. Hence, the binding of a small number of hormone molecules to cell-surface receptors leads to the release of a very large number of sugar units.

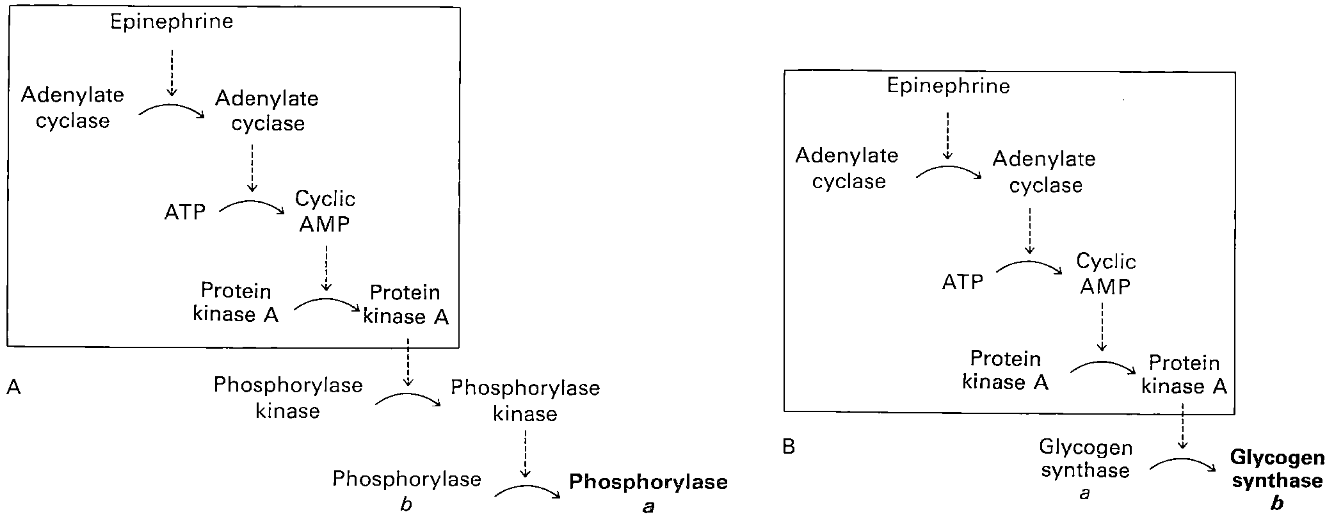

Figure 23-14
Coordinate control of glycogen metabolism by hormone-triggered cyclic AMP cascades. (A) glycogen degradation; (B) glycogen synthesis. Inactive forms are shown in red, and active ones in green. The sequence of reactions leading to the activation of protein kinase A is the same in the regulation of glycogen degradation and synthesis. Phosphorylase kinase also inactivates glycogen synthase.

PROTEIN PHOSPHATASE 1 REVERSES THE REGULATORY EFFECTS OF KINASES ON GLYCOGEN METABOLISM

The changes in enzymatic activity produced by protein kinases are reversed by protein phosphatases. The hydrolysis of nearly all phosphorylated serine and threonine residues in proteins is catalyzed by only four protein phosphatases. For historical reasons, they are known as *protein phosphatase 1, 2A, 2B,* and *2C (PP1, PP2A, PP2B,* and *PP2C).* Philip Cohen has found that *PP1 plays key roles in regulating glycogen metabolism.* PP1 inactivates phosphorylase kinase and phosphorylase *a* by dephosphorylating these enzymes.

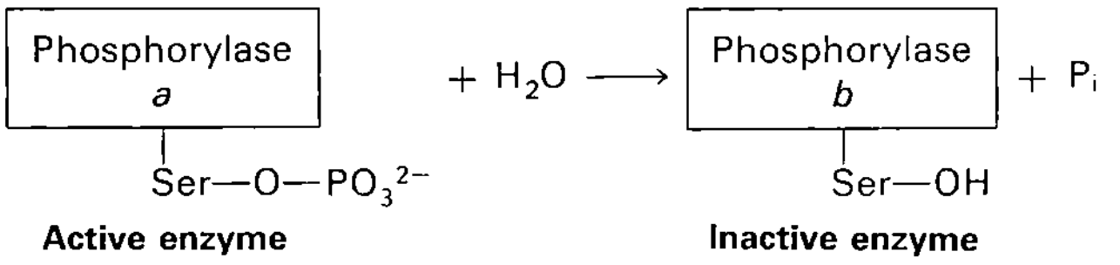

Thus, *PP1 decreases the rate of glycogen breakdown.* Moreover, PP1 also removes the phosphoryl group from glycogen synthase *b* to convert it into the much more active *a* form. Hence, *PP1 accelerates glycogen synthesis.* This is yet another molecular device for coordinating carbohydrate storage.

The phosphatase activity of PP1 is rigorously controlled. The 37-kd catalytic subunit of PP1 is ineffective by itself because it has low affinity for glycogen particles. High affinity is conferred by its association with a 160-kd G subunit (G stands for "glycogen binding," not "GTP binding"). *The G subunit draws PP1 into glycogen particles, where it can efficiently remove phosphoryl groups from glycogen synthase and phosphorylase kinase.* However, phosphorylation of the G subunit by protein kinase A prevents it from binding the catalytic subunit of PP1. Consequently, activation of the cAMP cascade by epinephrine leads to the inactivation of PP1 because it no longer binds its substrates.

The other key regulator of this protein phosphatase is *inhibitor 1.* Catalysis by PP1 is blocked by the phosphorylated but not the dephosphorylated form of inhibitor 1, a small protein. The degree of phosphorylation of inhibitor 1, in turn, is under hormonal control. cAMP, acting through protein kinase A, blocks PP1 by phosphorylating inhibitor 1. Thus, when glycogen degradation is switched on by cAMP, the accompanying phosphorylation of inhibitor 1 keeps phosphorylase in the active *a* form and glycogen synthase in the inactive *b* form. The epinephrine-induced phosphorylation of the G subunit and that of inhibitor 1 are complementary devices for sustaining glycogen degradation (Figure 23-15).

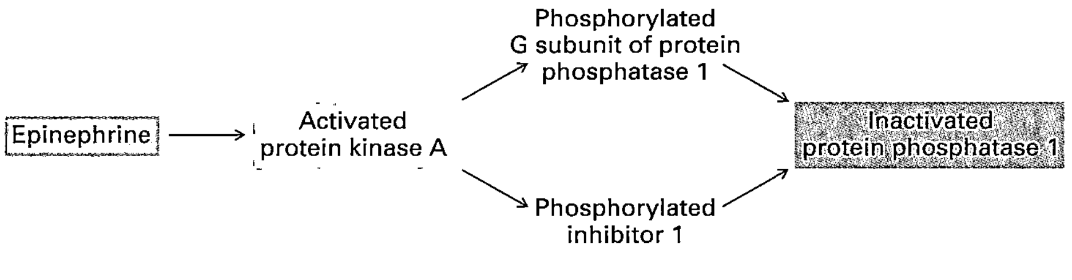

Figure 23-15
Epinephrine inactivates protein phosphatase 1 by phosphorylating its glycogen-targeting subunit (G subunit) and inhibitor 1. [After P. Cohen. *Ann. Rev. Biochem.* 58(1989):488.]

INSULIN STIMULATES GLYCOGEN SYNTHESIS BY ACTIVATING PROTEIN PHOSPHATASE 1

In times of plenty, insulin stimulates the synthesis of glycogen by triggering a pathway that dephosphorylates and thereby activates glycogen synthase. Glycogen degradation is simultaneously inhibited by the dephosphorylation of phosphorylase kinase and phosphorylase *a*. Recall that the first step in the action of insulin is its binding to a receptor tyrosine kinase in the plasma membrane (p. 351). How does activation of a membrane-bound tyrosine kinase change catalytic rates in glycogen granules? Multiple phosphorylations again serve as the regulatory mechanism. The binding of insulin to its receptor leads to the activation of an *insulin-sensitive protein kinase* that phosphorylates the G subunit of PP1 at a site *different* from that modified by protein kinase A (Figure 23-16). Insulin, unlike epinephrine, *makes protein phosphatase 1 more active.* The consequent dephosphorylation of glycogen synthase, phosphorylase kinase, and phosphorylase promotes glycogen synthesis and blocks its degradation. Once again we see that *glycogen synthesis and breakdown are coordinately controlled.*

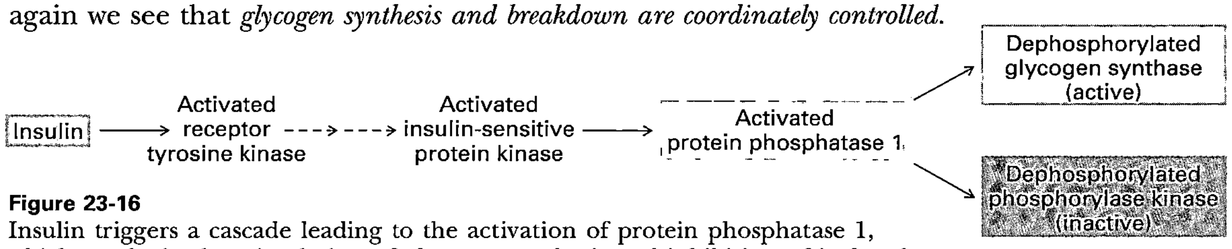

Figure 23-16
Insulin triggers a cascade leading to the activation of protein phosphatase 1, which results in the stimulation of glycogen synthesis and inhibition of its breakdown. The activated receptor tyrosine kinase switches on a putative master kinase that phosphorylates the insulin-sensitive protein kinase. In turn, the glycogen-targeting subunit (G subunit) of the phosphatase is phosphorylated, which activates the enzyme. [After P. Dent, A. Lavoinne, S. Nakielny, F.B. Caudwell, P. Watt, and P. Cohen. *Nature* 348(1990):306.]

GLYCOGEN METABOLISM IN THE LIVER REGULATES THE BLOOD GLUCOSE LEVEL

The control of the synthesis and degradation of glycogen in the liver is central to the regulation of the blood glucose level. The concentration of glucose in the blood normally ranges from about 80 to 120 mg per 100 ml (4.4 to 6.7 mM). The liver senses the concentration of glucose in the blood and takes up or releases glucose accordingly. The amount of liver phosphorylase *a* decreases rapidly when glucose is infused (Figure 23-17). After a lag period, the amount of glycogen synthase *a* increases, which results in the synthesis of glycogen. In fact, *phosphorylase* a *is the glucose sensor in liver cells.* The binding of glucose to phosphorylase *a* shifts its allosteric equilibrium from the active R form to the inactive T form (see Figure 23-8), which *exposes the phosphoryl group on serine 14 to hydrolysis by protein phosphatase 1.* It is significant that PP1 binds tightly to phosphorylase *a* but acts catalytically only when glucose induces the transition to the T form.

How does glucose activate the synthase? Phosphorylase *b*, in contrast with *a*, does not bind the phosphatase. Consequently, the conversion of *a* into *b* is accompanied by the *release of the PP1, which is then free to activate glycogen synthase.* Removal of the phosphoryl group of inactive synthase *b* converts it into the active *a* form. Initially, there are about 10 phosphorylase *a* molecules per phosphatase. Hence, *the activity of the synthase begins to increase only after most of phosphorylase* a *is converted into* b (see Figure 23-17).

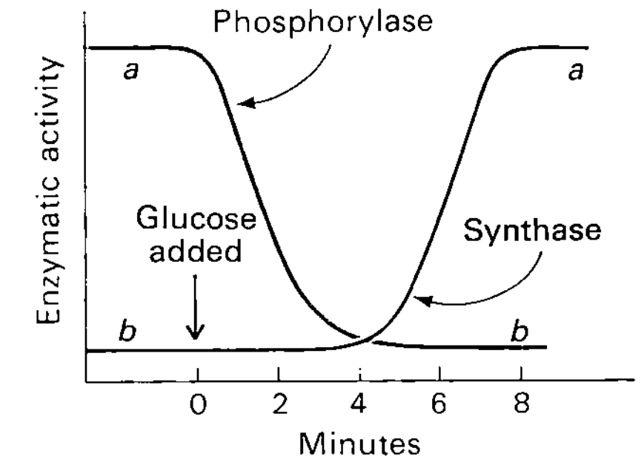

Figure 23-17
The infusion of glucose into the bloodstream leads to the inactivation of phosphorylase, followed by the activation of glycogen synthase, in liver. [After W. Stalmans, H. De Wulf, L. Hue, and H.-G. Hers. *Eur. J. Biochem.* 41(1974):127.]

This remarkable glucose-sensing system depends on three key elements: (1) communication between the serine phosphate and the allosteric site for glucose, (2) the use of PP1 to inactivate phosphorylase and activate the synthase, and (3) the binding of the phosphatase to phosphorylase *a* to prevent premature activation of the synthase.

GLYCOGEN-STORAGE DISEASES ARE PRODUCED BY A VARIETY OF GENETIC DEFECTS

The first glycogen-storage disease was described by Edgar von Gierke in 1929. A patient with this disease has a huge abdomen caused by a *massive enlargement of the liver.* There is a pronounced *hypoglycemia* between meals. Furthermore, the blood glucose level does not rise on administration of epinephrine and glucagon. An infant with this glycogen-storage disease may have convulsions because of the low blood glucose level.

The enzymatic defect in von Gierke's disease was elucidated by the Coris in 1952. They found that *glucose 6-phosphatase is missing from the liver of a patient with this disease.* This was the first demonstration of an inherited deficiency of a liver enzyme. The liver glycogen is normal in structure but present in abnormally large amounts. The absence of glucose 6-phosphatase in the liver causes hypoglycemia because glucose cannot be formed from glucose 6-phosphate. This phosphorylated sugar does not leave the liver because it cannot traverse the plasma membrane. A

Table 23-1
Glycogen-storage diseases

Type	*Defective enzyme*	*Organ affected*	*Glycogen in the affected organ*	*Clinical features*
I VON GIERKE'S DISEASE	Glucose 6-phosphatase or transport system	Liver and kidney	Increased amount; normal structure.	Massive enlargement of the liver. Failure to thrive. Severe hypoglycemia, ketosis, hyperuricemia, hyperlipemia.
II POMPE'S DISEASE	α-1,4-Glucosidase (lysosomal)	All organs	Massive increase in amount; normal structure.	Cardiorespiratory failure causes death, usually before age 2.
III CORI'S DISEASE	Amylo-1,6-glucosidase (debranching enzyme)	Muscle and liver	Increased amount; short outer branches.	Like type I, but milder course.
IV ANDERSEN'S DISEASE	Branching enzyme (α-1,4 $\longrightarrow$ α-1,6)	Liver and spleen	Normal amount; very long outer branches.	Progressive cirrhosis of the liver. Liver failure causes death, usually before age 2.
V McARDLE'S DISEASE	Phosphorylase	Muscle	Moderately increased amount; normal structure.	Limited ability to perform strenuous exercise because of painful muscle cramps. Otherwise patient is normal and well developed.
VI HERS' DISEASE	Phosphorylase	Liver	Increased amount.	Like type I, but milder course.
VII	Phosphofructokinase	Muscle	Increased amount; normal structure.	Like type V.
VIII	Phosphorylase kinase	Liver	Increased amount; normal structure.	Mild liver enlargement. Mild hypoglycemia.

Note: Types I through VII are inherited as autosomal recessives. Type VIII is sex linked.

compensatory increase in glycolysis in the liver leads to a high level of lactate and pyruvate in the blood. Patients who have von Gierke's disease also have an increased dependence on fat metabolism. This disease can also be produced by a mutation in the gene that encodes the *glucose 6-phosphate transporter.* Recall that glucose 6-phosphate must be transported into the lumen of the endoplasmic reticulum to be hydrolyzed by the phosphatase (p. 571). Mutations in the other three essential proteins of this system can likewise lead to von Gierke's disease.

Seven other glycogen-storage diseases have been characterized (Table 23-1). In Pompe's disease (type II), lysosomes become engorged with glycogen because they lack α-1,4-glucosidase, a hydrolytic enzyme confined to these organelles (Figure 23-18). The Coris elucidated the biochemical defect in another glycogen-storage disease (type III), which cannot be distinguished from von Gierke's disease (type I) by physical examination alone. In type III disease, the structure of liver and muscle glycogen is abnormal and the amount is markedly increased. Most striking, the outer branches of the glycogen are very short. *Patients having this type lack the debranching enzyme (α-1,6-glucosidase),* and so only the outermost branches of glycogen can be effectively utilized. Thus, only a small fraction of this abnormal glycogen is functionally active as an accessible store of glucose.

A defect in glycogen metabolism confined to muscle is found in McArdle's disease (type V). *Muscle phosphorylase activity is absent,* and the patient's capacity to perform strenuous exercise is limited because of painful muscle cramps. The patient is otherwise normal and well developed. Thus, effective utilization of muscle glycogen is not essential for life. Phosphorus-31 nuclear magnetic resonance studies of these patients have been very informative. The pH of skeletal muscle cells of normal people drops during strenuous exercise because of the production of lactic acid. In contrast, the muscle cells of patients with McArdle's disease become more alkaline during exercise because of the breakdown of creatine phosphate (p. 458). Lactate does not accumulate in these patients because the glycolytic rate of their muscle is much lower than normal; their glycogen cannot be mobilized. NMR studies have also shown that the painful cramps in this disease are correlated with high levels of ADP (Figure 23-19). NMR spectroscopy is a valuable, noninvasive technique for assessing dietary and exercise therapy for this disease.

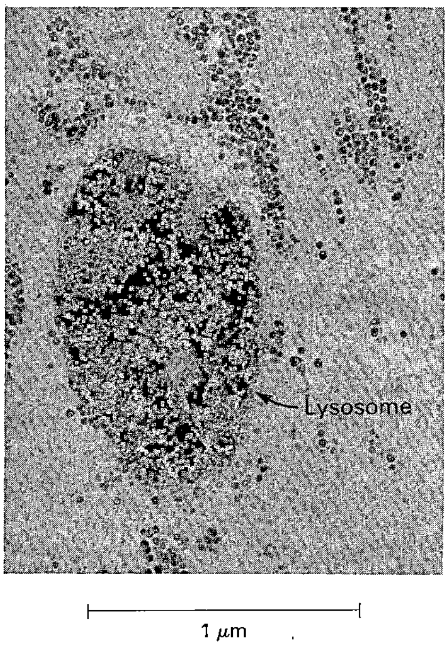

Figure 23-18
Electron micrograph of skeletal muscle from an infant with type II glycogen-storage disease (Pompe's disease). The lysosomes are engorged with glycogen because of a deficiency in the α-1,4-glucosidase, a hydrolytic enzyme confined to lysosomes. The amount of glycogen in the cytosol is normal. [From H.-G. Hers and F. Van Hoof (eds). *Lysosomes and Storage Diseases* (Academic Press, 1973), p. 205.]

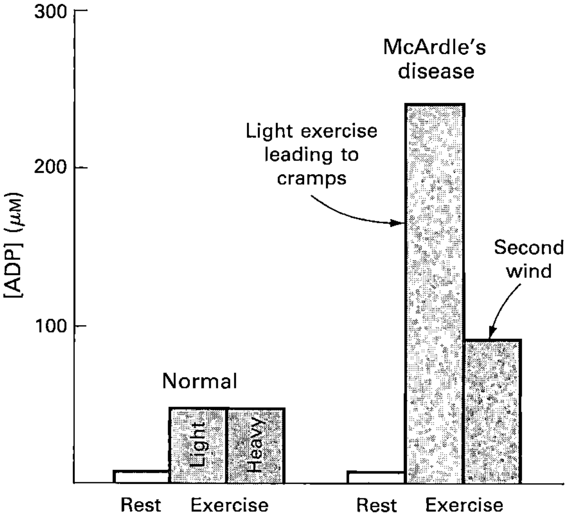

Figure 23-19
NMR study of human arm muscle. The level of ADP during exercise increases much more in a patient with McArdle's glycogen-storage disease (type V) than in normal controls. [After G.K. Radda. *Biochem. Soc. Trans.* 14(1986):522.]

SUMMARY

Glycogen, a readily mobilized fuel store, is a branched polymer of glucose residues. Most of the glucose units in glycogen are linked by α-1,4 glycosidic bonds. At about every tenth residue, a branch is created by an α-1,6 glycosidic bond. Glycogen is present in large amounts in muscle and liver, where it is stored in the cytoplasm in the form of hydrated granules. Most of the glycogen molecule is degraded to glucose 1-phosphate by the action of phosphorylase. The glycosidic linkage between C-1 of a terminal residue and C-4 of the adjacent one is split by orthophosphate to give glucose 1-phosphate, which can be reversibly converted into glucose 6-phosphate. Pyridoxal phosphate, a derivative of vitamin B_6, participates in the phosphorolytic cleavage of glycogen. Branch points are degraded by the concerted action of an oligosaccharide transferase and an α-1,6-glucosidase. The latter enzyme (also known as the debranching enzyme) catalyzes the hydrolysis of α-1,6 linkages, yielding free glucose.

Glycogen is synthesized by a different pathway. UDP-glucose, the activated intermediate in glycogen synthesis, is formed from glucose 1-phosphate and UTP. Glycogen synthase catalyzes the transfer of glucose from UDP-glucose to the C-4 hydroxyl group of a terminal residue in the growing glycogen molecule. Synthesis is primed by glycogenin, an autoglycosylating protein that contains a covalently attached oligosaccharide unit on a specific tyrosine residue. Glycogen synthase is active only when associated with glycogenin, which serves to limit the size of glycogen granules. A branching enzyme converts some of the α-1,4 linkages into α-1,6 linkages to increase the number of ends so that glycogen can be made and degraded more rapidly.

Phosphorylase, a dimeric enzyme, is regulated by allosteric effectors and reversible covalent modifications. Phosphorylase *b*, which is usually inactive, is converted into active phosphorylase *a* by the phosphorylation of a single serine residue in each subunit. This reaction is catalyzed by phosphorylase kinase. The *b* form in muscle can also be activated by the binding of AMP, an effect antagonized by ATP and glucose 6-phosphate. The *a* form in the liver is inhibited by glucose. The AMP binding sites and phosphorylation sites are located at the subunit interface. Conformational changes induced by AMP binding and phosphorylation are transmitted over distances of more than 30 Å to the catalytic sites and other binding sites. In muscle, phosphorylase is activated to generate sugar for use inside the cell as a fuel for contractile activity. By contrast, liver phosphorylase is activated to liberate glucose for export to other organs, such as skeletal muscle and brain.

Glycogen synthesis and degradation are coordinated by several amplifying reaction cascades. Glycogen synthase is inactive when phosphorylase is active, and vice versa. Epinephrine and glucagon stimulate glycogen breakdown and inhibit its synthesis by increasing the cytosolic level of cyclic AMP, which activates protein kinase A (PKA). Phosphorylase kinase becomes more active, whereas glycogen synthase becomes less active when phosphorylated by PKA. Elevated cytosolic Ca^{2+} levels directly activate phosphorylase kinase, which contains calmodulin as one of its subunits. Hence, muscle contraction and calcium-mobilizing hormones promote glycogen breakdown.

The glycogen-mobilizing actions of PKA are reversed by protein phosphatase 1, which is regulated by several hormones. Epinephrine inhibits this phosphatase by blocking its attachment to glycogen granules and by turning on an inhibitor; both effects are mediated by PKA-catalyzed phosphorylation. Insulin, by contrast, activates this phosphatase by triggering a cascade that phosphorylates the glycogen-targeting subunit of this enzyme. Hence, glycogen synthesis is decreased by epinephrine and increased by insulin. Glycogen synthase and phosphorylase are also regulated by noncovalent allosteric interactions. In fact, phosphorylase is a key part of the glucose-sensing system of liver cells. Glycogen metabolism exemplifies the power and precision of reversible phosphorylation in regulating biological processes.

SELECTED READINGS

Where to start

Krebs, E.G., 1993. Protein phosphorylation and cellular regulation I. *Biosci. Rep.* 13:127–142. [Nobel Lecture.]

Fischer, E.H., 1993. Protein phosphorylation and cellular regulation II. *Angew. Chem. Int. Ed.* 32:1130–1137. [Nobel Lecture.]

Johnson, L.N., 1992. Glycogen phosphorylase: Control by phosphorylation and allosteric effectors. *FASEB J.* 6:2274–2282.

Browner, M.F., and Fletterick, R.J., 1992. Phosphorylase: A biological transducer. *Trends Biochem. Sci.* 17:66–71.

Books and general reviews

Boyer, P.D., and Krebs, E.G. (eds.), 1986. *The Enzymes,* vol. 17. Academic Press. [This volume of the series, entitled *Control by Phosphorylation,* contains several excellent articles on glycogen metabolism. Also see vol. 18 of this series for critical reviews of control by reversible phosphorylation.]

Roach, P.J., Cao, Y., Corbett, C.A., DePaoli, R.A., Farkas, I., Fiol, C.J., Flotow, H., Graves, P.R., Hardy, T.A., and Hrubey, T.W., 1991. Glycogen metabolism and signal transduction in mammals and yeast. *Advan. Enzyme Regul.* 31:101–120.

Shulman, G.I., and Landau, B.R., 1992. Pathways of glycogen repletion. *Physiol. Rev.* 72:1019–1035.

Preiss, J., and Romeo, T., 1989. Physiology, biochemistry, and genetics of bacterial glycogen synthesis. *Advan. Microbiol. Physiol.* 30:183–238.

X-ray crystallographic studies

Barford, D., Hu, S.H., and Johnson, L.N., 1991. Structural mechanism for glycogen phosphorylase control by phosphorylation and AMP. *J. Mol. Biol.* 218:233–260.

Sprang, S.R., Withers, S.G., Goldsmith, E.J., Fletterick, R.J., and Madsen, N.B., 1991. Structural basis for the activation of glycogen phosphorylase *b* by adenosine monophosphate. *Science* 254:1367–1371.

Johnson, L.N., and Barford, D., 1990. Glycogen phosphorylase. The structural basis of the allosteric response and comparison with other allosteric proteins. *J. Biol. Chem.* 265:2409–2412.

Browner, M.F., Fauman, E.B., and Fletterick, R.J., 1992. Tracking conformational states in allosteric transitions of phosphorylase. *Biochemistry* 31:11297–11304.

Martin, J.L., Johnson, L.N., and Withers, S.G., 1990. Comparison of the binding of glucose and glucose 1-phosphate derivatives to T-state glycogen phosphorylase *b. Biochemistry* 29:10745–10757.

Johnson, L.N., Acharya, K.R., Jordan, M.D., and McLaughlin, P.J., 1990. Refined crystal structure of the phosphorylase-heptulose-2-phosphate-oligosaccharide-AMP complex. *J. Mol. Biol.* 211:645–661.

Priming of glycogen synthesis

Smythe, C., and Cohen, P., 1991. The discovery of glycogenin and the priming mechanism for glycogen biogenesis. *Eur. J. Biochem.* 200:625–631.

Lomako, J., Lomako, W.M., and Whelan, W.J., 1990. The nature of the primer for glycogen synthesis in muscle. *FEBS Lett.* 268:8–12.

Lomako, J., Lomako, W.M., Whelan, W.J., Domdro, R.S., Neary, J.T., and Norenberg, M.D., 1993. Glycogen synthesis in the astrocyte: from glycogenin to proglycogen to glycogen. *FASEB J.* 7:1386–1393.

Catalytic mechanisms

Palm, D., Klein, H.W., Schinzel, R., Buehner, M., and Helmreich, E.J.M., 1990. The role of pyridoxal 5′-phosphate in glycogen phosphorylase catalysis. *Biochemistry* 29:1099–1107.

Martin, J.L., Veluraja, K., Ross, K., Johnson, L.N., Fleet, G.W., Ramsden, N.G., Bruce, I., Orchard, M.G., Oikonomakos, N.G., and Papageorgiou, A.C., 1991. Glucose analogue inhibitors of glycogen phosphorylase: The design of potential drugs for diabetes. *Biochemistry* 30:10101–10116.

Regulation of glycogen metabolism

Roach, P.J., 1990. Control of glycogen synthase by hierarchal protein phosphorylation. *FASEB J.* 4:2961–2968.

Cohen, P., 1989. The structure and regulation of protein phosphatases. *Ann. Rev. Biochem.* 58:453–508.

Barford, D., 1991. Molecular mechanisms for the control of enzymic activity by protein phosphorylation. *Biochim. Biophys. Acta* 1133:55–62.

Dent, P., Lavoinne, A., Nakielny, S., Caudwell, F.B., Watt, P., and Cohen, P., 1990. The molecular mechanism by which insulin stimulates glycogen synthesis in mammalian skeletal muscle. *Nature* 348:302–308.

Nakielny, S., Campbell, D.G., and Cohen, P., 1991. The molecular mechanism by which adrenalin inhibits glycogen synthesis. *Eur. J. Biochem.* 199:713–722.

Genetic diseases

Hers, H.-G., Van Hoof, F., and de Barsy, T., 1989. Glycogen storage diseases. 1989. *In* Scriver, C.R., Beaudet, A.L., Sly, W.S., and Valle, D. (eds.), *The Metabolic Basis of Inherited Disease* (6th ed.), pp. 425–452. McGraw-Hill.

Burchell, A., and Waddell, I.D., 1991. The molecular basis of the hepatic microsomal glucose-6-phosphatase system. *Biochim. Biophys. Acta* 1092:129–137.

Lei, K.J., Shelley, L.L., Pan, C.J., Sidbury, J.B., and Chou, J.Y., 1993. Mutations in the glucose-6-phosphatase gene that cause glycogen storage disease type Ia. *Science* 262:580–583.

Ross, B.D., Radda, G.K., Gadian, D.G., Rocker, G., Esiri, M., and Falconer-Smith, J., 1981. Examination of a case of suspected McArdle's syndrome by ^{31}P NMR. *New Engl. J. Med.* 304:1338–1342.

Historical aspects

Larner, J., 1990. Insulin and the stimulation of glycogen synthesis. The road from glycogen structure to glycogen synthase to cyclic AMP-dependent protein kinase to insulin mediators. *Advan. Enzymol. Mol. Biol.* 63:173–231.

Cori, C.F., and Cori, G.T., 1947. Polysaccharide phosphorylase. In *Nobel Lectures: Physiology or Medicine (1942–1962)*, pp. 186–206. American Elsevier (1964).

Leloir, L.F., 1971. Two decades of research on the biosynthesis of saccharides. *Science* 172:1299–1302.

PROBLEMS

1. *Carbohydrate conversion.* Write a balanced equation for the formation of glycogen from galactose.

2. *Telltale products.* A sample of glycogen from a patient with liver disease is incubated with orthophosphate, phosphorylase, the transferase, and the debranching enzyme (α-1,6-glucosidase). The ratio of glucose 1-phosphate to glucose formed in this mixture is 100. What is the most likely enzymatic deficiency in this patient?

3. *Excessive storage.* Suggest an explanation for the fact that the amount of glycogen in type I glycogen-storage disease (von Gierke's disease) is increased.

4. *A shattering experience.* Crystals of phosphorylase *a* grown in the presence of glucose shatter when a substrate such as glucose 1-phosphate is added. Why?

5. *Recouping an essential phosphoryl.* The phosphoryl group on phosphoglucomutase is slowly lost by hydrolysis. Propose a mechanism for restoring this essential phosphoryl group that utilizes a known catalytic intermediate. How might this phosphoryl donor be formed?

6. *Hydrophobia.* Why is water excluded from the active site of phosphorylase? Predict the effect of a mutation that allows a water molecule to occasionally enter.

7. *Two in one.* A single polypeptide chain houses the transferase and debranching enzyme. Cite a potential advantage of this arrangement.

8. *Metabolic mutants.* Predict the major consequence of each of the following mutations:
   (a) Loss of the AMP binding site in muscle phosphorylase.
   (b) Mutation of Ser 14 to Ala 14 in liver phosphorylase.
   (c) Overexpression of phosphorylase kinase in liver.
   (d) Loss of the gene for inhibitor 1 of protein phosphatase 1.
   (e) Loss of the gene for the glycogen-targeting subunit of protein phosphatase 1.
   (f) Loss of the gene for glycogenin.

9. *Multiple phosphorylation.* Protein kinase A activates muscle phosphorylase kinase by rapidly phosphorylating its β subunits. The α subunits of phosphorylase kinase are then slowly phosphorylated, which makes α and β susceptible to the action of protein phosphatase 1. What is the functional significance of the slow phosphorylation of α?

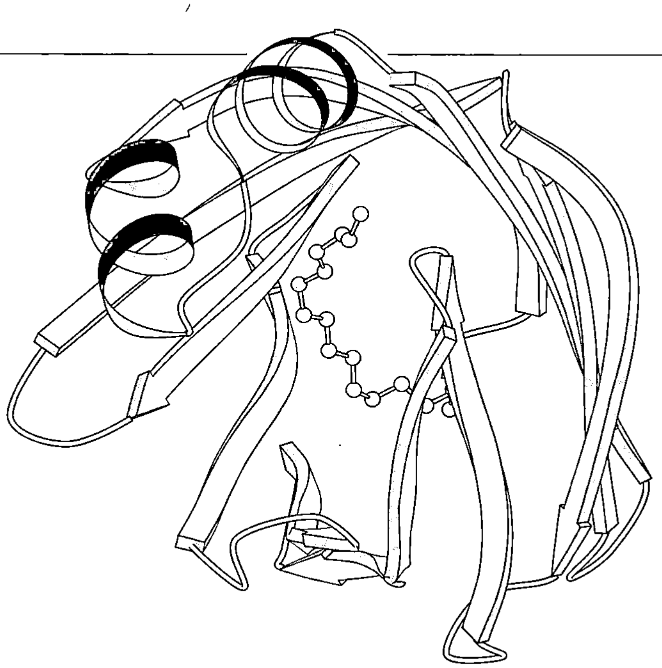

Fatty Acid Metabolism

We turn now from the metabolism of carbohydrates to that of fatty acids, a class of compounds containing a long hydrocarbon chain and a terminal carboxylate group. Fatty acids have four major physiologic roles. First, *they are building blocks of phospholipids and glycolipids.* These amphipathic molecules are important components of biological membranes, as was discussed in Chapter 11. Second, many proteins are modified by the *covalent attachment of fatty acids, which targets them to membrane locations* (p. 934). Third, *fatty acids are fuel molecules.* They are stored as *triacylglycerols,* which are uncharged esters of glycerol. Triacylglycerols are also called *neutral fats* or *triglycerides.* Fourth, fatty acid derivatives serve as *hormones* and *intracellular messengers.*

$$H_3C-(CH_2)_7-\overset{H}{C}=\overset{H}{C}-(CH_2)_7-\overset{O}{\underset{}{C}}-O-\overset{\displaystyle CH_2-O-\overset{O}{\underset{}{C}}-(CH_2)_{14}-CH_3}{\underset{\displaystyle CH_2-O-\underset{O}{\underset{}{C}}-(CH_2)_{16}-CH_3}{\overset{|}{CH}}}$$

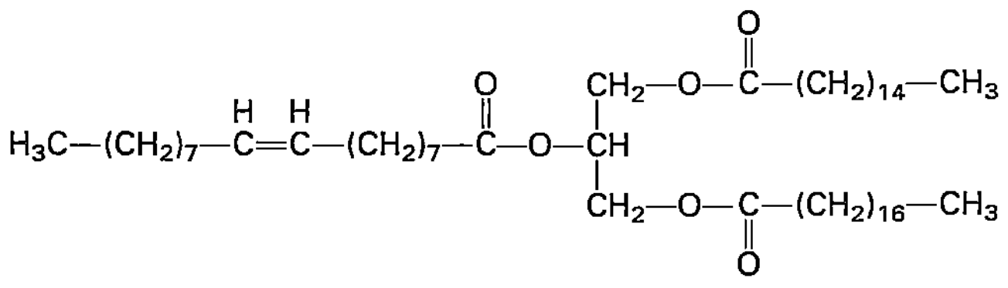

A triacylglycerol

NOMENCLATURE OF FATTY ACIDS

The systematic name for a fatty acid is derived from the name of its parent hydrocarbon by the substitution of *oic* for the final *e.* For example, the C_{18} saturated fatty acid is called *octadecanoic acid* because the parent hydrocarbon is octadecane. A C_{18} fatty acid with one double bond is called octadec*enoic* acid; with two double bonds, octadeca*dienoic* acid; and with

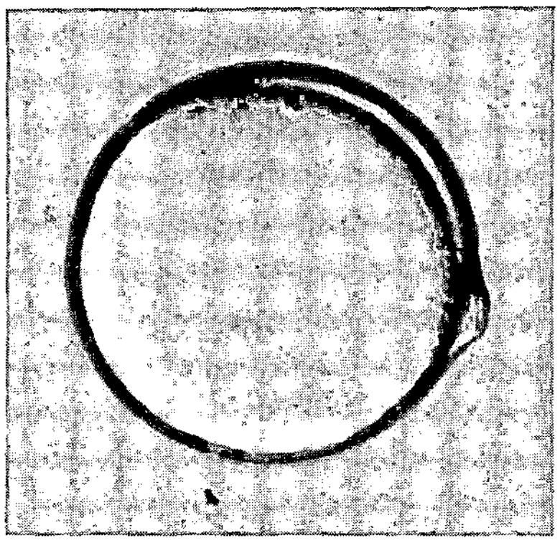

Figure 24-1
Photomicrograph of a fat cell. A large globule of fat is surrounded by a thin rim of cytoplasm and a bulging nucleus. [Courtesy of Dr. Pedro Cuatrecasas.]

Opening Image: Structure of palmitate (yellow) complexed to a fatty acid–binding protein. Fatty acids are usually bound to carrier proteins. β Strands are shown in green and α helices in red. The fatty acid is located in the hydrophobic center of this β barrel. [Drawn from 2ifb.pdb. J.C. Sacchettini, J.I. Gordon, and L.J. Banaszak. J. Mol. Biol. 208(1989):327.]

three double bonds, octadeca*trienoic* acid. The symbol $18:0$ denotes a C_{18} fatty acid with no double bonds, whereas $18:2$ signifies that there are two double bonds.

Fatty acid carbon atoms are numbered starting at the carboxy terminus.

$$H_3\underset{\omega}{C}-(CH_2)_n-\underset{\beta}{\overset{3}{C}H_2}-\underset{\alpha}{\overset{2}{C}H_2}-\overset{1}{C}\diagdown^{\displaystyle\nearrow O}_{\displaystyle OH}$$

ω-Carbon
atom

ω-3
double bond

$$\begin{array}{c} CH_3 \\ | \\ CH_2 \\ | \\ C-H \\ \| \\ C-H \\ | \\ CH_2 \\ | \\ R \\ | \\ COO^- \end{array}$$

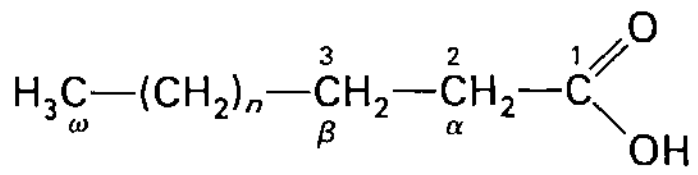

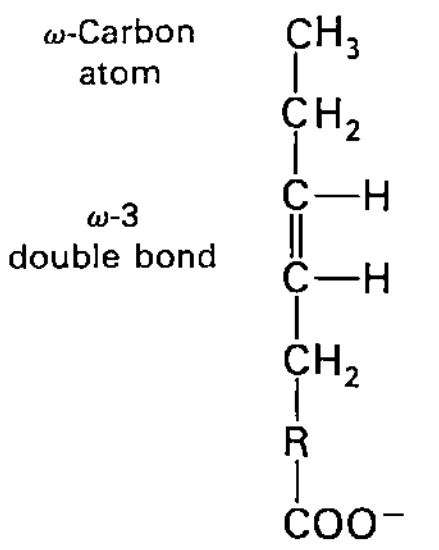

An ω-3 fatty acid

Carbon atoms 2 and 3 are often referred to as α and β, respectively. The methyl carbon atom at the distal end of the chain is called the *ω carbon*. The position of a double bond is represented by the symbol Δ followed by a superscript number. For example, *cis*-Δ^9 means that there is a *cis* double bond between carbon atoms 9 and 10; *trans*-Δ^2 means that there is a *trans* double bond between carbon atoms 2 and 3. Alternatively, the position of a double bond can be denoted by counting from the distal end, with the ω carbon atom (the methyl carbon) as number 1. An ω-3 fatty acid, for example, has the structure shown at the left. Fatty acids are ionized at physiologic pH, and so it is appropriate to refer to them according to their carboxylate form: for example, palmitate or hexadecanoate.

FATTY ACIDS VARY IN CHAIN LENGTH AND DEGREE OF UNSATURATION

Fatty acids in biological systems (Table 24-1) usually contain an even number of carbon atoms, typically between 14 and 24. The 16- and 18-carbon fatty acids are most common. The hydrocarbon chain is almost invariably unbranched in animal fatty acids. The alkyl chain may be saturated or it may contain one or more double bonds. The configuration of the double bonds in most unsaturated fatty acids is cis (p. 265). The double bonds in polyunsaturated fatty acids are separated by at least one methylene group.

Table 24-1
Some naturally occurring fatty acids in animals

Number of carbons	Number of double bonds	Common name	Systematic name	Formula
12	0	Laurate	*n*-Dodecanoate	$CH_3(CH_2)_{10}COO^-$
14	0	Myristate	*n*-Tetradecanoate	$CH_3(CH_2)_{12}COO^-$
16	0	Palmitate	*n*-Hexadecanoate	$CH_3(CH_2)_{14}COO^-$
18	0	Stearate	*n*-Octadecanoate	$CH_3(CH_2)_{16}COO^-$
20	0	Arachidate	*n*-Eicosanoate	$CH_3(CH_2)_{18}COO^-$
22	0	Behenate	*n*-Docosanoate	$CH_3(CH_2)_{20}COO^-$
24	0	Lignocerate	*n*-Tetracosanoate	$CH_3(CH_2)_{22}COO^-$
16	1	Palmitoleate	*cis*-Δ^9-Hexadecenoate	$CH_3(CH_2)_5CH{=}CH(CH_2)_7COO^-$
18	1	Oleate	*cis*-Δ^9-Octadecenoate	$CH_3(CH_2)_7CH{=}CH(CH_2)_7COO^-$
18	2	Linoleate	*cis,cis*-Δ^9, Δ^{12}-Octadecadienoate	$CH_3(CH_2)_4(CH{=}CHCH_2)_2(CH_2)_6COO^-$
18	3	Linolenate	all-*cis*-Δ^9, Δ^{12}, Δ^{15}-Octadecatrienoate	$CH_3CH_2(CH{=}CHCH_2)_3(CH_2)_6COO^-$
20	4	Arachidonate	all-*cis*-Δ^5, Δ^8, Δ^{11}, Δ^{14}-Eicosatetraenoate	$CH_3(CH_2)_4(CH{=}CHCH_2)_4(CH_2)_2COO^-$

The properties of fatty acids and of lipids derived from them are markedly dependent on their chain length and on their degree of saturation. Unsaturated fatty acids have a lower melting point than saturated fatty acids of the same length. For example, the melting point of stearic acid is 69.6°C, whereas that of oleic acid (which contains one cis double bond) is 13.4°C. The melting points of polyunsaturated fatty acids of the C_{18} series are even lower. Chain length also affects the melting point, as illustrated by the fact that the melting temperature of palmitic acid (C_{16}) is 6.5 degrees lower than that of stearic acid (C_{18}). Thus, *short chain length and unsaturation enhance the fluidity of fatty acids and of their derivatives* (p. 279).

TRIACYLGLYCEROLS ARE HIGHLY CONCENTRATED ENERGY STORES

Triacylglycerols are highly concentrated stores of metabolic energy because they are *reduced* and *anhydrous.* The yield from the complete oxidation of fatty acids is about 9 kcal/g, in contrast with about 4 kcal/g for carbohydrates and proteins. The basis of this large difference in caloric yield is that fatty acids are much more highly reduced. Furthermore, triacylglycerols are very nonpolar, and so they are stored in a nearly anhydrous form, whereas proteins and carbohydrates are much more polar and hence more highly hydrated. In fact, a gram of dry glycogen binds about two grams of water. Consequently, *a gram of nearly anhydrous fat stores more than six times as much energy as a gram of hydrated glycogen,* which is the reason that triacylglycerols rather than glycogen were selected in evolution as the major energy reservoir. Consider a typical 70-kg man, who has fuel reserves of 100,000 kcal in triacylglycerols, 25,000 kcal in protein (mostly in muscle), 600 kcal in glycogen, and 40 kcal in glucose. Triacylglycerols constitute about 11 kg of his total body weight. If this amount of energy were stored in glycogen, his total body weight would be 55 kg greater.

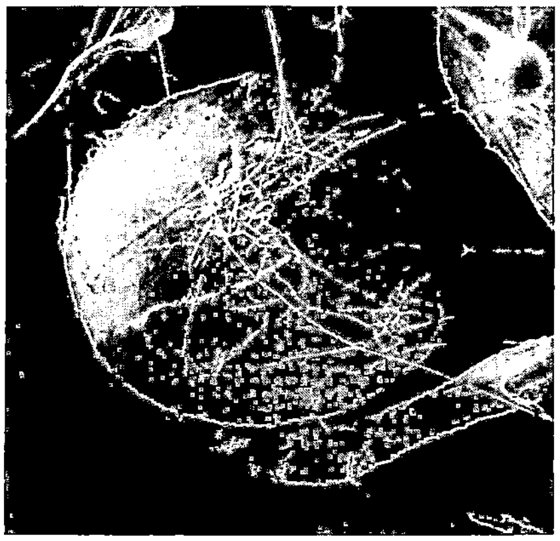

Figure 24-2
Scanning electron micrograph of a fat cell. [From *Tissues and Organs,* by Richard G. Kessel and Randy H. Kardon. Copyright © 1979. W.H. Freeman and Company.]

In mammals, the major site of accumulation of triacylglycerols is the cytoplasm of *adipose cells (fat cells). Droplets of triacylglycerol coalesce to form a large globule, which may occupy most of the cell volume. Adipose cells are specialized for the synthesis and storage of triacylglycerols and for their mobilization into fuel molecules that are transported to other tissues by the blood.*

TRIACYLGLYCEROLS ARE HYDROLYZED BY CYCLIC AMP-REGULATED LIPASES

The initial event in the utilization of fat as an energy source is the hydrolysis of triacylglycerol by lipases.

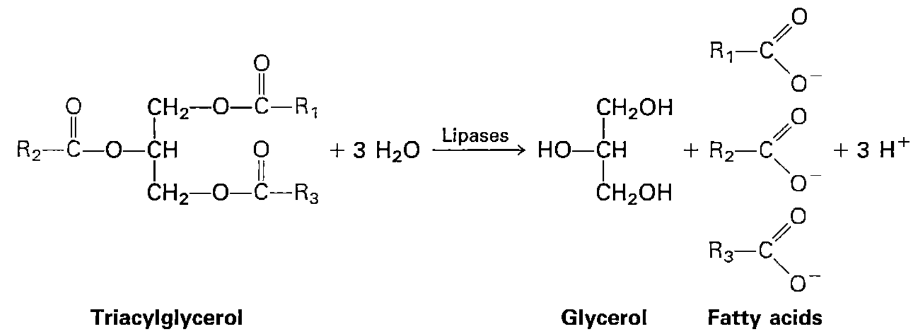

The activity of adipose-cell lipase is regulated by hormones. Epinephrine, norepinephrine, glucagon, and adrenocorticotropic hormone activate adenylate cyclase in adipose cells by triggering seven-helix receptors (p. 341). The increased level of cyclic AMP then stimulates protein kinase A, which activates the lipase by phosphorylating it. Thus, *epinephrine, norepinephrine, glucagon, and adrenocorticotropic hormone induce lipolysis.* Cyclic AMP is a messenger in the activation of lipolysis in adipose cells, as it is in the activation of glycogen breakdown (p. 594). In contrast, *insulin inhibits lipolysis.*

Glycerol formed by lipolysis is phosphorylated and oxidized to dihydroxyacetone phosphate, which in turn is isomerized to glyceraldehyde 3-phosphate. This intermediate is on both the glycolytic and the gluconeogenic pathways. Hence, glycerol can be converted into pyruvate or glucose in the liver, which contains the appropriate enzymes. The reverse process can occur by the reduction of dihydroxyacetone phosphate to glycerol 3-phosphate. Hydrolysis by a phosphatase then gives glycerol. Thus, glycerol and glycolytic intermediates are readily interconvertible.

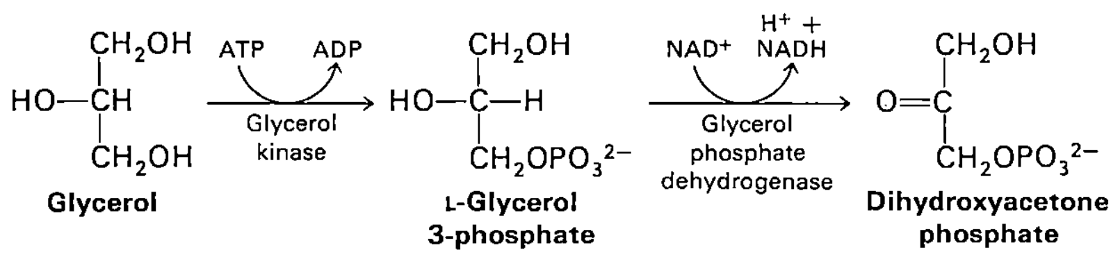

FATTY ACIDS ARE DEGRADED BY THE SEQUENTIAL REMOVAL OF TWO-CARBON UNITS

In 1904, Franz Knoop made a critical contribution to the elucidation of the mechanism of fatty acid oxidation. He fed dogs straight-chain fatty acids in which the ω carbon atom was joined to a phenyl group. Knoop found that the urine of these dogs contained a derivative of phenylacetic acid when they were fed phenylbutyrate. In contrast, a derivative of benzoic acid was formed when they were fed phenylpropionate. In fact, phenylacetic acid was produced whenever a fatty acid containing an even number of carbon atoms was fed, whereas benzoic acid was formed whenever a fatty acid containing an odd number was fed (Figure 24-3). Knoop deduced from these findings that *fatty acids are degraded by oxidation at the β carbon.* These experiments are a landmark in biochemistry because they were the first to use a synthetic label (the phenyl group) to elucidate reaction mechanisms. Deuterium and radioisotopes came into biochemistry several decades later.

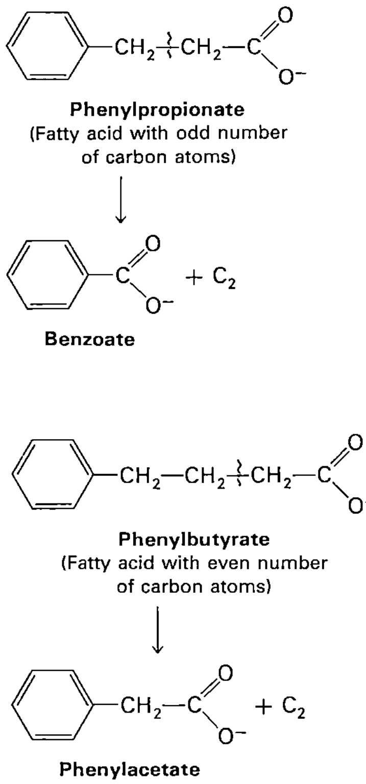

Figure 24-3
Knoop's experiment showing that fatty acids are degraded by the removal of two-carbon units.

FATTY ACIDS ARE LINKED TO COENZYME A BEFORE THEY ARE OXIDIZED

Eugene Kennedy and Albert Lehninger showed in 1949 that fatty acids are oxidized in mitochondria. Subsequent work demonstrated that they are activated before they enter the mitochondrial matrix. Adenosine triphosphate (ATP) drives the formation of a thioester linkage between the carboxyl group of a fatty acid and the sulfhydryl group of CoA. This activation reaction occurs on the outer mitochondrial membrane, where it is catalyzed by *acyl CoA synthetase* (also called *fatty acid thiokinase*).

$$R-C\overset{O}{\underset{O^-}{<}} + ATP + HS-CoA \rightleftharpoons R-\overset{O}{\overset{\|}{C}}-S-CoA + AMP + PP_i$$

Paul Berg showed that the activation of a fatty acid occurs in two steps. First, the fatty acid reacts with ATP to form an *acyl adenylate*. In this mixed anhydride, the carboxyl group of a fatty acid is bonded to the phosphoryl group of AMP. The other two phosphoryl groups of the ATP substrate are released as pyrophosphate. The sulfhydryl group of CoA then attacks the acyl adenylate, which is tightly bound to the enzyme, to form acyl CoA and AMP.

$$R-\overset{O}{\overset{\|}{C}}-O-\overset{O}{\underset{\underset{O^-}{|}}{\overset{\|}{P}}}-O-Ribose-Adenine$$

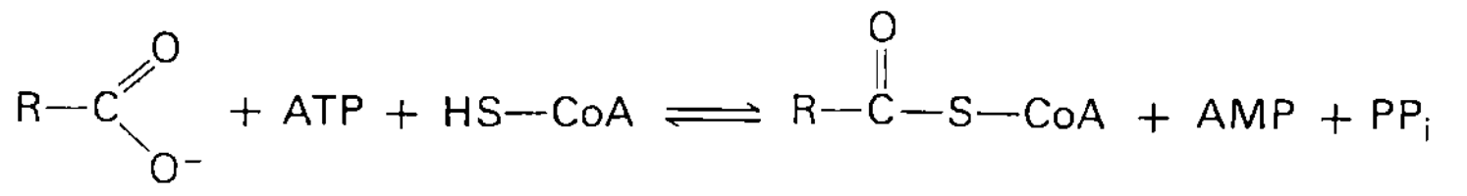

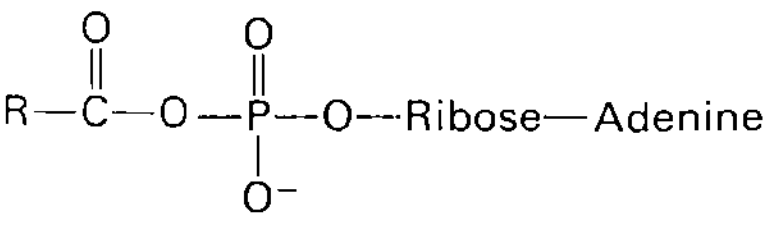

**Acyl adenylate
(Acyl AMP)**

$$R-C\overset{O}{\underset{O^-}{<}} + ATP \rightleftharpoons R-\overset{O}{\overset{\|}{C}}-AMP + PP_i \qquad (1)$$

Fatty acid **Acyl adenylate**

$$R-\overset{O}{\overset{\|}{C}}-AMP + HS-CoA \rightleftharpoons R-\overset{O}{\overset{\|}{C}}-S-CoA + AMP \qquad (2)$$

Acyl CoA

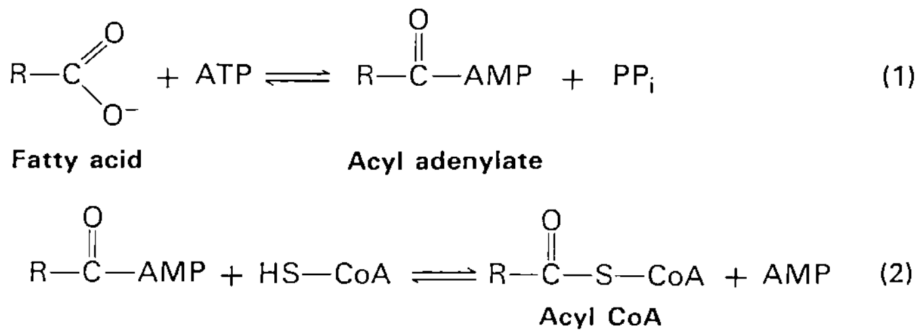

These partial reactions are freely reversible. In fact, the equilibrium constant for the sum of these reactions is close to 1.

$$R-COO^- + CoA + ATP \rightleftharpoons acyl\ CoA + AMP + PP_i$$

One high-energy bond is broken (between PP_i and AMP) and one high-energy bond is formed (the thioester linkage in acyl CoA). How is this reaction driven forward? The answer is that pyrophosphate is rapidly hydrolyzed by a pyrophosphatase.

$$R-COO^- + CoA + ATP + H_2O \longrightarrow acyl\ CoA + AMP + 2\ P_i + 2\ H^+$$

This makes the overall reaction irreversible because two high-energy bonds are consumed, whereas only one is formed. We see here another example of a recurring theme in biochemistry: *many biosynthetic reactions are made irreversible by the hydrolysis of inorganic pyrophosphate.*

Another motif recurs in this activation reaction. The enzyme-bound acyl adenylate intermediate is not unique to the synthesis of acyl CoA. *Acyl adenylates are frequently formed when carboxyl groups are activated in biochemical reactions.* For example, amino acids are activated for protein synthesis by a similar mechanism (p. 880).

CARNITINE CARRIES LONG-CHAIN ACTIVATED FATTY ACIDS INTO THE MITOCHONDRIAL MATRIX

Fatty acids are activated on the outer mitochondrial membrane, whereas they are oxidized in the mitochondrial matrix. Long-chain acyl CoA molecules do not readily traverse the inner mitochondrial membrane, and so a special transport mechanism is needed. Activated long-chain fatty acids are carried across the inner mitochondrial membrane by conjugating them to *carnitine*, a zwitterionic compound formed from lysine. The acyl group is transferred from the sulfur atom of CoA to the hydroxyl group of carnitine to form *acyl carnitine*. This reaction is catalyzed by *carnitine acyltransferase I*, which is bound to the outer mitochondrial membrane.

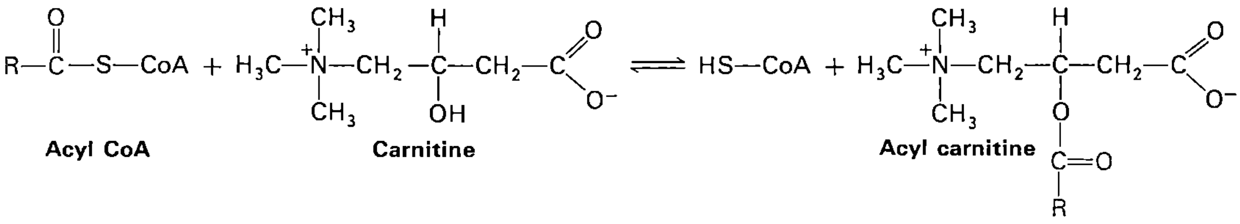

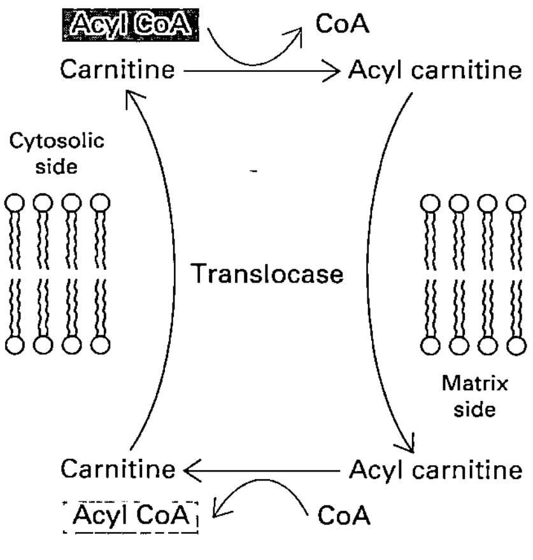

Figure 24-4
The entry of acyl carnitine into the mitochondrial matrix is mediated by a translocase. Carnitine returns to the cytosolic side of the inner mitochondrial membrane in exchange for acyl carnitine.

Acyl carnitine is then shuttled across the inner mitochondrial membrane by a translocase (Figure 24-4). The acyl group is transferred back to CoA on the matrix side of the membrane. This reaction, which is catalyzed by carnitine acyltransferase II, is thermodynamically feasible because the *O*-acyl link in carnitine has a high group transfer potential. Finally, carnitine is returned to the cytosolic side by the translocase, in exchange for an incoming acyl carnitine.

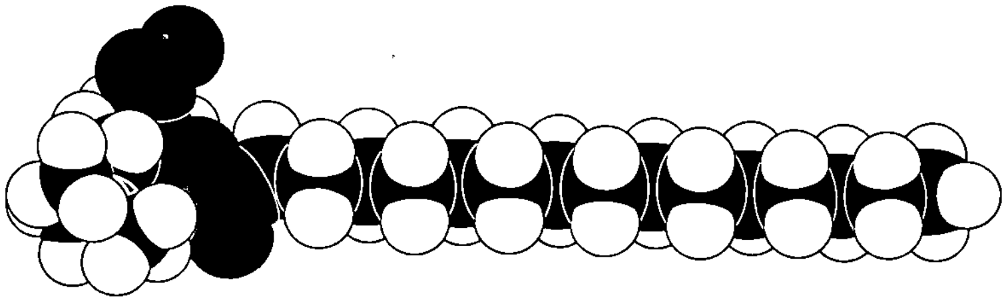

Molecular model of palmitoyl carnitine.

A defect in the transferase or translocase, or a deficiency of carnitine, might be expected to impair the oxidation of long-chain fatty acids. Such a disorder has in fact been found in identical twins who had had aching muscle cramps since early childhood. The aches were precipitated by fasting, exercise, or a high-fat diet; fatty acid oxidation is the major energy-yielding process in these three states. The enzymes of glycolysis and glycogenolysis were found to be normal. Lipolysis of triacylglycerols was normal, as evidenced by a rise in the concentration of unesterified fatty acids in the plasma after fasting. Assay of a muscle biopsy showed that the long-chain acyl CoA synthetase was fully active. Furthermore, medium-chain (C_8 and C_{10}) fatty acids were normally metabolized. It is known that carnitine is not required for the permeation of medium-chain acyl CoAs into the mitochondrial matrix. This case report demonstrates in a striking way that *the impaired flow of a metabolite from one compartment of a cell to another can cause disease.*

ACETYL CoA, NADH, AND $FADH_2$ ARE GENERATED IN EACH ROUND OF FATTY ACID OXIDATION

A saturated acyl CoA is degraded by a recurring sequence of four reactions: oxidation by flavin adenine dinucleotide (FAD), hydration, oxidation by NAD^+, and thiolysis by CoA (Figure 24-5). The fatty acyl chain is shortened by two carbon atoms as a result of these reactions, and $FADH_2$,

NADH, and acetyl CoA are generated. David Green, Severo Ochoa, and Feodor Lynen made important contributions to the elucidation of this series of reactions, which is called the *β-oxidation pathway*.

The first reaction in each round of degradation is the *oxidation* of acyl CoA by an *acyl CoA dehydrogenase* to give an enoyl CoA with a trans double bond between C-2 and C-3.

$$\text{Acyl CoA} + \text{E-FAD} \longrightarrow \text{trans-}\Delta^2\text{-enoyl CoA} + \text{E-FADH}_2$$

As in the dehydrogenation of succinate in the citric acid cycle, FAD rather than NAD^+ is the electron acceptor because the ΔG of this reaction is insufficient to drive the reduction of NAD^+. Electrons from the $FADH_2$ prosthetic group of the reduced acyl CoA dehydrogenase are transferred to a second flavoprotein called *electron-transferring flavoprotein (ETF)*. In turn, ETF donates electrons to *ETF:ubiquinone reductase*, an iron-sulfur protein. Ubiquinone is thereby reduced to ubiquinol, which delivers its high-potential electrons to the second proton-pumping site of the respiratory chain (p. 537). Consequently, 1.5 ATP are generated per molecule of $FADH_2$ formed in this dehydrogenation step, as in the oxidation of succinate to fumarate.

$$\begin{array}{ccccc}
\text{R—CH}_2\text{—CH}_2\text{—R'} & \text{E-FAD} & \text{ETF-FADH}_2 & \text{Fe-S (oxidized)} & \text{Ubiquinol (QH}_2) \\
\text{R—CH=CH—R'} & \text{E-FADH}_2 & \text{ETF-FAD} & \text{Fe-S (reduced)} & \text{Ubiquinone (Q)}
\end{array}$$

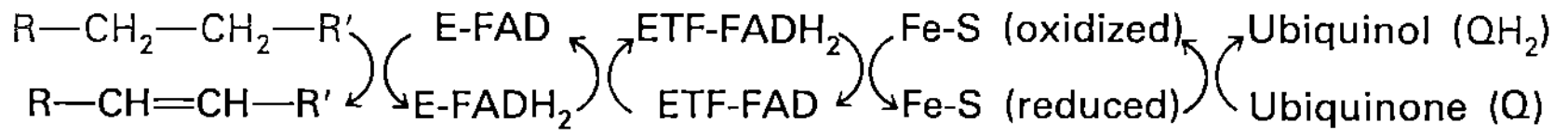

The next step is the *hydration* of the double bond between C-2 and C-3 by *enoyl CoA hydratase*.

$$\text{trans-}\Delta^2\text{-Enoyl CoA} + \text{H}_2\text{O} \rightleftharpoons \text{L-3-hydroxyacyl CoA}$$

The hydration of the enoyl CoA is stereospecific, as are the hydrations of fumarate and aconitate. Only the L isomer of 3-hydroxyacyl CoA is formed when the trans-Δ^2 double bond is hydrated. The enzyme also hydrates a cis-Δ^2 double bond, but the product then is the D isomer. We shall return to this point shortly in considering how unsaturated fatty acids are oxidized.

The hydration of enoyl CoA is the prelude to the second *oxidation* reaction, which converts the hydroxyl group at C-3 into a keto group and generates NADH. This oxidation is catalyzed by *L-3-hydroxyacyl CoA dehydrogenase*, which is absolutely specific for the L isomer of the hydroxyacyl substrate.

$$\text{L-3-Hydroxyacyl CoA} + \text{NAD}^+ \rightleftharpoons \text{3-ketoacyl CoA} + \text{NADH} + \text{H}^+$$

These three reactions in each round of fatty acid degradation closely resemble the last steps in the citric acid cycle:

$$\text{Acyl CoA} \longrightarrow \text{enoyl CoA} \longrightarrow \text{hydroxyacyl CoA} \longrightarrow \text{ketoacyl CoA}$$

$$\text{Succinate} \longrightarrow \text{fumarate} \longrightarrow \text{malate} \longrightarrow \text{oxaloacetate}$$

The preceding reactions have oxidized the methylene group at C-3 to a keto group. The final step is the *cleavage* of 3-ketoacyl CoA by the thiol group of a second molecule of CoA, which yields acetyl CoA and an acyl CoA shortened by two carbon atoms. This thiolytic cleavage is catalyzed by *β-ketothiolase*.

$$\text{3-Ketoacyl CoA} + \text{HS—CoA} \rightleftharpoons \text{acetyl CoA} + \quad \text{acyl CoA}$$
$$\quad\quad (n \text{ carbons}) \quad\quad\quad\quad\quad\quad\quad\quad\quad (n-2 \text{ carbons})$$

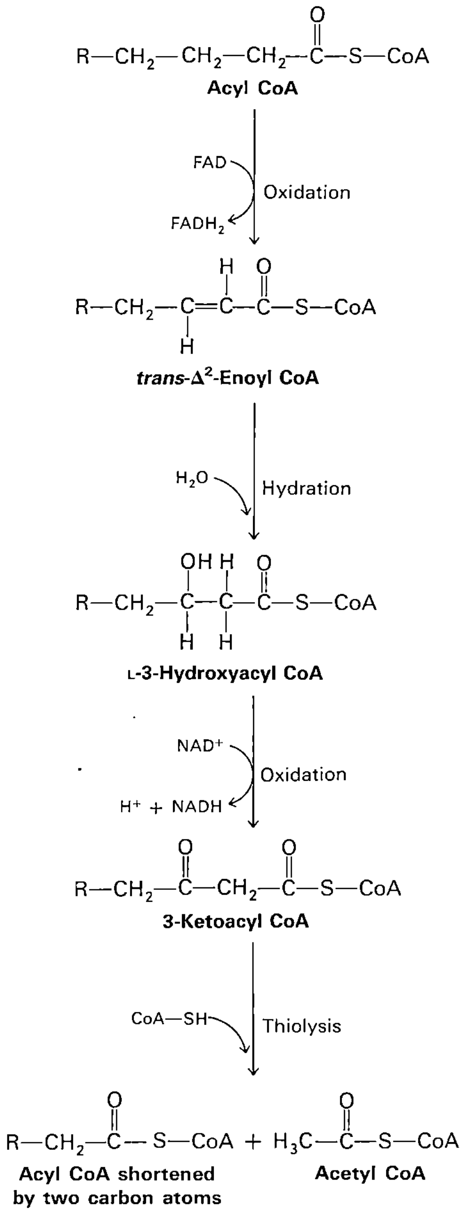

Figure 24-5
Reaction sequence in the degradation of fatty acids: oxidation, hydration, oxidation, and thiolysis.

The shortened acyl CoA then undergoes another cycle of oxidation, starting with the reaction catalyzed by acyl CoA dehydrogenase (Figure 24-6). Fatty acyl chains containing 12 to 18 carbons are oxidized by the long-chain acyl CoA dehydrogenase. The medium-chain acyl CoA dehydrogenase oxidizes fatty acyl chains having 14 to 4 carbons, whereas the short-chain acyl CoA dehydrogenase acts only on 4- and 6-carbon acyl chains. By contrast, β-ketothiolase, hydroxyacyl dehydrogenase, and enoyl CoA hydratase have broad specificity with respect to the length of the acyl group.

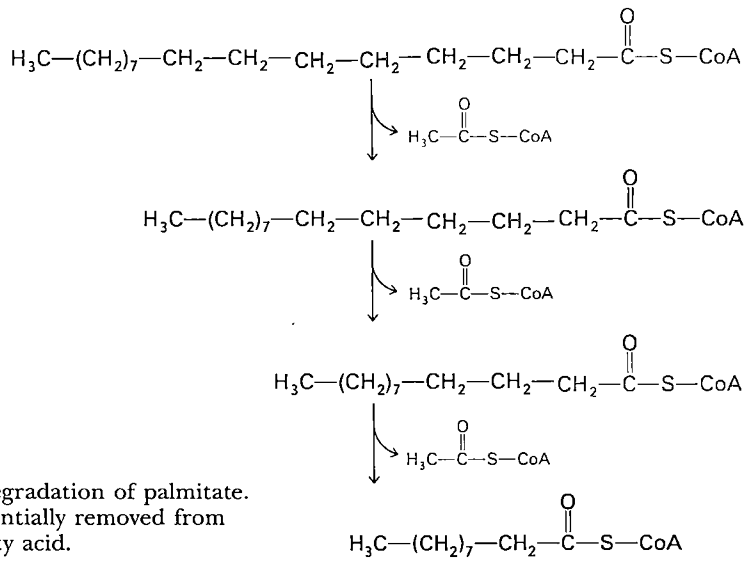

Figure 24-6
First three rounds in the degradation of palmitate. Two-carbon units are sequentially removed from the carboxyl end of the fatty acid.

Table 24-2
Principal reactions in fatty acid oxidation

Step	*Reaction*	*Enzyme*
1	Fatty acid + CoA + ATP $\rightleftharpoons$ acyl CoA + AMP + PP_i	Acyl CoA synthetase (also called fatty acid thiokinase and fatty acid:CoA ligase [AMP])
2	Carnitine + acyl CoA $\rightleftharpoons$ acyl carnitine + CoA	Carnitine acyltransferase
3	Acyl CoA + E-FAD $\longrightarrow$ *trans*-Δ^2-enoyl CoA + E-$FADH_2$	Acyl CoA dehydrogenases (several enzymes having different chain-length specificity)
4	*trans*-Δ^2-Enoyl CoA + H_2O $\rightleftharpoons$ L-3-hydroxyacyl CoA	Enoyl CoA hydratase (also called crotonase or 3-hydroxyacyl CoA hydrolyase)
5	L-3-Hydroxyacyl CoA + NAD^+ $\rightleftharpoons$ 3-ketoacyl CoA + NADH + H^+	L-3-Hydroxyacyl CoA dehydrogenase
6	3-Ketoacyl CoA + CoA $\rightleftharpoons$ acetyl CoA + acyl CoA (shortened by C_2)	β-Ketothiolase (also called thiolase)

THE COMPLETE OXIDATION OF PALMITATE YIELDS 106 ATP

We can calculate the energy yield derived from the oxidation of a fatty acid. In each reaction cycle, an acyl CoA is shortened by two carbons, and one $FADH_2$, NADH, and acetyl CoA are formed.

$$C_n\text{-acyl CoA} + FAD + NAD^+ + H_2O + CoA \longrightarrow C_{n-2}\text{-acyl CoA} + FADH_2 + NADH + \text{acetyl CoA} + H^+$$

The degradation of palmitoyl CoA (C_{16}-acyl CoA) requires seven reaction cycles. In the seventh cycle, the C_4-ketoacyl CoA is thiolyzed to two molecules of acetyl CoA. Hence, the stoichiometry of oxidation of palmitoyl CoA is

$$\text{Palmitoyl CoA} + 7\ \text{FAD} + 7\ \text{NAD}^+ + 7\ \text{CoA} + 7\ \text{H}_2\text{O} \longrightarrow 8\ \text{acetyl CoA} + 7\ \text{FADH}_2 + 7\ \text{NADH} + 7\ \text{H}^+$$

Two and a half ATP are generated when each of these NADH is oxidized by the respiratory chain, whereas 1.5 ATP are formed for each $FADH_2$, because their electrons enter the chain at the level of ubiquinol. Recall that the oxidation of acetyl CoA by the citric acid cycle yields 10 ATP. Hence, the number of ATP formed in the oxidation of palmitoyl CoA is 10.5 from the 7 $FADH_2$, 17.5 from the 7 NADH, and 80 from the 8 molecules of acetyl CoA, which gives a total of 108. Two high-energy phosphate bonds are consumed in the activation of palmitate, in which ATP is split into AMP and 2 P_i. Thus, *the complete oxidation of a molecule of palmitate yields 106 ATP.*

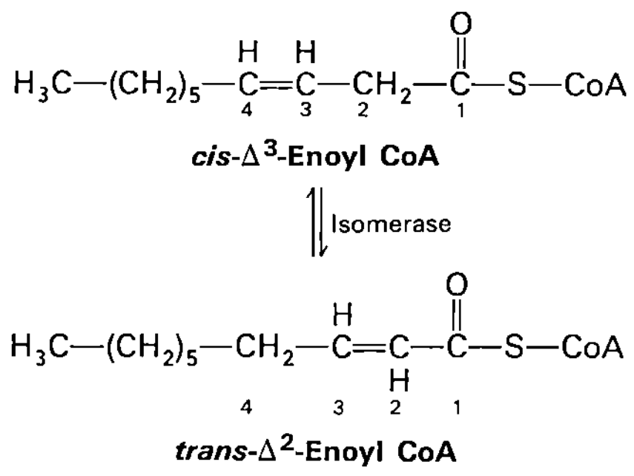

AN ISOMERASE AND A REDUCTASE ARE REQUIRED FOR THE OXIDATION OF UNSATURATED FATTY ACIDS

We turn now to the oxidation of unsaturated fatty acids. Many of the reactions are the same as those for saturated fatty acids. In fact, only two additional enzymes—an isomerase and a reductase—are needed to degrade a wide range of unsaturated fatty acids.

Consider the oxidation of palmitoleate. This C_{16} unsaturated fatty acid, which has one double bond between C-9 and C-10, is activated and transported across the inner mitochondrial membrane in the same way as saturated fatty acids. Palmitoleoyl CoA then undergoes three cycles of degradation, which are carried out by the same enzymes as in the oxidation of saturated fatty acids. However, the cis-Δ^3-enoyl CoA formed in the third round is not a substrate for acyl CoA dehydrogenase. The presence of a double bond between C-3 and C-4 prevents the formation of another double bond between C-2 and C-3. This impasse is resolved by a new reaction that shifts the position and configuration of the cis-Δ^3 double bond. *An isomerase converts this double bond into a trans-Δ^2 double bond.* The subsequent reactions are those of the saturated fatty acid oxidation pathway, in which the *trans*-Δ^2-enoyl CoA is a regular substrate.

A second accessory enzyme is needed for the oxidation of polyunsaturated fatty acids. Consider linoleate, a C_{18} polyunsaturated fatty acid with cis-Δ^9 and cis-Δ^{12} double bonds. The cis-Δ^3 double bond formed after three rounds of β-oxidation is converted into a trans-Δ^2 double bond by the isomerase mentioned above, as in the oxidation of palmitoleate. The cis-Δ^{12} double bond of linoleate poses a new problem. The acyl CoA produced by four rounds of β-oxidation contains a cis-Δ^4 double bond (Figure 24-7). Dehydrogenation of this species by acyl CoA dehydrogenase yields a *2,4-dienoyl intermediate,* which is not a substrate for the next enzyme in the β-oxidation pathway. This impasse is circumvented by *2,4-dienoyl CoA reductase,* an enzyme that uses NADPH to reduce the 2,4-dienoyl intermediate to *cis*-Δ^3-enoyl CoA. The isomerase mentioned above then converts *cis*-Δ^3-enoyl CoA to the trans form, a customary intermediate in the β-oxidation pathway. These catalytic strategies are elegant and economical. Only two extra enzymes are needed for the oxidation of *any* polyunsaturated fatty acid. *Odd-numbered double bonds are handled by the isomerase, and even-numbered ones by the reductase and the isomerase.*

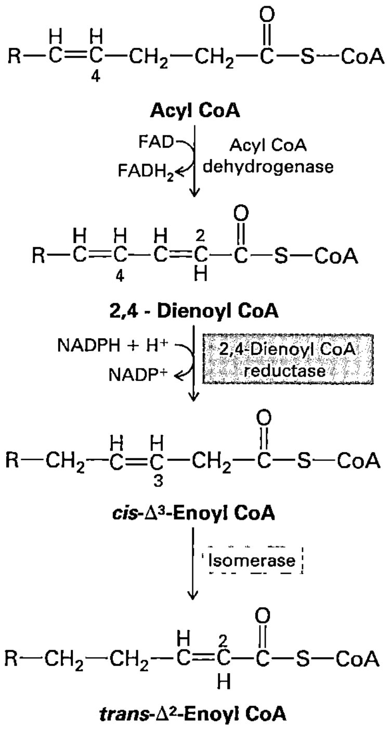

Figure 24-7
Two accessory enzymes—2,4-dienoyl CoA reductase and *cis*-Δ^3-enoyl isomerase—make possible the β-oxidation of polyunsaturated fatty acids containing a cis double bond at an even-numbered carbon atom.

ODD-CHAIN FATTY ACIDS YIELD PROPIONYL COENZYME A IN THE FINAL THIOLYSIS STEP

$$H_3C-CH_2-\overset{\overset{\displaystyle O}{\|}}{C}-S-CoA$$

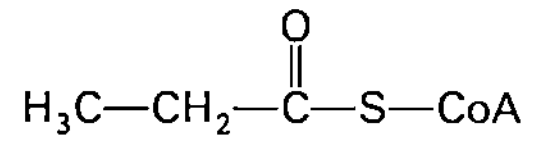

Propionyl CoA

Fatty acids having an odd number of carbon atoms are minor species. They are oxidized in the same way as fatty acids having an even number, except that propionyl CoA and acetyl CoA, rather than two molecules of acetyl CoA, are produced in the final round of degradation. The activated three-carbon unit in propionyl CoA enters the citric acid cycle after it is converted into succinyl CoA. The pathway from propionyl CoA to succinyl CoA will be discussed in the next chapter (p. 641) because propionyl CoA is also formed in the oxidation of several amino acids.

KETONE BODIES ARE FORMED FROM ACETYL COENZYME A WHEN FAT BREAKDOWN PREDOMINATES

The acetyl CoA formed in fatty acid oxidation enters the citric acid cycle only if fat and carbohydrate degradation are appropriately balanced. The reason is that the entry of acetyl CoA into the citric acid cycle depends on the availability of oxaloacetate for the formation of citrate, but the concentration of oxaloacetate is lowered if carbohydrate is unavailable or improperly utilized. Recall that oxaloacetate is normally formed from pyruvate, the product of glycolysis. The molecular basis of the adage that *fats burn in the flame of carbohydrates* is now evident.

In fasting or diabetes, oxaloacetate is consumed to form glucose by the gluconeogenic pathway (p. 571) and hence is unavailable for condensation with acetyl CoA. Under these conditions, acetyl CoA is diverted to the formation of acetoacetate and D-3-hydroxybutyrate. Acetoacetate, D-3-hydroxybutyrate, and acetone are sometimes referred to as *ketone bodies*. Abnormally high levels of ketone bodies are present in the blood of untreated diabetics (p. 780).

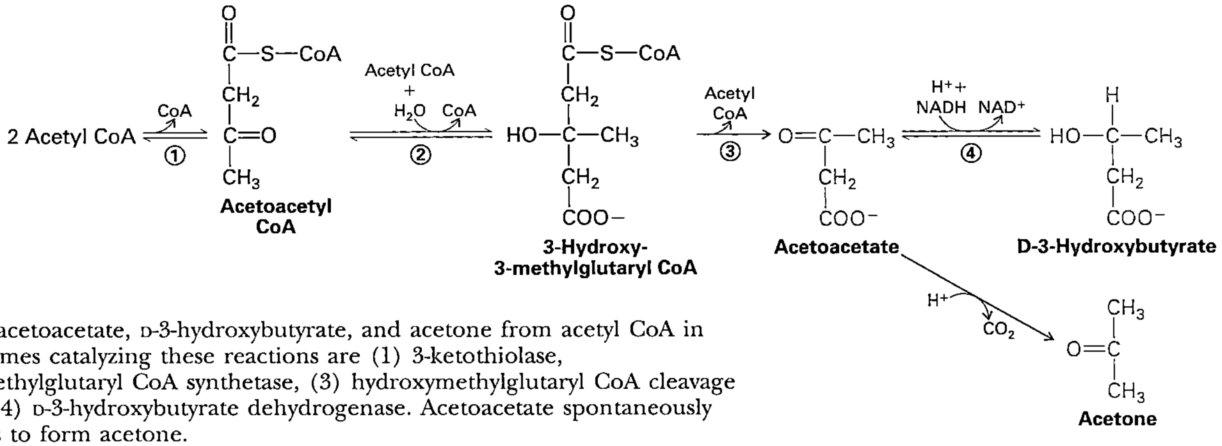

Figure 24-8
Formation of acetoacetate, D-3-hydroxybutyrate, and acetone from acetyl CoA in the liver. Enzymes catalyzing these reactions are (1) 3-ketothiolase, (2) hydroxymethylglutaryl CoA synthetase, (3) hydroxymethylglutaryl CoA cleavage enzyme, and (4) D-3-hydroxybutyrate dehydrogenase. Acetoacetate spontaneously decarboxylates to form acetone.

Acetoacetate is formed from acetyl CoA in three steps (Figure 24-8). Two molecules of acetyl CoA condense to form acetoacetyl CoA. This reaction, which is catalyzed by thiolase, is a reversal of the thiolysis step in the oxidation of fatty acids. Acetoacetyl CoA then reacts with acetyl CoA and water to give 3-hydroxy-3-methylglutaryl CoA (HMG-CoA) and CoA. This condensation resembles the one catalyzed by citrate synthase (p. 510). The unfavorable equilibrium in the formation of acetoacetyl

CoA is compensated for by this reaction, which has a favorable equilibrium owing to the hydrolysis of a thioester linkage. 3-Hydroxy-3-methylglutaryl CoA is then cleaved to acetyl CoA and acetoacetate. The sum of these reactions is

$$2\text{ Acetyl CoA} + H_2O \longrightarrow \text{acetoacetate} + 2\text{ CoA} + H^+$$

3-Hydroxybutyrate is formed by the reduction of acetoacetate in the mitochondrial matrix. The ratio of hydroxybutyrate to acetoacetate depends on the $NADH/NAD^+$ ratio inside mitochondria. Because it is a β-keto acid, acetoacetate also undergoes a slow, spontaneous decarboxylation to acetone. The odor of acetone may be detected in the breath of a person who has a high level of acetoacetate in the blood.

ACETOACETATE IS A MAJOR FUEL IN SOME TISSUES

The major site of production of acetoacetate and 3-hydroxybutyrate is the liver. These substances diffuse from the liver mitochondria into the blood and are transported to peripheral tissues. Until a few years ago, these ketone bodies were regarded as degradation products of little physiologic value. However, the studies of George Cahill and others have revealed that these derivatives of acetyl CoA are important molecules in energy metabolism. *Acetoacetate and 3-hydroxybutyrate are normal fuels of respiration and are quantitatively important as sources of energy.* Indeed, heart muscle and the renal cortex use acetoacetate in preference to glucose. In contrast, glucose is the major fuel for the brain and red blood cells in well-nourished people on a balanced diet. However, the brain adapts to the utilization of acetoacetate during starvation and diabetes (p. 776). In prolonged starvation, 75% of the fuel needs of the brain are met by acetoacetate.

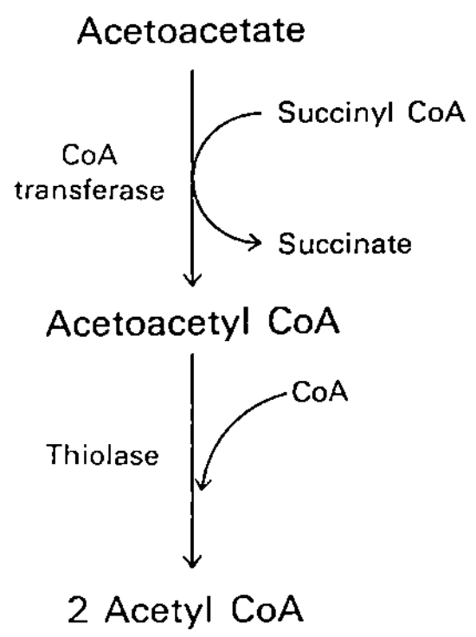

Figure 24-9
Utilization of acetoacetate as a fuel. Acetoacetate can be converted into two molecules of acetyl CoA, which then enter the citric acid cycle.

Acetoacetate can be activated by the transfer of CoA from succinyl CoA in a reaction catalyzed by a specific CoA transferase. Acetoacetyl CoA is then cleaved by thiolase to yield two molecules of acetyl CoA, which can then enter the citric acid cycle. The liver can supply acetoacetate to other organs because it lacks this particular CoA transferase.

Acetoacetate can be regarded as a water-soluble, transportable form of acetyl units. Fatty acids are released by adipose tissue and converted into acetyl units by the liver, which then exports them as acetoacetate. As might be expected, acetoacetate also has a regulatory role. *High levels of acetoacetate in the blood signify an abundance of acetyl units and lead to a decrease in the rate of lipolysis in adipose tissue.*

ANIMALS CANNOT CONVERT FATTY ACIDS INTO GLUCOSE

It is important to note that *animals are unable to convert fatty acids into glucose.* Specifically, acetyl CoA cannot be converted into pyruvate or oxaloacetate in animals. The two carbon atoms of the acetyl group of acetyl CoA enter the citric acid cycle, but two carbon atoms leave the cycle in the decarboxylations catalyzed by isocitrate dehydrogenase and α-ketoglutarate dehydrogenase. Consequently, oxaloacetate is regenerated but it is not formed de novo when the acetyl unit of acetyl CoA is oxidized by the citric acid cycle. In contrast, plants have two additional enzymes enabling them to convert the carbon atoms of acetyl CoA into oxaloacetate (p. 523).

FATTY ACIDS ARE SYNTHESIZED AND DEGRADED BY DIFFERENT PATHWAYS

Fatty acid synthesis is not simply a reversal of the degradative pathway. Rather, it consists of a new set of reactions, again exemplifying the principle that *synthetic and degradative pathways in biological systems are usually distinct.* Some important features of the pathway for the biosynthesis of fatty acids are:

1. Synthesis takes place in the *cytosol,* in contrast with degradation, which occurs in the mitochondrial matrix.

2. Intermediates in fatty acid synthesis are covalently linked to the sulf-hydryl groups of an *acyl carrier protein (ACP),* whereas intermediates in fatty acid breakdown are bonded to coenzyme A.

3. The enzymes of fatty acid synthesis in higher organisms are joined in a *single polypeptide chain* called *fatty acid synthase.* In contrast, the degradative enzymes do not seem to be associated.

4. The growing fatty acid chain is elongated by the *sequential addition of two-carbon units* derived from acetyl CoA. The activated donor of two-carbon units in the elongation step is *malonyl–ACP.* The elongation reaction is driven by the release of CO_2.

5. The reductant in fatty acid synthesis is *NADPH,* whereas the oxidants in fatty acid degradation are NAD^+ and *FAD.*

6. Elongation by the fatty acid synthase complex stops upon formation of *palmitate (C_{16}).* Further elongation and the insertion of double bonds are carried out by other enzyme systems.

THE FORMATION OF MALONYL COENZYME A IS THE COMMITTED STEP IN FATTY ACID SYNTHESIS

Salih Wakil's finding that bicarbonate is required for fatty acid biosynthesis was an important clue in the elucidation of this process. In fact, fatty acid synthesis starts with the carboxylation of acetyl CoA to *malonyl CoA.* This irreversible reaction is the committed step in fatty acid synthesis.

$$H_3C-\overset{\overset{O}{\|}}{C}-S-CoA \; + \; ATP \; + HCO_3^- \; \longrightarrow \; \overset{O}{\underset{-O}{\diagdown}}C-CH_2-\overset{\overset{O}{\|}}{C}-S-CoA \; + \; ADP \; + \; P_i \; + \; H^+$$

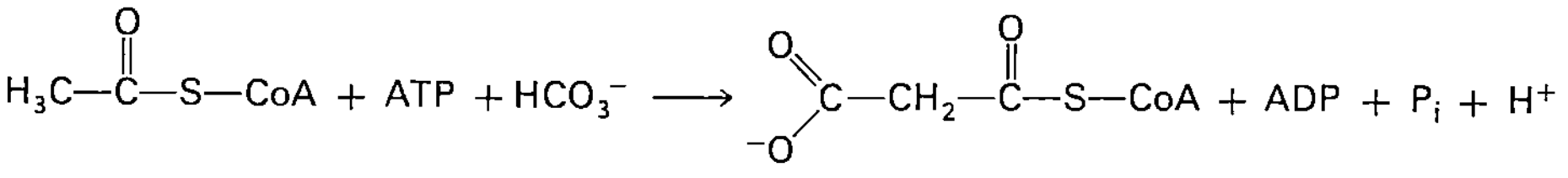

Acetyl CoA **Malonyl CoA**

The synthesis of malonyl CoA is catalyzed by *acetyl CoA carboxylase,* which contains a biotin prosthetic group. The carboxyl group of biotin is covalently attached to the *ε*-amino group of a lysine residue, as in pyruvate carboxylase (p. 572). Another similarity between acetyl CoA carboxylase and pyruvate carboxylase is that acetyl CoA is carboxylated in two stages. First, a carboxybiotin intermediate is formed at the expense of an ATP. The activated CO_2 group in this intermediate is then transferred to acetyl CoA to form malonyl CoA.

Biotin-enzyme + ATP + HCO_3^- $\rightleftharpoons$

$CO_2 \sim$ biotin-enzyme + ADP + P_i

$CO_2 \sim$ biotin-enzyme + acetyl CoA $\longrightarrow$

malonyl CoA + biotin-enzyme

Molecular model of the malonyl thio-ester unit of malonyl CoA.

Substrates are bound to this enzyme and products are released in a specific sequence (Figure 24-10). Acetyl CoA carboxylase exemplifies a *ping-pong reaction mechanism* in which one or more products are released before all the substrates are bound.

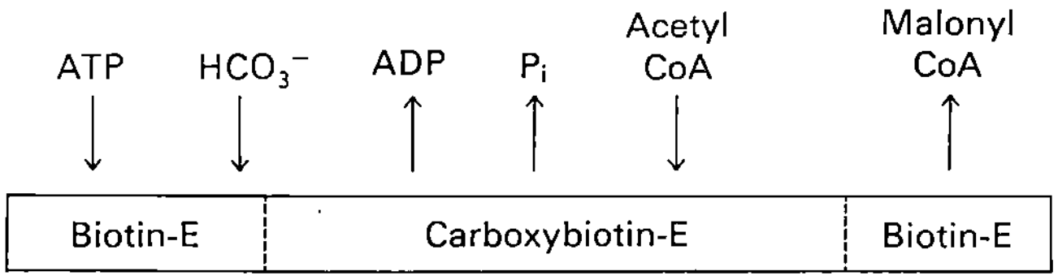

Figure 24-10
Reaction sequence of acetyl CoA carboxylase.

Acetyl CoA carboxylase from *Escherichia coli* has been separated into subunits that catalyze partial reactions. Biotin is covalently attached to a small protein (22 kd) called the *biotin carboxyl carrier protein.* The carboxylation of the biotin unit in this carrier protein is catalyzed by *biotin carboxylase,* a second subunit. The third component of the system is a *transcarboxylase,* which catalyzes the transfer of the activated CO_2 unit from carboxybiotin to acetyl CoA. The length and flexibility of the link between biotin and its carrier protein enable the activated carboxyl group to move from one active site to another in the enzyme complex (Figure 24-11), as in pyruvate carboxylase.

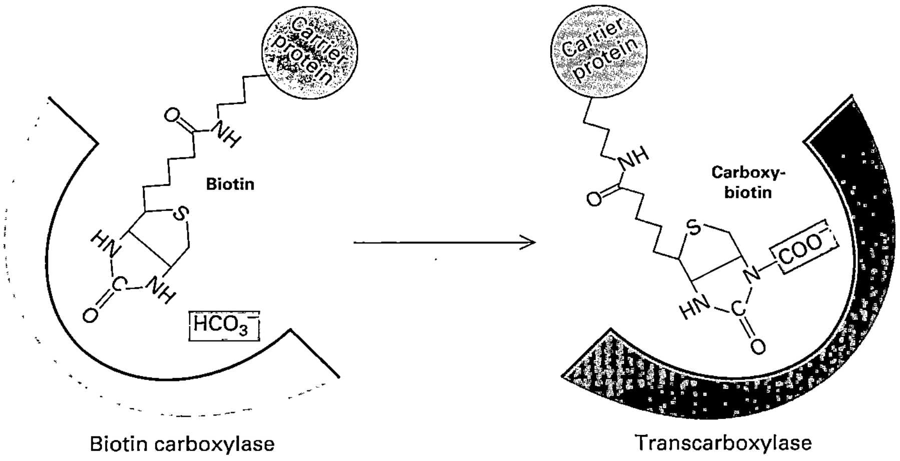

Figure 24-11
Schematic diagram showing the proposed movement of the biotin prosthetic group from the site where it acquires a carboxyl group from HCO_3^- to the site where it donates this group to acetyl CoA.

INTERMEDIATES IN FATTY ACID SYNTHESIS ARE ATTACHED TO AN ACYL CARRIER PROTEIN (ACP)

P. Roy Vagelos discovered that the intermediates in fatty acid synthesis in *E. coli* are linked to an acyl carrier protein. Specifically, they are linked to the sulfhydryl terminus of a phosphopantetheine group (Figure 24-12). In the degradation of fatty acids, this unit is part of CoA, whereas, in synthesis, it is attached to a serine residue of the ACP. This ACP, a single polypeptide chain of 77 residues, can be regarded as a giant prosthetic group, a "macro CoA."

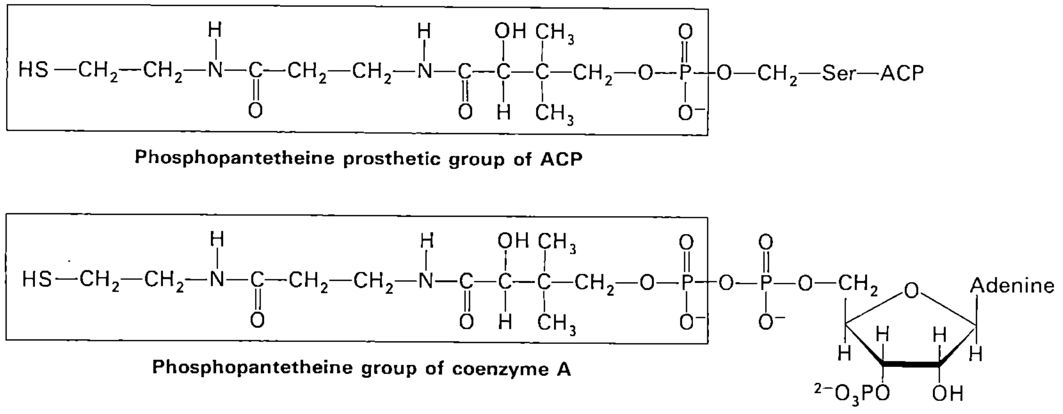

Figure 24-12
Phosphopantetheine is the reactive unit of ACP and CoA.

THE ELONGATION CYCLE IN FATTY ACID SYNTHESIS

The enzyme system that catalyzes the synthesis of saturated long-chain fatty acids from acetyl CoA, malonyl CoA, and NADPH is called the *fatty acid synthase.* The constituent enzymes of bacterial fatty acid synthases are dissociated when the cells are disrupted. The availability of these isolated enzymes has facilitated the elucidation of the steps in fatty acid synthesis (Table 24-3). In fact, the reactions leading to fatty acid synthesis in higher organisms are very much like those of bacteria.

Table 24-3
Principal reactions in fatty acid synthesis in bacteria

Step	*Reaction*	*Enzyme*
1	Acetyl CoA + HCO_3^- + ATP ⟶ malonyl CoA + ADP + P_i + H^+	Acetyl CoA carboxylase
2	Acetyl CoA + ACP ⇌ acetyl–ACP + CoA	Acetyl transacylase
3	Malonyl CoA + ACP ⇌ malonyl–ACP + CoA	Malonyl transacylase
4	Acetyl–ACP + malonyl–ACP ⟶ acetoacetyl–ACP + ACP + CO_2	Acyl-malonyl–ACP condensing enzyme
5	Acetoacetyl–ACP + NADPH + H^+ ⇌ D-3-hydroxybutyryl–ACP + $NADP^+$	β-Ketoacyl–ACP reductase
6	D-3-Hydroxybutyryl–ACP ⇌ crotonyl–ACP + H_2O	3-Hydroxyacyl–ACP dehydratase
7	Crotonyl–ACP + NADPH + H^+ ⟶ butyryl–ACP + $NADP^+$	Enoyl–ACP reductase

The elongation phase of fatty acid synthesis starts with the formation of acetyl–ACP and malonyl–ACP. *Acetyl transacylase* and *malonyl transacylase* catalyze these reactions.

$$\text{Acetyl CoA} + \text{ACP} \rightleftharpoons \text{acetyl–ACP} + \text{CoA}$$

$$\text{Malonyl CoA} + \text{ACP} \rightleftharpoons \text{malonyl–ACP} + \text{CoA}$$

Malonyl transacylase is highly specific, whereas acetyl transacylase can transfer acyl groups other than the acetyl unit, though at a much slower rate. Fatty acids with an odd number of carbon atoms are synthesized starting with propionyl–ACP, which is formed from propionyl CoA by acetyl transacylase.

Acetyl–ACP and malonyl–ACP react to form acetoacetyl–ACP. This condensation reaction is catalyzed by the *acyl-malonyl-ACP condensing enzyme.*

$$\text{Acetyl–ACP} + \text{malonyl–ACP} \longrightarrow \text{acetoacetyl–ACP} + \text{ACP} + CO_2$$

In the condensation reaction, a four-carbon unit is formed from a two-carbon unit and a three-carbon unit, and CO_2 is released. Why isn't the four-carbon unit formed from two two-carbon units? In other words, why are the reactants acetyl–ACP and malonyl–ACP rather than two molecules of acetyl–ACP? The answer is that the equilibrium for the synthesis of acetoacetyl–ACP from two molecules of acetyl–ACP is highly unfavorable. In contrast, *the equilibrium is favorable if malonyl–ACP is a reactant because its decarboxylation contributes a substantial decrease in free energy.* In effect, the condensation reaction is driven by ATP, though ATP does not directly participate in the condensation reaction. Rather, ATP is used to carboxylate acetyl CoA to malonyl CoA. The free energy thus stored in malonyl CoA is released in the decarboxylation accompanying the formation of acetoacetyl–ACP. Although HCO_3^- is required for fatty acid synthesis, its carbon atom does not appear in the product. Rather, *all the carbon atoms of fatty acids containing an even number are derived from acetyl CoA.*

The next three steps in fatty acid synthesis reduce the keto group at C-3 to a methylene group (Figure 24-13). First, acetoacetyl–ACP is reduced to D-3-hydroxybutyryl–ACP. This reaction differs from the corresponding one in fatty acid degradation in two respects: (1) The D rather than the L epimer is formed, and (2) NADPH is the reducing agent, whereas NAD^+ is the oxidizing agent in β-oxidation. This difference exemplifies the general principle that *NADPH is consumed in biosynthetic reactions, whereas NADH is generated in energy-yielding reactions.* Then D-3-hydroxybutyryl–ACP is *dehydrated* to form crotonyl–ACP, which is a *trans*-Δ^2-enoyl–ACP. The final step in the cycle *reduces* crotonyl–ACP to butyryl–ACP. NADPH is again the reductant, whereas FAD is the oxidant in the corresponding reaction in β-oxidation. These last three reactions—a reduction, a dehydration, and a second reduction—convert acetoacetyl–ACP into butyryl–ACP, which completes the first elongation cycle.

In the second round of fatty acid synthesis, butyryl–ACP condenses with malonyl–ACP to form a C_6-β-ketoacyl–ACP. This reaction is like the one in the first round, in which acetyl–ACP condenses with malonyl–ACP to form a C_4-β-ketoacyl–ACP. Reduction, dehydration, and a second reduction convert the C_6-β-ketoacyl–ACP into a C_6-acyl–ACP, which is ready for a third round of elongation. The elongation cycles continue until C_{16}-acyl–ACP is formed. This intermediate is not a substrate for the condensing enzyme. Rather, it is hydrolyzed to yield palmitate and ACP.

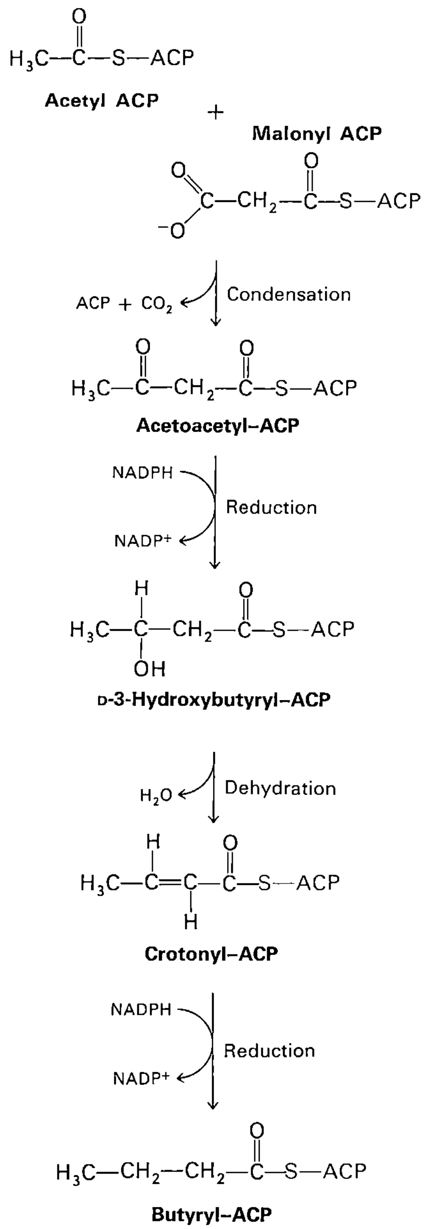

Figure 24-13
Reaction sequence in the synthesis of fatty acids in *E. coli:* condensation, reduction, dehydration, and reduction. The intermediates shown here are produced in the first round of synthesis.

STOICHIOMETRY OF FATTY ACID SYNTHESIS

The stoichiometry of the synthesis of palmitate is

$$\text{Acetyl CoA} + 7\ \text{malonyl CoA} + 14\ \text{NADPH} + 20\ H^+ \longrightarrow \text{palmitate} + 7\ CO_2 + 14\ NADP^+ + 8\ \text{CoA} + 6\ H_2O$$

The equation for the synthesis of the malonyl CoA used in the above reaction is

$$7\ \text{Acetyl CoA} + 7\ CO_2 + 7\ \text{ATP} \longrightarrow 7\ \text{malonyl CoA} + 7\ \text{ADP} + 7\ P_i + 14\ H^+$$

Hence, the overall stoichiometry for the synthesis of palmitate is

$$8 \text{ Acetyl CoA} + 7 \text{ ATP} + 14 \text{ NADPH} + 6 \text{ H}^+ \longrightarrow \text{palmitate} + 14 \text{ NADP}^+ + 8 \text{ CoA} + 6 \text{ H}_2\text{O} + 7 \text{ ADP} + 7 \text{ P}_i$$

FATTY ACIDS ARE SYNTHESIZED IN EUKARYOTES BY A MULTIFUNCTIONAL ENZYME COMPLEX

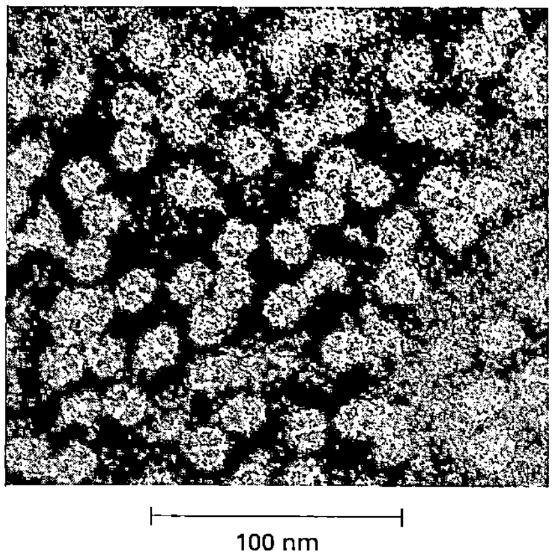

Figure 24-14
Electron micrograph of the fatty acid synthase complex from yeast. [Courtesy of Dr. Felix Wieland and Dr. Elmar A. Siess.]

The fatty acid synthases of eukaryotes, in contrast with those of *E. coli*, are well-defined *multienzyme complexes*. The one from yeast has a mass of 2400 kd and appears in electron micrographs as an ellipsoid with a length of 25 nm and a cross-sectional diameter of 21 nm (Figure 24-14). It consists of just two kinds of polypeptide chains and has the subunit composition $\alpha_6\beta_6$. The α chain contains the acyl carrier protein, the condensing enzyme, and the β-ketoacyl reductase, whereas the β chain contains acetyl transacylase, malonyl transacylase, β-hydroxyacyl dehydratase, and enoyl reductase.

Mammalian fatty acid synthase is a dimer of identical 260-kd subunits. Each chain is folded into three domains joined by flexible regions (Figure 24-15). *Domain 1, the substrate entry and condensation unit,* contains acetyl transferase, malonyl transferase, and β-ketoacyl synthase (condensing enzyme). *Domain 2, the reduction unit,* contains the acyl carrier protein, β-ketoacyl reductase, dehydratase, and enoyl reductase. *Domain 3, the palmitate release unit,* contains the thioesterase. Thus, *seven different catalytic sites are present on a single polypeptide chain.* It is noteworthy that many eukaryotic multienzyme complexes are multifunctional proteins in which different enzymes are linked covalently. An advantage of this arrangement is that the synthesis of different enzymes is coordinated. Also, a multienzyme complex consisting of covalently joined enzymes is more stable than one formed by noncovalent attractions. Furthermore, intermediates can be efficiently handed from one active site to another without leaving the assembly. The time for diffusion is markedly reduced and side reactions are minimized. It seems likely that multifunctional enzymes such as fatty acid synthase arose in eukaryotic evolution by exon shuffling (p. 114).

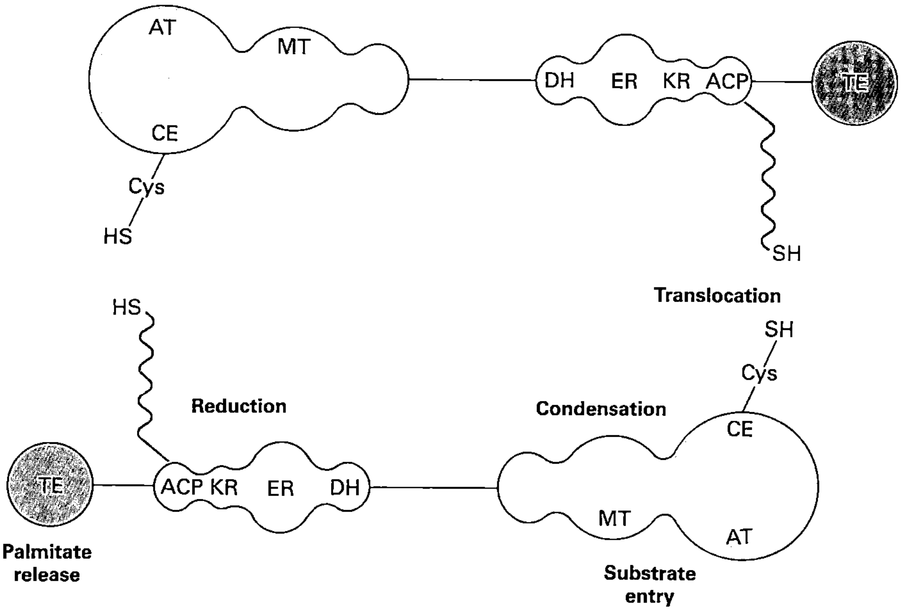

Figure 24-15
Schematic diagram of animal fatty acid synthase. Each of the identical chains in the dimer contains three domains. Domain 1 (blue) contains acetyl transferase (AT), malonyl transferase (MT), and condensing enzyme (CE). Domain 2 (yellow) contains acyl carrier protein (ACP), β-ketoacyl reductase (KR), dehydratase (DH), and enoyl reductase (ER). Domain 3 (red) contains thioesterase (TE). The flexible phosphopantetheinyl group (green) carries the fatty acyl chain from one catalytic site on a chain to another, and also between chains in the dimer. [After Y. Tsukamoto, H. Wong, J.S. Mattick, and S.J. Wakil. *J. Biol. Chem.* 258(1983):15312.]

THE FLEXIBLE PHOSPHOPANTETHEINYL UNIT OF ACP CARRIES SUBSTRATE FROM ONE ACTIVE SITE TO ANOTHER

Fatty acid synthesis in animals begins with the attachment of the acetyl group of acetyl CoA to the oxygen atom in the side chain of serine in the active site of acetyl transferase. The malonyl group of malonyl CoA likewise becomes *O*-linked to the active site of malonyl transferase. These entry reactions take place in domain 1 of the synthase (see Figure 24-15). The acetyl unit is then transferred to the cysteine sulfur in the active site of the condensing enzyme, and the malonyl unit is transferred to the sulfur atom of the phosphopantetheinyl group of the acyl carrier protein (ACP) of the *other* chain in the dimer. Domain 1 of each chain of this dimer interacts with domains 2 and 3 of the other chain. Thus, each of the two functional units of the synthase consists of domains formed by different chains. Indeed, the arenas of catalytic action are the interfaces between domains on opposite chains.

Elongation begins with the joining of the acetyl unit on the condensing enzyme to a two-carbon portion of the malonyl unit on ACP (Figure 24-16). CO_2 is released and an acetoacetyl-*S*-phosphopantetheinyl unit is formed on ACP. The active-site sulfhydryl on the condensing enzyme is restored. The acetoacetyl group is then delivered to three active sites in domain 2 of the opposite chain to reduce it to a butyryl unit. This saturated C_4 unit then migrates from the phosphopantetheinyl sulfur on ACP to the cysteine sulfur atom on the condensing enzyme. The synthase is now ready for another round of elongation. The butyryl unit on the condensing enzyme next becomes linked to a two-carbon part of the malonyl unit on ACP to form a C_6 unit on ACP, which undergoes reduction. Five more rounds of condensation and reduction produce a palmitoyl (C_{16}) chain on CE, which is hydrolyzed to palmitate by the thioesterase on domain 3 of the opposite chain. The migration of the growing fatty acyl chain back and forth between ACP and the condensing enzyme in each round of elongation is noteworthy. Analogous translocations of growing peptide chains take place in protein synthesis (p. 900).

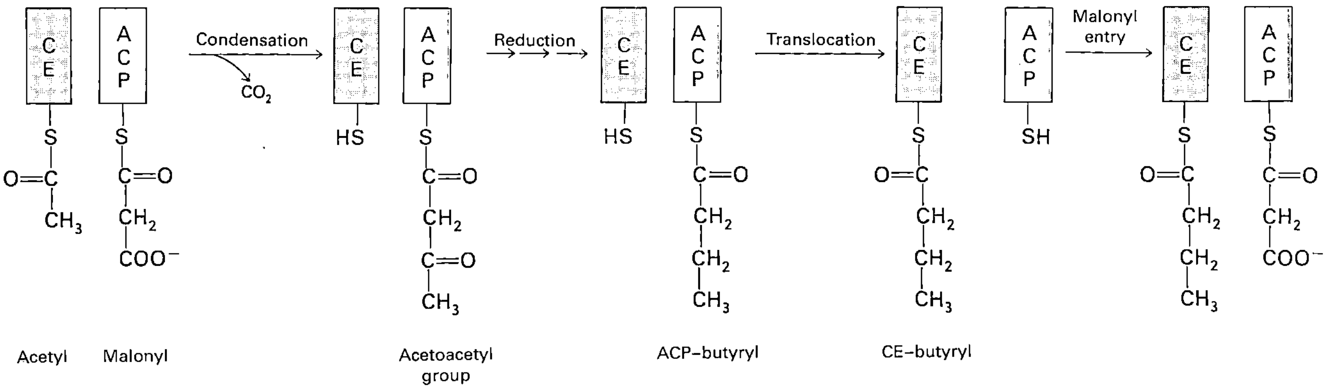

Figure 24-16
Translocations of the elongating fatty acyl chain between the cysteine sulfhydryl group of condensing enzyme (CE, blue) and the phosphopantetheine sulfhydryl group of acyl carrier protein (ACP, yellow).

The flexibility and 20-Å maximal length of the phosphopantetheinyl moiety are critical for the function of this multienzyme complex. The enzyme subunits need not undergo large structural rearrangements to interact with the substrate. Instead, the substrate is on a long, flexible arm that can reach each of the numerous active sites. Recall that biotin and lipoamide are also on long, flexible arms in their multienzyme complexes. The organization of the fatty acid synthases of yeast and higher organisms enhances the efficiency of the overall process because intermediates are directly transferred from one active site to the next. The reactants are not diluted in the cytosol. Moreover, they do not have to find each other by random diffusion. Another advantage of such a multienzyme complex is that covalently bound intermediates are sequestered and protected from competing reactions.

CITRATE CARRIES ACETYL GROUPS FROM MITOCHONDRIA TO THE CYTOSOL FOR FATTY ACID SYNTHESIS

The synthesis of palmitate requires the input of 8 molecules of acetyl CoA, 14 NADPH, and 7 ATP. Fatty acids are synthesized in the cytosol, whereas acetyl CoA is formed from pyruvate in mitochondria. Hence, acetyl CoA must be transferred from mitochondria to the cytosol. Mitochondria, however, are not readily permeable to acetyl CoA. Recall that carnitine carries only long-chain fatty acids. *The barrier to acetyl CoA is bypassed by citrate, which carries acetyl groups across the inner mitochondrial membrane.* Citrate is formed in the mitochondrial matrix by the condensation of acetyl CoA with oxaloacetate (Figure 24-17). When present at high levels, citrate is transported to the cytosol, where it is cleaved by *ATP-citrate lyase.*

> *Lyases*—
> Enzymes catalyzing the cleavage of C–C, C–O, or C–N bonds by elimination. A double bond is formed in these reactions.

$$\text{Citrate} + \text{ATP} + \text{CoA} + H_2O \longrightarrow \text{acetyl CoA} + \text{ADP} + P_i + \text{oxaloacetate}$$

Thus, acetyl CoA and oxaloacetate are transferred from mitochondria to the cytosol at the expense of an ATP.

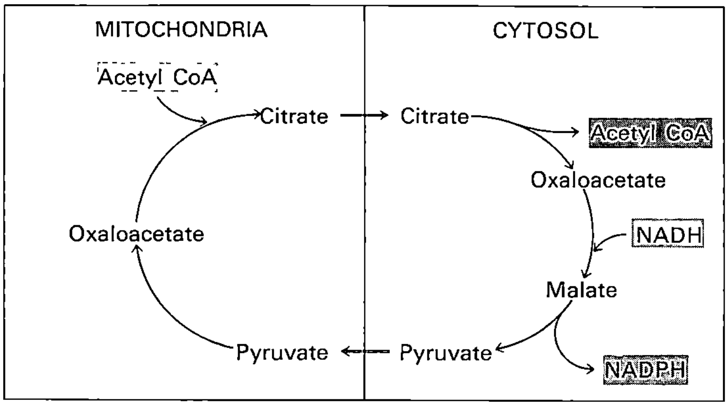

Figure 24-17
Acetyl CoA is transferred from mitochondria to the cytosol, and the reducing potential of NADH is concomitantly converted into that of NADPH by this series of reactions.

SOURCES OF NADPH FOR FATTY ACID SYNTHESIS

Oxaloacetate formed in the transfer of acetyl groups to the cytosol must now be returned to the mitochondria. The inner mitochondrial membrane is impermeable to oxaloacetate. Hence, a series of bypass reactions are needed. Most important, these reactions generate much of the NADPH needed for fatty acid synthesis. First, oxaloacetate is reduced to malate by NADH. This reaction is catalyzed by a *malate dehydrogenase* in the cytosol.

$$\text{Oxaloacetate} + \text{NADH} + H^+ \rightleftharpoons \text{malate} + \text{NAD}^+$$

Second, malate is oxidatively decarboxylated by an *$NADP^+$-linked malate enzyme* (also called *malic enzyme*).

$$\text{Malate} + NADP^+ \longrightarrow \text{pyruvate} + CO_2 + NADPH$$

The pyruvate formed in this reaction readily enters mitochondria, where it is carboxylated to oxaloacetate by pyruvate carboxylase.

$$\text{Pyruvate} + CO_2 + ATP + H_2O \longrightarrow \text{oxaloacetate} + ADP + P_i + 2\ H^+$$

The sum of these three reactions is

$$NADP^+ + NADH + ATP + H_2O \longrightarrow NADPH + NAD^+ + ADP + P_i + H^+$$

Thus, *one NADPH is generated for each acetyl CoA that is transferred from the mitochondria to the cytosol.* Hence, eight NADPH are formed when eight molecules of acetyl CoA are transferred to the cytosol for the synthesis of palmitate. *The additional six NADPH required for this process come from the pentose phosphate pathway* (p. 565).

ACETYL CoA CARBOXYLASE PLAYS A KEY ROLE IN CONTROLLING FATTY ACID METABOLISM

Fatty acid metabolism is stringently controlled so that synthesis and degradation are highly responsive to physiologic needs. Fatty acid synthesis is maximal when carbohydrate and energy are plentiful and when fatty acids are scarce. *Acetyl CoA carboxylase plays a key role in regulating fatty acid metabolism* (Figure 24-18). Recall that this enzyme catalyzes the committed step in fatty acid synthesis, the production of malonyl CoA (the activated two-carbon donor). The carboxylase is controlled by three global signals—glucagon, epinephrine, and insulin—that reflect the overall needs of the organism. *Insulin stimulates fatty acid synthesis by activating the carboxylase, whereas glucagon and epinephrine have the reverse effect.* Control is also exerted by the levels of citrate, palmitoyl CoA, and AMP within a cell. *Citrate,* a signal that building blocks and energy are abundant, activates the carboxylase. Palmitoyl CoA and AMP, in contrast, lead to the inhibition of the carboxylase. Thus, this key enzyme is subject to both global and local regulation.

Acetyl CoA carboxylase is switched off by phosphorylation (Figure 24-19). Modification of a single serine residue by an *AMP-activated protein kinase* converts the carboxylase into an inactive form. Cyclic AMP has no effect on this kinase, which is different from protein kinase A. Rather, it is stimulated by AMP and inhibited by ATP, which enables it to communicate the energy charge of the cell. The phosphoryl group on the inhibited carboxylase is removed by *protein phosphatase 2A,* one of the four major cellular phosphatases that act on proteins (p. 596).

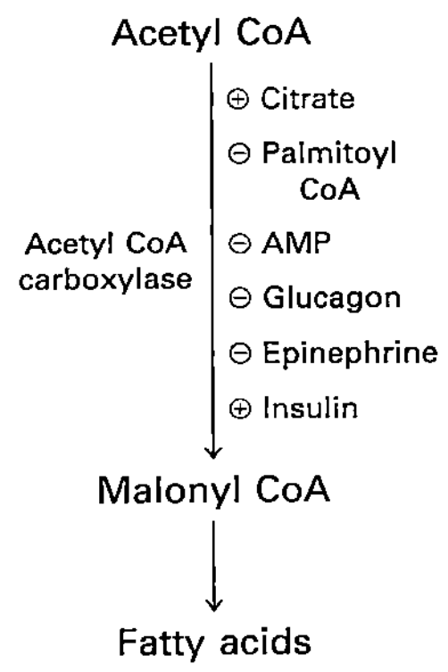

Figure 24-18
Acetyl CoA carboxylase, which catalyzes the committed step in fatty acid synthesis, is a key control site.

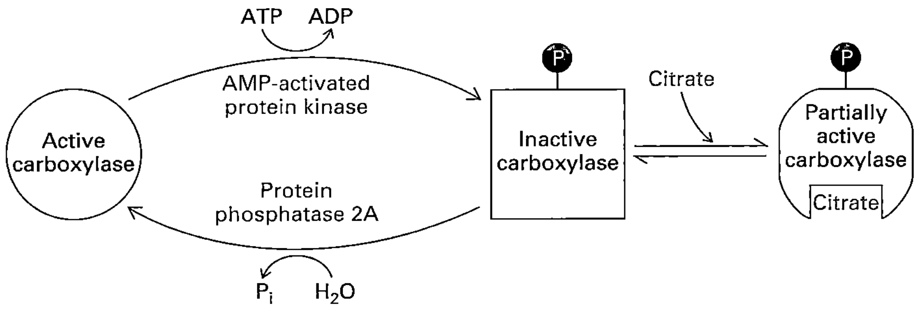

Figure 24-19
Acetyl CoA carboxylase is inhibited by phosphorylation and activated by the binding of citrate.

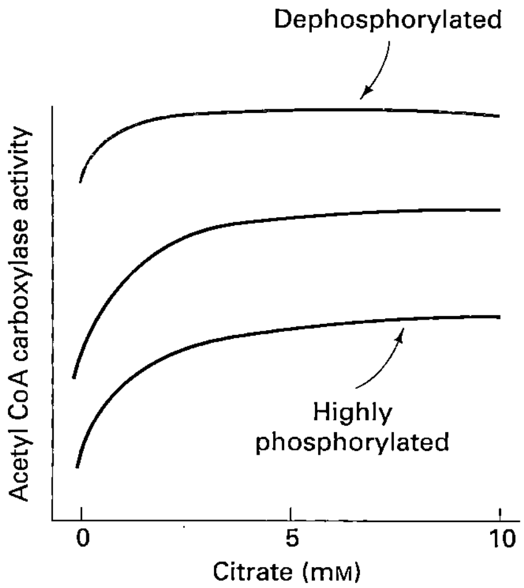

Figure 24-20
Dependence of the catalytic activity of acetyl CoA carboxylase on the concentration of citrate. The dephosphorylated form of the carboxylase is highly active even when citrate is absent. Citrate partially overcomes the inhibition produced by phosphorylation. [After G.M. Mabrouk, I.M. Helmy, K.G. Thampy, and S.J. Wakil. *J. Biol. Chem.* 265(1990):6330.]

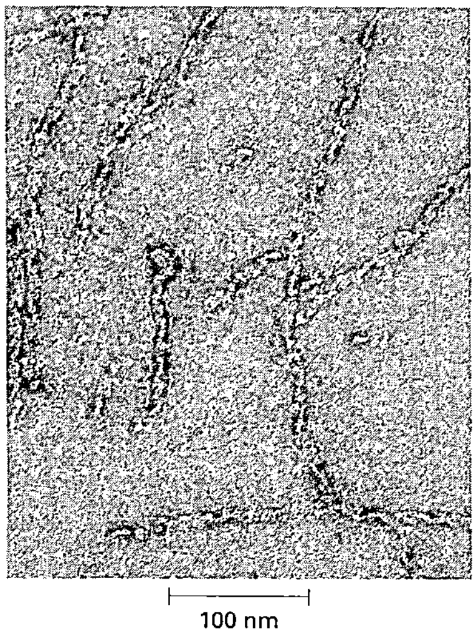

Figure 24-21
Electron micrograph of the enzymatically active filamentous form of acetyl CoA carboxylase from chicken liver. The inactive form is an octamer of 265-kd subunits. [Courtesy of Dr. M. Daniel Lane.]

The proportion of carboxylase in the active dephosphorylated form depends on the catalytic rates of these opposing enzymes. Epinephrine and glucagon activate protein kinase A, which in turn inhibits the phosphatase by phosphorylating it. Hence, *these catabolic hormones switch off fatty acid synthesis by keeping the carboxylase in the inactive phosphorylated state.* The inactive form also predominates when the energy charge in a cell is low because phosphorylation is stimulated by the high AMP level. *Insulin, by contrast, stimulates the carboxylase.* One possibility is that insulin activates protein phosphatase 2A to convert the carboxylase into the active dephosphorylated state. The hormonal control of acetyl CoA carboxylase is reminiscent of that of glycogen synthase (p. 593).

Acetyl CoA carboxylase is allosterically stimulated by citrate. Specifically, citrate partially reverses the inhibition produced by phosphorylation (Figure 24-20). The level of citrate is high when both acetyl CoA and ATP are abundant. Recall that mammalian isocitrate dehydrogenase is inhibited by a high energy charge (p. 525). Hence, a high level of citrate signifies that two-carbon units and ATP are available for fatty acid synthesis. The stimulatory effect of citrate on the carboxylase is antagonized by *palmitoyl CoA,* which is abundant when there is an excess of fatty acids. Palmitoyl CoA also inhibits the translocase that transports citrate from mitochondria to the cytosol, as well as glucose 6-phosphate dehydrogenase, which generates NADPH.

Fatty acid synthesis and degradation are reciprocally regulated so that both are not simultaneously active. *In starvation, the level of free fatty acids rises because adipose-cell lipase is stimulated by hormones such as epinephrine and glucagon* (p. 606). *Insulin, in contrast, inhibits lipolysis.* The entry of fatty acyl CoAs into the mitochondrial matrix is also regulated. Malonyl CoA, which is present at a high level when fuel molecules are abundant, inhibits carnitine acyltransferase I. Hence, fatty acyl CoAs do not have ready access to the mitochondrial matrix in times of plenty. Moreover, two enzymes in the β-oxidation pathway are markedly inhibited when the energy charge is high. NADH inhibits 3-hydroxyacyl CoA dehydrogenase, and acetyl CoA inhibits thiolase.

Long-term control is mediated by changes in the rates of synthesis and degradation of the enzymes participating in fatty acid synthesis. Animals that have fasted and then fed high-carbohydrate, low-fat diets show marked increases in their amounts of acetyl CoA carboxylase and fatty acid synthase within a few days. This type of regulation is known as *adaptive control.*

ELONGATION AND UNSATURATION OF FATTY ACIDS ARE CARRIED OUT BY ACCESSORY ENZYME SYSTEMS

The major product of the fatty acid synthase is palmitate. In eukaryotes, longer fatty acids are formed by elongation reactions catalyzed by enzymes on the cytosolic face of the *endoplasmic reticulum membrane.* For study of these elongation reactions, the membrane is fragmented into closed vesicles called *microsomes.* As in the in vivo reactions leading to the synthesis of palmitate, microsomes add two-carbon units sequentially to the carboxyl end of both saturated and unsaturated fatty acids. Malonyl CoA is the two-carbon donor in the elongation of fatty acyl CoAs. Again, condensation is driven by the decarboxylation of malonyl CoA.

Microsomal systems also introduce double bonds into long-chain acyl CoAs. For example, in the conversion of stearoyl CoA into oleoyl CoA, a cis-Δ^9 double bond is inserted by an oxidase that employs *molecular oxygen* and *NADH* (or *NADPH*).

$$\text{Stearoyl CoA} + \text{NADH} + H^+ + O_2 \longrightarrow \text{oleoyl CoA} + \text{NAD}^+ + 2\ H_2O$$

This reaction is catalyzed by a complex of three membrane-bound enzymes: *NADH-cytochrome* b5 *reductase, cytochrome* b5, and a *desaturase* (Figure 24-22). First, electrons are transferred from NADH to the FAD moiety of NADH-cytochrome *b*5 reductase.

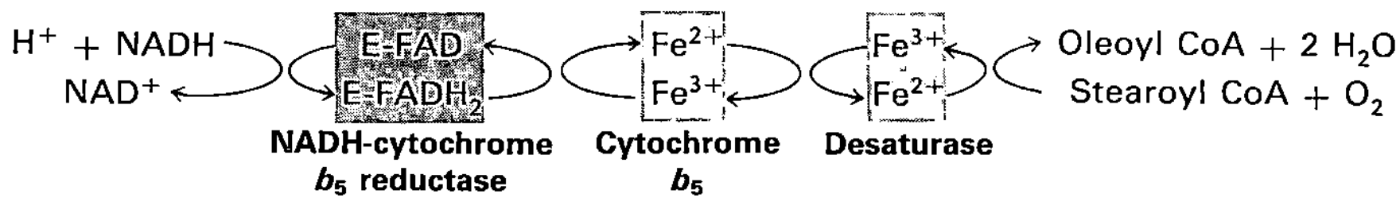

Figure 24-22
Electron-transport chain in the desaturation of fatty acids.

The heme iron atom of cytochrome b_5 is then reduced to the ferrous form. The nonheme iron atom of the desaturase is subsequently converted to the Fe^{2+} state, which enables it to interact with O_2 and the saturated fatty acyl CoA substrate. A double bond is formed and two molecules of H_2O are released. Two electrons come from NADH and two from the single bond of the fatty acyl substrate.

A variety of unsaturated fatty acids can be formed from oleate by a combination of elongation and desaturation reactions. For example, oleate can be elongated to a 20:1 cis-Δ^{11} fatty acid. Alternatively, a second double bond can be inserted to yield an 18:2 cis-Δ^6,Δ^9 fatty acid. Similarly, palmitate (16:0) can be oxidized to palmitoleate (16:1 cis-Δ^9), which can then be elongated to *cis*-vaccenate (18:1 cis-Δ^{11}).

Unsaturated fatty acids in mammals are derived from either palmitoleate (16:1), oleate (18:1), linoleate (18:2), or linolenate (18:3). The number of carbons from the ω end of a derived unsaturated fatty acid to the nearest double bond identifies its precursor.

Precursor	*Formula*
Linolenate (ω-3)	$CH_3—(CH_2)_1—CH{=}CH—R$
Linoleate (ω-6)	$CH_3—(CH_2)_4—CH{=}CH—R$
Palmitoleate (ω-7)	$CH_3—(CH_2)_5—CH{=}CH—R$
Oleate (ω-9)	$CH_3—(CH_2)_7—CH{=}CH—R$

Mammals lack the enzymes to introduce double bonds at carbon atoms beyond C-9 in the fatty acid chain. Hence, mammals cannot synthesize linoleate (18:2 cis-Δ^9, Δ^{12}) and linolenate (18:3 cis-Δ^9, Δ^{12}, Δ^{15}). *Linoleate and linolenate are the two essential fatty acids.* The term *essential* means that they must be supplied in the diet because they are required by the organism and cannot be endogenously synthesized. Linoleate and linolenate furnished by the diet are the starting points for the synthesis of a variety of other unsaturated fatty acids.

A

B

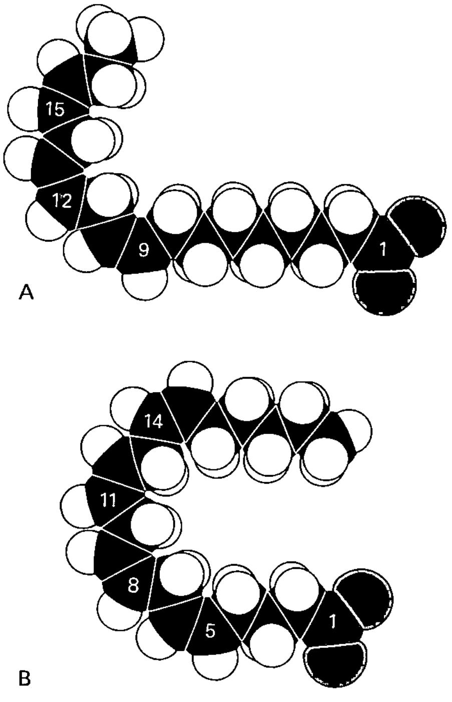

Figure 24-23
Models of polyunsaturated fatty acids. (A) Linoleate, a C_{18} fatty acid with three cis double bonds. (B) Arachidonate, a C_{20} fatty acid with four cis double bonds.

EICOSANOID HORMONES ARE DERIVED FROM POLYUNSATURATED FATTY ACIDS

Arachidonate, a 20:4 fatty acid derived from linoleate, is the major precursor of several classes of signal molecules—prostaglandins, prostacyclins, thromboxanes, and leukotrienes (Figure 24-24).

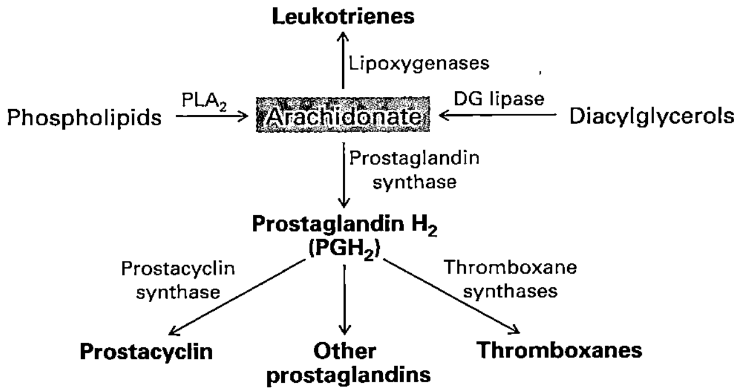

Figure 24-24
Arachidonate is the major precursor of eicosanoid hormones. Prostaglandin synthase catalyzes the first step in a pathway leading to prostaglandins, prostacyclins, and thromboxanes. Lipoxygenase catalyzes the initial step in a pathway leading to leukotrienes.

Prostaglandins are 20-carbon fatty acids containing a 5-carbon ring (Figure 24-25). They were discovered in the 1930s in secretions of the prostate gland but became prominent only in the 1960s, largely because of the pioneering work of Sune Bergstrom. A series of prostaglandins are fashioned by reductases and isomerases. The major classes are designated PGA through PGI; a subscript denotes the number of carbon-carbon double bonds outside the ring. Prostaglandins with two double bonds, such as PGE_2, are derived from arachidonate; the other two double bonds of this precursor are lost in forming a five-membered ring. *Prostacyclin and thromboxanes* are related compounds that arise from a nascent prostaglandin. They are generated by *prostacylin synthase* and *thromboxane synthase.* Alternatively, arachidonate can be converted into *leukotrienes* by the action of *lipoxygenase.* These compounds, first found in leukocytes, contain three conjugated double bonds—hence, the name. Prostaglandins, prostacyclin, thromboxanes, and leukotrienes are called *eicosanoids* because they contain 20 carbon atoms (*eikosi* is the Greek word for "twenty").

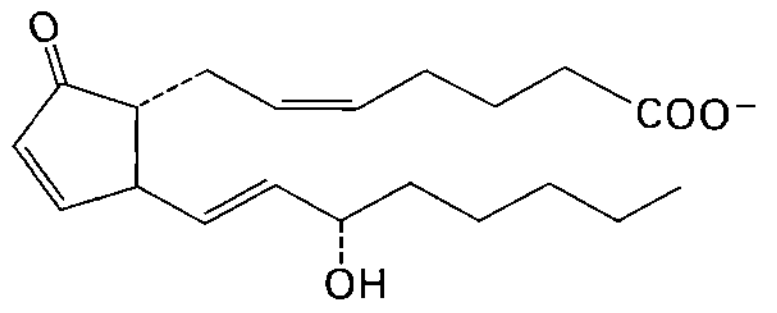

Prostaglandin A_2

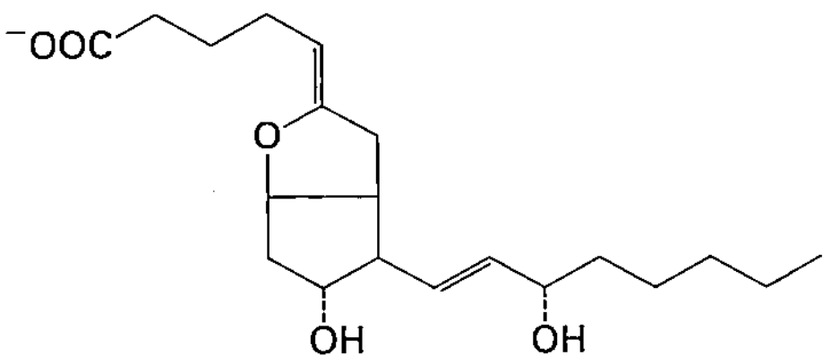

Prostacyclin (PGI_2)

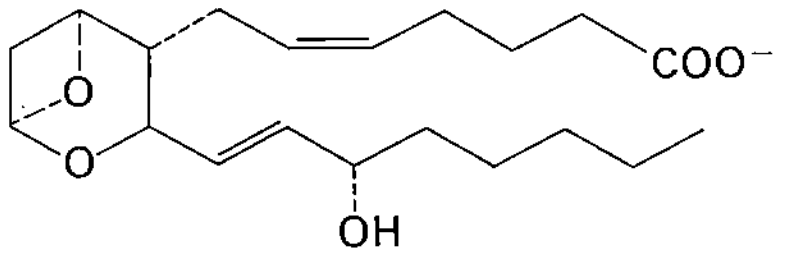

Thromboxane A_2 (TXA_2)

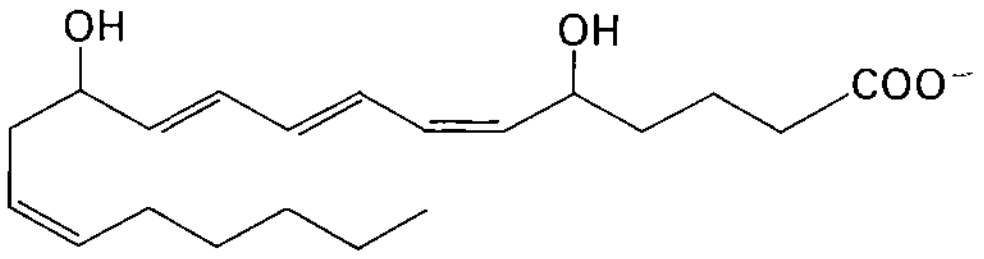

Leukotriene B_4

Figure 24-25
Structures of several eicosanoids.

Prostaglandins and other eicosanoids are *local hormones* because they are short-lived. They alter the activities of the cells in which they are synthesized and of adjoining cells. The nature of these effects may vary from one type of cell to another, in contrast with the more uniform actions of global hormones such as insulin and glucagon. Prostaglandins stimulate inflammation, regulate blood flow to particular organs, control ion transport across membranes, modulate synaptic transmission, and induce sleep.

ASPIRIN INHIBITS THE SYNTHESIS OF PROSTAGLANDINS BY ACETYLATING THE CYCLOOXYGENASE

Prostaglandins are synthesized in membranes from C_{20} polyunsaturated fatty acids containing at least three double bonds. These fatty acids are generated from phospholipids by action of phospholipase A_2, or from diacylglycerol by the action of a lipase. Recall that diacylglycerol is formed following activation of the phosphoinositide cascade (p. 344). The biosynthesis of most eicosanoids starts at arachidonate, which has four double bonds (Figure 24-26). A cyclopentane ring is formed and four oxygen atoms are introduced by the *cyclooxygenase* component of *prostaglandin synthase* (Figure 24-27). The four oxygen atoms introduced into PGG_2, the first intermediate, come from two molecules of O_2. The *hydroperoxidase* component of the synthase then catalyzes a two-electron reduction of the 15-hydroperoxy group to a 15-hydroxyl group. The heme-containing dioxygenase catalyzing these reactions is bound to the smooth endoplasmic reticulum. The highly unstable PGH_2 product is rapidly transformed into other prostaglandins, prostacylin, and thromboxanes.

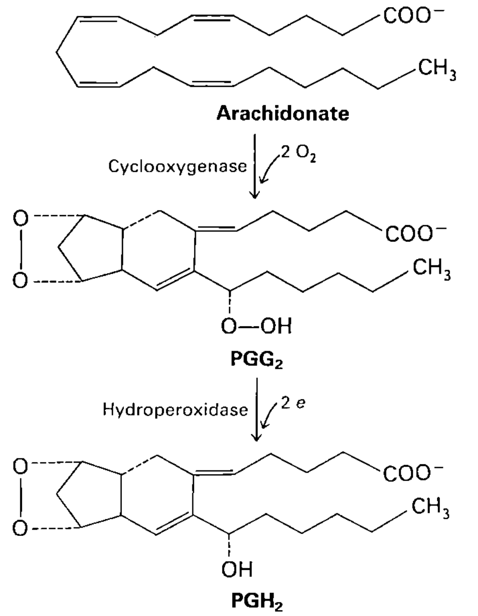

Figure 24-26
Synthesis of PGH_2, a key prostaglandin, from arachidonate. PGH_2 is the precursor of other prostaglandins, prostacylin, and thromboxanes.

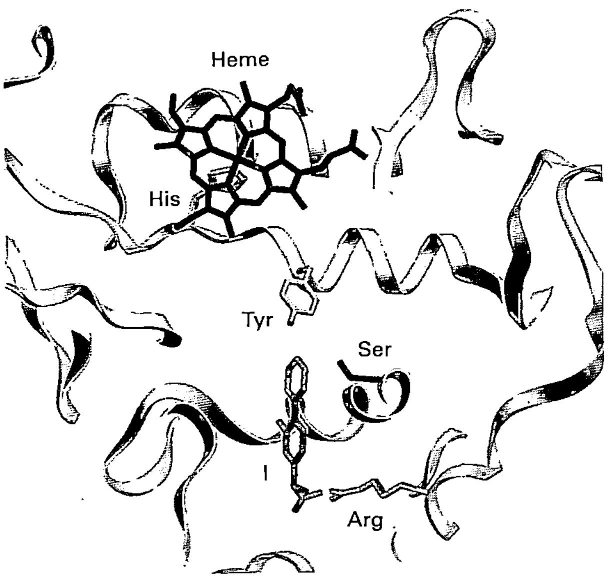

Figure 24-27
Structure of the active site of prostaglandin H_2 synthase. A competitive inhibitor (I) occupying the substrate site forms a salt bridge with an arginine side chain. Aspirin irreversibly inhibits the enzyme by acetylating the adjacent serine residue. [Courtesy of Dr. Michael Garavito. D. Picot, P.J. Loll, and R.M. Garavito. *Nature* 367(1994):243.]

Aspirin has been used for centuries to decrease inflammation, pain, and fever. Its mode of action was an enigma until 1974. John Vane then discovered that *aspirin inhibits the synthesis of prostaglandins by inactivating prostaglandin synthase.* Specifically, aspirin (acetylsalicylate) irreversibly inhibits the cyclooxygenase activity of this enzyme by acetylating a specific serine hydroxyl group. *Aspirin is a potent anti-inflammatory agent because it blocks the first step in the synthesis of prostaglandins.* This remarkably simple and effective drug is also widely used to prevent excessive blood clotting, which can lead to heart attacks and strokes. *Aspirin is antithrombotic because it blocks the formation of thromboxane A_2 (TXA_2), a potent aggregator of blood platelets.* Inhibition of the cyclooxygenase blocks the formation of PGH_2, the precursor of TXA_2 (see Figure 24-24). A small dose of aspirin taken once a day is effective in minimizing platelet aggregation because these enucleated cells are unable to synthesize new molecules of prostaglandin synthase.

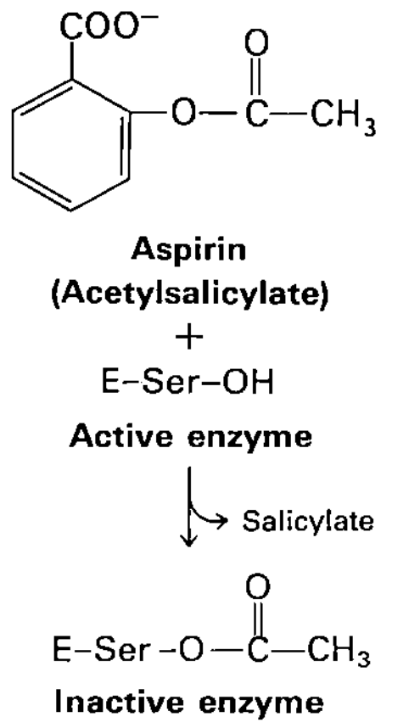

SUMMARY

Fatty acids are physiologically important as (1) components of phospholipids and glycolipids, (2) lipophilic modifiers of proteins, (3) fuel molecules, and (4) hormones and intracellular messengers. They are stored in adipose tissue as triacylglycerols (neutral fat), which can be mobilized by the hydrolytic action of lipases that are under hormonal control. Fatty acids are activated to acyl CoAs, transported across the inner mitochondrial membrane by carnitine, and degraded in the mitochondrial matrix by a recurring sequence of four reactions: oxidation linked to FAD, hydration, oxidation linked to NAD^+, and thiolysis by CoA. The $FADH_2$ and NADH formed in the oxidation steps transfer their electrons to O_2 by means of the respiratory chain, whereas the acetyl CoA formed in the thiolysis step normally enters the citric acid cycle by condensing with oxaloacetate. When the concentration of oxaloacetate is insufficient, acetyl CoA gives rise to acetoacetate and 3-hydroxybutyrate, which are normal fuel molecules. In starvation and in diabetes, large amounts of acetoacetate, 3-hydroxybutyrate, and acetone (collectively known as ketone bodies) are formed in the liver and accumulate in the blood. Mammals are unable to convert fatty acids into glucose because they lack a pathway for the net production of oxaloacetate, pyruvate, or other gluconeogenic intermediates from acetyl CoA.

Fatty acids are synthesized in the cytosol by a different pathway from that of β-oxidation. A reaction cycle based on the formation and cleavage of citrate carries acetyl groups from mitochondria to the cytosol. NADPH needed for synthesis is generated by the pentose phosphate pathway and in the transfer of reducing equivalents from mitochondria by the malate-pyruvate shuttle. Synthesis starts with the carboxylation of acetyl CoA to malonyl CoA, the committed step. This ATP-driven reaction is catalyzed by acetyl CoA carboxylase, a biotin enzyme. The intermediates in fatty acid synthesis are linked to an acyl carrier protein (ACP), specifically to the sulfhydryl terminus of its phosphopantetheine prosthetic group. Acetyl–ACP is formed from acetyl CoA, and malonyl–ACP is formed from malonyl CoA. Acetyl–ACP and malonyl–ACP condense to form acetoacetyl–ACP, a reaction driven by the release of CO_2 from the activated malonyl unit. This is followed by a reduction, a dehydration, and a second reduction. NADPH is the reductant in these steps. The butyryl–ACP formed in this way is ready for a second round of elongation, starting with the addition of a two-carbon unit from malonyl–ACP. Seven rounds of elongation yield palmitoyl–ACP, which is hydrolyzed to palmitate. The synthesis of palmitate requires 8 molecules of acetyl CoA, 14 NADPH, and 7 ATP; 7 HCO_3^- play a catalytic role. In higher organisms, the enzymes carrying out fatty acid synthesis are covalently linked in a multifunctional enzyme complex. The flexible phosphopantetheinyl unit of ACP carries the substrate from one active site to another in this complex.

Fatty acid synthesis and degradation are reciprocally regulated so that both are not simultaneously active. Acetyl CoA carboxylase, the key control site, is stimulated by insulin and inhibited by glucagon and epinephrine. These hormonal effects are mediated by changes in the amounts of the active dephosphorylated and inactive phosphorylated forms of the carboxylase. Citrate, which signals an abundance of building blocks and energy, allosterically stimulates the carboxylase. Glucagon and epinephrine stimulate triacylglycerol breakdown by activating the lipase. Insulin, by contrast, inhibits lipolysis. In times of plenty, fatty acyl CoAs do not enter the mitochondrial matrix because malonyl CoA inhibits carnitine acyltransferase I.

Fatty acids are elongated and desaturated by enzyme systems in the endoplasmic reticulum membrane. Desaturation requires NADH and O_2 and is carried out by a complex consisting of a flavoprotein, a cytochrome, and a nonheme iron protein. Mammals lack the enzymes to introduce double bonds distal to C-9, and so they require linoleate and linolenate in their diets.

Arachidonate, a key precursor of prostaglandins and other signal molecules, is derived from linoleate. This 20:4 polyunsaturated fatty acid is the precursor of several classes of signal molecules—prostaglandins, prostacyclins, thromboxanes, and leukotrienes—that act as messengers and local hormones because of their transience. They are called eicosanoids because they contain 20 carbon atoms. Prostaglandin synthase, the enzyme catalyzing the first step in the synthesis of all eicosanoids except for the leukotrienes, consists of a cyclooxygenase and a hydroperoxidase. Aspirin (acetylsalicylate), an anti-inflammatory and antithrombotic drug, irreversibly blocks the synthesis of these eicosanoids by acetylating a serine residue in the cyclooxygenase moiety of the synthase.

SELECTED READINGS

Where to start

Wakil, S.J., 1989. Fatty acid synthase, a proficient multifunctional enzyme. *Biochemistry* 28:4523–4530.

Lynen, F., 1972. The pathway from "activated acetic acid" to the terpenes and fatty acids. In *Nobel Lectures: Physiology or Medicine (1963–1970)*, pp. 103–138. American Elsevier (1973). [An account of the author's pioneering work on fatty acid degradation and synthesis and of the roles of acetyl CoA and biotin.]

Weissmann, G., 1991. Aspirin. *Sci. Amer.* 264(1):84–90.

Picot, D., Loll, P.J., and Garavito, R.M., 1994. The x-ray crystal structure of the membrane protein prostaglandin H_2 synthase-1. *Nature* 367:243–249. [Elucidation of the structure of aspirin's target.]

Books

Vance, D.E., and Vance, J.E. (eds.), 1991. *Biochemistry of Lipids, Lipoproteins, and Membranes.* Elsevier. [Contains many excellent articles on lipid structure, dynamics, metabolism, and assembly into membranes.]

Boyer, P.D. (ed.), 1983. *The Enzymes* (3rd ed.), vol. 16: *Lipid Enzymology.* Academic Press. [This volume contains a wealth of information about a broad range of topics in lipid metabolism.]

Numa, S. (ed.), 1984. *Fatty Acid Metabolism and Its Regulation.* Elsevier. [A concise and highly readable volume emphasizing control of fatty acid metabolism in bacteria, plants, and animals.]

Fatty acid oxidation

Foster, D.W., 1984. From glycogen to ketones—and back. *Diabetes* 33:1188–1199.

McGarry, J.D., and Foster, D.W., 1980. Regulation of hepatic fatty acid oxidation and ketone body production. *Ann. Rev. Biochem.* 49:395–420.

Garland, P.B., Shepherd, D., Nicholls, D.G., Yates, D.W., and Light, P.A., 1969. Interactions between fatty acid oxidation and the tricarboxylic acid cycle. *In* J.M. Lowenstein (ed.), *Citric Acid Cycle: Control and Compartmentation,* pp. 163–212. Dekker.

Carnitine

McGarry, J.D., Sen, A., Esser, V., Woeltje, K.F., Weis, B., and Foster, D.W., 1991. New insights into the mitochondrial carnitine palmitoyltransferase enzyme system. *Biochimie* 73:77–84.

Rebouche, C.J., 1992. Carnitine function and requirements during the life cycle. *FASEB J.* 6:3379–3386.

Fatty acid synthesis

Stoops, J.K., Kolodziej, S.J., Schroeter, J.P., Bretaudiere, J.P., and Wakil, S.J., 1992. Structure-function relationships of the yeast fatty acid synthase: Negative-stain, cryo-electron microscopy, and image analysis studies of the end views of the structure. *Proc. Nat. Acad. Sci.* 89:6585–6589.

Knowles, J.R., 1989. The mechanism of biotin-dependent enzymes. *Ann. Rev. Biochem.* 58:195–221.

Vagelos, P.R., 1973. Acyl group transfer (acyl carrier protein). *In* P.D. Boyer (ed.), *The Enzymes* (3rd ed.), vol. 8, pp. 155–199. Academic Press.

Iritani, N., 1992. Nutritional and hormonal regulation of lipogenic-enzyme gene expression in rat liver. *Eur. J. Biochem.* 205:433–442.

Acetyl CoA carboxylase

Mabrouk, G.M., Helmy, I.M., Thampy, K.G., and Wakil, S.J., 1990. Acute hormonal control of acetyl-CoA carboxylase: The roles of insulin, glucagon, and epinephrine. *J. Biol. Chem.* 265:6330–6338.

Kim, K.H., Lopez, C.F., Bai, D.H., Luo, X., and Pape, M.E., 1989. Role of reversible phosphorylation of acetyl-CoA carboxylase in long-chain fatty acid synthesis. *FASEB J.* 3:2250–2256.

Witters, L.A., and Kemp, B.E., 1992. Insulin activation of acetyl–CoA carboxylase accompanied by inhibition of the 5′-AMP-activated protein kinase. *J. Biol. Chem.* 267:2864–2867.

Cohen, P., and Hardie, D.G., 1991. The actions of cyclic AMP on biosynthetic processes are mediated indirectly by cyclic AMP-dependent protein kinases. *Biochim. Biophys. Acta* 1094:292–299.

Moore, F., Weekes, J., and Hardie, D.G., 1991. Evidence that AMP triggers phosphorylation as well as direct allosteric activation of rat liver AMP-activated protein kinase. A sensitive mechanism to protect the cell against ATP depletion. *Eur. J. Biochem.* 199:691–697.

Eicosanoid hormones and aspirin

Sigal, E., 1991. The molecular biology of mammalian arachidonic acid metabolism. *Amer. J. Physiol.* 260:L13–L28.

Lands, W.E., 1991. Biosynthesis of prostaglandins. *Ann. Rev. Nutrit.* 11:41–60.

Vane, J.R., Flower, R.J., and Botting, R.M., 1990. History of aspirin and its mechanism of action. *Stroke* 21 (Suppl. IV):12–23. [A fascinating account of the process of discovery. Many aspects of prostaglandin synthase are discussed in this incisive and interesting article.]

DeWitt, D.L., El-Harith, E.A., Kraemer, S.A., Andrews, M.J., Yao, E.F., Armstrong, R.L., and Smith, W.L., 1990. The aspirin and heme-binding sites of ovine and murine prostaglandin endoperoxide synthases. *J. Biol. Chem.* 265:5192–5198.

Genetic diseases

Roe, C.R., and Coates, P.M., 1989. Acyl-CoA dehydrogenase deficiencies. *In* Scriver, C.R., Beaudet, A.L., Sly, W.S., and Valle, D. (eds.), *The Metabolic Basis of Inherited Disease* (6th ed.), pp. 889–914. McGraw-Hill.

Blom, W., and Scholte, H.R., 1981. Acetyl CoA carboxylase deficiency: An inborn error of de novo fatty acid synthesis. *New Engl. J. Med.* 305:465–466.

PROBLEMS

1. *After lipolysis.* Write a balanced equation for the conversion of glycerol into pyruvate. Which enzymes are required in addition to those of the glycolytic pathway?

2. *From fatty acid to ketone body.* Write a balanced equation for the conversion of stearate into acetoacetate.

3. *Counterpoint.* Compare and contrast fatty acid oxidation and synthesis with respect to
   (a) Site of the process.
   (b) Acyl carrier.
   (c) Reductants and oxidants.
   (d) Stereochemistry of the intermediates.
   (e) Direction of synthesis or degradation.
   (f) Organization of the enzyme system.

4. *Sources.* For each of the following unsaturated fatty acids, indicate whether the biosynthetic precursor in animals is palmitoleate, oleate, linoleate, or linolenate.
   (a) 18:1 cis-Δ^{11}
   (b) 18:3 cis-Δ^{6}, Δ^{9}, Δ^{12}
   (c) 20:2 cis-Δ^{11}, Δ^{14}
   (d) 20:3 cis-Δ^{5}, Δ^{8}, Δ^{11}
   (e) 22:1 cis-Δ^{13}
   (f) 22:6 cis-Δ^{4}, Δ^{7}, Δ^{10}, Δ^{13}, Δ^{16}, Δ^{19}

5. *Tracing carbons.* Consider a cell extract that actively synthesizes palmitate. Suppose that a fatty acid synthase in this preparation forms one palmitate in about five minutes. A large amount of malonyl CoA labeled with ^{14}C in each carbon of its malonyl unit is suddenly added to this system, and fatty acid synthesis is stopped a minute later by altering the pH. The fatty acids in the supernatant are analyzed for radioactivity. Which carbon atom of the palmitate formed by this system is more radioactive—C-1 or C-14?

6. *Two plus three to make four.* Propose a reaction mechanism for the condensation of an acetyl unit with a malonyl unit to form an acetoacetyl unit in fatty acid synthesis.

7. *Driven by decarboxylation.* What is the role of decarboxylation in fatty acid synthesis? Name another key reaction in a metabolic pathway that employs this mechanistic motif.

8. *Kinase surfeit.* Suppose that a promoter mutation leads to the overproduction of protein kinase A in adipose cells. How would fatty acid metabolism be altered by this mutation?

9. *An unaccepting mutant.* The serine in acetyl CoA carboxylase that is the target of the AMP-dependent protein kinase is mutated to alanine. What is a likely consequence of this mutation?

10. *Blocked assets.* The presence of a fuel molecule in the cytosol does not assure that it can be effectively used. Give two examples of how impaired transport of metabolites between compartments leads to disease.

11. *Elegant inversion.* Peroxisomes have an alternative pathway for oxidizing polyunsaturated fatty acids. They possess a hydratase that converts D-3-hydroxyacyl CoA into *trans*-Δ^{2}-enoyl CoA. How can this enzyme be used to oxidize CoAs containing a cis double bond at an even-numbered carbon atom (e.g., the cis-Δ^{12} double bond of linoleate)?

12. *Covalent catastrophe.* What is a potential disadvantage of having many catalytic sites together on one very long polypeptide chain?

CHAPTER

25

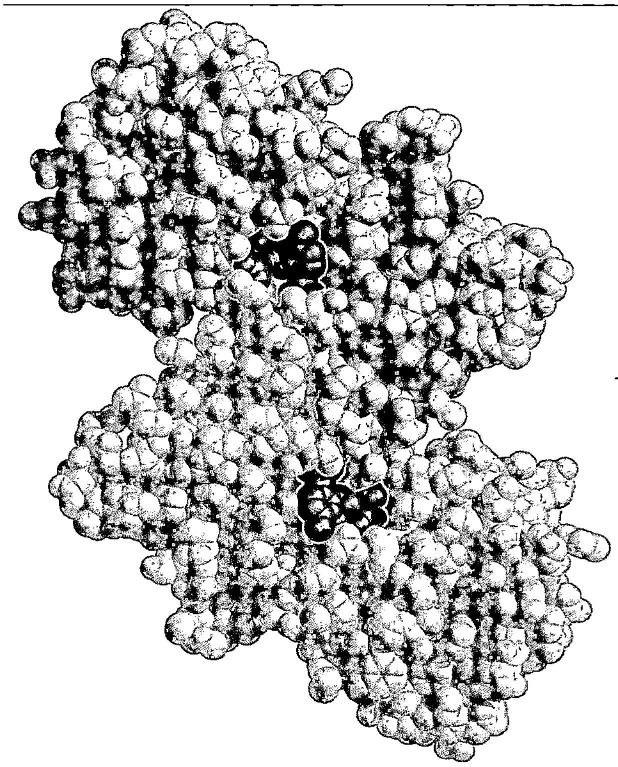

Amino Acid Degradation and the Urea Cycle

Amino acids in excess of those needed for the synthesis of proteins and other biomolecules cannot be stored, in contrast with fatty acids and glucose, nor are they excreted. Rather, surplus amino acids are used as metabolic fuel. *The α-amino group is removed and the resulting carbon skeleton is converted into a major metabolic intermediate.* Most of the amino groups of surplus amino acids are converted into urea, whereas their carbon skeletons are transformed into acetyl CoA, acetoacetyl CoA, pyruvate, or one of the intermediates of the citric acid cycle. Hence, *fatty acids, ketone bodies, and glucose can be formed from amino acids.*

Many coenzymes play key roles in amino acid degradation. *Pyridoxal phosphate,* for example, forms Schiff base intermediates that allow α-amino groups to be shuttled between amino acids and keto acids. *Vitamin B_{12} (cobalamin),* a cobalt-containing porphyrin-like cofactor, serves as a source of free radicals in the rearrangement of a carbon skeleton. We will also consider several genetic diseases that lead to brain damage and mental retardation unless remedial action is initiated soon after birth. *Phenylketonuria,* which is caused by a block in the conversion of phenylalanine to tyrosine, can readily be diagnosed and treated by removing phenylalanine from the diet. The study of amino acid metabolism is especially rewarding because it is rich in connections between basic biochemistry and clinical medicine.

Opening Image: *Structure of aspartate aminotransferase, a key enzyme in amino acid metabolism. The two identical subunits are shown in blue and yellow. A pyridoxal 5′-phosphate cofactor bound to each is shown in red. The front of the protein has been sliced away to expose the cofactors. [Drawn from 1asl.pdb, deposited by J. Jaeger, M. Moser, U. Sauder, and J.N. Jansonius.]*

α-AMINO GROUPS ARE CONVERTED INTO AMMONIUM ION BY OXIDATIVE DEAMINATION OF GLUTAMATE

The major site of amino acid degradation in mammals is the liver. The fate of the α-amino group will be considered first, followed by that of the carbon skeleton. The α-amino group of many amino acids is transferred to α-ketoglutarate to form *glutamate*, which is then oxidatively deaminated to yield NH_4^+.

$$\text{+H}_3\text{N}-\overset{\overset{\displaystyle H}{|}}{\underset{\underset{\displaystyle COO^-}{|}}{C}}-R \longrightarrow \text{+H}_3\text{N}-\overset{\overset{\displaystyle H}{|}}{\underset{\underset{\displaystyle COO^-}{|}}{C}}-CH_2-CH_2-COO^- \longrightarrow NH_4^+$$

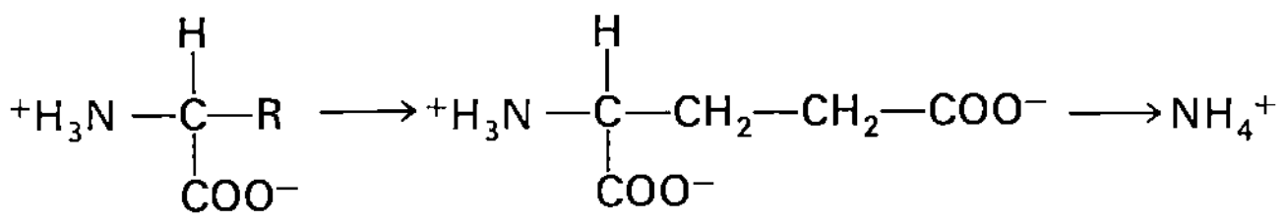

Amino acid **Glutamate**

Aminotransferases catalyze the transfer of an α-amino group from an α-amino acid to an α-keto acid. These enzymes, also called *transaminases*, generally funnel α-amino groups from a variety of amino acids to α-ketoglutarate for conversion into NH_4^+.

$$H-\overset{\overset{\displaystyle NH_3^+}{|}}{\underset{\underset{\displaystyle \boxed{R_1}}{|}}{C}}-COO^- + \overset{\overset{\displaystyle O}{\|}}{\underset{\underset{\displaystyle \textcircled{R_2}}{|}}{C}}-COO^- \rightleftharpoons \overset{\overset{\displaystyle O}{\|}}{\underset{\underset{\displaystyle \boxed{R_1}}{|}}{C}}-COO^- + H-\overset{\overset{\displaystyle NH_3^+}{|}}{\underset{\underset{\displaystyle \textcircled{R_2}}{|}}{C}}-COO^-$$

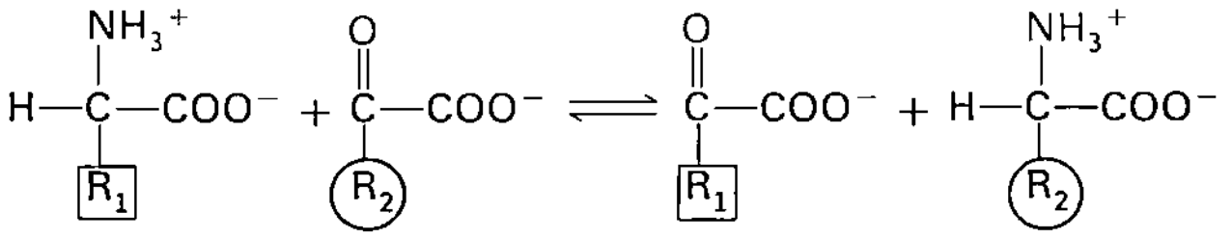

Aspartate aminotransferase, one of the most important of these enzymes, catalyzes the transfer of the amino group of aspartate to α-ketoglutarate.

Aspartate + α-ketoglutarate $\rightleftharpoons$ oxaloacetate + glutamate

Alanine aminotransferase, which is also prevalent in mammalian tissue, catalyzes the transfer of the amino group of alanine to α-ketoglutarate.

Alanine + α-ketoglutarate $\rightleftharpoons$ pyruvate + glutamate

Ammonium ion is formed from glutamate by oxidative deamination. This reaction is catalyzed by *glutamate dehydrogenase,* which is unusual in being able to utilize either NAD^+ or $NADP^+$.

$$H-\overset{\overset{\displaystyle NH_3^+}{|}}{\underset{\displaystyle \underset{\displaystyle \underset{\displaystyle \underset{\displaystyle COO^-}{|}}{CH_2}}{\underset{\displaystyle |}{CH_2}}}{C}}-COO^- + \begin{array}{c} NAD^+ \\ (\text{or } NADP^+) \end{array} + H_2O \rightleftharpoons NH_4^+ + \overset{\overset{\displaystyle O}{\|}}{\underset{\displaystyle \underset{\displaystyle \underset{\displaystyle \underset{\displaystyle COO^-}{|}}{CH_2}}{\underset{\displaystyle |}{CH_2}}}{C}}-COO^- + \begin{array}{c} NADH \\ (\text{or } NADPH) \end{array} + H^+$$

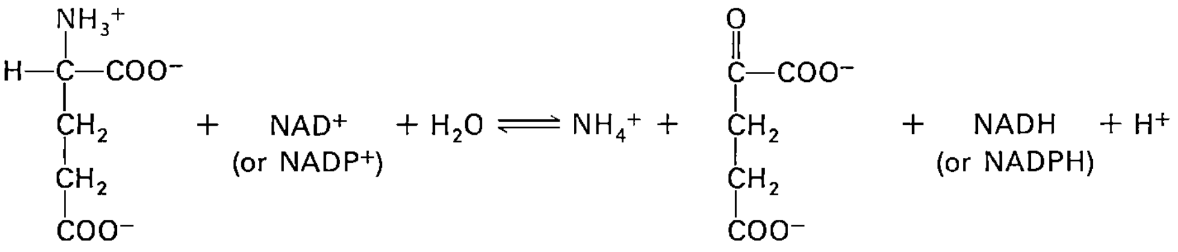

Glutamate **α-Ketoglutarate**

The activity of glutamate dehydrogenase is allosterically regulated. The vertebrate enzyme consists of six identical subunits. Guanosine triphosphate (GTP) and adenosine triphosphate (ATP) are allosteric inhibitors, whereas guanosine diphosphate (GDP) and adenosine diphosphate (ADP) are allosteric activators. Hence, *a lowering of the energy charge accelerates the oxidation of amino acids.*

The sum of the reactions catalyzed by aminotransferases and glutamate dehydrogenase is

α-Amino acid + $\begin{array}{c} NAD^+ \\ (\text{or } NADP^+) \end{array}$ + H_2O $\rightleftharpoons$

α-keto acid + NH_4^+ + $\begin{array}{c} NADH \\ (\text{or } NADPH) \end{array}$ + H^+

In most terrestrial vertebrates, NH_4^+ is converted into urea, which is excreted. The synthesis of urea will be discussed shortly.

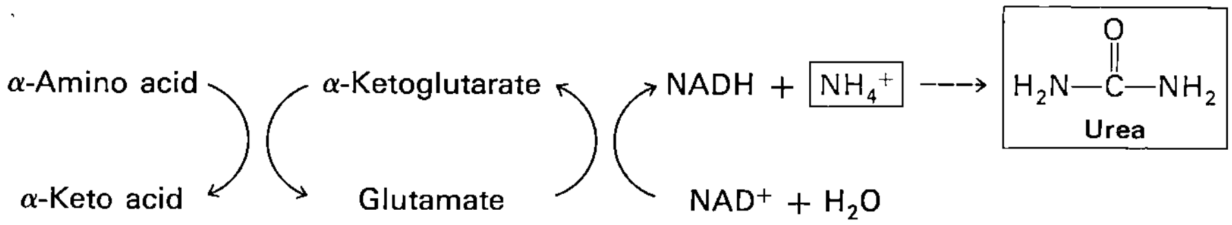

PYRIDOXAL PHOSPHATE FORMS SCHIFF BASE INTERMEDIATES IN AMINOTRANSFERASES

The prosthetic group of all aminotransferases is *pyridoxal phosphate (PLP)*, which is derived from *pyridoxine (vitamin B_6)*. During transamination, pyridoxal phosphate is transiently converted into *pyridoxamine phosphate (PMP)*.

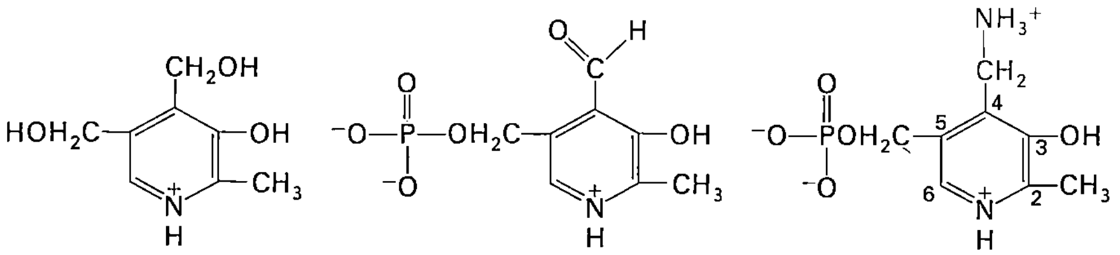

Pyridoxine (Vitamin B_6) | **Pyridoxal phosphate (PLP)** | **Pyridoxamine phosphate (PMP)**

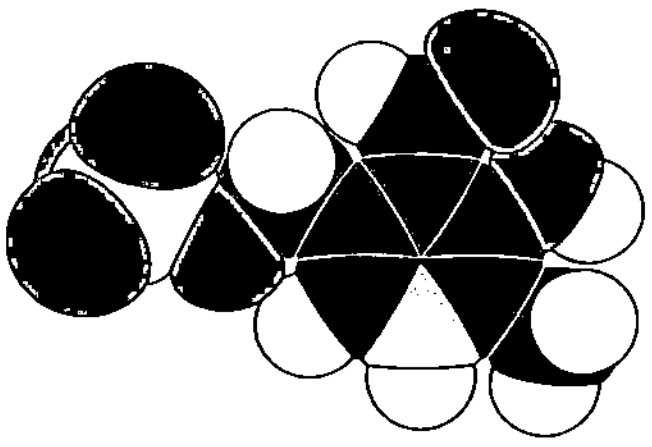

Figure 25-1
Space-filling model of pyridoxal 5-phosphate (PLP).

PLP enzymes form covalent Schiff base intermediates with their substrates. In the absence of substrate, the aldehyde group of PLP is in Schiff base linkage with the *ε-amino group of a specific lysine residue at the active site.* A new Schiff base linkage is formed on addition of an amino acid substrate. *The α-amino group of the amino acid substrate displaces the ε-amino group of the active-site lysine.* In other words, an *internal* aldimine becomes an *external* aldimine. The amino acid–PLP Schiff base that is formed remains tightly bound to the enzyme by multiple noncovalent interactions.

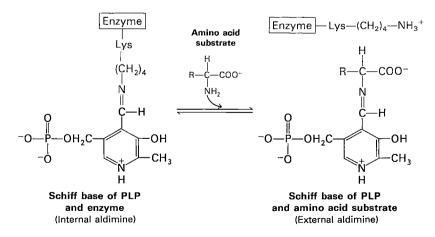

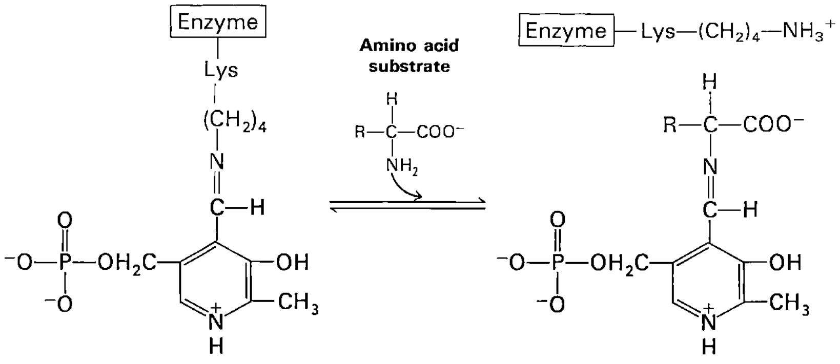

Schiff base of PLP and enzyme (Internal aldimine) | **Schiff base of PLP and amino acid substrate** (External aldimine)

Esmond Snell and Alexander Braunstein proposed a half century ago a reaction mechanism for transamination that has proved to be generally valid (Figure 25-2). The Schiff base between the amino acid substrate and PLP, the external *aldimine*, loses a proton from its α carbon to form a *quinonoid* intermediate. Reprotonation yields a *ketimine*, which contains a double bond between N and C_α of the substrate. In contrast, the aldimine contains a double bond between N and the carbonyl C of PLP.

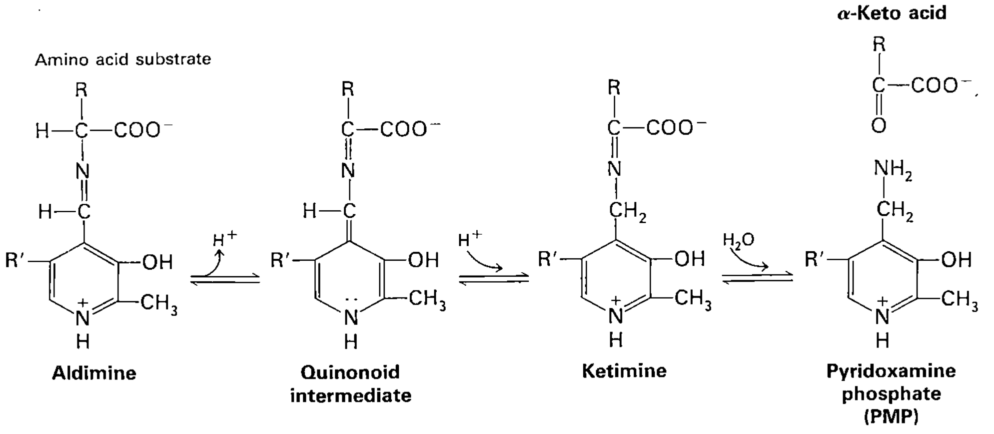

Figure 25-2
Proposed mechanism of transamination reactions. The steps shown constitute the first half of the overall reaction. In the second half, a different α-keto acid is converted into an amino acid by a reversal of this reaction sequence.

The ketimine is then hydrolyzed to an α-keto acid and pyridoxamine phosphate. These steps constitute half the transamination reaction.

$$\text{Amino acid}_1 + \text{E-PLP} \rightleftharpoons \alpha\text{-keto acid}_1 + \text{E-PMP}$$

The second half occurs by a reversal of the above pathway. A second α-keto acid reacts with the enzyme–pyridoxamine phosphate complex (E-PMP) to yield a second amino acid and regenerate the enzyme–pyridoxal phosphate complex (E-PLP).

$$\alpha\text{-Keto acid}_2 + \text{E-PMP} \rightleftharpoons \text{amino acid}_2 + \text{E-PLP}$$

The sum of these partial reactions is

$$\text{Amino acid}_1 + \alpha\text{-keto acid}_2 \rightleftharpoons \text{amino acid}_2 + \alpha\text{-keto acid}_1$$

THE ACTIVE-SITE CLEFT OF ASPARTATE AMINOTRANSFERASE CLOSES WHEN SUBSTRATE FORMS A SCHIFF BASE LINKAGE

X-ray crystallographic studies of mitochondrial aspartate aminotransferase have provided detailed views of how PLP and substrate are bound and have confirmed much of the proposed catalytic mechanism. Each of the identical 45-kd subunits of this dimer consists of a large domain and a small one. PLP is bound to the large domain, in a pocket near the subunit interface (Figure 25-3). The pyridine nitrogen atom of PLP is hydrogen bonded to an aspartate carboxylate, and the 2-methyl group is in van der Waals contact with several hydrophobic residues. The 3-hydroxyl group is ionized and hydrogen-bonded to the phenolic OH of a tyrosine. The 5-phosphate group of PLP is hydrogen-bonded to seven groups. Its two negative charges are balanced by an arginine side chain and by the posi-

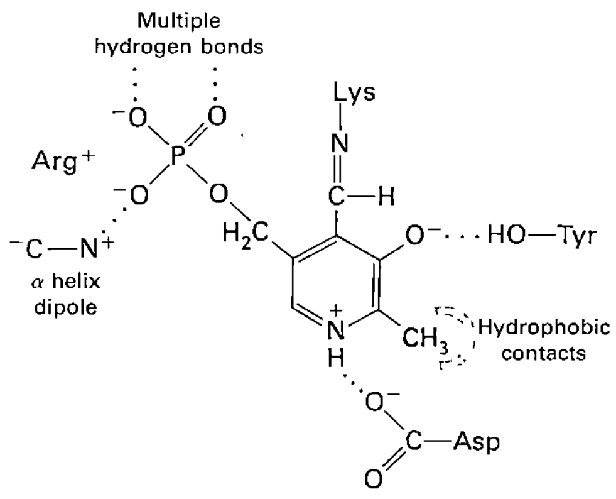

Figure 25-3
Mode of binding of PLP to aspartate aminotransferase. [After J.F. Kirsch, G. Eichele, G.F. Ford, M.G. Vincent, and J.N. Jansonius. *J. Mol. Biol.* 174(1984):510.]

tive dipole at the amino end of an α helix. As was postulated earlier on the basis of spectroscopic and chemical studies, the 4-aldehyde group of PLP has been shown by x-ray crystallography to be in Schiff base linkage with a lysine residue.

Each of the two carboxylates of the substrate, aspartate or glutamate, forms a salt bridge with an arginine residue of each subunit of the aminotransferase (Figure 25-4). These electrostatic interactions with guanidinium groups on different chains largely determine the substrate specificity of the enzyme. Moreover, they lead to a substantial movement of the small domain, which *closes the active-site crevice.* Recall that hexokinase (p. 499) and citrate synthase (p. 519) also undergo cleft closure during catalysis. The subsequent replacement of the ε-amino group of lysine by the α-amino group of the substrate induces a 30-degree tilting of the PLP ring, which promotes the subsequent conversions. Specifically, *the C_α–H bond becomes nearly perpendicular to the plane containing the PLP ring. This orientation facilitates the release of the α-H to form the quinonoid intermediate* (see Figure 25-2). Reprotonation gives the ketimine, which is hydrolyzed to give oxaloacetate, the α-keto acid. The lysine amino group that was initially in Schiff base linkage with PLP serves as a proton acceptor and donor in subsequent steps.

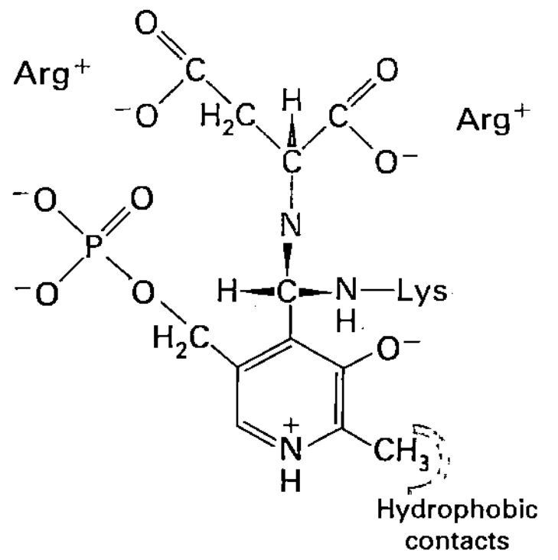

Figure 25-4
Mode of binding of aspartate in the tetrahedral intermediate of aspartate aminotransferase. [After J.F. Kirsch, G. Eichele, G.F. Ford, M.G. Vincent, J.N. Jansonius, H. Gehring, and P. Christen. *J. Mol. Biol.* 174(1984):510.]

PYRIDOXAL PHOSPHATE, A HIGHLY VERSATILE COENZYME, CATALYZES MANY REACTIONS OF AMINO ACIDS

Transamination is just one of a wide range of amino acid transformations that are catalyzed by PLP enzymes. The other reactions at the α carbon atom of amino acids are decarboxylations, deaminations, racemizations, and aldol cleavages (Figure 25-5). In addition, PLP enzymes catalyze elimination and replacement reactions at the β carbon atom (e.g., tryptophan synthetase, p. 726) and the γ carbon atom (e.g., cystathionase, p. 723) of amino acid substrates. Three common features of PLP catalysis underlie these diverse reactions. (1) A *Schiff base* is formed by the amino acid substrate (the amine component) and PLP (the carbonyl component). (2) The protonated form of PLP acts as an *electron sink* to stabilize catalytic intermediates that are negatively charged—the ring nitrogen of PLP attracts electrons from the amino acid substrate. In other words, PLP is an *electrophilic catalyst.* (3) The product Schiff base is then hydrolyzed.

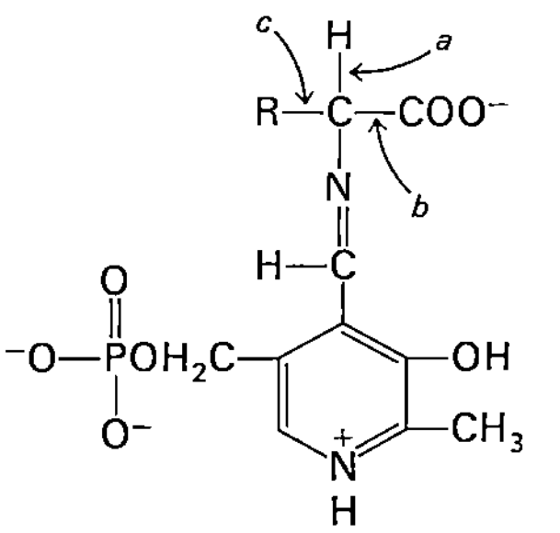

Figure 25-5
Pyridoxal phosphate enzymes labilize one of three bonds at the α carbon atom of an amino acid substrate. For example, bond *a* is labilized by aminotransferases, bond *b* by decarboxylases, and bond *c* by aldolases (such as threonine aldolases). PLP enzymes also catalyze reactions at the β and γ carbon atoms of amino acids.

How does an enzyme selectively break one of three bonds at the α carbon atom of an amino acid substrate? An important principle is that *the bond being broken must be perpendicular to the π-orbitals of the electron sink.* In an aminotransferase, for example, this is accomplished by binding the amino acid substrate so that the C_α–H bond is perpendicular to the PLP ring. This means of choosing one of several possible catalytic outcomes is called *stereoelectronic control.*

$$COO^-$$
$$^+H_3N—\overset{\displaystyle |}{\underset{\displaystyle |}{C}}—H$$
$$CH_2OH$$
Serine

↓ H₂O

$$COO^-$$
$$^+H_3N—\overset{\displaystyle |}{\underset{\displaystyle ||}{C}}$$
$$CH_2$$
Aminoacrylate

↓ H₂O

$$COO^-$$
$$\overset{\displaystyle |}{\underset{\displaystyle |}{C}}=O \quad + NH_4^+$$
$$CH_3$$
Pyruvate

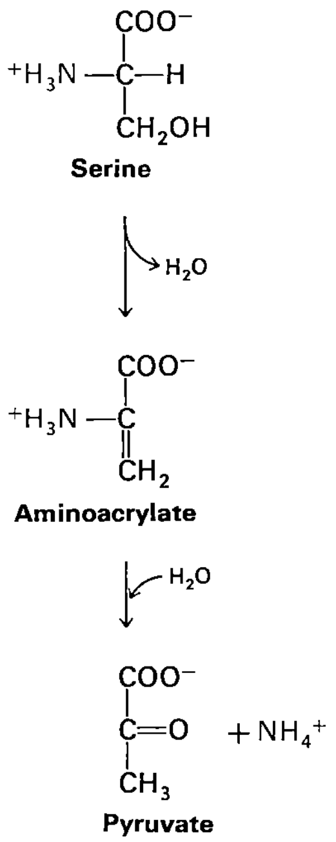

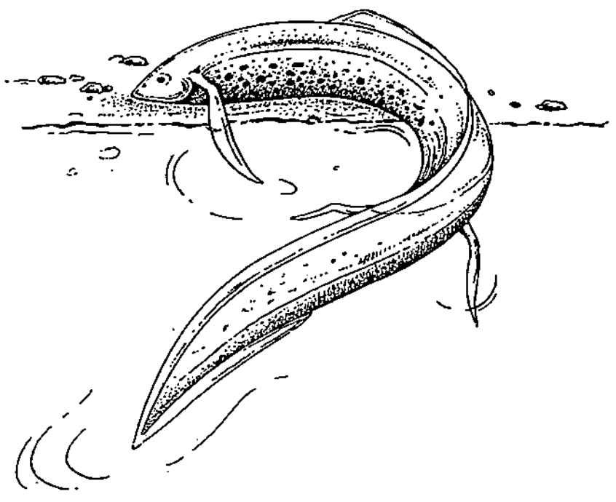

A lungfish. During droughts, lungfish become terrestrial and excrete urea instead of ammonia to conserve water.

SERINE AND THREONINE CAN BE DIRECTLY DEAMINATED

The α-amino groups of serine and threonine can be directly converted into NH_4^+ because each of these amino acids contains a hydroxyl group attached to its β carbon atom. These direct deaminations are catalyzed by *serine dehydratase* and *threonine dehydratase*, in which PLP is the prosthetic group.

$$\text{Serine} \longrightarrow \text{pyruvate} + NH_4^+$$

$$\text{Threonine} \longrightarrow \alpha\text{-ketobutyrate} + NH_4^+$$

These enzymes are called *dehydratases* because *dehydration precedes deamination*. Serine loses a hydrogen atom from its α carbon and a hydroxyl group from its β carbon to yield aminoacrylate. This unstable compound reacts with H_2O to give pyruvate and NH_4^+.

NH_4^+ IS CONVERTED INTO UREA IN MOST TERRESTRIAL VERTEBRATES AND THEN EXCRETED

Some of the NH_4^+ formed in the breakdown of amino acids is consumed in the biosynthesis of nitrogen compounds. In most terrestrial vertebrates, the excess NH_4^+ is converted into urea and then excreted. In birds and terrestrial reptiles, NH_4^+ is converted into uric acid for excretion, whereas in many aquatic animals, NH_4^+ itself is excreted. These three classes of organisms are called *ureotelic, uricotelic,* and *ammonotelic*.

In terrestrial vertebrates, urea is synthesized by the *urea cycle* (Figure 25-6). These reactions were proposed by Hans Krebs and Kurt Henseleit (a medical student) in 1932, five years before the elucidation of the citric

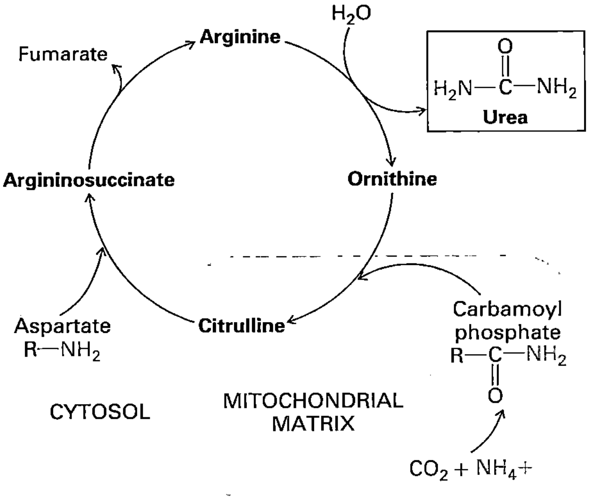

Figure 25-6
The urea cycle.

acid cycle. In fact, *the urea cycle was the first cyclic metabolic pathway to be discovered*. One of the nitrogen atoms of the urea synthesized by this pathway is transferred from an amino acid, aspartate. The other nitrogen atom is derived from NH_4^+, and the carbon atom comes from CO_2. *Ornithine*, an amino acid that is not a building block of proteins, is the carrier of these carbon and nitrogen atoms.

The immediate precursor of urea is *arginine*, which is hydrolyzed to urea and ornithine by *arginase*.

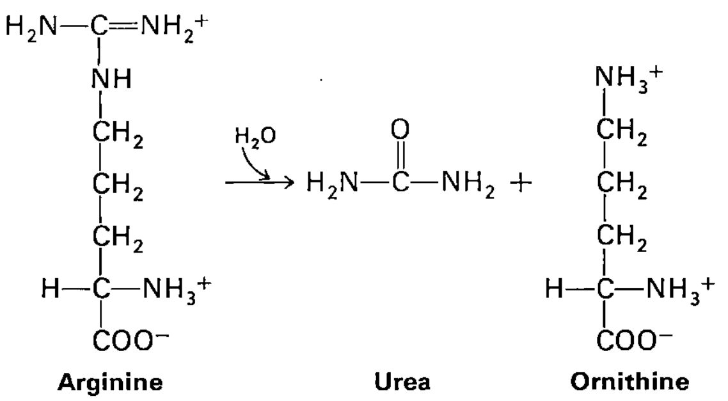

The other reactions of the urea cycle lead to the synthesis of arginine from ornithine. First, a carbamoyl group is transferred to ornithine to form *citrulline,* in a reaction catalyzed by *ornithine transcarbamoylase.* The carbamoyl donor in this reaction is *carbamoyl phosphate,* which has a high transfer potential because of its anhydride bond.

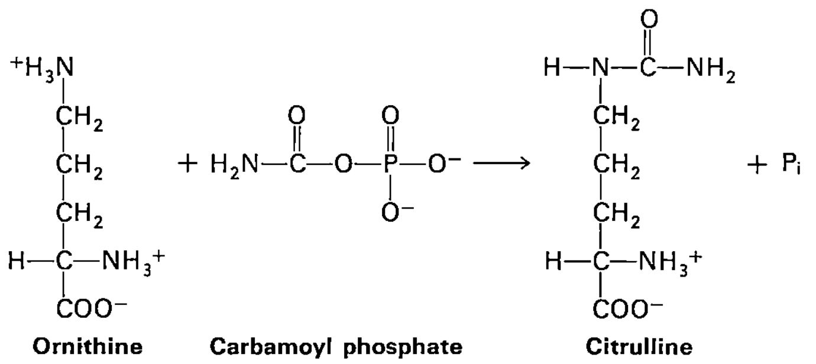

Argininosuccinate synthetase then catalyzes the condensation of citrulline and aspartate. This synthesis of *argininosuccinate* is driven by the cleavage of ATP into AMP and pyrophosphate and by the subsequent hydrolysis of pyrophosphate.

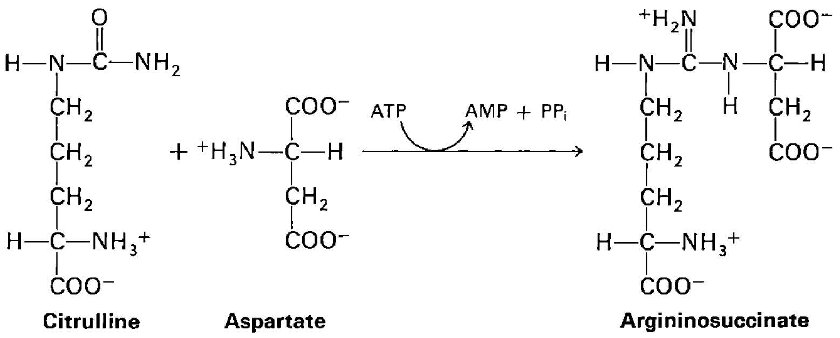

Finally, *argininosuccinase* cleaves argininosuccinate into *arginine* and *fumarate.* These two reactions, which transfer the amino group of aspartate to form arginine, preserve the carbon skeleton of aspartate in the form of fumarate.

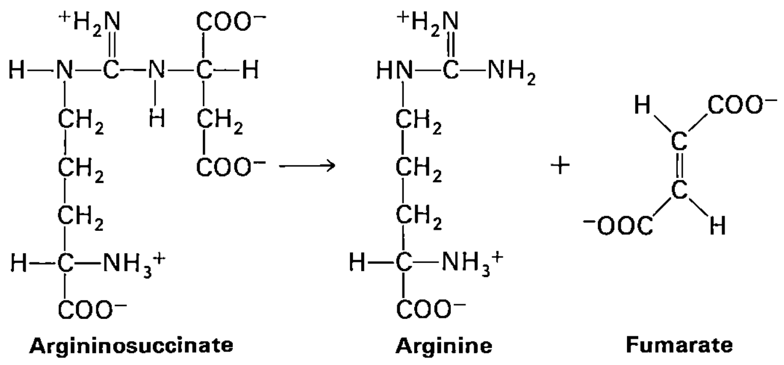

Carbamoyl phosphate is a simple molecule, but its synthesis is complex. This activated carbamoyl donor is formed from NH_4^+, CO_2, H_2O, and ATP in a reaction that is catalyzed by *carbamoyl phosphate synthetase*. An enigmatic feature of this enzyme is that it requires *N*-acetylglutamate for activity.

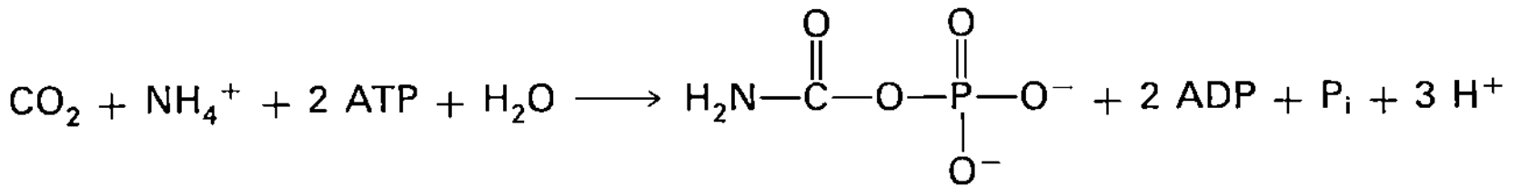

$$CO_2 + NH_4^+ + 2\ ATP + H_2O \longrightarrow H_2N-\overset{\overset{O}{\|}}{C}-O-\overset{\overset{O}{\|}}{\underset{\underset{O^-}{|}}{P}}-O^- + 2\ ADP + P_i + 3\ H^+$$

Carbamoyl phosphate

The consumption of two molecules of ATP makes this synthesis of carbamoyl phosphate essentially irreversible.

THE UREA CYCLE IS LINKED TO THE CITRIC ACID CYCLE

The stoichiometry of urea synthesis is

$$CO_2 + NH_4^+ + 3\ ATP + aspartate + 2\ H_2O \longrightarrow$$
$$urea + 2\ ADP + 2\ P_i + AMP + PP_i + fumarate$$

Pyrophosphate is rapidly hydrolyzed, and so four $\sim$P are consumed in these reactions to synthesize one molecule of urea. *The synthesis of fumarate by the urea cycle is important because it links the urea cycle and the citric acid cycle* (Figure 25-7). Fumarate is hydrated to malate, which is in turn oxidized to oxaloacetate. Oxaloacetate has several possible fates: (1) transamination to aspartate, (2) conversion into glucose by the gluconeogenic pathway, (3) condensation with acetyl CoA to form citrate, or (4) conversion into pyruvate.

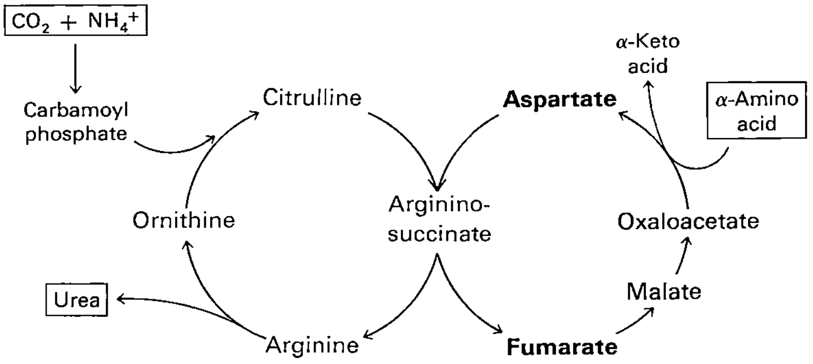

Figure 25-7
The urea cycle, the citric acid cycle, and the transamination of oxaloacetate are linked by fumarate and aspartate.

The compartmentation of the urea cycle and its associated reactions is also noteworthy. The formation of NH_4^+ by glutamate dehydrogenase, its incorporation into carbamoyl phosphate, and the subsequent synthesis of citrulline occur in the mitochondrial matrix. In contrast, the next three reactions of the urea cycle, which lead to the formation of urea, take place in the cytosol.

INHERITED DEFECTS OF THE UREA CYCLE CAUSE HYPERAMMONEMIA AND CAN LEAD TO BRAIN DAMAGE

The synthesis of urea in the liver is the major route of removal of NH_4^+. A blockage of carbamoyl phosphate synthesis or of any of the four steps of the urea cycle has devastating consequences because there is no alternative pathway for the synthesis of urea. *They all lead to an elevated level of NH_4^+ in the blood (hyperammonemia).* Some of these genetic defects become evident a day or two after birth, when the afflicted infant becomes lethargic and vomits periodically. Coma and irreversible brain damage may soon follow. Why are high levels of NH_4^+ toxic? The answer to this important question is not yet known. One possibility is that elevated levels of glutamine, formed from NH_4^+ and glutamate (p. 717), lead directly to brain damage.

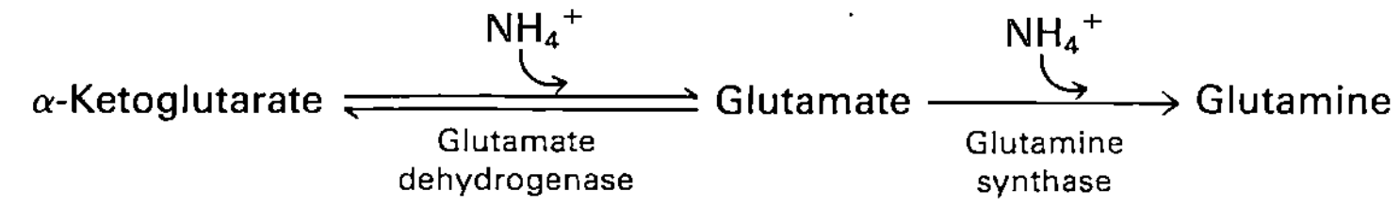

Ingenious strategies for coping with deficiencies in urea synthesis have been devised on the basis of a thorough understanding of the underlying biochemistry. Consider, for example, *argininosuccinase deficiency.* This defect can be partially bypassed by *providing a surplus of arginine in the diet and restricting the total protein intake.* In the liver, arginine is split into urea and ornithine, which then reacts with carbamoyl phosphate to form citrulline (Figure 25-8). This urea cycle intermediate condenses with aspartate to yield argininosuccinate, which is then excreted. Note that two nitrogen atoms—one from carbamoyl phosphate and one from aspartate—are eliminated from the body per molecule of arginine provided in the diet. In essence, *argininosuccinate substitutes for urea in carrying nitrogen out of the body.*

The treatment of *carbamoyl phosphate synthetase deficiency* or *ornithine transcarbamoylase deficiency* illustrates a different strategy for circumventing a metabolic block. Citrulline and argininosuccinate cannot be used to dispose of nitrogen atoms because their formation is impaired. Under these conditions, excess nitrogen accumulates in glycine and glutamine. The challenge then is to rid the body of these two amino acids. This is accomplished by supplementing a protein-restricted diet with *large amounts of benzoate and phenylacetate.* Benzoate is activated to benzoyl CoA, which reacts with glycine to form hippurate (Figure 25-9). Likewise, phenylacetate is activated to phenylacetyl CoA, which reacts with glutamine to form phenylacetylglutamine. These conjugates substitute for urea in the disposal of nitrogen. Thus, *latent biochemical pathways can be activated to partially bypass a genetic defect.*

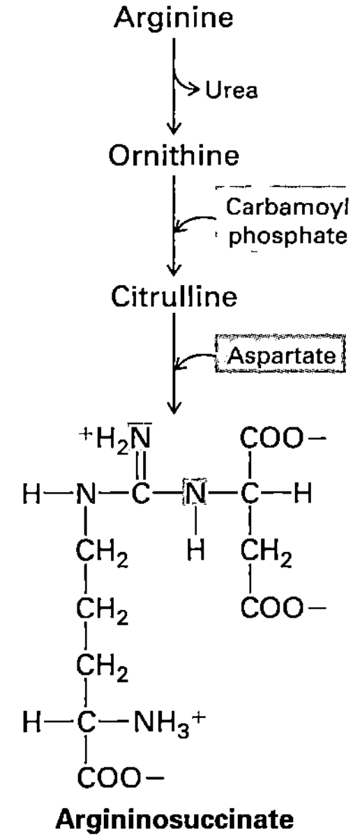

Figure 25-8
Argininosuccinase deficiency can be treated by supplementing the diet with arginine. Nitrogen is excreted in the form of argininosuccinate.

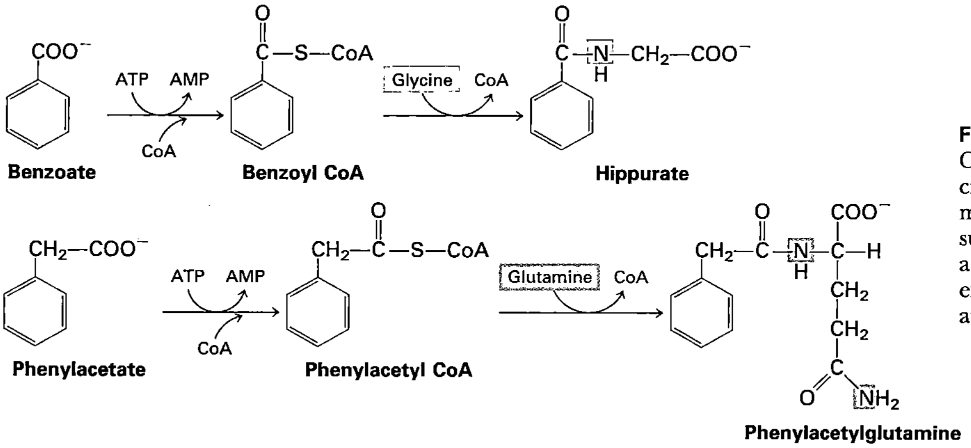

Figure 25-9
Carbamoyl phosphate synthetase deficiency and ornithine transcarbamoylase deficiency can be treated by supplementing the diet with benzoate and phenylacetate. Nitrogen is excreted in the form of hippurate and phenylacetylglutamine.

CARBON ATOMS OF DEGRADED AMINO ACIDS EMERGE IN MAJOR METABOLIC INTERMEDIATES

Thus far, we have considered a series of reactions that removes the α-amino group from amino acids and converts it into urea. We now turn to the fates of the carbon skeletons of amino acids. *The strategy of amino acid degradation is to form major metabolic intermediates that can be converted into glucose or be oxidized by the citric acid cycle.* In fact, the carbon skeletons of the diverse set of 20 fundamental amino acids are funneled into only seven molecules: *pyruvate, acetyl CoA, acetoacetyl CoA, α-ketoglutarate, succinyl CoA, fumarate,* and *oxaloacetate.* We see here a striking example of the remarkable economy of metabolic conversions.

Amino acids that are degraded to acetyl CoA or acetoacetyl CoA are termed *ketogenic* because they give rise to ketone bodies. In contrast, amino acids that are degraded to pyruvate, α-ketoglutarate, succinyl CoA, fumarate, or oxaloacetate are termed *glucogenic.* Net synthesis of glucose from these amino acids is feasible because these citric acid cycle intermediates and pyruvate can be converted into phosphoenolpyruvate and then into glucose (p. 570). Recall that mammals lack a pathway for the net synthesis of glucose from acetyl CoA or acetoacetyl CoA.

Of the basic set of 20 amino acids, only leucine and lysine are solely ketogenic (Figure 25-10). Isoleucine, phenylalanine, tryptophan, and tyrosine are both ketogenic and glucogenic. Some of their carbon atoms emerge in acetyl CoA or acetoacetyl CoA, whereas others appear in potential precursors of glucose. The other 14 amino acids are classed as solely glucogenic. This classification is not universally accepted because different quantitative criteria are applied. Whether an amino acid is regarded as being glucogenic, ketogenic, or both depends partly on the eye of the beholder.

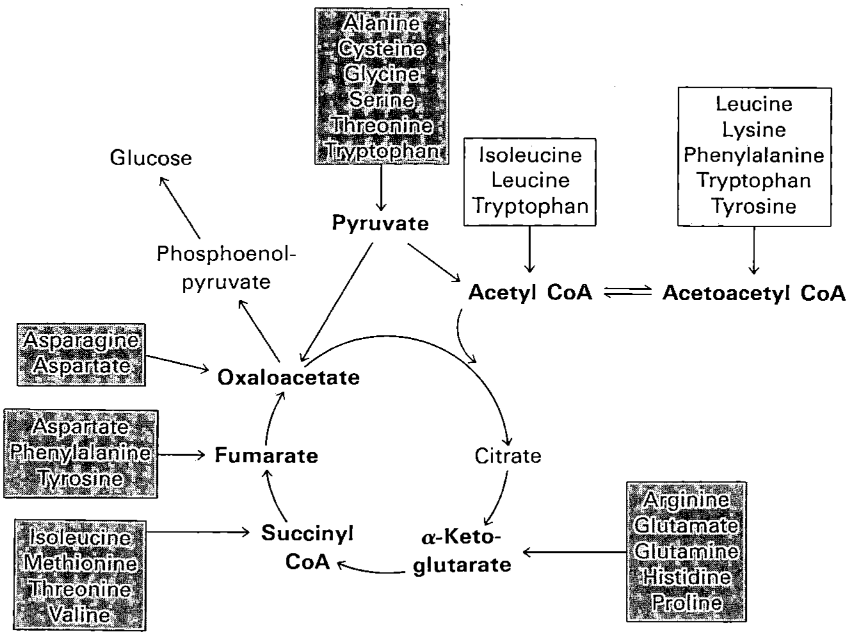

Figure 25-10
Fates of the carbon skeletons of amino acids. Glucogenic amino acids are shaded red, and ketogenic amino acids are shaded yellow. Some amino acids are both glucogenic and ketogenic.

THE C_3 FAMILY: ALANINE, SERINE, AND CYSTEINE ARE CONVERTED INTO PYRUVATE

Pyruvate is the entry point of the three-carbon amino acids—alanine, serine, and cysteine—into the metabolic mainstream (Figure 25-11). The transamination of alanine directly yields pyruvate:

$$\text{Alanine} + \alpha\text{-ketoglutarate} \rightleftharpoons \text{pyruvate} + \text{glutamate}$$

As mentioned previously (p. 630), glutamate is then oxidatively deaminated, yielding NH_4^+ and regenerating α-ketoglutarate. The sum of these reactions is

$$\text{Alanine} + NAD^+ + H_2O \longrightarrow \text{pyruvate} + NH_4^+ + NADH + H^+$$

Another simple reaction in the degradation of amino acids is the *deamination of serine to pyruvate* by *serine dehydratase* (p. 634).

$$\text{Serine} \longrightarrow \text{pyruvate} + NH_4^+$$

Cysteine can be converted into pyruvate by several pathways, with its sulfur atom emerging in H_2S, SO_3^{2-}, or SCN^-.

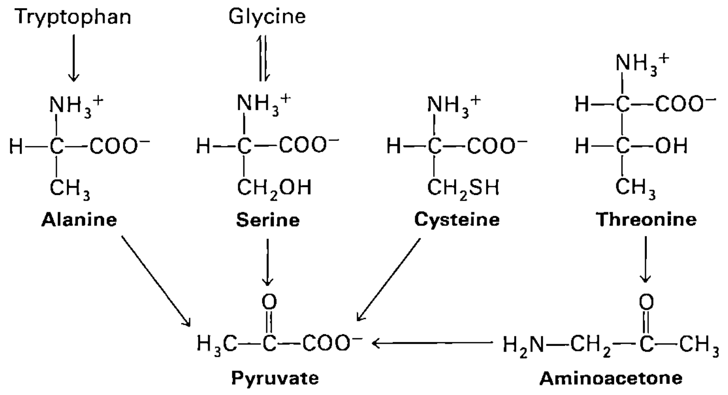

Figure 25-11
Pyruvate is the point of entry for alanine, serine, cysteine, glycine, threonine, and tryptophan.

The carbon atoms of three other amino acids can be converted into pyruvate. *Glycine* can be converted into serine by enzymatic addition of a hydroxymethyl group, or it can be cleaved to give CO_2, NH_4^+, and an activated one-carbon unit (p. 721). *Threonine* can give rise to pyruvate by way of aminoacetone. Three carbon atoms of *tryptophan* can emerge in alanine, which can be converted into pyruvate.

THE C_4 FAMILY: ASPARTATE AND ASPARAGINE ARE CONVERTED INTO OXALOACETATE

Aspartate, a four-carbon amino acid, is directly *transaminated to oxaloacetate,* a citric acid cycle intermediate.

$$\text{Aspartate} + \alpha\text{-ketoglutarate} \rightleftharpoons \text{oxaloacetate} + \text{glutamate}$$

Asparagine is hydrolyzed by *asparaginase* to NH_4^+ and aspartate, which is then transaminated.

Recall that aspartate can also be converted into *fumarate* by the urea cycle (p. 635). Fumarate is also a point of entry for half the carbon atoms of tyrosine and phenylalanine, as will be discussed shortly.

THE C_5 FAMILY: SEVERAL AMINO ACIDS ARE CONVERTED INTO α-KETOGLUTARATE THROUGH GLUTAMATE

The carbon skeletons of several five-carbon amino acids enter the citric acid cycle at *α-ketoglutarate.* These amino acids are first converted into *glutamate,* which is then oxidatively deaminated by glutamate dehydrogenase to yield α-ketoglutarate (Figure 25-12).

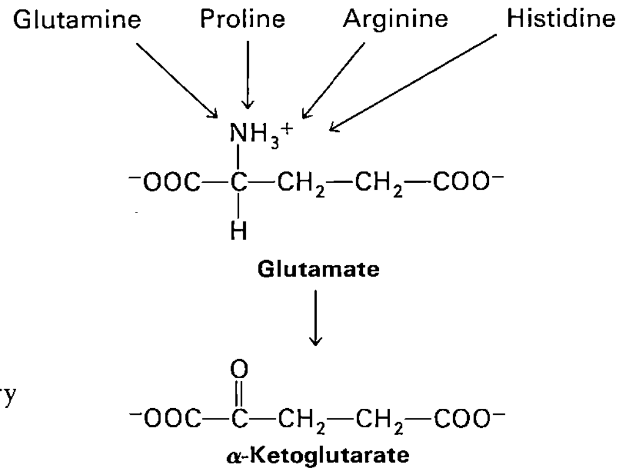

Figure 25-12
α-Ketoglutarate is the point of entry of several C_5 amino acids that are first converted into glutamate.

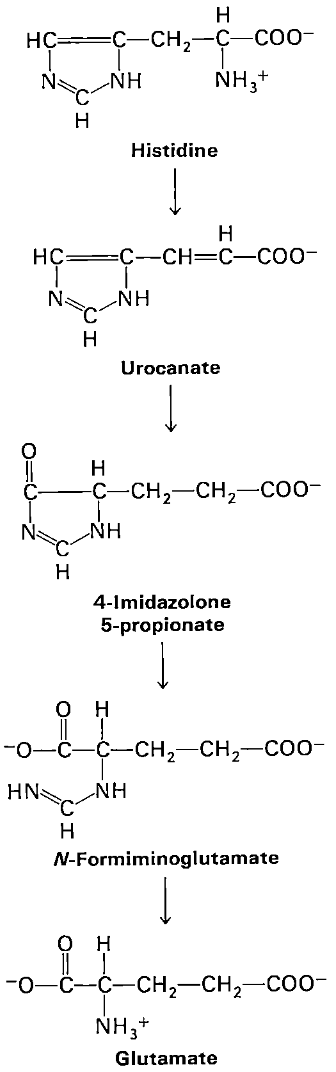

Figure 25-13
Conversion of histidine into glutamate.

Histidine is converted into 4-imidazolone 5-propionate (Figure 25-13). The amide bond in the ring of this intermediate is hydrolyzed to the *N*-formimino derivative of glutamate, which is then converted into glutamate by transfer of its formimino group to tetrahydrofolate, a carrier of activated one-carbon units (p. 720).

Glutamine is hydrolyzed to glutamate and NH_4^+ by *glutaminase. Proline* and *arginine* are converted into glutamate γ-semialdehyde, which is then oxidized to glutamate (Figure 25-14).

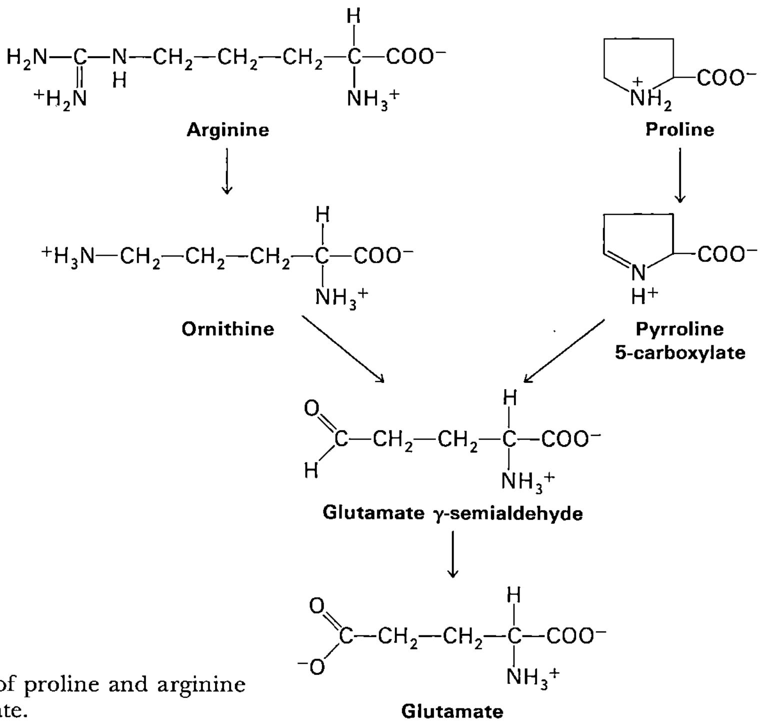

Figure 25-14
Conversion of proline and arginine into glutamate.

SUCCINYL COENZYME A IS A POINT OF ENTRY FOR SEVERAL NONPOLAR AMINO ACIDS

Succinyl CoA is the point of entry for some of the carbon atoms of methionine, isoleucine, and valine. Propionyl CoA and then methylmalonyl CoA are intermediates in the breakdown of these three nonpolar amino acids (Figure 25-15).

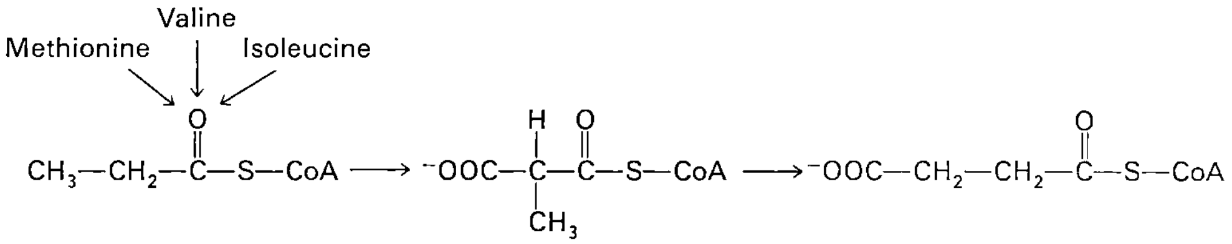

Propionyl CoA **Methylmalonyl CoA** **Succinyl CoA**

Figure 25-15
Conversion of methionine, isoleucine, and valine into succinyl CoA.

The pathway from propionyl CoA to succinyl CoA is especially interesting. Propionyl CoA is carboxylated at the expense of an ATP to yield the D isomer of methylmalonyl CoA. This carboxylation reaction is catalyzed by *propionyl CoA carboxylase,* a biotin enzyme that has a catalytic mechanism like that of acetyl CoA carboxylase and pyruvate carboxylase. The D isomer of methylmalonyl CoA is racemized to the L isomer, the substrate for the mutase that converts it into succinyl CoA.

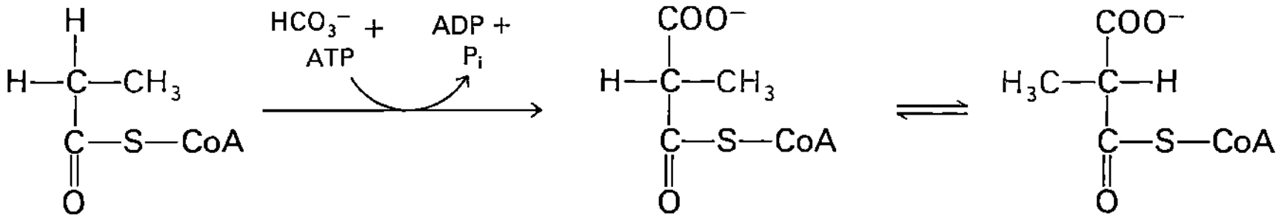

Propionyl CoA **D-Methylmalonyl CoA** **L-Methylmalonyl CoA**

Succinyl CoA is formed from L-methylmalonyl CoA by an *intramolecular rearrangement.* The –CO–S–CoA group migrates from C-2 to C-3 in exchange for a hydrogen atom. *This very unusual isomerization is catalyzed by* methylmalonyl CoA mutase, *one of the two mammalian enzymes known to contain a derivative of vitamin* B_{12} *as its coenzyme.*

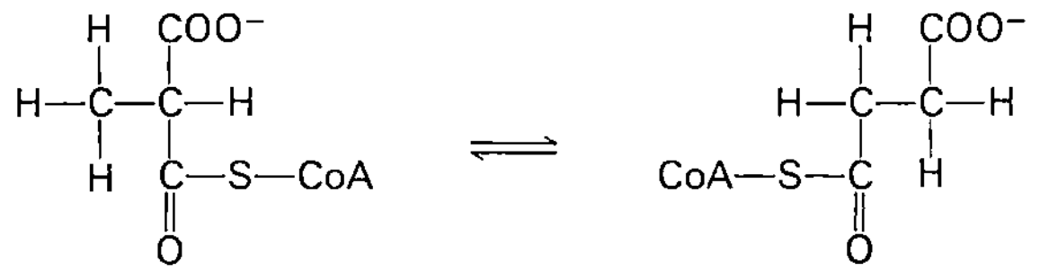

L-Methylmalonyl CoA **Succinyl CoA**

This pathway from propionyl CoA to succinyl CoA also participates in the oxidation of *fatty acids that have an odd number of carbon atoms.* The final thiolytic cleavage of an odd-numbered acyl CoA yields acetyl CoA and propionyl CoA (p. 612). Hence, odd-carbon fatty acids are partly glucogenic; three of their carbon atoms can emerge in glucose.

THE COBALT ATOM OF VITAMIN B_{12} IS BONDED TO THE 5′-CARBON OF DEOXYADENOSINE IN COENZYME B_{12}

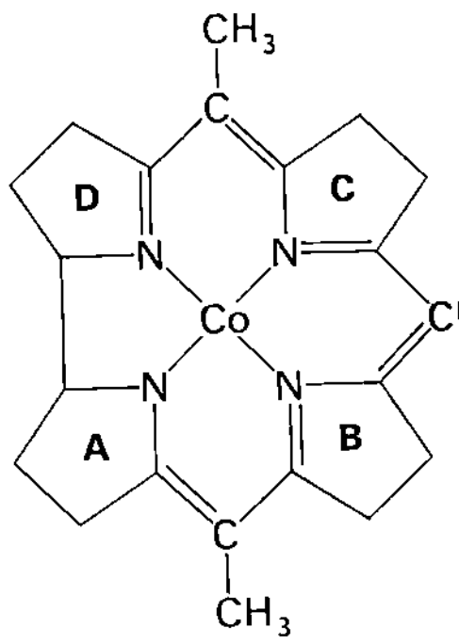

Figure 25-16
Corrin core of cobalamin (vitamin B_{12}). Substituents on the pyrroles and the other two ligands to cobalt are not shown in this diagram.

Cobalamin (vitamin B_{12}) has been a challenging problem in biochemistry and medicine since the discovery by George Minot and William Murphy in 1926 that pernicious anemia can be treated by feeding the patient large amounts of liver. Cobalamin was first purified in 1948, and it was crystallized then by Dorothy Hodgkin, who elucidated its complex three-dimensional structure in 1956. The core of cobalamin consists of a *corrin ring with a central cobalt atom* (Figure 25-16). The corrin ring, like a porphyrin, has *four pyrrole units.* Two of them (rings A and D) are directly bonded to each other, whereas the others are joined by methene bridges, as in porphyrins. The corrin ring is more reduced than that of porphyrins and the substituents are different.

A cobalt atom is bonded to the four pyrrole nitrogens. *The fifth substituent* (below the corrin plane in Figure 25-17) is a derivative of *dimethylbenzimidazole* that contains ribose 3-phosphate and aminoisopropanol. One of the nitrogen atoms of dimethylbenzimidazole is linked to cobalt. The amino group of aminoisopropanol is in amide linkage with a side chain. The *sixth substituent* of the cobalt atom (located above the corrin plane in Figure 25-17) can be $-CH_3$, OH^-, or a 5′-deoxyadenosyl unit.

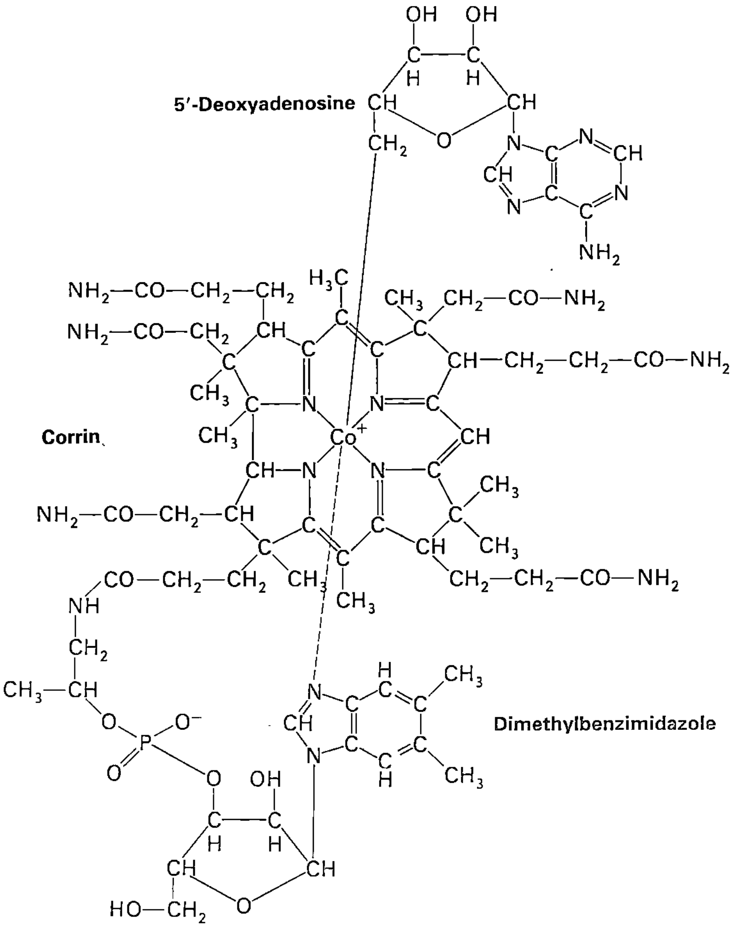

Figure 25-17
Structure of coenzyme B_{12} (5′-deoxyadenosylcobalamin).

The cobalt atom in cobalamin can be in the +1, +2, or +3 oxidation state. The cobalt atom is in the +3 state in hydroxocobalamin (where OH^- occupies the sixth coordination site). This form, called *B_{12a} (Co^{3+})*, is reduced to a divalent state, called *B_{12r} (Co^{2+})*, by a flavoprotein reductase. The B_{12r} (Co^{2+}) form is reduced by a second flavoprotein reductase to B_{12s} (Co^{+}). NADH is the reductant in both reactions.

$$B_{12a}\ (Co^{3+}) \longrightarrow B_{12r}\ (Co^{2+}) \longrightarrow B_{12s}\ (Co^{+})$$

The B_{12s} form is the substrate for the final enzymatic reaction that yields the active coenzyme. Co^{+} attacks the 5′-carbon atom of ATP and displaces the triphosphate group to form *5′-deoxyadenosylcobalamin,* also known as *coenzyme B_{12}* (Figure 25-18). This compound is remarkable in having a *carbon-metal bond,* the only one known in a biomolecule. Another unusual feature of this reaction is that the 5′-methylene carbon atom of ATP, rather than its α or β phosphorus atom, is the target of nucleophilic attack. The formation of *S*-adenosylmethionine (p. 721) is the only other biochemical reaction in which a nucleophile displaces the triphosphate group of ATP.

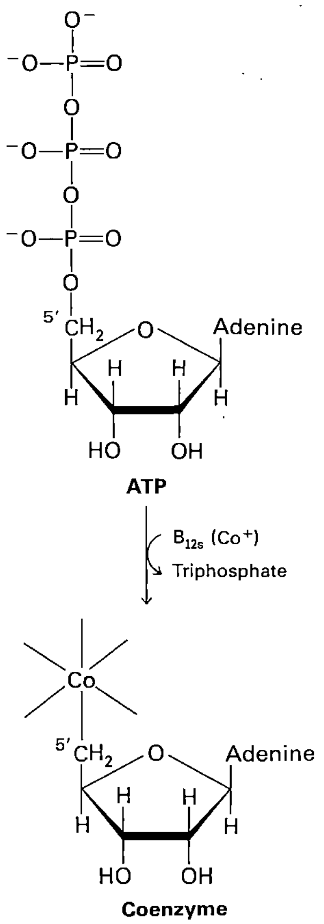

Figure 25-18
Formation of coenzyme B_{12} from cobalamin and ATP. The 5′-carbon of 5′-deoxyadenosine is coordinated to the cobalt atom in this coenzyme.

COENZYME B_{12} PROVIDES FREE RADICALS TO CATALYZE INTRAMOLECULAR MIGRATIONS INVOLVING HYDROGEN

Cobalamin enzymes catalyze three types of reactions: (1) *intramolecular rearrangements,* (2) *methylations,* as in the synthesis of methionine (p. 722), and (3) *reduction of ribonucleotides to deoxyribonucleotides* (p. 749). The conversion of L-methylmalonyl CoA into succinyl CoA (an intramolecular rearrangement) and the formation of methionine by methylation of homocysteine are the only reactions in mammals that require coenzyme B_{12}.

The rearrangement reactions catalyzed by coenzyme B_{12} are exchanges of two groups attached to adjacent carbon atoms (Figure 25-19). A hydrogen atom migrates from one carbon atom to the next, and an R group (such as the –CO–S–CoA group of methylmalonyl CoA) concomitantly moves in the reverse direction. The first step in these intramolecular rearrangements is the cleavage of the carbon–cobalt bond of 5′-deoxyadenosylcobalamin to form B_{12r} (Co^{2+}) and a 5′-deoxyadenosyl radical ($-CH_2\cdot$) (Figure 25-20). In this *homolytic cleavage reaction,* one electron of the Co–C bond stays with Co and the other with C, generating a free radical. In contrast, nearly all other cleavage reactions in biological systems are *heterolytic*—an electron pair is transferred to one of the two atoms that were bonded together.

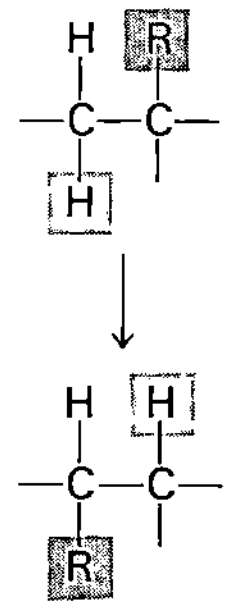

Figure 25-19
Rearrangement reaction catalyzed by cobalamin enzymes. The R group can be an amino group, a hydroxyl group, or a substituted carbon.

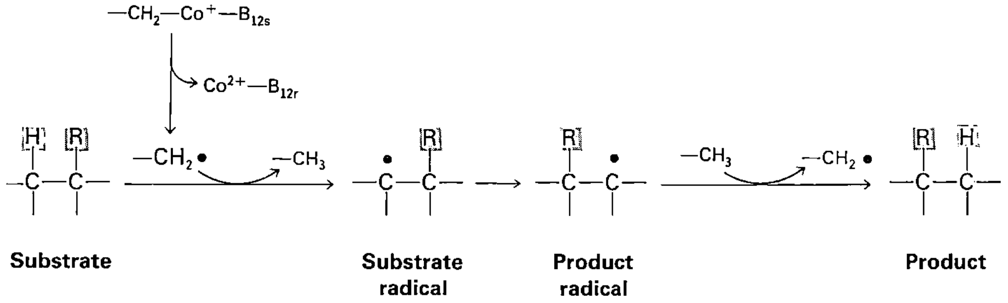

Figure 25-20
Coenzyme B_{12} provides the free radical that abstracts a hydrogen atom in rearrangement reactions.

What is the purpose of this very unusual $-CH_2\cdot$ radical? This highly reactive species abstracts a *hydrogen atom* from the substrate to form 5′-deoxyadenosine ($-CH_3$) and a substrate radical. This sets the stage for the migration of R to the position formerly occupied by H on the neighboring carbon atom. Finally, the product radical abstracts a hydrogen atom from the 5′-methyl group to complete the rearrangement and return the deoxyadenosyl unit to the radical form. *The role of B_{12} in such intramolecular migrations is to serve as a source of free radicals for the abstraction of hydrogen atoms.* A key property of coenzyme B_{12} is the weakness of its cobalt–carbon bond, whose facile cleavage generates a radical. Steric crowding around the cobalt atom prevents the formation of a stronger bond, which would make the coenzyme a less effective catalyst.

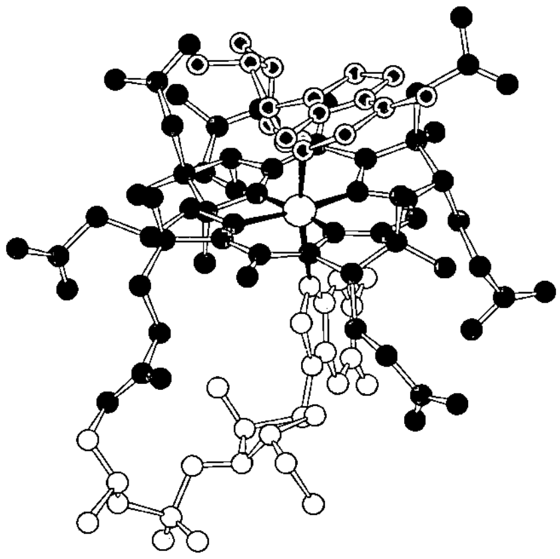

Figure 25-21
Model of coenzyme B_{12}. The cobalt atom is shown in yellow, the corrin unit in red, the 5′-deoxyadenosyl unit in blue, and the benzimidazole unit in green.

Dietary treatment of pernicious anemia—
"The dietary treatment of pernicious anemia may be of more importance than hitherto generally recognized. . . . Forty-five patients with pernicious anemia have continued to take a diet for from six weeks to two years composed especially of foods rich in complete proteins—particularly liver (120 to 240 gm. a day). The diet contains for each day at least 120 gm. of muscle meat. . . . Following the diet, all of the patients showed a prompt, rapid and distinct remission of their anemia coincident with marked symptomatic improvement, except for marked disorders due to spinal-cord degeneration. Improvement was often striking, so that where the red blood cell count averaged for all before starting the diet 1,470,000 per cubic millimeter, about one month afterward it averaged 3,400,000 per cubic millimeter. . . . "

G.R. Minot and W.P. Murphy
J. Amer. Med. Assoc. 87(1926):470

ABSORPTION OF COBALAMIN IS IMPAIRED IN PERNICIOUS ANEMIA

Animals and plants are unable to synthesize cobalamin. *This vitamin is unique in being synthesized only by microorganisms,* in particular, anaerobic bacteria. Mammals absorb cobalamin using a specialized transport system. The stomach secretes a 59-kd glycoprotein called *intrinsic factor (IF),* which binds cobalamin in the intestinal lumen. The cobalamin–IF complex, which is resistant to proteolysis by digestive enzymes in the gut, is taken up by a *specific receptor in the lining of the ileum.* The vitamin is released from IF inside the ileal epithelial cell and then transported across the opposite side of the plasma membrane into the bloodstream. Cobalamin then becomes bound to *transcobalamin II,* which enables it to enter the liver, the hematopoietic (blood-forming) system, and other cells by receptor-mediated endocytosis (p. 935). Recent cDNA cloning studies have shown that intrinsic factor and transcobalamin II contain homologous domains that bind the vitamin.

Pernicious anemia is caused by a deficiency of intrinsic factor, which leads to impaired absorption of cobalamin. Consequently, the synthesis of purines and of thymine, which are dependent on cobalamin, is impaired (p. 752). The hematopoietic system is especially vulnerable to a deficiency of this vitamin because blood cells turn over rapidly. This disease was originally treated by feeding patients large amounts of liver, a rich source of cobalamin, so that enough of the vitamin was absorbed even in the absence of intrinsic factor. The most reliable therapy is intramuscular injection of cobalamin at monthly intervals.

A normal person requires less than 10 μg of cobalamin per day. Pernicious anemia has a slow onset because several milligrams are normally stored in the liver and other organs. Nutritional deficiency of cobalamin is rare (except in strict vegetarians) because this vitamin is found in virtually all animal tissues. Pernicious anemia may be an autoimmune disease.

SEVERAL INHERITED DEFECTS OF METHYLMALONYL COENZYME A METABOLISM ARE KNOWN

Several inherited disorders of methylmalonyl CoA metabolism have been characterized. They usually become evident in the first year of life, when the striking symptom is *acidosis*. The pH of arterial blood is about 7.0, rather than the normal value of 7.4. *Large amounts of methylmalonate appear in the urine of patients who have these disorders.* A normal person excretes less than 5 mg of methylmalonate per day, whereas a patient with defective methylmalonyl CoA metabolism may excrete more than 1 g. About half the patients with methylmalonic aciduria improve markedly when large doses of cobalamin are administered intramuscularly. The arterial blood pH returns to normal, and there is a marked decrease in the excretion of methylmalonate. Responsive patients usually have a defect in the transferase that catalyzes the synthesis of coenzyme B_{12} from B_{12s} and ATP.

$$B_{12s}\ (Co^{+}) \longrightarrow\!\!|\,|\!\longrightarrow \text{deoxyadenosylcobalamin}$$

In contrast, other patients with impaired methylmalonyl CoA metabolism do not respond to large doses of cobalamin. Some of them have a defective methylmalonyl CoA mutase apoenzyme.

LEUCINE IS DEGRADED TO ACETYL COENZYME A AND ACETOACETATE

As was mentioned earlier, leucine and lysine are the only solely ketogenic amino acids of the common set of 20. First, leucine is transaminated to the corresponding α-keto acid, *α-ketoisocaproate.* This α-keto acid is then degraded by reactions akin to those occurring in the citric acid cycle and fatty acid oxidation. It is *oxidatively decarboxylated to isovaleryl CoA* by the *branched-chain α-keto dehydrogenase complex.* The α-keto acids of valine and

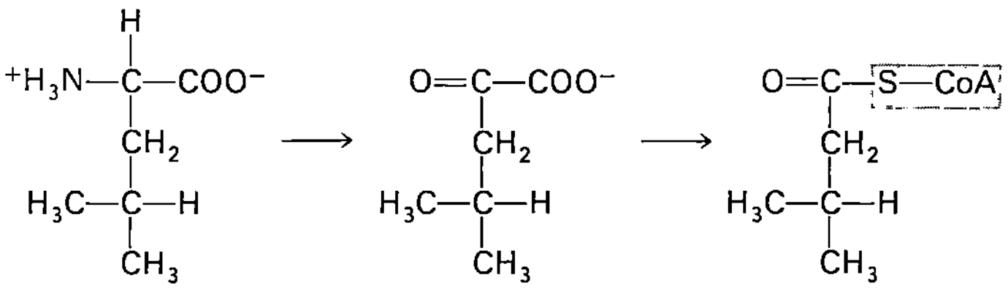

isoleucine, the other two branched-chain aliphatic amino acids, are also substrates. The oxidative decarboxylation of these α-keto acids is analogous to that of pyruvate to acetyl CoA and of α-ketoglutarate to succinyl CoA. The branched-chain α-keto dehydrogenase, a multienzyme complex, is a homolog of pyruvate dehydrogenase (p. 515) and α-ketoglutarate dehydrogenase (p. 517). Indeed, the E3 component of these enzymes, which regenerates the oxidized form of lipoamide, is identical.

Isovaleryl CoA is *dehydrogenated* to yield *β-methylcrotonyl CoA.* This oxidation is catalyzed by *isovaleryl CoA dehydrogenase.* The hydrogen acceptor is FAD, as in the analogous reaction in fatty acid oxidation that is catalyzed by acyl CoA dehydrogenase. *β-Methylglutaconyl CoA* is then formed by *carboxylation* of β-methylcrotonyl CoA at the expense of an ATP. As might be expected, the carboxylation mechanism of β-methylcrotonyl CoA carboxylase is very similar to that of pyruvate carboxylase and acetyl CoA carboxylase.

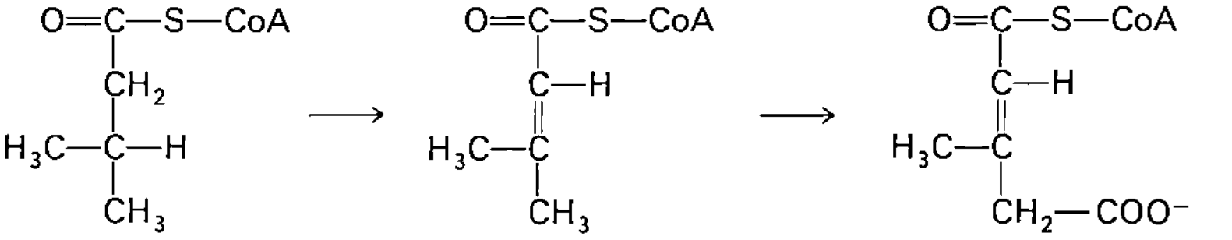

Isovaleryl CoA **β-Methylcrotonyl CoA** **β-Methylglutaconyl CoA**

β-Methylglutaconyl CoA is then *hydrated* to form *3-hydroxy-3-methylglutaryl CoA,* which is cleaved to *acetyl CoA* and *acetoacetate.* This reaction has already been discussed in regard to the formation of ketone bodies from fatty acids (p. 612).

Acetyl CoA

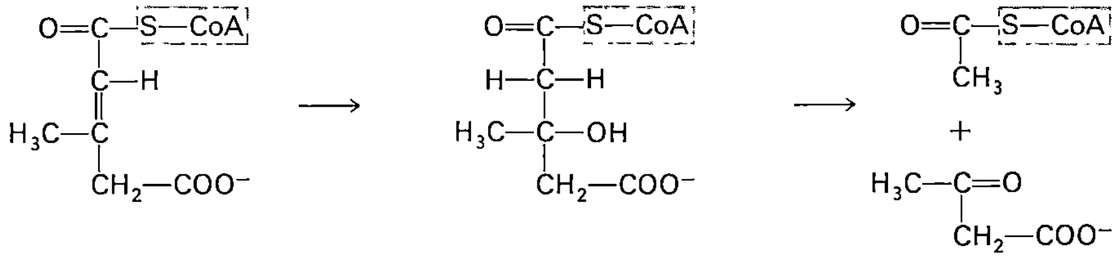

β-Methylglutaconyl CoA **3-Hydroxy-3-methylglutaryl CoA** **Acetoacetate**

It is interesting to note that seven coenzymes participate in the degradation of leucine to acetyl CoA and acetoacetate: *PLP* in transamination; *TPP, lipoate, FAD, and NAD⁺* in oxidative decarboxylation; FAD again in dehydrogenation; and *biotin* in carboxylation. *Coenzyme A* is the acyl carrier in these reactions.

The degradative pathways of valine and isoleucine resemble that of leucine. All three amino acids are initially transaminated to the corresponding α-keto acid, which is then oxidatively decarboxylated by the branched-chain α-keto dehydrogenase to yield a CoA derivative. The subsequent reactions are like those of fatty acid oxidation. Isoleucine yields acetyl CoA and propionyl CoA, whereas valine yields CO_2 and propionyl CoA. As was discussed earlier (p. 443), the number of reactions in metabolism is large, but the number of *kinds* of reactions is relatively small. The degradation of leucine, valine, and isoleucine provides a striking illustration of the underlying simplicity and elegance of metabolism.

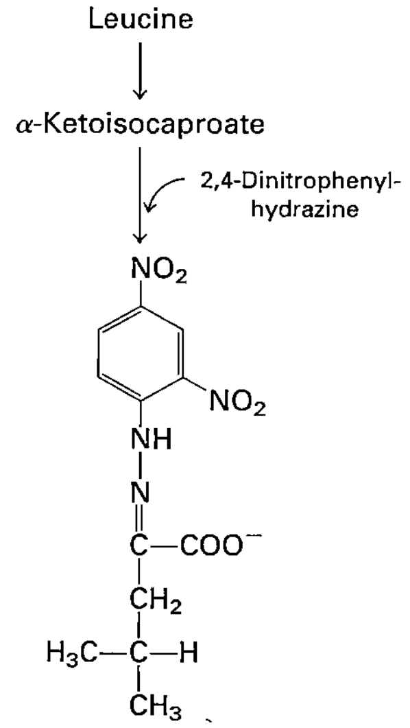

2,4-Dinitrophenylhydrazone derivative

In *maple syrup urine disease,* the oxidative decarboxylation of α-keto acids derived from valine, isoleucine, and leucine is blocked because the branched-chain dehydrogenase is missing or defective. Hence, the levels of these α-keto acids and the branched-chain amino acids that gave rise to them are markedly elevated in both blood and urine. Indeed, the urine of patients has the odor of maple syrup, hence the name of the disease (also called *branched-chain ketoaciduria*). Maple syrup urine disease usually leads to mental and physical retardation unless the patient is placed on a diet low in valine, isoleucine, and leucine early in life. The disease can readily be detected in newborns by screening urine samples with 2,4-dinitrophenylhydrazine, which reacts with α-keto acids to form 2,4-dinitrophenylhydrazone derivatives. A definitive diagnosis can be made by mass spectrometry.

PHENYLALANINE AND TYROSINE ARE DEGRADED BY OXYGENASES TO ACETOACETATE AND FUMARATE

The pathway for the degradation of phenylalanine and tyrosine has some very interesting features. This series of reactions shows how *molecular oxygen is used to break an aromatic ring.* The first step is the hydroxylation of phenylalanine to tyrosine, a reaction catalyzed by *phenylalanine hydroxylase.* This enzyme is called a *monooxygenase* (or *mixed-function oxygenase*) because *one atom of* O_2 appears in the product and the other in H_2O.

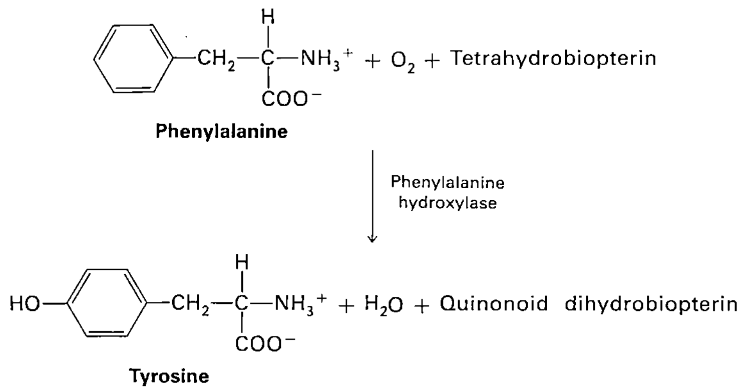

The reductant here is *tetrahydrobiopterin,* an electron carrier that has not been previously discussed. Tetrahydrobiopterin is initially formed by reduction of dihydrobiopterin by NADPH, in a reaction catalyzed by *dihydrofolate reductase* (Figure 25-22). The quinonoid form of dihydrobiopterin produced in the hydroxylation of phenylalanine is reduced back to tetrahydrobiopterin by NADH in a reaction catalyzed by *dihydropteridine reductase.* The sum of the reactions catalyzed by phenylalanine hydroxylase and dihydropteridine reductase is

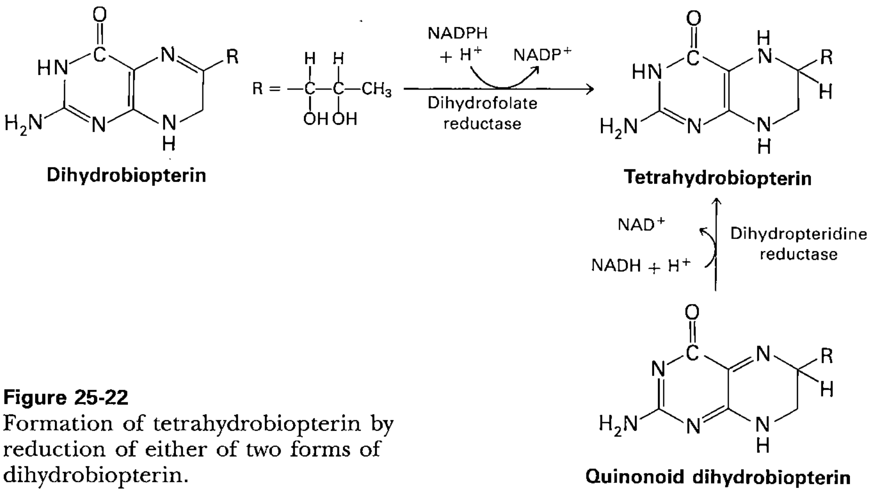

Figure 25-22
Formation of tetrahydrobiopterin by reduction of either of two forms of dihydrobiopterin.

$$\text{Phenylalanine} + O_2 + \text{NADH} + H^+ \longrightarrow \text{tyrosine} + \text{NAD}^+ + H_2O$$

The next step in the degradation of these aromatic amino acids is the transamination of tyrosine to p-*hydroxyphenylpyruvate* (Figure 25-23). This α-keto acid then reacts with O_2 to form *homogentisate.* The enzyme catalyzing this complex reaction, p-*hydroxyphenylpyruvate hydroxylase,* is called a

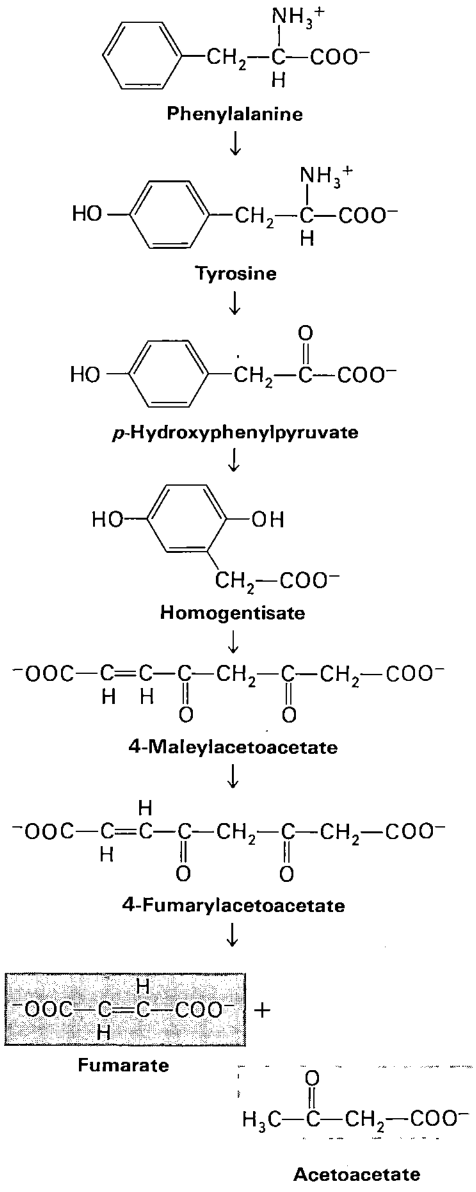

Figure 25-23
Pathway for the degradation of phenylalanine and tyrosine.

dioxygenase because *both atoms of* O_2 become incorporated into the product, one on the ring and one in the carboxyl group. The aromatic ring of homogentisate is then cleaved by O_2, which yields 4-maleylacetoacetate. This reaction is catalyzed by *homogentisate oxidase,* another dioxygenase. 4-Maleylacetoacetate is then isomerized to *4-fumarylacetoacetate* by an enzyme that uses glutathione as a cofactor. Finally, 4-fumarylacetoacetate is hydrolyzed to *fumarate* and *acetoacetate.*

We previously encountered a dioxygenase in the hydroxylation of proline (p. 454). Recall that α-ketoglutarate participates in this reaction and that one atom of O_2 emerges in hydroxyproline and the other in succinate. Prolyl hydroxylase and lysyl hydroxylase are *intermolecular dioxygenases* because they catalyze the incorporation of one atom of oxygen into each of *two* separate products. In contrast, *p*-hydroxyphenylpyruvate hydroxylase and homogentisate oxidase are *intramolecular dioxygenases* because both atoms of O_2 appear in the same product (Figure 25-24). The active sites of these enzymes contain iron that is not part of heme or an iron-sulfur cluster. *Nearly all cleavages of aromatic rings in biological systems are catalyzed by dioxygenases,* a class of enzymes discovered by Osamu Hayaishi.

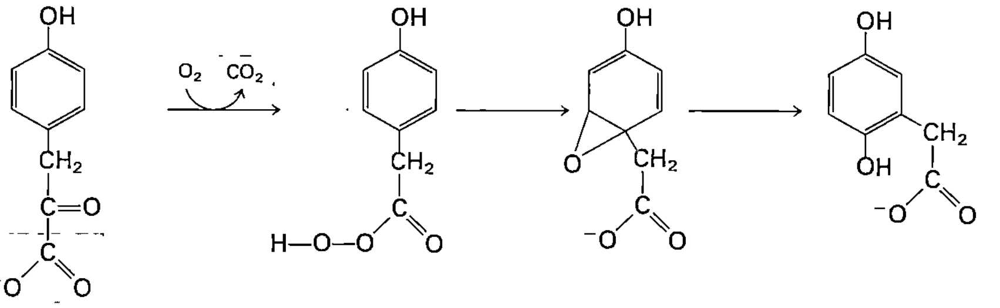

p-Hydroxyphenylpyruvate | Peracid | Epoxide | Homogentisate

Figure 25-24
Formation of homogentisate. *p*-Hydroxyphenylpyruvate hydroxylase, the enzyme catalyzing this reaction, is an intramolecular dioxygenase. Both atoms of O_2 emerge in homogentisate.

> *Urine black as ink*—
> "The patient was a boy who passed black urine and who, at the age of fourteen years, was submitted to a drastic course of treatment which had for its aim the subduing of the fiery heat of his viscera, which was supposed to bring about the condition in question by charring and blackening his bile. Among the measures prescribed were bleedings, purgation, baths, a cold and watery diet and drugs galore. None of these had any obvious effect, and eventually the patient, who tired of the futile and superfluous therapy, resolved to let things take their natural course. None of the predicted evils ensued, he married, begat a large family, and lived a long and healthy life, always passing urine black as ink."
>
> ZACUTUS LUSITANUS (1649)

GARROD'S DISCOVERY OF INBORN ERRORS OF METABOLISM

Alcaptonuria is an inherited metabolic disorder caused by the absence of homogentisate oxidase. Homogentisate accumulates and is excreted in the urine, which turns dark on standing as homogentisate is oxidized and polymerized to a melaninlike substance. Alcaptonuria is a relatively harmless condition. In 1902, Archibald Garrod showed that alcaptonuria is transmitted as a single recessive Mendelian trait. Furthermore, he recognized that homogentisate is a normal intermediate in the degradation of phenylalanine and tyrosine and that it accumulates in alcaptonuria because its degradation is blocked. He concluded that "the splitting of the benzene ring in normal metabolism is the work of a special enzyme, that in congenital alcaptonuria this enzyme is wanting." Garrod perceived the direct relationship between genes and enzymes, and he recognized the importance of chemical individuality. His imaginative and prescient book *Inborn Errors of Metabolism* opened new vistas in biology and medicine.

A BLOCK IN THE HYDROXYLATION OF PHENYLALANINE CAN LEAD TO SEVERE MENTAL RETARDATION

In 1934, Asbjörn Fölling examined two mentally retarded siblings and observed that their urine turned olive-green on addition of $FeCl_3$. Eight more positive reactions were found on testing the urine of 430 other mentally retarded patients. The occurrence of 4 pairs of siblings amongst these 10 reactive patients suggested that their disorder had a hereditary basis. The compound in urine responsible for the color reaction was identified as *phenylpyruvate*—hence, the disease was named *phenylketonuria.* Fölling proposed that *a genetic defect in phenylalanine metabolism was the cause of mental retardation in phenylketonuric individuals.*

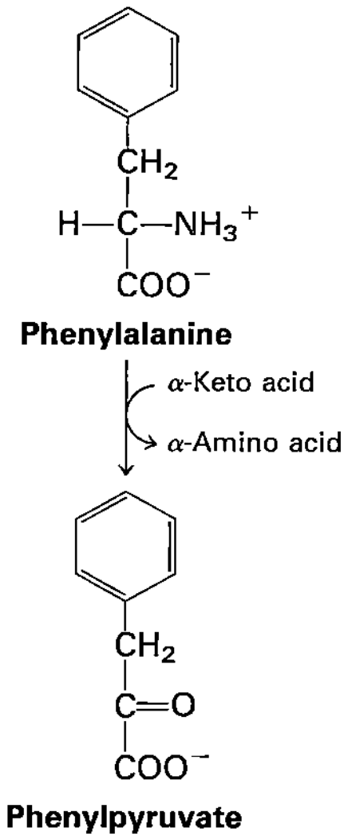

Indeed, phenylketonuria is caused by an *absence or deficiency of phenylalanine hydroxylase* or, more rarely, of its tetrahydrobiopterin cofactor. *Phenylalanine accumulates in all body fluids because it cannot be converted into tyrosine.* Normally, three-quarters of the phenylalanine is converted into tyrosine, and the other quarter becomes incorporated into proteins. Because the major outflow pathway is blocked in phenylketonuria, the blood level of phenylalanine is typically at least 20-fold higher than in normal individuals. Minor fates of phenylalanine in normal people, such as the formation of phenylpyruvate, become prominent in phenylketonurics. *Almost all untreated phenylketonurics are severely mentally retarded.* In fact, about 1% of patients in mental institutions have phenylketonuria. The brain weight of these people is below normal, myelination of their nerves is defective, and their reflexes are hyperactive. The life expectancy of untreated phenylketonurics is drastically shortened. Half are dead by age 20, and three-quarters by age 30. *The biochemical basis of their mental retardation is an enigma.*

Phenylketonurics appear normal at birth but are severely defective by age one if untreated. The therapy for phenylketonuria is a *low phenylalanine diet.* The aim is to provide just enough phenylalanine to meet the needs for growth and replacement. Proteins that have a low content of phenylalanine, such as casein from milk, are hydrolyzed and phenylalanine is removed by adsorption. A low phenylalanine diet must be started very soon after birth to prevent irreversible brain damage. In one study, the average IQ of phenylketonurics treated within a few weeks after birth was 93; a control group of siblings treated starting at age one had an average IQ of 53.

Early diagnosis of phenylketonuria is essential and has been accomplished by mass screening programs. The phenylalanine level in the blood is the preferred diagnostic criterion because it is more sensitive and reliable than the $FeCl_3$ test. Prenatal diagnosis of phenylketonuria with DNA probes has become feasible because the gene has been cloned and many mutations have been pinpointed. The incidence of phenylketonuria is

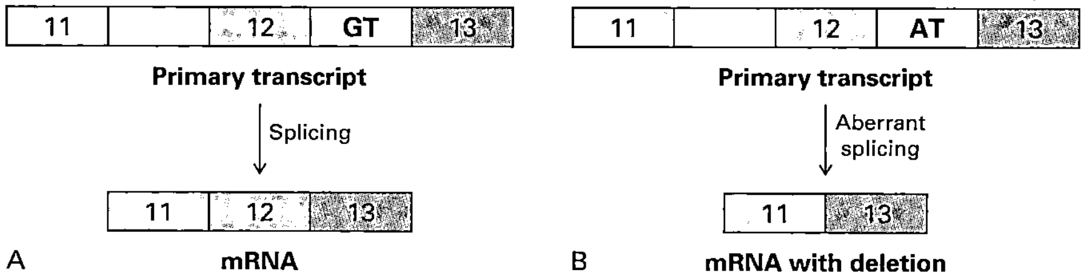

Figure 25-25
Phenylketonuria can be caused by an intron mutation that leads to aberrant splicing. (A) Normal primary transcript and mRNA. (B) Mutation of G to A in intron 12 results in the skipping of exon 12. Most mutations causing phenylketonuria occur in coding regions.

about 1 in 20,000 newborns. The disease is inherited as an *autosomal recessive,* as was proposed by Fölling. Heterozygotes, which make up about 1.5% of a typical population, appear normal. Carriers of the phenylketonuric gene have a reduced level of phenylalanine hydroxylase, as reflected in an increased level of phenylalanine in the blood. However, these criteria are not absolute, because the blood levels of phenylalanine in carriers and normal people overlap to some extent. The measurement of the kinetics of disappearance of intravenously administered phenylalanine is a more definitive test for the carrier state. It should be noted that a high blood level of phenylalanine in a pregnant woman can result in abnormal development of the fetus. This is a striking example of maternal-fetal relationships at the molecular level.

SUMMARY

Surplus amino acids are used as metabolic fuel. The degradation of most surplus amino acids starts with the removal of their α-amino groups by transamination to an α-keto acid. Pyridoxal phosphate is the coenzyme in all aminotransferases and in many other enzymes catalyzing transformations of amino acids. The α-amino groups funnel into α-ketoglutarate to form glutamate, which is then oxidatively deaminated by glutamate dehydrogenase to give NH_4^+ and α-ketoglutarate. NAD^+ or $NADP^+$ is the electron acceptor in this reaction. In most terrestrial vertebrates, NH_4^+ is converted into urea by the urea cycle. Urea is formed by the hydrolysis of arginine. The subsequent reactions of the urea cycle synthesize arginine from ornithine, the other product of the hydrolysis reaction. First, ornithine is carbamoylated to citrulline by carbamoyl phosphate. Citrulline then condenses with aspartate to form argininosuccinate, which is cleaved to arginine and fumarate. The carbon atom and one nitrogen atom of urea come from carbamoyl phosphate, which is synthesized from CO_2, NH_4^+, and ATP. The other nitrogen atom of urea comes from aspartate. Some enzymatic deficiencies of the urea cycle can be bypassed by supplementing the diet with arginine or compounds that form conjugates with glycine and glutamine.

The carbon atoms of degraded amino acids are converted into pyruvate, acetyl CoA, acetoacetate, or an intermediate of the citric acid cycle. Most amino acids are solely glucogenic, two are solely ketogenic, and a few are both ketogenic and glucogenic. Alanine, serine, cysteine, glycine, threonine, and tryptophan are degraded to pyruvate. Asparagine and aspartate are converted into oxaloacetate. α-Ketoglutarate is the point of entry for glutamate and four amino acids (glutamine, histidine, proline, and arginine) that can be converted into glutamate. Succinyl CoA is the point of entry for some of the carbon atoms of three amino acids (methionine, isoleucine, and valine) that are degraded by way of methylmalonyl CoA. 5′-Deoxyadenosylcobalamin, a coenzyme formed from vitamin B_{12} and ATP, is the free-radical source in the isomerization of methylmalonyl CoA to succinyl CoA. Leucine is degraded to acetoacetyl CoA and acetyl CoA. The breakdown of valine and isoleucine is like that of leucine. Their α-keto acid derivatives are oxidatively decarboxylated by the branched-chain dehydrogenase. This reaction is blocked in maple syrup urine disease, a genetic disorder that leads to brain damage.

The aromatic rings of tyrosine and phenylalanine are degraded by oxygenases. Phenylalanine hydroxylase, a monooxygenase, uses tetrahydrobiopterin as the reductant. One of the oxygen atoms of O_2 emerges in tyrosine and the other in water. Absence or inactivity of this enzyme causes phenylketonuria; mental retardation results unless a low phenylalanine diet is started in infancy. Subsequent steps in the degradation of these aromatic amino acids are catalyzed by intramolecular dioxygenases, which catalyze the insertion of both atoms of O_2 into a single product. Four of the carbon atoms of phenylalanine and tyrosine are converted into fumarate, and four emerge in acetoacetate.

SELECTED READINGS

Where to start

Torchinsky, Y.M., 1989. Transamination: Its discovery, biological and chemical aspects. *Trends Biochem. Sci.* 12:115–117.

Halpern, J., 1985. Mechanisms of coenzyme B_{12}-dependent rearrangements. *Science* 227:869–875.

Eisensmith, R.C., and Woo, S.L.C., 1991. Phenylketonuria and the phenylalanine hydroxylase gene. *Mol. Biol. Med.* 8:3–18.

Levy, H.L., 1989. Nutritional therapy for selected inborn errors of metabolism. *J. Amer. Coll. Nutrit.* 8:54S–60S.

Books

Bender, D.A., 1985. *Amino Acid Metabolism* (2nd ed.). Wiley.

Meister, A., 1965. *Biochemistry of the Amino Acids* (2nd ed.), vols. 1 and 2. Academic Press.

Lippard, S.J., and Berg, J.M., 1994. *Principles of Bioinorganic Chemistry.* University Science Books. [Chapters 5 and 11 deal with cobalamin-catalyzed reactions.]

Schauder, P., Wahren, J., Paoletti, R., Bernardi, R., and Rinetti, M. (eds.), 1992. *Branched-Chain Amino Acids: Biochemistry, Physiopathology, and Clinical Sciences.* Raven Press.

Grisolia, S., Báguena, R., and Mayor, F. (eds.), 1976. *The Urea Cycle.* Wiley.

Reviews

Barker, H.A., 1981. Amino acid degradation by anaerobic bacteria. *Ann. Rev. Biochem.* 50:23–40.

Cooper, A.J.L., 1983. Biochemistry of sulfur-containing amino acids. *Ann. Rev. Biochem.* 52:187–222.

Abeles, R., and Dolphin, D., 1976. The vitamin B_{12} coenzyme. *Acc. Chem. Res.* 9:114–120.

Nichol, C.A., Smith, G.K., and Duch, D.S., 1985. Biosynthesis and metabolism of tetrahydrobiopterin and molybdopterin. *Ann. Rev. Biochem.* 54:729–764.

Structure of aspartate aminotransferases

McPhalen, C.A., Vincent, M.G., and Jansonius, J.N., 1992. X-ray structure refinement and comparison of three forms of mitochondrial aspartate aminotransferase. *J. Mol. Biol.* 225:495–517.

McPhalen, C.A., Vincent, M.G., Picot, D., Jansonius, J.N., Lesk, A.M., and Chothia, C., 1992. Domain closure in mitochondrial aspartate aminotransferases. *J. Mol. Biol.* 227:197–213.

Reaction mechanisms

Walsh, C., 1979. *Enzymatic Reaction Mechanisms.* W.H. Freeman. [Contains excellent accounts of the catalytic mechanisms of pyridoxal phosphate enzymes, cobalamin enzymes, and oxygenases.]

Christen, P., and Metzler, D.E., 1985. *Transaminases.* Wiley.

Dakshinamurti, K. (ed.), 1990. *Vitamin B_6.* Annals of the New York Academy of Sciences. Vol. 585.

Nozaki, M., Yamamoto, S., Ishimura, Y., Coon, M.J., Ernster, L., and Estabrook, R.W. (eds.), 1982. *Oxygenases and Oxygen Metabolism. A Symposium in Honor of Osamu Hayaishi.* Academic Press.

Kirsch, J.F., Eichele, G., Ford, G.C., Vincent, M.G., Jansonius, J.N., Gehring, H., and Christen, P., 1984. Mechanism of action of aspartate amino transferase proposed on the basis of its spatial structure. *J. Mol. Biol.* 174:497–525.

Snell, E.E., and DiMari, S.J., 1970. Schiff base intermediates in enzyme catalysis. *In* Boyer, P.D. (ed.), *The Enzymes* (3rd ed.), vol. 2, pp. 335–370. Academic Press.

Barker, H.A., 1972. Coenzyme B_{12}-dependent mutases causing carbon chain rearrangements. *In* Boyer, P.D. (ed.), *The Enzymes* (3rd ed.), vol. 6, pp. 509–537. Academic Press.

Platica, O., Janeczko, R., Quadros, E.V., Regec, A., Romain, R., and Rothenberg, S.P., 1991. The cDNA sequence and the deduced amino acid sequence of human transcobalamin II show homology with rat intrinsic factor and human transcobalamin I. *J. Biol. Chem.* 266:7860–7863.

Genetic diseases

Scriver, C.R., Beaudet, A.L., Sly, W.S., and Valle, D. (eds.), 1989. *The Metabolic Basis of Inherited Disease* (6th ed.), McGraw-Hill. [Part 4, pp. 493-771, entitled "Amino Acids," contains fine articles on phenylketonuria, urea cycle disorders, and other genetic diseases of amino acid metabolism.]

Nyhan, W.L. (ed.), 1984. *Abnormalities in Amino Acid Metabolism in Clinical Medicine.* Appleton-Century-Crofts.

Historical aspects and the process of discovery

Cooper, A.J.L., and Meister, A., 1989. An appreciation of Professor Alexander E. Braunstein. The discovery and scope of enzymatic transamination. *Biochimie* 71:387–404.

Garrod, A.E., 1909. *Inborn Errors in Metabolism.* Oxford University Press (reprinted in 1963 with a supplement by H. Harris).

Childs, B., 1970. Sir Archibald Garrod's conception of chemical individuality: A modern appreciation. *New Engl. J. Med.* 282:71–78.

Holmes, F.L., 1980. Hans Krebs and the discovery of the ornithine cycle. *Fed. Proc.* 39:216–225.

PROBLEMS

1. *Keto counterparts.* Name the α-keto acid that is formed by transamination of each of the following amino acids:
   (a) Alanine. (d) Leucine.
   (b) Aspartate. (e) Phenylalanine.
   (c) Glutamate. (f) Tyrosine.

2. *A versatile building block.*
   (a) Write a balanced equation for the conversion of aspartate into glucose by way of oxaloacetate. Which coenzymes participate in this transformation?
   (b) Write a balanced equation for the conversion of aspartate into oxaloacetate by way of fumarate.

3. *Migratory route.* Consider the mechanism of conversion of L-methylmalonyl CoA into succinyl CoA by L-methylmalonyl CoA mutase.
   (a) Design an experiment to distinguish between the migration of the COO^- group and that of the –CO–S–CoA group in this reaction.
   (b) What is the significance of the finding that no tritium is incorporated into succinyl CoA when the mutase reaction is carried out in tritiated water?

4. *Effective electron sinks.* Pyridoxal phosphate stabilizes carbanionic intermediates by serving as an electron sink. Which other prosthetic group catalyzes reactions in this way?

5. *Helping hand.* Propose a role for the positively charged guanidinium nitrogen in the cleavage of argininosuccinate into arginine and fumarate.

6. *Telltale tag.* Methylmalonyl mutase is incubated with deuterated methylmalonyl CoA. Coenzyme B_{12} extracted from mutase in this reaction mixture is found to contain deuterium in its 5′-methylene group. Account for the transfer of label from substrate to coenzyme.

7. *Divergent pathways.* Heterolytic cleavage of a C–H bond can yield two types of products. What are they?

8. *Completing the cycle.* Four ~P are consumed in synthesizing urea according to the stoichiometry given on page 636. In this reaction, aspartate is converted into fumarate. Suppose that fumarate is converted back to aspartate. What is the resulting stoichiometry of urea synthesis? How many ~P are spent?

9. *A precise diagnosis.* The urine of an infant gives a positive reaction with 2,4-dinitrophenylhydrazine. Mass spectrometry then showed abnormally high blood levels of pyruvate, α-ketoglutarate, and the α-keto acids of valine, isoleucine, and leucine. Identify a likely molecular defect and propose a definitive test of your diagnosis.

10. *Therapeutic design.* How would you treat an infant who is deficient in argininosuccinate synthetase? Which molecules would carry nitrogen out of the body?

11. *Sweet hazard.* Why should phenylketonurics avoid using aspartame, an artificial sweetener? [Hint: Aspartame is L-aspartyl-L-phenylalanine methyl ester.]

12. *Déjà vu.* *N*-acetylglutamate is required as a cofactor in the synthesis of carbamoyl phosphate. How is it synthesized from glutamate?

CHAPTER

26

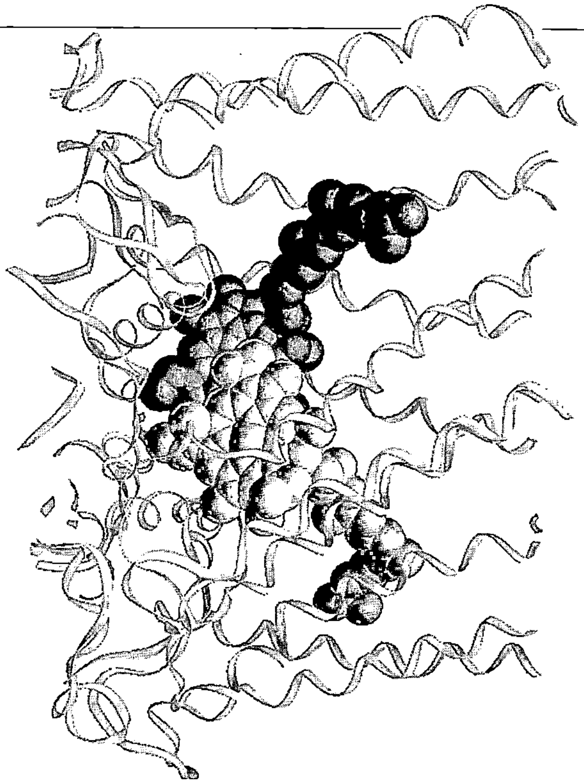

Photosynthesis

All free energy consumed by biological systems arises from solar energy that is trapped by the process of photosynthesis. The basic equation of photosynthesis is simple—indeed deceptively so:

$$H_2O + CO_2 \xrightarrow{\text{Light}} (CH_2O) + O_2$$

In this equation, (CH_2O) represents carbohydrate, primarily sucrose and starch. The mechanism of photosynthesis is complex and requires the interplay of many proteins and small molecules. Photosynthesis in green plants takes place in *chloroplasts.* The energy conversion apparatus is an integral part of the *thylakoid membrane system* of these organelles (Figure 26-1). The first step in photosynthesis is the absorption of light by *chlorophyll,* a porphyrin with a coordinated magnesium ion. The resulting electronic excitation passes from one chlorophyll molecule to another in a light-harvesting complex until the excitation is trapped by a chlorophyll pair with special properties. At such a *reaction center,* the energy of the excited electron is converted into a separation of charge. In essence, *light is used to create reducing potential.*

Photosynthesis in green plants is mediated by two kinds of light reactions. *Photosystem I* generates reducing power in the form of NADPH. *Photosystem II* transfers the electrons of water to a quinone and concomitantly evolves O_2. Electron flow within each photosystem and between them generates a transmembrane proton gradient that drives the synthe-

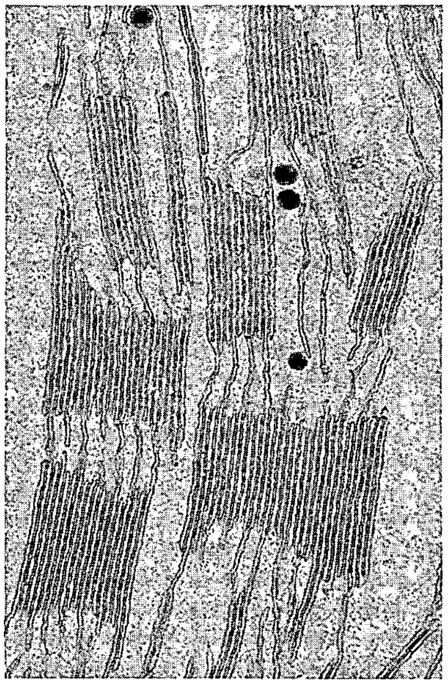

Figure 26-1
Electron micrograph of part of a chloroplast from a spinach leaf. The thylakoid membranes pack together to form grana. [Courtesy of Dr. Kenneth Miller.]

Opening Image: *The heart of a bacterial photosynthetic reaction center. The excitation of this pair of bacteriochlorophyll molecules leads to the separation of charge, the primary energy-harvesting step in photosynthesis. One member of this pair is shown in green, the other in red, and a small portion of the reaction center in gray. [Drawn from 1prc.pdb. J. Deisenhofer and H. Michel. Science 245(1989):1463.]*

sis of ATP, as in oxidative phosphorylation. Indeed, photosynthesis closely resembles oxidative phosphorylation. The principal difference between these energy transduction processes is the source of high-potential electrons. In oxidative phosphorylation, they come from the oxidation of fuels; in photosynthesis, they are produced by photoexcitation of chlorophyll. NADPH and ATP formed by the action of light then reduce CO_2 and convert it into *3-phosphoglycerate* by a series of dark reactions called the *Calvin cycle,* which occur in the stroma of chloroplasts. Hexoses are formed from 3-phosphoglycerate by the gluconeogenic pathway.

THE PRIMARY EVENTS OF PHOTOSYNTHESIS OCCUR IN THYLAKOID MEMBRANES

Chloroplasts, the organelles of photosynthesis, are typically 5 μm long. Like a mitochondrion, a chloroplast has an outer membrane and an inner membrane, with an intervening intermembrane space (Figure 26-2). The inner membrane surrounds a *stroma* containing soluble enzymes and membranous structures called *thylakoids,* which are flattened sacs. A pile of these sacs is called a *granum.* Different grana are connected by regions of thylakoid membrane called *stroma lamellae.* The thylakoid membranes separate the thylakoid space from the stroma space. Thus, chloroplasts have three different membranes (*outer, inner,* and *thylakoid membranes*) and three separate spaces (*intermembrane, stroma,* and *thylakoid spaces*). In developing chloroplasts, thylakoids arise from invaginations of the inner membrane, and so they are analogous to mitochondrial cristae.

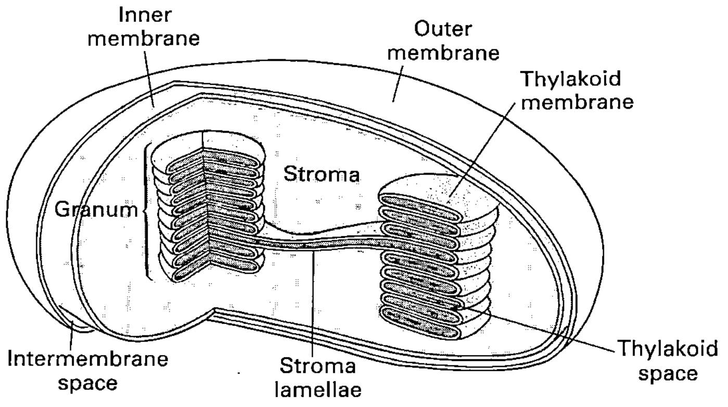

Figure 26-2
Diagram of a chloroplast. [After S.L. Wolfe, *Biology of the Cell,* p. 130. © 1972 by Wadsworth Publishing Company, Inc. Adapted by permission of the publisher.]

The thylakoid membranes contain the energy-transducing machinery: the light-harvesting proteins, reaction centers, electron transport chains, and ATP synthase. They have nearly equal amounts of lipids and proteins. The lipid composition is highly distinctive: about 40% of the total lipids are *galactolipids* and 4% are *sulfolipids,* whereas only 10% are phospholipids. The thylakoid membrane, like the inner mitochondrial membrane, is impermeable to most molecules and ions. The outer membrane of a chloroplast, like that of a mitochondrion, is highly permeable to small molecules and ions. The stroma contains the soluble enzymes that utilize the NADPH and ATP synthesized by the thylakoids to convert CO_2 into sugar. Chloroplasts contain their own DNA and the machinery for replicating and expressing it. However, chloroplasts (like mitochondria) are not autonomous: they also contain many proteins encoded by nuclear DNA.

DISCOVERY OF THE BASIC EQUATION OF PHOTOSYNTHESIS

Most of the basic equation of photosynthesis could have been written at the end of the eighteenth century. The production of oxygen in photosynthesis was discovered in 1780 by Joseph Priestley, an English chemist and Nonconformist minister. He found that plants could "restore air which has been injured by the burning of candles." He placed a sprig of mint in an inverted glass jar in a vessel of water and found several days later "that the air would neither extinguish a candle, nor was it all inconvenient to a mouse which I put into it" (Figure 26-3).

Figure 26-3
Priestley's classic experiment on photosynthesis. [After E.I. Rabinowitch. Photosynthesis. Copyright © 1948 by Scientific American, Inc. All rights reserved.]

The next major contribution to the elucidation of photosynthesis was made by Jan Ingenhousz, a Dutchman, who was court physician to the Austrian empress. Ingenhousz was a worldly man who liked to visit London. He once heard a discussion of Priestley's experiments on the restoration of air by plants and was so taken by it that he decided that he must do some experiments at the "earliest opportunity." This came six years later, when Ingenhousz rented a villa near London and spent a summer feverishly performing more than five hundred experiments. He discovered the role of light in photosynthesis:

> I observed that plants not only have the faculty to correct bad air in six or ten days, by growing in it, as the experiments of Dr. Priestley indicate, but that they perform this important office in a complete manner in a few hours; that this wonderful operation is by no means owing to the vegetation of the plant, but to the influence of light of the sun upon the plant.

Similar experiments were being carried out in Geneva by Jean Senebier, a Swiss pastor. His distinctive contribution was to show that "fixed air"—namely, CO_2—is taken up in photosynthesis. The role of water in photosynthesis was demonstrated by Theodore de Saussure, also a Genevan. He showed that the sum of the weights of organic matter produced by plants and of the oxygen evolved is much more than the weight of CO_2 consumed. From Lavoisier's law of the conservation of mass, de Saussure concluded that another substance was utilized. The only inputs in his system were CO_2, water, and light. Hence, de Saussure concluded that the other reactant must be water.

The final contribution to the basic equation of photosynthesis came nearly a half century later. Julius Robert Mayer, a German surgeon, discovered the law of conservation of energy in 1842. Mayer recognized that plants convert solar energy into chemical free energy:

> The plants take in one form of power, light; and produce another power, chemical difference.

The amount of energy stored by photosynthesis is enormous. More than 10^{17} kcal of free energy is stored annually by photosynthesis on earth, which corresponds to the assimilation of more than 10^{10} tons of carbon into carbohydrate and other forms of organic matter.

CHLOROPHYLLS TRAP SOLAR ENERGY

Mayer stated, "Nature has put itself the problem of how to catch in flight light streaming to the earth and to store the most elusive of all powers in rigid form." What is the mechanism of trapping this most elusive of all powers? The first step is the absorption of light by a photoreceptor molecule. The principal photoreceptor in the chloroplasts of most green plants is *chlorophyll* a, a substituted tetrapyrrole (Figure 26-4). The four

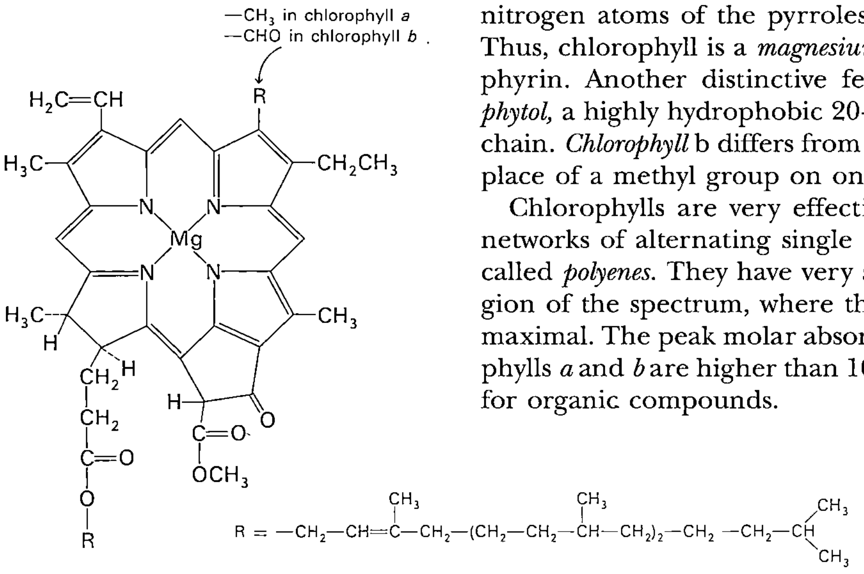

Figure 26-4
Formulas of chlorophylls *a* and *b*.

nitrogen atoms of the pyrroles are coordinated to a magnesium atom. Thus, chlorophyll is a *magnesium porphyrin,* whereas heme is an iron porphyrin. Another distinctive feature of chlorophyll is the presence of *phytol,* a highly hydrophobic 20-carbon alcohol, esterified to an acid side chain. *Chlorophyll* b differs from chlorophyll *a* in having a formyl group in place of a methyl group on one of its pyrroles.

Chlorophylls are very effective photoreceptors because they contain networks of alternating single and double bonds. Such compounds are called *polyenes.* They have very strong absorption bands in the visible region of the spectrum, where the solar output reaching the earth also is maximal. The peak molar absorption coefficients (ε, see p. 73) of chlorophylls *a* and *b* are higher than 10^5 cm^{-1} M^{-1}, among the highest observed for organic compounds.

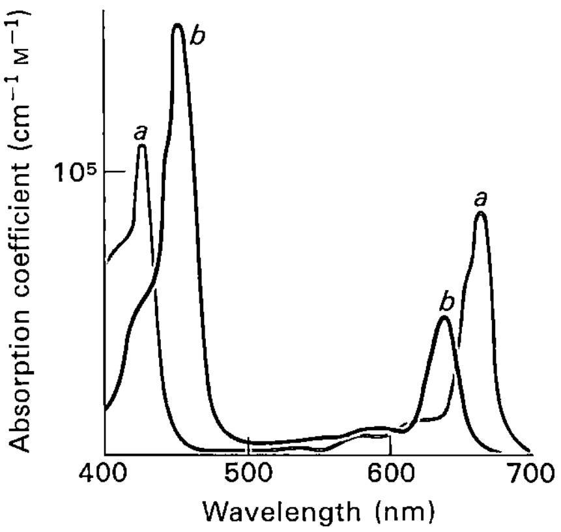

Figure 26-5
Absorption spectra of chlorophylls *a* and *b*.

The absorption spectra of chlorophylls *a* and *b* are different (Figure 26-5). Light that is not appreciably absorbed by chlorophyll *a*—at 460 nm, for example—is captured by chlorophyll *b,* which has intense absorption at that wavelength. Thus, *these two kinds of chlorophyll complement each other in absorbing the incident sunlight.* The spectral region from 500 to 600 nm is only weakly absorbed by these chlorophylls, but this does not pose a problem for most green plants. By contrast, light is often a limiting factor for cyanobacteria (blue-green algae) and red algae. They possess accessory light-harvesting pigments (p. 667) that enable them to trap light that is not absorbed strongly by the chlorophylls of photosynthetic organisms lying above them.

PHOTONS ABSORBED BY MANY CHLOROPHYLLS FUNNEL INTO A REACTION CENTER

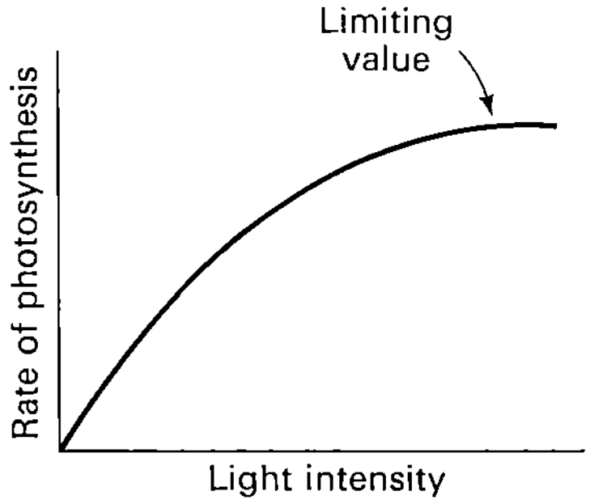

Figure 26-6
The rate of photosynthesis reaches a limiting value when the light intensity suffices to excite only a small fraction of the chlorophyll molecules.

Measurements of the dependence of the rate of photosynthesis on the intensity of illumination show that it increases linearly at low intensities and reaches a saturating value at high intensities (Figure 26-6). A saturating value is observed in strong light because chemical reactions utilizing the absorbed photons become rate-limiting. This experiment provided the first intimation that *photosynthesis can be separated into light reactions and dark reactions.* As will be discussed shortly, the light reactions generate NADPH and ATP, whereas the dark reactions use these energy-rich molecules to reduce CO_2.

In 1932, Robert Emerson and William Arnold measured the oxygen yield of photosynthesis when *Chlorella* cells (unicellular green algae) were exposed to light flashes lasting a few microseconds. They expected to find that the yield per flash would increase with the flash intensity until each chlorophyll molecule absorbed a photon, which would then be used in dark reactions. Their experimental observation was entirely unexpected: a saturating light flash led to the production of only one molecule of O_2 per 2500 chlorophyll molecules.

This experiment led to the concept of the *photosynthetic unit.* Hans Gaffron proposed that light is absorbed by hundreds of chlorophyll molecules, which then transfer their excitation energy to a site at which

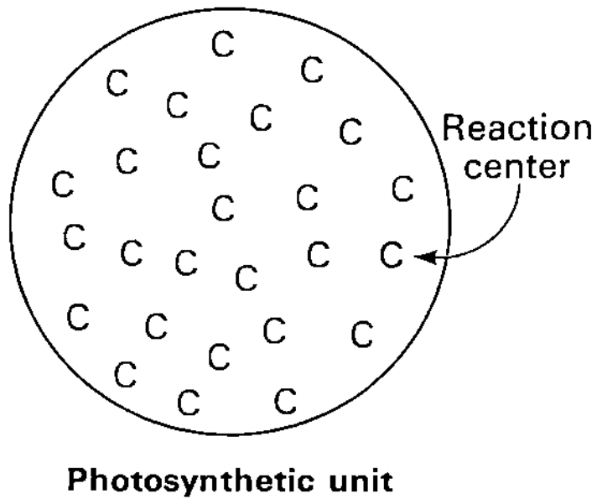

Figure 26-7
Diagram of a photosynthetic unit. The antenna chlorophyll molecules (denoted by green C's) transfer their excitation energy to a specialized chlorophyll at the reaction center (denoted by a red C).

chemical reactions occur (Figure 26-7). This site is called a *reaction center.* Thus, most chlorophyll molecules in the photosynthetic unit absorb light, but only a small portion of them, those at reaction centers, mediate the transformation of light into chemical energy. The energy level of chlorophylls at the reaction center is lower than that of other chlorophylls, which enables the reaction center to trap the excitation (Figure 26-8). The transfer of energy by direct electromagnetic interactions between chlorophylls and then to the reaction center is very rapid, occurring in picoseconds (10^{-12} s).

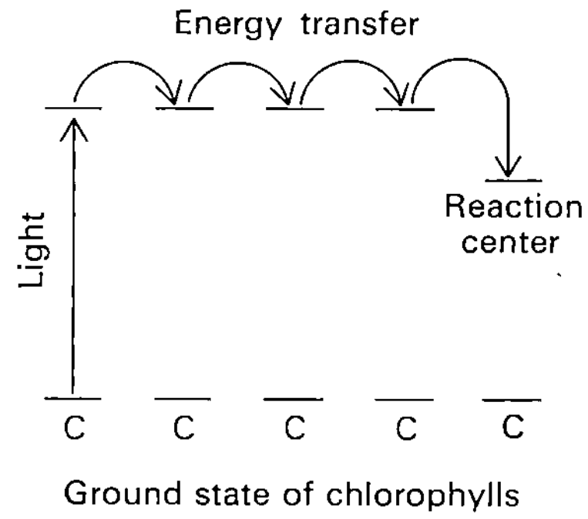

Figure 26-8
Diagram of the energy levels of the excited state of the antenna chlorophylls and of the reaction center.

O_2 EVOLVED IN PHOTOSYNTHESIS COMES FROM WATER

Let us turn now to the chemical changes in photosynthesis. The source of the oxygen evolved in green plants has important implications for the mechanism of photosynthesis. Comparative studies of photosynthesis in many organisms carried out as early as 1931 revealed the source. Some photosynthetic bacteria convert hydrogen sulfide into sulfur in the presence of light. Cornelis Van Niel perceived a common pattern in the overall reactions of photosynthesis carried out by green plants and by green sulfur bacteria.

$$CO_2 + 2\,H_2O \xrightarrow{\text{Light}} (CH_2O) + O_2 + H_2O$$

$$CO_2 + 2H_2S \xrightarrow{\text{Light}} (CH_2O) + 2S + H_2O$$

The sulfur that is formed by the photosynthetic bacteria is analogous to the oxygen that is evolved in plants. Van Niel proposed a general formula for photosynthesis.

$$\underset{\substack{\text{Hydrogen}\\\text{acceptor}}}{CO_2} + \underset{\substack{\text{Hydrogen}\\\text{donor}}}{2\,H_2A} \xrightarrow{\text{Light}} \underset{\substack{\text{Reduced}\\\text{acceptor}}}{(CH_2O)} + \underset{\substack{\text{Dehydrogenated}\\\text{donor}}}{2\,A} + H_2O$$

The hydrogen donor H_2A is H_2O in green plants and H_2S in the photosynthetic sulfur bacteria. Thus, photosynthesis in plants could be formulated as a reaction in which CO_2 is reduced by hydrogen derived from water. Oxygen evolution would then be the necessary consequence of this dehydrogenation process. The essence of this view of photosynthesis is that *water is split by light.*

The availability in 1941 of a heavy isotope of oxygen, ^{18}O, made it feasible to test this concept directly. In fact, ^{18}O appeared in the evolved oxygen when photosynthesis was carried out in water enriched in this isotope. This result confirmed the proposal that the O_2 formed in photosynthesis comes from water.

$$H_2{}^{18}O + CO_2 \xrightarrow{\text{Light}} (CH_2O) + {}^{18}O_2$$

HILL REACTION: ILLUMINATED CHLOROPLASTS EVOLVE O_2 AND REDUCE ELECTRON ACCEPTORS

In 1939, Robert Hill discovered that isolated chloroplasts evolve oxygen when they are illuminated in the presence of a suitable electron acceptor, such as ferricyanide. There is a concomitant reduction of ferricyanide to ferrocyanide. The Hill reaction is a landmark in the elucidation of the mechanism of photosynthesis for several reasons:

$$2\,H_2O + 4\,Fe^{3+} \xrightarrow{\text{Illuminated chloroplasts}} O_2 + 4\,H^+ + 4\,Fe^{2+}$$

1. It dissected photosynthesis by showing that oxygen evolution can occur without the reduction of CO_2. Artificial electron acceptors such as ferricyanide can substitute for CO_2.

2. It confirmed that the evolved oxygen comes from water rather than from CO_2, because no CO_2 was present.

3. It showed that isolated chloroplasts can perform a significant partial reaction of photosynthesis.

4. It revealed that a primary event in photosynthesis is the *light-driven transfer of an electron from one substance to another in the thermodynamically uphill direction.* The reduction of ferric to ferrous ion by light is a conversion of light into chemical energy.

TWO LIGHT REACTIONS INTERACT IN PHOTOSYNTHESIS

Investigations of the dependence of the rate of photosynthesis on the wavelength of incident light led to the discovery that chloroplasts contain two different photosystems. The photosynthetic rate (i.e., the rate of O_2 evolution) divided by the number of quanta absorbed gives the relative quantum efficiency of the process. For a single kind of photoreceptor, the quantum efficiency is expected to be independent of wavelength over its entire absorption band. This is not the case in photosynthesis: the quantum efficiency of photosynthesis drops sharply at wavelengths longer than 680 nm, although chlorophyll still absorbs light in the range from 680 to 700 nm (Figure 26-9). However, the rate of photosynthesis using long-wavelength light can be enhanced by adding light of a shorter wavelength, such as 600 nm. The photosynthetic rate in the presence of both 600-nm and 700-nm light is greater than the sum of the rates when the two wavelengths are given separately. These observations, called the *red drop* and the *enhancement phenomenon,* led Emerson to propose that *photosynthesis requires the interaction of two light reactions: both of them can be driven by light of wavelength less than 680 nm, but only one of them by light of longer wavelength.*

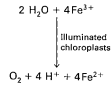

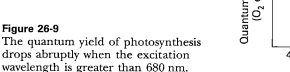

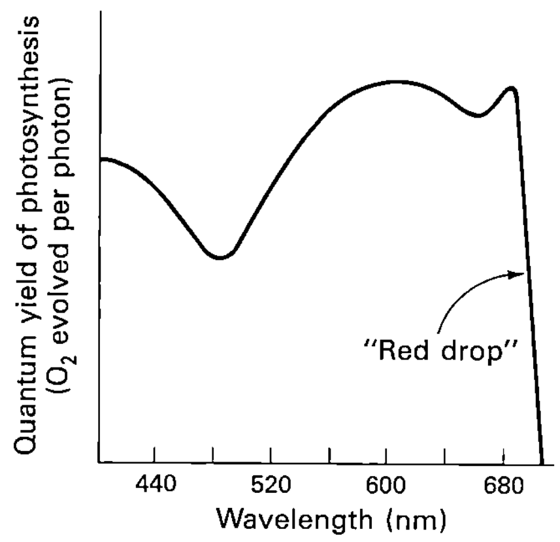

Figure 26-9
The quantum yield of photosynthesis drops abruptly when the excitation wavelength is greater than 680 nm.

PHOTOSYSTEMS I AND II HAVE COMPLEMENTARY ROLES

Indeed, photosynthesis by oxygen-evolving organisms depends on the interplay of two photosystems (Figure 26-10). Photosystem I, which can be excited by light of wavelength shorter than 700 nm, generates a strong reductant that leads to the formation of NADPH. Photosystem II, which requires light of wavelength shorter than 680 nm, produces a strong oxidant that leads to the formation of O_2. In addition, photosystem I produces a weak oxidant, whereas photosystem II produces a weak reductant. Electron flow from photosystem II to I, as well as electron flow within each photosystem, generates a transmembrane proton gradient that drives the formation of ATP. This part of photosynthesis, discovered by Daniel Arnon, is called *photosynthetic phosphorylation* or *photophosphorylation.*

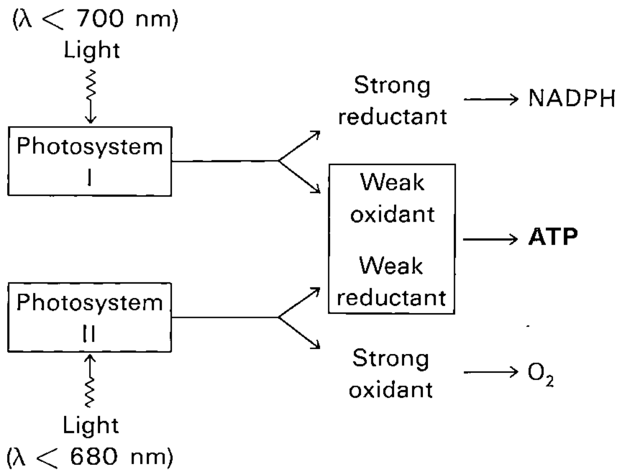

Figure 26-10
Interaction of photosystems I and II in photosynthesis by green plants.

PHOTOSYSTEM II TRANSFERS ELECTRONS FROM WATER TO PLASTOQUINONE AND GENERATES A PROTON GRADIENT

Photosystem II, a transmembrane assembly of some 10 polypeptide chains (>600 kd), catalyzes the light-driven transfer of electrons from water to *plastoquinone.* This electron acceptor closely resembles ubiquinone, a component of the electron transport chain of mitochondria (p. 536). Plastoquinone cycles between an oxidized form (Q) and a reduced form (QH_2, plastoquinol). The intermediate in this two-electron reduction is the free-radical semiquinone anion ($Q\cdot^-$). The net reaction catalyzed by photosystem II is

$$2\,Q + 2\,H_2O \xrightarrow{\text{Light}} O_2 + 2\,QH_2$$

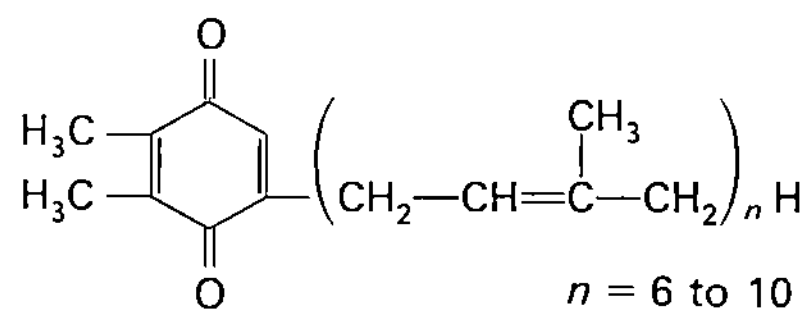

Plastoquinone (oxidized form, Q)

The electrons in QH_2 are at a higher potential than those in H_2O. Recall that in oxidative phosphorylation electrons flow from ubiquinol to O_2 rather than in the reverse direction (p. 534). Photosystem II drives the reaction in the thermodynamically uphill direction by utilizing the free energy of light. Our understanding of this membrane assembly of green plants has been greatly enriched by structural and mechanistic studies of a homologous photosystem of purple bacteria (p. 669).

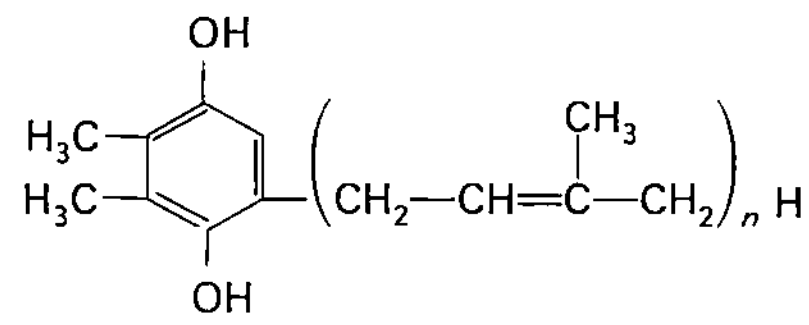

Plastoquinol (reduced form, QH_2)

Photosystem II consists of *a light-harvesting complex (LHC-II), a core with a reaction center, and an oxygen-evolving unit.* The 26-kd subunit of LHC-II, the most abundant membrane protein of chloroplasts, contains seven

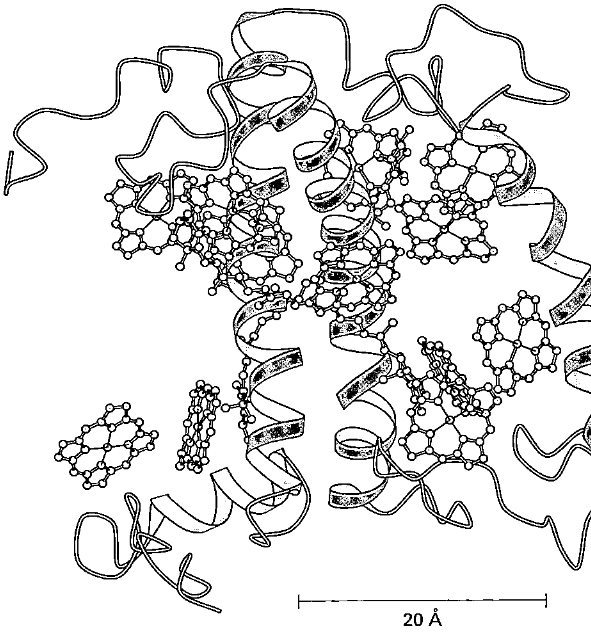

Figure 26-11
Schematic diagram of a subunit of light-harvesting complex II (LHC-II), the major antenna of photosystem II. Transmembrane helices are shown in red, chlorophyll *a* in green, chlorophyll *b* in blue, and carotenoids in yellow. The phytol tails of the chlorophylls are not shown. This structure was determined by electron crystallography. [Drawn from coordinates kindly provided by Dr. Werner Kühlbrandt. W. Kühlbrandt, D.-N. Wang, and Y. Fujiyoshi. *Nature* 367(1994):614.]

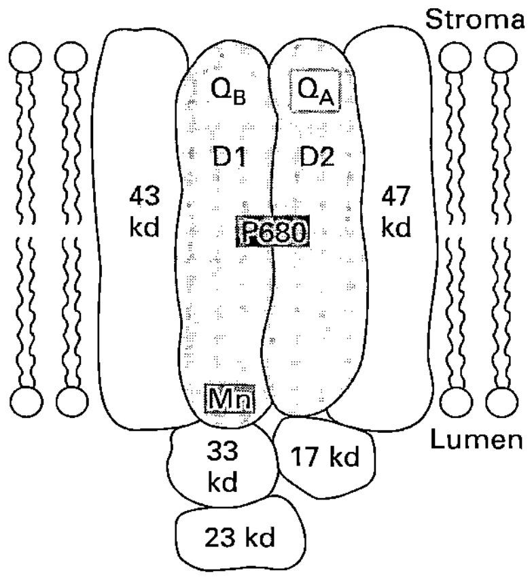

Figure 26-12
Schematic diagram of the core of photosystem II and the oxygen-evolving unit. The similar D1 and D2 subunits form the dimeric reaction center. Excitation of P680, a pair of chlorophyll molecules, induces a series of electron transfer reactions culminating in the reduction of a quinone bound to Q_B. The electrons donated to Q_B come from H_2O. The manganese cluster (labeled Mn) is the site of oxidation of H_2O to O_2. The 33-, 17-, and 23-kd proteins are associated with the Mn cluster. The 43- and 47-kd proteins of the core, as well as light-harvesting complex II, contain antenna chlorophylls that funnel energy into P680. [After G.W. Brudvig, W.F. Beck, and J.C. de Paula. *Ann. Rev. Biophys. Biophys. Chem.* 18(1989):25.]

chlorophyll *a* molecules, six chlorophyll *b* molecules, and two carotenoids (Figure 26-11). LHC-II, the major antenna of the system, is designed for highly efficient energy transfer from *b* to *a*. The energy from all chlorophylls in LHC-II funnels into the reaction center.

The core of photosystem II is formed by D1 and D2, a pair of similar 32-kd subunits that span the thylakoid membrane (Figure 26-12). *D1 and D2 contain the reaction center and the electron transfer chain.* Several additional polypeptides are required for oxygen evolution. Electronic excitation energy is funneled from antenna chlorophylls to a reaction-center chlorophyll pair called *P680* (*P* stands for the reaction-center *p*igment and *680* for the wavelength, in nm, of maximal absorption). The excited state of this reaction center, P680*, is a much stronger reductant than the ground state. Within picoseconds of excitation, an electron is transferred from P680* to bound *pheophytin (Ph),* a porphyrin identical with chlorophyll *a* except that it lacks Mg. The reaction center becomes a cation radical, $P680^+$. Much of the energy of the absorbed photon is conserved in this separation of charge.

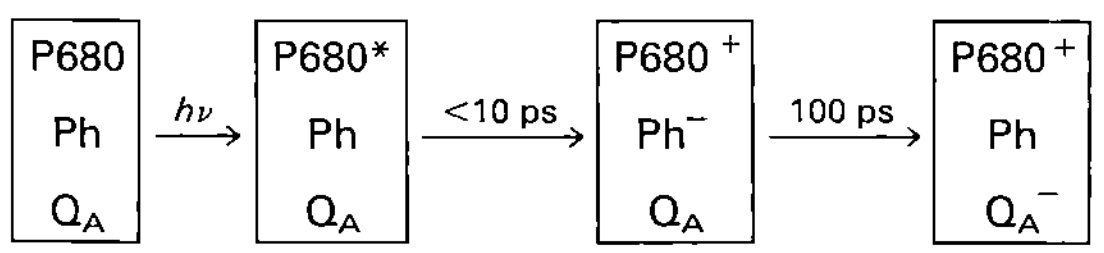

The electron then goes from reduced pheophytin to a permanently bound plastoquinone at the Q_A site on the D2 subunit (Figure 26-13). A mobile Q from the membrane pool binds to the Q_B site on the D1 subunit. The Q_A plastoquinone accepts an electron from reduced pheophytin and donates it to the Q_B plastoquinone to generate $Q\cdot^-$. This par-

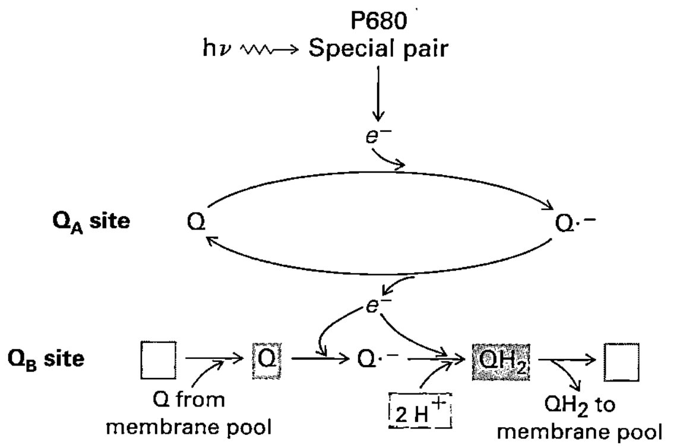

Figure 26-13
Interplay of the Q_A and Q_B sites of photosystem II in the two-electron reduction of Q to QH_2.

tially reduced species stays bound at Q_B until it acquires a second electron from the Q_A site. QH_2 is then released from Q_B into the membrane pool. At this point, *the energy of two photons has been safely and efficiently stored in the reducing potential of QH_2.* The interplay of the Q_A and Q_B sites enables a two-electron reduction (Q to QH_2) to be efficiently carried out with one-electron inputs (from the reaction center, via pheophytin). As will be discussed shortly, QH_2 feeds its electrons into a proton-pumping electron transport chain that is linked to photosystem I.

MANGANESE IONS PLAY A KEY ROLE IN EXTRACTING ELECTRONS FROM WATER TO FORM O_2

The other half of the task of photosystem II is to extract electrons from water. The $P680^+$ cation formed by photosystem II in the primary charge-separation step is a very strong oxidant. Indeed, its affinity for electrons is

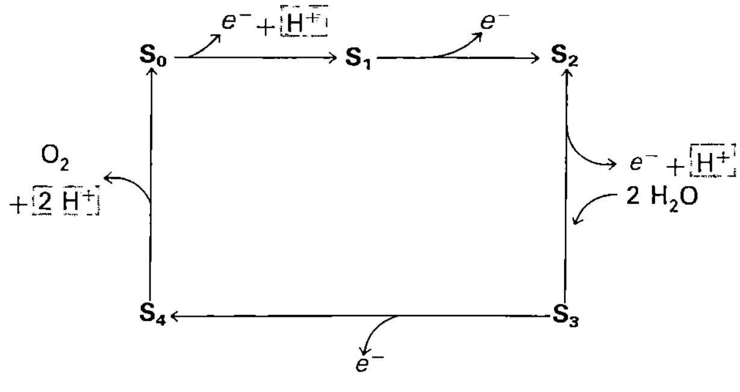

Figure 26-14
Charge-accumulator model for the splitting of water by the manganese center of photosystem II. The sequential withdrawal of four electrons by $P680^+$ drives the formation of O_2 from two molecules of H_2O. Four H^+ are released in each cycle.

even greater than that of O_2. *$P680^+$ extracts electrons from water, leading to the formation of O_2 and the return of the reaction center to the unexcited state.* This remarkable four-electron redox reaction is catalyzed by a cluster of four manganese ions. This Mn center cycles through a sequence of five oxidation states (Figure 26-14). The absorption of a photon by photosystem II leads to the removal of an electron from this cluster. Tyrosine 161 in the D1 protein, by forming a *tyrosyl radical,* serves as the intermediary in the transfer of an electron from the Mn center to $P680^+$. The loss of four electrons takes the Mn center from the S_0 to the S_4 state. The release of O_2 then brings the cluster back to the S_0 state. In essence, *the Mn center serves as a charge accumulator that enables O_2 to be formed without generating hazardous partly reduced intermediates.* Recall that a similar problem in oxidative phosphorylation is solved using the Fe^{2+}-Cu^+ center in cytochrome oxidase (p. 540).

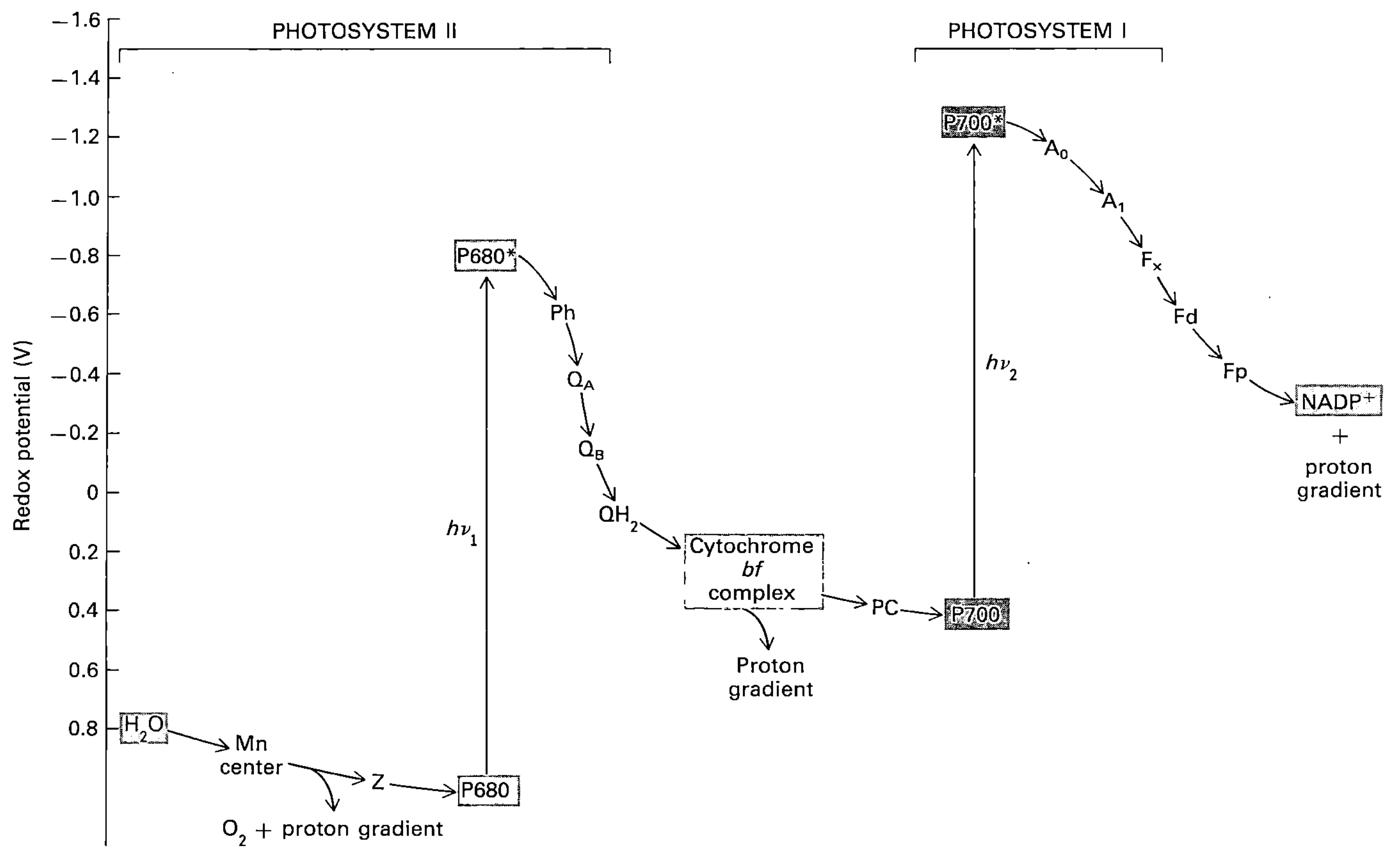

Figure 26-15
Pathway of electron flow from H_2O to $NADP^+$ in photosynthesis. This endergonic reaction is made possible by the absorption of light by photosystem II (P680) and photosystem I (P700). Reduced plastoquinone (QH_2) formed by photosystem II feeds electrons into the cytochrome *bf* complex. Reduced plastocyanin (PC) carries electrons to photosystem I, which generates reduced ferredoxin (Fd). This powerful reductant transfers its electrons to $NADP^+$ to form NADPH. A proton gradient across the thylakoid membrane (inside acidic) is formed when electrons flow through the cytochrome *bf* complex. The splitting of water and the reduction of $NADP^+$ on opposite sides of the thylakoid membrane also contribute to a proton gradient. Other abbreviations used: Z, a tyrosyl radical located between the Mn center and P680; Ph, pheophytin; Q_A and Q_B, plastoquinone-binding proteins; A_0 and A_1, acceptors of electrons from P700*; F_x, an iron-sulfur center; Fp, flavoprotein (ferredoxin-$NADP^+$ reductase). [After R.E. Blankenship and R.C. Prince. *Trends Biochem. Sci.* 10(1985):383.]

The energetics and pathway of electron flow from H_2O to QH_2 are most simply depicted in terms of redox potentials (Figure 26-15). QH_2 lies at a higher potential (0.1 V) than H_2O (0.82 V), which means that QH_2 is a stronger reductant (see p. 531 to review redox potentials). This uphill transfer of electrons is achieved using the energy of photons absorbed by photosystem II. A 680-nm photon has an energy of 1.82 electron volts (eV), which is more than sufficient to change the potential of an electron by 0.72 V (from 0.82 to 0.1 V) under standard conditions.

A PROTON GRADIENT IS FORMED AS ELECTRONS FLOW THROUGH CYTOCHROME *bf* FROM PHOTOSYSTEM II TO I

We turn now to the *cytochrome* bf *complex,* the next assembly mediating photosynthesis. Electrons flow through this complex from photosystem II to photosystem I. Cytochrome *bf* (also called *cytochrome* b_6f) catalyzes the transfer of electrons from plastoquinol (QH_2) to plastocyanin (PC) and concomitantly pumps protons across the thylakoid membrane.

$$QH_2 + 2\,PC(Cu^{2+}) \longrightarrow Q + 2\,PC(Cu^{+}) + 2\,H^{+}$$

The cytochrome *bf* complex contains four subunits: a 33-kd cytochrome *f*, a 23-kd cytochrome b_{563} with two hemes, a 20-kd Fe-S protein, and a 17-kd chain. This transmembrane complex closely resembles cytochrome reductase of mitochondria. Recall that cytochrome reductase pumps protons across the inner mitochondrial membrane in catalyzing the reduction of cytochrome *c*, a water-soluble protein, by ubiquinol (p. 537). The reaction mechanisms of these complexes also are similar. An Fe-S center participates directly in the reduction of plastocyanin, as it does in the reduction of cytochrome *c*. The cytochrome *b* component is again a recycling device that enables a two-electron carrier (plastoquinol) to interact with a one-electron carrier (Fe-S center). These electron transfers drive the pumping of protons from the stroma to the inner thylakoid space. Two protons are pumped per QH_2 oxidized.

Electrons flow from the Fe-S center of the cytochrome *bf* complex to ***plastocyanin***, an 11-kd water-soluble protein. The redox center of plastocyanin consists of a *copper ion* coordinated to the side chains of a cysteine, a methionine, and two histidine residues (Figure 26-16). These ligands distort the coordination geometry from the planar array characteristic of low-molecular-weight Cu^{2+} complexes. The localized strain at the Cu atom facilitates electron transfer as Cu alternates between +1 and +2 oxidation states.

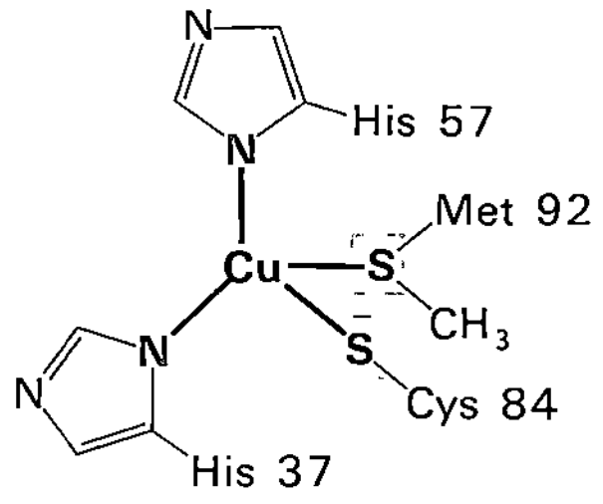

Figure 26-16
Structure of the coordinated copper ion in plastocyanin.

PHOTOSYSTEM I GENERATES NADPH BY FORMING REDUCED FERREDOXIN, A POWERFUL REDUCTANT

Plastocyanin is reduced and a proton gradient is generated by the actions of photosystem II and the cytochrome *bf* complex. The next stage of photosynthesis is mediated by ***photosystem I***, a transmembrane complex consisting of at least 13 polypeptide chains (>800 kd). The heart of photosystem I is a dimer of nearly identical psaA (83 kd) and psaB (82 kd) proteins (Figure 26-17). Light is funneled from antenna chlorophylls to *P700*, a pair of chlorophyll *a* molecules. As in photosystem II, the primary event at this reaction center is a light-induced separation of charge. An electron is transferred from $P700^*$, the excited state, to a monomeric acceptor chlorophyll called A_0 to form A_0^- and $P700^+$. A_0^- ($E_0' = -1.1$ V) is the most potent reductant in biological systems. Meanwhile, $P700^+$ captures an electron from reduced plastocyanin to return to P700, so that it can be excited again.

The very high potential electron of A_0^- is transferred to A_1, a quinone (vitamin K_1), and then to F_x, an iron-sulfur cluster. These three electron acceptors also are located in the psaA-psaB dimer core of photosystem I. The activated electron then hops to an iron-sulfur cluster in the adjacent 9-kd psaC protein. The final step is the reduction of *ferredoxin (Fd)*, a 12-kd water-soluble protein containing at least one iron-sulfur cluster. This reaction occurs on the stromal side of the thylakoid membrane. Thus, the net reaction catalyzed by photosystem I is

$$PC(Cu^+) + \text{ferredoxin}_{\text{oxidized}} \longrightarrow PC(Cu^{2+}) + \text{ferredoxin}_{\text{reduced}}$$

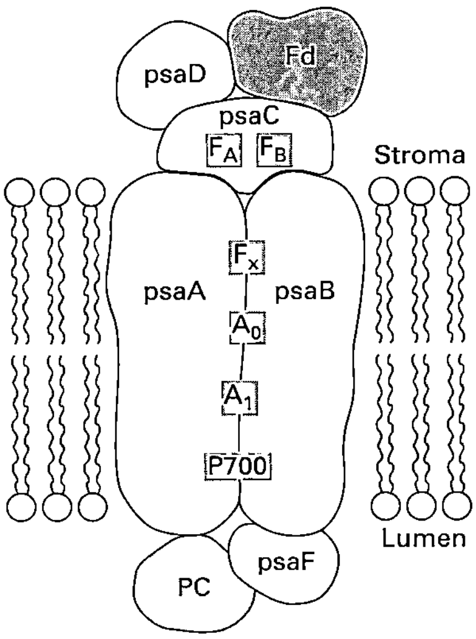

Figure 26-17
Schematic diagram of the core of photosystem I. The psaA and psaB proteins form the reaction center. Excited P700 transfers an electron to ferredoxin (Fd) via a series of redox sites shown in green. Reduced plastocyanin (PC) donates an electron to $P700^+$. The psaF and psaD proteins are needed for the binding of plastocyanin and ferredoxin to the lumenal and stromal side, respectively, of the reaction center.

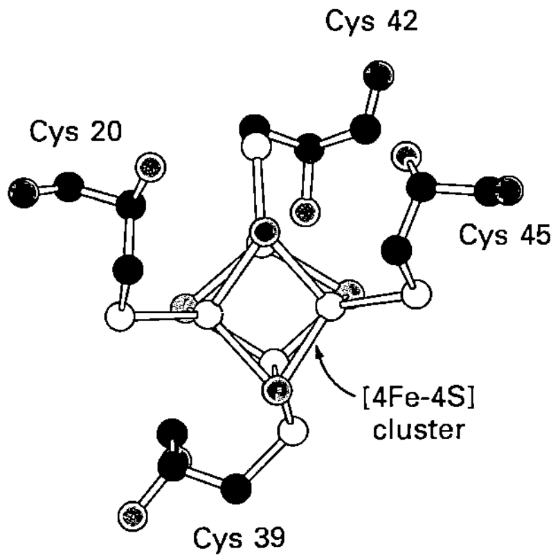

Figure 26-18
Structure of an iron-sulfur cluster in the oxidized form of ferredoxin, the electron acceptor in photosystem I. The protein contains a [3Fe-3S] cluster in addition to the [4Fe-4S] cluster depicted here. Iron atoms are shown in green, inorganic sulfurs in orange, and cysteine sulfurs in yellow. [Drawn from 5fdl.pdb. C.D. Stout. *J. Biol. Chem.* 268(1993):25920.]

The high-potential electrons of two molecules of ferredoxin are then transferred to $NADP^+$ to form NADPH. This reaction is catalyzed by *ferredoxin-NADP⁺ reductase,* a flavoprotein (Fp) with an FAD prosthetic group. The semiquinone form of the bound FAD is the intermediate in the convergence of two electrons from two molecules of reduced ferredoxin to one molecule of $NADP^+$. This reaction occurs on the stromal side of the membrane. Hence, the uptake of a proton in the reduction of $NADP^+$ further contributes to the generation of a proton gradient across the thylakoid membrane, with the inside acidic.

$$2\ \text{Ferredoxin}_{\text{reduced}} + H^+ + NADP^+ \xrightarrow{\text{Ferredoxin-NADP}^+\ \text{reductase}} 2\ \text{Ferredoxin}_{\text{oxidized}} + NADPH$$

The net reaction carried out by photosystem II, the cytochrome *bf* complex, and photosystem I is

$$2\,H_2O + 2\,NADP^+ \xrightarrow{\text{Light}} O_2 + 2\,NADPH + 2\,H^+$$

In essence, *light causes electrons to flow from H_2O to NADPH and leads to the generation of a proton-motive force* (see Figure 26-15). This pathway is called the *Z scheme of photosynthesis* because the redox diagram from P680 to P700* looks like a Z.

CYCLIC ELECTRON FLOW THROUGH PHOTOSYSTEM I LEADS TO THE PRODUCTION OF ATP INSTEAD OF NADPH

An alternative pathway for electrons arising from P700, the reaction center of photosystem I, contributes to the versatility of photosynthesis. The high-potential electron in ferredoxin can be transferred to the cytochrome *bf* complex rather than to $NADP^+$. This electron then flows back to the oxidized form of P700 through plastocyanin. The net outcome of this cyclic flow of electrons is the pumping of protons by the cytochrome *bf* complex. The resulting proton gradient then drives the synthesis of ATP. In this process, called *cyclic photophosphorylation* (Figure 26-19), *ATP is generated without the concomitant formation of NADPH.* Photosystem II does not participate in cyclic photophosphorylation, and so O_2 is not formed from H_2O. Cyclic photophosphorylation takes place when $NADP^+$ is unavailable to accept electrons from reduced ferredoxin because of a very high ratio of NADPH to $NADP^+$.

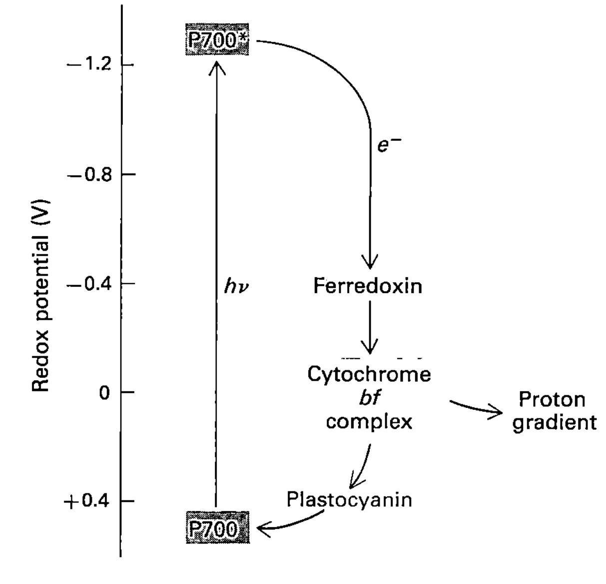

Figure 26-19
Electron flow in cyclic photophosphorylation. Electrons are transferred from P700*, the excited reaction center of photosystem I, to ferredoxin and then to the cytochrome *bf* complex. Protons are pumped by this complex as electrons return to the reaction center through plastocyanin.

ATP SYNTHESIS IS DRIVEN BY A PROTON GRADIENT ACROSS THE THYLAKOID MEMBRANE

In 1966, André Jagendorf showed that chloroplasts synthesize ATP in the dark when an artificial pH gradient is imposed across the thylakoid membrane. To create this transient pH gradient, chloroplasts first were soaked in a pH 4 buffer for several hours. These chloroplasts were then rapidly mixed with a pH 8 buffer containing ADP and P_i. The pH of the stroma suddenly increased to 8, whereas the pH of the thylakoid space remained at 4. *A burst of ATP synthesis then accompanied the disappearance of the pH gradient across the thylakoid membrane* (Figure 26-20). This incisive experiment was one of the first to unequivocally support Mitchell's hypothesis that ATP synthesis is driven by proton-motive force (p. 544).

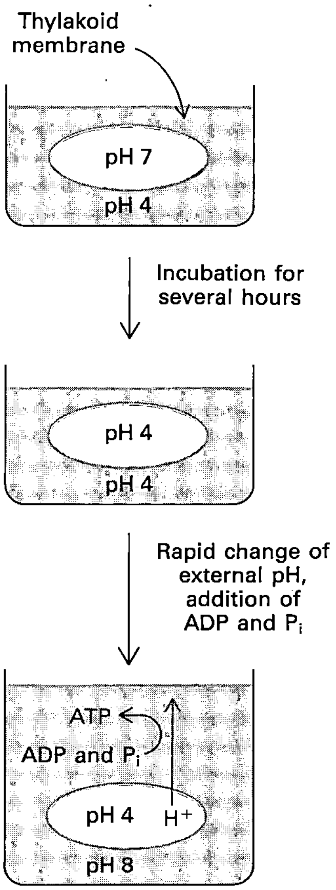

Figure 26-20
Synthesis of ATP by chloroplasts following the imposition of a pH gradient.

ATP SYNTHASE OF CHLOROPLASTS CLOSELY RESEMBLES THOSE OF BACTERIA AND MITOCHONDRIA

The mechanism of ATP synthesis in chloroplasts is very similar to that in mitochondria. *ATP formation is driven by a proton-motive force in both photophosphorylation and oxidative phosphorylation.* Furthermore, the enzyme assembly catalyzing ATP formation in chloroplasts is very similar to that of mitochondria and bacteria. The *ATP synthase* of chloroplasts, also called the *CF_1-CF_0 complex* (*C* stands for "chloroplast" and *F* for "factor"), closely resembles the F_1-F_0 complex in oxidative phosphorylation (p. 546). CF_0, which consists of four kinds of subunits, conducts protons across the thylakoid membrane. CF_1, like F_1, catalyzes the formation of ATP from ADP and P_i. The knobs on the external surface of thylakoid membranes are the CF_1 units of ATP synthase (Figure 26-21).

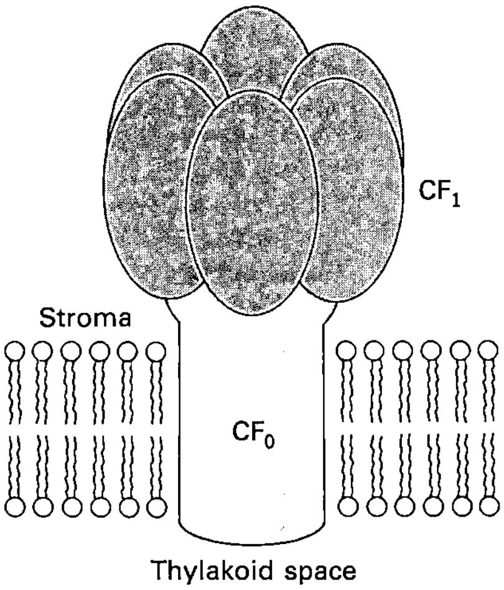

Figure 26-21
Schematic diagram of ATP synthase of chloroplasts. This assembly consists of a transmembrane CF_0 unit and a catalytic CF_1 unit on the stromal side of the thylakoid membrane.

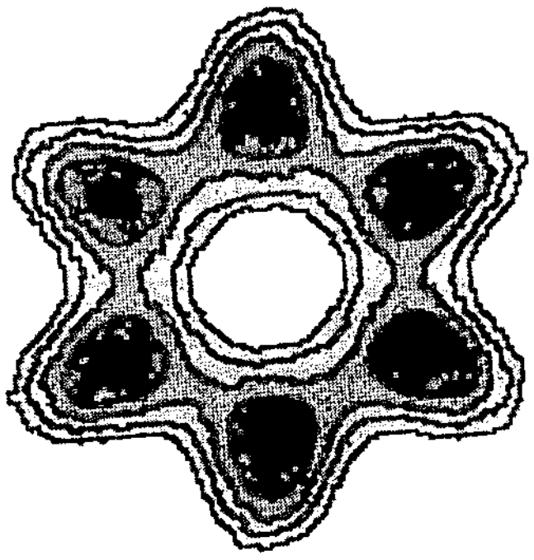

Figure 26-22
Two-dimensional image reconstruction of the CF_1 unit of chloroplast ATP synthase. This low-resolution image exhibits approximate sixfold symmetry, which implies that the three α and three β subunits are similar to each other. [From C.W. Ackey, R.H. Crepeau, S.D. Dunn, R.E. McCarty, and S.J. Edelstein. *EMBO J.* 2(1983):1412.]

CF_1 has the subunit composition $\alpha_3\beta_3\gamma\delta\varepsilon$. The α and β subunits contain the binding sites and catalytic sites for ATP and ADP. The δ subunit binds CF_1 to CF_0, and the γ subunit controls proton flow. The ε subunit inhibits the catalytic activity of the complex in the dark to block the wasteful hydrolysis of ATP.

Electron transfer through the asymmetrically oriented photosystems I and II and the cytochrome *bf* complex produces a large proton gradient across the thylakoid membrane (Figure 26-23). *The thylakoid space becomes markedly acidic, with the pH approaching 4. The light-induced transmembrane*

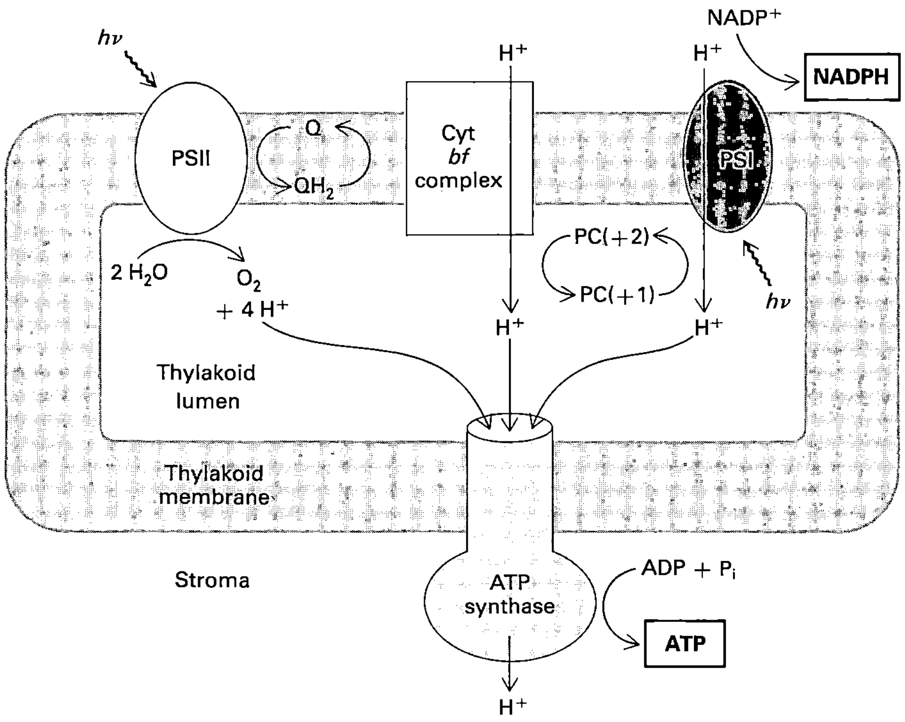

Figure 26-23
Vectorial arrangement of photosystems I and II, the cytochrome *bf* complex, and ATP synthase in the thylakoid membrane. Light-induced proton pumping makes the inner space acidic. The flow of protons through CF_0 to the stromal side leads to the synthesis of ATP by CF_1. NADPH is also formed on the stromal side. [After F.M. Harold, *The Vital Force: A Study of Bioenergetics* (W.H. Freeman, 1986), p. 271.]

proton gradient is about 3.5 pH units. As discussed previously (p. 545), the proton-motive force, Δp, consists of a pH-gradient contribution and a membrane-potential contribution. In chloroplasts, nearly all of Δp arises from the pH gradient, whereas in mitochondria the contribution from the membrane potential is larger. The reason for this difference is that the thylakoid membrane is quite permeable to Cl^- and Mg^{2+}. The light-induced transfer of H^+ into the thylakoid space is accompanied by the transfer of either Cl^- in the same direction or Mg^{2+} (1 per 2 H^+) in the opposite direction. Consequently, electrical neutrality is maintained and no membrane potential is generated. A pH gradient of 3.5 units across the thylakoid membrane corresponds to a proton-motive force of 0.2 V or a ΔG of -4.8 kcal/mol. *About three protons flow through the CF_1-CF_0 complex per ATP synthesized, which corresponds to a free-energy input of 14.4 kcal per mole of ATP.* No ATP is synthesized if the pH gradient is less than two units, because the driving force is then too small.

CF_1 is on the stromal surface of the thylakoid membrane, and so the newly synthesized ATP is released into the stromal space. Likewise, NADPH formed by photosystem I is released into the stromal space. Thus, *ATP and NADPH, the products of the light reactions of photosynthesis, are appropriately positioned for the subsequent dark reactions, in which CO_2 is converted into carbohydrate.*

PHOTOSYSTEM I AND ATP SYNTHASE ARE LOCATED IN UNSTACKED THYLAKOID MEMBRANES

Thylakoid membranes of most plants are differentiated into *stacked* (appressed) and *unstacked* (nonappressed) regions (see Figures 26-1 and 26-2). Stacking increases the amount of thylakoid membrane in a given chloroplast volume. Both regions surround a common internal thylakoid space, but only unstacked regions make direct contact with the chloroplast stroma. Stacked and unstacked regions differ in the nature of their photosynthetic assemblies (Figure 26-24). Photosystem I and ATP synthase are located almost exclusively in unstacked regions, whereas photosystem II is present mostly in stacked regions. The cytochrome *bf* complex is found in both regions. Indeed, this complex rapidly moves back and forth between the stacked and unstacked regions. Plastoquinone and plastocyanin are the mobile carriers of electrons between assemblies located in different regions of the thylakoid membrane. A common internal thylakoid space enables protons liberated by photosystem II in stacked membranes to be utilized by ATP synthase molecules that are located far away in unstacked membranes.

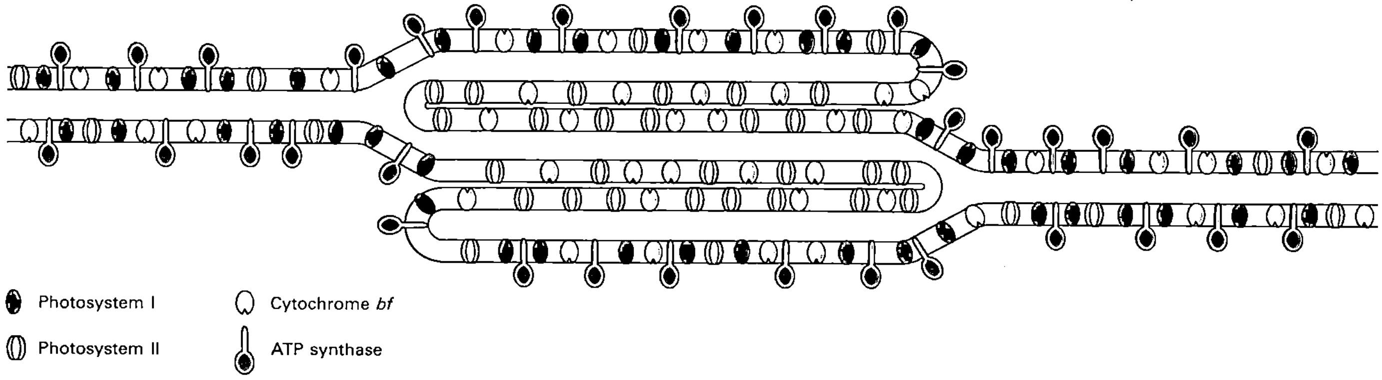

Figure 26-24
Photosynthetic assemblies are differentially distributed in the stacked (appressed) and unstacked (nonappressed) regions of thylakoid membranes. [After a drawing kindly provided by Dr. Jan M. Anderson and Dr. Bertil Andersson.]

What is the functional significance of this lateral differentiation of the thylakoid membrane system? The positioning of photosystem I in the unstacked membranes also gives it direct access to the stroma for the reduction of $NADP^+$. ATP synthase, too, is located in the unstacked region to provide space for its large CF_1 globule and to give access to ADP. In contrast, the tight quarters of the appressed region pose no problem for photosystem II, which interacts with a small polar electron donor (H_2O) and a highly lipid-soluble electron carrier (plastoquinone).

PHYCOBILISOMES SERVE AS MOLECULAR LIGHT PIPES IN CYANOBACTERIA AND RED ALGAE

Little blue or red light reaches algae living at a depth of a meter or more in seawater because such light is absorbed by water and by chlorophyll molecules in organisms lying above. Cyanobacteria (blue-green algae) and red algae contain large protein assemblies called *phycobilisomes* that enable them to harvest the green and yellow light that passes through. Phycobilisomes are bound to the outer face of thylakoid membranes, where they serve as light-absorbing antennas to funnel excitation energy into the reaction centers of photosystem II. They absorb maximally in the 470- to 650-nm region, in the valley between the blue and far-red absorption peaks of chlorophyll *a*. Phycobilisomes are very large assemblies (several million daltons) of many *phycobiliprotein* subunits, each containing many covalently attached *bilin* prosthetic groups, and linker polypeptides. Phycobilisomes contain hundreds of bilins. Phycocyanobilin and phycoerythrobilin are the two most common ones.

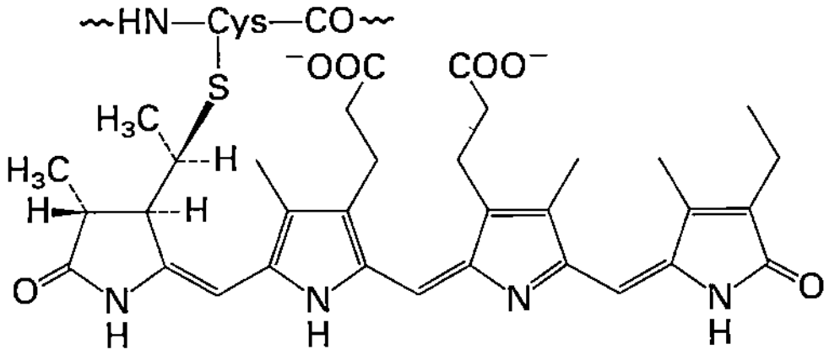

Peptide-linked phycocyanobilin

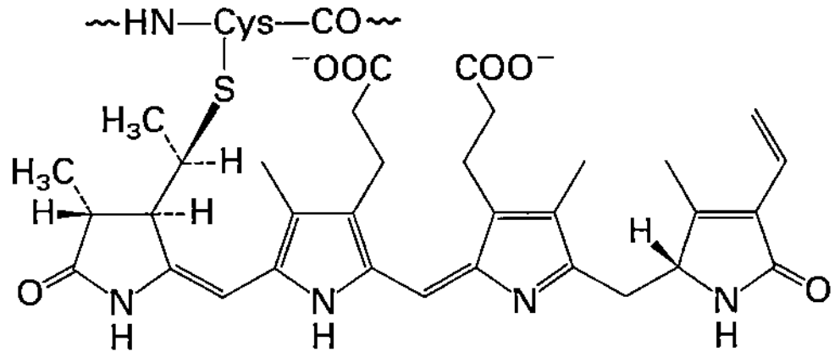

Peptide-linked phycoerythrobilin

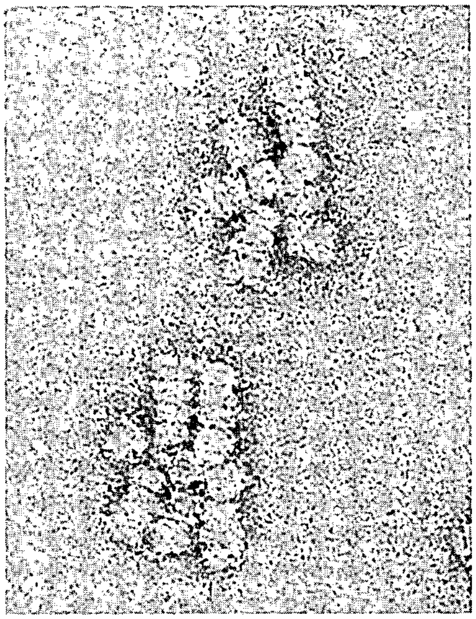

Figure 26-25
Electron micrograph of phycobilisomes from a cyanobacterium (*Synechocystis*). [Courtesy of Dr. Robley Williams and Dr. Alexander Glazer.]

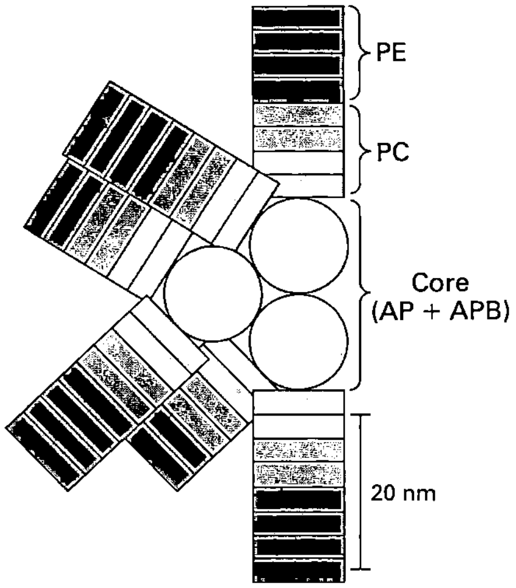

Figure 26-26
Schematic diagram of a phycobilisome from the cyanobacterium *Synechocystis* 6701. Rods containing phycoerythrin (PE) and phycocyanin (PC) emerge from a core made of allophycocyanin (AP) and allophycocyanin B (APB). The core region binds to the thylakoid membrane. [After a drawing kindly provided by Dr. Alexander Glazer.]

Phycobilisomes absorb light over a broad spectral region because they contain several phycobiliproteins with different spectral properties. Light energy collected by phycobiliproteins absorbing maximally at shorter wavelengths is transferred to phycobiliproteins absorbing maximally at longer wavelengths and then to the reaction-center chlorophyll of photosystem II. The energy is transferred by a direct electromagnetic interaction that requires overlap of the absorption spectrum of the energy acceptor and the emission spectrum of the energy donor. A suitably matched donor and acceptor as much as 70 Å apart can transfer energy efficiently. In the phycobilisomes of blue-green algae, for example, excitation energy is transferred from one phycobiliprotein to another in the following sequence:

$$\text{Phycoerythrin} \longrightarrow \text{phycocyanin} \longrightarrow \text{allophycocyanin} \longrightarrow \begin{matrix}\text{reaction}\\ \text{center}\end{matrix}$$

The geometrical arrangement of phycobiliproteins in phycobilisomes (Figure 26-26), as well as their spectral properties, contributes to the efficiency of energy transfer, which is greater than 95%. Excitation energy absorbed by phycoerythrin subunits at the periphery of these antennas appears at the reaction center in less than 100 ps. Phycobilisomes are elegantly designed light pipes that enable algae to occupy ecological niches that would not support organisms relying solely on chlorophyll for the trapping of light.

A BACTERIAL PHOTOSYNTHETIC REACTION CENTER HAS BEEN VISUALIZED AT ATOMIC RESOLUTION

Rhodopseudomonas viridis, a purple bacterium, contains a photosynthetic reaction center that is homologous to photosystem II of green plants. The x-ray crystallographic analysis of this bacterial photosynthetic assembly by Johann Deisenhofer, Hartmut Michel, and Robert Huber has been exceptionally rewarding. The elucidation of its structure was a tour de force because of both its size (more than 10,000 atoms) and membrane localization. The reaction center consists of four polypeptides: L (31 kd), M

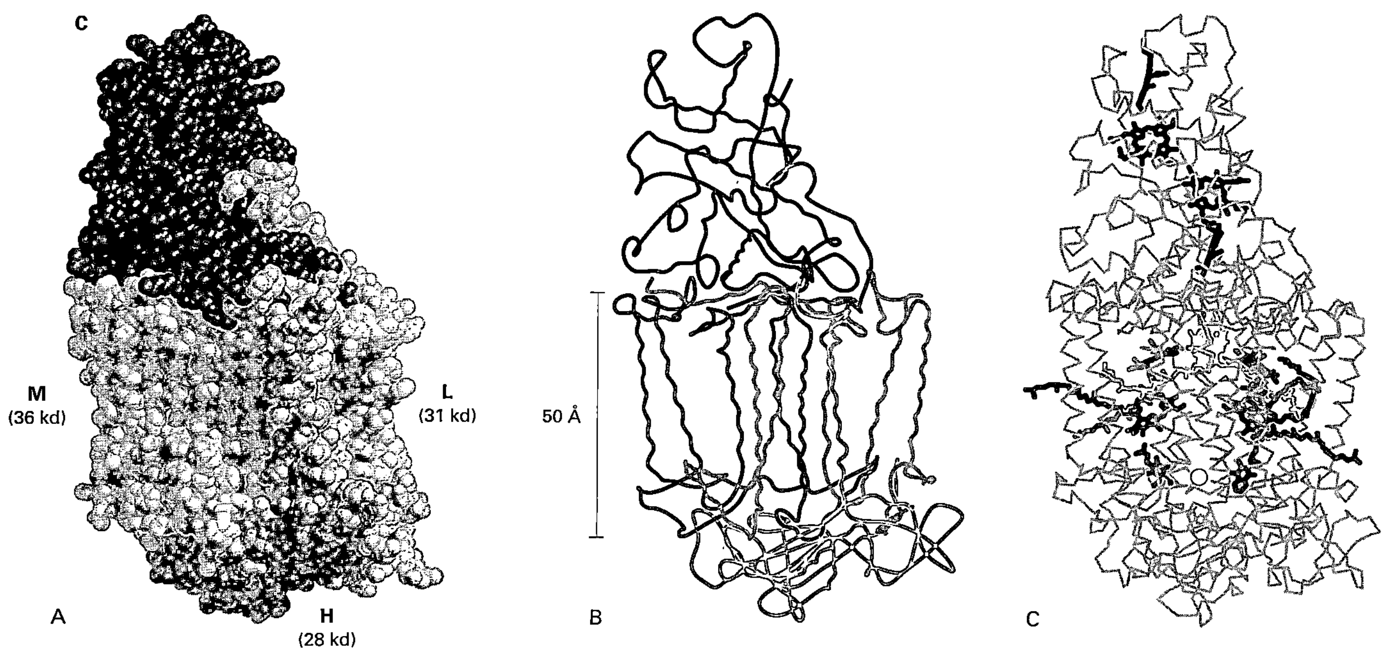

Figure 26-27
Structure of the photosynthetic reaction center of *Rhodopseudomonas viridis,* a purple bacterium. (A) Space-filling model. The reaction center consists of four subunits: C (red), L (yellow), M (blue), and H (green). Prosthetic groups are not shown. (B) Schematic diagram of the main-chain conformation. The L and M subunits each contain five transmembrane helices, and the H subunit contains one. (C) The electron transfer chain is depicted in color, and the polypeptide backbone in gray. The colors of the prosthetic groups are hemes (H), red; special-pair bacteriochlorophylls (BChl), green; the other two bacteriochlorophylls, orange; bacteriopheophytins (BP), blue; quinones (at sites Q_A and Q_B), purple; and iron (Fe), yellow. [Drawn from 1prc.pdb. J. Deisenhofer and H. Michel. *Science* 245(1989):1463.]

(36 kd), and H (28 kd) subunits and C, a *c*-type cytochrome (Figure 26-27). The cytochrome lies on the periplasmic side of the thylakoid membrane, and most of the H subunit on the opposite cytosolic side. The L and M subunits form the structural and functional core of the assembly. They are very similar, each containing five transmembrane helices, in contrast with the H subunit, which has just one. Most of the side chains of these helical segments are hydrophobic. The reaction center is 130 Å long; its core is 70 Å by 30 Å in cross section.

The entire electron transfer chain is evident in the atomic resolution map of the reaction center. Four bacteriochlorophyll *b* molecules (BChl-b), two bacteriopheophytin *b* molecules (BP), two quinones (Q_A and Q_B), and a ferrous ion are associated noncovalently with the L and M subunits (see Figure 26-27). Most striking, they are symmetrically positioned with respect to the twofold axis that relates the L and M subunits. Bacteriochlorophylls are similar to chlorophylls except for small modifications that shift their absorption maxima to the near infrared, to wavelengths as long as 1000 nm. In the reaction center of *R. viridis,* photons are captured by a dimer of BChl-b that absorbs maximally at 960 nm. This dimer is called the *special pair* because of its fundamental role in photosynthesis. Excitation of the special pair (P960) leads to the ejection of an electron, which is accepted by another BChl-b and then by pheophytin in the L subunit (Figure 26-28). Thus, the primary energy-conserving event is the separation of charge to form $P960^+$ and BP_L-. The activated electron is then transferred from pheophytin to Q_A, a quinone that is tightly

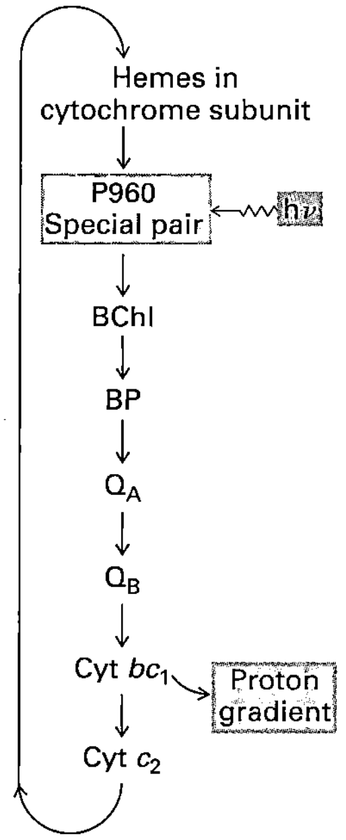

Figure 26-28
Light-induced electron flow in *R. viridis.* This cycle of electron transfer reactions generates a proton gradient and the consequent synthesis of ATP.

bound to the M subunit. Finally, the electron is transferred to Q_B, a mobile quinone that is transiently associated with the L subunit. Two photons need to be absorbed to drive the two-electron reduction of Q to QH_2. As in photosystem II, Q_B stays bound on receiving the first electron but is released into the hydrophobic core of the bilayer after accepting the second electron.

How does $P960^+$ regain an electron to return to the ground state? In photosystem II of green plants, the electron comes from water. Photosynthetic purple bacteria, however, do not contain a manganese center for the evolution of oxygen. Rather, the high-potential electrons of QH_2 return to the reaction center after flowing through the cytochrome bc_1 complex. Electron transfer through this assembly leads to the generation of a proton gradient, which is used to drive the synthesis of ATP. The electron is transferred from cytochrome bc_1 to cytochrome c_2, a water-soluble periplasmic protein, and then to the cytochrome subunit of the reaction center. Finally, the electron flows through the four covalently attached hemes of this subunit to return to $P960^+$ and restore the ground state of the special pair. Thus, R. viridis *carries out cyclic photophosphorylation. Its reaction center converts light into high-potential electrons, which are harvested by the proton-pumping cytochrome* bc_1 *complex.* Two protons are translocated to the periplasmic side of the membrane for each electron that is cycled. The proton-motive force drives the synthesis of NADH as well as that of ATP. The reduction of NAD^+ is carried out by NADH dehydrogenase, which operates in a direction opposite that in oxidative phosphorylation (p. 536).

MANY HERBICIDES INHIBIT PHOTOSYNTHESIS BY BLOCKING THE REDUCTION OF A QUINONE

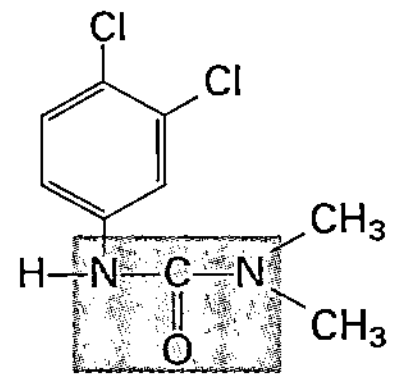

Diuron
(a urea derivative)

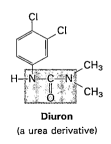

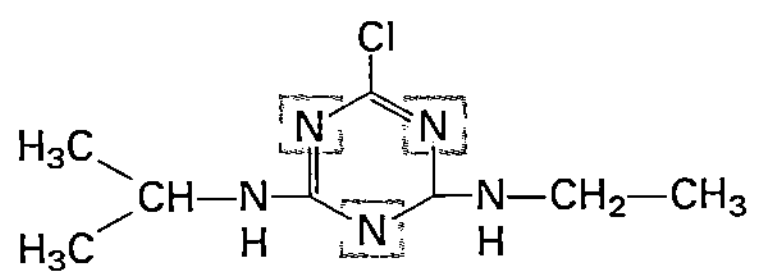

Atrazine
(a triazine derivative)

Many commercial herbicides kill weeds by interfering with the action of photosystem II. Urea derivatives such as *diuron* and triazine derivatives such as *atrazine* bind to the Q_B site of the D1 subunit of photosystem II and block the formation of plastoquinol (QH_2). Many substituted ureas and triazines also inhibit photosynthesis in purple bacteria by binding to the Q_B site in the L subunit of the reaction center. Indeed, *the similar action of triazine herbicides in green plants and purple bacteria is an expression of the evolutionary kinship of photosystem II and the bacterial reaction center.* The D1 and D2 subunits of photosystem II correspond to the L and M subunits, respectively, of the bacterial reaction center. The electron transfer chain from the special pair to Q_B is essentially the same in plants and purple bacteria.

RECURRING MOTIFS AND MECHANISMS IN PHOTOSYNTHETIC REACTION CENTERS

Studies of photosynthetic reaction centers of diverse prokaryotic and eukaryotic organisms have revealed a common set of principles:

1. *Chlorophyll (or bacteriochlorophyll) dimers serve as the special pair in all known reaction centers.* Why has nature chosen magnesium porphyrins rather than iron porphyrins to capture radiant energy? A likely reason is that excited iron porphyrins return to the ground state much more rapidly than do excited magnesium porphyrins. The substitution of magnesium for iron buys time: an electron can be transferred before the excitation is dissipated as heat.

2. *Separation of charge is the primary energy-conserving event in all known reaction centers.* The excited special pair donates an electron to a chlorophyll or pheophytin acceptor and becomes a radical cation.

3. *Reaction centers display approximate twofold symmetry.* The special pair is located at the interface of two similar subunits that serve as the structural and functional core of the reaction center.

4. *The initial electron transfer steps are extremely rapid.* Picosecond kinetics are essential for efficient photosynthesis because competing processes occur in times of nanoseconds. Very fast electron ejection by the special pair prevents the excitation from returning to the light-harvesting complex. Also, the ejected electron and the radical cation cannot recombine because the subsequent electron transfers also are extremely swift.

5. *Reaction centers contain multiple redox sites that are less than 15 Å apart.* The rate of electron transfer between two groups decreases exponentially with distance (Figure 26-29). In proteins, the rate of an energetically favorable electron transfer reaction decreases from about 10^{12} s^{-1} when the donor and acceptor are in van der Waals contact (at 3.6 Å) to 1 s^{-1} when their edges are 25 Å apart. Very rapid electron transfer is possible only if the electron donor and acceptor are in close proximity. The presence of relay stations between the special pair and the ultimate acceptor greatly accelerates electron transfer and thereby markedly increases the efficiency of photosynthesis.

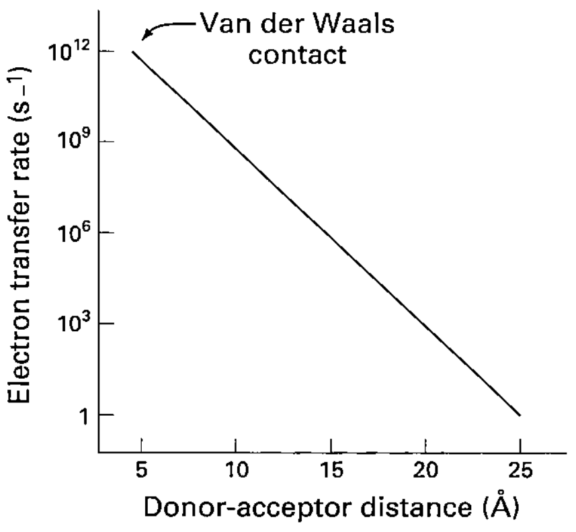

Figure 26-29
The rate of electron transfer decreases exponentially with distance between the donor and acceptor. For energetically favorable transfers, the rate is typically 10^{12} s^{-1} when the donor and acceptor are in van der Waals contact.

6. *Reaction centers generate transmembrane proton gradients.* Reaction centers span lipid bilayer membranes. Light-induced electron flow invariably leads to the translocation of protons across the membrane. In essence, light is converted first into electron-motive force and then into proton-motive force. In some systems, a stable reductant also is formed.

THE PATH OF CARBON IN PHOTOSYNTHESIS WAS TRACED BY PULSE LABELING WITH RADIOACTIVE CO_2

In 1945, Melvin Calvin and his colleagues started a series of investigations that resulted in the elucidation of the dark reactions of photosynthesis. They used the unicellular green alga *Chlorella* in their work because it was easy to culture this organism consistently. Their findings later proved to be pertinent to a wide variety of organisms ranging from photosynthetic bacteria to higher plants.

The aim of their work was to determine the pathway by which CO_2 becomes fixed into carbohydrate. The experimental strategy was to use radioactive carbon (^{14}C) as a tracer. $^{14}CO_2$ was injected into an illuminated suspension of algae that had been carrying out photosynthesis with normal CO_2. The algae were killed and enzymatic reactions stopped after a predetermined time by dropping the suspension into alcohol. Radioactive compounds in the algae were then separated by two-dimensional paper chromatography and visualized by autoradiography. In his Nobel Lecture, Calvin noted that their primary data resided "in the number, position, and intensity—that is, radioactivity—of the blackened areas. The paper ordinarily does not print out the names of these compounds, unfortunately, and our principal chore for the succeeding ten years was to properly label those blacked areas on the film."

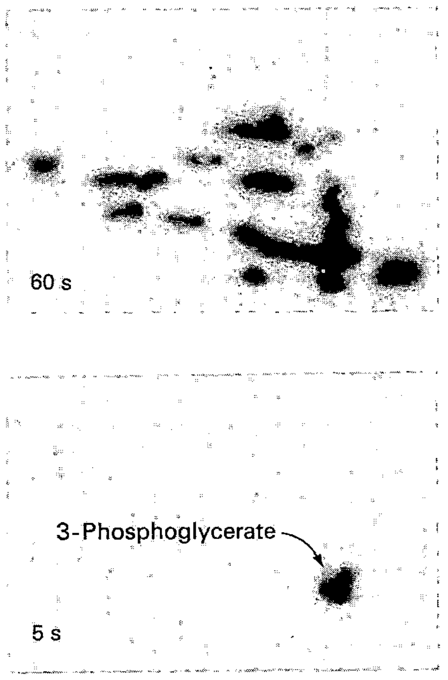

Figure 26-30
Radiochromatograms of illuminated suspensions of algae 60 s and 5 s after the injection of CO_2. [Courtesy of Dr. J.A. Bassham.]

CO_2 REACTS WITH RIBULOSE 1,5-BISPHOSPHATE TO FORM TWO MOLECULES OF 3-PHOSPHOGLYCERATE

The autoradiogram of the algal suspension after 60 s of illumination was so complex (Figure 26-30) that it was not feasible to detect the earliest intermediate in the fixation of CO_2. However, the pattern after only 5 s of illumination was much simpler. In fact, there was just one prominent radioactive spot, which proved to be *3-phosphoglycerate*. The formation of 3-phosphoglycerate as the first detectable radioactive intermediate suggested that a two-carbon compound is the acceptor for the CO_2. This proved not to be so. The actual reaction sequence is more complex.

$$C_5 \xrightarrow{CO_2} C_6 \xrightarrow{H_2O} C_3 + C_3$$

The CO_2 molecule condenses with ribulose 1,5-bisphosphate to form a transient six-carbon compound, which is rapidly hydrolyzed to two molecules of 3-phosphoglycerate (Figure 26-31). This highly exergonic reac-

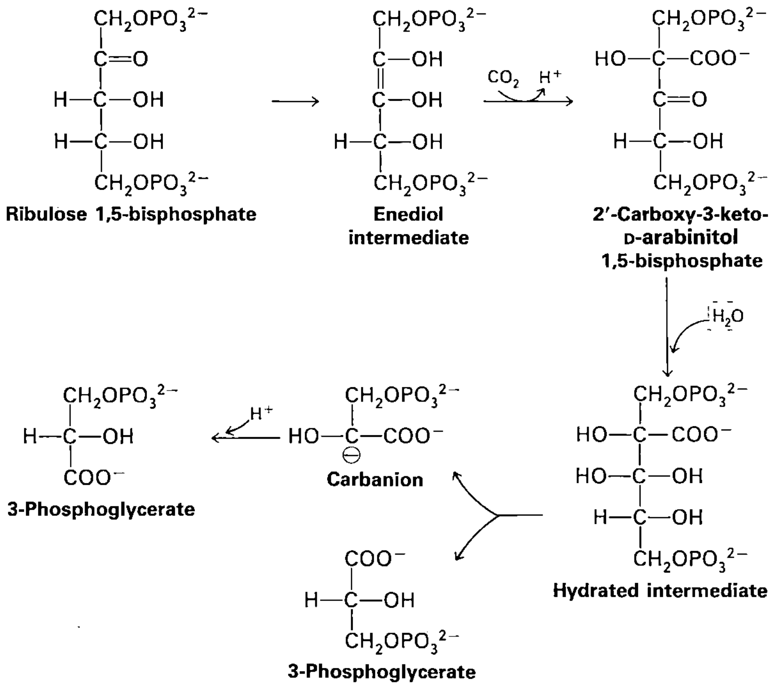

Figure 26-31
The carbon-fixation reaction catalyzed by rubisco (ribulose 1,5-bisphosphate carboxylase/oxygenase).

tion ($\Delta G^{\circ\prime} = -12.4$ kcal/mol) is catalyzed by *ribulose 1,5-bisphosphate carboxylase/oxygenase* (usually called *rubisco*), an enzyme located on the stromal surface of thylakoid membranes. Rubisco in chloroplasts consists of eight large subunits (L, 55 kd) and eight small ones (S, 13 kd). Each L chain contains a catalytic site and a regulatory site. The S chains enhance the catalytic activity of the L chains. This enzyme is very abundant in chloroplasts, comprising more than 16% of their total protein. In fact, rubisco is the most abundant enzyme, and probably the most abundant protein, in the biosphere. Large amounts are present because rubisco is a slow enzyme; its maximal catalytic rate is only 3 s^{-1}.

The enzyme is converted into a catalytically active form by the addition of CO_2 to the uncharged ε-amino group of a specific lysine residue to form a *carbamate.* This negatively charged adduct then binds a divalent metal ion (Mg^{2+} or Mn^{2+}) to form a positively charged center. The bound metal ion serves as an electron sink during catalysis. The activator CO_2 is distinct from the substrate CO_2.

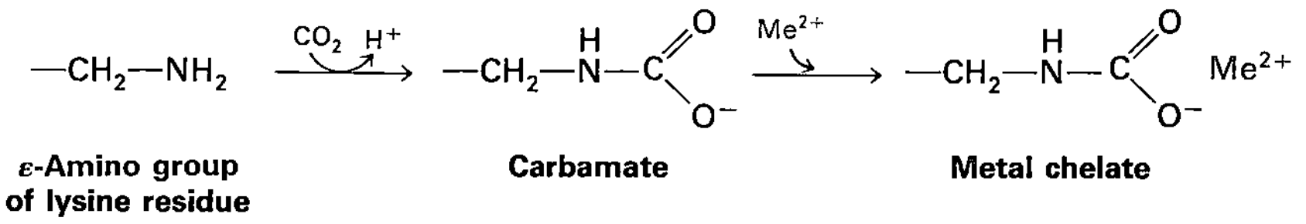

The first step in the reaction catalyzed by rubisco is the formation of an enediol intermediate, which reacts with CO_2 to produce a six-carbon intermediate, 2′-carboxy-3-keto-D-arabinitol 1,5-bisphosphate. Hydration of this C_6 species yields a diol at C-3. Cleavage of a C–C bond gives a molecule of 3-phosphoglycerate and the carbanion of a second one. The second molecule of 3-phosphoglycerate is then formed by protonation of the carbanion.

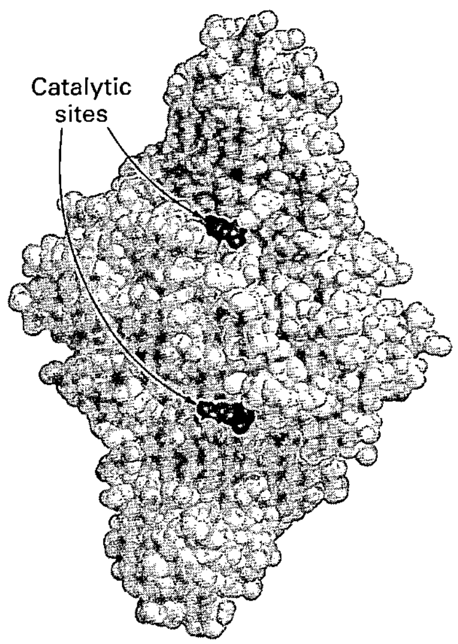

Figure 26-32
Structure of rubisco from *Rhodospirillum rubrum,* a purple bacterium. The enzyme is a dimer of identical subunits (blue and yellow), which resemble the L chains of plants. The active sites at the subunit interfaces are marked by the bound 3-phosphoglycerates (red), products of the carboxylation reaction. [Drawn from 1rus.pdb. T. Lundqvist and G. Schneider. *J. Biol. Chem.* 263(1988):3643.]

CATALYTIC IMPERFECTION: RUBISCO ALSO CATALYZES A WASTEFUL OXYGENASE REACTION

Ribulose 1,5-bisphosphate carboxylase is also an *oxygenase.* It catalyzes the addition of O_2 to ribulose 1,5-bisphosphate to form *phosphoglycolate* and 3-phosphoglycerate (Figure 26-33). The oxygenase and carboxylase reactions are carried out by the same active site and compete with each other. The rate of the carboxylase reaction is four times that of the oxygenase reaction under normal atmospheric conditions at 25°C; the stromal concentration of CO_2 is then 10 μM and that of O_2 is 250 μM. The oxygenase reaction, like the carboxylase reaction, requires that the same lysine be in the carbamate form with a bound divalent metal ion.

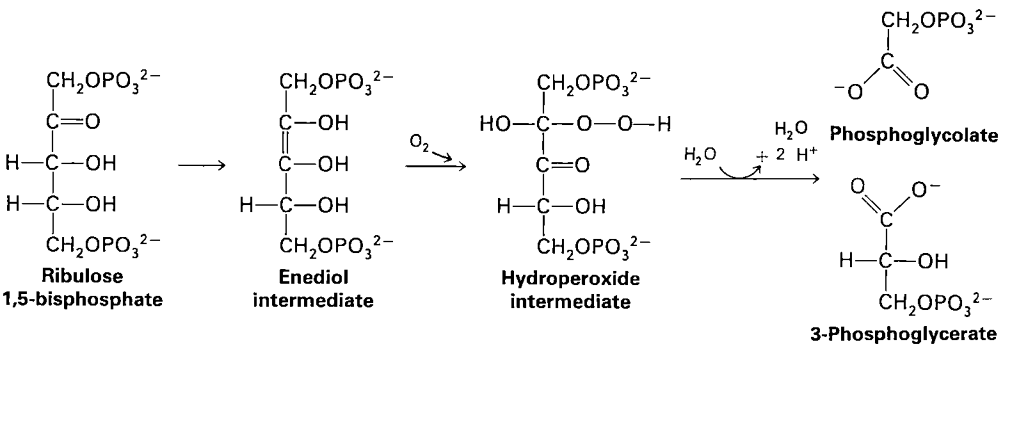

Figure 26-33
Oxygenase reaction catalyzed by rubisco.

Ribulose 1,5-bisphosphate

$$\downarrow^{O_2}_{\searrow 3\text{-Phosphoglycerate}}$$

$$\begin{array}{c} COO^- \\ | \\ CH_2OPO_3{}^{2-} \end{array}$$
Phosphoglycolate

$$\downarrow^{H_2O}_{\searrow P_i}$$

$$\begin{array}{c} COO^- \\ | \\ CH_2OH \end{array}$$
Glycolate

$$\downarrow^{O_2}_{\searrow H_2O_2}$$

$$\begin{array}{c} COO^- \\ | \\ C \\ O{\diagdown}\ \diagup H \end{array}$$
Glyoxylate

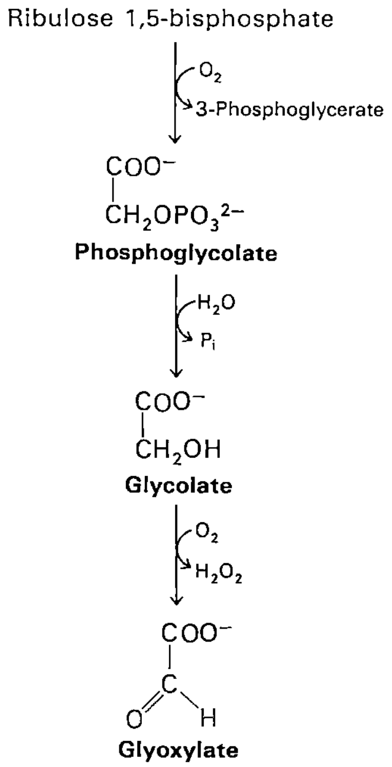

Figure 26-34
Formation and breakdown of glycolate.

Phosphoglycolate is not a versatile metabolite. A salvage pathway recovers part of its carbon skeleton (Figure 26-34). A specific phosphatase converts it into *glycolate*, which enters *peroxisomes* (also called *microbodies*) (Figure 26-35). Glycolate is then oxidized to *glyoxylate* by glycolate oxidase, an enzyme with a flavin mononucleotide prosthetic group. The H_2O_2 produced in this reaction is cleaved by catalase to H_2O and O_2. Transamination of glyoxylate then yields *glycine*. In mitochondria, serine, a potential precursor of glucose, is formed from two molecules of glycine, with the release of CO_2 and $NH_4{}^+$ (p. 721).

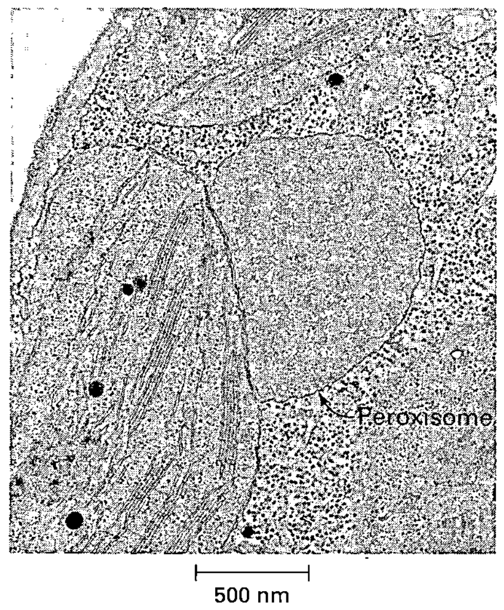

Figure 26-35
Electron micrograph of a peroxisome in a plant cell. [Courtesy of Dr. Sue Ellen Frederick.]

500 nm

This pathway serves to recycle three of the four carbon atoms of two molecules of glycolate. However, one of them is lost as CO_2, and one of two amino groups donated in transamination reactions is lost as $NH_4{}^+$. This process is called *photorespiration* because O_2 is consumed and CO_2 is released. Photorespiration is wasteful because organic carbon is converted into CO_2 without the production of ATP, NADPH, or another energy-rich metabolite. It is a consequence of the imperfection of rubisco. Efforts to increase the ratio of carboxylase to oxygenase activity of rubisco by mutagenesis have thus far not been fruitful.

HEXOSE PHOSPHATES ARE MADE FROM PHOSPHOGLYCERATE, AND RIBULOSE BISPHOSPHATE IS REGENERATED

The steps in the conversion of 3-phosphoglycerate into fructose 6-phosphate (Figure 26-36) are like those of the gluconeogenic pathway (p. 570), except that glyceraldehyde 3-phosphate dehydrogenase in chloroplasts is specific for NADPH, rather than NADH. These reactions and that catalyzed by rubisco bring CO_2 to the level of a hexose.

The remaining task is to regenerate ribulose 1,5 bisphosphate, the acceptor of CO_2 in the first dark step. The problem is to construct a five-carbon sugar from six-carbon and three-carbon sugars. In the pentose phosphate pathway (p. 561), this rearrangement is accomplished by the actions of transketolase and transaldolase. In photosynthesis, transketolase is used, and aldolase serves in place of transaldolase. Recall that *transketolase*, a thiamine pyrophosphate (TPP) enzyme, transfers a two-carbon unit ($CH_2OH-CO-$) from a ketose to an aldose. *Aldolase*, which we encountered in glycolysis (p. 487), catalyzes an aldol condensation between

Fructose 6-phosphate

$\uparrow$

Fructose 1,6-bisphosphate

$\uparrow$

Glyceraldehyde $\longrightarrow$ Dihydroxyacetone
3-phosphate $\rightleftharpoons$ phosphate

$\uparrow\!\!\nearrow$ NADP$^+$
$\quad\searrow$ NADPH

1,3-Bisphosphoglycerate

$\uparrow\!\!\nearrow$ ADP
$\quad\searrow$ ATP

3-Phosphoglycerate

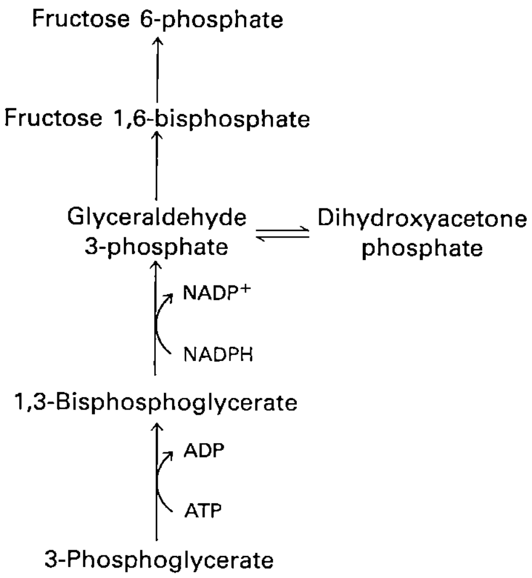

Figure 26-36
Pathway for the conversion of 3-phosphoglycerate into fructose 6-phosphate in chloroplasts.

dihydroxyacetone phosphate and an aldehyde. This enzyme is highly specific for dihydroxyacetone phosphate, but it accepts a wide variety of aldehydes. The specific dark reactions catalyzed by transketolase and aldolase are

$$\text{Fructose 6-phosphate} + \text{glyceraldehyde 3-phosphate} \xrightarrow{\text{Transketolase}} \text{xylulose 5-phosphate} + \text{erythrose 4-phosphate}$$

$$\text{Erythrose 4-phosphate} + \text{dihydroxyacetone phosphate} \xrightarrow{\text{Aldolase}} \text{sedoheptulose 1,7-bisphosphate}$$

$$\text{Sedoheptulose 7-phosphate} + \text{glyceraldehyde 3-phosphate} \xrightarrow{\text{Transketolase}} \text{ribose 5-phosphate} + \text{xylulose 5-phosphate}$$

$$C_6 + C_3 \xrightarrow{\text{Transketolase}} C_4 + C_5$$

$$C_4 + C_3 \xrightarrow{\text{Aldolase}} C_7$$

$$C_7 + C_3 \xrightarrow{\text{Transketolase}} C_5 + C_5$$

Four additional enzymes are needed for the dark reactions of photosynthesis. The hydrolytic action of *sedoheptulose 1,7-bisphosphate phosphatase* yields sedoheptulose 7-phosphate. *Phosphopentose epimerase* converts xylulose 5-phosphate into ribulose 5-phosphate, and *phosphopentose isomerase* converts ribose 5-phosphate into ribulose 5-phosphate. Recall that the epimerase and isomerase also participate in the pentose phosphate pathway (p. 560). The sum of these reactions is

$$\text{Fructose 6-phosphate} + 2\ \text{glyceraldehyde 3-phosphate} + \text{dihydroxyacetone phosphate} \longrightarrow 3\ \text{ribulose 5-phosphate}$$

Finally, *phosphoribulose kinase* catalyzes the phosphorylation of ribulose 5-phosphate to regenerate ribulose 1,5-bisphosphate, the acceptor of CO_2.

$$\text{Ribulose 5-phosphate} + \text{ATP} \longrightarrow \text{ribulose 1,5-bisphosphate} + \text{ADP} + H^+$$

This series of reactions is called the *Calvin cycle* (Figure 26-37).

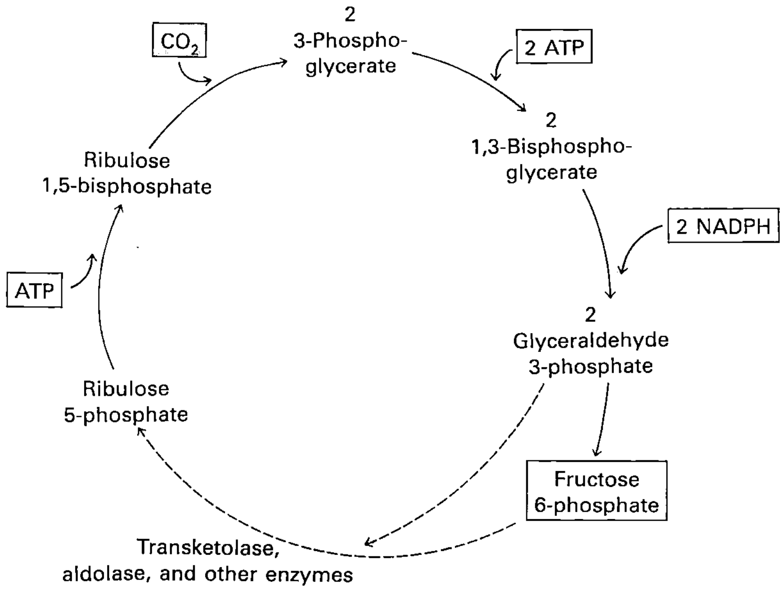

Figure 26-37
The Calvin cycle. The formation of ribulose 5-phosphate from three-carbon and six-carbon sugars is not explicitly shown in the diagram.

STARCH AND SUCROSE ARE THE MAJOR CARBOHYDRATE STORES IN PLANTS

Plants contain two major storage forms of sugar: *starch* and *sucrose.* Starch, like glycogen, is a polymer of glucose residues, but it is less branched because it contains a smaller proportion of α-1,6 glycosidic linkages (p. 581). Another difference is that ADP–glucose, CDP–glucose, or GDP–glucose, but not UDP–glucose, is the activated precursor. Starch is synthesized and stored in chloroplasts.

By contrast, sucrose (common table sugar), a disaccharide, is synthesized in the cytosol. An abundant phosphate translocator mediates the transport of triose phosphates from chloroplasts to the cytosol in exchange for phosphate. Fructose 6-phosphate formed from triose phosphates joins the glucose unit of UDP–glucose to form sucrose 6-phosphate (Figure 26-38). Hydrolysis of the phosphate ester yields sucrose, a readily transportable and mobilizable sugar that is stored in many plant cells, as in sugar beets and sugar cane.

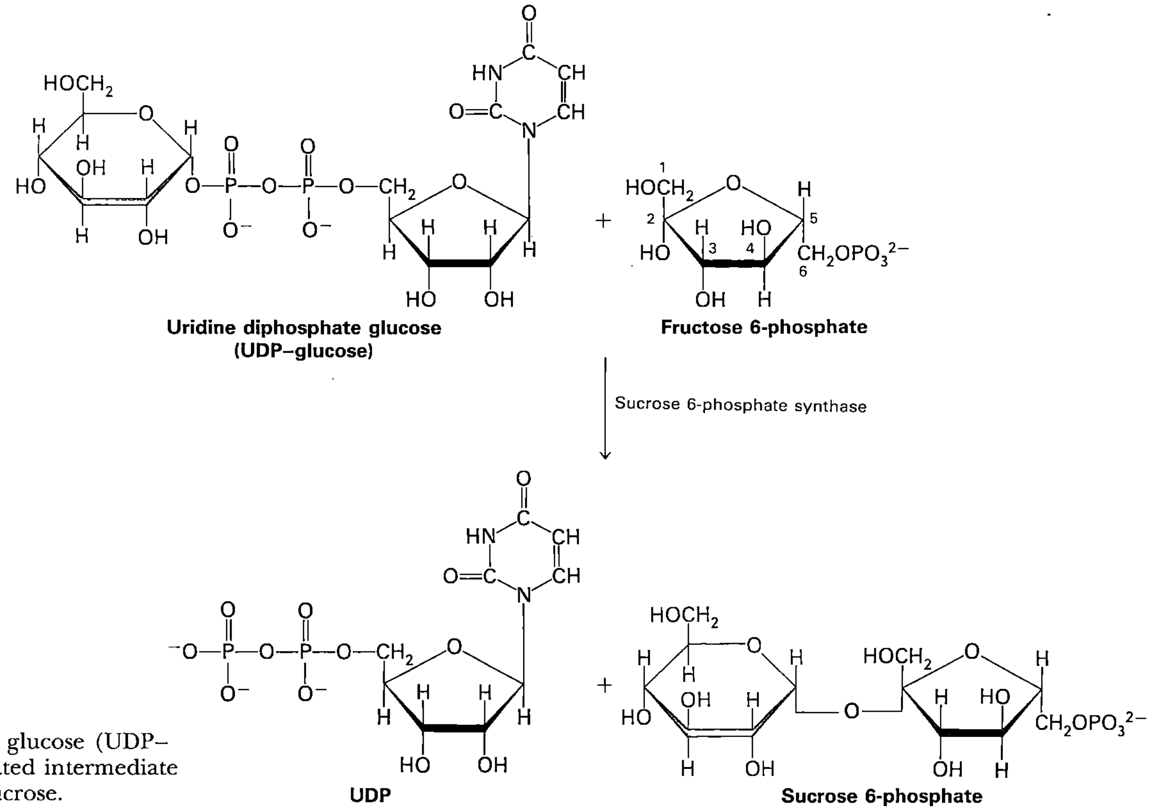

Figure 26-38
Uridine diphosphate glucose (UDP–glucose) is the activated intermediate in the synthesis of sucrose.

THREE ATP AND TWO NADPH ARE USED TO BRING CO_2 TO THE LEVEL OF A HEXOSE

What is the energy expenditure for synthesizing a hexose? Six rounds of the Calvin cycle are required, because one carbon atom is reduced in each (see Figure 26-37). Twelve ATP are expended in phosphorylating twelve molecules of 3-phosphoglycerate to 1,3-bisphosphoglycerate, and twelve NADPH are consumed in reducing twelve molecules of 1,3-bisphosphoglycerate to glyceraldehyde 3-phosphate. An additional six ATP are spent in regenerating ribulose 1,5-bisphosphate.

We can now write a balanced equation for the net reaction of the Calvin cycle.

$$6\,CO_2 + 18\,ATP + 12\,NADPH + 12\,H_2O \longrightarrow C_6H_{12}O_6 + 18\,ADP + 18\,P_i + 12\,NADP^+ + 6\,H^+$$

Thus, three molecules of ATP and two of NADPH are consumed in converting CO_2 into a hexose such as glucose or fructose.

The efficiency of photosynthesis can be estimated in the following way:

1. The $\Delta G^{\circ\prime}$ for the reduction of CO_2 to the level of hexose is +114 kcal/mol.

2. The reduction of $NADP^+$ is a two-electron process. Hence, the formation of two NADPH requires the pumping of four photons by photosystem I. The electrons given up by photosystem I are replenished by photosystem II, which needs to absorb an equal number of photons. Hence eight photons are needed to generate the required NADPH. The proton gradient generated in producing two NADPH is more than sufficient to drive the synthesis of three ATP.

3. A mole of 600-nm photons has an energy content of 47.6 kcal, and so the energy input of eight moles of photons is 381 kcal. Thus, the overall efficiency of photosynthesis under standard conditions is at least 114/381, or 30%.

THIOREDOXIN PLAYS A KEY ROLE IN COORDINATING THE LIGHT AND DARK REACTIONS OF PHOTOSYNTHESIS

Carbon dioxide assimilation by the Calvin cycle operates during the day, whereas carbohydrate degradation occurs primarily at night. How are synthesis and degradation coordinately controlled? Light leads to the generation of regulatory signals as well as of ATP and NADPH. The most important regulator is *thioredoxin*, a 12-kd protein containing neighboring cysteine residues that cycle between a reduced sulfhydryl and an oxidized disulfide form. The reduced form of thioredoxin, which is abundant in light, activates many biosynthetic enzymes by reducing disulfide bridges that control their activity (Figure 26-39). The catalytic activity of phosphoribulose kinase, for example, increases 100-fold on illumination. Conversely, several degradative enzymes are inhibited when their disulfide bridges are reduced by thioredoxin. In chloroplasts, oxidized thioredoxin is reduced by ferredoxin in a reaction catalyzed by *ferredoxin-thioredoxin reductase.* Thus, the activities of the light and dark reactions of photosynthesis are coordinated through the reducing potential of ferredoxin and then thioredoxin. We shall return to thioredoxin when we consider the reduction of ribonucleotides (p. 750).

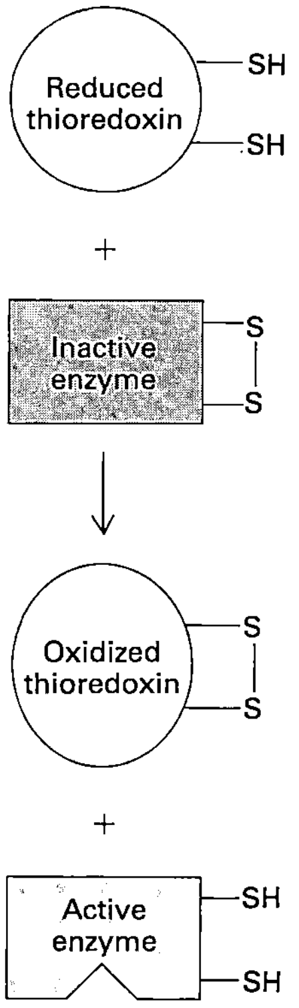

Figure 26-39
The reduced form of thioredoxin activates several enzymes in the Calvin cycle.

The rate-limiting step in the Calvin cycle is the carboxylation of ribulose 1,5-bisphosphate to form two molecules of 3-phosphoglycerate. *The activity of rubisco increases markedly on illumination.* In the stroma, the pH increases from 7 to 8, and the level of Mg^{2+} increases. Both effects are consequences of proton-pumping into the thylakoid space. The activity of the carboxylase increases under these conditions because carbamate formation is favored at alkaline pH. CO_2 adds to the deprotonated ε-amino group of the regulatory lysine residue, and Mg^{2+} binds to the carbamate to form an adduct essential for catalysis (p. 673).

THE C_4 PATHWAY OF TROPICAL PLANTS ACCELERATES PHOTOSYNTHESIS BY CONCENTRATING CO_2

The oxygenase activity of rubisco increases more rapidly with temperature than does its carboxylase activity. How then do tropical plants, such as sugar cane, avoid very high rates of wasteful photorespiration? Their solution to this problem is to achieve a high local concentration of CO_2 at the site of the Calvin cycle in their photosynthetic cells. The first clue to the existence of a CO_2-transport mechanism came from studies showing that radioactivity from a pulse of $^{14}CO_2$ appeared initially in oxaloacetate, malate, and other four-carbon compounds rather than in 3-phosphoglycerate. The essence of this pathway, which was elucidated by M.D. Hatch and C.R. Slack, is that *C_4 compounds carry CO_2 from mesophyll cells, which are in contact with air, to bundle-sheath cells, which are the major sites of photosynthesis* (Figure 26-40). Decarboxylation of the C_4 compound in the bundle-sheath cell maintains a high concentration of CO_2 at the site of the Calvin cycle. The C_3 compound returns to the mesophyll cell for another round of carboxylation.

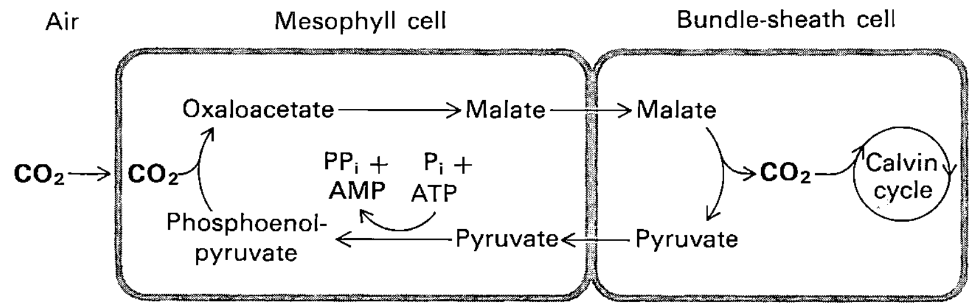

Figure 26-40
Schematic diagram of the essential features of the C_4 pathway. CO_2 is concentrated in bundle-sheath cells by the expenditure of ATP.

The C_4 pathway for the transport of CO_2 starts in the mesophyll cell with the condensation of CO_2 and phosphoenolpyruvate to form *oxaloacetate,* in a reaction catalyzed by *phosphoenolpyruvate carboxylase.* In some species, oxaloacetate is converted into *malate* by an $NADP^+$-linked malate dehydrogenase. Malate goes into the bundle-sheath cell and is decarboxylated within the chloroplasts by an $NADP^+$-linked malate enzyme. The released CO_2 enters the Calvin cycle in the usual way by condensing with ribulose 1,5-bisphosphate. Pyruvate formed in this decarboxylation reaction returns to the mesophyll cell. Finally, phosphoenolpyruvate is formed from pyruvate in an unusual reaction catalyzed by pyruvate-P_i dikinase (Figure 26-41).

$$\text{Pyruvate} + \text{ATP} + P_i \rightleftharpoons \text{phosphoenolpyruvate} + \text{AMP} + PP_i + H^+$$

ATP donates its γ phosphoryl group to orthophosphate, and its β phosphoryl group to a histidine residue of the enzyme. The phosphohistidine residue then reacts with pyruvate to form phosphoenolpyruvate. Why are two ~P consumed in this reaction rather than one? The reason for the phosphorylation of orthophosphate is that its subsequent hydrolysis makes the overall reaction irreversible. The net reaction of this C_4 pathway is

$$CO_2 \text{ (in mesophyll cell)} + \text{ATP} + H_2O \longrightarrow CO_2 \text{ (in bundle-sheath cell)} + \text{AMP} + 2\ P_i + H^+$$

Thus, *two high-energy phosphate bonds are consumed in transporting CO_2 to the chloroplasts of the bundle-sheath cells.*

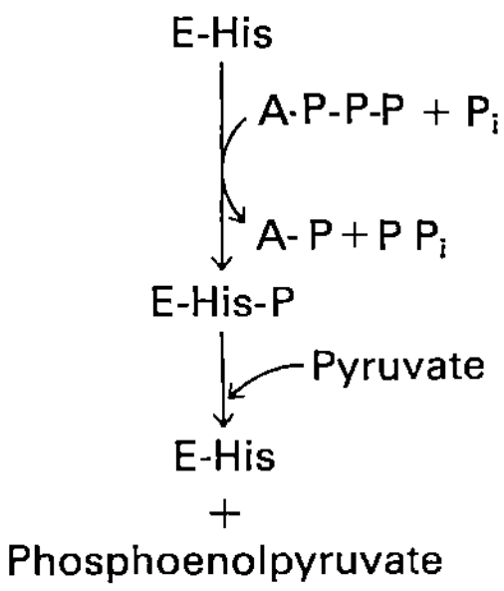

Figure 26-41
Reaction catalyzed by pyruvate-P_i dikinase (E-His). A-P-P-P denotes ATP, and A-P denotes AMP.

When the C_4 pathway and the Calvin cycle operate together, the net reaction is

$$6\,CO_2 + 30\,ATP + 12\,NADPH + 12\,H_2O \longrightarrow C_6H_{12}O_6 + 30\,ADP + 30\,P_i + 12\,NADP^+ + 18\,H^+$$

Note that 30 ATP are consumed per hexose formed when the C_4 pathway delivers CO_2 to the Calvin cycle, in contrast with 18 ATP per hexose in the absence of the C_4 pathway. The high concentration of CO_2 in the bundle-sheath cells of C_4 plants, which is due to the expenditure of the additional 12 ATP, is critical for their rapid photosynthetic rate, because CO_2 is limiting when light is abundant. A high CO_2 concentration also minimizes the energy loss caused by photorespiration.

Tropical plants with a C_4 pathway do little photorespiration because the high concentration of CO_2 in their bundle-sheath cells accelerates the carboxylase reaction relative to the oxygenase reaction. This effect is especially important at higher temperatures. The geographical distribution of plants having this pathway (C_4 plants) and those lacking it (C_3 plants) can now be understood in molecular terms. C_4 plants have the advantage in a hot environment and under high illumination, which accounts for their prevalence in the tropics. C_3 plants, which consume only 18 ATP per hexose formed in the absence of photorespiration (compared with 30 ATP for C_4 plants), are more efficient at temperatures of less than about 28°C, and so they predominate in temperate environments.

Rubisco emerged early in evolution, when the atmosphere was rich in CO_2 and almost devoid of O_2. The enzyme was not originally selected to operate in an environment rich in O_2 and almost devoid of CO_2. In essence, the C_4 pathway creates a microcosm in which CO_2 is abundant, which for photosynthesis is Paradise regained.

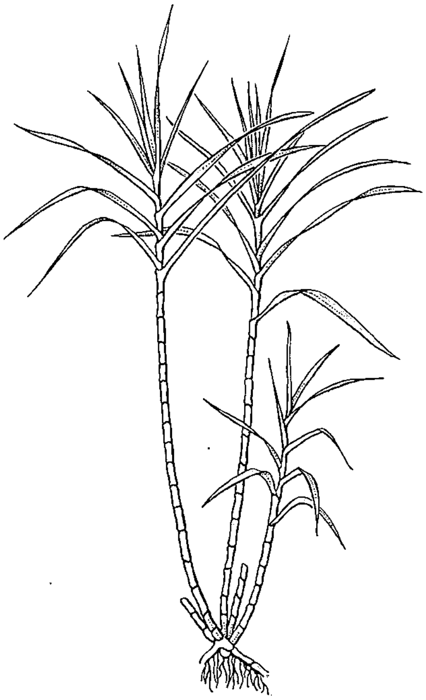

Figure 26-42
Sugar cane, a tropical plant, uses the C_4 pathway to concentrate CO_2.

SUMMARY

Photosynthesis in green plants is mediated by two photosystems located in the thylakoid membranes of chloroplasts. Illumination leads to (1) the generation of a transmembrane proton gradient for the formation of ATP and (2) the creation of reducing power for the production of NADPH. Light absorbed by chlorophylls in the light-harvesting complex of photosystem II funnels into P680, a special pair of chlorophylls located at the interface of two similar subunits that form the core of the reaction center. An electron is transferred from the excited P680* to pheophytin and then to plastoquinones attached to Q_A and Q_B to form reduced plastoquinone (QH_2). The reaction center regains electrons from water by the action of a manganese cluster that evolves O_2. Thus, the net reaction catalyzed by photosystem II is the light-induced transfer of electrons from water to plastoquinone. A transmembrane proton gradient is concomitantly generated. The photosynthetic reaction center of a purple bacterium, which is homologous to that of photosystem II of plants, has been visualized at atomic resolution. The critical event in all photosynthetic reaction centers is the light-induced transfer of an electron to an acceptor against an electrochemical potential gradient.

Electrons from photosystem II flow to photosystem I through the cytochrome *bf* complex. This transmembrane complex pumps protons into the thylakoid space as electrons are transferred from QH_2 to plastocyanin, a water-soluble protein. Photosystem I mediates the light-activated transfer of electrons from plastocyanin to P700 and then to ferredoxin, a powerful reductant. Ferredoxin-$NADP^+$ reductase, a flavoprotein located on the stromal side of the membrane, then catalyzes the formation of NADPH. Thus, the interplay of photosystems I and II leads to the transfer of electrons from H_2O to NADPH and the concomitant generation of a proton gradient for ATP synthesis. Alternatively, electrons from ferredoxin can flow back to photosystem I through the cytochrome *bf* complex; this mode of action of photosystem I, called cyclic photophosphorylation, leads to the generation of a proton gradient without the formation of NADPH. The ATP synthase of chloroplasts (also called CF_0-CF_1) closely resembles the ATP-synthesizing assemblies of bacteria and mitochondria (F_0-F_1). ATP synthesis is driven by the flow of protons from the thylakoid space through the CF_0 transmembrane channel into CF_1 on the stromal side of the membrane.

ATP and NADPH formed in the light reactions of photosynthesis are used to convert CO_2 into hexoses and other organic compounds. The dark phase of photosynthesis, called the Calvin cycle, starts with the reaction of CO_2 and ribulose 1,5-bisphosphate to form two molecules of 3-phosphoglycerate. The steps in the conversion of 3-phosphoglycerate into fructose 6-phosphate and glucose 6-phosphate are like those of gluconeogenesis, except that glyceraldehyde 3-phosphate dehydrogenase in chloroplasts is specific for NADPH rather than NADH. Ribulose 1,5-bisphosphate is regenerated from fructose 6-phosphate, glyceraldehyde 3-phosphate, and dihydroxyacetone phosphate by a complex series of reactions. Several of the steps in the regeneration of ribulose 1,5-bisphosphate are like those of the pentose phosphate pathway. Reduced thioredoxin formed by transfer of electrons from ferredoxin activates enzymes of the Calvin cycle by reducing disulfide bridges. The light-induced increase in pH and Mg^{2+} level of the stroma is also important in stimulating the carboxylation of ribulose 1,5-bisphosphate. Three ATP and two NADPH are consumed for each CO_2 converted into a hexose. Four photons are absorbed by photosystem I and another four by photosystem II to generate two NADPH and a proton gradient sufficient to drive the synthesis of three ATP. Starch in chloroplasts and sucrose in the cytosol are the major carbohydrate stores in plants.

Ribulose 1,5-bisphosphate carboxylase also catalyzes a competing oxygenase reaction, which produces phosphoglycolate and 3-phosphoglycerate. The recycling of phosphoglycolate leads to the release of CO_2 and further consumption of O_2, in a process called photorespiration. This wasteful side reaction is minimized in tropical plants, which have an accessory pathway for concentrating CO_2 at the site of the Calvin cycle. This C_4 pathway enables tropical plants to take advantage of high light levels and minimize the oxygenation of ribulose 1,5-bisphosphate.

SELECTED READINGS

Where to start

Huber, R., 1989. A structural basis of light energy and electron transfer in biology. *EMBO J.* 8:2125–2147. [Nobel Lecture.]

Deisenhofer, J., and Michel, H., 1989. The photosynthetic reaction centre from the purple bacterium *Rhodopseudomonas viridis. EMBO J.* 8:2149–2170. [Nobel Lecture.]

Glazer, A.N., 1989. Light guides. Directional energy transfer in a photosynthetic antenna. *J. Biol. Chem.* 264:1–4.

Youvan, D.C., and Marrs, B.L., 1987. Molecular mechanisms of photosynthesis. *Sci. Amer.* 256:42–48.

Barber, J., and Andersson, B., 1994. Revealing the blueprint of photosynthesis. *Nature* 370:31–34.

Books and general reviews

Andersson, B., and Franzén, L.-G., 1992. The two photosystems of oxygenic photosynthesis. *In* Ernster, L. (ed.), *Molecular Mechanisms in Biosynthesis,* pp. 121–143. Elsevier.

Amesz, J. (ed.), 1987. *Photosynthesis.* Elsevier. [A valuable collection of reviews.]

Cramer, W.A., and Knaff, D.B., 1991. *Energy Transduction in Biological Membranes. A Textbook of Bioenergetics.* Springer-Verlag. [Chapter 6 of this excellent book deals with photosynthesis.]

Nicholls, D.G., and Ferguson, S.J., 1992. *Bioenergetics* (2nd ed.). Academic Press. [Chapter 6 provides a lucid and concise account of photosynthetic generators of proton-motive force.]

Hoober, J.K., 1984. *Chloroplasts.* Plenum. [An excellent overview of the structure, function, and genetics of chloroplasts. Contains many informative and attractive illustrations.]

Halliwell, B., 1984. *Chloroplast Metabolism* (rev. ed.). Oxford University Press. [Emphasizes carbon dioxide fixation, photorespiration, and membrane lipid biosynthesis.]

Harold, F.M., 1986. *The Vital Force: A Study of Bioenergetics.* W.H. Freeman. [See Chapter 8, "Harvesting the Light," for a perceptive overview of photosynthesis.]

Miller, K.R., 1979. The photosynthetic membrane. *Sci. Amer.* 241(4):102–113.

Light-harvesting assemblies

Kühlbrandt, W., Wang, D.-N., and Fujiyoshi, Y., 1994. Atomic model of plant light-harvesting complex by electron crystallography. *Nature* 367:614–621.

Glazer, A.N., 1983. Comparative biochemistry of photosynthetic light-harvesting systems. *Ann. Rev. Biochem.* 52:125–157.

Zuber, H., 1986. Structure of light-harvesting antenna complexes of photosynthetic bacteria, cyanobacteria, and red algae. *Trends Biochem. Sci.* 11:414–419.

Electron transfer mechanisms

Moser, C.C., Keske, J.M., Warncke, K., Farid, R.S., and Dutton, P.L., 1992. Nature of biological electron transfer. *Nature* 355:796–802.

Boxer, S.G., 1990. Mechanisms of long-distance electron transfer in proteins: Lessons from photosynthetic reaction centers. *Ann. Rev. Biophys. Biophys. Chem.* 19:267–299.

Photosystem II and related reaction centers

Okamura, M.Y., and Feher, G., 1992. Proton transfer in reaction centers from photosynthetic bacteria. *Ann. Rev. Biochem.* 61:891–896.

Deisenhofer, J., and Michel, H., 1991. High-resolution structures of photosynthetic reaction centers. *Ann. Rev. Biophys. Biophys. Chem.* 20:247–266.

Sinning, I., 1992. Herbicide binding in the bacterial photosynthetic reaction center. *Trends Biochem. Sci.* 17:150–154.

Vermaas, W., 1993. Molecular-biological approaches to analyze photosystem II structure and function. *Ann. Rev. Plant Physiol. Plant Mol. Biol.* 44:457–481.

Oxygen evolution

Yamachandra, V.K., DeRose, V.J., Latimer, M.J., Mukerji, I., Sauer, K., and Klein, M.P., 1993. Where plants make oxygen: A structural model for the photosynthetic oxygen-evolving manganese complex. *Science* 260:675–679.

Brudvig, G.W., Beck, W.F., and de Paula, J.C., 1989. Mechanism of photosynthetic water oxidation. *Ann. Rev. Biophys. Biophys. Chem.* 18:25–46.

Debus, R.J., 1992. The manganese and calcium ions of photosynthetic oxygen evolution. *Biochim. Biophys. Acta* 1102:269–352.

Babcock, G.T., Barry, B.A., Debus, R.J., Hoganson, C.W., Atamian, M., McIntosh, L., Sithole, I., and Yocum, C.F., 1989. Water oxidation in photosystem II: From radical chemistry to multielectron chemistry. *Biochemistry* 28:9557–9565.

Photosystem I and the cytochrome *bf* complex

Blankenship, R.E., and Prince, R.C., 1985. Excited-state redox potentials and the Z scheme of photosynthesis. *Trends Biochem. Sci.* 10:382–383. [A concise and lucid statement of the redox properties of excited states.]

Malkin, R., 1992. Cytochrome bc_1 and b_6f complexes of photosynthetic membranes. *Photosyn. Res.* 33:121–136.

Krauss, N., Hinrichs, W., Witt, I., Fromme, P., Pritzkow, W., Dauter, Z., Betzel, C., Wilson, K.S., Witt, H.T., and Saenger, W., 1993. Three-dimensional structure of system I photosynthesis at 6 Å resolution. *Nature* 361:326–331.

Karplus, P.A., Daniels, M.J., and Herriott, J.R., 1991. Atomic structure of ferredoxin-$NADP^+$ reductase: Prototype for a structurally novel flavoenzyme family. *Science* 251:60–66.

ATP synthase

Walker, J.E., Fearnley, I.M., Lutter, R., Todd, R.J., and Runswick, M.J., 1990. Structural aspects of proton-pumping ATPases. *Phil. Trans. Roy. Soc. Lond. B* 326:367–378.

McCarty, R.E., and Moroney, J.V., 1985. Functions of the subunits and regulation of chloroplast coupling factor 1. *In* Martonosi, A. (ed.), *The Enzymes of Biological Membranes* (2nd ed.), vol. 4, pp. 383–413. Plenum.

Jagendorf, A.T., 1967. Acid-base transitions and phosphorylation by chloroplasts. *Fed. Proc.* 26:1361–1369.

Carbon dioxide fixation

Buchanan, B.B., 1992. Carbon dioxide assimilation in oxygenic and anoxygenic photosynthesis. *Photosyn. Res.* 33:147–162.

Schneider, G., Linqvist, Y., and Brändén, C.-I., 1992. Rubisco: Structure and mechanism. *Ann. Rev. Biophys. Biomol. Struct.* 21:119–143.

Spreitzer, R.J., 1993. Genetic dissection of rubisco structure and function. *Ann. Rev. Plant Physiol. Plant Mol. Biol.* 44:411–434.

Hatch, M.D., 1987. C_4 photosynthesis: A unique blend of modified biochemistry, anatomy, and ultrastructure. *Biochim. Biophys. Acta* 895:81–106.

Rawsthorne, S., 1992. Towards an understanding of C3-C4 photosynthesis. *Essays Biochem.* 27:135–146.

Regulation of photosynthesis

Allen, J.F., 1992. Protein phosphorylation in regulation of photosynthesis. *Biochim. Biophys. Acta* 1098:275–335.

Knaff, D.B., 1991. Regulatory phosphorylation of chloroplast antenna proteins. *Trends Biochem. Sci.* 16:82–83.

Anderson, J.M., 1992. Cytochrome b_6f complex: Dynamic molecular organization, function, and acclimation. *Photosyn. Res.* 34:341–357.

Barber, J., and Andersson, B., 1992. Too much of a good thing: Light can be bad for photosynthesis. *Trends Biochem. Sci.* 17:61–66.

Recurring motifs

Golbeck, J.H., 1993. Shared thematic elements in photochemical reaction centers. *Proc. Nat. Acad. Sci.* 90:1642–1646.

Nitschke, W., and Rutherford, A.W., 1991. Photosynthetic reaction centres: Variations on a common structural theme? *Trends Biochem. Sci.* 16:241–245.

Meyer, T.E., 1991. Evolution of cytochromes and photosynthesis. *Biochim. Biophys. Acta* 1058:31–34.

Historical aspects of photosynthesis

Arnon, D.I., 1991. Photosynthetic electron transport: Emergence of a concept, 1949–59. *Photosyn. Res.* 29:117–131.

Arnon, D. I., 1984. The discovery of photosynthetic phosphorylation. *Trends Biochem. Sci.* 9:258–262.

Bassham, J. A., 1962. The path of carbon in photosynthesis. *Sci. Amer.* 206(6):88–100.

PROBLEMS

1. *Electron transfer.* Calculate the $\Delta E_0'$ and $\Delta G^{\circ\prime}$ for the reduction of $NADP^+$ by ferredoxin. Use data given in Table 21-1 on p. 532.

2. *Weed killer.* Dichlorophenyldimethylurea (DCMU), an herbicide, interferes with photophosphorylation and O_2 evolution. However, it does not block the Hill reaction. Propose a site for the inhibitory action of DCMU.

3. *Infrared harvest.* Consider the relation between the energy of a photon and its wavelength.
   (a) Some bacteria are able to harvest 1000-nm light. What is the energy (in kcal) of a mole (also called an einstein) of 1000-nm photons?
   (b) What is the maximum increase in redox potential that can be induced by a 1000-nm photon?
   (c) What is the minimum number of 1000-nm photons needed to form ATP from ADP and P_i? Assume a ΔG of 12 kcal/mol for the phosphorylation reaction.

4. *Subito fortissimo.* A chloroplast suspension is exposed to a train of 1-μs light pulses. The dark interval between light pulses is 100 ms. A burst of O_2 evolution occurs after every fourth light pulse. Account for this periodicity.

5. *Variation on a theme.* Sedoheptulose 1,7-bisphosphate is an intermediate in the Calvin cycle but not in the pentose phosphate pathway. What is the enzymatic basis of this difference?

6. *Total eclipse.* An illuminated suspension of *Chlorella* is actively carrying out photosynthesis. Suppose that the light is suddenly switched off. How would the levels of 3-phosphoglycerate and ribulose 1,5-bisphosphate change during the next minute?

7. *CO_2 deprivation.* An illuminated suspension of *Chlorella* is actively carrying out photosynthesis in the presence of 1% CO_2. The concentration of CO_2 is abruptly reduced to 0.003%. What effect would this have on the levels of 3-phosphoglycerate and ribulose 1,5-bisphosphate during the next minute?

8. *Self-preservation.* Blue-green bacteria deprived of a source of nitrogen digest their least essential proteins. Which component of phycobilisomes is likely to be degraded first under starvation conditions?

9. *A potent analog.* 2-Carboxyarabinitol 1,5-bisphosphate (CABP) has been useful in studies of rubisco.
   (a) Write the structural formula of CABP.
   (b) Which catalytic intermediate does it resemble?
   (c) Predict the effect of CABP on rubisco.

10. *Salvage operation.* Write a balanced equation for the transamination of glyoxylate to yield glycine.

PART

IV

Biosynthesis of Building Blocks

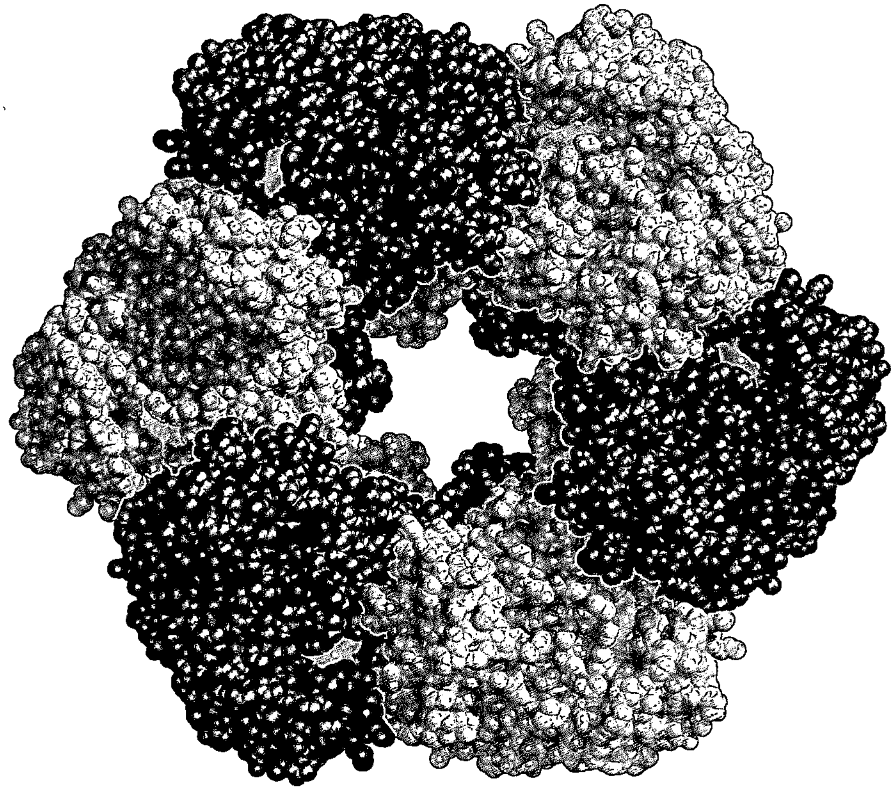

Glutamine synthetase catalyzes the ATP-powered synthesis of glutamine from glutamate and NH_4^+, a key step in the incorporation of nitrogen into biomolecules. The 12 identical subunits (red and blue) are arranged in two hexameric rings joined face to face. Catalytic sites are shown in yellow. [Drawn from 2gls.pdb. M.M. Yamashita, R.J. Almassy, C.A. Janson, and D. Cascio. J. Biol. Chem. 264(1989):17681.]

CHAPTER

27

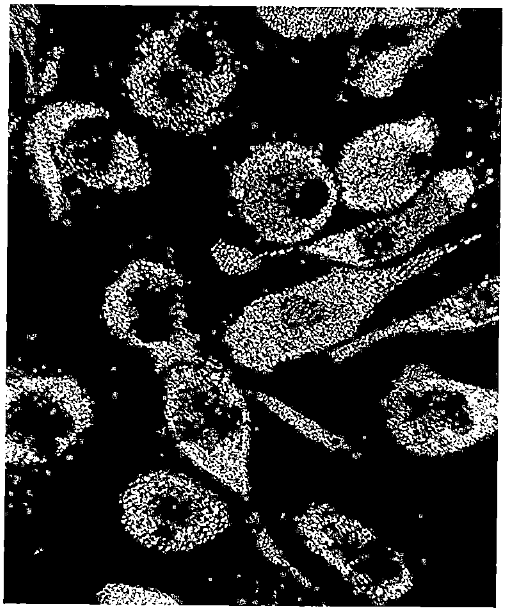

Biosynthesis of Membrane Lipids and Steroids

We turn now from the generation of metabolic energy to the biosynthesis of building blocks that serve as recurring units of membranes, proteins, DNA, and RNA. This chapter deals with the biosynthesis of three important components of biological membranes—phosphoglycerides, sphingolipids, and cholesterol (Chapter 11). Triacylglycerols are also considered here because the pathway for their synthesis overlaps that of phosphoglycerides. Cholesterol is of interest both as a membrane component and as a precursor of many signal molecules. Steroid hormones such as progesterone, testosterone, and cortisol are derived from cholesterol. The biosynthesis of cholesterol exemplifies a fundamental mechanism for the assembly of extended carbon skeletons from five-carbon units.

The transport of cholesterol in blood by the low-density lipoprotein and its uptake by a specific receptor on the cell surface are discussed in some detail because they vividly illustrate a recurring mechanism for the entry of metabolites and signal molecules into cells. The absence of this receptor in familial hypercholesterolemia, a genetic disease, leads to markedly elevated cholesterol levels in the blood, cholesterol deposits on blood vessels, and heart attacks in childhood.

Anabolism—
Biosynthetic processes.

Catabolism—
Degradative processes.

Derived from the Greek words *ana*, "up"; *kata*, "down"; *ballein*, "to throw."

PHOSPHATIDATE IS AN INTERMEDIATE IN THE SYNTHESIS OF PHOSPHOGLYCERIDES AND TRIACYLGLYCEROLS

Phosphatidate (diacylglycerol 3-phosphate) is a common intermediate in the synthesis of phosphoglycerides and triacylglycerols. The pathway begins with *glycerol 3-phosphate,* which is formed primarily by reduction of

Opening Image: *Polarized light micrograph of macrophages containing birefringent crystals of cholesterol esters. [Courtesy of Dr. Richard Anderson.]*

dihydroxyacetone phosphate and to a lesser extent by phosphorylation of glycerol. Glycerol 3-phosphate is acylated by acyl CoA to form *lysophosphatidate,* which is again acylated by acyl CoA to yield phosphatidate. These acylations are catalyzed by *glycerol phosphate acyltransferase.* In most phosphatidates, the fatty acyl chain attached to C_1 is saturated, whereas the one attached to C_2 is unsaturated.

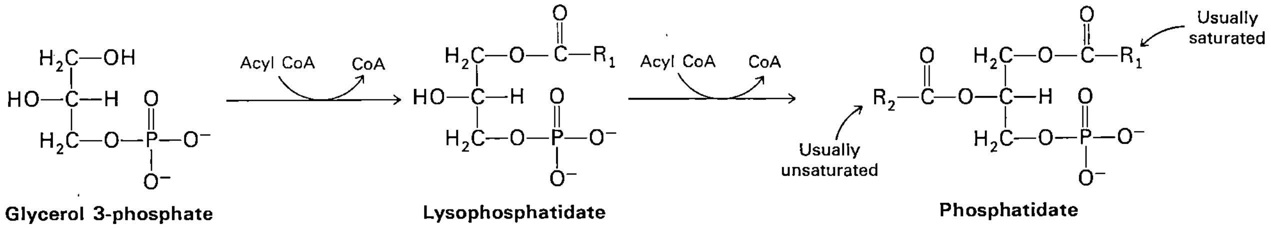

The pathways diverge at phosphatidate. In the synthesis of triacylglycerols, phosphatidate is hydrolyzed by a specific phosphatase to give a *diacylglycerol (DAG).* This intermediate is acylated to a *triacylglycerol* in a reaction that is catalyzed by *diglyceride acyltransferase.* These enzymes are associated in a *triacylglycerol synthetase complex* that is bound to the endoplasmic reticulum membrane.

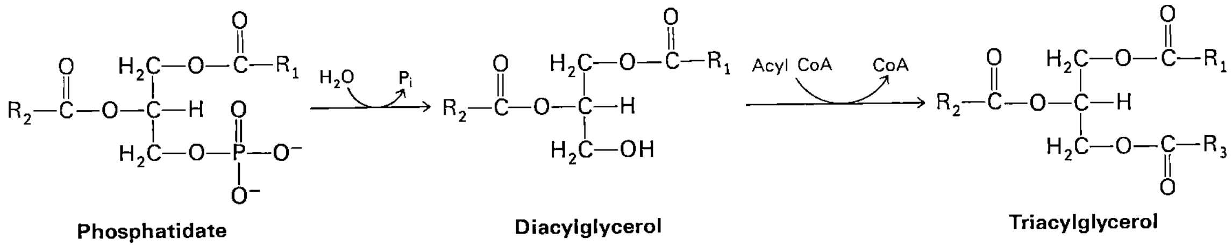

CDP-DIACYLGLYCEROL IS THE ACTIVATED INTERMEDIATE IN THE DE NOVO SYNTHESIS OF SOME PHOSPHOGLYCERIDES

Phosphoglycerides can be synthesized in several ways. One de novo pathway starts with the formation of *cytidine diphosphodiacylglycerol (CDP–diacylglycerol)* from phosphatidate and cytidine triphosphate (CTP). This reaction, like many biosyntheses, is driven forward by the hydrolysis of pyrophosphate. Recall that PP_i is formed in the synthesis of UDP–glucose from glucose 1-phosphate and UTP (p. 586).

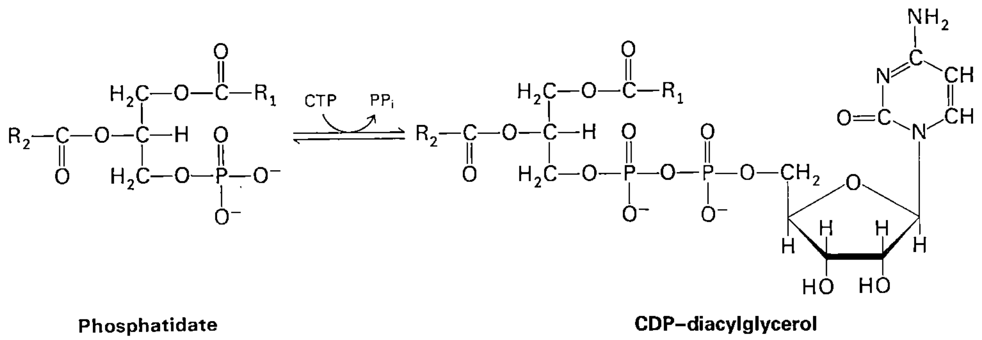

The activated phosphatidyl unit then reacts with the hydroxyl group of a polar alcohol. If the alcohol is serine, the products are *phosphatidyl serine* and cytidine monophosphate (CMP).

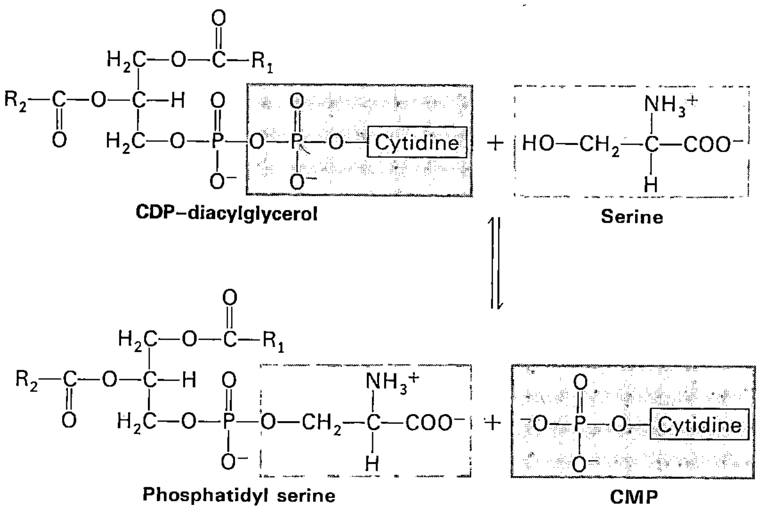

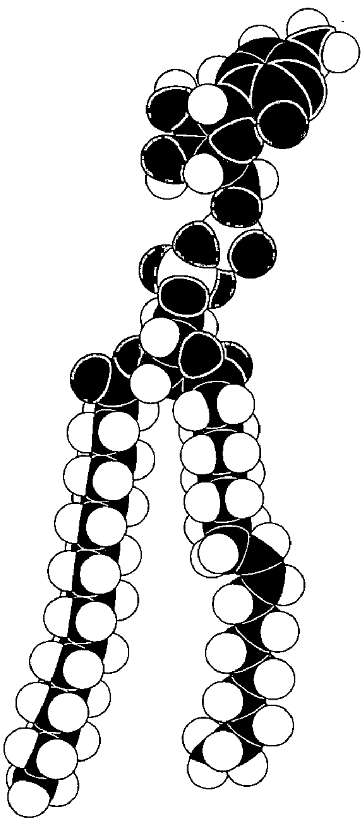

Figure 27-1
Model of CDP–diacylglycerol, a key intermediate in the synthesis of phospholipids.

Likewise, *phosphatidyl inositol* is formed by the transfer of a diacylglycerol phosphate unit from CDP–diacylglycerol to inositol. Subsequent phosphorylations catalyzed by specific kinases lead to the synthesis of *phosphatidyl inositol 4,5-bisphosphate,* a key molecule in signal transduction. Recall that hormonal and sensory stimuli activate phospholipase C, an enzyme that hydrolyzes this phospholipid into two intracellular messengers—diacylglycerol and inositol 1,4,5-trisphosphate (p. 344).

Thus, a cytidine nucleotide plays the same role in the synthesis of these phosphoglycerides as does a uridine nucleotide in the formation of glycogen (p. 586). In both biosyntheses, an activated intermediate (UDP–glucose or CDP–diacylglycerol) is formed from a phosphorylated substrate (glucose 1-phosphate or phosphatidate) and a nucleoside triphosphate (UTP or CTP). The activated intermediate then reacts with a hydroxyl group (the 4-OH terminus of glycogen or the hydroxyl side chain of serine).

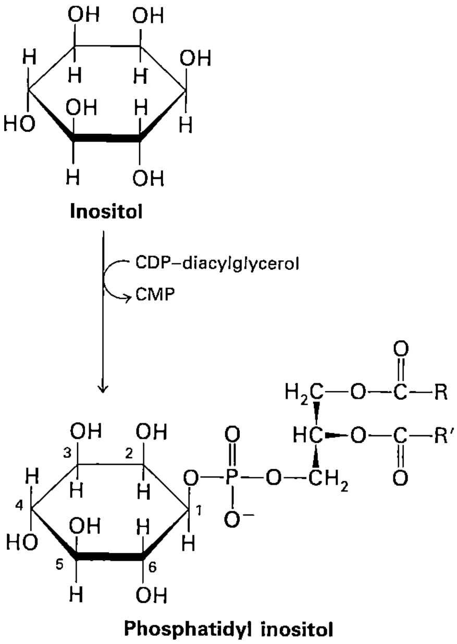

PHOSPHATIDYL ETHANOLAMINE AND PHOSPHATIDYL CHOLINE CAN BE FORMED FROM PHOSPHATIDYL SERINE

In bacteria, decarboxylation of phosphatidyl serine by a pyridoxal phosphate enzyme yields *phosphatidyl ethanolamine.* The amino group of this phosphoglyceride is then methylated three times to form *phosphatidyl choline.* S-*adenosylmethionine,* the methyl donor in this and many other methylations, will be discussed later (p. 721).

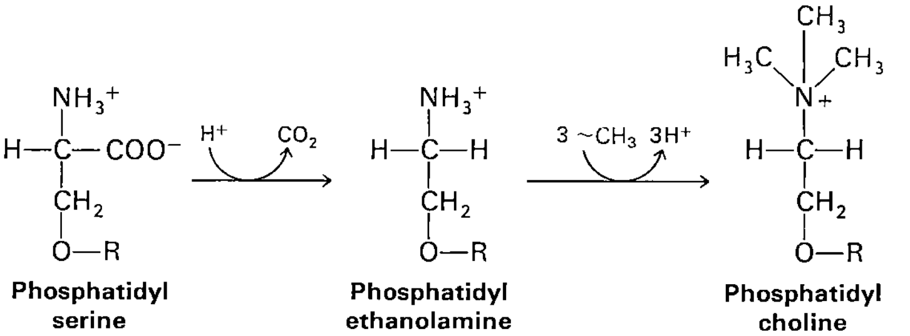

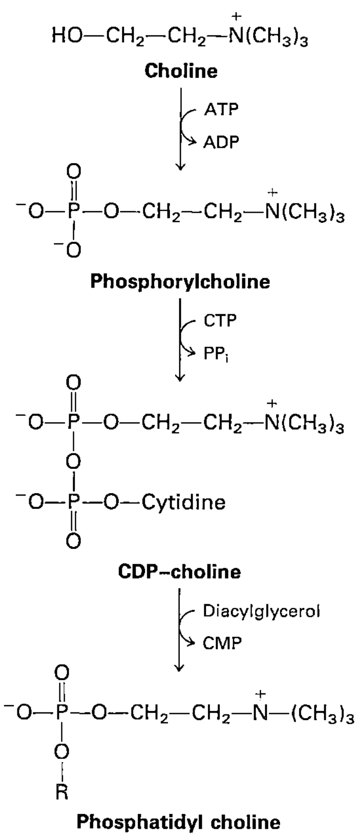

Figure 27-2
Synthesis of phosphatidyl choline.

PHOSPHOGLYCERIDES CAN ALSO BE SYNTHESIZED FROM A CDP-ALCOHOL INTERMEDIATE

In mammals, phosphatidyl choline is synthesized by a pathway that utilizes choline obtained from the diet (Figure 27-2). Choline is phosphorylated by ATP to *phosphorylcholine,* which then reacts with CTP to form *CDP–choline.* The phosphorylcholine unit of CDP–choline is then transferred to a diacylglycerol to form *phosphatidyl choline.* Note that the activated species in this pathway is the cytidine derivative of phosphorylcholine rather than of phosphatidate.

Likewise, *phosphatidyl ethanolamine* can be synthesized from ethanolamine by forming a CDP–ethanolamine intermediate by analogous reactions. Alternatively, phosphatidyl ethanolamine can be formed from phosphatidyl serine by the enzyme-catalyzed exchange of ethanolamine for the serine moiety of the phospholipid.

PLASMALOGENS AND OTHER ETHER PHOSPHOLIPIDS ARE FORMED FROM DIHYDROXYACETONE PHOSPHATE

Some phospholipids contain an ether unit instead of an acyl unit at C_1. *Glyceryl ether phospholipids* are synthesized starting with dihydroxyacetone phosphate (Figure 27-3). Acylation by a fatty acyl CoA yields a 1-acyl derivative that exchanges with a long-chain alcohol to form an ether at C_1. The keto group at C_2 is reduced by NADPH, and the resulting alcohol is acylated by a long-chain CoA. Removal of the 3-phosphate group yields 1-alkyl-2-acylglycerol, which reacts with CDP–choline to form the ether analog of phosphatidyl choline.

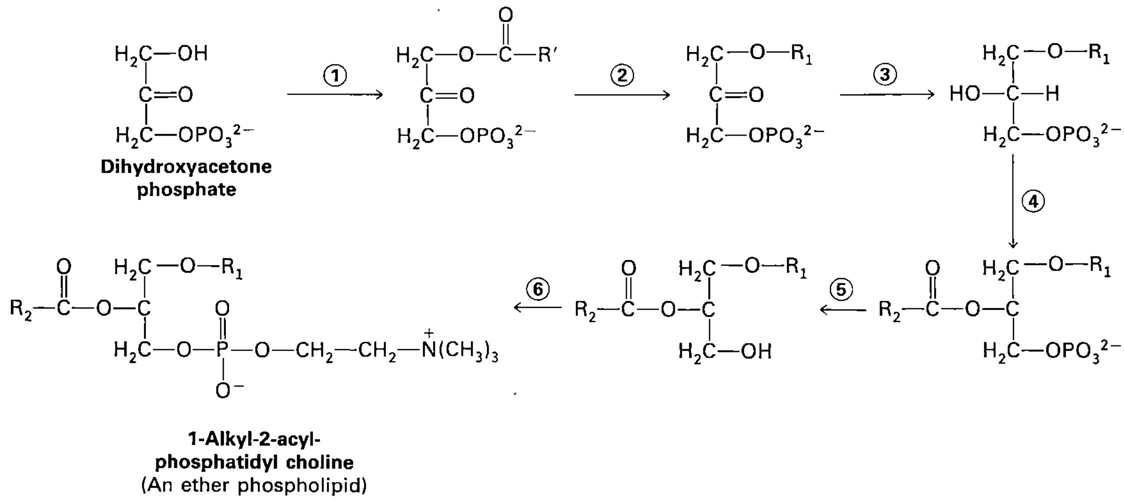

Figure 27-3
Synthesis of an ether phospholipid. The steps are (1) acylation by fatty acyl CoA, (2) exchange of an alcohol for the carboxylate moiety, (3) reduction by NADPH, (4) acylation by a second fatty acyl CoA, (5) hydrolysis of the phosphate ester, (6) transfer of a phosphocholine moiety.

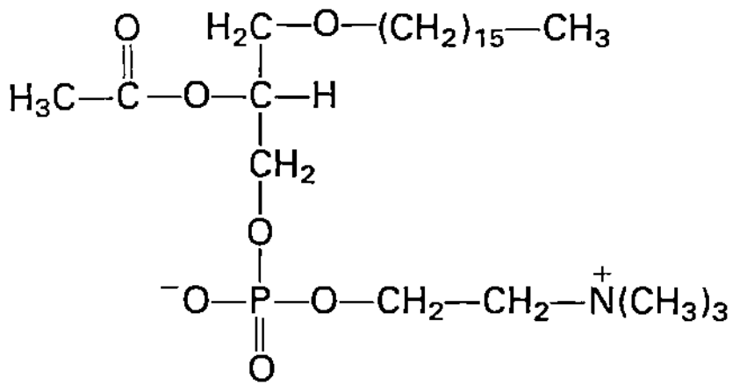

Platelet-activating factor is an ether phospholipid with striking activity. Subnanomolar concentrations of this 1-alkyl-2-acetyl ether analog of phosphatidyl choline induce the aggregation of blood platelets and the dilation of blood vessels. The presence of an acetyl group rather than a long-chain acyl group at C_2 increases the water-solubility of this lipid, enabling it to function in an aqueous environment.

Plasmalogens are phospholipids containing an α,β-unsaturated ether at C_1. Phosphatid*al* choline, the plasmalogen corresponding to phosphati*dyl* choline, is formed by desaturation of a 1-alkyl precursor. The desaturase catalyzing this final step in the synthesis of a plasmalogen is a microsomal enzyme akin to the one that introduces double bonds into long-chain fatty acyl CoAs: O_2 and NADH are reactants, and cytochrome b_5 participates in catalysis (p. 623).

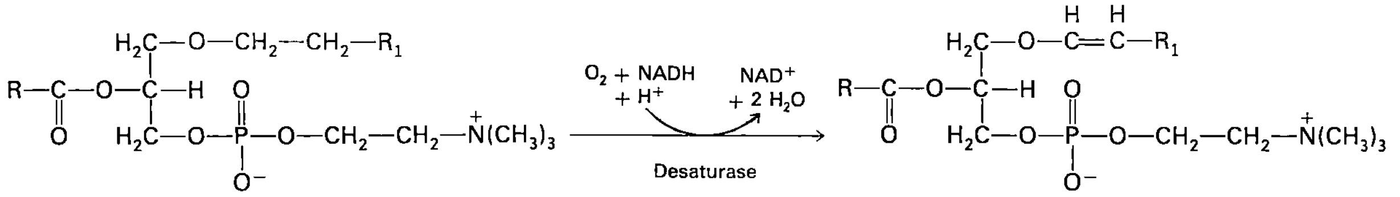
O2 + NADH
+ H+
NAD+
+ 2 H2O
Desaturase

1-Alkyl precursor

Phosphatidal choline (A plasmalogen)

PHOSPHOLIPASES SERVE AS DIGESTIVE ENZYMES AND AS GENERATORS OF SIGNAL MOLECULES

Phospholipases have two principal roles. Many of them are *digestive enzymes* present in high concentration in intestinal juices, bacterial secretions, and venoms. Phospholipases also *generate highly active signal molecules or their immediate precursors.* For example, phospholipase A_2 releases arachidonate, a precursor of prostaglandins, leukotrienes, and thromboxanes (p. 624), and phospholipase C unleashes two messengers in the phosphoinositide cascade (p. 344). Phospholipases are grouped according to their specificities. The bonds hydrolyzed by *phospholipases A_1, A_2, C,* and *D* are shown in Figure 27-5. Phospholipases preferentially hydrolyze substrates that are located in bilayer membranes, micelles, or lipoprotein particles. They carry out *interfacial catalysis* at the boundary of water and a condensed lipid phase.

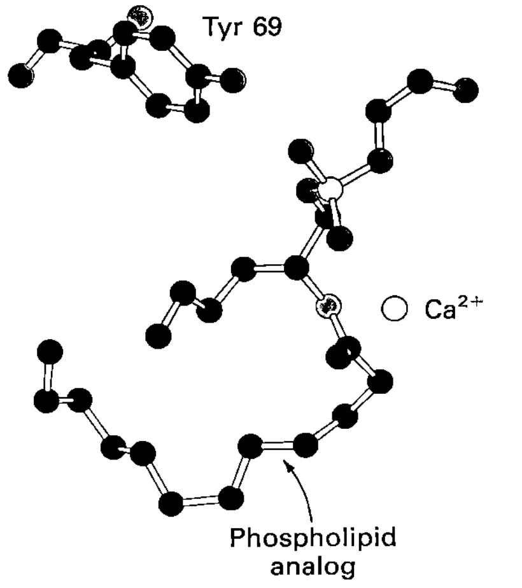
Tyr 69
Ca2+
Phospholipid analog

Figure 27-4
Structure of a phospholipid substrate analog bound to the active site of pancreatic venom phospholipase A_2. A calcium ion (green) is bound to the substrate analog. [Drawn from 5p2p.pdb. B.W. Dijkstra, M.M. Thunnissen, K.H. Kalk, and J. Drenth. *Nature* 347(1990):689.]

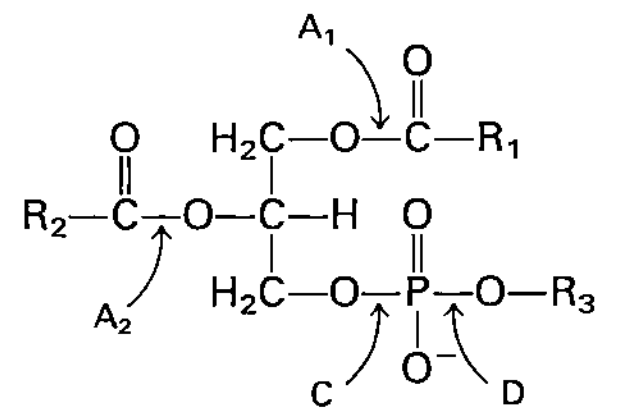

Figure 27-5
Specificity of phospholipases.

SYNTHESIS OF CERAMIDE, THE BASIC STRUCTURAL UNIT OF SPHINGOLIPIDS

We turn now from phosphoglycerides to sphingolipids. The backbone of a sphingolipid is *sphingosine,* rather than glycerol. Palmitoyl CoA and serine condense to form dehydrosphinganine, which is then converted to sphingosine. The enzyme catalyzing this reaction requires pyridoxal phosphate, a prominent cofactor in amino acid metabolism.

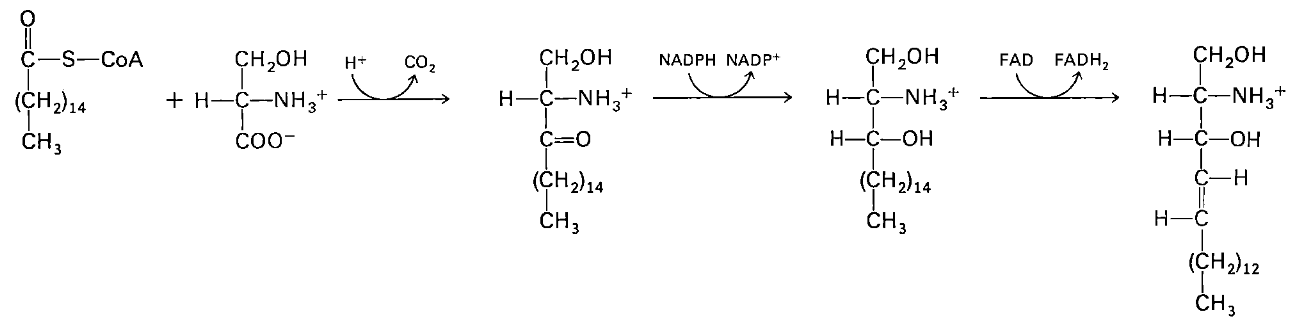
H+
CO2
NADPH
NADP+
FAD
FADH2

Palmitoyl CoA **Serine** **Dehydrosphinganine** **Dihydrosphingosine** **Sphingosine**

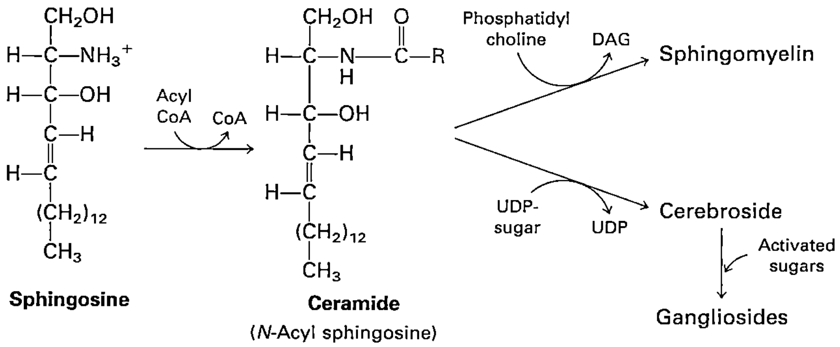

Figure 27-6
Ceramide is the precursor of sphingomyelin and of gangliosides.

In all sphingolipids, the amino group of sphingosine is acylated (Figure 27-6): a long-chain acyl CoA reacts with sphingosine to form *ceramide* (N-*acyl sphingosine*). The terminal hydroxyl group also is substituted. In *sphingomyelin*, the substituent is phosphorylcholine, which comes from phosphatidyl choline. In a *cerebroside*, glucose or galactose is linked to the terminal hydroxyl group of ceramide. UDP–glucose or UDP–galactose is the sugar donor in the synthesis of cerebrosides. In a *ganglioside*, an oligosaccharide is linked to ceramide by a glucose residue (Figure 27-7).

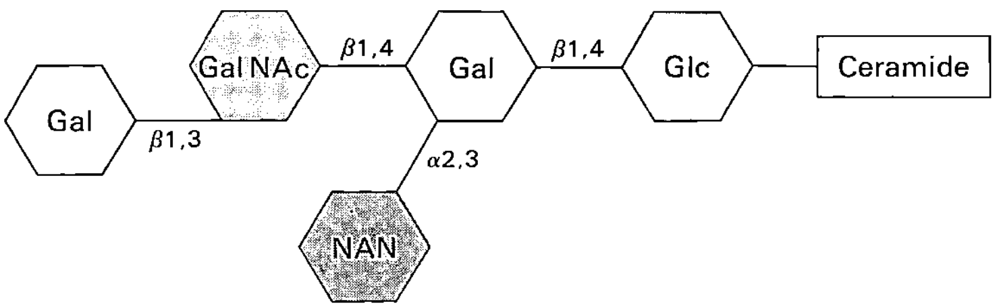

Figure 27-7
Structure of ganglioside G_{M1}. Abbreviations used: Gal, galactose; GalNAc, *N*-acetylgalactosamine; Glc, glucose; NAN, *N*-acetylneuraminate. The types of linkages between the sugars—such as β-1,4—are noted at the bonds joining them.

GANGLIOSIDES ARE CARBOHYDRATE-RICH SPHINGOLIPIDS THAT CONTAIN ACIDIC SUGARS

In gangliosides, the most complex sphingolipids, an *oligosaccharide chain* containing at least one acidic sugar is attached to ceramide. The acidic sugar is N-*acetylneuraminate* or N-*glycolylneuraminate*. These acidic sugars are called *sialic acids*. Their nine-carbon backbones are synthesized from phosphoenolpyruvate (a C_3 unit) and *N*-acetylmannosamine 6-phosphate (a C_6 unit).

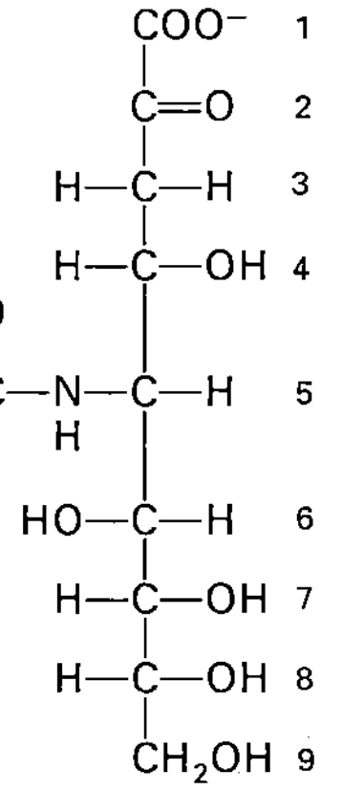

N-Acetylneuraminate
(Open-chain form)

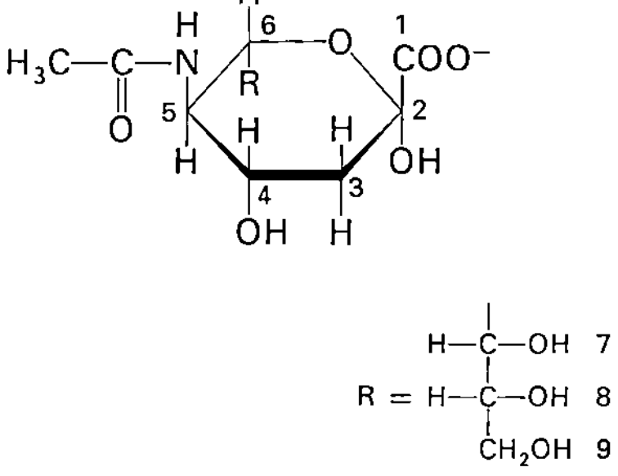

N-Acetylneuraminate
(Pyranose form)

Gangliosides are synthesized by the ordered, stepwise addition of sugar residues to ceramide. UDP–glucose, UDP–galactose, and UDP–*N*-acetylgalactosamine are activated donors of these sugar residues, as is the CMP derivative of *N*-acetylneuraminate. CMP–*N*-acetylneuraminate is synthesized from CTP and *N*-acetylneuraminate. The structure of the resulting ganglioside is determined by the specificity of the glycosyltransferases in the cell. More than 15 different gangliosides have been characterized (see Figure 27-7 for the structure of ganglioside G_{M1}).

TAY-SACHS DISEASE IS AN INHERITED DISORDER OF GANGLIOSIDE BREAKDOWN

Gangliosides are found in highest concentration in the nervous system, particularly in gray matter, where they constitute 6% of the lipids. Gangliosides are degraded inside lysosomes by the sequential removal of their terminal sugars. Disorders of ganglioside breakdown can have serious clinical consequences. In *Tay-Sachs disease,* the symptoms are usually evident before an affected infant is a year old. Weakness and retarded psychomotor development are typical early symptoms. The child is demented and blind by age two, and usually dead before age three. Striking pathological changes occur in the nervous system, where neurons become enormously swollen with lipid-filled lysosomes.

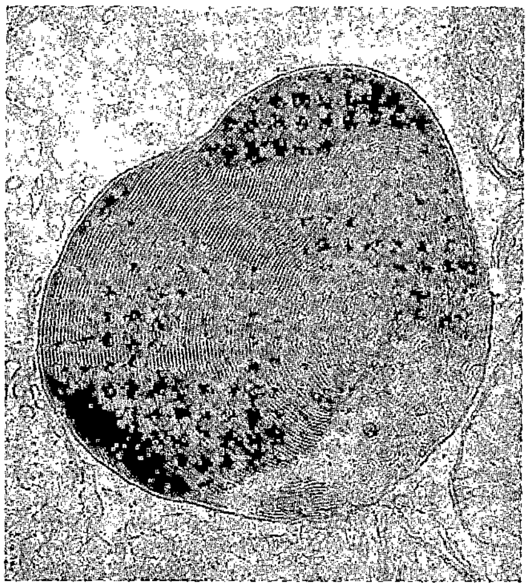

Figure 27-8
Electron micrograph of a lysosome containing an abnormal amount of lipid. [Courtesy of Dr. George Palade.]

The ganglioside content of the brain of an infant with Tay-Sachs disease is greatly elevated. *The concentration of ganglioside G_{M2} is many times higher than normal because its terminal* N-*acetylgalactosamine residue is removed very slowly or not at all.* The missing or deficient enzyme is a specific *β-*N-*acetylhexosaminidase.* Tay-Sachs disease is inherited as an autosomal recessive. The carrier rate is 1/30 in Jewish Americans originally from eastern Europe and 1/300 in non-Jewish Americans. Hence, the incidence of the disease is about 100 times higher in Jewish Americans. Tay-Sachs disease can be diagnosed during fetal development. Amniotic fluid is obtained by amniocentesis and assayed for *β-N*-acetylhexosaminidase activity.

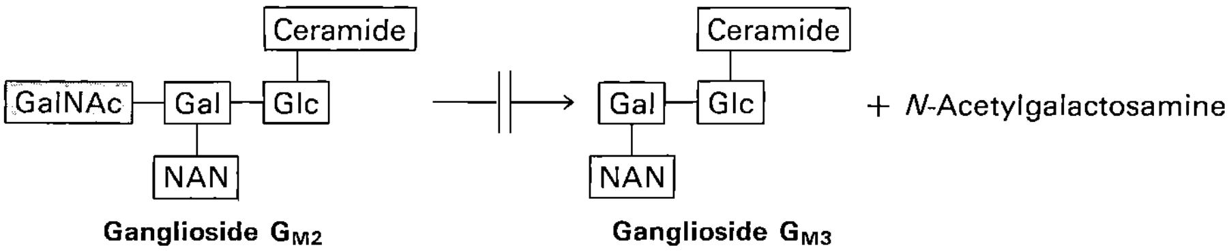

CHOLESTEROL IS SYNTHESIZED FROM ACETYL COENZYME A

We now turn to the synthesis of cholesterol, a steroid that modulates the fluidity of eukaryotic membranes (p. 279). Cholesterol is also the precursor of steroid hormones such as progesterone, testosterone, estradiol, and cortisol. An early and key clue concerning the synthesis of cholesterol

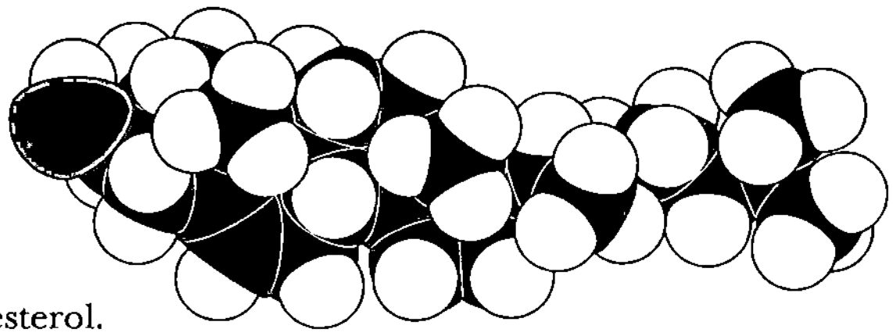

Figure 27-9
Model of cholesterol.

came from the work of Konrad Bloch in the 1940s. Acetate isotopically labeled in its carbon atoms was prepared and fed to rats. The cholesterol that was synthesized by these rats contained the isotopic label, which showed that acetate is a precursor of cholesterol. In fact, *all 27 carbon atoms of cholesterol are derived from acetyl CoA.* Further insight came from studies in which cholesterol was synthesized from acetate labeled in either its methyl or its carboxyl carbon. Degradation of cholesterol synthesized from such labeled acetate showed the origin of each atom of the cholesterol molecule (Figure 27-10). This pattern played a crucial role in the formation and testing of hypotheses concerning the pathway of cholesterol synthesis.

> "Cholesterol is the most highly decorated small molecule in biology. Thirteen Nobel Prizes have been awarded to scientists who devoted major parts of their careers to cholesterol. Ever since it was isolated from gallstones in 1784, cholesterol has exerted an almost hypnotic fascination for scientists from the most diverse areas of science and medicine. . . . Cholesterol is a Janus-faced molecule. The very property that makes it useful in cell membranes, namely its absolute insolubility in water, also makes it lethal."
>
> Michael Brown and Joseph Goldstein
> *Nobel Lectures (1985)*
> © The Nobel Foundation, 1985

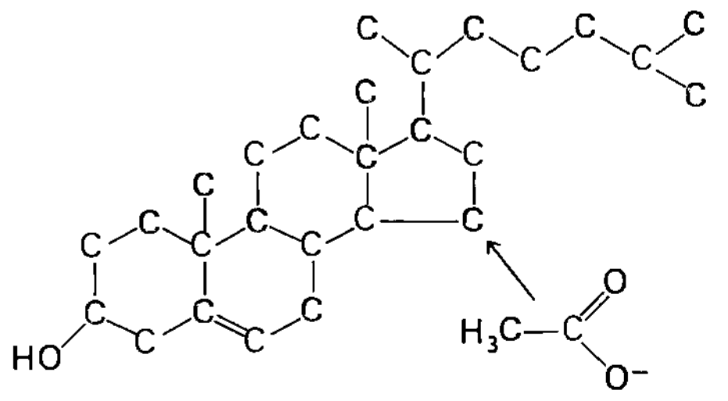

Figure 27-10
Isotope-labeling pattern of cholesterol synthesized from acetate labeled in its methyl (green) or carboxyl (red) carbon atom.

MEVALONATE AND SQUALENE ARE INTERMEDIATES IN THE SYNTHESIS OF CHOLESTEROL

The next major clue came from the discovery that *squalene,* a C_{30} hydrocarbon, is an intermediate in the synthesis of cholesterol. Squalene consists of six *isoprene* units.

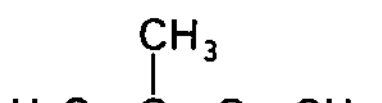

Isoprene

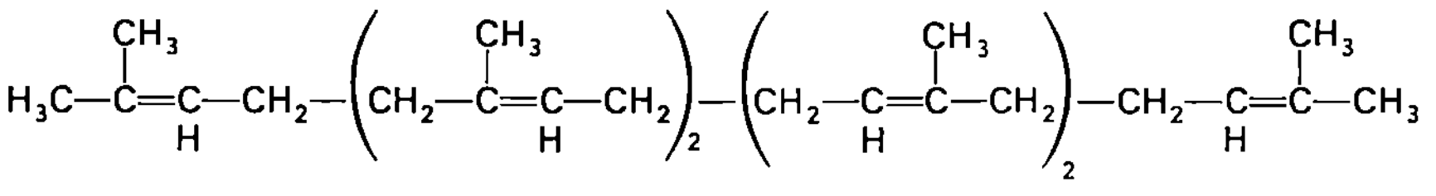

Squalene

This finding raised the question of how isoprene units are formed from acetate.

$$\underset{C_2}{\text{Acetate}} \longrightarrow \underset{C_5}{[\text{isoprene}]} \longrightarrow \underset{C_{30}}{\text{squalene}} \longrightarrow \underset{C_{27}}{\text{cholesterol}}$$

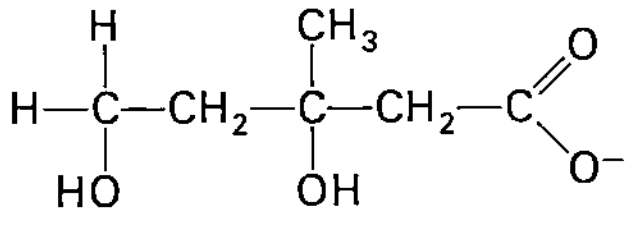

Mevalonate

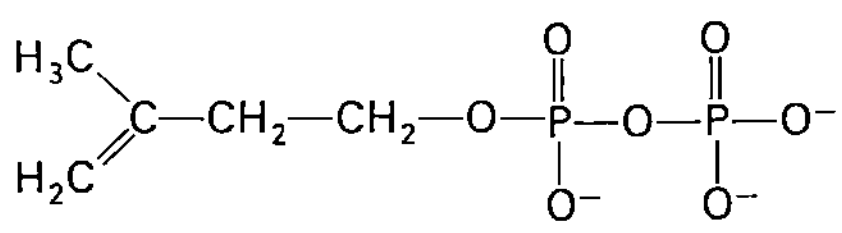

Isopentenyl pyrophosphate

The answer came unexpectedly from unrelated studies of bacterial mutants. It was found that *mevalonate* could substitute for acetate to meet the nutritional needs of acetate-requiring mutants. The discovery of mevalonate was a key step in elucidating the pathway of cholesterol biosynthesis because it was soon recognized that this C_6 acid could decarboxylate to yield the postulated C_5 isoprene intermediate. Isotopic labeling studies then showed that mevalonate is indeed a precursor of squalene and that it could be made from acetate. The activated isoprene intermediate proved to be *isopentenyl pyrophosphate,* which is formed by decarboxylation of a derivative of mevalonate. A pathway for the synthesis of cholesterol from acetate could then be outlined.

$$\underset{C_2}{\text{Acetate}} \longrightarrow \underset{C_6}{\text{mevalonate}} \longrightarrow$$
$$\underset{C_5}{\text{isopentenyl pyrophosphate}} \longrightarrow \underset{C_{30}}{\text{squalene}} \longrightarrow \underset{C_{27}}{\text{cholesterol}}$$

ACETYL CoA AND ACETOACETYL CoA CONDENSE TO FORM 3-HMG CoA, THE PRECURSOR OF MEVALONATE

The first stage in the synthesis of cholesterol is the formation of isopentenyl pyrophosphate from acetyl CoA. This set of reactions starts with the formation of 3-hydroxy-3-methylglutaryl CoA (3-HMG CoA) from acetyl CoA and acetoacetyl CoA. One of the fates of 3-hydroxy-3-methylglutaryl CoA, its cleavage to acetyl CoA and acetoacetate, was discussed previously in regard to the formation of ketone bodies (p. 612). Alternatively, it can

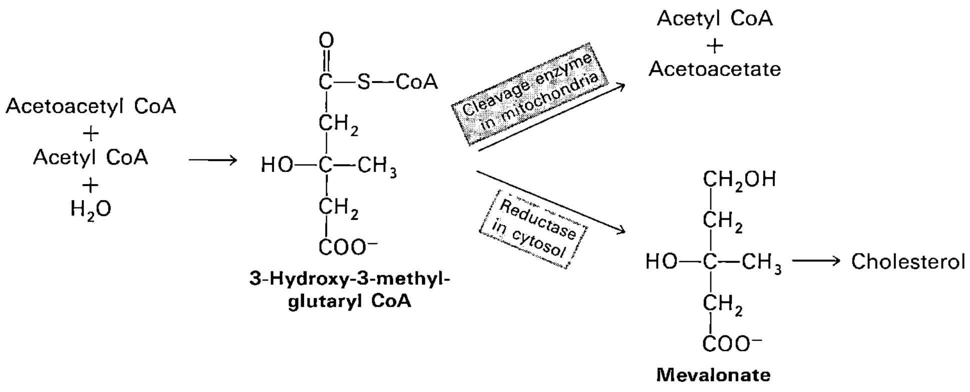

Figure 27-11
Synthesis and fates of 3-hydroxy-3-methylglutaryl CoA.

be reduced to mevalonate (Figure 27-11). 3-Hydroxy-3-methylglutaryl CoA is present both in the cytosol and in the mitochondria of liver cells. The mitochondrial pool of this intermediate is mainly a precursor of ketone bodies, whereas the cytoplasmic pool gives rise to mevalonate for the synthesis of cholesterol.

The synthesis of mevalonate is the committed step in cholesterol formation. The enzyme catalyzing this irreversible step, 3-hydroxy-3-methylglutaryl CoA reductase, is an important control site in cholesterol biosynthesis, as will be discussed shortly.

$$\text{3-Hydroxy-3-methylglutaryl CoA} + 2\ \text{NADPH} + 2\ \text{H}^+ \longrightarrow \text{mevalonate} + 2\ \text{NADP}^+ + \text{CoA}$$

Mevalonate is converted into *3-isopentenyl pyrophosphate* by three consecutive reactions involving ATP (Figure 27-12). In the last step, the release of CO_2 from 5-pyrophosphomevalonate occurs in concert with the hydrolysis of ATP to ADP and P_i.

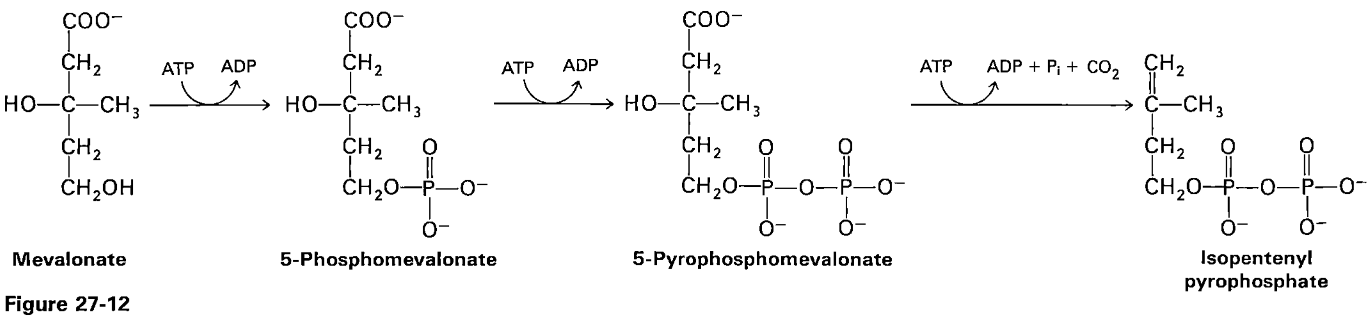

Figure 27-12
Synthesis of isopentenyl pyrophosphate from mevalonate.

SQUALENE (C_{30}) IS SYNTHESIZED FROM SIX MOLECULES OF ISOPENTENYL PYROPHOSPHATE (C_5)

Squalene is synthesized from isopentenyl pyrophosphate by the reaction sequence

$$C_5 \longrightarrow C_{10} \longrightarrow C_{15} \longrightarrow C_{30}$$

This stage in the synthesis of cholesterol starts with the isomerization of *isopentenyl pyrophosphate* to *dimethylallyl pyrophosphate.*

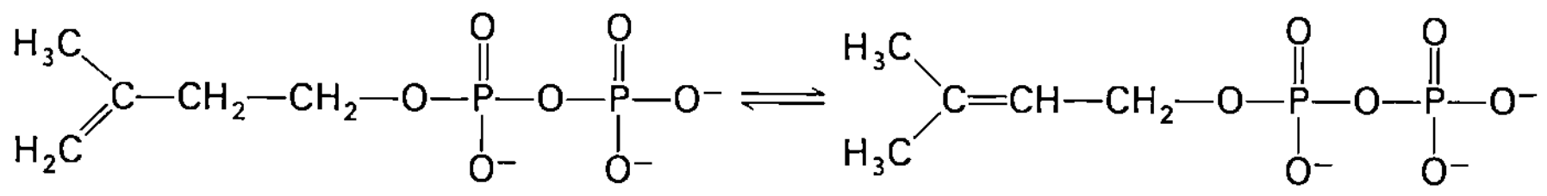

Isopentenyl pyrophosphate **Dimethylallyl pyrophosphate**

These isomeric C_5 units condense to form a C_{10} compound: an allylic carbonium ion formed from dimethylallyl pyrophosphate is attacked by isopentenyl pyrophosphate to form *geranyl pyrophosphate* (Figure 27-13).

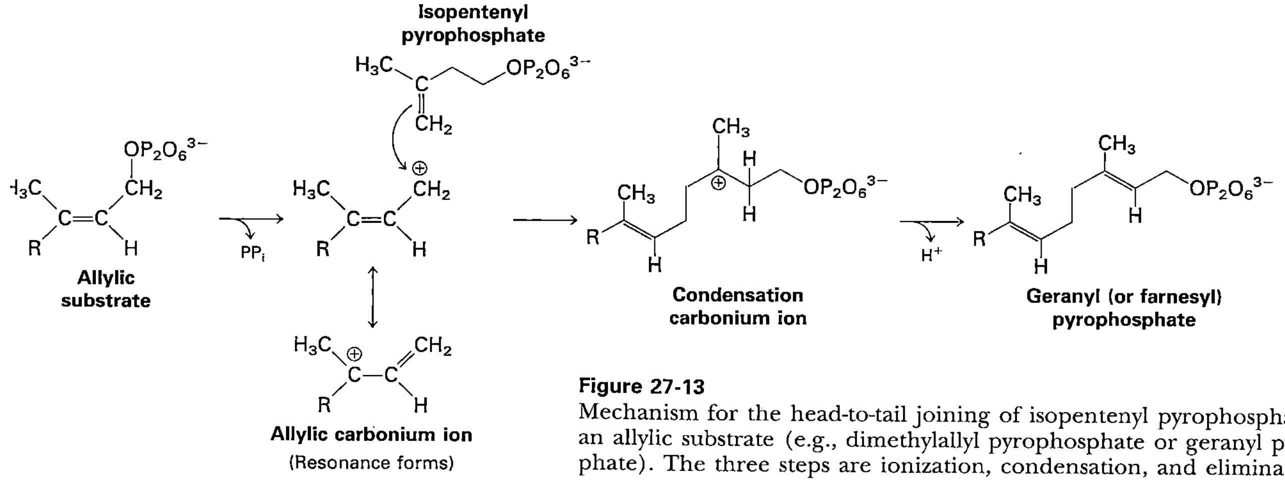

Figure 27-13
Mechanism for the head-to-tail joining of isopentenyl pyrophosphate to an allylic substrate (e.g., dimethylallyl pyrophosphate or geranyl pyrophosphate). The three steps are ionization, condensation, and elimination.

The same kind of reaction occurs again: geranyl pyrophosphate is converted into an allylic carbonium ion, which is attacked by isopentenyl pyrophosphate. The resulting C_{15} compound is called *farnesyl pyrophosphate.*

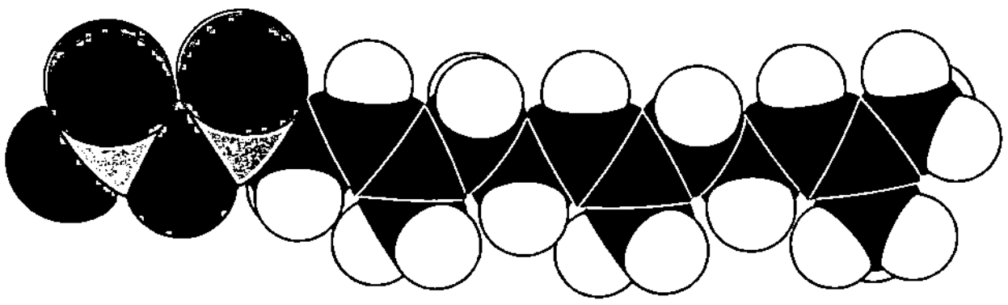

Figure 27-14
Model of farnesyl pyrophosphate.

The last step in the synthesis of *squalene* is a reductive condensation of two molecules of farnesyl pyrophosphate:

$$\underset{C_{15}}{2\ \text{Farnesyl pyrophosphate}} + \text{NADPH} \longrightarrow \underset{C_{30}}{\text{squalene}} + 2\ PP_i + NADP^+ + H^+$$

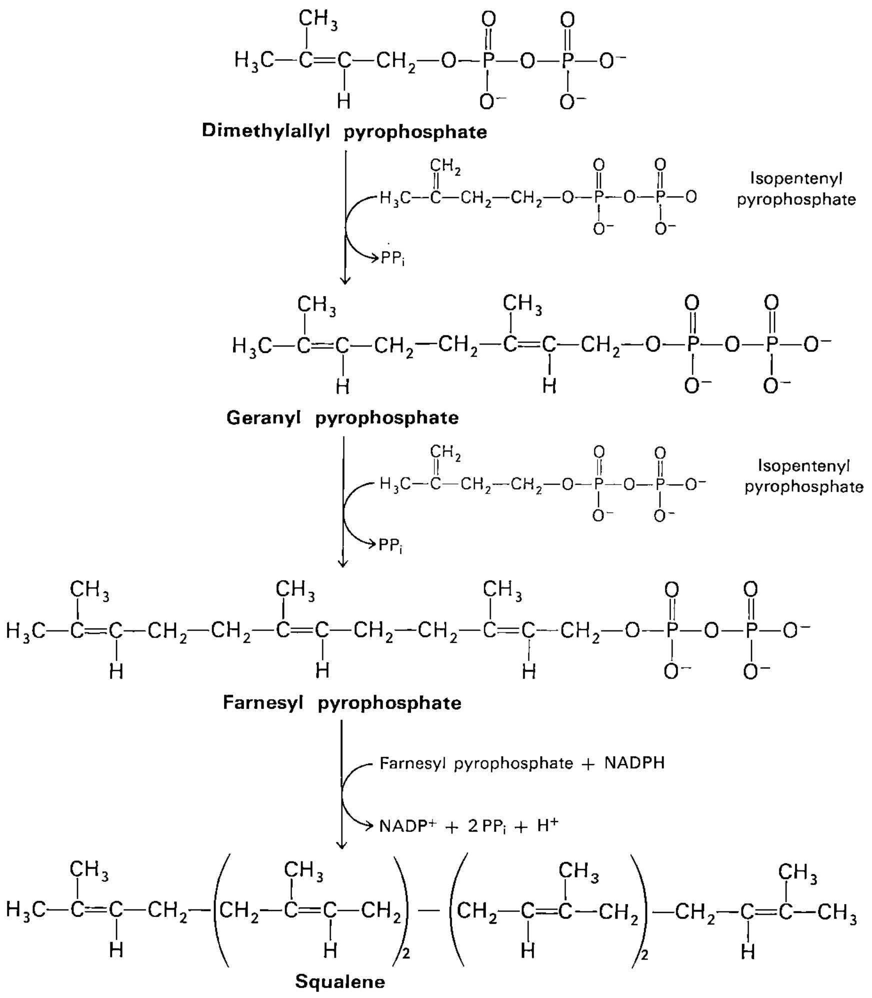

Figure 27-15
Synthesis of squalene from dimethylallyl pyrophosphate, an isomer of isopentenyl pyrophosphate. The joining of two C_{15} units to form squalene is a tail-to-tail condensation, in contrast with the preceding condensations, which are head-to-tail.

The reactions leading from C_5 units to squalene, a C_{30} isoprenoid, are shown in Figure 27-15.

SQUALENE EPOXIDE CYCLIZES TO LANOSTEROL, WHICH IS CONVERTED INTO CHOLESTEROL

The final stage of cholesterol biosynthesis starts with the cyclization of squalene (Figure 27-16). This stage, in contrast with the preceding ones, *requires molecular oxygen. Squalene epoxide,* the reactive intermediate, is formed in a reaction that uses O_2 and NADPH. Squalene epoxide is then cyclized to *lanosterol* by a cyclase. This remarkable closure is achieved by the concerted movement of electrons through four double bonds and the migration of two methyl groups. Finally, lanosterol is converted into *cholesterol* by the removal of three methyl groups, the reduction of one double bond by NADPH, and the migration of the other double bond. It is interesting to note that *cholesterol appeared only after the earth's atmosphere became aerobic.* Cholesterol is ubiquitous in eukaryotes but absent from most prokaryotes.

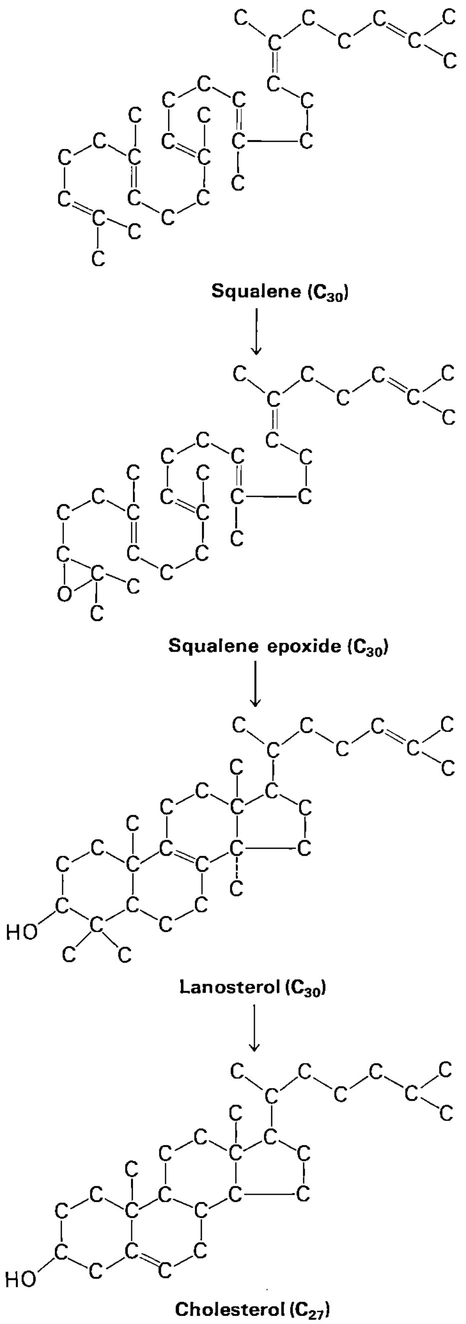

Figure 27-16
Synthesis of cholesterol from squalene.

BILE SALTS DERIVED FROM CHOLESTEROL FACILITATE THE DIGESTION OF LIPIDS

Bile salts are polar derivatives of cholesterol. These compounds are highly effective *detergents* because they contain both polar and nonpolar regions. Bile salts are synthesized in the liver, stored and concentrated in the gallbladder, and then released into the small intestine. Bile salts, the major constituent of bile, *solubilize dietary lipids.* The effective increase in the surface area of the lipids has two consequences: it promotes their hydrolysis by lipases and facilitates their absorption by the intestine. Bile salts are also the major breakdown products of cholesterol.

Cholesterol is converted into trihydroxycoprostanoate and then into *cholyl CoA,* the activated intermediate in the synthesis of most bile salts (Figure 27-17). The activated carboxyl carbon of cholyl CoA then reacts with the amino group of glycine to form *glycocholate,* or with the amino group of taurine ($H_2N{-}CH_2{-}CH_2{-}SO_3^-$), derived from cysteine, to form *taurocholate. Glycocholate is the major bile salt.*

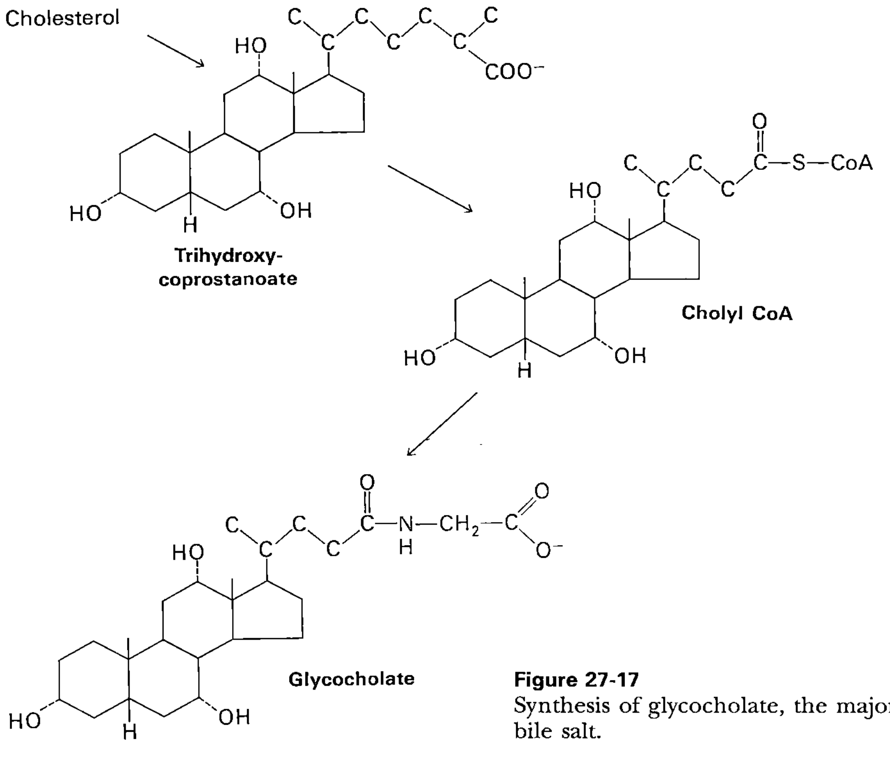

Figure 27-17
Synthesis of glycocholate, the major bile salt.

HMG CoA REDUCTASE PLAYS A KEY ROLE IN SETTING THE RATE OF CHOLESTEROL SYNTHESIS

Cholesterol can be obtained from the diet, or it can be synthesized de novo. The major site of cholesterol synthesis in mammals is the liver. Appreciable amounts of cholesterol are also formed by the intestine. An adult on a low-cholesterol diet typically synthesizes about 800 mg of cholesterol per day. The rate of cholesterol formation by these organs is highly responsive to the cellular level of cholesterol. *This feedback regulation is mediated primarily by changes in the amount and activity of 3-hydroxy-3-methylglutaryl CoA reductase (HMG CoA reductase).* As discussed previously, this enzyme catalyzes the formation of mevalonate, the committed step in cholesterol biosynthesis. HMG CoA reductase is controlled in multiple ways:

1. The rate of *synthesis of reductase mRNA* is controlled by *sterol regulatory element (SRE)*, a short DNA sequence on the 5′ side of the gene. In the presence of sterols, SRE markedly inhibits mRNA production.

2. The rate of *translation of reductase mRNA* is inhibited by nonsterol metabolites derived from mevalonate.

3. The degradation of the reductase is also stringently controlled. The enzyme is bipartite: its cytosolic domain carries out catalysis, and its membrane domain senses the level of derivatives of cholesterol and mevalonate. *A high level of these products leads to rapid degradation of the reductase* (Figure 27-18). Indeed, the amount of enzyme can be varied over a 200-fold range by a combination of these three regulatory devices.

4. *Phosphorylation decreases the activity of the reductase.* This enzyme, like acetyl CoA carboxylase (which catalyzes the committed step in fatty acid synthesis, p. 614), is switched off by an AMP-activated protein kinase. Thus, cholesterol synthesis ceases when the ATP level is low.

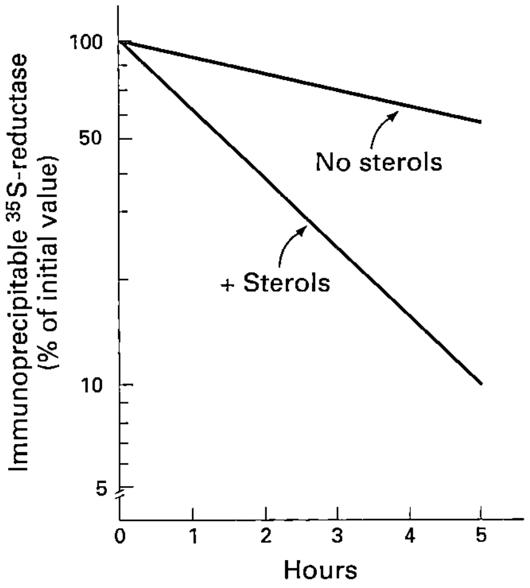

Figure 27-18
Sterols accelerate the degradation of HMG CoA reductase. Cells were pulse-labeled with ^{35}S-methionine and then placed in a nonradioactive medium. The amount of enzyme was measured by precipitating ^{35}S-labeled protein with a specific antibody. [After J.L. Goldstein and M.S. Brown. *Nature* 343(1990):428.]

CHOLESTEROL AND TRIACYLGLYCEROLS ARE TRANSPORTED TO TARGET CELLS BY LIPOPROTEINS

Cholesterol and triacylglycerols are transported in body fluids in the form of *lipoprotein particles.* The protein components of these macromolecular aggregates have two roles: *they solubilize hydrophobic lipids and contain cell-targeting signals.* Lipoproteins are classified according to increasing density: *chylomicrons, chylomicron remnants, very low density lipoproteins (VLDL), intermediate-density lipoproteins (IDL), low-density lipoproteins (LDL),* and *high-density lipoproteins (HDL)* (Table 27-1). Each is a particle consisting of a core of hydrophobic lipids surrounded by a shell of polar lipids and apoproteins. Ten principal apoproteins—A-1, A-2, A-4, B-48, B-100, C-1, C-2, C-3, D, and E—have been isolated and characterized. They are synthesized and secreted by the liver and the intestine.

Triacylglycerols, cholesterol, and other lipids obtained from the diet are carried away from the intestine in the form of large *chylomicrons* (180

Table 27-1
Properties of plasma lipoproteins

Lipoproteins	*Major core lipids*	*Apoproteins*	*Mechanism of lipid delivery*
Chylomicron	Dietary triacylglycerols	B-48, C, E	Hydrolysis by lipoprotein lipase
Chylomicron remnant	Dietary cholesterol esters	B-48, E	Receptor-mediated endocytosis by liver
Very low density lipoprotein (VLDL)	Endogenous triacylglycerols	B-100, C, E	Hydrolysis by lipoprotein lipase
Intermediate-density lipoprotein (IDL)	Endogenous cholesterol esters	B-100, E	Receptor-mediated endocytosis by liver and conversion to LDL
Low-density lipoprotein (LDL)	Endogenous cholesterol esters	B-100	Receptor-mediated endocytosis by liver and other tissues
High-density lipoprotein (HDL)	Endogenous cholesterol esters	A	Transfer of cholesterol esters to IDL and LDL

After M.S. Brown and J.L. Goldstein, *The Pharmacological Basis of Therapeutics,* 7th ed. (ed. by A.G. Gilman, L.S. Goodman, T.W. Rall, and F. Murad), Macmillan, 1985, p. 828.

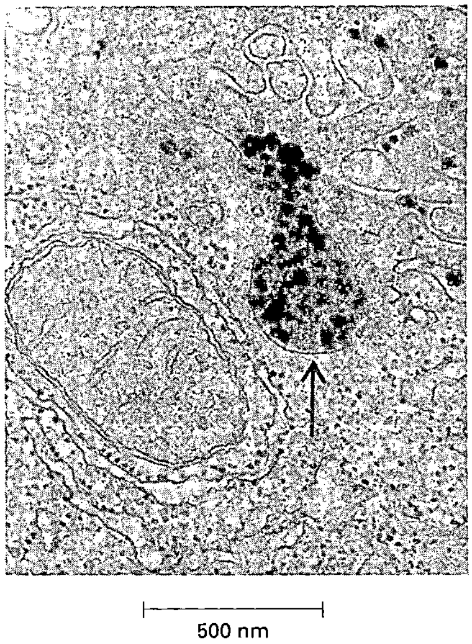

Figure 27-19
Electron micrograph of a part of a liver cell actively engaged in the synthesis and secretion of very low density lipoprotein (VLDL). The arrow points to a vesicle that is releasing its content of VLDL particles. [Courtesy of Dr. George Palade.]

to 500 nm in diameter). They have a very low density ($d < 0.94\ g/cm^3$) because triacylglycerols constitute ~99% of their content. Apolipoprotein B-48 (apo B-48), a large protein (240 kd), forms an amphipathic spherical shell around the fat globule; the external face of this shell is hydrophilic. The triacylglycerols in chylomicrons are hydrolyzed by *lipoprotein lipases* located on the lining of blood vessels in muscle and other tissues that use fatty acids as fuels and precursors of endogenously synthesized fat. The cholesterol-rich residues, known as *chylomicron remnants,* are then taken up by the liver.

Triacylglycerols and cholesterol in excess of the liver's own needs are exported into the blood in the form of *very low density lipoproteins (VLDL)* ($d < 1.006\ g/cm^3$) (Figure 27-19). These particles are stabilized by two lipoproteins, apo B-100 and apo E (34 kd). Apo B-100, one of the largest proteins known (513 kd), is a longer version of apo B-48. Both apo B proteins are encoded by the same gene and produced from the same initial RNA transcript. In the intestine, RNA editing (p. 860) modifies the transcript to generate the mRNA for apo B-48, the truncated form. Triacylglycerols in VLDL, as in chylomicrons, are hydrolyzed by lipases on capillary surfaces. The resulting remnants, which are rich in cholesterol esters, are called *intermediate-density lipoproteins (IDL)* ($1.006 < d < 1.019\ g/cm^3$). These particles have two fates. Half of them are taken up by the liver, whereas the other half are converted into *low-density lipoprotein (LDL)* ($1.019 < d < 1.063\ g/cm^3$).

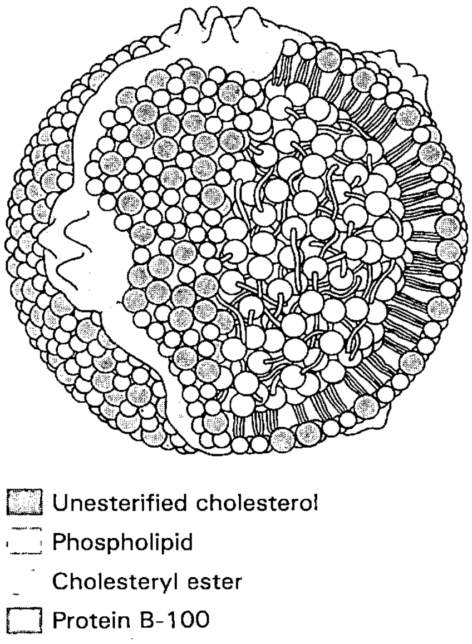

Figure 27-20
Schematic model of low-density lipoprotein (LDL). The conformation of the B-100 protein is not known. The diameter of the LDL particle is 22 nm (220 Å).

LDL, the major carrier of cholesterol in blood, has a diameter of 22 nm and a mass of about three million daltons. It contains a core of some 1500 esterified cholesterol molecules; the most common fatty acyl chain in these esters is linoleate, a polyunsaturated fatty acid. This highly hydrophobic core is surrounded by a shell of phospholipids and unesterified cholesterols. The shell also contains a single copy of apo B-100, which is recognized by target cells. *The role of LDL is to transport cholesterol to peripheral tissues and regulate de novo cholesterol synthesis at these sites,* as described below. A different purpose is served by *high-density lipoprotein (HDL)* ($1.063 < d < 1.21\ g/cm^3$), which picks up cholesterol released into the plasma from dying cells and from membranes undergoing turnover. An acyltransferase in HDL esterifies these cholesterols, which are then rapidly shuttled to VLDL or LDL by a transfer protein.

THE LOW-DENSITY-LIPOPROTEIN RECEPTOR PLAYS A KEY ROLE IN CONTROLLING CHOLESTEROL METABOLISM

Cholesterol, a component of all eukaryotic plasma membranes, is essential for the growth and viability of cells in higher organisms. However, high serum levels of cholesterol cause disease and death by contributing to the formation of atherosclerotic plaques in arteries throughout the body. It is evident that cholesterol metabolism must be precisely regulated. The mode of control in the liver, the primary site of cholesterol synthesis, has already been discussed: dietary cholesterol reduces the activity and amount of 3-hydroxy-3-methylglutaryl CoA reductase, the enzyme catalyzing the committed step. Studies of cultured human fibroblasts by Michael Brown and Joseph Goldstein have been sources of insight into the control of cholesterol metabolism in nonhepatic cells. In general, cells outside the liver and intestine obtain cholesterol from the plasma rather than by synthesizing it de novo. Specifically, *their primary source of cholesterol is the low-density lipoprotein.* The steps in the uptake of cholesterol by the *LDL pathway* (Figure 27-21) are:

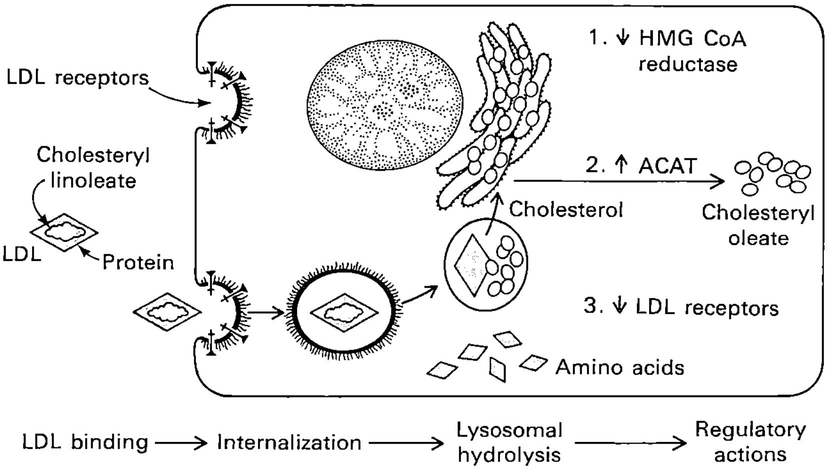

Figure 27-21
Steps in the low-density-lipoprotein pathway in cultured human fibroblasts. HMG CoA reductase denotes 3-hydroxy-3-methylglutaryl CoA reductase, and ACAT denotes acyl CoA:cholesterol acyltransferase. [After a drawing kindly provided by Dr. Michael Brown and Dr. Joseph Goldstein.]

1. Apolipoprotein B-100 on the surface of an LDL particle binds to a specific receptor protein on the plasma membrane of nonhepatic cells. The receptors for LDL are localized in specialized regions called *coated pits,* which contain *clathrin* (p. 936).

2. The receptor-LDL complex is internalized by *endocytosis*—that is, the plasma membrane in the vicinity of the complex invaginates and then fuses to form an endocytic vesicle (Figure 27-22).

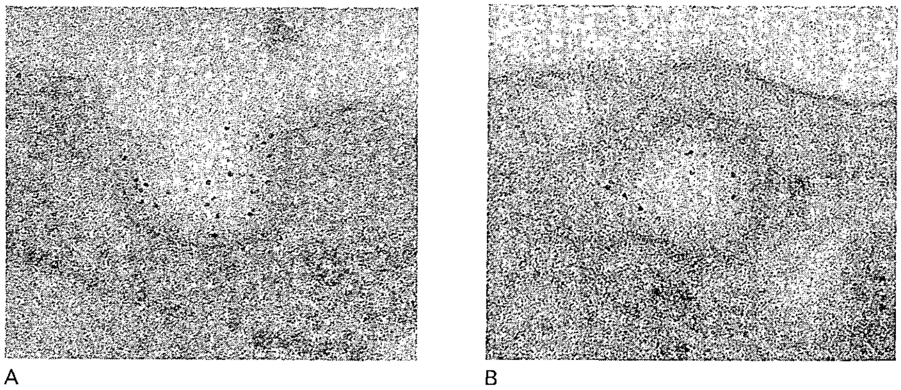

Figure 27-22
Endocytosis of LDL bound to its receptor on the surface of cultured human fibroblasts (LDL was visualized by conjugating it to ferritin): (A) electron micrograph showing LDL-ferritin (dark dots) bound to a coated-pit region of the cell surface; (B) this region invaginates and fuses to form an endocytic vesicle. [From R.G.W. Anderson, M.S. Brown, and J.L. Goldstein. *Cell* 10(1977): 351.]

3. These vesicles, containing LDL, subsequently fuse with *lysosomes*, which carry a wide array of degradative enzymes. The protein component of the LDL is hydrolyzed to free amino acids. The cholesterol esters in the LDL are hydrolyzed by a lysosomal acid lipase. The LDL receptor itself usually returns unscathed to the plasma membrane. The round-trip time for a receptor is about 10 minutes; in its lifetime of about a day, it may bring many LDL particles into the cell.

4. *The released unesterified cholesterol can then be used for membrane biosynthesis.* Alternatively, it can be *reesterified for storage inside the cell.* In fact, free cholesterol activates acyl CoA:cholesterol acyltransferase (ACAT), the enzyme catalyzing this reaction. Reesterified cholesterol contains mainly oleate and palmitoleate, which are monounsaturated fatty acids, in contrast with the cholesterol esters in LDL, which are rich in linoleate, a polyunsaturated fatty acid (see Table 24-1).

The LDL receptor is itself subject to feedback regulation. *When cholesterol is abundant inside the cell, new LDL receptors are not synthesized, and so the uptake of additional cholesterol from plasma LDL is blocked.* The gene for the LDL receptor, like that for the reductase, contains a sterol regulatory element that controls the rate of mRNA synthesis.

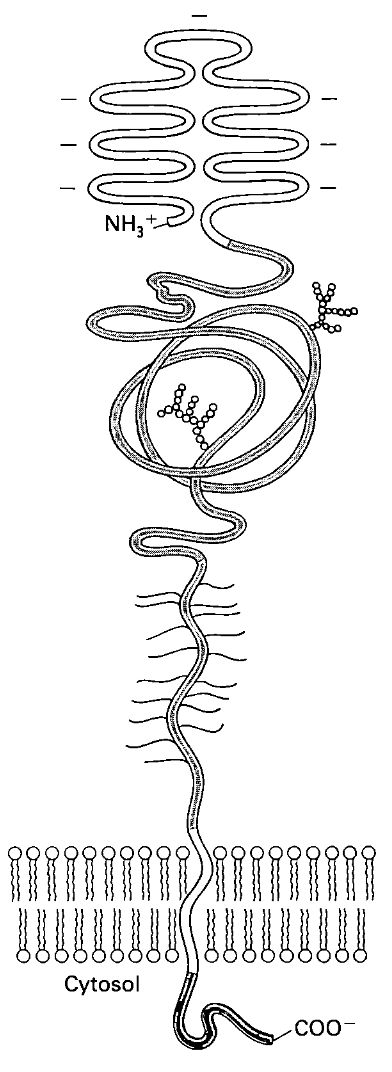

Figure 27-23
The LDL receptor consists of five domains with different functions: an LDL-binding domain, 292 residues (green); a domain bearing *N*-linked sugars, 350 residues (gray); a domain bearing *O*-linked sugars, 58 residues (blue); a membrane-spanning domain, 22 residues (yellow); and a cytosolic domain, 50 residues (red).

THE LDL RECEPTOR IS A TRANSMEMBRANE PROTEIN WITH FIVE DIFFERENT FUNCTIONAL DOMAINS

The cloning and sequencing of the cDNA for the human LDL receptor has revealed that this 115-kd protein consists of five domains (Figure 27-23). The amino-terminal domain of the mature receptor consists of a cysteine-rich sequence of about 40 residues that is repeated, with some variation, seven times. A cluster of negatively charged side chains in this *LDL-binding domain* interacts with a positively charged site on an apo B-100 molecule on the surface of an LDL particle. Protonation of the glutamate and aspartate side chains of the receptor in acidic endosomes (p. 939) leads to the release of LDL from its receptor. The second domain of the LDL receptor is homologous to part of the precursor of epidermal growth factor (EGF, p. 351). It contains two *N*-linked oligosaccharide chains. The third domain, which is very rich in serines and threonines, contains *O*-linked sugars. These oligosaccharides, like those in glycophorin (p. 283), may function as struts to keep the receptor extended from the membrane so that the amino-terminal domain is accessible to LDL. The fourth domain contains 22 hydrophobic residues that span the plasma membrane. The fifth domain, consisting of 50 residues, emerges on the cytosolic side of the membrane, where it controls the interaction of the receptor with coated pits and participates in endocytosis. The gene for the LDL receptor consists of 18 exons, which correspond closely to structural units of the protein. *The LDL receptor is a striking example of a mosaic protein encoded by a gene that was assembled by exon shuffling.*

ABSENCE OF THE LDL RECEPTOR LEADS TO HYPERCHOLESTEROLEMIA AND ATHEROSCLEROSIS

The physiologic importance of the LDL receptor was revealed by Brown and Goldstein's pioneering studies of *familial hypercholesterolemia (FH).* The total concentration of cholesterol and LDL in the plasma is markedly elevated in this genetic disorder, which results from a mutation at a single autosomal locus. The cholesterol level in the plasma of homozygotes is typically 680 mg/dl, compared with 300 mg/dl in heterozygotes (clinical assay results are often expressed in mg/dl, which is equal to mg per 100 ml). A value of 175 mg/dl is regarded as desirable, but most Americans have higher levels. *In familial hypercholesterolemia, cholesterol is deposited in various tissues because of the high concentration of LDL-cholesterol in the plasma.* Nodules of cholesterol called *xanthomas* are prominent in skin and tendons. More harmful is the deposition of cholesterol in atherosclerotic plaques, causing arterial narrowing and leading to heart attacks. In fact, *most homozygotes die of coronary artery disease in childhood.* The disease in heterozygotes has a milder and more variable clinical course.

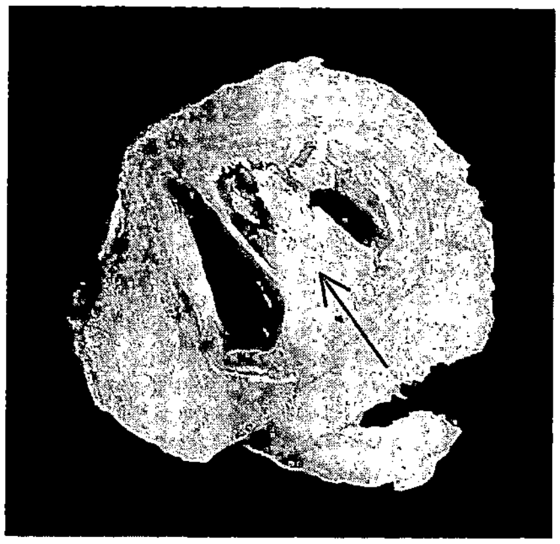

5 mm

Figure 27-24
An atherosclerotic plaque (marked by the arrow) blocks most of the lumen of this blood vessel. The plaque is rich in cholesterol. [Courtesy of Dr. Jeffrey Sklar.]

The molecular defect in most cases of familial hypercholesterolemia is an absence or deficiency of functional receptors for LDL. Homozygotes have almost no receptors for LDL, whereas heterozygotes have about half the normal number. Consequently, the entry of LDL into liver and other cells is impaired, leading to an increased plasma level of LDL. Furthermore, less IDL enters liver cells because IDL entry, too, is mediated by the LDL receptor. Consequently, IDL stays in the blood longer in FH, and more of it is converted into LDL than in normal people. All deleterious consequences of an absence or deficiency of the LDL receptor can be attributed to the ensuing elevated level of LDL-cholesterol in the blood.

LOVASTATIN LOWERS THE BLOOD CHOLESTEROL LEVEL BY INHIBITING HMG CoA REDUCTASE

Homozygous FH can be treated by transplanting a normal liver. A more generally applicable therapy is available for heterozygotes (1 in 500 people). *The goal is to stimulate the single normal gene to produce more than the customary number of LDL receptors.* Studies of cultured fibroblasts showed that the production of LDL receptors is controlled by the cell's need for cholesterol. When cholesterol is required, the amount of mRNA for LDL receptor rises and more receptor is found on the cell surface. This state can be induced by inhibiting the intestinal reabsorption of bile salts (which promote the absorption of dietary cholesterol, p. 696) and by blocking cholesterol synthesis. The reabsorption of bile is impeded by oral administration of positively charged polymers that bind negatively charged bile salts and are not themselves absorbed.

Cholesterol synthesis can be effectively blocked by *lovastatin* (also called *mevinolin*) (Figure 27-25), a potent competitive inhibitor (K_i of 1 nM) of HMG CoA reductase, the key control point in the biosynthetic pathway. The consequent increase in the number of LDL receptors on liver cells leads to a decrease in the LDL level in blood. Indeed, plasma cholesterol levels decrease by 50% in many patients given both lovastatin and inhibitors of bile salt reabsorption. There is much interest in drugs that lower cholesterol levels because atherosclerosis is the leading cause of death in industrialized societies. Lovastatin and other inhibitors of HMG CoA reductase are widely used to lower the plasma cholesterol level.

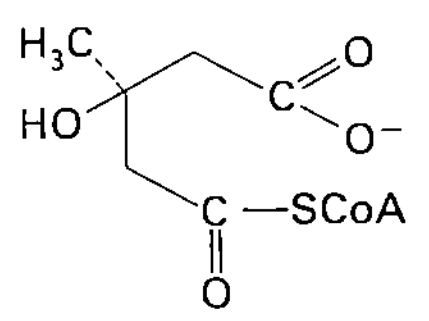

3-Hydroxy-3-methylglutaryl CoA

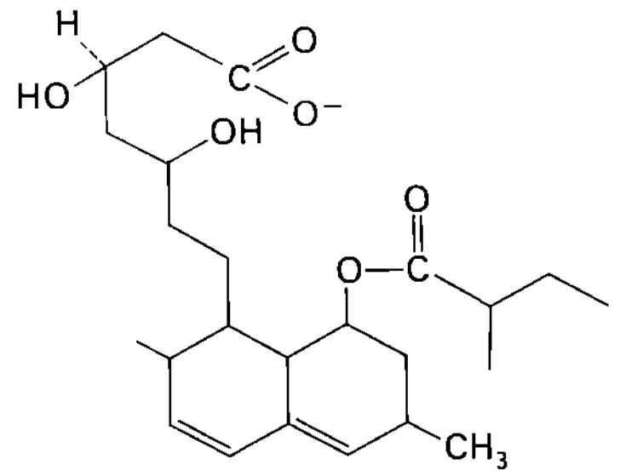

Lovastatin (Mevinolin)

Figure 27-25
Lovastatin (mevinolin), a potent competitive inhibitor of HMG CoA reductase, resembles 3-hydroxy-3-methylglutaryl CoA, the substrate. The active carboxylate form of lovastatin formed by hydrolysis of the lactone is shown here.

NOMENCLATURE OF STEROIDS

Carbon atoms in steroids are numbered as shown for cholesterol (Figure 27-26). The rings in steroids are named A, B, C, and D. Cholesterol contains two angular methyl groups: the C-19 methyl group is attached to C-10, and the C-18 methyl group is attached to C-13. By convention, the C-18 and C-19 methyl groups of cholesterol are *above* the plane containing the four rings. A substituent that is above the plane is termed *β-oriented* and is shown by a *solid* bond. In contrast, a substituent that is below the plane is *α-oriented* and denoted by a *dashed* or dotted line.

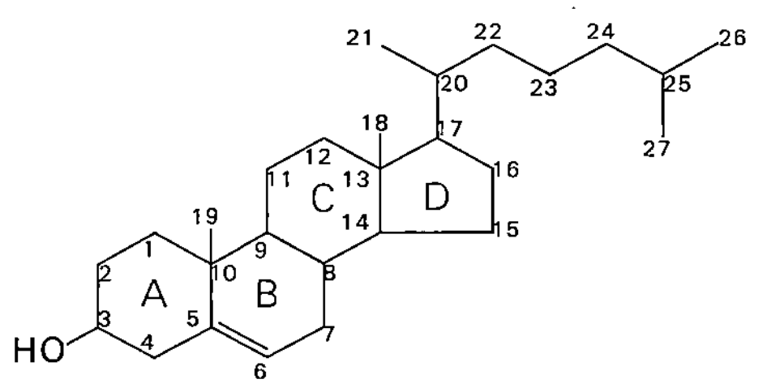

Figure 27-26
Numbering of the carbon atoms of cholesterol.

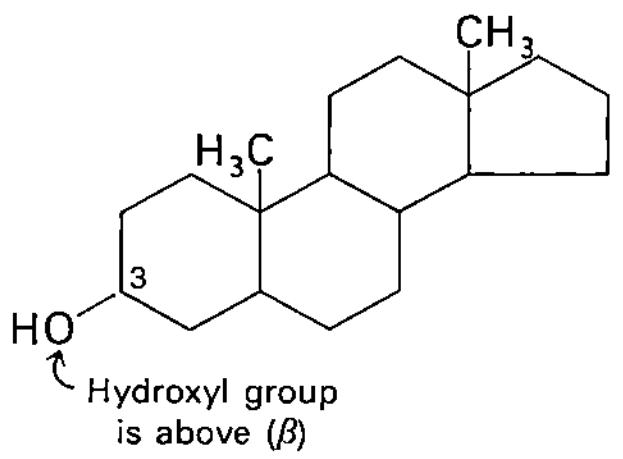

3β-Hydroxy

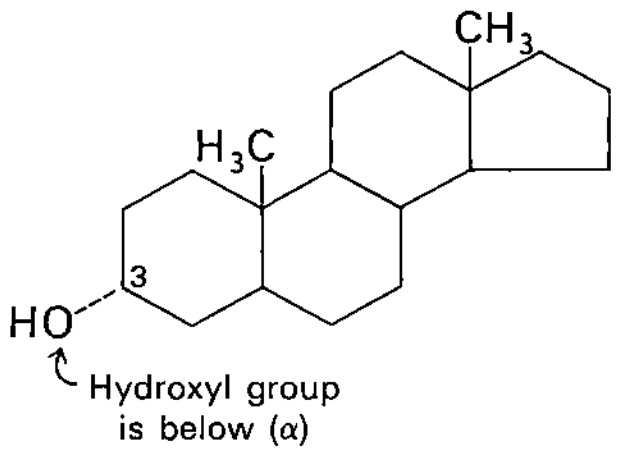

3α-Hydroxy

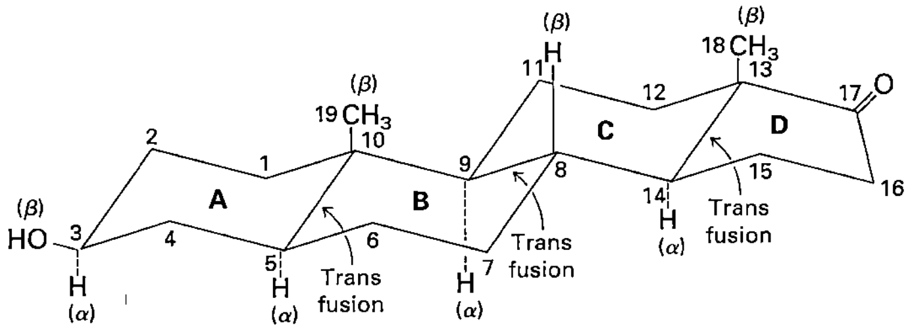

Stereochemical diagram of steroid nucleus

If a hydrogen atom is attached to C-5, it can be α- or β-oriented. The A and B steroid rings are fused in a *trans* conformation if the C-5 hydrogen is *α-oriented,* and *cis* if it is *β-oriented.* The absence of a symbol for the C-5 hydrogen atom implies a trans fusion. The C-5 hydrogen atom is α-oriented in all steroid hormones that contain a hydrogen atom in that position. In contrast, bile salts have a β-oriented hydrogen atom at C-5. Thus, *a cis fusion is characteristic of the bile salts, whereas a trans fusion is characteristic of all steroid hormones that possess a hydrogen atom at C-5.* A trans fusion yields a nearly planar structure, whereas a cis fusion gives a buckled structure.

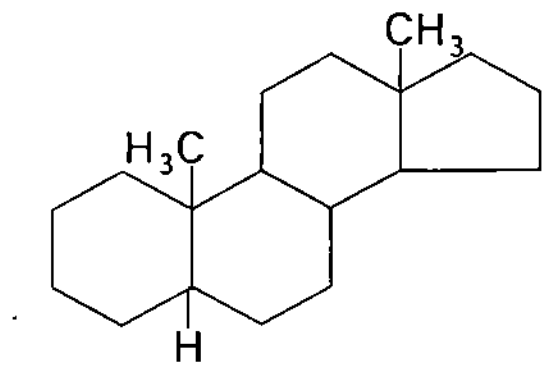

5β-Hydrogen
(A cis fusion)

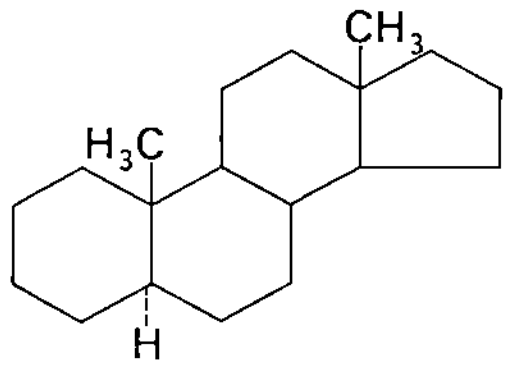

5α-Hydrogen
(A trans fusion)

STEROID HORMONES ARE DERIVED FROM CHOLESTEROL

Cholesterol is the precursor of the five major classes of steroid hormones: progestagens, glucocorticoids, mineralocorticoids, androgens, and estrogens (Figure 27-27). *Progesterone,* a *progestagen,* prepares the lining of the uterus for implantation of an ovum. Progesterone is also essential for the maintenance of pregnancy. *Androgens* (such as *testosterone*) are responsible for the development of male secondary sex characteristics, whereas *estrogens* (such as *estrone*) are required for the development of female secondary sex characteristics. Estrogens, along with progesterone, also participate in the ovarian cycle. *Glucocorticoids* (such as *cortisol*) promote gluconeogenesis and the formation of glycogen and enhance the degradation of fat and protein. They enable animals to respond to stress—indeed, the absence of glucocorticoids can be fatal. *Mineralocorticoids* (primarily *aldosterone*) act on the distal tubules of the kidney to increase the reabsorption of Na^+ and the excretion of K^+ and H^+, which lead to an increase in blood volume and blood pressure. The major sites of synthesis of these classes of hormones are the corpus luteum, progestagens; ovary, estrogens; testis, androgens; adrenal cortex, glucocorticoids and mineralocorticoids.

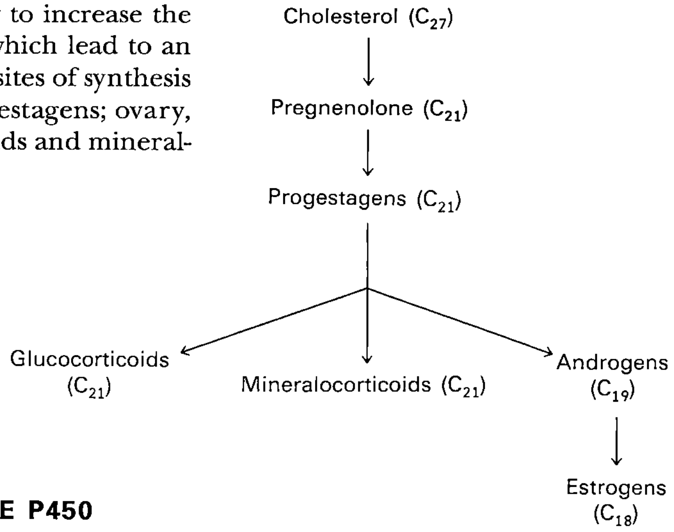

Figure 27-27
Biosynthetic relations of steroid hormones.

STEROIDS ARE HYDROXYLATED BY CYTOCHROME P450 MONOOXYGENASES THAT UTILIZE NADPH AND O_2

Hydroxylation reactions play a very important role in the synthesis of cholesterol from squalene and in the conversion of cholesterol into steroid hormones and bile salts. All these hydroxylations require *NADPH and* O_2. The oxygen atom of the incorporated hydroxyl group comes from O_2 rather than from H_2O, as shown by the use of ^{18}O-labeled O_2 and H_2O. One oxygen atom of the O_2 molecule goes into the substrate, whereas the other is reduced to water. The enzymes catalyzing these reactions are called *monooxygenases* (or *mixed-function oxygenases*). Recall that a monooxygenase also participates in the hydroxylation of phenylalanine to tyrosine (p. 647).

$$RH + O_2 + NADPH + H^+ \longrightarrow ROH + H_2O + NADP^+$$

Hydroxylation requires the activation of oxygen. In the synthesis of steroid hormones and bile salts, activation is accomplished by P450, a family of cytochromes that absorbs light maximally at 450 nm when complexed in vitro with exogenous CO. These membrane-anchored proteins (~50 kd) contain a heme prosthetic group. P450 is the terminal component of an *electron transport chain* in adrenal mitochondria and liver microsomes. The role of this assembly is hydroxylation rather than oxidative phosphorylation. NADPH transfers its high-potential electrons to a flavoprotein in this chain, which are then conveyed to *adrenodoxin*, a nonheme iron protein. Adrenodoxin transfers an electron to reduce the ferric form of P450 to the ferrous form (Figure 27-28). The binding of O_2 to the heme is followed by the acceptance of a second electron from adrenodoxin. At this point, the iron atom is in the +3 state, and the bound oxygen is in the −2 state. One of the bound oxygen atoms is then reduced to water. A hydrogen atom is abstracted from the substrate RH to form R·. This transient free radical acquires the other bound oxygen to form ROH, the hydroxylated product.

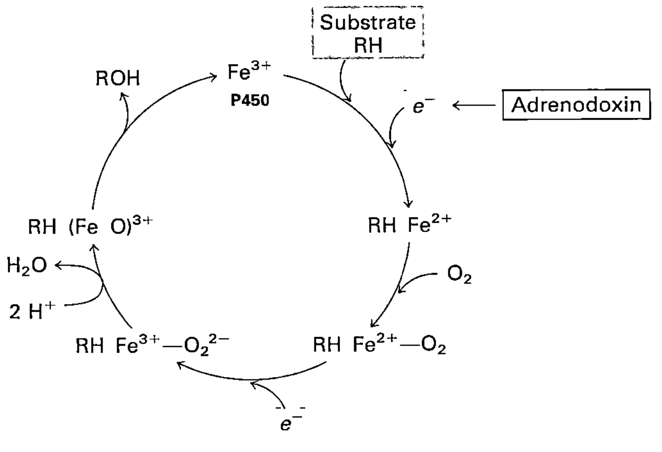

Figure 27-28
Proposed catalytic cycle of cytochrome P450. [After M.J. Coon, X. Ding, S.J. Pernecky, and A.D.N. Vaz. *FASEB J.* 6(1992):669.]

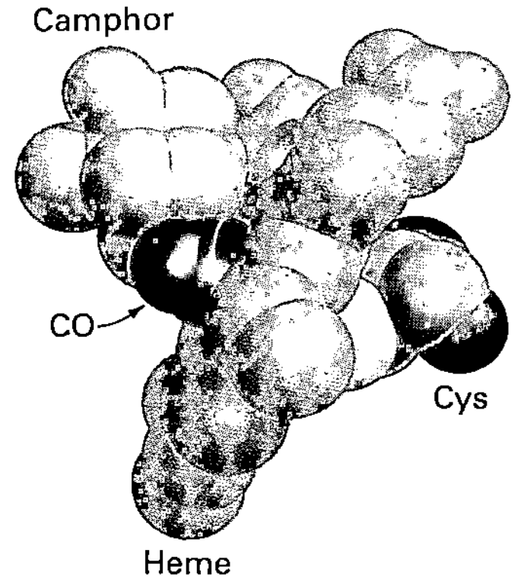

Figure 27-29
Structure of the active site of $P450_{CAM}$, a camphor-metabolizing, soluble P450 enzyme from *Pseudomonas putida.* The heme group is shown in light red. The sulfur atom (yellow) of a cysteine residue is coordinated to the heme iron. Carbon monoxide (blue) occupies the position that would normally be filled by O_2. Camphor (green) is hydroxylated by this enzyme. [Drawn from 3cpp.pdb. R. Raag and T.L. Poulos. *Biochemistry* 28(1989):7586.]

The cytochrome P450 system is also important in the *detoxification of foreign substances* (xenobiotic compounds). For example, the hydroxylation of phenobarbital, a barbiturate, *increases its solubility* and *facilitates its excretion.* Likewise, polycyclic aromatic hydrocarbons are hydroxylated by P450. The introduction of hydroxyl groups provides sites for conjugation with highly polar units (e.g., glucuronate or sulfate), which markedly increases the solubility of the modified aromatic molecule. However, the action of the P450 system is not always beneficial. *Some of the most powerful carcinogens are converted in vivo into a chemically reactive form.* This process of *metabolic activation* is usually carried out by the P450 system.

P450 enzymes in mammals are encoded by 10 gene families (their names begin with *CYP,* to recall *cy*tochrome *P*450). Indeed, humans have more than 100 genes that encode P450 enzymes with distinct specificities. Many P450 enzymes act on a broad range of substrates. The duration of action of many medications depends on their rate of inactivation by the P450 system. Prokaryotes too have these monooxygenases, as exemplified by $P450_{CAM}$ (Figure 27-29), a camphor-metabolizing enzyme in *Pseudomonas putida,* an aerobic microorganism that thrives on decaying organic matter.

PREGNENOLONE IS FORMED FROM CHOLESTEROL BY CLEAVAGE OF ITS SIDE CHAIN

Steroid hormones contain 21 or fewer carbon atoms, whereas cholesterol contains 27. The first stage in the synthesis of steroid hormones is the removal of a C_6 unit from the side chain of cholesterol to form *pregnenolone.* The side chain of cholesterol is hydroxylated at C-20 and then at C-22, and the bond between these carbon atoms is then cleaved by desmolase. Three NADPH and three O_2 are consumed in this remarkable six-electron oxidation.

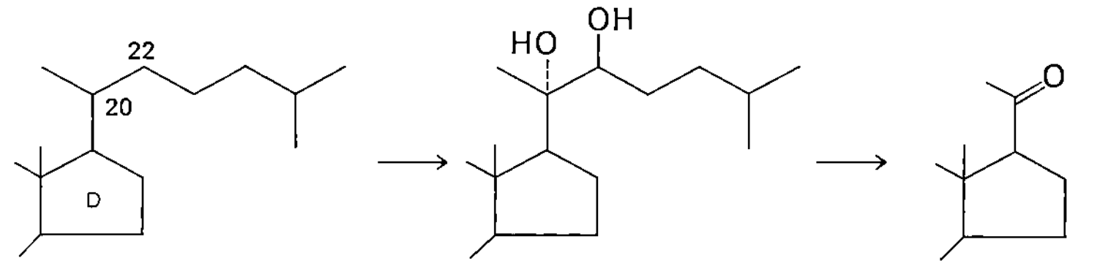

Adrenocorticotropic hormone (ACTH, or corticotropin), a polypeptide synthesized by the anterior pituitary gland, stimulates the conversion of cholesterol into pregnenolone, the precursor of all steroid hormones.

SYNTHESIS OF PROGESTERONE AND CORTICOIDS

Progesterone is synthesized from pregnenolone in two steps. The 3-hydroxyl group of pregnenolone is oxidized to a 3-keto group, and the Δ^5 double bond is isomerized to a Δ^4 double bond (Figure 27-30). *Cortisol,* the major glucocorticoid, is synthesized from progesterone by hydroxylations at C-17, C-21, and C-11; C-17 must be hydroxylated before C-21, whereas hydroxylation at C-11 may occur at any stage. The enzymes catalyzing these hydroxylations are highly specific, as shown by some inherited disorders. The initial step in the synthesis of *aldosterone,* the major mineral-

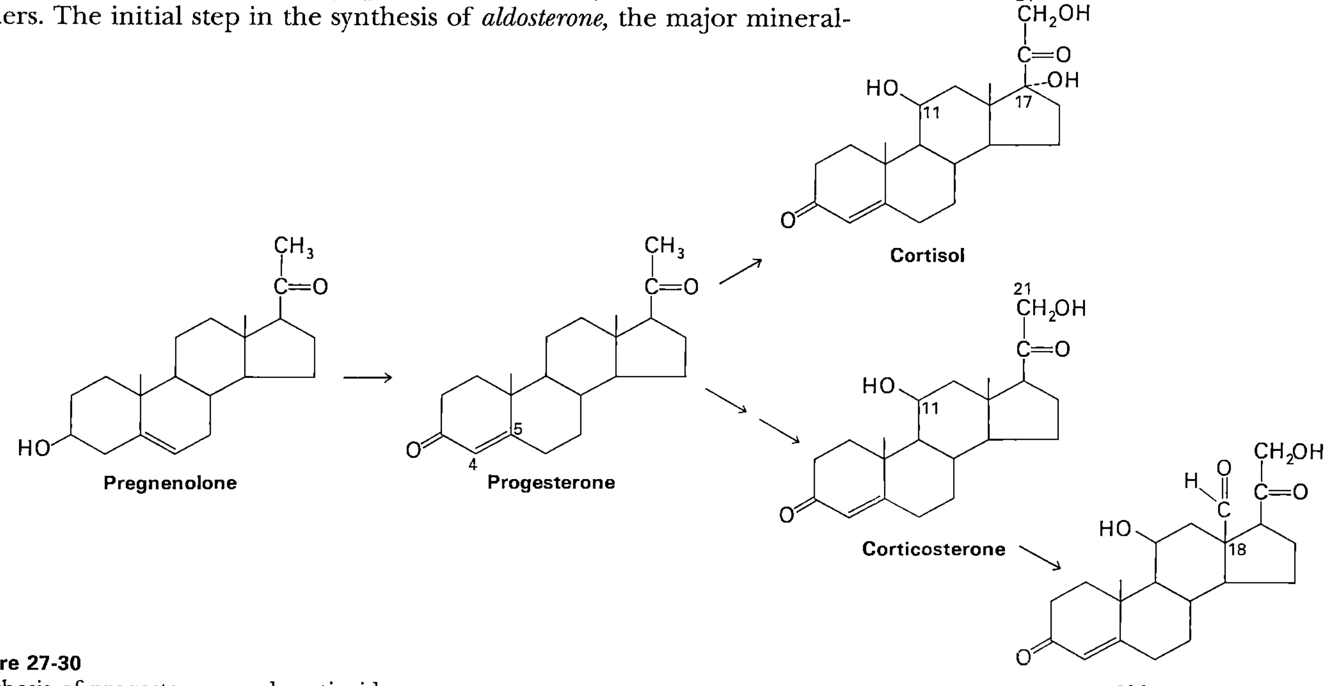

Figure 27-30
Synthesis of progesterone and corticoids.

ocorticoid, is the hydroxylation of progesterone at C-21. The resulting deoxycorticosterone is hydroxylated at C-11. The oxidation of the C-18 angular methyl group to an aldehyde then yields aldosterone.

SYNTHESIS OF ANDROGENS AND ESTROGENS

The synthesis of androgens (Figure 27-31) starts with the hydroxylation of progesterone at C-17. The side chain consisting of C-20 and C-21 is then cleaved to yield *androstenedione,* an androgen. *Testosterone,* another androgen, is formed by reduction of the 17-keto group of androstenedione. Androgens contain 19 carbon atoms. Estrogens are synthesized from androgens by the loss of the C-19 angular methyl group and the formation of an aromatic A ring. These reactions require NADPH and O_2. *Estrone,* an estrogen, is derived from androstenedione, whereas *estradiol,* another estrogen, is formed from testosterone.

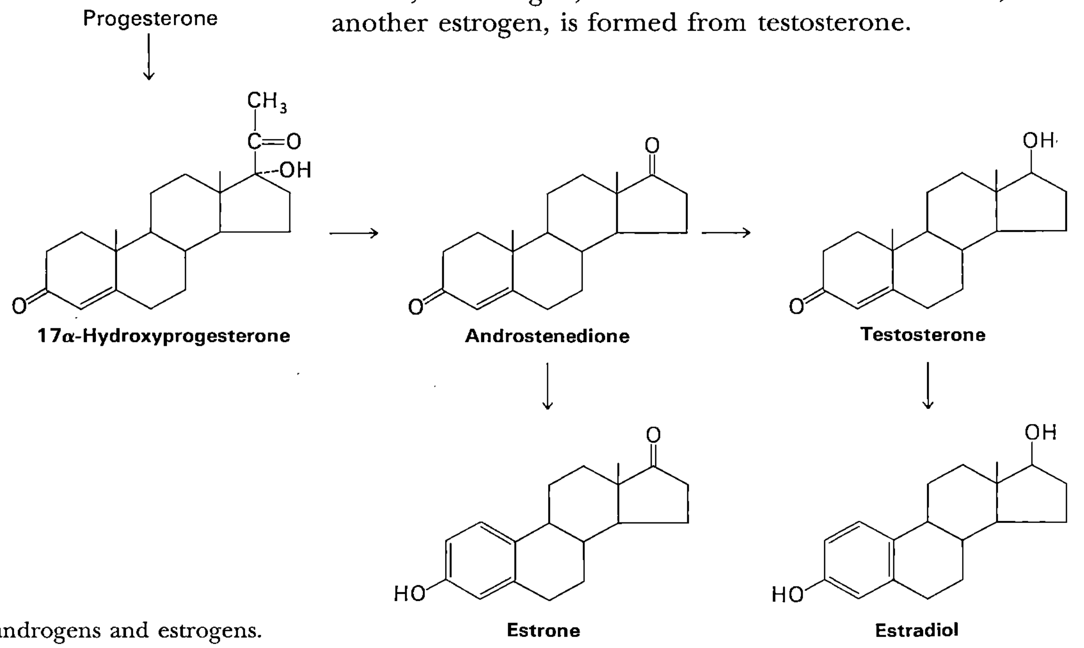

Figure 27-31
Synthesis of androgens and estrogens.

DEFICIENCY OF 21-HYDROXYLASE CAUSES VIRILIZATION AND ENLARGEMENT OF THE ADRENALS

The most common inherited disorder of steroid hormone synthesis is a deficiency of *21-hydroxylase,* an enzyme needed for the synthesis of glucocorticoids and mineralocorticoids. The diminished production of glucocorticoids leads to an increased secretion of ACTH by the anterior pituitary gland. This response is an expression of a normal feedback mechanism that controls adrenal cortical activity. *The adrenal glands enlarge because of the high level of ACTH in the blood. ACTH, acting through cyclic AMP, stimulates the conversion of cholesterol to pregnenolone.* Consequently, the concentrations of progesterone and 17α-hydroxyprogesterone increase. In turn, there is a *marked increase in the amount of androgens,* because they are derived from 17α-hydroxyprogesterone.

The striking clinical finding in 21-hydroxylase deficiency is *virilization* caused by high levels of androgens. In affected females, virilization is usually evident at birth. Androgens secreted during the development of the female fetus produce a masculinization of the external genitalia. In males, the sexual organs appear normal at birth. Sexual precocity becomes apparent several months later. There is accelerated growth and very early bone maturation so that short stature is the typical final result. Many patients with 21-hydroxylase deficiency *persistently lose Na^+ in the urine* because they have very low levels of aldosterone, the principal mineralocorticoid. Loss of salt leads to dehydration and hypotension, which may result in shock and sudden death.

Effective therapy is available for 21-hydroxylase deficiency. The administration of a glucocorticoid provides this needed hormone and concomitantly eliminates the excessive secretion of ACTH. Excessive formation of androgens is thereby stopped. A mineralocorticoid is also given to patients who lose salt. Some of the symptoms of 21-hydroxylase deficiency are reversed if therapy is started early in life.

VITAMIN D IS DERIVED FROM CHOLESTEROL BY THE RING-SPLITTING ACTION OF LIGHT

Cholesterol is also the precursor of vitamin D, which plays an essential role in the control of calcium and phosphorus metabolism. *7-Dehydrocholesterol (provitamin D_3) is photolyzed by ultraviolet light to previtamin D_3, which spontaneously isomerizes to vitamin D_3* (Figure 27-32). Vitamin D_3 (cholecalciferol) is converted into *calcitriol* (1,25-dihydroxycholecalciferol), the active hormone, by hydroxylation reactions occurring in the liver and kidneys. Vitamin D deficiency in childhood produces *rickets,* which is characterized by inadequate calcification of cartilage and bone.

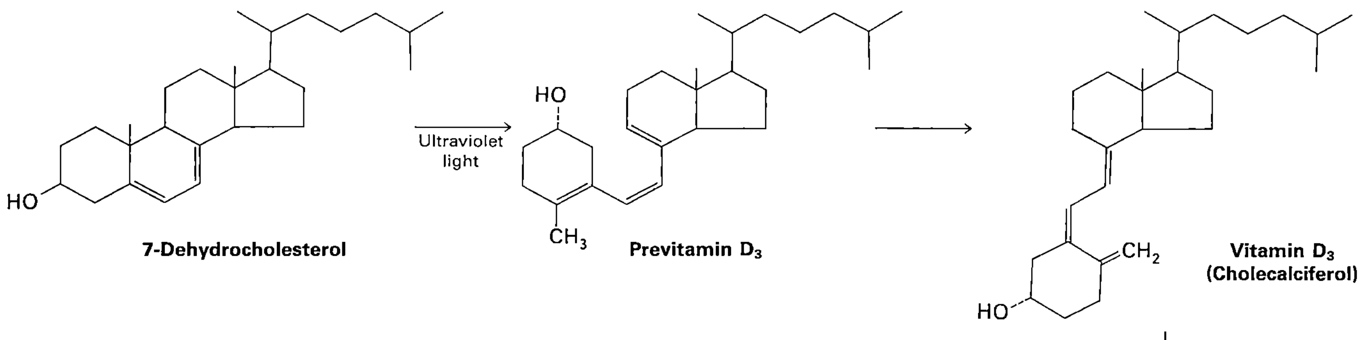

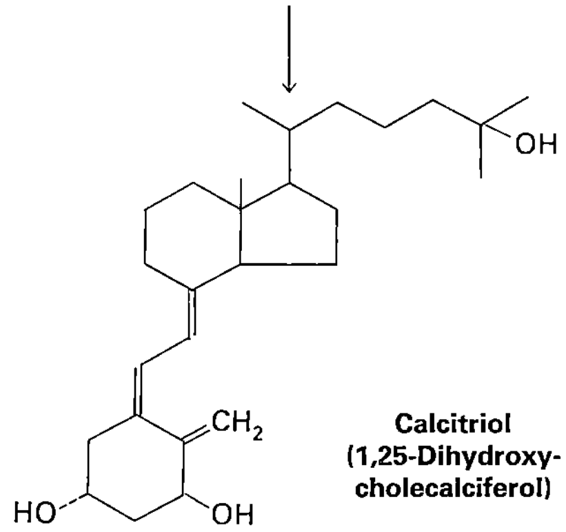

Figure 27-32
Conversion of 7-dehydrocholesterol into vitamin D_3 (cholecalciferol) and calcitriol, the active hormone. Vitamin D_2 (ergocalciferol) can be formed in a similar way starting with ergosterol, a plant sterol. Vitamin D_2 differs from D_3 in having a C-22-C-23 double bond and a C-24 methyl group.

Rickets was so common in the seventeenth century that it was called the "children's disease of the English." The 7-dehydrocholesterol in the skin of these children was not photolyzed to previtamin D_3 because there was little sunlight for many months of the year. Furthermore, their diets provided little vitamin D because most naturally occurring foods have a low content of this vitamin. Fish-liver oils are a notable exception. Cod-liver oil, abhorred by generations of children because of its unpleasant taste, was used in the past as a rich source of vitamin D. Today, the most reliable dietary sources of vitamin D are fortified foods. Milk, for example, is fortified to a level of 400 international units per quart (10 μg per quart). The recommended daily intake of vitamin D is 400 international units, irrespective of age. In adults, vitamin D deficiency leads to softening and weakening of bones, a condition called *osteomalacia.* The occurrence of osteomalacia in Bedouin Arab women who are clothed so that only their eyes are exposed to sunlight is a striking reminder that vitamin D is needed by adults as well as by children.

FIVE-CARBON UNITS ARE JOINED TO FORM A WIDE VARIETY OF BIOMOLECULES

The synthesis of squalene (C_{30}) from isopentenyl pyrophosphate (C_5) exemplifies a fundamental mechanism for the assembly of carbon skeletons of biomolecules. *A remarkable array of compounds are formed from isopentenyl pyrophosphate, the basic five-carbon building block.* The fragrances of many plants arise from volatile C_{10} and C_{15} compounds, which are called *terpenes.* For example, myrcene ($C_{10}H_{16}$) from bay leaves consists of two isoprene units, as does limonene ($C_{10}H_{15}$) from lemon oil (Figure 27-33). Zingiberene ($C_{15}H_{24}$), from the oil of ginger, is made up of three isoprene units. Some terpenes, such as geraniol from geraniums and menthol from peppermint oil, are alcohols, and others, such as citronellal, are aldehydes. *Natural rubber* is a linear polymer of *cis*-isoprene units.

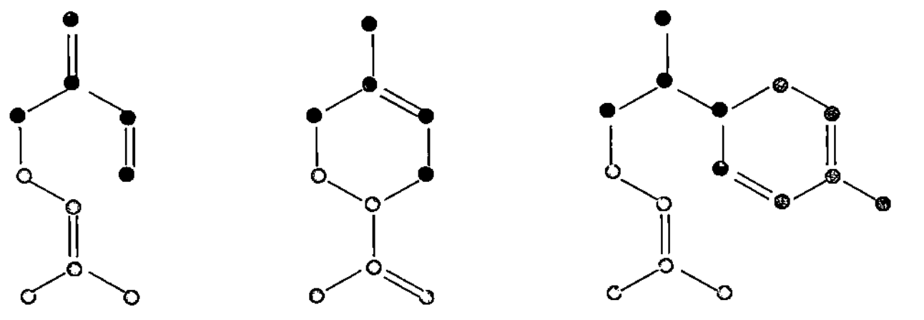

Myrcene Limonene Zingiberene

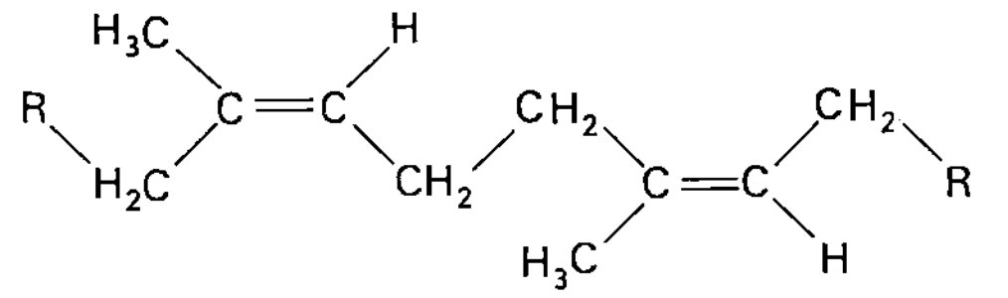

Natural rubber
(*cis*-Polyisoprene)

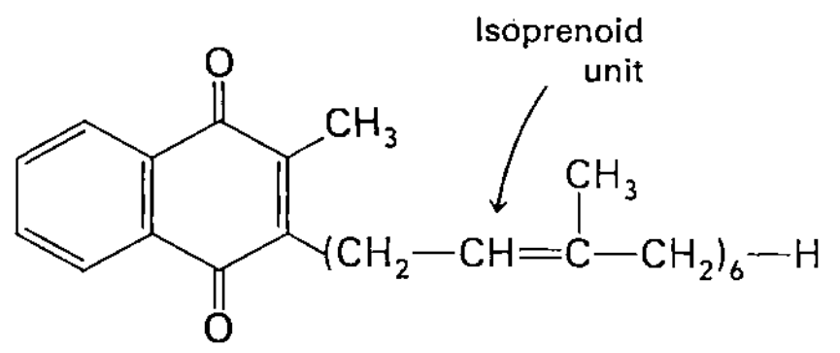

Vitamin K_2

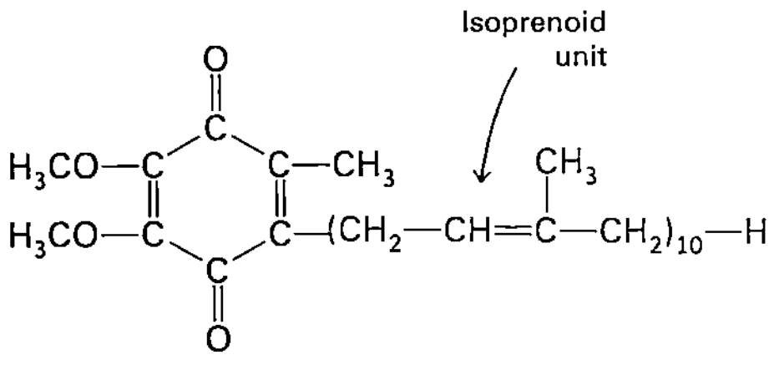

Ubiquinone
(Coenzyme Q_{10})

Figure 27-33
Formulas for some isoprenoids. The five-carbon building blocks are shown in color.

"Perfumes, colors, and sounds echo one another."

CHARLES BAUDELAIRE
Correspondances

We have already encountered several molecules that contain isoprenoid side chains. The C_{30} hydrocarbon *side chain of vitamin K_2*, a key molecule in clotting (p. 255), is built from six C_5 units. *Coenzyme Q_{10}* in the mitochondrial respiratory chain (p. 536) has a side chain made up of 10 isoprene units. Yet another example is the *phytol side chain of chlorophyll* (p. 656), which is formed from four isoprene units. As will be discussed in a subsequent chapter (p. 934), many proteins are targeted to membranes by the covalent attachment of a farnesyl (C_{15}) or a geranylgeranyl (C_{20}) unit to their carboxyl-terminal cysteine residue. *Prenylation confers hydrophobic character.*

Isoprenoids can delight by their color as well as by their fragrance. Indeed, isoprenoids can be regarded as the sensual molecules! The color of tomatoes and carrots comes from *carotenoids,* specifically from *lycopene* and *β-carotene,* respectively. These compounds absorb light because they contain extended networks of single and double bonds—that is, they are *polyenes.* Their C_{40} carbon skeletons are built by the successive addition of C_5 units to form *geranylgeranyl pyrophosphate, a C_{20} intermediate,* which then condenses tail-to-tail with another molecule of geranylgeranyl pyrophosphate. This biosynthetic pathway is like that of squalene except that C_{20} rather than C_{15} units are assembled and condensed.

$$C_5 \longrightarrow C_{10} \longrightarrow C_{15} \longrightarrow C_{30} \text{ (squalene)}$$

$$C_5 \longrightarrow C_{10} \longrightarrow C_{15} \longrightarrow C_{20} \longrightarrow C_{40} \text{ (phytoene)}$$

Phytoene, the C_{40} condensation product, is dehydrogenated to yield lycopene. Cyclization of both ends of lycopene gives β-carotene (Figure 27-34). Carotenoids serve as light-harvesting molecules in photosynthetic assemblies and also play a role in protecting prokaryotes from the deleterious effects of light. Carotenoids are also essential for vision. β-Carotene is the precursor of retinal, the chromophore in all known visual pigments (p. 333). *These examples illustrate the fundamental role of isopentenyl pyrophosphate in the assembly of extended carbon skeletons of biomolecules. It is also evident that isoprenoids are ubiquitous in nature and have diverse significant roles.*

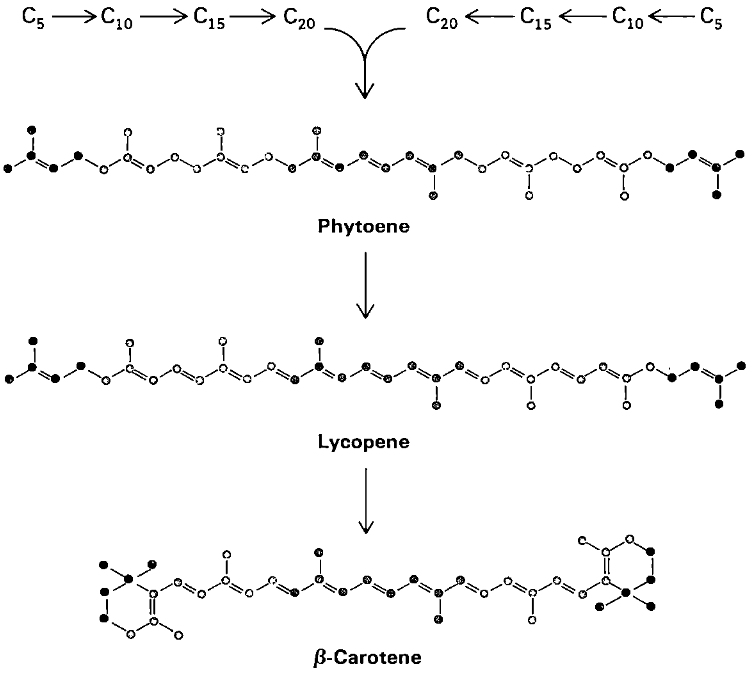

Figure 27-34
Synthesis of C_{40} carotenoids: phytoene, lycopene, and β-carotene.

SUMMARY

Phosphatidate, an intermediate in the synthesis of phosphoglycerides and triacylglycerols, is formed by successive acylations of glycerol 3-phosphate by acyl CoA. Hydrolysis of its phosphoryl group followed by acylation yields a triacylglycerol. CDP–diacylglycerol, the activated intermediate in the de novo synthesis of several phosphoglycerides, is formed from phosphatidate and CTP. The activated phosphatidyl unit is then transferred to the hydroxyl group of a polar alcohol, such as serine, to form phosphatidyl serine. In bacteria, decarboxylation of this phosphoglyceride yields phosphatidyl ethanolamine, which is methylated by *S*-adenosylmethionine to form phosphatidyl choline. In mammals, this phosphoglyceride is synthesized by a pathway that utilizes dietary choline. CDP–choline is the activated intermediate in this route. Sphingolipids are synthesized from ceramide, which is formed by the acylation of sphingosine. Gangliosides are sphingolipids that contain an oligosaccharide unit having at least one residue of *N*-acetylneuraminate or a related sialic acid. They are synthesized by the stepwise addition of activated sugars, such as UDP–glucose, to ceramide.

Cholesterol, a steroid component of eukaryotic membranes and a precursor of steroid hormones, is formed from acetyl CoA. The committed step in its synthesis is the formation of mevalonate from 3-hydroxy-3-methylglutaryl CoA (derived from acetyl CoA and acetoacetyl CoA). Mevalonate is converted into isopentenyl pyrophosphate (C_5), which condenses with its isomer, dimethylallyl pyrophosphate (C_5), to form geranyl pyrophosphate (C_{10}). Addition of a second molecule of isopentenyl pyrophosphate yields farnesyl pyrophosphate (C_{15}), which condenses with itself to form squalene (C_{30}). This intermediate cyclizes to lanosterol (C_{30}), which is modified to yield cholesterol (C_{27}). The synthesis of cholesterol by the liver is regulated by changes in the amount and activity of 3-hydroxy-3-methylglutaryl CoA reductase. Transcription of the gene, translation of the mRNA, and degradation of the protein are stringently controlled. In addition, the activity of the reductase is regulated by phosphorylation.

Cholesterol and other lipids are transported in the blood to specific targets by a series of lipoproteins. Triacylglycerols exported by the intestine are carried by chylomicrons and then hydrolyzed by lipases lining the capillaries of target tissues. Cholesterol and other lipids in excess of those needed by the liver are exported in the form of very low density lipoprotein (VLDL). After delivering its content of triacylglycerols to adipose tissue and other peripheral tissue, VLDL is converted into intermediate-density lipoprotein (IDL) and then into low-density lipoprotein (LDL). IDL and LDL carry cholesterol esters, primarily cholesterol linoleate. LDL is taken up by liver and peripheral tissue cells by receptor-mediated endocytosis. The LDL receptor, a protein spanning the plasma membrane of these target cells, binds LDL and mediates its entry into the cell. Absence of the LDL receptor in the homozygous form of familial hypercholesterolemia leads to a markedly elevated plasma level of LDL-cholesterol, deposition of cholesterol on blood vessel walls, and heart attacks in childhood. Apolipoprotein B, a very large protein, is a key structural component of chylomicrons, VLDL, and LDL.

Five major classes of steroid hormones are derived from cholesterol: progestagens, glucocorticoids, mineralocorticoids, androgens, and estrogens. Hydroxylations by P450 monooxygenases that use NADPH and O_2 play an important role in the synthesis of steroid hormones and bile salts from cholesterol. P450 enzymes, a large superfamily, also participate in the detoxification of drugs and other foreign substances. Pregnenolone (C_{21}), a key intermediate in the synthesis of steroid hormones, is formed by scission of the side chain of cholesterol. Progesterone (C_{21}), synthesized from pregnenolone, is the precursor of cortisol and aldosterone. Hydroxylation of progesterone and cleavage of its side chain yields androstenedione, an androgen (C_{19}). Estrogens (C_{18}) are synthesized from androgens by the loss of an angular methyl group and the formation of an aromatic A ring. Vitamin D, which is important in the control of calcium and phosphorus metabolism, is formed from a derivative of cholesterol by the action of light. A remarkable array of biomolecules in addition to cholesterol and its derivatives are synthesized from isopentenyl pyrophosphate, the basic five-carbon building block. The hydrocarbon side chains of vitamin K_2, coenzyme Q_{10}, and chlorophyll are extended chains constructed from this activated C_5 unit. Many proteins are targeted to membranes by prenyl groups that are derived from this activated intermediate.

SELECTED READINGS

Where to start

Brown, M.S., and Goldstein, J.L., 1986. A receptor-mediated pathway for cholesterol homeostasis. *Science* 232:34–47.

Goldstein, J.L., and Brown, M.S., 1990. Regulation of the mevalonate pathway. *Nature* 343:425–430.

Chan, L., 1992. Apolipoprotein B, the major protein component of triglyceride-rich and low density lipoproteins. *J. Biol. Chem.* 267:25621–25624.

Endo, A., 1992. The discovery and development of HMG-CoA reductase inhibitors. *J. Lipid Res.* 33:1569–1582. [An account of the discovery of drugs that lower blood cholesterol levels.]

Dennis, E.A., 1994. Diversity of group types, regulation, and function of phospholipase A_2. *J. Biol. Chem.* 269:13057–13060.

Books

Vance, D.E., and Vance, J. (eds.), 1991. *Biochemistry of Lipids, Lipoproteins and Membranes.* Elsevier. [A fine collection of articles on the metabolism, genetics, and assembly of lipids.]

Scriver, C.R., Beaudet, A.L., Sly, W.S., and Valle, D. (eds.), 1989. *The Metabolic Basis of Inherited Disease* (6th ed.). McGraw-Hill. [Contains many excellent articles on genetic disorders of sphingolipid, lipoprotein, cholesterol, and steroid metabolism.]

Hawthorne, J.N., and Ansell, G.B. (eds.), 1982. *Phospholipids.* Elsevier.

Walsh, C., 1979. *Enzymatic Reaction Mechanisms.* W.H. Freeman. [Contains excellent discussions of reaction mechanisms in the biosynthesis of cholesterol and other compounds derived from isopentenyl pyrophosphate.]

Phospholipids, sphingolipids, and phosphoinositides

Dennis, E.A., Rhee, S.G., Billah, M.M., and Hannun, Y.A., 1991. Role of phospholipase in generating lipid second messengers in signal transduction. *FASEB J.* 5:2068–2077.

Kent, C., Carman, G.M., Spence, M.W., and Dowhan, W., 1991. Regulation of eukaryotic phospholipid metabolism. *FASEB J.* 5:2258–2266.

Koval, M., and Pagano, R.E., 1991. Intracellular transport and metabolism of sphingomyelin. *Biochim. Biophys. Acta* 1082:113–125.

Majerus, P.W., 1992. Inositol phosphate biochemistry. *Ann. Rev. Biochem.* 61:225–250.

Raetz, C.R., and Dowhan, W., 1990. Biosynthesis and function of phospholipids in *Escherichia coli. J. Biol. Chem.* 265:1235–1238.

Vance, D.E., 1990. Phosphatidylcholine metabolism: Masochistic enzymology, metabolic regulation, and lipoprotein assembly. *Biochem. Cell Biol.* 68:1151–1165.

Vanden, B.T., and Cronan, J.J., 1989. Genetics and regulation of bacterial lipid metabolism. *Ann. Rev. Microbiol.* 43:317–343.

Hakomori, S., 1986. Glycosphingolipids. *Sci. Amer.* 254(5):44–53. [An interesting presentation of changes in these lipids during cell differentiation and oncogenesis.]

Hanahan, D.J., 1986. Platelet-activating factor: A biologically active phosphoglyceride. *Ann. Rev. Biochem.* 55:483–509.

Raetz, C.R., 1993. Bacterial endotoxins: Extraordinary lipids that activate eucaryotic signal transduction. *J. Bacteriol.* 175:5745–5753.

Biosynthesis of cholesterol and steroids

Bloch, K., 1965. The biological synthesis of cholesterol. *Science* 150:19–28.

Bloch, K., 1983. Sterol structure and membrane function. *Crit. Rev. Biochem.* 14:47–92.

Schroepfer, G.J., Jr., 1982. Sterol biosynthesis. *Ann. Rev. Biochem.* 51:555–585.

DeLuca, H.F., 1992. New concepts of vitamin D functions. *Ann. N.Y. Acad. Sci.* 669:59–68.

Lipoproteins and their receptors

Breslow, J.L., 1992. Apolipoprotein genes and atherosclerosis. *Clin. Invest.* 70:377–384.

Russell, D.W., 1992. Cholesterol biosynthesis and metabolism. *Cardiovasc. Drugs Therapy* 6:103–110.

Wilson, C., Wardell, M.R., Weisgraber, K.H., Mahley, R.W., and Agard, D.A., 1991. Three-dimensional structure of the LDL receptor-binding domain of human apolipoprotein E. *Science* 252:1817–1822.

Plump, A.S., Smith, J.D., Hayek, T., Aalto-Setälä, K., Walsh, A., Verstuyft, J.G., Rubin, E.M., and Breslow, J.L., 1992. Severe hypercholesterolemia and atherosclerosis in apolipoprotein E-deficient mice created by homologous recombination in ES cells. *Cell* 71:343–353.

Brown, M.S., and Goldstein, J.L., 1984. How LDL receptors influence cholesterol and atherosclerosis. *Sci. Amer.* 251(5):58–66.

Sudhof, T.C., Goldstein, J.L., Brown, M.S., and Russell, D.W., 1985. The LDL receptor gene: A mosaic of exons shared with different proteins. *Science* 228:815–822.

Ross, R., 1993. The pathogenesis of atherosclerosis: A perspective for the 1990s. *Nature* 362:801–809.

Oxygen activation and P450 catalysis

Coon, M.J., Ding, X., Pernecky, S.J., and Vaz, A.D.N., 1992. Cytochrome P450: Progress and predictions. *FASEB J.* 6:686–694. [This issue contains many excellent articles on the P450 superfamily.]

Vaz, A.D., and Coon, M.J., 1994. On the mechanism of action of cytochrome P450: Evaluation of hydrogen abstraction in oxygen-dependent alcohol oxidation. *Biochemistry* 33:6442–6449.

Gonzalez, F.J., and Nebert, D.W., 1990. Evolution of the P450 gene superfamily: Animal-plant "warfare," molecular drive and human genetic differences in drug oxidation. *Trends Genet.* 6:182–186.

Hayaishi, O. (ed.), 1974. *Molecular Mechanisms of Oxygen Activation.* Academic Press.

PROBLEMS

1. *Making fat.* Write a balanced equation for the synthesis of a triacylglycerol, starting from glycerol and fatty acids.

2. *Making phospholipid.* Write a balanced equation for the synthesis of phosphatidyl serine by the de novo pathway, starting from serine, glycerol, and fatty acids.

3. *Activated donors.* What is the activated reactant in each of these biosyntheses?
   (a) Phosphatidyl serine from serine.
   (b) Phosphatidyl ethanolamine from ethanolamine.
   (c) Ceramide from sphingosine.
   (d) Sphingomyelin from ceramide.
   (e) Cerebroside from ceramide.
   (f) Ganglioside G_{M1} from ganglioside G_{M2}.
   (g) Farnesyl pyrophosphate from geranyl pyrophosphate.

4. *Telltale labels.* What is the distribution of isotopic labeling in cholesterol synthesized from each of these precursors?
   (a) Mevalonate labeled with ^{14}C in its carboxyl carbon atom.
   (b) Malonyl CoA labeled with ^{14}C in its carboxyl carbon atom.

5. *Familial hypercholesterolemia.* Several classes of LDL receptor mutations have been identified as causes of this disease. Suppose that you were given cells from patients and antibody specific for the LDL receptor, and had access to an electron microscope. Which categories of mutations are likely to be found?

6. *RNA editing.* A shortened version (apo B-48) of apolipoprotein B is formed by intestine, whereas the full-length protein (apo B-100) is synthesized by liver. A Gln codon (CAA) is changed into a stop codon. Propose a simple mechanism for this change.

7. *Inspiration for drug design.* Some actions of androgens are mediated by dihydrotestosterone, which is formed by reduction of testosterone. This finishing touch is catalyzed by an NADPH-dependent 5α-reductase.

   Chromosomal XY males with a genetic deficiency of this reductase are born with a male internal urogenital tract but predominantly female external genitalia. These people are usually reared as girls. At puberty, they masculinize because the testosterone level rises. The testes of these reductase-deficient men are normal, whereas their prostate glands remain small. How might this information be used to design a drug to treat *benign prostatic hypertrophy,* a common consequence of the normal aging process in men? A majority of men over age 55 have some degree of prostatic enlargement, which often leads to urinary obstruction.

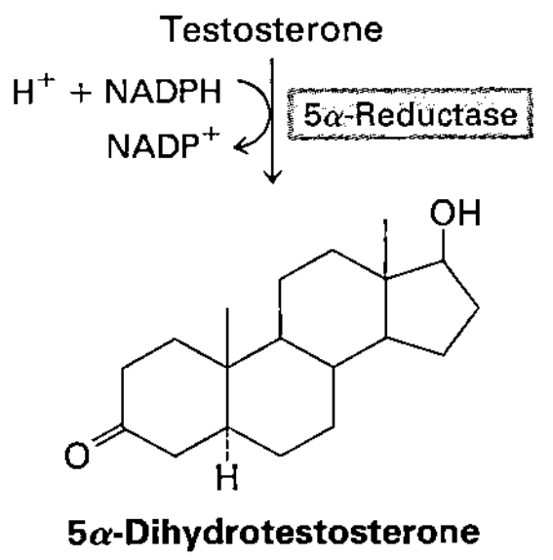

8. *Drug idiosyncracies.* Debrisoquine, a β-adrenergic blocking agent, has been used to treat hypertension. The optimal dose varies greatly (20 to 400 mg daily) in a population of patients. The urine of most patients taking the drug contains a high level of 4-hydroxydebrisoquine. However, those most sensitive to the drug (about 8% of the group studied) excrete debrisoquine and very little of the 4-hydroxy derivative. Propose a molecular basis for this drug idiosyncrasy. Why should caution be exercised in giving other drugs to patients who are very sensitive to debrisoquine?

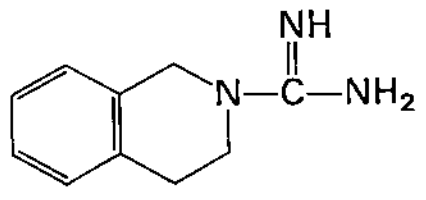

Debrisoquine

9. *Congenital adrenal hyperplasia.* As discussed earlier, a genetic deficiency of the 21-hydroxylase causes virilization and enlargement of the adrenals. Which other enzymatic defects in the synthesis of corticoids would lead to similar symptoms?

10. *Removal of odorants.* Many odorant molecules are highly hydrophobic and concentrate within the olfactory epithelium. They would give a persistent signal independent of their concentration in the environment if they were not rapidly modified. Propose a mechanism for converting hydrophobic odorants into water-soluble derivatives that can be rapidly eliminated.

CHAPTER

28

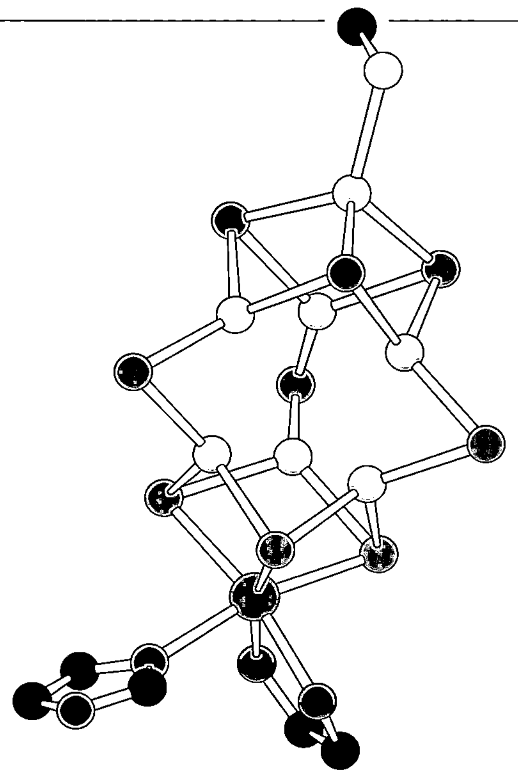

Biosynthesis of Amino Acids and Heme

This chapter deals with the biosynthesis of amino acids and some key derivatives, such as heme. The flow of nitrogen into amino acids will be considered first. This process starts with the reduction of N_2 to NH_3 by nitrogen-fixing microorganisms, one of the most important chemical transformations in the biosphere. A molybdenum-iron cofactor in a nitrogenase enzyme deftly carries out this demanding reaction. NH_4^+ is then assimilated into amino acids by way of glutamate and glutamine, two pivotal molecules in nitrogen metabolism. The master enzymes controlling the entry of nitrogen into amino acids are glutamate dehydrogenase and glutamine synthetase.

The assembly of the carbon skeletons of amino acids will then be discussed. Of the basic set of 20 amino acids, 11 are synthesized from intermediates of the citric acid cycle and other major metabolic pathways by quite simple reactions. We shall consider how they are formed and then examine the biosyntheses of aromatic amino acids as examples of more complex pathways. Humans are unable to synthesize 9 of the basic set of 20; these missing ones, called *essential amino acids,* must be obtained from dietary sources. Amino acid biosynthesis is regulated by control devices that are generally used to tune and coordinate biosynthetic pathways. Examination of their molecular logic is very rewarding. In particular, we shall focus on bacterial glutamine synthetase—a sentient enzyme that precisely adjusts its activity according to the nitrogen balance of the cell.

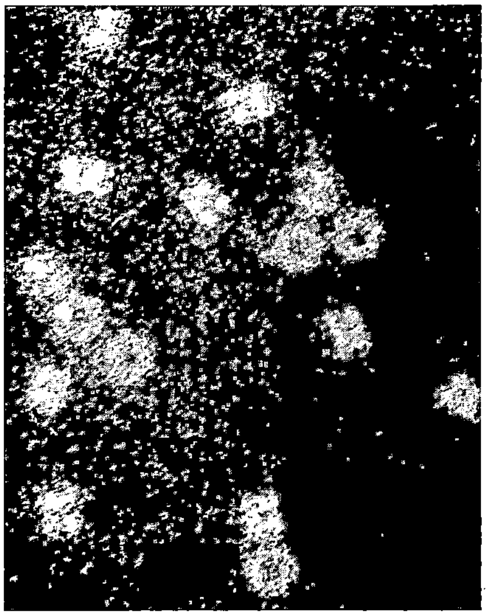

Figure 28-1
Electron micrograph of glutamine synthetase from *Escherichia coli.* This enzyme plays a key role in nitrogen metabolism. [Courtesy of Dr. Earl Stadtman.]

Opening Image: *The reduction of* N_2 *to* NH_3*, one of the most important reactions in the biosphere, is carried out in the central cavity of the iron-molybdenum cofactor of nitrogenase. Molybdenum is shown in purple, iron in green, and inorganic sulfur in orange. [Drawn from 1min.pdb. J. Kim and D.C. Rees. Nature 360(1992):553.]*

Two interesting carriers participate in amino acid metabolism. Tetrahydrofolate serves as a highly versatile shuttler of activated one-carbon units at three oxidation stages, and *S*-adenosylmethionine is the major methyl donor. An understanding of one-carbon interconversions prepares us for the study of nucleotide metabolism, which is presented in the next chapter. Many important biomolecules, including purines and pyrimidines, are derived from amino acids. For example, glutathione serves as a sulfhydryl buffer, thyroxine as a hormone, and nitric oxide as a short-lived signal molecule. The final section of this chapter is concerned with the synthesis and degradation of heme. Elegant isotope labeling studies unraveled the biosynthetic pathway of porphyrins and illustrated the power of this approach in revealing how biomolecules are constructed.

Figure 28-2
The nodules in the root system of the soybean are the sites of nitrogen fixation by *Rhizobium* bacteria. [© Dr. Jeremy Burgess, Science Source/Photo Researchers.]

NITROGEN FIXATION: MICROORGANISMS USE ATP AND A POWERFUL REDUCTANT TO REDUCE N_2 TO NH_3

The nitrogen in amino acids, purines, pyrimidines, and other biomolecules ultimately comes from atmospheric nitrogen. However, higher organisms are unable to reduce N_2 to NH_3. This conversion—called *nitrogen fixation*—is carried out by bacteria and blue-green algae (cyanobacteria). Some of these microorganisms—namely, the symbiotic *Rhizobium* bacteria—invade the roots of leguminous plants and form root nodules, where nitrogen fixation takes place, supplying both the bacteria and the plants (Figure 28-2). The amount of N_2 fixed by *diazotrophic microorganisms* has been estimated to be of the order of 10^{11} kg per year, about 60% of the earth's newly fixed nitrogen. Lightning and ultraviolet radiation fix another 15%; the other 25% comes from industrial processes.

The N≡N bond, which has a bond energy of 225 kcal/mol, is highly resistant to chemical attack. Indeed, Lavoisier named it "azote," meaning "without life," because it is quite unreactive. The industrial process for nitrogen fixation devised by Fritz Haber in 1910 is still being used in fertilizer factories.

$$N_2 + 3\ H_2 \rightleftharpoons 2\ NH_3$$

Fixation of N_2 is typically carried out over an iron catalyst at about 500°C and a pressure of 300 atm.

It is not surprising, then, that the biological process of nitrogen fixation requires a complex enzyme with multiple redox centers. The *nitrogenase complex,* which carries out this fundamental transformation, consists of two proteins: a *reductase,* which provides electrons with high reducing power, and a *nitrogenase,* which uses these electrons to reduce N_2 to NH_3 (Figure 28-3). The transfer of electrons from the reductase to the nitrogenase component is coupled to the hydrolysis of ATP by the reductase. The nitrogenase complex is exquisitely sensitive to inactivation by O_2. Leguminous plants maintain a very low concentration of free O_2 in their root nodules by binding O_2 to *leghemoglobin,* a homolog of hemoglobin.

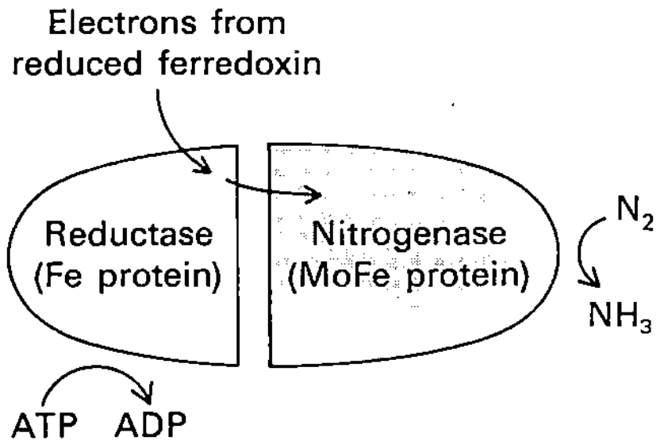

Figure 28-3
Schematic diagram of the nitrogenase complex. The reductase and nitrogenase components undergo cyclic association and dissociation in the conversion of N_2 into NH_3. The reductase is also known as the iron protein (Fe protein), and the nitrogenase as the molybdenum-iron protein (MoFe protein).

The reduction of N_2 to NH_3 is a six-electron process.

$$N_2 + 6\ e^- + 6\ H^+ \longrightarrow 2\ NH_3$$

However, the reductase is imperfect: H_2 is formed along with NH_3. Hence, an input of two additional electrons is required.

$$N_2 + 8\ e^- + 8\ H^+ \longrightarrow 2\ NH_3 + H_2$$

In most nitrogen-fixing microorganisms, *the eight high-potential electrons come from reduced ferredoxin.* Recall that reduced ferredoxin is produced in chloroplasts by the action of photosystem I (p. 663). Alternatively, reduced ferredoxin can be generated by oxidative processes. Nitrogen fixation is energetically very costly. *At least 16 ATP are hydrolyzed for each N_2 reduced.*

$$N_2 + 8\,e^- + 16\,ATP + 16\,H_2O \longrightarrow 2\,NH_3 + H_2 + 16\,ADP + 16\,P_i + 8\,H^+$$

THE IRON-MOLYBDENUM COFACTOR OF NITROGENASE BINDS AND REDUCES N_2

Both the reductase and nitrogenase components of the complex are *iron-sulfur proteins,* in which iron is bonded to the sulfur atom of a cysteine residue and to inorganic sulfide (p. 535). The *reductase* (also called the *iron protein* or *Fe protein*), a dimer of identical 30-kd subunits, contains a [4Fe-4S] cluster (Figure 28-4). An ATP/ADP binding site is also present at the subunit interface. The role of the reductase is to transfer electrons from a high-potential donor, such as reduced ferredoxin, to the nitrogenase component. Hydrolysis of ATP triggers electron transfer to the nitrogenase component and leads to the dissociation of the complex. The reductase has the shape of a butterfly, with the ATP/ADP site located at the head. The fluttering of its wings plays an essential role in electron transfer.

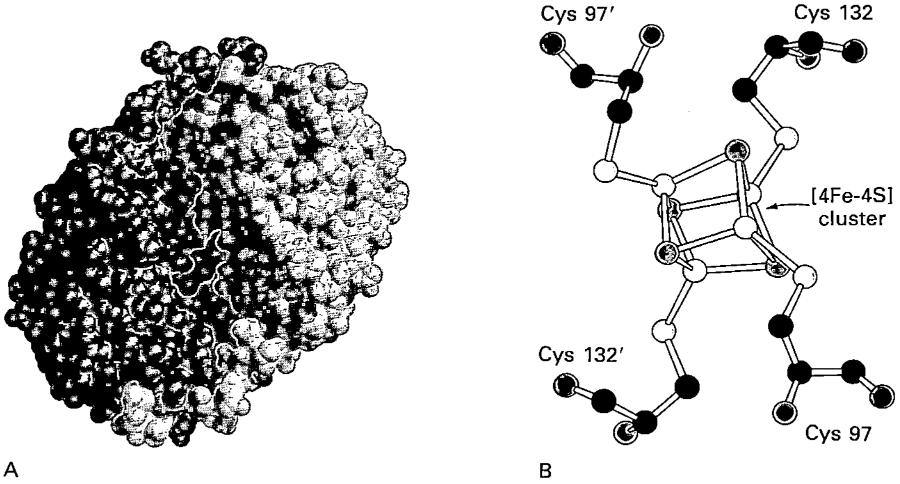

Figure 28-4
Structure of the reductase component (iron protein) of the nitrogenase complex. (A) ADP (red), formed by hydrolysis of bound ATP, is located at the interface of the two identical subunits (blue and yellow). (B) A [4Fe-4S] cluster, also located at the subunit interface, accepts electrons from ferredoxin and transfers them to the nitrogenase component. Fe is shown in green, inorganic sulfur in orange, and the other atoms in standard colors. [Drawn from coordinates kindly provided by Dr. Douglas Rees. M.M. Georgiadis, H. Komiya, P. Chakrabarti, D. Woo, J.J. Kornuc, and D.C. Rees. *Science* 257(1992):1653.]

The nitrogenase component, an $\alpha_2\beta_2$ tetramer (240 kd), also has a provocative structure (Figure 28-5A). Electrons enter at the *P-cluster,* which is located at an $\alpha\beta$ interface. Two [4Fe-4S] cubes are joined by sulfur atoms, which are in turn linked to cysteine residues (Figure 28-5B). Electrons flow from the P-cluster to the *FeMo-cofactor,* a very unusual

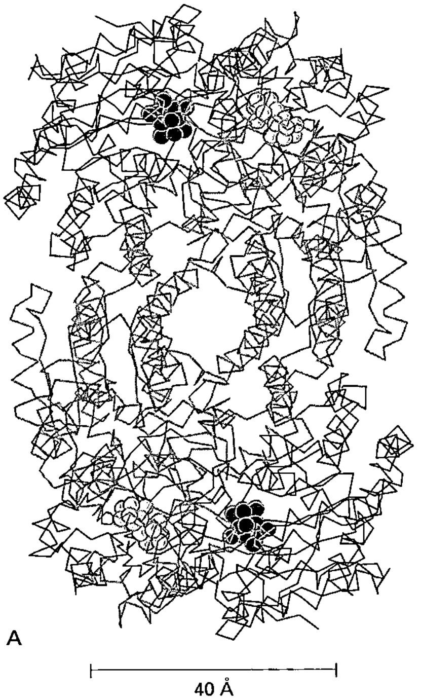

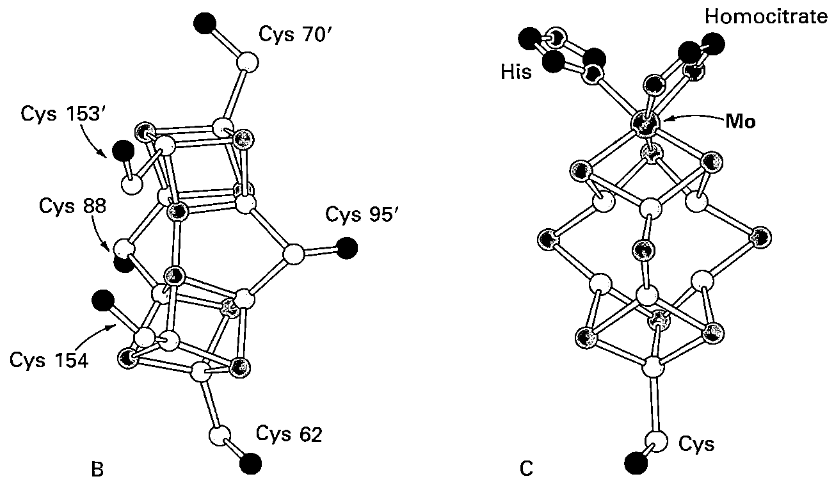

Figure 28-5
Structure of the nitrogenase component (molybdenum-iron protein). (A) The two α subunits of the tetramer are shown in blue, and the two β subunits in yellow in this α carbon trace. Electrons are transferred from the reductase to the P-clusters (green) and then to the FeMo-cofactor (red), the site of N_2 reduction. (B) The P-cluster consists of two linked [4Fe-4S] clusters. Fe is shown in green, inorganic sulfur in orange, and the other atoms in standard colors. Three of the cysteines come from the α chain and three from the β chain. (C) The molybdenum atom of the FeMo-cofactor is coordinated to homocitrate and a histidine side chain. The iron atom at the opposite end is coordinated to a cysteine residue. [Drawn from 1min.pdb. J. Kim and D.C. Rees. *Nature* 360(1992):553.]

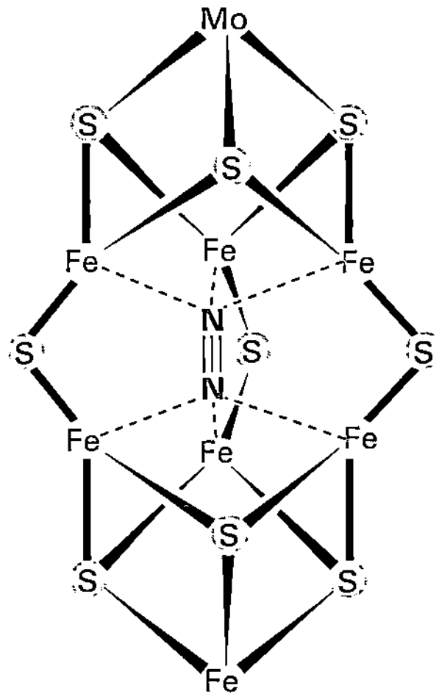

Figure 28-6
Proposed mode of binding of N_2 to the FeMo-cofactor, the catalytic center of nitrogenase. [After M.K. Chan, J. Kim, and D.C. Rees. *Science* 260(1993):782.]

redox center. Because molybdenum is present in this cluster, the nitrogenase component is also called the *molybdenum-iron protein (MoFe protein)*. The FeMo cofactor consists of two [M-3Fe-3S] clusters that are joined by three sulfur atoms (Figure 28-5C). Molybdenum occupies the M site in one cluster, and iron in the other. *The FeMo-cofactor is the site of nitrogen fixation.* It seems likely that N_2 binds in the central cavity of this cofactor (Figure 28-6). The formation of multiple Fe-N interactions in this complex weakens the N≡N bond and thereby lowers the activation barrier for reduction.

NH_4^+ IS ASSIMILATED INTO AMINO ACIDS BY WAY OF GLUTAMATE AND GLUTAMINE

The next step in the assimilation of nitrogen into biomolecules is the entry of NH_4^+ into amino acids. *Glutamate* and *glutamine* play pivotal roles in this regard. The α-amino group of most amino acids comes from the α-amino group of glutamate by transamination. Glutamine, the other major nitrogen donor, contributes its side-chain nitrogen in the biosynthesis of a wide range of important compounds.

Glutamate is synthesized from NH_4^+ and α-ketoglutarate, a citric acid cycle intermediate, by the action of *glutamate dehydrogenase.* This enzyme has already been encountered in the degradation of amino acids (p. 630). Recall that NAD^+ is the oxidant in catabolism. By contrast, NADPH is the reductant in biosyntheses.

$$NH_4^+ + \alpha\text{-ketoglutarate} + NADPH + H^+ \rightleftharpoons \text{glutamate} + NADP^+ + H_2O$$

Ammonium ion is incorporated into glutamine by the action of *glutamine synthetase* on glutamate. This amidation is driven by the hydrolysis of ATP.

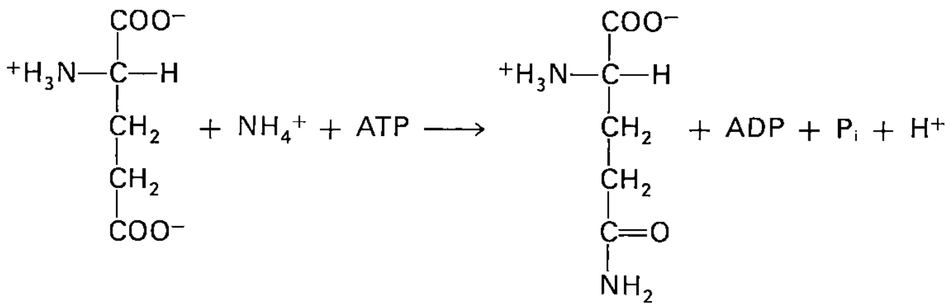

Glutamate **Glutamine**

The regulation of glutamine synthetase plays a critical role in controlling nitrogen metabolism, as will be discussed shortly.

Glutamate dehydrogenase and glutamine synthetase are present in all organisms. Most prokaryotes also contain *glutamate synthase,* which catalyzes the reductive amination of α-ketoglutarate. The nitrogen donor in this reaction is glutamine, and so two molecules of glutamate are formed.

$$\alpha\text{-Ketoglutarate} + \text{glutamine} + \text{NADPH} + H^+ \longrightarrow 2\ \text{glutamate} + \text{NADP}^+$$

When NH_4^+ is limiting, most of the glutamate is made by the sequential action of glutamine synthetase and glutamate synthase. The sum of these reactions is

$$NH_4^+ + \alpha\text{-ketoglutarate} + \text{NADPH} + \text{ATP} \longrightarrow \text{glutamate} + \text{NADP}^+ + \text{ADP} + P_i$$

Note that this stoichiometry differs from that of the glutamate dehydrogenase reaction in that an ATP is hydrolyzed. Why is this more expensive pathway sometimes used by bacteria? The answer is that the K_M of glutamate dehydrogenase for NH_4^+ is high (~1 mM), and so this enzyme is not saturated when NH_4^+ is limiting. In contrast, glutamine synthetase has very high affinity for NH_4^+.

AMINO ACIDS ARE MADE FROM INTERMEDIATES OF THE CITRIC ACID CYCLE AND OTHER MAJOR PATHWAYS

Thus far, we have considered the conversion of N_2 into NH_4^+ and the assimilation of NH_4^+ into glutamate and glutamine. We turn now to the biosynthesis of the other amino acids. Bacteria such as *Escherichia coli* can synthesize the entire basic set of 20 amino acids, whereas humans cannot make 9 of them. The amino acids that must be supplied in the diet are called *essential,* whereas the others are termed *nonessential* (Table 28-1). *These designations refer to the needs of an organism under a particular set of conditions.* For example, enough arginine is synthesized by the urea cycle to meet the needs of an adult but perhaps not those of a growing child. A deficiency of even one amino acid results in a *negative nitrogen balance.* In this state, more protein is degraded than is synthesized, and so more nitrogen is excreted than is ingested.

The pathways for the biosynthesis of amino acids are diverse. However, they have an important common feature: *their carbon skeletons come from intermediates of glycolysis, the pentose phosphate pathway, or the citric acid cycle.* A further simplification is that there are only *six biosynthetic families* (Figure 28-7).

Table 28-1
Basic set of 20 amino acids

Nonessential	*Essential*
Alanine	Histidine
Arginine	Isoleucine
Asparagine	Leucine
Aspartate	Lysine
Cysteine	Methionine
Glutamate	Phenylalanine
Glutamine	Threonine
Glycine	Tryptophan
Proline	Valine
Serine	
Tyrosine	

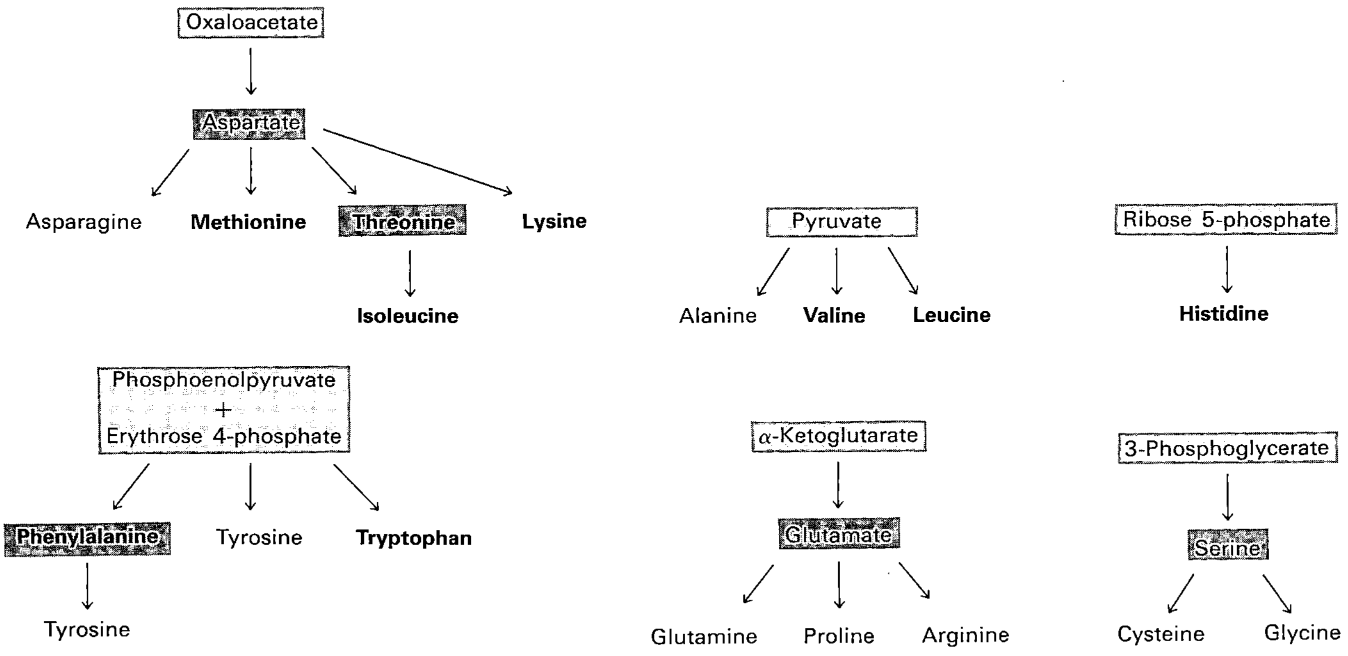

Figure 28-7
Biosynthetic families of amino acids in bacteria and plants. Major metabolic precursors are shaded blue. Amino acids that give rise to other amino acids are shaded red. Essential amino acids are shown in boldface.

The nonessential amino acids are synthesized by quite simple reactions, whereas the pathways for the formation of the essential amino acids are quite complex. For example, the nonessential amino acids *alanine* and *aspartate* are synthesized in a single step from pyruvate and oxaloacetate, respectively. Each acquires its amino group from glutamate in a transamination reaction in which pyridoxal phosphate is the cofactor (p. 631):

$$\text{Pyruvate} + \text{glutamate} \rightleftharpoons \text{alanine} + \alpha\text{-ketoglutarate}$$

$$\text{Oxaloacetate} + \text{glutamate} \rightleftharpoons \text{aspartate} + \alpha\text{-ketoglutarate}$$

Asparagine is then synthesized by the amidation of aspartate:

$$\text{Aspartate} + NH_4^+ + \text{ATP} \longrightarrow \text{asparagine} + \text{AMP} + PP_i + H^+$$

In mammals, the nitrogen donor in the synthesis of asparagine is glutamine rather than NH_4^+, as in bacteria. Ammonia generated at the active site of the enzyme is directly transferred to bound aspartate. Recall that high levels of NH_4^+ are toxic to humans (p. 637).

Another one-step synthesis of a nonessential amino acid in mammals is the hydroxylation of phenylalanine (an essential amino acid) to *tyrosine.*

$$\text{Phenylalanine} + O_2 + \text{NADPH} + H^+ \longrightarrow \text{tyrosine} + \text{NADP}^+ + H_2O$$

This reaction is catalyzed by *phenylalanine hydroxylase,* a monooxygenase discussed previously (p. 647). It is noteworthy that tyrosine is an essential amino acid in people lacking this enzyme.

GLUTAMATE IS THE PRECURSOR OF GLUTAMINE, PROLINE, AND ARGININE

The synthesis of glutamate by the reductive amination of α-ketoglutarate has already been discussed, as has the conversion of glutamate into *glutamine* (p. 717). Glutamate is the precursor of two other nonessential amino acids, *proline* and *arginine.* First, the γ-carboxyl group of glutamate reacts with ATP to form an acyl phosphate. This mixed anhydride is then reduced by NADPH to an aldehyde. Glutamic γ-semialdehyde cyclizes with a loss of H_2O to give Δ^1-pyrroline-5-carboxylate, which is reduced by NADPH to proline. Alternatively, the semialdehyde can be transaminated to ornithine, which is converted in several steps into arginine (p. 635).

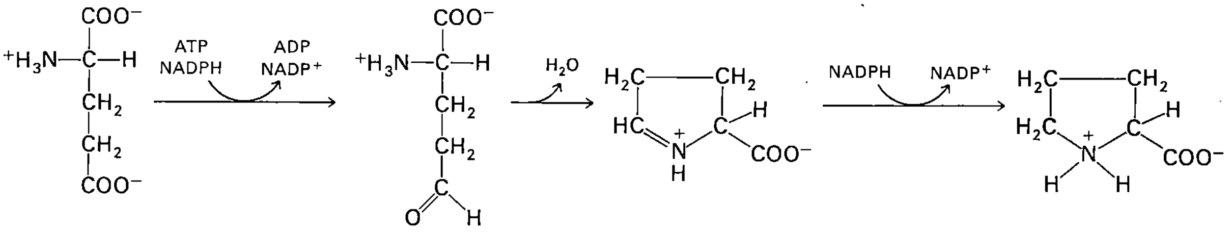

SERINE IS SYNTHESIZED FROM 3-PHOSPHOGLYCERATE

Serine is synthesized from 3-phosphoglycerate, an intermediate in glycolysis. The first step is an oxidation to 3-phosphohydroxypyruvate. This α-keto acid is transaminated to 3-phosphoserine, which is then hydrolyzed to serine.

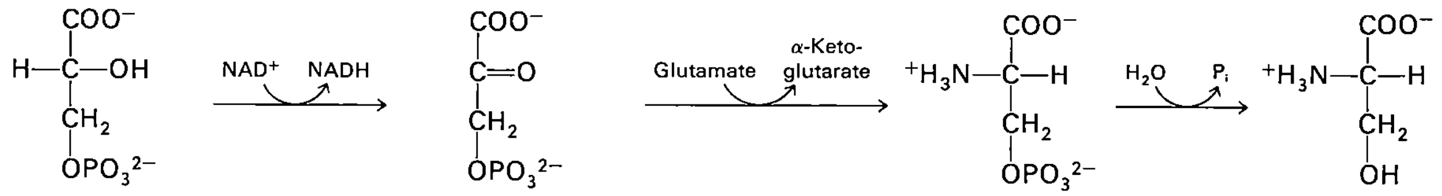

Serine is the precursor of *glycine* and *cysteine*. In the formation of glycine, the side-chain β carbon atom of serine is transferred to *tetrahydrofolate*, a carrier of one-carbon units that will be discussed shortly.

$$\text{Serine} + \text{tetrahydrofolate} \rightleftharpoons \text{glycine} + \text{methylenetetrahydrofolate} + H_2O$$

This interconversion is catalyzed by *serine transhydroxymethylase*, a pyridoxal phosphate (PLP) enzyme. The bond between the α and β carbon atoms of serine is labilized by the formation of a Schiff base between serine and PLP. The β carbon atom of serine is then transferred to tetrahydrofolate. The conversion of serine into cysteine requires the substitution of a sulfur atom derived from methionine for the side-chain oxygen atom (p. 723).

TETRAHYDROFOLATE CARRIES ACTIVATED ONE-CARBON UNITS AT SEVERAL OXIDATION LEVELS

Tetrahydrofolate (also called *tetrahydropteroylglutamate*), a highly versatile carrier of activated one-carbon units, consists of three groups: a substituted pteridine, *p*-aminobenzoate, and glutamate. Mammals can synthesize a pteridine ring but they are unable to conjugate it to the other two units. They obtain tetrahydrofolate from their diets or from microorganisms in their intestinal tracts.

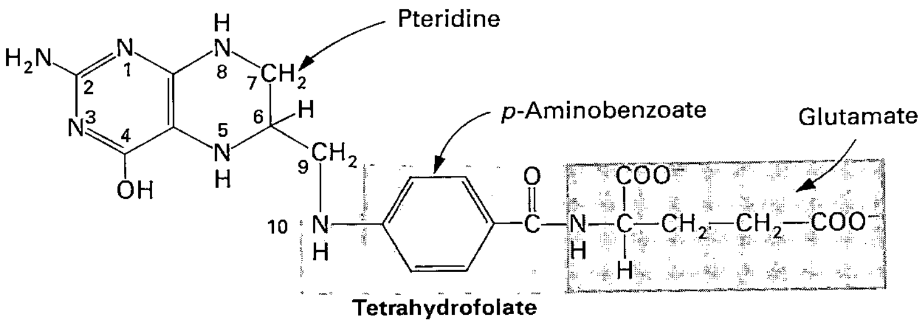

Table 28-2
One-carbon groups carried by tetrahydrofolate

Oxidation state		Group
Most reduced (= methanol)	$-CH_3$	Methyl
Intermediate (= formaldehyde)	$-CH_2-$	Methylene
Most oxidized (= formic acid)	$-CHO$	Formyl
	$-CHNH$	Formimino
	$-CH=$	Methenyl

The one-carbon group carried by tetrahydrofolate is bonded to its N-5 or N-10 nitrogen atom (denoted as N^5 and N^{10}) or to both. This unit can exist in three oxidation states (Table 28-2). The most reduced form carries a *methyl* group, whereas the intermediate form carries a *methylene* group. More oxidized forms carry a *formyl, formimino,* or *methenyl* group. The fully oxidized one-carbon unit, CO_2, is carried by biotin (p. 572) rather than by tetrahydrofolate.

The one-carbon units carried by tetrahydrofolate are interconvertible (Figure 28-8). N^5,N^{10}-*Methylene*tetrahydrofolate can be reduced to N^5-*methyl*tetrahydrofolate or oxidized to N^5,N^{10}-*methenyl*tetrahydrofolate. N^5,N^{10}-*Methenyl*tetrahydrofolate can be converted into N^5-*formimino*tetrahydrofolate or N^{10}-*formyl*tetrahydrofolate, which are at the same oxidation level. N^{10}-*Formyl*tetrahydrofolate can also be synthesized from

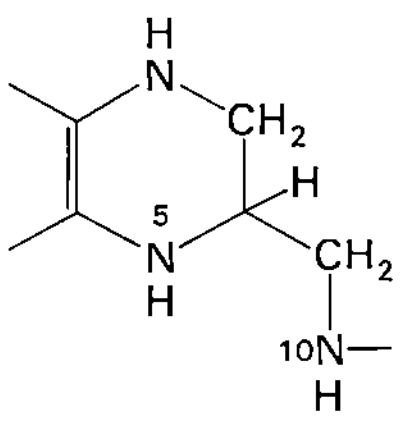

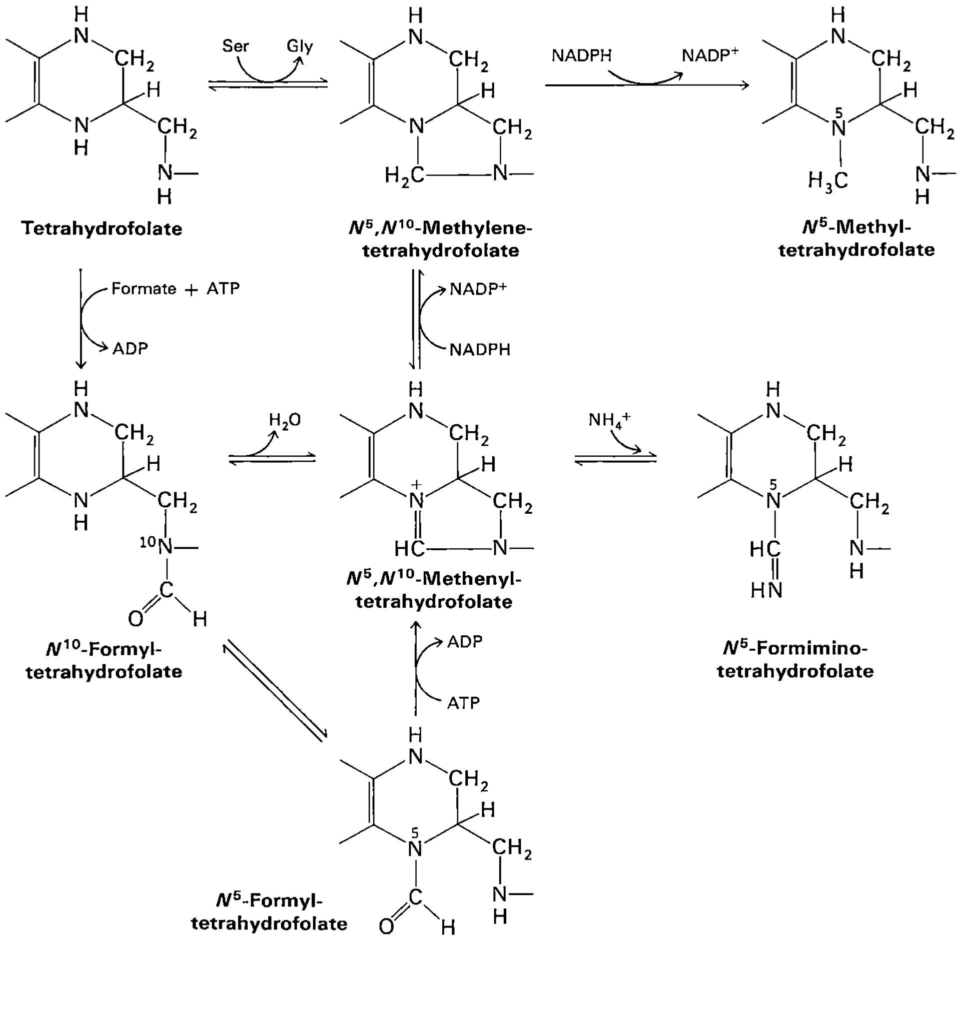

Figure 28-8
Conversions of one-carbon units attached to tetrahydrofolate.

tetrahydrofolate, formate, and ATP. N^5-*Formyl*tetrahydrofolate can be reversibly isomerized to N^{10}-*formyl*tetrahydrofolate, or it can be converted into N^5,N^{10}-*methenyl*tetrahydrofolate.

$$\text{Formate} + \text{ATP} + \text{tetrahydrofolate} \rightleftharpoons N^{10}\text{-formyltetrahydrofolate} + \text{ADP} + P_i$$

These tetrahydrofolate derivatives serve as donors of one-carbon units in a variety of biosyntheses. Methionine is regenerated from homocysteine by transfer of the methyl group of N^5-methyltetrahydrofolate, as will be discussed shortly. Some of the carbon atoms of *purines* are derived from the N^{10}-formyl derivatives of tetrahydrofolate. The methyl group of *thymine,* a pyrimidine, comes from N^5,N^{10}-methylenetetrahydrofolate. This tetrahydrofolate derivative also donates a one-carbon unit in the synthesis of *glycine* from CO_2 and NH_4^+, a reaction catalyzed by *glycine synthase* (called the *glycine cleavage enzyme* when it operates in the reverse direction).

$$CO_2 + NH_4^+ + N^5,N^{10}\text{-methylenetetrahydrofolate} + \text{NADH} \rightleftharpoons \text{glycine} + \text{tetrahydrofolate} + \text{NAD}^+$$

Thus, one-carbon units at each of the three oxidation levels are utilized in biosyntheses. Furthermore, *tetrahydrofolate serves as an acceptor of one-carbon units in degradative reactions.* The major source of one-carbon units is the facile conversion of serine into glycine, which yields N^5,N^{10}-methylenetetrahydrofolate. Serine can be derived from 3-phosphoglycerate (p. 719), and so *this pathway enables one-carbon units to be formed de novo from carbohydrate.* The breakdown of glycine by the *glycine cleavage enzyme* yields CO_2, NH_4^+, and a one-carbon fragment that becomes part of N^5,N^{10}-methylenetetrahydrofolate. Likewise, one-carbon units from *betaine* and *choline* funnel into tetrahydrofolate to form the N^5,N^{10}-methylene derivative. The breakdown of *histidine* yields *N*-formiminoglutamate (p. 640), which transfers its formimino group to tetrahydrofolate to give the N^5 derivative.

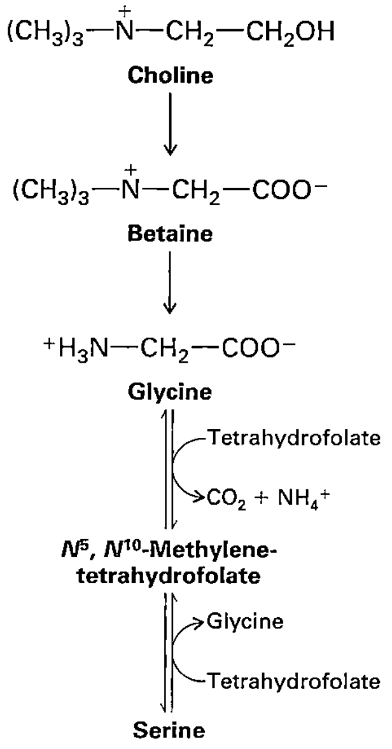

Figure 28-9
One-carbon fragments from glycine and serine funnel into N^5,N^{10}-methylenetetrahydrofolate. Tetrahydrofolate also accepts two of the methyl groups of betaine to form this methylene derivative. The third methyl group of betaine is taken up by homocysteine to form methionine.

S-ADENOSYLMETHIONINE IS THE MAJOR DONOR OF METHYL GROUPS

Tetrahydrofolate can carry a methyl group on its N^5 atom, but its transfer potential is not sufficiently high for most biosynthetic methylations. Rather, the activated methyl donor is usually *S-adenosylmethionine,* which we encountered earlier in the conversion of phosphatidyl ethanolamine into phosphatidyl choline (p. 687). *S*-Adenosylmethionine is synthesized by the transfer of an adenosyl group from ATP to the sulfur atom of methionine. The methyl group of the methionine unit is activated by the positive charge on the adjacent sulfur atom, which makes the molecule much more reactive than N^5-methyltetrahydrofolate.

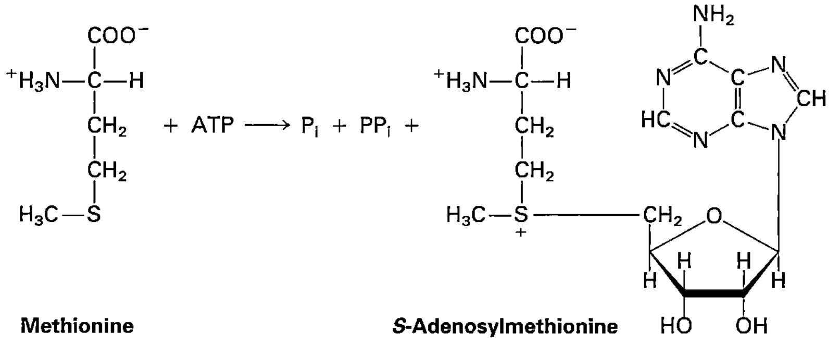

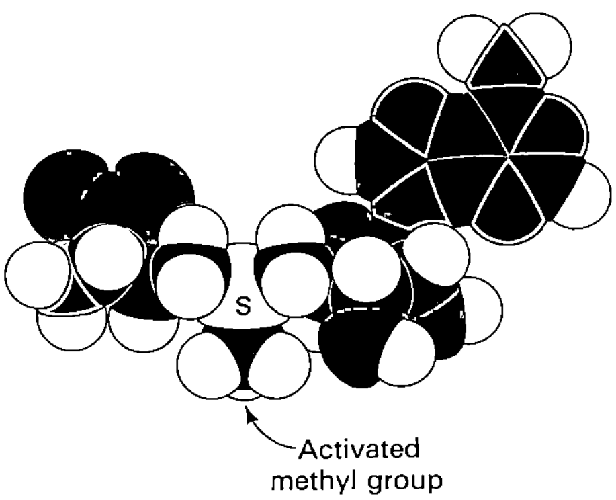

Figure 28-10
Model of S-adenosylmethionine.

The synthesis of S-adenosylmethionine is unusual in that the triphosphate group of ATP is split into pyrophosphate and orthophosphate. A total of three ~P are ultimately consumed because pyrophosphate is subsequently hydrolyzed to two P_i.

S-*Adenosylhomocysteine* is formed when the methyl group of S-adenosylmethionine is transferred to an acceptor such as phosphatidyl ethanolamine. S-Adenosylhomocysteine is then hydrolyzed to *homocysteine* and adenosine.

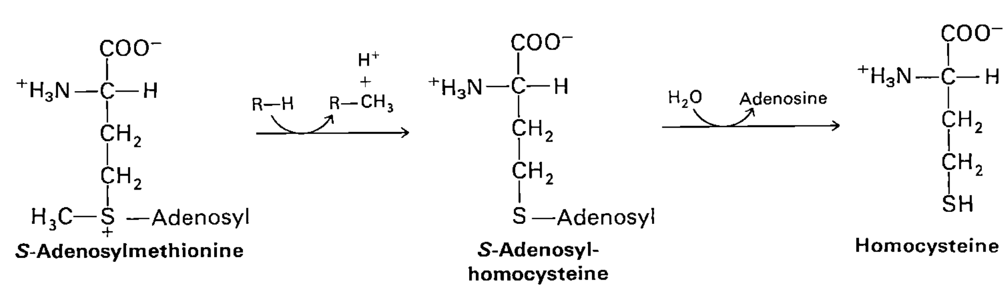

Methionine can be regenerated by the transfer of a methyl group to homocysteine from N^5-methyltetrahydrofolate, a reaction catalyzed by *homocysteine methyltransferase.*

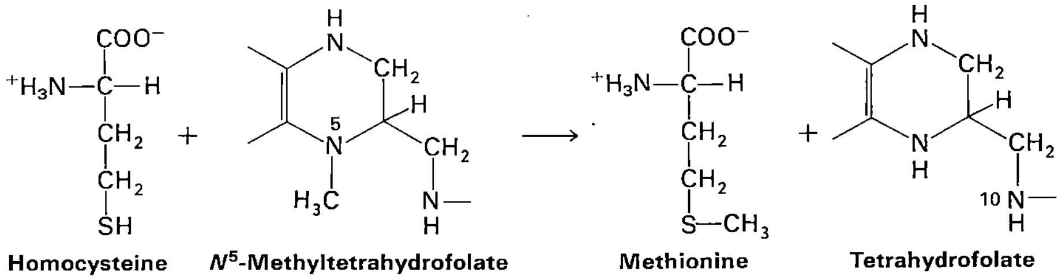

The coenzyme that mediates this transfer of a methyl group is *methylcobalamin,* derived from vitamin B_{12}. In fact, this reaction and the rearrangement of L-methylmalonyl CoA to succinyl CoA (p. 641) are the only two B_{12}-dependent reactions in mammals. Alternatively, homocysteine can be methylated to methionine by donors such as *betaine,* an oxidation product of choline (see Figure 28-9).

These reactions constitute the *activated methyl cycle* (Figure 28-11). Methyl groups enter the cycle in the conversion of homocysteine into methionine and are then made highly reactive by the insertion of an adenosyl group, which makes the sulfur atom positively charged. The

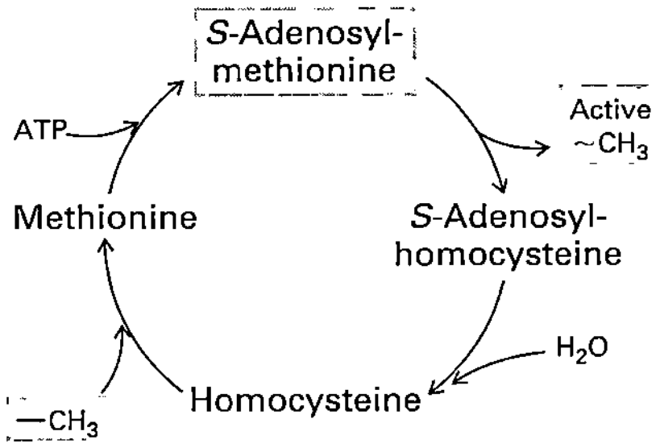

Figure 28-11
Activated methyl cycle. Humans derive methionine from the diet.

high transfer potential of this *S*-methyl group enables it to be transferred to a wide variety of acceptors, such as the amino group of norepinephrine (the precursor of epinephrine, p. 730) and specific glutamate residues of bacterial chemotaxis receptors (p. 331).

S-Adenosylmethionine serves not only as a methyl donor. Following decarboxylation, it donates a propylamine group in the synthesis of *spermidine* from *putrescine.* Donation of a second propylamine group results in the formation of *spermine.* These *polyamines* bind tightly to nucleic acids and are abundant in rapidly proliferating cells.

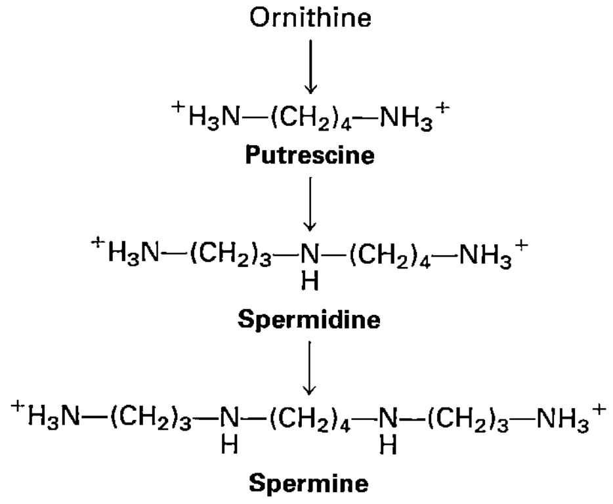

S-Adenosylmethionine is also the precursor of *ethylene,* a gaseous plant hormone that induces the ripening of fruit. The release of *S*-methylthioadenosine yields a cyclopropane derivative that is oxidized to ethylene, carbon dioxide, and hydrogen cyanide. The Greek philosopher Theophrastus recognized more than 2000 years ago that sycamore figs do not ripen unless they are scraped with an iron claw. The reason is now known: *wounding triggers ethylene production, which in turn induces ripening.* Much effort is being devoted to this biosynthetic pathway because ethylene is a culprit in the spoilage of fruit.

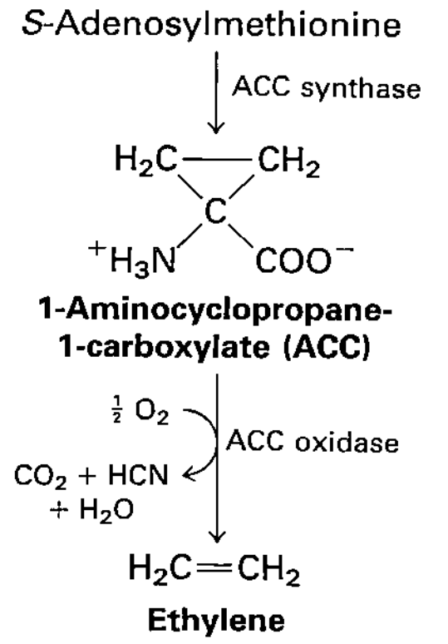

CYSTEINE IS SYNTHESIZED FROM SERINE AND HOMOCYSTEINE

In addition to being a precursor of methionine in the activated methyl cycle, homocysteine is an intermediate in the synthesis of cysteine. Serine and homocysteine condense to form *cystathionine* (Figure 28-12). This reaction is catalyzed by *cystathionine synthase,* a PLP enzyme. Cystathionine is then deaminated and cleaved to cysteine and α-ketobutyrate by *cystathioninase,* another PLP enzyme. The net reaction is

$$\text{Homocysteine} + \text{serine} \longrightarrow \text{cysteine} + \alpha\text{-ketobutyrate}$$

Note that the sulfur atom of cysteine is derived from homocysteine, whereas the carbon skeleton comes from serine.

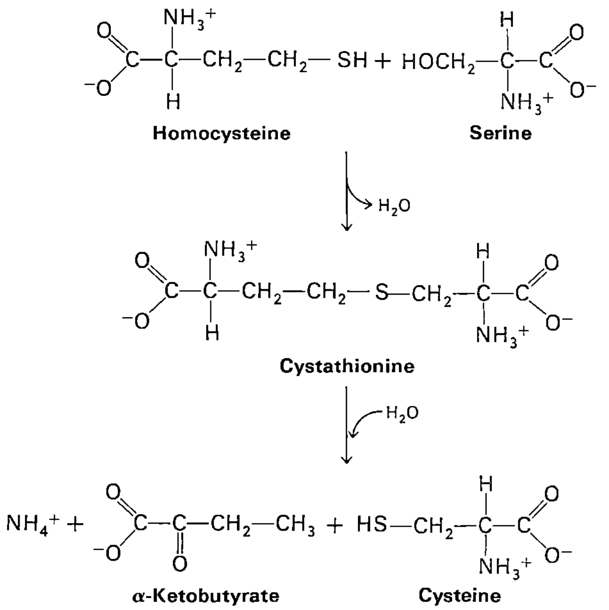

Figure 28-12
Synthesis of cysteine.

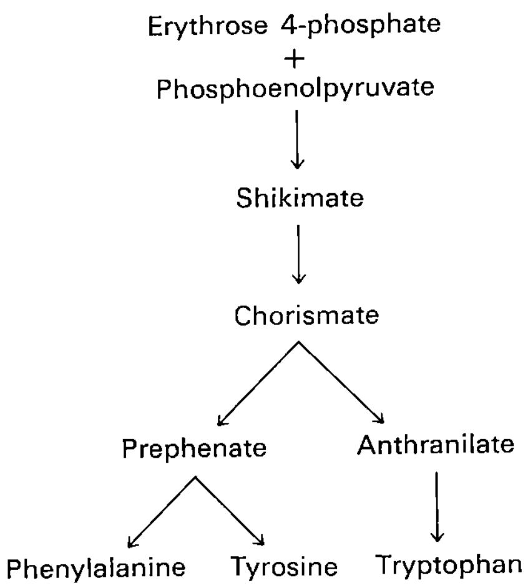

Figure 28-13
Pathway for the biosynthesis of aromatic amino acids in *E. coli.*

SHIKIMATE AND CHORISMATE ARE INTERMEDIATES IN THE BIOSYNTHESIS OF AROMATIC AMINO ACIDS

We turn now to the biosynthesis of essential amino acids, which are formed by much more complex routes than are the nonessential amino acids. Essential amino acids are synthesized by plants and microorganisms. The ones in the diet of humans are ultimately derived primarily from plants. The pathways for the synthesis of aromatic amino acids in bacteria have been selected for discussion because they are well understood and exemplify recurring regulatory motifs.

Phenylalanine, tyrosine, and tryptophan are synthesized by a common pathway in *E. coli* (Figure 28-13). The initial step is the condensation of phosphoenolpyruvate (a glycolytic intermediate) with erythrose 4-phosphate (a pentose phosphate pathway intermediate). The resulting C_7 open-chain sugar loses its phosphoryl group and cyclizes to 5-dehydroquinate. Dehydration then yields 5-dehydroshikimate, which is reduced by NADPH to *shikimate* (Figure 28-14). Phosphorylation of shikimate by ATP gives 5-phosphoshikimate, which condenses with a second molecule of phosphoenolpyruvate. This 3-enolpyruvyl intermediate loses its phosphoryl group, yielding *chorismate,* the common precursor of all three aromatic amino acids.

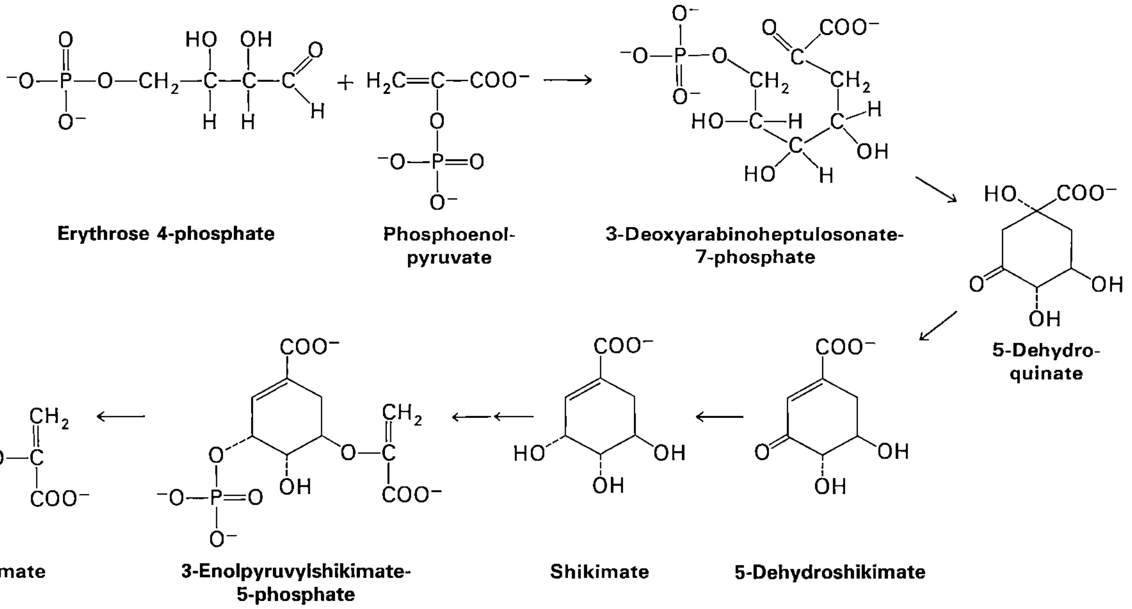

Figure 28-14
Synthesis of chorismate, an intermediate in the biosynthesis of phenylalanine, tyrosine, and tryptophan in *E. coli.*

The pathway bifurcates at chorismate. Let us first follow the *prephenate branch* (Figure 28-15). A mutase converts chorismate into prephenate, the immediate precursor of the aromatic ring of phenylalanine and tyrosine. Dehydration followed by decarboxylation yields *phenylpyruvate.* Alternatively, prephenate can be oxidatively decarboxylated to p-*hydroxyphenylpyruvate.* These α-keto acids are then transaminated to form *phenylalanine* and *tyrosine.*

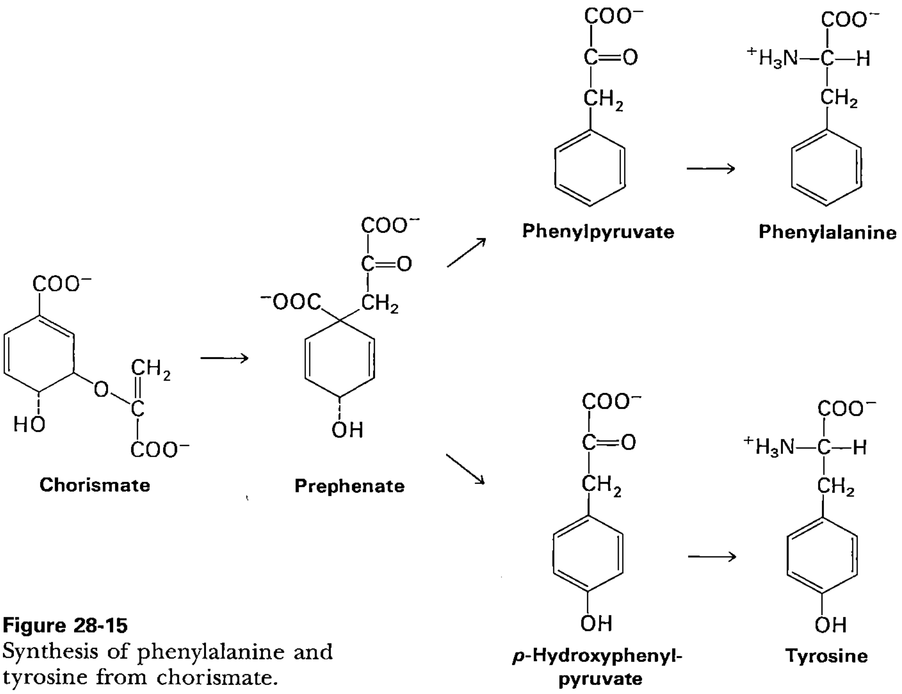

Figure 28-15
Synthesis of phenylalanine and tyrosine from chorismate.

The branch starting with *anthranilate* leads to the synthesis of *tryptophan* (Figure 28-16). Chorismate acquires an amino group from the side chain of glutamine and releases pyruvate to form anthranilate. In fact, *glutamine serves as an amino donor in many biosynthetic reactions.* Anthranilate then condenses with *phosphoribosylpyrophosphate (PRPP), an activated form of ribose phosphate.* PRPP is also a key intermediate in the synthesis of histidine, purine nucleotides, and pyrimidine nucleotides (p. 741). The C-1 atom of ribose 5-phosphate becomes bonded to the nitrogen atom of anthranilate in a reaction that is driven by the hydrolysis of pyrophosphate.

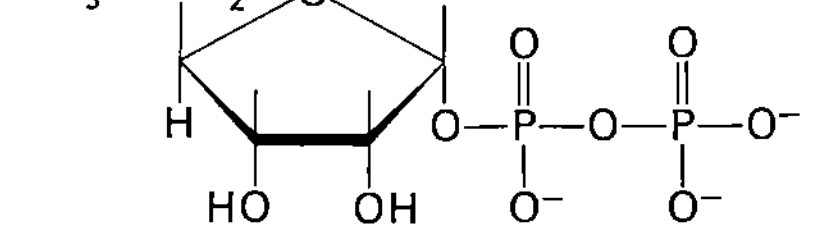

Phosphoribosylpyrophosphate (PRPP)

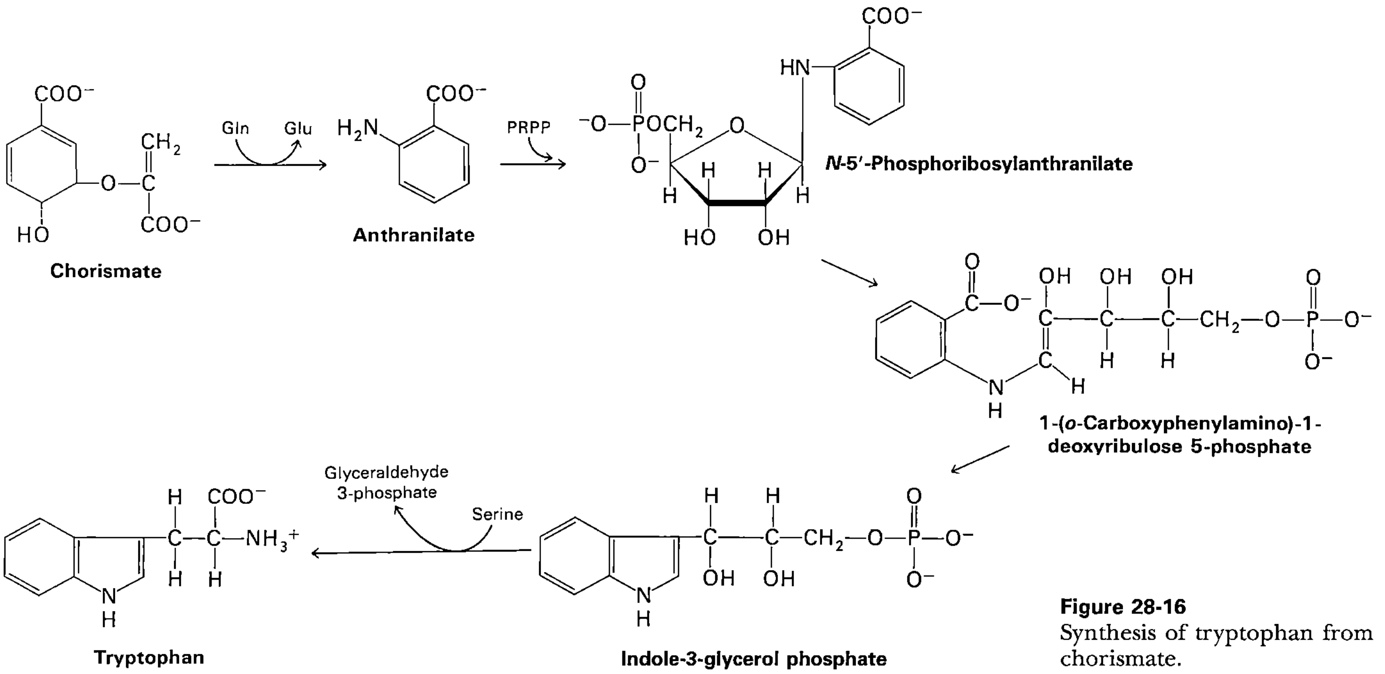

Figure 28-16
Synthesis of tryptophan from chorismate.

The ribose moiety of phosphoribosylanthranilate undergoes rearrangement to yield 1-(*o*-carboxyphenylamino)-1-deoxyribulose 5-phosphate. This intermediate is dehydrated and then decarboxylated to *indole-3-glycerol phosphate,* which reacts with serine to form *tryptophan.* In this final step, which is catalyzed by *tryptophan synthetase (tryptophan synthase),* the glycerol phosphate side chain of indole-3-glycerol phosphate is replaced by the carbon skeleton and amino group of serine.

TRYPTOPHAN SYNTHETASE ILLUSTRATES SUBSTRATE CHANNELING IN ENZYMATIC CATALYSIS

Tryptophan synthetase of *E. coli,* an $\alpha_2\beta_2$ tetramer, can be dissociated into two α subunits and a β_2 subunit. The α subunit catalyzes the formation of indole from indole-3-glycerol phosphate, whereas the β_2 subunit catalyzes the condensation of indole and serine to form tryptophan.

$$\text{Indole-3-glycerol phosphate} \xrightarrow{\alpha \text{ subunit}} \text{indole} + \text{glyceraldehyde 3-phosphate}$$

$$\text{Indole} + \text{serine} \xrightarrow{\beta_2 \text{ subunit}} \text{tryptophan} + H_2O$$

Each active site on the β_2 subunit contains a PLP prosthetic group. Serine forms a Schiff base with this PLP, which is then dehydrated to give the *Schiff base of aminoacrylate* (Figure 28-17). This reactive intermediate is attacked by indole to give tryptophan.

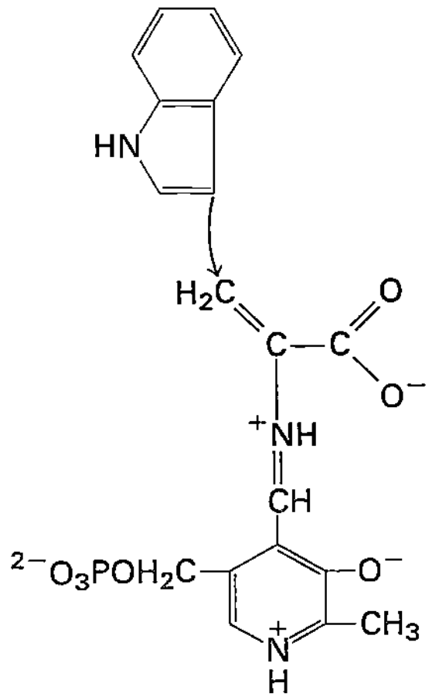

Figure 28-17
Intermediate in the synthesis of tryptophan. PLP (green) on a β chain of tryptophan synthetase forms a Schiff base with serine, which is then dehydrated to give the Schiff base of aminoacrylate (red). This enzyme-bound intermediate is attacked by indole, the product of the partial reaction catalyzed by the α subunit, to give tryptophan.

The synthesis of tryptophan poses a challenge. Indole, a hydrophobic molecule, readily traverses membranes and would be lost from the cell if it were allowed to diffuse away from the enzyme. This problem is solved in an ingenious way. A 25-Å-long channel connects the active site of the α subunit with that of the adjacent β subunit in the $\alpha_2\beta_2$ tetramer. Thus, indole can diffuse from one active site to the other without being released into bulk solvent. Indeed, isotopic labeling experiments have shown that indole formed by the α subunit does not leave the enzyme when serine is present. Furthermore, the two partial reactions are coordinated. Indole is not formed by the α subunit until the highly reactive aminoacrylate is ready and waiting in the β subunit. We see here a clear-cut example of *substrate channeling* in catalysis by a multienzyme complex. Channeling increases the overall catalytic rate more than tenfold. Furthermore, a deleterious side reaction—in this case, the potential loss of an intermediate—is avoided.

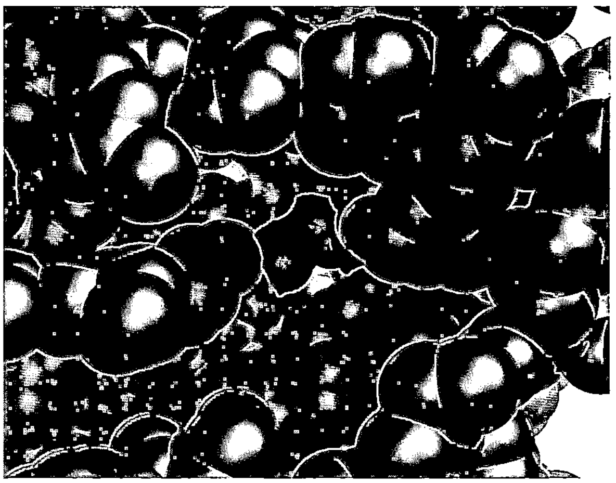

Figure 28-18
Tunnel in tryptophan synthetase leading from the active site of an α chain (blue) to that of a β chain (yellow). A pyridoxal phosphate (red) bound to β lies at the end of the tunnel. Only the main-chain atoms are shown. [Drawn from 1wsy.pdb. C.C. Hyde, S.A. Ahmed, E.A. Padlan, E.W. Miles, and D.R. Davies. *J. Biol. Chem.* 263(1988):17857.]

AMINO ACID BIOSYNTHESIS IS REGULATED
BY FEEDBACK INHIBITION

The rate of synthesis of amino acids depends mainly on the *amounts* of the biosynthetic enzymes and on their *activities*. We shall now consider the control of enzymatic activity. The regulation of enzyme synthesis will be discussed in Chapter 36.

The first irreversible reaction in a biosynthetic pathway, called the *committed step*, is usually an important regulatory site. *The final product of the pathway (Z) often inhibits the enzyme that catalyzes the committed step (A → B).*

$$A \longrightarrow | \;\; \overset{\text{Inhibited}}{\underset{\text{by Z}}{|}} \longrightarrow B \longrightarrow C \longrightarrow D \longrightarrow E \longrightarrow Z$$

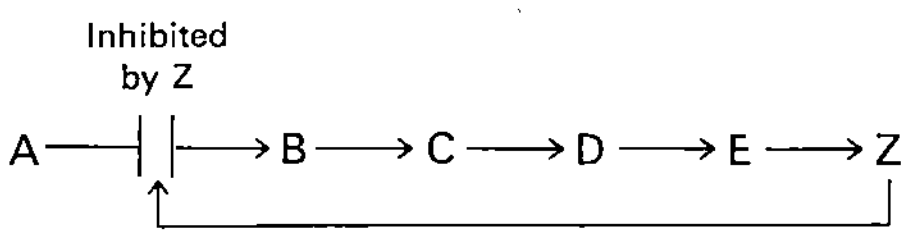

This kind of control is essential for the conservation of building blocks and metabolic energy. The first example was seen in studies of the biosynthesis of isoleucine in *E. coli*. The dehydration and deamination of threonine to α-ketobutyrate is the committed step. *Threonine deaminase,* the PLP enzyme that catalyzes this reaction, is allosterically inhibited by isoleucine (Figure 28-19). Likewise, tryptophan inhibits the enzyme complex that catalyzes the first two steps in the conversion of chorismate into tryptophan.

Consider a branched biosynthetic pathway in which Y and Z are the final products:

$$A \longrightarrow B \longrightarrow C \overset{\nearrow \; D \longrightarrow E \longrightarrow Y}{\searrow \; F \longrightarrow G \longrightarrow Z}$$

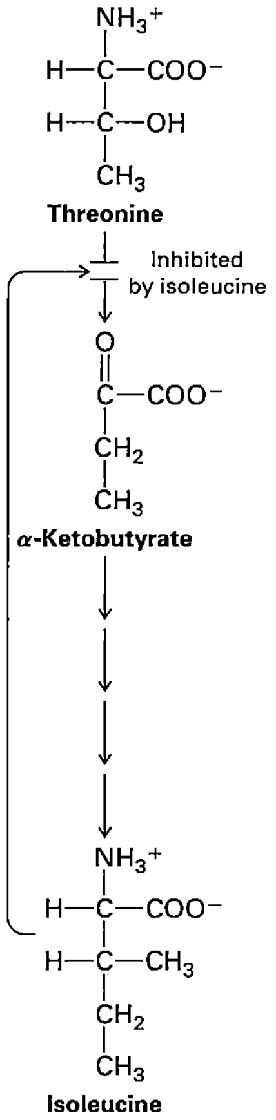

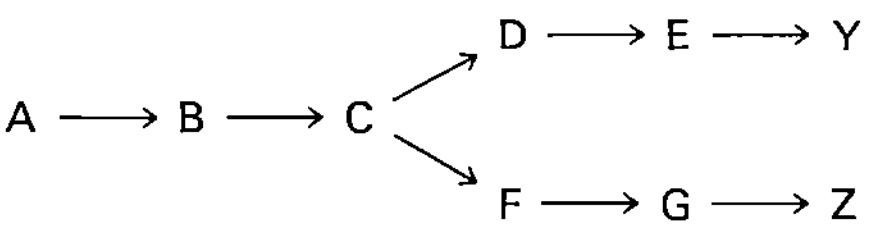

Suppose that high levels of Y *or* Z completely inhibit the first common step (A → B). Then, high levels of Y would prevent the synthesis of Z even if there were a deficiency of Z. Such a regulatory scheme is obviously not optimal. In fact, several intricate feedback inhibition mechanisms have been found in branched biosynthetic pathways:

1. *Sequential feedback inhibition.* The first common step (A → B) is not inhibited directly by Y or Z. Rather, these final products inhibit the reactions leading away from the point of branching: Y inhibits the C → D step, and Z inhibits the C → F step. In turn, high levels of C inhibit the A → B step. Thus, the first common reaction is blocked only if both final products are present in excess.

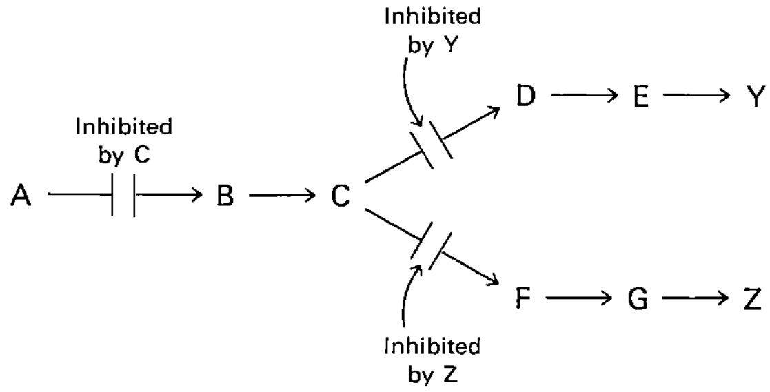

Sequential feedback control regulates the synthesis of aromatic amino acids in *Bacillus subtilis*. The first divergent steps in the synthesis of phenyl-

Figure 28-19
Feedback inhibition of threonine deaminase by isoleucine, the final product of the pathway.

alanine, tyrosine, and tryptophan are inhibited by their final products. If all three are present in excess, chorismate and prephenate accumulate. These branch-point intermediates in turn inhibit the first common step in the overall pathway, which is the condensation of phosphoenolpyruvate and erythrose 4-phosphate (see Figure 28-13).

2. *Enzyme multiplicity.* The distinguishing feature of this mechanism is that the first common step (A ⟶ B) is catalyzed by either of two different enzymes. One of them is directly inhibited by Y, and the other by Z. A high level of either Y or Z partially blocks the first step. Both Y and Z must be present at high levels to prevent the conversion of A into B completely. In the rest of this control scheme, as in sequential feedback control, Y inhibits the C ⟶ D step, and Z inhibits the C ⟶ F step.

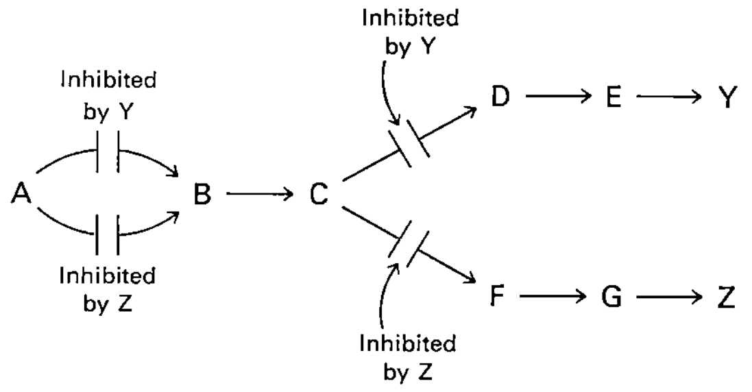

Differential inhibition of multiple enzymes controls a variety of biosynthetic pathways in microorganisms. In *E. coli,* the condensation of phosphoenolpyruvate and erythrose 4-phosphate is catalyzed by three different enzymes. One is inhibited by phenylalanine, another by tyrosine, and the third by tryptophan. Furthermore, there are two different mutases that convert chorismate into prephenate. One of them is inhibited by phenylalanine, the other by tyrosine.

3. *Concerted feedback inhibition.* The first common step (A ⟶ B) is inhibited only if high levels of Y *and* Z are simultaneously present. In contrast with enzyme multiplicity, a high level of either product alone does not appreciably inhibit the A ⟶ B step. Concerted feedback inhibition behaves as a logical *AND* gate. As in the two preceding control schemes , Y inhibits the C ⟶ D step and Z inhibits the C ⟶ F step.

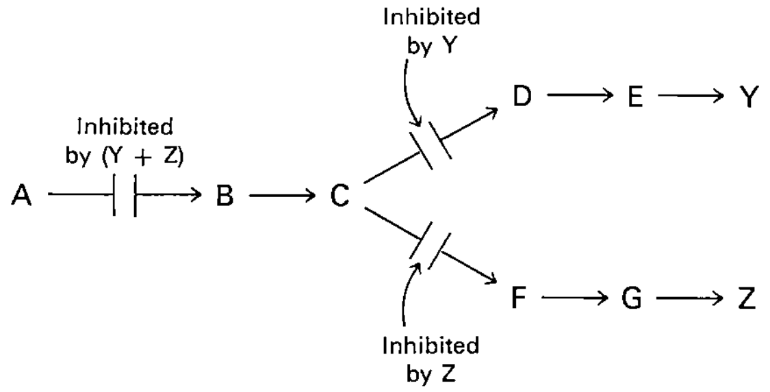

An example of concerted feedback control is the inhibition of bacterial aspartyl kinase by threonine and lysine, the final products.

4. *Cumulative feedback inhibition.* The first common step (A ⟶ B) is partly inhibited by each of the final products, both acting independently of the other. Suppose that a high level of Y decreased the rate of the A ⟶ B step from 100 to 60 s^{-1} and that Z alone decreased the rate from 100 to 40 s^{-1}. Then, the rate of the A ⟶ B step in the presence of high levels of Y and Z would be 24 s^{-1} ($0.6 \times 0.4 \times 100\ s^{-1}$).

THE ACTIVITY OF GLUTAMINE SYNTHETASE IS MODULATED BY AN ENZYMATIC CASCADE

The regulation of glutamine synthetase in *E. coli* is a striking example of *cumulative feedback inhibition.* Recall that glutamine is synthesized from glutamate, NH_4^+, and ATP (p. 717). Glutamine synthetase consists of 12 identical 50-kd subunits arranged in two hexagonal rings that face each other (Figure 28-20). Earl Stadtman showed that this enzyme regulates the flow of nitrogen and hence plays a key role in controlling bacterial metabolism. The amide group of glutamine is a source of nitrogen in the biosyntheses of a variety of compounds, such as tryptophan, histidine, carbamoyl phosphate, glucosamine 6-phosphate, CTP, and AMP. Glutamine synthetase is cumulatively inhibited by each of these final products of glutamine metabolism, as well as by alanine and glycine. The enzymatic activity of glutamine synthetase is switched off almost completely when all final products are bound to the enzyme.

The activity of glutamine synthetase is also controlled by *reversible covalent modification*—the attachment of an *AMP unit* by a phosphodiester bond to the hydroxyl group of a specific tyrosine residue in each subunit. *This adenylylated enzyme is more susceptible to cumulative feedback inhibition than is the deadenylylated form.* The covalently attached AMP unit is removed from the adenylylated enzyme by phosphorolysis. These reactions

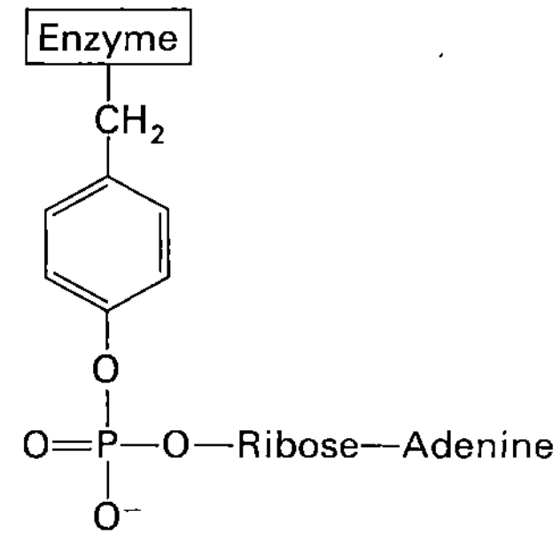

Adenylylated enzyme

are catalyzed by the same enzyme, *adenylyl transferase.* What determines whether it inserts or removes an AMP unit? The specificity of adenylyl transferase turns out to be controlled by a *regulatory protein* (designated P), which can exist in two forms, P_A and P_D (Figure 28-21). The complex of P_A and adenylyl transferase catalyzes the attachment of an AMP unit to glutamine synthetase, which reduces its activity. Conversely, the complex of P_D and adenylyl transferase removes AMP from the adenylylated enzyme. These opposing catalytic activities of adenylyl transferase are carried out by different catalytic sites on the same polypeptide chain—a motif encountered earlier in the control of the levels of fructose 2,6-bisphosphate (p. 494) and isocitrate dehydrogenase (p. 524).

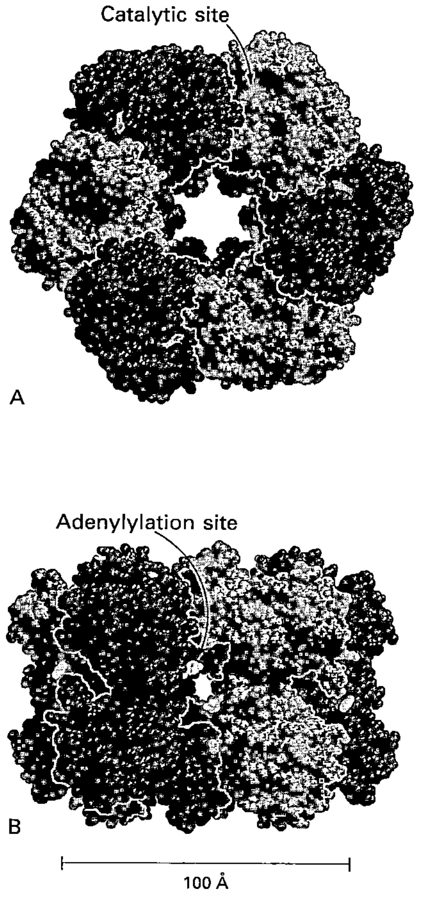

Figure 28-20
Structure of glutamine synthetase from *E. coli.* The 12 identical subunits (red and blue) are arranged in two hexameric rings joined face to face. Catalytic sites are shown in yellow, and adenylylation sites in white. (A) A view down the sixfold axis. (B) A view down a twofold axis (perpendicular to that in A). The adenylylation sites are more than 20 Å away from catalytic sites. [Drawn from 2gls.pdb. M.M. Yamashita, R.J. Almassy, C.A. Janson, and D. Cascio. *J. Biol. Chem.* 264(1989):17681.]

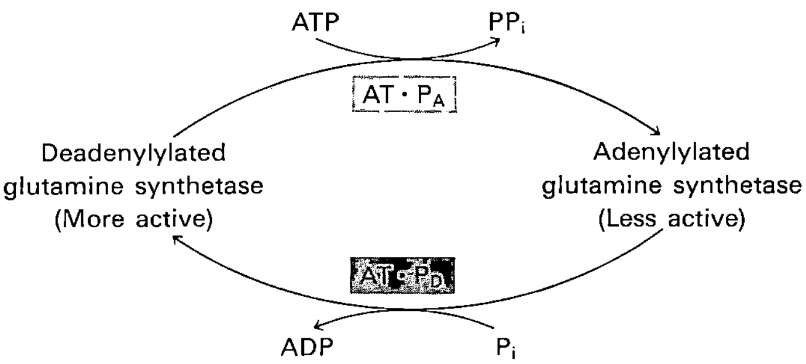

Figure 28-21
Control of the activity of glutamine synthetase by reversible covalent modification. Adenylylation is catalyzed by a complex of adenylyl transferase (AT) and one form of a regulatory protein (P_A). The same enzyme catalyzes deadenylylation when it is complexed with the other form (P_D) of the regulatory protein.

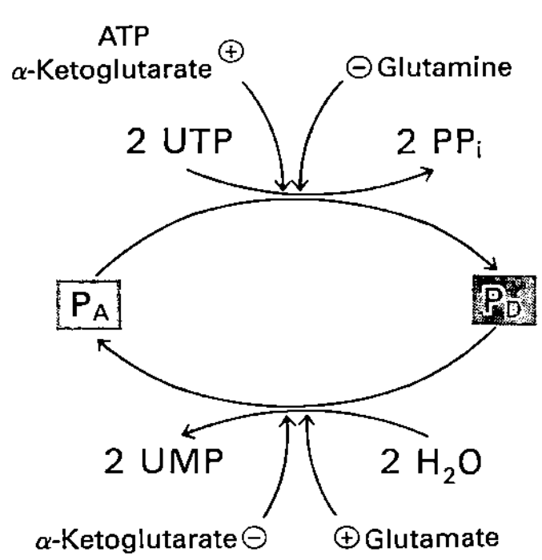

Figure 28-22
A higher level in the regulatory cascade of glutamine synthetase. P_A and P_D, the regulatory proteins that control the specificity of adenylyl transferase, are interconvertible. P_A is converted into P_D by uridylylation, which is reversed by hydrolysis. The enzymes catalyzing these reactions sense the concentrations of metabolic intermediates.

This brings us to another level of reversible covalent modification. P_A is converted into P_D by the attachment of uridine monophosphate (UMP) (Figure 28-22). This reaction, which is catalyzed by *uridylyl transferase,* is stimulated by ATP and α-ketoglutarate, whereas it is inhibited by glutamine. In turn, the two UMP units on P_D are removed by hydrolysis, a reaction promoted by glutamate and inhibited by α-ketoglutarate. These opposing catalytic activities also are present on a single polypeptide chain and are controlled so that the enzyme does not simultaneously catalyze uridylylation and hydrolysis.

Why is an enzymatic cascade used to regulate glutamine synthetase? One advantage of a cascade is that it *amplifies signals,* as in blood clotting (p. 253) and the control of glycogen metabolism (p. 595). Another reason is that the *potential for allosteric control is markedly increased when each enzyme in the cascade is an independent target for regulation.* The integration of nitrogen metabolism in a cell requires that a large number of input signals be detected and processed. There are limits to what a single protein can accomplish on its own—even a molecule as sentient as glutamine synthetase! The evolution of a cascade provided many more regulatory sites and made possible a finer tuning of the flow of nitrogen in the cell.

AMINO ACIDS ARE PRECURSORS OF MANY BIOMOLECULES

Amino acids are the building blocks of proteins and peptides. They also serve as precursors of many kinds of small molecules that have important and diverse biological roles. Let us briefly survey some of the biomolecules that are derived from amino acids (Figure 28-23). *Purines* and *pyrimidines* are derived in part from amino acids. The biosynthesis of these

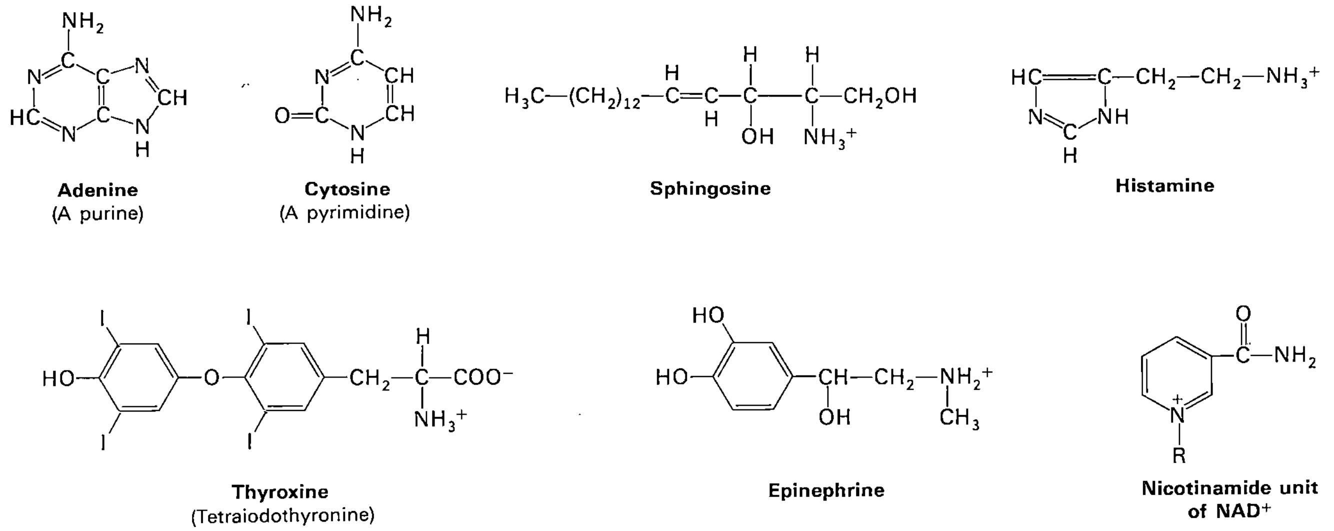

Figure 28-23
Biomolecules derived from amino acids. Atoms contributed by amino acids are shown in blue.

precursors of DNA, RNA, and numerous coenzymes is discussed in detail in the next chapter. Six of the nine atoms of the purine ring and four of the six atoms of the pyrimidine ring are derived from amino acids. The reactive terminus of *sphingosine,* an intermediate in the synthesis of sphingolipids, comes from serine. *Histamine,* a potent vasodilator, is derived from histidine by decarboxylation. Tyrosine is a precursor of the hormones *thyroxine* (tetraiodothyronine) and *epinephrine* and of *melanin,* a polymeric pigment. The neurotransmitter 5-hydroxytryptamine (*serotonin*) and the *nicotinamide ring* of NAD^+ are synthesized from tryptophan. Glutamine contributes the amide group of the nicotinamide moiety.

GLUTATHIONE, A γ-GLUTAMYL PEPTIDE, SERVES AS A SULFHYDRYL BUFFER AND AN ANTIOXIDANT

Glutathione, a tripeptide containing a sulfhydryl group, is a highly distinctive amino acid derivative with several important roles. For example, glutathione protects red cells from oxidative damage (p. 568). The first step in the synthesis of glutathione is the formation of a peptide linkage between the γ-carboxyl group of glutamate and the amino group of cysteine, in a reaction catalyzed by *γ-glutamylcysteine synthetase* (Figure 28-24). Formation of this peptide bond requires activation of the γ-carboxyl

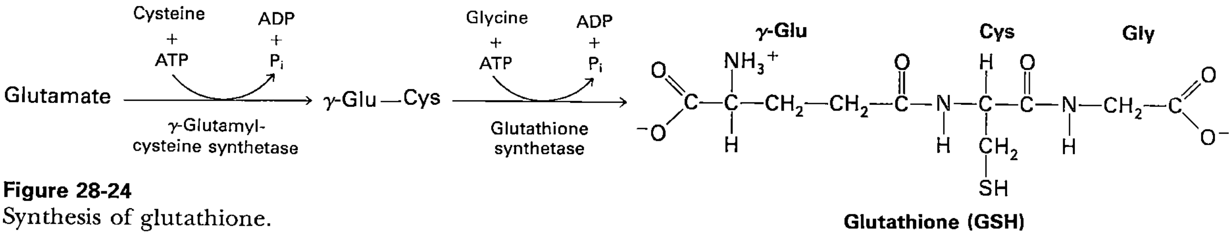

Figure 28-24
Synthesis of glutathione.

group, which is achieved by ATP. The resulting acyl-phosphate intermediate is then attacked by the amino group of cysteine. This reaction is feedback-inhibited by glutathione. In the second step, which is catalyzed by *glutathione synthetase,* ATP activates the carboxyl group of cysteine to enable it to condense with the amino group of glycine.

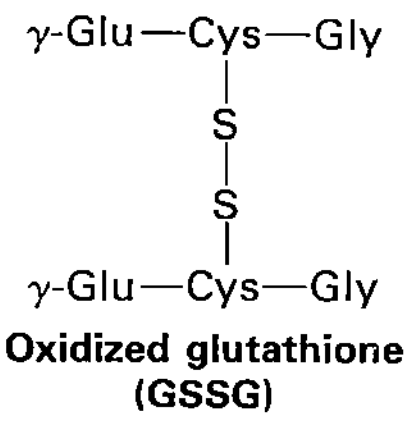

Glutathione, present at high levels (~5 mM) in animal cells, serves as a sulfhydryl buffer. It cycles between a reduced thiol form (GSH) and an oxidized form (GSSG) in which two tripeptides are linked by a disulfide bond. GSSG is reduced to GSH by *glutathione reductase,* a flavoprotein that uses NADPH as the electron source (p. 568). The ratio of GSH to GSSG in most cells is greater than 500.

Glutathione plays a key role in detoxification by reacting with hydrogen peroxide and organic peroxides, the harmful byproducts of aerobic life.

$$2\ \text{GSH} + \text{R—O—OH} \longrightarrow \text{GSSG} + H_2O + \text{ROH}$$

Glutathione peroxidase, the enzyme catalyzing this reaction, is remarkable in having a covalently attached *selenium* (Se) atom. Its active site contains the selenium analog of cysteine, in which Se has replaced S (Figure 28-25). The selenolate (E-Se$^-$) form of this residue reduces the peroxide substrate to an alcohol and is in turn oxidized to selenenic acid (E-SeOH) (Figure 28-26). Glutathione now comes into action by forming a selenosulfide adduct (E-Se-S-G). A second glutathione then regenerates the active form of the enzyme by attacking the selenosulfide to form oxidized glutathione.

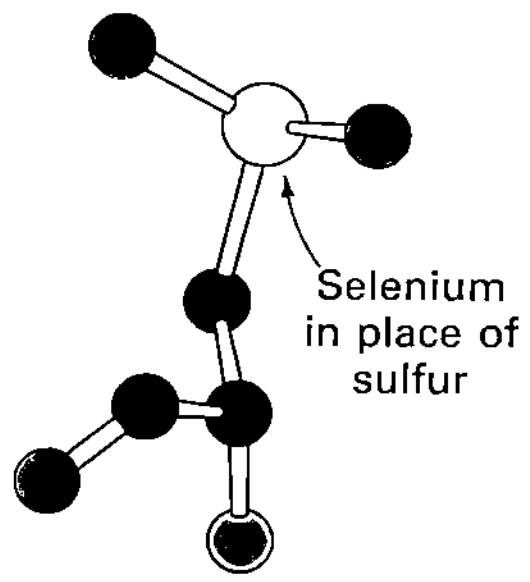

Figure 28-25
Model of the selenocysteine residue in the active site of glutathione peroxidase. The selenium atom is shown in green. The crystal contains the fully oxidized seleninate form of this modified amino acid. [Drawn from 1gpl.pdb. O. Epp, R. Ladenstein, and A. Wendel. *Eur. J. Biochem.* 133(1983):51.]

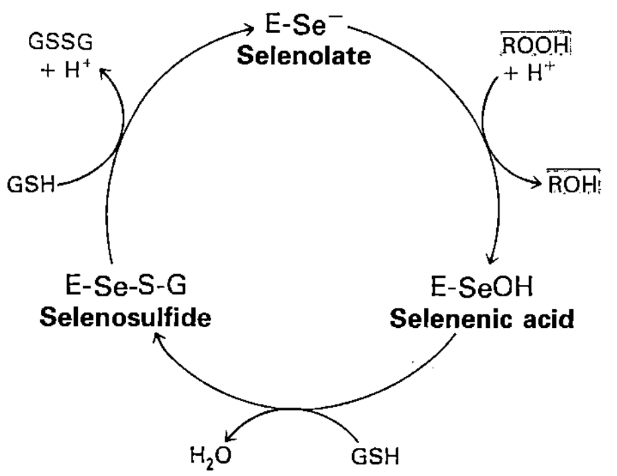

Figure 28-26
Proposed catalytic mechanism of glutathione peroxidase. [Based on O. Epp, R. Ladenstein, and A. Wendel. *Eur. J. Biochem.* 133(1983):51.]

NITRIC OXIDE (NO), A SHORT-LIVED SIGNAL MOLECULE, IS FORMED FROM ARGININE

Nitrous oxide (N_2O), a chemically stable oxide of nitrogen, has been used as an anesthetic for more than a century. *Nitric oxide (NO)*, a very different molecule formed from the same elements, has recently been shown to be an important messenger in many vertebrate signal transduction processes. This free-radical gas is produced endogenously from *arginine* in a complex reaction that is catalyzed by *nitric oxide synthase (NOS)*. Citrulline is the other product.

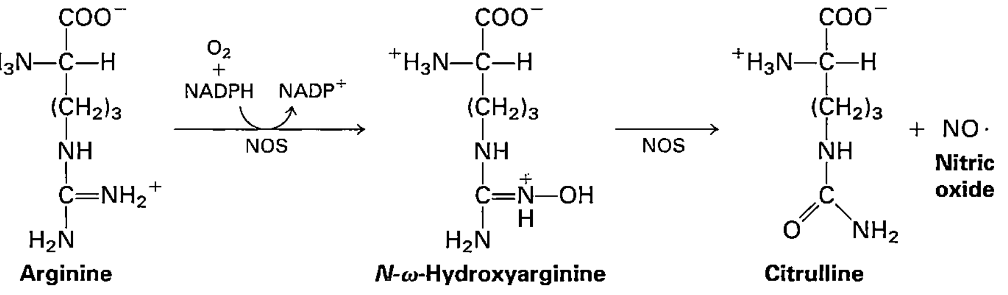

NADPH and O_2 are required for the synthesis of nitric oxide. This dependence suggests that NOS may be akin to cytochrome P450 enzymes that activate molecular oxygen in catalyzing hydroxylation reactions (p. 703). Indeed, the sequence of the carboxyl-terminal half of the cDNA for NOS resembles that of P450 oxygenases (Figure 28-27). In particular, two binding sites for flavins and one for NADPH are present in this moiety of NOS, as in P450 enzymes. Nitric oxide diffuses freely across membranes but has a short life, less than a few seconds, because it is highly reactive. Hence, *NO is well suited to serving as a transient signal molecule within cells and between adjacent cells.*

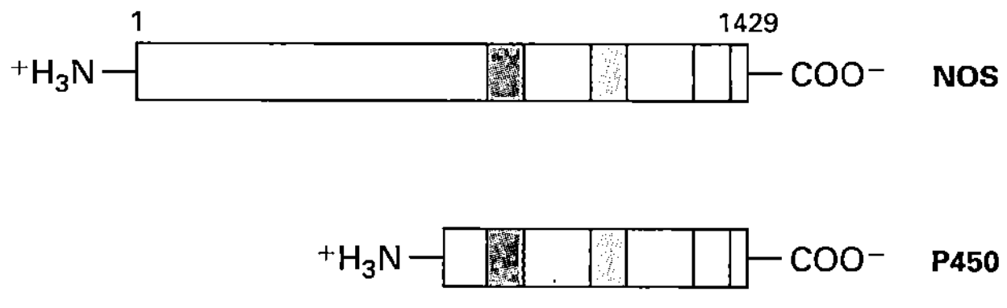

Figure 28-27
The carboxyl-terminal half of nitric oxide synthase (NOS) resembles cytochrome P450. The locations of binding sites for electron donors and acceptors are marked: FMN (red), FAD (blue), NADPH (green). [After D.S. Bredt, P.M. Hwang, C.E. Glatt, C. Lowenstein, R.R. Reed, and S.H. Snyder. *Nature* 351(1991):714.]

PORPHYRINS IN MAMMALS ARE SYNTHESIZED FROM GLYCINE AND SUCCINYL COENZYME A

The involvement of an amino acid in the biosynthesis of the porphyrin rings of hemes and chlorophylls was first revealed by isotopic labeling experiments carried out by David Shemin and his colleagues. In 1945, they showed that the nitrogen atoms of heme were labeled after the feeding of ^{15}N-glycine to human subjects (of which Shemin was the first), whereas ^{15}N-glutamate resulted in very little labeling. Using carbon-14, which had just become available, they discovered that eight of the carbon atoms of heme in nucleated duck erythrocytes are derived from the α carbon of glycine and none from the carboxyl carbon (Figure 28-28). Subsequent studies demonstrated that the other 26 carbon atoms of heme can arise from acetate. Moreover, the ^{14}C in methyl-labeled acetate emerged in 24 of these carbons, whereas the ^{14}C in carboxyl-labeled acetate appeared only in the other two.

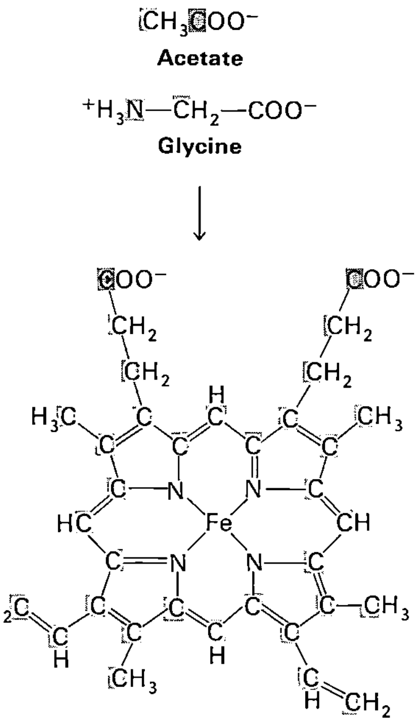

Figure 28-28
Labeling pattern of heme synthesized from glycine and acetate. The nitrogen atoms (blue) arise from the amino group of glycine. Colors denoting the origins of the carbon atoms are yellow, from the α carbon of glycine; green, mainly from the methyl carbon of acetate; and red, from the carboxyl carbon of acetate.

This highly distinctive labeling pattern led Shemin to propose that a heme precursor is formed by the condensation of glycine with an activated succinyl compound. In fact, *the first step in the biosynthesis of porphyrins in mammals is the condensation of glycine and succinyl CoA to form δ-aminolevulinate.*

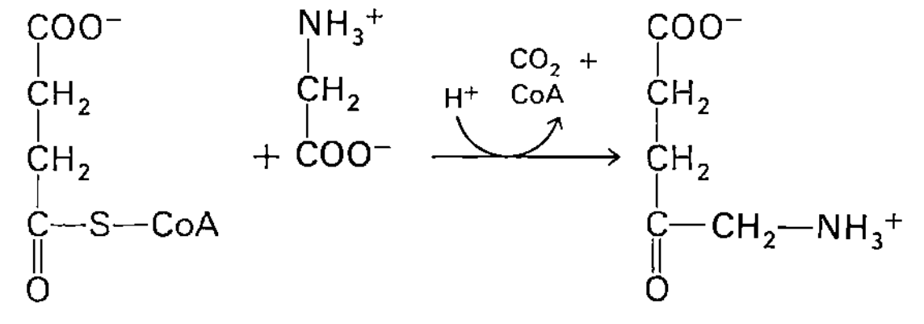

This reaction is catalyzed by *δ-aminolevulinate synthase,* a PLP enzyme in mitochondria. As might be expected, this committed step in the biosynthesis of porphyrins is regulated. *The translation of mRNA for this synthase is feedback-inhibited by heme.* The 5′ untranslated region of the mRNA contains a sequence that binds a heme-sensitive regulatory protein. Furthermore, the transport of the enzyme into mitochondria is blocked when heme is abundant.

Two molecules of δ-aminolevulinate condense to form *porphobilinogen,* the next intermediate. This dehydration reaction is catalyzed by *δ-aminolevulinate dehydrase.*

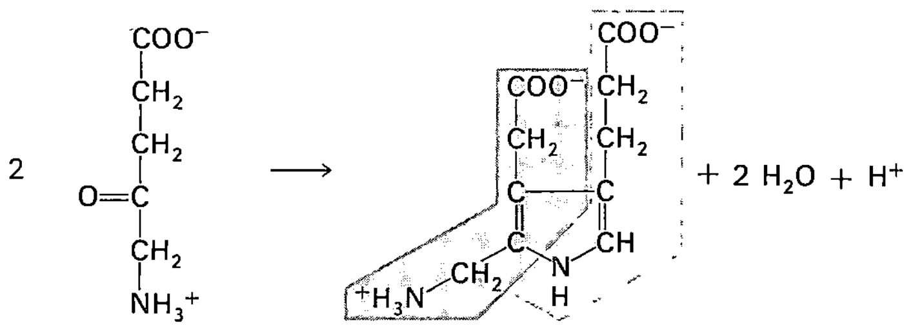

---

^{15}N labeling: A pioneer's account—

"Myself as a Guinea Pig

". . . in 1944, I undertook, together with David Rittenberg, an investigation on the turnover of blood proteins of man. To this end I synthesized 66 g of glycine labeled with 35 percent ^{15}N at a cost of $1000 for the ^{15}N. On 12 February 1945, I started the ingestion of the labeled glycine. Since we did not know the effect of relatively large doses of the stable isotope of nitrogen and since we believed that the maximum incorporation into the proteins could be achieved by the administration of glycine in some continual manner, I ingested 1 g samples of glycine at hourly intervals for the next 66 hours. . . . At stated intervals, blood was withdrawn and after proper preparation the ^{15}N concentrations of different blood proteins were determined."

David Shemin
BioEssays 10(1989):30

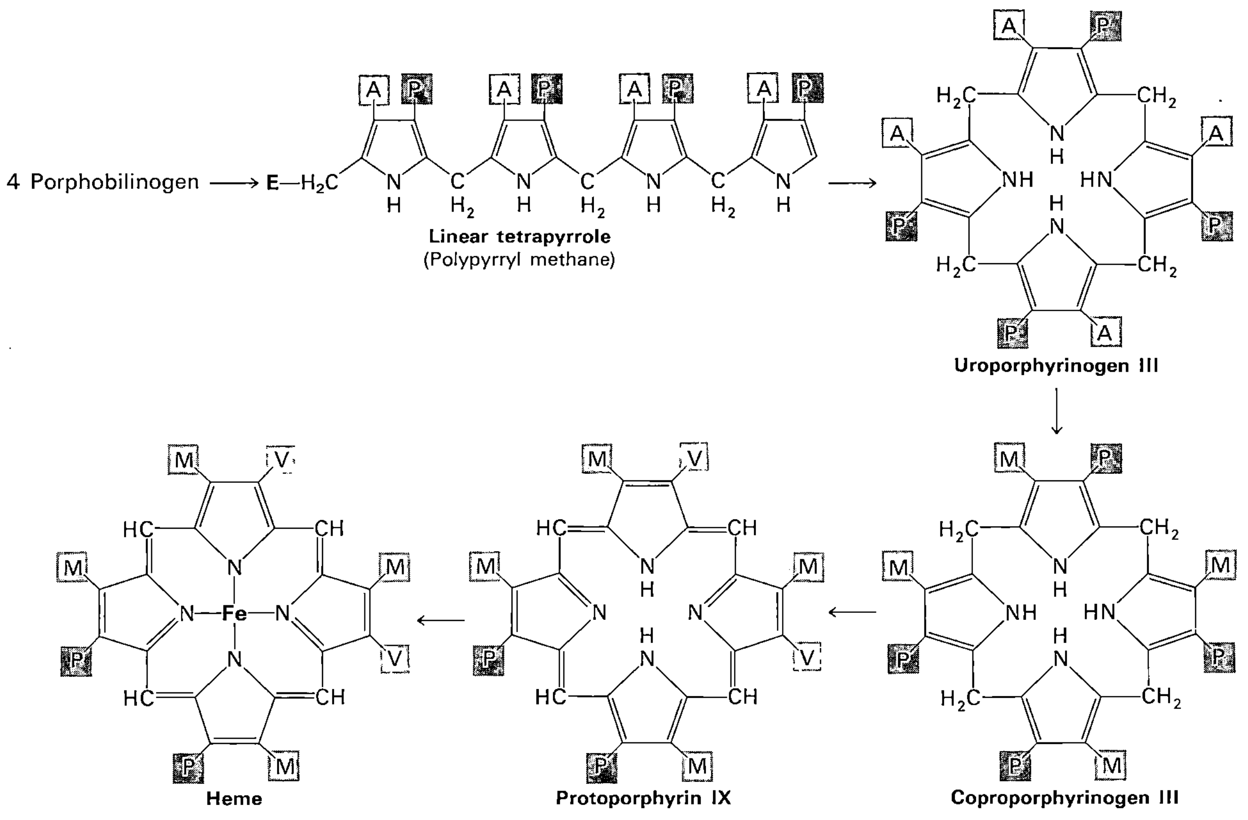

Figure 28-29
Pathway for the synthesis of heme from porphobilinogen. (Abbreviations: A, acetate; M, methyl; P, propionate; V, vinyl.)

Four porphobilinogens then condense head-to-tail to form a *linear tetrapyrrole* in a reaction catalyzed by *porphobilinogen deaminase* (Figure 28-29). An ammonium ion is released for each methylene bridge formed. The enzyme-bound linear tetrapyrrole then cyclizes to form *uroporphyrinogen III,* which has an asymmetric arrangement of side chains. This reaction requires a *cosynthase.* In the presence of synthase alone, uroporphyrinogen I, the nonphysiologic symmetric isomer, is produced. Uroporphyrinogen III is also a key intermediate in the synthesis of vitamin B_{12} by bacteria, and of chlorophyll by bacteria and plants.

The porphyrin skeleton is now formed. Subsequent reactions alter the side chains and the degree of saturation of the porphyrin ring (see Figure 28-29). *Coproporphyrinogen III* is formed by decarboxylation of the acetate side chains. Desaturation of the porphyrin ring and conversion of two of the propionate side chains into vinyl groups yield *protoporphyrin IX.* Chelation of iron finally gives *heme,* the prosthetic group of proteins such as myoglobin, hemoglobin, catalase, peroxidase, and cytochrome *c.* The insertion of the *ferrous* form of iron is catalyzed by *ferrochelatase.* Iron is transported in the plasma by *transferrin,* a protein that binds two ferric ions, and stored in tissues inside molecules of *ferritin.* The large internal cavity (~80-Å diameter) of ferritin can hold as many as 4500 ferric ions.

Figure 28-30
Model of protoporphyrin IX, the immediate precursor of heme.

PORPHYRINS ACCUMULATE IN SOME INHERITED DISORDERS OF PORPHYRIN METABOLISM

Porphyrias are inherited or acquired disorders caused by a deficiency of an enzyme in the heme biosynthetic pathway. *Congenital erythropoietic porphyria,* for example, results from insufficient cosynthase. In this porphyria, the synthesis of the required amount of uroporphyrinogen III is accompanied by the formation of very large quantities of uroporphyrinogen I, the useless symmetric isomer. Uroporphyrin I, coproporphyrin I, and other symmetric derivatives also accumulate. Erythrocytes are prematurely destroyed in this disease, which is transmitted as an autosomal recessive. The urine of patients having this disease is red because of the excretion of large amounts of uroporphyrin I. Their teeth exhibit a strong red fluorescence under ultraviolet light because of the deposition of porphyrins. Furthermore, their *skin is usually very sensitive to light* because photoexcited porphyrins are quite reactive.

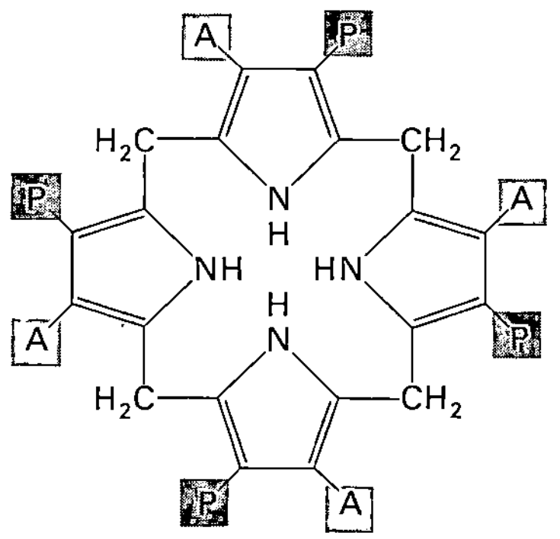

Uroporphyrinogen I
(Symmetric isomer)

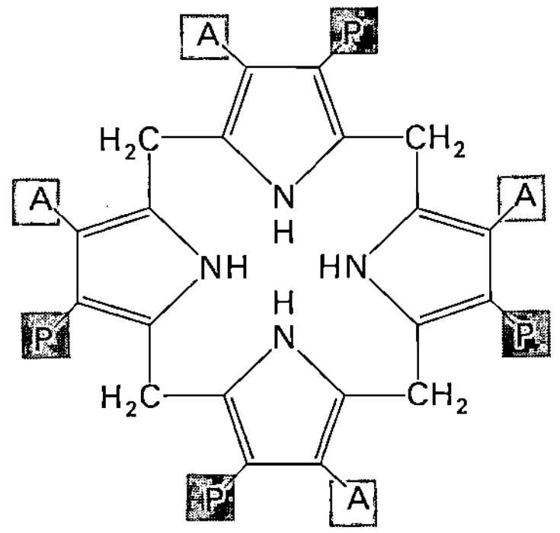

Uroporphyrinogen III
(Asymmetric isomer)

BILIVERDIN AND BILIRUBIN ARE INTERMEDIATES IN THE BREAKDOWN OF HEME

The normal human erythrocyte has a life span of about 120 days, as was first shown by the time course of ^{15}N in Shemin's own hemoglobin after he ingested ^{15}N-labeled glycine. Old cells are removed from the circulation and degraded by the spleen. The apoprotein of hemoglobin is hydrolyzed to its constituent amino acids. The first step in the degradation of the heme group to bilirubin is the cleavage of its α-methene bridge to form *biliverdin,* a linear tetrapyrrole (Figure 28-31). This reaction is catalyzed by *heme oxygenase,* a member of the cytochrome P450 superfamily. O_2 and NADPH are required for the cleavage reaction. It is noteworthy that a methene-bridge carbon is released as *carbon monoxide.* This endogenous production of CO posed a challenge in the evolution of oxygen carriers (p. 152). The central methene bridge of biliverdin is then reduced by *biliverdin reductase* to form *bilirubin.* Again, the reductant is NADPH. The changing color of a bruise is a highly graphic indicator of these degradative reactions.

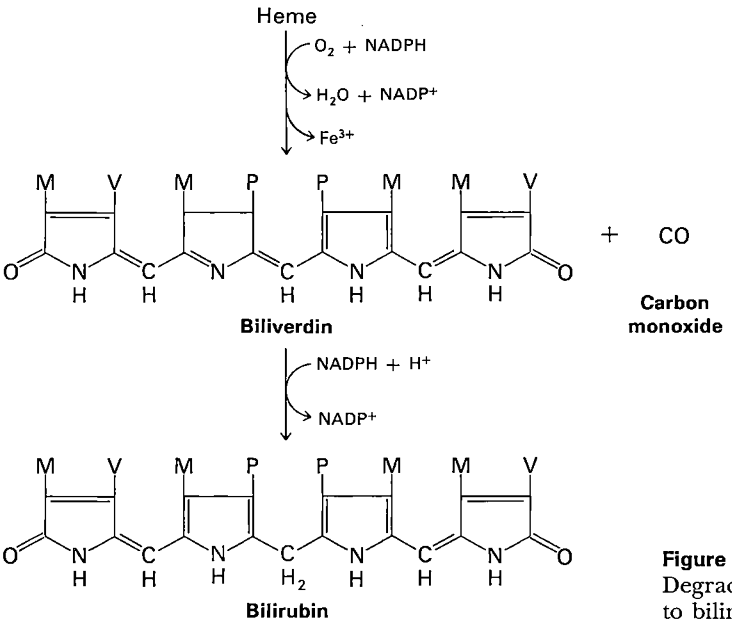

Figure 28-31
Degradation of heme to bilirubin.

Bilirubin complexed to serum albumin is transported to the liver, where it is rendered more soluble by the attachment of sugar residues to its propionate side chains. The solubilizing sugar is *glucuronate,* which differs from glucose in having a COO^- group at C-6 rather than a CH_2OH group. The conjugate of bilirubin and two glucuronates, called

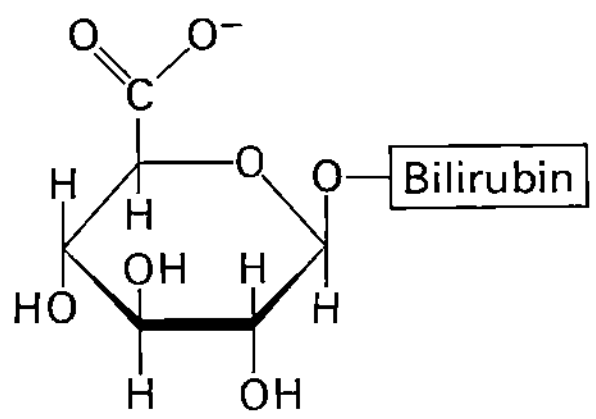

A glucuronate unit
in bilirubin diglucuronide.

bilirubin diglucuronide, is secreted into bile. *UDP–glucuronate, derived from the oxidation of UDP–glucose, is the activated intermediate in the synthesis of bilirubin diglucuronide.* Thus, whereas the iron atom of heme is recycled, the organic moiety is converted into a soluble, open-chain form that is excreted. *Jaundice,* the appearance of yellow pigmentation in the skin and white of the eye, points to an elevated level of bilirubin in the blood. Jaundice can be caused by excessive breakdown of red cells, impaired liver function, or mechanical obstruction of the bile duct. Measurement of the relative proportions of conjugated to unconjugated bilirubin is helpful in determining why a patient is jaundiced.

Bilirubin is much less soluble in aqueous media than is biliverdin. In reptiles and birds, the end product of heme catabolism is biliverdin rather than bilirubin. Why do mammals reduce biliverdin to bilirubin, a compound that poses a solubility problem? An intriguing clue is that *bilirubin is a very effective antioxidant,* whereas biliverdin is not. In scavenging two hydroperoxy radicals, bilirubin is oxidized to biliverdin, which is then rapidly reduced to again form bilirubin. On a molar basis, bilirubin bound to albumin is about a tenth as effective as ascorbate (vitamin C) in affording protection against water-soluble peroxides. In fact, bilirubin, urate (p. 758), and ascorbate (p. 455) are the three principal antioxidants in plasma. Most significant, bilirubin is an especially potent antioxidant in membranes, where it rivals vitamin E (p. 454). Thus, *the end product of a degradative pathway may be selected in evolution to exert a beneficial action.*

SUMMARY

Microorganisms use ATP and reduced ferredoxin, a powerful reductant, to reduce N_2 to NH_3. An iron-molybdenum cluster in nitrogenase deftly catalyzes the fixation of N_2, a very inert molecule. Higher organisms consume NH_4^+ to synthesize amino acids, nucleotides, and other nitrogen-containing biomolecules. The major points of entry of NH_4^+ into metabolism are glutamine, glutamate, and carbamoyl phosphate. Humans can synthesize 11 of the basic set of 20 amino acids. These amino acids are called nonessential, in contrast with the essential ones, which must be supplied in the diet. The pathways for the synthesis of nonessential amino acids are quite simple. Glutamate dehydrogenase catalyzes the reductive amination of α-ketoglutarate to glutamate. Alanine and aspartate are synthesized by transamination of pyruvate and oxaloacetate, respectively. Glutamine is synthesized from NH_4^+ and glutamate, and asparagine is synthesized similarly. Proline and arginine are derived from glutamate. Serine, formed from 3-phosphoglycerate, is the precursor of glycine and cysteine. Tyrosine is synthesized by the hydroxylation of phenylalanine, an essential amino acid.

The pathways for the biosynthesis of essential amino acids are much more complex than for the nonessential ones. Most of these pathways are regulated by feedback inhibition, in which the committed step is allosterically inhibited by the final product. Sequential feedback control, enzyme multiplicity, concerted feedback control, and cumulative feedback control are widely used regulatory devices in metabolism. The regulation of glutamine synthetase from *E. coli* provides a striking demonstration of cumulative feedback inhibition and of control by a cascade of reversible covalent modifications.

Tetrahydrofolate, a carrier of activated one-carbon units, plays an important role in the metabolism of amino acids and nucleotides. This coenzyme carries one-carbon units at three oxidation states, which are interconvertible: most reduced—methyl; intermediate—methylene; most oxidized—formyl, formimino, and methenyl. The major donor of activated methyl groups is *S*-adenosylmethionine, which is synthesized by the transfer of an adenosyl group from ATP to the sulfur atom of methionine. *S*-Adenosylhomocysteine is formed when the activated methyl group is transferred to an acceptor. It is hydrolyzed to adenosine and homocysteine, which is then methylated to methionine to complete the activated methyl cycle. *S*-Adenosylmethionine also serves as a donor of propylamine in the synthesis of polyamine, and is the precursor of ethylene, a plant hormone that induces ripening.

Amino acids are precursors of a variety of biomolecules. Glutathione (γ-Glu-Cys-Gly) serves as a sulfhydryl buffer and detoxifying agent. Glutathione peroxidase, a selenoenzyme, catalyzes the reduction of hydrogen peroxide and organic peroxides by glutathione. Nitric oxide (NO), a short-lived messenger, is formed from arginine. Porphyrins are synthesized from glycine and succinyl CoA, which condense to give δ-aminolevulinate. Two molecules of this intermediate become linked to form porphobilinogen. Four porphobilinogens combine to form a linear tetrapyrrole, which cyclizes to uroporphyrinogen III. Oxidation and side-chain modifications lead to the synthesis of protoporphyrin IX, which acquires an iron atom to form heme. The translation of mRNA for δ-aminolevulinate synthase, the enzyme catalyzing the committed step in this pathway, is inhibited by heme. This prosthetic group is degraded by a monooxygenase that converts it into biliverdin, a linear tetrapyrrole. Reduction of biliverdin by NADPH yields bilirubin, a potent scavenger of peroxides. Bilirubin is rendered soluble for excretion by the formation of the diglucuronide derivative.

SELECTED READINGS

Where to start

Kim, J., and Rees, D.C., 1994. Nitrogenase and biological nitrogen fixation. *Biochemistry* 33:389–397.

Rhee, S.G., Chock, P.B., and Stadtman, E.R., 1989. Regulation of *Escherichia coli* glutamine synthetase. *Advan. Enzymol. Mol. Biol.* 62:37–92.

Bredt, D.S., and Snyder, S.H., 1992. Nitric oxide, a novel neuronal messenger. *Neuron* 8:3–11.

Shemin, D., 1989. An illustration of the use of isotopes: The biosynthesis of porphyrins. *BioEssays* 10:30–35.

Battersby, A.R., 1994. How nature builds the pigments of life: The conquest of vitamin B_{12}. *Science* 264:1551–1557. [A stimulating account of how vitamin B_{12} is formed from uroporphyrinogen III, described as "the Everest of biosynthetic problems."]

Books

Bender, D.A., 1985. *Amino Acid Metabolism* (2nd ed.). Wiley.

Jordan, P.M. (ed.), 1991. *Biosynthesis of Tetrapyrroles.* Elsevier. [Contains excellent articles on the synthesis of heme, vitamin B_{12}, and chlorophyll.]

Scriver, C.R., Beaudet, A.L., Sly, W.S., and Valle, D. (eds.), 1989. *The Metabolic Basis of Inherited Disease* (6th ed.). McGraw-Hill. [Part 4 contains many excellent articles on disorders of amino acid metabolism, and Part 8, of porphyrins and heme.]

Meister, A., 1965. *Biochemistry of the Amino Acids* (2nd ed.), vols. 1 and 2. Academic Press. [A comprehensive and authoritative treatise on amino acid metabolism.]

Blakley, R.L., and Benkovic, S.J., 1985. *Folates and Pterins,* vol. 2. Wiley.

Walsh, C., 1979. *Enzymatic Reaction Mechanisms.* W.H. Freeman. [Chapter 25 provides an excellent account of reaction mechanisms in one-carbon metabolism.]

Nitrogen fixation

Burris, R.H., 1991. Nitrogenases. *J. Biol. Chem.* 266:9339–9342.

Postgate, J., 1989. Trends and perspectives in nitrogen fixation research. *Advan. Microbiol. Physiol.* 30:1–22.

Chan, M.K., Kim, J., and Rees, D.C., 1993. The nitrogenase FeMo-cofactor and P-cluster pair: 2.2 Å resolution studies. *Science* 260:792–794.

Georgiadis, M.M., Komiya, H., Chakrabarti, P., Woo, D., Kornuc, J.J., and Rees, D.C., 1992. Crystallographic structure of the nitrogenase iron protein from *Azotobacter vinelandii. Science* 257:1653–1659.

Regulation of amino acid biosynthesis

Cooper, A.J.L., 1983. Biochemistry of sulfur-containing amino acids. *Ann. Rev. Biochem.* 52:187–222.

Umbarger, H.E., 1978. Amino acid biosynthesis and its regulation. *Ann. Rev. Biochem.* 47:533–606. [An excellent review of amino acid biosynthesis in bacteria.]

Yamashita, M.M., Almassy, R.J., Janson, C.A., Cascio, D., and Eisenberg, D., 1989. Refined atomic model of glutamine synthetase at 3.5 Å resolution. *J. Biol. Chem.* 264:17681–17690.

Rhee, S.G., Park, R., Chock, P.B., and Stadtman, E.R., 1978. Allosteric regulation of monocyclic interconvertible enzyme cascade systems: Use of *Escherichia coli* glutamine synthetase as an experimental model. *Proc. Nat. Acad. Sci.* 75:3138–3142.

Aromatic amino acid biosynthesis

Crawford, I.P., 1989. Evolution of a biosynthetic pathway: The tryptophan paradigm. *Ann. Rev. Microbiol.* 43:567–600.

Miles, E.W., 1991. Structural basis for catalysis by tryptophan synthase. *Advan. Enzymol. Mol. Biol.* 64:93–172.

Anderson, K.S., Miles, E.W., and Johnson, K.A., 1991. Serine modulates substrate channeling in tryptophan synthase. A novel intersubunit triggering mechanism. *J. Biol. Chem.* 266:8020–8033.

Kishore, G.M., and Shah, D.M., 1988. Amino acid biosynthesis inhibitors as herbicides. *Ann. Rev. Biochem.* 57:627–663.

Glutathione

Meister, A., and Anderson, M.E., 1983. Glutathione. *Ann. Rev. Biochem.* 52:711–760.

Karplus, P.A., and Schulz, G.E., 1989. Substrate binding and catalysis by glutathione reductase as derived from refined enzyme: Substrate crystal structures at 2 Å resolution. *J. Mol. Biol.* 210:163–180.

Epp, O., Ladenstein, R., and Wendel, A., 1983. The refined structure of the selenoenzyme glutathione peroxidase at 0.2-nm resolution. *Eur. J. Biochem.* 133:51–69.

Stadtman, T.C., 1990. Selenium biochemistry. *Ann. Rev. Biochem.* 59:111–127.

Ethylene and nitric oxide

Theologis, A., 1992. One rotten apple spoils the whole bushel: The role of ethylene in fruit ripening. *Cell* 70:181–184.

Bredt, D.S., Hwang, P.M., Glatt, C.E., Lowenstein, C., Reed, R.R., and Snyder, S.H., 1991. Cloned and expressed nitric oxide synthase structurally resembles cytochrome P-450 reductase. *Nature* 351:714–718.

Nathan, C., 1992. Nitric oxide as a secretory product of mammalian cells. *FASEB J.* 6:3051–3064.

Biosynthesis of porphyrins

Leeper, F.J., 1989. The biosynthesis of porphyrins, chlorophylls, and vitamin B_{12}. *Nat. Prod. Rep.* 6:171–199.

Porra, R.J., and Meisch, H.-U., 1984. The biosynthesis of chlorophyll. *Trends Biochem. Sci.* 9:99–104.

Antioxidant role of bilirubin

Stocker, R., McDonagh, A.F., Glazer, A.N., and Ames, B.N., 1990. Antioxidant activities of bile pigments: Biliverdin and bilirubin. *Meth. Enzymol.* 186:301–309.

Stocker, R., Yamamoto, Y., McDonagh, A.F., Glazer, A.N., and Ames, B.N., 1987. Bilirubin is an antioxidant of possible physiologic importance. *Science* 235:1043–1046.

PROBLEMS

1. *From sugar to amino acid.* Write a balanced equation for the synthesis of alanine from glucose.

2. *From air to blood.* What are the intermediates in the flow of nitrogen from N_2 to heme?

3. *One-carbon transfers.* Which derivative of folate is a reactant in the conversion of
   (a) Glycine into serine?
   (b) Homocysteine into methionine?

4. *Telltale tag.* In the reaction catalyzed by glutamine synthetase, an oxygen atom is transferred from the side chain of glutamate to orthophosphate, as shown by ^{18}O-labeling studies. Account for this finding.

5. *Therapeutic glycine.* Isovaleric acidemia is an inherited disorder of leucine metabolism caused by a deficiency of isovaleryl CoA dehydrogenase. Many infants having this disease die in the first month of life. The administration of large amounts of glycine sometimes leads to marked clinical improvement. Propose a mechanism for the therapeutic action of glycine.

6. *Deprived algae.* Blue-green algae (cyanobacteria) form *heterocysts* when deprived of ammonia and nitrate. They lack nuclei and are attached to adjacent vegetative cells. Heterocysts have photosystem I activity but are entirely devoid of photosystem II activity. What is their role?

7. *Cysteine and cystine.* Most cytosolic proteins lack disulfide bonds, whereas extracellular proteins usually contain them. Why?

8. *To and fro.* The synthesis of δ-aminolevulinate occurs in the mitochondrial matrix, whereas the formation of porphobilinogen takes place in the cytosol. Propose a reason for the mitochondrial location of the first step in heme synthesis.

9. *A dangerous trap.* In pernicious anemia, much of the body's tetrahydrofolate is sequestered in the form of N^5-methyltetrahydrofolate. Why?

10. *Messengers in the wings.* Signal molecules such as epinephrine, nitric oxide, and ethylene are often generated in a small number of steps from a major biomolecule. Each year, new molecules that participate in signal transduction processes are being discovered. Propose two candidates that appeared in this chapter. What kind of receptor might be used to detect your putative messengers?

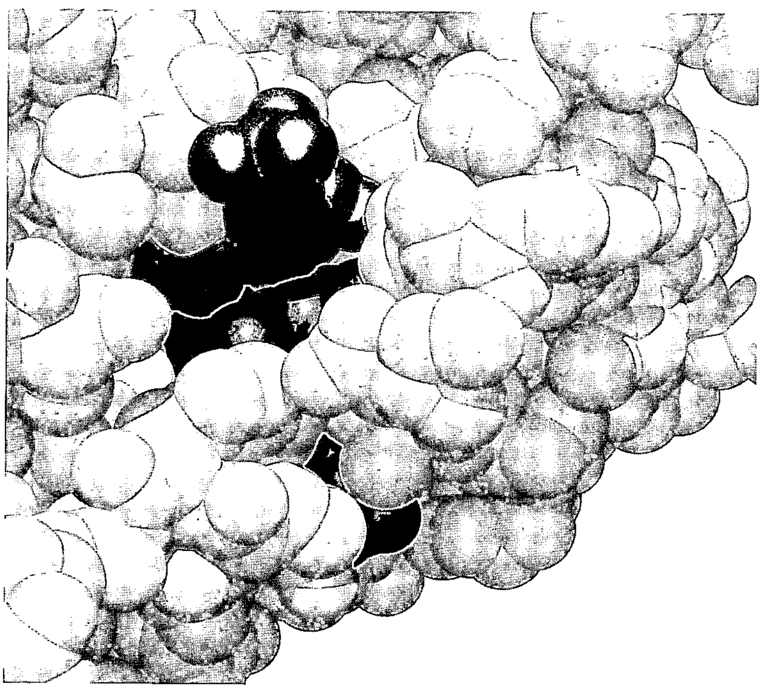

Biosynthesis of Nucleotides

This chapter deals with the biosynthesis of nucleotides. These compounds play key roles in nearly all biochemical processes:

1. They are the *activated precursors of DNA and RNA.*

2. Nucleotide derivatives are *activated intermediates in many biosyntheses.* For example, UDP–glucose and CDP–diacylglycerol are precursors of glycogen and phosphoglycerides, respectively. *S*-Adenosylmethionine carries an activated methyl group.

3. ATP, an adenine nucleotide, is a *universal currency of energy* in biological systems. GTP powers many movements of macromolecules, such as the translocation of nascent peptide chains on ribosomes and the activation of signal-coupling proteins.

4. Adenine nucleotides are *components of three major coenzymes:* NAD^+, FAD, and CoA.

5. Nucleotides also serve as *metabolic regulators.* Cyclic AMP is a ubiquitous mediator of the action of many hormones. Covalent modifications introduced by ATP alter the activities of many enzymes, as exemplified by the phosphorylation of glycogen synthase and the adenylylation of glutamine synthetase.

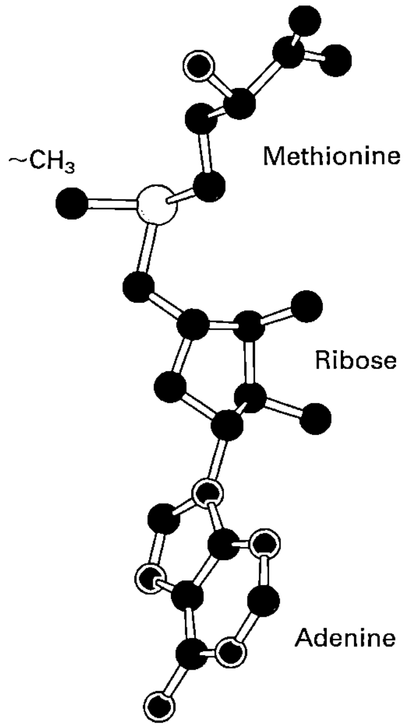

Figure 29-1
Model of *S*-adenosylmethionine. [Drawn from 1hmy.pdb. S. Kumar, X. Cheng, J.W. Pflugrath, and R.J. Roberts. *Cell* 74(1993):299.]

Opening Image: *Catalytic site of thymidylate synthase. This enzyme catalyzes the methylation of deoxyuridylate (green) to deoxythymidylate, a building block of DNA. Thymidylate synthase is a choice target in cancer chemotherapy because DNA is rapidly synthesized in proliferating tumor cells. A bound antifolate drug is shown in red. [Drawn from 2tsc.pdb. W.R. Montfort, K.M. Perry, E.B. Fauman, J.S. Finer-Moore, G.F. Maley, and R.M. Stroud. Biochemistry 29(1990):6964.]*

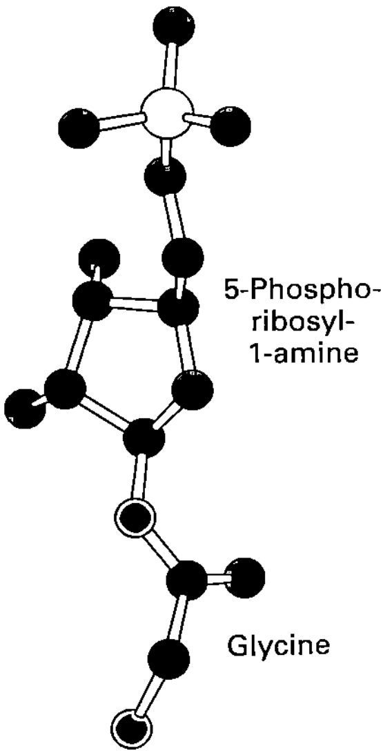

Figure 29-2
Model of glycinamide ribonucleotide, an early intermediate in the synthesis of purine ribonucleotides. Glycine is attached to the amino group of a phosphorylated ribose derivative. [Drawn from 1cde.pdb. R.J. Almassy, C.A. Janson, C.-C. Kan, and Z. Hostomska. *Proc. Nat. Acad. Sci.* 89(1992):6114.]

Nucleotides are synthesized from simple building blocks (*de novo synthesis*) or by the recycling of preformed bases (*salvage synthesis*). Purines and pyrimidines are built de novo from amino acids, tetrahydrofolate derivatives, NH_4^+, and CO_2. The sugar phosphate moiety of ribonucleotides comes from 5-phosphoribosyl-1-pyrophosphate, an activated donor. Deoxyribonucleotides are synthesized by reduction of ribonucleotides. Finally, deoxythymidylate (dTMP) is formed by methylation of deoxyuridylate (dUMP).

We shall see how diverse nucleotides are crafted from a simple set of building blocks. Equally interesting are the intricate regulatory circuits that provide for the balanced synthesis of the principal nucleotides. An understanding of nucleotide metabolism is also rewarding for the insights it provides into disease processes and their therapy. Nucleotide analogs are valuable drugs in the treatment of cancers, viral infections, autoimmune diseases, and genetic disorders such as gout.

NOMENCLATURE OF BASES, NUCLEOSIDES, AND NUCLEOTIDES

The nomenclature of nucleotides and their constituent units was presented earlier (p. 76). Recall that a *nucleoside* consists of a purine or pyrimidine base linked to a pentose, and that a *nucleotide* is a phosphate ester of a nucleoside. The names of the major bases of RNA and DNA, and of their nucleoside and nucleotide derivatives, are given in Table 29-1.

Table 29-1
Nomenclature of bases, nucleosides, and nucleotides

Base	*Ribonucleoside*	*Ribonucleotide (5′-monophosphate)*
Adenine (A)	Adenosine	Adenylate (AMP)
Guanine (G)	Guanosine	Guanylate (GMP)
Uracil (U)	Uridine	Uridylate (UMP)
Cytosine (C)	Cytidine	Cytidylate (CMP)
Base	*Deoxyribonucleoside*	*Deoxyribonucleotide (5′-monophosphate)*
Adenine (A)	Deoxyadenosine	Deoxyadenylate (dAMP)
Guanine (G)	Deoxyguanosine	Deoxyguanylate (dGMP)
Thymine (T)	Deoxythymidine	Deoxythymidylate (dTMP)
Cytosine (C)	Deoxycytidine	Deoxycytidylate (dCMP)

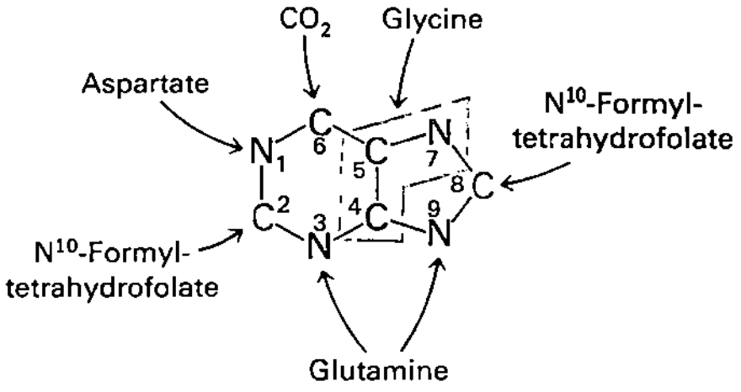

Figure 29-3
Origins of the atoms in the purine ring.

THE PURINE RING IS SYNTHESIZED FROM AMINO ACIDS, TETRAHYDROFOLATE DERIVATIVES, AND CO_2

Let us begin with the synthesis of purine nucleotides. The purine ring is assembled de novo from several simple precursors (Figure 29-3). *Glycine* provides C-4, C-5, and N-7. The N-1 atom comes from *aspartate.* The other two nitrogen atoms, N-3 and N-9, come from the amide group of the side chain of *glutamine.* Activated derivatives of *tetrahydrofolate* furnish C-2 and C-8, whereas CO_2 is the source of C-6.

PRPP IS THE DONOR OF THE RIBOSE PHOSPHATE UNIT OF NUCLEOTIDES

The ribose phosphate portion of purine and pyrimidine nucleotides comes from *5-phosphoribosyl-1-pyrophosphate* (PRPP), which we encountered earlier in the synthesis of tryptophan (p. 725). PRPP is synthesized

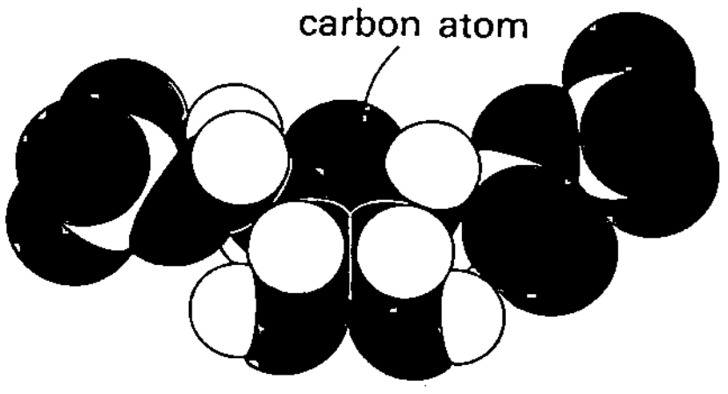

Figure 29-4
Model of 5-phosphoribosyl-1-pyrophosphate (PRPP), the activated donor of the sugar unit in the biosynthesis of nucleotides.

from ATP and ribose 5-phosphate, which is primarily formed by the pentose phosphate pathway (p. 559). *PRPP synthetase* catalyzes the transfer of the β-γ pyrophosphoryl group of ATP to C-1 of ribose 5-phosphate. PRPP has an α configuration at C-1, the activated carbon atom.

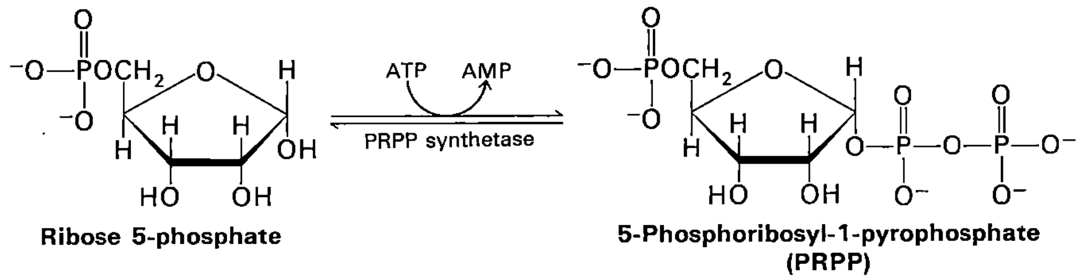

THE PURINE RING IS ASSEMBLED ON RIBOSE PHOSPHATE

The committed step in the de novo synthesis of purine nucleotides is the formation of *5-phosphoribosylamine* from PRPP and glutamine. The amide group from the side chain of glutamine displaces the pyrophosphate group attached to C-1 of PRPP. The configuration at C-1 is inverted from α to β in this reaction. The resulting C–N glycosidic bond has the β configuration that is characteristic of naturally occurring nucleotides. This reaction is driven forward by the hydrolysis of pyrophosphate.

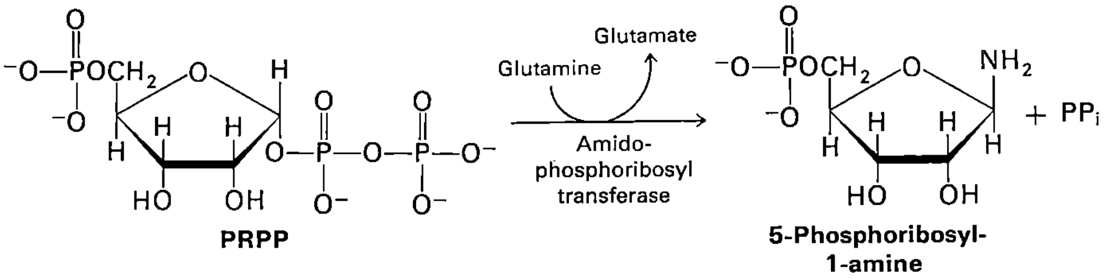

Glycine joins phosphoribosylamine to yield *glycinamide ribonucleotide* (Figure 29-5). An ATP is consumed in the formation of an amide bond between the carboxyl group of glycine and the amino group of phosphoribosylamine. An acyl phosphate intermediate is formed, as in the biosyntheses of glutamine (p. 717), glutathione (p. 731), and almost all

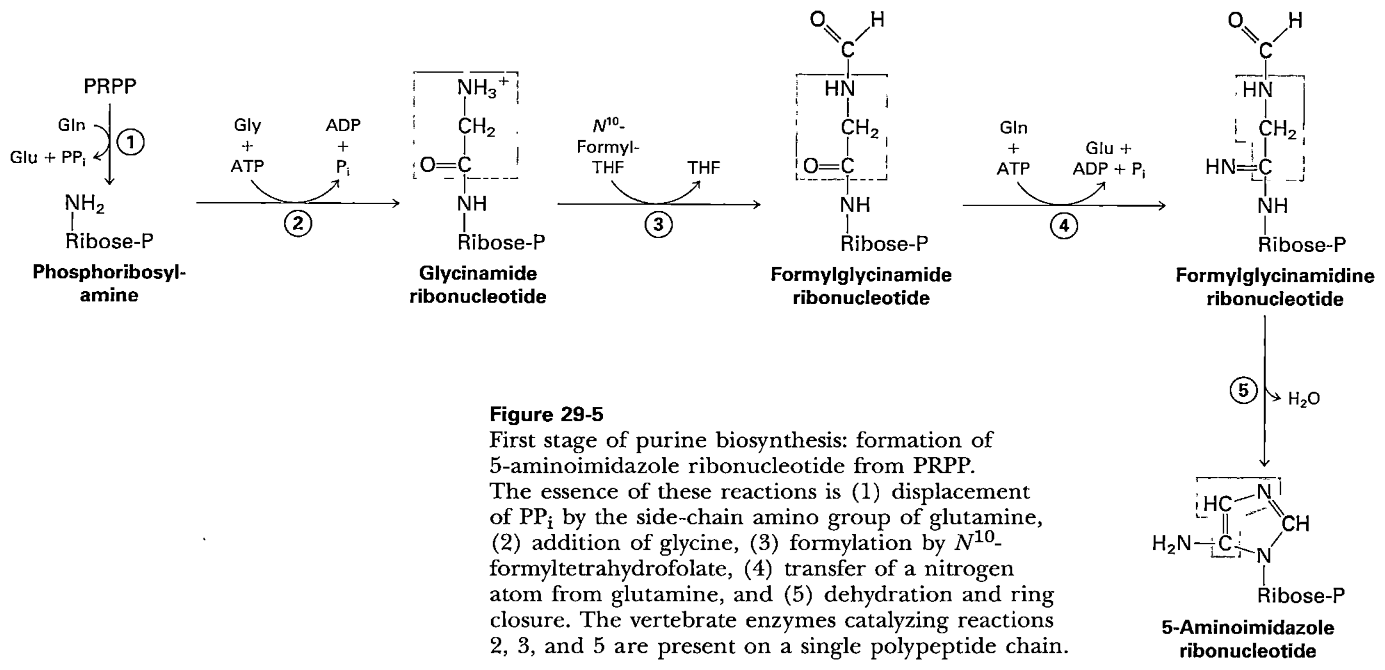

Figure 29-5
First stage of purine biosynthesis: formation of 5-aminoimidazole ribonucleotide from PRPP. The essence of these reactions is (1) displacement of PP_i by the side-chain amino group of glutamine, (2) addition of glycine, (3) formylation by N^{10}-formyltetrahydrofolate, (4) transfer of a nitrogen atom from glutamine, and (5) dehydration and ring closure. The vertebrate enzymes catalyzing reactions 2, 3, and 5 are present on a single polypeptide chain.

Figure 29-6
Second stage of purine biosynthesis: formation of inosinate from 5-aminoimidazole ribonucleotide. The essence of these reactions is (6) carboxylation, (7) addition of aspartate, (8) elimination of fumarate (leaving the amino group of aspartate), (9) formylation by N^{10}-formyltetrahydrofolate, and (10) dehydration and ring closure. The vertebrate enzymes catalyzing steps 6 and 7 are present on a single polypeptide chain, as are those catalyzing steps 9 and 10.

other amide bonds. The α-amino terminus of the glycine residue is then formylated by N^{10}-formyltetrahydrofolate to give *α-N-formylglycinamide ribonucleotide.* Folate derivatives serve as activated intermediates in several steps of nucleotide biosynthesis. The substitution of NH for O changes the amide group of this compound into an amidine group. The nitrogen atom is donated by the side chain of glutamine in a reaction that consumes an ATP. *Formylglycinamidine ribonucleotide then undergoes ring closure to form 5-aminoimidazole ribonucleotide.* This intermediate contains the complete five-membered ring of the purine skeleton.

The next phase in the synthesis of the purine skeleton, the formation of a six-membered ring, starts at this point (Figure 29-6). Three of the six atoms of this ring are already present in aminoimidazole ribonucleotide.

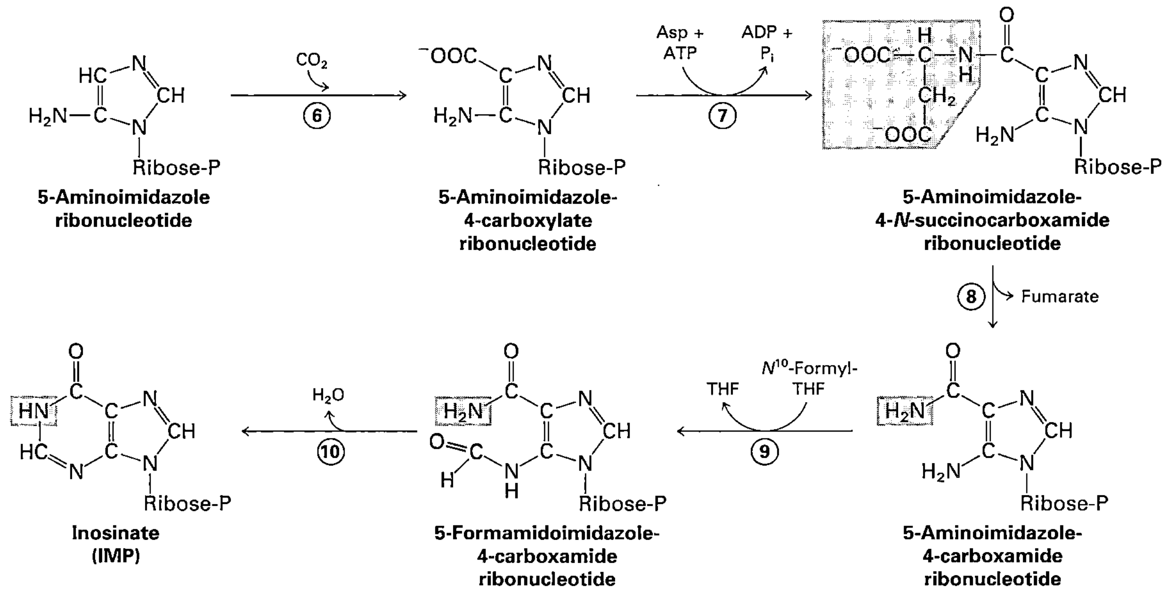

The other three come from CO_2, aspartate, and formyltetrahydrofolate. The next carbon atom in the six-membered ring is introduced by the carboxylation of aminoimidazole ribonucleotide, yielding *5-aminoimidazole-4-carboxylate ribonucleotide.* This carboxylation reaction is unusual in not utilizing biotin.

The amino group of aspartate then reacts with the carboxyl group of this intermediate to form *5-aminoimidazole-4-*N*-succinocarboxamide ribonucleotide.* An ATP is consumed in the formation of this amide bond. The carbon skeleton of the aspartate moiety comes off as fumarate in the next reaction to give *5-aminoimidazole-4-carboxamide ribonucleotide.* Note that the result of these reactions is the conversion of a carboxylate into an amide. Thus, *aspartate contributes only its nitrogen atom to the purine ring.* Recall that aspartate likewise donates only its nitrogen atom in the formation of arginine in the urea cycle (p. 635). The final atom of the purine ring is contributed by N^{10}-formyltetrahydrofolate. The resulting *5-formamidoimidazole-4-carboxamide ribonucleotide* undergoes dehydration and ring closure to form *inosinate* (IMP), which contains a complete purine ring. The purine base of inosinate is called *hypoxanthine.*

AMP AND GMP ARE FORMED FROM IMP

Inosinate, the product of the de novo pathway, is the precursor of AMP and GMP (Figure 29-7). *Adenylate* is synthesized from inosinate by the substitution of an amino group for the carbonyl oxygen atom at C-6. Again, the addition of aspartate followed by the elimination of fumarate contributes the amino group. GTP is the donor of a high-energy phosphate bond in the synthesis of the *adenylosuccinate* intermediate from inosinate and aspartate. The removal of fumarate from adenylosuccinate and from 5-aminoimidazole-4-*N*-succinocarboxamide ribonucleotide is catalyzed by the same enzyme.

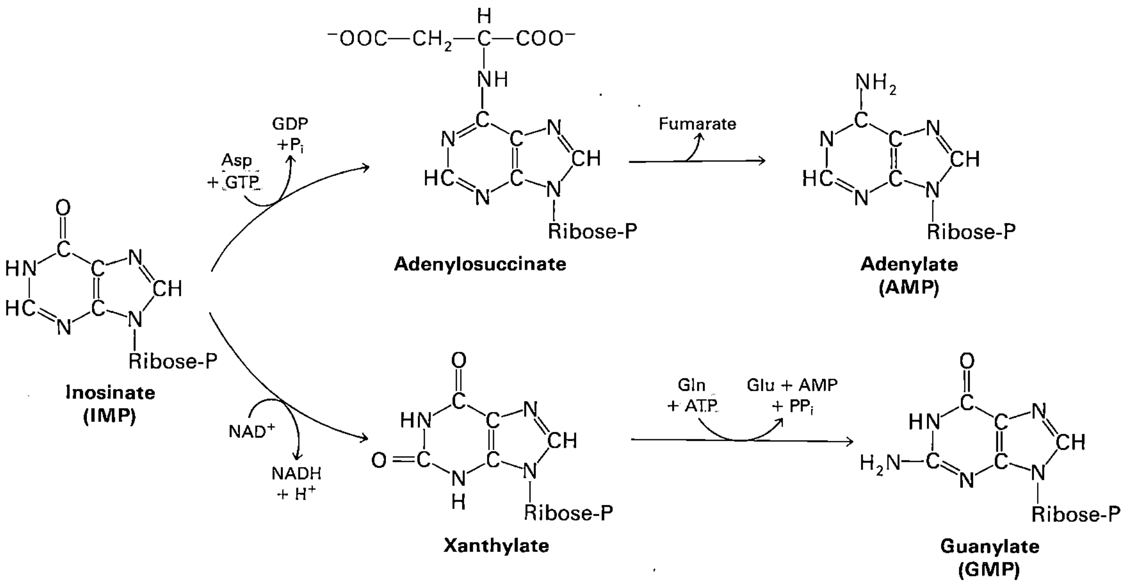

Figure 29-7
Adenylate (AMP) and guanylate (GMP) are synthesized from inosinate (IMP).

Guanylate (GMP) is synthesized by the oxidation of inosinate, followed by the insertion of an amino group at C-2. NAD^+ is the hydrogen acceptor in the oxidation of inosinate to xanthylate (XMP). The amide–NH_2 group of glutamine is then transferred to xanthylate. Two high-energy phosphate bonds are consumed in this reaction, because ATP is cleaved into AMP and PP_i, which is subsequently hydrolyzed.

In the conversions of inosinate into adenylate and guanylate, a carbonyl oxygen atom is replaced by an amino group. A similar change occurs in the synthesis of formylglycinamidine ribonucleotide from its amide precursor (step 4 on p. 742), in the formation of CTP from UTP (p. 748), and in the conversion of citrulline into arginine in the urea cycle (p. 635). *The common mechanistic theme of these reactions is the conversion of the carbonyl oxygen into a derivative that can be readily displaced by an amino group.* The tautomeric form of the carbonyl group reacts with ATP (or GTP) to form a phosphoryl ester, which is then nucleophilically attacked by a nitrogen atom (Figure 29-8). Inorganic phosphate is then expelled from this tetrahedral adduct to complete the reaction. The attacking nitrogen can be from NH_3, the side-chain amide group of glutamine, or the α-amino group of aspartate. The leaving group in this class of reactions can be P_i, PP_i, or the AMP moiety. In a related reaction, PP_i is displaced by the amino group of glutamine in the synthesis of 5-phosphoribosyl-1-amine from PRPP (p. 741).

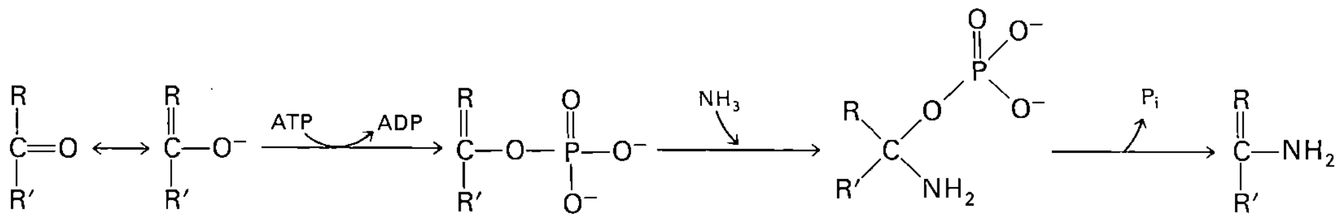

Figure 29-8
Reaction mechanism for the replacement of a carbonyl oxygen by an amino group.

PURINE BASES CAN BE RECYCLED BY SALVAGE REACTIONS THAT UTILIZE PRPP

Free purine bases are formed by the hydrolytic degradation of nucleic acids and nucleotides. Purine nucleotides can be synthesized from these preformed bases by a *salvage reaction,* which is simpler and much less costly than the reactions of the *de novo pathway* discussed above. In salvage reactions, the ribose phosphate moiety of PRPP is transferred to a purine to form the corresponding ribonucleotide:

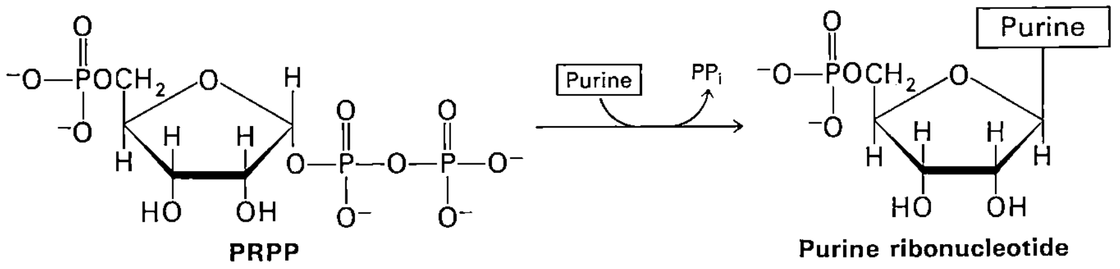

Two salvage enzymes with different specificities recover purine bases. *Adenine phosphoribosyl transferase* catalyzes the formation of adenylate:

$$\text{Adenine} + \text{PRPP} \longrightarrow \text{adenylate} + PP_i$$

whereas *hypoxanthine-guanine phosphoribosyl transferase* catalyzes the formation of inosinate and guanylate:

$$\text{Hypoxanthine} + \text{PRPP} \longrightarrow \text{inosinate} + PP_i$$
$$\text{Guanine} + \text{PRPP} \longrightarrow \text{guanylate} + PP_i$$

AMP, GMP, AND IMP ARE FEEDBACK INHIBITORS OF PURINE NUCLEOTIDE BIOSYNTHESIS

The synthesis of purine nucleotides is controlled by feedback inhibition and other regulatory mechanisms at several sites (Figure 29-9).

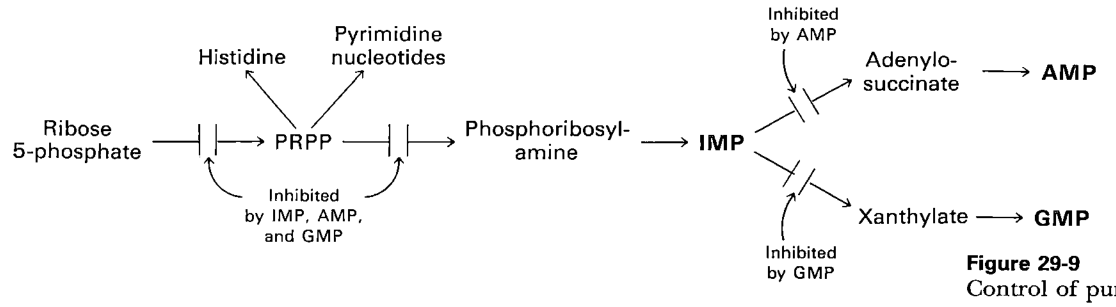

Figure 29-9
Control of purine biosynthesis.

1. *5-Phosphoribosyl-1-pyrophosphate synthetase,* the enzyme that synthesizes PRPP, is partially inhibited by high levels of purine nucleotides. The enzyme is not totally switched off when purines are abundant. PRPP is also a precursor of pyrimidines and of histidine. Hence, the synthetase is subject to cumulative feedback inhibition by compounds arising from multiple pathways.

2. The committed step in purine nucleotide biosynthesis is the conversion of PRPP into phosphoribosylamine by *glutamine-PRPP amidotransferase.* This key enzyme is feedback-inhibited by many purine ribonucleotides. It is noteworthy that AMP and GMP, the final products of the pathway, are synergistic in inhibiting the amidotransferase.

3. Inosinate is the branch point in the synthesis of AMP and GMP. *The reactions leading away from inosinate are sites of feedback inhibition.* AMP inhibits the conversion of inosinate into adenylosuccinate, its immediate precursor. Similarly, GMP inhibits the conversion of inosinate into xanthylate, its immediate precursor.

4. GTP is a substrate in the synthesis of AMP, whereas ATP is a substrate in the synthesis of GMP. This *reciprocal substrate relation* tends to balance the synthesis of adenine and guanine ribonucleotides.

5. In *Escherichia coli,* most of the genes encoding enzymes of the de novo pathway are coordinately regulated. Specifically, their transcription is blocked by the *purine repressor (PurR),* a DNA-binding protein, when hypoxanthine and guanine are abundant.

THE PYRIMIDINE RING IS SYNTHESIZED FROM CARBAMOYL PHOSPHATE AND ASPARTATE

We turn now from purines to pyrimidines. The pyrimidine ring is assembled first and then linked to ribose phosphate to form a pyrimidine nucleotide, in contrast with the reaction sequence in the de novo synthesis of purine nucleotides. The precursors of the pyrimidine ring are carbamoyl phosphate and aspartate (Figure 29-10).

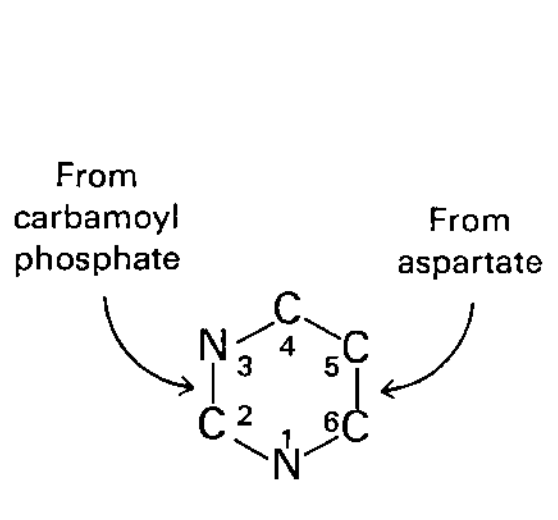

Figure 29-10
Origins of the atoms in the pyrimidine ring. C-2 and N-3 come from carbamoyl phosphate, whereas the other atoms of the ring come from aspartate.

Pyrimidine biosynthesis starts with the formation of *carbamoyl phosphate,* which is also an intermediate in the synthesis of urea (p. 635). The synthesis of this activated carbamoyl donor is compartmentalized in eukaryotes. Carbamoyl phosphate used to synthesize pyrimidines is formed in the cytosol, whereas that used to make urea is formed in mitochondria, by a

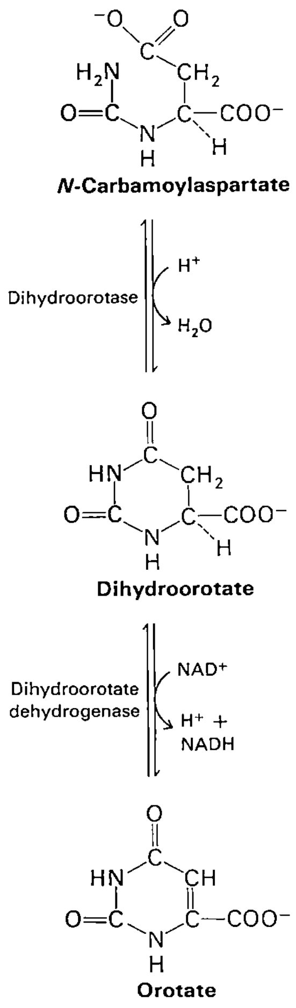

different carbamoyl phosphate synthetase (p. 636). Another noteworthy difference is that glutamine rather than NH_4^+ is the nitrogen donor in the cytosolic synthesis of carbamoyl phosphate. Also, *N*-acetylglutamate does not serve as an allosteric activator in the cytosolic synthesis.

$$\text{Glutamine} + 2\ \text{ATP} + HCO_3^- \longrightarrow \text{carbamoyl phosphate} + 2\ \text{ADP} + P_i + \text{glutamate}$$

The committed step in the biosynthesis of pyrimidines is the formation of N-*carbamoylaspartate from aspartate and carbamoyl phosphate.* This carbamoylation is catalyzed by *aspartate transcarbamoylase,* an especially interesting regulatory enzyme (p. 238).

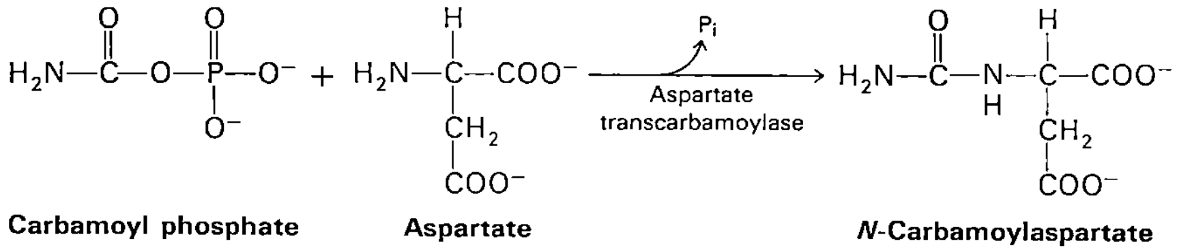

The pyrimidine ring is formed in the next reaction, in which carbamoylaspartate cyclizes with loss of water to yield *dihydroorotate. Orotate* is then formed by dehydrogenation of dihydroorotate.

OROTATE ACQUIRES A RIBOSE PHOSPHATE MOIETY FROM PRPP TO FORM A PYRIMIDINE NUCLEOTIDE

The next step in the synthesis of pyrimidine nucleotides is the *acquisition of a ribose phosphate group.* As in the synthesis of purines, the donor is PRPP. Orotate (a free pyrimidine) reacts with PRPP to form *orotidylate* (a pyrimidine nucleotide). This reaction, which is catalyzed by orotate phosphoribosyl transferase, is driven forward by the hydrolysis of pyrophosphate. Orotidylate is then decarboxylated to yield *uridylate* (UMP), a major pyrimidine nucleotide.

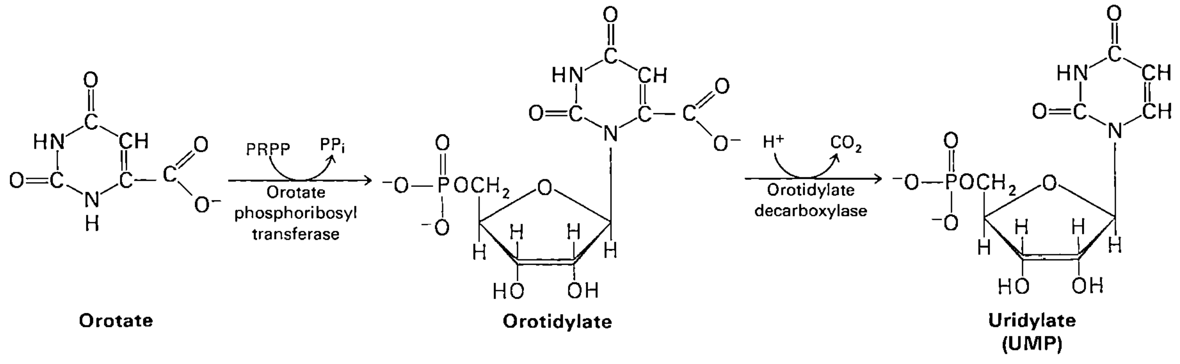

PYRIMIDINE BIOSYNTHESIS IN HIGHER ORGANISMS IS CATALYZED BY MULTIFUNCTIONAL ENZYMES

In *E. coli,* the six enzymes that synthesize UMP from simple precursors appear to be unassociated. In eukaryotes, by contrast, five of them are clustered in two complexes. One of these multifunctional enzymes was discovered when cultured mammalian cells were treated with N-*(phosphonacetyl)-*L*-aspartate* (*PALA*), a potent inhibitor of aspartate transcarbamoylase (ATCase). Recall that PALA binds tightly to ATCase (K_i = 10 nM) because it resembles the transition state in catalysis (p. 241). The surviving cells overcame the inhibitory effect of PALA by synthesizing 100-fold more ATCase than do normal cells. Quite unexpectedly, the concentrations of carbamoyl phosphate synthetase and of dihydroorotase were also elevated 100-fold. By contrast, the enzymes catalyzing subsequent steps in pyrimidine biosynthesis were unaffected by PALA. These observations led to the finding that *carbamoyl phosphate synthetase, aspartate transcarbamoylase, and dihydroorotase are covalently joined in a single 240-kd polypeptide chain.* This multifunctional enzyme is called *CAD.*

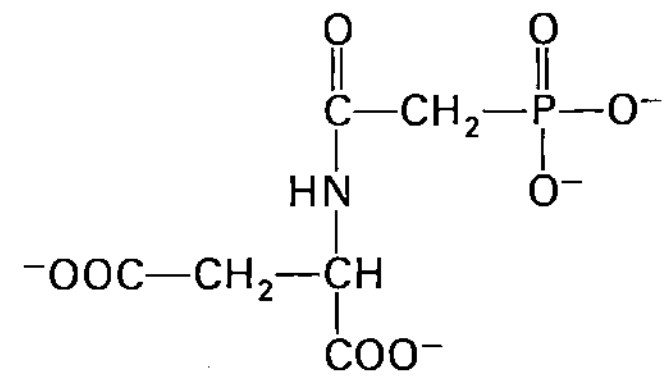

N-(Phosphonacetyl)-L-aspartate (PALA)

Orotate phosphoribosyl transferase and orotidylate decarboxylase, the enzymes catalyzing the last two steps in pyrimidine biosynthesis, are also associated in eukaryotes. Multifunctional enzymes also mediate the synthesis of purines in vertebrates (see Figures 29-5 and 29-6). Indeed, *the covalent linkage of functionally related enzymes occurs often in eukaryotes.* The mammalian fatty acid synthase, which contains seven enzymatic activities in each of two chains (p. 618), is another striking example. The clustering of enzymes catalyzing a reaction sequence has several potential advantages. Their synthesis is coordinated and their assembly into a coherent complex is easily assured. Side reactions are minimized as substrates are channeled from one catalytic site to the next. A covalently linked multifunctional complex is likely to be more stable than one formed by noncovalent interactions. *Multifunctional enzymes probably evolved by exon shuffling.*

NUCLEOSIDE MONO-, DI-, AND TRIPHOSPHATES ARE INTERCONVERTIBLE

The active forms of nucleotides in biosyntheses and energy conversions are the diphosphates and triphosphates. Nucleoside monophosphates are converted to the diphosphates by specific *nucleoside monophosphate kinases* that utilize ATP as the phosphoryl donor. For example, UMP is phosphorylated by *UMP kinase.*

$$\text{UMP} + \text{ATP} \rightleftharpoons \text{UDP} + \text{ADP}$$

AMP, ADP, and ATP are interconverted by *adenylate kinase.* The equilibrium constants of these reactions are close to 1.

$$\text{AMP} + \text{ATP} \rightleftharpoons 2\ \text{ADP}$$

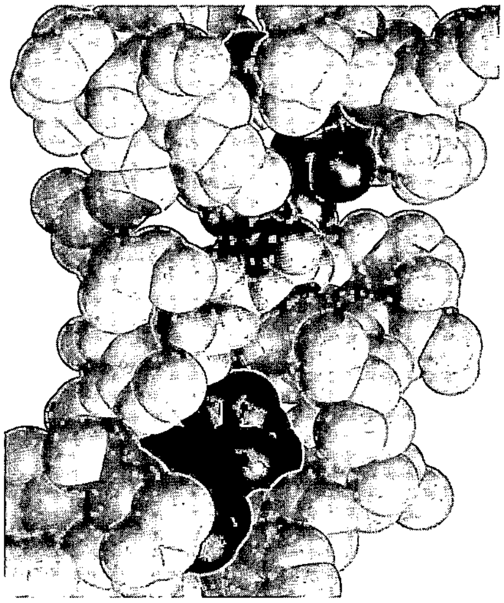

Figure 29-11
Adenylate kinase (gray) catalyzes the interconversion of ATP, ADP, and AMP. A bisubstrate analog is bound to the active site. The adenosine units of the analog are shown in blue and red, and the polyphosphate bridge in yellow. [Drawn from 1ake.pdb. C.W. Mueller and G.E. Schultz. *J. Mol. Biol.* 224(1992):159.]

Nucleoside diphosphates and triphosphates are interconverted by *nucleoside diphosphate kinase,* an enzyme that has broad specificity, in contrast with the monophosphate kinases. In the following equation, X and Y can be any of several ribonucleosides or deoxyribonucleosides.

$$\text{XDP} + \text{YTP} \rightleftharpoons \text{XTP} + \text{YDP}$$

For example,

$$\text{UDP} + \text{ATP} \rightleftharpoons \text{UTP} + \text{ADP}$$

CTP IS FORMED BY AMINATION OF UTP

Cytidine triphosphate (CTP) is derived from uridine triphosphate (UTP), the other major pyrimidine ribonucleotide. The carbonyl oxygen at C-4 is replaced by an amino group. In mammals, the amide group of glutamine is the amino donor, whereas in *E. coli* NH_4^+ is used in this reaction. Mammals avoid having a high level of NH_4^+ in the plasma by generating it in situ from a donor such as glutamine. ATP is consumed in both amination reactions. As in the conversions of inosinate to AMP and GMP, an acyl phosphate intermediate is nucleophilically attacked by a nitrogen atom (p. 743).

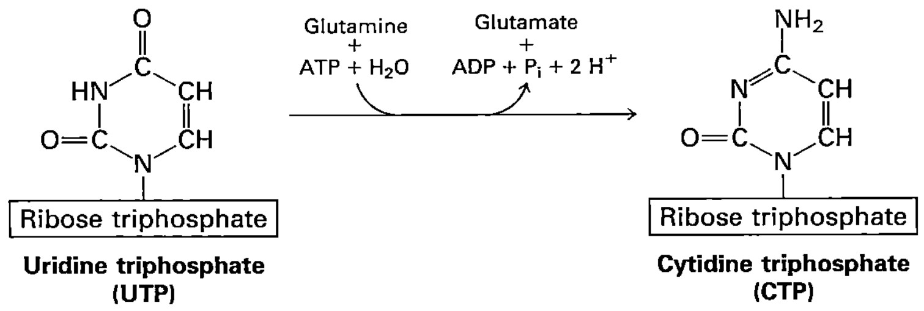

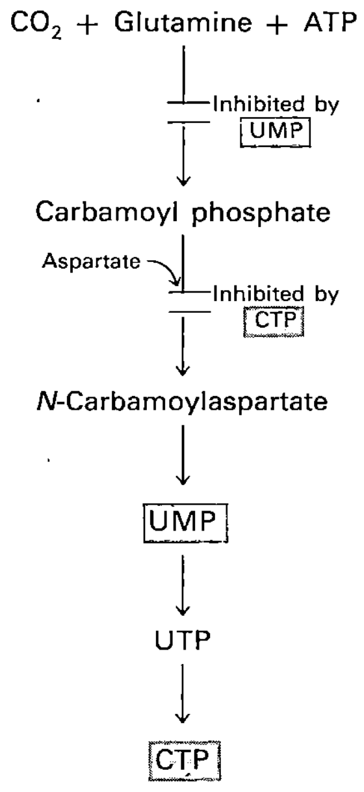

Figure 29-12
Control of pyrimidine biosynthesis in *E. coli.*

PYRIMIDINE NUCLEOTIDE BIOSYNTHESIS IN BACTERIA IS REGULATED BY FEEDBACK INHIBITION

The committed step in pyrimidine nucleotide biosynthesis in *E. coli* is the formation of *N*-carbamoylaspartate from aspartate and carbamoyl phosphate. *Aspartate transcarbamoylase (ATCase), the enzyme that catalyzes this reaction, is feedback-inhibited by CTP, the final product in the pathway.* A second control site is *carbamoyl phosphate synthetase,* which is feedback-inhibited by UMP (Figure 29-12). The structure and allosteric mechanism of *E. coli* ATCase were discussed in an earlier chapter (p. 238).

RIBONUCLEOTIDE REDUCTASE, A RADICAL ENZYME, CATALYZES THE SYNTHESIS OF DEOXYRIBONUCLEOTIDES

We turn now to the synthesis of deoxyribonucleotides. These precursors of DNA are formed by the reduction of ribonucleotides. The 2'-hydroxyl group on the ribose moiety is replaced by a hydrogen atom. The substrates are ribonucleoside diphosphates or triphosphates, and the ultimate reductant is NADPH. The same active site acts on all four ribonucleotides. The overall stoichiometry is

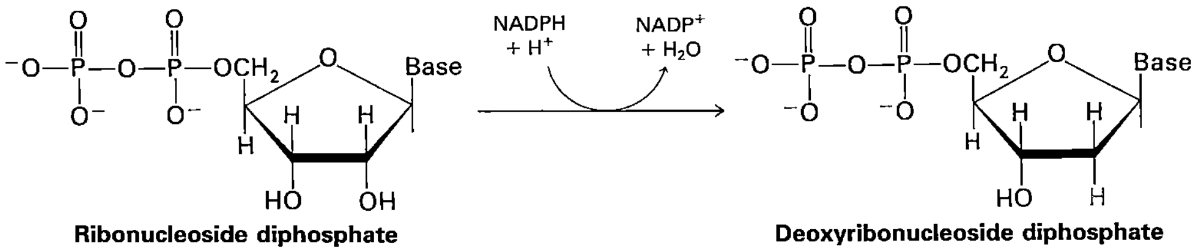

The actual reaction mechanism is more complex than implied by this equation. The best-understood system is that of *E. coli* living aerobically. Peter Reichard has shown that electrons from NADPH are transferred to the substrate through a series of carriers—a flavin, the sulfhydryls of a small protein, a pair of irons that generate a tyrosyl radical, and then another pair of sulfhydryls. *Ribonucleotide reductase* catalyzes the final stage, which has the stoichiometry

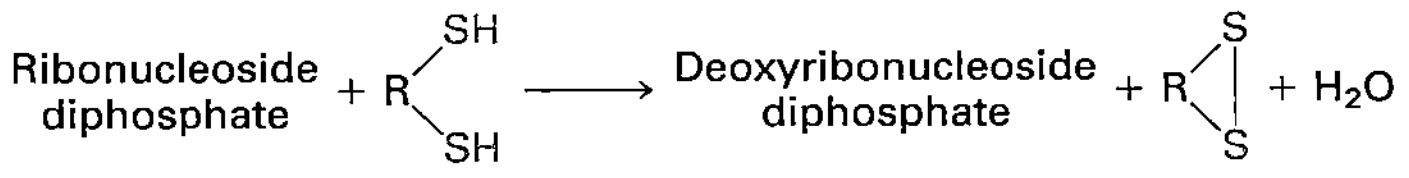

Ribonucleotide reductase consists of two subunits, B1 (a 172-kd dimer) and B2 (an 87-kd dimer). Each chain of B1 contains a substrate binding site, two allosteric control sites, and a sulfhydryl pair that serves as the immediate electron donor in the reduction of the ribose unit. B2 participates in catalysis by generating a remarkable free radical in each of its chains. The B1 and B2 subunits together form the catalytic sites of the enzyme (Figure 29-13).

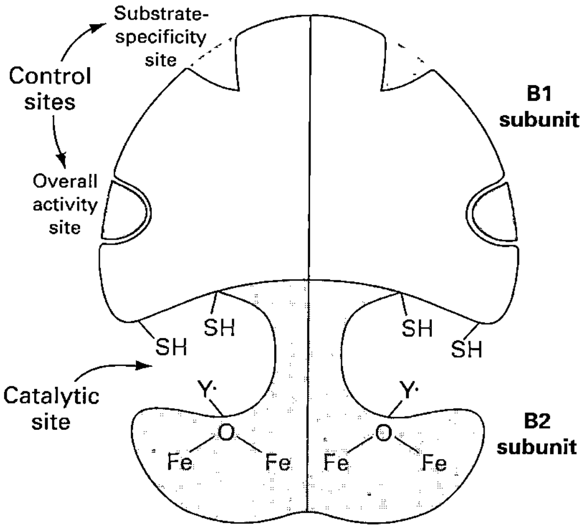

Figure 29-13
Model of ribonucleotide reductase from *E. coli.* [After L. Thelander and P. Reichard. *Ann. Rev. Biochem.* 48(1979): 136. ©1979 by Annual Reviews Inc.]

Each chain of the B2 subunit contains a stable *tyrosyl radical* (Y·) with an unpaired electron on its aromatic ring. This very unusual free radical is generated by a nearby *iron center* consisting of two Fe^{3+} ions connected by an oxygen atom in the −2 valence state. Active B2 is formed from the apoprotein and Fe^{2+} in the presence of O_2 and a thiol.

In the synthesis of a deoxyribonucleotide, the hydroxyl group bonded to C-2 of the ribose ring is stereospecifically replaced by H. No other hydrogen atoms are inserted or removed. The tyrosyl radical plays a catalytic rather than stoichiometric role in this reaction. The essence of the mechanism is a *transient transfer of radical properties from the enzyme to the substrate.* The unpaired electron is shuttled from a tyrosine on B2 to a cysteine on B1, which then abstracts a hydrogen atom from C-3 of the ribose unit (Figure 29-14). The presence of the radical at C-3 promotes the ejection of OH^- from C-2. Deoxyribose is then formed by the reduction of C-2 by thiols on the B1 subunit. The hydrogen atom abstracted by the radical is concomitantly returned to C-3.

The generation of the stable free radical requires O_2. How, then, does *E. coli* reduce ribonucleotides under anaerobic conditions? A different ribonucleotide reductase is at work when O_2 is absent. A stable free radical is again generated but in a different way. An iron-sulfur center, rather than an iron-oxygen center, forms the free radical. S-*Adenosylmethionine binds to the enzyme, and a 5′-deoxyadenosyl radical is produced.* The free electron on this radical then migrates to a site adjacent to the ribonucleoside *tri*phosphate substrate to initiate catalysis. In the ribonucleotide reductase of *Lactobacillus leichmannii,* an *adenosyl cobalamin free radical is produced from coenzyme* B_{12} (p. 642). Why do all known ribonucleotide reductases rely on a radical to achieve catalysis? The answer probably lies in the very demanding nature of the reaction. No change other than the exchange of −H for −OH is permitted in the conversion of a ribonucleotide into a deoxyribonucleotide.

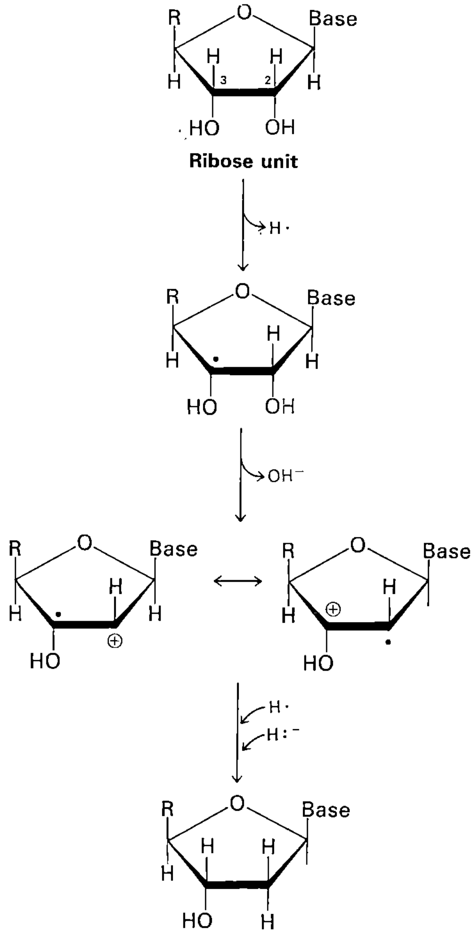

Figure 29-14
Proposed mechanism for the reduction of ribonucleotides by abstraction of a hydrogen atom and formation of free-radical intermediates. [After P. Reichard and A. Ehrenberg. *Science* 221(1983):514.]

THE SUBSTRATE SPECIFICITY AND CATALYTIC ACTIVITY OF RIBONUCLEOTIDE REDUCTASE ARE PRECISELY CONTROLLED

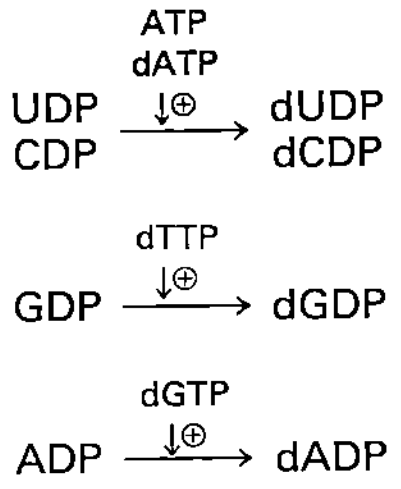

The reduction of ribonucleotides to deoxyribonucleotides is precisely controlled by allosteric interactions. Each polypeptide of the B1 subunit of the aerobic ribonucleotide reductase of *E. coli* contains two allosteric sites: one of them controls the *overall activity* of the enzyme, whereas the other regulates *substrate specificity* (see Figure 29-13). The overall catalytic activity of ribonucleotide reductase is diminished by the binding of dATP, which signals an abundance of deoxyribonucleotides. This feedback inhibition is reversed by the binding of ATP. The binding of dATP or ATP to the substrate-specificity control sites enhances the reduction of UDP and CDP, the pyrimidine nucleotides. The reduction of GDP is promoted by the binding of dTTP, which also inhibits the further reduction of pyrimidine ribonucleotides. The subsequent increase in the level of dGTP leads to a stimulation of ADP reduction. It is evident that ribonucleotide reductase has a variety of conformational states, each with different catalytic properties. This complex pattern of regulation provides the appropriate supply of the four deoxyribonucleotides needed for the synthesis of DNA.

$$\begin{array}{c} \text{ATP} \\ \text{dATP} \\ \text{UDP} \xrightarrow{\downarrow \oplus} \text{dUDP} \\ \text{CDP} \longrightarrow \text{dCDP} \end{array}$$

$$\begin{array}{c} \text{dTTP} \\ \text{GDP} \xrightarrow{\downarrow \oplus} \text{dGDP} \end{array}$$

$$\begin{array}{c} \text{dGTP} \\ \text{ADP} \xrightarrow{\downarrow \oplus} \text{dADP} \end{array}$$

THIOREDOXIN AND GLUTAREDOXIN CARRY ELECTRONS TO RIBONUCLEOTIDE REDUCTASE

How are electrons transferred from NADPH to the sulfhydryl groups at the catalytic site of aerobic ribonucleotide reductase? One carrier of reducing power is *thioredoxin*, a 12-kd protein with two exposed cysteine residues near each other (Figure 29-15). All thioredoxins from archaebacteria to humans have the sequence

-Trp-**Cys**-Gly-Pro-**Cys**-

These sulfhydryls are oxidized to a disulfide in the reaction catalyzed by ribonucleotide reductase. In turn, reduced thioredoxin is regenerated by electron flow from NADPH. This reaction is catalyzed by *thioredoxin reductase*, a flavoprotein. Electrons flow from NADPH to bound FAD of the reductase and then to the disulfide of oxidized thioredoxin.

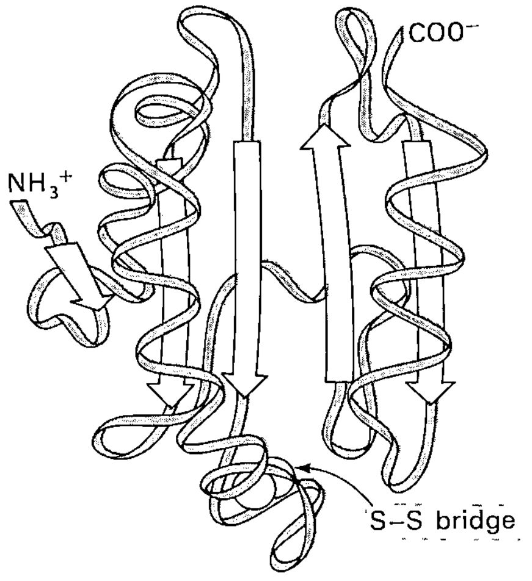

Figure 29-15
Schematic diagram of the main-chain conformation of oxidized thioredoxin from *E. coli*. The reactive disulfide is shown in yellow. [After a drawing kindly provided by Dr. Carl-Ivar Brändén.]

NADPH + H⁺ → FAD / FADH₂ (Thioredoxin reductase (TR)) → TR-SH/SH, TR-S/S (Thioredoxin (T)) → T-S/S, T-SH/SH → RR-SH/SH, RR-S/S (Ribonucleotide reductase (RR)) → Ribose unit / Deoxyribose unit

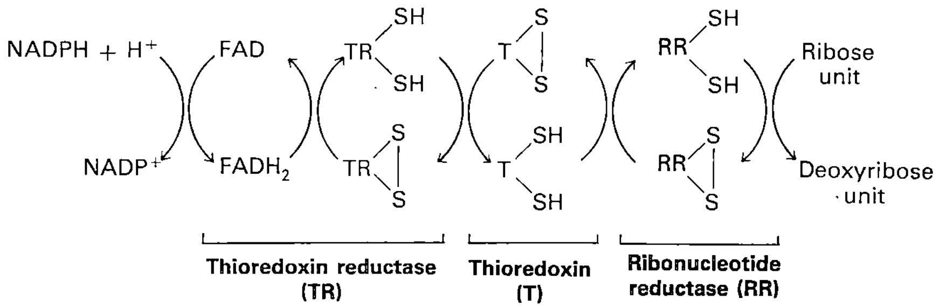

Thioredoxin is an electron donor not only in the reduction of ribonucleotides. Recall that thioredoxin plays an important role in controlling the dark reactions of photosynthesis (p. 677). In chloroplasts, thioredoxin is reduced by ferredoxin when photosystem I is active. Reduced thioredoxin regulates the activities of enzymes in many cells by reducing disulfides to thiols.

Thioredoxin is not the only carrier of reducing power to ribonucleotide reductase. A mutant of *E. coli* totally devoid of thioredoxin was found to form deoxyribonucleotides. This surprising observation led to the isolation of a second carrier system. The electron donor in this mutant proved to be *glutathione*, a cysteine-containing tripeptide. As discussed previously (p. 568), *glutathione reductase* catalyzes the reduction of oxidized glutathione (the disulfide form) by NADPH. A flavin participates in this redox reaction, as in the one catalyzed by thioredoxin reductase. *Glutaredoxin*, a hitherto unknown protein, transfers the reducing power of glutathione to ribonucleotide reductase. Glutaredoxin and thioredoxin are homologous proteins. In particular, the protruding segment containing the thiol pair is very similar in the two.

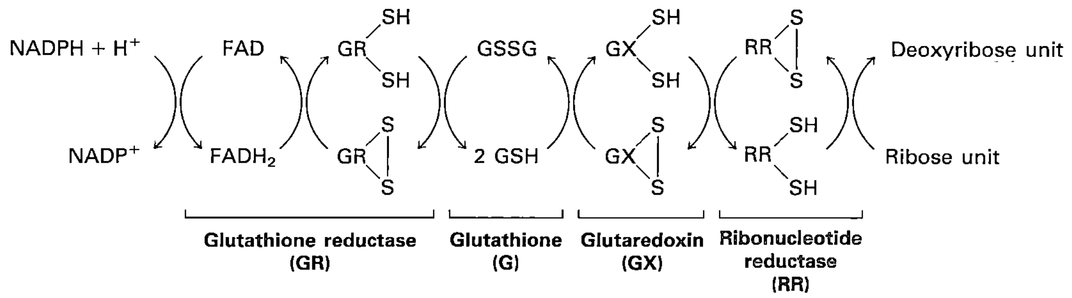

DEOXYTHYMIDYLATE IS FORMED BY METHYLATION OF DEOXYURIDYLATE

Uracil is not a component of DNA. Rather, DNA contains *thymine*, the methylated analog of uracil. *Thymidylate synthase* catalyzes this finishing touch: deoxyuridylate (dUMP) is methylated to deoxythymidylate (dTMP). The methyl donor in this reaction is a tetrahydrofolate derivative rather than *S*-adenosylmethionine. Specifically, the methyl carbon comes from N^5,N^{10}-methylenetetrahydrofolate (Figure 29-16).

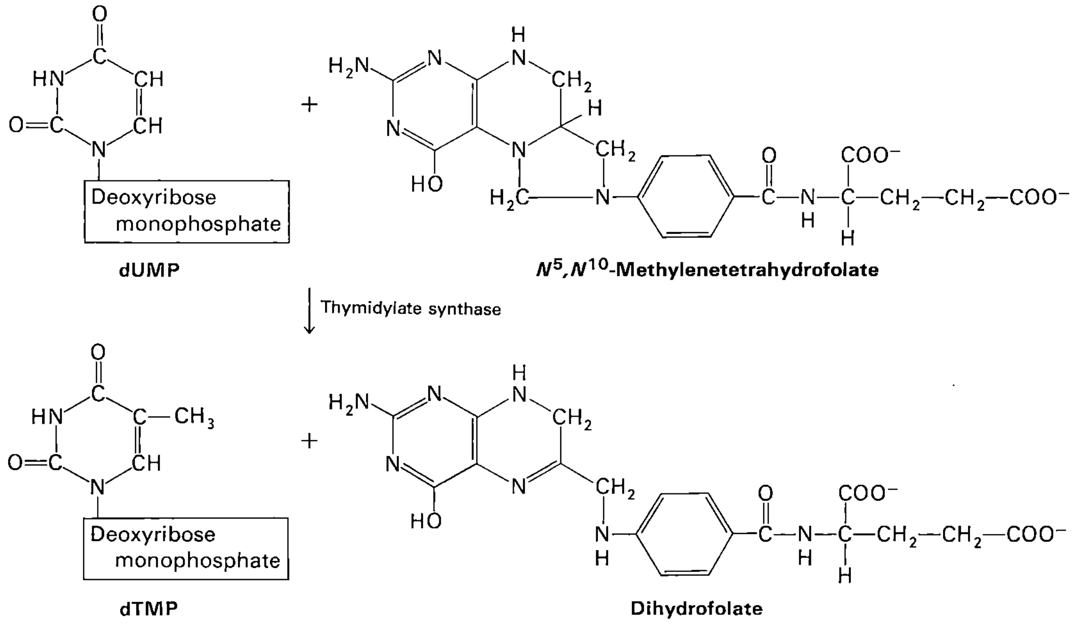

Figure 29-16
Thymidylate synthase catalyzes the methylation of dUMP to dTMP.

The methyl group inserted into deoxyuridylate is more reduced than the methylene group in the donor. What is the source of electrons for this reduction? *The two electrons come in the form of a hydride ion ($H:^-$) from the tetrahydrofolate moiety itself.* This hydrogen becomes part of the methyl group of dTMP. In this reaction, tetrahydrofolate is oxidized to dihydrofolate. Thus N^5,N^{10}-methylenetetrahydrofolate serves both as an *electron donor* and as a *one-carbon donor* in the methylation reaction. We see here, as in the synthesis of purines, the key role of tetrahydrofolate derivatives. *Indeed, nucleotide metabolism and amino acid metabolism are closely tied by one-carbon transfers.*

It is noteworthy that the deoxyribose and thymine units of DNA are formed by modification of ribonucleotides. In contrast, no known ribonucleotide is formed from a deoxyribonucleotide. These precursor-product relations strongly imply that ribonucleotides came first in evolution. *The reactions catalyzed by ribonucleotide reductase and thymidylate synthase are recapitulations of the transition from an RNA world to one in which DNA became the store of genetic information* (p. 115).

DIHYDROFOLATE REDUCTASE CATALYZES THE REGENERATION OF TETRAHYDROFOLATE, A ONE-CARBON CARRIER

Recall that one-carbon transfers are made by derivatives of *tetra*hydrofolate rather than of *di*hydrofolate (p. 719). Hence, tetrahydrofolate must be regenerated from the dihydrofolate that is produced in the synthesis of deoxythymidylate. This is accomplished by *dihydrofolate reductase* using NADPH as the reductant.

$$\text{Dihydrofolate} + \text{NADPH} + \text{H}^+ \longrightarrow \text{tetrahydrofolate} + \text{NADP}^+$$

A hydride ion is directly transferred from the nicotinamide ring of NADPH to the pteridine ring of dihydrofolate (Figure 29-17). The bound substrate and electron donor are in close contact. Indeed, they are separated by less than the normal van der Waals distance—this crowding promotes the formation of the transition state for electron transfer.

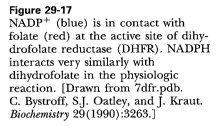

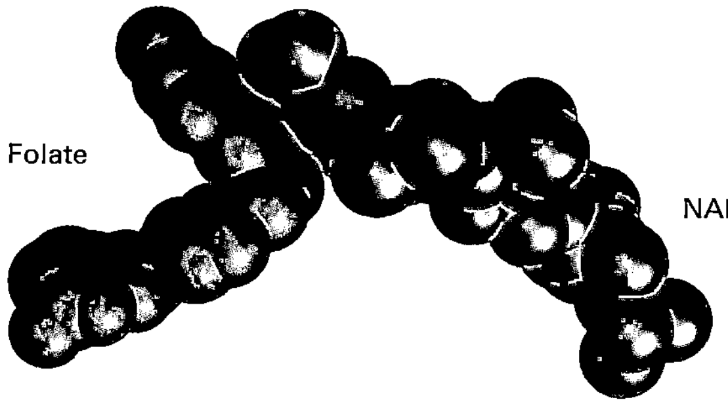

Figure 29-17
$NADP^+$ (blue) is in contact with folate (red) at the active site of dihydrofolate reductase (DHFR). NADPH interacts very similarly with dihydrofolate in the physiologic reaction. [Drawn from 7dfr.pdb. C. Bystroff, S.J. Oatley, and J. Kraut. *Biochemistry* 29(1990):3263.]

SEVERAL VALUABLE ANTICANCER DRUGS BLOCK THE SYNTHESIS OF DEOXYTHYMIDYLATE

Rapidly dividing cells require an abundant supply of deoxythymidylate for the synthesis of DNA. The vulnerability of these cells to the inhibition of dTMP synthesis has been exploited in cancer chemotherapy. Thymidylate synthase and dihydrofolate reductase are choice target enzymes (Figure 29-18).

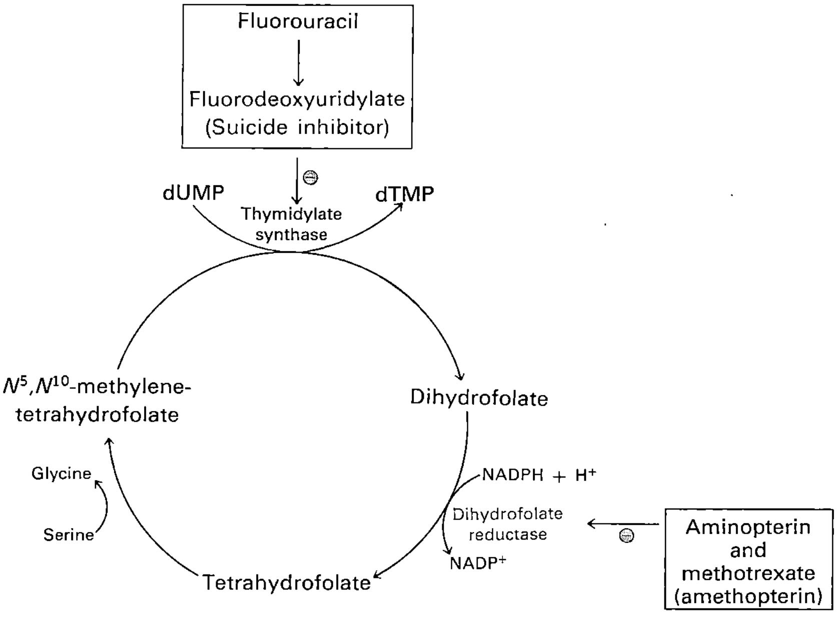

Figure 29-18
Thymidylate synthase and dihydrofolate reductase are choice target enzymes in cancer chemotherapy. Fluorodeoxyuridylate inhibits the methylation of dUMP. The folate analogs aminopterin and methotrexate block the regeneration of tetrahydrofolate.

Fluorouracil (or fluorodeoxyuridine), a clinically useful anticancer drug, is converted in vivo into *fluorodeoxyuridylate* (F-dUMP). This analog of dUMP irreversibly inhibits thymidylate synthase after acting as a normal substrate through part of the catalytic cycle. First, a sulfhydryl group of the enzyme adds to C-6 of the bound F-dUMP (Figure 29-19). Methylenetetrahydrofolate then adds to C-5 of this intermediate. In the case of dUMP, a hydride ion of the folate is subsequently shifted to the methylene group, and a proton is taken away from C-5 of the bound nucleotide. However, F^+ cannot be abstracted from F-dUMP by the enzyme, and so catalysis is blocked at the stage of the covalent complex formed by F-dUMP, methylenetetrahydrofolate, and the sulfhydryl group of the enzyme. We see here an example of *suicide inhibition,* in which an enzyme converts a substrate into a reactive inhibitor that immediately inactivates its catalytic activity.

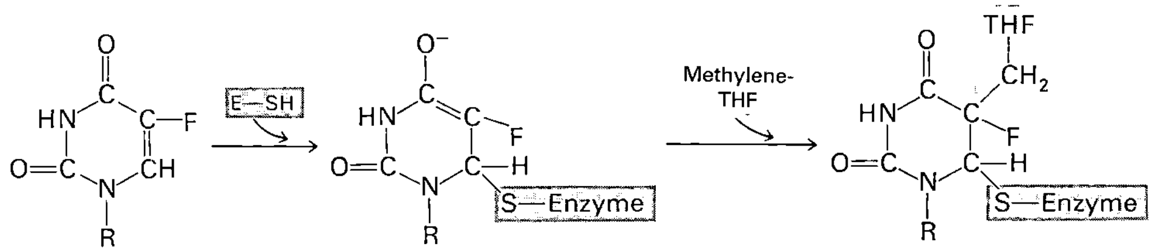

Figure 29-19
Thymidylate synthase is irreversibly inhibited by fluorodeoxyuridylate (F-dUMP). This analog forms a covalent complex with both a sulfhydryl residue of the enzyme (shown in blue) and methylenetetrahydrofolate (shown in yellow).

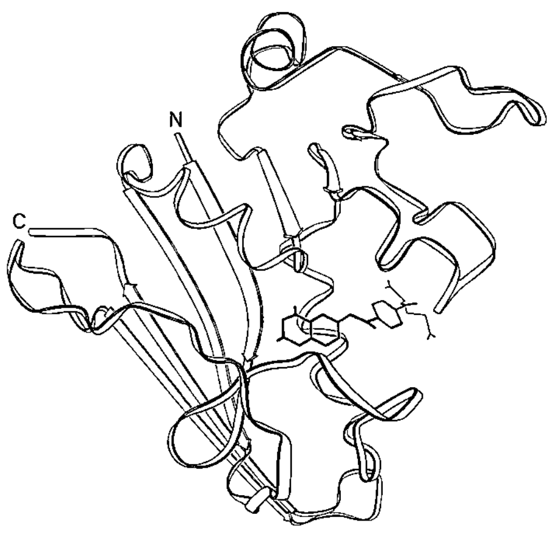

Three-dimensional structure of *E. coli* dihydrofolate reductase with a bound methotrexate. [Courtesy of Dr. Joseph Kraut.]

The synthesis of dTMP can also be blocked by inhibiting the regeneration of tetrahydrofolate. Analogs of dihydrofolate, such as *aminopterin* and *methotrexate* (amethopterin), are potent competitive inhibitors ($K_i <$ 1 nM) of dihydrofolate reductase. Methotrexate is a valuable drug in the treatment of many rapidly growing tumors, such as acute leukemia and choriocarcinoma. However, methotrexate is quite toxic because it kills rapidly replicating cells whether they are malignant or not. Stem cells in bone marrow, epithelial cells of the intestinal tract, and hair follicles are vulnerable to the action of this folate antagonist, accounting for many of its toxic side effects. Folate analogs such as *trimethoprim* have potent antibacterial and antiprotozoal activity. Trimethoprim binds 10^5-fold less tightly to mammalian dihydrofolate reductase than it does to reductases of susceptible microorganisms. Small differences in the active-site clefts of these enzymes account for its highly selective antimicrobial action. The combination of trimethoprim and sulfamethoxazole (an inhibitor of folate synthesis) is widely used to treat infections.

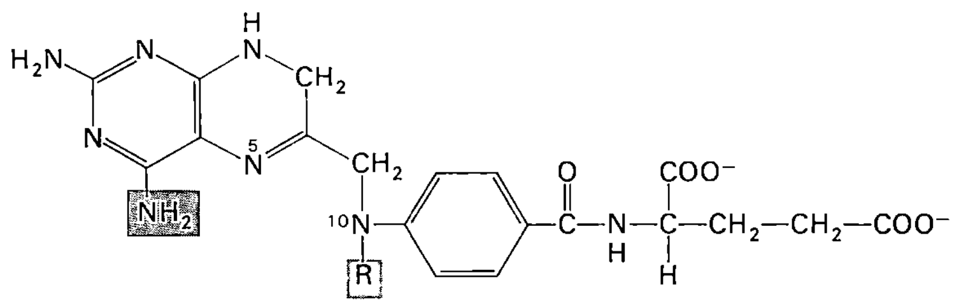

A Aminopterin (R = H) or methotrexate (R = CH_3)

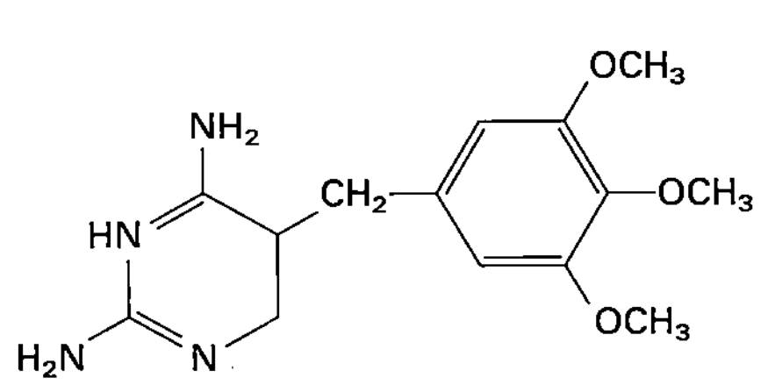

B Trimethoprim

Figure 29-20
Potent competitive inhibitors of dihydrofolate reductase. (A) The anticancer drugs aminopterin and methotrexate contain an $-NH_2$ group in place of the $-OH$ group of dihydrofolate. Methotrexate also differs in having a $-CH_3$ group instead of $-H$ at N^{10}. (B) Trimethoprim, an antibacterial folate analog.

Cultured mammalian cells grown in methotrexate become resistant to its inhibitory action. Some of these mutants do not take up methotrexate because of defects in transport systems. Others are resistant because of changes in the active site of dihydrofolate reductase leading to diminished affinity for methotrexate. A third mechanism of resistance is the *overproduction of dihydrofolate reductase arising from gene amplification* (p. 995). Robert Schimke showed that cells producing as much as 1000 times the normal level of enzyme are selected stepwise as the drug level in the medium is increased. Some overproducing cells contain hundreds of copies of the gene for dihydrofolate reductase. These studies of cultured cells strongly suggest that the dose of an anticancer drug should be sufficiently high to prevent the survival of mutants that become increasingly resistant by gene amplification.

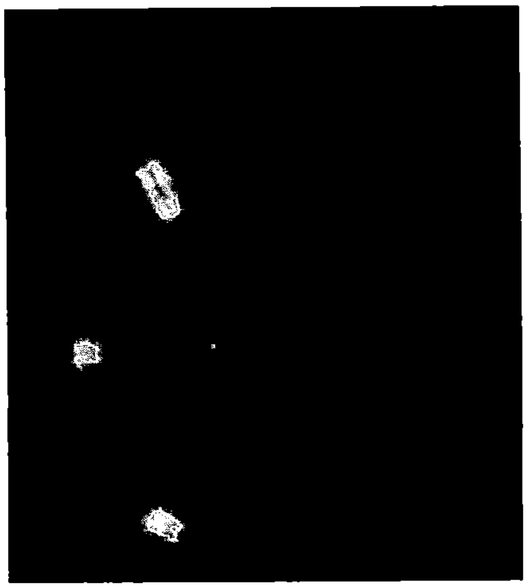

Figure 29-21
Fluorescence micrograph showing multiple copies of the gene for dihydrofolate reductase following selection of cultured cells for resistance to high concentrations of methotrexate. The bright yellow regions on three of the chromosomes each contain hundreds of copies of the gene. [Courtesy of Dr. Barbara Trask and Dr. Joyce Hamlin.]

NAD$^+$, FAD, AND COENZYME A ARE FORMED FROM ATP

The first step in the synthesis of *nicotinamide adenine dinucleotide* (NAD$^+$) is the formation of *nicotinate ribonucleotide* from nicotinate and PRPP. *Nicotinate* (also called *niacin*) is derived from tryptophan. Humans can synthesize the required amount of nicotinate if the supply of tryptophan in the diet is adequate. However, an exogenous supply of nicotinate is required if the dietary intake of tryptophan is low. A dietary deficiency of tryptophan and nicotinate can lead to *pellagra*, a disease characterized by dermatitis, diarrhea, and dementia. An endocrine tumor that consumes

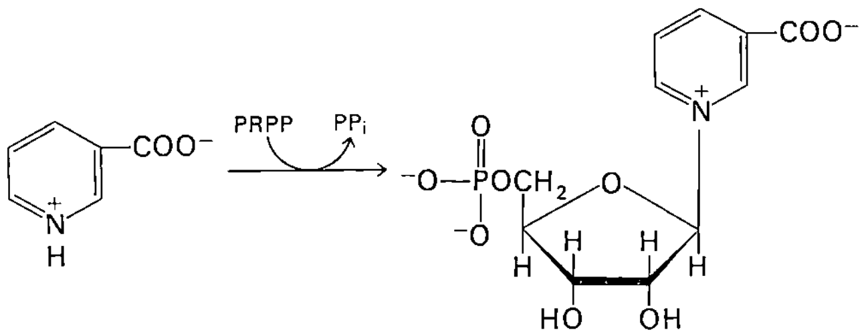

Nicotinate — Nicotinate ribonucleotide

large amounts of tryptophan in synthesizing serotonin (5-hydroxytryptamine), a hormone and neurotransmitter, leads to pellagralike symptoms.

An AMP moiety is then transferred from ATP to nicotinate ribonucleotide to form *desamido-NAD*$^+$. The final step is the transfer of the amide group of glutamine to the nicotinate carboxyl group to form NAD$^+$ (Figure 29-22). NADP$^+$ is derived from NAD$^+$ by phosphorylation of the 2′-hydroxyl group of the adenine ribose moiety. This transfer of a phosphoryl group from ATP is catalyzed by *NAD*$^+$ *kinase.*

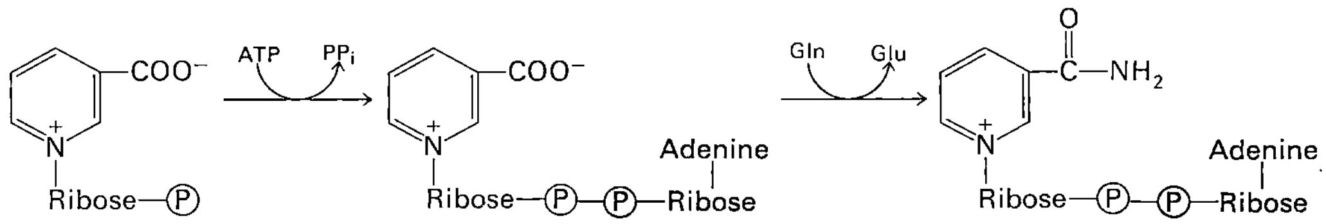

Nicotinate ribonucleotide — Desamido-NAD$^+$ — NAD$^+$

Figure 29-22
Synthesis of NAD$^+$ from nicotinate ribonucleotide.

Flavin adenine dinucleotide (*FAD,* p. 450) is synthesized from riboflavin and two molecules of ATP. Riboflavin is phosphorylated by ATP to give *riboflavin 5′-phosphate* (also called *flavin mononucleotide*). FAD is then formed by the transfer of an AMP moiety from a second molecule of ATP to riboflavin 5′-phosphate.

$$\text{Riboflavin} + \text{ATP} \longrightarrow \text{riboflavin 5'-phosphate} + \text{ADP}$$

$$\text{Riboflavin 5'-phosphate} + \text{ATP} \rightleftharpoons \text{flavin adenine dinucleotide} + \text{PP}_i$$

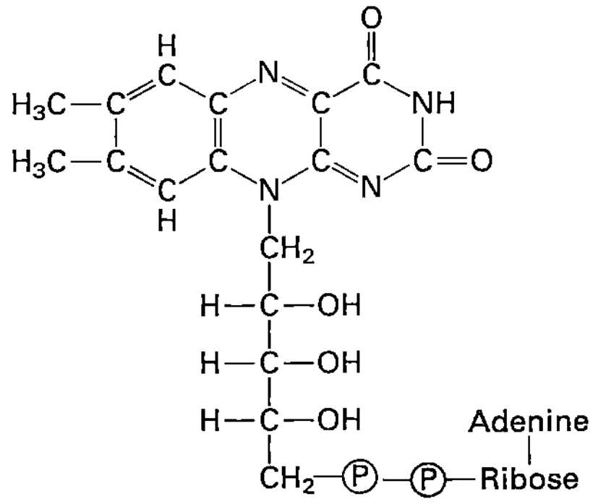

Flavin adenine dinucleotide (FAD)

The AMP moiety of coenzyme A (p. 451) also comes from ATP. A common feature of the biosyntheses of NAD$^+$, FAD, and CoA is the *transfer of the AMP moiety of ATP to the phosphate group of a phosphorylated intermediate.* The pyrophosphate formed in these condensations is then hydrolyzed to orthophosphate. As in many other biosyntheses, *much of the thermodynamic driving force comes from the hydrolysis of the released pyrophosphate.*

PURINES IN HUMANS ARE DEGRADED TO URATE

The nucleotides of a cell undergo continuous turnover. Nucleotides are hydrolytically degraded to nucleosides by *nucleotidases.* Phosphorolytic cleavage of nucleosides to free bases and ribose 1-phosphate (or deoxyribose 1-phosphate) is catalyzed by *nucleoside phosphorylases.* Ribose 1-phosphate is isomerized by *phosphoribomutase* to ribose 5-phosphate, a substrate in the synthesis of PRPP. Some of the bases are reused to form nucleotides by salvage pathways.

The pathway for the degradation of AMP (Figure 29-23) includes an additional step. AMP is deaminated to IMP by *adenylate deaminase.* The 5′-phosphate of IMP is hydrolytically removed, and the glycosidic bond is then split by P_i to yield hypoxanthine, the free base. *Xanthine oxidase,* a molybdenum- and iron-containing flavoprotein, oxidizes hypoxanthine to *xanthine* and then to *uric acid.* Molecular oxygen, the oxidant in both reactions, is reduced to H_2O_2, which is decomposed to H_2O and O_2 by catalase. The keto form of uric acid is in equilibrium with the enol form, which loses a proton at physiologic pH to form *urate. In humans, urate is the final product of purine degradation and is excreted in the urine.*

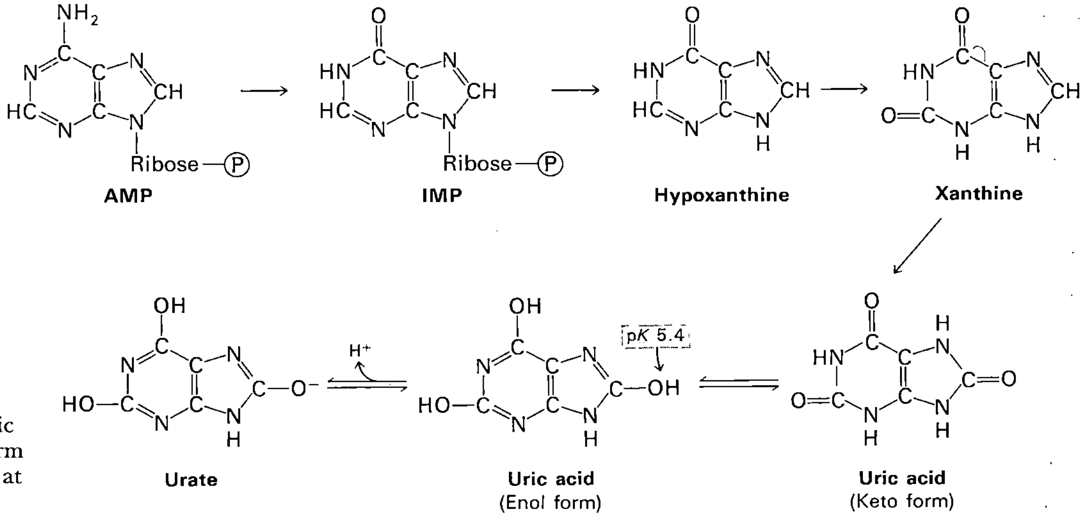

Figure 29-23
Degradation of AMP to uric acid. Urate, the ionized form of uric acid, predominates at physiologic pH.

GOUT IS INDUCED BY HIGH SERUM LEVELS OF URATE

Sodium urate, the predominant form at neutral pH, is more soluble than uric acid. However, its solubility limit of 7 mg/dl (7 mg per 100 ml) at 37°C poses a problem for people who have a high serum level of this purine breakdown product. *Hyperuricemia can induce gout, a disease that affects the joints and kidneys.* Inflammation of the joints is triggered by the *precipitation of sodium urate crystals.* The kidneys too may be damaged by the deposition of urate crystals. A vivid description of an acute attack of gout was given by Thomas Sydenham, an outstanding seventeenth-century English physician, who himself was afflicted with this disease:

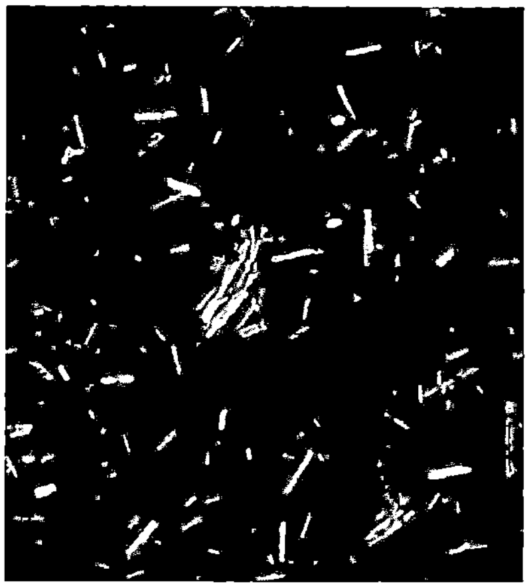

Figure 29-24
Micrograph of sodium urate crystals. Joints and kidneys are damaged by these crystals in gout. [Courtesy of Dr. James McGuire.]

> The victim goes to bed and sleeps in good health. About two o'clock in the morning he is awakened by a severe pain in the great toe; more rarely in the heel, ankle or instep. This pain is like that of a dislocation, and yet the parts feel as if cold water were poured over them. Then follow chills and shivers, and a little fever. The pain, which was at first moderate, becomes more intense. With its intensity the chills and shivers increase. After a time this comes to its height, accommodating itself to the bones and ligaments of the tarsus and metatarsus. Now it is a violent stretching and tearing of the ligaments—now it is a gnawing pain and now a pressure and tightening. So exquisite and lively meanwhile is the feeling of the part affected, that it cannot bear the weight of the bedclothes nor the jar of a person walking in the room. The night is passed in torture, sleeplessness, turning of the part affected, and perpetual change of posture; the tossing about of the body being as incessant as the pain of the tortured joint, and being worse as the fit comes on.

Gout is thought to be an inherited metabolic disease, but the biochemical lesion in most cases has not been elucidated. A small proportion of patients with gout have a partial deficiency of *hypoxanthine-guanine phosphoribosyl transferase* (HGPRT), the enzyme catalyzing the salvage synthesis of IMP and GMP (p. 744).

$$\text{Hypoxanthine} + \text{PRPP} \xrightarrow{\text{HGPRT}} \text{IMP} + \text{PP}_i$$

$$\text{Guanine} + \text{PRPP} \xrightarrow{\text{HGPRT}} \text{GMP} + \text{PP}_i$$

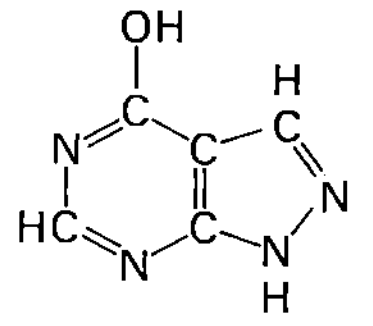

Allopurinol

A deficiency of HGPRT leads to reduced synthesis of GMP and IMP by the salvage pathway. *The consequent increase in the level of PRPP markedly accelerates purine biosynthesis by the de novo pathway.* The formation of 5-phosphoribosyl-1-amine, the first committed intermediate, is normally limited by the availability of PRPP. Excessive PRPP also interferes with feedback inhibition of the amidotransferase that catalyzes this step. Gout can also result from excess PRPP produced by a *hyperactive synthetase* having impaired allosteric regulation.

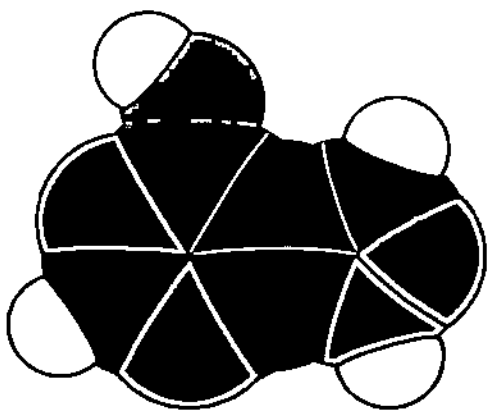

Model of allopurinol.

Allopurinol, an analog of hypoxanthine in which the N and C atoms at positions 7 and 8 are interchanged, is extensively used to treat gout. The mechanism of action of allopurinol is very interesting: it acts *first as a substrate* and *then as an inhibitor* of xanthine oxidase. The oxidase hydroxylates allopurinol to *alloxanthine* (*oxipurinol*), which then remains tightly bound to the active site. The molybdenum atom of xanthine oxidase is kept in the +4 oxidation state by the binding of alloxanthine instead of returning to the +6 oxidation state as in a normal catalytic cycle. We see here another example of *suicide inhibition.*

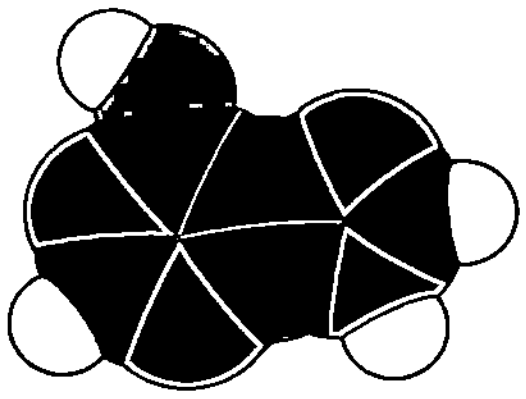

Model of hypoxanthine.

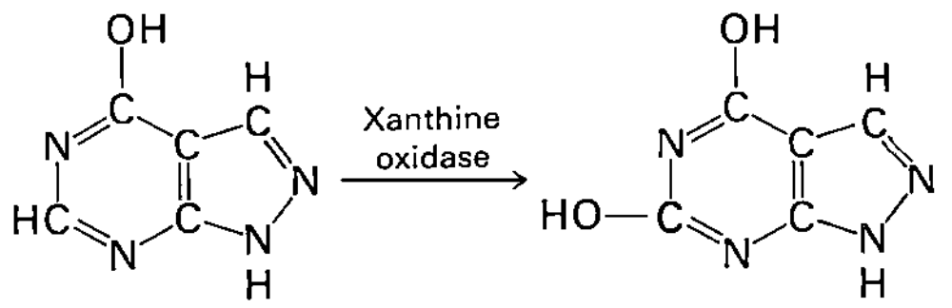

Allopurinol **Alloxanthine**

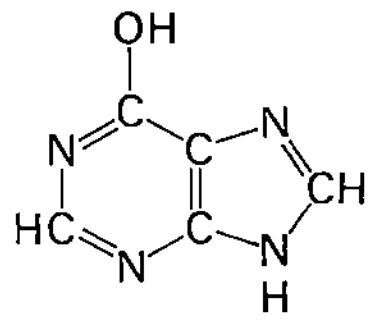

Hypoxanthine

The synthesis of urate from hypoxanthine and xanthine decreases soon after the administration of allopurinol.

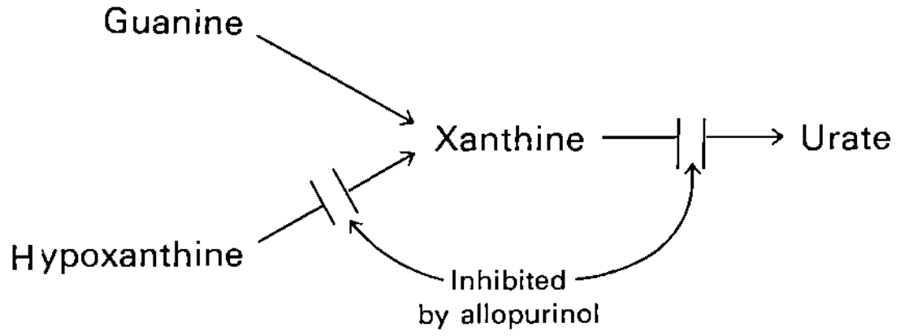

Hence, the serum concentrations of hypoxanthine and xanthine rise, whereas that of urate drops. The formation of uric acid stones is virtually abolished by allopurinol, and the arthritis becomes less severe. Also, *the rate of purine biosynthesis decreases because allopurinol sequesters PRPP by forming the ribonucleotide.* Furthermore, allopurinol ribonucleotide inhibits the conversion of PRPP into phosphoribosylamine.

URATE PLAYS A BENEFICIAL ROLE AS A POTENT ANTIOXIDANT

The average serum level of urate in humans is close to the solubility limit. In contrast, prosimians (such as lemurs) have tenfold lower levels. A striking increase in urate level occurred in the evolution of primates. What is the selective advantage of a urate level so high that it teeters on the brink of gout in many people? It turns out that urate has a markedly beneficial action. Urate is a very efficient scavenger of highly reactive and harmful oxygen species—namely, hydroxyl radicals, superoxide anion, singlet oxygen, and oxygenated heme intermediates in high Fe valence states (+4 and +5). Indeed, urate is about as effective as ascorbate as an antioxidant. The increased level of urate in humans compared with prosimians and other lower primates may contribute significantly to the longer life span of humans and to the lower incidence of human cancer. We see in urate, as in bilirubin (p. 736), an expression of the principle that *some end products of degradative metabolic pathways play important roles as protective agents.*

LESCH-NYHAN SYNDROME: SELF-MUTILATION, MENTAL RETARDATION, AND EXCESSIVE PRODUCTION OF URATE

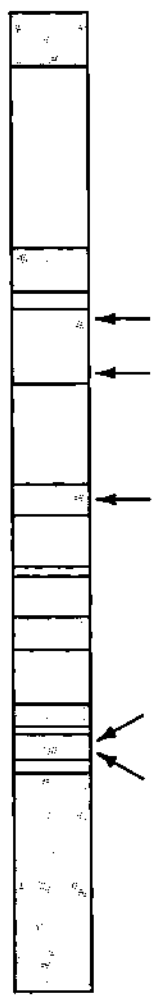

Figure 29-25
The human gene for hypoxanthine-guanine phosphoribosyl transferase (HGPRT) spans 44 kb of DNA and contains 9 exons (blue). Analyses of genomic DNA from patients with the Lesch-Nyhan syndrome have shown that the disease is caused by point mutations (red arrows) in three of the exons, partial and total gene deletions, an insertion, and a partial duplication. [After D.W. Melton, D.S. Konecki, J. Brennand, and C.T. Caskey. *Proc. Nat. Acad. Sci.* 81(1984)2147.]

A nearly total absence of hypoxanthine-guanine phosphoribosyltransferase has devastating consequences. The most striking expression of this inborn error of metabolism, called the *Lesch-Nyhan syndrome,* is *compulsive self-destructive behavior.* At age two or three, children with this disease begin to bite their fingers and lips. They also tend to be aggressive toward others. *Mental deficiency* and *spasticity* are other characteristics of the Lesch-Nyhan syndrome. Elevated levels of urate in the serum lead to the formation of kidney stones early in life, followed by the symptoms of gout years later. The disease is inherited as a sex-linked recessive disorder.

The biochemical consequences of the virtual absence of HGPRT are *an elevated concentration of PRPP, a marked increase in the rate of purine biosynthesis by the de novo pathway, and an overproduction of urate.* The relation between the absence of the transferase and the bizarre neurologic signs is an enigma. The brain may be very dependent on the salvage pathway for the synthesis of IMP and GMP. The level of HGPRT is normally higher in the brain than in any other tissue. In contrast, the activity of the amidotransferase that catalyzes the committed step in the de novo pathway is rather low in the brain.

Allopurinol is highly effective in diminishing urate synthesis in the Lesch-Nyhan syndrome. However, patients with this disease do not convert allopurinol into the ribonucleotide because they lack hypoxanthine-guanine phosphoribosyl transferase. Hence, the administration of allopurinol does not lower their level of PRPP and so de novo purine synthesis is not diminished. Furthermore, allopurinol fails to alleviate the neurologic symptoms.

The Lesch-Nyhan syndrome demonstrates that the salvage pathway for the synthesis of IMP and GMP is not gratuitous. The salvage pathway evidently serves a critical role that is not yet fully understood. Furthermore, the interplay between the de novo and salvage pathways of purine synthesis remains to be elucidated. Moreover, *the Lesch-Nyhan syndrome reveals that abnormal behavior such as self-mutilation and extreme hostility can be caused by the absence of a single enzyme.* Psychiatry will no doubt benefit from a better understanding of the molecular basis of such mental disorders.

SUMMARY

The purine ring is assembled from a variety of precursors: glutamine, glycine, aspartate, methylenetetrahydrofolate, N^{10}-formyltetrahydrofolate, and CO_2. The committed step in the de novo synthesis of purine nucleotides is the formation of 5-phosphoribosylamine from 5-phosphoribosyl-1-pyrophosphate (PRPP) and glutamine. The purine ring is assembled on ribose phosphate. The addition of glycine, followed by formylation, amination, and ring closure, yields 5-aminoimidazole ribonucleotide. This intermediate contains the completed five-membered ring of the purine skeleton. The addition of CO_2, the nitrogen atom of aspartate, and a formyl group, followed by ring closure, yields inosinate (IMP), a purine ribonucleotide. AMP and GMP are formed from IMP. Purine ribonucleotides can also be synthesized by a salvage pathway in which a preformed base reacts directly with PRPP. Feedback inhibition of 5-phosphoribosyl-1-pyrophosphate synthetase and of glutamine-PRPP amidotransferase by purine nucleotides is important in regulating their biosynthesis.

The pyrimidine ring is assembled first and then linked to ribose phosphate to form a pyrimidine nucleotide, in contrast with the sequence in the de novo synthesis of purine nucleotides. PRPP is again the donor of the ribose phosphate moiety. The synthesis of the pyrimidine ring starts with the formation of carbamoylaspartate from carbamoyl phosphate and aspartate, a reaction catalyzed by aspartate transcarbamoylase. Dehydration, cyclization, and oxidation yield orotate, which reacts with PRPP to give orotidylate. Decarboxylation of this pyrimidine nucleotide yields UMP. CTP is then formed by amination of UTP. Pyrimidine biosynthesis in *E. coli* is regulated by feedback inhibition of aspartate transcarbamoylase, the enzyme that catalyzes the committed step. CTP inhibits and ATP stimulates this enzyme. In higher organisms, the first three enzymes of pyrimidine biosynthesis (called CAD) are present on a single polypeptide chain.

Deoxyribonucleotides, the precursors of DNA, are formed in aerobic *E. coli* by the reduction of ribonucleoside diphosphates. These conversions are catalyzed by ribonucleotide reductase. Electrons are transferred from NADPH to sulfhydryl groups at the active sites of this enzyme by thioredoxin or glutaredoxin. A tyrosyl free radical that is generated by an iron center in the reductase participates in catalyzing the exchange of H for OH at C-2 of the sugar unit. dTMP is formed by methylation of dUMP. The one-carbon and electron donor in this reaction is N^5,N^{10}-methylenetetrahydrofolate, which is converted into dihydrofolate. Tetrahydrofolate is regenerated by the reduction of dihydrofolate by NADPH. Dihydrofolate reductase, which catalyzes this reaction, is inhibited by folate analogs such as aminopterin and methotrexate. These compounds and fluorouracil, an inhibitor of thymidylate synthase, are used as anticancer drugs.

Purines are degraded to urate in humans. Gout, a disease that affects joints and leads to arthritis, is associated with excessive production of urate. Allopurinol, a suicide inhibitor of xanthine oxidase, is used to treat gout. This analog of hypoxanthine blocks the formation of urate from hypoxanthine and xanthine. The Lesch-Nyhan syndrome, a genetic disease characterized by self-mutilation, mental deficiency, and gout, is caused by the absence of hypoxanthine-guanine phosphoribosyl transferase. This enzyme is essential for the synthesis of purine nucleotides by the salvage pathway.

SELECTED READINGS

Where to start

Zalkin, H., and Dixon, J.E., 1992. De novo purine nucleotide biosynthesis. *Prog. Nucleic Acid Res. Mol. Biol.* 42:259–287.

Reichard, P., 1993. From RNA to DNA, why so many ribonucleotide reductases? *Science* 260:1773–1777.

Uhlin, U., and Eklund, H., 1994. Structure of ribonucleotide reductase protein R1. *Nature* 370:533–539.

Elion, G.B., 1989. The purine path to chemotherapy. *Science* 244:41–47.

Seegmiller, J.E., 1989. Contributions of Lesch-Nyhan syndrome to the understanding of purine metabolism. *J. Inher. Metab. Dis.* 12:184–196.

Books on nucleotide metabolism

Kornberg, A., and Baker, T.A., 1992. *DNA Replication* (2nd ed.). W.H. Freeman. [Chapter 2 gives an excellent account of the biosynthesis of DNA precursors.]

Reese, C.B. (ed.), 1984. *Recent Aspects of the Chemistry of Nucleosides, Nucleotides, and Nucleic Acids.* Pergamon.

Walsh, C., 1979. *Enzymatic Reaction Mechanisms.* W. H. Freeman. [Several important reaction mechanisms in the biosynthesis of nucleotides are lucidly treated. See Chapter 5 for a discussion of amino transfers, Chapter 13 for xanthine oxidase, and Chapter 25 for one-carbon transfers.]

Blakley, R.L., and Benkovic, S.J., 1985. *Folates and Pterins,* vols. 1 and 2. Wiley.

Purine biosynthesis

He, B., Shiau, A., Choi, K.Y., Zalkin, H., and Smith, J.M., 1990. Genes of the *Escherichia coli* pur regulon are negatively controlled by a repressor-operator interaction. *J. Bacteriol.* 172:4555–4562.

Daubner, S.C., Schrimsher, J.L., Schendel, F.J., Young, M., Henikoff, S., Patterson, D., Stubbe, J., and Benkovic, S.J., 1985. A multifunctional protein possessing glycinamide ribonucleotide synthetase, glycinamide ribonucleotide transformylase, and aminoimidazole ribonucleotide synthetase activities in de novo purine biosynthesis. *Biochemistry* 24:7059–7062.

Schendel, F.J., Mueller, E., Stubbe, J., Shiau, A., and Smith, J.M., 1989. Formylglycinamide ribonucleotide synthetase from *Escherichia coli*: Cloning, sequencing, overproduction, isolation, and characterization. *Biochemistry* 28:2459–2471.

Weber, G., Nagai, M., Natsumeda, Y., Ichikawa, S., Nakamura, H., Eble, J.N., Jayaram, H.N., Zhen, W.N., Paulik, E., and Hoffman, R., 1991. Regulation of de novo and salvage pathways in chemotherapy. *Advan. Enzyme Regul.* 31:45–67.

Pyrimidine biosynthesis

Jones, M.E., 1980. Pyrimidine nucleotide biosynthesis in animals: Genes, enzymes, and regulation of UMP biosynthesis. *Ann. Rev. Biochem.* 49:253–279.

Jones, M.E., 1992. Orotidylate decarboxylase of yeast and man. *Curr. Top. Cell. Regul.* 33:331–342.

Lee, L., Kelly, R.E., Pastra-Landis, S.C., and Evans, D.R., 1985. Oligomeric structure of the multifunctional protein CAD that initiates pyrimidine biosynthesis in mammalian cells. *Proc. Nat. Acad. Sci.* 82:6802–6806.

Ribonucleotide reductases

Reichard, P., 1993. The anaerobic ribonucleotide reductase from *Escherichia coli. J. Biol. Chem.* 268:8383–8386.

Stubbe, J., 1990. Ribonucleotide reductases: Amazing and confusing. *J. Biol. Chem.* 265:5329–5332.

Bollinger, J.J., Edmondson, D.E., Huynh, B.H., Filley, J., Norton, J.R., and Stubbe, J., 1991. Mechanism of assembly of the tyrosyl radical-dinuclear iron cluster cofactor of ribonucleotide reductase. *Science* 253:292–298.

Stubbe, J.A., 1989. Protein radical involvement in biological catalysis? *Ann. Rev. Biochem.* 58:257–285.

Ashley, G.W., Harris, G., Stubbe, J., 1986. The mechanism of *Lactobacillus leichmannii* ribonucleotide reductase. Evidence for 3′ C-H bond cleavage and a unique role for coenzyme B_{12}. *J. Biol. Chem.* 261:3958–3964.

Thioredoxin and glutaredoxin

Holmgren, A., 1989. Thioredoxin and glutaredoxin systems. *J. Biol. Chem.* 264:13963–13966.

Eklund, H., Gleason, F.K., and Holmgren, A., 1991. Structural and functional relations among thioredoxins of different species. *Proteins* 11:13–28.

Dyson, H.J., Gippert, G.P., Case, D.A., Holmgren, A., and Wright, P.E., 1990. Three-dimensional solution structure of the reduced form of *Escherichia coli* thioredoxin determined by nuclear magnetic resonance spectroscopy. *Biochemistry* 29:4129–4136.

Thymidylate synthase and dihydrofolate reductase

Schweitzer, B.I., Dicker, A.P., and Bertino, J.R., 1990. Dihydrofolate reductase as a therapeutic target. *FASEB J.* 4:2441–2452.

Shoichet, B.K., Stroud, R.M., Santi, D.V., Kuntz, I.D., and Perry, K.M., 1993. Structure-based discovery of inhibitors of thymidylate synthase. *Science* 259:1445–1450.

Liu, L., and Santi, D.V., 1992. Mutation of asparagine 229 to aspartate in thymidylate synthase converts the enzyme to a deoxycytidylate methylase. *Biochemistry* 31:5100–5104.

Brown, K.A., and Kraut, J., 1992. Exploring the molecular mechanism of dihydrofolate reductase. *Faraday Discussions* 1992:217–224.

Bystroff, C., Oatley, S.J., and Kraut, J., 1990. Crystal structures of *Escherichia coli* dihydrofolate reductase: The $NADP^+$ holoenzyme and the folate $NADP^+$ ternary complex. Substrate binding and a model for the transition state. *Biochemistry* 29:3263–3277.

Schimke, R.T., 1992. Gene amplification; what are we learning? *Mutation Res.* 276:145–149.

Benkovic, S.J., 1980. On the mechanism of action of folate- and biopterin-requiring enzymes. *Ann. Rev. Biochem.* 49:227–251.

Uric acid as an antioxidant

Ames, B.N., Cathcart, R., Schwiers, E., and Hochstein, P., 1981. Uric acid provides an antioxidant defense in humans against oxidant- and radical-caused aging and cancer: A hypothesis. *Proc. Nat. Acad. Sci.* 78:6858–6862.

Genetic diseases

Scriver, C.R., Beaudet, A.L., Sly, W.S., and Valle, D. (eds.), *The Metabolic Basis of Inherited Disease* (6th ed.). McGraw-Hill.

[Part 6 contains a wealth of information on inborn errors of purine and pyrimidine metabolism.]

Davidson, B.L., Pashmforoush, M., Kelley, W.N., and Palella, T.D., 1989. Human hypoxanthine-guanine phosphoribosyltransferase deficiency. The molecular defect in a patient with gout (HPRTAshville). *J. Biol. Chem.* 264:520–525.

Sculley, D.G., Dawson, P.A., Emerson, B.T., and Gordon, R.B., 1992. A review of the molecular basis of hypoxanthine-guanine phosphoribosyltransferase (HPRT) deficiency. *Human Genet.* 90:195–207.

PROBLEMS

1. *Activated ribose phosphate.* Write a balanced equation for the synthesis of PRPP from glucose via the oxidative branch of the pentose phosphate pathway.

2. *Making a pyrimidine.* Write a balanced equation for the synthesis of orotate from glutamine, CO_2, and aspartate.

3. *Identifying the donor.* What is the activated reactant in the biosynthesis of each of these compounds?
 (a) Phosphoribosylamine.
 (b) Carbamoylaspartate.
 (c) Orotidylate (from orotate).
 (d) Nicotinate ribonucleotide.
 (e) Phosphoribosylanthranilate.

4. *Inhibiting purine biosynthesis.* Amidotransferases are inhibited by the antibiotic azaserine (*O*-diazoacetyl-L-serine), which is an analog of glutamine.

$$^-N{=}\overset{+}{N}{=}CH{-}\underset{\underset{O}{\|}}{C}{-}O{-}CH_2{-}\underset{\underset{NH_3{}^+}{|}}{CH}{-}COO^-$$

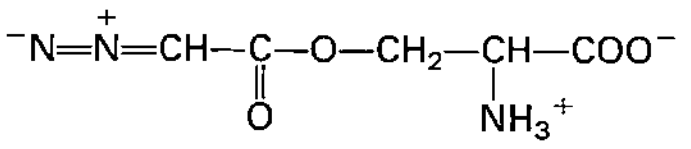

Azaserine

Which intermediates in purine biosynthesis would accumulate in cells treated with azaserine?

5. *The price of methylation.* Write a balanced equation for the synthesis of dTMP from dUMP that is coupled to the conversion of serine into glycine.

6. *Sulfa action.* Bacterial growth is inhibited by sulfanilamide and related sulfa drugs, and there is a concomitant accumulation of 5-aminoimidazole-4-carboxamide ribonucleotide. This inhibition is reversed by the addition of *p*-aminobenzoate.

$$H_2N{-}\boxed{}{-}SO_2NH_2$$

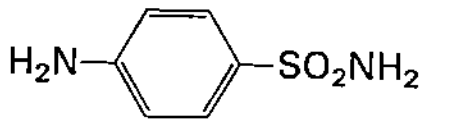

Sulfanilamide

Propose a mechanism for the inhibitory effect of sulfanilamide.

7. *A generous donor.* What are the major biosynthetic reactions that utilize PRPP?

8. *Catalytic duet.* Carbamoyl phosphate synthetase from *E. coli* consists of a small subunit (40 kd) and a large one (130 kd). The isolated small subunit has glutaminase activity, whereas the isolated large subunit can synthesize carbamoyl phosphate from NH_3 but not from glutamine. Furthermore, the isolated small subunit is an ATPase in the presence of bicarbonate. Propose a plausible reaction mechanism for the intact enzyme on the basis of the activities of its separated subunits. How might the holoenzyme be constructed to favor the interplay of subunits?

9. *Déjà vu.* In purine biosynthesis, aspartate becomes linked to the carboxylate group of an intermediate. In the next reaction, the carbon skeleton of the aspartate unit leaves in the form of fumarate. These reactions are reminiscent of a similar pair of reactions in the degradation of amino acids. Which ones?

10. *Pernicious anemia.* Purine biosynthesis is impaired in vitamin B_{12} deficiency. Why?

11. *A devastating deficiency.* Adenosine deaminase (ADA) catalyzes the irreversible deamination of adenosine and 2′-deoxyadenosine to inosine and 2′-deoxyinosine, respectively. Inherited defects of ADA lead to abnormalities in purine nucleoside metabolism that are selectively toxic to lymphocytes. The result is a *severe combined immunodeficiency disease* (*SCID*).
 (a) The intracellular level of *S*-adenosylhomocysteine is elevated in ADA-deficient patients. Why?
 (b) Predict the major consequence of an elevated level of *S*-adenosylhomocysteine.

12. *HAT medium.* Mutant cells unable to synthesize nucleotides by salvage pathways are very useful tools in molecular and cell biology. Suppose that cell A lacks thymidine kinase, the enzyme catalyzing the phosphorylation of thymidine to thymidylate, and that cell B lacks hypoxanthine-guanine phosphoribosyl transferase.
 (a) Cell A and cell B do not proliferate in a *HAT* medium containing *hy*poxanthine, *a*minopterin or

aminopterin (methotrexate), and thymine. However, cell C formed by the fusion of A and B grows in this medium. Why?

(b) Suppose that you wanted to introduce foreign genes into cell A. Devise a simple means of distinguishing between cells that have taken up foreign DNA and those that have not.

13. *Adjunct therapy.* Allopurinol is often given to patients with acute leukemia who are being treated with anticancer drugs. Why is allopurinol used?

14. *A hobbled enzyme.* Both side-chain oxygen atoms of aspartate 27 at the active site of dihydrofolate reductase form hydrogen bonds with the pteridine ring of folates (Figure 29-26). The importance of this interaction has been assessed by studying two mutants at this position, Asn 27 and Ser 27. The dissociation constant of methotrexate was 0.07 nM for the wild type, 1.9 nM for the Asn 27 mutant, and 210 nM for the Ser 27 mutant, at 25°C. Calculate the free energy of binding of methotrexate by these three proteins. What is the decrease in binding energy resulting from each mutation?

15. *Hyperuricemia.* Many patients with glucose 6-phosphatase deficiency have high serum levels of urate. Hyperuricemia can also be induced in normal people by the ingestion of alcohol or by strenuous exercise. Propose a common mechanism that accounts for these findings.

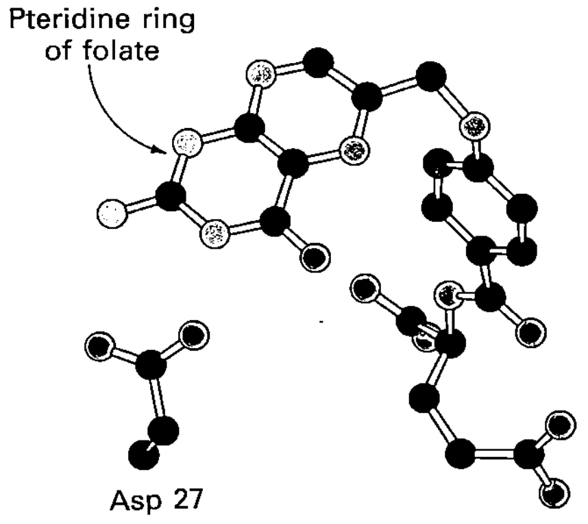

Figure 29-26
The carboxylate group of Asp 27 interacts with the pteridine ring of folate in the active site of dihydrofolate reductase. [Drawn from 7dfr.pdb. C. Bystroff, S.J. Oatley, and J. Kraut. *Biochemistry* 29(1990):3263.]

CHAPTER

30

Integration of Metabolism

How is the intricate network of reactions in metabolism coordinated to meet the needs of the whole organism? This chapter presents some of the principles underlying the integration of metabolism in mammals. It starts with a recapitulation of the strategy of metabolism and of recurring motifs in its regulation. The interplay of different pathways is then described in terms of the flow of molecules at three key crossroads: glucose 6-phosphate, pyruvate, and acetyl CoA. This view of major junctions is followed by a discussion of differences in the metabolic patterns of the brain, muscle, adipose tissue, and liver. The major hormonal regulators of fuel metabolism—insulin, glucagon, epinephrine, and norepinephrine—are considered next. We then turn to a most important aspect of the integration of metabolism—the control of the level of glucose in the blood. The last part of the chapter deals with the usage of fuels, as exemplified by the remarkable metabolic adaptations in prolonged starvation. The contrasting fuels used to power sprinting and marathon running are also instructive. The chapter closes with a discussion of diabetes mellitus, a serious and complex disease characterized by a grossly abnormal pattern of fuel usage.

> "To every thing there is a season, and a time to every purpose under the heaven:
> A time to be born, and a time to die; a time to plant, and a time to pluck up that which is planted;
> A time to kill, and a time to heal; a time to break down, and a time to build up."
>
> *Ecclesiastes 3:1–3*

Opening Image: *Detail of runners on a Greek amphora painted in the sixth century,* B.C. *The stadion, run over a distance of 192 meters, was the main sprint event at athletic festivals. The power output during this race approaches the maximum that can be attained by a human [Metropolitan Museum of Art, Rogers Fund, 1914 (14.130.12). Copyright © 1977 by the Metropolitan Museum of Art.]*

STRATEGY OF METABOLISM: A RECAPITULATION

As was explained in Chapter 17, the basic strategy of metabolism is to form ATP, reducing power, and building blocks for biosyntheses. Let us briefly review these central themes:

1. *ATP is the universal currency of energy.* The high phosphoryl transfer potential of ATP enables it to serve as the energy source in muscle contraction, active transport, signal amplification, and biosyntheses. The hydrolysis of an ATP molecule changes the equilibrium ratio of products to reactants in a coupled reaction by a factor of about 10^8. Hence, a thermodynamically unfavorable reaction sequence can be made highly favorable by coupling it to the hydrolysis of a sufficient number of ATP molecules. For example, three molecules of ATP are consumed in the conversion of mevalonate into isopentenyl pyrophosphate, the activated five-carbon unit in the synthesis of cholesterol.

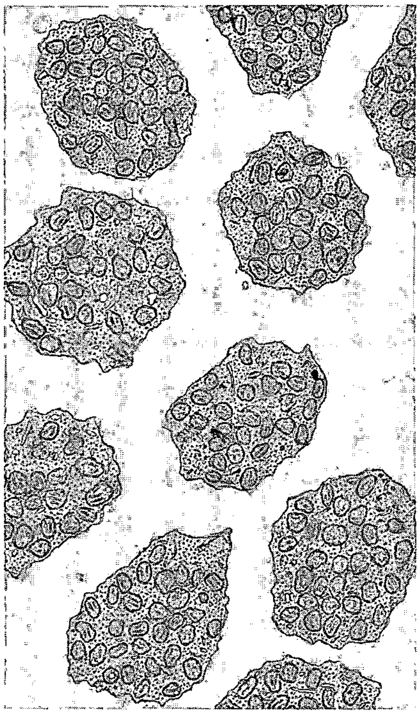

Figure 30-1
Electron micrograph showing numerous mitochondria in the inner segment of retinal rod cells. These photoreceptor cells generate large amounts of ATP and are highly dependent on a continuous supply of O_2. [Courtesy of Dr. Michael Hogan.]

2. *ATP is generated by the oxidation of fuel molecules such as glucose, fatty acids, and amino acids.* The common intermediate in most of these oxidations is acetyl CoA. The carbons of the acetyl unit are completely oxidized to CO_2 by the citric acid cycle with the concomitant formation of NADH and $FADH_2$. These electron carriers then transfer their high-potential electrons to the respiratory chain. The subsequent flow of electrons to O_2 leads to the pumping of protons across the inner mitochondrial membrane. This proton gradient is then used to synthesize ATP. Glycolysis also generates ATP, but the amount formed is much smaller than in oxidative phosphorylation. The oxidation of glucose to pyruvate yields only 2 ATP, whereas 30 (or 32) ATP are formed when glucose is completely oxidized to CO_2. However, glycolysis can proceed rapidly for a short time under anaerobic conditions, in contrast with oxidative phosphorylation, which requires a continuous supply of O_2.

3. *NADPH is the major electron donor in reductive biosyntheses.* In most biosyntheses, the products are more reduced than the precursors, and so reductive power is needed as well as ATP. The high-potential electrons required to drive these reactions are usually provided by NADPH. For example, in fatty acid biosynthesis, the keto group of an added two-carbon unit is reduced to a methylene group by the input of four electrons from two molecules of NADPH. The activation of O_2 by mixed-function P_{450} oxygenases in hydroxylation reactions also illustrates the pervasive role of NADPH as a reductant. The pentose phosphate pathway supplies much of the required NADPH. This reductant is also formed by the tandem operation of the mitochondrial citrate-pyruvate shuttle and the malate enzyme in the cytosol.

4. *Biomolecules are constructed from a relatively small set of building blocks.* The highly diverse molecules of life are synthesized from a much smaller number of precursors. The metabolic pathways that generate ATP and NADPH also provide building blocks for the biosynthesis of more complex molecules. For example, dihydroxyacetone phosphate formed in glycolysis gives rise to the glycerol backbone of phosphatidyl choline and other phosphoglycerides. Phosphoenolpyruvate, another glycolytic intermediate, provides part of the carbon skeleton of the aromatic amino acids. Acetyl CoA, the common intermediate in the breakdown of most fuels, supplies a two-carbon unit in a wide variety of biosyntheses. Succinyl CoA, formed in the citric acid cycle, is one of the precursors of porphyrins. Ribose 5-phosphate, which is formed in addition to NADPH by the pentose phosphate pathway, is the source of the sugar unit of nucleo-

tides. Many biosyntheses also require one-carbon units. Tetrahydrofolate is a source of these units at several oxidation levels. The formation of these derivatives and of *S*-adenosylmethionine, the major donor of methyl groups, is closely tied to amino acid metabolism. Thus, the central metabolic pathways have anabolic as well as catabolic roles.

5. *Biosynthetic and degradative pathways are almost always distinct.* For example, the pathway for the synthesis of fatty acids is different from that of their degradation. Likewise, glycogen is synthesized and degraded by different sets of reactions. This separation enables both biosynthetic and degradative pathways to be thermodynamically favorable at all times. A biosynthetic pathway is made exergonic by coupling it to the hydrolysis of a sufficient number of ATP. For example, four more ~P are spent in converting pyruvate into glucose in gluconeogenesis than are gained in converting glucose into pyruvate in glycolysis. The four additional ~P assure that gluconeogenesis, like glycolysis, is highly exergonic under all cellular conditions. *The essential point is that the rates of metabolic pathways are governed more by the activities of key enzymes than by mass action.* The separation of biosynthetic and degradative pathways contributes greatly to the effectiveness of metabolic control.

RECURRING MOTIFS IN METABOLIC REGULATION

The complex network of reactions in a cell is precisely regulated and coordinated. Metabolism is controlled in several ways:

1. *Allosteric interactions.* The flow of molecules in most metabolic pathways is determined primarily by the amounts and activities of certain enzymes rather than by the amount of substrate available. Essentially irreversible reactions are potential control sites. The first irreversible reaction in a pathway (the committed step) is nearly always tightly controlled. Enzymes catalyzing committed steps are allosterically regulated, as exemplified by phosphofructokinase in glycolysis and acetyl CoA carboxylase in fatty acid synthesis. Subsequent irreversible reactions in a pathway also may be controlled. Allosteric interactions enable such enzymes to rapidly detect diverse signals and to integrate this information.

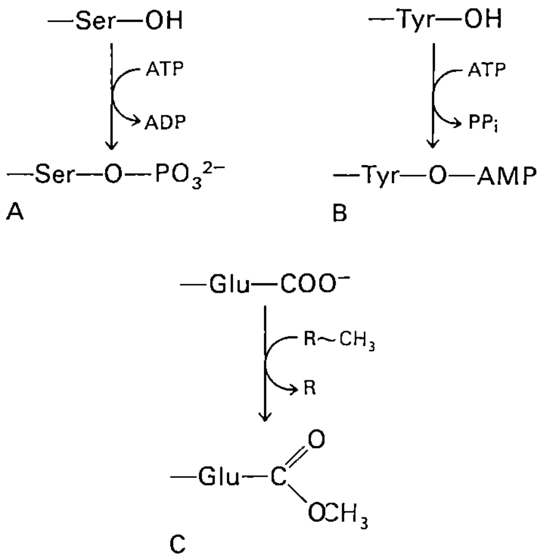

Figure 30-2
Examples of reversible covalent modifications of proteins: (A) phosphorylation, (B) adenylylation, and (C) carboxymethylation.

2. *Covalent modification.* Some regulatory enzymes are controlled by covalent modification in addition to allosteric interactions. For example, the catalytic activity of glycogen phosphorylase is enhanced by phosphorylation, whereas that of glycogen synthase is diminished. These covalent modifications are catalyzed by specific enzymes. Another example is glutamine synthetase, which is rendered less active by the covalent insertion of an AMP unit. Again, addition and removal of this modifying group are catalyzed by specific enzymes. Why is covalent modification used in addition to noncovalent allosteric control? Covalent modifications of key enzymes in metabolism are the final stage of amplifying cascades. Consequently, metabolic pathways can be rapidly switched on or off by very small triggering signals, as shown by the action of epinephrine in stimulating the breakdown of glycogen. Also, covalent modifications usually last longer (seconds to minutes) than do reversible allosteric interactions (milliseconds to seconds).

3. *Enzyme levels.* The amounts of enzymes, as well as their activities, are controlled. The rates of synthesis and degradation of many regulatory enzymes are altered by hormones.

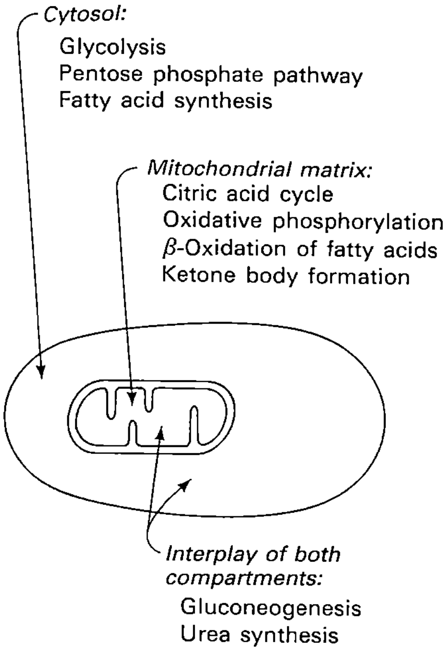

Figure 30-3
Compartmentation of the major pathways of metabolism.

4. *Compartmentation.* The metabolic patterns of eukaryotic cells are markedly affected by the presence of compartments. Glycolysis, the pentose phosphate pathway, and fatty acid synthesis take place in the cytosol, whereas fatty acid oxidation, the citric acid cycle, and oxidative phosphorylation are carried out in mitochondria. Some processes, such as gluconeogenesis and urea synthesis, depend on the interplay of reactions that occur in both compartments. The fates of certain molecules depend on whether they are in the cytosol or in mitochondria, and so their flow across the inner mitochondrial membrane is often regulated. For example, fatty acids transported into mitochondria are rapidly degraded, in contrast with fatty acids in the cytosol, which are esterified or exported. Recall that long-chain fatty acids are transported into the mitochondrial matrix as esters of carnitine, a carrier that enables these molecules to traverse the inner mitochondrial membrane.

5. *Metabolic specializations of organs.* Regulation in higher eukaryotes is profoundly affected and enhanced by the existence of organs with different metabolic roles. These interactions will be discussed shortly.

MAJOR METABOLIC PATHWAYS AND CONTROL SITES

Let us review the roles of the major pathways of metabolism and the principal sites for their control:

1. *Glycolysis.* This sequence of reactions in the cytosol converts glucose into two molecules of pyruvate with the concomitant generation of two ATP and two NADH. The NAD^+ consumed in the reaction catalyzed by glyceraldehyde 3-phosphate dehydrogenase must be regenerated for glycolysis to proceed. Under anaerobic conditions, as in highly active skeletal muscle, this is accomplished by the reduction of pyruvate to lactate. Alternatively, under aerobic conditions, NAD^+ is regenerated by the transfer of electrons from NADH to O_2 through the electron transport chain. Glycolysis serves two main purposes: it degrades glucose to generate ATP and it provides carbon skeletons for biosyntheses. The rate of conversion of glucose into pyruvate is regulated to meet these dual needs.

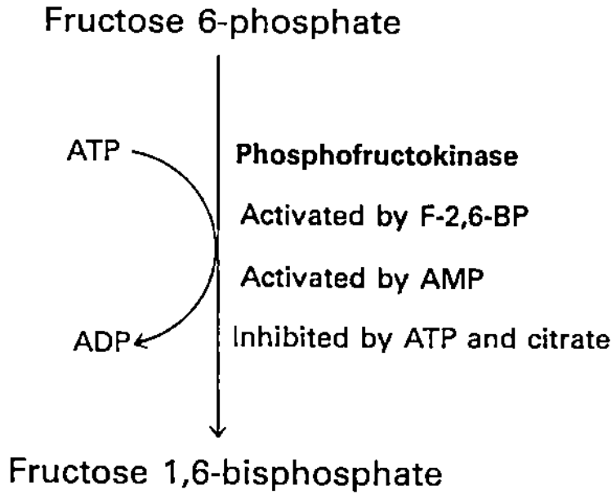

Figure 30-4
Phosphofructokinase is the key enzyme in the regulation of glycolysis.

Phosphofructokinase, which catalyzes the committed step in glycolysis, is the most important control site. A high level of ATP inhibits phosphofructokinase—the allosteric sites are distinct from the substrate sites and have a lower affinity for the nucleotide. This inhibitory effect is enhanced by citrate and reversed by AMP. Thus, the rate of glycolysis depends on the need for ATP, as signaled by the ATP/AMP ratio, and on the need for building blocks, as signaled by the level of citrate. *In liver, the most important regulator of phosphofructokinase activity is fructose 2,6-bisphosphate (F-2,6-BP).* Recall that the level of F-2,6-BP is determined by the activity of the kinase that forms it from fructose 6-phosphate and of the phosphatase that hydrolyzes the 2-phosphoryl group (see Figure 19-8 on p. 495). When the blood glucose level is low, a glucagon-triggered cascade leads to activation of this phosphatase and inhibition of this kinase in liver. The resulting decrease in the level of F-2,6-BP leads to the deactivation of phosphofructokinase, and hence glycolysis is slowed.

It is important to note that phosphofructokinase is controlled rather differently in muscle. The kinase catalyzing the synthesis of F-2,6-BP is stimulated, not inhibited, by cAMP-induced phosphorylation. Thus, epinephrine stimulates glycolysis in muscle but inhibits it in liver. The increased breakdown of liver glycogen induced by epinephrine serves to supply glucose to muscle, which rapidly consumes it to generate ATP for contractile activity.

Pasteur effect—
The inhibition of glycolysis by respiration, discovered by Louis Pasteur in studying fermentation by yeast. The consumption of carbohydrate is about sevenfold lower under aerobic conditions than under anaerobic ones. The inhibition of phosphofructokinase by citrate and ATP accounts for much of the Pasteur effect.

2. *Citric acid cycle.* This common pathway for the oxidation of fuel molecules—carbohydrates, amino acids, and fatty acids—takes place inside mitochondria. Most fuels enter the cycle as acetyl CoA. The complete oxidation of an acetyl unit generates one GTP, three NADH, and one $FADH_2$. The four pairs of electrons are then transferred to O_2 through the electron transport chain, which results in the formation of a proton gradient that drives the synthesis of nine ATP. NADH and $FADH_2$ are oxidized only if ADP is simultaneously phosphorylated to ATP. *This tight coupling, called respiratory control, ensures that the rate of the citric acid cycle matches the need for ATP.* An abundance of ATP also diminishes the activities of three enzymes in the cycle—citrate synthase, isocitrate dehydrogenase, and α-ketoglutarate dehydrogenase. The citric acid cycle also has an anabolic role. In concert with pyruvate carboxylase, the citric acid cycle provides intermediates for biosyntheses, such as succinyl CoA for the formation of porphyrins.

3. *Pentose phosphate pathway.* This series of reactions, which takes place in the cytosol, has two purposes: the generation of NADPH for reductive biosyntheses and the formation of ribose 5-phosphate for the synthesis of nucleotides. Two NADPH are generated in the conversion of glucose 6-phosphate into ribose 5-phosphate. The dehydrogenation of glucose 6-phosphate is the committed step in this pathway. This reaction is controlled by the level of $NADP^+$, the electron acceptor. The extra phosphoryl group in NADPH is a tag that distinguishes it from NADH. This differentiation makes it possible to have both a high $NADPH/NADP^+$ ratio and a high $NAD^+/NADH$ ratio in the same compartment. Consequently, reductive biosyntheses and glycolysis can proceed simultaneously at a high rate.

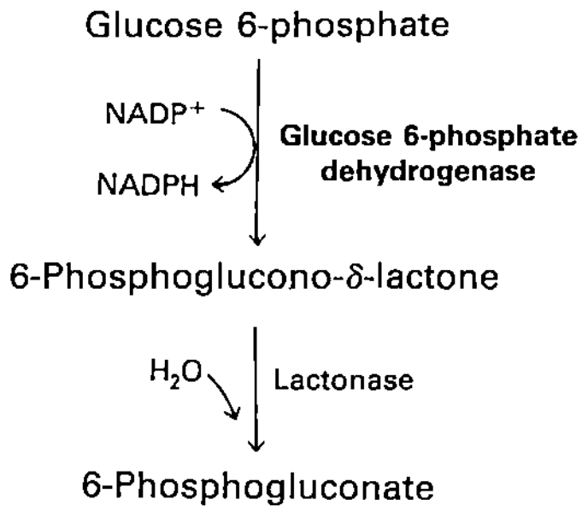

Figure 30-5
The dehydrogenation of glucose 6-phosphate is the committed step in the pentose phosphate pathway.

4. *Gluconeogenesis.* Glucose can be synthesized by the liver and kidneys from noncarbohydrate precursors such as lactate, glycerol, and amino acids. The major entry point of this pathway is pyruvate, which is carboxylated to oxaloacetate in mitochondria. Oxaloacetate is then decarboxylated and phosphorylated in the cytosol to form phosphoenolpyruvate. The other distinctive reactions of gluconeogenesis are two hydrolytic steps that bypass the irreversible reactions of glycolysis. *Gluconeogenesis and glycolysis are usually reciprocally regulated so that one pathway is quiescent while the other is highly active.* For example, AMP inhibits and citrate activates fructose 1,6-bisphosphatase, a key enzyme in gluconeogenesis, whereas these molecules have opposite effects on phosphofructokinase, the pacemaker of glycolysis. F-2,6-BP also coordinates these processes by inhibiting fructose 1,6-bisphosphatase. Hence, when glucose is abundant, the high level of F-2,6-BP inhibits gluconeogenesis and activates glycolysis.

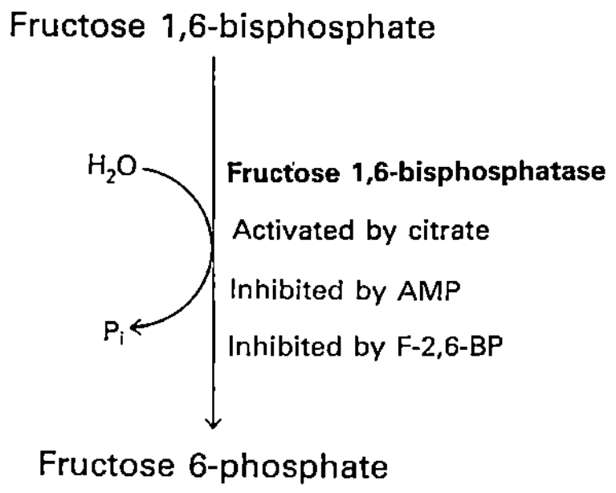

Figure 30-6
Fructose 1,6-bisphosphatase is the key control site in gluconeogenesis.

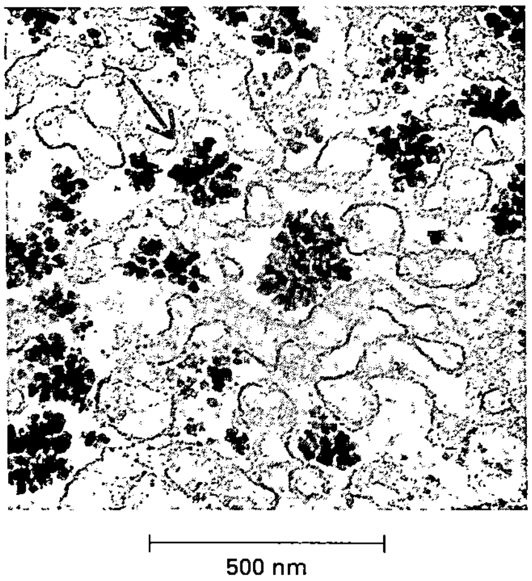

Figure 30-7
Electron micrograph of a portion of a liver cell showing glycogen particles. [Courtesy of Dr. George Palade.]

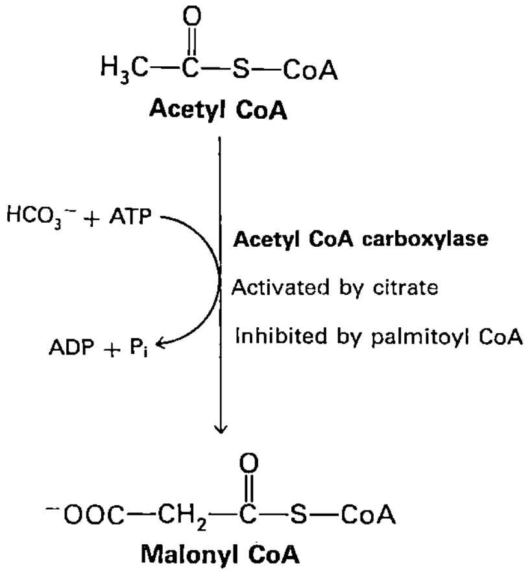

Figure 30-8
Acetyl CoA carboxylase is the key control site in fatty acid synthesis.

5. *Glycogen synthesis and degradation.* Glycogen, a readily mobilizable fuel store, is a branched polymer of glucose residues. The activated intermediate in its synthesis is UDP–glucose, which is formed from glucose 1-phosphate and UTP. Glycogen synthase catalyzes the transfer of glucose from UDP–glucose to the terminal hydroxyl residue of a growing strand. Glycogen is degraded by a different pathway. Phosphorylase catalyzes the cleavage of glycogen by orthophosphate to give glucose 1-phosphate. *Glycogen synthesis and degradation are coordinately controlled by a hormone-triggered amplifying cascade so that the synthase is inactive when phosphorylase is active, and vice versa.* These enzymes are regulated by phosphorylation and noncovalent allosteric interactions (p. 595).

6. *Fatty acid synthesis and degradation.* Fatty acids are synthesized in the cytosol by the addition of two-carbon units to a growing chain on an acyl carrier protein. Malonyl CoA, the activated intermediate, is formed by the carboxylation of acetyl CoA. Acetyl groups are carried from mitochondria to the cytosol by the citrate-malate shuttle. *Citrate in the cytosol stimulates acetyl CoA carboxylase, the enzyme catalyzing the committed step. When ATP and acetyl CoA are abundant, the level of citrate increases, which accelerates the rate of fatty acid synthesis.* Fatty acids are degraded by a different pathway in a different compartment. They are degraded to acetyl CoA in the mitochondrial matrix by β-oxidation. The acetyl CoA then enters the citric acid cycle if the supply of oxaloacetate is sufficient. Alternatively, acetyl CoA can give rise to ketone bodies. The $FADH_2$ and NADH formed in the β-oxidation pathway transfer their electrons to O_2 through the electron transport chain. Like the citric acid cycle, β-oxidation can continue only if NAD^+ and FAD are regenerated. Hence, *the rate of fatty acid degradation is also coupled to the need for ATP.*

KEY JUNCTIONS: GLUCOSE 6-PHOSPHATE, PYRUVATE, AND ACETYL CoA

The factors governing the flow of molecules in metabolism can be further understood by examining three key crossroads: glucose 6-phosphate, pyruvate, and acetyl CoA. Each of these molecules has several contrasting fates:

1. *Glucose 6-phosphate.* Glucose entering a cell is rapidly phosphorylated to glucose 6-phosphate, which can be stored as glycogen, degraded by way of pyruvate, or converted into ribose 5-phosphate (Figure 30-9). Glycogen is formed when glucose 6-phosphate and ATP are abundant. In contrast, glucose 6-phosphate flows into the glycolytic pathway when ATP or carbon skeletons for biosyntheses are required. Thus, the conversion

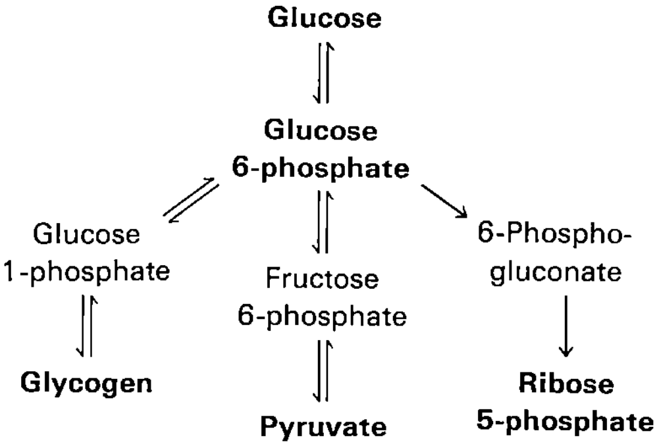

Figure 30-9
Metabolic fates of glucose 6-phosphate.

of glucose 6-phosphate into pyruvate can be anabolic as well as catabolic. The third major fate of glucose 6-phosphate, to flow through the pentose phosphate pathway, provides NADPH for reductive biosyntheses and ribose 5-phosphate for the synthesis of nucleotides. The relative amounts of these two outputs can be adjusted over a very broad range by this highly flexible series of reactions, as discussed previously (p. 563). Glucose 6-phosphate can be formed by the mobilization of glycogen, or it can be synthesized from pyruvate and glycogenic amino acids by the gluconeogenic pathway. As will be discussed shortly, a low level of glucose in the blood stimulates both glycogenolysis and gluconeogenesis in the *liver and kidneys. These organs are distinctive in possessing glucose 6-phosphatase, which enables glucose to be released into the blood.*

2. *Pyruvate.* This three-carbon α-keto acid is another major metabolic junction (Figure 30-10). Pyruvate is derived primarily from glucose 6-phosphate, alanine, and lactate (the reduced form of pyruvate). The facile reduction of pyruvate catalyzed by lactate dehydrogenase serves to regenerate NAD^+, enabling glycolysis to proceed transiently under anaerobic conditions. Lactate formed in active tissues such as contracting muscle is subsequently oxidized back to pyruvate, primarily in the liver. The essence of this interconversion is that it buys time and shifts part of the metabolic burden of active muscle to the liver. Another readily reversible reaction in the cytosol is the transamination of pyruvate, an α-keto acid, to alanine, the corresponding amino acid. Conversely, several amino acids can be converted into pyruvate. Thus, transamination is a major link between amino acid and carbohydrate metabolism.

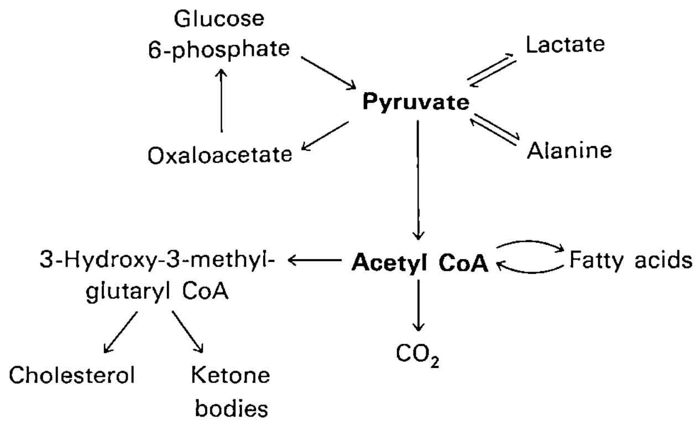

Figure 30-10
Major metabolic fates of pyruvate and acetyl CoA in mammals.

A third fate of pyruvate is its carboxylation to oxaloacetate inside mitochondria. This reaction and the subsequent conversion of oxaloacetate into phosphoenolpyruvate bypass an irreversible step of glycolysis and hence enable glucose to be synthesized from pyruvate. The carboxylation of pyruvate is also important for replenishing intermediates of the citric acid cycle. The activation of pyruvate carboxylase by acetyl CoA enhances the synthesis of oxaloacetate when the citric acid cycle is slowed by a paucity of this intermediate. On the other hand, oxaloacetate synthesized from pyruvate flows into the gluconeogenic pathway when the citric acid cycle is inhibited by an abundance of ATP.

A fourth fate of pyruvate is its oxidative decarboxylation to acetyl CoA. This irreversible reaction inside mitochondria is a decisive reaction in metabolism: it commits the carbon atoms of carbohydrates and amino acids to oxidation by the citric acid cycle or to the synthesis of lipids. The pyruvate dehydrogenase complex, which catalyzes this irreversible funneling, is stringently regulated by multiple allosteric interactions and covalent modifications. Pyruvate is rapidly converted into acetyl CoA only if ATP is needed or if two-carbon fragments are required for the synthesis of lipids.

3. *Acetyl CoA.* The major sources of this activated two-carbon unit are the oxidative decarboxylation of pyruvate and the β-oxidation of fatty acids (see Figure 30-10). Acetyl CoA is also derived from ketogenic amino acids. The fate of acetyl CoA, in contrast with many molecules in metabolism, is quite restricted. The acetyl unit can be completely oxidized to CO_2 by the citric acid cycle. Alternatively, 3-hydroxy-3-methylglutaryl CoA can be formed from three molecules of acetyl CoA. This six-carbon unit is a precursor of cholesterol and of ketone bodies, which are transport forms of acetyl units between liver and some peripheral tissues. A third major fate of acetyl CoA is its export to the cytosol in the form of citrate for the synthesis of fatty acids. It is important to stress again that acetyl CoA cannot be converted into pyruvate in mammals. Consequently, mammals are unable to convert lipid into carbohydrate.

METABOLIC PROFILES OF THE MAJOR ORGANS

The metabolic patterns of the brain, muscle, adipose tissue, and the liver are strikingly different. Let us consider how these organs differ in their use of fuels to meet their energy needs:

1. *Brain. Glucose is virtually the sole fuel for the human brain, except during prolonged starvation.* The brain lacks fuel stores and hence requires a continuous supply of glucose, which enters freely at all times. It consumes about 120 g daily, which corresponds to an energy input of about 420 kcal. The brain accounts for some 60% of the utilization of glucose by the whole body in the resting state. Noninvasive ^{13}C nuclear magnetic resonance measurements have shown that the concentration of glucose in brain is about 1 mM when the plasma level is 4.7 mM (84.7 mg/dl), a normal value (Figure 30-11). Glycolysis slows down when the glucose level approaches the K_M of hexokinase (~50 μM). This danger point occurs when the plasma glucose level drops below about 2.2 mM (39.6 mg/dl).

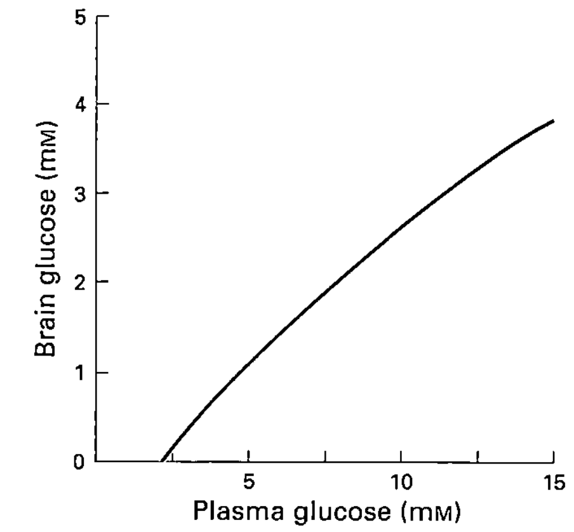

Figure 30-11
Dependence of the brain glucose level on the plasma glucose level. Glucose labeled with ^{13}C at C-1 was infused into six subjects. The concentration of glucose in the intact brain was measured by spatially localized ^{13}C nuclear magnetic resonance spectroscopy. [After R. Gruetter, E.J. Novotny, S.D. Boulware, D.L. Rothman, G.F. Mason, G.I. Shulman, R.G. Shulman, and W.V. Tamborlane. *Proc. Nat. Acad. Sci.* 89(1992):1109.]

During starvation, ketone bodies (acetoacetate and 3-hydroxybutyrate, its reduced counterpart) generated by the liver partly replace glucose as fuel for the brain. Acetoacetate is activated by the transfer of CoA from succinyl CoA to give acetoacetyl CoA (Figure 30-12). Cleavage by thiolase then yields two molecules of acetyl CoA, which enter the citric acid cycle. Fatty acids do not serve as fuel for the brain, because they are bound to albumin in plasma, and so they do not traverse the blood-brain barrier. In essence, *ketone bodies are transportable equivalents of fatty acids.* As will be discussed shortly, the shift in fuel usage from glucose to ketone bodies is critical in minimizing protein breakdown during starvation.

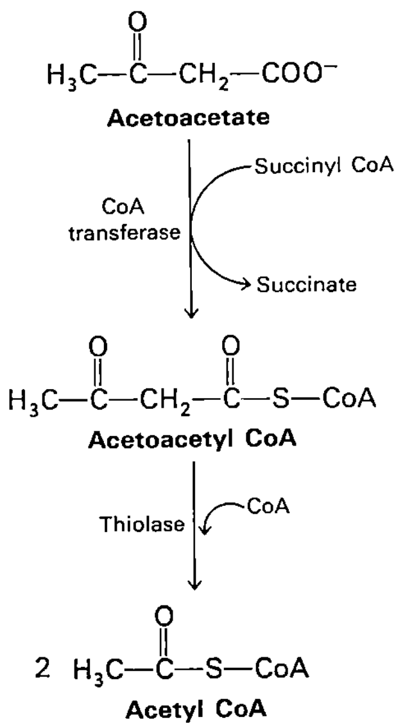

Figure 30-12
Entry of ketone bodies into the citric acid cycle.

2. *Muscle. The major fuels for muscle are glucose, fatty acids, and ketone bodies.* Muscle differs from brain in having a large store of glycogen (1200 kcal). In fact, about three-fourths of all the glycogen in the body is stored in muscle (Table 30-1). The glycogen content of muscle after a meal can be as high as 1%. This glycogen is readily converted into glucose 6-phosphate for use within muscle cells. Muscle, like the brain, lacks glucose 6-phosphatase, and so it does not export glucose. Rather, *muscle retains glucose, its preferred fuel for bursts of activity.*

Table 30-1
Fuel reserves in a typical 70-kg man

Organ	Available energy (kcal)		
	Glucose or glycogen	*Triacylglycerols*	*Mobilizable proteins*
Blood	60	45	0
Liver	400	450	400
Brain	8	0	0
Muscle	1200	450	24,000
Adipose tissue	80	135,000	40

After G.F. Cahill, Jr. *Clin. Endocrinol. Metab.* 5(1976):398.

In actively contracting skeletal muscle, the rate of glycolysis far exceeds that of the citric acid cycle. Much of the pyruvate formed under these conditions is reduced to lactate, which flows to the liver, where it is converted into glucose (Figure 30-13). These interchanges, known as the *Cori cycle* (p. 577), shift part of the metabolic burden of muscle to the liver. In addition, a large amount of alanine is formed in active muscle by the transamination of pyruvate. Alanine, like lactate, can be converted into glucose by the liver.

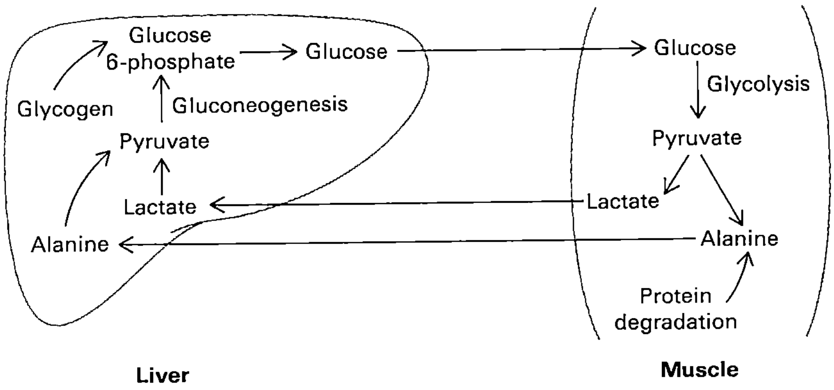

Figure 30-13
Metabolic interchanges between muscle and the liver.

The metabolic pattern of resting muscle is quite different. *In resting muscle, fatty acids are the major fuel.* Ketone bodies can also serve as fuel for heart muscle. In fact, heart muscle consumes acetoacetate in preference to glucose.

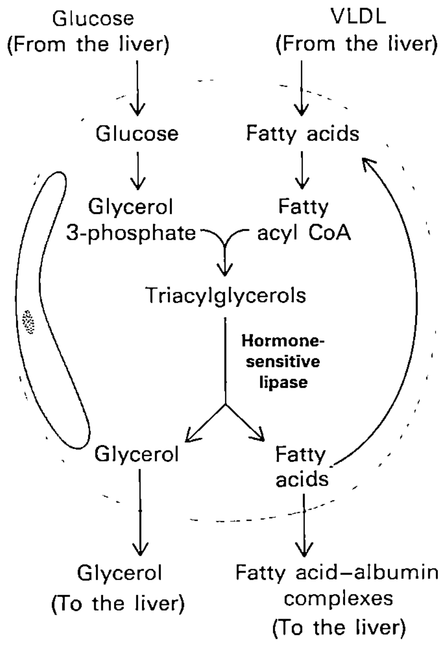

Figure 30-14
Synthesis and degradation of triacylglycerols by adipose tissue. Fatty acids are delivered to adipose cells in the form of triacylglycerols contained in very low density lipoproteins (VLDL).

3. *Adipose tissue. The triacylglycerols stored in adipose tissue are an enormous reservoir of metabolic fuel.* Their energy content is 135,000 kcal in a typical 70-kg man. Adipose tissue is specialized for the esterification of fatty acids and for their release from triacylglycerols. In humans, the liver is the major site of fatty acid synthesis, whereas the principal biosynthetic task of adipose tissue is to activate these fatty acids and transfer the resulting CoA derivatives to glycerol. Glycerol 3-phosphate, a key intermediate in this biosynthesis (p. 686), comes from the reduction of dihydroxyacetone phosphate, which is formed from glucose by the glycolytic pathway. Adipose cells are unable to phosphorylate endogeneous glycerol because they lack the kinase. Hence, *adipose cells need glucose for the synthesis of triacylglycerols.*

Triacylglycerols are hydrolyzed to fatty acids and glycerol by lipases. The release of the first fatty acid from a triacylglycerol, the rate-limiting step, is catalyzed by a hormone-sensitive lipase that is reversibly phosphorylated. Cyclic AMP, the intracellular messenger in this hormone-triggered amplifying cascade, activates a protein kinase—a recurring theme in hormone action (p. 606). Triacylglycerols in adipose cells are continually being hydrolyzed and resynthesized. Glycerol derived from their hydrolysis is exported to the liver. Most of the fatty acids formed on hydrolysis are reesterified if glycerol 3-phosphate is abundant. In contrast, they are released into the plasma if glycerol 3-phosphate is scarce because of a paucity of glucose. Thus, the glucose level inside adipose cells is a major factor in determining whether fatty acids are released into the blood.

4. *Liver. The metabolic activities of the liver are essential for providing fuel to the brain, muscle, and other peripheral organs.* Most compounds absorbed by the intestine pass through the liver, which enables it to regulate the level of many metabolites in the blood. The liver can take up large amounts of glucose and convert it into glycogen. As much as 400 kcal can be stored in this way. The liver can release glucose into the blood by breaking down its store of glycogen and by carrying out gluconeogenesis. The main precursors of glucose are lactate and alanine from muscle, glycerol from adipose tissue, and glucogenic amino acids from the diet.

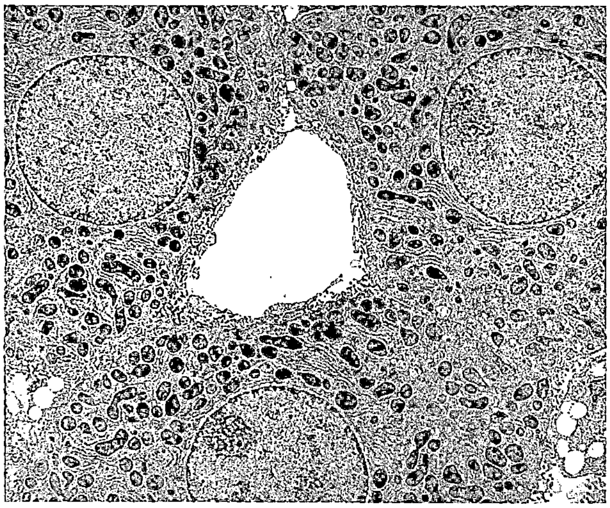

Figure 30-15
Electron micrograph of liver cells. The liver plays a key role in the integration of metabolism. [Courtesy of Dr. Ann Hubbard.]

The liver also plays a central role in the regulation of lipid metabolism. When fuels are abundant, fatty acids derived from the diet or synthesized by the liver are esterified and secreted into the blood in the form of very low density lipoprotein (VLDL) (Figure 30-16). This plasma lipoprotein is the main source of fatty acids used by adipose tissue to synthesize triacylglycerols. However, in the fasting state, the liver converts fatty acids into ketone bodies. How does a liver cell choose between these contrasting paths? The selection is made according to whether the fatty acid enters the mitochondrial matrix. Recall that long-chain fatty acids traverse the inner mitochondrial membrane only if they are esterified to carnitine (p. 608). Carnitine acyltransferase I, which catalyzes the formation of acyl carnitine, is inhibited by malonyl CoA, the committed intermediate in the synthesis of fatty acids. Thus, *when malonyl CoA is abundant, long-chain fatty acids are prevented from entering the mitochondrial matrix, the compartment of β-oxidation and ketone body formation. Instead, fatty acids are exported to adipose tissue for incorporation into triacylglycerols.* In contrast, the level of malonyl CoA is low when fuels are scarce. Under these conditions, fatty acids liberated from adipose tissues enter the mitochondrial matrix for conversion into ketone bodies.

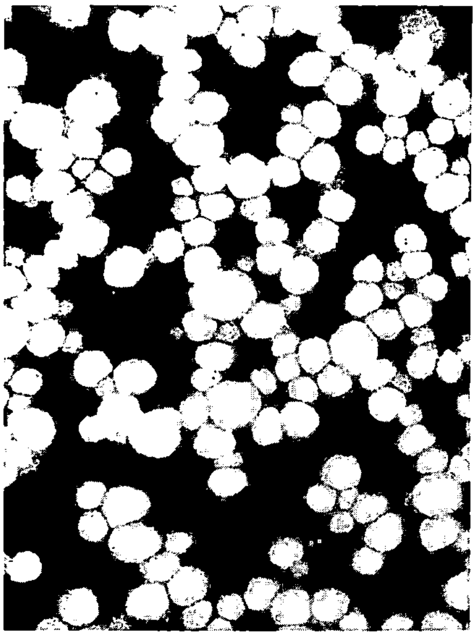

Figure 30-16
Electron micrograph of very low density lipoprotein (VLDL) particles. These particles, which have diameters ranging between 30 and 80 nm, carry triacylglycerols from the liver to adipose tissue and other peripheral tissues. [Courtesy of Dr. Robert Mahley.]

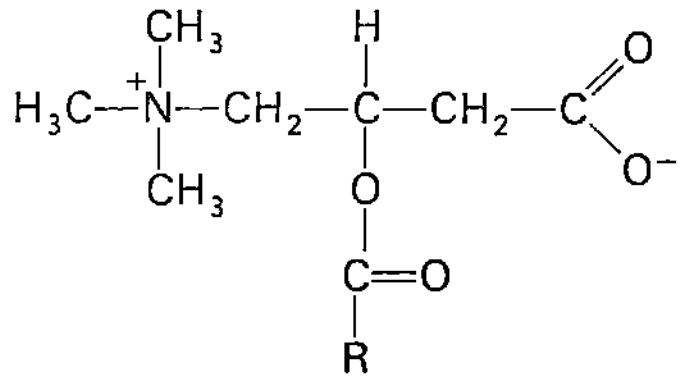

Fatty acyl carnitine

How does the liver meet its own energy needs? Keto acids derived from the degradation of amino acids are preferred to glucose as the liver's own fuel. In fact, the main purpose of glycolysis in the liver is to form building blocks for biosyntheses. Furthermore, the liver cannot use acetoacetate as a fuel because it has little of the transferase needed for its activation to acetyl CoA. Thus, the liver eschews the fuels it exports to muscle and the brain—an altruistic organ indeed!

HORMONAL REGULATORS OF FUEL METABOLISM

Hormones play a key role in integrating metabolism. In particular, *insulin, glucagon, epinephrine, and norepinephrine* have large effects on the storage and mobilization of fuels and on related facets of metabolism:

1. *Insulin.* This 6-kd protein (p. 25) and glucagon are the most important hormonal regulators of fuel metabolism. The secretion of insulin by the β cells of the pancreas is stimulated by glucose and the parasympathetic nervous system. *In essence, insulin signals the fed state: it stimulates the storage of fuels and the synthesis of proteins in a variety of ways.* Insulin acts on protein kinase cascades. It stimulates glycogen synthesis in both muscle and liver and suppresses gluconeogenesis by the liver. Insulin also accelerates glycolysis in the liver, which in turn increases the synthesis of fatty acids. The entry of glucose into muscle and adipose cells is promoted by insulin. The abundance of fatty acids and glucose in adipose tissue results in the synthesis and storage of triacylglycerols. The action of insulin also extends to amino acid and protein metabolism. Insulin promotes the uptake of branched-chain amino acids (valine, leucine, and isoleucine) by

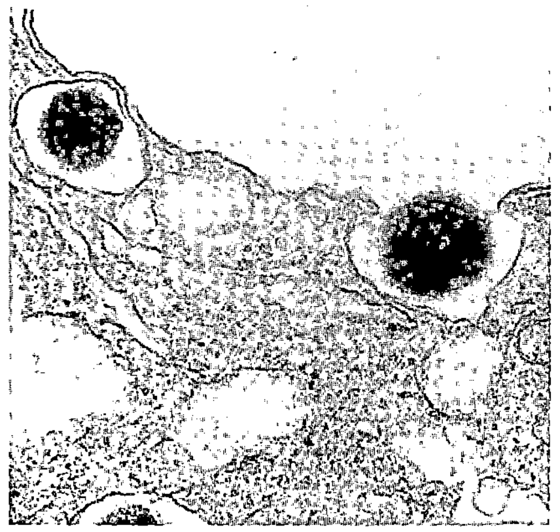

Figure 30-17
Electron micrograph showing the release of insulin from a pancreatic B cell. One secretory granule is on the verge of fusing with the plasma membrane and releasing insulin into the extracellular space, and the other has already released the hormone. [Courtesy of Dr. Lelio Orci. L. Orci, J.-D. Vassalli, and A. Perrelet. *Sci. Amer.* 259(September 1988):85–94.]

muscle, which favors a building up of muscle protein. Indeed, insulin has a general stimulating effect on protein synthesis. In addition, it inhibits the intracellular degradation of proteins.

2. *Glucagon.* This 3.5-kd polypeptide hormone (p. 594) is secreted by the α cells of the pancreas in response to a *low blood sugar level in the fasting state. The main target organ of glucagon is the liver.* Glucagon stimulates glycogen breakdown and inhibits glycogen synthesis by triggering the cAMP cascade leading to the phosphorylation of phosphorylase and glycogen synthase (p. 595). Glucagon also inhibits fatty acid synthesis by diminishing the production of pyruvate and by lowering the activity of acetyl CoA carboxylase. In addition, glucagon stimulates gluconeogenesis and blocks glycolysis by lowering the level of F-2,6-BP. All known actions of glucagon are mediated by protein kinases that are activated by cyclic AMP. The net result of these actions of glucagon is to *markedly increase the release of glucose by the liver.* Likewise, glucagon raises the level of cyclic AMP in adipose cells, which activates a lipase that mobilizes triacylgylcerols.

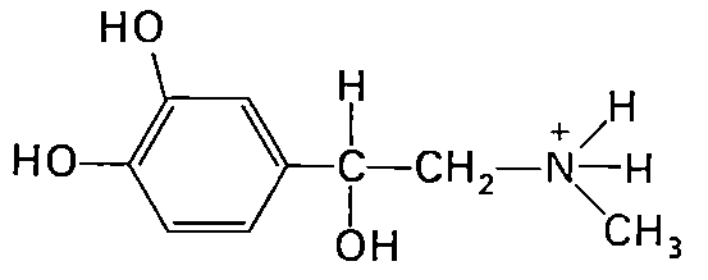

Epinephrine

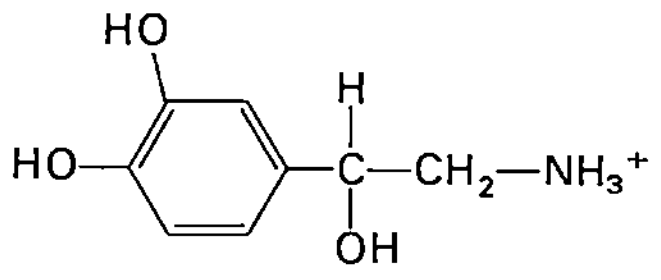

Norepinephrine

3. *Epinephrine and norepinephrine.* These catecholamine hormones are secreted by the adrenal medulla and sympathetic nerve endings in response to a *low blood glucose level.* Like glucagon, they stimulate the mobilization of glycogen and triacylglycerols by triggering the cAMP cascade. They differ from glucagon in that their glycogenolytic effect is greater in muscle than in the liver. Another action of the catecholamines is to inhibit the uptake of glucose by muscle. Instead, fatty acids released from adipose tissue are used as fuel. Epinephrine also stimulates the secretion of glucagon and inhibits the secretion of insulin. Thus, *the catecholamines increase the amount of glucose released into the blood by the liver and decrease the utilization of glucose by muscle.*

THE BLOOD GLUCOSE LEVEL IS BUFFERED BY THE LIVER

The blood glucose level in a typical person after an overnight fast is 80 mg/dl (80 mg/100 ml, or 4.4 mM). The blood glucose level during the day normally ranges from about 80 mg/dl before meals to about 120 mg/dl after meals. How is blood glucose maintained at a quite constant level despite large changes in the input and utilization of glucose? The major control elements have already been discussed, and so they need only be brought together here. *The blood glucose level is controlled primarily by the liver, which can take up or release large amounts of glucose in response to hormonal signals and the level of glucose itself* (Figure 30-18). After a

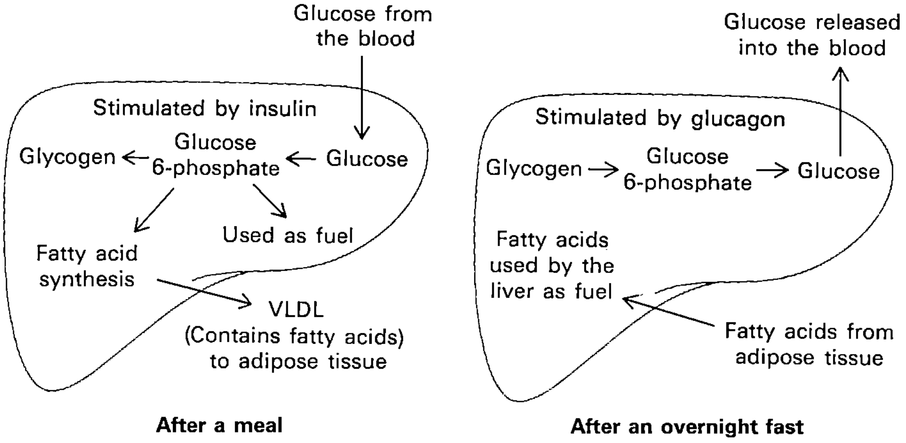

Figure 30-18
Control of the blood glucose level by the liver.

carbohydrate-containing meal, the markedly increased concentration of glucose in the blood leads to a higher level of glucose 6-phosphate in the liver because only then do the catalytic sites of glucokinase become filled with glucose. Recall that glucokinase, in contrast with hexokinase, has a high K_M for glucose (~10 mM, compared with a fasting blood glucose level of 4.4 mM) and is not inhibited by glucose 6-phosphate. Consequently, *glucose 6-phosphate is formed more rapidly by the liver as the blood glucose level rises.*

The fate of glucose 6-phosphate is then largely controlled by the opposing effects of glucagon and insulin. Glucagon triggers the cAMP cascade that results in the breakdown of glycogen (p. 594), whereas insulin antagonizes this action. *A high blood glucose level leads to a decreased secretion of glucagon and an increased secretion of insulin by the pancreas. Consequently, glycogen is rapidly synthesized when the blood glucose level is high.* These hormonal effects on glycogen synthesis and storage are reinforced by a direct action of glucose itself. As was discussed in an earlier chapter (p. 597), *phosphorylase* a *is a glucose sensor in addition to being the enzyme that cleaves glycogen.* When the glucose level is high, the binding of glucose to phosphorylase *a* renders it susceptible to the action of a phosphatase that converts it into phosphorylase *b*, which does not degrade glycogen. This conversion also releases the phosphatase, which enables it to activate glycogen synthase. Thus, *glucose allosterically shifts the glycogen system from a degradative to a synthetic mode.*

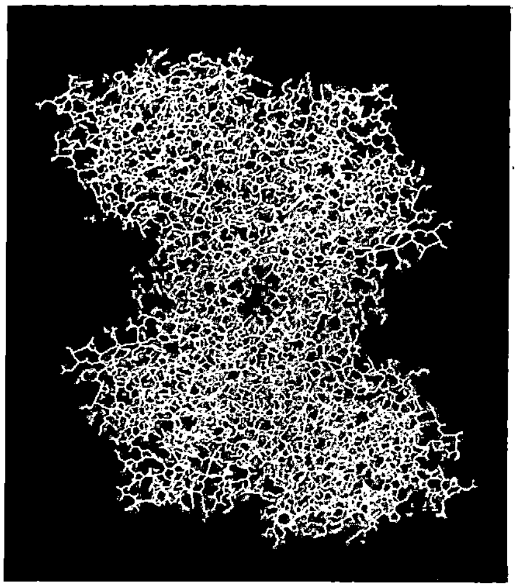

Figure 30-19
Model of phosphorylase *a*, a key enzyme in the regulation of the blood glucose level. [Courtesy of Dr. Robert Fletterick.]

The high insulin level in the fed state also promotes *the entry of glucose into muscle and adipose tissue.* The synthesis of glycogen by muscle as well as by liver is stimulated by insulin. In fact, muscle can store about three times as much glycogen as the liver because of its larger mass. The entry of glucose into adipose tissue provides glycerol 3-phosphate for the synthesis of triacylglycerols.

The blood glucose level begins to drop several hours after a meal, which leads to decreased insulin secretion and increased glucagon secretion. The effects just described are then reversed. The activation of the cAMP cascade results in a higher level of phosphorylase *a* and a lower level of glycogen synthase *a*. The effect of hormones on this cascade is reinforced by the diminished binding of glucose to phosphorylase *a*, which makes it less susceptible to the hydrolytic action of the phosphatase. Instead, the phosphatase remains bound to phosphorylase *a*, and so the synthase stays in the inactive phosphorylated form. Consequently, there is a rapid mobilization of glycogen. The large amount of glucose formed by the hydrolysis of glucose 6-phosphate derived from glycogen is then released from the liver into the blood. The diminished utilization of glucose by muscle and adipose tissue also contributes to the maintenance of the blood glucose level. The entry of glucose into muscle and adipose tissue decreases because of the low insulin level. Both muscle and liver use fatty acids as fuel when the blood glucose level drops. Thus, *the blood glucose level is kept at or above 80 mg/dl by three major factors: the mobilization of glycogen and release of glucose by the liver, the release of fatty acids by adipose tissue, and the shift in the fuel used from glucose to fatty acids by muscle and the liver.*

METABOLIC ADAPTATIONS IN PROLONGED STARVATION MINIMIZE PROTEIN DEGRADATION

We turn now to metabolic adaptations in prolonged starvation. A typical well-nourished 70-kg man has fuel reserves of some 1600 kcal in glycogen, 24,000 kcal in mobilizable protein, and 135,000 kcal in triacylglycerols (see Table 30-1). The energy need for a 24-hour period ranges from about 1600 kcal in the basal state to 6000 kcal, depending on the extent of activity. Thus, stored fuels suffice to meet caloric needs in starvation for one to three months. However, the carbohydrate reserves are exhausted in only a day. Even under these conditions the blood glucose level is maintained above 2.2 mM (40 mg/dl). The brain cannot tolerate appreciably lower glucose levels for even short periods. Hence, *the first priority of metabolism in starvation is to provide sufficient glucose to the brain and other tissues (such as red blood cells) that are absolutely dependent on this fuel.* However, precursors of glucose are not abundant. Most of the energy is stored in the fatty acyl moieties of triacylglycerols. Recall that fatty acids cannot be converted into glucose because acetyl CoA cannot be transformed into pyruvate (p. 613). The glycerol moiety of triacylglycerols can be converted into glucose, but only a limited amount is available. The only other potential source of glucose is amino acids derived from the breakdown of proteins. Muscle is the largest potential source of amino acids during starvation. However, survival for most animals depends on being able to move rapidly, which requires a large muscle mass. Thus, *the second priority of metabolism in starvation is to preserve protein. This is accomplished by shifting the fuel being used from glucose to fatty acids and ketone bodies* (Figure 30-20).

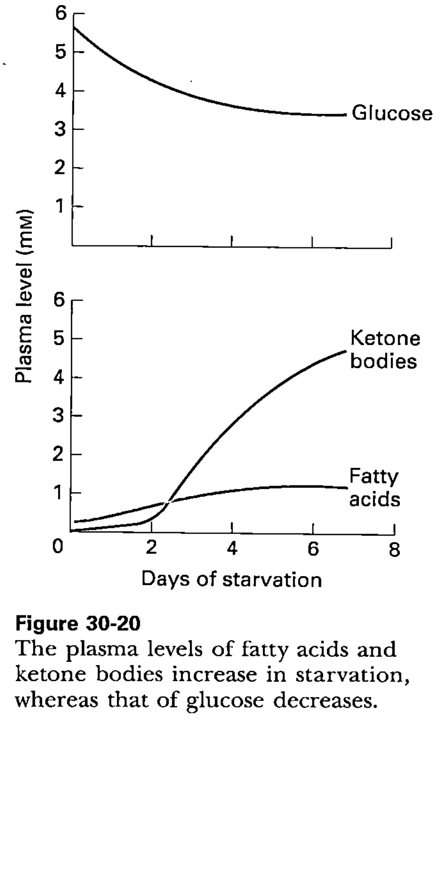

Figure 30-20
The plasma levels of fatty acids and ketone bodies increase in starvation, whereas that of glucose decreases.

The metabolic changes during the first day of starvation are like those after an overnight fast. The low blood sugar level leads to decreased secretion of insulin and increased secretion of glucagon. *The dominant metabolic processes are the mobilization of triacylglycerols in adipose tissue and gluconeogenesis by the liver. The liver obtains energy for its own needs by oxidizing fatty acids released from adipose tissue.* The concentrations of acetyl CoA and citrate consequently increase, which switches off glycolysis. The uptake of glucose by muscle is markedly diminished because of the low insulin level, whereas fatty acids enter freely. Consequently, *muscle shifts almost entirely from glucose to fatty acids for fuel.* The β-oxidation of fatty acids by muscle halts the conversion of pyruvate into acetyl CoA. Recall that acetyl CoA stimulates phosphorylation of the pyruvate dehydrogenase complex, which renders it inactive (p. 524). Hence, pyruvate, lactate, and alanine are exported to the liver for conversion into glucose. Proteolysis of muscle protein provides some of these three-carbon precursors of glucose. Glyc-

Table 30-2
Fuel metabolism in starvation

Fuel exchanges and consumption	*Amount formed or consumed in 24 hours (grams)*	
	3rd day	*40th day*
Fuel use by the brain		
Glucose	100	40
Ketone bodies	50	100
All other use of glucose	50	40
Fuel mobilization		
Adipose-tissue lipolysis	180	180
Muscle-protein degradation	75	20
Fuel output of the liver		
Glucose	150	80
Ketone bodies	150	150

erol derived from the cleavage of triacylglycerols is another raw material for the synthesis of glucose by the liver.

The most important change after about three days of starvation is that large amounts of acetoacetate and 3-hydroxybutyrate (ketone bodies) are formed by the liver (Figure 30-21). Their synthesis from acetyl CoA increases markedly because the citric acid cycle is unable to oxidize all the acetyl units generated by the degradation of fatty acids. Gluconeogenesis depletes the supply of oxaloacetate, which is essential for the entry of acetyl CoA into the citric acid cycle. Consequently, the liver produces large quantities of ketone bodies, which are released into the blood. At this time, the brain begins to consume appreciable amounts of acetoacetate in place of glucose. After three days of starvation, about a third of the energy needs of the brain are met by ketone bodies (Table 30-2). The heart also uses ketone bodies as fuel. These changes in fuel usage are accompanied by a rise in the level of ketone bodies in the plasma.

After several weeks of starvation, ketone bodies become the major fuel of the brain. Only 40 g of glucose is needed per day for the brain, compared with about 120 g in the first day of starvation. *The effective conversion of fatty acids into ketone bodies by the liver and their use by the brain markedly diminishes the need for glucose. Hence, less muscle is degraded than in the first days of starvation.* The breakdown of 20 g of muscle compared with 75 g early in starvation is most important for survival. The duration of starvation compatible with life is mainly determined by the size of the triacylglycerol depot.

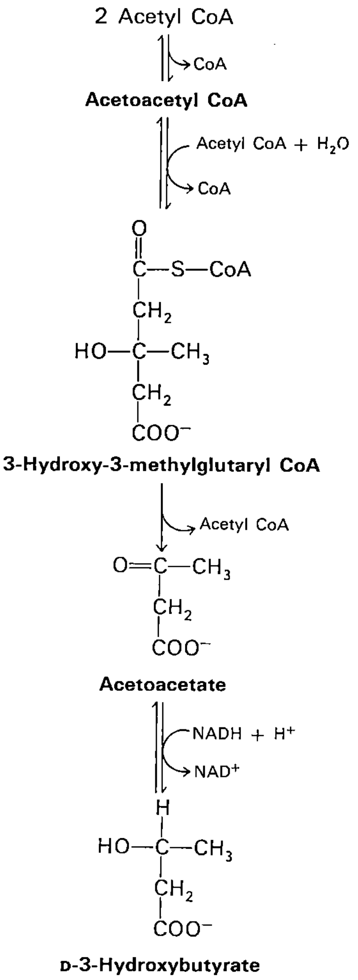

Figure 30-21
Synthesis of ketone bodies by the liver.

HUGE FAT DEPOTS ENABLE MIGRATORY BIRDS TO FLY LONG DISTANCES

Migratory birds provide a striking illustration of the biological value of triacylglycerols. Some small land birds fly in the autumn from their breeding grounds in New England to their wintering grounds in the West Indies and then back again in the spring. They fly nonstop over water for a distance of some 2400 kilometers. These birds sustain a velocity of 40 kilometers per hour for 60 hours. This remarkable feat is made possible by the existence of very large fat depots that are efficiently mobilized during the long flight. Birds that migrate for short distances or not at all are quite lean. They have fat indices of about 0.3. (The fat index is defined as the ratio of the dry weight of total body fat to that of nonfat.) In contrast, long-range migrants become moderately obese in preparing to travel over land and then become highly obese just before they begin their overseas journey. Indeed, their fat index is then close to 3. In the ruby-throated hummingbird, about 0.15 g of triacylglycerols is accumulated each day per gram of body weight. The comparable weight gain in a human would be 10 kg per day. The accumulated fat in migratory birds is stored under the skin, in the abdominal cavity, in muscle, and in the liver. About two-thirds of this fat store is consumed in the long flight over water. The transition to the use of fatty acids and ketone bodies as fuels must be very rapid because almost no protein is degraded during the 60-hour flight. Oxidation of fat also provides these birds with the water needed to make up for losses through respiratory passages. The high efficiency of triacylglycerols as a storage fuel is also noteworthy. Recall that *triacylglycerols store six times as much energy as does glycogen because they are anhydrous and more highly reduced* (p. 605). A migratory bird carrying the same amount of fuel in the form of glycogen would never get off the ground!

A ruby-throated hummingbird.

SPRINTING AND MARATHON RUNNING ARE POWERED BY VERY DIFFERENT FUELS

The selection of fuels during intense exercise illustrates many important facets of energy generation and metabolic integration. Myosin is directly powered by ATP, but the amount of ATP in muscle is relatively small. Hence, the power output and, in turn, the velocity of running depends on the rate of ATP production from other fuels. As shown in Table 30-3, *creatine phosphate* (phosphocreatine) can swiftly transfer its high-potential phosphoryl group to ADP to generate ATP (p. 447). However, the amount of creatine phosphate, like that of ATP itself, is quite limited. A good deal more ATP can be generated by converting *muscle glycogen* to lactate, but the rate is slower than from creatine phosphate. Complete oxidation of muscle glycogen to CO_2 substantially increases the energy yield, but this aerobic process is substantially slower than anaerobic glycolysis. Liver glycogen complements muscle glycogen as an energy store that can be tapped. Much larger quantities of ~P can be obtained by oxidation of fatty acids derived from the breakdown of *fat in adipose tissue*, but the maximal rate of ATP generation is more than tenfold slower than with creatine phosphate. Thus, *ATP is generated much more slowly from high-capacity stores than from limited ones.*

Table 30-3
Fuel sources for muscle contraction

Fuel source	*Maximal rate of ATP production (mmol/s)*	*Total ~P available (mmol)*
Muscle ATP		223
Creatine phosphate	73.3	446
Muscle glycogen to lactate	39.1	6,700
Muscle glycogen to CO_2	16.7	84,000
Liver glycogen to CO_2	6.2	19,000
Adipose-tissue fatty acids to CO_2	6.7	4,000,000

Note: Fuels stored are estimated for a 70-kg person having a muscle mass of 28 kg.
After E. Hultman and R.C. Harris. In *Principles of Exercise Biochemistry*, J.R. Poortmans, ed. (Karger, 1988), pp. 78–119.

A 100-meter sprint is powered by stored ATP, creatine phosphate, and anaerobic glycolysis of muscle glycogen. The ATP level in muscle drops from 5.2 to 3.7 mM, and that of creatine phosphate from 9.1 to 2.6 mM, during this ~10-s sprint. The key role of anaerobic glycolysis here is reflected in the elevation of the blood lactate level from 1.6 to 8.3 mM. The release of H^+ from the intensely active muscle concomitantly lowers the blood pH from 7.42 to 7.24. This pace cannot be sustained in a 1000-meter run (~132 s) for two reasons. First, creatine phosphate would be consumed within a few seconds. Second, anaerobic glycolysis could not persist for two minutes because the supply of NAD^+ would be exhausted far sooner. Also, too much acid would be produced. *Part of the ATP consumed in a 1000-meter run must come from oxidative phosphorylation.* Because ATP is produced more slowly by oxidative phosphorylation than by glycolysis (see Table 30-3), the pace is necessarily slower than in a 100-meter sprint. The championship velocity for the 1000-meter run is 7.6 m/s, compared with 10.1 m/s for the 100-meter event (Figure 30-22).

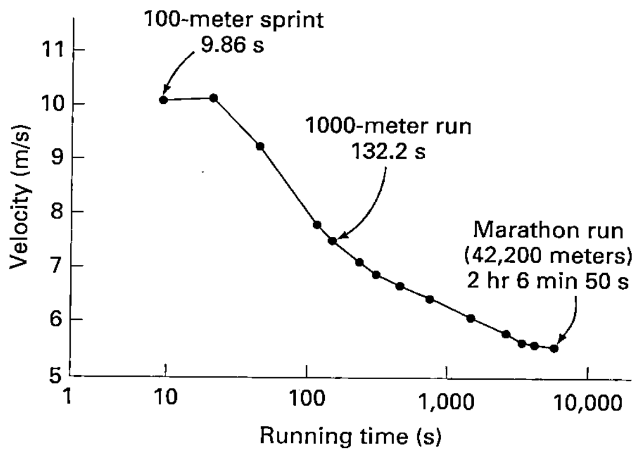

Figure 30-22
Dependence of the velocity of running on the duration of the race. The values shown are world track records (*World Almanac,* 1994, p. 922).

The running of a marathon (42,200 meters, or 26 miles 385 yards) requires a different selection of fuels. The total body glycogen stores (103 mol ATP at best) are insufficient to provide the 150 mol of ATP needed for this grueling ~2-h event. *ATP generation from fatty acids is essential.* However, a marathon would take about six hours to run if all the ATP came from fatty acid oxidation, which is much slower than glycogen oxidation (see Table 30-3). Elite runners consume about equal amounts of glycogen and fatty acids during a marathon to achieve a mean velocity of 5.5 m/s, about half that of a 100-meter sprint (see Figure 30-22). How is an optimal mix of these fuels achieved? *A low blood sugar level leads to a high glucagon/insulin ratio, which in turn mobilizes fatty acids from adipose tissue.* Fatty acids readily enter muscle, where they are degraded by β-oxidation to acetyl CoA and then to CO_2. The elevated acetyl CoA level decreases the activity of the pyruvate dehydrogenase complex to block the conversion of pyruvate to acetyl CoA. Hence, fatty acid oxidation decreases the funneling of sugar into the citric acid cycle and oxidative phosphorylation. Glucose is spared so that just enough remains available at the end of the marathon. The simultaneous use of both fuels gives a higher mean velocity than would be attained if glycogen were totally consumed before the start of fatty acid oxidation.

METABOLIC DERANGEMENTS IN DIABETES RESULT FROM RELATIVE INSULIN INSUFFICIENCY AND GLUCAGON EXCESS

We close this chapter with a consideration of *diabetes mellitus.* This complex disease is characterized by grossly abnormal fuel usage: *glucose is overproduced by the liver and underutilized by other organs.* The incidence of diabetes mellitus (usually referred to simply as *diabetes*) in industrialized countries is about 1%. Indeed, diabetes is the most common serious metabolic disease in the world; it affects hundreds of millions. Type I diabetes is caused by autoimmune destruction of the insulin-secreting β cells in the pancreas. Type II diabetes, by contrast, has a different cause. A genetic basis seems likely, but the molecular lesion has not been identified.

In an untreated diabetic, *the level of insulin is too low and that of glucagon is too high relative to the needs of the patient.* Because insulin is deficient, *the entry of glucose into cells is impaired.* The excessive level of glucagon relative to insulin leads to a decrease in the amount of F-2,6-BP in the liver. Hence, glycolysis is inhibited and gluconeogenesis is stimulated because of the opposite effects of F-2,6-BP on phosphofructokinase and fructose-1,6-bisphosphatase (p. 575). The high glucagon/insulin ratio in diabetes also promotes glycogen breakdown. Hence, *an excessive amount of glucose is*

Diabetes—
Named for the excessive urination in the disease. Aretaeus, a Cappadocian physician of the second century A.D., wrote: "The epithet diabetes has been assigned to the disorder, being something like passing of water by a siphon." He perceptively characterized diabetes as "being a melting-down of the flesh and limbs into urine."

Mellitus—
From Latin, meaning "sweetened with honey." Refers to the presence of sugar in the urine of patients having the disease.
Mellitus distinguishes this disease from diabetes *insipidus,* which is caused by impaired renal reabsorption of water.

produced by the liver and released into the blood. Glucose is excreted in the urine (hence the name *mellitus)* when its concentration in the blood exceeds the reabsorptive capacity of the renal tubules. Water accompanies the excreted glucose, and so an untreated diabetic in the acute phase of the disease is hungry and thirsty.

Impaired carbohydrate utilization in the diabetic leads to the breakdown of fat and protein. Fatty acids readily enter the mitochondrial matrix in diabetes because carnitine acyltransferase I is highly active as a consequence of the low level of malonyl CoA. Large amounts of acetyl CoA are then produced by β-oxidation. However, much of the acetyl CoA cannot enter the citric acid cycle because there is insufficient oxaloacetate for the condensation step. Recall that mammals can synthesize oxaloacetate from pyruvate, a product of glycolysis, but not from acetyl CoA—instead, they generate ketone bodies. *A striking feature of diabetes is the shift in fuel usage from carbohydrates to fats—glucose, more abundant than ever, is spurned.*

Most of the acid produced in normal metabolism is in the form of CO_2, which is readily excreted by the lungs. In contrast, ketone body acids cannot be excreted by the lungs. In high concentrations, they overwhelm the kidney's capacity to maintain acid-base balance. The untreated diabetic can go into a coma because of a lowered blood pH level and dehydration. *Accelerated ketone production leading to acidosis occurs in type I, or insulin-dependent, diabetes mellitus (IDDM),* which usually begins before age 20. The term *insulin-dependent* means that these patients require insulin to live. Most diabetics, by contrast, have a normal or even higher level of insulin in their blood, but they are quite unresponsive to the hormone. This form of the disease, known as *type II, or non-insulin-dependent, diabetes mellitus* (*NIDDM*), typically arises later in life than does the insulin-dependent form.

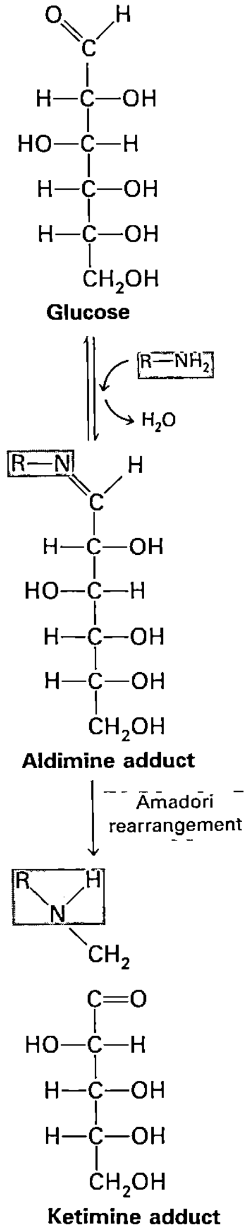

Figure 30-23
Formation of hemoglobin A_{Ic} by the nonenzymatic addition of glucose (or glucose 6-phosphate) to the α-amino group of each β chain. A stable adduct is formed by an Amadori rearrangement of the aldimine to a ketimine.

GLUCOSE REACTS WITH HEMOGLOBIN TO FORM A REVEALING INDICATOR OF BLOOD SUGAR LEVEL

Diabetes can have devastating long-term consequences. This metabolic disease causes retinal degeneration, which can lead to blindness. Kidney damage, nerve damage, and atherosclerosis are other frequent sequelae. It is suspected but not established that these degenerative complications arise from elevated blood sugar levels over a span of years. Hence, *a prime objective in the treatment of diabetes is to lower the blood glucose level.* A valuable indicator of blood sugar levels was unexpectedly found in a different field of inquiry—oxygen transport. During the 120-day lifetime of the red cell, glucose, glucose 6-phosphate, and other sugars react nonenzymatically to form stable conjugates with the α-amino group of the β chains of hemoglobin. The aldehyde group of the open-chain form of glucose condenses with this amino group to form a Schiff base (Figure 30-23). This reversible reaction is followed by a virtually irreversible Amadori rearrangement in which the double bond shifts to C-2 of the sugar to give a stable fructose derivative of hemoglobin, called hemoglobin A_{Ic}, having altered electrophoretic properties.

The red blood cells of all people contain a small proportion of hemoglobin A_{Ic}. The rate of its formation is proportional to the sugar level, and so diabetics have a higher proportion of hemoglobin A_{Ic} than do normal individuals (6% to 15% compared with 3% to 5%). *The level of hemoglobin A_{Ic} reveals the integral of the blood sugar concentration over a period of several weeks.* Hence, measurements of hemoglobin A_{Ic} levels every several weeks are very useful in determining whether the blood glucose levels of diabetic patients were adequately controlled. Before the discovery of hemoglobin A_{Ic}, more frequent monitoring of the blood glucose levels of diabetics was necessary.

Hemoglobin A_{Ic} is also interesting as a model of how proteins can be damaged by high levels of reducing sugars. Some of the late complications of diabetes may arise from the covalent attachment of glucose to vulnerable proteins, as in hemoglobin A_{Ic}. Sorbitol produced from excess glucose by aldol reductase, an NADPH-dependent enzyme, may also be deleterious. Extensive clinical studies are in progress to learn whether the long-term implications of diabetes are consequences of hyperglycemia or of other factors.

SUMMARY

The basic strategy of metabolism is to form ATP, reducing power, and building blocks for biosyntheses. This complex network of reactions is controlled by allosteric interactions and reversible covalent modifications of enzymes and changes in their amounts, by compartmentation, and by interactions between metabolically distinct organs. The enzyme catalyzing the committed step in a pathway is usually the most important control site, as exemplified by phosphofructokinase in glycolysis and acetyl CoA carboxylase in fatty acid synthesis. Opposing pathways such as gluconeogenesis and glycolysis are reciprocally regulated so that one pathway is usually quiescent while the other is highly active. Another pair of opposed reaction sequences, glycogen synthesis and degradation, are coordinately controlled by a hormone-triggered amplifying cascade that leads to the phosphorylation of glycogen synthase and phosphorylase. The role of compartmentation in control is illustrated by the contrasting fates of fatty acids in the cytosol and the mitochondrial matrix.

The metabolic patterns of brain, muscle, adipose tissue, and liver are very different. Glucose is essentially the sole fuel for the brain in a well-fed person. During starvation, ketone bodies (acetoacetate and 3-hydroxybutyrate) become the predominant fuel of the brain. Muscle uses glucose, fatty acids, and ketone bodies as fuel, and it synthesizes glycogen as a fuel reserve for its own needs. Adipose tissue is specialized for the synthesis, storage, and mobilization of triacylglycerols. The diverse metabolic activities of the liver support the other organs. The liver can rapidly mobilize glycogen and carry out gluconeogenesis to meet the glucose needs of other organs. It plays a central role in the regulation of lipid metabolism. When fuels are abundant, fatty acids are synthesized, esterified, and sent from the liver to adipose tissue in the form of very low density lipoprotein (VLDL). In the fasting state, however, fatty acids are converted into ketone bodies by the liver. The activities of these organs are integrated by hormones. Insulin signals the fed state: it stimulates the formation of glycogen and triacylglycerols and the synthesis of proteins. In contrast, glucagon signals a low blood glucose level: it stimulates glycogen breakdown and gluconeogenesis by the liver and triacylglycerol hydrolysis by adipose tissue. The effects of epinephrine and norepinephrine on fuels are like those of glucagon, except that muscle rather than the liver is their primary target.

The blood glucose level in a well-fed person typically ranges from 80 mg/dl (4.4 mM) to 120 mg/dl (6.7 mM). After a meal, the rise in the blood glucose level leads to increased secretion of insulin and decreased secretion of glucagon. Consequently, glycogen is synthesized in muscle and the liver. The increased entry of glucose into adipose tissue provides glycerol 3-phosphate for the synthesis of triacylglycerols. These effects are reversed when the blood glucose level drops several hours later. Glucose is then formed by the degradation of glycogen and by the gluconeogenic pathway, and fatty acids are released by the hydrolysis of triacylglycerols. Liver and muscle then use fatty acids instead of glucose to meet their own energy needs so that glucose is conserved for use by the brain and other tissues that are highly dependent on it.

The metabolic adaptations in starvation are designed to minimize protein degradation. Large amounts of ketone bodies are formed by the liver from fatty acids and released into the blood within a few days after the onset of starvation. After several weeks of starvation, ketone bodies become the major fuel of the brain. The diminished need for glucose decreases the rate of muscle breakdown, and so the likelihood of survival is enhanced. Maximal power output during intense exercise depends on the selection of appropriate fuels and the integration of their use. The 100-meter sprint is powered by stored ATP and creatine phosphate and by anaerobic glycolysis. By contrast, oxidation of both muscle glycogen and fatty acids derived from adipose tissue is essential in the running of a marathon, a highly aerobic process.

Diabetes mellitus, the most common serious metabolic disease, is produced by an insufficiency of insulin and an excess of glucagon relative to the needs of the patient. Insulin deficiency impairs the entry of glucose into cells and its utilization. Excess glucagon enhances glucose formation by the liver. The elevated blood glucose level leads to the excretion of a large volume of glucose-rich urine. Triacylglycerols are mobilized and ketone bodies are formed to an abnormal extent. A striking feature of diabetes is the shift in fuel usage from carbohydrates to fats. Accelerated ketone body formation can lead to acidosis, coma, and death in untreated insulin-dependent diabetics. Hemoglobin A_{Ic}, a conjugate formed by the nonenzymatic addition of glucose to terminal amino groups, is a valuable indicator of the blood sugar level over a period of weeks.

SELECTED READINGS

Where to start

Foster, D.W., 1984. From glycogen to ketones—and back. *Diabetes* 33:1188–1199. [A lucid account of how anabolism and catabolism are sensitively controlled by the insulin: glucagon ratio.]

Randle, P.J., 1986. Fuel selection in animals. *Biochem. Soc. Trans.* 14:799–806. [A fine presentation of the control of the pyruvate dehydrogenase complex and its significance in fuel selection.]

Cerami, A., Vlassara, H., and Brownlee, M., 1987. Glucose and aging. *Sci. Amer.* 256(3):90–96. [An interesting account of deleterious consequences arising from the reactivity of glucose.]

Shulman, R.G., Blamire, A.M., Rothman, D.L., and McCarthy, G., 1993. Nuclear magnetic resonance imaging and spectroscopy of human brain function. *Proc. Nat. Acad. Sci.* 90:3127–3133.

Lienhard, G.E., Slot, J.W., James, D.E., and Mueckler, M.M., 1992. How cells absorb glucose. *Sci. Amer.* 266(1):86–91.

Books

Newsholme, E.A., and Leech, A.R., 1983. *Biochemistry for the Medical Sciences.* Wiley.

Poortmans, J.R. (ed.), 1988. *Principles of Exercise Biochemistry.* Karger.

Ochs, R.S., Hanson, R.W., and Hall, J. (eds.), 1985. *Metabolic Regulation.* Elsevier. [A collection of many interesting articles on metabolic control in prokaryotes as well as higher organisms.]

Bondy, P.K., and Rosenberg, L.E. (eds.), 1980. *Metabolic Control and Disease.* Saunders.

NMR: A window on metabolism

Radda, G.K., 1992. Control, bioenergetics, and adaptation in health and disease: Noninvasive biochemistry from nuclear magnetic resonance. *FASEB J.* 6:3032–3038.

Gruetter, R., Novotny, E., Boulware, S.D., Rothman, D.L., Mason, G.F., Shulman, G.I., Shulman, R.G., and Tamborlane, W.V., 1992. Direct measurement of brain glucose concentrations in humans by ^{13}C NMR spectroscopy. *Proc. Nat. Acad. Sci.* 89:1109–1112.

Carbohydrate metabolism

McGarry, J.D., Kuwajima, M., Newgard, C.B., and Foster, D.W., 1987. From dietary glucose to liver glycogen—the full circle round. *Ann. Rev. Nutrit.* 7:51–73.

Nordlie, R.C., 1984. Fine tuning of blood glucose concentrations. *Trends Biochem. Sci.* 10:70–75.

Felig, P., 1980. Disorders of carbohydrate metabolism. *In* Bondy, P.K., and Rosenberg, L.E. (eds.), *Metabolic Control and Disease.* Saunders.

Cahill, G.F., Jr., and Owen, O.E., 1968. Some observations on carbohydrate metabolism in man. *In* Dickens, F., Randle, P.J., and Whelan, W.J. (eds.), *Carbohydrate Metabolism and Its Disorders,* vol. 1, pp. 497–522. Academic Press.

Kitakura, K., and Uyeda, K., 1987. The mechanism of activation of heart fructose 6-phosphate,2-kinase:fructose-2,6-bisphosphatase. *J. Biol. Chem.* 262:679–681. [Reports experiments showing that the enzyme in heart, in contrast to the one in liver, is activated by phosphorylation.]

Hers, H.-G., and Hue, L., 1983. Gluconeogenesis and related aspects of glycolysis. *Ann. Rev. Biochem.* 52:617–653.

Amino acid metabolism

Snell, K., 1979. Alanine as a gluconeogenic carrier. *Trends Biochem. Sci.* 4:124–128.

Felig, P., 1975. Amino acid metabolism in man. *Ann. Rev. Biochem.* 44:933–955.

Dice, J.F., Walker, C.D., Byrne, B., and Cardiel, A., 1978. General characteristics of protein degradation in diabetes and starvation. *Proc. Nat. Acad. Sci.* 75:2093–2097.

Lipid metabolism

McGarry, J.D., and Foster, D.W., 1980. Regulation of hepatic fatty acid oxidation and ketone body production. *Ann. Rev. Biochem.* 49:395–420.

Rebouche, C.J., 1992. Carnitine function and requirements during the life cycle. *FASEB J.* 6:3379–3386.

Williamson, D.H., 1979. Recent developments in ketone-body metabolism. *Biochem. Soc. Trans.* 7:1313–1321.

Metabolic adaptations in starvation

Cahill, G.F., Jr., 1976. Starvation in man. *Clin. Endocrinol. Metab.* 5:397–415.

Grey, N.J., Karl, I., and Kipnis, D.M., 1975. Physiologic mechanisms in the development of starvation ketosis in man. *Diabetes* 24:10–16.

Sugden, M.C., Holness, M.J., and Palmer, T.N., 1989. Fuel selection and carbon flux during the starved-to-fed transition. *Biochem. J.* 263:313–323.

Fuel selection in exercise

Hagerman, F.C., 1992. Energy metabolism and fuel utilization. *Med. Sci. Sports Exerc.* 24:S309–S314.

Felig, P., and Wahren, J., 1975. Fuel homeostasis in exercise. *New Engl. J. Med.* 293:1078–1084.

Newsholme, E., and Leech, T., 1983. *The Runner: Energy and Endurance.* Fitness Books. [A fascinating presentation of the biochemical and physiologic basis of energy utilization in running.]

Elia, M., and Livesey, G., 1992. Energy expenditure and fuel selection in biological systems: The theory and practice of calculations based on indirect calorimetry and tracer methods. *World Rev. Nutrit. Dietet.* 70:68–131.

Odum, E.P., 1965. Adipose tissue in migratory birds. *In* Renold, A.E., and Cahill, G.F. (eds.), *Handbook of Physiology,* sec. 5: *Adipose Tissue,* pp. 37–43. American Physiological Society.

Diabetes mellitus

Foster, D.W., 1989. Diabetes mellitus. *In* Scriver, C.R., Beaudet, A.L., Sly, W.S., and Valle, D. (eds.), *The Metabolic Basis of Inherited Disease* (6th ed.), pp. 375–397. McGraw-Hill.

Granner, D.K., and O'Brien, R.M., 1992. Molecular physiology and genetics of NIDDM. Importance of metabolic staging. *Diabetes Care* 15:369–395.

Steiner, D.F., Tager, H.S., Chan, S.J., Nanjo, K., Sanke, T., and Rubenstein, A.H., 1990. Lessons learned from molecular biology of insulin-gene mutations. *Diabetes Care* 13:600–609.

Foster, D.W., and McGarry, J.D., 1983. The metabolic derangements and treatment of diabetic ketoacidosis. *New Engl. J. Med.* 309:159–169.

Cohen, M.P., 1986. *Diabetes and Protein Glycosylation.* Springer-Verlag.

PROBLEMS

1. *Distinctive organs.* What are the key enzymatic differences between liver, muscle, and brain that account for their differing utilization of metabolic fuels?

2. *Missing enzymes.* Predict the major consequence of each of the following enzymatic deficiencies:
   (a) Hexokinase in adipose tissue.
   (b) Glucose 6-phosphatase in liver.
   (c) Carnitine acyltransferase I in skeletal muscle.
   (d) Glucokinase in liver.
   (e) Thiolase in brain.
   (f) Kinase in liver that synthesizes fructose 2,6-bisphosphate.

3. *Contrasting milieux.* Cerebrospinal fluid has a low content of albumin and other proteins compared with plasma.
   (a) What effect does this have on the concentration of fatty acids in the extracellular medium of the brain?

(b) Propose a plausible reason for the selection by brain of glucose rather than fatty acids as the prime fuel.
(c) How does the fuel preference of muscle complement that of the brain?

4. *Metabolic energy and power.* The rate of energy expenditure of a typical 70-kg person at rest is about 70 watts (W), like a light bulb.
(a) Express this rate in kilojoules per second and in kilocalories per second.
(b) How many electrons flow through the electron transport chain of mitochondria per second under these conditions?
(c) Estimate the corresponding rate of ATP production.
(d) The total ATP content of the body is about 50 grams. Estimate how often an ATP molecule turns over in a person at rest.

5. *Respiratory quotient (RQ).* This classic metabolic index is defined as the volume of CO_2 released divided by the volume of O_2 consumed.
(a) Calculate the RQ values for the complete oxidation of glucose and of tripalmitoylglycerol.
(b) What do RQ measurements reveal about the contributions of different energy sources during intense exercise? (Assume that protein degradation is negligible.)

6. *Camel's hump.* Compare the H_2O yield from the complete oxidation of 1 g glucose with that of 1 g tripalmitoylglycerol. Relate these values to the evolutionary selection of the contents in a camel's hump.

7. *The wages of sin.* How long does one have to jog to offset the calories obtained from eating 10 macadamia nuts? [Assume an incremental power consumption of 400 W.]

8. *Sweet hazard.* Ingesting large amounts of glucose before a marathon might seem to be a good way of increasing the fuel stores. However, experienced runners do not ingest glucose before a race. What is the biochemical reason for their avoidance of this potential fuel? [Hint: consider the effect of glucose ingestion on the level of insulin.]

PART

V

Genes: Replication and Expression

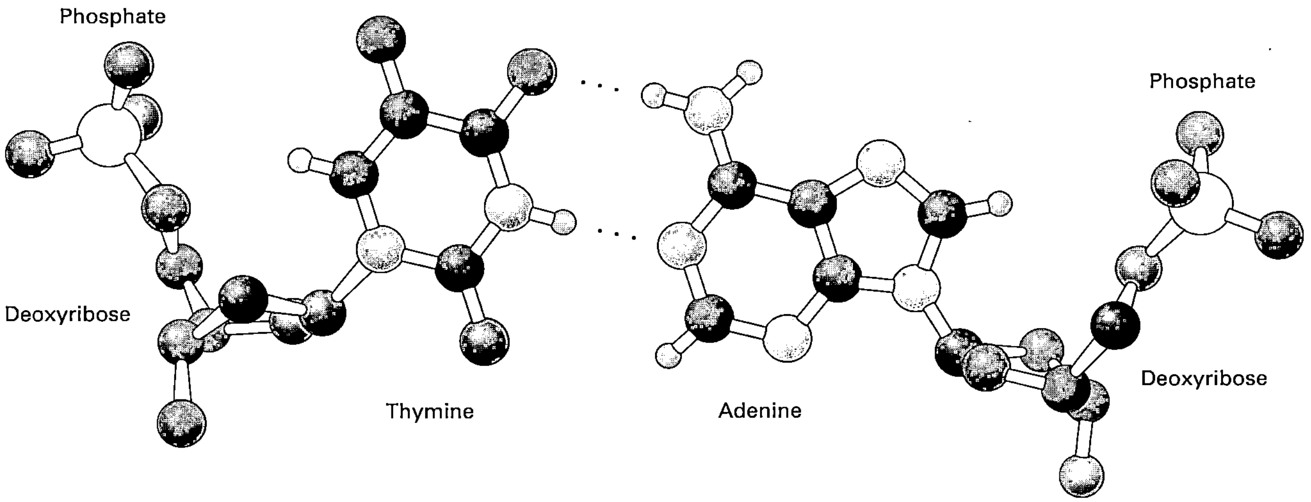

Model of a pair of deoxyribonucleotide units of a DNA double helix. Adenine is hydrogen-bonded to thymine in this base pair.

CHAPTER

31

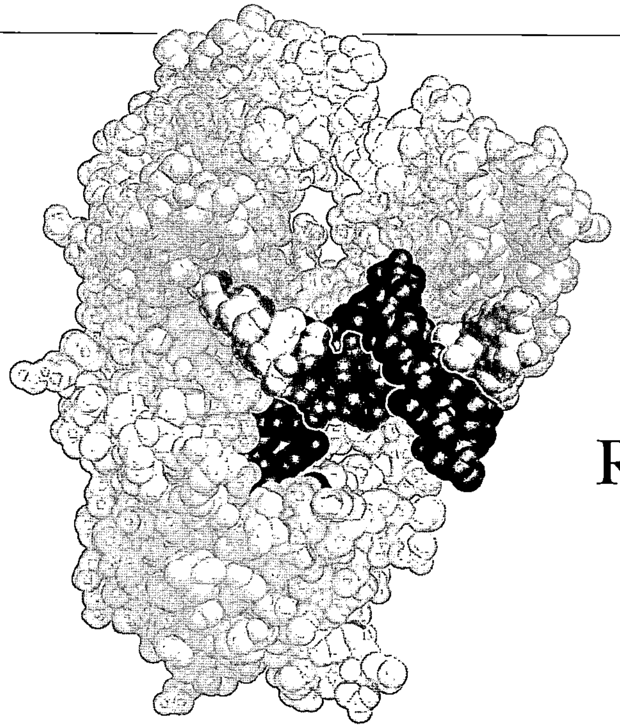

DNA Structure, Replication, and Repair

The storage, transmission, and expression of genetic information is the central theme of this part of the book. The genetic role and structure of DNA were introduced in Part I (Chapter 4), as was the flow of genetic information (Chapter 5). Our consideration of this theme now resumes, enriched by a knowledge of proteins and metabolism.

This chapter deals with the structure, replication, mutation, and repair of DNA. We shall focus on prokaryotic systems because they have been intensively studied and exemplify general principles. More complex eukaryotic systems will be discussed in a later chapter (p. 975). The structure of DNA is dynamic. The Watson-Crick double helix can be bent, kinked, and unwound. It can also adopt different helical forms. X-ray analyses of crystals of DNA oligomers have provided insight into these conformational properties. Enzymes that cut and join DNA are considered next. The structure of DNA bound to a restriction endonuclease reveals how this exquisitely specific enzyme finds and cleaves its target sequence. Proteins can recognize specific DNA sequences by scanning the pattern of hydrogen bond donors and acceptors in the major and minor grooves of the double helix. The next topic is the topology and supercoiling of DNA. Topoisomerases, a fascinating group of enzymes, catalyze the supercoiling of DNA molecules and their relaxation by cutting and rejoining DNA strands.

We shall then be ready to consider one of the most demanding and remarkable of all biological processes, the replication of DNA. How does DNA synthesis begin? How are parental strands unwound and separated

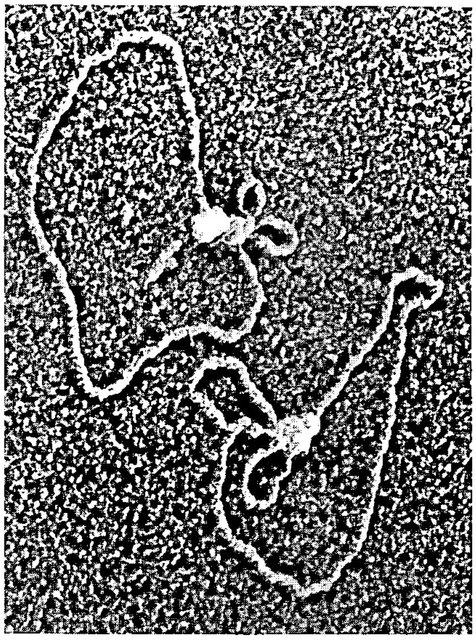

Figure 31-1
Electron micrograph showing the initiation of replication of ϕX174 viral DNA. Each of the two circular DNA molecules is bound to a primosome, a protein assembly that begins DNA replication. [Courtesy of Dr. Jack Griffith.]

Opening Image: *Structure of a catalytically active large fragment of DNA polymerase I. Polymerization is carried out by the domain shown in blue, and editing by the one in yellow. The bound duplex DNA (red and green strands) is bent. [Drawn from coordinates kindly provided by Dr. Thomas Steitz. L.S. Beese, V. Derbyshire, and T.A. Steitz. Science 260(1993):352.]*

to serve as templates? How is very high fidelity replication achieved? A striking property of DNA polymerases is their built-in proofreading capability. The last part of this chapter deals with the molecular nature of mutations and their repair. Lesions in DNA are continually being repaired by multiple processes that use information in the intact strand to direct the correction of the damaged strand. Defective repair of DNA markedly increases the likelihood of many cancers.

DNA IS STRUCTURALLY DYNAMIC AND CAN ASSUME A VARIETY OF FORMS

The double-helical structure of DNA deduced by Watson and Crick profoundly influenced the course of biology because it immediately suggested how genetic information is stored and replicated. As was discussed before (p. 80), the essential features of their model are:

1. Two polynucleotide chains running in opposite directions coil around a common axis to form a right-handed double helix.

2. The purine and pyrimidine bases are on the inside of the helix, whereas the phosphate and deoxyribose units are on the outside.

3. Adenine (A) is paired with thymine (T), and guanine (G) with cytosine (C). A · T base pairs are reinforced by two precisely directed hydrogen bonds, and G · C base pairs by three such bonds. The double helix is also stabilized by interactions between stacked bases on the same strand.

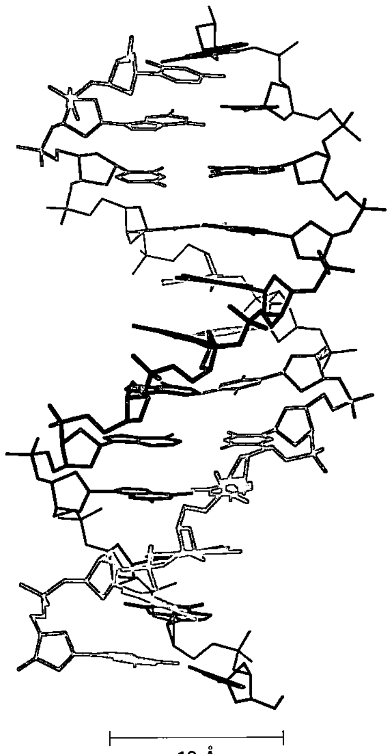

Figure 31-2
Skeletal model of a 12-bp DNA molecule in the B form. One of the strands is shown in green and the other in red. Note the large local deviations from the classic Watson-Crick structure and the curvature of the helix axis. [Drawn from 2bna.pdb. H.R. Drew, S. Samson, and R.E. Dickerson. *Proc. Nat. Acad. Sci.* 79(1982):4040.]

The model proposed by Watson and Crick (known as the *B-DNA helix*) was based on x-ray diffraction patterns of DNA *fibers,* which provide information about properties of the double helix that are *averaged* over its constituent residues. Much more structural information can be obtained from x-ray analyses of DNA *crystals.* However, such studies had to await the development of techniques for synthesizing large amounts of DNA oligomers with defined base sequences (p. 124). X-ray analyses of these crystals at atomic resolution have revealed that DNA exhibits much more structural variability and diversity than formerly envisaged. A DNA chain can be rotated about six bonds in each monomer—the glycosidic bond between the base and sugar, the $C_{4'}$–$C_{5'}$ bond of the sugar, and four bonds in the phosphodiester bridge joining $C_{3'}$ of one sugar and $C_{5'}$ of the next one—compared with just two for polypeptide chains (p. 28). As will be discussed shortly, the puckering of the ribose ring is an important structural determinant.

The x-ray analysis of a crystallized DNA dodecamer by Richard Dickerson and his co-workers revealed that its overall structure is very much like a Watson-Crick double helix (Figure 31-2). However, the dodecamer differed from the Watson-Crick model in not being uniform; there are rather large local deviations from the average structure. The Watson-Crick model has 10 residues per complete turn, and so a residue is related to the next along a chain by a rotation of 36 degrees. In Dickerson's dodecamer, the rotation angles range from 28 degrees (less tightly wound) to 42 degrees (more tightly wound). Furthermore, the two bases of some base pairs are not coplanar (Figure 31-3). Rather, they are twisted, like the blades of a propeller. This deviation from the idealized structure, called *propeller twisting,* enhances the stacking of bases along a strand. *These and other local variations of the double helix depend on base sequence. A protein searching for a specific target sequence in DNA may sense its presence through its effect on the precise shape of the double helix.*

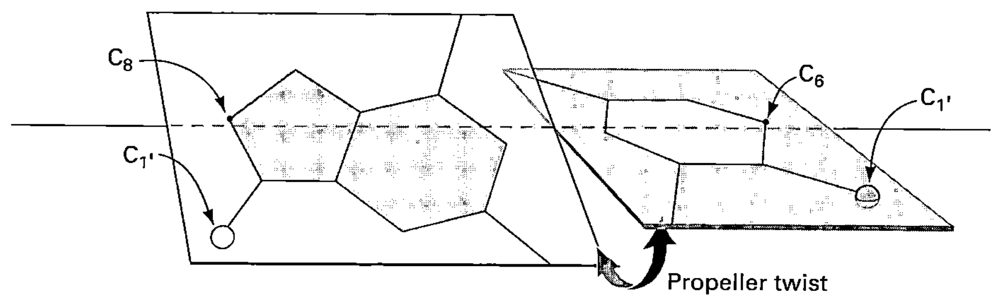

Figure 31-3
The bases of a DNA base pair are not perfectly coplanar. They are twisted with respect to each other, like the blades of a propeller.

One other feature of the B-DNA helix is noteworthy. *The helix can be smoothly bent into an arc or it can be supercoiled with rather little change in local structure.* This ease of deformation is biologically important because it enables circular DNA to be formed and allows DNA to be wrapped around proteins. The compacting of DNA to fit into a much smaller volume depends on its deformability. If DNA were constrained to be linear, it would not fit into a cell. DNA can also be *kinked*—that is, bent at discrete sites. Kinking can be induced by specific base sequences, such as a run of at least four adenine residues, or by the binding of a protein, as will be evident shortly.

THE MAJOR AND MINOR GROOVES ARE LINED BY SEQUENCE-SPECIFIC HYDROGEN-BONDING GROUPS

B-DNA contains two kinds of grooves, called the *major groove* (12 Å wide) and the *minor groove* (6 Å wide) (Figure 31-4). They arise because the glycosidic bonds of a base pair are not diametrically opposite each other (see Figures 4-9 and 4-10 on p. 81). The minor groove contains the pyrimidine O-2 and the purine N-3 of the base pair, and the major groove is on the opposite side of the pair (Figure 31-5). The major groove is slightly deeper than the minor one (8.5 versus 7.5 Å). Each groove is lined by potential hydrogen bond donor and acceptor atoms. In the minor groove, N-3 of adenine and guanine, and O-2 of thymine and cytosine, can serve as hydrogen acceptors, and the amino group attached to C-2 of guanine can be a hydrogen donor. Let us denote N-3 by *n*, O-2 by *o*, and an amino group hydrogen by *h*. Hence, the patterns of donors and acceptors in the minor groove are *no* (AT), *on* (TA), *nho* (GC), and *ohn* (CG). In the major groove, N-7 of guanine and adenine is a potential acceptor, as is O-4 of thymine and O-6 of guanine. The amino groups attached to C-6 of adenine and C-4 of cytosine can serve as hydrogen donors. Thus, the patterns exhibited by the major groove are *nho* (AT), *ohn* (TA), *noh* (GC), and *hon*

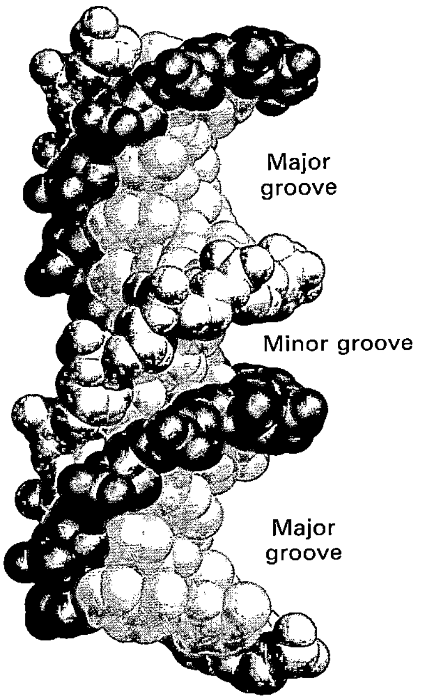

Figure 31-4
B-DNA contains a major groove and minor groove. This helix has been tilted to emphasize the markedly different sizes of these grooves.

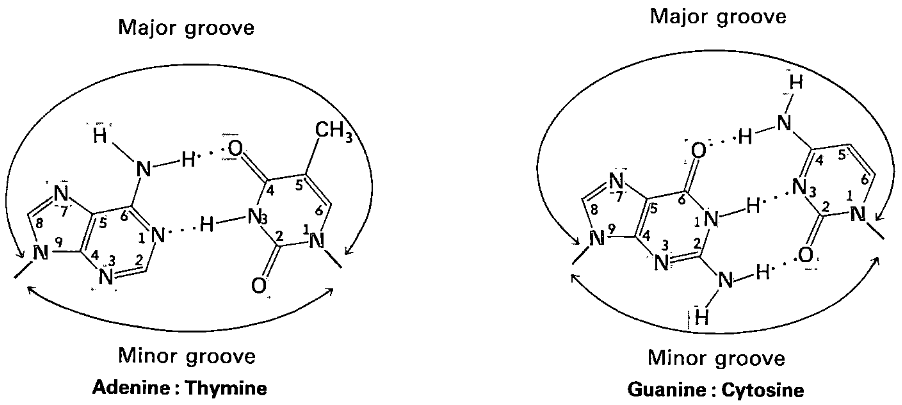

Figure 31-5
The A · T and G · C base pairs of DNA contain atoms that can form additional hydrogen bonds. The major and minor grooves are lined by potential hydrogen bond donors and acceptors (marked in yellow).

(CG), where *n* denotes N-7, and *o* denotes O-4 or O-6. Note that the major groove displays more distinctive features than does the minor groove. Also, the larger size of the major groove makes it more accessible for interactions with proteins that recognize specific DNA sequences.

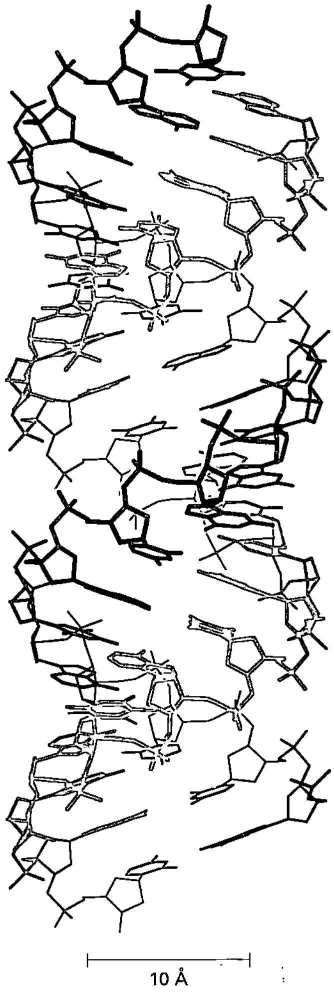

Figure 31-6
Skeletal model of an idealized A-DNA helix. [Drawn from coordinates reported in S. Arnott and D.W.L. Hukins. *Biochem. Biophys. Res. Comm.* 47(1972):1504.]

THE 2'-OH OF RNA FITS IN AN A-DNA HELIX WITH TILTED BASE PAIRS BUT NOT IN A B-DNA HELIX

X-ray diffraction studies of dehydrated DNA fibers revealed a different form called *A-DNA,* which appears when the relative humidity is reduced below about 75%. A-DNA, like B-DNA, is a right-handed double helix made up of antiparallel strands held together by Watson-Crick base pairing. The A helix is wider and shorter than the B helix, and its base pairs are tilted rather than normal to the helix axis (Figure 31-6). Many of the structural differences between them arise from different puckerings of their ribose units (Figure 31-7). In A-DNA, $C_{3'}$ is out of the plane (called $C_{3'}$-endo) formed by the other four atoms of the furanose ring; in B-DNA, C_2' is out of plane (called $C_{2'}$-endo). The $C_{3'}$-*endo* puckering in A-DNA leads to a 19-degree tilting of the base pairs away from the normal to the helix. Moreover, the minor groove nearly vanishes. The phosphate groups in the A helix bind fewer H_2O molecules than do phosphates in B-DNA. Hence, dehydration favors the A form.

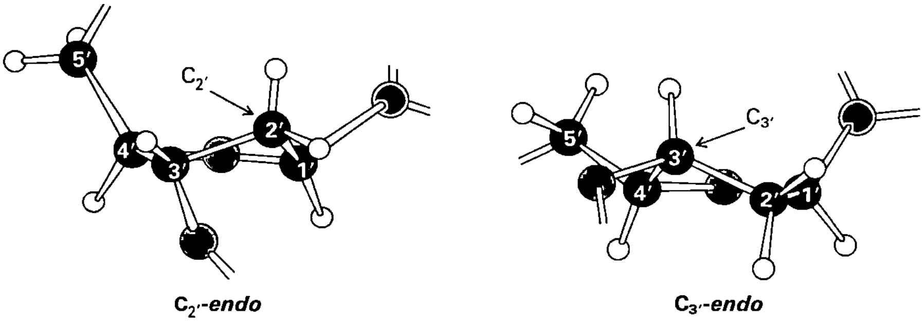

Figure 31-7
Sugar puckering markedly affects the orientation of the phosphodiester bridges and the glycosidic bond. $C_{2'}$-*endo* (found in B-DNA) and $C_{3'}$-*endo* (found in A-DNA) denote which atom of the ribose ring lies above the plane in the orientation shown here.

The A helix is not confined to dehydrated DNA. *Double-stranded regions of RNA* (as in hairpins, p. 102) *and RNA-DNA hybrids adopt a double-helical form very similar to that of A-DNA.* The 2'-OH of ribose prevents RNA from forming a classic Watson-Crick B helix because of steric hindrance. O-2' would come too close to three atoms of the adjoining phosphate group and one in the next base. In an A-type helix, by contrast, O-2' projects outward, away from other atoms.

Z-DNA IS A LEFT-HANDED DOUBLE HELIX IN WHICH BACKBONE PHOSPHATES ZIGZAG

A third type of DNA helix was discovered by Alexander Rich and his associates when they solved the structure of CGCGCG. They found that this hexanucleotide forms a duplex of antiparallel strands held together by Watson-Crick base pairing, as expected. What was surprising, however, was that this double helix was *left-handed,* in contrast with the *right-handed* screw sense of the A and B helices. Furthermore, the phosphates in the backbone *zigzagged;* hence, they called this new form *Z-DNA* (Figure 31-8). The zigzagging is a consequence of the fact that the repeating unit is a dinucleotide rather than a mononucleotide. Z-DNA also differs from the A and B forms in containing only one deep helical groove.

The Z-DNA form is adopted by short oligonucleotides that have *sequences of alternating pyrimidines and purines.* High salt concentrations are required to minimize electrostatic repulsion between the backbone phosphates, which are closer to each other than in A- and B-DNA. Under physiologic conditions, *most of the DNA in a bacterial or a eukaryotic genome is in the classic Watson-Crick B-DNA form.* While the biological role of Z-DNA is uncertain, its existence graphically shows that *DNA is a flexible, dynamic molecule.* A-, B-, and Z-DNA are compared in Table 31-1 and Figure 31-9.

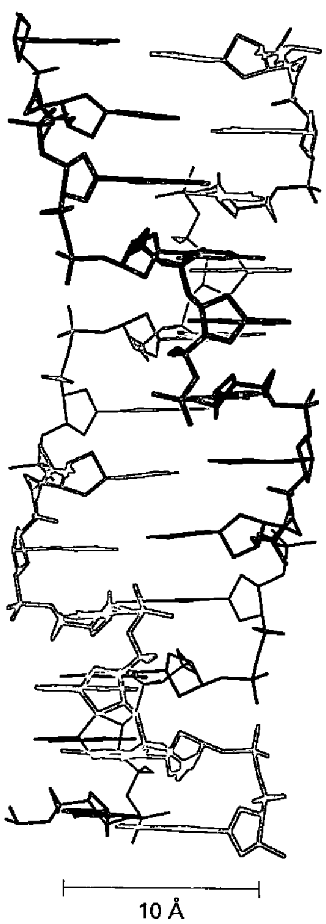

Figure 31-8
Skeletal model of Z-DNA, a left-handed helical form of DNA. The sugar-phosphate backbone zigzags in Z-DNA because the repeating unit is a dinucleotide. [Drawn from 3zna.pdb. A.H.-J. Wang, G.J. Quigley, F.J. Kolpak, G. Van der Marel, J.H. Van Boom, and A. Rich. *Science* 211(1981):171.]

Table 31-1
Comparison of A-, B-, and Z-DNA

	Helix type		
	A	*B*	*Z*
Shape	Broadest	Intermediate	Narrowest
Rise per base pair	2.3 Å	3.4 Å	3.8 Å
Helix diameter	25.5 Å	23.7 Å	18.4 Å
Screw sense	Right-handed	Right-handed	Left-handed
Glycosidic bond	*anti*	*anti*	*anti* for C, T *syn* for G
Base pairs per turn of helix	11	10.4	12
Pitch per turn of helix	25.3 Å	35.4 Å	45.6 Å
Tilt of base pairs from normal to helix axis	19°	1°	9°
Major groove	Narrow and very deep	Wide and quite deep	Flat
Minor groove	Very broad and shallow	Narrow and quite deep	Very narrow and deep

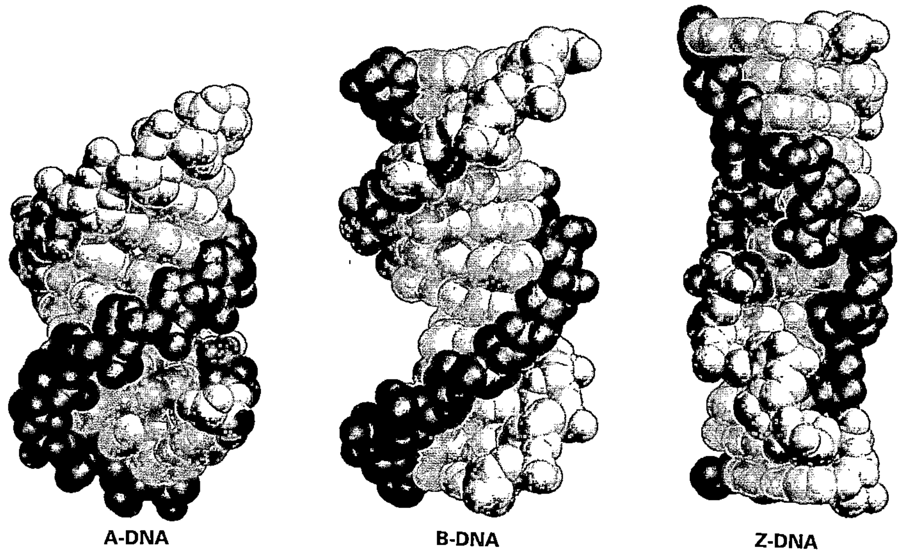

Figure 31-9
Space-filling models of idealized forms of A-DNA, B-DNA, and Z-DNA, each 12-bp in length. Note that the Z-DNA helix is left-handed, whereas A-DNA and B-DNA are right-handed. Nearly all DNA under physiologic conditions is in the B-DNA form.

NIΨONANOMHMATAMHMONANOΨIN
A palindromic inscription, in Greek, on the sacred font in the courtyard of the Hagia Sophia in Istanbul (Constantinople), meaning *"Wash your sins, not only your face."*

THE RESTRICTION ENDONUCLEASE EcoRV FINDS ITS TARGET BY SCANNING THE MAJOR GROOVE OF DNA

Bacteria can detect and destroy foreign DNA molecules such as those of invading viruses. They possess *restriction endonucleases* (*restriction enzymes*) that cleave foreign DNA but leave untouched their own DNA, which is methylated at potential recognition sites. Restriction endonucleases must be swift and precise: invading DNA must be rapidly degraded, but the bacterium's own genome must not be cut even once. Mechanistic and structural studies of the EcoRV endonuclease have revealed how exquisite specificity is achieved with retention of catalytic prowess. This restriction enzyme recognizes a specific hexanucleotide target sequence that is *palindromic*—it possesses *twofold rotational symmetry* (p. 120). Indeed, most targets of restriction enzymes are palindromic. Both strands are cleaved at identical sites symmetrically positioned with respect to the twofold axis. A change of a single base pair in the recognition sequence lowers the cleavage rate more than a millionfold. One would therefore expect that the enzyme, like its target, is symmetric. Indeed, EcoRV is a dimer of identical subunits, and binds DNA so that the twofold axis of the target site coincides with the twofold axis of the enzyme (Figure 31-10). Thus, *the symmetry of the endonuclease matches the symmetry of its targets.*

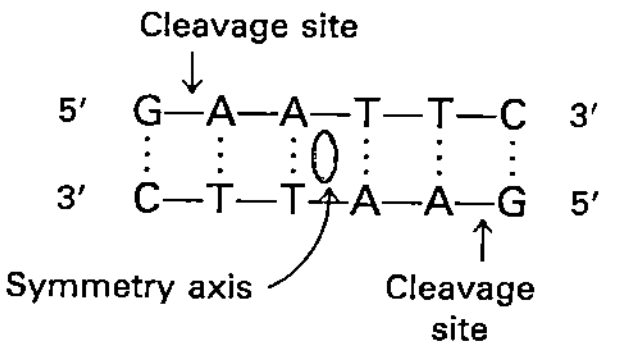

EcoRI target

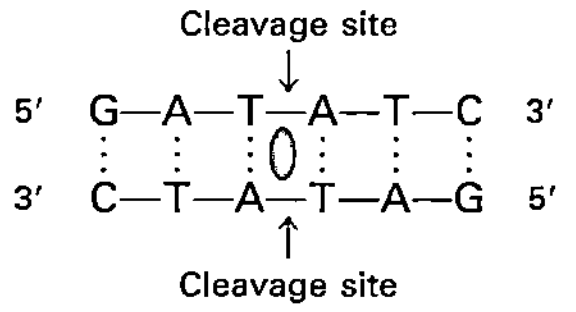

EcoRV target

The EcoRV endonuclease searches DNA for its GATATC target sequence by diffusing along its major groove. Specifically, a surface loop from a β turn of each subunit makes contact with the major groove. When the specific sequence is encountered, a large structural rearrangement occurs in both the enzyme and its DNA target (Figure 31-11). In this *induced fit,* DNA becomes kinked by 50 degrees at the center of the hexanucleotide recognition site. Each recognition loop forms six hydrogen bonds, all in the major groove, with the outer two base pairs of a GAT half

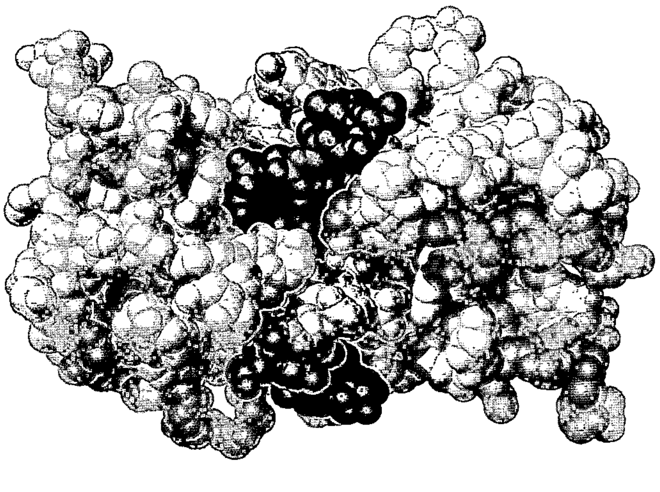

Figure 31-10
The twofold symmetry of EcoRV endonuclease matches that of its target DNA. The two identical subunits of the endonuclease are shown in blue and yellow, and the two identical strands of target DNA in green and red. [Drawn from 4rve.pdb. F.K. Winkler, D.W. Banner, C. Oefner, D. Tsernoglou, R.S. Brown, S.P. Heathman, R.K. Bryan, P.D. Martin, K. Petratos, and K.S. Wilson. *EMBO J.* 12(1993):1781.]

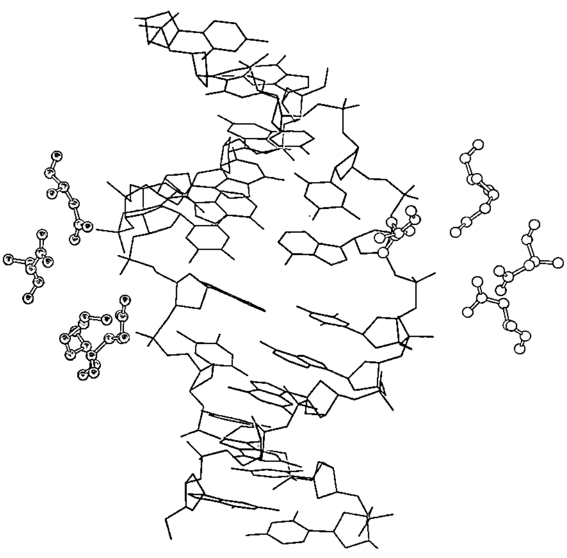

Figure 31-11
The EcoRV endonuclease bends duplex DNA by 50 degrees when it encounters the specific hexanucleotide sequence. One of the DNA strands is shown in red, and the other in green. Four key catalytic residues of one subunit of the dimer are shown in blue, and of the other subunit in yellow. [Drawn from 4rve.pdb. F.K. Winkler, D.W. Banner, C. Oefner, D. Tsernoglou, R.S. Brown, S.P. Heathman, R.K. Bryan, P.D. Martin, K. Petratos, and K.S. Wilson. *EMBO J.* 12(1993):1781.]

site. Most significant, Mg^{2+}, which is essential for hydrolysis, enters the catalytic site and becomes coordinated only when the target sequence is encountered. We see here the *highly dynamic character of protein-DNA recognition. The action of EcoRV shows that nucleotide sequences in DNA can be read swiftly and with high precision by a discerning protein that scans the major groove.*

DNA LIGASE JOINS ENDS OF DNA IN DUPLEX REGIONS

The finding of circular DNA (p. 88) pointed to the existence of an enzyme that connects the ends of DNA chains. In 1967, scientists in several laboratories simultaneously discovered *DNA ligase.* This enzyme *catalyzes the formation of a phosphodiester bond between the 3′-OH group at the end of one DNA chain and the 5′-phosphate group at the end of the other* (Figure 31-12). An energy source is required to drive this endergonic reaction. In *Escherichia coli* and other bacteria, *NAD*$^+$ serves this role; in animal cells and bacteriophage, *ATP* is the energy source. As will be discussed later, this joining process is essential for the *normal synthesis of DNA, the repair of damaged DNA, and the splicing of DNA chains in genetic recombination.*

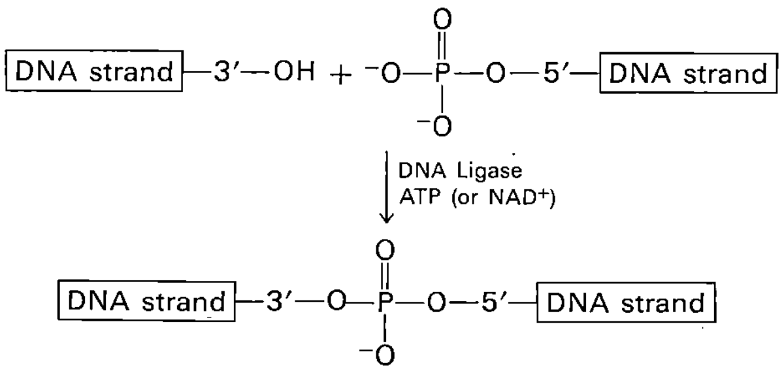

Figure 31-12
DNA ligase catalyzes the joining of DNA strands that are part of a double-helical molecule.

DNA ligase cannot link two molecules of single-stranded DNA or circularize single-stranded DNA. Rather, *ligase seals breaks in double-stranded DNA molecules.* The enzyme from *Escherichia coli* ordinarily forms a phosphodiester bridge only if there are at least several base pairs near this link. Ligase encoded by T4 bacteriophage can link two blunt-ended double-helical fragments, a capability that is exploited in recombinant DNA technology.

Let us look at the mechanism of joining, which was elucidated by I. Robert Lehman. ATP (or, in some ligases, NAD^+) donates its activated AMP unit to DNA ligase to form a *covalent enzyme-AMP (enzyme-adenylate) complex* in which AMP is linked to the ε-amino group of a lysine residue of the enzyme through a phosphoamide bond (Figure 31-13). Pyrophosphate (or nicotinamide mononucleotide, NMN) is concomitantly released. The activated AMP moiety is then transferred from the lysine residue to the phosphate group at the 5′ terminus of a DNA chain, forming a *DNA–adenylate complex.* The final step is a nucleophilic attack by the 3′-OH group on this activated 5′-phosphorus atom. This sequence of reactions (Figure 31-14) is driven by the hydrolysis of pyrophosphate re-

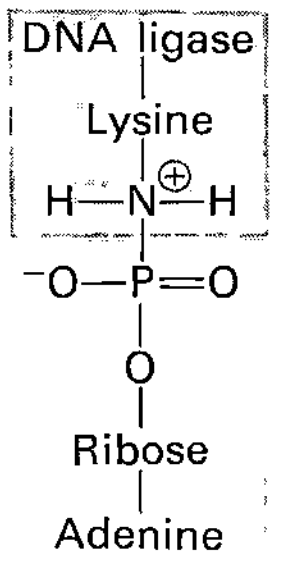

Figure 31-13
Covalent ligase-AMP intermediate formed by the reaction of NAD^+ with a lysine residue of the enzyme. The AMP moiety is activated in this intermediate.

E + ATP (or NAD$^+$) ⇌ E-AMP + PP$_i$ (or NMN)

E-AMP + Ⓟ—5′-DNA ⇌ E + AMP—Ⓟ—5′-DNA

DNA-3′-OH + AMP—Ⓟ—5′-DNA ⇌ DNA-3′-O—Ⓟ—5′-DNA + AMP

DNA-3′-OH + Ⓟ—5′-DNA + ATP (or NAD$^+$) ⇌ DNA-3′-O—Ⓟ—5′-DNA + AMP + PP$_i$ (or NMN)

Figure 31-14
Mechanism of the reaction catalyzed by DNA ligase.

leased in the formation of the enzyme-adenylate complex. Thus, two ~P are spent in constructing a phosphodiester bridge in the DNA backbone when ATP is the adenylate donor. Likewise, two ~P are spent in regenerating NAD^+ from NMN and ATP when NAD^+ is the adenylate donor.

THE LINKING NUMBER OF DNA, A TOPOLOGICAL PROPERTY, DETERMINES THE DEGREE OF SUPERCOILING

In 1963, Jerome Vinograd found that circular DNA from polyoma virus emerged in two bands when it was centrifuged. In pursuing this puzzle, he discovered an important property of circular DNA not possessed by linear DNA. Consider a 260-bp DNA duplex in the B-DNA form (Figure 31-15A). Because the number of residues per turn in an unstressed DNA

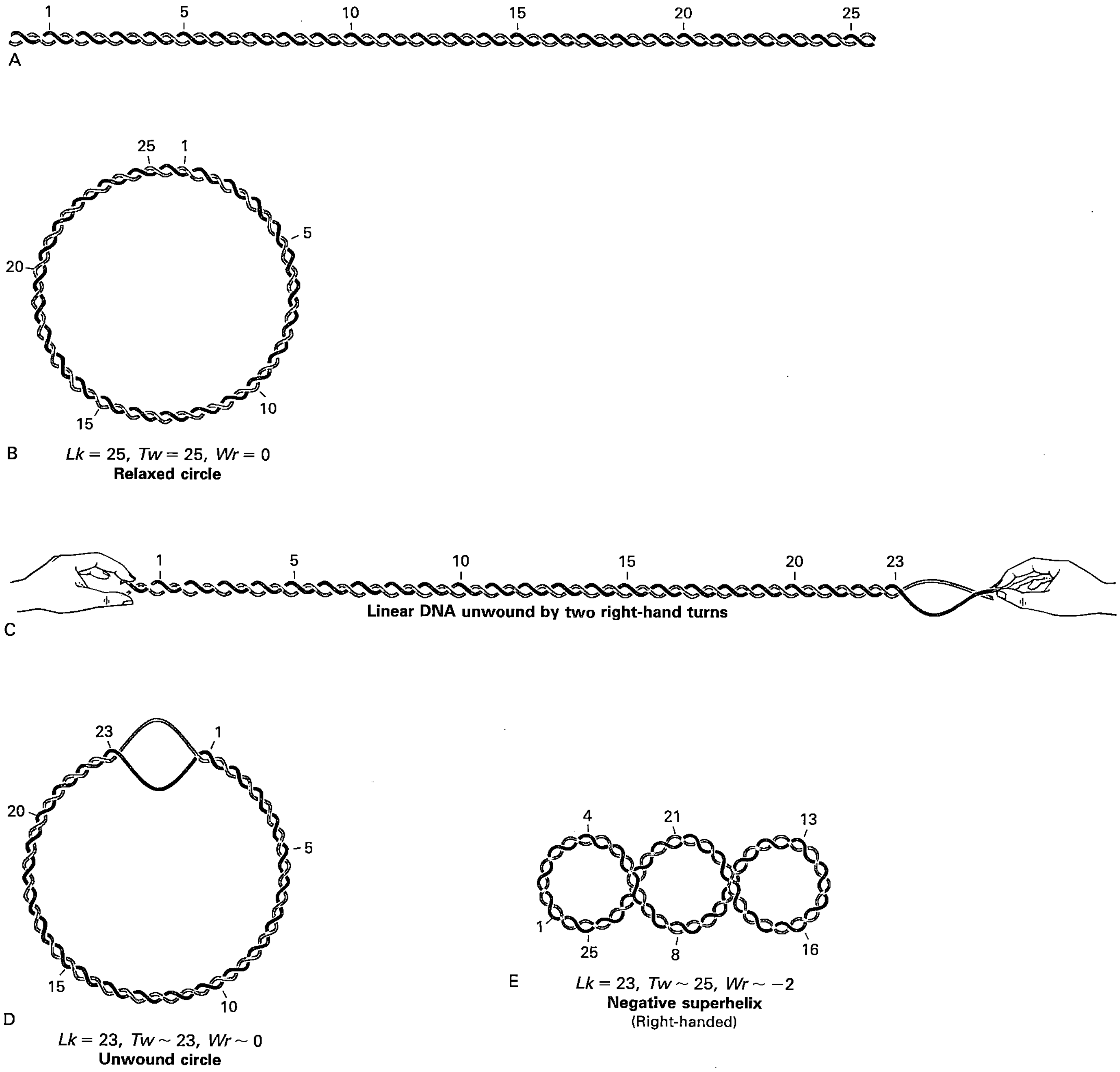

Figure 31-15
Schematic diagram depicting the linking number (*Lk*), twisting number (*Tw*), and writhing number (*Wr*) of circular DNA molecules. [After W. Saenger. *Principles of Nucleic Acid Structure* (Springer-Verlag, 1984), p. 452.]

molecule is 10.4, this linear DNA molecule has 25 (260/10.4) turns. Now let us join the ends of this helix to produce a *relaxed* circular DNA (Figure 31-15B). A different circular DNA can be formed by unwinding the linear duplex by two turns before joining its ends (Figure 31-15C). What is the structural consequence of unwinding prior to ligation? Two limiting conformations are possible. The DNA can fold into a structure containing 23 turns of B helix and an unwound loop (Figure 31-15D). Alternatively, it can adopt a *supercoiled* structure with 25 turns of B helix and 2 turns of *right-handed* (termed *negative*) superhelix (Figure 31-15E).

Supercoiling markedly alters the overall form of DNA. *A supercoiled DNA molecule is more compact than a relaxed DNA molecule of the same length* (Figure 31-16). Hence, supercoiled DNA moves faster than relaxed DNA when they are centrifuged or electrophoresed. The rapidly sedimenting DNA in Vinograd's experiment was supercoiled, whereas the slowly sedimenting DNA was relaxed because one of its strands was nicked.

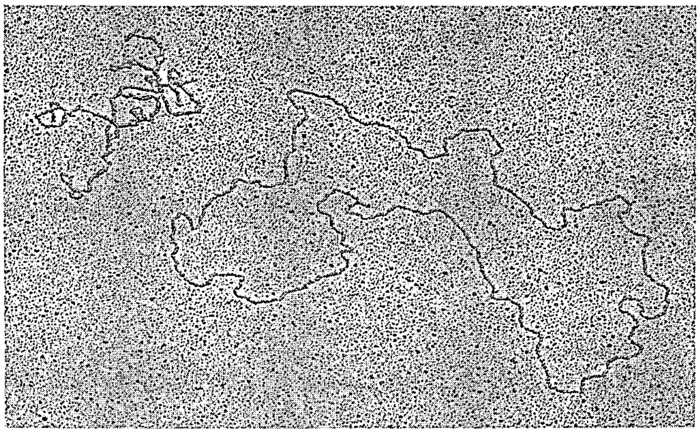

Figure 31-16
Electron micrograph of relaxed and negatively supercoiled DNA. [Courtesy of Dr. Jack Griffith.]

Our understanding of the conformation of DNA is enriched by concepts drawn from topology, a branch of mathematics dealing with structural properties that are unchanged by deformations such as stretching and bending. A key topological property of a circular DNA molecule is its *linking number* (Lk), which is equal to the number of times a strand of DNA winds in the right-handed direction around the helix axis when the axis is constrained to lie in a plane. For the relaxed DNA shown in Figure 31-15B, $Lk = 25$. For the unwound molecule shown in D and the supercoiled one shown in E, $Lk = 23$ because the linear duplex was unwound two complete turns *before* closure. Molecules differing only in linking number are *topological isomers (topoisomers)* of one another. *Topoisomers of DNA can be interconverted only by cutting one or both DNA strands and then rejoining them.*

The unwound DNA and supercoiled DNA shown in parts D and E of Figure 31-15 are topologically identical but geometrically different. They have the same value of *Lk* but differ in *Tw (twist)* and *Wr (writhe)*. Twist reflects the helical winding of the DNA strands around each other, whereas writhe is a measure of the coiling of the axis of the double helix (the rigorous definitions of *Tw* and *Wr* are complex). Is there a relation between *Tw* and *Wr*? Indeed, there is. Topology tells us that the sum of *Tw* and *Wr* is simply equal to *Lk*.

$$Lk = Tw + Wr$$

In Figure 31-15, the unwound circular DNA has $Tw \sim 23$ and $Wr \sim 0$, whereas the supercoiled DNA has $Tw \sim 25$ and $Wr \sim -2$. These forms can be interconverted without cleaving the DNA chain because they have the same value of *Lk*, namely, 23. The partitioning of *Lk* (which must be an integer) between *Tw* and *Wr* (which need not be integers) is determined by energetics. The free energy is minimal when about 70% of the change in *Lk* is expressed in *Wr* and 30% in *Tw*. Hence, the most stable form would be one with $Tw = 24.4$ and $Wr = -1.4$. Thus, *a lowering of* Lk *causes both right-handed (negative) supercoiling of the DNA axis and unwinding of the Watson-Crick duplex.* Topoisomers differing by just 1 in *Lk*, and consequently by 0.7 in *Wr*, can readily be separated by agarose gel electrophoresis because their hydrodynamic volumes are quite different—*supercoiling condenses DNA.*

MOST NATURALLY OCCURRING DNA MOLECULES ARE NEGATIVELY SUPERCOILED

The degree of supercoiling of a DNA molecule can be expressed in terms of the *supercoiling density* (σ), which is independent of length.

$$\sigma = (Lk - Lk_0)/Lk_0$$

Lk_0 is the linking number of the relaxed circular DNA molecule. For the supercoiled DNA in Figure 31-15E, $Lk = 23$, $Lk_0 = 25$, and so $\sigma = -0.08$. The supercoiling density of most natural DNA molecules is about -0.06. The negative sign means that DNA superhelices in nature, as in this example, are right-handed. In other words, they arise from *underwinding* or *unwinding*. *Negative supercoiling prepares DNA for processes requiring the separation of strands: replication, recombination, and transcription.* Positive supercoiling (Figure 31-17) would compact DNA as effectively as negative supercoiling does, but it would make strand separation much more difficult. We see here a compelling reason for the selection of negative supercoiling.

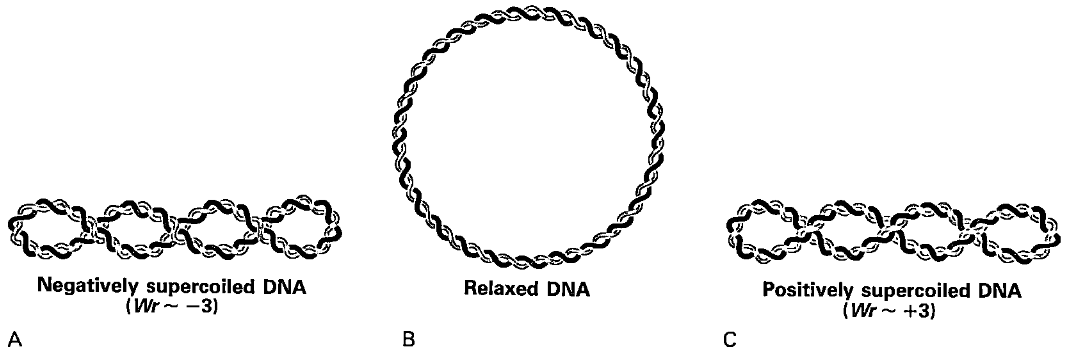

Figure 31-17
Schematic diagram of DNA molecules that differ in their supercoiling. (A) Negatively supercoiled ($Wr \sim -3$, right-handed). (B) Relaxed ($Wr = 0$). (C) Positively supercoiled ($Wr \sim +3$, left-handed). Most naturally occurring DNA molecules are negatively supercoiled.

TOPOISOMERASE I CATALYZES THE RELAXATION OF SUPERCOILED DNA

> "With thy sharp teeth this knot intrinsicate of life at once untie."
>
> William Shakespeare (1608)
> *Antony and Cleopatra* (V, ii)

The interconversion of topoisomers of DNA (Figure 31-18) is catalyzed by enzymes called *topoisomerases* that were discovered by James Wang and Martin Gellert. These enzymes alter the linking number of DNA by catalyzing a three-step process: (1) *cleavage* of one or both strands of DNA, (2) *passage* of a segment of DNA through this break, and (3) *resealing* of the DNA break. Type I topoisomerases cleave just one strand of DNA, whereas type II enzymes cleave both strands.

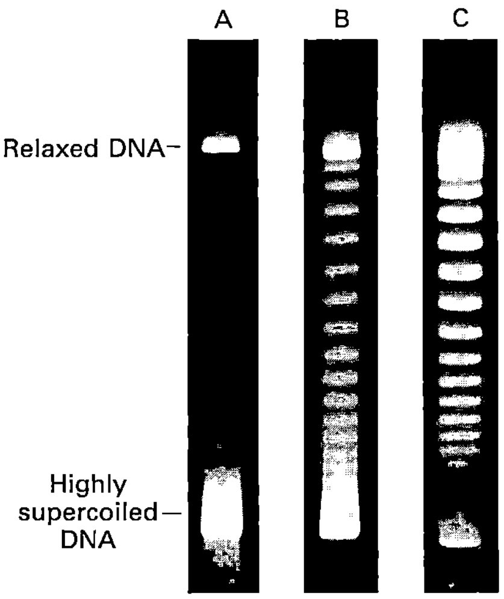

Figure 31-18
Gel patterns showing the relaxation of supercoiled SV40 viral DNA. Part A is a highly negatively supercoiled DNA. Incubation of the DNA with a topoisomerase for (B) 5 minutes and (C) 30 minutes leads to a series of bands that have less supercoiling. The *Lk* values of adjacent bands differ by 1, and average *Wr* values by about 0.7. [From W. Keller. *Proc. Nat. Acad. Sci.* 72(1975):2553.]

The type I topoisomerase of *E. coli* catalyzes the relaxation of negatively supercoiled DNA. One possibility a priori was that this enzyme acts first as a nuclease and then as a ligase. However, intact relaxed circles were formed in the absence of ATP, NAD^+, and other potential energy donors. This finding ruled out an initial hydrolytic event—the formation of a phosphodiester bridge between a free hydroxyl terminus and a free phosphoryl terminus would require an energy source. It suggested instead that *one end of the cleaved strand stays activated by being covalently attached to the enzyme.* In fact, a covalent enzyme-DNA complex has been isolated from a mixture in which sodium dodecyl sulfate denatured the topoisomerase in the midst of its catalytic cycle. The 5′-phosphate moiety of a DNA strand was found to be linked to a tyrosine hydroxyl group of this 97-kd enzyme (Figure 31-19). The 3′-OH at the other end of the cleaved strand nucleophilically attacks this activated intermediate to restore the continuity of the circle. It is noteworthy that the chemical bonds in the substrate and product of this reaction are identical. The only role of this topoisomerase is to create a transient break that allows the passage of a segment of DNA. The linking number of a negatively supercoiled DNA increases by +1 (relaxation) with each catalytic cycle. A striking feature of the structure of the topoisomerase is a hole large enough to accommodate a DNA double helix (Figure 31-20).

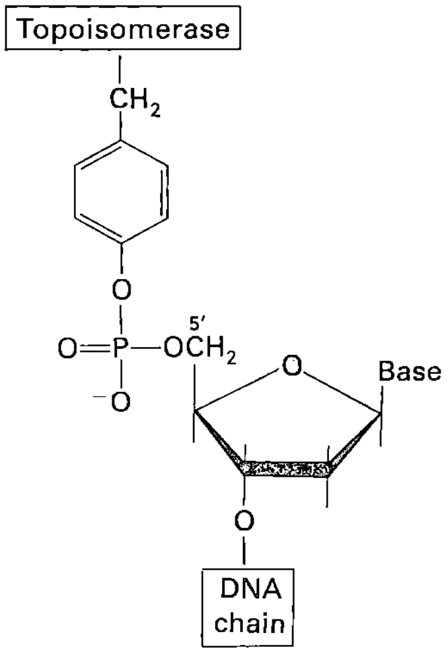

Figure 31-19
Covalent enzyme-substrate intermediate in the action of topoisomerases. The 5′-phosphate end of the cleaved DNA strand is covalently linked to the hydroxyl group of a specific tyrosine residue of the enzyme.

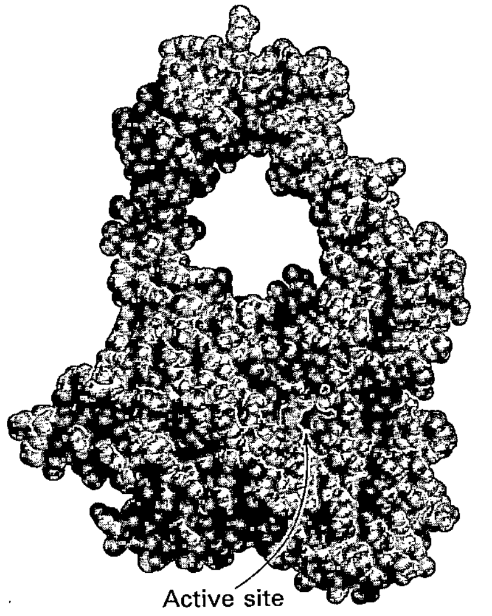

Figure 31-20
Structure of topoisomerase I from *E. coli.* This 67-kd amino-terminal fragment of the 97-kd monomeric enzyme contains a 27.5-Å-diameter hole, large enough to accommodate a DNA double helix. The tyrosine residue that becomes covalently bonded to the 5′-phosphoryl end of the cleaved strand is shown in red, and several other residues of the active site in yellow. The enzyme must undergo large structural changes to encircle double-helical DNA and allow it to pass through the gap between the ends of the cleaved strand. [Drawn from coordinates kindly provided by Dr. A. Mondragón. C.D. Lima, J.C. Wang, and A. Mondragón. *Nature* 367(1994):138.]

DNA GYRASE CATALYZES THE ATP-DRIVEN INTRODUCTION OF NEGATIVE SUPERCOILS INTO DNA

The reaction catalyzed by topoisomerase I, the relaxation of supercoiled DNA, is thermodynamically downhill. What about the introduction of supercoils? Supercoiling costs energy because a supercoiled molecule, in contrast with its relaxed counterpart, is torsionally stressed. The introduction of an additional supercoil into a DNA molecule with $\sigma = -0.06$ costs about 9 kcal/mol. Thus, a good deal of energy is stored in the supercoiling of naturally occurring DNA molecules.

Supercoiling in *E. coli* is catalyzed by *DNA gyrase,* a topoisomerase consisting of two 105-kd A chains and two 95-kd B chains. *DNA gyrase is an energy-transducing device: it converts the free energy of ATP into the torsional energy of supercoiling.* The reaction begins with the wrapping of about 200 bp of DNA around the enzyme (step 1 in Figure 31-21). The binding of ATP then triggers the cleavage of *both* strands of DNA (step 2); the cleavage site on one strand is staggered by four nucleotides from the site on the other strand. The 5′-phosphate terminus of each cleaved strand is linked to a specific tyrosine residue of an A subunit of the enzyme. The anchoring of the two ends of the cut DNA is essential for preventing their free rotation, which would quickly lead to the loss of supercoiling.

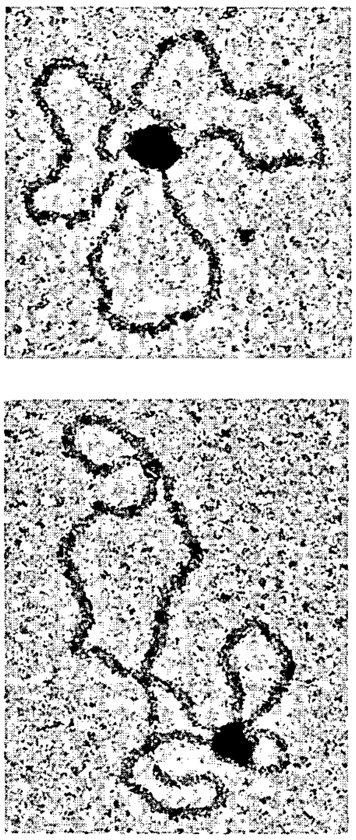

Electron micrographs showing DNA gyrase bound to negatively supercoiled DNA molecules.

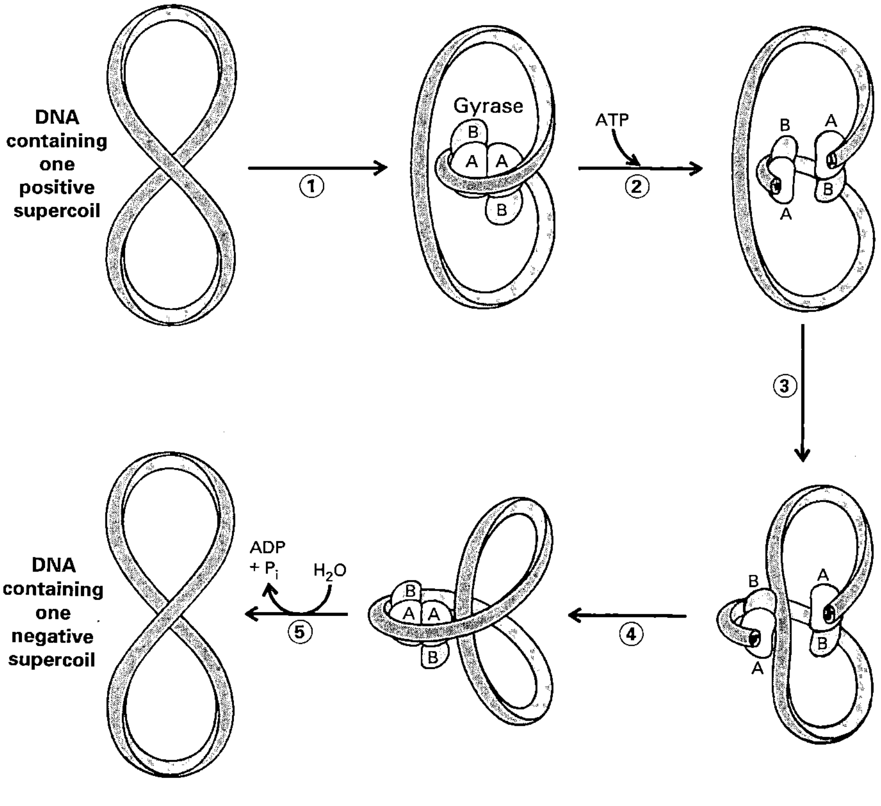

Figure 31-21
Schematic diagram depicting the catalytic action of DNA gyrase in changing the linking number of a duplex DNA molecule. The DNA initially has one positive supercoil. After both strands are cut and rejoined, the DNA has one negative supercoil. The linking number decreases by 2. [After a drawing kindly provided by Dr. Nicholas Cozzarelli.]

A segment of double-stranded DNA then passes through the gap between the fixed ends (step 3). This passage is *vectorial*—DNA gyrase allows facile movement only in the direction leading to negative supercoiling. The cleaved DNA ends then come together to reestablish the continuity of both strands of the duplex (step 4). Finally, hydrolysis of bound ATP leads to the release of the transported DNA segment, enabling the enzyme to begin another round of catalysis (step 5). DNA gyrase decreases the linking number of its substrate in steps of *two*, because duplex DNA passes through a break in *both* strands. About two negative supercoils are introduced per second. DNA gyrase can act repeatedly on the same DNA substrate without dissociating from it.

The degree of supercoiling of bacterial DNA is thus determined by the opposing actions of two enzymes. Negative supercoils are introduced by DNA gyrase and are removed by topoisomerase I. The amounts of these enzymes are regulated to maintain an appropriate degree of negative supercoiling. DNA gyrase is the target of several antibiotics that inhibit the prokaryotic enzyme much more than the eukaryotic one. *Novobiocin* blocks the binding of ATP to gyrase. *Nalidixic acid* and *ciprofloxacin*, by contrast, interfere with the breakage and rejoining of DNA chains. These two gyrase inhibitors are widely used to treat urinary tract and other infections.

DNA POLYMERASE I WAS THE FIRST TEMPLATE-DIRECTED ENZYME TO BE DISCOVERED

We turn now to the molecular events in the replication of DNA. The search for an enzyme that synthesizes DNA was initiated by Arthur Kornberg and his associates in 1955. Their search soon proved fruitful, largely because three inspired choices were made:

1. *What are the activated precursors of DNA?* They correctly deduced that *deoxyribonucleoside 5′-triphosphates* are the activated intermediates in DNA synthesis. This inference was based on two clues. First, pathways of purine and pyrimidine biosynthesis lead to nucleoside 5′-phosphates rather than to nucleoside 3′-phosphates (p. 741). Second, ATP is the activated intermediate in the synthesis of a pyrophosphate bond in coenzymes such as NAD^+, FAD, and CoA (p. 754).

2. *How should DNA synthesis be assayed?* It was surmised that the net amount of DNA synthesized might be very small and so a sensitive assay was essential. This was accomplished by using *radioactive precursor nucleotides.* The incorporation of these precursors into DNA was detected by measuring the radioactivity of an *acid precipitate of the incubation mixture.* (DNA is precipitated by trichloroacetic acid, whereas precursor nucleotides stay in solution.)

3. *Which kinds of cells should be analyzed?* Bacterial cells were used after initial experiments with animal-cell extracts were negative. *E. coli* was chosen because it has a generation time of only 20 minutes and can be harvested in large quantities. As expected, this bacterium is a choice source of enzymes that synthesize DNA.

An extract of *E. coli* was incubated with radioactive deoxythymidine. The level of radioactivity in this ^{14}C-labeled precursor was one million counts per minute. The acid precipitate from this incubation mixture gave just fifty counts. Only a few picomoles of DNA was synthesized, but it was a start. Kornberg wrote: "Although the amount of nucleotide incorporated into nucleic acid was miniscule, it was nonetheless significantly above the level of background noise. Through this tiny crack we tried to drive a wedge. The hammer was enzyme purification, a technique that had matured during the elucidation of alcoholic fermentation."

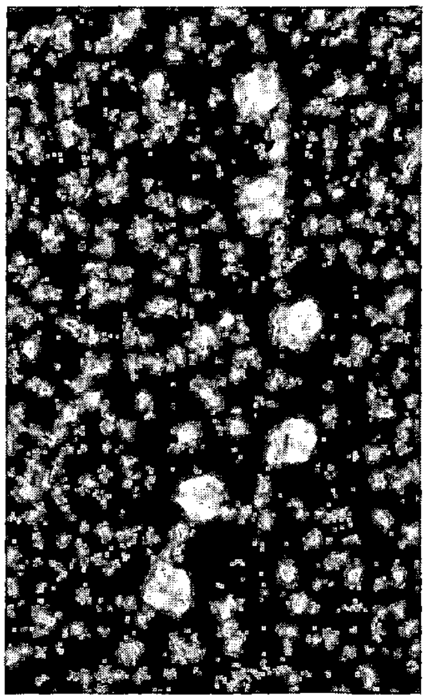

Figure 31-22
Electron micrograph of DNA polymerase I molecules bound to a synthetic DNA template. [Courtesy of Dr. Jack Griffith.]

This new enzyme was named *DNA polymerase*—it is now called *DNA polymerase I* because other DNA polymerases have since been isolated. After a decade of effort in Kornberg's laboratory, DNA polymerase I was

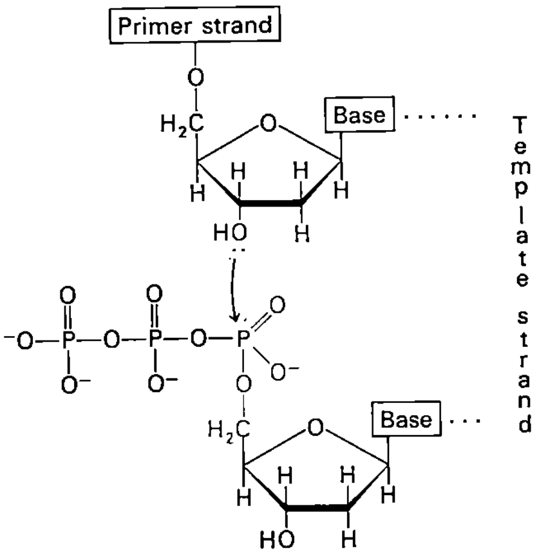

Figure 31-23
Chain-elongation reaction catalyzed by DNA polymerases.

> *Processive enzyme*—
> From the Latin *procedere,* "to go forward."
> An enzyme that catalyzes multiple rounds of elongation or digestion of a polymer while the polymer stays bound. A *distributive enzyme,* by contrast, releases its polymeric substrate between successive catalytic steps.

purified to homogeneity and characterized in detail. An appreciation of the magnitude of the task can be gained by noting that it took 100 kilograms of *E. coli* cells to produce 500 milligrams of pure enzyme. As will be discussed shortly, DNA polymerase I is not the enzyme that replicates most of the DNA in *E. coli.* However, it plays a critical role in replication and also in the repair of DNA. Moreover, it is the simplest DNA polymerase and the best understood. The study of this enzyme has been highly rewarding because it exemplifies many key principles of the action of both prokaryotic and eukaryotic DNA polymerases.

DNA polymerase I, a 103-kd monomer, catalyzes the *step-by-step addition of deoxyribonucleotide units to the 3′ end of a DNA chain:*

$$(\text{DNA})_{n\ \text{residues}} + \text{dNTP} \rightleftharpoons (\text{DNA})_{n+1} + \text{PP}_i$$

As was discussed in Chapter 4 (p. 88), DNA polymerase I requires *all four deoxyribonucleoside 5′-triphosphates—dATP, dGTP, dTTP, and dCTP—and* Mg^{2+} to synthesize DNA. The symbol dNTP denotes any deoxyribonucleoside 5′-triphosphate. The enzyme adds deoxyribonucleotides to the free 3′-OH of the chain undergoing elongation, *which proceeds in the 5′ → 3′ direction* (Figure 31-23). A *primer chain* with a free 3′-OH group is needed at the start. A *DNA template* containing a single-stranded region is also essential. The polymerase catalyzes the nucleophilic attack of the 3′-OH terminus of the primer on the innermost phosphorus atom of a dNTP. A phosphodiester bond is formed and pyrophosphate is released. The reaction is driven forward by the subsequent hydrolysis of PP_i by inorganic pyrophosphatase. DNA polymerase I is a *moderately processive* enzyme—it catalyzes the formation of about 20 phosphodiester bonds before dissociating from the template DNA.

A striking feature of the enzyme is that it takes instructions from its template. Indeed, DNA polymerase I was the first template-directed enzyme to be discovered. Polymerization is catalyzed by a single active site that can bind any of the four dNTPs. Which one binds depends on the corresponding base on the template strand. *The likelihood of binding and of making a phosphodiester bond is very low unless the incoming nucleotide forms a Watson-Crick base pair with the opposing nucleotide on the template.* The most compelling evidence for the high fidelity of this enzyme is that ϕX174 viral DNA replicated in vitro by DNA polymerase I is fully infectious.

DNA POLYMERASE I IS ALSO A PROOFREADING 3′ → 5′ EXONUCLEASE

DNA polymerase I can catalyze the hydrolysis of DNA chains as well as their polymerization. The enzyme catalyzes the hydrolysis of *unpaired* nucleotides, one at a time, at the 3′ end of DNA chains. In other words, *DNA polymerase I is a 3′ → 5′ exonuclease.* As will be seen shortly, the exonuclease site is distinct from the polymerase site. Experiments using synthetic DNA with a mismatched residue at the primer terminus revealed that the *3′ → 5′ nuclease activity has an editing function in polymerization.* The polymer shown in Figure 31-24 contains an unpaired C at the end of a sequence of T residues that form a double helix with a longer sequence of A residues. On addition of DNA polymerase I and dTTP, the unpaired C is first excised. dCMP is released and a 3′-OH group is exposed on the terminal T of the primer strand. Additional T are added only after removal of the unpaired C. In general, *DNA polymerase I removes mismatched residues at the primer terminus before proceeding with polymerization.*

This 3′ → 5′ exonuclease activity markedly enhances the accuracy of DNA replication by serving as a second test of the correctness of base pairing. Polymerization does not usually occur unless the base pair fits into a double helix. However, an error made at this stage is nearly always corrected before the next nucleotide is added. In effect, *DNA polymerase I examines the result of each polymerization it catalyzes before proceeding to the next.* An incorrectly paired nucleotide is nearly always removed. Indeed, about 1 in 10 *correctly* paired nucleotides is removed by the 3′ → 5′ exonuclease activity of DNA polymerase III, a price paid for having a very vigilant editor. The error frequency of the enzyme is estimated to be 10^{-7}, the product of the initial mispairing frequency of 10^{-4} and the proofreading error frequency of 10^{-3}. Three repair systems acting after replication further improve the fidelity to give an overall mutation rate of 10^{-10} per base pair per generation.

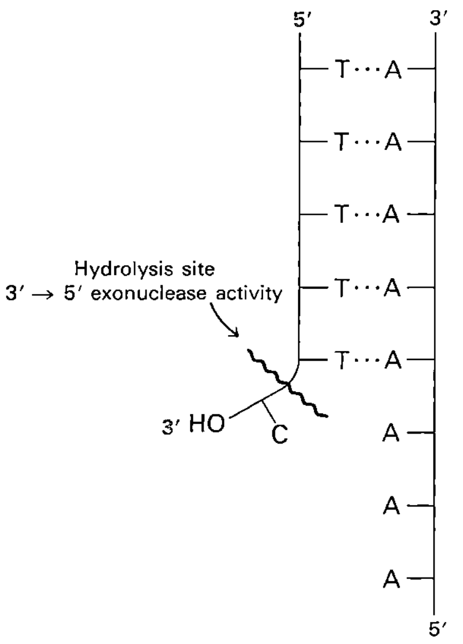

Figure 31-24
3′ → 5′ exonuclease action of DNA polymerase I.

DNA POLYMERASE I IS ALSO AN ERROR-CORRECTING 5′ → 3′ EXONUCLEASE

DNA polymerase I can also hydrolyze DNA starting from the 5′ end of a chain. This *5′ → 3′ nuclease activity* (Figure 31-25) is very different from the 3′ → 5′ exonuclease action discussed above. First, cleavage can occur at the terminal phosphodiester bond or at a bond several residues away from the 5′ terminus. The 5′ end can bear a free hydroxyl group or it can be phosphorylated. Second, the cleaved bond must be in a double-helical region. Third, 5′ → 3′ exonuclease activity is enhanced by concomitant DNA synthesis. Fourth, the active site for exonuclease action is clearly separate from the active sites for polymerization and 3′ → 5′ hydrolysis. Thus, *DNA polymerase I contains three different active sites on its single polypeptide chain.*

The 5′ → 3′ exonuclease activity plays a key role in DNA replication by removing RNA primer, as will be discussed shortly (p. 806). Moreover, the 5′ → 3′ exonuclease complements the 3′ → 5′ exonuclease activity by correcting errors of a different type. For example, the 5′ → 3′ exonuclease participates in the excision of pyrimidine dimers formed by exposure of DNA to ultraviolet light.

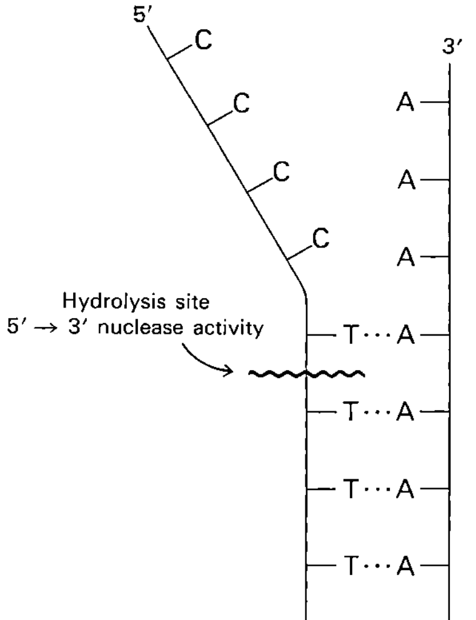

Figure 31-25
5′ → 3′ nuclease action of DNA polymerase I.

DNA POLYMERASE I CONTAINS A TEMPLATE-BINDING CLEFT AND A POLYMERASE ACTIVE-SITE CLEFT

This trifunctional enzyme can be split by proteases into a 36-kd *small fragment* with all the original 5′ → 3′ exonuclease activity and a 67-kd *large fragment* (also called the *Klenow fragment*) with all of the polymerase and 3′ → 5′ exonuclease activities (Figure 31-26).

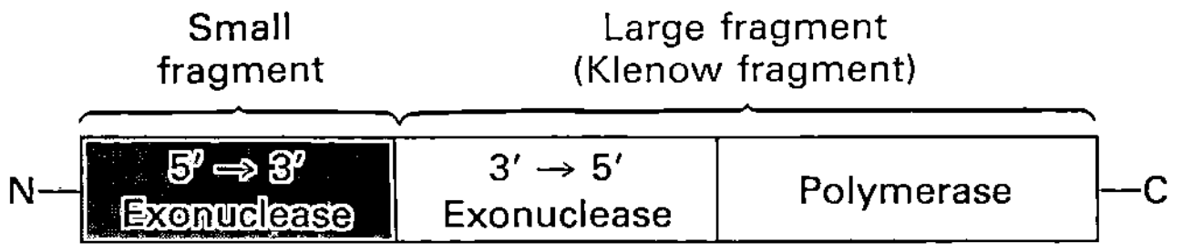

Figure 31-26
DNA polymerase I has three enzymatic activities in a single polypeptide chain.

X-ray crystallographic studies of the large fragment show that it is about 90 Å long. Two prominent clefts that are nearly perpendicular to each other are evident (Figure 31-27A). One cleft serves as the binding site for duplex DNA (Figure 31-27B). The other cleft is thought to contain the catalytic site for polymerization and the binding site for single-stranded template. The 3′ → 5′ exonuclease site is located sufficiently close to the polymerase site to allow the 3′ terminus of the growing chain to shuttle back and forth between them. Thus, polymerization and editing can proceed nearly simultaneously without release of DNA from the enzyme.

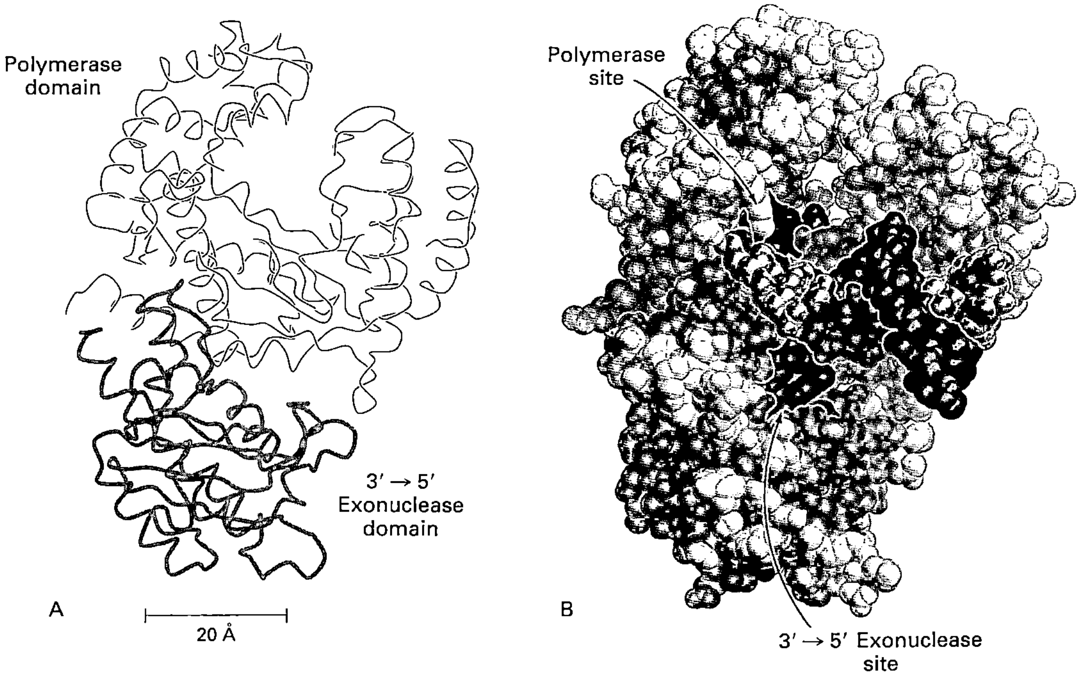

Figure 31-27
Structure of the large fragment (Klenow fragment) of DNA polymerase I. (A) Main-chain trace. The polymerase domain is shown in blue, and the 3′ → 5′ exonuclease domain in yellow. (B) Space-filling model. The primer strand of the bent DNA duplex is shown in red, and the template strand in green. The 3′ end of the primer strand is located in the exonuclease editing site. Several key residues in the polymerase catalytic site are shown in purple. The 3′ end of the primer strand shuttles between these sites, which are about 35 Å apart. [Drawn from coordinates kindly provided by Dr. Thomas Steitz. L.S. Beese, V. Derbyshire, and T.A. Steitz. *Science* 260(1993):352.]

DISCOVERY OF DNA POLYMERASES II AND III

DNA polymerase I can synthesize and repair DNA in vitro. Does it perform these functions in vivo? This question is pertinent because an enzyme need not carry out the same reaction in vivo as it does in vitro. The reaction conditions inside a cell may be different, and other enzymes may be present. In fact, *E. coli* contains two other DNA polymerases, named *II* and *III*, which were discovered some 15 years after DNA polymerase I. Why this lag? The reason is that the presence of polymerases II and III was masked by the high level of activity of polymerase I.

The breakthrough came in 1969, when Paula DeLucia and John Cairns isolated a mutant of *E. coli* that had only about 1% of the normal polymerizing activity of DNA polymerase I. This mutant (named *polA1*) multiplied at the same rate as the parent strain, but it was much more easily killed by ultraviolet light. DeLucia and Cairns inferred that DNA replication in *polA1* was normal, whereas DNA repair was markedly impaired. They suggested that a polymerase different from DNA polymerase I might be essential for DNA synthesis.

Indeed, two new polymerases were soon isolated and characterized in several laboratories. *DNA polymerases II and III are like polymerase I in several respects:*

1. They catalyze a template-directed synthesis of DNA from deoxyribonucleoside 5′-triphosphate precursors.
2. A primer with a free 3′-OH group is required.
3. Synthesis is in the 5′ → 3′ direction.
4. They possess 3′ → 5′ exonuclease activity.

What are the roles of the three polymerases in vivo? As will be discussed shortly, *a multisubunit assembly containing polymerase III synthesizes most new DNA, whereas polymerase I erases the primer and fills gaps.* DNA polymerase II participates in DNA repair but is not needed for DNA replication. Biochemical and genetic studies have revealed that more than 20 proteins in addition to DNA polymerases I and III are required for DNA replication in *E. coli*. Before turning to the interplay of these proteins with DNA, let us take a view of replication at the level of the whole chromosome.

PARENTAL DNA IS UNWOUND AND NEW DNA IS SYNTHESIZED AT REPLICATION FORKS

DNA in the midst of replication has been visualized by autoradiography and electron microscopy. Autoradiography has low resolution (~50 nm) but is attractive because only what is labeled is seen. DNA is made visible in autoradiographs by the incorporation of tritiated thymine or thymidine. Autoradiographs and electron micrographs show that replicating DNA from *E. coli* has the form of a closed circle with an internal loop (Figure 31-28). These forms, called *theta structures* because of their resem-

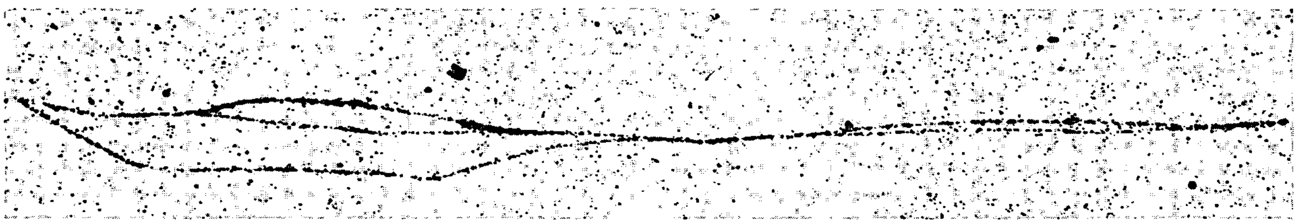

Figure 31-28
Autoradiograph of replicating DNA from *E. coli*. [Courtesy of Dr. John Cairns.]

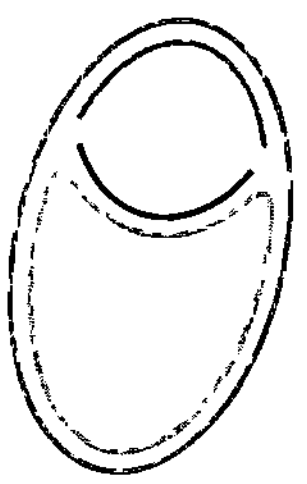

Figure 31-29
Schematic diagram of the circular chromosome of *E. coli* during replication. The green strands in this theta structure depict parental DNA, and the red strands represent newly synthesized DNA.

blance to the Greek letter θ (Figure 31-29), reveal that the *DNA molecule maintains its circular form while it is being replicated.* The resolution of neither technique suffices to display free ends. However, it is clear that long stretches of single-stranded DNA are absent. Thus, these pictures rule out a replication mechanism in which the parental DNA strands unwind completely before serving as templates for the synthesis of new DNA. Rather, *the synthesis of new DNA is closely coupled to the unwinding of parental DNA.* A site of simultaneous unwinding and synthesis is called a *replication fork.*

ONE STRAND OF DNA IS MADE IN FRAGMENTS AND THE OTHER STRAND IS SYNTHESIZED CONTINUOUSLY

At a replication fork, both strands of parental DNA serve as templates for the synthesis of new DNA. Recall that the parental strands are antiparallel (p. 80) and that DNA replication is semiconservative (p. 83). Hence, the *overall* direction of DNA synthesis must be $5' \rightarrow 3'$ for one daughter strand and $3' \rightarrow 5'$ for the other (Figure 31-30). However, all known DNA polymerases synthesize DNA in the $5' \rightarrow 3'$ direction but not in the $3' \rightarrow 5'$ direction. How then does one of the daughter DNA strands *appear at low resolution* to grow in the $3' \rightarrow 5'$ direction?

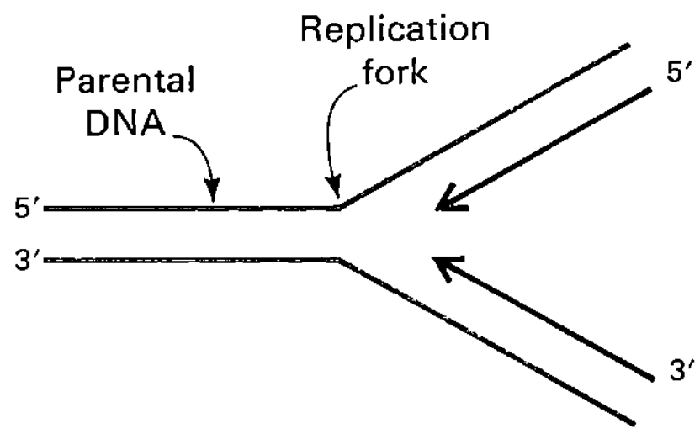

Figure 31-30
At *low resolution,* the *apparent* direction of DNA replication is $5' \rightarrow 3'$ for one daughter strand and $3' \rightarrow 5'$ for the other. Actually, both strands are synthesized in the $5' \rightarrow 3'$ direction, as shown in Figure 31-31.

This dilemma was resolved by Reiji Okazaki, who found that a *significant proportion of newly synthesized DNA exists as small fragments.* These units of about a thousand nucleotides (called *Okazaki fragments*) are present briefly in the vicinity of the replication fork. As replication proceeds, these fragments become covalently joined by DNA ligase to form one of the daughter strands (Figure 31-31). The other new strand is synthesized continuously. The strand formed from Okazaki fragments is termed the *lagging strand,* whereas the one synthesized without interruption is the *leading strand.* Both the Okazaki fragments and the leading strand are synthesized in the $5' \rightarrow 3'$ direction. *The discontinuous assembly of the lagging strand enables $5' \rightarrow 3'$ polymerization at the nucleotide level to give rise to overall growth in the $3' \rightarrow 5'$ direction.*

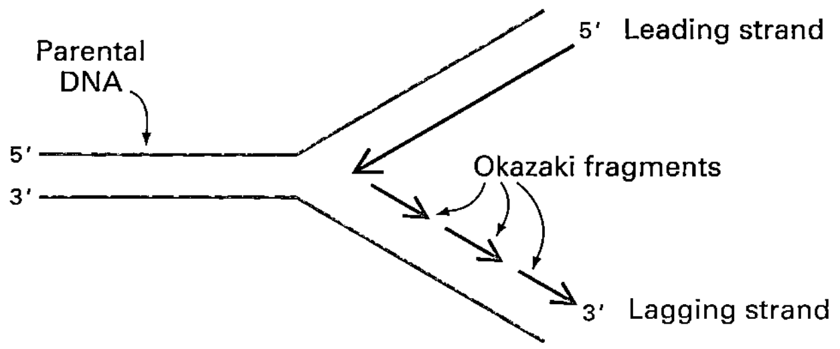

Figure 31-31
Schematic diagram of a replication fork. Both strands of DNA are synthesized in the $5' \rightarrow 3'$ direction. The leading strand is synthesized continuously, whereas the lagging strand is synthesized in the form of short fragments (Okazaki fragments).

REPLICATION BEGINS WITH THE UNWINDING OF THE *OriC* (ORIGIN) SITE

How does DNA synthesis begin? Plasmids bearing *oriC*, the unique origin of replication in *E. coli*, have been invaluable in unraveling the molecular mechanism of this key process. The *oriC* locus is a sequence of 245 bp in the *E. coli* chromosome, a DNA molecule containing 4×10^6 bp. The origin has some unusual features (Figure 31-32). It contains a tandem array of three 13-mers with nearly identical sequences. Each begins with GATC, a sequence that appears 11 times in *oriC*, many times more often than would be expected on a random basis. These 13-mers are rich in A·T base pairs, which facilitates the melting of the duplex to begin DNA synthesis. Methylation of adenine in GATC may be important in controlling when replication begins. The timing of DNA replication is critical because it must respond to signals of increased cell mass and be coordinated with cell division.

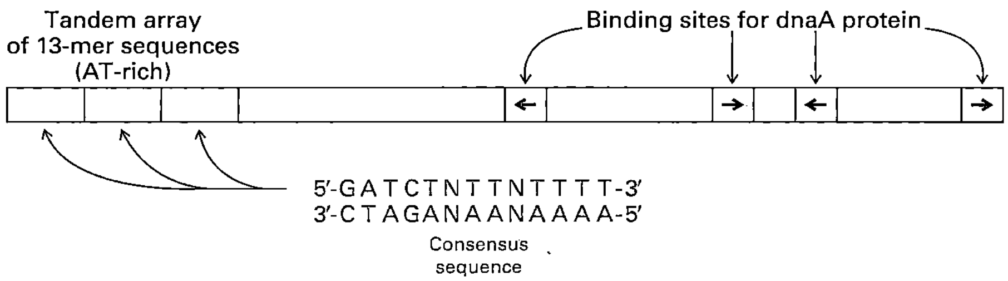

Figure 31-32
OriC, the origin of replication in *E. coli*, has a length of 245 bp. It contains a tandem array of three nearly identical 13-nucleotide sequences (green) and four binding sites (yellow) for dnaA protein. The orientations of the four dnaA sites are denoted by arrows.

The binding of the *dnaA protein* (named for its gene) to four sites on *oriC* near these AT-rich 13-mers initiates an intricate sequence of steps leading to the unwinding of the template DNA and the synthesis of a primer. The DNA must be negatively supercoiled to enable dnaA to bind. The dnaB and dnaC proteins join dnaA to bend and open the double helix. *DnaB protein is a helicase: it catalyzes the ATP-driven unwinding of double-helical DNA.* The unwound portion of DNA is then stabilized by *single-strand binding protein* (SSB), a tetramer of 19-kd subunits, which binds cooperatively to single-stranded DNA. The unwinding of circular DNA at the origin produces positive supercoiling, which must be relieved to allow replication to continue. Relief is provided by the compensatory action of *DNA gyrase* (p. 798), which introduces negative supercoils as it hydrolyzes ATP.

AN RNA PRIMER SYNTHESIZED BY PRIMASE ENABLES DNA SYNTHESIS TO BEGIN

The DNA template is now exposed, but new DNA cannot be synthesized until a primer is constructed. Recall that all known DNA polymerases require a primer with a free 3′-OH group for DNA synthesis. How is this primer formed? An important clue came from the observation that RNA synthesis is essential for the initiation of DNA synthesis. This finding, taken together with the fact that RNA polymerases can start chains de novo, suggested that RNA might prime the synthesis of DNA. Kornberg then found that *nascent DNA is covalently linked to a short stretch of RNA.*

In fact, *RNA primes the synthesis of DNA.* A specialized RNA polymerase called *primase* joins the prepriming complex in a multisubunit assembly called the *primosome.* Primase synthesizes a short stretch of RNA (~5 nu-

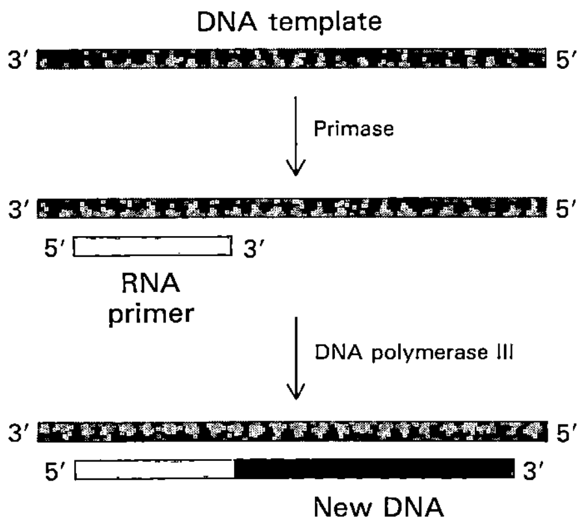

Figure 31-33
DNA replication is primed by a short stretch of RNA that is synthesized by primase, an RNA polymerase. The primer RNA is excised at a later stage of replication.

cleotides) that is complementary to one of the template DNA strands (Figure 31-33). This primer RNA is removed at the end of replication by the 5′ → 3′ exonuclease activity of DNA polymerase I. An RNA primer would be unnecessary if DNA polymerases could start chains de novo. However, such a property would be incompatible with the very high fidelity of DNA polymerases, which is due in part to their proofreading of nascent DNA. Recall that DNA polymerase I tests the correctness of the preceding base pair before forming a new phosphodiester bond. RNA polymerases can start chains de novo because they do not examine the preceding base pair. Consequently, their error rates are orders of magnitude higher than those of DNA polymerases. The ingenious solution is to start DNA synthesis with a low-fidelity stretch of polynucleotide but mark it "temporary" by placing ribonucleotides in it. The use of an RNA polymerase to initiate DNA synthesis is also plausible from an evolutionary viewpoint, because RNA was probably present long before DNA emerged as a more reliable and stable store of genetic information.

DNA POLYMERASE III HOLOENZYME, A HIGHLY PROCESSIVE AND PRECISE ENZYME, SYNTHESIZES MOST OF THE DNA

The stage is now set for the entry of *DNA polymerase III holoenzyme.* The hallmarks of this multisubunit assembly are its *very high processivity, catalytic potency, and fidelity.* The holoenzyme catalyzes the formation of many thousands of phosphodiester bonds before releasing its template, compared with only 20 for DNA polymerase I. The much lower degree of processivity of DNA polymerase I is well suited to its role as a gap filler (p. 809) and repair enzyme. DNA polymerase III holoenzyme, on the other hand, is designed to grasp its template and not let go until it has been completely replicated. A second distinctive feature of the holoenzyme is its catalytic prowess: 1000 nucleotides added per second compared with only 10 per second by DNA polymerase I. This acceleration is accomplished with no loss of accuracy. The greater catalytic prowess of polymerase III is largely due to its processivity; no time is lost in repeatedly stepping on and off the template.

These striking features of DNA polymerase III do not come cheaply. The holoenzyme consists of 10 kinds of polypeptide chains and has a mass of ~900 kd, nearly an order of magnitude larger than that of single-chain DNA polymerase I. This replication complex is an *asymmetric dimer* (Figure 31-34). The holoenzyme is a dimer because both strands of parental DNA must be replicated in the same place at the same time. How-

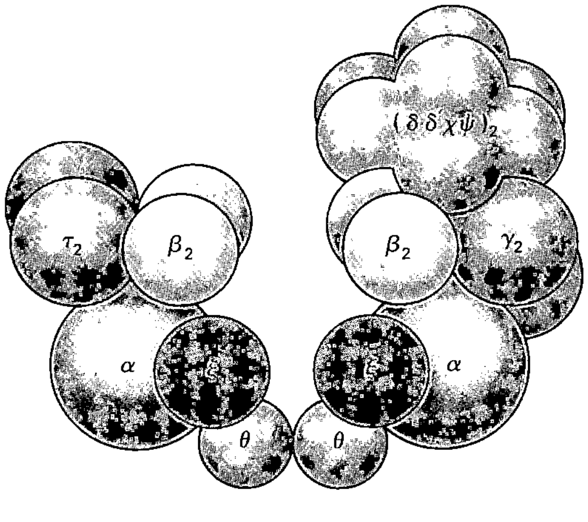

Figure 31-34
Proposed architecture of DNA polymerase III holoenzyme, an asymmetric dimer. The catalytic (α) subunits are shown in green, the $3' \rightarrow 5'$exonuclease subunits (ε) in red, and the sliding clamp subunits (β_2) in yellow. [After *DNA Replication* (2nd ed.), by A. Kornberg and T. Baker. W.H. Freeman and Company. Copyright © 1992.]

ever, the leading and lagging strands are synthesized differently—hence, the holoenzyme must also be asymmetric. A β_2 subunit is associated with one branch of the holoenzyme, and τ_2 and $(\gamma\delta\delta'\chi\psi)_2$ with the other. The core of each branch is the same, an $\alpha\varepsilon\theta$ complex. The α subunit is the polymerase and the ε subunit is the proofreading $3' \rightarrow 5'$ exonuclease. Each core is catalytically active but not processive. Processivity is conferred by β_2 and τ_2.

The β subunits make the holoenzyme highly processive by encircling the template DNA. The β_2 dimer has the form of a star-shaped ring (Figure 31-35A). The 35-Å-diameter hole in its center can readily accommodate a duplex

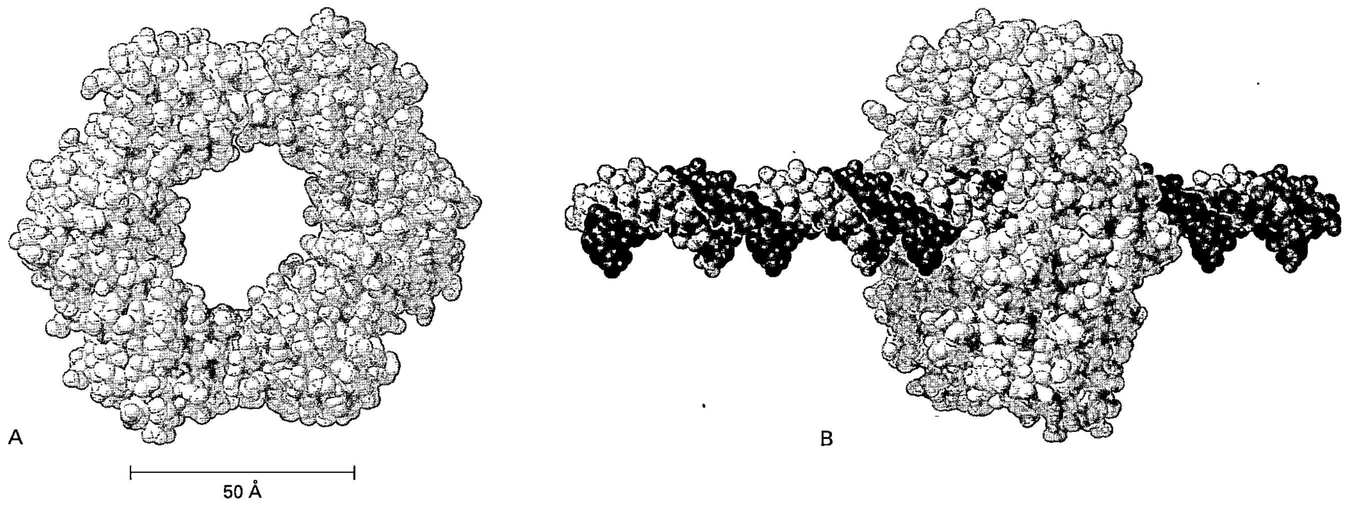

Figure 31-35
A sliding DNA clamp is critical for processivity. The dimeric β subunit (yellow and blue) of DNA polymerase III encircles the DNA template (red and green). (A) β_2 alone. (B) β_2 encircling duplex DNA. The structure of the protein was determined by x-ray crystallography; the position of DNA was inferred from model building. [Drawn from coordinates kindly provided by Dr. John Kuriyan. After X.-P. Kong, R. Onrust, M. O'Donnell, and J. Kuriyan. *Cell* 69(1992):425.]

DNA molecule (Figure 31-35B). *The template is trapped, but there is enough space between the DNA and protein to allow rapid sliding and turning during replication.* A catalytic rate of 1000 nucleotides polymerized per second requires the sliding of 100 turns of duplex DNA (a length of 3400 Å or 0.34 μm) through the central hole of β_2. Thus, β_2 plays a key role in replication by serving as a *sliding DNA clamp.*

THE LEADING AND LAGGING DAUGHTER STRANDS ARE SIMULTANEOUSLY SYNTHESIZED BY THE HOLOENZYME

DNA polymerase III holoenzyme is now positioned at the replication fork, where it begins the synthesis of the leading strand using the RNA primer formed by primase (Figure 31-36). The duplex DNA ahead of the polymerase is unwound by the ATP-driven helicase. Single-strand binding protein again keeps the unwound DNA extended and accessible so that both strands can serve as templates. The leading strand is synthesized continuously by polymerase III, which does not release the template until replication has been completed. DNA gyrase concurrently introduces right-handed (negative) supercoils to avert a topological crisis.

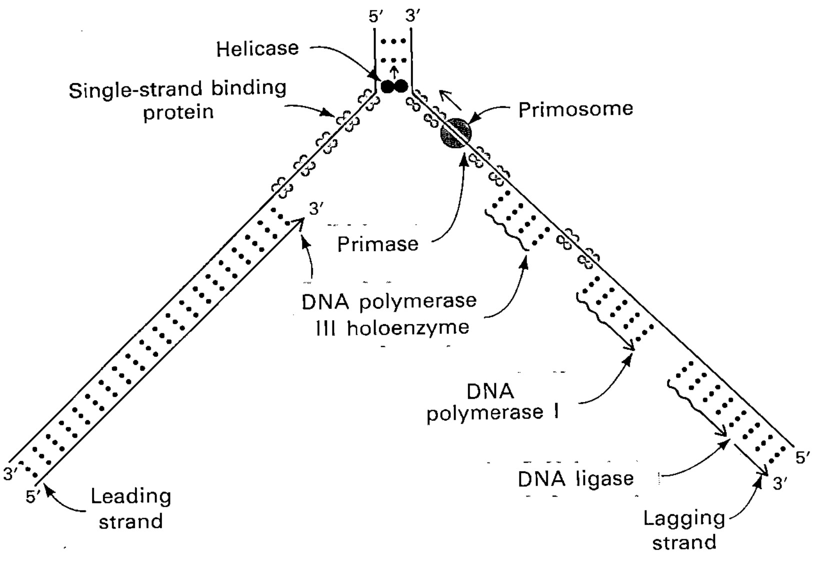

Figure 31-36
Schematic diagram of the enzymatic events at a replication fork of *E. coli.* Enzymes shaded yellow catalyze chain initiation, elongation, and ligation. The wavy lines on the lagging strand denote RNA primers. [After *DNA Replication* (2nd ed.), by A. Kornberg and T. Baker. W.H. Freeman and Company. Copyright © 1992.]

Table 31-2
DNA replication proteins of *E. coli*

Protein	*Role*	*Size (kd)*	*Molecules per cell*
Helicase (dnaB protein)	Unwinds the double helix	300	20
Primase	Synthesizes RNA primers	60	50
SSB	Stabilizes single-stranded regions	74	300
DNA gyrase	Introduces negative supercoils	400	250
DNA polymerase III holoenzyme	Synthesizes DNA	~900	20
DNA polymerase I	Erases primer and fills gaps	103	300
DNA ligase	Joins the ends of DNA	74	300

The mode of synthesis of the lagging strand is necessarily more complex. As was mentioned earlier, the lagging strand is synthesized in fragments so that $5' \rightarrow 3'$ polymerization leads to overall growth in the $3' \rightarrow$

5′ direction. *This may be accomplished by a looping of the template for the lagging strand* (Figure 31-37). *The lagging-strand template would then pass through the polymerase site in one subunit of a dimeric polymerase III in the same direction as the leading-strand template in the other subunit.* DNA polymerase III would have to let go of the lagging strand template after about 1000 nucleotides have been added to the lagging strand. A new loop would then be formed, and primase would again synthesize a short stretch of RNA primer to initiate the formation of another Okazaki fragment.

The gaps between fragments of the nascent lagging strand are then filled by DNA polymerase I. This essential enzyme also uses its 5′ → 3′ exonuclease activity to remove the RNA primer lying ahead of the polymerase site (Figure 31-38). The primer cannot be erased by DNA polymerase III because the enzyme lacks 5′ → 3′ editing capability. Finally, DNA ligase joins the fragments.

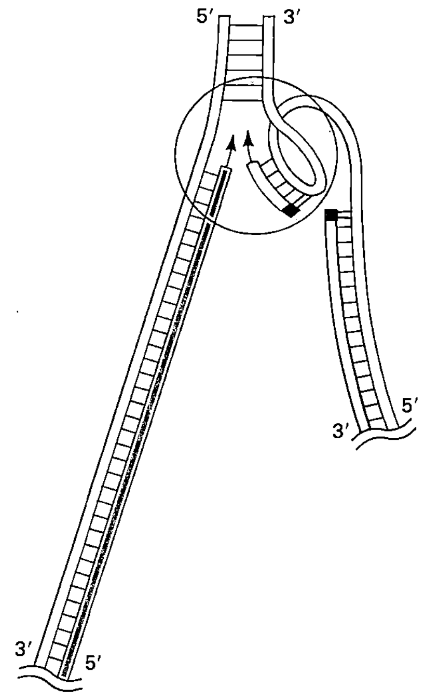

Figure 31-37
The looping of the template for the lagging strand enables a dimeric DNA polymerase III holoenzyme (colored yellow) at the replication fork to synthesize both daughter strands. The leading strand is shown in blue, the lagging strand in green, and the RNA primer in red. [Courtesy of Dr. Arthur Kornberg.]

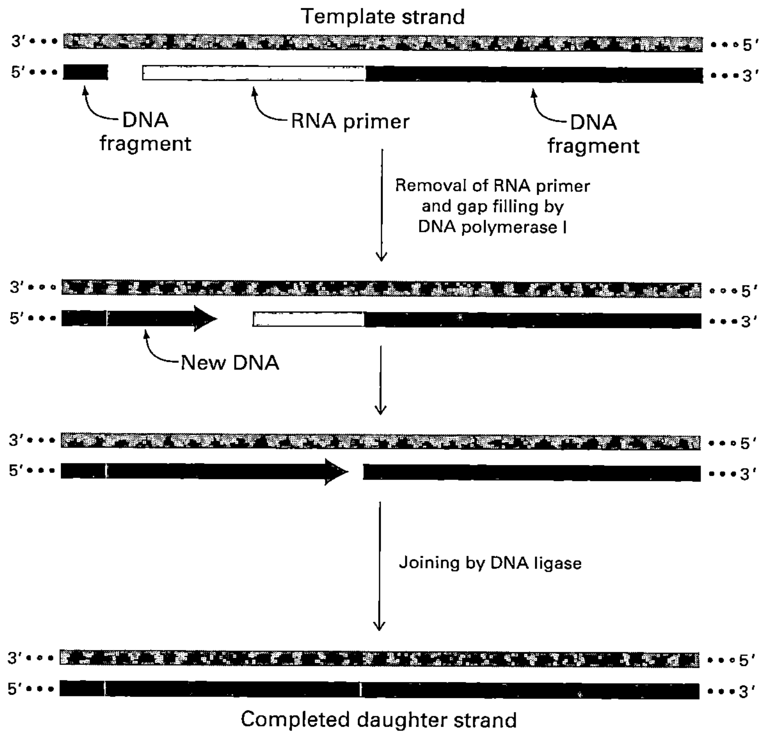

Figure 31-38
The RNA primer is removed by the 5′ → 3′ exonuclease activity of DNA polymerase I. The gap in the daughter strand is then filled by the polymerase action of this enzyme. DNA ligase catalyzes the final linking step.

MUTATIONS ARE PRODUCED BY SEVERAL TYPES OF CHANGES IN THE BASE SEQUENCE OF DNA

We turn now from DNA replication to DNA mutations and repair. Several types of mutations are known: (1) the *substitution* of one base pair for another, (2) the *deletion* of one or more base pairs, and (3) the *insertion* of one or more base pairs. The spontaneous mutation rate of T4 phage is about 10^{-7} per base per replication. *E. coli* and *Drosophila melanogaster* have much lower mutation rates, of the order of 10^{-10}.

The substitution of one base pair for another is the most common type of mutation. Two types are possible. A *transition* is the replacement of one purine by the other, or of one pyrimidine by the other. In contrast, a *transversion* is the replacement of a purine by a pyrimidine, or of a pyrimidine by a purine.

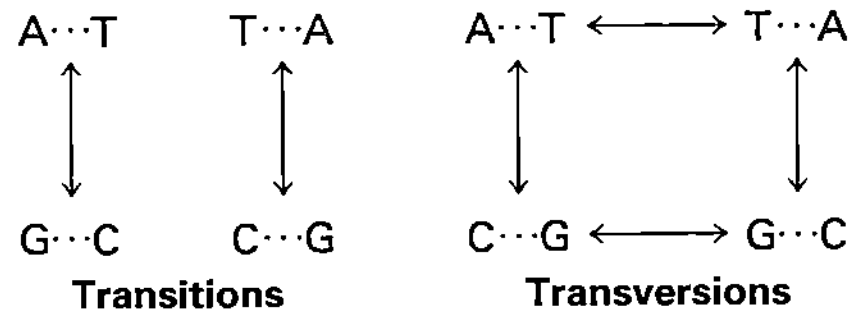

A mechanism for the spontaneous occurrence of transitions was suggested by Watson and Crick in their classic paper on the DNA double helix. They noted that some of the hydrogen atoms on each of the four bases can change their location to produce a tautomer. An *amino* group ($-NH_2$) can tautomerize to an *imino* form (=NH). Likewise, a *keto* group (−C=O) can tautomerize to an *enol* form (=C−OH). The fraction of each base in the form of these imino and enol tautomers is about 10^{-4}.

These transient tautomers can form nonstandard base pairs that fit into a double helix. For example, the imino tautomer of adenine can pair with cytosine (Figure 31-39). This $A^* \cdot C$ pairing (the asterisk denotes the imino tautomer) would allow C to become incorporated into a growing DNA strand where T was expected, and it would lead to a mutation if left uncorrected. In the next round of replication, A* will probably re-tautomerize to the standard form, which pairs as usual with thymine, but the cytosine will pair with guanine. Hence, one of the daughter DNA molecules will contain a G · C base pair in place of the normal A · T base pair (Figure 31-40).

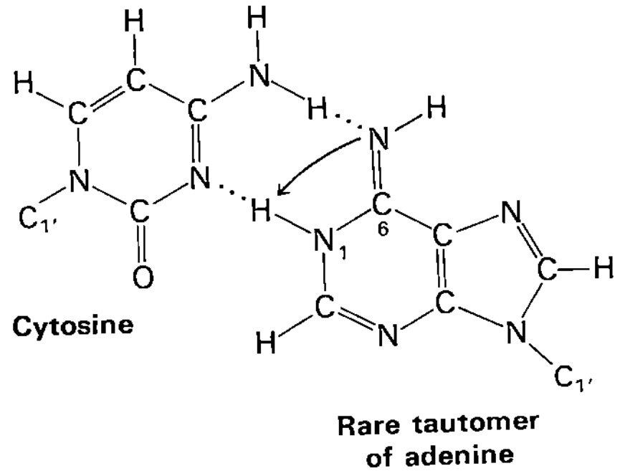

Figure 31-39
The rare tautomer of adenine pairs with cytosine instead of thymine. This tautomer is formed by the shift of a proton from the 6-amino group to N-1.

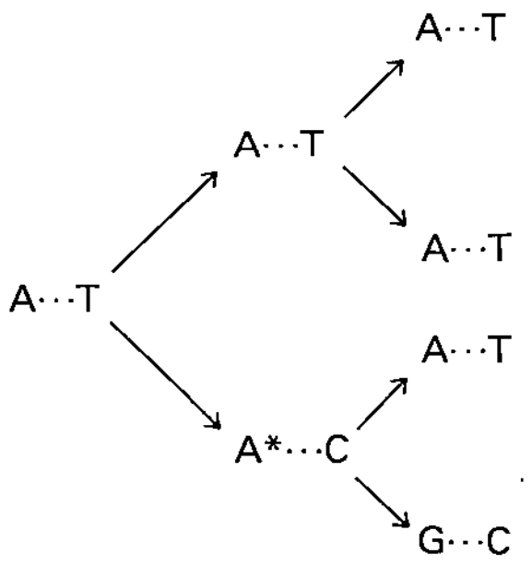

Figure 31-40
The pairing of the rare tautomer of adenine (A*) with cytosine leads to a G · C base pair in the next generation.

SOME CHEMICAL MUTAGENS ARE QUITE SPECIFIC

Base analogs such as 5-bromouracil and 2-aminopurine can be incorporated into DNA. They lead to transition mutations as a consequence of altered base pairing in a subsequent round of DNA replication (Figure 31-41). *5-Bromouracil*, an analog of thymine, normally pairs with adenine. However, the proportion of the enol tautomer of 5-bromouracil is higher than that of thymine, probably because the bromine atom is much more electronegative than a methyl group at C-5. The enol form of 5-bromouracil pairs with guanine, which causes AT ↔ GC transitions. *2-Aminopurine* normally pairs with thymine. In contrast with adenine, the normal tautomer of 2-aminopurine can form a single hydrogen bond with cytosine. Hence, 2-aminopurine can produce AT ↔ GC transitions.

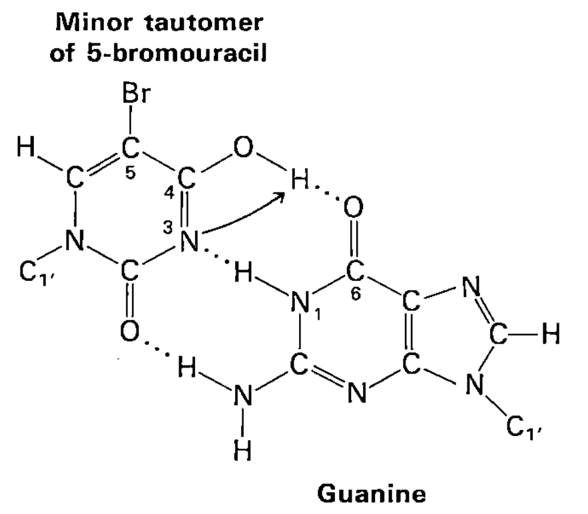

Figure 31-41
5-Bromouracil, an analog of thymine, occasionally pairs with guanine instead of adenine. The presence of bromine at C-5 increases the proportion of the rare tautomer formed by the shift of a proton from N-3 to the C-4 oxygen atom.

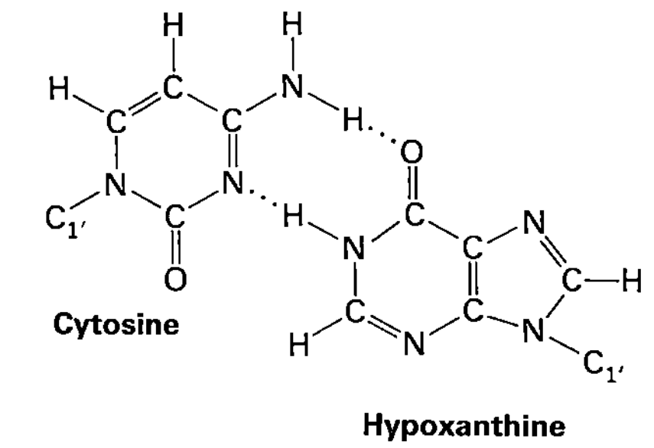

Figure 31-42
Base pairing of hypoxanthine with cytosine. Hypoxanthine is formed by oxidative deamination of adenine. Hence, an A · T base pair becomes a G · C base pair in a later round of replication.

Other mutagens act by chemically modifying the bases of DNA. For example, nitrous acid reacts with bases that contain amino groups. Adenine is oxidatively deaminated to hypoxanthine, cytosine to uracil, and guanine to xanthine. Hypoxanthine pairs with cytosine rather than with thymine (Figure 31-42). Uracil pairs with adenine rather than with guanine. Xanthine, like guanine, pairs with cytosine. Consequently, nitrous acid causes AT ⇔ GC transitions. *Hydroxylamine* (NH_2OH) is a highly specific mutagen. It reacts almost exclusively with cytosine to give a derivative that pairs with adenine rather than with guanine. Hydroxylamine produces a unidirectional transition of C · G to A · T.

A different kind of mutation is produced by flat aromatic molecules such as the acridines. These compounds *intercalate* in DNA—that is, they slip in between adjacent base pairs in the DNA double helix (p. 851). Consequently, *they lead to the insertion or deletion of one or more base pairs.* The effect of such mutations is to alter the reading frame in translation, unless an integral multiple of three base pairs is inserted or deleted. In fact, it was the analysis of such mutants that revealed the triplet nature of the genetic code.

LESIONS IN DNA SUCH AS PYRIMIDINE DIMERS FORMED BY ULTRAVIOLET LIGHT ARE CONTINUALLY REPAIRED

DNA is damaged by a variety of chemical and physical agents, and so all cells possess mechanisms for repair. Bases can be altered or lost, phosphodiester bonds in the backbone can be broken, and strands can become covalently cross-linked. These lesions are produced by ionizing radiation, ultraviolet light, and a variety of chemicals. *Much of the damage sustained by DNA can be repaired because genetic information is stored in both strands of the double helix, so that information lost by one strand can be retrieved from the other.*

One of the best-understood repair mechanisms is the excision of a *pyrimidine dimer, which is formed on exposure of DNA to ultraviolet light.* Adjacent pyrimidine residues on a DNA strand can become covalently linked under these conditions. Such a pyrimidine dimer cannot fit into a double helix, and so replication and gene expression are blocked until the lesion is removed. Three enzymatic activities are essential for this repair process in *E. coli* (Figure 31-43). First, an enzyme complex consisting of the proteins encoded by the *uvrABC* genes detects the distortion produced by the pyrimidine dimer. The *uvrABC enzyme* then cuts the damaged DNA strand at two sites, eight nucleotides away from the dimer on the 5′ side and four nucleotides away on the 3′ side. The 12-residue oligonucleotide excised by this highly specific *excinuclease* (*exci,* from the Latin words "to cut out")

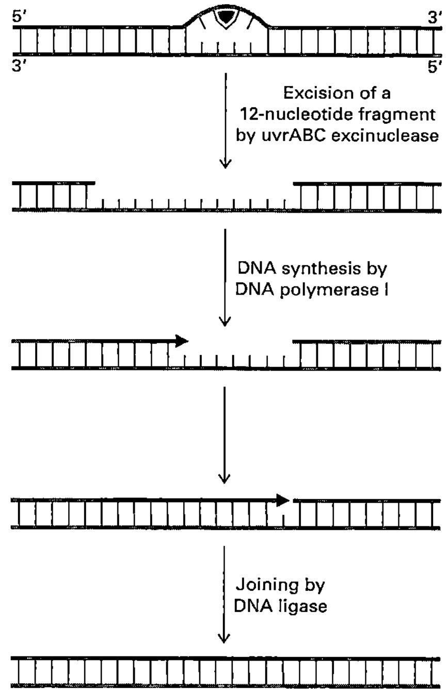

Figure 31-43
Repair of a region of DNA containing a thymine dimer by the sequential action of a specific excinuclease, a DNA polymerase, and a DNA ligase. The thymine dimer is shown in blue, and the new region of DNA in red. [After P.C. Hanawalt. *Endeavor* 31(1972):83.]

then diffuses away. DNA polymerase I enters the gap to carry out repair synthesis. The 3′ end of the nicked strand is the primer, and the intact complementary strand is the template. Finally, the 3′ end of the newly synthesized stretch of DNA and the original portion of the DNA chain are joined by *DNA ligase.*

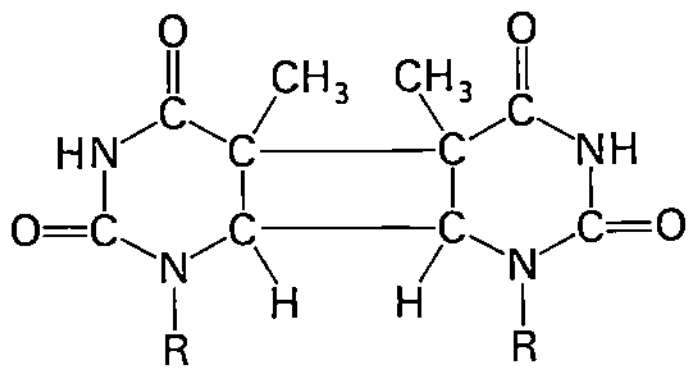

Thymine dimer

Alternatively, the pyrimidine dimer can be photochemically split. Nearly all cells contain a *photoreactivating enzyme* called *DNA photolyase.* The *E. coli* enzyme, a 35-kd protein, acquires an absorption band in the near-ultraviolet and blue spectral region on binding to the distorted region of DNA. The absorption of a photon creates an excited state that cleaves the dimer into its original bases.

In the repair of DNA, it is essential to identify the normal and aberrant strands. This information is sometimes contained in the defect itself, if it is a modified base (e.g., a pyrimidine dimer). Consider, on the other hand, an A · C base pair in a newly synthesized DNA molecule. Both bases are normal constituents of DNA. How does the repair machinery distinguish the parental strand (the authentic one) from the daughter strand (the one with the error)? *The tags that identify the strands are methyl groups on adenine residues in GATC sequences.* The parental strand is methylated at many of these sites, but the new daughter strand has not yet been modified in this way, because methylation takes time. Hence, the mismatch-correction enzyme cuts the unmethylated strand to remove the wrong nucleotide and leaves the parental strand intact so that it can again serve as a correct template.

THE PRESENCE OF THYMINE INSTEAD OF URACIL IN DNA PERMITS THE REPAIR OF DEAMINATED CYTOSINE

The presence in DNA of thymine rather than uracil was an enigma for many years. Both bases pair with adenine. The only difference between them is a methyl group in thymine in place of the C-5 hydrogen in uracil. Why is a methylated base employed in DNA and not in RNA? Recall that the methylation of deoxyuridylate to form deoxythymidylate is energetically expensive (p. 751). The existence of an active repair system to correct the deamination of cytosine provides a convincing solution to this puzzle.

Cytosine in DNA spontaneously deaminates at a perceptible rate to form uracil. The deamination of cytosine is potentially mutagenic because uracil pairs with adenine, and so one of the daughter strands will contain an A · U base pair rather than the original G · C base pair.

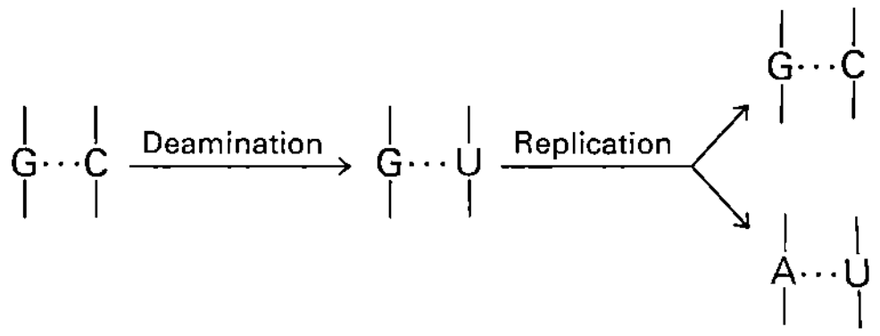

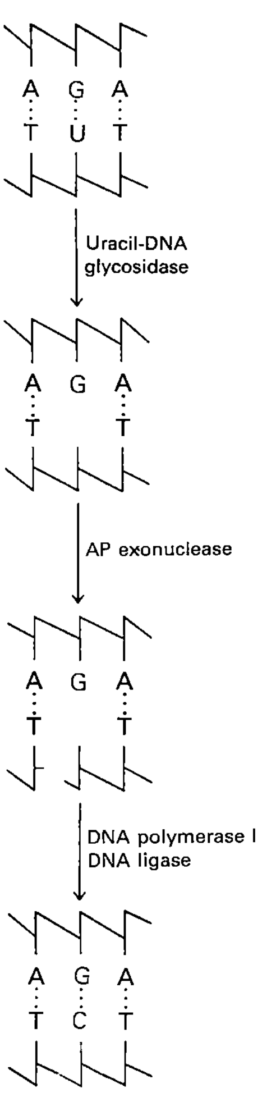

Figure 31-44
Uracil residues in DNA are excised and replaced with cytosine, the original base.

This mutation is prevented by a repair system that recognizes uracil to be foreign to DNA (Figure 31-44). First, a *uracil-DNA glycosidase* hydrolyzes the glycosidic bond between the uracil and deoxyribose moieties. At this stage, the DNA backbone is intact, but a base is missing. This hole is called an *AP site* because it is *ap*urinic (devoid of A or G) or *ap*yrimidinic (devoid of C or T). An *AP endonuclease* then recognizes this defect and nicks the backbone adjacent to the missing base. DNA polymerase I excises the residual deoxyribose phosphate unit and inserts cytosine, as dictated by the presence of guanine on the undamaged complementary strand. Finally, the repaired strand is sealed by DNA ligase. Uracil-DNA glycosidase does not remove thymine from DNA.

Thus, *the methyl group on thymine is a tag that distinguishes it from deaminated cytosine.* If thymine were not used in DNA, uracil correctly in place would be indistinguishable from uracil formed by deamination. The defect would persist unnoticed, and so a G · C base pair would necessarily be mutated to A · U in one of the daughter DNA molecules. This mutation is prevented by a repair system that searches for uracil and leaves thymine alone. It seems likely that *thymine is used instead of uracil in DNA to enhance the fidelity of the genetic message.* In contrast, RNA is not repaired, and so uracil is used in RNA because it is a less expensive building block.

MANY CANCERS ARE CAUSED BY DEFECTIVE REPAIR OF DNA

Xeroderma pigmentosum, a rare skin disease in humans, is genetically transmitted as an autosomal recessive trait. The skin in an affected homozygote is extremely sensitive to sunlight or ultraviolet light. In infancy, severe changes in the skin become evident and worsen with time. The skin becomes dry and there is a marked atrophy of the dermis. Keratoses appear, the eyelids become scarred, and the cornea ulcerates. Skin cancer usually develops at several sites. Many patients die before age 30 from metastases of these malignant skin tumors.

> *Xeroderma*—
> From the Greek words for "dry skin."
> The term was first used by F. Hebra and M. Kaposi in 1874 to describe the "parchment skin" and abnormal pigmentation seen in one of their patients.

Ultraviolet light produces pyrimidine dimers in human DNA, as it does in *E. coli* DNA. Furthermore, the repair mechanisms seem to be similar. Studies of skin fibroblasts from patients with xeroderma pigmentosum have revealed a biochemical defect in one form of this disease. In normal fibroblasts, half the pyrimidine dimers produced by ultraviolet radiation are excised in less than 24 hours. In contrast, almost no dimers are excised in this time interval in fibroblasts derived from patients with xeroderma pigmentosum. These studies show that *xeroderma pigmentosum can be produced by a defect in the excinuclease that hydrolyzes the DNA backbone near a pyrimidine dimer. The drastic clinical consequences of this enzymatic defect emphasize the critical importance of DNA repair processes.* The disease is also caused by mutations in eight other genes for DNA repair, which attests to the complexity of repair processes.

In 1895, Alfred Warthin's seamstress told him that she would die at an early age from cancer of her colon or female organs because "most of my family members die of these cancers." Her premonition was accurate: she died of carcinoma of the uterus. Warthin studied her family and found that they were indeed highly cancer-prone; many developed cancer of the colon, stomach, or uterus. Subsequent studies showed that this genetic disorder, called *hereditary nonpolyposis colorectal cancer (HPCC,* or *Lynch syndrome),* is not rare—as many as 1 in 200 people are affected. HPCC results from *defective DNA mismatch repair.* Mutations in two genes, called *hMSH2* and *hMLH1,* account for most cases of this hereditary predisposition to cancer. The striking finding is that these genes are the human counterparts of *MutS* and *MutL* of *E. coli.* The MutS protein binds to mismatched base pairs (e.g., G · T) in DNA. The MutH protein, together with MutL, participates in cleaving one of the DNA strands in the vicinity of this mismatch to initiate the repair process (Figure 31-45). It seems likely that mutations in *hMSH2* and *hMLH1* lead to the accumulation of mutations throughout the genome. In time, genes important in controlling cell proliferation become altered, resulting in the onset of cancer, and its eventual progression.

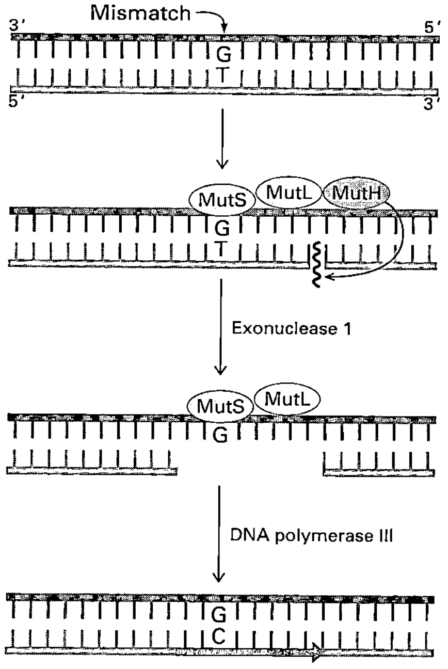

Figure 31-45
DNA mismatch repair in *E. coli* is initiated by the interplay of the MutS, MutL, and MutH proteins. A G · T mismatch is recognized by MutS. MutH cleaves the backbone in the vicinity of the mismatch. A segment of the DNA strand containing the erroneous T is removed by exonuclease 1, and synthesized anew by DNA polymerase III. [After R.F. Service. *Science* 263(1994):1559.]

MANY POTENTIAL CARCINOGENS CAN BE DETECTED BY THEIR MUTAGENIC ACTION ON BACTERIA

Many human cancers are caused by exposure to toxic chemicals. These chemical carcinogens are usually mutagenic, which suggests that *damage to DNA is a fundamental event in both carcinogenesis and mutagenesis.* It is important to identify these compounds and ascertain their potency so that human exposure to them can be minimized. Bruce Ames has devised a simple and sensitive test for detecting chemical mutagens. A thin layer of agar containing about 10^9 bacteria of a specially constructed tester strain of *Salmonella* is placed on a petri dish. These bacteria are unable to grow in the absence of histidine because of a mutation in one of their genes for the biosynthesis of this amino acid. The addition of a mutagen to the center of the plate results in many new mutations. A small proportion of them reverse the original mutation so that histidine is synthesized. These *revertants* multiply in the absence of exogenous histidine and appear as discrete colonies after the plate is incubated at 37°C for two days (Figure 31-46). For example, 0.5 μg of 2-aminoanthracene gives 11,000 revertant colonies compared with only 30 spontaneous revertants in its absence. A series of concentrations of a chemical can readily be tested to generate a dose-response curve. These curves are usually linear, which suggests that there is no threshold concentration for mutagenesis.

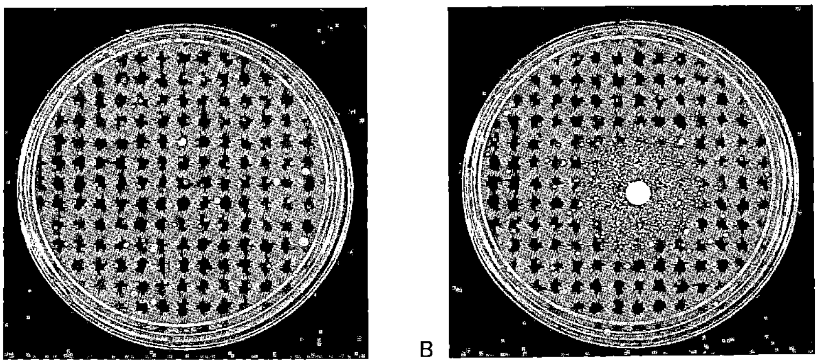

Figure 31-46
Salmonella test for mutagens: (A) petri plate containing about 10^9 bacteria that cannot synthesize histidine; (B) a plate containing a filter-paper disk with a mutagen, which produces a large number of revertants that can synthesize histidine. After two days, revertants appear as a ring of colonies around the disk. The small number of visible colonies in plate A are spontaneous revertants. [From B.N. Ames, J. McCann, and E. Yamasaki. *Mutation Res.* 31(1975):347.]

Some of the tester strains are responsive to *base-pair substitutions,* whereas others detect *deletions or additions of base pairs (frameshifts).* The sensitivity of these specially designed strains has been enhanced by genetically deleting their excision-repair systems. Also, the lipopolysaccharide barrier that normally coats the surface of *Salmonella* is incomplete in the tester strains, which facilitates the entry of potential mutagens.

A key feature of this detection system is the inclusion of a *mammalian liver homogenate.* Recall that some potential carcinogens are converted into their active forms by enzyme systems in the liver or other mammalian tissues (p. 704). Bacteria lack these enzymes, and so the test plate requires a few milligrams of a liver homogenate to activate this group of mutagens. The P450 system present in mammalian cells but not in bacteria increases the carcinogenicity of many compounds by enabling them to form covalent bonds with DNA.

The *Salmonella* test is extensively used to help evaluate the mutagenic and carcinogenic risks of a large number of chemicals. This rapid and inexpensive bacterial assay for mutagenicity complements epidemiologic surveys and animal tests that are necessarily slower, more laborious, and far more expensive. The *Salmonella* test for mutagenicity is an outgrowth of studies of gene-protein relationships in bacteria. It is a striking example of how fundamental research in molecular biology can lead directly to important advances in public health.

SUMMARY

DNA is a structurally dynamic molecule that can exist in a variety of helical forms: A-DNA, B-DNA (the classic Watson-Crick helix), and Z-DNA. DNA can be bent, kinked, and unwound. In A-, B-, and Z-DNA, two antiparallel chains are held together by Watson-Crick base pairs and stacking interactions between bases in the same strand. The sugar-phosphate backbone is on the outside, and the bases are inside the double helix. A- and B-DNA are right-handed helices in which the repeating unit is a mononucleotide. In B-DNA, the base pairs are nearly perpendicular to the helix axis, whereas in A-DNA they are tilted. Dehydration induces the transition from B- to A-DNA. The 2′-OH of ribose cannot fit into a B helix because of steric hindrance. Double-helical RNA hairpins and RNA–DNA hybrids adopt the A-helix structure. Z-DNA is a left-handed helix in which the repeating unit is a dinucleotide. It can be formed in regions of DNA in which purines alternate repeatedly with pyrimidines, as in CGCG or CACA. Most of the DNA in a genome is in the Watson-Crick B-helix form. An important structural feature of the B helix is the presence of major and minor grooves, which display different potential hydrogen bond acceptors and donors according to the base sequence.

Bacteria possess restriction endonucleases that enable them to destroy foreign DNA molecules. These exquisitely specific enzymes cleave both strands of foreign DNA at palindromic target sequences. The bacterium's own DNA is spared because potential target sequences are methylated soon after replication. The EcoRV endonuclease scans the major groove of DNA for GATATC. When this palindromic hexanucleotide sequence is encountered, the DNA becomes kinked by 50 degrees and Mg^{2+} becomes bound to the active site to enable cleavage to occur. The twofold symmetry of the dimeric enzyme matches that of its duplex DNA substrate.

A key topological property of DNA is its linking number (*Lk*), which is defined as the number of times one strand of DNA winds around the other in the right-hand direction when the DNA axis is constrained to lie in a plane. Molecules differing in linking number are topoisomers of one another and can be interconverted only by cutting one or both DNA strands; these reactions are catalyzed by topoisomerases. Changes in linking number generally lead to changes in both the number of turns of double helix and the number of turns of superhelix. DNA gyrase catalyzes the ATP-driven introduction of negative supercoils, which leads to the compaction of DNA and renders it more susceptible to unwinding. Supercoiled DNA is relaxed by topoisomerase I.

DNA polymerases are template-directed enzymes that catalyze the formation of phosphodiester bonds by the nucleophilic attack of a 3′-OH on the innermost phosphorus atom of a deoxyribonucleoside 5′-triphosphate. They cannot start chains de novo; a primer with a free 3′-OH is required. DNA polymerases proofread the nascent product; their $3' \rightarrow 5'$ exonuclease activity examines the outcome of each polymerization step. A mispaired nucleotide is excised before the next polymerization step. In *E. coli,* DNA polymerase I (a 103-kd single-chain protein) repairs DNA and participates in replication. Most of the DNA is synthesized by DNA polymerase III holoenzyme, a multisubunit assembly (900 kd) of 10 kinds of polypeptide chains. Polymerase III is a highly processive enzyme with a very low error rate. The β_2 subunit serves as a sliding clamp that encircles the DNA template.

DNA replication in *E. coli* starts at a unique origin (*oriC*) and proceeds sequentially in opposite directions. More than 20 proteins are required for replication. DnaB protein, an ATP-driven helicase, unwinds the *oriC* region to create a replication fork. At this fork, both strands of parental DNA serve as templates for the synthesis of new DNA. DNA synthesis is primed by a short stretch of RNA formed by primase, an RNA polymerase. One strand of DNA (the leading strand) is synthesized continuously, whereas the other strand (the lagging strand) is synthesized discontinuously, in the form of 1-kb fragments (Okazaki fragments). Both new strands are formed simultaneously by the concerted actions of DNA polymerase III holoenzyme, an asymmetric dimer. The discontinuous assembly of the lagging strand enables $5' \rightarrow 3'$ polymerization at the atomic level to give rise to overall growth of this strand in the $3' \rightarrow 5'$ direction. The RNA primer portion is hydrolyzed by the $5' \rightarrow 3'$ nuclease activity of DNA polymerase I, which also fills gaps. Finally, nascent DNA fragments are joined by DNA ligase in a reaction driven by ATP or NAD^+. The unwinding of DNA at the replication fork is catalyzed by an ATP-driven helicase. DNA gyrase also plays a key role by introducing negative supercoils and allowing the daughter duplex to separate from the parental one.

Mutations are produced by mistakes in base pairing, covalent modification of bases, and the deletion and insertion of bases. The $3' \rightarrow 5'$ exonuclease activity of DNA polymerases is critical in lowering the spontaneous mutation rate, which arises largely from mispairing by tautomeric bases. Lesions in DNA are continually being repaired. Multiple repair processes utilize information present in the intact strand to correct the damaged strand. For example, pyrimidine dimers formed by the action of ultraviolet light are excised by the uvrABC excinuclease, an enzyme that removes a 12-nucleotide region containing the dimer. Xeroderma pigmentosum, a genetically transmitted disease, is caused by defective repair of lesions in DNA, such as pyrimidine dimers; patients with this disease usually develop skin cancers. Many cancers of the colon are caused by defective DNA mismatch repair arising from mutations in human genes that have been highly conserved in evolution. Damage to DNA is a fundamental event in both carcinogenesis and mutagenesis. Many potential carcinogens can be detected by their mutagenic action on bacteria.

SELECTED READINGS

Where to begin

Kornberg, A., 1988. DNA replication. *J. Biol. Chem.* 263:1–4.

Dickerson, R.E., 1983. The DNA helix and how it is read. *Sci. Amer.* 249(6):94–111.

Wang, J.C., 1982. DNA topoisomerases. *Sci. Amer.* 247(1):94–109.

Lindahl, T., 1993. Instability and decay of the primary structure of DNA. *Nature* 362:709–715.

Books

Kornberg, A., and Baker, T.A., 1992. *DNA Replication* (2nd ed.). W.H. Freeman.

Saenger, W., 1984. *Principles of Nucleic Acid Structure.* Springer-Verlag.

Friedberg, E.C., 1985. *DNA Repair.* W.H. Freeman.

Cozzarelli, N.R., and Wang, J.C. (eds.), 1990. *DNA Topology and Its Biological Effects.* Cold Spring Harbor Laboratory Press.

DNA structure

Dickerson, R.E., 1992. DNA structure from A to Z. *Meth. Enzymol.* 211:67–111.

Dickerson, R.E., Drew, H.R., Conner, B.N., Wing, R.M., Fratini, A.V., and Kopka, M.L., 1982. The anatomy of A-, B-, and Z-DNA. *Science* 216:475–485.

Rich, A., Nordheim, A., and Wang, A.H.-J., 1984. The chemistry and biology of left-handed Z-DNA. *Ann. Rev. Biochem.* 53:791–846.

Quintana, J.R., Grzeskowiak, K., Yanagi, K., and Dickerson, R.E., 1992. Structure of a B-DNA decamer with a central T-A step: C-G-A-T-T-A-A-T-C-G. *J. Mol. Biol.* 225:379–395.

Verdaguer, N., Aymami, J., Fernandez, F.D., Fita, I., Coll, M., Huynh, D.T., Igolen, J., and Subirana, J.A., 1991. Molecular structure of a complete turn of A-DNA. *J. Mol. Biol.* 221:623–635.

Johnston, B.H., 1992. Generation and detection of Z-DNA. *Meth. Enzymol.* 211:127–158.

Restriction endonucleases and sequence recognition

Heitman, J., 1992. How the EcoRI endonuclease recognizes and cleaves DNA. *BioEssays* 14:445–454.

Kim, Y.C., Grable, J.C., Love, R., Greene, P.J., and Rosenberg, J.M., 1990. Refinement of EcoRI endonuclease crystal structure: A revised protein chain tracing. *Science* 249:1307–1309.

Taylor, J.D., and Halford, S.E., 1989. Discrimination between DNA sequences by the EcoRV restriction endonuclease. *Biochemistry* 28:6198–6207.

Winkler, F.K., Banner, D.W., Oefner, C., Tsernoglou, D., Brown, R.S., Heathman, S.P., Bryan, R.K., Martin, P.D., Petratos, K., and Wilson, K.S., 1993. The crystal structure of EcoRV endonuclease and of its complexes with cognate and non-cognate DNA fragments. *EMBO J.* 12:1781–1795.

Steitz, T.A., 1990. Structural studies of protein-nucleic acid interaction: The sources of sequence-specific binding. *Quart. Rev. Biophys.* 23:205–280.

DNA topology, topoisomerases, and gyrases

Bauer, W.R., Crick, F.H.C., and White, J.H., 1980. Supercoiled DNA. *Sci. Amer.* 243(1):118–133. [A lucid presentation of topological principles.]

Cozzarelli, N., Boles, T.C., and White, J.H., 1990. Primer on the topology and geometry of DNA supercoiling. *In* Cozzarelli, N.R., and Wang, J.C. (eds.), 1990. *DNA Topology and Its Biological Effects*, pp. 139–184. Cold Spring Harbor Laboratory Press.

Vologodskii, A.V., Levene, S.D., Klenin, K.V., Frank, K.M., and Cozzarelli, N.R., 1992. Conformational and thermodynamic properties of supercoiled DNA. *J. Mol. Biol.* 227:1224–1243.

Wang, J.C., 1991. DNA topoisomerases: why so many? *J. Biol. Chem.* 266:6659–6662.

Fisher, L.M., Austin, C.A., Hopewell, R., Margerrison, M., Oram, M., Patel, S., Plummer, K., Sng, J.-H., and Sreedharan, S., 1992. DNA supercoiling and relaxation by ATP-dependent DNA topoisomerases. *Phil. Trans. Royal Soc. Lond.* B 336:83–91.

Wigley, D.B., Davies, G.J., Dodson, E.J., Maxwell, A., and Dodson, G., 1991. Crystal structure of an N-terminal fragment of the DNA gyrase B protein. *Nature* 351:624–629.

Lindsley, J.E., and Wang, J.C., 1993. On the coupling between ATP usage and DNA transport by yeast DNA topoisomerase II. *J. Biol. Chem.* 268:8096–8104.

Liu, L.F., 1989. DNA topoisomerase poisons as antitumor drugs. *Ann. Rev. Biochem.* 58:351–375.

Mechanism of replication

Marians, K.J., 1992. Prokaryotic DNA replication. *Ann. Rev. Biochem.* 61:673–719.

McHenry, C.S., 1991. DNA polymerase III holoenzyme. Components, structure, and mechanism of a true replicative complex. *J. Biol. Chem.* 266:19127–19130.

Salas, M., 1991. Protein-priming of DNA replication. *Ann. Rev. Biochem.* 60:39–71.

Lohman, T.M., 1993. Helicase-catalyzed DNA unwinding. *J. Biol. Chem.* 268:2269–2272.

Echols, H., and Goodman, M.F., 1991. Fidelity mechanisms in DNA replication. *Ann. Rev. Biochem.* 60:477–511.

Kunkel, T.A., 1992. DNA replication fidelity. *J. Biol. Chem.* 267:18251–18254.

DNA polymerases and ligases

Steitz, T.A., 1993. DNA- and RNA-dependent DNA polymerases. *Curr. Opin. Struct. Biol.* 3:31–38.

Beese, L.S., Derbyshire, V., and Steitz, T.A., 1993. Structure of DNA polymerase I Klenow fragment bound to duplex DNA. *Science* 260:352–355.

Polesky, A.H., Steitz, T.A., Grindley, N.D., and Joyce, C.M., 1990. Identification of residues critical for the polymerase activity of the Klenow fragment of DNA polymerase I from *Escherichia coli*. *J. Biol. Chem.* 265:14579–14591.

Kong, X.P., Onrust, R., O'Donnell, M., and Kuriyan, J., 1992. Three-dimensional structure of the beta subunit of *E. coli* DNA polymerase III holoenzyme: A sliding DNA clamp. *Cell* 69:425–437.

Lindahl, T., and Barnes, D.E., 1992. Mammalian DNA ligases. *Ann. Rev. Biochem.* 61:251–281.

Willis, A.E., and Lindahl, T., 1987. DNA ligase I deficiency in Bloom's syndrome. *Nature* 325:355–357.

Mutations and DNA repair

Sancar, A., and Sancar, G.B., 1988. DNA repair enzymes. *Ann. Rev. Biochem.* 57:29–67.

Modrich, P., 1991. Mechanisms and biological effects of mismatch repair. *Ann. Rev. Genet.* 25:229–253.

Sancar, G.B., and Rupp, W.D., 1983. A novel repair enzyme: uvrABC excision nuclease of *E. coli* cuts a DNA strand on both sides of the damaged region. *Cell* 33:249–260.

Fang, W.-H., and Modrich, P., 1993. Human strand-specific mismatch repair occurs by a bidirectional mechanism similar to that of the bacterial reaction. *J. Biol. Chem.* 268:11838–11844.

Caskey, C.T., Pizzuti, A., Fu, Y.H., Fenwick, R.J., and Nelson, D.L., 1992. Triplet repeat mutations in human disease. *Science* 256:784–789.

Hayes, W., 1968. *The Genetics of Bacteria and Their Viruses* (2nd ed.). Wiley. [Chapter 13, on the nature of mutations, is lucid and concise.]

Defective DNA repair and cancer

Cleaver, J.E., and Kraemer, K.R., 1989. Xeroderma pigmentosum. *In* Scriver, C.R., Beaudet, A.L., Sly, W.S., and Valle, D. (eds.), *The Metabolic Basis of Inherited Disease* (6th ed.), pp. 2949–2971. McGraw-Hill.

Lynch, H.T., Smyrk, T.C., Watson, P., Lanspa, S.J., Lynch, J.F., Lynch, P.M., Cavalieri, R.J., and Boland, C.R., 1993. Genetics, natural history, tumor spectrum, and pathology of hereditary nonpolyposis colorectal cancer: An updated review. *Gastroenterology* 104:1535–1549.

Fishel, R., Lescoe, M.K., Rao, M.R.S., Copeland, N.G., Jenkins, N.A., Garber, J., Kane, M., and Kolodner, R., 1993. The human mutator gene homolog *MSH2* and its association with hereditary nonpolyposis colon cancer. *Cell* 75:1027–1038.

Papadopoulos, N., et al., 1994. Mutation of a *mutL* homolog in hereditary colon cancer. *Science* 263:1625–1629.

Ames, B.N., and Gold, L.S., 1991. Endogenous mutagens and the causes of aging and cancer. *Mutation Res.* 250:3–16.

Ames, B.N., 1979. Identifying environmental chemicals causing mutations and cancer. *Science* 204:587–593.

PROBLEMS

1. *Activated intermediates.* DNA polymerase I, DNA ligase, and topoisomerase I catalyze the formation of phosphodiester bonds. What is the activated intermediate in the linkage reaction catalyzed by each of these enzymes? What is the leaving group?

2. *Fuel for a new ligase.* Whether the joining of two DNA chains by known DNA ligases is driven by NAD^+ or ATP depends on the species. Suppose that a new DNA ligase requiring a different energy donor is found. Propose a plausible substitute for NAD^+ or ATP in this reaction.

3. *AMP-induced relaxation.* DNA ligase from *E. coli* relaxes supercoiled circular DNA in the presence of AMP but not in its absence. What is the mechanism of this reaction, and why is it dependent on AMP?

4. *Life in a hot tub.* An archaebacterium (*Sulfolobus acidocaldarius*) found in acidic hot springs contains a topoisomerase that catalyzes the ATP-driven introduction of positive supercoils into DNA. How might this enzyme be advantageous to this unusual bacterium?

5. *A cooperative transition.* The transition from B-DNA to Z-DNA occurs over a small change in the superhelix density, which shows that the transition is highly cooperative.
   (a) Consider a DNA molecule at the midpoint of this transition. Are B- and Z-DNA regions frequently intermingled or are there long stretches of each?
   (b) What does this finding reveal about the energetics of forming a junction between the two kinds of helices?
   (c) Would you expect the transition from B- to A-DNA to be more or less cooperative than the one from B- to Z-DNA? Why?

6. *A revealing analog.* AppNHp, the β,γ-imido analog of ATP, is hydrolyzed very slowly by most ATPases. The addition of AppNHp to DNA gyrase and circular DNA leads to the negative supercoiling of a single molecule of DNA per gyrase. DNA remains bound to gyrase in the presence of this analog. What does this finding reveal about the catalytic mechanism?

7. *Mutagenic trail.* Suppose that the single-stranded RNA from tobacco mosaic virus was treated with a chemical mutagen, that mutants were obtained having serine or leucine instead of proline at a specific position, and that further treatment of these mutants with the same mutagen yielded phenylalanine at this position.

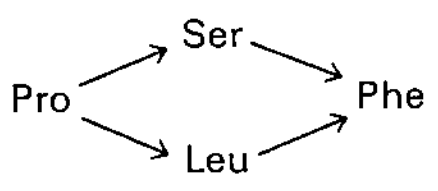

   (a) What are the plausible codon assignments for these four amino acids?
   (b) Was the mutagen 5-bromouracil, nitrous acid, or an acridine dye?

8. *Revealing tracks.* Suppose that replication is initiated in a medium containing *moderately* radioactive tritiated thymine. After a few minutes of incubation, the bacteria are transferred to a medium containing *highly* radioactive tritiated thymidine. Sketch the autoradiographic pattern that would be seen for (a) undirectional replication, and (b) bidirectional replication, from a single origin.

9. *Induced spectrum.* DNA photolyases convert the energy of light in the near UV or visible region (300 to 500 nm) into chemical energy to break the cyclobutane ring of pyrimidine dimers. In the absence of substrate, these photoreactivating enzymes do not absorb light of wavelengths longer than 300 nm. Why is the substrate-induced absorption band advantageous?

10. *Molecular motors in replication.*
   (a) How fast does template DNA spin (expressed in revolutions per second) at an *E. coli* replication fork?
   (b) What is the velocity of movement (in μm/s) of DNA polymerase III holoenzyme relative to the template?
   (c) Compare these rates with those of flagellar rotation in *E. coli* and sarcomere shortening in vertebrate skeletal muscle.

CHAPTER

32

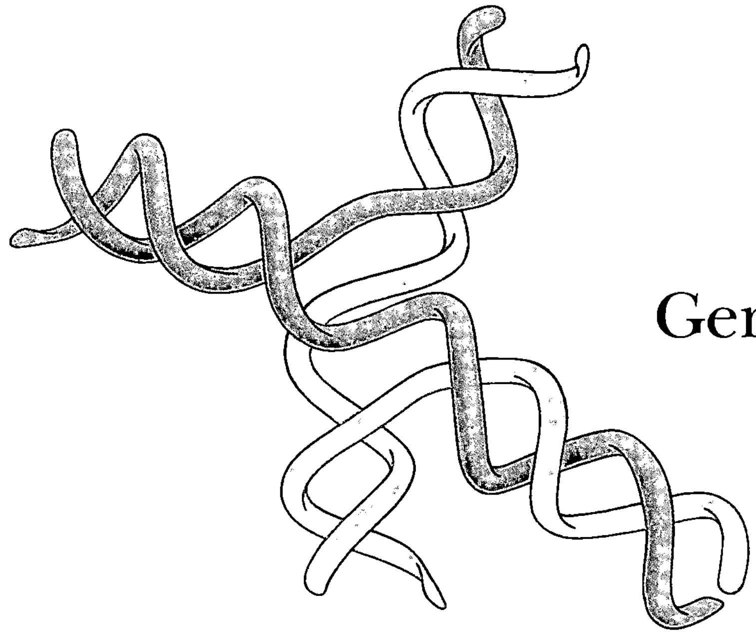

Gene Rearrangements

The theme of this chapter is the formation of new arrangements of genes by the movement of large blocks of DNA. In *homologous recombination,* parent DNA duplexes align at regions of sequence similarity, and new DNA molecules are formed by the breakage and joining of homologous segments. Intermediates in recombination have been isolated, and the actions of enzymes catalyzing the exchange of DNA strands have been delineated. The ATP-powered recA protein mediates homologous recombination by enabling single-stranded DNA to rapidly scan duplex DNA for an identical or similar sequence. RecA then catalyzes strand exchange at the site of homology. Homologous recombination plays a key role in the *repair of DNA.* When both strands of a duplex are damaged, information for repair must come from another DNA molecule. Recombination provides a template for the synthesis of DNA to fill the gap. The DNA of mutants incapable of undergoing recombination is highly vulnerable to irreversible damage by ultraviolet light and agents that react with DNA.

Transposition is the movement of a gene from one chromosome to another or from one site to a different one on the same chromosome. In contrast with homologous recombination, transposition does not require extensive sequence similarity. *Transposons* are mobile genetic elements that enable genes to move between nonhomologous sites in DNA. R factor plasmids are a medically important group of transposons that make bacteria resistant to multiple antibiotics (Figure 32-1). Recombination and transposition also *generate new combinations of genes.* These processes are immensely important in evolution because they markedly enlarge the genetic repertoire from which natural selection chooses.

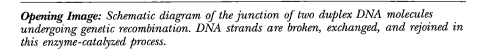

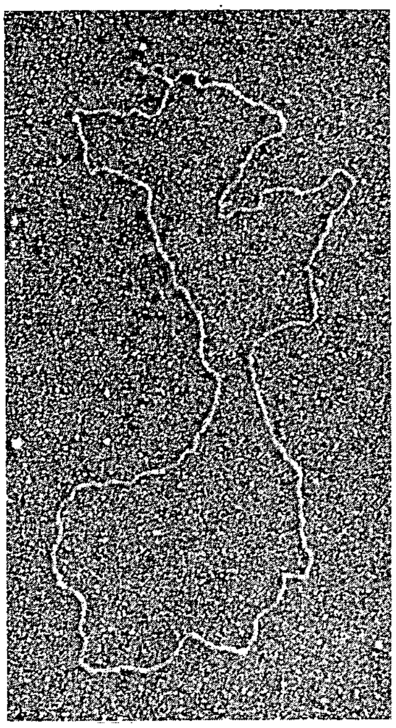

Figure 32-1
Electron micrograph of a circular DNA molecule containing genes that confer resistance to several antibiotics. Infectious drug resistance is a consequence of the transmission of such R (resistance) factor plasmids. [Courtesy of Dr. Stanley Cohen.]

Opening Image: *Schematic diagram of the junction of two duplex DNA molecules undergoing genetic recombination. DNA strands are broken, exchanged, and rejoined in this enzyme-catalyzed process.*

Site-specific recombination is the exchange of two specific (but not necessarily homologous) DNA sequences. Lambda (λ) phage can either replicate and destroy the infected bacterial cell, or its DNA can become integrated in quiescent form in the bacterial genome by site-specific recombination. The cutting and splicing of DNA duplexes at specific sites is also important in eukaryotes, as exemplified by the joining of gene segments that encode antibodies and T-cell receptors. The combinatorial association of *V*, *D*, and *J* segments creates a very large repertoire of sites that can recognize foreign molecules.

This chapter concludes with fascinating gene rearrangements that involve a reverse flow of genetic information, from RNA to DNA. The genomes of *retroviruses* such as human immunodeficiency virus-1 cycle between an RNA form in the virion and a DNA form in the host cell. Integration of viral DNA in the genome of the host, a necessary step in the life cycle of the virus, occurs by recombination that is catalyzed by a virus-encoded protein. This remarkable genetic strategy is also employed by *retroposons* (retrotransposons), which hop from one DNA site to another in their host by directing the synthesis of an RNA intermediate. The emerging picture is that prokaryotic and eukaryotic genomes are dynamic stores of information that continually undergo change.

HOLLIDAY MODEL FOR HOMOLOGOUS RECOMBINATION

In genetic recombination, a DNA molecule is formed with a nucleotide sequence derived partly from one parental DNA molecule and partly from another. A key study was carried out in 1961 using the density-gradient equilibrium sedimentation technique (p. 84). Analyses of hybrid phage DNA formed in infected *Escherichia coli* from bromouracil-labeled

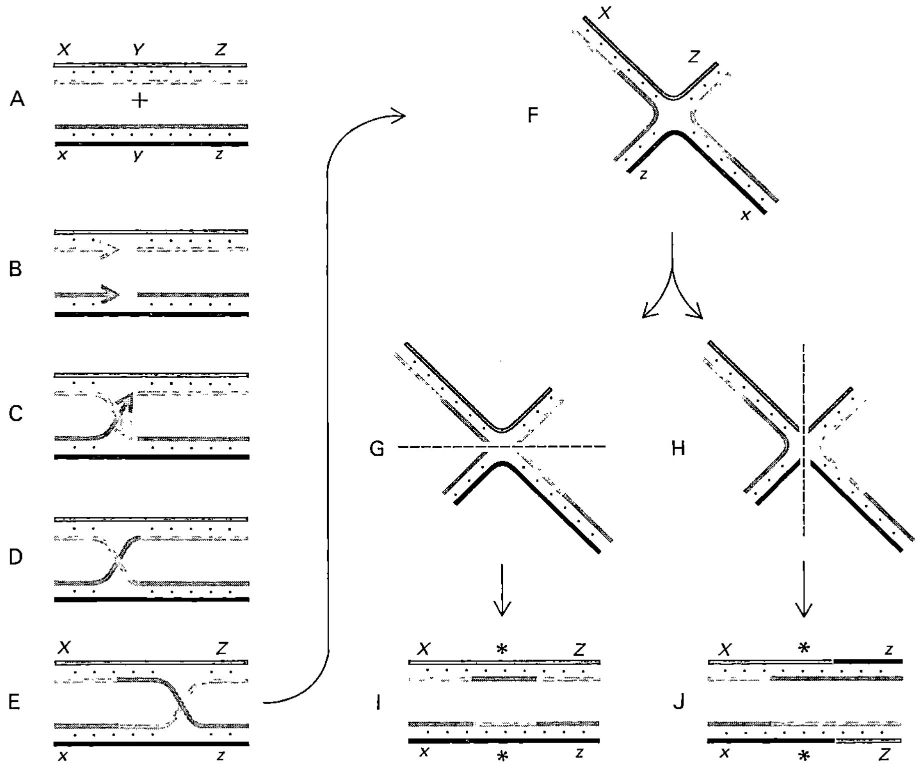

Figure 32-2
A model of homologous recombination proposed by Robin Holliday. One parental duplex is shown in green and the other in red. Either dark strand can base-pair with either light strand, and vice versa. *X*, *Y*, and *Z* denote three genes; *x*, *y*, and *z* are alleles. The steps are (A) alignment, (B) cleavage of one strand of each duplex, (C) invasion, (D) sealing, and (E) branch migration. Part F is an alternative representation of part E. The lower duplex of E is rotated 180 degrees about the vertical axis to give F. Note that F can be cleaved along the (G) horizontal or the (H) vertical axis, which corresponds to the cleavage of the invading or the noninvading strands, respectively. Rejoining of strands gives two different sets of recombinants (I and J). Regions containing one strand from each parental duplex, called heteroduplexes, are labeled with an asterisk. [After H. Potter and D. Dressler. *Cold Spring Harbor Symp. Quant. Biol.* 43(1979):973.]

and ^{32}P-labeled phage DNA revealed that *single-stranded regions of DNA are intermediates in recombination.* Furthermore, they suggested that *enzymes are intimately involved in the process.*

In 1964, Robin Holliday proposed a model of homologous recombination (also called *general recombination*) to account for the products of meiosis in fungi. In his scheme (Figure 32-2), the first step is the alignment of two homologous duplexes (A). A strand of one duplex and the corresponding strand of the other duplex are nicked by an endonuclease (B). An end of each nicked strand leaves its own duplex and invades the other duplex (C). Strands from different duplexes are then joined to each other to form a recombination intermediate that is dynamic (D). Strand exchange can continue, allowing the crossover point between duplexes to move (E), a process called *branch migration.* This recombinational intermediate can be cleaved and rejoined in two contrasting ways. If the two *invading* strands are cut (G), the products (I) are the same as the parent molecules except that each contains a heteroduplex (hybrid) region in the middle. Alternatively, if the two *noninvading* strands are cut (H), the products (J) are recombinant molecules in which the left half comes from one parental DNA and the right half from the other; the heteroduplex region is the site of crossover.

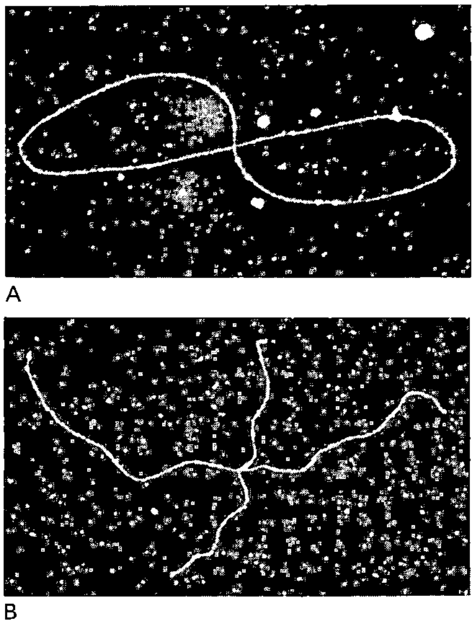

Figure 32-3
Electron micrographs of DNA molecules undergoing recombination: (A) a figure-eight intermediate consisting of two DNA molecules; (B) cleavage by a restriction endonuclease yields a chi form. [From H. Potter and D. Dressler. *Proc. Nat. Acad. Sci.* 76(1979):1086.]

HOMOLOGOUS DNA STRANDS PAIR IN RECOMBINATION TO FORM CHI INTERMEDIATES

The Holliday model has served as a stimulating conceptual framework for studies of molecular mechanisms of recombination. To test the model, intermediates in the homologous recombination of plasmids in *E. coli* were visualized by electron microscopy. Bacterial cells treated with chloramphenicol become filled with plasmids because this antibiotic inhibits replication of bacterial DNA but not that of plasmid DNA. These plasmids, like bacterial DNA, readily undergo recombination. When they were isolated from the treated cells, about a quarter of the plasmids were found to be dimers in the shape of a figure eight (Figure 32-3A), as predicted by the Holliday model (Figure 32-4).

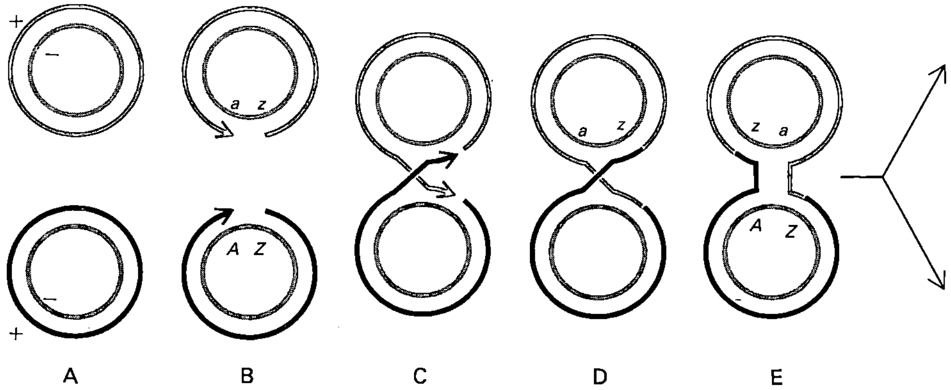

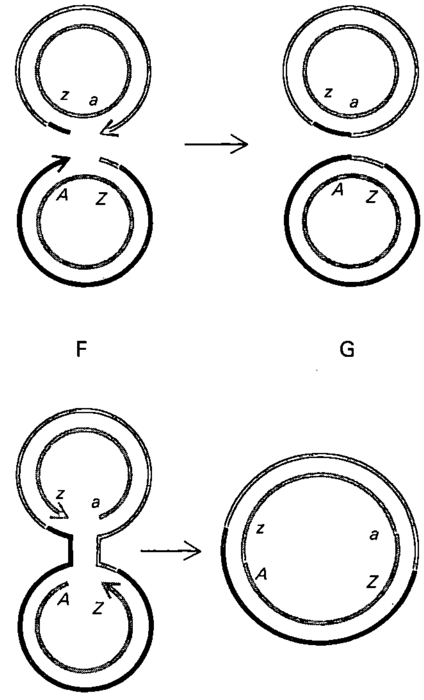

Figure 32-4
Recombination of circular DNA duplexes. The intermediate shown in (E) corresponds to the Holliday intermediate for the recombination of linear duplexes (F in Figure 32-2). Two kinds of products are formed, depending on how the Holliday intermediate is resolved. [After D. Dressler and H. Potter. Molecular mechanisms in genetic recombination. *Ann. Rev. Biochem.* 51:733. Copyright © 1982 by Annual Reviews, Inc. All rights reserved.]

These dimers were then cleaved by EcoRI endonuclease, which cuts the original plasmid at a unique site. If a figure-eight dimer were made of interlocked circles or if it were simply a twisted double-length circle, unit-size rods would be formed by this procedure. On the other hand, a structure with four arms resembling the Greek letter χ (chi) would be generated if a dimer consisted of two plasmid circles covalently joined at a single point. In fact, nearly all figure eights were converted into chi forms (Figure 32-3B), which strongly suggests that *figure-eight forms are intermediates in recombination.* This inference was supported by the finding that figure eights are not formed by *E. coli* mutants that are defective in recombination.

What is the nature of the crossover region of the two genomes in the figure-eight forms? The contact point in the chi forms always divides the structure into pairs of equal-length arms. This means that *the genomes are joined at a region of homology* (Figure 32-5). There would be no special

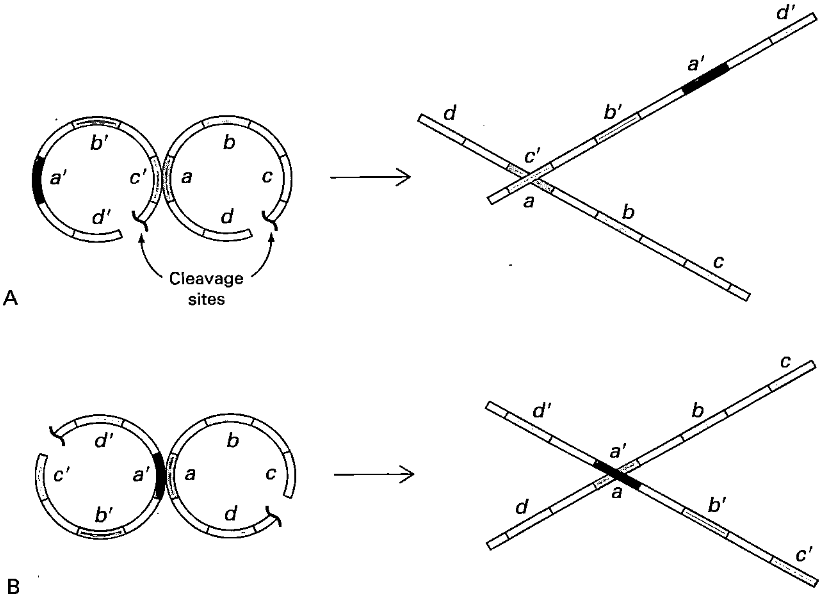

Figure 32-5
Expected cleavage patterns for two DNA molecules that are joined at (A) nonhomologous sites and (B) homologous sites. A restriction endonuclease cleaves each plasmid at a specific site. The observed symmetry of chi forms (as in Figure 32-3B) shows that DNA molecules in figure-eight forms are joined at homologous sites.

relations between the sizes of the arms if the plasmids were joined at unrelated sequences. Furthermore, the contact point occurs with nearly equal probability along the entire plasmid, which shows that *pairing can occur at many locations.* The connections at the crossover region have been visualized by electron microscopy after being exposed by selective denaturation. *The four duplexes emerge from a ring of connecting single strands at the junction between the two genomes* (Figure 32-6).

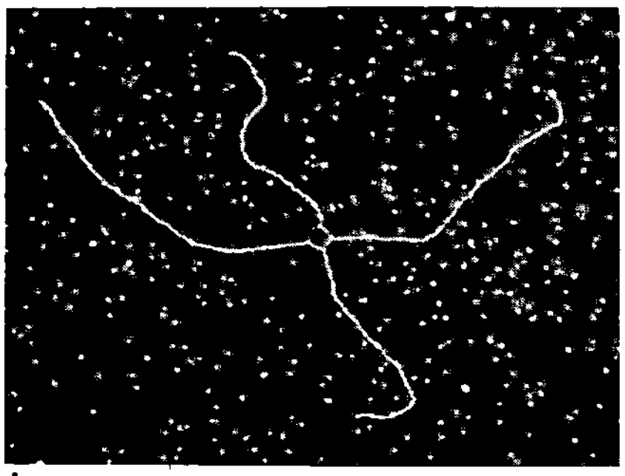
A

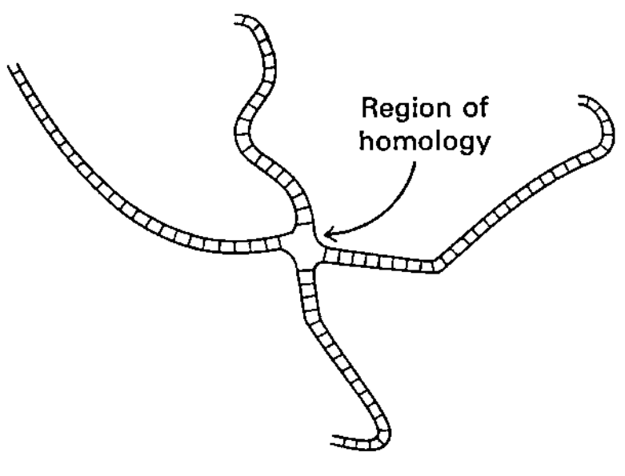

B

Figure 32-6
(A) Electron micrograph of a chi form. (B) Interpretive diagram. The region of homology (which is rich in A · T base pairs) was selectively denatured by formamide to show the strand connections in the crossover. [From H. Potter and D. Dressler. *Cold Spring Harbor Symp. Quant. Biol.* 43(1979):973.]

THE recA PROTEIN CATALYZES THE ATP-DRIVEN EXCHANGE OF DNA STRANDS IN HOMOLOGOUS RECOMBINATION

The process discussed thus far is called *homologous,* or *general, recombination* because exchange can occur between *any* pair of homologous sequences on the parental DNA molecules. In *E. coli,* homologous recombination is mediated by proteins encoded by the *rec* genes. The *recA protein* plays a central role, as evidenced by the finding that the frequency of recombination is 10^4-fold lower in *recA*$^-$ mutants than in the wild type. An early clue as to how recA protein acts was provided by the observation that *single-stranded DNA markedly stimulates its ATPase activity.* The breakthrough came when recA protein was shown to *catalyze the pairing of a single-stranded DNA molecule with the complementary region of a duplex DNA molecule* (Figure 32-7). The resulting three-stranded intermediate is called a *D loop* because the displaced strand has the appearance of the loop of the letter D.

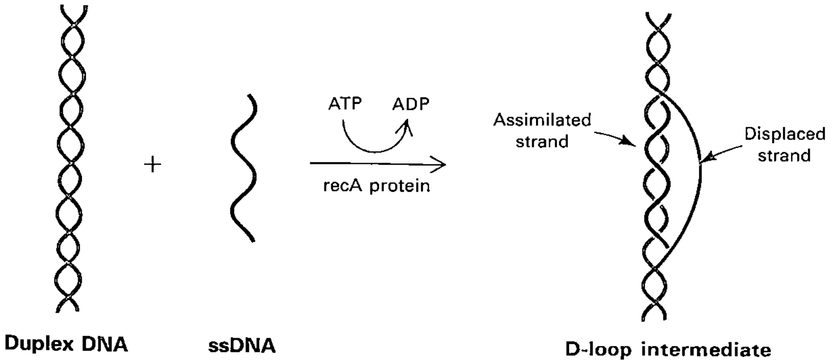

Figure 32-7
The pairing of a single-stranded DNA molecule with the complementary strand of a duplex is catalyzed by the recA protein. The resulting structure is called a D loop. ATP hydrolysis releases recA protein from DNA.

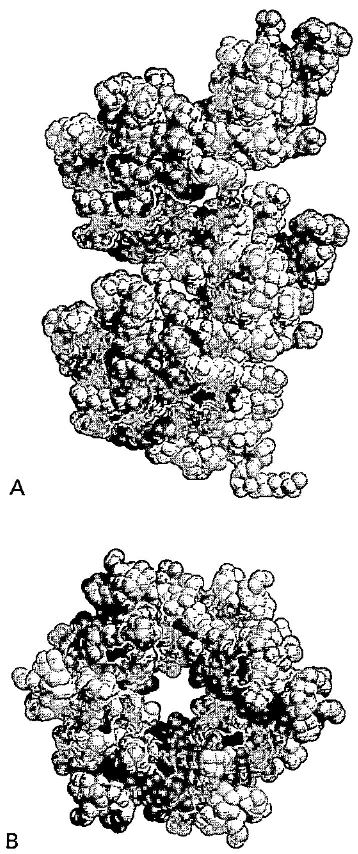

Figure 32-8
Structure of the recA helical filament seen in the crystal of the ADP complex. Monomer units are shown alternately in blue and yellow, and bound ADP in red. (A) Axial view. (B) Cross-sectional view. DNA molecules occupy the central cavity. [Drawn from 1rea.pdb. R.M. Story, I.T. Weber, and T.A. Steitz. *Nature* 355(1992):318.]

The first step in the action of recA protein is the formation of a filament around single-stranded DNA (ssDNA). This helical filament contains six 38-kd monomers per turn and has a diameter of 100 Å (Figure 32-8). The ssDNA in the center is stretched out—the axial distance between bases is 5 Å, compared with 3.4 Å in a Watson-Crick double helix. ATP is required for the formation of this extended complex, which enables ssDNA to search for regions of homology in duplex DNA. In vivo, ssDNA is present in regions of damaged DNA.

The recA–ssDNA filament then binds duplex DNA. The duplex DNA is partly unwound to facilitate the reading of its base sequence; a turn of double helix in this complex contains 18.6 bp, compared with 10.4 for B-DNA. ssDNA and duplex DNA are next to each other in a single binding site of recA protein (Figure 32-9). The recA–ssDNA filament then rapidly scans the duplex DNA for a sequence complementary to that of the ssDNA. When a complementary sequence is found, the duplex is further unwound to allow a switch in base pairing. *ssDNA now pairs with the complementary strand of the target duplex. Once started in this way, the exchange of strands (branch migration) continues.*

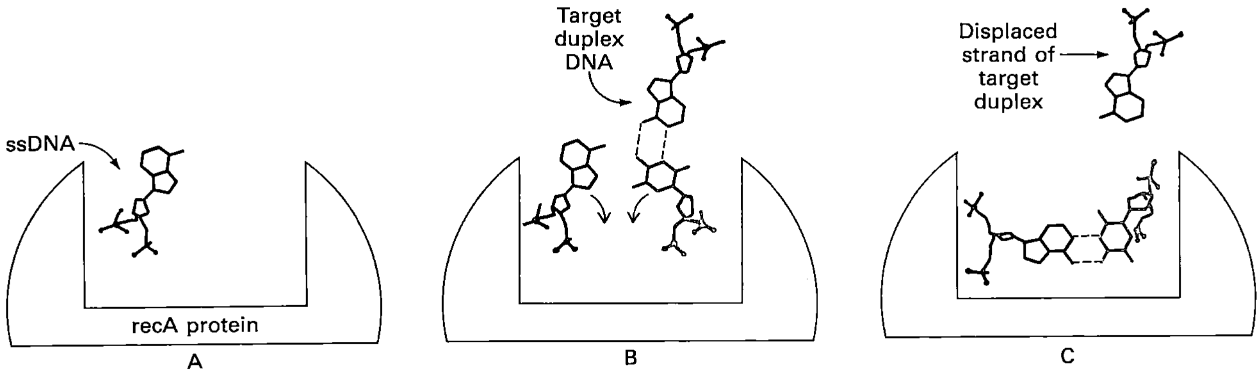

Figure 32-9
The binding of ssDNA to recA (A) leads to the binding of duplex DNA (B), which is scanned for a homologous region. The invading ssDNA becomes paired to the complementary strand of the target duplex (C). [After P. Howard-Flanders, S.C. West, and A. Stasiak. *Nature* 309(1984):217.]

The recombinational actions of recA protein are powered by the hydrolysis of ATP. ATP is required for the binding of DNA to recA protein. Strand exchange can occur without ATP hydrolysis. However, recA protein does not dissociate from DNA until bound ATP is hydrolyzed. Thus, ATP hydrolysis serves to release recA from DNA. Such a reaction cycle is reminiscent of the interplay of actin and myosin in muscle contraction (p. 399) and of the action of DNA gyrase (p. 798). Another pertinent finding is that the recA filament, like F-actin (p. 397), has polarity (see Figure 32-8). Moreover, recA monomers assemble unidirectionally ($5' \rightarrow 3'$) on ssDNA. These properties of recA enable it to catalyze unidirectional strand assimilation.

The driving force for coherent movement of the recA filament and the invading ssDNA is the hydrolysis of bound ATP. Without an input of free energy, the invading ssDNA would take a few steps forward and then a few steps back, an aimless random walk. Unidirectional, processive movement has another important function. It enables the invading DNA strand to bypass damaged regions of the target duplex, up to hundreds of base pairs, that are not complementary to it. ATP hydrolysis also promotes branch migration by rotating a duplex DNA molecule around the filament (Figure 32-10). The mechanochemistry of recombination is an intriguing and challenging area of inquiry.

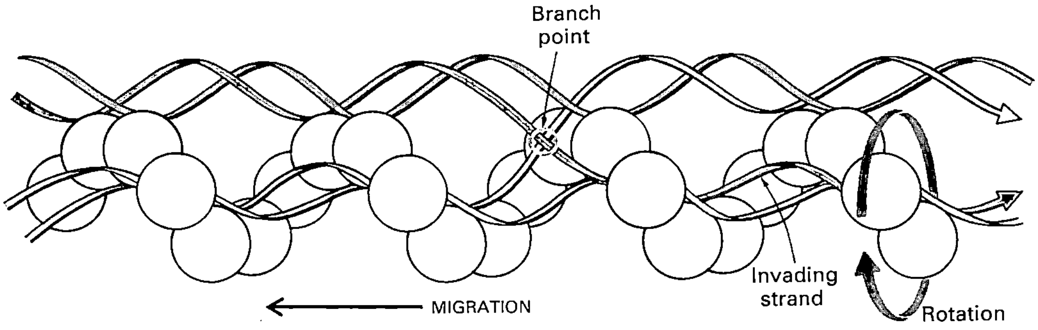

Figure 32-10
Model for branch migration, the movement of the crossover point between DNA duplexes. A recA filament containing one DNA molecule rotates a second DNA molecule joined to the first by a crossover junction. Branch migration is powered by the hydrolysis of ATP by recA. [After M.M. Cox and I.R. Lehman. Enzymes of general recombination. *Ann. Rev. Biochem.* 56:252. Copyright © 1987 by Annual Reviews, Inc. All rights reserved.]

THE HELICASE AND ENDONUCLEASE ACTIONS OF recBCD GENERATE SINGLE-STRANDED DNA FOR RECOMBINATION

Recombination requires that the invading DNA be single-stranded. How is ssDNA formed for this purpose? Genetic studies showed that the *recB*, *recC*, and *recD* genes are important in recombination. In fact, these genes encode proteins that form a 328-kd *recBCD complex* having both *helicase* and *nuclease* activity. RecB protein alone catalyzes the ATP-driven unwinding of duplex DNA, and recD protein by itself is a nuclease. The recBCD complex generates ssDNA from a duplex in an interesting way (Figure 32-11). The complex attaches to one end of a linear duplex and unwinds it. The hydrolysis of ATP powers the unwinding, which proceeds at about 300 nucleotides per second. Rewinding of DNA at the back end of the complex occurs more slowly than unwinding at the forward end. Hence, two single-stranded loops emerge from the complex, and they grow in size as the enzyme moves ahead on the DNA.

Cleavage occurs near a specific sequence called *chi*. The *E. coli* chromosome has about a thousand chi sites, about one per 4 kb. The sequence recognized is

5′-GCTGGTGG-3′

RecBCD cuts the backbone of this DNA strand between four and six nucleotides away from the 3′ end of this sequence (see Figure 32-11). The enzyme complex continues to move forward and unwind the DNA to produce single-stranded DNA that can invade a homologous duplex.

Several other proteins that are needed for homologous recombination have been identified. *Single-strand binding protein (SSB)* participates in this process as well as in DNA replication (p. 808). The cooperative binding of SSB to single-stranded DNA prevents the formation of hairpin loops. SSB stabilizes an unpaired DNA strand until recA protein forms a filament around it. *Topoisomerase I* (p. 796) also plays an important role in recombination by relieving torsional strain produced by the winding of one DNA molecule around another during branch migration. Branch migration, the movement of Holliday junctions, is driven by the ATP-powered *ruvB protein* acting in concert with the *ruvA protein*. These junctions are cut by the *ruvC protein*, a nuclease that recognizes cruciform structures, and are sealed by *DNA ligase*.

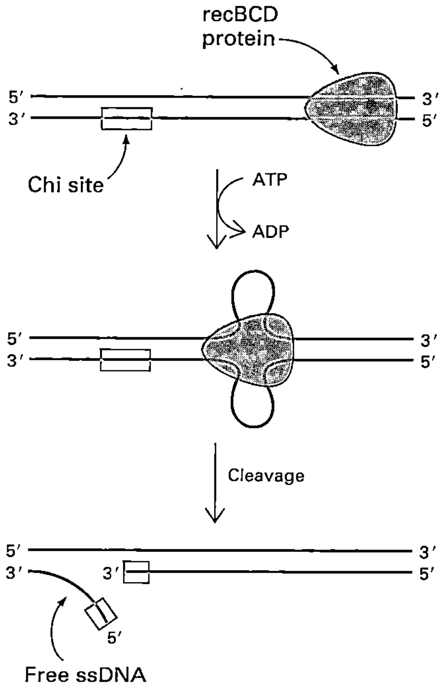

Figure 32-11
Model for the generation of single-stranded DNA by the recBCD complex (red), an ATP-driven helicase and nuclease. Two single-stranded loops are formed as the complex moves forward. The enzyme cleaves one of the strands when it encounters a chi sequence (yellow) to form single-stranded DNA with a free end. Recombination can then take place. [After A. Taylor and G.R. Smith. *Cell* 22(1980):447.]

DAMAGE TO DNA TRIGGERS AN SOS RESPONSE THAT IS INITIATED BY AUTOPROTEOLYSIS OF A REPRESSOR

An *E. coli* cell normally contains several hundred copies of the recA protein. The level of this key enzyme in recombination increases a hundredfold following damage to DNA. Indeed, damage rapidly induces the synthesis of more than 15 proteins that mediate the repair of DNA. The switching on of this repair machinery is called the *SOS response*. Under normal conditions, repair proteins are present at low levels because the synthesis of their messenger RNAs is blocked by a repressor protein called *lexA*. The presence of extensive single-stranded DNA signals that DNA has been damaged and reverses repression by lexA.

The SOS response is triggered by cleavage of the lexA protein. What is the protease that unleashes this response? Remarkably, *lexA protein digests itself when instructed to do so* (Figure 32-12). The directive is provided by the ssDNA–recA filament, which binds multiple copies of lexA. An activated

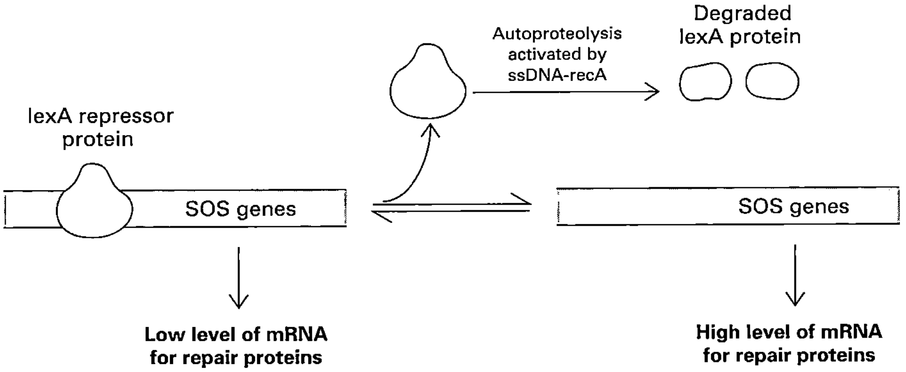

Figure 32-12
SOS genes for DNA repair are switched on by damage to DNA because the lexA repressor is then destroyed. The ssDNA–recA complex activates the protease activity of lexA. Digestion of a lexA molecule by another lexA on the filament lowers the concentration of this repressor, thereby activating expression of the SOS genes.

lexA protein on this filament then digests neighboring lexA molecules. Thus, *ssDNA–recA, a signal of DNA damage, serves as a coprotease.* The consequent lowering of the level of lexA repressor leads to the full expression of many genes that encode DNA repair proteins. For example, large amounts of recA, SSB, and the uvrABC complex (a nuclease that excises thymine dimers, p. 812) are produced following the destruction of lexA.

BACTERIA CONTAIN PLASMIDS AND OTHER MOBILE GENETIC ELEMENTS

Homologous recombination generates new combinations of specific *alleles,* but it does not readily change the arrangement of entire *loci.* In other words, *ABC′D′E′* can be readily formed by homologous recombination of *ABCDE* with *A′B′C′D′E′*, whereas *ABXYZCDE* and *ABE* cannot. These larger-scale genetic rearrangements are mediated by *mobile genetic elements,* which are also known as *transposable elements* (Table 32-1).

Alleles—
Alternative forms of a gene that can occupy a particular chromosomal site.
From the Greek word meaning "of one another."
For example, the β^S sickle gene is an allele of the normal β^A gene. In contrast, the genes for the α and β chains of hemoglobin are not alleles of each other.

Table 32-1
Mobile genetic elements in *E. coli*

Type	*Size (kb)*	*Characteristics*
Plasmids		
F (fertility) factor	93	Confers maleness; transmissible by conjugation.
F′ factors	>100	Carry *E. coli* genes in addition to F factor genes.
R (resistance) factors	4 to 117	Carry drug-resistance genes; some also carry genes for conjugation.
Colicinogenic factors	6 to 141	Carry genes for colicin (toxin) production; some also carry genes for conjugation.
Lysogenic phages		
Lambda	48	A small proportion carry *E. coli* genes (*gal* or *bio*) in addition to viral genes.
Mu	38	All mu carry a small piece of the *E. coli* genome.
Insertion sequences (IS)	0.8 to 1.4	Genes flanked by a pair of IS are translocatable within a cell.

Plasmids are a major class of mobile genetic elements. These circular duplex DNA molecules (see Figure 32-1), ranging in size from two to several hundred kilobases, carry genes for the inactivation of antibiotics (p. 128), the metabolism of natural products (p. 138), and the production of toxins. In essence, *plasmids are accessory chromosomes* that differ from the bacterial chromosome in being dispensable under certain conditions. *Plasmids are replicons; they contain their own origin of replication and hence can replicate autonomously.*

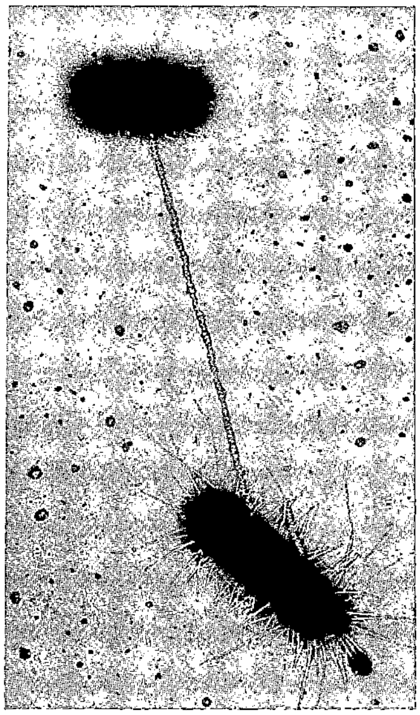

Figure 32-13
Transmission of genetic information from one *E. coli* cell to another. This electron micrograph shows two *E. coli* joined by a pilus during conjugation. DNA is transmitted through the pilus from the donor to the acceptor cell. [Courtesy of Dr. Charles Brinton and Dr. Judith Carnahan.]

THE F FACTOR ENABLES BACTERIA TO DONATE GENES TO RECIPIENTS BY CONJUGATION

Some plasmids enable bacteria to transfer genetic material to each other by forming a direct cell-cell contact. This process, called *conjugation,* was discovered in 1946 by Joshua Lederberg and Edward Tatum. During the conjugation of *E. coli,* one partner (the male) is the genetic donor and the other (the female) is the genetic recipient. Male bacteria contain special appendages called *sex pili* on their surfaces, whereas female cells contain receptor sites that bind pili. A male bacterium contains a plasmid called *F factor (fertility factor),* which contains genes for the formation of the sex pili and other components required for conjugation. A male and a female cell become joined by a pilus (Figure 32-13), which retracts to draw the cells into direct contact for the passage of DNA. One strand of the F factor plasmid gets nicked, and the duplex unwinds (Figure 32-14). The 5′ end of the nicked strand enters the recipient cell, and a complementary strand is synthesized to form a closed, circular duplex. *The presence of an F factor plasmid in the recipient cell (originally F^-) converts it into a male (F^+).* On the other hand, a male cell can spontaneously lose the F factor and thereby revert to F^-.

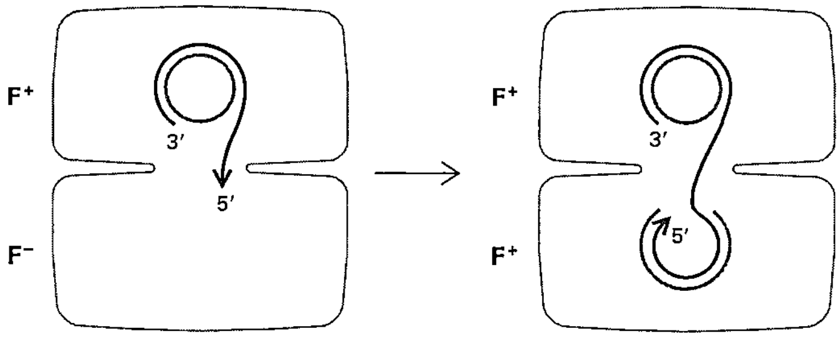

Figure 32-14
Proposed model for strand transfer in conjugation. The supercoiled F factor in the F^+ donor unwinds after being nicked. Transfer of one strand of the F factor to the F^- recipient is coupled to replication of that strand in the donor. A complementary strand is then synthesized in the recipient. The newly synthesized DNA in the donor is shown in blue, and in the recipient, in green. [After G.J. Warren, A.J. Twigg, and D.J. Sherratt. *Nature* 274(1978):260.]

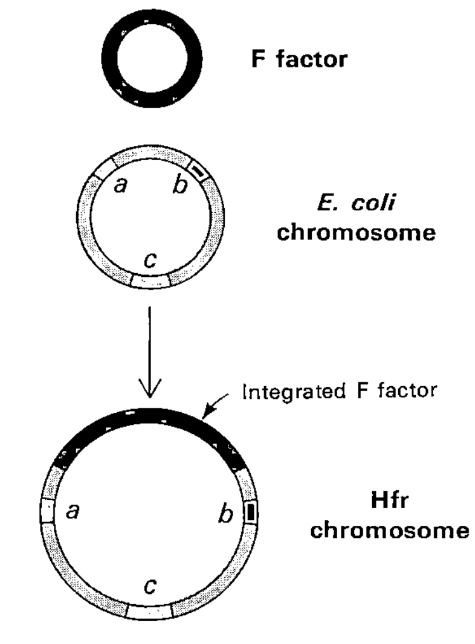

Figure 32-15
Schematic diagram showing the formation of an Hfr cell by integration of the F factor into the *E. coli* chromosome.

An F factor plasmid can become integrated into the bacterial chromosome (Figure 32-15). Integration occurs by crossing-over at one of a number of sites on the bacterial chromosome. The frequency of integration is about 10^{-5} per generation. Bacteria harboring F factors in their chromosomes are called *Hfr cells* because they exhibit a high frequency of recombination. Hfr cells, like F^+ cells, are donors in conjugation. The difference between them is that *an Hfr cell donates the entire bacterial chromosome (including the integrated F factor),* whereas an F^+ cell donates only the F factor. In both

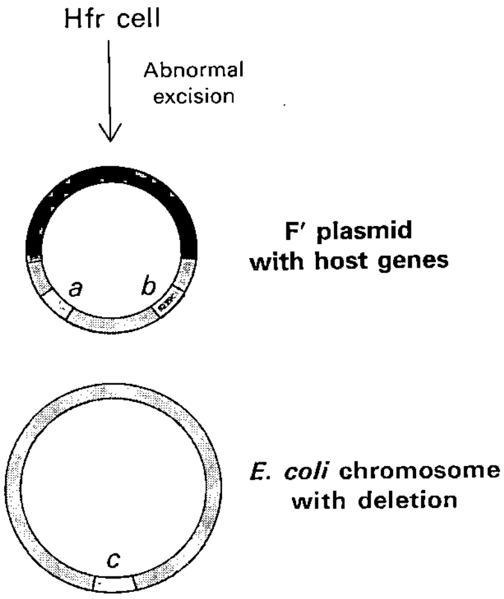

Figure 32-16
Abnormal excision leads to the formation of an F′ plasmid, which contains part of the *E. coli* chromosome.

cases, a single strand of DNA is transferred, and its complement is then synthesized. Transfer occurs in about 90 minutes in an Hfr × F^- cross. The order of entry of the donated genes into the recipient depends on the site of integration of the F factor and on its orientation. Hence, it is feasible to ascertain the order of genes in the donor's chromosome by interrupting the conjugation at different times and determining which markers have been transferred. The donated chromosome can recombine with the recipient's chromosome. The frequency of recombination is highest for genes that first entered the recipient because they are present for the longest time. Thus, genetic maps can be constructed by determining the *time of entry* and the *frequency of recombination* of markers contributed by the donor.

An Hfr cell reverts to the F^+ state by the excision of the F factor from its chromosome. This reversal of the integration of the F factor also has a frequency of about 10^{-5} per generation. In a small proportion of the revertants, excision occurs at a site different from that of integration, which results in the *formation of a plasmid that contains chromosomal genes in addition to the F factor genes* (Figure 32-16). Such a plasmid is called an *F′ factor,* the prime denoting the presence of chromosomal genes. Conjugation of an F′ cell with an F^- cell results in the transfer of these chromosomal genes from the donor to the recipient cell, which then becomes diploid for them.

Thus, bacteria have mechanisms for transferring sets of genes and even entire chromosomes from one cell to another. *The F factor can be regarded as a specially designed vector for the interchange of genetic material.* It is interesting to note that lysogenic phages can also mediate the exchange of host genes. For example, λ phage DNA can be inserted between the *gal* (galactose) and *bio* (biotin) genes on the *E. coli* chromosome (p. 832). When progeny virions are formed, excision is usually, but not invariably, precise. In about 1 of 10^5 virions, λ DNA contains either the *gal* operon or the *bio* gene. Infection by such phages, which are called λ*gal* or λ*bio,* introduces these *E. coli* genes in addition to λ genes. A related phage, called *φ80,* inserts near the *trp* operon and can carry genes for the synthesis of tryptophan from one infected cell to another. Bacteriophage *μ,* which inserts nearly anywhere on the *E. coli* chromosome, invariably emerges with a piece of the bacterial chromosome. *These transducing phages, like the F factor, are mobile genetic elements that promote the exchange of bacterial genes and hence accelerate bacterial evolution.*

> *Episome*—
> A genetic element that can exist as a separate piece of DNA or become integrated into the bacterial genome. Plasmids and F factors are episomes.

R FACTOR PLASMIDS MAKE BACTERIA RESISTANT TO ANTIBIOTICS

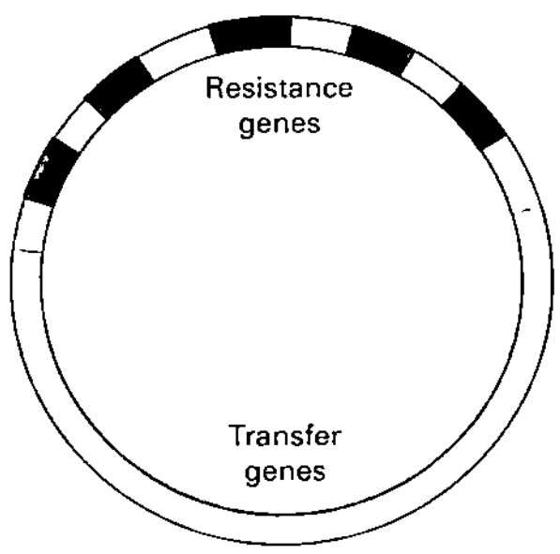

Figure 32-17
Schematic diagram of an R factor. The *RTF* genes (for conjugation and replication) are shown in green, and the *r*-genes (for drug resistance) in red. Insertion sequences are shown in yellow.

A striking example of very rapid bacterial evolution took place in 1955 during an epidemic of bacterial dysentery. A strain of *Shigella dysenteriae* became simultaneously resistant to chloramphenicol, streptomycin, sulfanilamide, and tetracycline. Multiple-drug resistance of this kind is now common among many pathogenic microorganisms. The genes conferring resistance to multiple antibiotics are linked together on *R factor (resistance factor) plasmids.* The larger of these plasmids contain a *resistance transfer factor (RTF)* in addition to several r-*genes* (Figure 32-17). The RTF region enables the plasmid to be transmitted to other bacteria by conjugation. In fact, the genes in the RTF region closely resemble those in F factors. The *r*-genes code for enzymes that inactivate specific drugs. R factors with an RTF region can be transmitted in mixed cultures even between different bacterial species. Thus, *multiple-drug resistance can be infectious.*

The small R factor plasmids are devoid of an RTF region and usually confer resistance to a single antibiotic. For example, the 8.2-kb pSC101 plasmid carries a gene for tetracycline resistance, but it cannot be transmitted by conjugation. A transmissible R plasmid is formed when this *r*-gene integrates into a plasmid containing an RTF region (Figure 32-18). Additional *r*-genes conferring resistance to different antibiotics can join this plasmid. Thus, *complex R factor plasmids are built from highly mobile modules that confer resistance to individual drugs.*

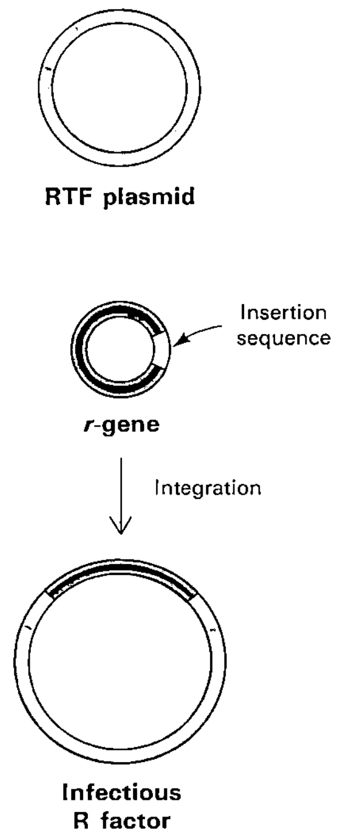

Figure 32-18
An infectious R factor is formed when an *r*-gene joins an RTF plasmid.

TRANSPOSONS ARE HIGHLY MOBILE DNA SEQUENCES THAT MOVE GENES TO NEW SITES IN BACTERIAL GENOMES

Mobile genetic elements such as R factor plasmids are called *transposons.* What is the structural basis of their high mobility? The simplest transposons are *insertion sequences (IS),* which are typically about a kilobase long (Table 32-2). They encode a highly specific nuclease called a *transposase.* The sequences of 20 or so base pairs at opposite ends of an IS are very similar but their order is reversed. These boundary sequences are known as *inverted terminal repeats.* An IS can move to virtually any site on a chromosome. The recipient site becomes duplicated during transposition (Figure 32-19). A sequence of 4 to 12 bp that was present once in the

Table 32-2
Properties of some transposons in *E. coli*

Type	*Length (bp)*	*Duplicate recipient length (bp)*	*Function (in addition to transposition)*
Insertion sequences			
IS1	768	9	—
IS2	1,327	5	—
IS4	1,426	12	—
IS5	1,195	4	—
Complex transposons			
Tn3	4,957	5	Ampicillin resistance
Tn5	5,700	9	Kanamycin resistance
Tn2571	23,000	9	Resistance to multiple antibodies and to Hg

Sources: J. Cullum, in *Genetics of Bacteria,* J. Scaife, D. Leach, and A. Galizzi, eds. (Academic Press, 1985), p. 86; N. Kleckner, *Ann. Rev. Genet.* 15(1981):341.

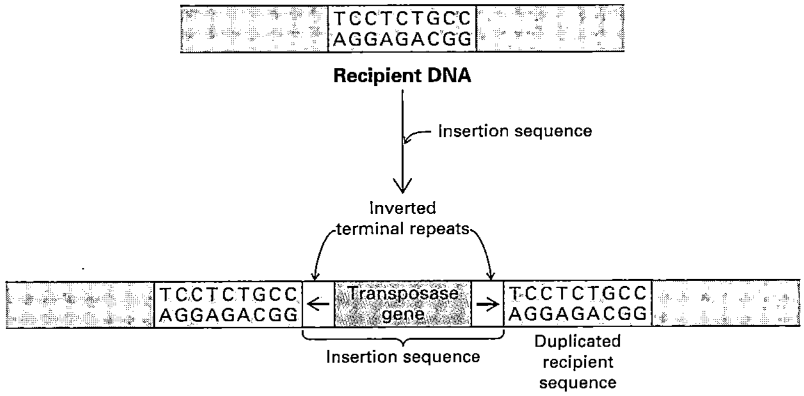

Figure 32-19
An insertion sequence consists of a gene for a transposase (red) bounded by inverted terminal repeats (yellow). Transposition leads to a duplication of a short sequence of the recipient DNA (blue).

target DNA before transposition occurs twice after transposition, one on each side of the IS. The *duplicated recipient sequence* is a noninverted repeat. There is no homology between these flanking sequences and the terminal sequences of the IS. Furthermore, the products of *rec* genes do not participate in transposition. Rather, ***the inverted terminal repeats of an IS bind its own transposase, which makes staggered cuts at both the donor site and recipient site.*** DNA polymerase I and DNA ligase are also essential for transposition, which occurs in bacteria at a frequency of the order of 10^{-6} per generation.

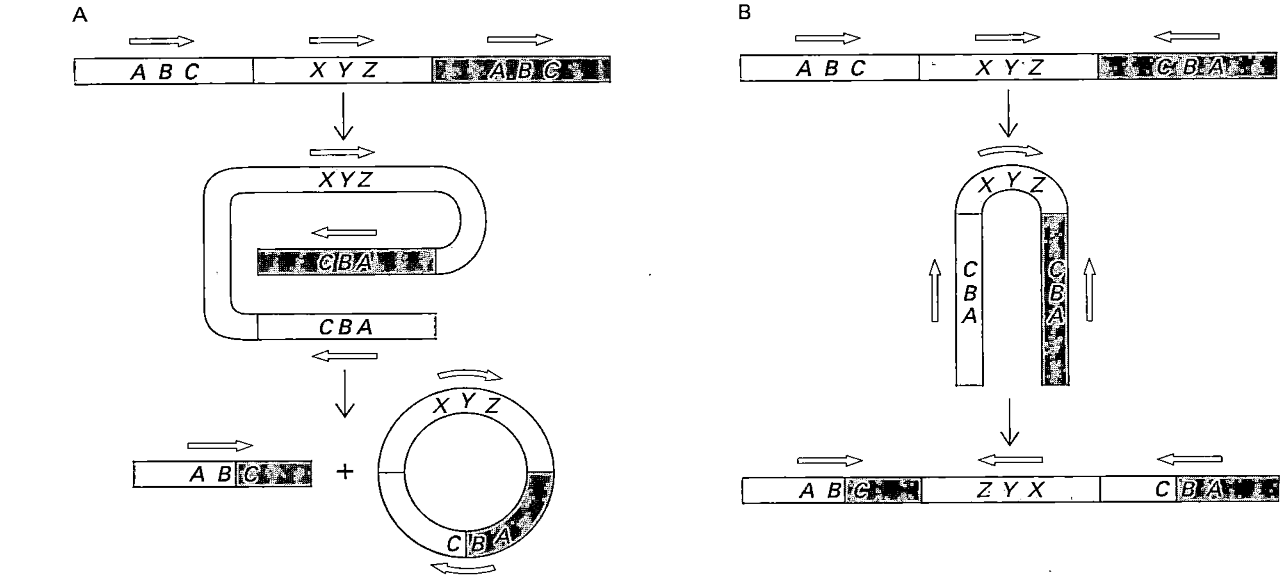

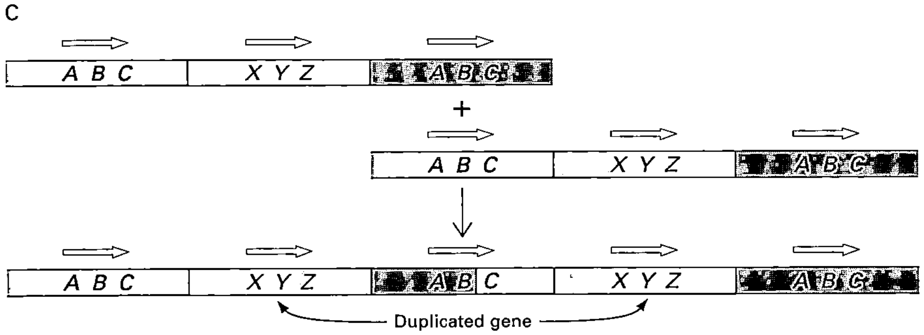

Figure 32-20
Recombination between identical insertion sequences (shown in red and green) leads to (A) the deletion or (B) the inversion of genes (yellow) located between them. The orientation of the insertion sequences determines the outcome. (C) Genes can be duplicated by crossing-over of insertion sequences at different loci on sister chromosomes.

Most genes are inactivated by the transposition of an insertion sequence into their midst. However, each insertion sequence carries its own promoter for the transcription of its transposase gene. This promoter sometimes enhances the expression of a neighboring gene. Moreover, insertion sequences can cause *deletions, inversions,* and *duplications* of genes (Figure 32-20). The *E. coli* chromosome contains multiple copies of several insertion sequences. Recombination between two identical insertion sequences oriented the same way leads to the excision of genes between them (see part A of Figure 32-20). In contrast, the genes between oppositely oriented insertion sequences are inverted by a reciprocal crossover (B). The expression of a gene can be profoundly altered by inversion, as exemplified by a switching from one type of flagellar protein to another in *Salmonella* (p. 969). In eukaryotes, genes bounded by insertion sequences can be tandemly duplicated by unequal crossing-over of sister chromosomes (C). Thus, the presence of transposons makes possible large-scale frequent rearrangements of the genome.

Complex transposons, in contrast with insertion sequences, contain genes mediating functions in addition to transposition. For example, Tn3 consists of three genes that are bounded by inverted terminal repeats (Figure 32-21). The gene for *β-lactamase,* a hydrolytic enzyme, confers resistance to the antibiotic ampicillin. Transposition is carried out by proteins encoded by the *transposase* and *resolvase* genes. Tn3, unlike most simple insertion sequences, is duplicated when it moves. This process is called *replicative transposition,* to distinguish it from *conservative replication,* which begins and ends with one copy of a transposon. In replicative transposition, a *cointegrate* intermediate that contains both transposons is formed. The role of resolvase is to cut and rejoin DNA strands to craft two DNA circles from the cointegrate. *Complex transposons such as Tn3 probably arose from cellular genes that became bounded by a pair of insertion sequences.* It seems likely that R factor plasmids conferring resistance to multiple antibiotics evolved by the coming together of several transposons.

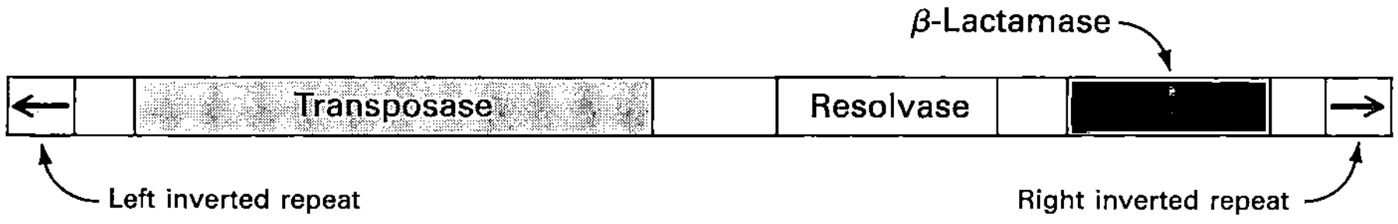

Figure 32-21
Tn3, a complex transposon, consists of genes for three proteins bounded by inverted terminal repeats.

Plasmids and phages can exchange blocks of genes with bacterial chromosomes and can recombine with each other. A tetracycline-resistance element has been shown to move from an R factor plasmid to a *Salmonella* phage and then to the *Salmonella* chromosome, and from there to λ phage and then to the *trp* operon of *E. coli* and back to λ. This remarkable journey emphasizes the highly mobile nature of prokaryotic genes, a theme that recurs in eukaryotes.

LAMBDA PHAGE DNA BECOMES INTEGRATED INTO THE BACTERIAL GENOME BY SITE-SPECIFIC RECOMBINATION

Some bacteriophages have a choice of life styles: they can multiply and then lyse an infected cell (*lytic pathway*), or their DNA can join the infected cell, retaining the capacity for multiplication and lysis (*lysogenic pathway*) (see Figure 6-15 on p. 128). Viruses that do not always kill their hosts are called *temperate* or *moderate,* of which *lambda* (λ) phage is the best understood. The DNA in the λ virion is a linear 48-kb duplex that contains single-stranded sequences 12 nucleotides long at each 5′ end. These sequences are called *cohesive ends* because they are mutually complementary and can base-pair to each other. In fact, the cohesive ends come together after infection of a bacterium. The 5′ phosphate terminus of each strand is then adjacent to the 3′ hydroxyl end of the same strand. A host DNA ligase seals these two gaps to yield a *circular λ DNA molecule* (Figure 32-22).

This circular λ DNA molecule can be expressed and replicated by λ proteins acting in concert with host proteins (p. 808). Alternatively, the λ

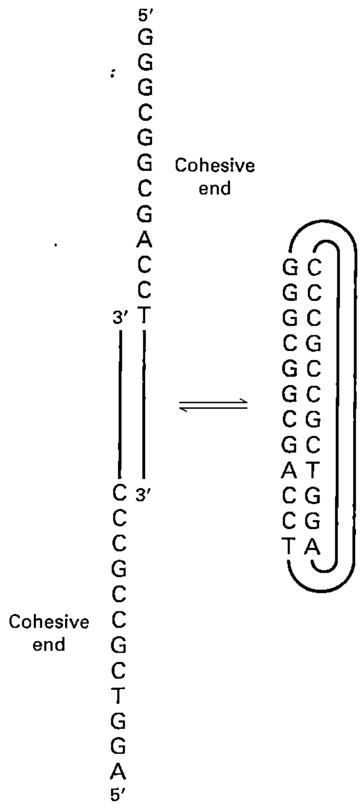

Figure 32-22
Conversion of linear λ DNA into the circular form. DNA ligase seals the ends.

DNA circle can be inserted into the bacterial chromosome by recombination between specific sites on both DNA molecules. This process is called *site-specific recombination* to distinguish it from homologous recombination, which can occur between any two long stretches of identical or similar sequences. *Site-specific recombination is guided primarily by proteins that recognize particular DNA sequences rather than by sequence homology.* The recA protein does not participate. Another difference is that single-stranded DNA is needed to initiate homologous recombination but not site-specific recombination.

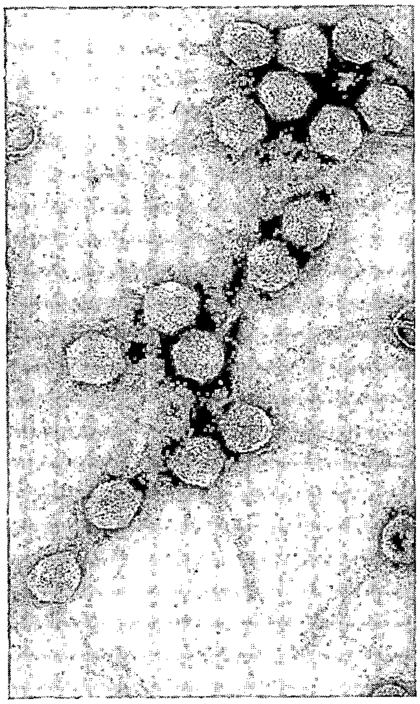

Electron micrograph of λ phage. [Courtesy of Dr. A. Dale Kaiser.]

The attachment site for λ DNA in *E. coli* DNA, called *attB*, is located between *galE* and *bioA*, genes in the galactose and biotin operons (Figure 32-23). The base sequence of *attB* is symbolized by B-O-B′ (*B* for "bacterial"). The specific attachment site on λ phage DNA, called *attP*, is located next to the genes *int* (for "integrate") and *xis* (for "excise"). The base sequence of *attP* is symbolized by P-O-P′ (*P* for "phage"). O (shown in yellow in the figure) denotes the 15-bp identical core sequences in the phage and bacterial segments that recombine. Insertion of λ DNA into *E. coli* DNA is mediated by *integrase*, a protein encoded by λ phage, and *integration host factor (IHF)*, an *E. coli* protein. Integrase recognizes the P-O-P′ sequence in the phage DNA and the B-O-B′ sequence in *E. coli* DNA. The enzyme makes staggered cuts with seven-nucleotide overhangs in both strands of the O sequences. IHF serves to bend the P-O-P′ sequence. After all four chains are cut, *P joins B′ and B joins P′ to form one DNA circle from two.* This scheme (Figures 32-23 and 32-24) was originally proposed by Allan Campbell on the basis of genetic evidence.

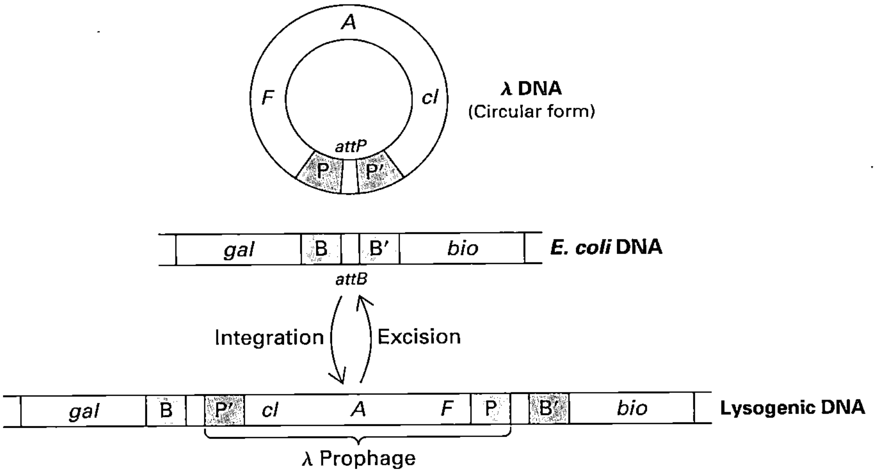

Figure 32-23
Schematic diagram of reciprocal recombination of λ DNA and *E. coli* DNA. The yellow regions mark identical 15-mer sequences (termed O) of the bacterial *attB* site and the phage *attP* site.

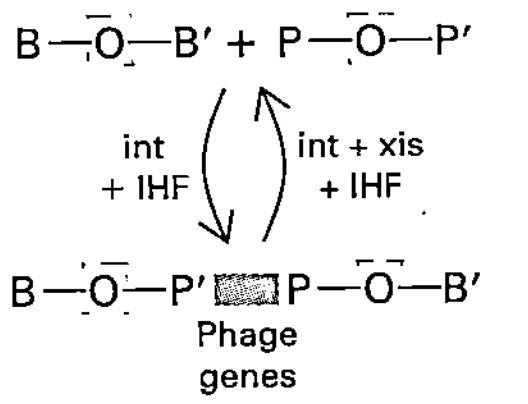

Figure 32-24
Integration and excision of λ DNA. Int denotes integrase and xis denotes excisionase. IHF (integration host factor) is a host protein, whereas int and xis are encoded by λ.

The λ DNA is now part of the *E. coli* DNA molecule. This form of λ is called the *prophage*, and the *E. coli* cell containing the prophage is called a *lysogenic bacterium*. The prophage is stable in the absence of *excisionase*, an enzyme encoded by the *xis* gene in λ DNA. As will be discussed in a later chapter (p. 958), transcription of *xis* is blocked by the λ repressor. When repression is released, excisionase, integrase, and IHF together catalyze the breaking of the B-P′ and P-B′ sequences, and again a transfer occurs (see Figure 32-24): P rejoins P′ and B rejoins B′, which recreates a circular molecule of λ DNA and a nonlysogenic *E. coli* chromosome. A key feature of this recombination system is that integrase alone cannot recognize the two new sequences at the ends of the prophage (B-P′ and P-B′), which makes the prophage DNA stable. Thus, insertion occurs when integrase alone is present, whereas excision occurs when both integrase and excisionase are present.

SITE-SPECIFIC RECOMBINATION OF *V, D,* AND *J* GENES GENERATES VAST DIVERSITY IN THE IMMUNE SYSTEM

Site-specific recombination is important in eukaryotes as well as prokaryotes. The development of the immune system and its response to antigens provide striking examples of how recombination is used to create an enormous number of different antibody molecules and T-cell receptors. Recall that the variable region of the genes for the light (L) chains of immunoglobulins are formed by the association of *V* and *J* genes (p. 373). In the case of heavy (H) chains, *V, D,* and *J* genes come together (see Figure 14-21 on p. 374). The combinatorial association of different gene segments greatly increases the diversity of the antibody repertoire. For example, 56,250 different kinds of H genes can be formed from 250 *V* genes, 15 *D* genes, and 5 *J* genes that can be joined in three frames. The genes for the variable region of the α and β chains of the T-cell receptor are also formed by the association of gene segments (p. 383).

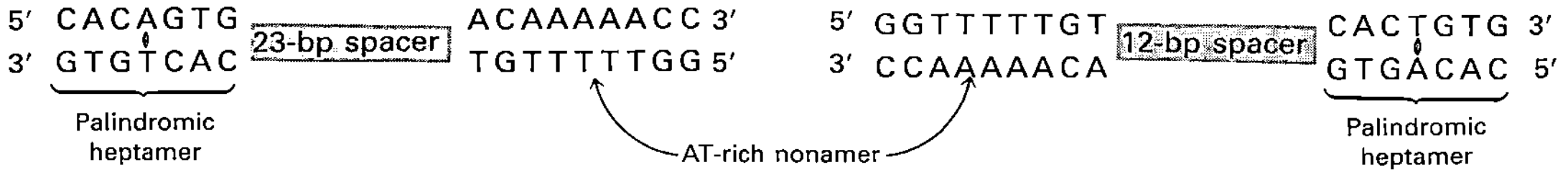

Figure 32-25
Recognition sites for the recombination of *V-*, *D-*, and *J*-segment genes. Sites are of two types: one with a 23-bp spacer (blue), and the other with a 12-bp spacer (red). Recombination can occur only between different types of sites.

The site-specific recombination of *V, D,* and *J* genes is mediated by proteins that recognize sequences flanking these segments. Each recognition sequence consists of a conserved palindromic heptamer that is separated from a conserved, AT-rich nonamer by a spacer (Figure 32-25). The spacer is either about 12 bp long or 23 bp long. All *V*-segment genes are *followed* by a recognition sequence containing a 23-bp spacer. All *D*-segment genes, by contrast, are flanked on *both* sides by recognition sequences of the 12-bp type. All *J*-segment genes are *preceded* by a 23-bp type of recognition sequence. The fundamental rule is that *recombination can occur between the 12-bp and 23-bp types but not between two 12-bp types or two 23-bp types.* Hence, *V-* and *J*-segment genes cannot recombine with each other. Likewise, *V* or *D* or *J* gene segments cannot recombine amongst themselves. The only allowed pairings are *V* with *D,* and *D* with *J.* Hence, the arrangement of the 12-bp and 23-bp types of recognition sequences assures that *VDJ* and not any other combination is produced by recombination of these gene segments (Figure 32-26).

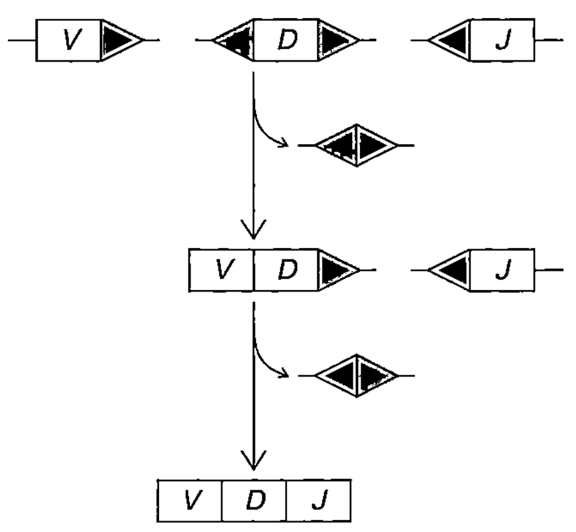

Figure 32-26
The joining of *V, D,* and *J* segments occurs by site-specific recombination between the two types of recognition sequences. The 23-bp spacer sequence is shown in blue, and the 12-bp spacer sequence in red.

The molecular machinery mediating these site-specific recombinations is coming into view. Two *recombination activating genes (RAG-1* and *RAG-2)* have been identified. Absence of a different gene product is responsible for *severe combined immunodeficiency disease (scid),* a genetic disorder of mice. In *scid* mice, VDJ joining and the repair of double-strand breaks anywhere in the DNA are markedly impaired. As noted earlier (p. 374), a few nucleotides may be lost or added to coding sequences in the normal association of *V, D,* and *J* gene segments. This imprecision in recombination markedly increases the diversity of the immune repertoire. It will be interesting to learn how nature abets slight disorder to achieve a beneficial end. Another challenge is to learn how site-specific recombination generates different classes of antibodies, a process called *class switching* (p. 377).

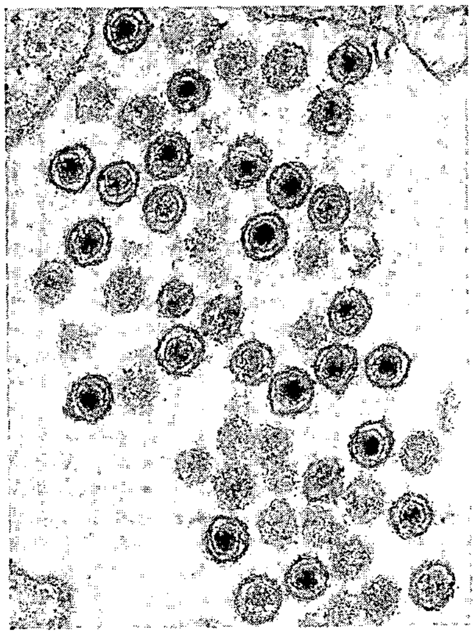

Figure 32-27
Electron micrograph of avian sarcoma virus (Rous sarcoma virus), a retrovirus. The densely stained virions are near the surface of an infected chicken cell. [Courtesy of Dr. Samuel Dales.]

RETROVIRAL GENOMES CYCLE BETWEEN AN RNA FORM IN THE VIRION AND A DNA FORM IN THE HOST CELL

Retroviruses contain an RNA genome but replicate through a double-stranded DNA intermediate that becomes integrated in the host-cell genome (p. 91). Recombination plays an essential role in their life cycle. In 1964, Howard Temin observed that infection by RNA tumor viruses such as avian sarcoma virus (also known as *Rous sarcoma virus,* p. 354) is blocked by inhibitors of DNA synthesis. Methotrexate, 5-fluorodeoxyuridine, and cytosine arabinoside block viral replication if present during the first 12 hours following the introduction of the virus. This finding suggested that *DNA synthesis is required for the growth of RNA tumor viruses.* Furthermore, the production of progeny virus particles is inhibited by actinomycin D, which is known to block the synthesis of RNA from DNA templates. Hence, *transcription of DNA seemed to be essential for the multiplication of RNA tumor viruses.* These unexpected results led Temin to propose that *a DNA provirus is an intermediate in the replication and oncogenic action of RNA tumor viruses.*

$$\text{RNA tumor virus} \longrightarrow \text{DNA provirus} \longrightarrow \text{RNA tumor virus}$$

Temin's bold hypothesis that genetic information can flow from RNA to DNA was initially greeted with little enthusiasm by most investigators. It required the existence of a then unknown enzyme—one that synthesizes DNA according to instructions given by an RNA template (an *RNA-directed DNA polymerase*). In 1970, Temin and David Baltimore independently discovered such an enzyme, which is called *reverse transcriptase,* in the virions of some RNA tumor viruses. All RNA tumor viruses subsequently studied have been found to contain a reverse transcriptase in their virions, and so they have become known as *retroviruses.* Not all retroviruses induce cancer. *Human immunodeficiency virus-1 (HIV-1),* for example, causes AIDS (p. 384).

The life cycle of a typical retrovirus starts when infecting virions bind to specific receptors on the surface of the host and enter the cell. Viruses are masters at gaining entry into cells by recognizing cell-surface receptors that are used for an entirely different purpose by the host. The viral (+) RNA is uncoated in the cytosol (by convention, the plus strand of viral RNA is the one that serves as mRNA). Reverse transcriptase (Figure 32-28) brought in by the virus particle then synthesizes the (−) strand of DNA and digests the viral (+) RNA. Finally, reverse transcriptase synthesizes the (+) strand of DNA. Thus, *reverse transcriptase carries out three kinds of reactions: RNA-directed DNA synthesis, DNA-directed DNA synthesis, and hydrolysis of RNA.* DNA synthesis is carried out by the polymerase moiety of reverse transcriptase, and RNA hydrolysis by the RNAse H moiety. DNA synthesis and RNA hydrolysis are coordinated so that RNA is removed after it has served as a template.

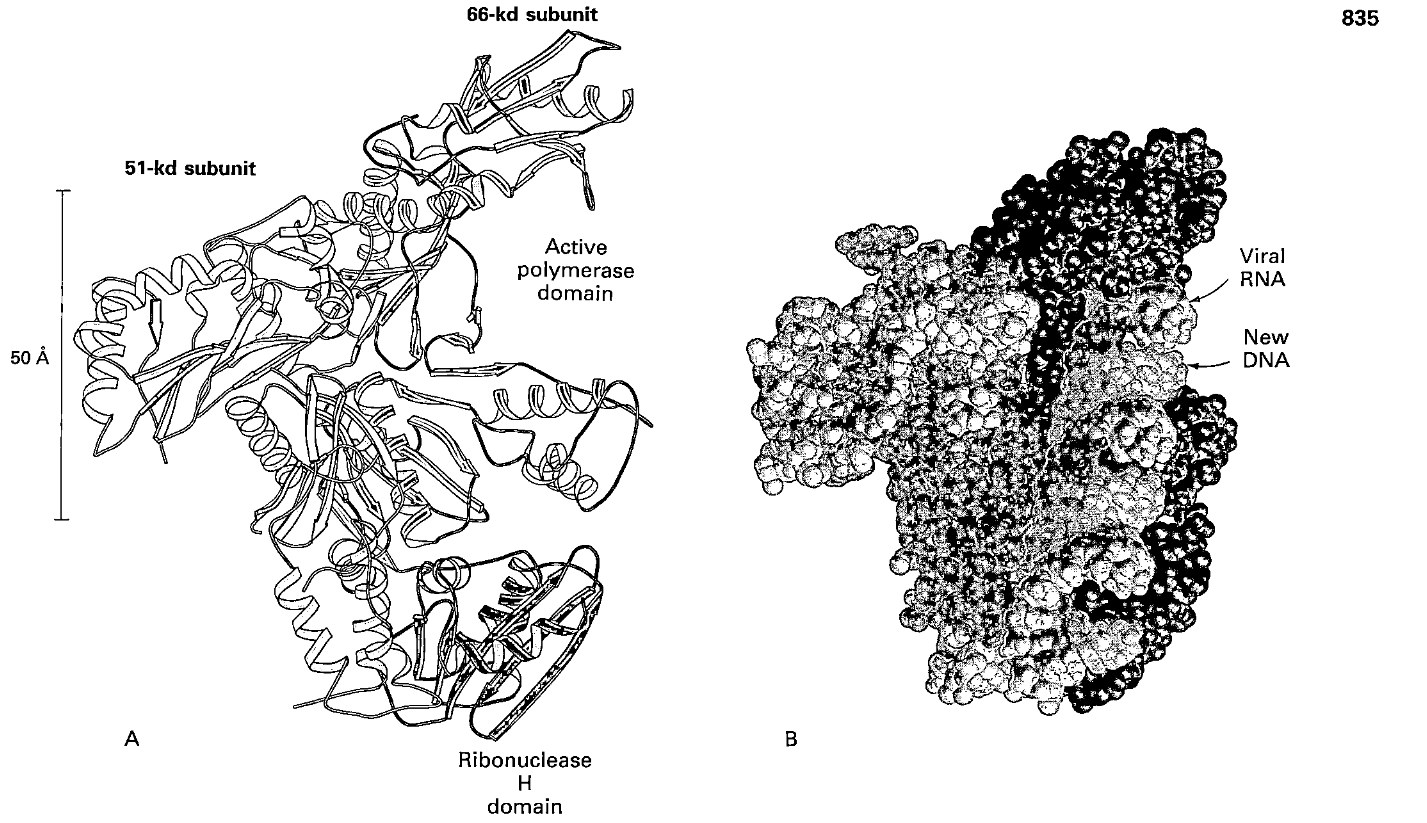

Figure 32-28
Structure of reverse transcriptase from HIV-1. The enzyme consists of a 51-kd subunit (p51) and a 66-kd subunit (p66). HIV-1 protease generates this heterodimer by cleaving one of the chains of the nascent p66-p66 homodimer. The polymerase domain of p66 but not of p51 is active. A RNase H domain is present in p66 but not in p51. (A) Schematic diagram of the heterodimer. p51 is shown in green, the polymerase domain of p66 in blue, and the RNAse H domain of p66 in red. The active polymerase domain resembles that of DNA polymerase I (see Figure 31-27). (B) Postulated mode of binding of an RNA–DNA duplex to the heterodimer. The newly synthesized DNA strand is shown in purple, and the viral RNA strand undergoing digestion is shown in yellow. [Drawn from 3hvt.pdb and coordinates kindly provided by Dr. Joseph Jaeger. L.A. Kohlstaedt, J. Wang, J.M. Friedman, P.A. Rice, and T.A. Steitz. *Science* 256(1992):1783.]

Reverse transcriptase, like other DNA polymerases, synthesizes DNA in the 5′ to 3′ direction and is unable to initiate chains de novo. How then is viral DNA synthesis primed? Initiation is accomplished very economically: the (+) RNA viral genome contains a noncovalently bound transfer RNA that was acquired from the host in the packaging of the virion. *The 3′ OH of this base-paired tRNA acts as the primer for DNA synthesis* (Figure 32-29). As will be discussed later (p. 984), a special problem is encountered in the replication of any *linear* DNA: the 5′ ends of daughter strands are initially incomplete. The same problem arises in the synthesis of DNA from an RNA template. Retroviruses have met this challenge in an ingenious way. Their genomic (+) RNA contains the same sequence (called R) at the 5′ and 3′ ends. This *terminal redundancy* plays a critical role in the synthesis of duplex DNA. Nascent DNA strands undergo two large shifts in base pairing before a complete duplex is formed. The resulting double-helical DNA contains identical ends called *long terminal repeats (LTR)*. LTRs are rich in signals for integration and transcription.

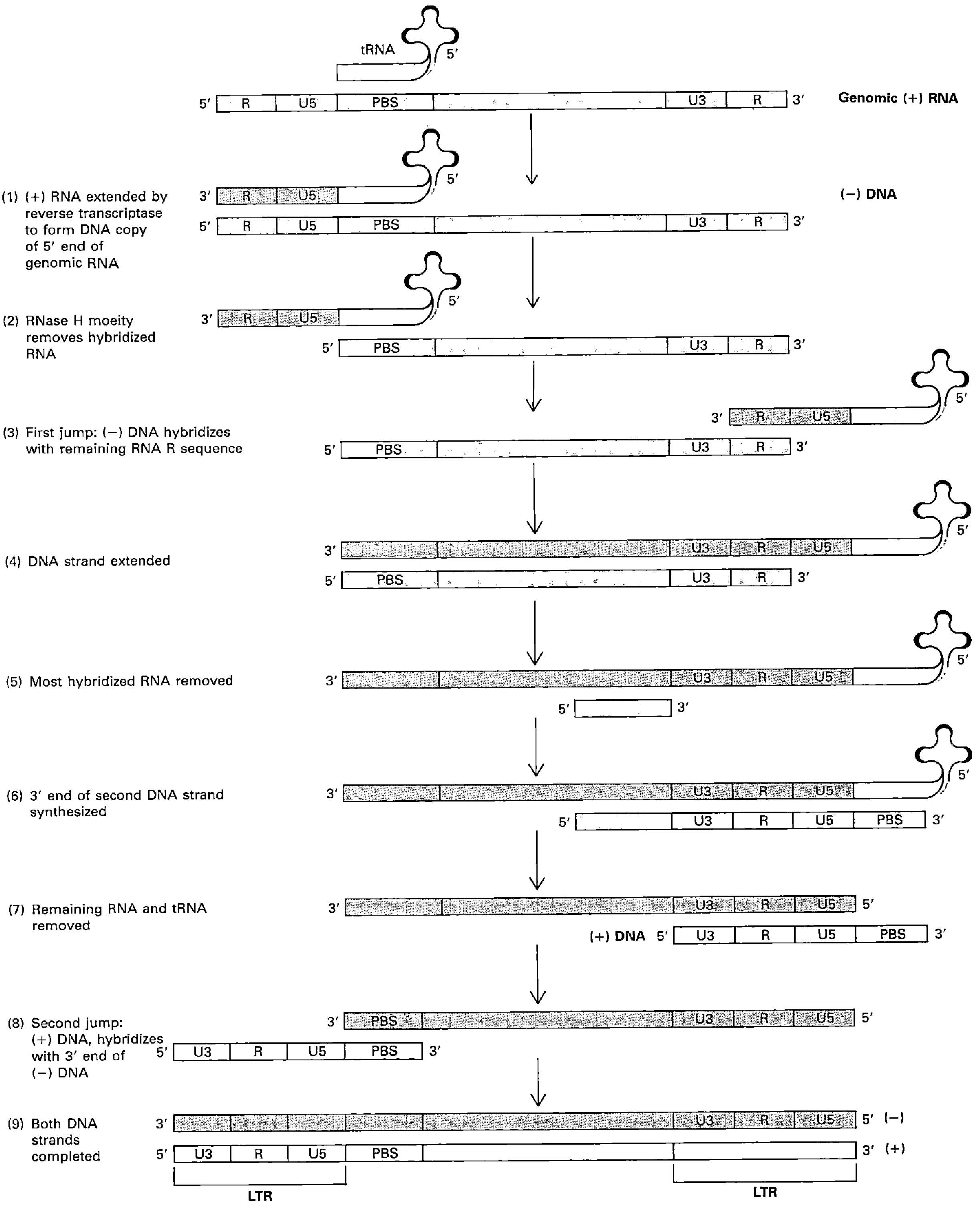

Figure 32-29
Reverse transcriptase converts the genomic RNA of a retrovirus into a double-stranded proviral DNA. (+) RNA is shown in blue, (−) DNA in red, (+) DNA in green, and the tRNA primer in yellow. PBS is the primer binding site, U5 and U3 are different sequences, and R is a repeated sequence. LTR denotes the long terminal repeat at each end of the proviral DNA. [After J. Darnell, H. Lodish, and D. Baltimore. *Molecular Cell Biology*, 2nd ed. (Scientific American Books, 1990), p. 972.]

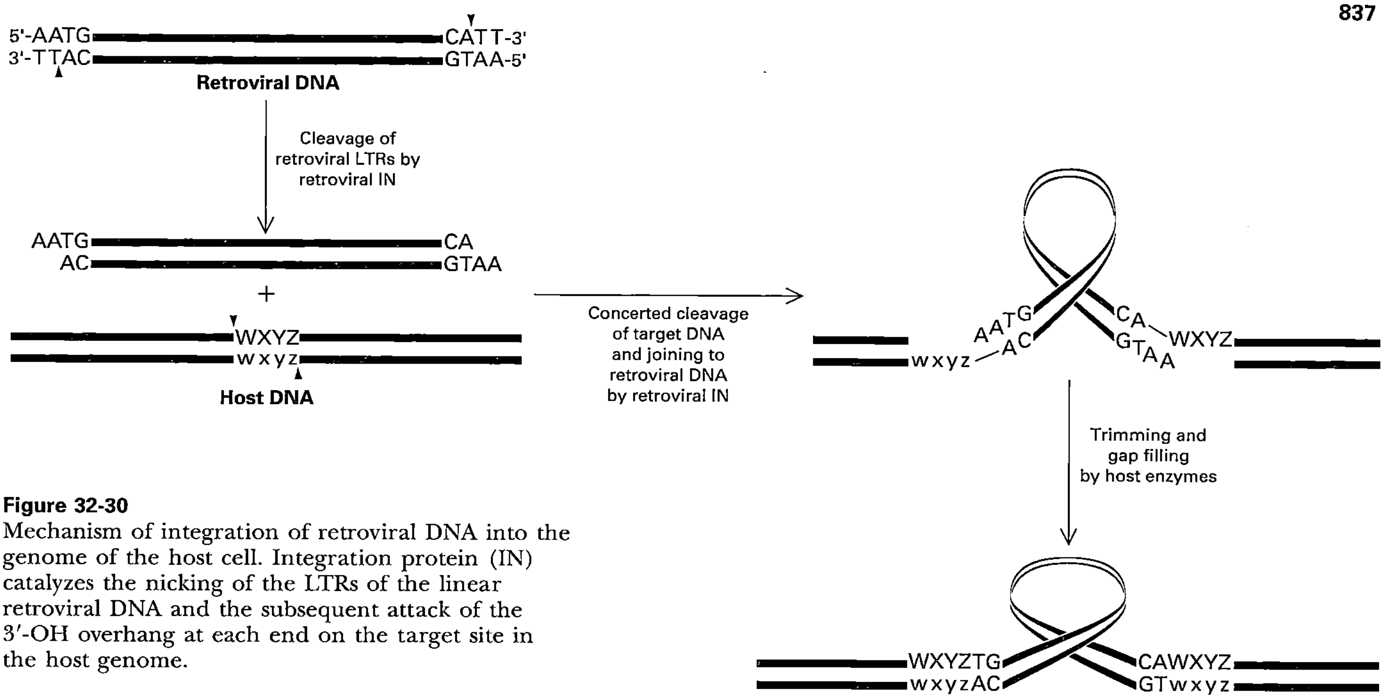

Figure 32-30
Mechanism of integration of retroviral DNA into the genome of the host cell. Integration protein (IN) catalyzes the nicking of the LTRs of the linear retroviral DNA and the subsequent attack of the 3'-OH overhang at each end on the target site in the host genome.

Transcription of retroviral DNA occurs only after it has been integrated into the host-cell DNA. Thus, *integration is an obligatory step in the life cycle of retroviruses.* Recombination of retroviral DNA with host DNA is mediated by *integration protein (IN),* which the virus brings into the cell along with reverse transcriptase. IN, an endonuclease, recognizes specific sequences at the ends of the LTRs. Two nucleotides are removed from the 3′ end of each DNA strand (Figure 32-30). The integrase-viral DNA complex then binds to the host-cell DNA. *Integration can occur nearly anywhere—no specific sequence is selected.* The insertion step is simple: each 3′-hydroxyl group of viral DNA directly attacks a phosphodiester bond in the target DNA. No energy source is needed because a phosphodiester bond is formed while another is cleaved. The target DNA sites on opposite strands are staggered by four to six nucleotides with 5′ overhangs. These gaps are filled, and the trimmed 5′ ends of viral DNA are then joined to target DNA strands. Hence, four to six nucleotides at the host integration site are duplicated, as in the movement of bacterial transposons.

The genetic strategy employed by retroviruses recurs in a variety of eukaryotic genetic elements. *Ty elements* are fascinating transposons found in yeast (*T* stands for transposon and *y* for yeast). Ty1, a 6-kb DNA integrated in the yeast genome, is bounded by LTRs called δ and flanked by short duplicated sequences. It encodes proteins akin to the reverse transcriptases and integrases of retroviruses. Furthermore, duplication and movement to another site in the yeast genome occurs through an RNA intermediate that is produced by transcription (Figure 32-31). Ty RNA then directs the synthesis of a linear DNA duplex, which is inserted into the genome. Thus, the flow of information of Ty—*from DNA to RNA back to DNA*—is like that of retroviruses. Genetic elements employing this strategy are called *retrotransposons* or *retroposons.* Another example is the *copia* elements of *Drosophila,* which generate copious amounts of RNA (hence their name). *Retroviruses can be regarded as retroposons that have evolved the capacity to move from one cell to another.*

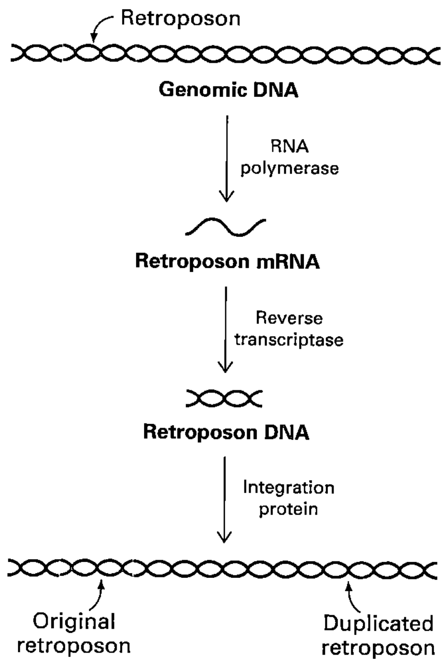

Figure 32-31
Retroposons (retrotransposons) replicate and move to another site on DNA through an RNA intermediate.

SUMMARY

New arrangements of genes are formed by the processes of recombination and transposition. In homologous (general) recombination, any pair of homologous sequences on parental DNA molecules can be exchanged. New DNA molecules are formed by the breakage and reunion of DNA strands at points of homology, as shown by the formation of chi intermediates. In *E. coli,* homologous recombination is mediated by products of the *rec* genes. The recBCD protein complex, a helicase and sequence-specific endonuclease, generates single-stranded DNA (ssDNA). RecA protein then forms a filament around ssDNA. The ssDNA–recA complex binds duplex DNA, partly unwinds it, and searches for a sequence complementary to that of the ssDNA. RecA protein then catalyzes a switch in base pairing; one of the strands of the duplex becomes paired with the invading ssDNA. The exchange of strands continues, a process called branch migration. These actions of recA protein, like the helicase activity of the recBCD complex, are powered by the hydrolysis of ATP.

Homologous recombination plays a key role also in repairing DNA. Recombination provides a new template for the synthesis of DNA when both strands are damaged. Damage to DNA triggers an SOS response leading to increased synthesis of many proteins that mediate repair. The first step in this adaptive response is the autoproteolysis of lexA repressor induced by the ssDNA–recA complex, a signal that damage has occurred.

Genetic rearrangements of nonhomologous loci are mediated by mobile genetic elements called transposons (transposable elements). Plasmids, which are small circular DNA duplexes, are an important class of mobile genetic elements. These accessory chromosomes carry genes for the inactivation of antibiotics, metabolism of natural products, and production of toxins. Plasmids can replicate autonomously or become integrated into the host chromosome. The F factor (fertility factor) plasmid enables bacteria to donate genes to other bacteria by conjugation, a process mediated by direct physical contact between the donor (F^+) and recipient (F^-) cells. R factor plasmids confer resistance to antibiotics. The larger of these plasmids contain a resistance transfer factor (RTF) that enables the plasmid to be transmitted to other bacteria by conjugation. Several resistance-conferring genes can become linked to an RTF region. Hence, multiple-drug resistance can be infectious.

The simplest transposons are insertion sequences. They consist of a transposase and one other gene (or none) bounded by inverted terminal repeats. Transposons can move to nearly any site on a chromosome. Transposition is accompanied by duplication of a short sequence of the recipient site. Genes located between a pair of transposons that recombine can be inverted, deleted, or duplicated. Complex transposons contain genes mediating functions in addition to transposition. R factor plasmids probably evolved by the coming together of several transposons.

The integration of lambda (λ) phage in the *E. coli* chromosome exemplifies site-specific recombination, which is guided primarily by proteins rather than by extensive sequence homology. Recombination between these two circular DNA molecules occurs at the *attP* site on phage DNA and the *attB* site on *E. coli* DNA. This reciprocal recombination is mediated by integrase (int), a phage-encoded nuclease that makes staggered cuts in both duplexes, and integration host factor (IHF), a bacterial protein. Phage DNA can remain in quiescent form in the bacterial DNA for many generations (lysogeny). Expression of phage excisionase (xis) activates recombination in the reverse direction, leading to excision of the phage DNA and activation of the lytic pathway.

Site-specific recombination is also important in eukaroytes, as illustrated by the joining of *V*, *D*, and *J* gene segments in the synthesis of immunoglobulins and T-cell receptors. Each recognition sequence consists of a conserved palindromic heptamer that is separated from a conserved AT-rich nonamer by a 12-bp or 23-bp spacer. Recombination can occur between 12-bp and 23-bp types but not between types of the same kind. The combinatorial association of gene segments, aided by imprecision in joining, generates vast recognition potentiality. Site-specific recombination is also the basis of class switching in the synthesis of antibodies having the same specificity (the same V regions) but different effector functions (different C regions).

Retroviral genomes, such as those of HIV-1 and avian sarcoma virus, cycle between a single-stranded RNA form in the virion and a duplex DNA form in the host cell. Reverse transcriptase brought into the cell by the virus synthesizes duplex DNA using viral RNA as the template. In this complex synthesis, the DNA strand is synthesized by the polymerase moiety of the p66 subunit of the p55-p66 heterodimer, and the RNA strand is concomitantly digested by the ribonuclease H moiety of p66. Two jumps by reverse transcriptase lead to the synthesis of a long terminal repeat (LTR) at each end of the retroviral DNA. LTRs are rich in signals for integration and transcription. Integration protein (IN), which also is carried by the virus into the cell, catalyzes the insertion of retroviral DNA into the host genome, which can occur at virtually any site. Integration is an obligatory step in the life cycle of all retroviruses. The genetic strategy employed by retroviruses is also used by genetic elements called retroposons (retrotransposons) that do not leave their host cell. An RNA intermediate is formed during transposition by retroposons such as the Ty elements of yeast and the copia element of *Drosophila.*

SELECTED READINGS

Where to start

Stahl, F.W., 1987. Genetic recombination. *Sci. Amer.* 256(2):90–101.

West, S.C., 1992. Enzymes and molecular mechanisms of genetic recombination. *Ann. Rev. Biochem.* 61:603–640.

Radding, C.M., 1991. Helical interactions in homologous pairing and strand exchange driven by RecA protein. *J. Biol. Chem.* 266:5355–5358.

Lederberg, J., 1987. Genetic recombination in bacteria: A discovery account. *Ann. Rev. Genet.* 21:23–46.

Books

Lewin, B., 1994. *Genes V.* Oxford University Press. [Chapters 33–36 on genetic recombination and transposition are rich in information and insight.]

Watson, J.D., Gilman, M., Witkowski, J., and Zoller, M., 1992. *Recombinant DNA* (2nd ed.). Scientific American Books. [Chapter 10, entitled Movable Genes, is lucid and well illustrated.]

Singer, M., and Berg, P., 1991. *Genes and Genomes.* [Chapter 10, entitled Genomic Rearrangements, contains an excellent account of retrotransposons, retrogenes, and programmed gene rearrangements.]

Griffiths, A.J.F., Miller, J.H., Suzuki, D.T., Lewontin, R.C., and Gelbart, W.M., 1993. *An Introduction to Genetic Analysis* (5th ed.). W.H. Freeman. [Chapters 19 and 20 give a concise account of recombination and transposition from a genetic viewpoint.]

Berg, D.E., and Howe, M.M. (eds.), 1989. *Mobile DNA.* American Society for Microbiology. [A rich source of information on transposons and other mobile genetic elements.]

Homologous recombination

Kowalczykowski, S.C., 1991. Biochemistry of genetic recombination: Energetics and mechanism of DNA strand exchange. *Ann. Rev. Biophys. Biophys. Chem.* 20:539–575.

Rao, B.J., Chiu, S.K., and Radding, C.M., 1993. Homologous recognition and triplex formation promoted by RecA protein between duplex oligonucleotides and single-stranded DNA. *J. Mol. Biol.* 229:328–343.

Honigberg, S.M., Rao, B.J., and Radding, C.M., 1986. Ability of recA protein to promote a search for rare sequences in duplex DNA. *Proc. Nat. Acad. Sci.* 83:9586–9590.

Dressler, D., and Potter, H., 1982. Molecular mechanisms in genetic recombination. *Ann. Rev. Biochem.* 51:727–761. [The Holliday model is lucidly presented in this review.]

Cox, M.M., and Lehman, I.R., 1987. Enzymes of general recombination. *Ann. Rev. Biochem.* 56:229–262.

Capecchi, M.R., 1989. Altering the genome by homologous recombination. *Science* 244:1288–1292.

Potter, H., and Dressler, D., 1978. In vitro system from *Esche-*

richia coli that catalyzes generalized genetic recombination. *Proc. Nat. Acad. Sci.* 75:3698–3702.

RecA protein

Roca, A.I., and Cox, M.M., 1990. The RecA protein: Structure and function. *Crit. Rev. Biochem. Molec. Biol.* 25:415–456.

Story, R.M., Weber, I.T., and Steitz, T.A., 1992. The structure of the *E. coli* recA protein monomer and polymer. *Nature* 355:318–325.

Egelman, E.H., and Stasiak, A., 1986. Structure of helical recA-DNA complexes. Complexes formed in the presence of ATPγS or ATP. *J. Mol. Biol.* 191:677–697.

DNA repair

Howard-Flanders, P., 1981. Inducible repair of DNA. *Sci. Amer.* 245(5):72–80.

Walker, G.C., 1985. Inducible DNA repair systems. *Ann. Rev. Biochem.* 54:425–457.

Kim, B., and Little, J., 1993. LexA and λ CI repressors as enzymes: Specific cleavage in an intermolecular reaction. *Cell* 73:1165–1173. [Demonstration that recA stimulates the proteolysis of LexA in the triggering of the SOS response in DNA repair.]

Friedberg, E.C., 1985. *DNA Repair.* W.H. Freeman. [SOS repair is discussed in Chapter 7.]

Transposition

Kleckner, N., 1990. Regulation of transposition in bacteria. *Ann. Rev. Cell Biol.* 6:297–327.

Mizuuchi, K., 1992. Transpositional recombination: Mechanistic insights from studies of mu and other elements. *Ann. Rev. Biochem.* 61:1011–1051.

Cohen, S.N., and Shapiro, J.A., 1980. Transposable genetic elements. *Sci. Amer.* 242(2):40–49.

Infectious drug resistance

Clewell, D.B., and Gawron-Burke, C., 1986. Conjugative transposons and the dissemination of antibiotic resistance in streptococci. *Ann. Rev. Microbiol.* 40:635–659.

Clowes, R.C., 1973. The molecule of infectious drug resistance. *Sci. Amer.* 228(4):18–27.

Site-specific recombination

Landy, A., 1989. Dynamic, structural, and regulatory aspects of lambda site-specific recombination. *Ann. Rev. Biochem.* 58:913–949.

Wasserman, S.A., and Cozzarelli, N.R., 1986. Biochemical topology: Applications to DNA recombination and replication. *Science* 232:951–960. [A fascinating account of topological factors in site-specific recombination.]

Lieber, M.R., 1991. Site-specific recombination in the immune system. *FASEB J.* 5:2934–2944.

Gellert, M., 1992. Molecular analysis of V(D)J recombination. *Ann. Rev. Genet.* 26:425–446.

Retroviral replication and integration

Whitcomb, J.M., and Hughes, S.H., 1992. Retroviral reverse transcription and integration: Progress and problems. *Ann. Rev. Cell Biol.* 8:275–306.

Kohlstaedt, L.A., Wang, J., Friedman, J.M., Rice, P.A., and Steitz, T.A., 1992. Crystal structure at 3.5 Å resolution of HIV-1 reverse transcriptase complexed with an inhibitor. *Science* 256:1783–1790.

Goff, S.P., 1992. Genetics of retroviral integration. *Ann. Rev. Genet.* 26:527–544.

PROBLEMS

1. *Unwinding DNA.* A variety of ATP-driven proteins unwind duplex DNA in *E. coli.* Name three of them and cite their functions.

2. *Linking, writhing, and twisting.* One of the strands of a 1040-bp circular duplex DNA is nicked. RecA protein is then added in an amount such that an average of 100 protein molecules are bound per DNA duplex. DNA ligase and NAD^+ are added to close the nicked circle. RecA protein is then removed.
   (a) Compare the linking number of the resulting DNA molecule with that of the corresponding relaxed duplex.
   (b) Compare the hydrodynamic properties of this DNA molecule with that of its relaxed counterpart.

3. *Autorepression.* The lexA protein represses its own synthesis. How does this control feature enable the SOS response to be reversed following repair of DNA?

4. *A revealing analog.* ATPγS is a hydrolysis-resistant analog of ATP in which a nonesterified oxygen atom of the terminal phosphoryl group is replaced by a sulfur atom. Predict the consequences of using ATPγS instead of ATP in the reactions catalyzed by (a) recA protein and (b) the recBCD complex.

5. *Prelude to recombination.* Why is recombination favored by negative supercoiling?

6. *Gene knockout.* Devise an experimental strategy for permanently inactivating a specific mammalian gene.

7. *Heterodimer from homodimer.* The polymerase domain of the p55 subunit of reverse transcriptase has a rather different conformation from that of the p66 subunit. Furthermore, only one chain of the nascent p66-p66 homodimer is cleaved to generate the functional p55-p66 heterodimer. Initially identical chains usually do not have such different fates. Why does HIV-1 craft the reverse transcriptase heterodimer from two initially identical chains rather than two different ones?

CHAPTER

33

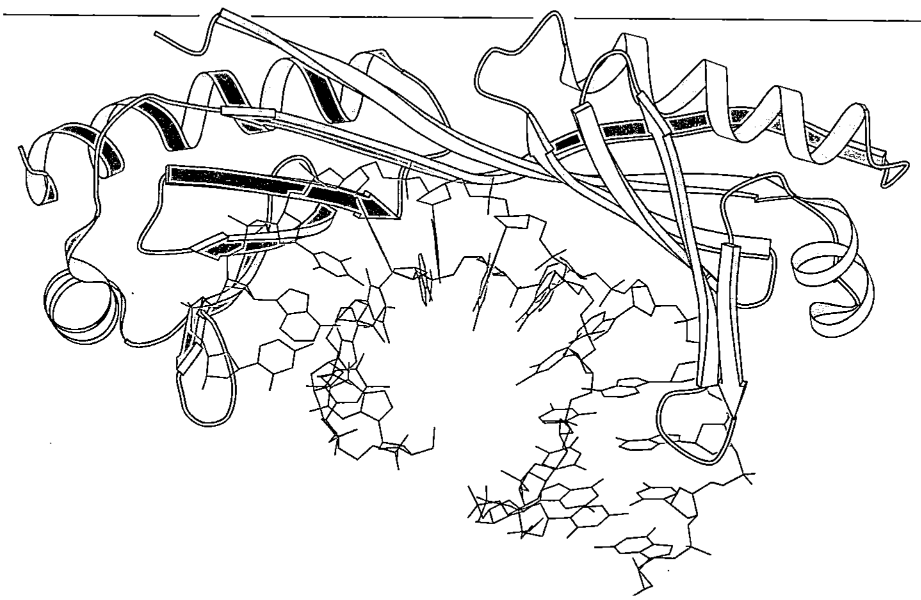

RNA Synthesis and Splicing

The flow of genetic information from DNA to RNA to protein was introduced in Chapter 5. An overview was given of the different kinds of RNA, the role of messenger RNA, the action of RNA polymerase, and the split nature of many eukaryotic genes. This chapter deals with how RNA is synthesized and spliced. We begin with transcription in *Escherichia coli,* where it is best understood, and focus on three questions: what are the properties of promoters, and how do they cause specific initiation? How do RNA polymerase, the DNA template, and the nascent RNA chain interact with one another? How is transcription terminated?

We shall then turn to transcription in eukaryotes, beginning with promoter sequences and the transcription-factor proteins that bind them. A distinctive feature of eukaryotic DNA templates is the presence of enhancer sequences that can stimulate initiation more than a thousand base pairs away. Primary transcripts in eukaryotes are extensively modified, as exemplified by the capping of the 5′ end of an mRNA precursor and the addition of a long poly(A) tail to its 3′ end. Most striking is the splicing of mRNA precursors, which is catalyzed by spliceosomes consisting of ribonucleoproteins (snRNPs). The small nuclear RNA molecules (snRNAs) in these complexes play a key role in directing the alignment of splice sites and in mediating catalysis. Indeed, some RNA molecules can splice themselves in the absence of protein. This landmark discovery by Thomas Cech revealed that RNA molecules can serve as catalysts and profoundly influenced our view of molecular evolution.

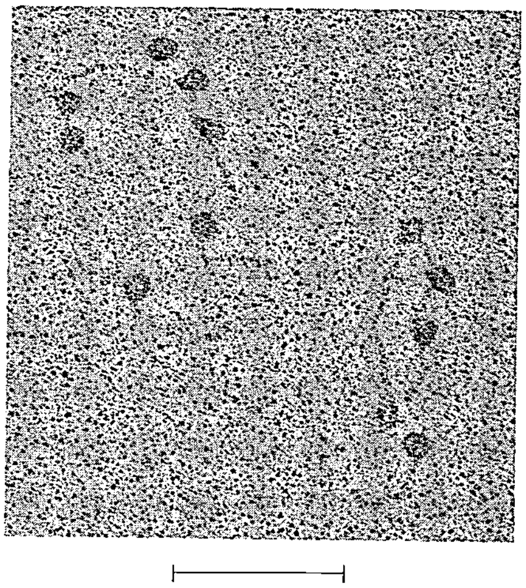

Figure 33-1
Electron micrograph of RNA polymerase holoenzyme molecules bound to several promoter sites on a fragment of T7 bacteriophage DNA. [From R.C. Williams. *Proc. Nat. Acad. Sci.* 74(1977):2313.]

Opening Image: *Structure of the heart of the initiation complex in the synthesis of messenger RNA. The saddle-shaped TATA box–binding protein (blue and yellow) is bound to a TATA sequence in DNA (red and green). The start site for RNA synthesis is defined by this complex. [Drawn from coordinates kindly provided by Dr. Youngchang Kim and Dr. Paul Sigler. Y. Kim, J.H. Geiger, S. Hahn, and P.B. Sigler. Nature 365(1993):512.]*

RNA POLYMERASE FROM *E. COLI* IS A MULTISUBUNIT ENZYME

RNA polymerase from *E. coli* is a very large (~500 kd) and complex enzyme consisting of four kinds of subunits (Table 33-1). The subunit composition of the entire enzyme, called the *holoenzyme,* is $\alpha_2\beta\beta'\sigma$. As will be discussed shortly, the σ subunit finds a promoter site where transcription begins, helps initiate RNA synthesis, and then dissociates from the rest of the enzyme. RNA polymerase without this subunit is called the *core enzyme* ($\alpha_2\beta\beta'$); it contains the catalytic site. The α subunits bind regulatory proteins, the β' subunit binds the DNA template, and the β subunit binds ribonucleoside triphosphate substrates.

Table 33-1
Subunits of RNA polymerase from *E. coli*

Subunit	*Gene*	*Number*	*Mass (kd)*	*Role*
α	*rpoA*	2	37	Binds regulatory sequences
β	*rpoB*	1	151	Forms phosphodiester bonds
β'	*rpoC*	1	155	Binds DNA template
σ^{70}	*rpoD*	1	70	Recognizes promoter and initiates synthesis

The synthesis of RNA by *E. coli* RNA polymerase, like nearly all biological polymerization reactions, takes place in three stages: *initiation, elongation,* and *termination.* RNA polymerase performs multiple functions in this process:

1. It searches DNA for initiation sites. *E. coli* DNA has about 2000 promoter sites in its 4×10^6 bp.

2. It unwinds a short stretch of double-helical DNA to produce a single-stranded DNA template from which it takes instructions.

3. It selects the correct ribonucleoside triphosphate and catalyzes the formation of a phosphodiester bond. This process is repeated many times as the enzyme moves unidirectionally along the DNA template. RNA polymerase is totally processive—a transcript is synthesized from start to end by a single RNA polymerase molecule.

4. It detects termination signals that specify where a transcript ends.

5. It interacts with activator and repressor proteins that modulate the rate of transcription over a wide dynamic range. Gene expression is controlled mainly at the level of transcription, as will be discussed in detail in Chapters 36 and 37.

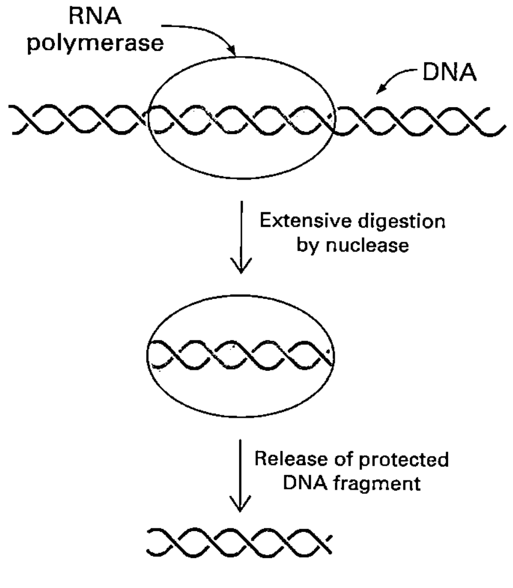

Figure 33-2
Binding sites on DNA for RNA polymerase and other proteins can be isolated by digesting DNA with nucleases. Fragments that are bound to protein are protected from digestion.

TRANSCRIPTION IS INITIATED AT PROMOTER SITES ON THE DNA TEMPLATE

Transcription starts at *promoters* on the DNA template. These initiation sites have been identified in several ways. One approach is to add RNA polymerase, and then a deoxyribonuclease, to DNA. In the absence of ribonucleoside triphosphates, RNA polymerase remains bound to promoters and protects them from digestion. Fragments of DNA about 60 bp long are produced after extensive digestion of the unprotected DNA (Figure 33-2). The position and sequence of the protected region can readily be identified by a complementary technique called *footprinting*. An end of one of the strands of a specific DNA fragment (produced by the action of a restriction enzyme, for example) is first labeled with ^{32}P. RNA polymerase is added to the labeled DNA, and the complex is digested with DNase I just long enough to make an average of one cut in each chain. An aliquot without RNA polymerase is treated in the same way to serve as a control. The resulting DNA fragments are separated according to size by electrophoresis. The gel pattern (Figure 33-3) is highly revealing: a series of bands present in the control sample is absent from the aliquot containing RNA polymerase. These bands are missing because RNA polymerase shields DNA from cleavages that would give rise to the corresponding fragments.

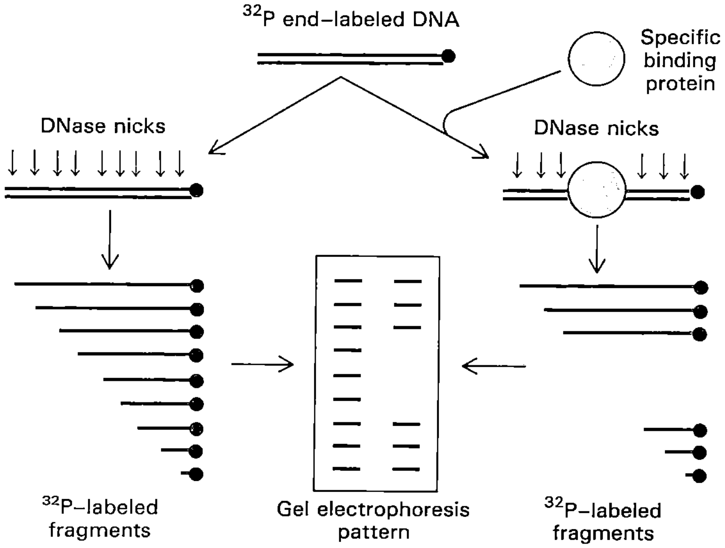

Figure 33-3
Footprinting technique. One end of a DNA chain is labeled with ^{32}P (shown by a red circle). This labeled DNA is then cut at a limited number of sites by DNase I. The same experiment is carried out in the presence of a protein that binds to specific sites on DNA. The bound protein protects a segment of DNA from the action of DNase I. Hence, certain fragments will be absent. The missing bands in the gel pattern identify the binding site on DNA.

A striking pattern is evident when the sequences of many prokaryotic promoters are compared. *Two common motifs are present on the 5′ (upstream) side of the start site.* They are known as the *−10 sequence* and the *−35 sequence* because they are centered at about 10 and 35 nucleotides upstream of the start site. These sequences are each 6 bp long. Their consensus sequences, deduced from analyses of many promoters, are

−35 −10 +1
5′~~TTGACA~~~~~~~~~~TATAAT~~~~ Start site

The first nucleotide (the start site) of a transcribed DNA sequence is denoted as +1, the second one as +2; the nucleotide preceding the start site is denoted as −1. These designations refer to the coding strand of DNA. Recall that the sequence of the *template strand of DNA* is the *complement* of that of the RNA transcript (see Figure 5-10 on p. 101). By contrast, the *coding strand of DNA* has the *same* sequence as that of the RNA transcript except for T in place of U. The coding strand is also known as the *sense (+) strand,* and the template strand as the *antisense (−) strand.*

Promoters differ markedly in their efficacy. Strong promoters cause frequent initiation of transcription, as often as every two seconds in *E. coli*. In contrast, genes with very weak promoters are transcribed about once in 10 minutes. The −10 and −35 regions of most strong promoters

Transcription starts here ↓

	5′ −10 3′
(A)	C G T A T G T T G T G T G G A
(B)	G C T A T G G T T A T T T C A
(C)	G T T A A C T A G T A C G C A
(D)	G T G A T A C T G A G C A C A
(E)	G T T T T C A T G C C T C C A
	T A T A A T

Figure 33-4
Prokaryotic promoter sequences showing homology in the −10 region. The sequences are from the (A) *lac*, (B) *gal*, and (C) *trp* operons of *E. coli*; (D) from λ phage, and (E) from φX174 phage. Homologies are shown in green and the −10 consensus sequence (deduced from a large number of promoter sequences) in red. The +1 nucleotides are shown in blue.

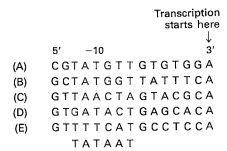

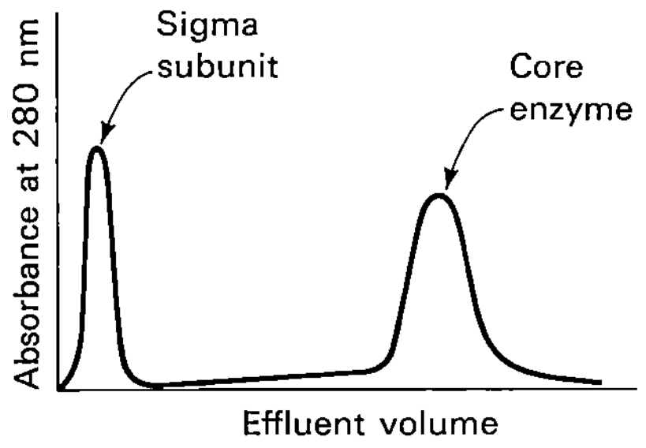

Figure 33-5
Resolution of RNA polymerase into the σ subunit and the core enzyme ($\alpha_2\beta\beta'$) on a phosphocellulose column. The separation of σ from the core made it possible to ascertain their roles. The core enzyme cannot recognize promoters. Specific initiation is restored by the addition of σ.

have sequences that correspond closely to the consensus ones, whereas weak promoters tend to have multiple substitutions at these sites. Indeed, mutation of a single base in either the −10 sequence or the −35 sequence can lead to a loss of promoter activity. The distance between these conserved sequences is also important; a separation of 17 nucleotides is optimal. The frequency of transcription of many genes is also markedly influenced by regulatory proteins that bind to specific sequences near promoter sites and interact with RNA polymerase (Chapters 36 and 37).

SIGMA SUBUNITS ENABLE RNA POLYMERASE TO RECOGNIZE PROMOTER SITES

The $\alpha_2\beta\beta'$ core of RNA polymerase is unable to start transcription at promoter sites. Rather, *the complete* $\alpha_2\beta\beta'\sigma$ holoenzyme is essential for specific initiation. The σ subunit contributes to specific initiation by *decreasing the affinity of RNA polymerase for general regions of DNA by a factor of* 10^4. Sigma also enables RNA polymerase to recognize promoters. The holoenzyme binds to duplex DNA and searches for a promoter site by forming transient hydrogen bonds with exposed hydrogen donor and acceptor groups of base pairs. The search is rapid for two reasons. First, RNA polymerase slides along DNA instead of repeatedly binding and dissociating from it. *The promoter site is encountered by a random walk in one dimension rather than in three dimensions.* The observed rate constant for the binding of RNA polymerase holoenzyme to promoter sequences is $10^{10}\ \text{M}^{-1}\ \text{s}^{-1}$, more than 100 times as fast as could be accomplished by repeated encounters on and off the DNA. Second, *RNA polymerase can detect the −10 and −35 sequences of DNA without unwinding the double helix.* Recall that the EcoRV endonuclease slides along DNA without distorting the double helix until the enzyme encounters its palindromic target (p. 792). A search requiring extensive DNA unwinding would be much slower than one exploring an essentially intact duplex.

PROMOTERS FOR HEAT-SHOCK GENES ARE RECOGNIZED BY A SPECIAL KIND OF SIGMA SUBUNIT

E. coli contains multiple σ factors. The type that recognizes the consensus sequences described earlier is called σ^{70} because it has a mass of 70 kd. A different σ comes into play when the temperature is raised abruptly. *E. coli* responds by synthesizing σ^{32}, which recognizes the promoters of *heat-shock genes.* These promoters exhibit −10 sequences that differ markedly from the one for standard promoters (Figure 33-6). The −35 region of heat-shock promoters resembles that of general promoters, but it also has some distinctive features. The increased transcription of heat-shock genes leads to the coordinated synthesis of a series of protective proteins. Nitrogen starvation leads to the synthesis of σ^{54}, which stimulates the expression of genes that assimilate nitrogen. A change of sigma subunit is employed by *Bacillus subtilis* as a developmental device. Different σ factors control sporulation in this bacterium by determining which battery of genes is expressed. Likewise, SPO1, a phage that infects *B. subtilis*, en-

Sequence (−35 … −10)	
5′~~T T G A C A~~~~~~T A T A A T~~~~3′	Standard promoter
5′~~T N N C N C N C T T G A A~~~~~~C C C A T N T~~~~3′	Heat-shock promoter
5′~~C T G G G N A~~~~~~T T G C A~~~~3′	Nitrogen-starvation promoter

Figure 33-6
Comparison of the consensus sequences of nitrogen-starvation promoters, heat-shock promoters, and standard promoters of *E. coli*. They are recognized by σ^{54}, σ^{32}, and σ^{70}, respectively.

codes several new σ factors that lead to sequential expression of viral genes. These findings demonstrate that *σ plays a key role in determining where RNA polymerase initiates transcription.*

RNA POLYMERASE UNWINDS NEARLY TWO TURNS OF TEMPLATE DNA BEFORE INITIATING RNA SYNTHESIS

In searching for promoter sites, RNA polymerase is bound to double-helical DNA. However, duplex DNA cannot serve directly as the template for RNA synthesis because its bases are paired. A region of duplex DNA must be unpaired so that nucleotides on one of its strands become accessible for base pairing with incoming ribonucleoside triphosphates. The DNA template strand selects the correct ribonucleoside triphosphate by forming a Watson-Crick base pair with it (p. 100), as in DNA synthesis. What is the extent of unwinding of DNA by RNA polymerase? This question was answered by analyzing the effect of the enzyme on the supercoiling of a circular duplex DNA. Different amounts of RNA polymerase were added to the DNA. Topoisomerase I, an enzyme catalyzing the concerted nicking and resealing of duplex DNA (p. 797), was then added to relax the portion of circular DNA not in contact with polymerase molecules. These DNA samples were analyzed by gel electrophoresis (see Figure 31-18 on p. 797) following removal of bound protein. *The degree of negative supercoiling increased in proportion to the number of RNA polymerases bound per template DNA* (Figure 33-7), *showing that the enzyme unwinds DNA. Each bound polymerase unwinds a 17-bp segment of DNA, which corresponds to 1.6 turns of B-DNA helix.*

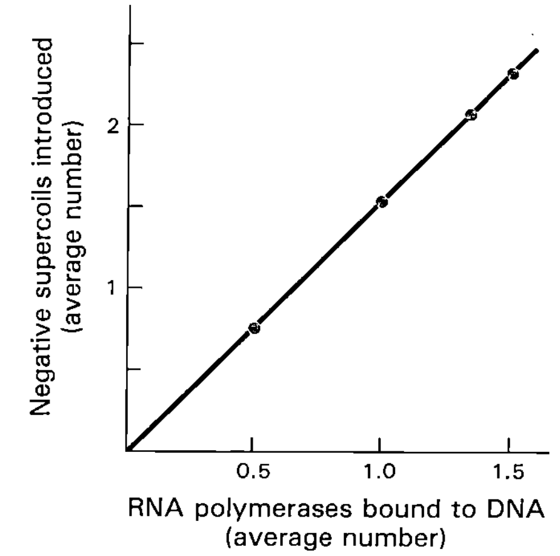

Figure 33-7
The unwinding of template DNA by RNA polymerase was analyzed by measuring the change in the number of supercoils produced by topoisomerase I in the presence of bound polymerase. [After H.B. Gamper and J.E. Hearst. *Cell* 29(1982):83.]

Negative supercoiling of circular DNA favors the transcription of many genes because it facilitates unwinding (p. 796). Thus, the pumping of negative supercoils into DNA by DNA gyrase can increase the efficiency of promoters located at distant sites. However, not all promoters are stimulated by negative supercoiling. The promoter for DNA gyrase is a noteworthy exception. The rate of transcription of this gene is decreased by negative supercoiling, a nice feedback control assuring that DNA does not become excessively supercoiled. Negative supercoiling could decrease the efficiency of this promoter by changing the steric relation of the −10 and −35 regions.

The transition from the *closed promoter complex* (in which DNA is double-helical) to the *open promoter complex* (in which a DNA segment is unwound) is a key event in transcription. The stage is now set for the formation of the first phosphodiester bond of the new RNA chain.

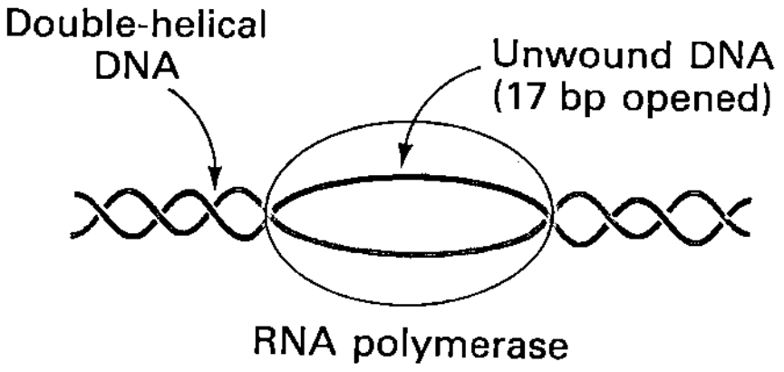

Figure 33-8
RNA polymerase unwinds about 17 bp of template DNA.

RNA CHAINS START WITH pppG OR pppA AND ARE SYNTHESIZED IN THE 5′ → 3′ DIRECTION

Most newly synthesized RNA chains carry a tag that reveals how RNA chains are started. The 5′ end of a new RNA chain is highly distinctive: the molecule starts with either *pppG* or *pppA*. In contrast with DNA synthe-

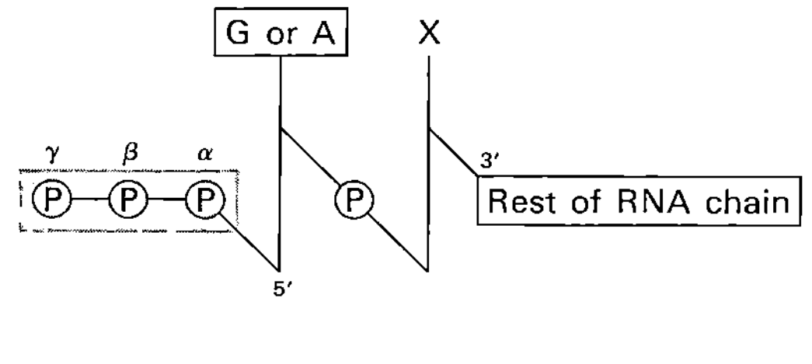

sis, a primer is not needed. *RNA chains can be formed de novo.* Labeling experiments with γ-^{32}P substrates revealed that RNA chains, like DNA chains, grow in the *5′ → 3′* direction.

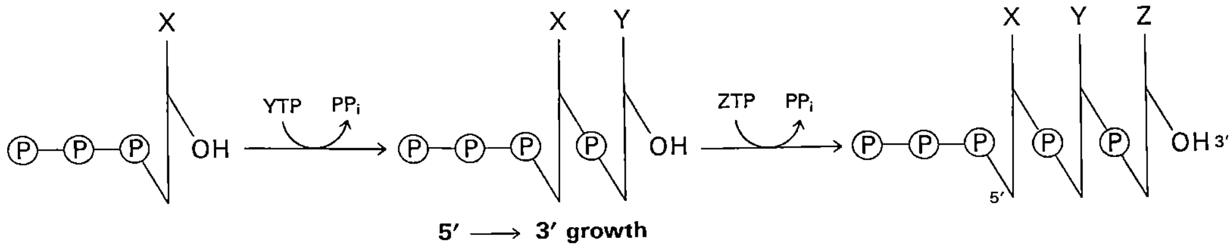

Figure 33-9
RNA chains are synthesized in the 5′ → 3′ direction.

ELONGATION TAKES PLACE AT TRANSCRIPTION BUBBLES THAT MOVE ALONG THE DNA TEMPLATE

The elongation phase of RNA synthesis begins after formation of the first phosphodiester bond. An important change is the loss of σ—the core enzyme without σ binds more strongly to the DNA template. Indeed, RNA polymerase stays bound to its template until a termination signal is reached. The region containing RNA polymerase, DNA, and nascent RNA is called a *transcription bubble* because it contains a locally melted "bubble" of DNA (Figure 33-10). The newly synthesized RNA forms a

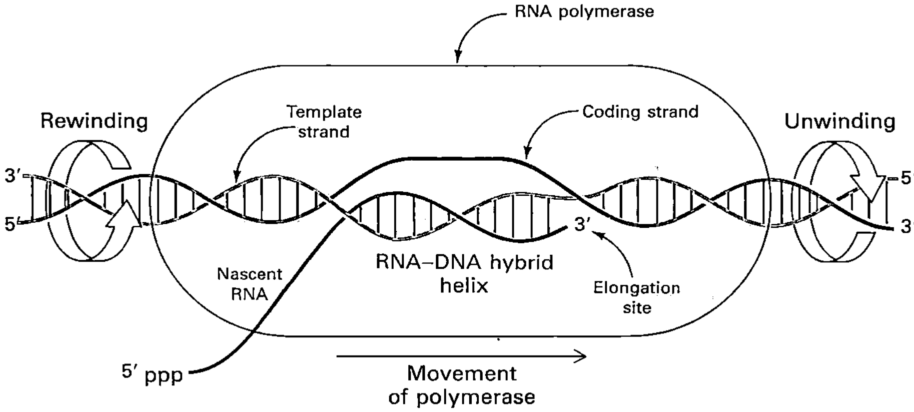

Figure 33-10
Model of a transcription bubble during elongation of the RNA transcript. Duplex DNA is unwound at the forward end of RNA polymerase and rewound at its rear end. The RNA–DNA hybrid helix rotates in synchrony.

hybrid helix with the template DNA strand. This RNA–DNA helix is thought to be about 12 bp long, which corresponds to nearly one turn of an A-DNA-type helix. The 3′-hydroxyl of the RNA in this hybrid helix is positioned so that it can attack the α phosphorus atom of an incoming ribonucleoside triphosphate. The core enzyme also contains a binding site for the other DNA strand. About 17 bp of DNA are unwound throughout the elongation phase, as in the initiation phase. The rate of elongation is about 50 nucleotides per second; the transcription bubble moves a distance of 170 Å (17 nm) in this time.

The length of the RNA–DNA hybrid and of the unwound region of DNA stays rather constant as RNA polymerase moves along the DNA template. This finding indicates that DNA is rewound at about the same rate at the rear of RNA polymerase as it is unwound at the front of the enzyme. The RNA–DNA hybrid must also rotate each time a nucleotide is added so that the 3′-OH end of the RNA stays at the catalytic site. The 12-bp length of the hybrid is just short of a full turn of helix to avoid a knotty problem. The RNA strand is bent sharply away from the template DNA strand just before a full turn is formed to prevent the 5′ end of the RNA from becoming intertwined with the DNA.

The nucleic acid framework of the elongation complex can be mimicked by a *synthetic RNA–DNA bubble* (Figure 33-11). This construct consists of (1) a DNA duplex containing an internal mismatched sequence that creates a bubble and (2) an RNA oligomer that is paired with one of the DNA strands in the bubble. The addition of core polymerase and ribonucleoside triphosphates to this construct leads to the processive elongation of the RNA primer. The efficient template-directed synthesis of RNA by this complex in the absence of sigma factor and a promoter shows that core polymerase is active once a bubble is formed.

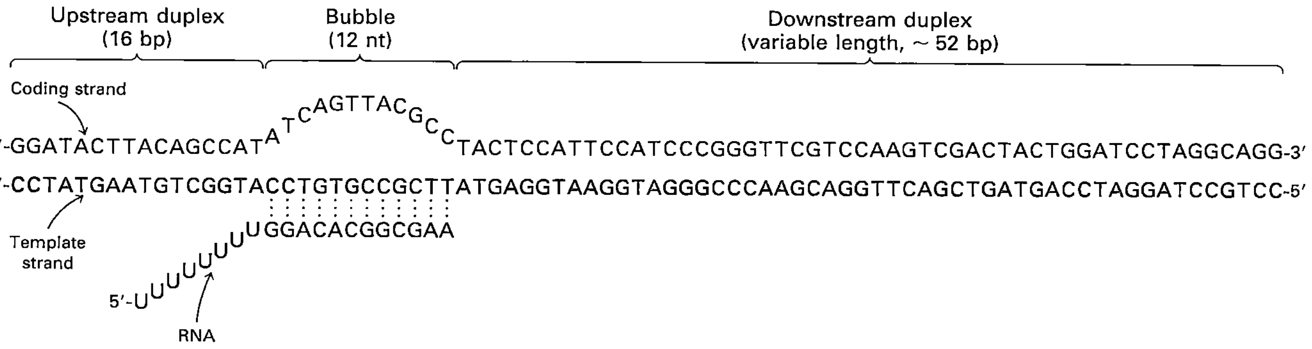

Figure 33-11
Synthetic RNA–DNA bubble. The template DNA strand is shown in green, the "coding" DNA strand in blue, and the nascent RNA chain in red. The bubble is generated by a 12-nucleotide (12-nt) mismatched region in the DNA duplex. The RNA is complementary to the template DNA in this region. [After S.S. Daube and P.H. von Hippel. *Science* 258(1992):1320.]

It is noteworthy that RNA polymerase lacks nuclease activity. *In contrast with DNA polymerase, it does not edit the nascent polynucleotide chain. Consequently, the fidelity of transcription is much lower than that of replication.* The error rate of RNA synthesis is of the order of 1 mistake per 10^4 or 10^5, about 10^5 times as high as that of DNA synthesis. The much lower fidelity of RNA synthesis can be tolerated because mistakes are not transmitted to progeny. For most genes, many RNA transcripts are synthesized each generation; a few defective transcripts are unlikely to be harmful.

AN RNA HAIRPIN FOLLOWED BY SEVERAL U RESIDUES LEADS TO THE TERMINATION OF TRANSCRIPTION

In the termination phase of transcription, the formation of phosphodiester bonds ceases, the RNA–DNA hybrid dissociates, the melted region of DNA rewinds, and RNA polymerase releases the DNA. The termination of transcription is as precisely controlled as its initiation. The transcribed regions of DNA templates contain stop signals. The simplest one is a *palindromic GC-rich region followed by an AT-rich region* (Figure 33-12). *The RNA transcript of this DNA palindrome is self-complementary.* Hence, its bases can pair to form a hairpin structure with a stem and loop, a structure favored by its high content of G · C residues (Figure 33-13). Recall that G · C base pairs are more stable than A · T pairs (p. 86). This stable hairpin is followed by a sequence of four or more U residues. The RNA transcript ends within or just after them.

How does this hairpin-oligo(U) structure terminate transcription? First, it seems likely that RNA polymerase pauses immediately after it has synthesized a stretch of RNA that folds into a hairpin. Furthermore, the RNA–DNA hybrid helix produced after the hairpin is unstable because of its content of rU · dA base pairs, which are the weakest of the four kinds.

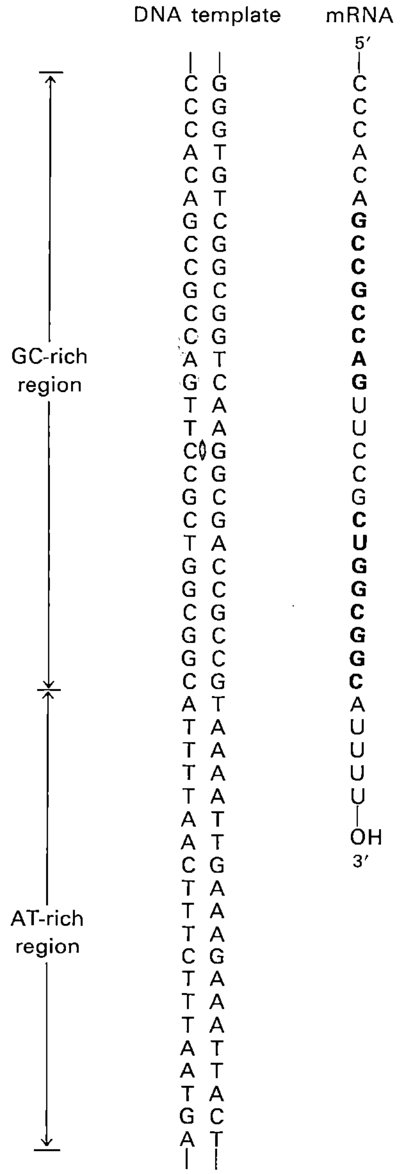

Figure 33-12
DNA sequence corresponding to the 3′ end of *trp* mRNA from *E. coli.* A twofold axis of symmetry (marked by the green symbol) relates the base pairs that are shaded yellow.

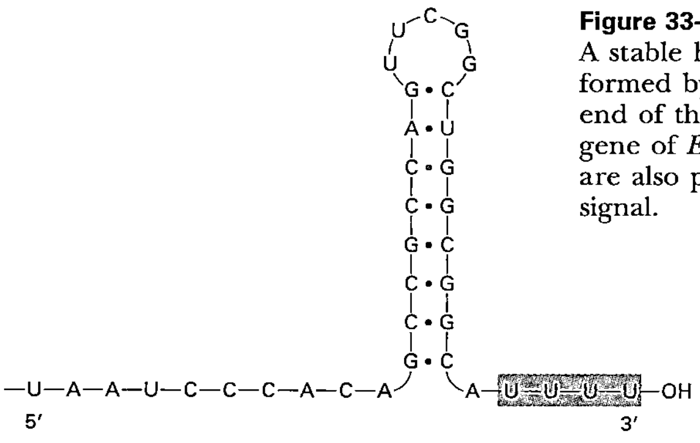

Figure 33-13
A stable hairpin structure can be formed by the base sequence of the 3′ end of the mRNA transcript of the *trp* gene of *E. coli.* The four U residues are also part of the termination signal.

Hence, *nascent RNA dissociates from the DNA template and then from the enzyme.* The solitary DNA template strand rejoins its partner to re-form the DNA duplex in the bubble region. The core enzyme (devoid of σ) has much less affinity for duplex DNA than for single-stranded DNA, and so the DNA is released. Sigma rejoins the core enzyme to form holoenzyme, which again searches for a promoter to initiate a new transcript.

RHO PROTEIN HELPS TERMINATE TRANSCRIPTION OF SOME GENES

RNA polymerase needs no help to terminate transcription at a hairpin followed by several U residues. At other sites, however, termination requires the participation of an additional factor. Some RNA molecules synthesized in vitro by RNA polymerase alone are *longer* than those made in vivo. The missing factor, a protein, was isolated and named rho (ρ). RNA synthesized from M13 filamentous phage DNA in the presence of ρ, for example, has a sedimentation coefficient of 10S, whereas RNA made in its absence is 23S. Additional information about the action of ρ was obtained by adding this termination factor to an incubation mixture at various times after the initiation of RNA synthesis (Figure 33-14). RNAs with sedimentation coefficients of 13S, 17S, and 23S were obtained when ρ was added a few seconds, two minutes, and ten minutes, respectively, after initiation. It is evident that the template contains at least three termination sites that respond to ρ (yielding 10S, 13S, and 17S RNA) and one termination site that does not (yielding 23S RNA). Thus, specific termination can occur in the absence of ρ. However, ρ detects additional termination signals that are not recognized by RNA polymerase alone.

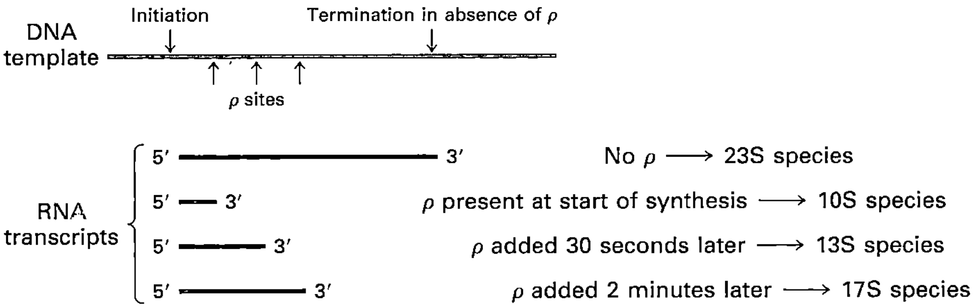

Figure 33-14
Effect of ρ protein on the size of RNA transcripts.

How does ρ provoke termination of RNA synthesis? *A key clue is the finding that ρ hydrolyzes ATP in the presence of single-stranded RNA but not of DNA or duplex RNA.* Hexameric ρ specifically binds single-stranded RNA;

a stretch of 72 nucleotides is bound, 12 per subunit. Rho is brought into action by sequences located in the nascent RNA—the absence of a simple consensus sequence implies that ρ detects noncontiguous structural features. The ATPase activity of ρ enables it to move unidirectionally along nascent RNA toward the transcription bubble. It then breaks the RNA–DNA hybrid helix by pulling RNA away (Figure 33-15). *A common feature of ρ-independent and ρ-dependent termination is that the active signals lie in newly synthesized RNA rather than in the DNA template.*

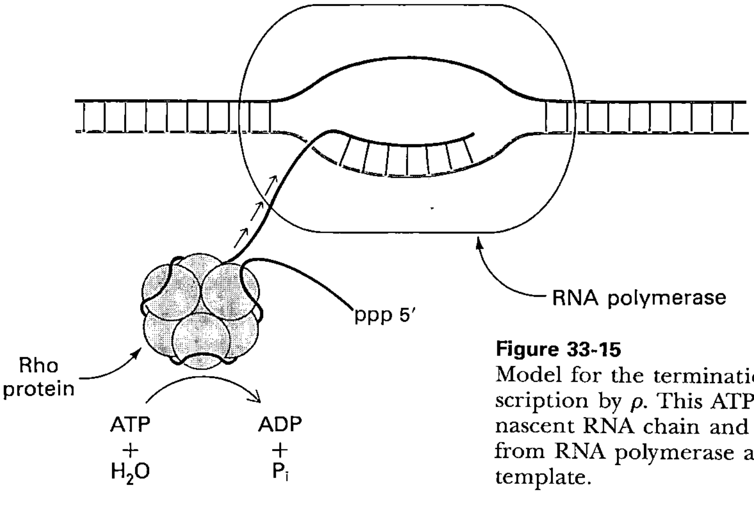

Figure 33-15
Model for the termination of transcription by ρ. This ATPase binds the nascent RNA chain and pulls it away from RNA polymerase and the DNA template.

Proteins in addition to ρ mediate and modulate termination. For example, the *nusA protein* enables RNA polymerase in *E. coli* to recognize a characteristic class of termination sites. As will be described later, λ phage synthesizes *antitermination proteins* that allow certain genes to be transcribed and expressed (p. 958). In *E. coli*, specialized termination signals called *attenuators* are regulated to meet the nutritional needs of the cell (p. 966). Again, the active signals are on nascent RNA.

PRECURSORS OF TRANSFER AND RIBOSOMAL RNA ARE CLEAVED AND CHEMICALLY MODIFIED AFTER TRANSCRIPTION

In prokaryotes, mRNA molecules undergo little or no modification following synthesis by RNA polymerase. Indeed, many of them are translated while they are being transcribed. In contrast, *transfer RNA and ribosomal RNA molecules are generated by cleavage and other modifications of nascent RNA chains.* For example, in *E. coli*, three kinds of ribosomal RNA molecules and a transfer RNA molecule are excised from a single primary RNA transcript that also contains spacer regions (Figure 33-16). Other transcripts contain arrays of several kinds of tRNAs or of several copies of the same tRNA. The nucleases that cleave and trim these precursors of rRNA and tRNA are highly precise. Ribonuclease P, for example, generates the correct 5′ terminus of all tRNA molecules in *E. coli*. This interesting enzyme contains a catalytically active RNA molecule. Ribonuclease III excises 5S, 16S, and 23S rRNA precursors from the primary transcript by cleaving double-helical hairpin regions at specific sites.

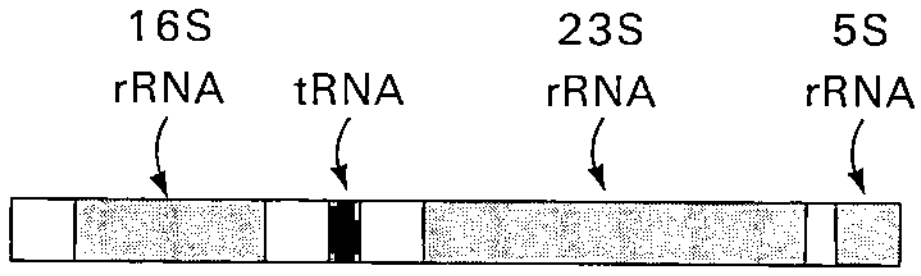

Figure 33-16
Cleavage of this primary transcript produces 5S, 16S, and 23S rRNA molecules and a tRNA molecule. Spacer regions are shown in yellow.

A second type of processing is the *addition of nucleotides to the termini of some RNA chains.* For example, CCA is added to the 3′ end of tRNA molecules that do not already possess this terminal sequence. *Modifications of*

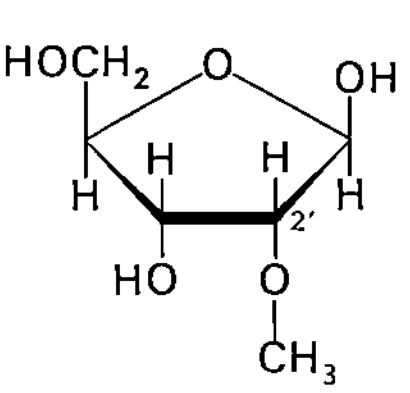

2′-*O*-Methylribose
(In eukaryotes)

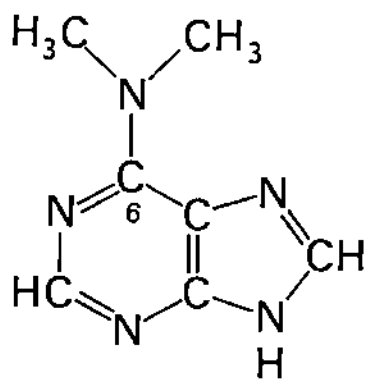

6-Dimethyladenine
(In prokaryotes)

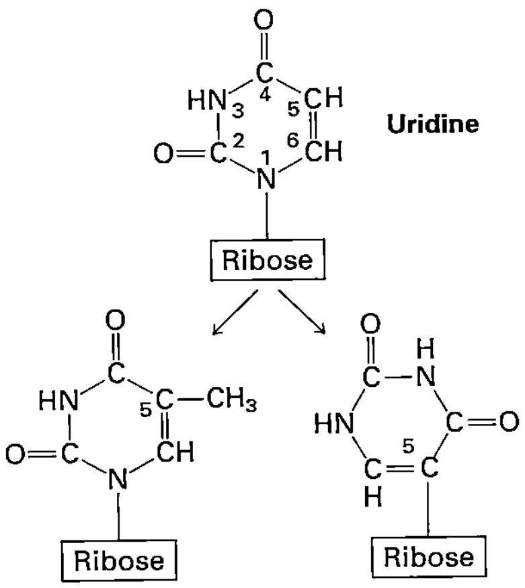

Uridine

Ribothymidine **Pseudouridine**

bases and ribose units of ribosomal RNAs are a third category. In prokaryotes, some bases are methylated, whereas in eukaryotes the 2′-hydroxyl group of about one in a hundred ribose units is methylated. Unusual bases are found in all tRNA molecules (p. 877). They are formed by enzymatic modification of a standard ribonucleotide in a tRNA precursor. For example, *ribothymidylate* and *pseudouridylate* are formed by modification of uridylate residues after transcription. These modifications generate diversity, as if tRNAs were trying to emulate proteins.

In eukaryotes, tRNA precursors are converted into mature tRNAs by a series of alterations: cleavage of a 5′ leader sequence, splicing to remove an intron, replacement of the 3′-terminal UU by CCA, and modification of several bases. The processing of yeast tRNA precursors has been investigated in detail (Figure 33-17). The 14-nucleotide intron next to the anticodon is excised by an endonuclease that generates a 2′,3′-cyclic phosphate at the end of the upstream exon and a 5′-OH at the end of the downstream exon. These two ends of the cleaved tRNA molecule stay together during the subsequent steps leading to their ligation.

The splicing of tRNAs is very similar in species that are evolutionarily far apart. Cloned genes for a yeast tRNA precursor were microinjected into *Xenopus* oocytes to determine whether a gene from a unicellular eukaryote can be transcribed and processed in an amphibian. The striking result is that the yeast gene is transcribed and properly processed in *Xenopus,* even though the genes for this tRNA are quite different in the two species. In particular, a 14-nucleotide intron was properly spliced out. Thus, the specificity of splicing enzymes has been conserved over a very long evolutionary period.

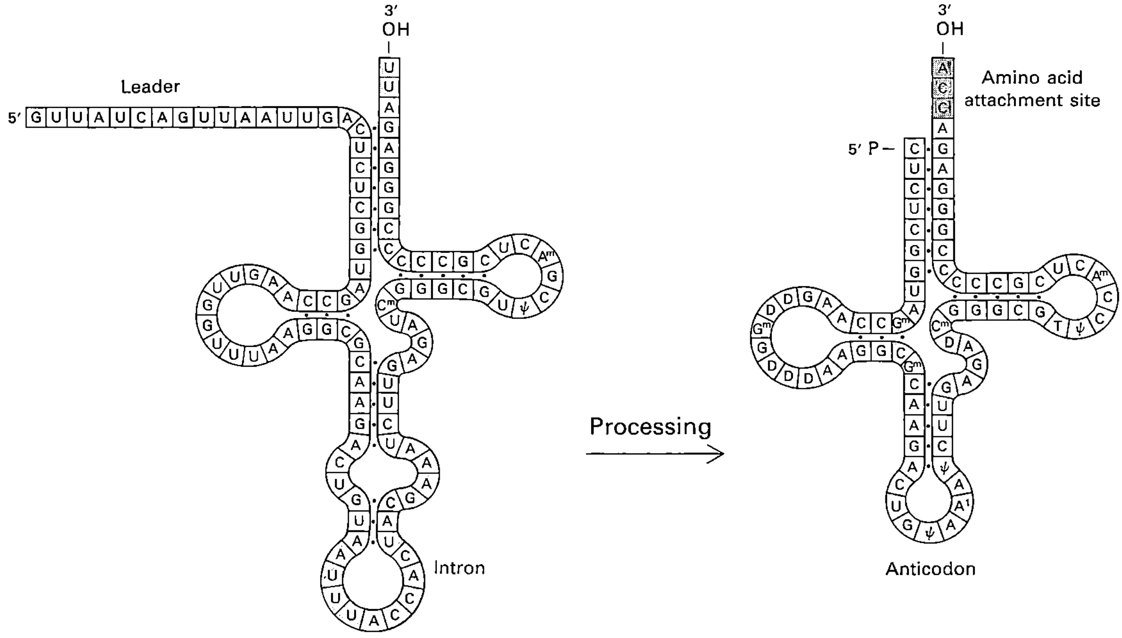

Figure 33-17
Processing of a precursor of yeast tyrosine tRNA. A 14-nucleotide intron (yellow) is removed, and a number of bases are modified. The 5′ leader (green) is cleaved, and the 3′ end is changed to CCA (red).

ANTIBIOTIC INHIBITORS OF TRANSCRIPTION: RIFAMPICIN AND ACTINOMYCIN D

Antibiotics are interesting molecules because many of them are highly specific inhibitors of biological processes. Rifampicin and actinomycin are two antibiotics that inhibit transcription in quite different ways. The *rifamycins,* which are derived from a strain of *Streptomyces,* and *rifampicin,* a semisynthetic derivative, *specifically inhibit the initiation of RNA synthesis.* Rifampicin does not block the binding of RNA polymerase to the DNA template; rather, it interferes with the formation of the first phosphodiester bond in the RNA chain. In contrast, chain elongation is not appreciably affected. This high degree of selectivity makes rifampicin a very useful experimental tool. For example, it can be used to block the initiation of new RNA chains without interfering with the transcription of chains that are already being synthesized. The site of action of rifampicin is the β *subunit of RNA polymerase.* Some mutants (called rif-r) having an altered β chain are resistant to rifampicin because they cannot bind the antibiotic.

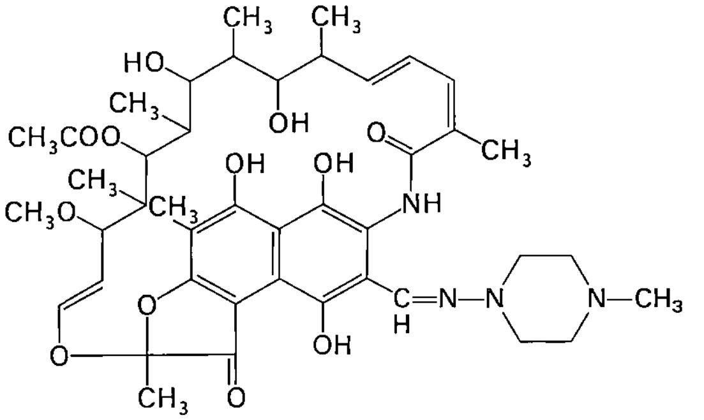

Rifampicin (Rifampin)

Actinomycin D, a polypeptide-containing antibiotic (Figure 33-18) from a different strain of *Streptomyces,* inhibits transcription by an entirely different mechanism. *Actinomycin D binds tightly and specifically to double-helical DNA and thereby prevents it from being an effective template for RNA synthesis.* It does not bind to single-stranded DNA or RNA, double-stranded RNA, or RNA–DNA hybrids. The binding of actinomycin to DNA is markedly enhanced by the presence of guanine residues. Spectroscopic and hydrodynamic studies of complexes of actinomycin D and DNA suggested that the phenoxazone ring of actinomycin slips in between neighboring base pairs in DNA. This mode of binding is called *intercalation.* At low concentrations, actinomycin D inhibits transcription without appreciably affecting DNA replication or protein synthesis. Hence, *actinomycin D has been extensively used as a highly specific inhibitor of the formation of new RNA in both prokaryotic and eukaryotic cells.* Its inhibition of the growth of rapidly dividing cells makes it an effective therapeutic agent in the treatment of some cancers.

The structure of a crystalline complex of one actinomycin molecule and two deoxyguanosine molecules is known at atomic resolution. The phenoxazone ring of actinomycin is sandwiched between two guanine rings. One of its two cyclic polypeptides extends above the phenoxazone ring, the other below. A significant feature of this complex is that it is nearly symmetric. There is a twofold axis of symmetry along the line joining the central O and N atoms of the phenoxazone ring. The conformation of this crystalline complex suggested that *actinomycin recognizes the base sequence GpC in DNA.* If the sequence along one strand of DNA is

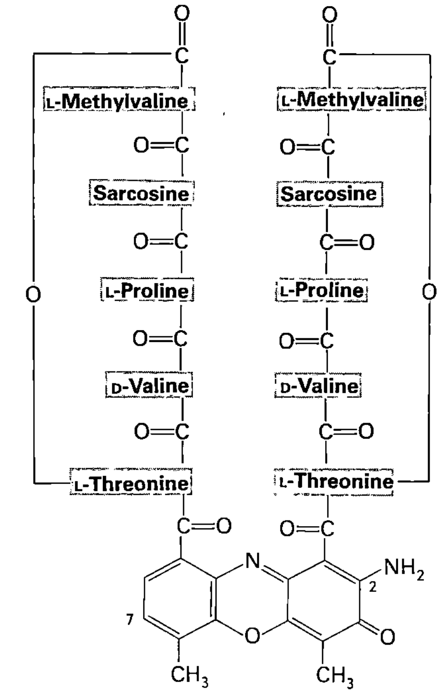

Figure 33-18
Structure of actinomycin D.

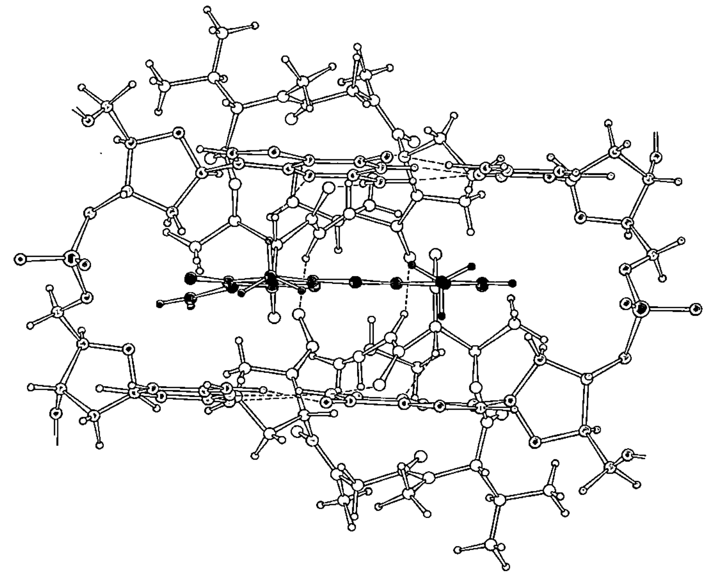

Figure 33-19
Structure of the complex of actinomycin D and DNA. The phenoxazone ring of actinomycin D (shown in red) is intercalated between two G · C base pairs of the DNA (shown in blue). The cyclic peptides of actinomycin D (shown in yellow) bind to the narrow groove of the DNA helix. The axis of symmetry of actinomycin coincides with the axis of symmetry of the two G · C base pairs. [After a drawing kindly provided by Dr. Henry Sobell.]

5′-GC-3′, the sequence of the complementary strand is 3′-CG-5′. Indeed, actinomycin intercalates between two G · C base pairs in DNA and interacts with the G residues in much the same way as it does in its complex with deoxyguanosine (Figure 33-19). Hence, *the symmetry of actinomycin matches the symmetry of a specific base sequence in DNA.* In fact, the symmetry of many DNA-binding proteins matches that of their palindromic target sites, as exemplified by restriction endonucleases (p. 792) and repressor proteins (p. 962).

TRANSCRIPTION AND TRANSLATION IN EUKARYOTES ARE SEPARATED IN SPACE AND TIME

We turn now to transcription in eukaryotes, a much more complex process than in prokaryotes. Eukaryotes, by definition, contain a membrane-bounded nucleus, where transcription occurs. In contrast, translation takes place outside the nucleus. Thus, transcription and translation in eukaryotes occur in different cellular compartments, whereas in prokaryotes they are closely coupled (Figure 33-20). Indeed, translation of bacterial mRNA begins while the transcript is still being synthesized. *The spatial and temporal separation of transcription and translation enables eukaryotes to*

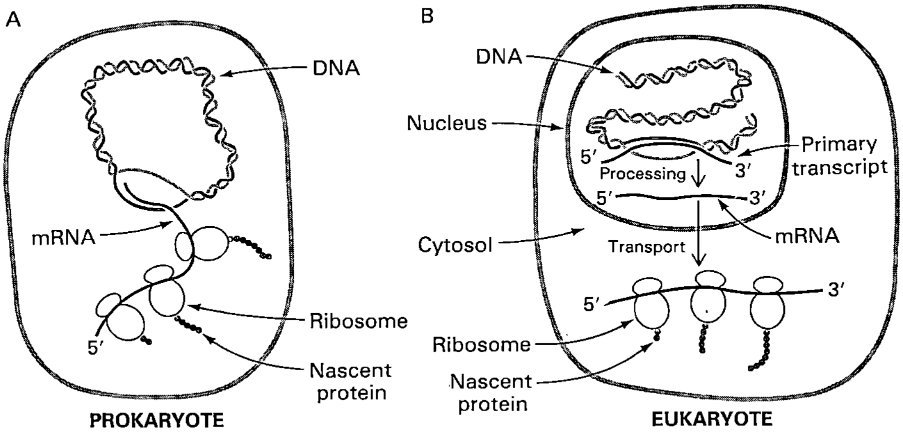

Figure 33-20
Transcription and translation are closely coupled in prokaryotes, whereas they are spatially and temporally separate in eukaryotes. (A) In prokaryotes, the primary transcript serves as mRNA and is used immediately as the template for protein synthesis. (B) In eukaryotes, mRNA precursors are processed and spliced in the nucleus before being transported to the cytosol. [After J. Darnell, H. Lodish, and D. Baltimore. *Molecular Cell Biology,* 2nd ed. (Scientific American Books, 1990), p. 230.]

regulate gene expression in much more intricate ways, contributing to the richness of eukaryotic form and function.

A second major difference between prokaryotes and eukaryotes is that *eukaryotes very extensively process nascent RNA destined to become mRNA.* Primary transcripts—the products of RNA polymerase action—acquire a cap at their 5′ end and a poly(A) tail at their 3′ end. Most importantly, *nearly all mRNA precursors in higher eukaryotes are spliced.* Introns are precisely excised from primary transcripts to form mature mRNAs with continuous messages. Some mRNAs are only a tenth the size of their precursors, which can be as large as 30 kb or more. The pattern of splicing can be developmentally regulated to generate variations on a theme, such as the membrane-bound and secreted forms of antibody molecules. Alternative splicing enlarges the repertoire of proteins in eukaryotes.

RNA IN EUKARYOTIC CELLS IS SYNTHESIZED BY THREE TYPES OF RNA POLYMERASES

In prokaryotes, RNA is synthesized by a single kind of polymerase. By contrast, the nucleus of eukaryotes contains three types of RNA polymerases differing in template specificity, localization, and susceptibility to inhibitors (Table 33-2). *RNA polymerase I* is located in nucleoli, where it transcribes the tandem array of genes for 18S, 5.8S, and 28S ribosomal RNA (p. 992). The other ribosomal RNA molecule (5S rRNA, p. 999) and all the transfer RNA molecules (p. 877) are synthesized by *RNA polymerase III,* which is located in the nucleoplasm rather than in nucleoli. The precursors of messenger RNA are synthesized by *RNA polymerase II,* which is also located in the nucleoplasm. Several small RNA molecules, such as the U1 snRNA of the splicing apparatus (p. 863), are also synthesized by polymerase II.

Table 33-2
Eukaryotic RNA polymerases

Type	*Localization*	*Cellular transcripts*	*Effect of α-amanitin*
I	Nucleolus	18S, 5.8S, and 28S rRNA	Insensitive
II	Nucleoplasm	mRNA precursors and snRNA	Strongly inhibited
III	Nucleoplasm	tRNA and 5S rRNA	Inhibited by high concentrations

Eukaryotic RNA polymerases are like prokaryotic ones in catalyzing the nucleophilic attack by the 3′-OH of the growing RNA chain on the α phosphorus atom of an incoming ribonucleoside triphosphate. Again, a primer is not needed, and synthesis proceeds in the 5′ → 3′ direction according to instructions given by an antiparallel DNA template strand. Eukaryotic RNA polymerases also lack nuclease activity, which means that errors in nascent RNA are not corrected.

Three features of type II eukaryotic RNA polymerases are conserved. First, they contain between 8 and 12 subunits. The one from yeast, which is the best characterized, consists of 11 subunits. Second, they contain two large subunits (~220 kd and ~140 kd), termed *RPB1* and *RPB2,* which correspond to the catalytic β and β' subunits in the core of the prokaryotic enzyme. The carboxyl-terminal domain (CTD) of the 220-kd subunit is remarkable in containing multiple repeats of a YSPTSPS consensus sequence. Twenty-seven of these CTD sequences are present in the yeast enzyme, and more than 50 in mammalian ones. The CTDs, which are regulated by phosphorylation, participate in the recognition of activating signals. The importance of this region is highlighted by the finding that yeasts containing mutant polymerase II with fewer than 10 CTD sequences are not viable. Third, all type II RNA polymerases share three subunits (RPB5, 6, and 8). Indeed, these subunits are also present in RNA polymerases I and III.

More than a hundred deaths result each year in the world from the ingestion of poisonous mushrooms, particularly *Amanita phalloides,* which is also called the *death cup* or the *destroying angel.* One of the toxins in this mushroom is *α-amanitin,* a cyclic octapeptide that contains several unusual amino acids. α-Amanitin binds very tightly ($K = 10$ nM) to RNA polymerase II and thereby blocks the formation of precursors of mRNA. Polymerase III is inhibited by higher concentrations of α-amanitin (1 μM), whereas polymerase I is insensitive to this toxin. α-Amanitin blocks the elongation phase of RNA synthesis.

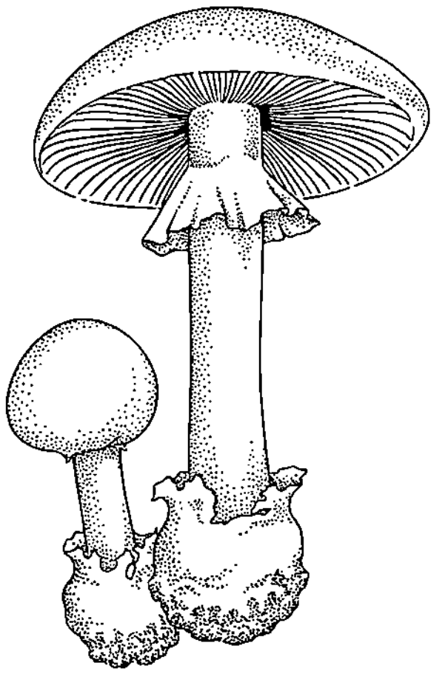

Figure 33-21
Amanita phalloides, a poisonous mushroom that contains α-amanitin. This mushroom also contains phalloidin, a potent inhibitor of cytoskeletal function. [After G. Lincoff and D.H. Mitchel. *Toxic and Hallucinogenic Mushroom Poisoning.* (Van Nostrand Reinhold, 1977), p. 30.]

5′ T_{82} A_{97} T_{93} A_{85} A_{63} A_{88} A_{50} 3′

A_7 T_2 A_7 T_{10} T_{37} T_{10} T_{33}

TATA box

Figure 33-22
TATA box in eukaryotic promoters for mRNA precursors. Comparisons of the sequences of more than a hundred promoters led to the consensus shown in the upper row. The second most frequently found bases are shown in the second row. The subscript denotes the frequency (in percent) of the base at that position. The TATA box is centered at about −25.

EUKARYOTIC PROMOTERS CONTAIN A TATA BOX NEAR THE START SITE FOR TRANSCRIPTION

Promoters for RNA polymerase II, like those for bacterial polymerases, are located on the 5′ side of the start site for transcription. Mutagenesis experiments, footprinting studies, and comparisons of many higher eukaryotic genes have demonstrated the importance of several upstream regions. The one closest to the start site, centered at about −25, is called the *TATA box.* The consensus sequence is a heptanucleotide of A and T residues (Figure 33-22). In yeast, a TATAAA sequence centered between −30 and −90 plays the same role. The TATA box is present in nearly all eukaryotic genes giving rise to mRNA. Note that *the TATA box of eukaryotes closely resembles the −10 sequence of prokaryotes (TATAAT) but is farther from the start site.* Mutation of a single base in the TATA box markedly impairs promoter activity. Likewise, interchanges of A and T bases lead to loss of activity, showing that the precise sequence, not just a high content of AT pairs, is essential. Most TATA boxes are flanked by GC-rich sequences.

The TATA box is necessary but not sufficient for strong promoter activity. Additional elements are located between −40 and −110. Many promoters contain a *CAAT box* and some contain a *GC box* (Figures 33-23 and 33-24). Constitutive genes (genes that are continuously expressed rather than regulated) tend to have GC boxes in their promoters. The positions of these upstream sequences vary from one promoter to another, in contrast with the quite constant location of the −35 region in prokaryotes. Another difference is that the CAAT box and the GC box can be effective when present on the template (antisense) strand, unlike the −35 region, which must be present on the coding (sense) strand.

5′ GGNCAATCT 3′
CAAT box

5′ GGGCGG 3′
GC box

Figure 33-23
Consensus sequences of CAAT and GC boxes of eukaryotic promoters for mRNA precursors. Most are located in the −40 to −110 region.

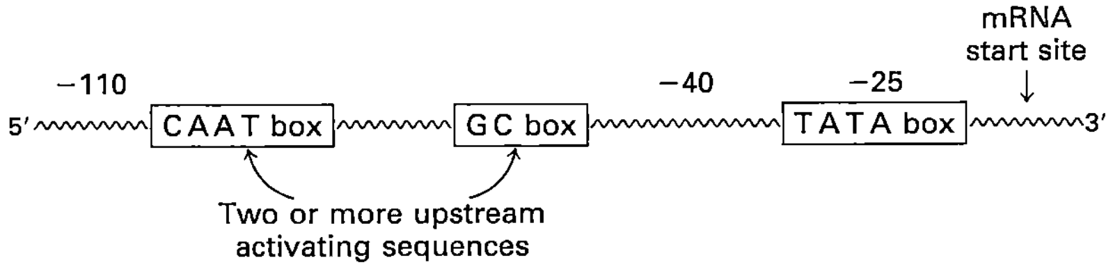

Figure 33-24
General features of promoters for mRNA precursors in higher eukaryotes.

THE TATA BOX–BINDING PROTEIN PLAYS A KEY ROLE IN ASSEMBLING ACTIVE TRANSCRIPTION COMPLEXES

RNA polymerase II is unable to initiate transcription on its own. Rather, it is guided to the start site by a set of transcription factors known collectively as *TFII* (*TF* stands for *T*ranscription *F*actor, and *II* refers to RNA polymerase *II*). Initiation begins with the binding of TFIID to the TATA box (Figure 33-25). TFIIA is recruited, followed by TFIIB. RNA polymerase II and then TFIIE join the other factors to form a complex called the *basal transcription apparatus.* This complex can transcribe DNA at a slow rate. However, additional transcription factors that bind to other sites are required for high-level mRNA synthesis and selectivity.

The key initial event is the recognition of the TATA box by the *TATA box-binding protein (TBP),* an ~30-kd component of the 700-kd TFIID. TBP binds 10^5-fold more tightly to the TATA box than to noncognate sequences; the dissociation constant of the specific complex is ~1 nM. The C-terminal 185 residues of TBP are highly conserved in eukaryotes. TBP is a saddle-shaped protein consisting of two similar domains (Figure 33-26). Each contains two α helices and a five-stranded antiparallel β sheet. All mutations affecting DNA binding and TATA box recognition map to the underside of the molecular saddle, which is formed by the curved antiparallel β sheets.

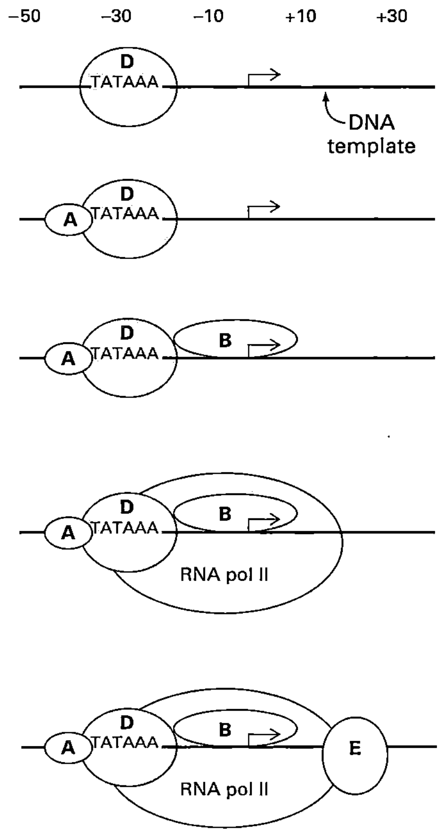

Figure 33-25
Transcription factors TFIIA, B, D, and E are essential in initiating transcription by RNA polymerase II. The stepwise assembly of these general transcription factors begins with the binding of TFIID (blue) to the TATA box. RNA polymerase II (yellow) joins the complex after TFIIA, B, and D are positioned on the DNA. The arrow marks the transcription start site. [After L. Guarente. *Trends Genet.* 8(1992):28.]

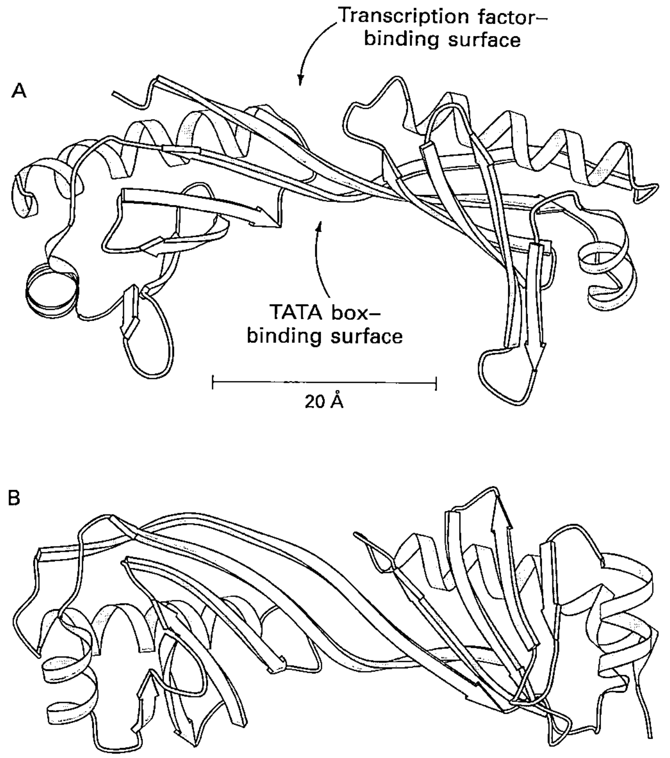

Figure 33-26
Three-dimensional structure of the TATA box–binding protein (TBP), the DNA recognition subunit of TFIID. Each of the two similar domains (yellow and blue) of the conserved C-terminal fragment of TBP contains two α helices and a five-stranded antiparallel β pleated sheet. Two views of the protein are shown in (A) and (B). [Drawn from coordinates kindly provided by Dr. Paul Sigler and Dr. Youngchang Kim. Y. Kim, J.H. Geiger, S. Hahn, and P.B. Sigler. *Nature* 365(1993):512.]

Indeed, *the TATA box of DNA binds to the concave surface of TBP* (Figure 33-27 and see p. 841). The saddle-shaped protein induces large conformational changes in the bound DNA. A sharp kink is introduced at each end of the seven-bp TATA box. Moreover, the double helix is partially unwound to widen its *minor groove,* enabling it to make extensive contact with the antiparallel β sheets on the concave side of TBP. Hydrophobic interactions are prominent at this interface. Four phenylalanines, for example, are intercalated between base pairs of the TATA box. The flexibility of AT-rich sequences generally is exploited here in bending the DNA. Immediately outside the TATA box, classical B-DNA resumes. This complex is distinctly asymmetric, in contrast with many protein-DNA complexes that display twofold symmetry (p. 792). *The asymmetry of the TBP–TATA box complex is crucial for specifying a unique start site and assuring that transcription proceeds unidirectionally.*

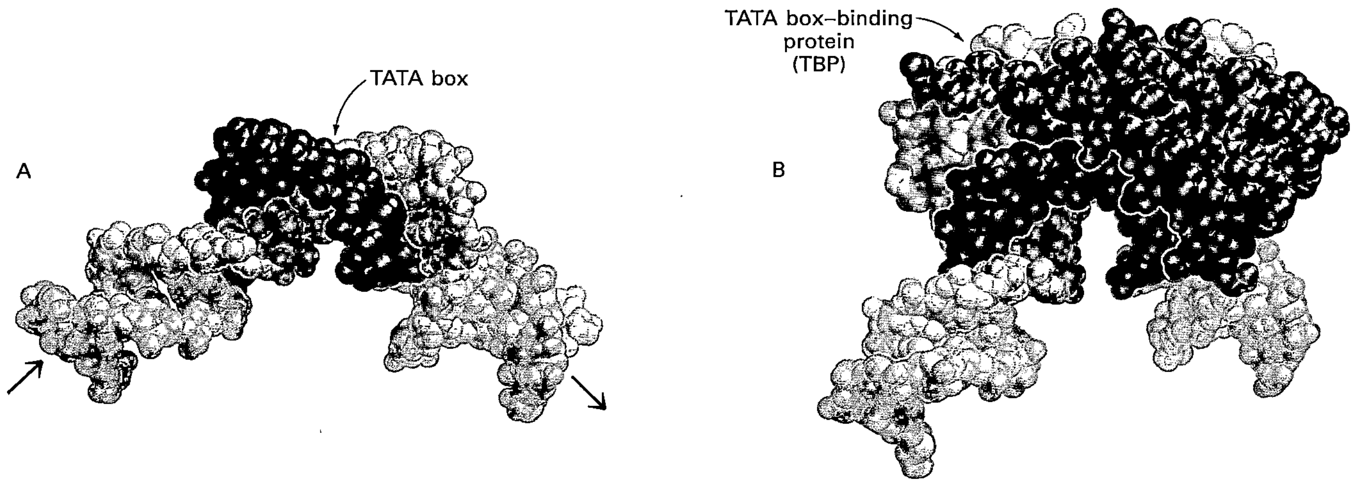

Figure 33-27
Three-dimensional structure of TBP bound to duplex DNA containing a TATA box. One of the domains of the protein is shown in yellow, and the other in blue. The DNA strands are shown in red and green. The 8-bp TATA box is shown in dark colors, and the adjoining segments of DNA in light colors. (A) DNA portion of the complex. Note that the DNA axis is bent about 90 degrees by the interaction of the TATA box with TBP. (B) TBP–DNA complex. The structure shown in A is rotated mainly about the vertical axis to provide this view. [Drawn from coordinates kindly provided by Dr. Paul Sigler and Dr. Youngchang Kim. Y. Kim, J.H. Geiger, S. Hahn, and P.B. Sigler. *Nature* 365(1993):512.]

MULTIPLE TRANSCRIPTION FACTORS INTERACT WITH EUKARYOTIC PROMOTERS

TBP bound to the TATA box is the heart of the initiation complex. The convex surface of the TBP saddle provides docking sites for the binding of other components of TFII (Figure 33-28). Additional transcription factors assemble on this nucleus. Moreover, DNA can be looped so that sites far apart in the linear sequence are brought into proximity.

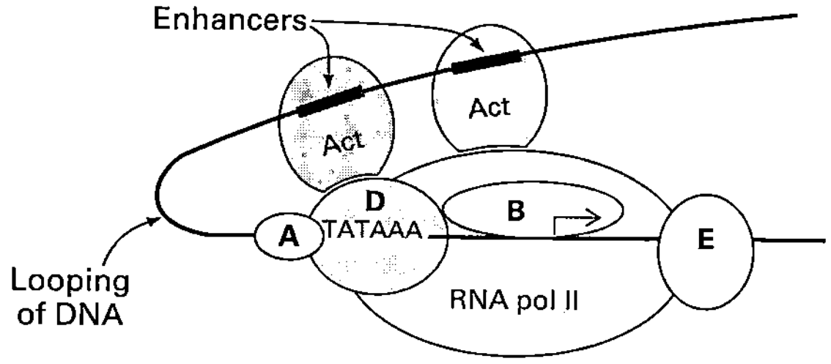

Figure 33-28
The looping of DNA enables activator proteins (Act) bound to distant enhancers to interact with TFIID (blue), other TFII components, and RNA polymerase (yellow). Looping enables proteins bound to distant enhancers to interact with the TATA box–binding protein and participate in the formation of a functional transcription complex. [After R.G. Roeder. *Trends Biochem. Sci.* 16(991):406.]

The diversity of upstream stimulatory sequences in eukaryotic genes and the variability of their position suggested that they are recognized by many different specific proteins. Indeed, many transcription factors in addition to TFII have been isolated, and their binding sites have been identified by footprinting experiments. For example, *Sp1,* an ~100-kd protein from mammalian cells, is required for transcription of genes whose promoters contain GC boxes. The duplex DNA of SV40 virus (a cancer-producing virus that infects monkey cells) contains five GC boxes 50 to 100 bp upstream or downstream from start sites (Figure 33-29A). DNase I digestion footprints showed that three of these GC boxes bind Sp1 tightly and two bind it weakly. The binding of a molecule of Sp1 protects about 20 bp (~68 Å) of DNA from digestion. A *heat-shock transcription factor (HSTF)* is expressed in *Drosophila* following an abrupt increase in temperature. This 93-kd DNA-binding protein binds to the consensus sequence

5′-CNNGAANNTCCNNG-3′

Several copies of this sequence, known as the *heat-shock response element,* are present starting at a site 15 bp upstream from the TATA box (Figure 33-29B). HSTF differs from σ^{32}, a heat-shock protein of *E. coli* (p. 844), in binding directly to response elements in target sites in heat-shock promoters rather than first becoming associated with RNA polymerase. The *CCAAT-binding transcription factor (CTF;* also called *NF1),* a 60-kd protein from mammalian cells, binds to the CAAT box (see Figure 33-23).

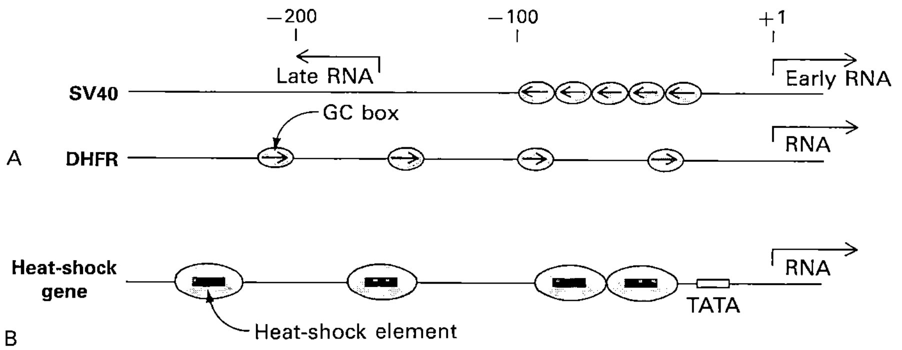

Figure 33-29
Multiple binding sites for transcription factors in eukaryotic promoters mapped by footprinting. (A) Binding of Sp1 (green) to the SV40 viral DNA promoter and the dihydrofolate reductase (DHFR) promoter.
(B) Binding of HSTF (blue) to a *Drosophila* heat-shock promoter. [After W.S. Dynan and R. Tjian. *Nature* 316(1985):774.]

ENHANCER SEQUENCES CAN STIMULATE TRANSCRIPTION AT START SITES THOUSANDS OF BASES AWAY

The activities of many promoters in higher eukaryotes are greatly increased by sequences called *enhancers* that have no promoter activity of their own. *Remarkably, enhancers can exert their stimulatory actions over distances of several thousand base pairs. They can be upstream, downstream, or even in the midst of a transcribed gene.* The SV40 viral enhancer, a tandem repeat of a 72-bp sequence, remains active when moved anywhere in the 5.2-kb circular viral genome. Moreover, enhancers are effective when present on *either DNA strand* (equivalently, in either orientation). Enhancers in yeast are known as *upstream activator sequences (UAS).*

A particular enhancer is effective only in certain cells. For example, the immunoglobulin enhancer functions in B lymphocytes but not elsewhere. Enhancers act only in cells that contain cognate stimulatory proteins.

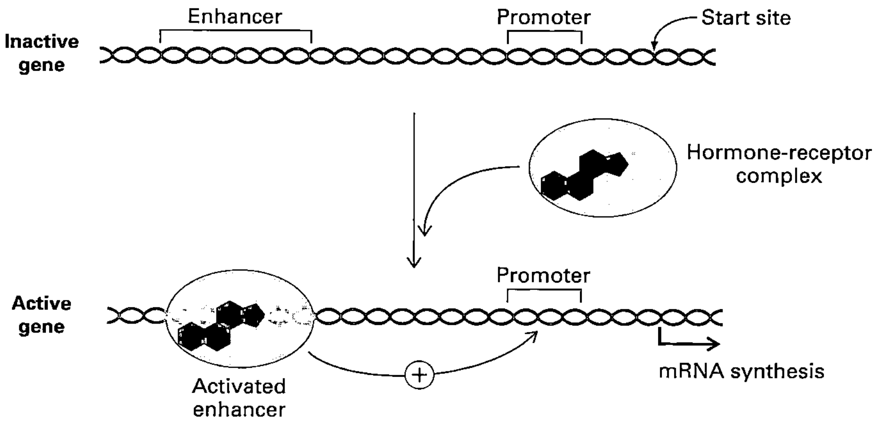

Figure 33-30
Enhancer sequences play a key role in mediating the action of steroid hormones. A glucocorticoid hormone binds to a soluble receptor protein. This complex binds to enhancer sequences, enabling them to stimulate the transcription of hormone-responsive genes.

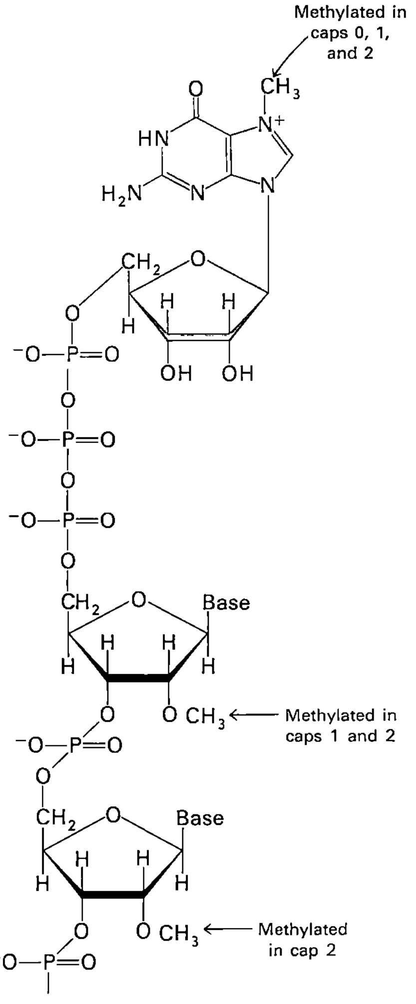

Figure 33-31
Structure of caps at the 5′ end of eukaryotic mRNAs. All caps contain 7-methylguanylate (shown in green) attached by a triphosphate linkage to the sugar at the 5′ end. None of the riboses are methylated in cap 0, one is methylated in cap 1, and two are methylated in cap 2.

This is well understood for the enhancer mediating the hormonal action of glucocorticoids. These steroids interact with a soluble receptor. The binding of the hormone-receptor complex to the glucocorticoid enhancer then leads to stimulation of transcription of a distinctive set of genes (p. 1001). As might be expected, the DNA of viruses infecting eukaryotic cells usually contains enhancers that are activated by host-cell proteins. For example, the DNA of mouse mammary tumor virus (MMTV) contains the glucocorticoid enhancer. Hence, MMTV thrives in cells that are normally stimulated by these steroids—namely, epithelial cells of the mammary gland—because they contain the hormone receptor that activates the enhancer. The restricted host range of viruses is partly a consequence of their having tissue-specific and species-specific enhancers.

Transcription factors and other proteins that bind to regulatory sites on DNA can be regarded as passwords that cooperatively open multiple locks, giving RNA polymerase access to specific genes. The discovery of promoters and enhancers has opened the door to understanding how genes are selectively expressed in eukaryotic cells. We shall return to this exciting area of inquiry in Chapter 37.

mRNA PRECURSORS ACQUIRE 5′ CAPS DURING TRANSCRIPTION

As in prokaryotes, eukaryotic transcription usually begins with A or G. However, the 5′ triphosphate end of the nascent RNA chain is immediately modified. A phosphate is released by hydrolysis. The diphosphate 5′ end then attacks the α phosphorus atom of GTP to form a very unusual 5′-5′ triphosphate linkage. This distinctive terminus is called a *cap*. The N-7 nitrogen of the terminal guanine is then methylated by *S*-adenosylmethionine to form *cap 0*. The adjacent riboses may be methylated to form *cap 1* or *cap 2* (Figure 33-31). Transfer RNA and ribosomal RNA molecules, in contrast with mRNAs and some snRNAs, do not have caps. Caps contribute to the stability of mRNAs by protecting their 5′ ends from phosphatases and nucleases. In addition, caps enhance the translation of mRNA by eukaryotic protein-synthesizing systems (p. 904).

A 3′ POLYADENYLATE TAIL IS ADDED TO MOST mRNA PRECURSORS AFTER CLEAVAGE BY AN ENDONUCLEASE

Rather little is known about eukaryotic termination signals. Some nascent RNAs contain a hairpin followed by a series of U residues, as in prokaryotic transcripts. *Most eukaryotic mRNAs contain a polyadenylate (poly(A)) tail at their 3′ end.* This poly(A) tail is not encoded by DNA. Indeed, the nucleotide preceding poly(A) is not the last nucleotide to be transcribed. Some primary transcripts contain hundreds of nucleotides beyond the 3′ end of the mature mRNA.

Eukaryotic primary transcripts are cleaved by a specific endonuclease that recognizes the sequence AAUAAA (Figure 33-32). Cleavage does not occur if this sequence or a segment of some 20 nucleotides on its 3′ side is deleted. The presence of internal AAUAAA sequences in some mature mRNAs indicates that AAUAAA is only part of the cleavage signal; its context is also important. After cleavage by the endonuclease, a *poly(A) polymerase* adds about 250 A residues to the 3′ end of the transcript; ATP is the donor in this reaction. This poly(A) tail then wraps around several copies of a 78-kd binding protein.

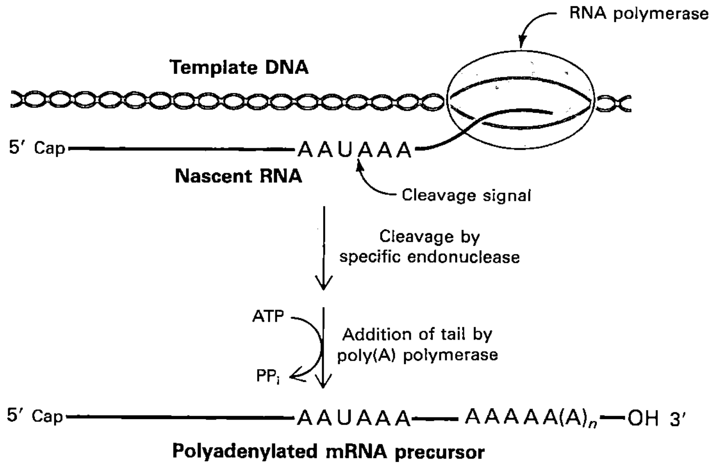

Figure 33-32
Cleavage and polyadenylation of a primary transcript. A specific endonuclease cleaves the RNA downstream of AAUAAA. Poly(A) polymerase then adds about 250 A residues.

The role of the poly(A) tail is becoming evident. Its synthesis can be blocked by *3′-deoxyadenosine (cordycepin),* which does not interfere with synthesis of the primary transcript. Messenger RNA devoid of a poly(A) tail can be transported out of the nucleus. However, an mRNA devoid of a poly(A) tail is usually a much less effective template for protein synthesis than is one with a poly(A) tail. Furthermore, a long tail protects an mRNA molecule from digestion by nucleases. The life of an mRNA molecule is determined in part by the rate of degradation of its poly(A) tail.

RNA EDITING CHANGES THE PROTEINS ENCODED BY mRNA

The information content of some mRNAs is altered following transcription. *Apolipoprotein B (apo B)* plays an important role in the transport of triacylglycerols and cholesterol by forming an amphipathic spherical shell

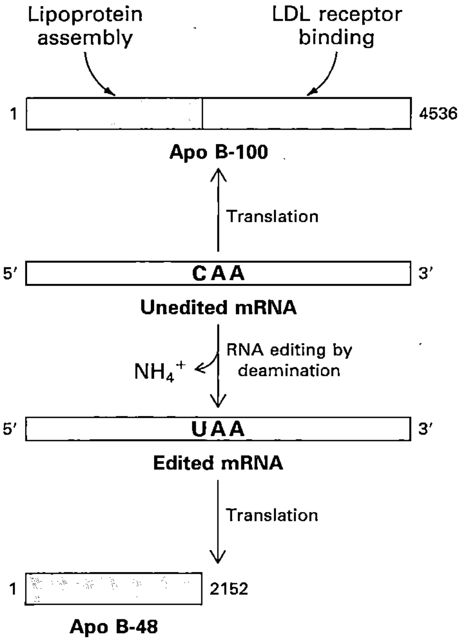

Figure 33-33
RNA editing of apolipoprotein B mRNA. Enzyme-catalyzed deamination of a specific C in the mRNA for apolipoprotein B-100 changes the codon for glutamine (CAA) to a stop signal (UAA). Apo B-48, a truncated version of the protein lacking the LDL receptor binding domain, is generated by this post-transcriptional change in mRNA sequence. [After P. Hodges and J. Scott. *Trends Biochem. Sci.* 17(1992):77.]

around the lipids carried in lipoprotein particles (p. 697). Apo B exists in two forms, a 512-kd *apo B-100* and a 240-kd *apo B-48*. The larger form, synthesized by the liver, participates in the transport of endogenously synthesized lipids. The smaller form, synthesized by the small intestine, carries dietary fat in the form of chylomicrons. Apo B-48 contains the 2152 N-terminal residues of the 4536-residue apo B-100. This truncated molecule can form lipoprotein particles but cannot bind to the LDL receptor on cell surfaces. What is the biosynthetic relation of these two forms of apo B? One possibility a priori is that apo B-48 is produced by proteolytic cleavage of apo B-100, and another is that the two forms arise from alternative splicing. Experiments showed neither occurs here. A totally unexpected and new mechanism for generating diversity is at work: *the changing of the nucleotide sequence of mRNA following its synthesis* (Figure 33-33). *A specific cytidine of the mRNA is deaminated to uridine, which changes the codon at residue 2153 from CAA (Gln) to UAA (Stop).* The deaminase that catalyzes this reaction is present in the small intestine but not the liver, and is developmentally regulated.

RNA editing—the change in the information content of RNA following transcription by processes other than RNA splicing—is not confined to apolipoprotein B. Glutamate opens cation-specific channels in the vertebrate central nervous system by binding to receptors in postsynaptic membranes. RNA editing changes a single glutamine codon (CAG) in the message for the glutamate receptor to one for arginine (CGG). The Gln to Arg substitution in the receptor prevents Ca^{2+} but not Na^+ from flowing through the channel. In trypanosomes, a different kind of RNA editing markedly changes several mitochondrial mRNAs. Nearly half the uridines in these mRNAs are inserted by RNA editing. A *guide RNA molecule* identifies the sequences to be modified, and a *poly(U) tail* on the guide donates uridines to the mRNAs undergoing editing. It is evident that DNA sequences do not always faithfully disclose the sequence of encoded proteins—functionally crucial changes can occur at the level of mRNA.

SPLICE SITES IN mRNA PRECURSORS ARE SPECIFIED BY SEQUENCES AT THE ENDS OF INTRONS

Introns are precisely spliced out of mRNA precursors. A one-nucleotide slippage in a splice point would shift the reading frame on the 3′ side of the splice to give an entirely different amino acid sequence. How does the splicing machinery process its targets with high fidelity? The base sequences of thousands of intron-exon junctions of RNA transcripts are known, and they are revealing (Table 33-3). These sequences in eukaryotes from yeast to mammals have a common structural motif: *the base*

Table 33-3
Base sequences of splice points in transcripts containing introns

Gene region	*Exon*																			*Exon*
Ovalbumin intron 2	U	A	A	G	G	U	G	A	G	C	∿∿∿	U	U	A	C	A	G	G	U	U	G
Ovalbumin intron 3	U	C	A	G	G	U	A	C	A	G	∿∿∿	A	U	U	C	A	G	U	C	U	G
β-Globin intron 1	G	C	A	G	G	U	U	G	G	U	∿∿∿	C	C	U	U	A	G	G	C	U	G
β-Globin intron 2	C	A	G	G	G	U	G	A	G	U	∿∿∿	C	C	A	C	A	G	U	C	U	C
Immunoglobulin λ_1 intron 1	U	C	A	G	G	U	C	A	G	C	∿∿∿	U	U	G	C	A	G	G	G	G	C
SV40 virus early T-antigen	U	A	A	G	G	U	A	A	A	U	∿∿∿	U	U	U	U	A	G	A	U	U	C

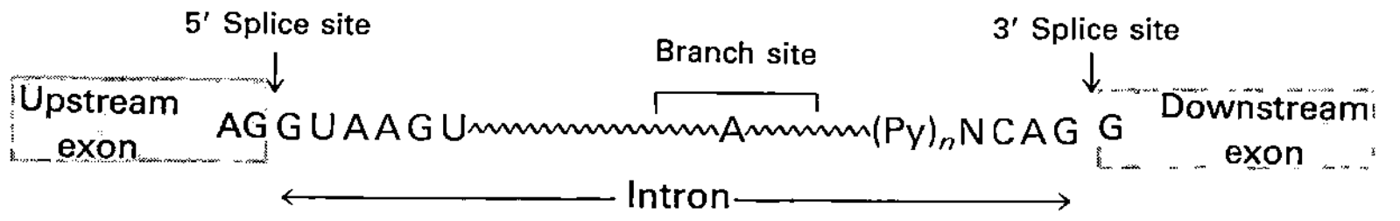

Figure 33-34
Splicing signals. Consensus sequences for the 5′ splice site and the 3′ splice site are shown.

sequence of an intron begins with GU and ends with AG. The consensus sequence at the 5′ splice site of vertebrate introns is AGGUAAGU (Figure 33-34). At the 3′ end of introns, the consensus sequence is a stretch of *10 pyrimidines* (U or C), followed by any base and then by C, and ending with the invariant AG. Introns also have an important internal site located between 20 and 50 nucleotides upstream of the 3′ splice site; it is called the *branch site* for reasons that will be evident shortly. In yeast, the branch site sequence is nearly always UACUAAC, whereas in mammals a variety of sequences are found.

Portions of introns other than the 5′ and 3′ splice sites and the branch site appear to be unimportant in determining where splicing occurs. The length of most introns ranges from 50 to 10,000 nucleotides. Much of an intron can be deleted without altering the site or efficiency of splicing. Likewise, splicing is unaffected by the insertion of long stretches of DNA into the introns of genes. Moreover, chimeric introns crafted by recombinant DNA methods from the 5′ end of one intron and the 3′ end of a very different intron are properly spliced, provided that the splice sites and branch site are unaltered. In contrast, mutations in each of these three critical regions lead to aberrant splicing. Mutations far from splice junctions can lead to abnormal splicing only if the change creates a consensus sequence for 5′ or 3′ splicing at a new location.

Some forms of thalassemia, a group of hereditary anemias characterized by defective synthesis of hemoglobin (p. 175), are caused by *aberrant splicing.* In one patient, a mutation of G to A 19 nucleotides away from the normal 3′ splice site of the first intron created a new 3′ splice site (Figure 33-35). The resulting mRNA contains a series of codons not normally present. The sixth codon after the splice is a stop signal for protein synthesis, and so the aberrant protein ends prematurely.

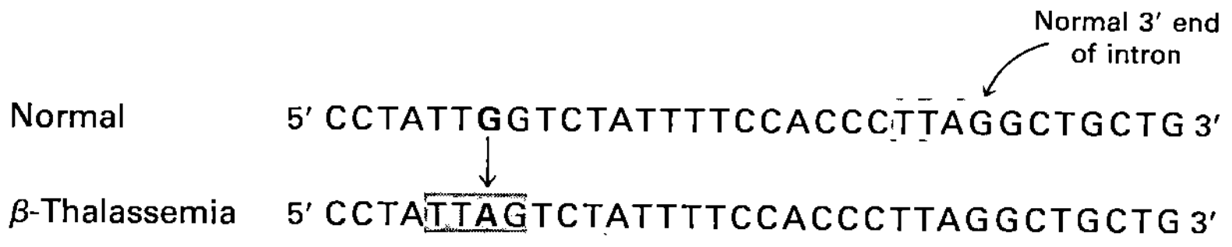

Figure 33-35
Mutation of a single base (G to A) in an intron of the β-globin gene leads to thalassemia. This mutation probably generates a new 3′ splice site (blue) akin to the normal one (yellow) farther downstream. [After R.A. Spritz, P. Jagedeeswaran, P.V. Choudary, P.A. Biro, J.T. Elder, J.K. deRiel, J.L. Manley, M.L. Gefter, B.G. Forget, and S.M. Weissman. *Proc. Nat. Acad. Sci.* 78(1981):2457.]

LARIAT INTERMEDIATES ARE FORMED IN THE SPLICING OF mRNA PRECURSORS

How does the end of one exon in an mRNA precursor become linked to the beginning of the adjacent exon? Splicing begins with the cleavage of the phosphodiester bond between the upstream exon (exon 1) and the 5′

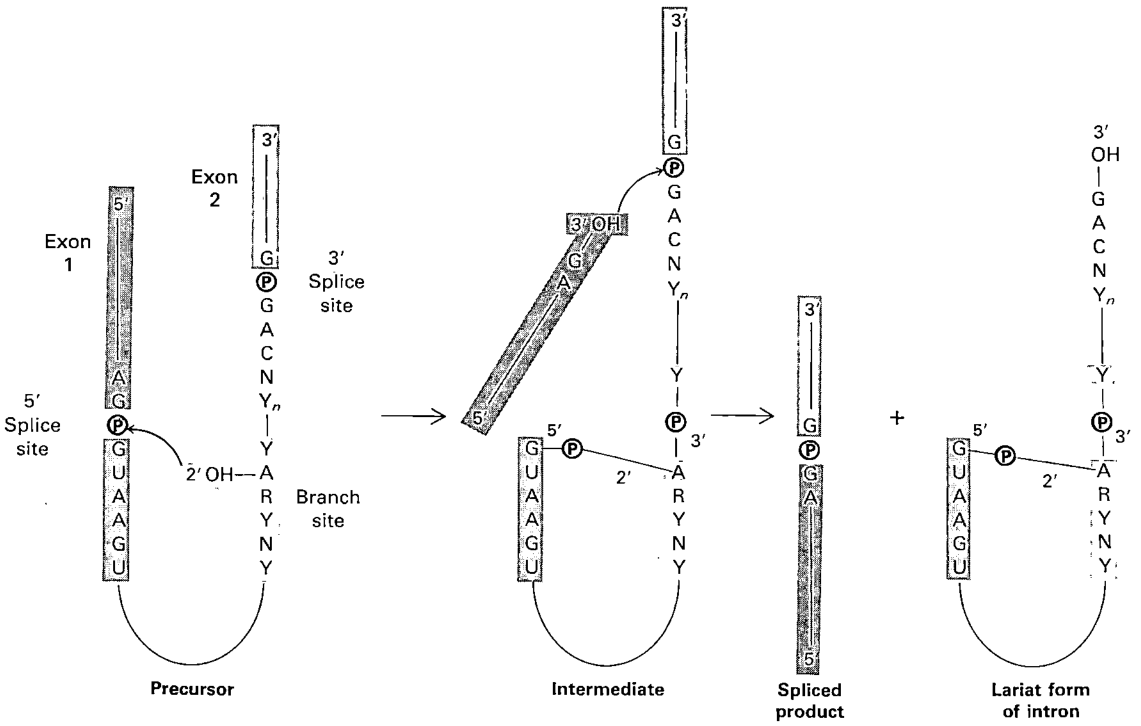

Figure 33-36
Mechanism of splicing of mRNA precursors in eukaryotic nuclei. The upstream (5′) exon is shown in red, the downstream (3′) exon in green, and the branch site in yellow. Y stands for a purine nucleotide, R for a pyrimidine nucleotide, and N for any nucleotide. The 5′ splice site is attacked by the 2′-OH of the branch site adenosine. The 3′ splice site is then attacked by the newly formed 3′-OH of the upstream exon. The exons are joined, and the intron is released in the form of a lariat. [After P.A. Sharp. *Cell* 2(1985):3980.]

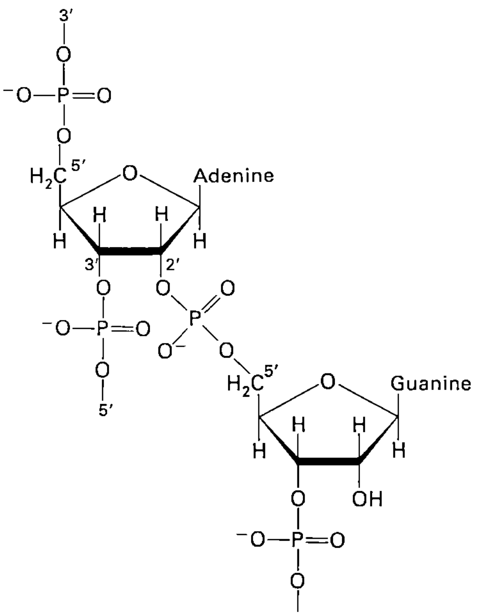

Figure 33-37
Structure of the branch point in the lariat intermediate in splicing. This adenylate residue is joined to three nucleotides by phosphodiester bonds. The new 2′ → 5′ phosphodiester bridge is shown in red and the usual 5′ → 3′ bridges in blue.

end of the intron (Figure 33-36). The attacking group in this reaction is the 2′-OH of an adenylate residue in the branch site. A 2′,5′-phosphodiester bond is formed between this A residue and the 5′ terminal phosphate of the intron. Note that this adenylate residue is also joined to two other nucleotides by normal 3′,5′-phosphodiester bonds (Figure 33-37). Hence a *branch* is generated at this site and a *lariat intermediate* is formed.

The 3′-OH terminus of exon 1 then attacks the phosphodiester bond between the intron and exon 2. Exons 1 and 2 become joined, and the intron is released in lariat form. Note that splicing is accomplished by two *transesterification reactions* rather than by hydrolysis followed by ligation. The first reaction generates a free 3′-OH at the 3′ end of exon 1, and the second reaction links this group to the 5′-phosphate of exon 2. *The number of phosphodiester bonds stays the same during these steps.* Until the two exons are joined, the products of the first reaction are held together by a *spliceosome,* an assembly of ribonucleoproteins that recognizes the 5′ splice site, 3′ splice site, and branch site of mRNA precursors.

SMALL NUCLEAR RNAs (snRNAs) IN SPLICEOSOMES CATALYZE THE SPLICING OF mRNA PRECURSORS

The nucleus and cytosol contain many types of small RNA molecules with fewer than 300 nucleotides. They are called *snRNAs (small nuclear RNAs)* and *scRNAs (small cytoplasmic RNAs).* These RNA molecules are associated

Table 33-4
Small nuclear ribonucleoprotein particles (snRNPs) involved in the splicing of mRNA precursors

snRNP	*Size of snRNA (nucleotides)*	*Role*
U1	165	Binds the 5′ splice site and then the 3′ splice site
U2	185	Binds the branch site and forms part of the catalytic center
U5	116	Binds the 5′ splice site
U4	145	Masks the catalytic activity of U6
U6	106	Catalyzes splicing

with specific proteins to form complexes termed *snRNPs* (*s*mall *n*uclear *r*ibo*n*ucleoprotein *p*articles) and *scRNPs* (*s*mall *c*ytoplasmic *p*articles); investigators often speak of them as "snurps" and "scurps." Spliceosomes are large (60S), dynamic assemblies of snRNPs and mRNA precursors (Table 33-4). The key role of snRNPs in splicing was revealed by the finding that splicing is blocked by specific antibodies obtained from patients with *systemic lupus erythematosus*. In this autoimmune disease, complexes of antibodies and nuclear components lead to arthritis and kidney damage.

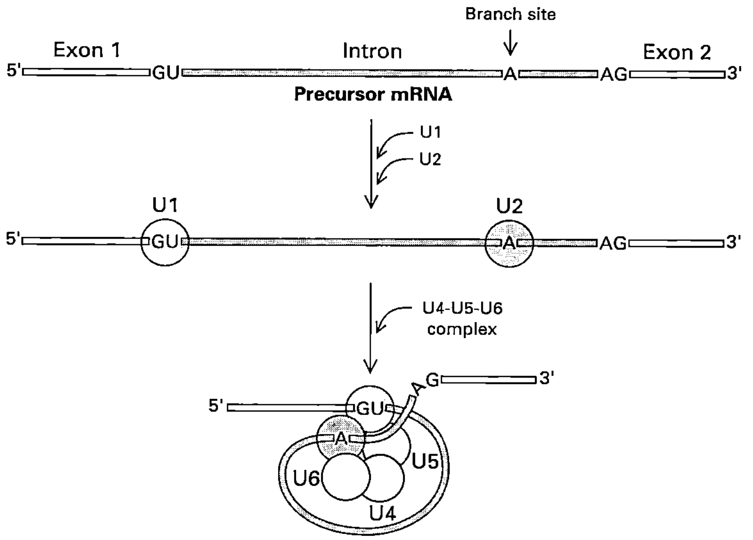

Figure 33-38
Pathway for spliceosome assembly. U1 (blue) binds to the 5′ splice site and U2 (red) to the branch site. A preformed U4-U5-U6 complex then joins this assembly to form a complete spliceosome.

In mammalian cells, splicing begins with the recognition of the 5′ splice site by U1 snRNP (Figure 33-38). In fact, U1 RNA contains a sequence that is complementary to this splice site, and the binding of U1 snRNP to an mRNA precursor protects a 15-nucleotide region at the 5′ splice site from digestion.

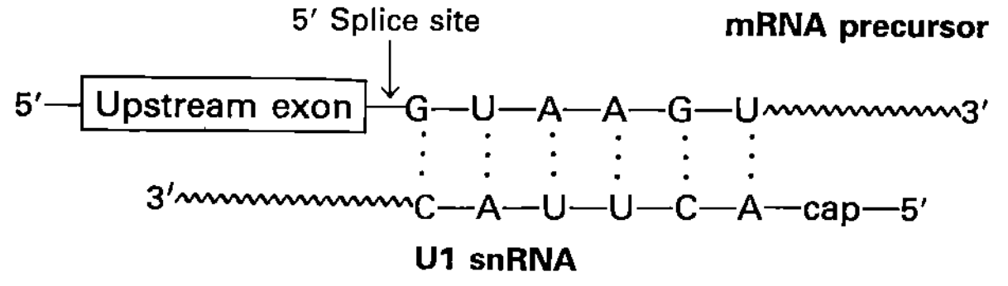

U2 snRNP then binds the branch site in the intron. The consequent association of U1 and U2 brings together the 5′ and 3′ ends of the intron, which enables U1 also to pair with the 3′ splice site. This complex of U1, U2, and the mRNA precursor is joined by a preassembled U4-U5-U6 complex to form a complete spliceosome.

A revealing view of the interplay of RNA molecules in this assembly came from the pattern of cross-links formed by *psoralen,* a photoactivable reagent that joins neighboring pyrimidines in base-paired regions. These cross-links suggest that splicing occurs in the following way. First, U5 interacts with exon sequences in the 5′ splice site. The next step is the ATP-driven melting of the base pairing between U1 and the 5′ splice site. U1 leaves the spliceosome, and U5 extends its pairing with the mRNA precursor to include the 5′ splice site. Disruption of the pairing between U4 and U6 then sets the stage for splicing by unleashing the catalytic activity of U6. U4 serves as an inhibitor that masks U6 until the specific splice sites are aligned.

U6, released from U4, now pairs with conserved intron sequences at the 5′ splice site. Furthermore, U6 becomes paired with U2, which in turn is paired with the branch point of the intron. *U2 and U6 snRNAs probably form the catalytic center of the spliceosome* (Figure 33-39). The 2′ hydroxyl of the branch point then attacks the 5′ splice site, and the newly formed 3′-OH of exon 1 attacks the 3′ splice site to join the two exons. U2, U5, and U6 bound to the excised lariat intron are released to complete the splicing reaction. Two features of this process are noteworthy. First, *RNA molecules play key roles in directing the alignment of splice sites and in carrying out catalysis.* Second, *ATP-powered proteins unwind RNA duplex intermediates and induce the release of ribonucleoproteins from mRNA precursors and products.*

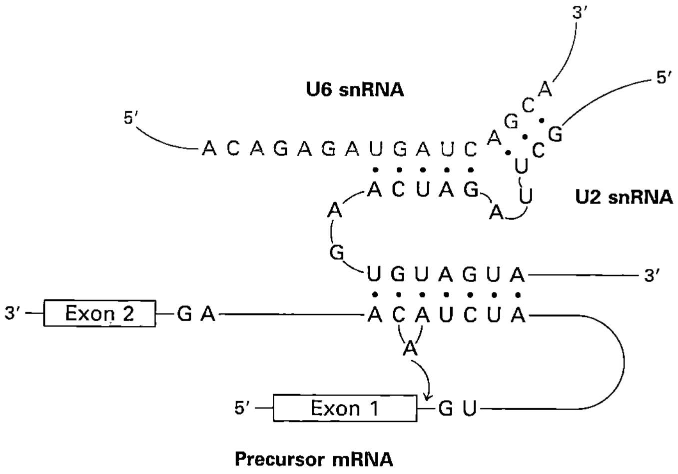

Figure 33-39
The catalytic center of the spliceosome is formed by U2 snRNA (red) and U6 snRNA (green), which are base-paired. U2 is also base-paired to the branch site of the mRNA precursor. [After H.D. Madhani and C. Guthrie. *Cell* 71(1992):803].

SELF-SPLICING RNA: THE DISCOVERY OF CATALYTIC RNA

Ribosomal RNA molecules in eukaryotes, like those of prokaryotes, are formed by the cleavage of primary transcripts. In *Tetrahymena* (a ciliated protozoan), a 414-nucleotide intron is then removed from a 6.4-kb precursor to yield the mature 26S rRNA molecule (Figure 33-40). Studies of this splicing reaction by Thomas Cech and his co-workers led to some very unexpected findings. They added cell extracts to the precursor RNA to determine the protein requirements for splicing. Surprisingly, they found that a control containing the precursor and nucleoside triphos-

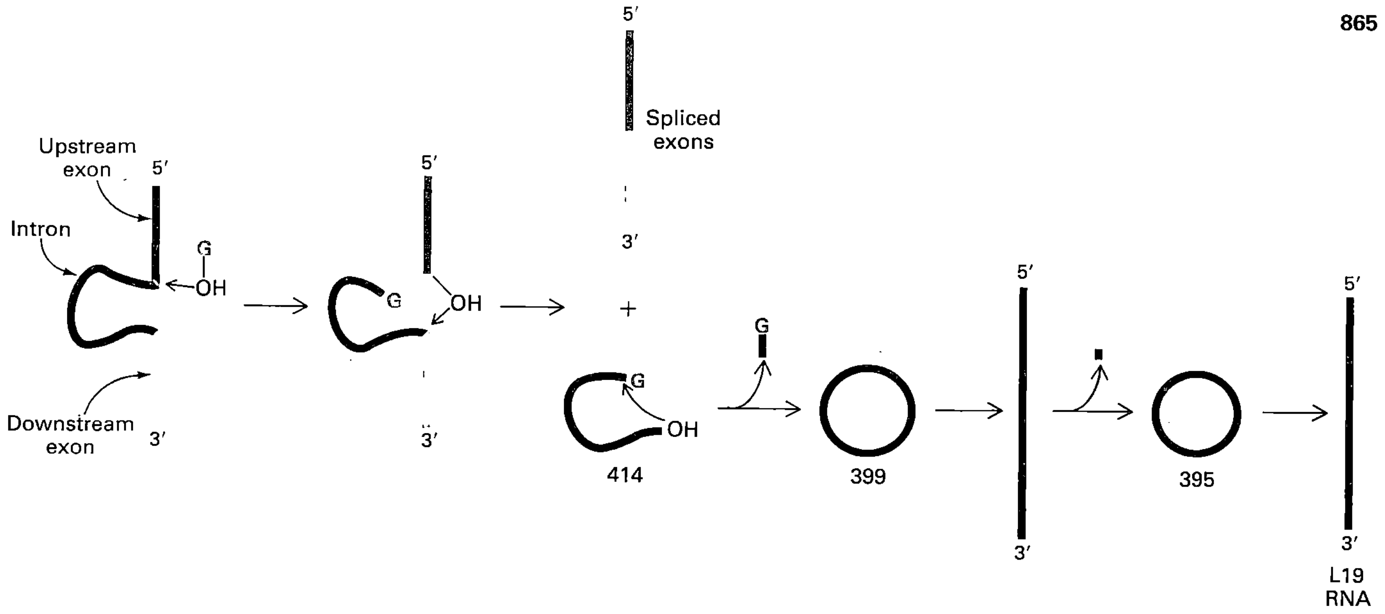

Figure 33-40
Self-splicing of a ribosomal RNA precursor from *Tetrahymena*. A 414-nucleotide intron (red) is released in the first splicing reaction in which guanosine or a guanyl nucleotide (green G) serves as a cofactor. This intron splices itself twice again to produce a linear RNA that has lost a total of 19 nucleotides. This L19 RNA is catalytically active. [After T. Cech. RNA as an enzyme. Copyright © 1986 by Scientific American, Inc. All rights reserved.]

phates but apparently devoid of protein underwent splicing. Could it be that an essential protein was not removed from the precursor RNA? This doubt was addressed by using recombinant DNA methods to prepare in *E. coli* the DNA corresponding to this 6.4-kb precursor (*E. coli* lacks this RNA molecule and cannot splice RNA). Purified DNA encoding this precursor was then transcribed in vitro to yield a synthetic RNA substrate for the splicing reaction. The result was the same: in the presence of nucleotides, the RNA spliced itself to precisely excise the 414-nucleotide intron. *This remarkable experiment demonstrated that an RNA molecule can have highly specific catalytic activity and splice itself in the absence of protein.*

Nucleotides were originally included in the reaction mixture because it was thought that ATP or GTP might be needed as an energy source. The required cofactor proved to be a guanosine unit, in the form of either guanosine, GMP, GDP, or GTP. G (denoting any one of these species) serves not as an energy source but as an attacking group that becomes transiently incorporated into the RNA (see Figure 33-40). G binds to the RNA and then attacks the 5′ splice site to form a phosphodiester bond with the 5′ end of the intron. This transesterification reaction generates a 3′-OH at the end of the upstream exon. The 3′ splice site is then attacked by the newly formed 3′-OH group of the upstream exon. This second transesterification reaction joins the two exons and leads to the release of the 414-nucleotide intron.

Two more rounds of self-splicing take place. The 3′-OH of the intron attacks a phosphodiester bond near the 5′ end to form a circle and a 15-nucleotide fragment containing the G that was incorporated in the first step. The 399-nucleotide circle opens into a linear molecule, which cyclizes to lose a 4-nucleotide fragment. This circle opens into a linear RNA called L − 19 IVS (*l*inear minus *19* *i*nter*v*ening *s*equence); we shall refer to it as L19 RNA.

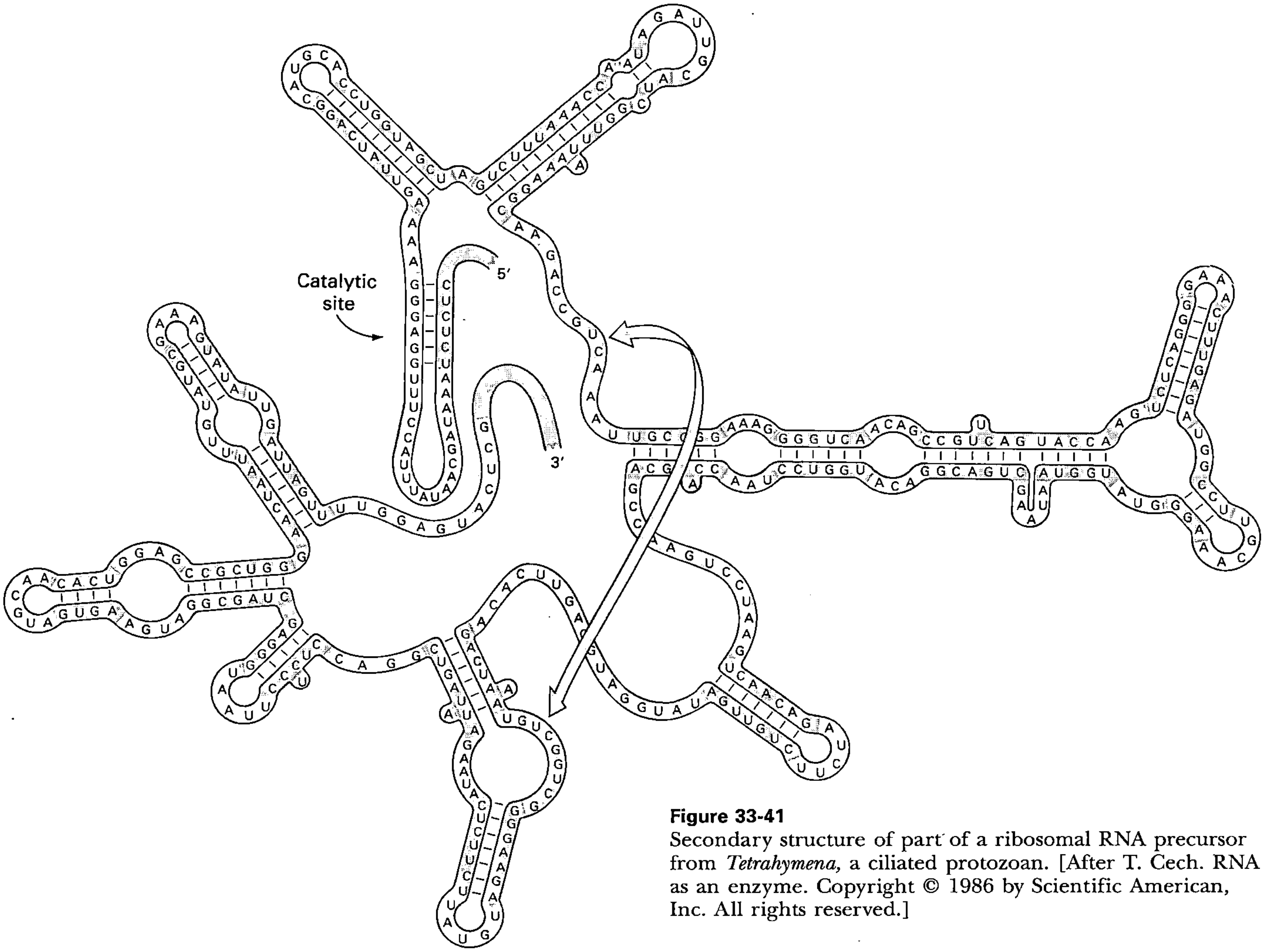

Figure 33-41
Secondary structure of part of a ribosomal RNA precursor from *Tetrahymena,* a ciliated protozoan. [After T. Cech. RNA as an enzyme. Copyright © 1986 by Scientific American, Inc. All rights reserved.]

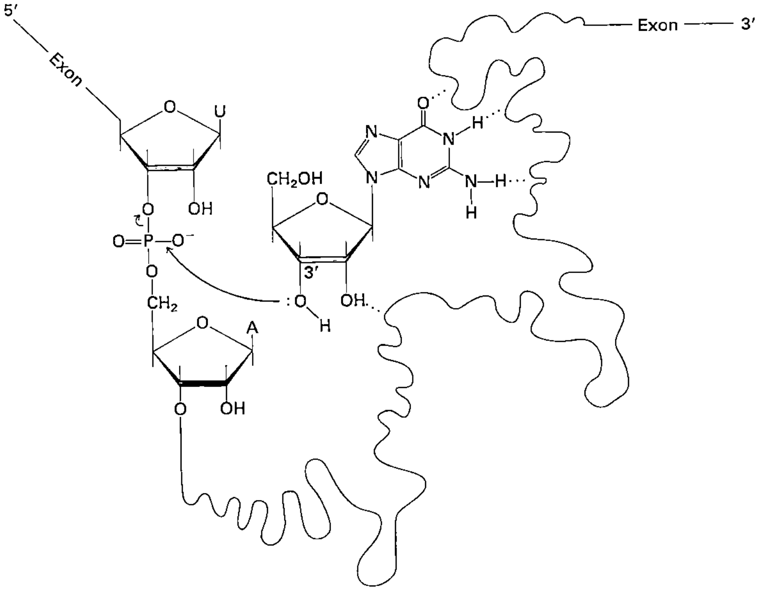

Figure 33-42
Proposed mode of binding of guanosine at the catalytic site of an RNA enzyme such as the self-splicing intron from *Tetrahymena.* [After B.L. Bass and T.R. Cech. *Nature* 308(1984):820.]

Self-splicing depends on the structural integrity of the rRNA precursor. Much of the intron is needed for self-splicing. This molecule, like many RNAs, has a folded structure formed by many double-helical stems and loops (Figure 33-41). The folded RNA contains weak GU base pairs in addition to the stronger AU and GC pairs; their approximate affinities are in the ratio of 1:100:1000. Splicing is blocked when secondary and tertiary structure are disrupted by the addition of denaturing agents such as dimethylformamide. Furthermore, the binding of G is saturable (K_M of 32 μM) and can be competitively inhibited. *These findings strongly imply that the precursor contains a specific binding pocket for G* (Figure 33-42).

Analysis of the base sequence of the rRNA precursor suggested that the 5′ splice site is aligned with the catalytic residues by base pairing between a *pyrimidine-rich region* (CUCUCU) of the upstream exon and a *purine-rich guide sequence* (GGGAGG) within the intron (Figure 33-43). The intron brings together the guanosine cofactor and 5′ splice site so that the 3′-OH of G can nucleophilically attack the phosphorus atom at this splice site. Another part of the intron then holds the downstream exon in position for attack by the newly formed 3′-OH of the upstream exon. A phosphodiester bond is formed between the two exons, and the intron is released as a linear molecule. Self-catalysis of bond formation and breakage in this rRNA precursor is highly stereospecific, like catalysis by protein enzymes.

The purine-rich guide sequence then binds another pyrimidine-rich sequence of the linear intron to form a circle and release a 15-nucleotide fragment. The circle opens, another pyrimidine-rich sequence binds to the guide sequence, and a new circle shortened by 4 nucleotides is produced. Finally, this 395-nucleotide intron opens to give L19 RNA, which is stable because no complementary pyrimidine-rich sequences are left.

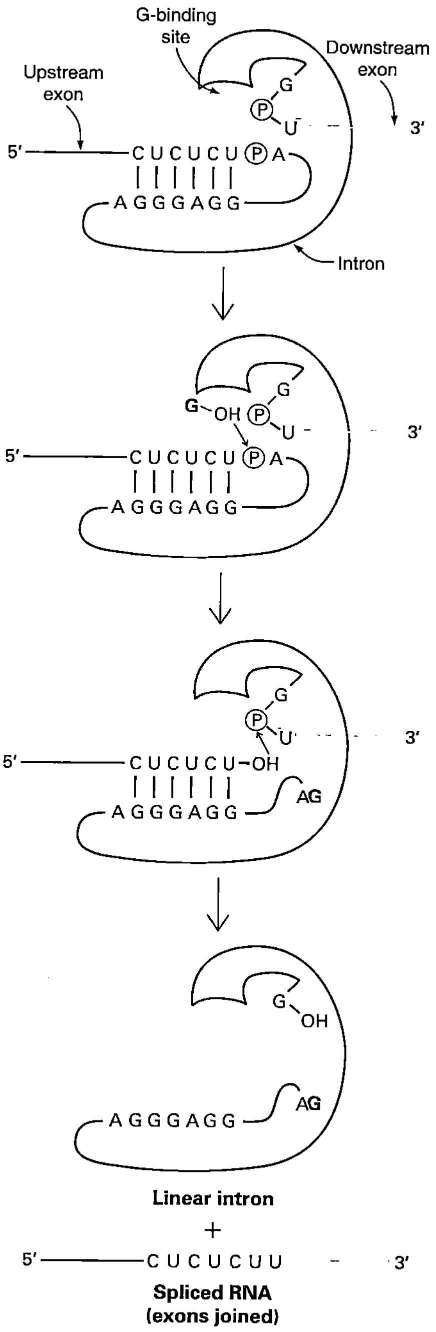

Figure 33-43
Postulated catalytic mechanism of the self-splicing intron from *Tetrahymena*. [After T. Cech. RNA as an enzyme. Copyright © 1986 by Scientific American, Inc. All rights reserved.]

L19 RNA IS BOTH A NUCLEASE AND A POLYMERASE

L19 RNA still possesses a G-binding site and a guide sequence. Cech reasoned that L19 RNA might act on external substrates. Indeed it does. As was discussed in an earlier chapter (p. 231), L19 RNA catalyzes the conversion of pentacytidylate (C_5) into longer and shorter oligomers. Thus, *L19 RNA is a true enzyme: it is both a nuclease and a polymerase.* The rate of hydrolysis of C_5 by this RNA enzyme (ribozyme) is about 10^{10} times the uncatalyzed rate, showing that RNA molecules can have great catalytic prowess. This discovery suggests that RNA at an early stage of evolution could have replicated itself without the participation of proteins.

How does a ribozyme cleave its substrate? Studies of the cleavage of a 13-mer RNA substrate have provided insight into the nature of the transition state (Figure 33-44).

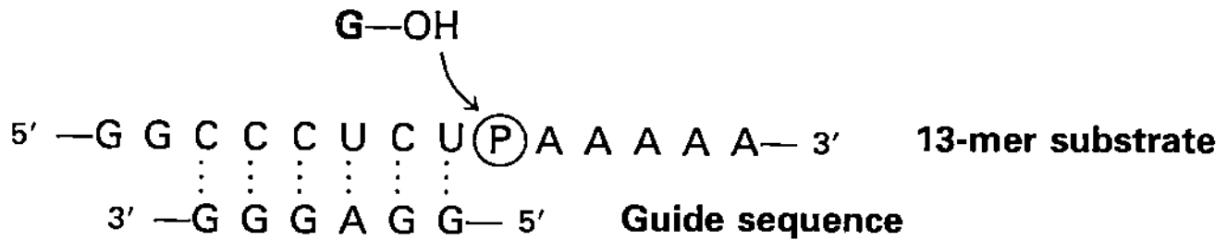

The attacking guanosine residue that becomes attached to A_5 is located in a pocket of the ribozyme, where it forms hydrogen bonds with two purines. In the transition state, *phosphorus is pentacovalent,* just as it is in most protein-catalyzed phosphoryl transfer reactions (e.g., hydrolysis of RNA by ribonuclease, see p. 218). One apex of the trigonal bipyramid is occu-

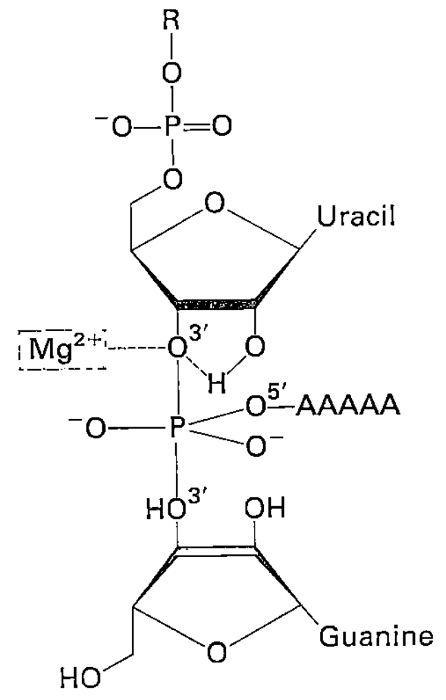

Figure 33-44
Proposed transition state in ribozyme catalysis. The pentacovalent phosphoryl group is shown in red, the attacking guanosine in green, and the leaving group in blue. Mg^{2+} (yellow) plays a key role in catalysis by stabilizing the negative charge on the 3′ oxygen atom of the leaving group. [After T.R. Cech, D. Herschlag, J.A. Piccirilli, and A.M. Pyle. *J. Biol. Chem.* 267(1992):17479.]

pied by the attacking oxygen atom, and the other by the departing oxygen atom. *Mg^{2+} plays an essential role in catalysis by this ribozyme by stabilizing the negative charge that develops on the 3′ oxygen atom of the leaving group.* Indeed, it seems likely that ribozymes generally draw on metal ions for catalysis because their functional groups are less suited for acid-base catalysis than are those of proteins.

HAMMERHEAD RNAs CONTAINING ONLY 43 NUCLEOTIDES ARE CATALYTICALLY ACTIVE

What is the minimum size of a catalytic RNA molecule? Studies of pathogenic RNAs from plants have led to the synthesis of ribozymes that are much smaller than the 395-nucleotide L19 RNA. *Viroids* are single-stranded circular RNA molecules about 300 nucleotides long that infect plants. The plant disease *cadang-cadang,* which has killed more than 30 million coconut palms in the Philippines, illustrates the destructive capacity of these small RNAs. Indeed, viroids are the smallest pathogens. They replicate by forming concatameric daughter strands in which multiple genomes are linked end to end. Plants do not contain an enzyme that can precisely cleave these tandem arrays to form monomer-length RNAs, and viroids do not appear to encode any proteins. How then are the concatamers cleaved? The answer is that *viroids undergo self-cleavage.* In this transesterification reaction, a 2′-3′-cyclic phosphodiester (rather than a 3′-OH, as in a hydrolytic reaction) is generated at one terminus, and a 5′-OH at the other.

Comparisons of nucleotide sequences in the vicinity of specific cleavage sites suggested that the active site has a *hammerhead* secondary structure consisting of three helical regions radiating from a central core of apparently unpaired nucleotides (Figure 33-45A). This model led to the synthesis of a hammerhead formed from a 19-nucleotide RNA and a 24-nucleotide RNA (Figure 33-45B). Indeed, the 24-nucleotide strand is cleaved at the expected site. Furthermore, multiple copies of the 24-nucleotide strand were cleaved by the 19-nucleotide strand when an excess of the 24-nucleotide RNA was added. Thus, *the 19-nucleotide strand of this synthetic hammerhead is the enzyme, and the 24-nucleotide strand is the sub-*

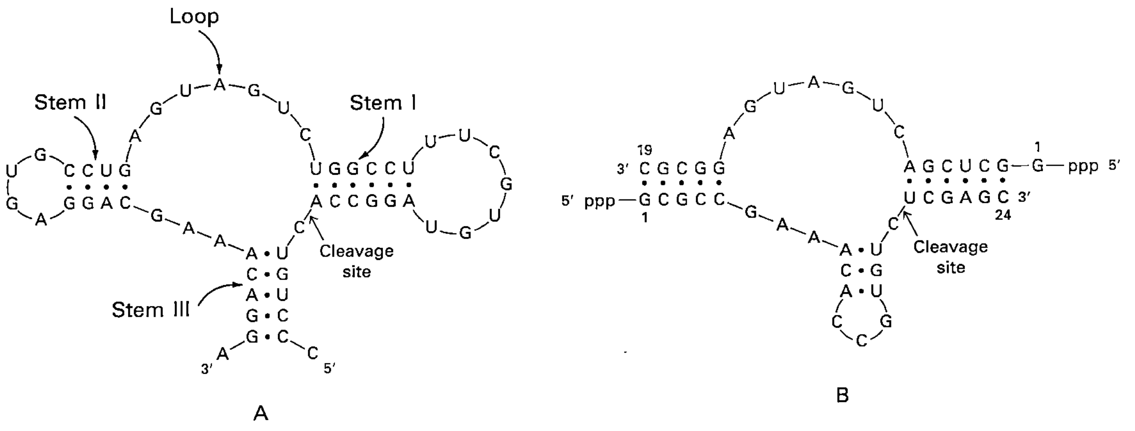

Figure 33-45
Many self-cleaving RNAs adopt a hammerhead conformation. (A) Self-cleaving sequence of tobacco ringspot virus satellite RNA. (B) A synthetic hammerhead catalytic RNA. The 19-nucleotide strand (green) is the enzyme, and the 24-nucleotide strand (red) is the substrate. The single cleavage site is marked. [After D.M. Long and O.C. Uhlenbeck. *FASEB J.* 25(1993):25.]

strate. The recent elucidation of the crystal structure of an RNA–DNA hammerhead (Figure 33-46) opens the door to a deeper understanding of these highly precise catalysts. It may be feasible to design ribozymes that specifically cut target RNAs to block the expression of particular genes or destroy the genomes of pathogenic RNA viruses.

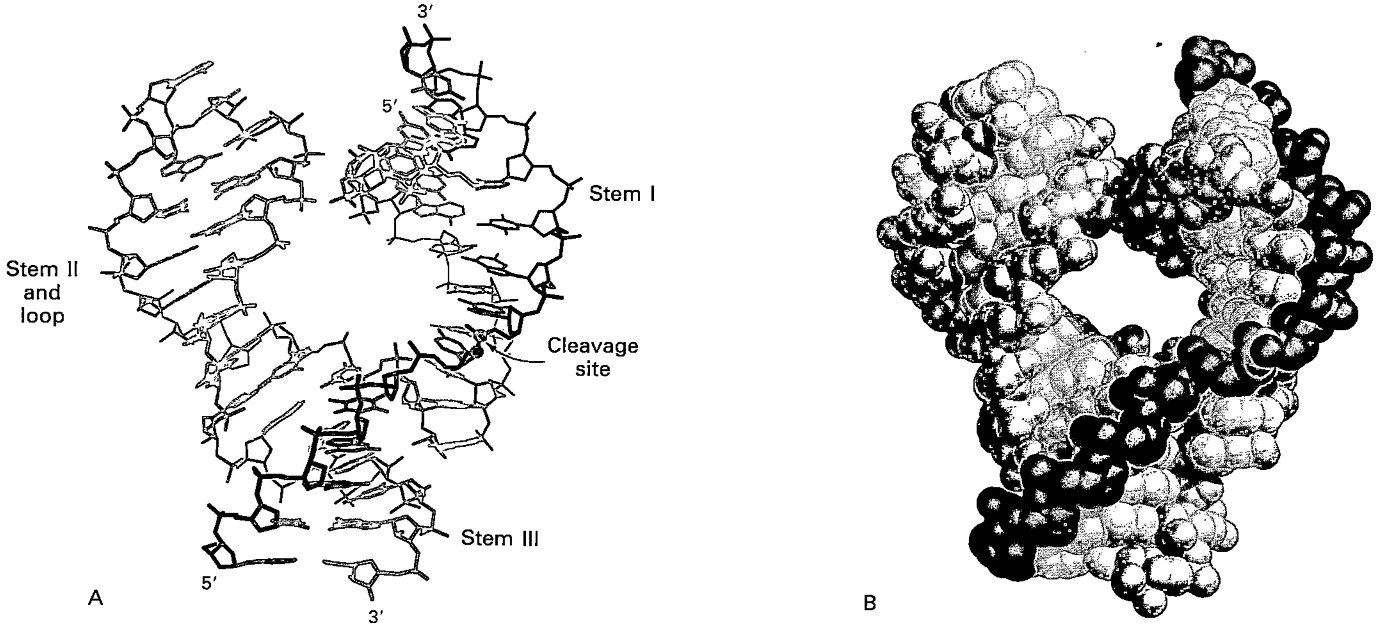

Figure 33-46
Three-dimensional structure of a hammerhead enzyme analog. The hammerhead does not undergo cleavage because the RNA substrate strand is replaced by a nonhydrolyzable DNA strand. The 34-nucleotide RNA strand is shown in green, and the 13-nucleotide DNA strand in red. (A) Skeletal model. (B) Space-filling model. [Courtesy of Heinz Pley, Kevin Flaherty, and Dr. David McKay.]

SPLICEOSOME-CATALYZED SPLICING MAY BE EVOLUTIONARILY DERIVED FROM SELF-SPLICING

Messenger RNA precursors in the mitochondria of yeast and fungi also undergo self-splicing, as do some RNA precursors in the chloroplasts of unicellular organisms such as *Chlamydomonas.* Self-splicing reactions can be classified according to the nature of the unit that attacks the upstream splice site. Group I self-splicing is mediated by a guanosine cofactor, as in *Tetrahymena.* The attacking moiety in group II splicing is the 2′-OH of a specific adenylate of the intron (Figure 33-47). Transfer RNA precursors are spliced by an entirely different mechanism, as was discussed earlier (p. 850).

Group I and group II self-splicing resemble spliceosome-catalyzed splicing in two respects. First, the initial step is an attack by a ribose hydroxyl group on the 5′ splice site. The newly formed 3′-OH terminus of the upstream exon then attacks the 3′ splice site to form a phosphodiester bond with the downstream exon. Second, both reactions are transesterifications in which the phosphate moieties at each splice site are retained in the products. The number of phosphodiester bonds stays constant. Group II splicing is like spliceosome-catalyzed splicing of mRNA precursors in two additional ways. The first cleavage is carried out by a part of the intron itself (the 2′-OH of A) rather than by an external cofactor (G). Furthermore, the intron is released in the form of a lariat.

Phillip Sharp has proposed that spliceosome-catalyzed splicing of mRNA precursors evolved from RNA-catalyzed self-splicing. Group II splicing may well be an intermediate between group I splicing and that occurring in the nuclei of higher eukaryotes. *A major step in this transition*

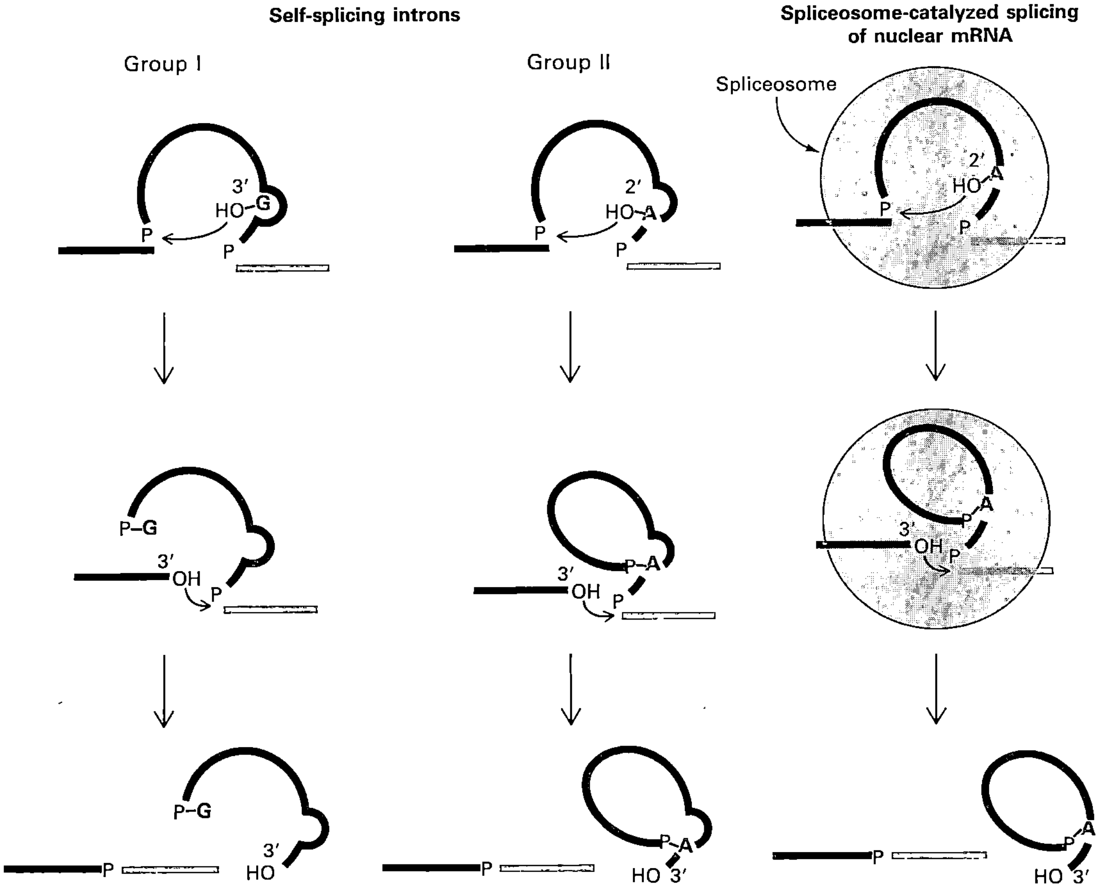

Figure 33-47
Comparison of self-splicing and spliceosome-catalyzed splicing. The exons being joined are shown in blue and yellow, and the attacking unit in green. The catalytic site is formed by the intron itself (red) in group I and II splicing. In contrast, the splicing of nuclear mRNA precursor is catalyzed by snRNPs in a spliceosome. [After P.A. Sharp. *Science* 235(1987):769.]

was the transfer of catalytic power from the intron itself to other molecules. The formation of spliceosomes gave introns a new freedom because they were no longer constrained to provide the catalytic center for splicing. Another advantage of external catalysts for splicing is that they can be more readily regulated. Our eyes have been opened to more of the underlying continuity between the early RNA world and contemporary life.

SUMMARY

All cellular RNA molecules are synthesized by RNA polymerases according to instructions given by DNA templates. The activated monomer substrates are ribonucleoside triphosphates. The direction of RNA synthesis is 5′ → 3′, as in DNA synthesis. RNA polymerases, unlike DNA polymerases, do not need a primer and do not possess proofreading nuclease activity. Another difference is that the DNA template is fully conserved in RNA synthesis, whereas it is semiconserved in DNA synthesis.

All RNA in *E. coli* is synthesized by one kind of RNA polymerase. The subunit composition of the ~500-kd holoenzyme is $\alpha_2\beta\beta'\sigma$ and that of the core enzyme is $\alpha_2\beta\beta'$. Transcription is initiated at promoter sites consisting of two sequences, one centered near −10 and the other near −35, that is, 10 and 35 nucleotides away from the start site in the 5′ (upstream) direction. The consensus sequence of the −10 region is TATAAT. The sigma subunit enables the holoenzyme to recognize promoter sites. When the temperature is raised, *E. coli* expresses a special σ that selectively binds the distinctive promoter of heat-shock genes. Likewise, *Bacillus subtilis* switches σ subunits when it sporulates. The binding of RNA polymerase to a promoter leads to a local unwinding of the bound DNA, which exposes some 17 bases on the template strand and sets the stage for the formation of the first phosphodiester bond. RNA chains usually start with pppG or pppA. The sigma subunit dissociates from the holoenzyme following initiation of the new chain. Elongation takes place at transcription bubbles that move along the DNA template at a rate of about 50 nucleotides per second. The nascent RNA chain contains stop signals that end transcription. One stop signal is an RNA hairpin followed by several U residues. A different stop signal is read by rho protein, an ATPase. In *E. coli,* transfer RNA and ribosomal RNA precursors are cleaved and chemically modified after transcription, whereas mRNA is used unchanged as a template for protein synthesis.

Eukaryotes contain in their nuclei three kinds of RNA polymerases: type I makes ribosomal RNA precursors, II makes mRNA precursors, and III makes transfer RNA precursors. Promoters for RNA polymerase II are located on the 5′ side of the start site for transcription. They consist of a TATA box centered at about −25 and additional upstream sequences usually located between −40 and −200. They are recognized by proteins called transcription factors rather than by RNA polymerase II. The binding sites of several transcription factors have been delineated by footprinting, a technique that identifies regions of template DNA that are protected from digestion or chemical modification. The activity of many promoters is greatly increased by enhancer sequences that have no promoter activity of their own. Enhancers can act over distances of several kilobases, and they can be located either upstream or downstream of a gene. The saddle-shaped TATA box–binding protein sharply bends DNA at TATA box sequences and serves as a focal point for the assembly of transcription complexes. The 5′ ends of mRNA precursors become capped by methylation during transcription. A 3′ poly(A) tail is added to most mRNA precursors after the nascent chain is cleaved by an endonuclease. RNA editing processes alter the nucleotide sequence of some mRNAs, such as the one for apolipoprotein B.

Splicing of mRNA precursors is carried out by spliceosomes, which consist of small nuclear ribonucleoprotein particles (snRNPs). Splice sites are specified by sequences at ends of introns and by a branch site near their 3′ end. The 2′-OH of an A in the branch site attacks the 5′ splice site to form a lariat intermediate. The newly generated 3′-OH terminus of the upstream exon then attacks the 3′ splice site to become joined to the downstream exon. The number of phosphodiester bonds stays constant during these two transesterification reactions. The active center of spliceosomes is formed by U2 and U6 small nuclear RNAs (snRNAs). Some RNA molecules, such as the 26S ribosomal RNA precursor from *Tetrahymena,* undergo self-splicing in the total absence of protein. Spliceosome-catalyzed splicing may have evolved from self-splicing. Some self-cleaved RNAs display a hammerhead secondary-structure pattern. Indeed, RNAs less than 50 nucleotides long can be effective enzymes. The discovery of catalytic RNA has opened new vistas in our exploration of molecular evolution.

SELECTED READINGS

Where to begin

Darnell, J.E., Jr., 1985. RNA. *Sci. Amer.* 253(4):68–78.

Sharp, P.A., 1994. Split genes and RNA splicing. *Cell* 77:805–815. [A lucid and beautifully illustrated Nobel Lecture.]

Cech, T.R., 1990. Self-splicing and enzymatic activity of an intervening sequence RNA from *Tetrahymena. Biosci. Rep.* 10:239–261. [A landmark discovery is recounted in this Nobel Lecture.]

Guthrie, C., 1991. Messenger RNA splicing in yeast: Clues to why the spliceosome is a ribonucleoprotein. *Science* 253:157–163.

Books

Lewin, B., 1994. *Genes* (5th ed.). Wiley. [Chapters 14, 16, and 30–32 give excellent accounts of RNA synthesis, processing, splicing, and catalysis.]

Kornberg, A., and Baker, T.A., 1992. *DNA Replication* (2nd ed.). W.H. Freeman. [Chapter 7 deals with RNA polymerases.]

Darnell, J., Lodish, H., and Baltimore, D., 1990. *Molecular Cell Biology* (2nd ed.). Scientific American Books. [Chapters 7 and 8 deal with RNA synthesis and processing. A very interesting discussion of evolutionary aspects of splicing is given in Chapter 26.]

Watson, J.D., Hopkins, N.H., Roberts, J.W., Steitz, J.A., Weiner, A.M., 1987. *Molecular Biology of the Gene* (4th ed.). Benjamin/Cummings. [Chapters 13, 18, and 21 deal with transcription and its control.]

Gesteland, R.F., and Atkins, J.F. (eds.), 1993. *The RNA World.* Cold Spring Harbor Laboratory Press. [This volume, subtitled *The Nature of Modern RNA Suggests a Prebiotic RNA World,* contains many informative and interesting articles on RNA splicing, catalytic RNA, and molecular evolution.]

RNA polymerases

Chamberlin, M., 1982. Bacterial DNA-dependent RNA polymerases. *In* Boyer, P.D. (ed.), *The Enzymes,* vol. 15, pp. 61–108. Academic Press.

Young, R.A., 1991. RNA polymerase II. *Ann. Rev. Biochem.* 60:689–715.

Lewis, M.K., and Burgess, R.R., 1982. Eukaryotic RNA polymerases. *In* Boyer, P.D. (ed.), *The Enzymes,* vol. 15, pp. 109–153. Academic Press.

Darst, S.A., Edwards, A.M., Kubalek, E.W., and Kornberg, R.D., 1991. Three-dimensional structure of yeast RNA polymerase II at 16 Å resolution. *Cell* 66:121–128.

Initiation and elongation

Conaway, R.C., and Conaway, J.W., 1993. General initiation factors for RNA polymerase II. *Ann. Rev. Biochem.* 62:161–190.

Erie, D.A., Yager, T.D., and von Hippel, H.P., 1992. The single-nucleotide addition cycle in transcription: A biophysical and biochemical perspective. *Ann. Rev. Biophys. Biomol. Struct.* 21:379–415.

Goodrich, J.A., and McClure, W.R., 1991. Competing promoters in prokaryotic transcription. *Trends Biochem. Sci.* 16:394–397.

Roeder, R.G., 1991. The complexities of eukaryotic transcription initiation: Regulation of preinitiation complex assembly. *Trends Biochem. Sci.* 16:402–408.

Daube, S.S., and von Hippel, H.P., 1992. Functional transcription elongation complexes from synthetic RNA–DNA bubble duplexes. *Science* 258:1320–1324.

Krummel, B., and Chamberlin, M.J., 1992. Structural analysis of ternary complexes of *Escherichia coli* RNA polymerase. Individual complexes halted along different transcription units have distinct and unexpected biochemical properties. *J. Mol. Biol.* 225:221–237.

Rice, G.A., Kane, C.M., and Chamberlin, M.J., 1991. Footprinting analysis of mammalian RNA polymerase II along its transcript: An alternative view of transcription elongation. *Proc. Nat. Acad. Sci.* 88:4245–4249.

Sobell, H.M., 1974. How actinomycin binds to DNA. *Sci. Amer.* 231(2):82–91.

Promoters, enhancers, and transcription factors

Kustu, S., North, A.K., and Weiss, D.S., 1991. Prokaryotic transcriptional enhancers and enhancer-binding proteins. *Trends Biochem. Sci.* 16:397–402.

Johnson, W., Moran, C., and Losick, R., 1983. Two RNA polymerase sigma factors from *Bacillus subtilis* discriminate between overlapping promoters for a developmentally regulated gene. *Nature* 302:800–804.

Greenblatt, J., 1991. RNA polymerase-associated transcription factors. *Trends Biochem. Sci.* 16:408–411.

Mitchell, P.J., and Tjian, R., 1989. Transcriptional regulation in mammalian cells by sequence-specific DNA binding proteins. *Science* 245:371–378.

Schleif, R., 1992. DNA looping. *Ann. Rev. Biochem.* 61:199–223.

Kim, Y., Geiger, J.H., Hahn, S., and Sigler, P.B., 1993. Crystal structure of a yeast TBP/TATA-box complex. *Nature* 365:512–520.

Kim, J.L., Nikolov, D.B., and Burley, S.K., 1993. Co-crystal structure of TBP recognizing the minor groove of a TATA element. *Nature* 365:520–527.

White, R.J., and Jackson, S.P., 1992. The TATA-binding protein: A central role in transcription by RNA polymerases I, II and III. *Trends Genet.* 8:284–288.

Termination and polyadenylation

von Hippel, H.P., and Yager, T.D., 1992. The elongation-termination decision in transcription. *Science* 255:809–812.

Das, A., 1993. Control of transcription termination by RNA-binding proteins. *Ann. Rev. Biochem.* 62:893–930.

Richardson, J.P., 1993. Transcription termination. *Crit. Rev. Biochem. Molec. Biol.* 28:1–30.

Sachs, A., and Wahle, E., 1993. Poly(A) tail metabolism and function in eucaryotes. *J. Biol. Chem.* 268:22955–22958.

RNA editing

Hodges, P., and Scott, J., 1992. Apolipoprotein B mRNA editing: A new tier for the control of gene expression. *Trends Biochem. Sci.* 17:77–81.

Hajduk, S.L., Harris, M.E., and Pollard, V.W., 1993. RNA editing in kinetoplastid mitochondria. *FASEB J.* 7:54–63.

Splicing of mRNA precursors

Nilsen, T.W., 1994. RNA–RNA interactions in the spliceosome: Unraveling the ties that bind. *Cell* 78:1–4.

Madhani, H.D., and Guthrie, C., 1992. A novel base-pairing interaction between U2 and U6 snRNAs suggests a mechanism for the catalytic activation of the spliceosome. *Cell* 71:803–817.

Wassarman, D.A., and Steitz, J., 1992. Interactions of small nuclear RNA's with precursor messenger RNA during in vitro splicing. *Science* 257:1918–1925.

Wolin, S.L., and Walther, P., 1991. Small ribonucleoproteins. *Curr. Opin. Struct. Biol.* 1:251–257.

Weiner, A.M., 1993. mRNA splicing and autocatalytic introns: Distant cousins or the products of chemical determinism? *Cell* 72:161–164.

Kim, S.H., and Lin, R.J., 1993. Pre-mRNA splicing within an assembled yeast spliceosome requires an RNA-dependent ATPase and ATP hydrolysis. *Proc. Nat. Acad. Sci.* 90:888–892.

Sharp, P.A., 1985. On the origin of RNA splicing and introns. *Cell* 42:397–400.

Green, M.R., 1991. Biochemical mechanisms of constitutive and regulated pre-mRNA splicing. *Ann. Rev. Cell Biol.* 7:559–599.

Self-splicing and RNA catalysis

Cech, T.R., 1986. RNA as an enzyme. *Sci. Amer.* 255(5):64–75.

Cech, T.R., 1990. Self-splicing of group I introns. *Ann. Rev. Biochem.* 59:543–568.

Uhlenbeck, O.C., and Long, D.M., 1993. Self-cleaving catalytic RNA. *FASEB J.* 7:25–30.

Symons, R.H., 1992. Small catalytic RNAs. *Ann. Rev. Biochem.* 61:641–671.

Pyle, A.M., 1993. Ribozymes: A distinct class of metalloenzymes. *Science* 261:709–714.

Cech, T.R., Herschlag, D., Piccirilli, J.A., and Pyle, A.M., 1992. RNA catalysis by a group I ribozyme. Developing a model for transition state stabilization. *J. Biol. Chem.* 267:17479–17482.

Herschlag, D., and Cech, T.R., 1990. Catalysis of RNA cleavage by the *Tetrahymena thermophila* ribozyme. 1. Kinetic description of the reaction of an RNA substrate complementary to the active site. *Biochemistry* 29:10159–10171.

Piccirilli, J.A., Vyle, J.S., Caruthers, M.H., and Cech, T.R., 1993. Metal ion catalysis in the *Tetrahymena* ribozyme reaction. *Nature* 361:85–88.

Christian, E.L., and Yarus, M., 1993. Metal coordination sites that contribute to structure and catalysis in the group I intron from *Tetrahymena. Biochemistry* 32:4475–4480.

Wang, J.F., Downs, W.D., and Cech, T.R., 1993. Movement of the guide sequence during RNA catalysis by a group I ribozyme. *Science* 260:504–508.

Pley, H.W., Flaherty, K.M., and McKay, D.B., 1994. Three-dimensional structure of a hammerhead ribozyme. *Nature* 372:68–74.

PROBLEMS

1. *Complements.* The sequence of part of an mRNA is

   5′-AUGGGGAACAGCAAGAGU
   GGGGCCCUGUCCAAGGAG-3′

   What is the sequence of the DNA coding strand? Of the DNA template strand?

2. *Potent inhibitor.* Heparin inhibits RNA polymerase by binding to its β' subunit. Why does heparin bind so effectively to β'?

3. *A loose cannon.* Sigma protein by itself does not bind to promoter sites. Predict the effect of a mutation enabling sigma to bind to the −10 and −35 regions in the absence of other subunits of RNA polymerase.

4. *Stuck sigma.* What would be the likely effect of a mutation that would prevent σ from dissociating from the RNA polymerase core?

5. *Transcription time.* What is the minimum length of time required for the synthesis by *E. coli* polymerase of an mRNA encoding a 100-kd protein?

6. *Between bubbles.* How far apart are transcription bubbles on *E. coli* genes that are being transcribed at a maximal rate?

7. *A revealing bubble.* Consider the synthetic RNA–DNA bubble shown in Figure 33-11. Let us refer to the coding DNA strand, the template strand, and the RNA strand as strands 1, 2, and 3, respectively.
   (a) Suppose that strand 3 is labeled with ^{32}P at its 5′ end and that polyacrylamide gel electrophoresis is carried out under nondenaturing conditions. Predict the autoradiographic pattern for (i) strand 3 alone, (ii) strands 1 and 3, (iii) strands 2 and 3, (iv) strands 1, 2, and 3, and (v) strands 1, 2, and 3, and core RNA polymerase.
   (b) What is the likely effect of rifampicin on RNA synthesis in this system?
   (c) Heparin blocks elongation of the RNA primer if it is added to core RNA polymerase prior to the onset of transcription but not if added after transcription starts. Account for this difference.

(d) Suppose that synthesis is carried out in the presence of ATP, CTP, and UTP. Compare the length of the longest product obtained with that expected when all four ribonucleoside triphosphates are present.

8. *An extra piece.* What is the amino acid sequence of the extra segment of protein synthesized in the thalassemic patient with a mutation leading to aberrant splicing (see Figure 33-35)? The reading frame after the splice site begins with TCT.

9. *A long-tailed messenger.* Another thalassemic patient had a mutation leading to the production of an mRNA for the β chain of hemoglobin that was 900 nucleotides longer than the normal one. The poly(A) tail of this mutant mRNA was located a few nucleotides after the only AAUAAA sequence in the additional sequence. Propose a mutation that would lead to the production of this altered mRNA.

10. *The right metal.* In the proposed transition state for catalysis by a ribozyme (see Figure 33-44), Mg^{2+} stabilizes the negative charge on the 3′ oxygen of the leaving group. If this oxygen is replaced by sulfur, Mg^{2+} is ineffective in catalyzing cleavage. However, cleavage readily occurs if Mn^{2+} is used in place of Mg^{2+}. Why?

11. *RNA editing.* Many U's are inserted into some mitochondrial mRNAs in trypanosomes. The uridines come from the poly(U) tail of a donor strand. Nucleoside triphosphates do not participate in this reaction. Propose a reaction mechanism that accounts for these findings. [Hint: relate RNA editing to RNA splicing.]

CHAPTER

34

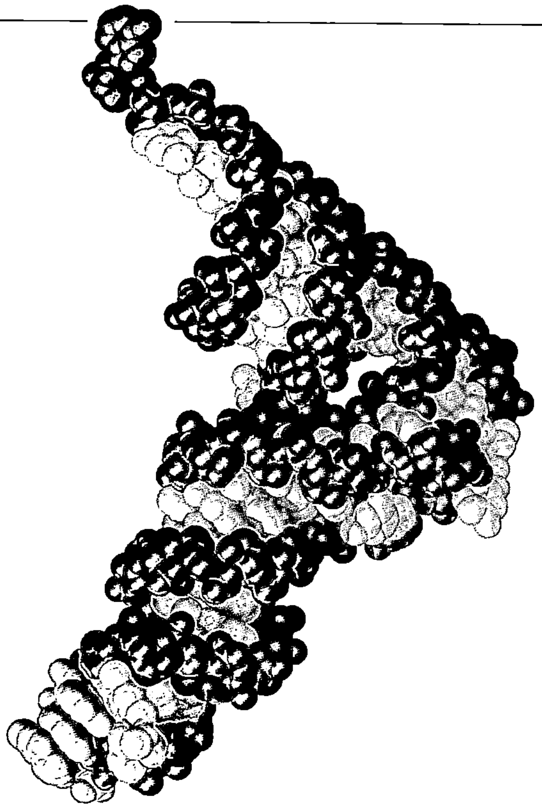

Protein Synthesis

In Chapter 5, we considered the flow of genetic information from DNA to RNA to protein and the nature of the genetic code. We turn now to the mechanism of protein synthesis, a process called *translation* because the four-letter alphabet of nucleic acids is translated into the entirely different alphabet of proteins. As might be expected, translation is a more complex process than either replication or transcription, which take place within the framework of a common base-pairing language. In fact, translation necessitates the coordinated interplay of more than a hundred macromolecules. Transfer RNA molecules, mRNA, and many proteins are required in addition to ribosomes. The focus of this chapter will be on protein synthesis in *Escherichia coli* because it illustrates many general principles and is well understood. Some distinctive features of protein synthesis in eukaryotes will also be presented.

Let us take an overview of protein synthesis before examining it in some detail. A protein is synthesized in the amino-to-carboxyl direction by the sequential addition of amino acids to the carboxyl end of the growing peptide chain. The activated precursors are aminoacyl-tRNAs, in which the carboxyl group of an amino acid is joined to the 3′-OH of a transfer RNA (tRNA). The linking of an amino acid to its corresponding

Opening Image: *Transfer RNAs, the central molecules of protein synthesis, serve as adaptors between the four-base language of nucleic acids and the twenty-amino acid language of proteins. The anticodon is shown in yellow and the attachment site for the amino acid in red. The RNA backbone is shown in dark blue and the bases in light blue. [Drawn from 6tna.pdb. J.L. Sussman, S.R. Holbrook, R.W. Warrant, G.M. Church, and S.-H. Kim. J. Mol. Biol. 123(1978):607.]*

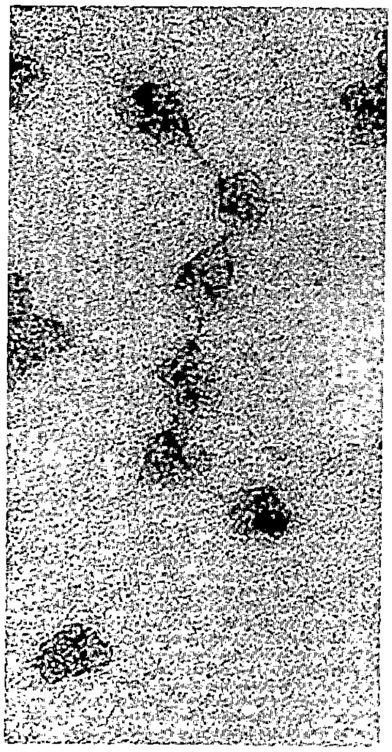

Figure 34-1
Electron micrograph of ribosomes on an mRNA molecule. Ribosomes are template-directed catalysts of peptide-bond formation. [Courtesy of Dr. Alex Rich.]

tRNA is catalyzed by an aminoacyl-tRNA synthetase. This activation reaction, which is analogous to the activation of fatty acids, is driven by ATP. For each amino acid, there is at least one kind of tRNA and activating enzyme.

Protein synthesis takes place in three stages: initiation, elongation, and termination. *Initiation* results in the binding of the initiator tRNA to the start signal of mRNA. The initiator tRNA occupies the P (peptidyl) site on a ribosome. *Elongation* starts with the binding of an aminoacyl-tRNA to the A (aminoacyl) site, a distinct tRNA-binding site on the ribosome. A peptide bond then forms between the amino group of the incoming aminoacyl-tRNA and the carboxyl group of the formylmethionine carried by the initiator tRNA. The resulting dipeptidyl-tRNA then moves from the A site to the P site, and the initiator tRNA molecule moves to the E (exit) site before leaving the ribosome. The binding of aminoacyl-tRNA, the movement of peptidyl-tRNA, and the associated movement of the ribosome to the next codon are powered by the hydrolysis of GTP. An aminoacyl-tRNA then binds to the vacant A site to start another round of elongation, which proceeds as described above. *Termination* occurs when a stop signal on the mRNA is read by a protein release factor, which leads to the release of the completed polypeptide chain from the ribosome. *Ribosomes, in essence, are enzymes that catalyze mRNA-directed formation of peptide bonds.*

Ribosomes are large ribonucleoprotein assemblies consisting of a small (30S) and a large (50S) subunit. They contain 55 different proteins and 3 RNA molecules. The name *ribosome* is apt because two-thirds of the mass is RNA. Indeed, ribosomal RNA molecules are information-rich molecules that play a central role in translation. Their action in contemporary ribosomes provides clues as to how primitive ribosomes consisting only of RNA may have carried out protein synthesis.

TRANSFER RNA (tRNA) MOLECULES HAVE A COMMON DESIGN

We begin our consideration of protein synthesis with tRNA because it serves as the adaptor molecule that recognizes both the enzyme that attaches the correct amino acid and the anticodon on mRNA. The base sequence of a tRNA molecule was first determined by Robert Holley in 1965, the culmination of seven years of effort. Indeed, his study of yeast alanine tRNA provided the first complete sequence of any nucleic acid and suggested how tRNA functions. This adaptor molecule is a single chain of 76 ribonucleotides (Figure 34-2). The 5′ terminus is phosphorylated (pG), whereas the 3′ terminus has a free hydroxyl group. A striking feature of this RNA is its high content of bases other than A, U, G, and C. Many unusual nucleosides are present: inosine (I, p. 887), pseudouridine (ψ, p. 850), dihydrouridine (UH_2, p. 878), ribothymidine (T, p. 850), and methylated derivatives of guanosine and inosine. The *amino acid attachment site* is the 3′-hydroxyl group of the adenosine residue at the 3′ terminus of the molecule. The sequence IGC in the middle of the molecule is the *anticodon.* It is complementary to GCC, one of the codons for alanine.

Anticodon

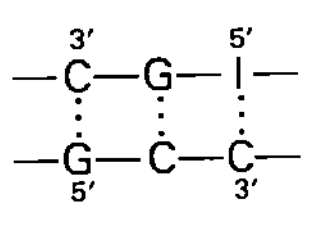

Codon

The sequences of several other tRNA molecules were determined a short time later. More than 100 sequences are now known. The striking finding is that all of them can be written in a cloverleaf pattern in which about half the residues are base-paired. Hence, *tRNA molecules have many common structural features.* This finding is not unexpected, because all

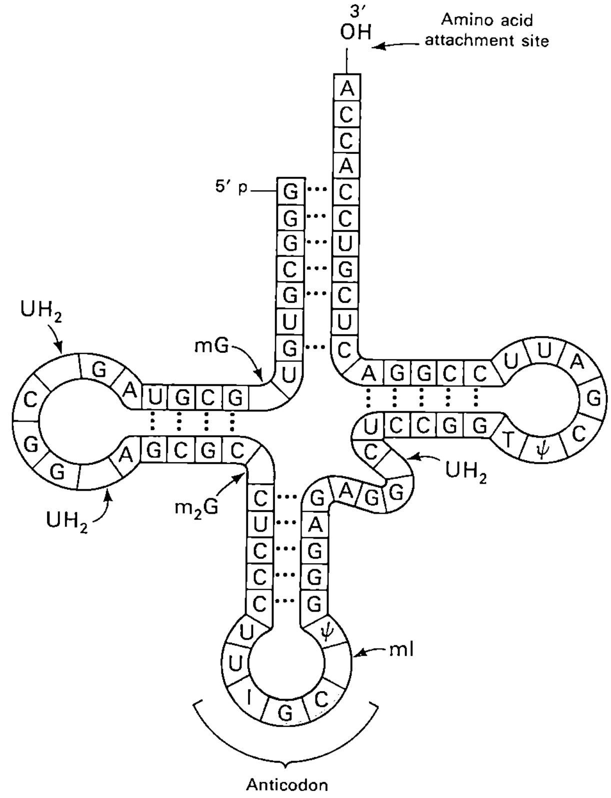

Figure 34-2
Base sequence of yeast alanine tRNA. Modified nucleosides (shown in yellow) are abbreviated as follows: methylinosine (mI), dihydrouridine (UH_2), ribothymidine (T), pseudouridine (ψ), methylguanosine (mG), and dimethylguanosine (m_2G). Inosine (I), another modified nucleoside, is part of the anticodon (green).

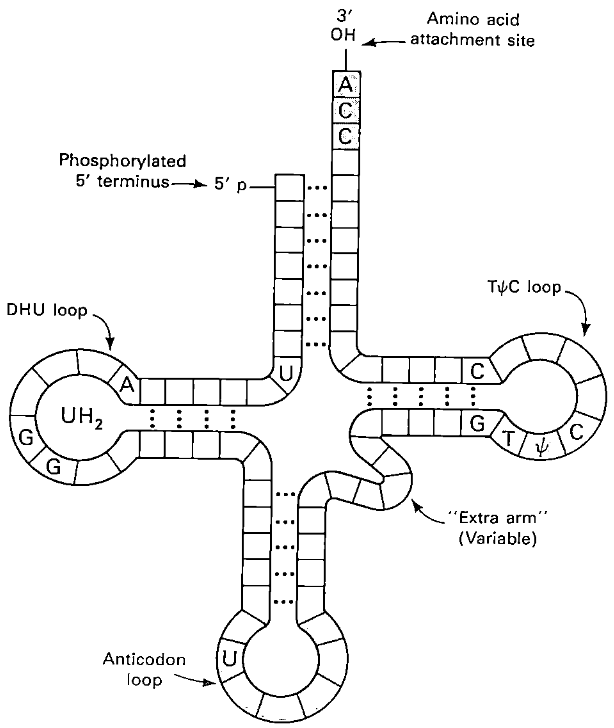

Figure 34-3
Common features of tRNA molecules.

tRNA molecules must be able to interact in nearly the same way with ribosomes, mRNAs, and elongation factors. For example, they must fit into the A, P, and E sites on the ribosome and interact with the enzymatic site that catalyzes peptide-bond formation.

All known transfer RNA molecules share the following features (Figure 34-3):

1. They are single chains containing between *73 and 93 ribonucleotides* (~25 kd).

2. They contain *many unusual bases,* typically between 7 and 15 per molecule. Some are methylated or dimethylated derivatives of A, U, C, and G that are formed by enzymatic modification of a precursor tRNA (p. 850). Methylation prevents the formation of certain base pairs, thereby rendering some of the bases accessible for other interactions. Also, methylation imparts a hydrophobic character to some regions of tRNAs, which may be important for their interaction with synthetases and ribosomal proteins and for their folding. Other modifications alter codon recognition, as will be discussed shortly.

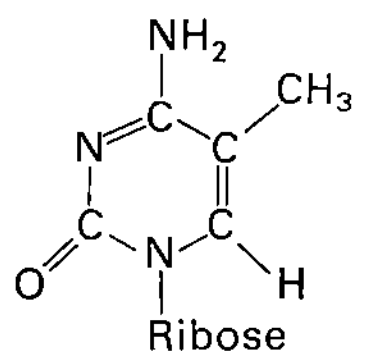

5-Methylcytidine

3. The 5′ end of tRNAs is phosphorylated. The 5′ terminal residue is usually pG.

4. The base sequence at the 3′ end of mature tRNAs is CCA. The activated amino acid is attached to the 3′-hydroxyl group of the terminal adenosine.

5. About half the nucleotides in tRNAs are base-paired to form double helices. Five groups of bases are not base paired: the 3′ *CCA terminal region;* the *TψC loop,* which acquired its name from the sequence ribothymine-pseudouracil-cytosine; the "*extra arm,*" which contains a variable number of residues; the *DHU loop,* which contains several dihydrouracil residues; and the *anticodon loop.*

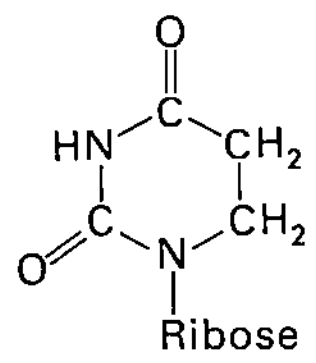

Dihydrouridine

6. The *anticodon loop* consists of seven bases, with the following sequence:

5′ —Pyrimidine—Pyrimidine—[X—Y—Z]—Modified purine—Variable base— 3′

THE ACTIVATED AMINO ACID AND ANTICODON OF tRNA ARE AT OPPOSITE ENDS OF THE L-SHAPED MOLECULE

The three-dimensional structure of a tRNA molecule was first solved in 1974. X-ray crystallographic studies of yeast phenylalanine tRNA carried out in the laboratories of Alexander Rich and Aaron Klug provided a wealth of structural information:

1. The molecule is *L-shaped* (Figure 34-4).

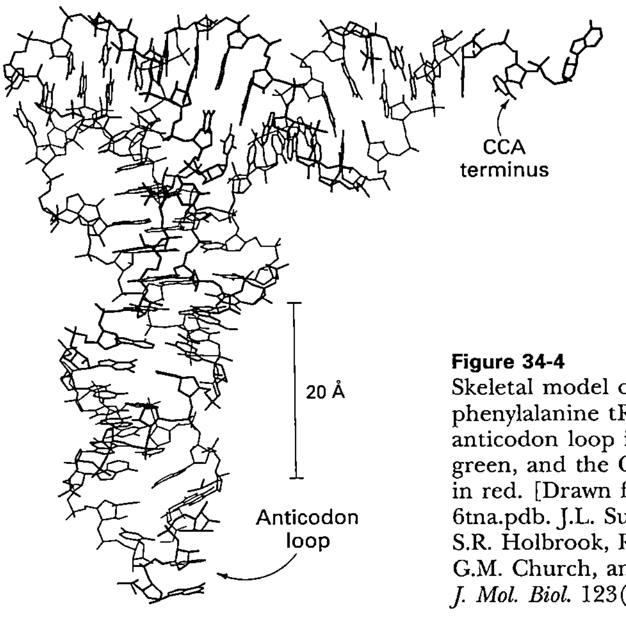

Figure 34-4
Skeletal model of yeast phenylalanine tRNA. The anticodon loop is shown in green, and the CCA terminus in red. [Drawn from 6tna.pdb. J.L. Sussman, S.R. Holbrook, R.W. Warrant, G.M. Church, and S.-H. Kim. *J. Mol. Biol.* 123(1978):607.]

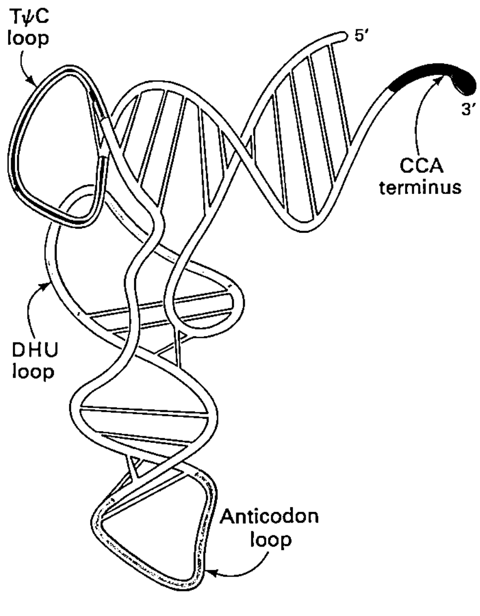

Figure 34-5
Schematic diagram of the three-dimensional structure of yeast phenylalanine tRNA. [After a drawing kindly provided by Dr. Sung-Hou Kim.]

2. There are *two segments of double helix.* They are like A-DNA, as expected for an RNA duplex (p. 790). Each of these helices contains about 10 base pairs, which correspond to one turn of helix. The helical segments are perpendicular to each other, giving the molecule its L shape (Figure 34-5). The base pairing in the cloverleaf model (p. 877), postulated on the basis of sequence studies, is correct.

3. Most of the bases in the nonhelical regions participate in unusual hydrogen-bonding interactions. These *tertiary interactions* are between bases that are not usually complementary (e.g., GG, AA, and AC). More-

over, the ribose-phosphate backbone interacts with some bases and even with another region of the backbone itself. The 2′–OH groups of the ribose units act as hydrogen bond donors or acceptors in many of these interactions. In addition, most bases are stacked. These hydrophobic interactions between adjacent aromatic rings play a major role in stabilizing the architecture of the molecule.

4. The CCA terminus containing the *amino acid attachment site* is at one end of the L. The other end of the L is the *anticodon loop*. Thus, *the amino acid in aminoacyl-tRNA is far from the anticodon* (about 80 Å). The DHU and TψC loops form the corner of the L.

5. The CCA terminus and the adjacent helical region do not interact strongly with the rest of the molecule. This part of the molecule may change its conformation during amino acid activation and also during protein synthesis on the ribosome.

Subsequent x-ray analyses of other prokaryotic and eukaryotic tRNAs have shown that their molecular architecture follows the same plan as that of yeast phenylalanine tRNA. As will be discussed shortly, crystallographic studies have also revealed how tRNAs interact with synthetases when amino acids become attached to their 3′ CCA terminus.

MULTIPLE TRANSFER RNA MOLECULES ARISE FROM CLEAVAGE OF A LARGE PRECURSOR BY RIBONUCLEASE P

The 60 genes for transfer RNA molecules in *E. coli* are clustered in 25 units that are transcribed into multimeric precursors. Some units encode ribosomal RNAs as well, whereas others contain only tRNAs. One of these transcripts is the precursor of seven tRNAs: one specific for leucine, two for internal methionines, and two each for two kinds of glutamine codons (Figure 34-6). The primary transcript of 950 nucleotides is cleaved by *ribonuclease P* (RNase P) on the 5′ side of the first nucleotide of each mature tRNA-to-be. *Ribonuclease D* (RNase D) then trims the exposed 3′ end of each precursor until it reaches the CCA sequence, which becomes the 3′ terminus of the mature molecule.

Ribonuclease P is a ribonucleoprotein enzyme comprising a 377-nucleotide M1 RNA molecule and a 20-kd protein. In 1983, Sidney Altman discovered that the RNA component alone possesses enzymatic activity. At higher than physiologic concentrations of Mg^{2+} (~60 mM), *M1 RNA by itself recognizes target sites in primary transcripts and cleaves them at an appreciable rate. This experiment showed that M1 RNA interacts specifically with the substrate and possesses the catalytic groups.* The role of the protein is subsidiary—it increases the hydrolytic rate and enables the reaction to occur at a much lower concentration of Mg^{2+}. We see here another example of *catalytically active RNA* and catch a glimpse of the early RNA world.

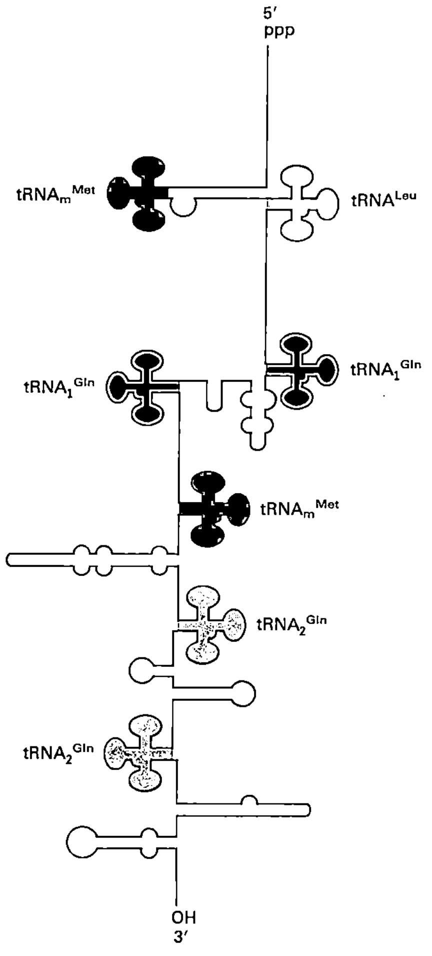

Figure 34-6
Seven tRNA molecules are formed by cleavage of this 950-nucleotide primary transcript. [After N. Nakajima, H. Ozeki, and Y. Shimura. *Cell* 23(1981):245.]

AMINO ACIDS ARE ACTIVATED AND LINKED TO PARTICULAR TRANSFER RNAs BY SPECIFIC SYNTHETASES

The formation of a peptide bond between the amino group of one amino acid and the carboxyl group of another is thermodynamically unfavorable. This barrier is overcome by activating the carboxyl group of the precursor amino acids. *The activated intermediates in protein synthesis are amino acid esters,* in which the carboxyl group of an amino acid is linked to

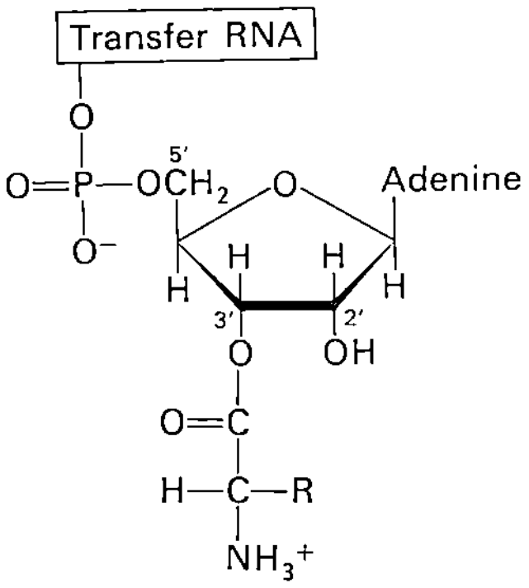

Aminoacyl-tRNA

Figure 34-7
An amino acid is esterified to the 2′- or the 3′-hydroxyl group of the terminal adenosine in an aminoacyl-tRNA.

either the 2′- or the 3′-hydroxyl group of the ribose unit at the 3′ end of tRNA. In some tRNAs, the activated amino acid migrates very rapidly between the 2′- and 3′-hydroxyl groups. An amino acid ester of tRNA is called an *aminoacyl-tRNA* (Figure 34-7); it is sometimes called a *charged tRNA*.

The attachment of an amino acid to a tRNA is important not only because it activates the carboxyl group, but also because amino acids by themselves cannot recognize the codons on mRNA. Rather, amino acids must be carried to the ribosomes by specific tRNAs, which do recognize codons on mRNA and thereby act as adaptor molecules.

In 1957, Paul Zamecnik and Mahlon Hoagland discovered that the activation of amino acids and their subsequent linkage to tRNAs are catalyzed by specific *aminoacyl-tRNA synthetases,* which are also called *activating enzymes.* The first step is the formation of an *aminoacyl-adenylate* from an amino acid and ATP. This activated species is a mixed anhydride in which the carboxyl group of the amino acid is linked to the phosphoryl group of AMP; hence, it is also known as *aminoacyl-AMP.*

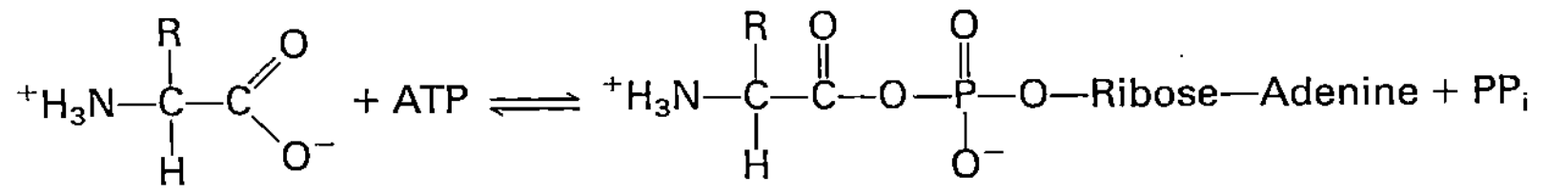

Aminoacyl adenylate
(Aminoacyl-AMP)

The next step is the transfer of the aminoacyl group of aminoacyl-AMP to a tRNA molecule to form *aminoacyl-tRNA.*

$$\text{Aminoacyl-AMP} + \text{tRNA} \rightleftharpoons \text{aminoacyl-tRNA} + \text{AMP}$$

The sum of these activation and transfer steps is

$$\text{Amino acid} + \text{ATP} + \text{tRNA} \rightleftharpoons \text{aminoacyl-tRNA} + \text{AMP} + \text{PP}_i$$

The $\Delta G^{\circ\prime}$ of this reaction is close to 0, because the free energy of hydrolysis of the ester bond of aminoacyl-tRNA is similar to that of the terminal phosphoryl group of ATP. What then drives the synthesis of aminoacyl-tRNA? As expected, the reaction is driven by the hydrolysis of pyrophosphate. The sum of these three reactions is highly exergonic:

$$\text{Amino acid} + \text{ATP} + \text{tRNA} + H_2O \longrightarrow \text{aminoacyl-tRNA} + \text{AMP} + 2\ \text{P}_i$$

Thus, *two ~P are consumed in the synthesis of an aminoacyl-tRNA.* One of them is consumed in forming the ester linkage of aminoacyl-tRNA, whereas the other is consumed in driving the reaction forward.

The activation and transfer steps for a particular amino acid are catalyzed by the same aminoacyl-tRNA synthetase. In fact, *the aminoacyl-AMP intermediate does not dissociate from the synthetase.* Rather, it is tightly bound to the active site of the enzyme by noncovalent interactions. The aminoacyl-AMP is normally a transient intermediate in the synthesis of aminoacyl-tRNA, but it is quite stable and readily isolated if tRNA is absent from the reaction mixture.

We have already encountered an acyl adenylate intermediate in fatty acid activation (p. 607). In fact, Paul Berg first discovered this intermediate in fatty acid activation, and then recognized that it is also formed in amino acid activation. The major difference between these reactions is that the acceptor of the acyl group is CoA in the former and tRNA in the latter. The energetics of these biosyntheses are very similar: both are made irreversible by the hydrolysis of pyrophosphate.

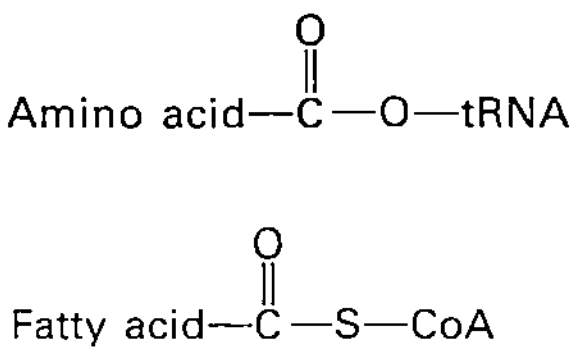

AMINOACYL-tRNA SYNTHETASES BELONG TO TWO STRUCTURAL CLASSES

At least one aminoacyl-tRNA synthetase exists for each amino acid. The diverse sizes, subunit composition, and sequences of these enzymes were bewildering for many years. However, recent studies have shown that synthetases can be grouped into two classes, termed *class I* and *class II*, according to the presence of short signature sequences. The synthetases for 10 of the basic set of 20 amino acids belong to class I enzymes, and those for the other 10 to class II (Table 34-1). It is interesting to note that the smaller amino acids are generally activated by class II synthetases, whereas the larger amino acids and also the more hydrophobic ones are activated by class I enzymes. The class II enzymes may be the more ancient ones. Another difference is that class I enzymes acylate the 2′-hydroxyl group of the terminal adenosine of tRNA, whereas class II enzymes (except the one for Phe) acylate the 3′-hydroxyl. The amino acid activating domain of the two classes is also different. Class I enzymes have a parallel β domain (the classical dinucleotide binding fold, p. 498), whereas class II enzymes have an antiparallel β domain. They also differ in how they recognize tRNA (p. 886).

Table 34-1
Classification and subunit structure of aminoacyl-tRNA synthetases in *E. coli*

Class I	*Class II*
Arg (α)	Ala (α_4)
Cys (α)	Asn (α_2)
Gln (α)	Asp (α_2)
Glu (α)	Gly ($\alpha_2\beta_2$)
Ile (α)	His (α_2)
Leu (α)	Lys (α_2)
Met (α)	Phe ($\alpha_2\beta_2$)
Trp (α_2)	Ser (α_2)
Tyr (α_2)	Pro (α_2)
Val (α)	Thr (α_2)

TYROSYL-AMP FORMATION IS GREATLY ACCELERATED BY THE BINDING OF γ PHOSPHATE IN THE TRANSITION STATE

X-ray crystallographic and protein engineering studies have provided insight into the catalytic action of *E. coli* tyrosyl-tRNA synthetase, a class I dimer of 47-kd subunits. The amino-terminal 320 residues are needed for the activation reaction, whereas the carboxyl-terminal 99 residues participate in the binding of tRNA and the formation of tyrosyl-tRNA. The crystal structure of the synthetase containing bound tyrosyl-AMP has been solved at high resolution. This activated intermediate is stable in the absence of the matching tRNA and is bound to the enzyme by some 12 hydrogen bonds (Figure 34-8).

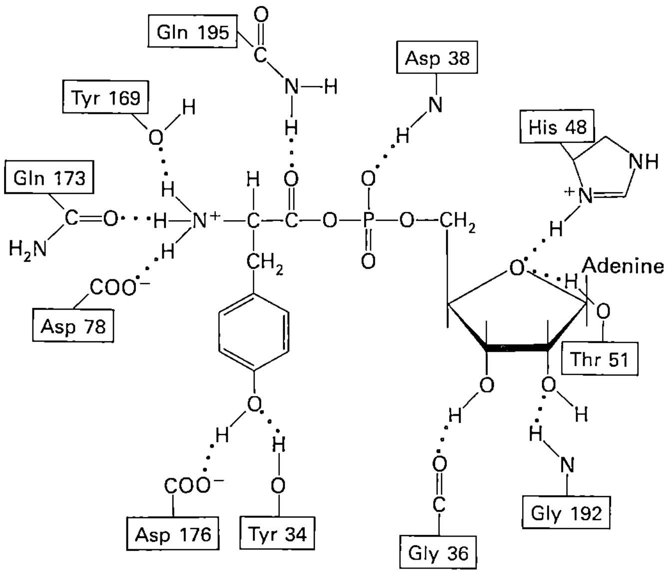

Figure 34-8
Tyrosyl-AMP (shown in red) is bound to its synthetase by multiple hydrogen bonds. [Courtesy of Dr. David Blow.]

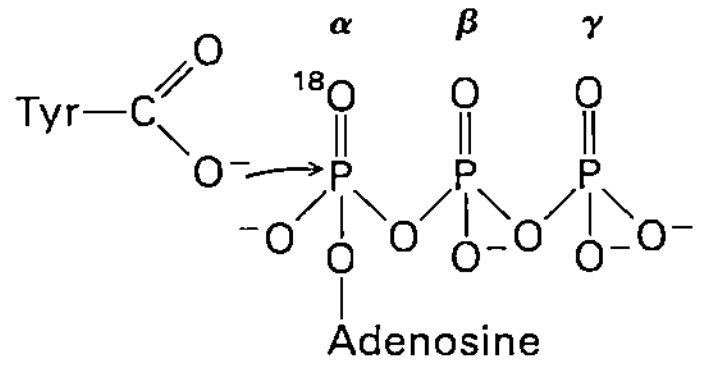

The formation of tyrosyl-AMP from tyrosine and ATP stereospecifically labeled with ^{18}O at the α phosphoryl group leads to an *inversion* of configuration. Hence, the reaction probably proceeds by an *in-line displacement* in which the tyrosyl carboxylate is the attacking nucleophile and pyrophosphate is the leaving group. The α phosphorus atom in the transition state is pentacovalent and has the geometry of a trigonal bipyramid, as in the hydrolytic reaction catalyzed by ribonuclease (p. 218) and the self-splicing of a ribozyme (p. 868).

A plausible structure of the transition state was deduced by model building (Figure 34-9). A key feature is the hydrogen bonding of the γ phosphate group to the side chains of threonine 40 and histidine 45.

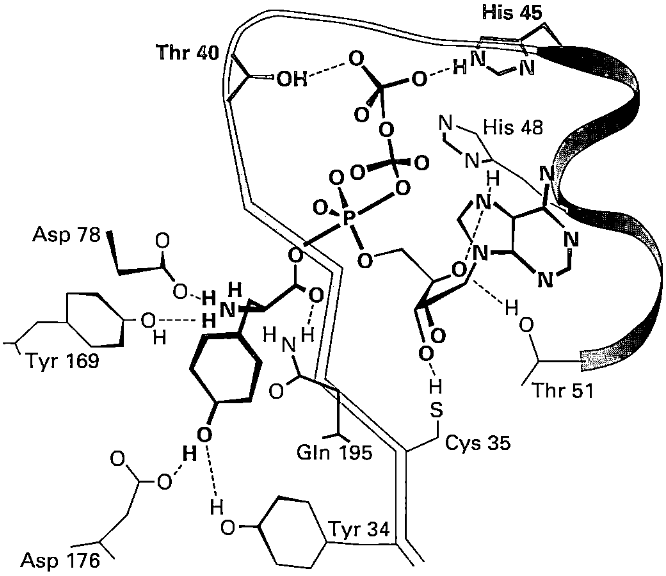

Figure 34-9
Proposed structure of the transition state in the formation of tyrosyl-AMP from tyrosine and ATP (shown in red). The side chains of threonine 40 and histidine 45 (shown in green) play a critical role in catalysis by selectively binding the γ phosphate in the transition state. [After R.J. Leatherbarrow, A.R. Fersht, and G. Winter. *Proc. Nat. Acad. Sci.* 82(1985):7841.]

Their importance was assessed by mutating both to alanine, to eliminate the possibility of hydrogen bonding. Indeed, the catalytic activity of the double mutant was less than that of the wild type by a factor of 3×10^6, whereas the binding affinities of the enzyme for ATP and tyrosine were altered relatively little. This experiment shows that threonine 40 and histidine 45 are essential for catalysis but not for substrate binding (Figure 34-10). They probably interact strongly with the γ phosphate in the transition state but not in the initial enzyme-substrate complex. The large change in position of the pyrophosphate unit accompanying the shift from a tetrahedral to a bipyramidal geometry triggers this selective binding. Recall that *the essence of catalysis is selective stabilization of the transition state* (p. 188).

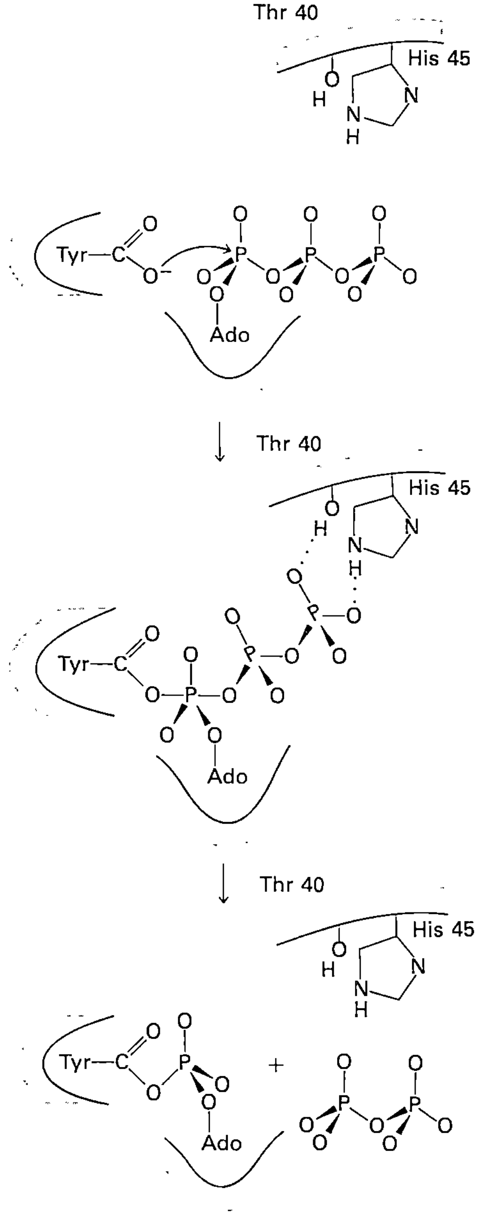

Figure 34-10
Proposed mechanism for the formation of tyrosyl-AMP by tyrosyl-tRNA synthetase. The α phosphorus atom of ATP is nucleophilically attacked by a carboxylate oxygen of tyrosine. The transition state is stabilized by hydrogen bonding of the γ phosphate to the side chains of a threonine and a histidine residue. Pyrophosphate bound to Mg^{2+} (not shown) leaves the pentacovalent transition state to give tyrosyl-AMP. Ado denotes adenosine. [After R.J. Leatherbarrow, A.R. Fersht, and G. Winter. *Proc. Nat. Acad. Sci.* 82(1985):7841.]

Which groups on the enzyme directly participate in making and breaking bonds? Probably none. The carboxylate group of tyrosine is an intrinsically effective nucleophile, ATP is already activated, and Mg^{2+}–pyrophosphate is a good leaving group. The enzyme accelerates catalysis by a factor of about 4×10^4 simply by bringing tyrosine and ATP together, and it gains another factor of 3×10^5 mainly by binding γ phosphate in the transition state. It is noteworthy that the new interactions in the transition state are at some distance from the α phosphorus atom, the reaction center. Thus, *catalysis can be delocalized*—what counts is selective binding of the transition state, however achieved.

PROOFREADING BY AMINOACYL-tRNA SYNTHETASES INCREASES THE FIDELITY OF PROTEIN SYNTHESIS

Aminoacyl-tRNA synthetases are highly selective in their recognition of both the amino acid to be activated and the prospective tRNA acceptor. As will be discussed shortly, tRNA molecules that accept different amino acids have different base sequences, and so they can be readily distinguished by their synthetases. A much more demanding task for these enzymes is to discriminate between similar amino acids. For example, the only difference between isoleucine and valine is that isoleucine contains a methylene group not present in valine (Figure 34-11). The additional binding energy (~−3 kcal/mol) contributed by this extra $-CH_2-$ group favors the activation of isoleucine over valine by isoleucyl-tRNA synthetase by a factor of about 200. Even so, the concentration of valine in vivo is about five times that of isoleucine, and so valine would be mistakenly incorporated into proteins in place of isoleucine 1 in 40 times. However, the observed error frequency in vivo is only 1 in 3000, indicating that there must be a subsequent editing step to enhance fidelity.

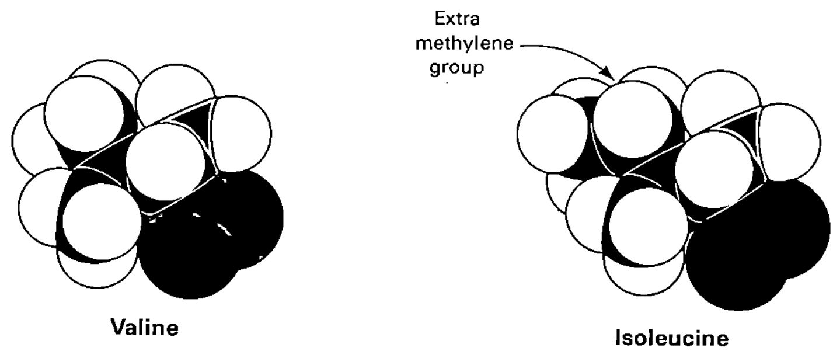

Figure 34-11
Molecular models of valine and isoleucine. The extra methylene group in isoleucine is marked. The synthetases specific for these amino acids are highly discerning.

In fact, *the synthetase corrects its own errors.* Mistakenly activated valine is not transferred to tRNA specific for isoleucine. Instead, *this tRNA promotes the hydrolysis of valine–AMP and thereby prevents its erroneous incorporation into proteins* (Figure 34-12). Furthermore, hydrolysis frees the synthetase to activate and transfer isoleucine, the correct amino acid. How does the synthetase avoid hydrolyzing isoleucine–AMP, the correct intermediate? Most likely, the hydrolytic site is just large enough to accommodate valine–AMP but too small to allow the entry of isoleucine–AMP.

How do synthetases distinguish between valine and threonine, a pair of amino acids that are very nearly identical in size? Choosing between these

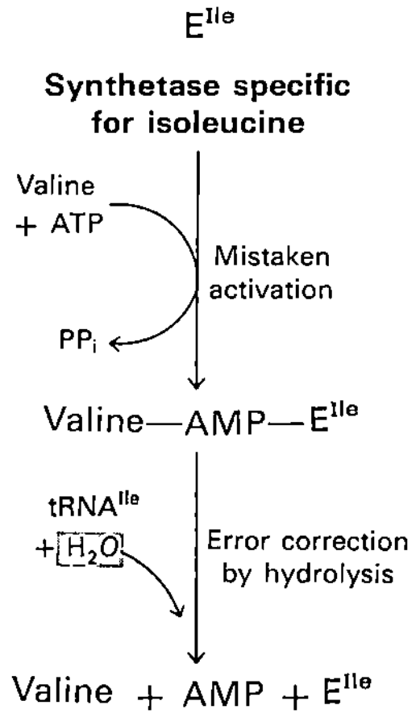

Figure 34-12
Proofreading by hydrolysis of an incorrect aminoacyl-AMP. The entry of tRNA specific for isoleucine induces hydrolysis of valyl-AMP.

amino acids is demanding because threonine differs from valine only in having an $-OH$ in place of a $-CH_3$ group. The aminoacyl-tRNA synthetase for valine contains two adjacent catalytic sites, one for the acylation of tRNA and the other for hydrolysis of incorrectly acylated tRNA (Figure 34-13). Valine is preferred over threonine in the acylation reaction because the acylation site is hydrophobic. By contrast, threonyl-tRNA is hydrolyzed much more rapidly than valyl-tRNA because the hydrolytic site is hydrophilic. The synthetase for valine does most of its editing at the level of aminoacyl-tRNA, whereas the one for isoleucine does so at the level of aminoacyl-AMP.

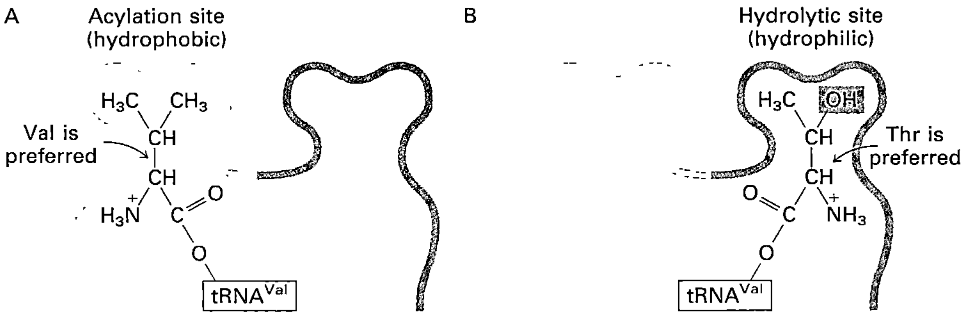

Figure 34-13
Valyl-tRNA synthetase rejects threonine at two stages. (A) The hydrophobic acylation site for the transfer of activated amino acid to tRNA prefers valine over threonine because valine is more hydrophobic. (B) The adjacent hydrolytic site, by contrast, is hydrophilic. Hence, threonine erroneously linked to the tRNA is preferentially hydrolyzed because the binding energy of the hydroxyl group of threonine is used to stabilize the transition state. [After A. Fersht. *Enzyme Structure and Mechanism,* 2nd ed. (W.H. Freeman, 1984), p. 356.]

Most aminoacyl-tRNA synthetases contain hydrolytic sites in addition to acylation sites. Complementary pairs of sites function as a *double sieve* to assure very high fidelity. The acylation site rejects amino acids that are *larger* than the correct one because there is insufficient room for them, whereas the hydrolytic site destroys activated intermediates that are *smaller* than the correct species. Hydrolytic proofreading is central to the fidelity of many aminoacyl-tRNA synthetases, as it is to DNA polymerases (p. 800). However, a few synthetases achieve high accuracy without editing their covalently attached intermediates. For example, tyrosyl-tRNA synthetase has no difficulty discriminating between tyrosine and phenylalanine; the hydroxyl group on the tyrosine ring enables it to be bound to the enzyme 10^4 times as strongly as phenylalanine. *Proofreading is costly in energy and time and hence is selected in the course of evolution only when fidelity must be enhanced beyond what can be obtained through an initial binding interaction.*

SYNTHETASES RECOGNIZE THE ANTICODON LOOP AND ACCEPTOR STEM OF TRANSFER RNA MOLECULES

How do synthetases choose their tRNA partners? Precise recognition of tRNAs is equally important for high-fidelity protein synthesis as is accurate selection of amino acids. A priori, the anticodon of tRNA would seem to be a good choice for establishing its identity because each tRNA has a different one. Indeed, *some synthetases recognize their tRNA partner primarily on the basis of its anticodon.* The most direct evidence comes from *identity swap* experiments. The CCA anticodon of tryptophanyl-tRNA (abbreviated $tRNA^{Trp}$) and the UAC anticodon of $tRNA^{Val}$ were replaced with CAU, the anticodon for methionine (Figure 34-14A). The genes for these altered tRNAs were transcribed in vitro, and the kinetics of acylation by the synthetase specific for methionine were measured. Changing the anticodon of $tRNA^{Trp}$ and $tRNA^{Val}$ to CAU increases the rate of aminoacylation with methionine more than 50,000-fold. In fact, $tRNA^{Val}$

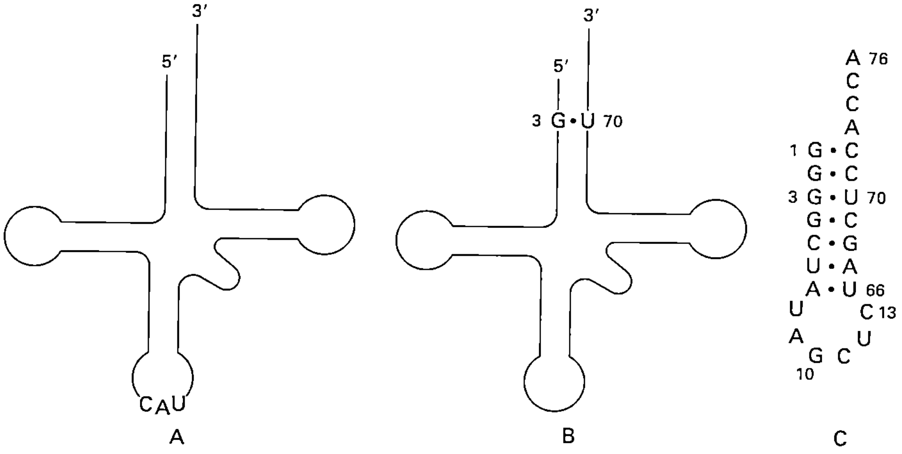

Figure 34-14
Determinants of the identity of tRNA molecules in recognition by synthetases. (A) The identity of $tRNA^{Val}$ is changed from Val to Met by mutating its anticodon from UAC to CAU. (B) The identity of $tRNA^{Cys}$ is changed from Cys to Ala by changing its 3:70 base pair from C · G to G · U. (C) A "microhelix" containing only 24 of the 76 nucleotides of $tRNA^{Ala}$ is recognized by the cognate synthetase.

bearing the CAU anticodon is aminoacylated by methionyl-tRNA synthetase at nearly the same rate as is $tRNA^{Met}$.

In contrast, the anticodon of $tRNA^{Ala}$ can be changed with no effect on the aminoacylation specificity. Rather, *the crucial determinant is the presence of a G · U base pair at the 3:70 position in the 3′ acceptor stem of this 76-nucleotide molecule* (Figure 34-14B). Cysteine tRNA differs from $tRNA^{Ala}$ at 40 positions and contains a C · G base pair at the 3:70 position. When this C · G base pair is changed to G · U, $tRNA^{Cys}$ is recognized by alanyl-tRNA synthetase as though it were $tRNA^{Ala}$. This finding raised the question as to whether a fragment of tRNA suffices for aminoacylation by alanine-tRNA synthetase. Indeed, *a "microhelix" containing just 24 of the 76 nucleotides of the native tRNA is specifically aminoacylated by the synthetase.* This microhelix contains only the 3′ acceptor stem and a hairpin loop (Figure 34-14C).

The identities of other tRNAs are conferred by multiple determinants. For example, *both the anticodon and acceptor stems of $tRNA^{Gln}$ are recognized by glutaminyl-tRNA synthetase.* X-ray crystallographic studies of a complex of this synthetase, $tRNA^{Gln}$, and ATP have provided a revealing view of how this enzyme recognizes its tRNA partner (Figure 34-15). This highly

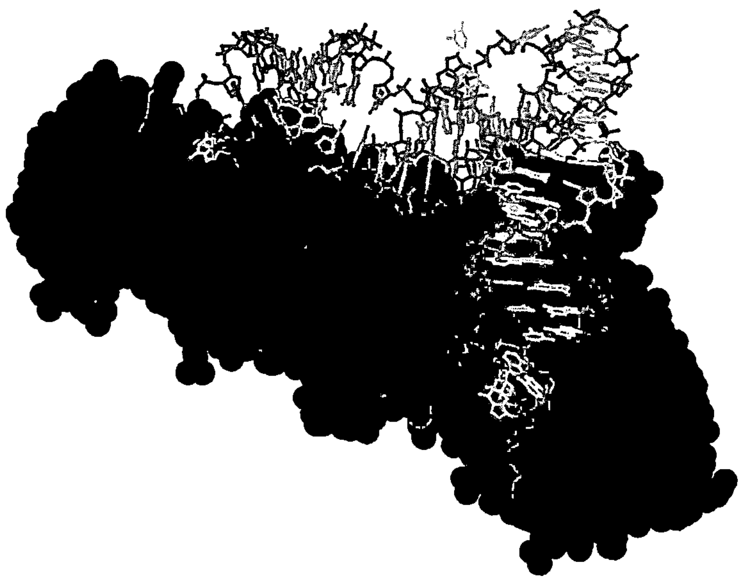

Figure 34-15
Structure of a tRNA bound to its cognate synthetase. The sugar-phosphate backbone of $tRNA^{Gln}$ is shown in red and the bases in yellow. The solvent-accessible surface of glutaminyl-tRNA synthetase is depicted in blue. ATP (green) is also bound to the synthetase. [After M.A. Rould, J.J. Perona, D. Söll, and T.A. Steitz. *Science* 246(1989):1135.]

elongated class I synthetase interacts with tRNA along the entire inside of the L, from the anticodon to the 3′ acceptor end. Yeast aspartyl-tRNA synthetase, a class II enzyme, also binds a face of tRNA extending from the anticodon to the 3′ end. However, different sides of tRNA are recognized by these enzymes.

THE CODON IS RECOGNIZED BY THE ANTICODON OF tRNA RATHER THAN BY THE ACTIVATED AMINO ACID

It has already been mentioned that the anticodon on tRNA is the recognition site for the codon on mRNA and that recognition occurs by base pairing. Does the amino acid attached to the tRNA play a role in this recognition process? This question was answered in the following way. First, cysteine was attached to its cognate tRNA. The attached cysteine unit was then converted into alanine by reacting Cys-tRNA^{Cys} with Raney nickel, which removed the sulfur atom from the activated cysteine residue without affecting its linkage to tRNA. Thus, a *hybrid aminoacyl-tRNA* was produced in which alanine was covalently attached to a tRNA specific for cysteine.

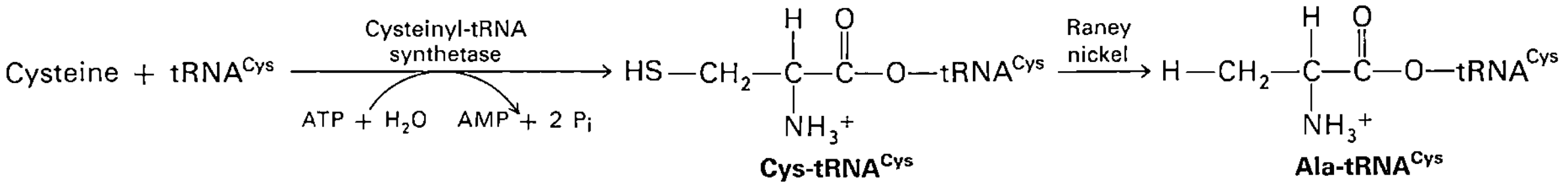

Does this hybrid tRNA recognize the codon for alanine or for cysteine? The answer came on adding the tRNA to a cell-free protein-synthesizing system. The template was a random copolymer of U and G in the ratio of 5:1, which normally leads to the incorporation of cysteine (encoded by UGU) but not of alanine (encoded by GCX). However, alanine was incorporated into a polypeptide when Ala-tRNA^{Cys} was added to the incubation mixture, because it was attached to the tRNA specific for cysteine. The same result was obtained when mRNA for hemoglobin served as the template and ^{14}C alanyl-tRNA^{Cys} was used as the hybrid aminoacyl-tRNA. The only radioactive tryptic peptide produced was one that normally contained cysteine but not alanine. On the other hand, peptides normally containing alanine but not cysteine were devoid of radioactivity. Thus, *the amino acid in aminoacyl-tRNA does not play a role in selecting a codon.*

SOME TRANSFER RNA MOLECULES RECOGNIZE MORE THAN ONE CODON BECAUSE OF WOBBLE IN BASE PAIRING

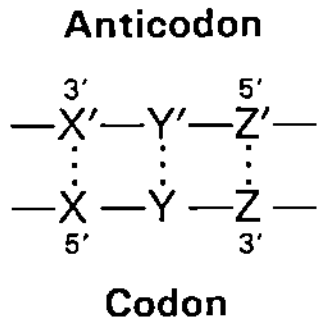

What are the rules that govern the recognition of a codon by the anticodon of a tRNA? A simple hypothesis is that each of the bases of the codon forms a Watson-Crick type of base pair with a complementary base on the anticodon. The codon and anticodon would then be lined up in an antiparallel fashion. In the accompanying diagram at the left, the prime denotes the complementary base. Thus X and X′ would be either A and U (or U and A) or G and C (or C and G). A specific prediction of this model is that a particular anticodon can recognize only one codon.

The facts are otherwise. *Some pure tRNA molecules can recognize more than one codon.* For example, the yeast alanine tRNA studied by Holley binds to *three* codons: GCU, GCC, and GCA. The first two bases of these codons are the same, whereas the third is different. Could it be that the recognition of the third base of a codon is sometimes less discriminating than recognition of the other two? The pattern of degeneracy of the genetic code indicates that this might be so. XYU and XYC always code for the same amino acid; XYA and XYG usually do. Crick surmised from these data that the steric criteria for pairing of the third base might be less stringent than for the other two. Models of various base pairs were built to determine which ones are similar to the standard A · U and G · C base pairs with regard to the distance and angle between the glycosidic bonds. Inosine was included in this study because it appeared in several anticodons.

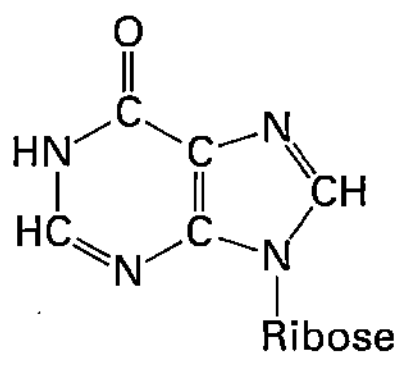

Inosine

Assuming some steric freedom ("*wobble*") in the pairing of the third base of the codon, the combinations shown in Table 34-2 seemed plausible.

The wobble hypothesis is now firmly established. The anticodons of tRNAs of known sequence bind to the codons predicted by this hypothesis. For example, the anticodon of yeast alanine tRNA is IGC. This tRNA recognizes the codons GCU, GCC, and GCA. Recall that, by convention, nucleotide sequences are written in the 5′ → 3′ direction unless otherwise noted. Hence, I (the 5′ base of this anticodon) pairs with U, C, or A (the 3′ base of the codon), as predicted.

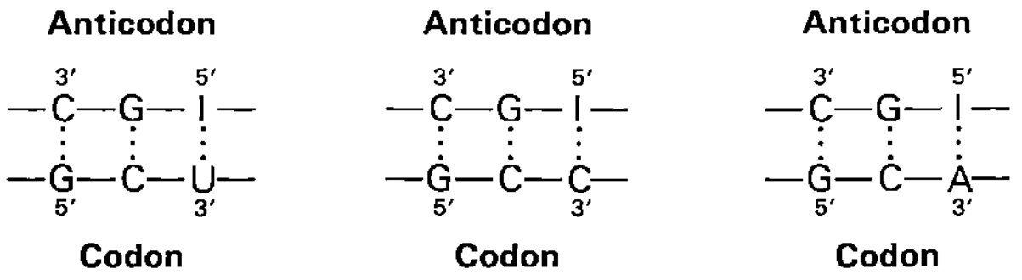

Phenylalanine tRNA, which has the anticodon GAA, recognizes the codons UUU and UUC but not UUA and UUG.

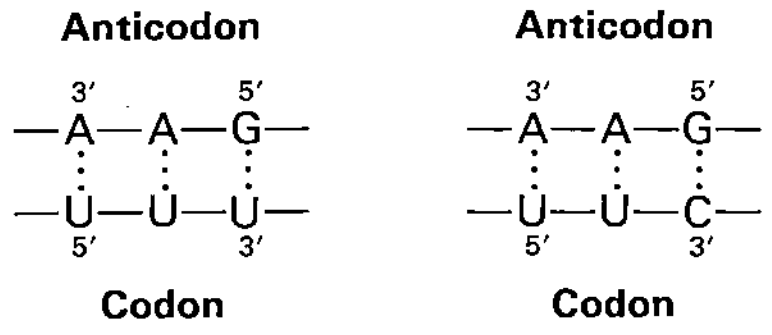

Thus, G pairs with either U or C in the third position of the codon, as predicted by the wobble hypothesis.

Two generalizations concerning the codon-anticodon interaction can be made:

1. The first two bases of a codon pair in the standard way. Recognition is precise. Hence, *codons that differ in either of their first two bases must be recognized by different tRNAs.* For example, both UUA and CUA code for leucine but are read by different tRNAs.

2. The first base of an anticodon determines whether a particular tRNA molecule reads one, two, or three kinds of codons: C or A (1

Table 34-2
Allowed pairings at the third base of the codon according to the wobble hypothesis

First base of anticodon	*Third base of codon*
C	G
A	U
U	A or G
G	U or C
I	U, C, or A

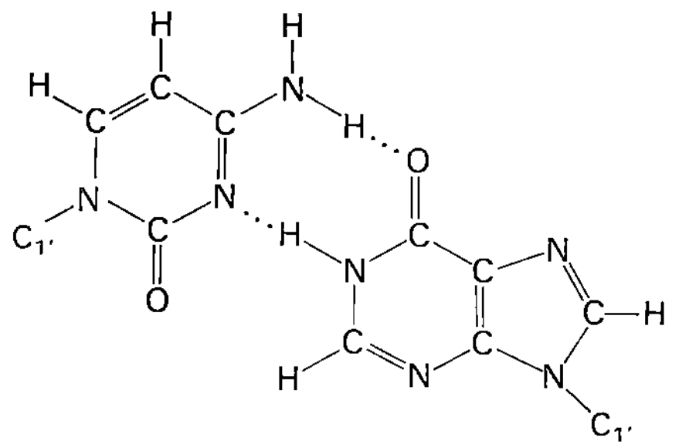

Cytidine-inosine base pair

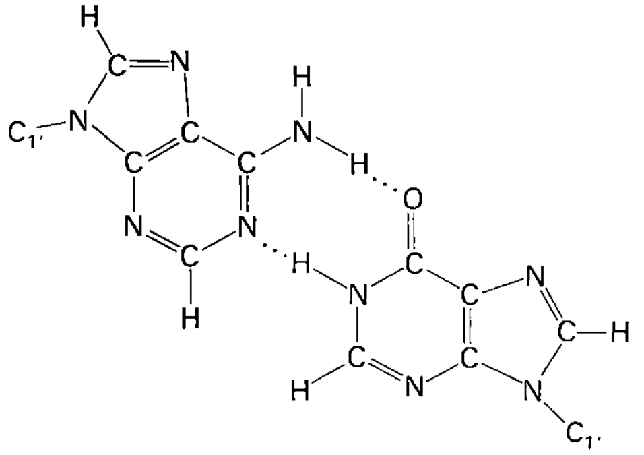

Adenosine-inosine base pair

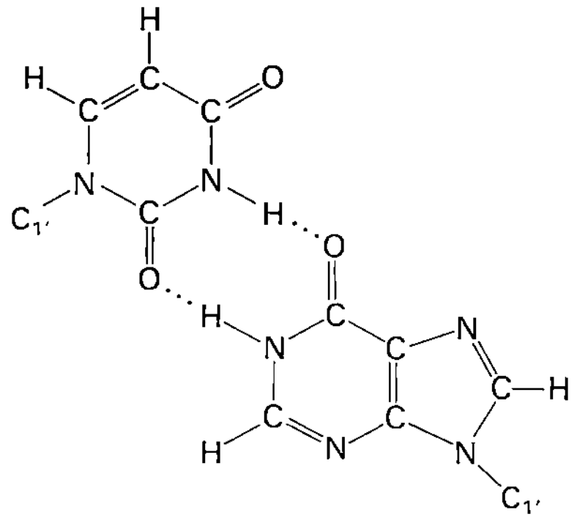

Uridine-inosine base pair

Figure 34-16
Inosine can base-pair with cytosine, adenine, or uracil because of wobble.

codon), U or G (2 codons), or I (3 codons). Thus, *part of the degeneracy of the genetic code arises from imprecision (wobble) in the pairing of the third base of the codon with the first base of the anticodon.* We see here a strong reason for the frequent appearance of inosine, one of the unusual nucleosides, in anticodons. *Inosine maximizes the number of codons that can be read by a particular tRNA molecule* (Figure 34-16). The inosines in tRNA are formed by deamination of adenosine following synthesis of the primary transcript.

RIBOSOMES ARE RIBONUCLEOPROTEIN PARTICLES (70S) MADE OF A SMALL (30S) AND A LARGE (50S) SUBUNIT

We have seen how transfer RNAs, the adaptor molecules between two fundamentally different alphabets, become specifically aminoacylated and, in turn, how their anticodons are recognized by codons. We turn now to ribosomes, the molecular machines that coordinate the interplay of tRNAs, mRNA, and proteins in this complex process and catalyze peptide-bond formation. An *E. coli* ribosome is a ribonucleoprotein assembly with a mass of about 2700 kd, a diameter of approximately 200 Å, and a

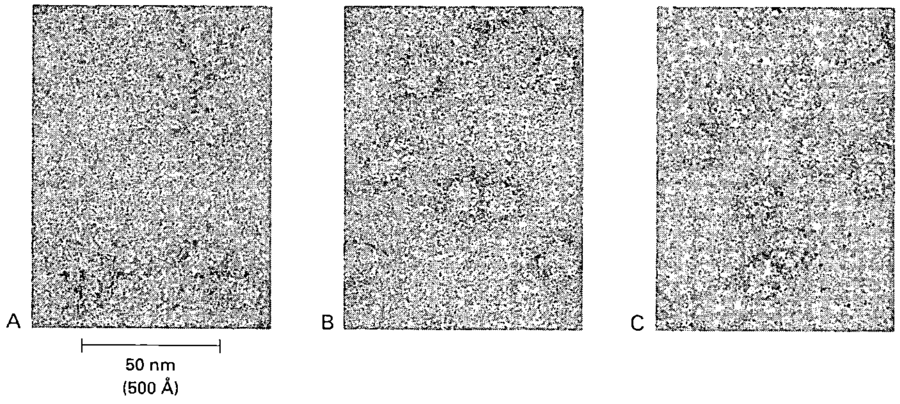

Figure 34-17
Electron micrographs of (A) 30S subunits, (B) 50S subunits, and (C) 70S ribosomes. [Courtesy of Dr. James Lake.]

sedimentation coefficient of 70S (Figure 34-17). The 20,000 ribosomes in a bacterial cell constitute nearly a fourth of its mass. A ribosome can be dissociated into a *large subunit* (*50S*) and a *small subunit* (*30S*) (Figure 34-18). These subunits can be further split into their constituent proteins

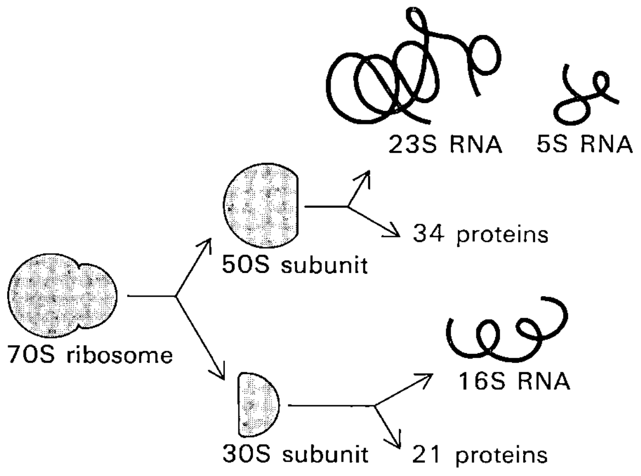

Figure 34-18
Ribosomes can be dissociated into 55 kinds of proteins and 3 RNA molecules.

and RNAs. The 30S subunit contains 21 different proteins (labeled S1 to S21) and a 16S RNA molecule. The 50S subunit contains 34 different proteins (labeled L1 to L34) and two RNA molecules, a 23S and a 5S species. A ribosome contains one copy of each RNA molecule, two of the L7 and L12 proteins, and one of each other protein. L7 is identical with L12 except that its amino terminus is acetylated. Both the 30S and 50S subunits can be reconstituted in vitro from their constituent proteins, as was first achieved by Masayasu Nomura in 1968.

RIBOSOMAL RNAs (5S, 16S, AND 23S rRNA) PLAY A CENTRAL ROLE IN PROTEIN SYNTHESIS

The prefix *ribo* in the name *ribosome* is apt, for RNA constitutes nearly two-thirds of the mass of these large molecular assemblies. The three RNAs present—5S, 16S, and 23S—are critical for ribosomal architecture and function. They are formed by cleavage of primary 30S transcripts and further processing. The base-pairing patterns of these molecules have been deduced by carrying out chemical modification and digestion experiments and by comparing nucleotide sequences of many species to detect conserved features (Figure 34-19). The striking finding is that *ribosomal RNAs (rRNAs) are folded into defined structures with many short duplex regions.*

For many years, it was presumed that ribosomal proteins orchestrate protein synthesis and that ribosomal RNAs serve primarily as a structural scaffold. The current view is almost the reverse. The discovery of catalytic RNA made us receptive to the possibility of a much more active role for RNA in ribosomal function. Indeed, several lines of evidence now suggest that *ribosomal RNAs have directive roles in protein synthesis and may indeed be dominant:*

1. Cleavage of a single bond in 16S rRNA by colicin E3, a nuclease secreted by some bacteria, abolishes protein synthesis.

2. The omission of single proteins in the in vitro reconstitution of 30S subunits leads to decreased ribosomal activity rather than total loss of function.

3. A sequence in 16S rRNA selects the start site in mRNA.

4. The wobble base of tRNA occupying the P site in a ribosome is paired with a base located in a 14-nucleotide sequence in 16S rRNA that is identical in more than a thousand molecules sequenced thus far. The identity of this sequence in archaebacteria, eubacteria, and eukaryotes indicates that it plays a critical role, one that has been conserved over several billion years of evolution.

5. The 3′ acceptor end of tRNA interacts with a conserved region of 23S rRNA.

6. Ribosomes virtually depleted of protein can still catalyze the formation of peptide bonds. 23S rRNA, by contrast, is essential for peptidyl transferase activity.

7. Most antibiotics that interfere with protein synthesis do so by interacting specifically with a ribosomal RNA rather than a ribosomal protein.

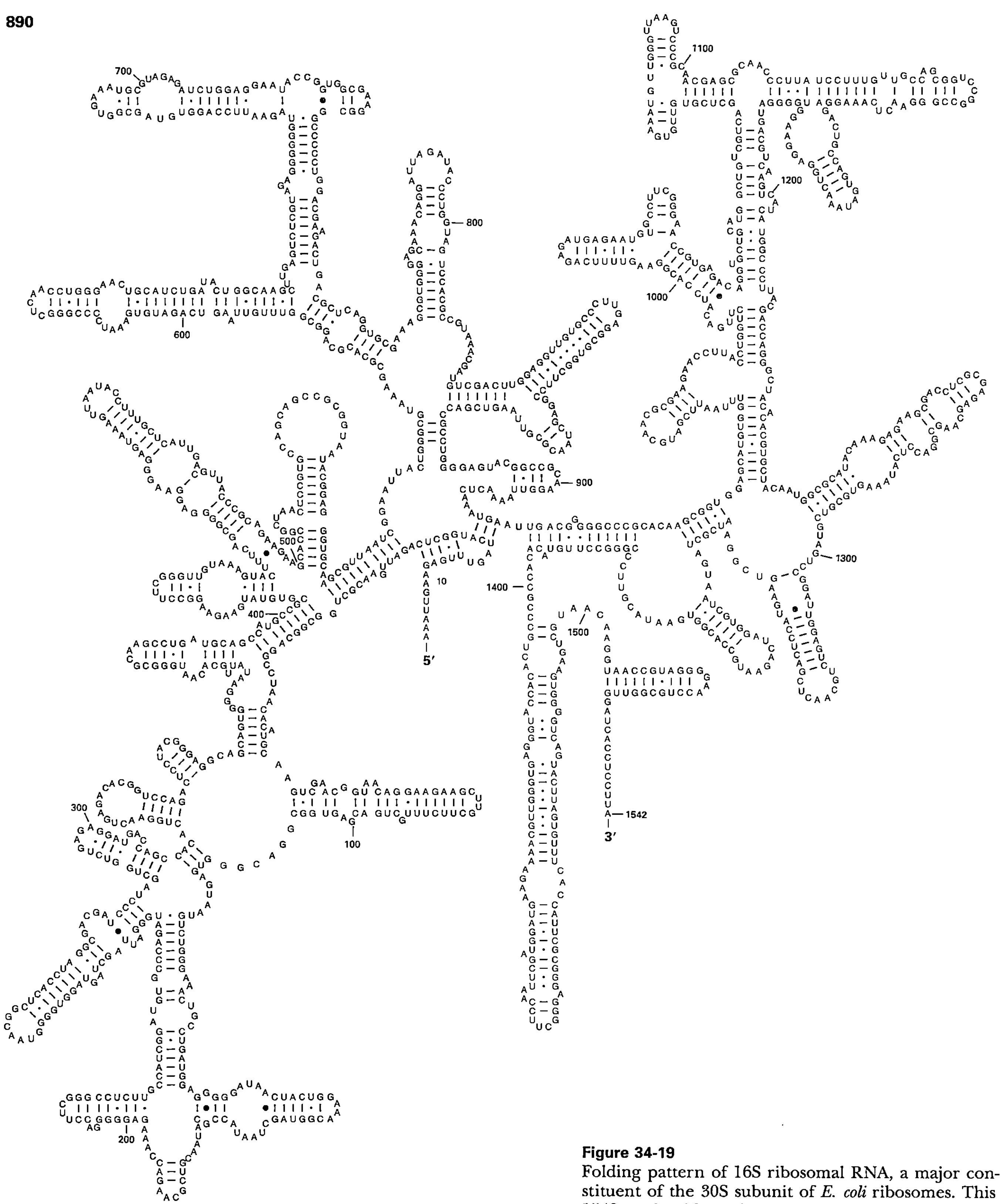

Figure 34-19
Folding pattern of 16S ribosomal RNA, a major constituent of the 30S subunit of *E. coli* ribosomes. This 1542-nucleotide molecule consists of a 5′-terminal domain (red), a central domain (green), and a 3′-terminal domain (blue). The high degree of conservation of this RNA makes it a choice document of evolutionary history. [Courtesy of Dr. Bryn Weiser and Dr. Harry Noller.]

RIBOSOMAL ARCHITECTURE, INTERACTIONS, AND DYNAMICS ARE BEING MAPPED BY MULTIPLE TECHNIQUES

Elucidating the relations between ribosome structure, dynamics, and function is a formidable challenge because of the large size (megadaltons) and complexity of these assemblies. Nevertheless, investigators have succeeded in delineating the overall shape of the ribosome, its surface topography, and the location of its protein and RNA constituents. This impressive progress comes from the application of a wide range of techniques by many laboratories. The shape of the ribosome and its 30S and 50S subunits has been reconstructed from a large number of electron-microscopic images. Immunoelectron microscopy using antibodies specific for particular proteins has revealed the identity of many surface features (Figure 34-20). The mRNA binding site and the 3′ end of 16S RNA are situated on a platform located between the upper and lower parts of the 30S subunit. In a cleft formed by this platform and the upper third of the subunit are the two tRNA binding sites. The 50S subunit contains three protuberances. The peptidyl transferase site that catalyzes peptide-bond formation is located in the valley between two of them; a fingerlike projection formed by a tetramer of L7 and L12 proteins contains the GTPase site that powers the movements of tRNAs and mRNA. The growing polypeptide chain emerges from the ribosome on the opposite side of the 50S subunit.

The locations of all 21 proteins of the 30S subunit have been determined by neutron diffraction analyses of concentrated solutions. Neutrons rather than x-rays were used because neutrons are scattered very differently by deuterium and hydrogen. The 30S subunit was reconstituted with two of its proteins deuterated—the subunits were obtained from bacteria grown in D_2O. Neutron scattering then revealed the distance between the centers of mass of the two deuterated proteins in the reconstituted particle. Many measurements of this kind on subunits containing different pairs of deuterated proteins led to an unequivocal map of their positions (Figure 34-21).

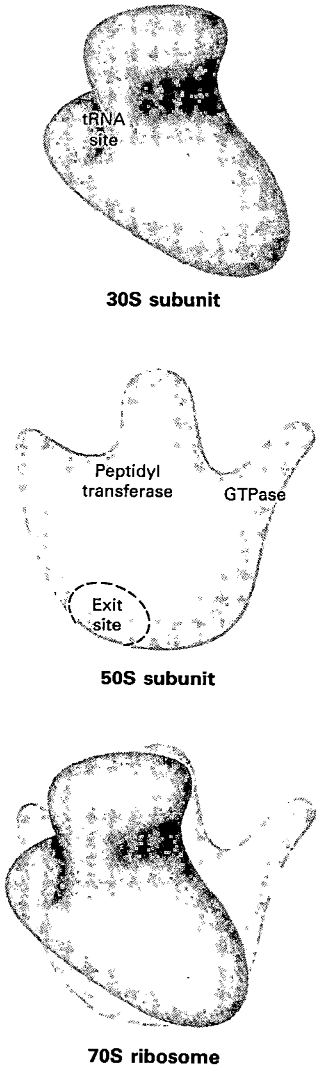

Figure 34-20
Surface topography and functional sites of the 30S subunit, 50S subunit, and intact 70S ribosome. [Courtesy of Dr. James Lake.]

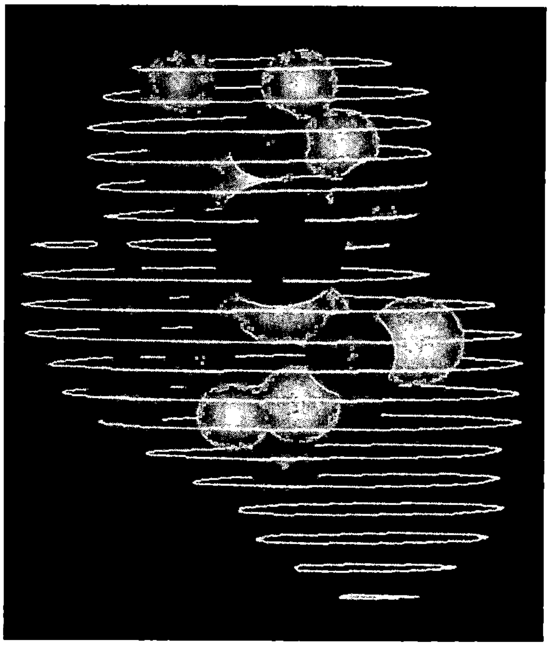

Figure 34-21
The location of all 21 proteins in the 30S ribosomal subunit has been mapped by neutron diffraction analyses of reconstituted ribosomes containing pairs of deuterated proteins. The white arcs depict the surface of the 30S particle, and the colored spheres show the positions of the proteins (the different colors serve only to help distinguish individual spheres). RNA occupies the unmarked volume of the particle. [Courtesy of Drs. M.S. Capel, D.M. Engelman, B.R. Freeborn, M. Kjeldgaard, J.A. Langer, V. Ramakrishnan, D.G. Schindler, D.K. Schneider, B.P. Schoenborn, I.-Y. Sillers, S. Yabuki, and P.B. Moore.]

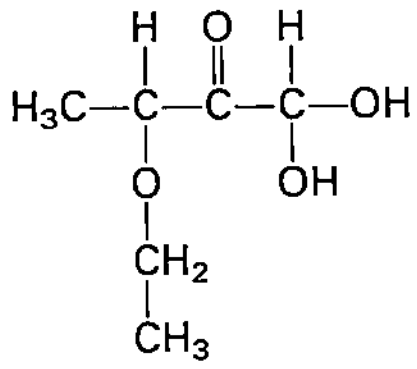

Kethoxal
(Modifies G)

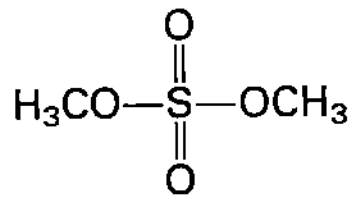

Dimethylsulfate
(Modifies A and C)

OH·
Hydroxyl radical
(Modifies ribose units)

Chemical modification studies of rRNAs have also been very informative. Naked RNA is vulnerable to attack by reagents such as *kethoxal, dimethylsulfate,* and *hydroxyl radical.* Nucleotides interacting with proteins and other RNA molecules are usually shielded from attack by these small chemical probes. Regions of 16S rRNA that bind each of the proteins in the 30S subunit have been mapped in this way. These *footprinting experiments* by Harry Noller have led to a model for the three-dimensional folding of 16S rRNA (Figure 34-22). Changes in footprints recorded at different steps in peptide elongation provide revealing glimpses of the dynamics of ribosome function. *Photocross-linking experiments* too have been valuable in identifying groups that are in close proximity. For example, a pyrimidine dimer is formed between the wobble base of the anticodon of tRNA occupying the P site and the cytosine at nucleotide 1400 in 16S rRNA.

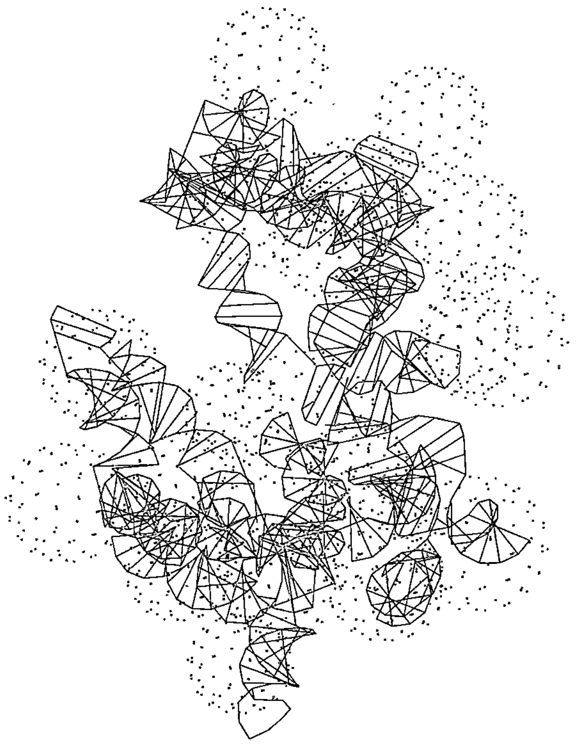

Figure 34-22
Model of the folding of 16S ribosomal RNA in the 30S subunit. The 5′, central, and 3′ domains of the RNA are shown in red, green, and blue, respectively. The dotted spheres mark the positions of individual proteins, as determined by neutron diffraction (see Figure 34-21). The proximity of particular regions of RNA to these proteins was determined by footprinting experiments. [After S. Stern, T. Powers, L.-M. Changchien, and H.F. Noller. *Science* 244(1989):783.]

PROTEINS ARE SYNTHESIZED IN THE AMINO-TO-CARBOXYL DIRECTION

One of the first questions asked about the mechanism of protein synthesis was whether proteins are synthesized in the amino-to-carboxyl direction or the reverse direction. Pulse-labeling studies by Howard Dintzis provided a clear-cut answer. Reticulocytes (young red blood cells) that were actively synthesizing hemoglobin were exposed to ^{3}H-leucine. Completed hemoglobin was sampled frequently during a period shorter than required to synthesize a complete chain. Each sample was separated into α and β chains and then treated with trypsin to fingerprint them (p. 171). In the earliest samples, only peptides from the carboxyl ends were labeled. Later samples yielded labeled peptides closer and closer to the amino ends. Over all the samples, *a gradient of radioactivity increasing from the amino to the carboxyl end of each chain was found* (Figure 34-23). This would be expected if the amino part of the sampled chains was already synthesized prior to the addition of radioactive leucine. If the carboxyl end was synthesized last, radioactive label would appear there first, in chains that were almost complete when label was added to the medium. This experiment demonstrated that *the direction of chain growth is from the amino to the carboxyl end.*

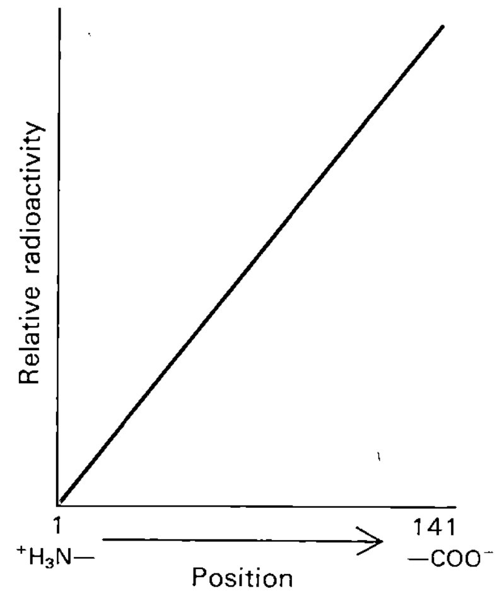

Figure 34-23
Distribution of ^{3}H-leucine in α chains of hemoglobin synthesized after exposure of reticulocytes to tritiated leucine. The higher radioactivity of the carboxyl ends relative to the amino ends indicates that the carboxyl end of each chain was synthesized last.

MESSENGER RNA IS TRANSLATED IN THE 5′ → 3′ DIRECTION

The direction of reading of mRNA was determined by using the synthetic polynucleotide

$$\overset{5'}{A}—A—A—(A—A—A)_n—A—A—\overset{3'}{C}$$

as the template in a cell-free protein-synthesizing system. AAA codes for lysine, whereas AAC codes for asparagine. The polypeptide product was

$$^{+}H_3N—Lys—(Lys)_n—Asn—C(=O)O^-$$

Because asparagine was the carboxyl-terminal residue, the codon AAC was the last to be read. Hence, *the direction of translation is 5′ → 3′.*

If the direction of translation were opposite to that of transcription, only fully synthesized mRNA could be translated. In contrast, if the directions were the same, mRNA could be translated while it is being synthesized. In fact, mRNA too is synthesized in the 5′ → 3′ direction (p. 845). In *E. coli,* almost no time is lost between transcription and translation. The 5′ end of mRNA interacts with ribosomes very soon after it is made, much before the 3′ end of mRNA is finished (Figure 34-24). *An important feature of prokaryotic gene expression is that translation and transcription are closely coupled in space and time.*

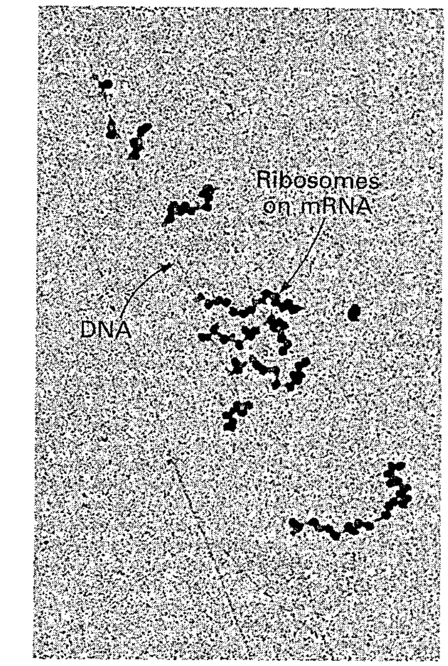

Figure 34-24
Transcription of a section of the DNA of *E. coli* and translation of the nascent mRNA. Only part of the chromosome is being transcribed. [From O.L. Miller, Jr., Barbara A. Hamkalo, and C.A. Thomas, Jr. Visualization of bacterial genes in action. *Science* 169(1970):392.]

SEVERAL RIBOSOMES SIMULTANEOUSLY TRANSLATE A MESSENGER RNA MOLECULE

Many ribosomes can simultaneously translate an mRNA molecule. This parallel synthesis markedly increases the efficiency of utilization of the mRNA. The group of ribosomes bound to an mRNA molecule is called a *polyribosome* or a *polysome.* The ribosomes in this unit operate independently, each synthesizing a complete polypeptide chain. The maximum density of ribosomes on mRNA is about 1 ribosome per 80 nucleotides. Polyribosomes synthesizing hemoglobin (which contains about 145 amino acids per chain, or 500 nucleotides per mRNA) typically consist of five ribosomes bound to an mRNA molecule. Ribosomes closest to the 5′ end of the messenger have the shortest polypeptide chains, whereas those nearest the 3′ end have almost finished chains (Figure 34-25). Ribosomes dissociate into 30S and 50S subunits soon after the polypeptide product is released.

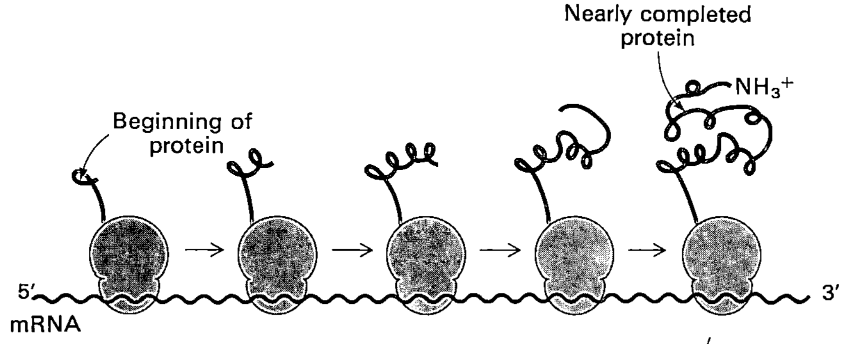

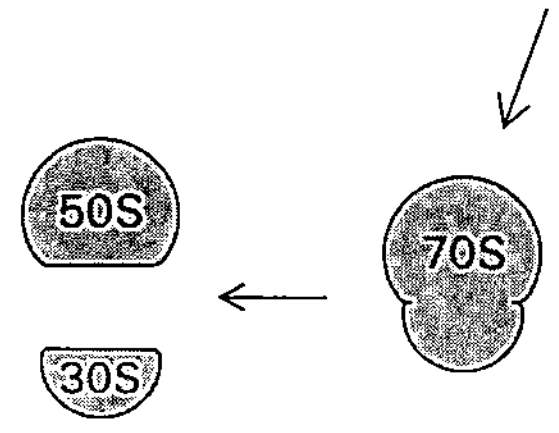

Figure 34-25
Diagram of a polyribosome (polysome). Ribosomes move along the mRNA in the 5′ → 3′ direction. They function independently of one another.

PROTEIN SYNTHESIS IN BACTERIA IS INITIATED BY FORMYLMETHIONYL TRANSFER RNA

How does protein synthesis start? The simplest possibility a priori is that the first three nucleotides of each mRNA serve as the first codon; no special start signal would then be needed. However, the experimental fact is that translation does not begin immediately at the 5′ terminus of mRNA. Indeed, the first translated codon is nearly always more than 25 nucleotides away from the 5′ end. Furthermore, many mRNA molecules in prokaryotes are *polycistronic*—that is, they code for two or more polypeptide chains. For example, a single mRNA molecule about 7000 nucleotides long specifies five enzymes in the biosynthetic pathway for tryptophan in *E. coli*. Each of these five proteins has its own start and stop signals on the mRNA. In fact, *all known mRNA molecules contain signals that define the beginning and end of each encoded polypeptide chain.*

A clue to the mechanism of initiation was the finding that nearly half the amino-terminal residues of proteins in *E. coli* are methionine, yet this amino acid is uncommon at other positions of the polypeptide chain. Furthermore, the amino terminus of nascent proteins is usually modified, which suggests that a derivative of methionine participates in initiation. In fact, *protein synthesis in bacteria starts with* N-*formylmethionine (fMet)*. A special tRNA brings formylmethionine to the ribosome to initiate protein synthesis. This *initiator tRNA* (abbreviated as $tRNA_f$) is different from the one that inserts methionine in internal positions (abbreviated as $tRNA_m$). The subscript *f* indicates that methionine attached to the initiator tRNA can be formylated, whereas it cannot be formylated when attached to $tRNA_m$.

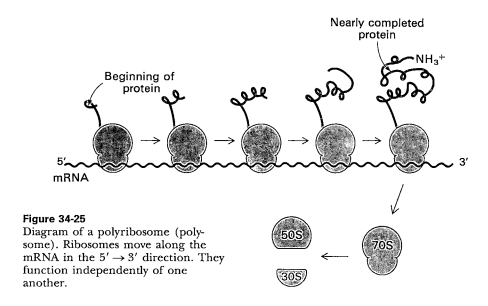

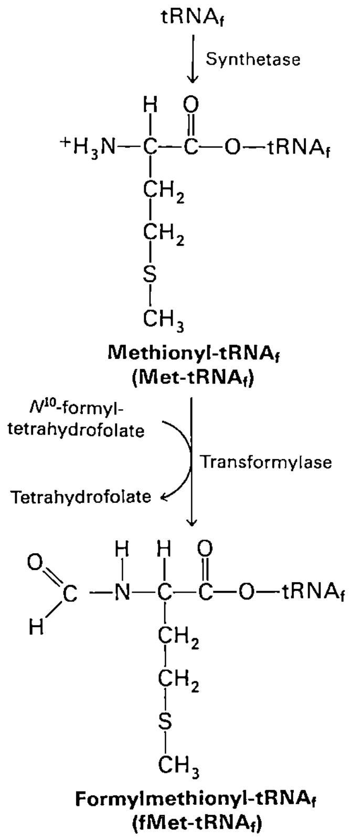

Figure 34-26
Formation of formylmethionyl-$tRNA_f$.

Methionine is linked to these two kinds of tRNAs by the same aminoacyl-tRNA synthetase (Figure 34-26). A specific enzyme then formylates the amino group of methionine that is attached to $tRNA_f$. The activated formyl donor in this reaction is N^{10}-formyltetrahydrofolate (p. 720). It is significant that free methionine and methionyl-$tRNA_m$ are not substrates for this transformylase.

THE START SIGNAL IS AUG (OR GUG) PRECEDED BY SEVERAL BASES THAT PAIR WITH 16S rRNA

The initiating codon in mRNA is AUG (methionine) or, much less frequently, GUG (valine). How is an initiating AUG or GUG distinguished from one that encodes an internal residue of a protein? The first step toward answering this question was the isolation of initiator regions from a number of mRNAs. This was accomplished by digesting mRNA-ribosome complexes (formed under conditions of chain initiation but not elongation) with pancreatic ribonuclease. In each case, a sequence of about 30 nucleotides was protected from digestion. As expected, each of these initiator regions displays an AUG (or GUG) codon (Figure 34-27). In addition, each initiator region contains a purine-rich sequence centered about 10 nucleotides on the 5′ side of the initiator codon.

5′ → 3′	
AGCACGAGGGGAAAUCUGAUGGAACGCUAC	*E. coli trpA*
UUUGGAUGGAGUGAAACGAUGGCGAUUGCA	*E. coli araB*
GGUAACCAGGUAACAACCAUGCGAGUGUUG	*E. coli thrA*
CAAUUCAGGGUGGUGAAUGUGAAACCAGUA	*E. coli lacI*
AAUCUUGGAGGCUUUUUUAUGGUUCGUUCU	φX174 phage A protein
UAACUAAGGAUGAAAUGCAUGUCUAAGACA	Qβ phage replicase
UCCUAGGAGGUUUGACCUAUGCGAGCUUUU	R17 phage A protein
AUGUACUAAGGAGGUUGUAUGGAACAACGC	λ phage *cro*

Pairs with 16S rRNA — Pairs with initiator tRNA

Figure 34-27
Sequences of initiation sites for protein synthesis in some bacterial and viral mRNA molecules.

The role of this purine-rich region (called the *Shine-Dalgarno sequence*) became evident when the sequence of 16S rRNA was elucidated. The 3′ end of this RNA component of the 30S subunit contains a sequence of several bases that is complementary to the purine-rich region in the initiator sites of mRNA. In fact, a complex of the 3′ end of 16S rRNA and the initiator region of mRNA has been isolated from an enzymatic digest of the initiation complex (Figure 34-28). The sequences of more than 70

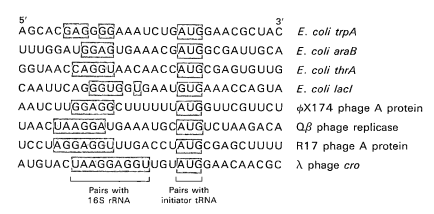

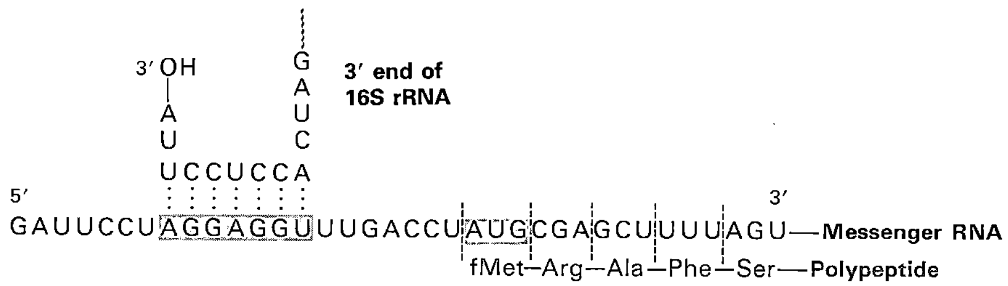

Figure 34-28
Base pairing between the purine-rich region (blue) in the initiator region of an mRNA and the 3′ end of 16S rRNA (red). The AUG codon (green) defines the start of the polypeptide chain. The mRNA shown here codes for the A protein of R17 phage. The purine-rich region is known as a Shine-Dalgarno sequence.

known initiator sites show that the number of base pairs between mRNA and 16S rRNA ranges from three to nine. Mutagenesis of the CCUCC sequence near the 3′ end of 16S rRNA to ACACA markedly interferes with the recognition of start sites in mRNA. Thus, *two kinds of interactions determine where protein synthesis starts: the pairing of mRNA bases with the 3′ end of 16S rRNA, and the pairing of the initiator codon on mRNA with the anticodon of fMet initiator tRNA.*

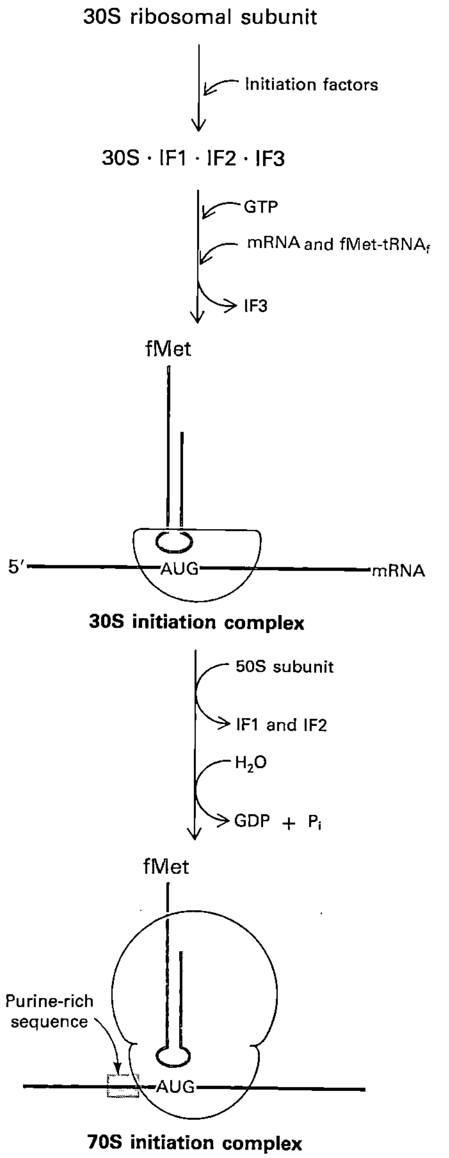

Figure 34-29
Initiation phase of protein synthesis: formation of a 30S initiation complex, followed by a 70S initiation complex.

FORMYLMETHIONYL-tRNA$_f$ IS PLACED IN THE P SITE OF THE RIBOSOME BY THE 70S INITIATION COMPLEX

How are mRNA and formylmethionyl-tRNA$_f$ brought to the ribosome to initiate protein synthesis? Three protein *initiation factors* (IF1, IF2, and IF3) are essential. The 30S ribosomal subunit first forms a complex with these three factors (Figure 34-29). The binding of GTP to IF2 enables mRNA and the initiator tRNA to join the complex as IF3 is released. Formylmethionyl-tRNA$_f$ is specifically recognized by IF2, and the release of IF3 allows the 50S subunit to join the complex. Hydrolysis of GTP bound to IF2 on entry of the 50S subunit leads to the release of IF1 and IF2. The result is a *70S initiation complex.* The L7 and nearly identical L12 proteins of the 50S subunit participate in GTP hydrolysis, which is essential for forming a productive initiation complex. The L7/L12 spike on the large subunit (see Figure 34-20 on p. 891) also participates in GTP hydrolysis during the elongation phase of protein synthesis.

Formation of the 70S initiation complex signifies that the ribosome is ready for the elongation phase of protein synthesis. *The fMet-tRNA$_f$ molecule occupies the P (peptidyl) site on the ribosome.* The other two sites for tRNA molecules—the *A (aminoacyl) site* and the *E (exit) site*—are empty. The existence of distinct P and A sites was inferred from studies of puromycin, an antibiotic to be discussed shortly (p. 902). The important point now is that fMet-tRNA$_f$ is positioned so that its anticodon pairs with the initiating AUG (or GUG) codon on mRNA. *The reading frame is defined by this interaction and by the pairing of the adjoining purine-rich sequence to a pyrimidine-rich sequence in 16S RNA* (see Figures 34-27 and 34-28). How were poly(U) and other synthetic polypeptides without start signals translated in the studies leading to the elucidation of the genetic code (p. 105)? Fortunately, nonspecific initiation occurred because the concentration of Mg^{2+} in the reaction mixture was much higher than it is in vivo.

THE GTP FORM OF ELONGATION FACTOR Tu DELIVERS AMINOACYL-tRNA TO THE A SITE OF THE RIBOSOME

The elongation cycle in protein synthesis begins with the insertion of an aminoacyl-tRNA into the empty A site on the ribosome. The particular species inserted depends on the mRNA codon that is positioned in the A site. The cognate aminoacyl-tRNA does not simply leave the synthetase and diffuse to the A site. Rather, it is delivered to the A site by a 43-kd protein called *elongation factor Tu* (*EF-Tu*). EF-Tu, like IF2, contains a bound guanyl nucleotide and cycles between a GTP and a GDP form (Figure 34-30). After EF-Tu has positioned the aminoacyl-tRNA in the A site, GTP is hydrolyzed. The GDP form of EF-Tu then dissociates from the ribosome. *EF-Ts*, a second elongation factor, joins the EF-Tu complex and induces the dissociation of GDP. Finally, GTP binds to EF-Tu, and EF-Ts is

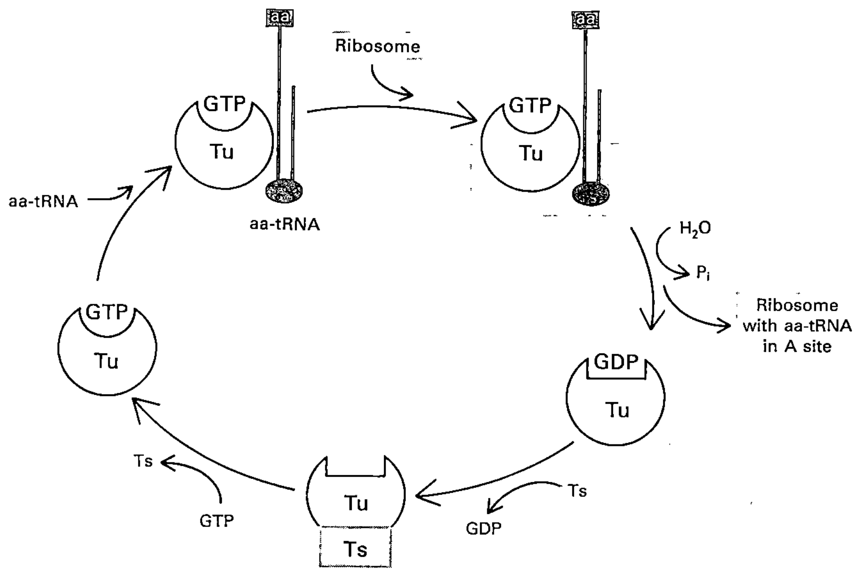

Figure 34-30
Reaction cycle of elongation factor Tu (EF-Tu).

concomitantly released. EF-Tu containing bound GTP is now ready to pick up another aminoacyl-tRNA and deliver it to the A site of the ribosome.

It is noteworthy that *EF-Tu does not interact with fMet-tRNA$_f$*. Hence, this initiator tRNA is not delivered to the A site. In contrast, Met-tRNA$_m$, like all other aminoacyl-tRNAs, does bind to EF-Tu. These findings account for the fact that *internal AUG codons are not read by the initiator tRNA*. Conversely, initiation factor 2 recognizes fMet-tRNA$_f$ but no other tRNA.

This GTP-GDP cycle of EF-Tu is reminiscent of that of transducin in vision (p. 336), the stimulatory G protein in hormone action (p. 342), and the ras protein in growth control (p. 355). Indeed, the amino-terminal domain (the G-domain) of EF-Tu is structurally similar to the GTP-binding subunit of these signal-transducing proteins. The other two do-

Figure 34-31
Structure of the GTP form of EF-Tu, a tripartite protein. The GTPase domain is shown in red, and the other two domains in blue and green. The bound hydrolysis-resistant analog of GTP is shown in yellow. The location of the binding site for aminoacyl-tRNA was inferred from labeling and mutagenesis studies. (A) Schematic diagram. (B) Space-filling model. [Drawn from left.pdb. M. Kjeldgaard, P. Nissen, S. Thirup, and J. Nyborg. *Structure* 1(1993):35.]

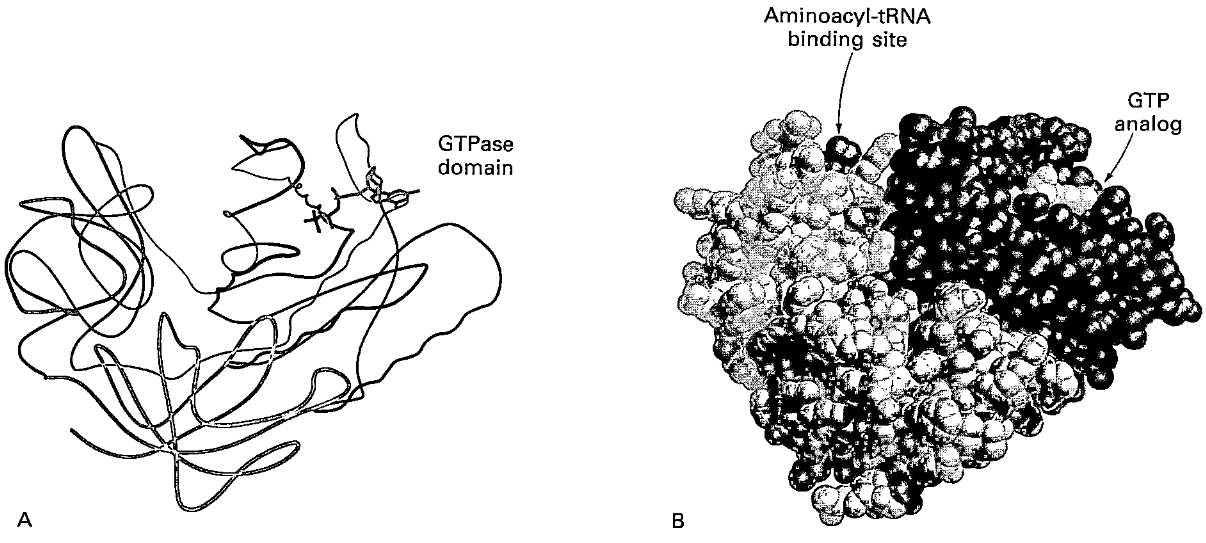

mains of the tripartite EF-Tu are distinctive—they mediate interactions with aminoacyl-tRNA and the ribosome. The hydrolysis of bound GTP to GDP triggers a change in the G-domain that is very much like that seen in the ras protein.

Why is EF-Tu needed in protein synthesis? The activated ester bond of aminoacyl-tRNA is quite labile. *The rapid binding of EF-Tu to a tRNA that has just been activated protects this linkage from hydrolysis by water.* The very high concentration of EF-Tu (5% of the total bacterial protein) enables it to sequester virtually all the aminoacyl-tRNA and do so rapidly. EF-Tu also escorts aminoacyl-tRNA to the right site on the ribosome. But EF-Tu must then be released from the delivered tRNA. Hydrolysis of GTP bound to EF-Tu markedly lowers its affinity for both the aminoacyl-tRNA and the ribosome. The loss of just one phosphoryl group loosens EF-Tu's grip on both tRNA and the ribosome. EF-Tu–GDP instead prefers EF-Ts, which triggers the release of GDP and the entry of another GTP to complete the cycle. Hence, *one role of the GTP-GDP cycle is to change the affinity of EF-Tu for its reaction partners.*

THE GTPase RATE OF EF-Tu SETS THE PACE OF PROTEIN SYNTHESIS AND DETERMINES ITS FIDELITY

The GTP-GDP cycle of EF-Tu plays a second key role. The fidelity of protein synthesis depends on having the correct aminoacyl-tRNA in the A site when the peptide bond is formed. This is the point of no return, for cells lack a means of excising an incorrect amino acid residue once the peptide bond is made. As might be expected, the incoming aminoacyl-tRNA is carefully scrutinized to make sure that its anticodon is complementary to the mRNA codon occupying the A site. An aminoacyl-tRNA that can pair with two of the three bases of a codon may bind transiently, but it will usually leave the A site before a peptide bond is formed. It takes time for the ribosome to decide whether the bound aminoacyl-tRNA is the right one. Protein synthesis would be much less accurate if a peptide bond were to form immediately on arrival of an aminoacyl-tRNA.

The timer that determines the period of scrutiny is the GTPase site of EF-Tu (Figure 34-32). A peptide bond cannot be formed until EF-Tu is released from the aminoacyl-tRNA; release requires that bound GTP be hydrolyzed to GDP. On average, several milliseconds pass before GTP is hydrolyzed, and another several milliseconds pass before EF-Tu–GDP leaves the ribosome. An incorrect aminoacyl-tRNA usually leaves the ribosome during one of these intervals, whereas the correct one stays bound. Furthermore, *the conformation of EF-Tu changes when it hydrolyzes GTP, and so the context of the codon-anticodon interaction is altered. The correct aminoacyl-tRNA*

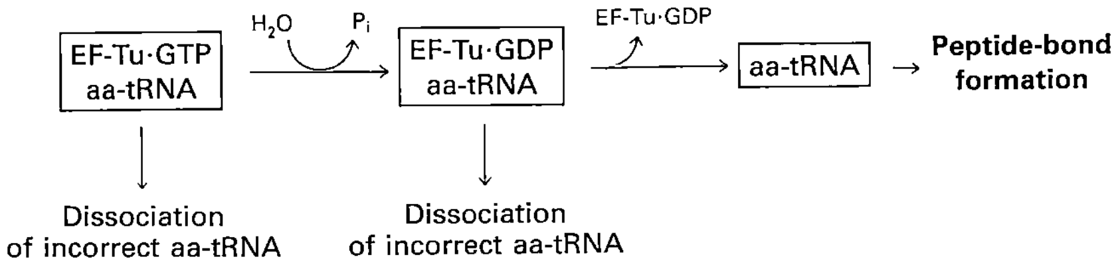

Figure 34-32
Proofreading of the aminoacyl-tRNA occupying the A site of the ribosome occurs both before and after hydrolysis of GTP bound to EF-Tu. Incorrect aminoacyl-tRNAs dissociate from the A site (steps shown in red), whereas the correct aminoacyl-tRNA stays bound and a peptide bond is formed (green arrows).

interacts strongly with mRNA in both states, but an incorrect one does not. In effect, the codon-anticodon interaction is scrutinized twice, in different ways, to achieve higher accuracy, just as proofreading of a manuscript by two readers for an hour each is much more effective than two hours of proofreading by one. This double-check mechanism for assuring that the right amino acid becomes attached to tRNA is another expression of the value of multiple views in testing authenticity.

How accurate must aminoacyl-tRNA selection be? The probability p of forming a protein with no errors depends on n, the number of amino acid residues, and ε, the frequency of inserting a wrong amino acid:

$$p = (1 - \varepsilon)^n$$

As Table 34-3 shows, an error frequency of 10^{-2} would be intolerable, even for quite small proteins. An ε of 10^{-3} would usually lead to the error-free synthesis of a 300-residue protein (~33 kd) but not of a 1000-residue protein (~110 kd). Thus, an error frequency of the order of no more than 10^{-4} per residue is needed to effectively produce the larger proteins. In fact, the observed values of ε are close to 10^{-4}. It is interesting to note that higher accuracy can be attained in vitro by using GTPγS, a GTP analog that is hydrolyzed by EF-Tu a thousand times less rapidly than is GTP. The longer residency time of EF-Tu on the ribosome enhances fidelity by giving more time for the departure of incorrect aminoacyl-tRNAs. However, higher accuracy achieved in this way carries a high kinetic price with little gain in the number of functional proteins synthesized. On the other hand, protein synthesis with an ε of 10^{-2} would be rapid, but nearly all the products would be inactive. *An error frequency of about 10^{-4} per amino acid residue has been selected in the course of evolution to produce the greatest number of functional proteins in the shortest time.*

Table 34-3
Accuracy of protein synthesis

Frequency of inserting an incorrect amino acid	*Probability of synthesizing an error-free protein*		
	Number of amino acid residues		
	100	*300*	*1000*
10^{-2}	0.366	0.049	0.000
10^{-3}	0.905	0.741	0.368
10^{-4}	0.990	0.970	0.905
10^{-5}	0.999	0.997	0.990

FORMATION OF A PEPTIDE BOND IS FOLLOWED BY GTP-DRIVEN TRANSLOCATION OF tRNAs AND mRNA

We now have a complex in which an aminoacyl-tRNA occupies the A site and fMet-tRNA occupies the P site. EF-Tu has left the ribosome. The stage is set for the *formation of a peptide bond* (Figure 34-33). This reaction is catalyzed by *peptidyl transferase,* an enzymatic activity of the 50S subunit. Recent experiments show that the 3′-CCA end of tRNA interacts with a highly conserved domain of 23S rRNA and suggest that *23S rRNA forms the peptidyl transferase active site.* The activated formylmethionine unit of fMet-$tRNA_f$ is transferred to the amino group of the aminoacyl-tRNA to form a dipeptidyl-tRNA. The attack of the amino group on the ester linkage to

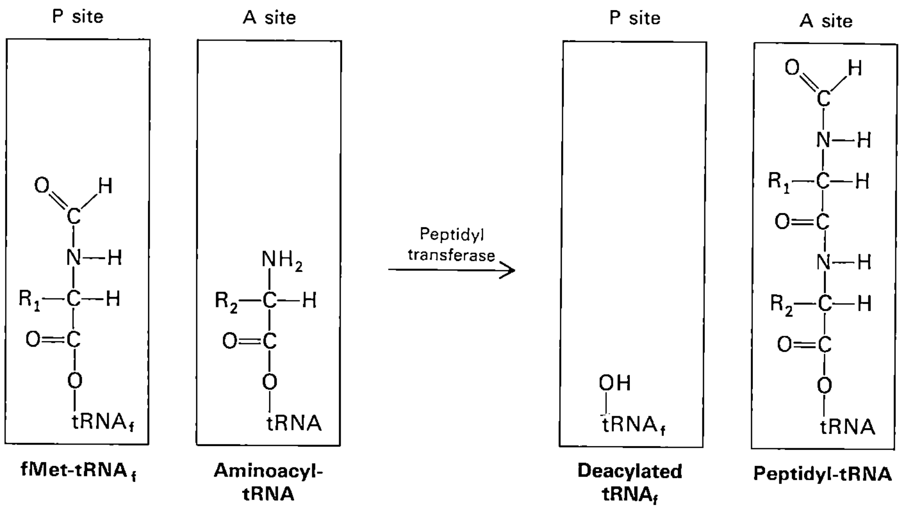

Figure 34-33
Peptide-bond formation is catalyzed by the peptidyl transferase site of the ribosome. The elongating chain is transferred to the amino group of the incoming aminoacyl-tRNA. This reaction proceeds by nucleophilic attack of the amino nitrogen atom of the aminoacyl-tRNA on the ester carbon atom of fMet-tRNA (or peptidyl-tRNA).

form a peptide bond is a thermodynamically favorable reaction; the free-energy cost of making a peptide bond was paid earlier in forming an aminoacyl-tRNA.

Peptide-bond formation is accompanied by a change in the interactions of both tRNAs with the 50S subunit but not with the 30S subunit (Figure 34-34). The deacylated tRNA now occupies the E (exit) site on the 50S subunit while staying in the P site on the 30S subunit. The new dipeptidyl-tRNA occupies the P site on the 50S subunit while staying in the A site on the 30S subunit. The next phase of the elongation cycle is *translocation.* Three movements occur: the deacylated tRNA moves out of the P site on the 30S subunit, the dipeptidyl-tRNA moves from the A site on the 30S subunit to the P site on the 30S subunit, and mRNA moves a distance of three nucleotides. The result is that the next codon is positioned for reading by the incoming aminoacyl-tRNA.

Translocation is mediated by *elongation factor G* (*EF-G,* also called *translocase*). EF-G, like IF2 and EF-Tu, cycles between a GTP and a GDP form. The GTP form of EF-G is the one that drives the translocation step. Several proteins in the 50S subunit (L11 and the L7/L12 tetramer) as well as 23S rRNA act in concert with EF-G to move peptidyl tRNA and mRNA. The hydrolysis of bound GTP enables EF-G to be released from the ribosome. L7/L12 may serve as a lever in this GTP-driven directed movement. After translocation, the A site is empty, ready to bind an aminoacyl-tRNA to start another round of elongation. The filling of the A site induces the release of deacylated tRNA from the E site; the A and E sites cannot be simultaneously occupied.

Why does a peptidyl-tRNA molecule move from the A site to the P site in *two* steps rather than *one?* One of the challenges facing the ribosome is to make sure that peptidyl-tRNA does not drift away before the polypeptide chain is completed. A peptidyl-tRNA that wanders into solution is very vulnerable to deacylation because its ester linkage is not shielded. *Premature termination would be highly deleterious because most truncated proteins are inactive. This potential catastrophe is averted by the anchoring of one end of the tRNA while the other moves during peptide-bond formation and translocation.* The movement of both tRNAs through a hybrid state in which they occupy one set of sites in the 50S subunit and a different set in the 30S subunit (see lower-right part of Figure 34-34) enhances the processivity of protein synthesis.

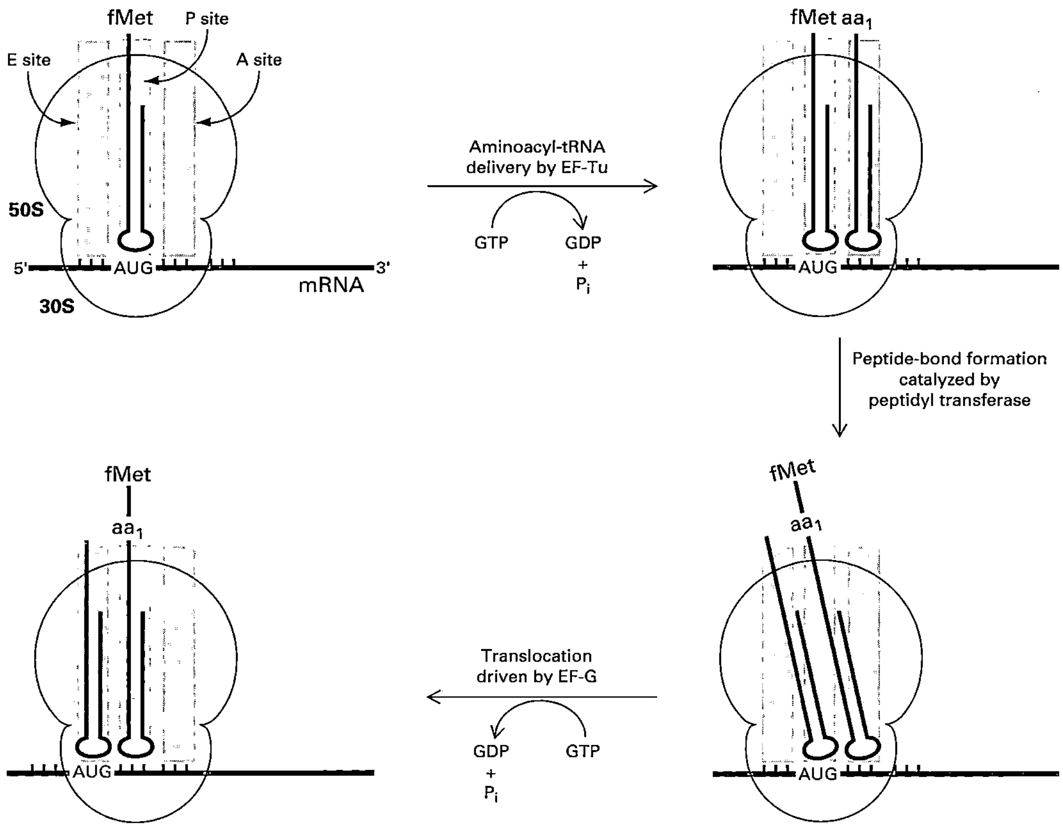

Figure 34-34
Elongation cycle of protein synthesis: binding of aminoacyl-tRNA, peptide-bond formation, and translocation. The A (aminoacyl) site is shown in green, the P (peptidyl) site in yellow, and the E (exit) site in blue. After peptide-bond formation but before translocation, the sites occupied by the two tRNAs in the 50S subunit are different from those in the 30S subunit. This hybrid state can be produced by a tilting of the tRNAs (as shown here), or alternatively, by a movement of the 50S subunit relative to the 30S subunit.

PROTEIN SYNTHESIS IS TERMINATED BY RELEASE FACTORS THAT READ STOP CODONS

How does the synthesis of a polypeptide chain come to an end when a stop codon is encountered? Aminoacyl-tRNA does not normally bind to the A site of a ribosome if the codon is UAA, UGA, or UAG. Normal cells do not contain tRNAs with anticodons complementary to these stop signals. Instead, *stop codons are recognized by release factors, which are proteins.* One of these release factors, RF1, recognizes UAA or UAG. A second factor, RF2, recognizes UAA or UGA. It is interesting to note that *16S rRNA also plays a key role in termination.* Deletion of a single 16S nucleotide, C1054, causes ribosomes to read through UGA stop codons. This nucleotide is near tandem UCA triplets that are complementary to UGA.

The binding of a release factor to a termination codon in the A site activates peptidyl transferase so that it hydrolyzes the bond between the polypeptide and the tRNA in the P site. *The specificity of peptidyl transferase is altered by the release factor so that water rather than an amino group is the acceptor of the activated peptidyl moiety.* The detached polypeptide chain leaves the ribosome, followed by tRNA and mRNA. Finally, the ribosome dissociates into 30S and 50S subunits as the prelude to the synthesis of another protein molecule. The binding of IF3 to the 30S subunit prevents it from prematurely joining the 50S subunit to form a dead-end 70S complex devoid of mRNA and fMet-$tRNA_f$.

PUROMYCIN CAUSES PREMATURE CHAIN TERMINATION BY MIMICKING AMINOACYL TRANSFER RNA

The antibiotic puromycin inhibits protein synthesis by releasing nascent polypeptide chains before their synthesis is completed. Puromycin is an analog of the terminal aminoacyl-adenosine portion of aminoacyl-tRNA (Figure 34-35). It binds to the A site on the ribosome and inhibits the entry of aminoacyl-tRNA. Furthermore, puromycin contains an α-amino group. This amino group, like the one on aminoacyl-tRNA, forms a peptide bond with the carboxyl group of the growing peptide chain in a reaction that is catalyzed by peptidyl transferase. The product is a peptide having a covalently attached puromycin residue at its carboxyl end. Peptidyl puromycin then dissociates from the ribosome. Puromycin has been used to probe the functional state of ribosomes. In fact, the concept of A and P sites arose from the use of puromycin to ascertain the location of peptidyl-tRNA. When peptidyl-tRNA is anchored in the A site of the 30S subunit (before translocation, as shown in Figure 34-34), it cannot react with puromycin.

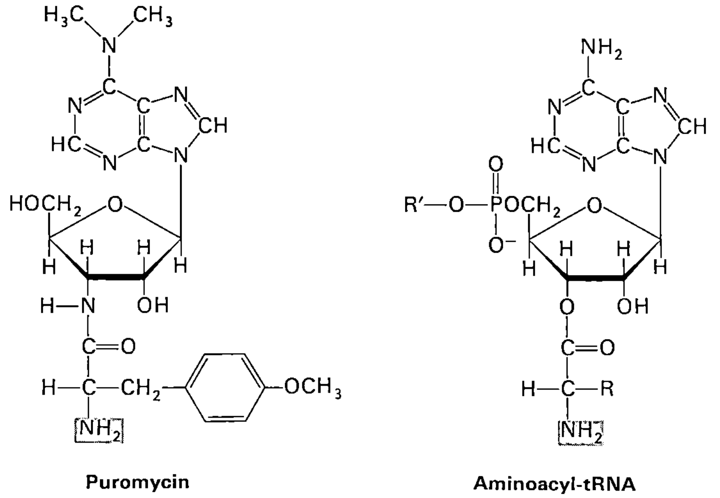

Figure 34-35
Puromycin resembles the aminoacyl terminus of an aminoacyl-tRNA. Its amino group becomes joined to the carboxyl group of the growing peptide chain to form an adduct that dissociates from the ribosome.

ANTIBIOTICS INDUCE MISCODING AND BLOCK PEPTIDE-BOND FORMATION AND TRANSLOCATION

Many other antibiotic inhibitors of protein synthesis are known, and the mechanisms of several have been elucidated (Table 34-4). Streptomycin, tetracycline, chloramphenicol, and erythromycin are highly potent antibacterial agents that act by blocking different steps in prokaryotic protein synthesis.

Table 34-4
Antibiotic inhibitors of protein synthesis

Antibiotic	*Action*
Streptomycin and other aminoglycosides	Inhibit initiation and cause misreading of mRNA (prokaryotes)
Tetracycline	Binds to the 30S subunit and inhibits binding of aminoacyl-tRNAs (prokaryotes)
Chloramphenicol	Inhibits the peptidyl transferase activity of the 50S ribosomal subunit (prokaryotes)
Cycloheximide	Inhibits the peptidyl transferase activity of the 60S ribosomal subunit (eukaryotes)
Erythromycin	Binds to the 50S subunit and inhibits translocation (prokaryotes)
Puromycin	Causes premature chain termination by acting as an analog of aminoacyl-tRNA (prokaryotes and eukaryotes)

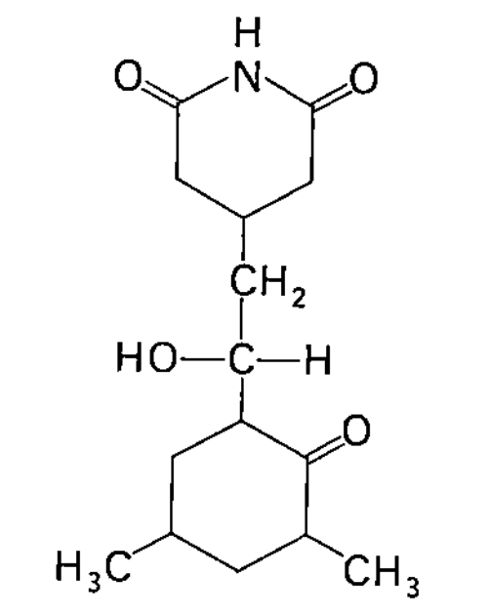

Cycloheximide

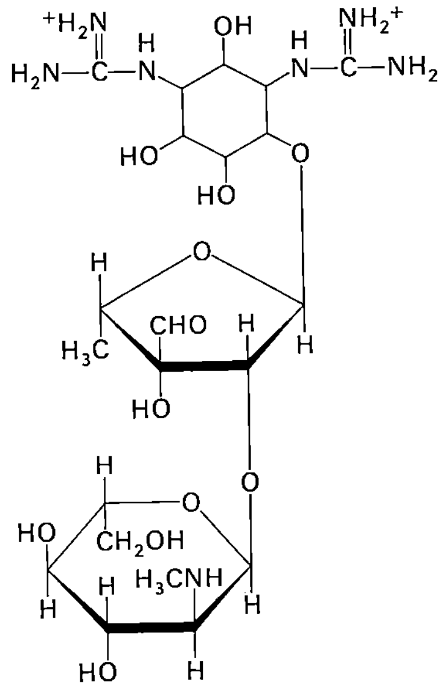

Streptomycin

Streptomycin, a highly basic trisaccharide, interferes with the binding of formylmethionyl-tRNA to ribosomes and thereby prevents the correct initiation of protein synthesis. Streptomycin also leads to a misreading of mRNA. For example, if poly(U) is the template, isoleucine (AUU) is incorporated in addition to phenylalanine (UUU). The site of action of streptomycin in the ribosome has been determined by forming hybrid ribosomes containing components from streptomycin-sensitive and streptomycin-resistant bacteria. These bacterial strains are related by mutation of a single gene. Hybrid ribosomes consisting of 50S subunits from resistant bacteria and 30S subunits from sensitive bacteria were found to be sensitive, whereas ribosomes prepared from the reverse combination were resistant. Subsequent experiments showed that mutations in a single protein (S12) of the 30S subunit can make bacteria resistant to streptomycin. Likewise, a mutation of nucleotide 523 in 16S rRNA leads to streptomycin resistance. Footprinting experiments have shown that this loop in 16S rRNA is protected from chemical modification by the binding of the S12 protein. Thus, streptomycin most likely acts at the interface of S12 and the loop containing nucleotide 523 in 16S rRNA.

Aminoglycoside antibiotics such as neomycin, kanamycin, and gentamycin interfere with the *decoding site* located in the vicinity of nucleotide 1400 in 16S rRNA of the 30S subunit. Recall that this region interacts with the wobble base in the anticodon of tRNA. These antibiotics also block the self-splicing of group I introns (p. 870), which raises the intriguing possibility that ribosomes and ribozymes may have a common evolutionary origin. *Chloramphenicol* blocks protein synthesis by inhibiting peptidyl transferase activity. It binds to the loop in 23S rRNA that interacts with the 3′-CCA end of tRNA. Mutation of a single base in this region, most likely the transferase catalytic site, confers resistance to chloramphenicol. *The sites of action of most antibiotic inhibitors of protein synthesis have been mapped to the RNA rather than the protein components of ribosomes.*

EUKARYOTIC AND PROKARYOTIC PROTEIN SYNTHESIS HAVE MANY COMMON STRUCTURAL AND MECHANISTIC MOTIFS

The basic plan of protein synthesis in eukaryotes is similar to that in prokaryotes. The major structural and mechanistic themes recur in higher forms. However, eukaryotic protein synthesis involves more pro-

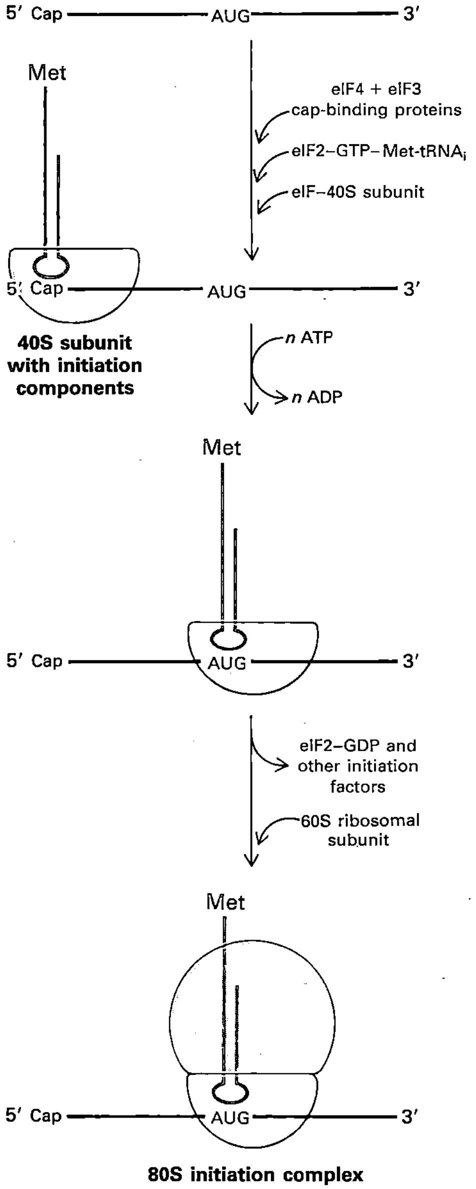

Figure 34-37
A 40S ribosomal subunit containing initiator tRNA and several initiation factors scans mRNA starting at the 5′ cap. The initiation factors dissociate when the first AUG codon is found. The 60S subunit then joins the 40S subunit to form an 80S initiation complex that is ready for the elongation phase.

tein components, and some steps are more intricate. Some noteworthy similarities and differences are listed below:

1. *Ribosomes.* Eukaryotic ribosomes are larger. They consist of a 60S large subunit and a 40S small subunit, which come together to form an 80S particle having a mass of 4200 kd (Figure 34-36), compared with 2700 kd for the prokaryotic 70S ribosome (see Figure 34-17). The 40S subunit contains an 18S RNA that is homologous to the prokaryotic 16S RNA. The 60S subunit contains three RNAs: the 5S and 28S RNAs are the counterparts of the prokaryotic 5S and 23S molecules; its 5.8S RNA is unique to eukaryotes.

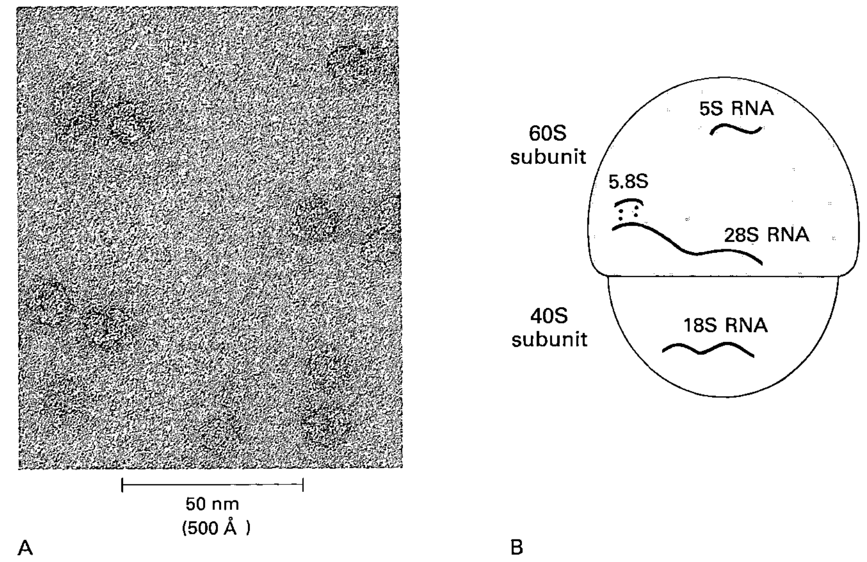

Figure 34-36
(A) Electron micrograph of eukaryotic ribosomes. (B) Schematic diagram of an 80S eukaryotic ribosome. [(A) Courtesy of Dr. Miloslav Bublik.]

2. *Initiator tRNA.* In eukaryotes, the initiating amino acid is methionine rather than *N*-formylmethionine. However, as in prokaryotes, a special tRNA participates in initiation. This aminoacyl-tRNA is called Met-tRNA$_f$ or Met-tRNA$_i$ (the subscript *f* indicates that it can be formylated in vitro, and *i* stands for initiation).

3. *Start signal.* The initiating codon in eukaryotes is always AUG. Eukaryotes, in contrast with prokaryotes (p. 895), do not use a purine-rich sequence on the 5′ side to distinguish initiator AUGs from internal ones. Instead, the AUG nearest the 5′ end of mRNA is usually selected as the start site. 40S ribosomes attach to the cap at the 5′ end of eukaryotic mRNAs and search for an AUG codon by moving stepwise in the 3′ direction. This scanning process in eukaryotic protein synthesis (Figure 34-37) is powered by the hydrolysis of ATP. Pairing of the anticodon of Met-tRNA$_i$ with the AUG codon of the mRNA signals that the target has been found. A eukaryotic mRNA has only one start site and hence is the template for a single protein. In contrast, a prokaryotic mRNA can have multiple start sites and it can serve as the template for the synthesis of several proteins.

4. *Initiation complexes.* Eukaryotes contain many more initiation factors than do prokaryotes, and their interplay is much more intricate. Nine are

known, and several consist of multiple subunits. The prefix *eIF* denotes a eukaryotic initiation factor. The GTP form of eIF2 (1000 kd) brings the initiator tRNA to the 40S subunit. Cap-binding proteins (CBPs) bind the cap of mRNA. They are joined by eIF3, which finds the AUG closest to the 5′ end; eIF4 is the ATP-driven engine in this search. The initiation factor eIF5 induces the release of eIF2 and eIF3 following the pairing of Met-$tRNA_i$ with the initiating AUG; eIF5 exerts these actions by triggering the hydrolysis of GTP bound to eIF2. Finally, the 60S subunit joins the complex of initiator tRNA, mRNA, and 40S subunit to form the 80S initiation complex.

5. *Elongation and termination factors.* Eukaryotic elongation factors EF1α and EF1$\beta\gamma$ are the counterparts of prokaryotic EF-Tu and EF-Ts. The GTP form of EF1α delivers aminoacyl-tRNA to the A site of the ribosome, and EF1$\beta\gamma$ catalyzes the exchange of GTP for bound GDP. Eukaryotic EF2 mediates GTP-driven translocation in much the same way as does prokaryotic EF-G. Termination in eukaryotes is carried out by a single release factor eRF, a GTP-driven protein, compared with two in prokaryotes. Finally, eIF3, like its prokaryotic counterpart IF3, prevents the reassociation of ribosomal subunits in the absence of an initiation complex.

TRANSLATION IN EUKARYOTES IS REGULATED BY PROTEIN KINASES THAT INACTIVATE AN INITIATION FACTOR

Initiation is usually the rate-limiting step in protein synthesis, which makes it an attractive target for regulation. One of the best-understood examples of translational control emerged from studies of the kinetics of hemoglobin synthesis. Extracts of reticulocytes synthesize hemoglobin subunits at a rapid rate until the supply of heme is depleted. In the absence of heme, protein synthesis halts because of the abrupt formation of an inhibitor of protein synthesis called *heme-controlled inhibitor* (HCI). The mechanism of action of HCI was an enigma until it was discovered that HCI is a kinase that phosphorylates the α subunit of eIF2, the initiation factor that brings Met-$tRNA_i$ to the 40S ribosomal subunit (see Figure 34-37). Phosphorylation blocks the recycling of eIF2 in the following way (Figure 34-38). Initiation factor eIF2 is in the inactive GDP form when it

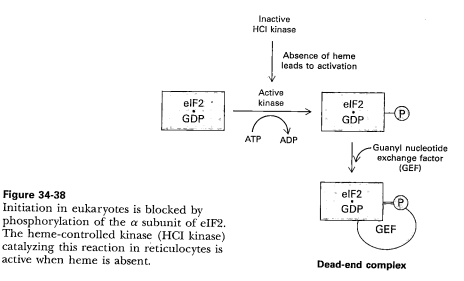

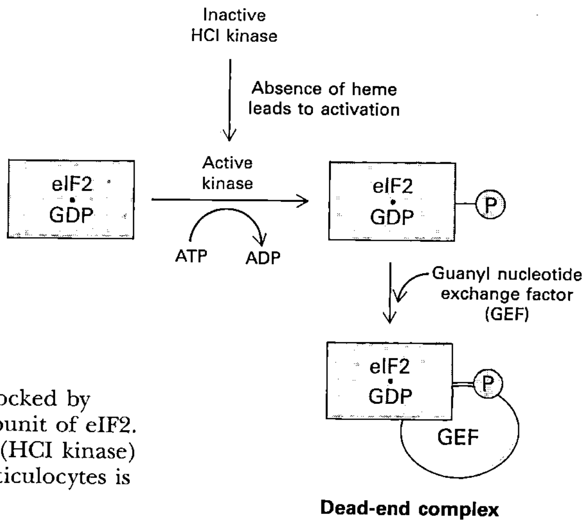

Figure 34-38
Initiation in eukaryotes is blocked by phosphorylation of the α subunit of eIF2. The heme-controlled kinase (HCI kinase) catalyzing this reaction in reticulocytes is active when heme is absent.

is released from the 40S subunit prior to the formation of the 80S initiation complex. Until it is converted into the GTP form, eIF2 cannot pick up a Met-$tRNA_i$ to initiate another round of protein synthesis. The requisite exchange of GTP for GDP bound to eIF2 is catalyzed by a protein known as GEF (*g*uanyl nucleotide *e*xchange *f*actor), or eIF2B. But if eIF2 is phosphorylated by the heme-controlled kinase, it has such high affinity for GEF that the complex is essentially irreversible. Hence, *phosphorylated eIF2 molecules are unable to initiate protein synthesis.* Moreover, they tie up GEF, which is present in lesser amounts than is eIF2. Consequently, the phosphorylation of only 30% of eIF2 leads to the complete cessation of protein synthesis. Phosphorylated eIF2 becomes functional again upon removal of its phosphate group by a specific phosphatase.

The physiologic role of the phosphorylation of eIF2 in reticulocytes is to coordinate the synthesis of heme and globin. When the level of heme is low, the synthesis of globins stops to prevent the formation of apohemoglobin, which is readily denatured. It seems likely that phosphorylation of eIF2 by different kinases is used to control translation in other cells too. The finding of an interferon-induced kinase that acts similarly in virus-infected cells suggests that this mechanism is widely used to control gene expression in eukaryotic cells.

DIPHTHERIA TOXIN BLOCKS PROTEIN SYNTHESIS IN EUKARYOTES BY INHIBITING TRANSLOCATION

Diphtheria was a major cause of death in childhood before the advent of effective immunization. The lethal effects of this disease are due mainly to a protein toxin produced by *Corynebacterium diphtheriae,* a bacterium that grows in the upper respiratory tract. The gene that encodes the toxin comes from a lysogenic phage that is harbored by some strains of *C. diphtheriae.* A few micrograms of diphtheria toxin is usually lethal in an unimmunized person because it inhibits protein synthesis. As will be discussed in the next chapter (p. 941), the toxin is cleaved shortly after entering a target cell into a 21-kd A fragment and a 40-kd B fragment. *The A fragment of the toxin catalyzes the covalent modification of a key component of the protein-synthesizing machinery, whereas the B fragment enables the A fragment to enter the cytosol of its target cell.*

A single A fragment of the toxin in the cytosol can kill a cell. Why is it so lethal? The target of the A fragment is EF2, the elongation factor catalyzing translocation in eukaryotic protein synthesis. EF2 contains *diphthamide,* an unusual amino acid residue of unknown function that is formed by post-translational modification of histidine. The A fragment catalyzes the transfer of the adenosine diphosphate ribose unit of NAD^+ to a nitrogen atom of the diphthamide ring (Figure 34-39). *This ADP-ribosylation of a single side chain of EF2 blocks its capacity to carry out translocation of the growing polypeptide chain.* Protein synthesis ceases, accounting for the remarkable toxicity of diphtheria toxin.

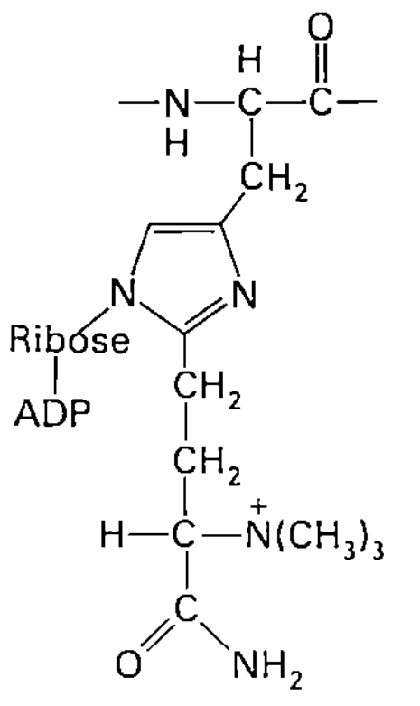

Figure 34-39
Diphtheria toxin blocks protein synthesis in eukaryotes by catalyzing the transfer of an ADP-ribose unit (red) from NAD^+ to diphthamide, a modified amino acid residue in elongation factor 2 (translocase). Diphthamide is formed by post-translational modification (blue) of a histidine residue.

SUMMARY

Protein synthesis (translation) is mediated by the coordinated interplay of more than a hundred macromolecules, including mRNA, tRNAs, activating enzymes, and protein factors, in addition to ribosomes. Protein synthesis starts with the ATP-driven activation of amino acids by aminoacyl-tRNA synthetases (activating enzymes), which link the carboxyl group of an amino acid to the 2′- or 3′-hydroxyl group of the adenosine unit at the 3′ end of tRNA. There is at least one specific activating enzyme for each amino acid. Also, there is at least one specific tRNA for each amino acid. Transfer RNAs of different specificity have a common structural design. They are single RNA chains of about 80 nucleotides, some of which are modified after transcription. The base sequences of all known tRNAs can be written in a cloverleaf pattern in which about half the nucleotides are base-paired. X-ray crystallographic studies have shown that tRNA is L-shaped. The amino acid attachment site at the 3′-CCA terminus is at one end of the L, whereas the anticodon is 80 Å away at the other end.

Synthetases recognize cognate tRNAs by their anticodons and acceptor stems. Messenger RNA recognizes the anticodon rather than the amino acid attached to a tRNA. A codon on mRNA forms base pairs with the anticodon of the tRNA. Some tRNAs are recognized by more than one codon because pairing of the third base of a codon is less discriminating than that of the other two (the wobble mechanism).

Protein synthesis takes place on ribosomes, which consist of large and small subunits, each about two-thirds RNA and one-third protein. In *E. coli,* the 70S ribosome (2700 kd) is made up of 30S and 50S subunits. The 30S subunit consists of 16S ribosomal RNA (16S rRNA) and 21 different proteins; the 50S subunit consists of 23S and 5S rRNA and 34 different proteins. Proteins are synthesized in the amino-to-carboxyl direction, and mRNA is translated in the 5′ → 3′ direction. In prokaryotes, transcription and translation are closely coupled. Several ribosomes can simultaneously translate an mRNA.

Protein synthesis takes place in three phases: initiation, elongation, and termination. In prokaryotes, mRNA, formylmethionyl-$tRNA_f$ (the special initiator tRNA), and a 30S ribosomal subunit come together to form a 30S initiation complex. The start signal on mRNA is AUG (or GUG) preceded by a purine-rich sequence that can base-pair with 16S rRNA. A 50S ribosomal subunit then joins this complex to form a 70S initiation complex, in which fMet-$tRNA_f$ occupies the P (peptidyl) site of the ribosome. EF-Tu delivers the appropriate aminoacyl-tRNA to the A (aminoacyl) site to begin the elongation phase. A peptide bond is then formed between the activated carboxyl group of the growing polypeptide and the amino group of the aminoacyl-tRNA. This reaction is catalyzed by the peptidyl transferase center of the 50S subunit. The tRNAs shift to new positions on the 50S subunit. Translocation induced by EF-G then alters their location on the 30S subunit. Deacylated tRNA moves from the P to the E (exit) site, the new peptidyl tRNA moves from the A to the P site, and mRNA moves a distance of three nucleotides. Both EF-Tu and EF-G are powered by GTP. Protein synthesis is terminated by release factors, which recognize the termination codons UAA, UGA, and UAG and cause the hydrolysis of the bond between the polypeptide and tRNA. Streptomycin induces misreading of mRNA, chloramphenicol blocks peptidyl transferase, and puromycin leads to premature release of the polypeptide chain.

The basic plan of protein synthesis in eukaryotes is similar to that of prokaryotes, but there are some significant differences between them. Eukaryotic ribosomes (80S) consist of a 40S small subunit and a 60S large subunit. The AUG closest to the 5′ end of mRNA is nearly always the start site. The initiating amino acid is again methionine, but it is not formylated. The initiation of protein synthesis in eukaryotes is more complex than in prokaryotes. eIF2, the eukaryotic initiation factor that delivers the initiator tRNA to the ribosome, is inactivated by phosphorylation. The kinases catalyzing this inactivation reaction are controlled. For example, a deficiency of heme in reticulocytes leads to the activation of this kinase and the consequent cessation of protein synthesis. Diphtheria toxin blocks protein synthesis in eukaryotes by catalyzing the ADP-ribosylation of EF2, the elongation factor mediating translocation.

The emerging picture is that ribosomal function is directed to a large extent by ribosomal RNAs rather than by ribosomal proteins. A sequence in 16S rRNA selects the start site in prokaryotic mRNA. The 3′ acceptor end of tRNA interacts with a conserved region of 23S RNA, which also forms the peptidyl transferase center. Most antibiotic inhibitors of protein synthesis interact specifically with rRNAs. Ribosomes early in evolution may have consisted entirely of RNA; proteins may have been recruited later to increase the fidelity and speed of translation.

SELECTED READINGS

Where to start

Noller, H.F., 1991. Ribosomal RNA and translation. *Ann. Rev. Biochem.* 60:191–227.

Nierhaus, K.H., 1990. The allosteric three-site model for the ribosomal elongation cycle: Features and future. *Biochemistry* 29:4997–5008.

Merrick, W.C., 1992. Mechanism and regulation of eukaryotic protein synthesis. *Microbiol. Rev.* 56:291–315.

Altman, S., 1990. Enzymatic cleavage of RNA by RNA. *Biosci. Rep.* 10:317–337. [Nobel Lecture account of the discovery that the RNA moiety of RNase P, the enzyme that cleaves tRNA precursors, is catalytically active in the absence of protein.]

Books

Gesteland, R.F., and Atkins, J.F. (eds.), 1993. *The RNA World.* Cold Spring Harbor Laboratory Press. [Many articles in this fine volume deal with ribosomes and translation.]

Hill, W.E., Dahlberg, A., Garrett, R.A., Moore, P.B., Schlessinger, D., and Warner, J.R. (eds.), 1990. *The Ribosome: Structure, Function, and Evolution.* American Society for Microbiology.

Clark, B.F.C., and Petersen, H.F. (eds.), 1984. *Gene Expression. The Translational Step and Its Control.* Munksgaard, Copenhagen.

Bermek, E. (ed.), 1985. *Mechanisms of Protein Synthesis. Structure-Function Relations, Control Mechanisms, and Evolutionary Aspects.* Springer-Verlag.

Aminoacyl-tRNA synthetases

Carter, C.W., Jr., 1993. Cognition, mechanism, and evolutionary relationships in aminoacyl-tRNA synthetases. *Ann. Rev. Biochem.* 62:715–748.

Cavarelli, J., and Moras, D., 1993. Recognition of tRNAs by aminoacyl-tRNA synthetases. *FASEB J.* 7:79–86.

Rould, M.A., Perona, J.J., Söll, D., and Steitz, T.A., 1989. Structure of *E. coli* glutaminyl-tRNA synthetase complexed with tRNA(Gln) and ATP at 2.8 Å resolution. *Science* 246:1135–1142.

Moras, D., 1992. Structural and functional relationships between aminoacyl-tRNA synthetases. *Trends Biochem. Sci.* 17:159–164.

Burbaum, J.J., and Schimmel, P., 1991. Structural relationships and the classification of aminoacyl-tRNA synthetases. *J. Biol. Chem.* 266:16965–16968.

Bedouelle, H., and Winter, G., 1986. A model of synthetase/transfer RNA interaction as deduced by protein engineering. *Nature* 320:371–373.

Transfer RNA

Normanly, J., and Abelson, J., 1989. Transfer RNA identity. *Ann. Rev. Biochem.* 58:1029–1049.

Rich, A., and Kim, S.H., 1978. The three-dimensional structure of transfer RNA. *Sci. Amer.* 238(1):52–62.

Basavappa, R., and Sigler, P.B., 1991. The 3 Å crystal structure of yeast initiator tRNA: Functional implications in initiator/elongator discrimination. *EMBO J.* 10:3105–3111.

Ribosomes and ribosomal RNAs

Moore, P., 1991. Ribosomes. *Curr. Opin. Struct. Biol.* 1:258–263.

Lake, J.A., 1981. The ribosome. *Sci. Amer.* 245(2):84–97.

Nomura, M., 1984. The control of ribosome synthesis. *Sci. Amer.* 250(1):102–114.

Stern, S., Powers, T., Changchien, L.M., and Noller, H.F., 1989. RNA-protein interactions in 30S ribosomal subunits: Folding and function of 16S rRNA. *Science* 244:783–790.

Olsen, G.J., and Woese, C.R., 1993. Ribosomal RNA: A key to phylogeny. *FASEB J.* 7:113–123.

Moller, W., 1991. Functional aspects of ribosomal proteins. *Biochimie* 73:1093–1100.

Dahlberg, A.E., 1989. The functional role of ribosomal RNA in protein synthesis. *Cell* 57:525–529.

Eisenstein, M., Sharon, R., Berkovitch, Y.Z., Gewitz, H.S., Weinstein, S., Pebay, P.E., Roth, M., and Yonath, A., 1991. The interplay between X-ray crystallography, neutron diffraction, image reconstruction, organo-metallic chemistry and biochemistry in structural studies of ribosomes. *Biochimie* 73:879–886.

Moazed, D., and Noller, H.F., 1991. Sites of interaction of the CCA end of peptidyl-tRNA with 23S rRNA. *Proc. Nat. Acad. Sci.* 88:3725–3728.

Elongation factors

Sprinzl, M., 1994. Elongation factor Tu: A regulatory GTPase with an integrated effector. *Trends Biochem. Sci.* 19:245–250.

Kjeldgaard, M., Nissen, P., Thirup, S., and Nyborg, J., 1993. The crystal structure of elongation factor EF-Tu from *Thermus aquaticus* in the GTP conformation. *Structure* 1:35–50.

Berchtold, H., Reshetnikova, L., Reiser, C.O.A., Schirmer, N.K., Sprinzl, M., and Hilgenfeld, R., 1993. Crystal structure of active elongation factor Tu reveals major domain rearrangements. *Nature* 365:126–132.

Powers, T., and Noller, H.F., 1993. Evidence for functional interaction between elongation factor Tu and 16S ribosomal RNA. *Proc. Nat. Acad. Sci.* 90:1364–1368.

Peptide bond formation and translocation

Noller, H.F., 1993. tRNA-rRNA interactions and peptidyl transferase. *FASEB J.* 7:87–89.

Noller, H.F., Hoffarth, V., and Zimniak, L., 1992. Unusual resistance of peptidyl transferase to protein extraction procedures. *Science* 256:1416–1419.

Moazed, D., and Noller, H.F., 1989. Intermediate states in the movement of transfer RNA in the ribosome. *Nature* 342:142–148.

Termination

Craigen, W.J., Lee, C.C., and Caskey, C.T., 1990. Recent advances in peptide chain termination. *Mol. Microbiol.* 4:861–865.

Prescott, C.D., and Goringer, H.U., 1990. A single mutation in 16S rRNA that affects mRNA binding and translation-termination. *Nucl. Acids Res.* 18:5381–5386.

Lee, C.C., Craigen, W.J., Muzny, D.M., Harlow, E., and Caskey, C.T., 1990. Cloning and expression of a mammalian peptide chain release factor with sequence similarity to tryptophanyl-tRNA synthetases. *Proc. Nat. Acad. Sci.* 87:3508–3512.

Brown, C.M., McCaughan, K.K., and Tate, W.P., 1993. Two regions of the *Escherichia coli* 16S ribosomal RNA are important for decoding stop signals in polypeptide chain termination. *Nucl. Acids Res.* 21:2109–2115.

Fidelity and proofreading

Kurland, C.G., 1992. Translational accuracy and the fitness of bacteria. *Ann. Rev. Genet.* 26:29–50.

Kirkwood, T.B.L., Rosenberger, R.F., and Galas, D.J. (eds.), 1986. *Accuracy in Molecular Processes. Its Control and Relevance to Living Systems.* Chapman and Hall. [A fascinating book. Chapters 4, 5, and 6 deal with the fidelity of protein synthesis, and Chapter 11 with the kinetic costs of proofreading.]

Fersht, A., 1985. *Enzyme Structure and Mechanism.* W. H. Freeman. [Chapter 13 on specificity and editing mechanisms is lucid and concise.]

Thompson, R.C., Dix, D.B., and Karim, A.M., 1986. The reaction of ribosomes with elongation factor Tu-GTP complexes. Aminoacyl-tRNA–independent reactions in the elongation cycle determine the accuracy of protein synthesis. *J. Biol. Chem.* 261:4868–4874.

Eukaryotic protein synthesis

Samuel, C.E., 1993. The eIF-2 alpha protein kinases, regulators of translation in eukaryotes from yeasts to humans. *J. Biol. Chem.* 268:7603–7606.

Proud, C.G., 1992. Protein phosphorylation in translational control. *Curr. Top. Cell. Regul.* 32:243–369.

Kozak, M., 1992. Regulation of translation in eukaryotic systems. *Ann. Rev. Cell Biol.* 8:197–225.

Riis, B., Rattan, S.I., Clark, B.F., and Merrick, W.C., 1990. Eukaryotic protein elongation factors. *Trends Biochem. Sci.* 15:420–424.

Shatkin, A.J., 1985. mRNA cap binding proteins: Essential factors for initiating translation. *Cell* 40:223–224.

Antibiotics and toxins

Tai, P.-C., Wallace, B.J., and Davis, B.D., 1978. Streptomycin causes misreading of natural messenger by interacting with ribosomes after initiation. *Proc. Nat. Acad. Sci.* 75:275–279.

Allen, P.N., and Noller, H.F., 1991. A single base substitution in 16S ribosomal RNA suppresses streptomycin dependence and increases the frequency of translational errors. *Cell* 66:141–148.

Vannuffel, P., Di, G.M., Morgan, E.A., and Cocito, C., 1992. Identification of a single base change in ribosomal RNA leading to erythromycin resistance. *J. Biol. Chem.* 267:8377–8382.

De Stasio, E.A., Moazed, D., Noller, H.F., and Dahlberg, A.E., 1989. Mutations in 16S ribosomal RNA disrupt antibiotic-RNA interactions. *EMBO J.* 8:1213–1216.

Choe, S., Bennett, M.J., Fujii, G., Curmi, P.M., Kantardjieff, K.A., Collier, R.J., and Eisenberg, D., 1992. The crystal structure of diphtheria toxin. *Nature* 357:216–222.

PROBLEMS

1. *Synthetase mechanism.* The formation of Ile-tRNA proceeds through an enzyme-bound Ile-AMP intermediate. Predict whether ^{32}P-labeled ATP is formed from $^{32}PP_i$ when each of the following sets of components is incubated with the specific activating enzyme:
   (a) ATP and $^{32}PP_i$.
   (b) tRNA, ATP, and $^{32}PP_i$.
   (c) Isoleucine, ATP, and $^{32}PP_i$.

2. *Light and heavy ribosomes.* Ribosomes were isolated from bacteria grown in a "heavy" medium (^{13}C and ^{15}N) and from bacteria grown in a "light" medium (^{12}C and ^{14}N). These 70S ribosomes were added to an in vitro system actively engaged in protein synthesis. An aliquot removed several hours later was analyzed by density-gradient centrifugation. How many bands of 70S ribosomes would you expect to see in the density gradient?

3. *The price of protein synthesis.* How many high-energy phosphate bonds are consumed in the synthesis of a 200-residue protein, starting from amino acids?

4. *Contrasting modes of elongation.* There are two basic mechanisms for the elongation of biomolecules (Figure 34-40). In type 1, the activating group (labeled X) is

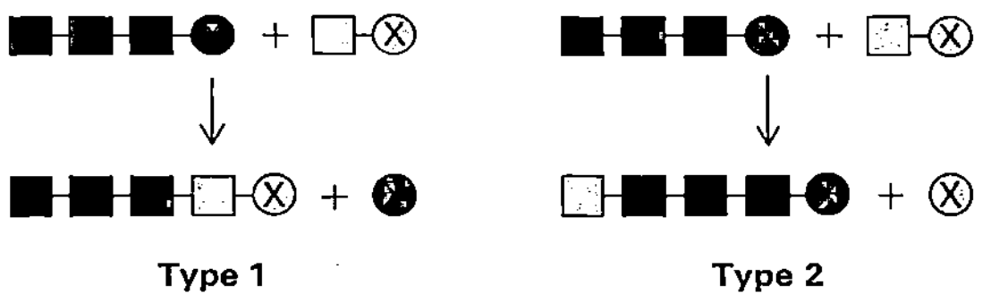

Figure 34-40
Two modes of elongation.

released from the growing chain. In type 2, the activating group is released from the incoming unit as it is added to the growing chain. Indicate whether each of the following biosyntheses occurs by means of a type 1 or a type 2 mechanism:
   (a) Glycogen synthesis.
   (b) Fatty acid synthesis.
   (c) $C_5 \rightarrow C_{10} \rightarrow C_{15}$ in cholesterol synthesis.
   (d) DNA synthesis.
   (e) RNA synthesis.
   (f) Protein synthesis.

5. *Suppressing frameshifts.* The insertion of a base in a coding sequence leads to a shift in the reading frame, which in most cases produces a nonfunctional protein. Propose a mutation in tRNA that would suppress frameshifting.

6. *Tagging a ribosomal site.* Design an affinity-labeling reagent for one of the tRNA binding sites in *E. coli* ribosomes. How would you synthesize such a reagent?

7. *Viral mutation.* An mRNA transcript of a T7 phage gene contains the base sequence

$$\downarrow$$
5′-A ACUGCACGA G GUAACACA AGAUGGCU-3′

Predict the effect of a mutation that changes the G marked by an arrow to A.

8. *Molecular attack.* What is the nucleophile in the reaction catalyzed by peptidyl transferase?

9. *Two synthetic modes.* Compare and contrast protein synthesis by ribosomes with that by the solid-phase method.

10. *Enhancing fidelity.* Compare the accuracy of (a) DNA replication, (b) RNA synthesis, and (c) protein synthesis. Which mechanisms are used to assure the fidelity of each of these processes?

11. *Triggered GTP hydrolysis.* Ribosomes markedly accelerate the hydrolysis of GTP bound to the complex of EF-Tu and aminoacyl-tRNA. What is the biological significance of this enhancement of GTPase activity by ribosomes?

12. *Déjà vu.* Which protein in G protein cascades plays a role akin to that of elongation factor Ts?

13. *Blocking translation.* Devise an experimental strategy for switching off the expression of a specific mRNA without changing the gene encoding the protein or the gene's control elements.

CHAPTER

35

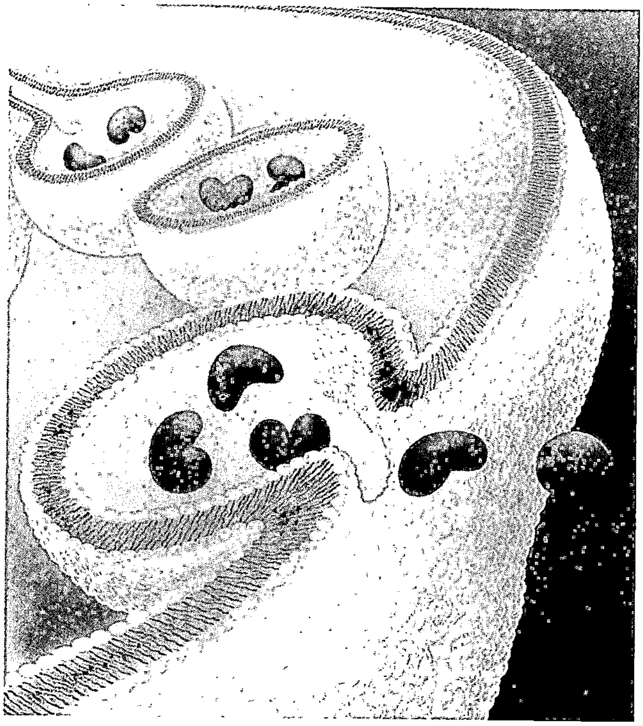

Protein Targeting

Nascent proteins contain signals that determine their ultimate destination. A newly synthesized protein in *Escherichia coli,* for example, can stay in the cytosol, or it can be sent to the plasma membrane, the outer membrane, the space between them, or the extracellular medium. Eukaryotic cells can direct proteins to internal sites such as lysosomes, mitochondria, chloroplasts, and the nucleus. How is sorting accomplished? In eukaryotes, a key choice is made soon after the synthesis of a protein begins. A ribosome remains free in the cytosol unless it is directed to the endoplasmic reticulum (ER) by a signal sequence in the protein being synthesized. Nascent polypeptide chains formed by such membrane-bound ribosomes are translocated across the ER membrane. In the lumen of the ER, many of them are glycosylated and modified in other ways. They are then transported to the Golgi complex, where they are further modified. Finally, they are sorted for delivery to lysosomes, secretory vesicles, and the plasma membrane. Transported proteins are carried by vesicles that bud from donor compartments and fuse with target compartments.

The signals used to target eukaryotic proteins for transfer across the endoplasmic reticulum membrane are ancient—bacteria employ similar sequences to send proteins to their plasma membrane and to secrete

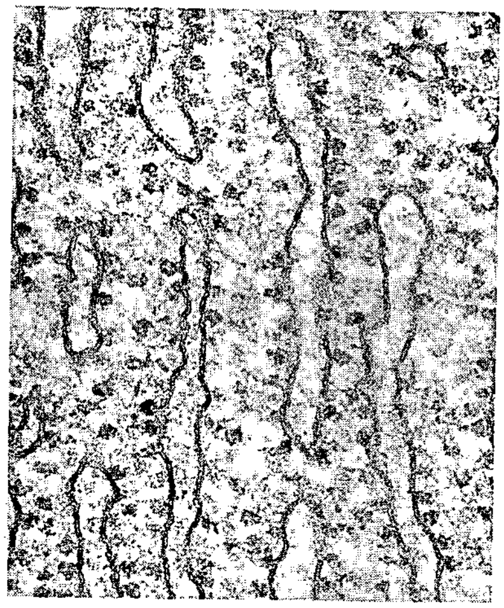

Figure 35-1
Electron micrograph of the rough endoplasmic reticulum. Ribosomes are bound to the cytosolic face of the membrane. [Courtesy of Dr. George Palade.]

Opening Image: *Proteins are sorted in the Golgi complex for delivery to secretory vesicles, lysosomes, and the plasma membrane. The budding and fusion of membranes are key events in protein targeting.*

them. Different strategies are used to send proteins synthesized by free ribosomes to the nucleus, peroxisomes, mitochondria, and chloroplasts of eukaryotic cells. The molecular machinery that recognizes these signals and translocates proteins bearing them across membranes is being unraveled by genetic and biochemical approaches. Indeed, many targeting processes can now be reconstituted in vitro.

A complementary aspect of protein targeting is the import of proteins into eukaryotic cells by *receptor-mediated endocytosis.* Proteins in the extracellular space bind to specific receptors in specialized regions of the plasma membrane called *coated pits.* Clathrin, a protein that forms polyhedral lattices, converts coated-pit regions into membrane-bound vesicles inside the cell. When these vesicles become acidified, the imported proteins dissociate from their receptors. Many viruses and toxins enter cells by mimicking molecules that are normally imported by endocytosis.

The final section of this chapter deals with the *programmed destruction of proteins.* The half-life of a protein is very much influenced by the nature of its amino-terminal residue. Ubiquitin, a protein found in all eukaryotes, becomes conjugated to proteins marked for degradation.

> "The choice of the pancreatic acinar cell, a very efficient protein producer, as the object of our studies reflected in part our training, and in part our environment. . . . Perhaps the most important factor in this selection was the appeal of the amazing organization of the pancreatic acinar cell, whose cytoplasm is packed with stacked endoplasmic reticulum cisternae studded with ribosomes. Its pictures had for me the effect of the song of a mermaid: irresistible and half transparent. Its meaning seemed to be buried only under a few years of work, and reasonable working hypotheses were already suggested by the structural organization itself."
>
> George Palade
> *Nobel Lectures (1975)*
> © 1975 The Nobel Foundation

RIBOSOMES BOUND TO THE ENDOPLASMIC RETICULUM MAKE SECRETORY AND MEMBRANE PROTEINS

In eukaryotic cells, some ribosomes are free in the cytosol, whereas others are bound to an extensive membrane system called the *endoplasmic reticulum (ER),* which comprises about half the total membrane of a cell. The region that binds ribosomes is called the *rough ER* because of its studded appearance, in contrast with the *smooth ER,* which is devoid of ribosomes. Membrane-bound ribosomes synthesize three major classes of proteins: *secretory proteins* (proteins exported by the cell), *lysosomal proteins,* and *proteins spanning the plasma membrane.* Indeed, virtually all integral membrane proteins of the cell, except those located in the membranes of mitochondria and chloroplasts, are formed by ribosomes bound to the ER. George Palade's pioneering studies of the mechanism of secretion of zymogens by the pancreatic acinar cell, a cell specialized for protein synthesis and export, opened a new area of inquiry—protein targeting—and delineated the pathway taken by secretory proteins.

Are membrane-bound ribosomes different from cytosolic ones, or are all ribosomes intrinsically the same? Studies of the protein-synthesizing activities of ribo-

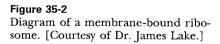

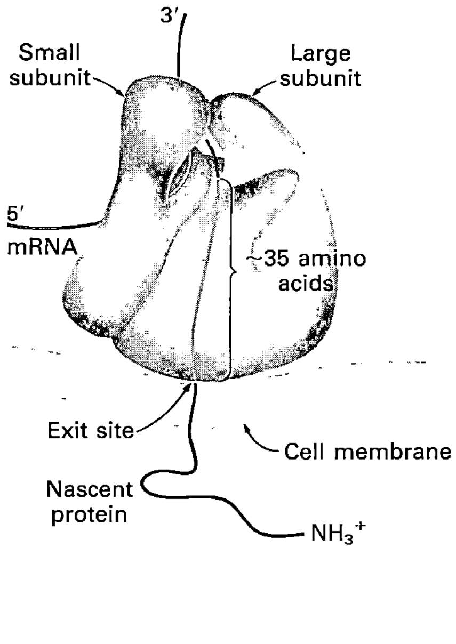

Figure 35-2
Diagram of a membrane-bound ribosome. [Courtesy of Dr. James Lake.]

somes in cell-free systems answered this important question. Free ribosomes from the cytosol were isolated and then added to rough ER membranes that had been stripped of their ribosomes. This reconstituted system actively synthesized secretory proteins when presented with the appropriate mRNAs and other soluble factors. Likewise, ribosomes isolated from the rough ER were fully active in synthesizing proteins that are normally released into the cytosol. Furthermore, there were no detectable structural differences between free ribosomes and ribosomes derived from the rough ER. Thus, *membrane-bound ribosomes and free ribosomes are intrinsically identical. Whether a particular ribosome is free or attached to the rough ER depends only on the kind of protein it is making.*

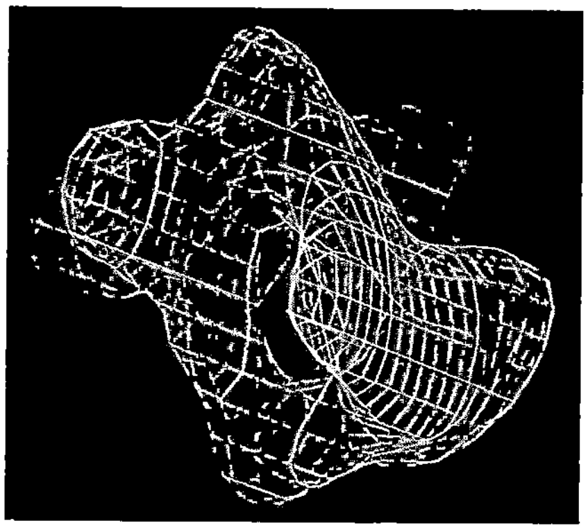

Tunnel in the large subunit of the *E. coli* ribosome. This image was obtained by analyses of electron micrographs of two-dimensional crystalline arrays. [From A. Yonath, K.R. Leonard, and H.G. Wittmann. *Science* 236(1987):815.]

SIGNAL SEQUENCES MARK PROTEINS FOR TRANSLOCATION ACROSS THE ENDOPLASMIC RETICULUM MEMBRANE

The attachment of an actively synthesizing ribosome to the ER membrane is a key event in translocating a protein across this membrane. What is it that directs the ribosome to the membrane? In 1970, David Sabatini and Günter Blobel postulated that *the signal for attachment is a sequence of amino acid residues near the amino terminus of the nascent polypeptide chain.* This *signal hypothesis* was soon supported; Cesar Milstein and George Brownlee found that an immunoglobulin chain synthesized in vitro by free ribosomes contained an amino-terminal sequence of 20 residues that was absent from the mature protein synthesized in vivo. Blobel then found that the major secretory proteins of the pancreas contain amino-terminal extensions of about 20 residues when synthesized in vitro by free ribosomes. These *signal sequences* are absent from normally secreted proteins because they are cleaved by a *signal peptidase* on the lumenal side of the ER membrane.

The cleaved amino-terminal signal sequences of more than a hundred secretory proteins from a wide variety of eukaryotic species are now known (Figure 35-3). A well-defined consensus sequence, such as the

		Cleavage site
Human growth hormone	MATGSRTSLLLAFGLLCLPWLQEGSA	FPT
Human proinsulin	MALWMRLLPLLALLALWGPDPAAA	FVN
Bovine proalbumin	MKWVTFISLLLFSSAYS	RGV
Mouse antibody H chain	MKVLSLLYLLTAIPHIMS	DVQ
Chicken lysozyme	MRSLLILVLCFLPKLAALG	KVF
Bee promellitin	MKFLVNVALVFMVVYISYIYA	APE
Drosophila glue protein	MKLLVVAVIACMLIGFADPASG	CKD
Zea maize protein 19	MAAKIFCLIMLLGLSASAATA	SIF
Yeast invertase	MLLQAFLFLLAGFAAKISA	SMT
Human influenza virus A	MKAKLLVLLYAFVAG	DQI

Figure 35-3
Amino-terminal signal sequences of some eukaryotic secretory and plasma membrane proteins. The hydrophobic core (yellow) is preceded by basic residues (blue) and followed by a cleavage site (red) for signal peptidase.

TATA box guiding the initiation of transcription, is not evident. However, signal sequences do exhibit several common features: (1) They range in length from 13 to 36 residues. (2) The amino-terminal part of the signal contains at least one positively charged residue. (3) A highly hydrophobic stretch, typically 10 to 15 residues long, forms the center of the signal sequence. Alanine, leucine, valine, isoleucine, and phenylalanine are common in this region. Substitution of a charged residue for a nonpolar one in this hydrophobic core usually destroys the directing activity of the signal sequence. (4) The residue on the amino-terminal side of the cleavage site usually has a small, neutral side chain. Alanine is most common.

Not all secretory and plasma membrane proteins contain an amino-terminal signal sequence that is cleaved following translocation across the ER membrane. Some proteins contain an *internal signal sequence* that serves the same role. The location of the internal signal sequence of ovalbumin, the major protein of hen egg white, has been determined. By recombinant DNA techniques, different segments of the ovalbumin gene were deleted to find out which amino acid residues are essential for targeting. This analysis revealed that the sequence between residues 22 and 41 is critical for the transfer of nascent ovalbumin across the endoplasmic reticulum membrane.

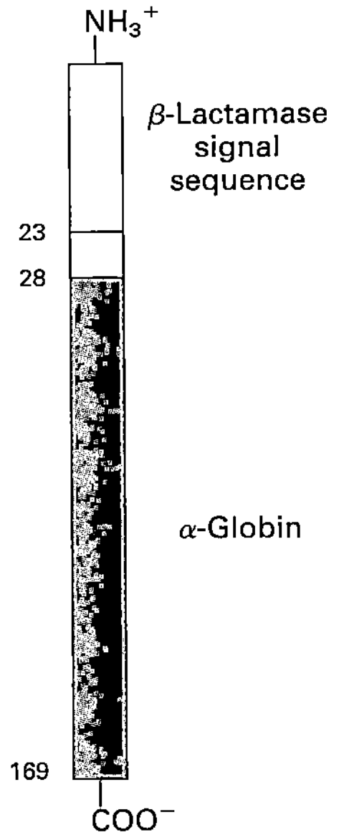

Figure 35-4
α-Globin (red) is redirected from the cytosol to the ER by joining a bacterial signal sequence (green) and five adjoining residues (yellow) to its amino terminus.

A CYTOSOLIC PROTEIN CAN BE REDIRECTED TO THE ER BY JOINING A SIGNAL SEQUENCE TO ITS AMINO TERMINUS

Chimeric proteins formed by recombinant DNA techniques have provided insight into mechanisms of protein targeting. This approach has been used to determine whether a cleavable amino-terminal signal sequence contains all the information needed to direct the translocation of a protein across the ER membrane. The signal sequence in this experiment came from β-lactamase, an *E. coli* enzyme that hydrolyzes penicillin and other β-lactams. As it is synthesized, this enzyme is secreted into the periplasmic space between the outer and plasma membranes of the bacterium. In the process, a 23-residue signal sequence is cleaved from the polypeptide. This sequence and five adjoining residues of β-lactamase were joined to the amino terminus of the α chain of hemoglobin by forming a hybrid gene (Figure 35-4). α-Globin was chosen as the test protein because it normally stays in the cytosol and is highly hydrophilic.

The mRNA for this chimeric protein was added to an in vitro protein-synthesizing system that can translocate secretory proteins across the ER membrane (Figure 35-5). This in vitro system, derived from a eukaryote,

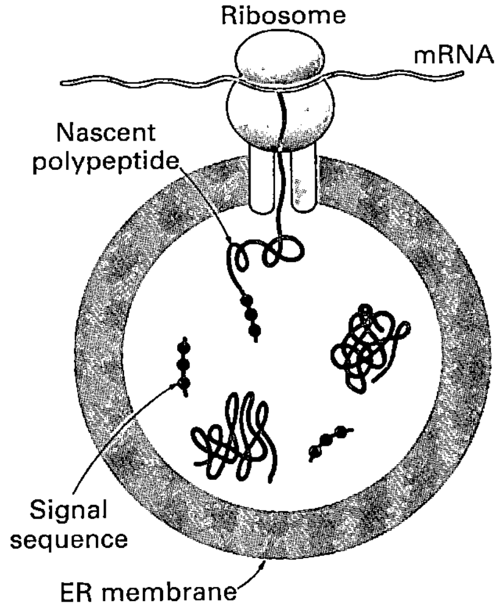

Figure 35-5
Membrane and secretory proteins can be synthesized and sequestered in vitro by using microsomal vesicles produced by fragmentation of the endoplasmic reticulum.

consisted of microsomes, ribosomes, tRNAs, and other soluble factors required for protein synthesis and translocation. (Microsomes are closed vesicles formed by the self-sealing of fragments of the endoplasmic reticulum membrane.) Ribosomes synthesizing α-globin with a signal sequence became bound to the microsomes. Most significant, the nascent polypeptide chain traversed the ER membrane, the signal sequence was cleaved, and α-globin was sequestered inside the vesicle. Thus, *the addition of a signal sequence to α-globin converted it from a cytosolic protein into a secretory protein. This experiment also demonstrated that bacterial and eukaryotic signal sequences are functionally similar.*

SIGNAL RECOGNITION PARTICLE (SRP) DETECTS SIGNAL SEQUENCES AND BRINGS RIBOSOMES TO THE ER MEMBRANE

How are signal sequences recognized and the proteins containing them translocated across the ER membrane? Blobel and Peter Walter showed that the adaptor coupling the protein-synthesizing machinery in the cytosol to the protein-translocating machinery in the ER membrane is a ribonucleoprotein called *signal recognition particle (SRP)*. This 325-kd assembly consists of a 300-nucleotide RNA molecule called *7SL RNA* and six different polypeptide chains. SRP binds tightly to ribosomes containing a nascent chain with a signal sequence but not to other ribosomes (Figure 35-6). This binding occurs soon after the emergence of the amino-terminal signal sequence from the ribosome. Elongation of the polypeptide chain is arrested or slowed while SRP is bound. The SRP-ribosome complex then diffuses to the ER membrane, where SRP binds to the *SRP receptor* (docking protein). A GTP-driven process discussed below then delivers the ribosome containing the nascent polypeptide chain to the translocation machinery. The concomitant release of SRP from the ribosome leads to the resumption of elongation. Thus, *SRP acts catalytically to deliver ribosomes with a signal sequence to the ER membrane.* SRP also prevents premature elongation and folding of the nascent chain, which would interfere with translocation.

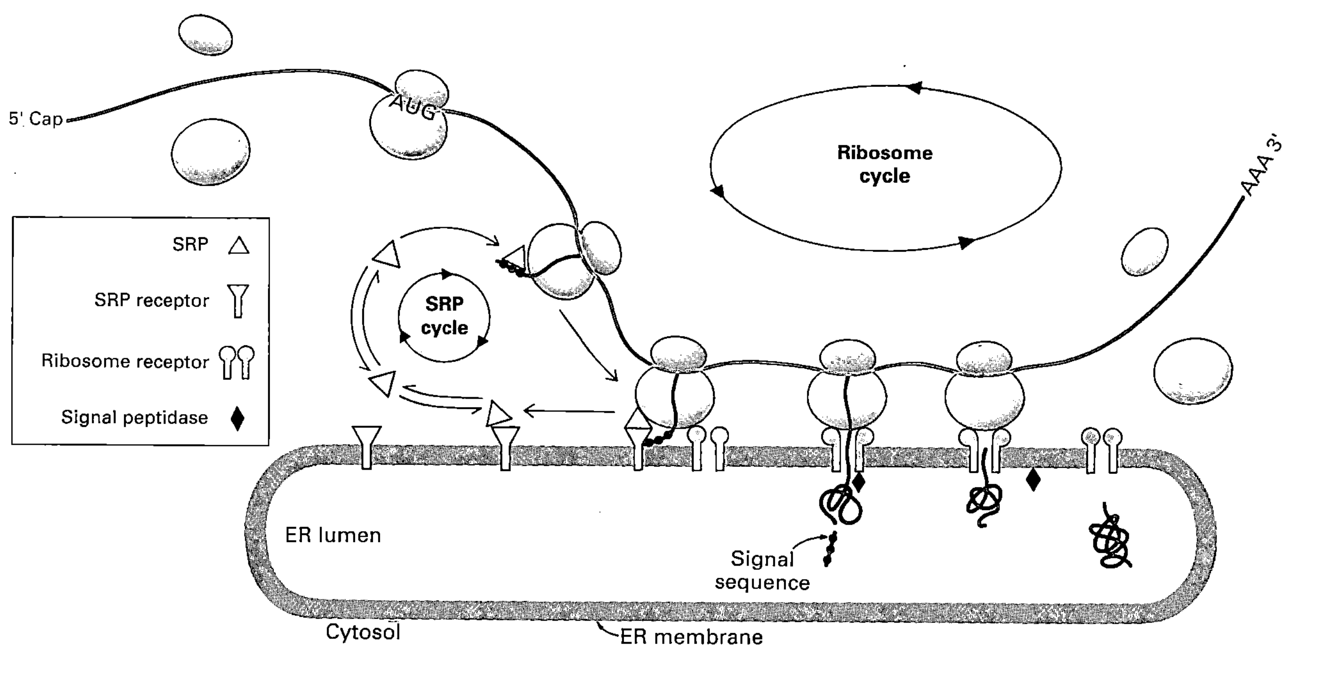

Figure 35-6
Signal recognition particle (SRP) participates in delivering ribosomes to the ER. The signal sequence of a nascent polypeptide chain is recognized by SRP. The complex consisting of SRP, the nascent chain, and the ribosome binds to the SRP receptor in the ER membrane. The ribosome is then transferred to the translocation machinery, which actively threads the polypeptide chain across the ER membrane. SRP released from its receptor is free to bind another emerging signal sequence. [After P. Walter, R. Gilmore, and G. Blobel. *Cell* 38(1984):6.]

A GTP-GDP CYCLE RELEASES THE SIGNAL SEQUENCE FROM SRP AND THEN DETACHES SRP FROM ITS RECEPTOR

The signal sequence on the nascent polypeptide must be sequestered by SRP until it is delivered to the translocation machinery on the ER membrane. How does SRP know when to let go? Precise timing is achieved by a GTP-GDP cycle in the *SRP receptor,* an integral membrane protein consisting of α (68 kd) and β (30 kd) subunits. The binding of SRP-signal peptide to the receptor triggers the exchange of GTP for GDP bound to the α subunit (Figure 35-7). *The GTP form of the receptor tenaciously binds SRP, which releases its grip on the signal peptide. The freed signal peptide swiftly binds to the translocation machinery.* The α subunit of the receptor then hydrolyzes its bound GTP to GDP, which releases SRP. The delay in GTP hydrolysis gives the signal peptide time to find its new partner. The interval is sufficiently long to ensure that the signal peptide is not recaptured by SRP. *Ribosomes bearing signal sequences are vectorially targeted to the ER membrane because of the unidirectionality of this GTP-GDP cycle.* We see here mechanistic motifs that are reminiscent of the action of seven-helix receptors in triggering G proteins (p. 341) and of elongation factor Tu in delivering aminoacyl-tRNAs to ribosomes (p. 897).

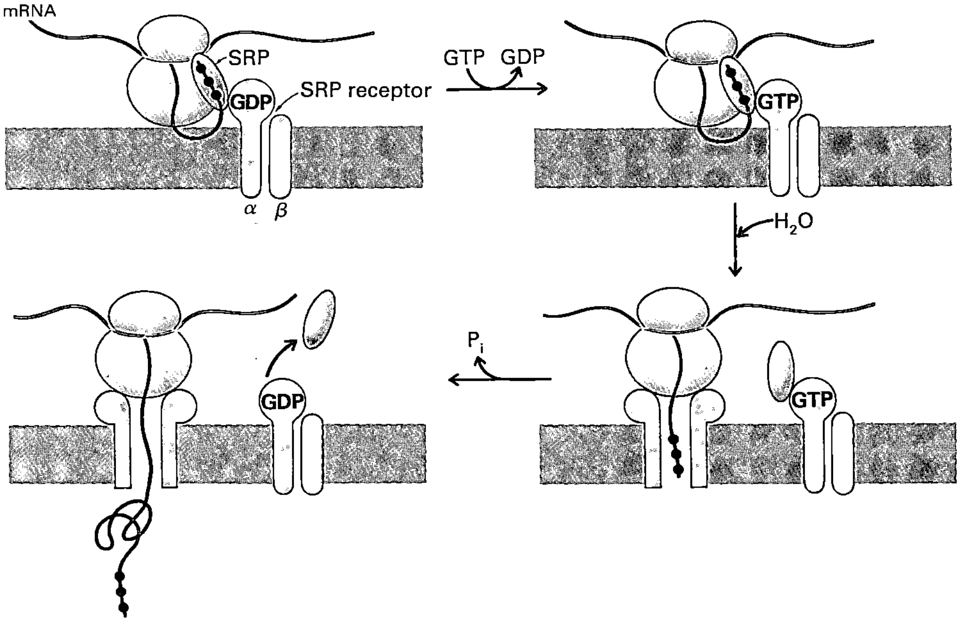

Figure 35-7
GTP-GDP cycle of the signal recognition particle receptor. This cycle drives the delivery of the signal sequence to the translocation machinery of the ER membrane. [After T.A. Rapoport. *Science* 258(1992):931.]

SIGNAL PEPTIDES OPEN PROTEIN-CONDUCTING CHANNELS

The translocation machinery, called the *translocon,* is a multisubunit assembly of integral and peripheral membrane proteins. Only a few components have been identified, and little is known about how nascent chains are threaded through the ER membrane. Recent electrophysiological studies demonstrate the existence of protein-conducting channels in the ER membrane, most likely formed by activated translocons (see p. 272 and p. 294 for discussions of electrophysiological methods). Microsomes derived from the rough ER of the pancreas were fused to one side of a planar bilayer membrane; membrane asymmetry is preserved in fusion.

The addition of puromycin (p. 902) resulted in a large increase in ionic conductance, most likely because puromycin cleared nascent protein chains from the lumen of the channel. Single channels having a conductance of 220 pS were detected when puromycin was added at low concentration. The magnitude of the conductance, about five times that of the acetylcholine receptor channel (p. 295), suggests that the diameter of the channel is about 15 Å. Most striking, *the addition of signal peptide to the cytosolic but not the periplasmic side of a planar bilayer membrane containing fused bacterial cell membrane led to the opening of channels* (Figure 35-8). The gating of protein-conducting channels by signal peptides is an economical device for assuring that they open only when nascent proteins are ready to be translocated. The proton-motive force across the membrane would be rapidly dissipated if the channels were open but unoccupied by proteins undergoing translocation.

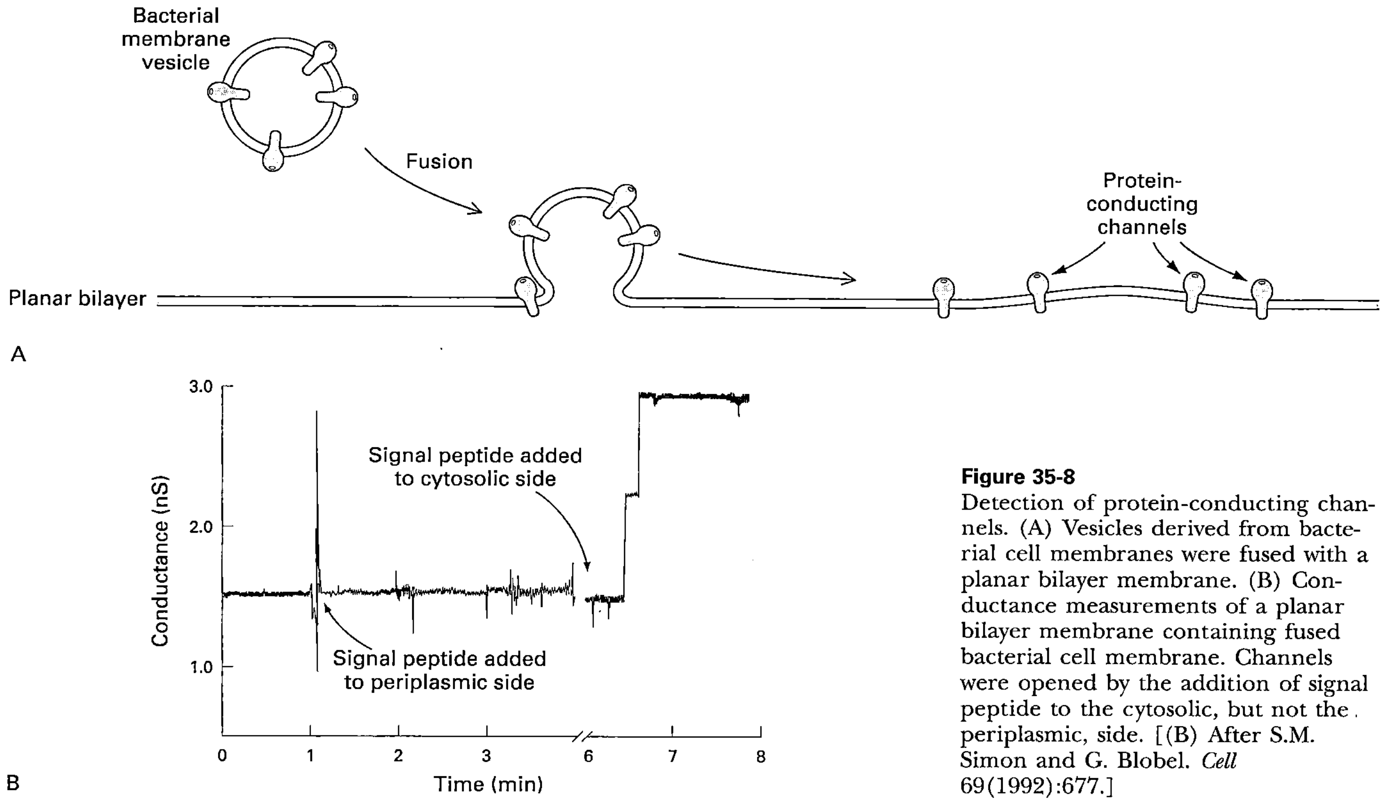

Figure 35-8
Detection of protein-conducting channels. (A) Vesicles derived from bacterial cell membranes were fused with a planar bilayer membrane. (B) Conductance measurements of a planar bilayer membrane containing fused bacterial cell membrane. Channels were opened by the addition of signal peptide to the cytosolic, but not the periplasmic, side. [(B) After S.M. Simon and G. Blobel. *Cell* 69(1992):677.]

TRANSLOCATION IS GUIDED BY SIGNAL SEQUENCES AND STOP-TRANSFER SEQUENCES

What is the relation between the elongation of a polypeptide chain and its translocation across the ER membrane? It was first thought that elongation drives translocation, that peptide synthesis pushes the nascent chain through a protein tunnel in the bilayer. However, some proteins, such as an 18-kd precursor of a yeast mating factor, can be translocated after they have been completely synthesized and released from the ribosome. Indeed, the glucose transporter of human erythrocytes, a large protein with 12 membrane-spanning α helices, can be slowly translocated in vitro after being fully synthesized. Thus, *translocation and elongation are mechanistically separate processes.*

Why then do they usually occur simultaneously? The answer is that most fully synthesized proteins cannot be efficiently translocated, most likely because they have folded. A pair of interacting α helices, for example, probably cannot fit in the protein-conducting channel. *Unfolded polypeptide chains are optimal substrates for translocation across the ER membrane.* As mentioned earlier, elongation arrest induced by the binding of SRP to ribosomes prevents premature folding of the nascent chain. The ribosome, too, participates in enabling the nascent chain to be translocated by keeping its most recently synthesized part fully stretched out in the narrow tunnel of the large subunit.

The translocation process for integral membrane proteins is more complex than for secretory and lysosomal proteins, which are threaded through in entirety. Some integral membrane proteins have one membrane-spanning helix, whereas others have several (Figure 35-9). Moreover, the amino and carboxy termini can be on either side of the membrane, depending on the particular protein. How is a polypeptide chain prevented from being translocated in entirety? The translocation machinery, like the broom the Sorcerer's Apprentice sought to control, is unrelenting in its action unless stopped by a specific instruction, which in this case is a *stop-transfer sequence* (also called a *membrane anchor sequence*) on the nascent polypeptide. A second signal sequence is needed to initiate another round of translocation of a chain that spans the membrane more than once. Furthermore, the translocation machinery must also be able to thread polypeptide chains in the reverse direction. Much remains to be learned about the assembly and placement of membrane proteins, a fascinating area of inquiry.

Figure 35-9
Different topological arrangements of integral membrane proteins. [After W.T. Wickner and H.F. Lodish. *Science* 230(1985):400.]

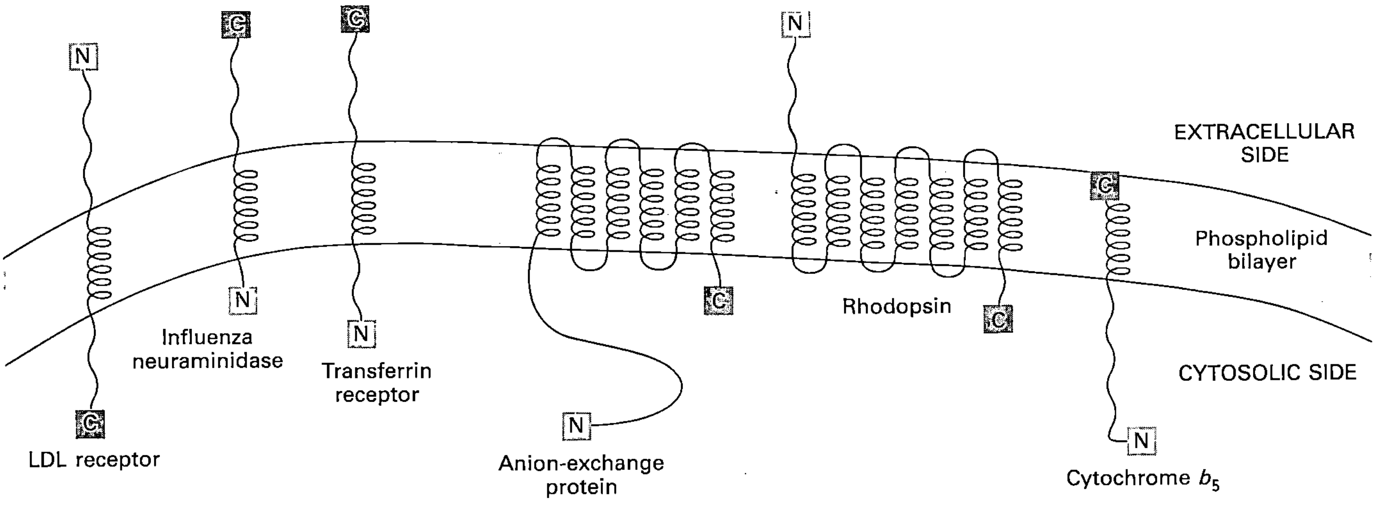

ATP-DRIVEN HEAT-SHOCK PROTEINS SERVE AS CHAPERONES THAT BIND NASCENT PROTEINS AND ASSIST THEIR FOLDING

Nascent polypeptide chains in the lumen of the endoplasmic reticulum do not immediately fold. Rather, they bind to *chaperone proteins* that keep them unfolded for minutes. At first glance, the need for a chaperone seems puzzling, given that the conformation of a protein is determined by its amino acid sequence (p. 37). However, the situation in vivo, in the endoplasmic reticulum, differs from that in vitro in two important ways. First, in vitro refolding experiments are usually carried out with whole proteins. In contrast, the amino-terminal end of a nascent protein emerges from the ER channel many seconds or even minutes before the carboxyl-terminal end does. Second, the protein concentration in the lumen of the ER is very high, of the order of 200 mg/ml (compared with less than 1 mg/ml for optimal in vitro refolding). Hence, nascent chains have many opportunities for promiscuous interactions that are diversionary. In the absence of chaperones, the multitude of nascent chains would become hopelessly entangled. Chaperones assist folding by blocking illicit intermolecular interactions. They hold in trust the emerging part of the nascent chain until synthesis is complete.

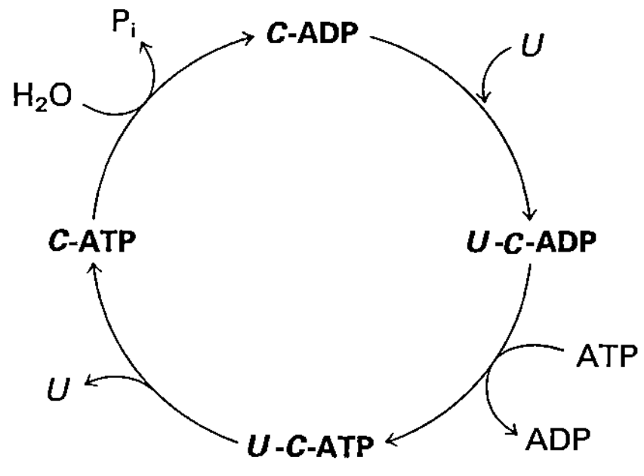

C = chaperone;
U = unfolded peptide segment

What is the timer that determines when a chaperone will release its bound nascent chain? *Chaperone proteins are slow ATPases.* The ADP-chaperone complex has high affinity for unfolded polypeptides but not for native proteins. The binding of an unfolded peptide segment to the chaperone stimulates the release of ADP and the entry of ATP into the chaperone's catalytic site. The ATP-chaperone complex releases the peptide segment. The subsequent hydrolysis of bound ATP enables the chaperone to again bind an unfolded segment; the interval between release and binding is determined by the hydrolytic rate. As folding proceeds, less and less of the nascent chain can be bound by the chaperone. In essence, *chaperones buy time*—proper folding is then a matter of thermodynamics.

The major chaperone in the lumen of the ER is *BiP* (*bi*nding *p*rotein), a 78-kd member of the *heat-shock protein (hsp) family.* All cells, prokaryotic and eukaryotic, respond to an increase in temperature (a heat shock) by synthesizing new proteins that belong to this family. Other stresses, such as free-radical damage, have a similar effect. The common denominator in the induction of heat-shock proteins is an abundance of *unfolded polypeptide chains.* It should be noted that many members of this family are not induced—rather, they are normally present in virtually all cellular compartments (Table 35-1).

Table 35-1
Heat-shock proteins

Hsp60 (chaperonin-60) family
GroEL (in bacterial cytosol)
Hsp60 (in mitochondrial matrix)
Rubisco binding protein (in chloroplasts)
Hsp70 (stress-70) proteins
Hsp70 (in mammalian cytosol)
BiP (in ER of eukaryotes)
Grp75 (in mitochondria)
DnaK (in bacterial cytosol)
Hsp90 (stress-90) proteins
Hsp83 (in eukaryotic cytosol)
Grp94 (in mammalian ER)
HtgP (in bacterial cytosol)

Source: M.-J. Gething and J. Sambrook. *Nature* 355(1992):33.

The hsp70 class of heat-shock proteins, which includes BiP, is highly conserved in evolution—the amino acid sequences of the E. coli *and human proteins are 50% identical.* Hsp70 proteins consist of an ATPase domain and a peptide-binding domain.

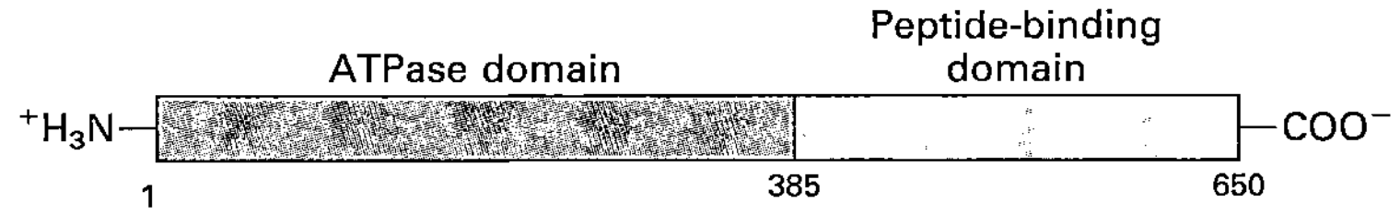

The ATPase moiety has two lobes with a cleft between them; the nucleotide is bound at the base of the cleft (Figure 35-10). We have already encountered this structural motif in actin, which also is a slow ATPase (see Figure 15-14 on p. 397).

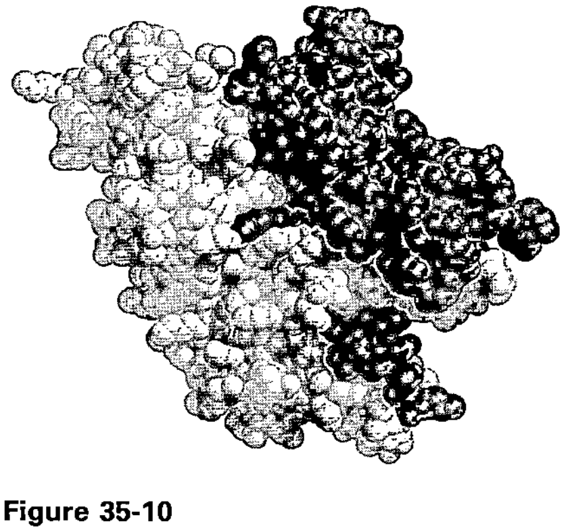

Figure 35-10
Structure of an ATPase fragment derived from hsc70, a cytosolic heat-shock protein. ADP (red) is bound in the cleft between the two domains (yellow and blue). [Drawn from 1ats.pdb. K.M. Flaherty, C. DeLuca-Flaherty, and D.B. McKay. *Nature* 346(1990):623.]

The lumen of the endoplasmic reticulum contains proteins in addition to chaperones that are essential for the folding and processing of newly synthesized polypeptide chains. *Protein disulfide isomerase* accelerates protein folding by catalyzing the formation of correct disulfide pairings (p. 429). This dimeric enzyme contains four thioredoxinlike domains; the thiol-disulfide active site is located in an exposed loop of each domain. *Peptidyl prolyl isomerases* are important too (p. 430). They accelerate protein folding by catalyzing the cis-trans isomerization of X-Pro bonds (*X* denotes any residue).

GLYCOPROTEINS ACQUIRE THEIR CORE SUGARS FROM DOLICHOL DONORS IN THE ENDOPLASMIC RETICULUM

Many secretory proteins and membrane proteins bear covalently attached carbohydrate moieties. As was discussed earlier (p. 280), the oligosaccharide units of glycoproteins are linked to either asparagine side chains by *N*-glycosidic bonds or to serine and threonine side chains by

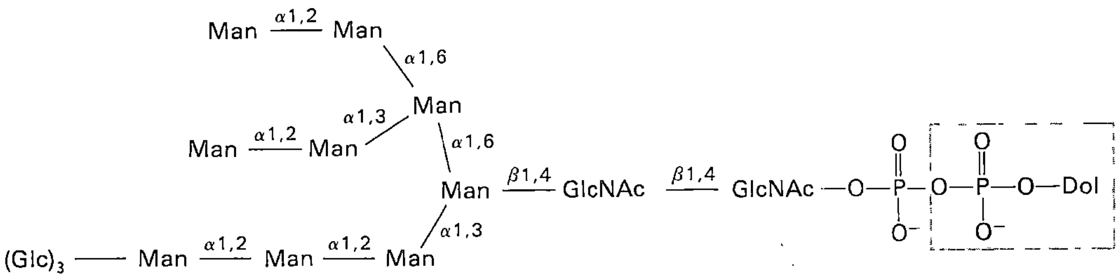

Figure 35-11
Structure of an activated core oligosaccharide (green). The carrier (yellow) is dolichol phosphate.

O-glycosidic bonds. *N*-linked oligosaccharides, though diverse, contain a common pentasaccharide core consisting of three mannose and two *N*-acetylglucosamine residues. This unifying motif expresses the mode of biosynthesis of such oligosaccharide units. A somewhat larger common oligosaccharide block can be constructed on an activated carrier, which then transfers it to a growing polypeptide chain on the lumenal side of the ER membrane (Figure 35-11). The carrier is *dolichol phosphate,* a very long lipid containing some 20 isoprene (C_5) units. The terminal phosphoryl group of this highly hydrophobic carrier is the site of attachment of the activated oligosaccharide.

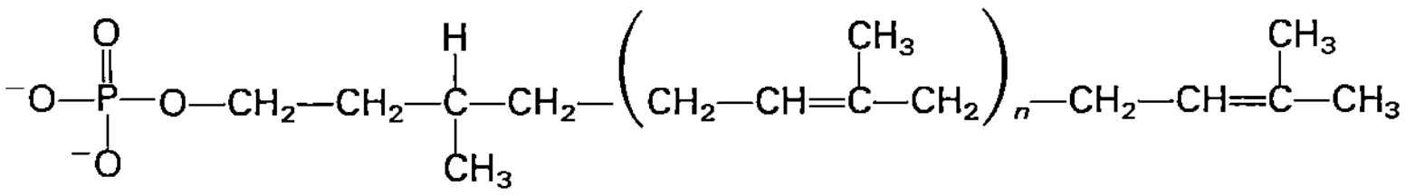

Dolichol phosphate
(n = 15–19)

Dolichol phosphate acquires an oligosaccharide unit consisting of two *N*-acetylglucosamines, nine mannoses, and three glucoses by the sequential addition of monosaccharides (Figure 35-12). The activated sugar donors in these reactions are derivatives of UDP, GDP, and dolichol pyrophosphate. A series of specific transferases catalyzes the synthesis of the activated core oligosaccharide. Dolichol phosphate, located in the ER membrane with its phosphate terminus on the cytosolic face, acquires two *N*-acetylglucosamine residues and five mannoses from activated intermediates in the cytosol. Dolichol bearing this growing chain is translocated across the membrane so that its sugar end is located on the lumenal side. It then acquires four mannose and three glucose residues from dolichol phosphate intermediates rather than from UDP or GDP. Dolichol glucose and dolichol mannose are regenerated on the cytosolic side of the endoplasmic reticulum membrane.

The 14 sugar residues in this dolichol phosphate intermediate are then transferred en bloc to a specific asparagine residue of the growing poly-

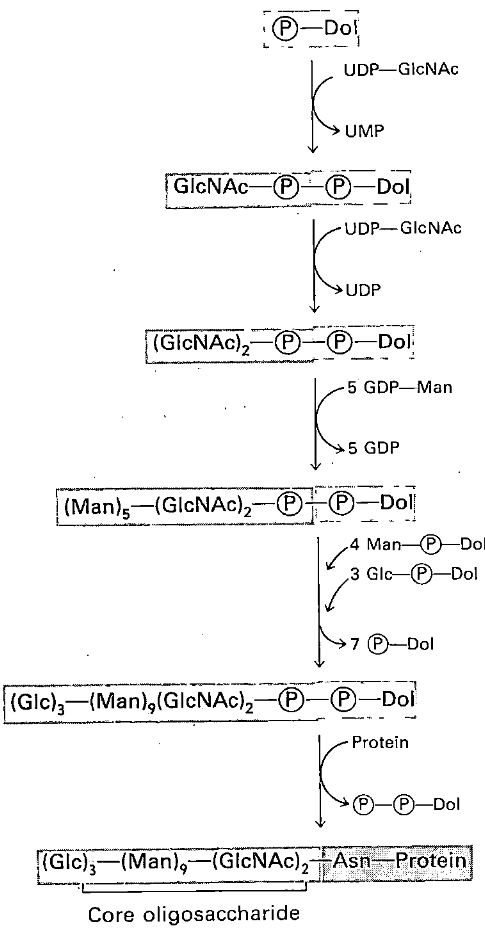

Figure 35-12
An activated core oligosaccharide is built by the sequential addition of single sugar units. The block is then transferred to an asparagine side chain of a nascent protein in the lumen of the ER.

peptide chain. The activated oligosaccharide and the requisite transferase are located on the lumenal side of the ER, accounting for the fact that proteins in the cytosol are not glycosylated. An asparagine residue can accept the oligosaccharide only if it is part of an Asn-X-Ser or Asn-X-Thr sequence, and if it is sterically accessible to the transferase.

Dolichol pyrophosphate released in the transfer of the oligosaccharide to the protein is recycled to dolichol phosphate by the action of a phosphatase. This hydrolysis is blocked by *bacitracin,* an antibiotic. Another interesting antibiotic inhibitor of *N*-glycosylation is *tunicamycin,* a hydrophobic analog of UDP–*N*-acetylglucosamine. Tunicamycin blocks the addition of *N*-acetylglucosamine to dolichol phosphate, the first step in the formation of the core oligosaccharide.

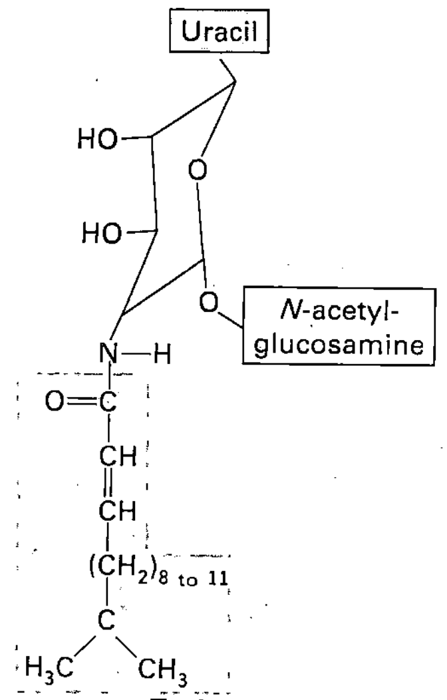

Tunicamycin

THE ABSENCE OF GLUCOSE SIGNALS THAT A GLYCOPROTEIN IS FULLY FOLDED AND READY FOR EXPORT TO THE GOLGI

The three glucose residues and one of the mannoses are rapidly trimmed from the *N*-linked oligosaccharide while the glycoprotein is still in the ER (Figure 35-13). *Totally unexpected, glucose is added back to the N-linked oligosaccharide if the protein is unfolded or misfolded.* Indeed, many rounds of glucose addition and trimming occur while the glycoprotein is in the ER. Why? This reaction cycle plays a key role in determining when a glycoprotein is properly folded and hence ready for export to the Golgi complex.

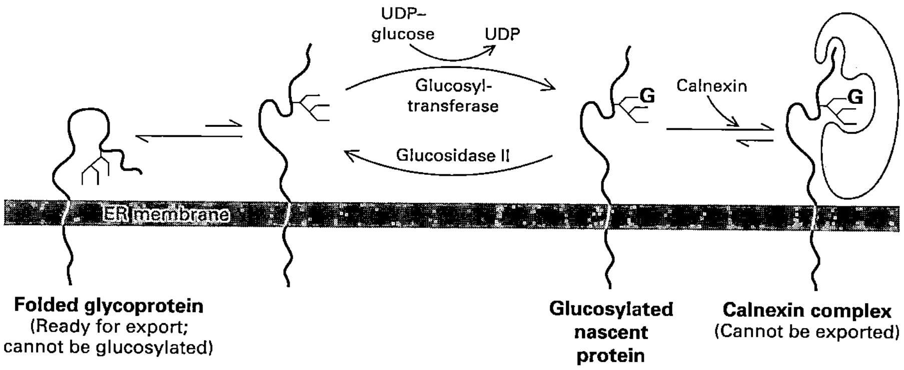

Figure 35-13
Quality control system in the ER for determining when a glycoprotein is ready for export to the Golgi complex. A glucosyltransferase that uses UDP–glucose as the donor adds glucose to the oligosaccharide of unfolded or misfolded, but not folded, proteins. Only glucosylated glycoproteins bind to calnexin, which serves as a chaperone to prevent the export of immature or defective proteins. [After C. Hammond, I. Braakman, and A. Helenius. *Proc. Nat. Acad. Sci.* 91(1994):913].

A glycoprotein bearing one or more glucose residues on the *N*-linked oligosaccharide stays bound to *calnexin,* a chaperone protein anchored in the ER membrane. When the glycoprotein is fully folded, glucosylation ceases. The rapid trimming of glucose in the absence of readdition leads to the release of the glycoprotein from calnexin. The chaperone lets go when convinced that its charge has reached maturity. Thus, *multiple cycles of glucosylation and trimming, and the binding only of glucosylated glycoproteins to calnexin, play key roles in controlling the quality of exports from the ER.*

TRANSPORT VESICLES CARRY PROTEINS FROM THE ER TO THE GOLGI FOR FURTHER GLYCOSYLATION AND SORTING

Proteins in the lumen of the ER and in the ER membrane are transported to the *Golgi complex,* which is a stack of flattened membranous sacs (Figure 35-14). The Golgi has two principal roles. First, *carbohydrate units of glycoproteins are altered and elaborated in the Golgi. O*-linked sugar units are fashioned there, and *N*-linked ones are modified in many different ways. Second, *the Golgi is the major sorting center of the cell.* It sends proteins to lysosomes, secretory granules, or the plasma membrane according to signals encoded by their three-dimensional structures. The transport of proteins between the ER and Golgi, and between the Golgi and subsequent destinations, is mediated by small (~50 to 100 nm in diameter) membrane-bound compartments called *transport vesicles (transfer vesicles).*

The Golgi of a typical mammalian cell has three or four membranous sacs (cisternae), and those of many plant cells have about twenty. The stack of cisternae is asymmetric. The Golgi is differentiated into (1) a *cis* compartment, the receiving end, which is closest to the ER; (2) *medial* compartments; and (3) a *trans* compartment, which exports proteins to a variety of destinations. These compartments contain different enzymes and mediate distinctive functions. Different vesicles transfer proteins from one Golgi compartment to another and then to lysosomes, secretory vesicles, and the plasma membrane (Figure 35-15). Many cells have more than one secretory pathway. *Constitutive secretion* is rapid and continuous, whereas *regulated secretion* is episodically triggered by specific signals. Soluble proteins in the lumen of the ER emerge in constitutive secretory vesicles and are exported unless special signals are present to direct them to other locations. Likewise, integral membrane proteins initially in the ER go to the plasma membrane unless they carry instructions to the contrary. In other words, *constitutive secretory vesicles and the plasma membrane are the default destinations.*

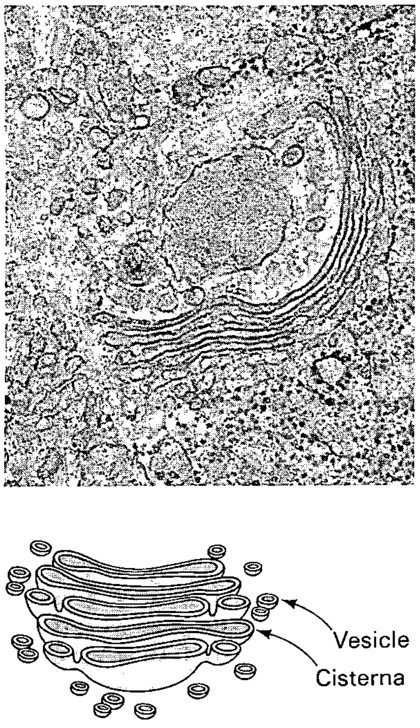

Figure 35-14
Electron micrograph and schematic diagram of the Golgi complex. [Micrograph courtesy of Lynne Mercer.]

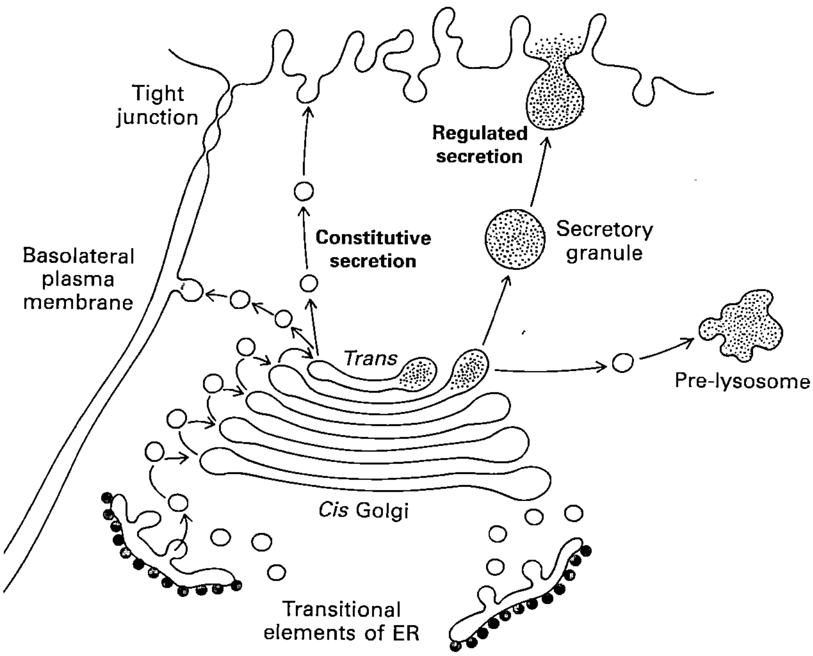

Figure 35-15
The Golgi complex is the sorting center in the targeting of proteins to lysosomes, secretory vesicles, and the plasma membrane. The cis face of the Golgi receives vesicles from the ER, and the trans face sends a different set of vesicles to different target sites. Proteins are also transferred by vesicles from one compartment of the Golgi to another. Polarized cells, such as those in epithelia, contain apical and basolateral regions separated by tight junctions that prevent intermixture of their contents. Different proteins are targeted to these respective regions. [Courtesy of Dr. Marilyn Farquhar.]

Carbohydrate units of glycoproteins are modified in each of the compartments of the Golgi (Figure 35-16). In the *cis* Golgi, three mannoses are removed from the oligosaccharide chains of proteins destined for secretion or for insertion in the plasma membrane. The carbohydrate units of glycoproteins targeted to the lumen of lysosomes are modified differently, as described below. In the *medial* Golgi of some cells, two more mannoses are removed, and two *N*-acetylglucosamines and a fucose are added. Finally, in the *trans* Golgi, another *N*-acetylglucosamine is added, followed by galactose and sialic acid, to form a complex oligosaccharide unit (p. 475). The sequence of *N*-linked oligosaccharide units of glycoproteins is determined both by (1) the sequence and conformation of the protein undergoing glycosylation and by (2) the glycosyl transferase composition of the Golgi in which they are processed. Carbohydrate processing in the Golgi is called *terminal glycosylation* to distinguish it from *core glycosylation,* which takes place in the ER.

Figure 35-16
Asparagine-linked oligosaccharides are processed in the ER and the three compartments of the Golgi. The enzymes catalyzing these steps are (1) glucosidase I, (2) glucosidase II, (3) ER α-1,2-mannosidase, (4) *N*-acetylglucosaminylphosphotransferase, (5) phosphodiester glycosidase, (6) Golgi mannosidase I, (7) GlcNAc transferase I, (8) mannosidase II, (9) GlcNAc transferase II, (10) fucosyl transferase, (11) GlcNAc transferase IV, (12) galactosyltransferase, (13) sialyltransferase. Steps 4 and 5 apply only to proteins destined for delivery to lysosomes (p. 927). [After D.E. Goldberg and S. Kornfeld. *J. Biol. Chem.* 258(1983):3160.]

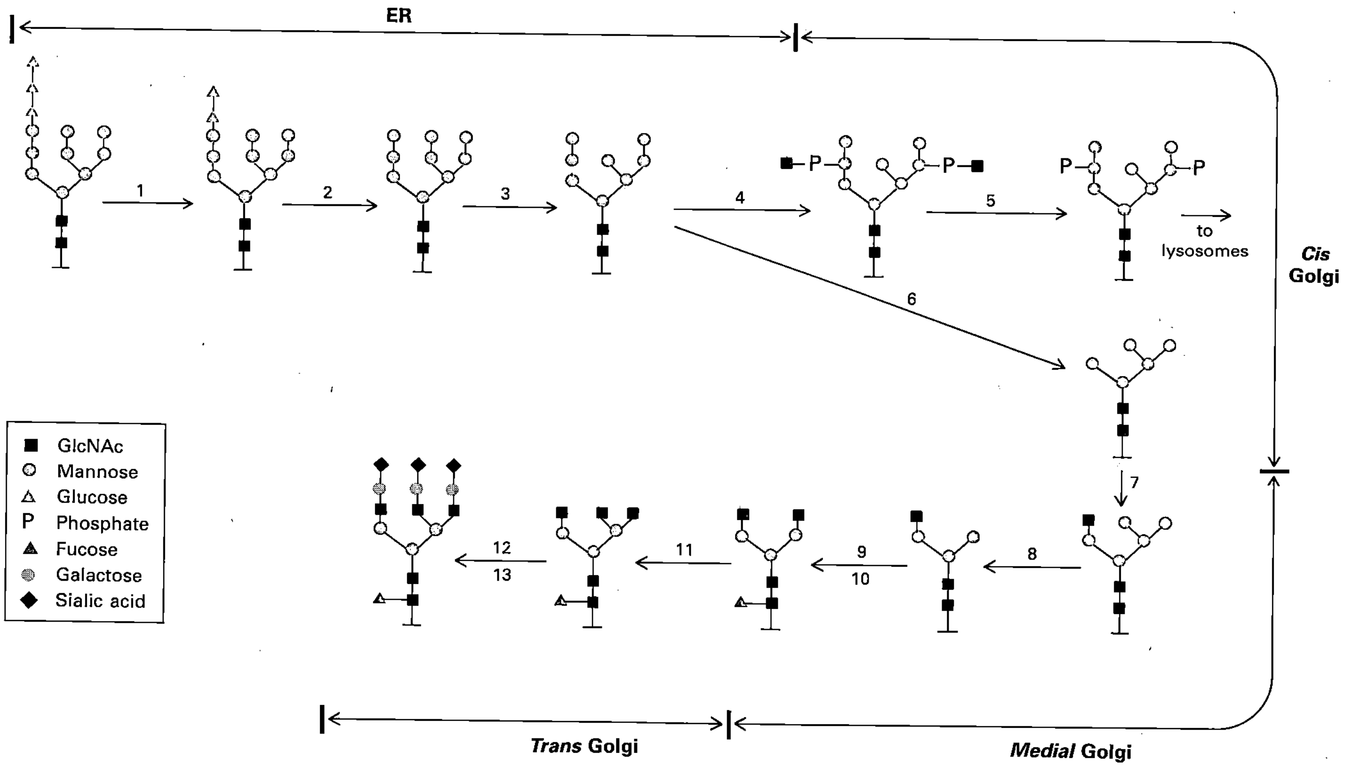

MEMBRANE ASYMMETRY IS PRESERVED IN THE BUDDING AND FUSION OF TRANSPORT VESICLES

The asymmetry of biological membranes is central to their function (p. 278). *In vesicular transport, membrane asymmetry is preserved* (Figure 35-17). The cytosolic face of a transport vesicle corresponds to the cytosolic face of the donor compartment. The topology of fusion mirrors that of budding. Following fusion, the cytosolic face of the transport vesicle becomes continuous with the cytosolic face of the target compartment. *The corollary is that the lumenal side of transport vesicles and the Golgi corresponds to*

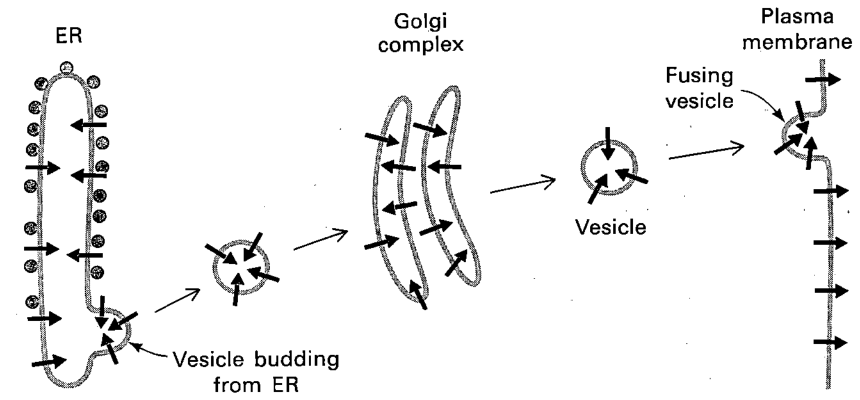

Figure 35-17
Vesicles transferred between membranes preserve their asymmetry. The lumenal face of the ER and other organelles corresponds to the extracellular face of the plasma membrane.

the lumenal side of the ER membrane. However, when a vesicle ultimately fuses with the plasma membrane, its lumenal surface becomes part of the extracellular surface of the plasma membrane. Hence, *the lumenal side of ER and other organelle membranes corresponds to the external face of the plasma membrane.* For this reason, the carbohydrate groups of glycoproteins in the plasma membrane are always on its extracellular surface (p. 280). It is also noteworthy that the lumenal contents of intracellular compartments are not released into the cytosol during budding or fusion. Entry into the ER is irreversible: *once a protein enters the ER, it never returns to the cytosol.*

SMALL GTP-BINDING PROTEINS, COAT PROTEINS, SNAPs, AND SNAREs PLAY KEY ROLES IN VESICULAR TRANSPORT

How does a vesicle bud from the donor compartment? What determines its cargo? How does the transport vesicle identify the target membrane? And, finally, how does fusion occur? The molecular mechanisms of budding, selectivity, and fusion were a total mystery until a few years ago. The veil is being lifted, and we are beginning to catch glimpses of how these complex processes take place and are regulated.

The reconstitution of Golgi transport in a cell-free system by James Rothman has provided insight into the molecular mechanism of these processes. When isolated Golgi stacks are incubated with cytosol and ATP, many ~75 nm diameter vesicles bud from their cisternae. These vesicles are initially covered by a set of *coat proteins* (α-, β-, γ-, and δ-*COPS)* that form a *coatomer* shell (Figure 35-18). A small GTP-binding protein called *ARF* (*A*DP-*r*ibosylation *f*actor) is also required. *The formation of a coat on a segment of the donor membrane increases its curvature so that it can be shaped into a vesicle.* When the coated vesicle finds its target membrane, GTP bound to ARF is hydrolyzed to GDP. The coat is then released from the vesicle, a key step in preparing the vesicle for fusion with the target membrane.

The finding that fusion following uncoating can be blocked by *N*-ethylmaleimide (NEM), which alkylates sulfhydryl groups, led to the isolation of an *NEM-sensitive fusion factor* called *NSF.* ATP, fatty acyl CoAs, and several other proteins are also needed for fusion. In fact, NSF is an ATPase—*ATP binding is necessary for membrane attachment, whereas ATP hydrolysis is required for release.*

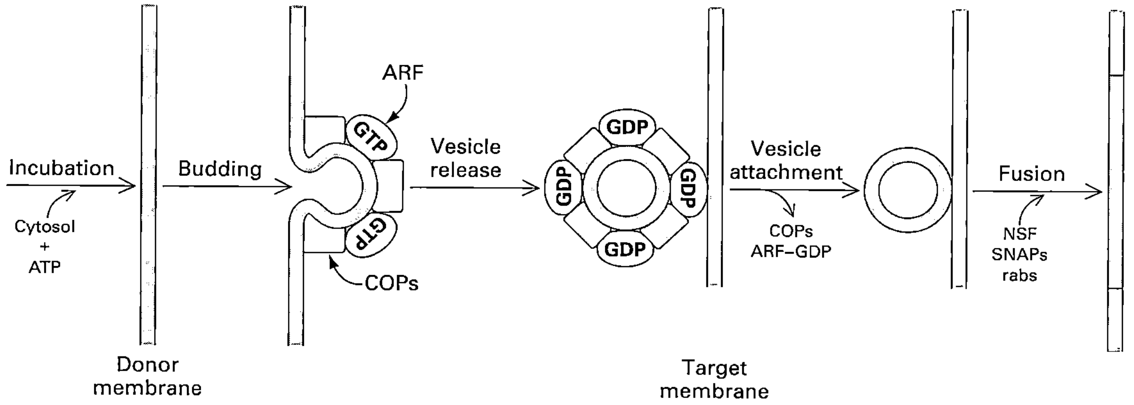

Figure 35-18
Proposed mechanism of vesicular transport between Golgi compartments. The binding of ARF and coatomers to the surface of a donor membrane (blue) leads to its budding. The resulting COP-coated vesicle binds to a target membrane (yellow), which triggers uncoating. ARF cycles between GTP and GDP forms. An *N*-ethylmaleimide-sensitive protein (NSF) and several other proteins are required for fusion. [After J.E. Rothman and L. Orci. *Nature* 355(1992):409.]

NSF requires additional cytosolic proteins to bind to Golgi membranes. These *soluble NSF attachment proteins,* termed *SNAPs,* recognize a membrane receptor. Several *SNAP receptors* (called *SNAREs*) from brain have been identified. *Synaptobrevin* is present on the surface of synaptic vesicles, which fuse with the plasma membrane to deliver neurotransmitters into the synaptic cleft. A single transmembrane helix anchors this receptor in the vesicle membrane; the remainder of the protein is located on the cytosolic side, where it is accessible for interaction with NSF and SNAP. Synaptobrevin is critical for synaptic vesicle fusion in vivo, as evidenced by the finding that cleavage of this receptor by *botulinum B toxin* blocks neurotransmitter release and causes paralysis. Botulinum toxin, produced by the anaerobic bacterium *Clostridium botulinum,* is one of the most potent toxins known; food-borne botulism has a mortality rate of 10%.

Syntaxin, a second SNARE, is present in the presynaptic plasma membrane, adjacent to synaptic vesicles that are ready to fuse. The trigger for fusion is a rise in the cytosolic Ca^{2+} level caused by depolarization of the nerve terminal. An attractive working hypothesis is that *vesicles find specific targets by the pairing of complementary receptors, a v-SNARE from the transport vesicle, and a t-SNARE from the target compartment* (Figure 35-19). The complementary v- and t-SNARES are thought to associate in a fusion particle that contains other recognition and regulatory elements, such as calcium sensors.

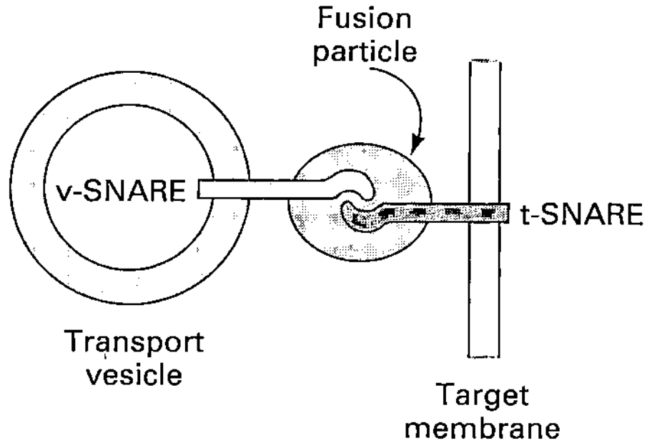

Figure 35-19
Membrane-anchored proteins called SNAREs may encode specificity in vesicle fusion. These proteins have been found in 20S particles that participate in membrane fusion. The binding of a donor vesicle v-SNARE (green) to a target membrane t-SNARE (red) in a postulated fusion particle (gray) could lead to specific membrane targeting.

A family of *rab proteins* has also been implicated in vesicle targeting and fusion. These mammalian proteins were discovered as an outgrowth of studies of yeast that are defective in secretion. *Sec4* mutants accumulate secretory vesicles, whereas ER to Golgi transport is blocked in *yptl* mutants. Sec4 and yptl are small GTP-binding proteins akin to mammalian ras proteins, which participate in growth-control cascades (p. 355). A search for the mammalian counterparts of these yeast proteins has led to the identification of more than 30 gene products. Rab proteins possess carboxyl-terminal C_{20} prenyl groups (p. 934) that enable them to become membrane-associated. The members of this family are specific in their membrane localization. Rab1, for example, is found on the ER and Golgi complex, whereas rab3a is present on regulated secretory vesicles. An intriguing possibility is that the *GTP-GDP cycle of rab proteins promotes the precise delivery of transport vesicles to target membranes.*

PROTEINS WITH A CARBOXYL-TERMINAL KDEL SEQUENCE ARE RETURNED TO THE ENDOPLASMIC RETICULUM

As was discussed earlier, the endoplasmic reticulum is rich in chaperones and other proteins that assist the folding of nascent polypeptide chains. What prevents the ER from losing these essential resident proteins? In principle, either the resident set or the secreted set could contain a distinguishing tag. However, many experiments have shown that no special tag is required for secretion: a cytosolic protein given a signal sequence will emerge in the ER and then be efficiently secreted. Rather, *residents of the ER lumen carry a retention signal.* More than 50 resident ER proteins from vertebrates, plants, insects, and nematodes have a C-terminal Lys-Asp-Glu-Leu (KDEL) or closely related tetrapeptide sequence. In yeast, HDEL plays the same role. Appending KDEL to the C terminus of a secretory protein keeps it in the ER of higher eukaryotes. Conversely, removing KDEL from the C terminus of a resident ER protein changes its fate: the mutant is secreted rather than retained in the ER. Thus, *KDEL sequences are both necessary and sufficient to keep resident proteins in the ER.*

A KDEL sequence does not actually block the departure of a protein from the ER. Rather, it serves as a *retrieval tag.* On reaching the Golgi complex, KDEL proteins bind to membrane receptors that recognize their C-terminal tail (Figure 35-20). These protein-receptor complexes

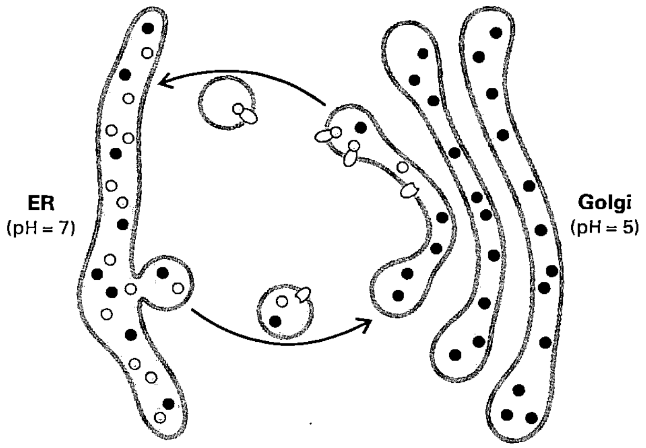

Figure 35-20
Recycling of resident proteins of the endoplasmic reticulum. Both resident proteins (yellow) and secretory proteins (black) are carried from the ER to the Golgi by transport vesicles. The KDEL sequence of resident proteins enables them to be retrieved by a membrane-bound protein that acts as a recycling receptor. Affinity of the receptors for KDEL proteins is controlled by the pH (the higher pH of the ER is shown in red, and the lower pH of the Golgi in blue). [After H.R.B. Pelham. *Proc. Roy. Soc. Lond. Ser. B* 250(1992):1.]

are then incorporated into vesicles that return them to the ER. *This recycling scheme requires that the receptor have high affinity for KDEL in the Golgi but low affinity in the ER.* What accounts for this large difference in binding strength? The affinity of the KDEL tail for its receptor increases more than tenfold when the pH is lowered from 7 to 5. Because the lumen of the Golgi is more acidic than that of the ER, KDEL proteins are membrane bound in the Golgi, whereas they are free in solution in the ER. *Tight binding in the Golgi assures efficient retrieval, whereas weak binding in the ER gives chaperones and other resident ER proteins the freedom to promote the folding and modification of nascent proteins.*

MANNOSE 6-PHOSPHATE TARGETS LYSOSOMAL ENZYMES TO THEIR DESTINATION

A different kind of molecular marker sends proteins from the Golgi to lysosomes. A key clue came from analyses of *I-cell disease* (also called *mucolipidosis II*), a lysosomal storage disease that is inherited as an autosomal recessive trait. Patients with I-cell disease have severe psychomotor retardation and skeletal deformities. Their lysosomes contain large *inclusions* of undigested glycosaminoglycans and glycolipids—hence the "I" in the name of the disease. These inclusions are present because at least eight acid hydrolases required for their degradation are missing from affected lysosomes. In contrast, very high levels of the enzymes are present in the blood and urine. Thus, active enzymes are synthesized, but they are exported instead of being sequestered in lysosomes. In other words, *a whole series of enzymes is mislocated in I-cell disease.* Normally these enzymes contain a mannose 6-phosphate residue, but in I-cell disease, the attached mannose is unmodified. *Mannose 6-phosphate is in fact the marker that normally directs many hydrolytic enzymes from the Golgi to lysosomes.*

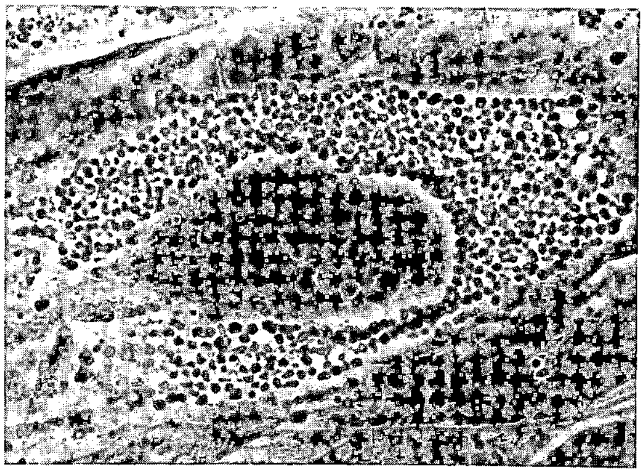

Phase-contrast micrograph of a cultured fibroblast from a patient with I-cell disease. The lysosomes are filled with undigested glycoaminoglycans and glycolipids because several lysosomal enzymes are missing. [Courtesy of Dr. George H. Thomas.]

A glycoprotein destined for delivery to lysosomes acquires a phosphoryl marker in the cis Golgi. A phosphotransferase adds a phospho-*N*-acetylglucosamine unit to the 6-OH group of a mannose, and a phosphodiesterase removes the added sugar to generate a mannose 6-phosphate residue in the core oligosaccharide (Figure 35-21). *I-cell patients are deficient in the phosphotransferase catalyzing the first step; the consequence is the mistargeting of eight essential enzymes.* This phosphotransferase is a highly discriminating enzyme. It does not act on a mannose residue in an unattached oligosaccharide or on a short peptide containing such a unit. Rather, *the transferase recognizes a three-dimensional motif that is present only in glycoproteins addressed to lysosomes.*

How does mannose 6-phosphate act as a marker for lysosomal targeting? The Golgi membrane contains a receptor that specifically recognizes the mannose 6-phosphate unit and binds proteins marked by it. Vesicles containing this protein-receptor complex bud off the rims of the trans Golgi (see Figure 35-15). These carriers then fuse with pre-lysosomal vesicles, which are more acidic than the Golgi. The lowered pH leads to the

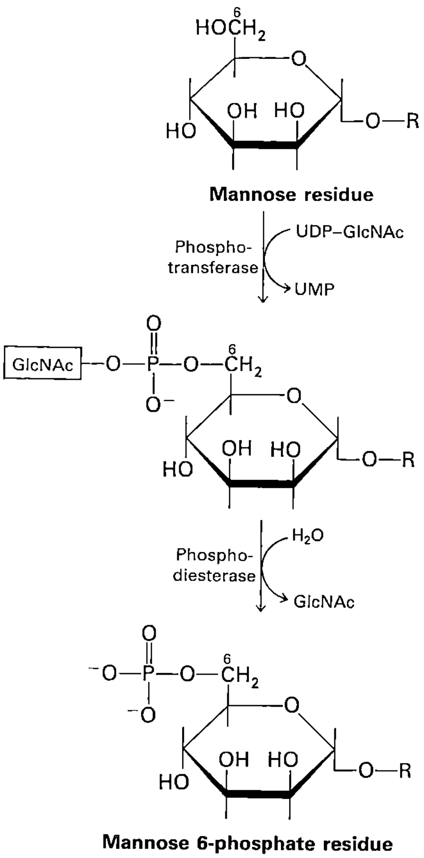

Figure 35-21
Formation of a mannose 6-phosphate marker for targeting to lysosomes. A phosphotransferase and a phosphodiesterase in the cis Golgi catalyze the addition of a phosphoryl tag.

dissociation of the marked glycoprotein from its receptor. At this stage, the receptor returns to the Golgi. Pre-lysosomes mature into lysosomes by fusing with lysosomes and receiving their enzymes.

The membrane-bound mannose 6-phosphate receptor returns to the Golgi by a different set of vesicles. This receptor, like many others, is recycled so that it can be used many times. The importance of acidification in this cycle is evidenced by the finding that *lysosomal targeting is blocked by agents that make sorting vesicles less acidic.* NH_4Cl and chloroquine, for example, raise the pH of sorting vesicles and lead to the export of lysosomal enzymes from the cell rather than to lysosomes. The reason is that, if the sorting vesicles are not sufficiently acidic, the enzyme-receptor complex does not dissociate and so the receptor does not return to the Golgi. In the absence of receptors, newly formed glycoproteins containing mannose 6-phosphate continue along the secretory pathway and are exported from the cell.

BACTERIA ALSO USE SIGNAL SEQUENCES TO TARGET PROTEINS

Protein targeting is not an invention of eukaryotes. Bacteria also target proteins to destinations encoded in their sequences. A gram-negative microorganism such as *E. coli* can send nascent proteins synthesized by ribosomes in the cytosol to the *cell membrane* (also called the plasma membrane or inner membrane), the *outer membrane,* the *periplasmic space* between membranes, or (rarely) the *extracellular medium* (Figure 35-22). The

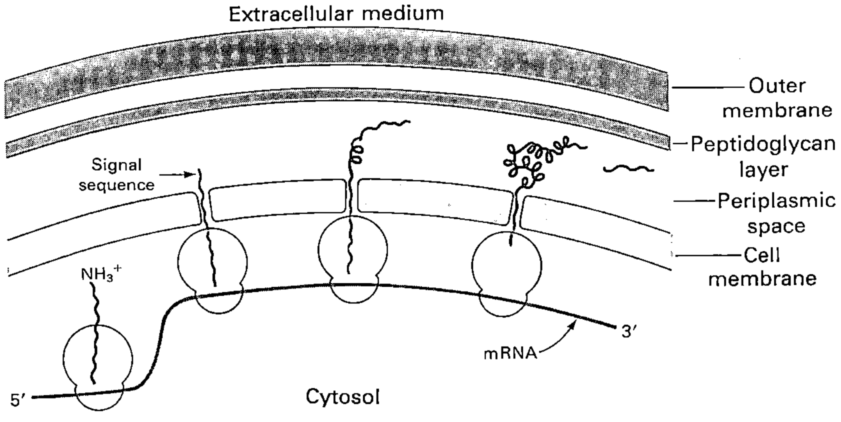

Figure 35-22
Prokaryotic proteins destined for locations other than the cytosol are synthesized by ribosomes bound to the plasma membrane. A signal sequence (red) on the nascent chain directs the ribosome to the plasma membrane and enables the protein to be translocated. As in eukaryotes, translocation is not mechanistically coupled to chain elongation. The translocation machinery is not depicted in this schematic diagram.

targeting of periplasmic and outer-membrane proteins is best understood. Amino-terminal signal sequences (also called *leader sequences*) very much like those of eukaryotic secretory proteins direct bacterial proteins across the cell and outer membranes. These signal sequences are usually 16 to 26 residues long. The one in prolipoprotein, the precursor of the most abundant outer membrane protein of *E. coli,* has the sequence

Cleavage site

^+H_3N-MKATKLVLGAVILGSTLLAG | CSSN

Prokaryotic signal sequences, though diverse, have a positively charged N-terminal region, a central hydrophobic region, and a helix-breaking segment. As in eukaryotes, the signal sequence is usually cleaved by a signal peptidase at the helix-breaking site. Another similarity is that polypeptide chain elongation and translocation usually occur at about the same time but are not mechanistically coupled.

Translocation of a polypeptide chain across the cell membrane of *E. coli* is catalyzed by a soluble chaperone and a membrane-bound, multisubunit translocase (Figure 35-23). *SecB protein,* the major chaperone for export, keeps nascent chains in an unfolded or partially folded state to enable them to traverse the membrane. SecB presents the nascent chain to SecA, a peripheral membrane component of the translocase. SecA works in concert with SecY and SecE, the membrane-embedded portion of the translocase. Two forms of free energy drive protein translocation in *E. coli*: ATP and proton-motive force. SecA is an ATPase—the ATP state has high affinity for the protein undergoing translocation, whereas the ADP state has low affinity. Segments of the nascent chain are successively handed from SecA to the SecY-SecE channel as a result of multiple rounds of ATP hydrolysis. The proton-motive force across the cell membrane then drives the threading of the nascent chain across the membrane. As was mentioned earlier, the conductance of single protein-conducting channels in the *E. coli* cell membrane has been measured (p. 917).

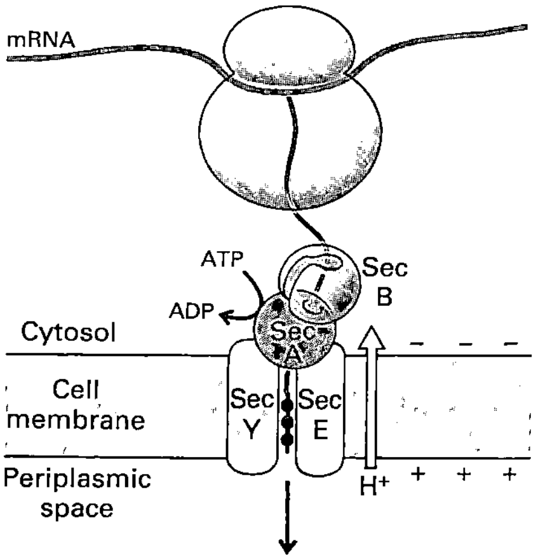

Figure 35-23
A highly schematic diagram showing the interplay of Sec proteins in protein translocation across the cell membrane of *E. coli.* The signal sequence of the nascent protein is shown in red, and the other part of the unfolded polypeptide in blue. Proton-motive force powers the unidirectional translocation of the polypeptide from the cytosolic to the periplasmic side of the membrane.

Many components of the protein targeting machinery are conserved in evolution. The signal sequence of a nascent prokaryotic secretory protein (β-lactamase) synthesized on plant ribosomes was correctly recognized by mammalian SRP. *E. coli* contains a 4.5S RNA that is similar to part of the 7S RNA of mammalian SRP. Furthermore, *E. coli* contains *Ffh* (*f*ifty-*f*our *h*omolog), a protein akin to the 54-kd subunit of mammalian SRP. Indeed, Ffh can substitute for this subunit of SRP in recognizing the signal sequence of a nascent mammalian secretory protein and in arresting its elongation. In addition, FtsY of *E. coli* is like the α subunit of the receptor for SRP. Thus, prokaryotes probably contain a homolog of the mammalian signal recognition particle. SecY and SecE of *E. coli* also have counterparts in eukaryotes. Finally, the specificities of bacterial and eukaryotic signal peptidases are very similar.

MOST MITOCHONDRIAL PROTEINS ARE SYNTHESIZED IN THE CYTOSOL AND IMPORTED INTO THE ORGANELLE

Mitochondrial DNA encodes all the RNA but only a few of the proteins in mitochondria. Most mitochondrial proteins are encoded by nuclear DNA and synthesized in the cytosol by free ribosomes. Indeed, about 10% of the proteins in a eukaryotic cell are imported into mitochondria. How do these proteins reach their mitochondrial destinations? The problem is intriguing because mitochondrial proteins reside in four locations: the outer membrane, the inner membrane, the intermembrane space, and the matrix. Gottfried Schatz discovered that *proteins are targeted to the mitochondrial matrix by their amino-terminal sequence. Matrix-targeting sequences* (also known as *presequences*) are typically 15 to 35 residues long. They are rich in positively charged residues and in serines and threonines. No consensus sequence has been found. In fact, matrix-targeting sequences, like prokaryotic and eukaryotic signal sequences, are highly degenerate—about 20% of randomly generated sequences allow proteins to enter mitochondria. The common denominator of matrix-targeting sequences is the capacity to form an amphipathic α helix or β sheet in which the basic and hydroxyl-containing residues are clustered on one face and the hydrophobic residues on the opposite face.

$^{+}H_3N$–MLRTSSLFTRRVQPSLFRNILRLQST

Figure 35-24
A mitochondrial matrix-targeting sequence. This presequence is recognized by receptors on the external face of the outer membrane of mitochondria and leads to the import of the protein bearing it into the matrix. Hydrophobic residues are shown in yellow, basic ones in blue, and serine and threonine in red.

Import of precursor proteins into the mitochondrial matrix is mediated by proteins in the cytosol, both mitochondrial membranes, and the matrix. *Fully folded proteins cannot enter mitochondria.* Hence, chaperones play a key role in maintaining precursor proteins in an unfolded or partially folded state. A cytosolic *hsp70 protein* hands the precursor to an *import receptor* in the outer mitochondrial membrane (Figure 35-25). The receptor-bound precursor then moves to a site where the outer and inner mitochondrial membranes make contact. The precursor protein threads through a channel formed by several subunits from each membrane. Translocation is driven by the membrane potential across the inner mitochondrial membrane and by the ATP-powered uptake of the nascent chain by *mhsp70,* a 70-kd heat-shock protein in the mitochondrial matrix. Finally, the amino-terminal targeting sequence is cleaved in the matrix by a soluble metalloprotease to produce the mature protein. The mechanism of protein targeting to other mitochondrial destinations is being intensively studied.

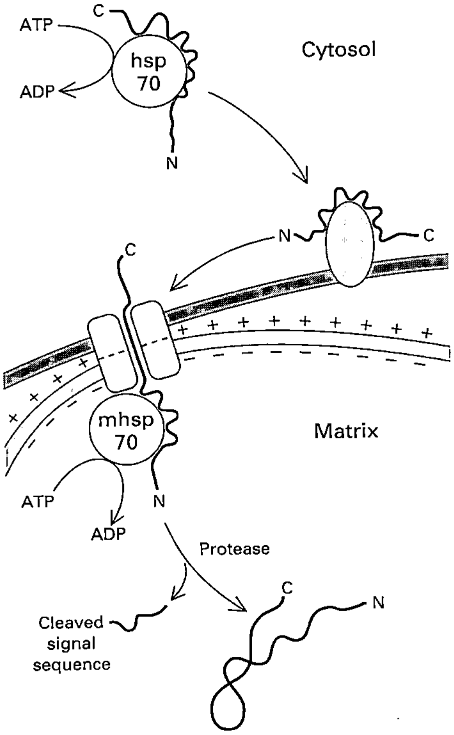

Figure 35-25
Proposed mechanism for the delivery of a protein from the cytosol to the mitochondrial matrix. The nascent chain (red) is delivered by a chaperone to a receptor (blue) in the outer mitochondrial membrane. The nascent protein threads through a channel (green) that traverses both membranes at a contact site. A transmembrane potential across the inner mitochondrial membrane is essential for transport. The matrix-targeting sequence is removed by a protease. [After G. Schatz. *Protein Science* 2(1993):141. Copyright © 1993 The Protein Society. Reprinted with the permission of Cambridge University Press.]

CHLOROPLASTS ALSO IMPORT MOST OF THEIR PROTEINS AND SORT THEM ACCORDING TO THEIR PRESEQUENCES

Most proteins of chloroplasts, like those of mitochondria, are synthesized by cytosolic ribosomes and imported. Proteins imported into chloroplasts have six potential destinations: the outer membrane, the inner membrane, the intermembrane space, the stroma, the thylakoid membrane, and the thylakoid lumen. The existence of thylakoid membranes separate from the inner membrane gives rise to two more destinations than are present in mitochondria. As in mitochondria, targeting is largely achieved by *amino-terminal presequences.* Chloroplast presequences (also called *transit sequences*) resemble mitochondrial presequences in being positively charged and rich in serine and threonine residues. Indeed, a chloroplast presequence derived from the small subunit of ribulose bisphosphate carboxylase, the plant enzyme that fixes CO_2, effectively targets a cytosolic protein to yeast mitochondria in vitro. It will be interesting to learn why the enzyme in vivo is imported into chloroplasts only.

The presequences of chloroplast proteins targeted to the stroma and the thylakoids are cleaved during the transport process. Transport across the outer and inner chloroplast membranes is powered by ATP hydrolysis, whereas transport across the thylakoid membrane is driven by the pH gradient. The presequences of proteins destined for the thylakoid lumen appear to contain two signals (Figure 35-26). The amino-terminal signal leads to the import of the precursor protein into the stroma of the chloroplast. This part of the presequence is cleaved in the stroma or en route to it, exposing a second signal that directs the translocation of the modified precursor across the thylakoid membrane. A protein targeted to the stroma lacks this second signal, which contains a hydrophobic core reminescent of bacterial and ER signal sequences. This mechanism is supported by the finding that plastocyanin (normally located in the thylakoid lumen, p. 666) is redirected to the stroma when its own presequence is replaced with that of ferredoxin (a stromal protein).

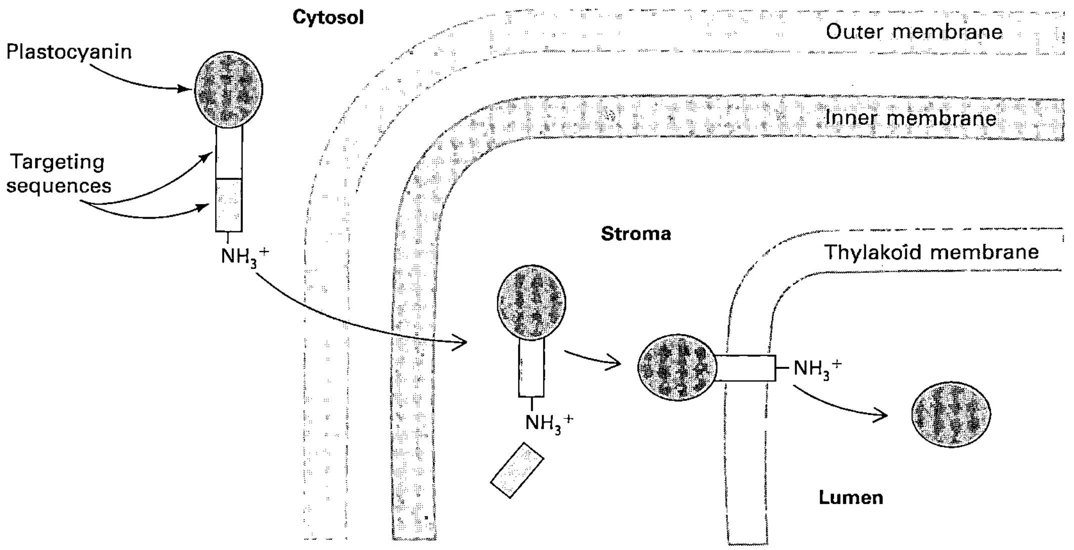

Figure 35-26
Plastocyanin is targeted to the thylakoid lumen of chloroplasts by the sequential action of two amino-terminal sequences. The first one enables it to enter the stroma, and the second (exposed by proteolysis) enables it to cross the thylakoid membrane. [After S. Smeckens, C. Bauerle, J. Hageman, K. Keegstra, and P. Weisbeck. *Cell* 46(1986):373.]

CYTOSOLIC PROTEINS ARE TARGETED TO PEROXISOMES BY CARBOXYL-TERMINAL SKF SEQUENCES

Peroxisomes are small membrane-bound compartments that are present in the cells of most eukaryotes. These organelles contain oxidases that generate H_2O_2, and catalase, which degrades H_2O_2 to water and O_2. It is now appreciated that peroxisomes have a much broader metabolic role than detoxification. They catalyze, for example, the first two steps in the

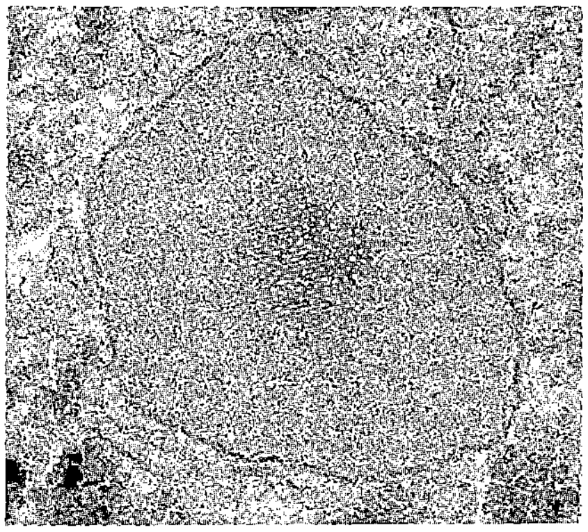

Figure 35-27
Electron micrograph showing a peroxisome in a liver cell. A crystal of urate oxidase is present inside the organelle, which is bounded by a single bilayer membrane. The dark granular structures outside the peroxisome are glycogen particles. [Courtesy of Dr. George Palade.]

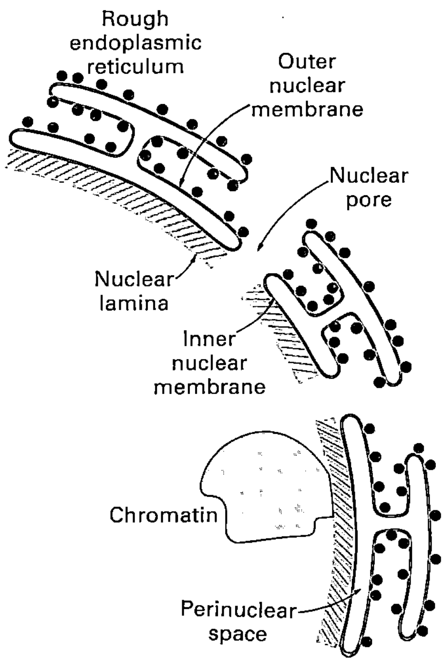

Figure 35-28
Diagram of a nuclear envelope. The outer nuclear membrane is continuous with the endoplasmic reticulum.

Figure 35-29
(A) Electron micrograph of a preparation of nuclear pores from *Xenopus laevis* oocytes. A nuclear pore is a very large protein assembly with eightfold symmetry. The mass of this assembly is 124 megadaltons, about 30 times that of a ribosome. (B) Working model of the structure of a nuclear pore. Only a few of the components of this complex structure are shown. [Courtesy of Dr. Nigel Unwin and Dr. Ronald Milligan.]

synthesis of plasmalogens (ether phospholipids, p. 688). In humans, the β-oxidation of fatty acids longer than C_{18} occurs primarily in peroxisomes rather than in mitochondria. Furthermore, bile salts (p. 735) are synthesized in peroxisomes. In plants, similar organelles (also called *glyoxysomes)* play a key role in the recycling of phosphoglycolate, which is produced by the oxygenase action of rubisco (p. 673). Peroxisomes and glyoxysomes are also known as *microbodies.*

Peroxisomes and glyoxysomes are devoid of DNA. The soluble proteins in their matrix are imported from the cytosol. *The targeting signal for many peroxisomal matrix proteins is simply a carboxyl-terminal Ser-Lys-Phe (SKF) tripeptide sequence.* Mutation of the SKF sequence of a cytosolic precursor protein blocks its import into peroxisomes. Moreover, a protein that normally resides in the cytosol can be redirected to peroxisomes just by adding an SKF sequence to its C terminus. This targeting signal, in contrast with mitochondrial and chloroplast import signals, is not cleaved. Two other differences are its brevity and C-terminal location.

NUCLEAR LOCALIZATION SIGNALS ENABLE PROTEINS TO RAPIDLY ENTER THE NUCLEUS THROUGH NUCLEAR PORES

The nucleus of eukaryotic cells is surrounded by a *nuclear envelope* consisting of two membranes—an *inner nuclear membrane* and an *outer nuclear membrane*—separated by a perinuclear space (Figure 35-28). The outer nuclear membrane is continuous with the rough endoplasmic reticulum, and the perinuclear space is continuous with the lumen of the ER. The nuclear lamina, composed of intermediate filament proteins called *lamins,* links the inner nuclear membrane to chromatin.

All proteins of the cell nucleus are synthesized in the cytosol by free ribosomes. How do DNA polymerases, RNA polymerases, histones (p. 977), and other nuclear proteins traverse the nuclear envelope? This selective barrier contains openings, called *nuclear pores* (Figure 35-29). Small proteins (~15 kd) such as histones enter readily, whereas large proteins (>90 kd) are excluded unless they contain specific signals. Such signals can also accelerate the entry of small proteins.

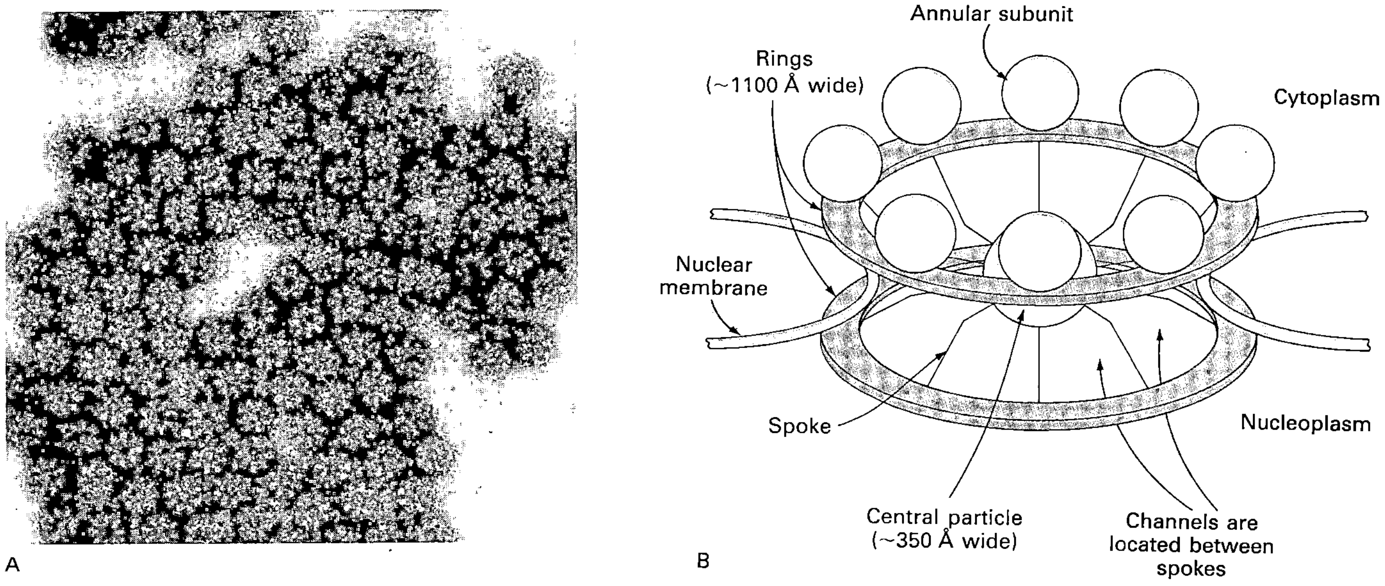

The T antigen of SV40 virus is a 92-kd protein that regulates transcription and replication of viral DNA. Studies of the accumulation of T antigen in nuclei have shown that the transport of this large protein into the nucleus depends on the presence of a *nuclear localization sequence* containing five consecutive positively charged residues:

-Pro-Lys-Lys-Lys-Arg-Lys-Val-
128

A change of a single amino acid residue can render this sequence inactive. For example, T antigen containing threonine or asparagine in place of lysine at residue 128 stays in the cytosol. The transport of large proteins into nuclei is powered by ATP hydrolysis.

Can a protein that normally resides in the cytosol be brought into the nucleus by attaching to it this viral heptapeptide sequence? A test was carried out on pyruvate kinase, a tetramer of 58-kd subunits that normally resides in the cytosol. By recombinant DNA techniques, cDNA encoding the targeting sequence was joined to cDNA encoding the kinase. A plasmid containing this chimeric cDNA was then microinjected into mammalian cells. Pyruvate kinase containing this SV40 nuclear location sequence was found almost exclusively in the cell nucleus (Figure 35-30). In contrast, pyruvate kinase bearing an altered nuclear localization sequence (threonine instead of lysine at position 128) stayed in the cytosol. The joining of this viral heptapeptide to a variety of other cytosolic proteins led in each case to their localization in the cell nucleus. Hence, *a short peptide can direct a protein into the nucleus.*

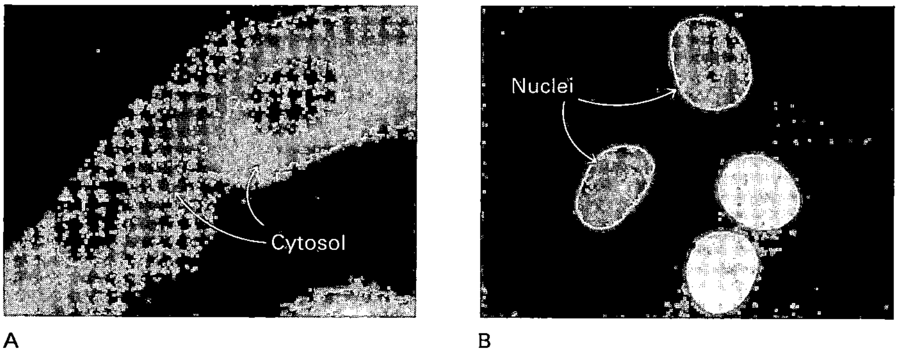

Figure 35-30
Localization of (A) unmodified pyruvate kinase, and (B) pyruvate kinase containing at its amino terminus a nuclear localization signal derived from the T antigen of SV40 virus. Pyruvate kinase was visualized by fluorescence microscopy after staining with a specific antibody. [From W.D. Richardson, B.L. Roberts, and A.E. Smith. *Cell* 44(1986):79.]

Several nuclear localization signals have been identified. None are cleaved on entry into the nucleus. It is important to note that fully folded proteins can be imported into nuclei but not into mitochondria or chloroplasts, which must maintain a tight permeability barrier to sustain a proton-motive force. No bilayer membrane is crossed on entering the nucleus—hence, unfolding is not essential. The nucleus can afford to be more relaxed about its border because the pH and ionic composition of the nucleoplasm is essentially the same as that of the cytosol.

MANY MEMBRANE-ASSOCIATED PROTEINS CONTAIN COVALENTLY ATTACHED FATTY ACYL OR PRENYL UNITS

We have thus far considered how proteins are transported from the cytosol into diverse membrane-bound compartments—the ER, Golgi, lysosomes, mitochondria, chloroplasts, peroxisomes, and the nucleus—and across the plasma membrane. Different strategies are used to send eukaryotic proteins to the *cytosolic* face of either the plasma membrane or a compartment membrane. *A soluble cytosolic protein can be given a membrane anchor by the covalent attachment of a fatty acyl or prenyl group.* Four kinds of modification are generally employed:

1. *Myristoylation at the N-terminus.* The amino terminus of a number of proteins is cotranslationally acylated with a myristoyl (C_{14}) or similar fatty acyl group. The N-terminal residue must be glycine, residue 5 is usually serine or threonine, and residues 6 and 7 are typically basic. The acyl donor in the reaction catalyzed by N-*myristoyl transferase* is myristoyl CoA. Myristoylation enables a modified protein to interact with a membrane receptor or the lipid bilayer itself. The essentiality of myristoylation for the function of certain proteins is vividly illustrated by the viral src protein (p. 354), which is rendered nononcogenic by mutations that block its myristoylation.

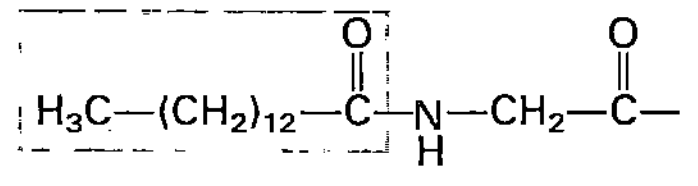

N-Myristoylglycine

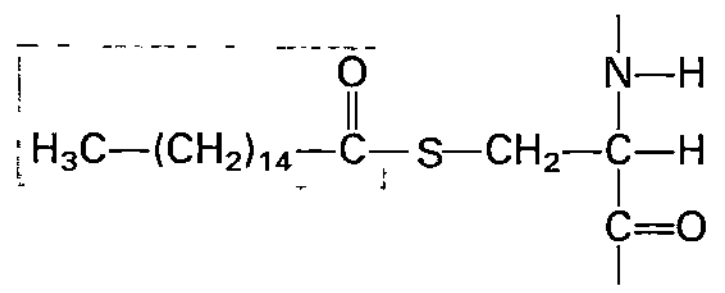

***S*-Palmitoylcysteine**

2. *Palmitoylation of cysteine residues.* The thiol group of certain cysteine residues in proteins can be acylated by palmitoyl CoA to form an *S*-palmitoyl derivative (C_{16}). Rhodopsin, for example, contains two adjacent *S*-palmitoyl groups that serve as membrane anchors.

3. *Farnesylation at the C terminus.* Many proteins that participate in signal transduction and protein targeting contain either a farnesyl (C_{15}) or a geranylgeranyl (C_{20}) unit at their C terminus.

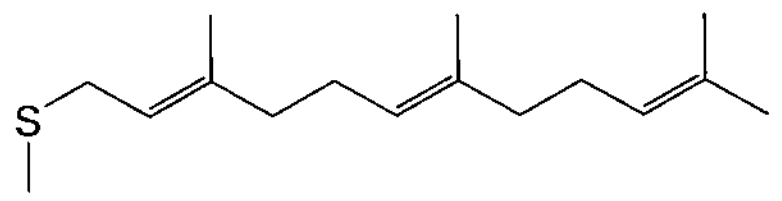

***S*-Farnesyl (C_{15}) unit**

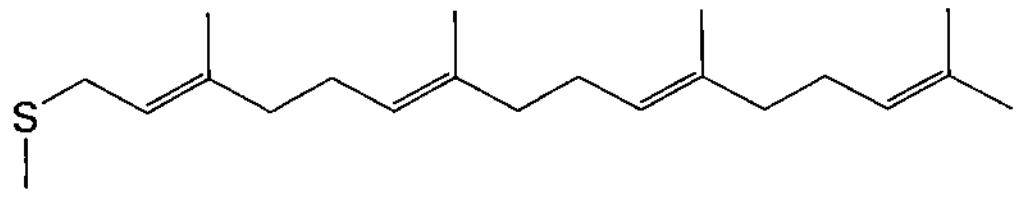

***S*-Geranylgeranyl (C_{20}) unit**

These *prenyl* groups are attached to C-terminal cysteine residues by thioether linkages. We have already encountered the activated donors—farnesyl pyrophosphate and geranylgeranyl pyrophosphate—in the biosynthesis of cholesterol and other biomolecules built from isoprene (C_5) units. Farnesylation occurs at CaaX sequences in which cysteine is followed by two aliphatic residues (a) and a carboxyl-terminal residue (X) (Figure 35-31). After attachment of the C_{15} unit to this cysteine, the aaX residues are proteolytically removed, and the new terminal carboxylate group is methylated. Thus, a highly hydrophobic C terminus is fashioned by a series of modifications. The ras protein does not insert in the plasma membrane unless it is farnesylated. Indeed, unfarnesylated ras is unable to transduce growth signals.

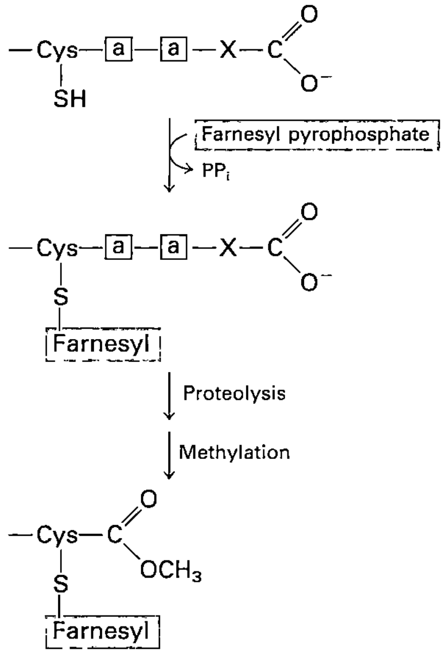

Figure 35-31
Cytosolic proteins can be given a hydrophobic C-terminal anchor by the covalent attachment of a farnesyl (C_{15}) unit. Farnesylation is followed by trimming of three C-terminal residues and methylation of the terminal carboxylate.

4. *Geranylgeranylation at the C terminus.* When the C-terminal sequence is CC, CXC, or CCXX, a geranylgeranyl (C_{20}) unit rather than a farnesyl unit becomes attached to one or both cysteines. The rab family of small GTP-binding proteins, which participate in membrane targeting (p. 925), are geranylgeranylated. The attachment of this highly hydrophobic prenyl unit is necessary for membrane binding, but it alone does not specify the target membrane—different rabs are present in different membranes.

GLYCOSYL PHOSPHATIDYL INOSITOL UNITS SERVE AS MEMBRANE ANCHORS FOR MANY CELL-SURFACE PROTEINS

A different molecular device is used to anchor a protein in the *outer* leaflet of the plasma membrane. *Glycosyl phosphatidyl inositol (GPI)* (Figure 35-32) attached to the carboxy terminus of a protein serves as a *flexible leash* that gives it freedom to act on molecules outside the cell. The entire protein except for this glycolipid anchor is located in the extracellular space. Many cell-surface hydrolytic enzymes and adhesins are tethered to the cell by a GPI unit.

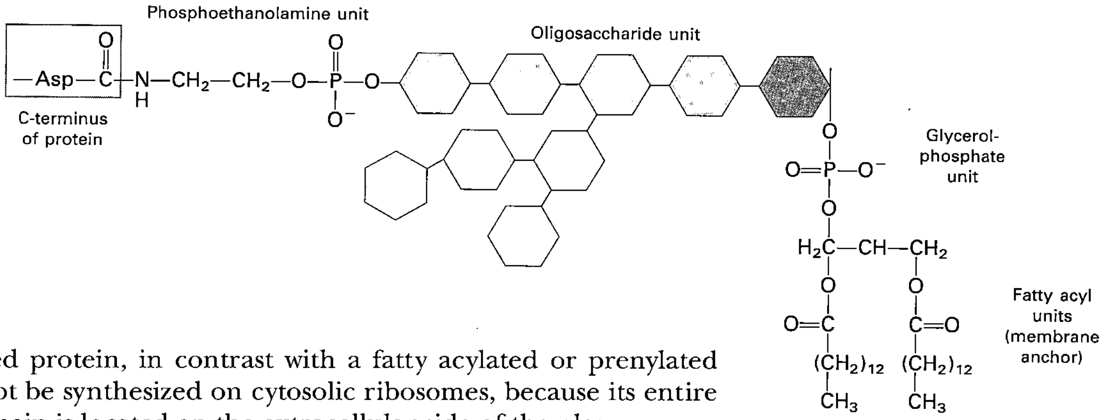

Figure 35-32
Glycosyl phosphatidyl inositol (GPI) units anchor many cell-surface proteins to the outer leaflet of the plasma membrane. The carboxy terminus of the protein is attached to the phosphoethanolamine unit of GPI. The oligosaccharide unit is schematically represented by a series of hexagons (mannose in green, galactose in yellow, *N*-acetylglucosamine in blue, and inositol in red). The two fatty acyl chains linked to the glycerolphosphate unit serve as membrane anchors. [After T.L. Doering, W.J. Masterson, G.W. Hart, and P.T. Englund. *J. Biol. Chem.* 265(1990):611.]

A GPI-linked protein, in contrast with a fatty acylated or prenylated protein, cannot be synthesized on cytosolic ribosomes, because its entire polypeptide chain is located on the extracellular side of the plasma membrane. Rather, GPI-anchored proteins initially contain an N-terminal signal sequence that directs the nascent chain to the ER, and a C-terminal hydrophobic sequence 20 to 30 residues long. This hydrophobic C-terminal region transiently ties the protein to the ER membrane. The ethanolamine end of a preformed GPI unit then attacks a peptide bond at the lumenal end of this C-terminal tether. The C-terminal peptide is released, and a GPI-anchored protein is generated. This protein then goes from the ER to the Golgi and then to the plasma membrane, the default destination.

SPECIFIC PROTEINS ARE IMPORTED INTO CELLS BY RECEPTOR-MEDIATED ENDOCYTOSIS

We turn now to a different facet of protein targeting—the import of specific proteins into a cell by their binding to receptors in the plasma membrane and their inclusion into vesicles. This process of *receptor-mediated endocytosis* (Figure 35-33) has broad biological significance. First, it is

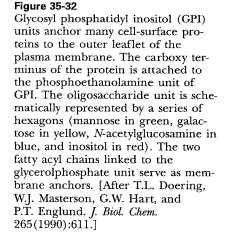

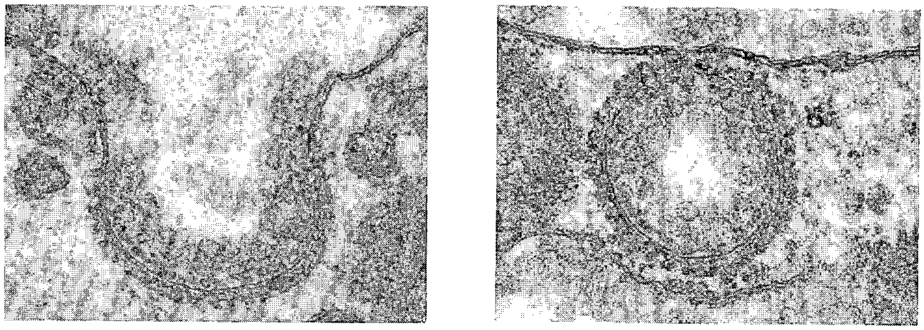

Figure 35-33
Receptor-mediated endocytosis takes place at coated pits in the plasma membrane. These electron micrographs show the uptake of vitellogenin, a lipoprotein, by hen oocytes. [Courtesy of Dr. M.M. Perry and Dr. A.A. Gilbert.]

a means of *delivering essential metabolites to cells.* As was discussed earlier (p. 699), low-density lipoprotein (LDL) carrying cholesterol is taken up by the LDL receptor in the plasma membrane and internalized. Likewise, the complexes of vitamin B_{12} bound to transcobalamin II and of iron bound to transferrin are recognized by cell-surface receptors and imported. Second, *endocytosis modulates responses to many protein hormones and growth factors.* Epidermal growth factor and nerve growth factor are taken into the cell and degraded together with their receptors. Endocytosis removes these hormones from the circulation and makes the cell temporarily less responsive to them by decreasing the number of receptors. Third, *proteins targeted for destruction are taken up and delivered to lysosomes for digestion.* For example, phagocytic cells have receptors that enable them to take up antigen-antibody complexes. Liver cells take up and destroy senescent plasma glycoproteins that are marked by terminal galactose residues; the loss of sialic acid residues from oligosaccharide chains exposes these markers (p. 477). Fourth, *receptor-mediated endocytosis is exploited by many viruses and toxins to gain entry into cells.* The ingenious mode of entry and departure of Semliki Forest virus will be considered shortly. Fifth, *disorders of receptor-mediated uptake can lead to disease,* as exemplified by some forms of familial hypercholesterolemia.

CLATHRIN PARTICIPATES IN ENDOCYTOSIS BY FORMING A POLYHEDRAL LATTICE AROUND COATED PITS

Most cell-surface receptors mediating endocytosis are *transmembrane glycoproteins.* They have a large extracellular domain, one or two transmembrane helices, and a small cytosolic region, as exemplified by the receptors for transferrin and asialoglycoproteins (Figure 35-34). Many of these

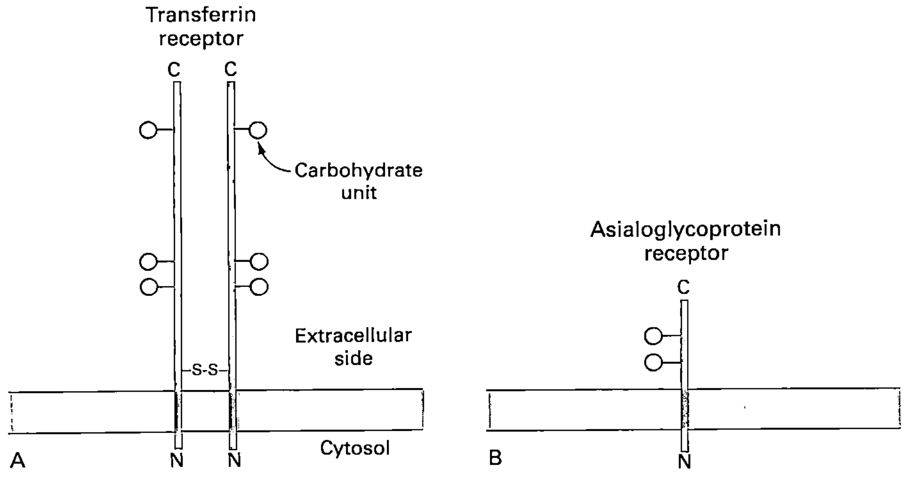

Figure 35-34
Schematic diagram of two receptors that are internalized at coated pits. (A) transferrin receptor and (B) the asialoglycoprotein receptor. Their short N-terminal tails are critical for internalization.

receptors are located in specialized regions of the plasma membrane called *coated pits.* The cytosolic side of these indentations has a thick coat of *clathrin,* a protein designed to form lattices around membranous vesicles (Figure 35-35). Coated pits occupy about 2% of the surface of a typical animal cell. A number of receptors (such as those for LDL, transferrin,

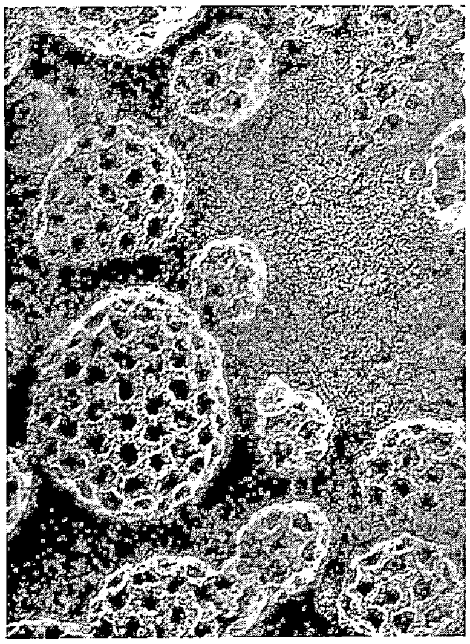

Figure 35-35
Electron micrograph of clathrin lattices on the cytosolic face of a plasma membrane. [Courtesy of Dr. John Heuser.]

asialoglycoproteins, and insulin) congregate in coated pits whether or not ligand is bound. Others (such as the receptor for epidermal growth factor) cluster there after binding their cognate protein.

Receptor-mediated endocytosis begins with the invagination of a coated pit (see Figure 35-33). Clathrin forms a lattice around the coated pit, excising it from the plasma membrane to form a *coated vesicle,* which has a diameter of about 80 nm. The coated vesicle rapidly loses its clathrin shell and fuses with an *endosome.* In turn, endosomes fuse with one another to form vesicles having diameters ranging from 200 to 600 nm. The conversion of a coated pit into an endosome takes about 20 seconds. An important characteristic of endosomes is their acidity, which dissociates most protein-receptor complexes, enabling the partners to have different fates. Hence, *sorting decisions are made in endosomes.*

What is the structural basis of clathrin's propensity to form closed polyhedral lattices? Clathrin consists of three heavy chains (H, 180 kd) and three light chains (L, ~35 kd). The $(HL)_3$ clathrin unit (8S, 650 kd), obtained by solubilizing coats at alkaline pH in the absence of divalent cations, has a striking appearance in electron micrographs (Figure 35-36). It is a three-legged structure, a *triskelion.* The carboxy termini of the three heavy chains, each about 500 Å long, come together at a vertex (Figure 35-37A). A bend in the heavy chain divides it into a proximal arm, closest to the vertex, and a distal arm. Each of the light chains is aligned with the proximal arm of a heavy chain. Purified trimers spontaneously reassemble into closed shells when dialyzed into a Mg^{2+}-containing buffer at pH 6. These closed polyhedra are made of pentagons and hexagons (Figure 35-37B). Both are used because an icosahedral shell

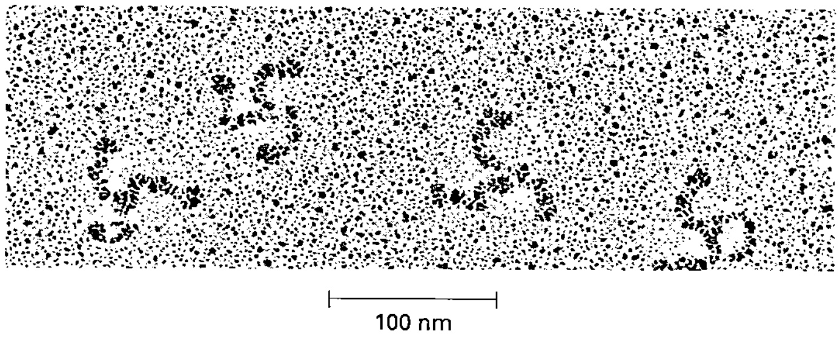

Figure 35-36
Electron micrograph of triskelions, the building blocks of the polyhedral shell surrounding coated vesicles. [Courtesy of Dr. Daniel Branton.]

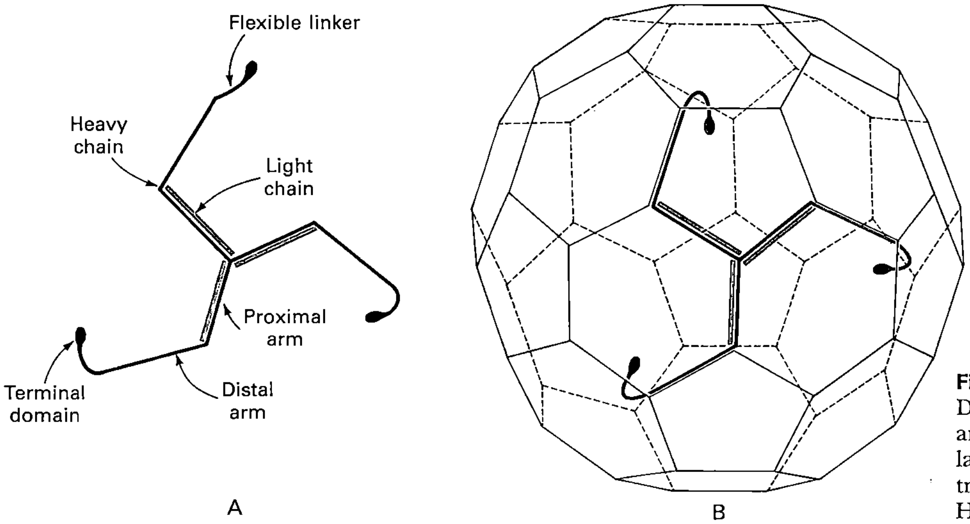

Figure 35-37
Diagram of (A) a triskelion molecule, and (B) its position in the polyhedral lattice formed by the association of triskelions. [Courtesy of Dr. Stephen Harrison.]

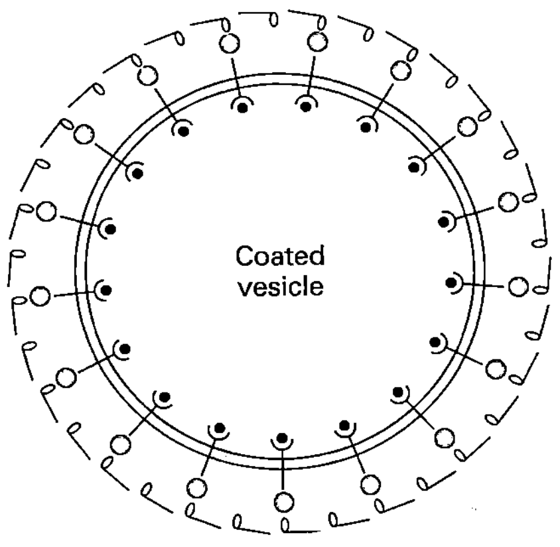

Figure 35-38
Adaptors play a key role in forming coated vesicles and in selecting their protein cargo. One face of an adaptor (green) binds clathrin (red), whereas the other face binds the cytosolic tail of a transmembrane protein (black). [After F.M. Brodsky, B.L. Hill, S.L. Acton, I. Näthke, D.H. Wong, S. Ponnambalam, and P. Parham. *Trends Biochem. Sci.* 16(1991):208.]

formed from more than 60 subunits cannot be constructed from pentagons alone or hexagons alone. *A single edge of a pentagon or hexagon is made of parts of four triskelions, two proximal arms and two distal ones.* The flexibility of a triskelion is important in enabling it to fit into a pentagon or a hexagon.

Triskelions polymerize into a polyhedral network on the cytosolic side of the plasma membrane. They interact with *adaptor proteins* that in turn select the cargo to be carried in the clathrin cage (Figure 35-38). Adaptors are assemblies of two *adaptin subunits* and two other proteins. They recognize a tyrosine-containing sequence in the cytosolic tail of transmembrane receptors that undergo internalization. The release of clathrin from fully formed coated vesicles is the next key step in endocytosis. Disassembly of the outer clathrin shell of coated vesicles is mediated by an *ATP-driven uncoating enzyme.* This member of the hsp70 heat-shock protein family hydrolyzes three ATP for each triskelion removed from the polyhedral lattice. It is evident that lattice formation through noncovalent binding interactions is thermodynamically favorable, whereas its disruption requires an input of free energy. Clathrin light chains regulate the dynamics of coated vesicles by controlling the timing of assembly and disassembly.

ENDOCYTOSED PROTEINS AND RECEPTORS ARE SORTED IN ACIDIC ENDOSOMES

The pathway from coated pits to coated vesicles to endosomes is common to all proteins that have undergone endocytosis. The fates of internalized proteins and receptors then diverge. Acidification of endosomes by ATP-driven proton pumps leads to the dissociation of protein-receptor complexes, a necessary prelude to their sorting and targeting. The pathway taken by transferrin and its receptor (Figure 35-39) illustrates one of four potential outcomes (Table 35-2). *Transferrin* transports iron from sites of

Table 35-2
Four modes of receptor-mediated endocytosis

Mode	*Fate of receptor*	*Fate of protein*	*Examples*
1	Recycled	Recycled	Transferrin, major histocompatibility proteins
2	Recycled	Degraded	Low-density lipoprotein, transcobalamin II
3	Degraded	Degraded	Epidermal growth factor, immune complexes
4	Transported	Transported	Maternal immunoglobulin G, immunoglobulin A

absorption and storage to sites of utilization. Two Fe^{3+} ions are bound per 77-kd protein, which contains two similar domains. The protein devoid of iron is called *apotransferrin.* The binding of Fe^{3+} to the protein involves HCO_3^- and a tyrosine side chain in the anionic form. Transferrin, but not apotransferrin, binds to a dimeric receptor (see Figure 35-34) in coated pits. On reaching an endosome, which is acidified to a pH between 5 and 6, Fe^{3+} dissociates from transferrin. The acidity lowers the

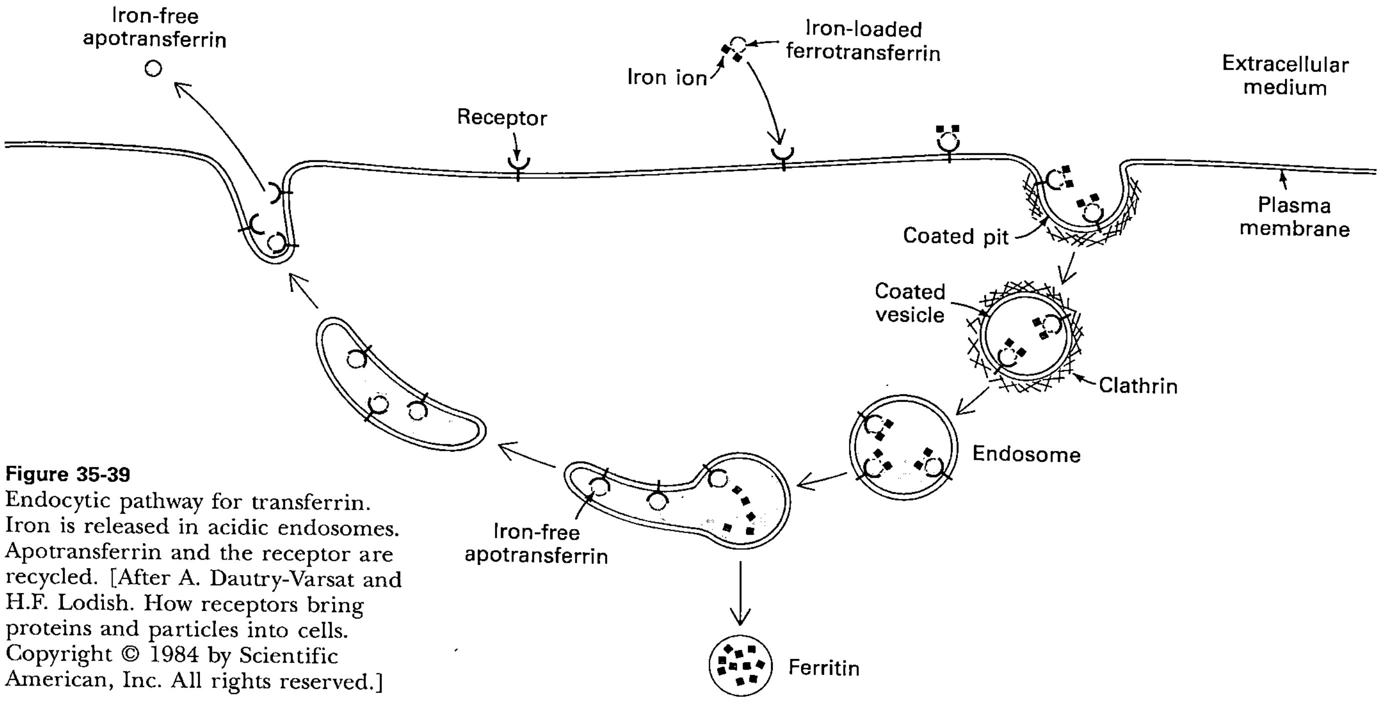

Figure 35-39
Endocytic pathway for transferrin. Iron is released in acidic endosomes. Apotransferrin and the receptor are recycled. [After A. Dautry-Varsat and H.F. Lodish. How receptors bring proteins and particles into cells. Copyright © 1984 by Scientific American, Inc. All rights reserved.]

affinity of transferrin for Fe^{3+} more than a millionfold. *Acidification leads to the release of Fe^{3+} by protonating bicarbonate, the phenolate $-O^-$ of tyrosine, and other groups that contribute to binding.* However, apotransferrin remains bound to the receptor.

Sorting then takes place: part of the vesicle bearing apotransferrin bound to the receptor pinches off and is directed to the plasma membrane, whereas the remaining Fe^{3+} is stored in ferritin in the cytosol. *When the pinched-off vesicle fuses with the plasma membrane, apotransferrin is released from the receptor because of the sudden shift in pH.* Apotransferrin has little affinity for the receptor at pH 7.4. Thus, pH changes are used twice to drive the transport cycle, first to release iron from transferrin in the endosome, and then to discharge apotransferrin into the extracellular fluid. Both the carrier of iron and the receptor are recycled with little loss. The cycle takes about 16 minutes—4 minutes for the binding of transferrin, 5 minutes for transport to endosomes, and 7 minutes for the return of the iron carrier and the receptor to the cell surface. A liver cell can take up some 20,000 iron atoms per minute in this way.

MANY MEMBRANE-ENVELOPED VIRUSES ENTER CELLS BY RECEPTOR-MEDIATED ENDOCYTOSIS

Membrane-enveloped viruses exploit endocytic pathways to enter cells and exocytic pathways to leave them. Studies of the life cycle of *Semliki Forest virus (SFV)* have provided insight into how membrane-enveloped viruses take advantage of the protein-targeting machinery of host cells. SFV, an RNA virus that infects mosquitoes, was named for a rain forest in Uganda. It is closely related to the virus causing yellow fever in humans. The RNA genome of SFV is surrounded by 180 copies of C protein, which form an icosahedral shell (Figure 35-40). This nucleoprotein capsid is

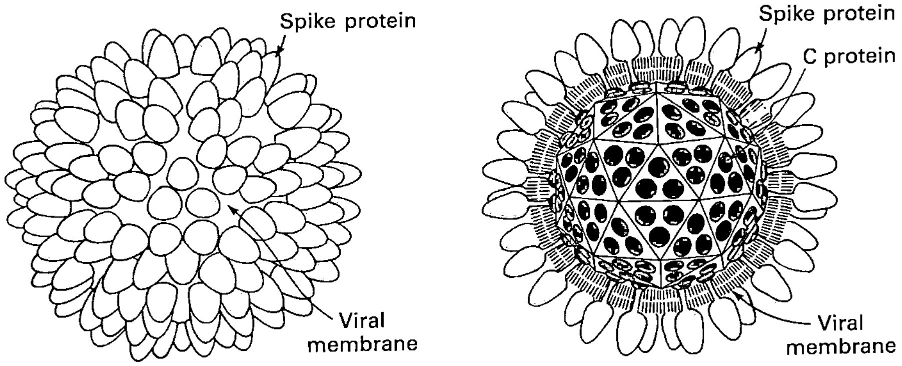

Figure 35-40
Diagram of the structure of Semliki Forest virus (SFV), a membrane-enveloped virus. [After K. Simons, H. Garoff, and A. Helenius. How an animal virus gets into and out of its host cell. Copyright © 1982 by Scientific American, Inc. All rights reserved.]

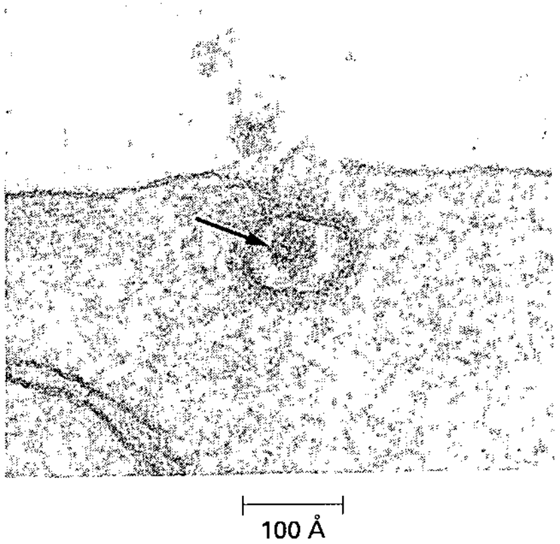

Figure 35-41
Electron micrograph showing the entry of Semliki Forest virus into an infected cell. [Courtesy of Dr. Kai Simons.]

enveloped by a lipid bilayer membrane containing 180 spike glycoproteins, each consisting of three subunits. The E1 and E2 subunits each include a large external domain, a single membrane-spanning helix, and a short carboxyl-terminal region that makes contact with the C protein of the nucleocapsid. The small E3 subunit is bound to the external domain of E2; these subunits arise by cleavage of a common precursor.

Electron micrographs of infected cells show virus particles in coated pits, coated vesicles, and endosomes. It is evident that SFV enters susceptible cells by binding to receptors in coated pits that are then endocytosed (Figures 35-41 and 35-42). *The acidic environment of the endosome leads to a conformational change that is necessary for infection.* The E1 spike protein changes shape and triggers the fusion of the viral membrane with the membrane of the endosome. Hence, the nucleocapsid is released into the cytosol. Inhibitors of acidification prevent infection by blocking this release step.

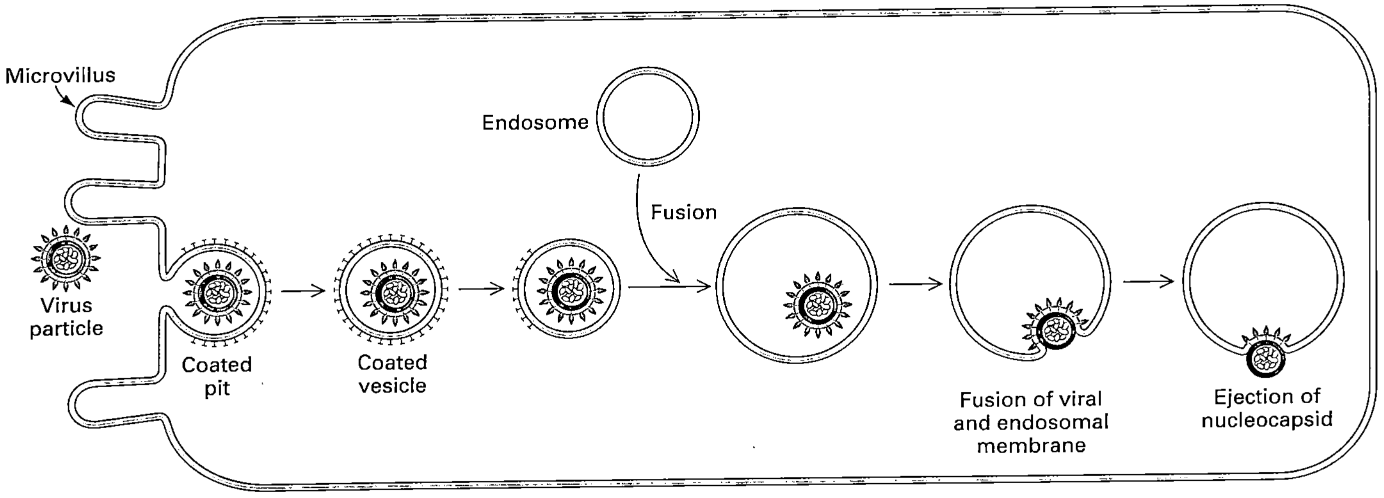

Figure 35-42
Entry of Semliki Forest virus into a susceptible cell by receptor-mediated endocytosis. The virus is internalized after it binds to specific cell-surface receptors. The acidic milieu of the endosome triggers the fusion of membranes and release of the viral nucleocapsid. [After K. Simons, H. Garoff, and A. Helenius. How an animal virus gets into and out of its host cell. Copyright © 1982 by Scientific American, Inc. All rights reserved.]

The viral RNA is then translated and replicated. The C protein is synthesized in the cytosol on free ribosomes. In contrast, the precursors of spike protein contain signal sequences, which direct ribosomes synthesizing them to the endoplasmic reticulum. These viral glycoproteins are processed by the Golgi and sent to the plasma membrane, where they cluster (Figure 35-43). Newly synthesized nucleocapsids consisting of RNA and C protein interact with these viral spike proteins. The plasma membrane curves around the nucleocapsid as successively more bonds are made between spikes and C proteins. Finally, when all 180 C proteins are bound to spikes, the membrane coat around the virus breaks off from the plasma membrane, releasing the virus particle into the extracellular space. It is interesting to note that *influenza virus* uses strategies similar to those employed by SFV to enter and leave target cells.

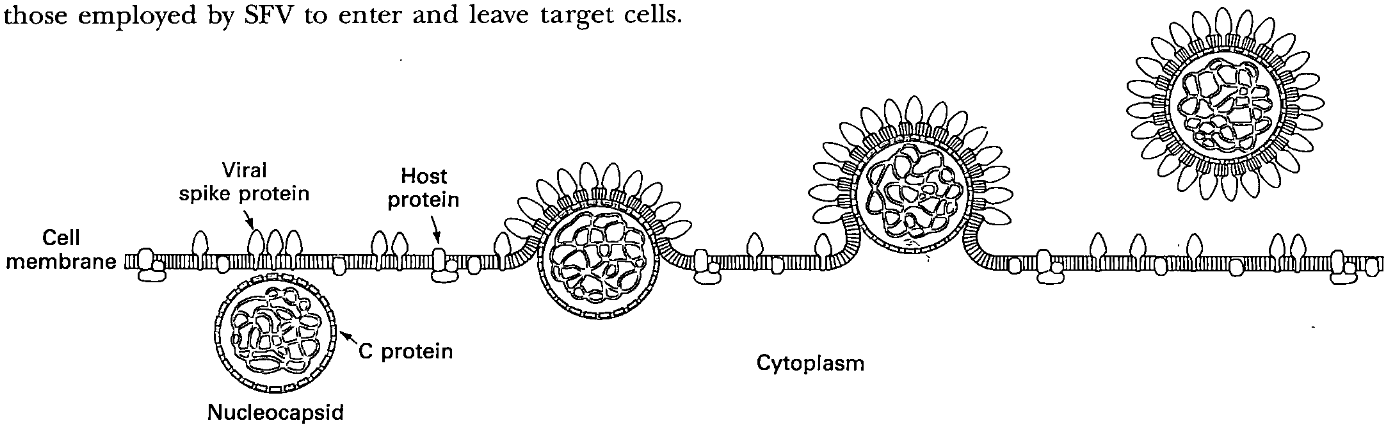

Figure 35-43
Membrane assembly and release of newly formed Semliki Forest virus particles. [After K. Simons, H. Garoff, and A. Helenius. How an animal virus gets into and out of its host cell. Copyright © 1982 by Scientific American, Inc. All rights reserved.]

DIPHTHERIA TOXIN AND CHOLERA TOXIN GAIN ENTRY INTO TARGET CELLS BY BINDING TO CELL-SURFACE RECEPTORS

Toxins as well as viruses enter cells by receptor-mediated pathways. Diphtheria toxin, produced by *Corynbacterium diphtheria,* is a tripartite protein containing a *catalytic domain,* a *membrane-insertion domain,* and a *receptor-binding domain* (Figure 35-44). Binding of the toxin to a growth factor precursor on the cell surface triggers the internalization of the complex by endocytosis. The 61-kd toxin is cleaved in endosomes into a 21-kd A fragment and a 40-kd B fragment. The membrane-insertion domain of the B fragment, activated by the acidic milieu of the endosome, enables the catalytic A fragment to enter the cytosol. As was discussed in the preceding chapter (p. 906), *a single A fragment of the toxin in the cytosol can kill a cell by catalyzing the ADP-ribosylation of elongation factor 2 (EF2), which leads to the cessation of protein synthesis.*

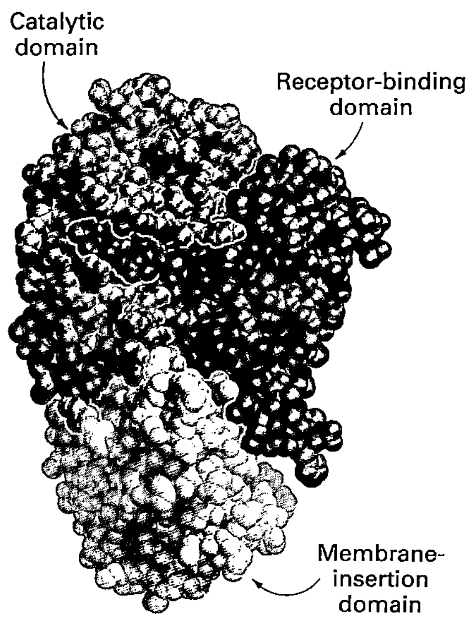

Figure 35-44
Structure of diphtheria toxin. This ADP-ribosylase consists of three domains: an amino-terminal catalytic domain (blue), a membrane-insertion domain (yellow), and a receptor-binding carboxyl-terminal domain (red). A dinucleotide occupying the NAD^+-binding site in the catalytic domain is shown in purple. [Drawn from coordinates kindly provided by Dr. David Eisenberg and Dr. Melanie Bennett. S. Choe, M.J. Bennett, G. Fujii, P.M.G. Curmi, K.A. Kantardjieff, R.J. Collier, and D. Eisenberg. *Nature* 357(1992):216.]

Figure 35-45
Structure of the *E. coli* heat-labile enterotoxin. Cholera toxin has a similar structure. (A) Axial view of the complex consisting of a catalytic A chain (red) and five membrane-targeting B chains. In the crystal, lactose (purple) occupies the ganglioside-binding site of each B chain. (B) Cross-sectional view showing the fivefold symmetry of the membrane-targeting subunit. The catalytic chain (not shown) probably enters the membrane through the hole in the pentamer. [Drawn from 1ltt.pdb. T.K. Sixma, S.E. Pronk, K.H. Kalk, B.A. van Zanten, A.M. Berghuis, and W.G. Hol. *Nature* 351(1992):561.]

Cholera toxin and several related toxins directly enter intestinal epithelial cells after binding to a cell-surface receptor. They consist of a *catalytic unit* (an A chain) and a *membrane-penetration unit* (five B chains) (Figure 35-45). The pentameteric B subunit binds to multiple molecules of ganglioside G_{M1}, a carbohydrate-rich sphingolipid (p. 690) in the outer leaflet. The A subunit then enters the cytosol, most likely through a membrane hole created by B_5, and ADP-ribosylates the α subunit of the stimulatory G protein ($G_{s\alpha}$) (see Figure 13-33 on p. 342). This covalent modification of $G_{s\alpha}$ blocks its capacity to hydrolyze GTP to GDP, which locks G_s in the active state. Adenylate cyclase becomes persistently active and cyclic AMP levels rise. The consequent activation of ion pumps leads to a very large efflux of Na^+ and water into the gut. Indeed, the fulminating diarrhea can be lethal if fluids are not swiftly replaced intravenously.

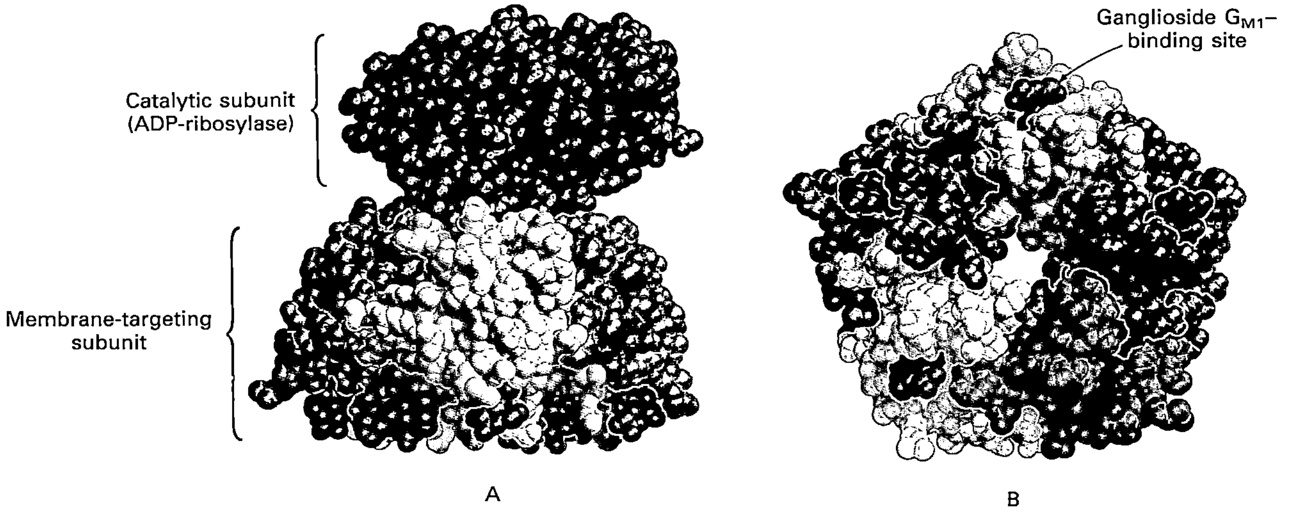

UBIQUITIN TAGS PROTEINS FOR DESTRUCTION

Proteins are also targeted for destruction. They differ markedly in their half-lives. Some are turned over very rapidly, whereas others are very stable. Most enzymes that are important in metabolic regulation have short lives. Metabolic patterns can be rapidly changed by altering the amounts of these potentially labile enzymes. Controlled degradation is also important in removing abnormal proteins. A significant proportion of newly synthesized protein molecules are defective because of errors in translation. Moreover, proteins undergo oxidative damage and are altered in other ways with the passage of time. Cells have mechanisms for detecting and removing damaged proteins.

Ubiquitin, a small (8.5 kd) protein present in all eukaryotic cells, plays an important role in tagging proteins for destruction. This protein is highly conserved in evolution: yeast and human ubiquitin differ at only 3 of 76 residues. The carboxyl-terminal glycine of ubiquitin becomes covalently attached to the ε-amino group of lysine residues of proteins destined to be degraded. The energy for the formation of these *isopeptide bonds* (*iso* because ε- rather than α-amino groups are partners) comes from ATP. Three enzymes (E_1, E_2, and E_3) participate in the conjugation of ubiquitin to proteins. First, the terminal carboxylate group of ubiquitin

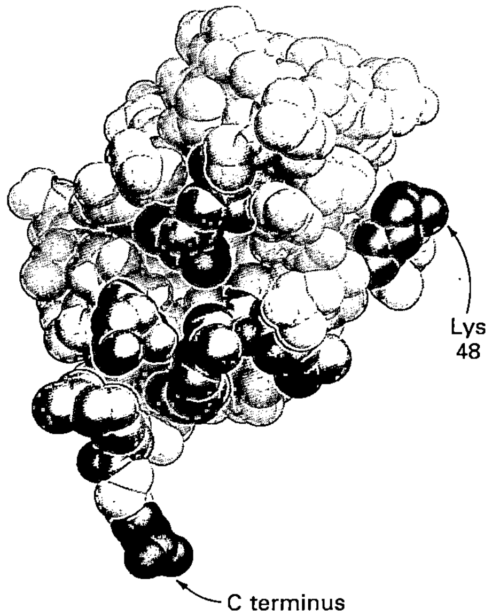

Figure 35-46
Three-dimensional structure of ubiquitin. The C-terminal carboxylate of residue 76 (red) becomes covalently linked to proteins that are targeted for destruction. The ε-amino group of Lys 48 (blue) can become joined to the C terminus of another ubiquitin molecule to form a polyubiquitinylated derivative. Hydrophilic residues that may interact with E_2, the ubiquitin conjugating enzyme, are shown in green. [Drawn from 1ubq.pdb. S. Vijay-Kumar, C.E. Bugg, and W.J. Cook. *J. Biol. Chem.* 262(1967):6396.]

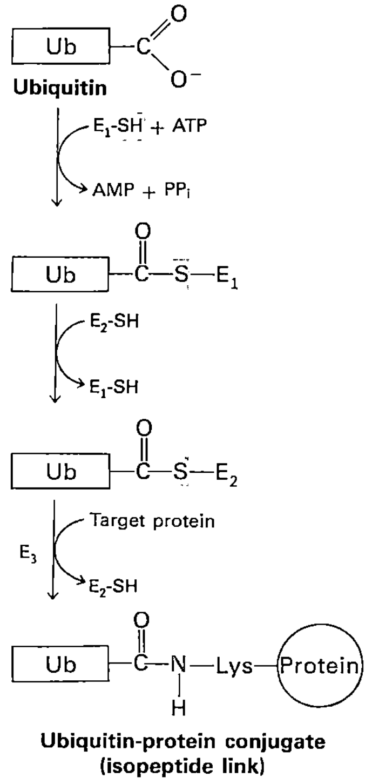

Figure 35-47
Activation and attachment of ubiquitin to a protein targeted for degradation.

becomes linked to a sulfhydryl group of E_1 by a thioester bond (Figure 35-47). This ATP-driven reaction is reminiscent of fatty acid activation (p. 607) and amino acid activation (p. 880). Activated ubiquitin is then shuttled to a sulfhydryl of E_2. Finally, E_3 catalyzes the transfer of ubiquitin from E_2 to the target protein. A protein tagged for destruction usually acquires several molecules of ubiquitin. The ε-amino group of a lysine residue of one ubiquitin molecule can become linked to the terminal carboxylate of another.

What determines whether a protein becomes ubiquitinylated? One signal turns out to be unexpectedly simple. *The half-life of a cytosolic protein is determined to a large extent by its amino-terminal residue* (Table 35-3). A yeast protein with methionine at its N terminus typically has a half-life of more than 20 hours, whereas one with arginine at this position has a half-life of about 2 minutes. A highly destabilizing N-terminal residue such as arginine or leucine favors rapid ubiquitinylation, whereas a stabilizing residue such as methionine or proline does not. Proteins with an N-terminal aspartate or glutamate residue react with arginyl-tRNA to acquire highly destabilizing arginine as their new N terminus. Likewise, asparagine and glutamine are destabilizing because they are deamidated to aspartate and glutamate. The E_3 enzyme in the conjugation reaction is the reader of N-terminal residues. *The ordering of amino-terminal residues according to whether they are stabilizing or destabilizing is similar for bacteria, yeast, and mammals.*

Finally, the ubiquitinylated protein is digested by a *26S protease complex.* This ATP-driven multisubunit protease spares ubiquitin, which is then recycled. Multiple rounds of ATP hydrolysis enable the protease to unfold the ubiquitinylated protein and processively digest it. It is evident that protein degradation is controlled and conducted by sophisticated molecular devices.

Table 35-3
Dependence of the half-lives of cytosolic yeast proteins on the nature of their amino-terminal residue

Highly stabilizing residues ($t_{1/2} > 20$ hours)			
Ala	Cys	Gly	Met
Pro	Ser	Thr	Val
Intrinsically destabilizing residues ($t_{1/2}$ = 2 to 30 minutes)			
Arg	His	Ile	Leu
Lys	Phe	Trp	Tyr
Destabilizing following chemical modification ($t_{1/2}$ = 3 to 30 minutes)			
Asn	Asp	Gln	Glu

Source: J.W. Tobias, T.E. Schrader, G. Rocap, and A. Varshavsky. *Science* 254(1991):1374.

SUMMARY

Proteins contain signals that determine their ultimate destination. The synthesis of all proteins (except those encoded by mitochondrial and chloroplast DNA) begins on free ribosomes in the cytosol. In eukaryotes, protein synthesis continues in the cytosol unless the nascent chain contains a signal sequence that directs the ribosome to the endoplasmic reticulum (ER). Amino-terminal signal sequences of prokaryotes as well as eukaryotes consist of a hydrophobic segment of 10 to 15 residues preceded by a basic region. Signal recognition particle (SRP), a ribonucleoprotein assembly, recognizes signal sequences and brings ribosomes bearing them to the ER. A GTP-GDP cycle releases the signal sequence from SRP and then detaches SRP from its receptor. The nascent chain is then translocated across the ER membrane. Translocation continues until a stop-transfer sequence is encountered. Multiple signal sequences and stop-transfer sequences give rise to proteins that weave back and forth across the bilayer. Most signal sequences are cleaved after translocation.

The lumen of the ER is a protein folding factory. ATP-driven heat-shock proteins serve as chaperones that bind nascent proteins and assist their folding. Glycosylation also begins in the ER. Glycoproteins acquire the core of their *N*-linked sugars from a dolichol pyrophosphate donor in the ER. A block of 14 sugar residues is transferred from this highly hydrophobic carrier to asparagine side chains. Three glucose residues and a mannose residue are rapidly trimmed. Transport vesicles carry proteins from the ER to the Golgi complex for further modification of carbohydrate units and for sorting. Resident ER proteins such as chaperones are recognized by their carboxyl-terminal KDEL sequence and retrieved.

The Golgi consists of a stack of membranous sacs that are differentiated into cis, medial, and trans compartments. The cis compartment is closest to the ER and receives vesicles from it. A different set of vesicles transfers proteins from the cis to the medial and then to the trans compartment of the Golgi. Carbohydrate units of glycoproteins are modified in each of these compartments. Enzymes destined to be delivered to lysosomes contain a conformational motif that leads to the addition of a mannose 6-phosphate unit. This phosphorylated sugar is recognized by a membrane-bound receptor that brings glycoproteins to pre-lysosomes, which then fuse with lysosomes. Sorting takes place primarily in the trans Golgi. The molecular mechanisms of budding and fusion, which are at the heart of membrane targeting, are being unraveled by biochemical and genetic approaches. A multitude of players—small GTP-binding proteins (ARF and rab proteins), coat proteins, SNAPs, and SNAREs—are coming into view.

Most mitochondrial proteins are encoded by nuclear DNA and are synthesized by free ribosomes in the cytosol. They contain amino-terminal sequences (called mitochondrial presequences) and other signals that direct their entry into mitochondria and specify whether they are to remain in the outer membrane or be translocated to the inner membrane, the intermembrane space, or the matrix. Transport to the mitochondrial matrix is driven by the potential across the inner mitochondrial membrane. A cytosolic protein can be directed to the mitochondrial matrix by joining a mitochondrial presequence to its amino terminus. Chloroplasts also import most of their proteins and send them to six sites according to the nature of their presequences (also called transit sequences). Proteins must be partially or completely unfolded to enter mitochondria and chloroplasts. Most of the presequences of proteins imported into these organelles are cleaved following entry.

Cytosolic proteins containing a carboxyl-terminal SKF sequence are directed to peroxisomes. Nuclear localization signals enable proteins to enter the cell nucleus through large pores in the nuclear envelope. Soluble cytosolic proteins can be given a membrane anchor by the covalent attachment of a myristoyl group to the N terminus, a palmitoyl group to an internal cysteine, or a farnesyl (C_{15}) or geranylgeranyl (C_{20}) unit to a C-terminal cysteine. Some cell-surface proteins are tethered to the outer leaflet of the bilayer by a glycosyl phosphatidyl inositol (GPI) unit.

Specific proteins are imported into eukaryotic cells by receptor-mediated endocytosis. Proteins bound to transmembrane receptors cluster in coated-pit regions of the plasma membrane. Clathrin participates in endocytosis by forming a polyhedral lattice around the coated pit. Adaptor proteins select the membrane proteins to be endocytosed. The resulting coated vesicle rapidly loses its clathrin shell by the action of an uncoating ATPase that belongs to the heat-shock family. The vesicle then fuses with an endosome. The acidity of endosomes (pH 5 to 6) induces dissociation of most protein-receptor complexes and leads to their sorting. Many viruses and toxins enter cells by receptor-mediated endocytosis. Acid-induced conformational changes permit their release from endosomes into the cytosol.

Proteins are also targeted for destruction. Ubiquitin, a protein present in all eukaryotes and highly conserved in evolution, becomes covalently linked to proteins destined to be degraded. The conjugation of the carboxy terminus of ubiquitin to ε-amino groups of proteins involves activation by ATP and recognition of a marker. The rate of ubiquitinylation of cytosolic proteins is determined to a large extent by the nature of their amino terminus. Methionine, for example, leads to a long half-life, whereas arginine leads to a very short half-life. Ubiquitinylated proteins are digested by an ATP-powered 26S protease complex.

SELECTED READINGS

Where to start

Palade, G., 1975. Intracellular aspects of the process of protein synthesis. *Science* 189:347–358. [This Nobel Lecture gives a lucid and beautiful account of the mechanism of secretion of zymogens by pancreatic acinar cells.]

Rothman, J.E., 1994. Mechanisms of intracellular protein transport. *Nature* 372:59–67.

Rapoport, T.A., 1992. Transport of proteins across the endoplasmic reticulum membrane. *Science* 258:931–936.

Schatz, G., 1993. The protein import machinery of mitochondria. *Protein Sci.* 2:141–146.

Orci, L., Vassalli, J.-D., and Perrelet, A., 1988. The insulin factory. *Sci. Amer.* 259(3):85–94.

Books

Alberts, B., Bray, D., Lewis, J., Raff, M., Roberts, K., and Watson, J.D., 1994. *Molecular Biology of the Cell* (3rd ed.). Garland. [Chapter 13 gives a lucid and beautifully illustrated account of vesicular traffic in the secretory and endocytic pathways.]

Neupert, W., and Hill, R. (eds.), 1992. *Membrane Biogenesis and Protein Targeting.* Elsevier. [Contains many excellent articles on prokaryotic and eukaryotic protein targeting.]

Rothman, J.E. (ed.), 1992. *Methods in Enzymology,* vol. 219: *Reconstitution of Intracellular Transport.* Academic Press. [A valuable source of experimental techniques used in studying protein targeting.]

Signal sequences, signal recognition, and translocation

Gilmore, R., and Kellaris, K.V., 1992. Translocation of proteins across and integration of membrane proteins into the rough endoplasmic reticulum. *Ann. N.Y. Acad. Sci.* 674:27–37.

Nunnari, J., and Walter, P., 1992. Protein targeting to and translocation across the membrane of the endoplasmic reticulum. *Curr. Opin. Cell. Biol.* 4:573–580.

Wolin, S.L., 1994. From the elephant to *E. coli*: SRP-dependent protein targeting. *Cell* 77:787–790.

Wickner, W., Driessen, A.J., and Hartl, F.U., 1991. The enzymology of protein translocation across the *Escherichia coli* plasma membrane. *Ann. Rev. Biochem.* 60:101–124.

Miller, J.D., Wilhelm, H., Gierasch, L., Gilmore, R., and Walter, P., 1993. GTP binding and hydrolysis by the signal recognition particle during initiation of protein translocation. *Nature* 366:351–354.

Görlich, D., Hartmann, E., Prehn, S., and Rapoport, T.A., 1992.

A protein of the endoplasmic reticulum involved early in polypeptide translocation. *Nature* 357:47–52.

Simon, S.M., and Blobel, G., 1991. A protein-conducting channel in the endoplasmic reticulum. *Cell* 65:371–380.

Doering, T.L., Masterson, W.J., Hart, G.W., and Englund, P.T., 1990. Biosynthesis of glycosyl phosphatidylinositol membrane anchors. *J. Biol. Chem.* 265:611–614.

Hartmann, E., Rapoport, T.A., and Lodish, H.F., 1989. Predicting the orientation of eukaryotic membrane-spanning proteins. *Proc. Nat. Acad. Sci.* 86:5786–5790.

ER, Golgi complex, and vesicular transport

Ferro-Novick, S., and Jahn, R., 1994. Vesicle fusion from yeast to man. *Nature* 370:191–193.

Mellman, I., and Simons, K., 1992. The Golgi complex: In vitro veritas? *Cell* 68:829–840.

Pfeffer, S.R., 1992. GTP-binding proteins in intracellular transport. *Trends Cell. Biol.* 2:41–46.

Nuoffer, C., and Balch, W.E., 1994. GTPases: Multifunctional molecular switches regulating vesicular traffic. *Ann. Rev. Biochem.* 63:949–990.

Bennett, M.K., and Scheller, R.H., 1994. A molecular description of synaptic vesicle membrane trafficking. *Ann. Rev. Biochem.* 63:63–100.

Hammond, C., Braakman, I., and Helenius, A., 1994. Role of *N*-linked oligosaccharide recognition, glucose trimming, and calnexin in glycoprotein folding and quality control. *Proc. Nat. Acad. Sci.* 91:913–917.

Pelham, H.R., 1992. The secretion of proteins by cells. *Proc. Roy. Soc. Lond. Ser. B* 250:1–10.

Rothman, J.E., 1985. The compartmental organization of the Golgi apparatus. *Sci. Amer.* 253(3):74–89.

Palmer, D.J., Helms, J.B., Beckers, C.J., Orci, L., and Rothman, J.E., 1993. Binding of coatomer to Golgi membranes requires ADP-ribosylation factor. *J. Biol. Chem.* 268:12083–12089.

Orci, L., Palmer, D.J., Ravazzola, M., Perrelet, A., Amherdt, M., and Rothman, J.E., 1993. Budding from Golgi membranes requires the coatomer complex of non-clathrin coat proteins. *Nature* 362:648–652.

Whiteheart, S.W., Griff, I.C., Brunner, M., Clary, D.O., Mayer, T., Buhrow, S.A., and Rothman, J.E., 1993. SNAP family of NSF attachment proteins includes a brain-specific isoform. *Nature* 362:353–355.

Sollner, T., Whiteheart, S.W., Brunner, M., Erdjument, B.H., Geromanos, S., Tempst, P., and Rothman, J.E., 1993. SNAP receptors implicated in vesicle targeting and fusion. *Nature* 362:318–324.

Heat-shock proteins and chaperones

Welch, W., and Georgopoulos, C., 1993. Heat-shock proteins or chaperones. *Ann. Rev. Cell Biol.* 9:601–634.

Hendrick, J.P., and Hartl, F.-U., 1993. Molecular chaperone functions of heat-shock proteins. *Ann. Rev. Biochem.* 62:349–384.

Gething, M.J., and Sambrook, J., 1992. Protein folding in the cell. *Nature* 355:33–45.

McKay, D.B., 1993. Structure and mechanism of 70-kDa heat-shock-related proteins. *Advan. Prot. Chem.* 44:67–98.

Fatty acylation and prenylation of proteins

Johnson, D.R., Bhatnager, R.S., Knoll, L.J., and Gordon, J.I., 1994. *Ann. Rev. Biochem.* 63:869–914.

Schafer, W.R., and Rine, J., 1992. Protein prenylation: Genes, enzymes, targets, and functions. *Ann. Rev. Genet.* 26:209–237.

Clarke, S., 1992. Protein isoprenylation and methylation at carboxyl-terminal cysteine residues. *Ann. Rev. Biochem.* 61:355–386.

Targeting to mitochondria and chloroplasts

Glick, B.S., Beasley, E.M., and Schatz, G., 1992. Protein sorting in mitochondria. *Trends Biochem. Sci.* 17:453–459.

Pfanner, N., Rassow, J., van der Klei, I.J., and Neupert, W., 1992. A dynamic model of the mitochondrial protein import machinery. *Cell* 68:999–1002.

Smeekens, S., Weisbeek, P., and Robinson, C., 1990. Protein transport into and within chloroplasts. *Trends Biochem. Sci.* 15:73–76.

Targeting to peroxisomes

Subramani, S., 1993. Protein import into peroxisomes and biogenesis of the organelle. *Ann. Rev. Cell Biol.* 9:445–478.

Targeting to nuclei

Silver, P.A., 1991. How proteins enter the nucleus. *Cell* 64:489–497.

Forbes, D.J., 1992. Structure and function of the nuclear pore complex. *Ann. Rev. Cell Biol.* 8:495–527.

Dingwall, C., 1991. Transport across the nuclear envelope: Enigmas and explanations. *BioEssays* 13:213–218.

Hinshaw, J.E., Carragher, B.O., and Milligan, R.A., 1992. Architecture and design of the nuclear pore complex. *Cell* 69:1133–1141.

Targeting in bacteria

Pugsley, A.P., 1993. The complete general secretory pathway in gram-negative bacteria. *Microbiol. Rev.* 57:50–108.

Bassilana, M., and Wickner, W., 1993. Purified *Escherichia coli* preprotein translocase catalyzes multiple cycles of precursor protein translocation. *Biochemistry* 32:2626–2630.

Simon, S.M., and Blobel, G., 1992. Signal peptides open protein-conducting channels in *E. coli*. *Cell* 69:677–684.

Hartmann, E., Sommer, T., Prehn, S., Gorlich, D., Jentsch, S., and Rapoport, T.A., 1994. Evolutionary conservation of components of the protein translocation complex. *Nature* 367:654–657.

Miller, J.D., Bernstein, H.D., and Walter, P., 1994. Interaction of E. coli Ffh/4.5S ribonucleoprotein and FtsY mimics that of mammalian signal recognition particle and its receptor. *Nature* 367:657–659.

Endocytosis, clathrin, and coated vesicles

Dautry-Varsat, A., and Lodish, H.F., 1984. How receptors bring proteins and particles into cells. *Sci. Amer.* 250(5):52–58.

Pastan, I., and Willingham, M.C. (eds.), 1985. *Endocytosis.* Plenum. [Contains excellent articles on topics such as receptor-mediated endocytosis, clathrin, asialoglycoprotein receptor, and toxins.]

Pearse, B.M., and Robinson, M.S., 1990. Clathrin, adaptors, and sorting. *Ann. Rev. Cell Biol.* 6:151–171.

Brodsky, F.M., Hill, B.L., Acton, S.L., Näthke, I., Wong, D.H., Ponnambalam, S., and Parham, P., 1991. Clathrin light chains: Arrays of protein motifs that regulate coated-vesicle dynamics. *Trends Biochem. Sci.* 16:208–213.

Entry of viruses and toxins

Simons, K., Garoff, H., and Helenius, A., 1982. How an animal virus gets into and out of its host cell. *Sci. Amer.* 246(2):58–66.

Sixma, T.K., Pronk, S.E., Kalk, K.H., Wartna, E.S., van Zanten, B.A.M., Witholt, B., and Hol, W.G., 1991. Crystal structure of a cholera toxin–related heat-labile enterotoxin from *E. coli.* *Nature* 351:371–377.

Mosser, G., Mallouh, V., and Brisson, A., 1992. A 9 Å two-dimensional projected structure of cholera toxin B-subunit-G_{M1} complexes determined by electron crystallography. *J. Mol. Biol.* 226:23–28.

Choe, S., Bennett, M.J., Fujii, G., Curmi, P.M., Kantardjieff, K.A., Collier, R.J., and Eisenberg, D., 1992. The crystal structure of diphtheria toxin. *Nature* 357:216–222.

Protein degradation

Ciechanover, A., 1994. The ubiquitin-proteasome proteolytic pathway. *Cell* 79:13–21.

Peters, J.-M., 1994. Proteosomes: Protein degradation machines of the cell. *Trends Biochem. Sci.* 19:377–382.

Goldberg, A.L., and Rock, K.L., 1992. Proteolysis, proteasomes and antigen presentation. *Nature* 357:375–379.

Hershko, A., and Ciechanover, A., 1992. The ubiquitin system for protein degradation. *Ann. Rev. Biochem.* 61:761–807.

Varshavsky, A., 1992. The N-end rule. *Cell* 69:725–735.

PROBLEMS

1. *Targeting signals.* What are the distinctive features of each of the following targeting sequences?
   (a) Eukaryotic signal directing a nascent protein to the ER.
   (b) Prokaryotic signal directing a nascent protein to the plasma membrane.
   (c) Signal directing a protein in the Golgi to lysosomes.
   (d) Signal directing a membrane protein in the Golgi to the plasma membrane.
   (e) Signal directing a protein in the cytosol to mitochondria.
   (f) Signal targeting a cytosolic protein for rapid destruction.

2. *Insouciance.* A mutant LDL receptor is found to be uniformly distributed in the plasma membrane rather than being concentrated in coated-pit regions. The mutant binds LDL normally but is not internalized. Which region of the receptor is likely to be altered?

3. *A different route.* Cells derived from patients with I-cell disease take up lysosomal enzymes purified from normal cells. The added enzymes appear in the lysosomes of these cells.
   (a) What is the likely pathway for the transport of these enzymes from the extracellular medium to the lysosomes of I-cells?
   (b) This experiment led to the hypothesis that lysosomal enzymes are normally secreted and taken up by this route. Which sugar phosphate would you add to the extracellular medium to test this hypothesis?

4. *Predicting the locations of new proteins.* Membrane-bound immunoglobulin and its secreted counterpart differ in the carboxyl-terminal regions of their heavy chains. As was noted in an earlier chapter (p. 115), these variations on a theme arise from alternative splicing of mRNA.
   (a) Suppose that the transmembrane sequence of the antibody is placed by recombinant DNA methods at the amino terminus of a cytosolic protein. Where is this chimeric protein likely to be located?
   (b) Suppose that this transmembrane sequence is placed at the carboxy terminus of chymotrypsinogen. Where is this chimeric protein likely to be located?

5. *Contrasting amino termini.* The amino-terminal residue of secreted eukaryotic proteins is usually different from that of cytosolic proteins. A majority of secretory proteins have leucine, phenylalanine, aspartate, lysine, or arginine as their amino-terminal residue. In contrast, these residues are rarely found in cytosolic proteins. What is a potential benefit of having different amino-terminal residues for these two classes of proteins?

6. *Defective peroxisomes.* A newborn with multiple developmental abnormalities and severe neurological damage was found to have peroxisomes that are unable to oxidize long-chain fatty acids, carry out steps in the synthesis of plasmalogens, and form bile acids. Catalase was found in the cytosol rather than in the peroxisomes of this infant's cells. Propose a molecular defect that would account for these findings.

7. *Redirected enzymes.* Dihydrofolate reductase (DHFR), a cytosolic enzyme, can be redirected into mitochondria by attaching a matrix-targeting sequence to its amino terminus. DHFR can be placed in the mitochondrial matrix by adding to its amino terminus the 27-residue presequence of an alcohol dehydrogenase isozyme that normally resides in the matrix. The import of this chimeric DHFR into the matrix is blocked by methotrexate. However, methotrexate does not block the transport of authentic mitochondrial proteins, nor does it interfere with the binding of the chimeric protein to receptors on the outer mitochondrial membrane. How does methotrexate interfere with the import of this presequence-DHFR chimeric protein?

8. *Cancer chemotherapy.* Mutant forms of ras that are persistently in the GTP state accelerate the progression of many tumors. Propose a new class of anticancer agents based on altered targeting of ras.

CHAPTER

36

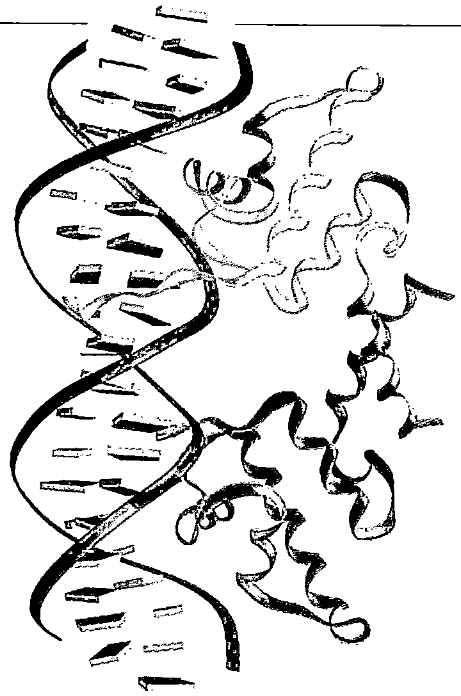

Control of Gene Expression in Prokaryotes

Bacteria are highly versatile and responsive organisms: they sense the levels of many metabolites and regulate their metabolic patterns accordingly by a wide variety of mechanisms. In earlier chapters, we saw that the activities of many key proteins are controlled by allosteric changes and reversible covalent modifications. Equally critical in determining the pattern of cellular processes is the control of gene expression. An *Escherichia coli* cell contains only 10 or so copies of scarce proteins and as many as 10^5 copies of highly abundant ones. Moreover, the rate of synthesis of some proteins varies over a 1000-fold range in response to the supply of nutrients or to environmental challenges.

Gene activity is regulated primarily at the level of transcription. In bacteria, many genes are clustered in units called *operons.* The coordinate transcription of genes in an operon is blocked by *repressor proteins* and activated by *stimulatory proteins.* We shall focus on the lactose and tryptophan operons of *E. coli* because much is known about the molecular basis of their control. Bacteriophage lambda (λ) is another rewarding object of inquiry because it illustrates how a choice is made between alternative developmental pathways. The elucidation of the three-dimensional structure of several control proteins has revealed a recurring *helix-turn-helix* motif in protein-DNA interactions (Figure 36-1). Success in exploring gene regulation in prokaryotes has inspired studies of more complex eukaryotic systems, the theme of the next chapter.

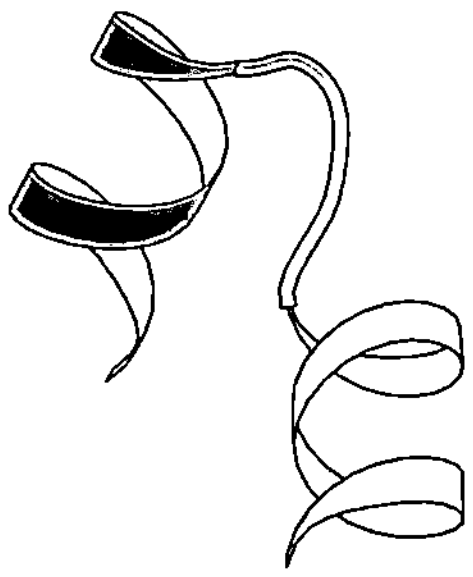

Figure 36-1
The helix-turn-helix motif occurs in many prokaryotic regulators of gene expression. The second helix of this motif (shown in green) fits neatly into the major groove of DNA.

Opening Image: *The λ repressor (blue and yellow) silences the expression of λ phage genes by binding to the operator site on DNA (green and red). The symmetry of the dimeric λ repressor matches that of its DNA target. [Drawn from 1lmb.pdb. L.J. Beamer and C.O. Pabo. J. Mol. Biol. 227(1992):177.]*

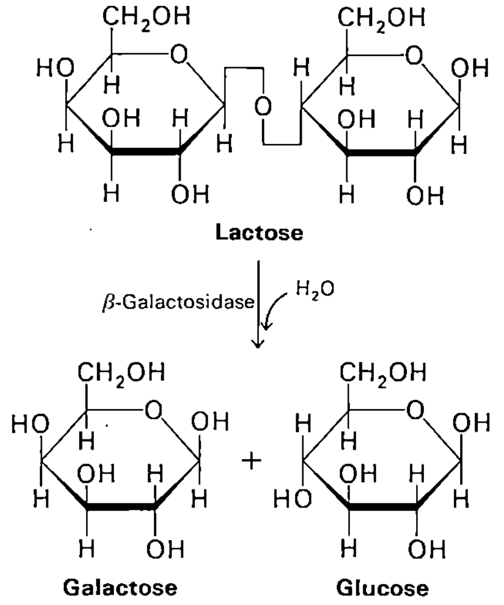

Figure 36-2
Lactose is hydrolyzed by β-galactosidase.

β-GALACTOSIDASE IS AN INDUCIBLE ENZYME

E. coli can use lactose as its sole source of carbon. An essential enzyme in the metabolism of this disaccharide is *β-galactosidase,* which hydrolyzes lactose to galactose and glucose (Figure 36-2). An *E. coli* cell growing on lactose contains several thousand molecules of β-galactosidase. In contrast, the number of β-galactosidase molecules per cell is fewer than 10 if *E. coli* is grown on other sources of carbon, such as glucose or glycerol. The presence of lactose in a culture medium induces a large increase in the amount of β-galactosidase in *E. coli* by eliciting the synthesis of new enzyme molecules rather than by activating a proenzyme (Figure 36-3). Hence, β-galactosidase is an *inducible enzyme.* Two other proteins are synthesized in concert with β-galactosidase—namely, *galactoside permease* and *thiogalactoside transacetylase.* The permease is required for the transport of lactose across the bacterial cell membrane. The transacetylase is not essential for lactose metabolism; its physiologic role is uncertain. In vitro, it catalyzes the transfer of an acetyl group from acetyl CoA to the C-6 hydroxyl group of a thiogalactoside.

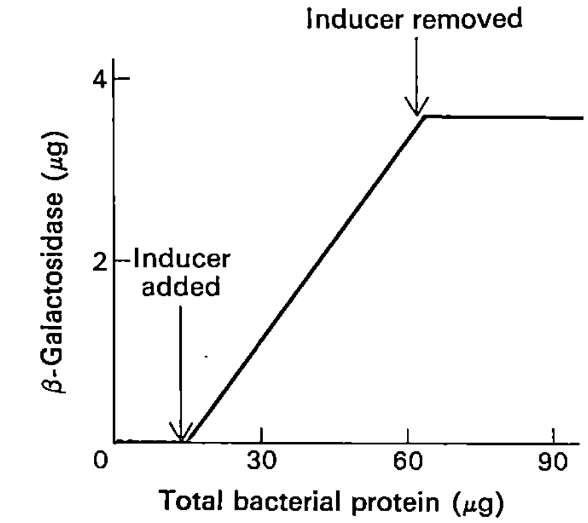

Figure 36-3
The increase in the amount of β-galactosidase parallels the increase in the number of cells in a growing culture of *E. coli.* The slope of this plot indicates that 6.6% of the protein synthesized is β-galactosidase.

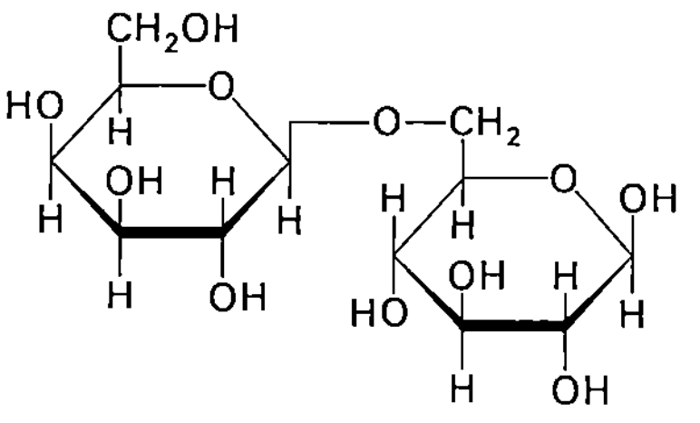

1,6-Allolactose

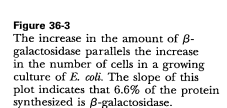

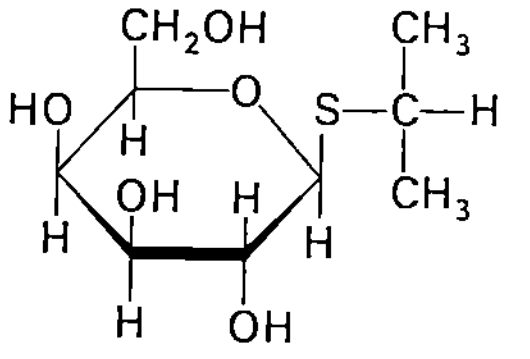

IPTG

Within an *E. coli* cell, the physiologic inducer is *allolactose,* which is formed from lactose by transglycosylation. The synthesis of allolactose is catalyzed by the few β-galactosidase molecules that are present prior to induction. Studies of synthetic inducers showed that some β-galactosides are inducers without being substrates of β-galactosidase, whereas other compounds such as lactose are substrates without being inducers. For example, *isopropylthiogalactoside (IPTG)* is a nonmetabolizable inducer.

DISCOVERY OF A REGULATORY GENE

An important clue concerning the nature of the induction process was the finding that the amounts of the permease and the transacetylase increased in direct proportion to that of β-galactosidase for all inducers tested. Further insight came from studies of mutants, which showed that β-galactosidase, the permease, and the transacetylase are encoded by three contiguous genes, called *z*, *y*, and *a*, respectively. Mutants defective in only one of these proteins were isolated. For example, $z^-y^+a^+$ denotes a mutant lacking β-galactosidase but having normal amounts of the permease and the transacetylase. A most interesting class of mutants affecting all three proteins was then isolated. These *constitutive mutants* synthesize large amounts of β-galactosidase, the permease, and the transacetylase whether or not inducer is present. Francois Jacob and Jacques Monod deduced that *the rate of synthesis of these three proteins is normally governed by a common element that is different from the genes specifying their structures.* The gene for this common regulatory element was named *i*. Wild-type inducible bacteria have the genotype $i^+z^+y^+a^+$, whereas the constitutive lactose mutants have the genotype $i^-z^+y^+a^+$.

How does the i^+ gene affect the rate of synthesis of the proteins encoded by the *z*, *y*, and *a* genes? The simplest hypothesis was that the i^+ gene determines the synthesis of a cytoplasmic substance called a *repressor*, which is missing or inactive in i^- mutants. This idea was tested in an ingenious series of genetic experiments involving partly diploid bacteria that contained two sets of genes for the lactose region (Figure 36-4). One

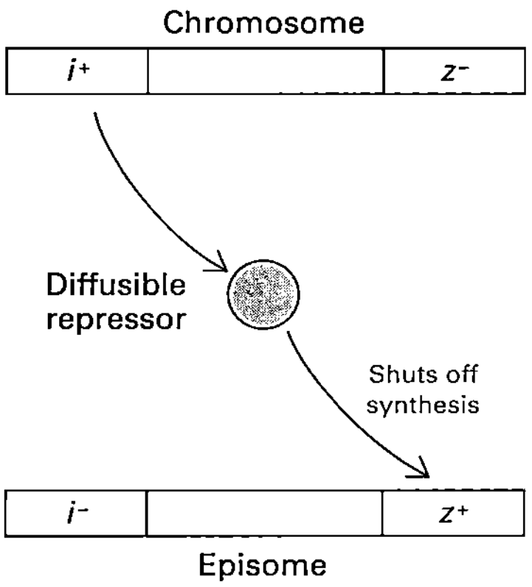

Figure 36-4
A diffusible repressor encoded by the i^+ gene on the chromosome switches off synthesis of β-galactosidase encoded by the z^+ gene on the episome. Other experiments showed that the repressor does not act directly on z^+.

set was on the bacterial chromosome, whereas the other was on an F′ sex factor (p. 827), an episome introduced by conjugation. For example, an i^+z^-/Fi^-z^+ diploid was isolated. In this diploid, i^+z^- is on the chromosome, whereas i^-z^+ is on the episome. Is this diploid inducible or constitutive for β-galactosidase? In other words, does i^+ on the bacterial chromosome repress the expression of z^+ on the episome? The experimental result was clear-cut: *the diploid was inducible rather than constitutive.* The same result was obtained for the diploid i^-z^+/Fi^+z^-. Hence, *a diffusible repressor is specified by the* i$^+$ *gene.* A diffusible repressor is an example of a *trans-acting factor,* one that can be encoded by a locus on a DNA molecule different from the one containing its target.

AN OPERON IS A COORDINATED UNIT OF GENE EXPRESSION

These and other trenchant experiments led Jacob and Monod to propose the *operon model* for the regulation of RNA synthesis. The genetic elements of the model are a *regulator gene,* an *operator site,* and a set of *structural genes* (Figure 36-5). The regulator gene produces a *repressor* that can interact with the operator. Subsequent work revealed that the repressor is a protein. The operator, by contrast, is a DNA segment adjacent to the

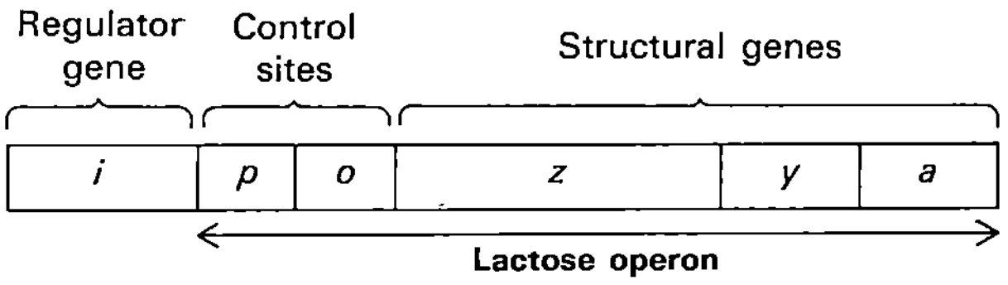

Figure 36-5
Map of the lactose operon and its regulator gene. (This map is not drawn to scale: the promoter (*p*) and operator (*o*) sites are actually much smaller than the other genes.)

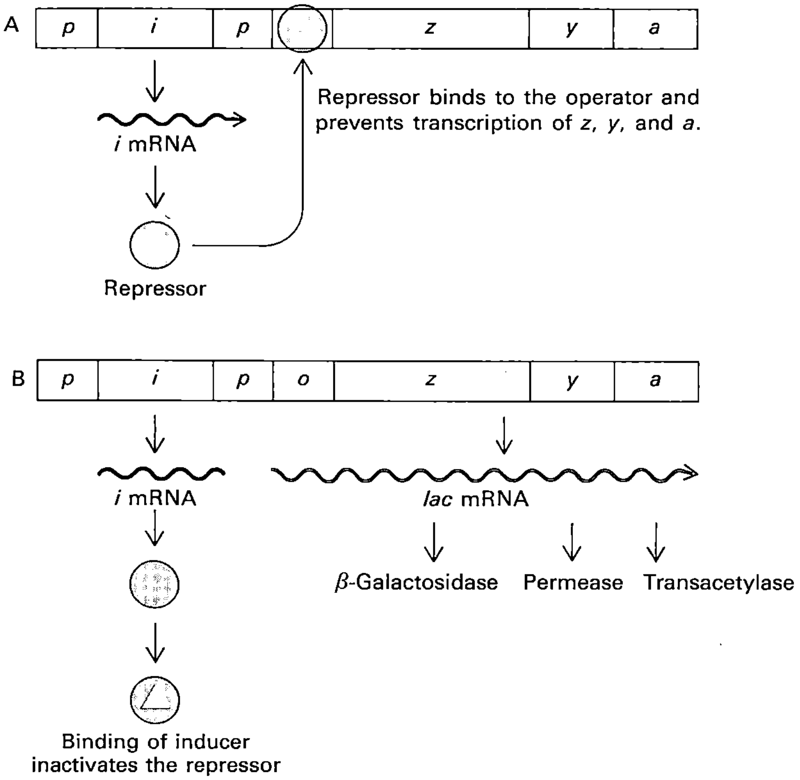

Figure 36-6
Diagram of the lactose operon in the (A) repressed and (B) induced states.

structural genes it controls. The binding of the repressor to the operator prevents the transcription of these genes. The operator and its associated structural genes are called an *operon.* For the lactose operon, the *i* gene is the regulator gene, *o* is the operator, and the *z*, *y*, and *a* genes are the structural genes. The operon also contains a *promoter site* (denoted by *p*) for the binding of RNA polymerase. This site for the initiation of transcription is next to the operator. An *inducer* such as IPTG binds to the repressor, which prevents it from interacting with the operator. The *z*, *y*, and *a* genes can then be transcribed to give a single mRNA molecule that codes for all three proteins (Figure 36-6). An mRNA molecule coding for more than one protein is known as a *polygenic* (or *polycistronic*) *transcript.*

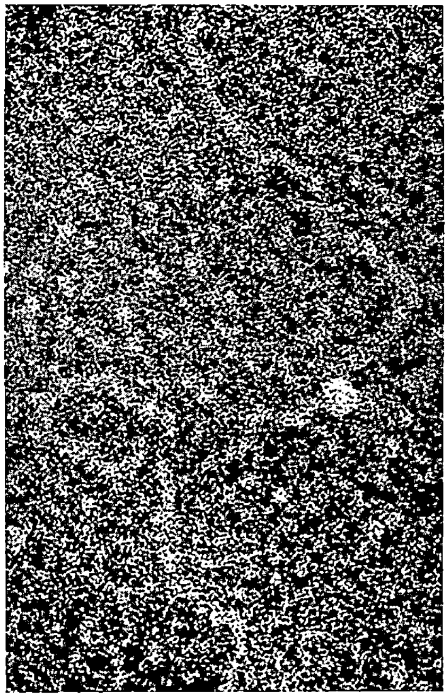

Figure 36-7
Electron micrograph of the *lac* repressor bound to DNA containing the *lac* operator. [Courtesy of Dr. Jack Griffith.]

LAC REPRESSOR PROTEIN IN THE ABSENCE OF INDUCER BINDS TO THE OPERATOR AND BLOCKS TRANSCRIPTION

The repressor of the lactose operon (called the *lac repressor*) was isolated on the basis of its binding of IPTG. Walter Gilbert and Benno Müller-Hill showed that the *lac* repressor is a protein that binds to DNA carrying the *lac* operon but not to other DNA molecules. As predicted, IPTG prevents the binding of *lac* repressor to *lac* operator DNA. A wild-type *E. coli* cell contains only about 10 molecules of the *lac* repressor, only 0.001% of the total protein, so that it was difficult to purify the repressor from these cells. However, much larger amounts of the *lac* repressor were made in mutants having a more efficient promoter for the *i* gene. The amount of *lac* repressor was increased further by infecting the bacteria with transducing phages carrying the *lac* region. Such an infected *E. coli* cell contains about 20,000 repressors (about 2% of the total protein), which made it a choice starting material for the purification of *lac* repressor.

The repressor is a tetramer of identical 37-kd subunits, each with a binding site for the inducer. Crystallographic studies have shown that the protein contains twofold axes of symmetry, as might be expected for a protein that binds specifically to DNA (p. 792). The repressor, in the absence of inducer, binds very tightly and rapidly to the operator. The dissociation constant of the repressor-operator complex is about 0.1 pM (10^{-13} M). The rate constant for association ($\sim 10^{10}$ M^{-1} s^{-1}) is strikingly high, indicating that *the repressor finds the operator site by diffusing along a DNA molecule (a one-dimensional search) rather than by encountering it from the aqueous medium (a three-dimensional search).* Recall that RNA polymerase finds promoter sites in a similar way (p. 844). The *lac* repressor binds 4×10^6 times as strongly to its operator as to other sites on the chromosome. This high degree of selectivity is necessary because the genome contains a vast excess (1.6×10^5) of competing sites.

THE *LAC* OPERATOR HAS A SYMMETRIC BASE SEQUENCE

The availability of pure *lac* repressor made it feasible to isolate the *lac* operator and determine its base sequence. Gilbert and his associates sonicated the DNA of a phage carrying the *lac* region into fragments that were approximately 1000 base pairs long. The *lac* repressor was added to the mixture of fragments, which was then filtered through a cellulose nitrate membrane. DNA fragments without bound *lac* repressor passed through this filter, whereas DNA-repressor complexes bound tightly to the filter. DNA fragments containing the *lac* operator were eluted from the filter by the addition of IPTG, which dissociates *lac* repressor-operator complexes. The free DNA fragments, following removal of IPTG, were treated with pancreatic deoxyribonuclease in the presence of repressor, which protected the operator region from digestion (Figure 36-8).

The base sequence of the protected operator region is very interesting: a total of 28 base pairs are related by a twofold axis of symmetry (Figure 36-9), just as the subunits of *lac* repressor are. Thus, *the symmetry of the repressor molecule matches that of its target site in DNA.* Recall that this recognition principle is also utilized in the binding of restriction endonucleases to palindromic cleavage sites (p. 120). *Symmetry matching is a recurring theme in protein-DNA interactions.*

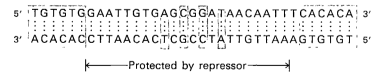

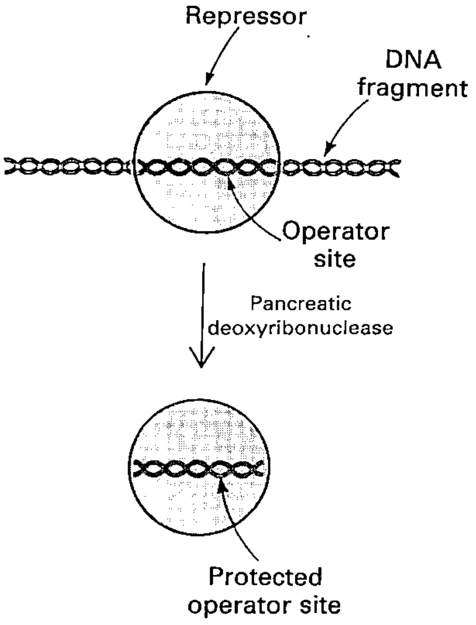

Figure 36-8
The *lac* repressor protects the *lac* operator from digestion by pancreatic deoxyribonuclease.

5′ TGTGTGGAATTGTGAGCGGATAACAATTTCACACA 3′
3′ ACACACCTTAACACTCGCCTATTGTTAAAGTGTGT 5′
|←——Protected by repressor——→|

Figure 36-9
Nucleotide sequence of the *lac* operator. The symmetrically related regions are shown in matching colors.

INDUCIBLE CATABOLIC OPERONS ARE GLOBALLY REGULATED BY CAP PROTEIN CONTAINING BOUND CYCLIC AMP

It has long been known that *E. coli* grown on glucose, a preferred energy source, have very low levels of catabolic enzymes, such as β-galactosidase, galactokinase, arabinose isomerase, and tryptophanase. Clearly, it would be wasteful to synthesize these enzymes when glucose is abundant. The

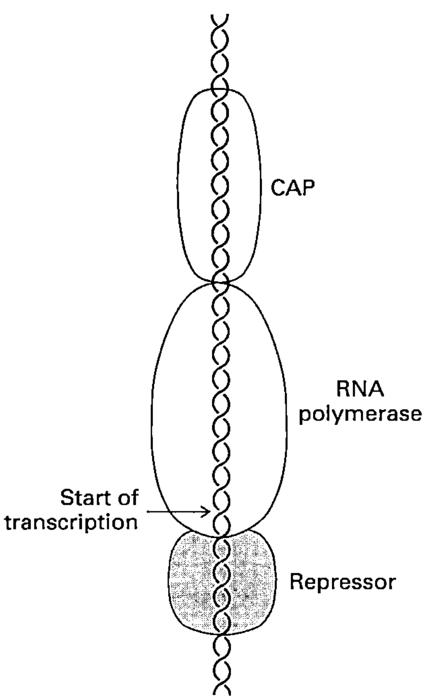

Figure 36-10
Schematic diagram of CAP, RNA polymerase, and *lac* repressor on a DNA template. The locations of these proteins were inferred from nuclease digestion studies. Both CAP and *lac* repressor enhance the binding of RNA polymerase. The repressor, in contrast with CAP, blocks transcription.

molecular basis of this inhibitory effect of glucose, called *catabolite repression,* has been elucidated. A key clue was the observation that glucose lowers the concentration of cyclic AMP in *E. coli.* It was then found that exogenous cyclic AMP can relieve the repression exerted by glucose. Subsequent biochemical and genetic studies revealed that *cyclic AMP stimulates the concerted initiation of transcription of many inducible operons.* It is interesting to note that cyclic AMP serves as a hunger signal both in bacteria and in mammals. Recall that a low level of blood sugar stimulates the secretion of glucagon, which leads to elevated cyclic AMP levels inside hormone-sensitive cells (p. 594).

In mammalian cells, cyclic AMP (cAMP) acts by stimulating a protein kinase that phosphorylates many target proteins, such as those controlling glycogen synthesis and breakdown (p. 595). The action of cAMP in bacteria is very different. cAMP binds to *CAP* (the *catabolite gene activator protein*), a dimer of identical 22-kd subunits. Proteolytic digestion experiments have shown that each subunit contains a DNA-binding domain and a cAMP-binding domain. *The complex of CAP and cAMP, but not CAP alone, stimulates transcription by binding to certain promoter sites.* In the *lac* operon, CAP binds next to the site for RNA polymerase, as shown by nuclease digestion studies. Specifically, CAP protects nucleotides −87 to −49 from digestion, whereas RNA polymerase protects nucleotides −48 to +5 (Figure 36-10). In this numbering system, the first transcribed nucleotide is +1. CAP exhibits twofold symmetry (Figure 36-11) that matches that of its DNA-binding site. *Lac* repressor binds on the other side of RNA polymerase, as shown by its protection of nucleotides −3 to +21. The binding of repressor enhances the binding of RNA polymerase but blocks the progression of the closed complex to the open one (p. 845).

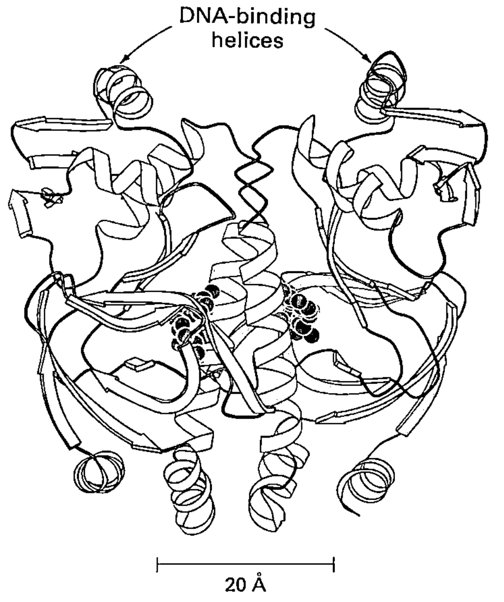

Figure 36-11
Structure of CAP protein. Each subunit of the dimer (yellow and blue chains) contains a bound cyclic AMP (purple). The DNA-binding helices of the helix-loop-helix motifs are marked. [Drawn from 3gap.pdb. I.T. Weber and T.A. Steitz. *J. Mol. Biol.* 198(1987):311.]

CAP stimulates the initiation of *lac* mRNA synthesis by a factor of 50. How? The contiguous and nonoverlapping arrangement of the binding sites for CAP and RNA polymerase suggested that *the binding of CAP to DNA creates an additional interaction site for RNA polymerase.* Indeed, the binding of RNA polymerase to the promoter is enhanced by its energetically favorable contacts with bound CAP. *CAP also stimulates transcription by bending DNA.* The three-dimensional structure of a CAP–DNA complex reveals that the protein bends DNA by 94 degrees (Figure 36-12). Specifi-

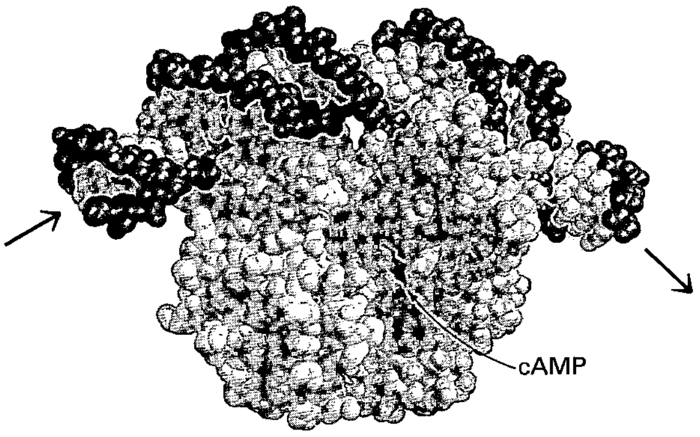

Figure 36-12
The binding of CAP protein bends DNA by about 90 degrees. The two subunits of CAP are shown in yellow and blue, cyclic AMP in purple, the DNA backbone in red, and the bases of DNA in green. [Drawn from 1cgp.pdb. S.C. Schultz, G.C. Shields, and T.A. Steitz. *Science* 253(1991):1001.]

cally, the two helix-loop-helix motifs (p. 963) of the dimer insert into successive major grooves of the DNA double helix. Two 43-degree kinks account for most of the right angle turn of the DNA axis. The binding of CAP and of RNA polymerase are mutually reinforcing because they bend DNA in the same direction.

The cAMP–CAP complex probably acts similarly at other inducible catabolic operons. All contain TGTGA upstream of the promoter site—in fact, TGTGA is part of the CAP-binding site. Another common feature is that their −35 and −10 sequences differ markedly from the consensus sequence of strong promoters (Figure 36-13). Evolution has probably weakened these promoters to make their operons dependent on a helper protein for efficient initiation of transcription. Thus, *inducible catabolic operons have been placed under dual control.* A high level of expression requires the simultaneous presence of cAMP and a specific inducer, such as a galactoside for the *lac* operon. The specific inducer acts on a single operon, whereas the cAMP–CAP complex affects many.

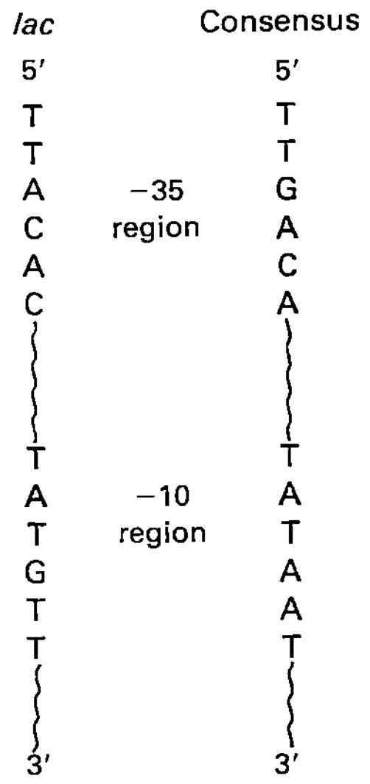

Figure 36-13
The base sequence of the *lac* promoter deviates considerably from the consensus sequence (green) of strong promoters. cAMP–CAP is required for optimal expression of the *lac* operon.

The operons controlled in concert by cyclic AMP are members of a *global regulatory circuit.* Another such network is made up of the heat-shock genes that are activated by a distinctive sigma subunit of RNA polymerase induced by a rise in temperature (p. 844). Yet another example of global control is the induction of SOS genes that encode repair enzymes following damage to DNA (p. 825). Each of these higher-level regulatory systems is specifically triggered and has built-in mechanisms for returning to the unstimulated state.

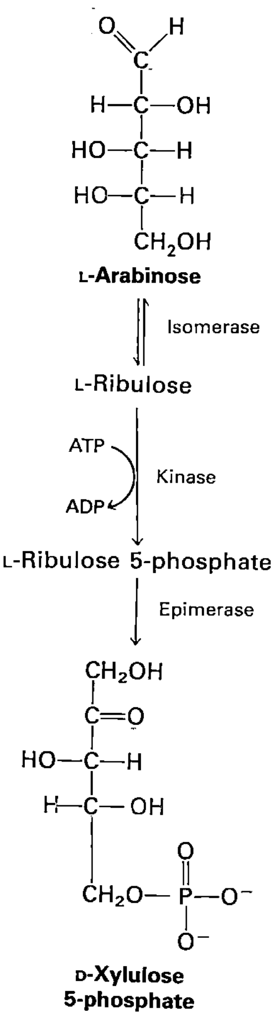

Figure 36-14
The conversion of arabinose to xylulose 5-phosphate is catalyzed by the three enzymes encoded by the arabinose operon.

DIFFERENT FORMS OF THE SAME PROTEIN ACTIVATE AND INHIBIT TRANSCRIPTION OF THE ARABINOSE OPERON

Bacteria can use arabinose as a fuel by converting it into xylulose 5-phosphate, a pentose phosphate that can be converted into glycolytic intermediates (p. 561). Xylulose 5-phosphate is formed from arabinose by the sequential action of arabinose isomerase, ribulokinase, and ribulose 5-phosphate epimerase (Figure 36-14). These enzymes are encoded by the *araA, araB,* and *araD* genes, respectively, which all belong to the arabinose operon. This region also contains a gene (*araC*) for a regulatory C protein, two operator sites (*ara*O_1 and *ara*O_2), a binding site for CAP, and another regulatory site (*araI*) (Figure 36-15).

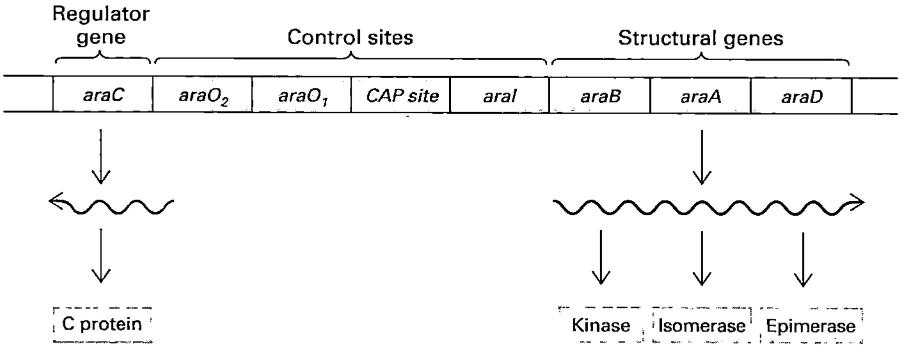

Figure 36-15
Map of the arabinose operon and its regulator gene.

The arabinose operon, like other inducible catabolic operons, is under dual control. Two signals—the cAMP–CAP complex and arabinose bound to the C protein—are required for efficient transcription. The C protein also serves as a negative regulator of *araC.* Transcription of *araC,* which proceeds in a direction opposite to that of *araBAD,* is controlled by the O_1 rather than the O_2 operator site. mRNA for C protein is formed when the level of this protein and of cAMP–CAP is low (Figure 36-16A). When C protein is abundant and cAMP–CAP is not, transcription of the *C* gene stops because a C protein binds to *ara*O_1. Thus, *the synthesis of C protein is autoregulated.* The binding of a second molecule of C to *ara*O_2 blocks the synthesis of *BAD* mRNA by forming a DNA loop and also binding to *araI,* which is adjacent to the promoter for the *BAD* genes (Figure 36-16B). Looping blocks the binding of RNA polymerase.

This loop is not formed when cAMP–CAP is abundant and arabinose is bound to the C protein, which alters its conformation. Instead, the presence of these positive regulatory factors enables RNA polymerase to bind to the promoter site for the *BAD* genes and to transcribe them (Figure 36-16C). In this case, mRNA for C protein is not formed because *ara*O_1 is occupied by C protein.

The arabinose operon illustrates several general principles of gene regulation. First, a protein can regulate its own synthesis by repressing the transcription of its gene. Second, the binding of a signal molecule to a protein can switch it from being an inhibitor of transcription to being an activator. These alternative conformations bind to different regulatory sites on DNA. Third, protein-binding regulatory sites on DNA need not be contiguous with the genes controlled by them. The arabinose operon provides a

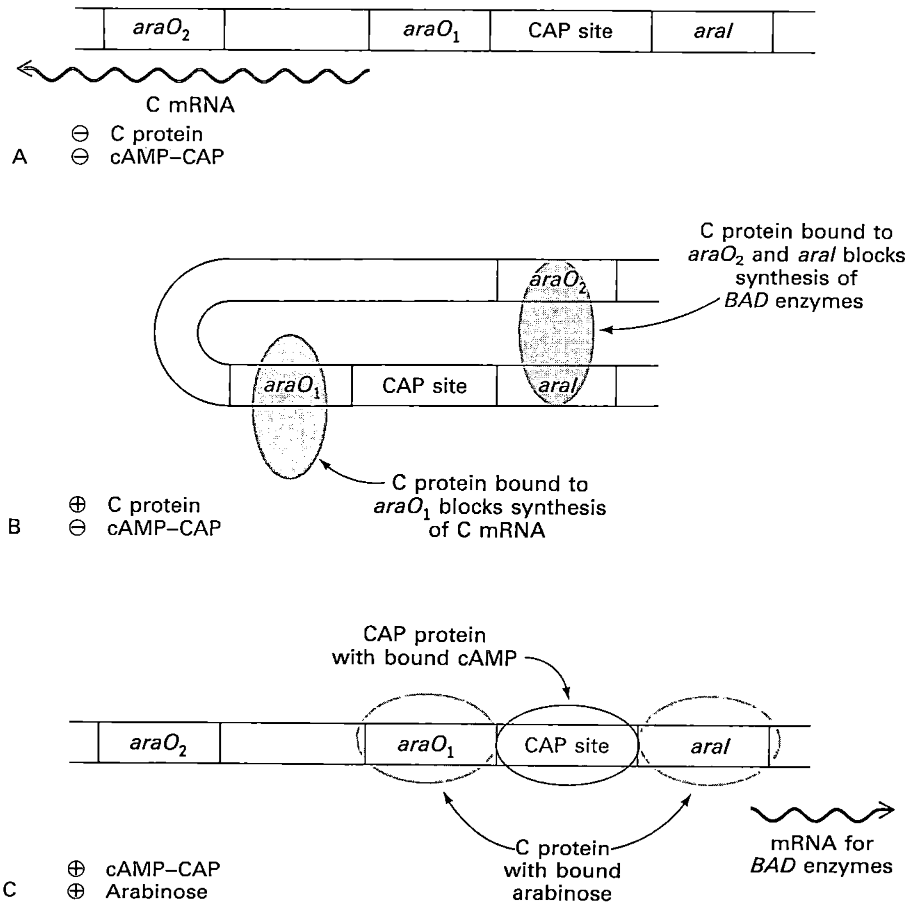

Figure 36-16
Three states of the arabinose operon. (A) mRNA for C protein is formed at low levels of this protein. (B) No mRNA is formed when the cAMP-CAP level is low and C protein is abundant, irrespective of the level of arabinose. (C) When both cAMP-CAP and arabinose are abundant, mRNA for enzymes catabolizing arabinose is formed.

concrete example of how transcription can be modulated by a site at some distance from the transcribed gene. DNA can readily be looped by the binding of a protein or a protein complex to noncontiguous sites on the DNA. DNA looping is widely used in gene regulation, especially in eukaryotes (p. 856). Fourth, the changes induced by signal molecules are readily reversed. The system responds continuously and rapidly to variations in the levels of metabolites.

REPRESSORS AND ACTIVATORS OF TRANSCRIPTION DETERMINE THE DEVELOPMENT OF TEMPERATE PHAGE

We turn now to the role of repressors and activators of transcription in regulating the life cycle of *lambda (λ) bacteriophage.* The mature virus particle consists of a linear double-helical DNA molecule (48 kb) surrounded by a protein coat. After it has infected a bacterium, two developmental pathways are open to this virus: destroy the bacterium or join it (hence the name *temperate*; see Figure 6-15 on p. 128). In the *lytic pathway,* the viral functions are fully expressed, leading to the lysis of the bacterium and the release of about 100 progeny virus particles. Alternatively, λ can enter the *lysogenic pathway,* in which its DNA becomes covalently inserted into the host-cell DNA at a specific site. This recombination process, involving a circular λ DNA molecule, has already been discussed (p. 832). Most of the phage functions are switched off when its DNA is

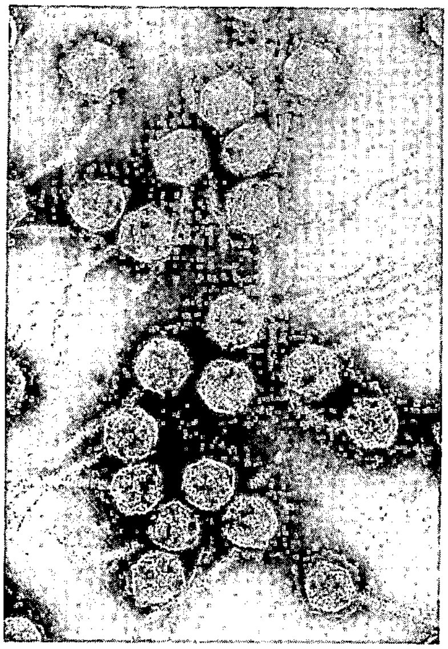

Figure 36-17
Electron micrograph of λ phages. [Courtesy of Dr. A. Dale Kaiser.]

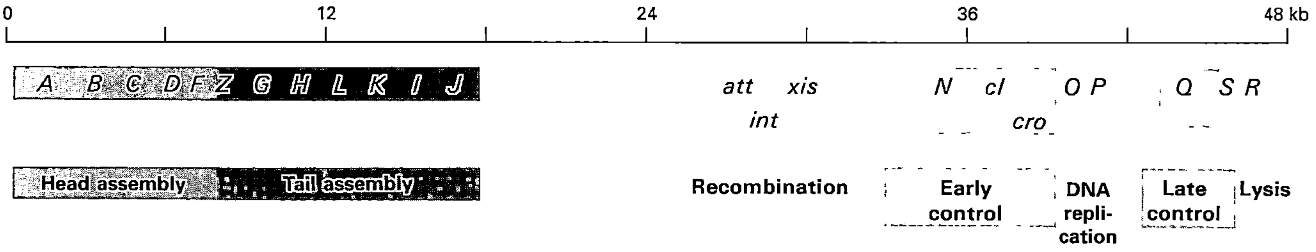

Figure 36-18
Genetic map of λ phage. Key processes are encoded by blocks of contiguous genes. Only a few genes are shown here. The linear duplex DNA is converted into the circular form after it enters the bacterial cell.

integrated in the host DNA. The viral DNA in this state is called a *prophage;* a host cell containing a prophage is called a *lysogenic bacterium.* The prophage replicates as part of the host chromosome in a lysogenic bacterium, usually for many generations. The lytic functions of λ are dormant but not lost in the lysogenic state. A variety of agents that damage the host DNA induce the prophage to undergo lytic development.

Let us first consider the pattern of gene expression of λ in the lytic pathway. The goal of producing a large number of progeny is achieved by the *sequential transcription of viral genes.* Proteins needed for DNA replication and recombination are made first, followed by the synthesis of the head and tail proteins of the virus particle and of the proteins required for lysis of the host cell. Timing is critical; premature destruction of the host would of course be disadvantageous to the virus. There are three stages of gene expression in the lytic pathway: immediate-early, delayed-early, and late (Figure 36-19).

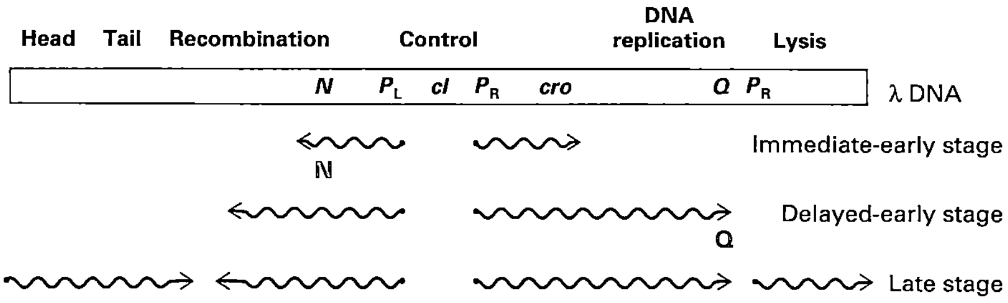

Figure 36-19
Three stages of transcription in the lytic growth of λ phage. The N protein formed in the immediate-early stage activates the delayed-early stage. In turn, the Q protein is formed, which activates the late stage. [After H. Echols. *The Bacteria* 8(1979):502.]

In the *immediate-early stage,* RNA synthesis starts at two promoter sites, P_L and P_R. The left transcript is the message for the N protein, which has a critical regulatory role. In the absence of N protein, the immediate-early transcripts end at either of two termination sites. The N protein antagonizes the termination of transcription at both sites and thereby permits further expression of λ genes. Thus, the N protein turns on the *delayed-early stage.* Proteins necessary for the replication of λ DNA and for recombination are made at this time. In addition, the *Q* gene is transcribed in the delayed-early stage. The Q protein is another critical regulator of gene expression in λ; it is required for the *late stage.* The genes for proteins that compose the head and tail of the phage and those that lyse the host are transcribed in the late stage. The Q protein, like the N protein, antagonizes the termination of transcription. In short, *the sequential regulation of lytic development is effected by two positive regulatory proteins, encoded by genes* N *and* Q, *which act by allowing transcription to proceed beyond several termination sites.*

Now let us consider lysogeny, the alternative developmental pathway for λ. The three stages of the lysogenic cycle are *establishment, maintenance,* and *release.* The establishment of the prophage state requires the integration of the viral DNA into the host DNA and the inactivation of the lytic functions of the virus. These processes are complex, whereas the maintenance of the prophage state is much simpler. Dale Kaiser showed that *only the* cI *gene is expressed in the prophage.* This gene codes for the λ repressor, which binds to two operator regions, O_L and O_R (Figure 36-20).

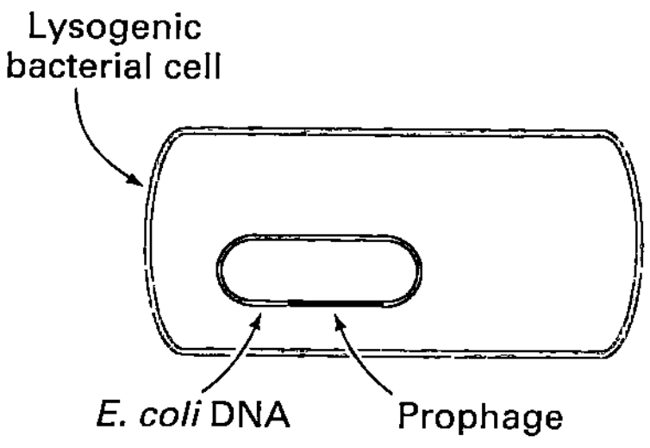

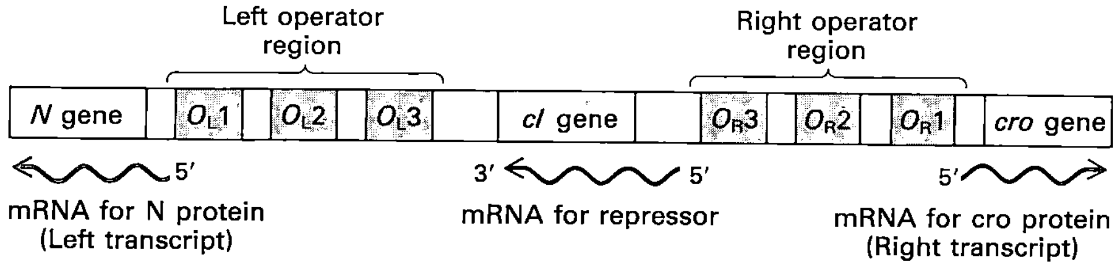

Figure 36-20
Diagram of the O_L and O_R operator regions and the adjacent genes. O_L1 and O_R1 have the highest affinity for the λ repressor. The repressor gene is *cI*. The left transcript starts with the *N* gene, whereas the right transcript starts with the *cro* gene.

Binding of the λ repressor to O_L directly prevents the leftward transcription of the immediate-early genes. In particular, the N protein is not synthesized, and so the lytic pathway is blocked. By binding to O_R, the λ repressor prevents the rightward expression of the *cro* and *Q* genes. Indeed, the binding of the λ repressor to O_L and O_R silences the whole λ genome except for the *cI* gene, which codes for the λ repressor. As will be discussed shortly, the λ repressor itself controls the *cI* gene, thereby regulating its own level. Inactivation of the λ repressor enables lytic genes to be transcribed. The prophage is then excised from the host chromosome, and the lytic functions are expressed.

TWO OPERATORS IN λ CONTAIN A SERIES OF BINDING SITES FOR THE REPRESSOR

The λ repressor, a dimer of identical 26-kd subunits, has been purified and studied in detail by Mark Ptashne. Each chain contains an amino-terminal domain that binds DNA and a carboxyl-terminal domain that interacts with its counterpart to hold the dimer together (Figure 36-21). The three-dimensional structure of the repressor and how it binds to an operator will be discussed shortly (p. 963). Two operator regions, O_L and O_R, are recognized by the same λ repressor. The *cI* gene that codes for the repressor is located between O_L and O_R (see Figure 36-20). Each of these operators contains three binding sites for the λ repressor. Nuclease digestion studies have shown that the binding sites are 17 base pairs long and that they are separated from each other by AT-rich regions that are from 3 to 7 base pairs in length. The base sequences of these operator sites are

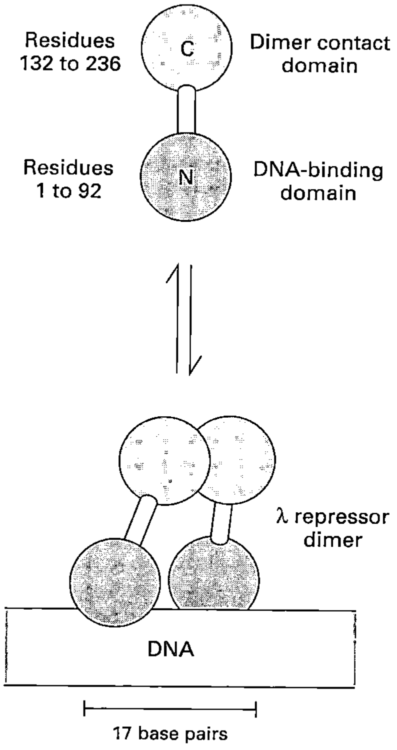

Figure 36-21
Domain structure of λ repressor.

similar but not identical (Figure 36-22), which accounts for their differing affinities for the λ repressor. The six operator sites, like the *lac* operator, exhibit partial twofold symmetry. Again, the symmetry of the binding sites in DNA matches that of the dimeric repressor molecule.

Figure 36-22
Sequences of one of the strands of the six binding sites for λ repressor. The base lying on the twofold axis is shown in red. Bases obeying twofold symmetry are shown in blue. The sequences are listed in the order of their binding affinities. The affinity for λ repressor is greatest for O_L1 and least for O_R3, the reverse of the affinities for Cro protein.

O_L1	T A T C A C C G C C A G T G G T A
O_R1	T A T C A C C G C C A G A G G T A
O_L2	T A T C T C T G G C G G T G T T G
O_L3	T A T C A C C G C A G A T G G T T
O_R2	T A A C A C C G T G C G T G T T G
O_R3	T A T C A C C G C A A G G G A T A

The strongest binding site for repressor in both O_L and O_R is the one closest to the start of the first structural gene in the operon—O_L1 and O_R1, respectively. The promoter site for the *N* gene lies within O_L1, and the promoter site for the *cro* gene lies within O_R1. As in the arabinose operon, the binding of repressor to these operators blocks the binding of RNA polymerase to the corresponding promoter, and so transcription of the *N* gene (left transcript) and of the *cro* gene (right transcript) is not initiated. In contrast, transcription of the *cI* gene itself is not blocked by the binding of the repressor to these high-affinity sites.

THE λ REPRESSOR REGULATES ITS OWN SYNTHESIS

The number of λ repressor molecules in a lysogenized *E. coli* cell is precisely regulated. Too little repressor, even briefly, would push the cell onto the lytic pathway. On the other hand, too much repressor would make it difficult for the phage to emerge should the bacterium become inhospitable. What controls the level of the repressor? The solution is elegantly simple: *the repressor regulates its own synthesis.* The different affinities of O_R1, O_R2, and O_R3 for repressor are at the heart of its autoregulation. Of these three, O_R1 has the highest intrinsic affinity for the repressor. At low levels of repressor, the O_R1 site is filled, which blocks the transcription of the *cro* gene and others lying to the right (Figure 36-23).

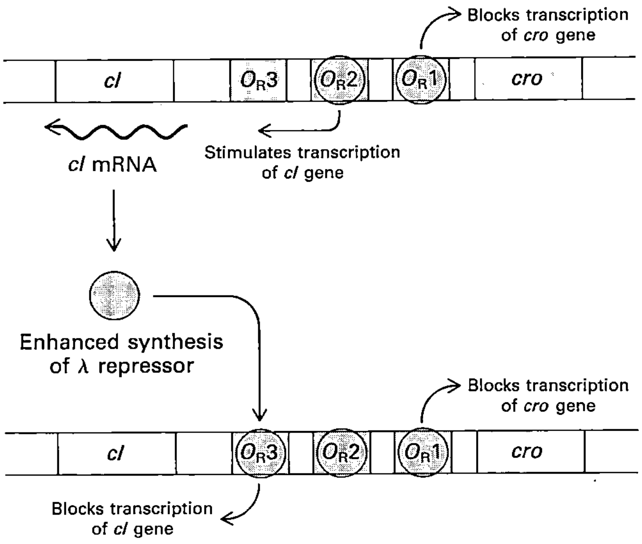

Figure 36-23
Self-regulation of the level of λ repressor. When its level is low, the repressor binds to O_R1 and then to O_R2. Occupancy of O_R2 stimulates transcription of *cI*. As the level of the repressor increases, it binds to O_R3, which inhibits further transcription of *cI*.

A repressor molecule bound to O_R1 favors the binding of another one to O_R2, which enhances the transcription of the *cI* gene and leads to the production of more repressor. As the concentration of repressor increases, O_R3 becomes filled. However, the occupancy of this site, the closest to the *cI* gene, blocks transcription of the gene. Thus, expression of the *cI* gene, like that of the *araC* gene (p. 957), is self-regulated.

LYSOGENY IS TERMINATED BY PROTEOLYSIS OF λ REPRESSOR AND SYNTHESIS OF CRO PROTEIN

The feedback circuit just described will tend to maintain the level of repressor so that the rest of the phage genome is not expressed. How then does the phage ever emerge from lysogeny? The critical trigger is a reduction in the number of λ repressor molecules just sufficient to enable the *cro* gene to be transcribed. The completed Cro protein (a dimer of 7-kd subunits) then binds to O_R3 to prevent the transcription of the *cI* gene (Figure 36-24). The important point is that O_R3 has a higher affinity for the Cro protein than does O_R1. *Low levels of Cro protein completely repress the synthesis of λ repressor without turning off the synthesis of Cro protein itself.* The λ repressor is now prevented from regaining the upper hand. At this point, a chain of events leading to lysis is irreversibly set in motion.

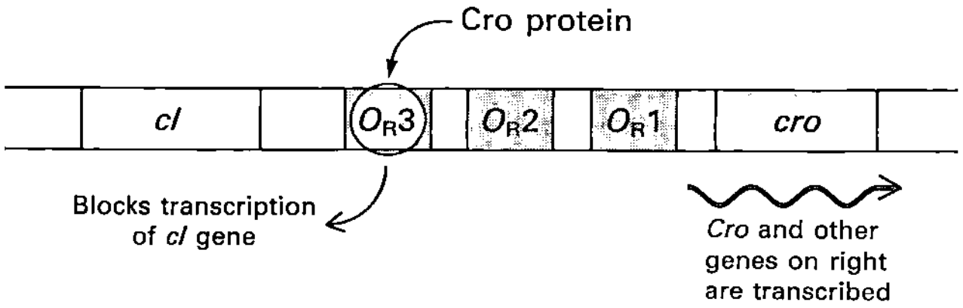

Figure 36-24
Cro protein at low levels blocks the expression of *cI* by binding to O_R3. Cro mRNA is formed because site O_R1 is empty.

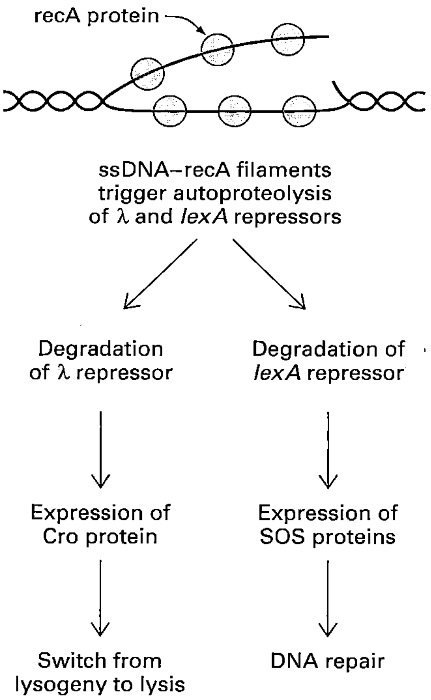

Figure 36-25
Autoproteolysis of λ repressor triggered by single-stranded DNA containing bound recA protein leads to the termination of lysogeny and the induction of the lytic pathway of λ phage.

The transition from lysogeny to lysis is induced by damage to DNA arising from ultraviolet radiation or chemical agents. What is the link between these events? *The key step is the proteolytic destruction of λ repressor* (Figure 36-25). Single-stranded DNA (ssDNA) containing bound recA protein (p. 825) activates a latent protease activity in λ repressor. One repressor molecule bound to an ssDNA–recA filament digests a neighboring bound repressor. Recall that the lexA repressor similarly digests itself in the SOS response to DNA damage (p. 826). Bacteriophage lambda takes advantage of *triggered proteolysis* to switch to its lytic mode so that it can flee from what is no longer a safe haven. The subtle behavior of λ arises from the remarkable interplay of just a few proteins and operator sites. The economy and elegance of its regulatory circuits are striking.

A HELIX-TURN-HELIX MOTIF MEDIATES THE BINDING OF MANY REGULATORY PROTEINS TO CONTROL SITES IN DNA

How do repressors and activators bind to DNA and recognize their target sequences? The elucidation of the three-dimensional structure of Cro protein by Brian Matthews and his co-workers immediately suggested how it functions. Each 66-residue subunit of dimeric Cro contains three α

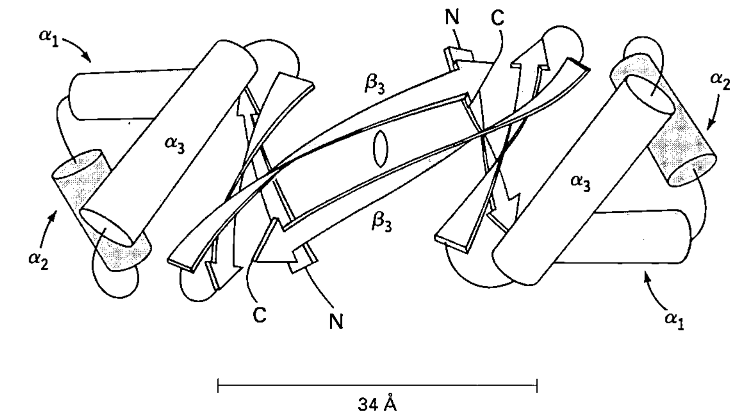

Figure 36-26
Three-dimensional structure of Cro protein, a dimer of identical subunits. [Courtesy of Dr. Brian Matthews.]

helices and three antiparallel β sheets (Figure 36-26). The dimer is held together by the association of two antiparallel β strands; the twofold axis of symmetry lies between them. Helix 3 in one subunit is separated from its counterpart in the other subunit by 34 Å, the same distance as the separation between neighboring major grooves of a Watson-Crick double helix. Furthermore, an α helix (including its side chains) has a diameter of about 12 Å, just the right size to fit into the major groove of DNA, which is about 12 Å wide and 7 Å deep.

Matthews was also struck by the match between the twofold symmetry of Cro and the approximate symmetry of the 17-bp operator sites recognized by it. *It therefore seemed likely that Cro fits into DNA by placing the third helix of each subunit into adjoining major grooves of DNA* (Figure 36-27). Specificity would arise from hydrogen bonding between the side chains on the exposed side of this helix and exposed portions of base pairs in the major groove. Subsequent x-ray studies showed that Cro indeed binds as predicted.

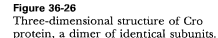

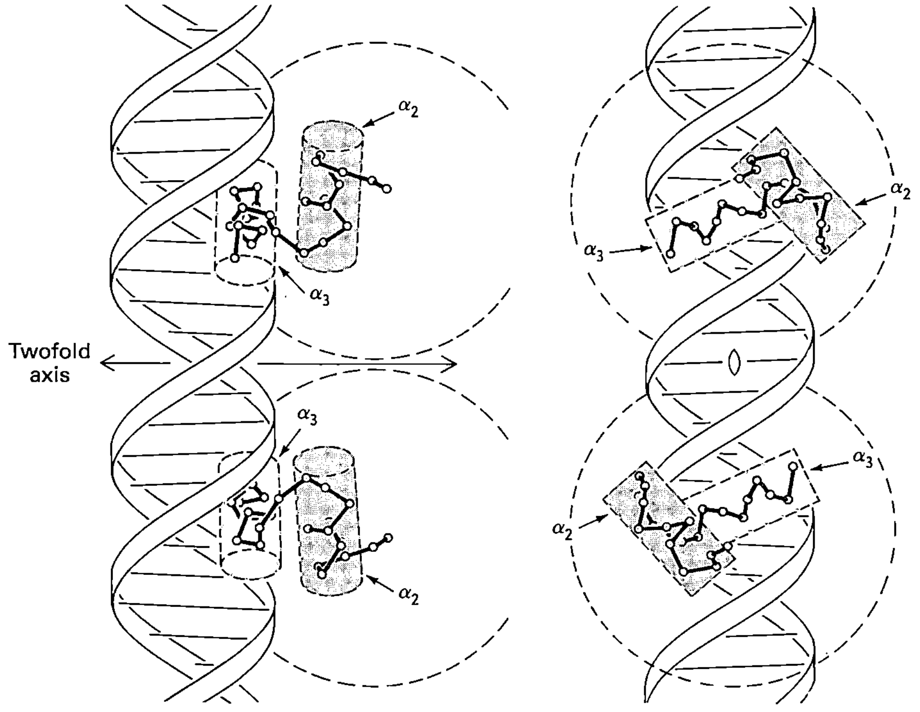

Figure 36-27
Two views of the predicted interaction of Cro with DNA. A pair of symmetry-related α helices of the dimer can be fit neatly into neighboring major grooves of DNA. Cro indeed interacts in this way, as shown by subsequent crystallographic studies. The DNA is bent on binding Cro. [Courtesy of Dr. Brian Matthews.]

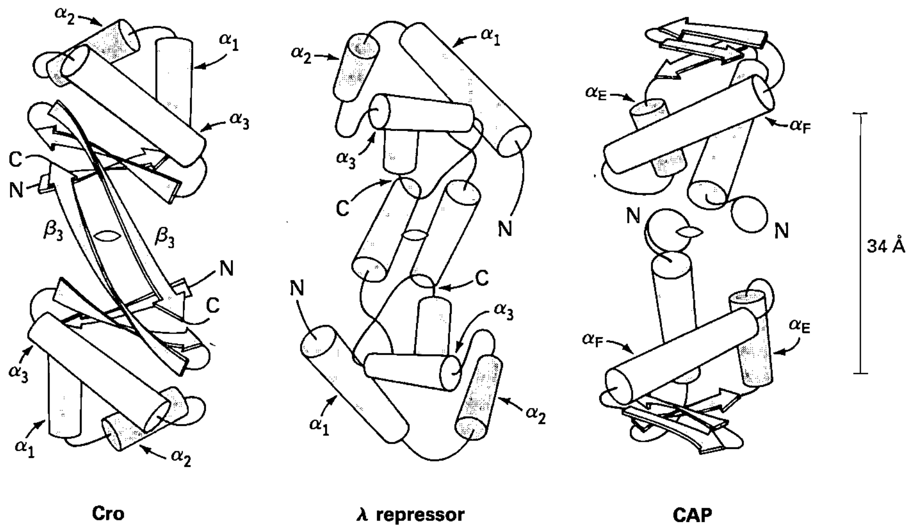

Figure 36-28
Comparison of the three-dimensional structures of Cro, λ repressor, and CAP protein. Each of these regulatory proteins contains a pair of helix-turn-helix (HTH) motifs. The C-terminal helix (green) of each motif binds to the major groove of DNA.

We now know that many DNA-binding proteins contain two α-helical recognition units related by a twofold axis of symmetry and separated by 34 Å, the pitch of the DNA helix. The structures of Cro, the DNA-binding domain of λ repressor, and CAP protein are compared in Figure 36-28. The common DNA-binding element of these regulatory proteins is a *helix-turn-helix (HTH)* motif (Figure 36-29). Residues 1 to 7 form the N-terminal helix, residues 8 to 11 form the turn, and residues 12 to 20 form the C-terminal helix of this recurring module. Residue 9 is usually glycine to allow a hairpin turn. The motif is stabilized by nonpolar interactions between side chains of the two helices.

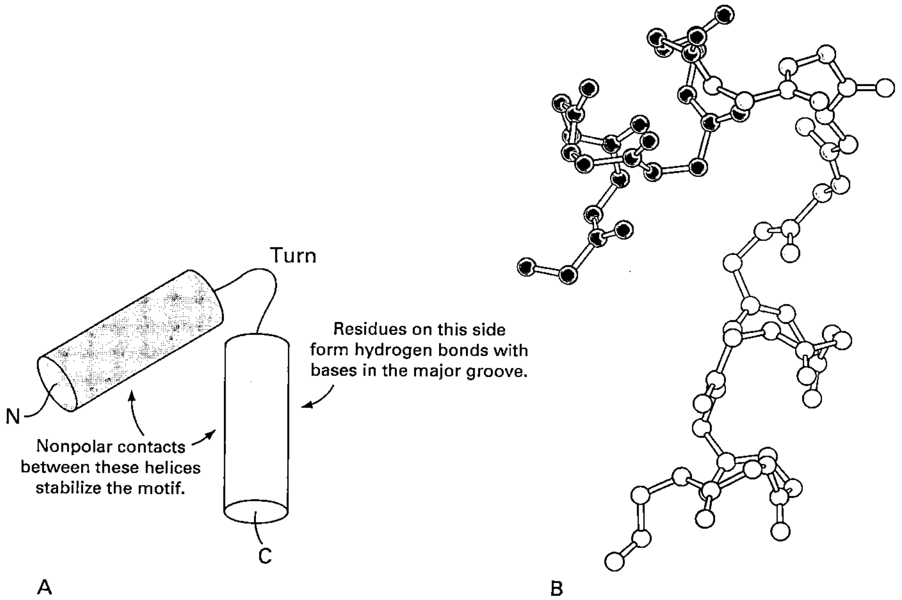

Figure 36-29
Helix-turn-helix motif of DNA-binding proteins. (A) Schematic diagram. (B) Ball-and-stick model of the HTH motif of λ repressor. The N-terminal helix is shown in red, the turn in yellow, and the C-terminal helix in green. [Part B is drawn from 1lmb.pdb. L.J. Beamer and C.O. Pabo. *J. Mol. Biol.* 227(1992):177.]

Several repressors containing the HTH motif have been crystallized together with their operator DNA. X-ray analyses of these crystals have shown that the second helix of the HTH motif indeed occupies the major groove of DNA (Figure 36-30 and the image on p. 949). Side chains emerging from this helix interact specifically with several bases in the major groove—hence, residues 12 to 20 are sometimes called the *recognition helix.* However, it should be noted that other residues of the HTH motif and of other parts of the protein do contribute to specific binding. Furthermore, no simple code relates interacting amino acid residues to the base sequence of the target site. α Helices are well suited for recognizing specific DNA sequences, but they do not have a monopoly on DNA recognition. DNA-binding modules crafted from β sheets have been found in several regulatory proteins, such as the *met* repressor, which regulates the transcription of genes for the synthesis of methionine and *S*-adenosylmethionine.

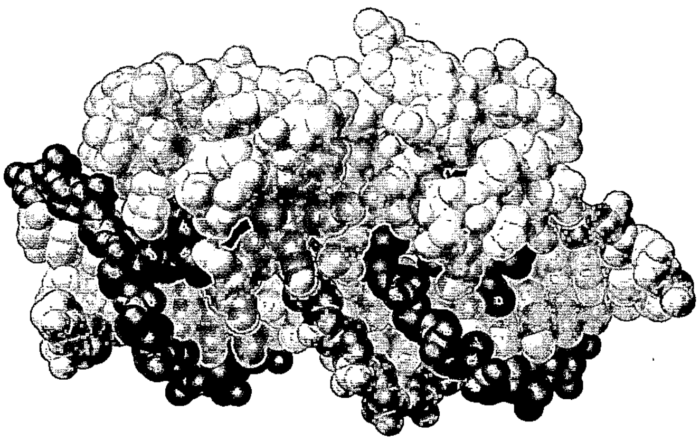

Figure 36-30
Structure of phage 434 repressor bound to DNA (red and green strands). One subunit of the amino-terminal domain of the dimeric repressor is shown in yellow, and the other in blue. The C-terminal helix (purple) of each HTH motif fits into the major groove of DNA. The sugar-phosphate backbone of DNA is shown in dark colors, and the bases in light colors. [Drawn from 2orl.pdb. A.K. Aggarwal, J.E. Anderson, D.W. Rodgers, and S.C. Harrison. *Science* 242(1988):899.]

TRANSCRIPTION OF THE *TRP* OPERON IS BLOCKED BY REPRESSOR CONTAINING BOUND TRYPTOPHAN

The 7-kb mRNA transcript of the tryptophan operon encodes the five enzymes in *E. coli* that convert chorismate into tryptophan (p. 725). These five proteins are synthesized sequentially, coordinately, and in equimolar amounts by translation of this polygenic (polycistronic) *trp* mRNA. Translation begins before completion of transcription. *Trp* mRNA is synthesized in about four minutes and then rapidly degraded. The short lifetime of *trp* mRNA, which is only about three minutes after its synthesis has been completed, enables bacteria to respond quickly to their changing needs for tryptophan. Indeed, *E. coli* can vary the rate of production of its biosynthetic enzymes for tryptophan over a 700-fold range.

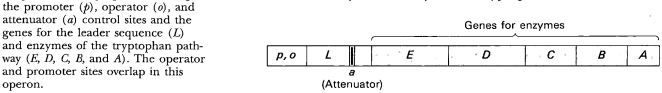

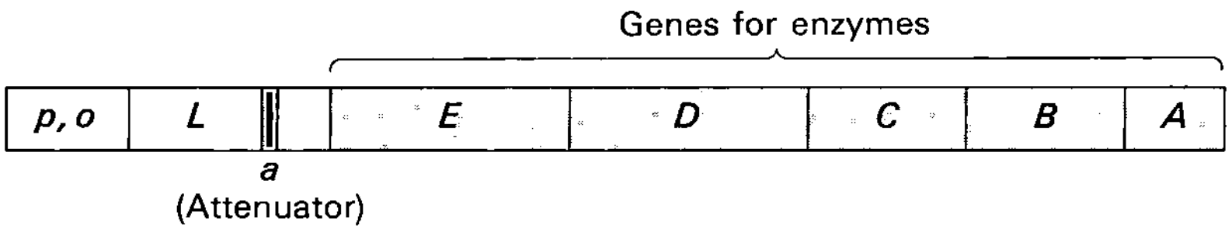

Figure 36-31
Diagram of the *trp* operon showing the promoter (*p*), operator (*o*), and attenuator (*a*) control sites and the genes for the leader sequence (*L*) and enzymes of the tryptophan pathway (*E, D, C, B,* and *A*). The operator and promoter sites overlap in this operon.

How is this regulation accomplished? One level of control is achieved by the interaction of a specific repressor with the *trp* operator site on the DNA. The *trp* repressor, a dimer of 107-residue subunits, is encoded by the *trpR* gene, which is far from the *trp* operon. *A complex of this repressor and tryptophan binds tightly to the operator, whereas the repressor alone does not.* In other words, tryptophan itself is a *corepressor*. The target for the tryptophan-repressor complex is a DNA sequence with twofold symmetry (Figure 36-32). Again, symmetry plays an important role in the interaction of a protein with DNA. This operator site overlaps the promoter site for the initiation of transcription. Hence, *binding of the trp repressor to the operator prevents RNA polymerase from binding to the trp promoter, so that the trp genes are not transcribed.*

```
                                   Start
                                     ↓
 −21              −10               +1
5'-CGAACTAGTTAACTAGTACGCAAG-3'
3'-GCTTGATCAATTGATCATGCGTTC-5'
```

Figure 36-32
Base sequence of the *trp* operator. The twofold axis of symmetry is denoted by the green symbol. The base pair labeled +1 is the start of the transcribed part of the operon.

The *trp* repressor consists of two extensively interlocked subunits (Figure 36-33). The dimer contains three domains: a central core, formed by the amino-terminal half of both subunits, and two flexible DNA-reading heads, each formed from the carboxyl-terminal half of a subunit. *A helix-turn-helix unit is present in each subunit, as in λ repressor, Cro, and CAP.* The central core serves as a spacer between the reading heads. In the aporepressor (no bound tryptophan), the reading heads are only 26 Å apart, much too close to fit into successive major grooves of DNA. The binding of tryptophan increases the distance between subunits by 8 Å, so that they fit readily into adjacent major grooves of DNA. In essence, *tryptophan acts as a wedge that pries the reading heads apart to match the distance between major grooves.* Furthermore, tryptophan molds the neighboring side chains of the binding pocket so that they form stabilizing contacts with DNA. In addition, the indole ring of tryptophan makes a hydrogen bond with a backbone phosphate of DNA. The target DNA as well as the repressor undergoes structural change in forming this complex. Another interesting feature of this complex is the role of structured water—several precisely positioned water molecules act as bridges between the DNA and protein.

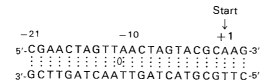

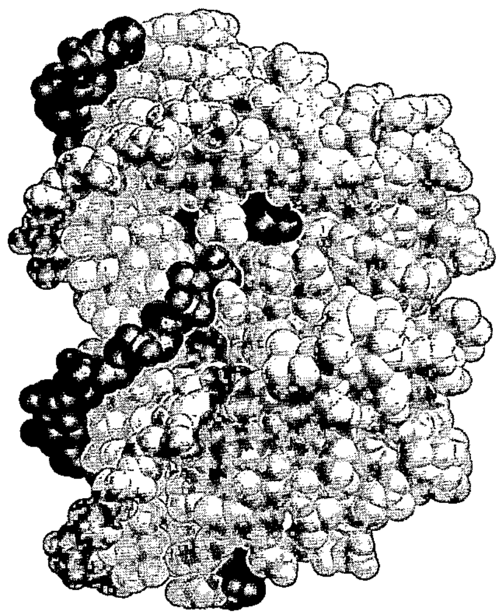

Figure 36-33
Structure of the *trp* repressor protein bound to DNA (red and green). The two subunits of the repressor (yellow and blue) are extensively interlocked. Tryptophan, the corepressor, is shown in purple. [Drawn from 1tro.pdb, deposited by Z. Otwinowski, R.G. Zhang, and P.B. Sigler.]

ATTENUATION IS A KEY MEANS OF CONTROLLING OPERONS THAT ENCODE ENZYMES FOR AMINO ACID BIOSYNTHESIS

A new means of controlling gene expression was discovered by Charles Yanofsky and his colleagues as a result of their studies of the tryptophan operon. For some time it was thought that end-product inhibition of the catalytic activity of the first enzyme complex in the tryptophan pathway (p. 727) and repressor-mediated inhibition of transcription accounted for most of the regulation of tryptophan biosynthesis. This view was abruptly overturned by the unexpected finding that certain mutants with *deletions* between the operator and the gene for the first enzyme (*trpE*) showed *increased* production of *trp* mRNA. Furthermore, analysis of the 5′ end of *trp* mRNA revealed the presence of a *leader sequence* of 162 nucleotides before the initiation codon of *trpE.* It was then found that deletions resulting in enhanced *trp* mRNA levels mapped within this leader region, some 30 to 60 nucleotides before the start of *trpE.* The next striking observation was that nonmutants produced a transcript consisting of only the first 130 nucleotides of the leader when the tryptophan level was high, but produced a 7000-nucleotide *trp* mRNA including the entire leader sequence when tryptophan was scarce. Hence, Yanofsky concluded that transcription of the *trp* operon must be regulated by a *controlled termination site,* called an *attenuator,* that is located between the operator and the gene for the first enzyme in the pathway. This physiologically regulated termination site, like the stop signals at the ends of some operons (p. 847), contains a GC-rich sequence followed by an AT-rich one. Each of these regions in the attenuator exhibits a twofold axis of symmetry (Figure 36-34). Moreover, the terminated leader transcript ends with a series of U residues.

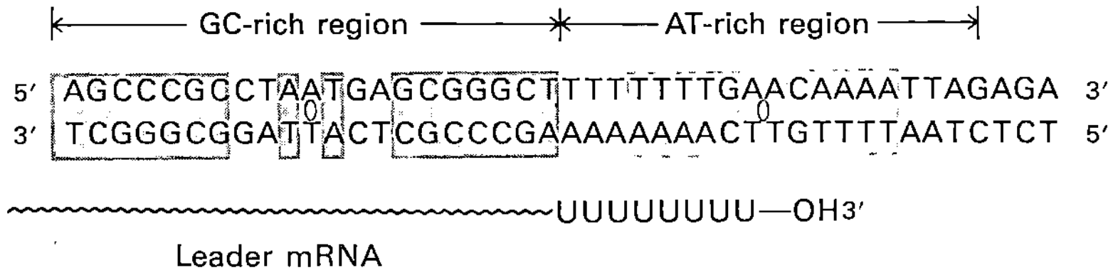

Figure 36-34
Base sequence of the *trp* attenuator site. Base pairs in the GC-rich region related by a twofold axis of symmetry are shown in blue, and those in the AT-rich region are shown in yellow.

ATTENUATION IS MEDIATED BY THE TIGHT COUPLING OF TRANSCRIPTION AND TRANSLATION

The attenuator site complements the operator site in regulating transcription of *trp* genes. When tryptophan is plentiful, initiation of transcription is blocked by the binding of the tryptophan-repressor complex to the operator. As the level of tryptophan in the cell decreases, repression is lifted and transcription begins. However, some of the RNA polymerase molecules dissociate from the template at the attenuator site, whereas others continue to synthesize the entire *trp* message. *The proportion of RNA polymerase molecules that proceed past the attenuator site increases as tryptophan becomes scarcer.*

How does the attenuator site in the *trp* operon sense the level of tryptophan in the cell? An important clue was the finding that part of the leader mRNA is translated. The presence of tryptophan residues at positions 10 and 11 of the 14-residue leader polypeptide (Figure 36-35) is highly significant. When tryptophan is abundant, this complete polypeptide is syn-

Met - Lys - Ala - Ile - Phe - Val - Leu - Lys - Gly - Trp(10) - Trp(11) - Arg - Thr - Ser - Stop

5′~AUG AAA GCA AUU UUC GUA CUG AAA GGU UGG UGG CGC ACU UCC UGA~3′

Figure 36-35
Amino acid sequence of the *trp* leader peptide and the base sequence of the corresponding leader mRNA.

thesized. However, *when tryptophan is scarce, the ribosome stalls at the tandem UGG codons because of a paucity of tryptophanyl-tRNA. The stalled ribosome alters the structure of the mRNA so that the RNA polymerase transcribing it proceeds beyond the attenuator site.* A key aspect of this mechanism is that translation and transcription are closely coupled. The ribosome translating the *trp* leader mRNA follows closely behind the RNA polymerase molecule that is transcribing the DNA template. The stalled ribosome switches the secondary structure of the mRNA from a base-paired arrangement that favors termination of transcription to a very different one that allows RNA polymerase to read through the attenuator site (Figure 36-36). RNA molecules, like proteins, can adopt alternative conformations that are regulated with far-reaching physiologic consequences.

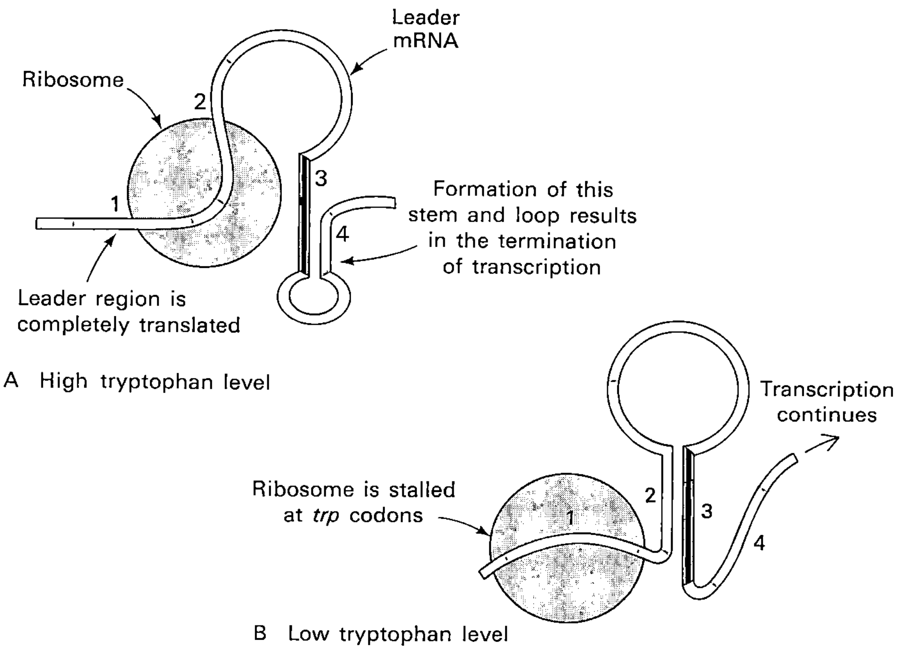

Figure 36-36
Model for attenuation in the *E. coli trp* operon. When tryptophan is abundant (A), the leader region (segment 1) of the *trp* mRNA is fully translated. Segment 2 enters the ribosome, which enables segments 3 and 4 to base-pair. This base-paired region signals RNA polymerase to terminate transcription. In contrast, when tryptophan is scarce (B), the ribosome is stalled at the *trp* codons of segment 1. Segment 2 interacts with 3 instead of being drawn into the ribosome and so segments 3 and 4 cannot pair. Consequently, transcription continues. [After D.L. Oxender, G. Zurawski, and C. Yanofsky. *Proc. Nat. Acad. Sci.* 76(1979):5524.]

THE ATTENUATOR SITE IN THE HISTIDINE OPERON CONTAINS SEVEN HISTIDINE CODONS IN A ROW

Several other operons for the biosynthesis of amino acids in *E. coli* are now known to have attenuator sites. The leader peptide of each contains an abundance of amino acid residues of the kind controlled by the operon. For example, the threonine operon encodes enzymes that synthesize both threonine and isoleucine; the leader peptide contains 8 threonines and 4 isoleucines in a 16-residue sequence (Figure 36-37). Seven phenylalanines are present in the 15-residue leader of the phenylalanine operon. Even more striking, the leader peptide of the histidine operon contains *seven histidine residues in a row.* Clearly, these leader mRNAs are

A

Met- Lys - Arg - Ile - Ser - Thr - Thr - Ile - Thr - Thr - Thr - Ile - Thr - Ile - Thr - Thr -

5′ AUG AAA CGC AUU AGC ACC ACC AUU ACC ACC ACC AUC ACC AUU ACC ACA 3′

B

Met- Lys - His - Ile - Pro - Phe - Phe - Phe - Ala - Phe - Phe - Phe - Thr - Phe - Pro - Stop

5′ AUG AAA CAC AUA CCG UUU UUC UUC GCA UUC UUU UUU ACC UUC CCC UGA 3′

C

Met- Thr - Arg - Val - Gln - Phe - Lys - His - His - His - His - His - His - His - Pro - Asp -

5′ AUG ACA CGC GUU CAA UUU AAA CAC CAC CAU CAU CAC CAU CAU CCU GAC 3′

Figure 36-37
Amino acid sequence of the leader peptide and base sequence of the corresponding portion of mRNA from the (A) threonine operon, (B) phenylalanine operon, and (C) histidine operon.

designed to sense the level of the amino acid synthesized by the encoded proteins. If the corresponding charged tRNA is scarce, translation of the leader is arrested. As in the *trp* operon, a stalled ribosome facilitates switching to an mRNA conformation that allows RNA polymerase to read through the attenuator site. The presence of seven consecutive codons for histidine in the histidine operon leader mRNA markedly heightens the sensitivity of this detection system. Indeed, the histidine operon is controlled solely by attenuation.

FREE RIBOSOMAL PROTEINS REPRESS THE TRANSLATION OF mRNA ENCODING THEM

How is the synthesis of more than 50 kinds of ribosomal proteins coordinated? Their genes are located in more than 20 operons, yet their synthesis is precisely balanced. Tight control is essential, for ribosomal components make up some 40% of the dry weight of a bacterial cell. Control of ribosomal protein synthesis is exerted primarily at the level of *translation* rather than of transcription, in contrast with that of most other proteins. At least one protein encoded by each ribosomal protein operon serves as a *translational repressor* (Figure 36-38). It binds to mRNA near the initiation site for its own synthesis and blocks the synthesis of several proteins

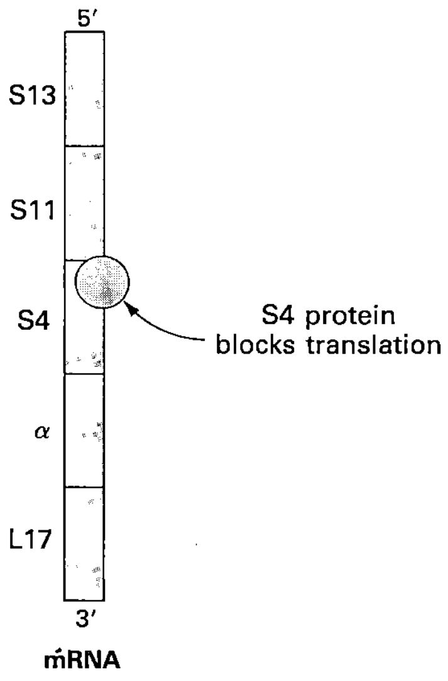

Figure 36-38
Feedback control by translational repression. A ribosomal protein (S4) represses the translation of an mRNA encoding components of the small (S) and large (L) ribosomal subunits and the α subunit of RNA polymerase.

encoded by that polygenic message. This binding to mRNA does not interfere with the assembly of ribosomes because ribosomal proteins bind more tightly to ribosomal RNA than to mRNA. Translational repression occurs only when the production of ribosomal proteins runs ahead of that of ribosomal RNAs or when these proteins are not formed in equimolar amounts.

The synthesis of ribosomal RNAs and transfer RNAs is also coordinated with the synthesis of nonribosomal proteins. When amino acids are scarce, the synthesis of these RNAs stops. This adaptation to adversity, called the *stringent response,* is mediated by an unusual nucleotide, originally called *magic spot* because of its mysterious emergence in chromatograms of certain cell extracts. The new compound formed under conditions of amino acid deprivation turned out to be *guanosine 5'-diphosphate 3'-diphosophate (ppGpp).* It can be formed by the transfer of a diphosphate unit from ATP to the 3'-OH of GDP. Alternatively, ppGpp can arise by hydrolysis of pppGpp formed from ATP and GTP. The formation of ppGpp and pppGpp, which occurs at the A site of the ribosome, is catalyzed by a synthetase called *stringent factor.* Synthesis occurs only if an unacylated tRNA is present in the A site. Hence, ppGpp and pppGpp are formed only in cells starved of amino acids because the A site in unstarved cells is nearly always occupied by an aminoacyl-tRNA. Once formed, ppGpp inhibits the initiation of transcription of operons encoding ribosomal RNAs and slows the elongation of many other transcripts. Its precise mechanism of action is an intriguing problem.

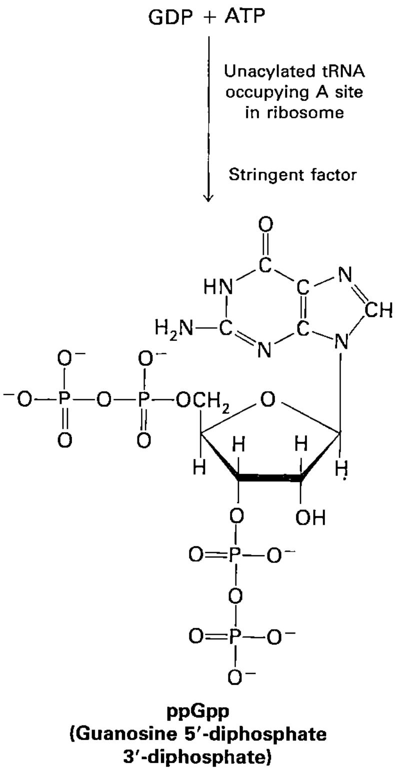

Figure 36-39
ppGpp is synthesized from ATP and GDP during amino acid starvation. This signal molecule blocks the formation of ribosomal and transfer RNAs in the stringent response.

DNA INVERSIONS LEAD TO ALTERNATING EXPRESSION OF A PAIR OF FLAGELLAR GENES

Salmonella, a bacterium living in the intestine of mammals, swims by rotating flagella that emerge from its surface (p. 411). Flagella are thin helical protein filaments built from many 53-kd flagellin subunits. *Salmonella* contains two genes for flagellins and expresses only one of them at a given time. The type expressed changes spontaneously an average of once in a thousand cell divisions. This switching between flagellins H1 and H2, called *phase variation,* helps the bacterium evade the immune response of its host. To be effective, the change must be complete. In phase 2, for example, the presence in the flagellum of even a few H1 molecules amidst many H2 would render this motile appendage vulnerable to inactivation by antibodies directed against H1, the type produced in phase 1. The switch must be both *absolute* and *reversible.*

These two objectives are met by an ingeniously simple means of switching back and forth between two genes (Figure 36-40). The mRNA for H2 is also the template for a repressor that silences the H1 gene. This finding explains phase 2, but how then can H1 be expressed to the exclusion of H2? In phase 1, the promoter for H2 and the repressor is ineffective because the DNA segment containing it has been inverted. In this phase, the promoter is in the wrong place and points in the wrong direction for expression of the genes for H2 and the repressor. The H1 gene containing its own promoter is now expressed because the repressor is absent.

What controls the inversion of the DNA segment bearing the promoter for H2 and the repressor? The ends of this 993-bp segment are a pair of 13-bp inverted terminal repeats; the middle of it encodes a recombinase, called *Hin,* that catalyzes site-specific recombination between the repeats (*hix* sites). Hin, a dimer of 22-kd subunits, is assisted by *Fis* (*f*actor for *i*nversion *s*timulation), a small protein that bends DNA to bring two *hix*

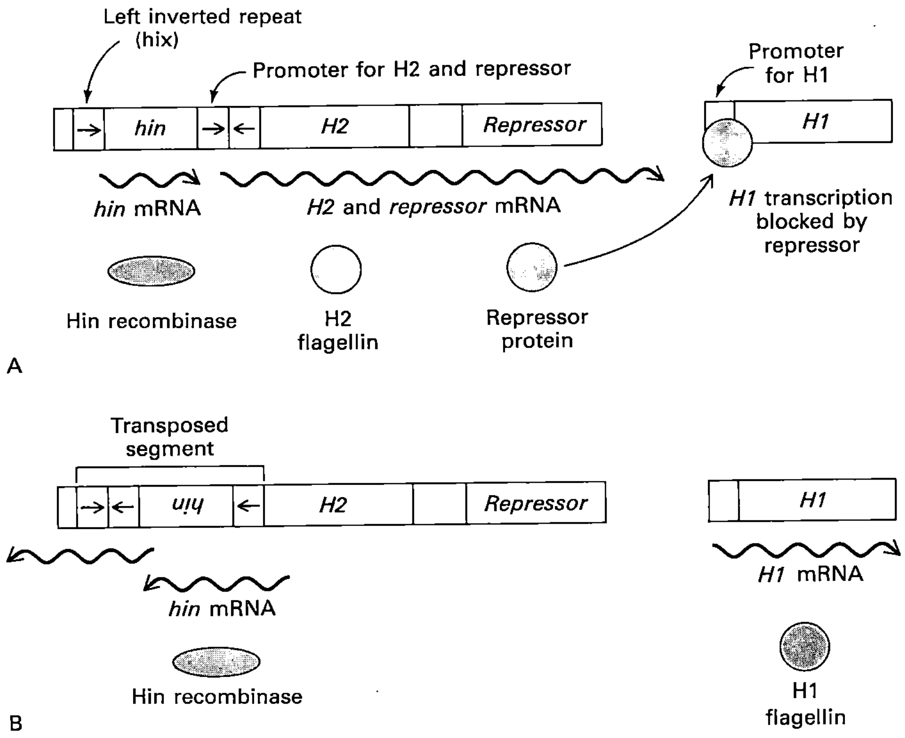

Figure 36-40
Phase variation in *Salmonella.* Flagellins H1 and H2 are expressed in a mutually exclusive manner. (A) In phase 2, the H1 gene is silenced by a repressor protein formed along with the H2 protein. (B) In phase 1, inversion of a DNA segment catalyzed by a recombinase (Hin) encoded by it leads to the loss of the promoter for H2 and the repressor. H1 is then expressed. Further inversions switch the system back and forth between these phases.

sites together. Cleavage and strand exchange takes place in the resulting *invertasome* (Figure 36-41). Recall that recombination between *oppositely* oriented sites leads to the *inversion* of the DNA segment located between them (see Figure 32-20B on p. 830). Thus, four genes—those for H1, H2, the repressor, and Hin—constitute a *flip-flop circuit that changes an exposed bacterial protein at a frequency determined by the activity of the recombinase.*

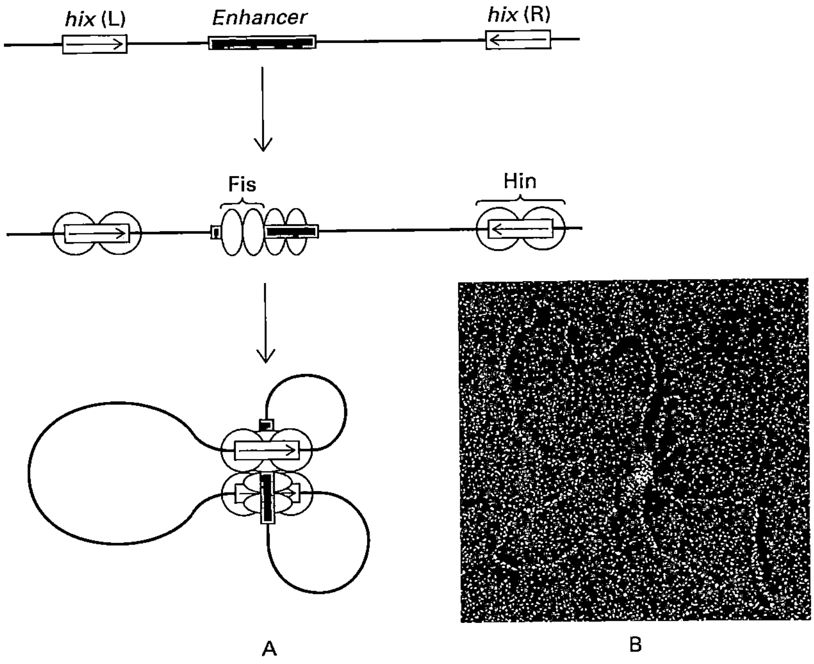

Figure 36-41
Mechanism of DNA inversion in *Salmonella* phase variation. A complex of proteins and DNA called an invertasome mediates this site-specific recombination reaction. (A) The invertasome consists of Hin protein bound to the two *hix* sequences and of Fis protein bound to an enhancer sequence between them. Cleavage of DNA and strand exchange occur in this complex. The DNA substrate must be negatively supercoiled.
(B) Electron micrograph of an invertasome. [After K.A. Heichman and R.C. Johnson. *Science* 249(1990):511.]

SUMMARY

Prokaryotes regulate the amounts of proteins synthesized in a variety of ways. The expression of most genes is regulated primarily at the level of transcription rather than translation. Many genes are clustered into operons, which are units of coordinated genetic expression. An operon consists of control sites (an operator and a promoter) and a set of structural genes. In addition, regulator genes encode proteins that interact with the operator and promoter sites to stimulate or inhibit transcription. In the absence of a galactoside inducer, *lac* repressor protein binds to the operator and blocks transcription of the genes encoding β-galactosidase and two other proteins of the lactose operon. The twofold symmetry of the repressor protein matches the nearly palindromic base sequence of the *lac* operator. Symmetry matching is an important general feature of protein-DNA interactions. Binding of an inducer to the *lac* repressor displaces it from the operator. RNA polymerase can then move through the operator to transcribe the *lac* operon.

Cyclic AMP, a hunger signal, stimulates the transcription of many catabolic operons by binding to the catabolite gene activator protein (CAP). The binding of cAMP–CAP to a specific site in the promoter region of these inducible catabolic operons enhances the binding of RNA polymerase and the initiation of transcription. Full expression of the *lac* operon requires both a galactoside inducer and cAMP, which is formed when glucose is scarce. Different forms of a control protein activate and inhibit transcription of the arabinose operon, another member of this global regulatory circuit controlled by cAMP. The looping of DNA enables the C protein to bind to noncontiguous target sites in the arabinose operon. DNA looping is widely used in gene regulation.

Repressors and activators of transcription determine whether λ phage follows the lysogenic or lytic pathway. The λ repressor prevents the expression of all viral genes except for the one (*cI*) encoding the repressor itself. The expression of the repressor is autoregulated; high concentrations block the transcription of *cI*. Damage to host DNA leads to formation of single-stranded DNA filaments containing bound recA protein. These filaments trigger the autoproteolysis of the λ repressor and the *lexA* repressor (the controller of the SOS genes for DNA repair). The consequent expression of Cro silences further formation of λ repressor and commits the virus to the lytic pathway. The switch between the lysogenic and lytic pathways is mediated by the interplay of λ repressor and Cro with six binding sites in the O_R and O_L operator sites of the viral DNA.

The λ repressor, Cro, CAP, *trp* repressor, and many other DNA-binding proteins are dimers of a common helix-turn-helix motif. One of the α helices of this recurring unit plays a key role in detecting specific sequences in DNA by binding to the major groove. The angular relation and distance between the two DNA-reading heads of these dimeric control proteins enable them to fit into adjoining major grooves of B-DNA. The binding of inducer or corepressor changes the form of these DNA-binding units and their spatial relationship.

The tryptophan operon and several others for the biosynthesis of amino acids are controlled by attenuation, a mechanism based on the coupling of transcription and translation. Transcription terminates at an attenuator site preceding the first structural gene if the amino acid end product is abundant. Attenuation is mediated by translation of a leader mRNA. A ribosome stalled by the absence of an aminoacyl-tRNA alters the structure of mRNA so that RNA polymerase transcribes the operon beyond the attenuator site.

Purely translational control is evident in the regulation of ribosomal protein synthesis. Free ribosomal proteins repress the translation of mRNAs encoding them by binding to initiation sites. The synthesis of ribosomal RNAs is inhibited by ppGpp, a signal molecule formed when amino acids are scarce.

A specialized mechanism of controlling gene expression is used by *Salmonella* to alternately express H1 or H2 flagellin in a mutually exclusive manner. This reversible switching is mediated by the inversion of a DNA segment containing the promoter for *H2* and the gene for a recombinase (*hin*) catalyzing this inversion.

SELECTED READINGS

Where to start

Maniatis, T., and Ptashne, M., 1976. A DNA operator-repressor system. *Sci. Amer.* 234(1):64–76.

Ptashne, M., Johnson, A.D., and Pabo, C.O., 1982. A genetic switch in a bacterial virus. *Sci. Amer.* 247(5):128–140.

Pabo, C.O., and Sauer, R.T., 1992. Transcription factors: Structural families and principles of DNA recognition. *Ann. Rev. Biochem.* 61:1053–1095.

Harrison, S.C., and Aggarwal, A.K., 1990. DNA recognition by proteins with the helix-turn-helix motif. *Ann. Rev. Biochem.* 59:933–969.

Books

McKnight, S.L., and Yamamoto, K.R. (eds.), 1992. *Transcriptional Regulation,* vols. 1 and 2. Cold Spring Harbor Laboratory Press. [Contains many excellent articles on both prokaryotic and eukaryotic gene regulation. Chapters 16 to 19 are especially pertinent to the topics covered here.]

Branden, C., and Tooze, J., 1991. *Introduction to Protein Structure.* Garland. [Chapter 7 provides a beautifully illustrated account of prokaryotic transcriptional regulators having a helix-turn-helix motif.]

Schleif, R., 1986. *Genetics and Molecular Biology.* Addison-Wesley. [Chapters 12, 13, and 14 provide a lucid account of the experimental basis of major concepts of gene regulation in prokaryotes.]

Ptashne, M., 1991. *A Genetic Switch. Phage λ and Higher Organisms* (2nd ed.). Cell Press and Blackwell Scientific Publications. [A superb account of how λ phage switches between two developmental pathways.]

Hendrix, R.W., Roberts, J.W., Stahl, F.W., and Weisberg, R.A. (eds.), 1983. *Lambda II.* Cold Spring Harbor Laboratory Press. [Contains a wealth of information on the structure, assembly, regulation, and evolution of λ phage.]

Miller, J.H., and Reznikoff, W.S. (eds.), 1978. *The Operon.* Cold Spring Harbor Laboratory Press. [A fine collection of articles about regulation in *E. coli* and λ phage. The lactose, tryptophan, arabinose, histidine, and galactose operons are discussed in detail.]

Lactose operon

Gilbert, W., and Müller-Hill, B., 1966. Isolation of the *lac* repressor. *Proc. Nat. Acad. Sci.* 56:1891–1898.

Dickson, R., Abelson, J., Barnes, W., and Reznikoff, W., 1975. Genetic regulation: The *lac* control region. *Science* 187:27–35.

Lee, J., and Goldfarb, A., 1991. Lac repressor acts by modifying the initial transcribing complex so that it cannot leave the promoter. *Cell* 66:793–798.

Cyclic AMP, CAP, and catabolite repression

Schultz, S.C., Shields, G.C., and Steitz, T.A., 1991. Crystal structure of a CAP–DNA complex: The DNA is bent by 90°. *Science* 253:1001–1007.

Kolb, A., Busby, S., Buc, H., Garges, S., and Adhya, S., 1993. Transcriptional regulation by cAMP and its receptor proteins. *Ann. Rev. Biochem.* 62:749–795.

Arabinose operon

Bustos, S.A., and Schleif, R.F., 1993. Functional domains of the AraC protein. *Proc. Nat. Acad. Sci.* 90:5638–5642.

Reeder, T., and Schleif, R., 1993. AraC protein can activate transcription from only one position and when pointed in only one direction. *J. Mol. Biol.* 231:205–218.

Wilcox, G., Meuris, P., Bass, P., and Englesberg, E., 1974. Regulation of the arabinose operon in vitro. *J. Biol. Chem.* 249:2946–2952.

Schleif, R., 1992. DNA looping. *Ann. Rev. Biochem.* 61:199–223.

Control of transcription in λ phage

Johnson, A.D., Poteete, A.R., Lauer, G., Sauer, R.T., Ackers, G.K., and Ptashne, M., 1981. Lambda repressor and cro—components of an efficient molecular switch. *Nature* 294:217–223.

Hendrix, R.W. (ed.), 1983. *Lambda II.* Cold Spring Harbor Laboratory Press.

Kim, B., and Little, J.W., 1993. LexA and λ C1 repressors as enzymes: Specific cleavage in an intermolecular reaction. *Cell* 73:1165–1173.

Griffith, J., Hochschild, A., and Ptashne, M., 1986. DNA loops induced by cooperative binding of λ repressor. *Nature* 322:750–752.

Structure of λ repressor and cro protein

Brennan, R.G., and Matthews, B.W., 1989. Structural basis of DNA-protein recognition. *Trends Biochem. Sci.* 14:286–290.

Brennan, R.G., Roderick, S.L., Takeda, Y., and Matthews, B.W., 1990. Protein-DNA conformational changes in the crystal structure of a lambda Cro-operator complex. *Proc. Nat. Acad. Sci.* 87:8165–8169.

Beamer, L.J., and Pabo, C.O., 1992. Refined 1.8 Å crystal structure of the lambda repressor-operator complex. *J. Mol. Biol.* 227:177–196.

Clarke, N.D., Beamer, L.J., Goldberg, H.R., Berkower, C., and Pabo, C.O., 1991. The DNA binding arm of lambda repressor: Critical contacts from a flexible region. *Science* 254:267–270.

Sauer, R.T., Jordan, S.R., and Pabo, C.O., 1990. Lambda repressor: A model system for understanding protein-DNA interactions and protein stability. *Advan. Prot. Chem.* 40:1–61.

Ohlendorf, D.H., Anderson, W.F., Lewis, M., Pabo, C.O., and Matthews, B.W., 1983. Comparison of the structures of cro and λ repressor proteins from bacteriophage λ. *J. Mol. Biol.* 169:757–769.

Tryptophan operon

Sigler, P.B., 1992. The molecular mechanism of *trp* repression. *In* McKnight, S.L., and Yamamoto, K.R. (eds.), *Transcriptional Regulation,* vol. 1, pp. 475–499. Cold Spring Harbor Laboratory Press.

Platt, T., 1978. Regulation of gene expression in the tryptophan operon of *Escherichia coli*. *In* Miller, J.H., and Reznikoff, W.S. (eds.), *The Operon,* pp. 213–302. Cold Spring Harbor Laboratory Press. [A lucid introduction to the *trp* operon.]

Landick, R., Carey, J., and Yanofsky, C., 1985. Translation activates the paused transcription complex and restores transcription of the trp operon leader region. *Proc. Nat. Acad. Sci.* 82:4663–4667.

Hurlburt, B.K., and Yanofsky, C., 1992. Trp repressor/trp operator interaction. Equilibrium and kinetic analysis of complex formation and stability. *J. Biol. Chem.* 267:16783–16789.

Control of ribosomal protein and RNA synthesis

Nomura, M., Gourse, R., and Baughman, G., 1984. Regulation of the synthesis of ribosomes and ribosomal components. *Ann. Rev. Biochem.* 53:75–117.

Portier, C., Dondon, L., and Grunberg, M.M., 1990. Translational autocontrol of the *Escherichia coli* ribosomal protein S15. *J. Mol. Biol.* 211:407–414.

Stephens, J.C., Artz, S.W., and Ames, B.N., 1975. Guanosine 5′-diphosphate 3′-diphosphate (ppGpp): Positive effector for histidine operon transcription and general signal for amino acid deficiency. *Proc. Nat. Acad. Sci.* 72:4389–4393.

DNA inversion and phase variation

van den Putte, P., and Goosen, N., 1992. DNA inversions in phages and bacteria. *Trends Genet.* 8:457–462.

Heichman, K.A., and Johnson, R.C., 1990. The Hin invertasome: Protein-mediated joining of distant recombination sites at the enhancer. *Science* 249:511–517.

Lim, H.M., and Simon, M.I., 1992. The role of negative supercoiling in Hin-mediated site-specific recombination. *J. Biol. Chem.* 267:11176–11182.

Discovery of operons and control elements

Yanofsky, C., 1992. Transcriptional regulation: Elegance in design and discovery. *In* McKnight, S.L., and Yamamoto, K.R. (eds.), *Transcriptional Regulation,* vol. 1, pp. 3–23. Cold Spring Harbor Laboratory Press.

Ptashne, M., and Gilbert, W., 1970. Genetic repressors. *Sci. Amer.* 222(6):36–44.

Jacob, F., and Monod, J., 1961. Genetic regulatory mechanisms in the synthesis of proteins. *J. Mol. Biol.* 3:318–356. [The operon model and the concept of messenger RNA were proposed in this landmark paper.]

Lwoff, A., and Ullmann, A. (eds.), 1979. *Origins of Molecular Biology. A Tribute to Jacques Monod.* Academic Press. [A wonderful collection of essays about the discovery of the operon concept and the scientists involved in this research. A book to be savored.]

PROBLEMS

1. *Missing genes.* What is likely to be the effect of each of these mutations?
   (a) Deletion of the *lac* repressor gene.
   (b) Deletion of the *trp* repressor gene.
   (c) Deletion of the *araC* protein gene.
   (d) Deletion of the *cI* gene of λ.
   (e) Deletion of the *N* gene of λ.

2. *Turned off.* Superrepressed mutants (i^s) of the *lac* operon behave as noninducible mutants. The i^s gene is dominant over i^+ in partial diploids. What is the molecular nature of this mutation likely to be?

3. *Uncontrolled synthesis.* A mutant of *E. coli* synthesizes large amounts of β-galactosidase whether or not inducer is present. A partial diploid formed from this mutant and $Fi^+o^+z^-$ also synthesizes large amounts of β-galactosidase whether or not inducer is present. What kind of mutation might give these results?

4. *A limited palette.* A mutant unable to grow on galactose, lactose, arabinose, and several other carbon sources is isolated from a wild-type culture. The cyclic AMP level in this mutant is normal. What kind of mutation might give these results?

5. *A grant of immunity.* An *E. coli* cell bearing a λ prophage is immune to lytic infection by λ. Why?

6. *Translational efficacy.* Transcription of *cI* can be initiated at p_{RE}, the promoter for repressor establishment, or at p_{RM}, the promoter for repressor maintenance. The p_{RM} transcript begins at its 5′ end with an AUG

codon for the λ repressor, whereas in the p_{RE} transcript this initiation codon is preceded by a sequence that is complementary to the 3′ end of 16S rRNA. Initiation at the p_{RE} site requires phage-encoded proteins that are not expressed in the lysogenic state.

(a) Which transcript will be translated more efficiently?

(b) What is the likely physiologic significance of this difference?

7. *The price of tenacity.* Predict the consequence of a mutation that increases by a factor of 100 the binding affinity of *lac* repressor for *lac* operator without changing the binding affinity for nonspecific sites on DNA.

8. *A kinetic difference.* The binding of most repressors to operator sites is much faster at low ionic strength than at high. In contrast, the affinity of the operator for the repressor is not markedly affected by ionic strength. Why?

9. *Shatterproof mutant.* Crystals of normal *trp* aporepressor shatter on addition of tryptophan. However, crystals of a mutant aporepressor are unaffected by the addition of tryptophan. The unit cell dimensions of this crystal are virtually identical with those of a crystal of normal repressor containing bound tryptophan. Predict the physiologic activity of this mutant repressor protein.

10. *Terminating a terminator.* The amount of termination factor RF2 is regulated by an unusual mechanism. Either a truncated polypeptide ending at residue 25 or a complete protein ending at residue 340 is produced from the mRNA for RF2. The amino acid sequence in the vicinity of residue 25 and the corresponding base sequence are

-Gly-Tyr-Leu-Asp-Tyr-Asp-
23 24 25 26 27 28

5′-GGGUAUCUUUGACUACGAC-3′

Propose a control mechanism based on these data. What is the outcome of this regulatory circuit?

11. *On the importance of being integral.* Lambda repressor binds cooperatively to operator sites that are separated by 5 or 6 turns of double helix. In contrast, binding is noncooperative when the operator sites are 4.5, 5.5, or 6.5 turns apart. Propose a molecular basis for this difference.

12. *A pregnant pause.* In transcribing the *trp* leader gene, RNA polymerase pauses at a discrete site immediately after it makes segments 1 and 2 (see Figure 36-36). Why?

13. *Helix swap.* Five amino acid residues on the outside of the C-terminal helix of the helix-turn-helix motif of the phage 434 repressor were replaced with those occupying the same position in the P22 repressor encoded by a phage that infects *Salmonella.* The DNA-binding specificity of this chimeric protein corresponded to that of the P22 repressor rather than the 434 repressor. What does this helix redesign experiment reveal about the recognition process?

CHAPTER

37

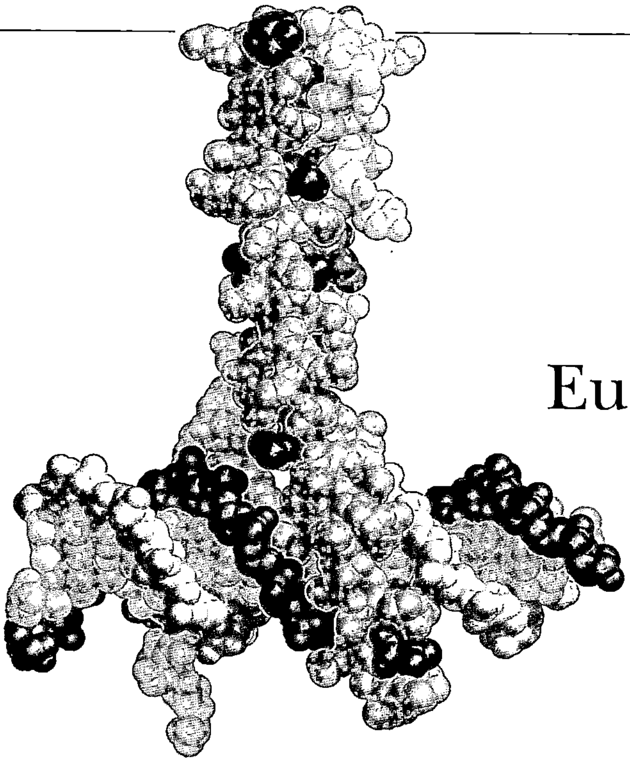

Eukaryotic Chromosomes and Gene Expression

Eukaryotic chromosomes are larger and have a higher degree of structural organization than those of prokaryotes. The genomes of yeast, fruit flies, and humans contain about 4, 40, and 1000 times as much DNA, respectively, as the genome of *Escherichia coli*. This abundance endows eukaryotes with potentialities that are absent from prokaryotes and poses additional challenges in replication and expression.

This chapter focuses on three aspects of eukaryotic chromosomes and gene expression: (1) How is eukaryotic DNA packaged and replicated? Basic proteins called *histones* play a key role in the compaction of DNA, which is wound around them to form *nucleosomes,* like beads on a string. The entire DNA-protein complex is called *chromatin.* (2) How are genes arranged in a eukaryotic chromosome? A striking feature of the genomes of higher eukaryotes is the *abundance of repeated sequences and the small proportion of protein-coding sequences.* (3) How is gene expression controlled in eukaryotes? As in prokaryotes, control is exerted primarily at the level of transcription. *In eukaryotes, however, most genes are silent unless they are turned on by the binding of transcription-factor proteins to multiple control sites on DNA.* These activators act cooperatively to promote the assembly of the transcriptional apparatus at the initiation site. The three-dimensional structure of several DNA-binding modules—*zinc fingers, nuclear receptors, leucine zippers, homeodomains*—will be considered because they recur in

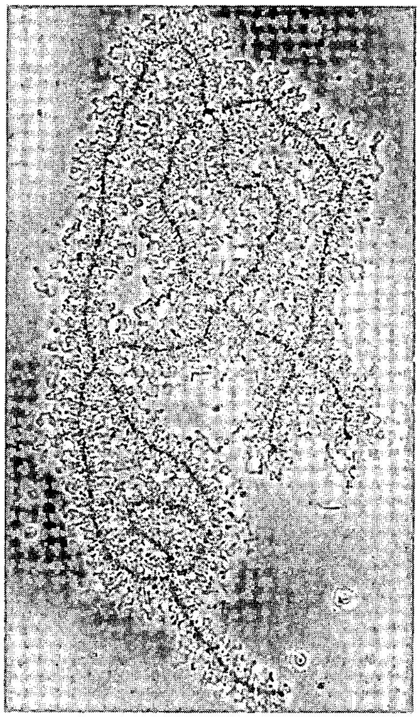

Figure 37-1
Phase-contrast light micrograph of a lampbrush chromosome from an amphibian oocyte. The lampbrush appearance comes from extended chromatin loops that are covered with nascent RNA bound to protein. [Courtesy of Dr. Joseph Gall.]

Opening Image: *Structure of a leucine zipper module (yellow and light blue) bound to DNA (red and green). Leucine zipper proteins play key roles in controlling eukaryotic gene expression. The α-helical coiled coil region of these transcriptional regulators is stabilized by interactions between regularly spaced leucine residues (dark blue and orange). [Drawn from 1ysa.pdb. T.E. Ellengerger, C.J. Brandl, K. Struhl, and S.C. Harrison. Cell 71(1992):1223.]*

many transcriptional regulators from yeast to humans. We are beginning to understand the molecular circuits that guide the development of multicellular organisms and enable their genomes to be dynamically controlled.

A EUKARYOTIC CHROMOSOME CONTAINS A SINGLE LINEAR MOLECULE OF DOUBLE-HELICAL DNA

How is DNA packed into chromosomes? This important question was difficult to answer for many years because very large DNA molecules are exquisitely sensitive to degradation by shearing forces. Bruno Zimm circumvented this problem by using the *viscoelastic technique,* which provides a measure of the size of the *largest DNA molecules in a mixture.* Cells from fruit flies were lysed in the measuring chamber to avoid the breakage of DNA that would be caused by transferring samples. Nucleases were inactivated by incubating the sample in detergent at 65°C. In addition, a mixture of proteases was added to digest proteins bound to DNA. The DNA molecules were then stretched out by flowing the solution and allowed to recoil to their normal form. The half-time for recoiling depends on the mass of the molecule.

The observed mass of the largest DNA molecules in this mixture was 41×10^6 kd, which agreed closely with the known value of the DNA content of the largest chromosome in *Drosophila melanogaster,* which is 43×10^6 kd. Excellent agreement also was obtained for a translocation mutant of the largest chromosome, which has an extra piece of DNA, giving it a calculated DNA content of 59×10^6 kd. The measured viscoelastic mass of this DNA molecule was 58×10^6 kd. Radioautographs of *D. melanogaster* DNA (Figure 37-2) confirmed the existence of very long DNA molecules. These images revealed that *a* Drosophila *chromosome contains a single uninterrupted molecule of DNA.* Furthermore, this DNA molecule is *linear and unbranched.* A mass of 43×10^6 kd corresponds to 76×10^6 base pairs (76 Mb). Thus, the largest *Drosophila* chromosome is about 20 times as large as the *E. coli* chromosome (Table 37-1).

Megabase (Mb)—
A length of DNA consisting of 10^6 base pairs (if double-stranded) or 10^6 bases (if single-stranded).

1 Mb = 10^3 kb = 10^6 bases

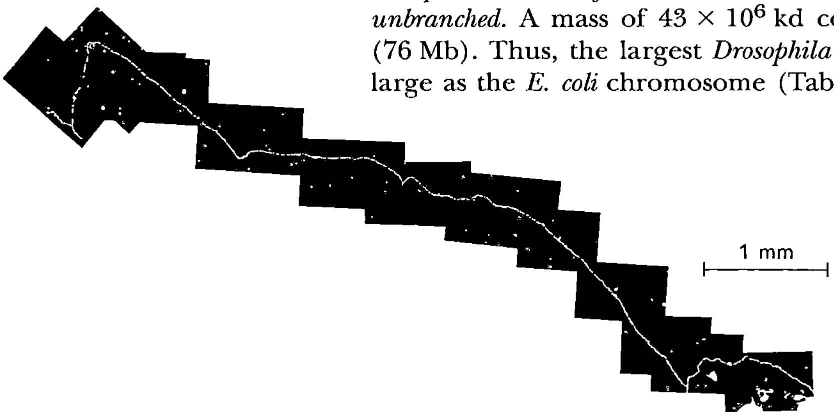

Figure 37-2
Autoradiograph of a DNA molecule from *Drosophila melanogaster.* [From R. Kavenoff, L.C. Klotz, and B.H. Zimm. *Cold Spring Harbor Symp. Quant. Biol.* 38(1974):4.]

Table 37-1
DNA contents of several prokaryotic and eukaryotic genomes

Organism	*Number of base pairs*	*Total DNA length (mm)*	*Number of chromosomes*
E. coli	4×10^6	1.4	1
Yeast *(S. cerevisiae)*	1.4×10^7	4.6	16
Fruit fly *(D. melanogaster)*	1.7×10^8	56	4
Human	3.9×10^9	1326	23

Note: The values given are for haploid genomes.

Baker's yeast (*Saccharomyces cerevisiae*) contains 16 chromosomes ranging in size from 0.2 to 2.2 Mb. These DNA molecules can be separated by *pulsed field electrophoresis* (Figure 37-3). Electric fields oriented 120 degrees apart in the plane of the agarose gel are applied in an alternating sequence. Large molecules reorient much more slowly than do small ones; they move much more slowly in the gel because they cannot follow the changing applied field. This powerful technique makes it possible to analyze and separate much larger DNA molecules than can be accomplished with a constant electric field in a single direction. Megabase DNA molecules, such as entire yeast chromosomes, can now be resolved.

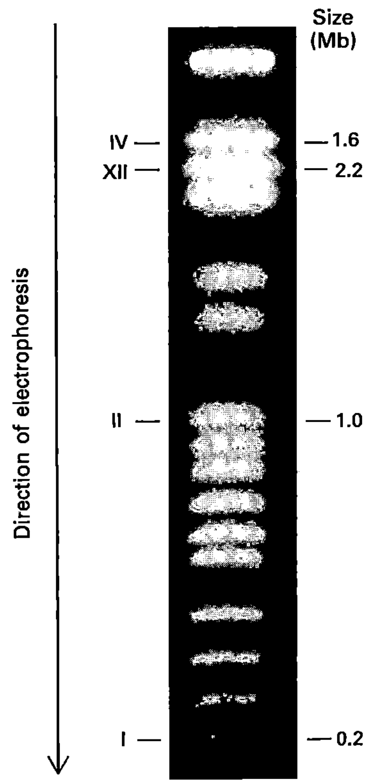

Figure 37-3
Separation of yeast chromosomes by pulsed field electrophoresis, which resolves megabase DNA molecules. The locations of chromosomes I, II, IV, and XII were determined by Southern blotting. [From G. Chu, D. Vollrath, and R.W. Davis. *Science* 234(1986):1583.]

DNA IN EUKARYOTIC CHROMOSOMES IS TIGHTLY BOUND TO BASIC PROTEINS CALLED HISTONES

The DNA in eukaryotic chromosomes is not bare. Rather, eukaryotic DNA is tightly bound to a group of small basic proteins called *histones.* In fact, histones constitute about half the mass of eukaryotic chromosomes, the other half being DNA. This nucleoprotein chromosomal material is termed *chromatin.* Histones can be dissociated from DNA by treating chromatin with salt or dilute acid. The resulting mixture can then be fractionated by ion-exchange chromatography (p. 50). The five types of histones are called *H1, H2A, H2B, H3,* and *H4.* They range in mass from about 11 kd to 21 kd (Table 37-2). A striking feature of histones is their *high content of positively charged side chains:* about one in four residues is either lysine or arginine.

Table 37-2
Types of histones

Type	*Lys/Arg ratio*	*Number of residues*	*Mass (kd)*	*Location*
H1	20.0	215	21.0	Linker
H2A	1.25	129	14.5	Core
H2B	2.5	125	13.8	Core
H3	0.72	135	15.3	Core
H4	0.79	102	11.3	Core

Each type of histone can exist in a variety of forms because of *post-translational modifications* of certain side chains. For example, lysine 16 in H4 is often acetylated. Histones can also be methylated, ADP-ribosylated, and phosphorylated. The modulation of the charge, hydrogen-bonding capabilities, and shape of histones by these reversible covalent modifications may be important in packaging DNA and in regulating its availability for replication and transcription.

Ser-Gly-Arg-Gly -Lys -Gly-Gly-Lys -Gly-Leu- 10
Gly -Lys -Gly-Gly -Ala -Lys-Arg-His -Arg-Lys- 20
Val -Leu-Arg-Asp-Asn-Ile -Gln-Gly -Ile -Thr- 30
Lys -Pro-Ala -Ile -Arg -Arg-Leu-Ala -Arg-Arg- 40
Gly -Gly-Val -Lys -Arg-Ile -Ser-Gly -Leu-Ile - 50
Tyr -Glu-Glu-Thr -Arg-Gly-Val -Leu-Lys-Val - 60
Phe-Leu-Glu-Asn-Val -Ile -Arg-Asp-Ala -Val - 70
Thr -Tyr -Thr-Glu -His -Ala -Lys-Arg -Lys -Thr- 80
Val -Thr-Ala -Met-Asp-Val -Val -Tyr -Ala -Leu- 90
Lys -Arg-Gln-Gly -Arg -Thr-Leu-Tyr -Gly -Phe- 100
Gly -Gly 102

Figure 37-4
Amino acid sequence of histone H4 from calf thymus. Several residues (shown in red) are modified. The N-terminal α-amino group and the ε-amino group of Lys 16 are acetylated. The ε-amino group of Lys 20 is methylated or dimethylated. Histone H4 from pea seedlings has the same amino acid sequence except that residue 60 is isoleucine and residue 77 is arginine.

THE AMINO ACID SEQUENCES OF H3 AND H4 ARE NEARLY THE SAME IN ALL PLANTS AND ANIMALS

The amino acid sequences of H4 from pea seedlings and calf thymus differ at only 2 sites out of 102 residues. The changes at these two sites are quite small: valine in place of isoleucine, and lysine in place of arginine. Thus, *the amino acid sequence of H4 has remained nearly constant in the 1.2 × 10^9 years since the divergence of plants and animals.* Likewise, H3 has changed little throughout this very long evolutionary period. The amino acid sequences of H3 from pea seedlings and calf thymus differ at just four positions. The remarkably conserved structure of H3 and H4 strongly suggests that they play a critical role that was established early in the evolution of eukaryotes and has remained nearly invariant since then.

NUCLEOSOMES ARE THE REPEATING UNITS OF CHROMATIN

How do histones combine with DNA to form the chromatin fiber? In 1974, Roger Kornberg proposed that *chromatin is made up of repeating units, each consisting of 200 bp of DNA and of two each of H2A, H2B, H3, and H4.* These repeating units are now known as *nucleosomes.* Most of the DNA is wound around the outside of a core of histones. The remainder of the DNA, called the *linker,* joins adjacent nucleosomes and contributes to the flexibility of the chromatin fiber. Thus, *a chromatin fiber is a flexibly jointed chain of nucleosomes,* rather like beads on a string.

This model for the structure of chromatin is supported by a wide range of experimental findings:

1. *Electron microscopy.* Linear arrays of 100-Å-diameter spheres connected by a thin strand are seen in electron micrographs of chromatin (Figure 37-5). The degree of extension of the chromatin fiber depends on the procedure used to prepare the specimen for electron microscopy.

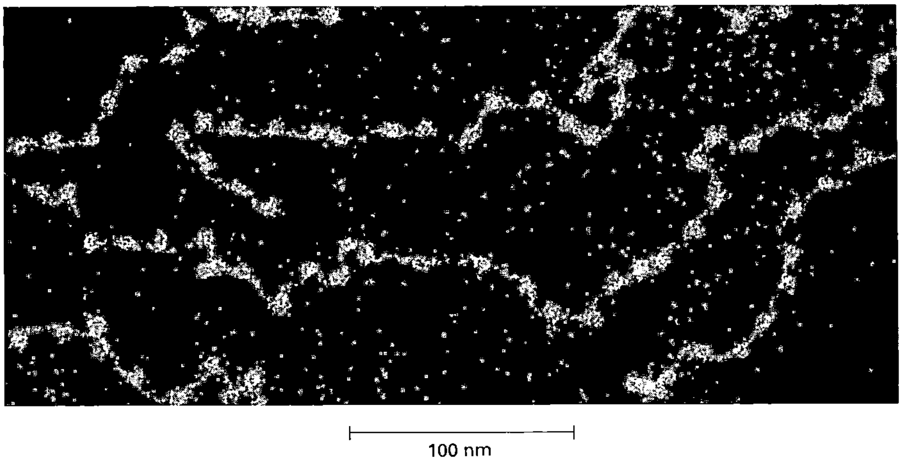

Figure 37-5
Electron micrograph of chromatin. The beadlike particles have diameters of nearly 100 Å. [Courtesy of Dr. Ada Olins and Dr. Donald Olins.]

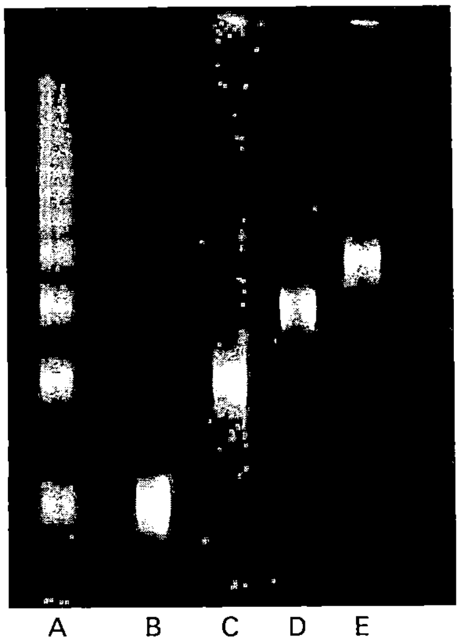

Figure 37-6
Gel electrophoresis patterns of defined lengths of chromatin produced by limited micrococcal nuclease digestion. Lane A shows an unfractionated digest. Fractionation by centrifugation on a sucrose gradient yielded monomer (B), dimer (C), trimer (D), and tetramer (E). [From J.T. Finch, M. Noll, and R.D. Kornberg. *Proc. Nat. Acad. Sci.* 72(1975):3321.]

2. *X-ray and neutron diffraction.* A 100-Å repetition is also seen in x-ray diffraction patterns of chromatin fibers. Neutron diffraction studies indicate that the DNA is located near the outside of the nucleosome.

3. *Nuclease digestion.* Free DNA in solution can be cleaved at any of its phosphodiester bonds by pancreatic deoxyribonuclease I (DNase I) or micrococcal nuclease. In contrast, DNA in chromatin is protected from digestion except at particular sites. The digestion pattern of chromatin is striking in its simplicity: it consists of a ladder of discrete bands (Figure 37-6). The DNA contents of these fragments are multiples of a basic unit of about 200 base pairs. Electron micrographs show that the number of

spherical particles in a fragment of chromatin is equal to the number of 200-bp units. For example, a fragment with 600 base pairs of DNA consists of three 100-Å-diameter particles. Thus, a bead seen in an electron micrograph corresponds to a nucleosome defined by nuclease digestion.

4. *Reconstitution.* A chromatinlike fiber can be formed in vitro by adding histones to DNA from adenovirus or simian virus 40 (SV40). The amount of DNA associated with a nucleosome in these reconstituted systems is close to 200 base pairs. Furthermore, equimolar amounts of H2A, H2B, H3, and H4 are required to form nucleosomes. Characteristic beads are not formed if any of the four histones is absent from the reconstitution mixture. X-ray diffraction studies also showed that H2A, H2B, H3, and H4 must be added to DNA to regain the diffraction pattern of chromatin. In contrast, H1 is not required, which fits the finding that H1 is not an integral part of the nucleosome.

Table 37-3
DNA content of nucleosomes

Cell type	*Number of base pairs*
Yeast	165
HeLa	183
Rat bone marrow	192
Rat liver	196
Rat kidney	196
Chick oviduct	196
Chicken erythrocyte	207
Sea urchin gastrula	218
Sea urchin sperm	241

A NUCLEOSOME CORE CONSISTS OF 140 BASE PAIRS OF DNA WOUND AROUND A HISTONE OCTAMER

The DNA content of nucleosomes from different organisms and cell types ranges from about 160 to 240 base pairs (Table 37-3). What is the structural basis of these differences? Again, nucleases have proved to be highly informative probes. Nucleosomes can be further digested with micrococcal nuclease to yield a *core particle* that contains 140 base pairs of DNA, irrespective of the initial DNA content of the particular nucleosome. The 140-nucleotide DNA is not further digested because the phosphodiester backbone is inaccessible to the nuclease. The nucleosome core is probably nearly the same in all eukaryotes. *This core consists of 140 base pairs of DNA bound to a histone octamer (two each of H2A, H2B, H3, and H4).*

The crystallization of nucleosome cores (Figure 37-7) demonstrated that the particles are quite homogeneous and opened the door to a deeper understanding of how eukaryotic DNA is packaged. X-ray and electron-microscopic analyses of these crystals by Aaron Klug and John Finch revealed that the core is a flat particle of dimensions 110 × 110 × 55 Å. The 140 base pairs of DNA are wound on the outside of the core to form 1.75 turns of a left-handed supercoil with a pitch of about 28 Å. A model of the nucleosome based on low-resolution (20 Å) data is shown in Figure 37-8. Further analysis suggested that a tetramer formed from two histone H3 and two H4 occupies the center of the nucleosome, flanked at

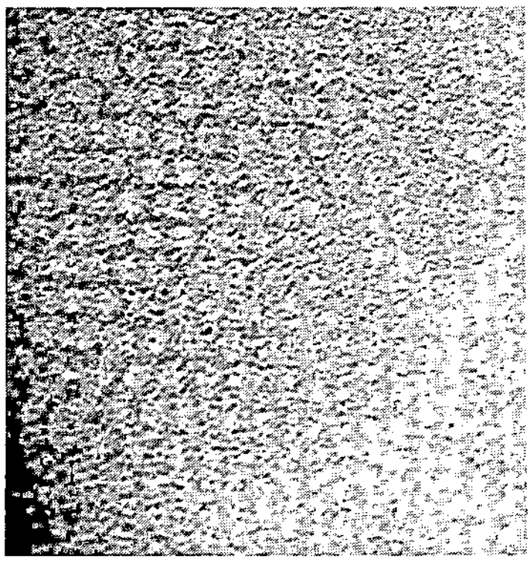

Figure 37-7
Electron micrograph of a crystal of nucleosome cores. Centers of adjacent nucleosome cores in this hexagonal array are 100 Å apart. [From J.T. Finch, L.C. Lutter, D. Rhodes, R.S. Brown, B. Rushton, M. Levitt, and A. Klug. *Nature* 269(1977):31.]

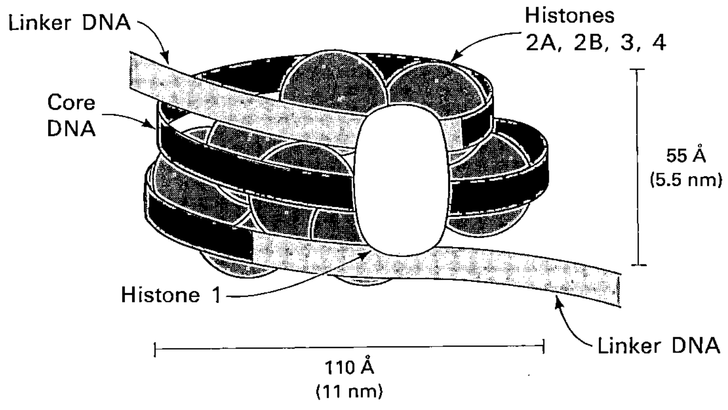

Figure 37-8
Schematic diagram of a region of chromatin containing a nucleosome. The DNA double helix (shown in red) is wound around an octamer of histones (two molecules each of H2A, H2B, H3, and H4, shown in blue). The DNA in the nucleosome core is shown in dark red. Histone 1 (shown in yellow) binds to the outside of this core particle and to the linker DNA (light red). [After A. Kornberg. *DNA Replication.* (W.H. Freeman, 1980), p. 294.]

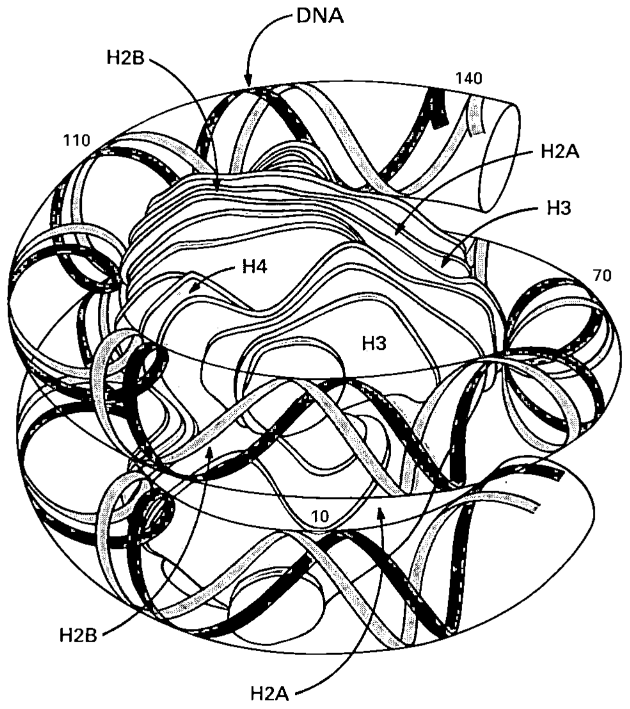

Figure 37-9
Model of a nucleosome core showing DNA wound in a left-handed superhelix around the histone octamer. The locations of the four kinds of histone subunits are marked. [After R. Kornberg and A. Klug. The nucleosome. Copyright © 1981 by Scientific American, Inc. All rights reserved.]

either end by an H2A-H2B dimer (Figure 37-9). The flexibility of DNA enables the formation of nucleosomes and higher-order structures in chromosomes.

DNA-histone contacts are made on nearly every turn of the DNA double helix and are confined to the inner surface of the superhelix. The histones do not embrace the DNA, nor do they protrude between the turns of the superhelix. The most substantial contacts are made between the H3-H4 tetramer and the central loop of the superhelix. α-Helical rods projecting from the H3 dimer fit snugly into the minor grooves of DNA, on either side of the twofold symmetry axis of the nucleosome. *Histones can interact with most DNA sequences, in keeping with their role as a device for packaging DNA.* An H2A-H2B dimer interacts with each exposed face of the H3-H4 tetramer. The binding of H2A-H2B to each last half turn of the the superhelix further stabilizes the nucleosome core. The disassembly of a nucleosome, as in DNA replication, is probably initiated by the dissociation or pulling away of an H2A-H2B dimer.

Histone H1 plays a key role at the next level of chromosome structure by serving as a bridge between adjacent nucleosomes. H1 is located on the outside of the nucleosome, where it binds linker DNA and interacts with H2A subunits of the core. It is released from nucleosomes when the DNA per particle is trimmed from 160 to 140 base pairs. H1 has a stoichiometry of one per nucleosome, compared with two for the other histones. It is also noteworthy that several types of H1 have been found, in contrast with the constancy of the other histones. Moreover, H1 is phosphorylated just before mitosis, and is dephosphorylated following mitosis, suggesting that this covalent modification regulates its capacity to make DNA compact.

NUCLEOSOMES ARE THE FIRST STAGE IN THE CONDENSATION OF DNA

The winding of DNA around a nucleosome core contributes to its packing by decreasing its linear extent. A 200-bp stretch of DNA would have a length of about 680 Å (68 nm) in solution. In contrast, this amount of DNA fits within the 100-Å (10-nm) diameter of the nucleosome. Thus, the *packing ratio* (degree of condensation) of the nucleosome is about 7. How does this value compare with the degree of condensation of DNA in chromosomes? Human metaphase chromosomes, which are highly condensed, contain a total of 7.8×10^9 base pairs, corresponding to a contour length of 2.6 m. The total length of these 46 metaphase chromosomes is 200 μm. Thus, *the packing ratio of human chromosomes in metaphase is about* 10^4. In interphase nuclei, where the chromatin is more dispersed, the same DNA has a packing ratio of about 10^2 to 10^3. Clearly, the nucleosome is just the first step in DNA compaction.

What then is the next level of organization of DNA? One possibility is that the nucleosomes themselves form a helical array. For example, a postulated solenoidal model of chromatin with a 360-Å (36-nm) diameter (Figure 37-10) would have a packing ratio of about 40. The folding of such solenoids into loops would provide additional condensation. It seems likely too that a series of nonhistone proteins stabilizes higher-order structures. For example, metaphase chromosomes stripped of histones display a central protein scaffold that is surrounded by many very long loops of DNA (Figure 37-11). These DNA loops are probably bound to the scaffold. The presence of topoisomerase II in the scaffold makes it

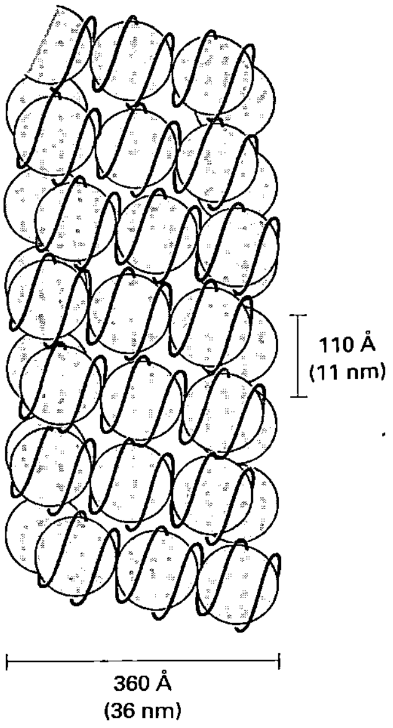

Figure 37-10
A proposed solenoidal model of chromatin consisting of six nucleosomes per turn of helix. The DNA double helix (shown in red) is wound around each histone octamer (shown in blue). [After J.T. Finch and A. Klug. *Proc. Nat. Acad. Sci.* 73(1976):1900.]

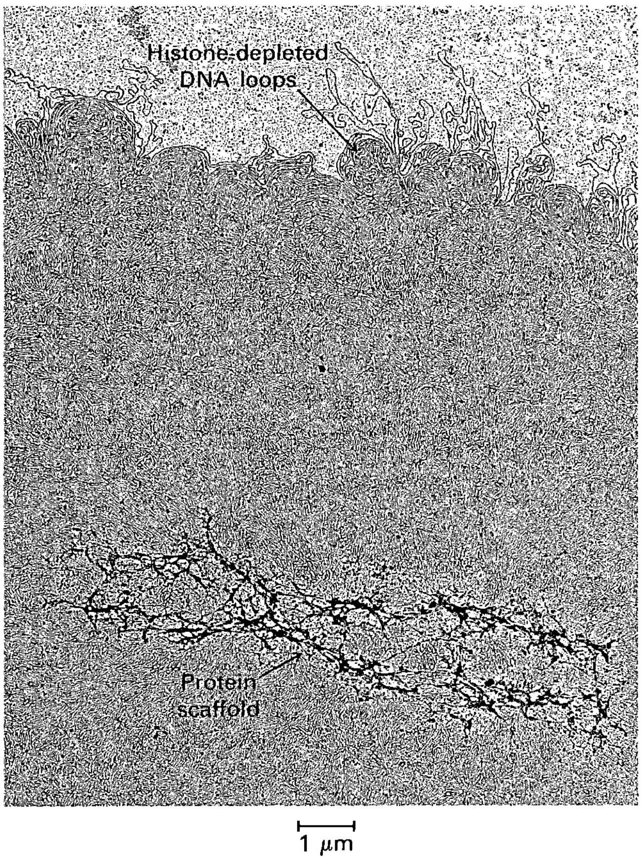

Figure 37-11
Electron micrograph showing histone-depleted DNA attached to a central protein scaffold. Histones were removed from metaphase chromosomes of HeLa cells by treating them with polyanions. [Courtesy of Dr. Ulrich Laemmli.]

likely that changes in supercoiling are important in altering the architecture and accessibility of large segments of DNA in mitosis and meiosis.

The most compacted DNA is found in sperm heads, in which histones are replaced by *protamines,* a series of arginine-rich proteins that become highly α-helical on binding to DNA. The α helices of protamine probably lie in the major grooves of DNA, where they neutralize the negatively charged phosphate backbone and so enable DNA duplexes to pack tightly together.

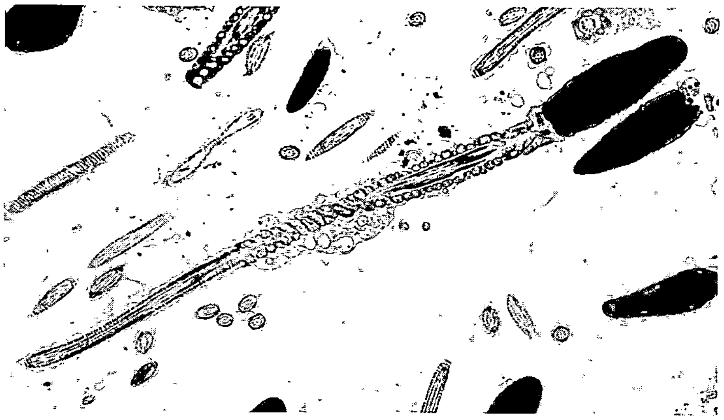

Electron micrograph of a sperm. The dense head of the sperm contains highly compacted DNA. [Courtesy of Lynne Mercer.]

EUKARYOTIC DNA IS REPLICATED BIDIRECTIONALLY FROM MANY ORIGINS

Eukaryotic DNA, like all other DNA molecules, is *replicated semiconservatively.* Electron-microscopic studies have also shown that eukaryotic DNA is replicated *bidirectionally from many origins.* The use of many initiation points is necessary for rapid replication because of the great length of eukaryotic DNAs. Consider, for example, a 62-Mb *Drosophila* chromosome. A DNA replication fork in *Drosophila* moves at the rate of 2.6 kb/min (0.16 Mb/h), compared with about 16 kb/min in *E. coli.* The replication of this *Drosophila* chromosome would take more than 16 days if there were only one origin for replication. The actual replication time of less than 3 minutes is achieved by *the cooperative action of more than 6000 replication forks per DNA molecule.* A DNA molecule from the cleavage nuclei of *Drosophila* exhibits a serial array of replicated regions, or "eye forms" (Figure 37-12). The activation of each initiation point generates two di-

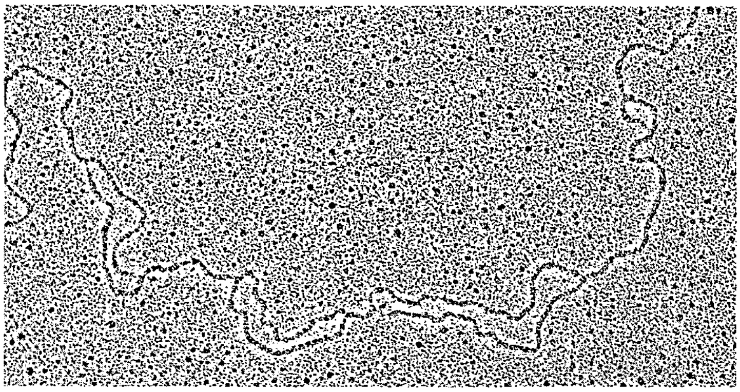

Figure 37-12
Electron micrograph of replicating chromosomal DNA from cleavage nuclei of *Drosophila.* The eye forms are the newly replicated regions. [Courtesy of Dr. David Hogness.]

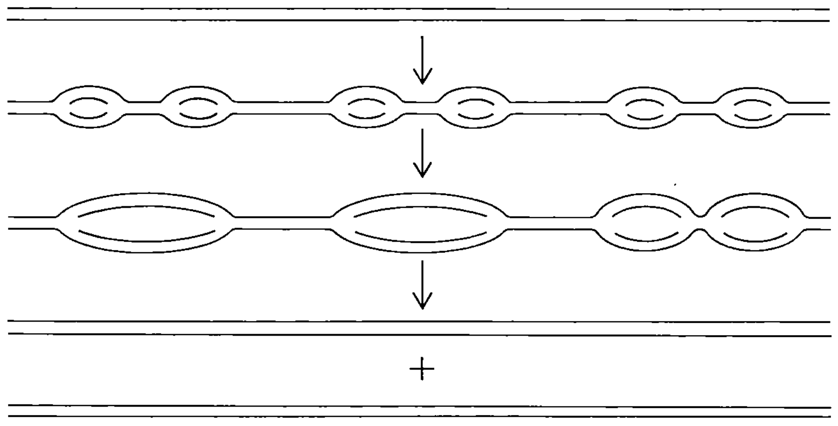

Figure 37-13
Schematic diagram of the replication of a eukaryotic chromosome segment. The parental DNA is shown in blue, and the newly replicated DNA in red.

verging replication forks. The eye forms expanding in both directions merge to form the two daughter DNA molecules (Figure 37-13). An eye form within an eye form has not been observed, indicating that an origin cannot be reactivated until after the entire DNA molecule is replicated.

A yeast chromosome contains about 400 initiation sites for DNA replication. These origins of replication, called *ARS* for *autonomous replication sequence,* share an 11-bp consensus sequence. Insertion of an ARS sequence into a bacterial plasmid enables it to replicate autonomously in yeast cells. ARS is recognized by a 250-kd assembly of eight proteins called the *origin recognition complex,* which initiates replication.

EUKARYOTIC DNA IS REPLICATED AND REPAIRED BY SEVERAL TYPES OF POLYMERASES

Eukaryotic cells contain several types of DNA polymerases (Table 37-4). The α and δ polymerases play a major role in chromosome replication, whereas the β and ε polymerases participate in DNA repair. The γ polymerase is responsible for the replication of mitochondrial DNA. The actions of these polymerases can be distinguished by their different susceptibilities to inhibitors. 2',3'-Dideoxyribonucleotides strongly inhibit β and γ, whereas aphidocolin (a fungal diterpenoid) blocks α, δ, and ε. These enzymes, like prokaryotic DNA polymerases, use deoxyribonucleoside triphosphates as activated intermediates and carry out template-directed elongation of a primer in the $5' \rightarrow 3'$ direction (p. 800). Also, the two daughter strands are synthesized differently. The discontinuous synthesis of the lagging strand is catalyzed by α, and the continuous synthesis of the leading strand by δ. The α complex carries a primase subunit

Table 37-4
Eukaryotic DNA polymerases

Type	*Location*	*Mass (kd)*	*Major role*
α	Nucleus	250	Replication of nuclear DNA
β	Nucleus	39	Repair of nuclear DNA
γ	Mitochondria	200	Replication of mitochondrial DNA
δ	Nucleus	170	Replication of nuclear DNA
ε	Nucleus	260	Repair of nuclear DNA

that makes temporary RNA stretches that serve as primer and are then erased (p. 805). The δ, ε, and γ enzymes possess $3' \rightarrow 5'$ exonuclease activity with editing capabilities akin to those of prokaryotic polymerases. An accessory protein probably proofreads the DNA synthesized by α.

The three-dimensional structure of DNA polymerase β has recently been solved. This 39-kd protein has a conformation like that of the large fragment (Klenow fragment) of DNA polymerase I from *E. coli* (p. 802). An 8-kd amino-terminal domain is followed by a 31-kd carboxyl-terminal domain arranged like the fingers, palm, and thumb of the right hand (Figure 37-14A). A duplex DNA consisting of a 7-nucleotide primer strand and an 11-nucleotide template strand with an overhang is bound in the cleft between the fingers and thumb (Figure 37-14B). When DNA is bound, the highly basic 8-kd domain of the enzyme shifts in position to interact with the template strand. The catalytic site contains two aspartates that are present in the same position in all DNA polymerases and reverse transcriptases studied thus far. These aspartates bind two Mg^{2+} that are critical for catalysis, and one of them accepts a proton from the attacking 3'-OH group.

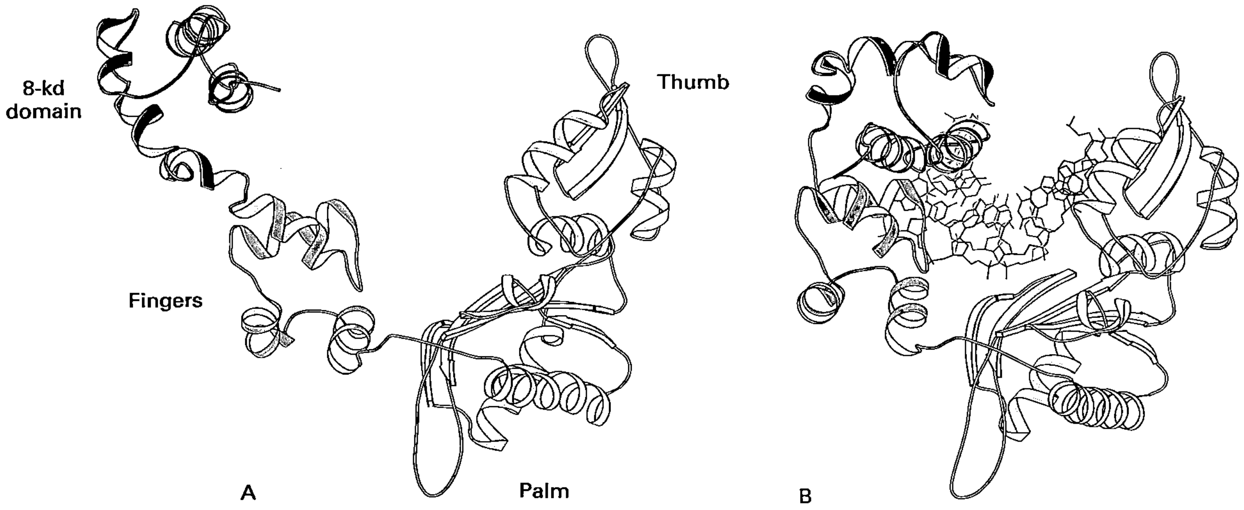

Figure 37-14
Structure of DNA polymerase β. The folding of this eukaryotic DNA repair enzyme is similar to that of the polymerase domain of prokaryotic DNA polymerase I. (A) Enzyme alone. (B) DNA-enzyme complex. The primer strand is shown in green, and the template strand in blue. The 8-kd domain (dark red) rotates to make contact with the elongating primer strand. [Drawn from coordinates kindly provided by Dr. Huguette Pelletier and Dr. Joseph Kraut. H. Pelletier, M.R. Sawaya, A. Kumar, S.H. Wilson, and J. Kraut. *Science* 264(1994):1891.]

THE ENDS OF CHROMOSOMES (TELOMERES) ARE REPLICATED BY A REVERSE TRANSCRIPTASE BEARING A TEMPLATE

> "There's a divinity that shapes our ends,
> Rough-hew them how we will."
>
> William Shakespeare (1601)
> *Hamlet* (V, ii)

The linearity of eukaryotic chromosomes poses a problem not encountered with circular DNA molecules such as those of prokaryotes. Eukaryotic DNA polymerases, like prokaryotic ones, are unable to synthesize in the $3' \rightarrow 5'$ direction or start chains de novo. Erasure of the RNA primer on the lagging strand (p. 806) produces a daughter DNA molecule with an incomplete 5' end (Figure 37-15). Shorter and shorter daughter DNA molecules would result from successive rounds of replication. How is this genetic catastrophe averted? The first clue came from sequence analysis of the ends of chromosomes, which are called *telomeres* (from the Greek word *telos,* meaning "an end"). Telomeric DNA contains hundreds of tandem repeats of a hexanucleotide sequence. One of the strands (the 3' end) is G-rich, whereas the complementary one (the 5' end) is C-rich. In *Tetrahymena* (a ciliated protozoan), the repeating G-rich sequence is GGGGTT, and in humans, AGGGTT.

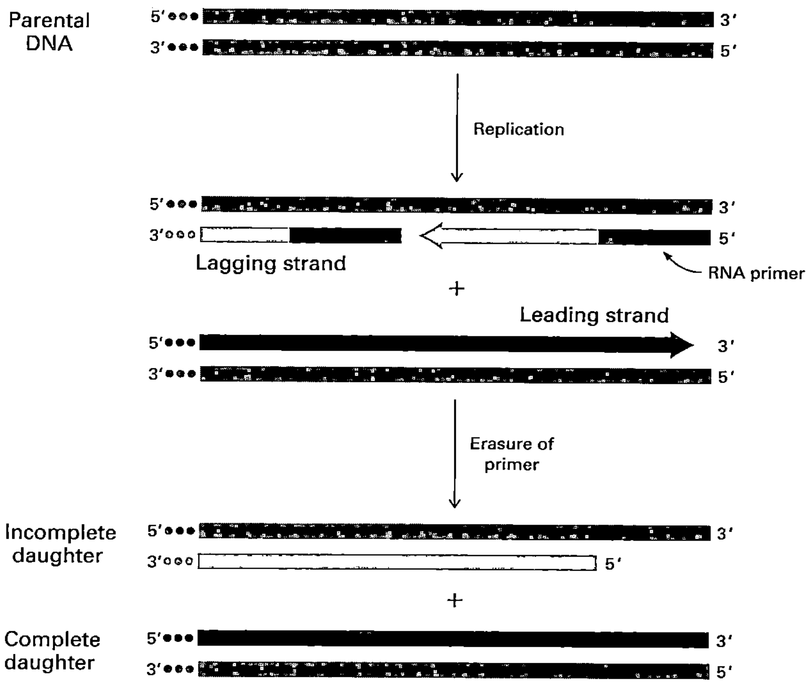

Figure 37-15
Problem posed in the replication of linear DNA. Erasure of the RNA primer (red) produces a daughter DNA molecule in which one strand is shortened. The parental DNA strands are shown in gray, the continuously synthesized leading daughter strand in blue, and the discontinuously synthesized lagging daughter strand in green. [After E.H. Blackburn. *Nature* 350(1991):569.]

The second clue came from in vitro studies of the action of *telomerase,* a large ribonucleoprotein enzyme catalyzing the elongation of the 3′-ending strand. When a primer ending in *GGTT* was added to the *Tetrahymena* enzyme in the presence of deoxyribonucleoside triphosphates, *GGTT*GGGGTT and longer products were formed. Human telomerase elongates the same primer to form *GGTT*AGGGTT and longer products. Thus, *the telomerase of a given species synthesizes the G-rich telomeric strand characteristic of that species.* Where does this specificity arise? Elizabeth Blackburn discovered that *telomerase contains an RNA molecule that serves as the template for the elongation of the G-rich strand.* The 159-nucleotide RNA in *Tetrahymena* telomerase contains a C-rich sequence that is the complement of the G-rich sequence it synthesizes.

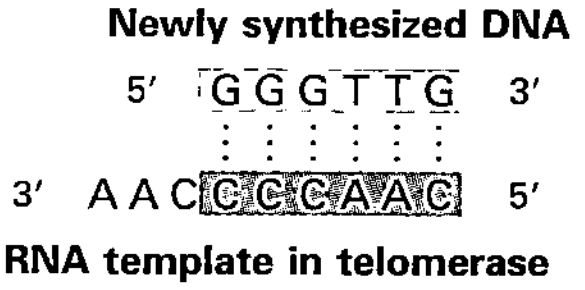

Thus, *telomerase is a template-bearing reverse transcriptase.* No other known polymerase carries a polynucleotide template. The enzyme binds to the 3′ end of the G-rich overhang of telomeric DNA (Figure 37-16). Several nucleotides of this overhang form base pairs with the template RNA of telomerase. Six deoxyribonucleotides are added to form an additional tandem repeat on the G-rich telomeric DNA. Telomerase then translocates a distance of six nucleotides on the elongated primer to begin another round of synthesis. The enzyme is *highly processive*—hundreds of nucleotides are added before telomerase dissociates from the elongated chain.

How is the complementary C-rich strand of telomeric DNA synthesized? The precise mechanism has not been elucidated. The overhanging G-rich strand with a free 3′-OH group could form a hairpin that would

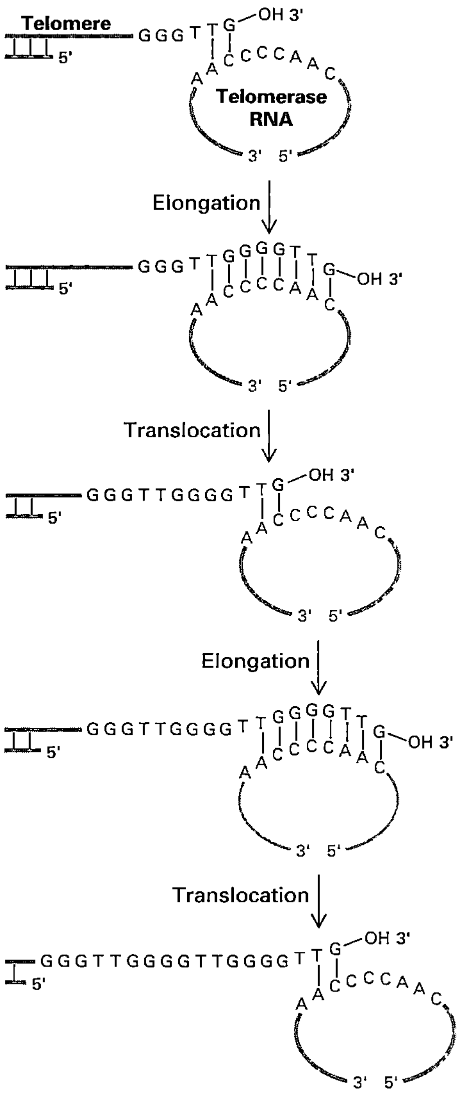

Figure 37-16
Mechanism of synthesis of the G-rich strand of telomeric DNA. The RNA template of telomerase is shown in green, and the nucleotides added to the G-rich strand of the primer are shown in red. [After E.H. Blackburn. *Nature* 350(1991):569]

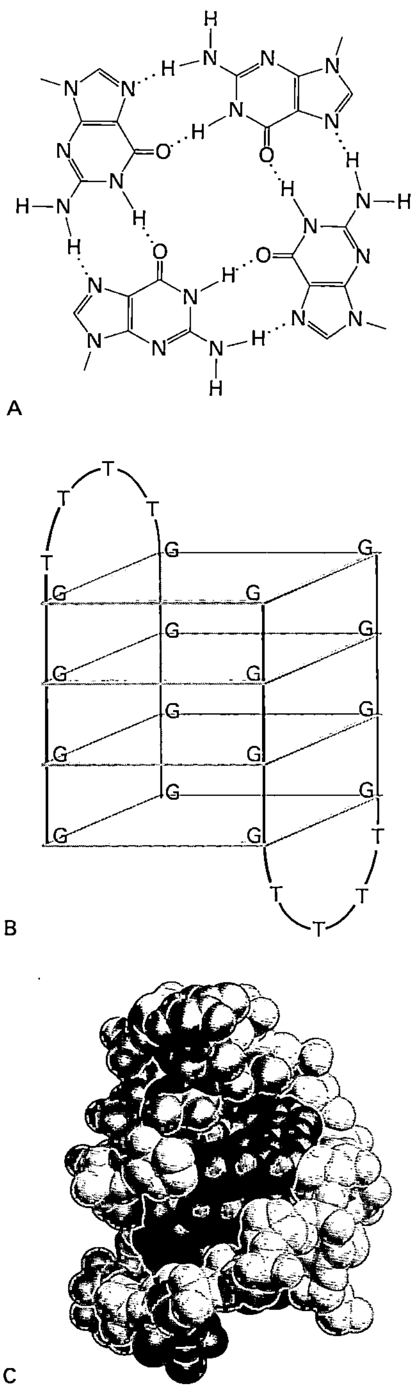

A B C

Figure 37-17
G-rich sequences can form distinctive tetrameric structures called G-quartets. (A) Four guanines can hydrogen-bond to form a ring. A cation (not shown) is bound at the center. (B) Schematic diagram and (C) space-filling model of the structure of d($G_4T_4G_4$). Guanines are shown in red, and thymines in blue. [Drawn from 1d59.pdb. C. Kang, X. Zhang, R. Ratliff, R. Moyzis, and A. Rich. *Nature* 356(1992):126.]

serve as a primer for the synthesis of the complementary strand. Polymerase α with its associated primase subunit is well suited for the completion of telomeres. Erasure of RNA at the 5′ end of the C-rich complement would pose no problem because of the high degree of redundancy of the terminal sequence; a few of the more than hundred repeats could be lost with impunity. Most important, *the information needed to generate tandem terminal repeats in future replications is contained in the RNA and protein components of telomerase, which are encoded by internal DNA sequences, rather than by DNA ends themselves.*

The conformation of telomeric DNA may be very different from that of classic B-DNA because of the abundance of G. Structural studies of G-rich oligonucleotides such as d($G_4T_4G_4$) have shown that they form *four-stranded structures* called *G-quartets.* Four guanosines can hydrogen-bond to one another in a cyclic fashion to form a layer of a G-quartet (Figure 37-17A). K^+ or another cation is bound at the center. These clusters stack on top of each other to form a four-stranded structure (Figure 37-17B and C).

OLD HISTONES ARE ACQUIRED BY THE LEADING DAUGHTER DNA DUPLEX

How are old and new histones distributed on daughter chromosomes? This question was answered by carrying out DNA synthesis in vitro in the presence of cycloheximide, an inhibitor of protein synthesis. Under these conditions, DNA synthesis continues for about 15 minutes. Half the newly synthesized DNA is completely degraded by DNase I, whereas the other half is split into fragments containing 200 base pairs. This experiment, as well as density-labeling studies, suggested that parental histones are associated with one of the daughter DNA duplexes, the other being bare because of the absence of new histones. This interpretation is directly supported by electron micrographs showing that one of the daughter duplexes at a replication fork is beaded, whereas the other is naked (Figure 37-18). In other words, *parental histones segregate conservatively during replication.*

This arrangement indicates that histones do not dissociate from DNA during replication. In fact, *old histones stay with the DNA duplex containing the leading strand, whereas new histones assemble on the DNA duplex containing the lagging strand.* A likely reason for this difference between the daughter DNA molecules is that histones bind much more strongly to double-stranded than to single-stranded DNA. Old histones probably do not follow the lagging duplex because it contains single-stranded regions before the joining of its Okazaki fragments.

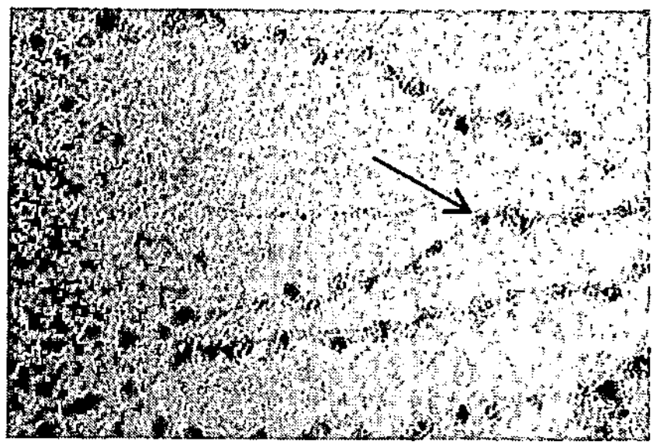

Figure 37-18
An asymmetric replication fork (arrow) generated by DNA synthesis in the presence of cycloheximide, which blocks the formation of new histones. One of the daughter duplexes is beaded, whereas the other is bare, which shows that parental histones are bound to only one daughter duplex. [From D. Riley and H. Weintraub. *Proc. Nat. Acad. Sci.* 76(1979):331.]

ENTRY OF CELLS INTO MITOSIS IS CONTROLLED BY A HIGHLY CONSERVED CYCLIN-DEPENDENT PROTEIN KINASE

In the cell cycle of eukaryotes, DNA is replicated during the S (*s*ynthesis) phase, and the replicated chromosomes are segregated into daughter nuclei during the M (*m*itosis) phase. A *g*ap phase G_1 intervenes between mitosis and DNA synthesis, and another gap, G_2, occurs between DNA synthesis and mitosis. The two key decisions are when to enter the S phase (the *start* point) and, subsequently, the M phase. Mitosis begins only after DNA synthesis has been completed and requires major changes in cell architecture: breakdown of the nuclear envelope, condensation of chromatin, and extensive reorganization of the cytoskeleton.

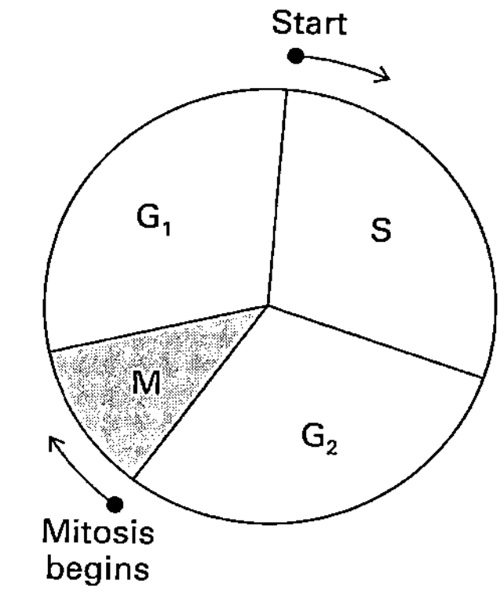

Eukaryotic cell cycle. M (*m*itosis), S (DNA *s*ynthesis); G_1 (*g*ap 1) and G_2 (*g*ap 2).

How is the entry of cells into mitosis regulated? The convergence of biochemical studies of frog oocytes and genetic analyses of the cell cycle of yeast revealed that eukaryotes employ a highly conserved mechanism to time and orchestrate this critical transition. The major controller is a 34-kd protein kinase known as the *cdc2 protein* because it is encoded by a *c*ell-*d*ivision-*c*ycle gene. Cdc2 kinase requires *cyclin B*, a 45-kd protein, for activity. Cyclin was discovered by a group of students taking a summer laboratory course at Woods Hole, Massachusetts. They observed that the concentration of a hitherto unknown protein dropped precipitously at the end of mitosis, and increased during interphase (G_1, S, and G_2)—hence the name *cyclin*.

During S phase, newly synthesized cyclin joins cdc2 kinase to form a complex called *maturation promoting factor (MPF)*. MPF is phosphorylated on Tyr 15, and then on Thr 161, by different enzymes (Figure 37-19). The

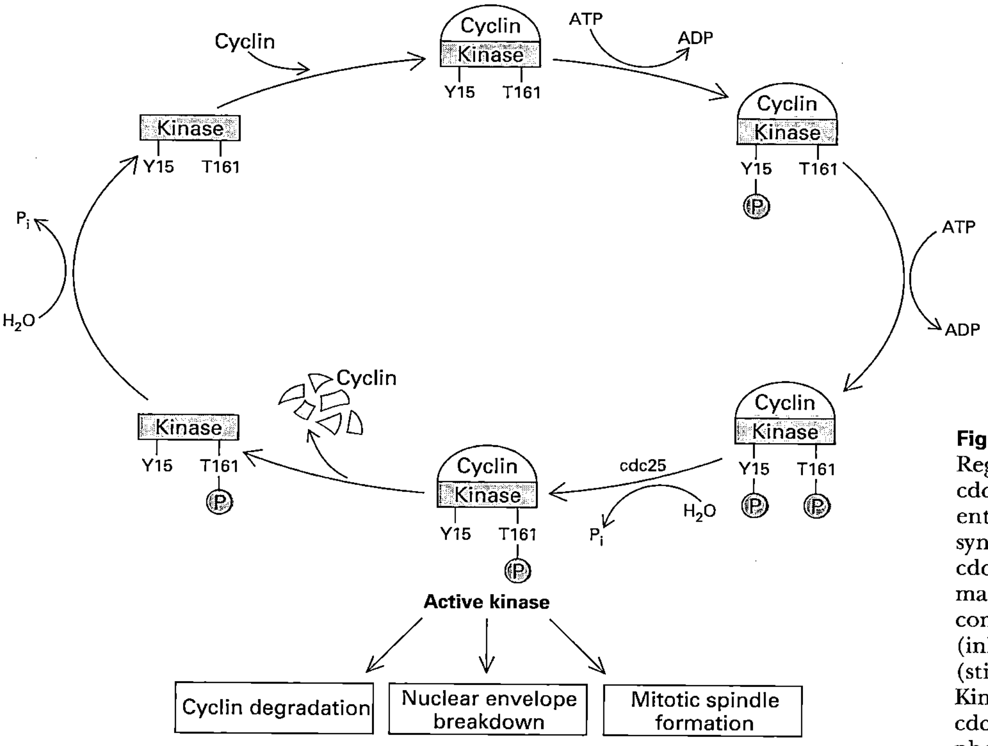

Figure 37-19
Regulation of the catalytic activity of cdc2 kinase, the major controller of entry into mitosis. Cyclin B (yellow) synthesized during interphase joins cdc2 kinase (blue) to form an inactive maturation promoting factor (MPF) complex. Phosphorylation of Tyr 15 (inhibitory site) and then Thr 161 (stimulatory site) keeps MPF inactive. Kinase activity is switched on when cdc25, a phosphatase, removes the phosphoryl group from Tyr 15. Activated MPF then phosphorylates many target proteins to initiate mitosis. The enzymatic activity of the complex is switched off by the destruction of cyclin, which also is triggered by the kinase.

doubly phosphorylated kinase is inactive but ready to be triggered. The key event is dephosphorylation of Tyr 15 by cdc25, a protein phosphatase that is switched on after DNA synthesis has been completed. *Activated cdc2 kinase then phosphorylates numerous proteins to initiate mitosis.* For example, phosphorylation of several kinds of lamins (filamentous proteins in the nuclear lamina) leads to the breakdown of the nuclear envelope. Phosphorylation also induces the reorganization of microtubules, culminating in the formation of the mitotic spindle. The condensation of chromatin may be aided by the phosphorylation of histone H1. *Cdc2 kinase also turns off its own activity by switching on an enzyme that degrades cyclin.* The generation of a new amino terminus leads to the conjugation of ubiquitin, which tags the protein for destruction (p. 943). After the kinase becomes quiescent, removal of phosphoryl groups from the lamins and other targets by phosphatases leads to the reassembly of the nuclear envelope and other structures of the interphase cell.

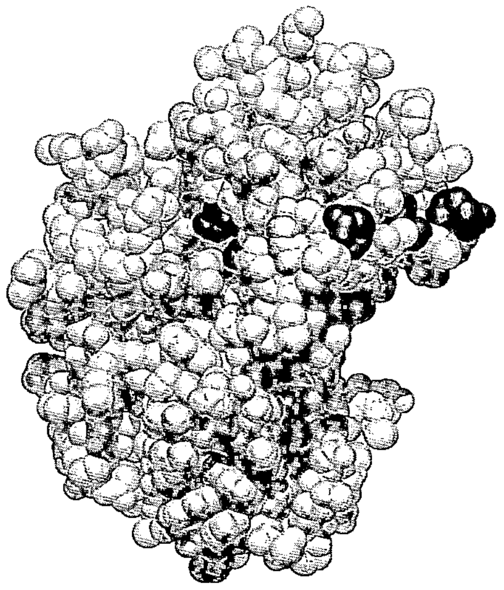

Model of human cyclin-dependent protein kinase 2 (CDK2), a regulator of the cell cycle. This kinase, which controls entry into S phase in mammals, is similar to the cdc2 kinase, which controls entry into M phase in all eukaryotes. The residues shown in blue are identical in CDK2 and cdc2. The stimulatory phosphorylation site (green) and three acidic residues (red) implicated in the binding of cyclin are also conserved. The catalytic site of this inactive form of the kinase is covered by an inhibitory domain (yellow). The bound ATP (purple) is almost completely buried. [Drawn from coordinates kindly provided by Dr. Sung-Hou Kim.]

Several features of cell-cycle regulation are noteworthy. First, *major control elements are highly conserved in eukaryotic evolution.* For example, the amino acid sequences of human and yeast cdc2 proteins are 65% identical—so similar, in fact, that a mutant yeast lacking cdc2 can be rescued by insertion of the homologous human gene. Second, cdc2 kinase plays a key role in yeast in initiating DNA synthesis as well as mitosis. In humans, a closely related protein called *cyclin-dependent protein kinase 2 (CDK2)* plays this role. Third, cdc2 kinase itself is regulated by *ordered multisite phosphorylation and dephosphorylation.* It is evident that tyrosine- and serine-threonine-specific protein kinases and phosphatases have essential roles in controlling the cell cycle. In particular, the cdc25 phosphatase assures that mitosis does not begin until all the parental DNA is replicated. Fourth, *controlled protein degradation by a ubiquitin-mediated pathway* brings mitosis to an end. Cdc2 kinase, by activating the destruction of cyclin, directs the termination of mitosis as well as its onset.

MITOCHONDRIA AND CHLOROPLASTS CONTAIN THEIR OWN DNA

Not all genetic information of eukaryotic cells is encoded by their nuclear chromosomal DNA. Genetic studies of yeast led to the discovery of a mitochondrial genome distinct from the nuclear one. In 1949, Boris Ephrussi found that some mutants of baker's yeast are unable to carry out oxidative phosphorylation. These respiration-deficient mutants grow slowly by fermentation. They are called *petites* because they form very small colonies. Genetic analyses led to the surprising finding that genes carrying petite mutations segregate independently of the nuclear genes, which suggested that mitochondria have their own genomes. Several years later, mitochondria were found to contain DNA. Furthermore, mitochondrial DNA from a petite strain had a different buoyant density from that of wild-type yeast; much of the mitochondrial genome was missing in this mutant. It was subsequently found that chloroplasts in photosynthetic eukaryotes likewise contain DNA that is replicated, transcribed, and translated. In fact, chloroplast DNA is larger and more complex than mitochondrial DNA. Mitochondria and chloroplasts import many proteins (p. 930) in addition to synthesizing proteins encoded by their own genomes. *Both organelles depend on the interplay of two distinct genomes for their formation and function.*

Human mitochondrial DNA is a double-helical circle containing 16,569 base pairs, which have been sequenced in entirety. This genome encodes 13 proteins, 22 tRNAs, and 2 rRNAs (Figure 37-20). About 60% of the protein-coding capacity of this DNA is used to specify seven subunits of NADH-Q reductase, the first of three proton-pumping complexes in the inner mitochondrial membrane (p. 534). This DNA mole-

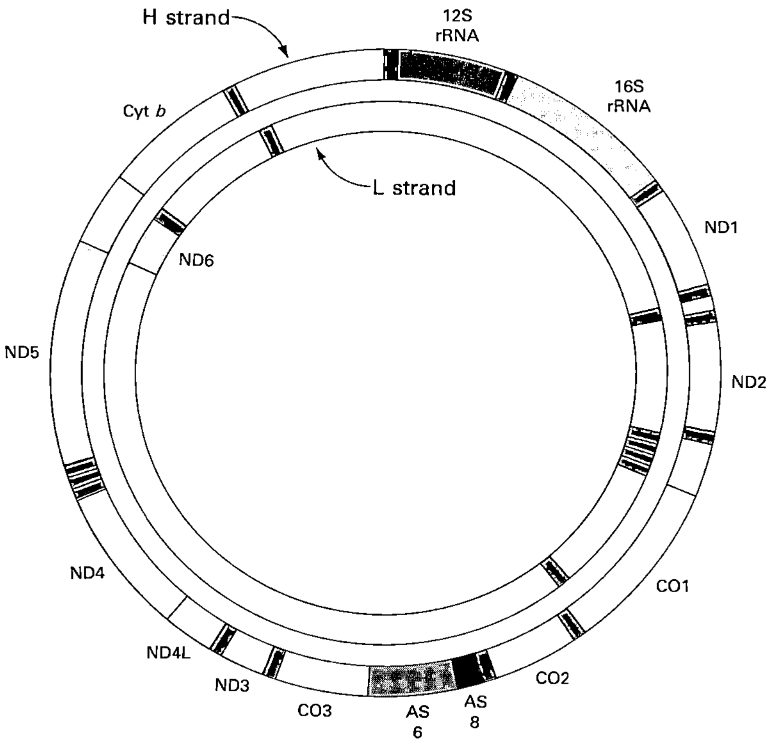

Figure 37-20
Map of the 16,569-bp human mitochondrial genome showing the genes encoded by the two strands (H and L). The ND genes (green) encode subunits of NADH-Q reductase (also called NADH dehydrogenase or Complex I); CO genes (yellow), of cytochrome oxidase; Cyt *b* gene (light yellow), the *b* component of cytochrome reductase; AS (black), of ATP synthase. The tRNA genes are shown in red, and the rRNA genes in blue.

cule also encodes a cytochrome reductase subunit, three cytochrome oxidase subunits, and two ATP synthase subunits. *A striking feature of the human mitochondrial genome is its extreme economy.* Nearly every base pair in human mitochondrial DNA encodes a protein or RNA product, and some even do double duty in that the last base of one gene serves as the first one of the next gene. The H strand (the denser of the two) encodes all RNA and protein products except for the 1 protein and 14 tRNAs that are specified by the L strand. A single primary transcript arises from the H strand, and another one from the L strand. These transcripts are cleaved on either side of tRNA sequences.

Another remarkable feature of gene expression in human mitochondria is the use of only 22 kinds of tRNAs, compared with 61 in the cytosol. Some mitochondrial tRNAs read four kinds of codons. As was mentioned in an earlier chapter (p. 111), mitochondria have a distinctive genetic code. Four codons have nonstandard meanings: AGA and AGG, which normally encode arginine, are stop codons; AUA codes for methionine instead of isoleucine; and UGA codes for tryptophan rather than being a stop signal.

HYBRIDIZATION STUDIES REVEALED THAT EUKARYOTIC DNA CONTAINS MANY REPEATED BASE SEQUENCES

Studies of the kinetics of reassociation of thermally denatured DNA by Roy Britten and his associates revealed that eukaryotic DNA, in contrast with prokaryotic DNA, contains many repeated base sequences. In these experiments, DNA was sheared into small fragments and then denatured by heating the solution above the melting temperature (T_m) of the DNA. This solution of single-stranded DNA was then cooled to about 25°C below the T_m, which is optimal for the reassociation of complementary strands. The kinetics of reassociation can be measured in a variety of ways. One technique is to follow the absorbance of the solution at 260 nm (p. 86). At this wavelength, the absorbance coefficient of double-stranded DNA is about 40% less than that of single-stranded DNA; this decrease is called *hypochromicity*. Another experimental approach is based on the fact that *double-stranded DNA binds to hydroxyapatite (calcium phosphate) columns, whereas single-stranded DNA passes through.* An attractive feature of this technique is that large amounts of DNA can be fractionated according to their rates of reassociation.

The observed kinetics of reassociation of DNA from *E. coli* or T4 phage followed the time course expected for the bimolecular reaction

$$S + S' \xrightarrow{k} D$$

in which S and S′ are complementary single-stranded molecules, D is the reassociated double helix, and k is the rate constant for association. For such a reaction, the fraction f of single-stranded molecules decreases with time according to the expression

$$f = \frac{1}{1 + kC_0t}$$

in which C_0 is the total concentration of DNA (expressed in moles of nucleotides per liter) and t is time (in seconds). For a particular DNA and set of experimental conditions (e.g., ionic strength, temperature, size of the DNA fragments), f depends only on C_0t, the product of DNA concentration and time. It is convenient to depict the kinetics of reassociation by plotting f versus the logarithm of C_0t. Such a C_0t *curve* has a sigmoidal shape (Figure 37-21).

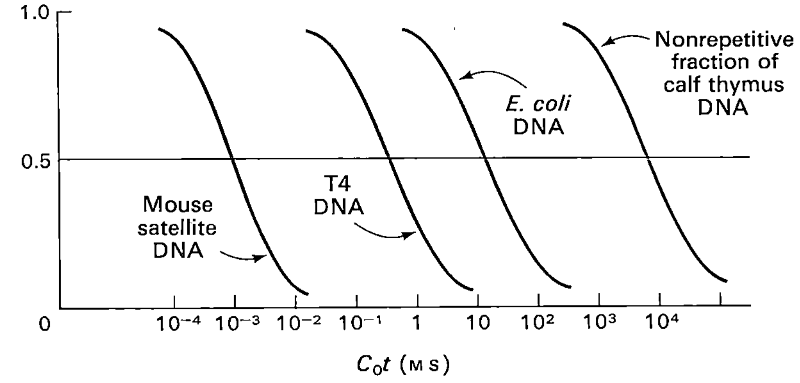

Figure 37-21
These C_0t curves depict the kinetics of reassociation of several kinds of thermally denatured DNA. The fraction of single-stranded molecules is plotted as a function of C_0t. The rapid reassociation of mouse satellite DNA shows that it contains very many repeated sequences. [After R.J. Britten and D.E. Kohne. *Science* 161(1968):530.]

An informative index of this plot is $C_0t_{0.5}$, the value of C_0t corresponding to reassociation of half the DNA ($f = 0.5$). For *E. coli* DNA, $C_0t_{0.5}$ is about 15 M s, whereas for T4 phage, $C_0t_{0.5}$ is 0.3 M s. These values indicate that *E. coli* DNA reassociates about 30 times as slowly as T4 phage DNA. The reason is that *E. coli* DNA is larger and so the number of kinds of fragments in a sheared sample is greater than for T4 DNA. Hence, *the concentration of complementary fragments is lower in a solution of sheared* E. coli *DNA compared with one of T4 DNA* (containing the same concentration of nucleotides) *and thus its rate of reassociation is slower.* Studies of a variety of prokaryotic DNA molecules showed that $C_0t_{0.5}$ is directly related to the size of the genome.

An unexpected result was obtained when mouse DNA was studied in this way. Because mammalian genomes are about three orders of magnitude larger than that of *E. coli*, a $C_0t_{0.5}$ value of the order of 10^4 M s seemed likely. A 10^{-4} M solution of DNA having such a $C_0t_{0.5}$ value would take 10^8 s (about 3 years) to reassociate halfway. It came as a surprise to find that 10% of the mouse DNA was half-reassociated in a few seconds. Indeed, *this fraction of mouse DNA reassociates more rapidly than even the smallest viral DNAs, which indicates that it contains many repeated sequences.* An analysis of the C_0t curve suggested that this rapid fraction contains *on the order of a million copies of a repeating sequence of about 300 base pairs.* About 20% of the mouse DNA renatured at an intermediate rate, pointing to the presence of moderately repetitive sequences. The other 70% renatured very slowly, suggesting that it consists of sequences that are unique or nearly so.

THE GENOMES OF HIGHER EUKARYOTES CONTAIN MANY REPETITIVE DNA SEQUENCES

More than 30% of human DNA consists of sequences repeated at least 20 times. Families of repeated DNA sequences 100 to 500 bp long that are interspersed in the genome are called SINES (*s*hort *in*terspersed repeats). Longer ones (several kb or more) are known as LINES (*l*ong *in*terspersed repeats). *Alu* sequences are especially abundant SINES. These 300-bp sequences recur nearly a million times in the human genome. In fact, most 20-kb fragments of human DNA contain an Alu sequence (the name comes from the *AluI* restriction endonuclease, which cuts at many of these sequences). The degree of sequence identity between any two members of this family is about 85%. An intriguing finding is that *Alu sequences are similar to 7SL RNA, the small cytoplasmic RNA molecule in signal recognition particle* (Figure 37-22). Recall that this assembly plays a key role in delivering nascent membrane and secretory proteins to the endoplasmic reticulum for translocation (p. 915). Alu sequences probably arose by reverse transcription of 7SL RNA and integration of this cDNA into the genome. Indeed, most SINES appear to have arisen from genes that encoded small cytosolic RNAs, such as tRNAs and 7SL RNA. The functions of SINES and LINES are an enigma.

Alu sequences—
A family of 300-bp sequences occurring nearly a million times in the human genome. Many of these sequences contain a target site for the AluI restriction endonuclease; hence their name.

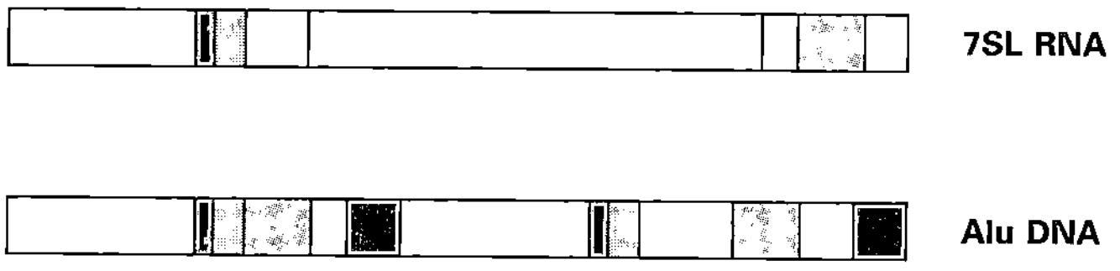

Figure 37-22
Alu sequences in human DNA are similar to portions of 7SL RNA, an abundant cytosolic RNA that is a component of signal recognition particle and participates in protein targeting. Homologous sequences are shown in red, yellow, and black. The distinctive central sequence of 7SL RNA is shown in green, and the AT-rich regions of Alu DNA in blue. Human Alu sequences are head-to-tail dimers of similar sequences. [After E. Ullu and C. Tschudi. *Nature* 312(1984):171.]

Some highly repetitive DNA is clustered rather than dispersed in the chromosome. Several such DNAs can be isolated by density-gradient centrifugation because fragments containing them have distinctive buoyant densities. For example, the sedimentation pattern of fragmented *Drosophila virilis* DNA exhibits a main band and three less dense satellite bands, each consisting exclusively of highly repetitive DNA. Each band has a distinctive repeating heptanucleotide sequence:

Satellite I	Satellite II	Satellite III
5′—ACAAACT—3′	5′—ATAAACT—3′	5′—ACAAATT—3′
3′—TGTTTGA—5′	3′—TATTTGA—5′	3′—TGTTTAA—5′

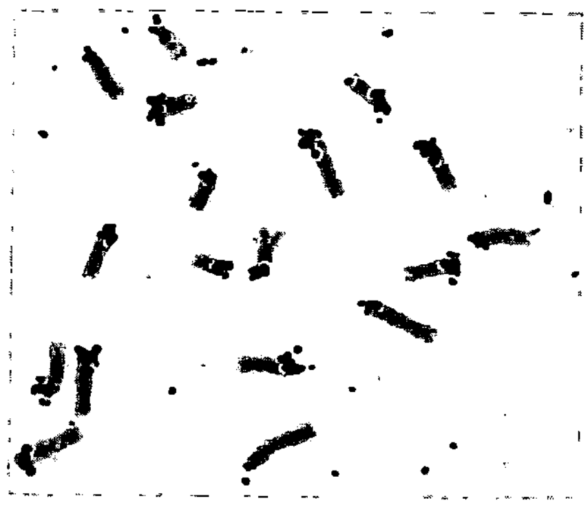

Figure 37-23
Autoradiograph of a mouse cell showing the localization of the satellite DNA at the centromeres of chromosomes. [Courtesy of Dr. Joseph Gall.]

The chromosomal localization of mouse *satellite DNA* has been determined by *in situ hybridization*, a technique devised by Joseph Gall and Mary Lou Pardue. Cells immobilized under a thin layer of agar were treated with alkali to denature their DNA. This preparation was then incubated with tritium-labeled RNA that had been transcribed in vitro from purified mouse satellite DNA. Hybrids of this radioactive RNA and chromosomal regions containing satellite DNA were detected by autoradiography. A clear-cut result was obtained: nearly all satellite DNA is located in *centromeres*, the attachment sites of mitotic spindles. It will be interesting to learn the role of this satellite DNA, which is repeated more than 10,000 times and comprises up to 5% of a chromosome.

GENES FOR RIBOSOMAL RNAs ARE TANDEMLY REPEATED SEVERAL HUNDRED TIMES

The genes coding for ribosomal RNA molecules are highly distinctive in two respects. First, they are *tandemly repeated.* Nearly all eukaryotes have more than 100 copies of these genes. Second, most rRNA genes are located in specialized regions of chromosomes that are associated with *nucleoli.* Mutants lacking nucleoli synthesize very little rRNA and hence are not viable. The genes for the four kinds of ribosomal RNA—18S, 5.8S, 28S, and 5S rRNA—have been purified from the DNA of the African clawed toads *Xenopus laevis* and *Xenopus mulleri.* These creatures were chosen because their oocytes contain many copies of these genes. Eukaryotic 18S, 28S, and 5S rRNA are homologous to prokaryotic 16S, 23S, and 5S rRNA, respectively. A homolog of eukaryotic 5.8S rRNA is present at the 5′ end of prokaryotic 23S rRNA.

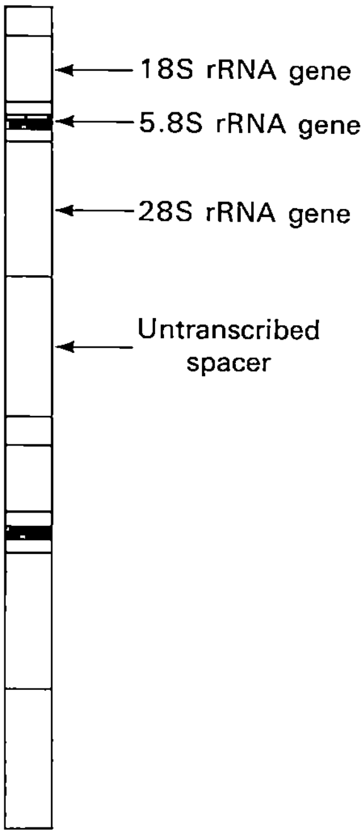

Figure 37-25
Organization of the genes for the 40S precursor of 18S, 5.8S, and 28S rRNAs in *Xenopus.* The tandemly repeated genes are separated by untranscribed spacers (yellow). The repeating unit is about 13 kb long.

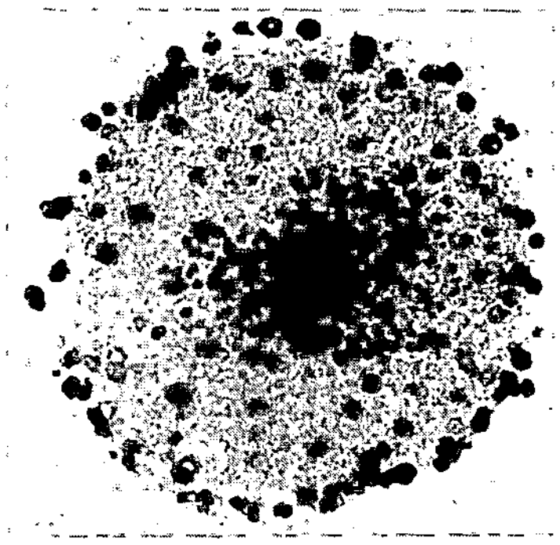

Figure 37-24
Micrograph of an isolated nucleus from a *Xenopus* oocyte stained to show the hundreds of nucleoli that are formed in the amplification of the ribosomal RNA genes. [From D.D. Brown and I.B. Dawid. *Science* 160(1968):272.]

The genes for 18S, 5.8S, and 28S rRNA are clustered and tandemly repeated, separate from the 5S rRNA genes. In situ hybridization studies showed that these three genes are located in nucleoli. Repeated clusters of them are separated by spacers that are not transcribed (Figures 37-25 and 37-26). Somatic cells of toads contain about 500 copies of this repeating unit, all in a tandem array. *In oogenesis, these genes are selectively replicated several thousand times to form some 2×10^6 copies.* Indeed, the genes for these rRNA molecules comprise nearly 75% of the total DNA in the oocyte. The amplified DNA is present as extrachromosomal circles that are bound to the very many new nucleoli. Furthermore, the 5S rRNA gene is also highly amplified; some 20,000 copies are present at the ends of several chromosomes. *Selective gene amplification* enables the oocyte to amass

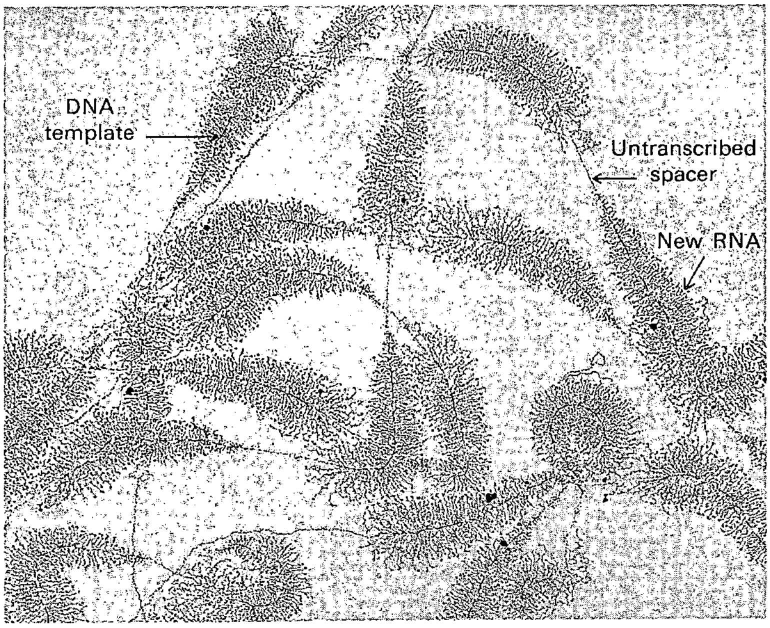

Figure 37-26
The tandem array of genes for 18S, 5.8S, and 28S rRNA is evident in this electron micrograph of nucleolar DNA. The dense axial fiber is DNA. The fine lateral fibers are newly synthesized RNA molecules with bound proteins. The tip of each RNA "arrowhead" is an initiation point for transcription. The bare regions between arrowheads are the untranscribed spacers. [From O.L. Miller, Jr., and B.R. Beatty. Portrait of a gene. *J. Cell Physiol.* 74(supp. 1, 1969):225.]

the 10^{12} ribosomes needed for the very rapid protein synthesis that occurs during cell cleavages. In the absence of gene amplification, it would take several centuries to assemble 10^{12} ribosomes!

HISTONE GENES ARE CLUSTERED AND TANDEMLY REPEATED MANY TIMES

What is the arrangement of eukaryotic genes that code for proteins? The histone genes were among the first to be isolated and characterized because of the abundance of histone mRNA in rapidly dividing sea urchin embryos. These marine invertebrates develop from a zygote to a 1000-cell blastula in about 10 hours, and so large amounts of histones must be synthesized for the assembly of new chromatin. Indeed, more than a quarter of the proteins synthesized during early embryogenesis are histones. Moreover, about 70% of the messengers synthesized at this stage are histone mRNAs, and so it was relatively easy to isolate them. The kinetics of hybridization of histone mRNA with sea urchin DNA was then measured to determine the number of copies of histone genes. *The rate of hybridization was several hundred times as fast as would be given by single-copy sequences, showing that histone genes are highly reiterated.* The number of copies of the histone genes ranges from 300 to 1000 in several species of sea urchin. In other organisms, the number is lower (Table 37-5). Yeast, for example, have only two copies of each histone gene.

The genes for the five major histones in the sea urchin are clustered in a basic 7-kb repeating unit (Figure 37-27). The five coding regions of this repeating unit alternate with five spacers. *This five-gene cluster is tandemly repeated many times.* The repetitions are very similar but not identical,

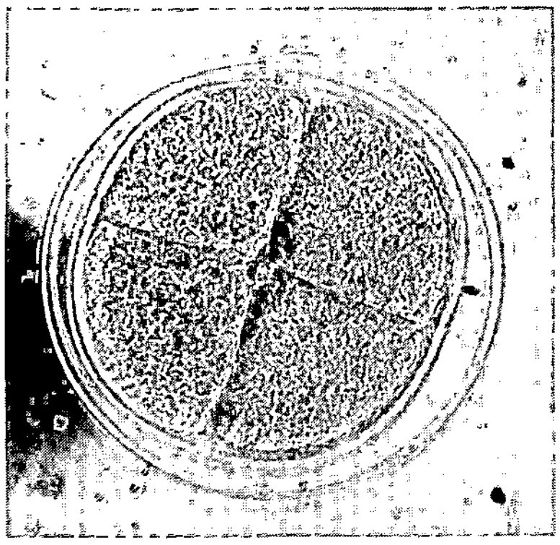

Light micrograph of a sea urchin embryo at the four-cell stage. [Courtesy of Dr. Annamma Spudich.]

Table 37-5
Reiteration frequency of the histone genes

Species	*Number of copies*
Sea urchin	300–1000
Fruit fly *(Drosophila melanogaster)*	110
Clawed toad *(Xenopus laevis)*	20–50
Human	30–40
Mouse	10–20
Chicken	10
Yeast	2

which fits the finding that H1, H2A, and H2B are groups of closely related proteins rather than unique species. Different sets of histone genes are expressed in different tissues and at different times in development. The genes for the five kinds of histones are also clustered together and tandemly repeated in the fruit fly. However, in some vertebrates, histone genes are dispersed, indicating that clustering is not essential for coordinated gene expression.

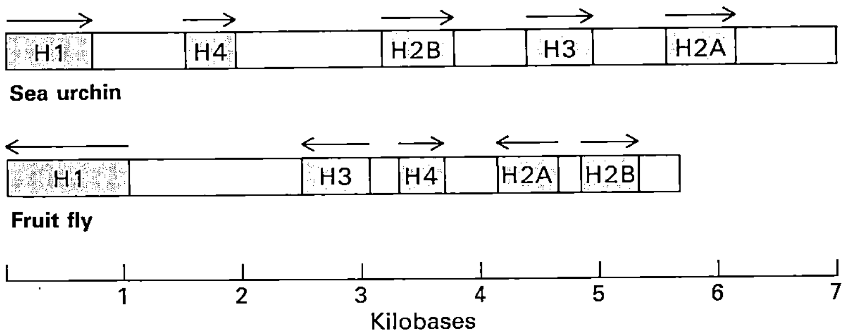

Figure 37-27
Maps of the histone gene cluster in a sea urchin (*Strongylocentrotus purpuratus*) and a fruit fly (*Drosophila melanogaster*). Coding regions are shown in blue and spacers in yellow. Arrows denote the direction of transcription.

Two additional features of histone gene expression are noteworthy. First, *histone genes lack introns.* Continuous genes are uncommon in vertebrates. Second, *histone mRNAs do not have poly(A) tails.* The absence of introns and poly(A) tails may enable histone mRNAs to be formed and transported to the cytosol very rapidly.

MANY MAJOR PROTEINS ARE ENCODED BY SINGLE-COPY GENES

A silkworm. [Courtesy of Dr. Karen Sprague.]

We have seen that the genes for ribosomal RNAs and histones are reiterated many times. Are there a large number of copies of other genes that code for abundant products? Donald Brown explored this question before the advent of recombinant DNA technology by isolating mRNA for silk fibroin (p. 30) from the silkworm *Bombyx mori.* This system was chosen because the mRNA for silk fibroin has distinctive chemical characteristics. It is the template for repeating amino acid sequences that are rich in glycine. Furthermore, very large amounts of silk fibroin are synthesized by a single type of giant cell at a particular stage of larval development. Fibroin mRNA was purified on the basis of its large size (9.1 kb) and its high content of G compared with that of rRNA and other RNA molecules. Hybridization of purified fibroin mRNA with genomic DNA then revealed that *there is only one gene for silk fibroin per haploid genome.*

This result shows that large amounts of a particular protein can be synthesized even if there is only one copy of its gene. The single gene for silk fibroin is the template for the synthesis of 10^4 mRNA molecules, which are stable for days. Each mRNA is the template for the synthesis of 10^5 proteins. Thus, *a single gene is sufficient for the synthesis of* 10^9 *protein molecules.*

The reiterated genes for ribosomal RNAs and histones are exceptions rather than the common pattern. *The single gene found for silk fibroin is much more typical of genes that code for proteins, even abundant ones.* For example, reticulocytes contain one or just a few copies of the genes for the subunits of hemoglobin (p. 996). Likewise, large amounts of ovalbumin, the major protein in egg white, are synthesized by the oviduct of a laying hen, which contains only one copy of this gene per haploid genome.

SINGLE-COPY GENES CAN BECOME HIGHLY AMPLIFIED UNDER SELECTIVE PRESSURE

Single-copy genes can become highly reiterated as a result of *gene amplification and selection.* For example, cells containing hundreds of copies of the gene for dihydrofolate reductase (DHFR), a key enzyme in the synthesis of deoxythymidylate, have been obtained by exposing them to sublethal doses of methotrexate (p. 754). The rate of *spontaneous* duplication of the DHFR gene is about 10^{-3} per division. Methotrexate selects for the few cells that contain multiple copies of this gene. In this way, a chromosomal region can acquire a tandem array of DHFR. More than 20 genes have been amplified in a similar way. The size of the repeating unit is typically 1 Mb, much larger than the gene undergoing amplification.

The discovery of gene amplification highlights the dynamic nature of the genome. Gene duplications occur frequently, and they can be stabilized by selective pressures. The work on dihydrofolate reductase has also influenced the strategy of cancer chemotherapy by revealing that drug resistance can arise by prolonged exposure of cells to sublethal doses of an inhibitory agent.

HEMOGLOBIN GENES IN TWO CLUSTERS ARE ARRANGED IN THE ORDER OF EXPRESSION DURING DEVELOPMENT

The study of hemoglobin has contributed a great deal to our understanding of protein structure and function (Chapter 7). Likewise, investigations of the genes for hemoglobin and their mode of expression have been sources of insight into eukaryotic gene function in general. For example, studies of the hybridization of β-globin mRNA with the β-globin gene contributed to the discovery that eukaryotic genes are interrupted by introns (p. 112).

As was discussed earlier (p. 154), a series of hemoglobins are synthesized in development: embryonic hemoglobins are followed by fetal and then by adult hemoglobins. Each of them consists of two α-type and two β-type chains. The α-type genes are clustered on one chromosome, and the β-type genes on another (Figure 37-28A). In the α cluster, the gene for the embryonic zeta (ζ) chain precedes two genes for α chains, which are components of both fetal and adult hemoglobins. In the β cluster, the gene for the embryonic epsilon (ε) chain is followed by two genes for fetal γ chains, and then by the genes for the adult δ and β chains. Thus, *the sequence of the human globin genes matches the order in which they are expressed in development.* It will be interesting to learn why these genes are clustered in developmental sequences. One possibility is that the $\zeta \rightarrow \alpha$ switch and the $\varepsilon \rightarrow \gamma \rightarrow \delta,\beta$ switch depend on the proximity and ordering of these genes.

The linkage of these genes also reflects their evolutionary history. The pattern of differences between their DNA sequences suggests that an ancestral hemoglobin gene duplicated and diverged to give α and β globin genes some 500 million years ago, early in the evolution of vertebrates. Myoglobin differs from both the α and β subunits of hemoglobin more than they differ from each other, indicating that myoglobin diverged before the α and β genes arose. Mammals, reptiles, birds, amphibians, and bony fish have distinct α and β subunits, whereas primitive vertebrates have only a single kind of subunit. The α and β genes later duplicated and diversified to give the two series of genes now present in humans (Figure 37-28B).

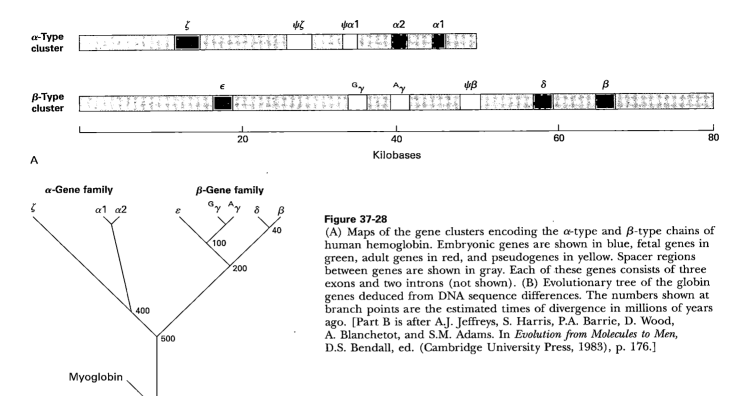

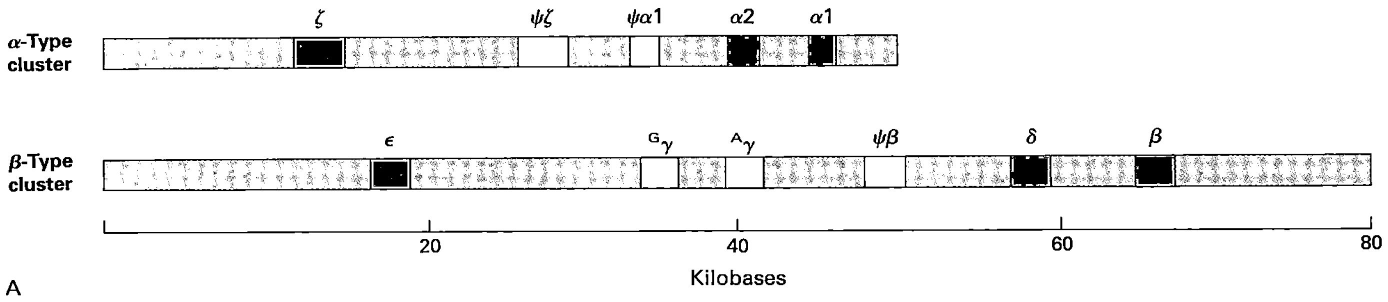

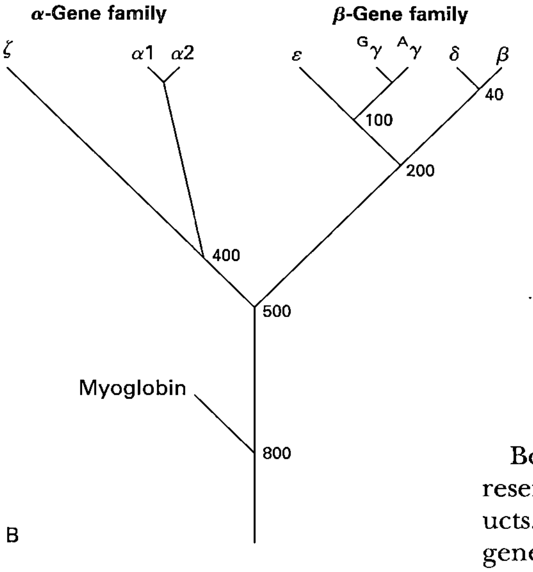

Figure 37-28
(A) Maps of the gene clusters encoding the α-type and β-type chains of human hemoglobin. Embryonic genes are shown in blue, fetal genes in green, adult genes in red, and pseudogenes in yellow. Spacer regions between genes are shown in gray. Each of these genes consists of three exons and two introns (not shown). (B) Evolutionary tree of the globin genes deduced from DNA sequence differences. The numbers shown at branch points are the estimated times of divergence in millions of years ago. [Part B is after A.J. Jeffreys, S. Harris, P.A. Barrie, D. Wood, A. Blanchetot, and S.M. Adams. In *Evolution from Molecules to Men*, D.S. Bendall, ed. (Cambridge University Press, 1983), p. 176.]

Both the α and β gene clusters contain *pseudogenes.* These sequences resemble the adjoining genes, but they do not code for functional products. The prefix ψ denotes a pseudogene; ψα1, for example, is a pseudogene resembling the α1 gene. Several structural features of this pseudogene account for its inactivity: (1) it lacks the 5′ consensus sequence for splicing; (2) the polyadenylation signal has mutated from AATAAA to AATGAA; (3) the normal ATG initiation codon is replaced by GTG; (4) a 20-nucleotide deletion starting at residue 38 generates three stop codons that prevent completion of the polypeptide chain. How were pseudogenes generated? Some are reverse transcripts of cDNAs that have made their way into the genome. Others are relics of genes that arose by duplication but have drifted in sequence and become inactive. We see in eukaryotic genomes both the mighty and the fallen.

Table 37-6
Classes of eukaryotic DNA

Protein-coding genes
Single-copy genes
Duplicated genes
RNA-coding genes
Most are tandemly duplicated
Pseudogenes
Repetitive DNA
Simple-sequence DNA (as in satellite DNA)
Dispersed repetitive DNA (includes mobile genetic elements)
Spacer DNA of unknown function

After J. Darnell, H. Lodish, and D. Baltimore. *Molecular Cell Biology*, 2nd ed. (Scientific American Books, 1990), p. 348.

ONLY A SMALL PART OF A MAMMALIAN GENOME CODES FOR PROTEINS

The functional globin genes in the α and β clusters are quite far from one another. The clusters occupy about 35 and 60 kb, respectively, which corresponds to about 12 kb per gene, but the amount of DNA needed to encode an ~150-residue globin chain is only 0.45 kb. Thus, less than 4% of these clusters codes for protein. Furthermore, the globin genes are far from the nearest gene encoding another protein. *Perhaps only 2% of mammalian DNA actually codes for proteins.* Thus, mammalian DNA is strikingly different from bacterial DNA, which is very compact and largely dedicated to specifying proteins. *The human genome is about a thousand times as large as the* E. coli *genome, but the number of genes is more likely to be fiftyfold rather than a thousandfold greater.* One of the great challenges in molecular biology today is to unravel the function of the vast majority of DNA, the more than three billion base pairs in the human genome that have no known encoding function.

TRANSCRIPTIONALLY ACTIVE REGIONS OF CHROMOSOMES ARE UNDERMETHYLATED AND HYPERSENSITIVE TO DNase I

Gene expression in eukaryotes, as in prokaryotes, is controlled to a large extent by the pattern of transcription. Some of the earliest evidence came from studies of the chromosomes of developing insects. The giant *polytene chromosomes* from *Drosophila* salivary glands contain more than a thousand replicated DNA molecules aligned side by side that have not separated. Each polytene chromosome has a characteristic series of bands (chromomeres) that can be seen under a light microscope. The *Drosophila* genome contains about 5000 bands, each containing, on average, fewer than 10 genes. In the development of a larva into a pupa, *certain bands become transiently enlarged (puffed) because the chromatin in that region has been converted from a condensed form into a dispersed one* (Figure 37-29).

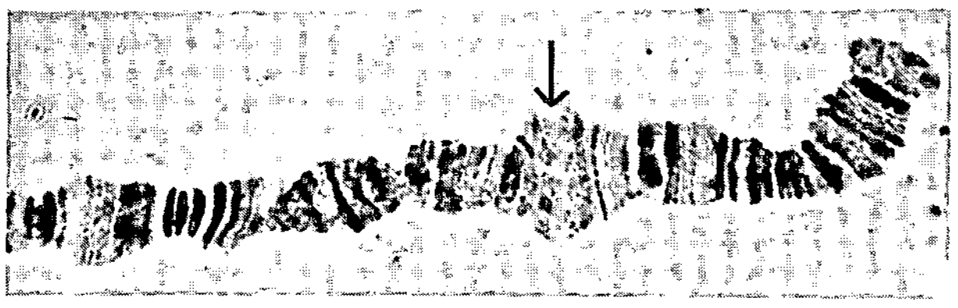

Figure 37-29
Puff formation in a polytene chromosome during development. The arrow marks a prominent puff in a *Drosophila* chromosome. [Courtesy of Dr. Joseph Gall.]

Puffs correspond to transcriptionally active regions. A striking finding is that puffs can be induced in isolated salivary glands by *ecdysone,* an insect steroid hormone. Specific bands enlarge and then contract in a precise temporal sequence, and there are concomitant changes in the population of mRNAs being synthesized. Some bands can be puffed either by developmental signals or by heat shock, but the batteries of genes activated by these stimuli are different. Puffing is necessary but not sufficient for gene activation.

Changes in the structure and accessibility of DNA induced by gene activation can be probed using nucleases. The susceptibility of DNA to nucleases is not uniform. DNA packaged in nucleosomes is digested by low concentrations of DNase I or micrococcal nuclease into fragments that are multiples of 200 base pairs (see Figure 37-6). In contrast, *transcriptionally active regions of DNA are digested by DNase I into much smaller and irregular pieces.* These susceptible regions are rather large. For example, in the globin genes they extend some 7 kb beyond the 5′ and 3′ boundaries of the α and β clusters. Enhanced susceptibility to digestion implies that transcriptionally active regions have few nucleosomes or nucleosomes that are altered in structure. Covalent modifications of histones, such as phosphorylation and ubiquitination, may be important in this regard because they alter the affinity of histones for DNA. Moreover, some sites on DNA become exquisitely sensitive to cleavage by DNase I. Most of these *hypersensitive sites* are located at the 5′ ends of transcribed genes. Naked DNA is not hypersensitive to digestion by DNase I. Rather, hypersensitivity is conferred by the binding of specific regulatory proteins to target sites.

Enhanced sensitivity and hypersensitivity to digestion are tissue-specific and developmentally regulated. For example, globin genes in the precursors of erythroid cells are insensitive to DNase I in 20-hour-old chicken embryos. However, when hemoglobin synthesis begins at 35 hours, these genes become highly susceptible to digestion. In contrast, globin genes in the brain, which are not transcribed, are resistant to DNase I throughout

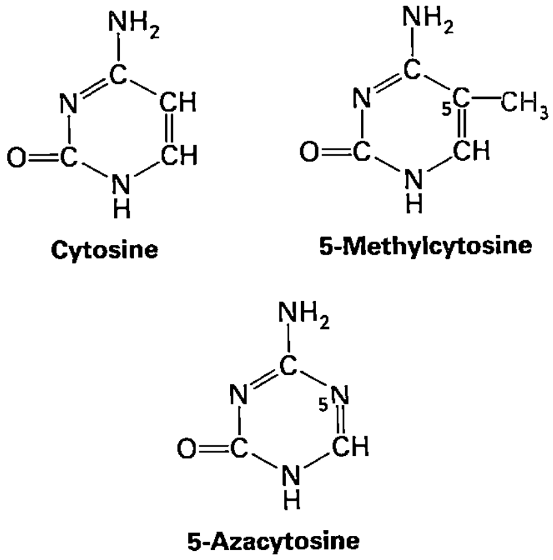

Figure 37-30
Formulas of cytosine, 5-methylcytosine, and 5-azacytosine (the pyrimidine base of 5-azacytidine). Transcriptionally inactive regions of DNA have a high content of 5-methylcytosine. 5-Azacytosine incorporated into DNA cannot be methylated at N-5 by the methyltransferase.

development and in adulthood. Hypersensitivity is also correlated with active expression of specific genes in an array. The 5′ ends of γ, δ, and β genes in fetal erythroid cells are hypersensitive to DNase I, whereas only the δ and β genes are hypersensitive in adult bone marrow cells.

The degree of methylation of cytosines in DNA is also correlated with gene activity. C-5 of cytosine can be methylated by specific methyltransferases (Figure 37-30). About 70% of C-G sequences in mammalian DNA are methylated. Methylation patterns can be assessed by comparing Southern blots of DNA cleaved by a pair of restriction endonucleases. HpaII cleaves CCGG but not C^{m}CGG sequences (^{m}C denotes a methylated cytosine), whereas MspI cleaves both. *Studies using these enzymes have shown that globin genes are extensively methylated in nonerythroid cells but much less so in erythroid cells.* The γ genes, for example, contain a CCGG at position −54, between the TATA box and an upstream promoter site. Methylation of a cytosine in this sequence places a protruding methyl group in the major groove of DNA, which could interfere with the binding of a factor that stimulates transcription. The importance of methylation is reinforced by the finding that methylation of a recombinant globin gene blocks its expression in transfected cells. Specifically, methylation of the −760 to +100 region, but not other parts of the gene, interferes with transcription.

Base substitution provides further evidence for the significance of methylation in gene regulation. Incorporating *5-azacytidine* (*5-azaC*, which contains N-5 in place of C-5) (see Figure 37-30) into DNA prevents methylation and induces differentiation—that is, activation of genes that would otherwise be quiescent. In general, *active mammalian genes are less methylated than inactive ones.* However, it should be noted that methylation of DNA is not a universal regulatory device in higher eukaryotes. *Drosophila* DNA, for example, is not methylated at all.

TRANSCRIPTION IN EUKARYOTES IS CONTROLLED BY THE COMBINATORIAL ASSOCIATION OF MULTIPLE PROTEINS

Gene expression in eukaryotes, as in prokaryotes, is controlled primarily at the level of transcription. Regulation in eukaryotes is necessarily more complex than in prokaryotes because their genomes are much larger and many more genes must be coordinately controlled. *Most genes in eukaryotes are silent unless they are specifically turned on. Nucleosomes must be cleared from the start site to allow transcription to begin.* Eukaryotic RNA polymerases, in contrast with prokaryotic ones, cannot transcribe DNA on their own. A multisubunit transcriptional apparatus must assemble first at the TATA box of the gene (p. 855). *Activators and repressors of eukaryotic gene expression act by altering the rate of formation of this transcriptional complex.* A second difference between prokaryotes and eukaryotes is that *most eukaryotic genes are controlled by multiple proteins rather than by just one or two.* The serum albumin gene in mice, for example, contains six binding sites for regulatory proteins near its TATA box. Additional control sites are present as far as several kilobases away.

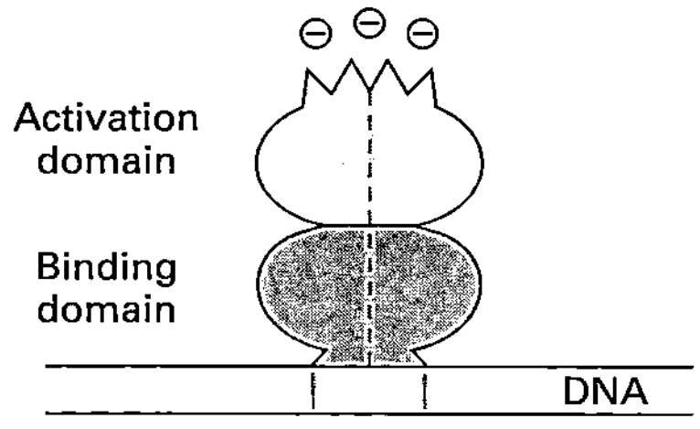

Figure 37-31
Transcriptional activators contain a DNA-binding domain (red) and an activation domain (blue). The specific DNA site is shown in green.

Transcriptional activators contain two essential regions: a *DNA-binding domain* that recognizes a particular DNA sequence (or set of closely related sequences) and an *activation domain* that aids the assembly of the transcription complex at the TATA box (Figure 37-31). The striking finding is that activation domains are functional when moved from one DNA-binding protein to another. Indeed, yeast activation domains are effective in human cells. A common feature of many activation domains is the presence of net negative charge; their targets probably contain a complementary positively charged region. Other activation domains are rich in glutamine or proline. Activation domains may interact with components of the transcription complex or with an adapter protein that links the two.

We have seen that proteins bound to DNA sequences several kilobases away from the TATA box can regulate transcription because DNA is a flexible molecule that can be looped (see Figure 33-28 on p. 856). Looping enables multiple regulatory proteins to interact with the transcriptional complex. Indeed, *most eukaryotic genes are switched on by the action of several activators rather than by a single one acting alone. Different combinations of activators switch on different sets of genes.*

TANDEM ZINC FINGERS REGULATE GENE EXPRESSION BY BINDING TO EXTENDED DNA SEQUENCES

Recall that the helix-turn-helix (HTH) motif is found in many proteins that control prokaryotic gene expression (p. 963). Likewise, recurring modules are present in eukaryotic gene regulators. The *zinc finger,* a major eukaryotic motif, was discovered in studies of the transcriptional control of 5S ribosomal RNA genes in *Xenopus.* A sequence essential for the transcription of 5S genes by RNA polymerase III was identified by deleting parts of the gene. The surprising finding was that deletions upstream of the start site and those downstream of the termination site had little effect on the number of transcripts formed. Rather, a 45-nucleotide region in the *center* of the gene proved to be critical for efficient transcription. This *internal control region* is recognized by a 40-kd protein called *transcription factor IIIA (TFIIIA).* One molecule of TFIIIA binds to a 5S RNA gene to form a complex that sequentially binds TFIIIC, TFIIIB, and RNA polymerase III. Factors B and C also serve to initiate transcription of tRNA genes, whereas TFIIIA is specific for 5S RNA genes.

TFIIIA has a remarkable structure. It contains nine similar domains, each about 30 residues long (Figure 37-32). Each repeat contains *two*

A

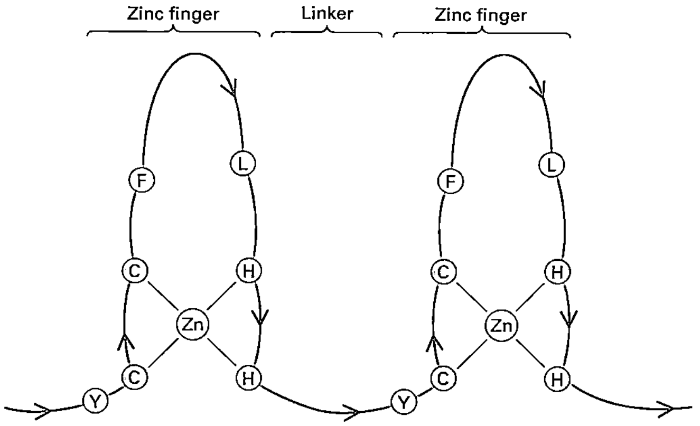

B

1 Y I C S F A D C G A A Y N K N W K L Q - A H L C - K H
2 T G E K - P F P C K E E G C E K G F T S L H H L T - R H S L - T H
3 T G E K - N F T C D S D G C D L R F T T K A N M K - K H F N R F H
4 N I K I C V Y V C H F E N C G K A F K K H N Q L K - V H Q F - S H
5 T Q Q L - P Y E C P H E G C D K R F S L P S R L K - R H E K - V H
6 A G - - - - Y P C K K D D S C S F V G K T W T L Y L K H V A E C H
7 Q D - - - L A V C - - D V C N R K F R H K D Y L R - D H Q K - T H
8 E K E R T V Y L C P R D G C D R S Y T T A F N L R - S H I Q S F H
9 E E Q R - P F V C E H A G C G K C F A M K K S L E - R H S V - V H

Figure 37-32
Zinc finger motif. (A) Schematic diagram showing a pair of fingers joined by a linker. (B) The amino acid sequence of transcription factor IIIA (TFIIIA) from *Xenopus* displays nine zinc finger sequences. Zinc is coordinated to a pair of cysteines (green) and histidines (blue). Three hydrophobic residues (yellow) in each repeat are also conserved. [After D. Rhodes and A. Klug. *Sci. Amer.* 268(Feb. 1993):58.]

cysteines and *two histidines* in identical locations. These invariant residues are tetrahedrally coordinated to a *zinc ion.* In addition, three hydrophobic residues are conserved. *Each 30-residue repeating unit is folded into an elongated zinc-binding domain having the shape of a finger*—hence the name *zinc finger.* An antiparallel β hairpin (residues 1 to 10) is followed by a turn and then by an α helix (residues 12 to 25) (Figure 37-33). The zinc atom, buried in the interior, stabilizes the module by binding a pair of cysteines from the β hairpin and a pair of histidines from the α helix.

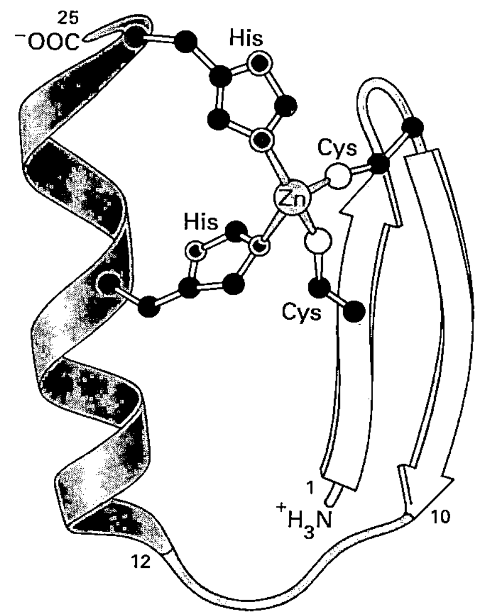

Figure 37-33
Schematic diagram of a zinc finger. An antiparallel β hairpin (green) and loop (gray) are followed by an α helix (red) that plays a key role in the recognition of DNA. A zinc ion is bound to two histidine residues from the helix and two cysteine residues from the β sheet. [After C. Branden and J. Tooze. *Introduction to Molecular Structure.* (Garland, 1991), p. 117.]

X-ray crystallographic studies have provided an informative view of how a zinc finger grips a specific region of DNA. The three zinc fingers of *Zif268,* a transcription factor that participates in the early development of mice, bind to the major groove of DNA and wrap part of the way around the double helix (Figure 37-34). Each finger makes contact with a three base-pair subsite of the DNA. Residues from the amino-terminal part of the α helix form hydrogen bonds with the exposed bases in the major groove. In particular, arginine-guanine interactions play a key role in the specific binding of Zif268 to DNA. Multiple residues of each finger also make contact with the sugar-phosphate backbone of DNA.

Zinc finger proteins have been found in all eukaryotes studied, from yeast to humans. They have the common sequence motif

$$X_3\text{-}\boxed{\text{Cys}}\text{-}X_{2\text{-}4}\text{-}\boxed{\text{Cys}}\text{-}X_{12}\text{-}\boxed{\text{His}}\text{-}X_{3\text{-}4}\text{-}\boxed{\text{His}}\text{-}X_4\text{-}$$

where X can be any amino acid. The conserved structural features are the tetrahedral coordination of zinc by two Cys and two His, and the presence of an antiparallel β hairpin followed by an α helix, the principal recognition element. Zinc fingers are autonomous folding units: an isolated zinc finger folds into a compact domain. Hence, this motif is highly versatile. Indeed, the number of zinc fingers in proteins analyzed thus far ranges from 1 to 37. Extended DNA sequences can be specifically recognized simply by bringing together a series of zinc fingers that have different residues in their α helices. Thus, *arrays of zinc fingers are well suited for combinatorial recognition of DNA sequences.*

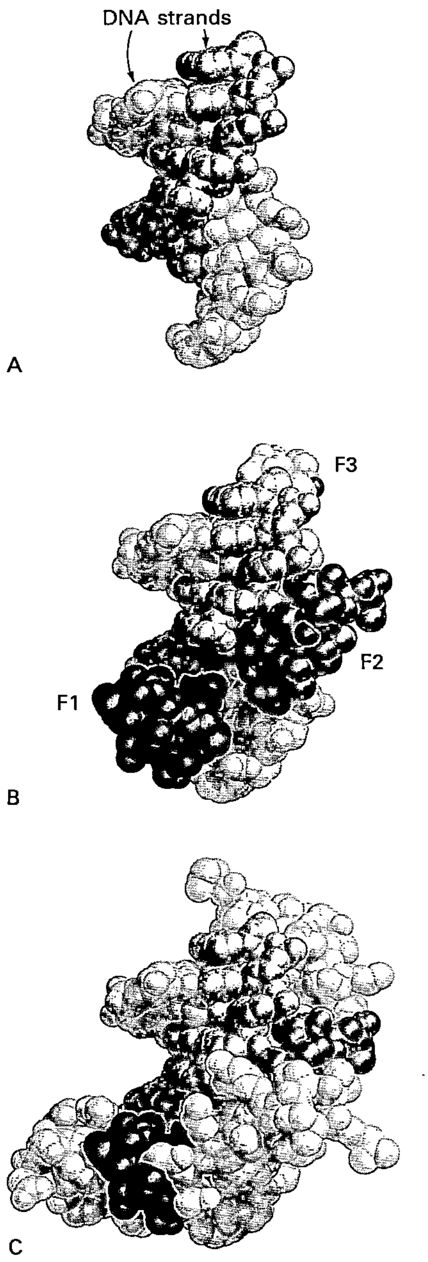

Figure 37-34
Three-dimensional structure of a zinc finger protein (mouse Zif268) bound to DNA. The two strands of DNA are shown in shades of green. The recognition helices of Zif268 are shown in red, blue, and yellow, the other parts of the protein in white, and the zinc ions in purple. (A) DNA alone.
(B) DNA with bound recognition helices and zinc ions. F1, F2, and F3 mark the three fingers. (C) Entire protein-DNA complex. The three zinc fingers of this protein are arranged in a semicircle that wraps partway around the DNA double helix by fitting snugly in the major groove. [Drawn from 1zaa.pdb. N.P. Pavletich and C.O. Pabo. *Science* 252(1991):809.]

STEROID HORMONES AND MORPHOGENS ACTIVATE DNA-BINDING PROTEINS THAT CONTROL TRANSCRIPTION

Many hormones and growth factors exert their biological effects by activating cell-surface receptors. Recall that epinephrine, for example, activates the adenylate cyclase cascade by binding to a seven-helix receptor in the plasma membrane (p. 341). Glucagon, insulin, and epidermal growth factor likewise trigger intracellular responses by switching on receptors in the plasma membrane. *Steroid hormones, by contrast, must enter target cells to act.* Hormones such as estradiol, progesterone, testosterone, and cortisol traverse the plasma membrane and bind first to specific receptor proteins in the cytosol. These hormone-receptor complexes then migrate to the nucleus, where they bind to specific sites on DNA. Each induces or represses the transcription of a particular set of 50 to 100 genes. *Steroid hormones primarily alter the pattern of gene expression rather than the activity of a particular enzyme or membrane transporter.* The full impact of steroids is evident in hours rather than in seconds or minutes because new mRNAs and new proteins must be synthesized.

Receptors for several steroid hormones were purified by taking advantage of their high affinity for ligand ($K_d \sim 1$ nM). Antibodies specific for these receptors were then used to screen cDNA expression libraries. The sequencing and expression of positive cDNAs revealed that steroid hormone receptors such as those for estrogens, progesterone, and glucocorticoids have common motifs (Figure 37-35). They contain a conserved 66-residue *DNA-binding domain* and a conserved 240-residue *hormone-binding domain.* An isolated DNA-binding domain will interact with cognate DNA sequences irrespective of whether or not hormone is present. *The role of the hormone-binding domain in an intact receptor is to prevent the DNA-binding domain from interacting with DNA unless hormone is bound.* The amino-terminal domain, which is not conserved, enables a receptor to interact with other transcriptional regulators.

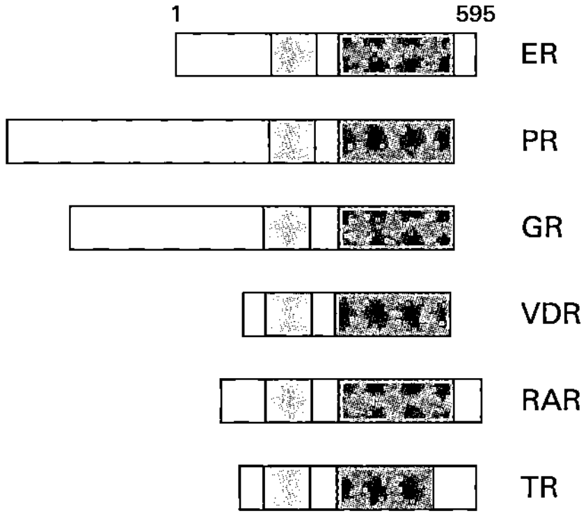

Figure 37-35
Domain structure of the nuclear receptor superfamily. The highly variable activation domain is shown in yellow, the highly conserved DNA-binding domain in blue, and the hormone-binding domain in red. Receptors for steroids, vitamin D_3, retinoic acid, and thyroxine belong to this superfamily of transcriptional regulators. Abbreviations used: ER, estrogen receptor; PR, progesterone receptor; GR, glucocorticoid receptor; VDR, vitamin D_3 receptor; RAR, retinoic acid receptor; TR, thyroxine receptor.

Activated receptors bind to specific DNA sequences called *hormone response elements (HREs).* These elements, like other eukaryotic transcriptional control sites, regulate the expression of nearby genes. HREs for steroid receptors are palindromes consisting of a pair of 6-bp sequences separated by a 3-bp spacer. The glucocorticoid response element differs from the estrogen response element in having TT instead of AC at positions +5 and +6 of the palindrome (Figure 37-36). The DNA-binding domains of the glucocorticoid and estrogen receptors are able to distinguish between these elements because they differ at three key residues.

5'—N A G A A C A N N N T G T T C T N—3'
3'—N T C T T G T N N N A C A A G A N—5'
Glucocorticoid response element (GRE)

5'—N A G G T C A N N N T G A C C T N—3'
3'—N T C C A G T N N N A C T G G A N—5'
Estrogen response element (ERE)

Figure 37-36
Target DNA sequences that are recognized by several nuclear receptors. These hormone response elements are palindromes. The twofold axis of symmetry relating 6-bp sequences is shown in green.

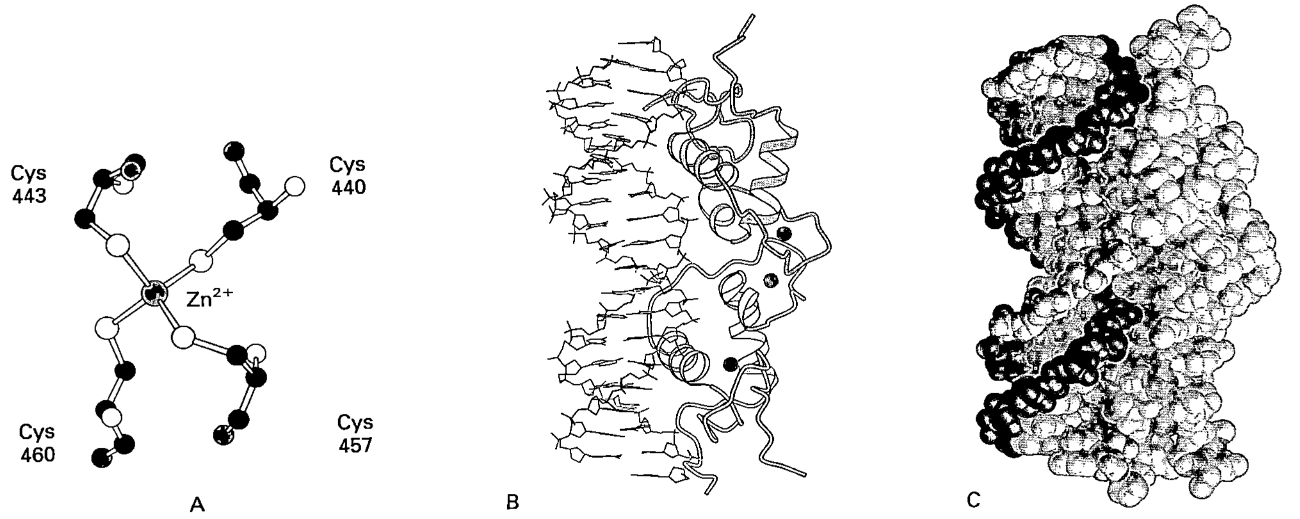

Figure 37-37
The dimeric glucocorticoid receptor recognizes a palindromic DNA sequence. (A) Each polypeptide chain of the receptor is stabilized by a pair of zinc clusters. One of them is depicted here. A Zn^{2+} ion is coordinated to four cysteines. (B) Skeletal model showing the binding of the DNA-binding modules of the dimeric protein (yellow and blue; Zn^{2+} in purple) to a palindromic DNA target sequence (red and green). (C) Space-filling model. Each recognition helix fits snugly in a widened major groove. [Drawn from 1glu.pdb B.F. Luisi, W.X. Xu, Z. Otwinski, L.P. Freedman, K.R. Yamamoto, and P.B. Sigler. *Nature* 352(1991):497.]

The DNA-binding domain of the glucocorticoid receptor complexed to a DNA target has been visualized by x-ray crystallography (Figure 37-37). The domain dimerizes on binding to the palindromic response element. As anticipated, the symmetry of the dimer matches that of its DNA target. Each protein subunit contains two *zinc clusters* in which a zinc ion is tetrahedrally coordinated to *four cysteine side chains.* The secondary and tertiary structures of this DNA-binding domain are different from those of a zinc finger. The domain is globular rather than elongated. *The two subunits of a steroid receptor recognize the same DNA sequence, whereas adjacent zinc fingers act independently and can read quite different DNA sequences.* The major groove of DNA widens by 2 Å to accommodate a recognition helix from each domain of the dimer.

Vitamin D_3, a steroid derivative (p. 707), acts by binding to a nuclear receptor that is similar to those for adrenal steroids and steroid sex hormones. Vitamin D_3 plays a key role in the control of calcium metabolism and the differentiation of bone. The receptors for *thyroxine* and *retinoic acid* also belong to this *nuclear receptor superfamily* (see Figure 37-35). This finding was totally unexpected because these hormones are structurally and biosynthetically unrelated to the steroids. Thyroxine is an iodinated derivative of tyrosine and retinoic acid is formed by oxidation of vitamin A (retinol, p. 333).

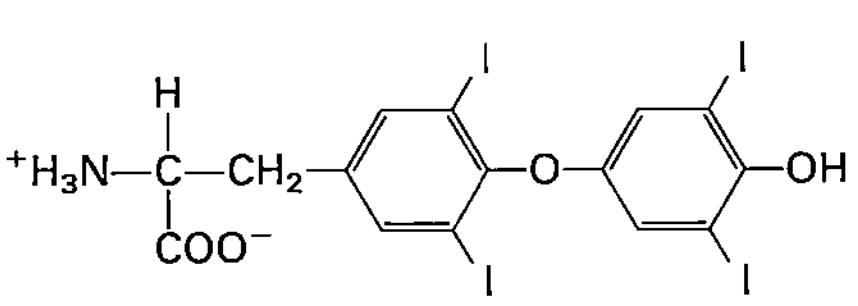

Thyroxine
(L-3,5,3′,5′-Tetraiodothyronine)

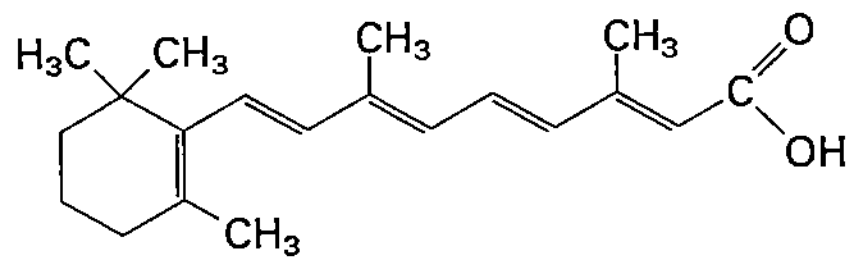

All-*trans*-retinoic acid

Model of thyroxine. The four iodine atoms of this thyroid hormone are shown in purple. Thyroxine is formed by the iodination and joining of tyrosine residues in thyroglobulin. Proteolysis of this precursor gives rise to thyroxine.

Both act as *morphogens*—substances that induce differentiation or control its pattern. Thyroxine induces the metamorphosis of tadpoles into frogs, and retinoic acid establishes the anterior-posterior axis in developing limbs. The nuclear receptor superfamily is present in insects and arthropods as well as vertebrates. In *Drosophila,* the steroid hormone ecdysone triggers the metamorphosis of larvae into adult flies. The ecdysone receptor of *Drosophila* is strikingly similar to vertebrate receptors for steroids, thyroxine, and retinoic acid.

LEUCINE ZIPPER PROTEINS CONTAIN AN α-HELICAL COILED COIL AND A PAIR OF DNA-BINDING DOMAINS

A third major class of eukaryotic transcriptional regulators are the *leucine zipper proteins.* The hallmark of these proteins is the presence of leucine at every seventh position in a stretch of ~35 residues. This regularity suggested that they form an α-helical coiled coil like that found in the rod region of myosin (p. 395). Indeed, *leucine zipper proteins form dimers held together by an α-helical coiled coil* (Figure 37-38). The two strands run in the same direction, as in myosin. A coiled coil has 3.5 residues per turn, which means that every seventh residue occupies an equivalent position with respect to the helix axis. The regular array of leucines inside the coiled coil stabilizes the structure by hydrophobic and van der Waals interactions, hence the name of this motif.

The sequences of these proteins revealed another common feature, the presence of an ~30-residue basic region at their amino-terminal ends. This region serves as a *DNA-binding module.* It becomes α-helical on binding to DNA. *The role of the leucine zipper region is to bring together a pair of DNA-binding modules to bind two adjacent DNA sequences.* The coiled coil forms the stem of a Y, and the basic regions are the arms of the Y (Figure 37-39). The α helix in each basic region is likely to be bent to enable it to wrap around the major groove of DNA.

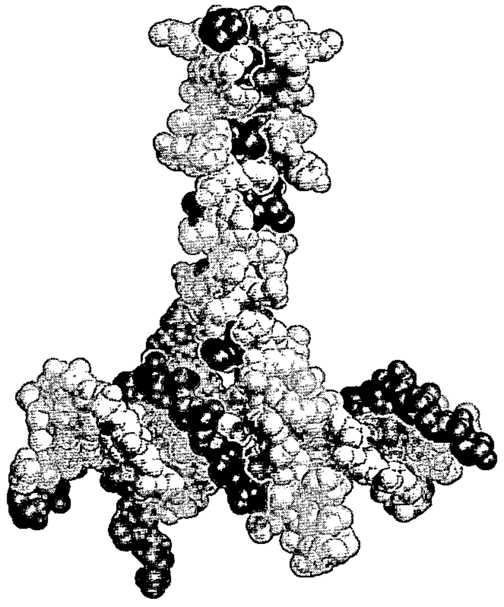

Figure 37-38
Structure of the α-helical coiled coil region of GCN4, a leucine zipper protein, bound to DNA. The zipper is stabilized by hydrophobic and van der Waals interactions between leucines (purple) at every seventh residue of the two parallel strands (blue and yellow). The two strands of the DNA are shown in red and green. [Drawn from 1ysa.pdb. T.E. Ellengerger, C.J. Brandl, K. Struhl, and S.C. Harrison. *Cell* 71(1992):1223.]

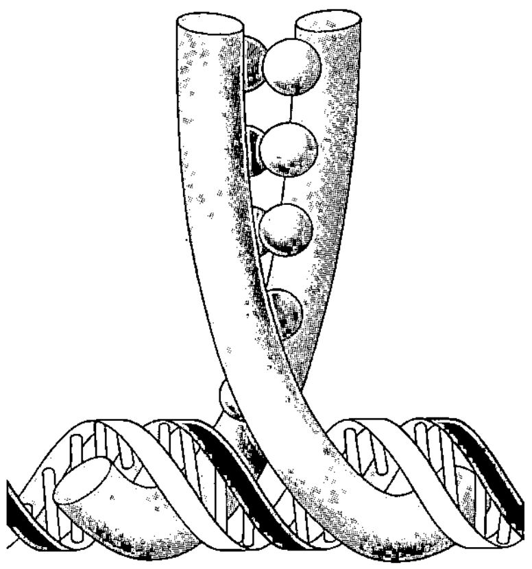

Figure 37-39
Schematic diagram showing the proposed mode of binding of a leucine zipper protein to a palindromic target DNA sequence. [After C.R. Vinson, P.B. Sigler, and S.L. McKnight. *Science* 246(1989):911.]

Yeast cells respond to amino acid starvation by synthesizing *GCN4,* a leucine zipper protein. *GCN4 coordinately activates the transcription of more than 40 genes that encode biosynthetic enzymes.* This dimeric activator binds to 9-bp palindromic sequences that are upstream of genes for amino acid biosynthesis. The acidic transcriptional activation domain of GCN4 is separate from its basic DNA-binding module and leucine zipper (Figure 37-40). When bound to DNA, the activation domain interacts with transcription factor IID (TFIID) at the TATA box to initiate transcription (p. 855).

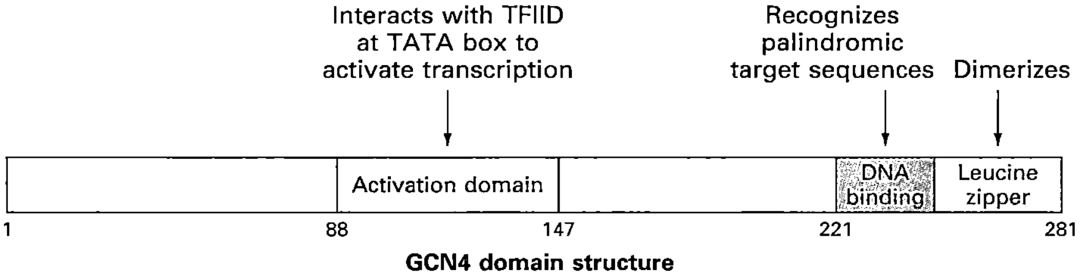

Figure 37-40
Domain structure of GCN4, a transcriptional activator in yeast. GCN4 stimulates the expression of genes that encode biosynthetic enzymes for amino acids.

5′—T G A C G T C A—3′
3′—A C T G C A G T—5′

Cyclic AMP response element (CRE)

In higher eukaryotes, *leucine zipper proteins mediate the effect of cyclic AMP (cAMP) on transcription.* Genes under the control of cAMP contain a cAMP response element (CRE), a palindromic 8-bp DNA sequence. A 43-kd protein called the *cAMP response element binding protein (CREB)* binds to this DNA target sequence. The dimerization of CREB through its C-terminal leucine zipper region brings together a pair of basic DNA-binding modules that are positioned to bind the palindromic CRE. How does cAMP control the transcriptional activity of CREB? Recall that cAMP, formed by the action of a G protein cascade, activates protein kinase A (PKA), a multifunctional kinase (p. 245). *Phosphorylation of CREB by PKA promotes dimerization and increases its efficacy as a transcriptional activator.* CREB is a target for the Ca^{2+}–calmodulin-activated CaM kinase II (p. 349), protein kinase C (p. 345), and other kinases, which suggests that it integrates many signals.

Leucine zippers can dimerize identical or nonidentical chains. GCN4 and CREB are examples of *homodimers.* The mammalian transcription factor *AP-1* (activator protein 1), by contrast, is a group of *heterodimers.* The Jun and Fos proteins, first identified as the products of the *jun* and *fos* oncogenes, can associate through their leucine zipper regions to form a potent transcriptional regulator that recognizes AP-1 target sites in DNA. In fact, the Jun-Fos heterodimer is much more stable than either the Jun-Jun or Fos-Fos homodimer. Why is the heterodimer preferred? The coiled coil is stabilized not only by the packing of leucines in the interior. Polar residues at positions *e* and *g* of the heptad repeat (*abcdefg*) are also important (see Figure 16-30 on p. 435). The heterodimer is stabilized by salt bridges between oppositely charged side chains at these external positions, whereas the homodimers are destabilized by electrostatic repulsions between side chains having the same charge. Indeed, *many different heterodimers of leucine zipper proteins are formed.* The ATF-CREB family, for example, contains 13 members that can dimerize with one another to form 13 homodimers and 78 heterodimers. The leucine zipper family graphically illustrates the use of *combinatorial association* in generating diverse regulatory proteins that can read many different DNA sequences.

THE HOMEO BOX IS A RECURRING MOTIF IN GENES CONTROLLING DEVELOPMENT IN INSECTS AND VERTEBRATES

Insect bodies are divided into segments that are formed early in embryonic development. *Drosophila* contains three thoracic segments and eight abdominal ones (Figure 37-41). Cells of different segments usually retain their identity and do not intermix. However, certain mutations, called *homeotic mutations,* transform one body part into another. The Antennapedia (Antp) mutation, for example, changes an antenna into a leg. It can

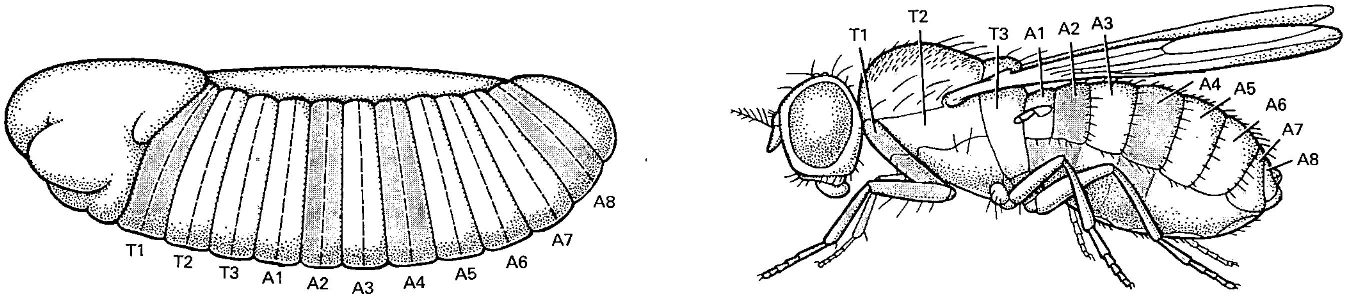

Figure 37-41
Segmentation pattern of a *Drosophila* embryo and adult fly. Segmentation is controlled by homeotic genes.

be inferred that the normal $Antp^{+}$ gene determines that legs are formed in the right place. In general, *homeotic genes control the fundamental architectural plan of the embryo.* The two main clusters of homeotic genes are the Antennapedia complex (ANT-C), which regulates the head and anterior thoracic segments, and the Bithorax complex (BX-C), which regulates the posterior thoracic segments and abdominal segments.

These large clusters of homeotic genes have been explored by David Hogness and Walter Gehring using the technique of chromosome walking (p. 132). ANT-C is longer than 100 kb, and BX-C is longer than 310 kb. Hybridization studies revealed that many genes in these clusters contain a common sequence. The recurring motif proved to be a 180-bp sequence, which was called the *homeo box.* It encodes a 60-residue protein *homeodomain* (Figure 37-42). The high content of basic residues suggested that the homeodomain binds to DNA. Proteins containing the homeodomain are in fact localized in the nucleus and bind to specific DNA sequences.

Figure 37-42
The homeodomain is a recurring motif encoded by genes controlling development. Sequences from mammals, amphibians, insects, and yeast are similar. The homologous sequences compared here are from the (A) mouse *MO*-10 gene, (B) frog *MM*3 gene, (C) *Drosophila Antp* gene, and (D) yeast *MAT*α*2* gene. Residues shaded green are highly conserved in all homeodomains.

Helix 1 Helix 2 Helix 3

A S K R G R T A Y T R P Q L V E L E K E F H F N R Y L M R P R R V E M A N L L C L T E R Q I K I W F Q N R R M K Y K K D N
B R K R G R Q T Y T R Y Q T L E L E K E F H F N R Y L T R R R R I E I A H V L C L T E R Q I K I W F Q N R R M K W K K E N
C R K R G R Q T Y T R Y Q T L E L E K E F H F N R Y L T R R R R I E I A H A L C L T E R Q I K I W F Q N R R M K W K K E N
D K P Y R G H R F T K E N V R I L E S W F A K N P Y L D T K G L E N L M K N T S L S R I Q I K N W V S N R R R K E K T I T

Indeed, *homeodomains serve as regulators of gene expression in all eukaryotes studied, from yeast to humans.* Nuclear magnetic resonance and x-ray crystallographic studies of homeodomains from *Drosophila* and yeast show that they contain an extended amino-terminal arm and three α helices (Figure 37-43A). Helices 1 and 2 are antiparallel to each other, and nearly perpendicular to helix 3. The hydrophobic face of helix 3 packs against helices 1 and 2 to form the interior of the protein. Homeodomains bind as monomers to specific sites on DNA. Their amino-terminal arm fits into the minor groove, whereas helix 3 fits into the major groove (Figure 37-43B). The arrangement of the second and third helices in the homeodomain is similar to that of the *helix-turn-helix (HTH) motif* seen in prokaryotic gene regulators such as the λ repressor (p. 963).

Figure 37-43
Structure of engrailed, a homeodomain protein. (A) Ribbon drawing of the homeodomain motif. (B) Schematic diagram showing the binding of engrailed to a 21-bp duplex DNA (red and green strands). Helix 3 (purple) of the homeodomain occupies the major groove of DNA, whereas the N-terminal arm (yellow) binds to the minor groove. (C) Space-filling model of the engrailed-DNA complex. [Drawn from 1hdd.pdb. C.R. Kissinger, B. Liu, E. Martin-Blanco, T.B. Kornberg, and C.O. Pabo. *Cell* 63(1990):579.]

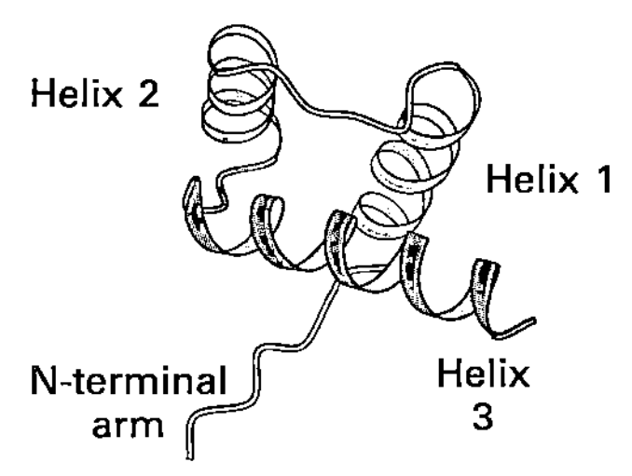

A

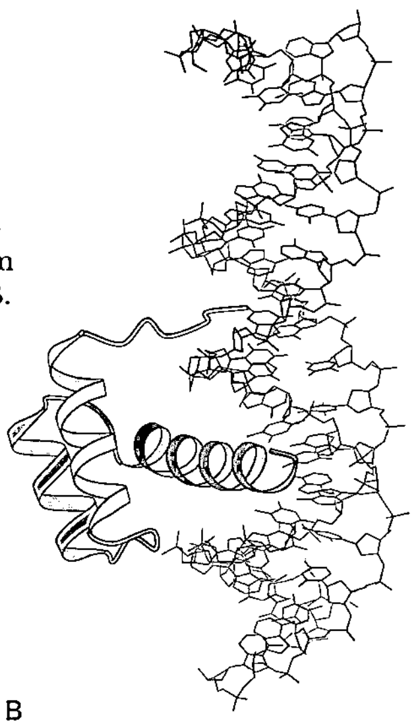

B

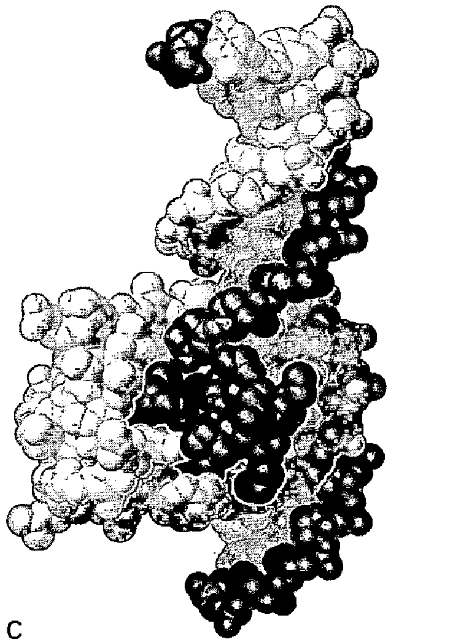

C

The conservation of homeodomains in evolution is striking. Vertebrate homeodomain genes, called *HOX genes*, are similar to their homologs in *Drosophila*. In mice and humans, homeo box genes are arranged in four clusters—*HOX1, HOX2, HOX3,* and *HOX4*—that are found on different chromosomes. The homeodomain encoded by the mouse HOX 1.1 gene differs from that encoded by fly *Antp* by just one amino acid residue. Furthermore, mammalian HOX genes are functional in flies. Even more remarkable, ***the sequence of homeo box genes in both flies and mammals corresponds precisely to their order of action along the anterior-posterior axis of the embryo*** (Figure 37-44). Homeo box genes specifying head development are at one end of the cluster, and those directing the formation of posterior structures (the abdomen in flies and lumbar vertebrae in mammals) are at the other end. It is evident that a fundamental mechanism of development, ***the specification of segment identity***, has been preserved over 600 million years of evolution. We are beginning to decipher how the one-dimensional information contained in the sequence of DNA is converted into the intricate and beautiful three-dimensional form of multicellular organisms.

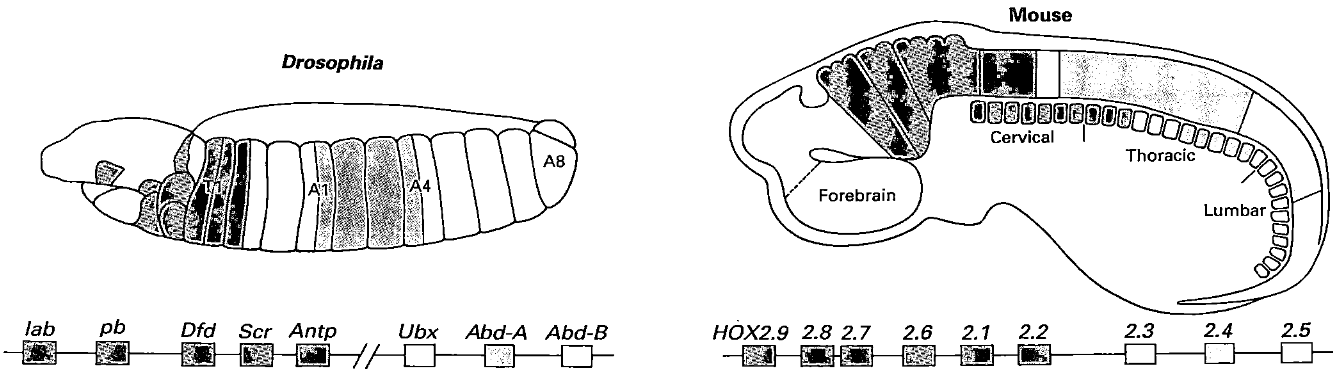

Figure 37-44
The arrangement of homeo box genes and the sequences of their encoded proteins are highly conserved in evolution. The left side of the figure shows the order of genes in the Antennapedia complex and the Bithorax complex of *Drosophila* and the anterior boundary of expression of each gene in the fly embryo. The right side shows the order of mouse HOX2 genes and their pattern of expression. The colors indicate the correspondences between fly and mouse genes and their patterns of expression. [After E.B. Lewis. *J. Amer. Med. Assoc.* 267(1992):1524.]

SUMMARY

A eukaryotic chromosome contains a single linear molecule of double-helical DNA. The largest chromosome of *Drosophila* contains 76×10^6 base pairs (76 Mb), 19 times the size of the entire *E. coli* genome. The human genome contains nearly a thousand times as much DNA as that of *E. coli.* Eukaryotic DNA is tightly bound to basic proteins called histones; the combination is called chromatin. The chromatin fiber is a flexibly jointed chain of nucleosomes. The core of these repeating units consists of 140 base pairs of DNA wound around the outside of a histone octamer made up of two each of histones H2A, H2B, H3, and H4. Nucleosome cores are joined by linker DNA, typically 60 base pairs long, which binds one molecule of H1. A sevenfold reduction in the length of DNA is achieved by organization into nucleosomes, which are the first stage in the condensation of DNA. Higher-order structures lead to the >100-fold compaction of interphase chromosomes, and 10^4-fold compaction of metaphase chromosomes.

Eukaryotic DNA is replicated semiconservatively and semidiscontinuously from several thousand origins. A large number of initiation points are necessary because of the great length of eukaryotic DNA. Old histones stay with the DNA duplex containing the leading strand, whereas new histones assemble on the DNA duplex containing the lagging strand. Telomeres (the ends of chromosomes) are replicated by a reverse transcriptase that carries an RNA template. Telomerase synthesizes tandem hexanucleotide repeats rich in G to complete the synthesis of the lagging strand. In the cell cycle of eukaryotes, DNA is replicated during the S (synthesis) phase, and the replicated chromosomes are segregated into daughter nuclei during the M (mitosis) phase. Entry into mitosis is controlled by a protein kinase (cdc2 kinase) that is activated by cyclin B, a protein that is alternately synthesized and degraded during the cell cycle. The catalytic activity of cdc2 kinase also depends on its phosphorylation state, which is controlled by growth factor stimuli.

Mitochondria and chloroplasts contain their own DNA, which encodes a small portion of the proteins in these organelles. The human mitochondrial genome is a 16.6-kb double-helical circular DNA molecule. Nearly all the DNA is used to encode proteins, tRNAs, and rRNAs. Several codons in mitochondria have meanings different from those in the cytosol.

The kinetics of reassociation of thermally denatured DNA revealed that eukaryotic DNA contains many repeated base sequences. The human genome, for example, contains nearly a million copies of highly homologous 300-bp Alu sequences. They are dispersed through the genome and may serve as initiation sites in DNA replication. A different class of highly repetitive DNA, called satellite DNA, is localized at centromeres. In *Drosophila,* this DNA consists of a heptanucleotide sequence repeated more than 10,000 times. The genes for 5.8S, 18S, and 28S rRNA are clustered and tandemly repeated many times. These ribosomal RNA genes are located in nucleoli. The genes for histones are also clustered and tandemly repeated many times. In contrast, other abundant proteins, such as silk fibroin, hemoglobin, and ovalbumin, are encoded by genes that are present once (or a few times) per genome. Single-copy genes, such as the one for dihydrofolate reductase, can become highly amplified under selective pressure, as when cells are exposed to an inhibitor of the gene product.

Our understanding of the fundamental architecture of eukaryotic genomes and the regulation of their expression is at an early stage. Only a small proportion (perhaps as little as 2%) of the human genome codes for proteins. The function of most of the DNA is an enigma. Intensive study of a number of gene clusters, such as the ones encoding the α-type and β-type genes of hemoglobin, is proving to be highly rewarding. Control is mediated largely at the level of transcription. Actively transcribed regions of DNA are more susceptible to digestion by DNase I than are inactive regions, and fewer of their cytosine bases are methylated. These differences are tissue-specific and developmentally regulated.

Most eukaryotic genes are not expressed unless they are activated by the binding of specific proteins, called transcription factors, to sites on DNA. These factors activate transcription by promoting the assembly of the large multisubunit initiation complex at the TATA box. The looping of DNA enables multiple transcription factors to act at distances up to several kilobases from the start site of transcription. Several DNA-binding motifs recur in a large number of transcription factors. The zinc finger, a 30-residue elongated unit held together by the coordination of a zinc ion to two cysteine and two histidine side chains, is ubiquitous in eukaryotes.

Tandem repeats of zinc fingers (nine in transcription factor IIIA, for example) recognize extended DNA sequences. A second major motif is found in nuclear receptors, which mediate the effects of hormones and morphogens such as steroids, vitamin D, thyroxine, retinoic acid, and ecdysone. The binding of hormone to the hormone-binding domain of these receptors allows their DNA-binding domain to interact with cognate DNA sequences. Recognition is largely mediated by the interaction of an α helix of each subunit of the dimer with the major groove of DNA. Helix–major groove interactions are important in the recognition of DNA by most transcriptional regulators.

A third major class of eukaryotic transcriptional regulators are the leucine zipper proteins, which are dimers held together by an α-helical coiled coil that is stabilized by the presence of leucine at every seventh residue. The function of the leucine zipper is to bring together a pair of basic DNA-binding modules. Many different homodimers and heterodimers of leucine zipper proteins can be formed from a smaller repertoire of subunits. Homeodomain proteins contain a different DNA-binding module—the extended amino-terminal arm and three α helices are reminiscent of the helix-loop-helix motif seen in prokaryotic gene regulators. Homeodomains, which are encoded by homeo boxes, were first found in *Drosophila,* where mutations in homeotic genes transform one body part to another. Indeed, the homeo box is a recurring motif in genes that specify segment identity in insects and vertebrates. The presence of the homeo box in the genes of both amphibians and mammals points to an underlying unity in the molecular basis of development.

SELECTED READINGS

Where to start

Kornberg, R.D., and Klug, A., 1981. The nucleosome. *Sci. Amer.* 244(2):52–64.

Ptashne, M., 1989. How gene activators work. *Sci. Amer.* 260(1):40–47.

Rhodes, D., and Klug, A., 1993. Zinc fingers. *Sci. Amer.* 268(2):56–65.

Gehring, W.J., Affolter, M., and Bürglin, T., 1994. Homeodomain proteins. *Ann. Rev. Biochem.* 63:487–526.

Murray, A.W., and Kirschner, M.W., 1991. What controls the cell cycle? *Sci. Amer.* 264(3):56–63.

Books

Cold Spring Harbor Laboratory Symposia on Quantitative Biology, 1993. *DNA and Chromosomes.* Volume 58. [A rich and stimulating series of articles presented at a symposium that celebrated the 40th anniversary of the discovery of the DNA double helix.]

Darnell, J., Lodish, H., and Baltimore, D., 1990. *Molecular Cell Biology* (2nd ed.). Scientific American Books. [Chapters 9, 10, and 11 deal with eukaryotic genomes and gene regulation.]

Singer, M., and Berg, P., 1991. *Genes and Genomes.* University Books. [Part III deals with the molecular anatomy and expression of eukaryotic genes.]

Watson, J.D., Gilman, M., Witkowski, J., and Zoller, M., 1992. *Recombinant DNA* (2nd ed.). Scientific American Books. [Chapters 18, 19, and 20 deal with oncogenes, cell-cycle control, and the regulation of development in *Drosophila.*]

Wolffe, A., 1992. *Chromatin Structure and Function.* Academic Press.

Van Holde, K.E., 1989. *Chromatin.* Springer-Verlag.

Murray, A., and Hunt, T., 1993. *The Cell Cycle.* W.H. Freeman.

McKnight, S.L., and Yamamoto, K.R. (eds.), 1992. *Transcriptional Regulation,* vols. 1 and 2. Cold Spring Harbor Laboratory Press. [Contains many excellent articles on eukaryotic gene regulation. Chapters 20 to 22 of vol. 1 and most of the contents of vol. 2 pertain to eukaryotes.]

Ptashne, M., 1991. *A Genetic Switch. Phage λ and Higher Organisms* (2nd ed.). Cell Press and Blackwell Scientific Publications. [A lucid and concise account of the principles of transcriptional regulation in eukaryotes is given in Chapter 6.]

Branden, C., and Tooze, J., 1991. *Introduction to Protein Structure.* Garland. [Chapter 8 contains many informative and attractive drawings of the structure of eukaryotic transcriptional regulators.]

Conaway, R.C., and Conaway, J.W. (eds.), 1994. *Transcription: Mechanisms and Regulation.* Raven Press.

Parker, M.G. (ed.), 1991. *Nuclear Hormone Receptors: Molecular Mechanism, Cell Function, and Clinical Abnormalities.* Academic Press.

Nucleosomes and chromatin

Arents, G., Burlingame, R.W., Wang, B.C., Love, W.E., and Moudrianakis, E.N., 1991. The nucleosomal core histone octamer at 3.1 Å resolution: A tripartite protein assembly and a left-handed superhelix. *Proc. Nat. Acad. Sci.* 88:10148–10152.

Stewart, A., 1990. The functional organization of chromosomes and the nucleus—a special issue. *Trends Genet.* 6:377–379.

Workman, J.L., and Buchman, A.R., 1993. Multiple functions of nucleosomes and regulatory factors in transcription. *Trends Biochem. Sci.* 18:90–95.

Selker, E.U., 1990. DNA methylation and chromatin structure: A view from below. *Trends Biochem. Sci.* 15:103–107.

Kornberg, R.D., and Lorch, Y., 1992. Chromatin structure and transcription. *Ann. Rev. Cell Biol.* 8:563–587.

Struck, M.M., Klug, A., and Richmond, T.J., 1992. Comparison of x-ray structures of the nucleosome core particle in two different hydration states. *J. Mol. Biol.* 224:253–264.

DNA replication in eukaryotes

Kornberg, A., and Baker, T.A., 1992. *DNA Replication* (2nd ed.). W.H. Freeman. [Chapter 6 deals with eukaryotic DNA polymerases.]

Coverley, D., and Laskey, R.A., 1994. Regulation of eukaryotic DNA replication. *Ann. Rev. Biochem.* 63:745–776.

DePamphilis, M.L., 1993. Origins of DNA replication in metazoan chromosomes. *J. Biol. Chem.* 268:1–4.

Pelletier, H., Sawaya, M.R., Kumar, A., Wilson, S.H., and Kraut, J., 1994. Structures of ternary complexes of rat DNA polymerase β, a DNA template-primer, and ddCTP. *Science* 264:1891–1903.

Blackburn, E.H., 1992. Telomerases. *Ann. Rev. Biochem.* 61:113–129.

Blackburn, E.H., 1991. Structure and function of telomeres. *Nature* 350:569–573.

Williamson, J.R., 1994. G-quartet structures in telomeric DNA. *Ann. Rev. Biophys. Biomol. Struct.* 23:703–730.

Cell-cycle control

Norbury, C., and Nurse, P., 1992. Animal cell cycles and their control. *Ann. Rev. Biochem.* 61:441–470.

Hunt, T., and Kirschner, M., 1993. Cell multiplication. *Curr. Opin. Cell. Biol.* 5:163–165.

De Bondt, H.L., Rosenblatt, J., Jancarik, J., Jones, H.D., Morgan, D.O., and Kim, S.-H., 1993. Crystal structure of cyclin-dependent kinase 2. *Nature* 363:595–602.

Glotzer, M., Murray, A.W., and Kirschner, M.W., 1991. Cyclin is degraded by the ubiquitin pathway. *Nature* 349:132–138.

Mitochondrial DNA

Grivell, L.A., 1983. Mitochondrial DNA. *Sci. Amer.* 248(3): 78–89.

Borst, P., and Grivell, L.A., 1981. Small is beautiful—portrait of a mitochondrial genome. *Nature* 290:443–444.

Wallace, D.C., 1992. Diseases of the mitochondrial DNA. *Ann. Rev. Biochem.* 61:1175–1212.

Mechanism of transcriptional regulation

Pabo, C.O., and Sauer, R.T., 1992. Transcription factors: Structural families and principles of DNA recognition. *Ann. Rev. Biochem.* 61:1053–1095.

Suzuki, M., 1993. Common features in DNA recognition helices of eukaryotic transcription factors. *EMBO J.* 12:3221–3226.

Harrison, S.C., 1991. A structural taxonomy of DNA-binding domains. *Nature* 353:715–719.

Pugh, B.F., and Tjian, R., 1992. Diverse transcriptional functions of the multisubunit eukaryotic TFIID complex. *J. Biol. Chem.* 267:679–682.

Lin, Y.S., Carey, M., Ptashne, M., and Green, M.R., 1990. How different eukaryotic transcriptional activators can cooperate promiscuously. *Nature* 345:359–361.

McKinney, J.D., and Heintz, N., 1991. Transcriptional regulation in the eukaryotic cell cycle. *Trends Biochem. Sci.* 16:430–435.

Zinc finger proteins

Berg, J.M., 1990. Zinc fingers and other metal-binding domains. *J. Biol. Chem.* 265:6513–6516.

Pavletich, N.P., and Pabo, C.O., 1991. Zinc finger-DNA recognition: Crystal structure of a Zif268-DNA complex at 2.1 Å. *Science* 252:809–817.

Schule, R., and Evans, R.M., 1991. Cross-coupling of signal transduction pathways: Zinc finger meets leucine zipper. *Trends Genet.* 7:377–381.

Nuclear receptors for hormones and morphogens

Tsai, M.-J., and O'Malley, B.W., 1994. Molecular mechanisms of action of steroid/thyroid receptor superfamily members. *Ann. Rev. Biochem.* 63:451–486.

Schwabe, J.W.R., and Rhodes, D., 1991. Beyond zinc fingers: Steroid hormone receptors have a novel structural motif for DNA recognition. *Trends Biochem. Sci.* 16:291–296.

Luisi, B.F., Xu, W.X., Otwinowski, Z., Freedman, L.P., Yamamoto, K.R., and Sigler, P.B., 1991. Crystallographic analysis of the interaction of the glucocorticoid receptor with DNA. *Nature* 352:497–505.

Koelle, M.R., Talbot, W.S., Segraves, W.A., Bender, M.T., Cherbas, P., and Hogness, D.S., 1991. The *Drosophila EcR* gene encodes an ecdysone receptor, a new member of the steroid receptor superfamily. *Cell* 67:59–77.

Pearce, D., and Yamamoto, K.R., 1993. Mineralocorticoid and glucocorticoid receptor activities distinguished by nonreceptor factors at a composite response element. *Science* 259:1161–1165.

Leucine zipper proteins

McKnight, S.L., 1991. Molecular zippers in gene regulation. *Sci. Amer.* 264(4):54–64.

Lamb, P., and McKnight, S.L., 1991. Diversity and specificity in transcriptional regulation: The benefits of heterotypic dimerization. *Trends Biochem. Sci.* 16:417–422.

Alber, T., 1992. Structure of the leucine zipper. *Curr. Opin. Genet. Develop.* 2:205–210.

O'Shea, E.K., Rutkowski, R., and Kim, P.S., 1992. Mechanism of specificity in the Fos-Jun oncoprotein heterodimer. *Cell* 68:699–708.

Cyclic AMP activation of transcription

Brindle, P.K., and Montminy, M.R., 1992. The CREB family of transcription activators. *Curr. Opin. Genet. Develop.* 2:199–204.

Foulkes, N.S., Borrelli, E., and Sassone, C.P., 1991. CREM gene: Use of alternative DNA-binding domains generates multiple antagonists of cAMP-induced transcription. *Cell* 64:739–749.

Sheng, M., Thompson, M.A., and Greenberg, M.E., 1991. CREB: A Ca^{2+}-regulated transcription factor phosphorylated by calmodulin-dependent kinases. *Science* 252:1427–1430.

Homeo box and homeotic genes

Wolberger, C., Vershon, A.K., Liu, B., Johnson, A.D., and Pabo, C.O., 1991. Crystal structure of a MAT alpha 2 homeodomain-operator complex suggests a general model for homeodomain-DNA interactions. *Cell* 67:517–528.

Kissinger, C.R., Liu, B.S., Martin, B.E., Kornberg, T.B., and Pabo, C.O., 1990. Crystal structure of an engrailed homeodomain-DNA complex at 2.8 Å resolution: A framework for understanding homeodomain-DNA interactions. *Cell* 63:579–590.

Furukubo, T.K., Flister, S., and Gehring, W.J., 1993. Functional specificity of the Antennapedia homeodomain. *Proc. Nat. Acad. Sci.* 90:6360–6364.

Andres, A.J., and Thummel, C.S., 1992. Hormones, puffs and flies: The molecular control of metamorphosis by ecdysone. *Trends Genet.* 8:132–138.

Lewis, E.B., 1991. Clusters of master control genes regulate the development of higher organisms. *J. Amer. Med. Assoc.* 267:1524–1531.

Tabin, C.J., 1992. Why we have (only) five fingers per hand: Hox genes and the evolution of paired limbs. *Development* 116:289–296.

Edelman, G.M., and Jones, F.S., 1993. Outside and downstream of the homeobox. *J. Biol. Chem.* 268:20683–20686.

PROBLEMS

1. *Eons apart.* Compare prokaryotic and eukaryotic gene expression in regard to:
   (a) Degree of coupling of transcription and translation.
   (b) Number of gene products on a primary transcript.
   (c) Number of proteins arising from the translation of a primary transcript.
   (d) Density of coding sequences in DNA.
   (e) Organization of genes into operons.

2. *YACs.* Synthetic yeast chromosomes can be constructed by the joining of three kinds of DNA elements. What are they?

3. *Tolerance.* The error rate of the mitochondrial DNA polymerase is much higher than that of the nuclear DNA polymerase. Why can lower fidelity be tolerated in the mitochondrial enzyme?

4. *Rapid evolution.* In the course of evolution, genes encoding mitochondrial proteins probably moved from the mitochondrial to the nuclear genome. Human mitochondrial DNA, for example, is fivefold smaller and simpler than yeast mitochondrial DNA. Different sets of mitochondrial proteins are encoded by mitochondrial DNA in the two species. Hence, there seems to be no structural reason why a mitochondrial protein must be synthesized within the organelle rather than be imported. Do you expect the human mitochondrial genome to be even shorter 10 million years from now? Which aspect of mitochondrial gene expression would impede the transfer of a mitochondrial gene to the nuclear genome?

5. *The sense of the supercoil.* Design an experiment to determine whether DNA coiled around histones forms a left-handed or a right-handed superhelix.

6. *Dual allegiance.* Transcription factor IIIA (TFIIIA) binds to 5S RNA as well as to the internal control region of the gene encoding it. Why? How might this binding of the gene product be used to regulate expression of the 5S genes?

7. *Tenacity.* TFIIIA stays on a 5S ribosomal RNA gene while it is being transcribed. How might TFIIIA remain bound in the face of an advancing RNA polymerase?

8. *Seven but not six.* Recall that yeast cells respond to amino acid starvation by synthesizing GCN4, a leucine zipper protein. The insertion of six amino acid residues at the amino-terminal junction of the zipper sequence destroys the capacity of GCN4 to activate transcription of amino acid biosynthetic genes. In contrast, seven residues can be inserted with retention of activity. Account for this difference. Predict the effects of inserting four and five residues.

9. *Replacing the leucine zipper.* Design an experiment to test the hypothesis that the sole purpose of the leucine zipper is to bring two basic DNA-binding regions together in an appropriate spatial relationship for interaction with the target DNA sequence.

10. *Crafting a repressor.* Describe a simple method for converting a transcriptional activator into a transcriptional repressor.

11. *Never nuclear.* A mutant estrogen receptor binds hormone but is unable to move from the cytosol to the nucleus. Propose a structural basis for this defect.

12. *Identifying target genes.* Suppose that you have cloned a new cDNA that encodes a protein resembling members of the nuclear receptor superfamily. The ligand that activates this new receptor is unknown. How would you identify the target of this receptor and the battery of genes that it activates?

Appendixes
Answers to Problems
Index

A

Physical Constants and Conversion of Units

Values of physical constants

Physical constant	*Symbol*	*Value*
Atomic mass unit (dalton)	amu	1.660×10^{-24} g
Avogadro's number	N	6.022×10^{23} mol^{-1}
Boltzmann's constant	k	1.381×10^{-23} J deg^{-1} 3.298×10^{-24} cal deg^{-1}
Electron volt	eV	1.602×10^{-19} J 3.828×10^{-20} cal
Faraday constant	F	9.649×10^{4} C mol^{-1} 2.306×10^{4} cal volt^{-1} eq^{-1}
Curie	Ci	3.70×10^{10} disintegrations s^{-1}
Gas constant	R	8.315 J mol^{-1} deg^{-1} 1.987 cal mol^{-1} deg^{-1}
Planck's constant	h	6.626×10^{-34} J s 1.584×10^{-34} cal s
Speed of light in a vacuum	c	2.998×10^{10} cm s^{-1}

Abbreviations: C, coulomb; cal, calorie; cm, centimeter; deg, degree Kelvin; eq, equivalent; g, gram; J, joule; mol, mole; s, second.

Mathematical constants

$\pi = 3.14159$

$e = 2.71828$

$\log_e x = 2.303 \log_{10} x$

Conversion factors

Physical quantity	*Equivalent*
Length	1 cm = 10^{-2} m = 10 mm = 10^{4} μm = 10^{7} nm 1 cm = 10^{8} Å=0.3937 inch
Mass	1 g = 10^{-3} kg = 10^{3} mg = 10^{6} μg 1 g = 3.527×10^{-2} ounce (avoirdupois)
Volume	1 cm^3 = 10^{-6} m^3 = 10^{3} mm^3 1 ml = 1 cm^3 = 10^{-3} l = 10^{3} μl 1 cm^3 = 6.1×10^{-2} in^3 = 3.53×10^{-5} ft^3
Temperature	K = °C + 273.15 °C = (5/9)(°F − 32)
Energy	1 J = 10^{7} erg = 0.239 cal = 1 watt s
Pressure	1 torr = 1 mm Hg (0°C) = 1.333×10^{2} newton/m^2 = 1.333×10^{2} pascal = 1.316×10^{-3} atmospheres

Standard prefixes

Prefix	*Symbol*	*Factor*
kilo	k	10^{3}
hecto	h	10^{2}
deca	da	10^{1}
deci	d	10^{-1}
centi	c	10^{-2}
milli	m	10^{-3}
micro	μ	10^{-6}
nano	n	10^{-9}
pico	p	10^{-12}

B

Atomic Numbers and Weights of the Elements

Element	Symbol	Atomic number	Atomic weight
Actinium	Ac	89	227.03
Aluminum	Al	13	26.98
Americium	Am	95	243.06
Antimony	Sb	51	121.75
Argon	Ar	18	39.95
Arsenic	As	33	74.92
Astatine	At	85	210.99
Barium	Ba	56	137.34
Berkelium	Bk	97	247.07
Beryllium	Be	4	9.01
Bismuth	Bi	83	208.98
Boron	B	5	10.81
Bromine	Br	35	79.90
Cadmium	Cd	48	112.40
Calcium	Ca	20	40.08
Californium	Cf	98	249.07
Carbon	C	6	12.01
Cerium	Ce	58	140.12
Cesium	Cs	55	132.91
Chlorine	Cl	17	35.45
Chromium	Cr	24	52.00
Cobalt	Co	27	58.93
Copper	Cu	29	63.55
Curium	Cm	96	245.07
Dysprosium	Dy	66	162.50
Einsteinium	Es	99	254.09
Erbium	Er	68	167.26
Europium	Eu	63	151.96
Fermium	Fm	100	252.08
Fluorine	F	9	18.99
Francium	Fr	87	223.02
Gadolinium	Gd	64	157.25
Gallium	Ga	31	69.72
Germanium	Ge	32	72.59
Gold	Au	79	196.97
Hafnium	Hf	72	178.49
Helium	He	2	4.00
Holmium	Ho	67	164.93
Hydrogen	H	1	1.01
Indium	In	49	114.82
Iodine	I	53	126.90
Iridium	Ir	77	192.22
Iron	Fe	26	55.85
Khurchatovium	Kh	104	260
Krypton	Kr	36	83.80
Lanthanum	La	57	138.91
Lawrencium	Lr	103	256
Lead	Pb	82	207.20
Lithium	Li	3	6.94
Lutetium	Lu	71	174.97
Magnesium	Mg	12	24.31
Manganese	Mn	25	54.94
Mendelevium	Md	101	255.09
Mercury	Hg	80	200.59
Molybdenum	Mo	42	95.94
Neodymium	Nd	60	144.24
Neon	Ne	10	20.18
Neptunium	Np	93	237.05
Nickel	Ni	28	58.71
Niobium	Nb	41	92.91
Nitrogen	N	7	14.01
Nobelium	No	102	255
Osmium	Os	76	190.20
Oxygen	O	8	16.00
Palladium	Pd	46	106.40
Phosphorus	P	15	30.97
Platinum	Pt	78	195.09
Plutonium	Pu	94	242.06
Polonium	Po	84	208.98
Potassium	K	19	39.10
Praseodymium	Pr	59	140.91
Promethium	Pm	61	145
Protactinium	Pa	91	231.04
Radium	Ra	88	226.03
Radon	Rn	86	222.02
Rhenium	Re	75	186.20
Rhodium	Rh	45	102.91
Rubidium	Rb	37	85.47
Ruthenium	Ru	44	101.07
Samarium	Sm	62	150.40
Scandium	Sc	21	44.96
Selenium	Se	34	78.96
Silicon	Si	14	28.09
Silver	Ag	47	107.87
Sodium	Na	11	22.99
Strontium	Sr	38	87.62
Sulfur	S	16	32.06
Tantalum	Ta	73	180.95
Technetium	Tc	43	98.91
Tellurium	Te	52	127.60
Terbium	Tb	65	158.93
Thallium	Tl	81	204.37
Thorium	Th	90	232.04
Thulium	Tm	69	168.93
Tin	Sn	50	118.69
Titanium	Ti	22	47.90
Tungsten	W	74	183.85
Uranium	U	92	238.03
Vanadium	V	23	50.94
Xenon	Xe	54	131.30
Ytterbium	Yb	70	173.04
Yttrium	Y	39	88.91
Zinc	Zn	30	65.37
Zirconium	Zr	40	91.22

APPENDIX

C

pK' Values of Some Acids

Acid	pK′ (at 25°C)
Acetic acid	4.76
Acetoacetic acid	3.58
Ammonium ion	9.25
Ascorbic acid, pK_1'	4.10
pK_2'	11.79
Benzoic acid	4.20
n-Butyric acid	4.81
Cacodylic acid	6.19
Carbonic acid, pK_1'	6.35
pK_2'	10.33
Citric acid, pK_1'	3.14
pK_2'	4.77
pK_3'	6.39
Ethylammonium ion	10.81
Formic acid	3.75
Glycine, pK_1'	2.35
pK_2'	9.78
Imidazolium ion	6.95

Acid	pK′ (at 25°C)
Lactic acid	3.86
Maleic acid, pK_1'	1.83
pK_2'	6.07
Malic acid, pK_1'	3.40
pK_2'	5.11
Phenol	9.89
Phosphoric acid, pK_1'	2.12
pK_2'	7.21
pK_3'	12.67
Pyridinium ion	5.25
Pyrophosphoric acid, pK_1'	0.85
pK_2'	1.49
pK_3'	5.77
pK_4'	8.22
Succinic acid, pK_1'	4.21
pK_2'	5.64
Trimethylammonium ion	9.79
Tris (hydroxymethyl) aminomethane	8.08
Water	14.0

APPENDIX

D

Standard Bond Lengths

Bond	*Structure*	*Length (Å)*
C—H	R_2CH_2	1.07
	Aromatic	1.08
	RCH_3	1.10
C—C	Hydrocarbon	1.54
	Aromatic	1.40
C=C	Ethylene	1.33
C≡C	Acetylene	1.20
C—N	RNH_2	1.47
	O=C—N	1.34
C—O	Alcohol	1.43
	Ester	1.36
C=O	Aldehyde	1.22
	Amide	1.24
C—S	R_2S	1.82
N—H	Amide	0.99
O—H	Alcohol	0.97
O—O	O_2	1.21
P—O	Ester	1.56
S—H	Thiol	1.33
S—S	Disulfide	2.05

Answers to Problems

Chapter 2

1. (a) Each strand is 35 kd and hence has about 318 residues (the mean residue mass is 110 daltons). Because the rise per residue in an α helix is 1.5 Å, the length is 477 Å.
   (b) Eighteen residues in each strand (40 minus 4 divided by 2) are in a β sheet conformation. Because the rise per residue is 3.5 Å, the length is 63 Å.
2. The methyl group attached to the β carbon of isoleucine sterically interferes with α helix formation. In leucine, this methyl group is attached to the γ carbon atom, which is farther from the main chain and hence does not interfere.
3. The first mutation destroys activity because valine occupies more space than alanine, and so the protein must take a different shape. The second mutation restores activity because of a compensatory reduction of volume; glycine is smaller than isoleucine.
4. The native conformation of insulin is not the thermodynamically most stable form. Indeed, insulin is formed from proinsulin, a single-chain precursor containing 33 additional residues. In proinsulin, residue 30 of the future B chain of insulin is linked to residue 1 of the future A chain. See D.F. Steiner, *Harvey Lectures* 78(1983):191, for a lucid account of the discovery of proinsulin and the molecular physiology of the biosynthetic process.
5. A segment of the main chain of the protease could hydrogen-bond to the main chain of the substrate to form an extended parallel or antiparallel pair of β strands.
6. Glycine has the smallest side chain of any amino acid. Its smallness often is critical in allowing polypeptide chains to make tight turns or to approach one another closely.
7. Glutamate, aspartate, and the terminal carboxylate can form salt bridges with the guanidinium group of arginine. In addition, this group can be a hydrogen bond donor to the side chains of glutamine, asparagine, serine, threonine, aspartate, and glutamate, and the main chain carbonyl.
8. Disulfide bonds in hair are broken by adding a thiol and applying gentle heat. The hair is curled, and an oxidizing agent is added to re-form disulfide bonds to stabilize the desired shape.

Chapter 3

1. (a) Phenyl isothiocyanate.
   (b) Dansyl chloride or dabsyl chloride.
   (c) Urea; β-mercaptoethanol to reduce disulfides.
   (d) Chymotrypsin.
   (e) CNBr.
   (f) Trypsin.
2. 0.01, 0.1, 1, 10, and 100.
3. Each amino acid residue, except the carboxyl-terminal one, gives rise to a hydrazide on reacting with hydrazine. The carboxyl-terminal residue can be identified because it yields a free amino acid.
4. The *S*-aminoethylcysteine side chain resembles that of lysine. The only difference is a sulfur atom in place of a methylene group.
5. A 1 mg/ml solution of myoglobin (17.8 kd) corresponds to 5.62×10^{-5} M. The absorbance of a 1-cm path length is 0.84, which corresponds to an I_0/I ratio of 6.96. Hence 14.4% of the incident light is transmitted.
6. Tropomyosin is rod shaped, whereas hemoglobin is approximately spherical.
7. The frictional coefficient f as well as the mass m determines S. Specifically, f is proportional to r (see eq. 2 on p. 46). Hence, f is proportional to $m^{1/3}$, and so S is proportional to $m^{2/3}$ (see eq. 5 on p. 51). An 80-kd spherical protein sediments 1.59 times as rapidly as a 40-kd spherical protein.
8. 50 kd.
9. The positions of disulfide bonds can be determined by diagonal electrophoresis (p. 58). The disulfide pairing is unaltered by the mutation if the off-diagonal peptides formed from the native and mutant proteins are the same.
10. (a) Electrostatic repulsion between positively charged ε-amino groups prevents α helix formation at pH 7. At pH 10, the side chains become deprotonated, allowing α helix formation.
    (b) Poly-L-glutamate is a random coil at pH 7 and becomes α-helical below pH 4.5 because the γ-carboxylate groups become protonated.
11. A fluorescent-labeled derivative of a bacterial degradation product (e.g., a formylmethionyl peptide) would bind to cells containing the receptor of interest.
12. The sedimentation and electrophoretic properties of the L enzyme and the mirror-image D form would be the same. The circular dichroism spectra would have the same magnitude but be of opposite sign because the two structures have opposite screw-sense. Peptide substrates that are mirror images of one another would be cleaved at the same rate by the L and D enzymes. See R.C. Milton, S.C. Milton, and S.B. Kent, *Science* 256(1992):1445, for the total chemical synthesis of the D and L enantiomers of the HIV-1 protease and demonstration of their reciprocal chiral substrate specificity.
13. Light was used to direct the synthesis of these peptides.

Each amino acid added to the solid support contained a photolabile protecting group instead of a *t*-Boc protecting group at its α-amino group. Illumination of selected regions of the solid support led to the release of the protecting group, which exposed the amino groups in these sites to make them reactive. The pattern of masks used in these illuminations and the sequence of reactants define the ultimate products and their locations. See S.P.A. Fodor, J.L. Read, M.C. Pirrung, L. Stryer, A.T. Lu, and D. Solas, *Science* 251(1991):767, for an account of light-activated, spatially addressable parallel chemical synthesis.

Chapter 4

1. (a) TTGATC; (b) GTTCGA; (c) ACGCGT; and (d) ATGGTA.
2. (a) [T] + [C] = 0.46.
   (b) [T] = 0.30, [C] = 0.24, and [A] + [G] = 0.46.
3. 5.71×10^3 base pairs.
4. In conservative replication, after 1.0 generation, one half of the molecules would be ^{15}N-^{15}N, the other half ^{14}N-^{14}N. After 2.0 generations, one-quarter of the molecules would be ^{15}N-^{15}N, the other three-quarters ^{14}N-^{14}N. Hybrid ^{14}N-^{15}N molecules would not be observed in conservative replication.
5. The DNA renatured when the heat-killed pneumococci were cooled before they were injected into mice.
6. Noncompetent strains may not be able to take up DNA. Alternatively, they may have potent deoxyribonucleases, or they may not be able to integrate fragments of DNA into their genome.
7. In the Hershey-Chase experiment, ^{35}S-labeled T2 viral proteins did not become incorporated into infected cells. The labeled viral proteins were found in the supernatant when infected cells were centrifuged. In contrast, M13 proteins become imbedded in the inner membrane of infected cells; they would appear in the pellet rather than the supernatant after centrifugation. Hershey and Chase would not have been able to separate M13 into genetic and nongenetic parts, as they did for T2.
8. (a) Tritiated thymine or tritiated thymidine.
   (b) dATP, dGTP, dCTP, and dTTP labeled with ^{32}P in the innermost (α) phosphorus atom.
9. Molecules (a) and (b) would not lead to DNA synthesis because they lack a 3′-OH group (a primer). Molecule (d) has a free 3′-OH at one end of each strand but no template strand beyond. Only (c) would lead to DNA synthesis.
10. A deoxythymidylate oligonucleotide should be used as the primer. The poly(rA) template specifies the incorporation of dT; hence, radioactive dTTP should be used in the assay.
11. The ribonuclease serves to degrade the RNA strand, a necessary step in forming duplex DNA from the RNA–DNA hybrid.
12. Treat one aliquot of the sample with ribonuclease and another with deoxyribonuclease. Test these nuclease-treated samples for infectivity.
13. Deamination changes the original G·C base pair into a G·U pair. After one round of replication, one daughter duplex will contain a G·C pair, and the other duplex an A·U pair. After two rounds of replication, there would be two G·C pairs, one A·U pair, and one A·T pair.
14. Hydrogen cyanide. Adenine can be viewed as a pentamer of HCN.
15. (a) $4^8 = 65{,}536$. In computer terminology, there are 64K 8-mers of DNA.
    (b) A bit specifies two bases (say, A and C) and a second bit specifies the other two (G and T). Hence, two bits are needed to specify a single nucleotide (or base pair) in DNA. For example, 00, 01, 10, and 11, could encode A, C, G, and T. An 8-mer stores 16 bits ($2^{16} = 65{,}536$), the *E. coli* genome (4×10^6 bp) stores 8×10^6 bits, and the human genome (2.9×10^9 bases) stores 5.8×10^9 bits of genetic information.
    (c) A high-density diskette stores about 1.5 megabytes, which is equal to 1.2×10^7 bits. A large number of 8-mer sequences could be stored on such a diskette. The DNA sequence of *E. coli*, once known, could be written on a single diskette. Nearly 500 diskettes would be needed to record the human DNA sequence.

Chapter 5

1. (a) Deoxyribonucleoside triphosphates versus ribonucleoside triphosphates.
   (b) 5′ → 3′ for both.
   (c) Semiconserved for DNA polymerase I, conserved for RNA polymerase.
   (d) DNA polymerase I needs a primer, whereas RNA polymerase does not.
2. (a) 5′-UAACGGUACGAU-3′.
   (b) Leu-Pro-Ser-Asp-Trp-Met.
   (c) Poly(Leu-Leu-Thr-Tyr).
3. The 2′-OH group in RNA acts as an intramolecular catalyst. In the alkaline hydrolysis of RNA, it forms a 2′-3′ cyclic intermediate.
4. Cordycepin terminates RNA synthesis. An RNA chain containing cordycepin lacks a 3′-OH group.
5. Only single-stranded RNA can serve as a template for protein synthesis.
6. Incubation with RNA polymerase and only UTP, ATP, and CTP led to the synthesis of only poly(UAC). Only poly(GUA) was formed when GTP was used in place of CTP.
7. (a) A codon for lysine cannot be changed to one for aspartate by the mutation of a single nucleotide.
   (b) Arg, Asn, Gln, Glu, Ile, Met, or Thr.
8. A peptide terminating with Lys (UGA is a stop codon), -Asn-Glu-, and -Met-Arg-.
9. Highly abundant amino acid residues have the most codons (e.g., Leu and Ser each have six), whereas the least abundant ones have the fewest (Met and Trp each have only one). Degeneracy allows (a) variation in base composition and (b) decreases the likelihood that a substitution of a base will change the encoded amino acid. If the degeneracy were equally distributed, each of the 20 amino acids would have three codons. Benefits (a) and (b) are maximized by assigning more codons to prevalent amino acids than to less frequently used ones.
10. Phe-Cys-His-Val-Ala-Ala.

11. GUG and GUC are likely to be used more by the alga from the hot springs to increase the melting temperature of its DNA (see p. 86).
12. The genetic code is degenerate. Eighteen of the twenty amino acids are specified by more than one codon. Hence, many nucleotide changes (especially in the third base of a codon) do not alter the nature of the encoded amino acid. Mutations leading to an altered amino acid are usually more deleterious than those that do not and hence are subject to more stringent selection.

Chapter 6

1. (a) 5′-GGCATAC-3′.
   (b) The Sanger dideoxy method of sequencing would give the gel pattern shown below.

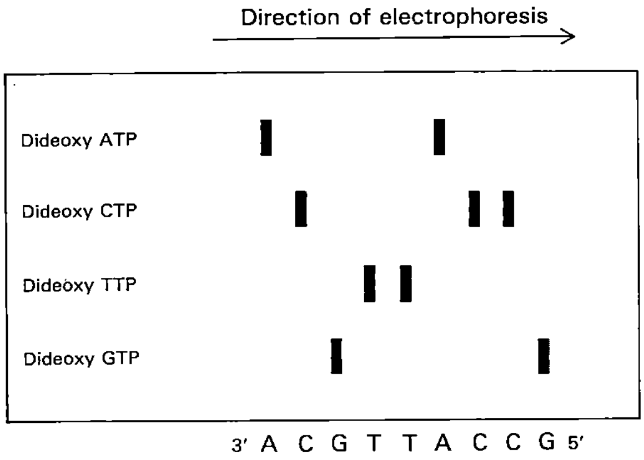

2. Ovalbumin cDNA should be used. *E. coli* lacks the machinery to splice the primary transcript arising from genomic DNA.
3. (a) No, because most human genes are much longer than 4 kb. One would obtain fragments containing only a small part of a complete gene.
   (b) No, chromosome walking depends on having *overlapping* fragments. Exhaustive digestion with a restriction enzyme produces nonoverlapping, short fragments.
4. Southern blotting of an MstII digest would distinguish between the normal and mutant genes. The loss of a restriction site would lead to the replacement of two fragments on the Southern blot by a single longer fragment (see p. 173). Such a finding would not prove that GTG replaced GAG; other sequence changes at the restriction site could yield the same result.
5. Cech replicated the recombinant DNA plasmid in *E. coli*, and then transcribed the DNA in vitro using bacterial RNA polymerase. He then found that this RNA underwent self-splicing in vitro in the complete absence of any proteins from *Tetrahymena*.
6. A simple strategy for generating many mutants is to synthesize a degenerate set of cassettes by using a mixture of activated nucleosides in particular rounds of oligonucleotide synthesis. Suppose that the 30-mer begins with GTT, which encodes Val. If a mixture of all four nucleotides is used in the first and second rounds of synthesis, the resulting oligonucleotides will begin with the sequence XYT (where X and Y denote A, C, G, or T). These 16 different versions of the cassette will encode proteins containing either Phe, Leu, Ile, Val, Ser, Pro, Thr, Ala, Tyr, His, Asn, Asp, Cys, Arg, or Gly at the first position. Likewise, one can make degenerate cassettes in which two or more codons are simultaneously varied.
7. Digest genomic DNA with a restriction enzyme and select the fragment that contains the known sequence. Circularize this fragment. Then carry out PCR using a pair of primers that serve as templates for the synthesis of DNA away from the known sequence.
8. The encoded protein contains four repeats of a specific sequence.
9. Sequence-tagged sites (STSs) can serve as a common framework for establishing the relation between clones. An STS is a sequence 200 to 500 bp long that occurs only once in the entire genome. These sequences and those of a pair of PCR primers that can generate the STS are stored in a database. Laboratory 1 states its YAC contains STS-34, STS-102, and STS-860. Laboratory 2 can then synthesize the corresponding PCR primers and learn whether their YAC contains any of these STSs. For an illuminating discussion of this strategy, see J.D. Watson, M. Gilman, J. Witkowski, and M. Zoller, *Recombinant DNA*, 2nd ed. (W.H. Freeman, 1992), pp. 610–612.

Chapter 7

1. (a) 2.96×10^{-11} g.
   (b) 2.71×10^{8} molecules.
   (c) No. There would be 3.22×10^{8} hemoglobin molecules in a red cell if they were packed in a cubic crystalline array. Hence, the actual packing density is about 84% of the maximum possible.
2. 2.65 g (or 4.75×10^{-2} moles) of Fe.
3. (a) In humans, 1.44×10^{-2} g (4.49×10^{-4} moles) of O_2 per kilogram of muscle. In sperm whale, 0.144 g (4.49×10^{-3} moles) of O_2 per kilogram.
   (b) 128.
4. (a) $k_{off} = k_{on}K = 20\ s^{-1}$.
   (b) Mean duration is 0.05 s (the reciprocal of k_{off}).
5. (a) Increased, (b) decreased, (c) decreased, and (d) increased oxygen affinity.
6. Inositol hexaphosphate.
7. The p*K* is (a) lowered, (b) raised, and (c) raised.
8. (a) Yes. $K_{AB} = K_{BA}(K_B/K_A) = 2 \times 10^{-5}$ M.
   (b) The presence of A enhances the binding of B; hence, the presence of B enhances the binding of A.
9. Carbon monoxide bound to one heme alters the oxygen affinity of the other hemes in the same hemoglobin molecule. Specifically, CO increases the oxygen affinity of hemoglobin and thereby decreases the amount of O_2 released in actively metabolizing tissues. Carbon monoxide

stabilizes the quaternary structure characteristic of oxyhemoglobin. In other words, CO mimics O_2 as an allosteric effector.

10. (a) For maximal transport, $K = 10^{-5}$ M. In general, maximal transport is achieved when $K = ([L_A][L_B])^{0.5}$.
    (b) For maximal transport, $P_{50} = 44.7$ torrs, which is considerably higher than the physiologic value of 26 torrs. However, it must be stressed that this calculation ignores cooperative binding and the Bohr effect.
11. (a) Lys or Arg at position 6.
    (b) GAG (Glu) to AAG (Lys).
    (c) This mutant hemoglobin moves more rapidly toward the anode than does Hb A and Hb S because it is more positively charged.
12. The fraction of molecules in the R form is 10^{-5}, 0.004 0.615, 0.998, and 1 when 0, 1, 2, 3, and 4 ligands, respectively, are bound.
13. Mutations in the α gene affect all three hemoglobins because their subunit structures are $\alpha_2\beta_2$, $\alpha_2\delta_2$, and $\alpha_2\gamma_2$. Mutations in the β, δ, or γ genes affect only one of them.
14. Deoxy Hb A contains a complementary site, and so it can add on to a fiber of deoxy Hb S. The fiber cannot then grow further because the terminal deoxy Hb A molecule lacks a sticky patch.
15. Carbamoylation of hemoglobin increases its oxygen affinity. Oxygenated Hb S does not sickle.
16. At very low fractional saturation, O_2 binds to hemoglobin molecules that are almost entirely in the T state. At very high fractional saturation, O_2 binds to molecules that are almost entirely in the R state. The slope is ~1 (nearly noncooperative) when binding is not accompanied by a significant switch from T to R.

Chapter 8

1. (a) 31.1 μmoles.
   (b) 0.05 μmoles.
   (c) 622 s^{-1}.
2. (a) Yes. $K_M = 5.2 \times 10^{-6}$ M.
   (b) $V_{max} = 6.84 \times 10^{-10}$ moles/min.
   (c) 337 s^{-1}.
3. Penicillinase, like glycopeptide transpeptidase, forms an acyl-enzyme intermediate with its substrate but transfers it to water rather than to the terminal glycine of the pentaglycine bridge.
4. (a) In the absence of inhibitor, V_{max} is 47.6 μmole/min and K_M is 1.1×10^{-5} M. In the presence of inhibitor, V_{max} is the same, and the apparent K_M is 3.1×10^{-5} M.
   (b) Competitive.
   (c) 1.1×10^{-3} M.
   (d) f_{ES} is 0.243 and f_{EI} is 0.488.
   (e) f_{ES} is 0.73 in the absence of inhibitor and 0.49 in the presence of 2×10^{-3} M inhibitor. The ratio of these values, 1.49, is the same as the ratio of the reaction velocities under these conditions.
5. (a) V_{max} is 9.5 μmole/min. K_M is 1.1×10^{-5} M, the same as without inhibitor.
   (b) Noncompetitive.
   (c) 2.5×10^{-5} M.
   (d) 0.73, in the presence or absence of this noncompetitive inhibitor.
6. (a) $V = V_{max} - (V/[S])\,K_M$.
   (b) Slope $= -K_M$, y-intercept $= V_{max}$, x-intercept $= V_{max}/K_M$.
   (c) An Eadie-Hofstee plot is shown below.

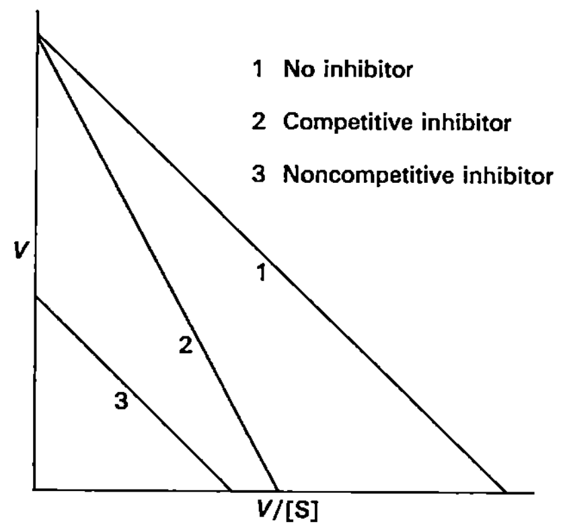

7. Potential hydrogen bond donors at pH 7 are the side chains of the following residues: arginine, asparagine, glutamine, histidine, lysine, serine, threonine, tryptophan, and tyrosine.
8. The rates of utilization of A and B are given by

$$V_A = \left(\frac{k_3}{K_M}\right)_A [E][A]$$

and

$$V_B = \left(\frac{k_3}{K_M}\right)_B [E][B]$$

Hence, the ratio of these rates is

$$V_A/V_B = \left(\frac{k_3}{K_M}\right)_A [A] \Big/ \left(\frac{k_3}{K_M}\right)_B [B]$$

Thus, an enzyme discriminates between competing substrates on the basis of their values of k_3/K_M rather than of K_M alone.

9. The mutation slows the reaction by a factor of 100 because the activation free energy is increased by +2.72 kcal/mol (2.303 RT log 100). Strong binding of the substrate relative to the transition state slows catalysis.

Chapter 9

1. The fastest is (b) and the slowest is (a).
2. (a) B-C, (b) A-B or E-F, and (c) A-B-C (one sugar residue does not interact with the enzyme, thus avoiding site D, which is energetically unfavorable).

3. The ^{18}O emerges in the C-4 hydroxyl of di-NAG (residues E-F).
4. This analog lacks a bulky substituent at C-5, and so it can probably bind to site D without being strained. Consequently, the binding of residue D of this analog is likely to be energetically favorable, whereas the binding of residue D of tetra-NAG costs free energy. See P. van Eikeren and D.M. Chipman, *J. Am. Chem. Soc.* 94(1972):4788.
5. (a) In oxymyoglobin, Fe is bonded to five nitrogens and one oxygen. In carboxypeptidase A, Zn is bonded to two nitrogens and two oxygens.
   (b) In oxymyoglobin, one of the nitrogens bonded to Fe comes from the proximal histidine residue, whereas the other four come from the heme. The oxygen atom linked to Fe is that of O_2. In carboxypeptidase A, the two nitrogen atoms coordinated to Zn come from histidine residues. One of the oxygen ligands is from a glutamate side chain, the other from a water molecule.
   (c) Aspartate, cysteine, and methionine.
6. If tyrosine 248 were essential for catalysis, a mutant of this enzyme containing phenylalanine in its place would be inactive. The tyrosine 248 codon (TAT) was converted to that of phenylalanine (TTT) by oligonucleotide-directed mutagenesis (p. 139). The recombinant plasmid containing this gene was inserted into yeast and expressed. The striking finding was that the mutant enzyme had the same k_{cat} value as did the native enzyme but a K_M value that was sixfold higher. This experiment showed that tyrosine 248 participates in the binding of substrate but is not essential for catalysis. Moreover, it graphically illustrated the power of site-specific mutagenesis in delineating the function of a particular residue in a protein. See S.J. Gardell, C.S. Craik, D. Hilvert, M.S. Urdea, and W.J. Rutter, *Nature* 317(1985):551.
7. The essential sulfhydryl in its active site should be highly susceptible to chemical modification. The enzyme should be inactivated by reacting it with iodoacetamide, *N*-ethylmaleimide, or another sulfhydryl-specific reagent. Furthermore, bound substrate is likely to protect the key sulfhydryl from alkylation.
8. (a) Yes.
   (b) Histidine 119 in ribonuclease A, histidine 57 in chymotrypsin, glutamic acid 35 in lysozyme, and a zinc-bound water molecule in carboxypeptidase A.
9. (a) Tosyl-L-lysine chloromethyl ketone (TLCK).
   (b) First, determine whether substrates protect trypsin from inactivation by TLCK and, second, ascertain whether the D isomer of TLCK inactivates trypsin.
10. (a) Serine.
    (b) A hemiacetal between the aldehyde of the inhibitor and the hydroxyl group of the active-site serine.
11. A dimer of identical subunits is advantageous for two reasons. First, less RNA is needed to encode the protease. A small genome can be more readily replicated and packaged than a large one. Second, the protease does not become active until a critical concentration is attained. Premature excision of viral subunits from the polyprotein is thereby prevented.
12. Inhibitors of protease subunit dimerization should be effective in blocking protease action.
13. Its sulfhydryl group probably binds to the zinc ion at the active site, and its carboxylate group very likely occupies the recognition site for the C terminus of the substrate.
14. Precise positioning of catalytic residues and substrates, geometrical strain (distortion of the substrate), electronic polarization, and desolvation of the substrate.

Chapter 10

1. The protonated form of histidine probably stabilizes the negatively charged carbonyl oxygen atom of the scissile bond in the transition state. Deprotonation would lead to a loss of activity. Hence, the rate is expected to be half-maximal at a pH of about 6.5 (the pK of an unperturbed histidine side chain in a protein) and decrease as the pH is raised.
2. (a) 100. The change in the [R]/[T] ratio on binding one substrate molecule must be the same as the ratio of the substrate affinities of the two forms (see equation 12 on p. 167).
   (b) 10. The binding of four substrate molecules changes the [R]/[T] by a factor of $100^4 = 10^8$. The ratio in the absence of substrate is 10^{-7}. Hence, the ratio in the fully liganded molecule is $10^8 \times 10^{-7} = 10$.
3. The sequential model can account for negative cooperativity, whereas the concerted model cannot. Homotropic allosteric interactions must be cooperative if the concerted model holds.
4. The binding of PALA switches ATCase from the T to the R state because it acts as a substrate analog. An enzyme molecule containing bound PALA has fewer free catalytic sites than does an unoccupied enzyme molecule. However, the PALA-containing enzyme will be in the R state and hence have higher affinity for the substrates. The dependence of the degree of activation on the concentration of PALA is a complex function of the allosteric constant L_0, and of the binding affinities of the R and T states for the analog and the substrates. For an account of this experiment, see J. Foote and H.K. Schachman, *J. Mol. Biol.* 186(1985):175.
5. Activation is independent of zymogen concentration because the reaction is intramolecular (p. 250).
6. Add blood from the second patient to a sample from the first. If the mixture clots, the second patient has a defect different from that of the first. This type of assay is called a complementation test.
7. Activated factor X remains bound to blood platelet membranes, which accelerates its activation of prothrombin.
8. Antithrombin III is a very slowly hydrolyzed substrate of thrombin. Hence, its interaction with thrombin requires a fully formed active site on the enzyme.
9. Residues *a* and *d* are located in the interior of an α-helical coiled coil, near the axis of the superhelix. Hydrophobic interactions between these side chains contribute to the stability of the coiled coil.
10. Leucine would be a good choice. It is much more stable than methionine but has nearly the same volume and also is very hydrophobic.

Chapter 11

1. 2.86×10^6 molecules, because each leaflet of the bilayer contains 1.43×10^6 molecules.
2. 2×10^{-7} cm, 6.32×10^{-6} cm, and 2×10^{-4} cm.
3. The radius of this molecule is 3.08×10^{-7} cm and its diffusion coefficient is 7.37×10^{-9} cm^2/s. The average distances traversed are 1.72×10^{-7} cm in 1 μs, 5.42×10^{-6} in 1 ms, and 1.72×10^{-4} cm in 1 s.
4. The membrane underwent a phase transition from a highly fluid to a nearly frozen state when the temperature was lowered. A carrier can shuttle ions across a membrane only when the bilayer is highly fluid. A channel former, by contrast, allows ions to traverse its pore even when the bilayer is quite rigid.
5. The initial decrease in the amplitude of the paramagnetic resonance spectrum results from the reduction of spin-labeled phosphatidyl cholines in the outer leaflet of the bilayer. Ascorbate does not traverse the membrane under these experimental conditions; hence, it does not reduce the phospholipids in the inner leaflet. The slow decay of the residual spectrum is due to the reduction of phospholipids that have flipped over to the outer leaflet of the bilayer. See R.D. Kornberg and H.M. McConnell, *Biochemistry* 10(1971):1111.

Chapter 12

1. The ratio of closed to open forms of the channel is 10^5, 5000, 250, 12.5, and 0.625 when 0, 1, 2, 3, and 4 ligands, respectively, are bound. Hence, the fraction of open channels is 10^{-5}, 2×10^{-4}, 3.98×10^{-3}, 7.41×10^{-2}, and 0.615.
2. These organic phosphates inhibit acetylcholinesterase by reacting with the active site serine to form a stable phosphorylated derivative. They cause respiratory paralysis by blocking synaptic transmission at cholinergic synapses.
3. (a) The binding of the first acetylcholine increases the open/closed ratio by a factor of 240, and binding of the second by a factor of 11,700.
   (b) Hence, the free energy contributions are 3.3 kcal/mol and 5.6 kcal/mol, respectively.
   (c) No. The MCW model predicts that the binding of each ligand will have the same effect on the open/closed ratio. The acetylcholine receptor channel is not perfectly symmetric. The two α chains are not in identical environments. Also, the presence of desensitized states in addition to the open and closed ones indicates that a more complex model is required.
4. (a) -22 mV; (b) $+4.5$; (c) 5.2 kcal/mol.
5. Batrachotoxin blocks the transition from the open to the inactivated state.
6. (a) Chloride ions flow into the cell.
   (b) Chloride flux is inhibitory because it hyperpolarizes the membrane.
   (c) The channel consists of five subunits.
7. The free-energy cost is 7.6 kcal/mol (4.88 kcal/mol for the chemical work performed and 2.76 kcal/mol for the electrical work performed).
8. (a) The conductance g of the channel is 100 pS.
   (b) 3.12×10^4 ions flow through the channel during its mean open time of 1 ms.
   (c) The mean transit time for an ion is 32 ns.
9. Membrane vesicles containing a high concentration of lactose in their inner volume are formed. The binding of lactose to the inner face of the permease will be followed by the binding of a proton. Both sites will then evert. Because the lactose concentration on the outside is low, lactose and the proton will dissociate from the permease. The downhill flux of lactose will drive the uphill flux of protons in this in vitro system.

Chapter 13

1. Phosphorylation mediates the excitatory response in chemotaxis and receptor tyrosine kinase cascades, whereas it participates in recovery and adaptation in vision. Autophosphorylation of cheA is essential in chemotaxis. Likewise, autophosphorylation of receptor tyrosine kinases is crucial in growth-control cascades. By contrast, phosphorylation of photoexcited rhodopsin leads to its deactivation.
2. The binding of an attractant rapidly leads to the formation of less phosphorylated cheY, and hence the probability of tumbling decreases. On a slower time scale, the binding of attractant leads to the formation of less phosphorylated cheB (the more potent form of the methylesterase). Also, the rate of methylation is increased. Consequently, the methylation level of the receptor increases, which speeds the production of phosphorylated cheY to bring the tumbling probability back to the initial value. Hence, excitation is followed by adaptation.
3. (a) Loss of *cheB* (the gene for the methylesterase) would give receptors that are fully methylated. The bacterium would persistently tumble, always convinced that it is heading in the wrong direction.
   (b) Loss of *cheR* (the gene for methyltransferase) would give receptors that are completely demethylated. The bacterium would swim smoothly all the time, confident that it is heading for greener pastures.
4. (a) The photoresponse would be prolonged because deactivation of R* would be inhibited.
   (b) The amplitude of the photoresponse would be decreased because less cGMP would be hydrolyzed.
   (c) EGTA would lower the calcium level, which would increase guanylate cyclase activity. The photoresponse would be diminished by the higher cGMP level.
   (d) The photoresponse would be markedly prolonged because GTPγS bound to transducin cannot be hydrolyzed. The phosphodiesterase would be persistently activated.
5. The absorption maximum is likely to be intermediate between that of the red and green receptors. The capacity to distinguish hues between red and green depends on having receptors with well-separated absorption maxima. The separation between the hybrid and green receptors is less than in the normal case. Hence, color vision is impaired.
6. cGMP phosphodiesterase, protein kinase A, and protein

kinase C are activated by removal of an inhibitory constraint.

7. ADP-ribosylation by cholera toxin leads to persistent activation of adenylate cyclase because G_s–GTP is very slowly hydrolyzed to G_s–GDP. The abnormally high cAMP level leads to complete activation of protein kinase A, which then phosphorylates chloride channels. The proportion of open channels is much higher than normal, which leads to excessive loss of salt and water.
8. The Ca^{2+}-ATPase pump contains a region (D) that binds this internal calmodulinlike sequence (C). The catalytic site is blocked by the formation of the $C \cdot D$ complex. Ca^{2+}-calmodulin (C') disrupts this internal complex and activates the pump by binding D. Thus, calcium-calmodulin reverses the effect of an autoinhibitory region of its target. Once again we see that stimulation of one protein by another can be mediated by relief of a built-in inhibitory constraint.
9. One strategy is to search for compounds that block the dimerization of a receptor tyrosine kinase. Inhibitors of dimerization should block autophosphorylation.

Chapter 14

1. (a) $\Delta G^{\circ\prime} = -8.9$ kcal/mol.
   (b) $K_a = 3.3 \times 10^6\ \text{M}^{-1}$.
   (c) $k_{on} = 4 \times 10^8\ \text{M}^{-1}\text{s}^{-1}$. This value is close to the diffusion-controlled limit for combination of a small molecule with a protein (p. 196). Hence, the extent of structural change is likely to be small; extensive conformational transitions take time.
2. The active site in lysozyme also accommodates six sugar residues (p. 211). The range of sizes of the combining sites of antibodies is like that of active sites of enzymes.
3. The fluorescence enhancement and shift to the blue indicate that water is largely excluded from the combining site when the hapten is bound. Hydrophobic interactions contribute significantly to the formation of most antigen-antibody complexes.
4. (a) 7.1 μM.
   (b) $\Delta G^{\circ\prime}$ is equal to $2 \times -7 + 3$ kcal/mol or -11 kcal/mol, which corresponds to an apparent dissociation constant of 8 nM. The avidity (apparent affinity) of bivalent binding in this case is 888 times as much as the affinity of the univalent interaction.
5. (a) An antibody combining site is formed by CDRs from both the H and L chains. The V_H and V_L domains are essential. A small proportion of F_{ab} fragments can be further digested to produce F_v, a fragment that contains just these two domains. C_H1 and C_L contribute to the stability of F_{ab} but not to antigen binding.
   (b) A synthetic F_v analog 248 residues long was prepared by expressing a synthetic gene consisting of a V_H gene joined to a V_L gene through a linker. See J.S. Huston et al., *Proc. Nat. Acad. Sci.* 85(1988):5879.
6. (a) Multivalent antigens lead to the dimerization or oligomerization of transmembrane immunoglobulins, an essential step in their activation. This mode of activation is reminiscent of that of receptor tyrosine kinases (p. 351).
   (b) An antibody specific for a transmembrane immunoglobulin will activate a B cell by cross-linking these receptors. This experiment can be carried out using, for example, a goat antibody to cross-link receptors on a mouse B cell.
7. μ_s and μ_m are formed by alternative splicing of the same primary transcript (p. 115). The mRNA for μ_m encodes a C-terminal transmembrane anchor that is not present in μ_s.
8. B cells do not express T-cell receptors. Hybridization of T-cell cDNAs with B-cell mRNAs removes cDNAs that are expressed in both cells. Hence, the mixture of cDNAs following this hybridization are enriched in those encoding T-cell receptors. This procedure, called subtractive hybridization, is generally useful in isolating low-abundance cDNAs. Hybridization should be carried out using mRNAs from a closely related cell that does not express the gene of interest. See S.M. Hedrick, M.M. Davis, D.I. Cohen, E.A. Nielsen, and M.M. Davis, *Nature* 308(1984):149, for an interesting account of how this method was used to obtain genes for T-cell receptors.

Chapter 15

1. (a) ATP for skeletal muscle and eukaryotic cilia, and proton-motive force for bacterial flagella.
   (b) Two in each system: myosin-actin, dynein-microtubule, and motA-motB, respectively.
   (c) Thick and thin filaments of muscle are located in the cytosol. The axoneme of cilia is enveloped by the plasma membrane; the contents are continuous with the cytosol. The rotatory motor that drives bacterial flagella is located in the cytoplasmic membrane of the cell; the flagella are extracellular appendages.
2. A large load resists the movement of the thin filament relative to the thick filament. ADP release is slower with a large load than a small one because the power stroke takes longer to complete.
3. The rapid decrease in the level of ATP following death has two consequences. First, the cytosolic level of calcium rises rapidly because the Ca^{2+}-ATPase pumps in the plasma membrane and sarcoplasmic reticulum membrane no longer operate. High Ca^{2+}, through troponin and tropomyosin, enables myosin to interact with actin. Second, a large proportion of S1 heads will be associated with actin. Recall that ATP is required to dissociate the actomyosin complex. In the absence of ATP, skeletal muscle is locked in the contracted (rigor) state.
4. The dissociation constant K sets an upper limit on k_{off}, the rate constant for calcium release from the binding protein. Because $k_{on} < 10^9\ \text{M}^{-1}\text{s}^{-1}$ (the diffusion-controlled limit), $k_{off} < K\ 10^9\ \text{M}^{-1}\text{s}^{-1}$. For calmodulin, $k_{off} < 10^3\ \text{s}^{-1}$, whereas for troponin C, $k_{off} < 10^4\ \text{s}^{-1}$. Indeed, as predicted, troponin C is a faster switch than calmodulin, which enables skeletal muscle to rapidly contract and relax. The price paid is that a higher concentration of Ca^{2+} is needed to activate troponin C, and hence more ATP must be spent to pump Ca^{2+} out of the cytosol to return to the basal state.
5. The effect is mediated through Ca^{2+}-calmodulin, which

stimulates myosin light-chain kinase (MLCK) (see Figure 13-47 on p. 350). Phosphoryl groups introduced by MLCK are removed from myosin by a Ca^{2+}-independent phosphatase.

6. In principle, phalloidin could (a) block the assembly of microfilaments by capping the plus (growing) end, (b) sever already formed microfilaments, (c) prevent the disassembly of microfilaments, and (d) interfere with the binding of microfilaments to membranes and other cellular structures. In fact, phalloidin binds to F-actin and blocks depolymerization.
7. (a) $F = 9.4 \times 10^{-10}$ dyne.
   (b) The work performed is 4.7×10^{-14} erg.
   (c) The energy content of ATP (based on a ΔG of -12 kcal/mol under typical cellular conditions) is 8.3×10^{-13} erg per molecule. Hence, 40 ATP could yield 3.3×10^{-11} erg, far more than the actual work performed in moving the 2-μm-diameter bead. Thus, the hydrolysis of ATP by a single kinesin motor provides more than enough free energy to power the transport of micrometer-size cargoes at micrometer-per-second velocities.

Chapter 16

1. EF hands can be detected by scanning amino acid sequences for the 29-residue motif shown in Figure 16-28 (p. 434). The sequences of the four EF hands in calmodulin are

12	EFKEAFSLFDKDGDGTITTKELGTVMRSL	40
48	ELQDMINEVDADGNGTIDFPEFLTMMARK	76
85	EIREAFRVFDKDGNGYISAAELRHVMTNL	113
121	EVDEMIREADIDGDGQVNYEEFVQMMTAK	149

2. (a) This experiment shows that there is considerable redundancy of information in the amino acid sequence of phage lysozyme.
   (b) Alanine is much smaller than leucine. The absence of three carbon atoms creates a cavity in the hydrophobic core and diminishes the strength of van der Waals interactions.
   (c) Serine is a helix breaker, whereas alanine is a helix former.
   (d) The substitution of noncritical residues with alanine focuses attention on positions that are important for folding and stability. See D.W. Heinz, W.A. Baase, and B.W. Matthews, *Proc. Nat. Acad. Sci.* 89(1992):3751.
3. The order of synthesis of the polypeptide chain is not important in determining the final folded structure. See K. Luger, U. Hommel, M. Herold, J. Hofsteenge, and K. Kirschner, *Science* 243(1989):206.
4. (a) The gene for each heavy chain is altered so that it encodes a coiled-coil sequence following the C_H1 domain. The F_{ab} units will then associate by forming an α-helical coiled coil.
   (b) A bispecific antibody has been used to increase the destruction of target cells by killer T cells. One of the F_{ab} units of this antibody was specific for a cell-surface protein on killer T cells, and the other was specific for an interleukin receptor on a target cell. See S.A. Kostelny, M.S. Cole, and J.Y. Tso, *J. Immunol.* 148 (1992):1547.
5. One possibility a priori is that cyclophilin catalyzes the cis-trans isomerization of a peptide bond in CsA. The newly formed isomer would then dissociate from cyclophilin, bind to calcineurin, and block its phosphatase activity. Alternatively, calcineurin might be inhibited by CsA only when the cyclic peptide is bound to cyclophilin. In fact, the inhibitory species is a complex of CsA and cyclophilin. We see here that the simplest mechanism is not always the one utilized by nature. The research that elucidated this subtle mechanism of action of an important drug is reviewed in S.L. Schreiber, *Cell* 70(1992):365.
6. A four-helix bundle in which the pattern of hydrophobic and hydrophilic residues is the reverse of that shown in Figure 16-31A (p. 436) might fold in a membrane and conduct ions. The length of each helical segment should be about 21 residues to span the bilayer.

Chapter 17

1. Reactions (a) and (c), to the left; reactions (b) and (d), to the right.
2. None whatsoever.
3. (a) $\Delta G^{\circ\prime} = +7.5$ kcal/mol and $K'_{eq} = 3.06 \times 10^{-6}$.
   (b) 3.28×10^4.
4. $\Delta G^{\circ\prime} = 1.7$ kcal/mol. The equilibrium ratio is 17.8.
5. (a) +0.2 kcal/mol.
   (b) −7.8 kcal/mol. The hydrolysis of PP_i drives the reaction toward the formation of acetyl CoA.
6. (a) $\Delta G^{\circ} = 2.303\ RT\ \text{p}K$.
   (b) 6.53 kcal/mol at 25°C.
7. The activated form of sulfate in most organisms is 3′-phosphoadenosine 5′-phosphosulfate. See P.W. Robbins and F. Lipmann. *J. Biol. Chem.* 229(1957):837.
8. (a) 310 Hz.
   (b) Both the protonation and deprotonation rates must be faster than 310 s^{-1}.
   (c) The pK' for the equilibrium of $H_2PO_4^-$ and HPO_4^{2-} is 7.21 (see Appendix C on p. 1014). Hence, the dissociation constant K is 6.16×10^{-8} M. The rate constant for association k_{on} is equal to k_{off}/K. Because k_{off} is greater than 310 s^{-1}, k_{on} must be greater than 5×10^9 $M^{-1}\ s^{-1}$.
9. Arginine phosphate in invertebrate muscle, like creatine phosphate in vertebrate muscle, serves as a reservoir of high-potential phosphoryl groups. Arginine phosphate maintains a high level of ATP in muscular exertion. NMR is a choice technique for monitoring the levels of arginine phosphate and ATP in contracting muscle.
10. An ADP unit (or a closely related derivative, in the case of CoA).

Chapter 18

1. Carbohydrates were originally regarded as *hydrates* of *carbon* because the empirical formula of many of them is $(CH_2O)_n$.

2. (a) Aldose-ketose; (b) epimers; (c) aldose-ketose; (d) anomers; (e) aldose-ketose; and (f) epimers.
3. Aldoses are converted into aldonic acids; the aldehyde group of the sugar is oxidized to a carboxylate.
4. The proportion of α anomer is 0.36, and that of the β anomer is 0.64.
5. Glucose is reactive because of the presence of an aldehyde group in its open-chain form. The aldehyde group slowly condenses with amino groups to form Schiff base adducts (see p. 780).
6. A pyranoside reacts with two molecules of periodate; formate is one of the products. A furanoside reacts with only one molecule of periodate; formate is not formed.
7. From methanol.
8. (a) β-D-Mannose; (b) β-D-galactose; (c) β-D-fructose; (d) β-D-glucosamine.
9. The trisaccharide itself should be a competitive inhibitor of cell adhesion if the trisaccharide unit of the glycoprotein is critical for the interaction.

Chapter 19

1. Glucose is reactive because its open-chain form contains an aldehyde group (see p. 467).
2. (a) The label is in the methyl carbon of pyruvate.
   (b) 5 mCi/mM. The specific activity is halved because the number of moles of product (pyruvate) is twice that of the labeled substrate (glucose).
3. (a) Glucose + 2 P_i + 2 ADP $\longrightarrow$ 2 lactate + 2 ATP
   (b) $\Delta G' = -27.2$ kcal/mol.
4. 3.06×10^{-5}.
5. The equilibrium concentrations of fructose 1,6-bisphosphate, dihydroxyacetone phosphate, and glyceraldehyde 3-phosphate are 7.76×10^{-4} M, 2.24×10^{-4} M, and 2.24×10^{-4} M, respectively.
6. All three carbon atoms of 2,3-BPG are ^{14}C-labeled. The phosphorus atom attached to the C-2 hydroxyl is ^{32}P-labeled.
7. Hexokinase has a low ATPase activity in the absence of a sugar because it is in a catalytically inactive conformation (p. 499). The addition of xylose closes the cleft between the two lobes of the enzyme. However, xylose lacks a hydroxymethyl group, and so it cannot be phosphorylated. Instead, a water molecule at the site normally occupied by the C-6 hydroxymethyl group acts as the phosphoryl acceptor from ATP.
8. (a) 2,3-Bisphosphoglycerate (BPG) lowers the oxygen affinity of hemoglobin (p. 160). The rate of synthesis of 2,3-BPG is controlled by the level of 1,3-BPG, a glycolytic intermediate.
   (b) The lowered level of glycolytic intermediates leads to less 2,3-BPG and, hence, a higher oxygen affinity.
   (c) Glycolytic intermediates are present at a higher than normal level. The level of 2,3-BPG is increased, which makes the oxygen affinity lower than normal.
9. (a) The fructose 1-phosphate pathway (p. 491) forms glyceraldehyde 3-phosphate. Phosphofructokinase, a key control enzyme, is bypassed. Furthermore, fructose 1-phosphate stimulates pyruvate kinase.
   (b) The rapid, unregulated production of lactate can lead to metabolic acidosis.
10. The metal ion serves as an electron sink, as does the protonated Schiff base in animal aldolases.
11. EDTA removes the metal ion from the catalytic site of prokaryotic aldolases, whereas sodium borohydride reduces the Schiff base intermediate in catalysis by animal aldolases.
12. (a) Increased; (b) increased; (c) increased; (d) decreased.

Chapter 20

1. (a) After one round of the citric acid cycle, the label emerges in C-2 and C-3 of oxaloacetate.
   (b) The label emerges in CO_2 in the formation of acetyl CoA from pyruvate.
   (c) After one round of the citric acid cycle, the label emerges in C-1 and C-4 of oxaloacetate.
   (d and e) Same fate as in (a).
2. (a) Isocitrate lyase and malate synthase are required in addition to the enzymes of the citric acid cycle.
   (b) 2 Acetyl CoA + 2 NAD^+ + FAD + 3 H_2O $\longrightarrow$ oxaloacetate + 2 CoA + 2 NADH + $FADH_2$ + 3 H^+
   (c) No. Hence, mammals cannot carry out the net synthesis of oxaloacetate from acetyl CoA.
3. −9.8 kcal/mol.
4. The coenzyme stereospecificity of glyceraldehyde 3-phosphate dehydrogenase is the opposite of that of alcohol dehydrogenase (type B versus type A, respectively.)
5. Thiamine thiazolone pyrophosphate is a transition state analog. The sulfur-containing ring of this analog is uncharged, and so it closely resembles the transition state of the normal coenzyme in thiamine-catalyzed reactions (e.g., the uncharged resonance form of hydroxyethyl-TPP, p. 516). See J.A. Gutowski and G.E. Lienhard, *J. Biol. Chem.* 251(1976):2863, for a discussion of this analog.
6. (a) The steady-state concentrations of the products are low compared with those of the substrates.
   (b) The ratio of malate to oxaloacetate must be greater than 1.75×10^4 for oxaloacetate to be formed.
7. The enol intermediate of acetyl CoA attacks the carbonyl carbon atom of glyoxylate to form a C–C bond. This reaction is like the condensation of oxaloacetate with the enol intermediate of acetyl CoA in the reaction catalyzed by citrate synthase (p. 519). Glyoxylate contains a hydrogen atom in place of the $-CH_2COO^-$ of oxaloacetate; the reactions are otherwise nearly identical.
8. The phosphoryl group on Ser 113 is transferred to bound ADP and then to water, rather than directly to water. See D.C. Laporte, *J. Cell. Biochem.* 51(1993):14.
9. The kinase that synthesizes fructose 2,6-bisphosphate (F-2,6,-BP) and the phosphatase that hydrolyzes F-2,6-BP to fructose 6-phosphate are present on a single polypeptide chain (p. 494). However, in contrast with the bifunctional enzyme that modifies bacterial isocitrate dehydrogenase, the kinase and phosphatase active sites are on different domains.

Chapter 21

1. (a) 12.5; (b) 14; (c) 32; (d) 13.5; (e) 30; (f) 16.
2. Biochemists use E_0', the value at pH 7, whereas chemists use E_0, the value in 1 M H^+. The prime denotes that pH 7 is the standard state.
3. (a) Blocks electron transport and proton pumping at site 3.
   (b) Blocks electron transport and ATP synthesis by inhibiting the exchange of ATP and ADP across the inner mitochondrial membrane.
   (c) Blocks electron transport and proton pumping at site 1.
   (d) Blocks ATP synthesis without inhibiting electron transport by dissipating the proton gradient.
   (e) Blocks electron transport and proton pumping at site 3.
   (f) Blocks electron transport and proton pumping at site 2.
4. (a) The P:O ratio is equal to the product of ($H^+/2e^-$) and ($\sim P/H^+$). Note that the P:O ratio is identical with the ($P{:}2e^-$) ratio. (b) 2.5 and 1.5, respectively.
5. $\Delta G^{\circ\prime}$ is +16.1 kcal/mol for oxidation by NAD^+ and +1.4 kcal/mol for oxidation by FAD. The oxidation of succinate by NAD^+ is not thermodynamically feasible.
6. Cyanide can be lethal because it binds to the ferric form of cytochrome oxidase and thereby inhibits oxidative phosphorylation. Nitrite converts ferrohemoglobin to ferrihemoglobin, which also binds cyanide. Thus, ferrihemoglobin competes with cytochrome oxidase for cyanide. This competition is therapeutically effective because the amount of ferrihemoglobin that can be formed without impairing oxygen transport is much greater than the amount of cytochrome oxidase.
7. The absolute configuration of thiophosphate is opposite to that of ATP in the reaction catalyzed by ATP synthase. This result is consistent with an in-line phosphoryl transfer reaction occurring in a single step. The retention of configuration in the Ca^{2+}-ATPase reaction points to two phosphoryl transfer reactions—inversion by the first, and a return to the starting configuration by the second. The Ca^{2+}-ATPase reaction proceeds by a phosphorylated enzyme intermediate. See M.R. Webb, C. Grubmeyer, H.S. Penefsky, and D.R. Trentham, *J. Biol. Chem.* 255(1980):255.
8. The available free energy from the translocation of 2, 3, and 4 protons is −9.23, −13.8, and −18.5 kcal, respectively. The free energy consumed in synthesizing a mole of ATP under standard conditions is 7.3 kcal. Hence, the residual free energy of −1.93, −6.5, and −11.2 kcal can drive the synthesis of ATP until the $[ATP]/[ADP][P_i]$ ratio is 26.2, 6.51×10^4, and 1.62×10^8, respectively. Suspensions of isolated mitochondria synthesize ATP until this ratio is greater than 10^4, which shows that the number of protons translocated per ATP synthesized is at least 3.
9. Such a defect (called Luft's syndrome) was found in a 38-year-old woman who was incapable of performing prolonged physical work. Her basal metabolic rate was more than twice normal, but her thyroid function was normal. A muscle biopsy showed that her mitochondria were highly variable and atypical in structure. Biochemical studies then revealed that oxidation and phosphorylation were not tightly coupled in these mitochondria. In this patient, much of the energy of fuel molecules was converted into heat rather than ATP. The development of mitochondrial medicine is lucidly reviewed in R. Luft, *Proc. Nat. Acad. Sci.* 91(1994):8731.
10. Dicylohexylcarbodiimide reacts readily with carboxyl groups, as was discussed earlier in regard to its use in peptide synthesis (p. 69). Hence, the most likely targets are aspartate and glutamate side chains. In fact, aspartate 61 of subunit c of *E. coli* F_0 is specifically modified by this reagent. Conversion of this aspartate to an asparagine by site-specific mutagenesis also eliminated proton conduction. See A.E. Senior, *Biochim. Biophys. Acta* 726(1983):81.
11. Triose phosphate isomerase converts dihydroxyacetone phosphate (a potential dead end) into glyceraldehyde 3-phosphate (a mainstream glycolytic intermediate).
12. This inhibitor (like antimycin A) blocks the reduction of c_1 by QH_2, the crossover point.

Chapter 22

1. The label emerges at C-5 of ribulose 5-phosphate.
2. Oxidative decarboxylation of isocitrate to α-ketoglutarate. A β-keto acid intermediate is formed in both reactions.
3. C-1 and C-3 of fructose 6-phosphate are labeled, whereas erythrose 4-phosphate is not labeled.
4. (a) 5 Glucose 6-phosphate + ATP ⟶ 6 ribose 5-phosphate + ADP + H^+
   (b) Glucose 6-phosphate + 12 $NADP^+$ + 7 H_2O ⟶ 6 CO_2 + 12 NADPH + 12 H^+ + P_i
5. The lactate level in the maternal circulation, and hence in the fetal circulation, increases during pregnancy because the mother is carrying a growing fetus with its own metabolic demands. The shift to H_4 in the fetal heart enables the fetus to use lactate as a fuel. Consequently, there is less need for gluconeogenesis by the mother.
6. Form a Schiff base between a ketose substrate and transaldolase, reduce it with tritiated $NaBH_4$, and fingerprint the labeled enzyme.
7. $\Delta E_0'$ for the reduction of glutathione by NADPH is +0.09 V. Hence, $\Delta G^{\circ\prime}$ is −4.15 kcal/mol, which corresponds to an equilibrium constant of 1126. The required $[NADPH]/[NADP^+]$ ratio is 8.9×10^{-2}.
8. Fructose 2,6-bisphosphate, present at a high concentration when glucose is abundant, normally inhibits gluconeogenesis by blocking fructose 1,6-bisphosphatase. In this genetic disorder, the phosphatase is active irrespective of the glucose level. Hence, substrate cycling is increased, which generates heat. The level of fructose 1,6-bisphosphate is consequently lower than normal. Less pyruvate is formed, resulting in less acetyl CoA. In turn, less ATP is formed by the citric acid cycle and oxidative phosphorylation.
9. The lipoic acid is covalently attached to a lysine side chain in the dihydrolipoyl dehydrogenase component (E_2) of pyruvate dehydrogenase (p. 516). This flexible unit carries acetyl groups from bound thiamine pyrophosphate to coenzyme A.
10. Reactions (b) and (e) would be blocked.

Chapter 23

1. Galactose + ATP + UTP + H_2O + glycogen$_n$ $\longrightarrow$
   glycogen$_{n+1}$ + ADP + UDP + 2 P_i + H^+
2. There is a deficiency of the branching enzyme.
3. The high level of glucose 6-phosphate in von Gierke's disease, resulting from the absence of glucose 6-phosphatase or the transporter, shifts the allosteric equilibrium of phosphorylated glycogen synthase toward the active form.
4. Glucose is an allosteric inhibitor of phosphorylase *a*. Hence, crystals grown in its presence are in the T state. The addition of glucose 1-phosphate, a substrate, shifts the $R \rightarrow T$ equilibrium toward the R state. The conformational differences between these states are sufficiently large that the crystal shatters unless it is stabilized by chemical cross-links. The shattering of a crystal caused by an allosteric transition was first observed by Haurowitz in the oxygenation of crystals of deoxyhemoglobin.
5. The phosphoryl donor is glucose 1,6-bisphosphate, which is formed from glucose 1-phosphate and ATP in a reaction catalyzed by phosphoglucokinase.
6. Water is excluded from the active site to prevent hydrolysis. The entry of water could lead to the formation of glucose rather than glucose 1-phosphate. A site-specific mutagenesis experiment is revealing in this regard. In phosphorylase, Tyr 573 is hydrogen bonded to the 2′-OH of a glucose residue. The ratio of glucose 1-phosphate to glucose product is 9000:1 for the wild-type enzyme, and 500:1 for the Phe 573 mutant. Model building suggests that a water molecule occupies the site normally filled by the phenolic OH of tyrosine and occasionally attacks the oxocarbonium ion intermediate to form glucose. See D. Palm, H.W. Klein, R. Schinzel, M. Buehner, and E.J.M. Helmreich, *Biochemistry* 29(1990):1099.
7. The substrate can be handed directly from the transferase site to the debranching site.
8. (a) Muscle phosphorylase *b* will be inactive even when the AMP level is high. Hence, glycogen will not be degraded unless phosphorylase is converted into the *a* form by hormone-induced or Ca^{2+}-induced phosphorylation.
   (b) Phosphorylase *b* cannot be converted into the much more active *a* form. Hence, the mobilization of liver glycogen will be markedly impaired.
   (c) The elevated level of the kinase will lead to the phosphorylation and activation of glycogen phosphorylase. Little glycogen will be present in the liver because it will be persistently degraded.
   (d) Protein phosphatase 1 will be continually active. Hence, the level of phosphorylase *b* will be higher than normal, and glycogen will be less readily degraded.
   (e) Protein phosphatase 1 will be much less effective in dephosphorylating glycogen synthase and glycogen phosphorylase. Consequently, the synthase will stay in the less active *b* form, and the phosphorylase will stay in the more active *a* form. Both changes will lead to increased degradation of glycogen.
   (f) The absence of glycogenin will block the initiation of glycogen synthesis. Very little glycogen will be synthesized in its absence.
9. The slow phosphorylation of the α subunits of phosphorylase kinase serves to prolong the degradation of glycogen. The kinase cannot be deactivated until its α subunits are phosphorylated. The slow phosphorylation of α assures that the kinase and, in turn, phosphorylase stay active for a defined interval. See H.G. Hers, *Ann. Rev. Biochem.* 45 (1976):167.

Chapter 24

1. (a) Glycerol + 2 NAD^+ + P_i + ADP $\longrightarrow$
   pyruvate + ATP + H_2O + 2 NADH + H^+
   (b) Glycerol kinase and glycerol phosphate dehydrogenase.
2. Stearate + ATP + 13½ H_2O + 8 FAD +
   8 NAD^+ $\longrightarrow$ 4½ acetoacetate + 14½ H^+ +
   8 $FADH_2$ + 8 NADH + AMP + 2 P_i
3. (a) Oxidation in mitochondria, synthesis in the cytosol.
   (b) Acetyl CoA in oxidation, acyl carrier protein for synthesis.
   (c) FAD and NAD^+ in oxidation, NADPH for synthesis.
   (d) L-Isomer of 3-hydroxyacyl CoA in oxidation, D-isomer in synthesis.
   (e) Carboxyl to methyl in oxidation, methyl to carboxyl in synthesis.
   (f) The enzymes of fatty acid synthesis, but not those of oxidation, are organized in a multienzyme complex.
4. (a) Palmitoleate; (b) linoleate; (c) linoleate; (d) oleate; (e) oleate; and (f) linolenate.
5. C-1 is more radioactive (see p. 892 for a discussion of this experimental approach in the elucidation of the direction of synthesis of a polypeptide chain).
6. The enolate anion of one thioester attacks the carbonyl carbon atom of the other thioester to form a C–C bond.
7. Decarboxylation drives the condensation of malonyl-ACP and acetyl-ACP. In contrast, the condensation of two molecules of acetyl-ACP is energetically unfavorable. In gluconeogenesis, decarboxylation drives the formation of phosphoenolpyruvate from oxaloacetate (p. 573).
8. Adipose-cell lipase is activated by phosphorylation. Hence, overproduction of the cAMP-activated kinase will lead to accelerated breakdown of triacylglycerols and depletion of fat stores.
9. The mutant enzyme would be persistently active because it could not be inhibited by phosphorylation. Fatty acid synthesis would be abnormally active. Such a mutation might lead to obesity.
10. Carnitine translocase deficiency and glucose 6-phosphate transporter deficiency.
11. In the fifth round of β-oxidation, *cis*-Δ^2-enoyl CoA is formed. Dehydration by the classic hydratase yields D-3-hydroxyacyl CoA, the wrong isomer for the next enzyme in β-oxidation. This dead-end is circumvented by a second hydratase that removes water to give *trans*-Δ^2-enoyl CoA. Addition of water by the classic hydratase then yields L-3-hydroxyacyl CoA, the appropriate isomer. Thus, hydratases of opposite stereospecificities serve to *epimerize* (invert the configuration of) the 3-hydroxyl group of the acyl

CoA intermediate. See J.K. Hiltunen, P.M. Palosaari, and W.-H. Kunau, *J. Biol. Chem.* 264(1989):13536.

12. The probability of synthesizing an error-free polypeptide chain decreases as the length of the chain increases. A single mistake can make the entire polypeptide ineffective. In contrast, a defective subunit can be spurned in forming a noncovalent multienzyme complex; the good subunits are not wasted.

Chapter 25

1. (a) Pyruvate; (b) oxaloacetate; (c) α-ketoglutarate; (d) α-ketoisocaproate; (e) phenylpyruvate; and (f) hydroxyphenylpyruvate.
2. (a) Aspartate + α-ketoglutarate + GTP + ATP + $2\ H_2O$ + NADH + H^+ $\longrightarrow$ ½ glucose + glutamate + CO_2 + ADP + GDP + NAD^+ + 2 P_i
   (b) Aspartate + CO_2 + NH_4^+ + 3 ATP + NAD^+ + $4\ H_2O$ $\longrightarrow$ oxaloacetate + urea + 2 ADP + 4 P_i + AMP + NADH + H^+
3. (a) Label the methyl carbon atom of L-methylmalonyl CoA with ^{14}C. Determine the location of ^{14}C in succinyl CoA. The group transferred is the one bonded to the labeled carbon atom.
   (b) The proton that is abstracted from the methyl group of L-methylmalonyl CoA is directly transferred to the adjacent carbon atom.
4. Thiamine pyrophosphate.
5. It acts as an electron sink. See C. Walsh, *Enzymatic Reaction Mechanisms* (W.H. Freeman, 1979), p. 178.
6. Deuterium is abstracted by the radical form of the coenzyme. The methyl group rotates before hydrogen is returned to the product radical.
7. A carbanion or a carbonium ion.
8. CO_2 + NH_4^+ + 3 ATP + NAD^+ + aspartate + 3 H_2O $\longrightarrow$ urea + 2 ADP + 2 P_i + AMP + PP_i + NADH + H^+ + oxaloacetate

   Four ~P are spent.
9. The mass spectrometric analysis strongly suggests that three enzymes—pyruvate dehydrogenase, α-ketoglutarate dehydrogenase, and the branched-chain α-keto dehydrogenase—are deficient. Most likely, the common E3 component of these enzymes (p. 517) is missing or defective. This proposal could be tested by purifying these three enzymes and assaying their capacity to catalyze the regeneration of lipoamide.
10. Benzoate, phenylacetate, and arginine would be given to supply a protein-restricted diet. Nitrogen would emerge in hippurate, phenylacetylglutamine, and citrulline. See S.W. Brusilow and A.L. Horwich in C.R. Scriver, A.L. Beaudet, W.S. Sly, and D. Valle (eds.), *The Metabolic Basis of Inherited Disease,* 6th ed. (McGraw-Hill, 1989), pp. 629–663, for a detailed discussion of the therapeutic rationale.
11. Aspartame, a dipeptide ester (aspartyl-phenylalanine methyl ester), is hydrolyzed to L-aspartate and L-phenylalanine. High levels of phenylalanine are harmful in phenylketonurics.
12. *N*-acetylglutamate is synthesized from acetyl CoA and glutamate. Once again, acetyl CoA serves as an activated acetyl donor. This reaction is catalyzed by *N*-acetylglutamate synthase.

Chapter 26

1. $\Delta E_0' = +0.11$ V and $\Delta G^{\circ\prime} = -5.1$ kcal/mol.
2. DCMU inhibits electron transfer in the link between photosystems II and I. O_2 evolution can occur in the presence of DCMU if an artificial electron acceptor such as ferricyanide can accept electrons from Q.
3. (a) 28.7 kcal/einstein (or 120 kJ/einstein).
   (b) 1.24 V.
   (c) One 1000-nm photon has the free energy content of 2.39 ATP. A minimum of 0.42 photon is needed to drive the synthesis of an ATP.
4. The formation of O_2 is a four-electron process. Each brief light pulse provides one electron for reduction, which is accumulated by the manganese center.
5. Aldolase participates in the Calvin cycle, whereas transaldolase participates in the pentose phosphate pathway.
6. The concentration of 3-phosphoglycerate would increase, whereas that of ribulose 1,5-bisphosphate would decrease.
7. The concentration of 3-phosphoglycerate would decrease, whereas that of ribulose 1,5-bisphosphate would increase.
8. Phycoerythrin, the most peripheral protein in the phycobilisome.
9. (a)

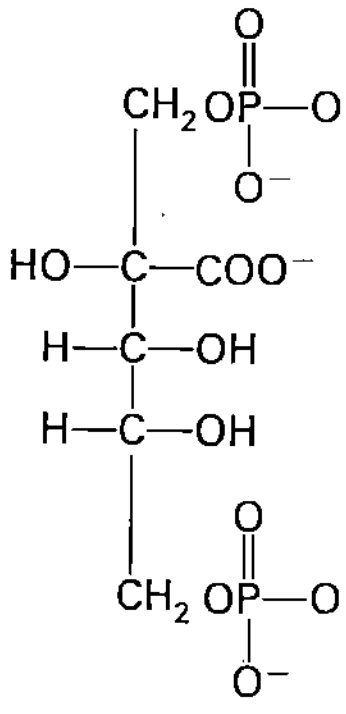

2-Carboxyarabinitol 1,5-bisphosphate (CABP)

   (b) CABP resembles the addition compound formed in the reaction of CO_2 and ribulose 1,5-bisphosphate (p. 672).
   (c) As predicted, CABP is a potent inhibitor of rubisco.
10. Aspartate + glyoxylate $\longrightarrow$ oxaloacetate + glycine.

Chapter 27

1. Glycerol + 4 ATP + 3 fatty acids + 4 H_2O $\longrightarrow$ triacylglycerol + ADP + 3 AMP + 7 P_i + 4 H^+
2. Glycerol + 3 ATP + 2 fatty acids + 2 H_2O + CTP + serine $\longrightarrow$ phosphatidyl serine + CMP + ADP + 2 AMP + 6 P_i + 3 H^+

3. (a) CDP–diacylglycerol; (b) CDP–ethanolamine; (c) acyl CoA; (d) CDP–choline; (e) UDP–glucose or UDP–galactose; (f) UDP–galactose; and (g) geranyl pyrophosphate.
4. (a and b) None, because the label is lost as CO_2.
5. (a) No receptor is synthesized.
   (b) Receptors are synthesized but do not reach the plasma membrane because they lack signals for intracellular transport or do not fold properly.
   (c) Receptors reach the cell surface, but they fail to bind LDL normally because of a defect in the LDL-binding domain.
   (d) Receptors reach the cell surface and bind LDL, but they fail to cluster in coated pits because of a defect in their carboxyl-terminal region.
6. Deamination of cytidine to uridine changes CAA (Gln) into UAA (stop).
7. Benign prostatic hypertrophy can be treated by inhibiting the 5α-reductase. *Finasteride,* the 4-aza steroid analog of dihydrotestosterone, competitively inhibits the reductase but does not act on androgen receptors. Patients taking finasteride have a markedly lower plasma level of dihydrotestosterone and a nearly normal level of testosterone. The prostate becomes smaller, whereas testosterone-dependent processes such as fertility, libido, and muscle strength appear to be unaffected (see E. Stoner, *Steroid Biochem. Molec. Biol.* 37(1990):375–378). Genetic deficiencies of 5α-reductase are discussed by J.E. Griffin and J.D. Wilson, in C.R. Scriver, A.L. Beaudet, W.S. Sly, and D. Valle (eds.), *The Metabolic Basis of Inherited Disease,* 6th ed., (1989, McGraw-Hill), pp. 1919–1944.

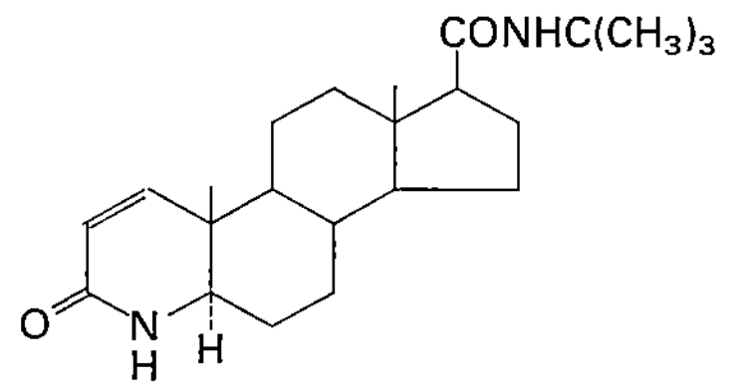

Finasteride

8. Patients who are most sensitive to debrisoquine have a deficiency of a liver P450 enzyme encoded by a member of the *CYP*2 subfamily. This characteristic is inherited as an autosomal recessive trait. The capacity to degrade other drugs may be impaired in people who hydroxylate debrisoquine at a slow rate because a single P450 enzyme usually handles a broad range of substrates. See W.B. Pratt and P. Taylor, *Principles of Drug Action: The Basis of Pharmacology,* 3rd ed. (Churchill Livingstone, 1990), pp. 496–500; and F.J. Gonzalez and D.W. Nebert, *Trends Genet.* 6(1990):182.
9. Deficiencies of 11β-hydroxylase, 17α-hydroxylase, 3β-dehydrogenase, and desmolase also lead to adrenal hyperplasia. See M.I. New, P.C. White, S. Pang, B. Dupont, and P.W. Speiser in C.R. Scriver, A.L. Beaudet, W.S. Sly, and D. Valle (eds.), *The Metabolic Basis of Inherited Disease,* 6th ed. (McGraw-Hill, 1989), pp. 1181–1917.
10. Many hydrophobic odorants are deactivated by hydroxylation. O_2 is activated by a cytochrome P450 monooxygenase. NADPH serves as the reductant. One oxygen atom of O_2 goes into the odorant substrate, whereas the other is reduced to water.

Chapter 28

1. Glucose + 2 ADP + 2 P_i + 2 NAD^+ + 2 glutamate
   $\longrightarrow$ 2 alanine + 2 α-ketoglutarate + 2 ATP + 2 NADH + 2 H_2O + 2 H^+
2. $N_2 \longrightarrow NH_4^+ \longrightarrow$ glutamate $\longrightarrow$ serine $\longrightarrow$ glycine $\longrightarrow$ δ-aminolevulinate $\longrightarrow$ porphobilinogen $\longrightarrow$ heme
3. (a) N^5;N^{10}-Methylenetetrahydrofolate; (b) N^5-methyltetrahydrofolate.
4. γ-Glutamyl phosphate is a likely reaction intermediate.
5. The administration of glycine leads to the formation of isovalerylglycine. This water-soluble conjugate, in contrast with isovaleric acid, is excreted very rapidly by the kidneys. See R.M. Cohn, M. Yudkoff, R. Rothman, and S. Segal, *New Engl. J. Med.* 299(1978):996.
6. They carry out nitrogen fixation. The absence of photosystem II provides an environment in which O_2 is not produced. Recall that the nitrogenase is very rapidly inactivated by O_2. See R.Y. Stanier, J.L. Ingraham, M.L. Wheelis, and P.R. Painter, *The Microbial World,* 5th ed. (Prentice-Hall, 1986), pp. 356–359, for a discussion of heterocysts.
7. The cytosol is a reducing environment, whereas the extracellular milieu is an oxidizing environment.
8. Succinyl CoA is formed in the mitochondrial matrix.
9. The reduction of N^5,N^{10}-methylenetetrahydrofolate to N^5-methyltetrahydrofolate is essentially irreversible. Tetrahydrofolate is normally recovered in the conversion of homocysteine to methionine, a methylcobalamin-dependent reaction. In pernicious anemia, vitamin B_{12} deficiency blocks this methyl transfer reaction. Hence, the folate pool is trapped in the form of N^5-methyltetrahydrofolate. Most one-carbon transfers are thereby blocked.
10. CO, which is formed in the breakdown of heme, and HCN, which is formed in the synthesis of ethylene, are two plausible candidates. Both bind to heme, CO to the ferrous form and CN^- to the ferric form. A heme protein could serve as a receptor for either. It will be interesting to learn whether CO and HCN have messenger roles.

Chapter 29

1. Glucose + 2 ATP + 2 $NADP^+$ + H_2O $\longrightarrow$
   PRPP + CO_2 + ADP + AMP + 2 NADPH + 3 H^+
2. Glutamine + aspartate + CO_2 + 2 ATP + NAD^+ $\longrightarrow$
   orotate + 2 ADP + 2 P_i + glutamate + NADH + H^+
3. (a, c, d, and e) PRPP; (b) carbamoyl phosphate.
4. PRPP and formylglycinamide ribonucleotide.
5. dUMP + serine + NADPH + H^+ $\longrightarrow$
   dTMP + $NADP^+$ + glycine
6. There is a deficiency of N^{10}-formyltetrahydrofolate. Sulfa-

nilamide inhibits the synthesis of folate by acting as an analog of p-aminobenzoate, one of the precursors of folate.

7. PRPP is the activated intermediate in the synthesis of (a) phosphoribosylamine in the de novo pathway of purine formation, (b) purine nucleotides from free bases by the salvage pathway, (c) orotidylate in the formation of pyrimidines, (d) nicotinate ribonucleotide, (e) phosphoribosyl-ATP in the pathway leading to histidine, and (f) phosphoribosylanthranilate in the pathway leading to tryptophan.
8. The small subunit releases ammonia from glutamine. The nascent ammonia then reacts with an activated form of CO_2 that is formed by the large subunit. The bicarbonate-dependent ATPase activity implies that this activated species is carbonyl phosphate. Reaction of this carbonic-phosphoric mixed anhydride with NH_3 yields carbamate, which would then react with ATP to give carbamoyl phosphate. See C. Walsh, *Enzymatic Reaction Mechanisms* (W.H. Freeman, 1979), pp. 150–154. The holoenzyme probably contains a channel for the diffusion of nascent NH_3 from the small subunit to the large one. Recall that tryptophan synthetase (p. 726) contains a channel for the diffusion of indole from the α to the β subunit. An intriguing possibility is that the catalytic activities of the subunits of carbamoyl phosphate synthetase are coordinated: NH_3 production by the small subunit is stimulated when carbonyl phosphate is formed by the large subunit.
9. Analogous reactions occur in the urea cycle—from citrulline to argininosuccinate, and then to arginine (p. 635).
10. In vitamin B_{12} deficiency, methyltetrahydrofolate cannot donate its methyl group to homocysteine to regenerate methionine. Because the synthesis of methyltetrahydrofolate is irreversible (p. 720), the cell's tetrahydrofolate will ultimately be converted into this form. No formyl or methylene tetrahydrofolate will be left for nucleotide synthesis. Pernicious anemia illustrates the intimate connection between amino acid and nucleotide metabolism.
11. (a) *S*-Adenosylhomocysteine hydrolase activity is markedly diminished in ADA-deficient patients because of reversible inhibition by adenosine and suicide inactivation by 2′-deoxyadenosine.
    (b) The activated methyl cycle (p. 722) is blocked in these patients. *S*-Adenosylhomocysteine is a potent inhibitor of methyl transfer reactions involving *S*-adenosylmethionine. For a discussion of adenosine deaminase deficiency, see the article by N.M. Kredich and M.S. Hershfield in C.R. Scriver, A.L. Beaudet, W.S. Sly, and D. Valle (eds.), *The Metabolic Basis of Inherited Disease,* 6th ed. (McGraw-Hill, 1989), pp. 1045–1075.
12. (a) Cell A cannot grow in a HAT medium because it cannot synthesize dTMP either from thymidine or dUMP. Cell B cannot grow in this medium because it cannot synthesize purines by either the de novo pathway or the salvage pathway. Cell C can grow in a HAT medium because it contains active thymidine kinase from cell B (enabling it to phosphorylate thymidine to dTMP) and hypoxanthine-guanine phosphoribosyl transferase from cell A (enabling it to synthesize purines from hypoxanthine by the salvage pathway).
    (b) Transform cell A with a plasmid containing foreign genes of interest and a functional thymidine kinase gene. The only cells that will grow in a *HAT* medium are those that have acquired a thymidylate kinase gene; nearly all of these transformed cells will also contain the other genes on the plasmid.
13. These patients have a high level of urate because of the breakdown of nucleic acids. Allopurinol prevents the formation of kidney stones and blocks other deleterious consequences of hyperuricemia by preventing the formation of urate (p.621).
14. The free energies of binding are −13.8 (wild type), −11.9 (Asn 27), and −9.1 (Ser 27) kcal/mol. The loss in binding energy is 1.9 and 4.7 kcal/mol. See E.H. Howell, J.E. Villafranca, M.S. Warren, S.J. Oatley, and J. Kraut, *Science* 231(1986):1125.
15. The cytosolic level of ATP in liver falls and that of AMP rises above normal in all three conditions. The excess AMP is degraded to urate. See C.R. Scriver, A.L. Beaudet, W.S. Sly, and D. Valle (eds.), *The Metabolic Basis of Inherited Disease,* 6th ed. (McGraw Hill, 1989), pp. 984–988, for an illuminating discussion.

Chapter 30

1. Liver contains glucose 6-phosphatase, whereas muscle and brain do not. Hence, muscle and brain, in contrast with liver, do not release glucose. Another key enzymatic difference is that liver has little of the transferase needed to activate acetoacetate to acetoacetyl CoA. Consequently, acetoacetate and 3-hydroxybutyrate are exported by the liver for use by heart muscle, skeletal muscle, and brain.
2. (a) Adipose cells normally convert glucose to glycerol 3-phosphate for the formation of triacylglycerols. A deficiency of hexokinase would interfere with the synthesis of triacylglycerols.
    (b) A deficiency of glucose 6-phosphatase would block the export of glucose from liver following glycogenolysis. This disorder (called von Gierke's disease) is characterized by an abnormally high content of glycogen in the liver and a low blood glucose level.
    (c) A deficiency of carnitine acyltransferase I impairs the oxidation of long-chain fatty acids. Fasting and exercise precipitate muscle cramps in these people.
    (d) Glucokinase enables the liver to phosphorylate glucose even in the presence of a high level of glucose 6-phosphate. A deficiency of glucokinase would interfere with the synthesis of glycogen.
    (e) Thiolase catalyzes the formation of two molecules of acetyl CoA from acetoacetyl CoA and CoA. A deficiency of thiolase would interfere with the utilization of acetoacetate as a fuel when the blood sugar level is low.
    (f) Phosphofructokinase will be less active than normal because of the lowered level of F-2,6-BP. Hence, glycolysis will be much slower than normal.
3. (a) A high proportion of fatty acids in the blood are bound to albumin. Cerebrospinal fluid has a low content of fatty acids because it has little albumin.
    (b) Glucose is highly hydrophilic and soluble in aqueous

media, in contrast with fatty acids, which must be carried by transport proteins such as albumin. Micelles of fatty acids would disrupt membrane structure.

(c) Fatty acids, not glucose, are the major fuel of resting muscle.

4. (a) A watt is equal to 1 joule (J) per second (0.239 calorie per second). Hence, 70 W is equivalent to 0.07 kJ/s or 0.017 kcal/s.

(b) A watt is a current of 1 ampere (A) across a potential of 1 volt (V). For simplicity, let us assume that all the electron flow is from NADH to O_2 (a potential drop of 1.14 V). Hence, the current is 61.4 A, which corresponds to 3.86×10^{20} electrons per second (1 A = 1 coulomb/s = 6.28×10^{18} charges/s).

(c) Three ATP are formed per NADH oxidized (two electrons). Hence, one ATP is formed per 0.67 electrons transferred. A flow of 3.86×10^{20} electrons per second therefore leads to the generation of 5.8×10^{20} ATP per second or 0.96 mmol per second.

(d) The molecular weight of ATP is 507. The total body content of ATP of 50 g is equal to 0.099 mol. Hence, ATP turns over about once per 100 seconds when the body is at rest.

5. (a) The stoichiometry of complete oxidation of glucose is

$$C_6H_{12}O_6 + 6\ O_2 \longrightarrow 6\ CO_2 + 6\ H_2O$$

and that of tripalmitoylglycerol is

$$C_{51}H_{98}O_6 + 72.5\ O_2 \longrightarrow 51\ CO_2 + 49\ H_2O$$

Hence, the RQ values are 1.0 and 0.703, respectively.

(b) An RQ value reveals the relative usage of carbohydrate and fats as fuels. The RQ of a marathon runner typically decreases from 0.97 to 0.77 during the race. The lowering of the RQ reflects the shift in fuel from carbohydrate to fat.

6. One gram of glucose (molecular weight 180.2) is equal to 5.55 mmol, and one gram of tripalmitoylglycerol (molecular weight 807.3) is equal to 1.24 mmol. The reaction stoichiometries (see problem 5) indicate that 6 mol of H_2O are produced per mol of glucose oxidized, and 49 mol of H_2O per mol of tripalmitoylglycerol oxidized. Hence, the H_2O yields per gram of fuel are 33.3 mmol (0.6 g) for glucose, and 60.8 mmol (1.09 g) for tripalmitoylglycerol. Thus, complete oxidation of this fat gives 1.82 times as much water as does glucose. Another advantage of triacylglycerols is that they can be stored in essentially anhydrous form, whereas glucose is stored as glycogen, a highly hydrated polymer (p. 605). A hump consisting mainly of glycogen would be an intolerable burden—far more than the straw that broke the camel's back!

7. A typical macadamia nut has a mass of about 2 g. Because it consists mainly of fats (~9 kcal/g), a nut has a value of about 18 kcal. The ingestion of 10 nuts results in an intake of about 180 kcal. As was discussed in problem 4, a power consumption of 1 W corresponds to 0.239 cal per second, and so 400-W running requires 95.6 cal/s, or .0956 kcal/s. Hence, one would have to run 1882 s, or about 31 min, to spend the calories provided by 10 nuts.

8. A high blood glucose level would trigger the secretion of insulin, which would stimulate the synthesis of glycogen and triacylglycerols. A high insulin level would impede the mobilization of fuel reserves during the marathon.

Chapter 31

1. DNA polymerase I uses deoxyribonucleoside triphosphates; pyrophosphate is the leaving group. DNA ligase uses a DNA-adenylate (AMP joined to the 5′-phosphate) as a reaction partner; AMP is the leaving group. Topoisomerase I uses a DNA-tyrosyl intermediate (5′-phosphate linked to the phenolic OH); the tyrosine residue of the enzyme is the leaving group.
2. FAD, CoA, and $NADP^+$ are plausible alternatives.
3. DNA ligase relaxes supercoiled DNA by catalyzing the cleavage of a phosphodiester bond in a DNA strand. The attacking group is AMP, which becomes attached to the 5′-phosphoryl group at the site of scission. AMP is required because this reaction is the reverse of the final step in the joining of pieces of DNA (see Figure 31-14 on p. 793).
4. Positive supercoiling resists the unwinding of DNA. The melting temperature of DNA increases in going from negatively supercoiled to relaxed to positively supercoiled DNA. Positive supercoiling is probably an adaptation to high temperature.
5. (a) Long stretches of each occur because the transition is highly cooperative.

(b) B–Z junctions are energetically highly unfavorable.

(c) A–B transitions are less cooperative than B–Z transitions because the helix stays right-handed at an A–B junction but not at a B–Z junction.

6. ATP hydrolysis is required to release DNA gyrase after it has acted on its DNA substrate. Negative supercoiling requires only the binding of ATP, not its hydrolysis.
7. (a) Pro (CCC), Ser (UCC), Leu (CUC) and Phe (UUC). Alternatively, the last base of each of these codons could be U.

(b) These C $\longrightarrow$ U mutations were produced by nitrous acid.

8. If replication were unidirectional, tracks with a low grain density at one end and a high grain density at the other end would be seen. On the other hand, if replication were bidirectional, the middle of a track would have a low density, as shown below. For *E. coli*, the grain tracks are denser on both ends than in the middle, indicating that replication is bidirectional.

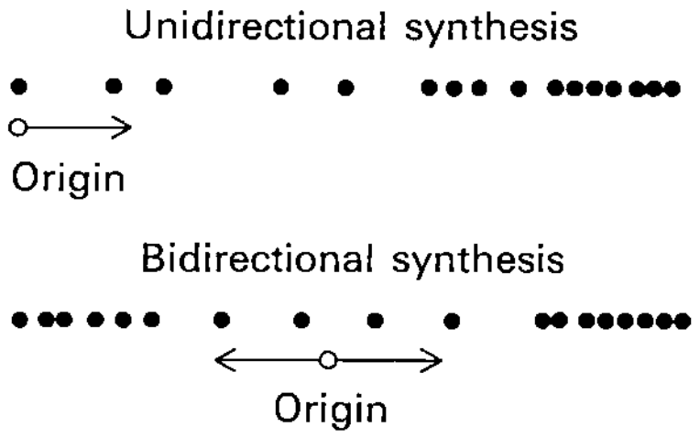

9. Potentially deleterious side reactions are avoided. The enzyme itself might be damaged by light if it could be acti-

vated by light in the absence of bound DNA harboring a pyrimidine dimer. The DNA-induced absorption band is reminiscent of the glucose-induced activation of the phosphotransferase activity of hexokinase (p. 499).

10. (a) 96.2 revolutions per second (1000 nucleotides/s divided by 10.4 nucleotides/turn for B-DNA gives 96.2 rps).
    (b) 0.34 μm/s (1000 nucleotides/s corresponds to 3400 Å/s because the axial distance between nucleotides in B-DNA is 3.4 Å).
    (c) The rate of rotation of DNA at a replication fork (96.2 rps) is nearly the same as that of a bacterial flagellum (100 rps). Skeletal muscle sarcomere shortening is about 15-fold as rapid as polymerase movement at a replication fork (5 μm/s, compared with 0.34 μm/s). Smooth muscle sarcomere shortening (0.4 μm/s) occurs at about the same speed as replication fork movement.

Chapter 32

1. Helicase participates in DNA replication. The recB component of the recBCD complex generates single-stranded DNA in general recombination. The recA protein binds to single-stranded DNA and catalyzes its invasion of duplex DNA and a switch in base pairing.
2. (a) The recA-DNA complex contains 18.6 base pairs per turn, compared with 10.4 for B-DNA. Hence, the linking number (*Lk*) is 56, compared with 100 for the relaxed circle formed in the absence of recA protein.
   (b) Recall that about 70% of the change in *Lk* in an isolated circular DNA molecule is expressed in *Wr* (writhing) and 30% in *Tw* (twisting) (p. 796). Hence, after removal of recA, *Wr* is 30.8 less than that of the relaxed circle, which means that the duplex is negatively (right-handed) supercoiled. The degree of supercoiling σ is $-.44$, compared with $-.06$ for most naturally occurring DNA. Thus, closure in the presence of recA generates a highly supercoiled molecule, which is much more compact than its relaxed counterpart and will sediment much more rapidly.
3. Proteolysis of the lexA protein at the onset of the SOS response relieves the repression of synthesis of lexA mRNA. DNA repair then leads to a decrease in the amount of the ssDNA–recA complex, the coprotease. Hence, the level of intact lexA increases, which in turn terminates the SOS response.
4. (a) In the presence of ATPγS, recA protein binds to DNA but its dissociation is blocked because this analog is not readily hydrolyzed. Strand exchange does not take place.
   (b) RecB protein binds to DNA in the presence of this analog but its release is blocked. Consequently, single-stranded DNA is not formed.
5. Negative supercoiling promotes recombination by inducing strand separation (negative *Wr* can be changed into negative *Tw*). Furthermore, negatively supercoiled DNA occupies a much smaller volume than does relaxed DNA. Hence, the effective concentration of sites undergoing recombination is increased by negative supercoiling. For an illuminating discussion, see R. Kanaar and N.R. Cozzarelli, *Curr. Opin. Struct. Biol.* 2(1992):369.
6. Homologous recombination between a genomic DNA sequence and a similar cloned DNA sequence introduced into a cell makes it possible to transfer any modification of the cloned gene into the genome of a cell. For example, a stop codon could be placed early in a sequence encoding a protein of interest. See M.R. Capecchi, *Science* 244 (1989):1288.
7. Viruses make the most of a small genome. The encoding of a different second subunit would require about 2 additional kb of RNA. Replication would be slower and packaging of the nucleic acid more difficult.

Chapter 33

1. The sequence of the coding (+, sense) strand is

   5′-ATGGGGAACAGCAAGAGTGGGGCCCTGTCCAAGGAG-3′

   and the sequence of template (−, antisense) strand is

   3′-TACCCCTTGTCGTTCTCACCCCGGGACAGGTTCCTC-5′
2. Heparin, a glycosaminoglycan (p. 474), is highly anionic. Its negative charges, like the phosphodiester bridges of DNA templates, bind to lysine and arginine residues of β'.
3. This mutant sigma would competitively inhibit the binding of holoenzyme and prevent the specific initiation of RNA chains at promoter sites.
4. The core enzyme without sigma binds more tightly to the DNA template than does the holoenzyme. The retention of sigma after chain initiation would make the mutant RNA polymerase less processive. Hence, RNA synthesis would be much slower than normal.
5. A 100-kd protein contains about 910 residues, which are encoded by 2730 nucleotides. At a maximal transcription rate of 50 nucleotides per second, the protein would be synthesized in 54.6 seconds.
6. Initiation at strong promoters occurs every two seconds. In this interval, 100 nucleotides are transcribed. Hence, centers of transcription bubbles are 34 nm (340 Å) apart.
7. (a) The lowest band on the gel will be that of (i), whereas the highest will be that of (v). Band (ii) will be at the same position as (i) because the RNA is not complementary to the nontemplate strand, whereas band (iii) will be higher because a complex is formed between RNA and the template strand. Band (iv) will be higher than the others because strand 1 is complexed to 2, and strand 2 to 3. Band (v) is the highest because core polymerase associates with the three strands.
   (b) None, because rifampicin acts prior to the formation of the open complex.
   (c) RNA polymerase is processive. Once the template is bound, heparin cannot enter the DNA-binding site.
   (d) The longest RNA product formed is a 36-mer rather than the full 72-mer. Because GTP is absent, synthesis stops when the first C downstream of the bubble is encountered in the template strand.
8. Ser-Ile-Phe-His-Pro-Stop.

9. A mutation that disrupted the normal AAUAA recognition sequence for the endonuclease could account for this finding. In fact, a change from U to C in this sequence caused this defect in a thalassemic patient. Cleavage occurred at the AAUAAA 900 nucleotides downstream from this mutant AACAAA site. See S.H. Orkin, T.-C. Cheng, S.E. Antonarakis, and H.H. Kazazian, Jr., *EMBO J.* 4(1985):453.
10. Mg^{2+} coordinates strongly to oxygen but not sulfur, whereas Mn^{2+} coordinates strongly to sulfur but not oxygen. See E.K. Jaffe, and M. Cohn, *J. Biol. Chem.* 253 (1978):4823.
11. One possibility is that the 3′ of the poly(U) donor strand cleaves the phosphodiester bond on the 5′ side of the insertion site. The newly formed 3′ terminus of the acceptor strand then cleaves the poly(U) strand on the 5′ side of the nucleotide that initiated the attack. In other words, a U could be added by two transesterification reactions. This postulated mechanism is akin to the one in RNA splicing. See T.R. Cech, *Cell* 64(1991):667.

Chapter 34

1. (a) No; (b) no; and (c) yes.
2. Four bands: light, heavy, a hybrid of light 30S and heavy 50S, and a hybrid of heavy 30S and light 50S.
3. About 799 high-energy phosphate bonds are consumed—400 to activate the 200 amino acids, 1 for initiation, and 398 to form 199 peptide bonds.
4. Type I: b, c, and f; type 2: a, d, and e.
5. A mutation caused by the insertion of an extra base can be suppressed by a tRNA that contains a fourth base in its anticodon. For example, UUUC rather than UUU is read as the codon for phenylalanine by a tRNA that contains 3′-AAAG-5′ as its anticodon.
6. One approach is to synthesize a tRNA that is acylated with a reactive amino acid analog. For example, bromoacetylphenylalanyl-tRNA is an affinity-labeling reagent for the P site of *E. coli* ribosomes. See H. Oen, M. Pellegrini, D. Eilat, and C.R. Cantor, *Proc. Nat. Acad. Sci.* 70(1973):2799.
7. The sequence GAGGU is complementary to a sequence of five bases at the 3′ end of 16S rRNA and is located several bases on the 5′ side of an AUG codon. Hence this region is a start signal for protein synthesis. The replacement of G by A would be expected to weaken the interaction of this mRNA with the 16S rRNA and thereby diminish its effectiveness as an initiation signal. In fact, this mutation results in a tenfold decrease in the rate of synthesis of the protein specified by this mRNA. See J.J. Dunn, E. Buzash-Pollert, and F.W. Studier, *Proc. Nat. Acad. Sci.* 75(1978):2741, for a discussion of this informative mutant.
8. The nitrogen atom of the deprotonated α-amino group of aminoacyl-tRNA is the nucleophile in peptide-bond formation.
9. Proteins are synthesized from the amino to the carboxyl end on ribosomes, and in the reverse direction in the solid-phase method. The activated intermediate in ribosomal synthesis is an aminoacyl-tRNA; in the solid-phase method, it is the adduct of the amino acid and dicyclohexylcarbodiimide.
10. The error rates of DNA, RNA, and protein synthesis are of the order of 10^{-10}, 10^{-5}, and 10^{-4} per nucleotide (or amino acid) incorporated. The fidelity of all three processes depends on the precision of base pairing to the DNA or mRNA template. No error correction occurs in RNA synthesis. In contrast, the fidelity of DNA synthesis is markedly increased by the 3′ → 5′ proofreading nuclease activity and by postreplicative repair. In protein synthesis, the mischarging of some tRNAs is corrected by the hydrolytic action of the aminoacyl-tRNA synthetase. Proofreading also takes place when aminoacyl-tRNA occupies the A site on the ribosome; the GTPase activity of EF-Tu sets the pace of this final stage of editing.
11. GTP is not hydrolyzed until aminoacyl-tRNA is delivered to the A site of the ribosome. An earlier hydrolysis of GTP would be wasteful because EF-Tu-GDP has little affinity for aminoacyl-tRNA.
12. EF-Ts catalyzes the exchange of GTP for GDP bound to EF-Tu. In G protein cascades, an activated seven-helix receptor catalyzes GTP-GDP exchange in a G protein. For example, photoexcited rhodopsin triggers GTP-GDP exchange in transducin (p. 336).
13. The translation of an mRNA molecule can be blocked by antisense RNA, an RNA molecule with the complementary sequence. The antisense-sense RNA duplex cannot serve as a template for translation; single-stranded mRNA is required. Furthermore, the antisense-sense duplex is degraded by nucleases. Antisense RNA added to the external medium is spontaneously taken up by many cells. A precise quantity can be delivered by microinjection. Alternatively, a plasmid encoding the antisense RNA can be introduced into target cells. For an interesting discussion of antisense RNA and DNA as research tools and drug candidates, see H.M. Weintraub, *Sci. Amer.* 262(January 1990):40.

Chapter 35

1. (a) Cleavable amino-terminal signal sequences are usually 13 to 36 residues long and contain a highly hydrophobic central region 10 to 15 residues long. The amino-terminal part has at least one basic residue. The cleavage site is preceded by small neutral residues.
   (b) Prokaryotic signal sequences are similar to eukaryotic ones. In addition, a stop-transfer sequence is needed to keep the protein in the plasma membrane.
   (c) Mannose 6-phosphate residues direct proteins to lysosomes.
   (d) Integral membrane proteins initially in the ER go to the plasma membrane unless they carry instructions to the contrary. No specific signal is needed.
   (e) An amino-terminal sequence containing positively charged residues, serine, and threonine, in addition to hydrophobic residues.
   (f) An amino-terminal arginine, histidine, isoleucine, leucine, lysine, phenylalanine, or tryptophan.
2. The cytosolic portion of the receptor, which enables it to interact with coated pits, is likely to be altered.
3. (a) Endocytosis mediated by a receptor specific for mannose 6-phosphate residues. Endocytic vesicles containing the added lysosomal enzymes then fuse with lysosomes.

(b) The addition of mannose 6-phosphate to the extracellular medium should prevent lysosomal enzymes from reaching their destination if the normal pathway involves secretion outside the cell and import back into the cell. However, mannose 6-phosphate does not inhibit normal lysosomal targeting, showing that lysosomal enzymes reach their destination without leaving the cell.

4. (a) The chimeric protein will probably be found in the cytosol. The transmembrane sequence of a membrane-bound immunoglobulin functions as a stop-transfer sequence and not as a signal sequence.
   (b) The chimeric protein will probably be found in the plasma membrane because chymotrypsinogen is synthesized with a signal sequence.
5. Secretory proteins that are erroneously targeted to the cytosol will be rapidly degraded because their amino termini mark them for rapid destruction (Table 35-3, on p. 913). Rapid elimination of mistargeted secretory proteins is important because some of them (e.g., trypsinogen) could wreak havoc in the cytosol.
6. This genetic disorder, known as Zellweger syndrome, is caused by defective import of peroxisomal matrix proteins bearing SKL sequences. The mutation is probably in the gene for the SKL receptor or the associated translocase. See P.B. Lazarow and H.W. Moser in C.R. Scriver, A.L. Beaudet, W.S. Sly, and D. Valle (eds.), *The Metabolic Basis of Inherited Disease*, 6th ed. (McGraw-Hill, 1989), pp. 1479–1509.
7. Methotrexate binds very tightly to DHFR (p. 753) and prevents it from unfolding. Recall that proteins must be partly or completely unfolded during their translocation across the inner mitochondrial membrane as well as other membranes.
8. Mutant ras proteins are oncogenic only if they are membrane-anchored. Farnesylation of ras is essential for membrane anchoring. Hence, specific inhibitors of farnesyl transferase are potentially valuable anticancer agents. See N.E. Kohl, S.D. Mosser, S.J. deSolms, E.A. Giuliani, D.L. Pompliano, S.L. Graham, R.L. Smith, E.M. Scolnick, A. Oliff, and J.B. Gibbs, *Science* 260(1993):1934.

Chapter 36

1. (a) The *lac* repressor is missing in an i^- mutant. Hence, this mutant is constitutive for the proteins of the *lac* operon.
   (b) This mutant is constitutive for the proteins of the *trp* operon because the *trp* repressor is missing.
   (c) The arabinose operon is not expressed in this mutant because the *araC* protein is needed to activate transcription.
   (d) This mutant is lytic but not lysogenic, because it cannot synthesize the λ repressor.
   (e) This mutant is lysogenic but not lytic because it cannot synthesize the N protein, a positive control factor in transcription.
2. One possibility is that an i^s mutant produces an altered *lac* repressor that has almost no affinity for inducer but normal affinity for the operator. Such a *lac* repressor would bind to the operator and block transcription even in the presence of inducer.
3. This mutant has an altered *lac* operator that fails to bind the repressor. Such a mutator is called O^c (operator constitutive).
4. The cyclic AMP binding protein (CAP) is probably defective or absent in this mutant.
5. An *E. coli* cell bearing a λ prophage contains λ repressor molecules, which also block the transcription of the immediate-early genes of invading λ DNA.
6. (a) Translation of the p_{RE} transcript is from 5 to 10 times as rapid as that of the p_{RM} transcript because it contains the full protein synthesis initiation signal (as discussed on p. 895).
   (b) The more effective translation of the p_{RE} transcript provides a burst of λ repressor molecules needed to establish the lysogenic state. See M. Ptashe, K. Backman, Z. Humagun, A. Jeffrey, R. Maurer, B. Meyer, and R.T. Sauer, *Science* 194(1976):156.
7. The rate constant for association of wild-type *lac* repressor with the operator site is near the diffusion-controlled limit. Hence, a 100-fold increase in binding affinity implies that the mutant dissociates from the operator site about 100-fold more slowly than does the wildtype. The lag between addition of inducer and the initiation of transcription of the *lac* operon will be much longer in the mutant than in the wild type.
8. A repressor finds its target site by first binding to a nonspecific site anywhere in the DNA molecule and then diffusing along the DNA to reach the specific site. The repressor binds less tightly to nonspecific DNA at low ionic strength than at high ionic strength because of increased electrostatic repulsion. In contrast, binding to the operator is nearly independent of ionic strength because the interaction is mediated by hydrogen-bond and van der Waals interactions with the bases of the operator site.
9. The mutant protein acts as a repressor even when tryptophan is not bound. Expression of the *trp* operon is permanently repressed in this mutant bacterium.
10. When RF2 is present, UGA is read as a stop codon, leading to the formation of a truncated 25-residue polypeptide.

```
- Gly - Tyr - Leu - Stop
                          Read by
GGG UAU CUU UGA           RF2
```

When the level of RF2 is very low, termination at UGA does not occur. Instead, a shift of the reading frame occurs, resulting in continued protein synthesis with Asp 26 as the next residue. See W. Craigen, R. Cook, W. Tate, and C. Caskey, *Proc. Nat. Acad. Sci.* 82(1985):3619.

```
- Gly - Tyr - Leu -  Asp - Tyr - Asp -

GGG UAU CUU UGAC UAC GAC
```

Frameshift in absence of RF2. U is skipped. GAC is first codon of new reading frame.

11. Operator sites separated by five or six turns of double helix are oriented so that they can bind to both sites of the dimeric receptor. Only one operator site can bind when the other is 4.5, 5.5, or 6.5 turns away because the second site faces the opposite direction (180 degrees out of phase).
12. Attenuation requires that transcription and translation be tightly coupled. Pausing is a means of synchronizing these two processes. Transcription is halted until a ribosome initiates translation and comes near the paused polymerase. The outcome then depends on whether or not the leader is completely translated (p. 967).
13. The second helix of the HTH motif of both repressors is the major determinant of their specificity. Hence, it is often called the recognition helix. See R.P. Wharton and M. Ptashne, *Nature* 316(1985):601.

Chapter 37

1. (a) In prokaryotes, translation begins while transcription is still in progress. In eukaryotes, these processes are separate in space and time.
   (b) Eukaryotic transcripts are monogenic, whereas prokaryotic transcripts are typically polygenic.
   (c) In prokaryotes, several proteins can be specified by a single primary transcript, one for each gene encoding the mRNA. In eukaryotes, a primary transcript encoded by a single gene can give rise to multiple proteins through alternative splicing. Cleavage of a polyprotein can also yield multiple proteins in eukaryotes.
   (d) Most of prokaryotic DNA encodes proteins and functional RNAs. In contrast, most of eukaryotic DNA does not encode functional macromolecules.
   (e) Most prokaryotic genes are clustered in operons. Eukaryotes do not have operons.
2. Centromeres, telomeres, and *ars* (autonomous replicating sequences serving as origins of replication) are needed to fashion a synthetic yeast chromosome. See A. Murray and J.W. Szostak, *Nature* 305(1983):189.
3. The mitochondrial genome is much smaller than the nuclear genome. Furthermore, a cell contains a large number of mitochondria. A substantial portion of them, say 5%, could be nonfunctional without injuring the cell.
4. Mitochondrial DNA is evolving at a very rapid rate (see R.L. Cann, M. Stoneking, and A.C. Wilson, *Nature* 325(1986):31). It is possible but not likely that a human mitochondrial gene will be transferred to the nuclear genome in the next 10 million years. The existence of different genetic codes in the mitochondrion and the cytosol is a formidable barrier to gene transfer. In particular, UGA specifies tryptophan in mitochondrial proteins but is a stop signal in cytosolic protein synthesis.
5. One experimental approach is to add histones to linear duplex DNA and then form closed circles using DNA ligase. The histones are then removed from the supercoiled circles. Topoisomerase I is added to an aliquot to form the relaxed counterpart. The melting temperature of the supercoiled DNA is compared with that of the relaxed circle. A left-handed superhelix has a lower melting temperature than the relaxed circle, whereas a right-handed superhelix has a higher melting temperature. Recall that left-handed supercoiled DNA (negatively supercoiled; *Wr* is negative) is poised to be unwound (p. 796).
6. 5S RNA has the same base sequence (except for U in place of T) as the coding strand of its gene, the one that strongly binds TFIIIA. Hence, 5S RNA competes with its gene for the binding of TFIIIA. When 5S RNA is abundant, little TFIIIA is bound to the internal control region, and so transcription is slowed. The binding of TFIIIA to 5S RNA as well as the internal control region of the gene is the basis of a feedback loop that regulates the amount of 5S RNA.
7. TFIIIA has nine zinc fingers. Some of its zinc fingers could stay bound to the 5S RNA gene and others could be released to allow RNA polymerase to pass.
8. The leucine zipper symmetrically orients the two basic DNA-binding regions at the half sites of the palindromic DNA target sequence. The insertion of six residues changes the angular relationship of the pair of basic sites. In contrast, the correct angle is preserved when seven residues are inserted because this number corresponds to a heptad repeat. The insertion of four or five residues also destroys activity. See W. P. Pu and K. Struhl, *Proc. Nat. Acad. Sci.* 88(1991):6901.
9. Covalently cross-link a pair of basic DNA-binding regions by forming a flexible disulfide at their C-terminal end. The resulting protein, devoid of a coiled coil, serves as a transcriptional activator. See R.V. Talanian, C.J. McKnight, and P.S. Kim, *Science* 249(1990):769.
10. Retain the DNA-binding domain but delete the activation domain to block the formation of a functional transcription complex.
11. The mutant receptor lacks a nuclear localization signal, which is required for passage through nuclear pores (p. 932).
12. One approach is to synthesize a truncated cDNA that encodes the DNA-binding domain but not the hormone-binding domain of the new nuclear receptor. The DNA-binding domain produced by expressing this cDNA in *E. coli* should then bind with high affinity to target DNA sequences. A complementary approach is to carry out a domain swap experiment. The target cell could be transfected with a gene for a hybrid receptor containing the hormone-binding domain of the glucocorticoid receptor and the DNA-binding domain of the new receptor. The addition of glucocorticoid to the transfected cell would then activate genes normally stimulated by the unknown ligand.

Index

References to structural formulas are given in **boldface** type.

ISBN 0-7167-2009-4
90000
EAN
9 780716 720096

COMMON ABBREVIATIONS IN BIOCHEMISTRY

A	adenine
ACP	acyl carrier protein
ADP	adenosine diphosphate
Ala	alanine
AMP	adenosine monophosphate
cAMP	cyclic AMP
Arg	arginine
Asn	asparagine
Asp	aspartate
ATP	adenosine triphosphate
ATPase	adenosine triphosphatase
C	cytosine
CDP	cytidine diphosphate
CMP	cytidine monophosphate
CoA	coenzyme A
CoQ	coenzyme Q (ubiquinone)
CTP	cytidine triphosphate
Cyclic AMP	adenosine 3′,5′-cyclic monophosphate
Cyclic GMP	guanosine 3′,5′-cyclic monophosphate
Cys	cysteine
Cyt	cytochrome
d	2′-deoxyribo-
DNA	deoxyribonucleic acid
cDNA	complementary DNA
DNase	deoxyribonuclease
EcoRI	EcoRI restriction endonuclease
EF	elongation factor
FAD	flavin adenine dinucleotide (oxidized form)
$FADH_2$	flavin adenine dinucleotide (reduced form)
fMet	formylmethionine
FMN	flavin mononucleotide (oxidized form)
$FMNH_2$	flavin mononucleotide (reduced form)
G	guanine
Gln	glutamine
Glu	glutamate
Gly	glycine
GDP	guanosine diphosphate
GMP	guanosine monophosphate
cGMP	cyclic GMP
GSH	reduced glutathione
GSSG	oxidized glutathione
GTP	guanosine triphosphate
GTPase	guanosine triphosphatase
Hb	hemoglobin
HDL	high-density lipoprotein
HGPRT	hypoxanthine-guanine phosphoribosyl transferase
His	histidine
Hyp	hydroxyproline
IgG	immunoglobulin G
Ile	isoleucine
IP_3	inositol trisphosphate
ITP	inosine triphosphate
LDL	low-density lipoprotein
Leu	leucine
Lys	lysine
Met	methionine
NAD^+	nicotinamide adenine dinucleotide (oxidized form)
NADH	nicotinamide adenine dinucleotide (reduced form)
$NADP^+$	nicotinamide adenine dinucleotide phosphate (oxidized form)
NADPH	nicotinamide adenine dinucleotide phosphate (reduced form)
PFK	phosphofructokinase
Phe	phenylalanine
P_i	inorganic orthophosphate
PLP	pyridoxal phosphate
PP_i	inorganic pyrophosphate
Pro	proline
PRPP	phosphoribosylpyrophosphate
Q	ubiquinone (or plastoquinone)
QH_2	ubiquinol (or plastoquinol)
RNA	ribonucleic acid
mRNA	messenger RNA
rRNA	ribosomal RNA
scRNA	small cytoplasmic RNA
snRNA	small nuclear RNA
tRNA	transfer RNA
RNase	ribonuclease
Ser	serine
T	thymine
Thr	threonine
TPP	thiamine pyrophosphate
Trp	tryptophan
TTP	thymidine triphosphate
Tyr	tyrosine
U	uracil
UDP	uridine diphosphate
UDP–galactose	uridine diphosphate galactose
UDP–glucose	uridine diphosphate glucose
UMP	uridine monophosphate
UTP	uridine triphosphate
Val	valine
VLDL	very low density lipoprotein